3rd International Photonics & OptoElectronics Meetings 2010

(POEM 2010)

AA003012

Journal of Physics: Conference Series Volume 276

Wuhan, China
2-5 November 2010

Volume 1 of 2

ISBN: 978-1-61782-660-3
ISSN: 1742-6588

Printed from e-media with permission by:

Curran Associates, Inc.
57 Morehouse Lane
Red Hook, NY 12571

Some format issues inherent in the e-media version may also appear in this print version.

Printed by Curran Associates, Inc. (2011)

Institute of Physics
Dirac House, Temple Back
Bristol BS1 6BE UK

Phone: 44 1 17 929 7481
Fax: 44 1 17 920 0979

techtracking@iop.org

Additional copies of this publication are available from:

Curran Associates, Inc.
57 Morehouse Lane
Red Hook, NY 12571 USA
Phone: 845-758-0400
Fax: 845-758-2634
Email: curran@proceedings.com
Web: www.proceedings.com

TABLE OF CONTENTS

Volume 1

INVITED PAPERS

LASER TECHNOLOGY AND APPLICATIONS (LTA)

NANO-ENABLED ENERGY TECHNOLOGIES AND MATERIALS (NETM)

OPTOELECTRONIC DEVICES AND INTEGRATION (OEDI)

Volume 2

OPTOELECTRONIC SENSING AND IMAGING (OSI)

SOLAR CELLS, SOLID STATE LIGHTING AND INFORMATION DISPLAY TECHNOLOGIES (SSID)

TERA-HERTZ SCIENCE AND TECHNOLOGY (THST)

PREFACE

The 3rd International Photonics and OptoElectronics Meeting (POEM 2010) was held from November 2–5, 2011, in Wuhan, China.

POEM takes place annually, usually in November, with the aim of focusing on the key techniques of scientific frontiers and industry in the field of optoelectronics, understanding future trends as well as making the most of the industrial advantages of Wuhan - Optics Valley of China (OVC).

POEM 2010 presented a plenary session and six parallel sessions. The latter comprised Laser Technology and Applications; Nano-enabled Energy Technologies and Materials; Optoelectronic Devices and Integration; Optoelectronic Sensing and Imaging; Solar Cells, Solid State Lighting and Information Display Technologies; and Tera-Hertz Science and Technology. 700 delegates from the field of optoelectronics – including world-famous experts, researchers, investors and entrepreneurs from more than 20 countries – attended the conference, among whom were 160 invited speakers.

POEM 2010 once again received extensive praise for its intricate planning, rich contents, and the high-level and influential invited speakers which it attracted. Participants remarked that the presentations by the invited experts, the 'hot topic' discussions, students' posters, and the awards for papers were very engaging. They appreciated this valuable and beneficial opportunity for exchanging ideas with top photonics and optoelectronics experts.

Our thanks are extended to the Conference Secretariat and Local Organizing Committee, who have been completely dedicated to their work, and who made the conference such a great success. We are also grateful for the financial support from 111 Project (B07038), and for the help with organization and coordination from Wuhan National Laboratory for Optoelectronics and Huazhong University of Science and Technology.

Proceedings of POEM 2010
234 papers were selected out of the 343 manuscripts submitted. The organizers of POEM 2010 are grateful to all the authors whose papers are being published in this volume of the *Journal of Physics: Conference Series*. The proceedings are divided into six sections according to different technical areas:

- Laser Technology and Applications (LTA)
- Nano-enabled Energy Technologies and Materials (NETM)
- Optoelectronic Devices and Integration (OEDI)
- Optoelectronic Sensing and Imaging (OSI)
- Solar Cells, Solid State Lighting and Information Display Technologies (SSID)
- Tera-Hertz Science and Technology (THST)

Wuhan, PR China
December, 2010

Chaohui Ye
Zhong Lin Wang
Bingkun Zhou
Conference Chairs

The 3rd International Photonics and OptoElectronics Meeting (POEM 2010)
November 2–5, 2010
Wuhan, China

Supporters:
Ministry of Education of China (MOE)
State Administration of Foreign Experts Affairs (SAFEA)
National Natural Science Foundation of China (NSFC)

Sponsors:
Huazhong University of Science and Technology (HUST)
China Hubei Provincial Science Technology Department (HBSTD)
Wuhan East Lake National Innovation Model Park

Co-operating Societies:
Institute of Physics (IOP)
American Institute of Physics (AIP)
International Biomedical Optics Society (IBOS)
Laser Institute of America (LIA)
Optical Society of America (OSA)
IEEE Photonics Society (Singapore and Hongkong Chapters)
Chinese Optical Society (COS)

Organizer:
Wuhan National Laboratory for Optoelectronics (WNLO)

1. LASER TECHNOLOGY AND APPLICATIONS (LTA)
Editors:
Peixiang Lu, Wuhan National Laboratory for Optoelectronics (China)
Katsumi Midorikawa, Extreme Photonics Research Group, RIKEN (Japan)
Bernd Wilhelmi, Jenoptik AG, Jena (Germany)

2. NANO-ENABLED ENERGY TECHNOLOGIES AND MATERIALS (NETM)
Editors:
Zhong Lin Wang, Wuhan National Laboratory for Optoelectronics (China) and Georgia Institute of Technology (USA)
Guozhen Shen, Wuhan National Laboratory for Optoelectronics (China)

3. OPTOELECTRONIC DEVICES AND INTEGRATION (OEDI)

Editors:

Chinlon Lin, Bell Laboratory (USA)
Jesper Moerk, Technical University of Denmark (Denmark)
Xun Li, McMaster University (Canada)
Xinliang Zhang, Wuhan National Laboratory for Optoelectronics (China)
Junqiang Sun, Wuhan National Laboratory for Optoelectronics (China)

4. OPTOELECTRONIC SENSING AND IMAGING (OSI)

Editors:

Kecheng Yang, Wuhan National Laboratory for Optoelectronics (China)
Pengcheng Li, Wuhan National Laboratory for Optoelectronics (China)

5. SOLAR CELLS, SOLID-STATE LIGHTING AND INFORMATION DISPLAY TECHNOLOGIES (SSID)

Editors:

Hiroshi Amano, Meijo University (Japan)
Yibing Cheng, Monash University (Australia)
Jinzhong Yu, Institute of Semiconductor, CAS (China)
Changqing Chen, Wuhan National Laboratory for Optoelectronics (China)
Hongwei Han, Wuhan National Laboratory for Optoelectronics (China)
Guoli Tu, Wuhan National Laboratory for Optoelectronics (China)

6. TERA-HERTZ SCIENCE AND TECHNOLOGY (THST)

Editors:

Jianquan Yao, Tianjin University (China)
Shenggang Liu, University of Electronic Science and Technology of China (China)
X C Zhang, Rensselaer Polytechnic Institute (USA)
Jinsong Liu, Wuhan National Laboratory for Optoelectronics (China)

International Advisory Committee:

Yibing Cheng, Monash University (Australia)
Stephen Z D Cheng, University of Akron (USA)
Min Gu, Swinburne University of Technology (Australia)
Andrew B Holmes, the University of Melbourne (Australia)
Chinlon Lin, Bell Laboratory (retired, USA)
Xun Li, McMaster University (Canada)
Shenggang Liu, University of Electronic Science and Technology of China (China)
Jesper Moerk, Technical University of Denmark (Denmark)
Dennis L Matthews, University of California, Davis (USA)
Jiacong Shen, Jilin University (China)
Ping Shum, Nanyang Technological University (Singapore)
Chester C T Shu, Chinese University of Hong Kong (China)
Valery V Tuchin, Saratov State University (Russia)

Bruce Tromberg, University of California/Irvine (USA)
Peiheng Wu, University of Nanjing (China)
Alan Willner, University of Southern California (USA)
Lihong Wang, Washington University in St. Louis (USA)
C P Wong, Georgia Institute of Technology (USA)
Jianquan Yao, Tianjin University (China)
Xi Zhang, Tsinghua University (China)
X C Zhang, Rensselaer Polytechnic Institute (USA)

Program Committee:
Qingming Luo, Wuhan National Laboratory for Optoelectronics (China) – Chair
Hiroshi Amano, Meijo University (Japan)
Yibing Cheng, Monash University (Australia)
Peixiang Lu, Wuhan National Laboratory for Optoelectronics (China)
Ruxin Li, Shanghai Institute of Optics and Fine Mechanics (China)
Chinlon Lin, Bell Laboratory (USA)
Xun Li, McMaster University (Canada)
Shenggang Liu, University of Electronic Science and Technology of China (China)
Katsumi Midorikawa, Extreme Photonics Research Group, RIKEN (Japan)
Jesper Moerk, Technical University of Denmark (Denmark)
Valery V Tuchin, Saratov State University (Russia)
Lihong Wang, Washington University in St. Louis (USA)
Zhong Lin Wang, Georgia Institute of Technology(USA)
Jinzhong Yu, Institute of Semiconductor, CAS (China)
Jianquan Yao, Tianjin University (China)
X C Zhang, Rensselaer Polytechnic Institute (USA)

Local Organizing committee:
Lin Lin, Wuhan National Laboratory for Optoelectronics (China) – Chair
Sheng Lu, Administration Committee of Wuhan East Lake Hi-tech Development Zone (China) – Chair
Changqing Chen, Wuhan National Laboratory for Optoelectronics (China)
Ling Fu, Wuhan National Laboratory for Optoelectronics (China)
Hongwei Han, Wuhan National Laboratory for Optoelectronics (China)
Peixiang Lu, Wuhan National Laboratory for Optoelectronics (China)
Pengcheng Li, Wuhan National Laboratory for Optoelectronics (China)
Jinsong Liu, Wuhan National Laboratory for Optoelectronics (China)
Junqiang Sun, Wuhan National Laboratory for Optoelectronics (China)
Guozhen Shen, Wuhan National Laboratory for Optoelectronics (China)
Guoli Tu, Wuhan National Laboratory for Optoelectronics (China)
Kecheng Yang, Wuhan National Laboratory for Optoelectronics (China)
Xinliang Zhang, Wuhan National Laboratory for Optoelectronics (China)
Yuandi Zhao, Wuhan National Laboratory for Optoelectronics (China)

Local Secretariat:
Xiaochun Xiao, Huazhong University of Science and Technology (China)
Weiwei Dong, Huazhong University of Science and Technology (China)

All papers published in this volume of *Journal of Physics: Conference Series* have been peer reviewed through processes administered by the proceedings Editors. Reviews were conducted by expert referees to the professional and scientific standards expected of a proceedings journal published by IOP Publishing.

This is a co-publication with *Journal of Physics: Conference Series* volume 277, The 9th International Conference on Photonics and Imaging in Biology and Medicine (PIBM 2010). These conferences were held together from November 2–5, 2010, at Wuhan Science & Technology Convention & Exhibition Center, Wuhan, PR China.

Journal of Physics: Conference Series **276** (2011) 012001 doi:10.1088/1742-6596/276/1/012001

Monolithic Integration of GaN-based LEDs

Jin-Ping Ao[1]

Institute of Technology and Science, The University of Tokushima
2-1 Minami-Josanjima, Tokushima 770-8506, Japan

jpao@ee.tokushima-u.ac.jp

Abstract. The technology of monolithically integrated GaN-based light-emitting diodes (LEDs) is reported. First, the technology details to realize monolithic integration are described, including the circuit design for high-voltage and alternating current (AC) operation and the technologies for device isolation. The performances of the fabricated monolithic LED arrays are then demonstrated. A monolithic series array with totally 40 LEDs exhibited expected operation function under AC bias. The operation voltage of the array is 72 V when 20 LEDs were connected in series. Some modified circuit designs for high-voltage operation and other monolithic LED arrays are finally reviewed.

1. Introduction

Recently the development and application of GaN-based light-emitting diodes (LEDs) are greatly accelerating in the fields of automotive, liquid crystal display (LCD) television backlighting and solid-state lighting. Especially in the solid-state lighting field, it has been expected that extremely high-brightness illuminating LEDs can significantly replace the current standard incandescent and fluorescent lamps owing to their low energy consumption and enhanced lifetime. Obviously it will be convenient if the LED component can be driven directly by the recent alternating current (AC) high-voltage power source. In this case, the power lose in the rectification circuits can be cut. To realize high-voltage operation, a numbers of LEDs connected in series are necessary to disperse the voltage. Monolithic integration of LEDs can offer not only compact, low-cost and reliable components for high-voltage operation, but also new smart components such as microdisplays and microarray biochips. In this presentation, the fundamental technologies of monolithic integration of GaN-based LEDs will be described. Some application cases, such as a compact monolithic blue LED series array that can operate under high AC voltage will also be reviewed.

2. Circuit design for high-voltage operation

To achieve high-voltage operation, numbers of LEDs should be connected in series according to the operation voltage and the working bias of the LEDs, as shown in Figure 1. Owing to the series connection, the operation voltage will become a multiplication of the forward voltage of a single LED. Figure 2 shows a circuit diagram of a LED array with series and parallel connection to meet AC operation, in which half of the LEDs are connected in series to form one group and then connected reversely with another group in parallel [1]. Shunts between the series LEDs are placed to protect the diodes from reverse bias breakdown, when the breakdown characteristics are not uniform among the

[1] To whom any correspondence should be addressed.

LEDs. Thanks to this design, a rectification part which converts AC power to DC power can be omitted.

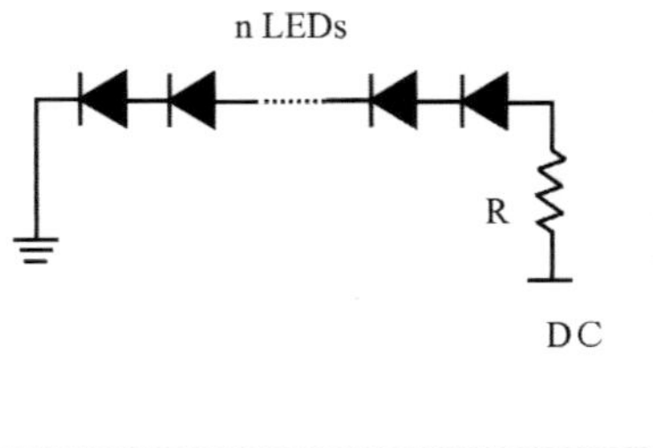

Figure 1. A circuit diagram of a LED series for DC operation.

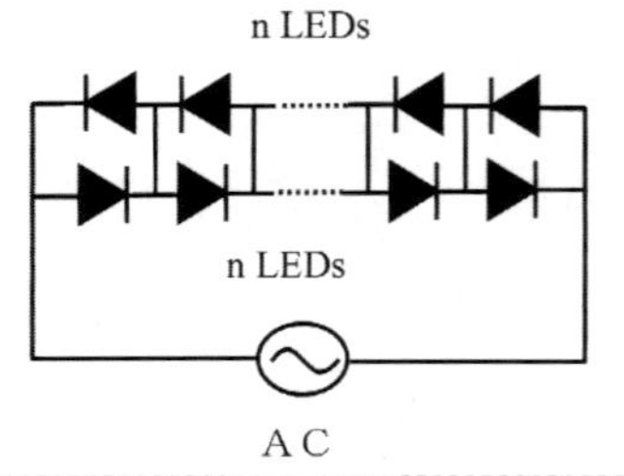

Figure 2. A circuit diagram of a series LED array for AC operation [1].

3. Fabrication technologies for monolithic integration

3.1. Device isolation technologies

The GaN LED structure used for monolithic integration is a popular structure which consists of p-type anode layers and n-type cathode layers with an active layer between them. To realize monolithic integration of LEDs, a key technology is device isolation process. If the substrate is a insulator or a semi-insulating substrate, such as sapphire and semi-insulating SiC, the device isolation process can be completed by deeply etching to the substrate forming a trench. For GaN-based materials, dry etching technique is popular in which reactive ion etching (RIE) or inductively coupled plasma (ICP) based on Chlorine-based plasmas is used. Figure 3 is a schematic structure of a monolithically integrated LED array with air-bridge interconnection.

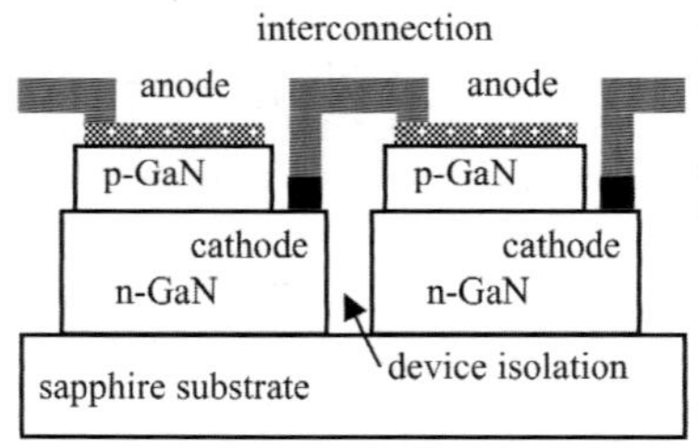

Figure 3. A schematic structure of a monolithically integrated LED series array.

Another possible technology for device isolation is heavy metal diffusion. Theoretically if a deep acceptor impurity is introduced, an n-type semiconductor where shallow dopant impurity exists will become semi-insulating. Figure 4 shows the calculation results of electron concentration and resistivity of an n-GaN with dopant impurity concentration of 2×10^{16} cm^{-3} and impurity level of 30 meV below the conduction band. When a deep acceptor with concentration of 2×10^{17} cm^{-3} and impurity level of 1.0 eV above the valence band is introduced, both the electron density and the hole density become extremely low and the resistivity of the semiconductor film becomes 9.63×10^{13} Ωcm while the Fermi level is 1.0 eV above the valence band.

To confirm the effect of heavy metal diffusion for GaN device isolation, Ni was adopted to diffuse into a n-GaN layer with doping concentration of 1×10^{17} cm^{-3} and thickness of 5 μm. The measurement pattern is a ring-type resistor with two Ohmic contacts in both sides and a space of 20 μm. When Ni was introduced into the space area, we found that the current between the two Ohmic pads decreased according to the diffusion condition, as shown in Figure 5. Compared with that of the as-grown sample, the current decreased if Ni was diffused under diffusion temperature of 1000 °C and time of 30 minutes. If the diffusion temperature and time were raised to 1050 °C and 180 minutes, respectively, the current decreased further by three orders. The diffusion depth of the latter condition was estimated to be 4 μm. This indicates that heavy metal diffusion is also a possible method for GaN device deep isolation. Further work is still necessary to investigate the reproductivity and estimate the energy level of Ni in GaN.

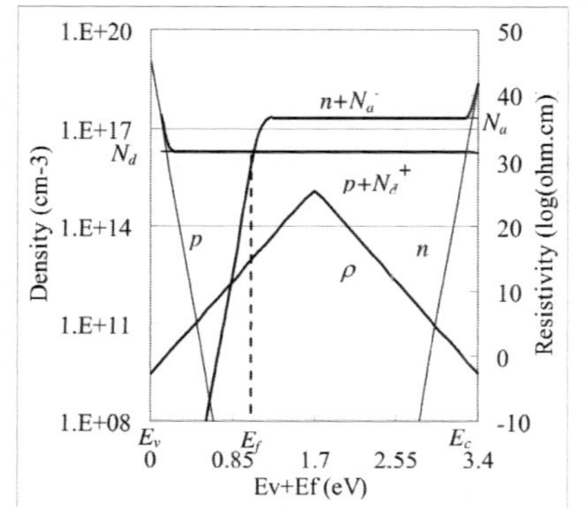

Figure 4. The calculation results of electron concentration and resistivity of an n-GaN when a deep acceptor is introduced.

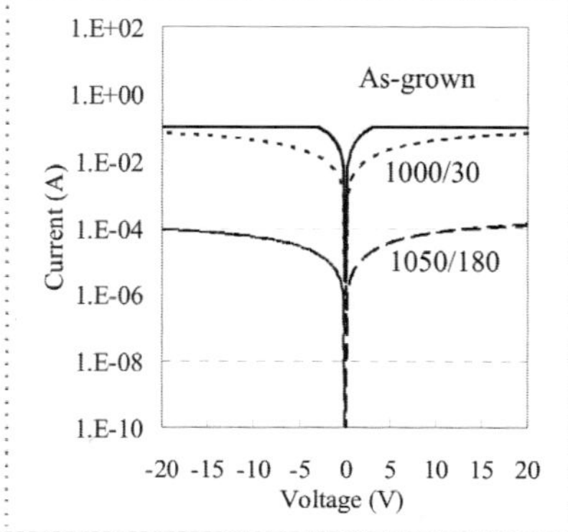

Figure 5. The current change of a n-GaN layer between two Ohmic pads under different diffusion conditions.

3.2. Fabrication of monolithically integrated LED array

After the device isolation process, the next step was a second dry etching by which the n-GaN layer was exposed for preparing the cathode Ohmic contact. Transparent anode electrode was then formed utilizing Ni/Au. Ti/Al/Ti/Au multi-player metal was adopted as the cathode electrode. The cathode contact metal also worked as the first local layer for interconnection. Finally, a gold electroplating process was utilized to form air-bridges, second layer interconnection lines and pads. Figure 6 shows a photography of a fabricated monolithic LED array with 5 LEDs in series.

Figure 6. A photography of a fabricated monolithic LED array with 5 LEDs in series.

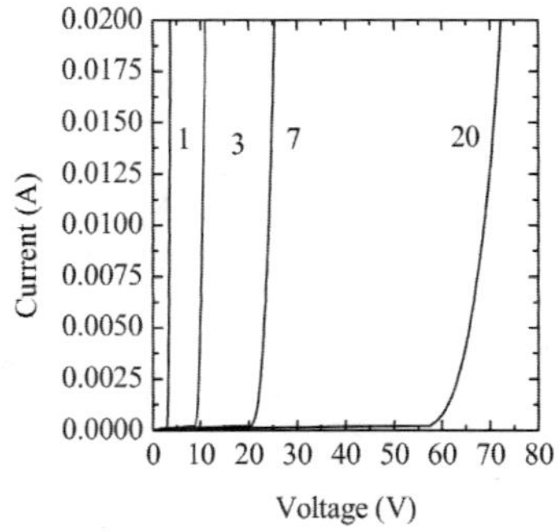

Figure 7. The forward current-voltage characteristics of 4 LED arrays with 1, 3, 7 and 20 LEDs in series.

3.3. Evaluation of monolithically integrated LED arrays

Figure 7 shows the forward current-voltage characteristics of 4 LED arrays with 1, 3, 7 and 20 LEDs in series. The forward voltages were almost proportional to the LED number connected in series. A monolithic series array with totally 40 LEDs exhibited expected operation function under AC bias. The operation voltage of the above array is 72 V when 20 LEDs were connected in series [1]. The output powers were also proportional to the LED number connected in series. This work also demonstrated a new method to achieve high output power. Based on this technology, a series of commercial products with operating voltage from 100 to 230 V has been developed [2].

4. Modified Circuit designs for High-Voltage Operation

In the above design, actually only 50% of the LEDs emit simultaneously, resulting low active layer utility ratio. Thermal burden might also be high due to the compact layout design. In order to improve the active layer utility ratio and to improve the thermal burden, there is still much space to optimize the circuit and layout design. A modified circuit configuration was reported in which a part of LEDs were designed in series at branch sides acting as Wheatstone bridge for rectification [3]. In this design, 10×4 LEDs are used for rectification and 15 LEDs are placed in the center of the bridge. Under operation, 35 LEDs will emit among totally 55 LEDs, resulting emitting ratio of 63.6%. Another circuit design with ladder type was also investigated to improve the uniformity and reliability [4]. In this design, 39×2 LEDs are used in both sides connected back-to-back and 36 LEDs are placed in the center. Under operation, 75 LEDs will emit among totally 114 LEDs, resulting emit ting ratio of 65.8%. Operation up to AC voltage of 220 V was demonstrated.

5. Other Integrated GaN LEDs Arrays

Monolithic integration technology can be also used to improve the emission efficiency of GaN LEDs. It was reported that when hundreds of LEDs were connected in parallel, the overall emission efficiency increased over the conventional LEDs for the same device area [5, 6]. Monolithic integration technology can be also expanded to realize GaN LED-based self-emission micro-displays [7] and GaN LED-based micro-array biochips [8].

6. Conclusion

To achieve high-voltage operation, numbers of LEDs can be monolithically integrated in series if the substrate is insulating or semi-insulating. A monolithic series array with totally 40 LEDs exhibited expected operation function under AC bias. The operation voltage of the above array is 72 V when 20 LEDs were connected in series. The active layer utility ratio and the thermal burden can be improved by carefully optimizing the circuit and layout design. Monolithic integration technology can be also used to improve the emission efficiency of GaN LEDs, to realize GaN LED-based self-emission micro-displays and GaN LED-based micro-array biochips. In conclusion, monolithic integration is a smart method to realize various functional integrated arrays by using

References

[1] Ao J P, Sato H, Mizobuchi T, Morioka K, Kawano S, Muramoto Y, Lee Y B, Sato D, Ohno Y and Sakai S 2002 *Phys. Stat. Sol.* (a) **194** 376
[2] http://www.acriche.com
[3] Yen H H, Kuo H C and Yeh W Y 2008 *Jpn. J. Appl. Phys.* **47** 8808
[4] Onushkin G A, Lee Y J, Yang J J, Kim H K, Son J K, Park G H and Park Y J 2009 *IEEE Photon. Technol. Lett.* **21** 33
[5] Jin S X, Li J, Lin J Y and Jiang H X 2000 *Appl. Phys. Lett.* **77** 3236-8
[6] Choi H W, Jeon C W and Dawson M D 2003 *Phys. Stat. Sol.* c **0** 2185-8
[7] Choi H W, Jeon C W and Dawson M D 2004 *IEEE Eelectron Dev. Lett.* **25** pp. 277-9
[8] McGuinness C D, Sagoo K, McLoskey D and Birch D J S 2005 *Appl. Phys. Lett.* **86** 261911

Journal of Physics: Conference Series **276** (2011) 012002 doi:10.1088/1742-6596/276/1/012002

High Performance Large Mode-Area Ytterbium-doped Photonic Crystal Fiber for Fiber Lasers

Wei Chen [1,2,*], Shiyu Li[2], Peixiang Lu[1] and Dongxiang Wang[2]

[1] Wuhan National Laboratory for Optoelectronics, Huazhong University of Science and Technology, Wuhan,P.R.China 430074

[2] State Key Laboratory of Optical Communication Technologies and Networks, Fiberhome Telecommunication Technologies Co. Ltd, P.R.China, 430074

E-mail: chenwei@fiberhome.com.cn

Abstract. In this letter, large-mode-area double-cladding ytterbium-doped photonic crystal fiber was designed in theory and fabricated in practice. This fiber we have fabricated successfully has endless single mode operation performance and large inner-cladding numerical aperture of more than 0.75. The struts width between large air-holes in the outer-cladding is about 0.22 μm. The photonic crystal fiber has a mode-area about 1465.7μm^2. Due to the material being pure silica and air, such structures have excellent capacity to with-stand high temperature. The laser light can have very good beam quality, even diffraction-limited beam quality because of the single-mode core. This fabrication technical breakthrough of novelty high performance double-cladding ytterbium-doped photonic crystal fibers will give contributions to the high power fiber lasers and promote the progress of technology in the fields of high power lasers.

1. Introduction

High power fiber lasers and fiber amplifiers are the research frontier in the fields of lasers at present. They are nearly always realized with rare-earth-doped double-cladding optical fibers [1-4], which are pumped with fiber-coupled high-power diode bars or other kinds of laser diodes. The pump light is launched into an inner cladding rather than into the much smaller fiber core, in which the laser light is generated. Conventional double-cladding optical fibers are based on step-index type of core designs and polymer-based outer-cladding materials to create an inner cladding with a high numerical aperture (NA). These fibers provide core size of up to around 10 μm to 30 μm for single mode to few modes operation. With large core sizes, these kinds of active fibers can handle power levels to the kW level. However, the laser beam quality and laser brightness with these fibers is not ideal, and the bending loss of these active fibers is big. Besides, the use of polymers limits the heat-handling capacities of the active fibers for high power fields.

To tackle these problems, many solutions have been taken. Among them, bending methods were taken to improve beam quality. The double cladding active fiber was bent with a diameter of about 20 cm to reduce bending loss for higher-order modes without affecting the fundamental mode [5]. But the bending leads some power leaking into the polymer outer-cladding, which results the softening and failing of the polymer cladding. P. Wang et al [6] demonstrated a approach to achieve nearly single-mode operation in a cladding-pumped multimode-core fiber laser based on the use of a fiber with helical-core geometry. All these methods cannot maintain robust and efficient single-mode operation.

J. Limpert et al [7-12] made laser experiments for robust single-transverse-mode propagation with double-cladding ytterbium-doped photonic crystal fibers (PCFs). In this letter, double-cladding ytterbium-doped photonic crystal fiber was designed and fabricated. This fiber has ultra low core numerical aperture (NA) for making very large, strictly single mode core, and very high inner-cladding NAs for making high coupling efficiency.

2. Theoretical Design

For PCFs, the Micro-hole arrays are distributed regularly in the PCFs, these periodical micro-hole arrays make great action on the transmission light. The shorter the transmission light wavelength is, the more centralized to the higher index domain the light intensity is, so the effective refractive index of the cladding will be raised. The cladding effective refractive index is decided by the Fundamental Space-filling Modes (FSMs) in the microstructure optical fiber, β_{FSM} is defined as the transmission constant of the FSMs. Therefore, the effective refractive index of fiber cladding $n_{eff}=\beta_{FSM}/k$, where $k=2\pi/\lambda$. The normalized transmission frequency v_{pcf} could be written as [13,14]:

$$v_{pcf} = \frac{2\pi \cdot \Lambda}{\lambda}\left(n_{core}^2 - n_{neff}^2\right)^{0.5} \quad (1)$$

Where Λ is the micro-hole pitch. When $\lambda \to 0$, the mode field distributions of the transmission light don't change basically, they don't depend on the light wavelength and the hole's distance, but are related to the relative dimension between the diameter of micro-hole (d) and transmission wavelength ( ). When $v_{pcf} < \pi$, light could transmit in the photonic crystal fibers in the single mode way [15,16]. The relations between normalized transmission frequency (v_{pcf}) of the photonic crystal fibers at different d/Λ and the Λ/π were theoretically calculated. The Figure 1 shows the theoretical results.

Light could transmit along the photonic crystal fibers in the single mode way, when the d/ •0.30 and is as big as fifteen times of . While d/ •0.45, the maximum ratio of to is 1.538, light can be also guided in the state of single-mode.

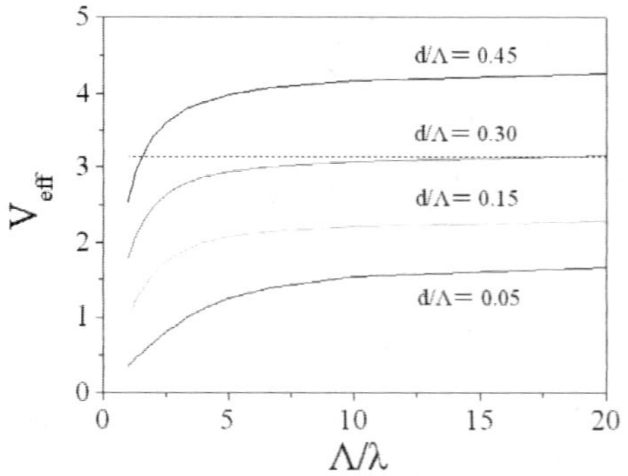

Figure 1. The relation between v_{eff} and / .

The numerical aperture of the inner cladding (NA_{iclad}) is relative to the refractive index of inner cladding and the refractive index of fundamental bridge mode in the air-silica compounded outer cladding (n_b). The NA_{iclad} can be calculated from the following equation.

$$NA_{iclad} = \sqrt{n_{iclad}^2 - n_b^2} \quad (2)$$

Figure 2 shows the air-clad structure, it includes air-holes and silica bridges. The n_b is dependent on the width of the bridge (b) and the radial-width of the large air-holes (w) in the air-silica compounded outer-cladding. The direction of power propagation in bridge modes has radial and longitudinal components. In the limit of w/ $\to\infty$, a continuum of these modes exists with effective indices in the range between 1 and n_b. The effective index n_b is therefore an upper bound for all such localized bridge modes that have a radial component to power propagation. In order to separate the leakage mechanisms of loss via

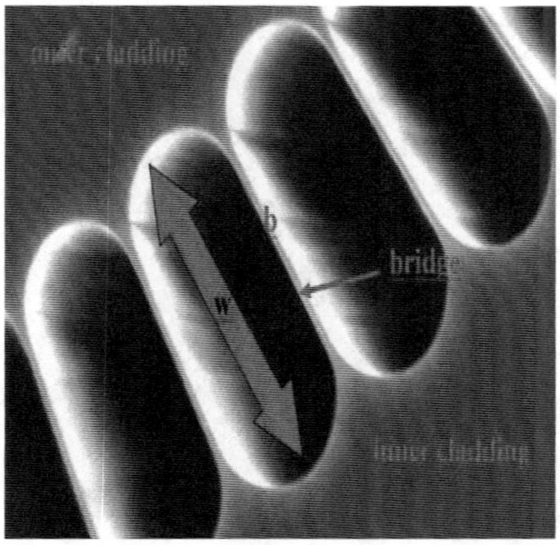

Figure 2. The air-clad structure of the PCF.

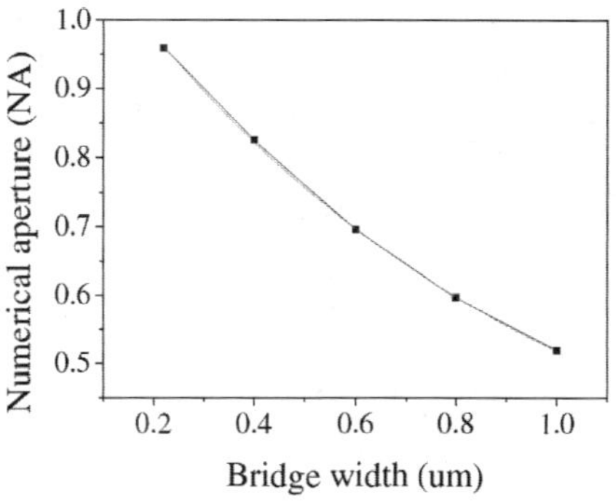

Figure 3. The relation between bridge width and NA of inner cladding.

the bridges from that of optical tunneling, the extreme situation of w/ >>1 is assumed. The upper bounds of NA_{iclad} are valuable for choosing the radial width of air-clad designs. It suggests that, when w/ >4, the role of optical tunneling in influencing the NA can be neglected for all practical lengths [17]. In our simulation, w/ =10 and =915nm are assumed, and the results are shown in Fig.3. The smaller the bridge width is, the bigger the NA_{iclad} is. When the b/ <<1, the NA_{iclad} is close to 1. The relation between the bridge width and the NA_{iclad} is shown in Fig.3 by simulation.

3. Fabrication and Results

By the method of Stack & Drawing, the double-cladding ytterbium-doped photonic crystal fiber (DY-PCF) was fabricated successfully, which was shown in the Fig.4. The fiber has a 43.22-µm diameter ytterbium-doped core with a numerical aperture of 0.06. In the inner cladding, the air-filling fraction is 30%. In the outer cladding, the radial width of the large hole size is 19.5µm, and the bridge thickness between the adjacent air-holes is 0.22µm, which is shown in Fig.5. At the wavelength of 1.06µm, the w/ =18.4>>1, b/ =0.21<<1. Therefore, the structure parameters meet the conditions of single-mode operation and big NA_{iclad}. The fiber has a 1465.7-µm^2 mode area and a NA_{iclad} of 0.75 by our measurements. This measuring NA of the inner-cladding is 20% smaller than the theoretical value of 0.95, the difference is due to test windage of the Far-field light distribution.

Figure 6 shows the simulation results of the power distribution in the DY-PCF. From the figure, we can see that the fundamental mode light is restricted tightly in the ytterbium-doped core.

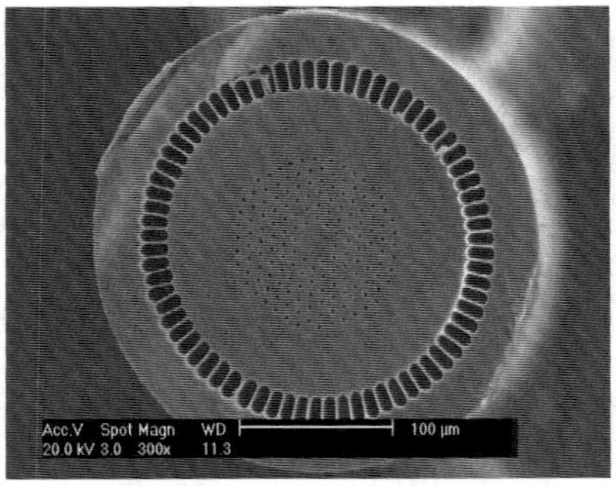

Figure 4. The SEM photo of DY-PCF.

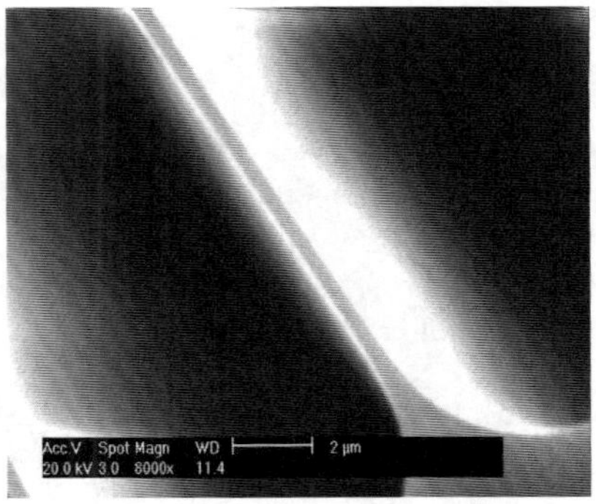

Figure5. The SEM photo of silica bridge.

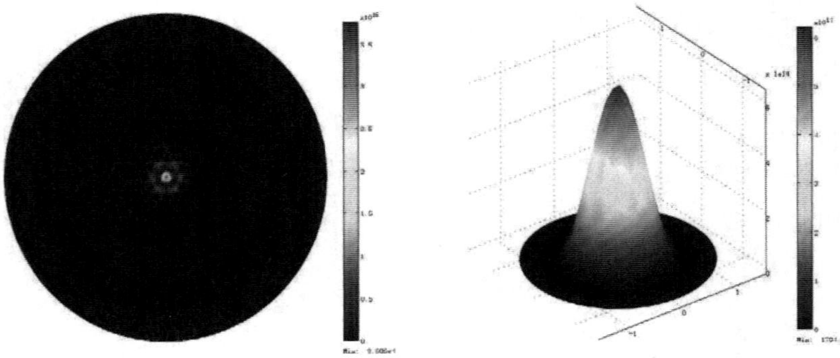

Figure 6. The power distribution in the DY-PCF.

4. Conclusions

In this letter, the large-mode-area double-cladding ytterbium-doped photonic crystal fiber was designed in theory and fabricated in practice. This fiber has endless single mode operation performance and large inner-cladding numerical aperture of more than 0.75. The photonic crystal fiber has a mode-area about 1465.7μm^2. Due to the material being pure silica and air, such structures have excellent capacity to withstand high temperature. The laser light can have very good beam quality, even diffraction-limited beam quality because of the single-mode core. This fabrication technical breakthrough of novelty high performance double-cladding ytterbium-doped photonic crystal fibers will give contributions to the high power fiber lasers and promote the progress of technology in the fields of high power lasers.

Acknowledgments

This work is supported by the National Basic Research Program of China (Grant No. 2003CB314905 and 2010CB327606), the National Key Basic Research Special Foundation (Grant No. 2006CB806006) and the National High Technology Research and Development Program of China (Grant No. 2007AA03Z447 and 2008AA03Z405).

References

[1] Dawson J W, Messerly M J, Beach R J, Shverdin M Y, Stappaerts E A, Sridharan A K, Pax P H, Heebner J E, Siders C W, and Barty C P J, 2008 *Opt. Express*, **16** 3240
[2] Minelly J, Teodoro F D, Brooks C, Desmoulins S, 2008 *Winter Topical Meeting Series, IEEE/LEOS*, **14** 148
[3] Chen Z L, Hou J and Jiang Z F, 2007 *Chin. Phys. Lett.*, **5** 178
[4] Limpert J, Roser F, Klingebiel S, Schreiber T, Wirth C,Peschel T, Eberhardt R, Tiinnermann A. 2007 *Selected Topics in Quantum Electronics, IEEE Journal,* **13** 537
[5] Jeong Y, Sahu J K, Payne D N, and Nilsson J, 2004 *Opt. Express*, **12** 6088
[6] Wang P, Cooper L J, Sahu J K, and Clarkson W A, 2006 *Opt. Lett.* **31** 226
[7] Limpert J, Schmidt O, Rothhardt J, Röser F, Schreiber T, and Tünnermann A, 2006 *Opt. Express,* **14** 2715
[8] Fomin V, Mashkin A, Abramov M, Ferin A, Gapontsev V, 2006 *3 rd International Symposium on High-Power Fiber Lasers and Their Applications*, session HPFL-3
[9] Boullet J, Zaouter Y, Desmarchelier R, Cazaux M, Saby J, Eric Cormier R B, 2008 *Opt. Express* **16** 17891
[10]Chen W, Li J Y, Lu P X, Li S Y, Ji L L, Jiang Z W, Zhang J H, Peng J G,2008 *Chin. Phys. Lett.* **25** 960
[11] Y J Song, M L Hu, Q W Liu, J Y Li, W Chen, L Chai, Q Y Wang. 2008 *Acta Phys. Sin.* **57** 5045.
[12] Liu Q W, WangQ Y, Xue Y H, Hu M L, Song Y J, Li J Y, Chen W, Chai L, 2008 *Chinese J. Lasers.* **35** 819

[13] Chen W, Li J Y, Li S Y, Li H Q, Jiang Z W, Peng J G,2007 *Chin. Opt. Lett.* **5** 83
[14] Yang S G, Zhang Y J, Peng X Z, Lu Y, Xie S Z, Li J Y, Chen W, Jiang Z W, Peng J G, Li H Q, 2006 *Opt. Express,* **14** 3015
[15] Mortensen N A, Folkenberg J R,2003 *Opt. Lett.*, **28** 1879
[16] Birks T A, Knight J C and Russell P St J,1997 Opt. Lett. **22** 961
[17] Åslund M, Canning J, Jackson S D, Teixeira A and Lyytikäinen K, 2005 *Opt. Express*, **13** 5227

Coherent Polarization Control of THz Waves Generated from Asymmetrically Ionized Gases

Jianming Dai[a)], Nicholas Karpowicz[b)], and X.-C. Zhang[a) 1]
a) Center for Terahertz Research, Rensselaer Polytechnic Institute, Troy, New York 12180, USA
b) Max-Planck Institute for Quantum Optics, Garching, Germany

E-mail: zhangxc@rpi.edu

Abstract. Unlike polarization control of optical waves, lossless control over the polarization of broadband terahertz waves remained challenging. We recently found that the polarization of terahertz waves generated from gas plasma excited by femtosecond fundamental pulse (ω) and its second harmonic (2ω) could be coherently controlled by changing the relative phase between the ω and 2ω pulses. In particular, when the ω and 2ω pulses are both circularly polarized (or close to it), the photo-excited electrons exhibit different trajectories as the relative phase between the two optical pulses changes, and subsequently terahertz polarization angle can be controlled arbitrarily through the relative phase while the intensity of the emitted terahertz wave is kept constant. This new finding may enable fast terahertz wave modulation and coherent control of nonlinear responses excited by intense terahertz waves with controllable polarization.

1. Introduction

As a promising nonlinear optical medium, femtosecond-laser-induced plasma in gases has been studied intensively over the past several years. It is not only used to generate tunable ultrashort laser pulses and coherent, extreme, ultraviolet radiation via phase-matched high-order difference-frequency mixing [1, 2], but also used to generate terahertz (THz) waves [3-8]. Such an air-plasma-based THz source provides very high THz electric field, which is comparable to those obtained with large-scale facilities only accessible in some national laboratories [9]. Moreover, recent results show that femtosecond laser-induced gas plasma exhibits a remarkable ability to detect pulsed (broadband) THz waves through nonlinear optical processes [10, 11]. The use of air (or selected gases) as both the broadband THz wave emitter and THz wave sensor provides superior bandwidth covering the entire THz gap and well-beyond [12]. Potential applications include nonlinear spectroscopy and imaging, as well as remote sensing and identification. The air-plasma-based THz source has already become a common tool in research laboratories for fundamental scientific research, such as nonlinear THz response of different materials [13, 14].

Coherent control of THz wave generation from gas plasma has been demonstrated in 2006 [15] with an interferometric phase compensator, which has been used to phenomenologically determine the mechanism for THz wave generation from gas plasma. More recently, quantum mechanical models have been employed to describe the physical mechanism for THz generation from gas plasma in cases

[1] To whom any correspondence should be addressed.

using phase-controlled femtosecond ω and 2ω pulses and few-cycle single-color pulses, respectively, both treated with linearly polarized excitation optical pulses [16, 17]. It has been shown that the full THz emission process takes place in two steps: first, a broadband pulse is produced through the asymmetric ionization due to the laser-atom interaction, and then a second step, an "echo" is produced by the interaction of the ionized wave packets with the surrounding gas and plasma.

Using the quantum mechanical model described in our aforementioned work [16], we calculated the electron expectation value trajectories in the case of circularly and elliptically polarized ω and 2ω beams. When linearly polarized optical excitation beams are used, the problem essentially reduces to two dimensions, and one expects the THz radiation to share the polarization of the pump pulses.

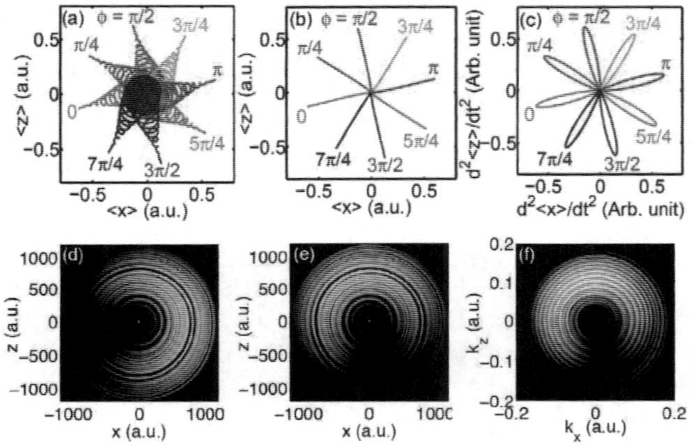

Figure 1. Three-dimensional, quantum-mechanical simulation of the effects of changing the phase ϕ between the circularly-polarized fundamental and second harmonic pulses. (a) Electron expectation value trajectories in the dual-color field. (b) Electron trajectories with the laser-driven quiver motion removed. (c) Second time-derivative of the trajectories, showing the effective polarization of the emitted radiation. (d) and (e) Final electron density distribution in the z-x plane (scaled logarithmically) for the phases π and π/2 respectively. (f) Momentum space electron density distribution for phase π/2 (scaled logarithmically, bound states removed).

When circularly or elliptically polarized optical fields are applied, the laser-atom interaction requires three dimensions, since the optical field is capable of coupling states with differing values of the z-projection of the angular momentum (m) in addition to the angular momentum ℓ. This was calculated by representing the electron wave function as a series of partial waves in spherical coordinates, with a spatial radial dimension and momentum-space angular dimensions, and numerically solving the time-dependent Schrödinger equation. The simulations were performed using hydrogen for simplicity. The system of coordinates was rotated dynamically such that the vector potential of the laser was always aligned with the z-axis (and the laser Poynting vector was along the y-axis), which allowed the m-coupling to be confined to a single operation, $\exp(i\theta L_y)$. The exact (real, dense) operator was used rather than the infinitesimal or Padé approximants so that arbitrary ellipticities could be utilized without the build-up of rotation errors. The laser-induced coupling between the ℓ partial waves was performed in the velocity gauge [18]. The two-dimensional electron polarization was continuously monitored throughout the simulation by calculating the expectation values <z> and <x> at each time step. This polarization completely describes the THz radiation produced by the first step of the emission process (ionization), and also determines the direction of the remaining emission processes.

The results of such a simulation are shown in Fig. 1, which plots the consequence of changing the relative phase between the fundamental and second harmonic carrier waves, ϕ, when the optical pulses are both right-circularly polarized. We can see that instead of the intensity modulation observed with linearly polarized excitation, the THz intensity remains constant, but the polarization angle rotates with ϕ. When left-circular excitation is used, the situation is similar, but the THz field rotation is counter-clockwise.

Motivated by the above physical picture and some preliminary experimental results indicating that the THz electric-field detected by polarization-sensitive electric-optic sampling does not increase when the THz power detected with a pyroelectric detector is tripled during the optimization of the THz emission, we performed systematic experiments to test the polarization behavior of THz waves generated from gas plasma. Specifically, focusing on the configuration with two-color (ω and 2ω) optical pulse excitation, we presented both theoretical and experimental investigations of the THz polarization characteristics as the relative phase between ω and 2ω pulses changes, with different polarization combinations of the two pulses. We found that the polarization of the THz waves is coherently controllable through the phase when at least one of the optical pulses (ω or 2ω) is elliptically polarized. In particular, when both ω and 2ω beams are circularly polarized (or close to it), the THz polarization angle can be rotated arbitrarily simply by changing the phase, with the THz amplitude kept unchanged. Our results not only give a clearer picture about the behavior of the THz emission from gas plasma but also add to the THz air source a more attractive feature that may lead to fast THz wave modulation devices and enable coherent control of nonlinear responses excited by intense THz waves.

2. Experimental investigation and analysis

Experimentally, a stable phase control mechanism with sufficient scan range is necessary. Instead of using the phase compensator described in reference [19] or a phase plate [20], a new phase compensator in an in-line configuration with attosecond phase-control accuracy is employed, as shown inside the dashed line in Fig. 2. A femtosecond pulse at 800 nm (ω) generates a second harmonic pulse at 400 nm (2ω) while passing through a type-I Beta Barium Borate (β-BBO) crystal. The ω and 2ω beams, which have polarizations perpendicular to each other, pass through an x-cut birefringent plate (BP, here we use a linear crystal, α-BBO) with its slow axis aligned with the ω beam polarization (o-ray) and the fast axis aligned with the 2ω beam (e-ray) so that right after this plate the 2ω pulse is leading to the ω pulse, as shown in the figure. A fused silica wedge pair with a small wedge angle of 3.93° is used to finely control the relative phase delay between the ω and 2ω pulses through the relationship $\Delta\tau = \Delta l(n_{2\omega} - n_{\omega})\tan(\theta_w)$, where Δl is the step size of the mechanical translation stage, n_{ω} and $n_{2\omega}$ is the refractive index of the fused silica at 800 nm and 400 nm, respectively, θ_w is the wedge angle, and $\Delta\tau$ is the resulting optical delay step. Finally, a tunable dual-band, zero-order waveplate is used to control polarizations of the ω and 2ω beams. In the experiments, additional quarter waveplates at 800 and 400 nm are necessary to further control the polarization of the two optical pulses.

The advantage of this in-line phase compensator is that it combines the minimal lateral displacement of our previous phase compensator and the minimal phase fluctuation of the phase plate as described in references [19, 20]. The collimated ω and 2ω beams are focused simultaneously to ionize the gas and emit THz waves with an off-axis parabolic mirror (M2). A polyethylene or silicon lens is used to collect the emitted THz waves. A broadband THz polarizer is used to analyze the polarization of the emitted THz waves, and a pyroelectric detector is used to monitor the transmitted THz power through the THz polarizer as it rotates.

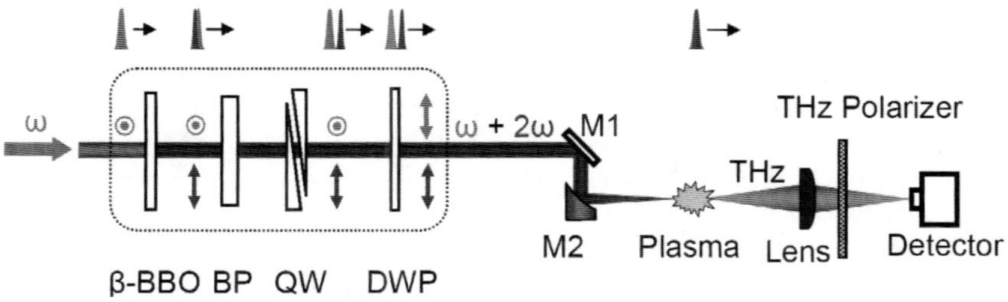

Figure 2. Schematic illustration of the experimental setup. Inside the dashed line is the in-line phase compensator. β-BBO, Beta Barium Borate crystal; BP, Birefringent Plate (α-BBO); QW, Quartz Wedges; DWP, Dual-wavelength Waveplate; the red and blue arrows indicate the polarization of the ω and 2ω beams, respectively; M1, dual-wavelength high-reflection mirror (HR at both 800 and 400 nm); M2, off-axis parabolic mirror used to focus optical beams; Lens, polyethylene or silicon lens.

Fig. 3 shows a typical phase curve obtained by changing relative phase between ω and 2ω pulses through the translation of one of the wedges while monitoring at the THz average power with a pyroelectric detector when ω and 2ω pulses are linearly polarized and parallel to each other.

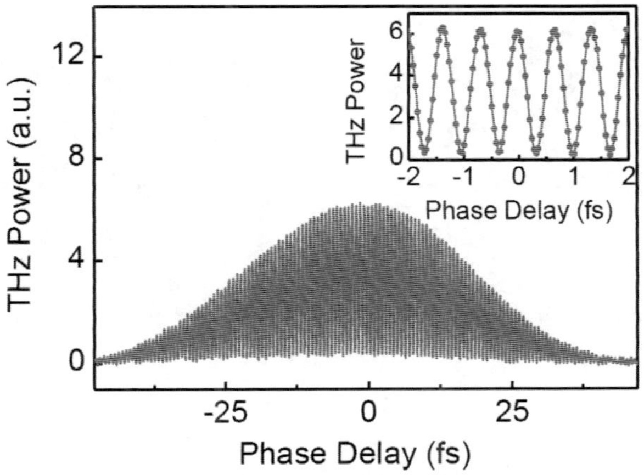

Figure 3. Schematic illustration of the experimental setup. Inside the dashed line is the in-line phase compensator. β-BBO, Beta Barium Borate crystal; BP, Birefringent Plate (α-BBO); QW, Quartz Wedges; DWP, Dual-wavelength Waveplate; the red and blue arrows indicate the polarization of the ω and 2ω beams, respectively; M1, dual-wavelength high-reflection mirror (HR at both 800 and 400 nm); M2, off-axis parabolic mirror used to focus optical beams; Lens, polyethylene or silicon lens.

Initially, the THz polarization effect is tested with linearly polarized ω and 2ω optical beams. In this case, the electron energies also change but the overall orientation does not change. Figs. 4(a) and 4(c) show the change in measured THz intensity versus THz polarizer angle and the relative phase between ω and 2ω pulses with the ω and 2ω beams parallel and orthogonally polarized (the 2ω beam is kept vertically polarized), respectively. Figs. 4(b) and 4(d) are the corresponding simulation results using the quantum mechanical model. Both experimental results and theoretical simulation show the THz polarization in this case essentially follows the polarization of the 2ω beam. However, the THz emission efficiency is about one order of magnitude higher when ω and 2ω are parallel polarized than when they are orthogonally polarized. When the polarizations of the two pulses are aligned, the second harmonic has the effect of causing destructive interference of the electron wave packets in one direction and constructive interference in the other, resulting in one of the electron "beams" emerging from the atom being effectively switched off [16]. When the two pulses have orthogonal polarizations, the relatively weak second harmonic can only "pull" the symmetric pair of beams to one side or the other, resulting in a much smaller net polarization.

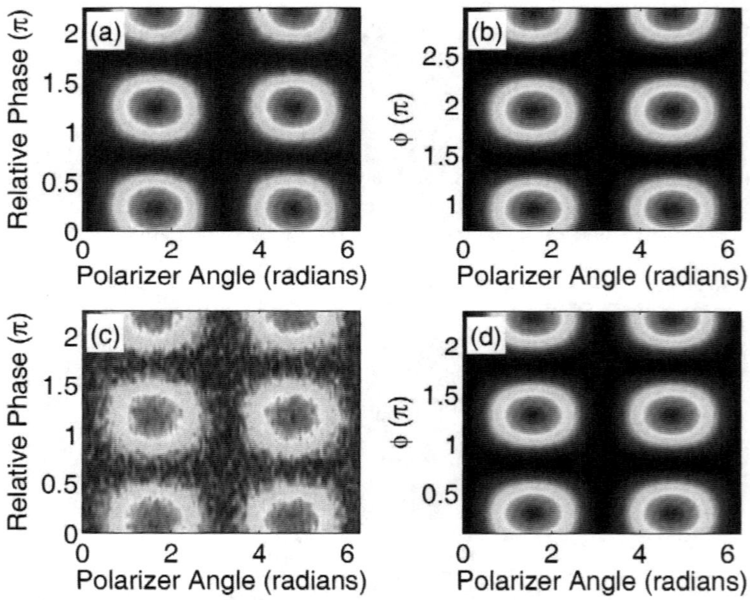

Figure 4. THz intensity versus THz polarizer angle and the relative phase between the ω and 2ω pulses with linear polarization. (a) and (c) are experimental results with the two pulses parallel and orthogonally polarized, respectively; (b) and (d) are the corresponding simulation results.

When at least one of the optical beams is circularly or elliptically polarized, the electron trajectories will change their orientation as the phase between ω and 2ω changes. As a consequence, the polarization of the emitted THz wave changes its orientation. Figs. 5(a) and 5(c) show the experimental results when the ω beam is left-handed and right-handed circularly polarized while the

Journal of Physics: Conference Series **276** (2011) 012003 doi:10.1088/1742-6596/276/1/012003

2ω beam is elliptically polarized with the ratio between the minor axis and major axis of the ellipse (defined here as ellipticity) of about 1/11 in terms of THz intensity, respectively. Figs. 4(b) and 4(d) are the corresponding simulation results using the quantum mechanical model, which reproduces the experimental results. We can see that when ω is right-handed circularly polarized the THz polarization rotates in a right-handed manner.

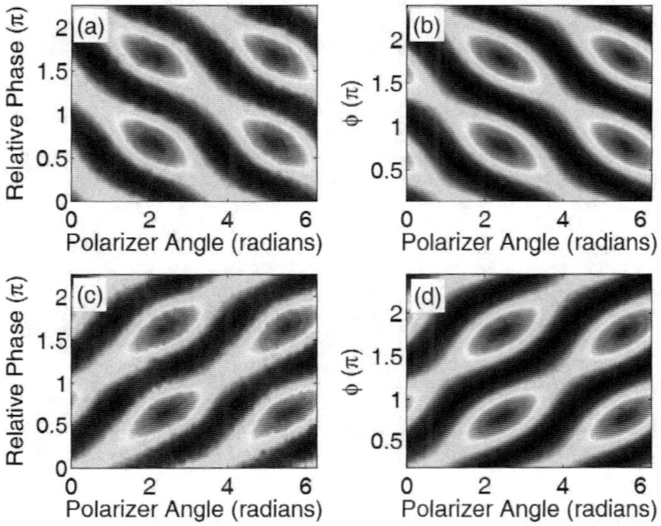

Figure 5. THz intensity versus THz polarizer angle and the relative phase between the ω and 2ω pulses with left- or right-handed circularly polarized ω pulse and with elliptically polarized 2ω pulse (with an ellipticity of about 1/11 in terms of optical intensity). (a) and (c) are the experimental results with left- and right-handed circularly polarized ω pulses, respectively, and (b) and (d) are the corresponding simulation results.

Figs. 6(a) and 6(c) show the experimental results when the 2ω beam is left-handed and right-handed circularly polarized while the ω beam is elliptically polarized with the ratio between the minor axis and major axis of the ellipse (defined here as ellipticity) of about 1/7 in terms of THz intensity, respectively. Figs. 6(b) and 6(d) are the corresponding simulation results using the quantum mechanical model, which also reproduces the experimental results. We can see that when 2ω is right-handed circularly polarized the THz polarization rotates in a left-handed manner.

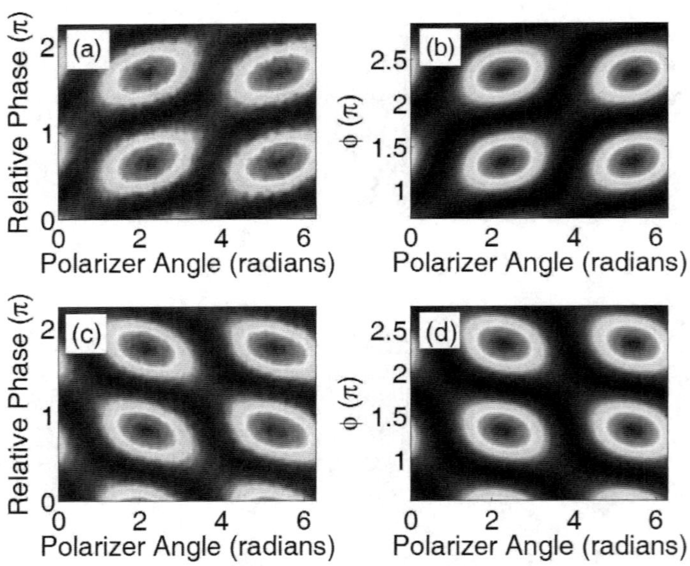

Figure 6. THz intensity versus THz polarizer angle and the relative phase between the ω and 2ω pulses with left- or right-handed circularly polarized 2ω pulse and with elliptically polarized ω pulse (with an ellipticity of about 1/7 in terms of optical intensity). (a) and (c) are the experimental results with left- and right-handed circularly polarized ω pulses, respectively, and (b) and (d) are the corresponding simulation results.

As has been simulated above, the case of both circularly-polarized ω and 2ω beams leads to the result that when the relative phase between the pulses changes the polarization of the emitted THz beam rotates while the intensity or the electric field of the THz wave is kept unchanged. This particular situation is very important for some applications, such as THz modulation devices. Fig. 6(a) shows the experimental results when both ω and 2ω beams are right-handed elliptically polarized with their ellipticities both higher than 0.8 (which means that both the ω and 2ω beams are close to circular polarization). Fig. 6(b) shows the simulation result for comparison.

Both the experimental and theoretical results indicate that the emitted THz waves are linearly polarized when the ω and 2ω beams are both linearly polarized. They also indicate that when the ω and 2ω beams are circularly or elliptically polarized the polarization of THz waves can be slightly elliptical but is very close to linear polarization with ellipticity less than 1/20 in terms of THz intensity. This result is basically consistent with the recent result reported by S.L. Chin's group [21]. Furthermore, our results indicate that, as the relative optical phase changes by 2π, the polarization direction of the THz wave rotates one complete circle accordingly when at least one optical pulse is elliptically polarized. With the above theoretical and experimental analysis, we have a much clearer picture about the THz generation during the gas ionization process. Our demonstration further verified the validity of the quantum mechanical model in our previous work [16].

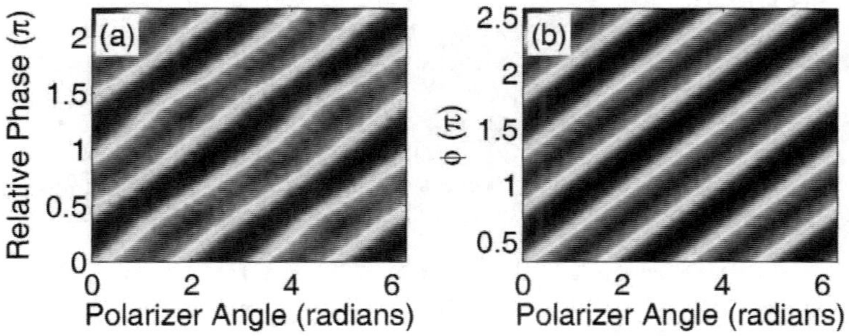

Figure 7. THz intensity versus THz polarizer angle and relative phase between ω and 2ω pulses. (a) the experimental result when both ω and 2ω beams are right-handed elliptically polarized with their ellipticities both higher than 0.8 (which means that both of the ω and 2ω beams are close to circular polarization), and (b) the simulation result with ideally circularly polarized optical beams. Both indicate the THz polarization can be rotated while keeping the amplitude of the THz wave constant.

3. Summary

In conclusion, we presented both theoretical and experimental investigations of the polarization characteristics of the THz waves generated from the gas plasma excited by dual-color optical pulses (ω and 2ω). We found that the polarization of the THz waves can be coherently controlled by changing the phase between the ω and 2ω pulses when at least one of the optical pulses is elliptically polarized. In particular, when both ω and 2ω beams are circularly polarized, the THz polarization angle can be rotated arbitrarily by simply changing the phase between the two optical pulses, with the THz amplitude kept constant. Similar results have been observed independently by two research groups [22, 23]. The demonstration adds a new feature into this novel THz gas source and should enable fast THz wave modulation and coherent control of nonlinear responses excited by intense THz waves with controllable polarization.

Acknowledgement

This work was supported in part by the U.S. Office of Naval Research (ONR), the Defense Threat Reduction Agency (DTRA), National Science Foundation (NSF), and the U.S. Department of Homeland Security through the DHS-ALERT Center under Award No. 2008-ST-061-ED0001. The views and conclusions contained in this document are those of the authors and should not be interpreted as necessarily representing the official policies, either expressed or implied, of the U.S. Department of Homeland Security.

References

[1] F. Théberge, N. Aközbek, W. Liu, A. Becker, and S. L. Chin, "Tunable ultrashort laser pulses generated through filamentation in gases", Phys. Rev. Lett. 97, 023904 (2006).
[2] S. Meyer, H. Eichmann, T. Menzel, S. Nolte, and B. Wellegehausen, "Phase-matched high-order difference-frequency mixing in plasmas", Phys. Rev. Lett. 76, 3336 (1996).
[3] T. Bartel, P. Gaal, K. Reimann, M. Woerner, and T. Elsaesser, "Generation of single-cycle THz transients with high electric-field amplitudes", Opt. Lett. 30, 2805-2807 (2005).

[4] H. Hamster, A. Sullivan, S. Gordon, W. White, and R. W. Falcone, "Subpicosecond, electromagnetic pulse from intense laser-plasma interaction", Phys. Rev. Lett. 71, 2725 (1993).
[5] S. Tzortzakis, G. Méchain, G. Patalano, Y.-B. André, B. Prade, M. Franco, A. Mysyrowicz, J.-M. Munier, M. Gheudin, G. Beaudin, and P. Encrenaz, "Coherent subterahertz radiation from femtosecond infrared filaments in air", Opt. Lett. 27, 1944-1946 (2002).
[6] D. J. Cook, and R. M. Hochstrasser, "Intense terahertz pulses by four-wave rectification in air", Opt. Lett. 25, 1210-1212 (2000).
[7] M. Kress, T. Töffler, S. Eden, M. Thomson, and H. G. Roskos, "Terahertz-pulse generation by photoionization of air with laser pulses composed of both fundamental and second-harmonic waves", Opt. Lett. 29, 1120-1122 (2004).
[8] T. Löffler, M. Kress, M. Thomson, and H. G. Roskos, "Efficient Terahertz Pulse Generation in Laser-Induced Gas Plasmas", Acta Phys. Pol. A 107, 99 (2005).
[9]Y. Shen, T.Watanabe, D. A. Arena, C.-C. Kao, J. B. Murphy, T. Y. Tsang, X. J. Wang, and G. L. Carr, "Nonlinear cross-phase modulation with intense single-cycle terahertz pulses," Phys. Rev. Lett. 99, 043901 (2007).
[10] J. Dai, X. Xie, and X.-C. Zhang, "Detection of broadband terahertz waves with a laser-induced plasma in gases",Phys. Rev. Lett. 97, 103903 (2006).
[11] N. Karpowicz et al., "Coherent heterodyne time-domain spectrometry covering the entire "Terahertz gap"", Appl. Phys. Lett. 92, 011131 (2008).
[12] I. Ho, X. Guo, X.-C. Zhang, "Design and performance of reflective terahertz air-biased-coherent-detection for time-domain spectroscopy", Opt. Express 18, 2872-2883 (2010).
[13] H. Wen, M. Wiczer, and A. M. Lindenberg, "Ultrafast electron cascades in semiconductors driven by intense femtosecond terahertz pulses", Phys. Rev. B 78, 125203 (2008).
[14] P. Gaal et al., "Nonlinear terahertz response of n-type GaAs", Phys. Rev. Lett. 96, 187402 (2006).
[15] X. Xie, J. Dai, and X.-C. Zhang, "Coherent control of THz wave generation in ambient air",Phys. Rev. Lett. 96, 075005 (2006).
[16] N. Karpowicz and X.-C. Zhang, "Coherent terahertz echo of tunnel ionization in gases", Phys. Rev. Lett. 102, 093001 (2009).
[17] A.A. Silaev and N.V. Vvedenskii, "Residual-current excitation in plasmas produced by few-cycle laser pulses" Phys. Rev. Lett. 102, 115005 (2009).
[18] H.G. Muller, "An Efficient Propagation Scheme for the Time-Dependent Schrödinger Equation in the Velocity Gauge", Laser Phys. 9, 138 (1999).
[19] J. Dai and X.-C. Zhang, "Terahertz wave generation from gas plasma using a phase compensator with attosecond phase-control accuracy", Appl. Phys. Lett. 94, 021117 (2009).
[20] Yu Oishi et al., "Generation of extreme ultraviolet continuum radiation driven by a sub-10-fs two-color field", Optics Express 14, 7230 (2006).
[21] Y. Zhang et al., "Non-radially polarized THz pulse emitted from femtosecond laser filament in air", Opt. Express 16, 15483 (2008).
[22] J. Dai, N. Karpowicz, and X.-C. Zhang, "Coherent polarization control of terahertz waves generated from two-color laser-induced gas plasma", Phys. Rev. Lett. 103, 023001 (2009).
[23] H. Wen and A. Lindenberg, "Coherent Terahertz Polarization Control through Manipulation of Electron Trajectories", Phys. Rev. Lett. 103, 023902 (2009).

Electron correlation dynamics of strong-field double ionization of atoms below recollision threshold

Yunquan Liu[1,2], Difa Ye[3,4], Jie Liu [3,4], A Rudenko[2,5], S Tschuch[2], M Dürr[2], M Siegel[6], U Morgner[6], Qihuang Gong[1], R Moshammer[2] and J Ullrich[2]

[1]Department of Physics and State Key Laboratory for Mesoscopic Physics, Peking University, Beijing 100871, P. R. China
[2]Max-Planck-Institut für Kernphysik, D-69117 Heidelberg, Germany
[3]Center for Applied Physics and Technology, Peking University, 100084 Beijing, P. R. China
[4]Institute of Applied Physics and Computational Mathematics, 100088 Beijing, P. R. China
[5]Max-Planck Advanced Study Group at CFEL, 22607 Hamburg, Germany
[6]Leibniz Universität Hannover, Welfengarten 1, D-30167 Hannover, Germany

E-mail: yunquan.liu@pku.edu.cn

Abstract. In recent combined experimental and theoretical study we have explored non-sequential double ionization of neon and argon atoms in the infrared light field (800nm) below the recollision threshold. We find that the two-electron correlation dynamics depends on atomic structure– "side-by-side emission" (correlation) for Ne and "back-to-back emission" (anticorrelation) for argon atoms. This can be explained theoretically within our three dimensional classical model calculation including tunnelling effect. The multiple recollisions as well as recollision-induced-excitation-tunnelling (RIET) effect dominate the anticorrelation of argon, whereas the laser-assisted instantaneous recollision dominates the correlation of neon.

1. Introduction

Within the past three decades extensive works on the interaction of intense laser fields with atoms and molecules have resulted in a profound understanding of various strong-field phenomena, e.g., above-threshold ionization [1], high-harmonic generation [2] and nonsequential double ionization (NSDI) [3,4]. Few-electron correlation is most important in NSDI. Until the present day, it still resists any comprehensive modelling. NSDI not only manifested its importance in the enhancement of double charged ions at moderate laser intensities [4] and but, represents an outstanding subtleties of electron-electron correlations in strong fields. Strong-field ionization mechanisms of single ionization are often classified by the Keldysh [5] parameter $=(I_p/2U_p)^{1/2}$ (I_p: ionization potential; $U_p=E_0^2/\omega^2$:pondermotive potential with E_0 the laser field amplitude and ω its frequency; atomic units (a.u.) are used throughout the paper). For < 1, at moderately large intensities, tunnelling ionization is dominant. Eventually, for $\ll 1$ and very strong fields the potential might be such strongly suppressed that the electrons can just escape "over the barrier". Vice versa, multi-photon processes are expected to dominate at smaller intensities and > 1. NSDI essentially occurs in the tunnelling ionization regime, below the over barrier ionization regime and is usually explained within the quasiclassical re-collision model [6-8].

Here, the first electron is set free near the maximum of laser field, accumulates energy (up to 3.17 U_p) during about half a laser cycle, is thrown back on its parent ion near a zero-crossing of the laser field and, in a inelastical collision, kicks out the second electron. As shown in many experiments and calculations, both emitted electrons prefer to move into the same hemisphere along the laser polarization axis [9-14]. Even though this general scenario is well established, distinct differences do occur for different target species, i.e., significant structure dependence is observed for NSDI. Up to now, this is qualitatively explained via target dependent cross sections for (re-colliding) electron-impact excitation of the ionic core: If this cross section is small, direct re-collision ionization dominates with both electrons being mainly ejected into the same hemisphere. Is it large, as, e.g. for Ar 3s to 3p excitation, re-collision-induced excitation with subsequent field ionization [10] becomes important, with both electrons being equally likely emitted in the same or opposite hemisphere. One should note, however, that this discussion cannot be considered completely settled, since other ionization mechanisms leading to anti-correlation at high intensities, close to the over barrier ionization regime, have been discussed recently [15].

One intriguing though still widely unexplored question did arise early on within above recollision picture. What does happen when the energy of the recolliding electron of up to 3.17 U_p is not sufficient to directly ionize the parent ion or, at lower energies, not even to reach its first excited state? This is a situation, where previous classical considerations do not hold any more and where one enters the fundamentally important non-perturbative multi-photon – multi-electron regime. Still, multiple ionization rates were found to be dramatically enhanced due to electron-electron correlation. At low intensities below the re-collision threshold, where one enters the multi-photon ionization regime, anti-correlation between the electrons from NSDI of argon has been observed recently [16] and new ionization mechanisms have been suggested [17]. In this contribution, we investigate electron correlation in NSDI and its target-species dependence for Ne and Ar at lowest intensities ever addressed, below the recollision threshold, even closing to the double ionization threshold. In kinematically complete measurements we find opposite correlation behaviour in the two-electron momentum distributions along the polarization direction, dominating emission into opposite hemispheres (back-to-back emission) for Ar (so-called anti-correlation) and into the same hemisphere (side-by-side emission) for Ne (so-called correlation), which we partly explain within three-dimensional classical calculations including tunnelling of the second electron. Inspecting the transverse momentum distributions we shed light on the dynamics, investigating Coulomb focusing effects, and find non-classical dynamics for the NSDI of Ne in threshold regime.

Dedicated classical model calculation [18] on the other hand, considering the contributions from multiple re-collisions and, in addition, implementing tunnelling of the second electron [17], clearly revealed the back-to-back emission characteristics observed in the experiment. Whereas in [21], the anti-correlated electrons were found to be mainly produced via multiple collisions in the first or second field maximum after re-collision, the analysis in [17] yields a more subtle picture. Here, the contributions to anti-correlated electron emission due to classical multiple recollisions induced direct-ionization can be distinguished from those due to recollision induced excitation plus tunneling (RIET) of the second electron. Whereas the recollision induced direct ionization contribution to correlated emission decreases with decreasing intensity, the RIET part is expected to increases. For anti-correlated events, instead, both mechanisms contribute when lowering the intensity making anticorrelation to dominate over correlation deep in the multiphoton ionization regime in agreement with experiment. Currently, quantum calculations are widely lacking below the recollision threshold and the dependence on the target structure, expected to be very important for the case of excitation-tunnelling ionization, has not been investigated at all.

2. Experimental setup

The coincident detection of electrons and ions of NSDI in the multiphoton ionization regime is challenging due to the exceedingly low double ionization yields. The major breakthrough enabling such measurements was via the demonstration of a high-power (peak power >16MW) femtosecond laser system approaching microjoule with unprecedented repetition rate (~6 MHz) [19,20]. The laser pulse duration is about 46fs. The laser pulse was tightly focused to about 1μm diameter and a maximum intensity more than $2{\cdot}10^{14}$ W/cm^2 on the target could be achieved. Charged fragments, i.e., ions and electrons, were guided onto two position-sensitive channel plate detectors by weak electric (2V/cm) and magnetic (4.5 G) fields applied along the laser polarization axis. The full momentum vectors of the coincident recoil ions and electrons were calculated from their times of flight (TOF) and hitting positions on the detectors. The gas-jet density of reaction system is about $1{\cdot}10^{12}$/cm^3. The coincidence method, the especially designed reaction microscope, the momentum resolutions achieved for the various fragments as well as the intensity calibration procedure and accuracy are described in detail elsewhere [21, 22]. Under present settings, data collection for each target took several weeks in order to achieve reasonable statistical significance, very much at the limit of what is technically feasible. In the experiments, we recorded a total of 1780 double coincidence events (one ion and one electron) for Ne^{2+} with a ratio of Ne^{2+}/Ne^{+} ~ $1.5{\cdot}10^{-4}$ at an intensity of 1.5×10^{14} W/cm^2 and 1160 counts for Ar at $3{\cdot}10^{13}$ W/cm^2 with Ar^{2+}/Ar^{+} ~ $8{\cdot}10^{-4}$.

3. Experimental results and model calculations

The electron longitudinal momentum distributions from double ionization of argon and neon are illustrated in Figs. 2(a) and (c). Both of them show double-hump structure, but they are resulted from different mechanism. Figures 2b and 2d show the momentum distributions of double charged argon and neon. At higher intensities, the longitudinal momentum spectra of double charged ions show so pronounced double-hump structure [12,13]. The longitudinal momentum distributions of Ar^{2+} and Ne^{2+} show very different feature in the NSDI threshold regime. It is found that the longitudinal momentum distribution of Ar^{2+} shows a sharp cusp-like structure. This is because large amount of the coincident electrons moving in opposite direction. The pronounced "double-hump" structure of the longitudinal momentum distribution at higher laser intensity completely disappears. Thus, the momentum distribution of double charged argon can not be predicted from simple re-collision picture in the multi-photon ionization regime. In contrast to argon, the momentum distribution of double charged neon is much broad because the two ionized electrons still prefer to move into the same hemisphere. However the double hump structure is not pronounced for the longitudinal momentum distribution of Ne^{2+} in the multi-photon ionization regime.

The momentum cut-off is about $\pm(10Up)^{1/2}$ (~ 5Up for energy cut-off) for argon. This is consistent with the observation that the high-energy part of single ionization and the sum of the energies of both electrons from double ionization of argon show similar shape [9]. As to double ionization of neon, there is very little excess energy during the recollsion process in the threshold regime. The electron momentum from double ionization of neon gives a $\pm(2Up)^{1/2}$-cut-off (~2Up for energy cut-off) and is consistent with "momentum constraint" [10].

In order to understand the influence of atomic structure on NSDI below recollision threshold, we have performed three-dimensional semi-classical model calculations, in which the recollision-induced excitation-tunneling (RIET) effect has been taken into account within the Wentzel-Kramers-Brillouin (WKB) approach [17]. Considering the cross sections for (re-colliding) electron-impact excitation of the ionic core, the recollision-induced excitation effect should be reduced more for Ne than that of Ar. Thus, RIET effect is less important in the case of Ne. The energy level of the first excitation state of Ne^{+} and the field-free double ionization threshold are much higher than those of argon. It is thus much more difficult to resonantly excite the second electron during re-collision in a more reasonable quantum-mechanical description. To take this effect approximately into account in our classical model, we have abandoned all double ionization trajectories for Ne if the inner electron is excited to an energy that is lower than the first excited state.

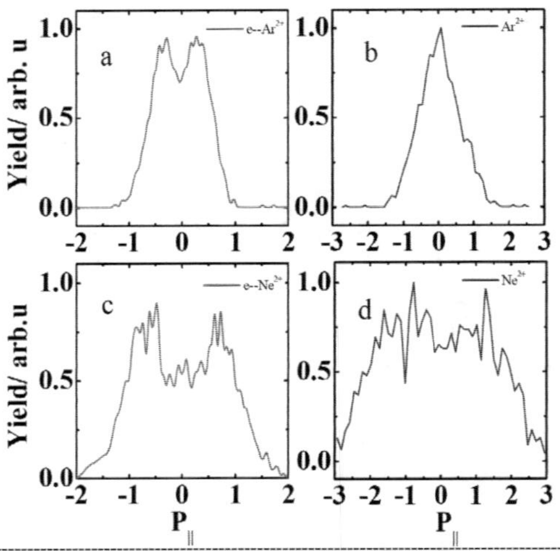

Figure 1._(color online)_ The electron momentum distribution in the laser polarization plane from NSDI of argon (a) (0.3×10^{14} W/cm^2) and neon (c) (1.5×10^{14} W/cm^2); The longitudinal momentum distributions of Ar^{2+} (b) (0.3×10^{14} W/cm2) and Ne^{2+} (d) (1.5×10^{14}W/cm$^{2)}$.

The calculated longitudinal momentum distributions of electrons and double charged ions are shown in Fig. 2. All of concerned results agree with the experimental data very well. The classical ensemble model can present the explanation of the production of significant numbers of double ionization electrons with energy much above $2Up$ for argon. In almost all trajectories leading to high energy, at least one electron remains bound for at least a portion of a laser cycle after recollision. During the interaction the electron which will achieve high energy experiences a sign change of momentum in the laser polarization plane. The anticipated process of nuclear scattering of a free electron at recollision is involved. The another process is that the bounded electron changes sign of momentum after the recollision, escapes over a suppressed barrier, and achieves high energy in the complex interaction. In this case, both electrons prefer to emit into opposite direction. In both processes, there is a change of phase of electron oscillation relative to the laser phase, thus electron achieve high energy beyond $2Up$.

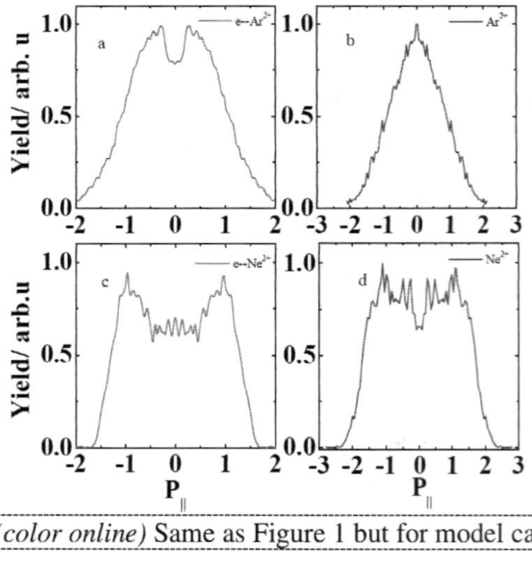

Figure 2. _(color online)_ Same as Figure 1 but for model calculations

The correlation behaviours of Ar and Ne are also quite different in the threshold regime [23]. Even though the absolute single ionization count rate is rather comparable for both targets, the correlation maps as well as $P_{\perp}$-distributions show distinctly different shapes, as shown in Fig. 3. For argon a substantial amount of anti-correlated electrons are observed in line with our previous measurements at slightly higher intensity [5] and the correlated emission is found to dominate in case of neon!

Differences in the dynamics become especially obvious if one inspects the transverse momenta of electrons coincident to double ionization (red curve, right row) with a cusp-like structure for Ar that is substantially broadened in case of Ne. Even taking into account an at least 30 % uncertainty in the absolute determination of the intensity and, thus, thinking we might be much deeper in the multiphoton ionization regime for Ar than for Ne, we still find similarly distinct differences if we compare the Ne spectrum to the one for Ar at $7 \cdot 10^{13}$ W/cm^2 (dashed red line in Fig. 3b).

Inspecting the electron transverse momentum distributions for single ionization and double ionization for both target species, as shown in Figs 3(b) and 3(d) for the experiment as well as in Figs 4(b) and 4(d) for our model calculations we find in general pronounced "cusp-like" structures as observed before. It has been proposed that Coulomb focusing [24] or the Coulomb singularity [25] are responsible for their formation for single ionization in accordance with the present data (blue curves in Figs 3(b), 3(d), 4(b) and 4(d)). Whereas the experimental momentum distribution of electrons coincident with Ar^{2+} ions (red line) also shows a clear cusp-like shape, only slightly broader than for single ionization and in good agreement with theory the respective distribution for Ne is significantly broader than predicted. In general, the cusp-like shape is obtained as a result of the singularity in the electron continuum wavefunction at zero energy in the presence of a Coulomb potential [21]. Thus, we conclude that for Ar double ionization both electrons have very little energy in the final state. This goes very much along with what would be expected within the RIET picture, where the second electron tunnels, being essentially set into the continuum with zero momentum and independently from the first electron, at another maximum in the laser field. Accordingly, our model calculations (Fig. 4 (b)) largely reproduce the experimental data in Fig. 3(b).

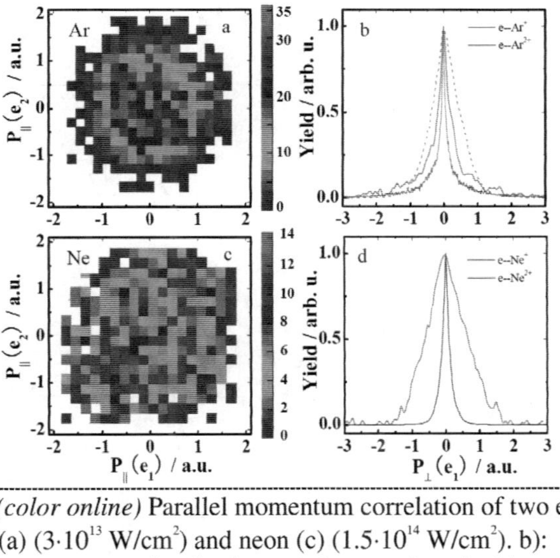

Figure 3.*(color online)* Parallel momentum correlation of two electrons for argon (a) ($3 \cdot 10^{13}$ W/cm^2) and neon (c) ($1.5 \cdot 10^{14}$ W/cm^2). b): Transverse momentum distributions of electrons in coincidence with Ar^{1+} (blue line) and Ar^{2+} at $3 \cdot 10^{13}$ W/cm^2 (solid red line) and at $7 \cdot 10^{13}$ W/cm^2 (dashed solid line) as well as d) for Ne^{1+} and Ne^{2+} at $1.5 \cdot 10^{14}$ W/cm^2.(experiment)

On the other hand electrons emerging coincident to Ne^{2+} production are observed to receive much larger transverse momenta and show a more Gaussian-like shape. This indicates that the recollision mechanism still dominates, where the re-colliding electron could be scattered to a larger angle with it's longitudinal momentum being transferred to the transverse direction. Surprisingly, such a feature is not found in our classical model calculations where the transverse momentum distribution of an electron coincident to the simulated Ne^{2+} ion, excluding trajectories that lead to unphysical low excitation energies, is shown in Fig. 4 (d). Obviously such a model does not capture the full quantum dynamics of the system in all of its aspects correctly, underlining the quest for quantum calculations.

In order to confirm the different correlation behaviour and to possibly even further penetrate into the multiphoton ionization regime, during the experiments, we have continued to decrease the laser intensity to about $1.5\cdot10^{13}$ W/cm^2 for Ar and $8\cdot10^{13}$ W/cm^2 for Ne. Here, however, we could not find any indication of double ionization events while running the measurements for several weeks for each species. This means that double ionization rates decrease by at least three orders of magnitude between $1.5\cdot10^{14}(3\cdot10^{13})$ W/cm^2 and $8\cdot10^{13}(1.5\cdot10^{13})$ W/cm^2 for neon (argon). The anticorrelation for neon was not able to be observed because of its extremely narrow transition region from correlation to anticorrelation, as discussed in [23].

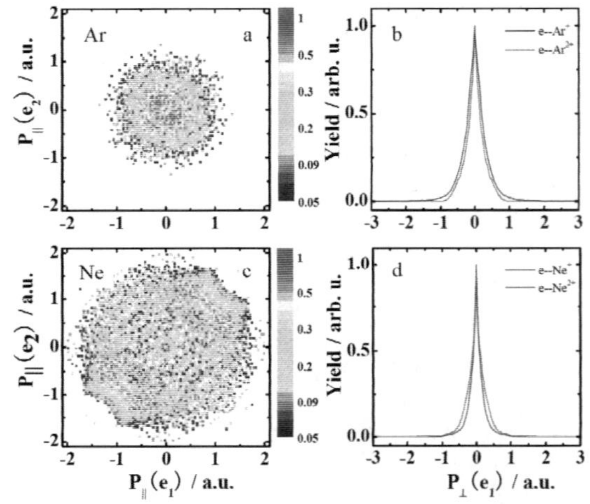

Figure 4. *(color online)* Same as Figure 3 but for model calculations

4. Conclusion

In this contribution, we present kinematically complete measurements and model calculations for NSDI of Ar and Ne in infrared laser fields below recollision threshold, at the lowest intensities ever investigated. Whereas for Ar anti-correlation (back-to-back emission) in the longitudinal two-electron momentum spectra is observed in agreement with previous measurements at slightly higher intensity, correlation (side-by-side emission) dominates for the Ne target similarly close to the threshold. Further lowering the intensity led to the complete absence of any double ionization event over weeks of data taking time. We can reproduce that salient feature in the correlation plot with 3D classical calculations including tunnelling for the second electron if, for the Ne target, all trajectories are rejected with re-collision energies below the first excitation level of Ne^{+}. This provides strong evidence that the target structure, i.e. the low-lying excited levels in Ar^{+} along with the large electron-impact excitation cross section are the reason for this observation.

The electrons from of Ne^{2+} can obtain larger transverse momentum and show more Gaussian shape. Under condition that the recollision mechanism dominates, the re-colliding electron could be in-elastically scattered to larger angle and obtain larger momentum. The transverse electron momentum of from Ar^{2+} shows cusp-like structure. The effect of Coulomb potential of the parent ion is responsible to this observation. The Coulomb interaction between the electron and the ion reduces the transverse spread of the electronic wave packet and shows strong focusing effect for argon. The Coulomb focusing effect leads to a small deflection of large impact parameter electrons whenever they approach the parent ion and enhance the probability for the electron to make a low impact parameter inelastic collision with the ionic core after multiple recollisions. This could be the intrinsic mechanism of sub-threshold re-collision process.

We thank the support by the Scientific Research Foundation for the Returned Overseas Chinese Scholars and the program for New Century Excellent Talents (NCET) in University (Ministry of Education), the National Fundamental Research Program of China (2006CB806007), the National Science Foundation of China (61078025, 10821062 and 10634020) and the Key lab funding of Shanxi Province (09JS077).

5. References:

[1] Agostini P, Fabre F, Mainfray G, *et al.* 1979 *Phys. Rev. Lett.* **42** 1127
[2] Huillier AL and Balcou Ph 1993 *Phys Rev Lett.* **70** 774
[3] Fittinghoff D N, Bolton P R, Chang B, *et al.* 1992 *Phys. Rev. Lett.* **69** 2642
[4] Walker B, Sheehy B, DiMauro L F, *et al.* 1994 *Phys. Rev. Lett.* **73** 1227
[5] Keldysh V 1965 *Sov. Phys JETP* **20** 1307
[6] Kuchiev M Yu 1987 *JETP Lett* **45** 404
[7] Schafer K, Yang B, DiMauro L F, *et al.* 1993 *Phys. Rev. Lett.* **70** 1599
[8] Corkum P 1993 *Phys. Rev. Lett.* **71** 1994
[9] Weber Th, Giessen H, Weckenbrock M, *et al.* 2000 *Nature* **405** 658
[10] Feuerstein B, Moshammer R, Fischer D, *et al.* 2001 *Phys. Rev. Lett.* **87** 043003.
[11]Weckenbrock M, Zeidler D, Staudte A, *et al.* 2004 *Phys. Rev. Lett.* **92** 213002.
[12] Staudte A, Ruiz C, Schoffler M, *et al.* 2007 *Phys. Rev. Lett.* **99** 263002.
[13] Rudenko A, Jesus V L B de, Ergler Th, *et al.* 2007 *Phys. Rev. Lett.* **99** 263003
[14] Eremina E, Liu X, Rottke H, *et al.* 2003 *J. Phys. B* **36** 3269
[15] Emmanouilidou A and Staudte A arXiv:0909.3409v1
[16] Liu Y, Tschuch S, Rudenko A, *et al.* 2008 *Phys. Rev. Lett.* **101** 053001.
[17] Ye D F and Liu J 2010 *Phys Rev A* **81** 043402
[18] Haan S L, Smith Z S, Shomsky K N, *et al. J Phys B*, 2008, **41** 211002.
[19] Dewald S, Lang T, Schroter C D, *et al.* 2006 *Opt. Lett.* **31** 2072.
[20 Liu Y, Tshuch S, Durr M, *et al.* 2007 Opt. Expr. **15** 18103.
[21] Ullrich J, Moshammer R, Dorn A, *et al.* 2003 *Rep Rog Phys.* **66** 1463.
[22] Jesus V L B de, Feuerstein B, Zrost K, *et al.* 2004 *J Phys B* **37** L161.
[23] Liu Y, Ye D F, Liu J, *et al.* 2010 *Phys. Rev. Lett.* **104** 173002.
[24] **Comtois D, Zeidler D, Pepin H, *et al.* 2005** *J Phys B* **38** 1923.
[25] Rudenko A, Zrost K, Ergler Th, *et al.* 2005 *J Phys B* **38** L191.

Studying circulation times of liver cancer cells by in vivo flow cytometry

G Liu[1], Y Li[1], Z Fan[1], J Guo[1], X Tan[1] and X Wei[1]
[1] Institutes of Biomedical Sciences, Fudan University, 138 Yi Xue Yuan Road, Shanghai, China 200032
E-mail: xwei@fudan.edu.cn

Abstract. Hepatocellular carcinoma (HCC) may metastasize to lung, bones, kidney, and many other organs. The survival rate is almost zero for metastatic HCC patients. Molecular mechanisms of HCC metastasis need to be understood better and new therapies must be developed. A recently developed "in vivo flow cytometer" combined with real-time confocal fluorescence imaging are used to assess spreading and the circulation kinetics of liver tumor cells. The in vivo flow cytometer has the capability to detect and quantify continuously the number and flow characteristics of fluorescently labeled cells in vivo in real time without extracting blood sample. We have measured the depletion kinetics of two related human HCC cell lines, high-metastatic HCCLM3 cells and low-metastatic HepG2 cells, which were from the same origin and obtained by repetitive screenings in mice. >60% HCCLM3 cells are depleted within the first hour. Interestingly, the low-metastatic HepG2 cells possess noticeably slower depletion kinetics. In comparison, <40% HepG2 cells are depleted within the first hour. The differences in depletion kinetics might provide insights into early metastasis processes.

1. Introduction

Liver cancer is one of the most common malignancies in the world, with approximately 1,000,000 cases reported every year. Hepatocellular carcinoma may metastasize to lung, bones, kidney, and many other organs [1-5]. Surgical resection, liver transplantation, chemotherapy and radiation therapy are the foundation of current HCC therapies. However the outcomes are poor: the survival rate is almost zero for metastatic HCC patients.

Metastasis is a very complicated process that still has yet to be completely understood. In metastasis, the cancer cells that travel through the body are capable of establishing new tumor sites in locations remote from the site of the original disease. To metastasize, a cancer cell must break away from its tumor, invade either the circulatory or lymph system, which will carry it to a new location, and establish itself in the new site [6]. Molecular mechanisms of HCC metastasis need to be understood better and new therapies must be developed to selectively target to unique characteristics of HCC cell growth and metastasis.

Conventionally, researchers extract samples of tumor tissue or blood from patients or animal models to study the mechanisms of HCC metastasis with biochemical analysis, along with extremely careful protocols to reduce the changes induced by sample extraction [7]. In order to reveal the exact mechanisms in certain circumstances, there is a great demand to monitor these molecular events in vivo, where modern optical techniques became more and more powerful.

Existing optical techniques including confocal microscopy, multi-photon microscopy [8], optical coherence tomography (OCT) [9], and so on are mainly imaging methods that give insight into real-time biological reactions or phenomenon. However, quantitative evaluations of such biological processes are crucial to discover mechanisms, for example cancer metastasis.

Flow cytometer was developed to analyze quantitative characteristics about cells in vitro, by flowing single cells through fine tube [10]. Cells are suspended in a stream of fluid and one by one pass the focus of a laser light beam or several laser light beams. On the excitation, fluorescent signals and other scattering characteristics will be immediately collected by the various detectors and later processed to analyze the numbers of different cells groups. Inspired from this technique and principles of confocal microscopy, scientists invented "in vivo flow cytometer" and applied it to various biological studies [11-13]. The blood flow within micro-vessel in the ear (or other accessible part) of small animals is used similarly as the stream of fluid in flow cytometer. Labelled cells passing through linearly focused laser beam are excited by certain laser and then emit specific fluorescence to be detected. Up to date, there are a number types of "in vivo flow cytometers" developed, i.e. two-photon [14-16], multi-color [11, 12, 17], photoacoustic [17-20] in vivo flow cytometer, for the biological studies varying from cancer cells to immune cells for quantitatively analysing kinetics of cells in either the blood or lymphatic circulation system.

In order to study characteristics of circulating HCC cells, here we apply the "in vivo flow cytometry" combined with optical imaging to study if there is any relationship between kinetics of circulating tumor cells (CTCs) and its metastatic potential.

2. Materials and methods

2.1. In vivo flow cytometry

The in vivo flow cytometer has the capability to detect and quantify continuously the number and flow characteristics of fluorescently labeled cells in vivo [11-13, 21-24]. The in vivo flow cytometer allows researchers to acquire cytometric information from the circulation in live animals without extracting blood sample (Figure 1). Previous researchers demonstrate that using this technique, the number of fluorescently-labeled circulating cells can be quantified in a real-time and reproducible manner in live animal. We set up our in vivo flow cytometer to detect fluorescence signal from a given circulating tumor cell (CTC) population in a confocal geometry based on previous groups' experiences [11, 13].

Briefly, fluorescence signal from a given circulating cell population is recorded as the cells pass through the slit of light. Confocal detection of the excited fluorescence enables continuous monitoring of labeled cells in the upper layers of scattering tissue, such as the skin of a mouse ear. The size of the slit at the focal plane of the sample is approximately 5x72 μm. The depth of focus (i.e. the full width at half maximum of the light slit onto the sample in the axial direction) is approximately 50 μm, a value chosen to match the vessels of interest. The sample is positioned so that the long dimension of the slit traverses the width of the blood vessel; thus, fluorescence is excited as the labeled cells in circulation pass through the slit. Fluorescence is detected with a photomultiplier tube placed directly behind the mechanical slit, sampled at a rate of 5 kHz with a data acquisition card, and displayed and stored on a computer. The device has been used to characterize the in vivo kinetics of red and white blood cells circulating in the mouse ear vasculature, as well as in the circulating liver cancer cells.

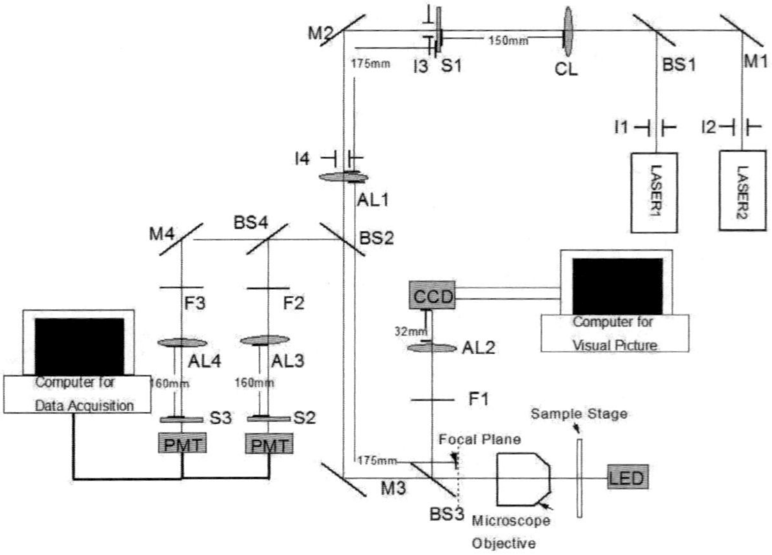

Figure 1. Schematic of the two-color in vivo flow cytometer experimental setup. Laser light (488nm or 635nm) is focused into a slit by a cylindrical lens (CL) and imaged across the selected blood vessel with a microscope objective lens (40X, 0.6 NA). The fluorescence is collected by the same microscope objective, directed through the dichroic beam splitter BS3 (reflection 25%, transmission 75%, Edmund Optics), reflected by a mirror, a second splitter BS2 (transmission • 90% for 488nm and 635nm; reflection• 90% for 499-622nm, and 652-755nm, Semrock) and a third dichroic beam splitter BS4 (edge wavelength 516nm, reflection band>90% for 490-510nm, transmission band>90% for 520-700nm, Semrock), and imaged onto a 200μm x 3000μm mechanical slit, which is confocal with the excitation slit. F1-3: bandpass filter (F1: 509-552nm; F2: 500-520nm; F3: 640-690nm; Semrock). BS1: beam splitter, edge wavelength 505nm, reflection band 513-725nm, transmission band 446-500nm, Semrock). AL1-3: achromatic lenses.

2.2. Animal experiments

In experiments, the BALB/c nude mice were anesthetized and injected intravenously with DiD-fluorescently-labeled HCCLM3 human liver tumor cells. The HCCLM3 is a hepatocellular carcinoma (HCC) cell line with high metastatic potential and comparatively easier to metastasize

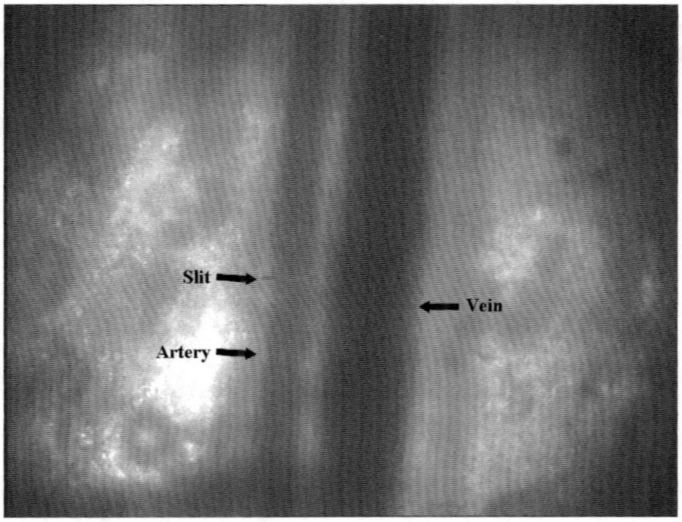

Figure 2. The mouse is anesthetized and placed on a heated stage. An artery is chosen in the ear for the measurement. The fluorescence signal from DiD-labeled circulating cells is recorded when the cells pass through a slit of light focused across the artery. DiD is a fluorescent dye to label the membrane lipid (excitation peak: 640nm; emission peak: 660-670nm; Molecular Probes, USA).

to the lung in BALB/c nude mice than other cell lines [25]. Studying the depletion kinetics of HCCLM3 cells in the circulation could help understand the early metastatic process. After the injection, mice were then placed on a heated stage and an artery in the ear was chosen for obtaining measurements. Fluorescence signal was excited as the labeled circulating cells passed through a laser slit focused across the blood vessel (Figure 2). Detecting the excited fluorescence confocally enables researchers to monitor labeled cells in the animal circulation system continuously. Signal was recorded at a rapid rate (5 Khz) to ensure the measurement of fast-flowing cells (Figure 3). The number of fluorescent peaks, along with the height and full width at half maximum of each peak, was determined using algorithms developed in-house. The device has been used to characterize the in vivo kinetics of red and white blood cells circulating in the mouse ear vasculature as well as the circulating prostate cancer cells [11-13, 26].

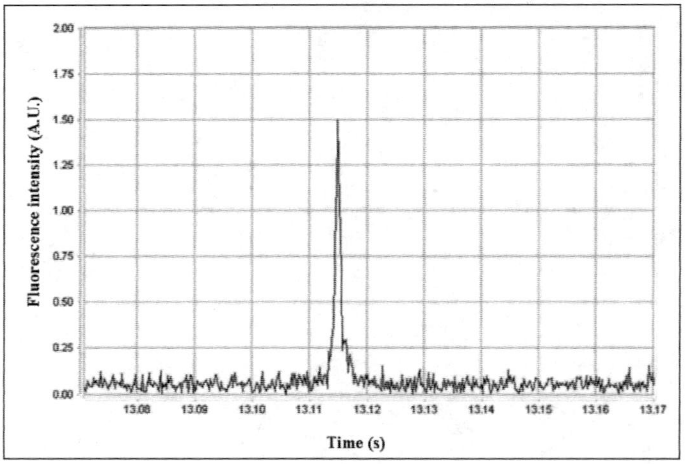

Figure 3. A trace of labeled circulating HCCLM3 hepatocellular carcinoma cell measured by the in vivo flow cytometer, after intravenously introduced into a mouse. The peak within the trace indicates that a DiD-labeled HCCLM3 cell passes through the slit of light and thus gives a burst of fluorescence.

The sample isolation and preparation steps here are bypassed altogether in contrast to the measurement of a typical conventional (in-vitro) flow cytometer. Multiple measurements in the in-vivo-flow-cytometer are carried out noninvasively. We have measured the depletion kinetics of two related human HCC cell lines, high-metastatic HCCLM3 cells and low-metastatic HepG2 cells, which were from the same HCC patient and obtained by repetitive screenings in mice [25].

3. Results

The depletion kinetics of circulating HCCLM3 cells in BALB/c nude mice during the first 12 hours following injection of the fluorescently labeled cells illustrates the depletion process (Figure 4). >60% cells are depleted within the first hour. After the initial depletion, there is a re-appearance in the number of circulating cells, quickly followed by a second depletion. This phenomenon was also observed in previous study on prostate cancer cells [26]. By 12 hours, ~94% HCCLM3 cells are depleted from the circulation. Interestingly, the low-metastatic HepG2 cells possess noticeably slower depletion kinetics. In comparison, <40% HepG2 cells are depleted within the first hour. By 12 hours, ~55% HepG2 cells are depleted from the circulation. When present in circulation for extensively long time, cancer cell might undergo cell death due to lack of survival signal from cell adhesion and the harsh environment imparted by the sheer stress [27, 28]. Therefore, the differences in depletion kinetics might provide insights into early metastasis processes.

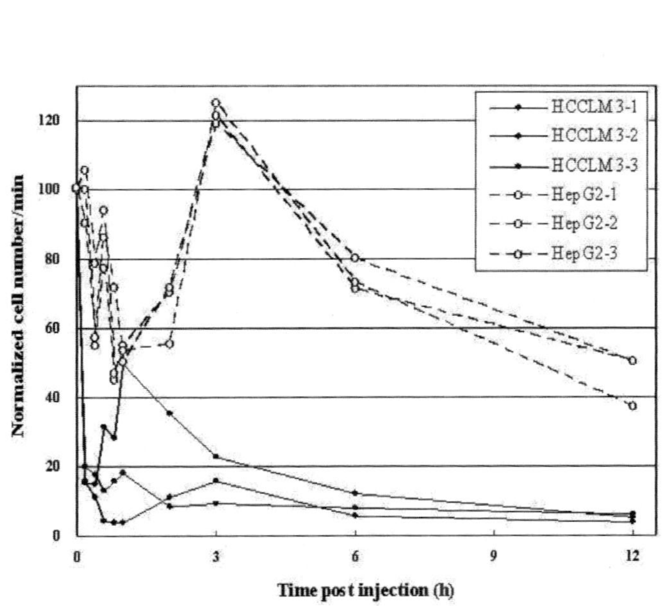

Figure 4. Depletion kinetics of circulating high-metastatic HCCLM3 cells and low-metastatic HepG2 cells in BALB/c nude mice. The normalized numbers of circulating cells per min are shown for 12 hours following injection of the DiD-fluorescently-labeled cells to illustrate the depletion process. >60% HCCLM3 cells are depleted within the first hour. By 12 hours, ~94% HCCLM3 cells are depleted from the circulation. In comparison, the low-metastatic HepG2 cells possess noticeably slower depletion kinetics. <40% HepG2 cells are depleted within the first hour. By 12 hours, ~55% HepG2 cells are depleted from the circulation.

4. Discussions and conclusions

The in-vivo-flow-cytometry is particularly useful for studying small animal models of tumors. For example, it is often difficult to obtain sufficient blood volume from a single mouse, whose total blood volume is only ~2 ml. This is particularly true when multiple blood samples are needed. With this in vivo technique, multiple measurements can be performed on a single animal over time, minimizing the number of animals and also eliminating variation between individuals.

Detection of green fluorescence signal on circulating tumor cells could be technically more difficult. It is particularly true for detection through blood because of the strong absorption below ~590 nm [15, 29]. Both excitation and emission will be attenuated respectively on the way in and on the way out. It might be a potential problem for in vivo flow cytometer to pick up the green fluorescence signal on circulating cells, depending on the green fluorescence level, although we have detected green fluorescence signal from different cell population in circulation (unpublished results). One method to enhance the sensitivity is to utilize the multi-photon detection, although with significantly higher cost. Low et al. developed a multiphoton intravital flow cytometer to quantifity rare circulating tumor cells in vivo and achieved moderately higher sensitivity than confocal detection used in our current in-vivo-flow-cytometer [15, 16, 23, 29-31]. Another method is to use the photothermal/photoacoustic detection, which also improves the detection depth. Tuchin et al. developed a photothermal image flow cytometer to detect target cells in blood and lymph flow in vivo

[20, 32, 33]. They used nanoparticles to enrich the rare cells and achieved significantly higher sensitivity [18, 19, 34].

The study here focuses on measuring circulating liver tumor cells. However the methods developed here shall be applicable to various potential biomarkers in circulation for detecting cancer, characterizing pathologic malignant tumors, assessing disease prognosis, and for predicting and measuring response to treatments, thus providing invaluable information about detecting and treating cancer. Furthermore, the developed methods here will be useful to monitor the circulating cells in general.

5. Acknowledgement

We thank Charles P. Lin for collaboration. We also thank Guiying Wang, Zenghui Zhou, Li Wang for helpful discussions. Guangda Liu and Yan Li contribute equally to this paper. This work is partially supported by the National Basic Research Program of China (973 Program, no. 2011CB910400), China National Key Projects for Infectious Disease (2008ZX10002-021) and partially supported by grant 30770524 and 20975027 of the Chinese National Science Foundation, grant 109056 of the Chinese Ministry of Education, "Shuguang Scholarship" from the Education Commission of Shanghai Municipality and Program for New Century Excellent Talents in University Award (NCET-08-0131).

6. References

[1] Massoud T F and Gambhir S S 2003 *Genes Dev* **17** 545-80

[2] Nery S, Wichterle H and Fishell G 2001 *Development* **128** 527-40

[3] Ntziachristos V, Schellenberger E A, Ripoll J, Yessayan D, Graves E, Bogdanov A J, Josephson L and Weissleder R 2004 *Proc Natl Acad Sci U S A* **101** 12294-9

[4] Osella-Abate S, Zaccagna A, Savoia P, Quaglino P, Salomone B and Bernengo M G 2001 *J Am Acad Dermatol* **44** 40-7

[5] Reed J C 1999 *J Clin Oncol* **17** 2941-53

[6] Bedner E, Halicka H D, Cheng W, Salomon T, Deptala A, Gorczyca W, Melamed M R and Darzynkiewicz Z 1999 *Cytometry* **36** 77-82

[7] Bayon L G, Izquierdo M A, Sirovich I, van Rooijen N, Beelen R H and Meijer S 1996 *Hepatology* **23** 1224-31

[8] Sipkins D A, Wei X, Wu J W, Runnels J M, Cote D, Means T K, Luster A D, Scadden D T and Lin C P 2005 *Nature* **435** 969-73

[9] Leung C K, Ye C, Weinreb R N, Cheung C Y, Qiu Q, Liu S, Xu G and Lam D S 2010 *Ophthalmology* **117** 267-74

[10] Maecker H T and Trotter J 2006 *Cytometry A* **69** 1037-42

[11] Boutrus S, Greiner C, Hwu D, Chan M, Kuperwasser C, Lin C P and Georgakoudi I 2007 *J Biomed Opt* **12** 20507

[12] Novak J and Puoris'Haag M 2007 *Opt Lett* **32** 2993-5

[13] Novak J, Georgakoudi I, Wei X, Prossin A and Lin C P 2004 *Opt Lett* **29** 77-9

[14] Tkaczyk E R, Zhong C F, Ye J Y, Myc A, Thomas T, Cao Z, Duran-Struuck R, Luker K E, Luker G D, Norris T B and Baker J R 2008 *Opt Commun* **281** 888-94

[15] Tkaczyk E R, Zhong C F, Ye J Y, Katnik S, Myc A, Thomas T, Luker K E, Luker G D, Baker J R and Norris T B 2007 *Proc SPIE* **6631** T6310

[16] Zhong C F, Ye J Y, Myc A and Thomas T 2005 *Proc SPIE* **5700** 78-89

[17] Galanzha E I, Shashkov E V, Spring P M, Suen J Y and Zharov V P 2009 *Cancer Res* **69** 7926-34

[18] Galanzha E I, Shashkov E V, Kelly T, Kim J W, Yang L and Zharov V P 2009 *Nat Nanotechnol* **4** 855-60

[19] Galanzha E I, Shashkov E V, Tuchin V V and Zharov V P 2008 *Cytometry A* **73** 884-94

[20] Zharov V P, Galanzha E I, Shashkov E V, Khlebtsov N G and Tuchin V V 2006 *Opt Lett* **31** 3623-5

[21] Sipkins D A, Wei X, Wu J W, Runnels J M, Cote D, Means T K, Luster A D, Scadden D T and Lin C P 2005 *Nature* **435** 969-73

[22] He W, Wang H, Hartmann L C, Cheng J X and Low P S 2007 *Proc Natl Acad Sci U S A* **104** 11760-5

[23] Zhong C F, Tkaczyk E R, Thomas T, Ye J Y, Myc A, Bielinska A U, Cao Z, Majoros I, Keszler B, Baker J R and Norris T B 2008 *J Biomed Opt* **13** 34008

[24] Fan Z, Spencer J A, Lu Y, Pitsillides C M, Singh G, Kim P, Yun S H, Toxavidis V, Strom T B, Lin C P and Koulmanda M 2010 *Nat Med* **16** 718-22

[25] Li Y, Tang Y, Ye L, Liu B, Liu K, Chen J and Xue Q 2003 *J Cancer Res Clin Oncol* **129** 43-51

[26] Georgakoudi I, Solban N, Novak J, Rice W L, Wei X, Hasan T and Lin C P 2004 *Cancer Res* **64** 5044-7

[27] Gaiano N, Nye J S and Fishell G 2000 *Neuron* **26** 395-404

[28] Wei X, Sipkins D A, Pitsillides C M, Novak J, Georgakoudi I and Lin C P 2005 *Mol Imaging* **4** 415-6

[29] Tkaczyk E R, Tkaczyk A H, Katnik S, Ye J Y, Luker K E, Luker G D, Myc A, Baker J J and Norris T B 2008 *J Biomed Opt* **13** 41319

[30] Tkaczyk E R, Zhong C F, Ye J Y, Myc A, Thomas T, Cao Z, Duran-Struuck R, Luker K E, Luker G D, Norris T B and Baker J R 2008 *Opt Commun* **281** 888-94

[31] Chang Y C, Ye J Y, Thomas T P, Cao Z, Kotlyar A, Tkaczyk E R, Baker J J and Norris T B 2010 *J Biomed Opt* **15** 47004

[32] Zharov V P, Galanzha E I and Tuchin V V 2005 *Opt Lett* **30** 628-30

[33] Zharov V P, Galanzha E I and Tuchin V V 2006 *J Cell Biochem* **97** 916-32

[34] Galanzha E I, Kokoska M S, Shashkov E V, Kim J W, Tuchin V V and Zharov V P 2009 *J Biophotonics* **2** 528-39

Journal of Physics: Conference Series **276** (2011) 012006 doi:10.1088/1742-6596/276/1/012006

Ultrafast broadband laser spectroscopy reveals energy and charge transfer in novel donor-acceptor triads for photovoltaic applications

T. Roland, G. Hernandez Ramirez, J. Léonard, S. Méry, and S. Haacke

Institut de Physique et Chimie des Matériaux de Strasbourg, Strasbourg University – CNRS UMR 7504, F-67034 Strasbourg, France

Corresponding author: stefan.haacke@ipcms.u-strasbg.fr

Abstract. Triggered by the quest for new organic materials and micro-structures for photovoltaic applications, a novel class of donor-acceptor-donor (DAD) triads extended with siloxane chains has been synthesized in our labs. Because of the siloxane chains, the molecules self-organize into a smectic liquid crystal phase, resulting in a stacking of the DAD cores.We report here a preliminary study of the ultrafast dynamics of energy and charge transfer studied by femtosecond broadband transient absorption experiments on isolated triads in chloroform.

1. Introduction

Solar cells made of organic broadband absorbing materials are coming of age, with the highest power conversion efficiency exceeding 7% [1]. Research is very active both with regard to the development of new electron donor D and acceptor A molecules featuring high absorption cross sections and improved frontier orbital properties [2] and with respect to the mesoscopic organization of D and A for optimized charge separation and collection.

Donor-acceptor dyads or triads exhibit very fast and efficient formation of intramolecular charge transfer (CT) states [3, 4], unlike polymer blends where excitons must be transported to the hetero-interface before the CT is formed. Arranging the DA or DAD molecules in a liquid crystalline phase [5] is promising in view of improved charge mobility and collection efficiencies. In this context, a novel class of donor-acceptor-donor (DAD) molecules, made of thiophene-based donors and a perylenediimide (PID) acceptor, has been synthesized in our labs. Siloxane chain substitution at both ends of the DAD core are found to strongly stabilize the formation of a smectic liquid crystal phase (SmA type) in a wide temperature range, including room temperature. As a result, an intermolecular DAD core stacking is obtained with the formation of D and A sublayers. The present paper reports a first study of the ultrafast photo-induced energy and charge transfer processes in this DAD molecule in solution, performed by femtosecond broadband transient absorption (TA) laser spectroscopy. This approach not only identifies the nature of the photo-chemically generated species, but also measures the kinetics and the efficiencies of their generation and decay [6].

2. Experiment and methods

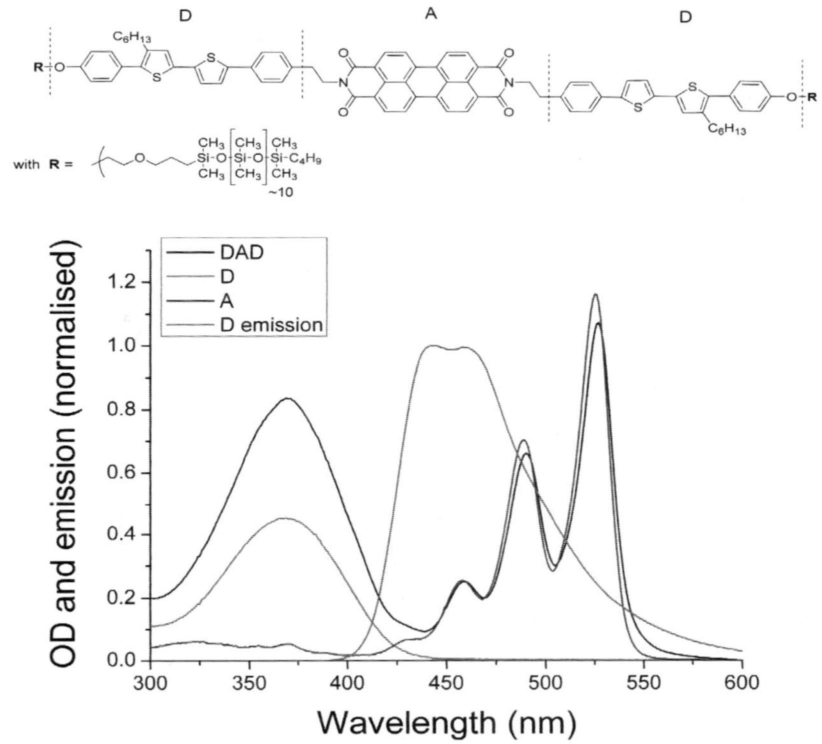

Figure 1: Absorption spectra of the donor (D, red) acceptor (A, blue) molecules and of the DAD triad (black). Absorbances per 1mm path length. The 400 nm pump pulses selectively excite the D moiety. The steady-state fluorescence of D (green) has large spectral overlap with the absorbance of A, facilitating resonance energy transfer in DAD.

Figure 1 shows the chemical structure of the DAD triad supplemented with terminal siloxane chains. Details regarding the synthesis and structural characterization will be published elsewhere. Optical excitation occurs on the conjugated parts, either in the thiophene-based donor (D) or the perylenediimide acceptor moiety (A). The absorption spectra of D and of DAD are given in figure 1. Note that the DAD absorption spectrum is in good approximation the sum of the D and A absorbances. Small deviations, as the broadening and peak shift in the low-energy part of DAD are indicative of excitonic interactions.
The CT state formation speed and efficiency has been studied for D and DAD in chloroform by femtosecond transient absorption (TA) spectroscopy. Details of the home-built experimental set-up can be found in [7]. Briefly, 400-nm, 50 fs pulses at 2.5 kHz repetition rate selectively excite the donor molecules. The absorption of A is almost zero at that wavelength (fig. 1). Molecules are pumped through a 0.5 mm quartz capillary so as to refresh the excited sample portion in between to laser shots. Care has been taken to work in a regime of linear dependence of TA signal versus excitation intensity. This means that typically less than 10% of the molecules are excited. Absorption changes are monitored in the range of 420-750 nm using a white-light supercontinuum generated in a 2-mm CaF_2 plate. The instrument response function is 80 fs, given by the FWHM of stimulated Raman scattering

of the pump pulses in the solvent. After 1h data collection and averaging, the experiment has a sensitivity of $\Delta A_{min} \approx 5\times10^{-5}$.

3. Results and discussion

The present contribution focuses on the differences in excited state lifetime and photochemistry of D when isolated or bound in the DAD triad.

3.1. The donor molecule alone

Figure 2 shows the transient absorption changes of isolated thiophene-based molecules. The overall evolution can be divided in three time regimes: a) excited state equilibration in less than 1 ps, b) excited state decay due to intersystem crossing, on a 0.2-0.3 ns time scale, and c) quasi-stationnary triplet-to-ground state difference absorbance spectrum. The latter is best seen in the 1 ns spectrum of fig. 2. With its maximum at 570 nm and relatively broad shape (FWHM > 100 nm) it is qualitatively in good agreement with the triplet-to-ground state difference spectrum determined for similar bi-thiophene compounds [8].

At times earlier than 0.3 ns, the molecule is in excited singlet state S_1 (phases a) and b)). This is readily deduced from the negative ΔA signal at shorter wavelengths with a maximum in the 460-475 nm range. Due to its similarity with the steady-state fluorescence spectrum of D (not shown), it is unambiguously attributed to stimulated emission along the $S_1 \rightarrow S_0$ transition. Bleach of the $S_0 \rightarrow S_1$ ground state absorption transition gives rise to a similar negative signal, but is located at wavelengths shorter than 400 nm (not shown). The positive ΔA at $\lambda > 520$ nm is then due to excited state absorption (ESA, $S_1 \rightarrow S_n$). During the first picosecond (phase a)), important changes in the spectral shape of ESA are occurring, together with a decay of the peak-like feature at 585 nm. Note that the SE also decays by almost 50 % on that time scale. We attribute these changes to intra- and inter-molecular vibrational energy relaxation. Ultrafast planarization of the bi-thiophene rings, as invoked by other authors [8], cannot be excluded. The subsequent S_1 decay during phase b) leads to the formation of the triplet-to-ground state difference spectrum. A global multi-exponential fitting procedure yields two excited state decay times of 2.0 ± 0.5 ps and 220 ps, for S_1 equilibration and decay respectively. The photochemical reactions of D can thus be summarized as follows

$$S_0 \xrightarrow{hv} S_1^* \xrightarrow{\approx 2\,ps} S_1^{eq} \xrightarrow{220\,ps} T_1$$

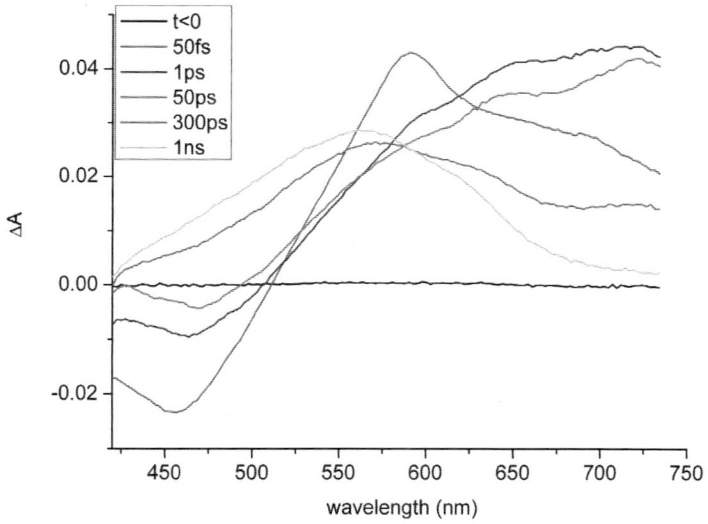

Figure 2: Time-resolved transient absorption spectra of the isolated D molecules, displaying intersystem crossing on a 220 ps time scale. The initial sub-ps spectral relaxation is due to excited state structural equilibration of the molecule and the solvent.

3.2. Energy and charge transfer dynamics in the DAD triad

Figure 3A shows the primary dynamics in the first 400 fs after optical excitation of the D moiety in the triads. The initial excited state species is D*AD. The negative signals at $\lambda < 570$ nm displays the characteristic modulations of the perylene-diimide (A) absorbance (cf. fig. 1) and is therefore due to bleach of the ground state of A. As the 400 nm pump beam is absorbed by the D but not by A, this indicates that A is excited indirectly through Förster resonant fluorescence energy transfer from D to A, forming the excited state species DA*D. The evolution of ΔA at longer wavelengths is consistent with this process. Indeed, the early 67 fs spectrum bears a lot of similarity of the ESA signal observed for the isolated D (cf. 50-fs spectrum in fig. 2). Spectra at later times indicate additional signals due to stim. emission along the $S_1 \rightarrow S_0$ transition in A, most obvious by the negative dips at 576 and 620 nm, and the build-up of A's excited state absorption at 705 nm (cf. fig. 3 b with the steady state fluorescence spectrum of A). These observations are consistent with those reported for a DAD triad with the same A and a different but similar thiophene-based donor [3].

The subsequent events related to the DA*D decay are revealed by the evolution of ΔA at longer delay times (fig. 3 B). The ground state bleach and stim. emission signals decay on a 5-10 ps time scale, and new absorption features at 633 and 712 nm are observed. Comparison with literature data [9, 10] allows to identify these as the absorption due to the cationic forms of D and the anions of A, meaning that charge transfer (CT) has occurred and the D+A-D species is formed. Additional experiments extending the present probe wavelength range into the near-IR and comparison with electro-absorption spectroscopy confirm these assignments (to be published). The D+A-D species is not stable however. Rather we observe the system returning to its DAD ground state after 200-300 ps. Clearly, the molecular orbital properties and the proximity of D and A make the charge transfer state

unstable, and very fast recombination occurs. This issue will be a major concern for functional materials and thin films when photovoltaic applications are targeted.
A detailed quantitative analysis using "global analysis strategies" with multi-exponential fitting of the reaction kinetics yields the following reaction scenario and lifetimes.

$$DAD \xrightarrow{h\nu} D^*AD \xrightarrow{120\,fs} DA^*D \xrightarrow{1.2/4.4\,ps} D^+A^-D \xrightarrow{46\,ps} DAD$$

The charge transfer process DA*D→D+A-D seems to involve two different reaction times (1.2 and 4.4 ps) possibly related to structural heterogeneity of the triads. In addition, the decay of A's SE is complete, indicating that the charge transfer build-up occurs with close to 100% quantum efficiency. This is not the case for the initial energy transfer process. As a matter of fact, recent time-resolved fluorescence experiments indicate that ≈ 10% of the triads are non-reactive, as they display the characteristic fluorescence of isolated donor molecules.

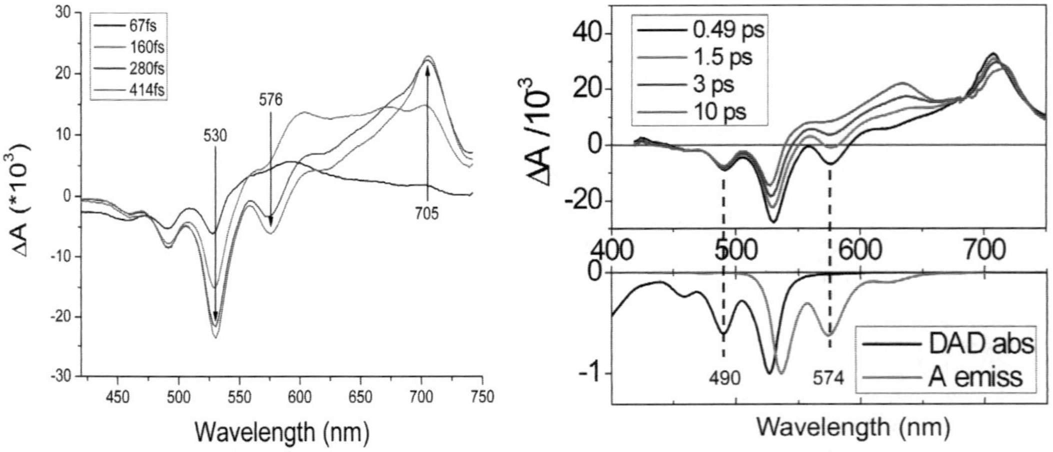

Figure 3. Time-resolved transient absorption spectra of the DAD triad. **Left:** The first 300 fs are characterized by excitation of the A via resonant energy transfer from D*. Spectra show the progressive build-up of A's ground state bleach and stim. emission (<600 nm) and the crossover from excited state absorption due to D* (160 fs) to the one of A* (410 fs) at longer wavelengths. **Right:** Decay of A's SE and formation of D+ and A- with the absorption bands at 633 and 712 nm (at 10 ps).

4. Conclusions

Femtosecond transient absorption has been used to reveal the ultrafast photochemistry at work in the molecular building blocks of novel photosensitive materials. When combined with ultrafast fluorescence techniques both the reaction kinetics and the quantum efficiencies of each reaction step can be determined. For the present DAD triad, we show that the large majority of the thiophene-based donor transfers its charge within less than 5 ps to the PDI acceptor. The data show that the non-radiative energy occurring on a 120 fs time scale is not detrimental to the charge transfer in this particular system, indicating that the electron transfer occurs between the HOMO levels of D and A.

Further studies will address the challenge of stabilizing the CT state, which in the isolated triads has only a very short lifetime (46 ps). Preliminary experiments on thin films of DAD indeed show a prolonged CT lifetime, most probably due to interchain delocalization. The application of an external bias should further contribute to reduction of CT recombination. The results of the photovoltaic property investigation of these novel DAD triads in their smectic phase will be reported in a subsequent paper.

Acknowledgements
The authors are grateful to G. Dekyndt, E. Couzignet and O. Crégut for important technical assistance. Work at IPCMS is funded by the Strasburg University and the "Centre National pour la Recherche Scientifique" (CNRS).

References

References

[1] Liang Y, Xu Z, Xia J, Tsai S T, Wu Y, Li G, Ray C and Yu L 2010 For the bright future—bulk heterojunction polymer solar cells with power conversion efficiency of 7.4% *Adv. Materials* **22** E135 - E8
[2] Scharber M, Mühlbacher D, Koppe M, Denk P, Waldauf C, Heeger A and Brabec C 2006 Design rules for donors in bulk-heterojunction solar cells—towards 10 % energy-conversion efficiency *Adv. Materials* **18** 789-94
[3] Fron E, Lor M, Pilot R, Schweitzer G, Dincalp H, De Feyter S, Cremer J, Bauerle P, Mullen K, Van der Auweraer M and De Schryver F C 2005 Photophysical study of photoinduced electron transfer in a bis-thiophene substituted peryleneimide *Photochem. & Photobiol. Sci.* **4** 61-8
[4] Marcos Ramos A, Beckers E H A, Offermans T, Meskers S C J and Janssen R A J 2004 Photoinduced multistep electron transfer in an oligoaniline,àíoligo(p-phenylene vinylene),àíperylene diimide molecular array *J. Phys. Chem. A* **108** 8201-11
[5] Bu L, Guo X, Yu B, Qu Y, Xie Z, Yan D, Geng Y and Wang F 2009 Monodisperse co-oligomer approach toward nanostructured films with alternating donor,àíacceptor lamellae *J. Am. Chem. Soc.* **131** 13242-3
[6] Ohkita H, Cook S, Astuti Y, Duffy W, Tierney S, Zhang W, Heeney M, McCulloch I, Nelson J, Bradley D D C and Durrant J R 2008 Charge carrier formation in polythiophene/fullerene blend films studied by transient absorption spectroscopy *J. Am. Chem. Soc.* **130** 3030-42
[7] Sinicropi A, Martin E, Ryazantsev M, Helbing J, Briand J, Sharma D, Léonard J, Haacke S, Cannizzo A, Chergui M, Zanirato V, Fusi S, Santoro F, Basosi R, Ferré N and Olivucci M 2008 An artificial molecular switch that mimics the visual pigment and completes its photocycle in picoseconds *Proc. Nat. Acad. Sc. USA* **105** 17642-7
[8] Paa W, Yang J P and Rentsch S 2001 Ultrafast intersystem crossing in thiophene oligomers investigated by fs-pump-probe spectroscopy *Synthetic Metals* **119** 525-6
[9] Evans C H and Scaiano J C 1990 Photochemical generation of radical cations from α-terthienyl and related thiophenes: Kinetic behavior and magnetic field effects on radical ion pairs in micellar solution *J. Am. Chem. Soc.* **112** 2694-701
[10] Gosztola D, Niemczyk M P, Svec W, Lukas A S and Wasielewski M R 2000 Excited doublet states of electrochemically generated aromatic imide and diimide radical anions *J. Phys. Chem. A* **104** 6545-51

Journal of Physics: Conference Series **276** (2011) 012007 doi:10.1088/1742-6596/276/1/012007

Transform-limited, achromatic injection-seeded terahertz-wave parametric generator

R Guo, H Minamide and H Ito*

RIKEN Advanced Science Institute, 519-1399, Aramaki Aoba, Aoba-ku, Sendai 980-0845, Japan.

* Email: h-ito@riken. jp

Abstract. A review to our effort on developing the transform-limited, frequency-agile terahertz-wave parametric generator (TPG) is presented. A frequency-agile THz-wave generation is realized by introducing the injection-seeding method and the optical design for the stationary dispersion-compensation. The purity of the THz-wave frequency was dramatically improved to $\Delta\nu/\nu<10^{-4}$. Simultaneously, the THz-wave output was several hundred times higher than that of a conventional TPG. In addition, a wide frequency tuning with fast tuning speed were realized. The THz-wave frequency can be set randomly or scanned continuously over a frequency range from 0.6 THz to 2.4 THz with narrow linewidth of sub 100 MHz. Furthermore, a tabletop, high-performance THz-wave gas spectrometer based on this achromatic injection-seeded TPG was developed. To demonstrate the potential of this system, we performed the measurement of the absorption line due to rotational transitions of the water molecules and determined their pressure-broadening coefficient.

1. Introduction

Frequency conversion in a nonlinear optical material is an effective method of achieving coherent THz-wave radiation due to its wide tunability and room-temperature operation[1-8]. We have focused on the development of THz-wave parametric oscillator (TPO)[9-13] and THz-wave parametric generator[14-19] (TPG) based on the laser light scattering from the A_1-symmetry polariton mode of $MgO:LiNbO_3$ nonlinear crystal, which generated widely tunable, coherent THz-wave radiations with compact and room-temperature operation.

The difference between a TPO and a TPG is that the former has an idler cavity while the latter does not. The THz-wave linewidth of a conventional TPG exceeds 500 GHz and the output magnitude is much smaller than that from a TPO. The linewidth of a conventional TPO is about several tens GHz and is continuously tunable in the 100 to 300 μm (1 to 3 THz) range in a single operation. It can emit peak power of up to several tenths mW.

Injection-seeding by external narrow-band laser radiation at low power levels can considerably reduce the linewidth of an OPO down to the Fourier transform limit. Since the TPO and TPG are pumped by a nanosecond pump laser, a spectral linewidth of less than 100 MHz is possible under the Fourier-transform-limited condition.

The injection-seeded TPO (is-TPO) [12] and the injection-seeded TPG (is-TPG) [14-19] has been demonstrated to generate Fourier transform-limited THz-wave radiation. TPGs, as compared to TPOs, are especially preferred for external seeding by narrow linewidth light sources such as a CW tunable

SLM laser for two reasons. Firstly, a TPG has no cavity, therefore, there is no need to actively control the cavity length to match the frequency of the seeding laser. The closed-loop serving control for frequency tuning is also not necessary, which greatly simplifies the techniques for scanning the wavelength of the TPG. Secondly, since the TPG operated in a single-pass fashion, there is no feedback from the TPG back toward the external seeding laser. This is important for the CW SLM lasers, because even a small amount of feedback can disturb the SLM operation, and under certain circumstances can even damage the gain material. The no feedback is particularly important when using diode lasers as seeding sources. Therefore, for developing the sub-100MHz narrow linewidth THz-wave sources, we concentrated our efforts on the development of an injection-seeded TPG system. We developed an achromatical injection-seeded TPG (ais-TPG)[17,18] that guarantees a fast and smooth frequency tuning. Furthermore, a tabletop, high-performance THz-wave gas spectrometer based on this ais-TPG was realized[19]. Using the frequency-agile characteristics, such as narrow linewidth, wide tunability, and rapid frequency control in this spectrometer, high-resolution THz-wave gas sensing is demonstrated.

In this article, we summarize our results on developing the Transform-limited, achromatic is-TPG system.

2. The principle for THz-wave generation

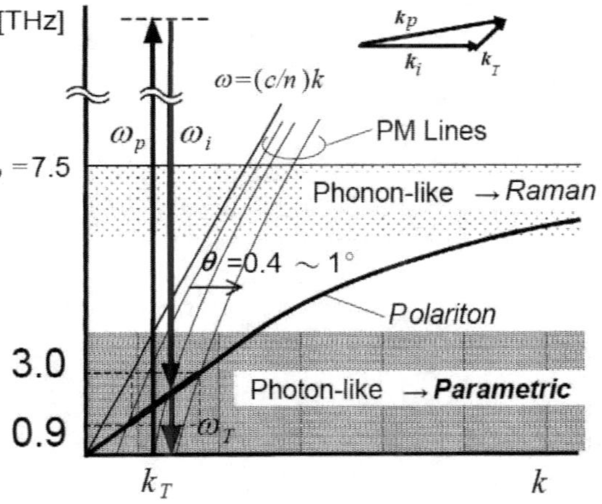

Figure 1 Dispersion relation of the polariton. The inset shows the noncollinear phase-matching condition.

The efficient parametric scattering of laser light via a polariton (stimulated polariton scattering)[1] generates coherent THz waves. The scattering process involves both the second-order and the third-order nonlinear processes. $LiNbO_3$ is one of the most suitable materials for generating the THz waves efficiently because of its large nonlinear coefficient (d_{33}=25.2pmV^{-1} at λ=1.064μm[20] and d_{33}=165pmV^{-1} at THz-wave region[21]) and its transparency over a wide wavelength range (0.4–5.5μm). $LiNbO_3$ has four infrared–active and Raman–active transverse optical (TO) phonon modes, called the A_1-symmetry modes. The lowest mode (ω_0~ 250cm^{-1}) is useful for efficient tunable THz-wave generation. Figure 1 shows the dispersion relation of the polariton of the $LiNbO_3$ crystal. Polaritons exhibit phonon-like behavior in the resonant frequency region (near the TO-phonon frequency ω_{TO}). However, they behave like photons in the nonresonant low-frequency region, where a signal photon at THz frequency (ω_T) and a near-infrared idler photon (ω_i) are created parametrically from a near-infrared pump photon (ω_p), according to the energy conservation law $\omega_p=\omega_T+\omega_i$ (p: pump; T: THz; i: idler). In the stimulated scattering process, the momentum conservation law $k_p=k_i+k_T$ (noncollinear phase-matching condition; see the insets of Figure 1), also holds. This leads to the angle-dispersive characteristics of the idler and THz waves.

By introducing injection-seed to the idler beam, the line width of THz-wave wave was narrowed to close to the linewidth of the pump beam. Simultaneously, the output was several tens times higher than that of a Terahertz-wave parametric generation without the injection-seed. In additional, wide tunability can be realized using a tunable seed by changing both the seed wavelength and the seed incident angle.

3. Injection-seeded THz-wave parametric generator (is-TPG)

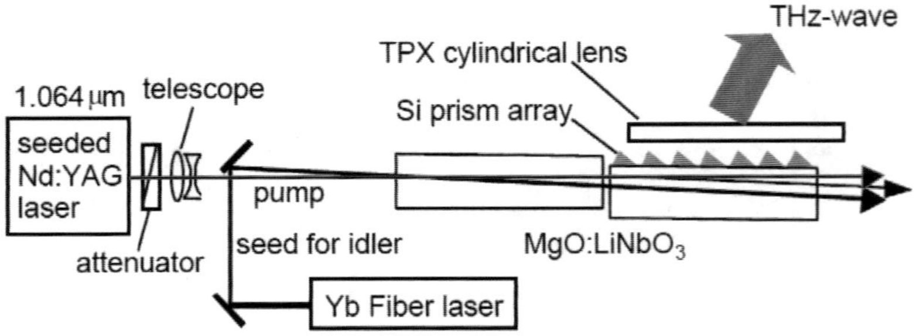

Figure 2 Experimental set-up of is-TPG. The pump was a single longitudinal mode Q-switched Nd:YAG laser (1.064μm), and the seed for the idler was a CW Yb-fibre laser (1.070μm).

Figure 2 shows the setup of our experimental is-TPG. The pump used was a single longitudinal mode (SLM) Q-switched Nd:YAG laser (Spectron SL404T, wavelength: 1.064μm; energy: <50mJ/pulse; pulse width: 15ns; beam profile: TEM_{00}, repetition rate: 10 Hz). The pump beam diameter was decreased to 0.8 mm using a telescope in order to increase the power density. The pump power density was < 530 MW/cm^2 at the crystal surface and could be varied with an attenuator. The pump beam was almost normal to the crystal surfaces as it entered the crystals and passed through the crystal close to the y-surface. A continuous-wave (CW) SLM Yb-fibre laser (wavelength: 1.070 μm fixed; power: < 300m W) was used as an injection-seed for the idler. Arrangements were tested using one, two, and three nonlinear crystals, 65mm in length. The maximum THz-wave output was obtained when two crystals (5mol% MgO-doped $LiNbO_3$) were used in series. Both crystals were cut into 65×6×5mm (x×y×z-axis) pieces. The x-surfaces were polished and antireflection-coated. An array of seven Si-prism couplers[22] was placed on the y-surface of the secondary $MgO:LiNbO_3$ crystal for efficient coupling of the THz wave as shown in Figure 2. The polarizations of the pump, seed, idler, and THz waves were all parallel to the z-axis of the crystals. The transmission direction of the seed beam with different wavelength was carefully adjusted to overlap with that of the idler beam in $MgO:LiNbO_3$. The THz-wave output and temporal waveform were measured with a 4K Si bolometer and a Schottky barrier diode detector, respectively.

Both of energies of the THz and idler waves were greatly enhancement by injection seeding. With increasing the seed power from 0 and 200mW, THz-wave and idler energy increased by factors of nearly 300 and 500, respectively. The maximum conversion efficiency was achieved when the pump and seed beams almost fully overlapped at the incident surface of the first $MgO:LiNbO_3$ crystal, as shown in Figure 2.

Figure 3 shows the measured THz-wave output energy from the is-TPG and a conventional TPG. The maximum THz-wave output of 900pJ/pulse (peak>100mW) was obtained with a pump of 45mJ/pulse and a seed power of 250mW. As the minimum sensitivity of the Si bolometer was almost

1fJ/pulse, the dynamic range of the injection-seeded TPG system was 900pJ to 1fJ~60dB, which is sufficient for most applications.

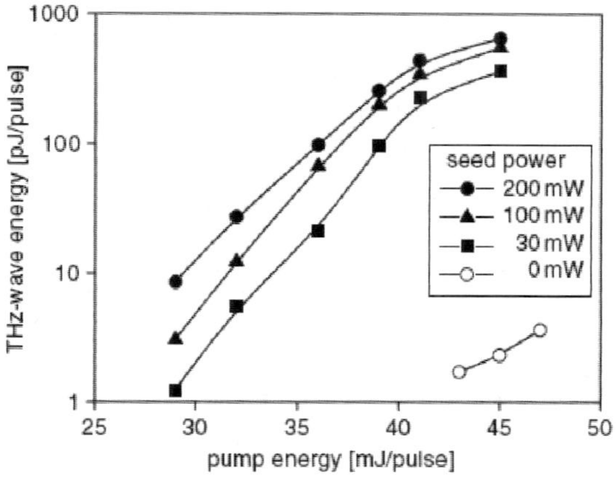

Figure 3 Input–output characteristics of an is-TPG, showing THz-wave (190μm) energy enhancement by a factor of 300 hundreds with injection seeding.

Figure 4 shows the effect of injection seeding on idler spectrum narrowing. The dotted line indicates the idler spectrum of a conventional TPG without injection seeding and the solid line indicates the idler spectrum of an injection-seeded TPG. The resolution limit of the spectrum analyzer used was 0.2nm, so the idler spectrum in Figure 4 shows only the spectrometer resolution of the spectrum analyzer, the real idler spectrum was much narrower than that shown in this figure.

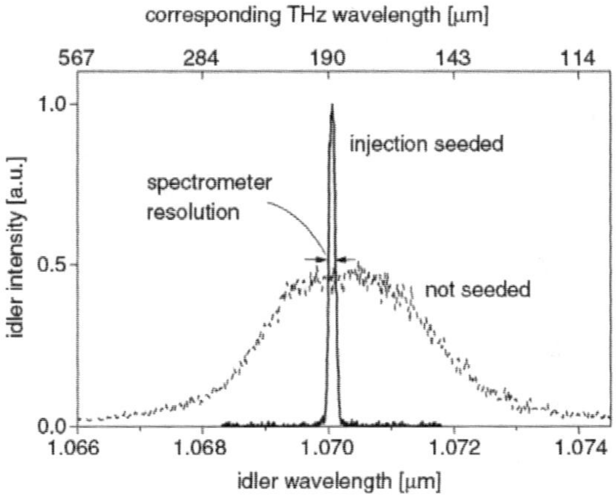

Figure 4 Narrowing of the idler (1.07μm) spectrum by injection seeding. The dotted and solid lines indicate the idler spectrum of a conventional TPG and an is-TPG, respectively.

The linewidth of THz-wave should be close to the linewidth of the idler. It can be determined from a scanning Fabry–Perot etalon consisting of two Ni metal meshes or the absorption spectrum of low pressure water vapour. We will discuss it at the following context.

4. An achromatic is-TPG (ais-TPG) for frequency-agile THz-wave generation

Noncollinear phase-matching condition (the inset in Figure 1) must be satisfied in the stimulated scattering process for THz-wave generation in $LiNbO_3$. This leads to the angle-dispersive characteristics of the idler (seed) and THz waves. If we tune the incident angle as the wavelength of the seed is changed in Figure 2, we may realize a tunable THz-wave radiation. However, rapid tuning is difficult using this setup. An achromatic phase matching in the seed beam was introduced to automatically compensate for the dispersion in the $LiNbO_3$ crystals, we realized a frequency-agile THz-wave generator with tuning the wavelength of seed beam only.

4.1. The optical design in seed beam for the stationary dispersion-compensation

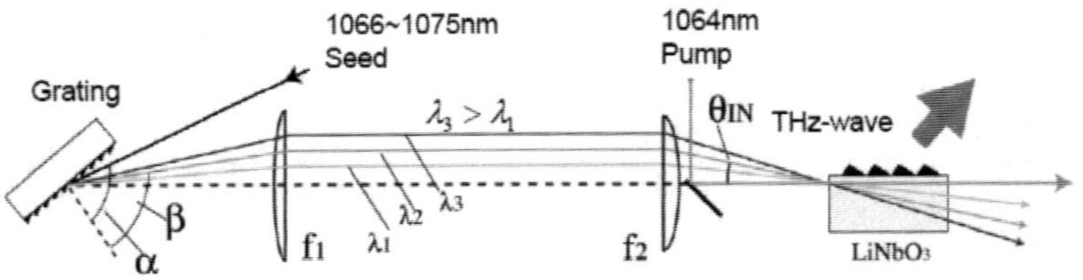

Figure 5 shows the optical arrangement in seed beam for the stationary dispersion-compensation.

An achromatic phase matching to compensate for the dispersion caused by the $LiNbO_3$ crystal is designed as shown in Figure 5. It consisted of a grating and two focus lenses that formed a 1:3 telescope device. The incident angle of the seed beam on the grating is α, and the first-order diffraction angle β(λ) is expressed by the grating equation $\lambda = d(\sin\alpha + \sin\beta)$ where d is the spacing between grooves. The grating disperses wavelengths laterally ($\lambda_1 < \lambda_3$), and the lenses (focal lengths f_1 and f_2) form a telescope to expand the dispersion of the grating and to converge the dispersed seed beam onto the input surface of the $LiNbO_3$ crystal. The telescope magnifies the angular dispersion by f_1/f_2.

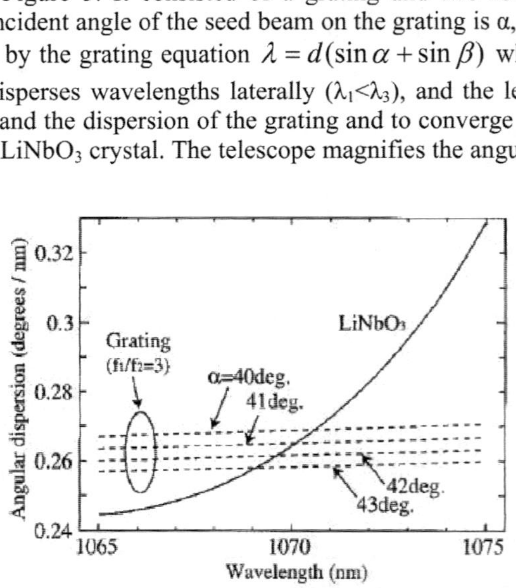

Figure 6 Calculated linear dispersion of the $LiNbO_3$ crystal (solid curve) and the1200-groove/mm gratings followed by a telescope that magnifies the angular dispersion by $3[3\partial\theta_{IN}/\partial\lambda)]$ (dashed lines).

The solid curve in Figure 6 shows the calculated first-order coefficient $\partial\theta_{IN}/\partial\lambda$ of $LiNbO_3$ crystal with Sellmeier equations [23, 24] at the optical and THz frequency regions. The dashed lines in figure 6 show the calculated angular dispersion of a 1200-groove/mm grating followed by a telescope that magnifies the angular dispersion by $3[3\partial\theta_{IN}/\partial\lambda)]$ for incident angles α of 40°–43°.

The measured Half-Width at Half-Maximum (HWHM) external angle tolerance for THz-wave radiation without dispersion compensation was 0.16° at 1070nm and corresponds to the full bandwidth of seed beam of 1.07 nm and a THz-wave frequency bandwidth of ~310GHz.

Using the achromatic phase matching device, the first order dispersion $\partial\theta_{IN}/\partial\lambda$ can be compensated. The calculated full bandwidth of the second order dispersion in the frequency tuning range of the seed was 12 nm. (The second derivative, $\partial^2\theta_{IN}/\partial\lambda^2$ is (~0.0093°/nm)/nm).

For THz-wave frequency tuning 0.6 to 2.7 THz at a pump wavelength of 1064nm, the required seed wavelength tuning range is 1066–1075 nm (bandwidth: 9nm). Therefore, matching the first derivative at an appropriate central wavelength is sufficient for the dispersion compensation of 9nm bandwidth in the $LiNbO_3$ crystal.

4.2. Experimental setup for ais-TPG

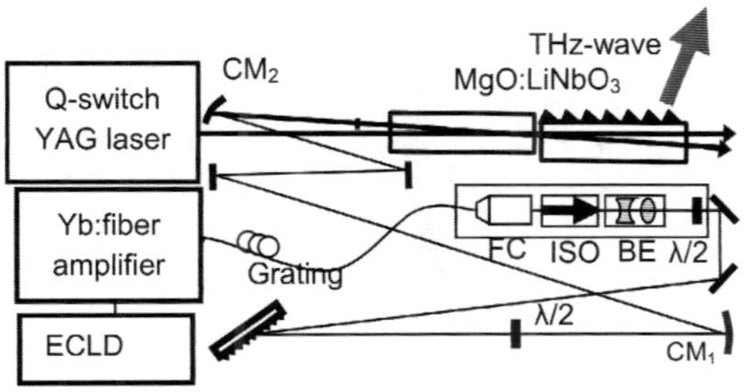

Figure 7 Experimental set-up of ais-TPG. CM_1, CM_2: Concave mirrors. The radius of curvature is 1.5m and 0.5 m, respectively. FC: Fiber collimator. ISO: Isolator. BE: Beam expander. λ/2: half-wave plate. Gratings: 1200-grooves/mm.

Figure 7 shows the experimental setup of the ais-TPG. It included a pump source, a seed source, the optics for achromatic phase-matching and the nonlinear crystals. The pump source was a specially designed injection-seeded Q-switch YAG laser (developed by Megaopto Co., Ltd), which generated single-longitudinal-mode pulses at the wavelength of 1064 nm. Its pulse width, repetition rate and maximum output energy were 15 ns, 500 Hz and 21 mJ/pulse, respectively. The repetition rate was 50 times higher than that of the flash-lamp-pumped Nd:YAG laser that we used in the previous is-TPG. The seed source was a continuous-wave external cavity laser diode (ECLD) (NEW FOCUS, Inc) amplified up to 1 W by an Yb:fiber amplifier. The wavelength was coarsely tuned by DC motor drive screw from 1056nm to 1083 nm without mode-hopping and finely tuned by a piezoelectric transducer over a range of approximately 20 GHz. The scan speed of the DC motor can be set to a value between 0.01~15.7nm/sec, which is corresponding to the frequency between 3 GHz~47 THz/sec.

The TPG gain media is the MgO:$LiNbO_3$ crystal, same as that we have used in Figure 2. The seed from the Yb:fiber amplifier was collimated and expanded to 3.0 mm in diameter. The beam divergence was smaller than 0.3 mrad. Then, the seed beam was diffracted using a 1200-groove/mm grating with an angle of incidence of ~40° for 1071 nm. The diffracted seed beam was reflected by a pair of concave mirrors, CM_1 and CM_2, and injected into the MgO:$LiNbO_3$ crystal. The radius of curvature of the two mirrors was 1.5m and 0.5 m, respectively. The pair of concave mirrors formed a telescope that expanded the dispersion of the grating and converged the dispersed seed beam onto the input surface of the MgO:$LiNbO_3$ crystal with a beam diameter of about 0.6 mm (FWHM). The center of rotation was fixed on the input surface of the first MgO:$LiNbO_3$ crystal. Owing to the design of the stationary dispersion-compensated optical arrangement, the seed beam automatically can enter the $MgO:LiNbO_3$ crystal at the appropriate phase-matching angle, by changing the frequency of the seed beam alone.

4.3. Frequency-agile THz-wave generation

The threshold energy of the pump beam for generating THz-wave radiation was 13mJ/pulse with the waist of the pump beam of 0.7mm (FWHM) at the first facet of the MgO: $LiNbO_3$ crystal. THz-wave radiation was detected by a He-cooled Si bolometer. The THz-wave frequency was determined by the frequency difference between the pump beam and the idler beam based on the energy conservation of parametric processes in nonlinear optics. The voltage signal from the bolometer was sent to an analogue signal-processing circuit, which was the synchronized with the repetition frequency of the pump laser, including a peak-hold circuit and a delayed-sampling circuit. Peak intensity of every pulse envelope of the THz-wave signal can be capture with this device. The data from this signal-processing circuit was transferred to the computer. The frequency tuning and data acquisition were controlled by the computer using the Labview intermediate program.

Figure 8 shows the THz-wave output energy in the THz-wave frequency region of 0.6~2.4 THz obtained by smoothly tuning the seed beam wavelength from 1066.4 nm to 1073.3 nm. The pump energy and the seed beam power were 19 mJ/pulse and 250 mW, respectively. It takes about 1 minute for one scanning.

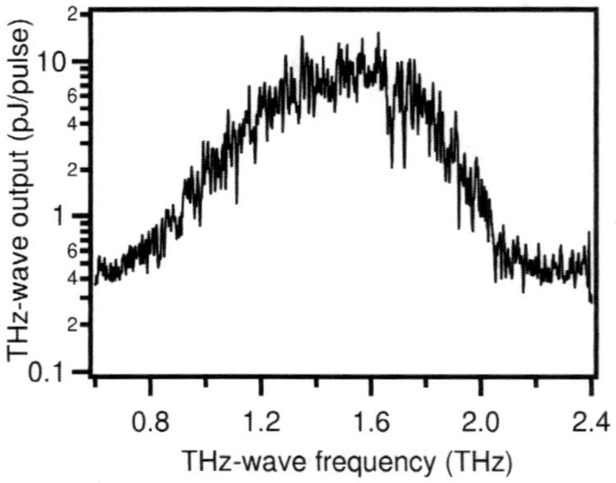

Figure 8 THz-wave output energy in the tuning region of 0.6 to 2.4 THz with the pump input energy of 19 mJ/pulse and seed beam power of 250 mW.

5. A frequency-agile high-resolution terahertz-wave spectrometer for gas sensing

The use of far-infrared or terahertz radiation for gas sensing is very much complementary to the well established microwave and mid-IR spectroscopy technique, and will greatly expand the number of gas species that can be detected. However, unlike gas sensing in both the microwave (<0.1 THz) and mid-infrared (>20 THz) frequency regions, far-infrared (IR) and terahertz sensing is relatively new, and its development has been hindered by a lack of suitable THz radiation sources and detectors. We have realized a narrow linewidth and tunable THz-wave generator, the establishment of a practical high resolution terahertz-wave spectrometer based on this ais-TPG is highly attractive for THz-wave gas-sensing.

5.1. Terahertz-wave spectrometer based on ais-TPG

Figure 9 shows the picture of the spectrometer. The spectrometer system is packaged into three chambers with a total volume of about 900 × 450 × 130 mm: the first chamber (I) contains the gas cells for THz-wave spectroscopy, the second (II) contains the nonlinear $MgO:LiNbO_3$ crystals for THz-wave generation, and the third (III) has the optics for steering both the seed beam (the thin white line in the picture) and the pump beam (black line) to the nonlinear crystals. To eliminate atmospheric water vapor, the first two chambers were designed to operate under a flow of pure nitrogen gas. The THz-wave radiation (the thick white line in the picture) from the $MgO:LiNbO_3$ was collimated by a cylindrical lens and then steered into the chamber (I). A standing wire-grid beam splitter divided the THz-wave radiation into reference and signal beams. The signal beam was steered into a portable gas cell with two gas inlets controlled by ultra-high vacuum leak valves to facilitate sample processing, and controlling and monitoring its pressure. The cell was built with two quartz wedge windows for eliminating interference fluctuations due to parallel reflection surfaces. Pressure was measured with a capacitance manometer. A two-channel liquid He-cooled Si bolometer was used for dual-beam balance detection to cancel the noise in the THz-wave source. Both ports of the bolometer were connected to the chamber directly to prevent further corruption by ambient water over the optical path.

The frequency scanning rate of this spectrometer can be set from 3 to 47 GHz/s in intervals of 6–100 MHz. Each absorption profile consists of 20–100 resolution elements. The broader the line, the larger the chosen step. Spectral measurements for high-quality line shape fits were performed using the fine-tuning mode, which can cover a continuous tuning range up to 50 GHz.

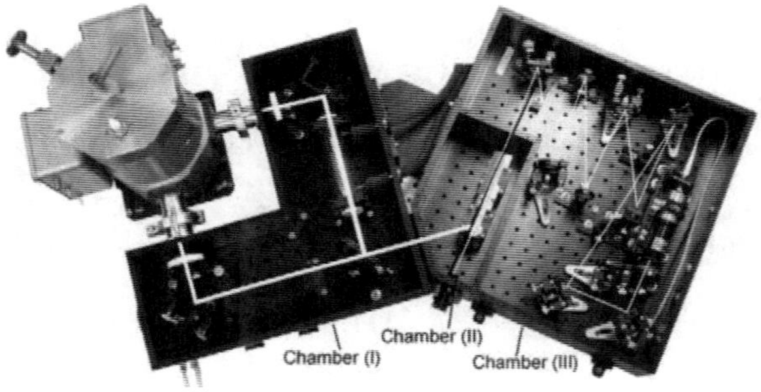

Figure 9 the picture of the tabletop THz-wave spectrometer.

5.2. Measuring the absorption spectroscopy of low pressure water vapor

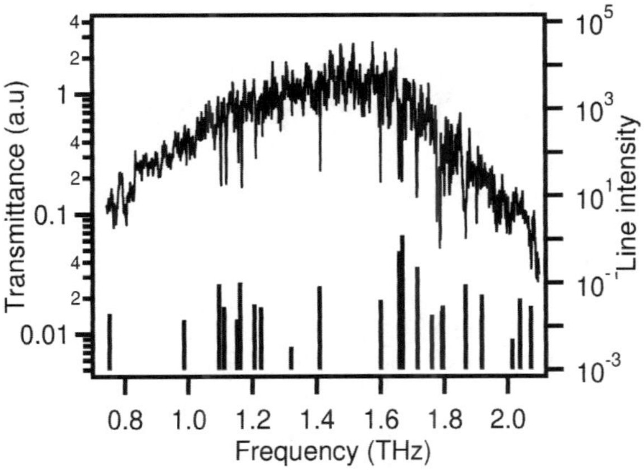

Figure 10 Accurate and rapid measurement of the water vapor spectrum in a wide frequency range from 0.8 to 2.1 THz with a 30MHz frequency step. The stick line shows the spectrum data from the NASA database.

We steer the total THz-wave radiation into the gas sample cell, then detect the THz-wave transmission spectroscopy using one of the two channels in Bolometer. Figure 10 shows the direct measured water vapor absorption spectrum from 0.8 THz to 2.1 THz. It takes only 3 minutes for scanning this wide frequency range with the frequency step 12 MHz. The pressure of water vapor in gas cell was 0.1 kPa. The absorption lines measured by this is-TPG spectrometer are in good agreement with the spectrum data from the NASA database[25] showed in Figure 10 with a stick spectrum. The serious background noise was observed in Figure 10, which was arising from the amplitude fluctuation in THz-wave source and the interference fringe due to the quartz parallel windows in the gas cell. The balance detection and digital filter were employed for eliminating those background fluctuations.

Figure 11 shows the transmittance spectra of water vapor from 1.2 to 1.9 THz at 0.1 kPa using dual-channel balanced detection. It takes about 4 min for the continuous frequency scan of 700 GHz in 6-MHz frequency steps. A broader frequency step can be chosen for faster scans. Twelve purely rotational transitions were observed. The measured absorption lines agree well with the spectrum data from the NASA database, shown as a stick spectrum in Figure 11. Special attention to four transitions: 2_{21}-2_{12}, 2_{12}-1_{01}, 6_{24}-6_{15}, and 7_{34}-7_{25} was given in Figure 11.

Because the balanced detection requires dividing the THz-wave into two paths, and the brightness of the signal beam decreases, the merit of the balanced detection for improving signal-to-noise ratio is lost at the low power region (0.6–1 THz and 2.0–2.4 THz). Single-channel direct detection offers the wider frequency region detection at the cost of a poorer signal-to-noise ratio.

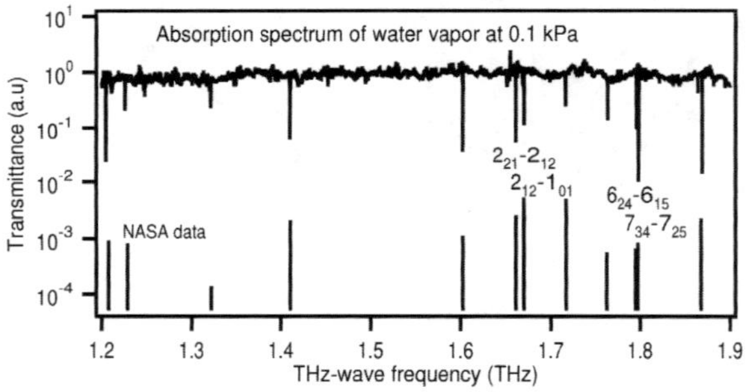

Figure 11 Accurate, rapid measurements of the absorption spectrum of the 12 purely rotational transitions of water vapor. The stick line shows the spectral data from the NASA database for comparison.

5.3. Pressure broadening detection

We observed pressure broadening of the absorption lines using an inert buffer gas (pure N_2). The atmosphere in the gas cell was pumped down to a pressure of 0.2 kPa, and then the N_2 gas was added until the desired final pressure was reached. No attempt was made to keep the pure water vapor in gas cell before the broadening gas was added. Any residual contribution to the total linewidth of water vapor due to the collision broadening from the 0.2 kPa atmosphere was mitigated by the fact that the foreign-gas collision-broadening coefficient is measured at a constant pressure atmosphere by varying the foreign gas pressure and computing the slope of the total linewidths as a function of this foreign-gas pressure. Further, the collision-broadening of a 0.2 kPa atmosphere is far narrower than the resolution of the THz-wave spectrometer; therefore, it can be ignored.

Absorption spectra of the 6_{24}-6_{15} and 7_{34}-7_{25} rotational transitions of water vapor were recorded while the pressure of the N_2 gas was varied between 0.2 and 10 kPa. The frequency interval between the 6_{24}-6_{15} and 7_{34}-7_{25} transitions is 2.4 GHz. Figure 12 shows the set of the two absorption lines under N_2 pressures of 0.4, 1.5, 2.5, 3.5, 4.5, 6, and 7 kPa (circles in Fig. 12). The nonlinear least-squares-fitted Voigt profiles were added to all the measured lines as solid lines in Figure 12. The spectral line half width at half maximum (HWHM) at each foreign-gas pressure was obtained from a nonlinear least-squares analysis.

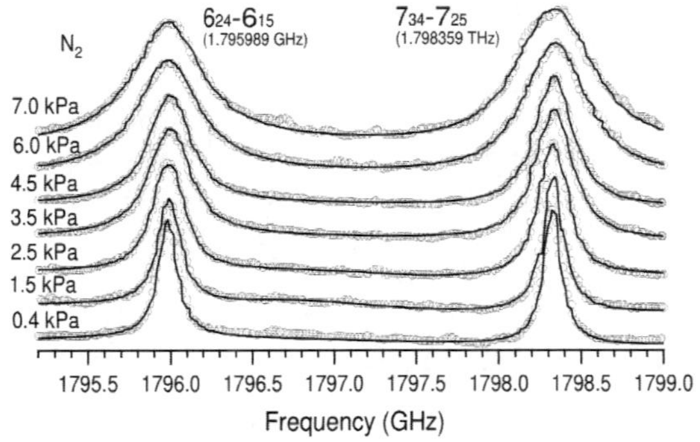

Figure 12 Set of absorption lines of the 6_{24}-6_{15} and 7_{34}-7_{25} rotational transitions of water vapor under N_2 pressures of 0.4, 1.5, 2.5, 3.5, 4.5, 6, and 7 kPa (circles).The nonlinear least-squares-fitted Voigt profiles of all the measured lines were added in solid line. The frequency interval between the 6_{24}-6_{15} and 7_{34}-7_{25} transitions is 2.4 GHz.

Figure 13 shows the linewidth of the 6_{24}-6_{15} transition as a function of the N_2 pressure. The linear least-squares fit to the HWHM data above a pressure of 2.5 kPa is given by the straight line. The linear fit slopes give pressure-broadening coefficients of 32.66±0.52 MHz/kPa and 33.93±0.56 MHz/kPa at the 6_{24}-$6_{15}$and 7_{34}-7_{25} transitions, respectively.

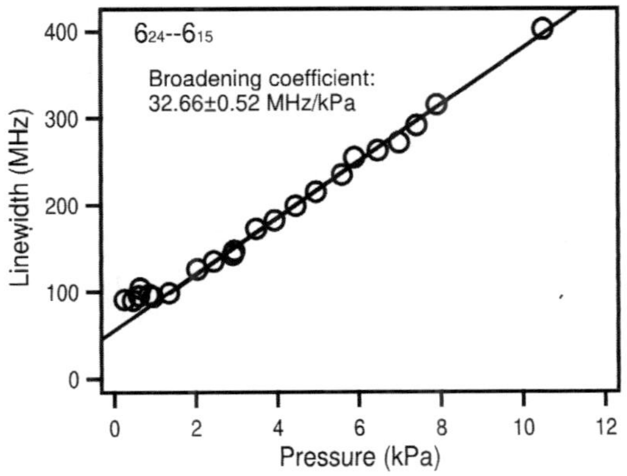

Figure 13 The linewidth of the 6_{24}-6_{15} transition as a function of N_2 pressure. The straight line shows the linear least-squares fit to the HWHM data above a pressure of 2.5 kPa. The slope gives a pressure-broadening coefficient of 32.66±0.52 MHz/kPa.

Table I compares our experimental measurements with theoretical predictions and other experimental results. The theoretical predications are taken from the work of Gamache and Davies[26]; the other experimental data in Table I are from Fourier-transform spectrophotometer (FTS) measurements taken by Townes et al[27]. They measured 17 N_2 broadened pure rotational transitions of water vapor using a commercial FTS in the spectral range of 0.75 to 3.3 THz with a gap in values between 1.5 and 2.2 THz. They give a weighted average for the experimental N_2 broadening coefficient in the spectral range of 0.75 to 3.3 THz (shown in Table I). No experimental data measuring the same rotational transitions as we do using FTS are currently available. However, our data are in good agreement with this weighted average experimental data. Dr. Townes has argued that the theory may be biased toward the narrower linewidth transitions and therefore predicts smaller linewidth broadening coefficients than are observed experimentally.

Our experiment demonstrates the possibility of obtaining information on the pressure broadening in a broad frequency region with high resolution in the THz-wave region. This information is very important to the calculation of atmospheric transmission, where requires high quality spectral parameters for many thousands of water vapor transitions.

	6_{24}–6_{15} (MHz/kPa)	7_{34}–7_{25} (MHz/kPa)
This work	32.66±0.52	33.93±0.56
FTS	32.25	
Theoretical	29.85	29.93

Table I. Comparison of our experimental measurements with theoretical predictions and the typical results from the commercial FTS.

5.4. The linewidth of THz-wave radiation

The typical pulse width of the THz wave was 3.4 ns measured by a Schottky barrier diode, and was almost identical to that of the idler, which was measured with a high-speed photodetector.[16] The calculated Fourier transform-limited linewidth was 136 MHz. Theoretically, the linewidth of the THz-wave generated by is-TPG is Fourier transform-limited by the linewidth of the pump beam (in our case, the 17-ns single-longitudinal-mode pulse width means a 30-MHz linewidth). These facts go to confirm that the linewidth of the THz wave was possible to be narrowed to near the Fourier transform-limited. However, the THz-wave linewidth was still more than 10GHz when an injection-seeded TPG is pumped by a multi-frequency Nd:YAG laser. Thus, in order to obtain the transform-limited THz wave with a TPG, both the pump and seed must be SLM lasers.

The THz linewidth can be measured using a scanning Fabry–Perot etalon consisting of two Ni metal meshes. But it is not easy for such a narrow-linewidth THz source to demonstrate a resolution better than 100 MHz (0.003 cm^{-1}) because of the instability of the scanning mirror for several meters.

The absorption spectrum of the low-pressure (<1.5kPa) water vapour in Fig. 12 clearly revealed that the resolution of THz-wave was less than 100 MHz. As shown in Fig. 13, the measured pressure-broadening linewidth does not decrease linearly below 1.5 kPa; furthermore, the linear fit line offset (Lorentzian width at a pressure of 0 Torr) does not correspond to the Doppler linewidth (~2 MHz), but rather reaches a linewidth of 50 MHz (HWHM). The linewidth of the THz-wave source is responsible for the linear bias at low pressure (the calculated values of the pressure broadening are not affected by the bias). It demonstrates that the linewidth of THz-wave generated by this is-TPG was 50 MHz or so, nearly the Fourier transform-limited by the pulse width of the pump beam.

6. Conclusion

We reviewed the widely tunable injection-seeded THz-wave generator based on the optical parametric conversion in the nonlinear crystal $MgO:LiNbO_3$. In comparison with a conventional TPG without injection seeding, the output was increased around 300 times and the linewidth was decreased to near the Fourier-transform-limitation. The stationary dispersion-compensation optics design in the seeded beam made a rapid and easy THz-wave frequency tuning. We realized an achromatic injection-seeded THz-wave generator that can realize THz-wave frequency tuning from 1-3 THz by changing the frequency of the seed beam alone.

We developed a frequency-agile THz-wave spectrometer for high-resolution gas sensing using this achromatic injection-seeded THz-wave generator. This THz-wave spectrometer demonstrates the capability of obtaining THz-wave gas phase spectral information over a wide frequency region at very fast data-acquisition times and high resolution. Its advantages include compactness, wide tunability, high resolution, rapid data acquisition, room-temperature operation, and ease of use.

7. Acknowledgment

The authors are grateful of Dr. K. Kawase, Nagoya Univ., Dr. K. Imai, Optical comb, Inc., the previous members of Tera-photonics group, RIKEN, Dr. J. Shikata, Tohoku Univ. and Dr. K. Akiyama, Mitsubishi Elextronic Corporation, for their valuable contribution in developing is-TPG. The authors thank Dr. T. Ikari, Dr. Y. Ishikawa and Dr. Seigo Ohno, RIKEN ASI, for useful discussions; Dr. Balasigamani Devaraj in New Focus Co. Ltd helps us for the development of the ECLD working in the wavelength range of 1056~1083 nm. Mega Opto Co., Ltd developed the special designed all-solid-state Q-switch Nd:YAG laser. Polishing and dielectric coating of crystals were done by technicians C.Takyu and T.Shoji of Tohoku University; Mechanical process was done by A.Kudo; we are thankful to all for their helpful cooperation.

References

[1] Henry C H and Garrett C G B, 1968 *Phys. Rev.*, **171**, 1058-64.
[2] Yarborough M , Sussman S S , Puthoff H E , Pantell R H and Johnson B C, 1969 *Appl. Phys. Lett.* **15,** 102-5.
[3] Johnson B C, Puthoff H E, Soo Hoo J M and Sussman S S, 1971 *Appl.Phys.Lett.***18**, 181-3.
[4] Piestrup M A, Fleming R N and Pantell R H, 1975 *Appl.Phys.Lett.*, **26,** 418-21.
[5] Kawase K, Sato M, Taniuchi T and Ito H, 1996 *Appl. Phys. Lett.* **68**, 2483-85.
[6] Taniuchi T, Shikata J and Ito H, 2000 Electr. Lett. **36**, 1409-11.
[7] Shi W and Ding Y J, 2003 Appl. Phys. Lett. **84**, 1635-7.
[8] Tanabe T, Suto K, Nishizawa J, Saito K and Kimura T, 2003 *Appl. Phys. Lett.* **83**, 237-39.
[9] Shikata J, Kawase K, Karino K, Taniuchi T and Ito H 2000 *IEEE Trans. Microwave Theory Tech.* **48**, 653-661.
[10] Shikata J, Sato M, Taniuchi T, Ito H and Kawase K, 1999 *Opt. Lett.* **24**, 202-04.
[11] Kawase K, Sato M, Nagamura K, Taniuchi T and Ito H, 1997 *Appl. Phys. Lett.* **71**, 753-759.
[12] Imai K, Kawase K , Shikata J, Minamide H and Ito H, 2001 *Appl. Phys. Lett.***78**, 1026-28.
[13] Ikari T, Zhang X B, Minamide H and H. Ito, 2006, *Opt. Expr.* **14**, 1604-10.
[14] Kawase K, Shikata J, Imai K and Ito H, 2001 *Appl. Phys. Lett.* **78**, 2819-21.
[15] Kawase K, Minamide H, Imai K, Shikata J and Ito H, 2002 *Appl. Phys. Lett.***80**, 195-7.
[16] Kawase K, Shikata J and Ito H, 2002 *J.Phys. D:Appl. Phys.* 35, R1-14.
[17] Imai K, Kawase K , Minamide H and Ito H, 2002 Opt. *Lett.* **27**, 2173-75.
[18] Guo R, Akiyama K, Minamide H, and Ito H, 2006 *Appl. Phys. Lett.***88**, 091120.
[19] Guo R, Akiyama K, Minamide H, and Ito H, 2007 *Appl. Phys. Lett.***90**, 121127.
[20] Shoji I, Kondo T, Kitamoto, Shirane M, Ito R, 1997. *J. Opt. Soc. Am.* **B**14, 2268-94.
[21] Avetisyan Y, Sasaki Y, Ito H, 2001 *Appl. Phys.***B73**, 511-4.
[22] Kawase K, Shikata J, Minamide H, Imai K and Ito H, 2001 *Appl. Opt.* **40**, 1423-26.
[23] Bosomworth D R, 1966 *Appl. Phys. Lett.* **9**, 330-332.
[24] Hobden H V and Warner J, 1966 *Phys. Lett.* **22**, 243.
[25] Johns J W, 1985 *J. Opt. Soc. Am. B,* **2**, 1340.
[26] Gamache R R and Davies R W, 1983 *Appl. Opt.* **22**, 4013-18.
[27] Gasster S D, Townes C H, Goorvitch D and Valero, 1988 *J. Opt. Soc. Am.* **B5**, 593-601.

Journal of Physics: Conference Series **276** (2011) 012008 doi:10.1088/1742-6596/276/1/012008

Solutions of kW Continuous-wave All-fiber Laser

Yan Dapeng[1], Li Libo[1], Liu Xiaoxu[1], Min Dayong[2]

[1]Wuhan Raycus Fiber Laser Technologies Co., Ltd, Wuhan, China;
[2]Wuhan HuaGong Laser Engineering Co.,Ltd, Wuhan, China

E-mail: dyan@raycuslaser.com

Abstract. Solutions of kW continuous-wave (CW) all-fiber laser are proposed. In our solutions, master oscillator power amplifier (MOPA) configuration is applied. Output power of master oscillator is 10W, and then is amplified to 70W with 1st pre-amplifier and next scaled up to 400W. Finally, 400W fiber laser is used as a basic power unit, and 1000W all-fiber laser can be achieved by means of beam combining with large core double clad fiber (DCF) combiner. In this solution, fiber laser has good stability and reliability for dispersion coupling of pump source and inhibition of photon darkening effect in the fiber. In addition, this solution assures us realize a 1000W all-fiber laser product easily, and the cost is low.

1. Introduction

Fiber laser hold a number of advantages, such as small volume, high efficiency, energy saving and low using cost, etc. It is an ideal light source for industrial applications and be considered as the third generation of the most advanced industrial processing laser and arouses great interest all over the world. In the last decade, output power of fiber lasers increase dramatically. In the laboratory, several kilowatts can be achieved with a single fiber[1] and mature products of kW fiber laser can be offered in the field of industry[2].

In consideration of reliability and maintenance-free, all-fiber laser, which means all parts of the laser are optical fiber components literally is preferred in applications of industry. By now, only IPG company can provide mature product of all-fiber laser exceed 1kW. kW all-fiber laser can be used for laser cutting, laser welding, etc of metal and non-metallic materials as an alternative to conventional high power CO_2 laser and solid-state laser. And applications of it expand to the field of automobile making, shipbuilding and aircraft industry [3, 4], etc.

2. Solutions of kW CW all-fiber laser

Scheme of kW-class fiber laser of IPG company can be seen in the following Figure 1. Pump power of LDs is injected into the Yb-doped double clad fiber (YDCF) with fiber combiners and the laser oscillates within the cavity resonator consists of Fiber Bragg Gratings (FBGs). One of FBGs possesses high reflective (HR) for laser and the other one used as output coupler (OC), which has 4%~10% optical transmissivity for laser.

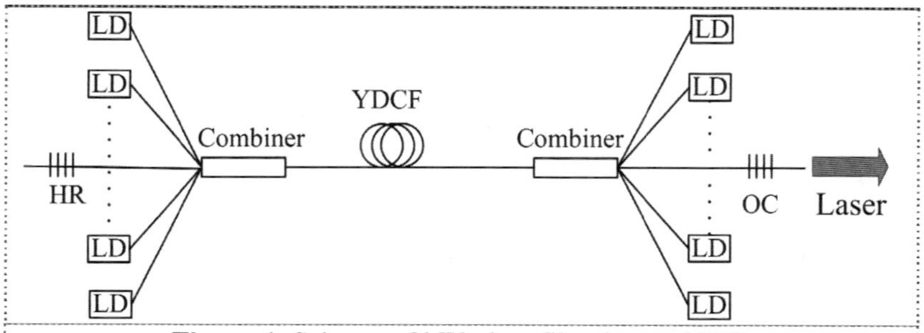

Figure 1. Scheme of kW-class fiber laser of IPG.

In this scheme, (N+1)×1 combiners are employed. Multi-mode fibers of combiners are spliced with LDs to get high pump power and input DCF fibers are spliced with FBGs. Served as cavity resonator, FBGs endure very high power. And above all, insertion loss of signal laser of combiners must be very low, always less than 0.1dB or they will burn out.

Different from this scheme, we present a kW fiber laser with MOPA configuration. In this solution, a 10W master oscillator is employed, and then is amplified to 70W with 1st pre-amplifier and next scaled up to 400W. Finally, 400W fiber laser is used as a basic power unit, and 1000W all-fiber laser can be achieved by means of beam combining with large core double clad fiber (DCF) combiner. Scheme of us can be seen in Figure2 and Figure3.

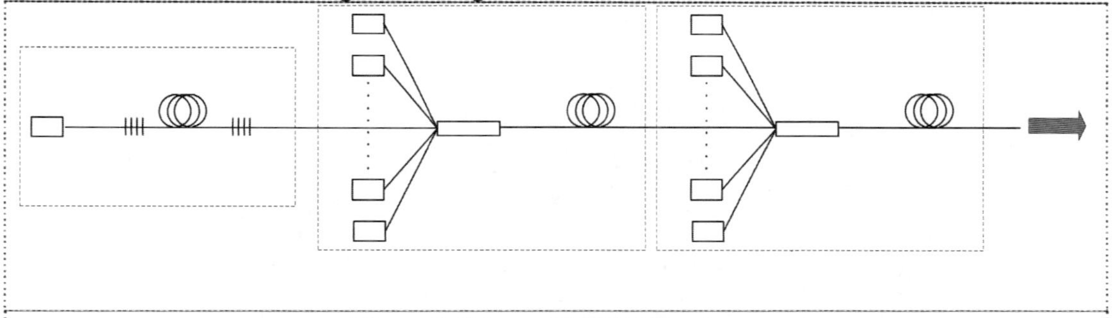

Figure 2. Scheme of 400W fiber laser.

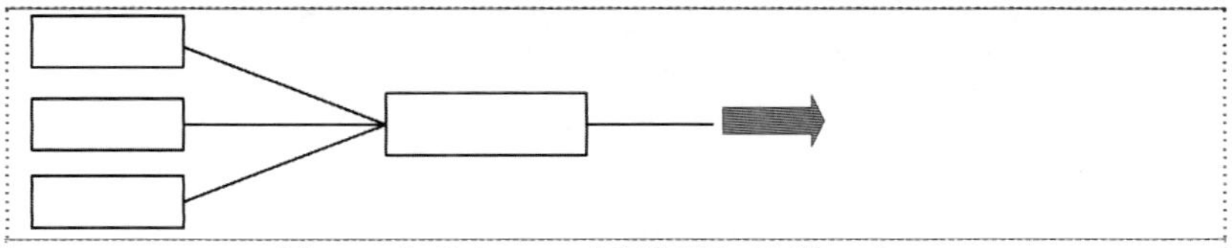

Figure 3. Scheme of 1000W all-fiber laser.

2.1. 10W Master Oscillator

6/125μm SM-YDCF is used as gain media and FBGs are also inscribed with 6/125μm photosensitive fiber used as the cavity resonator in the master oscillator. Figure4 shows reflection and transmission spectrums of HR FBG, whose reflectivity is higher than 99% for 1064nm. With this configuration, SM-DCF Bragg gratings can be inscribed in China and the cost is much lower than LMA-DCF Bragg gratings.

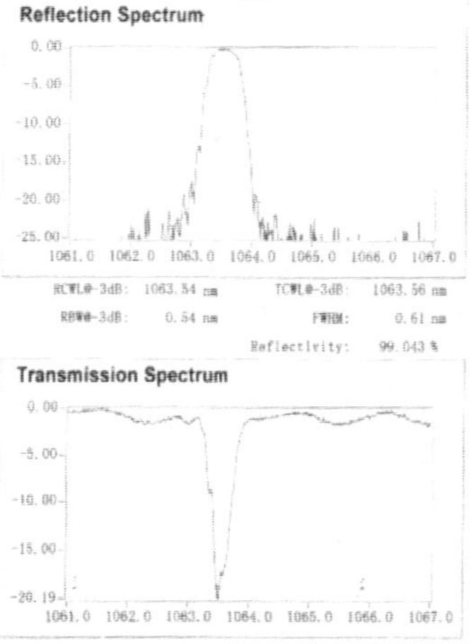

Figure 4. Spectrum of HR FBG.

2.2. 70W Pre-amplifier

10W single mode laser with 1064nm from the master oscillator is then injected into the 12/125μm YDCF with a (N+1)×1 side-coupler. And pump power of LDs can be coupled into the YDCF with multimode pigtails of the (N+1)×1 side-coupler. On the basis of our invented patent, we develop and manufacture side-coupling fiber coupler collaborated with leading domestic fiber-optic enterprises. Coupling efficiency for pump power of the coupler is higher than 90% and signal insertion loss is less than 0.2dB. Moreover, it is worth noting that the cost of it is much lower compared with similar foreign products. Figure5 shows side-coupler we used.

Figure 5. Appearance of side-coupler.

2.3. 400W Power Amplifier

Similar with the pre-amplifier, another (N+1)×1 side-coupler is used in the power amplifier. Signal of the amplifier is 70W laser from the pre-amplifier with 1064nm. In this stage, output of the laser is scaled to 400W then be used as a power unit in our solutions. Especially, 400W fiber laser can be applied in material processing, such as laser cutting and welding. Figure6 shows 400W fiber laser of us.

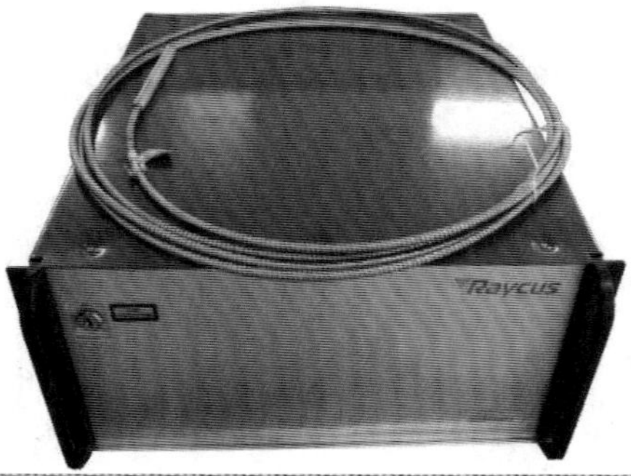

Figure 6. Appearance of 400W fiber laser of Raycus.

To achieve 400W power unit, a three-stage MOPA configuration is employed. As we know, photo darkening effect of active fiber can not be ignored for high power fiber laser. In our solution, pump powers are dispersed injected into the active fiber and photo darkening effect can be inhibited efficiently, which can assure the unit operates reliably. Furthermore, mode field of the multi-stage can match well each other and this leads its beam quality is better than one-stage fiber laser.

2.4. DCF Combiner

By means of DCF combiner, three power units of 400W fiber laser are combined into 1000W all-fiber laser. Unlike usual combiner, all pigtails of the DCF combiner are double clad fibers, several input ports and one output port. Laser from power unit transmits in the core of input DCFs separately and then combined as one laser beam.

To make the DCF combiner, double clad fiber is corroded and only inner clad and core of the fiber is left. Three fibers have small size core and one has large. Then ends of four such fibers are polished and placed tightly, three fibers with small core in and one fiber with large core out. Joint of fibers is fused and tapered and then recoated. Coupling efficiency of this DCF combiner is always higher than 98% for laser and water cooling must be employed for the combiner during the operation.

3. Conclusion

High power all-fiber laser is a direction of fiber laser and it is the only one mature technology in commercial field.With these solutions, we can realize a 1000W all-fiber laser product easily, and the cost is low. Furthermore, fiber laser based on these solutions has good stability and can be used in industrial laser processing reliably.

Acknowledgement

This work is supported by the National Key Technology R&D Program (2007BAF11B02) of China.

References

[1] Jeong Y, Sahu J K, Payne D N and Nilsson J. 2004 *Opt. Express* **12** 6088-6092
[2] http://www.ipgphotonics.com/products_1micron_lasers_cw_ylr-hpseries.htm
[3] Ozden H. 2007 *Welding Journal* **5** 26-29
[4] Quintino L, Costa A, Miranda R, Yapp D, Kumar V and Kong C J.2007 *Materials & Design* **28** 1231-1237

Photonic Crystal Waveguides in Terahertz Regime (invited)

Zhang Huaiwu*

State Key Laboratory of Electronic Films and Integrated Devices, University of Electronic Science and Technology of China, Chengdu, 610054

* email address: hwzhang@uestc.edu.cn; Tel: 028-83207063

Abstract: Using the finite difference time domain method, the electromagnetic field distribution of THz waves in photonic crystals (PCs) T-splitters and Y-splitters had been simulated. The simulation results show that those different T-splitters and Y-splitters can divide the power in an input wave guide equally between two output waveguides. By the improved T-splitter with a rod in the junction, we achieved the 84% amplitude- frequency characteristics consistency of pass-band from 1.12 THz to 1.22 THz, and surpass the 76% consistency of common T-splitter. The improved Y-splitter with a rod in the junction and without rod in the corners has widest -3db bandwidth 0.224 THz, and the amplitude reaches 1655.727. The improved Y-splitter has better performance than other Y-splitters. Introducing the photonic band gap structure with L-type defect composed of three defects. Three high-Q resonant frequencies appeared simultaneously in some monitor coordinates. The wavelength-add-drop properties of L-type defects may be used in multi-carrier communication and multi-frequency-monitoring for the THz regime. Also, a carefully designed PCs can be used as high Q narrowband filter in THz band. These results provide a useful guide and a theoretical basis for the developments of THz functional components.

1. Introduction

Terahertz (THz) radiation is a part of the electromagnetic spectrum. THz frequencies cannot clearly be classified to be either on the "electronic" side or on the "optics" side, commonly referred to as the "terahertz gap" (0.1-10 THz, λ=3mm-30μm). The THz frequency radiation has been proved as a fertile region in the electromagnetic spectrum and a powerful tool in scientific research and applications. Although enormous efforts have focused on the search for "THz" materials or alternative novel techniques to enable the construction of device components, much work remains [1]. THz waves have significant transmission loss in the atmosphere, so waveguide-based terahertz devices have become an important foundation for the THz transmission, as well as the bottle-neck for its practical applications.

One solution to the problem of THz manipulation is the application of photonic crystals (PCs), which is a low-loss periodic dielectric medium. With special design and construct the PCs can control the propagation of THz wave in certain directions with specified frequencies [2]. In this presentation, we introduce three kinds of PCS based THz devices: splitter, add-drop filter and High Q narrow-band filter.

2. Splitters

A useful waveguide device is a splitter, which divides the power in an input wave guide equally between two output waveguides [3]. Like the bend waveguide, the photonic band gap eliminates radiation loss and we need only deal with the possibility of reflection. Unlike the bend waveguide, it turns out that we cannot eliminate reflections by a symmetry argument, and must do something counterintuitive. We need to obstruct the output waveguides in order to increase transmission. In this paragraph, we demonstrate an improved method for guiding wave around T-splitter and Y-splitter, using photonic crystal waveguides.

For simplicity, we choose to study a 2-D photonic crystal of long dielectric rods in air on a square array with lattice constant of a. The lattice constant a is set to be 100μm, since it remains unchanged under rescaling for a photonic band gap (PBG) material, it can easily assure that the guided light will be in the infrared or THz region [3]. For example, if we choose a lattice constant a of 0.58 μm, the wavelength corresponding to the mid gap frequency will be 1.55 μm. Their 2D photonic crystals composed of long dielectric rods in air can be realized in a practicable form by incorporating the same lattice pattern in the thin space between the two metal plates of a parallel-plate waveguide, as discussed and demonstrated in References [2, 4]. According to Grischkowsky's work [5], High-resistivity and high-purity Silicon (Si) with the index of refraction (n) of 3.4176 is excellent for THz transmission due to its non-dispersive nature and low absorption in THz region. So here we choose Si rods for simulation. We had calculated the TE band gap structure for the square PCs with R = 18 μm for relatively wilder transverse-electric (TE) gap [6], two TE band gaps namely Ⅰ and Ⅱ exist in two different frequency regions. Gap Ⅰ is wider with a frequency range from 0.91 to 1.33 THz, while gap Ⅱ is much narrower with a frequency range from 2.23 to 2.28 THz. So we mainly focus on the transmission properties in Gap Ⅰ.

In Figure 1, it shows four different T-splitter configurations, (a) is the common T-splitter called T0, (b) is the T-splitter that has one rod in junction called T1, (c) is the T-splitter that has four rods in junction called T2 and (d) is the T-splitter that has nine rods in junction called T3. The field pattern of the propagating mode can be observed by a continuous wave (CW) excitation of the guided mode. According to the previous results [6], we set the CW excitation frequency to be 1.12 THz, and the corresponding electric field distribution pattern was simulated and shown in Figure 1. As can be clearly seen that, the mode is completely confined inside the guides, and the wave travels smoothly around the sharp bend, even though the radius of curvature of the bend is on the order of the wavelength [3].

In order to monitor the field amplitude that travels after T-splitter, we set a monitor point after the T-splitters at the position of (-400, 0, 0), labeled as point A in Figure 1. We study the transmission properties of waveguide splitters using a vector finite-difference time-domain program with quartic perfectly matched layer boundaries [7]. In our simulation, a dipole located at the entrance of the waveguide creates a pulse with a Gaussian envelope in time. Figure 1 shows the frequency dependence of field amplitude, which is monitored inside the T-splitters guides at point A. As can be seen that, four curves have the similar features that, field amplitude increases sharply at 0.843 THz and decreases quickly at 1.50 THz. The amplitude reaches its highest point near the frequency 1.12 THz and 1.22 THz, those features are decided by the structure of photonic crystals. But obviously the four curves are different. For example, T0 firstly reach to its sub-peak value of 785 at 1.12 THz and increase to its peak value of 1035 at 1.22 THz. For T1 its sub-peak is 810 at 1.12 THz and increasing to the highest point of 965 at 1.22 THz. T2 and T3 are more special than T0 and T1 curves. That is, T2 firstly reaches to its highest point of 905 at 1.12 THz and then decreases to the sub-peak point of 835 at 1.22 THz. Similarly, T3 firstly reaches to its highest point of 755 at 1.12 THz and decreases to the sub-peak value of 465 at 1.22 THz. Those differences are decided by the bend structure of T-splitters [8].

From figure 1 we can also see that the pass-band of T-splitters is from 1.12 THz to 1.22 THz. We define the amplitude-frequency characteristics consistency of pass-band as δ, its value is the ratio of the sub-peak amplitude comparing with that of the highest peak. So the value of δ implies the consistency of the pass-band. Considering the practical fact in signal propagation, that the loss at higher frequency is a little larger than that at lower frequency, higher transparency at higher frequency is prefect. This is the case of T1 curve. Thus considering the consistency quality as well as the demand in signal propagation, T1 is the best choice comparing with other configuration [6].

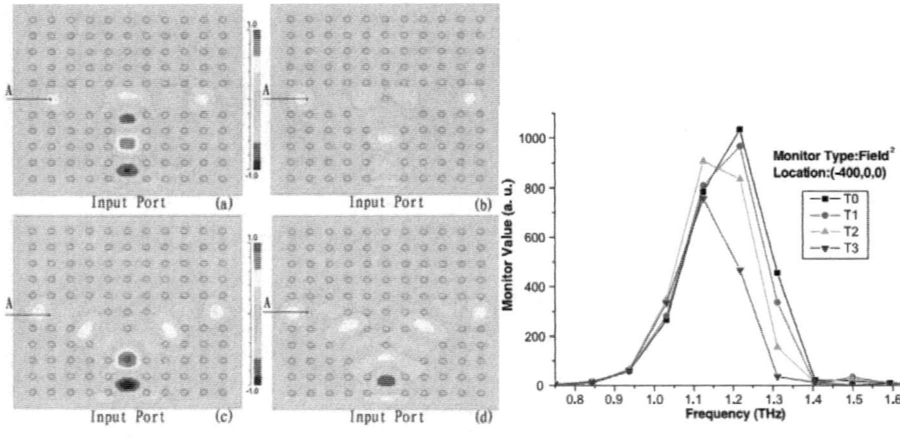

Fig.1 The electric field pattern in the vicinity of the T-splitter at frequency of 1.12 THz and field amplitude is monitored inside the T-splitters waveguides.

The amplitude-frequency characteristics of improved T-splitter were superior to the common T-splitter's. But the output and input direction of T-splitter are vertical position, and the T-splitter can not output the EM wave parallel to the input direction. In this paragraph, the different Y-splitters with a junction and two corners are studied.

Y-splitters are composed of a junction and two corners. The different Y-splitters with same junction structure and different corners structure are investigated. In Figure 2, it shows four different Y-splitter configurations, (a) is the conventional Y-splitter called Y00 without rod in the junction and corners, (b) is the Y-splitter that has no rod in the junction and one rod in the corners called Y01, (c) is the Y-splitter that has no rod in the junction and three rods in the corners called Y02 and (d) is the Y-splitter that has no rod in junction and six rods in the corners called Y03. The field pattern of the propagating mode can be observed by a CW excitation of the guided mode. According to the previous results [6], we set the CW excitation frequency to be 1.12 THz, and the corresponding electric field distribution pattern was simulated and shown in Figure 2. As can be clearly seen that, the mode is completely confined inside the guides, and the wave travels smoothly around the junction and corner, even though the radius of curvature of the bend is on the order of the wavelength. The observation can be explicated by the PCs' basic feature that preventing light from propagating in certain directions with specified frequencies [3].

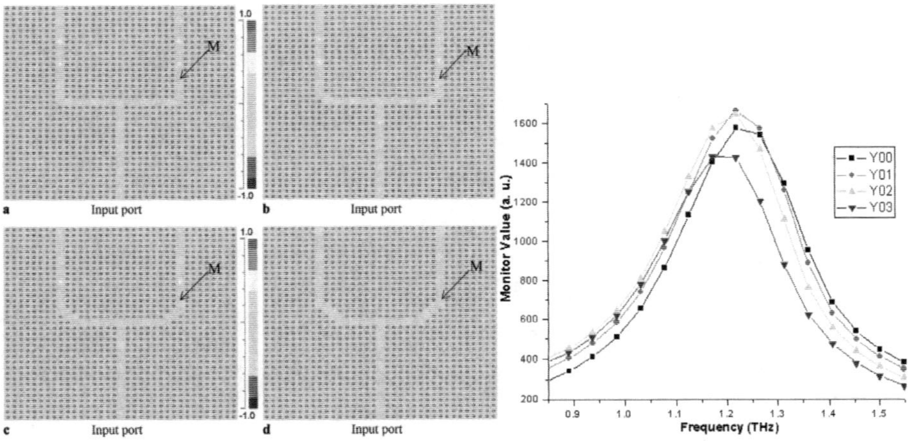

Fig. 2 The electric field pattern in the vicinity of the Y-splitters for frequency 1.12 THz. The field amplitude was monitored inside the Y00, Y01, Y02 and Y03 Y-splitters waveguides.

In order to monitor the field amplitude that travels after those Y-splitters, a monitor point after the Y-splitters was set at the position of (900, 400), labeled as point M in Figure 2. We study the amplitude-frequency transmission properties of waveguide splitters using a vector finite-difference time-domain program with quartic perfectly matched layer boundaries [7]. The frequency dependence of field amplitude which is monitored inside the Y-splitters guides at point M was shown in Figure 2. The curves of Y00, Y01, Y02 and Y03 reach the peak of 1577.247 at 1.219 THz, 1665.585 at 1.219 THz, 1645.622 at 1.219 THz and 1431.75 at 1.172 THz, respectively. The Y02 has highest peak amplitude and the Y03 has lowest peak amplitude. The -3db bandwidth is a key and useful evaluation index for bandpass properties. The -3db bandwidth of Y00, Y01, Y02 and Y03 are 0.216 THz, 0.211 THz, 0.209 THz and 0.215 THz, respectively. The Y00 has widest -3db bandwidth. Those differences are determined by the different corners structure of Y-splitters [9]. The effective lengths of corners bend waveguide variation lead to the amplitude frequency characteristics change [8].

In Figure 3, it shows four different Y-splitter configurations, (a) is the Y-splitter called Y10 with one rod in the junction and without rod in the corners, (b) is the Y-splitter that has one rod in the junction and corners called Y11, (c) is the Y-splitter that has one rod in the junction and three rods in the corners called Y12 and (d) is the Y-splitter that has one rod in junction and six rods in the corners called Y13. The field pattern of the propagating mode also can be observed by a CW excitation of the guided mode. The CW excitation frequency was also designated to be 1.12 THz, and the corresponding electric field distribution pattern was simulated and shown in Figure 3. As can also be clearly observation that, the mode is completely confined inside the guides, and the wave travels smoothly around the junction and corner [9].

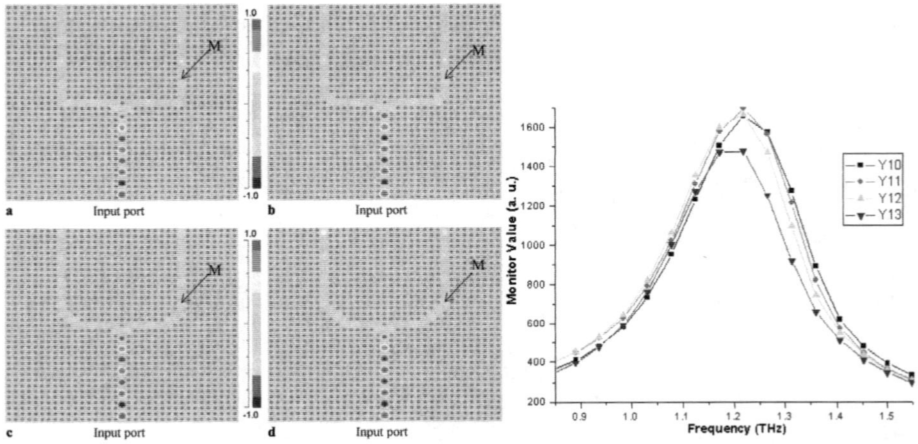

Fig. 3 The electric field pattern in the vicinity of the Y-splitters for frequency 1.12 THz. The field amplitude was monitored inside the Y10, Y11, Y12 and Y13 Y-splitters waveguides.

In order to monitor the field amplitude that travels after those Y-splitters, we also set a monitor point after the Y-splitters at the position of (900, 400), labeled as point M in Figure 3. The monitored conditions of the amplitude-frequency properties of the waveguide are same as the former. The frequency dependence of field amplitude which is monitored inside the Y-splitters guides at point M was shown in Figure 3. The curves of Y10, Y11, Y12 and Y13 reach the peak of 1655.727, 1696.228, 1663.664 and 1476.106 at 1.219 THz, respectively. The Y11 has highest peak magnitude and the Y13 has lowest peak magnitude. The -3db bandwidth of Y10, Y11, Y12 and Y13 are 0.224 THz, 0.208 THz, 0.207 THz and 0.211 THz, respectively. The Y10 has the widest -3db bandwidth. Those differences are determined by the different corners structure of Y-splitters [9].

Figure 2, 3 show that the transmission properties of the different Y-splitters. Y-splitters can divide the power in an input wave guide equally between two output waveguides. Those figures show that the Y-splitters with three rods in the corners have the highest peak amplitude at 1.219 THz. The Y-splitters with six rods in the corners have a feature that the amplitude at 1.172 THz has been increased relatively to the amplitude at 1.219 THz. The Y-splitters without rod in the corners have the widest -3db bandwidth. Those differences are determined by the different corners structure of Y-splitters. The effective lengths of corners bend waveguide variation lead to the amplitude frequency characteristics change [8].

In comparison with the property of different Y-splitters, the frequency and normalized amplitude of peak and the -3db bandwidth had been listed in Table 1. Table 1 evidently shows that only Y10, a rod in the junction and without rod in the corners, has the higher amplitude of peak and wider -3db bandwidth than the conventional Y-splitter Y00 [9].

Table 1 The comparison of different Y-splitters

Y-splitter	Peak		-3db bandwidth
	Freq.(THz)	Amp. (a. u.)	(THz)
Y00	1.219	1577.247	0.216
Y01	1.219	1665.585	0.211
Y02	1.219	1645.622	0.209
Y03	1.172	1431.75	0.215
Y10	1.219	1655.727	0.224
Y11	1.219	1696.228	0.208
Y12	1.219	1663.664	0.207
Y13	1.219	1476.106	0.211

Using the finite difference time domain method, the electromagnetic field distribution of terahertz waves in different photonic crystals Y-splitters with a junction and two corners were simulated. The simulation results show that those different Y-splitters can divide the power in an input wave guide equally between two output waveguides. The improved Y-splitter with a rod in the junction and without rod in the corners has widest -3db bandwidth 0.224 THz, and the amplitude reaches 1655.727. The improved Y-splitter has better amplitude frequency performance than other Y-splitters.

3. Dropped/Added filters

Terahertz waves could handle broadband signals and are expected to be applied to ultrafast wireless communications [1]. In a communication system, one of the most important issues is that the work frequency or narrowband signal should be selectively upload and download. Conventional components can not perform well in THz regime, but PCS is an excellent candidate [10]. In this work, PCs with both line- and cavity-defects are designed. By control the shape of the cavity defects, one, two, or three upload/download frequencies are realized, rendering this PCS filter a good candidate for multi-carrier communication and multi-frequency-monitoring in the THz regime.

In Figure 4, line-shaped defects designed as a straight waveguide is formed by removing one column of rods from the square PCs array, and with a single defect in the vicinity of the waveguide as illustrated in each figure. The coordinate of every single-defect are assigned as (300, 0), (300,100), (400, 0), (300,-100) and (200, 0), respectively. Relative to the coordinate of (300, 0), those single defect are defined as center, upward, rightward, downward and leftward defect, respectively. The wavelength-add-drop properties of the waveguide are investigated using a vector finite-difference time-domain program with quartic perfectly matched layer boundaries [7]. In our simulation, a dipole located at the entrance of the waveguide creates a pulse with a Gaussian envelope. The field amplitude of monitored inside the corresponding waveguides are shown in Figure 4 as M.

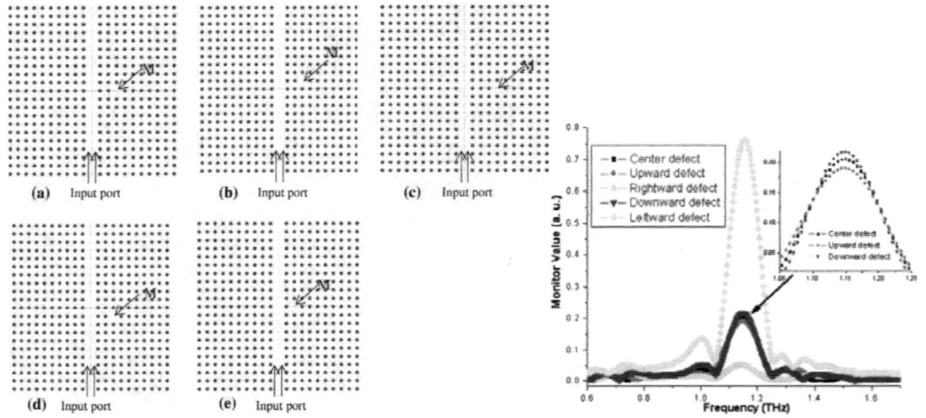

Fig.4 Schematics of 2D PBG with typical square lattices composed of the Si rods in air background with a=100 um and r = 18 μm. The coordinate of single-defects are (a) (300, 0), (b) (300,100), (c) (400, 0), (d) (300,-100) and (e) (200, 0), respectively. The field amplitude of monitored of different single-defect inside PBG

The field amplitude of monitored of different single-defect inside PBG are shown in Figure 4. Figure 4 shows that the input pulses propagating through the linear waveguide are trapped by the single-defect. The reason is that the single-defect has strong localization state. The single-defect acted as resonator with special frequency and quality factor. The resonance peak of the center, upward, rightward, downward and leftward single-defect are 0.206 at 1.149 THZ, 0.191 at 1.149 THZ, 0.05 at 1.14 THZ, 0.217 at 1.149 THZ, and 0.764 at 1.155 THZ, respectively. It shows that the upward, center and downward single-defect have same resonance frequency. But the filed amplitude increase from 0.191 to 0.217, the obvious reason is that the much closer of the single-defect to the input port, the more energy can be obtained from line-shaped defects waveguides. Meanwhile, with the single-defect position vary from the leftward, to rightward, the resonance frequency decrease from 1.155 THz to 1.14 THz and the field amplitude decrease from 0.764 to 0.05, as shown in Figure 4. This can be understood by the fact that much more energy would be obtained from the leftward single defect nearest to the line-shaped defects. In comparison, the resonance frequency and amplitude of the single-defect in vertical direction relative to the line-shaped defects waveguide are more sensitive to its position than that of the single-defect in parallel direction, and the difference are caused by the propagation properties of the line-shaped defects waveguides [11]. According to Figure 4, the sharpest curve of leftward defect shows most quality factor Q, and the descending trends of factor Q are the downward, center, upward and rightward single defect, respectively. The reason of most quality factor Q is that more energy can be obtained from the single-defect which is closer to the line-shaped defects.

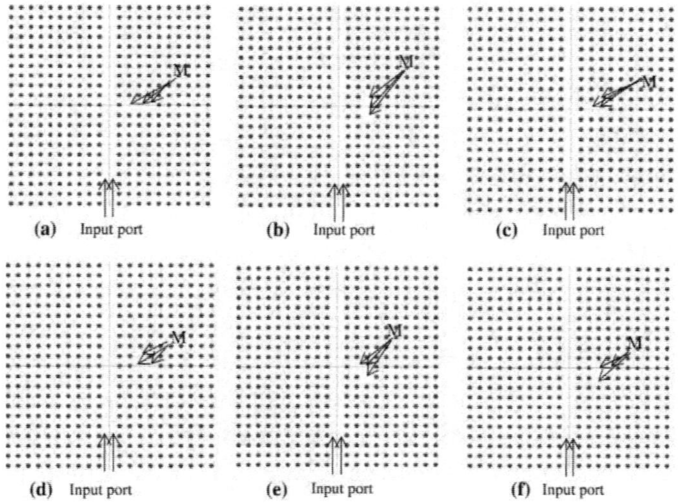

Fig. 5 Schematics of 2D PBG. (a) and (b) are vertical and parallel defects relatively to the line-shaped waveguides in the middle of PCs. (c), (d), (e) and (f) are L-type defects of upper-left, upper-right, lower-left, and lower-right, respectively.

The structures with three defects are shown in Figure 5. Two kinds line-defects as shown in Figure 5 (a) and (b), the line defects are named as vertical and parallel line defects relatively to the line-shaped waveguides in the middle of PCs, and four kinds L-type defects as shown in Figure 5 c-f, the four kinds L-type defects are named as upper-left, upper-right, lower-left and lower-right L-type defects, respectively. The field amplitude of monitored inside the corresponding waveguides are shown in Figure 5 as M. For comparing the amplitude of three single-defects, every single-defect was monitored. Other monitored conditions of the wavelength-add-drop properties of the waveguide are same as the former.

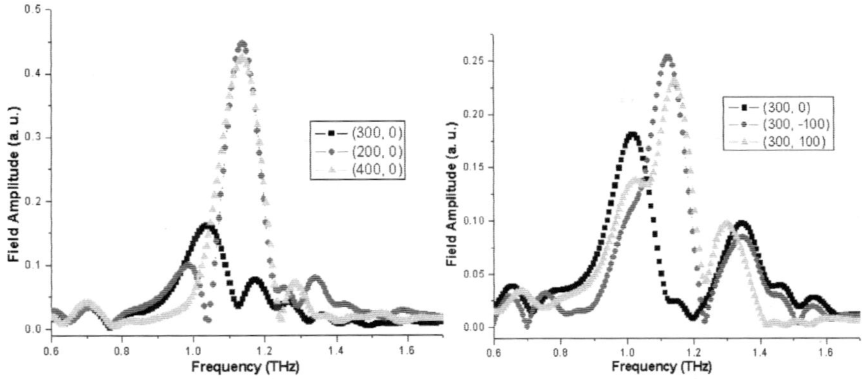

Fig. 6 The field amplitude of monitored inside the vertical and parallel line-shape defects

Figure 6 shows the field amplitude of monitored inside the vertical line-shape

Journal of Physics: Conference Series **276** (2011) 012009 doi:10.1088/1742-6596/276/1/012009

defects relatively to the line-shaped waveguides in the middle of PCs. The amplitude of coordinate at (300, 0) have two little difference of the resonance peak of 0.162 at 1.038 THz and 0.077 at 1.173 THz. The amplitude of (200, 0) have two large difference of the resonance peak of 0.099 at 0.987 THz and 0.448 at 1.137 THz. The amplitude of (400, 0) also have two large difference of the resonance peak of 0.424 at 1.137 THz and 0.072 at 1.284 THz. Figure 6 shows that the endpoint of the vertical line-shape defects have the high amplitude and the middle of the vertical defects has little different amplitude of two resonant frequency [11].

Figure 6 also shows the field amplitude of monitored inside the parallel line-shape defects relatively to the line-shaped waveguides in the middle of PCs. The amplitude of (300, 0) have two little difference of the resonance peak of 0.181 at 1.017 THz and 0.098 at 1.347 THz. The amplitude of (300, -100) also have two difference of the resonance peak of 0.254 at 1.122 THz and 0.085 at1.347 THz. The amplitude of (300, 100) also have two major of the resonance peak of 0.231 at 1.14 THz and 0.097 at 1.299 THz, meanwhile, an unremarkable resonance peaks of 0.138 exist in 1.023 THz. Figure 6 shows that the endpoint of the parallel line-shape defects have the high amplitude and the middle of the parallel defects has little different amplitude of two resonant frequency [11].

Figure 6 shows that the vertical or parallel line-shape defects, composed of three single-defects, have more than one resonance frequency. The defects structure change cause the localization state change, the result is the resonant state and mode change. The different field patterns measured could be explained by the eigenmodes of the defects and the excitations can be understood considering how the waveguide modes couple to each of the eigenmodes. The different resonant structure between the vertical and parallel line-shape defects had obtained different energy from the straight line-shape waveguide, and caused the difference peak location between Figure 6 in the same coordinate (300, 0). Figure 6 also show that the line-shape with three single-defects have the similar amplitude feature that two endpoint large and middle point small. The feature can be interpreted in accordance with Reference 3. The resonance cavity, the vertical or parallel line-shape three single-defects, that is effectively surrounded by reflected walls. The endpoint directly access to energy from three reflected walls, conversely, the middle-point directly access to energy from two reflected walls. The direct reflected walls number difference lead to the magnitude difference. From Figure 4, the amplitude peak of (200, 0) single-defect has more than (400, 0)'s, the similar reason cause the amplitude peak of (200, 0) has more than (400, 0)'s in the vertical line-shape defects. And the similar phenomenon and reason appear in the parallel line-shape defects.

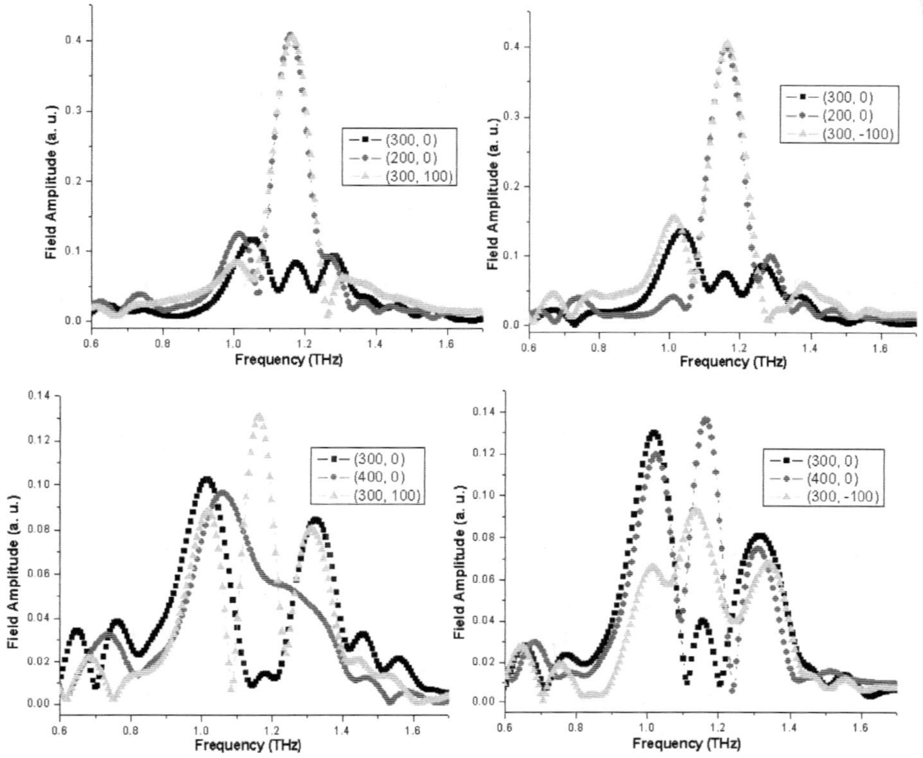

Fig. 7 (color) The field amplitude of monitored inside the upper-left, lower-left, upper-right and lower-right L-type defects.

Figure 7 shows the field amplitude of monitored inside the upper-left L-type defects. The amplitude of (300, 0) have three little difference of the resonance peak of 0.117 at 1.053 THz, 0.083 at 1.176 THz and 0.094 at 1.281 THz. The amplitude of (200, 0) have two large difference of the resonance peak of 0.126 at 1.014 THz and 0.409 at 1.158 THz. The amplitude of (300, 100) also have two large difference of the resonance peak of 0.084 at 1.008 THz and 0.405 at 1.161 THz. Figure 4 shows that three resonant frequency appeared simultaneously in the (300, 0) of upper-left L-type defects, and the main resonant frequency appeared in the other two monitored coordinates.

Figure 7 also shows the field amplitude of monitored inside the lower-left L-type defects. The amplitude of (300, 0) have three little difference of the resonance peak of 0.136 at 1.035 THz, 0.075 at 1.158 THz and 0.087 at 1.266 THz. The amplitude of (200, 0) have two large difference of the resonance peak of 0.399 at 1.161 THz and 0.100 at 1.287 THz. The amplitude of (300, -100) have two large difference of the resonance peak of 0.156 at 1.011 THz and 0.405 at 1.161 THz.

In Figure 7, the field amplitude of monitored inside the upper-right L-type defects is shown. The amplitude of (300, 0) have two little difference of the resonance peak of 0.102 at 1.014 THz and 0.085 at 1.326 THz. The amplitude of (400, 0) have main resonance peak of 0.097 at 1.059 THz. The amplitude of (300, 100) have three little difference of the resonance peak of 0.089 at 1.017 THz, 0.131 at 1.161 THz and

0.081 at 1.311 THz. Figure 7 shows that three resonant frequency appeared simultaneously in the (300, 100) of upper-right L-type defects, two resonant frequency appeared in the (300, 0), and one low-Q factor resonant frequency in the (400, 0).

As to the case of lower-right L-type defects, the amplitude of (300, 0) have three little difference of the resonance peak of 0.131 at 1.017 THz, 0.040 at 1.155 THz and 0.081 at 1.317 THz. The amplitude of (400, 0) also have three little difference of the resonance peak of 0.120 at 1.023 THz, 0.136 at 1.164 THz and 0.075 at 1.311 THz. The amplitude of (300, -100) also have three little difference of the resonance peak of 0.067 at 1.014 THz, 0.093 at 1.134 THz and 0.068 at 1.338 THz. Figure 7 shown that three resonant frequency appeared simultaneously in each monitor point of the lower-right L-type defects, the high-Q and small difference of amplitude of resonant appeared in (400, 0), the high-Q and large difference of amplitude of resonant appeared in (300, 0), and lower-Q resonant appeared in (300, -100).

Figure 7 show that the different L-type defects structures, composed of three single-defects, have different resonant frequency, amplitude and peak shape. But there have the similar feature that three resonant frequencies simultaneously appear in some monitored coordinates. The specific feature possible is interpreted below [3, 11, 12]. Firstly the three single-defects obtain energy from the straight line-shape waveguide, and then there exchange energy and resonate in the L-type cavity. The energy exchange and resonant inevitable had been effected by the corner with the existence of bound states localized in the vicinity of the corner. The multi-reflection will appear in the course of energy of one of single-defect reach the other two single-defect and return to the beginning coordinates. The complex reciprocity cause the resonance split to three points. In other words, the corner in the L-type defects lead to the resonant frequency split to three in some coordinates. The different field patterns measured also could be explained by the eigenmodes of the defects. The different three single-defects composed different L-type defects. Figure 4 shows the different single-defects with different resonances, and causes the different L-type defects have different resonant appeared Figure 7.

Our result provide a new way, introducing L-type defects composed of three defects, to develop the wavelength-add-drop device that three high-Q resonant frequencies appear simultaneously in one defect coordinates. The wavelength-add-drop properties of L-type defects may be used in multi-carrier communication and multi-frequency-monitoring for the THz regime.

4. Hgh-Q narrow-band filters

Also, a carefully designed PCs can be used as high Q narrowband filter in THz band. We introduce a waveguide–cavity–waveguide structure to the PC, as shown in Fig. 8, and it is found that the number and structure of cavity has very important influences on the characteristic of the filter. Our simulation results show that by control the number of the cavity we can achieve single THz narrow-band signals.

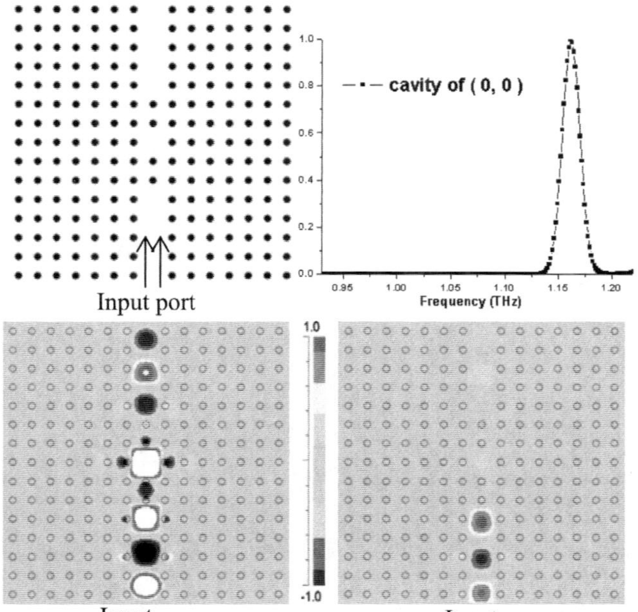

Fig. 8 (Top) the narrow-band filters structure and properties of the single-defect cavity. (Bottom) the electric field pattern for frequency 1.161 THz and 1.151THz

The narrow-band filter properties of the central cavity were shown in Figure 2. The transmission spectrum shows that 100% peak at cavity resonance frequency 1.161 THz, and Δf at half-maximum (50% transmission) is about 0.0193 THz with Q≈60.15. The fractional width $\Delta f/f_0$ at half-maximum (50% transmission) is precisely equal to 1/Q, where Q is the quality factor of the cavity mode when excited internally. This sharp peak means that the device acts as a narrow-band filter. The light is transmitted for frequencies near the resonant frequency of the cavity, and is reflected for somewhat lower or higher frequencies. The existence of the resonance peak conforms to intuition: near the resonant frequency, light from the input waveguide can couple into the cavity, and the cavity in turn can couple into the output waveguide [3].

The field pattern for transmission at resonance is shown in Figure 8. If we shift the frequency by only 0.01 THz, the transmission drops to less than 0.3, corresponding to the fields in the middle panel. The sharp dip in the transmission at around 1.13 THz corresponds to the zero-slope band edge of the waveguide mode, where coupling light through the device is especially difficult. The oscillations at high and low frequencies correspond to frequencies outside the band gap, where energy propagates through the crystal instead of being confined to the waveguide and cavity.

5. Conclusion

The application of photonic crystal, which is a low-loss periodic dielectric medium, can solve to the problem of THz manipulation. With special design and construct the PCs can control the propagation of THz wave in certain directions with specified frequencies. Three kinds of splitter, add-drop filter and High Q narrow-band filter had been introduced for the application of THz. These results provide a useful guide and a theoretical basis for the developments of THz functional components.

Acknowledgments

This work was partly supported by the National Basic Research Program (973) of China, under Grant No. 2007CB310407, Foundation for Innovative Research Group of the NSFC under Grant No.60721001, the International S&T Cooperation Program of China under Grant No.2006DFA53410 and 2007DFR10250.

References:

1. Hosako, I., Sekine, N., Patrashin, M., et al. At the Dawn of a New Era in Terahertz Technology, Proceedings of the IEEE , 95, 8,1611 – 1623, (2007)
2. A. Bingham, Yuguang Zhao, and D. Grischkowsky, THz Parallel Plate Photonic Waveguides, Appl. Phys. Lett. 87, 051101(2005)
3. J.D. Joannopoulos, R.D. Meade, and J.N. Winn, Photonic Crystals: Molding the Flow of Light (second edition) (Princeton, New York, 2007)
4. Yuguang Zhao and D. Grischkowsky, "Terahertz demonstrations of effectively two dimensional photonic bandgap structures," Optics Letters, 31, 1534-1536 (2006)
5. D. Grischkowsky, Søren Keiding, Martin van Exter and Ch Fattinger, "Far-infrared time-domain spectroscopy with terahertz beams of dielectrics and semiconductors", J. Opt. Soc. Am. B, 7, 2006-2015 (1990)
6. Sheng LI, Huai-Wu ZHANG, Qi-Ye WEN, et al. Improved Amplitude- Frequency Characteristics for T-splitter Photonic Crystal Waveguides in Terahertz Regime. Applied Physics B:Lasers and Optics, 2009, 95:745-749
7. J. C. Chen and K. Li, Quartic Perfectly Matched Layers for Dielectric Waveguides and Gratings, Microwave Opt. Technol. Lett. 10, 319–323 (1995).
8. Attila Mekis, J. C. Chen, I. Kurland, et al., High Transmission through Sharp Bends in Photonic Crystal Waveguides, Phys. Rev. Lett., 77,3787,(1996)
9. Sheng LI, Huai-Wu ZHANG, Qi-Ye WEN, et al. Improved Y-splitter Photonic Crystal Waveguides in Terahertz Regime. Applied Physics B:Lasers and Optics, 2010, 99:709-716
10. Susumu Noda, Alongkarn Chutinan and Masahiro Imada, Trapping and emission of photons by a single defect in a photonic bandgap structure, Nature, 407, 608 (2000)
11. Sheng LI, Huai-Wu ZHANG, Qi-Ye WEN, et al. Wavelength-drop properties of L-type defects in photonic bandgap structure for the terahertz regime. Opt. Quant. Electron. 2009, 41:159–168
12. Attila Mekis, Shanhui Fan, and J. D. Joannopoulos, Bound states in photonic crystal waveguides and waveguide bends, Phys. Rev. B, 58, 4809 (1998)

Review of self-focusing of high power lasers in large-mode-area optical fibers

Chujun Zhao, Ying Li, Dajun Lei, Jianguo Wen, Hua Yang, Shuangchun Wen[†], Dianyuan Fan

Key Laboratory for Micro/Nano Optoelectronic Devices of Ministry of Education, School of Computer and Communication, Hunan University, Changsha 410082, China

E-mail: scwen@vip.sina.com

Abstract. The main progress about the self-focusing of high power lasers in large-mode-area optical fiber has been reviewed. The theoretical models including the self-focusing effects have been discussed. Some different views on the whole beam self focusing and small scale self-focusing effects in optical fiber have been introduced. Moreover, the possible methods exceeding the bulk-media self-focusing threshold have been discussed and explored.

Self-focusing is a consequence of the dependence of the refractive index in dielectric materials on the light intensity [1]. The phase shift resulting from the nonlinearity leads to whole beam self-focusing and small scale self-focusing. Whole beam self-focusing extends over the whole wave front and leads to a narrowing of the beam profile. Small scale self-focusing increases the contrast across the beam profile, and small intensity ripples or irregularities will grow quickly in amplitude and lead to individual damage tracks. Therefore, large beams tend to break up into a number of filaments. Self-focusing of laser beams in Kerr media was recognized to be the primary mechanism limiting the maximum power that could be extracted from the glass amplifier rods [2].

In recent years, with tremendous progress in fiber lasers [3], nonlinear effects become important and can limit the power scalability in the fiber [4]. Some well-known nonlinear effects, such as stimulated Brillouin scattering, stimulated Raman scattering, four-wave mixing, self phase modulation, optical damage and etc., can be alleviated by using large-mode-area (LMA) fibers or increasing the signal bandwidth. Self-focusing is independent of the mode area and will determine the power scaling of the fiber lasers ultimately. Furthermore, there have also been recent demonstrations where peak powers in optical fibers have approached the self-focus limit [5-9]. Consequently, it is becoming more urgent to have a better understanding of transmission behavior in optical fibers near the limit of nonlinear self-focusing.

It is essential to develop comprehensive theoretical models to simulate the self-focusing effect. Self-focusing happens in bulk optical media due to the interaction between the Kerr nonlinearity and diffraction [10]. While for the optical fiber, the waveguide effect has a strong impact on the effect of the Kerr nonlinearity and the self-focusing is determined by the interaction between the Kerr nonlinearity, diffraction and waveguide effects. To solve the problem, Researchers from Sandia

[†] The authors are with Key Laboratory of Micro/Nano Optoelectronic Devices of Ministry of Education and School of Computer and Communication, Hunan University, Changsha 410082, China.
E-mail: scwen@vip.sina.com

National Laboratories have developed a family of fiber-amplifier models to design optimized fiber amplifiers and to analyze their performance [11]. However, the sophisticated models have not considered the mode competition yet. University of Michigan have studied nonlinear beam propagation in a step-index core fiber for pure fundamental mode input using both BPM numerical model and coupled-mode analysis [12]. The detailed results, however, have not been reported. We have also modeled the self-focusing in the large-mode-area optical fiber using wide-angle BPM method, and the model considering the time-dependent nonlinear effects are being validated. It is hard to get a complete physical insight of the problem using purely numerical simulations. To simplify the work, some researchers have adopted some approximations to solve the problems, such as Gaussian approximation [10] and perturbation method [13].

Despite the fact that nonlinear self-focusing has been well studied and reasonably well understood in bulk media, there has been some disputes regarding nonlinear self-focus in fibers. A notable one is on how to determine the critical power for self-focusing (P_{crit}). G. Fibich and A. L. Gaeta adopted a phenomenological definition for P_{crit} as the minimum power at which the focused peak irradiance reached 10^4 times the initial peak irradiance [16]. R. L. Farrow et al defined P_{crit} as the lowest power at which the beam radius decreased to less than 1 μm, corresponding to loss of resolution by the calculation mesh [17]. For more realistic consideration, the intrinsic damage threshold irradiance of material and beam narrowing should be considered in together when determining the P_{crit}. For the beam size variation below the self-focus limit, M. Igarash et al. have found oscillatory behaviors [12], again contradicted by some recent work [16-17]. Besides, there are some reports on the propagation characteristics below or near the limit of critical power for self-focusing [16-20].

Another problem we need to consider is that whether the small scale self-focusing can occur in large-mode-area optical fiber. It is well known that if a mode satisfies the waveguide boundary conditions, it can propagate forward with negligible loss, or the mode will cut off. For the modulated beams propagating in the optical fiber, if the mode excited by the modulations can not satisfy the waveguide condition, the mode will cut off and the small scale self-focusing can not happen. However, the intense light propagating in the optical fiber makes the refractive index distribution change, and thus the boundary conditions. With beam propagating, the mode will perhaps evolve to satisfy the waveguide structure. For the complicated relations among nonlinear effects and waveguide boundary condition, whether the small scale self-focusing can happen needs us to explore.

Up to now, there are not convincing experimental results to clarify the self-focusing in optical fiber. In 1987, P. L. Baldeck et al. observed the ring structure with picosecond laser pulses propagating in a large-core optical fiber of 100 μm diameter [21]. They attributed it to self-focusing. Later, some other researchers argued that the ring structure of self-focusing light field is attributed to the intrinsic refractive index profile of the fiber used in experiment [22, 23] or mode competition in a multimode fiber used [24]. We can do some calculation to see whether the self-focusing can happen using the data given in ref. [21]. The result shows that the nonlinear change of refractive index in the fiber core is about five orders of magnitude smaller than the index difference between the core and the cladding. Such a small change of refractive index can hardly modify the field distribution in the fiber, and thus self-focusing is unlikely to occur. By the calculations in Ref. [19], self-focusing plays a major role only if the modal area at low power exceeds 500 μm^2, or in straight fiber, the core radius exceeds 20 μm. For smaller modal areas, the damage threshold is reached at powers well below the self-focusing power.

Suppression of beam self-focusing may be regarded as one of the main problems in the design of high-peak-power laser. For the fiber laser system, there are few reports on how to suppress the self-focusing. Enlightened by the idea used in bulk system, the simple method is to use material with small n_2 and improve the uniformity. Also, we can suppress the self-focusing or elevate the self-focusing threshold by fiber design. A. D. Yablon et al. have presented optical fiber with a strong central dip at the center of a strictly axisymmetric refractive index profile can guide stable fundamental modes at more than 10 times the bulk media critical power for self-focusing [25]. Also, we can design mode filtering optical fiber to act as spatial filter. Since laser beams tend to be more intense at the center than at the edge, the beam is slowed at the center with respect to the edge and converges. We can design optical fiber to compensate the variation of refractive index.

Self-focusing in large-mode-area optical fiber will limit the power scaling of the fiber laser and amplifier ultimately. Although self-focusing in bulk solid-state laser have been well studied, there are some disputes on propagation characteristics of self-focusing in large-mode-area optical fiber below or near the self-focusing limit. Moreover, there are not convincing experimental results to validate the theoretical results. So it is imperative to carry out theoretical and experimental work to further understand the nonlinear characteristics of large-mode-area optical fiber.

References

[1] Koechner W 2006 *Solid State LaserEngineering* (Berlin: Springer)
[2] Boyd R W, Lukishova S G, Shen Y R 2008 *Self-focusing: Past and Present* (Berlin: Springer)
[3] Hecht J 2009 *Laser Focus World* **12** 45
[4] Tünnermann A, Schreiber T, Röser F, Liem A, Höfer S, Zellmer H, Nolte S and Limpert J 2005 *J. Phys. B: At. Mol. Opt. Phys.* **38** S681
[5] Cheng M Y, Chang Y C, A. Galvanauskas, Mamidipudi P, Changkakoti R, and Gatchell P 2005 *Opt. Lett.* **30** 58
[6] Teodoro F D and Brooks C D 2005 *Opt. Lett.* **30** 3299
[7] Brooks C D and Teodoro F D 2007 *Opt. Commun* **280** 424
[8] Schmidt O, Rothardt J, Roser F, Linke S, Schreiber T, Rademacher K, Limpert J, Ermeneux S, Yvernault P, Salin F, and Tunnermann A 2007 *Opt. Lett.* **32** 1551
[9] Desmoulins S and Teodoro F D 2008 *Opt. Express* **16** 2431
[10] Dong L 2008 *J. Lightwave Technology* **26** 3476
[11] Kliner D A V, Bambha R P, Do B T, Farrow RL, Fèv J P, Fox B P, Hadley G R, Hansen A, Hoffman H J, Hotoleanu M, Hoops A A, Hsu W L, Koplow J P, Koponen J, Moore S W, Schmitt R L, Schrader P E, Simmons J H, Simmons-Potter K, Smith A V, Soderlund M, Thomes W J, and Wien G 2008 *Proc. of SPIE* **6952** 695202
[12] Igarashi M and Galvanauskas A 2004 *OSA Trends Opt. Photon.* **96** 513
[13] Liao S Y 2009 Doctoral thesis of Tsinghua University (Beijing, Tsinghua University)
[14] Sun L, Marciante J R 2007 *Journal of the Optical Society of America. B* **24** 2321
[15] Tempea G and Brabec T 1998 *Opt. Lett.* **23** 762
[16] Fibich G and Gaeta A L 2000 *Opt. Lett.* **25** 335
[17] Farrow R L, Kliner D A V, Hadley G R, Smith A V 2006 *Opt. Lett.* **31** 3423
[18] Farrow R, Hadley G R, Smith A V, and Kliner D 2007 *Proc. SPIE* **6453** 645309.
[19] Smith A V, Do B T, Hadley G R, Farrow R L 2009 *IEEE Journal of Selected Topics in Quantum Electronics* **15** 153
[20] Hadley G R, Farrow R L, Smith A V 2007 *Proc. of SPIE* **6453** 64531B
[21] Baldeck P L, Raccah F, and Alfano R R 1987 *Opt. Lett.* **12** 588
[22] Hua Y M, Li Q, Chen Y L, Chen Y X 1990 *Opt. Commun.* **79** 459
[23] Chen Z H, Chen X Y and He C X 1992 *Opt. Commun.* **90** 39
[24] Chiang K S 1993 *Opt. Commun.* **95** 235
[25] Yablon A D and Jasapara J 2007 *Proc. of SPIE* **6453** 64531D

Infrared Laser Therapy using IR absorption of biomolecules

K Awazu[1,2,3], **K Ishii**[1] **and H Hazama**[1]

[1] Medical Beam Physics Lab., Graduate School of Engineering, Osaka University, 2-1 Yamadaoka, Suita, Osaka, 565-0871, Japan
[2] Research Institute of Nuclear Engineering, University of Fukui, 3-9-1 Bunkyo, Fukui, Fukui, 910-8507, Japan
[3] Institute for Chemical Research, Kyoto University, Gokasho, Uji, Kyoto, 611-0011, Japan

E-mail: awazu@see.eng.osaka-u.ac.jp

Abstract. Since numerous characteristic absorption lines caused by molecular vibration exist in the mid-infrared (MIR) wavelength region, selective excitation or selective dissociation of molecules is possible by tuning the laser wavelength to the characteristic absorption lines of target molecules. By applying this feature to the medical fields, less-invasive treatment and non-destructive diagnosis with absorption spectroscopy are possible using tunable MIR lasers. A high-energy nanosecond pulsed MIR tunable laser was obtained with difference-frequency generation (DFG) between a Nd:YAG and a tunable Cr:forsterite lasers. The MIR-DFG laser was tunable in a wavelength range of 5.5–10 μm and generated a laser pulses with an energy of up to 1.4 mJ, a pulse width of 5 ns, and a pulse repetition rate of 10 Hz. Selective removal of atherosclerotic lesion was successfully demonstrated with the MIR-DFG laser tuned at a wavelength of 5.75 μm, which corresponds to the characteristic absorption of the ester bond in cholesterol esters in the atherosclerotic lesions. We have developed a non-destructive diagnostic probe with an attenuated total reflection (ATR) prism and two hollow optical fibres. An absorption spectrum of cholesterol was measured with the ATR probe by scanning the wavelength of the MIR-DFG laser, and the spectrum was in good agreement with that measured with a commercial Fourier transform infrared spectrometer.

1. Introduction

Since numerous characteristic absorption lines due to the molecular vibration exist in the mid-infrared (MIR) wavelength range, this wavelength range is often termed the molecular fingerprint region. Selective excitation of a molecule is possible by tuning the laser wavelength to the characteristic absorption line of the target molecule. MIR absorption spectroscopy is frequently used to component analysis and structural analysis. By applying these characteristic absorption lines in the MIR wavelength range for the medical field, selective and less-invasive treatments and non-destructive diagnosis are expected to be possible using tunable MIR lasers.

For example, cholesteryl esters, cholesterol bound to a fatty acid such as oleic acid via an ester bond, are a principal component of atherosclerotic lesions. Cholesteryl esters have a characteristic absorption line at a wavelength of 5.75 μm originated from the C=O stretching vibration of ester bonds. It has been reported that cholesteryl esters were selectively removed from atherosclerotic lesions by using a free-electron laser (FEL) tuned to a wavelength of 5.75 μm with a pulse width of microseconds [1–3]. Since FEL can generate high-power laser in the MIR wavelength range and widely tunable, it has been used to various medical applications [4] such as soft tissue ablation [5], cardiovascular surgery [1–3], neurosurgery, preventive dentistry [6], lithotripsy [7], ophthalmology, and orthopaedic surgery. However, extremely large-size and expensive equipments have prevented FEL from the practical applications.

Recently, a tabletop MIR tunable laser using difference-frequency generation (MIR-DFG laser) has been developed [8, 9]. The MIR-DFG laser is tunable within a MIR wavelength range of 5.5–10 μm and generates laser pulses with a pulse energy up to 1.4 mJ, pulse width of 5 ns, and repetition rate of 10 Hz. In general, short pulse lasers in the nanosecond range are more suitable to induce mechanical ablation with suppressing thermal effects compared with pulsed lasers in the microsecond or longer range [10]. In this research, we have evaluated the effectiveness of irradiating the MIR-DFG laser with a wavelength of 5.75 μm for selective and less-invasive treatment of atherosclerotic lesions.

On the other hand, the Fourier transform infrared spectrometer (FT-IR) is one of the most popular spectrometer in the MIR wavelength range. In recent years, it is also applied to the analysis of remote samples by combining attenuated total reflection (ATR) method with optical fibres. This technique can be applied to diagnose a human body and to determine the optimum wavelength for the less-invasive treatment using MIR tunable lasers before the treatment. However, the types of the optical fibre are limited in the MIR wavelength range and the transmittance of the optical fibres for the MIR rays is low compared with those for the visible and near-infrared rays [11]. Since the brightness of the light source used in FT-IR is low, the signal intensity becomes weak by using a long optical fibres. In addition, the size of commercial ATR probe is too large to introduce into a human body. In recent years, hollow optical fibres which can transmit MIR rays have been developed [12, 13]. By using the hollow optical fibres and the MIR-DFG laser, higher signal intensity is expected compared with FT-IR, and the single laser source can be used for both treatments and diagnosis.

In this paper, our preliminary results of novel therapeutic and diagnostic applications using a tabletop MIR tunable laser were reported. That is, results of *in vitro* experiments for less-invasive laser angioplasty with the MIR-DFG laser at a wavelength of 5.75 μm and ATR spectroscopy of biomolecules performed by scanning the wavelength of the MIR-DFG laser are shown.

2. Experimental apparatus

2.1. Mid-infrared tunable laser

Figure 1(a) schematically depicts the tabletop mid-infrared tunable laser used in this research. The tunable MIR laser in a wavelength range of 5.5–10 μm was obtained by DFG between a *Q*-switched Nd:YAG laser with a wavelength of 1064 nm (Tempest 10, New Wave Research, Inc., USA) and the Cr:forsterite laser tunable within a wavelength range of 1190–1320 nm. The wavelength of the Cr:forsterite laser was varied by rotating the rear mirror of the optical resonator and was measured with a wavelength meter (WS6-IR, HighFinesse GmbH, Germany). The wavelength of the MIR output λ_3 was calculated from the wavelengths of the Nd:YAG laser λ_1 and the Cr:forsterite laser λ_2 using the following relationship:

$$\lambda_3^{-1} = \lambda_1^{-1} - \lambda_2^{-1}.$$

The Cr:forsterite laser can be pumped by the fundamental wavelength of the Nd:YAG laser. However, since the buildup time of the Cr:forsterite laser varies from 70 to 200 ns depending on its oscillating wavelength, it is difficult to synchronize the Nd:YAG and the Cr:forsterite laser pulses at the $AgGaS_2$ crystals by using an optical delay. Therefore, a *Q*-switched Nd:YAG laser different from that used for DFG (INDI-40-10, Spectra-Physics, inc., USA or Tempest 300, New Wave Research, Inc., USA) was used for pumping the Cr:forsterite laser. A digital delay/pulse generator (DG535, Stanford Research Systems, Inc., USA) was used to synchronize the Nd:YAG and the Cr:forsterite laser pulses.

Two $AgGaS_2$ crystals with the same dimensions and the same cutting angles were used to obtain high output energy and to compensate a displacement of the optical axis. The height, width, and length of

the $AgGaS_2$ crystals were 9, 12, and 24 mm, respectively. The type II phase-matching condition was used to obtain a higher effective nonlinearity compared with the type I phase-matching [14].
The prototypes of the tunable MIR laser and the tunable Cr:forsterite laser as a part of the MIR laser were developed by RIKEN and Kawasaki Heavy Industries, Ltd. (KHI). Then, full automation of the wavelength tuning and stabilization of the output energy have been carried out by KHI.

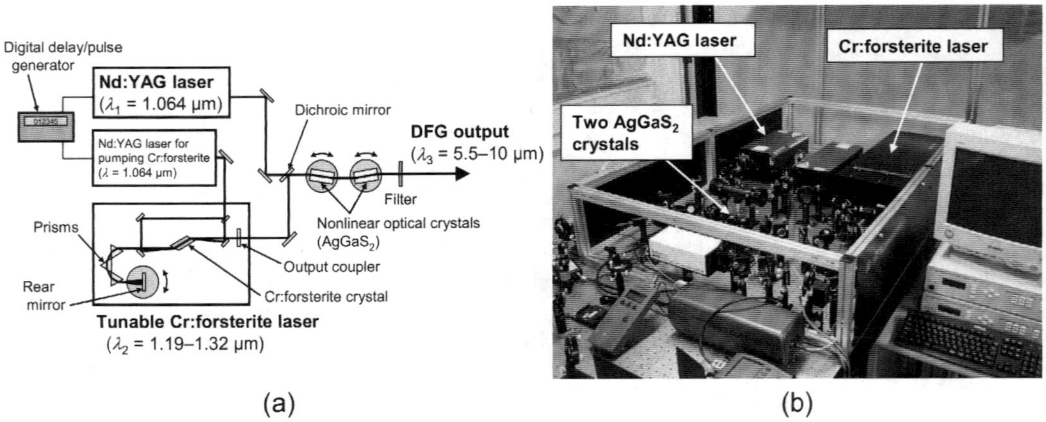

Figure 1. Schematic drawing (a) and photograph (b) of the MIR tunable laser. Pulses from two near infrared lasers are combined in two $AgGaS_2$ crystals to produce MIR tunable pulses.

2.2. ATR probe

As the diameter of commercial ATR probes is large (> 10 mm), we have designed and manufactured an ATR probe. Figure 2 shows a schematic drawing and photograph of the ATR probe. A prism was made of diamond which is chemically stable and not harmful in a human body. Type I b diamond which has weak internal absorption within the wavelength ranges of about 4–6 μm and 7–15 μm was used [15]. Two slanted surfaces of the prism do not satisfy the condition for the total internal reflection, so that the surfaces were coated by nickel to enhance the reflectance. The inner and outer diameter of the hollow optical fibres were 700 μm and 850 μm, respectively, and the length of each fibre was up to 2 m. The hollow optical fibre can also transmit a visible laser as a guide laser [12, 13]. The housing at the tip of the ATR probe holding the diamond ATR prism and the hollow optical fibres was made with polyether ether ketone (PEEK) and the cross-sectional area of the housing was 3.55 mm × 1.05 mm.

The MIR-DFG laser was focused by an off-axis parabolic mirror with an effective focal length of 150 mm and introduced into one of the two hollow optical fibres attached to the diamond ATR prism. Laser energy came back from the diamond ATR prism via another hollow optical fibre was measured with a laser energy meter with a detection limit of 0.2 μJ (PE9, Ophir Optronics, Israel).

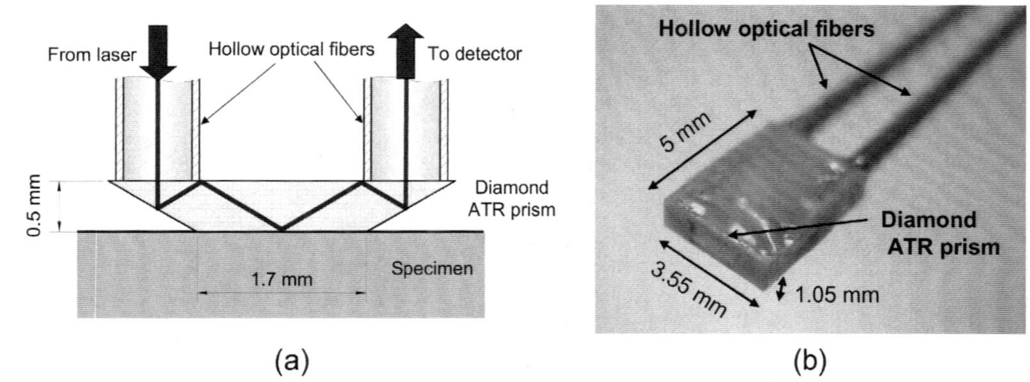

Figure 2. Schematic drawing (a) and photograph (b) of the ATR probe comprised of an diamond prism and two hollow optical fibres.

3. Materials and methods

3.1. Animals and sample preparations

An animal model for spontaneous hypercholesterolemia, coronary atherosclerosis, and myocardial infarction named myocardial infarction-prone Watanabe heritable hyperlipidemic (WHHLMI) rabbits [16, 17] were used as an atherosclerotic model, and Japanese white rabbits were used as a normal model. WHHLMI rabbits (females, 24 months old) were provided from Institute for Experimental Animals, Kobe University Graduates School of Medicine. Atherosclerotic and normal rabbits were sacrificed by an intravenous injection of pentobarbital sodium (50 mg/kg) (Nembutal, Dainippon Sumitomo Pharma Co., Ltd., Japan). Thoracic aortas of them were extracted and rinsed with saline. This study was approved by the Institutional Animal Experiments Committee and conducted in accordance with the guidelines of animal experimentation at Osaka University.

3.2. Laser irradiation and histological analysis

The MIR-DFG laser was focused using a ZnSe lens with a focal length of 100 mm to a diameter of about 140 μm. Then the laser was irradiated normal to the surface of the thoracic aorta samples from the intimal side. After laser irradiation, the samples were embedded by Tissue-Tek O.C.T. Compound (Sakura Finetechnical Co., Ltd., Japan). The embedded samples were frozen and sliced by using a cryostat microtome (Leica CM-1850, Leica Microsystems GmbH, Germany) for histological evaluation. Sections were cut vertically to the tissue surface at 10 μm intervals and attached onto slide glasses. The sections with a crater produced by laser irradiation were photographed using an optical microscope (DM IRBE, Leica Microsystems) with a cooled colour CCD camera (Nebula QICAM, Q Image).

3.3. Sample preparation for ATR spectroscopy

Cholesterol (C8667, Sigma-Aldrich, USA) was dissolved in carbon tetrachloride at a concentration of 6×10^{-2} mol/L and the solution with a volume of 5 μL was dropped onto a BaF_2 window with a thickness of 1 mm and diameter of 13 mm. After drying the droplet, an absorption spectrum was measured with ATR spectroscopy and was compared with that measured with a microscopic Fourier transform infrared spectrometer (FT-IR) (FT-520, HORIBA, Japan).

4. Results and discussion

4.1. Irradiation effects of the MIR-DFG laser for aorta samples

Figure 3 shows typical absorption spectra of the atherosclerotic and normal aortas measured using the FT-IR. The absorption peak originated from the C=O stretching vibration of the ester bond in cholesterol esters was observed at the wavelength of 5.75 μm only from atherosclerotic lesions.

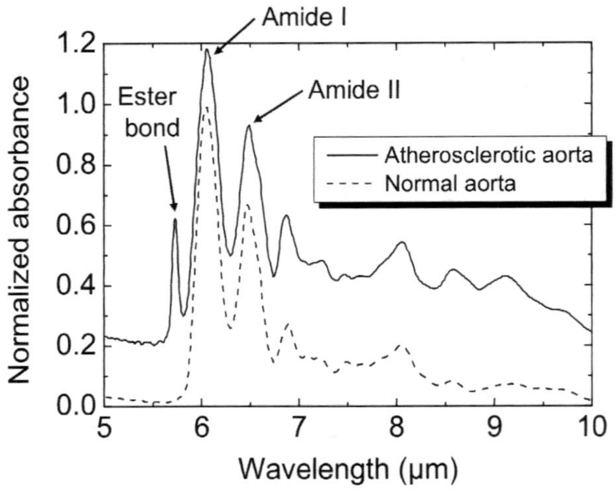

Figure 3. Typical infrared absorption spectra for the intimal layers of atherosclerotic and normal aortas. The absorption peak originated from C=O stretching vibration of the ester bond in cholesteryl esters was observed at 5.75 μm only from atherosclerotic lesions. (The spectrum for the atherosclerotic aorta is shifted for easier viewing.)

Figure 4 shows photomicrographs of the sections of the atherosclerotic and normal aortas after laser irradiation in a dry condition. Tuning the laser to 5.75 μm allowed selective removal of an atherosclerotic lesion without damaging normal tissue (Fig. 4(a) and (c)). On the other hand, tuning the laser to 6.09 μm which corresponds to the absorption peak of the amide I band resulted in perforation of both aortas (Fig. 4(b) and (d)). Therefore, it is suggested that the selective and less-invasive removal of atherosclerotic lesion is possible using the MIR-DFG laser tuned at the wavelength of 5.75 μm.

To evaluate the effectiveness of the irradiation of the MIR-DFG laser with a wavelength of 5.75 μm in more practical case, aorta samples were irradiated in a wet condition. Figure 5(a–d) shows photomicrographs of the atherosclerotic thoracic aorta after laser irradiation in a wet condition with the wavelength of 5.75 μm, where the average power density was 80 W/cm^2. Ablation was observed at 1 s and longer irradiation times. Figure 5(e–h) shows the photomicrographs of normal thoracic aorta after laser irradiation in a wet condition with the wavelength of 5.75 μm, where the average power density was 80 W/cm^2. Figure 6 shows the ablation depth as a function of the laser irradiation time. The experimental conditions were same as those for the results shown in Fig. 5. For all irradiation times, the ablation depth of the atherosclerotic aortas was larger than that of the normal aorta, and the maximum ablation depth for the normal aortas remained to be less than 100 μm. In the wet condition, laser power density is attenuated due to the absorption by water on the sample surface. Although the power densities required to ablate the aortas were higher compared to those required to ablate dry aortas, selective and less-invasive ablation using the MIR laser with a wavelength of 5.75 μm was also successfully demonstrated for wet aortas. The photomicrographs shown in Fig. 5 suggest that the short pulse laser with a pulse width of 5 ns is able to induce mechanical ablation with less-thermal damage.

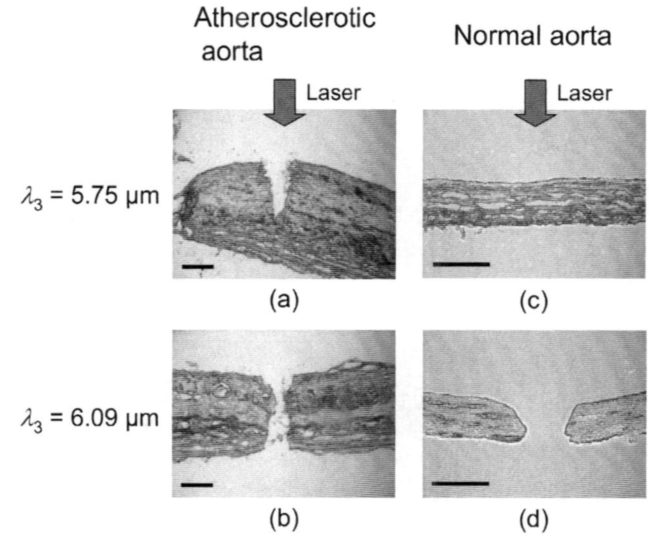

Figure 4. Photomicrographs of aorta cross-sections after laser irradiation. The top of each image is the intimal layer. When the MIR-DFG laser was tuned to 5.75 μm, the atherosclerotic lesion was removed without damaging normal tissue. Laser exposure time was 3 s (30 pulses) and average power densities were 50 and 40 W/cm^2 for atherosclerotic and normal aortas, respectively. Scale bars at the bottom of each image are 200 μm long.

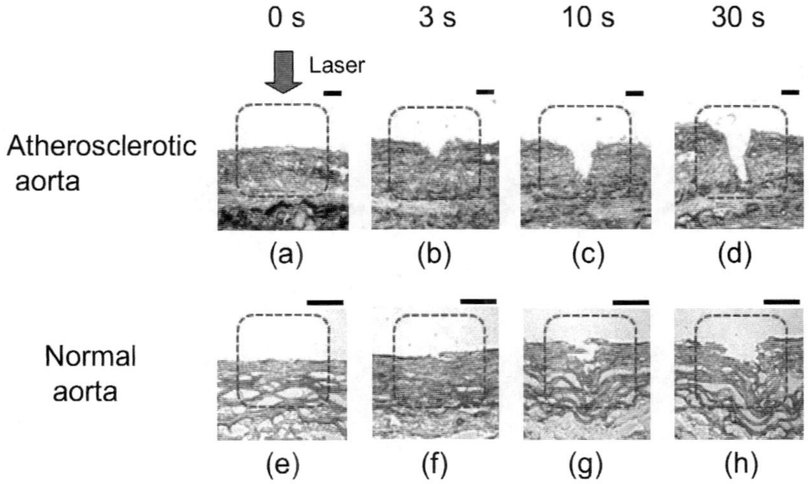

Figure 5. Photomicrographs of the sections of atherosclerotic and normal thoracic aortas after laser irradiation at a wavelength of 5.75 μm in a wet condition, where the average power density was 80 W/cm^2 and irradiation time was 0–30 s. The top of each image is the intimal layer. Scale bars at the top of each image are 100 μm long.

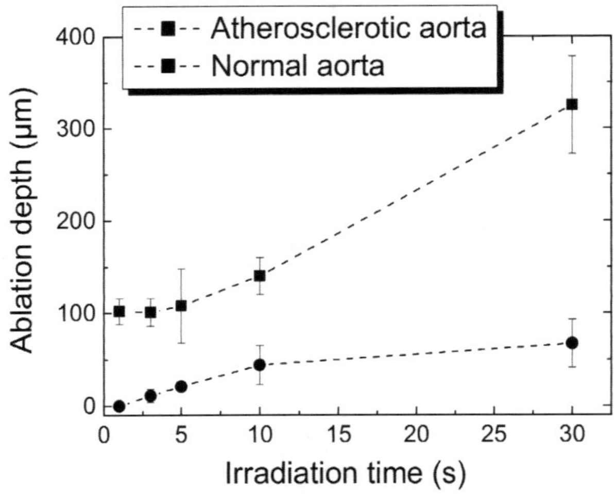

Figure 6. Ablation depth after laser irradiation with a wavelength of 5.75 μm in a wet condition as a function of the laser irradiation time, where the average power density was 80 W/cm^2. Dots and error bars are the averaged values and the standard deviations from ten samples, respectively.

To perform *in vivo* treatments using MIR lasers, an irradiation fibre was designed and manufactured as shown in Fig. 7. A hollow optical fibre with an inner and outer diameters of 700 and 850 μm, respectively, was used. A diamond lens with a focal length of 0.6 mm and a diameter of 1.3 mm was attached to the hollow optical fibre using a housing made with PEEK. Typical transmittance of the hollow optical fibre with a length of 2 m at a wavelength of 5.75 μm was about 70% in a straight condition. The maximum diameter of the irradiation fibre was 1.8 mm, so that the irradiation fibre can be introduced into blood vessels, digestive organs and so on via a catheter or an endoscope for angioplasty, lithotripsy, etc.

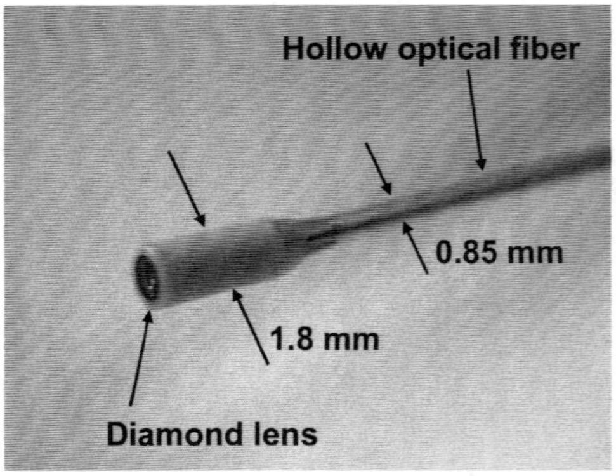

Figure 7. Photograph of the irradiation fibre developed for less-invasive MIR laser therapy such as angioplasty, lithotripsy, etc. The focal length of the diamond lens was 0.6 mm.

4.2. ATR spectroscopy

Figure 5 shows the comparison of absorption spectrum of cholesterol measured with the ATR probe by scanning the wavelength of the MIR-DFG laser and that measured with an FT-IR. In this case, the lengths of the inlet and outlet hollow optical fibres were 85 cm and 35 cm, respectively. Two spectral patterns were in good agreement as shown in Fig. 5. In both spectra, a strong absorption peak caused by the C–H bending vibration was observed at the wavelength of 6.83 μm. Typical transmittance of the ATR probe with a pair of 2 m long hollow optical fibres was about 8% at a wavelength of 6 μm. The tiny ATR probe was made with harmless materials and can be introduced into a human body via a catheter or endoscope. It is supposed that the combination of the ATR probe and the MIR-DFG laser is useful for diagnosis with MIR absorption spectroscopy.

The present control software of the MIR-DFG laser is designed to obtain the best performance for each wavelength by optimizing every parameters, i.e., the angles of the rear mirror of the Cr:forsterite laser and two $AgGaS_2$ crystals and the delay time between the Nd:YAG and Cr:forsterite laser pulses. Under the present circumstances, it takes a few minutes after inputting a wavelength until laser irradiation becomes possible. Therefore, a long time is required to perform the spectroscopic measurement over a wide wavelength range. For example, the measurement of the absorption spectrum in Fig. 8 spent about 2 hours and it was necessary to continuously operate the equipments during the measurement. Thus, we are now developing a new control software to automatically scan the wavelength of the MIR-DFG laser. By using this software, the time required to measure the absorption spectrum in Fig. 8 is estimated to be a few minutes and the whole measurement is performed automatically.

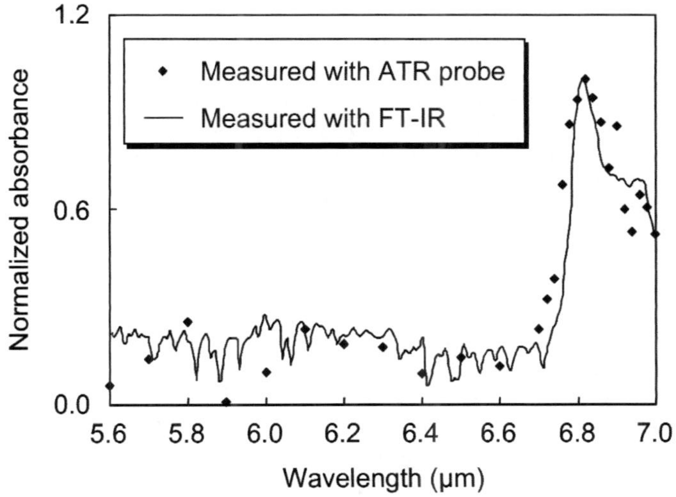

Figure 8. Infrared absorption spectra of cholesterol measured with the ATR probe and the FT-IR.

5. Conclusion

The objective of this research was to demonstrate the effectiveness of nanosecond pulsed laser at 5.75 μm irradiation for selective and less-invasive removal of atherosclerotic lesions. *In-vitro* experiments using rabbit aortas in both dry and wet condition showed that the irradiation of nanosecond pulsed laser with the wavelength of 5.75 μm could remove atherosclerotic lesions effectively. In addition, the irradiation effect to normal aortas at the wavelength of 5.75 μm was small. As the result, it was confirmed that less-invasive interaction to normal thoracic aortas could be induced by the wavelength of 5.75 μm, the average power densities of 60–80 W/cm^2 and the irradiation time shorter than 30 s.

We have developed a system for non-destructive and less-invasive diagnosis with the ATR spectroscopy. A tiny ATR probe was developed using a diamond ATR prism and two hollow optical fibres. The absorption spectra of cholesterol was measured with the ATR probe by scanning the wavelength of the tunable MIR laser, and it was in good agreement with that measured with the FT-IR. The ATR probe can be introduced into a human body via a catheter or endoscope. This fibre-based ATR spectroscopic technique using a high-energy MIR tunable laser realizes the non-labelling chemical diagnosis inside the body, and subsequent selective and less-invasive treatment.

Acknowledgements
This work was supported by Takeda Science Foundation, Japanese Foundation for Research and Promotion of Endoscopy, and Grants-in-Aid for Scientific Research (KAKENHI). The authors thank Dr. Minoru Yokoyama of KHI for his technical cooperation in the experiments using the MIR-DFG laser, Prof. Masashi Shiomi of Institute for Experimental Animals, Kobe University Graduate School of Medicine for providing us WHHLMI rabbits, and Mr. Shinya Yamada and Mr. Hideki Tsukimoto for their assistance in the experiments.

References
[1] Awazu K, Nagai A and Aizawa K 1998 *Lasers Surg. Med.* **23** 233
[2] Awazu K and Fukami Y 2001 *Nucl. Instrum. Methods Phys. Res.* A **475** 650
[3] Nakajima Y, Iwatsuki K, Ishii K, Suzuki S, Fujinaka T, Yoshimine T and Awazu K 2006 *J. Neurosurg.* **104** 426
[4] Edwards G S, Allen S J, Haglund R F, Nemanich R J, Redlich B, Simon J D and Yang W-C 2005 *Photochem. Photobiol.* **81** 711
[5] Edwards G, Logan R, Copeland M, Reinisch L, Davidson J, Johnson B, Maciunas R, Mendenhall M, Ossoff R, Tribble J, Werkhaven J and O'Day D 1994 *Nature* **371** 416
[6] Heya M, Sano S, Takagi N, Fukami Y and Awazu K 2003 *Lasers Surg. Med.* **32** 349
[7] Watanabe M, Kajiwara H, Awazu K and Aizawa K 2001 *Surg. Today* **31** 626
[8] Hazama H, Takatani Y and Awazu K 2007 *Proc. SPIE* **6455**, 645507
[9] Hazama H, Ishii K, Tsukimoto H and Awazu K 2008 *SPIE Newsroom*, 10.1117/2. 1200801.1000
[10] Niemz M H 2003 *Laser-tissue interactions: fundamentals and applications 3rd Enlarged Ed.* (Berlin: Springer)
[11] Harrington J A 2004 *Infrared fibers and their applications* (Washington: SPIE press)
[12] Shi Y W, Wang Y, Abe Y, Matsuura Y, Miyagi M, Sato S, Taniwaki M and Uyama H 1998 *Appl. Opt.* **37** 7758
[13] Shi Y W, Ito K, Matsuura Y and Miyagi M 2005 *Opt. Lett.* **30** 2867
[14] Dmitriev V G, Gurzadyan G G and Nikogosyan D N 1999 *Handbook of Nonlinear Optical Crystals 3rd Rev. Ed.* (Berlin: Springer)
[15] Walker J 1979 *Reports Progress Phys.* **42** 1605
[16] Shiomi M, Ito T, Yamada S, Kawashima S and Fan J 2003 *Arterioscler. Thromb. Vasc. Biol.* **23** 1239
[17] Ito T, Yamada S, Tamura T and Shiomi M 2005 *Exp. Anim.* **54** 413

Journal of Physics: Conference Series **276** (2011) 012012 doi:10.1088/1742-6596/276/1/012012

Photoelectron angular distributions from photodetachment of negative ions in strong laser fields

Lihua Bai[1], Xianghe Ren[2], Yuheng Liu[1], Tingting Cui[1], Yan Wang[1], Huifang Zhang[1] and Dongmei Deng[1]

[1]Department of Physics, Shanghai University, Shanghai 200444, People's Republic of China

[2]Institute of Advance Energy, Kyoto University, Gokasho, Uji, Kyoto, 611-0011, Japan

E-mail: lhbai@163.com

Abstract. Using a nonperturbative scattering theory, the photoelectron angular distributions (PADs) of negative ions irradiated by intense laser fields are studied. Various PADs are obtained. Similar to that of atoms, PADs of negative ions show main lobes and jet-like structures. Here, the main lobe means the formation of the detached photoelectrons around the direction of laser polarization, while the jet-like structure means a peaked-out formation of photoelectrons emitted from the waist between the two main lobes. For a set of above-threshold-detachment peaks, with one-more-photon absorption, the number of the jet-like structures is not always increased by one, which verifies that the jet-like structures are irrelevant to photoelectron angular momentum.

1. Introduction

Photoionization of atoms in intense laser fields is an important nonlinear effect in the domain of strong field physics. Recently, more and more interests have been arisen in the study of photoionization, among which the study of photoelectron angular distributions (PADs) is a hot topic. Both in experimental researches and in the comparison of experimental observations with theoretical studies, PADs are an important tool. With rapid developments in experimental techniques, it is possible to measure photoelectron kinetic-energy spectra with highly resolved angles [1-5]. Nandor *et al.* [1] and Schyja *et al.* [5] observed a jet structure sticking out from the main lobes in the PADs of low-energy above-threshold ionization peaks of xenon atoms. Here, the main lobes in PADs are the formations of photoelectrons emitted in and around the direction of laser polarization, while the jets are the formations of photoelectrons emitted from the waist between the two main lobes. It was revealed that when the number of absorbed photons increases by one, the number of jets (we denote the jet number throughout this paper, here the jet number denotes the number of jets on one side of PADs) also increases by one. Since in traditional perturbation theory, a photon absorption will alter the electron angular momentum state, it is likely to attribute the number of jets in PADs to the quantum number of the electron angular momentum state. Recently, we have studied the PADs of a hydrogen atom irradiated by linearly polarized laser light [6] by using a nonperturbative quantum scattering theory proposed by Guo *et al.* [7]. Our previous study [6] has shown that the calculated PADs show main lobes and jet-like structures. Our study has confirmed the experimental observation that when the absorbed-photon number increases by one, the jet number also increases by one. And has further predicted that in some cases three or other odd-number more jets may appear with just one-more-photon absorption. The jet-like structure of PADs is due to the maxima of generalized phased Bessel functions, not an indication of the quantum number of photoelectron angular momentum states.

Then do the predicted phenomena be confirmed by other strong field theories? Do the predicted phenomena only exist in photoionization of atoms? These questions will be answered in this paper. If the predicted phenomena are confirmed by other strong field theories, if the predicted phenomena exist in ionization of other matters with laser fields, they will provide a strong support to our statement. In this paper we investigate the PADs of negative ion H^- irradiated by linearly polarized laser light by using a nonperturbative scattering theory proposed by Reiss [8]. Our study shows that the calculated PADs from above-threshold detachment of H^- also show main lobes and jet-like structures. Similar to that of atoms, our study confirms the experimental observation [1] that when the absorbed-photon number increases by one, the jet number also increases by one. And our study also verifies the predicted phenomena that in the PADs of a set of above-threshold detachment peaks, three more jets may appear with just one-more-photon absorption. This study shows that the jet-like structures have no relation with the angular momentum of photoelectrons and provides a strong support to the prediction and the nonpertubative quantum scattering theory used in our previous study [7].

2. Theory

The vector potential for a monochromatic linearly polarized plane wave in long-wavelength approximation is $A(t)=a\vec{\varepsilon}\cos\omega t$, where $\vec{\varepsilon}$ is the polarization vector, $\vec{\varepsilon}$ is real and normalized ($\vec{\varepsilon}^{\,2}=1$). The total transition probability per unit time for photodetachment arising from monochromatic linearly polarized electromagnetic fields is given by [8]:

$$dW/d\Omega=(2m^3\omega^5)^{1/2}(2\pi)^{-2}\Sigma(n-z)^2(n-z-\varepsilon_B)^{1/2}\,|\phi_i(\vec{p})|^2 J^2_n(z^{1/2}\alpha,-z/2) \qquad (1)$$

where n is bounded from below, since $n \geqslant z+\varepsilon_B$, in terms of integer relationships, $n \geqslant n_0$, $n_0 \equiv [z+\varepsilon_B]$, $\varepsilon_B = E_B/\omega$, E_B is the initial binding energy of the electron. Since n_0 is the smallest index in the sum over n, the range of n is now confined to $n_0 \leq n \leq \infty$. z is a fundamental intensity parameter defined by $z \equiv e^2a^2/4m\omega$ [9], where a is the amplitude of the radiation-gauge vector potential of the applied plane-wave field of linear frequency ω. Energy of the photodetached electron is given by the energy contributed by an nth order interaction with the field, less the energy which must be invested in overcoming E_B. In the general case, where $z\neq 0$, the energy $z\omega$ is a minimal interaction energy of the charged particle with the electromagnetic field. $\phi_i(\vec{p})$ is the Fourier transform of $\phi_i(\vec{r})$, $\phi_i(\vec{p})=\int d^3r e^{-i\vec{p}\cdot\vec{r}}\phi_i(\vec{r})$, the initial-state wave function $\phi_i(\vec{r},t)=\phi_i(\vec{r})e^{-iE_it}$. $J_n(u, v)$ is a generalized Bessel function

$$J_n(u, v) = J_0(u)J_{n/2}(v)+\Sigma J_{2k}(u)[J_{k+n/2}(v)+J_{-k+n/2}(v)], \qquad (2)$$

where $\alpha=8^{1/2}(n-z-\varepsilon_B)^{1/2}\cos\theta$, the angle θ is the polar angle in spherical polar coordinates, but in linear polarization case, the polar axis is taken to be along the polarization vector, $\vec{\varepsilon}$.

3. Results and analysis

In this paper, by using the nonperturbative scattering theory [8], PADs of negative ion H^- irradiated by linearly polarized intense laser fields are studied. PADs for odd values of n and even values of n are obtained. Similar to that of atoms, PADs of negative ion H^- show main lobes and jet-like structures. We find that for odd-n PADs, a maximum locates at the direction of laser polarization, and a zero is at the direction perpendicular to the laser polarized direction. We also find that for even-n PADs, a peak exists in the direction perpendicular to the laser polarization, which is in consistent with the conclusion of Reiss [8]. We further find that for a set of above-threshold detachment peaks, when one-more-photon is absorbed, the number of the jet-like structures in PADs is increased by one and three. It verifies the prediction that the jet-like structures are irrelevant to photoelectron angular momentum.

According to equation (1), PADs of above-threshold detachment for negative ion H^- in linearly polarized laser fields are obtained. In our calculations, the binding energy of H^- is chosen as 0.75421 eV [10] and the initial momentum-space wave function of a negative ion H^- in its ground state is given as [8, 9] $\phi_i(\vec{p})=2f(2\pi\beta)^{1/2}(\beta^2+p^2)^{-1}$, where $\beta=(2mE_B)^{1/2}$, and f is an empirical constant with the value $f^2=2.65$. In this paper, we investigate PADs with shorter laser wavelength comparing to the study of Reiss [8], and different character of PADs are found. For longer laser wavelength and lower laser intensity [8], when the absorbed-photon number increases by one, the jet number also increases by one, as shown in figure 4 of reference [8]. Our calculation shows that for shorter laser wavelength and higher laser intensity, when one-more-photon is absorbed, the number of the jet-like structures in PADs is increased by one and three.

For shorter laser wavelength and lower laser intensity, when one-more-photon is absorbed, the number of the jet-like structures in PADs is increased by one, which is in good agreement with the study of Reiss [8].

We first calculate the PADs of H^- irradiated by the laser light of wavelength 2100 nm and the laser light of intensity 1×10^{12} W/cm^2, the threshold order is $n_0=2$. PADs for n=2, 3, 4 are depicted in figure 1. We can see from figure 1 that each plot shows the main lobes along the direction of laser polarization and several jets sticking out from the waist between the main lobes. For PADs of n=2, there are two jets, and for PADs of n=3 and n=4, there are three and four jets, respectively. Figure 1 shows that when the absorbed-photon number increases by one, the jet number also increases by one, which is in agreement with the observation of Nandor *et al.* [1] and the study of Reiss for PADs with shorter laser wavelength and lower laser intensity [8]. We also find that for odd-n PADs, as shown in plot (b) for three photons absorbed, the maximum locates at the direction of laser polarization and zero in the direction perpendicular to the laser polarization, which is in agreement with the study of Reiss [8]. In plot (a), more detached electrons are emitted to the direction perpendicular to the direction of laser polarization, which is the so-called threshold effect, which has been studied experimentally and theoretically [11, 12]. In plot (c) for four photons absorbed, we also find a peak in the direction perpendicular to the laser polarization. For even-n PADs, there is always a peak in the direction perpendicular to the laser polarization, as shown in plot (a) and plot (c), For odd-n PADs, no peak appears in the direction perpendicular to the laser polarization, as shown in plot (b), which is in agreement with the study of Zhang *et al.* [13].

We then calculate the PADs of H^- irradiated by laser fields with much higher laser intensity. Figure 2 depicts the PADs of H^- irradiated by laser fields with laser intensity at 1×10^{13} W/cm^2 and laser wavelength at 1600 nm, the threshold order is $n_0=5$, plots (a)-(c) depict PADs for n=5, 6, 7, respectively. For odd-n PADs, as shown in plot (a) and plot (c), there are almost no photons emitted along the direction perpendicular to the laser polarization. For even-n PADs, as shown in plot (b), there is a peak in the PAD along the direction perpendicular to the laser polarized direction, which is in agreement with the study of Zhang *et al.* [13]. For PADs of n=5, there are two jets, and for PADs of n=6 and n=7, there are five and six jets, respectively. By comparing an n-photon PAD with its neighboring (n+1)-photon PAD, figure 2 shows that the jet number increases by three and one, when the absorbed-photon number increases by one, the jet number is not always increase by one.

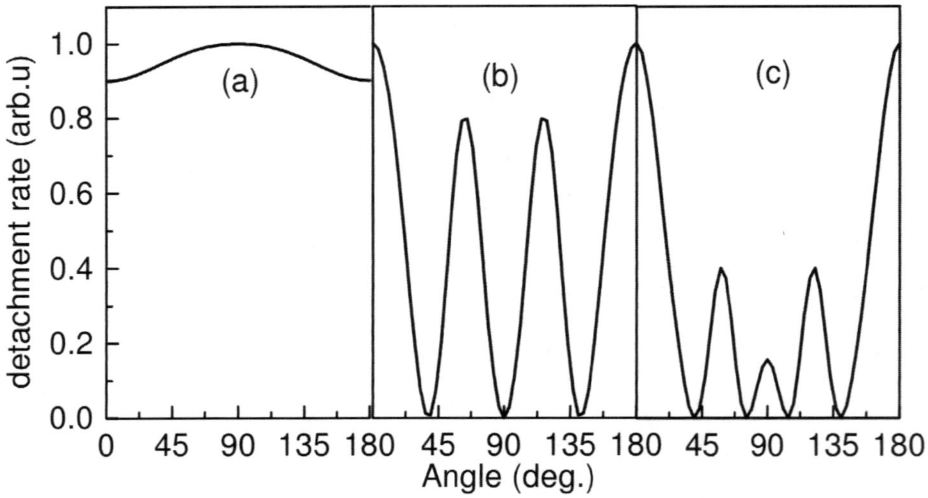

Figure 1. PADs of H^- by intense linearly polarized laser fields with laser wavelength at 2100 nm and laser intensity at 1×10^{12} W/cm^2. Plot (a) is the contribution from the lowest order n_0=2. Plot (b) and plot (c) are for the PADs from orders n=3 and n=4, respectively.

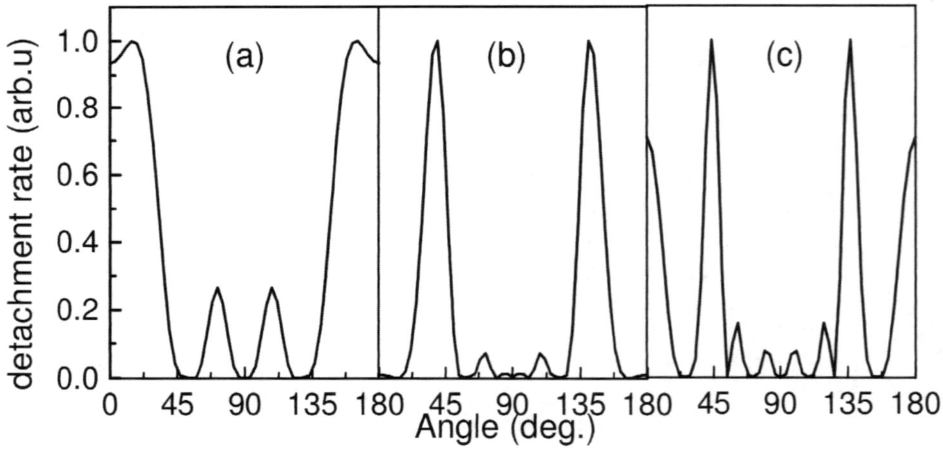

Figure 2. PADs of H^- by intense linearly polarized laser fields with laser wavelength at 1600 nm and laser intensity at 1×10^{13} W/cm^2. Plot (a) is the contribution from the lowest order n_0=5. Plot (b) and plot (c) are for the PADs from orders n=6 and n=7, respectively.

Similar feature is also found in figure 3. Figure 3 depicts PADs for n=6, 7, 8 with laser intensity at 1×10^{13} W/cm^2 and laser wavelength at 1800 nm, the threshold order is n_0=6. For PADs of n=6, there are three jets, and for PADs of n=7 and n=8, there are four and seven jets, respectively. We also find that for odd-n PADs, as shown in plot (b), the maximum locates at the direction of laser polarization, for even-n PADs, as shown in plot (a) and plot (c), there is a maximum in the direction perpendicular to the laser polarization. By comparing an n-photon PAD with its neighboring (n+1)-photon PAD, figure 3 shows that the jet number increases by one and three.

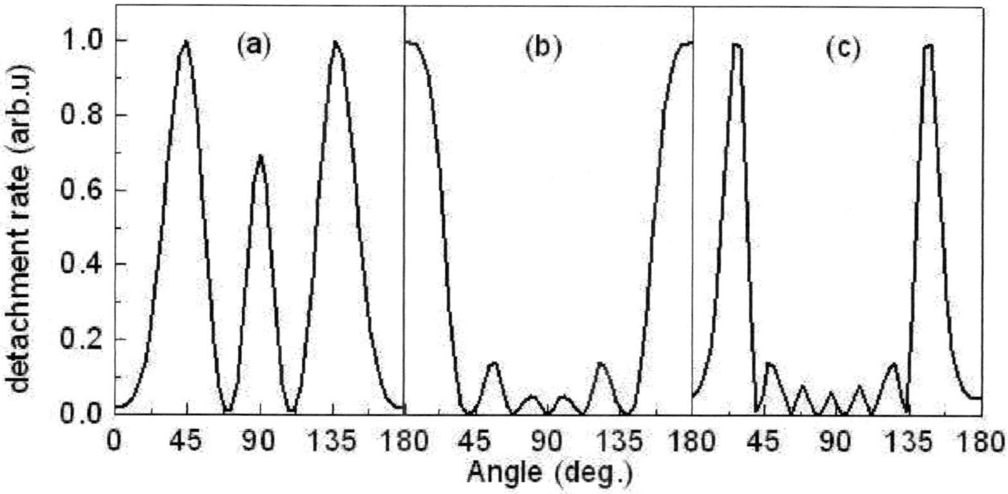

Figure 3. PADs of H^- by intense linearly polarized laser fields with laser wavelength at 1800 nm and laser intensity at 1×10^{13} W/cm^2. Plot (a) is the contribution from the lowest order n_0=6. Plot (b) and plot (c) are for the PADs from orders n=7 and n=8, respectively.

From the above calculation, we see that one more jet may not indicate one-more-photon absorption. The three-more-jet case shows that the number of jets does not indicate the angular momentum change: the jet-like structures are irrelevant to photoelectron angular momentum.

4. Conclusion

In conclusion, using a nonpertubative scattering theory, PADs from above-threshold photodetachment of negative ion H^- in linearly polarized laser fields are investigated. Similar to that of atoms, PADs of negative ions show main lobes and jet-like structures. For a set of above-threshold-detachment peaks, with one-more-photon absorption, the number of the jet-like structures is not always increased by one, which verifies that the jet-like structures are irrelevant to photoelectron angular momentum. For odd-n PADs, a maximum locates at the direction of laser polarization, and a zero is at the direction perpendicular to the laser polarization. We also find that for even-n PADs, a peak exists in the direction perpendicular to the laser polarization, which is in consistent with the investigation of Reiss [8].

In our previous study [7], based on the nonperturbative quantum scattering theory of photoionization, we have predicted that for PADs of above-threshold ionization of atoms in intense linearly polarized laser fields , the jet number may increase by one, three, or other odd numbers when one-more-photon is absorbed, which is confirmed by the nonperturbative scattering theory proposed by Reiss [8]. Good agreement is obtained by these two quite different treatments. And we find our prediction exists in above-threshold detachment of intense laser fields with negative ions, which provides a strong support to our prediction.

Acknowledgments

The authors acknowledge the National Natural Science Foundation of China for support under grants 60908006, 10804067, and 60407007.

References

[1] Nandor M J, Walker M A and Woerkom L D Van 1998 *J. Phys. B* **31** 4617
[2] Yang B, Schafer K J, Walker B, Kulander K C, Agostini P and Dimauro L F 1993 *Phys. Rev.*

Lett. **71** 3770
[3] Lewenstein M, Kulander K C, Schafer K J and Bucksbaum P H 1995 *Phys. Rev. A* **51** 1495
[4] Dionissopoulou S, Mercouris T, Lyras A and Nicolaides C A 1997 *Phys. Rev. A* **55** 4397
[5] Schyja V, Lang T and Helm H 1998 *Phys. Rev. A* **57** 3692
[6] Guo D S, Aberg T and Crasemann B 1989 *Phys. Rev. A* **40** 4997
[7] Bai L, Zhang J, Xu Z and Guo D 2006 *Phys. Rev. Lett.* **97** 193002
[8] Reiss H R 1980 *Phys. Rev. A* **22** 1788
[9] Reiss H R 1979 *Phys. Rev. A* **19** 1140
[10] Andersen T, Haugen H and Hotop H 1999 *J. Phys. Chem. Ref. Data* **28** 1511
[11] Reichle R, Helm H and Kiyan I Y 2001 *Phys. Rev. Lett.* **87** 243001
[12] Bai L, Zhang J, Zhang X and Xu Z 2006 *Phys. Rev. A* **74** 025402
[13] Zhang J, Zhang W, Xu Z, Li X, Fu P, Guo D and Freeman R R 2002 *J. Phys. B* **35** 4809

Journal of Physics: Conference Series **276** (2011) 012013 doi:10.1088/1742-6596/276/1/012013

Research Thinking of Low-intensity laser For the Treatment of Menopausal Syndrome

G Z Chen[1], Y X Xu[2*], X Y Wang[1], S H Liu[3] and L J Li[2]

[1] The Second Clinical Medical College, Guangzhou University of Chinese Medicine, Guangzhou, 510405 China;
[2] Acupuncture and Massage College, Guangzhou University of Chinese Medicine, Guangzhou, 510405 China;
[3] South China Normal University, Guangzhou, 510631 China

E-mail: xuyx1968@163.com

Abstract. Female climacteric syndrome is a clinical syndrome due to autonomic nerve dysfunction occurring in women during climacteric period, which may affect their physical and mental health. Therefore, how to pass climacteric period for women without any problems, avoid or reduce the occurrence of climacteric syndrome, prevent geriatric diseases and improve life quality is a key issue now for great attention. Looking for a convenient, effective, and safer method without toxic-side effects to control the disease is a modern medical problem. By analyzing the relationship between laser technology and traditional acupuncture and moxibustion, the advantage and the existing problems on acupuncture and moxibustion for the treatment of menopausal syndrome, the application of laser methods for the mechanism research on TCM diagnosis and treatment of menopausal syndrome was discussed. It's pointed out that the laser acupuncture is safe and effective to treat menopausal syndrome. Breakthrough will be achieved from the research of the selection of the acupoint prescription and mechanism of Acupuncture and Moxibustion for the treatment of menopausal syndrome by utilizing the advantage of interdisciplinary intersection. Laser technology will make the development of acupuncture and moxibustion science possess an unprecedented field.

1. Introduction

With the problems of side-effects of drugs and drugs induced diseases, the world voice of "Return to Nature" is stronger and stronger. So it is an aspiration for Medicine to seek a therapeutic method, which is effective in the clinic and without toxic and side-effect.

Acupuncture, an important component of Traditional Chinese Medicine (TCM), is widely used to treat various kinds of diseases and it played an important role in the medical practice over the centuries in China. Especially at the present day, with the increasing of drug-induced diseases, the strengths of acupuncture therapy are more outstanding and valuable. Therefore, more and more countries begin to pay attention to it and apply and study it in the clinic. However, in recent years because of the problem of ideas, institutions and policies, further more the lack of integration between acupuncture system and modern natural science, the medicine shares of acupuncture and moxibustion are declining and its development becomes more and more difficult. So, only if there is the innovation in theory and technology can acupuncture and moxibustion get out of the trouble and make a due contribution in the Medicare system. Laser acupuncture is the achievement in scientific research on the combination of the traditional acupuncture and the modern laser technology. In recent years, researches on using laser-irradiating acupionts instead of needling to ease pain and treat disease have drawn great attention from the biology community. This article will simply analyze and probe the application of low-intensity laser for the treatment of menopausal syndrome.

2. Prevalence and treatment of female menopausal syndrome

Menopause is a physiological period from middle age to old age. At the same time, it's a turning point for body physiological function from exuberance to recession. The climacteric syndrome (CS) appears at female menopause and assumes a group of syndromes by vegetative nerve functional disturbance, such as hot flash, red cheek bone, sweating, paramenia and emotional instability. Hence CS can influence women's health and work, how to avoid or relieve the symptoms of CS and make women spend menopause more smoothly is a key problem at present. That is why the World Menopause Society assigns Oct.18 for "the World Menopause Day".

According to the statistics, the age of menopause is often 45~55yrs.The average age in American and British is 50yrs, but 47.81+-3.97yrs in China according to the investigation on 3364 women above 40yrs in Guangdong province. There are about 75~88% menopause women having some menopausal symptoms, but only 15% go to see a doctor because of their severe symptoms [1]. The morbidity is 89.5% in menopausal women, 85.9% in premenopausal women and $P<0.05$ [2].

For CS, the modern medicine mainly adopts hormone replacement therapy (HRT) and symptomatic treatment [3-4]. However, the necessity and safety of HRT are always the discussion point and at the same time HRT has too many contraindications and adverse reactions [5]. A large study by American women's health program (WHIMS) in 2004 found that the application of HRT in a long time could increase the risk of stroke, thrombus and breast cancer, carcinoma of endometrium, ovarian cancer, etc. A successive therapeutic method is bound to normalize the deficit in the metabolic network and not to disturb those normal. Therefore, it is still one of problems we are facing to find a safer and more effective method to prevent and treat CS.

For CS, the TCM mainly starts with the theory "Kidney Governing Reproduction" to improve symptoms and make quality of life better on the basis of the syndrome differentiation and treatment. In recent years, acupuncture is under observation in medical community for its exact effect and little side-effect. Acupuncture prevents and treats diseases through regulating action of human body and it also has weak-points, for examples, patients fear to needle, it's easy to relapse due to too short time of needling, and the treatment course is too long. So, only if there is innovation in theory and technology can make acupuncture and moxibustion get out of the trouble and bring a big effect into play.

Laser acupuncture is a therapeutic method to prevent and treat diseases by irradiating acupoints with low-intensity laser under the guidance of TCM meridian theory. The core of laser acupuncture is using laser irradiation instead of needle or electricity. In some sense, it is an improvement of needle and acupuncture method. Laser acupuncture uses laser as stimulation so it is safe, painless, wide-use. Therefore, it is easily accepted by patients especially children and those who fear needle comparing with the traditional acupuncture which is painful and easier to infect or have an accident. So it is worth to be popularized in clinic. The study showed that low-intensity laser could defer senility and regulate endocrine function which lay the foundation for using it to treat CS. However, the clinical mechanism of laser acupuncture is not studied too much at present.

3. Principles of Selecting Acupoints for Low-level Laser Irradiation Treatment of CS

TCM holds that the disorder of zangfu organs and body metabolism caused by the recession of kidney qi is the root cause for CS, and the kidney is the origin of congenital constitution, the spleen is the root of acquired constitution. So the therapeutic principle of laser acupuncture for the treatment of CS should be strengthening spleen and kidney, regulating zang and fu organs, balancing yin and yang. About the acupoints prescription, those meridians related to CS or points having specific function should be chosen. Shenshu (BL 23) is a specific point at back where the qi of the kidney meridian pour into. The clinical research shows that this point has relative specificity for CS [6-7] and has a good effect. Sanyinjiao (SP 6) belonging to the spleen meridian is a key point to regulate reproductive endocrine system [8]. Guanyuan (CV 4) belonging to Ren meridian is good at strengthening primordial Qi. Our pilot study found that catgut implantation at BL 23, SP6 and CV4 in ovariectomized rats could not only change the weight of rats, but also enhance the estrogen level and lower the level of GnRH, FSH, LH of rats [9-11], and it had got a very good effect in clinical [12-13]. So we choose BL 23, SP6 and CV4 as the chief points to establish the prescription for the treatment of CS.

4. Investigations on the Effects of Low-level Laser Irradiation for the treatment of CS

Laser acupuncture has been widely used in the treatment of clinical conditions, and the type of laser used in this therapy is safe when its power output is below 1 W [14]. Recent studies show that laser acupuncture has useful healing effects on visceral postmenopausal obesity [15]. Negative results have been reported in a investigation examining laser acupuncture for the treatment of menopausal woes [16]; however, that particular study contained some limitations in parameters setting. Overall, the clinical efficacy of laser acupuncture for the treatment of female menopausal syndrome remains controversial.

Breakthrough might be obtained from the animal experimental research. Therefore, the aim of our study is to investigate the effects of laser acupuncture treatment using an animal model in which uterus and pituitary glands were also observed. The research showed that the semiconductor laser irradiation could exert beneficial effects on OVX rats through reducing body weight and increasing pituitary ER α expression, and 30 J/cm2 was the most effective among the doses used.

A critical analysis of some well-known negative studied on low-level laser therapy showed that it is all in the parameters: low doses, inclusion criteria, lack of control group, treatment technique, systematic effect, tissue condition, power density, mixed parameters, influence of ambient light, meta-analysis [17]. It indicated that the parameters must be carefully concerned to evaluate the efficacy of laser acupuncture for the treatment of CS either clinically or experimentally.

5. Thinking on the Research of Therapeutic Mechanism of low-level Laser Irradiation for CS

5.1. Mechanism research thought

Recently, it is widely believed that the cause of CS is the disorder of Hypothalamic-pituitary-ovarian axis (H-P-O-A) and hypothalamic-pituitary-adrenal axis (H-P-A-A) resulted from low-level estrogen. However the exact pathological mechanism is still unknown. Estrogen is the important component of internal environment of body and the estrogen receptor (ER) exists in almost all kinds of tissues studied not only in uterus and vagina. When the estrogen and ER combined, a wild physiological effect would be produced. So when the estrogen reducing, these tissues or organs will retrogress or change in mechanism and then appear a series of symptoms. So the reducing of the number and the function of ER may make the tissues inactive to estrogen and play a role in the occurrence of CS.

One kind of body function is influenced by many factors at the same time, so if you want to study the function of one specific factor to the metabolism course you must keep other factors stationary. Due to the restriction of its own condition, the clinical study of needling is unable to be blinded, so it is inevitable to be influenced by individual, society and mentality. However, the laser acupuncture could be double blinded. Further more, animal model can avoid ethical issues and social restriction, even some methods, which are not allowed to apply to human, can be adopted. So, animal experiment is always used when we want to research the mechanism of one stimulating method or drug in order to cover the shortage of clinical study. For this research, we can choose ovariectomized SD rats or natural aging SD rats as the animal model.

Laser acupuncture uses laser as stimulus so it is safe, painless, adjustable and has the function of both needling and moxibustion at the same time. In recent years, some researchers [17-19] who studied the biological effect of laser found that 808nm semiconductor laser could reach hypoderm through corium and get a good effect without pain. Therefore, we could choose 808nm semiconductor laser beam instead of needle to stimulate points, which also can improve patient compliance, reduce bias from researchers and subjects [20-22], and objectify evaluation therapeutic effect of acupoints.

5.2. Complement and Explanation

For CS, the western medicine has side-effects and the traditional acupuncture exists defects too. So the World Federation of Acupuncture and Moxibustion Societies indicate that it is the important direction in 21st century to find a long-acting, painless acupuncture therapy. Since 1970s the laser acupuncture with painlessness and asepsis has been a development in acupuncture therapy history. However, most of recent researches are limited to traditional acupuncture and moxibustion methods and few related to the mechanism of laser acupuncture. For this reason, several points should be noticed in the future research as follows:

- Give close attention to triumphs of modern Physics, Optics and their interdiscipline subjects and apply them to clinical studies which can keep the research of laser acupuncture for CS at a high starting point and level;
- Adopt recognized diagnostic and response criteria, and design experiment according to the Clinical research design, measurement and evaluation (DME);
- Research animal models which are closer to clinical situation and much more correspondent with syndrome differentiation and treatment of TCM. Ovariotomize rats model through imitating other influencing factors and observe behavior science with modern analytical testing techniques to approximate animal models with CS symptoms;
- The development of modern life science speeds up the fusion of multiple disciplines and then produces lots of new subjects and technology platforms, such as genomics, proteomics, metabonomics, bioinformatics, etc. With the advantage of multiple disciplines such as TTM, HPLC, NMR, etc., observe the physiological and pathological changes of menopause women in vivo, strive for discussing the pathological mechanism of CS and the mechanism of action of acupuncture for CS from multi-angle of multi-levels (whole, cell, molecular and gene level), multi-criteria (direct and indirect indicator) and multiple disciplines, and finally improve the theory of acupuncture, optimize the acupoints prescription and enhance the clinical effect.

Acknowledgments
This work was supported by the Post-doctor Foundation of China under grant No. 20070410242 and the Guangdong Natural Science Foundation under grant No. 8451063101001809, and Post-doctor Foundation of Guangzhou University of Chinese Medicine under grant No.B3YH1005.

References

[1] Zhang B P, Gao X J and Xia H H 1997 Diagnosis of Health Services and Demand Conditions in Female Community-Dwelling of 40-60yrs *Maternal and Child Health Care of China*.**12** 356
[2] Liu BR and Wu KM 1999 *The World Traditional Gynecology* Scientific and Technical Publishers 235
[3] Duzenli M A, Ozdemir K, Sokmen A et al. 2010 The Effects of Hormone Replacement Therapy on Myocardial Performance in Early Postmenopausal Women. *Climacteric* **13** 157-170
[4] Freedman DM, Rajaraman P, Fuhrman B et al. 2010 Sunlight, Hormone Replacement Status and Colorectal Cancer Risk in Postmenopausal Women. *Journal International Du Cancer* **126** 1997-2001
[5] Yang S Z and Yang Z L 2008 The Pros and Cons of Hormone Replacement Therapy (HRT) in Treating Perimenopausal Syndrome. *Guiding Journal of Traditional Chinese Medicine and Pharmacy* **5** 174-175.
[6] CHEN G Z and XU Y X 2008 Brief Discussion on Shenshu. *World Journal of Acupuncture-Moxibustion* **18** 61-63
[7] Zhu D, Ma Q, Li C, et al. 2000 Effect of Stimulation of Shenshu Point on the Aging Process of Genital System in Aged Female Rats and the Role of Monoamine Neurotransmitters. *J. Tradit. Chin. Med.* **20** 59
[8] Xia X H, Hu L and Qin Z Y 2008 The Multicenter Randomized Control Study on Treating Perimenopausal Syndrome with Electroacupuncture on San Yin Jiao. *Acupuncture Research* **33** 262-266

[9] CHEN G Z and XU Y X 2008 Application of Radioimmunoassay on the Study of Catgut Embedding in Points on Ovariectomized Rats *Advances in Biomedical Photonics and Imaging:Proc. Int. Conf. on Photonics and Imaging in Biology and Medicine(Wuhan,China , 4-6 November 2007)* ed Q luo(New York: World Scientific)pp 360-364

[10] CHEN G Z and XU Y X. 2007 Effect of Catgut Embedding in Points on Reproductive Endocrine of Ovariectomized Rats *The Journal of Alternative and Complementary Medicine* **13** 908

[11] Chen G Z and Xu Y X 2008 The Effects of Catgut Implantation Acupoint on Genital Endocrine Hormone and Weight Change of Ovariotomized Rats. *Journal of New Traditional Chinese Medicine* **40** 97-99

[12] Chen G Z and Xu Y X 2010 The Influence of Acupoint Catgut-Embedding on the Quality of Life, Reproductive Endocrine and Bone Metabolism of Postmenopausal Women. *Chinese Journal of Integrative Medicine* **16** 1-6

[13] Chen G Z and Xu Y X 2010 The Effects and Safety Evaluation of Catgut Implantation Acupoint on Bone Metabolism and the Level of Free Radicals of Postmenopausal Women. *Chinese Journal of Acupuncture &Moxibustion* **30** 177-181

[14] Karu I T 2003 Low-power laser therapy In: Biomedical Photonics Handbook. T. VoDinh (ed.). CRC Press, BocaRaton FL 1-2-5

[15] Piotr W, Grzegorz S, Agnieszka P S and Przemyslaw O 2003 Laser acupuncture andlow-calorie diet during visceral obesity therapy after menopause. *Acta Obstet Gynecol Scand* **82** 69-73

[16] Paul A 2010 Laser acupuncture fails to ease menopausal woes. *Menopause*, published online January 8

[17] Jun T, Lars H 1998 It is all in the parameters: A critical analysis of some well-known negative studied on low-level laser therapy. *Journal of Clinical Laser Medicine and Surgery* **16** 245-248

[18] Wei H J, Xing D and Wu G Y 2003 The Photometric Characteristics of 532 nm and 808 nm L aser on Irradiating the Normal Bladder Tissues and Neoplastic Tissues. *Progress in Biochemistry and Biophysics* **30** 209-215

[19] Wei H D, Xing D and Wu G Y 2004 The Light-transfer Characteristics of About 808 nm Laser on Irradiating Three Different in Vitro Human Tissues. *Chinese Journal of lasers* **31** 305-309

[20] Pan Z H, Wang X and Li C 2008 The Effects of 630/808 nm Double-wave length, Low-intensity Laser on the Wound Repair of Skin of Rats. *Journal of Clinical Rehabilitative Tissue Engineering Research* **12** 10254-58

[21] G. Litscher, D. Rachbauer, S. Ropele et al. 2004 Acupuncture using laser needles modulates brain function: first evidence from functional transcranial Doppler sonography and functional magnetic resonance imaging. *Lasers in Medical Science* **19** 6-11

[22] Ebru I, Aysegul C, Rian D et al. 2004 Comparison of Laser, Dry Needling, and Placebo Laser Treatments in Myofascial Pain Syndrome. *Photomedicine and Laser Surgery* **22** 306-311

[23] Christian M, Siedentopf, Florian K I et al. 2005 Laser acupuncture induced specific cerebral cortical and subcortical activations in humans. *Lasers in Medical Science* **20** 68-73

Study on focusing performance of the twice reflecting laser focusing system

Fu Qiang Cheng[1] and Zhi Guo Dou[2]

[1] Company of Postgraduate Management, the Academy of Equipment Command & Technology, Beijing 101416, China
[2] Department of Basic Theories, the Academy of Equipment Command & Technology, Beijing 101416, China

E-mail: nazis76@163.com

Abstract. Laser thrusters characterized by feasible application perspective all possess the twice reflecting laser focusing system. Nonetheless, studies on its focusing performance are imperative for the research of flight route of the thruster and haven't been developed yet. Under three different focusing design modes, assisted with optical design software ZEMAX which employing Monte Carlo ray tracing, performance of twice reflecting laser focusing system including focusing performance parameters, radiation intensity distribution and the details of ignition region on the focusing plane are studied comparatively, and the studied two main angular aberrations are beam out-coming aberration from laser source and flying aberration from thruster itself. Studies show even slight aberration will result in steep falling focusing performance and strong deviation of beam spot for any focusing design mode, while different aberrations bring in distinctive falling tendencies and the evolutions of ignition region. It strongly demands the directional precision of laser beam, and the attitude control of laser thruster is indispensable. And it's not recommended that one focusing design mode is superior to another. Which kind of aberration is dominant should be taken into account when choosing focusing mode.

1. Introduction

The advantages of laser thrusters like ASLPE[1] and LITA[2] which possess the twice reflecting laser focusing system are considered as: a) vital separation of optical units and jet nozzle, b) lack of effects of the exhaust jet on the laser beam delivered to the vehicle, c) independence of the vehicle motion control on mutual orientation relative to laser, capturing the thruster vector control. Apparently, all the three points suits ASLPE and LITA to thruster's demands of feasible application perspective. Consequentially, study on the focusing performance of focusing system is imperative, however, previous works were all carried around once reflecting focusing system incorporating just a parabolic reflector simultaneously playing the role of nozzle which may arouse many defects. Programs using RMT (Ray Tracing Method) in order to explore ray path have been used for focusing performance analysis of parabolic reflector for years[3]~[5], but due to the relative complexity of the twice reflecting focusing system, works on its performance haven't yet been developed. Meanwhile, different thrusters employing different twice reflecting focusing modes that are classified by geometrical type of reflecting surfaces result in which focusing mode is better and the contrast of their focusing performance is necessary. Now, with flexible non-sequential ray tracing created by optical design software also providing plentiful reflecting surface types, performance analyses of complex focusing system in high precision become simpler and more practicable.

Based on ZEMAX (a non-sequential Monte Carlo ray tracing program), we present works on focusing performance of twice reflecting laser focusing system under three different focusing modes comparatively. Focusing performance parameters and radiation intensity distributions on focusing plane were computed and analyzed under different angular aberrations.

2. Primary physical model

2.1 Geometry Model of Twice Reflecting Laser Focusing System

Schematic diagram of the twice reflecting focusing system, which is toroidal-symmetric and consists of two reflectors with known shape, is shown in figure 1. Z is the incident direction of laser beam and symmetric axis of the system, while axis X, Y, Z satisfy right-hand rule and plane XOY is defined as focusing plane. The laser beam is directed to the mirror R1, after the reflection of R1 to the mirror R2, then reflected to the O (zero point). α and β are focusing orientation meaning deviation degree of laser beam relative to the incident direction and focusing convergence angle the degree of beam converging. In common use, according to the different generatrix of the reflecting surfaces, integration modes of R1 and R2 could be paraboloid to hyperboloid, cone to paraboloid and paraboloid to cone, which are marked as focusing mode A, B, C respectively in the following text.

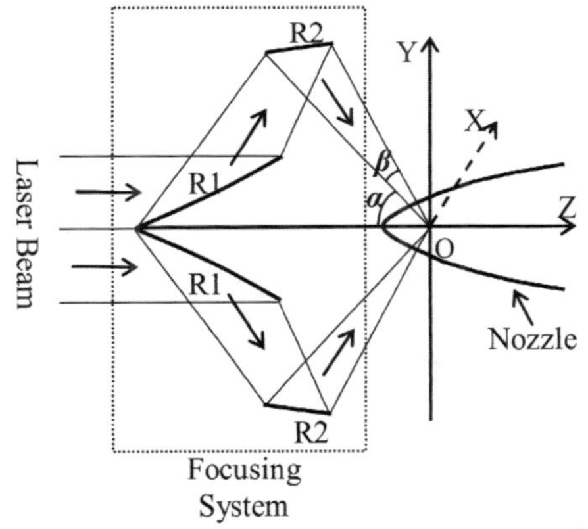

Figure 1. Schematic diagram of the twice reflecting laser focusing system

Based on the coordinate above, three groups of reflecting surfaces are selected as object and the geometrical details are shown in table1. For the comparability of the three focusing modes, $\alpha\approx35°$, $\beta\approx6°$ and the radius of focusing system of almost 100mm are specified.

Table 1. geometrical details of three groups of reflecting surfaces (unit: mm)

Focusing mode	Reflecting surfaces	Generatrix equation of reflecting surface	Start point (z, y)	End point (z, y)
A	R1:paraboloid	$(y-550)^2=-500(z-350)$	(−255, 0)	(−201.3, 25)
	R2:hyperboloid	$\sqrt{(z-225)^2+(y-550)^2}-\sqrt{z^2-y^2}=404.26$	(−138.4,102.5)	(−169.6,97.9)
B	R1:cone	$y=z+180$	(−180, 0)	(−162.3, 25)
	R2:paraboloid	$z^2=-580(y-145)$	(−180, 89.1)	(−162.3,99.6)
C	R1:paraboloid	$(y-270)^2=-500(z+18)$	(−163.8, 0)	(−138.1, 25)
	R2:cone	$y-135=0.52963(z+143)$	(−156.9,89.8)	(−139.6,98.9)

Surface type of paraboloid in focusing mode A, cone in B and C are already afforded in ZEMAX corresponding to the type of Odd Sphere, Cone and Cylinder Pipe. For the other surface types, we need to create file of IGS in CAD then imported to ZEMAX for use. Shaded models of the three focusing modes in ZEMAX are shown in figure 2 in which rays displayed are at random.

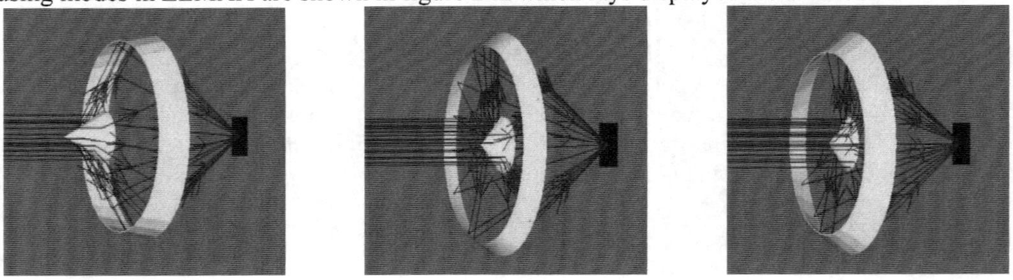

Focusing mode A Focusing mode B Focusing mode C

Figure 2. Shaded model of the three groups of reflecting surfaces in ZEMAX.

The two angular aberrations are: a) $\Delta\theta_L$, beam out-coming aberration (inclination of beam relative to the axis Z at the spot of laser source); b) $\Delta\theta_T$, flying aberration of thruster (inclination of focusing system relative to axis Z at the head of the focusing system). d_L is the distance between laser source and the focusing system in figure 3(a).

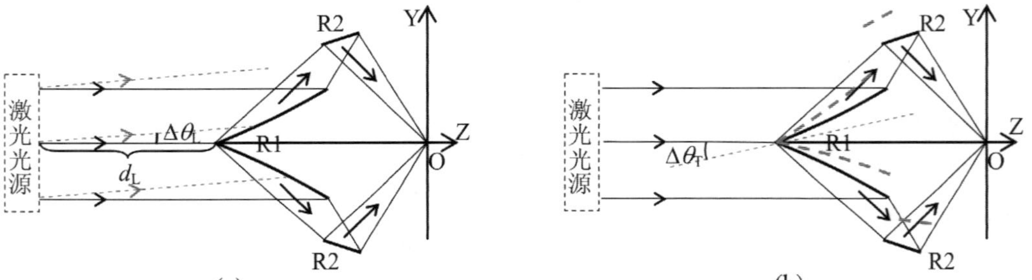

(a) (b)

Figure 3. Schematic diagram of the two angular aberrations: beam out-coming aberration (a) and flying aberration of thruster (b).

2.2 Laser Source Model and Monte Carlo Ray Tracing

Intensity distribution of laser beam is complex in practice, herein we use model of Gaussian beam of basic mode in which the distribution is axis-symmetric. In the polar coordinate, intensity distribution of laser beam on the plane vertical to axis Z is specified by

$$I(r) = I_0 \exp\left(-\frac{2r^2}{\omega_0^2}\right) \tag{1}$$

Where r is the distance from center of the beam, I_0 is the maximum intensity on the cross-section of the beam, ω_0 is radius of laser beam which represents the r where the intensity is $1/e^2$ of I_0, as shown in figure 3. In the near field, ω_0 can be assumed to be constant. If the power of the laser source is P, $I_0=2P/(\pi\omega_0^2)$.

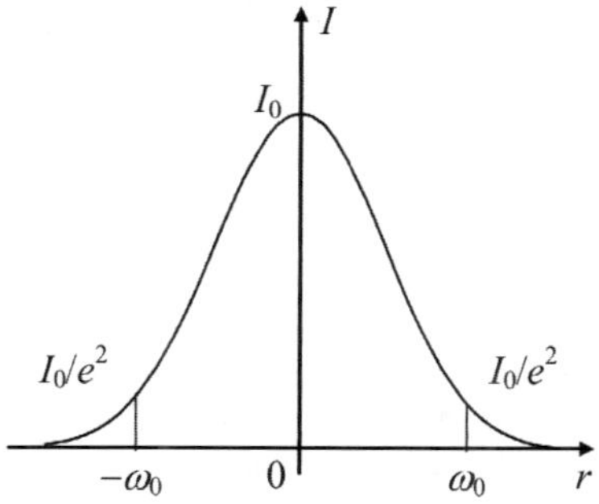

Figure 4. intensity distribution of laser radiation on cross-section

The ray tracing in ZEMAX adopts Monte Carlo (MC) approach, which is a kind of stochastic simulation method. Fundamental thought of MC approach is assuming the laser source as consisted of large quantity of mutually independent rays, while the emitting position or direction of rays, whether the ray is absorbed on the reflecting surface and the propagation process such as diffraction are all determined by corresponding probability model. If the amount of rays is N, the average power of each ray is P/N. Through tracing a ray till it is absorbed or escapes from the system then another ray launched, and with a large amount of rays repeating this process, stable statistic results would be gained. Herein, total reflection will occur when the traced ray reaches reflecting surface.

According to the axis-symmetry of intensity distribution of Gaussian beam, the probability model of ray emitting at the polar angle θ is

$$\theta \sim \frac{1}{2\pi} \tag{2}$$

At r, the probability model is obtained through normalizing equation (1), then is

$$r \sim \frac{2}{\pi\omega_0^2}\exp\left(-\frac{2r^2}{\omega_0^2}\right) \tag{3}$$

θ and r are mutually independent random variables, so the ray emitting probability $p(r_i, \theta_i)$ at $r=r_i$, $\theta=\theta_i$ is

$$p(r_i, \theta_i) = \frac{1}{\pi^2\omega_0^2}\exp\left(-\frac{2r_i^2}{\omega_0^2}\right) \tag{4}$$

For the computation of intensity distribution on the focusing plane, the plane is partitioned to a number of square grids, and then the sum of ray power absorbed in a grid divided by the grid area yields the result to the corresponding grid.

2.3 Focusing Performance Parameters

I_{max} and R_{RMS} are defined to assess the focusing performance of focusing system, where I_{max} is the peak intensity on the focusing plane representing whether the radiation intensity reaches breakdown threshold of the air (1×10^7W/cm^2); R_{RMS} is

$$R_{RMS} = \sqrt{\frac{\sum_{i=1}^{j} R_i^2}{j}} \tag{5}$$

Where, j is the total number of rays absorbed by focusing plane through twice reflecting, and R_i is the distance between the ray of No.i and the ideal focus ($x=y=0$) on the plane. R_{RMS} roughly represents deviation degree of laser spot from the ideal focus.

3. Simulation results and analyses

ω_0=25mm, number of random rays N=10^5 which can yields little fluctuating results, and the power of pulsed laser 3MW were set in ZEMAX. The focusing plane is set in absorption mode and its size is changing with the laser spot area while 25×25 grids was guaranteed in every square millimeter. Intensity distribution on the focusing plane is continuous in reality, but what the MC tracing obtained is discrete and inconvenient for observation and analysis. So the results of computing were smoothened by 5 times.

3.1 Influence of flying aberration of thruster on focusing performance

When $\Delta\theta_T$ increases, variation of I_{max} and R_{RMS} are shown in figure 5 and figure 6. I_{max} in the three focusing modes all decrease rapidly, but the tendencies of decreasing are apparently different. In focusing mode A, I_{max} drops monotonically with $\Delta\theta_T$: $\Delta\theta_T$=1°, I_{max} falls to 2.94×10^8W/cm^2 that is 4 times less than $\Delta\theta_T$=0°; I_{max} falls to 7.9×10^6W/cm^2 that is below the breakdown threshold of the air at $\Delta\theta_T$=5°. As for B, the decreasing tendency is less steep than A: $\Delta\theta_T$=1°, I_{max} falls almost 10 times less than $\Delta\theta_T$=0° drastically; but between $\Delta\theta_T$=1°~9°, the curve goes through a slow decreasing process, and falls below 1×10^7W/cm^2 at $\Delta\theta_T$=10°. In C, the decreasing amplitude of I_{max} is steeper than the other two modes: I_{max} is less than 1×10^7W/cm^2 at just $\Delta\theta_T$=3° with 9.7×10^6W/cm^2, then the decreasing tendency is slowed down gradually. Inversely, for the three modes, R_{RMS} all increase linearly along with $\Delta\theta_T$ and the increasing slopes from more to less are A, B, C orderly. At $\Delta\theta_T$=0°, R_{RMS} are all around zero point reasonably; $\Delta\theta_T$=10°, 83mm、52mm and 24mm for focusing mode A, B and C.

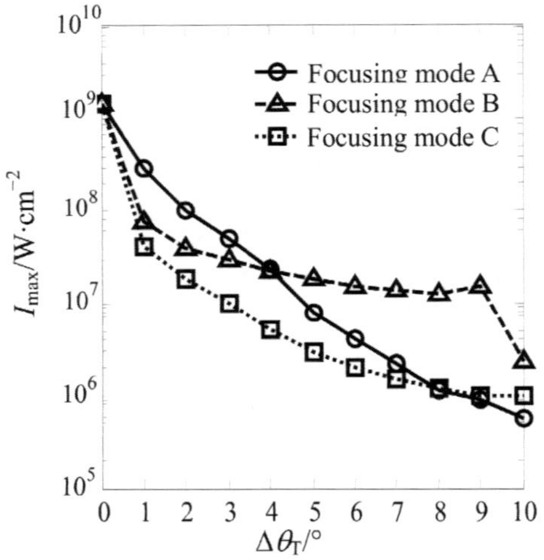

Figure 5. I_{max} changing with $\Delta\theta_T$

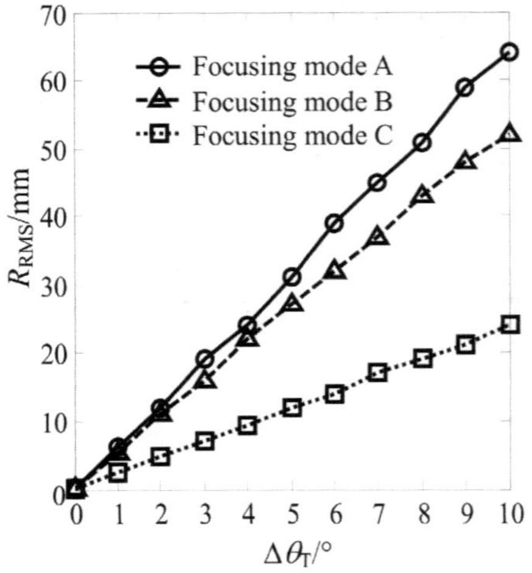

Figure 6. R_{RMS} changing with $\Delta\theta_T$

On the focusing plane, focusing spot is defined as the area where the intensity is larger than 0 and the ignition region is the area of larger than 1×10^7W/cm^2. Relative intensity is the value of practical intensity divided by 1×10^7W/cm^2. For the research of flight route of the thruster, the position and shape of ignition region are very meaningful.

Relative intensity distributions on focusing plane with $\Delta\theta_T$ are shown in figure 7. We can see distinctly dissimilar evolution of focusing spot and ignition region under different $\Delta\theta_T$ for different focusing mode. Generally, a tendency of division of ignition region is shown in both focusing mode A and B, while in C exhibits a region of steadily expanding round. When $\Delta\theta_T$=0°, there is no aberration and a round ignition region with center at the focus and radius of nearly 0.4mm is displayed in the three modes uniformly. At $\Delta\theta_T$=1°, a ignition region shaped with two joined annuluses and a round of high intensity in the center has a deviation of almost 6mm from the focus in focusing mode A; in B, an upside-wide annulus ignition region with radius about 1.1mm has a deviation of nearly 5mm; in C, the ignition region is also an annulus with radius about 2mm, but its center is at the focus. At $\Delta\theta_T$=3°, the two joined annuluses are stretched to two separated tilting elliptical loops at the bottom of which the intensity is much higher while lower at the outsides and the deviation is almost 16mm in mode A; in B, deviation of nearly 11mm, two inter-tilting annuluses are also shown and the intensity at their overlapped bottom is relative high; in C, the ignition region can not be formed and the focusing spot is still a center-in-focus annulus but the area is broader with radius of nearly 7mm.

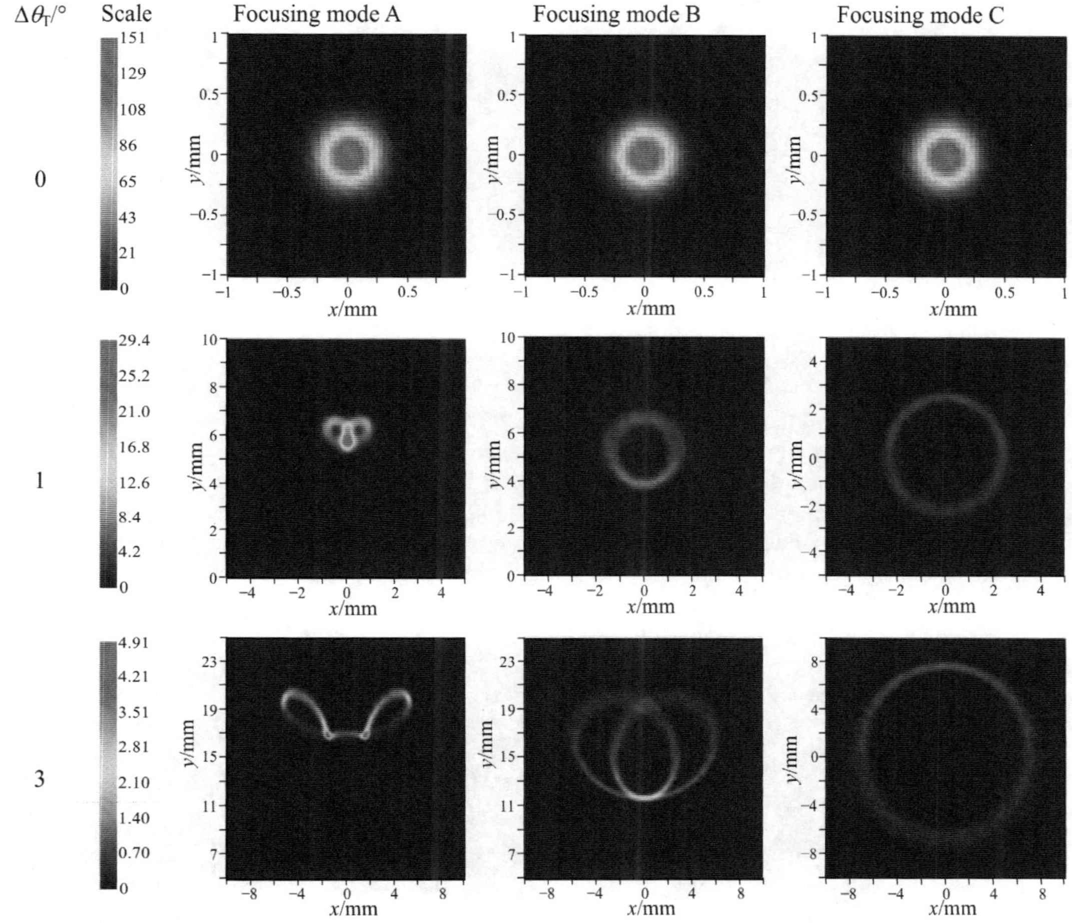

Figure 7. distribution of relative intensity on the focusing plane at certain $\Delta\theta_T$

From the results above, we can conclude that distinctive falling tendencies of the focusing performance parameters and evolutions of the ignition region mark different focusing modes when $\Delta\theta_T$ exits. In focusing mode C, the R_{RMS} is the least and the shape of ignition region is focus-symmetric all along with $\Delta\theta_T$, which could be beneficial for the stabilization of thruster.

3.2 Influence of beam out-coming aberration on focusing performance

The variations of I_{max} and R_{RMS} along with the $\Delta\theta_L$ are shown in figure 8 and figure 9 under d_L=1000mm. On the whole, the falling tendencies in the three modes are nearly coincident, and so are their focusing performances. I_{max} all decrease steeply by an order-of-magnitude or more and the decreasing slopes from more to less are C, A, B with relatively slight contrast. At θ_L=0.5°, I_{max} falls almost ten times than there is no aberration in the three modes and I_{max} in B is a little bigger; θ_L=2°, the I_{max} in A and C are 8.9×10^6W/cm^2 and 9.3×10^6W/cm^2 all below the breakdown threshold of the air, while in B it is 1.57×10^7W/cm^2; θ_L=2.5°, I_{max} are all below 1×10^7W/cm^2. The ascending slopes of R_{RMS} from more to less are C, A, B and all rise from zero. At θ_L=2.5°, R_{RMS} are 9.2mm, 10mm, and 12mm for focusing mode B, A and C.

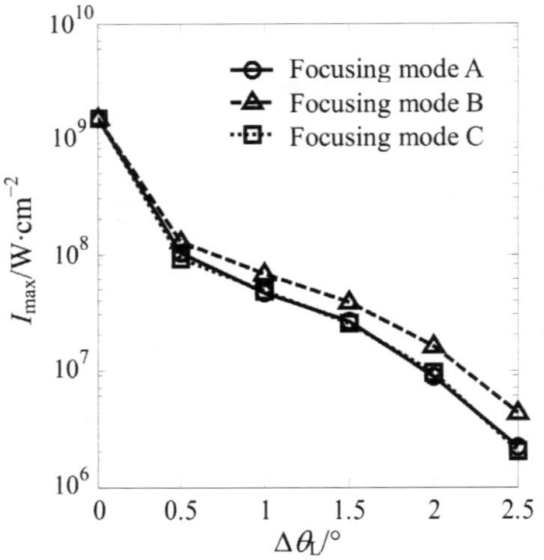

Figure 8. d_L=1m, I_{max} changing with $\Delta\theta_L$

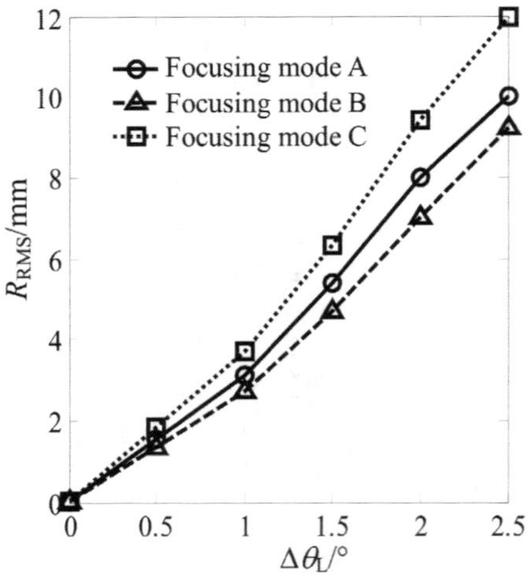

Figure 9. d_L=1m, R_{RMS} changing with $\Delta\theta_L$

Intensity distribution on the focusing plane when $\Delta\theta_L$ exits and d_L=1m is shown in figure 10. We can see that the focusing spots are all shaped of annulus in the three focusing modes and the intensity in upper part which is along the direction of aberration is stronger than the bottom, while the area of focusing spot in focusing mode B is the least corresponding to R_{RMS} of C in figure 9. At $\Delta\theta_L$=0.5°, the ignition regions are all annulus and intensity distributions are symmetric along axis Y; the radii of the annulus regions from bigger to smaller is C, A, B orderly; the intensity in B is apparently stronger than the other two. At $\Delta\theta_L$=1° and $\Delta\theta_L$=2°, the characteristic of intensity distributions is like the $\Delta\theta_L$=0.5°, but the ignition regions have degraded to shape of circular arc and the focusing spots expand to broader annuluses.

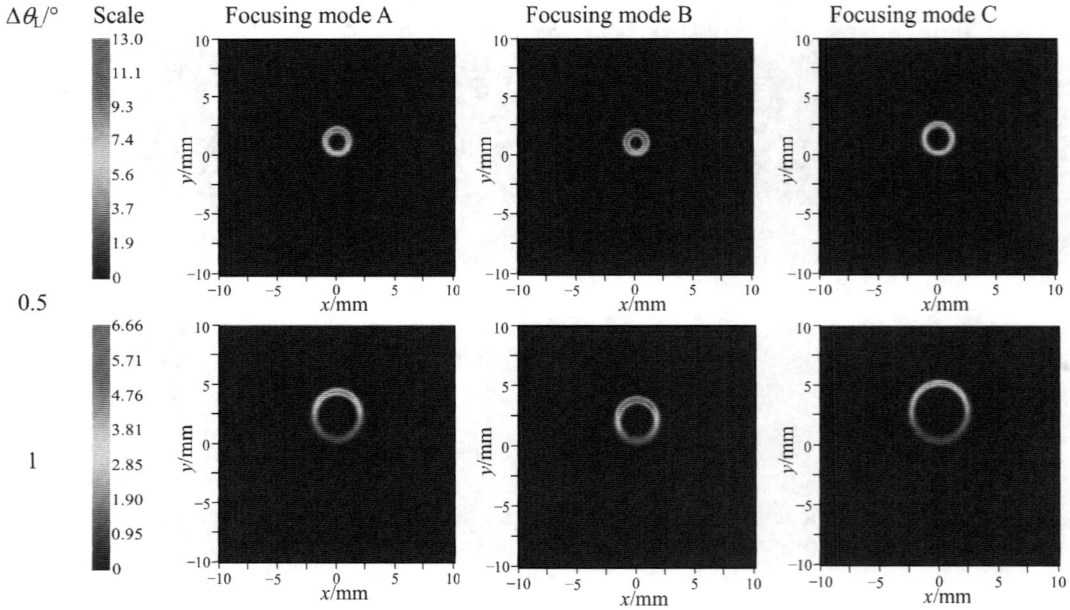

2

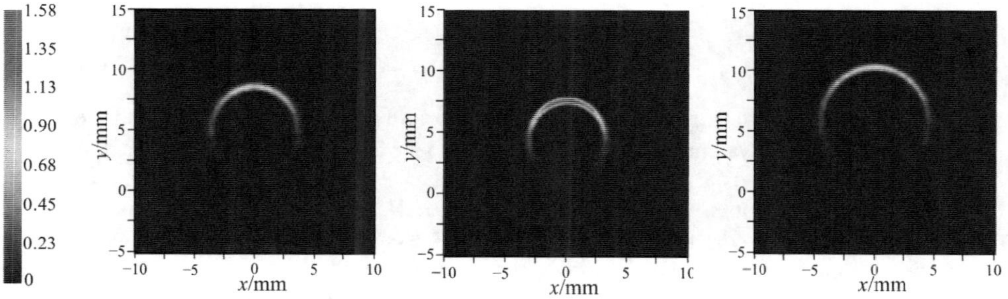

Figure 10. distribution of relative intensity on the focusing plane at certain $\Delta\theta_L$ (d_L=1m)

When $\Delta\theta_L$ exists, the intensity distributions in the three focusing modes are quite similar: the focusing spots all expands roundly and intensities decrease; the ignition regions evolve from the shape of annulus to arc with the area reduced. Coupled with the results of figure 8 and figure 9 that in focusing mode B the ignition region will disappear in the last and the spot deviation is the least, focusing performance of B is a little better. It should be noted that the existence of $\Delta\theta_L$ will bring in $\Delta\theta_T$, because the straying ignition area result in unbalanced force on the nozzle that will incline the thruster.

3.3 Influence of the distance of laser source on focusing performance

Figure 11 and figure 12 are the dependences of I_{max} and R_{RMS} on the d_L, when $\Delta\theta_L$=1°. Mainly, as the d_L increases, I_{max} in the three focusing modes all ascend within a narrow range at first, and then decrease rapidly. When d_L increases from 0 to 0.5m, the I_{max} in focusing mode B rises from 6.1×10^7W/cm^2 to 7.04×10^7W/cm^2; while I_{max} in A and C rise from 3.96×10^7W/cm^2 and 3.82×10^7W/cm^2 to 5.04×10^7W/cm^2 and 4.78×10^7W/cm^2 respectively at d_L=0~1m. At the descending phase when d_L=1.5m~2.5m, I_{max} in the three modes are close and fall below the breakdown threshold of the air at d_L=2.5m. As for R_{RMS}, the figure shows a stationary process at d_L=0~1m with the value around 3.1mm, 2.7mm and 3.6mm for focusing mode A, B and C, after d_L=1m the R_{RMS} all rise uniformly with the rising slope decreasing gradually. At d_L=2.5m, R_{RMS} in focusing mode A, B, and C are 4.21mm, 3.71mm and 5.04mm respectively.

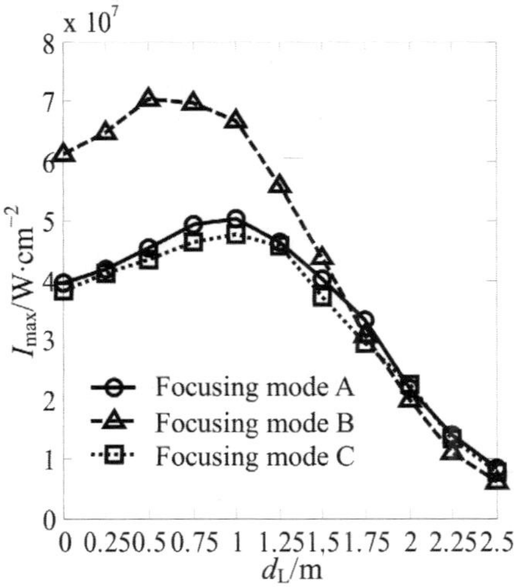

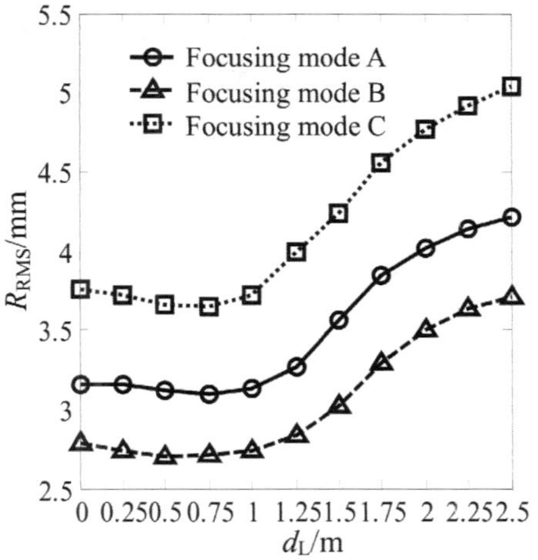

Figure 11. $\Delta\theta_L$=1°, I_{max} changing with d_L **Figure 12.** $\Delta\theta_L$=1°, R_{RMS} changing with d_L

We can see that the distance of laser source in a definitive range won't influence the focusing performance when the beam out-coming aberration exists, but escaping the very range will result in remarkable falling focusing performance.

4. Conclusions

Simulating the laser focusing systems by available software is a convenient and effective way. Modeling of focusing systems is a significant diagnostic and design development manner. Though optical design software is capable of simulating these systems, model construction is not trivial. Simulation accuracy counts on several factors: 1) understanding of the optical problem, 2) accurate computation of the physical model of the reflecting surfaces, 3) the amount of random tested rays, which means more rays tested helps more accurate results, but huge amount of rays being tested will consume computer resource highly and cost a very long time.

Studies show even insignificant aberrations will result in steep falling performance for any focusing mode and strong deviation of beam that directly result in asymmetry ignition area and unbalanced thruster simultaneously. So it strongly demands the directional precision of laser beam, and the attitude control system is indispensable for laser thrusters. While for the three modes, different aberrations cause distinctive falling tendencies of focusing performance: flying aberration of thruster shows absolutely different falling tendencies of focusing performance parameters and evolutions of ignition region; while out-coming aberration of laser beam brings similar falling tendencies and the ignition region; the distance of laser source bounded in certain range won't influence the focusing performance when the previous aberration exists.

With only two angular aberrations investigated, it is not safe to recommend one focusing mode superior to another. Other aberrations such as manufacture errors, thermal deformations and wavefront aberrations of the beam are not considered. These aberrations could couple and will arouse greater falling focusing performance. The eventual conclusion is each focusing mode has its points, and in practice the choice of focusing mode should take into account which aberration is prominent.

References

[1] Ageichik A A , Egorov M S, Rezunkov Y A, Safronov A L and Stepanov V V 2004 Experimental study on thrust characteristics of airspace laser propulsion engine *Sec. Int. Symp. on Beamed Energy Propulsion(Sendai, Japan, 20-23 Oct. 2003) (AIP Conf. Proc.* vol 702*)* ed Kimiya Komurasaki (New York: Proc. AIP) pp 49-60

[2] Sasoh A, Yu X, Ohtani T, Kim S and Jeung I S 2004 In-Tube Laser propulsion: performance and application prospects *Sec. Int. Symp. on Beamed Energy Propulsion(Sendai, Japan, 20-23 Oct. 2003) (AIP Conf. Proc.* vol 702*)* ed Kimiya Komurasaki (New York: Proc. AIP) pp 61-67

[3] Shi H B, Cheng Z G, Jiang J Li X Q, Xu G L and Xia J A 2000 Study on misalignment of parabolic mirror in lightcraft vehicle *High-Power Laser Ablation III (Proc. SPIE* vol 4065*)* ed Phipps C R pp 371-378

[4] Feikema D 2000 Analysis of the laser propelled lightcraft vehicle *31st AIAA Plasma dynamics and Lasers Conf. (Denver CO, US, 19-22 June, 2000)* AIAA-2000-2348 pp 1-14

[5] Wen M, Hong Y J and Li Q 2003 Study of the focus characteristics of the laser lightcraft *J. Equipment Command & Technol.* **14**(4) 93-7

Journal of Physics: Conference Series **276** (2011) 012015 doi:10.1088/1742-6596/276/1/012015

Surface-enhanced Raman scattering in femtosecond laser-nanostructured Ag substrate

Ye Dai[1], Min He[1], Bo Lu[2], Xiaona Yan[1], and Guohong Ma[1]

[1]Department of Physics, Shanghai University, Shanghai 200444, China
[2]Instrumental Analysis & Research Center, Shanghai University, Shanghai 200444, China

E-mail: yedai@shu.edu.cn

Abstract. We demonstrate that a surface-enhanced Raman scattering (SERS) substrate could be directly fabricated on the surface of Ag film by femtosecond laser micromachining. According to the morphology observation by SEM, an amount of nanoparticles, nanoprotrusions, and nanospikes were found to form in the ablation region and the density and size distribution of these Ag nanoparticles depended possibly on the incident laser intensity. Additionally, a large area of nanostructured region was produced by fast line scanning, and an enhancement factor of $\sim 10^5$ was obtained in this region after the sample was soaked in the rhodamine 6G solution for 30 min.

1. Introduction

In the past few decades, femtosecond laser micromachining, based on its ultrashort interaction time and ultrahigh power density, has demonstrated unique advantages in the field of micr- or nanostructuring on various material surfaces [1~3]. In the case of metals, the femtosecond laser micromachined areas have already ranged from surface ablation, drilling and cutting [4] to manufacture functional microstructures for microfluidics, biological sensor or chemical catalysts [5~8]. Recent research results show that some induced nanostructures on the metal surface could significantly enhance the resonant excitation of localized surface plasmons [9~11]. For example, Guo *et al.* reported a new technique to control surface reflectivity of metal from UV to terahertz band by femtosecond laser direct patterning [9]. In that process, femtosecond laser irradiation could produce periodic metal species in form of nanoparticle, nanosphere, nanoprotrusion, and nanocavity on the surface. As a result, these novel phenomena greatly richen the optical properties of metal. In addition, it is well known that some noble metal nanoparticles absorbed by individual molecules could be active sites for surface enhanced Raman scattering (SERS), so we try to use femtosecond laser to fabricate an SERS substrate on the surface of Ag film.

Since SERS has very high sensitivity in individual molecular detection, it has been extensively studied since its discovery in 1977 [12, 13]. Up to date, the fundamental mechanism of SERS is still far to understand, but most of the SERS results include a resonant scattering process in the rhodamine (R6G)/Ag system. In this paper, we used rhodamine solution to overlay a femtosecond laser-nanostructured Ag film, and an SERS enhancement factor (EF) of $\sim 10^5$ at a 514nm excitation wavelength was obtained. We think that this is an easy femtosecond laser writing technique to pattern large areas Ag substrate for SERS.

2. Experimental

The Ag films used in this experiment were coated by magnetron sputtering method (JGP-450a made by SKY Technology Development Co., Ltd), and the film substrates were borosilicate glass. The target material was Ag with purity of 99.99%. The sputtering time was 20 minutes.

A regenerative amplified Ti: Sapphire mode-locked pulse laser (Spitfire, Spectra-Physics Co.) emitted a train pulse of 1 kHz, 120 fs and 800 nm. A Gaussian profile laser beam was guided onto the Ag film by a 20× objective lens (N.A. = 0.3), and the diameter of the focused point was about 10 μm in the focal point. By using a computer controlled 5-axis motion stage with minimum resolution of 1μm, we could select the irradiated region through moving the laser focal point. The average power of incident laser could be controlled by adjusting a neutral density filter.

Two kinds of rhodamine aqueous solution with respective concentrations of 10^{-7}M and 10^{-3}M were prepared for SERS measurement. Firstly, the rhodamine powders were dissolved in distilled water, and then an amount of NaCl was added into the solution so that the Cl^- ion can stabilize the rhodamine's absorption on Ag nanoparticles. After the two solutions were stirred by magnetic stirrer for 1 hour respectively, the femtosecond laser-nanostructured Ag film sample was soaked in the 10^{-7}M rhodamine solution for 30 minutes. And a referenced Ag film sample without femtosecond laser irradiation was soaked in 10^{-3}M rhodamine solution for 30 minutes as well. The Micro-Raman spectrometer (Renishaw Invia) was used for SERS measurement with 514 nm excitation wavelength.

3. Results and discussion

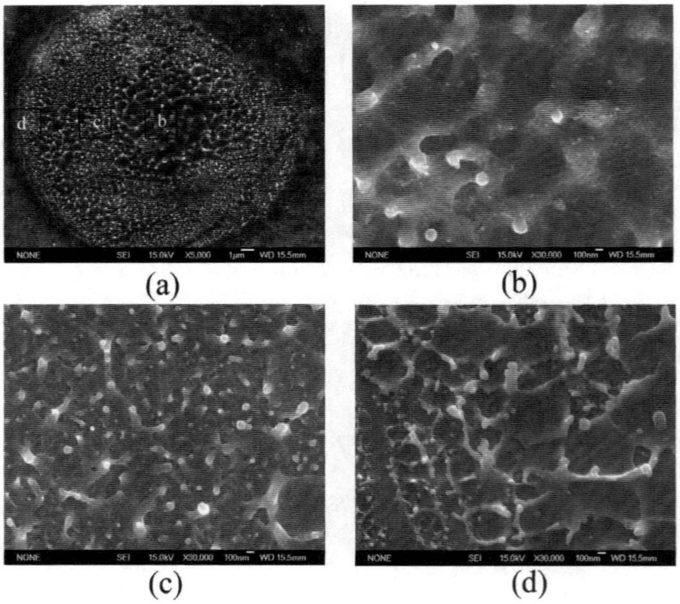

Figure 1. (a) SEM images of surface morphology of the nanostructed Ag film after one pulse irradiation at fluence of 1.27 J/cm^2. (b) ~ (d) The nanoscale structures in the respective regions in (a).

After the as-deposited Ag film was cleaned by de-ionized water, a femtosecond laser pulse with 4 μJ energy was irradiated to the surface through 20X objective lens. To avoid the optical breakdown of Ag film, the focused plane was set to be 100 μm above the sample surface. Figure 1 shows the SEM images of surface features of the nanostructured Ag film after femosecond laser irradiated at fluence of 1.27 J/cm^2. In the laser irradiated region, an obvious nano-scale web structure like a coral could be observed. Besides that, a few of nanoparticles, nanoprotrusions and nanospikes formed on the end of each branch of this structure, and they looked rather random and sparse. As the Gaussian distribution of the laser profile, the size and density of the formed Ag nanoparticles showed a little discrepancy [Figure 1. (b)~(d)]. The Ag nanopartilces (~ 150nm) in the center region was slightly larger than the ones (~50nm) in the surroundings, but their densities were contrast to their sizes. The proposal mechanism was that these larger structures were associated with surface defects or laser beam intensity inhomogeneities [10].

Since the Ag nanoparticles could be formed by femtosecond laser ablating, a large area of nanostructured region was produced by line-scanning the Ag film surface. The stage moving speed was about 50 μm/s. As shown in figure 2 (a), a series of parallel lines with uniform nanostructures occurred after femtosecond laser ablation. Under the magnified observation [Figure 2 (b) and (c)], we could found some microgrooves formed in the ablated region. In addition, many nanoparticles and nanoprotrusions resided on these concave-convex structures. In this condition, these generated Ag nanoparticles constituted an electric-coupling system in which the giant electromagnetic enhancements could be achieved through interacting their electric fields between any two adjacent ones under the resonance excitation [11]. By this means, these nanoparticles became the so-call "hot spots" to support the SERS.

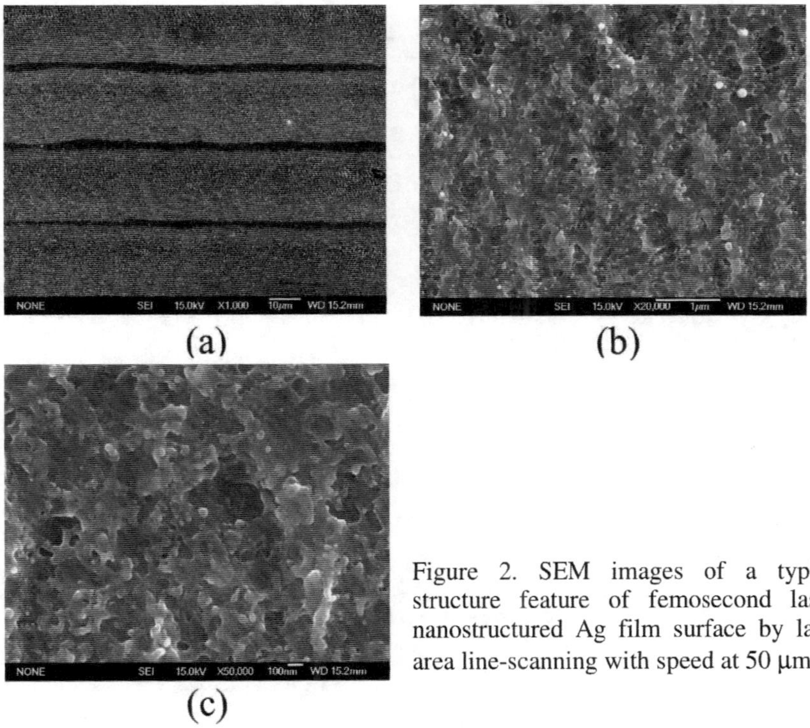

Figure 2. SEM images of a typical structure feature of femosecond laser-nanostructured Ag film surface by large area line-scanning with speed at 50 μm/s.

Figure 3. shows the SERS spectra of a femotosecond laser nanostructed Ag film with 10^{-7}M R6G solution (red line) and the normal spectra of the as-deposited Ag film with 10^{-3}M R6G solution (black line). For the two spectra, the 514 nm excitation powers are 10mw and 1mw, respectively. The accumulation times is 10s for both ones. The apparent enhancement factor was experimentally measured by direct comparison using the following relation: $EF = \left(RS^{ENH} / RS^{REF}\right) \times \left(C^{REF} / C^{ENH}\right)$, where RS^{ENH} and RS^{REF} are the measured Raman intensities, and C^{REF} and C^{ENH} are the solution's concentrations for normal and enhanced samples [14]. According to this equation, we calculated that their EFs at different Raman peaks 610 cm^{-1}, 770 cm^{-1}, 1358 cm^{-1}, 1507 cm^{-1}, and 1647 cm^{-1} were 5.4×10^5, 5.3×10^5, 7.3×10^5, 6.2×10^5, and 9.3×10^5, respectively. These EFs proved substantially that the femtosecond laser-nanostructured Ag film could be used to enhance signal sensitivity for the SERS measurement. Generally, there existed two enhancement processes derived from physical and chemical mechanisms in the SERS [15, 16]. In this experiment, the formed Ag nanoparticles (<200nm) created an electromagnetic enhancement because they could support surface plasmon resonance at visible wavelengths, which is in term of a physical enhancement process. Simultaneously, most of the Ag nanoparticles combined with the rhodamine molecules also yielded a resonance Raman enhancement due to a metal-to-molecule or molecule-to-metal charge-transfer electronic transition, which is a chemical enhancement process. Due to the working together of these two enhancement processes, the rhodamine Raman scattering signals on the surface of the nanostructured Ag film exhibited a very large cross-section enhancement factor of ~10^5.

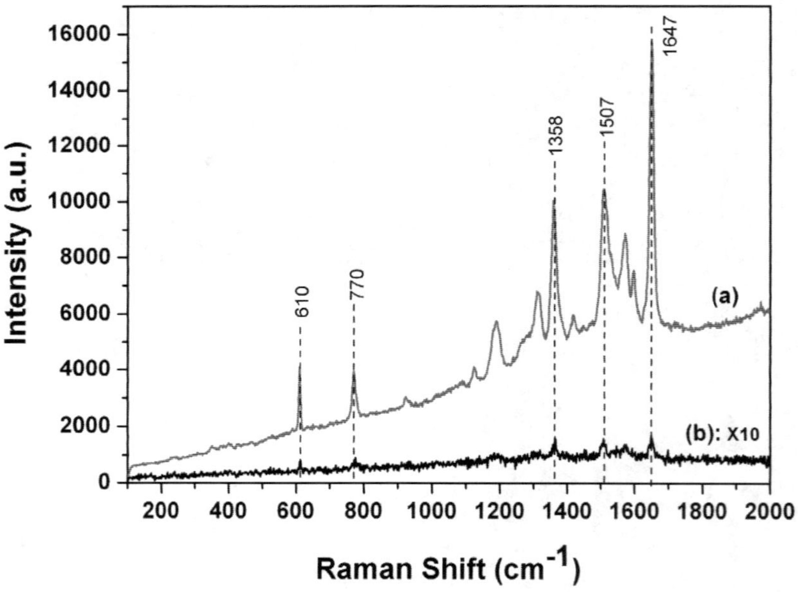

Figure 3. (a) The SERS spectrum of the nanostructured Ag film in 10^{-7} M R6G solution (red line) and (b) the Raman spectrum of the as-deposited Ag film in 10^{-3} M R6G solution (black line: $\times10$).

4. Conclusions

We fabricated a kind of SERS substrate by using femotsecond laser ablation on the surface of Ag film. After femtosecond laser irradiation, a textured nanostructure surface with an amount of Ag nanoparticles could be obtained. Due to the laser intensity varying, the density and size of these formed Ag nanoparticles showed a little discrepancy in the different parts of the ablation region. However, by means of a femtosecond laser line-scanning technique, a uniform nanostructure in large-scale could be produced, which was used to the substrate for the SERS measurement. In the current rhodamine/Ag system, the enhancement factor of the Raman signal is found to be as high as ~10^5. We suggest that this SERS substrate on the surface of Ag film can be applied for detecting micro quantity analyte in chemistry and biomedicine fields.

Acknowledgement
This work was financially supported by National Natural Science Foundation of China (Grants No. 60908007, 10774099), the Program for Professor of a Special Appointment (Eastern Scholar) at Shanghai Institutions of Higher Learning, Science and Technology Commission of Shanghai municipal (09530501100) and Shanghai Leading Academic Discipline Project (S30105).

References

[1] Chichkov B N, Momma C, Nolte S, Von Alvensleben F and Tunnermann A 1996 *Appl. Phys. A* **63** 109
[2] Perry M D, Stuart B C, Banks P S, Feit M D, Yanovsky V and Rubenchik A M 1996 *J. Appl. Phys.* **85** 6803
[3] Borowiec A and Haugen H K 2003 *Appl. Phys. Lett.* **82** 4462
[4] Kamlage G, Bauer T, Ostendorf A and Chichkov B N 2003 *Appl. Phys. A* **77** 307
[5] Kabashin A V, Meunier M, Kingston C and Luong J H T 2003 *J. Phys. Chem. B* **19** 107
[6] Lin C H, Jiang L, Chai Y H, Xiao H, Chen S J and Tsai H L 2009 *Opt. Express* **17** 21581
[7] Han Y, Lan X, Wei T, Tsai H L, Xiao H 2009 *Appl. Phys. A* **97** 721
[8] Zhou Z, Xu J, Liao Y, Cheng Y, Xu Z, Sugioka K, Midorikawa K 2009 *Opt. Comm.* **282** 1370
[9] Vorobyev A Y, Topkov A N, Gurin O V, Svich V A and Guo C 2009 *Appl. Phys. Lett.* **95** 121106
[10] Vorobyev A Y, Guo C 2006 *Opt. Express* **14** 2164
[11] Hwang T Y, Vorobyev A Y and Guo C 2009 *Appl. Phys. Lett.* **95** 123111
[12] Jeanmaire D L, Van Duyne R P 1977 *J. Electroanal. Chem.* **84** 1
[13] Albrecht M G, Creighton J A 1977 *J. Am. Chem. Soc.* **99** 5215
[14] Jackson J B, Halas H J 2004 *Proc. Natl. Acad. Sci.* **101** 17930
[15] Michaels A M, Nirmal M and Brus L E 1999 *J. Am. Chem. Soc.* **121** 9932
[16] Diebold E D, Mack N H, Doorn S K and E. Mazur E 2009 *Langmuir* **25** 1790

Development a Cu-based Metal Powder for Selective Laser Micro Sintering

C. Dai[1], H.H. Zhu1, L. D. Ke1, W.J. Lei1, B.J. Chen[2]*

[1]Division of Laser Science and Technology, Wuhan National Laboratory for Optoelectronics, Institute of Optoelectronics Science and Engineering,
[2]State Key Laboratory of Material Processing and Die & Mould Technology, Huazhong University of Science and Technology, Wuhan, Hubei, 430074, P. R. of China

E-mail: chenbaijin@mail.hust.edu.cn

Abstract. A Cu-based metal powder which consists of Cu and Cu-P alloy for selective micro laser sintering has been developed based on the theoretical analysis of the characteristics of the laser micro sintering metal powder. The characteristics of the wetting, capillary force and viscosity have been considered. The preliminary experimental investigation on the selective laser micro sintering Cu-based metal powder has been performed. A 50W CW Nd:YAG laser was employed to sinter the developed metal powder mixture. The sintering mechanism and the effect of the process parameters on the characteristics of the sintering samples have been preliminary investigated. The results show that the mechanism of laser micro sintering this developed metal powder is liquid-phase sintering and Cu-P alloy powder plays an effectively binder in the sintering process. The process parameter has significant effects on the characteristics of the sintering parts. From the SEM image, two different microstructures of samples with different scan spacing parameters were compared and a better binding effect was obtained at a parameter of 0.05mm scan spacing.

1. Introduction

As the mechanical engineering industry confront a growing demand of components with um-sized resolution for an increasing range of applications, miniaturization of components and moulds get more and more attractive in mechanical fabrication. Selective laser micro sintering (μ-SLS) can meet these requirements, which is the development from a modified selective laser sintering. Same as SLS process, μ-SLS can fabricate micro mould directly from CAD data. Comparing to SLS process, μ-SLS has the advantages of higher resolution [1].

As the μ-SLS developed, the process attractive more and more attentions, there are more and more researches in recent years. Y.P. Kathuria used a pulsed Nd:YAG laser to achieve a line width of about 221μm with one metal component and a line width of about 470μm with two different metal component in1999[2]. Horst Exner et. employed a Q-switched Nd:YAG laser with mixtures of copper and tungsten successfully obtain a micro part with a spacing resolution of 30μm in ambient atmosphere in 2003[3-4]. Since the beginning of 2005 oxide and non-oxide ceramics were investigated but the characteristics such as aggregation, dissociation, recombination and further reactions of this type material make it difficult to sinter than metals [5]. Although the μ-SLS has been introduced for several years, the technology of producing 3-D micro metal parts using μ-SLS process is still in its early stage. The investigation of workable materials is crucial for this technology which decides the feasibility and sinter ability of the process. Balling is a severe problem in laser sintering metallic powder. The previous research shows that the two-phase method is an effective method to overcome the balling effect. In this process, the material system consists of at least two phases with different melting points, which are called binder and structural metal powders. During the process, the binder melts, fills the pores and binds the solid particles together. The liquid flows and the structural particles rearrange under the influence of capillary force, leading to densification of the layered powder material. This paper focused on the development of a Cu-based metal powder for the μ-SLS.

2. Material development

For a superior performance in μ-SLS via liquid phase sintering binding mechanism, there are three significant physical aspects of the materials need to be considered such as wetting, capillary force and viscosity. So find a material system with outstanding characters on these physical aspects will be proper for μ-SLS.

2.1 Wetting

Wetting is the first requirement in laser sintering process, which determines the sinter ability, densification and microstructure. Wetting is expressed by the contact angle and dihedral angle depending on the phases in the system. Contact angle expresses the equilibrium of the interfacial energies in the system including liquid, solid and vapor. Assuming the solid surface is ideal and the curvature of L/V surface can be negligible, the contact angle θ can be expressed as the following classical Young equation [6]:

$$\cos\theta=(\gamma_{SV}-\gamma_{SL})/\gamma_{LV} \tag{1.1}$$

where γ_{SV}, γ_{LV} and γ_{SL} are the surface tensions of the solid-vapor, liquid-vapor and solid-liquid , respectively. The range of the contact angle can be from 0° to 180°. A low contact angle (<90°) means good wetting (Figure 1(a)) and a high contact angle (>90°) means poor wetting (Figure 1 (b)). Zero contact angle indicates a perfect wetting.

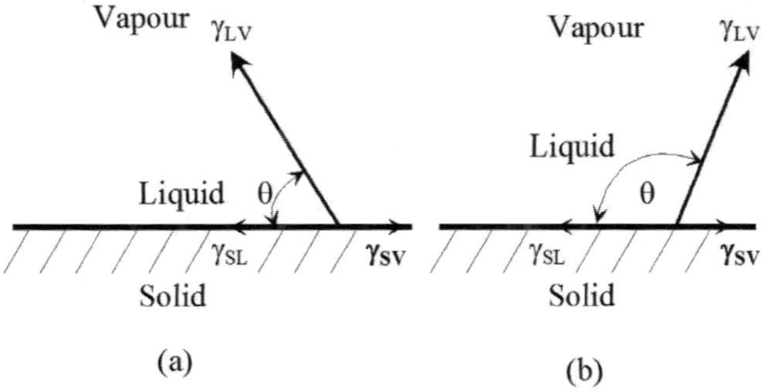

Figure 1. The schematic diagram of contact angle in solid-liquid-vapour system.
(a) good wetting (b) poor wetting

In systems B/A with negligible mutual solubility, the contact angle θ can be found from: [7]:

$$\cos\theta=1-0.28(T_{m1}/T_{m2}-1) \tag{1.2}$$

where T_{m1} and T_{m2} are the melting points of solid metal A and liquid metal B, respectively. From Eq. (1.2), it can be seen that same metal $T_{m1}=T_{m2}$, can wet perfectly since $\theta=0°$; if $T_{m1}/T_{m2}<3$, for instance Fe/Ag, it also has good wetting; if $T_{m1}/T_{m2}>4.2$, the insoluble system will not be wetting. Therefore, in liquid phase sintering, the difference of metal melting points must be suitable.

2.2 Capillary Force

Capillary force is the driving force for densification in laser sintering process. Using the two spheres model shown in Figure2.[6], the capillary force F_c on one of the particles depends on the meniscus size G and surface energy of the liquid γ_{LV} as shown in the following equation:

$$F_c=\pi G2\Delta P+2\pi G\gamma_{LV}\cos\theta \quad (1.3)$$

Where ΔP is proportional to the surface energy γ_{LV} and is inversely dependent on the radii of curvature, and can be obtained from Eq. (1.4):

$$\Delta P = \gamma_{LV}\,(1/r_1+1/r_2)=2\gamma_{LV}\cos\theta/d \quad (1.4)$$

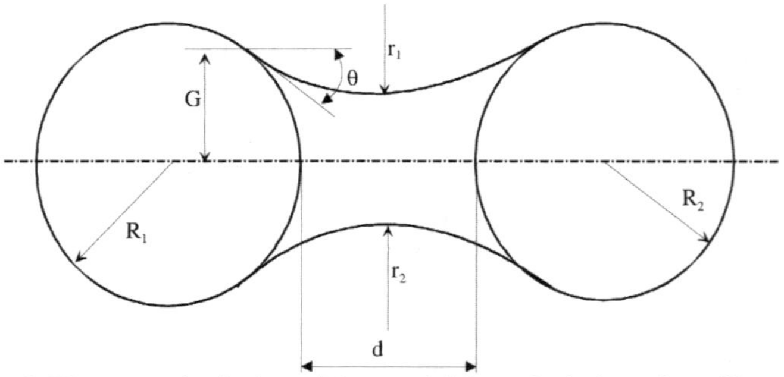

Figure 2. The two spherical particles model for calculation of capillary force.

The capillary force is attractive for good wetting, leading to the densification of the system, while the force is repulsive for poor wetting, leading to the swelling of the system. From the Eqs. (1.3) and (1.4), it can be found that the wetting force increases with the decrease of the contact angle and of particle separate distance. Thus using smaller binder particle size can lead to a higher capillary force and the pores caused by the melting binder can be easily filled.

2.3 Viscosity

In laser sintering, viscosity acts as a resistant force for the liquid spreading. The reciprocal of viscosity is known as fluidity. Therefore, low viscosity of the molten liquid is another requirement for the successful direct laser sintering processing. The viscosities of liquid metals are in the range of 0.5~ 8 mPas and decrease quickly with the increase of the temperature. Eq. (1.5) is a typical equation for calculating the viscosity of simple (monoatomic) liquids near the melting point [8]:

$$\mu_m=1.6\times10^{-4}(MT_m)^{1/2}/V_m^{2/3} \quad (1.5)$$

where V_m is the atomic volume at the melting point and M is the atomic weight. From Eq. (1.5), it can be found that the viscosities of metals decrease with decreasing melting point. For most liquid metals the variation of viscosity μ with the temperature T(K) may be written as Eq. (1.6):

$$\mu = \mu_l \exp(E/RT) \quad (1.6)$$

Where μ_l and E are constants, and R is the gas constant. Eq. (1.6) shows that the viscosity of a liquid metal decreases with the increase of the temperature. Therefore, increasing the sintering temperature can lower the viscosity, and hence promote densification.

In the mixture of solid/liquid system, the viscosity of the system is modified because the solid particles act as obstacles. Thus, the solid particles impede the spreading of the liquid. The viscosity of a solid-liquid mixture μ is related to the viscosity of the pure liquid phase μ_2, and the volume fraction of solid in liquid Φ as in:

$$\mu=\mu_2\,(1-\Phi/\Phi_m)^{-2} \quad (1.7)$$

where Φ_m is the critical volume fraction above which the mixture has essentially infinite viscosity. Particle bonding is driven by the liquid viscosity μ if the materials are suitably chosen for the occurrence of the inter-particle wetting. In laser sintering of metal powders, the volume fraction of solid particles is generally in the range of 30~70%. From Eq. (1.7), the viscosity of the solid/liquid mixture in direct laser sintering can be calculated to be about 2~11 times that of the corresponding pure liquid metal, resulting in slow liquid flow.
However, the viscosity can be decreased by the use of a low melting point binder, the right solid-liquid ratio, addition of alloy elements and the control of the degree of superheating of the liquid phase. On the other hand, the viscosity cannot be too low because the tendency for balling increases with the decrease of the viscosity[9].

2.4 Design of Cu-based Metal Powder
A Cu-based metal powder has been developed according to the above requirements in this paper. It mainly comprises of two powders: Cu powder and pre-alloyed Cu-P powder. The melting point of the "binary" eutectics Cu-P is 710°C. The ratio of melting point of Cu (1083°C) to melting point Cu-P (710°C) is approximately 1.52, which is lower than 3. Therefore the metal powder system has good wetting. The contact angle θ calculated according to Eq. (1.2) is approximately 31.46°. Except for the formation of eutectic, another role of phosphorus is as a fluxing agent to improve the wetting characteristics in the system. It decreases the surface tension of the Cu powder, reacts with the copper oxides from the powder surfaces during the laser sintering and protects the metal surfaces from re-oxidation during heating, which permits good wetting and formation of a sound bond. Firstly, the metal powder system consists of two metal powders with dispersive melting points. The melting point of Cu-P at 710°C is rather low, which hence does not require a high laser power to melt it. Secondly, the melting point difference is more than 300°C, which is beneficial to the liquid flow ability. Thirdly, the material system has good wetting and suitable mutual solubility. Lastly, Cu powder is easy to obtain in the market and the Cu-P is easy to produce.

3. Experimental

3.1 Powder preparation
Gas atomic 99%purity Cu_1 powder with spherical shape and a mean particle size of 150um and another Cu_2 powder with also spherical shape and a mean particle size of 75um were firstly mixed with mass ratio Cu_1: Cu_2 of 1: 0.2. Then the mixed Cu powder is mixed with Cu-P alloy (phosphor 7.25wt%, copper 92.75wt %) powder according to mass ratio of 60: 40. The shape of the Cu-P alloy powder is also spherical, and its mean particle size is about 45um. The mixed powders was loaded into the processing cylinder and levered by the blade for obtaining a flat processing surface.

3.2 Laser processing
In the experiments, a 50W CW Nd:YAG laser controlled by a galvanometer scanner was employed for sintering, powders processed in the ambient atmosphere at room temperature, the scan speed was fixed on 10mm/s. The figure3 shows the schematic set-up for laser micro-sintering. The experiment was divided to 3 groups and different scan spacing was conducted in every group to see varied surface morphology of the single layer sample. A 20mm×20mm single layer square CAD profile was created by the controlling computer for sintering blends powders. Every different process parameter were conducted to make a 20mm×20mm square sample for subsequent assessments.

3.3 Characterization
The microstructure and surface morphology analysis of the individual layer samples were characterized using an FEI Sirion 200 Scanning Electron Microscope, the composition of the sintered sample was analysed by an Energy Dispersive X-ray analysis (EDX).

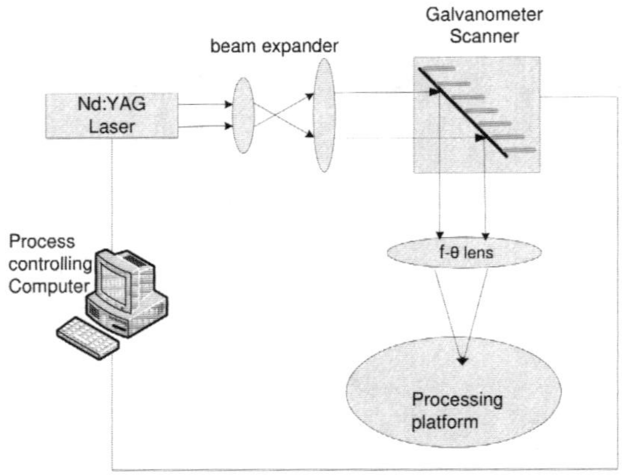

Figure 3 . Schematic diagram of the set-up for laser micro sintering.

4. Result and discussion

4.1 Sintering Mechanism

Figure4 shows the SEM image of surface morphology of laser-sintered sample at a scan spacing of 0.05mm and laser power of 10W. From Figure4, it can clear find many spherical powders with size around 100um, and some bridges connect the spherical powders. It can also find that many holes in the microstructure, suggesting non-density of the sample.

Figure 4. SEM image of surface morphology of laser-sintered sample.

From the image, it can see that the particles with size of about 100um keep there shapes as sphere very well, which suggest that some powders are not molten during the process. The "bridges" have different shapes, which are significantly differently from the spherical powder. This means that the "bridge" is solidified from a molten powder. The melting point of Cu-P is 710°C and the melting point of Cu is 1083°C. Therefore, melting Cu-P is much easy than that of Cu for same laser energy absorption.

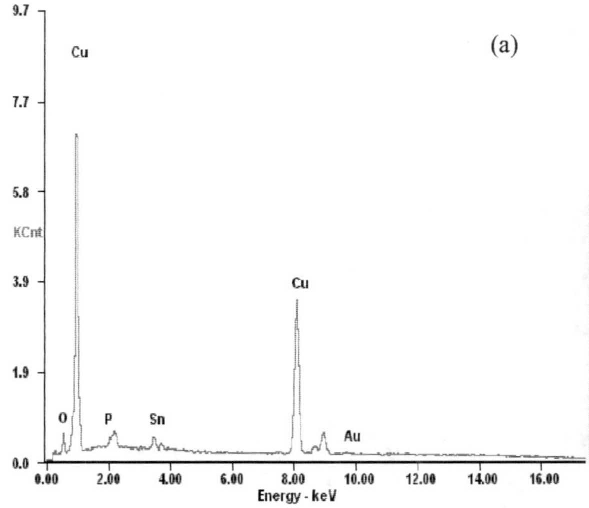

Element	*Wt%*	*At%*
OK	03.35	12.68
PK	01.31	02.55
CuK	84.56	80.60
Matrix	Correction	ZAF

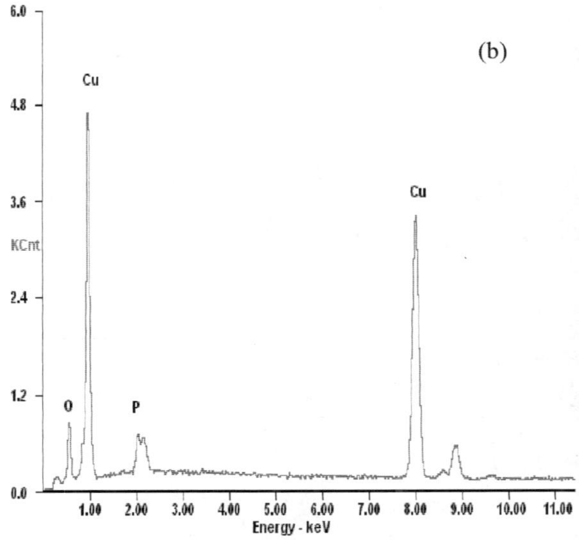

Element	*Wt%*	*At%*
OK	05.32	17.77
PK	03.04	05.24
CuK	91.63	76.99
Matrix	Correction	ZAF

Figure 5. EDX analyses showing chemical compositions of the non-melting powder (a) and binder (b).

The EDX results approve such conclusion. Figure5 shows the typical EDX analysis results of a spherical powder and the "bridge". Figure5(b) shows the components of the bridge. From Figure(b), it can be found that the components of the "bridge" are mainly P and Cu, and the content P is more than 3 wt%, which is much higher than that of the spherical powder, approving that the "bridge" is solidified by liquid Cu-P pre-alloy powder. Figure5(a) is the EDX analysis of the spherical powder. The EDX analysis shows that there are two elements in the surface of the powder: P and Cu, but the content of P element is only about 1 wt% and the content of Cu is about 99%, suggesting that Cu powder is not molten during the process. From Figure5(a), it can also see that it is relatively rough at the surface of the spherical powder, this is because the wetting between the liquid Cu-P and Cu is very good,. During laser sintering, the liquid Cu-P coats the Cu powder, after laser removing, the liquid Cu-P solidifies and brings a rough surface of the Cu powder and a minority phosphorous content.

From the above analysis, it can found that this kind of sintering is base on the mechanism of liquid phase sintering. During the process, the structure powder with higher melting point is not molten and another powder with lower melting point is complete molten [10]. In this metal powder system, Cu-P alloy component is melted and plays an effect of binder in the sintering processing and Cu powder has been connected by binder and rearranged to a network, so this Cu-based powder system can sintering by laser irradiation. Amount of Cu powders has been coated by the binder confirm that the Cu-P alloy has an excellent fluidity in this processing temperature.

In the mixed powders, Cu-P alloy as the binder due to it lower melting point (710℃)and higher absorption than Cu (1083℃) component. So when during the laser irradiation, Cu-P alloy can easily be melted and plays like a binder, which connects two adjacent Cu spherical powders. Cu-P alloy materials have a unique characteristic called the Flow Point, the Flow Point is defined as the temperature at which the filler metal is fluid enough to capillary through a joint even though not completely liquid. Another effect phosphorus component can provide in the processing is that the element phosphorus can act as flux to prevent Cu oxidization[11-12].

4.2 Effect of the scan spacing

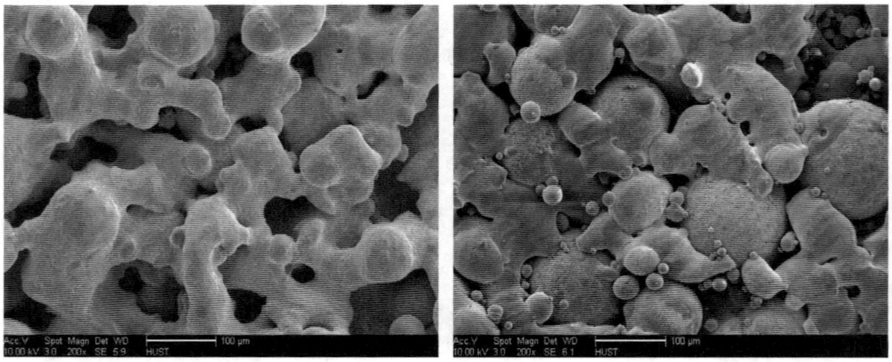

(a) 0.05mm (b) 0.1mm

Figure 6. The surface morphology of the sample using different scan spacing.

(a) (b)

Figure 7. (a) (b) showing the microstructure of Cu ball with a neck of Cu-P respectively at scan spacing 0.05mm (a) and 0.1mm (b).

In the experiments, the laser power was fixed on 10W and the scan speed was kept on 10mm/s, scan spacing changes at the range from 0.1mm to 0.05 mm.

Figure6 (a) is the surface morphology of the sample was obtained using scan spacing of 0.05mm and figure6 (b) is the surface morphology of the sample was obtained using scan spacing of 0.1mm. It clear see from figure 6(b) that many Cu-P alloy powders are not melted under 0.1mm scan spacing and the neck between two Cu powders are not strength enough. It seems that the liquid Cu-P alloy powder cannot flow and coat on Cu powders enough before re-solidify. But Cu-P alloy powders are melted absolutely and coat the Cu powders under 0.05mm scan spacing (figure6(a) and form a denser microstructure than that of using 0.1mm scan spacing. This may because it can have a higher energy injection lead to a higher temperature and capillary force.

Figure7 is the magnification of Figure 6. In the figure7, the microstructure of two Cu spherical powders with a neck under 0.05mm scan spacing (a) and 0.1mm scan spacing (b) are showing. From the (b), in clear see a boundary between the Cu ball and the neck, Cu powders are only a part of spherical surface coated by the binder, but in the (a), the spherical surface of Cu powder are absolutely coated by the binder, so it can not found the boundary between the Cu ball and the neck.

5. Conclusion

A Cu-based metal powder for selective micro laser sintering has been development, the physical character of this material system accord with the theoretical calculation and base on the experiment and analysis above, following conclusion can be given:

- The developed metal powder system is suitable for laser sintering, the mechanism of this laser sintering is liquid-phase sintering.
- Cu-P alloy material plays binder in u-SLS process, while Cu powder plays the structure, the wetting between them is very well.
- Small scan spacing leads to a dense and pretty connection microstructure.

Acknowledgements

This work is supported by the National Natural Science Foundation of China (60878022), the fund of Key Laboratory of Chemical Laser (KLCL-HT-200907) and Sponsored by the Scientific Research Foundation for the Returned Overseas Chinese Scholars, State Education Ministry. The Authors also thank the helping of Analytical and Testing Center of HUST for their providing SEM analysis.

Reference

[1] Kumar, S 2003 Selective Laser Sintering:A Qualitative and Objective Approach[J]. *Jom-Journal of the Minerals Metals & Materials Society* **55** 43

[2] Kathuria YP 1999 Microstructuring by selective laser sintering of metallic powder[J] *Surface & Coatings Technology* **116** 643

[3] Exner H, Regenfuss P, Hartwig L, Klotzer S and Ebert R 2003 Selective laser micro sintering with a novel process[C] *Fourth International Symposium on Laser Precision Microfabrication* **5063** 145

[4] *Regenfuss P, Hartwig L, Klotzer S,* Ebert R, Brabant T, Petsch T and Exner H 2005 Industrial freeform generation of microtools by laser micro sintering[J] *RPJ* **11** 18

[5] Regenfuss P, Streek A, Hartwig L, et 2007 Principles of laser micro sintering[J] *RPJ* **13** 204

[6] German, R. M. 1985 Liquid Phase Sintering. *New York: Plenum Press*. 13-94

[7] Guo, S. J. 1998 Theory of Powder Sintering. *Peiking: Metallurgy Industrial Press.* 308

[8] Iida, T. and R. L. Guthrie. 1988 The Physical Properties of Liquid Metals. *Oxford: Oxford University Press* 134-192

[9] Lu, L., Fuh, J. and Y. -S. Wong. 2001 Laser-Induced Materials and Processes for Rapid Prototyping. *Norwell: Kluwer Academic Publishers* 89-142

[10] Nikolay Tolochko Sregei Mozzharov Tahar Laoui and Ludo Froyen 2003 Selective laser sintering of single and two-component metal powders[J] *RPJ* **9** 68-78

[11] Dongdong Gu, Yifu Shen, Shangqing Fang and Jun Xiao 2007 Metallurgical mechanisms in direct laser sintering of Cu–CuSn–CuP mixed powder[J] *Journal of Alloy and Compounds* **438** 184

[12] Zhu HH, Lu L and Fuh JYH 2003 Development and characterisation of direct laser sintering Cu-based metal powder *6th APCMP* **140** 314

Ultra Stable, Industrial Green Tailored Pulse Fiber Laser with Diffraction-limited Beam Quality for Advanced Micromachining

P Deladurantaye[1], V Roy, L Desbiens, M Drolet, Y Taillon and P Galarneau

INO, 2740 rue Einstein, Quebec City, QC, Canada, G1P 4S4

E-mail: pascal.deladurantaye@ino.ca

Abstract. We report on a novel pulsed fiber laser platform providing pulse shaping agility at high repetition rates and at a wavelength of 532 nm. The oscillator is based on the direct modulation of a seed laser diode followed by a chain of fiber amplifiers. Advanced Large Mode Area (LMA) fiber designs as well as proprietary techniques to mitigate non-linear effects enable output energy per pulse up to 100 μJ at 1064 nm with diffraction-limited beam quality and narrow line widths suitable for efficient frequency conversion. Ultra stable pulses with tailored pulse shapes were demonstrated in the green region of the spectrum at repetition rates higher than 200 kHz. Pulse durations between 2.5 ns and 640 ns are available, as well as pulse to pulse dynamic shape selection at repetition rates up to 1 MHz. The pulse energy stability at 532 nm is better than ± 1.5%, 3σ, over 10 000 pulses. Excellent beam characteristics were obtained. The M^2 parameter is lower than 1.05, the beam waist astigmatism and beam waist asymmetry are below 10% and below 8% respectively, with high stability over time. We foresee that the small spot size, high repetition rate and pulse tailoring capability of this platform will provide advantages to practitioners who are developing novel, advanced processes in many industrially important applications.

1. Introduction

Pulsed fiber lasers are of growing interest to many practitioners since they offer capabilities and performances that are difficult to achieve with traditional lasers such as Diode Pumped Solid State (DPSS) Q-switched lasers. High repetition rates, excellent beam quality, high average power and industrial robustness are a few of the key attributes that explain why pulsed fiber lasers are becoming prime candidates to develop advanced processes in many industrially important applications such as laser repair of dynamic random access memory (DRAM), laser scribing of photovoltaic cells, laser drilling of microvias in semiconductors, flexible interconnects, IC packages, dielectrics, including glass, and metals [1].

The new Master Oscillator with Programmable Amplitude Waveform (MOPAW) Green fiber laser platform developed by INO and presented here is primarily intended to advanced laser material processes for production of small feature sizes at high speed and with high throughput of such processed features. Those processes are highly desirable in the microelectronic industry, where the

[1] To whom any correspondence should be addressed.

Published under licence by IOP Publishing Ltd

progressive miniaturization translates into demands for smaller spot sizes, higher repetition rates and increased laser flexibility. The ability to tailor the pulse characteristics of our MOPAW fiber laser in a straightforward manner, especially the pulse shape, can provide substantial advantages in terms of process yield, throughput and reliability. Producing smaller spot sizes often implies working at harmonic wavelengths which typically requires more energy per pulse than working at the fundamental. The need for high pulse energy, combined with operation at increasingly high repetition rates, call for increased average output power levels without sacrificing the laser beam quality nor the spectral purity of the emission.

A further commonly required laser critical characteristic in advanced laser material processing is the stability of the laser performances over time and with respect to environmental changes. For example, in semiconductor memory processing applications where a single pulse is used to process a given structure, such as severing conductive links for memory repair, it is important to keep the pulse energy within a well defined energy per pulse process window. If the pulse energy is too low, then the link may be incompletely removed. In the case where the energy per pulse exceeds the allowable energy process window, excess pulse energy may be coupled into adjacent or underlying link structures, or the substrate itself, thereby causing highly undesirable damage to the device. Short-term and long-term stability of the laser critical performances, including beam characteristics, average power, energy per pulse, pulse amplitude and beam pointing, have been extensively addressed during the course of the development of the MOPAW Green fiber laser presented here.

2. MOPAW Green laser platform description

A top level representation of the MOPAW Green fiber laser architecture is shown in figure 1. It is based on a directly modulated semiconductor laser diode seeding a chain of two fiber amplifiers in a Master Oscillator, Power Amplifier (MOPA) configuration.

2.1. High speed electronics

The high speed electronics includes a Digital Control Unit (DCU) and a high-speed digital to analog converter (DAC) used in conjunction to drive the seed laser diode. The DCU generates a waveform by concatenation of consecutive temporal "bins", the amplitude of each "bin" being programmed independently with a 10 bit resolution. In the current version of the MOPAW, each "bin" has a temporal width of 2.5 ns and 1 to 256 "bins" can be addressed, thus allowing pulse durations in the range of 2.5 ns to 640 ns. The waveform generated by the DCU is converted to an analog signal by the high-speed DAC. An RF amplifier is used to drive the seed laser diode with the generated waveform.

2.2. Seed stage

The first optical stage of the oscillator is the seed stage, as shown in figure 1. It includes the seed laser diode which is typically chosen to emit at 1064 nm while other wavelengths are possible. Additionally, the seed stage incorporates a seed laser beam conditioning unit to tailor the spectrum of the optical pulses for mitigating fiber nonlinearities such as Stimulated Brillouin Scattering (SBS) and Self-Phase Modulation (SPM) in the following fiber amplification stages. Finally, the seed stage includes a photodiode for optical power monitoring purposes.

2.3. Preamplifier stage

The preamplifier stage follows the seed stage. It is based on a highly efficient fiber amplifier design incorporating an optimized ytterbium-doped, single clad, polarization maintaining fiber from INO. A net gain of 30 dB is generated by the preamplifier stage, with a pump power budget of 750 mW at 976 nm. An optical filter whose transmission spectrum is centered at the seed laser emission wavelength is employed at the output of the preamplifier for minimizing the amount of ASE injected in the power amplification stage to maximize the extractable energy in that stage. Finally, as in the seed stage, a photodiode has been included in the preamplifier unit for monitoring purposes.

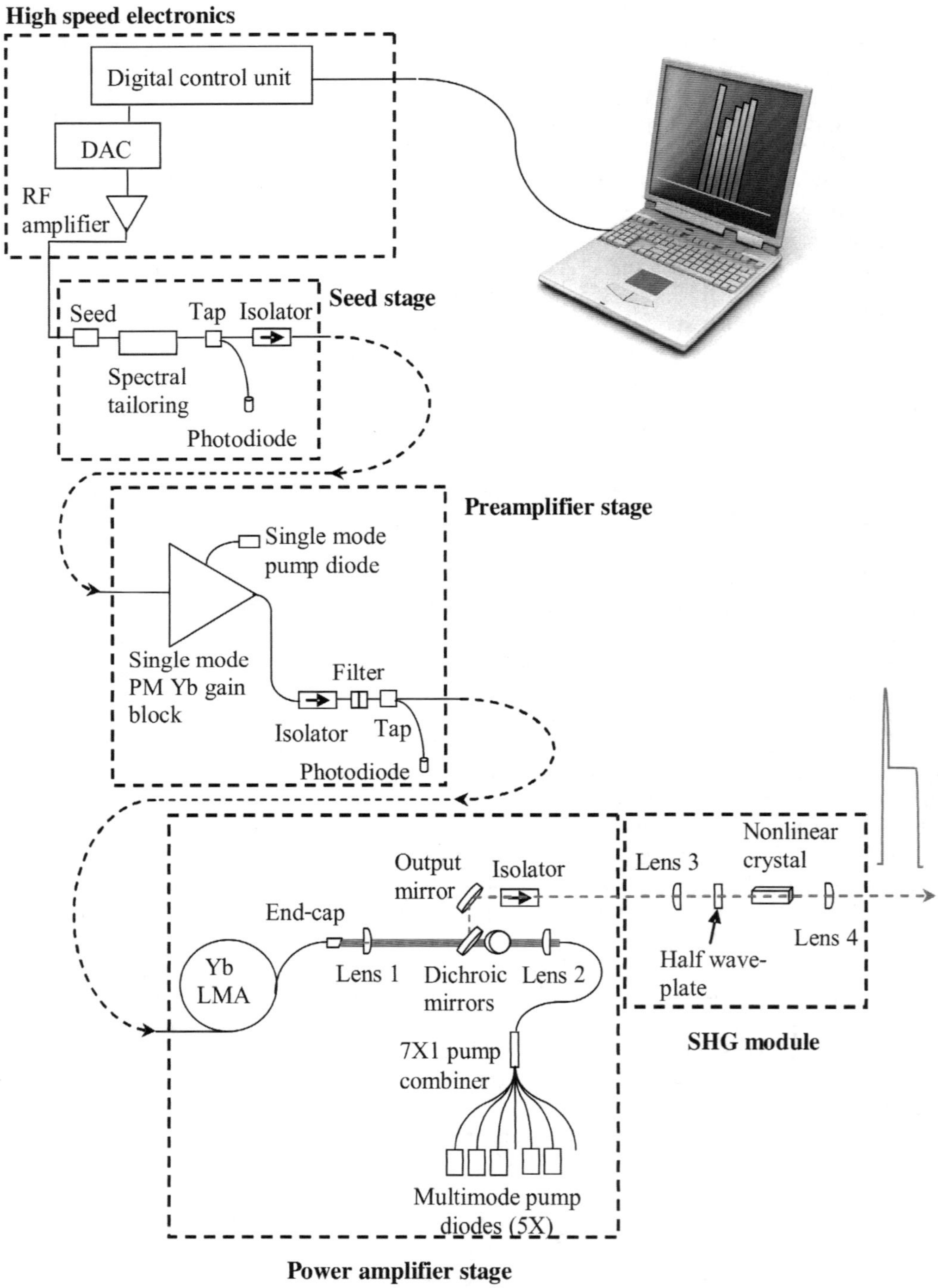

Figure 1. Top-level representation of the laser architecture.

2.4. Power amplifier stage

The power amplifier stage incorporates a photodarkening-resistant, polarization maintaining, triple clad ytterbium-doped Large Mode Area fiber from INO [2-4]. The fiber MFD is 16 μm at 1064 nm and the pump absorption is typically 4 dB/m at 976 nm. The amplifier employs a counter-propagating pumping scheme, mainly for optimizing the output power with respect to the threshold power levels associated with SBS and SPM [5]. Five fiber-coupled pump modules, each delivering 5W at 976 nm and combined with a commercially available pump combiner, constitute the pump source of the amplifier. The pump is injected through a pair of dichroic mirrors. One dichroic mirror is used to output the amplified 1064 nm laser beam while transmitting the pump, as shown in figure 1. The second dichroic mirror is used to further isolate the pump diodes from the residual ASE and from the 1064 nm signal leakage transmitted through the first dichroic mirror. The output of the LMA fiber is terminated by an end-cap to avoid surface damage at the air-silica interface, given the emitted fluences of ~50 J/cm^2 at the output of the fiber.

2.5. Second Harmonic Generation (SHG) module

The final stage of the laser is the Second Harmonic Generation (SHG) module. Frequency doubling is accomplished using type I non critical phase matching in a temperature-controlled lithium triborate (LBO) crystal having a length of 20 mm.

3. Laser performances and capabilities

3.1. Average power

The MOPAW Green laser has been designed to deliver an average power of up to 10 watts at 1.064 μm yielding up to 6 watts at 532 nm. Obviously, for a given average power in the infrared, the energy per pulse in the second harmonic is dependent on the pulse repetition rate but also on the pulse shape, as both characteristics determine the peak power in the infrared and therefore the conversion efficiency of the nonlinear frequency conversion process. Table 1 illustrates this, where different output power levels have been obtained at 532 nm from different pulse shapes at a constant infrared average power of 6 watts and at a fixed repetition rate of 200 kHz.

Table 1. Green output power obtained with different pulse shapes (all with an output power of 6 W at 1064 nm and a repetition rate of 200 kHz).

Green pulse shape	Pulse duration (ns)	Average power at 532 nm (W)
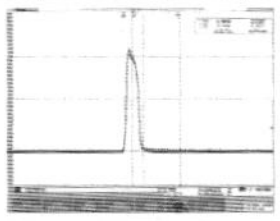	5	2.9
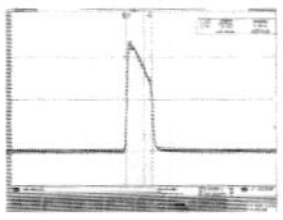	10	2.1
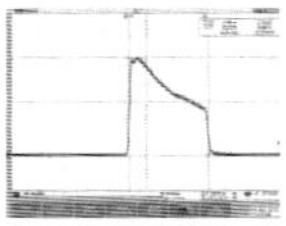	30	1.0
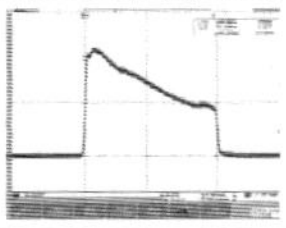	50	0.6

Figure 2 shows average power stability data at 532 nm over more than 300 hours in a typical laboratory environment with ambient temperature variations of ± 1˚C. For this experiment the laser output power was set at 1.4 W at 532 nm and the pulse duration was 17 ns at a repetition rate of 200 kHz. Figure 2(a) shows the evolution of the stability measured over 20 minutes whereas the stability over 24 hours appears in figure 2(b). The sampling rate was 1 point every 30 seconds, and the variation was calculated by taking the difference between the highest and the lowest power levels recorded during the considered interval (either 20 minutes or 24 hours). The average power variations remained lower than ±2.0% and ±2.5 % for the 20 minutes and the 24 hours measurement intervals respectively.

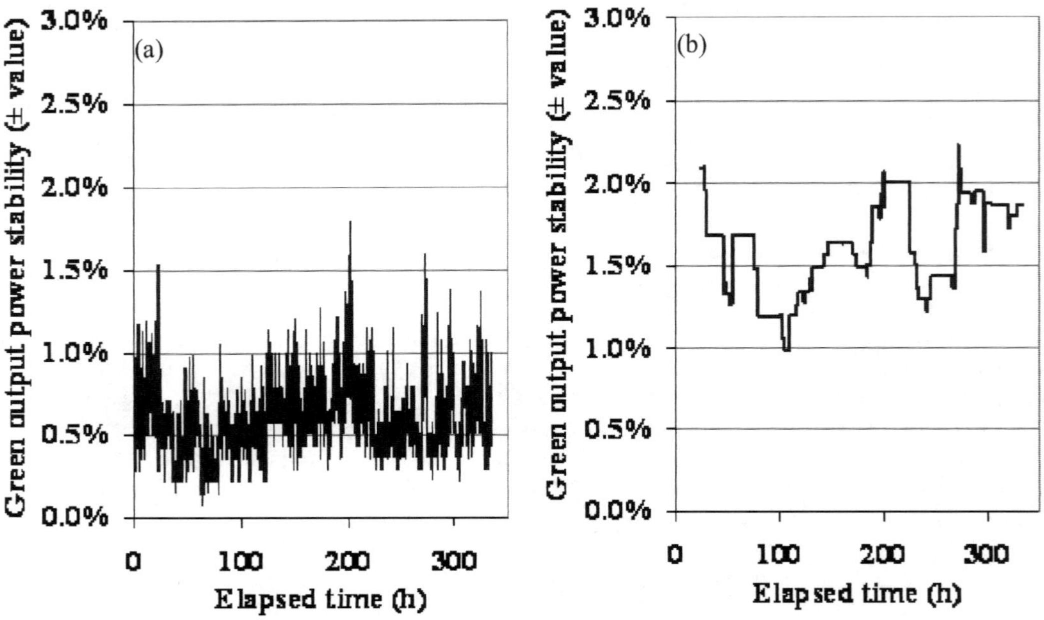

Figure 2. Evolution of the average output power stability at 532 nm for (a) 20 minutes and (b) 24 hours intervals.

3.2. Pulse characteristics

Figure 3 presents oscilloscope screen shots of four different optical pulse shapes from the MOPAW Green fiber laser. The amplitude profile can be accurately controlled to compensate nonlinear optical transfer functions such as fiber amplifier gain saturation and nonlinear frequency conversion so as to obtain virtually any suitable pulse shape, including single or multiple pulses. Up to 32 pulses can be stored in a non-volatile memory of the laser control electronics and recalled when needed. Any two of these 32 shapes can be dynamically selected from pulse to pulse, at repetition rates up to 1 MHz. This feature is particularly well-adapted for processing workpieces including different types of structures requiring different pulse amplitude profiles.

The typical rise time and fall time of the optical pulses are 800 ps and 1.5 ns respectively. Measurements of the pulse energy stability at 532 nm yielded maximum variations of ± 1.5% over 10 000 pulses (3σ value). Maximum pulse amplitude variations of ± 4.5% were recorded over 10 000 pulses. These measurements were carried out using a photodiode with a rise time of 35 ps and a digital oscilloscope having a bandwidth of 1 GHz.

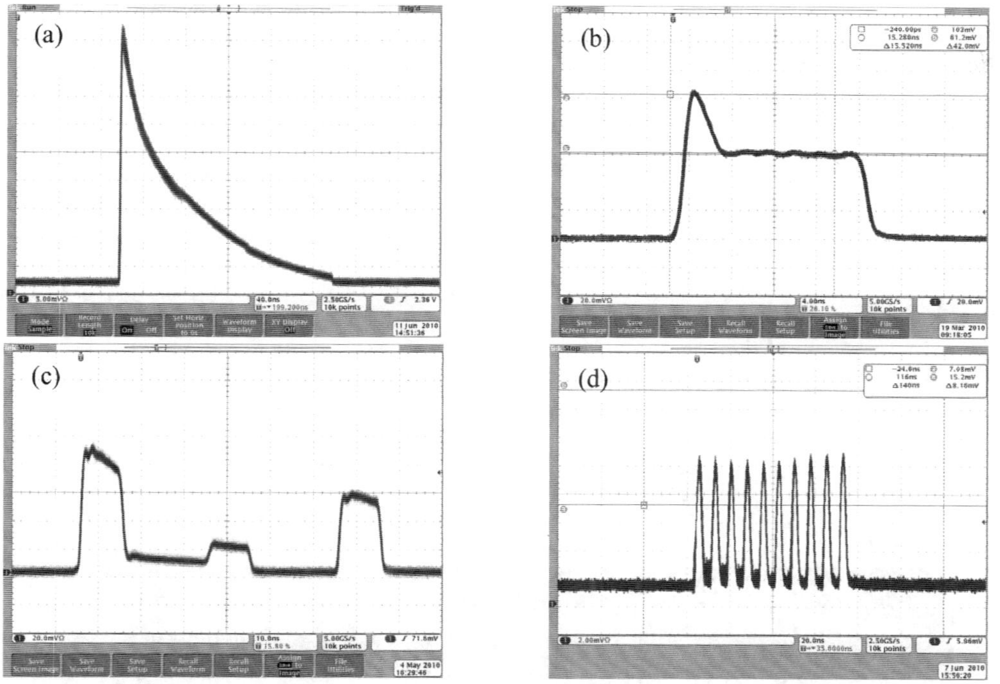

Figure 4. Examples of optical pulse shapes: (a) 200 ns exponential-like decay, (b) 17 ns chair-like, (c) 80 ns with three pulses and a plateau, (d) train of 10 pulses separated by 7.5 ns, each pulse having a duration of 3 ns.

3.3. Beam characteristics

The beam characterization setup was based on a Nanoscan™ beam profiler from Photon Inc. mounted on a rail for measuring the beam width along two perpendicular axes (denoted x and y) as a function of distance from the laser head output aperture along the beam propagation axis, denoted as the z axis. The x-y plane is by definition perpendicular to the z axis. Figure 5 depicts the setup. A plano-convex lens with 200 mm focal length was used to create a waist in the region accessible for sweeping the beam along the z axis. A typical data set included x and y beam width measurements for at least 20 points taken in the waist region along the z axis, over a total distance corresponding to about 6 times the Rayleigh range. A curve fit on the experimental data is performed thereafter to extract the most important beam characteristics such as the M^2, the beam waist astigmatism and the beam waist asymmetry. The setup was automated in order to track the evolution of these characteristics over time. Typical results at 532 nm are shown in figure 6. Values for the laser waist radius, waist asymmetry, waist astigmatism and M^2 appear in the inset of the figure. With a M^2 of 1.04 the beam is considered truly diffraction-limited. The very small waist astigmatism and asymmetry (both below 4%), yielded a beam roundness better than 95% everywhere along the z axis. The worst case values obtained when extensively testing three different MOPAW Green fiber lasers in various conditions of operation are 1.05 for the M^2, 10% for the astigmatism and 8% for the asymmetry. The stability of the beam characteristics was also tested, a typical result for the astigmatism and the asymmetry is shown in figure 7. Variations below ± 0.7 % where recorded for both characteristics over 40 hours. The robustness of the beam quality with respect to ambient temperature changes has also been measured. For example, at t ≅ 4 hours in figure 7, we can see that a sudden ambient temperature increase of 3°C does not have any impact on the beam characteristics.

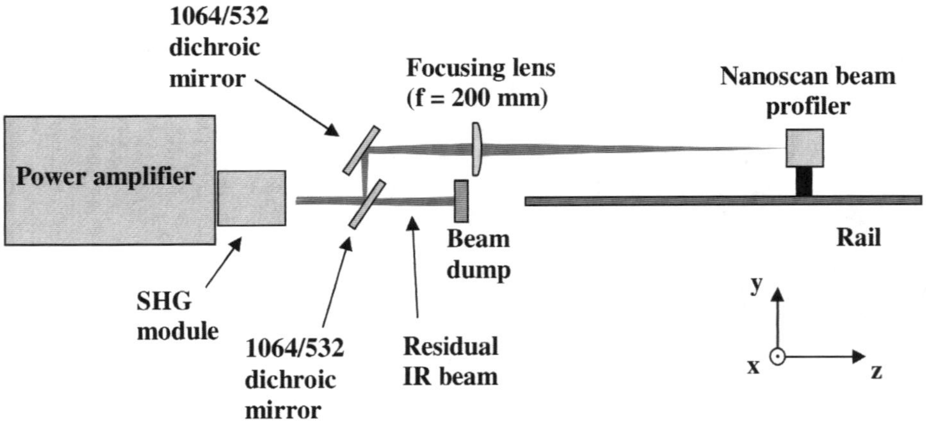

Figure 5. Beam characterization setup.

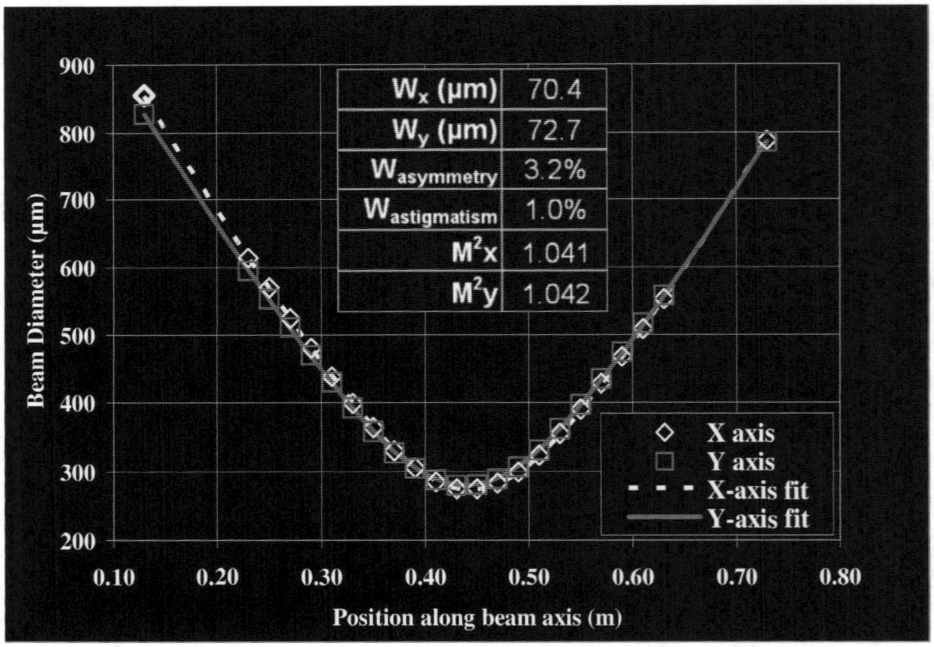

Figure 6. Typical beam characteristics.

3.4. Beam pointing stability

Beam pointing stability is essential for advanced micromachining on dense microelectronic structures. It was characterized with the setup of figure 5, except that the focusing lens was removed. The beam centroid position was measured every 30 seconds with the beam profiler. The centroid displacement divided by the propagation distance from the laser head output aperture then gives the angular beam displacement. The beam pointing stability is defined here as the maximum angular displacement (full angle) recorded over one hour. Figure 8 shows the evolution of the beam pointing stability over 140 hours. It remained very stable with a value below 20 µrad.

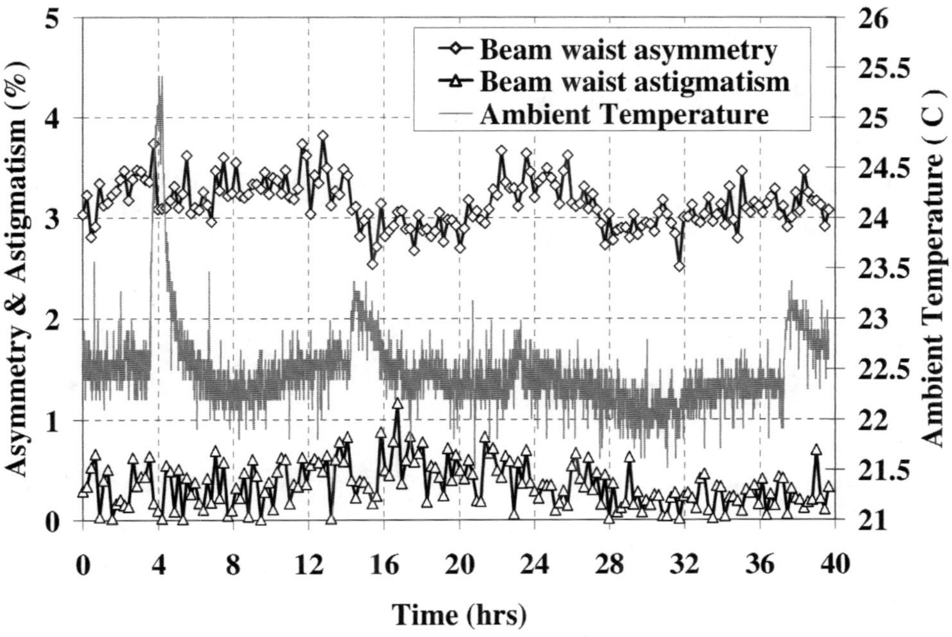

Figure 7. Typical stability of the beam characteristics over time and over ambient temperature.

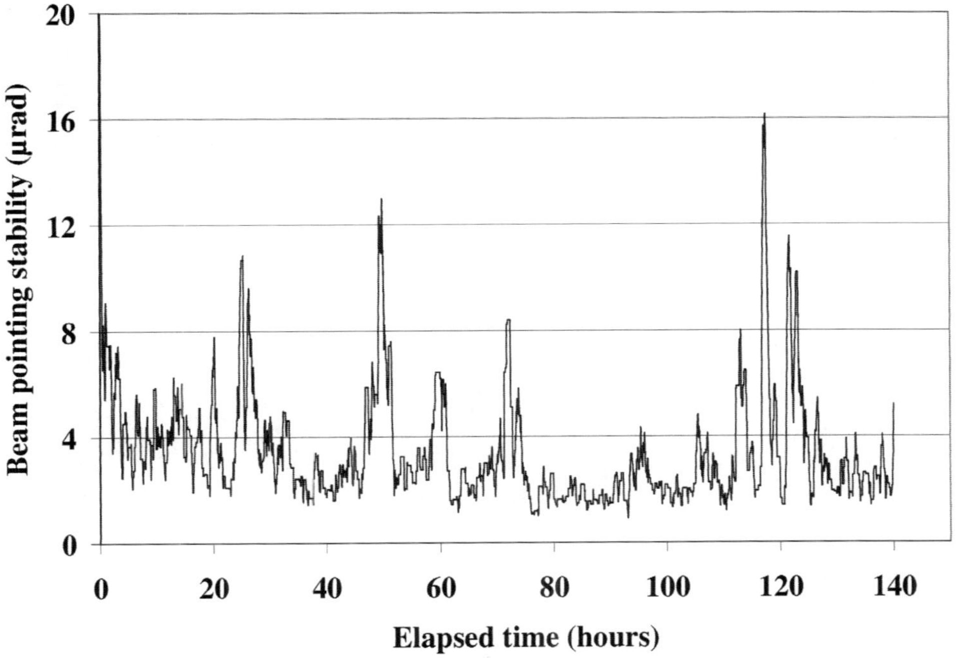

Figure 8. Beam pointing stability over 140 hours.

4. Conclusion

An ultra stable industrial green tailored pulse nanosecond fiber laser platform has been developed for advanced micromachining. This platform provides a fine control over the pulse shape, on both amplitude and pulse width, and can operate at high repetition rates. The pulse shapes can be programmed very easily using a computer. The platform possesses unique agility attributes, like the capability of selecting pulse shapes on the fly, from pulse to pulse, at repetition rates as high as 1 MHz. Proprietary LMA fiber designs and nonlinear effect mitigation methods have been employed to scale up the infrared output pulse energy to up to 100 μJ at 1064 nm giving a SHG output pulse energy of 60 μJ at 532 nm, while preserving diffraction-limited beam characteristics. It is foreseen that the very stable laser output, excellent beam quality and great flexibility of the platform will be beneficial for numerous industrial, military and medical applications.

References

[1] Hogan H Adding fiber pays off for industry 2010 *Photonics Spectra* **7** 44-46

[2] Lauzon et al. *Triple-clad rare-earth doped optical fiber and applications*, U.S. Patent No. 6,941,053 (issued Sept. 6, 2005)

[3] Proulx et al. *Polarization-maintaining optical fiber and method for manufacturing the same,* US. Patent Application No. 12/350,571, Publication No.20090181842 (Published Jul. 16, 2009)

[4] Laperle P et al. 2009 Optimization of signal gain and core composition for low photodegradation in Yb-doped fiber amplifiers *Proc. SPIE,* **7386**, 73860P (2009); doi:10.1117/12.839323

[5] Agrawal G.P. 1995 *Nonlinear Fiber Optics* (San Diego-Academic Press)

Acknowledgments

The authors want to thank Louis Martin, Marc Deladurantaye, Dany Lemieux, Nichola Desnoyers, Sebastien Deshaies, Philippe Goyette, Michel Jacob, François Duchesne, Pierre Laperle, Louis-Philippe Tremblay, Patrick Couture, François Lagacé, Julie Lambert, François Roy-Moisan, Geneviève Anctil, Mélanie Leclerc, Michel Doucet and Patrick Paradis for their engagement and effort in the development of the laser platform.

3rd International Photonics & OptoElectronics Meetings (POEM 2010) IOP Publishing
Journal of Physics: Conference Series **276** (2011) 012018 doi:10.1088/1742-6596/276/1/012018

Dependence of mis-alignment sensitivity of ring laser gyro cavity on cavity parameters

Feng Sun*, Xi Zhang, Hong-Bo Zhang and Chang-cheng Yang

Huazhong Institute of Electro-Optics - Wuhan National Lab for Optoelectronics, Wuhan, Hubei, China

E-mail: sunok1234@sohu.com

Abstract. The ring laser gyroscope (RLG), as a rotation sensor, has been widely used for navigation and guidance on vehicles and missiles. The environment of strong random-vibration and large acceleration may deteriorate the performance of the RLG due to the vibration-induced tilting of the mirrors. In this paper the RLG performance is theoretically analyzed and the parameters such as the beam diameter at the aperture, cavity mirror alignment sensitivities and power loss due to the mirror tilting are calculated. It is concluded that by carefully choosing the parameters, the significant loss in laser power can be avoided.

1. Introduction

Ring laser gyroscope (RLG) is an angular rate sensor based on Sagnac effect. The typical RLG is fabricated by boring or drilling connected optical cavities through a body block, which is made of ultra-low expansion materials. Flat or curved mirrors are attached to the corner of the cavities through the optical bonding process which is applied on the corresponding surface of the block to reflect the light and form a triangular or square shaped loop-path. In an ideal laser gyro, the mirrors can reflect the light to the perfect direction as designed. However, practically the angle may have a mismatch problem [1,2]. This mismatch may be caused by the tolerance of the manufacture or fabrication, or the distortion of the block when the RLG is placed in a circumstance of large acceleration.

The mismatch may deteriorate the performance of the RLG. In this paper, the power loss of a triangular RLG due to the mirrors mismatch is studied theoretically using the ABCD matrix [3]. For a given triangular RLG, the optimized parameters of the mirrors are obtained to reduce the influence of the mirror mismatch.

2. The triangular RLG

Published under licence by IOP Publishing Ltd

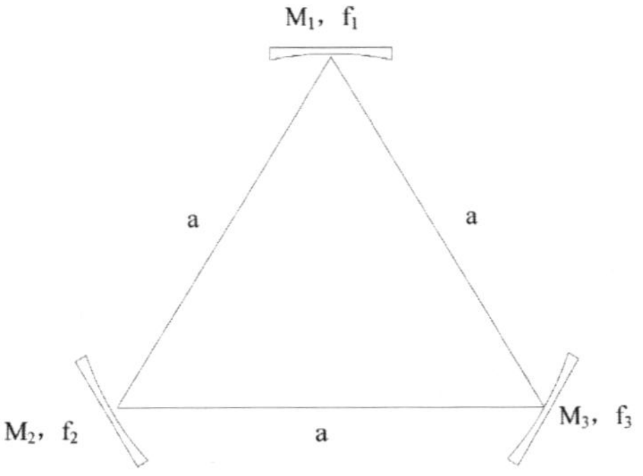

Figure 1. The RLG with a triangular optical path

The RLG with a triangular optical path are presented in Fig.1 schematically. The optical resonator is composed of 3 mirrors, M_1, M_2 and M_3, and the corresponding focus and curvature radii are defined as f_1, f_2, f_3 and R_1, R_2, R_3 respectively. The lateral of the triangle is α, so the perimeter is 3α. Since the difference of the laser beam radii among the different positions is typically less than 5%, the following approximation can be made in the calculation: the beam energy distribution is Gaussian in the cross section of the beam, and the beam is propagating as plane wave. It is assumed that a diaphragm is placed on the surface of M_3, and each mirror has an angular mismatch θ. The beam which starts from M_3 with the coordinate vector $\boldsymbol{R}_0$, travels one loop of the optical path via the mirrors M_i (i=1,2,3) with the coordinate vector $\boldsymbol{R}_i$, and reaches M_3 with the coordinate vector $\boldsymbol{R}_f$. The equation $\boldsymbol{R}_f$=$\boldsymbol{R}_0$ must be satisfied to meet the requirement of the laser oscillation. The following equations can be deduced to describe this process.

$$\left.\begin{aligned}
&\mathbf{R}_1 = \mathbf{M}_1\mathbf{R}_o \\
&\mathbf{R}_i = \mathbf{M}_i(\mathbf{R}_{i-1} + \mathbf{\Theta}), \quad \text{where} \quad i = 2,3 \\
&\mathbf{R}_f = \mathbf{R}_3 + \mathbf{\Theta} = \mathbf{R}_o \\
&\mathbf{\Theta} = \begin{bmatrix} 0 \\ 2\theta \end{bmatrix}, \quad \mathbf{R}_j = \begin{bmatrix} r_j \\ r_j' \end{bmatrix}, \quad \text{where} \quad j = 0,1,2,3,\mathrm{f} \\
&\mathbf{M}_k = \mathbf{T}_{f,k}\mathbf{T}_a, \mathbf{T}_{f,k} = \begin{bmatrix} 1 & 0 \\ -f_k^{-1} & 1 \end{bmatrix}, \quad \text{where} \quad k = 1,2,3 \\
&\mathbf{T}_a = \begin{bmatrix} 1 & a \\ 0 & 1 \end{bmatrix}
\end{aligned}\right\} \tag{1}$$

Where r_j and r'_j denotes the position and angular part of the position vector $\boldsymbol{R}_j$. The gain medium is treated as vacuum approximately in Eq. (1). The ABCD matrix which describes the light travels through a distance α in vacuum is denoted as $\mathbf{T}_a$. $\mathbf{T}_{f,k}$ denotes the ABCD matrix, describing the light being reflected by the mirror M_k. For meridian ray and sagittal ray, the relation between the focus and the radius of the mirror can be written as:

$$f_k = \begin{cases} f_{k,meridian} = \dfrac{R}{2}\cos\phi \\ f_{k,sagittal} = \dfrac{R}{2}\sec\phi \end{cases} \tag{2}$$

Where, ϕ is the angle of incidence, and ϕ=30° for the equilateral triangular optical path. It can be deduced from Eq. (1) that the original coordinate vector satisfies the linear equation:

$$\mathbf{M}\,\mathbf{R}_o = \mathbf{U} \tag{3}$$

Where

$$\mathbf{M} = \mathbf{M}_3\mathbf{M}_2\mathbf{M}_1 - \mathbf{I}$$
$$\mathbf{U} = -(\mathbf{M}_3\mathbf{M}_2 + \mathbf{M}_3 + \mathbf{I})\boldsymbol{\Theta} \tag{4}$$

Where **I** is the 2×2 identity matrix. By solving Eq. (3), the shift of the laser beam due to the mismatch of the mirrors can be obtained:

$$r_0 = \frac{2\theta(a-3f_1)(a-3f_2)f_3}{a^2 - 2a(f_1+f_2+f_3) + 3(f_1f_2+f_2f_3+f_3f_1)} \tag{5}$$

$$r_0' = \frac{2\theta(L-3f_2)(f_1-f_3)}{a^2 - 2a(f_1+f_2+f_3) + 3(f_1f_2+f_2f_3+f_3f_1)} \tag{6}$$

By ignoring the influence of the mismatch, the transform matrix which describes the beam traveling around the optical path can be written as:

$$\mathbf{T}_3 = \mathbf{M}_3\mathbf{M}_2\mathbf{M}_1 = \begin{bmatrix} A & B \\ C & D \end{bmatrix} \tag{7}$$

The following condition should be satisfied to meet the stable resonator requirement:

$$-1 < \frac{1}{2}(A+D) < 1 \tag{8}$$

It can be obtained from Eq. (7) and (8):

$$-1 < 2 - 3a\left(\frac{1}{f_1}+\frac{1}{f_2}+\frac{1}{f_3}\right) + 2a^2\left(\frac{1}{f_1f_2}+\frac{1}{f_2f_3}+\frac{1}{f_3f_1}\right) - \frac{a^3}{f_1f_2f_3} < 1 \tag{9}$$

Since the variety of the beam width is very small at different positions of the optical path, for the readout system of the RLG, the position shift part r'_0 is a significant factor of the power loss due to the mirror mismatch. For given resonator, the influence of the mirror mismatch can be calculated using Eq.(5) and (9). The following discussions are based on a typical triangular resonator of a RLG which is composed of 2 concave mirrors and 1 flat mirror, or 2 flat mirrors and 1 concave mirror. The detail of the beam shift due to the mirror mismatch for a triangular resonator is discussed as follows:

(1) The resonator is composed of 2 concave mirrors and 1 flat mirror.

If the diaphragm is placed on the surface of one concave mirror, $f_2=f_3=f$, $f\to\infty$, then

$$r_0^{(1a)} = 3f\theta \tag{10}$$

If the diaphragm is placed on the surface of the flat mirror, $f_1=f_2=f$, $f_3\to\infty$, we have

$$r_0^{(1b)} = (3f-a)\theta \tag{11}$$

Since $r_0^{(1a)} > r_0^{(1b)}$, the result of placing 3 diaphragms on 3 mirrors is equivalent to placing only one mirror on the curved mirror. When estimating the influence of the mirror mismatch, the larger beam shift is used. Thus, $r_0^{(1a)}$ is introduced, and the discussion is generally based on sagittal ray. However, when building the practical RLG, the diaphragm is placed on the flat mirror to reduce the influence of mirror mismatch.

The stable condition requirement should be met for both meridian ray and sagittal ray:

$$\begin{cases} R > \dfrac{4\sqrt{3}}{3}a \\ \text{or } \dfrac{\sqrt{3}}{2}l > R > \dfrac{4}{9}\sqrt{3}l \end{cases} \tag{12}$$

It is difficult to restrict the radius in the second region to keep the resonator stable [4], so the following region is chosen as the stable region.

$$R > \frac{4\sqrt{3}}{3}a \tag{13}$$

(2) The resonator is composed of 2 flat mirrors and 1 concave mirror.

If the diaphragm is placed on the surface of one flat mirrors, $f_1=f, f_2=f_3 \rightarrow \infty$, then

$$r_0^{(2a)} = (6f - 2a)\theta \tag{14}$$

If the diaphragm is placed on the surface of the concave mirror, $f_3=f, f_1=f_2 \rightarrow \infty$, then

$$r_0^{(2b)} = 6f\theta \tag{15}$$

The stable condition is

$$R > \sqrt{3}a \tag{16}$$

According to the aforementioned approximation, the beam has a Gaussian energy distribution on the cross section and the variety of the beam at different location along the propagating direction is neglected. The two concave mirrors have the same radius R. The beam width of the fundamental mode laser is determined if the radius of the concave mirrors is fixed.

$$w^2 = \frac{2\lambda B}{\pi\sqrt{4-(A+D)^2}} \tag{17}$$

Where, A, B and D are the ABCD matrix element in Eq. (7).

3. Power loss due to the mirror mismatch

To get the high fundamental mode volume, the diagram of the diaphragm is chosen to be Da=3.6w. If there is no mirror angle mismatch, the energy power which can be received should be written as

$$P_0 = \int_{\sigma_1+\sigma_2} I(r)2\pi r\,\mathrm{d}r = \frac{\pi w^2}{2}[1-\exp(-\frac{2\mathrm{Da}^2}{w^2})] \tag{18}$$

If each mirror has a angle mismatch θ, and the beam has a shift d, the energy power can be written as

$$P_0 = \int_{\sigma_1} I(r)2\pi r\,\mathrm{d}r \tag{19}$$

Where σ_1 and σ_2 are the regions which are displayed in Fig. 2, and show the cross section of the shifted beam.

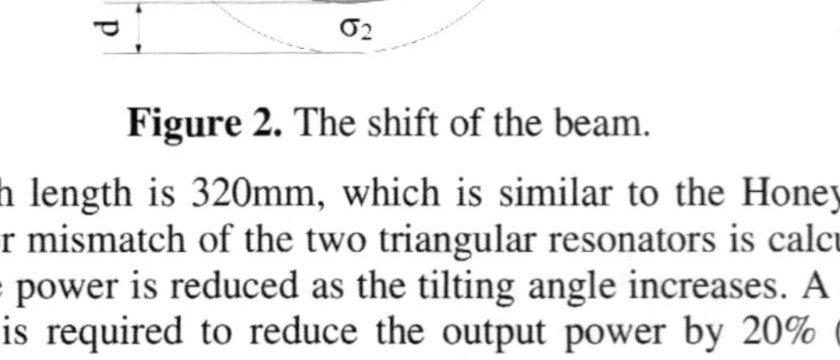

Figure 2. The shift of the beam.

Suppose the optical path length is 320mm, which is similar to the Honeywell GG1342 [5]. The power loss due to the mirror mismatch of the two triangular resonators is calculated. And the result is demonstrated in Fig. 3. The power is reduced as the tilting angle increases. A critical angle is defined as the tilting angle which is required to reduce the output power by 20% (the horizontal line). A vertical line of 50 micro radians defines the maximum expected mirror tilt due to the flexure of the

RLG cavity block in a 40g acceleration extreme environment [6]. If the critical angle is less than 50 micro radians, there may be a problem for the RLG in the extreme environment. It can be seen that for resonator (a) the radius should be smaller than 5 meters so that the RLG performance won't be deteriorated significantly in the extreme environment. However, for the resonator (b), this radius should be smaller than 1 meter. It is revealed that the influence of the mirror tilting for resonator (a) is much smaller than that for resonator (b).

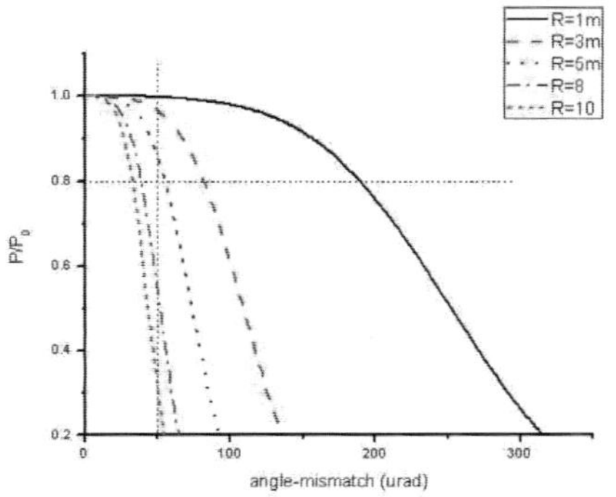

(a)

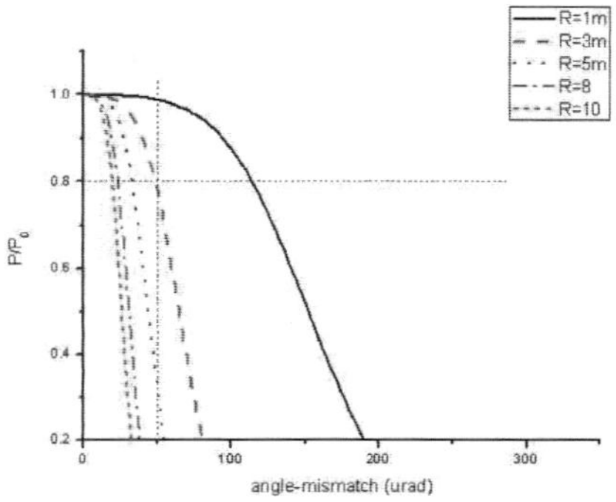

(b)

Figure 3. Power loss due to the mirror mismatch of the triangular resonator (a) composed of 1 flat mirror and 2 concave mirrors, (b) composed of 2 flat mirrors and 1 concave mirror.

4. Conclusion

In this paper, the mirror-tilting induced power loss of the triangular RLG is calculated. It is revealed that the large radius may be helpful to increase the laser mode volume and increase the efficiency of mode selection, while it may depress the performance of the RLG in extreme environment with very large acceleration. Besides, it can be seen that the triangular resonator which is compose of 2 concave mirrors and 1 flat mirror has a better capability to survive in the extreme environment than the triangular resonator composed of 2 flat mirrors and 1 concave mirror.

References

[1] S-L. Jin, X-W. Long, X-H. Li, etc, Research on New Type Control Mirror of Light Path and Optical Path Length for Ring Laser Gyroscope. ACTA Optica Sinica, Vol. 26, No. 4 (2006)
[2] J. Yuan, Research on the Slant Measuring system for the Cavity Translational Mirror Used by Laser Gyro, Ph. D. Dissertation: National University of Defence Technology (2003)
[3] B-D. Lu, Laser Opticas: Laser Beam Propagation and Beam Quality Control. Sichuan University Press (1992)
[4] Y-N Jiang, Ring Laser Gyro, Tsinghua University Press (1985)
[5] Z-W. Cao, Research on Key partial Technique of Mechanically Dithered Ring Laser Gyro Compass. Master dissertation: National University of Defense Technology (2006)
[6] R. Chodzko, Ring Laser Gyro Resonator Design, DTIC (1994)

The impacts of temperature on the absorption spectral lines of carbon monoxide

GUO Jian-Qiang, XU Yuan-Ze, Gao Xiao-Rong, WANG Li and WANG Ze-Yong

College of Physical Science and Technology, Southwest Jiaotong University, Chengdu, Sichuan, P. R. China 610031

E-mail: jianqguo@home.swjtu.edu.cn

Abstract. In order to study the change of temperature on the effects of carbon monoxide absorption spectral lines, first of all proceed from the principle of absorption spectra, using theoretical analysis method, and the transmission and absorption database of the high-resolution molecular educed the carbon monoxide absorption spectrum intensity of spectral lines, integrated widening line type function and absorption coefficient concerned with temperature, then we got the change curve between carbon monoxide absorption spectrum intensity of spectral lines, integrated widening line type function and absorption coefficient with temperature by the numerical simulation of MATLAB, and analyzed and discussed the relationship between the temperature and them .The results showed that the temperature on the effects of carbon monoxide absorption spectral lines, especially on an Integrated widening line type function is complex, and different laser frequencies will also affect the relationship of the line type function and the absorption coefficient change with temperature, which has important reference value for the absorption and measurement of carbon in practical application.

1. Introduction

At macro level, a temperature is one of factors, which affect the absorption coefficient of CO gas. However, the specific laws of the absorption coefficient of CO gas with temperature are seldom seen. In particular, the regular reports of the absorption spectra and comprehensive widening spectrum line type function with the impact of temperature are lacking. It is important to measure CO concentration in combustion furnace smelting with standard atmospheric [1]. In this paper, from the theory of absorption spectroscopy, by the use of the HITRAN molecular spectroscopy database, it is analyzed that the intensity of absorption spectrum line and widening line type function of spectrum line change with temperature, to obtain the dependent relation between absorption coefficient of spectrum line and temperature.

2. Theoretical basis

2.1. Absorption lines intensity

The absorption line intensity *S(T)* indicates the absorption intensity of a certain frequency ($cm^{-1} \cdot (mol \cdot cm^{-2})^{-1}$), it is a temperature-related function, and it is also affected by the type of molecular isotopes. The intensity of CO absorption line *S(T)* can be queried by use of the high-resolution transmission molecular absorption database (HITRAN), or by the analysis software to calculate. But in practice, in

Published under licence by IOP Publishing Ltd

order to get the intensity of the absorption line at any temperature, the formula (1) will be used to calculate the intensity [1, 2]:

$$S(T)=S(T_0)\frac{Q(T_0)}{Q(T)}\frac{\exp\left(-hcE_i''/kT\right)}{\exp\left(-hcE_i''/kT_0\right)}\frac{1-\exp\left(-hcv_{0,i}/kT\right)}{1-\exp\left(-hcv_{0,i}/kT_0\right)} \tag{1}$$

Where, T_0 is the reference temperature (K), $S(T_0)$ is the intensity of the absorption spectrum gotten by the HITRAN database with the temperature T_0, Q is the total molecular partition function, E_i'' is lower energy transition state (cm^{-1}), $_{0,i}$ is the transition frequency (cm^{-1}), h is Planck's constant, k is the Boltzmann constant, c is the speed of light. In this formula, the third item is the Boltzmann distribution, the fourth item is the excitation radiation. The total molecular partition function Q is relative to the molecular isotopic species. The molecule value of different isotopic partition function is different, its value can be query form the HITRAN molecular spectroscopy software, but in the HITRAN molecular database [3], the data provided with the temperature is not continuous, it is shown in table 1.

Table 1. The total internal partition function look-up table for $^{12}C^{16}O$.

Temperature T/K	296	297	298	299	300	301
Q(T)	107.1169	107.4785	107.8400	108.2016	108.5632	108.9249

Table 1 shows the total molecular partition function values of the isotope $^{12}C^{16}O$ with the temperature range of 296K to 301K, it is seen by the table that the temperature can not query in decimal. So, the second subdivision interpolation method will be taken to calculate $Q(T)$ as following:

$$Q(T_i)=Q(T_1)\frac{(T-T_2)(T-T_3)}{(T_1-T_2)(T_1-T_3)}+Q(T_2)\frac{(T-T_1)(T-T_3)}{(T_2-T_1)(T_2-T_3)}+Q(T_3)\frac{(T-T_1)(T-T_2)}{(T_1-T_2)(T_1-T_3)} \tag{2}$$

In the formula (2), the temperature T_i is between the temperature T_l and T_2, $T_l<T_i<T_3$, $Q(T_l)$ is the partition function of the temperature T_l, $Q(T_2)$ is the molecular partition function of the temperature T_2, $Q(T_3)$ is the molecular partition function of the temperature T_3, $Q(T_i)$ is the molecular partition function of the temperature T_i.

2.2. Comprehensive widen line type function

In general, CO gas is always appears in spectrum line broadening. When the gas is in pressure and temperature simultaneously, its spectrum line will is widened with homogeneous broadening and Doppler non-uniform broadening caused by collisions. Then the widening line type function is no longer a simple line type function, such as simple Lorenz or Gauss line type function [4, 5], but it is decided by both of them, that is Voigt line type function [6, 7], expressing as:

$$\tilde{f}_V(v,v_0)=\int_{-\infty}^{+\infty}\tilde{f}_G(v_0',v_0)\tilde{f}_L(v,v_0')dv_0' \tag{3}$$

In the formula (3), v_0' is atomic center frequency (cm^{-1}), v_0 is the center frequency of non-uniform broadening, $\tilde{f}_G(v_0',v_0)$ is the line type function of Doppler broadening, $\tilde{f}_L(v,v_0')$ is the line type function of collision broadening, their expressions are such as equation (4) and (5):

$$f_G(\nu,\nu_0)=\frac{2}{\Delta\nu_G}\sqrt{\frac{\ln 2}{\pi}}\exp\left[-4\ln 2\left(\frac{\nu-\nu_0}{\Delta\nu_G}\right)^2\right] \tag{4}$$

$$\Delta\nu_G=7.16\times10^{-7}\sqrt{\frac{T}{M}}\nu_0 \tag{5}$$

$$f_L(\nu,\nu_0)=\frac{\left(\frac{\Delta\nu_L}{2\pi}\right)}{(\nu-\nu_0)^2+\left(\frac{\Delta\nu_L}{2}\right)^2} \tag{6}$$

$$\Delta\nu_L=2\gamma_{air}\left(\frac{296}{T}\right)^n P \tag{7}$$

In the equation (4) and (6), v is the frequency of the laser (cm^{-1}), M is the molar mass of CO molecules g.mol^{-1}, $_{air}$ is the air-broadening coefficients, n is the collision broadening index, P is the pressure of the CO.

In the equation (3), when $_L$<< $_G$, the homogeneous broadening caused by the collision is much smaller than the non-uniform broadening caused by the Doppler effect, the above-mentioned integral of equation (3) has the non-zero value near $_0$'• in only a small range. In this context, the function $\tilde{f}_G(\nu_0',\nu_0)$ can be replaced with a constant $\tilde{f}_G(\nu,\nu_0)$, therefore:

$$\tilde{f}_V(\nu,\nu_0)=\tilde{f}_G(\nu,\nu_0)\int_{-\infty}^{+\infty}\tilde{f}_L(\nu,\nu_0')d\nu_0'=\tilde{f}_G(\nu,\nu_0) \tag{8}$$

That is, the Voigt line type function is approximately equal to Gauss line type function when $_L$<< $_G$. As the same, when $_G$<< $_L$, the non-uniform broadening caused by the Doppler effect is much smaller than the collision-induced homogeneous broadening, the spectral function is expressed as:

$$\tilde{f}_V(\nu,\nu_0)=\tilde{f}_L(\nu,\nu_0) \tag{9}$$

That is, the Voigt line type function is approximately equal to Lorenz line type function. From equations (7) and (9), it is known that the Lorenz line type function and the Gauss line type function are only the two extreme forms of the Voigt line type function. They are consistent with the physical models. For Voigt line type function, it can be solved numerically by look-up table or by polynomial to match the curve [8].

2.3. Absorption coefficient

The mathematical expression of the absorption coefficient is [9, 10]:

$$a(\nu)=S(T)\tilde{f}_v(\nu,\nu_0) \tag{10}$$

In the formula (10), *S(T)* is the intensity of molecular absorption spectrum lines, $\tilde{f}_v(\nu,\nu_0)$ is the integrated broadening line type function of the absorption spectrum lines (cm^{-1}), the frequency is the frequency of the laser output (cm^{-1}). The absorption coefficient here is also named as the monochromatic absorption coefficient (cm^2.mol^{-1}) and it represents the absorption ability of the point of the frequency . After obtaining the absorption coefficient, the concentration of CO can be calculated by using differential absorption methods and so on.

3. Simulation analysis

In order to study the absorption coefficient changed with temperature, in the measurement gas chamber, an atmospheric pressure is set and the parameter values of the expression are shown in the table 2. Here is the wave number instead of the frequency. CO has been found that it has six isotopic molecules. In them, $^{12}C^{16}O$ is for 98.654 percent, so it is a main element.

Table 2. The value of each parameter for CO.

Parameter	Meaning	Value	Units
$S(T_0)$	Line intensity at *To*=296K	2.179e-23	$cm \cdot mol^{-1}$
$Q(T_0)$	Total internal partition at T_0=296K	107.116880	/
γ_{air}	Air-broadened half width at 296K	0.0599	$cm^{-1} \cdot atm^{-1}$
n	Collision broadening index	0.5	/
E_i	Lower state energy	107.6424	cm^{-1}
T_0	Reference temperature (Room temperature)	296	K
v_0	Wave number (absorption peak)	6369	cm^{-1}
M	Molar Mass of CO	28	$g \cdot mol^{-1}$
C	The speed of light	2.99792458e+8	$m \cdot s^{-1}$
h	Planck constant	6.626176e-34	J·S
K	Boltzmann constant	1.380662e-23	$J \cdot K^{-1}$

3.1 The impacts of temperature to the intensity of absorption spectrum line

According to equation (1) and the reference temperature T_0=296K, the intensity of the spectrum line can be calculated under temperature T. Figure 1 shows the relative intensity of spectrum line at the different temperature:

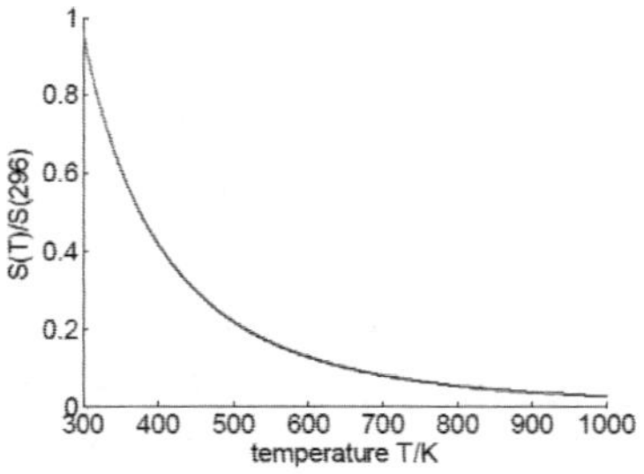

Figure 1. The relationship between the intensity of absorption spectrum line and temperature T at transition wave number 6369 cm^{-1}.

In figure 1, as the temperature is increased, the relative intensity of spectrum line is gradually reduced, due to *S(296)* is a constant, so the corresponding intensity of absorption spectrum line *S(T)* is gradually reduced.

3.2. The impacts of temperature on the integrated widening line type function

Selecting spectrum line 6369(cm^{-1}) as the center of CO absorption lines (absorption peak), it is corresponding to the 1.57um wavelength located to the optical fiber transmission window. When the pressure in CO chamber is 1 atmospheric pressure, the comprehensive widening line type function is shown as figure 2.

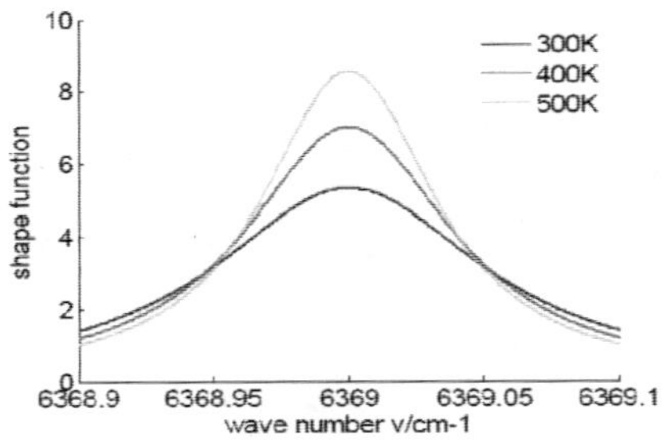

Figure 2. The relationship between comprehensive widening line type function and wave number at different temperature.

In Figure 2, using wave number as horizontal axis, the center absorption spectrum line is also 6369 (cm^{-1}), it can be seen that as the temperature increases, the peak value of widening line type function will increase. As the wave number off center position is far, the relative intensity of the absorption spectrum line will gradually decrease. The relative intensity value with high temperature will faster decrease than low temperature. With different temperature, the maximum value of relative intensity is located to the center position of the CO absorption spectrum line.

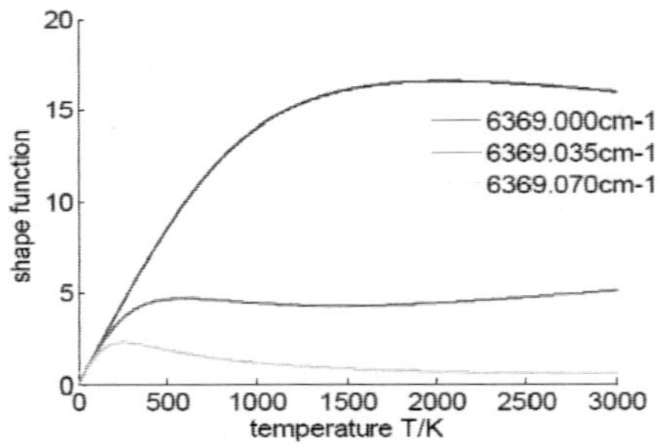

Figure 3. The relationship between widening line type function and temperature at absorption peak of 6369cm^{-1}.

Figure 3 reflects the relative intensity of peak nearby of the comprehensive widening line type function changes with the different temperature. As the temperature increases, the change trends of the comprehensive widening line type function have two directions. When the temperature is increasing from low temperature, the relative intensity value of the comprehensive widening line type function will increases, then when the temperature reaches high temperature, it finally increases to a saturation value. But to the wave number 6369(cm^{-1}) and 6369.07(cm^{-1}), at first, it increases with temperature, then, it reaches to saturation and decreases. Moreover to wave number 6369.035(cm^{-1}), at first, it increases with temperature, then, it reaches to saturation and begins decreasing and as temperature is gradually increases, it begins gradually increases. The relative intensity value of the comprehensive widening line type function is the greatest when the wave number is located to the central peak wave 6369 cm^{-1} peak.

3.3. The impacts of temperature on absorption coefficient

After obtaining the relative intensity of absorption lines and the relationship in which the comprehensive widening line type function changes with the temperature, the absorption coefficient

( ) can be described by equation (8). Figure 4 shows the relative absorption coefficient *( )/($_0$)* changes with temperature. This *($_0$)* is the absorption coefficient of the peak $_0$=6369.000cm^{-1} when the temperature is 296K. From the analysis in the previous, the relationship of the comprehensive widening line type function with temperature is not same at different wave number, so selecting the central absorption peak of the CO absorption line to 6369 cm^{-1}, the relation curves of absorption coefficient with temperature is shown as figure 4.

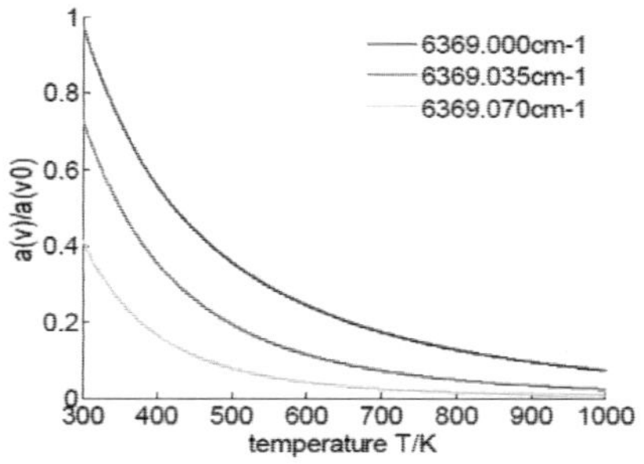

Figure 4. The relationship between the absorption coefficient and temperature at different laser wave number.

From Figure 4, as the temperature is increased, the absorption coefficient will decrease. And the absorption coefficient at low temperature is larger than high temperature. Meanwhile, the wave number is closer to the absorption peak, absorption coefficient is larger. In practice, it is best to make the laser frequency align the central peak frequency of absorption spectrum line to obtain a better measurement effect. And from the relationship of the absorption coefficient with the temperature, the intensity of absorption line has much more effect on the absorption coefficient than the comprehensive widening line type function.

4. Conclusion

The variation character of CO absorption spectral line with temperature can be analyzed by using the theory of absorption spectrum. The intensity of absorption lines and comprehensive widening line type function are the main the factors to determine spectral absorption coefficient, and the intensity of absorption lines makes a greater contribution to the spectral absorption coefficient than the comprehensive widening line type function. The change of temperature will give rise to the changes of the intensity of absorption spectrum line and comprehensive widening line type function, and then they will lead to change of absorption coefficient. In lower temperature and laser frequency closed to the central peak position of the absorption spectrum line, the absorption coefficient is the largest. It is good to measure absorption character of CO gas and its concentration.

References

[1] Liu W Q, Cui Z C and Dong F Z 2002 *J. Optoelectronic Technology & information (in Chinese)* **15** 1
[2] Wang X M, Zhang Y J, Xia H, et al. 2007 *J. ATMOSPHERIC AND ENVIRONMENTAL OPTICS (in Chinese)* **2** 290
[3] Li N, Wang F, Yan J H, et al. 2005 *J. Proceedings of the CSEE (in Chinese)* **25** 121
[4] Rothman L S, Jacquemart D, Barbe A, et al. 2005 *J. Quantitative Spectroscopy and Radiative Transfer* **96** 139
[5] Ma W G, Yin W B, Huang T, et al. 2004 *J. Spectroscopy and Spectral Analysis (in Chinese)* **24** 135

[6] Gao Z H. 2000 *J. LASER JOURNAL (in Chinese)* **21** 36
[7] Zhou B K, Gao Y Z, Cheng T R, et al. 2004 *Principles of Laser (in Chinese)* (Beijing: National Defense Industry Press) P 139-140
[8] Jian He and Qingguo Zhang 2006 J. Phys. Lett. A **359** 256
[9] Zhang G Q, You J H and He J. 2008 *Journal of Henan University of Science and Technology: Natural Science (in Chinese)* **29** 84
[10] Tong W H, Jang D, Zhou D F, et al. 2007 *J. Laser Technololy (in Chinese)* **31** 479

Investigation on a new local control technology with near-infrared spectroscopy for laser induced interstitial thermotherapy

Guo-ran Hua[1], Hua Zhang[*,1], Ai-ping Qian[1] and Zhi-yu Qian[2]
1 School of Mechanical Engineering, Nantong University, Nantong, Jiangsu 226019, China
2 Department of Biomedical Engineering, Nanjing University of Aeronautics and Astronautics, Nanjing, Jiangsu 210016, China
E-mail: ntuzhh@ntu.edu.cn

Abstract. In order to control the treatment of laser-induced interstitial thermotherapy, a new local control technique is put forward, which use a functional near infrared spectroscopy (fNIRS) system. The temperature of biological tissue is regarded as the most important indicator for treatment of LITT. So, based on the experiments of LITT being performed on a liver cancer, the temperature and reducing scattering coefficient have been obtained in real time, respectively, with probe-type thermometer and fNIRS system. The results of the two factors have the same trend with different power of laser. It is confirmed that the fNIRS can be a useful local control technology for LITT.

Introduction

Laser-induced interstitial thermotherapy (LITT) has proved to be an effective and minimally invasive surgical technique for the treatment of numerous types of vascular malformations, hemangiomas and different kinds of soft tissue tumours and metastases [1-3]. LITT involves the percutaneous introduction of an optical fiber through a rigid steel cannula. The laser fiber is located interstitially and is not visible in contrast to open surgical procedures. Control of treatment is a key issue for LITT. Previous studies have developed some ways that using open magnetic resonance imaging (MRI) or ultrasound (US) systems which offer an accurate image-guided placement of the optical fiber [1,3]. However, open MRI techniques have often been limited due to slow image acquisition sequences and high additional time-consuming and cost-intensive expense. US systems provide a good visual control of both, trajectory and placement of the cannulation needle. However, ultrasound technology cannot be used in particular anatomic areas as for example the deep oropharyngeal region.

The objects of LITT all are biological tissues. Biological tissues are composed of cells and interstitial cells with different sizes or components. There is light absorption and scattering during light transmits in the tissues. The temperature range of tissue solidification is 55-95℃. Under the temperature conditions, the composition and structure of tissue would change, which lead to changes of the optical properties, especially the light scattering properties are more significant changed [4-9].

* Corresponding author. Tel: +86051385012672.
E-mail addresses: ntuzhh@ntu.edu.cn, nuaazhh@126.com (Hua Zhang)

Therefore, the scattering coefficient could be used as the important basis during the control of treatment for LITT.

Biological tissue is relative transparency for near infrared light at wavelength range 700-1200nm. According to the phenomena, functional near infrared spectroscopy (fNIRS) technology uses near-infrared light being propagated with biological tissue. After many elastic scattering and absorption, the reflect light and scattering light could carry a large number of information about the structure and composition of biological tissues. Therefore, the organization's internal information can be obtained by analyzing the detection of infrared light. The reduced scattering coefficient (μ'_s) is an important optical coefficient of biological tissue and closely related to the temperature of the tissue.

Based on the above analysis, a new local control technology with near-infrared spectroscopy for laser induced interstitial thermotherapy has been developed in this paper. The technology uses an fNIRS system, which has the function of the temperature measurement and the reduced scattering coefficient. By comparing the changed trend of the two testing parameters under different laser power, it is confirmed that the fNIRS technology is useful for the control treatment during LITT.

Experimental procedures

1.1. Experimental system

Fig.1 shows the experimental system. It is consisted of LITT system and fNIRS monitoring system. LITT system includes: laser (808nm, NL-FC-2.0-763 LASER Light); optical fiber (built-in three 200μm fibers, one is for laser input, the others are for near infrared light input and output); thermoscope (WSY-4T, diameter 300μm). fNIRS monitoring system included: USB2000 spectrometer (Ocean Optics) and halogen light source (Mikropack, HL-2000) and monitoring software. fNIRS monitoring software was developed by department of biomedical engineering of Nanjing University of Aeronautics and Astronautics based on Labview7.1 platform, which could real-time record tissue optical properties, blood oxygen parameters and flow parameters in front of the probe [10].

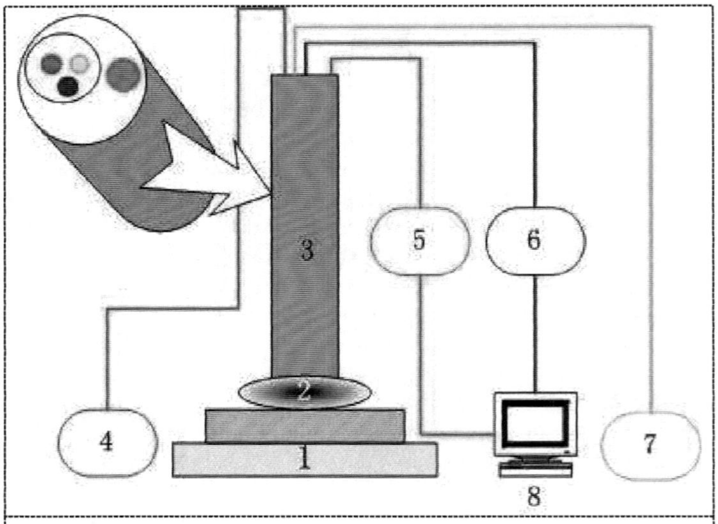

Figure 1. Illustrative diagram of experimental system.

1-CNC table 2-Tested tissue 3-Optical fiber and temperature probe 4-Laser 5-Thermoscope 6-Spectrometer 7- Input of NIR

1.2. Materials and methods

Fresh porcine liver tissue in vitro was used for this experiment. The liver tissue has been cut into five specimens (3×3×2.5cm) under room temperature (25•). The experiment includes five steps:
(1) The specimen was fixed on the experimental CNC table.
(2) The optical fiber and the temperature probe were penetrated the specimen vertically. The insertion depth was about 1mm.
(3) The specimen was treated with LITT, and the temperature of central zone and reduced scattering coefficient were recorded in real-time.
(4) Based on the compilation of these experimental data, the trend of temperature and reduced scattering coefficient under different laser power were analyses.
(5) At last, using MRI and Histopathological examination to validate the treatment of LITT.
The experimental parameters of LITT are listed in Table 1. The experiment with the same parameters has been repeated five times. The final experimental data was taken from the average of five tested data. In order to ensure the accuracy of tested value, the positions of optical fiber and temperature probe should be changed for the next thermotherapy on one sample after one test was end.

Table 1. Experimental parameters of LITT.

Laser Power(W)	Action time(s)
0.95	
1.23	600
1.42	

Results and discussion
The initial temperature of tissue is 25• at the beginning of LITT. The temperature of issue increases rapidly within 100s, and gradually reaches the stable state in the next 500s. The temperature rises faster when the laser is greater. The minimal final value of tissue temperature has been achieved 52• with laser power 0.95W. The maximal final value is 73• with laser power 1.42W. The initial value of μ's is 6cm-1. Then μ's continuously rise with increasing coagulation temperatures. μ's increases fast at the beginning and than rise slightly. The upward tendency of μ's at different laser powers was similitude, but the rising velocities were different. μ's rises fast when the laser power was greater. The minimal final value of μ's is about $12.6cm^{-1}$ with laser power 0.95W. The maximal final value is about $17.6cm^{-1}$ with 1.42W. The relationships between these two parameters with time at 1.42W are shown in Fig.2. The left y-axis and the x-axis is μ's versus time, the right y-axis and the x-axis is temperature versus time. As shown in Fig.2, it is obviously that the changes of μ's and temperature with time have the same trend.

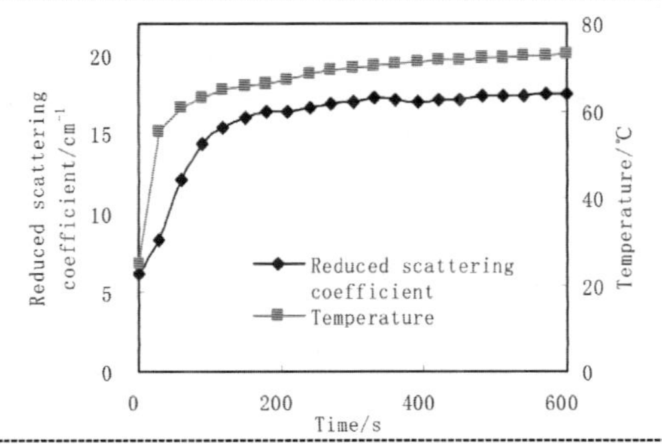

Figure 2. Change of temperature and μ'_s with time (1.42W).

According to the above analysis, the fNIRS can be used to control the treatment of LITT in real time.

At last, the treatment of LITT is validated by MRI and Histopathological examination.Fig.3 and Fig.4, respectively, shows the result of MRI and Histopathological examination. As shown in Fig.3 with white arrow, the destroyed region of tissue is obvious. As shown in Fig.4 with black arrow, the edge between the damage area and normal tissue is very clear. MRI and Histopathological examination both proved the effectiveness of the above experiments of LITT.

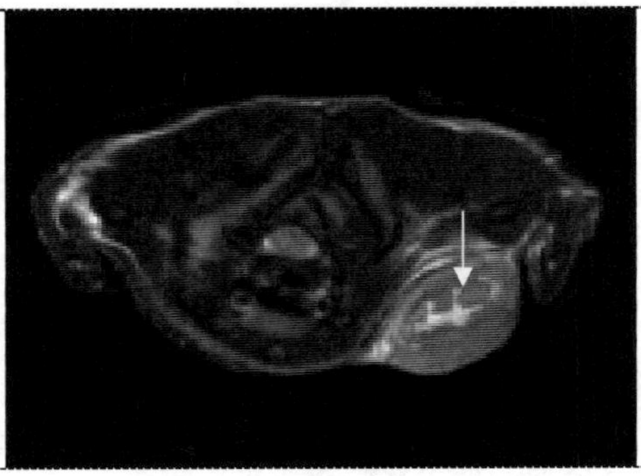

Figure 3. MRI of experimental result (1.42W, 300s).

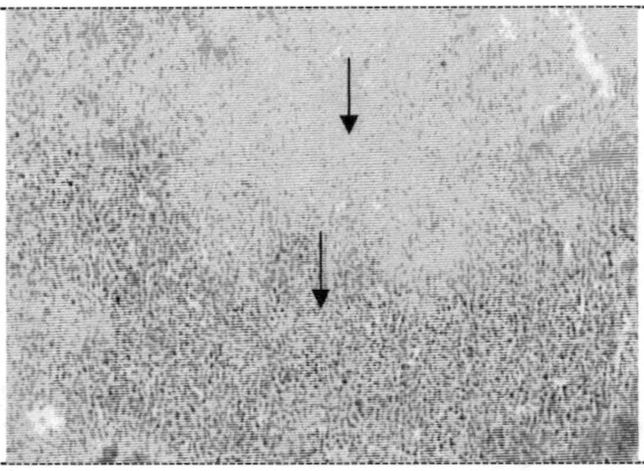

Figure 4. Histopathological examination result (1.42W, 300s).

Conclusion

This work has developed a new a new local control technology with near-infrared spectroscopy for laser induced interstitial thermotherapy and investigated the feasibility of the technology. The main conclusions can be summarized as follows:

(1) The changes of μ's and temperature with time have the same trend under different laser power.

(2) It is confirmed that the fNIRS can be a useful local control technology for LITT.

Acknowledgments

The author would like to acknowledge the support provided by the Natural Science Foundation of Jiangsu Province (BK2008184).

References

[1] Cholewa D 1998 *Lasers Surg. Med.* **23** 250
[2] Offergeld C 2003 *HNO* **51** 46
[3] Werner J A 1998 *Laryngoscope* **108** 463
[4] Zhu D, Luo Q M and Zhu G M 2002 *Lasers Surg. Med.* **31** 313
[5] Lin W C, Buttemere C and Jansen A M 2003 *IEEE J. Sel. Topl Quantum Electron* **9** 162

[6] Hans J S, Thomas G 1995 *SPIE* **2391** 451
[7] Cilesiz I F, Welch A J 1993 *Appl. Opt.* **32** 477
[8] Xie S S, Li H and Li B H 2003 *Chin. Opt. Lett.* **1** 44
[9] Zhu D, Luo Q M and Zeng S Q 2002 *Chinese J. Lasers* **29** 667
[10] Qian Z Y, Gu Y Q and Liu H L 2005 *Chinese J. Med. Phy.* **22** 463

Hypersonic wave drag reduction performance of cylinders with repetitive laser energy depositions

J Fang*, Y J Hong, Q Li and H Huang
Academy of Equipment Command & Technology, Post Box 3380-86, Huairou Dis. Beijing 101416, China

Email: fangjuan314@163.com

Abstract. It has been widely research that wave drag reduction on hypersonic vehicle by laser energy depositions. Using laser energy to reduce wave drag can improve vehicle performance. A second order accurate scheme based on finite-difference method and domain decomposition of structural grid is used to compute the drag performance of cylinders in a hypersonic flow of Mach number 2 at altitude of 15km with repetitive energy depositions. The effects of frequency on drag reduction are studied. The calculated results show: the recirculation zone is generated due to the interaction between bow shock over the cylinder and blast wave produced by energy deposition, and a virtual spike which is supported by an axis-symmetric recirculation, is formed in front of the cylinder. By increasing the repetitive frequency, the drag is reduced and the oscillation of the drag is decreased; however, the energy efficiency decreases by increasing the frequency.

1. Introduction

In recent years, due to the rapid development of hypersonic vehicles [1-3], drag reduction of demand on more and more strongly, which attracted researchers' attention. Drag reduction using laser energy deposition is a new way of drag reduction. America, Russia, Japan and other countries have carried out research on laser energy deposition drag reduction using numerical simulation and experiment methods [4-6]. Riggins team studied the influence of laser parameters on drag reduction by numerical simulation on the condition of continuous wave laser [7-8], Knight team simulated the interaction between the low density region produced by single pulse laser and bow shock over the blunt body [9-10], Sasoh team further studied the effect of laser energy deposition on different shapes of blunt body [11-12].

Literature results show that laser energy deposition is a very effective way to reduce hypersonic wave drag and pulsed laser is usually choose as energy source. So the choice of frequency is an important component of the research. Based on previous work, this paper studies the interaction between high-repetitive laser energy deposition and bow shock over the cylinder, and the effect of frequency on drag reduction. The condition of stable drag reduction is proposed under the conditions of this research.

2 Numerical method

Two-dimensional axisymmetric inviscid flow over a cylinders is assumed to be a idea gas with the constant specific heat ratio of 1.4. Mass, momentum and energy conservation equations are discretized using a finite-difference method. The MUSCL approach is employed with a min-mod limiter, and computed flow properties are second-order accurate in space. The discretized equations are

Published under licence by IOP Publishing Ltd

numerically integrated in time by using a second-order Runge-Kutta method. From reference [13] we know that this CFD method can describe the process of laser plasma reducing wave drag.

Two-dimensional planar graph of the blunt body used in the present study is in figure 1. The diameter of cylinder D is 2cm, the distance L between the stationary point of the body and energy deposition position is equal 2D, that is L=4cm. For all cases in the present study, the Mach number is set to 2, and the static pressure and temperature of the freestream flow are set to 1.2×10^4Pa and 216.65K respectively. The computational zone is divided into two zones-zone 1 and zone 2 which can be seen in figure 1. The grids of the two zones are 200×200 and 100×70 respectively. A heating source term for a periodic laser energy deposition is added to energy equation, the energy is deposited during the period of the order of the laser pulse duration at constant volume and that the deposited energy is injected instantaneously. The energy of single laser is 3mJ.

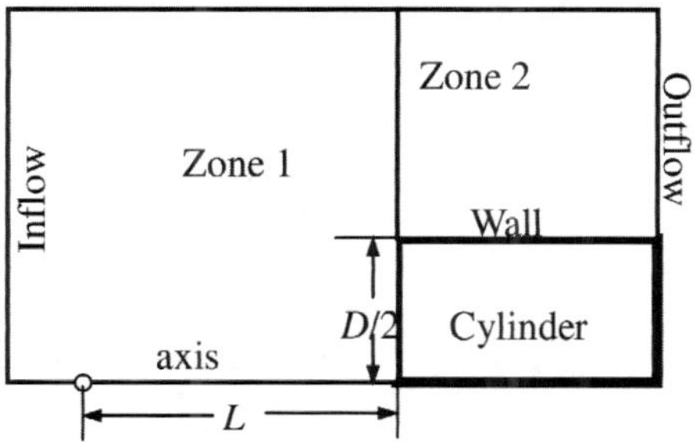

Figure 1. Two-dimensional planar graph of blunt body.

3 Results and discussion

Hypersonic wave drag reduction performances of cylinder at frequency f=10kHz, 20kHz, 50kHz, 100kHz and 200kHz are studied in this paper.

3.1 Unsteady flowfield at f=10kHz

Figure 2 shows the density contours and instantaneous streamlines between t=410μs and t=549μs. The blast wave and low density region produced by laser energy deposition are appear in figure2(a1). From the streamline shown in figure 2(a2), one can see vortices near the frontal surface of the body. This implies the low density region created by the previous energy deposition still stays near the corner of the cylinder surface.

We can see from the density contour in figure 2(b1) at t=449μs that the blast wave reaches at the frontal surface of the cylinder. This feature makes the drag increase. The low density region starts to interact with the bow shock at t=464μs and the bow shock is deformed one can see in figure 1(c1). This distortion is called lens effect [14].

The streamline at t=449μs shown in figure 2(b2) indicates that the vortices began to break down. By judging with the density contour in figure 2(b1), one can find that the vortex flow structure cannot keep in the frontal region over the body until the low density region produced at the next deposition reaches near the bow shock under this repetitive frequency condition. As shown in figure 2(c2), new vortices are produced in the region where the low density region interaction with the bow shock. The deformed low density region moves downstream as shown in figure 2(d2) and (e2). In figure 2(f1), the density contour at t=549μs is returns to the contour at t=449μs. One can know that at f=10kHz, the interaction between low density region and bow shock is independence in different laser depositions.

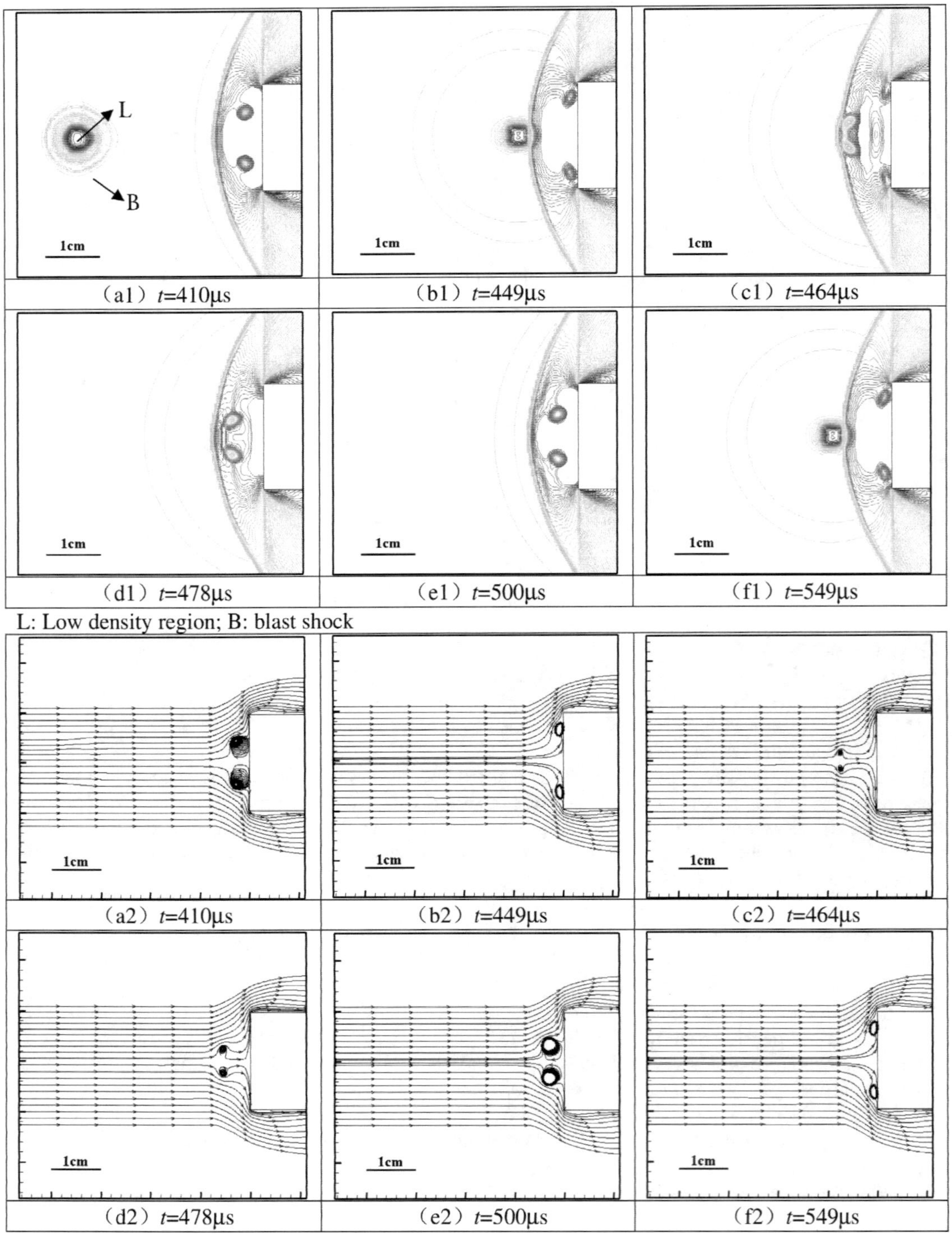

Figure 2. The density contour and streamlines at f=10kHz.

3.2 Quasi-steady state flowfield at different frequencies

The calculated quasi-steady flowfields at different frequencies are presented in figure 3. The flowfields are different between the lower repetitive frequency at *f*=10kHz and 20kHz, and the higher frequencies at *f*=50kHz, 100kHz and 200kHz. In figure 3(c), one can see an oblique shock wave in the frontal region of the cylinder. The oblique shock wave is formed through the interaction of the low density region with the bow shock. In contrast to the lower repetitive frequency, the oblique shock wave is strongly due to the multiple density regions incoming into the forebody region of the body. A large vortex flow region is produced due to the successive interaction of the low density regions with the bow shock. The flowfield shows a recirculation zone, as shown in the streamlines in figure 3(c) to 3(e), the recirculation zone is maintained in the frontal region of the body, the recirculation zone serves as virtual conical geometry in the flowfield, blocking the hypersonic flow incoming to the body [12]. Because the low density region flows through the recirculation zone in the forebody region, the pressure on the surface in kept lower as compared with the case at *f*=10kHz and 20kHz. From the comparison of the density contours between *f*=100kHz and 200kHz, as shown in figure 3(d) and 3(e), the angle of the oblique shock wave for *f*=200kHz is shallower that that for *f*=100kHz. This trend is due to the fact that the size of the recirculation zone is larger for *f*=200kHz as compared with the one for *f*=100kHz. As a result, the forebody region over cylinder is maintained at a lower pressure level, and the collision of the blast wave contributes to the increase of the drag only slightly.

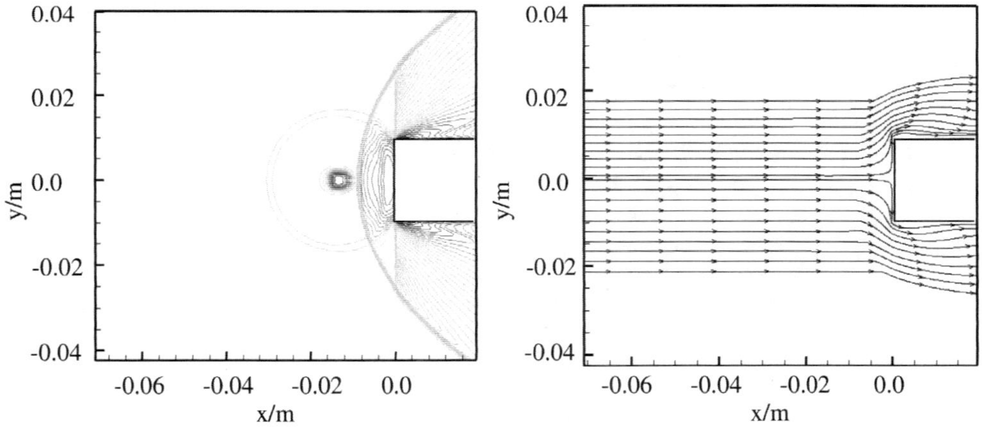

(a)10kHz

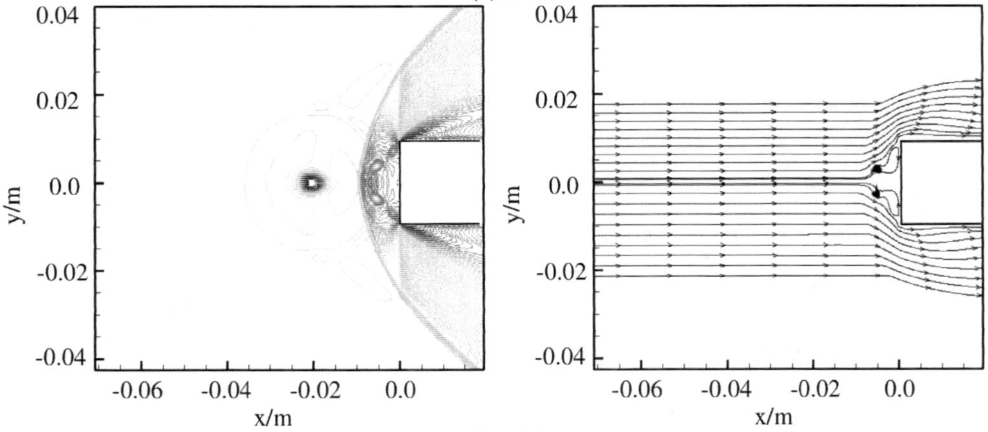

(b)20kHz

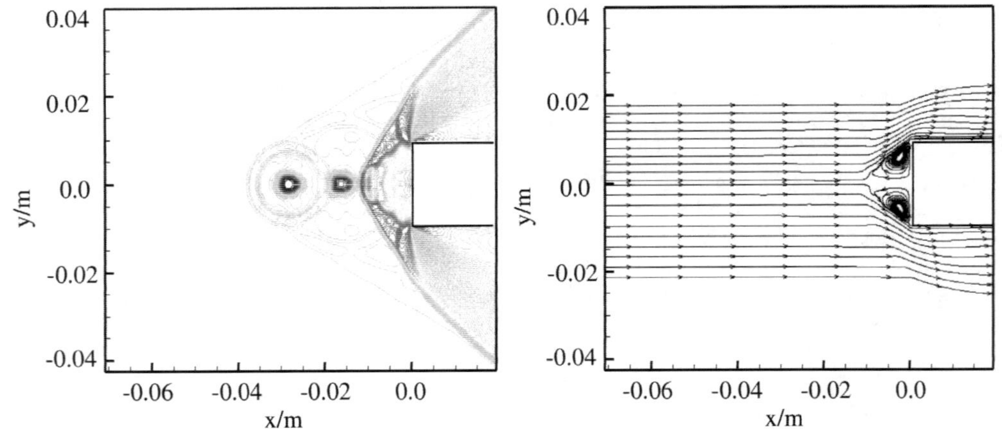

(c)50kHz

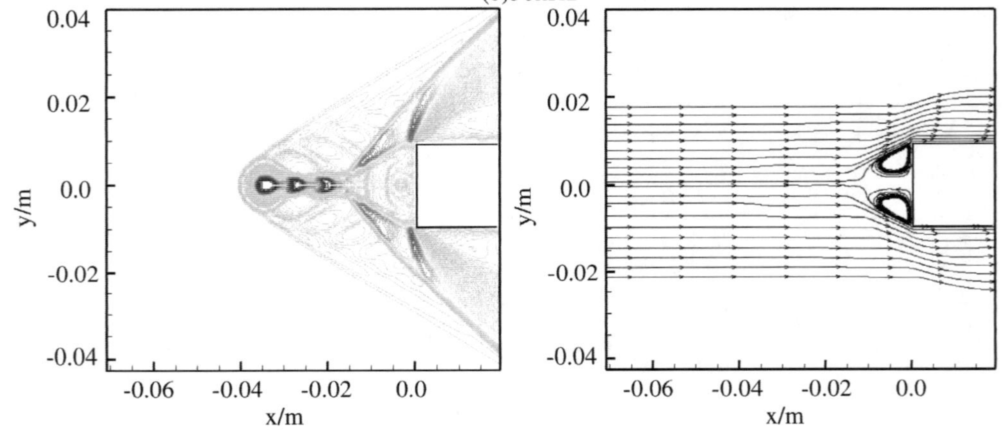

(d)100kHz

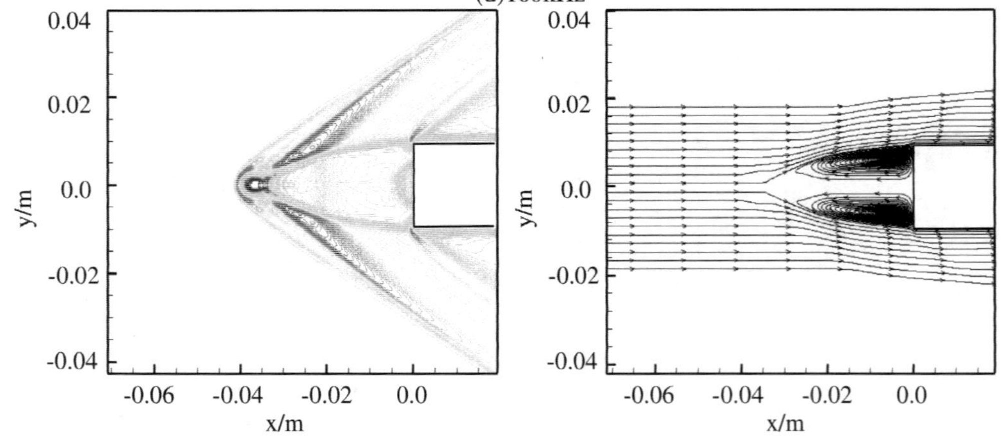

(e)200kHz

Figure 3. Quasi-steady state flowfield at different frequencies.

3.3 The influence of frequency on drag reduction

The wave drag curves at different repetitive frequencies of f=10kHz, 20kHz, 50kHz, 100kHz and 200kHz are given in figure 4. In each of the cases, the drag asymptotically approaches to a quasisteady state, showing periodical fluctuations. The averaged drag value is 13.4N, 13.2N, 9.4N, 6.1N and 2.5N, for the case of *f*=10kHz, 20kHz, 50kHz, 100kHz and 200kHz, respectively. The normalize drag which the ratio of averaged drag value to the one without energy deposition is about 91.7%, 90.9%, 64.8%, 42% and 17%. The oscillation becomes weaker with the repetitive frequencies increasing. Figure 5 shows the trend of the power efficiency *S* with frequencies *f*. One can see that *S* first increase and then decrease, the peak of *S* appear at *f*=50kHz and *S* is about 20. At *f*=200kHz, *S* is the minimum, but $S >> 1$.

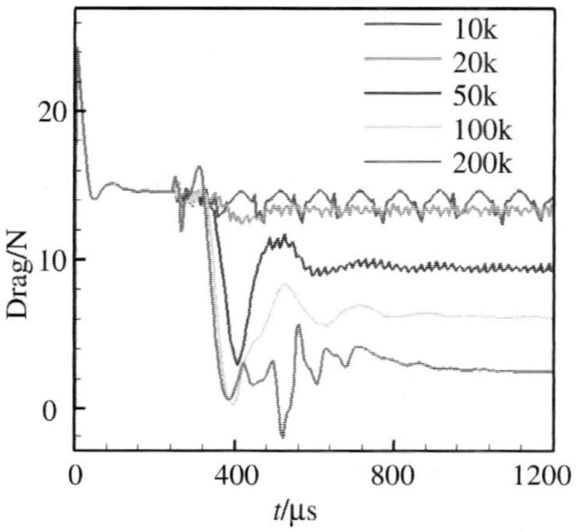

Figure 4. Drag curves at different frequencies.

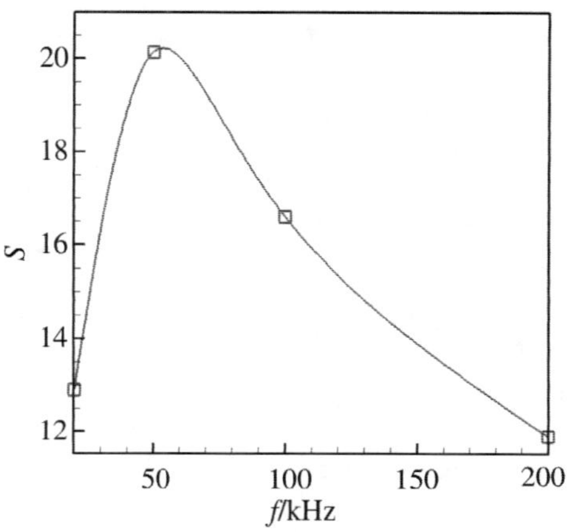

Figure 5. power efficiency vs. frequency.

4 Conclusions

In this paper, the influence of repetitive frequency on drag reduction is studied by numerical simulation. The numerical simulation results show that recirculation zone over blunt body is formed due to laser energy deposition, and the formation of recirculation zone reduces the wave drag of hypersonic vehicle. Frequency is an important parameter of influencing drag reduction, the higher frequency, the more drag reduction. When the frequency increases to 200kHz, the drag is least and only about 2.5N. However, power efficiency is not the best at *f*=200kHz, the maximum power efficiency is obtained at *f*=50kHz. Therefore, we should choose appropriate frequency according to the need of task. By increasing the repetitive frequency, the oscillation of the drag is decreased; At *f*=100kHz, the drag is steady.

Acknowledgement

This project is supported by National Natural Science Foundation of China(90916015).

References

[1] Salas M D. A review of hypersonic aerodynamics, aerothermodynamics and plasmadynamics activities within NASA's fundamental aeronautics program[R]. *AIAA*-2007-4264

[2] Hollingsworth P, Baets D and Ignacio E. Peregrine hypersonic strike fighter weapons system[R]. *The 1999/2000 AIAA foundation graduate team aircraft design competition*

[3] Okazaki M, Miyazawa K and Ishizawa K 1999 Engineering research for super/hyper-sonic transport propulsion system[C] *International Society for air breathing engines-ISOABE*, 13-20

[4] Maull D J 1960 Hypersonic flow over axially symmetric spiked bodies. *Journal of Fluid Mechanics*, **8** 584-592

[5] Hartley C S, Portwood T W, Filippelli M V et al 2005 Experimental and computational investigation of drag reduction by electric-arc airspikes at Mach 10[C]. *Beamed energy propulsion: third international symposium on beamed energy propulsion*, 499-513
[6] Riggins D, Barnett J T, Taylor T. Drag reduction and heat transfer mitigation for blunt bodies in hypersonic flight-A survey of techniques[R]. *AIAA* 2003-6968, 2003
[7] Riggins D, Nelson H F and Eric Johnson 1999 Blunt-body wave drag reduction using focused energy deposition[J]. *AIAA Journal*, **37** 460-467
[8] Khamooshi A, Taylor T, and Riggins D W 2007 Drag and heat transfer reductions in high-speed flows[J]. *AIAA Journal*, **45** 2401-2413
[9] Farzan F and Knight D. Interaction of microwave filament and blunt body in supersonic flows[R]. *AIAA* 2008-1356
[10] Knight D, Kolesnichenko Y F, Brovkin V, et al. Interaction of microwave-generated plasma with hemisphere-cone-cylinder[R]. *AIAA* 2010-1005
[11] Sakai T, Sekiya Y, Rosli M R B, et al. Unsteady interaction of blunt bodies with laser induced plasma in a supersonic flow[R]. *AIAA* 2008-3794
[12] Sakai T 2009 Supersonic drag performance of truncated cones with repetitive energy depositions[J]. *International journal of aerospace innovations*, **1** 31-43
[13] Li Qian, Jin Xin, Cao Zhengrui, et al 2010 Effects on aerodynamic drag of incident laser energy in technology of laser plasma point source drag reduction[J]. *Journal of propulsion technology*, **31** 377-380 (in Chinese)
[14] Georgievskii P Y and Levin V A 1993 Unsteady interaction of a sphere with atmospheric temperature inhomogeneity at supersonic speed[J], *Fluid Dynamics*, **28** 568-5

Spatial-Resolution Improvement in Optical Frequency Domain Reflectometry System Based on Tunable Linear Fiber Laser

Guoyu Li[1], **Peng Zhao**[2], **Yan Li**[1]
[1]Institute of Information Engineering, Handan College, Handan, 056005, P. R. China
[2]School of Physics & Optoelectronic Engineering, Dalian University of Technology, Dalian 116024, P. R. China

E-mail: guoyu_li@yahoo.cn

Abstract. In optical frequency domain reflectometry (OFDR) system, the spatial resolution is obtained by using the total frequency-sweep span of the tunable laser. However, in practice, the spatial resolution is severely limited by nonlinearity in the lightwave-frequency sweep of the tunable laser. A closed-loop PZT modulated DBR linear fiber laser is proposed to improve the spatial resolution of the OFDR system. Experimental results show that the spatial resolution of OFDR system has improved greatly. When the frequency sweep excursion is 66GHz and the fiber under test (FUT) is 7 m, the OFDR system has a spatial resolution of 1.5 m with open-loop PZT modulated laser. But the spatial resolution increases to 35 cm with closed-loop PZT modulated laser.

1. Introduction

Optical reflectometry is a powerful technique for the characterization of optical fibers and components. Different techniques have been developed, depending on the desired spatial range, resolution, dynamic range and sensitivity [1]. Now, Time domain techniques based on Optical time-domain reflectometry (OTDR) have maximum measurable fiber spans of a few tens kilometers, with resolutions of a few meters. However, the OTDR technique generally has to suffer from a tradeoff between dynamic range and spatial resolution. OFDR based on frequency-modulated continuous-wave (FMCW) reflectometry promises many advantages over the OTDR. It offers the possibilities of much higher sensitivity and larger dynamic range combined with higher spatial resolution than that of the OTDR [2].

In optical frequency domain reflectometry (OFDR) system, the spatial resolution is obtained by using the total frequency-sweep span of the tunable laser. That is, a wide frequency-sweep span provides a high spatial resolution. However, in practice, the spatial resolution is severely limited by nonlinearity in the lightwave-frequency sweep of the tunable laser [3].

In this paper, A way is proposed to improve the spatial resolution of the OFDR system by introducing a closed-loop PZT modulated DBR fiber laser. The PZT directly stretches the cavity length of the DBR fiber laser, so the frequency-modulation excursion is as much as 100GHz.

[1] Institute of Information Engineering, Handan College, Handan, 056005, P. R. China. E-mail: guoyu_li@yahoo.cn

2. Experimental setup

Figure 1 is a schematic diagram of our experimental setup for OFDR measurement. The active fiber used in the DBR laser was Er/Yb co-doped fiber, the fiber has a B-Ge-Si annulus with a diameter of 16μm around the 4-μm diameter Er-Yb phosphosilicate core. The DBR laser was fabricated by directly inscribing grating pair in active fiber with 193nm excimer laser and phase mask method. The laser cavity was fixed by the supports, and the PZT could control the laser cavity to stretch. The output optical power of the DBR fiber laser was kept 4 mW during the frequency tuning process.

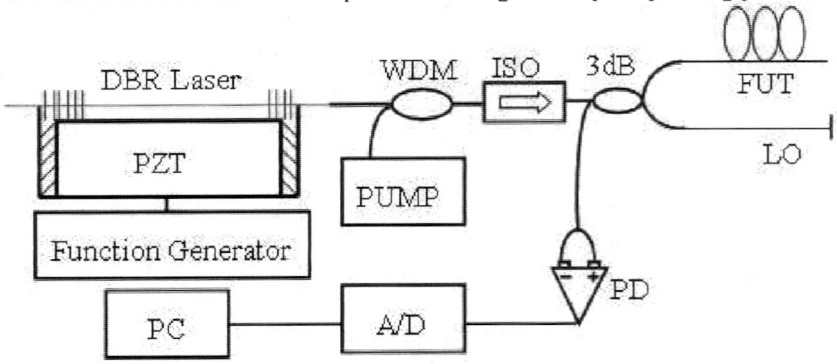

Figure 1. Experiment setup for OFDR measurement.

The ISO avoids the reflected light back to the DBR fiber laser. Two optical waves, which are derived from the DBR fiber laser whose frequency is modulated with a function generator but have traveled along different paths, are recombined at a point in the fiber [4,5]. The main interferometer is built with a 3dB coupler that is used for launching the light wave into the FUT and receiving the reflected light. The frequency-swept light from DBR laser is split using a 3dB coupler, 50% of the optical power goes into the FUT, and the remaining optical power goes into reference arm. Finally the reflected lights generate interference signal at a photodetector (PD), and the beat signals are acquired using the data acquisition card with a triggering signal generated each time when the frequency sweep begins.

3. Experimental results and discussion

3.1. tunable DBR fiber laser based on PZT modulation

In order to improve the spatial resolution of OFDR measurement system, the nonlinearity of the tunable fiber laser must be improved. But the PZT has the hysteresis characteristic, which leads to nonlinearity of the tunable fiber laser. A way is proposed to improve the linearity of the tunable DBR fiber laser based on closed-loop PZT modulation. That is, the fiber laser is modulated by a closed-loop PZT instead of open-loop PZT. The theory of closed-loop PZT modulation is as follow: the PZT stretches with the driving voltage increases, but the stretch of PZT has displacement difference because of hysteresis character. The displacement sensor attached to the PZT will monitor the displacement difference of the stretch, and the displacement difference is converted into the voltage signal, then the voltage signal modifies the driving voltage of the PZT, which assures the linearity of the PZT stretch. Fig.2 shows the nonlinear curve of the tunable DBR fiber laser based on open-loop PZT modulation.

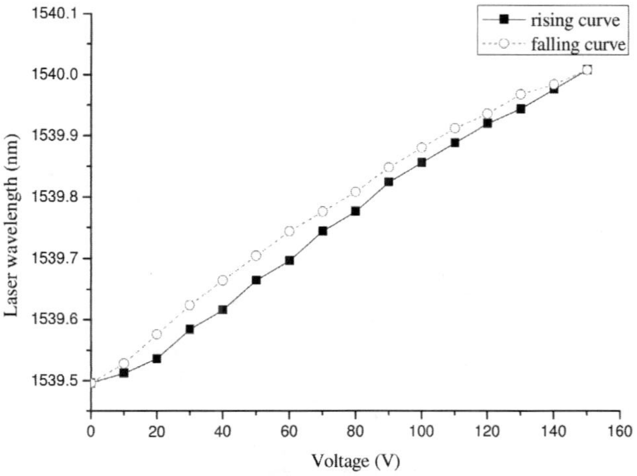

Figure 2. The tunable curve of DBR fiber laser based on open-loop PZT modulation.

The solid rectangles represent the tunable laser peak wavelength with driving voltage increases, and the hollow circles represent the tunable laser peak wavelength with driving voltage decreases. The solid line and the dashed line show the fit curves of rising and falling peak wavelength data, and they don't coincide with each other. From the figure 2, the open-loop PZT has serious hysteresis character, so the tunable curve of the DBR fiber laser based on open-loop PZT modulation is nonlinear apparently.

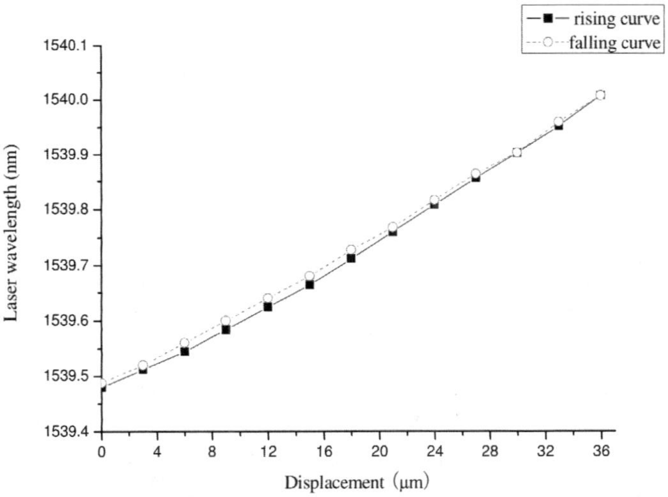

Figure 3. The tunable curve of DBR fiber laser based on closed-loop PZT modulation.

The solid rectangles represent the tunable laser peak wavelength with displacement increases, and the hollow circles represent the tunable laser peak wavelength with displacement decreases. The solid line and the dashed line show the fit curves of rising and falling peak wavelength data, and they coincide with each other. From the figure 3, the tunable curve of the DBR fiber laser based on closed-loop PZT modulation is linear apparently.

3.2. Spatial resolution improvement in OFDR system

In OFDR system measurement, the frequency sweep excursion is 66GHz, the FUT is 7m, and the DBR fiber laser is modulated by open-loop PZT, the spatial resolution is 1.5m, as shown in figure 4. The sweep nonlinearity of the tunable fiber laser based on open-loop PZT modulation causes its frequency to fluctuate, and this frequency fluctuation is converted to an amplitude change in an electric signal. So the OFDR waveform in figure 4 is indistinct with many sidebands, and the signal noise ratio (SNR) is low.

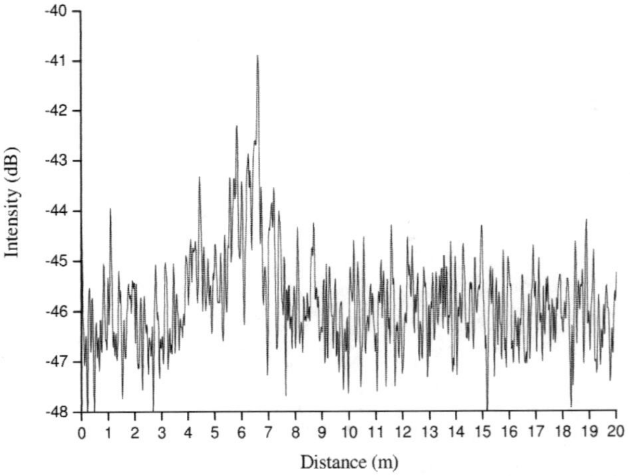

Figure 4. OFDR measurement based on open-loop PZT modulation.

In the figure 5, the frequency sweep excursion and the FUT are same as that in figure 4, but the tunable DBR fiber laser is modulated by closed-loop PZT, experimental results show that the spatial resolution of OFDR system increases to 35cm. The sweep linearity of the tunable fiber laser based on closed-loop PZT modulation decreases its frequency fluctuation, so the OFDR waveform in figure 5 is distinct with few sidebands, and the SNR is higher than SNR in figure 4.

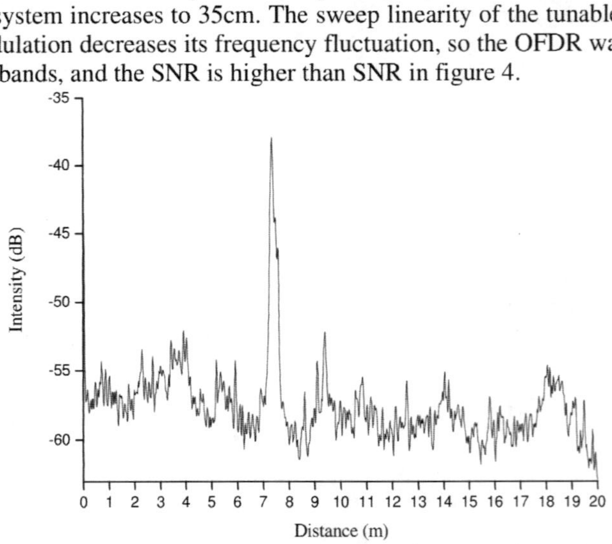

Figure 5. OFDR measurement based on closed-loop PZT modulation.

4. Conclusion

The spatial resolution improvement in OFDR system based on tunable linear fiber laser is studied. Experimental results show that the spatial resolution of OFDR system has improved greatly. When the frequency sweep excursion is 66GHz and the FUT is 7 m, the OFDR system has a spatial resolution of 1.5 m with open-loop PZT modulated nonlinear laser. But the spatial resolution increases to 35 cm with closed-loop PZT modulated linear laser.

5. References

[1] Oberson P Huttner B Guinnard O 2000 Optical frequency domain reflectometry with a narrow linewidth fiber laser *IEEE Photonics Technology Letters* Vol 12 No 7 pp 867-869

[2] Jihong Geng Christine Spiegelberg Shibin Jiang 2005 Narrow linewidth fiber laser for 100-km optical frequency domain reflectometry *IEEE Photonics Technology Letters* Vol 17 No 9 pp 1827-1829

[3] Tsuji K Shimizu K Horiguchi T 1997 Spatial-resolution improvement in long-range coherent optical frequency domain reflectometry by frequency-sweep linearization *Electronics Letters*, Vol 33, No 5 pp

[4] Jesse Zheng 2004 Analysis of optical frequency-modulated continuous-wave interference *Applied Optics* Vol 43 No 21 pp 4189-4197

[5] Jesse Zheng 2006 Optical frequency-modulated continuous-wave interferometers *Applied Optics* Vol 45 No 12 pp 2723-2730

Acknowledgments
This work was supported by the Project of National Natural Science Foundation of China (No.60807033).

Laser immunotherapy for treatment of patients with advanced breast cancer and melanoma

Xiaosong Li[1,2,3], Tomas Hode[4], Maria C. Guerra[4], Gabriela L. Ferrel[5], Robert E. Nordquist[6] and Wei R. Chen[3,*]

[1] Department of Oncology, the First Affiliated Hospital of Chinese PLA General Hospital, Beijing, China

[2] Department of Laser Medicines; Chinese PLA General Hospital, Beijing, China

[3] Department of Engineering and Physics, University of Central Oklahoma, Edmond, Oklahoma, United States

[4] Immunophotonics Inc., 1601 South Providence Road, Columbia, Missouri 65211, United States

[5] Hospital Nacional Edgardo Rebagliati Martins, Av. Edgardo Rebagliati #490 – Jesus Maria, Lima, Peru

[6] Wound Healing of Oklahoma, Inc., Oklahoma City, Oklahoma, United States

E-mail: wchen@uco.edu

Abstract Laser immunotherapy (LIT) was developed for the treatment of metastatic tumors. It combines local selective photothermal interaction and active immunological stimulation to induce a long-term, systemic anti-tumor immunity. During the past sixteen years, LIT has been advanced from bench-top to bedside, with promising outcomes. In our pre-clinical and preliminary clinical studies, LIT has demonstrated the capability in inducing immunological responses, which not only can eradicate the treated primary tumors, but also can eliminate untreated metastases at distant sites. Specifically, LIT has been used to treat advanced melanoma and breast cancer patients during the past five years. LIT was shown to be effective in controlling both primary tumors and distant metastases in late-stage patients, who have failed conventional therapies such as surgery, chemotherapy, radiation, and other more advanced approaches. The methodology and the development of LIT are presented in this paper. The patients' responses to LIT are also reported in this paper. The preliminary results obtained in these studies indicated that LIT could be an effective modality for the treatment of patients with late-stage, metastatic cancers, who are facing severely limited options.

Keywords: Laser immunotherapy, imiquimod, glycated chitosan, photothermal effect, systemic immune responses, metastasis, melanoma, breast cancer.

---

* To whom any correspondence should be addressed.

1. Introduction

Laser immunotherapy (LIT) is a new approach for cancer treatment using host immune system to fight the disease. LIT was proposed in 1997[1], which provides a convenient, efficient means to directly destroy the tumors at the treatment site and to induce tumor-specific host immune responses. LIT utilizes a near-infrared laser, a light-absorbing agent, and an immunoadjuvant[2,3,4] for the selective photothermal interaction and the targeted immunological stimulation. Our pre-clinical studies and preliminary clinical trials using LIT for the treatment for late-stage melanoma patients have shown promising outcomes[5].

Photothermal interaction using a near-infrared laser and a light-absorbing agent can induce a high temperature increase in the target tissue, which creates a selective tissue destruction zone covering the target tumor mass. Although this thermal reaction usually does not result in complete destruction or total acute eradication of target tumors, it causes tumor cells swell and break into pieces, allowing the release of antigens with the increase of temperature[6]. These antigens include tumor-associated antigens, thermally induced heat shock proteins (HSPs), and a large number of self-antigens. Antigen presenting cells (APCs), particularly dendritic cells (DCs), can capture these antigens and migrate to lymph nodes. They present the antigens to T cells to induce an immune response that can be effective against specific tumor cells. The LIT-treated tumors in patients serve as the sources of cancer antigens *in situ* and all the cancer antigens come from the patients' own tumor cells, as in the case of autologous vaccination. In addition, without *in vitro* procedure for pre-selection for specific cancer antigens, the whole cell serves as the vaccine in LIT.

The immune systems of late-stage cancer patients are often severely compromised and unable to respond vigorously to the traditional vaccination[7]. Therefore, two different immunoadjuvants, imiquimod and glycated chitosan, were applied to further enhance the immune response induced by the photothermal effect. Imiquimod, a special immunological modulator, activates immune cells through a toll-like receptor (TLR7), which are commonly involved in pathogen recognition[8]. Imiquimod can also induce localization and activation of Langerhans cells in treated skin, allowing them to subsequently migrate to local lymph nodes to activate the adaptive immune system[9]. Natural killer cells, macrophages, and B-lymphocytes are also activated by imiquimod. Glycated chitosan (GC) is a newly invented immunoadjuvant. According to our ongoing study (unpublished data), GC can directly activate DCs, and enhance antigen presentation to DCs, causing proliferation of T cells. The combined thermal treatment and GC stimulation can achieve even higher levels of T cell proliferation.

The present study is designed to investigate the clinical effects of LIT in the treatment of patients with late-stage melanoma and breast cancer.

2. Materials and Methods

2.1. Lasers

An 805-nm diode laser (AngioDynamics, Queensbury, NY) was used for the melanoma study. The light was delivered to the melanoma treatment surface through an optical fiber with a micro-diffusion lens. For the breast cancer study, an 805-nm laser (ImmunoPhotonics, Inc., Columbia, MO) was used and the laser light was delivered by an optical fiber to the breast cancer surface with a specially designed hand-piece (Figure 1). The laser power density at the surface of the skin was 1 W/cm^2 for both studies.

2.2. Indocyanine green (ICG)

ICG, obtained from Arkon Inc. (Buffalo Grove, IL), is a kind of cyanine dye used in medical diagnostics, which has a peak spectral absorption at about 800 nm. ICG can strongly absorb the light energy of the 805-nm laser. The interaction between ICG and 805-nm laser can induce the photothermal effect, which can increase the temperature in the tissue of treatment area. A solution of 0.25% ICG was injected at a dose of 0.5 mL/cm^3 prior to laser irradiation.

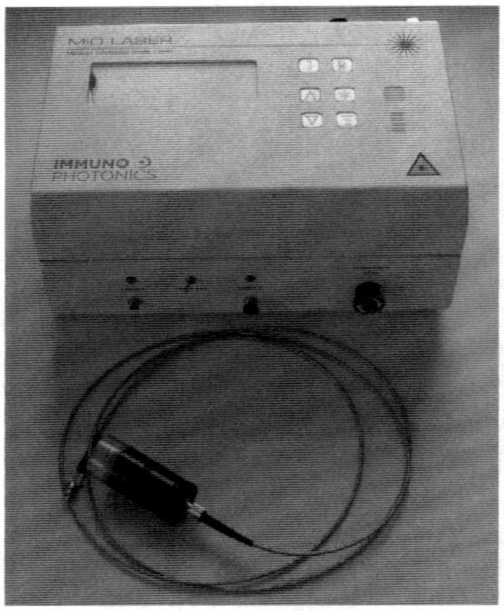

Figure 1. The 805-nm laser and the optical fiber with the hand-piece from by ImmunoPhotonics, Inc.

2.2. Indocyanine green (ICG)

ICG, obtained from Arkon Inc. (Buffalo Grove, IL), is a kind of cyanine dye used in medical diagnostics, which has a peak spectral absorption at about 800 nm. ICG can strongly absorb the light energy of the 805-nm laser. The interaction between ICG and 805-nm laser can induce the photothermal effect, which can increase the temperature in the tissue of treatment area. A solution of 0.25% ICG was injected at a dose of 0.5 mL/cm^3 prior to laser irradiation.

2.3. Imiquimod

Imiquimod (Aldara™) was purchased from 3M Pharmaceuticals (St. Paul, Minnesota, 5% cream under plastic occlusion). For melanoma patients, imiquimod was applied twice daily for two weeks before laser irradiation.

2.4. Glycated chitosan (GC)

GC was developed by our research group[2]. GC of 1% concentration was used for the treatment of breast cancer patients. GC was injected around the tumor for immunological stimulation immediately after laser irradiation.

2.5. Laser immunotherapy

For melanoma treatment, a 6-week cycle of LIT was carried out for each designated 200-cm^2 treatment area with the following five steps. (1) Topical application of imiquimod for 2 weeks; (2) Asepsis followed by local administration of anesthetic (lidocaine 2% with adrenaline); (3) Local injection of ICG; (4) Two times of laser irradiation, the first one was applied at the beginning of week 2, and the second one at the beginning of week 4; (5) Two weeks of imiquimod application after second laser session.

For breast cancer treatment, a 4-week cycle of LIT was carried out with the following steps: (1) Asepsis followed by local administration of anesthetic; (2) Local injection of ICG; (3) Laser irradiation; and (4) Local injection of GC.

2.6. Assessment of clinical outcomes

Toxicity was assessed according to National Cancer Institute Common Toxicity Criteria, version 3.0. The primary efficacy parameter was the best overall response by investigator's assessment using the Response Evaluation Criteria in Solid Tumors (RECIST).

3. Results

3.1. Patients

Eleven patients were enrolled in the melanoma study using laser immunotherapy with imiquimod from 2004 to 2010. Ten patients were enrolled in our breast cancer study using laser immunotherapy with glycated chitosan since September 2009. All the patients were more than 18 years old and had histologically confirmed stage III or IV melanoma or breast cancer according to the criteria of modified American Joint Commission on Cancer (AJCC) staging system[10,11]. All the patients are not available for traditional modalities.

3.2. Safety

LIT was generally well tolerated. Only local irritation was seen in all the patients treated by LIT. The most frequently reported adverse events were rash, pruritus, pain and edema. These side effects are induced mainly by the photothermal effects of LIT. The most severe side effects generally occurred during the first treatment cycle. The severity of these side effects was correlated with the previous irradiation therapy history of the treatment area. No grade 4 AEs were noted in our trials.

3.3. Efficacy

All the lesions at the treatment sites of the enrolled patients responded to LIT. In melanoma patients treated by LIT, complete response (CR) and partial response (PR) were observed in 5 patients. The objective response rate was 45.5%. One patient achieved stable disease (SD) and one patient had progression disease (PD).

In breast cancer patients treated by LIT, CR was observed in 1 patient, PR in 6 patients, SD in 2 patient and PD in 1 patient. The objective response rate was 70.0%, and the clinical beneficial response rate was 90.0%. All the patients with advanced breast cancer are still alive at the time of this report.

Figures 2 and 3 show the volume changes of the primary lesions and lung, liver and brain metastases for two stage-IV breast cancer patients. The data presented in Figure 2 is for a 40-year old patient. Before LIT, the diameter of primary lesion on the left breast was 4.0 cm. It completely disappeared after two LIT treatments. This patient had three liver metastatic nodules with diameters of 6.0 cm, 2.5 cm, and 2.0 cm, respectively, before LIT treatment. Five months after LIT treatment, two liver nodules were stable, and one metastasis disappeared (Figure 2).

The data presented in Figure 3 are for an 85-year old patient, who was diagnosed with stage IV triple-negative breast cancer, whose breast tumor does not express the genes for estrogen receptor (ER), progesterone receptor (PR), or Her2/neu. This subtype of breast cancer is clinically characterized as more aggressive and less responsive to standard treatments and with extremely poor overall prognosis. This patient had a 3.8-cm tumor in the right breast and a 3.6-cm tumor in the left breast before LIT treatment. Furthermore, this patient also developed metastases in the lungs and the brain. During the course of nine months, this patient received five LIT treatments. Her tumor in the right breast regressed completely while only a very small residual tumor was seen on the left breast (Figure 3). Brain metastasis with a diameter of 0.9 cm disappeared eight months after LIT treatment. The diameter of metastasis in the left lung decreased from 5 cm to 3.2 cm.

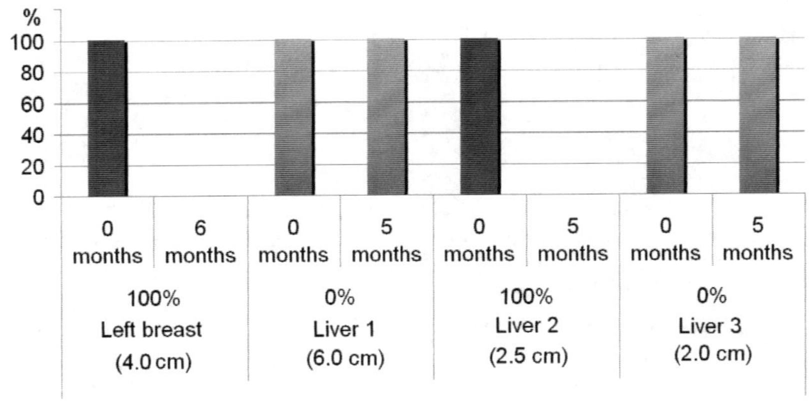

Figure 2. Percent change of tumor volumes (primary lesion and metastases) of a 40-year patient. The tumor diameter indicated in the figure is the size of the tumor before LIT treatment.

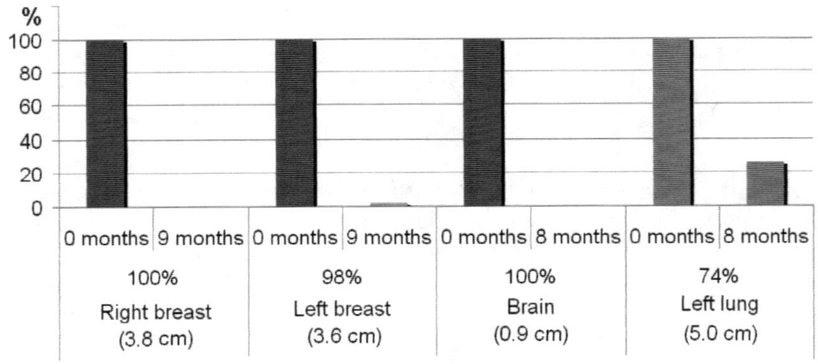

Figure 3. Percent change of tumor volumes (primary lesion and metastases) of an 85-year patient. The tumor diameter indicated in the figure is the size of the tumor before LIT treatment.

4. Conclusion

As a newly invented technology, LIT faces many challenges and many questions need to be further investigated. This report presented the preliminary clinical outcomes of LIT. Eleven advanced melanoma patients and ten advanced breast cancer patients were enrolled in our clinical trials. Each patient received at least one cycle of LIT treatment. The side effect is very limited, and it is well tolerated. After LIT treatment, not only the primary tumors responded well in the majority of the patients, the metastases at distant sites, such as the lungs, liver, and brain, were also responded positively in most stage IV patients. The metastases in most cases were either reduced in size or became stable. Our results show that LIT is not only a useful palliative modality for local primary tumors, but also a powerful method to control distant metastases. With further understanding of its fundamental mechanisms and continued research, it is believed that LIT will become a mature and necessary method for late-stage cancer treatment.

Acknowledgements

This research is supported by US National Institutes of Health (P20 RR016478 from the INBRE Program of the National Center for Research Resources) and National Natural Science Foundation of China (No. 81000994).

References:

[1] Chen WR, Adams RL, Carubelli R and Nordquist RE 1997 *Cancer Lett.* **115** 25
[2] Chen W R, Carubelli R, Liu H and Nordquist R E 2003 *Mol. Biotechnol.* **25** 37
[3] Chen W R, Liu H, Ritchey J W, Bartels K E, Lucroy M D and Nordquist R E 2002 *Cancer Res.* **62** 4295
[4] Chen W R, Jeong S W, Lucroy M D, Wolf R F, Howard E W, Liu H and Nordquist R E 2003 *Int. J. Cancer* **107** 1053
[5] Naylor M F, Chen W R, Teague T K, Perry L A and Nordquist R E 2006 *Br. J. Dermatol.* **155** 1287
[6] den Brok M H, Sutmuller R P, van der Voort R, Bennink E J, Figdor C G, Ruers T J and Adema G J 2004 *Cancer Res.* **64** 4024
[7] Cohen E P, Chopra A, O-Sullivan I and Kim T S 2009 *Immunotherapy* **1** 495
[8] Hemmi H, Kaisho T, Takeuchi O, Sato S, Sanjo H, Hoshino K, Horiuchi T, Tomizawa H, Takeda K and Akira S 2002 *Nat. Immunol.* **3** 196
[9] Gerster J F, Owens M L, Slade H B, Slade H B and Tomai M A 1999 *Int. J. Immunopharmacol.* **21** 1
[10] Singletary S E and Connolly J L 2006 *CA Cancer J. Clin.* **56** 37
[11] Balch C M, Buzaid A C, Soong S J, Atkins M B, Cascinelli N, Coit D G, Fleming I D, Gershenwald J E, Houghton A Jr, Kirkwood J M, McMasters K M, Mihm M F, Morton D L, Reintgen D S, Ross M I, Sober A, Thompson J A and Thompson J F 2001 *J. Clin. Oncol.* **19** 3635

Study of Linearly Polarized Light Measuring Fiber Diameter

Hong Liang
School of science, Communication University of China 100024

E-mail: lianghong@cuc.edu.cn

Abstract. The letter discusses a new way to measure fiber diameters, studies the theory of measuring fiber radius using this way, carries out numerical simulation based on this and provides experiment method as well.

1. Introduction

The ways to measure fiber diameter are contact and non-contact. The contact measurement is to use caliper and micrometer, which is easy to cause damage to fiber and poor in accuracy, as well as can't finish the real-time measurement during the progress of fiber drawing. There are two kinds of non-contact measurement instruments [1]. One is a device which forward or backward scattered light is used to measure fiber diameter, such measurement device is to work out the fiber radius mainly through analyzing electrode time of spot in forward or backward scattered image; the accuracy of this device is very high, but measurement needs optical collimator, and it is also very difficult to operate. A ray of collimated laser irradiates the measured fiber directly, high -frequency interference spot caused is modulated slowly due to inconsistent light-wave vibration directions, and the superposition of such slow signal makes it difficult to recognize local spot. Meanwhile, as the scattered angel is increasing, the amplitude of spot is becoming smaller and smaller, thus signal-to-noise ratio is becoming lower and lower, which is easy to be affected by interference. Therefore, it is not suitable to be used to measure real-time diameter [2]. The other one is fiber loss and mode field diameter measuring instrument [3]. This way is to measure mode field diameter between fibers, but can't carry out the real-time measurement during the progress of fiber drawing.

So far, determination of fiber radius is worked out by the known control parameters during making progress, however, if fiber radius is compressed to sub-millimeter magnitude, the above measurement way can't take effect, thus new calculation or measurement way is necessary. Non-contact way aforesaid seems very inapplicable due to tedious experiment device [4].

The Thesis discusses a new way to measure fiber diameters, studies the theory of measuring fiber radius using this way, carries out numerical simulation based on this and provides experiment method as well.

2. Theoretical Analysis

We shall discuss the isotropic column dielectric body and phase distribution of gauss wave packet scattered amplitude, and obtain the expression of light intensity.

For simplicity, we consider the two dimensional case: such as shown in figure 1. In such case, the wave equation met by the electric field component of the incident beam can be obtained through the following equation:

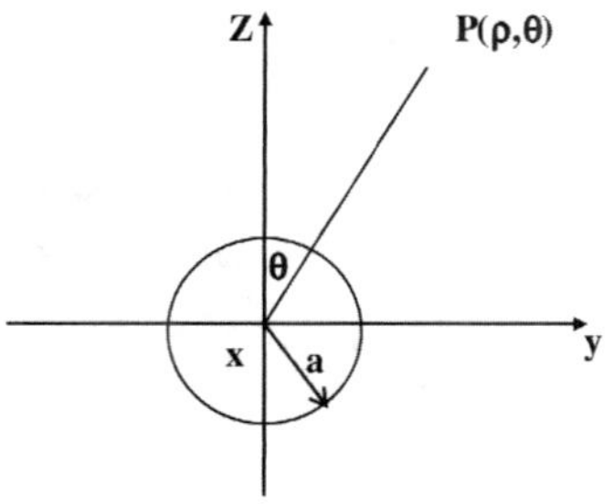

Figure 1. Two dimensional case.

$$\frac{\partial^2 E_x^i}{\partial y^2} + \frac{\partial^2 E_x^i}{\partial z^2} + k_0^2 E_x^i = 0 \tag{1}$$

Where $k_0^2 = \omega^2 \mu_0 \varepsilon_0$, ω is the angular frequency, μ_0, ε_0 are the magnetic permeability and dielectric coefficient in vacuum.

In order to solve the equation (1), we should return to the following anti-fourier Integral [5],

$$E_x^i(y,z) = \frac{1}{2\pi} \int_{-\infty}^{+\infty} E_i(\alpha) e^{-j(z+z_0)(k_0^2-\alpha^2)^{1/2} - j\alpha y} d\alpha \tag{2}$$

Where, $E_i(\alpha)$ is the fourier Integral of luminous source

$$E_i(\alpha) = \int_{-\infty}^{+\infty} E_i^{\,i}(y, -z_0) e^{j\alpha y} dy \tag{3}$$

When the luminous source is in $z = z_0$, the far field generated by the luminous source is expressed as

$$E_x^i(y, -z_0) = E_0 e^{-\beta^2 y^2} \tag{4}$$

Where $\beta^2 = a^2 + jb^2$, $1/a^2$ is the beam width;

When the origin of wave is assumed as the collimation laser beam, $\beta = \alpha$ is a real number. In the thesis, imaginary part of β is also considered, insert equation (4) into equation (3) and (2), we can get

$$E_x^i(y,z) = \frac{1}{2\pi\sqrt{\pi\beta}} \int_{-\infty}^{+\infty} E_i(\alpha) e^{-(a^2/4\beta^2) - j(z+z_0)(k_0^2-\alpha^2)^{1/2} - j\alpha y} d\alpha \tag{5}$$

In order to strictly figure out the integral of the above equation, it is difficult to obtain the analytic solution [6]. But we can get the approximate integral value. It is assumed that $|\beta\lambda|$ is little, and λ is the wavelength of the free space. Main contribution from the integral in equation (5) is derived from the small area of α. When α is little, the phase function in the integral expression can be spread into the Taylor series expression of α

$$(k_0^2 - \alpha^2)^{1/2} \approx k_0 - \frac{1}{2}\frac{\alpha^2}{k_0} \tag{6}$$

The integral in equation (5) can be strictly figured out, the following result can be get with the integral formula,

$$\frac{E_x^i(y,z)}{E_0} \approx \frac{1}{(1-jz)^{1/2}} e^{-\left[\beta y^2/(1-jz)\right]-jk_0(z+z_0)} \tag{7}$$

Where $z = 2\beta^2(1/k_0)(z+z_0)$.

In order to meet the boundary condition, the polar coordinate is changed from the cartesian coordinate, the integration variable is changed into

$$z = \rho\cos\theta, \quad y = \rho\sin\theta, \quad \alpha = k_0 \sin\gamma \tag{8}$$

The variable of exponential function in the integrated expression is turned out to be

$$z(k_0^2 - \alpha^2)^{1/2} + y\alpha = k_0\rho\cos[\theta - \gamma(\alpha)] \tag{9}$$

Therefore, the exponential function can be expressed as

$$e^{-jk_r\rho\cos[\theta-\gamma(\theta)]} = \sum j^{-n} e^{jn[\theta-\gamma(\theta)]} J_n(k_0\rho) \tag{10}$$

Where $J_n(k_0\rho)$ is the first kind of bessel function, insert equation (5) into equation (10), we can get

$$\frac{E_x^i(y,z)}{E_0} = \sum_{n=-\infty}^{\infty} j^{-n} e^{jn\theta} J_n(k_0\rho) A_n \tag{11}$$

Where,

$$A_n = \frac{1}{2\pi\sqrt{\pi\beta}} \int_{-\infty}^{+\infty} e^{-(a^2/4\beta^2)-jz_0(k_0^2-\alpha^2)^{1/2}-jn\gamma(\alpha)} d\alpha \tag{12}$$

In order to present the wave spread outwards, shape of the fringe field shall be as,

$$\frac{E_x^s(\rho,\theta)}{E_0} = \sum_{n=-\infty}^{\infty} j^{-n} e^{jn\theta} H_n^{(2)}(k_0\rho) B_n \tag{13}$$

Where $J_n^{(2)}(k_0\rho)$ is the second Hankel function, B_n is the undetermined coefficient determined by the boundary condition.

On the other hand, the light through the optical fiber cannot be neglected [7], and the wave equation in the dielectric circular cylinder is obtained from the following equation (in the polar coordinate)

$$\frac{\partial^2 E_x^i}{\partial\rho^2} + \frac{1}{\rho}\frac{\partial E_x^i}{\partial\rho} + \frac{1}{\rho^2}\frac{\partial^2 E_x^i}{\partial\theta^2} + k_0^2 \varepsilon E_x^i = 0 \tag{14}$$

Where, ε is the dielectric coefficient of dielectric circular cylinder, let the transmission wave in the cylinder be

$$\frac{E_x^s(\rho,\theta)}{E_0} = \sum_{n=-\infty}^{\infty} j^{-n} e^{jn\theta} R_n(\rho) C_n \tag{15}$$

Insert equation (15) into equation (14), we can get the differential equation related to the bessel function, the result is [8]:

$$R_n(\rho) = J_n(\sqrt{\varepsilon} k_0 \rho) \tag{16}$$

Two conditions must be in succession by using the boundary condition $\rho = a_0$ and tangential component of the field to determine two coefficients B_n and C_n. Therefore, the scattered beam and transmitted light beam are:

$$\frac{E_x^s(\rho,\theta)}{E_0} = \sum_{n=-\infty}^{\infty} j^{-n} e^{jn\theta} \left(\frac{J_n(k_0a_0) - \frac{J_n(\varepsilon k_0 a_0)}{\varepsilon J_n'(\varepsilon k_0 a_0)} J_n'(k_0 a_0)}{\frac{J_n(\varepsilon k_0 a_0)}{\varepsilon J_n'(\varepsilon k_0 a_0)} H_n^{(2)}(k_0 a_0)} \right) H_n^{(2)}(k_0\rho) A_n \tag{17}$$

$$\frac{E_x^t(\rho,\theta)}{E_0} = \frac{2}{\pi k_0 a_0} \sum_{n=-\infty}^{\infty} \frac{J^{-n+1} e^{jn\theta} J_n(\varepsilon k_0 a_0) A_n}{\varepsilon H_n^{(2)}(k_0 a_0) J_n'(\varepsilon k_0 a_0) - H_n^{(2)}(k_0 a_0) J_n(\varepsilon k_0 a_0)} \tag{18}$$

When the electric vector is parallel with the optical fiber,

$$\gamma_m = \frac{n_2 J^m(n_1 k_0 a_0) \left[\frac{J_m'(n_2 k_0 a_0)}{J_m(n_2 k_0 a_0)} \right] - n_1 J_m'(n_1 k_0 a)}{n_2 H_m^{(1)}(n_1 k_0 a_0) \left[\frac{J_m'(n_2 k_0 a_0)}{J_m(n_2 k_0 a_0)} \right] - n_1 H_m^{(1)}(n_1 k_0 a_0)} \tag{19}$$

When the magnetic-field vector is parallel with the optical fiber,

$$\gamma_m = \frac{n_1 J^m(n_1 k_0 a_0) \left[\frac{J_m'(n_2 k_0 a_0)}{J_m(n_2 k_0 a_0)} \right] - n_2 J_m'(n_1 k_0 a_0)}{n_1 H_m^{(1)}(n_1 k_0 a_0) \left[\frac{J_m'(n_2 k_0 a_0)}{J_m(n_2 k_0 a_0)} \right] - n_2 H_m^{(1)}(n_1 k_0 a_0)} \tag{20}$$

Where ω_0 is the Gaussian beam waist radius, k_0 is wave vector. J_m, H_m are order Bessel function and Hankel function.
The electric field is expressed as the following equation by the amplitude u:

$$E = jn_2 k_0 \hat{y} \quad \text{(E is parallel with the),} \tag{21}$$

$$E = -\frac{jmu}{r}\hat{r} - \frac{\partial u}{\partial r}\hat{\theta} \quad \text{(E is perpendicular to the optical fiber),} \tag{22}$$

Where $\hat{y}$ is the unit vector under the rectangular coordinate system (x,y). For these two conditions, the complex amplitude is in proportion with square of the electric field intensity [9],

$$|E|^2 \propto |u|^2 \tag{23}$$

In order to figure out the field intensity in the computer, the infinite summation function in equation (17) must be converted to the finitude summation function. With order number m increase, the summation function tends towards 0. For a great z, the bessel function (not in coefficient A_m and γ_m) in equation (17) can be substituted by the following equation [10]:

$$J_m(z) \to \left(\frac{2}{\pi z}\right)^{1/2} \cos(z - m\pi/2 - \pi/4) \tag{24}$$

$$Y_m(z) \to \left(\frac{2}{\pi z}\right)^{1/2} \sin(z - m\pi/2 - \pi/4) \qquad (25)$$

The light intensity passing through the optical fiber is superposition of scattered beam and transmitted beam [11], we can get the light intensity expression:

$$I(\theta, r) = \left| I_0 \sum_0^m \exp(-im\theta) i^m A_m(\omega_0, k_0) \times [J_m(n_1 k_0 r)] - H_m^{(1)}(n_1 k_0 r) \gamma_m(k_0, r, n) \right|^2 \qquad (26)$$

We shall establish a model with the mathematical software to simulate the infection of various parameters in the light intensity curve.

3. Numerical Simulation and test system

We can observe from equation (26) that after the optical fiber refractive index and optical fiber radius are determined, the light intensity is only the function of scattering angle θ. We can draw the relation curve between the scattering angle and light intensity through the mathematical software. For an assigned optical fiber radius, the light intensity corresponding to the scattering angle extreme value is regular. The minimal value and maximal value alternatively distribute and correspond to a set of scattering angle extreme value. Different optical fiber radius shall correspond to different extreme value of scattering angle. Curve shown in Fig. 2 is the light intensity curve when the radius is respectively r=0.6μm and r=0.3μm.

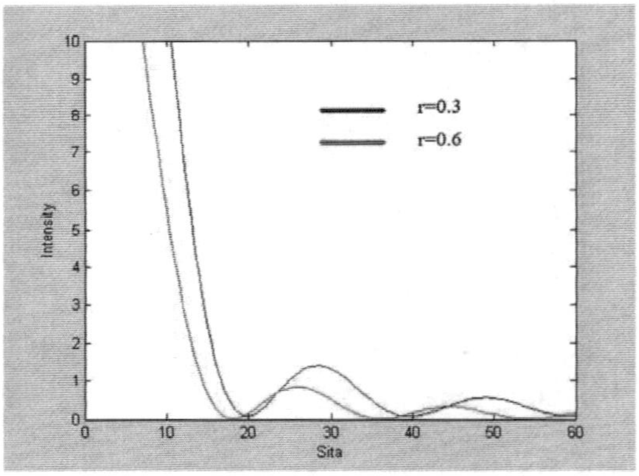

Figure 2. Light intensity curve when the radius, r, is 0.3μm and 0.6μm, respectively.

Let radius varies in a certain range, calculate the scattering angle corresponding to the light intensity extreme value within certain radius span with the mathematical software through equation 3-26, repeat the operation, you can draw a radius~angle relation curve.

Devices include fiber test system composed of laser, polarizer, condenser, magnifier, shadow shield and receiving screen and image acquisition and data analysis system composed of CCD camera and computer [12], shown in Fig. 3. Laser, polarizer, condenser, magnifier, shadow shield, receiving screen and CCD camera are fixed and installed in order within the shell. There is central gap left in the middle of shadow shield. Pre-position of measured fiber is set in the hollow of shell. Semiconductor monochromatic laser is used; laser beam sent out by it becomes polarized light through polarizer; condenser and magnifier are adjusted to make distance between two lens become the sum of focal distance of two lens; location of shadow shield is adjusted to make central gap level to align measured fire; CCD camera is adjusted to make lens to align receiving screen; CCD camera and computer are connected with data wire. CCD camera adopts interference diffraction fringe map on receiving screen; the interference diffraction fringe map is transmitted to computer through data wire; computer reads

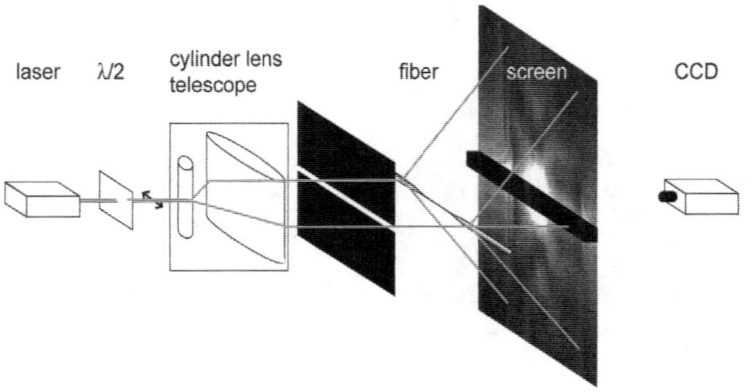

Figure 3. Test system.

strong maximum of interference diffraction fringe map and comparison of computer memory database; diameter value of fiber measured is output; measurement results is displayed by indicator finally. CCD camera takes photos in high frequency and transmits the map to the computer continuously, and the computer treat data in real time and outputs measurement result; real-time, rapid and accurate measurement is finished during the progress of fiber drawing.

4. Conclusion

We discuss a kind of new method in measuring the diameter of the fiber based on the forward scattering method, further study the theory on measuring the optical fiber radius with the method, and establish a set of integrated database on the foundation. For the tapered fiber with unknown radius, at first, measure the scattering intensity, draw the strength distribution curve, and then you can find out the optical fiber radius in the relation storage in accordance with the angle corresponding to the maximal and minimal value in the curve. The method is rapid and convenient in measuring the radius of the fiber, and can be taken as a real time method in measuring the fiber diameter.

References

[1] Birks T A, Wadsworth W J and Russell P St J. 2000 *Opt. Lett.* 25, 1415
[2] Udem T, Holzwarth R and Hansch T W, 2002 *Nature* 416 233
[3] Wu J B 1998 *Measuring technoiogy* No. 11
[4] Teipel J and Giessen H, 2007 *CLEO* p.58
[5] Teipel J, Franke I C, Turke D, Warken F, Meiser D, Leuschner M and Giessen H, 2006 *Appl. Phys. B* 77, 245
[6] Birks T A, Knight J C and Dimmick T E, 2000 *IEEE Photon. Technol. Lett.* 12, 182
[7] Butler D J and Forbes G W, 1998 *Appl.Opt.* 37, 2598
[8] Kozaki S, 1982 *J. Appl. Phys.* 53, 7195
[9] Butler D J and Forbes G W, 1998 *Appl. Opt.* 37, 2598
[10] Glass M, Dabbs T B and Chudleih P W, 1995, J65, 85-94
[11] Lentz W J 1976 *Appl.Apt.*15,668-671
[12] Warken F and Giessen H, 2004 *Optics Letters,* 29, 1727

Research on Image processing in laser triangulation system

Kai Liu, Qianqian Wang, Yang Wang and Chenrui Liu

School of Optoelectronics, Beijing Institute of Technology, 100081 Beijing, China

qqwang@bit.edu.cn

Abstract. Laser Triangulation Ranging is a kind of displacement distance measurement method which is based on the principle of optical triangulation using laser as the light source. It is superior in simple structure, high-speed, high-accuracy, anti-jamming capability and adaptability laser triangulation ranging. Therefore it is widely used in various fields such as industrial production, road test, three-dimensional face detection, and so on. In current study the features of the spot images achieved by CCD in laser triangulation system were analyzed, and the appropriate algorithms for spot images were discussed. Experimental results showed that the precision and stability of the spot location were enhanced significantly after applying these image processing algorithms.

1. Introduction

Laser triangle range-measurement technology is a crucial approach for the non-contact position measurement. The method can quickly give abundant information, such as displacement, thickness, vibration, velocity, acceleration, topography and outline of the target [1]. It is widely used in the field of modern industry production with the broadly advantages of high measuring speed, high accuracy, powerful anti-jamming performance, small measuring points, wide application range, etc [2]. During the process with the triangle range-measurement principle, the quality of the CCD image formation in the laser spot will be the primary factor for accuracy improving. In this article, the image processing algorithms (filter, image segmentation, morphology) were applied in the image which created in the laser triangle range-measurement system, so the higher pointing accuracy and the good stability would be achieved.

2. Laser spot image processing

2.1. Noise filtering

In the laser triangulation sensor system, the best laser spot got in the CCD should be a single sleek dot or single line. But in the actual measurement, many deviations appear in the image. These measurement errors come from system noise, just like CCD thermal noise, CCD drive noise, preamplifier noise, lights noise, stray light interference, and so on. To get smooth curves, filtering is applied to get rid of the noises. Generally speaking, the median filter, mean filter and Gaussian filter are frequently used [3-5].

2.2. Image segmentation

The filtering was applied to remove noise in the image. But there still exist some invalid areas around the spot. These regions have significant gray values, which are lower than the gray values in the spot zone. So they can be taken as background areas. The laser spot will be effectively discriminated from the background. As a result, the image in the laser triangulation system will be divided into two areas, the laser spot and the background. They are discrete in gray scale. So the threshold segmentation technique was chosen to divide the laser spot image. The valley between two peaks in the image histogram was set as the threshold for image segmentation.

2.3 .Morphological processing

After the noise filtering and segmentation, the image was a circular but irregular area. Its shape was unable to match for the primary spot. To get the desirable spot image, the morphological image processing method was utilized for further operation. Actually the ideal laser spot should be a circular Gauss area. So another circular structure was built to separate the image. With the erosion and Open-Close operations, the desirable shape emerged.

2.4. Spot centroid

In the laser triangulation range-measurement, the spot position was used to locate the site of target. But the spot got by CCD is just an irregular area, rather than an ideal point. After processing described above, the spot in the image was locked roughly in a small region. To satisfy the high positioning accuracy, some methods were brought in at gray level to obtain the precise spot centroid, such as the centroid, the surface fitting and so on [6]. The gray centroid and its modified form were analyzed in the aspects of position accuracy.

Supposing the spot focusing on the CCD used about $n \times n$ pixel, and the output signal envelop was similar to normal distribution or Gaussian distribution [7]. Consequently, the centroid method would be the first moment of the binary image. And then the gray scale value of the spot was given by the following formula:

$$X_C = \sum_{i=1}^{n} x_i \cdot g(x_i, y_i) / \sum_{i=1}^{n} g(x_i, y_i) + x_0 \tag{1}$$

$$Y_C = \sum_{i=1}^{n} y_i \cdot g(x_i, y_i) / \sum_{i=1}^{n} g(x_i, y_i) + y_0 \tag{2}$$

In the formula, X_C and Y_C were the spot centroid coordinates; x_i and y_i corresponded to the coordinates of the point (x_i, y_i); x_0 and y_0 were the coordinates for the starting point; $g(x_i, y_i)$ was the gray scale values of point (x_i, y_i). Before the computation, the bilinearity interpolation was applied to the selected laser spot region. It could effectively enhance the precision and the stability of spot positioning. If $0 < x < 1$, $0 < y < 1$ then the gray scale value of interpolation point $(i + x, j + y)$ was

$$g(i + x, j) = g(i, j) + x[g(i+1, j) - g(i, j)] \tag{3}$$

$$g(i + y, j) = g(i, j) + y[g(i, j+1) - g(i, j)] \tag{4}$$

$$\begin{aligned} g(i + x, j + y) &= x[g(i+1, j) - g(i, j)] + y[g(i, j+1) - g(i, j)] \\ &+ xy[g(i+1, j+1) + g(i, j) - g(i+1, j) - g(i, j+1)] + g(i, j) \end{aligned} \tag{5}$$

In the actual experiment, the value of x, y was set as 0.5. In other words, the interpolation nodal point was set in the middle of two pixels. The present study has discovered the low efficiency of traditional processing method based on gray centroid. It is vulnerable to the influence of system noise. And the low anti-jamming ability and the controversial localization accuracy restrain the positioning accuracy. Therefore the improved centroid method is presented. The new method takes the square of gray scale values, not the grey level, as the metrics. The pixel which has a large gray value plays a more important part in the processing. And the proportion of the white noise and system noise reduces simultaneously [8]. The improved method's formula expression is

$$\bar{x} = \frac{\sum x_i \cdot g_i^2}{\sum g_i^2} \tag{6}$$

$$\bar{y} = \frac{\sum y_i \cdot g_i^2}{\sum g_i^2} \tag{7}$$

g_i^2 is the square of gray scale value corresponding to the point (x_i, y_i).

3. Result and analysis

The original spot image is shown in Figure 1. And the result images after the median filtering are shown in Figure 2. After the median filtering, the noise in the image was weakening gradually.

As the filtering window enlarging, the image becomes smoother. But the edge of the spot becomes fuzzier. Therefore the best size of the filtering window should be able to remove the noise effectively and maintain the spot's contours to maximize.

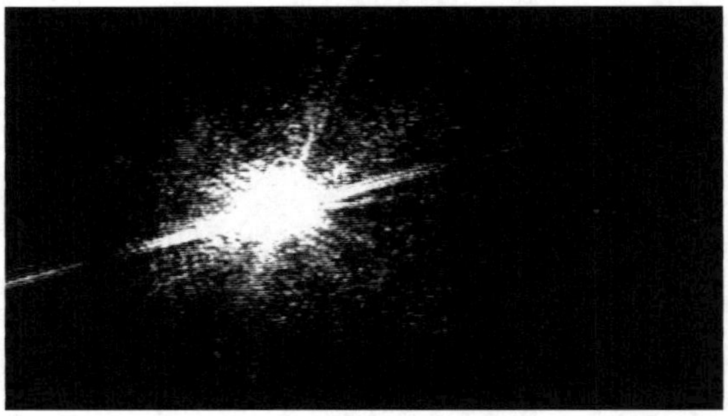

Figure 1. The original gray image of laser spot.

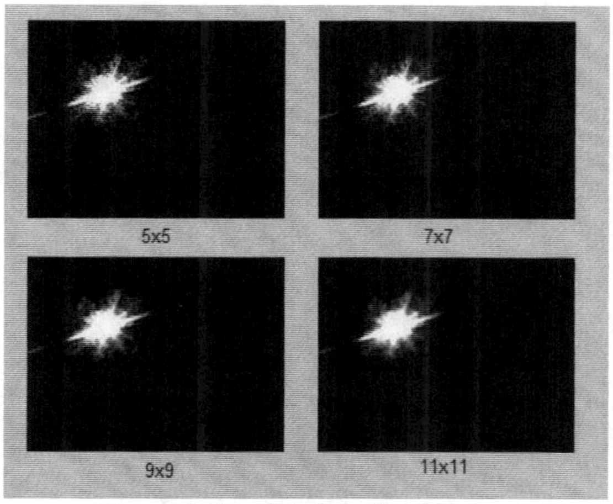

Figure 2. Results of median filtering.

The result of the threshold segmentation of target area is shown Figure 3. A 7×7 window was selected in the median filtering method, and the maximum value in grayscale was taken as the threshold T. As shown in Figure 3, after the processing, the spot shape was an irregular circular region owing to the existence of linear noise. To get the desirable spot image, the morphological image processing method was utilized for further operation.

Figure 3. Result of threshold segmentation processing.

Figure 4 illustrates morphological image processing results of the Figure 3.A circular structure with a 4 radius was used to complete the corrosion of figure 3. And then another circular structure with a 20 radius was utilized for open operations. After processing described above, the spot image has a higher anti-jamming capability, better shape and symmetry, smoother and clearer spot edge. It is beneficial to the accuracy and stability of the image positioning.

Figure 4. Morphological processing method.

The improved gray scale centroid method was applied to position the spot centroid (Figure 4). The result (as shown in Figure 5) of the operation was [234.1108 206.5719]. The image characteristics of Figure 4 have proven the effectiveness of the improved method. The spot is more symmetric horizontally, and asymmetric vertically. So the improved algorithm used for image processing is useful. And the horizontal and vertical coordinates are more correlated accurately with actual situation.

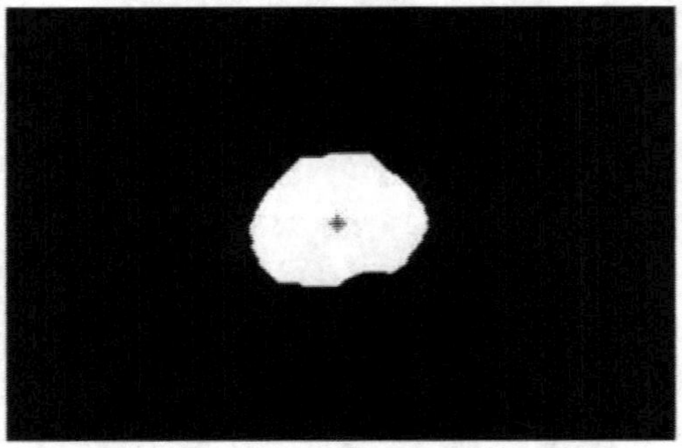

Figure 5. The result of the centroid method.

4. Conclusion

The image processing algorithms in triangle range-measurement system were discussed in this paper. After filtering, segmentation and morphology processing, the spot image has a high anti-jamming capability, good shape and ideal symmetry, glossy and clear spot edge. The result of the spot is more accurate and stable. In particular, an improved gray scale centroid method was applied for the positioning of the laser spot. The method enhanced the effective use of the spot information around the peak, and reduced the interference of white noise and other noises. The results demonstrated the accuracy and effectiveness of the improved image processing algorithm.

References

[1] Zhang Baojun and Cao Yonghua 2006 *J. Journal Of Zhanjiang Normal College*. **27** 41-43
[2] Lu Jinren and Huang Yuanqing 2004 *J. Journal of Xiamen University (Natural Science)*. **43** 50-53
[3] Wang Xiaojia, Gao Jun and Wang Lei 2004 *J. Chinese Journal of Scientific Instrument*. **8** 601-608
[4] Feng Junyan, Feng Qibo and Kuang Cuifang 2004 *J. Applied Optics*. **25** 33-36
[5] Pei S C and Tam I K 2000 *IEEE Int. Conf. Image Processing*. (Vancouver) P 488-491.

Journal of Physics: Conference Series **276** (2011) 012025 doi:10.1088/1742-6596/276/1/012025

[6] Su D and Willis P 2003 *Theory and Practice of Computer Graphics* (Birmingham: IEEE Computer Society Press) P 16

[7] Liang Zhongwei, Ye Bangyan, Peng Ruitao, et al 2007 *J. Aviation Precision Manufacturing Technology.* **43** 31-33

[8] Xu Yaming, Xing Cheng, Liu Guanlan, Tang Feng, Yu Zhiqi. 2007 *J. Hydrographic Surveying And Charting.* **27** 74-76

Numerical simulation for the influence of injected laser power on plasma drag reduction

Z Liu [1], Z G Dou[2], H Huang[2] and J Fang[1]

[1]Department of Postgraduates, the Academy of Equipment Command and Technology, 3380 Post box, Huairou Beijing 101416, China
[2]Department of Basic Theories, the Academy of Equipment Command and Technology, 3380 Post box, Huairou Beijing 101416, China

E-mail: liuzhun0@gmail.com

Abstract. Laser plasma drag reduction is a new method to reduce the wave drag of hypersonic flight. Inject laser power is an important parameter. An appropriate laser power should be chosen when laser power was injected to achieve the best drag reduction effect via the minimum laser power. The effect of inject laser power on the performance of laser plasma drag reduction when incoming flight Mach number is 6.5 and at 30km altitude was simulated numerically. The result indicates that the drag can be effectively reduced by energy injection in the upstream flow. The larger the inject power is, the smaller the drag of the blunt body obtained. The energy injection can also influence the pressure and temperature on the surface of blunt body. When laser energy injected, high pressure region on the surface moves to the back of the hemisphere, the pressure of stagnation point decreased. There are two peaks of temperature on the blunt surface, one is the stagnation point and the other is the high pressure region. Temperature of the surface after high pressure region is lower comparison to the condition that no energy injected.

1. Introduction

The wave drag of aircraft can be reduced effectively by inject laser energy into the hypersonic incoming flow. Myrabo, et al., first suggested that add energy to support the Directed-Energy Air-Spike (DEAS) ahead of the craft to reduce wave drag[1]. The drag of the aircraft may be greatly reduced is mainly due to that energy deposition could break the structure of bow shock on the craft surface and weak oblique shock formed, aerodynamic resistance due to the pressure difference between front and back of the craft. Besides, it changes the velocity direction of incoming flow, guide airflow to the periphery of aircraft, also made the drag reduced.

Laser plasma drag reduction has been investigated in some countries, and mainly concentrated in Russia, the United States and so on. Myrabo, et al., had experimental and computational investigation about the drag of blunt body in different inject energy[2]; David Riggins, et al., investigated blunt body drag reduction using focused energy deposition numerically[3]; Meiliang Mao, et al., had study the relation between positions of laser energy injection and the drag the blunt body suffered by solve viscous Navier-Stokes equations numerically[4].

Using laser plasma to reduce the drag of aircraft is an efficiency method, but related research is not extensive developed in our country, especially for drag reduction performance optimization. The inject power of laser is an important parameter among performance optimization. When laser energy

Published under licence by IOP Publishing Ltd

injected, we need to choose appropriate power, and expect to achieve best drag reduction performance by inject minimum energy. Thereby, base on previous work by others, this paper numerical simulated the effect of laser energy injection to the blunt body drag when *Ma*=6.5, the relation between injected laser power and laser plasma drag reduction performance has been researched. Through comparing the results, we can find that higher injection laser power made the blunt body suffered less resistance. But, if the power is too high, the efficiency of the laser energy significantly dropped. We expect that this work can provide assistance for the future experimental study to choose appropriate inject laser power.

2. Computation Methodology and Parameter Definition

2.1. Parameter Definition

Two parameters can be defined to provide insight into the usefulness of the energy deposition concept[5]. The first one is *R*, which is the normalized drag; the second one is *S*, which is power effectiveness.

$$R = \frac{D_{\mathrm{mod}}}{D_{\mathrm{ref}}} \quad (1)$$

$$S = \frac{(D_{\mathrm{ref}} - D_{\mathrm{mod}})V_{\infty}}{\dot{Q}} = \frac{D_{\mathrm{ref}}V_{\infty}}{\dot{Q}}(1-R) \quad (2)$$

In these expressions, $\dot{Q}$ is the injected power, V_{∞} is the velocity of the incoming flow. *R* is D_{ref} divided by D_{mod}, which means the actual drag obtained with the flowfield modification divided by the drag of the body without any flowfield modification (the reference case). *S* is the ratio of the propulsive power savings due to flowfield modification divided by the power required to modify the flow, or the energy deposition rate.

2.2. Computation Methodology

Aerodynamic was described by Euler gasdynamic equations and viscosity was not considered. The gas equation of state was under the air model of high-temperature equilibrium gas[6]. The equations were solved by using a Roe scheme of second order space accuracy and a MUSCL approach.

The energy is deposited upstream of the blunt body on the centerline axis. Assume that Laser energy deposited instantaneously, laser focus in the fire region, a high temperature and pressure region was formed. Laser energy was equal distributed to each computation grid. When the injected energy in the fire region grids exceeds the breakdown threshold of air, the air in this region was breakdown, and laser plasma generated.

2.3. Computation Condition

The blunt body used for computation is composed by a hemispherical nose which diameter is *D*=10cm and a cylinder which diameter is 10cm. The solutions use a 400×300 grid. The computation configuration and boundary conditions was show in Figure 1. "Inflow" is the supersonic inflow boundary condition; "Outflow" is the supersonic outflow boundary condition. Incoming flow comes from left side of x-axis to the right side. Fire region was in the upstream of the blunt body, the distance between blunt body and center of the fire region is 2*D*. The fire region is rectangular, which length is 0.1*D*, width is 0.05*D*. Because the computation model is axisymmetrical, the fire region is a cylindrical region essentially.

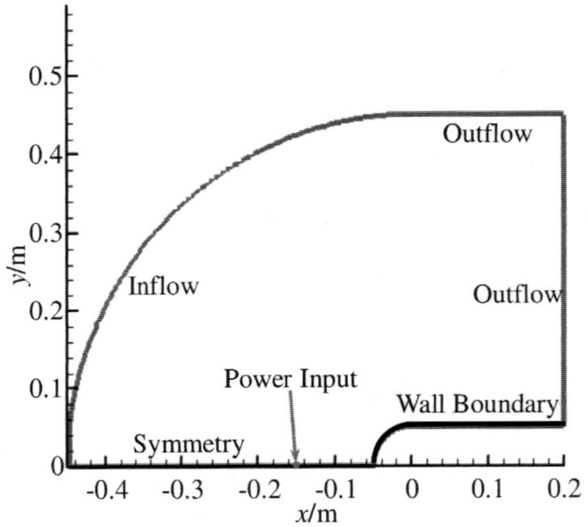

Figure 1. Computation configuration and boundary conditions.

The periphery conditions of the blunt body correspond to a standard altitude of 30km. In this condition, the air density is 1.841×10^{-2}kg/m^3, pressure is 1.197×10^3 Pa, sound velocity of unperturbed flow is 301.71m/s. The Mach number of incoming flow, *Ma*=6.5, the Reynolds number is 2.447×10^6.

3. Results and Discussion

3.1. Flowfield without laser energy injection

According to the shock wave theory, for each incoming Mach number *Ma*, there is an maximum flow turning angle α_{max}. Because the model is a blunt body, when the hypersonic flow near its surface, the turning angle approach to 90°, and this angle exceed α_{max}, the shock wave detach the blunt body surface. We could consider that the front of the detached bow shock as a normal shock. When the incoming flow penetrated normal shock, its velocity decelerated to subsonic, and the pressure increased sharply, the blunt body suffered great wave drag.

The computation result indicates that intensive bow shock will be formed ahead the nose of the blunt body, as shown in Figure 2. In order to obtain steady state field and stable drag of the blunt body, compute time was set to 250μs. The drag of blunt body suffered was about D_{ref}=247.17N. This drag was the reference drag. The distance between the bow shock and the surface of blunt body is about 0.0075m. The highest pressure, in the stagnation point, was about 6.50×10^4Pa, about 54 times of the environmental pressure.

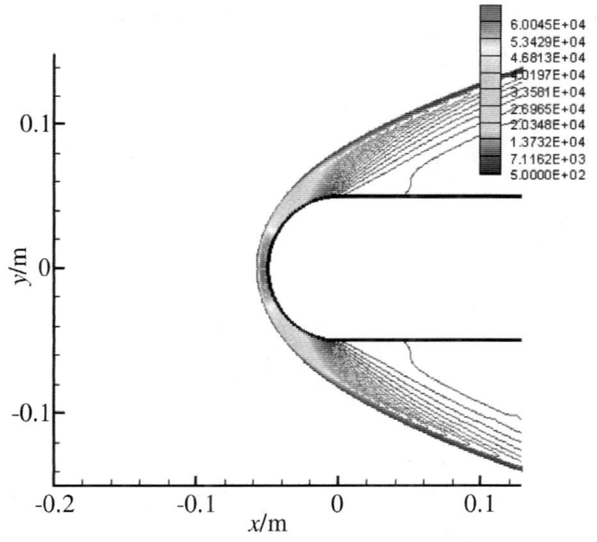

Figure 2. Pressure Contours of steady flow field at mach 6.5.

Lobb had investigated the relationship between the front face of the shock wave and the stagnation point distance in stable condition[7]:

$$\Delta = 0.41D\frac{(\gamma-1)M_\infty^2+2}{(\gamma+1)M_\infty^2} \tag{3}$$

in the equation, Δ is the distance between the front face of the shock wave and the stagnation point, D is diameter of craft, γ is specific heat ratio, M_∞ is Mach number of incoming flow.

According to the expression, substitute the computation conditions, we can obtain that $\Delta = 0.41\times 0.1\frac{(1.4-1)\times 6.5^2+2}{(1.4+1)\times 6.5^2} = 0.007642\text{m}$. Compare to this result, the numerical simulation result differ approximately 1.9%. So, it is feasible to simulate the aircraft wave drag in hypersonic stream by this program.

3.2. The effect of inject power to the drag of the blunt body

After the steady flow field build, the laser energy was input into the incoming stream ahead the blunt body. The computation compared the effect to the blunt body drag when inject energy power is 1kW, 2kW, 5kW, 10kW, 20kW, 50kW. Energy was continuous injected after steady flowfield has been established, that is, after 250μs. Compare to the condition when no energy injection, the drag of the blunt body suffered obvious decreased, the decreased level related to inject power.

Table 1 provides a comprehensive listing of 6 energy power performance with values *R* and *S*:

Table 1. Performance of different inject power.

P/kW	D_{mod}/N	*R*	*S*
1	207.94	0.841	76.93
2	202.80	0.820	43.51
5	188.70	0.763	22.93
10	153.71	0.622	18.33
20	102.66	0.415	14.17
50	67.90	0.275	7.03

If the inject energy power increase, the blunt body decreased. When inject power is 50kW, the drag of the blunt body suffered is 27.5% of the base drag. Figure 3 shows the normalized drag, power effectiveness when in different inject power.

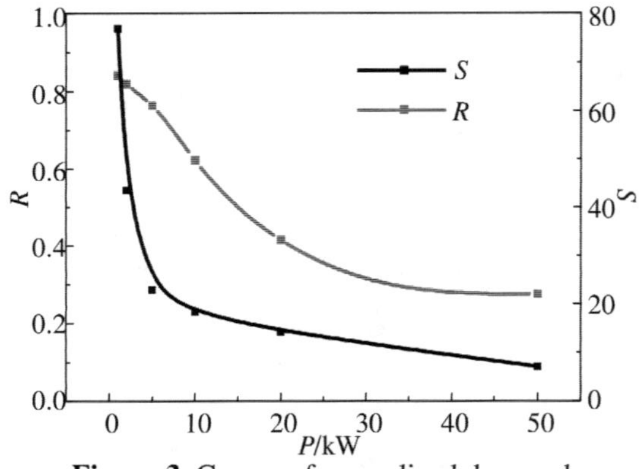

Figure 3. Curves of normalized drag and power effectiveness to inject power.

Fig 3 indicates that the drag of the blunt body dropped as the inject energy power increased; if the power is higher than 20kW, the tendency of drag decreasing is gently. As the inject power increase from 1kW to 10 kW, power effectiveness dropped sharply; if the power is higher than 10kW, power effectiveness dropping tend slowly. So, lower power is expected to get higher power effectiveness, but if the inject energy power is too low, the decrease of blunt body drag is inconspicuous.

3.3. Analysis of the computation result

The curves of drag vs. time under different inject energy power is shown is Figure 4.

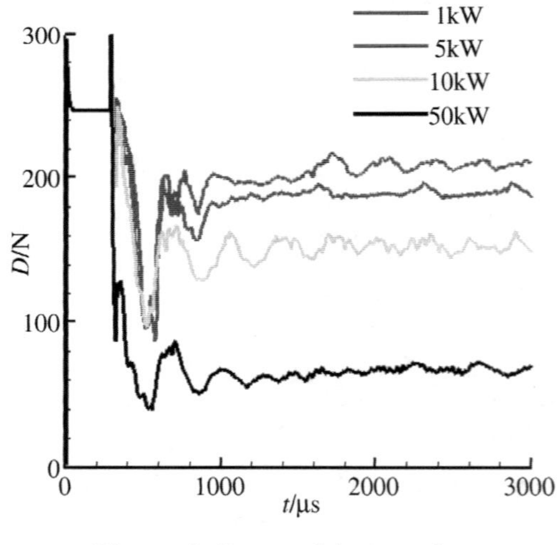

Figure 4. Curves of drag vs. time.

When no energy added, the drag of the blunt body kept stable (1~250μs). As energy begin to add in, an inverse impulse act on the blunt body surface, the drag increase at first, and then decrease rapidly. The decrease speed depends on the input power, if higher power is injected (50 kW for example), the drag drop almost vertically; if lower power is injected (10 kW for example), the decline of the drag is a little slowly. And if lower power is added in, the drag will reach a minimum value, and then resume to a higher level, for an example, the inject power is 1kW, the blunt body drag will get the minimum value at 577μs, and the value is 87.0N, then, it resume to 207.9N; if the inject power is 50kW, the drag continuous decrease, and will get the minimum value at 557μs, the minimum value is 39.7N, and then reach a stable drag near 67.9N. So, if low power is injected, a minimum drag will be reached, but it can't maintain a better drag reduction performance.

The flowfield of the blunt body periphery under 5kW injected power is shown in Figure 5. In the laser energy focus region, laser supported denotation wave (LSDW) is generated, as shown in Figure 5 a. The LSDW propagate to the surface of blunt body surface, and then penetrate the bow shock, an inverse thrust acted on the blunt body, the drag increase. Then, because the high temperature and low density region of the denotation wave interior, the pressure of the stagnation point decrease, the top of the bow shock move ahead, until the structure of the bow shock is destroyed (Figure 5 b). At last, the flowfield of the blunt body periphery get stable again, and a structure like oblique shock is formed (Figure 5 c).

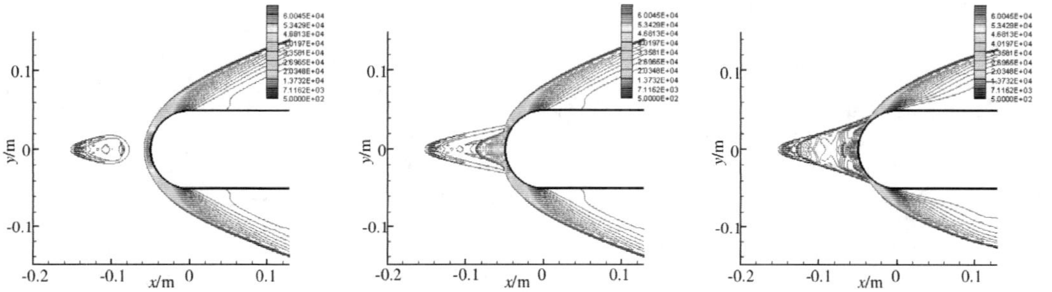

(a) 280μs (b) 320μs (c) 700μs

Figure 5. Pressure contours of flow field in different times.

Figure 6 shows the stable flow field under different inject power. The laser focus region is like a small "wedge" in the flow field. The high temperature and low density region formed by the LSDW cause the top of the bow shock move ahead, and transform to oblique shock (main shock). A triangle front separation region formed in the oblique shock, its top is behind the fire region. The velocity direction of incoming stream which far from the axis is changed, but the incoming stream which near the axis will penetrate the oblique shock into the separation region, and a second shock will be formed in the oblique shock (secondary shock). When the inject power increase, the main shock get more intensive, the front separation region get larger, and the secondary shock weakened.

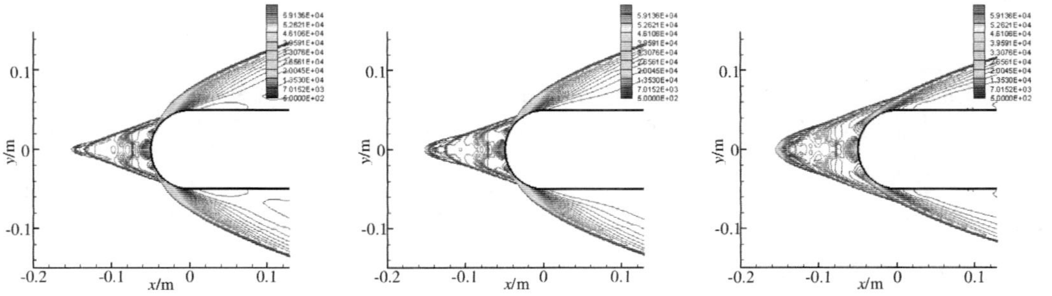

(a) 1kW (b) 5kW (c) 50kW

Figure 6. Pressure contours of flow field under different inject power.

The pressure and temperature in axis and blunt body surface under different inject power is shown in Figure 7.

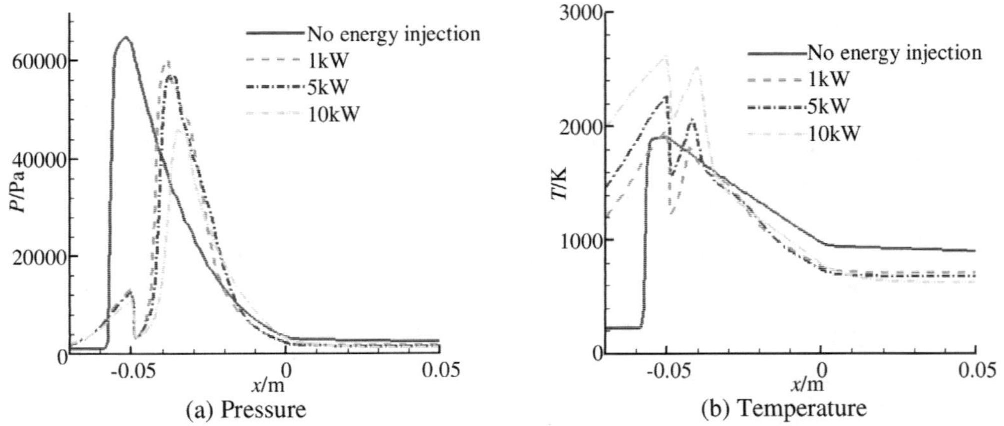

Figure 7. Pressure and temperature in axis and blunt body surface under different inject power.

When laser energy add in, the high pressure region on the blunt body surface move toward back of the hemisphere (Fig 5, Fig 6), the pressure of stagnation point decrease, from greater than 60kPa drop to near 10kPa. The highest pressure in the blunt body surface decreased as the inject energy power increased. In the back part of the blunt body, the influence of the inject power to surface pressure is much less than in the nose of the blunt body. There are two peaks of the temperature of body surface, one peak is the temperature of the stagnation point, higher than no energy added in, and increased as the inject power increased; the other peak is the region which pressure is the highest in the surface. The surface temperature drop rapidly after the second temperature peak and it is lower than no energy added in.

Myrabo et al. give a formula of the optimization laser energy power when radius of the LSDW is constant[8]:

$$P = 2(\gamma + 1)\alpha^{-4} p c_0 R^2 Ma \qquad (4)$$

In the equation, γ is specific heat ratio, α is numerical coefficient, α=0.937 for γ=4/3, p is pressure, c_0 is local sound velocity, R is radius of the denotation wave, Ma is Mach number of the incoming flow. From Figure 5 we can see that when the denotation wave propagates to the blunt body surface, its radius is about 0.035m. From the formula, the theoretical optimization power is about 15kW. From Figure 3 we can see that when inject laser power is 15kW, the normalized drag is small and the power effectiveness can hold in about 15.

4. Conclusions

The drag of the blunt body obvious dropped by injecting energy into the hypersonic incoming flow. This paper investigates the variation of the blunt body drag under different injected laser power by numerical simulation, and summarize as follows:

- Higher inject power can get better drag reduction effectiveness, when injected power is 50kW, the drag of the blunt body drag is 27.5% of the base drag;
- Less injected power can get higher power effectiveness, when injected power is 1kW, the power effectiveness reach 76.93;
- When the hypersonic incoming inflow penetrates the oblique shock, the pressure and temperature of the blunt body surface changed obviously. High pressure region moves towards the back of the blunt body nose, and the highest pressure decrease as the injected energy power increase. The temperature of the stagnation point will be higher compare to no energy add in, but the temperature behind the high pressure is lower than the temperature when no energy add in.

So, in the experimental investigation, the inject power should be appropriate chosen between normalized drag and power effectiveness to expect use least injected energy to save most propellant. At the same time, the variety of the pressure and temperature should also be considered.

References

[1] Myrabo L N 1978 Solar-Powered Global Air Transportation *AIAA Paper* No.78-0689
[2] Bracken R M, Hartley C S, Mann G, Myrabo L N, Nagamatsu H T, Shneider M N and Raizer Y P 2003 *Proc. of 1st Int. Symp. on Beamed Energy Propulsion*, 2003 485-96
[3] Riggins D W, Nelson H F and Johnson E 1999 *AIAA Journal* **37** 460-7
[4] Mao M L, Dong W Z and Deng X G 2001 *ACTA Aerodynamica Sinica* **19**(2) 172-6
[5] Riggins D W, Barnett J T and Taylor T 2003 *12th AIAA Int. Space Planes and Hypersonic Systems and Technologies (Norfolk Virginia 15-19 December 2003)*
[6] Li Q, Hong Y J and Cao Z R 2006 *Explosion and Shock Waves* **26**(6) 550-5
[7] Lobb R K 1964 *The High Temperature Aspects of Hypersonic Flow* 519-527
[8] Myrabo L N and Raizer Y P 1994 *25th AIAA Plasmadynamics and lasers Conf. (Colorado 20-23 June 1994)*

Measurement of the angular distributions of high-order harmonic generations from aligned CO_2

H Lu, S T Zhao, Z X Zhang, P Liu, Z N Zeng, R X Li, and Z Z Xu
Shanghai Institute of Optics and Fine Mechanics, Chinese Academy of Sciences, No. 390, Qinghe Road, Jiading District, Shanghai 201800, China

E-mail: peng@siom.ac.cn， ruxinli@mail.shcnc.ac.cn

Abstract: In this study, the angular distributions of harmonics emission from aligned CO_2 are explored experimentally and theoretically, and the validity of Strong Field Approximation (SFA) model in the molecular high harmonic generation is therefore studied. The study shows that for describing the angle distribution of high harmonic generation from molecules, SFA is roughly consistent with the qualitative analysis, while the quantitative analysis is different.

High-order harmonic generation (HHG) from molecules attracts much attention due to the increased freedom of molecules that reveals a rich set of phenomena[1-6]. In the well understood three-step process[7-9], in which the tunneling ionized electrons are accelerated by the strong laser field and then return to combine with the parent ions to eject high energy photons, the ionization and recombination steps are all dependent on the symmetry of the highest occupied molecular orbital (HOMO) of molecules[10,11]. Consequently, the harmonic emission from molecules irradiated by strong laser field shows angle dependence between the molecular axis and laser polarization direction[12]. This feature provides a means to shape the characteristics of harmonic emissions. The recent measurement of polarization state of harmonic emission from N_2, O_2 and CO_2 molecules shows a strong dependence on the molecular alignment and the orbital structures[13].

Moreover, the HHG from molecules shows potential application in probing molecular dynamics. From an ensemble of aligned nitrogen molecules, the harmonic intensities as a function of the angle θ between molecular axis and laser polarization is able to be utilized to rebuild the tomographic image of the highest occupied molecular orbital (HOMO) [14]. The recent measurement of high harmonic generation from aligned organic molecules[15] shows the imaging method can also extend to more complex molecules based on the validity of strong field approximation (SFA). This shows the potential application of harmonic generation process in extracting the molecular structure information from the angle distribution of harmonic emissions.

The structural retrieval is also possible by examining the quantum interference signatures in spectrum measurement. The CO_2 molecule provides good conditions for observing the quantum interference in harmonic generation, due to its characteristic linear structure. Its highest occupied molecular orbital (HOMO) is an antisymmetric π_g orbital dominated by the combination of the p_y orbitals of the two oxygen atoms, and the O - O distance (R_{O-O} = 0.232 nm) corresponds to an electron wavelength typical of the harmonic generation process. According to the two-center interference model[16,17], destructive interference appears as $R\cos\theta = n\lambda_e$, where R is the distance of the two ionization sites that reveals the bond length of molecule, θ is the angle of molecular axis and electron momentum, and $n\lambda_e$ is the integer times of electron de Broglie wavelength. Since the molecules are partially aligned in experiments and harmonic emission from molecules are θ dependent, one needs take into account not only the alignment distribution of molecules, P(θ), but also the harmonic intensity distributions, S(θ), in order to investigate the quantum interference effect.

Therefore, the angle dependence of harmonic emission - S(θ) - plays a key role in the study of probing molecular structures and shaping high harmonic emissions. Accurate measurement of the angle distributions of harmonics and theoretical apprehension are critical steps towards the tomographic imaging technique for larger molecules such as CO_2. So far the angle dependence for CO_2 molecule has not been measured directly yet and the comparison of SFA calculation with experimental observations is needed.

In this study we investigate the angular distributions of harmonics produced for aligned CO_2 molecules experimentally and theoretically. The calculated result by SFA agrees the experimental results qualitatively. From numerical simulations, the angular profiles of the harmonic emission have been discussed.

Due to the limit of partially aligned molecules, the angle distribution S(θ) in molecular frame can not be measured directly. Experimentally we can only measure the harmonic intensities as a function of the angle Θ between pump laser polarization and probe laser polarization in laboratory frame. Our experiment was performed using a Ti:sapphire based chirped pulse amplification laser system (Spectral-Physics, TSA-25), which produce 50 fs laser pulses of 10 Hz at 800 nm center wavelength. The output is split into two beams, one used as the pump pulse (for aligning molecules) and the other as the probe pulse (for driving HHG from molecules). The polarization of the pump laser beam is fixed as p polarization state, while the probe beam polarization is adjusted by a zero-order air filled wave plate from s to p state. The two beams are combined by a normal incidence beam splitter in the angle of < 3° and then were collinearly focused (with a 30 cm focal length lens) onto a pulsed supersonic molecular beam located in a high vacuum interaction chamber. The laser focus was about 1 mm downstream of a 0.25 mm diameter nozzle orifice. Stagnation pressure of CO_2 gas (99.998%) was 2 bars. Through the calculation to a rotational temperature of several tens of Kelvin. The HHG spectra were detected by a home-made flat-field grating spectrometer equipped with a soft-x-ray CCD camera (Princeton Instruments, PI: SX 400).

First the CO_2 molecules are first aligned in an optimized condition. The highest alignment degree $< \cos^2\theta >$ is normally improved by increasing the pump laser pulse energy, but the applied laser pulse energy is limited by the ionization due to increased laser intensity. In experiments we tune the laser field just below the value that produces detectable high harmonics. The field intensity is estimated to be 4×10^{13} W/cm^2. The intensity of the 23th harmonic as a function of the delay time after pump pulse is shown in figure 1(a) (solid line). As can be seen the alignment revivals are achieved at the full, half, quarter of molecular rotational period. Under the experimental conditions, the pump laser pulse duration (τ_{on} = 50 fs) is much shorter than the molecular rotational period, $\tau_{on} << 2\pi/\omega = 42.7ps$ for CO_2, nonadiabatic field-free alignment is achieved by the excitation of a rotational wavepacket $\psi(t) = \sum_{J,M} A_{J,M}(t)|J,M\rangle$ [18-20]. The time evolution of the wavepacket can be calculated through solving the time-dependent Schrödinger equation (TDSE). The time dependent alignment parameter defined as $< \cos^2\theta > (t) = \langle\psi(t)|\cos^2\theta|\psi(t)\rangle$ is calculated shown in figure 1(a) (dashed line, right axis). In our calculation the initial rotational temperature is taken to be 80 K estimated based jet kinetic models. It is evident that the modulation of harmonic signal is reversely matched with that of the molecular alignment parameters $< \cos^2\theta >$ exactly. This is consistent with the experimental results of Kanai et al. [11] and Vozzi et al. [21] that reveals quantum interference effect.

At the delay time t = 21.1 ps where CO_2 are aligned at the half revival, harmonic spectra are recorded by tuning the angle Θ between pump laser polarization and probe laser polarization from 0°

to 90°. Harmonic spectra are also taken by only introducing probe laser pulse, which generates the harmonic emission from isotropic molecules. The ratios of harmonic intensities (the 19th, 23rd, and 27th harmonic) from aligned and isotropic molecules are plotted in figure 1(b). It is seen that for all the observed harmonics, the normalized intensities all increase from a value lower than 1 at $\Theta = 0°$ to larger than 1 at $\Theta = 90°$. This is the inversion of the harmonic emission resulted from quantum interference effect, which is consistent with the previous report. It is also noted that the slopes of the increasing harmonic intensity is larger when the harmonic order is larger, for the 19th harmonic emission, the normalized intensity increase from 0.65 to 1.1, while for the 27th harmonics, the intensity increase from 0.3 to 1.4. Since the molecular alignment distribution is fixed, this indicates the variation of angle dependence of harmonic emission $S(\theta)$ for the observed harmonic orders.

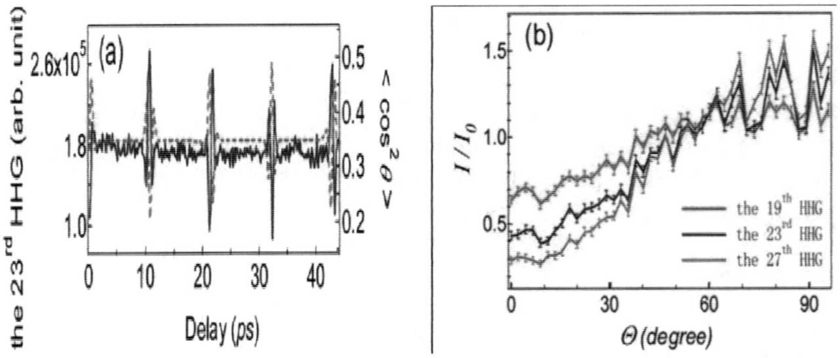

Figure 1. (a) Measured 23rd harmonic yield (solid line, left axis) and calculated alignment parameter $< \cos^2\theta >$ (dashed line, right axis) as functions of the pump-probe delay time; (b) the ratio of harmonic intensities (19th, 23th, 27th harmonic) from aligned and isotropic molecules.

The measured angular dependence in laboratory frame is related to the angular dependence of harmonic generation by

$$I_n(\Theta) = \int_{\omega_{n-1/2}}^{\omega_{n+1/2}} d\omega \left| \int \tilde{\mathbf{R}} P(\theta) \cdot \tilde{\mathbf{R}} S(\theta) \cdot d\theta \right|^2 \tag{1}$$

where $P(\theta)$ is the alignment distribution of the molecular axis in molecular frame, $S(\theta)$ is the amplitude and phase part of the harmonic emission, and $\tilde{\mathbf{R}}$ is the rotation operator that transform the distribution function in molecular frame to laboratory frame, which includes a rotation of polar angle θ and a integral of azimuthal angle χ. To simulate the measured result, $P(\theta)$ and $S(\theta)$ need be evaluated, then the harmonic intensities are calculated from Eq. (1) and finally the intensities are normalized by calculated harmonic emission from an isotropic ensemble of molecules.

In order to simulate the experimental results, we use the Lewenstein's strong field approximation model [22] to calculate the angular distribtution $S(\theta)$. In the model, the harmonic spectrum from molecular dipole moment in the time domain is calculated by using

$$\begin{aligned} \boldsymbol{x}(t) = i\int_0^{\infty} d\tau \left(\frac{\pi}{\in + i\tau/2}\right)^{3/2} \boldsymbol{d}^*[\boldsymbol{p}_{st} - \boldsymbol{A}(t)] e^{-iS(\boldsymbol{p}_{st},t,\tau)} \\ \times E(t-\tau)\cdot \boldsymbol{d}[\boldsymbol{p}_{st} - \boldsymbol{A}(t-\tau)] a^*(t) a(t-\tau) + c.c. \end{aligned} \tag{2}$$

where $\boldsymbol{d}^*[\boldsymbol{p}_{st} - \boldsymbol{A}(t)]$ and $\boldsymbol{d}[\boldsymbol{p}_{st} - \boldsymbol{A}(t-\tau)]$ are the transition dipole moments between the ground state and the continuum state, $\boldsymbol{p}_{st} = \int_{t-\tau}^{t} A(t')dt'/\tau$ is the canonical momentum at the stationary points with vector potential A, and the exponential term takes account of the action of the electron trajectories within the laser field. We consider the depletion of ground state molecules by a factor of $a(t)$, which is approximated by $a(t) = \exp[-\int_{-\infty}^{t} dt' W(t')/2]$ [23], and W(t') is the tunneling ionization rate obtained by MO-ADK theory [24].

To calculate the transition dipole moments the ground state of CO_2 is expressed by atomic $2p_y$ orbital $\Phi_{2p_{y'}}$ as [25-27]

$$\Psi_{\pi_g}(\boldsymbol{x}) \propto [\Phi_{2p_{y'}}(\boldsymbol{x}+\boldsymbol{R}/2) - \Phi_{2p_{y'}}(\boldsymbol{x}-\boldsymbol{R}/2)], \quad (3)$$

where R is the distance between oxygen atoms, and the $2p_{y'}$ orbital was expressed by $\sim \exp(-\alpha|x|)$ type basis function. The time dependent dipole transition moment is given by

$$\boldsymbol{d}_{\pi_g}(\boldsymbol{p}) \propto [2i\sin(\boldsymbol{p}\cdot\boldsymbol{R}/2) d_{2p_{y'}}(\boldsymbol{p}) - \cos(\boldsymbol{p}\cdot\boldsymbol{R}/2)\widetilde{\Phi}_{2p_{y'}}(\boldsymbol{p})\boldsymbol{R}], \quad (4)$$

in which the $d_{2p_{y'}}(\boldsymbol{p})$ is the atomic dipole moment from the $2p_{y'}$ orbital and $\widetilde{\Phi}_{2p_{y'}}(\boldsymbol{p})$ is the $2p_{y'}$ wavefunction in the momentum space. Finally the harmonic spectra as a function of the angle between molecular axis and laser polarization, $S(\theta)$, are obtained through the Fourier transform of the transition dipole moment.

The time-dependent alignment distribution of molecules $P(\theta, t = 21.1ps)$ is calculated from the $\psi(t)$ by solving the TDSE. The Θ dependence of harmonic intensity is then given by integrating the product of $S(\theta)$ and $P(\theta)$ as Eq. (1).

The calculated Θ dependence of 19th to 27th harmonics is plotted in figure 2(a). It can be seen that, the calculated harmonic intensities represent the same major features as observed in experiments. From $\Theta = 0°$ to $\Theta = 90°$ the intensities all increases from a value less than 1 to one larger, and the 27th harmonic emission has the largest increase while the 19th harmonic emission has the lest increment. The consistence shows the calculation of angle distribution $S(\theta)$ by SFA agrees with the observation generally.

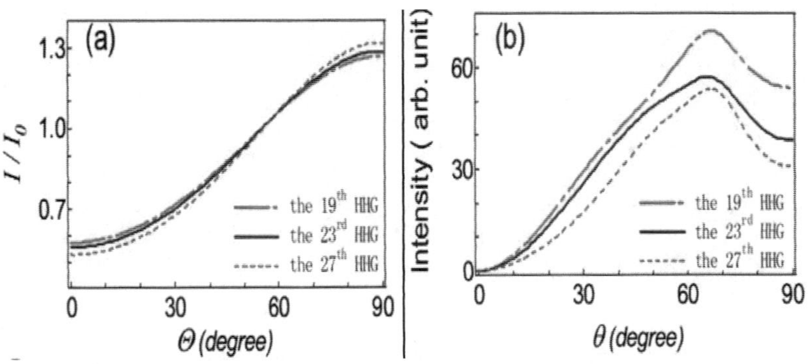

Figure 2. (a) The calculated Θ dependence of the normalized harmonic intensities for 19th, 23rd and 27th harmonics; (b) the calculated angle distribution $S(\theta)$ of 19th, 23rd and 27th harmonics.

In figure 2(b) the calculated angle distribution $S(\theta)$ of 19th, 23rd, and 27th harmonics are shown. Evidently the 19th harmonic emission has a broader angle distribution than the others and it is the different angle distributions that result in the change of Θ dependence for harmonic orders.

It is also noted that in figure 2(a), the values of normalized harmonic intensities does not match with the experimental results, the 19th, 23rd and 27th harmonic emission are all between 0.5 – 0.6 for Θ

= 0°, and they increase to 1.2 – 1.3 for Θ = 90°. One possibility is the uncertainty of estimated rotational temperature. To clarify the influence of temperature, we calculate the Θ dependence of harmonic intensities for an ensemble of aligned molecules at 120K. The normalized intensities are all within 0.6 – 0.7 for Θ = 0°, and increases to around 1.2 for Θ = 90°. This means the rotational temperature of molecules is not a critical influence of the inconsistence of experimental and calculated results.

To further discuss how the angle distribution $S(\theta)$ influences the observed Θ distribution of harmonic intensities, we carry out the following simulations. We assume the profile of $S(\theta)$ as a Gaussian function that is centered at 66° (about the peak positions calculated by SFA), then calculate the Θ distributions by set the full width at the half maximum (FWHM) of the function from 10° to 50° with a step size of 10°. The result is shown in figure 3(a), which shows that at Θ = 0° the harmonic intensity decreases to 0.5 by using a FWHM of 10°. In our second simulation, we choose the FWHM of Gaussian function as 20° and change the peak positions of $S(\theta)$ from 50° to 90° with a step size of 10°. The calculated Θ distributions of harmonic intensities are plotted in figure 3(b). As can be seen, when the peak position of $S(\theta)$ in molecular frame shift from 60° to 90°, the resulted harmonic emission changes from 0.6 to 0.3 for Θ = 0°. This shows that, the peak position and width of angle distribution $S(\theta)$ both significantly influence the Θ distributions of harmonic intensities in laboratory frame. By the peak position located at near 90°, the harmonic emission decreases to be about 0.3 for Θ = 0°, as observed in experiments for the 27th harmonics.

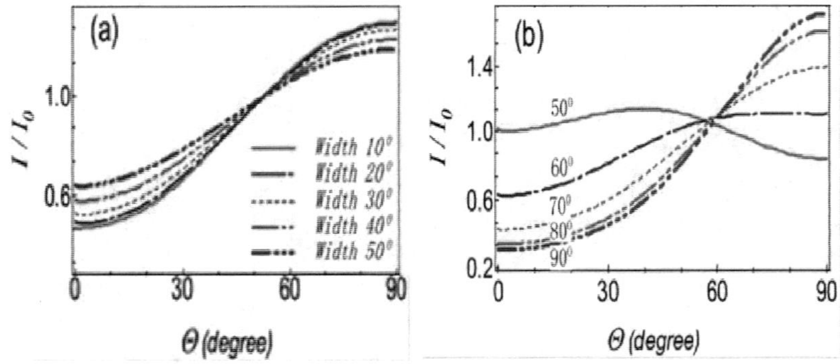

Figure 3. (a) The dependence of the normalized harmonic intensities for 19th, 23rd and 27th harmonics on the harmonic angle distribution $S(\theta)$: (a) dependence on the profile width; (b) dependence on the profile center.

The above analysis indicates that SFA is roughly consistent with the qualitative analysis but the quantitative analysis is different. The recent progress in tomographic imaging of molecular orbitals[28,29] shows that besides HOMO contribution to HHG, the lower bounds HOMO-n may also play a part in the process, which may lead to deviation of experimental results from calculation results from SFA. While this holds true for N_2, for CO_2, the conclusion is not settled yet.

In conclusion, we investigated experimentally the angular distributions of high harmonic generation from aligned CO2 molecules by tuning the angle between pump laser polarization and probe laser polarization from 0° to 90° while fixing the pump-probe delay time at the half revival of alignment. Although the calculated results by Strong Field Approximation (SFA) under the same conditions agree with the experimental results qualitatively, further discussion on the quantity difference between experimental results and calculation results indicate that modification is needed in the SFA model for accurate description of angle distribution $S(\theta)$.

This work is supported by Chinese Academy of Science, Chinese Ministry of Science and Technology, NSF of China (Grant Nos. 10734080, 10523003, 60921004, and 60978012), 973 Program of China (2006CB806000), and Shanghai Commission of Science and Technology (Nos. 06DZ22015 and 07pj14091).

References

[1] Dooley P W, Litvinyuk I V, Lee K F, Rayner D M, Spanner M, Villeneuve D M and Corkum P B 2003 *Phys. Rev. A* **68** 023406-1-12
[2] Xu N, Wu C Y, Huang J, Wu Z F, Liang QQ and Gong QH 2006 *Opt. Express* **14** 4992-7
[3] Ledingham K W D and Singhal R P 1997 *Int. J. Mass Spect. Ion Proc.* **163** 149-68
[4] Zavriyev A, Bucksbaum P H, Squier J and Saline F 1993 *Phys. Rev. Lett.* **70** 1077-80
[5] Velotta R, Hay N, Mason M B, Castillejo M and Marangos J P 2001 *Phys. Rev. Lett.* **87** 183901-1-4
[6] Miyazaki K, Kaku M, Miyaji G, Abdurrouf A and Faisal F H M 2005 *Phys. Rev. Lett.* **95** 243903-1-4
[7] Corkum, P B 1993 *Phys. Rev. Lett.* **71** 1994−7
[8] Krause J L, Schafer K J and Kulander K C 1992 *Phys. Rev. Lett.* 1992 **68** 3535−38
[9] Lewenstein M, Balcou P, Ivanov M Y, L'Huillier A and Corkum P B 1994 *Phys. Rev. A* **49** 2117−32
[10] Alnaser A S, Maharjan C, Tong X M, Ulrich B, Ranitovic P, Shan B, Chang Z, Lin C D, Cocke C L and Litvinyuk I V 2005 *Phys. Rev. A* **71** 031403-1-4
[11] Kanai T, Minemoto S and Sakai H 2005 *Nature* **435** 470−4
[12] Zhou X X, Tong X M, Zhao Z X and Lin C D 2005 *Phys. Rev. A* **71** 061801-1-4
[13] Mairesse Y, Levesque J, Dudovich N, Corkum P B and Villeneuve D M 2008 *Journal of Modern Optics* **55** 2591-602
[14] Itatani J, Levesque J, Zeidler, Niikura H, Pepin H, P B Corkum and Villeneuve D M 2004 *Nature (London)* **432** 867-71
[15] Torres R, Kajumba N, Underwood J G, Robinson J S, Baker S, Tisch J W G, de Nalda R, Bryan W A, Velotta R, Altucci C, Turcu I C E and Marangos J P 2007 *Phys. Rev. Lett.* **98** 203007-1-4
[16] Lein M, Hay N, Velotta R, Marangos J P and Knight P L 2002 *Phys. Rev. Lett.* **88** 183903-1-4
[17] Lein M, Hay N, Velotta R, Marangos J P and Knight P L 2002 *Phys. Rev. A* **66** 023805-1-6
[18] Stapelfeldt H and Seideman H 2003 *Rev. Mod. Phys.* **75** 543-57
[19] Miyazaki K, Kaku M, Miyaji G, Abdurrouf A and Faisal F H M 2005 *Phys. Rev. Lett.* **95** 243903-1-4
[20] Seideman T and Hamilton E 2005 Adv. At., *Mol., Opt. Phys.* **52** 289-329
[21] Vozzi C, Calegari F, Benedetti E, Caumes J P, Sansone G, Stagira S, Nisoli M, Torres R, Heesel E, Kajumba N, Marangos J P, Altucci C and Velotta R 2005 Phys. Rev. Lett. **95** 153902-1-4
[22] Lewenstein M, Balcou P, Ivanov M Y, L'Huillier A, and Corkum P B 1994 *Phys. Rev. A* **49** 2117-32
[23] Le A T, Tong X M and Lin C D 2006 *Phys. Rev. A* **73** 041402-1-4
[24] Tong X M, Zhao Z X, and Lin C D 2002 *Phys. Rev. A* **66** 033402-1-11
[25] Kanai T, Minemoto S and Sakai H 2007 *Phys. Rev. Lett.* **98** 053002-1-4
[26] Zhou X X, Tong X M, Zhao Z X and Lin C D 2005 *Phys. Rev. A* **72** 033412-1-4
[27] Zhou X X, Tong X M, Zhao Z X and Lin C D 2005 *Phys. Rev. A* **71** 061801-1-4
[28] McFarland B K, Farrell J P, Bucksbaum P H and Gühr M 2008 *Science* **322** 1232-5
[29] Smirnova O, Mairesse Y, Patchkovskii S, Dudovich N, Villeneuve D, Corkum P and Ivanov M Y 2009 *Nature* **460** 972-7

Journal of Physics: Conference Series **276** (2011) 012028 doi:10.1088/1742-6596/276/1/012028

Exploration of in vivo Effect Assessment Factor Monitoring by Near-infrared Spectroscopy during LITT

Ai-ping Qian[1], Guo-ran Hua[*,1], Hua Zhang[1] and Zhi-yu Qian[2]

1 School of Mechanical Engineering, Nantong University, Nantong, Jiangsu 226019, China

2 Department of Biomedical Engineering, Nanjing University of Aeronautics and Astronautics, Nanjing, Jiangsu 210016, China

E-mail: huagr@ntu.edu.cn

Abstract: By studying the variation trends of the absorption coefficient (μ_a) and the reduced scattering coefficient (μ'_s), which were monitored in vivo by functional near infrared spectroscopy (fNIRS) system in real time during laser induced interstitial thermotherapy (LITT), the optimized near infrared effect assessment factor would be explored. In vivo measurements of the absorption coefficient (u_a) and the reduced scattering coefficient (u'_s) were performed with a functional near infrared spectroscopy system during LITT. Fresh porcine liver tissue samples in vitro and the subcutaneous implanted rat liver cancers were examined in different laser doses and define heating times. The absorption coefficient obtained by the fNIRS increased in the pork liver experiments, but decreased in the rat liver cancer experiments. The reduced scattering coefficient increased in the pork liver experiments and the rat liver cancer experiments, it increased quickly at beginning, and gradually reached the stable state. Therefore, the reduced scattering coefficient is more suitable for reflecting the progress of damage during different biological tissues' LITT than the absorption coefficient. This conclusion will effectively guide the study of suitable therapy effect assessment system during LITT in real time.

1. Introduction

LITT is a minimally invasive technique involving the placement of laser fiber into the tumors, followed by thermal ablation of targeted lesions by the application of laser energy. Since its first description by Bown [1] in 1983, LITT has developed into a promising local procedure for the treatment of liver tumors. During the past two decades, basic theoretical research and clinical application of LITT have been developed quickly [2]. But the therapy effect assessment in real time during LITT is still a technical problem. The current conventional practice [2-4] is local temperature measurement by using thermocouple, thermal resistor or fiber optic thermometer, another way is to get dynamic temperature by reconstruction method through MRI (magnetic resonance imaging),

* Corresponding author. Tel: +13776918718.

E-mail addresses: huagr@ntu.edu.cn (Guo-ran Hua)

CT(computer tomography), US(ultrasound) and other imaging equipment. But these methods are difficult to popularize, due to the usual tissue infection, or technical difficulties and expensive equipments [5]. Previous experiments [6-13] have proved that the optical properties of biological tissue are affected significantly under the influence of thermal coagulation. And tissue damage degree is reflected through the changes of the optical parameters. The aim of the present study was to analyse the variety trends of the absorption coefficient and the reduced scattering coefficient on fresh porcine liver in vitro and the subcutaneous implanted rat liver cancers during LITT. Discover of these changes will play an important role in on line monitoring during LITT by using the near infrared optical methods.

2. Experimental procedures

2.1. Experimental Set-up

The experiment system was consisted of LITT thermotherapy system and fNIRS monitoring system (Figure 1). LITT thermotherapy system includes: laser (808nm, NL-FC-2.0-763 Laser Light,); optical fiber (built-in three 200μm fibers, one is for laser input, the others are for near infrared light input and output); thermoscope (WSY-4T, diameter 300μm). fNIRS monitoring system includes: USB2000 spectrometer (Ocean Optics) and halogen light source (Mikropack, HL-2000) and monitoring software. fNIRS monitoring software was developed by department of biomedical engineering of Nanjing University of Aeronautics and Astronautics based on Labview7.1 platform, which could real-time record tissue optical properties, blood oxygen parameters and flow parameters in front of the probe [14-15].

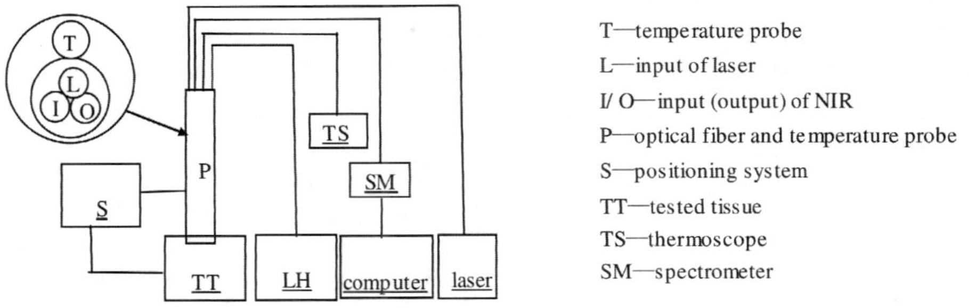

Figure 1. Illustrative diagram of experiment system.

2.2. Preparation and measurements

2.2.1 Pork liver experiments in vitro

Fresh porcine liver tissue in vitro was used for the experiment. The liver tissue would be cut into five tissue specimens (3*3*2.5cm) under room temperature (25•). After put the sample on the experiment platform horizontally, insert the optical fiber and temperature needle vertically into liver tissue about 1cm, adjust their position and fix them. Turn on the laser power switch, preheat for 15 minutes and then begin the experiment. The optical parameters (the absorption coefficient and the reduced scattering coefficient) were recorded in real time.

2.2.2. Rat liver cancer experiments in vivo

Twelve subcutaneous implanted H22 rat cancer models were used for the experiments. Establishing the subcutaneous implanted cancer model: Cultivate the tumor cell H22 of mouse in vitro, and make out solo cell suspension with phosphate buffered solution after centrifugalization, which is about 107/ml. Then take the injection (about 0.2ml) into the mice at the right axillary fossa. After three weeks, LITT will be performed when the subcutaneous tumors have grown to a diameter about 1.5cm. Fix the rat on the experiment platform after it was dopey, and expose the subcutaneous tumor. Then insert the optical fiber and temperature needle vertically into the tumor, adjust the their position and fix them

3. Results

3.1. Pork liver experiments in vitro

Groupings of pork liver were shown in table 1, which was the combination of four laser doses (0.76W, 0.95W, 1.23W, 1.42W) and three heating times (450s, 600s, 750s). Each group would be tested five times and test data of each group was given as mean value of the 5 measurement values. In order to ensure an accurate value, the position of optical fiber and temperature noodle should be changed for the next thermotherapy on one specimen after one test was end. The continuous changes in the absorption coefficient and the reduced scattering coefficient of porcine liver in vitro were obtained during LITT (Figure 2).

3.2. Rat liver cancer experiments in vivo

Groupings of rat model were shown in table 2. Each rat would be tested four times in different positions, test data of each tumor was given as mean value of the 4 measurement values. The continuous changes in the absorption coefficient and the reduced scattering coefficient of rat liver cancer in vivo were obtained during LITT (Figure 3).

Table 1: Grouping for pork liver LITT experiment.
(Power: output power from fiber cable)

Laser power(W)	0.76	0.95	1.23	1.42
Heating time(s)	450s 600s 750s	450s 600s 750s	450s 600s 750s	450s 600s 750s

Table 2: Rat grouping in LITT experiment.

Laser power(W)	1.23(2)	1.42(2)	1.88(4)	2.1(4)
Heating time(s)	600(1) 800(1)	600(1) 800(1)	600(2) 800(2)	600(2) 800(2)

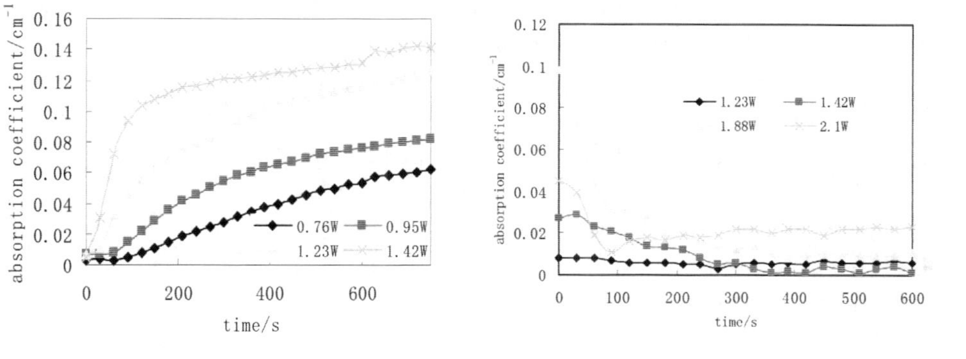

(a) (b)

Figure 2. Continuous changes in the absorption coefficient during LITT.
(a) pork liver experiments in vitro. (b)rat liver cancer experiments in vivo.

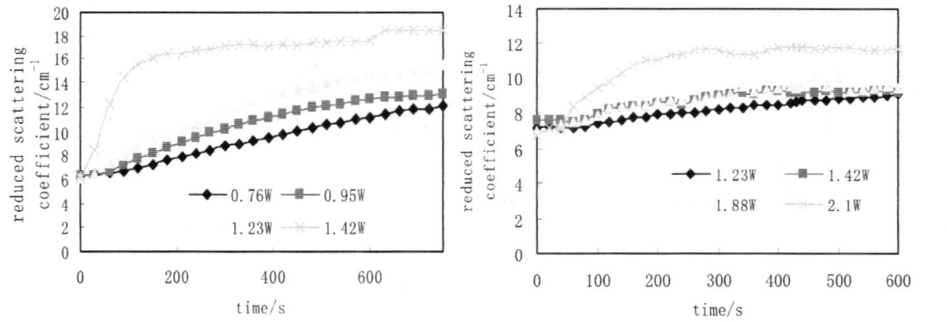

(a) (b)

Figure 3. Continuous changes in the reduced scattering coefficient during LITT.
(a) pork liver experiments in vitro. (b)rat liver cancer experiments in vivo.

4. Discussion

LITT is a promising therapeutic option for treating liver tumors. But the suitable on line monitoring procedure is still a key technique problem. Thus, it is necessary to have precise knowledge about optical property changes during LITT, which delivers information about degree of damage in tissue during therapy [16]. The thermal coagulation process of biological tissue is mainly due to protein thermal denaturation, which causes the changes of optical properties of biological tissue [17]. Existing studies [8-13] have shown that the scattering coefficient changes greatly with the cumulative damage, usually 2-4 times than before coagulation, however, in contrast to the scattering coefficient, the absorption changes slightly. And differences of t biological organization components led to different changes of their optical properties before and after coagulation progress. The goal of this study was to focus on that these two parameters, which are more suitable used for reflecting the progress of damage during different tissues' LITT.

The increased changes in the absorption coefficient of porcine liver in vitro were shown in Figure 2a..

The initial value of porcine liver absorption coefficient was about $0.01cm^{-1}$. Then μ_a continuously rose with increasing coagulation temperature. μ_a increased fast at the beginning, and than rose slowly. μ_a rose faster when the laser power was greater. The increased changes in the absorption coefficient of rat cancer in vivo were shown in Figure 2b. In contrast to the porcine liver experiment, the continuous changes in the absorption coefficient of rat cancer decreased during LITT. The initial values of rat tumors were different, due to the different component of tumor samples.

Figure 3a shows the increased changes in the reduced scattering coefficient of porcine liver in vitro. The initial value of μ'_s was $6cm^{-1}$. Then μ'_s continuously rose with increasing coagulation temperature. μ'_s increased fast at the beginning, and then rose slowly, finally it approached a stable state. μ'_s increased by 195% over the entire progress and by 165 % in the first 200s(1.42W). The upward tendency of μ'_s at different laser powers was similitude, but the rising velocities were different. μ'_s rose faster when the laser power was greater. The minimal final value of μ'_s was achieved at 0.95W about $12.6cm^{-1}$, and the maximal final value was achieved at 1.42W about $18.5cm^{-1}$.

Resemblance to the porcine liver in vitro, the reduced scattering coefficient of rat liver cancers in vivo increased during LITT (Figure 3b). The initial value of μ'_s was $7cm^{-1}$, and μ'_s continuously rose with increasing coagulation temperatures. It increased fast at the beginning, and then rose slowly, finally it approached a stable state. μ'_s increased by 73.6% over the entire progress and by 68.8% in the first 200s. The upward tendency of μ'_s at different laser powers was similitude, but μ'_s rose faster when the laser power was greater. The minimal final value of μ'_s was achieved at 1.23W about $8.9cm^{-1}$, and the maximal final value of μ'_s was achieved at 1.42W about $11.7cm^{-1}$.

5. Conclusion

The continuous changes of optical property(the absorption coefficient and the reduced scattering coefficient) on fresh porcine liver in vitro and the subcutaneous implanted rat liver cancers in vivo which got through the fNIRS system during the LITT have been analyzed in this paper. The absorption coefficient increased in the pork liver experiments, but decreased in the rat liver cancer experiments. The reduced scattering coefficient all increased in the pork liver experiments and the rat liver cancer experiments. It increased quickly at the beginning, and gradually approached a stable state. In summary, the reduced scattering coefficient is more suitable for reflecting the progress of damage during different biological tissues' LITT than the absorption coefficient. This conclusion will effectively guide the study of suitable therapy effect assessment system during LITT in real time.

6. Acknowledgments

The author would like to acknowledge the support provided by the Natural Science Foundation of Jiangsu Province (BK2008184).

References

[1] Bown S G 1983 *World J.Surg.* **7** 700
[2] Hans J S, Frank E, Volkhard U F and Frank U 2002 *Med.Laser Appl.* **17** 147
[3] Thomas J V, Thomas L and Katrin E 2007 *Eur Radiol* **17** 2020
[4] Jiang S C, Zhang X X 2005 *Lasers Med Sci.* **20** 122
[5] Liu J, Deng Z S 2008 Physics of Tumor Hyperthermia (Beijing:Scince Press)

[6] Ritz J P, Roggan A and Iabert C 2001 *Lasers Surg Med* **29** 205
[7] Germer C T, Roggan A and Ritz JP 1998 *Lasers Surg Med* **23** 194
[8] Zhu D, Luo Q M,Ceng J,Yu J S and Zeng S Q 2003 *Optics & Optoelectronic Technolog* **1** 51
[9] Zhu D, Luo Q M and Zhu G M 2002 *Lasers Surg. Med.* **31** 313
[10] Zhu D, Luo Q M and Zeng S Q 2002 *Chinese J. Lasers* **29** 667
[11] Wei H J, Xing D and He B H 2007 Spectroscopy and Spectral Analysis **27** 868
[12] Wei H J, Xing D and Wu G Y 2006 *Chinese J. Lasers* **33** 852
[13] Ao H L, Xing D and Wei H J 2008 *ChineseJ. Lasers* **35** 792
[14] Li R,Qian Z Y and Dai L J 2007 *Computer Measurement & Control* **15** 154
[15] Qian Z Y, Gu Y Q and Liu H L 2005 *Chinese J. Med. Phy.* **22** 463
[16] Ritz J P, Roggan A and Germer C 2001 *Lasers Surg Med* **28** 307
[17] Wei H J, Xing D and Xie S S 2006 *Spectroscopy and Spectral Analysis* **26** 1757

Prediction of drilling micro-hole in CO_2 laser irradiated sticking plaster

Zhiming Rao [1,2*], Tao Wu [3], Jianqiang Du [2] and Yanzhao Lu [1]
[1] Wuhan National Laboratory for Optoelectronic, the College of Optoelectronic Science and Engineering, Huazhong University of Science and Technology, Wuhan 430074, Hubei, PR China
[2] Depart of Computer Science, Jiangxi University of Traditional Chinese Medicine, Nanchang 330004, Jiangxi, PR China
[3] School of Science, Wuhan Institute of Technology, Wuhan 430073, PR China

Email: raozm24@163.com

Abstract. This paper reports a simulation model of drilling micro-hole in sticking plaster heated with a 1064 nm continuous CO_2 laser beam. Laser spot sizes ranged from 0.1 to 0.2mm diameter with axial irradiance power levels of 25-100W. To apply software Ansys, the measured steady-state surface temperature is calculated to rise with both increasing beam power and incident laser irradiance. For temperatures above 450 °C, sticking plaster vaporized into ventilation hole, and the size of ventilation hole 0.15mm diameter spent 1.7ms heated with laser power lever of 100W with the size of spot 0.15mm diameter, in good accordance with reported in earlier experiments studies. Similarly, the size of ventilation holes changed with beam power and laser spot diameter. These results show that software Ansys can be used to predict drilling micro-hole in CO_2 laser irradiated sticking plaster and the result of simulation can guide to laser drilling experiments.

1. Introduction

With the rapid development of laser technology, laser applications in the field of pharmaceutical production more and more widely in recent years. Drilling micro-hole in sticking plaster heated with a laser beam is an important application of laser technology [1-5]. Laser drilling micro-hole technology improved production efficiency and quality of medicines due to its more high-speed and more precision characteristics than that of conventional drilling micro-hole technology in the pharmaceutical production fields.

FEA (finite element analysis) software Ansys is good to predict the time evolution of temperature material properties, and it can calculate the distribution of the temperature field after laser irradiation materials [6-7]. It can be applied to a birth-and-death-element technique of Ansys to remove the part of the material and the formation of small holes if the temperature of the material is beyond the melting value. Materials can be simulated the size of the hole after laser irradiation.

This paper studies a simulation model of drilling micro-hole in sticking plaster heated with a 1064 nm continuous CO_2 laser beam. To apply software ANSYS, the measured steady-state surface temperature is calculated to rise with both increasing beam power and incident laser irradiance.

2. Theoretical approaches

The dimension of sticking plaster is 1mm×1mm×0.3mm and the sample is irradiated by a continuous CO_2 laser beam, which is shown in Figure 1. In order to analyze the thermal behavior numerically, the FEA software Ansys was used in this study. Before the mathematical model is established, some assumptions should be made:

1. All the physical parameters of sticking plaster are temperature-independent.
2. The CO_2 laser energy is fully absorbed by sticking plaster.
3. The sticking plaster sample is treated as the black body.
4. In this study, the energy loss is negligible in the phase change.
5. Heat transfer is not affected by thermal expansion.
6. The CO_2 laser beam is regarded as a surface-heating source.

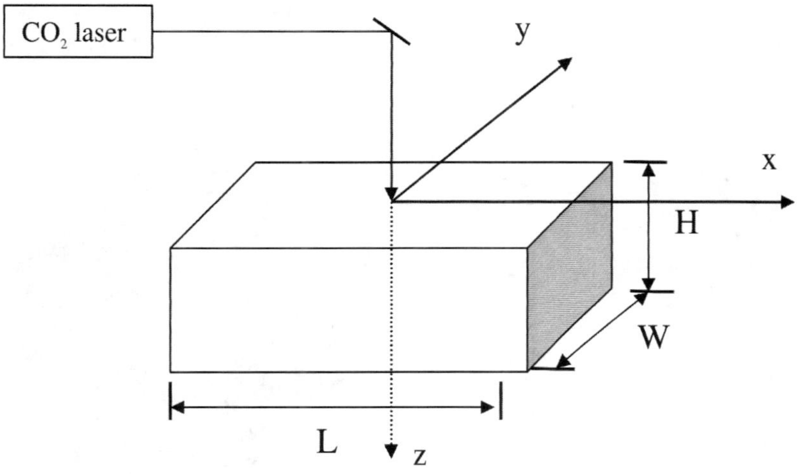

Figure 1. The setup of model with laser beam.

The grid structure of the sticking plaster is shown in Fig.2. A fine meshing was applied on the laser path to analyze the thermal behavior around the heated zone. Based on the above assumptions, the mathematical heat transfer model can be established as follows [8]

$$k\nabla^2 T = \rho c \frac{\partial T}{\partial t} \tag{1}$$

Where, k is the thermal conductivity. c and are the heat capacity and the density, respectively. A laser beam irradiated on the top of sticking plaster contains TEM_{00} mode, the intensity can be expressed in Eq.2. The CO_2 laser is absorbed fully by the sticking plaster and the absorbing depth is less than 15μm, so the laser beam is treated as a surface energy and the impulse function *(z)* is applied in Eq.2.

$$I(x, y, z, t) = \frac{P}{\pi r^2} \exp\left(-\frac{x^2 + y^2}{r^2}\right) \delta(z) \tag{2}$$

Where P and r are the power and the radius of the CO_2 laser beam, respectively. The initial temperature T_0=20•, which is the same as the ambient temperature. The boundary conditions of the FEA model illustrated in Fig.2 were carried out as Eq.3 and Eq.4.

$$-k\frac{\partial T}{\partial z} + h(T_s - T_0) + B\varepsilon(T_s^4 - T_0^4) = \alpha I(x, y, z, t) \text{ , at } z = 0 \tag{3}$$

$$-k\frac{\partial T}{\partial n} = h(T_n - T_0) \text{, at } z = H, x = \pm\frac{L}{2}, y = \pm\frac{W}{2} \tag{4}$$

Where, T_s denotes the temperature of the heated zone and T_n denotes the temperature of the area without laser heating, h=10Wm^{-2}K^{-1} is the convection heat-transfer coefficient, B= 5.670×10^{-8}Wm^{-2}K^{-4}is the Stefan-Boltzmann constant, *I(x, y, z, t)* is the density of the laser power, and n is the direction cosine of boundary. The surface emissivity is assumed as =1 and the absorption coefficient is assumed as =1.

The sticking plaster material is chosen as the thermal conductivity $k = 0.21W \cdot m^{-1} \cdot K^{-1}$, the heat capacity $c = 1890J \cdot kg^{-1} \cdot K^{-1}$ and the density $\rho = 1200kg \cdot m^{-3}$. For temperatures above 450•, sticking plaster will be vaporized. Based on the mathematical model of heat transfer, the process of drilling sticking plaster by laser beams can be simulated by Ansys. In this study, a birth-and-death-element technique was applied to simulating the process of drilling sticking plaster [9]. The principle of this technique is that the element will be killed and disappear in the model if the temperature of this element is beyond the melting value (450•) of sticking plaster. The drilling hole is illustrated in Figure3 which is simulated by Ansys.

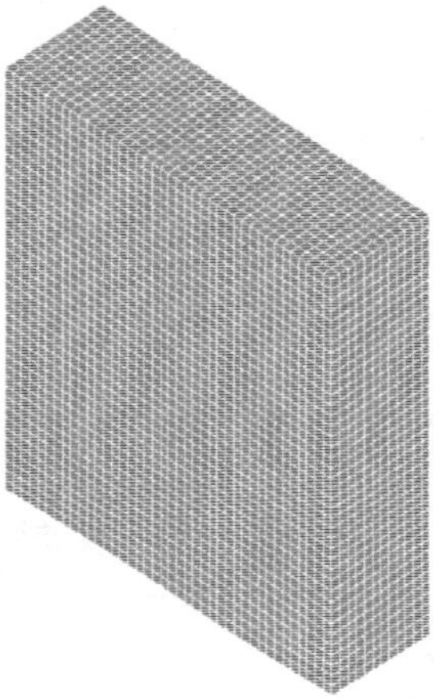

Figure 2. FEA model.

Figure 3. The mode of the hole.

3. Results and discussion

The continuous CO_2 laser is chosen as P=100 W, and the diameter is D=0.15mm, which calculated by Ansys. The time history of the depth of the hole is shown in Fig.4. It can be seen that the depth of the hole increases with the time of the laser irradiance. When sticking plaster heated with laser beam last time about 0.4ms, hole is produced. When t=1.5ms, the depth of the hole arrive at 0.3mm. When sticking plaster heated with laser bean spent on 1.7ms of time, the shape and size of the micro-hole is shown in Fig.5. As shown in Fig.5, the size of ventilation hole 0.15mm diameter spent 1.7ms heated with laser power lever of 100w with the size of spot 0.15mm diameter, in good accordance with reported in earlier experiments studies [2].

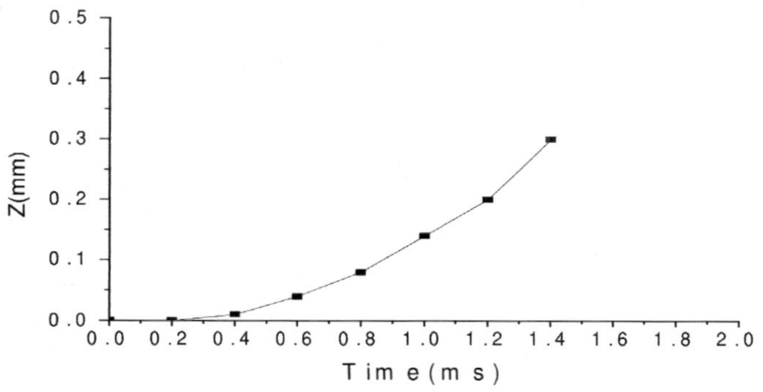

Figure 4. The time history of the depth of the hole.

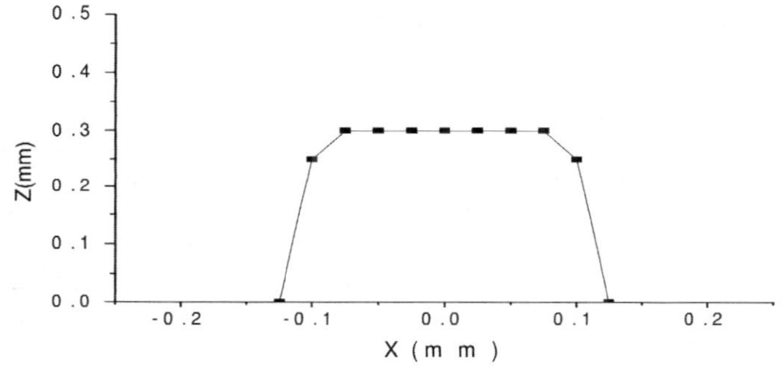

Figure 5. The depth and width of the hole.

Fig.6 shows the profile of the hole with different laser diameter but the same laser power (P=50 W, t=1.5ms). The depth of the hole increases with the decreasing laser diameter. On the other hand, the width of the hole is decreases with the decreasing laser diameter.
Fig.7 shows the profile of the hole with different laser power but the same laser diameter (D=0.15mm, t=1.5ms). We can see that the depth and width of the hole increases with the increasing laser power.

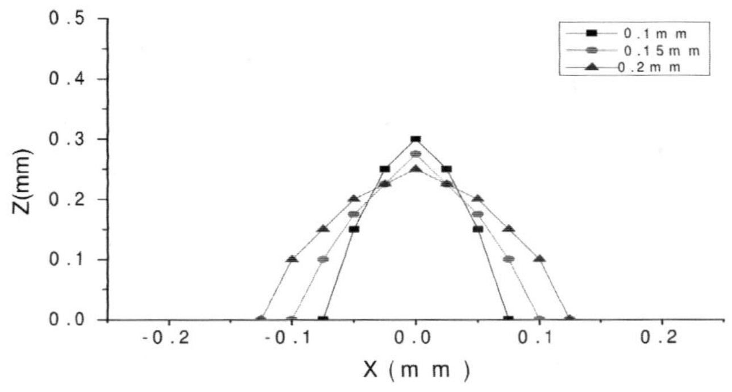

Figure 6. The depth and width of the hole with the different laser diameter.

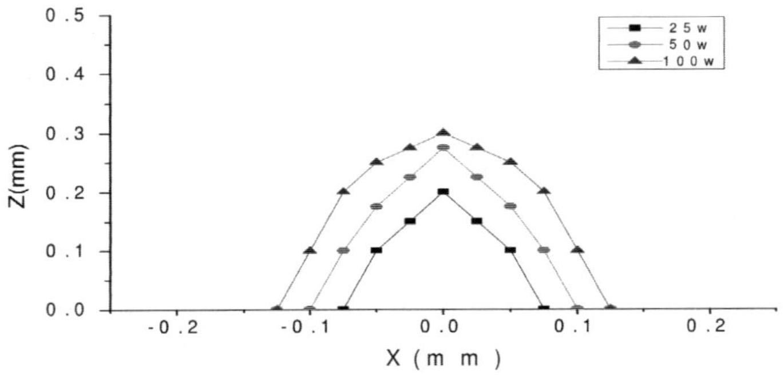

Figure 7. The depth and width of the hole with the different laser power.

4. Conclusion

The mathematical model of micro-hole by continuous CO_2 laser was illustrated and the micro-hole was simulated by FEA software Ansys. For temperatures above 450•, sticking plaster vaporized into ventilation hole, and the size of ventilation hole 0.15mm diameter spent 1.7ms heated with laser power lever of 100w with the size of spot 0.15mm diameter, in good accordance with reported in earlier experiments studies. Similarly, the size of ventilation holes changed with beam power and laser spot diameter. These results show that software Ansys can be used to predict drilling micro-hole in CO_2 laser irradiated sticking plaster and the result of simulation can guide to laser drilling experiments.

References

[1] Zhao XM, He YG, Wang SL, Wang XD, Yang H, Zhang ZG, Zhou Q and Li ZQ 2005 Laser drilling on plaster. *Laser Journal* **26** 81-82

[2] Zhao XM, Wang SL, Yang H, Wang XD, Zhou Q, Li ZQ and Zhang ZG 2006 Study on dynamic laser drilling for sticking plaster. *Laser Technology* **30** 308-310

[3] Chen PF, Wang Y, Li C and Hu CH 2007 Study on high power laser high speed drilling for sticking plaster. *Laser Journal* **28** 85-86

[4] Wang X B, Wang H, Pan C et al 2005 Laser on-line measuring the content of ointment [J]. *Applied Laser* **25** 7-9

[5] Chen P F, Zhou L, Zu M Z et al 2009 Study on optical system for high speed laser drilling [J]. *Laser Technology* **33** 12-14

[6] Jiao Junke, Wang Xinbing, Zhang Xueling 2007 Numerical analysis of cutting glass by dual CO_2 laser beams *Proceedings of SPIE* **6825** 68250G1-8

[7] Jiao Junke, Wang Xinbing 2008 A numerical simulation of machining glass by a dual CO_2-laser beam. *Opt. Laser Technol.* **40** 297-301

[8] Shen ZH, Zhang SY, Lu J, Ni XW 2001 Mathematical modeling of laser induced heating and melting in solid. *Opt. Laser Technol.* **33** 533-537

[9] ANSYS 10.0 Tutorials 2000 *Ansys Inc*

Numerical investigation on the effects of the laser energy and focal position on the multi-pulses laser propulsion

Song Junling[1], Hong Yanji[2] and Wen Ming[2]
[1] College of Postgraduate, the Academy of Equipment Command and Technology, Post Box 3380-86, Huairou Dis. Beijing, 101416, China
[2] Department of Basic theories, the Academy of Equipment Command and Technology,
Post Box 3380-86, Huairou Dis. Beijing, 101416, China

E-mail: songjl@mail.ustc.edu.cn, hongyanji@vip.sina.com and wujiewm@hotmail.com

Abstract. A detailed parametric study on the air-breathing laser propulsive performance is carried out for multi-pulses. Based on the finite volume scheme, the detailed evolving process of the inner and outer flow fluids is simulated. The numerical models with different focal positions and laser energies are employed to analyze the parameters effect on the multi-pulses impulse coupling coefficient C_m. Moreover, the laser frequency is discussed and compared with those calculations. The simulation results indicate that the focal position is one of the main factors to influence the multi-pulses C_m at low frequency. For high frequency, it is beneficial to replenish the air in the nozzle when the focal position locates near the nozzle exit. The influence of the laser energy is similar to the single pulse at low frequency, but at high frequency, the partial filling air in the nozzle causes low C_m by high laser energy. The multi-pulses C_m is lower than that of a single pulse. In the same calculative time, the higher the laser frequency, the higher the impulse value, but the lower the C_m.

1. Introduction

Laser propulsion is one of the most innovative propulsion for its high efficiency and specific impulse. To propel the small spacecraft, it must work at a sequence of multi-pulses on certain repeating rate in the air-breathing mode. Several experiments have been done on laser propulsion for spacecraft [1-5]. In 1998, the "Laser Lightcraft" had been flown successfully by air-breathing mode using a pulsed laser at 10Hz [6]. In another indoor experiment, it had been found the impulse coupling coefficients decrease when the laser repetition rate increases [7]. By using the numerical method, the non-equilibrium thermodynamics, unstructured-grid is developed to simulate the flow fluids for the single pulse in the nozzle [8]. A dynamic method was employed to calculate the multi-pulse performance [9]. The detailed parameters such as the nozzle configuration, focal systems laser energy etc. show impact on the multi-pulse propulsive performance. However, more systematic and detailed work is needed to be done.

Published under licence by IOP Publishing Ltd

This paper focused on the effects of the focal position and laser energy on multi-pulse propulsion. The detailed evolving process of the inner and outer flow fluids is simulated to compare with the differences among pulses. Moreover, three repetition rates are employed to analyze the frequency influence. The qualitative results gained in this paper will pave the way for designing the optimal laser thrusters for multi-pulses.

2. Computation method

2.1. The nozzle configuration
An expending cylinder nozzle is chosen which is shown in figure 1. The laser beam is focused to the site *B* mm away from the front of the nozzle using a lens. In the simulation, the incident laser pulse width is 7.57μs, the laser beam radio is 45mm and the pulse repetition rate is 20Hz, 50Hz and 100Hz respectively.

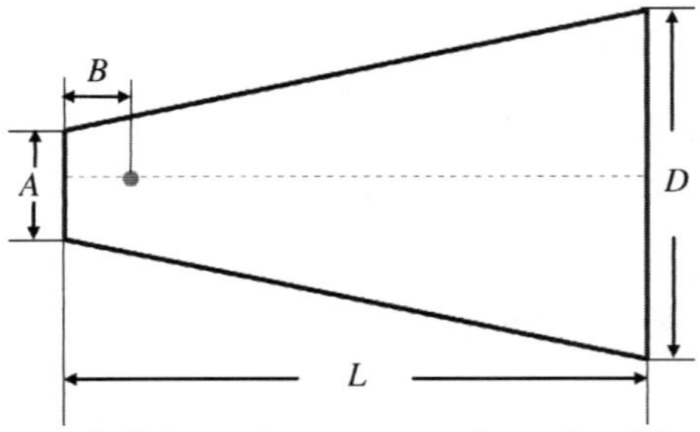

Figure 1. Schematic structure of nozzle. *A* is the diameter of the nozzle front wall, *D* is the diameter of the nozzle exit, and *L* is the total length of the nozzle. It is assumed that *A*=20mm, *D*=50mm, *L*=120mm.

2.2. Computation method
The cell-centered finite volume scheme is used. A Roe scheme is adopted to estimate numerical flux and the space accuracy is extended to 2nd-order by MUSCL (Monotone Upwind Scheme of Conservation Law) approach with Minmode flux limiter. The time integration performed with the predictor correction method.

2.3. Initial value and boundary condition
The gas is standard, and its physical parameters are as follows, atmospheric pressure p_0 is 1.01325×10^5Pa, temperature T is 288.15K, density ρ_0 is 1.225kg/m^3.

A two-dimensional axisymmetric computational model, as shown in figure 2, is created to describe the propagation of the inner and outer flow fluids. Structure grid is constructed and high grid density is used in the inner flow region for the actual computation. The wall is defined as a slip boundary and the interface is defined as a continuous overlapping type. The outer boundary is set far enough to suppress the influence of non-physical waves reflected from the boundary and is defined as a subsonic, characteristic condition.

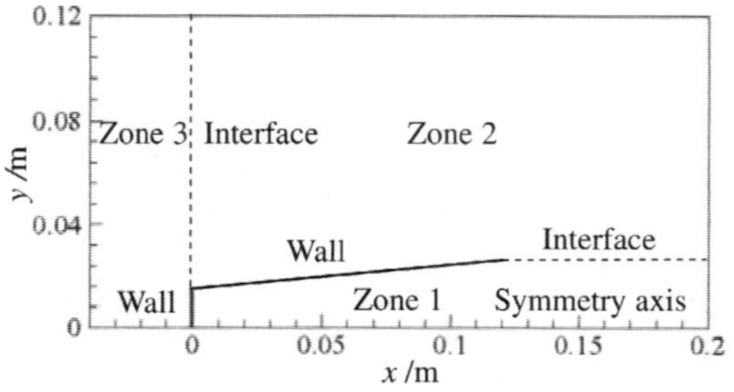

Figure 2. Computational zones near the nozzle. Zone 1 is the inner flow zone; zone 2 and zone 3 are the outer ones.

3. Results and discussion

A series of computations have been performed for different focused positions and laser energies. The nozzle numbers are listed in table 1.

Table 1. Nozzle number

(a) Focused positions

Nozzle number	B1	B2	B3	B4	B5	B6
B/mm	5	10	15	20	30	40

(b) Laser energy

Nozzle number	E1	E2	E3	E4	E5
E_{in}/J	14	28	42	56	70

3.1. Impulse coupling coefficient for different focused position

The laser energy is 14J. Figure 3 shows the thrust histories for different focused positions. The thrusts are formed at time of 0.6μs, 1.6μs, 3.7μs, 8.8μs, 15.1μs and 18.3μs respectively. The laser energy is deposited at the focused position and plasma is produced near the laser focus. As the blast wave expends, the portion of the wave in contact with the nozzle surface and the impulse is delivered to the nozzle. Therefore, the nearer the focused position to the front wall, the earlier the appearance thrust.

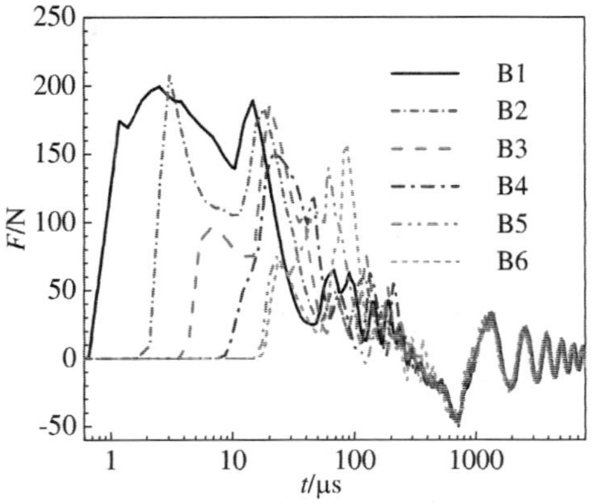

Figure 3. Thrust histories for different focused positions

Impulse coupling coefficient C_m is one of the important parameters of the propulsive characteristics in pulsed laser propulsion. Here, the multi-pulses is C_m defined as

$$C_m = \frac{I}{nE_{in}} = \frac{\int_0^{\frac{n+1}{f}} F\mathrm{d}t}{nE_{in}} \tag{1}$$

where I, the total impulse of the n pulses, is achieved through temporal force integrations, f is the laser repetition rate. For n pulses, C_m is the average evaluation of n pulses.

Figure 4 and Figure 5 show the C_m versus pulse numbers in 20Hz and 100Hz. The C_m decreases with the number of the laser pulses. In figure 4, the nozzle B2 gets higher C_m than B4 and B6 after 4 pulses in 20Hz. In contrast, the C_m of the nozzle B2, B4 and B6 get closed after the action of 10 pulses in 100Hz.

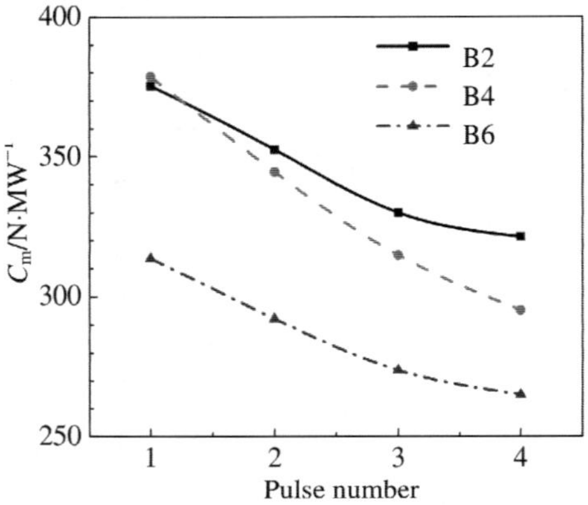

Figure 4. C_m versus number of pulses in 20Hz

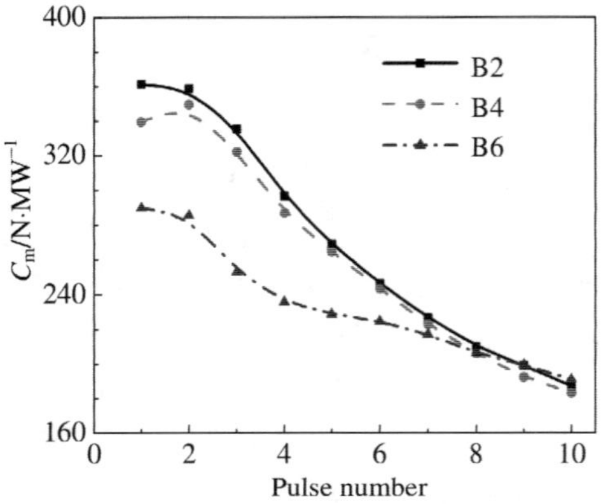

Figure 5. C_m versus number of pulses in 100Hz

Figure 6 shows the multi-pulses C_m versus the focal position. The C_m deceases with the increasing B in 20Hz, and the C_m varies slightly for laser repetition rate of 50Hz and 100Hz. As discussed in our previous work [10], the partial filling rate of the air has a critical influence on the pulse propulsive performance. It is beneficial to breathe the air in nozzle when the repetition rate is low. A strong shock wave is generated in the focus region, and its intensity decays rapidly as the wave expends. At 20Hz, the main factor for the C_m is focus position. So the nearer the focus position to the front wall the higher C_m will be got. The pulse separation becomes shorter as the repetition rate increases, the density of the air in the nozzle decreases with the increasing number of the pulses. When the focus position locates far away from the front wall that means it is near the exit, it is beneficial to breathe the air. Therefore, at 100Hz, as the increasing focus position, the C_m increases slightly.

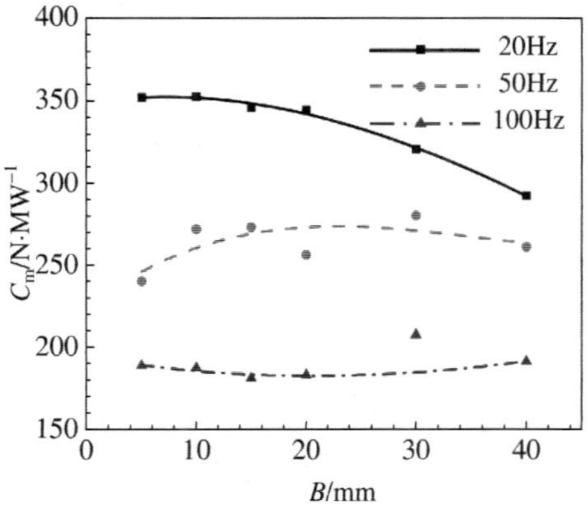

Figure 6. Multi-pulses C_m vs. focused position

3.2. Impulse coupling coefficient for different laser energies

Figure 7 shows the thrust histories of the first pulse for different laser energies. As shown, the higher the laser pulse energy, the earlier the generation of a pulse thrust and the higher the pick value. At around 550μs to 600μs, when the shock wave arrives at the exit of the nozzle, the density of the air in the nozzle is low and the shock attaching the outside of the wall generates a negative thrust. After the shock slips away from the exit, the air is breathed into the nozzle and the thrust oscillations

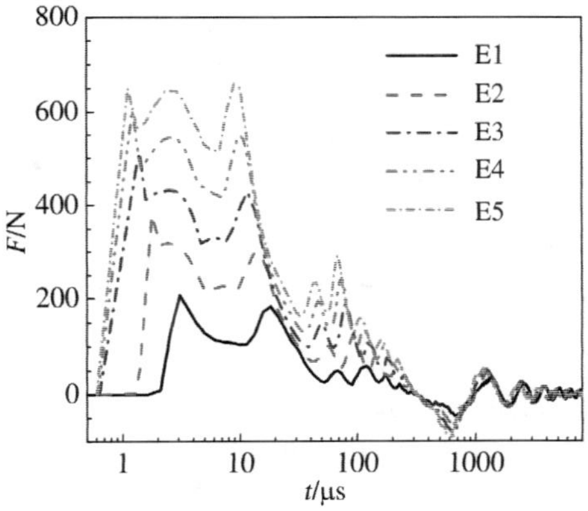

Figure 7. Thrust histories for different laser energies

In multi-pulses, the density of the air in the nozzle is important for initiating the air breakdown process and may have a significant impact on the numerical value of impulse coupling coefficient. As the laser repetition rate is 50Hz, the computational times to gain the density information for the 2nd, 3rd and 4th pulse start are at 20ms, 40ms and 60ms respectively. The initial density contours for the 2nd, 3rd and 4th pulse starts in the laser energies of 14J, 42J and 70J are shown in figure 8.The density of the air in the nozzle becomes larger with the number of the laser pulses. In the same pulse, the larger laser energy has the lower density area. That is because the air is exhausted more by higher laser energy and it takes a longer time to replenish the air in the nozzle.

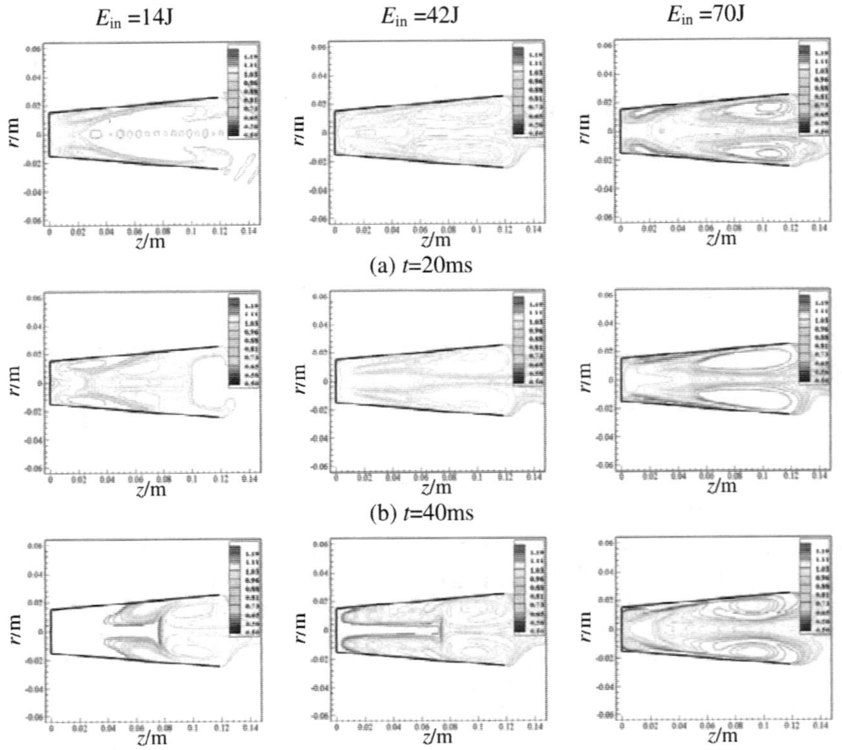

(a) t=20ms
(b) t=40ms
(c) t=60ms

Figure 8. Density contours for laser energies of 14J, 42J and 70J

The influence of the laser energy on impulse and C_m for single pulse has been studied widely [2, 5, 8]. Here, we simulate the laser energy impact on I and C_m with laser repetition rate of 20Hz, 50Hz and 100Hz for multi-pulses. The calculation time is 10ms.

Figure 9 and figure 10 show the calculation data of I and C_m for different repetition rates. The impulse is found to be a linearly increasing function at the pulse energy of 14-70J as shown in figure 9. The simulation results are similar to those results on single pulse in reference [2]. Also shown, the "f=100Hz" line gets highest rate of slope. One of the reasons is the higher repetition rate obtains more pulses in the same calculation time.

In figure 10, for the single pulse, the C_m reaches maximum at pulse energy of 40J, and does not grow even decrease slightly at subsequent increasing of the pulse energy. For the multi-pulses, the "f=20Hz" achieves higher C_m than the other repetition rates in the same laser energy, and its curve is similar to that of the single pulse. At 50Hz, the C_m has max peak at pulse energy of 32J, and then decreases at pulse energy of 40-70J. In contrast, the C_m grows slightly with the increasing laser energy at 100Hz. For low laser frequency, the influence of the energy is nearly the same with the single pulse. From the density contours in figure 8, it is found that higher energy causes larger low density area in the nozzle. For high frequency, the time between two pulses is short, so the air in the nozzle is partial filling. Hence, it impacts on the intensity of a pressure force upon the nozzle surface, and further influences the C_m. Also from figure 9 and figure 10, it indicates that the higher the repetition rate, the higher the impulse, but the lower the C_m.

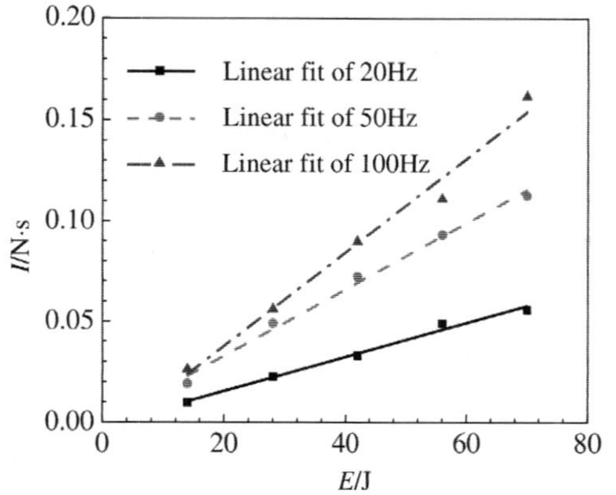

Figure 9. Multi-pulses I versus E_{in} in 20Hz, 50Hz and 100Hz

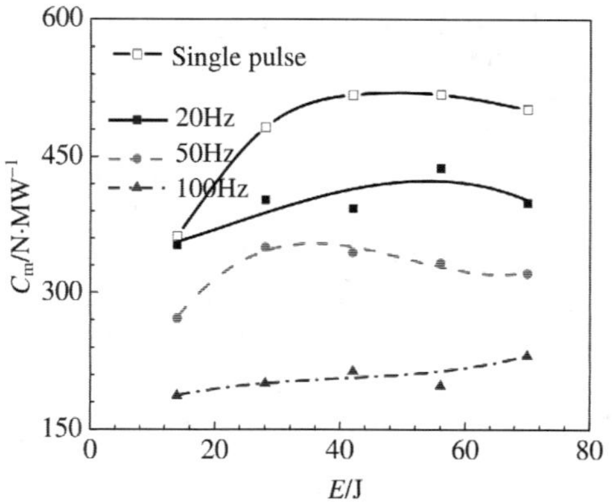

Figure 10. Multi-pulses C_m versus E_{in} in 20Hz, 50Hz and 100Hz

4. Conclusions

The focal position and laser energy impact on the propulsive performance of the air-breathing laser propulsion for multi-pulses have been studied. To contrast, three different repetition rates have been modelled to simulate the laser frequency affect.

The C_m decreases with the number of the laser pulses in multi-pulses propulsion. At the low frequency, the focal position is one of the main factors to influence the multi-pulses C_m. When the focal position locates near to the front nozzle wall, the high C_m is achieved. As increasing the frequency, the time to replenish air becomes short, so it is beneficial to breathe the air when the focal position near the exit.

The higher laser energy will exhaust more air in the nozzle, and it needs a longer time to recovery the density of the air. The reaches maximum at the energy of 32J, and then decreases at the energy of 40-70J at 50Hz. The curve of C_m at 20Hz is similar to that of the single pulse, but its value is lower than the single one. In the same calculative time, the 100Hz has the most pulse numbers, so it gets the highest impulse, but the lowest multi-pulse C_m.

References

[1] Myrabo L N 2003 Brief history of the lightcraft technology demonstrator (LTD) project In: *1st Int. Symp. on Beamed Energy Propulsion* ed Pakhomov A V (Troy: Proc. AIP) pp 49-60

[2] Bohn W L and Schall W O 2003 Laser propulsion activities in Germany In: *1st Int. Symp. on Beamed Energy Propulsion* ed Pakhomov A V (Troy: Proc. AIP) pp 79-91

[3] Katsurayama H, Komurasaki K, Hirooka Y, Mori K and Arakawa Y 2008 Numerical analyses of exhaust and refill processes of a laser pulse jet *J. Propul. Power* **24** 999-1006

[4] Wang T S, Chen Y S, Liu J, Myrabo L N and Franklin B M Jr 2001 Advanced Performance Modeling of Experimental Laser Lightcrafts In: *39th AIAA Aerospace Science Meeting & Exhibit* pp 1-12

[5] Rezunkov Y A, Safronov A L, Ageichik A A, Egorovl M S and Stepanov V V 2006 Performance characteristics of laser propulsion engine operating both in CW and in repetitively-pulesd modes. In: *4th Int. Symp. on Beamed Energy Propulsion* ed Komurasaki K, Yabe T *et al*(Troy:Proc. AIP) pp 3-13

[6] Myrabo L N, Messitt D G and Franklin B M Jr 1998 Ground and flight tests of a laser propelled vehicle AIAA paper: 98-1001

[7] Zheng Y, Tan R, Wang D, Zhenng G, Ke C, Zhang K, Wan C and Wu J 2005 Effect of laser pulse repetition rate on momentum coupling coefficients *High power laser and particle beams* **17** 972-82

[8] Chen Y S, Liu J and Wang T S 2003 Numerical modeling of laser supported propulsion with an aluminum surface breakdown model. In: *1st Int. Symp. on Beamed Energy Propulsion* ed Pakhomov A V (Troy: Proc. AIP) pp 138-48

[9] Xu R and Tang Z 2007 Numerical analyses of exhaust and refill processes of a laser propulsion of air breathing mode considering multi-pulses *High power laser and particle beams* **19** 370-2

[10] Song J, Hong Y J and Wen M 2010 Influence of laser pulse repetition rate on impulse coupling coefficient of air-breathing propulsive performance for multi-pulses *J. Aced. Equip. Comma. Tech.* (*Preprint* gr-qc/2010043)

Deposition and sputtering yields on EUV collector mirror from Laser Plasma Extreme Ultraviolet Sources

Tao Wu[1,2*], Zhiming Rao[3], Shifang Wang[4]
[1] Wuhan National Laboratory for Optoelectronics, School of Optoelectronic Science and Engineering, Huazhong University of Science and Technology, Wuhan 430074, P.R. China
[2] School of Science, Wuhan Institute of Technology, Wuhan 430073, P.R. China
[3] Depart of Computer Science, Jiangxi University of Traditional Chinese Medicine, Nanchang 330004, Jiangxi, P.R. China
[4] School of Physics and Electric Information, Hubei University of Education 1 Nanhuan Road, Wuhan East High-Tech. Zone, Wuhan 430205, Hubei, P.R. China

Email: flatime@163.com

Abstract. Based on the self-similar solution of gas dynamic equations, spherical expansion of the highly ionized plasma with limited mass into a vacuum is investigated for the droplet target laser-produced plasma extreme ultraviolet (LPP-EUV) sources. Using partially numerical and partially analytical technology, the velocity, the temperature and the density profiles in the plume versus ionization degree, adiabatic index and initial conditions are presented. Furthermore, the spatial thickness variations of the deposited substrate witness and ion sputtering yields for Ru, Mo, and Si under Sn ion bombardment are theoretically calculated, which can be useful to enable LPP-EUV sources suppliers to estimate collector lifetime and improve debris mitigation systems.

1. Introduction

Extreme ultraviolet lithography (EUVL) is currently a leading candidate for the manufacture of integrated circuits at the 32 nm node and below [1]. Recent theoretical and experimental data clearly demonstrate the advantage of the combination of a CO_2 laser with a Sn for the high conversion efficiency [2]. However, generation of debris in EUVL sources is an inherent and real threat to the lifetime of collection optics [3]. In order to mitigate the problem of debris from the Sn plasma, a mass-limited Sn-based target such as Sn droplet target with a diameter in the range of tens of μm, which has a minimum mass that can still deliver the sufficient EUV power, is effective [4,5]. Much theoretical work, using self-similar theory, has been done to understand the many complex physical processes in the plume dynamics and to reproduce the temporal and spatial evolution of the profiles in the pulsed laser deposition (PLD) expansion [6,7,8]. Nevertheless, Most of such previous works are described in planar geometry and the relevant analytical work on the droplet target LPP-EUV source plume expansion is rare. Moreover, most of existing self-similar models describing the behavior of expanding laser produced plasma plume are based on the neutral gases assumption. Recently, X.Y. Tan et.al. [9] and D.Bennaceur-Doumaz et.al. [10], using the Saha equation to estimate the ionization fraction, studied the single ionization effect in the expanding plume. However, the species emitting extreme ultraviolet light are highly ionized ions. In the present work, we deal with a theoretical investigation of highly ionized plasma expansion, limited to very small temporal and spatial domains just after the shot pulse, valid in vacuum, especially for LPP-EUV micro-droplet target sources. A simple analytical model is presented for hydrodynamic expansion of laser-produced plasma with a limited mass, which expands isothermally during laser irradiation and adiabatically after turning off the laser under the spherical geometrical conditions. Based on the gas dynamic equations, using the self-similar solution assumption and taking into account the high ionization degree of the plume, the ionization degree, adiabatic index and initial conditions are investigated to find their influence on the expansion features. The following spatial thickness variations and energetic ion-induced normal incidence sputtering yields of the Mo/Si multilayer materials are calculated.

2. Description of the model

Hydrodynamic spherical expansion of plasma plume into vacuum from a droplet target produced by pulsed laser irradiation can be divided into two stages before particles reached the mirror surface. The first stage is isothermal expansion during laser irradiation. After the cessation of the pulsed laser radiation, there is no additional input in the number of particles injected into the plasma from the target, nor is there any absorption of laser energy. Thus an adiabatic expansion occurs where the temperature can be related to the dimension of the plasma by a well-known thermodynamic relation given by

$$T\left(\frac{4}{3}\pi R^3\right)^{\gamma-1} = const. \tag{1}$$

In this equation, $=c_p/c_v$ is the adiabatic exponent and R refer to a time-dependent spherical radius of the leading edge of the plasma. The two successive expansions are found to be connected smoothly with each other in time and space and each of the two stages can be considered separately [11]. The later stage leads to the characteristic nature for the laser deposition and sputtering yields on collector mirror process.

2.1 Plasma expansion dynamics

During the initial stage of plasma expansion, the particle density is sufficiently high and the mean free path of the particles is short allowing the plasma to behave as a continuum fluid. Many collisions take place between the various particles so the plasma can be considered in local thermal equilibrium (LTE) [10]. This means that in a sufficiently small region, the electrons, ions and neutrals can be characterized with a common temperature. The plasma plume motion dynamics were considered as a compressible, viscous ideal gas. Therefore, the plume expansion is described by solution of gas-dynamic Euler equations in spherically symmetric case (r is the radial coordinate) and they are [12]

$$\frac{\partial \rho}{\partial t} + \frac{1}{r^2}\frac{\partial}{\partial r}(r^2 \rho u) = 0 \tag{2}$$

$$\frac{\partial u}{\partial t} + u\frac{\partial u}{\partial r} + \frac{1}{\rho}\frac{\partial P}{\partial r} = 0 \tag{3}$$

$$\frac{\partial T}{\partial t} + u\frac{\partial T}{\partial r} + (\gamma-1)\frac{T}{r^2}\frac{\partial}{\partial r}\left(ur^2\right) = 0 \tag{4}$$

where , u, P and T are the plasma vapor density, mass velocity, pressure and temperature, respectively.

When matter, especially medium and/or high Z element, is heated to high temperature, it will be ionized many times. The degree of ionization has a strong effect on many plasma properties. In LTE case, an approximate method to calculate mean ionization degree which is fitted to the scaling law and numerical results of the ionization potential of Thomas-Fermi statistical model is given in the form [13]

$$\eta = Z\{1-\exp[-4.115(\frac{T}{Z^{4/3}}\ln\frac{A\cdot T^{3/2}}{\eta\cdot n_i} - 1.16\times10^{-4})^{0.625}]\} - 0.5 \tag{5}$$

where A=4.8×10^{24}(cm)$^{-3}$(10^{6}K)$^{-3/2}$, is the mean ionization degree of the plume, n_i is the ion number density, Z=50 for Sn element. Eq. (5) can be iterative calculated to determine η for the particular ion number density and temperature. Under the condition of quasi-neutrality assumption $n_e=\eta n_i$, the governing equations (2)-(4) are closed by an ideal gas equation.

Equations (2)-(4) can be constructed for a self-similar isentropic solution in the following form:

$$u=\frac{r}{R}\frac{dR}{dt} \tag{6}$$

$$\rho=\frac{M}{J_1R^3}(1-\frac{r^2}{R^2})^{\frac{1}{\gamma-1}} \tag{7}$$

$$P=(1+\eta)\frac{E}{J_2R^3}(\frac{R_0}{R})^{3(\gamma-1)}(1-\frac{r^2}{R^2})^{\frac{\gamma}{\gamma-1}} \tag{8}$$

$$T=\frac{\mu}{R_g}\frac{E}{M}\frac{J_1}{J_2}\left(\frac{R_0}{R}\right)^{3(\gamma-1)}\left(1-\frac{r^2}{R^2}\right) \tag{9}$$

where R_0 is the initial size of the plume, r is the radial coordinate, R_g is the universal gas constant, μ is the molar mass of the fuel material. The values of the total mass M and the initial energy of the plasma plume E in expressions (7)-(9) are defined as the integrals over the volume of the plume. J_1 and J_2 are the normalization constants determined from the conditions of mass and energy conservation.

Substituting self-similar solutions (6)-(9) into the set of gas dynamic equations, one can easily find that the continuity Eq. (2), as well as the Eq. (4), are fulfilled identically. The Euler's Eq. (3) is transformed into an ordinary differential equation which describes the motion law for expanding plume boundary. The derived equation is similar to the formula in Ref. [9], which can be reduced to the normalized form as follows:

$$\frac{d^2R^*}{dt^{*2}}=(5\gamma-3)(1+\eta)\frac{1}{R^*}\left(\frac{1}{R^*}\right)^{3(\gamma-1)} \tag{10}$$

$$R^*\big|_{t^*=0}=1,\ \frac{dR^*}{dt^*}\Big|_{t^*=0}=u_0^*.$$

Here, we used the dimensionless variables $t_0=R_0\sqrt{M/E}$, $t^*=t/t_0$, $R^*=R/R_0$, $u_0^*=u_0t_0/R_0$.

The Eq. (10) has the solution in the form of inverse function through further integration

$$t^*=[u_0^{*2}+\frac{2}{3}(\gamma-1)]^{-\frac{1}{2}}\{R^*\cdot F\left(a,b;c;\frac{(-3+5\gamma)(1+\eta)}{1+\frac{3}{2}(\gamma-1)u_0^{*2}}R^{*-3(\gamma-1)}\right)-F\left(a,b;c;\frac{(-3+5\gamma)(1+\eta)}{1+\frac{3}{2}(\gamma-1)u_0^{*2}}\right)\}, \tag{11}$$

where $F(a,b;c;z)$ is the hypergeometric function with parameters a=1/2, b=-1/3(-1), c=1+b.

The normalized density, pressure and temperature profiles within the plume are given by

$$\rho^*=\frac{\rho}{\rho_0}=\frac{1}{R^{*3}}\left(1-\xi^2\right)^{\frac{1}{\gamma-1}} \tag{12}$$

$$P^*=\frac{P}{P_0}=(1+\eta)\frac{1}{R^{*3\gamma}}\left(1-\xi^2\right)^{\frac{\gamma}{\gamma-1}} \tag{13}$$

$$T^*=\frac{T}{T_0}=\frac{1}{R^{*3(\gamma-1)}}\left(1-\xi^2\right) \tag{14}$$

where $\rho_0=\frac{M}{J_1R_0^{\,3}}$, $P_0=\frac{1}{J_2}\frac{E}{R_0^{\,3}}$, $T_0=\frac{J_1}{J_2}\frac{\mu}{R_g}\frac{E}{M}$ and $\xi=\frac{r}{R}\left(0\le\xi\le1\right)$ is the Lagrangian coordinate.

2.2 Deposition and sputtering EUV mirror

In the case of Sn sources, mirror lifetime presents great challenge. Apart from erosion due to high energetic ions, Sn also condenses and deposits on mirror surfaces. The EUV reflectivity in this case is affected by erosion as well as roughness caused by deposition. A Mo/Si witness sample is exposed to a Sn EUV sources (see Fig.1). The plasma plume material contaminates the sample and causes non-uniform thickness profiles *h*( ). The mass flux to the substrate surface is as follows[6]:

$$j(r,R_s,t)=\rho(r,R_s,t)u(R_s,t)=\begin{cases}\dfrac{MR_s}{J_1R^4}\dfrac{dR}{dt}[1-\dfrac{r^2}{R^2}]^{\frac{1}{\gamma-1}}, & t\geq t_s(r)\\ 0 \quad, & t\leq t_s(r)\end{cases} \tag{15}$$

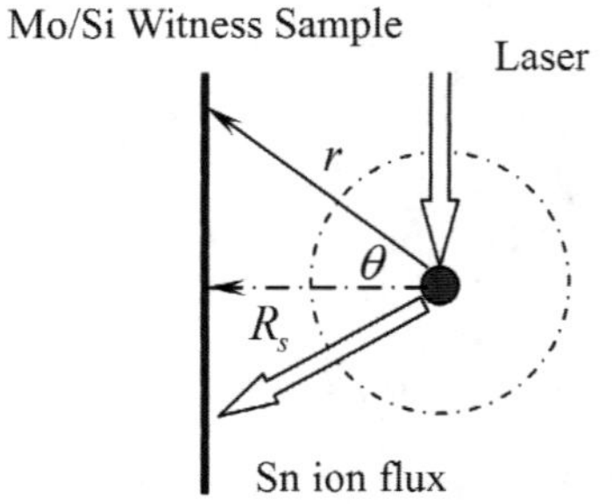

Figure 1. Schematic diagram of the spherical expansion plume deposition on witness sample.

Here $t_s(r)$ is the time when the outer boundary of the expanding cloud reaches the witness surface $r=R_s$sec $=R(t_s)$, where *r* is the radial coordinate and is the radial angle. Integrating the mass flux (15) over the time from $t_s(r)$ to infinity and dividing the result by the density of the deposited material $_s$, we obtain the thickness profile:

$$h(\theta)=\frac{MR_s}{\rho_sJ_1}\int_{t_s}^{\infty}\frac{1}{R^4}\frac{dR}{dt}\left(1-\frac{R_s^{\,2}}{R^2\cos^2\theta}\right)^{\frac{1}{\gamma-1}}dt \tag{16}$$

Sputtering is the erosion of a solid surface by energetic particles bombardment. Sputtering yield is defined as average number of ejected target atoms per incident ion. For a simple physical sputtering system, the yield is described in terms of $Y(E, ,M_1,M_2,Z_1,Z_2)$, which is dependent on the incident ion energy *E*, incident angle relative to the target surface normal, Z_1 and Z_2 are the atomic numbers, M_1 and M_2 are the masses of the projectile and the target atom, respectively. Based on the nuclear stopping power for the Thomas-Fermi potential $S_n^{TF}(E)$, Yamamura and Tawara derived a new universal equation as follows [14]:

$$Y(E)=0.042\frac{Q(Z_2)\alpha^*}{U_s}\frac{S_n^{TF}(E)}{1+\Gamma k_e\varepsilon^{0.3}}\left(1-\sqrt{\frac{E_{th}}{E}}\right)^s \tag{17}$$

where * and $Q(Z_2)$ are the parameters fitting to the data available, E_{th} is the sputtering threshold energy, k_e is the Lindhard electronic stopping coefficient, is the reduced energy, U_s is the surface binding energy of the target solid and $=W(Z_2)/[1+(M_1/7)^3]$ is a dimensionless factor. The detailed calculation method is given by Ref.[14].

3. Results and discussions

3.1 The calculation of the velocity evolution of the plume edge

With such a time-dependent homogeneous ionization degree assumption, equation (10) can be numerically solved coupling with equation (5). Fig.2 shows the evolution of ionization degree of monatomic gas (=1.67) versus normalized expanding plume length under the initial conditions given in table 1. We have noticed the ionization degree decreasing very fast with the expansion as a consequence of recombination processes and decrease of the temperature. The further integration of Eq. (10) yields $u_\infty^2=u_0^2+2(5\gamma-3)(1+\eta)E/3(\gamma-1)M$, which is consistent to the results from Fig.3. It is easy to see directly from the derived equation above that the velocity of expansion tends to constant $dR/dt \to u_\infty$ for sufficiently extended time, i.e. inertial stage of expansion. The results of calculations show that the maximum expansion velocity of the ionized plasma is increasing with the ionization degree. This shows that ionization strongly affects the dynamics evolvement of plasma, which is also verified by the Ref. [9-10]. In summary, the self-similar behavior of all the expansion profiles when ionization effect is taken into account, are more extended than that of a neutral gas vapor. We can see from Fig.4, the lower value of would result in higher terminal velocities, but at the early stage of adiabatic expansion vice versa. For laser-generated ionized species, the value of is much lower than monatomic gases, probably in the range of 1.1-1.3 when the degrees of freedom due to excitation and ionization are taken into account. We assume that the external source of energy is removed, so that the sum of the internal energy and the kinetic energy is conserved. This behavior is due to the adiabatic expansion where thermal energy is converted into directed kinetic energy. The plume dynamic behavior is also affected by different initial conditions, including initial plasma edge velocity and radius, i.e. the absorbed CO_2 pulsed laser energy and the entire disintegrated mass of the droplet target by the end of isothermal expansion stage.

Table 1. Initial parameters of the plume which have been used in calculations.

Initial parameters of the plume	value
Initial size, R_0[μm]	300
Plume mass, M[μg]	2
Plume total energy, E[mJ]	50
Initial velocity of expansion, u_0[km/s]	6
Initial temperature, T_0[eV]	33

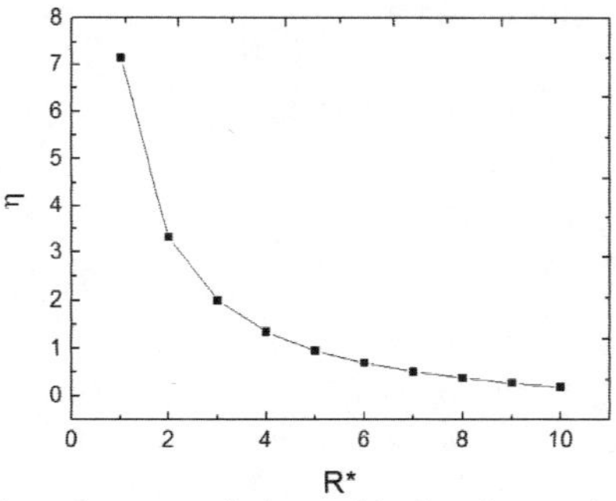

Figure 2. The averaged ion charge variation with the plume size under the assumption of homogeneous ionization degree. The initial parameters are listed in table 1.

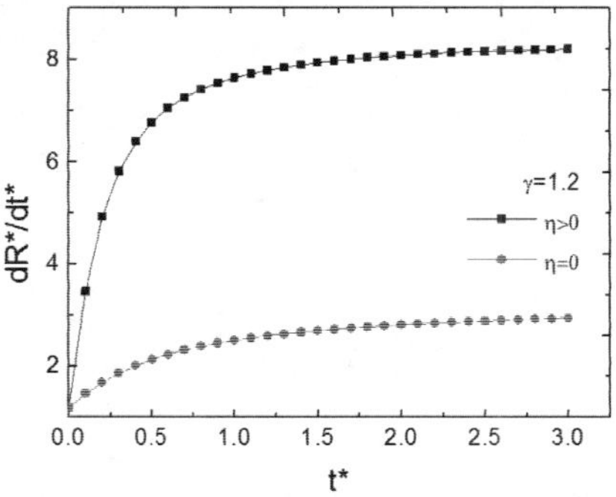

Figure 3. The time history of the normalized velocity at the edge of the plume for >0 and =0.

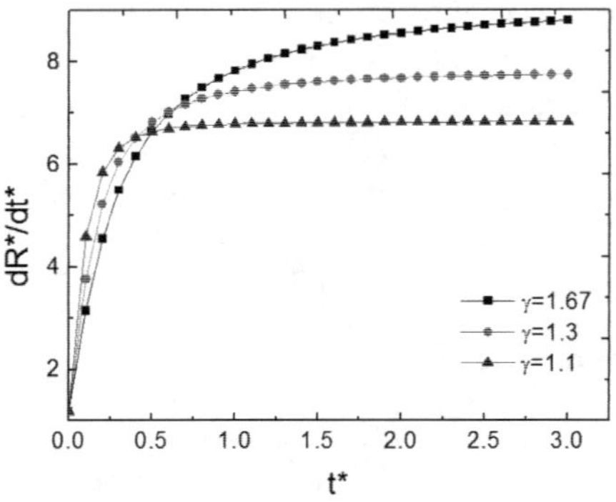

Figure 4. The normalized velocity evolvement at the edge of the plume considering different adiabatic index for >0.

3.2 Density, temperature and pressure profiles as a function of time and space

The normalized plasma plume density, temperature and pressure distributions versus time and space are given by the Fig. (5) and Fig. (6). In this free-expansion regime, we see that the density, temperature and pressure decrease with time and space. The results are just compatible with the analytical expressions mentioned above. Note that the density, temperature and pressure profiles describe a plume with a sharp external edge. This is the consequence of adiabatic assumption, which leads to the relation (1). Thus, the density, temperature and pressure approach zero at the front, in contrast to an isothermal expansion, where the velocity of sound at the outer edge of the plume remains finite and generates density and pressure tails.

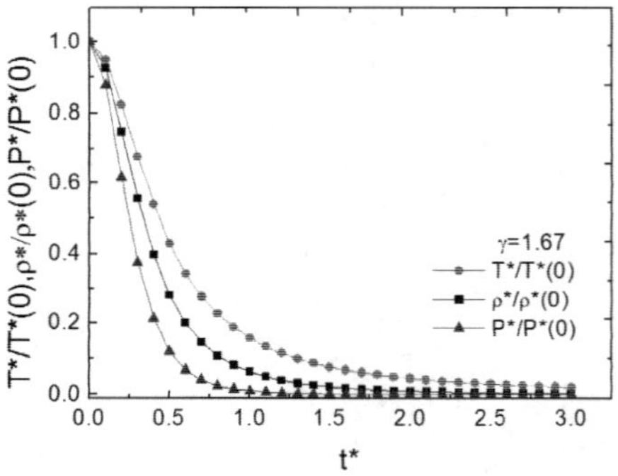

Figure 5. The density, temperature and pressure evolvements normalized by that at the centre.

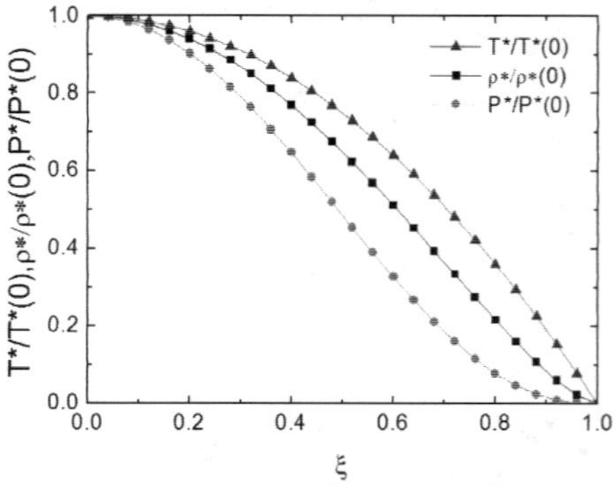

Figure 6. The density, temperature and pressure profiles normalized by that at the centre in the case of =1.2.

3.3 Plasma deposition thickness profiles and sputtering yields on the EUV mirror

EUVL technology, which is based on reflective optics for 13.5nm, by using Mo/Si reflective coatings with protecting capping layers of Ru, it is possible to achieve reflectivity of as high as 70% [15-16]. However, reflectivity degradation of the collector optics is an effect of debris interaction with the collector optics. For nearly normal incidence mirrors, sputter-induced erosion, in addition to roughness and contamination, plays a critical role in reflectivity degradation. A simple analytical expression for *h*( )•cos^3 can be obtained from equation (16) for the case of spherical expansion, which is satisfied in many experimental situations. The thickness variation controlled by adiabatic index is shown in Fig.7. It can be observed that the deposition thickness *h*( ) increases as adiabatic index and/or substrate-target distance decreases under the same conditions.

As mentioned above, in spite of mirror surface contamination, the energetic species bombardment may induce sputtering damage to the topmost surface. The sputtering yields of Ru, Mo and Si are theoretically calculated and plotted in Fig.8. It can be shown that the lightest element Si has the highest sputter threshold E_{th}. When the incident energy is in the range of 10keV, the sputter yields approach an asymptotic maximum value more than 10. Since energetic ion sputtering is found to be responsible for the degradation of reflectivity of the multilayer (ML), it is critical for longer lifetime to reduce energetic species flux to the EUV mirror surfaces.

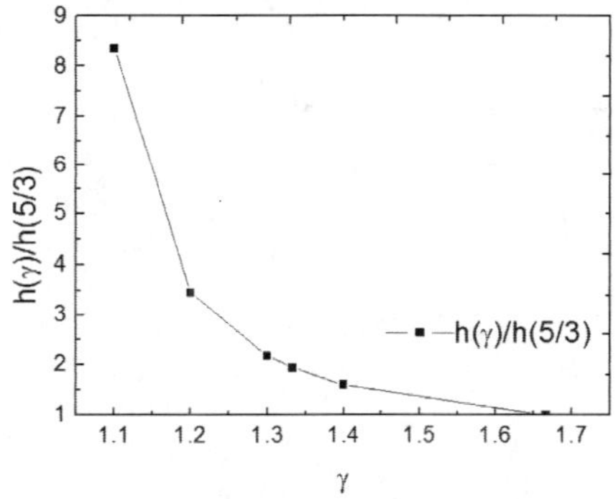

Figure 7. The thickness *h*( ) normalized by that of monatomic gas under the initial conditions listed in Table 1.

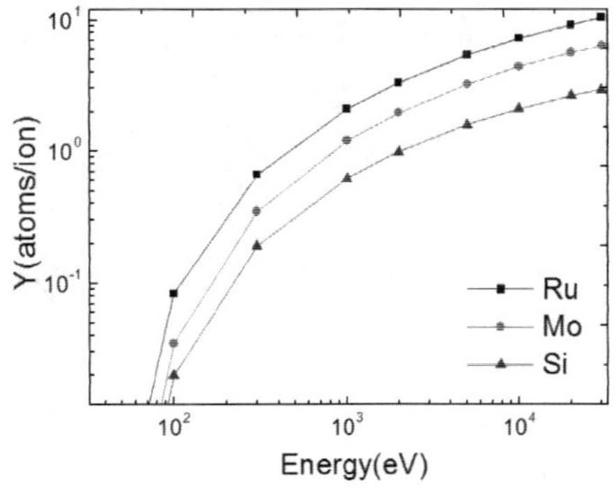

Figure 8. Energy dependence of the sputtering yields of Si, Mo and Ru with Sn using Yamamura and Tawara's formula.

4. Conclusion

Mathematically, the present self-similar solution under the physics of the one-fluid plasma model exists only for certain specific values of the adiabatic index depending on the spherical geometry in a very small temporal and spatial domain. The initial conditions required for the calculation above are chosen so as to be relevant to the EUV plasma. In addition to tin deposition, a major factor that determines the lifetime is ion sputtering of the material at the collector surface. Controlling or

mitigating the debris has been attempted by various groups using different methods, including cavity confined plasma[17], mass limited droplet targets, foil trap [18], the application of ambient gas, magnetic fields and combined effects [19].
In summary, based on the consideration of ionization degree and conservations of mass, momentum and energy of the plasma, plasma spherical expansion into vacuum has been studied, in which the plume evolvement profiles are obtained after the laser radiation. The influences of ionization degree, adiabatic index and initial conditions on plume expansion are discussed. In order to estimate the collector mirror lifetime and investigate the ML damage mechanism for EUV light source development, the non-uniform collector surface contamination and the sputtering yields of Si, Mo and Ru under the energetic Sn debris bombardment are also calculated theoretically.

References

[1] Silverman P J 2005 *J. Microlithogr., Microfabr., Microsyst.* **4** 011006
[2] Tao Y, Tillack M S, Harilal S S, Sequoia K L, Burdt R A and Najmabadi F 2007 *Opt. Lett.* **32** 1338
[3] Tanaka H, Hashimoto Y, Tamaru K, Takahashi A and Okada T 2006 *Appl. Phys. Lett.* **89** 181109
[4] Shinsuke F, Masashi S, Yoshinori S, Shinsuke M, Hirokazu S, Yuk N, Tatsuya A, Hiroaki N, Norimasa O, Atsushi S, Katsunobu N, Noriaki M, Yasukazu I and Kunioki M 2008 *Appl. Phys. Lett.* **92** 241502
[5] Keyser C, Schriever G, Richardson M and Turcu E 2003 *Appl. Phys.* A **77** 217
[6] Anisimov S I, Luk'yanchuk B S and Luches A 1996 *Appl. Surf. Sci.* **96** 24
[7] Singh R K, Holland OW and Narayan J 1990 *J. Appl. Phys.* **68** 233
[8] Alsonso C, Sema R, Catalina F and Bermejo D 1990 *Appl. Surf. Sci.* **46** 249
[9] Tan XY, Zhang D M, Li Z H, Liu G B, Li L and Fang R R 2007 *Phys. Lett.* A **370** 64
[10] Bennaceur-Doumaz D and Djibli M 2009 *Appl. Surf. Sci.* **255** 4595
[11] Murakami M, Kang Y G, Nishihara K, Fujioka S and Nishimura H 2005 *Phys. Plasmas* **12** 062706
[12] Luk'yanchuk B S, Marine W and Anisimov S I 1998 *Laser Phys.* **8** 291
[13] Zhang J and Gu P J 1987 *Chin. J. Comput. Phys.* **4** 1(In Chinese)
[14] Yamamura Y and Tawara H 1996 *Atomic data and nulcear data tables* **62** 149
[15] Wu S M, van de Kruijs R, Zoethout E and Bijkerk F 2009 *J.Appl.Phys.* **106** 054902
[16] Srivastava S N, Thompson K C, Antonsen E L, Qiu H, Spencer J B, Papke D and Ruzic D N 2007 *J.Appl.Phys.* **102** 023301
[17] Tomie T, Aota T, Ueno Y, Niimi G, Yashiro H, Lin J Q, Matsushima I, Komiyama K, Lee D H, Nishigori K and Yokota H 2003 *Proc. SPIE* **147** 5037
[18] Pankert J, Apetz R, Bergmann K, List A, Loeken M, Metzmacher C, Neff W and Probst S, 2005 *Proc. SPIE* **260** 5751
[19] Harilal S S, O'Shay B, Tao Y and Tillack M S 2007 *Appl. Phys.* B **86** 547

Numerical Simulation and Analysis on 3D Temperature Field of the Metal Ablated with Femtosecond Pulse Laser

Li Yang[1,2], Ching-Yue Wang[2], Wei Yang[1]

[1]Science College, Civil Aviation University of China, Tianjin 300300, China;

[2]Key Laboratory of Opto-electronic Information Science and Technology, Ministry of Education, Tianjin University, Tianjin 300072, China

e-mail address：y_lyang@yahoo.com.cn

Abstract. To describe femtosecond laser ablation on the metal, numerical simulation on the basis of the double-temperature equation for three-dimension temperature field of the copper ablated with femtosecond pulse laser was performed by finite-difference method. Based on imbalance of the electronic and lattice's temperatures, the calefactive process of the electron and the lattice was obtained, respectively. The dependence of the electron-lattice coupling time on irradiated laser fluence was studied. The ablation depth and the ablation radius of the copper for single pulse fluence were calculated. The dependence of the start ablation (phase explosion arises) time of the copper on irradiated laser fluence was studied. The results indicate that the material jet due to phase explosion is earlier and the duration of ablation is longer with the increase of the laser fluence. When the laser fluence is higher than 1.5J/cm^2, the ablation start time is about 2-3 ps.

1. Introduction

In recent years, femtosecond pulsed laser have shown great potential in the micro-nano machining field. Many issues group and researchers made a lot of research results in this area. Research on the interaction of ultrashort intense laser and materials had been attended more and more by the research scholars [1-3]. Femtosecond laser ablation on the materials is a fast and complex process. For understanding the properties and accumulating theoretical basis of femtosecond laser ablate the material, investigate on the temperature field of the material ablated with femtosecond pulse laser is a important issue. The double-temperatures model for metal was first established by the Soviet scholars S.I.Anisimov *et al* according to the non-steady state heat conduction equation [4]. The electronic and the lattice are divided into two systems in the model. And the temperature changes of the electronic and the lattice were described respectively. The model was built on the basis of reasonable physical analysis. And scholars have a series theoretical and experimental research on the interaction of ultra short intense laser and metal by one-dimensional double-temperature equation (DTE) [2,3,5-8]. The heat affected zone had been simulated when a rectangular pulse laser ablate the aluminium by French scholar S. Valette *et al* used the DTE and finite difference method (FDM) [9]. The numerical result was consistent with the literature [10].

In this paper, we report numerical simulation for 3D temperature field of the metal ablated with femtosecond pulse laser by the DTE and FDM. Based on imbalance of the electronic and the lattice temperatures, the calefactive process of the electron and lattice was obtained, respectively. The dependence of the electron-lattice energy coupling time on irradiated laser pulse energy was studied. The threshold fluence of single pulse in the copper had been obtained. The ablation depth and radius of the copper for single pulse fluence were calculated. The dependence of the start ablation (phase explosion arises) time of the copper on irradiated laser pulse fluence was studied.

2. Theoretical model

For the metal material, when the laser pulse duration was shorter than the electron-lattice energy coupling time, the time evolution of the electron and lattice temperatures can be described by two coupled nonlinear differential equation, i.e. the double-temperature equation:

$$C_e \frac{\partial T_e}{\partial t} = \nabla(K_e \nabla T_e) - g(T_e - T_i) + S \tag{1}$$

$$C_i \frac{\partial T_i}{\partial t} = g(T_e - T_i) \tag{2}$$

In these equations, C_e and C_i are the specific heats of the electron and lattice, respectively. T_e and T_i are the electron and lattice temperatures, respectively. K_e is the electron thermal conductivity. g is electron-lattice coupling coefficient. C_e and K_e vary with the changing electron temperature: $C_e = C_e' T_e$, $K_e = K_{e0} T_e / T_i$. S is the laser source items and is written as follows:

$$S(z, r, t) = I_0 \alpha (1 - R) \exp[-4\ln 2(\frac{t}{T_0})^2] \exp[-(\frac{r}{r_0})^2] \exp(-\alpha z) \tag{3}$$

Where: z is the laser direction. The laser direction is perpendicular to the material surface. r is the radius direction of laser beam. r_0 is the radius of the laser beam at the focusing point. R is the reflectance of the material surface. is the absorption coefficient of the material. I_0 is the peak power. T_0 is the laser pulse duration.

3. Results and discussions

In the present work, it is assumed that the material is a half of infinity, T_0=150fs, $r_0 = 8\mu m$, the initial temperatures of the electron and lattice both are 300K. Table 1 gives all the numerical data used to the system in the case of a copper sample.

Table 1. Physical constant used for calculations

metal	g ($Wm^{-3}K^{-1}$)	C'_e ($Jm^{-3}K^{-2}$)	K_{e0} ($Wm^{1-}K^{-1}$)	C_i($Jm^{-3}K^{-1}$)	R	α (/nm)	T_{tc}(K)
Cu	12×10^{16}	96.6	401	3.43×10^{6}	0.329	0.089	5890

To describe femtosecond laser ablation on the copper, numerical simulation on the basis of the DTE for 3D temperature field of the copper ablated with femtosecond pulse laser is performed by FDM. Based on imbalance of the electronic and the lattice's temperature, the calefactive process of the electron and the lattice was obtained, respectively. To femtosecond laser, the ablation of the material mainly is caused by phase explosion [11-13]. Because the energy injection time is very short, leading to that no time to spread, the temperature of the irradiated material rises rapidly, until the temperature of the material reached to $0.9T_{tc}$（T_{tc} is the thermal dynamic equilibrium critical temperature）, and then it is ejected from the metal surface. Therefore, in this simulation, the metal material can be ablated when the lattice's temperature is increased to $0.9T_{tc}$.

Figure 1 represents the evolution of the temperature of the electron and lattice with different incident laser pulse fluence. It can be observed that the electron temperature rises rapidly to the maximum and the lattice temperature keeps fixed in the duration of the femtosecond laser. After the laser pulse finishes, the electron-lattice energies exchange start off and the lattice temperature rises. The electron-lattice energy coupling time is relative to the incident laser fluence. As the laser fluence is greater, the electron-lattice energy coupling time goes into longer and the lattice temperature goes into higher after coupling. The incident fluence that makes the equilibrium temperature of the electron and lattice in accord with critical temperature is called the threshold fluence. When the laser fluence is increased to the threshold fluence, the center of material was ablated justly. The threshold fluence of single pulse in the copper is 0.76J/cm^2 by calculations, which is approximately consistent with the literature [14] (In the literature [14], the threshold fluence is 0.7 J/cm^2). And electron-lattice coupling time by calculations is about 17.06ps, accordingly.

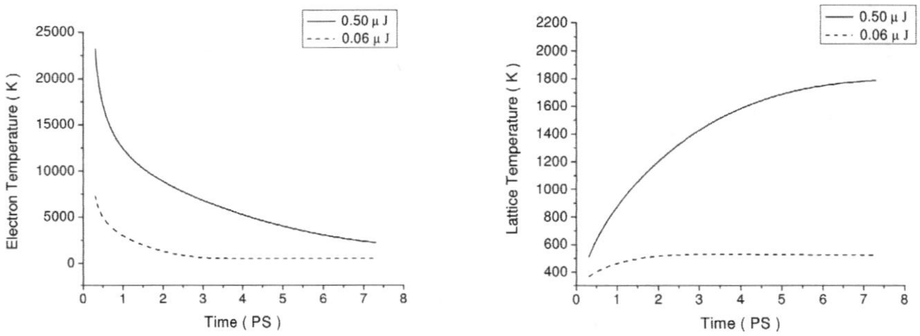

Figure 1. The time evolution of electron and lattice temperatures in the surface layer center of the copper with different incident laser pulse energies.

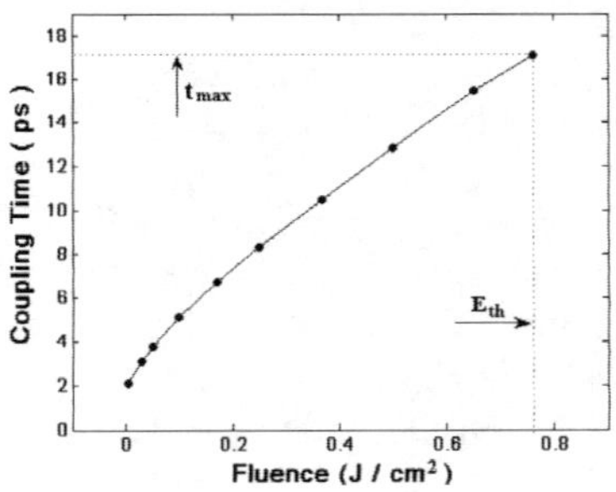

Figure 2. The change of the electron-lattice energy coupling time in the surface layer center of copper versus the pulse fluence.

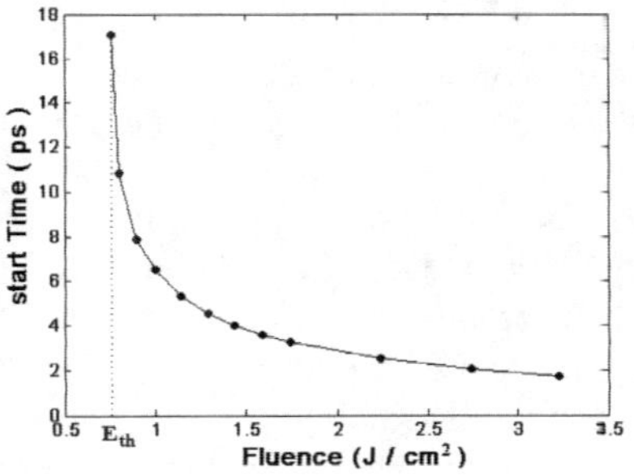

Figure 3. The change of the ablation start time in the surface layer center of copper versus the pulse Fluence.

Figure 2 represents the change of the electron-lattice energy coupling time in the center of copper surface versus the single pulse fluence. It can be seen that the electron-lattice energy coupling time is nearly relative to the incident fluence when it is below the threshold fluence. The electron-lattice energy coupling time becomes longer with the increasing laser fluence. When the incident fluence increases to the threshold fluence, the coupling time reaches maximum.

Figure 3 represents the change of the ablation start (phase explosion arises) time versus the laser energy for copper. The so-called the ablation start time of the metal material, that is the time that the lattice's temperature is increased to 0.9T_{tc} and then the material is ejected from the metal surface. From Figure 3, it can be seen that the higher laser pulse fluence, the earlier ablation start time. When the laser fluence is higher than 1.5 J/cm^2, the ablation start time is about 2-3 ps. With the further increasing laser fluence, the ablation start time finally becomes changeless.

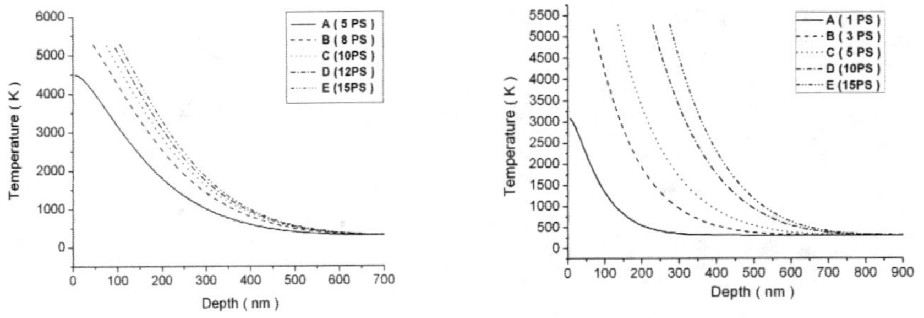

(a) pulse fluence: 1J/cm^2 (b) pulse fluence: 3.28 J/cm^2

Figure 4. Lattice temperature evolution for propagation direction of the beam, leading to an estimation of the ablated depth with different single pulse fluence for copper.

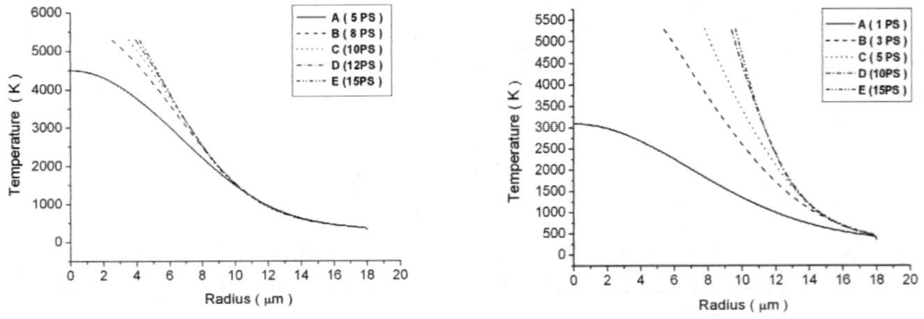

(a) pulse fluence: 1J/cm^2 (b) pulse fluence: 3.28 J/cm^2

Figure 5. Lattice temperature evolution for radial direction of the beam, leading to an estimation of the ablated radius with different single pulse fluence for copper.

Figure 4 represents the lattice temperature evolution for propagation direction of the beam, leading to an estimation of the ablated depth. Figure 5 represents the lattice temperature evolution for radial direction of the beam, leading to an estimation of the ablated radius. From Figure 4 and 5, we can see that the ablation velocity and the thermal effects zone relatively are larger when the ablation fluence is higher in the copper. The ablation in the copper can be mainly finished in 15-20 ps. The numerical results also show that the ablation depth and radius are gradually increasing along with the increase of the laser fluence and finally become changeless. It can be due to the injection depth and spot size of incident laser.

4. Summary

In short, we have reported numerical simulation for 3D temperature field of the metal ablated with femtosecond pulse laser by the DTE and FDM. The threshold fluence of single pulse in the copper had been obtained. The numerical results show that the electron-lattice energy coupling time presents a steady increase with the increase of the laser fluence. The copper jet due to phase explosion is earlier and the duration of ablation is longer with the increase of pulse fluence. When the laser fluence is higher than 1.5J/cm^2, the ablation start time is about 2-3 ps. With the further increasing laser fluence, the ablation start time finally becomes changeless. The ablation velocity and the thermal effects zone relatively are larger when the ablation fluence is higher in the copper. The ablation in the copper can be mainly finished in 15-20 ps.

Acknowledgements

This work was supported by the Key Grant Project of the Ministry of Education of the People's Republic of China (No.10410), and by the Basic Research Operating Expenses of Central University, the Ministry of Education of the People's Republic of China (No. ZXH2010C003)

References

[1] Pronko P P , Dutta S K, Squier J, Rudd J V, Du D and Mourou G 1995 *Opt .Commun* **114** 106
[2] Stuart B C , Feit M D, Herman S Rubenchik A M, Shore B W and Perry M D 1996 *Phys.Rev . B* **53**(4) 1749
[3] Xiaochang N and Chingyue W 2004 *Chinese Journal of Lasers* **31**(3) 277
[4] Anisimov S I, Kapeliovich B L and Perelman T L. 1974 *Sov . Phys. –JETP* **39** 375
[5] Elsayed-Ali H E, Norris T B, Pessot M A and Mourou G A 1987 *Phys. Rev . Let t.* **58** (12) 1212
[6] Zhu X 2000 Applied Surface Science, **167** 230
[7] Tzou D Y and Chiu K S 2001 *Int . J . Heat Mass Transf er* **44** 172
[8] Chen J K, Beraun J E and Grimes L E 2002 *International Journal of Solids and St ruct ures* **39** 3199
[9] Valette S, LeHarzic R and Huot N 2005 *Appl. Surf. Sci.* **247** 238
[10] LeHarzic R, Huot N, Audouard E, Jonin C, Laporte P, Valette S, Fraczkiewicz A and Fortunier R. 2002 *Appl. Phys. Lett.* **80**(21) 3886
[11] Miotello A and Kelly R 1999 *Appl. Phys. A* **69** 67
[12] Kelly R, Miotello A, Mele A Giardani G A, Hastie J W, Schenck P K and Okabe H 1998 *Appl. Surf. Sci.* **133** 251
[13] Bulgakova N M and Bulgakov A V 2001 *Appl. Phys. A* **73** 199
[14] Margetic V, Pakulev A, Stockhaus A, Bolshov M, Niemax K and Hergenr¨oder R 2000 *Spectrochimica Acta Part B* **55** 1771

Journal of Physics: Conference Series **276** (2011) 012033 doi:10.1088/1742-6596/276/1/012033

Temperature Fields of Al_2O_3/MgF_2 HR Coatings Prepared at Different Working Pressure

MeiQiong Zhan and ZhongLin Wu
Shanghai second polytechnic university, No.2360 Jinhai Road, Shanghai, P.R. China
E-mail: hbzmq@126.com

Abstract. A series of samples were deposited by electron beam evaporation using the same deposition process at different working pressure. Transmittance of the HR coatings was measured by Lambda 900 spectrometer. Laser-induced damage threshold (LIDT) was measured by a 355 nm Nd:YAG laser with a pulse width of 7 ns. The LIDT results range from 2.69 J/cm^2 to 11.03 J/cm^2 with the working pressure changing. It was found that working pressure has important effects on the absorption and LIDT of 355nm Al_2O_3/MgF_2 HR coatings. The temperature rise of HR coatings under laser irradiation was calculated by interfaces absorption model based on the theory of temperature fields. The results of temperature fields agree with the results of LIDT.

1. Introduction

High-power laser damage to optical devices seriously restricts the development of high-power laser systems. It is of great significance to have research on the laser damage and hardening of the optical devices [1]. Moreover, LIDT of 355 nm HR coatings is a more severe problem than that of 1064 nm HR coatings [2]. One of the continuing challenges facing the high energy laser field is to develop 355nm HR coatings that can withstand higher laser energy densities. the understanding of the origin and mechanism of LIDT in 355 nm HR coatings is very important for improving the LIDT [3].

For applications in the UV spectral range, the number of useful coating material is very limited due to the high absorption at wavelengths near to the electronic band-gap of the materials. Magnesium fluoride (MgF_2) thin film has transparency in the far ultraviolet (FUV) down to 115 nm[5], so it is widely used for optical applications in the UV and deep UV range [6]. Alumina (Al_2O_3) is one of the most important oxide thin film materials for the manufacture of interference coatings in the UV spectral region, down to 200 nm [7]. As a consequence, they are widely used to realize optical coatings for high performance and laser-damage resistant applications [8-9]. In this paper the temperature rise of Al_2O_3/MgF_2 HR coatings under laser irradiation was calculated by interfaces absorption model based on the theory of temperature fields. It is helpful to elucidate the damage mechanism of Al_2O_3/MgF_2 HR coatings prepared at different working pressure.

2. Experiment details

All the samples were deposited by electron beam evaporation in the same coating chamber with different process condition. The fused silica and UBK7 substrates were high quality super-polished with surface and cleaned by ultrasonic wave in alcoholic solution before deposition. The details of coating stack and working pressure are listed in Table 1, where H stands for high index material,

Al_2O_3, with one QWOT (Quarter Wavelength Optical Thickness) and L stands for low index material, MgF_2, with one QWOT, respectively. All the samples were annealed in the coating chamber for 2 hours after deposition.

Table 1. The details of three-group samples.

Sample ID	Coating stack	Working pressure
A	13 (hl) h	6.0×10^{-3} Pa
B	10 (hl) h	6.0×10^{-3} Pa
C	10 (hl) h	2.0×10^{-2} Pa

3. Results and discussion

Transmittance of the HR coatings was measured by Lambda 900 spectrometer. The LIDT test was performed in the ''1-on-1''mode, according to ISO standard 11254-1.2, by a 355nm Nd: YAG pulse laser with a pulse width of 7 ns. The LIDT (J/cm2) was defined as the incident pulse's energy density when the damage occured at 0% damage possibility, and it could be obtained by linear extrapolation of the damage probability data to zero damage probability.

The transmittance results of the samples in UV region are given in Figure 1. It can be found that the transmittance results of the sample A and C are the lowest and highest among three samples, respectively. Lower transmittance means higher absorption. The sample C, which have higher working pressure, have higher transmittance and lower absorption in the UV region. The reason is that increasing working pressure can make the samples more fuller oxidation during deposition, which leads to lower absorption in sample C.

Figure 2 illustrates the LIDT results of the samples. It shows that the sample C has the highest LIDT among three samples. Sample A, which is 27 layers, has the lowest LIDT. Compared with sample B and C, it can be found that increasing working pressure can improve LIDT effectively.

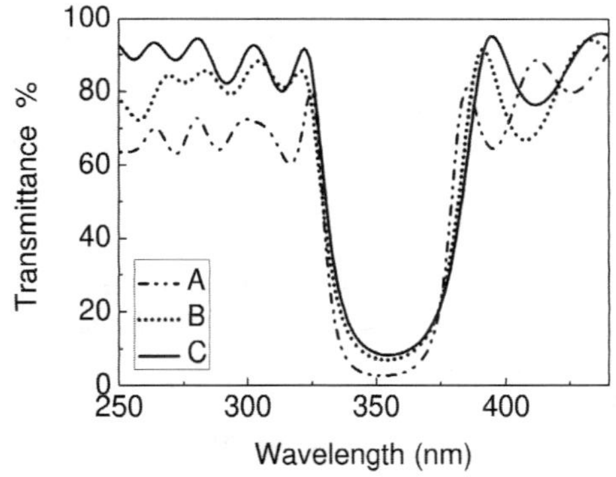

Figure 1. Transmittance of the samples.

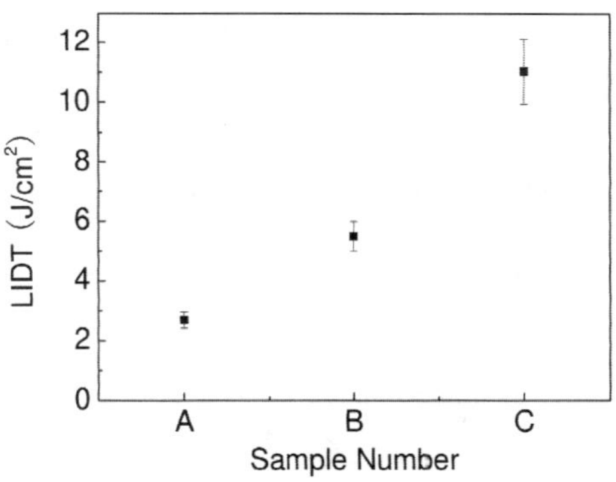

Figure 2. LIDT of the samples.

As far as the LIDT and transmittance are concerned in Figure 2 and 1, it can be found that sample A with the lowest transmittance and highest absorption in UV region has lowest LIDT among three samples. That is low LIDT emerging simultaneously with high absorption, which is agreed with the results in the reference [10].

According to the experimental results presented above, the most relevant damage mechanism associated with absorption is observed on 355nm Al_2O_3/MgF_2 HR coatings. In order to facilitate analysis of the damage mechanism, the temperature rise of different absorption in the HR coatings under laser irradiation can be calculated by interfaces absorption model based on the theory of temperature fields.

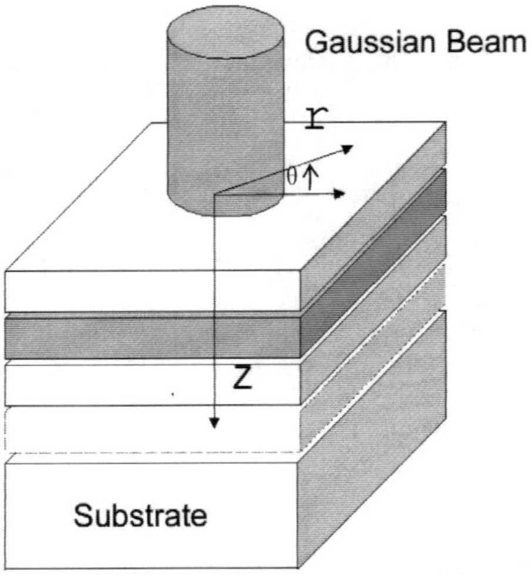

Figure 3. The HR coatings under laser radiation.

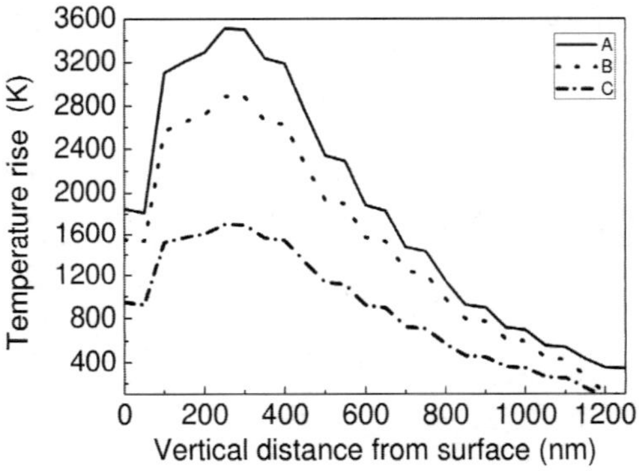

Figure 4. The temperature rise of Sample A, B and C.

Theory of temperature fields is based on the reference [11]. Figure 3 shows HR coatings with N-layer structure, which is illuminated on the surface by a Gaussian beam of light. The following Maxwell equations apply to the electric and magnetic field components in the kth layer:

$$\left\{\left(d^2/dz^2\right)+\left[2\pi n^{(k)}/\lambda_0\right]^2\right\}E_k(z)=0 \tag{1}$$

$$(d/dz)E_k(z)+i(2\pi/\lambda_0)H_k(z)=0 \tag{2}$$

The general solution to these equations can be written:

$$E_k(z)=A_1^{(k)}\left\{\exp\left[\frac{-i2\pi n^{(k)}}{\lambda_0}\left(z-\sum_{j=k=1}^{N+1}z_j\right)\right]\right\}+A_2^{(k)}\left\{\exp\left[\frac{+i2\pi n^{(k)}}{\lambda_0}\left(z-\sum_{j=k=1}^{N+1}z_j\right)\right]\right\} \tag{3}$$

$$H_k(z)=n^{(k)}A_1^{(k)}\left\{\exp\left[\frac{-i2\pi n^{(k)}}{\lambda_0}\left(z-\sum_{j=k=1}^{N+1}z_j\right)\right]\right\}-A_2^{(k)}\left\{\exp\left[\frac{+i2\pi n^{(k)}}{\lambda_0}\left(z-\sum_{j=k=1}^{N+1}z_j\right)\right]\right\} \tag{4}$$

With E and H determined, the rate of flow of energy through the multilayer can be calculated by invoking the Poynting vector theorem. The average amount of power per unit area that crosses the plane perpendicular to the Z axis at z can now be written

$$Y(z)=\mathrm{Re}\left[E(z)\cdot H^{*}(z)/2\right] \tag{5}$$

Starting at time t = 0, a narrow, circularly symmetric, Gaussian beam of light propagating in the z direction illuminates the surface. The incident intensity distribution can be written:

$$I(r,t)=\left[P_0(t)/(\pi r_0^2)\right]\exp\left[-(r/r_0)^2\right] \tag{6}$$

Where, r0 is the exp(-1) radius of the Gaussian beam and P0(t) is the instantaneous output power of the laser. From that equation the following equation can be obtained:

$$g(r,z,t)=\frac{d\left[I(r,t)Y(z)\right]}{dt\cdot dr\cdot dz} \tag{7}$$

g(r,z,t) is the power delivered to the unit volume by the laser.

Based on the theory of temperature fields mentioned above, the temperature rise of different absorption in the HR coatings under laser irradiation can be calculated by interfaces absorption model [12]. For HR coatings, we assume that the laser energy is $10J/cm^2$ (355nm, 7ns).

Figure 4 gives the result of temperature rise of Sample A, B and C. The highest temperature is occurred at about 300nm from the surface of samples. Under the same laser irradiation, Sample A, which has the highest absorption, also has the maximal temperature rise about 3500k. The maximal temperature rise of the sample C is about half of the sample A. correspondingly the LIDT and the absorption of the sample C is the highest and lowest, respectively. It is illustrated that the temperature rise of the HR coatings shows good correspondence with the LIDT and absorption results, which means that the damage mechanism for 355nm Al_2O_3/MgF_2 HR coatings is confirmed with temperature fields and interfaces absorption model.

4. Conclusions

As far as the temperature field of Al_2O_3/MgF_2 HR coatings deposited at different working pressure are concerned, it is found that the Al_2O_3/MgF_2 HR coatings with higher working pressure have higher transmittance, lower absorption in the UV region. Sample C has the highest LIDT among the samples. The damage mechanism for 355nm Al_2O_3/MgF_2 HR coatings is confirmed with temperature fields.

Acknowledgments

The work was supported by Innovation Program of Shanghai Municipal Education Commission (No. 09YZ452) and by youth foundation of Shanghai Municipal Education Commission (No. egd-07012).

References

[1] Ping W W et al 2003 *Opt. Laser. Technol.* **35** 303
[2] Feit M D et al 1998 *Proc. SPIE.* **3244** 350
[3] Li B C et al 2000 *Proc. SPIE.* **3902** 145
[4] Zhan M Q et al 2006 *Opt. Eng.* **45** 063801-1
[5] Larruquert J I et al 2003 *Opt. Commun.* **215** 93
[6] Jacob D et al 2000 *Thin Solid Films* **360** 133
[7] Carniglia C K et al 1985 *Nat. Bur. Stand. (U.S.) Spec. Publ.* **688** 347
[8] Kaiser N et al 1995 *Thin Solid Films* **260** 86
[9] Dijon J et al 1998 *Proc. SPIE.* **3244** 406

[10] Zhan M Q et al 2005 *Chin. Phys. Lett.* **22** 1246
[11] Mansuripur M et al 1982 *Appl. Opt.* **21** 1106
[12] Zhan M Q et al 2008 *J. Phys. D: Appl. Phys.* **41** 1

Journal of Physics: Conference Series **276** (2011) 012034 doi:10.1088/1742-6596/276/1/012034

Deflection tomography of a complex flow field based on the visualization of projection array

Bin Zhang and Zhanli Miao

College of Electromechanical Engineering, Qingdao University of Science and Technology, Qingdao, Shandong 266061, PR China

E-mail: zb-sh@163.com

Abstract. Tomographic techniques are used for the investigation of complex flow fields by means of deflectometric methods. A new deflection tomographic setup for obtaining an array of multidirectional deflectograms is presented. Deflection projections in different angles of view can be captured synchronously in same optical path condition and arranged on the camera in two rows with three views in each row. Tikhonov regularization method is used to reconstruct temperature distribution from deflectometric projection data. The conjugate gradient method is used to compute the regularized solution for the least-square equations. The asymmetric flame temperature distribution in the horizontal section was reconstructed from limited view angle projections. The experimental results of reconstruction from real projection data were satisfactory when compared with the direct thermocouple measurements.

1. Introduction

Tomography is a technique for determining the local distribution of a parameter in a system from path integrated or line-of-sigh measurements. In fluid flow systems line-of-sight measurements are usually based on refraction of a light beam. Refractive techniques involve measurements of the optical phase by interferometry and holographic interferometry. The gradient in the refractive index is measured by deflectometry [1]. Compared with interferometric methods applied to flow visualization and measurement [2], the deflection technique has been shown to be simpler and easier to use, and has been successfully applied to the mapping of two-dimensional phase objects [3-5]. It should be noted that because of its large dynamic range and applicability to unfavourable circumstances, the moire deflectometric method is widely utilized for measuring complex flows, such as rocket jet plumes [6] and supersonic flows in wind tunnels [7].

Deflection tomography comprises two major steps in reconstructing three-dimensional refractive-index fields. First, multidirectional deflection projections have to be experimentally captured. Second, the data are numerically processed with computational tomographic algorithms for reconstruction from projections.

To data, various experimental configurations have been reported for obtaining field projections. Typical of these can be rotation of the object field or probing beam, and spatial or temporal cascade division and redirection of the probing beam [5]. These approaches exhibit both merits and shortcomings. The object or beam rotation scheme is simple, but it dose not allow truly instantaneous capture of a field. The beam division approach can provide only a limited number of projections as a result of the complexity that arises because the setup has to be duplicated to obtain each projection.

Computational reconstruction techniques can be classified into transform methods and series expansion methods [8]. Transform algorithms are direct and computationally efficient, but require axisymmetric measured distribution or well-conditioned data that cover an angular range of 180° with a reasonable number of projections in the absence of an opaque object in the field. Series expansion methods, known as algebraic reconstruction technique (ART) [9], allow direct computational reconstruction from a limited data set, which may provide a restricted view, a limited number of projections, or incomplete projections as a result of the presence of an opaque object. Recently, a modified ART has been applied to moire deflection tomography [7, 8]. The algorithm was derived from the basic deflection formula and the deflection angles were used directly in iteration. Although the modified ART is proved to be more accurate than classic ART in simulated reconstruction using both limited and incomplete projection data, this algorithm has to be performed a large number of iterations.

In this paper, we introduce a new deflection tomographic setup that can obtain an array of multidirectional deflectograms. Deflection projections in different angles of view can be captured synchronously in same optical path condition. In reconstruction, the Tikhonov regularization method is used to calculate asymmetry temperature distributions. The technique is demonstrated by measuring a flame temperature field.

2. Deflection tomography apparatus

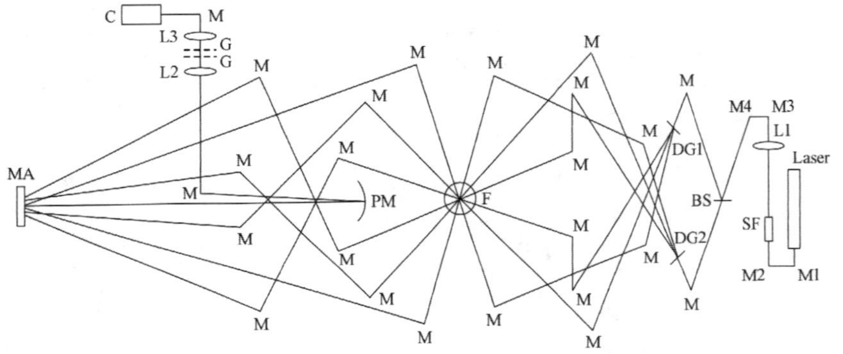

Figure 1. Tomographic setup: Mi, mirror; SF, spatial filter; Li, lens; BS, beam splitter; DGi, diffraction grating; F, flow field; PM, parabolic mirror; MA, mirror array; BS, beam splitter; C, CCD camera.

The optical configuration used in experiment is shown in figure 1. A helium-neon continuous wave laser (λ=632.8nm) is used as the light source. Two mirrors (M1, M2) and beam expanding optics (SF and L1) are used to produce a collimated beam. Mirrors M3 and M4 direct the ~40-mm diameter plane wave to a 50-50 beam splitter. The two resulting beams are directed onto separate diffraction gratings (DG1 and DG2). Each diffraction grating produces three beams of equal intensity. The resulting six probe beams intersect in the test volume with viewing directions of 0°, 26°, 52°, 88°, 114°, and 140°. After passing the apertures, the beams are directed onto an array of mirrors (two rows, each with three 30 × 30 mm mirrors). The mirror array is used to direct all the probe beams along the same optical axis. These parallel beams are then collected onto a 200-mm diameter parabolic mirror (PM), which was used to form a demagnified image of all views in front of lens L2. The probe beams are made parallel and imaged through two Ronchi gratings onto a CCD camera. The six deflectograms, where each deflectogram represents a tomographic viewing direction, are captured simultaneously on the camera with pixel resolution 600 (H) × 800 (W). The views are arranged on the camera in two rows with three views in each row and do not overlap.

3. Regularized reconstruction algorithm for deflection tomography

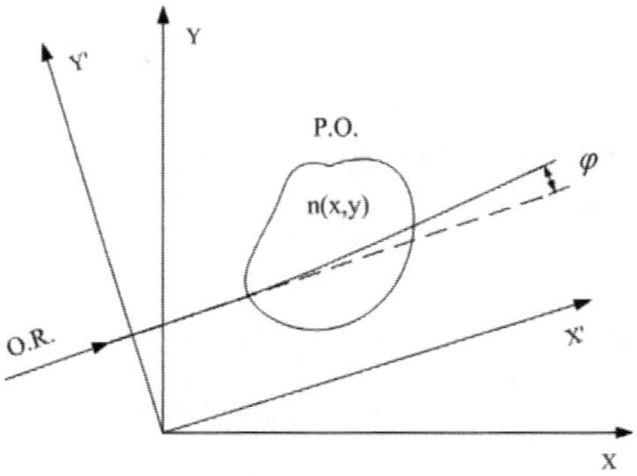

Figure 2. Schematic diagram of the coordinate system. is the ray deflection angle. O.R. represents a typical ray passing through the phase object P.O.

When a ray passes through a phase object, it is bent due to refraction associated with the gradients of the refractive-index within the object. In the coordinate system shown in figure 2, the ray deflection angle φ in the y' direction is [2]

$$\varphi \approx \int_{-\infty}^{\infty} \frac{1}{n_0} \frac{\partial n(x,y)}{\partial y'} dx', \tag{1}$$

Where $n(x, y)$ is the refractive-index associated with the two-dimensional phase object at a point (x, y) and n_0 is the refractive-index of the surroundings. In deriving equation (1) the in-plane deflection approximation and paraxial approximation have been used. The beam deflection angle obtained form deflectogram can be transformed into the path difference p as follow

$$p(x', y') = \int_{-\infty}^{\infty} n_0 \cdot \varphi(x', y') dy'. \tag{2}$$

A set of linear algebraic equations can be formed. That is

$$\mathbf{p} = \mathbf{A}\mathbf{n}, \tag{3}$$

where **p** is the measurement vector, **A** is the basis function projection matrix, and **n** is the object vector.

The basic principle of Tikhonov regularization method [10] is to find a **n** that minimizes

$$J_\alpha(\mathbf{n}) = \|\mathbf{An} - \mathbf{p}\|^2 + \alpha \|\mathbf{Dn}\|^2, \tag{4}$$

where **D** is a regularization operator and α is a regularization parameter, which plays an important role in reconstruction. This regularization method is effective for the kind of reconstruction problems where the unknown quantity to be reconstructed is continuously distributed in space. An approximation for α is suggested by [10] as

$$\alpha \approx \frac{2\|\mathbf{p} - \mathbf{An}(0)\|^2}{\|\mathbf{Dn}(0)\|^2}, \tag{5}$$

and **n**(0) is calculated by

$$\mathbf{n}(0) = (\mathbf{A}^T \mathbf{A})^{-1} \mathbf{A}^T \mathbf{p}. \tag{6}$$

The solution of equation (3) under the minimization of equation (4) can be obtained from normalized equation (7)

$$(\mathbf{A}^T \mathbf{A} + \alpha \mathbf{D}^T \mathbf{D}) \mathbf{n}_\alpha = \mathbf{A}^T \mathbf{p}. \tag{7}$$

A regularization operator for 2-D reconstruction [11] is used as follows

$$D_{0s} = n_0 - (1/8)(n_1 + n_2 + n_3 + n_4 + n_5 + n_6 + n_7 + n_8) \qquad s = 1, 2, \cdots k. \tag{8}$$

Three kinds of elements having different relationships with their neighbor elements are shown in figure 3. D_{0s} in equation (8) is the element in the line matching n_0.

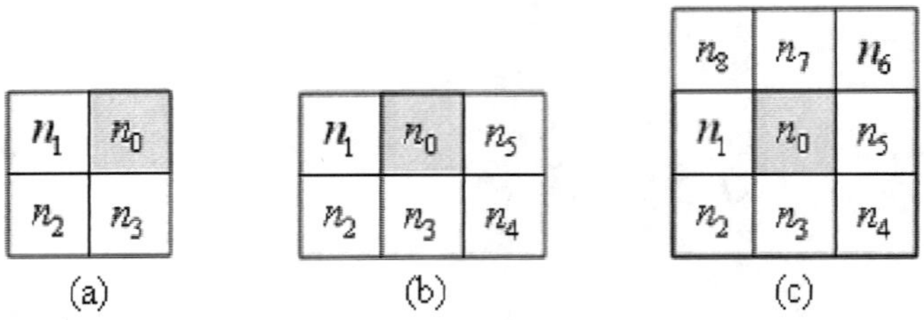

Figure 3. Three kinds of elements having different relationships with their neighbor elements. (a) the elements in the crank points; (b) the elements along the edge; (c) all the rest of the elements.

A refractive index field is calculated with the conjugate gradient method. The detail steps of the method are as follows.

(1) Initialize values.

$$\mathbf{E} = \mathbf{A}^T\mathbf{p}\,,$$
$$\mathbf{F} = \mathbf{A}^T\mathbf{A} + \alpha\mathbf{D}^T\mathbf{D}\,,$$
$$\mathbf{P}^{(1)} = \mathbf{Q}^{(0)} = \mathbf{E} - \mathbf{F}\mathbf{n}\,,$$
$$\gamma = 1\,,$$

where γ is the iteration number.

(2) Find values for a new iteration.

$$\begin{cases} \beta_r = (\mathbf{Q}^{(r-1)}, \mathbf{Q}^{(r-1)})/(\mathbf{P}^{(r)}, \mathbf{F}\mathbf{P}^{(r)}) \\ \mathbf{n}^{(r)} = \mathbf{n}^{(r-1)} + \beta_t \mathbf{P}^{(r)} \\ \mathbf{Q}^{(r)} = \mathbf{Q}^{(r-1)} - \beta_t \mathbf{F}\mathbf{P}^{(r)} \\ \chi_{r+1} = (\mathbf{Q}^{(r)}, \mathbf{Q}^{(r)})/(\mathbf{Q}^{(r-1)}, \mathbf{Q}^{(r-1)}) \\ \mathbf{P}^{(r+1)} = \mathbf{Q}^{(r)} + \chi_{r+1}\mathbf{P}^{(r)} \end{cases}$$

(3) Test the convergence.

If $\mathbf{P}^{(\gamma+1)} / \mathbf{P}^{(\gamma)} \geq \varepsilon$, where ε is a precision parameter in estimation, return to step (2); otherwise, stop the iteration estimation.

4. Experimental results

The temperature field was generated by flames of three candles, which were fixed at each vertex of an equilateral triangle. The six deflectograms at view angles of 0°, 26°, 52°, 88°, 114°, and 140°, respectively, which were captured by the deflectometric system, are shown in figure 4. With the use of automatic deflectogram-processing software, the deflectograms were processed with a PC-based image-processing system. From the processed data, the refractive-index field was reconstructed by applying the Tikhonov regularization method. One 30×30 mm^2 square region, which was in the horizontal plane 1.0 cm above the candles, was selected to be reconstructed. The reconstructed area was performed over a 30×30 grid array. Each projection consisted of 150 rays. The temperature distribution was then calculated by using the Glastone-Dale formula with an ambient temperature of 30 •. The cross-sectional temperature distribution is shown in figure 5. For cross-evaluation, the thermocouple measurements were also conducted. At the three points just above three candles, the construction result was in an average error of ~6% when compared with the thermocouple measurements.

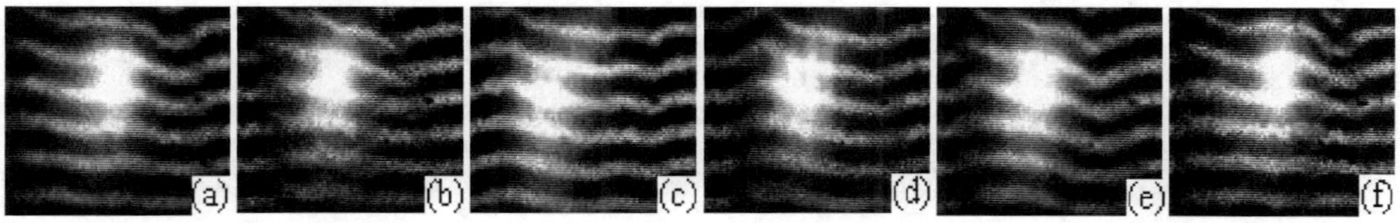

Figure 4. Multidirectional moire deflectograms at different view angles. (a) 0°; (b) 26°; (c) 52°; (d) 88°; (e) 114°; and (f) 140°.

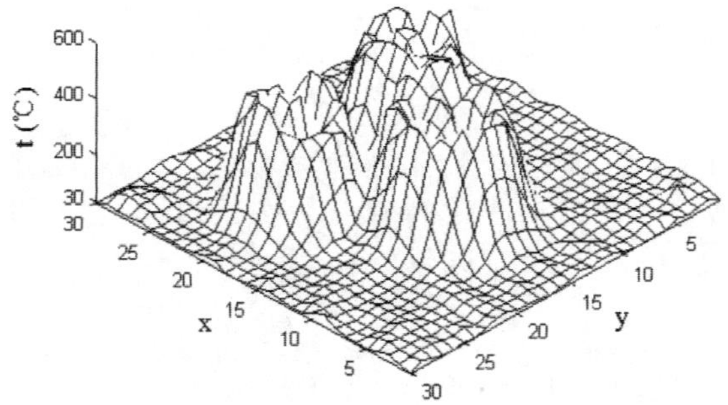

Figure 5. Reconstructed temperature distribution at a designated cross section.

5. Conclusions

A system for obtaining a large ensemble of scalar measurements using deflectometric tomography is presented. Six deflectograms, where each deflectogram represents a tomographic viewing direction, are captured simultaneously on the camera. The views are arranged on the camera in two rows with three views in each row and do not overlap. Tikhonov regularization method for deflection tomography is used to reconstruct the refractive-index field. A flame temperature distribution was reconstructed. In future research, some improvements to the current system will be performed, and some more complex flow fields, such as high temperature jet, will be measured with deflection tomography.

Acknowledgments

This work was supported in part by a grant from the Ph.D. Foundation of Department of Science and Technology of Shandong Province (No. 2008BS01004).

References

[1] Ajay K. Agrawal, Burt W. Albers and DeVon W. Griffin 1999 *Appl. Opt.* **38** 3394
[2] Vest C M and Prikryl I 1984 *Appl. Opt.* **23** 2433
[3] Kafri O 1985 *Opt. Eng.* **24** 944
[4] Chen Y Y, Song Y and He A Z 2009 *Appl. Opt.* **48** 489
[5] Song Y Z and Zhao Z M 2009 *Proc.SPIE* **7293** 72831Y
[6] Zhang B, Song Y, Song Y Z and He A Z 2006 *Opt. Eng.* **45** 117002
[7] Zhang B, He Y, Song Y and He A Z 2009 *Opt. Laser in Eng.* **47** 1183
[8] Yan D P and Cha S S 1998 *Appl. Opt.* **37** 1159
[9] Mishra D, Longtin JP, Singh RP and Prasad V 2004 *Appl. Opt.* **43** 1522
[10] Reginska T 1996 *SIAM J. Sci. Comput.* **17** 740
[11] Zhou H C, Lou C, Cheng Q and Jiang Z W 2005 *Proc. Comb. Inst.* **30** 1699

Approach dealing with transversely spatial profile of pump laser in Z-scan technique

Y Zhang[1], S Y WANG[1], P D Zhao[1, 3], D Y Wang[1], M Liu[1] and Z D Zhang[1]
[1]School of Science, Hebei University of Technology, 300130, Tianjin PRC

E-mail: pdzhao@eyou.com; pd_zhao@126.com

Abstract. The spatial pulse profile makes experimental values of Two–and three photon absorption (TPA and 3PA) coefficient deviate from its true value. We report an approach taking account of the influence of pump laser pulse profile on the nonlinear absorption coefficients (NAC) in Z-scan technique. We developed a new approach in which the factor (f factor) introduced to describe the influence coming from the spatial profile of laser on two and three photon absorption coefficient. The approximation of the approach is also estimated quantitatively. With Gaussian beams the numerically related results show that, compared with NAC based on the way dealing with the pulse in usual Z-scan, the relative differences of 2-photon and 3-photon absorption coefficients obtained based on the traditional one are less than 4.2% and 16.7%. The results suggest that the factor may become useful and simple parameters in dealing with the NAC deviation resulting from the pump laser pulse envelopes for the purpose of shortcutting the Z-scan datum process.

1. Introduction
It is well known that Z-scan technique has become a very popular tool for determining both magnitude and sing of nonlinear refractive index and nonlinear absorption coefficient of various media. This technique is based on the principles of spatial beam distortion but offers simplicity as well as very high sensitivity [1-4]. However, the temporal and spatial profiles of pump pulse laser may make experimental value of NAC deviate from its true value [5, 6].Consider the spatial-temporal profiles independent of each other, we report a new method for measuring the nonlinear absorption coefficients. In which a simple factor relative to the pulse spatial profile of pump laser was introduced to express the pulse influence on β or γ of the measured medium in nonlinear transmission. Here, we describe this method in detail and demonstrate it may be applied in Z-scan technique for the measurements of both β and γ [7].

2. Theory
According to the principle of 2PA and 3PA processes, the beam intensity change along the propagation direction (z axis) can be described as [8, 9]

$$dI / dz = -\alpha I - \beta I^2 - \gamma I^3 \quad (1)$$

where α is the attenuation coefficient due to linear absorption, β and γ are the nonlinear absorption coefficients due to 2PA and 3PA, respectively. The excitation irradiance I=I(r, z, t) is a function of time t and position r, z as well as beam penetrating depth z′ of sample. All of the four variables are

[3] To whom any correspondence should be addressed.

independent of each other.

For a spatially Gaussian pulse laser, we can write E as

$$E_0(z,r,t)=E_0(t)\frac{w_0}{w_z}\cdot\exp(-\frac{r^2}{w^2(z)}-\frac{ikr}{2R(z)})e^{-i\phi(z,t)} \tag{2}$$

Where $w^2(z)=w_0^2(1+z^2/z_0^2)$ is the beam radius, $R(z)=z(1+z^2/z_0^2)$ is the radius of curvature of the wave-front at z, $z_0 = kw_0^2/2$ is the diffraction length of the beam, $k=2\pi/\lambda$ is the wave vector, and λ is the laser wavelength, all in free space.

After consideration of its spatial-temporal influence on 2PA coefficient, the nonlinear normalized energy transmittance can be written as [5]

$$T_{2PA}=\frac{\int_{-\infty}^{+\infty}\ln[1+\beta LI_0 g(t)/(1+z^2/z_0^2)]dt}{\beta LI_0/(1+z^2/z_0^2)\int_{-\infty}^{+\infty}g(t)dt} \tag{3}$$

the 3PA coefficient, the nonlinear normalized energy transmittance can be written as

$$T_{3PA}=\frac{\int_{-\infty}^{+\infty}\ln(\sqrt{2L\gamma I_0^2 g^2(t)/(1+z^2/z_0^2)^2}+\sqrt{1+2L\gamma I_0^2 g^2(t)/(1+z^2/z_0^2)^2})dt}{\int_{-\infty}^{+\infty}\sqrt{2L\gamma I_0^2 g^2(t)/(1+z^2/z_0^2)^2}dt} \tag{4}$$

where The function g (t) represents the temporal envelope of pump laser pulse. For the Gaussian pulse model,

$$g(t)=e^{-2(\frac{t}{\tau_G})^2} \tag{5}$$

We assume that the laser intensity in media can be expressed as

$$I=I(r,z,t)=I_0 w(z)f(r)g(t) \tag{6}$$

where f(r) and g(t) represent, respectively, the spatial and the pulse-temporal profiles of I(r, z, t). I_0 is the incident peak intensity of pump pulse laser. Thus, nonlinear absorption equation (1) can be rewritten as

$$\frac{d}{dz}[I_0 w(z)f(r)g(t)]=\alpha[I_0 w(z)f(r)g(t)]+\beta[I_0 w(z)f(r)g(t)]^2+\gamma[I_0 w(z)f(r)g(t)]^3 \tag{7}$$

In order to obtain the expression of measurable pulse energy, we first integrate equation (7) over the time meanwhile ordering [10].

$$g_{2PA}=\frac{\int_{-\infty}^{+\infty}g^2(t)dt}{\int_{-\infty}^{+\infty}g(t)dt}=\frac{\int_{-\infty}^{+\infty}\int_0^{+\infty}I_0 w(z)f(r)g(t)g(t)2\pi rdrdt}{\int_{-\infty}^{+\infty}\int_0^{+\infty}I_0 w(z)f(r)g(t)2\pi rdrdt}$$

$$=\frac{\int_{-\infty}^{+\infty}\int_0^{+\infty}I(r,z,t)g(t)2\pi rdrdt}{\int_{-\infty}^{+\infty}\int_0^{+\infty}I(r,z,t)2\pi rdrdt}=<g(t)> \tag{8}$$

$$g_{3PA}=\frac{\int_{-\infty}^{+\infty}g^3(t)dt}{\int_{-\infty}^{+\infty}g(t)dt}=\frac{\int_{-\infty}^{+\infty}\int_0^{+\infty}I(r,z,t)g^2(t)2\pi rdrdt}{\int_{-\infty}^{+\infty}\int_0^{+\infty}I(r,z,t)2\pi rdrdt}=<g^2(t)> \tag{9}$$

and

$$f_{2PA}=\frac{\int_{-\infty}^{+\infty}f^2(r)2\pi rdr}{\int_{-\infty}^{+\infty}f(r)2\pi rdr} \tag{10}$$

$$f_{3PA} = \frac{\int_{-\infty}^{+\infty} f^3(r) 2\pi r dr}{\int_{-\infty}^{+\infty} f(r) 2\pi r dr} \tag{11}$$

The spatial part of equation (7) can be then expressed as

$$\frac{d}{dz'}[I_0 w(z) f(r)] = \alpha[I_0 w(z) f(r)] + g_{2PA} f_{2PA} \beta [I_0 w(z)]^2 + g_{3PA} f_{3PA} \gamma [I_0 w(z)]^3 \tag{12}$$

Usually equation (12) can be solved exactly to give the transmitted irradiance I(r, z, L) in terms of the incident irradiance I(r, z, 0), here, L is the thickness of measured medium. For the spatially Gaussian distribution of laser in Z-scan measurement, we can assume that

$$I_0 f_G(r, z, 0) = I_0 [1 + z^2 / z_0^2]^{-1} \exp[-\frac{2r^2}{w_0^2 (1 + z^2 / z_0^2)}] \tag{13}$$

Considering that the actual quantity typically measured in a pulsed experiment is the transmitted energy, this energy can be expressed, through the temporal and transversely spatial integral at a given position z of the sample, as

$$E_L = \int_0^{+\infty} \int_{-\infty}^{+\infty} I(r, z, L, t) dt 2\pi r dr \tag{14}$$

Similarly, the initial pulse energy can also be expressed as

$$E_0 = \int_0^{+\infty} \int_{-\infty}^{+\infty} I(r, z, 0, t) dt 2\pi r dr \tag{15}$$

In pure 2PA measurement of a Z-scan with a fully open aperture, γ can be assumed to be about zero in equation (12).Thus we can obtain an expression of nonlinear energy transmission for 2PA as following

$$T_{2PA} = \frac{1}{\beta g_{2PA} f_{2PA} L I_0} \tag{16}$$

We can also express the nonlinear energy transmission for only 3PA measurement of some medium by a fully open aperture Z-scan as

$$T_{3PA} = \frac{1}{\sqrt{1 + 2\gamma g_{3PA} f_{3PA} L I_0^2}} \tag{17}$$

According to equations (8) - (11), we can easily obtain Gaussian pulse influence factors as $g_{2PA} = \sqrt{2}/2$, $g_{3PA} = \sqrt{3}/3$; $f_{2PA} = 1/2(1 + z^2/z_0^2)$, $f_{3PA} = 1/3(1 + z^2/z_0^2)^2$.

Thus, the nonlinear Z-scan energy transmission relative to Gaussian pulse laser can be simple expressed as

$$T_{2PA}^G = \frac{1}{1 + \sqrt{2}\beta L I_0 / [4(1 + z^2 / z_0^2)]} \tag{18}$$

and

$$T_{3PA}^G = \frac{1}{\sqrt{1 + 2L\gamma\sqrt{3} I_0^2 / [9(1 + z^2 / z_0^2)^2]}} \tag{19}$$

Thus, the pulse influence on the β and γ may be simple described in terms of the two factors in equation (18) and equation (19).

3. Results and discussion

For Gaussian pulse model, the change curves of equation (3) and equation (18) with sample position z are figured out in Figure 1 at a fixed β_0 and I_0. Figure 2 shows the change curves of equation (4) and equation (19) with sample position z are figured out at a fixed γ_0 and I_0.

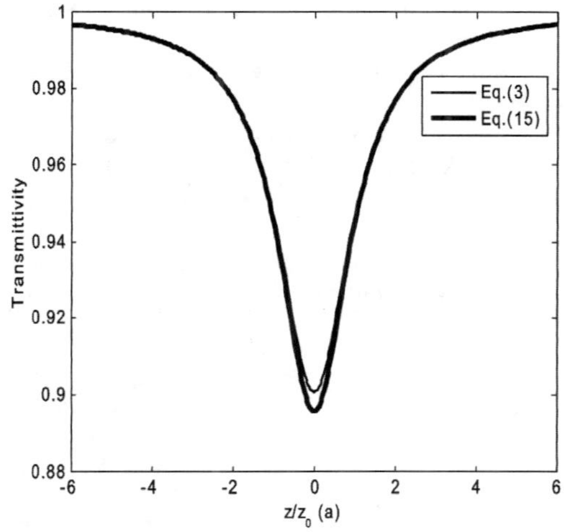

Figure 1. Z-scan transmittance curves for two–and three photon absorption
λ =532 nm , β_0 =5.8 cm / GW, I_0=0.21 GW/ cm^2, L=0.27cm.

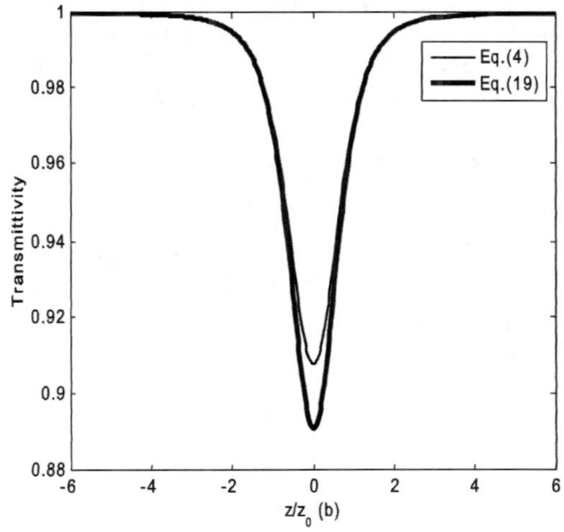

Figure 2. Z-scan transmittance curves for two–and three photon absorption
γ_0=2.7×10^{-6}cm^3/GW^2,I_0=500GW/cm^2, L=1cm.

For a given nonlinear energy transmission (usually in the range of 0.9 to 1.0), the approximation may be evaluated by the relative differences (δ) between β_0 (or γ_0) based on equation (3) (or equation (4)) and β (or γ) based on equation (18) (or equation (19)) obtained by way of numerical simulation. According to the usual error theory, the δ values for β and γ can be expressed as

$$\delta_\beta = \frac{\beta_0 - \beta}{\beta} \tag{20}$$

and

$$\delta_\gamma = \frac{\gamma_0 - \gamma}{\gamma} \tag{21}$$

Figure 1 shows that a slight difference among the two curves. Our numerical simulation results indicate that the relative difference between β_0 and β is less than 4.20%. and the relative difference between γ_0 and γ is less than 16.70%.We find out that the relative difference between β_0 (γ_0) and β (γ) based on Equation (3)(equation (4)) and based on equation (18) (equation (19)). In the typical range of 0.9 to 1.0, nonlinear normalized energy transmission with a fully open aperture Z-scan. One may find that this difference can be almost neglected. In other word, equation (3) may be substituted by Equation (18) and equation (4) may be substituted by Equation (19).The details are as following:

Table 1: Pulse influence factors of Gaussian laser pulse model and the relative differences of β and γ based the factors in the range of 0.9 ~ 1.0 of nonlinear energy transmission in Z-scan

Pulse model	g_{2PA}	f_{2PA}	$\delta_{\beta,\max}(\%)$	f_{3PA}	g_{3PA}	$\delta_{\gamma,\max}(\%)$
Gaussian	$\sqrt{2}/2$	$1/2(1+z^2/z_0^2)$ 1.1.	4.20	$1/3(1+z^2/z_0^2)^2$	$\sqrt{3}/3$	16.70

4. Conclusions

Based on the results mentioned above, one may conclude that the β and γ deviations from pump laser pulse profile can be simple described by the pulse influence factors if the pulse envelope of excitation laser is known (or assumed) in the NAC measurement of some medium by way of Z-scan technique. It may be acceptable more easily for experimenters. The simple relations of equations (18) and (19) allow the nonlinear absorption coefficients (β and γ) to be obtained directly from the Z-scan data without resorting to numerical integral. Especially, for theoretical analysis, the two equations may be very helpful to discuss the pulse profile influence of pump laser on β and γ. It is expected that this approach will be a valuable tool for experimenters searching for highly nonlinear absorption materials.

Acknowledgements

Project supported by Key Subject Construction of Hebei Provincial University and the Doctoral Foundation of Hebei University of Technology.

References

[1] M Sheik-Bahae, A A Said and E W Van Stryland 1990 *IEEE J. Quant. Electr.* **26** 760
[2] Zhao W and Muhoray P Palffy 1993 *Appl. Phys. Lett.* **63** 1613
[3] Hughes S and Burzler J M 1997 *Phys. Rev. A* **56** 1003
[4] Zhang W and Kuzyk M G 2006 *Appl. Phys. Lett.* **89** 101
[5] Gibson A F, Hatch C B, Maggs P N D, Tilley D R and Walker A C 1976 *J. Phys.* **C9** 3259
[6] Sutherland R L 1996 *Handbook of Nonlinear Optics* Marcel Dekker, Inc., New York
[7] Xu C and Webb W W 1996 *J. Opt. Soc. Am. B* **13** 481
[8] Boggess T F, Bohnert K M , Mansour K, Moss S C , Boyd I W and Smirl A L 1986 *Quant. Electr.* **22** 360
[9] Tutt L W and Boggess T F 1993 Prog. *Quant. Electr.* **17** 299
[10] Zhao Peide, Zhang Zhidong, Bao Wenxia, Yu Qing, Zhang Yong, Zhang Zhongya, Wang Dayun and Liu Ming 2010 *Opt.Comm.* **283** 1924

Detection of Atmospheric Composition Based on Lidar

Jinye Zhang[1,2], Yala Tong[1], Xiaoling Yang[1], Jiaoli Gong[1] and Wei Gong[2]
1 School of science, Hubei University of Technology, Wuhan 430068, China
2 State Key Laboratory for Information Engineering in Surveying, Mapping and Remote Sensing, Wuhan University, Wuhan 430079, China

E-mail: yezi.zh@163.com

Abstract. A summary overview about the types of lidar and their own applications on atmosphere detection is presented. Measurement of atmospheric aerosols by Mie lidar and Raman lidar is focused. The vertical profiles of aerosols in the atmosphere are retrieved. And at the same time, through analyzing aerosol vertical content distribution, the atmosphere boundary layer and the cloud are also observed. All the results show that the lidar has good performance on detecting the atmospheric composition.

1. Introduction

In recent years, natural disasters happened frequently. More and more awareness is attracted to the environment around us. The atmosphere is so important that we human being breath it. Various atmospheric constituents such as water vapor, aerosols and ozone, and their contents in the atmosphere directly affect our health. Therefore, atmosphere detection is one of the most concerns in the atmospheric science.

There are many detecting means such as satellite sensors, the various weather stations on the ground, aircraft, and balloons, and so on. There are blind spots by satellite sensor, because it is susceptible to clouds. Most of the equipments in the ground weather stations only detect the atmospheric parameters on the ground surface or about the whole atmosphere layer. Aircraft and balloon can not do continuous, long-term detection [1].

Lidar, as an active detection tool, because of its high spatial and spectral resolution, excellent direction and coherence, large vertical span and high detection accuracy, it is widely used in atmospheric measurements.

2. Lidar Applications on Atmosphere Detection

At present, there are several main kinds of lidar for atmosphere composition detection, such as Mie Lidar, Rayleigh Lidar, Raman Lidar, Differential Absorption Lidar and Resonance Fluorescence Lidar. The Mie backscatter signals are strong and the technical requirements for Mie lidar are relatively simple. So, Mie lidar is widely used to detect large particles in the atmosphere. The atmospheric aerosols and clouds are usually retrieved. According to the vertical distribution of the back signals, aerosols' content and distribution, the cloud height, cloud thickness and cloud shape can easily be judged. Because of its strong returns, the detection range is large. For example, the volcano eruption may produce an aerosol layer in the stratosphere, and the aerosol layer can be observed by Mie lidar [2].

Rayleigh lidar is based on the interaction of air molecules and laser. Rayleigh backscatters have the same wavelength with the out laser. Because there are a large number of aerosol particles and clouds in the lower atmosphere, Mie scattering and Rayleigh scattering are mixed together. But in the middle atmosphere, there are few aerosol particles and the main atmosphere composition is air molecules. So the lidar signal is only contributed by molecules and it is Rayleigh scattering. The temperature profiles of the middle atmosphere and the characteristics of gravity waves can be analyzed by Rayleigh lidar [3-4].

Raman lidar is different with the above two kinds. Raman scattering is inelastic and the scatter signals have a shifted wavelength. And also the backscatter cross section is much small. So, the echoes are weak. Raman lidar is mainly used to detect and identify molecular species that have high concentration in the atmosphere, such as nitrogen, water vapor, carbon monoxide, sulfur dioxide [5].

Absorption principle is used in Differential absorption lidar. There are two lasers out beams with similar wavelengths. The detected object will absorb one beam much and absorb another little. Then, through calculating the ratio of the two back signals and comparing the difference of the absorption cross-section on two wavelengths, the density distribution of the detected element will be retrieved. The Differential absorption lidar system needs more techniques and the structure is more complicated. The molecular ozone [6] and air pollutions such as sulfur dioxide, nitrogen dioxide, nitric oxide, can be detected.

Resonance fluorescence lidar is to receive the molecular and atomic resonance fluorescence.However, fluorescence is easy to be interfered and often disappear in the collision. So, it is difficult to play in the low atmosphere. But at high altitude, the collision decreases significantly and the Resonance fluorescence lidar shows its advantages. It is mainly used to detect the atoms and ions in the middle and top atmosphere [7].

Currently, a number of lidar network have been established. They are Micro Pulse Lidar Network (MPLNET) [8], Regional East Atmospheric Lidar Mesonet (REALM) [9], Asian Dust Network (AD-Net) [10] and European Aerosol Research Lidar Network (EARLINET) [11]. The EARLINET covers the largest area in Europe to do atmosphere observation. The observations include aerosol extinction coefficient, backscatter coefficient, lidar ratio, particle size, cloud, atmospheric boundary layer and other atmospheric parameters [12-14]. In particular, the transmission and movement of Sahara Desert dust have been observed over past several years [15-16]. The AD-Net is initiated by Japan and the lidar sites are located in Japan [17-18].

But in china, the lidar business is still relatively young. "Lidar Development Seminar for Atmospheric Research in China" sponsored by "Key Laboratory of Atmospheric Composition and Optical Radiation" was held on 5-8 November 2009 in Hefei. All the scientists studying on lidar, atmosphere and climate weather in china took part in it. The aim was to talk about establishing lidar network in china for atmosphere observation.

3. Observations of aerosol, cloud, nitrogen and boundary layer

A Lidar system has been developed and the lidar site is at the latitude 30°32•N and longitude114°21•E in Wuhan [19]. Wuhan is the largest city in central China. The lidar consists of laser, telescope and signal recorder. The lidar out wavelengths are 532nm and 1064nm. The nitrogen and water Raman scattering signals, aerosol Mie backscatters and molecular Raleigh returns can be received. This information will be used to retrieve the vertical distribution of the atmospheric aerosols, clouds, atmospheric boundary layer, and atmospheric humidity.

First, we receive aerosol Mie backscatter signals at 532nm. The vertical distribution of Mie backscatters reflects the vertical distribution of aerosol contents. Formula one is Mie lidar equation:

$$P(\lambda_L, r) = C_0 P_0(\lambda_L) \frac{\varepsilon(r)}{r^2} \Delta r \beta(\lambda_L, r) \exp[-2\int_0^r \alpha(\lambda_L, r) dr] \quad (1)$$

The laser output wavelength is λ_L. $P(\lambda_L, r)$ is the Mie returns received at distance r, $P_0(\lambda_L, r)$is the laser output power. $\varepsilon(r)$ is the overlap function between the laser beam and the field of view of the receiver. $\beta(\lambda_L, r)$ is the backscatter coefficient of aerosol, $\alpha(\lambda_L, r)$ atmosphere extinction coefficients at wavelength λ_L, C_0 contains all other distance independent parameters.

Figure 1 shows a series of aerosol Mie backscatters at the night on September 12, 2009, from the local time 21:00 to 21:50. As we know that the strength of Mie back signals represents the content of aerosols in the lower atmosphere. So, we can get the vertical aerosol density distribution from the profiles of lidar data. For example, from the time 21:00 to 21:20, the maximum of the Mie signal is about in the range of 0 to 1.3 km, and during the time 21:20 to 21:50, the maximum falls below 1 km. And the maximum of the signal strength becomes larger. And at same time, we can observe that there is a cloud appearing during the time at height 2.2km. From the changes of these signal profiles, we see that the atmospheric aerosols fall down as the deep night coming. Because people's activities deduce in the night and the temperature becomes low.

The aerosols can be observed by Mie scatters, and also can be retrieved by Raman scatters. Raman lidar equation can be written as:

$$P(\lambda_X,\lambda_L,r)=CP_0(\lambda_L)\frac{\varepsilon(r)}{r^2}\Delta r\beta_x(r)$$
$$\exp\left\{-\int_0^r\{[\alpha_{mol}(\lambda_L,r)+\alpha_{mol}(\lambda_X,r)]+[\alpha_{aer}(\lambda_L,r)+\alpha_{aer}(\lambda_X,r)]\}dr\right\} \tag{2}$$

$P(r,\lambda_X,\lambda_L)$ is the power received from distance r at the Raman-shifted wavelength λ_X if the laser output wavelength is λ_L. $\alpha_{mol}(\lambda_L, r)$ and $\alpha_{mol}(\lambda_X, r)$ are the molecular extinction coefficients at wavelength λ_L, λ_X. $\alpha_{aer}(\lambda_L, r)$ and $\alpha_{aer}(\lambda_X, r)$ are the aerosol extinction coefficients. Because the nitrogen has the largest density in the atmosphere, the Raman scatter is strong. The X molecular is nitrogen. The laser λ_L is 532nm and nitrogen Raman scatter wavelength λ_X is 607nm. Then the aerosol extinction profile can be expressed by [20]

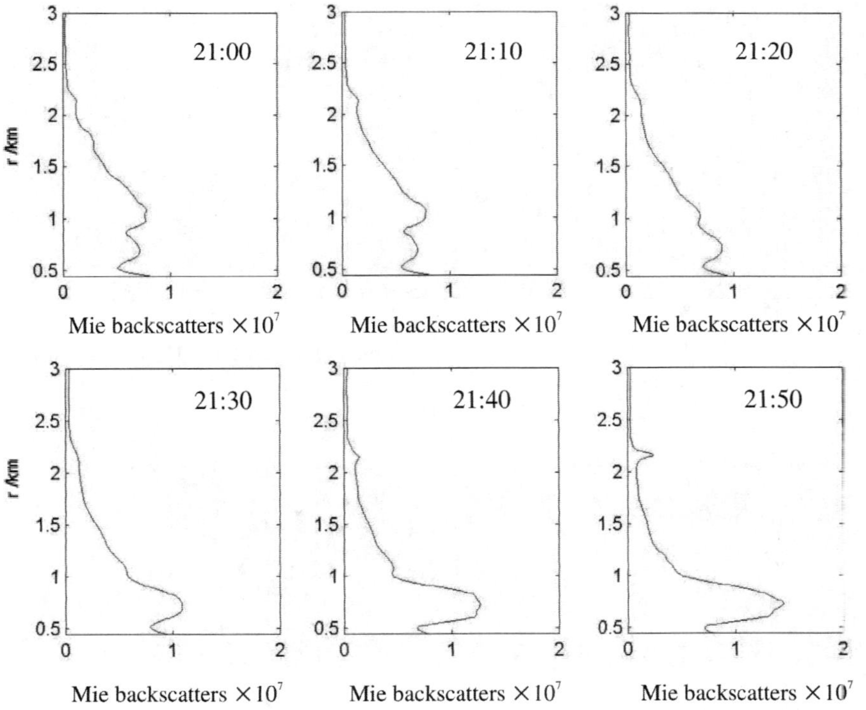

Figure 1. A series of aerosol Mie backscatters at the night on September 12, 2009.

$$\alpha_{aer}(\lambda_L, r) = \frac{\frac{d}{dr}\left[\ln \frac{n_x(r)}{r^2 P(r, \lambda_X, \lambda_L)}\right] - \alpha_{mol}(\lambda_L, r) - \alpha_{mol}(\lambda_X, r)}{1 + \frac{\lambda_L}{\lambda_X}} \tag{3}$$

Figure 2 shows the series of nitrogen Raman backscatters at the night on September 12, 2009, from the local time 21:00 to 21:50. As we know that the nitrogen is stable in the atmosphere and its percentage is 78%, so the backscatter profiles are stable too, not vary much.

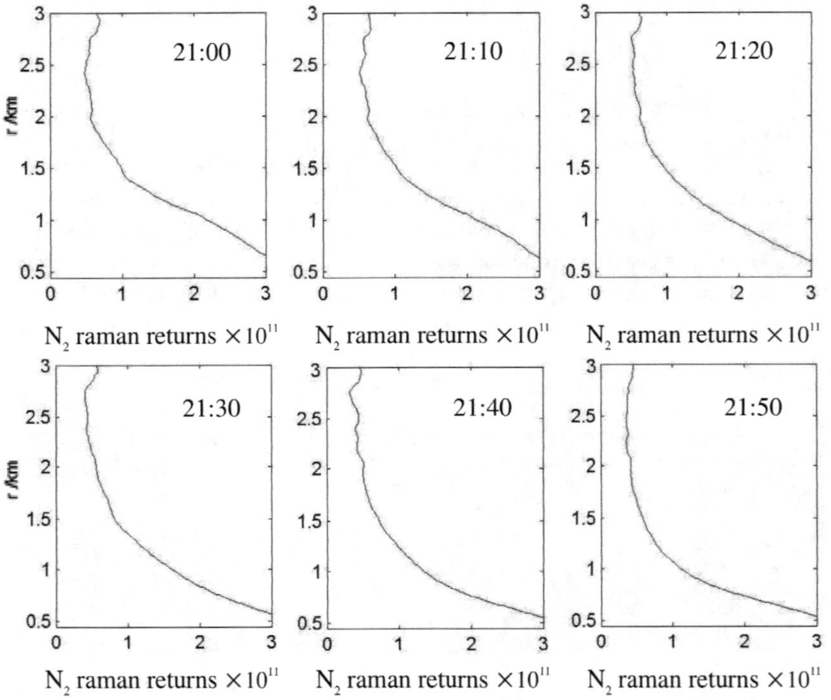

Figure 2. A series of N2 Raman backscatters at the night on September 12, 2009.

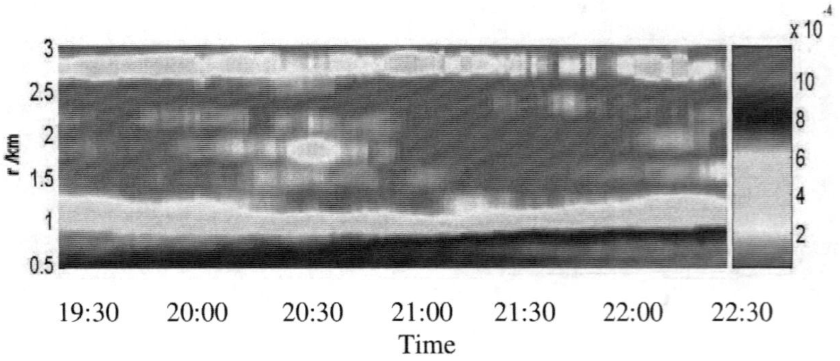

Figure 3. Atmosphere aerosol extinction coefficient at 532nm observed at 19:30-22:30 local time in the night on October 12, 2009.

Figure 4. A street scene on 9th October 2009.

Figure 3 shows the vertical distribution and continuous change of atmospheric aerosols below 3 km. This atmospheric aerosol observation is during the local time 19:30 to 22:30 in the night on October 12th in 2009. It is clear that there is a very stable atmospheric aerosol layer below 1.3 km. Because in early October, there has been no rain and wind for several weeks and there is an inversion layer over Wuhan. The dust is accumulated and difficult to disperse. Atmosphere boundary layer is characterized by the high density of aerosol particles, so, the stable aerosol layer between 0 and 1.3 km can be seen as the atmosphere boundary layer.

Figure 4 is a photograph of Wuhan taken by a reporter on October 9th. The photo shows that the air quality and visibility are very poor.

4. Conclusion

A Mie and Raman lidar has been used to detect the vertical distributions and changes of atmospheric aerosols and nitrogen during the night over Wuhan. The cloud and the atmosphere boundary layer have also been observed. These meteorological data are very important for knowing and forecasting the weather change rules. The lidar system can do real time, continuous and long-term observation of their vertical distribution profiles.

Acknowledgments

This work has been sponsored by Wuhan Programs for Science and Technology Development (Grant No. 201010621218).

References:

[1] Wilczak J M, Gossard E E, Neff W D and Eberhard W L 1996 *Bound .Layer Meteor.* **78** 321
[2] Northan 1973 *Appl. Opt.* **13** 2416
[3] Keckhut P and Haochecocne A 1993 *J.Atoms.Ocean.Tech.* **10** 850
[4] Cardner C S and Miller M S 1989 *J. Atmos. Sci.* **46** 1838
[5] Inaba H 1976 *laser monitoring of the Atmosphere* (Berlin: Spinger)
[6] Browell E V 1982 *Opt.Eng* **21** 128
[7] Yang G T, Liu B M and Wang J M 2003 *Chin. J. Geophy.s* **46** 576
[8] Welton E, Voss K, Quinn P, Flatau P, Markowicz K, Campbell J, Spinhirne J, Gordon H, and Johnson J 2002 *J. Geophys. Res.* **107** 8019
[9] Hoff R, Engel-Cox J, Krotkov N, Palm S, Rogers R, Mc-Cann K, Sparling L, Jordan N, Torres O and Spinhirne J 2004 *ESA Spec. Publ.* SP-561 683
[10] Yumimoto K, Uno I, Sugimoto N, Shimizu A and Satake S 2007 *Geophys. Res. Lett.* **34** L08806
[11] Murayama T and Sugimoto N, 2001 *J. Geophys. Res.* **106** 18346
[12] Böckmann C 2001 *Appl. Opt.* **40** 1329
[13] Wang X, Boselli A, Avino L D, Velotta R, Spinelli N, Bruscaglioni P, Ismaelli A and Zaccanti G 2005 *Appl. Phys.B* **80** 609
[14] Tomasi F D and Perrone M R 2006 *Atm. Res.* **80** 86

Journal of Physics: Conference Series **276** (2011) 012036 doi:10.1088/1742-6596/276/1/012036

[15] Matthias V and Bösenberg J 2002 *Atm. Res.* **63** 221
[16] Tafuro A M, Barnaba F, Tomasi F D, Perrone M R and Gobbi G P 2006 *Atm. Res.* **81** 67
[17] Yumimoto K, Uno I, Sugimoto N, Shimizu A, Liu Z and Winker D M 2007 *Atmos. Chem. Phys. Discuss.* **7** 15955
[18] Shimizu A, Sugimoto N, Matsui I, Arao K, Uno I, Murayama T, Kagawa N, Aoki K, Uchiyama A and Yamazaki A 2004 *J. Geophys. Res.* **109** D19S17
[19] Gong W, Zhang J Y, Mao F Y and Li J 2010 *Chin. Opt. Lett.* **8** 533
[20] Whiteman D N 1999 *Appl. Opt.* **38** 3360

Journal of Physics: Conference Series **276** (2011) 012037 doi:10.1088/1742-6596/276/1/012037

Laser pumping Cs atom magnetometer of theory research based on gradient tensor measuring

Zhang Yang, Kang Chong, Qingtao Wang, Cheng Lei and Caiping Zheng
College of Science, Harbin Engineering University, Harbin 150001, China
E-mail: zhangyang@hrbeu.edu.cn

Abstract. At present, due to space exploration, military technology, geological exploration, magnetic navigation, medical diagnosis and biological magnetic fields study of the needs of research and development, the magnetometer is given strong driving force. In this paper, it will discuss the theoretical analysis and system design of laser pumping cesium magnetometer, cesium atomic energy level formed hyperfine structure with the I-J coupling, the hyperfine structure has been further split into Zeeman sublevels for the effects of magnetic field. To use laser pump and RF magnetic field make electrons transition in the hyperfine structure to produce the results of magneto-optical double resonance, and ultimately through the resonant frequency will be able to achieve accurate value of the external magnetic field. On this basis, we further have a discussion about magnetic gradient tensor measuring method. To a large extent, it increases the magnetic field measurement of information.

1. Introduction

Magnetometer is an instrument used to measure the magnetic field, mainly in the air, land, and marine, biomedical and other fields to detect magnetic field. As the important role of the magnetometer in the national economy, space exploration, military technology and navigation field, it has been a hot issue in all countries. From the Development of magnetometer, could be divided into three stages [1], the first stage, adopt flux-gate magnetometer, proton magnetometer and optical pumping magnetometer of helium to measure total magnetic field intensity; the second stage, in addition to measurement of the total magnetic field intensity, but also make use of alkali metals (cesium, potassium, rubidium) laser optical pumping magnetometer measuring the horizontal and vertical magnetic field gradient which be paid more and more attention; the third stage, is just started, atomic magnetometer and superconducting magnetometer with very high resolution and sensitivity. From the magnetic detection technology perspective, now China is still in the first stage, with a big gap between the world advanced level, and high-performance magnetometer is under an embargo, included in the export control list of goods, thus developing laser optical pumping magnetometer of alkali metal is of great significance.

2. Theoretical analysis of Cesium (Cs) laser pumping magnetometer

Cesium laser optical pumping magnetometer is based on the Zeeman effect generated by energy levels of cesium atoms in a magnetic field , be made by the principle of magneto-optical double resonance caused by laser pump and RF magnetic field.

2.1. Cs atomic energy levels

Cesium is monobasic alkali metal, atomic number is 55, valence electrons are in the 6th shell, the main quantum number $n = 6$, the same principal quantum number n there

$$l = 0,1,\cdots,(n-1) \tag{1}$$

l is electron orbital angular momentum quantum number, in addition to orbital motion, the electron also do spin motion, the electron spin angular momentum quantum number s = 1/2, the interaction between spin and orbital motion of the electron (LS coupling)would produce energy levels splitting, known as the fine structure [2]. Therefore electronic orbital angular momentum and spin angular momentum composition of electronic total angular momentum, electronic total angular momentum quantum number is j, there

$$j = l+s, l+s-1,\cdots,|l-s| \tag{2}$$

Since s = 1 / 2, so the value of a set l,

$$j = l+1/2, |l-1/2| \tag{3}$$

Similar with electron, the atomic nucleus also has angular momentum, commonly called nuclear spin angular momentum, represent by I. The interaction between nuclear spin angular momentum and electronic total angular momentum (*I -J* coupling), producing hyperfine structure [3], using *F* to represent, *F* is the atomic total angular momentum quantum number, there

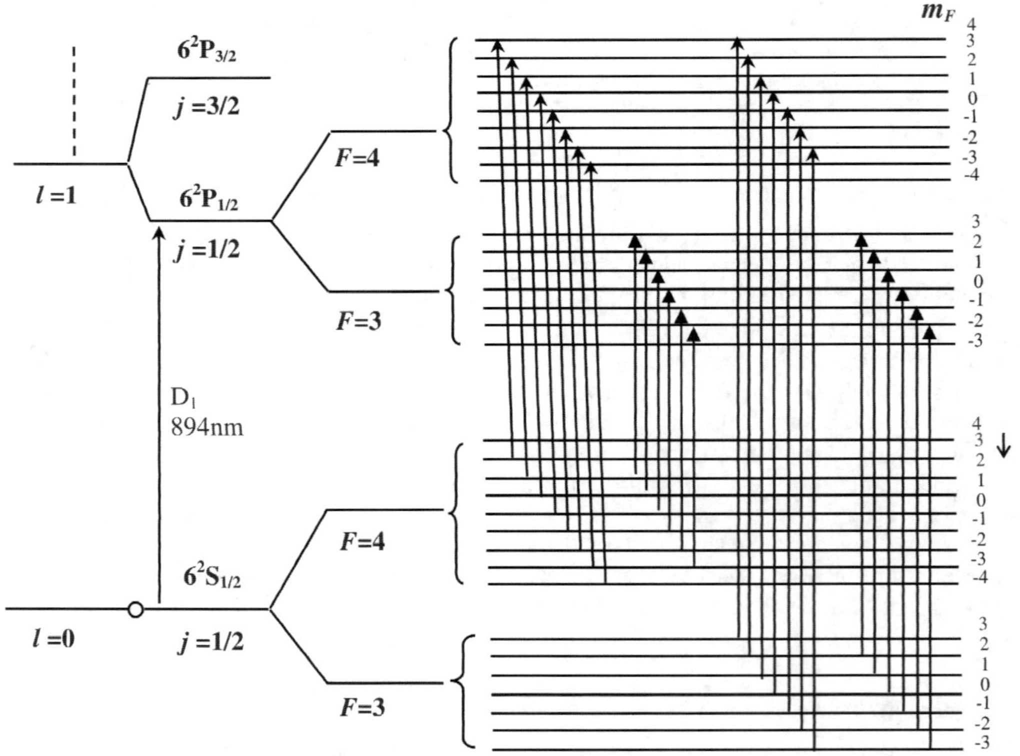

Figure 1. Cesium atomic energy level diagram with quantum number $n = 6$

$$F = I + j, I + j - 1, \cdots |I - j| \tag{4}$$

For the cesium atom, atomic nucleus spin angular momentum I = 7/2, when j = 1/2， with F = 4 and F = 3 two states; When j = 3/2, with F = 5, F = 4, F = 3 and F = 2 four states. When external magnetic field exists, the interaction between atomic total angular momentum and external magnetic field, hyperfine structure could produce Zeeman splitting further and form Zeeman sub-level, with the magnetic quantum number m_F to indicate that

$$m_F = F, F - 1, \cdots, -F \tag{5}$$

That is， split into $2F$ +1 Zeeman sub-levels which are equal level spacing. Cesium atomic-level structure talked above is shown in Figure 1.The condition of electronic transitions between energy levels [4] is shown

$$\begin{aligned} &\Delta L = \pm 1, \Delta J = 0, \pm 1, \Delta F = \pm 1, \Delta m_F = 0(\pi) \\ &or\ \Delta L = \pm 1, \Delta J = 0, \pm 1, \Delta F = 0, \pm 1, \Delta m_F = \pm 1(\sigma) \end{aligned} \tag{6}$$

2.2. *System design and principles analysis*

Cesium (Cs) laser optical pumping magnetometer system [5] is shown in Figure 2， the work process: the laser emit light of 894nm wavelength (D1 line), the light goes through the lens into parallel light, after a polarizer and 1/4 wave plate into laevogyrate circularly polarized light σ^+， Irradiate cesium absorption chamber (absorption chamber is made of the airtight glass, cesium elements of absorption chamber is gaseous), while adding a RF magnetic field in the vertical direction of the external magnetic field. When it produces magneto-optical double resonance, light through the absorption chamber is weakest, the moment could be detected by the optical detector, and moreover the value of the external magnetic field can be obtained through the resonance frequency.

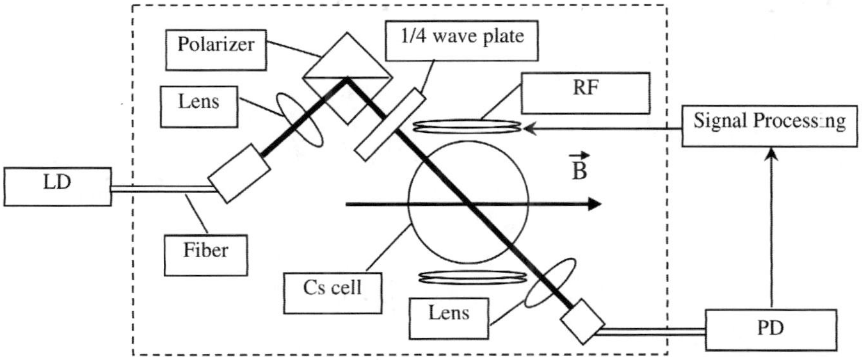

Figure 2. Cesium optical pumping magnetometer system diagram

Work principle: Laser emits wavelength of 894nm light(D1 line), known as the Figure 1, the electron can transit between the $6^2S_{1/2}$ state and $6^2P_{1/2}$ state, the formula (6) gives the transition of energy level conditions, according to Zeeman effect theory[6] , $\Delta m_F = +1$ is laevogyrate circularly polarized light σ^+ , $\Delta m_F = -1$ is dextrogyrate circularly polarized light σ^-， we adopt laevogyrate circularly polarized light σ^+ here , then only produce transition of $\Delta m_F = +1$， for example, the electron of ground state $6^2S_{1/2}$ with $m_F = +3$ energy level can transit to $6^2P_{1/2}$ with $m_F = +4$ energy level. From figure 1 we can find that $6^2P_{1/2}$ state does not exist $m_F = +5$ energy level, so the electron transition probabilities of ground state $6^2S1/2$ with $m_F = +4$ energy level is zero, that is, laevogyrate circularly polarized light could excitated each level on the ground state except $m_F = +4$ of electron to the states of $6^2P_{1/2}$. At the same time, electrons which transit to the $6^2P_{1/2}$ states after about 10^{-8} seconds return to the ground through state spontaneous transition, as the probability of return to the ground state of each level is basic equal, as a result electrons of $m_F = +4$ energy level will continue to increase. When the light irradiates continuously, it will reach dynamic equilibrium, and electrons of $m_F = +4$ energy level will greatly increase, that means electrons of ground state except $m_F = +4$ will be pumped empty. The optical pumping cause the uneven distribution of electrons of Zeeman energy levels of ground state, do not obey the Boltzmann distribution, that is, polarization [7]. The process can be seen from detect signal of the light detector, the moment when the light irradiation, equivalent of the total number of 7/8 of the electrons absorb σ^+ light, light through the absorption chamber is weak, as electronic continuous transition, light absorption decreased continuously, light through absorption chamber gradually increase, to the dynamic equilibrium, light intensity through the absorption chamber reach maximum and no longer changes. Now add a RF magnetic field in the vertical direction of the external magnetic field, when the RF magnetic field frequency f_0 is equal to adjacent Zeeman sub-level transition frequency, will produce magnetic resonance, resulting to the orientation effect, thus the electrons of Zeeman sub-level produce stimulated transition. As Figure 1, electrons of ground state with $m_F = +4$ energy level stimulated transition to $m_F = +3$ energy level, electrons of ground state with $m_F = +3$ energy level stimulated transition to $m_F = +2$ energy level, transition by order, So that the distribution of atoms obey the Boltzmann distribution law, then the orientation of atomic magnetic moment has been disrupted, meanwhile optical pumping goes simultaneously , electrons of ground state except $m_F = +4$ will be pumped to $m_F = +4$ once again, stimulated transition and optical pumping will reach a new dynamic equilibrium, at this time maximum absorption of light, the light intensity that light detector detects is weakest. The adjacent Zeeman sub-level transitions frequency is directly proportional to the measured external magnetic field, the Formula is:

$$f_0 = \frac{\Delta E}{h} = \frac{g_F \mu_B}{h} B = \gamma_p \frac{B}{2\pi} \tag{7}$$

g_F is Legendre factor, μ_B is Bohr magneton, γ_p is magnetogyric ratio, B is external magnetic field. From (7), we know, as long as measure the frequency of the RF magnetic field when the light detector detects the weakest light intensity, the value of the external magnetic field can be obtained.

3. Gradient measurement theory based on cesium optical pumping magnetometer

In order to understand the distribution characteristics and rules of the magnetic field better, in addition to measuring the magnetic field modulus and vectors, also need to measure space rate of change of the three components modulus and vectors, that is, gradient measurement[8].The advantage of gradient measurement can eliminate the effects of regional magnetic field and time-varying magnetic field. Gradient measurement is different from total magnetic field measurement, to obtain the gradient value of some direction at a point, requires cesium pumping vector magnetometer loaded with two three-axis coils measuring simultaneity. Set down the distance between the two magnetometer is ΔZ , the measurement value of magnetic above is B_1 , the measurement value of magnetic below is B_2 , the vertical gradient

$$GradB = \frac{B_1 - B_2}{\Delta Z} \tag{8}$$

By (8), can be seen that the two magnetometers closer, the higher resolution of gradient values, while the closer requires higher precision magnetometer, so system design should take overall consideration ,not only consider the precision of magnetometer, but also consider the resolution of gradient. Similarly longitudinal horizontal gradient and lateral horizontal gradient. According to the three components of the space rate of change above, could get magnetic gradient tensor [9], the complete magnetic field gradient tensor can be expressed as:

$$G = \begin{bmatrix} \frac{\partial B_x}{\partial x} & \frac{\partial B_x}{\partial y} & \frac{\partial B_x}{\partial z} \\ \frac{\partial B_y}{\partial x} & \frac{\partial B_y}{\partial y} & \frac{\partial B_y}{\partial z} \\ \frac{\partial B_z}{\partial x} & \frac{\partial B_z}{\partial y} & \frac{\partial B_z}{\partial z} \end{bmatrix} = \begin{bmatrix} g_{xx} & g_{xy} & g_{xz} \\ g_{yx} & g_{yy} & g_{yz} \\ g_{zx} & g_{zy} & g_{zz} \end{bmatrix} \tag{9}$$

B_x, B_y, B_z are the three components of the magnetic field B, magnetic gradient tensor with nine elements, as in the passive space, divergence and curl of B are zero, therefore

$$\nabla \cdot B = \frac{\partial B_x}{\partial x} + \frac{\partial B_y}{\partial y} + \frac{\partial B_z}{\partial z} = 0 \tag{10}$$

$$\nabla \times B = \begin{bmatrix} i & j & k \\ \frac{\partial}{\partial x} & \frac{\partial}{\partial y} & \frac{\partial}{\partial z} \\ B_x & B_y & B_z \end{bmatrix} = 0 \tag{11}$$

$B_{xx} + B_{yy} + B_{zz} = 0, B_{xy} - B_{yx} = 0, B_{xz} - B_{zx} = 0, B_{yz} - B_{zy} = 0$, Therefore, only five elements are independent in nine elements. And the magnetic gradient tensor is a real symmetric matrix, can be carried out real eigenvalue diagonalization,define eigenvalue of the magnetic gradient tensor as $\lambda_1, \lambda_2, \lambda_3$, the magnetic gradient tensor eigenvalue equation can be expressed as:

$$\lambda^3 - I_0\lambda^2 + I_1\lambda - I_2 = 0 \tag{12}$$

The coefficient I_0, I_1, I_2 are magnetic gradient tensor invariants, are respectively expressed as:

$$\begin{aligned} I_0 &= g_{xx} + g_{yy} + g_{zz} = 0 \\ I_1 &= g_{xx}g_{yy} + g_{yy}g_{zz} + g_{zz}g_{xx} - g_{xx}^2 - g_{xy}^2 - g_{xz}^2 \\ I_2 &= g_{xx}(g_{yy}g_{zz} - g_{yz}^2) + g_{xy}(g_{yz}g_{xz} - g_{xy}g_{zz}) + g_{xz}(g_{xy}g_{yz} - g_{xz}g_{yy}) \end{aligned} \tag{13}$$

Magnetic gradient tensor invariant itself has many nice properties [9], invariant contour map not affected by the direction of magnetic field, can draw a good field source boundary. When there are several field sources closely, the magnetic gradient tensor invariant can better distinguish field source compared to the total magnetic field. As the magnetic gradient measurement system's measurement object is the gradient of magnetic field vector component, unlimited by total field measurement, the measurement results can reflect information on vector magnetic moment of the objective body , geomagnetic field's obliquity and declination have a small effect on tensor elements, the calculated tensor invariants do not require additional processing can be a good description of the magnetic field source [10],and the magnetic gradient tensor inversion can accurately describe the field source body's magnetization direction and geometry, improving resolution of the magnetic source body.

4. Conclusions

This paper presents theoretical analysis and laser cesium optical pumping magnetometer and system design, explains the magneto-optical double resonance theory and the work process of cesium optical pumping magnetometer through cesium atomic hyperfine level structure diagram. On this basis, we discuss on the measuring method of the magnetic gradient tensor, which increase information of magnetic field measurement to a large extent. At present, our country's magnetic measurement technology is still at the first stage, with the great gap between the world advanced level, and the important role of the magnetometer in many key areas, research of laser optical alkali metal pumping magnetometer will have a very bright future.

Acknowledgements
The authors would like to acknowledge the support of International Science and Technology Cooperation Funds (2008DFR20420) and Fundamental Research Funds for the Central Universities(HEUCFL20101110).

References

[1] Weis A and R.Wynands 2005*J*. Optics and Lasers in Engineering. **43** 387
[2] Groeger S and Bison G 2006 *J*. Sensors and Actuators A. **129** 15
[3] Groeger S， Pazgalev A and Weis A 2005*J*.Applied PhysicsB. **80** 645
[4] Chung-Hoon Lee, Hang Guo and Shankar Radhakrishnan 2004*J*.Solid-State Sensor, Actuator and Mi-crosystems. **20** 23
[5] Chen Jing-biao, Wang Feng-zhi, Wang Yi-dao and Yang Dong-hai 2005*J*. Optoelectronics laser. **11** 563
[6] Harris M, Adams C and Cornish S 2006*J*.Physical Review A. **24** 64
[7] Smith D and Hughes I 2004 *J*. Am. J. Phys.**35** 631
[8] Harris M, Adams C and Cornish S 2006 *J*. Physical Review A. **17** 112
[9] Luo Yao and Changli Yao 2007 *J*. Journal of China University of Geosciences. **11** 431

Narrow linewidth Yb-doped double-cladding fiber laser utilizing fiber Bragg gratings inscribed by femtosecond pulses

Zhinan Zhang, Jiawei Shi, Jihuang Zhang, Haiyan Wang, Yuhua Li* and Peixiang Lu*

Wuhan National Laboratory for Optoelectronics, Huazhong University of Science and Technology, Wuhan 430074, China

E-mail: oeyhli@gmail.com; lupeixiang@mail.hust.edu.cn

Abstract. A narrow-linewidth high power laser in all fiber format at 1064 nm is demonstrated. The resonant cavity is composed of two distributed Bragg reflector (DBR) fiber gratings, which were inscribed into the core of the double-cladding fiber by use of 800 nm femtosecond laser pulses and a phase mask. The spectrum of the laser exhibited a narrow linewidth of 21 pm at the output power of 0.8 W. The wavelength and power of the laser featured long term stability.

1. Introduction

Rare-earth-doped fiber lasers have become an important tool in many industries with applications ranging from material processing to telecommunications [1]. High slope efficiencies, power scalability, broad wavelength tunability, and diffraction-limited beam qualities at high power are the prominent advantages of the fiber laser [2].

Narrow linewidth fiber laser, as an important branch of fiber laser, has excellent time coherence and low phase noise thus obtain widely application and great prospect in many fields such as sensor, high precision spectral analysis and large capacity optical fiber communication [3]. The construction of all-fiber narrow linewidth fiber laser utilizing fiber gratings makes the structure of fiber laser more compact and the output of laser coupling into fiber more convenient and efficient. It greatly expanded the application of the narrow linewidth fiber laser, as a new generation light source of WDM optical communication, thus promoted the optical fiber communication to the development of more channels and larger capacity. In addition, narrow linewidth fiber laser is widely used in optical fiber sensing because of its small size, low weight, and high sensitivity.

The resonant cavity structures of the narrow linewidth laser using FBG are divided into two categories: distributed feedback (DFB) and distributed Bragg reflector (DBR). The former one can only achieve the laser output at several mW due to its short cavity length [4]. Therefore, it cannot achieve higher power of narrow line width laser output.

However, using 800nm femtosecond laser pulses and phase mask technique to produce fiber Bragg gratings (FBGs) directly in the core of the common double-cladding fiber could achieve extremely narrow linewidth and excellent thermal stability. Hence, the structure of DBR, two FBGs as resonator mirrors, could help to achieve high power, narrow linewidth laser output.

2. The laser setup

Here we present a high-power (0.8 W) narrow linewidth (21 pm) fiber laser operating at 1064 nm. The laser cavity is established by two spectrally narrow passive-double-cladding fiber Bragg gratings (FBGs) that are fusion spliced to a 1.5 m long Yb-doped double-cladding fiber. The active fiber was an Yb-doped silica fiber with a 10 μm core diameter and a 125 μm hexagonally shaped inner cladding. The passive double-cladding fiber's structure is the same. Figure 1 shows the schematic of the laser.

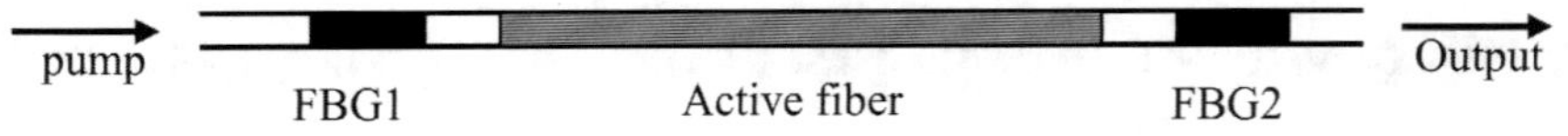

Figure 1. The schematic of the laser.

Two FBGs, the input coupler and the output coupler, were inscribed into the core of the passive fiber in order to create the frequency selective feedback for the cavity. To fabricate the FBGs, the outer polymer cladding of the passive fiber was stripped away, and the femtosecond laser beam was focused using a cylindrical lens before passing through a phase mask into the core of the passive fiber. The pulses had 50 fs pulse duration at 800 nm, 1kHz repetition rate and 800 nJ pulse energy. The induced FBG was 6 mm long and had the grating period of about 1.1 μm, corresponding to the 2nd order resonant reflection at about 1064 nm. Figure 2 shows the reflection spectrum of the FBG.

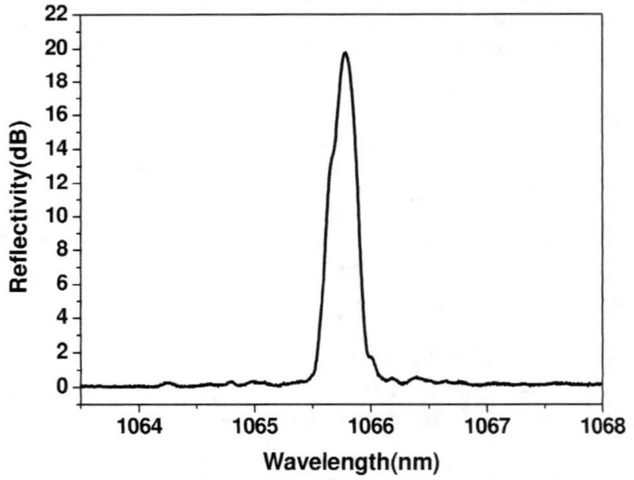

Figure 2. The reflection spectrum of the FBG.

3. Results and discussion

The laser is excited by a 915 nm LD pump laser with a maximum output power of 5W. Figure 3. shows the optical output power at 1064nm versus the launched power. More than 0.8W output power has been obtained for the maximum launched power of about 1.8w. The slope efficiency was 48.8% with a threshold of about 0.2 W launched power.

Figure 4 shows the typical output laser spectrum. The center wavelength of the laser was 1069.55nm with a bandwidth of about 21.9 pm for an output power of 0.8W, which is significantly narrow at these power levels [2]. The optical signal to noise ratio was measured to be about 40 dB. It can be observed that there is a greater amount of ASE, which is almost the same with the previous results [3]. This is due to the ASE being double passed through the cavity after reflecting off the high reflector.

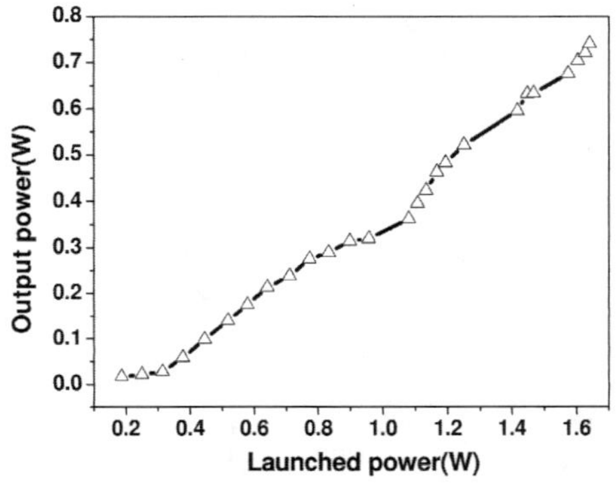

Figure 3. The laser power as a function of the launched power.

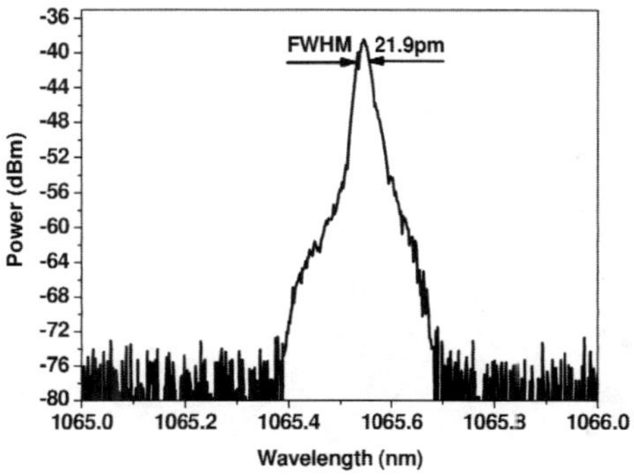

Figure 4. The output laser spectrum.

The stability of the center wavelength and laser spectrum was investigated at full power over a 3 h period. Figure 5 shows the stability of the center wavelength when the FBGs use water-cooling and no cooling. It can be seen that the stability of the center wavelength can be greatly improved by using water-cooling for FBGs.

Figure 6 shows the center wavelength variation with the launched power. The center wavelength shift toward longer wavelength as the launched power was increased because of the thermo-optical effect in the grating.

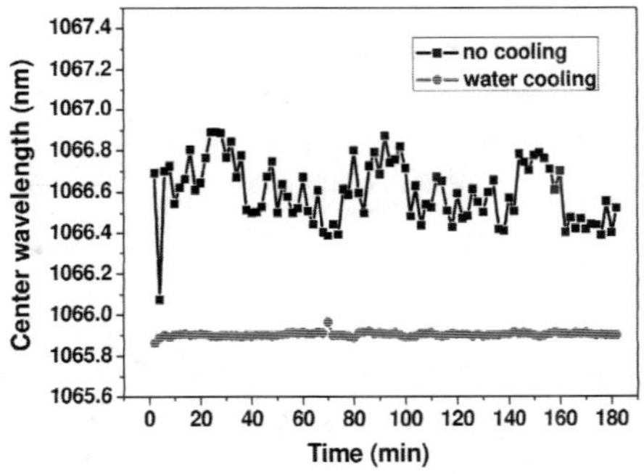

Figure 5. The stability of the center wavelength with and without water-cooling.

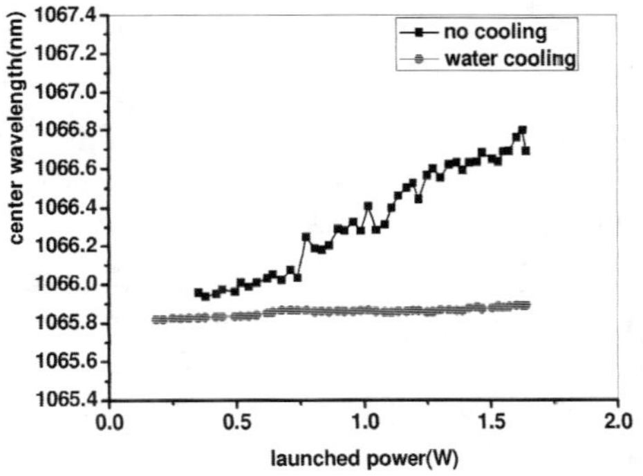

Figure 6. The center wavelength variation versus the launched power.

So we researched the thermo-optical effect of FBG and validated it through the experiment. Figure 7 shows the resonant wavelength of the FBG as the function of the temperature. The peak wavelength shift toward longer wavelength as the temperature was increased. This result is in line with our previous experiment.

Figure 8 and figure 9 show the laser spectrum variation over 3 h period with and without water-cooling.

So we find that FBGs cooling make the performance of the fiber laser more stable and reliable.

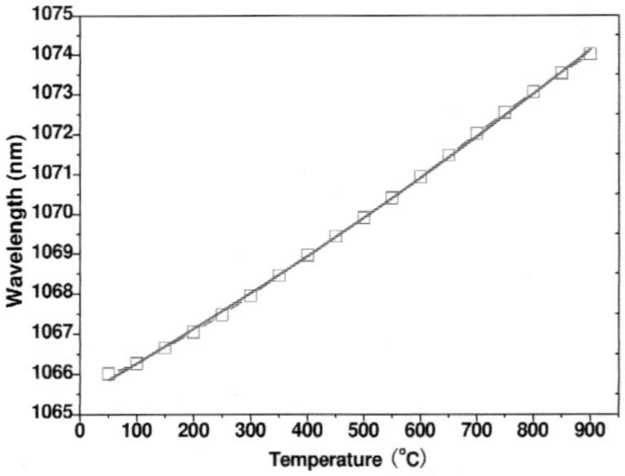

Figure 7. The peak wavelength of the FBG reflection spectrum variation versus the temperature.

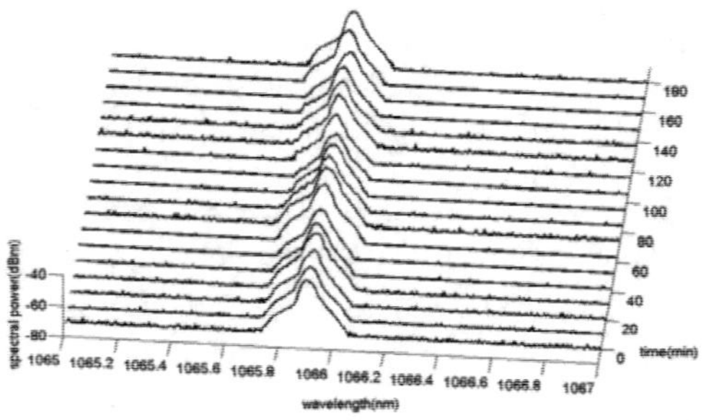

Figure 8. The laser spectrum variation with water-cooling.

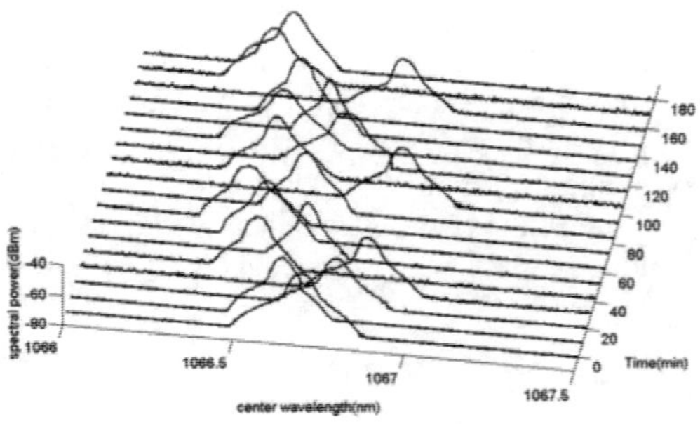

Figure 9. The laser spectrum variation without water-cooling.

4. Summary

In summary, we have presented a high power laser with more than 0.8W output, narrow linewidth of 21pm Yb-doped double-cladding fiber laser that utilizes two FBGs inscribed directly into the double-cladding fiber core using femtosecond laser pulses. The laser works at 1064nm, which exhibits extremely good stability by temperature control of the FBGs. The output power of the laser doesn't have any saturation at present power level. So we predict further power scaling to higher power by using a more powerful pump laser.

5. Acknowledge

The programme was supported by the National Natural Science Foundation of Hubei Province

References

[1] Jovanovic N, Aslund M, Fuerbach A, Jackson S D, Marshall G D and Withford M J 2007 *Optics Letters.* **32** 2804

[2] Jovanovic N, Fuerbach A, Marshall G D, Withford M J and Jackson S D 2007 *Optics Letters.* **32** 1486

[3] Spiegelberg C, Geng J, Hu Y, Kaneda Y, Jiang S, Peyghambarian N 2004 *J. Lightwave Technology.* **22** 57

[4] Loh W H, Samson B N, Dong L, Cowle G J and Hsu K 1998 *J. Lightwave Technology.* **16** 114

The Experimental Research and the Quantum and Thermotics Theory Analysis of Lethal Effect of CO_2 Laser Irradiation on Yeast Suspended in Liquid

ZHANG Hong-wei [1,2,*], ZHANG Li-li [1,2], ZHANG Can-bang [1], XU Lin [2] and ZHOU Ling-yun [1,3]

[1] Department of Physics, Honghe University, Mengzi, 661100, China
[2] Department of Physics, Yunnan Normal University, Kunming, 650092, China
[3] Kunming University of Science and Technology, Kunming, 650093, China

E-mail: zhhw7272@126.com

Abstract. In this paper, a lethal effect experiment was carried out, in which the biological cells of suspension-industrial saccharomyces cervisiac Sachromyce 2.1189 in liquid was irradiated by CO_2 laser. The experiment results were analyzed by means of quanta theory and thermotics theory. The theoretic analysis results can explain well the experimental results of the lethal effect.

1. Introduction

Laser breeding has been widely used. But in breeding experiments, it is difficult to control the radiation dose. Shorter laser irradiation time and lower irradiation laser power can't lead to variations and reach breeding purposes. However, overdose cause their genetic destructive denaturation, or simply to make living biological cells dead (lethal effects) and lead to breeding failure. In breeding experiment of CO_2 laser irradiation on yeast, we found bacteria had more deaths at the irradiation time of 26s, but closing to the irradiation time of 26s there is a high mutation rate[1]. Therefore, the lethal effect of laser irradiation research is extremely important in breeding and laser medical applications. In laser breeding microbial genetic mutation, the common method is to make a suspension with microbial cells, and then use the laser irradiation, in which CO_2 laser is used. As the wavelength of CO_2 laser is in infrared wave band, the major laser effect should be thermal effect. Repeated experiments showed that CO_2 laser irradiation could result in higher temperature in suspension, and longer radiation time, there is cell death. Through evaluating and testing some dates of lethal effect, we have a further understanding of the lethal effect of laser irradiation on yeast and its mechanism. We make microscopic mechanism analysis of experimental results of the lethal effect with quantum and thermotics theory.

2. Experiment

Experimental material is industrial alcohol from sugar cane molasses sacchromyces cerevisiae 2.1189, and culture media are YPD and $12B_c$ diluted sugar cane molasses. The $V=0.2ml$ yeast suspension is in sterile conditions of amperes bottle irradiation. The CO_2 laser power is 10W, the expanded beam spot diameter is 0.8cm. We set up six parallel time periods , which are 5s、10s、15s、20s、25s、30s.The irradiation time of comparison group is (CK) 0s. The bacterial suspension, which is irradiated by CO_2 laser, is cultivated by spread utensil diluting with normal saline. The comparison groups are set up. Bacterial numbers are counted after 3 days cultivating at 28 •, the results are shown in Table 1. The results of experiments will be analysis theoretically. We can obtain the relation of lethal rate of the cells and the irradiation time of CO_2 shown in the figure 1.

Table 1. The statistical results of lethal effect of CO_2 laser irradiation on the Sacchromyces cerevisiac (The average results of three experiments).

Group	Irradiation time/s	Tem/•	Number of yeast cells/ ml	Lethal rate/%
I	5	48	8.86×10^{7}	32.37
II	10	54	5.74×10^{7}	56.18
III	15	58	3.72×10^{7}	71.60
IV	20	61.5	1.66×10^{7}	87.32
V	25	63	3.76×10^{4}	99.79
VI	30	63.5	0.00	100.0
CK	0		1.31×10^{8}	0.00

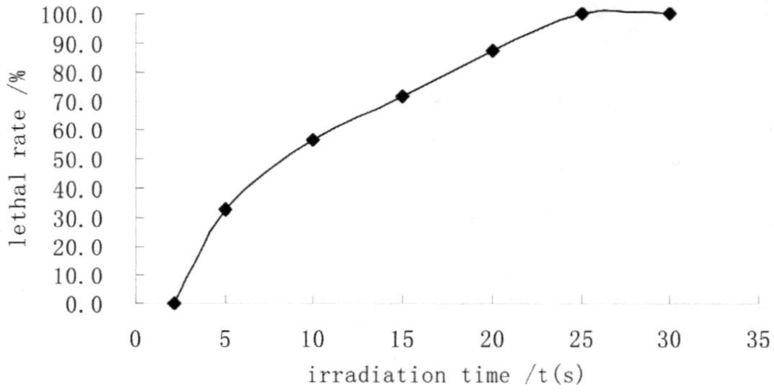

Figure1. The relation of lethal rate and irradiation time by CO_2 Laser.

3. The macroscopic themology theory of the experimental results

CO_2 laser irradiates the suspension from top to bottom, which energy is absorbed in the suspension. The relationship between laser energy flux density I and the depth of the laser entering into the suspension is given through Lambert's law [2] (simple terms, omitted nonlinear effect)

$$I = I_0 e^{-\gamma y} \tag{1}$$

where I_0 is the energy density of the laser at the suspension surface, and γ is the coefficient of cell body's absorbing CO_2 laser. If the reflector bottom is taken into account, then (1) should be

$$I = I_0(e^{-\gamma y} + \eta e^{\gamma y}) \tag{2}$$

where $\eta=\beta e^{-2\gamma H}$, β is the reflection coefficient of the bottom and H is the depth from cell layer to the bottle bottom. Taking one layer of cells of y-$y+dy$, then the laser energy of the layer by cells absorbing in the unit time is

$$dQ_1 = -sdI = s\gamma I_0(e^{-\gamma y} + \eta e^{\gamma y}) \tag{3}$$

where s is the cross-sectional square of the liquid column. As the liquid absorb the laser energy, the temperature distribution of the liquid changes. Thus the heat flow appeared in liquid from top to bottom, and due to heat flow the net heat enhancement in the liquid layer y-$y+dy$ is

$$dQ_2 = -s\frac{\partial q}{\partial y}dy \tag{4}$$

where q is the heat flux intensity. By the Fourier theorem (Strictly speaking, the application of Chester revising theory of heat conduction, Fourier heat conduction experiment is only an approximation [3]) it has

$$q = -\kappa\frac{\partial T}{\partial y} \tag{5}$$

In equation (5), κ is the coefficient of the liquid thermal conductivity, and T is the temperature of the liquid. Therefore

$$dQ_2 = \kappa s\frac{\partial^2 T}{\partial y^2}dy \tag{6}$$

Considering (3), (6), the increased energy per unit time in the liquid layer y-$y+dy$ should be

$$dQ = dQ_1 + dQ_2 = s[I_0\gamma(e^{-\gamma y} + \eta e^{\gamma y}) + \kappa\frac{\partial^2 T}{\partial y^2}]dy \tag{7}$$

On the other hand, the relationship between the rate of temperature change to the time in the liquid y-$y+dy$ layer and dQ is

$$dQ = c\rho s\frac{\partial T}{\partial t}dy \tag{8}$$

where c and ρ are respectively, the specific heat and density of liquid. Considering (7), (8),

$$\frac{\partial T}{\partial t} - a^2\frac{\partial^2 T}{\partial y^2} = b(e^{-\gamma y} - \eta e^{\gamma y}) \tag{9}$$

where $a^2=\kappa/c\rho$, $b=I_0/c\rho$. Equation (9) is the heat conduction equation that CO_2 laser irradiated to the suspension effect .The role of the right side of the equation (9) shown is equivalent to the heat source that irradiating action of suspension. Given that the temperature T_0 of the suspension in front of laser irradiation is uniform, the following initial conditions is

$$T(0, y) = T_0 \tag{10}$$

Generally speaking, the irradiation time of CO_2 laser acting to the suspension is short. The impact of suspension of the border can be neglected. Thus, according to (9) and (10) two equations relation that the temperature distribution of the suspension changes with time.

Neglecting the case of the boundary effect, the equation (9) in the following initial conditions of (10) has the following solution [4].

$$T(y,t) = \int_{-\infty}^{\infty}\frac{T_0}{2a\sqrt{\pi t}}e^{-\frac{(y-\zeta)^2}{4a^2 t}}d\zeta + \int_0^t\int_{-\infty}^{\infty}\frac{b}{2a\sqrt{\pi(t-\tau)}}e^{-\frac{(y-\zeta)^2}{4a^2(t-\tau)}}(e^{-\gamma\zeta} - \eta e^{\gamma\zeta})d\zeta d\tau \tag{11}$$

The first integrality of the equation

$$\int_{-\infty}^{\infty}\frac{T_0}{2a\sqrt{\pi t}}e^{-\frac{(y-\zeta)^2}{4a^2 t}}d\zeta \xrightarrow{x=y-\zeta} T_0\int_{-\infty}^{\infty}\frac{1}{\sqrt{2\pi}\sqrt{2t}a}e^{-\frac{x^2}{2(a\sqrt{2t})^2}}dx = T_0 \tag{12}$$

The second integrality of the equation

$$\int_0^t \int_{-\infty}^{\infty} \frac{b}{2a\sqrt{\pi(t-\tau)}} e^{-\frac{(y-\zeta)^2}{4a^2(t-\tau)}} (e^{-\gamma\zeta} - \eta e^{\gamma\zeta}) d\zeta d\tau$$
$$= b\int_0^t \int_{-\infty}^{\infty} \frac{1}{2a\sqrt{\pi(t-\tau)}} [e^{-\frac{(y-\zeta)^2+4a^2(t-\tau)\gamma\zeta}{4a^2(t-\tau)}} - \eta e^{-\frac{(y-\zeta)^2+4a^2(t-\tau)\gamma\zeta}{4a^2(t-\tau)}}] d\zeta d\tau \quad (13)$$
$$= b\int_0^t e^{a^2\gamma^2(t-\tau)} (e^{-\gamma y} - \eta e^{\gamma y}) d\tau$$
$$= \frac{b}{a^2\gamma^2} (e^{-\gamma y} - \eta e^{\gamma y})(e^{a^2\gamma^2 t} - 1)$$

Thus it can be solved that

$$T(y,t) = T_0 + \frac{b}{a^2\gamma^2}(e^{-\gamma y} - \eta e^{\gamma y})(e^{a^2\gamma^2 t} - 1) \quad (14)$$

In fact, the absorption of suspension by the CO_2 laser is very strong, while the reflection of the glass bottom for the CO_2 laser is not too strong. Thus η is small, and therefore $\eta e^{\gamma y}$ can be neglected in comparison with the experiment, namely

$$T(y,t) = T_0 + \frac{b}{a^2\gamma^2} e^{-\gamma y}(e^{a^2\gamma^2 t} - 1) \quad (15)$$

Considering (15), it shows that the temperature becomes higher along with the increasing of the laser irradiation time, and the temperature becomes lower and lower along with the increasing of the liquid depth.

In the suspended state, the yeast cells distribute in the liquid is uniform. Supposing T_c is the lethal temperature. Along the increasing of the laser irradiation time, T_c is moving from top to bottom. And suppose H is the total depth of the liquid, y_c is the liquid depth corresponding T_c, then when the irradiation time is t, considering (15), the death rate could be obtained

$$\frac{y_c}{H} = \frac{1}{\gamma H} \ln \frac{b(e^{a^2\gamma^2 t} - 1)}{(T_c - T_0)a^2\gamma^2} \quad (16)$$

Using the following data [4],

$s = 5.0\times10^{-5} m^2$; $v = 2.0\times10^{-7} m^3$; $H = 3.98\times10^{-3} m$; $\rho = 1.07\times10^3 Kg/m^3$; $\kappa = 0.403 J/mKs$;
$c = 4.22\times10^3 J/Kg$; $s = 5.0\times10^{-5} m^2$; $\gamma = 1.058\times10^3 m^{-1}$; $T_0 = 25°C$; $b = I_0\gamma/c\rho = 23.43(ks^{-1})$;
$a^2 = k/c\rho = 8.925\times10^2 (m^2 s^{-1})$

Using the above data and in a set of experimental values in table 1into (16),

$$T_c = 64°C = 337K \quad (17)$$

T_c is the lethal temperature of yeast cells, and it is consistent with our experimental results above. In order to further analyze the mechanism of our results, we will address below with the quantum theory.

4. Analysis of experimental results with the quantum theory

From the microcosmic view, the heating effect of biological molecules is, in fact, the chain of key biological molecules thermal vibration effects. In accordance with *Fröhlich* 's theory [5], the living biology molecular system is a very high coherence and high ordered system, hence it can be regarded as a related form of galvanic oscillator, while the oscillator frequency is in infrared wave band. So it is easy to be prone to resonance absorption for CO_2 laser and result in a strong warming effect to make the biological cells dead as CO_2 laser irradiation. We will make the microscopic mechanism of the lethal effect of CO_2 laser irradiation on yeast for specific analysis as follows. According to what is mentioned analyses, we can take the bimolecular system as a harmonic oscillation system. The partition function of the bimolecular system is [6]

$$Q = [1 - \exp(-\hbar\omega / kT)]^{s} \tag{18}$$

Where s is the freedom of bimolecular vibration, k is Boltzmann constant, T is the temperature of system. The average absorption energy that each bimolecular absorbs laser photon is

$$E = < n > \hbar\omega \tag{19}$$

Where $<n>$ is the average amount of absorbed laser photon by each molecule, ω is laser frequency (notice: resonance absorption is major). According to statistical physics, the average energy of each harmonic oscillation system is:

$$E = kT^{2} \frac{\partial \ln Q}{\partial T} \tag{20}$$

From (18) (19) and (20) we can obtain the temperature-rising formula of bimolecular system by laser heating.

$$\Delta T = \frac{s\hbar\omega}{k[\ln(1 + \frac{s}{< n >})]^{2} (< n > + s)} \tag{21}$$

From the above formula we can get that the temperature rising ΔT increases easily when the freedom s is little, but ΔT increases when absorption photos number of bimolecular $<n>$ increases. The $<n>$ is related to the laser intensity and laser-bimolecular interaction. Actually $<n>$ is related to the absorption coefficient of bimolecular for the laser photon. Owing to the complexity of the interaction of laser-bimolecular and of bimolecular self, the absorption coefficient of bimolecular is decided by many factors. We will analysis from tow respects.

The research of biophysics demonstrated that the bimolecular (e.g. DNA and protein) has semiconductor property. Hence we discussed the absorption coefficient of bimolecular for laser photon by means of band theory in solid quantum theory. The absorption coefficient η:

$$\eta = < n >_{s} \hbar\omega / I \tag{22}$$

Where $<n>_s$ is absorption photon number in unit time and unit volume, obviously related to $<n>$; I is the laser energy density of incident, obviously relate to laser intensity; η is a macroscopic value which can be macro-examined; $<n>$ is microscopic physical quantity. Now we analyze $<n>_s$ from micro-mechanism. According to band theory [9]

$$< s >_{s} = \frac{2\pi\hbar e^{2} A_{0}^{2}}{m^{2}} \sigma_{R} \tag{23}$$

Where σ_R is the quantity related to scattering cross-section. Then we can obtain the absorption coefficient η:

$$\eta = A\sqrt{(2\mu)^{3}(\hbar\omega - E_{g})} / \hbar\omega, \ A = \frac{2e^{2}}{\varepsilon_{0} v m^{2} \pi} \left|\langle f, k | e \cdot \nabla | i, k \rangle\right|^{2} \tag{24}$$

Where E_g of formula (24) is width of forbidden band, E_g of biomolecule (DNA and protein) compute by means of the methods of HMO, Hiickel, etc(in general $E_g \sim 10eV$). According to (24) and the under help of the relevant formulas in references [7], irradiation time of heating unit $\Delta T / T_0$ can be obtained through a series of calculations. Temperature ratio has the relationship with the laser power,

$$\Delta T = [A(\alpha)P + B(\alpha)P^{-1} - \frac{\hbar\omega}{k}]\delta(P - rp_{0}) \tag{25}$$

Where the order of magnitude of α take 10^{-1} or 10^{-2} when molecule size take 1×10^{-10}m, $\hbar\omega \sim 10^{-22}$, $P \sim 10W$. That is to say, when α is a determined value, $A(\alpha)$ and $B(\alpha)$ are two related constant, while $\hbar\omega / k$ is small which could be omitted, $\delta(P - rp_0)$ is the δ function,

$$\delta(P - rp_{0}) = \begin{cases} 1, & p = rp_{0} \\ 0, & p \neq rp_{0} \end{cases}, (r > 1) \tag{26}$$

According to (25), it could be seen that the increase of temperature with the laser power promoted the overall non-linear relationship. When P is big, by the first part of (25), ΔT and P overall linear relationship, while α's value is related to the thermal motion of molecules entropy, while ΔT is the increased temperature in unit irradiation time, with time t increasing, the increased temperature is raised, when the temperature increases, thermal motion intensify, the corresponding entropy increase. α's value becomes smaller. It could explain in table 1 when the temperature is higher, the temperature rise ration becomes slower. Thus we use quantum theory to explain the microscopic mechanism of our experimental results well.

5. Conclusions

1. CO_2 Laser irradiation on yeast lethal effect of the laser-induced breeding is significant.
2. Our theoretical analysis and experimental results are consistent. Here it should be stated that since the bimolecular interaction of the laser and the complexity, our research is only preliminary exploration. In addition, in heat transfer theory; we still applied the Fourier heat equation, which is slightly higher than the first. One of the authors in the reference [3] pointed out the limitations of the Fourier heat equation, and the process used in solving a number of similar, their approximate level is still to be improved [8], we make a special explanation here. In addition, using the Quantum theory to analyze the thermal effects of biology of laser is only an exploration; the method is worthy of discussion. In short, there are many areas worthy of further study about theoretical analysis of this article.

Acknowledgement

This work has been supported by the National Natural Science Foundation of China (Nos.60068001 and 60748002), the Natural Science Foundation of Yunnan Province (Nos.2000A0021M, 2006E0091M and 2008ZC044M), and the Natural Science Foundation of Education Department of Yunnan Province in China (No.07y40499).

Reference

[1] Zhang C B, Xu L and Zhou L Y 2008 *SPIE.* **72802H-I** 7280
[2] Hecht E and Zajac A 2003 *Optics* (Massachusetlsi: Audison–Wesley) p 890
[3] Zhou L Y, Zhang C B and Zhou Y Y 2003 *Chinese Optics Lett.* **1** 597
[4] Hlman J P 2002 *Heat Transfer* (New York: McGraw–Hill Companies) p 582
[5] *Fröhlich* H 1983 *Coherence in Biology* (Berlin: Springer) p 5
[6] Zhou Y Y and Zhou L Y 1999 *SPIE.* **423** 3863
[7] Zhou L Y 1996 *J. of Kunming Uni. of Sci. and Techno.* **21** 97
[8] Lisityn B M *et al.* 1989 *Approocimate solution of Heat–Conduction and Thermoelas Prob. Lems* (New York: Pleum Pub. Corp.)
[9] Medenlung O 1965 *Intoduction of Solid-State theory* (New York: JohnWiley) p 66

Journal of Physics: Conference Series **276** (2011) 012040 doi:10.1088/1742-6596/276/1/012040

Propagation stability of a chirped soliton in birefringent fibers

Hongjun Zheng[1,2*] **Shanliang Liu**[2] **Xin Li**[2] **Chongqing Wu**[1]

[1]School of Science, Beijing Jiaotong University, Beijing 100044, P. R. China.

[2]Institute of Optical Communication, Liaocheng University, Liaocheng Shandong, 252059, P. R. China.

zhj@lcu.edu.cn; hjzheng@yahoo.com

Abstract. The propagation stability of a chirped soliton at anomalous dispersion region in birefringent fibers is numerically studied by using the split-step Fourier-method. It is found that initial linear chirp can change obviously the threshold value A_{th} above which soliton stably propagates in birefringent fibers, the A_{th} increases with the decrease of the polarization angle | /4- |. The positive chirp makes obviously the A_{th} smaller for group velocity mismatch parameter >0.5, the negative one makes the A_{th} larger for <0.5. The effect of initial positive chirp on the A_{th} is greater than that of negative chirp for high birefringent fibers, is less than that of negative chirp for low birefringent fibers.

1. Introduction

The soliton propagation in birefringent fibers has been investigating extensively for its importance in science and practical applications. It is governed by a set of coupled nonlinear Schrödinger (CNLS) equations which have been receiving a great deal of attention in recent years [1-9]. The previous works on soliton propagation in birefringent fibers have been studied in the case of unchirped pulse [3-9]. However, optical pulses generally have frequency chirps which have great effects on characteristics of the pulses [10-16]. The frequency chirp can be controlled by changing the input current of laser or changing the length of input fiber, and so on. In this paper, propagation stability of chirped soliton at anomalous dispersion region in birefringent fibers is investigated numerically by using the split-step Fourier-method (SSFM), is compared with that of unchirped pulses.

2. Theoretical model of soliton propagation in birefringent fibers

The soliton propagation in birefringent fibers is described by the normalized coupled nonlinear Schrödinger equations (CNLS) [1, 2]

$$i(\frac{\partial u}{\partial \xi}+\delta\frac{\partial u}{\partial \tau})+\frac{1}{2}\frac{\partial^2 u}{\partial \tau^2}+(|u|^2+B|v|^2)u=0, \quad (1a)$$

$$i(\frac{\partial v}{\partial \xi}-\delta\frac{\partial v}{\partial \tau})+\frac{1}{2}\frac{\partial^2 v}{\partial \tau^2}+(|v|^2+B|u|^2)v=0, \quad (1b)$$

* 1) No.3 Shang Yuan Cun, Hai Dian District, Beijing 100044, P. R. China; 2) Culture Road 34, Liaocheng City, Shandong province 252059, P. R. China

where u and v are the normalized complex amplitudes of the two orthogonal polarization components in CNLS system, ξ is the propagation distance which normalized to the soliton period, τ is the time which normalized to T_0 (half-width of the input pulse at 1/e intensity), δ is the group velocity mismatch parameter and B is the cross-phase modulation (CPM) coefficient where 0•B•1. u is slow component and v is fast one for δ >0. The fiber loss is neglected because the propagation distance is short for all-optical switch application. Eqs.(1) reduce to a uncoupled NLS system for B=0 [1], reduce to the Manakov equations and have analytical solutions for B=1 [1,5-9]. However, Eqs.(1) are unintegrable for B=2/3 in linear birefringent fibers [1-4]. We numerically study the propagation characteristics of the chirped soliton by using the SSFM according to Eqs.(1) for B=2/3. The two orthogonal polarization components of an input soliton with initial linear chirp are, respectively,

$$u(\xi = 0, \tau) = A\cos\alpha\,\mathrm{sech}(\tau)\exp(-\frac{iC\tau^2}{2}) \tag{2a}$$

and

$$v(\xi = 0, \tau) = A\sin\alpha\,\mathrm{sech}(\tau)\exp(-\frac{iC\tau^2}{2}), \tag{2b}$$

where A is the amplitude, C is the linear chirp parameter, α determines the relative strengths of input soliton in each of the two polarizations. The dispersion and the nonlinearity effects can be taken into account respectively in propagating form ξ to ξ+h (small distance h) according to the SSFM algorithm. It becomes simple to deal with the pulse propagation by using the SSFM. To deal with the edge effects we enlarge the computational region to $\tau = (-80,80)$, set the number of sampling 2^{13} and use the edge damping method in the Ref.[4]. The initial amplitudes of input soliton in each polarizations are equal (= /4) in the section 3, are unequal (• /4) in section 4.

3. Effects of initial chirp on soliton propagation for equal amplitudes

3.1 Variation of the interval of the two components with propagation distance

The interval of the two polarization components is defined as

$$\Delta\tau(\xi) = \tau^u_{\max}(\xi) - \tau^v_{\max}(\xi), \tag{3}$$

where $\tau^u_{\max}(\xi)$ and $\tau^v_{\max}(\xi)$ are the positions above which $|u(\xi,\tau)|$ and $|v(\xi,\tau)|$ reach maximum value, respectively. We can obtain $\tau^u_{\max}(\xi) = -\tau^v_{\max}(\xi)$ from Eqs.(1) and (2). The slow component is before the fast one when $\Delta\tau(\xi)$<0. The two components will separate when $\Delta\tau(\xi)$ always increases with the increase of propagation distance. If $\Delta\tau(\xi)$ varies slower and slower with propagation distance and becomes finally unchanged at A=Ath, Ath is called the threshold value above which soliton stably propagates in birefringent fibers. The two polarization components can be mutually bound together for A•Ath, separate each other for A<Ath. The threshold value Ath can be determined by variation of the interval with propagation distance. We firstly assume $\delta = 0.5$. Fig.1 shows that the intervals of two polarization components vary with propagation distance for $\delta = 0.5$. The dashed curve is for A=0.845 and C=0.5, solid curve is for A=0.996 and C=0, dot-dashed curve is for A=1.02 and C=-0.5. It is found from Fig.1 that the intervals of two components for different cases are almost unchanged when $\xi > 16$. Therefore, Ath=0.996 which is more accurate than that in the Ref.[4] for C=0, Ath=0.845 for C=0.5 and Ath=1.02 for C=-0.5. It shows that initial chirp changes the threshold value Ath of soliton bound state in birefringent fibers. The positive chirp enhances the nonlinear effect and makes the threshold value Ath smaller, the negative one makes the threshold value Ath larger. It is important to the applications in optical communication and its device.

Journal of Physics: Conference Series **276** (2011) 012040 doi:10.1088/1742-6596/276/1/012040

Fig.2 shows that the intervals vary with propagation distance for A=1.1 and $\delta = 0.5$. The curves (1-5) are for C=-0.7, -0.5, 0, 0.5 and 0.7, respectively. It is found that the intervals oscillate with the increase of propagation distance for A>A_{th}. It shows that the two polarization components are mutually bound together and stably propagates as a single unit. For a given magnitude of A, the larger C (positive chirp) is, the earlier two components are bound; the larger |C| (negative chirp) is, the later two components are bound.

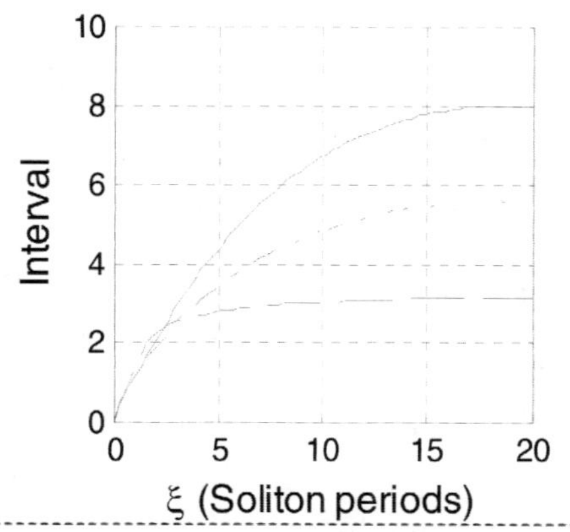

Figure 1. Variation of the intervals with propagation distance for $\delta = 0.5$. The dashed curve is for A=0.845 and C=0.5, solid curve is for A=0.996 and C=0, dot-dashed curve is for A=1.02 and C=-0.5.

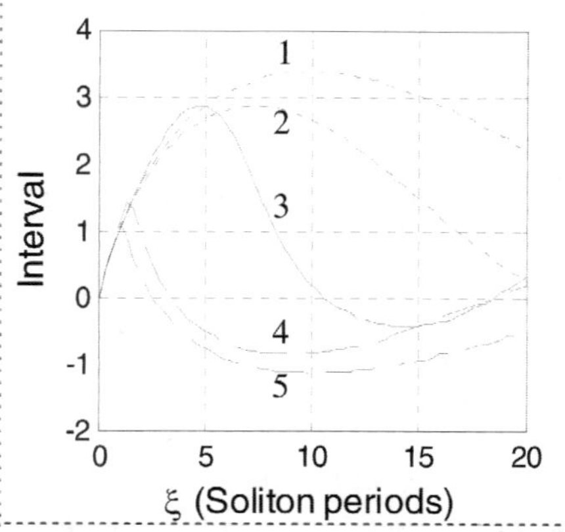

Figure 2. Variation of the intervals with propagation distance for A=1.1 and $\delta = 0.5$. The curves (1-5) are for C=-0.7, -0.5, 0, 0.5 and 0.7, respectively.

3.2 Effect of linear chirp on the relationship between A_{th} and δ

Fig.3 shows the effect of linear chirp on the relationship between A_{th} and δ. The solid curve is for C=0, dashed curve is for C=0.5, dotted curve is for C=-0.5. It is found from Fig.3 that the initial chirp changes the threshold value A_{th} which increases with the increase of δ. For example, the A_{th} is 1.608 at δ=0.8 for negative chirp, is 1.58 for unchirp case, is 1.34 for positive chirp; the A_{th} is 0.88 at δ=0.3 for negative chirp, is 0.76 for unchirp case, is 0.743 for positive chirp. The threshold value A_{th} of positive chirp is less than that of unchirped case for δ •0.3, is almost consistent with that of unchirped case for 0.15•δ<0.3; the difference of A_{th} between positive chirp and unchirped case increases with increase of δ for 0.3•δ<0.6, remains about 0.24 for 0.6•δ•1. The A_{th} of negative chirp is slightly larger than that of unchirped case for δ •0.5, is much larger than that of unchirped case for δ<0.5; the difference of A_{th} between negative chirp and unchirped case decreases with the increase of δ for 0.3<δ<0.5, remains about 0.13 for 0.15•δ•0.3.

It shows the effect of initial positive chirp on the threshold value A_{th} is greater than that of negative chirp for high birefringent fiber (about δ>0.5), is less than that of negative chirp for low birefringent fiber (about δ<0.5). One should consider the effect of initial chirp on the threshold value A_{th} changing in the practical applications.

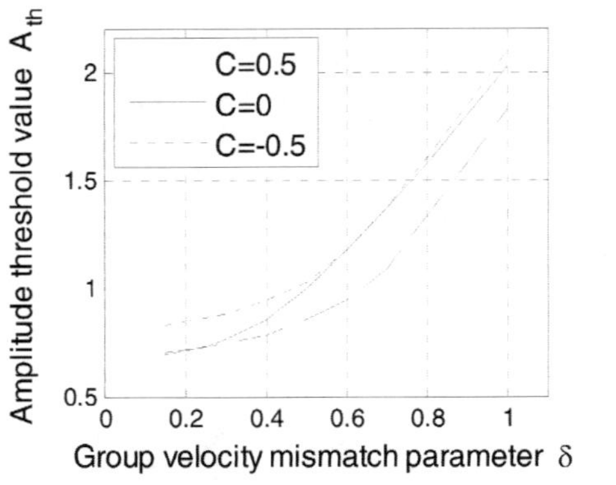

Figure 3. Effect of linear chirp on the relationship between A_{th} and δ. The solid curve is for C=0, dashed curve is for C=0.5, dotted curve is for C=-0.5.

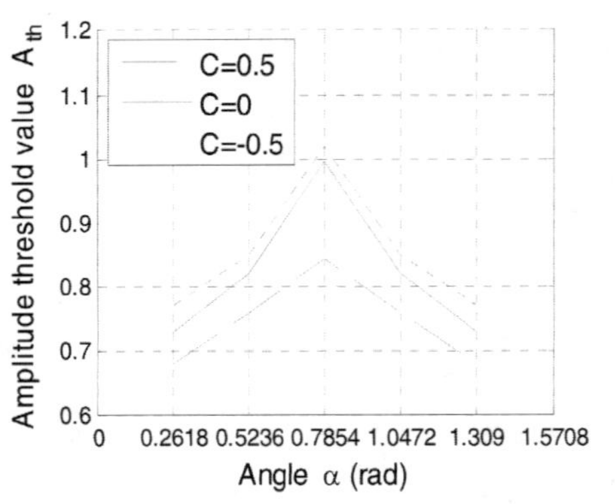

Figure 4. Variation of amplitude threshold value A_{th} with the angle for =0.5. The solid curve is for C=0, dashed curve is for C=0.5, dotted curve is for C=-0.5.

4 Effects of initial chirp on soliton propagation for unequal amplitudes

4.1 Variation of the threshold value A_{th} with the polarization angle

It is found from numerical results that the amplitude threshold value A_{th} increases with the decrease of the polarization angle | /4- |. Fig.4 shows that the A_{th} varies with the angle (0< < /2) in the case of =0.5. The solid curve is for C=0, dashed curve is for C=0.5, dotted curve is for C=-0.5. The threshold value A_{th} increases obviously with the decrease of the angle | /4- | and C, the maximum threshold value A_{th} is at = /4 (0.7854). The difference of A_{th} between positive chirp and unchirped case is lager than that between negative chirp and unchirped case. It shows that effect of positive chirp on variation of A_{th} with angle | /4- | is greater than that of negative chirp.

4.2 Variation of the threshold value A_{th} with parameter

We assume = /6 (0.5236). Fig.5 shows that the amplitude threshold value A_{th} varies with group velocity mismatch parameter in the case of = /6. The solid curve is for C=0, dashed curve is for C=0.5, dotted curve is for C=-0.5. It is found from Fig.5 that the initial chirp changes the threshold value A_{th} which increases with the increase of δ. The A_{th} of positive chirp is larger than that of unchirped case for 0.15• •0.4; is much less than that of unchirped case for 0.4< •1. The A_{th} of negative chirp is larger than that of unchirped case for 0.15•δ•0.75, is slightly less than that of unchirped case for 0.75<δ<1; the difference of A_{th} between negative chirp and unchirped case decreases with the increase of δ for 0.15•δ•0.75, remains about 0.05 for 0.75<δ<1. For a given magnitude of δ and C, the threshold value for = /6 is less than that for = /4. Comparing the difference of A_{th} between positive chirp and unchirped case with that between negative chirp and unchirped case, it shows that effect of initial positive chirp on the threshold value A_{th} is still greater than that of negative chirp for high birefringent fiber (about δ>0.5), is still less than that of negative chirp for low birefringent fiber (about δ<0.5).

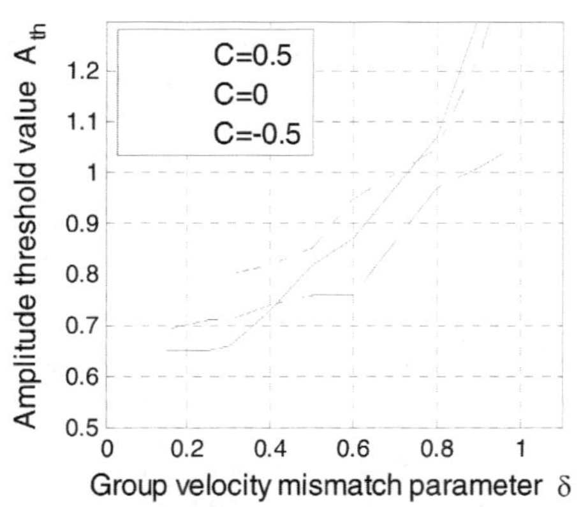

Figure 5. Variation of amplitude threshold value A_{th} with group velocity mismatch parameter for = /6. The solid curve is for C=0, dashed curve is for C=0.5, dot-dashed curve is for C=-0.5.

5 Conclusion

It is found that initial chirp can change obviously the threshold value A_{th} which increases with the decrease of the polarization angle | /4- |. The positive chirp obviously makes the A_{th} smaller for >0.5, the negative one makes the A_{th} larger for <0.5. The effect of initial positive chirp on the threshold value A_{th} is greater than that of negative chirp for high birefringent fiber, is less than that of negative chirp for low birefringent fiber. The threshold value A_{th} increases with the decrease of the angle | /4- | and C. The effect of positive chirp on variation of A_{th} with angle | /4- | is greater than that of negative chirp.

Acknowledgement

Project is supported in part by the National Natural Science Foundation of China (Grant No. 60778017 and 60877057), the Research Foundation of Education Department of Shandong Province, China(Grant No J05C09) and the Research Foundation of Liaocheng University.

References

[1] Agrawal G P 2001 *Nonlinear Fiber Optics 3E*, (New York: Academic press) pp 203-253
[2] Menyuk C R 1987 Nonlinear Pulse Propagation in Birefringent Optical Fibers *IEEE Journal of Quantum Electronics* **23** 174
[3] Curtis R Menyuk 1987 Stability of solitons in birefringent optical fibers. I: Equal propagation amplitudes *Opt. Lett.* **12** 614
[4] Curtis R Menyuk 1988 Stability of solitons in birefringent optical fibers. II. Arbitrary amplitudes *J. Opt. Soc. Am. B* **5** 392
[5] Manakov S V 1974 On the theory of two-dimensional stationary self-focusing of electromagnetic waves *Sov. Phys. JETP* **38** 248
[6] Smyth N F and Kath W L 2001 Radiative losses due to pulse interactions in birefringent nonlinear optical fibers *Phys. Rev. E* **63** 036614-1
[7] Kanna T and Lakshmanan M 2003 Exact soliton solutions of coupled nonlinear Schrödinger equations: Shape-changing collisions, logic gates, and partially coherent solitons *Phys. Rev. E* **67** 046617-1
[8] Solja i M, Steiglitz K, Sears S M, Segev M, Jakubowski M H and Squier R 2003 Collisions of Two Solitons in an Arbitrary Number of Coupled Nonlinear Schrödinger Equations *Phys.*

Rev. Lett. **90** 254102-1
[9] Horikis T P and Elgin J N 2004 Nonlinear optics in a birefringent optical fiber *Phys. Rev. E* **69** 016603-1
[10] Desaix M, Helczynski L, Anderson D and Lisak M 2002 Propagation properties of chirped soliton pulses in optical nonlinear Kerr media P*hys. Rev. E* **65** 056602-1
[11] Li Z H, Li L, Tian H P, Zhou G S and Spatschek K H 2002 Chirped Femtosecond Solitonlike Laser Pulse Form with Self-Frequency Shift *Physical Review Letters* **89** 263901-1
[12] Liu S L, Zheng H J 2006 Experimental research on the characteristic measurement of the short pulses before and after propagating in dispersion flattened fiber *Chin. J. Lasers* **33**: 199 (in Chinese)
[13] Zheng H J, Liu S L, Tian Z, Li X 2008 Effects of Raman Amplification on Propagation Characteristics of the Soliton *Chin. J. Lasers* **35** 861 (in Chinese)
[14] Liu S L, Zheng H J 2008 Measurement of nonlinear coefficient of optical fiber based on small chirped soliton transmission *Chin. Opt. Lett.* **6** 533
[15] Zheng H J, Liu S L, Li X, Tian Z 2007 Temporal characteristics of an optical soliton with distributed Raman amplification *J. Appl. Phys.* **102** 103106-1
[16] Zheng H J and Liu S L 2006 Effects of initial frequency chirp on the linear propagation characteristics of the exponential optical pulse *Chinese Phys.* **15** 1831

Localized modes in orientation-disordered uniaxial medium

Zhenzhu Wan[1], Yanling Han[1] and Hong Wang[2,3]

[1]School of Mathematics and Physics, China University of Geosciences, Wuhan 430074 China

[2]Center of Information and Laboratory, China University of Geosciences, Wuhan 430074 China

E-mail: w_pearl2008@163.com

Abstract. Using a 4×4 transfer matrix method, localized modes in orientation-disordered uniaxial medium have been investigated. We confirm that localized modes origin from the randomness of spatial orientation of optical axes. The misalignment of the optical axe provides the opportunity for a rearrangement of the localized modes. The number of localize mode also closely relative to the spatial orientation of optical axis. Numerical results indicate that it is possible to adjust the localized modes through altering the relative orientation of the optical axes of scatterers. This study is an importance for well understanding of localization of light wave and lasing action in anisotropic random media.

1. Introduction

Since Letokhov predicted that combination of multiple scattering and light amplification would lead to a lasing phenomenon (random lasers) [1], the diffusion and transport of light waves in complex dielectric structures have spurred a vast range of experimental and theoretical work during the past decade. Lawandy's group observed a narrowing of the spontaneous emission spectrum from a methanol solution of rhodamine 640 perchlorate dye and TiO_2 microparticles when excitation pump light intersity exceeded a threshold [2]; Wiersma's group reported the coherent backscattering measurement from amplifying random media using optically pumped Ti: sapphire powders [3]; Cao's group extensively studied the lasing process in disordered semiconductor polycrystalline films, powers and clusters [4,5]. In all the previous study, localization and interference effects which survive to multiple scattering events have been invoked to explain the random lasing observed in many exotic and complex systems. It is confirmed that the random laser origin from the amplification of such localized modes.

All the above study is relative to the isotropic scatterers; as a result, the spatial disorder of scatterers is the unique factor to yield the localization of lightwave. However, recently, coherent backscattering experiments performed with high accuracy apparatus, manifested weak localization of light even in tensorial systems characterized by high optical anisotropy, like nematic liquid crystals

[3] Corresponding author.
E-mail: wanghong745@sohu.com

[6, 7]. In such a system, the randomness of spatial orientation of optical axes of uniaxial scatterers should be another factor to lead to localization of lightwave.

In this paper, by using the 4×4 transfer matrix method, we investigate wave localization behaviour in one-dimensional random systems made of the orientation-disordered uniaxial scatterers, in which uniaxial scatterers are periodic in spatial location but disordered in spatial orientation of the optical axes.

Our results show that the localization behaviour in a propagating band is in general similar to that found in the random systems made of the isotropic scatterers and can be described by the standard localization theory developed for isotropic materials. However, the wave transport mechanism of such system is very different from that of isotropic materials. In the case of an anisotropic model, the localized modes origin from disorder in spatial orientation of the optical axes which are sharp contrast to the case of isotropic random system. The number of localized modes closely relative to the misalignment of the optical axial of scatterers. This study is an importance for well understanding of localization of lightwave and lasing action in anisotropic random media.

2. Theoretical models

We consider the propagation of electromagnetic (EM) waves through a one-dimensional random sample consisting of the uniaxial layers. The anisotropic layers have the same dielectric tensor $\bar{\varepsilon}$ and thickness L, but not the same orientation of optical axis. Randomness is introduced into system by angle of optical axes $\alpha=\theta(1+wr)$, where r is random number distributed uniformly between -0.5 and 0.5, $0\le w\le1$ gives the amplitude of randomness, and θ is angle between optical axes.

Suppose that optical axis is oriented along the X direction, the dielectric tensor of the uniaxial layer is defined as

$$\begin{bmatrix}\varepsilon_\Pi & 0 & 0\\ 0 & \varepsilon_\perp & 0\\ 0 & 0 & \varepsilon_\perp\end{bmatrix}. \tag{1}$$

As optical axis now is rotated in the XY plane, the dielectric tensor, is immediately written as

$$\begin{bmatrix}\varepsilon_\Pi\cos^2\alpha+\varepsilon_\perp\sin^2\alpha & (\varepsilon_\Pi-\varepsilon_\perp)\cos\alpha\sin\alpha & 0\\ (\varepsilon_\Pi-\varepsilon_\perp)\cos\alpha\sin\varepsilon & \varepsilon_\perp\cos^2\alpha+\varepsilon_\Pi\sin\alpha & 0\\ 0 & 0 & \varepsilon_\perp\end{bmatrix}. \tag{2}$$

Here α is the angle between the optical axis and the X direction. The random structure is constructed by stacking these uniaxial layers with random α.

Assume that the EM wave is incident on one surface of the sample. The transmission coefficient T of a layered system can be calculated by using the standard 4×4 transfer matrix method [8.9].

The field in the ith layer can be described by a vector $\psi(i)=[E_x(i),\ E_y(i),\ H_x(i),\ H_y(i)]^T$. The transfer matrix relates the field in the $(i+1)th$ layer to ith layer through $\psi(i+1)=M_i\psi(i)$, where M_i can be calculated by following a similar approach in [9] and T can be obtained by $M^{(N)}=\Pi_{i=1}^{N}M_i$ and N is the total number of layers [10].

3. Results and discuss

For compare with random structure, we first study a periodic structure in which two kinds of anisotropic layers with α_1 and α_2 are stacked in order. The band structures of such a periodical structure can be obtained by using 4×4 transfer matrix method. We find band structure. The band structure is shown in Fig.1 (a), where we have chosen $n_\perp=3$, $n_\Pi=9$, $\alpha_1=0$, $\alpha_2=\pi/2$ and $\Lambda=120nm$. We investigate the effect of the angle of optical axis on the band structure while the optical axis in the first layer is kept constant $\alpha_1=0$ and rotated optical axis in the second layer. Fig.1 show the progressive change in the band structure as the angle α_2 is made less. We find that band structure closely relative to

Journal of Physics: Conference Series **276** (2011) 012041 doi:10.1088/1742-6596/276/1/012041

the misalignment of optical axes that can be described by $\Delta\alpha = \alpha_2 - \alpha_1$. We obtain a wider stopband as $\alpha_2 = \pi/2$.

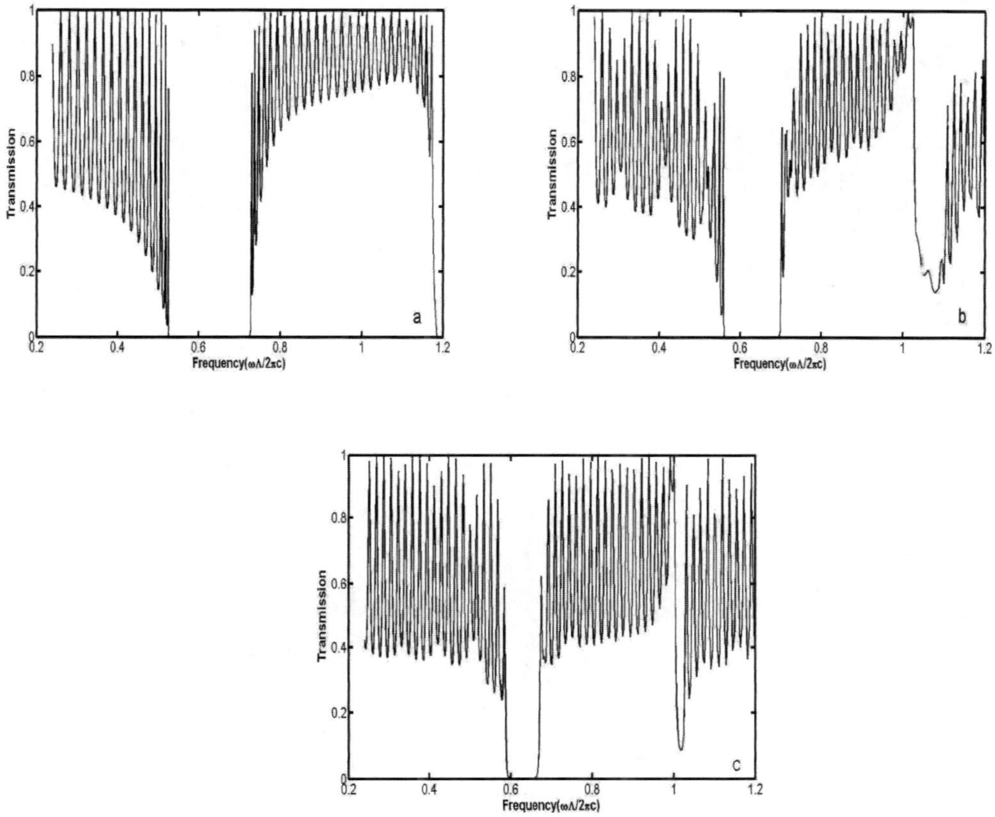

Figure 1. Transmission spectrum for different angle α, (a) $\alpha = 90$, (b) $\alpha = 40$ and (c) $\alpha = 20$.

When the above systems are random, we are interested in studying wave localization behaviours. For this purpose, we consider a finite system with any given number of periods. To introduce the randomness, we rearrange angle α of the uniaxial layers in a random manner as mentioned in Sec.2. By using 4×4 transfer matrix method, we calculate the transmission coefficient T for each random configuration. For any chosen frequency and sample thickness L, T is calculated for 500 configurations to study the localization behaviour.

Figure 2 shows the transmission spectra with various degrees of randomness, where we have chosen $n_{\perp} = 3$, $n_{\Pi} = 9$, $\theta = \pi$ and $\Lambda = 120nm$. The introduction of randomness to a perfectly ordered structure decreased transmissivity in the passbands slightly and changed drastically the transmissivity in the stopbands. Transmission peaks appear in the stopbands as angle increase. This peak indicates the frequency of localized modes and its linewidth reflects the decay rate of the localized mode. It is cleanly seen that number of localized mode increased as we increase the angle. As $w = 1.0$, the localization behavior in a propagating band is in general similar to that found in the random systems made of the isotropic scatterers.

Although the localization of waves for 1D system is well known, there are some interesting differences between the random isotropic and anisotropic system. We find all passbands still exist even w=0.9 as shown in Figure 2. This is different from the situation found in isotropic disordered medium. It is interesting to see the behaviour of our anisotropic disordered medium like a partially ordered system. This property may be useful for effectively pumping of random media, which is one of the practical issues of random laser.

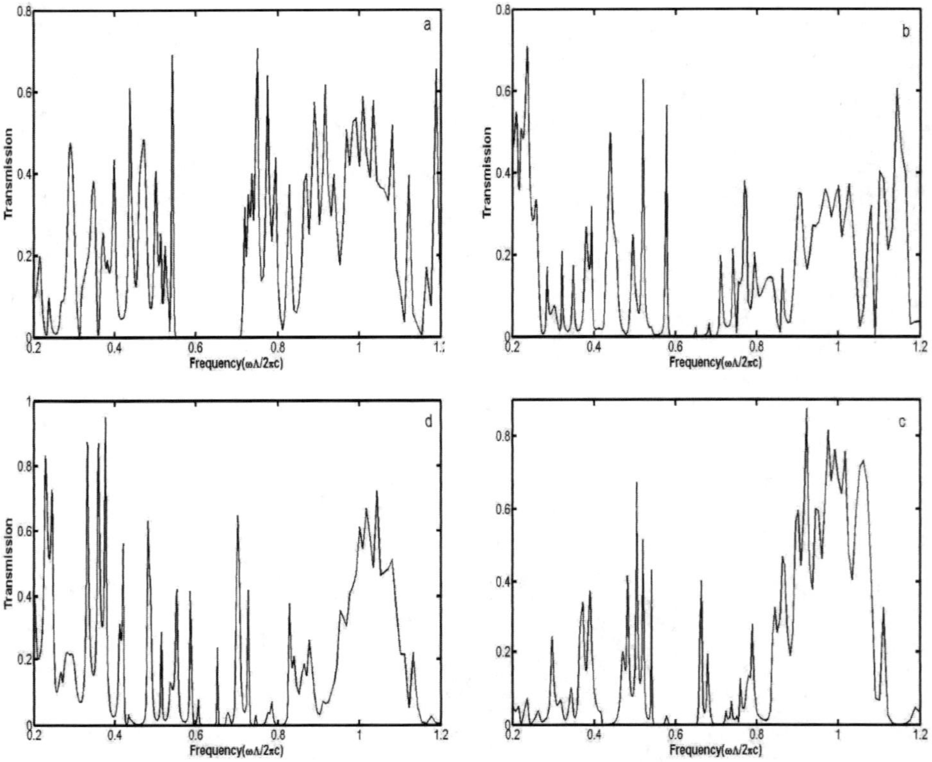

Figure 2. Transmission spectrum for different randomness strength w for (a) $w = 0.5$, (b) $w = 0.7$, (c) $w = 0.8$ and (d) $w = 1.0$

4. Summary and conclusions

We have investigated the wave localization behaviour in one-dimensional random systems made of the orientation-disordered uniaxial scatterers. We find the localization behaviour in a propagating band is in general similar to that found in the random systems made of the isotropic scatterers. However, the wave transport mechanism of such system is very different from that of isotropic materials. We confirm that localized modes origin from randomness of optical axe. The misalignment of the optical axes provides the opportunity for a rearrangement of the localized modes. This study is an importance for well understanding of localization of lightwave and lasing action in anisotropic random media.

Acknowledgment
This research is supported by the Fundamental Research Funds for the Central Universities (Grant N0:CUGL090250).

References

[1] Letokhov V S 1968 Sov. Phys.-JETP **26** 1246
[2] Lawandy N M Balachandran R M Gomes A S L and Sauvain E 1994 *Nature* **368** 436
[3] Wiersma D S Bartolini P Lagendijk A and Righini R 1997 *Nature* **390** 671
[4] Cao H Zhao Y G Ho S T Seelig E W Wang Q H and Chang R P H 1999 *Phys. Rev. Lett.* **82**

2278
[5] Cao H Xu J Y Seelig E W and Chang R P H 2000 *Appl. Phys.Lett.* **76** 2997
[6] Sapienza R Mujumdar S Cheung C Yodh A G and Wiersma D 2005 *Phys. Rev. Lett.* **92** 033903
[7] Kao M H Jeter K A Yodh A G and Colling P J 1996 *Phys. Rev. Lett.* **77** 2233
[8] Yeh. P 1979 *Opt. Soc. Am.* **69** 742
[9] Berreman D W 1971 *J. Opt. Soc. Am.* **62** 502
[10] Chen C J Lien A and Nathan M I 1997 *J. Opt. Soc. Am.* **14** 11

Photoassisted electrolysis of water for hydrogen generation with TiO_2 aggregate film

Guanying Cao[1,2], Qifeng Zhang[2], Dawei Liu[2], Junting Xi[2], Guozhong Cao[2*]

1. Research Institute of Photonics, Dalian Polytechnic University, Dalian, China, 116034
2. Department of Materials Science and Engineering, University of Washington, Seattle, USA, 98195

gzcao@uw.edu; gycao@dlpu.edu.cn

Abstract. In this paper, the nanocrystallite aggregates of TiO_2 were synthesized and characterized by X-Ray Diffraction (XRD), Scanning Electron Microscopy (SEM) and Brunauer-Emmett-Teller (BET). The aggregates are of submicron size, formed by nano-sized crystallites and able to offer both a large specific surface area and desirable size comparable to the wavelength of visible light. Therefore, the TiO_2 aggregates were also studied as photoelectrode in photoelectrochemical cell for hydrogen generation. The results show that the hydrogen generation rates are 0.47 ml/h*cm^2 and 0.27 ml/h*cm^2 during the first test with and without illumination, respectively. The current density also presented continually increasing during the light-on period. This was attributed to the photogenerated current, which benefited from the TiO_2 aggregates and may significantly enhance the electrolysis rate of water.

1. Introduction

Hydrogen has been becoming more and more attractive due to its high energy density as a possible alternative to the fossil fuel for solving the energy problems in future. Since Fujishima and Honda reported photoelectrochemical water splitting using titanium dioxide (TiO_2) photoelectrodes in 1972[1], many studies have been dedicated to the development of photocatalytic materials or photoelectrochemical cells for the conversion of solar energy into hydrogen[2-4]. The TiO_2 different in morphology, such as nanotubes, nanowires arrays and nanoparticles, has been widely investigated. Besides these nanostructures, a novel nanocrystallite aggregates was recently synthesized in our group and studied for hydrogen generation. The aggregates are of submicron size, formed by nano-sized crystallites and, therefore, able to offer both a large specific surface area and desirable size comparable to the wavelength of visible light. The aggregates have already been used for a photoelectrode in a dye-sensitized solar cell and achieved very high power conversion efficiencies[5]. The aggregates were employed to generate effective light scattering and thus extend the traveling distance of light within the photoelectrode film. This may result in an enhancement in the light harvesting efficiency of the photoelectrode. In this regard, the TiO_2 aggregates were also used as photoelectrode in photoelectrochemical cell for hydrogen generation.

In this paper, the nanocrystallite aggregates of TiO_2 were synthesized and characterized by XRD, SEM and BET. Then the hydrogen generation rates and current changes were investigated by using our lab-made electrochemical photocell. The results show that TiO_2 aggregate film has the potential application in hydrogen generation.

2. Experimental Section

2.1 Synthesis of TiO_2 aggregates and preparation of photoelectrode films

The TiO_2 aggregates were fabricated by firstly preparing TiO_2 spheres and then etching the spheres to be of porous structure. Specifically, for the preparation of TiO_2 spheres, a mixture of 1 mL of titanium isopropoxide (TTIP) and 30 mL of ethylene glycol (EG) was added into 400 mL of acetone containing 1 mL of DI-water under vigorous stirring. The reaction resulted in the yield of TiO_2 spheres about several hundred nanometers in diameter. The precipitate of the TiO_2 spheres was collected and then treated with 500 mL of DI-water containing 0.5 mL of acetate acid in a reflux at 120 °C for 1.5 h. Such a treatment leads to the formation of porously structured aggregates through an etching process on the TiO_2 spheres. The product was then washed with ethanol several times, dried at 100 °C, and finally ground into fine powder for use.

The photoelectrode films were prepared on a FTO glass substrate with a screen printing method using a paste of TiO_2 aggregates dispersed in ethanol. The films were annealed at 450 °C for 30 min. The final films were about 10 μm in thickness.

2.2 Characterization of TiO_2 aggregates

The film structure and morphology were characterized by X-ray diffraction (XRD) and scanning electron microscopy (SEM). The specific surface area was characterized by Brunauer-Emmett-Teller (BET) technique.

2.3 Photocell for hydrogen generation

Our lab-made electrochemical photocell for hydrogen generation was a two-electrode configuration with two separate compartments connected by a fine glass frit. The anode was made of TiO_2 aggregate film with a geometric area of 0.9 cm^2 on FTO. The cathode was a platinum wire spiral in a glass burette assembled with a syringe to raise the electrolyte level before start of the experiment. The Pt-stainless rod electrode was inserted through the burette such that all the hydrogen gas evolving at the cathode could be collected in the burette displacing the electrolyte level. The volume of hydrogen gas was measured by directly reading the variation of electrolyte level in the burette.1 M KOH (Fisher Scientific) solution with 10 vol% ethylene glycol was used as electrolyte. The light source was AM 1.5 illumination.

3. Results and discussion

Figure 1 shows the XRD patterns of films. It shows that the aggregates mainly contain anatase TiO_2 and, in addition, a small fraction of brookite. Figure 2 shows the SEM images of films that consist of TiO_2 aggregates. It can be seen that the TiO_2 aggregates are spherical in shape with submicron dimensions. A very rough surface was observed for these aggregates, as shown in the inset of Figure 2. Through a characterization of the specific surface area using the BET technique, these aggregates were evidenced to be high in porosity, approximately 100 m^2/g.

Hydrogen gas was collected using TiO_2 aggregates film at 2.0 V external voltage under AM 1.5 light in 1 M KOH with 10 vol% ethylene glycol. The hydrogen generation rates are 0.47 ml/h*cm^2 and 0.27 ml/h*cm^2 during the first test with and without illumination respectively. Figure 3 shows the relationship between the current density and operation time. During the light-on period, the current density presented continually increasing. The current density, however, showed to decrease gradually during the light-off period. This was attributed to the photogenerated current, which benefited from the TiO_2 aggregates being effective in light scattering and may thus significantly enhance the electrolysis rate of water.

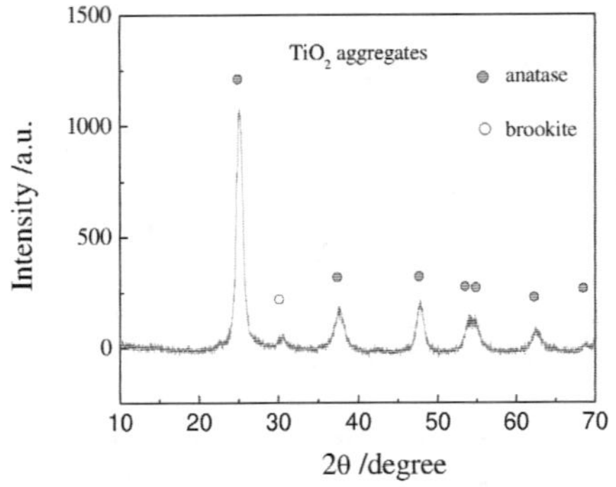

Fig. 1 XRD pattern of TiO_2 aggregates.

Fig. 2 SEM images of TiO_2 aggregates.

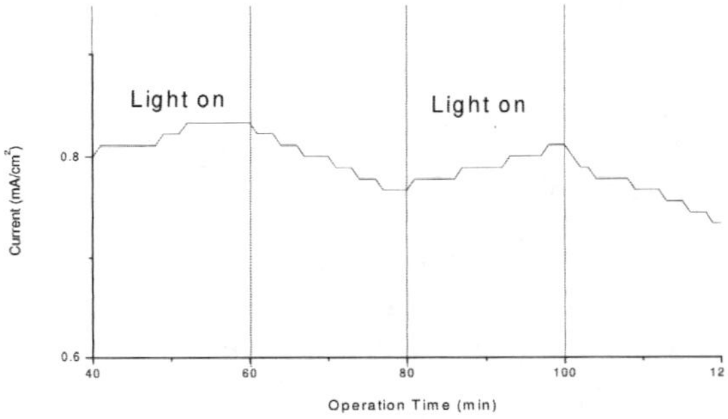

Fig. 3 The current density and the operation time during light on and off.

4. Conclusions

In summary, we have empolyed the nanocrystallite aggregates of TiO_2 for hydrogen generation primarily. The hydrogen generation rates are 0.47 ml/h*cm^2 and 0.27 ml/h*cm^2 during the first test with and without illumination respectively at 2.0 V external voltage under AM 1.5 light in 1 M KOH with 10 vol% ethylene glycol. Oxide nanocrystallite aggregates possess both large specific surface area and suitable size comparable to the wavelength of light. The oxide nanocrystallite aggregates are therefore a class of effective light scatterers for hydrogen generation. Future experiments involve increasing the intensity of the incident light, lowering the bias voltage so as to improve the hydrogen generation rate.

Acknowledgment

Guanying Cao gratefully acknowledges the fellowship from the Chinese Scholarship Council and partly supporting by the project of Education Department of Liaoning Province with number of 20060167.

References

[1] Fujishima A, Honda K. Electrochemical photolysis of water at a semiconductor electrode. Nature 1972;238:37–8.

[2] Mishra PR, Shukla PK, Singh AK, Srivastava ON. Investigation and optimization of nanostructured TiO2 photoelectrode in regard to hydrogenproductionthroughphoto-electrochemical process. Int J Hydrogen Energy 2003;28:1089–94.

[3] Park JH, Kim S, Bard AJ. Novel carbon-doped TiO2 nanotube arrays with high aspect ratios for efficient solar water splitting. Nano Lett 2006;6:24–8.

[4] Zhaoyue Liu, Batric Pesic, Krishnan S. Raja, Raghu R. Rangaraju, Mano Misra, Hydrogen generation under sunlight by self ordered TiO2 nanotube arrays, Int J Hydrogen Energy, 2 0 0 9; 3 4:3 2 5 0 – 3 2 5 7.

[5] Qifeng Zhang, Christopher S. Dandeneau,Kwangsuk Park, Dawei Liu, Xiaoyuan Zhou, Yoon-Ha Jeong, and Guozhong Cao, Light scattering with oxide nanocrystallite aggregates for dye-sensitized solar cell application, Journal of Nanophotonics, 2010, Vol. 4, 041540.

3rd International Photonics & OptoElectronics Meetings (POEM 2010) IOP Publishing
Journal of Physics: Conference Series **276** (2011) 012043 doi:10.1088/1742-6596/276/1/012043

Thermoelectric Properties of Silicon Microchannel Plates Structures

P L Ci[1], J Shi[1], F Wang[1], L Sun[1], S H Xu[1], P X Yang[1], L W Wang[1,3] and Paul K Chu[2]

[1] Laboratory of Polar Materials and Devices, Ministry of Education, and Department of Electronic Engineering, East China Normal University, Shanghai 200241, China

[2] Department of Physics and Material Sciences, City University of Hong Kong, Tat Chee Avenue, Kowloon, Hong Kong, China

[3] E-mail: lwwang@ee.ecnu.edu.cn

Abstract. We have fabricated silicon microchannel plates (MCPs) by photo-assisted electrochemical etching (PAECE) and determined the thermoelectric properties by measuring the Seebeck coefficient of the samples. The samples are composed of regular arrayed lattices with a width of about 5 μm and spacing of about 1 μm. The Seebeck coefficient along the edge of the lattice is 466 μV/K. The silicon MCPs are potential materials for power generation and refrigeration. After oxidation from 30 minutes to 70 minutes and removing the silicon dioxide layer by buffered hydrofluoric acid, the samples show an improved coefficient as high as 1019 μV/K after repeating oxidation and etching 5 times. Our results show that the Seebeck coefficient increases when the wall of the silicon MCPs is thinned.

1. Introduction

There are needs to explore new energy sources with high efficiency and low pollution and thermoelectric materials have attracted increasing interest because they are able to convert heat into electricity directly and have simple structures without any moving parts [1]. Therefore, they are used to generate power in industrial heating and refrigeration products. Silicon is the most abundant and widely used semiconductor which also benefits from the well-developed microfabrication technology [2-4]. Silicon microchannel plates (MCPs) are one of the most hopeful silicon-based thermoelectric materials. Silicon MCPs offer good thermoelectric properties in initial measurements. In addition, silicon MCPs have a porosity (about 69%) similar to that of porous silicon (PS), and the thermal conductivity of PS can be as low as 0.7~1.3 $Wm^{-1}K^{-1}$ [5]. Thus, a low conductivity is expected from silicon MCPs.

Sun *et al.* [6] have reported different Seebeck coefficients along different orientations in the silicon MCPs. We are interested in the change in the Seebeck coefficient when the structure of the silicon MCPs is changed. This paper reports our study of the thermoelectric effect of silicon MCPs fabricated by photo-assisted electrochemical etching (PAECE). The Seebeck coefficients are measured in order to evaluate the thermoelectric properties. In order to clarify the relationship between the Seebeck coefficient and thickness of pore wall, oxidation and etching are performed to thin the pore wall before determining the Seebeck coefficient.

2. Experiment details

Silicon MCPs were fabricated by the PAECE process. A p-type <100> single-side polished silicon wafer 500 μm thick with a resistivity of 2-9 Ω·cm was used.

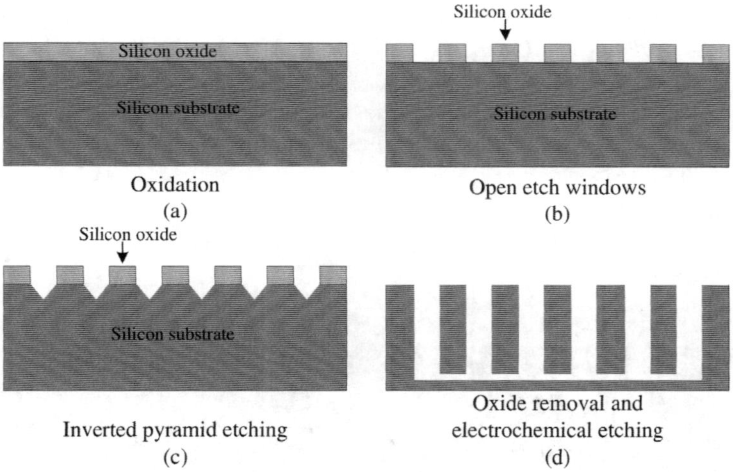

Figure 1. Schematic illustrating the photo-assisted electrochemical etching (PAECE) process: (a) oxidation; (b) open etch windows; (c) inverted pyramid etching; (d) oxide removal and electrochemical etching.

The process flow in electrochemical etching is illustrated in Figure 1. First of all, a silicon dioxide layer (3000 Å) was grown on the wafer by thermal oxidation at 1080°C and the oxide layer on the front side was patterned by a standard photolithographic process and wet etching in a buffered hydrofluoric acid which was used to open the etch windows. In order to fabricate the initial reaction interface, the pits with the shape of an inverted pyramid were formed by etching the samples in 25 wt% tetramethyl ammonium hydroxide (TMAH) at 85°C for a few minutes. The dimensions of the pore were designed to be 3 μm x 3 μm. The silicon dioxide layer was then removed and the PAECE process was performed at 20°C. The electrolyte was a solution of HF(40%):H_2O:DMF (Dimethylformamide: DMF) = 1:4:5. The etching current density was 15 mA/cm^2 and etching voltage was 12 V. A halogen lamp (150 W, 15 V OSRAM) was used to provide backside illumination. A stable etching current was achieved by adjusting the illumination which provided the holes during etching. A computer was used to monitor the etching current and adjust the illumination automatically. After 8 hours, the silicon microchannel plate layer with square hole arrays was automatically broken away from the substrate [7]. The hole had dimensions of about 5 μm x 5μm, and the space between two holes nearby was about 1 μm. In addition, the thickness of the silicon MCPs was about 200 μm.

The samples were then divided into five groups. The first group underwent dry thermal oxidation for 30 minutes at 1080°C. The second group was oxidized for 40 minutes at 1080°C, including dry oxidation of 15 minutes, wet oxidation of 10 minutes, and dry oxidation of 15 minutes. To keep the time of dry oxidation the same and increase the time of wet oxidation to 20 minutes, 30 minutes and 40 minutes, the total oxidation time was increased to 50 minutes, 60 minutes, and 70 minutes, respectively. The oxidized samples were then dipped in a buffered hydrofluoric acid for a few minutes to remove the oxide layer. Using this procedure, pore walls with different thicknesses would be fabricated. The microstructure of the samples was characterized by scanning electron microscopy (SEM).

3. Results and discussion

Figure 2 shows SEM images of the surface and cross-section of the silicon MCPs fabricated by PAECE process.

(a) (b)

Figure 2. SEM images of surface (a) and cross-section (b) of sample by PAECE.

The samples were first cut into pieces (2 cm long and 1 cm wide) by a laser beam for subsequent studies. In order to characterize the samples, they were placed on a thermally insulated substrate. The copper sheets serving as both the thermal and electrical conducting layers were spring loaded onto the sample by probes. This made the ohmic contact between the copper sheets and silicon MCPs. The schematic diagram is depicted in Figure 3.

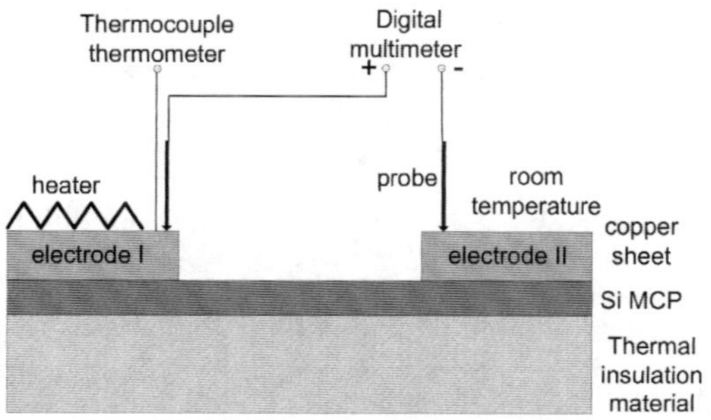

Figure 3. Schematic diagram of the apparatus for Seebeck coefficient measurement. All the measurements are performed along one axis of the square silicon MCP.

Electrode I shown in Figure 3 was heated from room temperature (25 °C) to 60 °C in the heater while Electrode II was kept at room temperature. The electrodes were parallel to the edge of the square lattice.

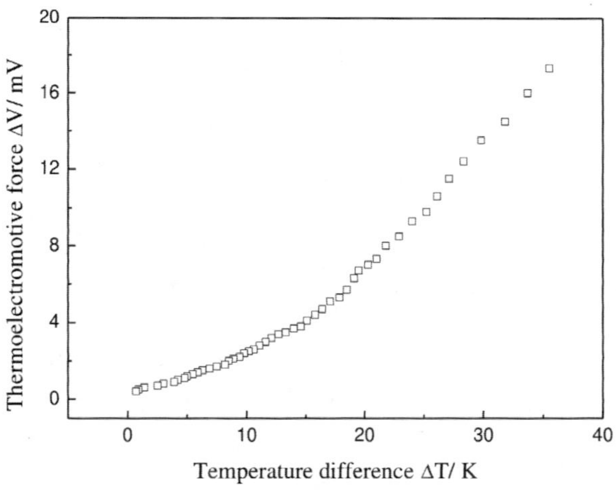

Figure 4. ΔT-ΔV plots of sample without thinning.

The Seebeck coefficient can be determined from the thermoelectromotive force (ΔV) and temperature difference (ΔT) by $S=\Delta V/\Delta T$. In other words, the slope of the ΔT-ΔV plots of samples is the Seebeck coefficient [8]. It must be emphasized that the distance between the electrodes has no influence on the data, and the key factor which affects the data is the temperature difference between the two electrodes.

Figure 4 shows a Seebeck coefficient of 466 μV/K along the edge of the lattice according to the slope. Meanwhile, the positive Seebeck coefficient reveals that the silicon MCPs fabricated by the PAECE process have exhausted the holes and behave like intrinsic semiconductor materials because a p-type semiconductor should have a negative Seebeck coefficient. The plot shows that the thermoelectromotive force (ΔV) increase quickly when the temperature difference (ΔT) (compared to room temperature) increases.

(a) (b)

Figure 5. SEM images of (a) surface and (b) and cross-section of the sample after undergoing oxidation for 70 minutes.

To investigate the relationship between the thickness of the pore wall and Seebeck coefficient, oxidation is conducted to control the wall thickness. A different oxidation time can lead to various degree of silicon consumption. The longer the oxidation time, the smaller is the thickness of the silicon layer on the pore wall. The samples with small pore thickness can be obtained after the oxidation layer is etched by buffered hydrofluoric acid. The oxidation time is 30 minutes, 40 minutes, 50 minutes, 60 minutes, or 70 minutes. Figure 5 shows the SEM image of the surface and cross-section of sample after oxidation for 70 minutes. This sample has a silicon dioxide layer of about 500 nm. It can also be observed that the pore walls on both sides are oxidized and the silicon dioxide layer has different color. The difference is caused by dry oxidation and wet oxidation in terms of density.

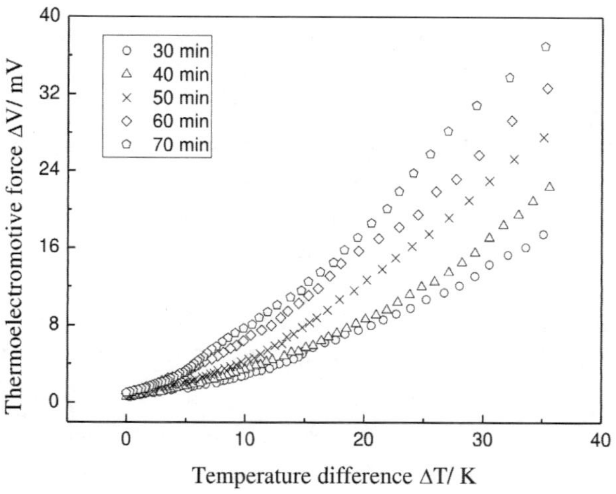

Figure 6. ΔT-ΔV plots of sample with different oxidation time.

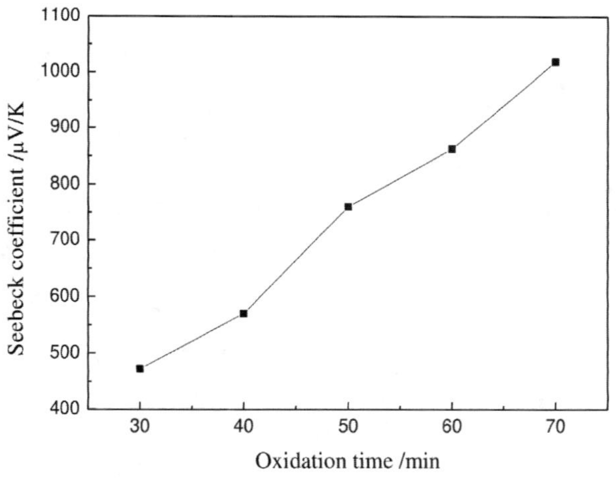

Figure 7. Relationship between the Seebeck coefficient and oxidation time.

The Seebeck coefficient is determined from the samples with thinned pore wall, and the curves which can be described by the ΔT-ΔV plot are shown in Figure 6. The Seebeck coefficient can be determined from the slope of each curve. Based on the Seebeck coefficient and corresponding oxidation time, another curve is presented in Figure 7. The first thinning process has little influence on the Seebeck coefficient (472 μV/K), because dry oxidation for 30 minutes only consumes a thickness of about 30 nm of silicon which is negligible compared to the thickness of the pore wall. When wet oxidation is added to the oxidation process, the pore wall is thinned greatly, and subsequently an improved coefficient appears. After thinning for 5 times, the coefficient reaches a value of 1019 μV/K. Under this condition, the sample has a fragile structure which is no longer suitable practically. The enhancement in the Seebeck coefficient is due to the change in the energy band while the pore wall becomes thinner. Therefore, Figure 7 indicates that a smaller thickness of the pore wall will increase the Seebeck coefficient notably. It should be noted that all the samples have the same order of magnitude as intrinsic silicon in terms of resistivity. It is attributed to the exhaustion of holes in the PAECE process and is very disadvantageous to application as thermal materials. However, doping can reduce the resistivity and more work is underway in our laboratory.

4. Conclusion

We have investigated the Seebeck coefficient of silicon MCPs fabricated by PAECE. The results show remarkable thermoelectric characteristics with a Seebeck coefficient of 466 μV/K. An enhanced Seebeck coefficient is observed when the silicon MCPs are thinned by thermal oxidation and etching in buffered hydrofluoric acid. The Seebeck coefficient is proportional to the thickness of the pore wall. This phenomenon is believed to be related to the change in the energy band which affects the effective mass of carriers. In addition, the thermal conductivity of silicon MCPs which has the similar porosity to PS is very low. Owing to these advantages, silicon MCPs are potential thermoelectric materials in application such as power generation, refrigeration, and sensing.

Acknowledgments

This work was supported by Shanghai Fundamental Key Project under contract number 08JC1408900, 10JC1404600 and international collaboration project under grant 10520704400. This work was also financially supported by the China NSFC support under the contract of Grant No 50672027 and No. 60990312, Hong Kong Research Grants Council (RGC) General Research Funds (GRF) No. CityU 112608, and City University of Hong Kong Strategic Research Grant (SRG) No. 7008009.

References

[1] Francis J. DiSalvo. 1999 Thermoelectric Cooling and Power Generation *Science* **285** 703-706

[2] Arun Majumdar 2004 Thermoelectricity in Semiconductor Nanostructures *Science* **303** 777-778

[3] Allon I. Hochbaum, Renkun Chen, Raul Diaz Delgado, Wenjie Liang, Erik C. Garnett, Mark Najarian, Arun Majumdar and Peidong Yang 2008 Enhanced thermoelectric performance of rough silicon nanowires *Nature* **451** 163-167

[4] Sabah K. Bux, Richard G. Blair, Pawan K. Gogna, Hohyun Lee, Gang Chen, Mildred S. Dresselhaus, Richard B. Kaner, and Jean-Pierre Fleurial 2009 Nanostructured Bulk Silicon as an Effective Thermoelectric Material *Adv. Funct. Mater.* **19** 2445–2452

[5] G Gesele, J Linsmeier, V Drach, J Fricke and R Arens-Fischer 1997 Temperature-dependent thermal conductivity of porous silicon *J. Phys. D: Appl. Phys.* **30** 2911-2916

[6] Li Sun, Fengjuan Miao, Lianwei Wang, Paul. K. Chu and P.M. Sarro 2010 Thermoelectric behavior of microchannel plates fabricated by photo-assisted electrochemical etching *23rd IEEE International Conference on Micro Electro Mechanical Systems* 576-579

[7] Xiaoming Chen, Jilei Lin, Ding Yuan, Pengliang Ci, Peisheng Xin, Shaohui Xu and LianweiWang 2008 Obtaining a high area ratio free-standing silicon microchannel plate via a modified electrochemical procedure *J. Micromech. Microeng.* **18** 037003
[8] Faiz Salleh, Kiyosumi Asai, Akihiro Ishida, and Hiroya Ikeda 2009 Seebeck Coefficient of Ultrathin Silicon-on-Insulator Layers *Applied Physics Express* **2** 071203

The study of electronic structures and optical properties of Al-doped GaN

Enling Li[1], Liping Hou[1], Lisha Li[2], Mancang Liu[1], Meng Xi[1], Xiqiang Wang[1] and Yuanbin Dai[1]

[1] Sciences School, Xi'an University of Technology, Xi'an, China 710054
[2] Physics, Northwest University, Xi'an, China 710068

E-mail:lienling@xaut.edu.cn, ping4917305@163.com

Abstract: The electronic structures and optical properties of undoped and Al-doped GaN ($Al_xGa_{1-x}N$, x=0.0625, 0.125, 0.25) have been studied based on generalized gradient approximation (GGA) method of density functional theory (DFT). The differences of the electronic structures and optical properties of undoped and Al-doped GaN have been discussed in detail. The result shows: according to total density of state of undoped and Al-doped GaN, the conduction band becomes width and moves to high energy level with gradual increase concentration of Al impurity. Impurity energy band isn't found in energy band structures of $Al_xGa_{1-x}N$, the same as energy band structures of undoped GaN, but the band gaps gradually become wide with increase of Al impurity. Absorption spectra of undoped and Al-doped GaN of main absorption peak moves to high energy level with increase of Al impurity.
Kew word: GaN super cell, Al-doped, DFT, optical property

0. Introduction

Wide band gap of compound semiconductor materials, such as GaN, SiC, AlN, which are the third generation semiconductor materials after the first generation of the Ge and Si semiconductor materials and the second generation compound semiconductor materials of GaAs and InP, have been developed. It is one of currently the global semiconductor researched fronts and hot spots in research and application. GaN-based semiconductor materials have wide band-gap, large electron drift velocity, high thermal conductivity, resistance to high voltage, high temperature, corrosion and radiation, and so on. Especially, it can be used to produce high-frequency, high efficiency, high temperature, high pressure, high power microwave devices. Ⅲ-nitride semiconductor materials represented by GaN are the direct band gap materials, including AlN, GaN, InN and the ternary alloy (AlGaN[1], InGaN), quaternary alloy[2] (Al/InGaN) materials. They have a lager rang of band gap from 1.9eV of the InN to 3.4eV of GaN, then to 6.2eV of AlN, equivalent to wavelength light-emitting area from 650nm to 200nm. This band covers the lager spectrum from visible light (red, yellow, green, blue) to the near UV range, so that the material system has a wide and continuous tunable direct band gap. Thus according to use Ⅲ-nitride material system, blue, green LEDs, blue-violet, UV LDs, ultraviolet (UV) light detectors and high power electronic devices can be prepared. GaN-based AlGaN/GaN HEMT devices have output power density, high temperature, resistance to radiation, etc. They can meet the next generation of electronic equipment demands on the microwave power device of higher power, higher frequency, smaller size and work under more adverse conditions. So they can be widely used in microwave millimeter sophisticated electronic equipment, civilian communications base stations. As the band structure of AlGaN band gap and the corresponding bending coefficient (bowing parameter)[3-6] are directly related to the emission wavelength, particularly in the design of bending coefficient emission wavelength of the light-emitting components play a vital role. Therefore there are many researchers at home and abroad to experiment or theoretical study. Now, Guo et al[7] studied the Al, Mg doped GaN (sphalerite structure) electronic structures and optical properties. Xing et al[8] studied Mn doped GaN (sphalerite structure) of the electronic structures and optical properties. Bo et al[9] studied the $Al_xGa_{1-x}N$, $In_xGa_{1-x}N$ and $Al_xIn_{1-x}N$ of the electronic structures and the bending coefficient. From then on, theoretical study of optical properties of Al doped wurtzite GaN crystals currently are not reported, and the use of plane wave pseudopotential method (PWP) of theoretical research has not been reported yet.

In order to study the optical properties of undoped and Al-doped the GaN crystals, this paper used plane wave pseudopotential method. The first, its different impurity concentration of the energy band structure, electronic densities of state were calculated and analysed. Then the optical properties of undoped and doped of the GaN crystals were studied under different impurity concentrations and the absorption spectra were obtained. Finally, the results of optical properties and electronic properties of undoped and Al-doped the GaN crystals were compared.

1. Theoretical model and calculation method

1.1 Theoretical Model

The most stable hexagonal wurtzite structure belongs to $P6_3mc$ space group. It has C^4_{6v} symmetry, lattice parameters α=β=90°, γ=120°, a=b=0.3189nm, c= 0.5185 nm, c/a is 1.626, which is smaller compared with the ideal hexagonal close packed structure of 1.633. The unit cell is made of the Ga and N of the hexagonal close-packing. In this paper, according to the ideal structural parameters of GaN, the super cell structure was obtained. Then based on super-cell model, 32, 16, 8 atoms of supercell systems were selected in term of different Al doping concentrations x. One Al atom was doped in one unit cell. In the super cell, one Al atom substituted for one Ga of lattice positions, thus we obtained 32, 16, 8 atoms of Al-doped super cell system of model structure.

1.2 Calculation Method

The calculation is completed using ab initio quantum mechanics program CASTEP based on DFT. Using periodic boundary conditions, the electronic wave functions are extended by plane wave basis vector. The ion core and valence electrons of the interaction is described by ultra-soft pseudo potentials [10] in order to reduce the number of plane wave basis vector. The selected Ga, N, Al of valence electron configuration are Ga: $3d^{10}4s^24p^1$, N: $2s^22p^3$, Al: $3s^23p^1$, respectively. In the reciprocal k-space, plane wave cutoff energy E_{cut} is 440eV, the exchange related energy makes use of the generalized gradient approximation (GGA) of PBE approximation method. Systematic total energy and charge density make use of Monkhorst-Pack program for the Brillouin zone of integral calculations. 32, 16 and 8 atom super-cell systems select 4×4×2, 2×4×2, 2×2×2 of k grid point, respectively, in order to ensure the systematic energy of convergence in the quasi-plane wave basis of level, convergence accuracy is 2×10^{-6}eV/atom. Internal atomic force of convergence criteria is 0.003eV/nm, single atomic energy of convergence criteria is 1.0×10^{-6}eV/ atom. To make the results stability and accuracy, the cell structures are firstly optimized and then internal coordinates are optimized in terms of the cell parameters. Finally the electronic structure and optical properties [11] are calculated.

2. Results and discussion

2.1. Undoped and Al-doped GaN of Electronic Structure

2.1.1 Optimized Structure of Properties. In order to compare conveniently with Al-doped GaN of electronic structure and optical properties, Undoped and Al-doped GaN super cells were optimized respectively. The experimental values and the optimized structural parameters of GaN lattice constants were shown in Table 1.

From Table 1, we can see that the undoped GaN of lattice constant a and c are consistent with the experimental values. c/a is equal to 1.629, which are consistent with the experimental value 1.626, deviated of only 0.18%. So it shows this method is reliable. Table 1 show that different Al-doped GaN of lattice parameters is different.

Table 1. GaN and $Al_xGa_{1-x}N$ of lattice constants

		lattice constants		
		a/nm	c/nm	c/a
experiment		0.3189	0.5185	1.626
theory	GaN	0.3225	0.5254	1.629
	$Al_{1/4}Ga_{3/4}N$	0.3198	0.5200	1.626
	$Al_{1/8}Ga_{7/8}N$	0.3211	0.5226	1.628
	$Al_{1/16}Ga_{15/16}N$	0.3221	0.5250	1.629

Lattice constants slightly increase with decrease of Al-doped concentration. Because the Al^{3+} ionic of radius is smaller than the Ga^{3+}ionic of radius. When one Al^{3+} replace with one Ga^{3+}, the volume will become smaller, meanwhile the lattice parameter becomes smaller. This change can be seen from bond length (Ga-N bond is 0.196nm, Al-N bond is 0.187nm).

2.1.2 The Density of States. Figure.1 shows the total density of states of the undoped and Al-doped GaN super cell. From Figure.1, we can see that the undoped GaN (a) and Al-doped GaN (b), (c), (d) super cell have three electronic states, including band 1, band 2 and band 3. In the band 1 and band 2, the electronic density of states of (a), (b), (c), (d) super cell are complete consistency. Band 1 has three peaks, left and right peaks are not high. Because the electrons are few in this energy, the middle of peak is very high and sharp. This shows that the density of electronic state is lager in this energy. It

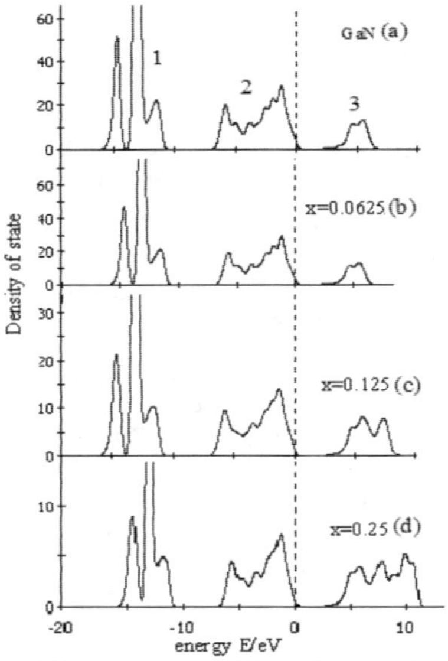

Figure 1. Densities of state of GaN and $Al_xGa_{1-x}N$

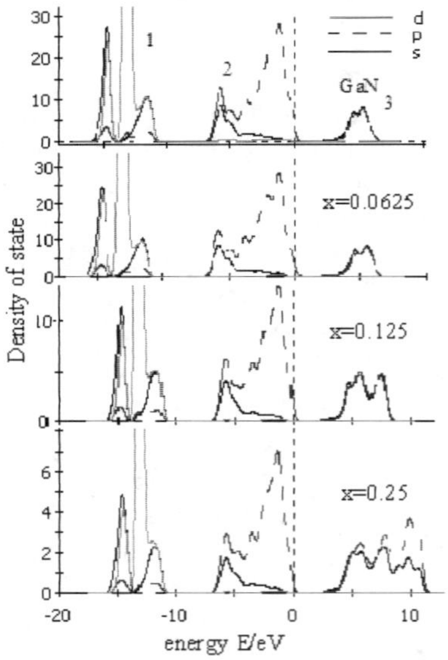

Figure 2. Partial densities of state of GaN and $Al_xGa_{1-x}N$

shows a strong localization and is no significant interaction compared with the other valence band. In band 2, the peak is not high, but relative wide, about 7eV. This shows that a wide range of electronic state is presented. In the band 3, the electron density changes significantly with the increase of Al doping concentration. The distribution of the electron density of states becomes wide, and the conduction band moves to the high energy area, resulting in wider optical band gap. This is because the Al impurity is doped, its 3p state electron is contributed.

Figure.2 shows the partial density of states of the undoped and Al-doped GaN, super cell. From Figure.2, we can see that the left peak is the Ga 3d states and the N 2s state of the electron distribution and the right sharp peak is the Ga 3d states and a litter amount of the Ga 4s, N 2s states of the electron distribution in band 1. Band 2 of width is about 7.28eV, including mainly the N 2p state and Ga 4s states of the electron distribution. The N 2p electronic state determines the top of valence band of positions. Band 3 is absolutely different from band 1 and band 2. Its band width is about 7.68eV and becomes wide with the increase of Al-doped concentration. Meanwhile it moves to high energy zone. The main reason is contribution of Al 3p [12] state of the electron distribution. Finally, band gap width is wider.

2.2 $Al_xGa_{1-x}N$ Optical Properties

2.2.1 $Al_xGa_{1-x}N$ Optical Band Gap. Figure 3 shows the energy band structures of GaN and $Al_xGa_{1-x}N$

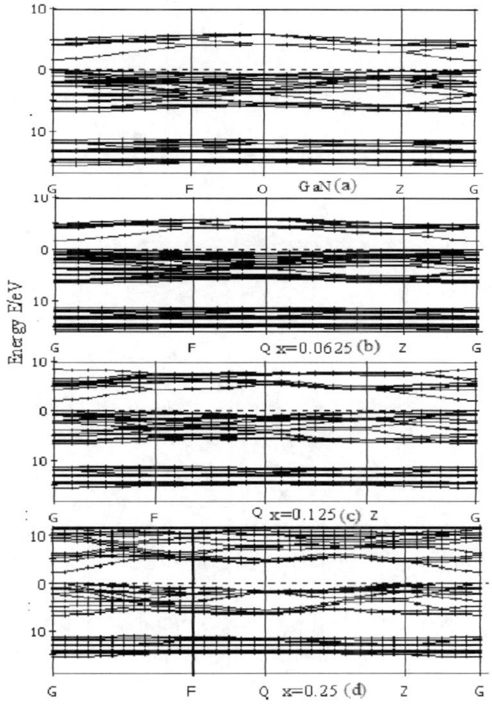

Figure 3. Energy band structures of Al $Al_xGa_{1-x}N$

(x=0.0626, 0.125, 0.25). From Figure.3 (a), calculated band gap value is 1.72eV. Compared with the experimental value Eg=3.4eV, it is quite lower. The reason is that used DFT calculation theory is the ground state theory, while the energy gap belongs to the properties of excited states. Therefore, the obtained result is very low and that is the prevailing theory of the phenomenon in the calculation. However, this phenomenon does not affect the theoretical analysis of electronic structure of GaN. From Fig. 3(b), (c), (d), we can see that Al-doped GaN super cells are still a direct band gap structure semiconductor, because the bottom of conduction band and the top of valence band lie at G-point of Brillouin zone (the point) Department. Different concentrations of Al doped GaN is not impurity energy level. Because Al atom is equivalence impurities， and different from other impurities. There are not the donor and the acceptor energy level between the top of valence band and bottom of conduction band. For different concentrations of Al-doped GaN, the energy gaps are different. Al_xGa_{1-}

$_x$N (x=0.25, 0.125, 0.0626) super cells of the energy gaps are 2.19eV, 1.95eV, 1.75eV, respectively. Their band gaps (Eg) are wider than the undoped GaN of band gap (Eg=1.72eV). The impurities concentration increase along with the wide of the energy gap. The reason is that the band gap of GaN is 3.4eV, while the band gap AlN is 6.2eV, Al-doped GaN of band gap should range from 3.4eV to 6.2eV.

2.2.2 The Absorption Spectrum of $Al_xGa_{1-x}N$. The medium of light absorption properties can be described by the virtual part of relative dielectric functions. Generally speaking, the imaginary part of relative dielectric function caused the absorption of electromagnetic wave, light absorption coefficient and relative dielectric function [13] are direct proportionality.

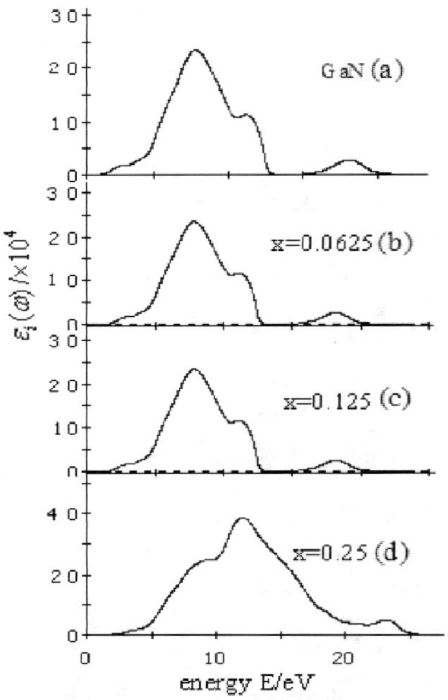

Figure 4. Absorption spectrum of $Al_xGa_{1-x}N$

Figure 4 shows the absorption spectrum of GaN and $Al_xGa_{1-x}N$, that is the dispersion curve of imaginary part of dielectric function [14]. The peaks come from the optical transitions of different critical point between valence band and conduction band. From Figure 4, we can see that when Al-doped concentration is equal to 0.0625 (b) and 0.125 (c), the absorption spectrum has three peaks and between the second peak and third peak there are the complete not-absorption state. When the Al doping concentration is equal to 0.25 (d), the absorption spectrum has three peaks which lie in completely the absorption state. This is because the absorption peaks come from the optical transition of different critical point between valence band and conduction band. The larger the doping concentration, the greater the valence band density of states, the more optical transitions different critical points between valence band and conduction band, so the not-absorb states disappear. Meanwhile, with Al doped concentration of increase, the main absorption peaks shift to higher energy[15]. The reason is that the band gap GaN is 3.4eV, while the band gap AlN is 6.2eV, Al-doped GaN of band gap ranges from 3.4eV to 6.2eV. Different concentrations have a different band gap. Of course, the greater the concentration of Al doping, the wider the band gap, that is consistent with the band structure of conclusions. According to the relationship of the band gap and wavelength, we obtained that the wavelength becomes small along with increases of the band gap, then the frequency increases. Therefore the absorption peaks move to the higher energy zone.

3. Conclusion

In this paper, the geometric structures of undoped and Al-doped GaN ($Al_xGa_{1-x}N$, x=0.0625, 0.125, 0.25) have been optimized based on the plane wave pseudopotential method and generalized gradient approximation of density functional theory(DFT). The structure of the lattice parameters were obtained and compared with experimental parameters, therefore this calculation method has good reliability. The electronic structure and optical properties of pure and Al-doped GaN ($Al_xGa_{1-x}N$, x=0.0625, 0.125, 0.25) have been studied. The result shows the total densities of states of pure and Al-doped GaN have obvious change with the increase of Al-doped concentration. That means the conduction band energy levels become wide and move to high energy. Band structure of calculations show that the Al-doped GaN do not appear donor and acceptor impurity levels and are similar the energy band of the ideal structure of GaN, but the optical band gap is wider than the ideal structure. Optical band gap continuously becomes wider with the increase of Al-doped concentration. These results provide some theoretical basics for the further design and application at the molecular and atomic scale.

Acknowledgements

This work was supported by Xi'an Applied Materials Innovation Fund and Xi'an Development Research Project Application.

4. References

[1] Miwa K and Fukumoto A 1993 *Physical Review B* **11** 7897
[2] Zhixiang Yangn and Zhizhen Ye 2006 *Zhejiang University*
[3] Yen Kuang Kuo and Wen Wei Lin 2002 *J. Appl. Phys* **8** 73
[4] Dridi Z, Bouhafs B and Ruterana P 2003 *Semicond. Sci. Techno* **5** 850
[5] Yun F, Reshchikov M A, He L and King T 2002 *J. Appl. Phys* **92** 4837
[6] Kuo Y K and Lin W W 2002 *J. Appl. Phys* **24** 73
[7] Jianyun Guo, Guang Zheng and Kaihua He 2008 *Acta Phys.Sin* **57** 3740
[8] Haiying Hing, GuanghanFan and Tianming Zhou 2008 *Acta Phys.Sin* **24** 1432
[9] Bo Ting Liou, Sheng Horng Yen and Yen Kuang Kuo 2005 *Proc. Of Spie* **56** 296
[10] J. Zenneck, T. Niermann, D. Mai and M. Roever 2007 *Appl.Phys* **101** 3504
[11] ZiLi Xie, Rong Zhang and RuoLian Jiang 2009 *Journal of electronic materials* **52** 332
[12] Fu jia Yang. 2005 *Atomic Physic* (Beijing: Higher education)
[13] Jie ping Fan, Liu hui Ming and Qiang Tian 2009 *College physics* **38** 24
[14] Rong chuan Fang 2005 *Solid spectroscopy* (Hefei: China science and technology university)
[15] Guomin He, Renzhi Wang and Yongmei Zheng 1998 *Journal of Xiamen University* **37** 300

3rd International Photonics & OptoElectronics Meetings (POEM 2010) IOP Publishing
Journal of Physics: Conference Series **276** (2011) 012045 doi:10.1088/1742-6596/276/1/012045

Enhanced Transmission through the Nanoparticle Coat

G Song [1, 2], L Yu [1, 2] , B Liu [1, 2] and T Li [1, 2]

[1]School of Science, Beijing University of Posts and Telecommunications, Beijing, 100876
[2]Key Laboratory of Information Photonics and Optical Communications (BUPT), Ministry of Education, Beijing 100876

E-mail: candlesinwind@gmail.com

Abstract. By using the full-vectorial three-dimensional finite-difference time-domain method and the perfect electronic conductor as the materials, we studied the enhanced transmission spectra through a substrate with the perfect electronic conductor nanoparticle coat. Single metal nanoparticle exhibited characteristic localized surface plasmon modes. By placing more plasmonic nanoparticles close to each other to make up of a plane, it was possible to observe the interaction between the modes of each individual nanostructure. The results showed that the enhanced transmission spectra through a substrate with different nanoparticle coats which were made by different numbers of the layers or by different metals exhibited different peaks whose values could reach 1. The number of the nanoparticle coat layers influenced the number and the positions of the enhanced transmission peaks.

1. Introduction

Since the phenomenon of the enhanced transmission through the subwavelength array holes is discovered by Ebessen and co-workers [1], it is focused by many scientists around the world. After that time, there are many researches on the enhanced transmission with both the experiments and theories. In the experiments, there are many researches on the enhanced transmission by using the noble metal materials such as the gold [2] and the silver [3] and the semiconductor for example the silicon [4]. Many kinds of subwavelength holes such as rectangular holes [5], cylinder holes [2] and ellipsoid holes [6] and nanoslits [7] were researched to investigate the enhanced transmission spectra. The nanoparticles also attract the scientists' eyes. The metal nanoparticles mixed in the dielectric were deposited on the dielectric substrate or a multiple substrates such as metal-insulator-metal structure [8] in experiments. At the same time, there are many researches on theories about the interaction between particles with different shapes and materials and substrates with metal-insulator-metal or insulator-metal-insulator structures. The single nanoparticle coupled with a metal film was just given in [9]. The different shapes of the particles on the single substrate were studied for the scattering [10-12]. The nanoparticle coat deposited on the substrate made by glass (SiO_2) has low reflection phenomena [13]. They studied the collective resonances of the whole nanoparticle coat. Single metal nanoparticle exhibited characteristic localized surface plasmon (LSP) modes. It exhibits the feature of enhanced electronic field. When they get together, the resonance of the nanoparticles is formed to be a bridge to guide the electromagnetic field passing by. They show the phenomenon of enhanced transmission. In another word, the reflective wave is reduced and the transmission wave is enhanced.

In this Letter, we use the nanoparticles as the elements and study the transmission spectrum to find out the position of the enhanced frequency and the effective of enhancement. We research a new phenomenon of enhanced transmission through such a structure with the perfect electronic conductor (PEC) nanoparticle coats deposited on a substrate. The results showed that when nanoparticle coats were made by different numbers of the layers or by different noble metals deposited on a insulator substrate, the enhanced transmission spectra peaks' positions and intensities were varied with the space between two nanoparticle lines, numbers of layers and the gold or silver nanoparticle coats. The PEC could be naturally extended to the real metal. In addition, we also take the gold and silver as real material not as PEC to discover the difference of the enhanced transmission. This kind of structure we researched might be as a new kind of filter in the future and also used in controlling light in nanosize.

2. Models and Results

We use the PEC as the material to make up of the nanoparticle. The radius of the nanoparticle is 105nm. The particles deposited on the SiO_2 substrate are arranged next to next in a line and the lines are made up of a plane which the area is about 1.995μm×1.995μm shown in Fig. 1. We use the FDTD method to get the transmission spectra.

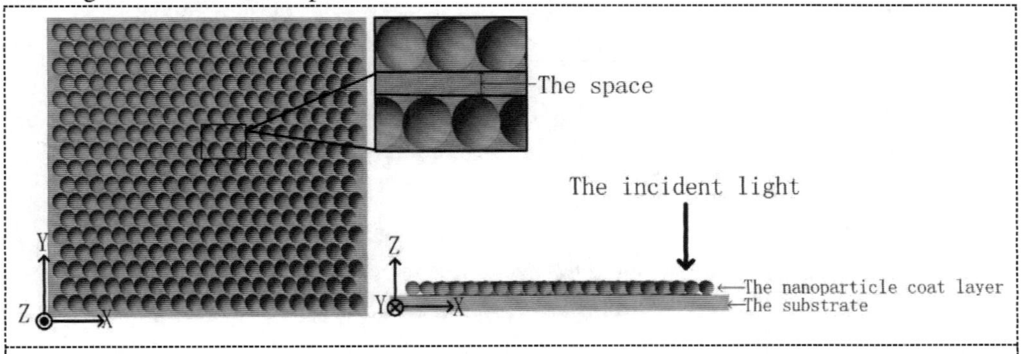

Figure 1. The schematic of the substrate with nanoparticle coat.

Firstly, a single nanoparticle coat layer is discussed and the transmission spectrum is shown in Fig. 2. The space between two nanoparticle lines is varied from 0 to 120 nm. All the spectra make up of a map. From this map, we can know that the transmission is changed acutely and brightly near the space at 27nm.Whereas, the effective of transmission at the same frequency always reaches the maximum with the space of two lines about 88nm. Compared with other's experiment work which the maximum transmission is 95% at about 600nm [13], it could be near 97% in the same wavelength in our simulation. Obviously they are in accordance with each other very well.

The value of transmission is all about 10^{-4} with the space at 0. The particles are so close to prevent the light. These electrons at the surface of the nanoparticles resonate to make up of the surface plasmon. The narrow space limits the resonance. With the space at about 27nm, the values of transmission change greatly. The space between two lines of the sample influences the collective resonance of the electrons at the surface. It also influences the scatting of the light. Single nanoparticle appears the LSP mode. When they are set in a line closed to each other and these lines are made up of a plane, the nanoparticle line exhibits a collective action. When the space becomes larger, the surface plasmon of the nanoparticles is enhanced. The surface plasmon course the enhanced transmission.

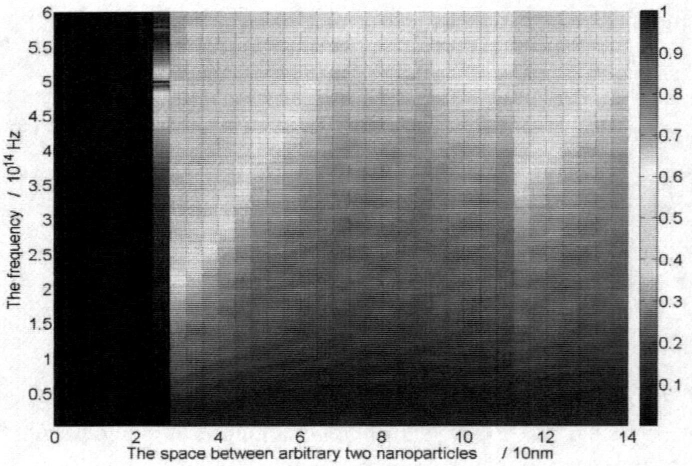

Figure 2. The spectra of a single nanoparticle layer: the space between two nanoparticle lines is varied from 0nm to 120nm.

The electrons at the surface of the nanoparticles resonate to form surface plasmons on metal nanoparticles (SPN). We take an area to investigate the electric field distributions of the samples and reflection spectrum. We choose two kinds of areas to keep the same numbers of particles: one is about 320nm×300nm for the space=0nm and the other is about 220nm×500nm for the space=110nm, 112nm and 114nm. The distributions of $|\boldsymbol{E}|^2$ at middle of the coat layer are shown in Fig. 3. Considering the effect of the periodic boundary conditions in the simulation, we should observe the field distribution the area of the around centre. For the case of space=0nm, we find the distribution of the electric intensity is weak. The reason why electric intensity excited by surface plasmon so weak in this case is that nanoparticles set next to next shown in Fig. 1 lead to the electric field decreased as shown in Fig. 3 (e). The excited surface plasmon of the nanoparticles is coupled to each other along the direction of the polarition. The electric field of two half spheres in one nanoparticle is opposite seen in Fig. 3 (f). When the particles are set next to next, the fields are decreased at the contact surface shown in Fig. 3 (g). The field near the surface of the particles is weak and the transmission is prevented. When the space of two nanoparticle lines increases linearly, the value of the transmission presents changing with periodic and increasing. From (b), (c) and (d), we can see that the average amplitude in (c) is more than the others and the value of the transmission is also larger than the others. Obviously, the surface plasmon excited lead to the strong intensity distribution. The coupled SPN helps to enhance the transmission light.

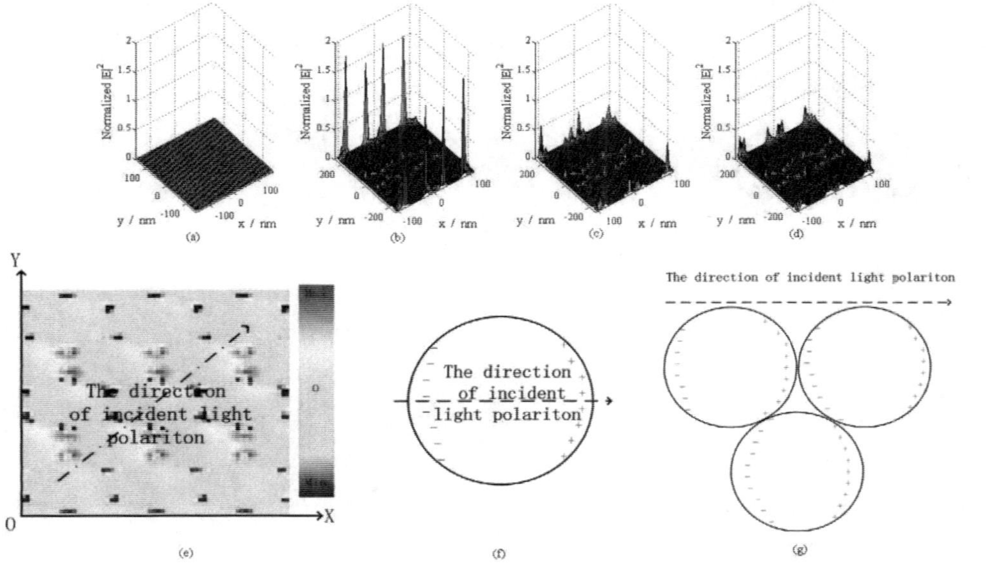

Figure 3. The electric field and amplitude distributions and the illustration of the coupling mode: (a) the space=0nm, (b) the space=110nm, (c) the space=112nm, (d) the space=114nm, (e) the field distribution at space=0nm, (f) one nanoparticle polariton, (g) the coupling of the nanoparticles set next to next.

When we put the nanoparticle lines in a plane perpendicular to the direction of the incident light, the nanoparticle coat exhibits the character of the enhanced transmission. If we put the nanoparticle lines in a plane parallel to the incident light, there is a new phenomenon. Then, we change the number of the nanoparticle layers. Comparing with one layer, the number of the layers which we considered is two, three, four and five, respectively. We choose the space of two nanoparticle lines is 28nm and get the transmission spectrum and show in Fig. 4.

From Fig. 4, we can see that there is more than one peak in each spectrum except (b) in Fig. 4. Except the first peak in the spectra, the other peak frequencies are all among the 1×10^{14}Hz~7.6×10^{14}Hz. In the spectrum of two, three, four and five coat layers, the peak number is accidently equal to the number of the layers. The reflection light and the transmission light couples and then the mixed light goes from one layer to another. The final result is that there are several peaks in the spectra. Comparing with the lattice of 28nm, we also choose the lattice of 88nm and get the same phenomenon. The difference between the two kinds of lattice is the values of the transmission. The values of the transmission are all over 0.8 with the lattice of 88nm. If we use this structure as a filter, the space of 88nm could not prevent the light which we do not need. If the space we used is too small, the transmission coefficients are too small. So we must choose the space of about 28nm.When the nanoparticle lines are set in a plane along the incident light, the plane is formed as a channel to guided the light. The incident light is *p*-polariton and the polariton is perpendicular to the coat and parallel to the direction of the layer. The space influences the width of the spectrum. The larger the space is, the wider the width is. When the particles were set in a line and parallel to the polariton of the incident light, the surface plasmon of nanoparticles strengthens the exciting field at the position of the other particle [14]. Influenced by the space and the polariton, the spectra exhibit several sharp peaks.

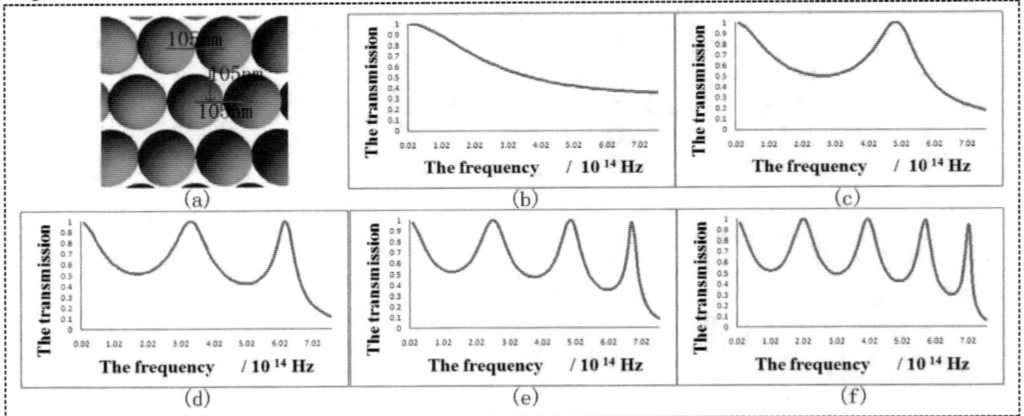

Figure 4. The transmission spectrum with the one (b), two (c), three (d), four (e) and five (f) coat layers.

3. Discussion

Similar behavior can also be expected in the optical regime for a real metal. Here we also calculate the transmission spectra through Ag and Au nanoparticle coat layers by the 3D FDTD method. The radius of the nanoparticles is 105nm. The space is 28nm. The frequency-dependent permittivity of Ag and the permittivity of Au are referred to the literature [15]. The incident light is *p*-polarition. We just use five different kinds of layers and get the transmission spectra shown in Fig. 4. The spectra of one layer, two layers, three layers and five layers in Fig. 5 are similar with Fig. 4. The intensities of some peaks in Fig. 5 are less than the ones in Fig. 4. The positions of the peaks in Fig. 5 have red-shifts. The electromagnetic field exists in the real metal nanoparticles. It influences the collective resonance of all the nanoparticles. The dielectric functions of metals also influence the surface plasmon of particles. Both the two reasons change the frequency of the enhanced transmission peak.

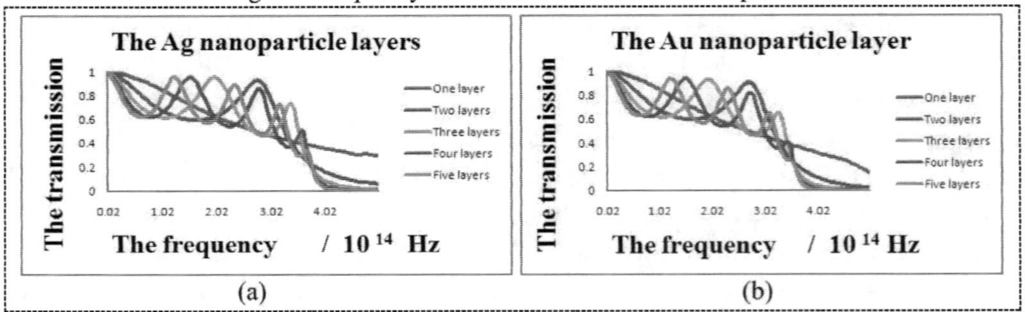

Figure 5. The transmission spectra of the Ag and Au nanoparticle coat layers with one, two, three, four and five layers: (a) is all the spectra of Ag nanoparticle coat layers; (b) is all the spectra of Au nanoparticle coat layers.

4. Conclusion

In conclusion, we discuss the enhanced transmission spectra through the substrate with nanoparticle coat layers. We change the number of the layer and get different transmission spectra. The single nanoparticle coat layer is perpendicular to the polariton of incident light to exhibit the enhancement. By changing the space of the samples, we know that the space constant influences the amplitude of the enhanced transmission spectra. The more the space is, the higher the amplitude is. By changing the number of the layers, we know that the direction of the nanoparticle layers which is parallel to the polarition of incident light exhibits the electromagnetic field and the energy propagating and the number of the layers influences the number of the enhanced transmission peaks. The more layers the more peaks. All of these, we know that the space between two lines, the number of the layers and the materials decide the frequency and the amplitude of the enhanced transmission peaks together.

Acknowledgement

This work has been supported by the National Basic Research Program (973) of China under Project No.2010CB923202.

References

[1] Ebbesen T W, Lezec H J, Ghaemi H, Thio T and Wolf P A, 1998 *Nature (London)* **391**,667

[2] Bai B F, Li L F and Zeng L J 2005 *Optics Letter* **Vol. 30, No. 18 182630**

[3] Haraguchi M, Okamoto T, Inoue T, Nakagaki M, Koizumi H, Yamaguchi K, Lai C, Fukui M, Kamano M and Fujii M 2008 *IEEE Journal of Selected Topics in Quantum Electronics* **Vol. 14, No. 6**

[4] Hendry E, Garcia-Vidal F J, Martin-Moreno L, Go´mez Rivas J, Bonn M, Hibbins A P and Lockyear M J 2008 *Phys. Rev. Lett.* **100, 123901**

[5] van der Molen K L, Klein Koerkamp K J, Enoch S, Segerink F B, van Hulst N F and Kuipers L 2005 *Phys. Rev. B* **72, 045421**

[6] Tammy K L, Alan D B, Hübner J and van Driel H M 2006 *J. Opt. Soc. Am. B* **Vol. 23, No. 10 2142-2147**

[7] Porto J A, Garcı´a-Vidal F J and Pendry J B 1999 *Phys. Rev. Lett.* **Vol. 83, No. 14 2845**

[8] Kume T, Hayashi S and Yamamoto K 1997 *Phys. Rev. B* **Vol. 55, No. 7, 0163-1829**

[9] Sondergaard T and Bozhevolnyi S I 2004 *Phys. Rev. B* **69, 045422**

[10] Evlyukhin A B 2005 *Phys. Rev. B* **71, 134304**

[11] Evlyukhin A B, Brucoli G and Martín-Moreno L, Bozhevolnyi S I and García-Vidal F J 2007 *Phys. Rev. B* **76, 075426**

[12] Yang T and Crozier K B 2008 *Optics Express* **Vol. 16, No. 17, 99120**

[13] Zhao Y, Wang J S and Mao G J 2009 *Optics Letter* **Vol. 30, No. 14 141885**

[14] Brongersma M. L. and Pieter G. K. *Surface Plasmon Nanophotonics* (Springer, USA, 2007)

[15] E. D. Palik, *Handbook of Optical Constants in Solids* (Academic, Boston, 1991), Vol.1

Study on the Electronic Transport Properties of Zigzag GaN Nanotubes

Enling Li[1),+], Xiqiang Wang[1),+], Liping Hou[1)], Danna Zhao[1)], Yuanbin Dai[1)], Xuewen Wang[2)]

[1)] Sciences School, Xi'an University of Technology, Xi'an, China 710054

[2)] Electronic Information Science and Technology, Northwest University, Xi'an, China 710068

Abstract: The electronic transport properties of zigzag GaN nanotubes (n, 0) ($4 \leq n \leq 9$) have been calculated using the density functional theory and non-equilibrium Green's functions method. Firstly, the density functional theory (DFT) is used to optimize and calculate the electronic structure of GaNNTs (n, 0) ($4 \leq n \leq 9$). Secondly, DFT and non-equilibrium Green function (NEGF) method are also used to predict the electronic transport properties of GaNNTs two-probe system. The results showed: there is a corresponding relation between the electronic transport properties and the valley of state density of each GaNNT. In addition, the volt-ampere curve of GaNNT is approximately linear.

1. IntroductionGaN

GaN has attracted much attention as a wide band gap semiconductor material [1]. Up to now, GaN bulk structures and GaN films have been studied extensively. In theoretical calculation, the main work of these research are calculations about some properties of one-dimentional GaN nanomaterials such as mechanics, electronic structure, light absorption, lattice vibration, electronic transportation characteristic and so on. However , in experiment, the prime task is only related to the preparation of one-dimentional GaN nanomaterials [2-3].

2. Calculation model and method

The structural optimization and the electronic structure calculation of the GaNNTs have been studied using first-principles method which is based on density function theory (DFT). During structural relaxation and electronic structure calculation, the generalized gradient approximation (GGA-PW91) is used to deal with the exchange of relevant potential energy, K-point sampling of Brillouin is $1 \times 1 \times 100$ corresponded to simple Brillouin zone.

In this paper, the electronic transport properties of zigzag GaN nanotubes (n, 0) ($4 \leq n \leq 9$) have been calculated using ATK software based on DFT and NEGF. In this work, GGA is used as the exchange correlation function, with a $1 \times 1 \times 100$ K-point sampling corresponding to simple Brillsouin zone. ATK software can fully complete self-consistent process of the atomic-scale system and the semi-infinite electrodes coupling systems, the voltage across the conductor directly get involved in self-consistent calculation. System can be divided into three parts: central scattering region, left electrode and right electrode areas. The model is shown in Fig.1 and the current through this system can be obtained from Landauer-Biittiker formula according to the corresponding Green function [4]:

$$I=\frac{2e}{h}\int T(E,V)\left[f(E-\mu_L)-f(E-\mu_R)\right]dE \quad (1)$$

Here, μ_L and μ_R are Chemical potential of Left and right electrode respectively. $(\mu_R-\mu_L)/e=V$ is the electrical potential difference between left and right ends, $\mu_R=E_f-\frac{V}{2}$, $\mu_L=E_f+\frac{V}{2}$, $[\mu_L,\mu_R]$ is energy integral interval, $f(E-\mu_L)$ and $f(E-\mu_R)$ are electronic fermi distribution function of the left and right electrodes respectively, $T(E,V)$ is transmission coefficient when energy is E and biasing is V. The transmission coefficient of Fig.1 system has been obtained using ATK.

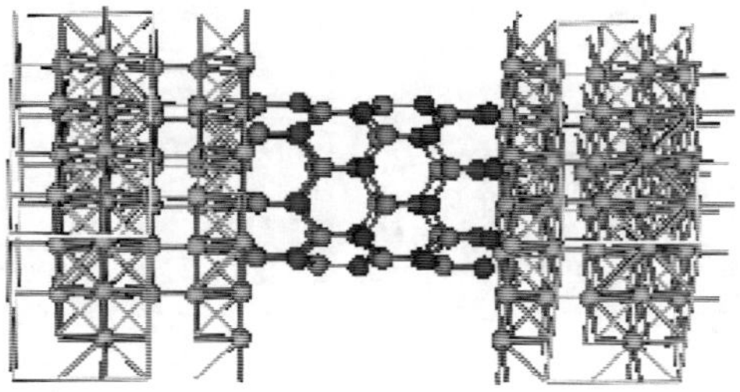

Fig.1the electron transport model

3. Simulation results and analysis

3.1. Optimization analysis

The optimized structures of zigzag GaNNTs (4 ≤ n ≤ 9) are shown in Fig.2, dark indicate N atoms, light indicate Ga atoms. The positions of atoms in the model are the results of optimization. The result shows that the Ga–N bonds display a buckling with the N atoms moving slightly outwards and the Ga atoms moving slightly inwards, zigzag GaNNTs are a buckling of the Ga–N bond. The reason for buckling is that surface layer atom has dangling bonds, in order to reduce surface energy, chemistry bonding needs to readjust structure adjustment, this result is consistent with Lee's [5] results.

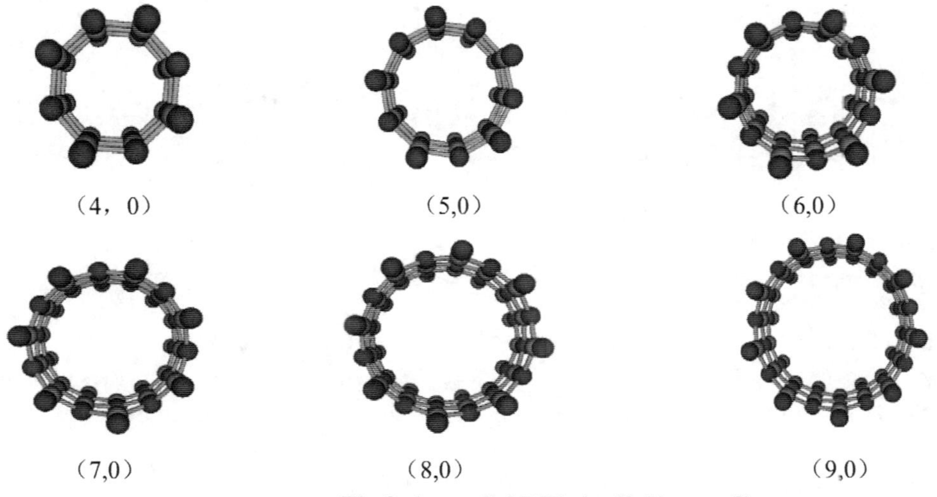

Fig.2 zigzag GaNNTs (n, 0) (4 ≤ n ≤ 9)

The results electronic band structures of zigzag GaNNTs (n, 0) ($4 \leq n \leq 9$) are shown in Fig.3. (3, 0) GaNNT has a direct-band-gap semiconductors [5-6], whereas the (4, 0) GaNNT has an indirect-band-gap. We think that the (4, 0) GaNNT has an indirect-band-gap because of a strong $\pi - \pi$ hybrid. When n>4, like（5，0），（6，0），（7，0），（8，0），（9，0）, zigzag GaNNTs are direct-band-gap semiconductors, this result is consistent with other literatures. The band -gap of GaNNTs are essentially unchanged. This is different from other articles. The reason is that the density functional theory is based on ground state theory in the calculations, but the energy gap is based on excited state, so the obtained results may have tolerance errors.

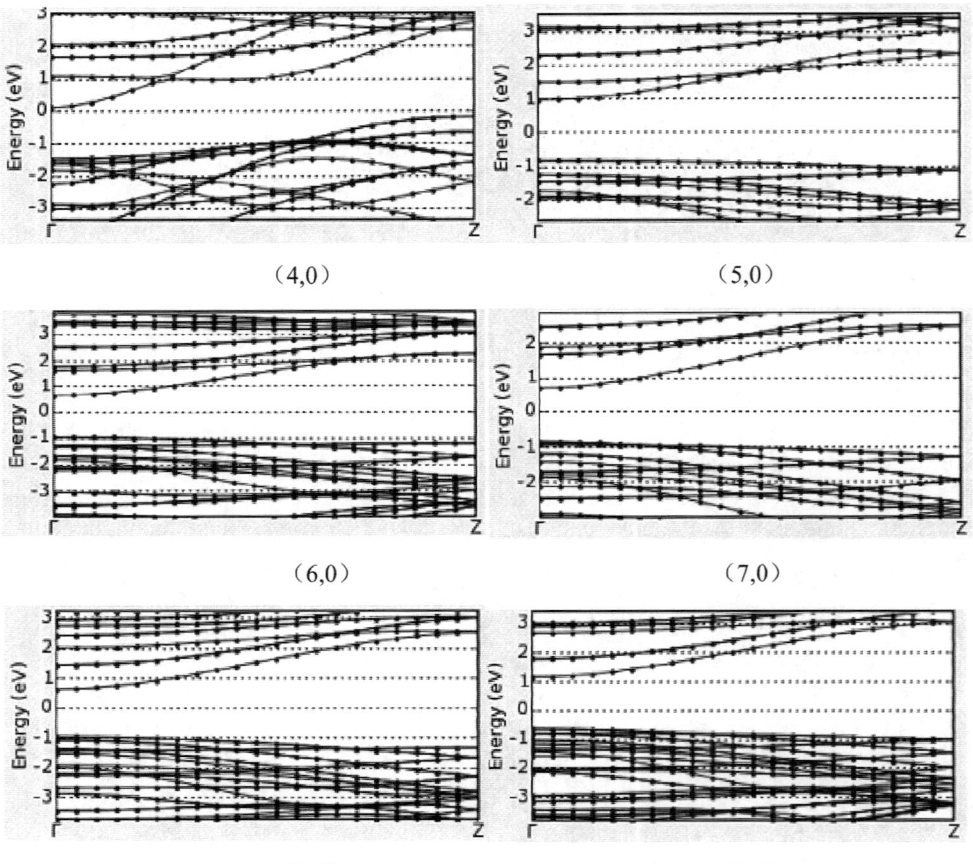

Fig.3 the band diagram of the zigzag GaNNTs (n, 0) ($4 \leq n \leq 9$)

3.2. Transport analysis

The transport properties of the two-probe system be using Au (001) as electrodes, which is exactly same with zigzag GaNNTs crystal face, we calculated the electronic transport properties of optimized zigzag GaNNTs [11]. In Fig.4, the density of states of zigzag GaNNTs under equilibrium conditions and the transmission spectrum are obtained when the electrons pass through nanotubes. The results clearly show that the probability that the electronics pass through zigzag GaNNTs are closely related to the electron energy. With the change in the absolute value of the energy, the probability that the electronics pass through zigzag GaNNTs presents the shake trend; it is completely different from the condition when the electrons go through a conductor. The transmission probability reaches high, the main reason is that when the electronics resonate with the covalent π orbital of the atomic [12].

In this paper, the voltage was 0V to 2V, unit of changes is 0.2V. We found that in this voltage range, the transmission spectrum of the system is almost unchanged, the transmission spectrum in the equilibrium state could be used to analyze qualitatively. As the molecular orbital density of states (DOS) peaks has a significant contribution to the system eigen state and the interaction between molecules and electrodes can cause peak broadening and level shift, the transmission coefficient peak is corresponding to the resonance passage when the electronics passed through the molecular state. From Fig.4, the computation has obtained the transmission spectrum and the density of states. The transmission spectrum has a series of peaks and valleys, and the peaks and valleys of transmission spectrum and the density state curves have a similar trend.

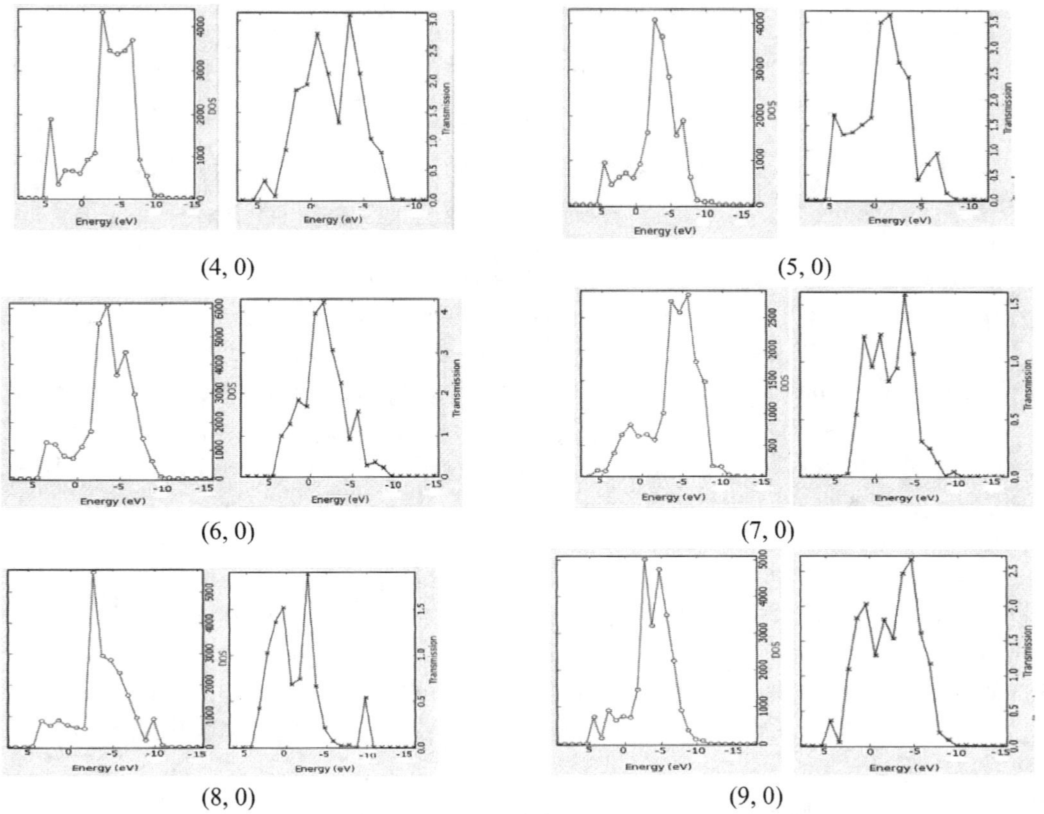

Fig.4 the density of states and transmission spectra of the zigzag GaNNTs (n, 0) ($4 \le n \le 9$)

Zigzag GaNNTs ($4 \le n \le 9$) I-V is characteristic as the curve in Fig.5 (dark behalf odd light behalf even).When voltage was 0V to 2V, performance of current and voltage has certain linear characteristics. In Fig.4, when the voltage was 0V to 2V zigzag GaNNTs($4 \le n \le 9$) I-V curve is approximately linear, the reason is that the zigzag GaNNTs density of state and the transmission spectrum and electronic density of states have the linear variation region in the small energy range (-1eV to 1eV). It is also observed that, when n is an odd number, under the same voltage, the electric current increased with the increasing of n value, when n is an even number the case is just opposite to the previous one. But, in the odd number tube (5, 0) the GaNNT does not follow the rule above, we thought that π-π track's hybridization is advantageous with compared to the electronic transportation. The rehybridization of the π-π orbits of (5, 0) GaNNT is prominent compared to that of (7, 0) GaNNT

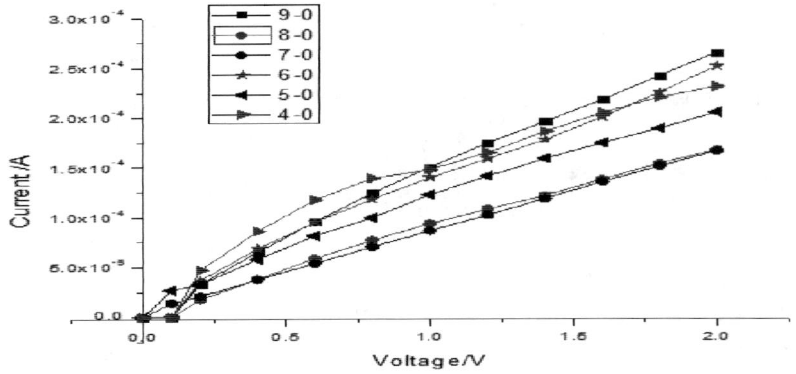

Fig.5 the current and voltage graph of the zigzag GaNNTs (n, 0) ($4 \leq n \leq 9$)

4.Conclusion

In this paper, we calculated the electronic structures of zigzag GaNNTs (4, 0) - (9, 0) using GGA method of the density functional theory, and then obtained their electronic energy structures. In the calculation results, we find that (4, 0) GaNNT is indirect band gap while the band gaps of (5, 0), (6, 0), (7, 0), (8, 0), (9, 0) GaNNTs are direct. By using the density functional theory and the non-equilibrium green's function method ,the electronic transportation characteristic of zigzag GaNNTs (4,0) - (9,0) are predicted, it can be seen that there are certain corresponding relationships between electricity transportation characteristic and the valleys of density states; within small energy range, the I –V curves of Zigzag GaNNTs show linear features

Reference

[1] Bing-She Xu, Lei-Ying Zhai,Jian Liang, Shu-Fang Ma,Hu-Sheng Jia, Xu-Guang Liu. Synthesis and characterization of high purity GaN nanowires [J].Journal of Crystal Growth, 2006, 291: 34-39

[2] Sohrab Ismail-Beigi. Electronic excitations in single-walled GaN nanotubes from first principles: Dark excitons and unconventional diameter dependences [J].PHYSICAL REVIEW, B, 2008, 77: 035306.

[3] Q. Wang,Q. Sun,P. Jena,Y. Kawazoe.Clustering of Cr in GaN nanotubes and the onset of ferrimagnetic order [J]. PHYSICAL REVIEW, B, 2006, 73: 205320.

[4] Mads Brandbyge, Jose´-Luis Mozos, Pablo Ordejo´n, Jeremy Taylor, Kurt Stokbro. Density-functional method for nonequilibrium electron transport [J].PHYSICAL REVIEW B, VOLUME, 2002, 65:165401.

[5] Jeong Won Kang, Ho Jung Hwang, Ki Oh Song, Won Young Choi, Ki-Ryang Byun. Structures, Nanomechanics and Disintegration of Single-Walled GaN Nanotubes: Atomistic Simulations [J].

Journal of the Korean Physical Society, 2003, 43(3):372-380.

[6] Sohrab Ismail-Beigi. Electronic excitations in single-walled GaN nanotubes from first principles: Dark excitons and unconventional diameter dependences [J].PHYSICAL REVIEW, B, 2008, 77:035306.

[7] Zhongchang Wang,Tingkun Gu,Tomofumi Tada,Satoshi Watanabe. Excess-silver induced bridge formation in a silver sulfide atomic switch [J].APPLIED PHYSICS LETTERS, 2008, 93:152106.

[8] M. L. Colussi, R.J. Baierle, R. H. Miwa. Stability and electronic properties of native defects and substitution alimpurities in GaN nanotubes [J].JOURNAL OF APPLIED PHYSICS, 2008, 104: 033712.

[9] Yu F Zhukovskii1, A I Popov, C Balasubramanian S Bellcci. Structural and electronic properties of single-walled AlN nanotubes of different chiralities and sizes [J].J.Phys. Condens, Matter, 2006, 18:2045–2054.

[10] Nicolas A.Bruque, Rajeev R.Pandey, Roger K.Lake. Electron transport through a conjugated molecule with carbon nanotube leads [J]. PHYSICAL REVIEW, B, 2007,76: 205322.

[11] Meng-Qiu Long, Ke-Qiu Chen, Lingling Wang, B. S.Zou, Z.Shuai. Negative differential resistance induced by intermolecular interaction in abimolecular device [J].APPLIED PHYSICS LETTERS, 2007, 91: 233512.

[12]Rostam Moradian, Sam Azadi, S.Vasheghani Farahani. Structure and electronic properties of native and defected gallium nitride nanotubes [J].Physics Letters, A, 2008 372:6935–6939.

[13] Yanhua Guo, Xiaohong Yan, Yurong Yang. First-principles study of narrow single-walled GaN nanotubes [J].Physics Letters, A, 2009, 373: 367–370.

A facile route to accelerate the formation of TiO_2 nanotube arrays

Yan Wang[1], Yucheng Wu[1, *], Gaobin Xu[2], Yongqiang Qin[1], Hongmei Zheng[1], Jiewu Cui[1], Yu Hong[1], Liang Liu[1], Xia Shu[1], Yuchun Zheng[1], Xinmin Huang[1]

1. School of Materials Sciences and Engineering, Hefei University of Technology, Hefei, 230009, P. R. China

2. School of Electronic Science and Applied Physics, Hefei University of Technology, Hefei, 230009, P. R. China

E-mail: ycwu@hfut.edu.cn

Abstract. Highly ordered TiO_2 nanotube arrays fabricated by electrochemical anodization of titanium have attracted significant attention due to their splendid promising applications. One of the factors limiting the application of TiO_2 nanotube arrays was their long sustaining reaction time by anodic oxidation, usually lasting 6 – 12 h and even longer when systhesizing thicker nanotubular layers. In this paper, we reported for the first time a facile and effective route to accelerate the formation of TiO_2 nanotube arrays by proper proportional addition of sodium carbonate(Na_2CO_3) into the anodization electrolyte. In our experiments, we adopted the 0.3 wt% NH_4F + EG (ethylene glycol) + 3.0 vol% H_2O electrolyte and we added Na_2CO_3 with the proportion n(NH_4F) : n(Na_2CO_3) = 1:1, 2:1, 3:1, 4:1 and 5:1. The field-emission scanning electron microscope (FESEM) characterization results suggested the Na_2CO_3 additives accelerated the growth rate of the TiO_2 nanotubes with the quickest growth rate 1100 nm/min when n(NH_4F) : n(Na_2CO_3) = 2:1. Finally, we investigated the mechanism of the Na_2CO_3 additives accelerating the growth rate of the TiO_2 nanotubes. It was believed that the hydrolyzation of the Na_2CO_3 additives in the electrolytes accelerated the formation of the TiO_2 nanotubes and at the same time restrained the chemical dissolution of the formed TiO_2 nanotubes.

1. Introduction

Nanostructured TiO_2 is one of the most widely studied materials due to its unique and excellent properties in optics, electronics, photochemistry and biology, as well as its

* Corresponding author: E-mail ycwu@hfut.edu.cn ;Tel.: +86-551-2901012, Fax: +86-551-2904517

applications in photovoltaic cells, photocatalysis, and sensors [1-4]. Among the various forms of nanostructural titania, TiO_2 nanotube had attracted increasing interest because of its highly ordered structure and the convenient controlling of the tube size. And since 2001[5], highly ordered TiO_2 nanotube arrays fabricated by electrochemical anodization of pure titanium foil had became the research focus due to their promising application in areas ranging from photocatalysis for hydrogen generation [6], photovoltaics [7, 8], gas sensing [9] to drug delivering [10, 11]. To the best of our knowledge, during the electrochemical synthesis of TiO_2 nanotube arrays, the reaction time of anodization generally lasts 6-12 hours, as in our previous work [12-14], and even longer when synthesizing thicker nanotubular layers [15, 16]. To explain the slow growth of nanotubes during the anodization, the chemical dissolution related to the acidification of the electrolyte ought to be firstly taken into account [17, 18]. Hence, in the present work we demonstrate how TiO_2 nanotube arrays can be achieved by rapid anodization with the growth rate up to 1100 nm per minute. And this outbreaking style formation of nanotubes is actualized by dominating the H^+ concentration at the pore tip and bottom with a certain amount of sodium carbonate (Na_2CO_3) additives into the anodization electrolyte.

2. Experimental details

In this experiment, Ti foils (0.1 mm thick foils with 99.6% purity) were firstly polished by various abrasive papers. After polishing, polished Ti substrates were degreased in acetone with ultrasonic treatment for 15 minutes, then chemical polished and pre-oxidized in 0.1 M NH_4F aqueous solution for 5 minutes. Finally, the pre-treated Ti were rinsed in ethanol and dried in oven. Titanium samples were anodized using the equipment consisting of a two-electrode configuration with a piece of highly pure graphite (50 × 30 × 5 mm) as cathode. All the experiments were carried out at room temperature about 295K. Electrolytes were 0.3M $NH_4F + x$ M Na_2CO_3 + 3.0 vol % H_2O in ethylene glycol and all the chemicals are AR grade. We controlled the value x with the proportion $n(NH_4F) : n(Na_2CO_3)$ = 1:1, 2:1, 3:1, 4:1 and 5:1. The anodization treatment processed under constant direct current potential 80V, which was achieved by a ramp from open-circuit potential to 80V with a sweep rate of 0.5V/s. The anodization time period adopted in this experiment was 60 minutes.

After the electrochemical treatment, the samples were rinsed in distilled water and dried in the oven. Samples were also annealed in order to obtain the crystalline phases of TiO_2. Field-emission scanning electron microscope (FESEM, FEI Sirion 200) was employed for the structural and morphological characterization of the TiO_2 nanotube arrays. The TiO_2 nanotube layers were taken off the titanium substrates by mechanical bending and then put directly on the electric adhesive tape in order to gain the thickness information from the cross-sectional FESEM characterization. X-ray diffraction analysis (XRD, Rigaku D/Max-2500VL/PC, Cu K_a) was performed for crystal phase identification.

3. Results and discussion

Fig. 1 shows the FESEM images of the as-prepared TiO_2 nanotubes with the top view ×20000 a), ×80000 a') magnifications and cross-sectional characteristic. From the FESEM images, it is evident that the self-organized highly ordered nanotube arrays consist of uniform pore arrays with pore diameter approximately 110 nm. Furthermore, only from the cross-sectional images of samples with series of anodization time can the thickness information of the porous arrays be obtained (Fig.1 b)-f) and Fig. 2). It can be clearly seen from Fig. 2 that the thickness of the as-prepared TiO_2 nanotube layers changed gradually with the addition of

Na_2CO_3 into the anodization electrolytes and the maximum thickness was achieved when the Na_2CO_3 additives concentration was 0.15 M ($n(NH_4F)$: $n(Na_2CO_3)$ = 2:1). The average growth rate of the whole rapid anodization process was 1100 nm per minute, much quicker than that in our previous work [12-14]. Finally, we annealed the TiO_2 nanotube arrays at 300℃ for 2h and the XRD patterns shown us anatase phase only (Fig. 3).

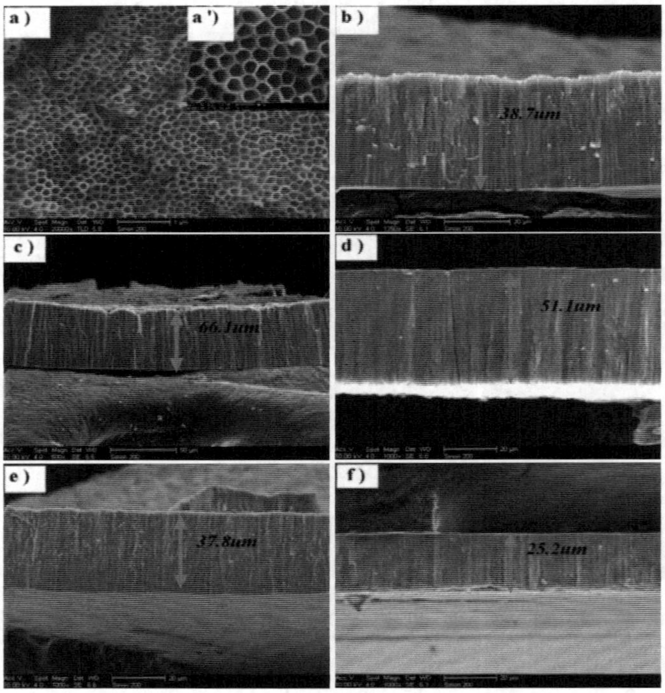

Fig. 1 The top FESEM view images of TiO_2 nanotubes with ×20000 a), ×80000 a'), and cross-sectional view of nanotubes arrays with gradually changed nanotube layers thickness produced in series of anodization electrolytes for 60 min time periods. The electrolytes and the thickness of the nanotube layers are b) 1:1, 38.7μm; c) 2:1, 66.1μm; d) 3:1, 51.1μm; e) 4:1, 37.8μm; f) 5:1, 25.2μm.

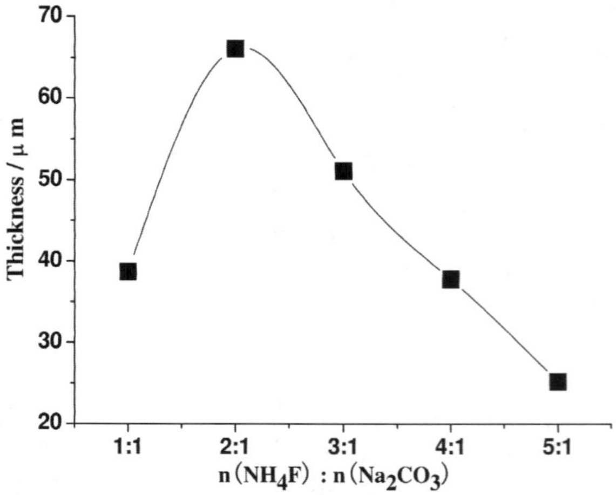

Fig. 2 Thickness of TiO_2 nanotube arrays in series of anodization electrolytes for 60 min

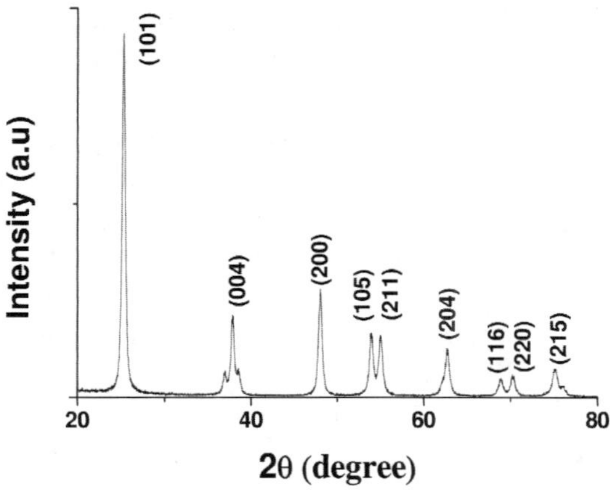

Fig. 3 XRD pattern of the TiO_2 nanotube arrays annealed at 300℃ for 2h.

In previous work the length of TiO_2 nanotubes has been recognized as the equilibrium of the electrochemical formation of TiO_2 at the pore tip and the chemical dissolution of the formed TiO_2 by fluorides at the mouth of the tubes (nanotube arrays surface) in the F^--containing electrolyte [1, 17, 18], shown as Eq. (1) and (2):

$$Ti + 2H_2O \rightarrow TiO_2 + 4H^+ + 4e^- \quad (1)$$

$$TiO_2 + 6F^- + 4H^+ \rightarrow TiF_6^{2-} + 2H_2O \quad (2)$$

From the TiO_2 electrochemical formation [Eq. (1)] and chemical dissolution [Eq. (2)] process, we discerned that the H^+ played an egregious role. We hypothesized Eq. (1) and (2) as separate dynamic chemical equilibrium during the anodization process. Whenever the H^+ concentration was lowered down by additive alkaline materials, the dynamic chemical balance of Eq. (1) would move forward which would accelerate the formation of the oxide layer locating at the tip of TiO_2 nanotube. Contemporarily, the H^+ concentration reduction would promote the equilibrium of Eq. (2) to move backward, restraining the chemical dissolution of formed TiO_2 in the electrolyte. Hence, timely removal of the H^+ generated during the anodization process would expedite the fabrication of TiO_2 nanotube arrays.

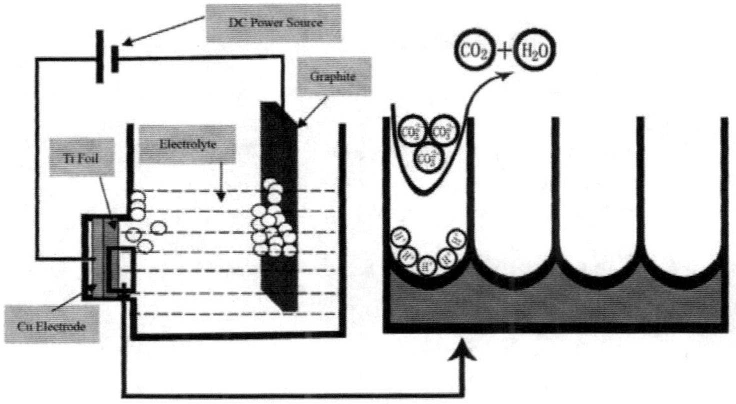

Fig. 4 Schematic representation of the acceleration mechanism

During the anodization in the electrolyte, a great deal of bubbles could be observed at the graphite cathode. And at the same time a few bubbles with little smaller diameters appeared at the Ti foil anode. We convinced this phenomenon consistent with the Eq. (3), shown as,

$$CO_3^{2-} + 2H^+ \rightarrow H_2O + CO_2\uparrow \qquad (3)$$

When the anodization was conducted in the electrolyte containing Na_2CO_3, the H^+ generated during the Ti oxidation at the oxide layer/Ti substrate would move quickly towards the graphite cathode under the assistance of the electric field. Hence, the chemical reaction between CO_3^{2-} and H^+ mainly occurred around the cathode (exhibiting as the bubbles) and only a small part of H^+ reacted with CO_3^{2-} in the vicinity of anode and elsewhere. The parallel experiment was taken in the electrolyte without Na_2CO_3 and almost no bubbles could be observed anywhere. Fig. 4 showed the mechanisms of the Na_2CO_3 addition accelerating the formation of TiO_2 nanotube arrays. Timely elimination of H^+ by the CO_3^{2-} (forming the gas CO_2) resulted in the drastic reduction of the H^+ concentration situated at the interface of oxide layer/Ti substrate and would in turn accelerate the Ti substrate oxidation, causing the forward movement of the interface into the substrate. Therefore, TiO_2 nanotube arrays could grow with a straightforward rate.

4. Conclusions

In summary, in this paper we report a facile route to accelerate the formation of the TiO_2 nanotube arrays by proper proportional addition of Na_2CO_3 into the anodization electrolyte. And the maximum growth rate achieved in our experiment was 1100nm/min. Finally, the accelerating mechanisms were discussed and it was believed that the hydrolyzation of the Na_2CO_3 additives in the electrolytes accelerated the formation of the TiO_2 nanotubes.

5. Acknowledgements

This work was supported by the National Natural Science Foundation of China (N.O. 91023030 and 51072044), International Scientific and Technological Cooperation Projects of Anhui Province (N. O. 10080703017), the 43rd China Postdoctoral Fund (N. O. 20080430758) and Major Science and Technology Projects of Education Department of Anhui Province (N. O. 2009AJZR0081).We would like to thank the Analysis and Testing Center of Hefei University of Technology. And we also would like to thank the reviewers for their kindness, patience and scientific evaluations.

References

[1] C. A. Grimes, G. K. Mor, TiO_2 Nanotube Arrays: Synthesis, Properties, and Applications; Springer: New York (2009), pp 1- 42;

[2] H. D. Jiang, S. K. Kim, S. J. Kim. J Nanopart Res, 3(2001), 141–147;

[3] O. K. Varghese, X. P. Yang, J. Kendig. Sens Lett, 4(2006), 120–128;

[4] G. K. Mor, O. K. Varghese, M. Paulose, K. Shankar, C. A. Grimes. Sol Energy Mater Sol Cells, 90(2006), 2011–2075;

[5] D. Gong, C. A. Grimes, O. K. Varghese, W. Hu, R. S. Singh, Z. Chen, E. C. Dickey, J. Mater. Res. 16(2001), 3331-3334.

[6] A. Fujishima, K. Honda, Nature 238(1972), 37-38 ;

[7] B. O'Regan, M. Gratzel, Nature 353(1991), 737-740 ;

[8] M. Gratzel, Nature 414(2001), 338-344;

[9] O. K. Varghese, D. Gong, M. Paulose, K. G. Ong, E. C. Dickey, C. A. Crimes, Adv. Mater. 15 (2003), 624-627;

[10] K. C. Popat, M. Eltgroth, T. J. LaTempa, C. A. Grimes, T. A. Desai, Small 3(2007), 1878 -1881;
[11] L. Peng, A. D. Mendelsohn, T. J. LaTempa, S. Yoriya, C. A. Grimes, T. A. Desai, Nano Lett. 9(2009), 1932-1936.
[12] X. Hu, T. Zhang, Z. Jin, S. Huang, M. Fang, Y. Wu, L. Zhang, Cryst. Grow. Des. 9(2009), 2324 - 2328.
[13] X. Hu, T. Zhang, Z. Jin, J. Zhang, W. Xu, J. Yan, J. Zhang, L. Zhang, Y. Wu, Materials Letters. 62 (2008), 4579 -4581.
[14] Rongjun Pan, Yucheng Wu and Kongyong Liew, Appl. Surf. Sci. 256 (2010) 6564 - 6568.
[15] M. Paulose, H. E. Prakasam, O. K. Varghese, L. Peng, K. C. Popat, G. K.Mor, T. A. Desai, C. A. Grimes, J. Phys.Chem. C. 111(2007), 14992-14997.
[16] Wen Zhu, Xi Liu, Huiqiong Liu, Dali Tong, Junyou Yang, and Jiangying Peng, J. Am. Chem. Soc. xxx, xxx, 000
[17] J. M. Macak, H.Tsuchiya, P. Schmuki, Angew. Chem. Int. Ed. 44 (2005), 2100-2102.
[18] J. M. Macak, H.Tsuchiya, L. Taveira, S. Aldabergerova, P. Schmuki, Angew. Chem. Int. Ed. 44 (2005), 7463-7465.

A study of color modulation of porous alumina processed by physical vapor deposition

Xulongqi Wang, Dongxian Zhang[*] and Haijun Zhang
State Key Laboratory of Modern Optical Instrumentation,
Zhejiang University, Hangzhou, 310027, China

Email: zhangdx@zju.edu.cn

Abstract. With the development of the porous alumina（PA）fabrication technology, more and more scholars plough into the research of its properties, especially optical properties. Recently, we observed an interesting phenomenon that the PA templates processed by Physical Vapor Deposition (PVD) show color differences related to light path difference. Our work attempts to make the principle clear and to find an effective method to modulate the color of PA samples. This article describes the details of our experimental and theoretical results. We successfully prepared some PA templates with different pore-depth by controlling the time of anodization in oxalic acid solution. In order to enhance the reflectivity of air-PA interface, a layer of TiO_2 film of 18 nm is coated with PVD technique, which makes PA templates display quite distinct colors with different hole-depth. By modelling and analyzing PA samples, we make the interpretation of this optical property by taking the PA sample with 150 nm pore-depth as an example, and then put forward a way to simulate sample's color within its hole-depth and material refraction-index. The results are in good agreement with our theoretical analysis, which proves the feasibility of our simulation mode.

1. Introduction

In recent decades, as the porous alumina (PA) gradually plays an important role in nanotechnology, a lot of researchers are interested in PA fabrication technique and its properties, including optical properties. Hideki Masuda [1][2] and Choi J [3] proposed different preparation methods of highly ordered PA membrane; B. Wang and W. J. Zheng, etc. successfully prepared PA samples with different pore-depth through controlling electrolyte temperature [4], timing in acid solution [5] and voltage of anodic oxidation [6] respectively; H. M. Chen used PA as the templates for growing of Ni

---

[*]To whom any correspondence should be addressed.

Published under licence by IOP Publishing Ltd

nanowires and nanotubes in pores [7]. Worth to mention, all of them have investigated certain particular optical properties of PA and also done some work involving the interpretation to this phenomenon, but still not sufficient.

In this work, PA template is observed distinct color after physical vapor deposition (PVD) processed, which is much more intuitive phenomenon. We attempt to make the principle clear and to find an effective method to modulate the color of PA templates. This article describes the details of our experimental and theoretical results.

2. Experiment

2.1. Preparation of PA templates

PA templates were prepared using high-purity aluminum foils (99.999%). Prior to anodization, the foils were degreased in acetone and ethanol, and then annealed at 500 ℃ under vacuum ambient (about 2×10-5 torr) for 5h to remove the mechanical stress, at last electrochemically polished by a mixture of HClO4 and C_2H_5OH (the volume ratio is 1:9). Orderly, the process of electrochemical anodization could be divided into two steps.

In the first step, anodization lasted for 2 h under a constant anodization voltage of 40 V and electrolyte temperature of 3 ◦C. The specimens were then immersed in a mixture of 6.0 wt% H_3PO_4 and 1.8 wt% CrO_3 at 60 ◦C for 4h to remove oxide layer on the surface.

The second step of anodization was actually the repeat of the first one. For obtaining PA templates with certain pore-depth precisely, only anodization time needs to be strictly controlled.

The morphology of PA template is characterized by a scanning electron microscopy (SEM, Carl zeiss, ULTRA55). Figure1 shows the typical top view of the SEM image of PA template. The ordered pore arrays as seen in figure1 have identical pore diameter of about 37.5 nm and inter-pores difference of about 100nm, which depend on the applied voltage.

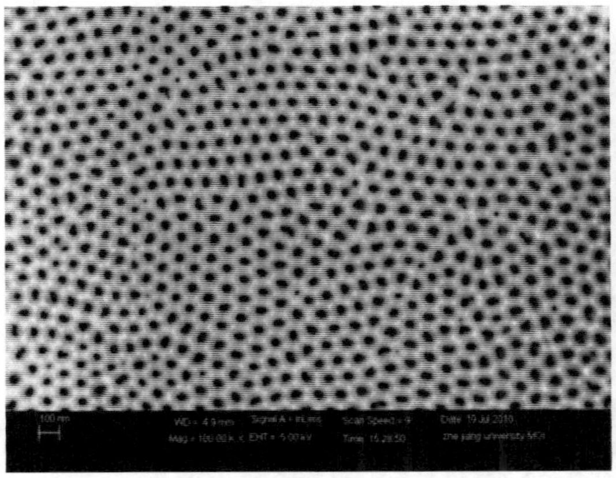

Figure 1. SEM image of PA template with 37.5nm pore-depth and 100nm inter-pores distance.

Figure 2. SEM image of the same PA template with PVD processed.

2.2. PVD processing

It is clear that PA is almost transparent in the range of visible spectrum, hence, majority of incident light reflects on the interface of PA membrane and Al base. In order to enhance the reflectivity of the interface of PA membrane and air, also make sure to avoid masking the properties of PA itself at the same time, a nano-scale layer with high reflective index needs to be coated. For this reason, TiO_2, this conventional colourless material, whose reflective index is about 2.4, is chosen to deposit on the PA surface with the help of PVD technique.

Figure 2 shows the SEM image of the same PA template PVD-deposited a TiO_2 film with the thickness round 18nm. In figure 2, surface morphology of PA template is obviously changed, which indicates that the TiO_2 film evenly overlays on the PA surface. Specially, some part of TiO_2 film above PA-pores collapses, so that the TiO_2 film is also pores-structure with 30nm pore-diameter and 100nm inter-pores distance, which is specified in figure3.

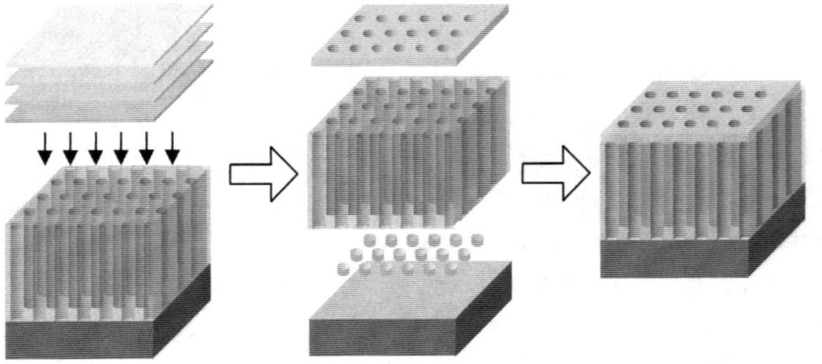

(a) (b) (c)

Figure 3. Simple flowchart for illustrating the inner and outer structure of PA processed by PVD. (a) PVD procedure. (b) The layered view of PA template processed by PVD. (c) The integral view of PA template processed by PVD.

2.3. Experimental results

Figure 4(a) and (b) are the contrast photos of PA templates before and after PVD processing in irradiation of white light with 20 and 60 degree incident angle. From figure 4, the original PA template seems a little dim which looks like metallic, while the template processed by PVD strikingly displays vivid green and purple respectively.

(a) (b)

Figure 4. Contrast photos of PA templates before (up) and after (below) PVD processing with (a) 20 degree and (b) 60 degree incidence angle respectively.

3. Theoretical analysis

3.1. Color Modulation

Figure 5(a) and (b) are physical models corresponding to the two samples in the figure 4(a) and (b), respectively.

As shown in figure 5(a), as PA is almost transparent in the range of visible spectra, the reflect light from Air-Al_2O_3 interface is much weaker compared to that from Al_2O_3-Al interface, which causes effective interference is not formed, consequently, PA template reveals colorless. With PVD technique, a thin TiO_2 film is evenly deposited on the PA surface that balances two parts of reflect light as seen in figure 5(b).

According to Bragg reflector principle,

$$2nd\cos\theta = m\lambda \tag{1}$$

where n, d, θ, m and λ is reflective index, thickness of PA membrane, refractive angle, order number and light wavelength. These two coherent light sources will be enhanced in several certain frequencies, which depend on the unity of $ndcos\theta$ defined as light path difference (LPD) involved pore-depth of PA templates and incidence angle.

Actually, the LPD contains two parts, one is the original part from PA template impact, the other is affected by the TiO_2 film. Hence, equation (1) may be rewritten as

$$2(n_{TiO2eff}d_{TiO2}\cos\theta_{TiO2} + n_{eff}d_{pore}\cos\theta_{PA}) = m\lambda \tag{2}$$

where $n_{TiO2eff}$, d_{TiO2}, d_{pore}, θ_{TiO2} and θ_{PA} are effective reflective index of TiO_2 film, thickness of TiO_2 film, pore-depth, refractive angle in TiO_2 film and refractive angle in PA membrane, respectively. It indicates that all of the wavelength fit equation (2) will be enhance mostly, but others will be weakened in different extent. As the result of that, the color according to the enhanced wavelength is observed on the PA surface.

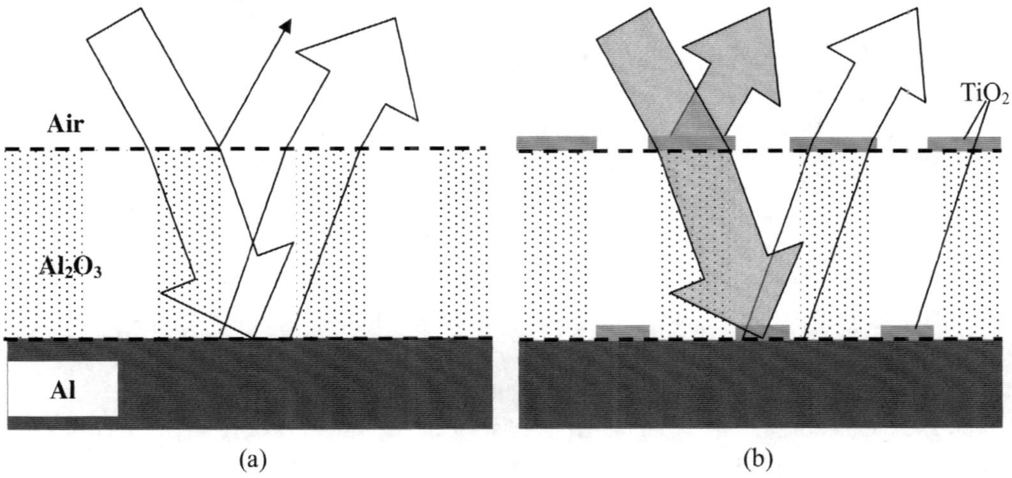

Figure 5. Model of light path diagram of PA template (a) before and (b) after PVD processing.

3.2. Simulation

In order to simulate the modulated color of PA templates with certain pore-depth, it is clear that the effective reflective indexes of layers with pores-structure are necessary to obtain, as equation (2) illuminated. According to the Maxwell-Garnett theory (Garnett, 1904) [8], which is a good approximation for isolated inclusions, if we assume that

$$\varepsilon = n^2 \tag{3}$$

effective dielectric constant ε_{eff} is determined by

$$\varepsilon_{eff} = \varepsilon_{mert} \frac{\varepsilon_{pore} + \varepsilon_{mert} + 2f(\varepsilon_{pore} - \varepsilon_{mert})}{\varepsilon_{pore} + \varepsilon_{mert} - f(\varepsilon_{pore} - \varepsilon_{mert})} \tag{4}$$

Where ε_{mert}, ε_{pore} is equal to dielectric coefficient of material and air respectively, and f is the filling fraction of air, here it is equal to the porosity p [9]. The porosity p of a hexagonal structure is given by

$$f = p = \frac{\pi}{2\sqrt{3}} \left(\frac{D}{D_{\text{int}}} \right)^2 \tag{5}$$

where D is diameter of the pores, D_{int} is inter-pores distance, which can be measured from SEM image. As shown in figure2, the value of D and D_{int} before PVD processing is 37.5 nm and 100 nm, consequently, the value of f is 0.145 according to equation (5). Furthermore, put f and the value of n_{Al2O3}=1.7 and n_{pore}=1into equation (4), we can calculate the effective dielectric constant of PA template is 1.52. Similarly, the effective dielectric constant of TiO_2 film is 2.32 by n_{TiO2}=2.4.

Aiming at 150nm pore-depth PA template with 18nm thickness TiO2 film coated by PVD in our experiment, we simulated its prospective observed color, and the specific values of various parameters are listed in Table1. As pore-depth is defined in the scale of wavelength or sub-wavelength, in addition of concerning visible light spectra limiting, the combinations between m and λ sometimes are single one or two if the light path difference is a certain constant.

Table 1. The simulation of PA template of 150nm pore-depth with different incident angle.

θ(°)	n_{eff}	d_{pore}(nm)	θ_{PA}(°)	$n_{TiO2eff}$	d_{TiO2}(nm)	θ_{TiO2}(°)	m	λ(nm)	*color*
20	1.52	150	13.0	2.32	18	8.5	1	526.9	green
60	1.52	150	34.7	2.32	18	21.9	1	452.4	purple

The "λ" column in Table 1 indicates corresponding wavelength to the interference frequency, concretely, that light of wavelength of 526.9 nm has enhanced mostly in 20 degree incidence angle, but 452.4 nm in 60 degree, moreover, the "*color*" column shows corresponding color in the chromaticity diagram, which is in good agreement with experimental results.

According to equation (3) ~ equation (5), it is clear that the effective reflective index of pores-structure layer has no relationship to pore-depth, in other words, if the pore-diameter and inter-pores difference have maintained the same, the effective reflective index is unchangeable.

Therefore, in the situation of identical PVD processing, PA templates with different pore-depth is supposed to reveal distinct colors.

Conversely, if a certain color is attempted to modulate on the surface of PA template, we could first calculate the corresponding pore-depth and then obtain such a depth by controlling anodization time strictly. The PVD technique can be finally employed to deposit a layer of TiO_2 as to acquire the modulated colors.

4. Conclusion

We have successfully prepared PA templates with different pore-depth by strictly controlling anodization time, and then processed them with PVD technique. To take the PA template with 150nm pore-depth as the given example, the according theoretical analysis has done, which has well illuminated the principle of color appearance on the surface of PA template. Furthermore, on the base of that, a method to modulate the colour of PA is proposed.

References

[1] Hideki Masuda, Kenji Fukuda 1995 Science **268** 1466
[2] Hideki Masuda, Haruki Yamada, Masahiro Satoh, Hidetaka Asoh 1997 J. Applied Physics Letter **71** 2770
[3] Choi J, Schilling J, Nielsch K, Hillebrand R, Reiche M, Wehrspohn R B, Gosele U 2002 J.Materials Research Society **722** L5.2.1
[4] Zheng W J, Fei G T , Wang B, Zhang L D 2009 J. Nanoscale Research Letters **4** 665
[5] Wang B, Fei G T, Wang M, Kong M G, Zhang L D 2007 J. Nanotechnology **18** 365601
[6] Zheng W J, Fei G T, Wang B, Jin Z, Zhang L D 2009 J. Materials Letters **63** 706
[7] Chen H M, Hsin C F, Liu R S, Hu S F, Huang C Y 2006 Journal of The Electrochemical Society **154** 11
[8] Hornyak G L, Patrissi C J, Oberhauser E B, Martin C R, Valmalette J-C, Lemaire L, Dutta J, Hofmann H 1997 J. Nanostructured Materials **9** 571
[9] Zhang D X, Zhang H J, He Y L 2006 J. Microscopy Research & Technique **69** 267

Research on Spectral Reflection Characteristics of Nanostructures in Morpho Butterfly Wing Scale

Wenjun Wu[1,2], Tielin Shi[2], Guanglan Liao[2], Haibo zuo[2]
1. Key Lab of Modern Manufacture Quality Engineering, Hubei University of Technology, Wuhan, Hubei, P. R. China 430068
2. State Key Laboratory of Digital Manufacturing Equipment and Technology, Huazhong University of Science and Technology, Wuhan, Hubei, P. R. China 430074

E-mail: guanglan.liao@mail.hust.edu.cn

Abstract. The intricate nanostructure in the scales of *Morpho*, which is composed of transparent cuticle protein, achieves an extremely high reflectivity in the range of visible light. The brilliant iridescent blue color is not produced by blue pigment but nanostructures. In order to investigate which structural parameters influenced the spectral reflection characteristics and formed the striking brilliance of blue color, a vector diffraction theoretical structural model was established, and simulation using rigorous coupled-wave analysis was carried out. The complex nanostructure was assumed as the diffraction grating structure of arbitrary configuration. The shape and size of the model was set according to the TEM photos of *Morpho* scale. The structure with irregular asymmetric multilayer lamellae ridge-like grating possessed best capability in reflectivity and color matching. The influence of every structural parameter to spectral reflectivity was cognized by comparing with the original spectrum. The results have revealed the nature of iridescent blue colors and high reflectivity, and enable us to control color and reflectivity by manufacturing nanostructure with specific structural parameter.

Keywords: *Morpho* butterfly, micro/nano structure, structural color, rigorous coupled-wave analysis.

1. Introduction

Three centuries of research, beginning with Hooke and Newton, have revealed a diversity of structural color at the submicrometre scale in nature [1-2]. Many birds, insects (particularly butterflies and beetles), fishes and lesser-known marine animals exploit structural colors on their surfaces to make their color changed with viewing angle (iridescence) or appear 'metallic' [3-7]. A tropical butterfly, *Morpho Rhetenor,* is famous for structural color. The striking brilliance of the iridescence blue color is so bright that it can reportedly be seen from "a quarter of a mile off" [8]. The blue color is not generated from blue pigment but nanostructures [9].The intricate nanostructure achieves an extremely high reflection in the short wavelength range of the visible spectrum [10-11]. These optical characters depend strongly on the micro/nano structural properties, such as the 2D-mode, the minimum size, periodicity and type of arrangement [12]. There are many techniques such as FDTD employed to investigate which structural parameters influenced the spectral reflection characteristics. It is found that the rigorous coupled-wave analysis (RCWA) has a good performance in calculating the sub-wavelength diffractive structures in terms of the vectorial wave diffraction theory. In this paper the complex nanostructure was modeled and simulated by the RCWA [13].

The rest of the paper is arranged as follows. Section 2 depicts the spectral reflectivity of *Morpho Rhetenor* scales and the experimental optical system. Section 3 describes the nanostructure of *Morpho Rhetenor* and the two-dimensional simulation model. The RCWA algorithm is also introduced briefly in Section 3. The simulation results of different structural parameters are presented in Section 4. Finally, Section 5 comes to the conclusions.

2. Spectral reflectivity of *Morpho Rhetenor*

We worked on a highly iridescent butterfly species, Morpho Rhetenor, as shown in Fig.1. Its nanostructures achieved an extremely high reflection, which depended strongly on the wavelength, polarization state, angle of incidence, shape of illuminating wave-front and the index of ambient medium [14-15].

To measure the reflective spectrum, an optical system was designed (as shown in Fig.2). In the experiments, the light was emitted from a HL-2000 Tungsten Halogen Light Source, went through the optical fiber, and irradiated on the reference plate (WS-1-SL, a white reflectance standard from Labsphere made from Spectralon) or samples. Then the light was reflected, captured by the reflection probe QR400-7-UV/VIS, and inputted into an Ocean Optics HR2000+ spectrometer. Here the high-resolution spectrometer with a response range of 200-1100 *nm* was employed to capture visible spectra.

Figure 1. The Morpho Rhetenor butterfly

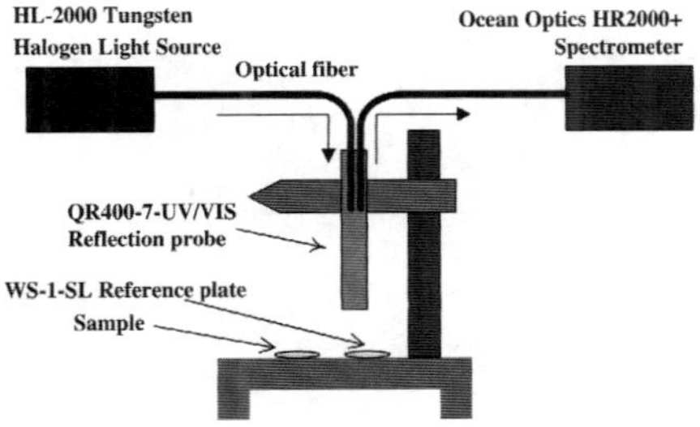

Figure 2. The schematic diagram of the experimental optical system

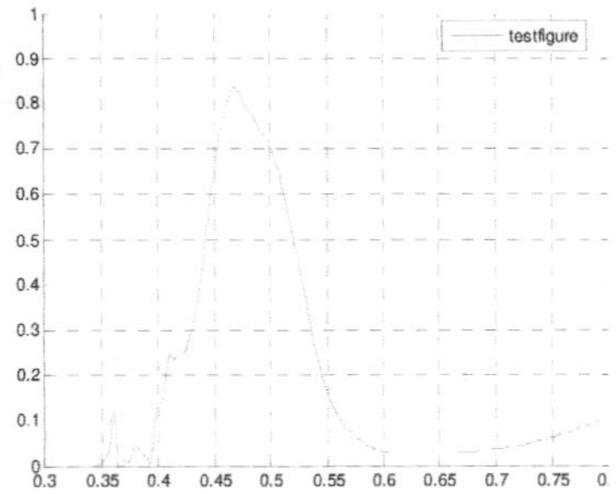

Figure 3. Reflective spectrum of *Morpho* Rhetenor scales

Here we only focused on the reflective spectrum of *Morpho* Rhetenor in the visible spectra range from 0.38 to 0.78 *μm*. As shown in Fig.3, the x-coordinate was wavelength and the Y-coordinate was reflectivity efficiency. It was found the peak was located in the range of 450 ~ 460nm with the value of 83%, which corresponds to blue light. The reflectivity was more than 60% from 448nm to 520nm, that is, the bandwidth was nearly 72nm, which was relevant to homochromatism.

3. Modeling the scale structure

Morpho Rhetenor has single layer of scales, as shown in Fig.4a, which contains little pigment. The reflectivity peak of *Morpho Rhetenor* reaches 80%.

It can be found from Fig.4b, the intricate ridges arranged periodically and composed a grating structure [16-17]. It was the grating structure who produces the iridescence and high reflectivity [18]. As shown in Fig.4c, the transverse section of ridge was just like a tree with many branches. Each ridge had 11 layers of lamella (the "branches" of the tree-like structure). Lamellas nearly paralleled the base of ridge. The topper lamella was narrow than lower lamella. Ridge and basement were not vertical, but angled.

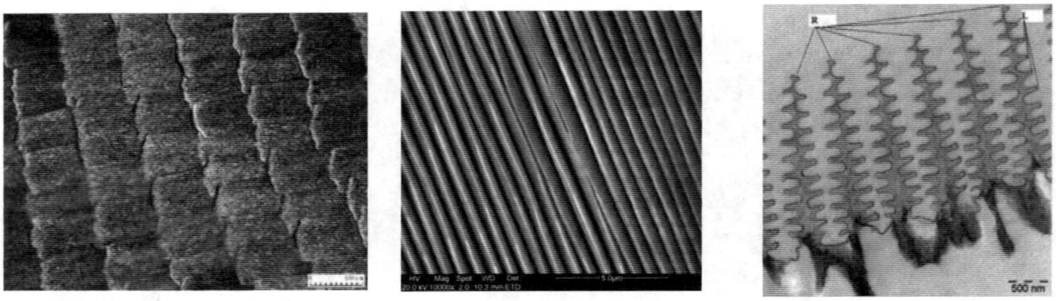

a. The OM image of scales b. The SEM image of one scale c. The TEM image of transverse section of one scale.

Figure 4.The structure of *Morpho Rhetenor* scale

According to the measurements reported in Vukusic at al [19], the complex optical index of this tree-like structure would be approximated by the constant value n = 1.55 + 0.05i in the whole optical range. The model had an x-invariant structure, in which there were infinite periods along x axis [20-21]. Here only 4 periods were set as an example (as shown in Fig 5). The asymmetric lamellae-ridges structure had an x-invariant period 800nm. The distance between two lamellae-ridges structure was 30nm. In each ridge, the width of trunk was 50nm. The widths of the isosceles trapezoid branches were different and increase equidifferently from 220nm to 400nm along the z axis. The distance between lamellae was 110nm.

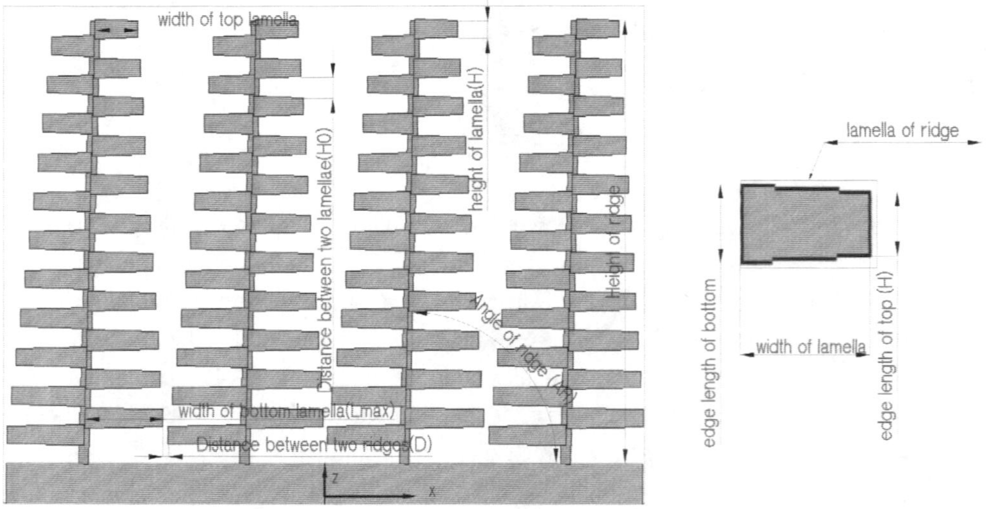

Figure 5. The original model

The two-dimensional model was constructed by DiffractMOD, a general design tool based on the RCWA for diffractive optical structures. For the electromagnetic diffraction by grating structures, the

RCWA was a relatively straightforward technique to obtain the exact solution of Maxwell's equations. It can be used accurately to analyze the diffraction of electromagnetic waves by periodic structures and even arbitrarily profiled dielectric-metallic surface-relief gratings. The technique was non-iterative and fundamentally stable, thus numerically stable and convergent results can be achieved [22-23].

4. The Simulation Analysis

We attempted to find the influence of structural parameters on the optical characteristics. The spectral reflectivity varied with the change of nanostructural parameters. It was very significant to find out the principles. Specific color can be achieved by designing specific structure. The reflectivity can be set as increase or decrease to meet actual demand.

The model shown in Fig.5 is employed to calculate the spectral reflectivity for both polarizations, monochromatic radiation of wavelengths ranging from 0.3 to 0.8μm at a regular interval of 5nm. Normal incidence was assumed.

4.1. The spectral reflectivity of original model

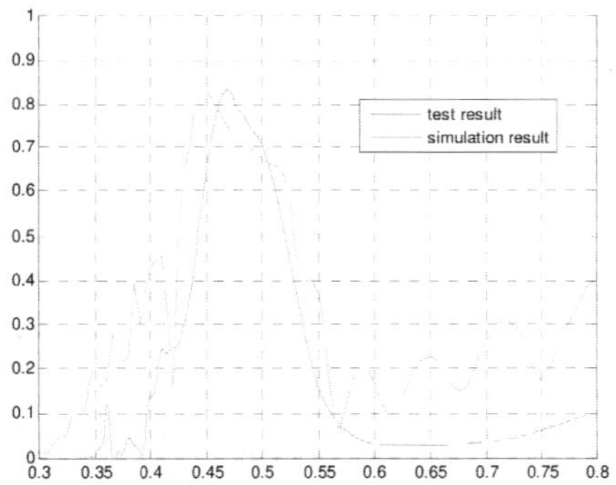

Figure 6. The spectral reflectivity of the *Morpho Rhetenor* scales and the original model

As shown in Fig.6, the spectral reflectivity of original model almost coincided with the test spectrum of *Morpho Rhetenor* scales at the wave crest in 420-570nm range. But the simulation spectrum was more fluctuant than test spectrum in 350-420nm and 570-780nm. Possible reasons were as follows: firstly, the shape of real nanostructure in scales is irregular curve. It was simplified as regular shape, such as isosceles trapezoid and rectangle; secondly, the real nanostructure has randomness in each size, which was simplified as a fixed value in model.

4.2. Influence of distance between two ridges

The original distance between two ridges (*D*) was 30nm. The reflectivity peak varied obviously with the distance, as shown in Fig.7.

The peak value declined obviously while bandwidth decreased slightly and peak position had tiny move along with the increase of *D*. In particular, the reflectivity decreased in visible range from 380 to 780nm, which indicated a decrease in whole brightness. It was clear that the peak reached 85% of 0-distance, and then dropped to 83% of 30nm-distance, and then 78% of 300nm-distance, and 50% of 1200nm-distance. The variations were non-linear.

The influence of ridge distance was significant to reflectivity in all visible range. Reflective efficiency can be raised by reducing distance between two ridges, and the color generated from nanostructure would remain as before.

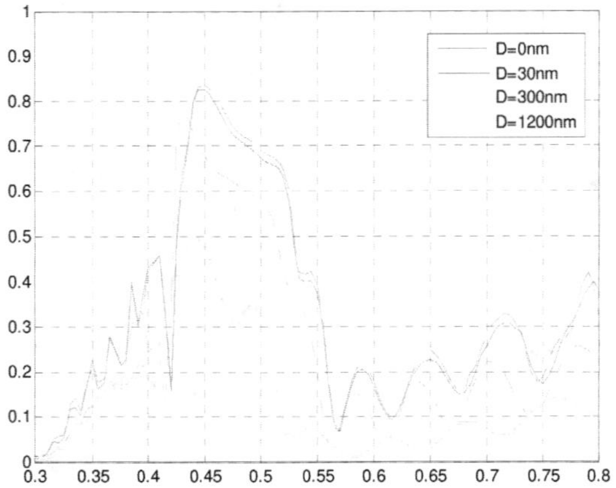

Figure 7. The spectral reflectivity changing with the distance between two ridges (D)

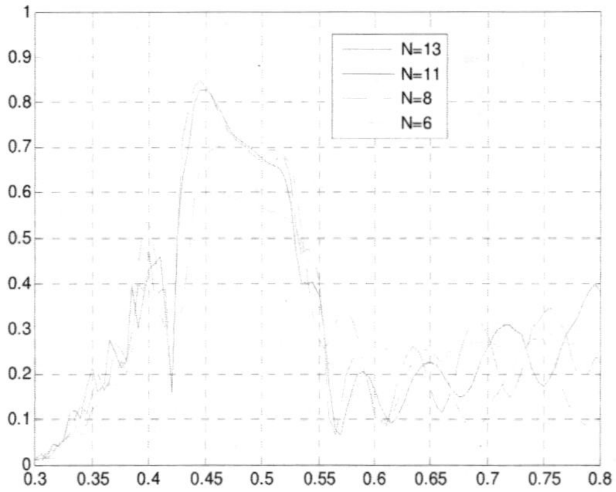

Figure 8. The spectral reflectivity changing with lamellas number (N)

4.3. *Influence of lamellae number*

The original lamellas number (N) was 11. The height of ridge increased with N.

As shown in Fig.8, the peak value decreased sharply and the peak position shifted to longer wavelengths slightly along with the decrease of N. The peak reached 85% of 13 layers in 450 ~ 460nm range. It was dropped to 60% of 6 layers in 460 ~ 470nm range. And the decrease occurred only in the range of wave crest from 430 to 460nm.

The influence of lamellae number was significant to reflectivity just in the range of main wave crest. We can improve reflective efficiency by increasing lamellae number, and the color generated from nanostructure would remain as before.

4.4. *Influence of lamellae height*

Lamella was simplified as isosceles trapezoid in model. So the height of lamella can be represented by edge length of top (H) and the ratio of bottom/top(R). The R was 1.2 and H was 85nm in original model.

Lamella was rectangle when R is1.

As Shown in Fig.9, reduction of R led to a small decline in peak and a decrease in gradient of wave crest. The peak position moved toward the shorter wavelength. The peak reached 90% as R=2. And it was declined to 82% as R=1. The difficulty of manufacture increased with R.

It can be found from Fig.10, the peak position of spectral reflectivity moved to longer wavelength and the bandwidth increased significantly with the increase of H. But the peak value varied merely in a narrow range of 10%. As H was more than 120nm, there was no longer single peak in visible range of 380-780nm. When H was more than 200nm, the anterior peak reached 80% in 390-400nm of blue light range, and the posterior peak was 85% in 740-750nm of red light range. That led to the nanostructure presented a mixed color of blue and red.

The influence of H was significant to color that generated from nanostructure.

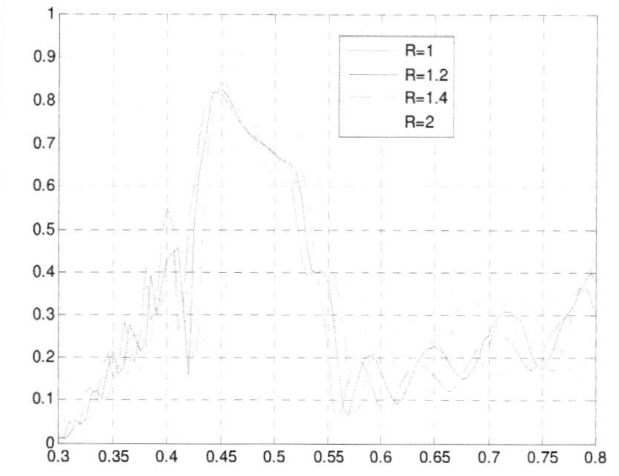

Figure 9. Spectral reflectivity changing with R

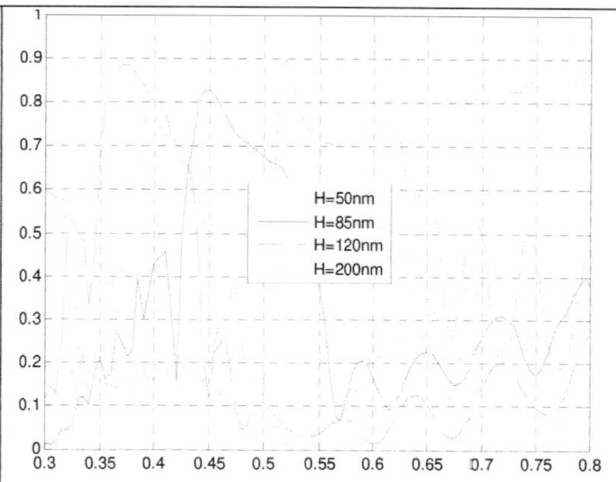

Figure 10. Spectral reflectivity changing with H

4.5. Influence of distance between two lamellae

Distance between two lamellae (H_0) was 110nm in original model. The increment of H_0 heightened the ridge.

As shown in Fig.11, the peak position shifted to longer wavelength and the peak value increased obviously while the bandwidth increased sharply along with the increase of H_0. When H_0=80nm, the peak value was just 74% in 410 ~ 420nm of purple light range, and the bandwidth was 90nm from 380nm to 470nm. When H_0=200nm, the peak value reached 95% in 570 ~ 580nm of yellow light range, and the reflectivity were more than 70% from 550nm to 760nm.

The influence of H_0 was significant to color that generated from nanostructure.

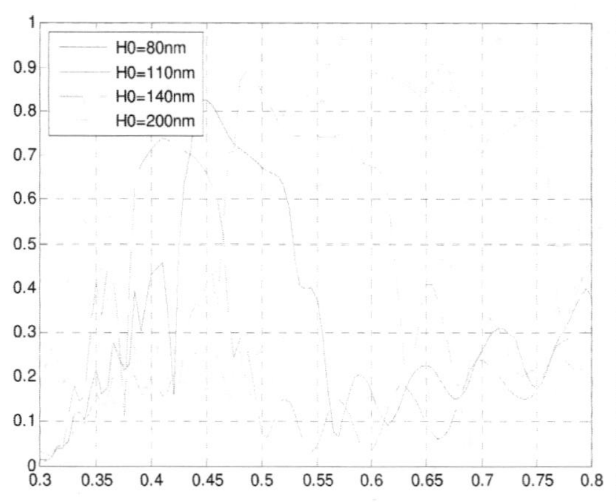

Figure 11. Spectral reflectivity changing with H_0

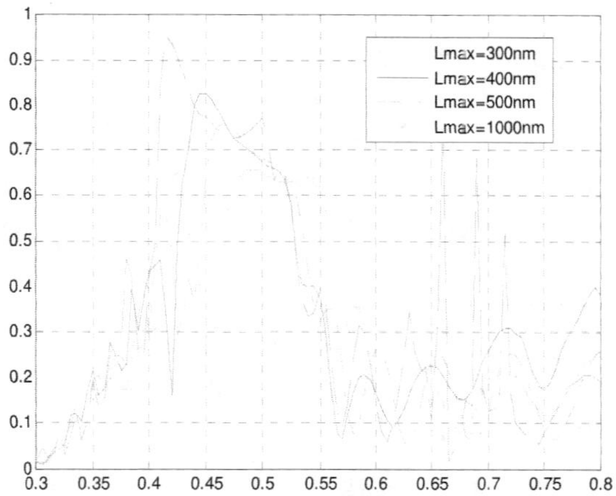

Figure 12. Spectral reflectivity changing with L_{max}

4.6. Influence of the width of lamella

In real scales, lamellae are not equal in width. The top layers are shorter than bottom layers, as shown in Fig.4c. That were simplify as decreasing with layer in model. And lamella was simplified as isosceles trapezoid. So the width of lamella was represented by height of isosceles trapezoid at bottom, which was the maximum width of lamella, and the height ratio of top-layer/bottom-layer. The maximum width of lamella (L_{max}) was 400nm and the height ratio of top-layer/bottom-layer (R_L) was 0.55 in original model.

It can be found in Fig.12, peak value declined obviously while peak position moved to longer wavelength and the gradient declined sharply along with increase of L_{max}. When L_{max}=300nm, peak value was nearly 95%. While L_{max} was 1000nm, peak value dropped to 65%.

The influence of L_{max} was significant to reflectivity just in the range of main wave crest. But if we enhanced reflective efficiency by increasing L_{max}, the color generated from nanostructure would be changed as well.

As shown in Fig.13, the gradient declined sharply along with the increase of R_L. Wave crest was nearly flat-roofed as R_L=1, that means width of every lamella was identical.

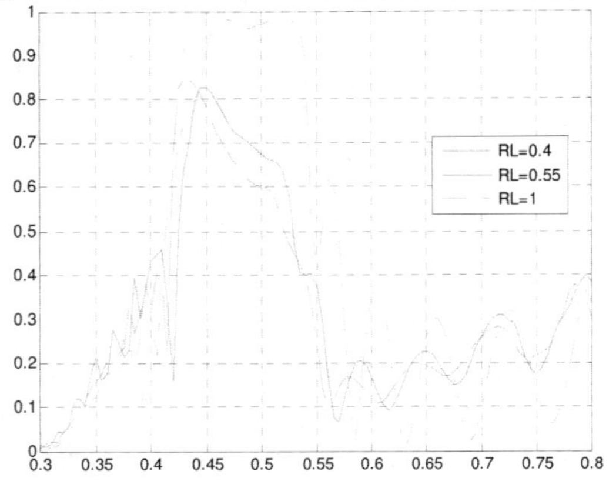

Figure 13. Spectral reflectivity changing with *RL*

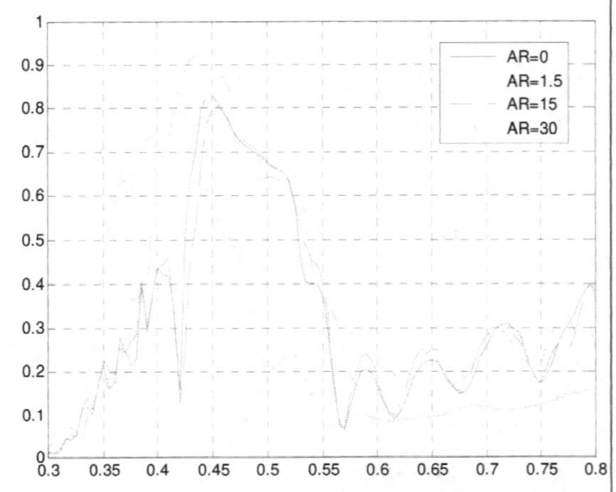

Figure 14. Spectral reflectivity changing with A_R

4.7. Influence of ridge angle.

Ridge angle (A_R) was angle between ridge and base. A_R was 1.5 degree in original model. Ridge height declined along with increase of A_R.

It can be found in Fig.14, the peak value increased from 80% to 92% as A_R varied from 0 to 15, and the gradient increased. While A_R more than 15, the peak declined with the increment.

5. Conclusion

The experiments and simulations were accomplished to investigate the reflection properties of *Morpho* butterfly wing scales. Some principles on the specific structure color can be depicted.

1) The blue color is produced by interference in ridge-lamellae structure, which is composed of alternating layers of high and low refractive index.

2) A narrow gap between the ridges and multi-layers lamellae generates high reflectivity. If the gap is much wider than 1000nm or the lamellae less than 4 layers, the incident light would penetrate the bottom of the multilayer and almost absorb or transmit through the substrate.

3) The ridge-lamellae structure can generate different colors by changing the height of lamellae and the height between lamellae. These two parameters must be less than 200 nm. Otherwise a mixed color is presented.

4) To obtain homochromous on condition of no color change, these parameters, the titling angle of

ridge, height ratio of one lamella and width ratio of lamellae, are significant.

The investigation gives us an opportunity to acknowledge the relation between structural parameters and the reflective color of the *Morpho* wing scales. It enables us to control color and reflectivity by manufacturing nanostructure with specific structural parameter. That should yield a wide variety of applications.

ACKNOWLEDGMENTS

This project was funded by National Key Basic Research Special Fund of China (Grant No. 2009CB724204) and National Natural Science Foundation of China (Grant No. 50775091).

REFERENCES

[1] R. Hooke, 1665 Micrographia , London: *Martyn and Allestry* .

[2] E. Newton, 1730 Opticks, 4th edn. Reprinted New York: Dover .

[3]A. A. Michelson, 1911, On metallic colouring in birds and insects. *Phil. Mag.* **21**, 554–566 .

[4] J. W. S. Rayleigh, 1923 Studies of iridescent colour, and the structure producing it. *IV. Iridescent beetles. Proc. R. Soc. Lond* **A 103** 233–239.

[5] C. W. Mason, 1923 Structural colors in feathers, *part I. J.Phys. Chem.* **27**, 201–251.

[6] E. Merritt, 1925 A spectrophotometric study of certain cases of structural color. Rev. Sci. Instrum. 11, 93–98.

[7] A. R. Parker, H. E. Townley 2007 Biomimetics of photonic nanostructures, *Nature* **2** 347-353

[8]H. W. Bates, 1864 The naturalist on the river Amazons (2nd ed.) London.

[9] P. Vukusic, & J. R. Sambles, 2003 Photonic structures in biology. *Nature* **424** 852-855

[10] S. Berthier , E.Charron, J.Boulenguez, 2006 Morphological structure and optical properties of the wings of Morphidae. *Insect Sci.* **13** 145–157

[11]T. Haruna, S. E. Mann, I. N. Miaoulis and P. Y. Wong, 1999 Effects of a butterfly scale microstructure on the iridescent color observed at different angles. *Optics Express* **5** 87-92.

[12] O. Deparis, C. Vandenbem, M. Rassart, V. L. Welch, J. P. Vigneron, 2006 Color-selecting reflectors inspired from biological periodic multilayer structures. *Optics Express* **14** 3547-3555

[13] C. E. Reuter，R. M. JosePh，E. T. Thiele，et al，Oct.1994 Ultrawideband Aborbing Boundary Condition for Termination of Waveguiding Struectrues in FDTD Simulations，*IEID*，**Vol.4, No.10**，344—346

[14] H. bata, K. Kumazawa, M. Funakawa, J. Takimoto and M. Akimoto, 1996 Microstructures and optical properties of scales of butterfly wings, *Opt. Rev.* **3** 139–149

[15] H. Tabata, K. Kumazawa, M. Funakawa, J. Takimoto and M. Akimoto 1996 Microstructures and optical properties of scales of butterfly wings, *Opt. Rev.* **3** 139–149

[16] H. Ghiradelia, D. Aneshansley, T. Eisner, R. E. Silberglied, and H. E. HInton, 1972 Ultraviolet reflection of a male butterfly: interference color caused by thin-layer elaboration of wing scales, *Science*, **178** 1214–1217.

[17] Gérard Tayeb, Boris Gralak and Stefan Enoch, 2003, Structural Colors in Nature and Butterfly-Wing Modeling, Optics & Photonics News, 41-49

[18] P. Vukusic, J. R. Sambles, C. R. Lawrence, and R. J. Wootton 2002 Limited-view iridescence in the butterfly Ancyluris meliboeus *Proc. R. Soc. Lond* **B 269** 7-14

[19] P. Vukusic, J. R. Sambles and C. R. Lawrence, 2000 Colour mixing in wing scales of a butterfly, *Nature*, 404- 457.

[20] S.E. Mann, I.N. Miaoulis, P.Y. Wong 2001 Spectral imaging, reflectivity measurements, and modeling of iridescent butterfly scale structures *Opt. Eng.* **40(10)** 2061–2068

[21] Boris Gralak, Gérard Tayeb and Stefan Enoch 2001 Morpho butterflies wings color modeled with lamellar grating theory, *OPTICS EXPRESS* **Vol. 9, No. 11** 567-578

[22] M. G. Moharam, E. B. Grann, D. A. Pommet, T. K. Gaylord, 1995 Formulation for stable and efficient implementation of the rigorous coupled-wave analysis of binary gratings *J. Opt. Soc.* Am **A 12** 1068-1076

[23] M. G. Moharam, D. A. Pommet, E. B. Grann, et al. 1995 Stable implementation of the rigorous coupled-wave analysis for surface-relief gratings: enhanced transmittance matrix approach *J Opt* **A 12(5)** 1077—1086

Thinking on the Application of Nanotechnology in the Mechanism Research on the Traditional Chinese Medicine Diagnosis and Treatment of Diabetes Mellitus

YUNXIANG XU[1], LI ZHANG[1], GUIZHEN CHEN (*)[2], PENGDIAN CHEN[1]

[1]Acupuncture and Massage College, Guangzhou University of Chinese Medicine, Guangzhou, 510405 China;
[2]The Second Clinical Medical College, Guangzhou University of Chinese Medicine, Guangzhou, 510405 China

E-mail: cgzhen2000@163.com

Abstract. Nanotechnology is an advanced scientific technique in the 21st century and diabetes mellitus (DM) is a commonly seen chronic disease, which seriously threatens the health of human beings. By analyzing the relationship between nanotechnology and biological medicine, nanotechnology and traditional Chinese medicine (TCM) and the advances and the existing problems of TCM diagnosing and treating DM, the application of nanotechnological methods for the mechanism research on TCM diagnosis and treatment of DM was discussed. It is indicated that nanotechnology is one of the fastest developmental, the most potential and the far-reaching high and new technologies in current world, and it greatly promotes the development of biological medicine and TCM. With the application of nanotechnology of medical diagnostics and medical materials, it will make the development of TCM possess an unprecedented field, which consequently could integrate the macroscopical and microscopical syndrome differentiation. It's pointed out that breakthrough will be achieved from the research of the administration route, the improvement of medical biological availability and the selection of the acupoint prescriptions on mechanism research on TCM for the diagnosis and treatment of diabetes mellitus.

Key words. Nanotechnology; Biological Medicine; Traditional Chinese Medicine (TCM); Diabetes Mellitus; Mechanism Research

1. Introduction

Diabetes mellitus (DM) is a metabolic disorder characterized by chronic hyperglycemia induced by multiple complicated reasons, which has seriously threatened the health of human beings. With the development of people's living standard, the incidence of DM grows year by year. According to a WHO report [1], the number of global DM patients was 120 million in 1994, 175 million in 2000 and may rise above 300 million by 2025. This has drawn much attention from scientists. Complications are the cause and pathological mechanism of DM, so the conventional research methods are time-consuming and labor-intensive with low therapeutic effects. In clinic, traditional Chinese medicine (TCM) has its own advantages and characteristics in the treatment of DM and its complications. By introducing advanced scientific techniques such as nanotechnology, it could provide thinking on the application of nanotechnology on TCM diagnosis and treatment of DM hence further understanding of the mechanism.

2. Nanotechnology and Biomedicine

Nanotechnology refers to an advanced new technique used to study the specificity and interactions of materials such as the management of atoms and molecules, which is a new scientific technology integrated with high techniques and multiple disciplinary cross [2]. Since the basic thoughts and methods of nanotechnology have been introduced into all the subjects, diverse newly-developing fields related to nanotechnology have grown, which demonstrates its great potential and superiority.

In the beginning, nanotechnology was just utilized in the techniques of micro-electron and micro-motor. However, at present, it has been penetrated to all the fields of bio-technology and biomedicine. Compared with the traditional electronic equipment based on Silicon, it could be applied to bio-materials, fluids and chemical materials as raw materials for biochips, which determines the diversity of the biochips. On the other hand, with the reduction of the area and size of the micro-bio-system, it could improve the functions, increase the production and save the cost of the system [3]. The researches and application of nanotechnology in biomedicine could be classified into the following three aspects.

2.1. Nano-biological Materials

Nano-materials have great impacts on biomedicine. To some extent, the development of nano-materials determines the future (or richness?) of nano-medicine. At present, nano-materials have been successfully applied in the medical field. Among them, drug carrier is the most attractive, which could be safely and effectively utilized to the target organs and gene therapy, or to be processed into the human biomedical materials such as artificial kidneys, artificial joints, etc.

2.2. Diagnostic Techniques of Nano-medicine

2.2.1. Optical Coherence Tomography (OCT).

With the resolution of a micron and 1000 times more accurate than CT and MRI, OCT is also called Molecular Radar by scientists, which could make the dynamic imaging of the living cells in the body 200 times per second, observe the dynamic state of living cells and detect the morbidity of single cell without killing living cells like X-ray, CT or MRI. Based on this, people could diagnose the cases in advance and avoid detecting the carcinoma of the late phase.

2.2.2. Laboratory Test.

Laser monatomic molecule detection is hypersensitive, which could be applied to obtain one of the one hundred billion atoms or molecules in the gaseous substance. Based on this, scientists who wish to can now detect bio-molecules in the human body to find out clues which influence human health. With the saliva, blood, stools and the gas breathed out, it could detect numerous kinds of pathogenic molecules or marks among the one hundred billion ones in time [4].

2.2.3. Pathologic Diagnosis.

Generally the sizes of nanoparticles are much smaller than cells and erythrocytes in living organisms, which provides a new research direction for biological study. That is, to make cell separation and cellular staining, or to process into some special drugs or antibody-free with nanoparticles and local-oriented therapy.

2.2.4. Diagnosis of Implantable Sensor Array.

Taking advantages of the nano-probe technique, it could be embedded in the body. According to different diagnosis and detections, it could locate in different regions of the body. With the blood flowing, it could move in vitro and produce feedbacks of biological information onto the in vitro record device [5].

2.2.5. Genetic Diagnosis.
Due to the great analytical skills of the gene chips, few samples applied, convenience, hypersensitivity and specialty, it could obtain abundant information, which is significant in the diagnosis of gene mutation and some genetic diseases with the copy number variations.

2.3. Therapeutic Techniques of Nano-medicine

2.3.1. Nanoscale Robot.
With the nanoparticles, scientists have produced the nanoscale robot, which could be injected in the vessels and makes the general physical examination possible. It could move in the microcosm of the organism, clear any harmful substances and activate the cell energy, which consequently keeps human beings healthy and long live.

2.3.2. Gene Therapy.
Gene therapy is thought to be the only method in treating genetic diseases. Nowadays, the critical challenge is, firstly find out the DNA chains of the abnormal cells; then with the nanotechnology, transmit the DNA chains applied into the abnormal cells and replace the abnormal DNA chains. With the nanotechnology, it could deliver the new-style genetic materials into the existing DNA without creating any immune reactions.

2.3.3. Drug Delivery and Nano-controlled Release System.
With the nanotechnology, the drug delivery could easily transmit to any regions of the body, release the drugs, kill the pathogens and eliminate the focus [6], which could apparently save the dosage, prevent the adverse effects after general administration and improve the therapeutic specialty. Nano-controlled system has great potential in the aspect of drug delivery, which could release drugs and prolong the drug effects, reduce the dosage, decrease or avoid the side-effects and improve the stability of the drugs. In addition, it could be applied to build up some new routes of administration, including in vivo local drug delivery, administration with mucous membrane absorption, oral administration with polypeptide drugs, etc [7].

3. Nanotechnology and Traditional Chinese Medicine

3.1. Nano-Chinese Materia Medica
In 1998, some scholars firstly came up with the definition of Nano-Chinese Materia Medica, which refers to the particles of the effective components, regions, raw materials and compound preparation of Chinese Materia Medica with nanotechnology, with diameter smaller than 100 nm [8]. Compared with the traditional Chinese Materia Medica, it has the following five advantages: (1) Improve the biological availability and save the limited resources of the Materia Medica; (2) Strengthen the target-oriented therapeutic effects; (3) Add up more pharmaceutical preparation choices; (4) Promote the standardization and internationalization of the Chinese drug preparation.

3.2. Application of Nanotechnology in the Acupuncture & Tuina
Acupuncture & Tuina is the specialty of TCM. With the nanotechnology, it could further study and understand the essence of Meridians and Collaterals, which consequently provides the theoretical and technical support for them and enlarges the research fields and indications.

3.2.1. Application of Nano-sensing Needle in Acupuncture & Tuina.
The research and development of needling sensor is a key technique in the mechanism study of Acupuncture & Tuina. Nano-sensing needle is based on the integrated techniques of acupoints people understand and the modern biomedical sensor, which has the functions of needling, delivering the bioinformation and detecting the changes of substances in the acupoints during the meridian qi transmission, the changing rules of the environment in the acupoints after taking the needling techniques and the relationship between human biorhythm and the meridian qi transmission, etc [9].

3.2.2. Nanotechnology and Acupoint Embedding.
Traditional catguts have the advantages of being absorbed without foreign matters. However, due to the individuals, after embedding excessive and thick catguts, the local regions would create inflammation and the pain obviously appears. Hence, taking advantages of drug-loaded nano-controlled release techniques, or finding new types of nano-materials to replace traditional catguts provide new methods and directions for the existing problems of application of catguts.

3.2.3. Nanotechnology and Acupoint Application.
Nano-Chinese Materia Medica could change the disadvantages of traditional acupoint application, which is to put the western or Chinese medicine onto the pathway of meridians, or to process them into various types of aerosols and penetrate directly into the body through the skin. On one hand, it could enlarge the drug administration methods of acupoint application; on the other hand, it could be more suitable for more groups of people.

3.2.4. Application of Nanotechnology in the Drug Meridian Distribution and Qi Meridian Transmission.
Thanks to the good magnet-controlled location of the magnetic nano-particles, it could trace the in vivo movement of nanoparticles with the photosensitive detecting device and the external magnetic field. Based on this, we could take the drug meridian transmission with the combination of drug and magnetic nano-particles. Nano-magnetic fluid materials could be applied in the mechanism research of meridian transmission of the basic subjects of acupuncture & Tuina. If we could process the solution applied in the acupoint injection into the nano-magnetic solution, it could provide a new research method for the mechanism study of meridian transmission.

4. Advantages and Existing Problems of TCM Diagnosing and Treating DM and its Complications

4.1. The advantages mainly indicate as follows.

- *Holistic Bi-directional Regulation.* Under the theoretical guidance of TCM, the effective components of herbs could treat DM by the entire regulation to the body hence apparently improve the clinical symptoms. Meanwhile, it could bi-directionally regulate endocrine functions and maintain the homeostatic equilibrium.
- *Individualized Diagnosis and Treatment.* With the syndrome differentiation to individuals of TCM and according to different constitutions, symptoms, even different characteristics, seasons and climates, select different suitable prescriptions and adopt different treatments and herbs, which could consequently improve the clinical symptoms and the quality of life.
- *Multiple Therapies of TCM.* Besides Chinese Materia Medica, other therapies such as acupuncture, Tuina, Qigong, physical therapy, psychotherapy, mental suggestion, etc, could be applied. Among them, acupuncture has many therapies such as filiform needling therapy, electro-acupuncture, moxibustion, acupoint catgut embedding, acupoint injection, acupoint application, integrated acupuncture and herbs, etc, all of which could be applied to treat type 2 DM and the complications caused by the diabetic neurovasculopathy, with definite therapeutic effects, convenient acupoint location and simple manipulation. Acupuncture combined with decreasing dosages of western medicine could satisfactorily control the glycemia and blood fat of the patients. As a result, it could reduce the adverse reactions caused by the drugs and prevent the complications caused by the neurovasculopathy, which has been widely applied in clinic and further studied [10].
- *Advantages of Chinese Materia Medica.* (1) Due to the diversity of prescriptions and herbs, it provides more choices for new-drug exploitation and treatments. (2) As most of the herbs are organics, which elements approach the human food, the toxic-side effects are lower than that of chemotherapy. (3) The herbs could strengthen the constitutions and improve the immune system. (4) Some herbs may control the glycemia from multiple perspectives without the

occurrence of hypoglycemia reactions. (5) The herbs could directly complement the Vitamin, microelements, electrolytes, mineral substances, cellulose, etc, and rectify the complications caused by the nutrient deficiency. (6) It could bi-directionally regulate the body, keep the homestatic equilibrium, maintain the osmotic pressure and protect the membrane functions. (7) Modern researches have found that there are numerous microelements such as Zinc, Co, etc, in the herbs, which participate in the synthesis of insulin and the decrease of glucose. There are rich vegetable fibers in many herbs, which could prolong the absorption of the glucose. Besides that, some herbs may stimulate the secretion of insulin.

4.2. There are also some Existing Problems as follows.

- In clinic, it's difficult for TCM diagnosing DM and its complications at the initial stage, especially some type 2 DM patients don't have the typical symptoms and are just diagnosed when take the physical examination. Before the occurrence of the complications, the organism of many patients could have the pathogenic changes, which affect the heart, brain, kidney, eyes, peripheral nerves, vessels, etc. So it's difficult to find out diabetic complications just with the simple TCM diagnosis.
- There are no standards for the syndrome types and the therapeutic effect evaluation. With the TCM macroscopical syndrome differentiation, the symptoms, signs and the evaluation couldn't be objective and accurate, which makes it difficult to be standardized.
- There are many herbs applied to treat DM and its complications. However, the actions of target glycemia-decreasing of herbs couldn't be swift and strong like that of western medicine, even the therapeutic effects couldn't be stable like insulin.
- Most of the Chinese herbs are compound decoction or Chinese patent drugs. Due to the slow effects and weak dependence of the patients, it brings much inconvenience for the DM patients.
- The mechanism study of traditional Chinese Materia Medica, acupunctutre, Tuina and other traditional therapies in treating DM and its complications is still at the initial stage, which focuses more on the clinical observation and mechanism speculation and needs further deeper researches.

5. Thinking on the Application of Nanotechnology in the Mechanism Research of TCM Diagnosis and Treatment of DM

5.1. Application of Nanotechnology in the TCM Diagnosis of DM and its complications

According to the clinical symptoms of DM and its complications at early phase, emphasize more on the detection to the groups of people with high incidence and take the biochemical indexes and clinical detections. With the techniques of nano-diagnosis such as optical coherence tomography, laser monatomic molecule detection, nano-particles, etc, we could detect the subtle situation of the cells and molecular metabolism in the body of DM patients and the substance metabolism before the occurrence of the complications hence provides the basis for the mechanism study of DM and its clinical diagnosis at early phase. Under the theoretical guidance of Surface Plasmon Resonance (SPR), the biosensor could detect the interactions among the protein, nuclein and other biological macromolecules without labels and stain. With the nano-particles such as polyactice acid (PLA), etc, some Japanese scientists have produced the detecting device, which could be applied to detect the signal substance neurotransmission and test the blood glucose and lactate levels related with the fatigue of the body and contributes to the prevention and diagnosis of DM. What's more, we could study the relationship between TCM syndrome types of DM and the objective biochemical indexes with the techniques of nano-diagnosis, which combines the traditional macroscopical syndrome differentiation with the modern microscopical nanotechnology and hopes to achieve the standards of syndrome types and therapeutic effect evaluation.

GAO, et al [11], studied the gene expression profile related with the types of kidney deficiency (dual yin-yang deficiency) and blood stasis of a DM family with the techniques of gene chips and made the detection and comparison between the 3 DM patients and the other 2 healthy family members. The result showed that there were totally 446 pieces of different genes. Among them, there were 8 pieces of up-regulated expressed genes and 438 pieces of down-regulated expressed genes, which indicated the close relationship between the deficiency type of DM and the down-regulated expression of genes. With the techniques of gene chips, the combination of TCM macroscopical syndrome differentiation with the microscopical gene functions provides a new research method in the modern study of TCM diagnosis of DM.

5.2. Application of Nanotechnology in the mechanism research of TCM treatment of DM and its complications

In the clinic, there are many Chinese herbs effectively treating DM. However, due to the slow actions and absorption, inconvenient administration, etc, they just play an auxiliary role in the treatment. With the advanced nanotechnology, the nano-Chinese Materia Medica could improve the biological availability and the therapeutic effects, take the convenient target-oriented administration, release the functions, reduce the dosage, etc, which would apparently improve the therapeutic effects, avoid the adverse effects caused by the long-term drug administration, enhance the quality of life and more importantly explore more prosperous markets for the Chinese Materia Medica treatment of DM and its complications. With the drug-loaded nanoparticles, deliver the herbs (mainly the effective components, regions or the extracts) to the focus, which could accelerate the solubility of the herbs and improve the effects and the amphipathic property of the surface of the drug-loaded nanoparticles and enhance the biological effectiveness, target orientation and specialty of Materia Medica. With the integration or interactions of nanoparticles and the human protein, it could prevent the progress of DM or treat its complications.

The pharmacological actions of Chinese Materia Medica have the characteristics of multiple approaches and targets. Hence, with the techniques of gene chips, it's significant to set up the technological platform for the herbal study, evaluate the therapeutic effects of the herbal treatment of DM, select the herbs and access the metabolism and the toxic-side effects. LI, et al [12], studied the gene expression changes of Gliguidone Capsule on the diabetic rats with the techniques of gene chips and set up the difference profiles of the gene expression for the pharmacological actions of Gliguidone. The techniques of gene chips integrate TCM, the related pathogenic gene expression of modern genomics with the pharmacological actions of modern chemical substances, which link TCM with the gene expression and explain the TCM pharmacological theories and the mechanism for the TCM treatment of DM and its complications from the perspective of gene expression.

The development of acupuncture promotes the further research of needling sensors, which was proposed with the requirement of modernization of traditional Materia Medica. Compared with the biomedical sensors at home and abroad, the sensing needle could be utilized in vivo at fixed positions, points and time with dynamic measurement, which could be applied as the sensor and the actuator [13]. With the techniques of Langmuir-Blodgett Films, it could assemble the bioactive molecules into the ultrathin membrane steadily and could be processed into the bio-sensor with special recognition functions. For instance, with the diffusion-absorption techniques, it could absorb the glucose oxidase onto the lipoid Langmuir-Blodgett Films. With the membrane modified onto the golden matrix of the electrode of H_2O_2, it could be processed into the glucose sensor [14]. Hence, with the techniques, we could absorb the glucose oxidase, insulin receptors, etc, onto the hypersensitive Langmuir-Blodgett Films and fix them onto the golden-plating acupuncture needles, they could be processed into the glucose sensing needle, insulin receptor sensing needle, etc. With the sensing needles applied in the treatment of DM and its complications, it could play an important role in the selection of acupoint prescriptions and the mechanism study of acupuncture therapeutics.

Acupoint catgut embedding belongs to one of the acupuncture therapies, even with better therapeutic effects and long-lasting actions. The retention of the catguts in the acupoints could continuously stimulate the acupoints during the process of being softened, resolved and absorbed hence finally cure the chronic diseases [15]. Acupoint catgut embedding has satisfactory therapeutic effects in the prevention of DM treatment and its complications. CHEN, et al[16], found that early intervention of this therapy could reduce the contents of endothelin (ET) and increase the contents of

NO, which indicates that it could protect the vascular endothelial cells (VEC), improve the functional disorders, protect the peripheral nerves (vagi) and reduce the injury and plays an important role in the early prevention of gastrointestinal dysfunction complicated with DM. Studies[17] indicated that Nano-controlled release system is promising, which is to put the drug powder or solution into the fine particles whose diameter size is measured in nanometer (nm). As carriers, the nanoparticles improve the curative effects and reduce the adverse effects [18]. With the nanotechnology, the catguts could effectively prolong the therapeutic effects in the body and the actions of drug target-orientation, which could remarkably improve the therapeutic effects of acupuncture treatment of DM and its complications.

Acupoint application is also one of the special therapies of acupuncture. SHAO, et al[19], found that to the cases of type 2 DM complicated with hyperlipemia, this therapy could improve the clinical symptoms, reduce the content of glycemia and apparently improve the lipid metabolism disorders, which could decrease the levels of cholesterol, triglyceride, low density lipoprotein (LDL) and increase the content of high density lipoprotein (HDL). With the advantages of convenient, safe, cheap and non-toxic-side effects, this therapy could be applied to the patients in the longer term. At present, one of the challenges of it is how to penetrate more herbs through the skin and into the blood circulation hence improve the therapeutic effects. For this, we could take advantages of the nano-Chinese Materia Medica or the nanoparticles as the herbal carriers hence improve the clinical effects of acupoint application in the prevention and treatment of DM and its complications.

Diverse studies have indicated that insulin resistance (IR) is the basic link of the etiology of DM due to the defect of actions of the insulin onto the receptors. As a result, the ability of muscles and fatty tissues in absorbing the glucose decreases. At the same time, the increase of the content of glucose in the liver shows the decrease of the sensitivity and response of the insulin. Quantum dot is a kind of nanoparticle made of materials of semiconductor crystal with the diameter smaller than 10 nm, which size is smaller than one thousandth of a normal cell and has the characteristics of absorbing large ranges of wavelengths and transmitting small ranges of wavelengths. It is a fluorescent laber indicator. After covalently bonding with the biological macromolecules, it could make the hypersensitive biological detections [20]. In addition, the quantum dot could couple various kinds of molecules such as Protein A, Antibody, Streptavidin, etc, into the fluorescent probes, which could detect the distributions and functions of the specific target molecules [21]. Based on the above mentioned, we could combine the quantum dot probes with the insulin receptors, which could trace the metabolism state in the body hence further explore the mechanism of TCM treatment of DM.

6. Concluding Remarks

With the development of science and technology, the fields of life sciences, information sciences and nanoscineces have been drawn much attention [22]. The history of the technological development indicates that multiple interdisciplinary cross could apparently promote the development of modern science and technology, and the achievements nanotechnology have obtained are the convincing instances. At present, the development of world medicine has the tendency of emphasis on the pharmacological study, exploitation and application of TCM and the TCM research with nanotechnology in the 21st century would become a new fashion, which would become a turning point for TCM to modernization. The etiology of DM is somewhat complicated, which mechanism exploitation and therapeutic targets from the perspective of molecular level are still the emphases. With the rapid development of the nanotechnology in the medical field, it has provided the thinking for the mechanism researches of TCM controlling and treating DM. The authors firmly believe that the mechanism of TCM delaying or checking the progress of DM and its complications would be further well developed with the nanotechnology widely applied in clinic in the longer term.

References

[1] 2004 *Guidebook for Prevention and Treatment of Diabetes Mellitus in China* (Beijing: Peking University Medical Press) P 2

[2] BAI C L 2002 Current Situation and Thinking of Chinese Nanotechnological Study *Physics* **31** 65-70

[3] Hoch HC, Jelinski LW, Craighead HG 1996 *Nanofabrication and Bio-systems: Integrating Materials Science, Engineering and Biology* (Cambridge: Cambridge University Press) P 202
[4] HE D R, XIE J P 2000 Nanotechnology Affecting Medical Development in 21st Century *Medicine and Philosophy* vol **21** 31-32
[5] GAO H M, LIU Y Q 2006 *Nanomedicine* (Beijing: Military Medical Science Press) PP 8-65
[6] QIN C Y, MU L M, ZHANG Y 2001 Application Prospect of Nanotechnology in Medicine *People's Military Surgeon* vol **44** 40
[7] ZHU Z F, YANG J 1998 The Latest Study Development of Nanometer Drug Release-controlled System *Foreign Medical Sciences Biomedical Engineering* **21** 327
[8] WANG Y M, WU Z Y, FAN H M 2003 *Consideration on Nanotechnology Applying to Chinese Materia Medica* (Chinese Traditional and Herbal Drugs vol 3) PP 193-196
[9] WANG H 1996 Application of TCM Sensing Needle in the Study of Meridians and Acupoints *Chinese Journal of Information on Traditional Chinese Medicine* **3** 10-11
[10] MA X L, LIU Y Y, WANG N 2009 Progress of the Clinical Studies of the Effect of Acupuncture and Moxibustion on Diabetes Mellitus *Shanghai Journal of Acupuncture and Moxibustion* **28** 249-252
[11] GAO Y, GAO H, WANG M Q 2005 Different Expressing Gene Analyses about both Kidney-Yang and Kidney-Yin Deficiency and Blood Stasis Syndromes of Diabetes in a Diabetes Family *Modern Journal of Integrated Traditional Chinese and Western Medicine* **14** 1-4
[12] LI P H, WANG X, YAN Z Y 2003 The Effects of Tangshen Ping Capsules on the Gene Expression of Diabetic Mice *West China Journal of Pharmaceutical Sicences* **18** 253-255
[13] REN S, ZHOU Y K, SHEN J 2004 Sensing Needle-Original Developed Sensor *Sensor World* **5** 6-10
[14] ZHAO P, FENG Q J 2001 Application of Langmuir-Blodgett Techniques in the Study of Biomedicine and TCM Prescriptions *Journal of Shanxi College of Traditional Chinese Medicine* **2** 28-30
[15] LI X Y, GE B H 2006 Clinical Study of Acupint Catgut Embedding in the Treatment of Obesity *Chinese Journal of Traditional Medical Science and Technology* **13** 70-72
[16] CHEN Y L, CHEN S H, CHEN R Z 2010 Impact of Acupoint Catgut Embedment on Contents of ET and NO in Diabetic Rats *World Chinese Medicine* **5** 69-71
[17] LIU H F, CHANG J, YAO K D 2001 Application of Polymer Nanoparticles in Medical Carriers *Chemistry* **6** 332-338
[18] Zhu Z F, Yang J 1998 The Latest Study Development of Nanometer Drug Release-controlled System *Foreign Medical Sciences Biomedical Engineering* **21** 327
[19] SHAO M, WEN L J, SUN R 2000 Influence of Acupoint Application on the Blood Fat Level of the Type 2 Diabetics *Shanghai Journal of Acupuncture and Moxibustion* **10** 5-6
[20] Pathak S, Choi S K, Amheim N 2001 Hydroxylated quantum dots as luminescent probes for in situ hybridization *J. Am. Chem. Soc.* **123** 4103-4104
[21] Goldman ER, Clapp AR, Anderson GP 2004 Multiplexed toxinan analysis using four colors of quantum dot fluororeagents *Anal Chem* **76** 684-688
[22] XU H B, YANG X L 2004 *Nano-medicine* (Beijing: Tsinghua University Press) P 268

Influence of SiO_2 buffer layer on the crystalline quality and photoluminescence of ZnO thin films

L H Xu[1,4], X Y Li[2], S Hua[3], Y L Chen[1] and F Xu[1]
[1]College of Math & Physics, Nanjing University of Information Science & Technology, Nanjing, 210044, China
[2]Department of Applied Physics, Nanjing University of Science and Technology, Nanjing, 210094, China
[3]Institute of Electronic Engineering and Photoelectric Technology, Nanjing University of Science and Technology, Nanjing, 210094, China

E-mail: congyu3256@sina.com

Abstract. In this work, a SiO_2 buffer layer was first grown on Si substrate by thermal oxidation, and then ZnO thin films were deposited on SiO_2 buffer layer and Si substrate by electron beam evaporation and sol-gel method. The influence of SiO_2 buffer layer on the crystalline quality and photoluminescence of the films was investigated. The analyses of X-ray diffraction (XRD) showed that all the ZnO thin films had a hexagonal wurtzite structure and were preferentially oriented along the c-axis perpendicular to the substrate surface. The SiO_2 buffer layer improved the crystalline quality and decreased the stress in ZnO thin films. The surface morphology analyses of the samples indicated that ZnO thin films deposited on SiO_2 buffer layers had densely packed grains which obviously increased compared with those grown on bare Si substrate. The photoluminescence spectra of the samples showed that the ZnO thin films deposited on SiO_2 buffer layers had stronger ultraviolet emission performance. The results suggest that SiO_2 buffer layer can improve the crystalline quality and ultraviolet emission of ZnO thin films.

1. Introduction

ZnO is an important compound semiconductor with excellent chemical and thermal stability. It has a wide band gap of 3.37 eV and a large exciton binding energy of 60 meV at room temperature. Lots of research results show that ZnO thin films and nanostructures have high excitonic emission efficiency. So far, ZnO materials are still studied mainly in the form of thin films. ZnO thin films can be used to fabricate many optoelectronic devices such as ultraviolet light-emitting diodes [1], ultraviolet lasers [2], nonvolatile memories [3], optical waveguides [4], ultraviolet photoconductive detectors [5], etc. Many research results indicate that the optical and electrical properties of ZnO thin films are directly connected with their crystallization quality. Therefore, if one wants to fabricate ZnO-based devices with good optoelectronic performance, he has to prepare high-quality ZnO thin films first. It is well-known that the crystalline quality of ZnO thin films has a close connection with the substrates. Many materials have been used as ZnO film substrates such as sapphire, MgO, GaN, Si, quartz, glass, etc. Compared with other materials, Si has some advantages like plentiful raw materials and mature

[4] To whom any correspondence should be addressed.

growth technique of single crystal. Especially, Si material is the cornerstone of the current semiconductor microelectronics industry. If high-quality ZnO thin films can be grown on Si substrate, it will be beneficial for effective integration of optoelectronic devices with Si IC technology. However, the mismatch of lattice constants and thermal expansion coefficients between ZnO and Si are large. What is more, when ZnO thin films are directly deposited on bare Si substrate or these films are annealed at high temperatures, Si atoms on the substrate surface are easy to "capture" oxygen atoms from ZnO thin films [6, 7], which will result in many oxygen vacancies in the films. The above-mentioned factors all lead to the decline of crystalline quality of ZnO thin films. Therefore, many researchers use buffer layers between ZnO thin films and Si substrates in order to improve the crystalline quality of the films. Some groups used ZnO homogenous buffer layer [8], others applied heterogeneous buffer layers such as GaN [9], Y_2O_3 [10], AlN [11], Ti [12], etc.

So far, ZnO thin films have been prepared by many techniques such as pulsed laser deposition, electron beam evaporation, magnetron sputtering, chemical vapor deposition, molecular beam epitaxy, sol-gel method, etc. Among these techniques, sol-gel method has attracted much attention due to many advantages like low cost, simple deposition procedures, no need of vacuum deposition conditions, easier fabrication of large-area film, easier control of composition, being able to deposit thin films on complex-shaped substrates, etc. Moreover, electron beam evaporation is also an important technique for fabrication of ZnO thin films. In this work, ZnO thin films were deposited on SiO_2 buffer layer and Si substrate by E-beam evaporation and sol-gel method. We studied the influence of SiO_2 buffer layer on the crystalline quality and photoluminescence of ZnO thin films and compared the difference of ZnO/ SiO_2 films prepared by the two techniques.

2. Experimental

2.1. Cleaning of Si substrate and formation of SiO_2 buffer layer

One-side polished Si (100) chips were used as film substrates. Si chips were first rinsed in acetone and ethanol, respectively, with ultrasonic vibration for 10 min. Then the Si chips were put into diluted HF solution for 3 min to remove native SiO_2 layer and subsequently rinsed in deionized water. At last, the Si chips were naturally dried in a clean glass chamber. An amorphous SiO_2 buffer layer of ~50 nm thickness was formed on the Si chip surface by thermal oxidation.

2.2. Deposition of ZnO thin films by electron beam evaporation

ZnO thin films were deposited on a bare Si substrate and a SiO_2 buffer layer, respectively, by electron beam evaporation (PMC90S, Protech Korea Ltd.). High purity (99.999%) ZnO particles were used as source material. When ZnO thin films were deposited, the Si substrates were placed on a sample holder which was about 1.5 m from evaporation source and rotated at 40 rpm. The substrate temperature was 200 ℃; the working pressure (Ar+O_2) was 2.3×10^{-2} Pa. A quartz crystal controller (IC5) was used to control the thickness of the films. The deposited film thicknesses were 150 nm. The film deposited on a bare Si substrate was labelled as sample A; the other one deposited on a SiO_2 buffer layer was labelled as sample B. The two samples were annealed in air at 600℃ for an hour.

2.3. Deposition of ZnO thin film by sol-gel method

Zinc acetate, ethanol and monoethanolamine were used to prepare ZnO sol. In this solution, the concentration of zinc acetate was 0.40 mol/L. After the ZnO sol was aged for 24 h at room temperature, ZnO thin films were deposited on SiO_2 buffer layer by spin-coating method. The spin-coating time was 30 s. In the beginning 10 s, the spin-speed was 1500 rpm; in the latter 20 s, the spin-speed was 2500 rpm. After each spin-coating, the sample was put into a furnace kept at 350℃ to be dried and given a pre-heating treatment for 10 min. The procedure from spin-coating to pre-heating treatment was repeated several times in order to increase the film thickness to ~150 nm. This sample was annealed in air at 600℃ for an hour and labelled as sample C.

2.4. *Characterization of structure and luminescence property of ZnO thin films*

The crystal phase and crystalline orientation of the samples were investigated by an X-ray diffractometer (Bruker D8 Advance). The surface morphology was analyzed by an atomic force microscope (CSPM4000) in contact mode. The photoluminescence spectra were used to analyze the luminescence behavior of the samples. The excitation source was a Xe lamp and the excitation wavelength was 325 nm. All the measurements were performed at room temperature in air.

3. Results and discussion

3.1. *Influence of SiO_2 on the structures of ZnO thin films*

Fig.1 shows the XRD patterns of the three samples. It can be seen from Fig.1 that all the samples have a diffraction peak at 34.5° or so, which corresponds to the diffraction of (002) plane of wurtzite-structured ZnO. This indicates that all the ZnO thin films have a hexagonal wurtzite structure and are preferentially oriented along the c-axis perpendicular to the substrate surface. For sample A, B and C, the (002) peak lies at 34.65°, 34.50° and 34.45°, respectively. The position of the (002) peak is associated with the stress in the ZnO thin films. As for the stress-free ZnO powder, its (002) peak lies at 34.43°. From the above data, it is obvious that the (002) peak position of ZnO thin films grown on SiO_2 buffer layer is closer to that of stress-free ZnO powder. This suggests that the usage of SiO_2 buffer layer decreases the stress in the film. For the stress in the ZnO thin film deposited on the bare Si substrate, it mainly results form the mismatch of lattice constants and thermal expansion coefficients between ZnO and Si substrate. Fig.2 shows the crystallite size and film stress derived from XRD data. The crystallite size is calculated by Scherrer formula and film stress is calculated according to the formula in the Ref. [13]. It can be seen from Fig.2 that due to the decrease of the film stress, the crystallites further grow up. The above results suggest that SiO_2 buffer layer improves the crystalline quality of ZnO thin films.

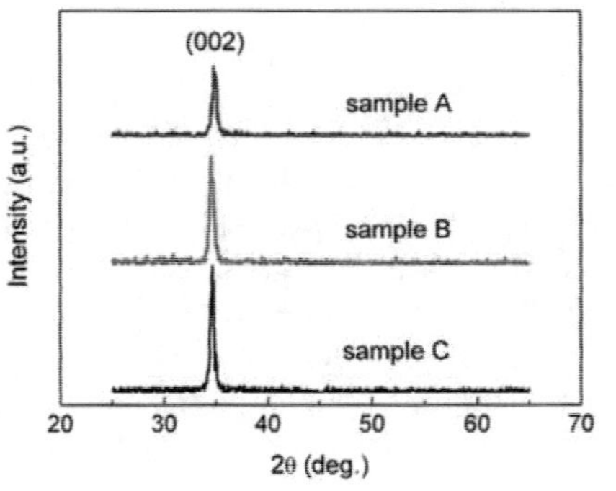

Figure 1. XRD patterns of the ZnO thin films.

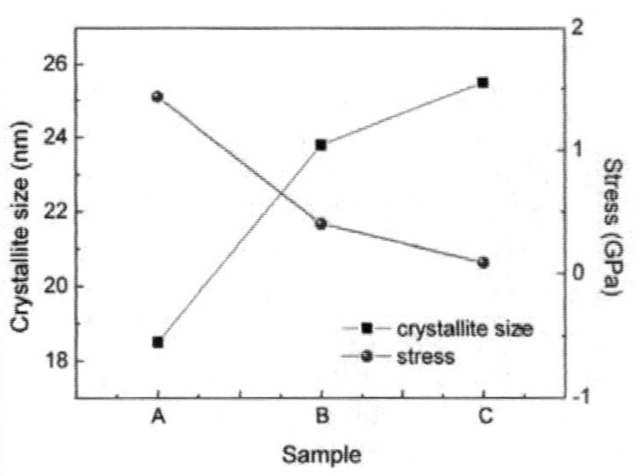

Figure 2. The crystallite sizes and stress of the ZnO thin films.

Fig.3 displays the surface morphology images of the samples. The scanning area is $4\mu m \times 4\mu m$. From Fig.3, it is obvious that all the samples have uniform grains and dense structure. Compared with sample A, the grains of sample B and C apparently increased. This is in agreement with the results derived from XRD data. However, the grains observed by AFM are bigger than that derived from XRD data. It is likely that a grain consists of several crystallites. The similar results have been also observed in many studies [14, 15].

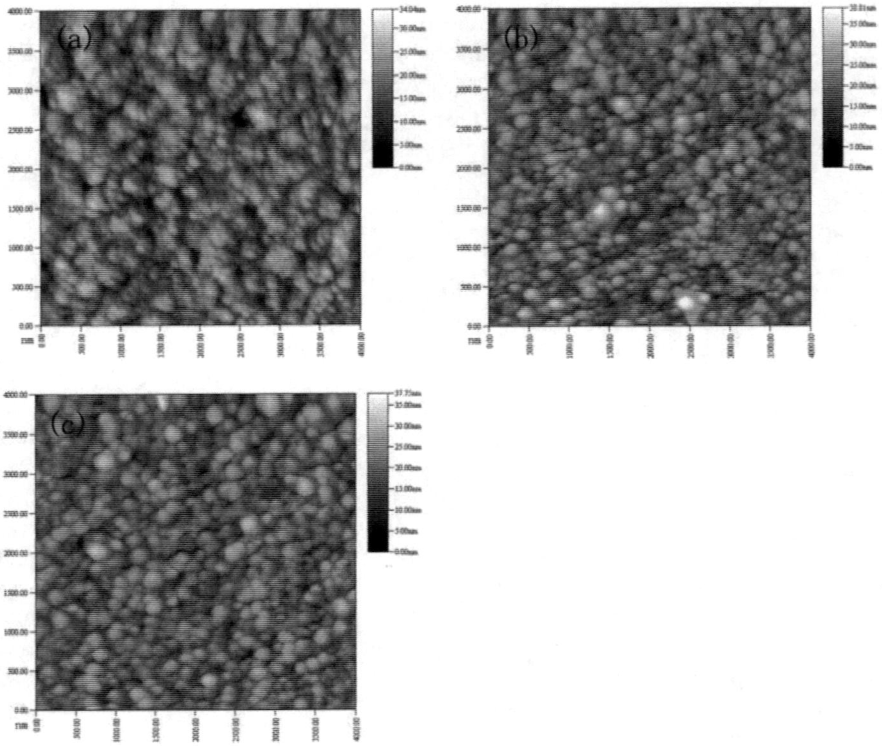

Figure 3. Surface morphology images of sample A (a), B (b), and C (c).

3.2. Influence of SiO_2 buffer layer on the photoluminescence of ZnO thin films

Fig.4 shows the photoluminescence spectra of the samples. As for sample A, besides an ultraviolet emission peak centered at 382 nm or so, it has a weak and wide green emission band. Compared with sample A, sample B and C have a stronger ultraviolet emission peak and weaker visible emissions; the full width at half maximum of the ultraviolet emission peaks is decreased.

In regard to the ultraviolet emission centered at 382 nm, it is generally accepted that it results from the recombination of free excitons [16, 17]. It is well-known that the density of free excitons is closely connected with the crystalline quality of ZnO thin films. Only the films with high crystalline quality and low defect density could have high density of free excitons. From the photoluminescence spectra, it can be seen that ZnO thin films deposited on SiO_2 buffer layers have stronger ultraviolet emission and lower visible emissions, which means that SiO_2 buffer layer improved the crystalline quality of ZnO thin films and reduced the density of native defects. Although both sample B and C are deposited on SiO_2 buffer layer, the ultraviolet emission of sample C seems stronger than that of sample B. Previously, Taschuk et al. [18] reported that the ultraviolet emission efficiency of the ZnO thin film deposited by sol-gel method was higher than those of ZnO thin films deposited by electron beam evaporation, pulsed laser deposition and magnetron sputtering. It is probably connected with the unintentionally doped hydrogen in the sample C deposited by the sol-gel process [19]. As for the green emission of ZnO thin films, it is often observed in the photoluminescence spectra. However, the green emission mechanism is unclear up to now. Some researchers ascribe the green emission to oxygen vacancy defects in ZnO [20, 21], others speculate that the green emission probably results from Zn vacancy defects [22] or antisite oxygen [23, 24]. We tend to think the green emission originates from oxygen vacancy defects. For the ZnO thin film deposited on the bare Si substrate by electron beam evaporation, the oxygen vacancies in the film is mainly connected with the two following factors: (1) the film growth environment is oxygen-deficient, which easily leads to oxygen vacancies in ZnO thin films; (2) the Si atoms on the substrate surface easily "capture" oxygen atoms from ZnO thin films. After the SiO_2 buffer layer is used, it effectively prevents Si atoms to "capture" oxygen from ZnO thin films, decreasing the density of oxygen vacancy. Accordingly, the green emission is reduced. For sample B, its growth environment is oxygen-deficient, so there are still some oxygen vacancies in the films. However, the sol-gel process offers the film an oxygen-surplus growth environment. As a result, sample C has almost no green emission related to oxygen vacancy defects.

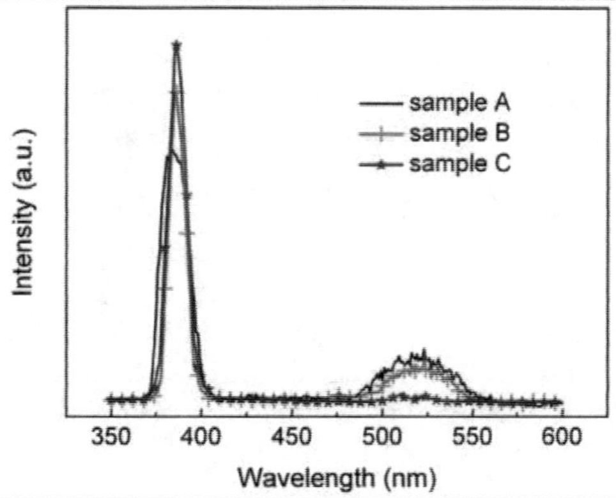

Figure 4. Room-temperature photoluminescence spectra of the ZnO thin films.

4. Conclusions

In this work, ZnO thin films were deposited on a SiO_2 buffer layer and a bare Si substrate by E-beam evaporation and sol-gel method. The influence of SiO_2 buffer layer on the structures and photoluminescence of the samples were studied. All the samples were wurtzite-structured and ZnO grains grew preferentially along the c-axis perpendicular to the substrate surface. ZnO thin films deposited on SiO_2 buffer layer had better crystalline quality. Photoluminescence spectra showed that the ultraviolet emission of sample B and C was increased and visible emission was reduced. The above results suggest that SiO_2 buffer layer can improve the crystalline quality and enhance the excitonic emission of ZnO thin films. The growth of SiO_2 buffer layer by thermal oxidation and ZnO thin films by sol-gel method is simple and low-cost, but the prepared film has good crystallization and excitonic emission. Therefore, it is a good method to combine sol-gel method with thermal oxidation to prepare high-quality ZnO thin films on Si substrates.

References

[1] Han W S, Kim Y Y, Kong B H, et al. Ultraviolet light emitting diode with n-ZnO:Ga/i-ZnO/p-GaN:Mg heterojunction. Thin Solid Films, 2009, **517**: 5106

[2] Ma Xiangyang, Chen Peiliang, Li Dongsheng, et al. Electrically pumped ZnO film ultraviolet random lasers on silicon substrate. Appl Phys Lett, 2007, **91**: 251109

[3] Chang Wen-Yuan, Lai Yen-Chao, Wu Tai-Bor, et al. Unipolar resistive switching characteristics of ZnO thin films for nonvolatile memory applications. Appl Phys Lett, 2008, **92**: 022110

[4] Jia Chuan-Lei, Wang Ke-Ming, Wang Xue-Lin, et al. Formation of c-axis oriented ZnO optical waveguides by radio-frequency magnetron sputtering. Opt Express, 2005, **13**: 5093

[5] Xu Zi-Qiang, Deng Hong, Xie Juan, et al. Ultraviolet photoconductive detector based on Al doped ZnO films prepared by sol-gel method. Appl Surf Sci, 2006, **253**: 476

[6] Fan X M, Lian J S, Guo Z X. Microstructure and photoluminescence properties of ZnO thin films grown by PLD on Si (111) substrates. Appl Surf Sci, 2005, **239**: 176

[7] Liu Z W, Fu W J, Liu M. Scaling behavior and coarsening transition of annealed ZnO films on Si substrate. Surf Coat Technol, 2008, **202**: 5410

[8] Park D J, Lee J Y, Park T E, et al. Improved microstructural properties of a ZnO thin film using a buffer layer in-situ annealed in argon ambient. Thin Solid Films, 2007, **515**: 6721

[9] Kim Il-Soo, Kim Tae-Hwan, Kim Sang Sub, et al. Magnetron sputtering growth and characterization of single crystal ZnO thin films on Si using GaN interlayer. J Cryst Growth, 2007, **299**: 295

[10] Liu W R, Li Y H, Hsieh W F, et al. Domain matching epitaxial growth of high-quality ZnO film using a Y2O3 buffer layer on Si (111). Cryst Growth Design, 2009, **9**: 239

[11] Wang L, Pu Y, Chen Y F, et al. MOVCD growth of ZnO films on Si (111) substrate using a thin AlN buffer layer. J Cryst Growth, 2005, **284**: 459

[12] Li Fan, Li Dongmei, Dai Jiangnan, et al. Effect of the initial thin Ti buffer layers on the quality of ZnO thin films grown on Si (111) substrates by MOVCD. Superlatti Microstru, 2006, **40**: 56

[13] Wang Y G, Lau S P, Lee H W, et al. Comprehensive study of ZnO films prepared by filtered cathodic vacuum arc at room temperature. J Appl Phys, 2003, **94**: 1597

[14] Sharma Ruchika, Sehrawat Kiran, Wakahara Akihiro, et al. Epitaxial growth of Sc-doped ZnO films on Si by sol-gel route. Appl Surf Sci, 2009, **255**: 5781

[15] Tan S T, Sun X W, Zhang X H, et al. Cluster coarsening in zinc oxide thin films by postgrowth annealing. J Appl Phys, 2006, **100**: 033502

[16] Chen S L, Lee S K, Chen W M, et al. On the origin of suppression of free exciton no-phonon emission in ZnO tetrapods. Appl Phys Lett, 2010, **96**: 033108

[17] Zhu Xiaming, Wu Huizhen, Wang Shuangjiang, et al. Optical and electrical properties of N-doped ZnO and fabrication of thin film transistors. Journal of Semiconductors, 2009, **30**: 033001

[18] Taschuk M T, Laforge J M, Nguyen H, et al. Absolute characterization of photoluminescence from pulsed-laser deposited, sol-gel, sputtered and evaporated ZnO thin films. Phys Status Solidi C, 2009, **6**: S131

[19] Dev A, Niepelt R, Richters J P, et al. Stable enhancement of near-band-edge emission of ZnO nanowires by hydrogen incorporation. Nanotechnology, 2010, **21**: 065709

[20] Peng Yu-Yun, Hsieh Tsung-Eong, Hsu Chia-Hung. White-light emitting ZnO-SiO2 nanocomposite thin films prepared by the target-attached sputtering method. Nanotechnology, 2006, **17**: 174

[21] Kurbanov S S, Panin G N, Kim T W, et al. Impact of visible light illumination on ultraviolet emission from ZnO nanocrystals. Phys Rev B, 2008, **78**: 045311

[22] Heo Y W, Norton D P, Pearton S J. Origin of green luminescence in ZnO thin film grown by molecular-beam epitaxy. J Appl Phys, 2005, **98**: 073502

[23] Lin Bixia, Fu Zhuxi, Jia Yunbo. Green luminescent center in undoped zinc oxide films deposited on silicon substrates. Appl Phys Lett, 2001, **79**: 943

[24] Sagar P, Shishodia P K, Mehra R M, et al. Photoluminescence and absorption in sol-gel-derived ZnO films. J Lumin, 2007, **126**: 800

3rd International Photonics & OptoElectronics Meetings (POEM 2010) IOP Publishing
Journal of Physics: Conference Series **276** (2011) 012052 doi:10.1088/1742-6596/276/1/012052

3-5μm F-P Tunable Filter Array based on MEMS technology

Wei Xu[1], Si-hai Chen [1,2], Shi-wei Bao[1], Yi-fan Zhou [1], Sang-zhi Zhao [1], Qing-hua Meng [3] and Huan Luo [1]

[1]College of Optoelectronic Science and Engineering, Huazhong University of Science and Technology, Wuhan 430074, China

[2]Wuhan National Laboratory for Optoelectronics, Wuhan430074, China

[3]Guangxi Normal University, Guilin541002, China

E-mail: xw198877@gmail.com

Abstract. In this paper the design, fabrication and test of a MEMS F-P tunable filter operating in 3-5μm range is presented. The initial design of micro-bridge structure was achieved through COMSOL simulation software. In the simulation, the micro-bridge could achieve a displacement of 0.6μm under very low applied voltage of 30V, while still preserving good mirror parallelism. By using surface micromachining techniques, a 200 × 200 array of F-P tunable micro-bridge was fabricated. SEM photographs of the fabricated device and method of optical performance test are also introduced. This MEMS F-P tunable filter can be a potential application in spectroscopic sensing and optical communication system.

1. Introduction

As an important multiple-beam interference structure, Micro Fabry-Perot cavity tunable filters are widely used in the area of hyper spectral imaging and DWDM (Dense Wavelength Division Multiplexing) optical communication system [1-2]. In recent years, a large effort has been directed towards development of miniaturized and hence cost-effective, portable micro-spectrometers systems. By depositing metal membrane on silicon chip based on MEMS surface process, a new micro F-P cavity has been achieved by us, which can be actuated by electrostatic force and realize the function of optical filter.

2. Design of micro-bridge structure

Micro F-P cavity is driven by static electricity to change the cavity length and achieve wavelength selective. The mechanical and electrical properties simulation is performed by finite element method (using software COMSOL) to establish and optimize the basic micro-bridge design.

The structure proposed by us is shown below in figure 1: the center membrane is a 90μm × 90μm square supported by four cantilever beams (15μm wide and 115μm long); and the optical window is a circular with 50μm diameter.

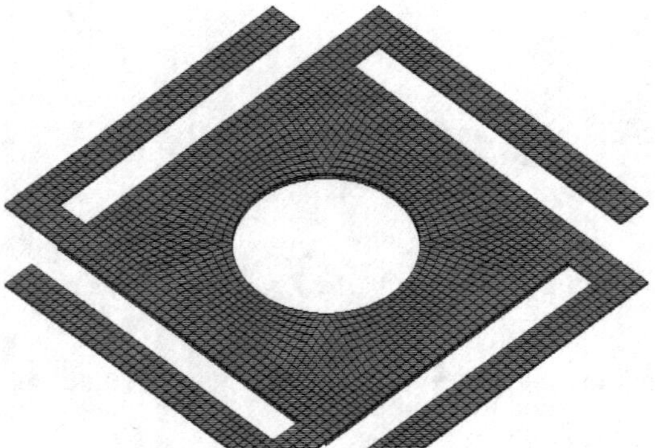

Figure 1. Schematic diagram of micro-bridge.

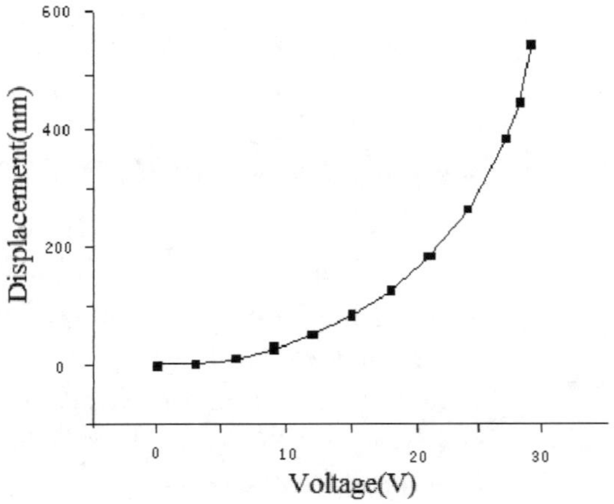

Figure 2. Voltage-displacement curve of micro-bridge.

This bridge could achieve a displacement of 0.6μm under very low applied voltage of 30V (figure 2) and be in good parallelism in the event of displacement. If the displacement of the bridge is d, and the transverse distance between two different places is Δx (figure 3), we can get the included angle between the top and bottom plane is:

$$\theta \approx \tan\theta \approx \frac{d_{\max} - d_{\min}}{\Delta x} \tag{1}$$

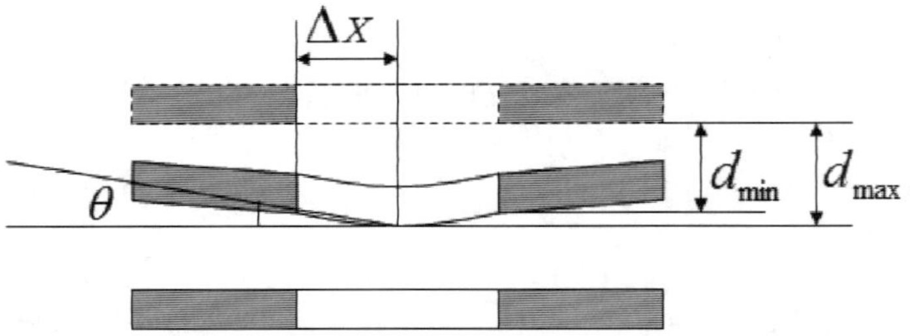

Figure 3. The included angle between two planes.

When applied tunable voltage U=29V,

$$\theta \approx \frac{(531.4 - 529.96) \times 10^{-9}}{3.5 \times 10^{-5}} = 4.11 \times 10^{-5}\,\text{rad} = 0.00235^{\circ} \tag{2}$$

The calculation shows that the value of θ is quite small, which demonstrates the micro bridge has a good parallelism performance in the tuning range.

3. Micro fabrication process

Journal of Physics: Conference Series **276** (2011) 012052 doi:10.1088/1742-6596/276/1/012052

The tunable F-P filter is fabricated by using surface micro-processing technology. The specific micro fabrication process is shown in figure 4. A resonant F-P cavity is designed to be formed between two $Ge/Al_2O_3/Ge$ distributed Bragg reflector (DBR) mirrors with the top DBR freely suspended about 2μm above the bottom plane.

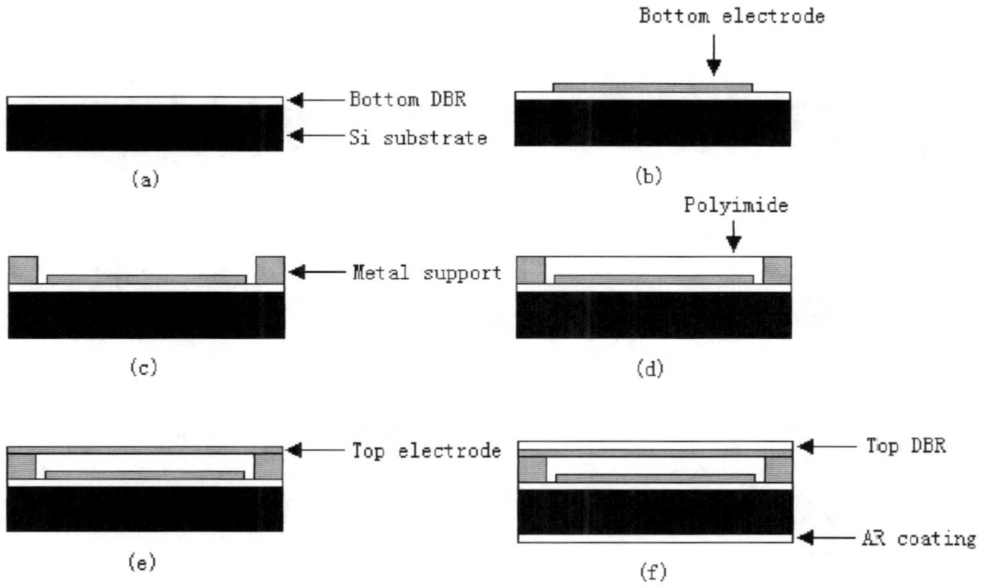

Figure 4. Flowchart of process.

Initially, the bottom DBR mirrors, which contain $Ge/Al_2O_3/Ge$ (250nm/629nm/250nm), were deposited on a bare, 500μm thick silicon substrate. Next, a bottom electrode (Ni-Cr/200nm) and metal support (Ni-Cr/2μm) were respectively formed on the bottom DBR mirrors, which is shown in figure 4 (b) and (c) above. Subsequently, polyimide was spun onto the required thickness of 2μm and defined the separation between the two DBR mirrors of the F-P cavity. The polyimide was formed as a sacrificial layer and baked at 250℃ for 2 hours to reduce the shrinkage effect and enhance its chemical stability to undergo the further fabrication process.

Then deposit a 200nm thick (Ni-Cr) bridge with the graphic structure already given before, shown in figure 4(e) above. The top DBR mirror consisting $Ge/Al_2O_3/Ge$ with a shape of circle (diameter 60μm) was deposited again by thermal evaporation. Finally, coat an antireflective film at the bottom, and release the polyimide sacrificial layer to suspend the micro-filter by using an O_2 plasma etching machine under high pressure.

In ordter to fabricate a suspended micro-bridge structure, the sacrifice layer between the bottom electorde and the top micro-brige must meet the several following requirements: 1) the thickness of the layer can reach 2 μm, and just a little lower than the metal support; 2) great surface flatness; 3) good chemistry stability and compatible with the following process.

A reactive ion etching (RIE) was applied to improve the qulity of polyimide layer surface(figure 5). Firstly, a 2μm polyimide layer was spun on the bottom electrode. However, this process would produce a 1~1.5μm peak. Spin another photoresist(ENPI) and follow with a reactive ion etching (RIE) to smooth the surface of polyimide layer. By using a acetone dissolution process, the flatness of the etched surface could be increased rapidly（figure 6）. This is because baked polyimide is unsolvable in acetone.

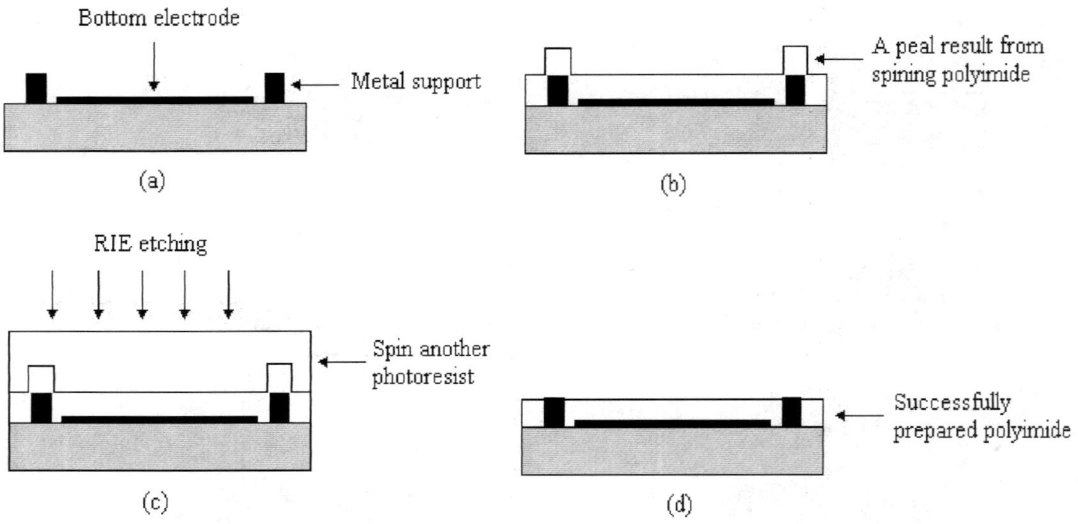

Figure 5. RIE etching process.

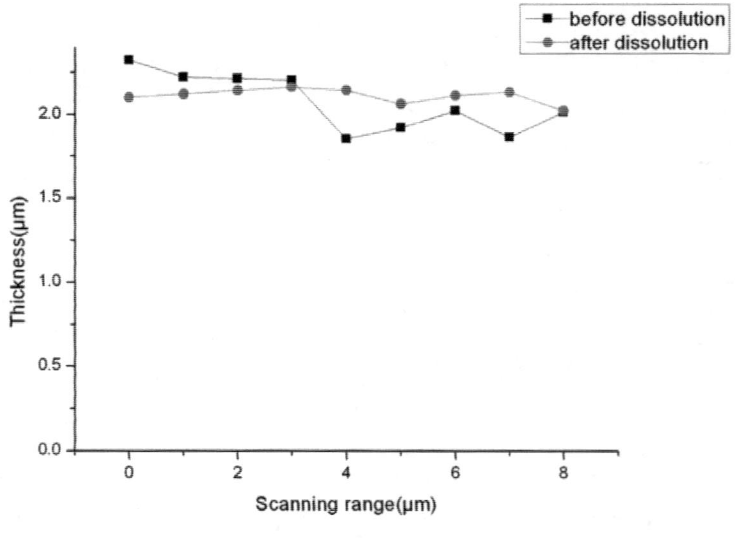

Figure 6. The thickness of etched polyimide.

Journal of Physics: Conference Series **276** (2011) 012052 doi:10.1088/1742-6596/276/1/012052

3. Results and test

The scanning electron microscopy (SEM) photographs of fabricated micro-bridge were shown in figure 7 and 8. It can be found in the cross-section image (figure 7) that a suspended structure has already been fabricated with no obvious ruptures and defects.

Figure 7. Cross-section SEM photograph of fabricated micro-bridge.

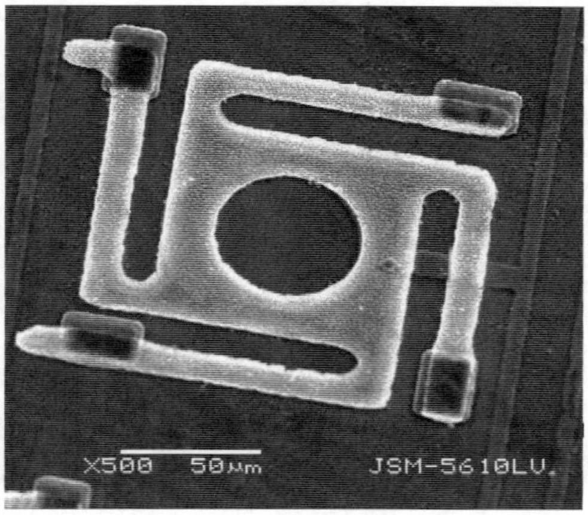

Figure 8. SEM photograph of fabricated micro-bridge.

Two main aspects of the optical performance would be characterized after the two DBR mirrors are added: static response and dynamic response. The static response was characterized by applying stable voltage between the top and bottom electrode to observe the tuning range and transmitted spectrum shifting with voltage. Next, the dynamic response was tested by using an optical method (figure 9) to measure the response time.

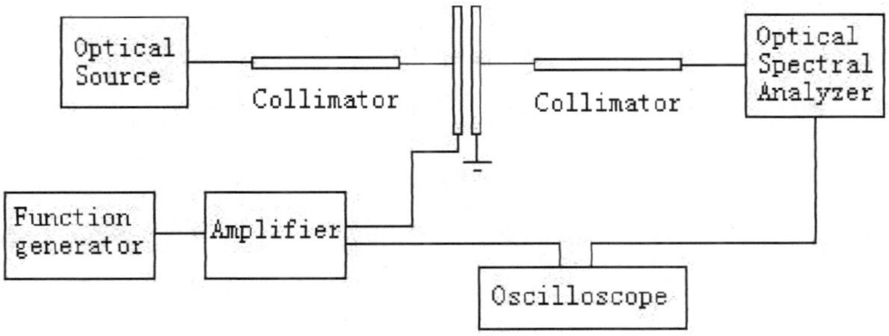

Figure 9. Set-up for dynamic response characterization.

Factual optical performance test will be discussed in another paper. And the simulated transmittance of the film is shown in figure 10. Peak transmission of the tunable filter has risen from 46.44% to 89.14% after depositing the anti-reflection film of Al_2O_3 (629nm) on both bottom and top of the device. And the FWHM (Full Wave at Half Maximum) is measured to be nearly 130nm.

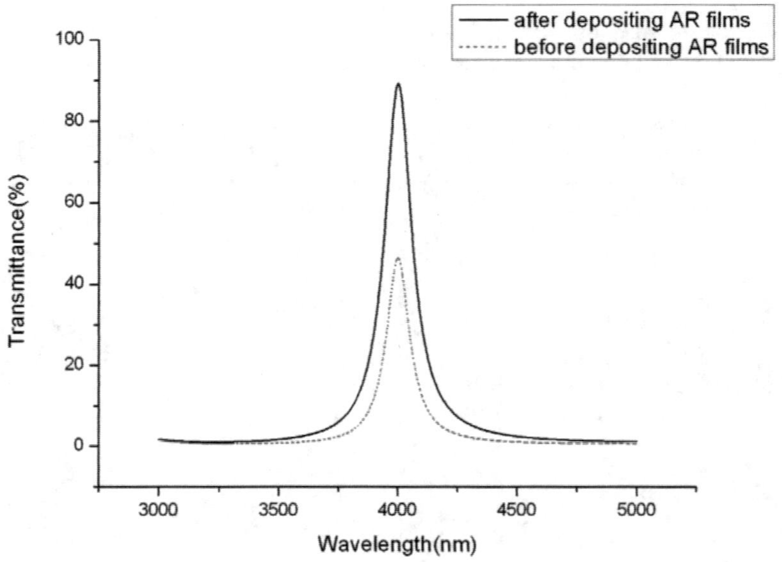

Figure 10. The device transmission before and after depositing anti-reflection film coating.

4. Conclusion

A novel structure F-P tunable micro-bridge operating in 3-5μm has been fabricated by surface micro-processing technology. The surface qulity of polyimide sarcrificial layer could be improved greatly to meet the further process requirement by applying reactive ion etching (RIE) and acetone dissolution. SEM photographs of the suspended structure and fabricated micro-bridge are given. And the tunable filter is designed to have a 89.14% peak transmission and 130nm linewidths after adding DBR mirrors.

Acknowledge

The work is supported by Chinese Aviation Science Foundation (No.2008ZC12002).

References

[1] ZuoYuhua *et al.* 2003 1.3μm Si-based MOEMS optical tunable filter with a tuning range of 90nm *Chinese Journal of Semiconductors* **24** 1140–1144

[2] Adrian J. Keating *et al.* 2007 Fabry-Perot MEMS microspectrometers spanning the SWIR and MWIR *Proc. of SPIE* **6542** 3G1–8

[3] Jaroslaw Antoszewski *et al.* 2006 Tunable Fabry-Perot filters operating in the 3 to 5 μm range for infrared micro-spectrometer applications *Proc. of SPIE* **6186** 081–8

[4] Xiang Min, Cai Yanmin, Wu Yaming, Yang Jianyi and Wang Yuelin 2004 A novel method of fabricating Fabry-Perot cavity employing MEMS wet-etching process *Proc. of SPIE* **5279** 312-318

[5] Xiang Min, Cai Yanmin, Wu Yaming, Yang Jianyi and Wang Yuelin 2003 Experimental study of the free spectral range (FSR) in FPI with a small plate gap *Optics Express* **11** 3147-52

Journal of Physics: Conference Series **276** (2011) 012053 doi:10.1088/1742-6596/276/1/012053

Increase The Absorptive Capacity Of Light Of The Photocells By Embedded In Bow-tie Antenna

Chenguang Zhu [1], Peilin Lang[2], Ru Zhang
[1]School of Science, Beijing University of Posts and Telecommunications, Beijing, 100876
[2]Key Laboratory of Information Photonics and Optical Communications (BUPT), Ministry of Education, Beijing 100876

E-mail: napolles@sohu.com

Abstract. The application of silicon photocells has been widely used in biological and energy field, how to improve the efficiency of silicon photocells has become the research hot spots. The light absorption efficiency is not ideal, only 10% to 20% of solar energy can be transformed into electricity, the paper embeds metal bow-tie antenna in the crystals of silicon, by the field enhancement of the surface plasma, it highly increase the absorptive capacity of light of the silicon photocells.

Key Words: silicon photocell, surface plasma, bow-tie antenna

1. Introduction

With the deterioration of the Earth's natural environment and human awareness of environmental protection, solar, wind, geothermal, biomass and other new kinds of energy are payed more and more attention, they are green and easy to use, without pollution, oil, coal and other traditional fossil energy sources will soon dry up, there are wide space for the research and use of these new kind of energy.

The silicon photocells is a new kind of green energy sources by useing solar energy, it is a kind of semiconductor devices that can transform the solar energy into electrical energy, its basic structure is a PIN knot. The silicon between the two electrodes are mainly consist of monocrystalline silicon and polycrystalline silicon, when the sunlight stimulate on the silicon photocells, the photoelectric effect will generate inside of the silicon,

[1] School of Science, Beijing University of Posts and Telecommunications, Beijing 100876

it will form a lot of electron-hole pairs, they will make directional movement to the two electrodes by the influence of the electrodes. Some of them reach the electrodes, this is the working principle of silicon photocells.

At first crystalline silicon is applied to make silicon photovoltaic, crystalline silicon is an indirect band gap semiconductors, its absorptive capacity of light is weak, in order to increase the efficiency of the photocells, the thickness of this kind of photocells has to reach about 200 μ m ~ 300 μ m; In order to reduce the cost of crystalline silicon photocells, people uses a thickness of 5 μ m ~ 10 μ m of polycrystalline silicon photocells. It can be very difficult to guarantee the efficiency as people reduces the costs at the same time. Additional, in case of the large number of defects in polysilicon, the reaching probability of the electron-hole pairs will be much lower. In order to improve light absorption efficiency, people use the polysilicon thin films with rough surface, however, it will bring more defects. Therefore, the key point to solve these problems is how to increase the efficiency as far as to compress the thickness the photocell.

It was suggested that the surface plasma wave can be used to resolve this problem by its field enhancement effect, surface plasma research is a hot topic today, when metal surface exposed in specific frequency of light, free electrons in the metal surface cause collective coherent shocks, electromagnetic wave will be formed between the metal and dielectric's surface. It has spatial locality, local field enhancement and other features, it is widely used in the field of antenna technology, bio-technology and so on. Some scholars doped nanometer-scale metal particles in the polysilicon films, it formed surface plasma waves on the surface of these metal particles, around the metal particles there forms local field enhancement effect, it enhances the ability of generating electron-hole pairs, this can guarantee the absorptive capacity when people compress the thickness of the crystals。

In this paper, the method is embedding bow-tie antenna into a polycrystalline silicon thin-film, simulate the electric field distribution when the sunlight irratiate throughout the silicon film, estimate the number of electron-hole pairs generated, and by the contrast, try to prove that with the same efficiency ,the model embedded in bow-tie antenna can be effectively compressed the film thickness.

2. Bow-tie antenna

One important purpose of the surface plasma is making nano-antenna, bow-tie antenna is a kind of antenna, the model is shown in Figure 1, two fence-type metal sheet are placed face to face, when light of certain frequency and wave vector perpendicularly irradiates to the bow-tie antenna, there will forms surface plasma waves on the surface of metal, this kind of electromagnetic wave transmit along the metal surface to the metal tip, as the cutting-edge field enhancement effect, in the vicinity there forms an enhanced electric field. The two metal tips are positive face to face, therefore, in the center of the antenna, it will form a broader and better region of field electric enhancement , which is the electric field enhancement of the bow-tie antenna.

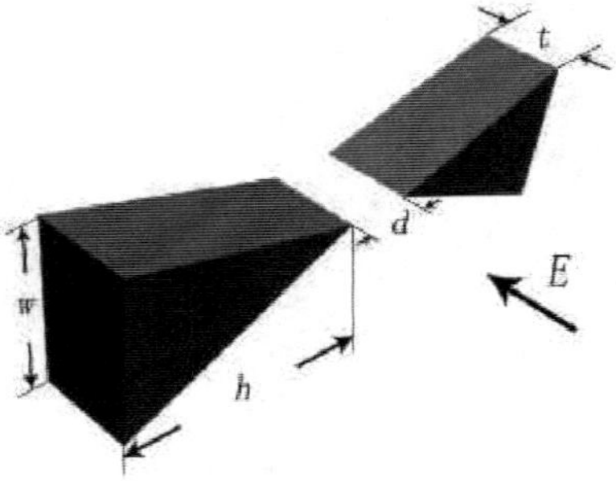

Fig.1. Bow-tie antenna, 4 important element (*h*,*w*,*d* and *t*). Light of the direction of E forms surface plasma waves on the surface of metal, this kind of electromagnetic wave transmit along the metal surface to the metal tip, as the cutting-edge field enhancement effect, in the center of the antenna, there forms a higher electric enhancement

3. Results and discussion.

First of all, we need to identify a right kind of the bow-tie antenna element, noble metals have relatively stable physical properties and chemical properties, so select the more commonly used gold as forming the antenna.

In Figure 1, on the assumption that the thickness t is in infinite case, the size of the bow-tie antenna is decided by the three parameters, the metal film height h, width w, and the distance of these two pieces of metal cutting-edge d. During the producing process of the silicon photocells, we want to reduce the producing costs as much as possible, so, first of all, this article selects h = 150nm, so there remains w and d to be determine.

Supposing w = 60nm, assuming the inputting wavelength is 750nm, the sine amplitude is 1V/m of sunlight, calculating by the method of FDTD(Finite Difference Time Domain), to the central point O, the relation between the electric field strength E and the distance d is shown in Figure 2 (1) .

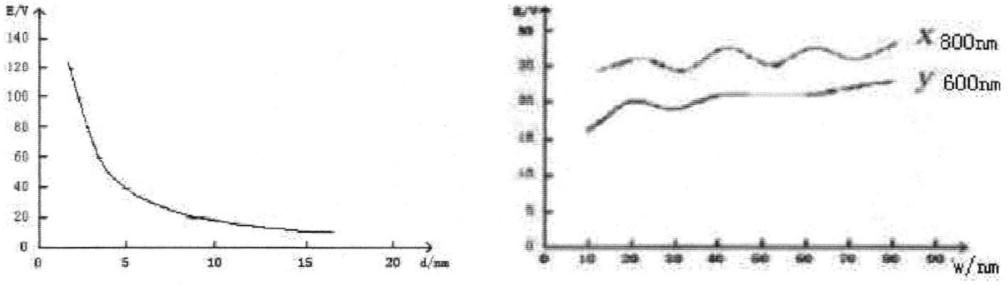

Fig.2. Analyse diagram. Analysis of Figure 2 (1) , we could get the results as follows, as d increases, the center field strength E is getting smaller and smaller, it is because the coherence coefficient between the two metal cutting-edge electric field in the O point is gradually decreasing, causing the electric field strength generated getting weaker and weaker, but considering the scope of the electric enhance field is getting smaller with the decreasing of d, this is not good for the improvement of the efficiency of photocells, so selecting the distance of d = 10nm comprehensiving these two factors .

After determining d = 10nm, h = 150nm, with the same assumption that the input is 1V/m amplitude sine wave, through the FDTD method, we can get the relationship between the electric field strength E of point O and the metal width w as it shown in Figure 2 (2). The upper line is about the sunlight with the wavelength of 800nm, the lower line is about the sunlight with the wavelength of 600nm. We can drew a conclusion that as w increase E appeared a few extreme values, whose reason is the effect of SPR（Surface Plasmon Resonance）between the electronics and the metal particles, according to the curve in the diagram, we choose one of the extreme value w = 60nm for the antenna.

In this way, we get a gold bow-tie antenna with the size of d = 10nm, h = 150nm, w = 50nm, in this structure, the antenna produces the best electric field enhancement when exposed in the sunlight.

4. Simulation of light absorption capacity calculation and comparison

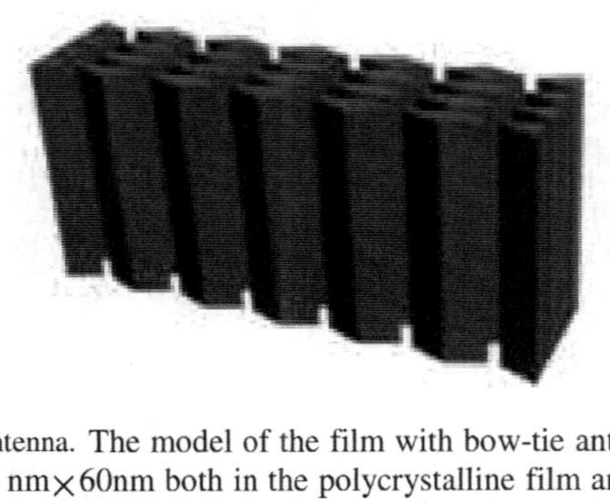

Fig.3. Film of bow-tie antenna. The model of the film with bow-tie antenna is shown in Figure 3, simultaneously select a cylinder of 310 nm$\times$60nm both in the polycrystalline film and the model consisted of bow-tie antenna, about the transverse section ,consider the absorption capacity of the 5 μm polysilicon , that is, to calculate the number of the electron-hole pairs, select 600nm, amplitude 1V/m sunlight as the source, irradiate on the surface of the transverse section, as the result of FDTD calculation, the electric in different locations of the film shows no great difference , the mean value is 0.984V, set light intensity as I, the number of the electron-hole pairs as Q, we can get this:

$$Q = kSI = kS\,E^2$$
$$= k \times 1.86 \times 10^{-14} \times 0.984^2$$
$$= 1.8\,k \times 10^{-14}$$

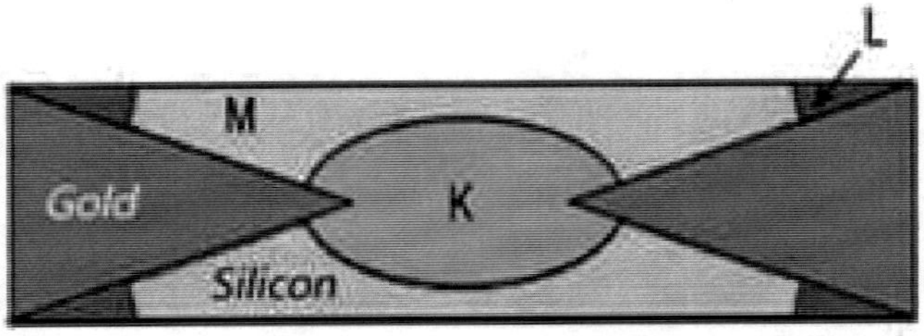

	S (nm²)	E(V/m)
K	3127	4.62
L	896	3.16
M	6577	0.993

Fig.4. electric field strength. Now we calculate the absorption capacity of the film with this d = 10nm, h = 150nm, w = 60nm bow-tie antenna, the wavelength is also 600nm , 1V/m sine wave ,the result is shown in figure 4, it can be seen that around the bow-tie antenna (two pieces of metal around the edge, and two around the corner) has been significantly enhanced of the electric field effect, in other field, the electric field strength did not change significantly, in the entire transverse section ,it's the semiconductor silicon where generate electron-hole pairs in , but gold parts not, so we divided the whole transverse section into six regions, as the figure shown, the electric field strength is shown in figure 4.

we can calculate the number of electron-hole pairs generated as this:

$$Q = k\sum SI = k\sum S\,E^2$$
$$= k\ (3127\times10^{-18}\times4.62^2 + 896\times10^{-18}\times3.16^2 + 6577\times10^{-18}\times0.99^2)$$
$$= 8.218\,k\times10^{-14}$$

By comparison, we can see that with the same thickness， the light absorption capacity of the film with bow-tie antenna is 4.56 times to the polycrystalline silicon.it means if we want to have two photocells of the same efficiency, one is a 5 μm polycrystalline silicon ,the other only has to be a film of 1.09 μm with bow-tie antenna embedded in, this significantly reduced costs.

5. Conclusion

This article used bow-tie antenna embedded in the polycrystalline photocell, through the method of finite-difference time-domain simulation, for the internal electric field distribution, under the sunlight, we calculated the number of the electron-hole pairs generated as a way to assess this structure, especially the absorption capacity of light. Though the comparing form the two kinds of photocells, we found that, the kind with bow-tie antenna has a higher efficiency, as it has a lower cost of production, this method has a broad space for developing. But as the calculating process in this paper, the author ignore the problems caused by optical nonlinear and some other complex factors, it is necessary to conduct further simulation, computing and improved.

6. Acknowledgments

This work is supported by the National Basic Research Program (973) of China under Project No.2010CB923202.

7. References

[1] Peterson A F 1988 Microwave Opt.Tech.Lett. 1 62
[2] Rocca J J,Shlyaptsev V,Tomasel F G,Cortazar O D,Hartshorn D,Chilla J L A 1994 Phys.Rev.Lett.73 2192
[3] D. M. Schaadt, B. Feng, and E. T. Yu, "Enhanced semiconductor optical absorption via surface plasmon excitation in metal nanoparticles," Appl. Phys. Lett. 86, 063106 (2005).
[4] R. B. Konda, R. Mundle, H. Mustafa, O. Bamiduro, A. K. Pradhan, U. N. Roy, Y. Cui, and A. Burger, "Surface plasmon excitation via Au nanoparticles in n-CdSe/p-Si heterojunction diodes," Appl. Phys. Lett. 91, 191111 (2007).
[5] U. Kreibig and M. Vollmer, Optical properties of metal clusters, Springer Series in Materials Science (Springer-Verlag, Berlin, 1995).

Comprehensive Assessment of New Modulation Techniques in 40Gb/s Optical Communication Systems

Bing Linlin[1], Lei Jianming[1 *], Lu Li[1, 2], Zou Xuecheng[1]

1 Research Center for VLSI and Systems, Department of Electronic Science and Technology, Huazhong University of Science and Technology, Wuhan 430074, China

2 Air Force Radar Academy, Wuhan 430019, China

*Email: leijianming@mail.hust.edu.cn

Abstract. In this paper, comprehensive assessments are proposed for intensity modulations and phase modulations respectively. Several key performance features, like receiver sensitivity, chromatic dispersion tolerance, nonlinearity tolerance, and noise sensitivity were compared among these modulation techniques. Results show that 66% duty cycle RZ signal so-called the carrier-suppressed return-to-zero (CSRZ) format has the best performance in the intensity modulations. In phase modulations, the most outstanding modulation is DPSK. DQPSK modulation is also a very promising choice if the effect of noise could be resolved. The combination of intensity modulation and phase modulation will compromise the performance advantages of both.

1. Introduction

In modern optical communication systems, modulation format has a great impact on system performance. NRZ (non-return-to-zero) modulation format is widely used in the 10G and lower rate optical communication system. Compared with other modulation formats, NRZ modulation is simple and convenient, and requires lower cost. But when it comes to the 40Gb/s and higher speed, transmission capacity is restricted because of the optical signal-to-noise ratio (OSNR) tolerance and nonlinear effects of NRZ. Therefore NRZ modulation is no longer suitable for the application of 40Gb/s or higher rate long-distance transmission systems. 40Gb/s long-distance transmission systems need spectrum-efficient modulation formats. For the current 40Gb/s commercial wavelength division multiplexing (DWDM) systems, special modulation formats are used, such as optical duo binary (ODB), return-to-zero (RZ), carrier-suppressed return-to-zero (CSRZ), differential phase shift keying(DPSK), and differential quadrature reference phase shift keying (DQPSK), etc. New modulation techniques have become a research focus in high-speed long haul transmission systems. Compared to NRZ, the most significant advantage of new modulation formats is to improve the N × 40Gb/s WDM transmission system for the most limiting factors (such as dispersion, nonlinear effects, back to back OSNR, etc.), and significantly expand the system's transmission distance[1].

This paper combines simulation contrast and theoretical analysis to reveal the features and application conditions for each intensity modulation and phase modulation. Moving a further step, a best combination of them is concluded.

2. Simulation Setup

All the simulations are performed using the commercial simulation software package OptiSystem Software. 40Gb/s optical communication systems with diverse modulation formats discussed are established and shown in Figure 1, Figure 2, Figure 3, and Figure 4, respectively. In these diagrams, PRBS represents a pseudo random bit sequence binary generator, which generates 2^9 (512 bites) long sequences. NRZ is a converter changing binary to electrical NRZ signal. OBPF is an optical band pass filter, as it is used as a second order Gaussian filter, the full width at half maximum (FWHM) bandwidth is optimised for individual modulation formats. V-Att is a variable attenuator. ELPF represents an electric low pass filter, which is fifth order and the -3 dB bandwidth is optimised for individual modulation formats. The frequency and power of the CW laser diode are set at 193.1 THz and 1 mW, respectively. Extinction ratio of Mach-Zehnder intensity modulators (MZM) is set at 30 dB. The link consists of a 100km span of standard single mode fiber (SSMF) with group velocity dispersion (GVD) of 17ps/nm/km, an attenuation coefficient of 0.2dB/km, an effective area Aeff of $70\mu m^2$ and a nonlinear index n2 of $2.6\times10^{-20} m^2/W$ .The chromic dispersion compensation is performed by dispersion compensation fiber (DCF). Erbium doped fiber amplifier (EDFA) is used to compensate the attenuation generated by SSMF fibers. A PIN photodetector with dark current of 10 nA and thermal noise of $1.0\times e^{-21} W/Hz$ is assumed.

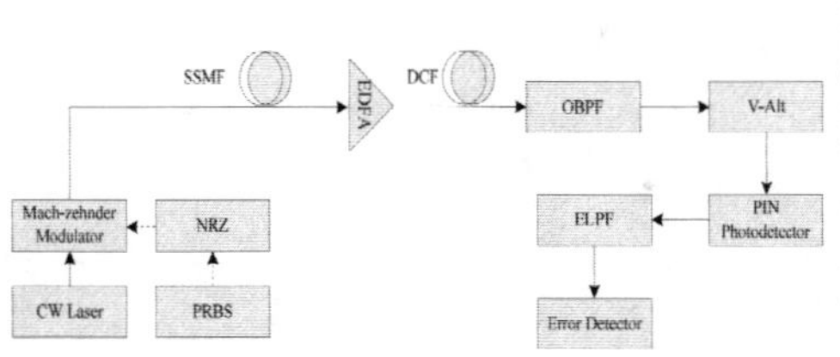

Figure 1. Simulation of 40 Gb/s NRZ transmission.

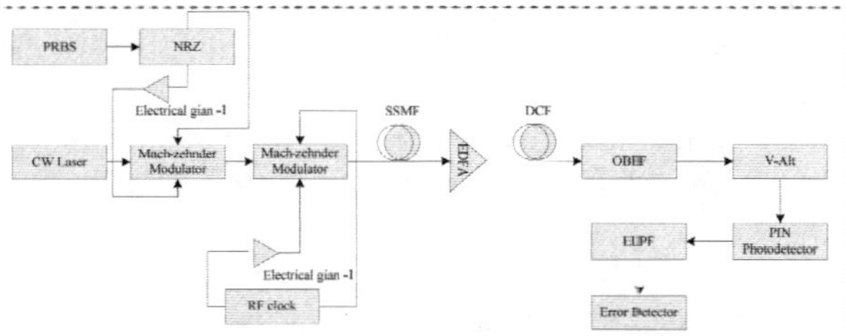

Figure 2. Simulation of 40 Gb/s RZ (cf 33%, 50% and 66% duty cycles) transmission.

In order to generate RZ signals with different duty cycles, one MZM is used to generate an NRZ format first, then another MZM is applied on top of the NRZ signal to obtain the corresponding RZ signals with different duty cycles (like 33%, 50% and 66% RZ signals), using RF clock signals with different amplitudes and frequencies.

The realization of DQPSK is a little complex, which comprises one phase modulator (PM) and one MZM in series, with 90 and 180 degree phase shift, respectively.

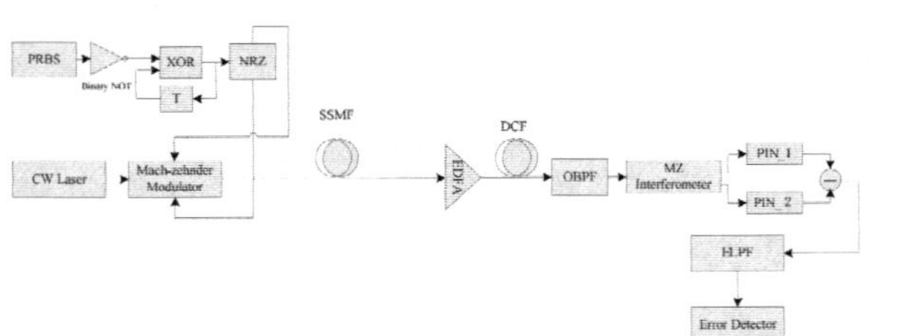

Figure 3. Simulation of 40 Gb/s DPSK transmission.

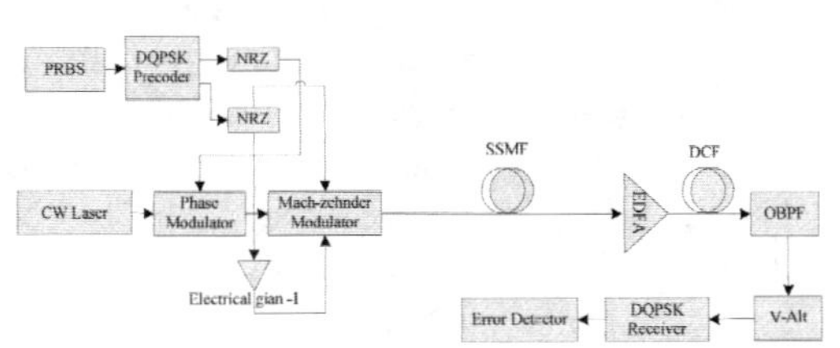

Figure 4. Simulation of 40 Gb/s DQPSK transmission.

3. Comparison of Intensity Modulations

3.1. Nonlinearity Tolerance

High-speed and long-distance transmission requires higher launch power of the transmitter, which will result in more severe nonlinear effects[2]. Therefore, the nonlinear effects become an important constraint to 40Gb/s or higher rate long-distance transmission.

The changes of nonlinear effects are achieved by controlling the launch power. Figure 5(a) reflects how the nonlinear effects impact bit error rate (BER) using different intensity modulation formats.

The four curves in Figure 5(a) all decrease at first and then increase in the simulation process. If the input optical power is small, the system will reach a high BER because the OSNR is too low. With the rising of input power, the BER continues to fall and finally arrives at the lowest point, where the system has the optimal performance. After that, the BER increases again, because the nonlinear effects become the dominant influence factor in the system. Hence, the lowest point represents the best launch power. We can see from the figure that best launch power of NRZ is 6dBm, while the RZ formats with 66%, 50% and 33% duty cycle have the best launch power of 8dBm, 9.3dBm, and 10.5dBm, respectively.

3.2. Dispersion Tolerance

The dispersion means that the phase velocity of a wave depends on its frequency, or alternatively, the group velocity depends on the frequency. It will cause inter-symbol interference and result in error bits during transmission [3].

Ignoring the polarization mode dispersion (PMD) effects and nonlinear effects, we set the launch power to the 4dBm. By adjusting the length of DCF, the chromatic dispersion (CD) can be changed from less compensation to over compensation. When the length of DCF is 10KM, chromatic dispersion is fully compensated, while less than 10KM means less compensation. The optical attenuator is adjusted to make the BER comparable.

The simulation results are shown in Figure 5(b). When the BER reaches at $10e^{-9}$, the residual dispersion that the system can tolerate is calculated as follow:

$$D = |(L - 10\text{km}) * (\text{-}80\text{ps}/\text{nm}/\text{km})| \qquad (1)$$

Where L is the length of DCF in different modulation systems.

It can be calculated that dispersion tolerance of NRZ is 32ps/nm, while that of RZ format of 66% duty cycle is 20ps/nm. RZ formats of 50% and 33% duty cycle have approximately equal dispersion tolerance, which is 12ps/nm and has become an important constraint.

3.3. Receiver Sensitivity

Receiver sensitivity is defined as the minimum average optical power required for the optical receiver to satisfy the communication quality conditions (BER or OSNR), and plays an important role in the measurement of receiver performance[4].

The thermal noise of PIN photodetector is set to $1.0\times e^{-21} W/Hz$. The simulation results are shown in Figure 5(c). When the BER reaches at $10e^{-9}$, the receiver sensitivity of NRZ modulation format is -9.2dBm, while that of RZ format of 66%,50% and 33% duty cycle are -10.5dBm, -11dBm and -11.7dBm, respectively.

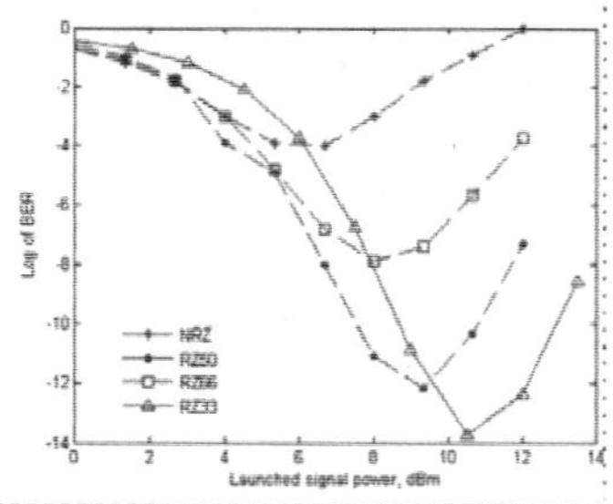

Figure 5(a). Nonlinear tolerance Comparison in intensity modulations.

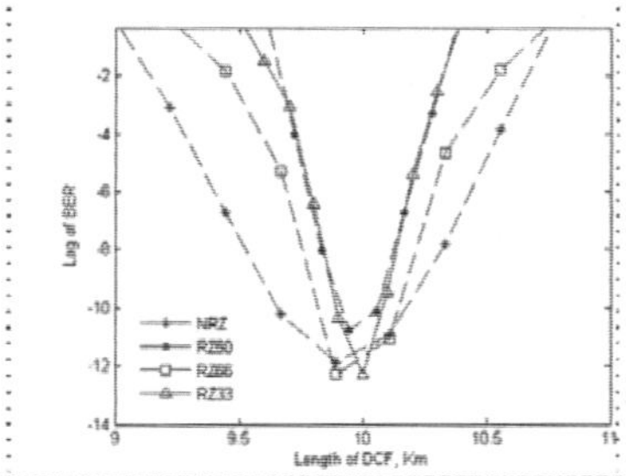

Figure 5(b). Dispersion tolerance Comparison in intensity modulations

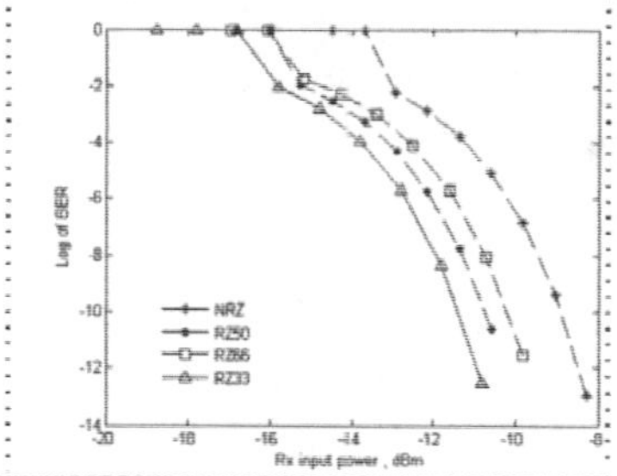

Figure 5(c). Comparison of receiver sensitivity in intensity modulations

Table 1 shows the comparison of intensity modulations. It can be seen that a smaller duty cycle of the intensity modulation brings about a chain of positive reactions, such as more intensive energy, stronger nonlinear tolerance and higher receiver sensitivity. However, on the downside, the poorer dispersion tolerance will arise. The RZ signal with 66% duty cycle which is called CSRZ is proposed to be the best trade-off among all the above performances in various intensity modulations.

Table 1 Comparison of intensity modulations

	best launch power	dispersion tolerance	receiver sensitivity
NRZ	6dBm	32ps/nm	-9.2dBm
RZ50	8dBm	20ps/nm	-10.5dBm
RZ66(CSRZ)	9.3dBm	12ps/nm	-11dBm
RZ33	10.5dBm	12ps/nm	-11.7dBm

4. Comparison of Phase Modulations

4.1. Noise Tolerance

Noise tolerance is an important performance index of phase modulations[5]. Considering the condition of full compensation and no nonlinear effects, we set the optical launch power to 4dBm, and the optical attenuator to 8.43db. By changing the thermal noise of PIN photodetector, the simulation results are shown in Figure 6(a).

We can see from the figure that, the BER of DPSK modulation is much lower than that of OOK(On-Off Keying) modulation system. When the BER reaches at $10e^{-9}$, the noise that can be tolerated is $1.5e^{-21}W/HZ$ in OOK system and $2.5e^{-21}W/HZ$ in DPSK system. However, the noise can be tolerated in DQPSK system is nearly zero in the simulation condition we set.

4.2. Nonlinear Tolerance

The thermal noise of PIN photodetector is set to $1.0e^{-21}W/HZ$ and the optical attenuator is adjusted to allow the system BER comparable. Figure 6(b) reflects how the nonlinear effects impact on BER in different phase modulation formats.

Similarly to the intensity modulations, the trends of the three curves in Figure 5 all decrease at first and then increase in the process. The best launch power of DPSK is 10dBm, while that non-linear tolerance of OOK and DQPSK are both 12dBm.

4.3. Dispersion Tolerance

The simulation results are shown in Figure 6(c). According to the formula (1), we can calculate that the dispersion tolerance of OOK is 8ps/nm, while that of phase modulation DPSK and DQPSK are improved to 13.6ps/nm and 32ps/ns, respectively.

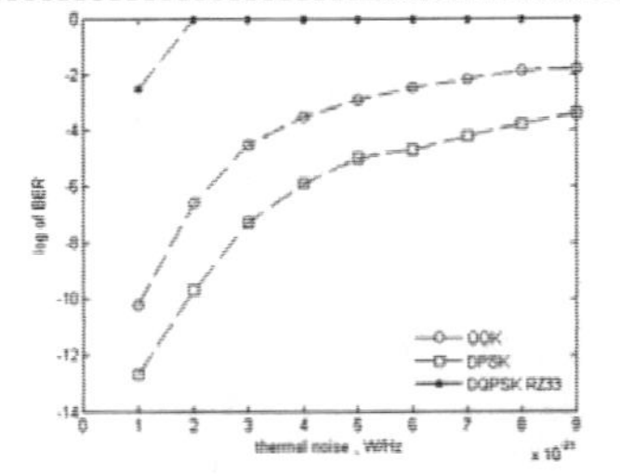

Figure 6 (a). Comparison of noise tolerance in phase modulations.

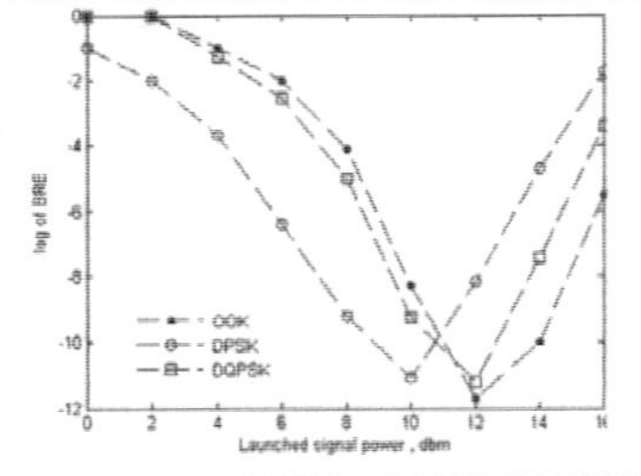

Figure 6 (b). Nonlinear tolerance Comparison in phase modulations.

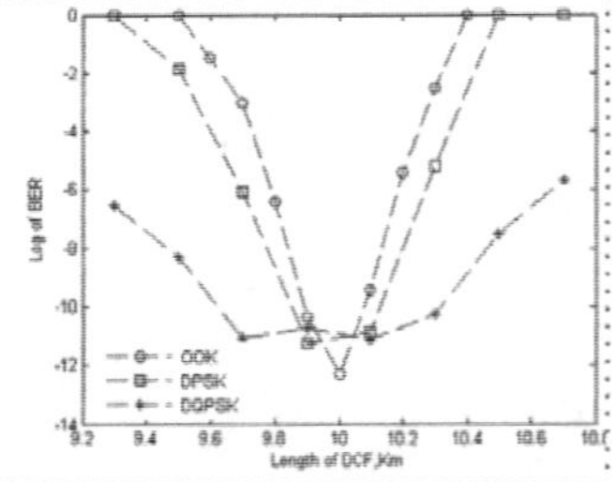

Figure 6 (c). Dispersion tolerance comparison in phase modulations.

Table 2 shows the comparison of phase modulation. Compared with the intensity modulations, the performance of phase modulations does not have strong nonlinear tolerance, but performs well in dispersion tolerance, especially DQPSK, for its half-reduced symbol rates. Compared with OOK, DPSK's noise tolerance enhanced greatly for the reason that the amplitude's variation range is expanded to from -1 to 1. DQPSK is particularly sensitive to noise as its phase is equally divided into four parts. Using more phases in phase modulation is accompanied with a lower symbol rate, stronger anti-dispersion ability, but higher noise sensitivity. The system noise can be reduced by coordinating with Raman amplifier[6].

Table 2 Comparison of phase modulations

	noise tolerance	best launch power	dispersion tolerance
OOK	$1.5e^{-21}W/HZ$	12dBm	8ps/nm
DPSK	$2.5e^{-21}W/HZ$	10dBm	13.6ps/nm
DQPSK	0	12dBm	32ps/nm

Compromising all the performance indexes mentioned above, DPSK is the most outstanding one in various phase modulations. If the impact of noise can be effectively reduced, multi-system phase modulation, such as DQPSK, is also very promising for its two positive characteristics: a performance enhancement and a low requirement of the electronic devices' speed, though its implementation is with high complexity.

5. Conclusions

Comprehensive assessment has been done in this article. Considering all performance indicators that have been simulated, the most outstanding intensity modulation format is CSRZ, while DPSK is the best in the phase modulations. If the effect of noise can be effectively solved, DQPSK modulation is also very promising. The combination of intensity modulation and phase modulation combines their performance advantages and will be the trend of modulation used in 40Gb/s optical communication systems.

References

[1] G. Keiser, Optical Fiber Communications, 3rd ed., New York: McGraw-Hill, 2000.
[2] S. Savory, "Digital Signal Processing Options in Long Haul Transmission", OFC2008, Paper OTuO3.

[3] G. Proakis, Digital Communications, 4th ed., New York: McGraw-Hill, 2001.

[4] Van Den Borne, D., Jansen, S.L., Khoe, G.D., De Waardt, H., and Gottwald, "Line optimization in long-haul transmission systems with 42.8 Gbit/s RZ-DQPSK modulation". Proc. Optical Fiber Conf., Anaheim, CA, USA, 2006, paper OFD2.

[5] Wickham, L.K., Essiambre, R.-J., Gnauck, A.H., Winzer, P.J., and Chraplyvy, "Bit pattern length dependence of intrachannel nonlinearties in pseudolinear transmission", IEEE Photonics Technol. Lett., 2004, 16, (6), pp. 1591–1593.

[6] Forestieri, "Evaluating the error probabilty in lightwave systems with chromatic dispersion, arbitrary pulse shape and pre- and postdetection filtering", J. Lightwave Technol., 2000, 18, (11), pp. 1493–1503.

Journal of Physics: Conference Series **276** (2011) 012055 doi:10.1088/1742-6596/276/1/012055

Low crosstalk Arrayed Waveguide Grating with Cascaded Waveguide Grating Filter

Yang Deng, Yuan Liu, Dingshan Gao*
Wuhan National Laboratory for Optoelectronics, Huazhong University of Science and Technology, Wuhan 430074, China

E-mail address: dsgao@mail.hust.edu.cn

Abstract We propose a highly compact and low crosstalk arrayed waveguide grating (AWG) with cascaded waveguide grating (CWGF). The side lobes of the silicon nanowire AWG, which are normally introduced by fabrication errors, can be effectively suppressed by the CWGF. And the crosstalk can be improved about 15dB.

1. Introduction

In recent years, silicon based on arrayed waveguide grating (AWG) as wavelength division multiplexer/demultiplexer (MUX/ DeMUX) has drawn great interests for its low insertion loss, high stability and low cost [1-2]. And it has great potential in chip-scale optical interconnection to enhance the data traffic capacity [3]. With the development of silicon nanowire technique, shrinking the device size can be achieved. E-beam lithography and dry etching can be used to fabricate the silicon nanowire AWGs [4]. However due to the rapid growing phase error in nanowire waveguides brought by fabrication errors, measured crosstalk is normally larger than 15dB [2, 5]. Suppressing the crosstalk of nano sized AWG has been a very big challenge. In this paper, a novel cascaded waveguide grating filter (CWGF) is proposed to decrease the crosstalk of AWG caused by fabrication errors.

2. Structure design

The initial design of structure is N×N (N can be 8 or 16 in our initial design) AWG with channel spacing of 3.2nm cascaded with N waveguide grating filters. Figure1 (a) shows a schematic layout of an 8×8 AWG with 8 CWGFs. The AWG consists of three main parts, the multiple input and output waveguides, two free propagation regions (FPR), and a waveguide array with equal length difference between adjacent waveguides connecting two FPRs. The AWG is designed on the silicon-on-insulator (SOI) wafer with silica thickness of 2μm and top silicon thickness of 340nm. The widths of input and output waveguides are $W_{inout} = 450\text{nm}$, the adjacent waveguide spacing is $d_{inout} = 2.5\mu\text{m}$. The length of FPR, or the diameter of Rowland circle is $D = 130.55\mu\text{m}$. The number of arrayed waveguides is $N_{array} = 39$. In order to reduce fabrication induced phase errors, the arrayed waveguides are widen to 800nm. However, higher order modes will be excited in the 800nm wide rectangle waveguide, which will confuse the focusing on spot at the second FPR. Thus tapers from 800nm to 450nm are applied in all bending parts to remove the higher order modes excited in the 800nm wide waveguides. And another original design is to add waveguide grating filter with two phase shift regions at every output port of AWG. It provides a slightly wider spectrum than AWG, which could maintain the 3dB bandwidth in the last spectrum.

The CWGF consists of a coupling cavity with two equivalent phase shift regions, and the left and right lateral Bragg reflectors as shown in Figure1 (b). The Bragg reflector comprises alternative layers of narrow and wide waveguide segments, each with light path of λ/4 to fulfill the Bragg condition. The widths of the narrow and wide segments are 450nm and 800nm respectively. For the channel of central wavelength of 1550nm, the effective indices of narrow and wide waveguides are 2.5933 and 2.9342 respectively, which determines their lengths to be 149.42nm and 131.79nm. And thus the grating period is 281.21nm. The left and right reflectors both include 11 periods, while the central reflector between two phase shift regions has 21 periods. The length and width of phase shift region are 1.9769μm and 800nm.

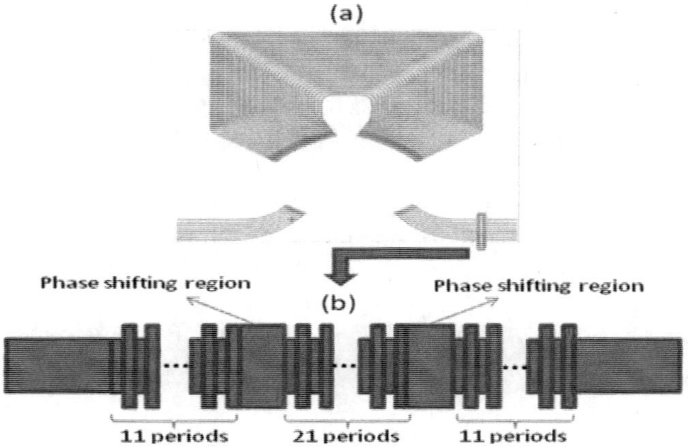

Figure1. Schematic layout of (a) AWG with CWGF and (b) the magnifying view of CWGF.

All the parameters above are for the channel of central wavelength of 1550nm. The principle and design method of phase shifting region in this filter has been described in reference [6] and the other channels' parameters can be got with the same method. Here we simply give the main parameters, which have been optimized.

3. Results of simulation and discussion.

Kirchhoff diffraction theory was used to calculate the transmission spectra of AWG. The calculated spectra for the central channel of AWG (with or without errors) and filter are shown in Figure2. The crosstalk between the adjacent output channels of AWG is about -30.7dB. When we consider practical fabrication errors of random waveguide width variation of 4nm, the crosstalk will be degraded by about -15dB. This is due to the emergence of large side lobes brought by phase errors in arrayed waveguides. In order to improve the crosstalk level, CWGFs at the outputs of AWG are introduced to cut off the side lobes, as shown in Figure1. The extinction ratio of the CWGF at two adjacent wavelengths of 1550nm and 1553.2nm is about -15.3dB. This high extinction ratio is utilized to suppress the side lobes of AWG.

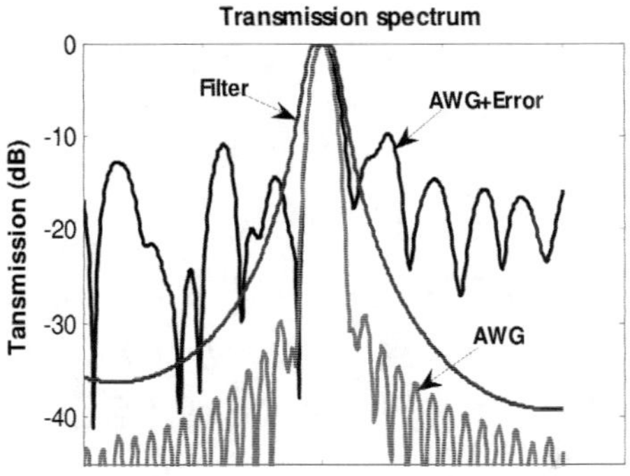

Figure2. Transmission spectrum of AWG (with or without errors) and CWGF for central wavelength of 1550nm.

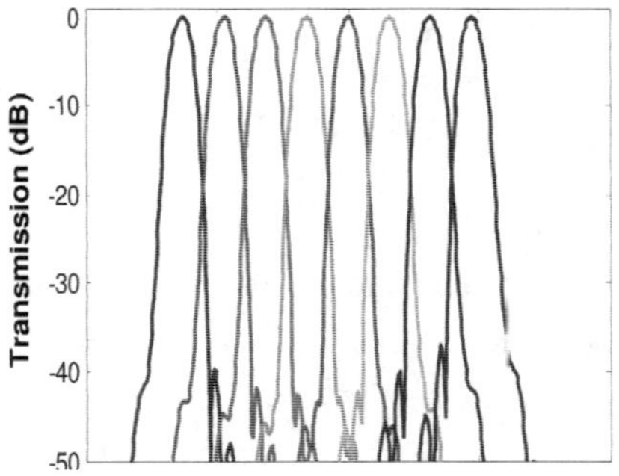

Figure3. Theoretical calculated transmission spectrum of AWG cascaded with grating filter.

Here we design eight different CWGFs. Every CWGF matches well in central wavelength with the corresponding output channel of AWG. Through tuning the central coupled cavity structure in the CWGF, we can get flat-top spectrum to maintain 3dB bandwidth. The transmission spectrum of AWG with CGFs was calculated as shown in Figure3. The crosstalk at central wavelength of 1550nm is -43.3dB, which matches well with the estimation of -30dB for AWG plus -12dB for filter from Figure2.

The above analysis just refers to the theoretical calculations. In order to estimate the impact of fabrication errors, we simplified the waveguide width error with normal distributed random error model. According to the real fabrication errors in our experiments, we set the maximum waveguide width variation to be 4nm. Then we can numerically simulate the random width errors to estimate the "experimental" transmission spectrum, as shown in Figure4. From the spectrum, we can find the side lobes brought by the fabrication errors is very high. The crosstalk level is degraded from -30dB to -15dB. But the crosstalk level can be improved and the side lobes can be suppressed efficiently with the help of CWGFs. And the crosstalk level can be recovered again to below -30dB. It should be pointed out that the fabrication errors are involved into not only AWG but also CWGFs. Furthermore, the length of the grating filter is only 18μm, which is much more compact compared with the traditional AWG cascaded with AWGs design in Ref [7].

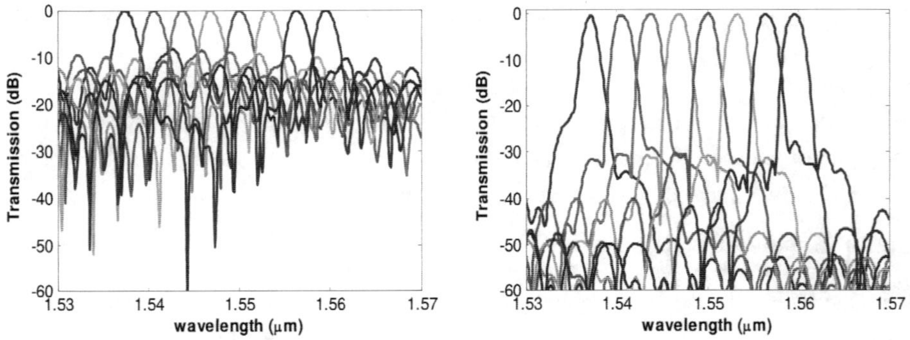

Figure4 (a) **Figure4 (b)**

Figure4. Transmission spectra of AWGs with maximum random waveguide width error of 4nm (a) single AWG (b) AWG with CWGF.

4. Conclusions

In summary, an ultra compact AWG with low crosstalk has been achieved by introducing CWGFs at the output of the AWG. The fabrication errors induced crosstalk in silicon nanowire AWG is efficiently suppressed and the crosstalk improvement is as high as 15dB.At the same time, the size of CWGF is only 18μm, which is very compact and easy to fabricate. Now we are experimentally fabricating this device and measuring the crosstalk performance.

Acknowledgements

The authors thank the members of the Silicon Photonics group in Wuhan National Laboratory for optoelectronics (WNLO). This work is financial supported by WNLO and also partly supported by the National Natural Science Foundation of China under the Grant No.60706013, Independent Creative Research Foundation of Huazhong University of Science and Technology under the Grant No.M2009026 and Creative Foundation of Wuhan National Laboratory for Optoelectronics under the Grant No.P080003.

References

[1] W.H. WONG and E.Y.B Pun, *Optics Express*, **Vol.13**, No.25, Dec 2005.

[2] K. Okamoto: *Fundamentals of Optical Waveguides* (Academic Press, New York, 2000).

[3] Bahram Jalali and Sasan Fathpour, *J. Lightwave Technol.* **24**, 4600-4615 (2006)

[4] K. Sasaki, F. Ohno, A. Motegi, and T. Baba, *Electron. Lett.* **41**, 801 (2005).

[5] W. Bogaerts, R. Baets, P. Dumon, V.Wiaux, S. Beckx, D. Taillaert, B. Luyssaert , J. V. Campenhout, P. Bienstman, and P. Jaenen, D. V Thourhout, *J.Lightwave Technol.* **23**.401-412(2005).

[6] Melloni, M. Floridi, F. Morichetti, and M. Martinelli, *J. Opt. Soc. A*, **20**, 273 (2003).

[7] Shin Kamei, Akimasa Kaneko, Motohaya Ishii, Tomohiro Shibata, Yasuyuki Inoue, *Member, IEEE,* and Yoshinori Hibino, *Member, IEEE, J. Lightwave Technol*. **23**, NO.5, MAY 2005.

Influence of wind speed on free space optical communication performance for Gaussian beam propagation through non Kolmogorov strong turbulence

Peng Deng, Xiuhua Yuan[1],Yanan Zeng, Ming Zhao, and Hanjun Luo

Wuhan National Laboratory for Optoelectronics, Huazhong University of Science & Technology, 1037 Luoyu Road, Wuhan, Hubei, P.R.China

**Email: yuanxh@mail.hust.edu.cn*

Abstract. In free-space optical communication links, atmospheric turbulence causes fluctuations in both the intensity and the phase of the received signal, affecting link performance. Most theoretical treatments have been described by Kolmogorov's power spectral density model through weak turbulence with constant wind speed. However, several experiments showed that Kolmogorov theory is sometimes incomplete to describe atmospheric turbulence properly, especially through the strong turbulence with variable wind speed, which is known to contribute significantly to the turbulence in the atmosphere. We present an optical turbulence model that incorporates into variable wind speed instead of constant value, a non-Kolmogorov power spectrum that uses a generalized exponent instead of constant standard exponent value 11/3, and a generalized amplitude factor instead of constant value 0.033. The free space optical communication performance for a Gaussian beam wave of scintillation index, mean signal-to-noise ratio<SNR>, and mean bit error rate <BER>, have been derived by extended Rytov theory in non-Kolmogorov strong turbulence. And then the influence of wind speed variations on free space optical communication performance has been analyzed under different atmospheric turbulence intensities. The results suggest that the effects of wind speed variation through non-Kolmogorov turbulence on communication performance are more severe in many situations and need to be taken into account in free space optical communication. It is anticipated that this work is helpful to the investigations of free space optical communication performance considering wind speed under severe weather condition in the strong atmospheric turbulence.

1. Introduction

As a technology providing high bandwidth wireless communication links over a distance of several kilometers using unlicensed optical wavelengths, free space optical (FSO) communication has increasingly attracted much attention in the past decade for a variety of applications, such as used for satellite-to-satellite cross-links, up-and-down links between space platforms and aircraft, ships, and other ground platforms, and among mobile or stationary terminals to solve the *last mile* problem through the atmosphere[1]. However, random spatial and temporal fluctuations in the refractive index of the atmosphere- referred to as *atmospheric turbulence* – give rise to random variations in the signal carrying laser beam intensity known as *atmospheric scintillation*[2-4]. Related turbulence induced

[1] Email: yuanxh@mail.hust.edu.cn.

effects including beam spreading and beam wander can result in power loss and increased bit error rates due to signal fading in communication channels that severely degrade the link performance [5-6].

Atmospheric turbulence occurs when turbulent winds in the atmosphere mix the always-present vertical moisture and temperature gradients caused by the Sun's heating of the Earth's surface. A quantitative measure of the intensity of atmospheric turbulence is the refractive index structure parameter, C_n^2, where averaged C_n^2 should be determined as a function of local differences in temperature, moisture, and wind speed at discrete points. Several atmospheric turbulence profile theoretical models for C_n^2 have been developed from experimental measurements made at a variety of locations [7-10]. One of the most popular models is the Hufnagel–Valley (HV) model[11] developed for inland sites and daytime viewing conditions. However, they are inadequate to represent the microphysical influences[12] on refractive index structure under severe environment conditions because they do not account for specific processes such as variations of near ground wind speed for local atmospheric turbulence strength. When the refractive index structure parameter C_n^2 increases due to the increasing fluctuation of wind speed and temperature, intensity of atmospheric turbulence would become stronger based on the Rytov variances $\sigma_R^2=1.23C_n^2k^{7/6}L^{11/6}$. This situation of strong turbulence would be very normal in the application of long-distance free space optical communication for ground-ground links along horizontal path or ground-satellite links along slant path under severe weather condition[6] [13]. Thus, It is necessary to develop a scintillation theory model that is valid from weak fluctuations through strong fluctuations, including the deep saturation regime.

To establish highly reliable optical communications links, quantitative estimates of various statistical quantities that are associated with atmospheric turbulence-induced scintillations are necessary. To date, all calculations relating to optical wave propagation in the atmosphere are based on the Kolmogorov model of the refractive index fluctuation spectrum. However, the significant deviations from the Kolmogorov model have been obtained in optical wave propagation through certain portions of the atmosphere, which is supported by numerous experimental evidences[14-15] and some results of theoretical investigation [16-19]. This has prompted the investigation of optical wave propagation through the atmospheric turbulence exhibiting non-Kolmogorov statistics. A.Zilberman suggested a model of turbulence spectra (Kolmogorov and non-Kolmogorov) changing with altitude on the basis of obtained experimental and theoretical data for turbulence profile in the troposphere and lower stratosphere[20]. Stribling et al defined the turbulence as non-Kolmogorov turbulence and presented an analysis of optical propagation in non-Kolmogorov atmospheric turbulence [21]. Gurvich and Belen'kii introduced a model for the power spectrum of stratospheric non-Kolmogorov turbulence and investigated the stratospheric turbulence on the scintillation and the degradation of star image[22]. Recently, Tosellia [23]et al presented a non-Kolmogorov theoretical power spectrum model and estimated free space optical system performance for laser beam propagating horizontally through non-Kolmogorov weak atmospheric turbulence. So far all of theoretical investigations for optical propagation in non-Kolmogorov atmospheric turbulence have focused on the unbounded plane or spherical wave models under the weak turbulence conditions. However, in many applications such as inter-satellite laser communication links[6, 24-25], the plane and spherical-wave approximations do not suffice to describe the propagation properties of optical wave, and the optical performance suffers from the strong atmospheric turbulence [19, 26]. Therefore, it is very necessary to extend the investigation of the plane and spherical-wave propagation in non-Kolmogorov weak atmospheric turbulence to Gaussian-beam wave propagation in non-Kolmogorov strong atmospheric turbulence.

In this paper, a non-Kolmogorov theoretical power spectrum is considered, which has a generalized power law that takes all the values ranging from 3 to 4. As the power law α is set to the standard Kolmogorov value 11/3, the spectrum reduces to the conventional Kolmogorov one. Based on this spectrum and the refractive index structure parameter C_n^2 model[12] within the surface boundary layer, taking account into the fluctuations of wind speed, free space optical communication performance for a Gaussian beam wave of scintillation index, mean signal-to-noise ratio<SNR>, and mean bit error rate <BER>, have been derived by extended Rytov theory in non-Kolmogorov strong turbulence. Then the influence of wind speed on free space optical communication performance has been analyzed.

2. Non-Kolmogorov Spectrum

The theoretical power spectrum model that describes non-Kolmogorov optical turbulence [23, 27] is considered, which obeys a more general power law and in which the power-law exponents can take all the values ranging from 3 to 4,

$$\Phi_n(\kappa,\alpha) = A(\alpha)\tilde{C}_n^2\kappa^{-\alpha}, 2\pi/L_0 \le \kappa \le 2\pi/l_0, 3<\alpha<4 \tag{1}$$

where κ is the magnitude of three dimensional wave number vector (in units of rad/m), α is the spectral power-law exponent, $\tilde{C}_n^2 = \beta C_n^2(v)$ is a generalized refractive-index structure parameter that describes the strength of the turbulence along the path, l_0 and L_0 denote the inner and outer scales of turbulence, respectively, and $A(\alpha)$ is a function defined by

$$A(\alpha) = \frac{1}{4\pi^2}\Gamma(\alpha-1)\cos(\frac{\pi\alpha}{2}) \tag{2}$$

where the symbol $\Gamma(x)$ represents the gamma function. When the power law α is equal to 11/3, $A(11/3)=0.033$, $\tilde{C}_n^2 = C_n^2(v)$, and the spectrum reduces to the conventional Kolmogorov spectrum

$$\Phi_n(\kappa,v) = 0.33C_n^2(v)\kappa^{-11/3} \tag{3}$$

where $C_n^2(v)$ represents the refractive-index structure parameter as function of wind speed v.

Based on local micrometeorological information derived through application of Monin–Obukhov turbulence theory[9, 12], a useful expression for the refractive index structure parameter as function of wind speed v can be given as

$$C_n^2(v) = \frac{bK_h(v)}{\varepsilon(v)^{1/3}}(\frac{\partial n}{\partial h})^2 \tag{4}$$

where b is a constant, $K_h(v)$ is the turbulent exchange coefficient for heat diffusion[9], $\varepsilon(v)$ is the energy dissipation rate due to wind shear and buoyancy[12], and $\partial n/\partial h$ is the vertical gradient of the index of refraction (n). For the hypothesis of universal equilibrium under steady-state conditions, the energy dissipation rate equals the sum of the rate of energy production due to wind shear M and buoyancy B

$$\varepsilon = M + B = K_m\left|\frac{\partial \bar{v}}{\partial h}\right|^2 + K_h(\frac{g}{\theta})(-\frac{\partial\theta}{\partial h}) \tag{5}$$

where K_m is the turbulent exchange coefficient for momentum, and $\bar{v}$ is average wind speed. Here g is acceleration due to the Earth's gravity (9.8m/s), K_h is the turbulent exchange coefficient for heat, and θ is mean potential temperature $\theta=T+0.0098h$.The flux profile relationships for the vertical gradients of average wind velocity and mean potential temperature have similar forms, and are given by

$$\frac{\partial\bar{v}}{\partial h} = \frac{u_*}{k_v h}\Phi_m(\frac{h}{L}), \frac{\partial\theta}{\partial h} = \frac{T_*}{k_v h}\Phi_h(\frac{h}{L}) \tag{6}$$

where k_v is von Karman's constant taken to be 0.4, $u*$ is friction velocity, T^* is characteristic temperature, $\Phi_m(h/L)$ is dimensionless wind shear, $\Phi_h(h/L)$ is dimensionless temperature gradient, L is the Monin-Obukhov length that is fundamental in describing the onset of turbulence and the vertical structure of turbulence-induced flux profiles in the Earth's surface boundary layer[9, 12].

With knowledge of the friction velocity u^*, it is possible to estimate the turbulent exchange coefficient for heat K_h and the turbulent exchange coefficient for momentum K_m.

$$K_h = k_v u_* h/\Phi_h(\frac{h}{L}),\quad K_m = k_v u_* h/\Phi_m(\frac{h}{L}) \tag{7}$$

By ignoring small contribution to the total differential from fluctuations in atmospheric pressure, optical turbulence-induced vertical gradient of the index of refraction can be approximated by

$$\frac{\partial n}{\partial h} \cong \frac{\partial n}{\partial\theta}\frac{\partial\theta}{\partial h} = \left(-77.6\times10^{-6}Pa\right)\frac{T^*}{k_v h T^2}\Phi_h(\frac{h}{L}) \tag{8}$$

3. Scintillation Index

The scintillation index is one of the most important parameters that is usually used to characterize laser beam propagation through atmospheric turbulence. For the propagation of Gaussian beam wave, the scintillation index is usually expressed as the sum of the radial component and the longitudinal component[2],

$$\sigma_I^2(\mathbf{r},L)=\sigma_{I,r}^2(\mathbf{r},L)+\sigma_{I,l}^2(L) \tag{9}$$

The radial component physically denotes the off-axis contribution to the scintillation index and vanishes at the beam center line (ρ=0) or as Λ=0 (corresponding to an infinite wave such as a plane or spherical wave), whereas the longitudinal component is constant through the beam cross section in any transverse plane. To describe the beam characteristics at the input plane and output plane of the transceivers, we use a sets of non-dimensional beam parameters. The curvature parameter Θ and the Fresnel ratio Λ at the output plane are defined by $\Theta=1-L/R$ and $\Lambda=\lambda L/\pi W^2$,respectively.Here, W and R denote the radius of the beam size and the radius of curvature of phase front at the receiver in the distance of L from the transmitter.

Our analysis evolved from the extend Rytov theory[2] by taking into account the role of decreasing spatial coherence of the optical wave as it propagates through strong atmospheric turbulence. Based on the extended Rytov theory, the irradiance of the field can be expressed as $I=X.Y$, where X represents large scale (refraction properties) and Y denotes small scale (diffraction properties) fluctuations. Assumed that both of the fluctuations are statistically independent from each other. the implied scintillation index takes the following form[2]

$$\begin{aligned}\sigma_I^2(\alpha)&=\langle X\rangle\langle Y\rangle-1\\&=\sigma_X^2(\alpha)+\sigma_Y^2(\alpha)+\sigma_X^2(\alpha)\sigma_Y^2(\alpha)\end{aligned} \tag{10}$$

Based on the relationship between the log-amplitude variance and the scintillation index, the longitudinal component of the total scintillation index then takes the form

$$\sigma_{I,l}^2(\alpha)=\exp(\sigma_{\ln X,l}^2(\alpha)+\sigma_{\ln Y,l}^2(\alpha))-1 \tag{11}$$

Based on the Non-Kolmogorov spectrum model and the one-dimensional quantities $\xi=1-z/L;\eta=L\kappa^2/k$, the large-scale and small-scale log-irradiance variance for a Gaussian beam wave are defined by

$$\begin{aligned}\sigma_{\ln X,l}^2(L,\alpha)=16\pi^2k^2L\int_0^1\int_0^\infty \kappa\Phi_n(\kappa,\alpha)G_X(\kappa,\alpha)\sin^2\left[\frac{L\kappa^2}{2k}\xi(1-\bar{\Theta}\xi)\right]\\ \times\exp(-\frac{\Lambda L\kappa^2\xi^2}{k})d\kappa d\xi\end{aligned} \tag{12}$$

$$\begin{aligned}\sigma_{\ln Y,l}^2(L,\alpha)=8\pi^2k^2L\int_0^1\int_0^\infty \kappa\Phi_n(\kappa,\alpha)G_Y(\kappa,\alpha)\exp(-\frac{\Lambda L\kappa^2\xi^2}{k})\\ \times\{1-\cos[\frac{L\kappa^2}{k}\xi(1-\bar{\Theta}\xi)]\}d\kappa d\xi\end{aligned} \tag{13}$$

The large-scale and small-scale filter function[2] are defined, respectively, by

$$G_X(\kappa,\alpha)=\exp(-\frac{\kappa^2}{\kappa_X^2}) \tag{14}$$

$$G_Y(\kappa,\kappa_Y)=\frac{\kappa^\alpha}{\left(\kappa^2+\kappa_Y^2\right)^{\alpha/2}}\exp[\frac{\Lambda L\kappa^2(z/L)^2}{k}] \tag{15}$$

The quantities κ_X and κ_Y represent the low-pass and high-pass spatial frequency cutoffs that related to the correlation width and scattering disk of the propagating optical wave.

Therefore, using the non-Kolmogorov power spectrum in Eq.(1) and following the same procedure as discussed in Ref. [2, 23] for the standard Kolmogorov spectrum, the scintillation index for a Gaussian beam wave propagating in non-Kolmogorov strong atmospheric turbulence are analyzed.

In strong fluctuation regimes where inner scale and outer scale effects are negligible, the longitudinal component of scintillation index is derived by,

$$\sigma_{I,G}^2(L,\alpha)=\exp\left(\frac{0.49\tilde{\sigma}_B^2}{(1+f_X(\alpha,\bar{\Theta})\tilde{\sigma}_R^{\frac{4}{\alpha-2}})^{3-\frac{\alpha}{2}}}+\frac{0.51\tilde{\sigma}_B^2}{(1+f_Y(\alpha)\tilde{\sigma}_B^{\frac{4}{\alpha-2}})^{\frac{\alpha}{2}-1}}\right)-1,\quad 0\le\sigma_R^2\le\infty \quad (16)$$

where $\tilde{\sigma}_B^2$ denotes the longitudinal component of the scintillation index for Gaussian beam wave under non-Kolmogorov weak turbulence can be expressed as[16]

$$\tilde{\sigma}_B^2=\frac{1}{\alpha-1}A(\alpha)\tilde{C}_n^2\pi^2k^{3-\frac{\alpha}{2}}L^{\alpha/2}\Gamma\left(\frac{2-\alpha}{2}\right)\{\Lambda^{\frac{\alpha}{2}-1}-\mathrm{Re}[\frac{2\alpha-2}{\alpha}i^{\frac{\alpha-2}{\alpha}}\,{}_2F_1(\frac{2-\alpha}{2},\frac{\alpha}{2};\frac{2+\alpha}{2};\bar{\Theta}+i\Lambda)]\} \quad (17)$$

where $\tilde{\sigma}_R^2$ is a non-Kolmogorov Rytov variance for plane wave[27] in weak turbulence defined by

$$\tilde{\sigma}_R^2=R(\alpha)\tilde{\sigma}_{Rytov}^2(\alpha)=-6.5\pi^2A(\alpha)\Gamma(1-\frac{\alpha}{2})\frac{1}{\alpha}\sin(\alpha\frac{\pi}{4})\times1.23\tilde{C}_n^2k^{3-\frac{\alpha}{2}}L^{\frac{\alpha}{2}} \quad (18)$$

where

$$\tilde{\sigma}_{Rytov}^2(\alpha)=1.23\tilde{C}_n^2k^{3-\frac{\alpha}{2}}L^{\frac{\alpha}{2}} \quad (19)$$

$$R(\alpha)=-6.5\pi^2A(\alpha)\Gamma(1-\frac{\alpha}{2})\frac{1}{\alpha}\sin(\alpha\frac{\pi}{4}) \quad (20)$$

Note that for Kolmogorov case *α=11/3, R(11/3)=1.*

The large scale and small scale parameters $f_X(\alpha)$ and $f_Y(\alpha)$ can be expressed by

$$f_x(\alpha,\bar{\Theta})=\left(\frac{V(\alpha)Z(\alpha,\bar{\Theta})}{0.98}\times\frac{\tilde{\sigma}_R^2}{\tilde{\sigma}_B^2}\right)^{\frac{2}{\alpha-6}} \quad (21)$$

$$f_Y(\alpha)=\left(\frac{\ln 2}{0.51}\right)^{\frac{2}{\alpha-2}} \quad (22)$$

where

$$Z(\alpha)=\int_0^1\frac{\xi^{\alpha-4}(1-\bar{\Theta}\xi)^2}{[(1-\bar{\Theta}\xi)^{\alpha-2}-(1-\bar{\Theta})^{\alpha-1}]^{\frac{6-\alpha}{\alpha-2}}}d\xi \quad (23)$$

$$B(\alpha)=-\frac{\pi^2}{1.23}4^{2-\frac{\alpha}{2}}\frac{\Gamma(1-\alpha/2)}{\Gamma(\alpha/2)}\frac{A(\alpha)}{R(\alpha)}\frac{1}{\bar{\Theta}(\alpha-1)} \quad (24)$$

$$V(\alpha)=\frac{8\pi^2}{1.23R(\alpha)(\alpha-2)}\Gamma\left(\frac{6-\alpha}{\alpha-2}\right)B(\alpha)^{\frac{\alpha-6}{\alpha-2}} \quad (25)$$

For the radial component, our analysis results in

$$\sigma_{I,r}^2(\mathbf{r},L,\alpha)_{untracked}=4.42\tilde{\sigma}_R^2\Lambda_e^{\frac{\alpha}{2}-1}\left(\frac{\sigma_{pe}}{W_{LT}}\right)^2+4.42\tilde{\sigma}_R^2\Lambda_e^{\frac{\alpha}{2}-1}\left(\frac{r-\sigma_{pe}}{W_{LT}}\right)^2;\quad \sigma_{pe}\le r\le W. \quad (26)$$

where σ_{pe} denotes an effective pointing error in the presence of beam wander effects[6], $W_{LT}^2(\alpha)$ is long term spot size caused by large scale induced beam wander defined by

$$W_{LT}^2(\alpha)=W^2(1+T(\alpha)),T(\alpha)=-16A(\alpha)\frac{1}{\alpha-1}\Gamma(1-\frac{\alpha}{2})\Lambda^{\frac{\alpha}{2}-1}\tilde{\sigma}_R^2(\alpha) \quad (27)$$

The long term spot size[2] is characterized by the effective beam parameter $\Lambda_e(\alpha) = 2L / kW_{LT}(\alpha)$

Therefore, the total scintillation index for Gaussian beam wave in non Kolmogorov strong turbulence and in absence of inner scale and outer scale effects is given by

$$\sigma_I^2(\mathbf{r}, L, \alpha)_{untracked} = \exp\left(\frac{0.49\tilde{\sigma}_B^2}{(1+f_X(\alpha,\overline{\Theta})\tilde{\sigma}_R^{\frac{4}{\alpha-2}})^{3-\frac{\alpha}{2}}} + \frac{0.51\tilde{\sigma}_B^2}{(1+f_Y(\alpha)\tilde{\sigma}_B^{\frac{4}{\alpha-2}})^{\frac{\alpha}{2}-1}}\right) - 1$$
$$+4.42\tilde{\sigma}_R^2\Lambda_e^{\frac{\alpha}{2}-1}\left(\frac{\sigma_{pe}}{W_{LT}}\right)^2 + 4.42\tilde{\sigma}_R^2\Lambda_e^{\frac{\alpha}{2}-1}\left(\frac{r-\sigma_{pe}}{W_{LT}}\right)^2, 3 < \alpha < 4, 0 \le \sigma_R^2 \le \infty, \sigma_{pe} \le r \le W. \tag{28}$$

Using scintillation index model, the comparison of scintillation index as function of wind speed for simulated model with that for the Hufnagel-Valley model[11] and BKB model[7] is shown in figure Figure **1**, taking *L = 1 km; T=300 K; λ= 1.55 nm; h= 4 m*. We deduce from figure Figure **1** that for wind speed lower than 11 m/s, excluding wind speed values close to 3, there is an increase of scintillation for the simulated Gaussian wave model, but for H-V model and BKB model it is constant and reduced, respectively. For wind speed greater than 11 m/s, the scintillation indexes for several models approach to the similar values. The physical interpretation of difference is that the power spectrum contains fewer eddies of high wave numbers; therefore, scintillation effects are reduced.

In addition, the longitude component and radial component of scintillation index are plotted as function of Rytov index in figure Figure **2**. We deduce from figure 2 that, for Rytov index lower than 1 under weak turbulence, scintillation index remarkable increases with respect to Rytov index; instead for Rytov index values greater than 1 in moderate to strong turbulence, scintillation index decreases slightly with respect to the Rytov index. That is the saturation effect of the scintillation index under strong irradiance fluctuations.

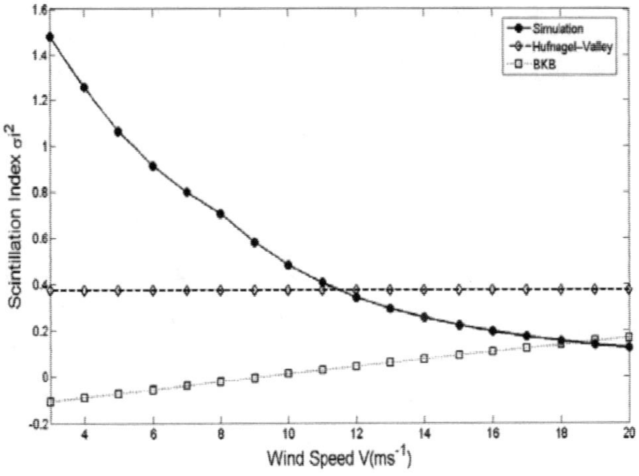

Figure 1. The scintillation index as function of wind speeds for various turbulence models.

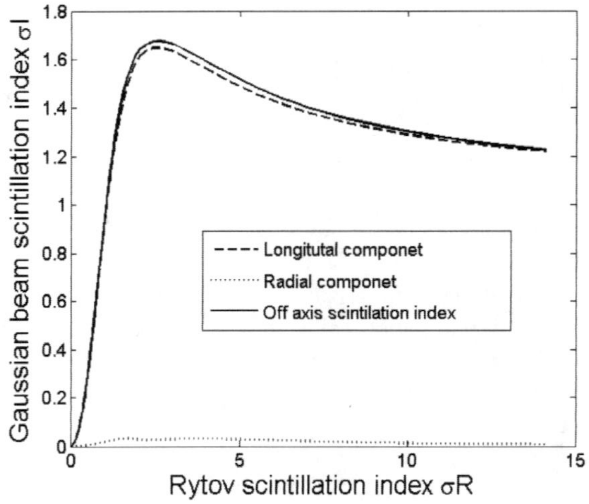

Figure 2 .The longitude component and radial component of scintillation index as function of Rytov scintillation index

4. Performance analysis of FSO system in non-Kolmogorov strong atmospheric turbulence

The performance of FSO system can be seriously deteriorated by turbulence –induced scintillation resulting from beam propagation through the atmospheric turbulence. Specifically, the moderate-strong scintillation can lead to power loss at the receiver and eventually to fading of the received signal below a prescribed threshold. The reliability of a FSO system operating in such an environment can be deduced from the theoretical model for the probability density function (PDF) of the randomly fading irradiance signal. From knowledge of the PDF model, we can calculate the mean signal to noise ratio (SNR) and the bit-error-rate (BER) of FSO system in non-Kolmogorov strong turbulence.

The probability density function (PDF) used under strong irradiance fluctuations for the case of a point receiver is the Gamma-Gamma distribution model. The Gamma-Gamma PDF[2, 26] is given by

$$p_I(I)=\frac{2(\alpha\beta)^{(\alpha+\beta)/2}}{\Gamma(\alpha)\Gamma(\beta)}\left(\frac{I}{\langle I(0,L)\rangle}\right)^{(\alpha+\beta)/2}K_{\alpha-\beta}\left(2\sqrt{\frac{\alpha\beta I}{\langle I(0,L)\rangle}}\right), I>0 \tag{29}$$

where we are taking the general case of mean irradiance $\langle I(0,L)\rangle \neq 1$ and $K_v(x)$ is a modified Bessel function. α and β are positive parameters directly related to the large-scale and small-scale scintillation index of the Gaussian wave according to

$$\alpha=\frac{1}{\sigma_X^2}=\frac{1}{\exp\left(\frac{0.49\tilde{\sigma}_B^2}{\left(1+f_X(\alpha,\overline{\Theta})\tilde{\sigma}_R^{\frac{4}{\alpha-2}}\right)^{3-\frac{\alpha}{2}}}\right)-1},\quad \beta=\frac{1}{\sigma_Y^2}=\frac{1}{\mathrm{e}\exp\left(\frac{0.51\tilde{\sigma}_B^2}{\left(1+f_Y(\alpha)\tilde{\sigma}_B^{\frac{4}{\alpha-2}}\right)^{\frac{\alpha}{2}-1}}\right)-1} \tag{30}$$

In the presence of atmospheric turbulence, the received signal exhibits additional power losses and random fluctuations. We examine the mean signal-to-noise ratio in the presence of atmospheric turbulence using a non- Kolmogorov power spectrum. Based on Refs.[2, 27], the mean signal-to-noise ratio <SNR> at the output of the detector in the case of a shot-noise limited system assumes the form

$$\langle SNR\rangle=\frac{SNR_0}{\sqrt{1+\sigma_I^2(\mathbf{r},L,\alpha)SNR_0^{\ 2}}} \tag{31}$$

where SNR_0 is the signal-to-noise ratio in the absence of turbulence.

Given the PDF model for irradiance fluctuation and the mean signal-to-noise ratio through atmospheric turbulence, the probability of error is considered a conditional probability that must be averaged over the PDF of random signal to determine the unconditional mean BER. In terms of a normalized signal with unit mean, this leads to the mean BER expression[2, 27]

$$\Pr(E) = \langle BER \rangle = \frac{1}{2}\int_0^{\infty} p_I(u) erfc(\frac{\langle SNR \rangle u}{2\sqrt{2}}) du \tag{32}$$

where $p_I(u)$ is the gamma-gamma distribution with unit mean and the threshold $i_T = 0.5s$.

Using the scintillation index for the Gaussian beam wave model, the probability density function under strong atmospheric turbulence are plotted as function of log intensity for several wind speeds in figure Figure **3**, taking *L = 1 km; T=300 K; λ= 1.55 nm; h= 4 m.* It is clear now that wind speed variation has a serious impact on the probability density function for Gaussian wave; in particular, for log intensity lower than zero, the probability density function increases with an increase in log intensity; while for log intensity greater than zero, it decreases rapidly with the increase of log intensity. Also for this diagram, there is a maximum point where the curves change their slope, because the scintillation begins to decrease to zero. With the increase in wind speed, the slope of the curves becomes steeper.

The mean signal-to-noise ratio<SNR> is plotted in decibel units as a function of signal-to-noise ratio without turbulence SNR_0 for several wind speed values in figure 8, using the Gaussian wave model for strong scintillation. We take the following parameters: *L = 1 km; T=300 K; λ= 1.55 nm; h= 4 m; α=3.3* for non-Kolmogorov spectrum, compared with Kolmogorov spectrum *α=11/3, v=11 m/s.* As shown in figure Figure **4**, illustrates the impact of the wind speed variation on the <SNR> performance. With the increase in wind speed values from 3 m/s to 20 m/s, excluding the values close to 3, there is a translation of the curves toward greater values, or, in other words, there is a gain on the system performance. That because the refractive index parameters reach to very low values with an increase in wind speed and consequently the scintillation reported before approaches smaller values.

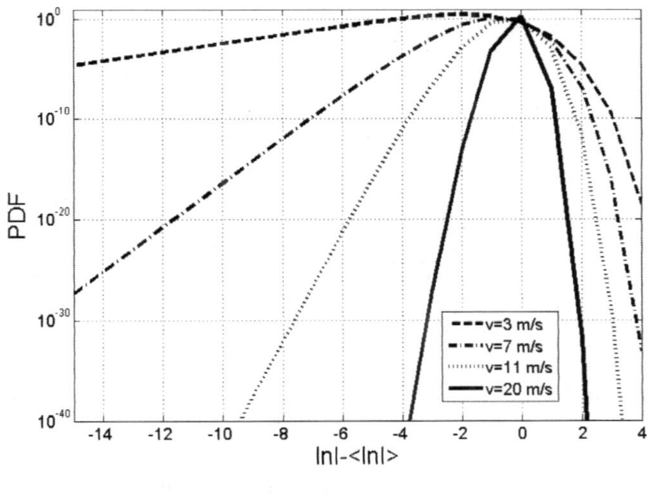

Figure 3. The probability density functions under strong atmospheric turbulence as function of log intensity for several wind speeds.

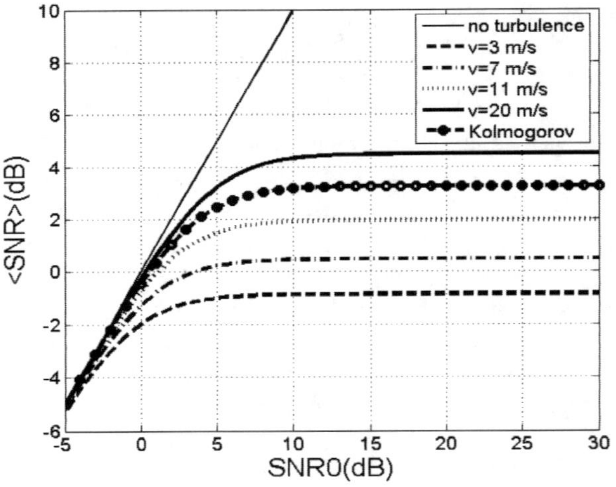

Figure 4. Mean Signal to Noise Ratio as a function of Signal to Noise Ratio without turbulence for several wind speeds, using Gaussian beam wave model for strong scintillation index.

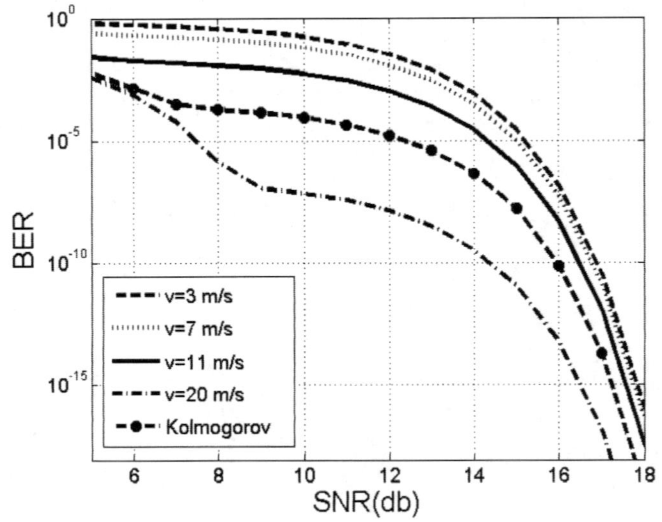

Figure 5 .Bit error rate as function of SNR for various wind speeds, using the Gaussian beam wave model through non-Kolmogorov strong turbulence

The mean bit error rate as a function of <SNR> for several wind speed values using the Gaussian beam wave model for strong scintillation in figure Figure . We take the same parameters, $L = 1\ km$; $T=300\ K$; $\lambda = 1.55\ nm$; $h= 4\ m$; $\alpha=3.3$ for non-Kolmogorov spectrum, compared with Kolmogorov spectrum $\alpha=11/3$, $v=11\ m/s$. The plot in figure Figure shows the impact of the wind speed variation on <BER> performance. Also in this analysis, when wind speed increases from 3 m/s to 20 m/s, excluding alpha values close to 3, there is a translation of the curves toward smaller values. In other words, there is an improvement on the system performance. It results from that the refractive index parameters reach to very low values with an increase in wind speed and consequently the scintillation reported before approaches smaller values. To the best of our knowledge, the physical mechanism of transition function of the turbulence spectral exponent α form between wind speeds v in turbulent atmospheric layers need be investigated.

5. Conclusion

The influence of wind speed on free space optical communication performance for a Gaussian beam wave of scintillation index, mean signal-to-noise ratio<SNR>, and mean bit error rate <BER>, have been analyzed by extended Rytov theory in non-Kolmogorov strong turbulence. It has been shown in strong turbulence, for a horizontal link, the scintillation index as variations depending on wind speed lead to results somewhat different than obtained with the standard value of Kolmogorov with constant wind speed. For wind speed lower than 11 m/s, excluding wind speed values close to 3, there is an increase of scintillation for the simulated Gaussian wave model. For wind speed greater than 11 m/s, the scintillation indexes for several models approach to the smaller values. With the increase in wind speed values from 3 m/s to 20 m/s, excluding the values close to 3, there is an improvement on the system performance. Specifically, the low wind speed gets more impact from a power law spectrum variation and it leads the most contribution on the scintillation index variation. Although our final expressions for the scintillation have been obtained by extended Rytov theory, which is necessary to adopt in strong turbulence conditions, they reduce to the proper results also in weak turbulence.

Acknowledgments

This research was financially supported by National High Technology Research and Development Program of China (No. 2008AA01Z207). The authors are grateful for a grant from National High Technology Research and Development Program of China.

References

[1] V. W. S. Chan, "Free-Space Optical Communications," JOURNAL OF LIGHTWAVE TECHNOLOGY **24**, 4750-4762 (2006).

[2] L. C. Andrews, and R. L. Phillips, *Laser Beam Propagation through Random Media* (SPIE Optical Engineering Press, Bellingham, 2005).

[3] A. Ishimaru, *Wave Propagation and Scattering in Random Media* (Academic, 1978).

[4] V. I. Tatarski, *Wave Propagation in a Turbulent Medium* (Nauka, 1967).

[5] J. C. Ricklin, S. M. Hammel, F. D. Eaton, and S. L.Lachinova, *Atmospheric channel effects on free-space laser communication* (J. Opt. Fiber. Commun. Rep., 2006).

[6] H. Guo, B. Luo, Y. Ren, S. Zhao, and A. Dang, "Influence of beam wander on uplink of ground-to-satellite laser communication and optimization for transmitter beam radius," Opt. Lett. **35**, 1977-1979 (2010).

[7] S. Bendersky, N. S. Kopeika, and N. Blaunstein, "Atmospheric Optical Turbulence Over Land in Middle East Coastal Environments: Prediction Modeling and Measurements," Appl. Opt. **43**, 4070-4079 (2004).

[8] T. T. Leclerc, R. L. Phillips, L. C. Andrews, D. T. Wayne, P. Sauer, and R. Crabbs, "Prediction of the ground-level refractive index structure parameter from the measurement of atmospheric conditions," L. M. W. Thomas, and E. J. Spillar, eds. (SPIE, Orlando, Florida, USA, 2010), pp. 76850A-76858.

[9] A. Tunick, "Modeling microphysical influences on optical turbulence in complex areas," Meteorology and Atmospheric Physics **96**, 293-304 (2007).

[10] A. Tunick, "Optical turbulence parameters characterized via optical measurements over a 2.33 km free-space laser path," Opt. Express **19**, 14645-14654 (2008).

[11] R. E. Hufnagel, "Variations of atmospheric turbulence," in *Digest of topical meeting on optical propagation through turbulence*(Optical Society of America, Washington, D.C., 1974), p. WA1.

[12] A. S. Monin, and A. M. Obukhov, "Basic law of turbulent mixing near the graund," Trans. Geophys. Inst. Akad. Nauk USSR **24**, 163-187 (1954).

[13] T. Weyrauch, and M. Vorontsov, "Atmospheric compensation with a speckle beacon in strong scintillation conditions: directed energy and laser communication applications," Appl. Opt. **44**, 6388-6401 (2005).

[14] M.S.Belen'kii, S.J.Karis, J.M.Brown, and R.Q.Fugate, "Experimental evidence of the effects of non-Kolmogorov turbulence and anisotropy of turbulence," Proc. SPIE **3749**, 50-51 (1999).
[15] A. Zilberman, E. Golbraikh, N. S. Kopeika, A. Virtser, I. Kupershmidt, and Y. Shtemler, "Lidar study of aerosol turbulence characteristics in the troposphere: Kolmogorov and non-Kolmogorov turbulence," Atmospheric Research **88**, 66-77 (2008).
[16] L. Tan, W. Du, J. Ma, S. Yu, and Q. Han, "Log-amplitude variance for a Gaussian-beam wave propagating through non-Kolmogorov turbulence," Opt. Express **18**, 451-462 (2010).
[17] W. Du, L. Tan, J. Ma, and Y. Jiang, "Temporal-frequency spectra for optical wave propagating through non-Kolmogorov turbulence," Opt. Express **18**, 5763-5775 (2010).
[18] A. Zilberman, E. Golbraikh, and N. S. Kopeika, "Some limitations on optical communication reliability through Kolmogorov and non-Kolmogorov turbulence," Optics Communications **283**, 1229-1235 (2010).
[19] A. Garcia-Zambrana, C. Castillo-Vazquez, and B. Castillo-Vazquez, "Space-time trellis coding with transmit laser selection for FSO links over strong atmospheric turbulence channels," OPTICS EXPRESS **18**, 5356-5366 (2010).
[20] A. Zilberman, E. Golbraikh, and N. S. Kopeika, "Propagation of electromagnetic waves in Kolmogorov and non-Kolmogorov atmospheric turbulence: three-layer altitude model," Appl. Opt. **47**, 6385-6391 (2008).
[21] B. E. Stribling, B. M. Welsh, and M. C. Roggemann, "Optical propagation in non-Kolmogorov atmospheric turbulence," Proc. SPIE **2471**, 181-196 (1995).
[22] A. S. Gurvich, and M.S.Belen'kii, "Influnce of stratospheric turbulence on infrared imaging," J. Opt. Soc. Am. A **12**, 2517-2522 (1995).
[23] I. Toselli, L. C. Andrews, R. L. Phillips, and V. Ferrero, "Free Space Optical System Performance for a Gaussian Beam Propagating Through Non-Kolmogorov Weak Turbulence," Ieee T Antenn Propag **57**, 1783-1788 (2009).
[24] L. Tan, Y. Yang, J. Ma, and J. Yu, "Pointing and tracking errors due to localized deformation in inter-satellite laser communication links," OPTICS EXPRESS **16**, 13372-13380 (2008).
[25] J. Ma, Y. Jiang, L. Tan, S. Yu, and W. Du, "Influence of beam wander on bit-error rate in a ground-to-satellite laser uplink communication system," Opt. Lett. **33**, 2611-2613 (2008).
[26] W. Lim, C. Yun, and K. Kim, "BER performance analysis of radio over free-space optical systems considering laser phase noise under Gamma-Gamma turbulence channels," Opt. Express **17**, 4479-4484 (2009).
[27] I. Toselli, L. C. Andrews, R. L. Phillips, and V. Ferrero, "Free-space optical system performance for laser beam propagation through non-Kolmogorov turbulence," Optical Engineering **47**, 026003-026009 (2008).

Multi-hole Optical Fiber Surface Plasmon Resonance Sensor

Chunying Guan [1] **, Yang Wang, Libo Yuan**
College of Science, Harbin Engineering University, Harbin 150001, China

E-mail: cyguan@163.com

Abstract A microstructured-fiber containing six large air holes is proposed to construct the surface plasmon resonance (SPR) sensor. The finite element method is used to analyze characteristics of the surface plasmon resonance sensor. The effects of the thickness of metal films, pitch between air holes, diameter of air hole, and refractive index of liquid on the resonance wavelength are elucidated. The results show that the resonance wavelength is sensitive to the thickness of metal film and refractive index of liquid, while the resonance wavelength doesn't change basically when the pitch between air holes and diameter of air holes vary. The proposed surface plasmon resonance sensor exhibits high sensitivity up to 10^{-4}.

1. Introduction

Surface plasmon resonance (SPR) biosensor was driven by modern physics and biology. Optical fiber SPR biosensors are undoubtedly required for the future development of miniaturized and compact SPR sensor systems. Fiber SPR sensors have some advantages over the conventional bulk SPR sensors, such as small size, light weight, remote sensing, multiplexing possibilities, and high capacity matched with the optical fiber communication system. As a result, optical fiber SPR sensors have generated considerable interests in the field of SPR sensors [1-4]. Optical fibers led to improve analysis method and signal measurement in long distance real-time multichannel. The detection and identification of biological analytes and biophysical analysis of biomolecular interactions are obtained by measuring resonant wavelength shift or change of light intensity from the output at a fixed wavelength or half-width of the power spectrum. Recently, wide applications of microstructured fiber in optical sensor and communication have been rapidly propelled by its mature fabrication process [5-7]. In 2006, Sazio [8] proposed a solution to fabricate homogeneous metallic films on inner surface of air holes in microstructured fibers using high-pressure microfluidic chemical deposition technique, which made it possible to design SPR sensor in air-hole microstructured fiber. Subsequently, many theoretical and experimental results were reported in the literatures [9-12]. M. Hautakorpi proposed a SPR sensor configuration based on coating the holes of a three-hole microstructured optical fiber, and added a low refractive index dielectric layer around the fiber core to confine most light within the fiber core, and obtained 10^{-4}/RIU sensitivity for aqueous analytes [11]. A. Hassani proposed a photonic crystal fiber (PCF) SPR sensor, in which phase matching between a surface plamon mode and a core mode can be enforced by introducing air-filled microstructure into the fiber core.

Simple multi-hole optical fibers were proposed to construct SPR sensors in the present paper. The effects of fiber parameters on the properties of SPR sensor were analyzed using the finite element method.

[1] To whom any correspondence should be addressed.

Published under licence by IOP Publishing Ltd

2. Sensor structure

It is well-known that only at higher frequencies plasmon refractive index becomes high enough as to match that of a waveguide core mode and consequently simple structured microstructured fibers can also be applied to be SPR sensor. Similar to optical fiber structure in Ref. [13], we designed one type of multi-hole fiber-based SPR sensor, illustrated in figure 1. The solid silica core was surrounded by six large air holes, where the diameter of air holes is d , the distance between two adjacent air holes is Λ and the thickness of gold films covered on inner surfaces of air holes is d_m . Next, we consider the performance of multi-hole optical fiber SPR sensor, launching the measured liquid with the refractive index n into air holes.

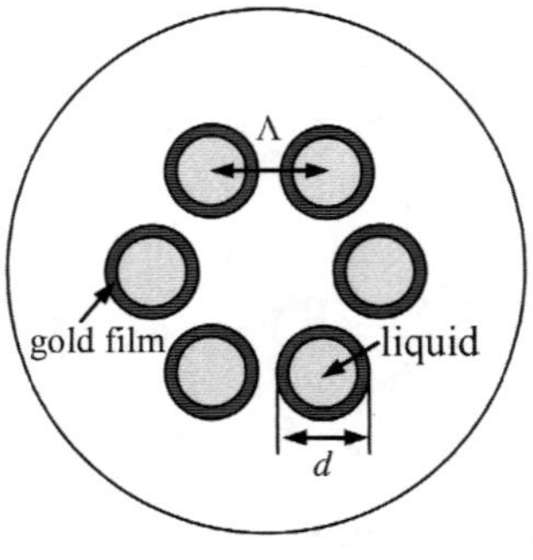

Figure 1 Structure of multi-hole fiber-based SPR sensor.

3. Numerical results and analysis

Considerable numerical methods have been adopted to conduct theoretical research of optical fiber and optical components. The finite element method is pretty powerful to research arbitrarily shaped optical fiber. Therefore, the properties of multi-hole optical fiber SPR sensors were calculated with the finite element method. Sellmeier equation described the dispersion of silica [14] and the relative permittivity of gold was found in Ref. 15.

Firstly, let us consider how the SPR resonant peak depends on the thickness of gold film. Supposed that the refractive index of liquid is 1.33, the distance of two air holes is $\Lambda = 5\mu m$ and the diameter of air holes is $d = 4.3\mu m$, the loss spectrum of the fundamental mode in the fiber is shown in figure 2 for different thick gold films with 30nm, 40m and 50nm. Several resonances occur in the whole wavelength range duo to asymmetrical feature of gold films to the fiber core, which is in good agreement with the conclusion in Ref. [16]. The resonant peak in the visible range moves to long wavelength as the thickness of gold film increases while the resonant peak in the near-infrared range moves to short wavelength. It is also observed that relative large loss in the nonresonant near-infrared

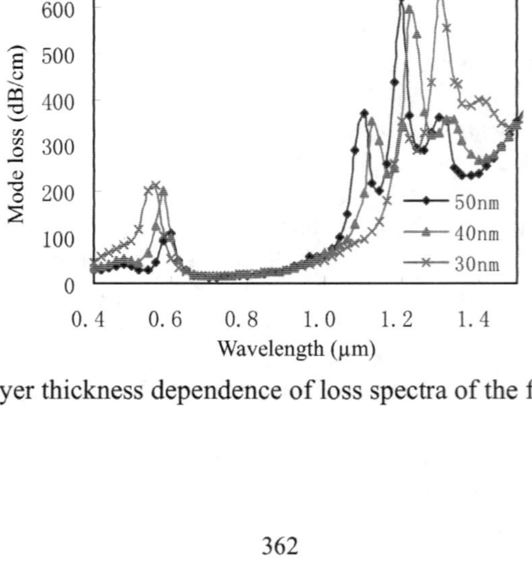

Figure 2 Gold layer thickness dependence of loss spectra of the fundamental mode.

range can not been ignored. Though the resonance occurs in the near-infrared range, they can hardly be detected duo to large loss in the nonresonant range. In the following parts, the properties of SPR in the visible range are mainly investigated. The loss spectrum of multi-hole fiber without gold films in figure 3 exhibits that losses in the whole visible range are low and light can mostly be confined into the fiber core. When the thickness of gold films is 50nm, the field distribution of resonant wavelength 0.59μm is described in figure 4. The field attenuates away from the interface between the liquid and gold film, however, the field is enhanced on one side of gold film close to the liquid owing to SPR.

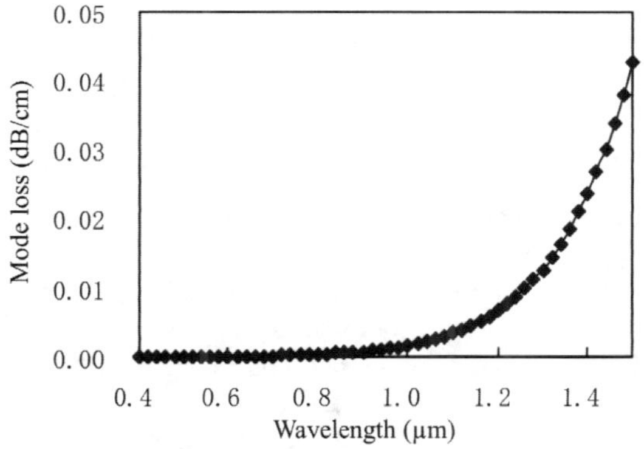

Figure 3 Loss spectra of the fundamental mode for the fiber without the gold film.

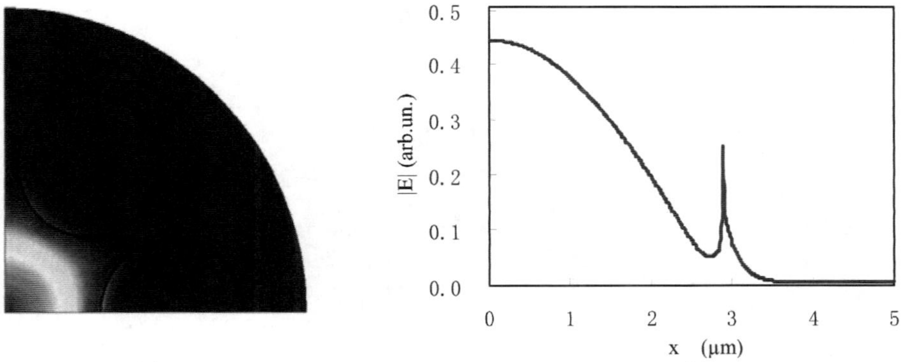

(a) (b)

Figure 4 Mode profile at 0.69μm wavelength (a) and |E| on the x-axis (b).

The responses of SPR to different measured liquid are also considered. Here, the thickness of gold film is 50nm, $\Lambda = 5\mu m, d = 4.3\mu m$. Loss spectra of the fundamental mode are shown in figure 5 when air holes are filled with different refractive index liquid samples. The resonant wavelength moves upwards and losses at the resonant peak increase and the half-width of the power spectrum is broadened resulting from the increasing refractive index. The increasing refractive index of measured liquids leads to small difference between fiber core and ambient medium, resulting in decreasing effective refractive index of fundamental modes. Thus, the resonant peak shifts to long wavelength. Therefore, the refractive index change of ambient medium can be measured by monitoring the power at a fixed wavelength or shift of resonant wavelength.

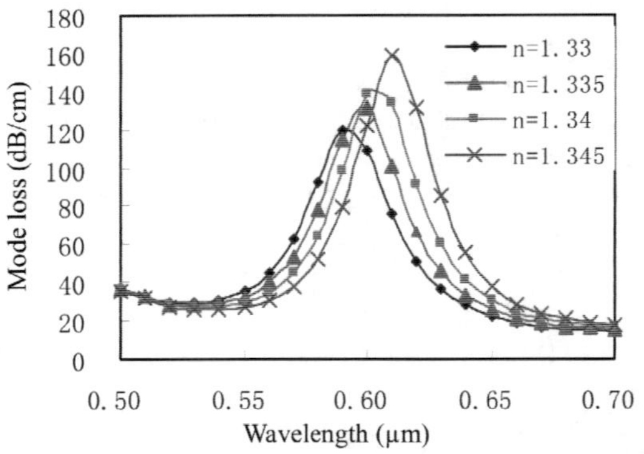

Figure 5 Loss spectra of the fundamental mode for different liquid refractive index.

The air holes will exert effects on SPR and the calculated results are shown in figure 6. The thickness of gold film is 50nm, the refractive index of liquid is 1.33. The diameter of air holes and the distance between two air holes are analyzed separately. Large diameter of air holes widens the resonant peak and brings about increasing loss, but the resonant peaks do not change their position. The effect of the distance between two air holes on the resonant peak is similar to that of air hole diameter. When the distance increases, the resonant peak is also kept unmoved, however, the loss at the resonance decreases.

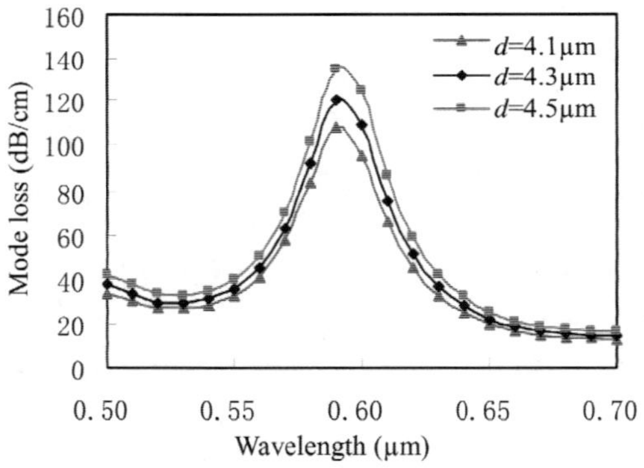

(a)

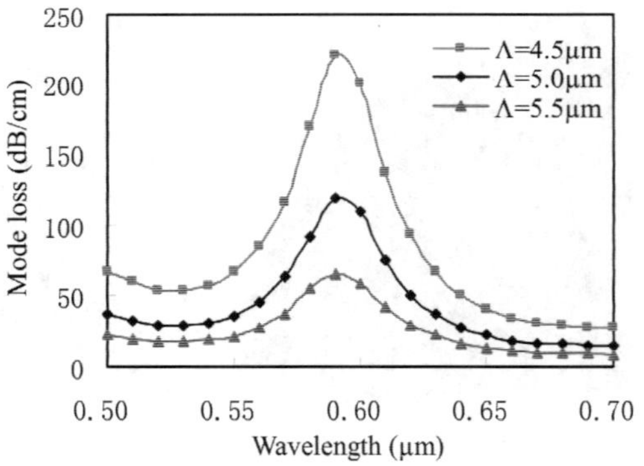

(b)

Figure 6 Effects of diameter of air holes (a) and pitch between air holes (b) on resonant peak.

One of the most important parameters of SPR sensors is its sensitivity. When the power at a fixed wavelength is measured to evaluate the performance of sensors, the sensitivity parameter can be expressed as [12]

$$S = (\Delta\alpha / \Delta n) / \alpha_a \qquad (1)$$

where α_a denotes the loss of optical fiber at $n = a$. For the sensor of length $1/\alpha_a$, Eq(1) is a good approximation when the refractive index of liquid in air holes changes slightly. At $n = a = 1.33$, the sensitivity curve is plotted in Figure7. If one assumes that 1% change in the transmitted power can reliably be detected, 1.4×10^{-4} change of refractive index (for $S = 70$) could be detected with this sensor. The proposed multi-hole optical fiber SPR sensors can treat weak signal detection indicating that its sensitivity is comparable to that of conventional SPR sensors.

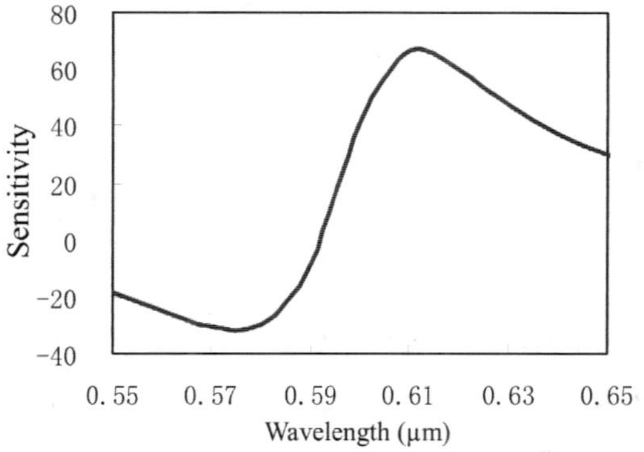

Figure 7 Sensitivity parameter for SPR sensor.

4. Conclusions

The SPR sensor was proposed on basis of multi-hole optical fiber containing six air holes. The finite element method was utilized to analyze effects of the geometrical and physical parameters on the resonance. The results reveal that the resonant wavelength is sensitive to the thickness of gold film and the refractive index of ambient medium while the peak wavelength of the resonance is insensitive to the diameter of air holes and distance between two air holes. As a result, the measurement of refractive index can be achieved based on monitoring the power in the resonant vicinity and the wavelength change of the resonant peak. The microstructured fiber SPR sensor proposed has high sensitivity up to 10^{-4}.

This work was supported by the Special Foundation for Harbin Young Scientists under grant 2008RFQXG031, and in part by the Special Foundation for Basic Scientific Research of Harbin Engineering University under grant HEUCF20101102.

Reference

[1] Jorgenson R C, Yee S S 1993 *Sens. Actuators B* **12** 213
[2] Fontana E, Dulman H D, Doggett D E, and Pantell R H 1998 *IEEE Trans. Instrum. Measure.* 4 168
[3] Iga M, Seki A, and Watanabe K 2004 *Sens. Actuators B* **101** 368
[4] Kim Y, Peng W, Banerji S, and Booksh K S 2005 *Opt. Lett.* **30** 2218
[5] Ranka J K, Windeler R S, and Stentz A J 2000 *Opt. Lett.* **25** 25
[6] Eggleton B, Kerbage C, Westbrook P, Windeler R, and Hale A 2001 *Opt. Express* **9** 698
[7] Mach P, Dolinski M, Baldwin K W, Rogers J A, Kerbage C, Windeler R S, and Eggleton B J 2002 *Applied Physics Letters* **80** 4294
[8] Sazio P J A, Amezcua-Correa A, Finlayson C E, Hayes J R, Scheidemantel T J, Baril N F, Jackson B R, Won D-J, Zhang F, Margine E R, Gopalan V, Crespi V H, and Badding J V, 2006 *Science* **311** 1583
[9] Zhang X, Wang R, Cox F, Kuhlmey B T, and Large M C J 2006 *Opt. Express* **15** 16270.
[10] Amezcua-Correa A, Yang J, Finlayson C E, Peacock A C, Hayes J R, Sazio P J A, Baumberg J J, and Howdle S M 2007 *Adv. Funct. Mater.* **17** 2024
[11] Hautakorpi M, Mattinen M, and Ludvigsen H 2008 *Opt. Express* **16** 8427
[12] Hassani A and Skorobogatiy M 2006 *Opt. Express* **14** 11616
[13] Ranka J K, Windeler R S, and Stentz A J 2000 *Opt. Lett.* **25** 796
[14] Tan C Z. J 1998 *Non-Cryst Solids.* **223** 158
[15] Johnson P B and Christy R W 1972 *Physical review B* **6** 4370
[16] Monzón-Hernández D and Villatoro J 2006 *Sensors and Actuators B: Chemical* **115** 227

Harmonic expressions for the eigen mode fields of step and graded index fibers

Fuyuan Guo, Lianhuang Li, Hua Zheng, Yi Wang, Tiangui Dai and Jinrui Ke
Key Laboratory of Optoelectronic Science and Technology for Medicine of Ministry of Education, Fujian Provincial Key Laboratory of Photonic Technology, Institute of Laser and Optoelectronics Technology, Fujian Normal University, 32 Shangsan Road, Cangshan, Fuzhou 350007, P. R. China

E-mail: guofy@fjnu.edu.cn

Abstract. For clarifying the mode field characteristics of step and graded index fibers, the Helmholtz's equation and local planar wave approximation are engaged, a novel idea that the eigenfunction of the traveling wave field in radial grade index dielectric is similar to positive or negative half order Hankel function is advanced. Then, the eigenfunction of core layer of graded index fiber is similar to the combination of half order Bessel function and half order Neumann function, the eigenfunction of cladding layer of graded index fiber is similar to the first kind half order imaginary parameter Hankel function, are suggested. The phase shift for total reflection effect at the turning point of grade fiber is proposed. The angular phase shift pi for the ray propagating along the fiber circumference is given. The eigen equation of new improved WKB method for analysing the mode field characteristics of graded index fiber is presented. Moreover, the eigen equation of new improved WKB method for analyzing the mode field characteristics of graded index fiber is held good for analysing the mode field characteristics of step index fiber too, it indicates that, the eigen equation of new improved WKB method is of universal significance.

1. Introduction

The fiber mode field theory is established creatively by Snitzer E in 1961 [1], and its simple description in weakly guiding approximation is described by Gloge D in 1971 [2]. In the classical theoretic system, the integral order Bessel function is the eigenfunction of fiber core layer, the integral order imaginary parameter Hankel function is the eigenfunction of fiber cladding layer, the hypothesis of false total reflection or caustic surface in fiber core layer is used for explaining the phase shifts π in the ray method of fiber characteristic analysis [3, 4]. A verdict of mode field theory for step index fiber appears to be formed [1-12]. But some questions of the fiber characteristics are presented in some literatures [13-16], several new ideas are presented for clarifying the essence of fiber, the theories of the electromagnetic adjoint transformation and the Sommerfeld's formula for spherical wave are brought forward to describe the characteristics of step index fiber by Yu S M et al [13, 14], the zero and first order Bessel functions, Neumann functions which modified by Dirac function and the imaginary parameter Hankel functions are put forward to describe the eigenfunctions of mode field distribution in step index fiber, the combination of half order Bessel function and half order Neumann function, the first kind half order imaginary parameter Hankel function are advanced to bewrite the eigenfunctions of mode field distribution in step index fiber by Guo F Y et al recently [15, 16].

Published under licence by IOP Publishing Ltd

For analysing the characteristics of graded index fiber, the classical and modified WKB methods are applied to analyse the mode field distribution and propagation constant frequently [12, 17-23]. The hypothesis of false total reflection in the core layer is engaged, the phase shifts of the inner knee point and outer turn point are equal to $\pi/2$ is adopted, and the angular phase shift π isn't explained. However, in the graded index planar waveguide domain, the calculating precision of the propagation constant by the classical WKB method is doubted by Cao Z Q et al, and the transfer matrix method was used for improving the calculating precision of the propagation constant [24-26], nevertheless, the physical significance of WKB method is unaccountability yet [26]. As a matter of fact, the classical WKB method, which stemmed from the solution of the Schrödinger equation in the quantum mechanics domain [27], is fit for solving the scalar issue only. As the light wave in the graded index dielectric is vector wave, if the classical WKB method is engaged to analyse the traveling wave in graded index dielectric, some questions would arise at the turning point, so, a new improved WKB method for analysing the symmetrical planar waveguide is suggested by Guo F Y et al lately [28, 29].

As a result of that the local planar wave approximation [3] is engaged to analyse the traveling wave in graded index dielectric, a novel expression of the traveling wave field in radial grade index dielectric is suggested. In succession, a new improved WKB method for analysing the mode field characteristics of the graded index fiber is presented. Moreover, when the index profile of grade index fiber degenerate into step form, the eigen equation of new improved WKB method for solving the characteristics of grade index fiber is effective for solving the characteristics of step index fiber also.

2. The traveling wave in radial graded index dielectric

In the radial graded index dielectric, the refractive index n change with the radial coordinate r solely, the schematic of the TE traveling wave in the meridian plane of the radial graded index dielectric is shown in figure 1, and the schematic of the TM traveling wave is as the same form as the TE wave, but interchange the symbols of the electric field E and the magnetic field H.

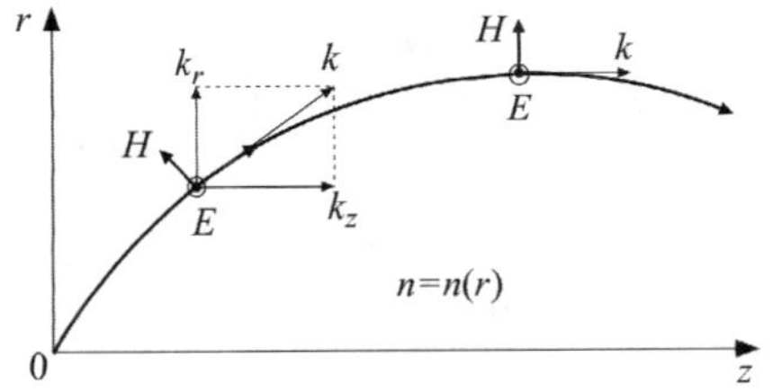

Figure 1. The traveling wave in the meridian plane of radial graded index dielectric.

In general, the electric field intensity of traveling wave is expressed by the electric components E_x, E_y and E_z, the magnetic field intensity of traveling wave is expressed by the magnetic components H_x, H_y and H_z, they are the solutions of the Helmholtz's equation [7, 8], and they are the functions of radial coordinate r, angular coordinate φ and axial coordinate z in cylindrical coordinate system, and time parameter t. Based on the Helmholtz's equation and the method of separation of variables, the expressions of electric components and magnetic components are expressed by the identical form $\Psi(r,\varphi,z,t)$, it is suggested as follows:

$$\Psi(r,\varphi,z,t) = C\eta_j(r)R(r)\exp(\pm \mathrm{i}m\varphi)\exp(\pm \mathrm{i}k_z z - \mathrm{i}\omega t) \quad (1)$$

where, C is a constant, $\eta_j(r)$ is a coefficient, subscripts $j=E$, H, here, $j=E$ labels for the coefficient of electric components E_x, E_y and E_z, $\eta_E(r)=1$, $j=H$ labels for the coefficient of is used for magnetic components H_x, H_y and H_z, $\eta_H(r)=k_0n(r)/(\omega\mu_0)$, respectively. k_0 is the wave number of the light wave with wavelength λ_0 in vacuum, $k_0=2\pi/\lambda_0$, $n(r)$ is the distribution function of refractive index in radial graded index dielectric, ω is the angular frequency of light wave, μ_0 is the permeability of light wave in vacuum, i is the square root of -1, namely, $\mathrm{i}=(-1)^{1/2}$, m is the order number of angular parameter, k_z is the component of wave number k in the z-axis direction, $k=k_0n(r)$, $R(r)$ is radial function.

As the Helmholtz's equation and equation (1) are engaged, the radial function $R(r)$ satisfies the following equation approximately:

$$\frac{\mathrm{d}^2[R(r)r^{1/2}]}{\mathrm{d}r^2}+Q(r)R(r)r^{1/2}=0 \tag{2}$$

where, $Q(r)$ is a parameter.

If the distribution function of the refractive index in radial graded index dielectric $n(r)$ is slowly varying function, $dn(r)/dr\to 0$, the parameter $Q(r)$ is a slowly varying real function also, it is given by the approximate expression [21], as follows:

$$Q(r)=k_0^2n^2(r)-\beta^2-\frac{m^2-1/4}{r^2} \tag{3}$$

According to the classical WKB theory [21], the approximate eigenfunction of the traveling wave field in radial grade index dielectric is expressed as follows:

$$\Psi(r,\varphi,z,t)=\frac{C}{\sqrt[4]{Q(r)}}\frac{1}{\sqrt{r}}\exp\left[\pm\mathrm{i}\int_0^r\sqrt{Q(\xi)}\mathrm{d}\xi\right]\exp(\pm\mathrm{i}m\varphi)\exp(\pm\mathrm{i}k_z z-\mathrm{i}\omega t) \tag{4}$$

where, m is integer, i.e. m=0, 1, 2, ····.

In the classical WKB theory [21], the eigenfunction of grade index fiber core layer is expressed by the standing wave form of equation (4), and the eigenfunction of grade index fiber cladding layer is expressed by the evanescent wave form of equation (4).

From equation (3), for a given radial coordinate r, the parameter $Q(r)$ may tend to zero, it means that, the amplitude of electromagnetic component $\Psi(r,\varphi,z,t)$ in equation (4) may close to the infinity, it breaks the law of energy conversation, so, equation (4) is a false expression of the electromagnet field distribution of traveling wave in the radial graded index dielectric.

For the expression of field distribution of traveling wave in the radial graded index dielectric is significative, the local planar wave approximation is employed, namely, the light wave in the fraction of the radial graded index dielectric is deemed as local planar wave [3] approximately, and the novel idea that the order number m is the positive or negative half is suggested, i.e. $m=\pm1/2$, then, the parameter $Q(r)$ is taken on a clear physical significance, it is the square of the component of the wave number k in the radial direction, namely, $Q(r)=k_r^2(r)$, and, the component $k_r(r)$ is expressed as following equation:

$$k_r(r)=\sqrt{k_0^2n^2(r)-k_z^2} \tag{5}$$

As radiant energy of traveling wave in the radial graded index dielectric is in need of satisfying the energy conversation law, from equation (2), equation (3) and $m=\pm1/2$, the radial function $R(r)$ is described by the harmonic expression approximately, it is suggested as follows:

$$R(r)=\frac{1}{\sqrt{k_0n(r)r}}\exp\left[\pm\mathrm{i}\int_0^r\sqrt{k_0^2n^2(\xi)-k_z^2}\mathrm{d}\xi\right] \tag{6}$$

Then, the eigenfunction of the traveling wave field in radial grade index dielectric is suggested as harmonic expression approximately [29]:

$$\Psi(r,\varphi,z,t)=\frac{C\eta_j(r)}{\sqrt{k_0n(r)r}}\exp\left[\pm\mathrm{i}\int_0^r\sqrt{k_0^2n^2(\xi)-k_z^2}\mathrm{d}\xi\right]\exp(\pm\mathrm{i}\frac{\varphi}{2})\exp(\pm\mathrm{i}k_z z-\mathrm{i}\omega t) \tag{7}$$

According to the characteristics of the Hankel functions, the first kind and second kind positive half order Hankel functions are expressed by: $H_{1/2}^{(1,2)}(x)=\pm(-\mathrm{i})[2/(\pi x)]^{1/2}\exp(\pm\mathrm{i}x)$, the first kind and second kind negative half order Hankel functions are expressed by: $H_{-1/2}^{(1,2)}(x)=[2/(\pi x)]^{1/2}\exp(\pm ix)$, the eigenfunctions of the travelling wave field in radial grade index dielectric $\Psi(r,\varphi,z,t)$ is similar to the positive or negative half order Hankel function.

3. A new improved WKB method for the radial graded index fiber

A given graded index fiber for example, the schematic of refractive index profile of fiber is shown as solid line in figure 2. In figure 2, n_1 is the maximal refractive index of fiber core layer for r=0, n_2 is the minimal refractive index of fiber cladding layer for $r\rightarrow\infty$, β is the propagation constant of the fiber, namely, it is the component of wave number k in the z-axis direction, $\beta=k_0n_{\text{eff}}$, here, n_{eff} is the equivalent index of fiber, a is the coordinate of turning point, a_{m} is the potential ultimate coordinate of turning point.

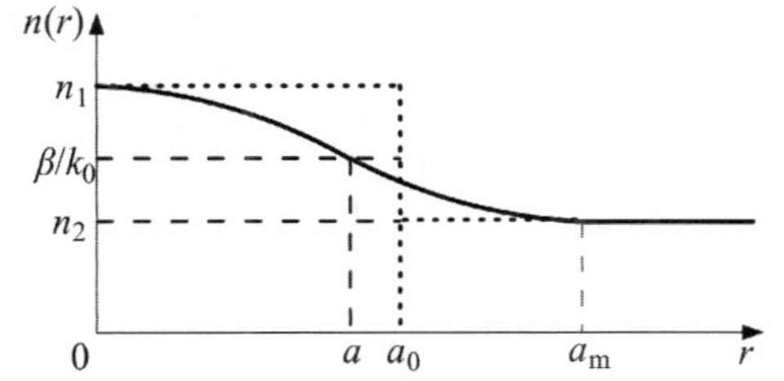

Figure 2. Schematic of the refractive index profile of grade index and step index fibers

In the graded index fiber, the standing wave is affirmed for the condition $k_0n(r)\geq\beta$, and evanescent wave is validated for the condition $k_0n(r)<\beta$ also. According to characteristics of the Bessel functions, Neumann functions and first kind imaginary parameter Hankel functions, the positive and negative half order Bessel functions are expressed by: $J_{1/2}(x)=[2/(\pi x)]^{1/2}\sin(x)$ and $J_{-1/2}(x)=[2/(\pi x)]^{1/2}\cos(x)$, the positive and negative half order Neumann functions are expressed by: $Y_{1/2}(x)=-[2/(\pi x)]^{1/2}\cos(x)$ and $Y_{-1/2}(x)=[2/(\pi x)]^{1/2}\sin(x)$, the positive and negative half order first kind imaginary parameter Hankel functions are expressed by: $K_{\pm 1/2}(x)=[2/(\pi x)]^{1/2}\exp(-x)$, and from equation (7), the eigenfunction of core layer of graded index fiber is similar to the combination of half order Bessel function and half order Neumann function, the eigenfunction of cladding layer of graded index fiber is similar to the first kind half order imaginary parameter Hankel function.

If the distribution function of the refractive index in graded index fiber $n(r)$ is slowly varying function, the difference between maximal refractive index n_1 and minimal refractive index n_2 is far less than minimal refractive index n_2, namely, n_1-n_2<<n_2, then, the weakly guiding approximation is engaged, and the scalar linear polarization (LP) mode theory of step index fiber [7, 8] is used for reference, the electromagnetic components of fiber core layer is equal to the electromagnetic components of fiber cladding layer in the turning point $r=a$ with $k_0n(r)=\beta$, so, the transverse eigen field distribution of core layer of graded index fiber $\Psi_1(r,\varphi,z,t)$ and the transverse eigen field distribution of cladding layer of graded index fiber $\Psi_2(r,\varphi,z,t)$ are expressed by following expressions approximately [29]:

$$\Psi_1(r,\varphi,z,t)=\frac{C\eta_j(r)}{\sqrt{k_0n(r)r}}\cos\left[\int_0^r\sqrt{k_0^2n^2(\xi)-\beta^2}\,\mathrm{d}\xi-\varphi_N\right]\exp(\pm \mathrm{i}\frac{\varphi}{2})\exp(\pm \mathrm{i}\beta z-\mathrm{i}\omega t) \quad (8)$$

$$\Psi_2(r,\varphi,z,t)=\frac{C\eta_j(r)\cos(U-\varphi_N)}{\sqrt{k_0n(r)r}}\exp\left[-\int_a^r\sqrt{\beta^2-k_0^2n^2(\xi)}\,\mathrm{d}\xi\right]\exp(\pm \mathrm{i}\frac{\varphi}{2})\exp(\pm \mathrm{i}\beta z-\mathrm{i}\omega t) \quad (9)$$

where, φ_N is the initial phase of eigen field distribution for the N order mode of graded index fiber in fiber center (z axis), N is the mode order of graded index fiber, U is the normalized standing wave parameter of graded index fiber.

For N order mode of graded index fiber, the initial phase of transverse eigen field distribution in fiber center φ_N is suggested as follows [16, 29]:

$$\varphi_N = (2N+1)\pi/4 \tag{10}$$

where, N is integer, i.e. N=0, 1, 2, ···.

As the concept of the normalized standing wave parameter of the step index fiber [7, 8] is used for reference, the normalized standing wave parameter of the graded index fiber [29] is suggested by following equation:

$$U = \int_0^a \sqrt{k_0^2 n^2(r) - \beta^2}\,\mathrm{d}r \tag{11}$$

In the interest of deducing the eigen equation of the WKB method, another equation is in need of establishing. As a example, $\Psi_1(r,\varphi,z,t)$ and $\Psi_2(r,\varphi,z,t)$ represent the transverse electric components E_{y1} and E_{y2} respectively, then, the longitudinal electric components of fiber core layer is equal to the longitudinal electric components of fiber core layer, namely, equation E_{z1}=E_{z2}, is a condition of deducing the eigen equation, According to the Maxwell's equations, the equation ∇E=0 is engaged, a special equation is presented as follows:

$$\sqrt{\beta^2 - k_0^2 n^2(r)} = \sqrt{k_0^2 n^2(r) - \beta^2}\,\tan\left[\int_0^a \sqrt{k_0^2 n^2(r) - \beta^2}\,\mathrm{d}r - \varphi_N\right]. \tag{12}$$

But, this equation can not work well.

According to the classical WKB method, the eigen equation for analysing the radial graded index fiber is expressed by:

$$4\int_0^a \sqrt{k_0^2 n^2(r) - \beta^2}\,\mathrm{d}r + 2\Phi(a) = 2N\pi \tag{13}$$

where, $\Phi(a)$ is the phase shift for total reflection effect at the turning point of the graded index fiber.

The value of phase shift $\Phi(a)$ is calculated by the formula of the Goos-Hänchen shift [30] with the reference points $r=a-\tau$ and $r=a+\tau$, As the parameter $\tau\to 0$ is considered in the classical WKB method, the classical value of the phase shift in turn point $\Phi(a)$ is $\pi/2$. i.e., $\Phi(a)=\pi/2$. If the angular phase shift π is ignored, namely, the $\varphi_N=\pi/2$ is a hypothesis for equation (12), then, equation (12) is consistent with equation (13). Whereas, the value of phase shift $\Phi(a)$ for planar graded index waveguide is doubted by Cao Z Q et al [24-26] and Guo F Y et al [28], namely, the phase shift $\Phi(a)=\pi/2$ can not work well.

As equation (8) and equation (9) are the exact solutions of the Helmholtz's equation for the points r=0 and $r\to\infty$ with $dn(r)/dr=0$, and approximate solutions for others points with $dn(r)/dr\neq 0$ merely, so, in the new improved WKB method for analysing the characteristics of radial graded index fiber, the reference points for calculating phase shift $\Phi(a)$ are suggested as r=0 and $r\to\infty$, then, the approximate computing formula of phase shift for total reflection effect at the turning point is recommended as follows [29]:

$$\Phi(a) = -2\arctan\left[\sqrt{\beta^2 - k_0^2 n_2^2}\Big/\sqrt{k_0^2 n_1^2 - \beta^2}\right] \tag{14}$$

As the ray method of fiber characteristic analysis [3, 4] is used for reference, the angular phase shift which caused by the light ray propagating along the circumference is considered, the eigen equation of new improved WKB method for analysing the characteristics of graded index fiber is suggested by following equation:

$$4\int_0^a k_r(r)\mathrm{d}r + 2\Phi(a) + \Phi_\varphi = 2N\pi\,, \tag{15}$$

In above equation, the angular phase shift which caused by the light ray propagating along the circumference Φ_φ is calculated by:

$$\Phi_\varphi = \int_0^{2\pi} k_\varphi r \mathrm{d}\varphi \tag{16}$$

where, k_φ is the component of the wave number k in the angular direction.

As $k_\varphi = m/r$ and $m = \pm 1/2$ are employed, then, $\Phi_\varphi = \pm\pi$. This implies that the value of phase shift which caused by the ray propagating along the circumference [29] is $\pm\pi$.

From equation (14) to equation (16), the approximate eigen equation of new improved WKB method [29] for analysing the graded index fiber is suggested as follows,

$$\sqrt{\beta^2 - k_0^2 n_2^2} = \sqrt{k_0^2 n_1^2 - \beta^2} \tan\left[\int_0^a \sqrt{k_0^2 n^2(r) - \beta^2}\mathrm{d}r - \varphi_N\right] \tag{17}$$

The normalized frequency V, it is called the normalized structure parameter also, is one of the main parameters of fiber. For clarifying the normalized meaning of the normalized parameters of the radial graded index fiber, the concepts of the normalized frequency of the step index fiber [7, 8] is used for reference. As the value of normalized frequency V is the ultimate value of normalized standing wave parameter U, from equation (11), the normalized frequency of the graded index fiber [29] is suggested by following equation:

$$V = k_0 \int_0^\infty \sqrt{n^2(r) - n_2^2}\mathrm{d}r \tag{18}$$

From equation (17) and equation (18), the condition for N order mode appeared in the grade index fiber [29] is commended as follows:

$$V \geq (2N+1)\pi/4 \tag{19}$$

Namely, the cut-off normalized frequency of N order mode of graded index fiber is $(2N+1)\pi/4$. It means that, the value of cut-off normalized frequency of nonzero order of graded index fiber is close to the corresponding value of the LP mode of step index fiber [29], but the cut-off normalized frequency of fundamental mode of graded index fiber is $\pi/4$, only the value of $\pi/4$ is an inconvincible solution, it is in need of making further discussion.

As the weakly guiding approximation is engaged for analysing the characteristics of radial graded index fiber, from equation (19), the mode number of multimode graded index fiber [29] is suggested as follows:

$$M = \mathrm{Round}(2V/\pi + 1/2) \tag{20}$$

where, Round is the function of round off operation.

4. New expressions for analysing the homogeneous dielectric and step index fiber

In the homogeneous dielectric, the distribution function of refractive index $n(r)$ is a constant, $n(r)=n_0$, as the Helmholtz's equation and the method of separation of variables are engaged, equation (2) is transformed to the Bessel's equation, the exact solutions of this equation are the first kind or second kind Hankel functions, i.e. the radial function $R(\mathrm{r})=H_m^{(1,2)}(k_r r)$, here, m is the order number.

In the classical fiber theory, the order number m is integer, the integral order Bessel function and integral order first kind imaginary parameter Hankel function are selected as the eigenfunctions of core layer and cladding layer of step index fiber. As the source function of integral order Bessel function and integral order first kind imaginary parameter Hankel function is integral order Hankel function, so, the integral order Hankel function ought to be the eigenfunction of radial traveling wave field in homogeneous dielectric. Owing to the Hankel function isn't the harmonic function, it isn't used for expressing radial traveling wave field in literatures.

A special expression is presented for expressing cylindrical traveling wave field in the homogeneous dielectric by Iizuka K [31], it is as follows:

$$\Psi(r,\varphi,z,t)=C\frac{1}{\sqrt{r}}\exp(\pm ikr-i\omega t) \tag{21}$$

Although the expression in the right side of equation (21) is a harmonic expression, equation (21) isn't a solution of the Helmholtz's equation.

As the distribution function of refractive index in homogeneous dielectric is a constant, i.e., $n(r)=n_0$, equation (7) is degenerated into the following equation [16, 29]:

$$\Psi(r,\varphi,z,t)=\frac{C\eta_j}{\sqrt{k_0n_0r}}\exp(\pm ik_rr)\exp(\pm i\frac{\varphi}{2})\exp(\pm ik_zz-i\omega t) \tag{22}$$

where, $k_r=(k_0^2n_0^2-k_z^2)^{1/2}$, it is a degenerated form of equation (5), the coefficient of electric components $\eta_E=1$, the coefficient of magnetic component $\eta_H=k_0n_0/(\omega\mu_0)$.

The expression in the right side of equation (22) is a harmonic expression, and equation (22) is an exact solution of the Helmholtz's equation. So, the eigenfunction of the radial traveling wave field in homogeneous dielectric is expressed by positive or negative half order Hankel function accurately.

A given step index fiber for example, the schematic of refractive index profile of fiber is shown as dotted line in figure 2. In figure 2, n_1 and n_2 are the refractive indexes of core layer and cladding layer respectively, a_0 is the radius of fiber core layer.

For the step index fiber, if the condition n_1-n_2<<n_2 is satisfied, the weakly guiding approximation is engaged, the scalar LP mode theory [7, 8] is used for reference, then, the transverse field distribution of fiber core layer $\Psi_1(r,\varphi,z,t)$ and the transverse field distribution of fiber cladding layer $\Psi_2(r,\varphi,z,t)$ are degenerated from equation (8) and equation (9), they are expressed as follows [16, 29]:

$$\Psi_1(r,\varphi,z,t)=\frac{C\eta_j}{\sqrt{k_0n_0r}}\cos(\frac{Ur}{a_0}-\varphi_N)\exp(\pm i\frac{\varphi}{2})\exp(\pm i\beta z-i\omega t) \tag{23}$$

$$\Psi_2(r,\varphi,z,t)=\frac{C\eta_j\cos(U-\varphi_N)}{\sqrt{k_0n_0r}}\exp[-W(\frac{r}{a_0}-1)]\exp(\pm i\frac{\varphi}{2})\exp(\pm i\beta z-i\omega t) \tag{24}$$

where, $U=(k_0^2n_1^2-\beta^2)^{1/2}a_0$, it is a degenerated form of equation (11), W is the normalized evanescent wave parameter of fiber cladding layer, $W=(\beta^2-k_0^2n_2^2)^{1/2}a_0$,

From equation (17), the eigen equation for analysing the characteristic of step index fiber [16, 29] is become as following equation:

$$W=U\tan(U-\varphi_N) \tag{25}$$

It is similar to the form of the eigen equation for analysing the characteristic of TE mode in planar waveguide, but the value φ_N is different, the value of φ_N for analysing the characteristic of step index fiber is more than the corresponding value for analysing the characteristic of planar waveguide, the difference between them is $\pi/4$.

For the step index fiber, the normalized frequency $V=k_0(n_1^2-n_2^2)^{1/2}a_0$, it is a degenerative form of equation (18). The existential condition for N order mode [16, 29] is calculated by equation (19), the cut-off normalized frequency of fundamental mode is $\pi/4$, this value is still an inconvincible solution, and the mode number M of multimode step fiber [16, 29] is calculated by equation (20) also.

5. Conclusions

The approximate expressions of eigenfunction of the traveling wave field in radial grade index dielectric and eigenfunctions of core layer and cladding layer in grade index fiber are suggested, then, a new improved WKB method for analysing the mode characteristics of grade index fiber is presented. Moreover, the eigen equation of new improved WKB method for solving the characteristics of grade index fiber is effective for solving the characteristics of step index fiber also.

Acknowledgments
This project was supported by the Natural Science Foundation of Fujian Province of China (No. 2009J01275), the program of Science and Technology Department of Fujian Province of China (No. 2008F3038) and the program of Education Department of Fujian Province of China (No. JB09069).

References
[1] Snitzer E 1961 *J. Opt. Soc. Am.* **51** 491
[2] Gloge D 1971 *Appl. Opt.* **10** 2252
[3] Ye P D and Wu Y Z 1981 *Foundation Theory for Waveguide Technology* (Beijing: People's Post and Telecommunications Publishing House)
[4] Xu S L, Ling S D and Loulergue 1990 *Optical Waveguide and its Application* (Hangzhou: Zhejiang University Press)
[5] Daly J C 1984 *Fiber Optics* (Florida: CRC Press)
[6] Yeh C 1989 *Handbook of Fiber Optics, Theory and Application* (California: Academic Press)83
[7] Wu C Q 2000 *Optical Waveguide Theory* (Beijing: Tsinghua University Press)
[8] Liao T B 2000 *Fiber Optics* (Beijing: Tsinghua University Press)
[9] Marcuse D 1974 *Theory of Dielectric Optical Waveguide* (New York: Academic Press)
[10] Owyuang G H 1981 *Foundation of Optical Waveguides* (London: Edward Arnold)
[11] Cheo P K 1990 *Fiber Optics and Optoelectronics, 2nd Edition* (New Jersey: Prentice-Hall)
[12] Vassallo C 1991 *Optical Waveguide Concepts* (Netherlands: Elsevier Science)
[13] Yu S M and Yu T 2001 *Acta Physica Sinica* **50** 1097
[14] Yu S M and Yu T 2001 *Acta Physica Sinica* **50** 2179
[15] Guo F Y, Chen Y Q, Lin B, Cao X Q and Shangguan C M S 2003 *Acta Photonica Sinica* **32** 1074
[16] Guo F Y, Chen Y Q and Lin B 2004 *Fujian Opt. Technol.* **27** 28
[17] Kim C and Chung M 1999 *Proc. Pacific Rim Conf. Lasers Electro-Opt.* (New Jersey: IEEE Service center) p 1042
[18] Kim C and Chung M 1999 *J. Lightwave Technol.* **17** 2534
[19] Paiano G and Pellicoro M 2003 *J. Lightwave Technol.* **21** 286
[20] Oyamada K and Okoshi T 1980 *IEEE Trans. Microwave Theory Tech.* **28** 839
[21] Qin B K and Sun Y N 1991 *Dielectric Optical Waveguide and Applications* (Beijing:Beijing Institute of Technology Press)
[22] North D 1979 *IEEE J. Quant. Electron.* **15** 17
[23] Liu K and Yu R J 2002 *Acta Photonica Sinica* **31** 819
[24] Liu Q, Cao Z Q, Shen Q S, Dou X M and Chen Y L 2001 *Opt. Quant. Electron.* **33** 675
[25] Cao Z Q, Zhan L and Chen Y L 1996 *Proc. SPIE* **2891** 289
[26] Cao Z Q 2000 *Transfer Matrix Method in Guided Optics* (Shanghai: Shanghai Jiao Tong University Press)
[27] Zeng J Y 2000 *Quantum Mechanics* vol. 2 3rd edition (Beijing: Science Press)
[28] Guo F Y, Li L H and Wang M H 2007 *Proc. SPIE* **6781** 67812G-1
[29] Guo F Y 2003 *Studies on the Theories of Optical Waveguides and their Applications in Wavelength Division Multiplexers* (Dissertation Submitted to Zhejiang University for the Degree of Doctor of Engineering)
[30] Chen J 1996 *Modern Optics and Technology* (*Electromagnetics*) (Hangzhou: Zhejiang University Press)
[31] Iizuka K 1991 *Engineering Optics* (Translated by Hu S Q, Cheng W Q and Wang D, Chongqing: Southwest China Normal University Press)

Journal of Physics: Conference Series **276** (2011) 012059 doi:10.1088/1742-6596/276/1/012059

Simulation and Comparison of Advanced Modulation Formats for Wavelength Reuse in High-SpeedWDM-PON System

Han Chenyang, Hong Wei
Room B209,Wuhan National Lab.for Optoelectronics,Huazhong Universityof Science andTechnology,LoyuRoad 1037, Wuhan, Hubei Province,China

E-mail:younghan@live.cn

Abstract. In recent years, the innovation in devices and schemes of physical layer plays an important role in promoting the development of WDM-PON (Wavelength Division MultiplexingPassive Optical Network).Among the key technologies adopted in the WDM-PON system, advanced modulation formats at downstream and remodulations at upstream attract the most attention. DPSK (Differential Phase Shift Keying Modulation), FSK(Frequency-Shift Keying), IRZ(Inverted Return-to-Zero) and Manchester are four of the most common formats and each of them was studied before, but, however, nobody compared and analyzed them together.So, we will do the job by numeralsimulation in this paper. During our simulation, the 40-Gbit/s FSK is generated through the demodulation of two DPSK signals and 40-Gbit/s Manchester signal bases of employing a delay interferometer (DI) and an optical delay line as format conversion device to convert RZ-DPSK(Return to Zero-Differential Phase Shift Keying Modulation) coding to Manchester coding, both of the two format generation schemes are novel.Meanwhile,this paper emphasizes the input optical power and dispersion tolerance, which both related to the sensitivity of the system. Analyze the sensitivity with DPSK, FSK, IRZ and Manchester, and then compare the four advanced modulation formats.

1. Introduction

In the 21th century, with the popularity of the Internet worldwide and the rapid development of the Microscale Integrated Manufacturing Technology, the High-bandwidth multimedia applications which use video signal as the major data load, including the three-dimensional video games, the network TV, the Multi-point video conference system and so on, are gradually become the main content of the network communications. At the same time, due to the optical properties of low-loss window and high bandwidth, the transmission system based on fiber has already standed out and becomes irreplaceabletechnology in the

Information transmission system, especially in the high-speed, large-capacity, long-distance transmission system. So, the facts that whether the network can meet the transmission requirementsof ultra-high speed and large capacity or not, will inevitably become the important development symbol of the next generation Optical Network in the future. In addition, with the continuous expansion and in depth of business forms,the requirements of network bandwidth and signal processing will be constantly strengthened[1].

In the Optical fiber transmission system, the common factors causing signal injury include: loss, the ASE of amplifiers, the nonlinear effects of fiber, the Chromatic dispersion, the Membrane polarization dispersion, the filter cascade, the crosstalk caused by multiplex channel bifurcation in DWDM, and so on[2,3]. With the transmission speed less than 10Gb/s, these factors can be easily overcome. However, when the transmission speed goes above 40Gb/s or even 100Gb/s, the traditional modulation formats has been unable to resist the transmission loss caused by the signal rate increase. Therefore, advanced modulation formats with good nonlinear resistance, high chromatic dispersion tolerance, high polarizing coating chromatic dispersion tolerance and high frequency spectrum utilization rate has been presented and deeply studied which had overcome many transmission limitations caused by transmission speed enhancement and channel narrowing, and it has been one of the revolutionary technologies which propelled the optical network transmission speed enhance quickly. The research objectives are as follows: firstly, obtain high optical spectrum efficiency, cut down unit bit transmission cost and can adjust to the new environment when the intervals between channels narrow down. Secondly, reduce the impact caused by the system loss (like chromatic dispersion, polarizing coating chromatic dispersion, optical fiber nonlinear and channel interference etc.). Lastly, support special user requirements (confidential transmission etc.). What's more, use advanced modulation format can update the transmission speed and tolerance without changing the original optical transmission link during practical applications, this will reduce the cost to a large extent.

With the growth of the correspondence demand of people, meanwhile, the FTTx has developed very fast. The FTTx technology based on the PON has widely spread all over the world in many different shapes. Moreover, with the combination of service operation and network in our country, high-bandwidth applications, such as video conferences, real-time games and IPTV, are emerging unceasingly, which not only set even higher demand on the bandwidth of the access network, but also impel the new FTTx technologies to emerge one after another continuously. Besides, some new technologies, like the integration of telecommunications networks, cable TV networks and the internet, the Internet of things, the cloud computing, the mobile network and the FMC, will also improve the applications of high-bandwidth and the related network. As a result, the FTTx technology will develop in the direction of higher speed, more transmission wavelengths, broader covering scope in the future. Thanks to its ability of full use of fiber resources, completely transparent speed and service, high security, easy management, long transmitting distance and good extension, the WDM-PON system is able to save the cost of branch optical fiber and other light-attribute-network largely, while it can also expand user covering scope and the quantity of the users, and WDM-PON system has provided an ideal solution for the operating agenciesto build a new-light-network with bigger capacity, higher transmission speed and facing entire service operation.

In summary, either advanced modulation formats or WDM-PON system is the most potential technology in the future; therefore, this paper aims to the applications of four advanced modulation formats (DPSK, FSK, IRZ and Manchester) in high-speed WDM-PON access network system is meaningful. Firstly, introduce schemes of the four advanced modulation formats. Then, give system simulation chart and results, which focus mainly on the input optical-fiber power influence and chromatic dispersion management, and then, compare and analyze their transmission performances. Lastly, summarize the job.

2. Numeral simulation

In this section, we will describe our scheme of WDM-PON system based on RSOA adopted in the numeral simulation and the schemes we take to generate the advanced modulation formats.

2.1. Scheme of WDM-PON system based on RSOA

The scheme adopted in our simulation is presented in Figure 1

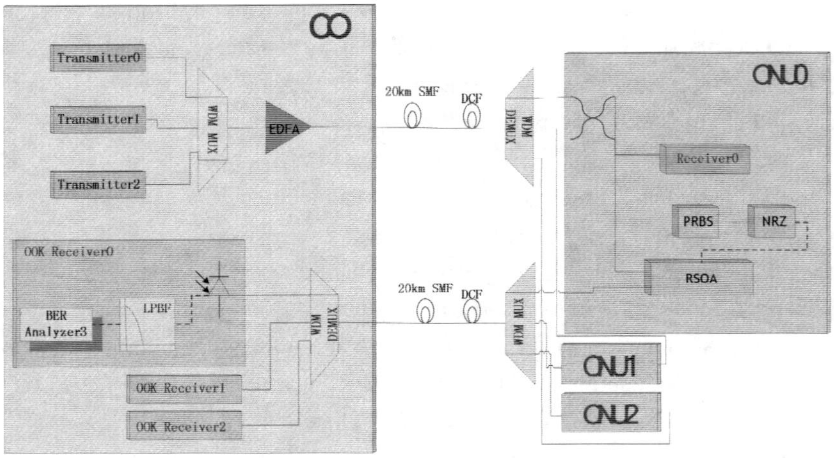

Figure 1. Construction of simulation system

In our system, the downstream bit rate is 40-Gbit/s, and the upstream is 2.5-Gbit/s; the four advanced modulation formats mentioned are used to modulate our downstream signals because of their constant power during transmission, and RSOA is employed to remodulate signals because of its advantage of modulating upstream signals and amplifying the power simultaneously. During our simulations for different formats, we remain all the parameters of the system the same value.

For one advanced format, the corresponding transmitters generate signals, which are carried by different wavelengths carriers. Firstly, the three channelsare combined by the multiplexer,and then enter into EDFA to amplify the power in the Central Office (CO) before getting to the fiber network. The optical fiber adopted in our simulation contains two parts: 20km SMF and 5km DCF, which aims to offset the chromatic dispersion phenomenon. After finishing the fiber travel, signals arrive at demultiplexer, where signals are separated respectively and sent into corresponding channel. Following the channel, they get to the ONU, where the signal are divided into two parts, one is received as the downstream signal from CO; meanwhile, the other flows into RSOA, which remodulate the downstream signals with 2.5-Gbit/s NRZ electrical signals to generate upstream signal. So, the upstream travel begins.

Similarly, the three channels entre into the multiplexer first, then the optical fiber, then the demultiplexer, and the receiver of the CO at last.In this paper, we make the intermediate channel as our study object.

We need to note that: the typical WDM-PON structure is a single bi-directionalfiber, the whole transmission will be done with the help of circulators. However, considering the scheme cannot avoid signal degradation caused by the parasitic reflection radically, we take the solution of two-fiber bi-directional. Moreover, the cost of optical fiber is low, and the two-fiber scheme may be more suitable for WDM-PON system.

2.2. schemes of advanced modulation formats

The four advanced modulation formats generation schemes are introduced in this section, meanwhile,as both of the FSK's and Manchester's are novel, we will emphasize to describe them.

2.2.1. DPSK scheme

NRZ (nonreturn-to-zero) data is differentiated by the encoderfirstly, and then modulate the continuous wave generated by the laser source in a phase modulator, and the DPSK signal is generated[4,5].

2.2.2. FSK scheme

Utilizing MZM suppressed carrier modulation to obtain the two waves needed by the 40Gb/s FSK, and the advantage of making use of single light source is the avoidance of wave and power's precise control of the two light sources. Besides, suppressed carrier modulation can be operated in multi-channel at the same time. It helps to save cost in the WDM system. MZ modulator works in the push-pull state, setting the DC bias at the "zero" spot of the transmission curve, the RF driving voltage amplitude was one half of the half-wave voltage($V_\pi/2$), and at this time, we can realize the suppressed carrier modulation, the interval between the two generated frequencies is the double of RF signal frequency. Then, through the phase modulator,40Gb/s data signal was modulated into this two carrier frequency so as to generate this two NRZ-DPSK signals and was sent to a 1-bit time-lapse demodulator. When this two carrier frequencies locate at the maximum and minimum value of the DI transmission spectrum, we can get two different signals with same amplitude but reverse logic, namely NRZ-FSK signal. And the interval between the two carrier frequencies can be modified by modulating the RF signal. When receiving, we can add a band pass filter which can make a frequency discrimination to the new FSK signal (Figure2)[6].

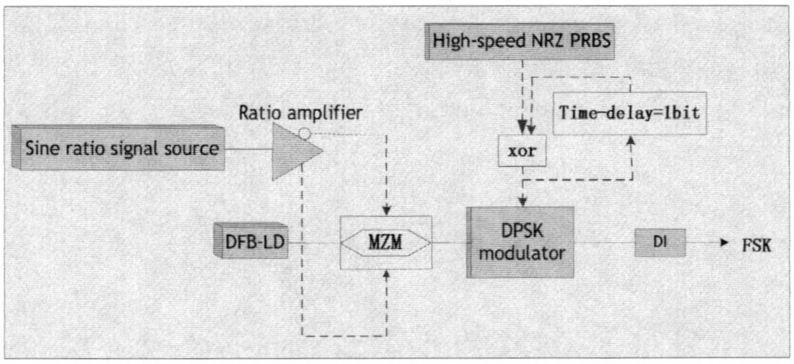

Figure 2. Scheme of FSK transmitter

2.2.3. IRZ scheme

The RZ data electrical signal will be reversed by a reversor firstly, and then, drive the amplify modulator so as to generate the IRZ signal by making it consistent with the continuous optical carrier wave generated by the light source.

2.2.4. Manchester scheme

In fact, Manchester encoding is a kind of redundance encoding, utilizing 1-bit signal from high level to low level and from low level to high level to signify binary character "1" and "0". Since the bit stream contains clock and data, the clock signal can be extracted easily. We can differential demodulate the RZ-DPSK signal to generate the phase-aided Manchester signal: the center frequency of the signal was setted at 193.1 THZ, the two cascaded MZM were drived by the 40Gb/s differential encoding data and 20GHz sinusoidal clock, making phase modulation to input optical signal, and then, generate RZ-DPSK signal with 4040Gb/s and 50%'s duty cycle . A DI with one bit delay will differentiate and demodulate the amplified RZ-DPSK signal, then get two-way RZ signal with reversed logic at the output terminal. One optical chronotron with 0.5 bit delay will be connected to one of the DI output terminal, And then, combine the delayed optical signal with the other one via a 3dB coupler and then, obtain the phase-aided Manchester signal consequently (Figure 3)[7,8].

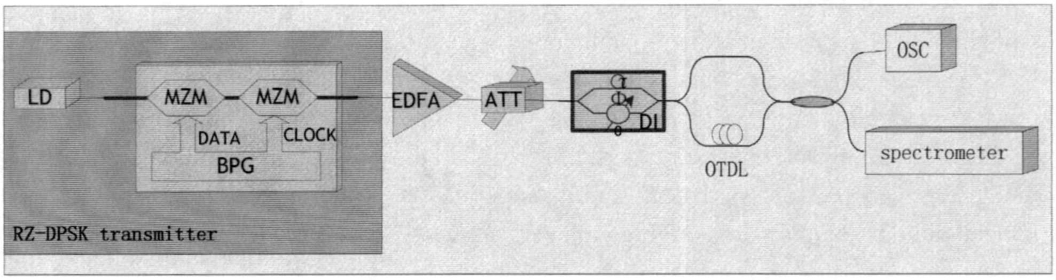

Figure 3. Scheme of Manchester's experimental system

3. Comparisons of simulation results

In this section, we will mainly compare the simulation results obtained in the schemes mentioned above.The comparisons emphasize two parts: input optical power and dispersion tolerance.

3.1. Effects from the input optical power

Retained DCF in the situation of well proper compensation to 20kmSMF;meanwhile, changed the value of input optical power of the WDM-PON system, and then we got the sensitivity curves of the four advanced modulation formats as follows (Figure4,Figure5.):

What we have to point out is that the values of power-into-fiber in the figure were measured after getting through EDFA, the power amplified before flowing into the optical fiber.

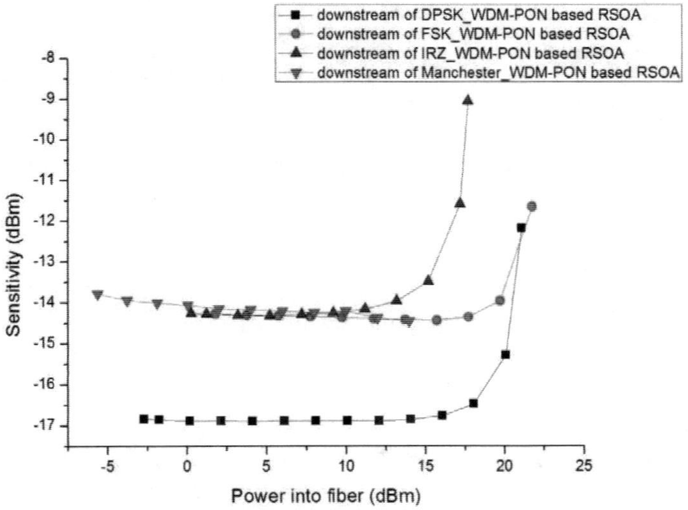

Figure 4. Measured sensitivity to power-into-fiber curves of the four advanced modulation formats for downstream in WDM-PON system based on RSOA

During thedownstream transmission, DPSK holds the highest sensitivity, which is nearly 3 dBm higher than that of the other three advanced modulation formats. While FSK, IRZ and Manchester go in a similar trace, that is they hold the nearly same sensitivity.

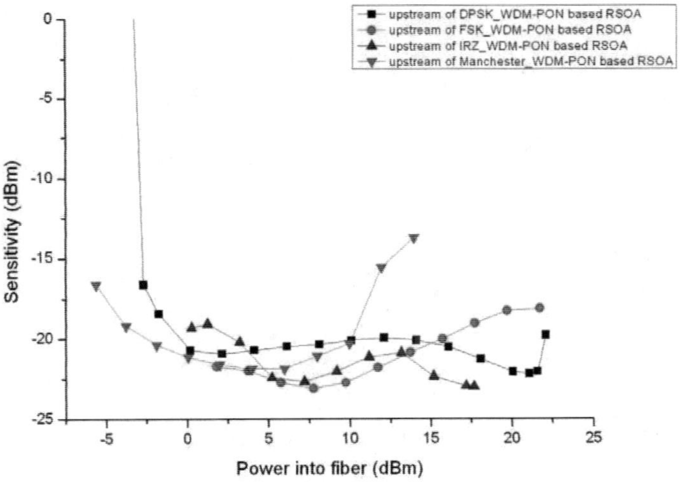

Figure 5. Measured sensitivity to power-into-fiber curves of the four advanced modulation formats for upstream in WDM-PON system based on RSOA

During the upstream transmission, differencesamong the four modulation formats arelittle.All of them possessthe sensitivity at about 21.8dBm.

3.2. Effects from dispersion

Keep the input optical-fiber power constant and change the length of DCF, the chromatic dispersion managements of the four advanced modulation formats are displayed as follows (Figure 6,Figure 7.):

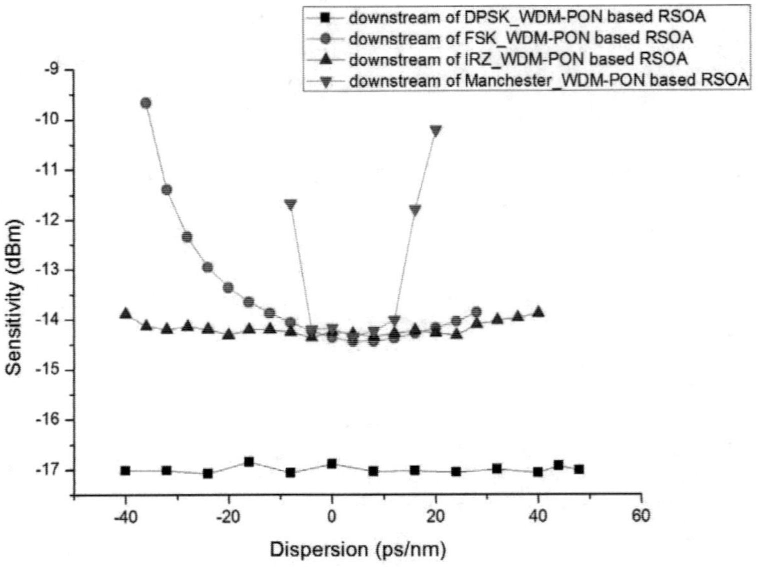

Figure 6. Measured sensitivity to dispersion curves of the four advanced modulation formats for upstream in WDM-PON system based on RSOA

During the downstream transmission, for the four advanced modulation formats, we can easily see that effects from dispersionare close to that from input power.DPSK also holds the highest sensitivity, and the other three ones share the nearly same sensitivity. However, for the chromatic dispersion management, Manchester's dynamic range is much smaller than the others.

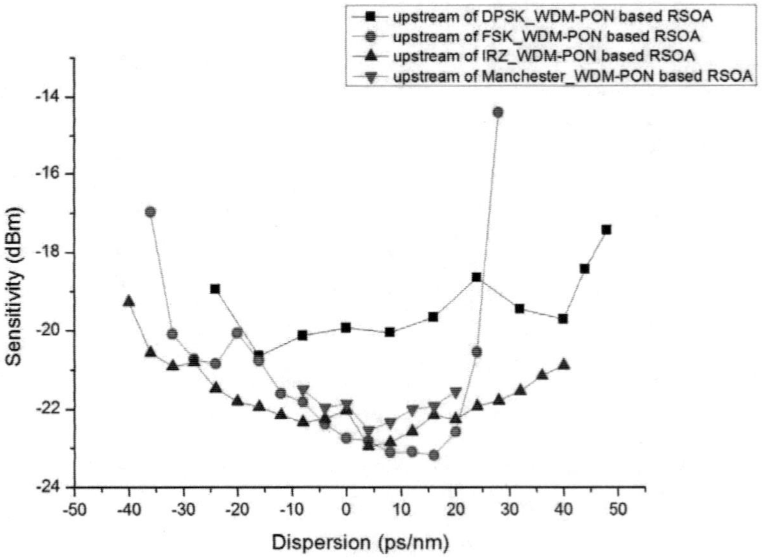

Figure 7. Measured sensitivity to dispersion curves of the four advanced modulation formats for upstream in WDM-PON system based on RSOA

During the upstream transmission, the sensitivity of DPSK modulation format is lower than the other three formats, about 1.5dBm. In comparison, the dynamic range of Manchester is very small, which is corresponding with its downstream transmission. As for the IRZ and FSK code, they perform equally with each other.

3.3. Analysis and discussion

For the downstream, due to their characteristic of constant power, NPR-DPSK, FSK, IRZ and Manchester possess the abilities of tolerance of nonlinearity. Therefore, the dynamic range of input power corresponding to the 1-dB receivers' sensitivity is large. However, since the IRZ with 50% duty-cycle, FSKand Manchester is not actuallypower constant after all.when the value of input power keeps increased, NRZ-DPSK holds the lowest peak power for the equal power. So, DPSK performs better than the other ones at tolerance of nonlinearity.

With regard to the chromatic dispersion, for the same transmission speed, DPSK holds the narrowest optical spectrum; therefore, the chromatic dispersion affects DPSK lest.While, IRZ, FSK and Manchester signal are affected by dispersion easily because of their large bandwidth. Especially is FSK affected, due to the special generation scheme of FSK adopted in this paper, the frequency interval reaches 60GHz. Additionally, considering that the generation of IRZ signal has no phase characteristic; whereas, either FSK or Manchester is generated by demodulating the DPSK signals, and consequently, their data has some phase characteristic,we can explain that although optical spectrum of FSK and Manchester signal is large relatively, the difference of the results between them and IRZ signal is very little. And it's the fact that the transmission speed of Manchester signal was doubled, which is Manchester signal at 40Gb/s transmit at 80Gb/s actually, that's why its dynamic range of chromatic dispersiontolerance is much narrower.

For the upstream, when the power entering into the network fiber becomes lower, the power entering into reflective SOA becomes lower, too, which results in not good remodulation of signals. On the other hand, if the power is too high, reflective SOA will be in gain saturation, which cannot modulate well. Both of these will consequently lead to wave distortion, meanwhile, the recovery time of carrier also effect the wave. Due to the large differences between the higher rate entering into reflective SOA and carriers' recovery time, Manchester signals go to bad quickly with its power being higher.

When itcomes to upstream dispersion, actually, it makes a little impact on 2.5Gb/s NRZ signals with 20km transmission distance. And considering about affect from phase, undulation is natural.

4. Conclusion

In summary, in the system of WDM-PON based on RSOA, among the four advanced modulation formats, IRZ and FSK performs equally, whatever in the situation of input power influence or chromatic dispersion management, either for the 40Gb/s downstream transmission or for the 2.5Gb/s upstream transmission. With regard to Manchester, its sensitivity is close to IRZ and FSK's in any situation, while its dynamic range is much smaller than the other three modulation formats. During the downstream transmission, DPSK's performance is the best.

On the whole, due to their own good characteristics, the advanced modulation formats have been regarded as one of the most potential technology in future fiber-optic communication systems.At the same time, WDM-PON system is regarded as the most promising plan for the passive optical-network access scheme. Only select the proper scheme according to specific project in practical operation, can we meet the requirements from all sides. The job this paper finished may be a good piece of material.

5. References

[1] Berthold, J., Saleh, A. A. M., Blair, L., et al. Optical networking: Past, present, and future. Journal of Lightwave Technology. 2008, 26(9-12): 1104-1118

[2] Essiambre, R. J., Foschini, G. J., Kramer, G., et al. Capacity Limits of Information Transmission in Optically-Routed Fiber Networks. Bell Labs Technical Journal. 2010, 14(4): 149-162

[3] Essiambre, R. J., Kramer, G., Winzer, P. J., et al. Capacity Limits of Optical Fiber Networks. Journal of Lightwave Technology. 2010,28(4):662-701

[4] Hong, W., M.H. Li, X.L. Zhang, et al., Dynamic Analysis of All-Optical Wavelength Optical Amplifier Mach-Zehnder Interferometer. Journal of Lightwave Technology, 2009.27(24): p. 5580-5589

[5] Lu, G. W., Abedin, K. S., Miyazaki, T. All-optical RZ-DPSK WDM to RZ-DQPSK phase multiplexing using four-wave mixing in highly nonlinear fiber. Ieee Photonics Technology Letters. 2007, 19(21-24): 1699-1701.

[6] Li, M.H., N. Chi, W. Hong, et al., Investigation of high-speed optical FSK generation scheme based on carrier suppression and phase modulation. Optics Communications, 2009.282(4): p.508-517

[7] Xu, J., Zhang, X. L., Zhang, Y., et al. Reconfigurable All-Optical Logic Gates forMulti-Input Differential Phase-Shift Keying Signals: Design and Experiments. Journal of Lightwave Technology. 2009, 27(23): 5268-5275.

[8] Song, Y. W., Dinh, K. T., Yamashita, S. Generation of all-fiber optical Manchester code using nonlinear polarization rotation. Optics Express. 2006, 14(1): 63-68.

PAPR Reduction in All-optical OFDM Systems Based on Phase Pre-emphasis

Zhou He[1,3], Wei Li[1], Zhiyong Tao[3], Ji ng Shao[1], Xiaojun Liang[1], Zhuanhua Deng[2], Dexiu Huang[1]

1.Wuhan National Lab for Optoelectronics, Department of Optoelectronics Engineering, Huazhong University of Science and Technology, Wuhan 430074, P.R.China.

2. School of Computer Science and Technology, Hubei University of Economics, Wuhan 430205, P.R.China.

3. State Key Laboratory of Optical Communication Technologies and Networks , Wuhan Research Institute of Posts and Telecommunications, Wuhan 430074, P.R.China.

Email: hezhou@wri.com.cn, weilee@mail.hust.edu.cn

Abstract. This paper investigates the peak-to-average power ratio (PAPR) theory in all-optical orthogonal frequency division multiplexing (OFDM) optical fibre communication systems. We find out that phase pre-emphasis could effectively reduce PAPR in all-optical OFDM communication systems which employ intensity modulation-direct detection (IM-DD) method. An equation is developed and proposed to calculate suitable phasing values for pre-emphasis. Furthermore, we find out that phase pre-emphasis cannot reduce PAPR effectively in all-optical OFDM systems that employ Phase Shift Keying (PSK) or Quadracture Amplitude Modulation (QAM) method.

1. Introduction

Optical OFDM has become a promising technique in long-haul and high-speed optical transmission systems, for its high spectral efficiency, relatively low bit rate and advanced robustness against chromatic dispersion and polarization mode dispersion [1-3]. All-optical OFDM has attracted much attention recently, for its advantage of eliminating processing speed limitation set by electronics [4-7].

High PAPR is a serious intrinsic defect in all-optical OFDM systems, deteriorating nonlinear impairment in optical fibers. However, hardly any investigations are focusing on the PAPR characteristic in all-optical OFDM systems. In conventional optical OFDM systems, PAPR reduction is implemented electronically by introducing phase pre-emphasis before modulation and post-processed by DSP at the receiver [9,10]. We have recently proposed a PAPR reduction scheme based on phase pre-emphasis in all-optical IM-DD OFDM systems, where we used numerical method rather than equations to obtain suitable phasing values [8].

This paper studies the PAPR theory in all-optical OFDM systems. We find out that phase pre-emphasis can effectively reduce PAPR in all-optical IM-DD OFDM systems; while in PSK or QAM OFDM systems, it is not as effective any more. An equation is developed to calculate suitable phasing values for all-optical IM-DD OFDM systems. The PAPR theory in all-optical OFDM systems is analyzed in Section 2. Section 3 and Section 4 evaluate the effectiveness of PAPR reduction based on phase pre-emphasis in all-optical IM-DD OFDM systems and in PSK or QAM OFDM systems, respectively. Section 5 is the conclusions.

2. PAPR theory in all-optical OFDM systems

Figure 1 depicts the configuration of an all-optical OFDM system. At the transmitter, a serial data stream is converted into multiple parallel data streams by the serial to parallel convertor (S/P). An electro-absorption modulator (EAM) generates a phase-locked optical pulse train, which is then split into multiple identical pulse trains. The subcarriers are modulated before entering an optical inverse discrete Fourier transformer (OIDFT), which implements the optical inverse Fourier transformation to generate all optical OFDM symbols. The modulation here can be Intension modulation or PSK modulation. After transmission, the optical discrete Fourier transformer (ODFT) rebuilds optical signals in every sub-carrier at the receiver. Detectors demodulate data in every sub-carrier after EAM sampling. The EAMs work as optical time gates, whose on-state is 1/N of an OFDM symbol's duration and off-state is the rest of an OFDM symbol's duration [4]. The parallel to serial processor (P/S) combines multiple parallel electronic data streams into a serial data stream. Because all optical OFDM systems need no electronic FFT or D/A circuits, the limitation of processing speed and high cost of electronic circuits are eliminated. The main difference between the all-optical OFDM system shown in Fig.1 and the optical coherent-OFDM systems [1,2] is that in the all-optical OFDM system, subcarriers are modulated before entering an OFDM multiplexer (the OIDFT), while in coherent-OFDM systems, OFDM symbols are generated by electronics and then modulated on optical light.

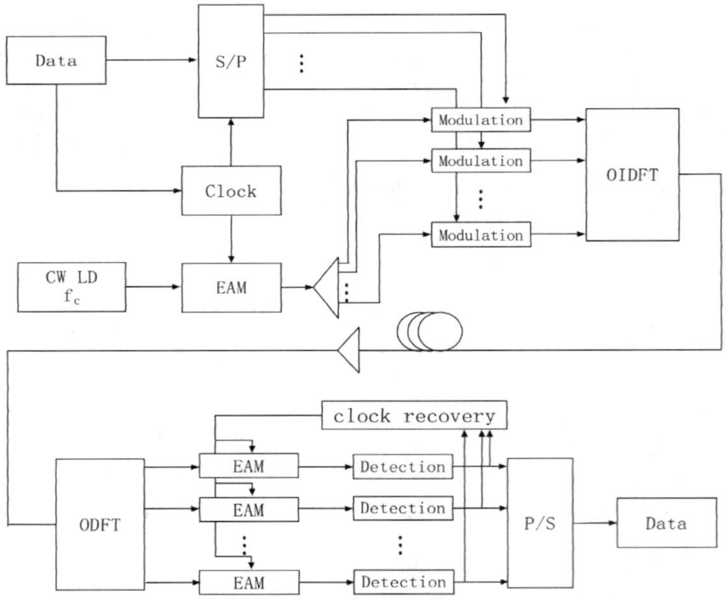

Figure 1 Configuration of an all-optical OFDM system

In optical OFDM systems, data stream is typically modulated by PSK or QAM, before entering an inverse fast Fourier transformation (IFFT) processor. The base band equivalent time-domain signal X_n can be expressed by Eq. (1).

$$x_n = \mathrm{IFFT}\{X_k\} = \frac{1}{\sqrt{N}}\sum_{k=0}^{N-1} X_k \exp(j\frac{2\pi}{N}nk) \qquad (n = 0,1,...,N-1) \tag{1}$$

where N is the number of sub-carriers, and X_k denotes the kth modulated PSK or QAM symbol. X_k can be written as $X_k = a_k + jb_k$, where X_k and b_k indicate the real component and imaginary component of X_k, respectively. The PAPR in discrete time OFDM systems is generally defined in Eq. (2).

$$PAPR_{dB} = 10\lg\left[\max_{0\le n\le N-1}|x_n|^2 / mean|x_n|^2\right] \tag{2}$$

The complementary cumulative distribution function (CCDF) of PAPR can be written as Eq.(3):

$$P(PAPR > PAPR0) = 1 - (1 - e^{-PAPR0})^N \tag{3}$$

Eq.(3) depicts the PAPR characteristics in conventional optical OFDM systems. Nevertheless, Eq.(3) is only applicable to all-optical OFDM systems with PSK or QAM mapping. Eq.(3) will be unestablished in all-optical IM-DD OFDM systems so that modifications are needed for PAPR investigation. Unfortunately, there are no analytical probability distribution functions so far. Numerical simulation is the only way to investigate PAPR characteristic under this condition [8].

3. Phase pre-emphasis in all-optical IM-DD OFDM systems

Eq. (1) implies that every OFDM symbol contains N components (X_0, X_1, $\cdots X_{n-1}$), lining up in sequence in time domain. In binary IM-DD systems shown in fig .1, X_k follows 0-1 distribution, it's clear that $|x_n| \le \frac{1}{\sqrt{N}}\sum_{k=0}^{N-1} X_k$ and $|X_n|$ reaches a maximum when n=0. Because the value of $|X_n|$ always reaches its maximum when n=0, PAPR will be reduced if the in-phase condition (when n=0) is eliminated. In our PAPR reduction scheme, different phase pre-emphases are introduced and Eq.(1) is converted to Eq.(4) and (5).

$$x_n = \mathrm{IFFT}\{X_k'\} = \frac{1}{\sqrt{N}}\sum_{k=0}^{N-1} X_k \exp(j\varphi_k)\exp(j\frac{2\pi}{N}nk) \qquad (n = 0,1,...,N-1) \tag{4}$$

$$X_k' = X_k \exp(j\varphi_k) \quad (k = 0,1,...,N-1) \tag{5}$$

where φ_k is the kth phase pre-emphasis. Chosen suitable values for φ_k, the in-phase condition (when n=0) can be eliminated and PAPR will be reduced.

From the PAPR definition Eq. (2), it is clear that the distribution of $|X_n|^2$ determines PAPR characteristics. Using equation (4), $|X_n|^2$ can be written as:

$$|x_n|^2 = \frac{1}{N}\{\sum_{k=0}^{N-1} X_k^2 + 2\sum_{k=0}^{N-2}\sum_{l=k+1}^{N-1} X_l X_k \exp\{j[\frac{2\pi}{N}n(l-k) + \varphi_l - \varphi_k]\}\} \tag{6}$$

Set

$$Q_n = \sum_{k=0}^{N-2}\sum_{l=k+1}^{N-1} X_l X_k \exp\{j[\frac{2\pi}{N}n(l-k) + \varphi_l - \varphi_k]\} \tag{7}$$

To reduce PAPR, suitable values of φ_k should be chosen to weaken Q_n fluctuation.

$$
\begin{aligned}
Q_n &= \sum_{l=1}^{N-1} X_0 X_l \exp(j\frac{2\pi}{N} l + \varphi_l - \varphi_0) + \sum_{l=2}^{N-1} X_1 X_l \exp(j\frac{2\pi}{N}(l-1) + \varphi_l - \varphi_1) + \ldots + X_{N-2} X_{N-1} \exp(j\frac{2\pi}{N} + \varphi_{N-1} - \varphi_{N-2}) \\
&= X_0 X_1 \exp(j\frac{2\pi}{N} + \varphi_1 - \varphi_0) + X_0 X_2 \exp(j\frac{2\pi}{N}\cdot 2 + \varphi_2 - \varphi_0) + \ldots + X_0 X_{N-1} \exp(j\frac{2\pi}{N}\cdot (N-1) + \varphi_{N-1} - \varphi_0) \\
&\quad + X_1 X_2 \exp(j\frac{2\pi}{N} + \varphi_2 - \varphi_1) + X_1 X_3 \exp(j\frac{2\pi}{N}\cdot 2 + \varphi_3 - \varphi_1) + \ldots + X_1 X_{N-1} \exp(j\frac{2\pi}{N}\cdot (N-2) + \varphi_{N-1} - \varphi_1) \\
&\quad + \ldots \\
&\quad + X_{N-2} X_{N-1} \exp(j\frac{2\pi}{N} + \varphi_{N-1} - \varphi_{N-2})
\end{aligned}
\tag{8}
$$

Q_n can be seen as the summation of vectors with a length of zero or one, because X_k follows 0-1 distribution. The vectors can be separated into N parts, N lines in Eq.(8). X_k is a random variable, this means that the vectors vary randomly. So that each line of Eq.(8) represents the summation of vectors and the number of the vectors vary randomly according to the values of X_k. Therefore, the summation will be small if all the vectors are symmetrically distributed on a circle. It is effective to choose suitable values of φ_k to make the vectors in the first or second line in Eq.(8) arranged symmetrically, in order to reduce PAPR. It should be mentioned that it is hard to calculate φ_k when more lines in Eq.(8) are considered. Considering the first line in Eq.(8), suitable values of φ_k following:

$$
(\varphi_{k+1} - \varphi_k) - (\varphi_k - \varphi_{k-1}) = \frac{2\pi}{N} \tag{9}
$$

A solution of Eq.(9) is:

$$
\varphi_k = \frac{k(k-1)\pi}{N-1} \tag{10}
$$

Similarly, when considering the second line in Eq.(8), another solution is:

$$
\varphi_k = \begin{cases} \dfrac{k(k-2)\pi}{2(N-2)} & k = 0,2,4,\ldots N-2 \\[2ex] \dfrac{(k-1)\pi}{N-1} + \dfrac{(k-1)^2\pi}{2(N-2)} & k = 1,3,5,\ldots,N-1 \end{cases} \tag{11}
$$

We calculate the average of Eq.(18) and (19) to get the values of φ_k.

$$
\varphi_k = \begin{cases} \dfrac{k(k-2)\pi}{4(N-2)} + \dfrac{k(k-1)\pi}{2(N-1)} & k = 0,2,4,\ldots N-2 \\[2ex] \dfrac{k(k-1)\pi}{2(N-1)} + \dfrac{(k-1)^2\pi}{4(N-2)} & k = 1,3,5,\ldots,N-1 \end{cases} \tag{12}
$$

Phasing values calculated by Eq.(12) are used to simulate the PAPR reduction effectiveness in a binary IM-DD OFDM system with 16 subcarriers (N=16). We develop a Matlab package to calculate PAPR. During each simulation, 40000 sample points are used. Figure 2 depicts the comparison of the probability distributions of PAPR between all-optical IM-DD OFDM systems with and without phase pre-emphasis, where X-axis and Y-axis represent the PAPR value and the corresponding possibility, respectively.

It can be seen that the possibility of high PAPRs become much smaller while phase pre-emphases are used. And the most possible PAPRs reduce for several dBs. Figure 3 depicts the differences between CCDFs, where X-axis and Y-axis represent PAPR0 value and the possibility when PAPR is larger than PAPR0, respectively. It is obvious that PAPRs have been reduced effectively. While the possibility that

PAPR is larger than PAPR0 is 0.0032, PAPR has been reduced by 3.46dB (N=16), 5.28dB (N=32) and 7.41dB (N=64). This means that phase pre-emphasis will be even more effective to reduce PAPR in OFDM systems with a large number of subcarriers. For the effectual reduce of PAPR, it will reduce the nonlinear effects in optical fibers greatly [8].

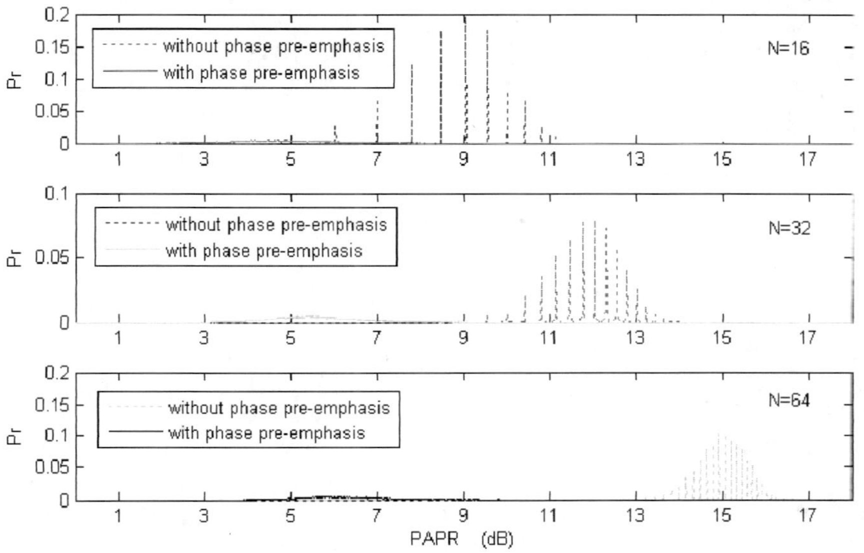

Figure 2 Probability distribution of PAPR

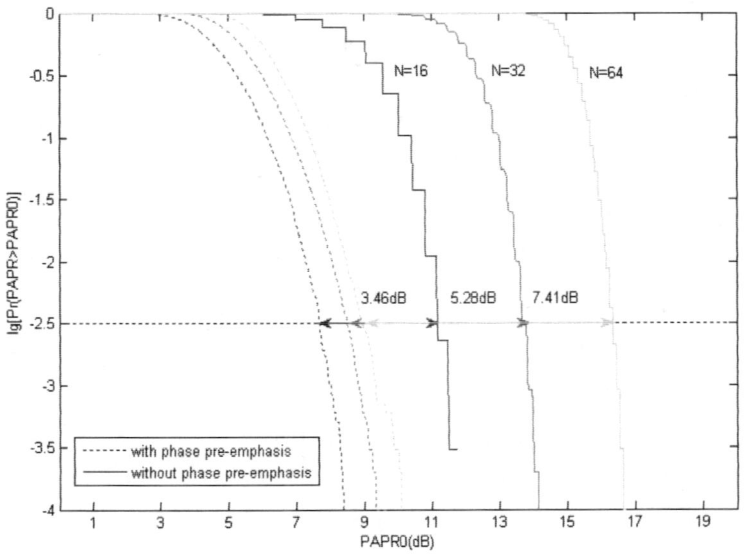

Figure 3 The CCDF of PAPR

4. Phase pre-emphasis in all-optical PSK or QAM OFDM systems

In all-optical PSK or QAM OFDM systems, phase pre-emphasis is ineffective for PAPR reduction. First of all, phase pre-emphases are a kind of phase noise. In order not to deteriorate the detection seriously, only small values of φ_k are acceptable. Second, because the aforementioned in-phase condition (when n=0) is not satisfied any longer, phase pre-emphasis which is aimed to destroy the in-phase condition cannot reduce PAPR noticeably. To illustrate the point clearly, a simple example will be discussed.

Assume a 4-subcarrier all-optical OFDM system that employs binary PSK modulation scheme, so X_k can only be 1 or -1. Eq.(1) can be written in a matrix form, as shown in Eq.(13). The operation can be seen as summations of vectors. Figure 4 depicts that X_k (k=0,1,2,3) is the summation of four vectors, indicated by the red arrows.

$$\begin{bmatrix} x_0 \\ x_1 \\ x_2 \\ x_3 \end{bmatrix} = \frac{1}{2}\begin{bmatrix} 1 & 1 & 1 & 1 \\ 1 & i & -1 & -i \\ 1 & -1 & 1 & -1 \\ 1 & -i & -1 & i \end{bmatrix}\begin{bmatrix} X_0 \\ X_1 \\ X_2 \\ X_3 \end{bmatrix} \tag{13}$$

As an example, assume the input values are [-1,-1,1,1]. Using the vector summation, 1 indicates the vector itself, while -1 indicates a vector with an opposite pointing direction. $[X_0,X_1,X_2,X_3]$ can be calculated out as [0,-2+2i, 0,-2-2i]. Figure 5 illustrates that $|X_n|^2$and $|X_n|^3$ reach the maximum and PAPR is 3dB, where the blue arrow represents the vectors' summation. Because X_1 is vertical to X_3, when phase pre-emphasis is introduced to reduce $|X_3|^2$ then $|X_1|^2$will be enhanced. For example, if φ_k are chosen as $[\pi/8,-\pi/8,\pi/8,-\pi/8]$, $|X_3|^2$will reduce from 2 to 0.5858, while $|X_1|^2$ increase from 2 to 3.4142, with the PAPR increases from 3dB to 5.33dB. Therefore, phase pre-emphasis cannot reduce PAPR here. What is more, when PSK or QAM are used, the phase of X_k is variable according to different input data, making the calculation of φ_k complicated."

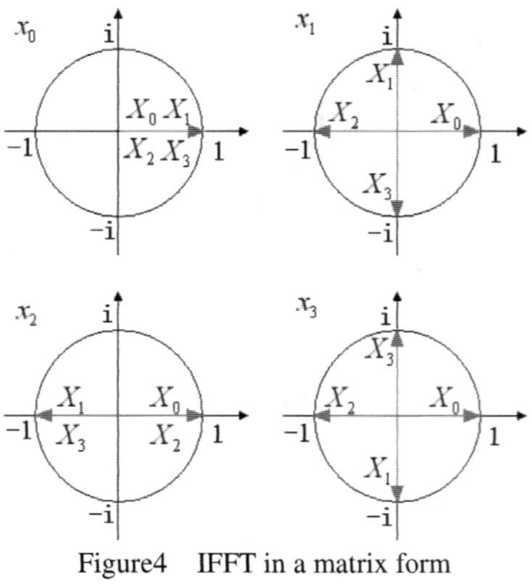

Figure4 IFFT in a matrix form

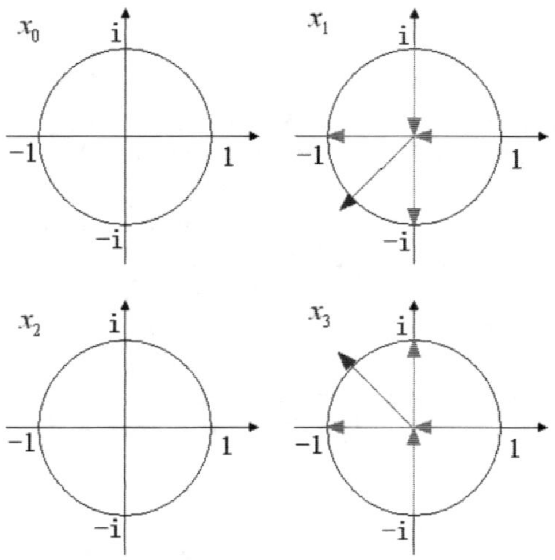

Figure 5 Vector summations ([-1,-1,1,1])

5. Conclusions

This paper investigates the PAPR theory in all-optical OFDM systems. We find out that phase pre-emphasis can effectively reduce PAPR in all-optical IM-DD OFDM systems. An equation is developed to calculate suitable phasing values. Phase pre-emphasis can reduce PAPR effectively in all-optical IM-DD OFDM systems, especially in those systems with a large number of subcarriers. Therefore, it will reduce nonlinear effects in optical fibers greatly. We also find out that phase pre-emphasis cannot reduce PAPR noticeably in all-optical PSK or QAM OFDM systems.

Acknowledgements

This work was supported by the National Basic Research Program of China (973 Program:2010CB328300, 2010CB328305), China National Science Foundation Project (under granted:60772013).

References

[1] J. Armstrong, "OFDM for Optical Communications," *J. Lightwave. Technol.*. **27**, 189-204 (2009).

[2] W. Shieh, H. Bao, and Y. Tang, "Coherent optical OFDM: theory and design," *Opt. Exp.* **16**, 841-859 (2006).

[3] J. Armstrong, "OFDM: From Copper and Wireless to Optical," in Optical Fiber Communication Conference, OSA Technical Digest (CD) (Optical Society of America, 2008), paper OMM1.

[4] K. Lee, T. T. Chan and J. K. Rhee, "All optical discrete Fourier transform processor for 100 Gbps OFDM transmission," *Opt. Express* .**16**, 4023-4028 (2008).

[5] E. Yamada, A. Sano, H. Masuda, T. Kobayashi, E. Yoshida, Y. Miyamoto, Y. Hibino, K. Ishihara, Y. Takatori1, K. Okada, K. Hagimoto, T. Yamada, and H. Yamazaki, "Novel No-Guard-Interval PDM CO-OFDM Transmission in 4.1Tb/s (50 x 88.8-Gb/s) DWDM Link over 800 km SMF Including 50-GHz Spaced ROADM Nodes," in National Fiber Optic Engineers Conference, OSA Technical Digest (CD) (Optical Society of America, 2008), paper PDP8.

[6] Y. Huang, D. Qian, R. E. Saperstein, P. N. Ji, N. Cvijetic, L. Xu, and T. Wang, "Dual-Polarization 2x2 IFFT/FFT Optical Signal Processing for 100-Gb/s QPSK-PDM All-Optical OFDM," in Optical Fiber Communication Conference, OSA Technical Digest (CD) (Optical Society of America, 2009), paper OTuM4.

[7] K. Takiguchi, M. Oguma, T. Shibata, and H. Takahashi, "Optical OFDM Demultiplexer Using Silica PLC Based Optical FFT Circuit," in Optical Fiber Communication Conference, OSA Technical Digest (CD) (Optical Society of America, 2009), paper OWO3.

[8] Xiaojun Liang, Wei Li, Weidong Ma and Kai Wang, "A simple peak-to-average power ratio reduction scheme for all optical orthogonal frequency division multiplexing systems with intensity modulation and direct detection", *Opt. Express*. **17**, 15614-15622 (2009).

[9] S. Chandrasekhar and Xiang Liu, "Experimental investigation on the performance of closely spaced multi-carrier PDM-QPSK with digital coherent detection," *Opt. Express*. **17**, 21350-21361 (2009)

[10] Jianjun Yu, Xiang Zhou, Ming-Fang Huang, Dayou Qian, Philip N. Ji, Ting Wang, and Peter Magill, "400Gb/s (4 × 100Gb/s) orthogonal PDM-RZ-QPSK DWDM Signal Transmission over 1040km SMF-28" *Optics Express*. Vol. **17**, Issue 20, pp. 17928-17933, 2009.

[11] J. Armstrong, "New OFDM peak-to-average power reduction scheme," in Proceedings of IEEE on Vehicular Technology, (IEEE, 2001), pp 756-760.A. N. Shiryaev, Probability, (New York, springer-verlag, 1996).

Journal of Physics: Conference Series **276** (2011) 012061 doi:10.1088/1742-6596/276/1/012061

Optimization of Wireless Optical Communication System Based on Augmented Lagrange Algorithm

Suxiang He[1], Hongchao Meng[1], Hui Wang[1], Yanli Zhao[2*]
[1]School of Science, Wuhan University of Technology, Wuhan 430070, China
[2]Wuhan National Laboratory for Optoelectronics, School of Optoelectronics Science and Engineering, Huazhong University of Science and Technology, Wuhan 430074, China

E-mail: yanlizhao@mail.hust.edu.cn

Abstract. The optimal model for wireless optical communication system with Gaussian pointing loss factor is studied, in which the value of bit error probability (BEP) is prespecified and the optimal system parameters is to be found. For the superiority of augmented Lagrange method, the model considered is solved by using a classical quadratic augmented Lagrange algorithm. The detailed numerical results are reported. Accordingly, the optimal system parameters such as transmitter power, transmitter wavelength, transmitter telescope gain and receiver telescope gain can be established, which provide a scheme for efficient operation of the wireless optical communication system.

1. Introduction

Wireless optical communication (WOC) technology, or free space optical communication (FSO) is a kind of broadband access way, which is the combination of wireless and optical communications product and transmits signal using the beam signal through the atmospheric space, rather than through fiber. For the feasibility for applications in: (i) last-mile access, (ii) service acceleration, (iii) metro network extensions, (iv) enterprise connectivity, (v) fiber backup, and (vi) backhaul, many researchers have been interested in WOC, see e.g. [1], [3], and [6]-[12] . Since the WOC must operate under the condition of line-of-sight, one of the critical affects is the transmitter and receivers sway caused by building sway due to strong wind, weak earthquake, and other reasons, which will make a significant impact on the bit error probability (BEP) of communication. Thus it has been an important issue that how to operate and design of WOC systems under the given BEP value. The model for wireless optical communication system with Gaussian pointing loss factor was studied and the optimal system parameters were reported by solving a system of nonlinear equations in [8]. However, the system of nonlinear equations involves the first-order derivative of BEP with respect to some variable in the model and the second-order derivative of BEP with respect to the same variable is concerned during the numerical implementation, which brings a heavy burden to calculation procedures.

The augmented Lagrange method is one kind of methods for solving optimization problems with equality constraints in mathematical programming, in which the augmented Lagrange function has the same smoothness with the primal objective function and constrained function [3]. Compared with the penalty method, the augmented Lagrange method can obtain the optimal solution without the penalty parameter being infinitely large, but with the penalty parameter being larger than a threshold. And the optimal Lagrange multipliers have special significance in the corresponding practical problem. In this paper, we conduct the analysis on the optimal model for wireless optical communication system with

Gaussian pointing loss factor under the prespecified value of BEP. Based on the superiority of augmented Lagrange method, the optimal solutions to the model are obtained by using a classical quadratic augmented Lagrange algorithm. Then the optimal transmitter power, transmitter wavelength, transmitter telescope gain and receiver telescope gain are determined, which means that the WOC systems can operate well still by appropriately adjusting these system parameters to alleviate the unfavorable factor of building sway.

2. Background

We list the variables and parameters used in this paper in the following.

d:: distance between the transmitter and the receiver (meter);
D_R/D_T : receiver/transmitter aperture diameter (meter);
G_R/G_T: receiver/transmitter antenna gain;
L_A: atmospheric loss factor;
R_h: detector's responsivity factor (ampere/watt/meter) ;
/ : receiver/transmitter radial pointing error angle (radian);
$L_R(G_R,\ )/L_T(G_T,\ )$: receiver/transmitter pointing loss factor;
P_R/P_T: received/transmitted optical power (watt);
$_R/_T$: optics efficiency of the receiver/transmitter;
: wavelength (meter);
$_H/_V$: azimuth/elevation pointing error deviation (radian);
$_{2N}$: covariance of the additive white Gaussian noise.

From the basic form of received optical power as follows

$$P_R = P_T G_T G_R \left(\frac{\lambda}{4\pi d}\right)^2 \eta_T \eta_R L_A L_T(G_T,\theta) L_R(G_R,\gamma),$$

for the assumptions that $L_T(G_T,\theta) = \exp\left(-G_T\theta^2\right)$ in the Gaussian beam case and $L_R(G_R,\gamma) \approx 1$ if the receiver telescope with a sufficiently wide field of view (FOV) is used, the BEP with the Gaussian pointing loss factor was reported in [8]:

$$BEP = \frac{1}{2}\int_0^\infty erfc\,[\alpha_0 \lambda P_T G_T \exp\left(-2G_T\sigma^2 x\right)] \exp(-x)dx\ ,$$

where $x = \dfrac{\theta^2}{2\sigma^2}, \alpha_0 = \left(\dfrac{R_h}{2\sqrt{2}\sigma_N}\right)\eta_T\eta_R\left(\dfrac{\lambda}{4\pi d}\right)^2 G_R L_A$. Furthermore, for convenience of computation, the normalized transmitter power and the normalized square-root transmitter gain are defined as $v = (\alpha_0 \pi D_T P_T)/\sigma$, $z = (\pi D_T \sigma)/\lambda$, then the BEP is simplified as

$$BEP = f(v,z) = \frac{1}{2}\int_0^\infty erfc\,[vz\ \exp\left(-2xz^2\right)] \exp(-x)dx\ . \qquad (1)$$

Under the prespecified value of BEP, i.e., BEP=B, the mathematical model to find the optimal system parameters is established in [8]:

$$\begin{aligned} &Min\ v \\ &st.\ g(v,z) = 0, \end{aligned} \qquad (2)$$

where $g(v,z) = f(v,z) - B$, and $\alpha_0, D_T, \theta, \sigma$ are taken as constants. The optimal solution to the model (2) was reported in [8] by solving the following system of nonlinear equations

$$\partial f(v,z)/\partial z=0, \quad g(v,z)=0. \tag{3}$$

And then the optimal system parameters were established.

However, the concrete algorithm for (3) was not given in [8]. Furthermore, during (3) is solved by some algorithm, not only first-order derivative of $f(v,z)$, in which the integrands concerns erfc(z), with respect to the variable z is involved, i.e., $\partial f(v,z)/\partial z$, but also $\partial^2 f(v,z)/\partial^2 z$ need to be calculated, which make the optimal solution difficulty to obtain numerically. For the superiority of augmented Lagrange method, the model (2) will be solved by using a classical quadratic augmented Lagrange algorithm in this paper. The detailed augmented Lagrange algorithm and the numerical results for the model (2) are given in the next section.

3. The Numerical Experiments

We will list the augmented Lagrange algorithm for model (2) and then report the detailed optimal solutions to the model (2) in this section.

Let $x=(x_1,x_2)^T=(v,z)^T \in R^2$, then the model (2) is written as the general optimization model with equality constraints

$$\begin{aligned} &Min\ x_1 \\ &s.t.\ \ g(x)=0. \end{aligned} \tag{4}$$

The quadratic augmented Lagrange function corresponding to (4) is constructed as follows

$$F(x,\lambda,M)=x_1+ug(x)+\frac{M}{2}g^2(x), \tag{5}$$

where $M>0$ is the penalty parameter, u is the Lagrange multiplier of constraint $g(x)=0$.

3.1. The Augmented Lagrange Algorithm

Based on the quadratic augmented Lagrange function (5), the concrete algorithm is listed in the following.

Step 1 Choose $M^{(0)}=1$, $u^{(0)}=1$, $\varepsilon=10^{-6}$ and let k:=0;

Step 2 Solve $x^{(k)}=\arg\min_x F(x,u^{(k)},M^{(k)})$;

Step 3 If $|g(x^{(k)})|\le\varepsilon$, then stop and $x^{(k)}$ is the approximate optimal solution to the model (2); else go to Step 4;

Step 4 Modify the Lagrange multiplier $u^{(k)}$ and penalty parameter $M^{(k)}$

$$u^{(k+1)}=u^{(k)}+M^{(k)}\times g(x^{(k)}),$$

$$M^{(k+1)}=2^k;$$

Let k: =k+1, go to Step 2.

3.2. Numerical Results

Using the above augmented Lagrange algorithm, we perform the numerical experiments for model (2) in which $B=10^{-3}$, $B=10^{-4}$, $B=10^{-5}$, $B=10^{-6}$, $B=10^{-7}$, $B=10^{-8}$, $B=10^{-9}$ and $B=10^{-10}$, respectively. The numerical experiments are implemented in MATLAB 7.1 runtime environment. The minimum point $x^{(k)}$ in Step 2 is obtained by the FMINUNC function from the Optimization Toolbox in MATLAB 7.1. And during this experiment, the BEP function $f(v,z)$ is not directly implemented in the form of (1) but in the following equivalent expression derived in [8]:

$$f(v,z)=\left(\frac{1}{4z^2}\right)\times\left(\frac{1}{vz}\right)^{\frac{1}{2z^2}}\times\int_0^{vz} erfc(s)s^{1/(2z^2)-1}ds \qquad (6)$$

in order to improve the efficiency of numerical computation.

The following table shows the detailed numerical results, *B* column denotes the prespecified values of BEP in model (2), v^* column and z^* column are the optimal solution to the model (2) with the prespecified value *B* respectively, u^* column is the optimal Lagrange multiplier and M^* column is the terminal value of *M* in the algorithm.

Table 3.1 Numerical Results

B	v^*	z^*	u^*	M^*
10^{-3}	10.787100818018	0.275174063651	1655.42937007216	16777216
10^{-4}	14.59452777894	0.23732142485	16494.660964832	1073741824
10^{-5}	18.386612180739	0.211654307102	159961.191264993	34359738368
10^{-6}	20.822819609920	0.198940748580	539481.109173006	68719476736
10^{-7}	25.962724172528	0.176462406865	1539431.34615309	4.5036e+015
10^{-8}	29.7915785238	0.16643574064	5374812.6071930	1.1259e+015
10^{-9}	33.58793484773	0.156756326365	15236881.1771831	1.8014e+016
10^{-10}	37.36576720047	0.148572096776	25794584.1061635	4.5036e+015

Remark 3.1 Table1shows that the values of v^* range from approximately 37.36576720047 at BEP = 1e-10 to about 10.787100818018 at BEP = 1e-3.

Remark 3.2 Table1shows that the values of z^* range from approximately 0.148572096776 at BEP = 1e-10 to about 0.275174063651 at BEP = 1e-3.

Remark 3.3 For the transmitter with D_T = 10 cm and = 2 μrad, from $\lambda=\frac{\pi D_T\sigma}{z}$, when BEP = 10^{-10}, the optimal wavelength *opt* • 4.2290480325 μm and when BEP =10^{-3}, the optimal wavelength *opt* • 2.2833493912 μm.

Remark 3.4 For the transmitter with D_T = 10 cm, = 2 μrad and $\alpha_0=1$, from $P_T=\frac{v\sigma}{\alpha_0\pi D_T}$, when BEP = 10^{-10}, the optimal transmitter power $P_T opt$ • 237.8778623292 μw and when BEP =10^{-3}, the optimal wavelength $P_T opt$ • 68.6728167402μw.

Remark 3.5 For the transmitter with = 2 μrad, from $G_T=\frac{z^2}{\sigma^2}$, when BEP = 10^{-10}, the optimal transmitter power $G_T opt$ • 0.0055184170 and when BEP =10^{-3}, the optimal wavelength $G_T opt$ • 0.01893019133.

Remark 3.6 For the receiver with D_R = 20 cm, and the transmitter D_T = 10 cm, = 2 μrad, from $G_R=\left(\frac{D_R z}{D_T\sigma}\right)^2$, when BEP = 10^{-10}, the optimal transmitter power $G_R opt$ • 0.022073668 and when BEP =10^{-3}, the optimal wavelength $G_R opt$ • 0.0757207653.

Remark 3.7 The change relationships between the optimal transmitter power and the optimal wavelength, and the corresponding prespecified BEP value respectively, are revealed in the following Fig. 1, in which the BEP values are put on the abscissa axis, v^* (represented by data1) and $1/z^*$ (represented by data1) are put on the ordinate axis.

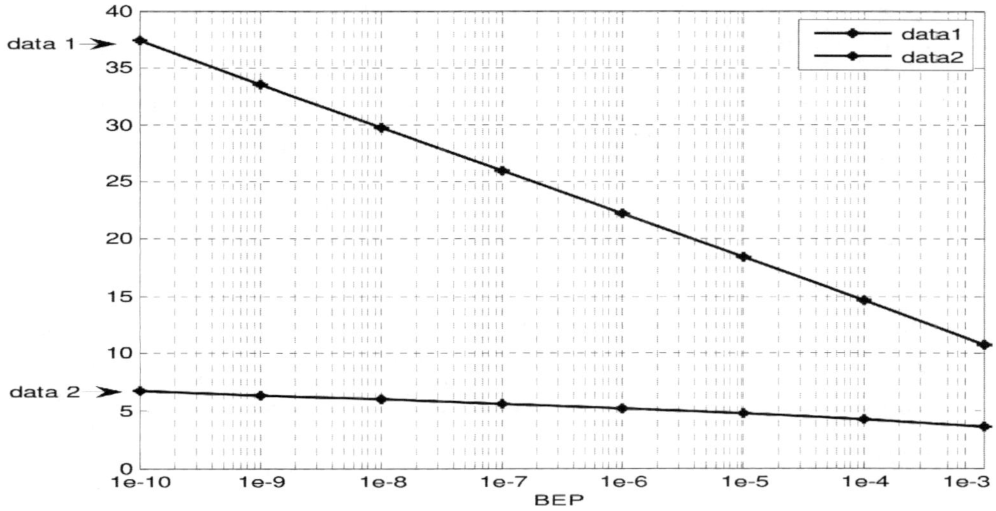

Fig. 1 Relationships between v^*, $1/z^*$ and BEP value respectively

The above figure shows that the BEP value increases as v^* or $1/z^*$ decreases at the optimal points, which implies that under the given D_T, σ and α_0, if the BEP value is larger, both the optimal transmitter power and the optimal wavelength are less.

4. Conclusions

In this paper, we present the optimal system parameters for the wireless optical communication system with Gaussian pointing loss factor by using the classical quadratic augmented Lagrange algorithm to solve the corresponding mathematical model with the BEP value being prespecified. The detailed numerical results such as transmitter power, transmitter wavelength, transmitter telescope gain and receiver telescope gain are reported. The presented solutions demonstrate that the BEP value decreases as the transmitter power or the wavelength increases at the optimal points.

Acknowledgements: This work was supported by the Fundamental Research Funds for the Central Universities (No.2010-Ia-040) and the National High-Tech Research and Development Plan of China (Grant No. 2008AA01Z207).

References

[1] Arnon S 2003 *IEEE Trans. wireless commun.* **2** 626
[2] Bertsekas B P 1982 Constrained Optimization and Lagrange Multiplier Methods. Academic Press, New York
[3] Blaser S, Hofstetter D, Beck M, and Faist J 2001 *IEEE Electron. Lett.* **37** 778
[4] Chen C C and Gardner C S 1989 *IEEE Trans. Commun.* **37** 252

[5] Gagliardi R M and Karp S 1976 *Optical Communications* New York: Wiley
[6] Kedar D and Arnon S 2004 *IEEE Commun. Mag.* **42** S2
[7] Liu X 2007 *Proc. IEEE ICC.* **67** 1247
[8] Liu X 2009 *IEEE transactions on wireless communication* **8** 492
[9] Martini R, Bethea C, Capasso F, Gmachl C, Paiella R, Whittaker E A, Hwang H Y, Sivco D L, Baillargeon J N, and Cho A Y 2002 *IEEE Electron. Lett.* **38** 181
[10] Martini R and Whittaker E A **2005** *J. Optical Fiber Commun. Reports.* **2** 279
[11] Simon M K and Vilnrotter V A 2005 *IEEE Trans. Wireless Commun.* **4** 35
[12] Wilson S G, Brandt-Pearce M, Cao Q, and Leveque J H 2005 *IEEE Trans.Commun.* **53** 1402

Journal of Physics: Conference Series **276** (2011) 012062 doi:10.1088/1742-6596/276/1/012062

A full-duplex radio-over-fiber system with differential phase-shift keying signals

Jing He, Dong Yang, Lin Chen

School of Computer and Communication, Hunan University, Changsha 410082, China

Email: hnu_jhe@hotmail.com

Abstract. We propose a full-duplex radio-over-fiber (ROF) system transmitting 2.5 Gb/s differential phase-shift keying (DPSK) signals with 40GHz optical millimeter-wave as downlink. Meanwhile it can be reused central wavelength as uplink connection for transmitting 2.5 Gb/s on-off keying (OOK) signals. The experimental and simulation results show that the downstream 2.5Gb/s DPSK data and the upstream 2.5Gb/s OOK data can transmit 40km single-mode fiber successfully.

1. Introduction

In the future wired and wireless networks, access bandwidth which used to provide for high-speed data and video services will increase to gigabit per-second. The combination of differential phase-shift keying (DPSK) signal and radio-over-fiber (ROF) techniques have become attractive solutions in realizing future broadband wireless networks because ROF technique can be used for the distribution of wireless signals. And DPSK signal has high spectrum efficiency and the tolerance to chromatic dispersion, which can increase the bandwidth and extend transmission distance over fiber. Previous investigation reported the generation of high cost-effective millimeter-wave signal, such as the frequency doubling using an optical carrier suppression modulation and frequency quadrupling[1-4], which can reduce the required bandwidth of optoelectronic devices and easy to achieve using commercial optoelectronic devices. At present, most millimeter-wave signals can support on-off keying (OOK) format, but seldom support vector modulation formats, such as phase-shift keying signal and quadrature amplitude modulation format, which is of importance for future broadband wireless applications.

In the paper, we have proposed a full-duplex ROF system transmitting 2.5 Gb/s DPSK signals with 40GHz optical millimeter-wave as downlink, Meanwhile it can be reused central wavelength as uplink connection for transmitting 2.5 Gb/s OOK signals, which can effectively utilize optical power. At the same time, we have simulated the novel duplex ROF system that it can be successfully transmitted 40km standard single-mode fiber (SMF).

2. Principle

Figure 1 shows the principle of full-duplex ROF architecture, which can generate optical millimeter-wave by using a differential Mazh-Zehnder modulator (DMZM) along with a FBG and realize wavelength reuse for upstream connection. In the central station (CS), we use a small RF signal to drive the DMZM. As DMZM is biased at the minimal output power, the odd-order optical sidebands can be suppressed. Then a circulator and a FBG are used to filter out the optical carrier. Two second-order optical sidebands are obtained at the output of FBG. In this way, the optical millimeter-wave with four times the frequency of the microwave drive signal is generated when the two second-order optical sideband signals are detected by a photodiode. However the remained optical carrier will be reflect from the other output of FBG. After the optical millimeter-wave is modulated by another DMZM driven by the downstream DPSK data, we use an optical coupler to combine the remained optical carrier and the optical millimeter-wave signal before they are transmitted to the base station (BS). At the BS, the optical millimeter-wave signal and the remained optical carrier are separated by another circulator and FBG as that one used in the central office. Since photo detection is inherently insensitive to the optical phase, DPSK cannot directly be received using square-law detection of photodiode, a delay interferometer (DI) is inserted in the optical path at the receiver to convert the differential phase modulation into intensity modulation. This optical preprocessing is necessary to accomplish demodulation [5]. The optical millimeter-wave signal will be broadcast by an antenna after optical-electrical (O/E) conversion, while the reflected optical carrier is acted as the continuous wave (CW) and modulated by an intensity modulator driven by the upstream OOK data. The upstream optical data will be sent back to the CS, where a low-cost low-frequency receiver detects the upstream data.

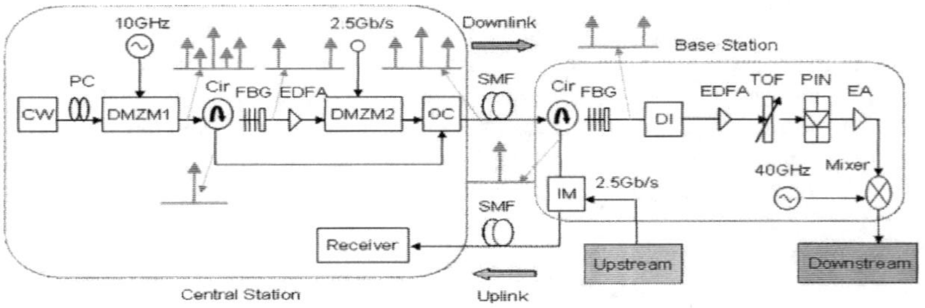

Figure 1. Principle of full-duplex ROF system

3. Experiment and simulation results

The experiment setup for optical millimeter-wave generation, transmission, and optical carrier reuse for uplink connection is shown in Figure 2. At the CS, a CW lightwave is generated by a tunable laser at 1543.7 nm. The DMZM is driven by a 10GHz RF sinusoidal wave with a peak-to-peak voltage of 1 V. The optical spectrum after modulation is shown as inset (a) in Figure 2.Since the DMZM is biased at the minimal output power, where it can suppress odd-order sidebands; therefore, two second-order sidebands are obtained after using a circulator and a FBG to filter out optical carrier. In this way, the optical millimeter-wave with two second-order sidebands has four times the RF frequency. The spectrum of the optical millimeter-wave signal is shown as inset (b) in Figure 2. Then the generated optical millimeter-wave is amplified by an erbium-doped fiber amplifier (EDFA) to get a power and modulated by another DMZM driven by a downstream 2.5Gb/s DPSK data. The remained optical carrier is reflected from the other output of FBG, and the optical spectrum is shown in Figure 2 as insets (c). After being combined by a 3dB optical coupler, the optical millimeter-wave signal and the remained optical carrier are transmitted 40km standard SMF before reaching the base station. The optical spectrum of the optical millimeter-wave signal and the remained optical carrier is shown in Figure 2 as insets (d).

At the BS, another circulator and FBG are used to separate the optical millimeter-wave signal and the remained optical carrier. The FBG filter has a 3dB reflection bandwidth of 0.15 nm and reflection ratio larger than 25 dB at the reflection peak wavelength, which is to filter the carrier to generate optical millimeter-wave for the downlink and to reflect the carrier acting as the CW for the uplink. The optical spectrum is shown in Figure 2 as insets (e) and (f), respectively. For the downlink, a delay interferometer (DI) is inserted in the optical path at the receiver to convert the differential phase modulation into intensity modulation. The DI lets two adjacent bits interfere with each other its output ports. This interference leads to the presence (absence) of power at a DI output port if two adjacent bits interfere constructively (destructively) with each other. Thus, the preceding bit in a DPSK encoded bit stream acts as the phase reference for demodulating the current bit. Then the optical millimeter-wave signal can be detected after preamplifier by a regular EDFA with a small-signal gain of 30dB, and then filtered by a tunable optical filter with the bandwidth of 0.5nm to suppress amplified spontaneous emission noise before optical-electrical down-conversion via a PIN photodiode with a 3dB bandwidth of 60GHz. The converted electrical signal is amplified by an electrical amplifier with a bandwidth of 10GHz centered at 40GHz. For the uplink, the reflected optical carrier from the FBG is modulated by an intensity modulator driven by the upstream 2.5Gb/s OOK data. The upstream optical signal is injected into the second 40km SMF and transmitted back to the CS.

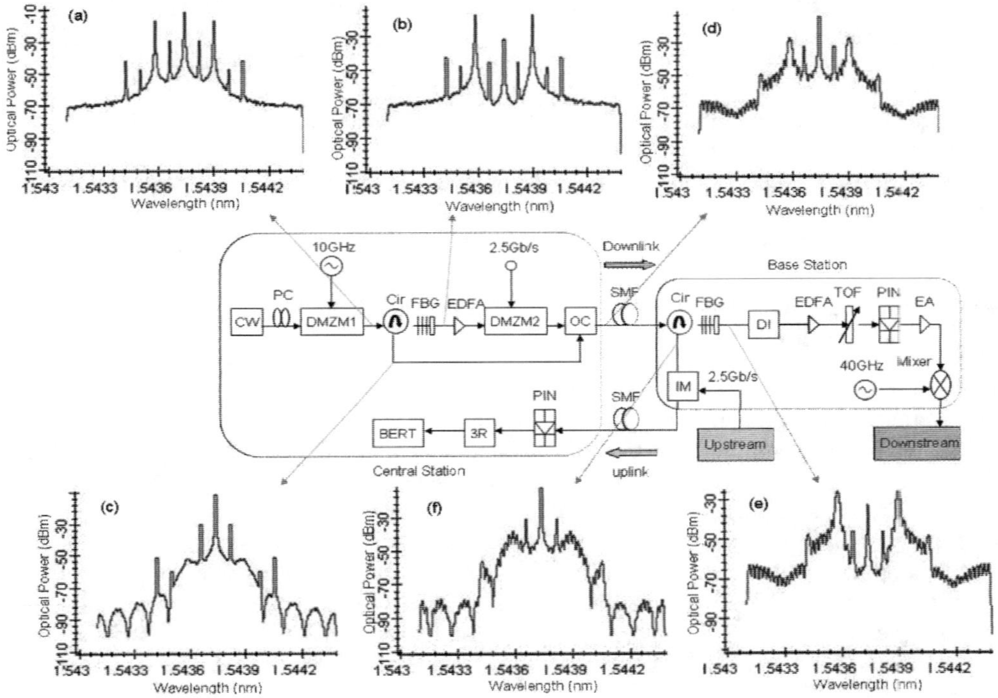

Figure 2. Experimental setup for full-duplex ROF system

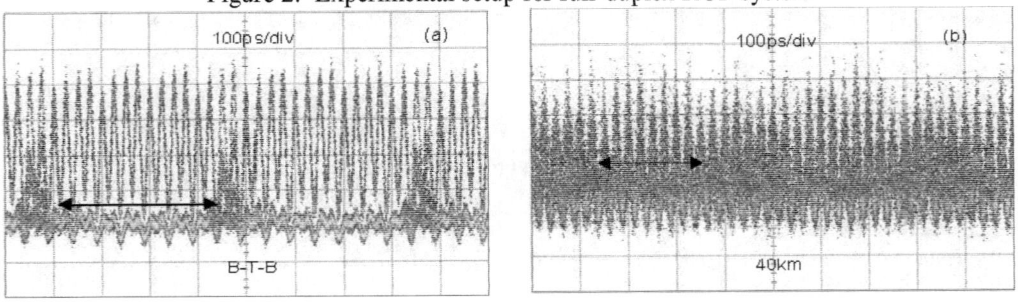

Figure 3. Measured eye diagrams of optical millimeter-wave signals. (a) back-to-back (b) transmission 40km SMF

For the downlink, the measured eye diagram of optical millimeter-wave signals at back-to-back and after transmission 40km in the SMF are shown in Figure 3. Figure 3(a) is the optical millimeter-wave signal before combining by the 3dB optical coupler. We can see that the 2.5Gb/s DPSK signal is carried by the 40GHz optical millimeter-wave. Figure 3(b) is the optical millimeter-wave after transmission 40km SMF. Fiber dispersion changed the shape of the optical millimeter-wave signal, which leads a different shape compared with the original millimeter-wave signal.

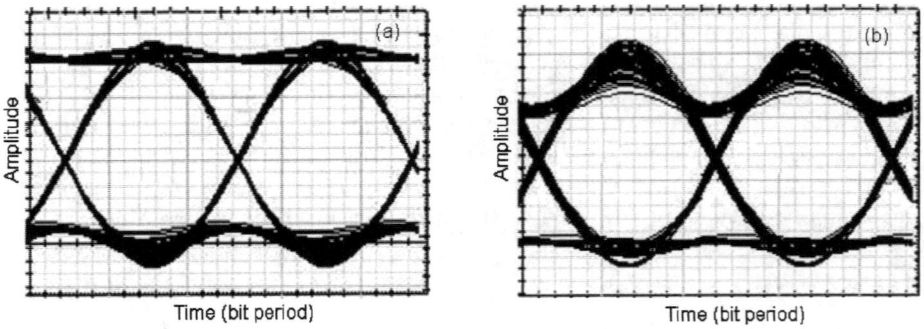

Figure 4. Measured eye diagrams of DPSK signal after demodulation. (a) back-to-back (b)transmission 40km SMF

Because of the limited experimental condition, the demodulation of DPSK signal for the downlink is used by simulation. The eye diagrams of demodulated DPSK signal at back-to-back and after transmission 40km in the SMF are shown in Figure 4. We can see that the 2.5Gb/s DPSK signal is carried by the 40GHz optical millimeter-wave. And the eye still keeps open despite transmission over 40km SMF, which is long enough for access network converge. It means that DPSK signal has good tolerance to chromatic dispersion.

4. Conclusions

We have proposed a full-duplex ROF system with optical carrier reuse for uplink connection. In this ROF system, the optical millimeter-wave is generated by using a DMZM along with a FBG. The optical millimeter-wave signal was generated with four times the frequency of the microwave drive signal, which can reduce the bandwidth of microwave component and modulator. The remained optical carrier is delivered to the base station for wavelength reuse for carrying the upstream data; it can effectively utilize optical power. From the experimental and simulation results, it can be seen that the downstream 2.5Gb/s DPSK data and the upstream 2.5Gb/s OOK data can transmit 40km single-mode fiber successfully.

Acknowledgments

Our thanks to the Open Fund of Key Laboratory of Optical Communication and Lightwave Technologies (Beijing University of Posts and Telecommunications, Ministry of Education , P. R. China).

References

[1] Kaszubowska A, Hu L, Barry, L P 2006 IEEE Photon. Technol. Lett., 18 562
[2] Chen L, Lu J, He J, et al., 2009 Opt. Fiber. Technology. 15 222
[3] Jia Z, Yu J J, Chang G K 2006 IEEE Photon. Technol. Lett., 18 1726
[4] Bakaul M, Nirmalathas A, Lim C 2005 IEEE Photon. Technol. Lett., 17 2718
[5] Winzer P J, Kim H 2003 IEEE Photon. Techonl. Lett., 15 1282

Journal of Physics: Conference Series **276** (2011) 012063 doi:10.1088/1742-6596/276/1/012063

Temperature dependence of active photonic band gap in bragg-spaced quantum wells

Zhiqiang Hu, Tao Wang, Chunchao Yu and Wei Xu
Wuhan National Laboratory for Optoelectronics; College of Optoelectronic Science and Engineering, Huazhong University of Science and Technology, Wuhan 430074 ,Hubei, P. R. China.

E-mail: huzhiqianghzq@163.com

Abstract. A novel all-optical polarization switch of active photonic band gap structure based on non-resonant optical Stark effect bragg-spaced quantum wells was investigated and it could be compatible with the optical communication system. The theory is based on InGaAsP/InP Bragg-spaced quantum wells (BSQWs). Mainly through the design of the InGaAsP well layer component and InP barrier thickness to make the quantum-period cycle meet the bragg condition and the bragg frequency is equal to re-hole exciton resonance frequency. When a spectrally narrow control pulse is tuned within the forbidden gap, such BSQWs have been shown to exhibit large optical nonlinearities and ps recovery times, which can form T hz switch. However, the exciton binding energy of InGaAsP will be automatically separate at room temperature, so the effect of all-optical polarization switching of active photonic band gap bragg structure quantum wells can only be studied at low temperature. By a large number of experiments, we tested part of the material parameters of BSQWs in the temperature range 10-300K. On this basis, the InGaAsP and InP refractive index changes with wavelength, InP thermal expansion coefficient are studied and a relationship equation is established. Experimental results show that the bragg reflection spectra with temperature mainly is effected by InP refractive index changes with temperature. Our theoretical study and experiment are an instruction as a reference in the designs and experiments of future practical optical switches.

1.Introduction

As the rapid development of the optical communication, the demand for ultrafast switching is increasingly urgent. However the response time of traditional switches are limited by a variety of inevitable factors. Whereas all-optical switches have attracted more and more interests because of their favorable optical characteristics and much faster switching speed. All-optical polarization switches that make use of nonlinear optical effects in semiconductor multiple quantum wells have exhibited potentials for the application in many fields[1].

The InGaAsP/InP system is the essential for the efficient operation of a number of optoelectronic and transport devices[2]. The all-optical polarization switches adopting InGaAsP/InP bragg spaced quantum wells (BSQWs) is investigated because it can be compatible with the optical communication system. The exciton binding energy of InGaAsP is 3meV in 20K, but it is automatically separated in room temperature, so the optical switch can only be achieved at low temperatures. But, a study of the temperature behavior of the InGaAsP /InP system is lacking.

In this article we report a detailed study of 20 periods InGaAsP/InP BSQWs fabricated by low pressure metal-organic chemical vapor deposition (LP-MOCVD). We have measured the photoluminescence (PL) spectra of InGaAsP/InP BSQWs in the temperature range10-300K. In order to understand the temperature behavior of the quantized transitions, we propose a model that takes into account the temperature-induced modification of the exciton energy. And also we have measured the reflection spectra of InGaAsP/InP BSQWs in the temperature range10-300K. Therefore, through our work we can change some parameters to promote the peak of exciton spectra to equal to the reflection spectra of the BSQWs, which can form T hz switch.

2. Experiment details

The BSQWs was fabricated by the LP-MOCVD method in a conventional, horizontal reactor, using trimethylindium, triethylgallium, AsH3,PH3. PH3 diluted to 25% and AsH3 diluted to 5% in H2 as source materials. The 20 periods InGaAsP/InP BSQWs sample was grown on undoped (001) InP substrate. Growth began with a 150nm InP buffer layer, followed by the 8nm In0.76Ga0.24As0.3P0.7 quantum well and 174nm InP barrier layer with all layers undoped. The structure schematic of BSQWs is shown in Figure 1. The LP-MOCVD growth conditions: the chamber pressure is 150mbar, the growth temperature is 620 °C, V/III ratio is 240, the growth rate is 0.3nm/s and 30-seconds growth stop at the quantum well interfaces. Nominal InGaAsP alloy composition was estimated from PL measurements and double x-ray diffraction measurements of the thick layer which were grown under the same conditions. The same thick layer was also used to measure the InGaAsP growth rate. The growth conditions indicated well width 174±0.4nm.

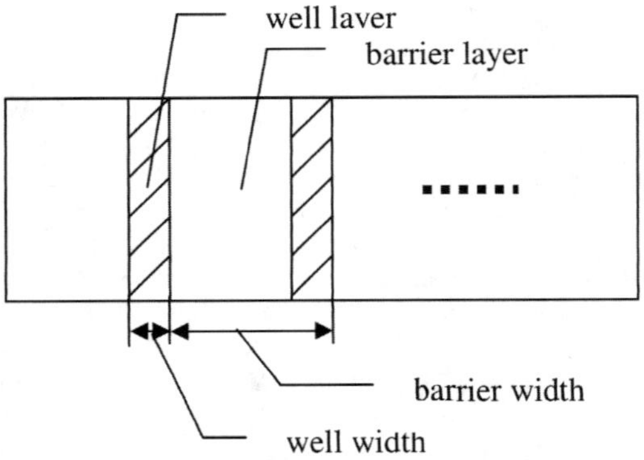

Figure 1. The structure schematic of BSQWs

For forming resonance frequency, the width of well and barrier have to meet the modified bragg condition:

$$n_b d_b + \frac{1+\rho}{1-\rho} n_w d_w = \frac{\lambda}{2} \tag{1}$$

Where, n_b , n_w are the refractive indices of barrier and well, d_b , d_w are the width of barrier and well,, ρ is the Fresnel coefficient. λ is quantum well excitonic resonance wavelength[3].

This switch operated by a circularly polarized control pulse to induce circular dichroism and birefringence in multiple quantum wells, and the absorption coefficient and index of refraction for right and left circularly polarized light are different. So this circular anisotropy changes the polarization state of the linearly polarized signal pulse. The change in the signal polarization is subsequently converted to a modulation of the signal amplitude by additional polarization sensitive elements.

The PL measurements were performed with an Ar+ laser(λ =514.5nm) as the excitation source. The signal was dispersed by a 1 m spectrometer and detected by a liquid-nitrogen-cooled Ge p-i-n photodiode which using conventional lock-in techniques[4]. The sample temperature was maintained in a continuous flow helium cryostat operating in the 10-300 K range.

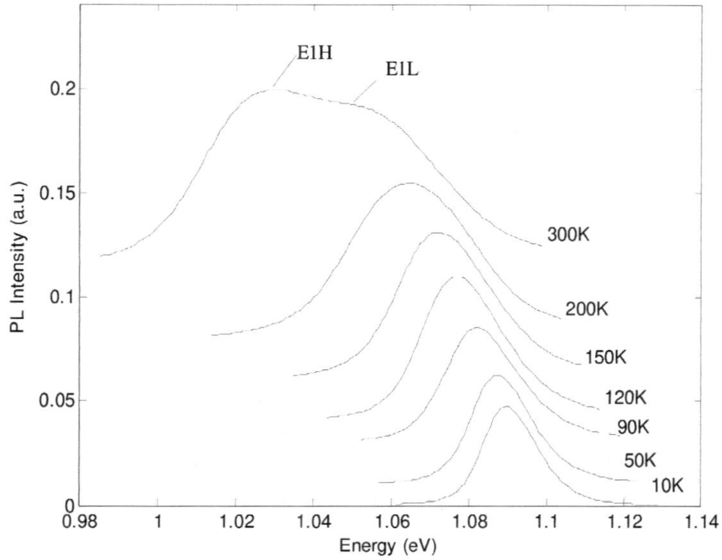

Figure 2. Photoluminescence spectra for the sample at various temperatures. The transition between the n =l electron subband to the n =l heavy flight hole subband is labeled E_{1H}.

Displayed in Figure 2 are the PL spectra of the BSQWs for the Photoluminescence spectra at 10, 50, 90, 120, 150, 200, and 300 K. The dominant PL peak positions shift to lower energies and the linewidths broaden with increasing temperature. The maximum of the PL spectra is due to the E_{1H} transition between the n =1 electron subband and the n =1 heavy-hole subband. At room temperature the light-hole exciton peak E_{1L} can also be seen and its intensity falls with the decrease of the temperature. In 10K, E_{1L} reduction disappeared and just have a sharp perfect spectra, which is the E_{1H} transition only be excited and that is what we want.

To facilitate comparison of PL spectra at different temperatures, the measured peak of each temperature PL spectra are drawn in Figure 3. Below 80K a small shift is observed with decreasing temperature. Using Varshni's equation to fit the experimental data

$$E_{1H}(T) = E_{1H}(0) - \alpha T^2 /(\beta + T) \quad (2)$$

where T is the absolute temperature measured in K.

$E_{1H}(0)$ is the n =l electron subband to the n =l heavy energy gap at 0 K, and α and β are adjustable parameters, Using Varshni's equation to fit the experimental data, the obtained values of α =-0.0008579eV/K and β =952.2K.

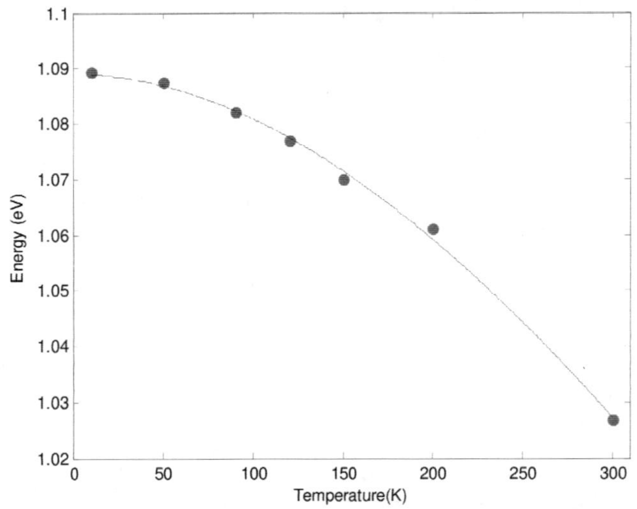

Figure 3. Temperature dependence of the excitonic transition energies for the InGaAsP/InP BSQWs. The dots are experimental test value. The curve is fits to Eq.2.

Described in the front of this paper, the BSQW structure is a periodically spaced sequence of quantum wells forming a one-dimensional photonic crystal, thus forming a strong reflection at the range of bragg spaced resonance frequency. We have measured the reflection spectra of BSQWs for the spectral range at 10, 50, 90, 120, 150, 200, and 300 K, which are Shown in Figure 3. The dominant reflection spectra peak positions shift to lower energies and the linewidths broaden with increasing temperature.

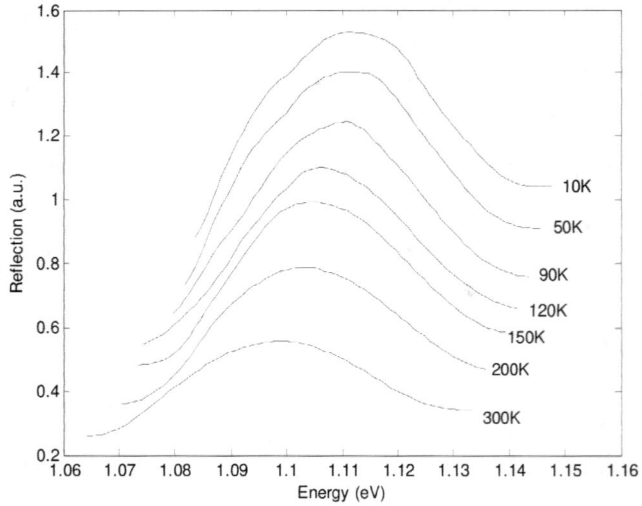

Figure 4. Reflection spectra for the BSQWs at various temperatures.

The refractive indices and thermal expansion changes of InP and InGaAsP in the infrared have been studied by many researchers. There is generally good agreement among most of the studies on the values of the room-temperature refractive indices, but at some temperatures and wavelengths, no values are available. In our optical switch we can achieved in a temperature range, not just a temperature dot. We can use the infrared laser interferometric thermometry (IRLIT) to measure the temperature of semiconductor wafers. In IRLIT, a low-power infrared laser which the photon energy is below the band-gap energy of a semiconductor is directed at the wafer of known thickness h which is must polished on both sides. The intensity of an infrared laser beam is transmitted through, or reflected off, and the sample of semiconductor wafer was measured as the sample temperature was slowly varied. Either the reflected or transmitted intensity is measured by a same photodiode. As the temperature of the wafer is increased or decreased the optical path length nh within the wafer increases or decreases, that can cause interference between the reflections in the front and back surfaces. The direction of temperature change can be measured by exploiting variations in wafer thickness, or by modulating the laser wavelength. The change of the optical path length in the semiconductor is dominated by the temperature dependence of n,

$$\beta(T) = \frac{1}{n}\frac{dn}{dT} \tag{3}$$

with a smaller contribution from thermal expansion,

$$\alpha(T) = \frac{1}{h}\frac{dh}{dT} \tag{4}$$

where n is the sample refractive indices, h is the sample thickness.

Because the interferograms are dominated by changes in the refracion index, $\beta(T)$ can be obtained by subtracting the contribution of $\alpha(T)$ from the interferometric data[5]. Since $\alpha(T)$ is small relative to $\beta(T)$, and well known $\beta(T)$ can be obtained with good accuracy. Using the experimental data in Ref. 19 we have calculated the coefficients of the polynomial expressions for $\alpha(T)$ and $\beta(T)$.
The peak of bragg reflection spectra and the previous calculation on the same graph are shown in Figure 4.

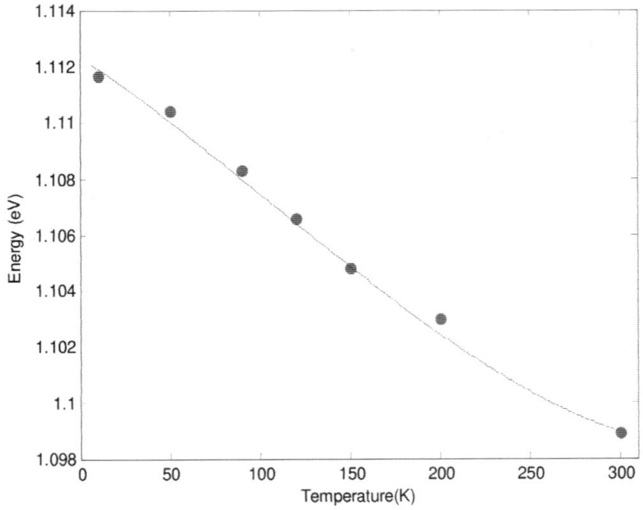

Figure 5. Temperature dependence of the Bragg reflection spectra energies for the InGaAsP/InP BSQWs. The dots are experimental test value. The curve is Simulation to Eq.1.

Because $\alpha(T)$ and $\beta(T)$ are increase with increasing temperature[6], result in the dominant bragg reflection spectra positions shift to lower energies and the linewidths broaden with increasing temperature. Computation show that the bragg reflection spectra with temperature mainly is effected by InP refractive index changes with temperature. Through the graph we can see that the bias of theoretical calculations and actual test results is small, within the experimental requirements. Our theoretical study and experiment are an instruction as a reference in the designs and experiments of future practical optical switches.

3. Conclusions

In conclusion, The temperature dependence of PL and reflection spectra in InGaAsP/InP BSQWs has been investigated experimentally and theoretically. The theoretical calculation is in good agreement with experimental results and it is shown that the dominant PL and reflection spectra peak positions shift to lower energies and the linewidths broaden with increasing temperature. However, the changes slope of PL spectra is much faster than that of reflection spectra. So we can design the components of BSQWs, reach the PL spectra peak and the bragg reflection spectra peak intersect within the range at 10 - 80K. The value of this experiment is that it is proved to be the possibility of all-optical polarization switches based on the InGaAsP/InP BSQWs structure, mainly through the design of the InGaAsP well layer component and InP barrier thickness to make the quantum-period cycle meet the bragg condition and the bragg frequency is equal to re-hole exciton resonance frequency. This study can be used as a basis of experimental research of all-optical spin-dependent polarization switching in BSQWs.

Acknowledgements

This work was Supported by the National Natural Science Foundation of China (Grant No. 60877040), National High Technology Research and Development Program of China (Grant No. 2008AA03Z404) and National Basic Research program of China (Grant. 2010CB923204).

References

[1] J. Prineas, J. Zhou, J. Kuhl, H.M. Gibbs, G. Khitrova, S.W. Koch, and A. Knorr 2002 *J.Appl. Phys.Lett.* **81** 4332
[2] Tao Wang, Gang Li, and Zheng Chen 2007 *Optics Express.* **16** 127
[3] Y. -G. Zhao, R. A. Masut, J. L. Brebner, C. A. Tran, and J. T. Graham 1995 *J.Appl. Phys.* **76** 5923
[4] C. F. Li, D. Y. Lin, and Y. S. Huanga 1997 *J.Appl. Phys.* **81** 401
[5] J. A. McCaulley,• V. M. Donnelly, M. Vernon, and I. Tahat 1994 *Amercian Physical Society.* **49** 7408
[6] W. J. Johnston, J. P. Prineas, and Arthur L. Smirl 2007 *J. Appl. Phys.* **101** 046101.

Journal of Physics: Conference Series **276** (2011) 012064 doi:10.1088/1742-6596/276/1/012064

Tuneable frequency up-conversion based on biased asymmetric coupled quantum well structure

Zhenhua Hu[1] and Dexiu Huang[2]
[1] Department of Physics, Science College, Wuhan University of Technology,, 430063, Wuhan, Hubei, China
[2] Wuhan national laboratory for optoelectronics, Huazhong University of Science and Technology, 430074, Wuhan, Hubei,China

E-mail: hzh267@sohu.com

Abstract. The behaviours of the optical nonlinear susceptibility $\chi^{(3)}$ responsible for the phase-conjugate beam of frequency conversion in non-degenerate four-wave mixing (NDFWM) are studied for a biased asymmetric coupled quantum well (ACQW) structure. It is shown that the frequency up-conversion peak position determined by $\chi^{(3)}$ is very sensitive to the external inverse electric field strength applied among the grown direction of quantum well but its value is insensitive to that. In other words, the frequency up-conversion peak has a large shift but its value maintains a constant when the electric field strength increases in a small bias range. The characteristics of the tuneable-frequency and the power balance of the ACQW structure may provide the high-efficient wavelength conversion in the optical communication system.

1. Introduction

Asymmetric coupled quantum well (ACQW) structures [1, 2] have been an interesting issue for using optoelectronic devices because the coupling effect between the two wells can not only provide a large quantum-confined Stark shift of the transition energy [3], a large refractive index change and a large second-order optical nonlinearity [4], but also allow for the nonlinear resonances beyond terahertz detuning rates in non-degenerate four-wave mixing (NDFWM) [5].

In general, the NDFWM mechanism is physically Bragg scattering from the index gratings results from the beating between the input pump- and probe-beams for bulk semiconductor materials [6]. Such gratings are associated with the modulation of both carrier density occupation probability in each (sub)-band at the pump-probe detuning rate $\delta=|\nu_p-\nu_s|$, where ν_p and ν_s are the frequencies of pump- and probe-beams, respectively, carrier-density pulsations [7] provide the most efficient scattering mechanism; however, they are unfortunately limited by the recombination lifetime τ_s [8], which is typically on the order of 100ps, so that the effectiveness of a Bragg reflection signal of large detuning rate via resulting gratings rapidly decreases at detuning rates in excess of several tens gigahertz [9]. In this sense, the ACQW structure is propitious to use as transparent all-optical-wavelength converters [10] and phase-conjugate generators [11]. In this paper, it is discussed that the third-order optical nonlinear susceptibility $\chi^{(3)}$ responsible for the phase-conjugate beam when the external inverse electric field strength along the grown direction of quantum well changes. It is shown that the resonant peak position of $\chi^{(3)}$ in the frequency domain is not only controlled by the inverse electric field but also its peak value may maintain a constant in a

[1] To whom any correspondence should be addressed.

small bias range. From a view of application, the ACQW structure as a high-efficient wavelength converter in the optical communication system can realize tuneable-frequency and power balance for converted signal.

2. Coupled quantum well structure and wave functions

ACQW structure considered consists of a wide well (WW), a narrow well (NW) and a thin barrier, as shown in figure 1(a). The widths of WW, NW are L_w and L_n, respectively,

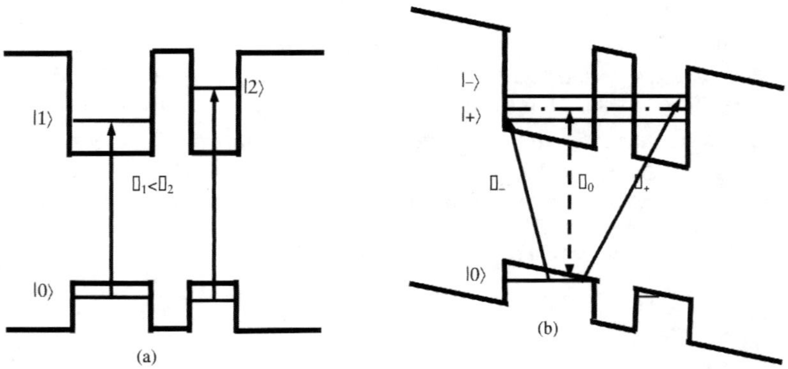

Figure1. Level diagrams of coupled quantum well structure: (a) is in uncoupled case, a double well structure, where ω_1 and ω_2 are respectively the transition angular frequencies between the first heavy-hole sublevel in the NW and that in the WW, and (b) is in coupled case, a ACQW, as an inverse static electric field is applied along the grown direction of quantum well, where ω_- and ω_+ are the transition angular frequencies from $|0\rangle$ to $|+\rangle$ and $|-\rangle$, respectively, and $\omega_0=(\omega_++\omega_-)/2$ is the central angular frequency of the upper two delocalized states $|+\rangle$ and $|-\rangle$ relative to a lower localized state $|0\rangle$.

and the thickness of the barrier layer between the two wells and L_b. If the L_b is thin enough, so that the significant overlap of electronic wave functions in the two wells may take place at bias, the lowest electronic states $|1\rangle$ and $|2\rangle$ are mixed into the new stationary states $|+\rangle$ and $|-\rangle$, and the heavy-hole may be viewed as a ground state due to large effective mass, as shown in figure 1(b). As a good approximation, the two delocalized states $|+\rangle$ and $|-\rangle$ in the conduction band are written, respectively, as [12]

$$|+\rangle = \frac{1}{\sqrt{2}}(|1\rangle + |2\rangle),\ |-\rangle = \frac{1}{\sqrt{2}}(|1\rangle - |2\rangle) \qquad (1)$$

with energy eigenvalues as E_- and E_+. The corresponding angular frequency splitting $\omega_c = (E_- - E_+)/\hbar$ may be written as

$$\omega_c = (\hbar\kappa / mL)\exp(-\kappa L), \qquad (2)$$

where $\kappa = 2\pi/\lambda\!\!\!^{-}, \lambda\!\!\!^{-} = 2\pi\hbar/\sqrt{2m(U-\bar{E})}$ is the de Broglie wavelength of the electron tunneling, $\bar{E} = (E_{1e}^W + E_{1e}^N)/2$ is the average quantum-confined energy of the electronic ground state in the two wells at zero bias and U is the barrier height depending on the biased electric field strength F along the growth direction of quantum well. The well depth at bias may be approximately written as $U = \Delta E_{CB} + eFL_b$, where ΔE_{CB} denotes the well depth at zero bias and eFL_b is a triangular potential, as a result of the external electric field applied along the growth direction of quantum well.

In the single particle state approximation, the lowest one-dimensionless quantum-confined wave function $\varphi_\alpha(z)$ for electron ($\alpha = e$) or heavy-hole ($\alpha = h$) in the two isolated wells satisfies the stationary Schrödinger equation [13]

$$\left\{\frac{-\hbar^2}{2m_{z\alpha}}\frac{\partial^2}{\partial z^2}+\left[V(z)\pm eFz\right]\right\}\varphi_\alpha(z)=E_\alpha\varphi_\alpha(z), \tag{3}$$

where $+$, $-$ correspond to electron and heavy hole, respectively, and $V(z)$ is the potential of quantum well. One assumes the electronic or heavy-hole effective mass in the well layer is the same that in the barrier layer and $V(z)=0$. Solving Eq.(3), $\varphi_\alpha(z)$ is given by linear combinations of the Airy function Ai(…) and Bi(…)

$$\varphi_\alpha(z)=c_\alpha\mathrm{Ai}(\eta_\alpha)+\tilde{c}_\alpha\mathrm{Bi}(\eta_\alpha) \tag{4}$$

with the argument as η_α linearly depending on the energy eigenvalue E_α and the position coordinate z at the quantum central position as zero:

$$\eta_z^\alpha=-\left[\frac{2m_\alpha}{(e\hbar F)^2}\right]^{1/3}(E_\alpha\mp eFz). \tag{5}$$

c_σ^α and $\tilde{c}_\sigma^\alpha$ are determined by the boundary condition of the single quantum well as shown in Figure 1.

3. Third-Order Nonlinear Optical Susceptibility

In the basis of both the density matrix approach and the rotting-wave approximation [14], the third-order optical nonlinear susceptibility $\chi^{(3)}$ responsible for the phase-conjugate beam via NDFWM is

$$\chi^{(3)}(\omega_s;-\omega_1,-\omega_1,\omega_2)=-\frac{i|\mu_{-0}|^4}{\varepsilon_0\hbar^3V}\frac{1}{\gamma_{-0}+i(\omega_0+\omega_c/2-\omega_1)}$$
$$\times\left[\frac{1}{\gamma_{-0}-i(\omega_0+\omega_c/2-\omega_1)}\frac{1}{\gamma_{00}+i(\omega_1-\omega_2)}-\frac{1}{\gamma_{-0}-i(\omega_0-\omega_c/2-\omega_2)}\frac{1}{\gamma_{--}+i(\omega_1-\omega_2)}\right.$$
$$\left.+\frac{1}{\gamma_{-0}+i(\omega_0+\omega_c/2-\omega_1)}\frac{1}{\gamma_{00}+i(\omega_1-\omega_2)}-\frac{1}{\gamma_{-0}+i(\omega_0+\omega_c/2+\omega_1)}\frac{1}{\gamma_{--}+i(\omega_1-\omega_2)}\right]$$
$$-\frac{i|\mu_{+0}|^4}{\varepsilon_0\hbar^3V}\frac{1}{\gamma_{+0}+i(\omega_0-\omega_c/2-\omega_1)}$$
$$\times\left[\frac{1}{\gamma_{+0}-i(\omega_0+\omega_c/2-\omega_2)}\frac{1}{\gamma_{00}+i(\omega_1-\omega_2)}-\frac{1}{\gamma_{+0}-i(\omega_0-\omega_c/2-\omega_2)}\frac{1}{\gamma_{++}+i(\omega_1-\omega_2)}\right.$$

$$+\frac{1}{\gamma_{+0}+i(\omega_0+\omega_c/2-\omega_1)}\frac{1}{\gamma_{00}+i(\omega_1-\omega_2)}-\frac{1}{\gamma_{+0}+i(\omega_0-\omega_c/2-\omega_1)}\frac{1}{\gamma_{++}+i(\omega_1-\omega_2)}\Bigg]$$
$$-\frac{i|\mu_{+0}|^2|\mu_{-0}|^2}{\varepsilon_0\hbar^3V}\frac{1}{\gamma_{+0}+i(\omega_0-\omega_c/2-\omega_1)}\frac{1}{\gamma_{00}+i(\omega_1-\omega_2)}$$
$$\times\left[\frac{1}{\gamma_{-0}-i(\omega_0+\omega_c/2-\omega_1)}+\frac{1}{\gamma_{-0}+i(\omega_0+\omega_c/2-\omega_1)}\right]. \quad (6)$$

Here ω_1 and ω_2 are the angular frequencies of pump- and probe-beam, respectively; ε_0 is the vacuum dielectric constant, V is the effective volume of well and barrier layers, γ_{ii} is the energy relaxation rate of the ith sub-band $|i\rangle$ and γ_{ij} is the dephasing rate between the sub-levels $|i\rangle$ and $|j\rangle$, μ_{0-} and μ_{0+} are the optical dipole matrix elements from the heavy-hole state $|0\rangle$ to the two delocalized electronic states $|-\rangle$ and $|+\rangle$, respectively; the terms associated with $|\mu_{-0}|^4$ and $|\mu_{+0}|^4$ are the nonlinear optical absorption and dispersion arising from the transitions $|0\rangle\rightarrow|-\rangle$ and $|0\rangle\rightarrow|+\rangle$, respectively, and the terms with $|\mu_{+0}|^2$ and $|\mu_{-0}|^2$ make an extra contribution to the optical nonlinearity because of the coupling between the two wells.

Making use of representation transformation, the optical dipole transition elements μ_{0-} and μ_{0+} in the representation $\{|0\rangle,|+\rangle,|-\rangle\}$ is transformed as μ_{10} and μ_{20} in the single quantum well representation $\{|0\rangle,|1\rangle,|2\rangle\}$. It is

$$\mu_{+0}=\frac{1}{\sqrt{2}}(\mu_{10}+\mu_{20}),\ \mu_{-0}=\frac{1}{\sqrt{2}}(\mu_{10}-\mu_{20}), \quad (7)$$

where μ_{10} and μ_{20} represent the optical dipole matrix element of the transitions $|0\rangle\rightarrow|1\rangle$ and $|0\rangle\rightarrow|2\rangle$, respectively. Thus μ_{0-} and μ_{0+} are attributed to the calculation of μ_{10} and μ_{20} in the single quantum well representation.

3. Analysis of results

From expression (4), it is known that $\chi^{(3)}$ depends not only on γ_{ii} and γ_{ij} but also depends on ω_c, μ_{+0} and μ_{-0}. In general, the values of γ_{ii} and γ_{ij} are treated by the screening of the Coulomb and Fröhlich interactions in the fully dynamic random-phase approximation [15], which is related to the temperature of the sample, the impurities, surface roughness and concentration fluctuation. One assumes $\gamma_{ii}\approx 2.0\text{ps}^{-1}$ and $\gamma_{ij}\approx 2.4\text{ps}^{-1}$ for $In_{1-x}Ga_xP$–$Ga_xIn_{1-x}As$ ACQW structure in the condition of low temperature. By using the band-gap formulae $E_g^{In_{1-x}Ga_xP}=1.35+0.668x+0.758x^2$ and $E_g^{In_{1-x}Ga_xAs}=0.35+0.629x+0.436x^2$, we have $E_g^{In_{0.532}Ga_{0.468}P}=1.829\text{ eV}$ and $E_g^{Ga_{0.468}In_{0.532}As}=0.74\text{ eV}$. Correspondingly, the effective electronic and heavy-hole masses are $m_e\approx 0.08m_0$ and $m_{hh}\approx 0.609m_0$, respectively; the spin-orbit splitting is $\Delta\approx 89\text{meV}$. The band-gap offset satisfies $\Delta E_g = E_g^{In_{0.532}Ga_{0.468}P}-E_g^{Ga_{0.468}In_{0.532}As}$

$= \Delta E_{CB} + \Delta E_{VB}$. If the ratio of conduction band offset to valence one is 6:4, the $\Delta E_{CB} \approx$ 635meV and the $\Delta E_{VB} \approx 436$meV. The lowest quantum-confined energies of electron are $E_e^W \approx 21$meV for the WW with $L^W = 15$ nm and $E_e^N \approx 47$meV for the NW with $L^N = 10$ nm, respectively. Similarly, the energies of heavy-hole are $E_h^W \approx 2.8$meV for the WW and $E_h^N \approx$ 3.7meV for the NW. Thus, the energy difference between the electronic states in coupling and the heavy-hole state is obtained as $E_0 \approx (E_{1e}^W + E_{1e}^W)/2 + (E_{1h}^W + E_{1h}^N)/2 \approx 0.78$eV, corresponding to the transition frequency $\nu_0 = \omega_0 / 2\pi =$ 188.3THz.

Our interest is that the frequency up-conversion peak changes with external electric field strength. Figure 2(a) denotes the dots of the frequency up-conversion peak determined by $|\chi^{(3)}|$ with respect to electric field strength and Figure 2(b) is the peak value changes with electric field strength. As seen

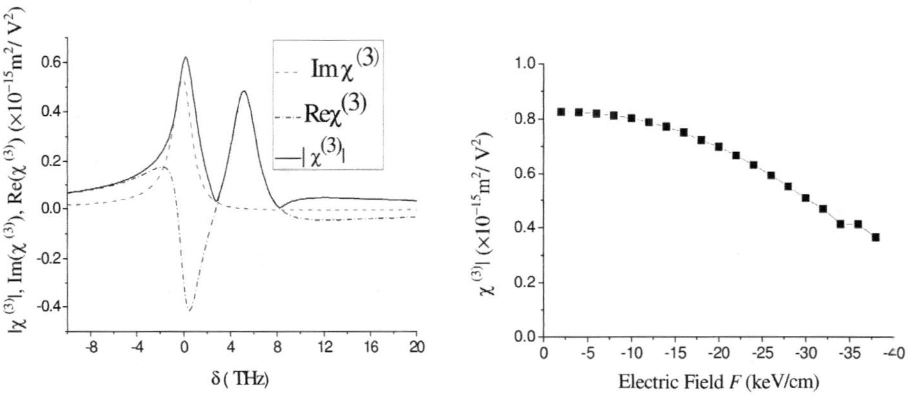

Figure 2. (a) is the relations of the third-order optical susceptibilities with respect to the detuning δ for the frequency up–conversion; (b) is the peak value changes of the third-order optical susceptibility $|\chi^{(3)}|$ with respect to an inverse electric field strength in the up-conversion. In the calculation, the decimal of inverse electric field is −2 keV/cm.

easily, the peak intensity of the large detuning signal via NDFWM maintains approximately a constant for a small bias range from 0 to -15keV/cm. Apparently, the ACQW structure as a tuneable up-frequency converter has the characteristic of output power balance for the phase-conjugate beam.

4. Conclusions

The third-order optical nonlinear susceptibility responsible for the phase- conjugate beam in NDFWM is analyzed for a biased ACQW structure. It is shown that the detuning of the frequency up-conversion may be in excess of 5THz based on the $In_{0.532}Ga_{0.468}P$-$Ga_{0.468}In_{0.532}$ As ACQW structure consisting of a wide well of 15nm width, a narrow well of 10nm width and a barrier of 4nm width. Also, the resonance peak of the frequency up-conversion maintains approximately a constant and goes with the tuneable frequency shift in excess of gigahertz when the inverse electric field strength changes from 0 to 15keV/cm. This means that the ACQW structure as a tuneable all-optical wavelength converter of output power balance has a potential application in the optical communication system.

References

[1] Stroscio, M. Kisin M A, Belenky G, and Luryi S 1999 *Appl. Phys. Lett.* **75** 3258.
[2] Yang W-X and Lee R-K 2008 Opt. Express **16** 17161
[3] Hu Z H and Huang D X 2004 *Acta Physica sinica* **53** 1195.
[4] Sa'ar A and Kapon R 1997 *IEEE J. Quantum. Electron.* **33** 1517.
[5] Paiella R and Vahala K J 1996 *IEEE J. Quantum. Electron.* **32** 721.
[6] Soto H and Erasme D 1999 *IEEE J. Quantum Electron.* **34** 2211.
[7] Zacharopoulos I , Tomkos I , Syvridis D, Sphicopoulos T, Caroubalos C , and Roditi E 1998 *IEEE Photon. Technol. Lett.* **10** 352.
[8] Agrawal G P, J. Opt. Soc. Am. 1988 **B 5** 147.
[9] D'Ottavi A, Iannone E, Mecozzi A, Scotti S, Spano P, Landreau J, Ougazzaden A, and Bouley J C, 1994 *Appl. Phys. Lett.* **64** 2492.
[10] Feiste U, Ludwig R, Schmidt C, Dietrich E, Diez S, Ehrke H J, Patzak E, Weber H G, and Merker T 1999 *IEEE Photon. Technol. Lett.* **11** 1063.
[11] Kim Y, Lee H, Kim S, Ko J, and Jeong J 1999 *IEEE J. Quantum Electron.* Select. Topic. **5** 873.
[12] Luryi S 1991 *IEEE J. Quantum. Electron.* **27** 54.
[13] Park T , Cumbs G and Chen Y C 1999 *J. Appl. Phys.* **86** 1467.
[14] Hu Z H and Huang D X 2005 Chin Phys., **14** 812.
[15] Kuwatsuka H, Simoyama T and Ishikawa H 1999 *IEEE J. Quantum Electron.* **35** 1817

Journal of Physics: Conference Series **276** (2011) 012065 doi:10.1088/1742-6596/276/1/012065

Transmission and Distribution of Optical Field in Prism Coupler

C Q Huang [1], J Liu[2], P Tian [3], Z M Wan [2], Z M Luo[2] and M Chen [1]

[1] College of Physics & Electronics Information, Hunan Institute of Science and Technology, Yueyang 414006, China
[2] College of Information & Communication Engineering, Hunan Institute of Science and Technology, Yueyang 414006, China
[3] College of Optoelectronic Science and Engineering, Wuhan National Laboratory for Optoelectronics, Huazhong University of Science and Technology, Wuhan, 430074, China

E-mail: namecqh@yahoo.com.cn

Abstract. Transmission and distribution characteristics of optical field in prism coupler are studies, and the phase matching function of prism coupler is deduced based on coupled wave theory. It is shown that the stable light field distribution and mode pattern are determined by its own geometric and dielectric parameters, but have nothing to do with the categories of incident light sources. It is also found that the coupling effect would generate between waveguides through evanescent field. Our numerical simulation is based on the finite difference time domain method with perfectly matched layer absorbing boundary condition. The simulation program is compiled in MATLAB. The simulation results are analyzed carefully.

1. Introduction

The necessary condition for realizing the integrated optical circuit is that optical energy can be coupled from one waveguide to the other [1]. At present, although there are several devices for realizing the coupling, most of them are either low efficient, or difficult to fabricate. For example, end-fire coupler has been shown to have a very low efficiency because of not having a finely-adjustable experimental setup. Grating coupler is a very promising optical device with high extraction efficiency of the light, but it is not well-suited for optical integrated circuits due to difficulty in mode matching, sensitivity to wavelengths, complexity of design and fabrication [2-4]. For these reasons, prism coupler is proposed for coupling power into or out of a planar optical circuit by us.

This letter is devoted to the study of energy transmission and distribution of guided modes in prism coupler. Because of the considerable mathematical complexity of the coupler construction, numerical simulation based on the finite difference time domain (FDTD) [5] technique with perfectly matched layer absorbing boundary condition [6,7] is employed. By using this technique, the optical propagation problems can be formulated systematically in terms of the well-known Maxwell's equations without using special mathematic knowledge. To the authors' knowledge, it is for the first time that the FDTD technique has been systematically performed to clarify the transmission and distribution of optical field in prism coupler.

2. Physical model and theoretical method

A schematic diagram of grating couplers which we discuss is shown in Figure 1. This can be seen that the prism coupler of our consideration is composed of four coupled media: a prism, an air gap, a thin-film, and a substrate. The prism is placed above the thin-film guide and is separated from it by the air gap of low refractive index. The refractive indices of air, prism, thin-film, and substrate are expressed by (n_a=1), n_p, n_f, and n_s ,respectively. The thickness of thin-film is h_f, the distance between the prism and thin-film (air gap) is h_a, the width of incident beam is $2W$. The vacuum wavelength of incident beam is λ_0, it refracts to bottom of the prime, and its direction of propagation is along the z axis. The bottom incident angle is α , and the critical angle is α_c.

Assumed that the optical wave refracted to bottom of the prime is given by the following [8]

$$A_0 \exp[jk_p(x\cos\alpha - z\sin\alpha)]$$

Where A_0 is the amplitude, $k_p(=k_p n_p/\lambda_0)$ is wave vector.

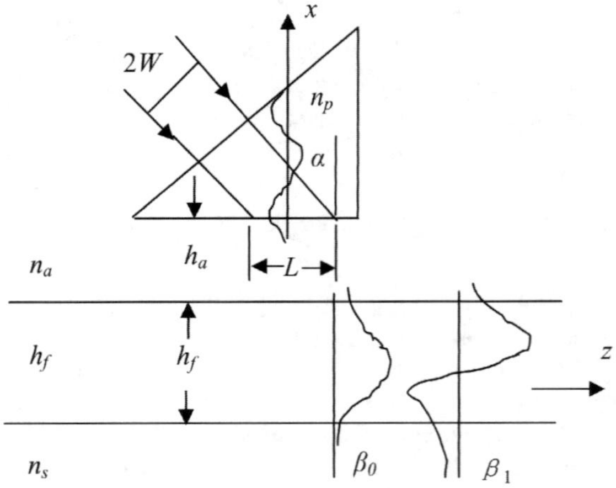

Figure 1. Basic structure of prism coupler

The reflected light can be expressed as

$$rA_0 \exp[jk_p(-x\cos\alpha - z\sin\alpha)]$$

Thus the total field in the prime can be expressed as

$$A_0 \exp(-jk_p z\sin\alpha)[\exp(jk_p x\cos\alpha) + r\exp(-jk_p x\cos\alpha)]$$

While total internal reflection occurs, the relation $r = 1$ is satisfied, a stronger standing wave field is localized in x direction of prime

$$2A_0 \cos(k_p x\cos\alpha)\exp(-jk_p z\sin\alpha)$$

It can be deduced that, the optical field of refractive wave is

$$A_0 t\exp[jk_p\cos\alpha' - z\sin\alpha')] = A_0 t\exp(jk_\alpha x\cos\alpha')\exp(-jk_\alpha x\cos\alpha')\exp(-jk_\alpha z\sin\alpha') \quad (1)$$

Where $k_\alpha = \frac{2\pi}{\lambda_0} n_\alpha$, $\sin\alpha' = \frac{n_p}{n_\alpha}\sin\alpha$

Obviously, $\cos\alpha' = \frac{n_p}{n_\alpha}\sqrt{\frac{n_\alpha^2}{n_p^2} - \sin^2\alpha}$

While $n_\alpha > n_p$, $\cos\alpha'$ is a imaginary number , which is given by the equation $\cos\alpha' = j(n_p/n_a)\sqrt{\sin^2\alpha - n_a^2/n_p^2}$. The distribution of refractive wave in air gap is then expressed as

$$A_0 t\exp\left[-k_\alpha \frac{n_p}{n_\alpha}\sqrt{\sin^2\alpha - \frac{n_\alpha^2}{n_p^2}}x\right]\exp\left(-jk_\alpha \frac{n_p}{n_\alpha} z\sin\alpha\right) \quad (2)$$

It demonstrates that, the optical field is exponentially decayed in x direction (called as elapsed wave), however the optical field in z direction is traveling wave field. Simultaneously, the wave field is coupled into air gap and thin-film.

According to the coupled wave theory, the phase matching condition must be satisfied in this coupler. That is to say, the change of optical-wave phase of prime in z direction must be equal to that of guided-mode in wave

$$2\pi n_p \sin\alpha / \lambda_0 = \beta_m \tag{3}$$

Where β_m is the propagation constant of preplanned guided-mode which is planned to couple. With different β_m of different order (m), the bottom incident angle α is different.

The condition for the optimal couple is

$$K\,L = \pi/2 \tag{4}$$

Where K is coupling coefficient, L is length of interaction which represent the intercept at the bottom of prime.

From Figure 1, it can be seen that

$$L = 2W/\cos\alpha \tag{5}$$

Thus

$$K = \pi\cos\beta/(4W) \tag{6}$$

From the aforementioned discussions, we can see that, in order to match the phase and obtain the high coupling efficiency, proper incident angle α and light-beam width W must be chosen, the edge of the incident angle is just intersected with corner angle of prime, and the spacing interval between the prime and the film is not suitable too big. Generally, chosen $h_a = (1/4 \sim 1/8)\lambda_0$.

Our research is based on FDTD technique with perfectly matched layer absorbing boundary condition. For a detailed description, see for example [6-8]. In what follows, only essential steps are outlined.

There are six vector field components in the system of Maxwell equations (E_x, E_y, E_z, H_x, H_y, H_z). From Figure 1, it can see that our chosen structure has nothing to do with variable y. For the sake of simplicity, Only TE_0 mode is considered in this paper. In these cases, it is easily proved that there is only three field components (E_y, H_x, H_z), and H_x and H_z can be expressed by E_y . Thus the field distribution of electric-field intensity E_y can be regarded as the total field distribution. To simplify solution procedure, we start with scalar wave equation for E_y

$$\frac{\partial^2 E_y}{\partial x^2} + \frac{\partial^2 E_y}{\partial z^2} - \mu\varepsilon\frac{\partial^2 E_y}{\partial t^2} = 0 \tag{7}$$

Where μ is magnetic permeability, ε is electrical permittivity.

According to central difference formula, at the conditions of satisfying the stabilization and avoiding the numerical dispersion, the Equation 7 becomes

$$\begin{aligned} E_y^{N+1}(i,k) = {} & p(i)[E_y^N(i-1,k)+E_y^N(i+1,k) + E_y^N(i,k-1)+E_y^N(i,k+1)-4E_y^N(i,k)] \\ & +2E_y^N(i,k)-E_y^{N-1}(i,k) \end{aligned} \tag{8}$$

Where $p(i) = 0.25\left(n_{min}/n(i,k)\right)^2$, n_{min} is the minimum refractive index of waveguide mediums, and $n(i,k)$ is the refractive index of the researched mesh.

The difference scheme of the lower boundary can be expressed as

$$\begin{aligned} E_y^{N+1}(1,k) = {} & -\frac{1}{3}E_y^{N+1}(2,k)+\frac{7}{6}[E_y^N(1,k) + E_y^N(2,k)] + [E_y^N(1,k-1)+E_y^N(1,k+1) \\ & +E_y^N(2,k-1)+ E_y^N(2,k+1)]/12 - E_y^{N-1}(1,k)/3 - E_y^{N-1}(2,k) \end{aligned} \tag{9}$$

Other difference schemes of the otherwise boundaries can be deduced in the same way. The procedure for calculating is compiled in MATLAB.

3. Numerical results

In the course of the numerical calculation, the geometric and dielectric parameters we have chosen are as follows: $h_a = 0.5\mu m$, $h_f = 1\mu m$, $h_s = 0.5\mu m$, n_p =2.192, , n_f =1.563, n_s =1.55, $\alpha = 45^0$, and the data of space step Δs and time step Δt are 0.0416 μm and 0.1074 ns , respectively. The incident wave is vertical to the sloped edge of prime with $\lambda_0 = 1.3\mu m$, and the edge of the incident beam is just overlapped each other in the right-angle of prime. Three kinds of optical sources are used as stimulants. They are Guass optical source, spherical optical source, and plane optical source.

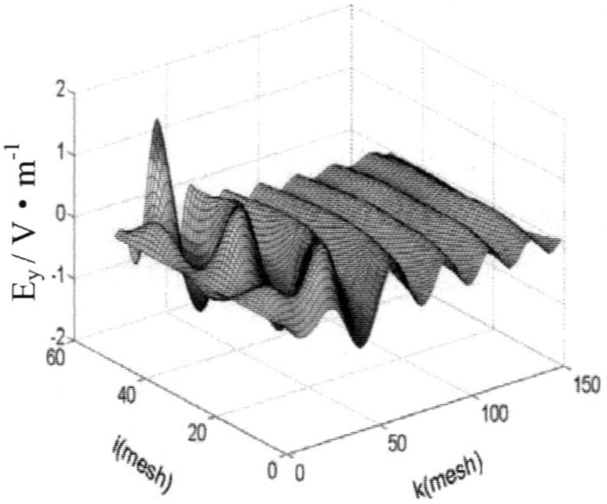

Figure 2. Field distribution stimulated by plane wave

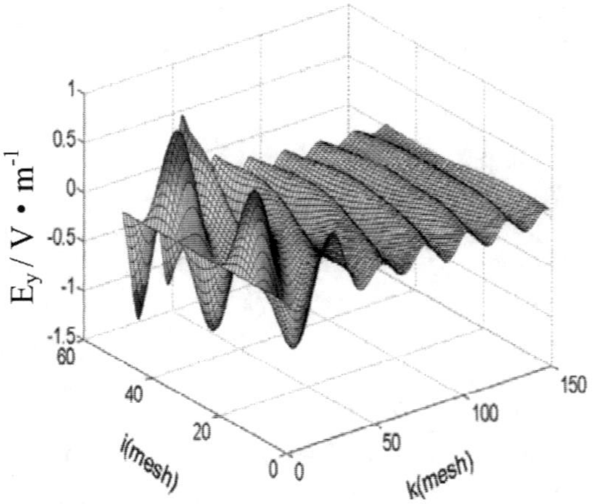

Figure 3. Field distribution stimulated by Guass wave

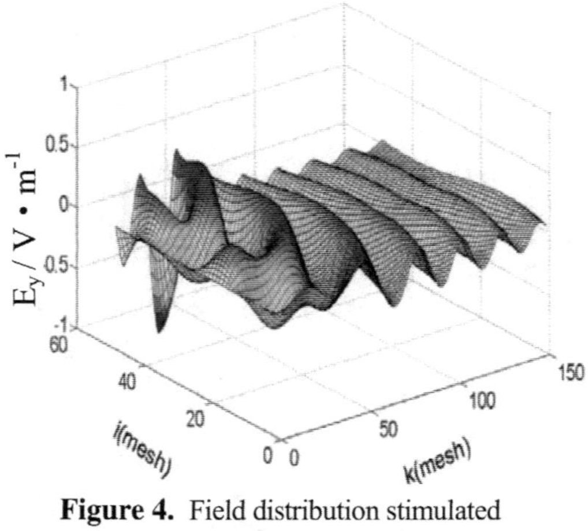

Figure 4. Field distribution stimulated by spherical wave

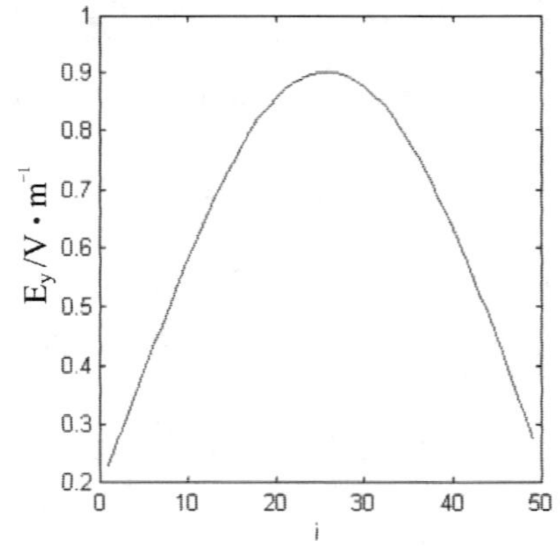

Figure 5. The stable mode of TE_0 for different incident waves

The transmissions and distributions of optical field profiles of TE_0 stimulated by different optical sources are shown in Figures 2, 3, 4, respectively. From these figures, it can be seen that the energy of incident light beam could be coupled from prime to waveguide film, and the guided-wave could be stimulated in the film. This fact can be explained that, while the incident light enters the prime, the high refractive index causes it to suffer total internal reflections, leading to the form of standing wave field. At the same time, part of the energy of standing wave field penetrates down into the air gap. Through the coupling effect of evanescent fields, the transition energy could transfer from the air gap to film, and could excite any modes existed in the film. It is clear that, at the beginning of prime coupler, there are not formed the stable modes. But with the increasing of propagation distance, the stable light field distributions are gradually formed. The stable mode pattern of TE_0 is illustrated in Figure 5, it is the same with analytic solutions.

From the aforementioned discussions, we know that a stable light field mode is determined by the geometric and dielectric parameters rather than the driving sources. Based on the profiles of transmissions and distributions, the transmission processes of energy in the prime coupler can be seen clearly, the accurate coupling length through evanescent field can be obtained, the coupling efficiency can be forecasted, and thus highly-efficient waveguide couplers which are of practical use in integrated optical circuits and optoelectronic devices can be designed conveniently.

4. Conclusion

This letter is devoted to the study of energy transmission and distribution of guided modes in prism coupler based on the FDTD technique with perfectly matched layer absorbing boundary condition, and the phase matching function of prism coupler is deduced based on coupled wave theory. It is found that a stable field distribution and mode pattern are determined by dielectric and geometric parameters of devices, but have nothing to do with the categories of incident light. According to the field distributions, the characteristic coupling length, detuning, resonance amplitudes of the fields can be also found. It is shown that the coupling is strongest if the components of wave vectors parallel to the air gap are equal to the propagation constant, of one of the film modes, and optical energy can be transferred from the prism to the film effectively. In these ways, the coupling efficiency of prism coupler can be predicted.

Acknowledgement. This work was supported by the Scientific Research Program of Education Bureau of Hunan Province (08C368) and Hunan Provincial Natural Science Foundation of China （10JJ6095）.

References

[1] Chaganti K, Salakhutdinov I, Avrutsky I and Auner G W 2006 Opt. Express **14** 4064-4072
[2] Liu X M, Zhu P F, Cao Z Q, Shen Q S and Chen J L 2006 JOSA B **23** 353-357
[3] Fick J and Vitrant G 1995 Optics Letters **20** 1462-1464
[4] Apiratikul P, Rossi A M and Murphy T E 2009 Opt. Express **17** 3396-3406
[5] Yee K S 1966 IEEE Trans. Antennas Propagat **14** 302-307
[6] Engquist B and Majda A 1975 Mathematics of Computation **31** 629-651
[7] Taflove A and Hagness S C 2000 Computational Electrodynamics: The Finite- Difference Time-Domain Method (Boston. London: Artech House)
[8] Ming H, Zhang G P and Xie J P 1998 Photoelectronic Technique(Hefei: China University of Science and Technology Press) pp 66-70

Journal of Physics: Conference Series **276** (2011) 012066 doi:10.1088/1742-6596/276/1/012066

Design of Super Narrowband DWDM Filters Based on the Effect of Spectral Splitting

C Q Huang [1], J Liu[2], P Tian [3], Z M Wan [2], Z M Luo[2] and M Chen [1]

[1] College of Physics & Electronics, Hunan Institute of Science and Technology, Yueyang 414006, China
[2] College of Information & Communication Engineering, Hunan Institute of Science and Technology, Yueyang 414006, China
[3] College of Optoelectronic Science and Engineering, Wuhan National Laboratory for Optoelectronics, Huazhong University of Science and Technology, Wuhan, 430074, China

E-mail: namecqh@yahoo.com.cn

Abstract. A novel approach is proposed to design super narrowband DWDM Filters consisting of multiple quantum wells (MQWs) by employing photonic crystals. Numerical investigations prove that the closed-cavity MQWs are more suitable for DWDM systems compared with the open-cavity MQWs. It is shown that different confined states could emerge from photonic band gap, which can be used as high-frequency carriers one-to-one. It is also found that these proposed MQWs could split the single spectral lines into multiples based on the effect of spectral splitting, and the number of the splitting is just equal to the number of the wells. In this way, the density of carriers can be increased multiplicatively in the same wave band, and thus the spectral efficiency can be improved multiplicatively. These results provide the prospects of channel density maximization and effective bandwidth optimization for optical communication.

1. Introduction

With rapid developments of computer and the global telecommunication market, especially the emergence of a series of new services such as the Internet, high-quality videoconference system and multimedia, the demand for bandwidth and capacity increases rapidly. Thus the spectral source is getting more and more lacked. The appearance of DWDM technology have avoided or relieved this situation to a certain extent.

In DWDM systems, MultiPlexers (MUX) and demultiPlexers (DEMUX) are key devices [1]. At the present time, thin film filter (TFF) and array waveguide grating (AWG) have been widely used as MUX /DEMUX devices in commercial DWDM systems due to their excellent temperature stability, low polarization dependent loss, high channel isolation and mature technology [2], etc.. However, they have also some drawbacks. For TFF, the insertion loss is big with non-uniform distribution; the price of its commercial application is shortening transmission distance, sacrificing communication quality, and reducing output power. For AWG, additional control devices and stable units are needed, leading to big volume of device, high trouble rate and maintaining difficulty [3]. In this case, it is urgent to explore new devices for satisfying the requirement of bandwidth and capacity with high quality and efficiency.

In this paper, we design the novel DWDM devices constructed by multiple-quantum-well (MQW) structures consisting of photonic crystal (PC), a series of new optical characteristics are discovered and analyzed. Because of the difficulty in fabricating the three dimensional PC, two dimension (2D) PC in reasonable arrangement is used as an alternative. The finite difference time domain method is used for numerical analysis.

2. Physical model and theoretical method

The most reported QW structure model composed by two-dimension PC in the literatures is shown in Figure 1 (a). Both potential barriers are consisted of the same two-dimension PC having a square lattice with a lattice constant *a*. The lattice is made of dielectric rods on an air background. These rods are placed along the *z* direction. For each barrier thickness, it is determined by $v = k \times a$, where k is the line number of rods in the x direction in the barrier region. The potential well is formed with air. It is obvious that the vacuum thickness in the $\pm y$ directions is infinite, which are called the side directions of QW structures; but it is finite in the $\pm x$ directions, which can be expressed as $w = n \times a$, where *n* is the line number of the removed rods in the x direction. P_1 and P_2 are the same detectors placed at each side of the QW, which can be used to measure the incident power and the transmission power, respectively. Thus, the transmissivity can be calculated readily as the ratio of transmitted power to the incident power.

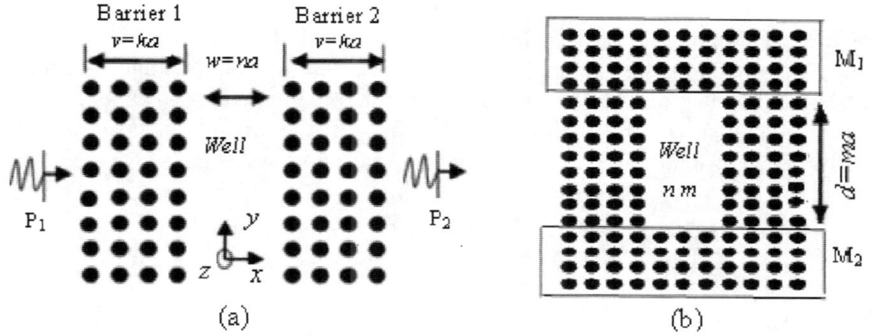

Figure 1. The schematics of the SQW structures (a) open-cavity SQW (b) closed-cavity SQW.

Apparently, the well region structure of Figure 1 (a) is open, we call this schematic diagram as "open-cavity single quantum well (SQW)". But we think that, the transmission efficiency of photon captured in the well region is impossible to be too high, it is because that the majority of them would radiate and leak out from the well as they are not restricted by any confinement effect in the $\pm y$ directions, resulting in lower transmissivity. For this reason, we propose a new type of two-dimensional quantum well structure named as closed-cavity SQW, as shown in Figure 1 (b). The difference is that two periodic structures $M_1$and M_2 is added in the side directions based on the Figure 1 (a). The geometric and dielectric parameters of M_1 and M_2 are the same of the case of potential barriers. The well region area is $m \times n$, where *m* is the number of rods removed from each line in the $\pm y$ directions, the lateral length of the well is $d = m \times a$. We make a conjecture that, as the existence of band gaps caused by M_1 and M_2, the propagation of waves is also inhibited at the side directions, just like that at $\pm y$ directions, so that the motion space of photons is reduced down. Once photons are trapped in the well, the probability for them to escape from the well would be very low, since the path of energy liberation is shut off, and the photon density and energy would be both increased, leading the transmissivity to nearly 1. The numerical calculation results have confirmed the correctness of our conjecture.

Stimulated by the high transmissivity of closed-cavity SQW aforementioned discussions, we further design the closed-cavity MQW structures, as shown in Figure 2 (a) and Figure 2 (b). To our surprise, a series of excellent optical characteristics are found. For the sake of simplicity, we only introduce the closed-cavity double QW (DQW) structure and the closed-cavity triple QW (TQW) structure with narrow barrier thickness, the size of the well regions are all the same. It is obvious that MQW can be regard as the cascade connection of multiple SQW structures, and we can obtain the open-cavity multiple QW (MQW) structures by only removing the all dielectric rods at the side directions.

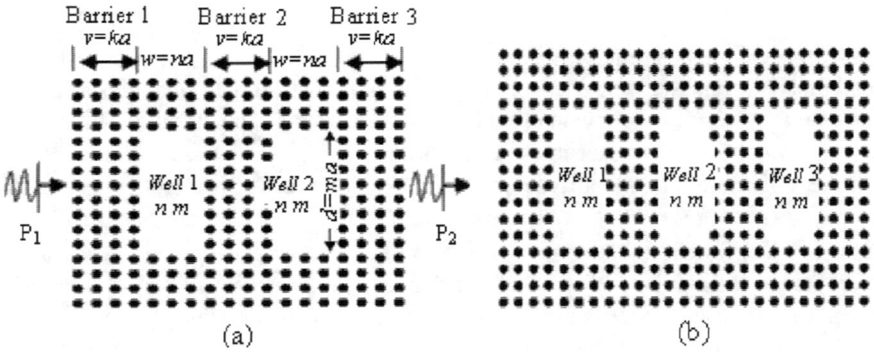

Figure 2. The schematics of MQW structures with narrower thickness for (a) closed-cavity DQW (b) closed-cavity TQW.

We use the finite-difference time-domain (FDTD) method [4] with PML [5] absorbing boundary conditions to truncate the computational region and minimize the reflections from the outer boundary as research tools. In the following discussion, we only consider the transverse-magnetic (TM) polarization since its omni-direction total reflection is always defined by the omni-directional photonic band gap (PBG). The difference formulas used in computational procedures are as follows

$$E_z^{N+1}(i,j) = E_z^N(i,j) + \frac{\Delta t}{\varepsilon(i,j)\Delta x}[H_y^{N+1/2}(i+\frac{1}{2},j) - H_y^{N+1/2}(i-\frac{1}{2},j)] - \frac{\Delta t}{\varepsilon(i,j)\Delta y}[H_y^{N+1/2}(i,j+\frac{1}{2}) - H_y^{N+1/2}(i,j-\frac{1}{2})] \quad (1)$$

$$H_y^{N+1/2}(i+\frac{1}{2},j) = H_y^{N-1/2}(i+\frac{1}{2},j) + \frac{\Delta t}{\mu(i,j)\Delta x}[E_z^N(i+1,j) - E_z^N(i,j)] \quad (2)$$

$$H_x^{N+1/2}(i,j+\frac{1}{2}) = H_y^{N-1/2}(i,j+\frac{1}{2}) - \frac{\Delta t}{\mu(i,j)\Delta x}[E_z^N(i,j+1) - E_z^N(i,j)] \quad (3)$$

Where N indicates the discrete time step, the subscripts i and j indicate the position of the grid point in the x and y directions, respectively. $E_z^{N+1}(i,j)$ represents the electric field E_z at the corresponding position (i,j) for time step N+1. $\varepsilon(i,j)$ and $\mu(i,j)$ are the position-dependent dielectric constant and magnetic permeability of the material, respectively. Δt is the time increment, and $\Delta x = \Delta y = \Delta s$ is the spatial increment in neighboring grid points along the x and y directions.

The stability condition for the FDTD is determined by choosing a suitable Δt to ensure the solutions with purely real frequencies for all possible wave vectors k. If the following condition is satisfied, the FDTD time stepping equations are stable numerically

$$\Delta t \le \Delta s/(c\sqrt{2}) \quad (4)$$

Where c is the speed of light in vacuum.

For simplicity of analyses, we use modulation Gaussian pulse, which can be given by

$$E_z(t) = \cos(2\pi f_0 t) \times \exp\left[-4\pi\frac{(t-t_0)^2}{T^2}\right] \quad (5)$$

Where f_0 is center frequency of fundamental wave $\cos(2\pi f_0 t)$, t_0 is center position of the Gaussian pulse, and T is a time constant. In this letter, we choose $t_0 = 20\Delta t$, $T = 2\Delta t$. In favor of the observation, in the calculated diagrammatic curves of transmission spectra, the frequencies described in X-coordinate are in reduced units which take ($\omega a/2\pi c$) as their units, and ω is the angular frequency.

The *Y*-coordinate values which represent the transmission coefficients can be calculated as the ratio of transmitted power to the incident power.

These difference formulas are simple and easy to be programmed. The time-varying electric fields at the input and output ports can be recorded in digit groups. By taking the fast Fourier transform (FFT) of these recorded fields at a large range of frequencies [6], the transmission spectra can be obtained. From the above equations, we can see that for a fixed total number of time step the computational time is proportional to the number of discretization points in the computation domain, which means that the FDTD algorithm is one of order N (where N is also the number of computational nodes). Thus, the FDTD method reduces significantly both memory requirement and CPU time. That is the very reason why we use the FDTD method for our research tool.

3. Numerical results

In the model of the Figure 1 or Figure 2, the parameter choices are as follows: the barrier width $v = 5a$, the well width w=10a , the lateral length of the well $d = 8a$ for the closed-cavity MQW structures, relative dielectric of dielectric rods $\varepsilon_r = 4.55$, and the radius of rods $r = 0.25a$. For simplicity of analyses, only the TM wave is analyzed, and the Gaussian pulse is in the case of normal incidence on the entrance face. The computational space have a spatial sampling rate of ($\lambda_0/40$), where λ_0 is the wavelength of fundamental wave in vacuum.

We first calculate the transmission spectrum of potential barriers (perfect 2D PC). We can clearly observe that the first PBG is from 0.325 to 0.465, in which light possessing certain values of wave vector is not allowed to propagate. In order to save the space of a whole page, we have not given the chart of transmission spectrum.

Figure 3 (a), 3 (b) and 3 (c) are the transmission spectra of open-cavity SQW, open-cavity DQW, and open-cavity TQW structures. Figure 4 (a), 4(b) and 4(c) are the transmission spectra of closed-cavity SQW, closed -cavity DQW, and closed -cavity TQW structures. It is clearly seen many defect modes (sharp peaks) with different frequencies emerge in the PBG. But for the open and closed QW structures, the amplitudes for transmission spectra of closed-cavity QW structures are bigger than that for open-cavity QW.

The formation of these sharp peaks embodies the new-fashioned dispersion relation which is caused by spatially defective arrangement of the well. While the energy of incident photons is matched with the confined state located in the well regions, they will fall into the potential wells and involve in the wells inescapably, just like falling into a deep water well. As long as the thickness of the potential barriers is not very thin, the moving photons will reflect basically after striking against the potential well walls. If the energy is not enough big, the photon can only limit in the wells level. Because of the quantum closeness of PBG, the effective state density of photons is concentrated in a certain specific narrow energy range, leading to narrow energy distribution. This is the sharpened phenomenon of the density of states. Based on the effect of the resonant tunneling, the tunneling probability of photons may achieve 1. On the contrary, if the energy of incident photons is matched with the confined state, the tunneling probability would be very close to 0.

The reason why the transmissibility of open-cavity MQW is that, for the open MQWs with vacuum well region, photons are not restricted by any confinement effect (not only the refractive index, but also the band gap) at the side directions of the MQWs. These photons cannot transmit effectively along the designed direction, and majority of them will radiate and leak out from the well, resulting in lower transmittance. While the closed-cavity MQW is just right opposite. This explanation just conforms to our conjecture above mentioned. That is the very reason we project the closed-cavity MQWs in this paper.

From Figure 3 and Figure 4, we can observe that, each sharp peak which appears in the SQW could split into two or three for DQW or TQW structures ······. Further researches indicate that the number of sharp peaks is just equal to that of the wells in MQW structures. This phenomenon is called as spectral splitting. It is noted that, these sharp peaks can be assigned to the center wavelength of the DWDM filters or the guide waves one-to-one for optical communication system. In this way, the number of the center wavelengths can be multiplied in a limited PBG. Thus the spectral efficiency can be improved multiplicatively in a fixed frequency band.

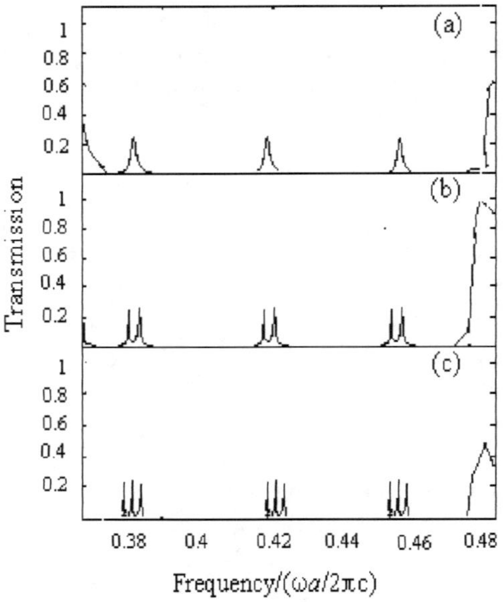

Figure 3. Transmission spectra of (a) open-cavity SQW (b) open-cavity DQW (c) open-cavity.

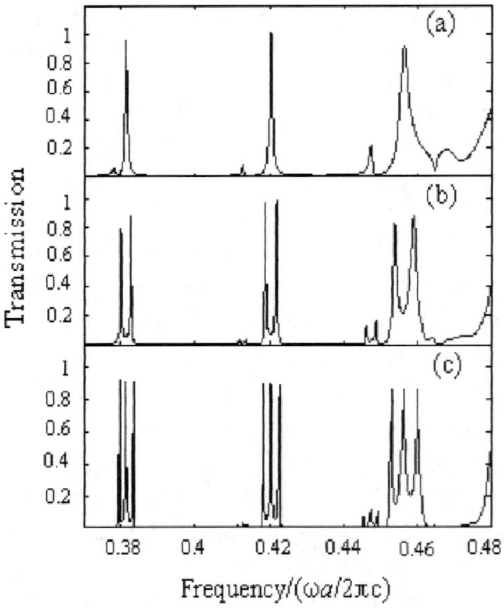

Figure 4. Transmission spectra of (a) closed -cavity SQW (b) closed -cavity DQW (c) closed –cavity.

In order to further improve the spectral utilization, we have been seeking for the new ways unceasingly. We find that, while the barrier thickness is changed, the frequency interval between two sharp peaks has a remarkable change. Figure 5 (a), (b) and (c) represent transmission spectra with the same well width w=10a but different barrier width v=3 a ,5 a , 7 a , respectively. It can obvious that, the narrower the barrier thickness is, the bigger the distance between adjacent sharp peaks is. This phenomenon indicates that centre wavelength of the DWDM filters can be fine adjusted by only changing the thickness of the barriers. There are no intersection and overlapping but PBG between neighboring sharp peaks, so that the signal crosstalk is reduced, and the resource of spectrum is effectively used. In DWDM systems, as it is used as MUX, it combines optical signals of different wavelengths before transmitting them; as it is used as DEMUX, it separates the combined optical signals in the optical fiber and then sends them to different communication terminals. By using this kind of technology, the request of the wavelength interval for DWDM devices is reduced, and the transmission capacity of the DWDM system is improved.

4. Conclusion

In this paper, 2D open-cavity and closed-cavity MQW structures are designed, the differences of transmission spectra for these two kinds of QW structures are compared and analyzed. The research results indicate that there are some superiorities for closed-cavity MQW structures compared with open-cavity MQW structures. It is also found that different confined states could emerge from photonic band gap (PBG). These confined states can be used as high-frequency carriers one-to-one for optical communication systems. It is also found that the single spectral lines split into several equidistant lines separated by band gap, and the number of the splitting is just equal to the number of the wells. In this way, the density of carriers can be increased multiplicatively in the same wave band, and thus the spectral efficiency can be improved multiplicatively.

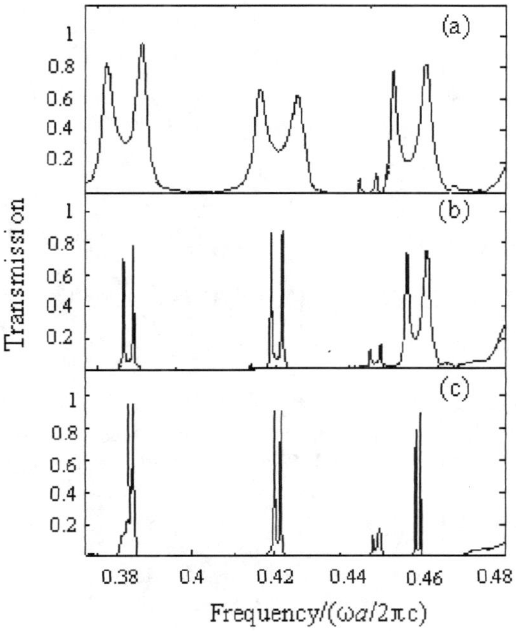

Figure 5. spectra of dual QW with w=10a but different barrier widths (a) $v=3a$ (b) $v=5a$ (c) $v=7a$.

Acknowledgement. This work was supported by the Scientific Research Program of Education Bureau of Hunan Province (08C368) and Hunan Provincial Natural Science Foundation of China (10JJ6095).

References

[1] Li C L, Sheng O Q , 2006 *Acta Photonica Sinica* **35** 734-737
[2] Guo H P, Yu C C and Gong R L 2003 *Quantum Electronics* **39** 160-165
[3] Li C L and Sheng Q Q 2006 *Acta Photonica Sinica* **35** 734-737
[4] Yee K S 1966 IEEE Trans. Antennas Propagat **14** 302-307
[5] Taflove A, Hagness S C 2000 Computational Electrodynamics: The Finite- Difference Time-Domain Method (Boston. London: Artech House)
[6] Yates S J C, Baryshev A M, Baselmans J J A, Klein B and Güsten R 2009 Applied Physics Letters **95** 042504

Tunable Microwave Photonic Notch Filter Based on a high-birefringence linearly chirped fiber Bragg grating

Yongxing Jin[1], Xinyong Dong[1], Junqiang Zhou[2] Jianfeng wang[1]
[1]Institute of Optoelectronic Technology, China Jiliang University, Hangzhou, China
[2]Network Technology Research Centre, Nanyang Technological University, Singapore

E-mail: phyjyxin@gmail.com

Abstract. In this paper, a continuously tunable microwave photonic notch filter is proposed and experimentally demonstrated. This filter is based on the differential group delay generated by a high-birefringence linearly chirped fiber Bragg grating. This microwave photonic filter belongs to the orthogonal polarization approach, polarization maintaining structure ensures the filter free from the random optical interference problem. Its response is induced by the differential group delay (DGD) of the Hi-Bi LCFBG and it can be varied by tuning the grating through adding gradient strength to the grating. Free spectral range tuning by 9.27 GHz with more than 35 dB notch rejection is achieved.

1. Introduction

Optical processing of microwave and millimetre-wave signals using microwave photonic filters (MPFs) has attracted significant interest because of the advantages of high time-bandwidth product, low loss, immunity to electromagnetic interference (EMI), and tunable and adaptive functions [1, 2]. In order to achieve stable operation, the incoherent MPF is more attractive in practice. There are mainly two approaches to satisfy the requirement of incoherent operation: keeping the laser coherence time smaller than the optical delay time, and making the state of polarization (SOP) of the optical signals to be orthogonal [4]. For the first approach, fulfilling the coherence time condition restricts the free spectral range (FSR) of the MPF [3]. If laser array [4], sliced wideband source [5], or multi-wavelength lasers [6] are used, the system complexity and cost will be high. For the applications which require low cost and compact MPFs, the second approach is a good choice. The MPF based on orthogonal polarization has been proposed in [7], but it has only step tuning ability. Continuously tunability has been achieved using a high-birefringence linearly chirped grating (Hi-Bi LCFBG), a polarization beam splitter (PBS) and a Hi-Bi coupler [8]. However, the free spectral range (FSR) tuning is relative small because of the uniform strength applied to the grating. The notch rejection may also be affected with a LCFBG inside a fibre loop because the limited reflectivity of the grating causes signal crosstalk.

In this paper, a continuously tunable MPF is proposed which is based on a Hi-Bi LCFBG and a polarization maintaining (PM) circulator. This MPF belongs to the orthogonal polarization approach, polarization maintaining structure ensures the MPF free from the random optical interference problem. The MPF response is induced by the differential group delay (DGD) of the Hi-Bi LCFBG and it can

[1] To whom any correspondence should be addressed.

Published under licence by IOP Publishing Ltd

be varied by tuning the grating through adding gradient strength to the grating. FSR tunability of 1.11 GHz and 40 dB notch rejection are achieved.

2. Configuration and operation principle

The configuration of the proposed MPF is shown in Figure 1. Continuous wave (CW) light from a laser diode (LD) is modulated by a Mach-Zehnder modulator (MZM) which is driven by an RF signal from a vector network analyzer (VNA). The output of the MZM is fed to a Hi-Bi LCFBG through a half waveplate and a PM circulator. The half waveplate is used to excite two orthogonal and equal power linearly polarized optical components. Since the whole structure is polarization maintaining, the two orthogonal components propagate along the slow and fast axes, get reflected by the Hi-Bi LCFBG, detected by a photodetector (PD) and the frequency response is measured by a network analyzer.

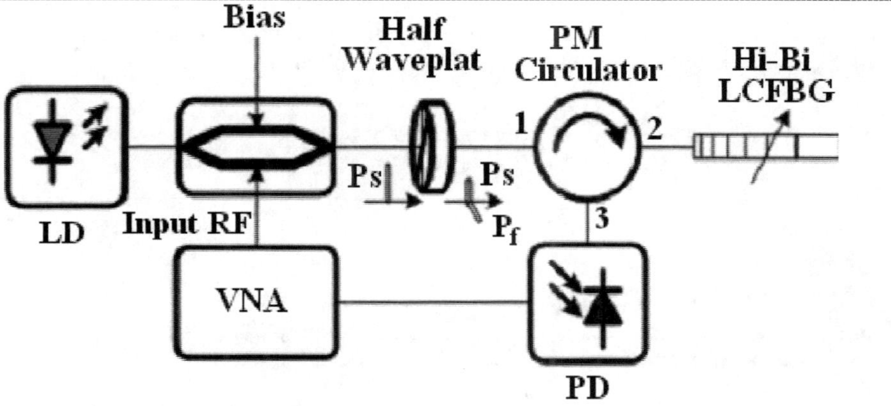

Figure 1. Experimental setup. (LD: laser diode, VNA: vector network analyzer, PD: photodetector)

The key component of our system is the Hi-Bi LCFBG. For a fiber Bragg grating, the strongest mode coupling occurs at the wavelength λ_B that can be back reflected from the grating. The Bragg wavelength is a function of grating pitch $\Lambda(z)$ [9]

$$\lambda_B = 2n_{eff}\Lambda(z) \tag{1}$$

where n_{eff} is the effective refractive index of the fibre. A linearly chirped FBG has a variable pitch which is a linear function of distance along the grating so that the Bragg wavelength and reflection location along the grating forms a linear relation. As a result, when the optical signals with different wavelengths input into a FBG, they are reflected differently at different locations along the grating so that the time delay of the LCFBG is also a linear function of the wavelength. When fabricating LCFBG into a Hi-Bi fiber, because of the refractive index difference between the x and y axes, the grating spectra for the two orthogonal polarization axes have a shift in wavelength domain. So, when the two intensity modulated orthogonal components reflect by the Hi-Bi LCFBG, they are delayed differently and a certain amount of differential group delay (DGD) occurs between the fast and slow axes. Since there are two orthogonal components, the MPF is a two-tap transversal filter and its normalized transfer function is

$$|H(f)| = |\cos(\pi f \Delta T)| \tag{2}$$

where f is the RF frequency and ΔT is the DGD of the system.
The FSR of the filter is described by

$$FSR = \frac{1}{\Delta T} \tag{3}$$

The DGD in Equation (2) and (3) actually contains two parts: Hi-Bi LCFBG DGD (τ_g), and fixed DGD (τ_0) which comes from the polarization maintaining fiber (PMF) pigtails of the modulator, circulator and Hi-Bi LCFBG. The Hi-Bi LCFBG DGD can be tuned by applying gradient strength to the grating to cause a change of the pitch. Figure 2 (a) shows the cantilever beam structure used for this purpose [10]. By tuning the FBG using this structure, the reflection spectrum broadens or narrows with the DGD reduced or increased a certain amount ($\Delta\tau_g$) as indicated in Figure 2 (b). However, the centre wavelength remains fixed. By using this property, the FSR tuning is achieved though varying the DGD at a fixed wavelength. Then Equation (3) can be rewritten as

$$FSR = \frac{1}{\tau_0 + \tau_g + \Delta\tau_g} \tag{4}$$

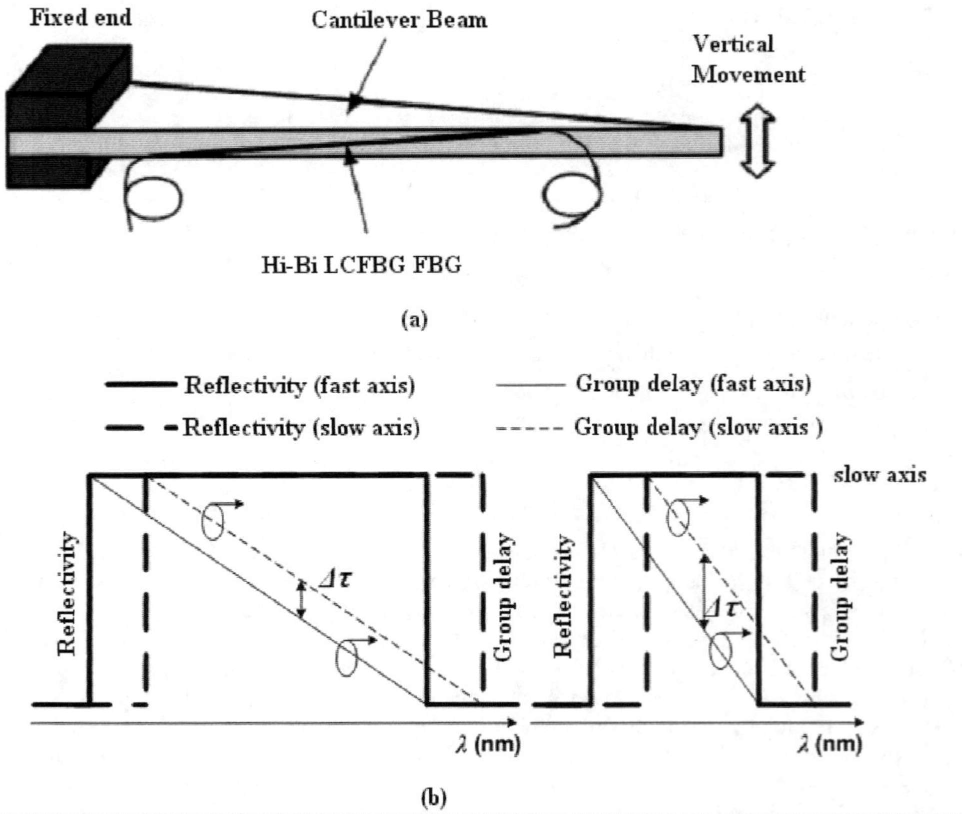

Figure 2 (a) Cantilever beam for tuning the Hi-Bi LCFBG and (b) schematic reflectivity and group delay

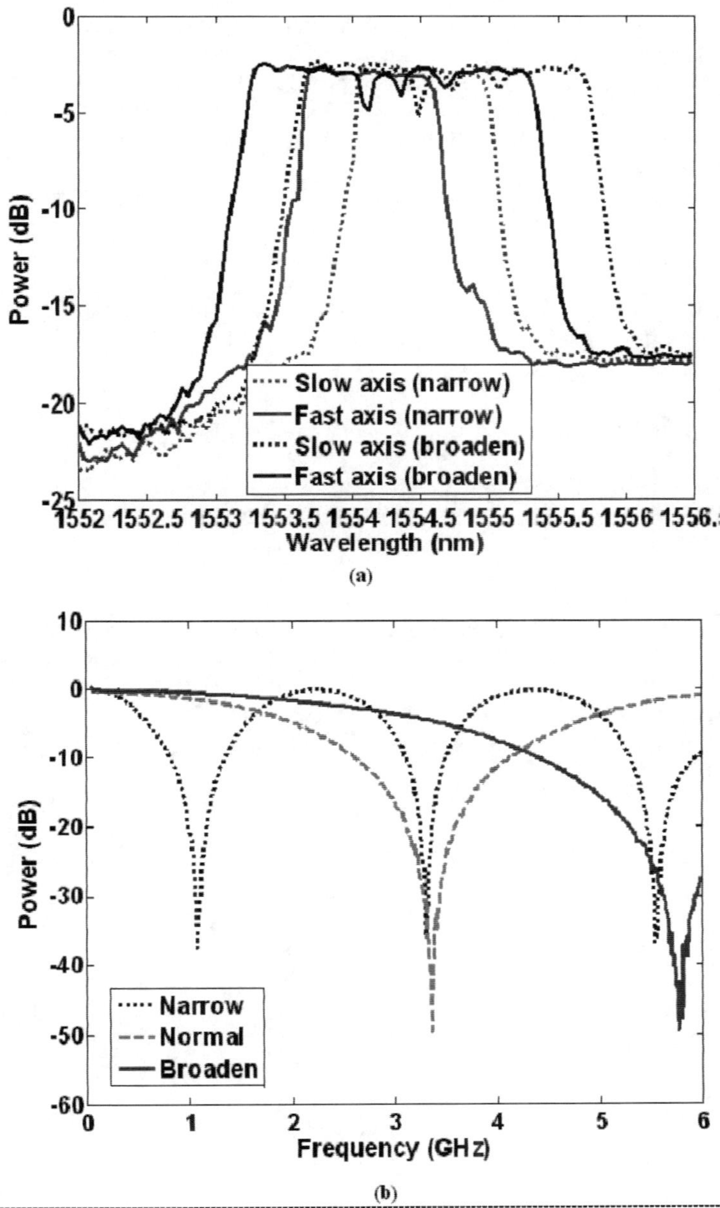

Figure 3. (a) Hi-Bi grating reflection spectrum after tuning and (b) Measured frequency responses

3. Results and discussion

The Hi-Bi LCFBG used in the experiments is fabricated by exposing a hydrogen-loaded PM fibre to a 244 nm UV laser beam through a linearly chirped phase mask. Figure 3 (a) shows the measured reflection spectra for the fast and slow axes. It is clearly seen that the spectra width can be varied up to 0.3 nm. According to Eq. (4), in order to achieve bigger FSR tuning range, τ_0 needs to be small. To get a smaller τ_0, the short wavelength port of the Hi-Bi LCFBG was connected to the PM circulator in our experiment and the working wavelength was the shortest wavelength which can be reflected by both slow and fast axes, e.g. 1554.3 nm. The measured filter responses are shown in Figure 3 (b), the FSRs when the FBG spectrum compressed, at normal position, or broadened are 2.21 GHz, 6.75 GHz and 11.48 GHz, respectively. An FSR tuning range of 9.27 GHz was achieved. Actually, the FSR tuning range can be further increased through compensating the fixed DGD by adding another section of PMF which has similar length as the PMF pigtails with its slow axis align with the fast axis of the PMF pigtails (90° rotation).

4. Conclusion

We have demonstrated a simple continuously tunable microwave photonic notch filter which is based on a Hi-Bi linearly chirped fibre Bragg grating as tuning element. The polarization maintaining structure is free from the problem of random optical interference. FSR tunability of 9.27 GHz with more than 35 dB notch rejection has been achieved.

References

[1] R. A. Minasian; 2006 *IEEE Trans Microwave Theory and Tech* 54, 832.
[2] J. Capmany and D. Novak; 2007 *Nature Photonics* 1, 319.
[3] W. Zhang, J.A.R. Williams, L.A. Everall and I. Bennion; 1998 *Electronics Letters* 34, 1770.
[4] V. Polo, B. Vidal, J. L. Corral, and J. Marti; 2003 *IEEE Photon Technol Let,* 15, 584.
[5] J. Mora, M. V. Andres, J. L. Cruz, B. Ortega, J. Capmany, D. Pastor, and S. Sales; 2003 *Opt Lett.*28 1308.
[6] D. Liu, N. Q. Ngo, and S. C. Tjin; 2007 *Opt Eng* 46, 054304.
[7] W. Zhang, J. A. R. Williams, and I. Bennion; 2001 *IEEE Photon. Technol Lett* 13, 523.
[8] X. K. Yi, C. Lu, X. F. Yang, W. D. Zhong, F. Wei, L. Ding, and Y. X. Wang,; 2003 *IEEE Photon Technol Lett* 15, 754.
[9] T. Erdogan; 1997 *J. Lightwave Technol* 15 1277.
[10] X. Y. Dong, P. Shum, N. Q. Ngo, C. C. Chan, J. H. Ng, and C. L. Zhao; 2003 *Opt Express* 11, 2970.

Cascaded passive silicon microrings for large bandwidth slow light device

Yuntao Li, Yingtao Hu, Xi Xiao, Zhiyong Li, Yude Yu ,Jinzhong Yu
State Key Laboratory of integrated Optoelectronics, Institute of Semiconductors, Chinese Academy of Sciences, P. O. Box 912, Beijing 100083, P. R. China
Email: ytli@semi.ac.cn

Abstract. Slow light devices have important applications in the areas of data buffering, signal processing, and phased array antenna. Cascaded microring resonators structure can obtain large delay and also enhance the bandwidth, which was considered as a potential approach for future on-chip optical buffer. In this paper, we demonstrated a large bandwidth slow light device using cascaded Silicon-on-insulator (SOI) based microring resonators. With carefully designed the gap between the bus and the ring waveguides and the distances between the adjacent rings, a 57 ps group delay was observed and 83 Gbps maximum allowable bit rate is suggested according the measured 3 dB spectral bandwidth in the 8-stage cascaded microrings.
Keywords: slow light; cascade microring; silicon-on-insulator

1. Introduction

Slow light, which means reducing the light propagation velocity in media and structures, has become a rapidly growing field with great scientific value and a lot of potential applications. Especially, the recently research indicated that slow light device has good potential applications in optical buffers of high-capacity communication networks, optical pulse synchronization and reshaping, ultrafast all-optical information processing, true-time delay (TTD) in a phased-array antenna (PAA) [1-4], etc. There are several effects and structures can be used to obtain the slow light, including Bose–Einstein condensates, low-pressure metal vapors, solid crystal materials, optical fibers, semiconductor quantum wells and quantum dots, photonic band-gap structures and so on.[5-10]. Among these, silicon based microring structure is regarded as the ideal structure for the on-chip optical delay line or optical buffer for the merits of small footprint, fabrication compatibility with the CMOS technology and so on.

For a single SOI microring resonator, the delay time and the bandwidth is in contradiction. In order to get an optical delay line with large delay time and bandwidth simultaneously, cascaded microring structures, such as coupled resonators optical waveguide (CROW) and side-coupled integrated spaced sequence of resonators (SCISSOR) structure, are usually introduced for SOI-based devices. In this paper, we demonstrate a SOI 8-microrings optical delay line with a SCISSOR structure. A 57ps delay time is measured at 3Gbps; however, the 3dB spectral bandwidth shows that the device's theoretical maximum allowable bit rate is 83 Gbps.

2. Design and fabrication

Among the factors affecting the performance of a SOI cascaded microring slow light device, the gaps between the bus and the ring waveguides and the distances between the adjacent rings are much more important because they determine the spectrum shape and therefore the delay time and wavelength of the device. Transfer matrix is employed here to get the transmission characteristics of the structure and the coupling coefficients can be, optimized.

The device is fabricated on a commercial SOI wafer with a 340-nm-thick silicon and 2 μm buried silica layer. Grating couplers are introduced to decrease the coupling loss on the input and output ports.

Published under licence by IOP Publishing Ltd

Journal of Physics: Conference Series **276** (2011) 012068 doi:10.1088/1742-6596/276/1/012068

Electron-beam-lithography (EBL, RAITH150) process is used to transfer the device pattern into the PMMA mask and the pattern is etched into the silicon layer by inductively-coupled-plasma (ICP) etching process.

The SEM images of the 8-stage single channel SCISSOR device are shown in Fig. 1. The radii of the microrings are 6 μm with the rib width of the bus waveguide and ring waveguide 460 nm and 590 nm, respectively. The grating's period is 620 nm with a filling factor of 0.5.

Fig.1 The SEM image of the 8-stage single channel SCISSOR.

3. Results and discussions

The optical spectrum is measured by coupling a TE-polarized light from a tunable laser into the bus waveguide and the output light was coupling into an optical spectrum analyzer at the same time. Fig.2 shows the spectra response of the 8-stage cascaded microring. There is an apparently broaden spectrum than a single microring, while there is also a spectra split. The spectra splits and resonance broaden can be explained by deviation of the coupling, the width of the waveguides, and the resonators' perimeters of different resonances. The maximum on-resonance insertion loss is about 24.7 dB at 1554.5nm and the 3 dB bandwidth is 1.326 nm, which theoretically means an 83 Gbps maximum allowable bit rate for the device.

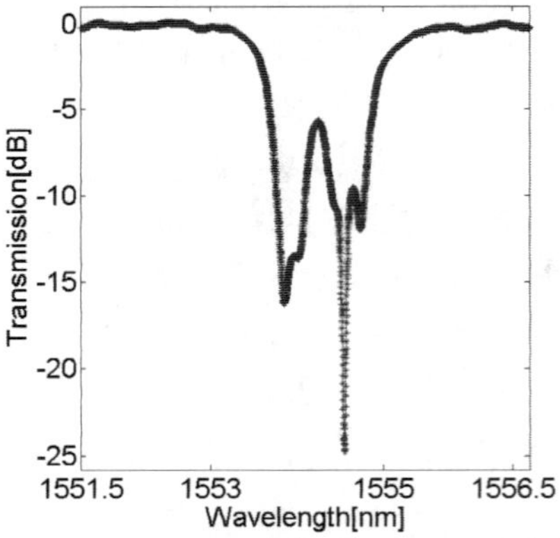

Fig.2 The spectra response of the 8-stage cascaded microring

The delay time is measured in a system shown in Fig.3. At first, light at a off-resonant wavelength from a tunable laser is modulated by a commercial $LiNbO_3$ modulator and then is detected by the detector and oscilloscope as a reference. Then, the wavelength of the input light is tuned to one of the broadened resonances, so the light will transmit through the microring resonantors and be detected. By comparing the two waveforms, the delay time is obtained. As shown in Fig.4, a 57ps optical delay is measured when the modulation frequency is 3Gbps.

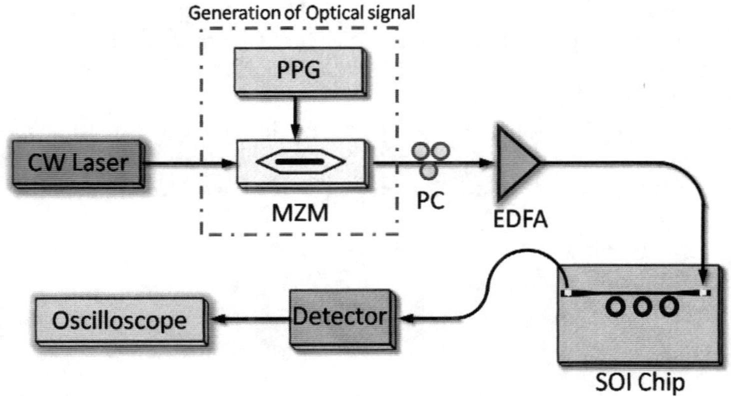

Fig.3 The experimental system of optical delay line

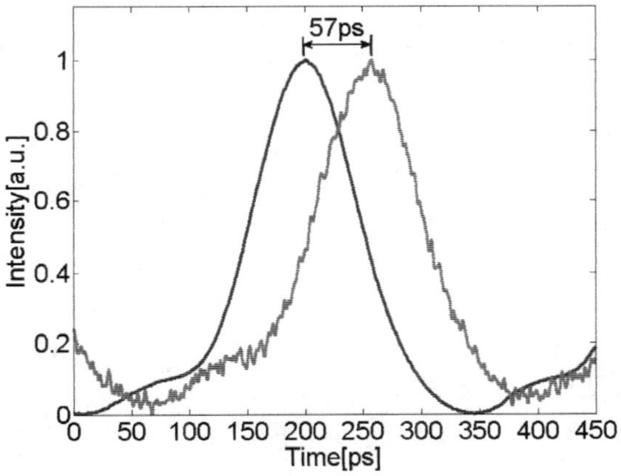

Fig.4 Group delay of 8-stage cascaded microring

4. Conclusion

We have experimentally fabricated optical delay lines based on SOI cascaded microrings and demonstrated their pulse delay performances. The cascaded microrings devices are provided with both broader bandwidth and larger group delay than single microring. The maximum delay we have measured is 57 ps at 3Gbps bit rate with 8-stage cascaded microrings, and the 3 dB spectral bandwidth suggested maximum allowable bit rate is large as 83 Gbps.

Acknowledgement

This work was supported by the National Natural Science Foundation of China (Grant No.60877036), the National Basic Research Program of China (Grant No.2006CB302803), State Key Laboratory of Advanced Optical Communication Systems and Networks, China (Grant No.2008SH02), Knowledge Innovation Program of ISCAS(Grant No.ISCAS2008T10).

References

[1] O'Mahony, M., et al., *The application of optical packet switching in future communication networks.* IEEE Communications Magazine, 2001. **39**(3): p. 128-135.

[2] Willner, A., et al., *Optical Signal Processing Using Tunable Delay Elements Based on Slow Light.* IEEE Journal of selected topics in quantum electronics, 2008. **14**(3): p. 691.

[3] Julsgaard, B., et al., *Experimental demonstration of quantum memory for light.* Nature, 2004. **432**(7016): p. 482-486.

[4] Liu, Z., et al., X-band continuously variable true-time delay lines using air-guiding photonic bandgap fibers and a broadband light source. Optics Letters, 2006. **31**(18): p. 2789-2791.

[5] Xia, F., L. Sekaric, and Y. Vlasov, *Ultracompact optical buffers on a silicon chip.* Nature Photonics, 2007. **1**(1): p. 65-71.

[6] Khurgin, J. and P. Morton, Tunable wideband optical delay line based on balanced coupled resonator structures. Optics Letters, 2009. **34**(17): p. 2655-2657.

Controlling the optical properties of gold nanoparticles periodic arrays by changing their topologic shapes and substrate properties

LI Ting(1) **, YU Li**(1)**, LU Zhi-Xin**(1)**, SONG Gang**(1)**, LIU Bin**(1)
(1) School of Science, Beijing University of Posts and Telecommunications, Beijing, China
E-mail:litinggirl39@gmail.com

Abstract. We investigated the influence of extinction of gold nanoparticles periodic arrays by varying the substrate properties and the shapes of nanoparticles with the full vectorial three dimensional finite difference time domain method. Substrates of different thicknesses and dielectric constants and ten topologically different gold nanostructures including diamond, cycle ring, rectangle ring, pentagon ring, five-pointed star, flower shape, L, Y, T and X shapes are considered. The results show that substrate properties have a significant impact on the extinction spectrum due to coupling of the modes excited in substrates and the one excited by localized surface plasmon. The extinction spectra are changed with the different shapes of nanoparticles periodic array. However, with similar structure particles periodic arrays, the extinction spectra appear similar in the visible regime. Therefore we can find an eligible shape and substrate which can be used in integrated devises.

Keywords: extinction spectra, gold nanoparticles, substrate properties, topologic shape

1. Introduction

Metallic nanoparticles usually exhibit localized surface plasmon resonance (LSPR) which is caused by the collective oscillation of electrons upon light excitation [1-2]. Excitation of LSPR results in the occurrence of pronounced extinction band [3] and local field enhancement effects [4-6] which has important applications in the potential chemical or biological nanosensors [7-8]. Different applications require the specific LSPR wavelength and extinction spectra. Therefore, extinction spectra could be tune by the different periodic arrays of nanoparticles. Reference [9] explain coupling between two adjacent nanoparticles using the dipolar interaction mechanism. And for the gold nanoparticles arrays, the light energy through the coupling between particles can be transport along the arrays due to the interaction between adjacent particles, which have the function of waveguide [10-15]. Many experimental investigations and theoretical calculations on both Au and Ag nanoparticles periodic arrays have shown that the characteristics of the resonance such as central frequency, bandwidth, amplitude and line shape depend on the particle size [16], shape [17-19], surrounding media [20-21], lattice constant [22-23], as well as the arrangement of the particle periodic arrays [24]. Therefore it is possible within certain limits to design the extinction spectra according by controlling these parameters.

In this letter, we investigated the extinction of gold nanoparticles periodic arrays on substrates of different dielectric constants and thicknesses. And then we investigated ten topologic shape different gold nanoparticles periodic arrays including diamond, cycle ring, rectangle ring, pentagon ring, five-pointed star, flower shape, L, Y, T and X shapes are considered. FDTD simulations that solve the full set of Maxwell's equations on a grid are a powerful method to model light-matter interactions in complex systems. All the results are obtained from the FDTD simulations.

In all the simulations, the wavelength of incident light is varied from 400 to 900 nm. The permittivity function of gold could be well described by a Drude-Lorentz model. These parameters used in our simulations are determined based on the experimental data of Johnson and Christy [25]. In our work, we implemented full 3-D FDTD simulation techniques based on rectangle uniform mesh. The grid size was set at around 2 nm. Two cases are considered. The first one, we adjusted the dielectric constant and thickness of substrate to investigate the influence on extinction spectra. The other, we changed the topologic shapes. Ten topologically different gold nanoparticles periodic arrays were considered, we divided them to four groups according to the similar topologic shapes. The group A is diamond, the group B are rectangle ring, pentagon ring, cycle ring, the group C are five-pointed star, flower shape and the group D are L, Y, T and X shapes.

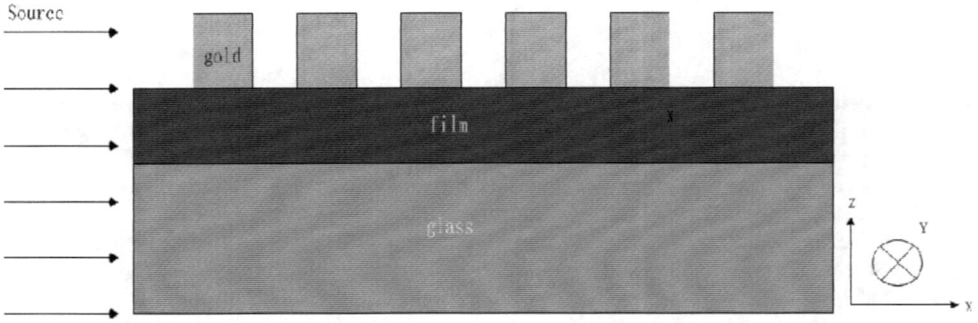

Figure 1. Sketch of gold nanoparticles periodic arrays on the quartz-glass substrates covered with film

2. Effect of dielectric constant and thickness of substrate

We simulated the extinction of cylinder gold nanoparticles periodic arrays deposed on the multiple substrates shown in Fig.1. The radius of gold nanoparticles is 50 nm, height is 100 nm. In our simulation, the X direction is the same direction as the source injection, therefore we use the absorbing boundaries (PML). We use six gold nanoparticles in X direction in simulation, which are enough to simulate the properties of periodic arrays [26]. Periodic boundaries are used in Y direction .The lattice constant kept the constant (d=450 nm). TE mode is adopted in order to observe extinction effectively. The extinction spectrum of gold nanoparticles periodic arrays is calculated by the transmissivity.

First we considered the thickness of the film influenced on extinction spectra. For experimental consideration, indium tin oxide (ITO) is usually chosen as the film to cover glass substrate [27]. The thicknesses of the ITO are 20 nm and 140 nm respectively. For the further research on extinction influenced by the thickness of substrate, we also considered the structure without any substrate, though it is not true for the real situation. The simulated results are shown in Fig.2(b). Due to extinction spectra completed differences among them, it is obvious to know that the thicknesses of substrates exert on significant influence. According to the optics waveguide theory [28],

$$\lambda_{cut} = \frac{2\pi * t\sqrt{\varepsilon_{film} - \varepsilon_{sub}}}{m\pi + ar\tan(\sqrt{\frac{\varepsilon_{sub} - \varepsilon_{air}}{\varepsilon_{film} - \varepsilon_{sub}}})} \quad (1)$$

Where t is the thickness of the film, ε_{film} is the dielectric constant of film, ε_{sub} is the dielectric constant of substrate, ε_{air} is the dielectric constant of air. $\varepsilon_{air} = 1$, $\varepsilon_{film} = \varepsilon_{ITO} = 3.61$, $\varepsilon_{sub} = 2.28$ .There will be TE_0 mode when the ITO film thickness 140nm and there is non- waveguide mode for 20 nm. For 140nm thickness case, the extinction spectrum appears two obvious narrow peaks compared with others. Their centre wavelengths are about 600 nm and 800 nm respectively. The reason might be that evanescent field of TE_0 mode existing in the ITO film could couple with the LSP mode excited by nanoparticles. Their interference could lead to the obvious narrow peaks. For 20 nm case, there is no waveguide mode in the ITO film. But radiation field excited by ITO film can also couple with the scattering light excited by nanoparticles. We also observed peaks, though the interference is not stronger than that of 140 nm case. Moreover, the width of peaks became wider than that of 140 nm case. For the non-substrate case, the extinction spectrum is determined by the superposition of scattering light of nanoparticles. Therefore, the width of spectrum get wider and the peak of extinction spectra get weaker.

Figure 2(a). Schematics of FDTD simulation. The above is gold nanoparticles periodic arrays without substrate. The middle and below are gold nanoparticles periodic arrays on the quartz-glass substrates covered with ITO film, the thickness of ITO film are 20 nm and 140 nm respectively

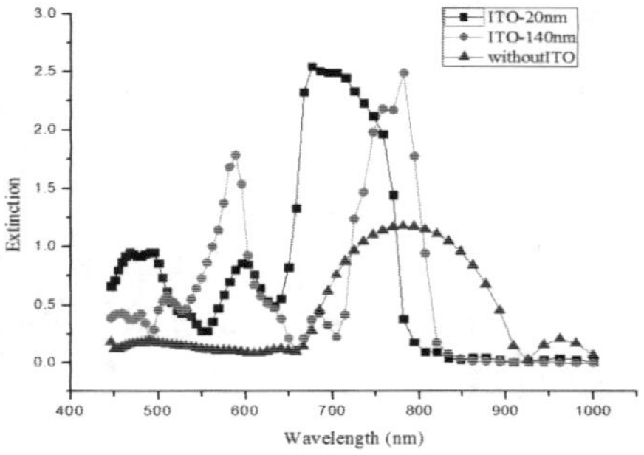

Figure 2(b). extinction of gold nanoparticles periodic arrays without ITO film (blue) , on a 20 nm thick ITO film (red), and 140 nm thick ITO film(black)

In order to more comprehensive investigated the effect of substrate properties. On one hand, the thickness of film kept 20 nm, we only varied the dielectric constants of film to 1.5, 2.5, 3.6, 10, 20 respectively. We noticed that the film never act as waveguide due to dielectric constant of film is smaller than the glass when the dielectric constant of the film is 1.5. But for dielectric constants of film are 2.5 or 3.6 respectively, no waveguide mode can be excited though it can be act waveguide. When the dielectric constant are 10 or 20, it can support one waveguide mode existed. On the other hand, we kept the thickness of film was 140 nm, changed the dielectric constant to 3.6 and 10 respectively. In this case, the film can support TE_0 mode for dielectric constant 3.6 and both TE_0 and TE_1 modes respectively for dielectric constant 10. All the results of simulations are shown in Fig. 3. We arrange the curve in graphics according to numbers of waveguide mode existed in films. Fig.3 (a) presents the extinction spectra of different dielectric constant of film which support non-mode. Fig.3 (b) and 3 (c) give the situations that the film can supports one and two modes. We found that the numbers of narrow peaks increasing with modes excited by film. The reason is obviously to explain by using the wave couple theory we mentioned above. We notice in Fig.3 (c) that there exist very narrow peak around 550 nm wavelength, which the width of spectrum is only 10 nm much smaller other any cases. It results from the couple of two waveguide modes and interference of evanescent field of waveguide modes and scattering light coming from nanoparticles. We could infer that when the thickness of films increasing so high that it can support many waveguide modes, the width of extinction spectra will be wide and due to many peaks. Finally we simulated the structure that nanoparticles periodic arrays on semi-infinite thick glass substrate without ITO film. The result is shown in Fig.3 (d), the extinction of it is similar to the case of film which support non-mode, which also can understood by the waveguide mode theory we mentioned above.

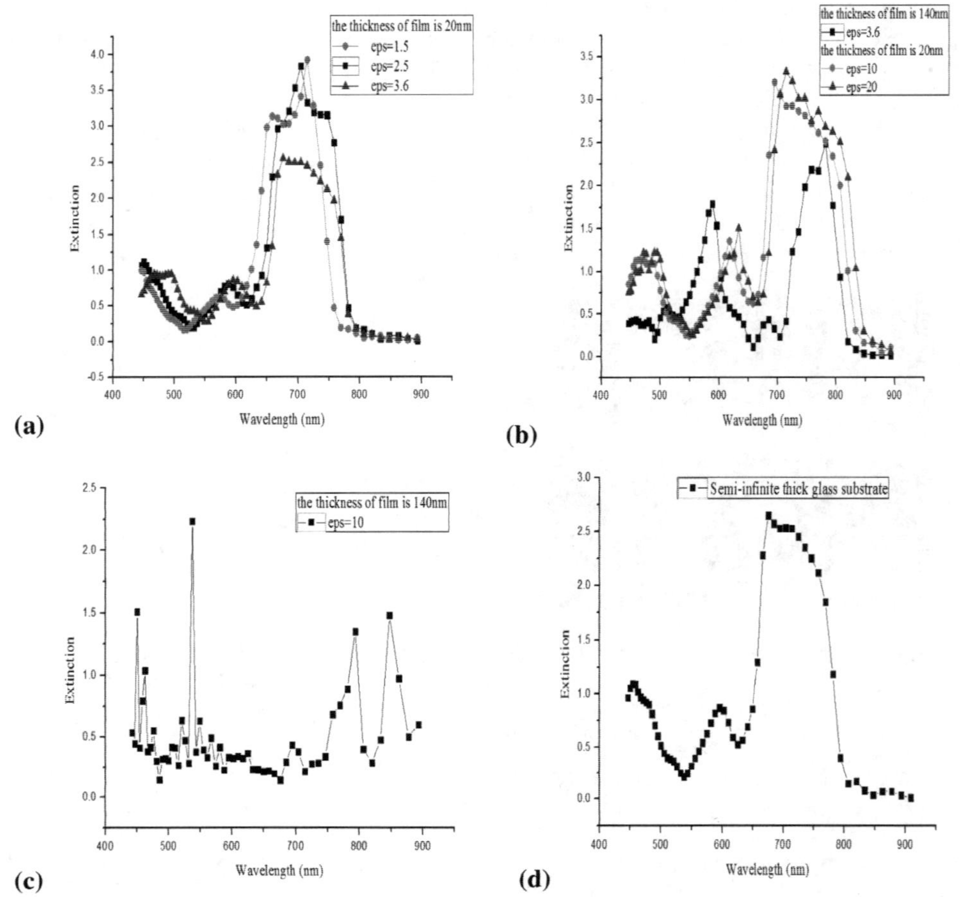

Figure 3(a) Extinction of gold nanoparticles periodic arrays on the quartz-glass substrates covered with film, t=20nm, the dielectric constants of the film are 1.5 (red) and 2.5 (black), t=140nm, the dielectric constant of the film is 10 (blue). **(b)**Extinction of gold nanoparticles arrays on the quartz-glass substrates covered with film, t=140nm the dielectric constant of the film is 3.6 (black), t=20nm (red), is 20, t=20nm the dielectric constant of the film are 10 (red) and 20 (black) (blue). **(c)**Extinction of gold nanoparticles periodic arrays on the quartz-glass substrates covered with film, t=140nm, the dielectric constant of the film is 10. **(d)** Extinction of gold nanoparticles arrays on the semi-infinite thick quartz-glass substrates.

3. Effect of nanoparticles topologic shape

In order to investigate shape effects on the extinction of the nanoparticles periodic arrays .We varied the topologic shape of nanoparticles periodic arrays while the substrate character kept unchanged. The substrate was the quartz-glass substrates covered with 100 nm-ITO film. The structure of system is the same as the Fig.1. In following simulations, the size of all samples is in the range of 50-100 nm and the height is 40 nm. The group A is diamond, we change arrangement of the nanoparticles periodic arrays. One is long symmetry axis vertical to X axis, the other is long symmetry axis parallel to X axis as shown in Fig.4a. The extinction spectra of both arrangement is similar each other (Fig.4b), Therefore the two arrangement of nanoparticles has little impact on the peak of extinction.

(a)
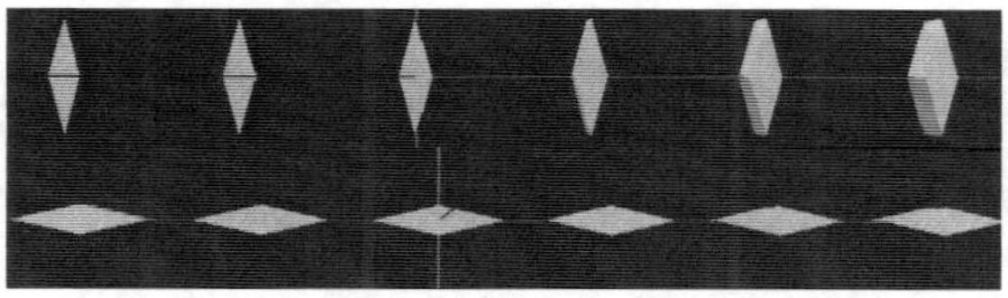

(b)
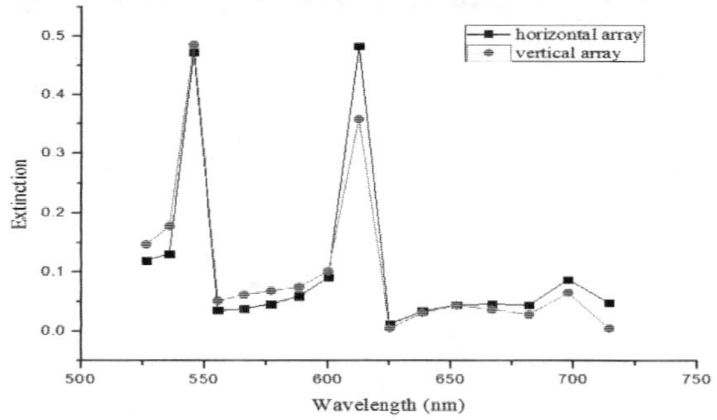

Figure 4(a). Schematic of periodic arrays of diamond nanoparticles with horizontal and vertical arrangement on quartz-glass substrates covered with ITO film. The above is the field distribution of them **(b).** Extinction of diamond of horizontal (black) and vertical (red) arrangement

The group B are including rectangle ring, pentagon ring, cycle ring, which are similar ring shape shown in Fig.5 (a).The outside radius and inside radius of them are all the same. The extinction spectra of them are shown in Fig.5 (b). The peaks and values of extinction spectra of particles of similar ring shape very close. It is may caused by the same coupling effect of adjacent nanoparticles. One of the peaks is sharp, but the other peak is wider than it. That may due to the closed shapes.

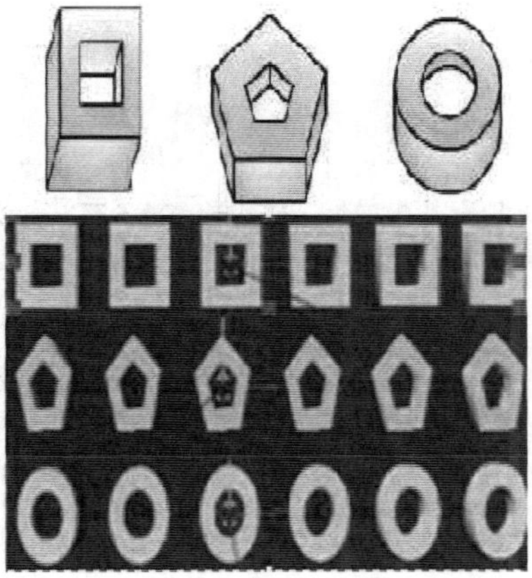

Figure 5(a). Schematic of periodic arrays of ring shape nanoparticles(rectangle ring, pentagon ring, cycle ring) on quartz-glass substrates covered with ITO film

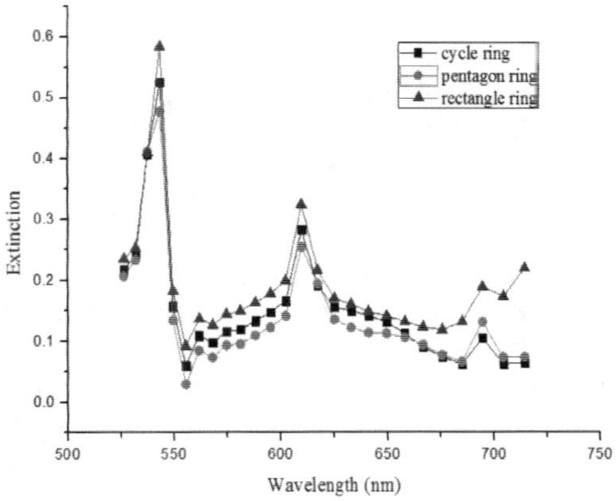

Figure 5(b). Extinction of rectangle ring(blue),cycle ring (black), pentagon ring(red).

The group C are including five-pointed star, flower shape, which are show in Fig.6 (a). We used TM and TE polarization wave illuminate. The flower shape has four leaves which are smoother than the five angles of five-pointed star. The extinction of group C are show in Fig.6 (b) (TM mode excitation), Fig.6 (c) (TE mode excitation). The extinction of different polarization wave incoming are very different, therefore, the optic properties of nanoparticles arrays can control by different polarization direction. For TM excitation, there are two extinction peaks of flower shape, one peak around 619 nm for both five-pointed star and flower shape, another peak at 700 nm for flower shape. For TE excitation, there is one peak around 610 nm for flower shape with slight concussion. However, there is so strong concussion of the peak for five-pointed star that we can see four weak peaks exist. It might caused by complex coupling among the angles of five-pointed star.

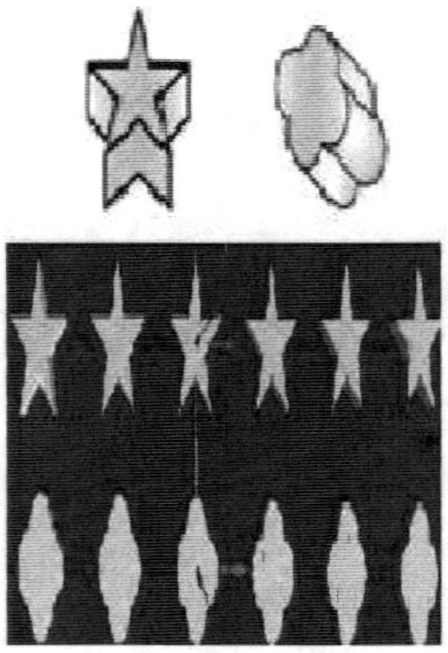

Figure 6(a). Schematic of periodic arrays of five-pointed star and flower shapes on quartz-glass substrates covered with ITO film.

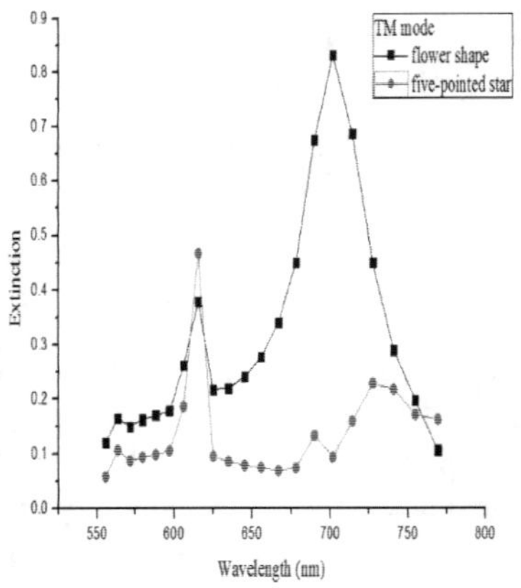

Figure 6(b). Extinction of five-pointed star (red), flower shapes (black) by TM excitation

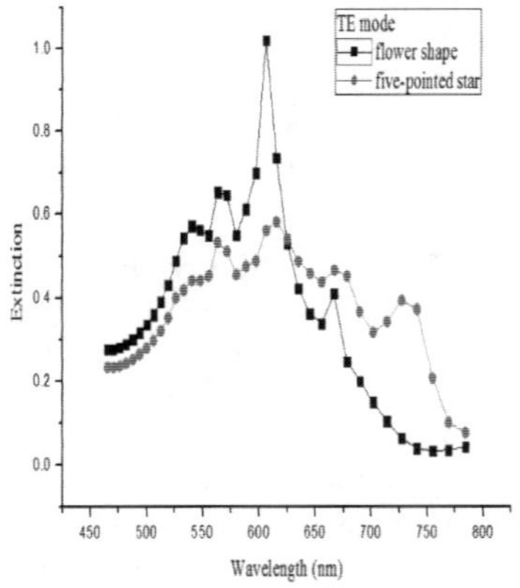

Figure 6(c). Extinction of five-pointed star (black), flower shapes (red) by TE excitation.

Finally, we choose a group of non-closed shapes. The group D including L, Y, T and X letter shapes (Fig.7(a)), The bending angle of L shape is 90^0,The two arms and the angle between the two arms of Y shape is same as the L shape, and the length of subjacent arm is the same as the above two arms. Two L shapes are constitute X shape. The length of upper horizontal nanorod of T shape is 60 nm, the length of subjacent vertical nanorod is the same as the arm of Y. Diameters of all arms are 20nm. The results are shown in Fig.7(b) ,the extinction spectra of them are also close. The third peak of X shape is larger than others might due to its volume is bigger than others. They all have two similar sharp peakS, It is may due to their two similar scale arms.

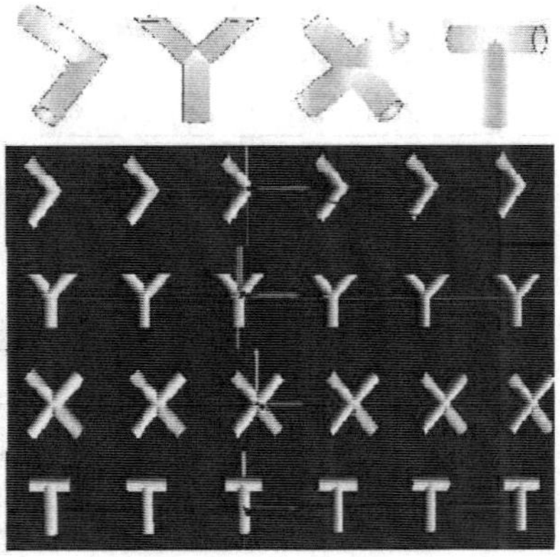

Figure 7(a). Schematic of periodic arrays of L, Y, X and T letter shapes on quartz-glass substrates covered with ITO film

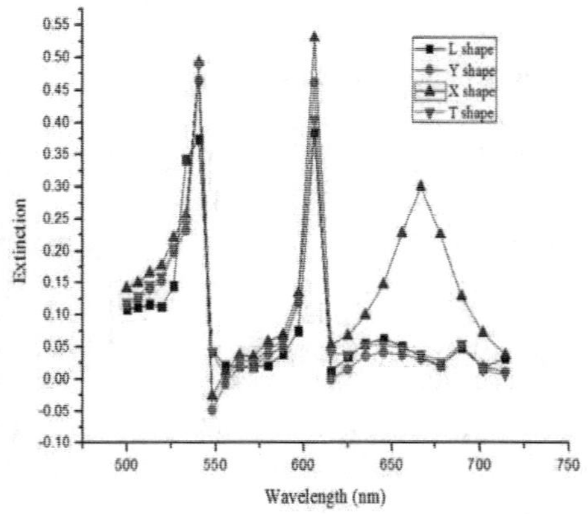

Figure 7(b). Extinction of periodic arrays of L (black), Y (red), X (blue) and T (green) letter shapes on quartz-glass substrates covered with ITO film

The four groups above are all the similar shape. Comparing extinction among the different shapes of the four groups, we can find spectra of structure of different shapes are largely different, such as closed (group B) and non-closed shapes (group D), horned(group A) and without angle (group B) ,their extinction are grate different. Which might due to the coupling among the arms and angles of them.

4. Conclusion

In conclusion, we have presented influence of extinction spectra by changing thicknesses and dielectric constants of substrates and topologic shapes of gold nanoparticles periodic arrays. The effect of different thicknesses of substrates has been investigated by variation of thickness film for a fixed dielectric constant. The results showed the spectra became narrower with the increasing of thickness of substrate, which due to the different waveguide modes excited film which can be considered a dielectric waveguide and the coupling of modes in film and scattering light excited by particles. The influence of extinction by dielectric constants of substrates were also investigated, we explained the reason by using the waveguide mode theory and wave couple theory. Gold nanoparticles periodic arrays of ten different topologic shape on the quartz-glass substrates covered with ITO film were simulated, we grouped them to four groups, the results showed that the peaks and values of extinction spectra of similar close structure particles very close and the two arrangement we considered of nanoparticles has little impact on the peak of extinction. Spectra of closed and non-closed structure were largely different might due to the coupling among the arms and angles of them in the visible regime. For the five-pointed star and flower shape, the extinction spectra of nanoparticles arrays were different by the different polarization direction wave incoming. Therefore we find optical properties of gold nanoparticles periodic arrays can control by changing their topologic shapes and substrate properties.

Acknowledgments
This work is supported by the National Basic Research Program (973) of China under Project No.2010CB923202.

5.References

[1] U.Kreibig, M.Vollmer, "Optical Properties of Metal Clusters," Springer Berlin(1995)

[2] Katherine A.Willets and Richard P.Van Duyne, "Localized Surface Plasmon Resonance Spectroscopy and Sensing," Annu. Rev. Phys. Chem. 58:267 – 97(2007)
[3] Baptiste Auguie and William L. Barnes, "Collective Resonances in Gold Nanoparticle Arrays," PRL 101, 143902(2008)
[4] Ping Chu and D. L. Mills, "Electromagnetic response of nanosphere pairs: Collective plasmon resonances, enhanced fields, and laser-induced forces," PHYSICAL REVIEW B 77, 045416(2008)
[5] Shengli Zou, George C. Schatz, lver, "nanoparticle array structures that produce giant nhancements in electromagnetic fields," CHEMICAL PHYSICS LETTERS 403,62(2005)
[6] L. A. Sweatlock, S. A. Maier and H. A. Atwater, "Highly confined electromagnetic fields in arrays of strongly coupled Ag nanoparticles," PHYSICAL REVIEW B.71,235408(2005)
[7] Yizhuo Chu, Ethan Schonbrun, Tian Yang and Kenneth B.Crozier, "Experimental observation of narrow surface plasmon resonances in gold nanoparticle arrays," APPLIED PHYSICS LETTERS 93, 181108(2008)
[8] Jonas Beermann, Sergey M.Novikov, Kristjan Leosson, and Sergey I. Bozhevolnyi, "Surface enhanced Raman imaging: periodic arrays and individual metal nanoparticles," OPTICS EXPRESS 17, 12698(2009)
[9] W. Rechberger, A. Hohenau, A. Leitner, J.R. Krenn, B. Lamprecht, F.R. Aussenegg, "Optical properties of two interacting gold nanoparticles," Optics Communications.220,137(2003)
[10] M.Quinten, A. Leitner, J. R. Krenn, and F. R. Aussenegg, "Electromagnetic energy transport via linear chains of silver nanoparticles," OPTICS LETTERS,23,17(1998)
[11] Stefan A. Maier, Pieter G. Kik, Luke A. Sweatlock, and Harry A. Atwater, "Energy transport in metal nanoparticle plasmon waveguides," Mat. Res. Soc. Symp. Proc. Vol. 777(2003)
[12] D.S. Citrin, "Coherent Excitation Transport in Metal-Nanoparticle Chains American Chemical Society,"4,9(2004)
[13] Stefan A.Maier, Pieter G. Kik, and Harry A. Atwater, "Observation of coupled plasmon-polariton modes in Au nanoparticle chain waveguides of different lengths: Estimation of waveguide loss," APPLIED PHYSICS LETTERS81,9(2002)
[14] Howard R. Stuart and Dennis G. Hall, "Enhanced Dipole-Dipole Interaction between Elementary Radiators Near a Surface," Phys.Rev.Lett.80,25(1998)
[15] Stefan A. Maier, Pieter G. Kik, Harry A. Atwater, Sheffer Meltzer, Elad Harel, Bruce E.koel, and Ari A.G. Requicha, "Local detection of electromagnetic energy transport below the diffraction limit in metal nanoparticle plasmon waveguides," nature, 10,1038(2003)
[16] Christopher Tabor, Raghunath Murali, Mahmoud Mahmoud, and Mostafa A. El-Sayed, "On the Use of Plasmonic Nanoparticle Pairs As a Plasmon Ruler: The Dependence of the Near-Field Dipole Plasmon Coupling on Nanoparticle Size and Shape," J. Phys. Chem. A, 113, 1946–1953(2009)
[17] Peng Yang, Herve´ Portale`s, and Marie-Paule Pileni, "Identification of Multipolar Surface Plasmon Resonances in Triangular Silver Nanoprisms with Very High Aspect Ratios Using the DDA Method," J. Phys. Chem. C, 113, 11597–11604(2009)
[18] Daeha Seo, Choong Il Yoo, Im Sik Chung, Seung Min Park,Seol Ryu and Hyunjoon Song, "Shape Adjustment between Multiply Twinned and Single-Crystalline Polyhedral Gold Nanocrystals: Decahedra, Icosahedra, and Truncated Tetrahedra," J. Phys. Chem. C, 112, 2469-2475(2008)
[19] Z.-Y. Zhang and Y.-P. Zhao, "Tuning the optical absorption properties of Ag nanorods by their topologic shapes: A discrete dipole approximation calculation," APPLIED PHYSICS LETTERS 89, 023110(2006)
[20] Patrick Rooney, Asad Rezaee, Songbo Xu, Touraj Manifar and Abdollah Hassanzadeh, "Control of surface plasmon resonances in dielectrically coated proximate gold nanoparticles immobilized on a substrate," PHYSICAL REVIEW B 77, 235446(2008)

Journal of Physics: Conference Series **276** (2011) 012069 doi:10.1088/1742-6596/276/1/012069

[21] K. Lance Kelly, Eduardo Coronado, Lin Lin Zhao, and George C. Schatz, "The Optical Properties of Metal Nanoparticles: The Influence of Size, Shape, and Dielectric Environment," J. Phys. Chem. B , 107, 668-677(2003)

[22] Andreas Hohenau, and Joachim R. Krenn, "Spectroscopy and nonlinear microscopy of Au nanoparticle arrays: Experiment and theory," PHYSICAL REVIEW B 73, 155404(2006)

[23] N. Félidj, G. Laurent, J. Aubard, and G. Lévi, "Grating-induced plasmon mode in gold nanoparticle arrays," THE JOURNAL OF CHEMICAL PHYSICS,123, 221103(2005)

[24] Kin Hung Fung, and C. T. Chan, "Analytical study of the plasmonic modes of a metal nanoparticle circular array," PHYSICAL REVIEW B 77, 205423 (2008)

[25] P. B. Johnson and R. W. Christy, "Optical constants of the noble metals," Phys. Rev. B, vol. 6, no. 12, pp. 4370–4379(1972)

[26] Hao Jiang, Jayshri Sabarinathan, Touraj Manifar, and Silvia Mittler, "3-D FDTD Analysis of Gold-Nanoparticle-Based Photonic Crystal on Slab Waveguide," JOURNAL OF LIGHTWAVE TECHNOLOGY, 27, 13(2009)

[27] V.M.shalaev and S.Kawata, "Nanophotonics with surface plasmons," Chapter 2(2007)

[28] H. Kogelnik and V. Ramaswamy, " Scaling rules for thin-film optical waveguides," Appl. Opt. vol. 13, pp. 1857-1862(1974)

Dependencey of resonant wavelength on the width of subwavelength metallic slits: analysis and modulation

Xu-Sheng Lin [1, 2*], **Jun-Hu Yan** [1], **Yun-Bao Zheng** [1], **Li-Jun Wu** [2], **Sheng Lan** [2]
[1]School of Electronic and Information Engineering, Guangdong Polytechnic Normal University, Guangzhou 510665
[2]Laboratory of Photonic Information Technology, School for Information and Optoelectronic Science and Engineering, South China Normal University, Guangzhou 510006

E-mail: xslin64@yahoo.cn

Abstract The variation trend of the resonant wavelength of a subwavelength metallic slit with respect to its width has been investigated by using the Fabry-Perot model. It is determined by two antagonistic factors related to the effective refractive index and the phase of reflection. We show that the trend can be changed by directly modulating the dielectric constant of the structure. Our analysis is supported by the numerical simulation based on the finite-difference time-domain technique.

1. Introduction

Surface plasmon polaritons (SPPs) are electromagnetic waves coupled to the oscillation of the collective electrons at metal surfaces. As one of the SPP-based waveguides, metal-dielectric-metal (MDM) structures recently attract much attention due to their subwavelength characteristic, i.e., guiding of light below the diffraction limit that is essential for nanochip-scale circuits [1-4]. In such waveguides, the SPP modes are called "X-ray waves" with optical frequencies because of their much shorter wavelengths than those in vacuum. Keeping this unique property in mind, researchers are now interested in developing MDM waveguide-based devices for manipulating light on the nanoscale, such as optical couplers across the nano-micro interfaces [5, 6], ultro-compact waveguide bends and splitters [7, 8], high resolution tunable filters [9, 10], and subwavelength all-optical switchers [11-14].

It is noted that many of the above device structures are based on the truncated MDM waveguides, which are also called slit resonators. Due to the complicated optical properties of metals, it is difficult to express explicitly the relationship between the resonance properties (the resonant wavelength, the peak transmission, and the quality factor of resonance, etc.) and the geometric and material parameters of the structures. Thus, intuitive models that reflect or describe the properties of the resonators are very helpful. One of them is the Fabry-Perot model that is widely used and proved to be effective [15-24], especially the corrected version that adopts the effective refractive index and the phase of reflection [18, 19, 22-24]. Due to the drop of the effective refractive index, the resonant wavelength undergoes a blue-shift with the increase of the slit width. However, as the slit width increases further, it can turn to a red-shift, showing a V-shape dependence [17, 23]. Although many optical properties of single slit resonators have been investigated, this phenomenon is still not explained properly. As the variation pattern of resonant wavelength is an important aspect of the SPP resonator design, it is

* To whom any correspondence should be addressed.

Published under licence by IOP Publishing Ltd

worthy of being investigated in detail. Here, we show that the phenomenon can be explained by using the concepts of the effective refractive index and the phase of reflection. As the phase of reflection cannot be expressed in an intuitive way such as the direct use of Fresnel's formula, people turn to rely on numerical calculations or simulations for better understanding [18, 22, 23]. We find that with the assistance of the results from the perfect electric conductor (PEC) cases, one can easily find out the influence of the phase of reflection and understand the variation pattern of the resonant wavelength.

2. Analysis and Simulation

The MDM resonator studied in this paper is shown in the inset of Fig. 1. It consists of a free-standing silver film with one isolated subwavelength slit that is filled with a dielectric. We use *t* and *w* to denote the thickness and width of the slit, while the permittivity of silver and the dielectric are represented by $_{Ag}$ and $_{d}$ respectively.

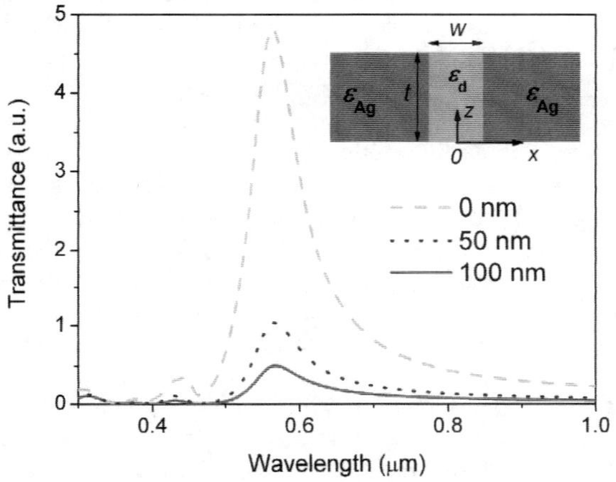

Figure 1. Output transmission spectra of the Ag slit measured at 0nm, 50nm and 100nm above the outlet of the slit. Geometrical and material parameters are: *t*=100nm, *w*=20nm, $_{d}$ =1. Inset is the structure of the slit resonator studied.

Normally, both the surface plasmonic and conventional optical modes can be excited in MDM waveguides. However, when the slit width drops to deep subwavelength, only the fundamental SPP modes with even magnetic distribution can be accessed [3]. They are TM modes whose magnetic field parallel to y-axis. If we neglect the imaginary part of $_{Ag}$, the dispersion of the fundamental SPP modes can be written as [3]

$$\tanh\frac{k_0 w\left(n_{eff}^2-\varepsilon_d\right)^{1/2}}{2}=\left(\frac{\varepsilon_d}{-\varepsilon_{Ag}}\right)\cdot\left(\frac{n_{eff}^2-\varepsilon_{Ag}}{n_{eff}^2-\varepsilon_d}\right)^{1/2} \tag{1}$$

where $k_0 = 2\ /\ _0$ and $_0$ is the wavelength of light in vacuum, n_{eff} is the effective refractive index of the SPP mode. Once n_{eff} is obtained from Eq. 1 and Fabry-Perot resonant condition is fulfilled, we have

$$\lambda_0=\frac{2n_{eff}t}{m-\delta} \tag{2}$$

where m is the resonance order and is the phase of reflection (unit:) obtained at the slit entrance. It is noted that $<m$ is valid for physical sense. Although the width of slit, *w*, does not appear in Eq. 2, it affects the location of $_0$ through n_{eff} and . To estimate the variation trend of $_0$ with respect to *w*, it would be more convenient when the derivative of Eq. 2 is used. That is

$$\frac{1}{\lambda_0}\cdot\frac{d\lambda_0}{dw}=\frac{1}{n_{eff}}\cdot\frac{dn_{eff}}{dw}+\frac{1}{1-\delta}\cdot\frac{d\delta}{dw}. \tag{3}$$

Here, we focus on the subwavelength resonators possessing only one resonance (m=1). It is expected that dn_{eff}/dw<0 and d /dw>0, so the variation of $_0$ with respect to w is decided by the "bout" of the two antagonistic factors, i.e., $n_{eff}^{-1}dn_{eff}/dw$ and $(1-)^{-1}$d /dw. Thus, the balance of them leads to d $_0/dw$=0. We consider it the reason why a maximum resonant frequency or a minimum resonant wavelength can be obtained [17, 23].

To verify the above assumption, we carry out numerical simulations of Maxwell equations based on the finite-difference time-domain method. Since recent investigation has shown that the length of metallic slit has little influence on the resonant wavelength positions [4], we simulate the two-dimensional slit structures instead of the three-dimensional ones for saving the computational time. The simulation domain is 3μm×3μm surrounded by perfectly matched layers of 0.2μm width. Spatial grid and temporal interval are set to be 2nm and (1/c)ns, respectively, where c is the speed of light in vacuum. As for the permittivity of Ag, we employ the Drude model plus two-pole Lorentzian form that shows a good description of empirical data over the wavelength range of 250nm to 1000nm [8]. Very short TM Gaussian pulse with centre wavelength at 1μm is launched normally from the position of z=–0.1μm (below the slit). By setting monitors at the outlet of the slit, we can numerically measure the transmission spectrum and obtain the resonant wavelength of the metallic slit. One example in which t=100nm, w=20nm, and $_d$=1.00 is shown in Fig. 1. The three spectra are measured at different positions (0nm, 50nm and 100nm) above the outlet. Since w is in deep subwavelength, only one resonant peak corresponding to the fundamental mode is excited. The peaks that correspond to the resonant wavelength appear at about 0.565μm, exhibiting weak dependence on the measuring position.

By repeating the above simulation for different slit widths, one can obtain the relationship between $_0$ and w for a given thickness of slit. When the slit is filled with air, i.e., n_d=1.00, the simulation results are depicted in Fig. 2 with solid circles. We can see that $_0$ first drops and reaches its minimum at w=50nm. Then, it begins to rise when w increases further. The trend is the same as those shown in Refs. [17, 23].

It has been suggested that two opposite factors are in action, as manifested in Eq. 3. Here we present a more detailed discussion. When the condition of (k_0w) •0.5 is fulfilled, which is valid for the case we study, one can easily derive the following relation from Eq. 1

$$n_{eff}\frac{dn_{eff}}{dw}=-\frac{4}{k_0^2w^3}\cdot\left(\frac{\varepsilon_d}{-\varepsilon_{Ag}}\right)^2. \tag{4}$$

Obviously, it shows that dn_{eff}/dw <0 and n_{eff} is very sensitive to w when w is small. One the other hand, in spite of the difficulty of deriving an analytical relation between and w, we find it helpful to consider the perfect electric conductor (PEC) case, i.e., $_{Ag}$•• •. From Eq. 4, one finds dn_{eff}/dw=0 for the PEC case. Thus, according to Eq. 3, we have

$$\frac{d\delta}{dw}=\frac{2n_{eff}}{\lambda_0^2}\cdot\frac{d\lambda_0}{dw}. \tag{5}$$

In Fig. 2, the relationship between $_0$ and w obtained by FDTD simulation for the PEC case of n_d=1.00 is depicted by dashed line. It shows that $_0$ increases monotonously with w. Actually, $_0$ is almost proportional to w, i.e., d $_0/dw$ is a constant, which is in good agreement with the results given in Ref. [19].Therefore, d /dw>0 and shows weak dependence on w according to Eq. 5. We can conclude that dn_{eff}/dw plays a dominant role so $_0$ decreases with w when w is very small. With increasing w, dn_{eff}/dw begins to decrease and its action is balanced gradually by d /dw. d /dw becomes dominant at larger w and as a result the dependence of $_0$ on w is changed. In Fig. 2, it is true that the dependence of $_0$ on w exhibits similar trend to the PEC case, implying that it is governed by the phase of reflection.

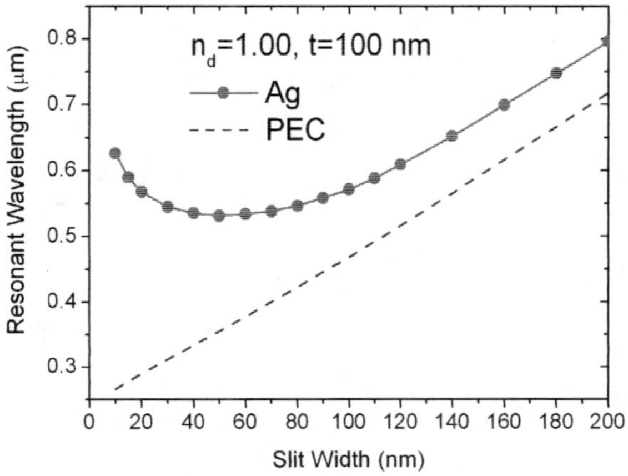

Figure 2. Dependence of the resonant wavelength on the slit width for the Ag and PEC cases.

From the above discussion, we find that the dependence of $_0$ on *w* can be engineered by modifying d /d*w* and dn_{eff}/d*w*, which have completely different contributions. Two extreme cases are $_0$ increases or decreases monotonously with *w*. The former corresponds to the situation of large (− $_{Ag}$), similar to the PEC case shown in Fig. 2. The latter occurs for small d /d*w*. We find that one simple method to reduce d /d*w* is to fill the slit with an insulator possessing a large refractive index. Fig. 3 shows the simulation results for n_d=2.00. Comparing with the case of n_d=1.00, $_0$ reaches its minimum $_0$ at a larger *w*(=100nm) and it rises much more slowly after the minimum. This can be attributed to the smaller d /d*w* caused by the larger n_d. Similarly, we can make a better understanding by comparing the PEC case, which is also depicted in Fig. 3. Apparently, d $_0$/d*w* (thus d /d*w*) is much smaller than that in the case of n_d=1.00.

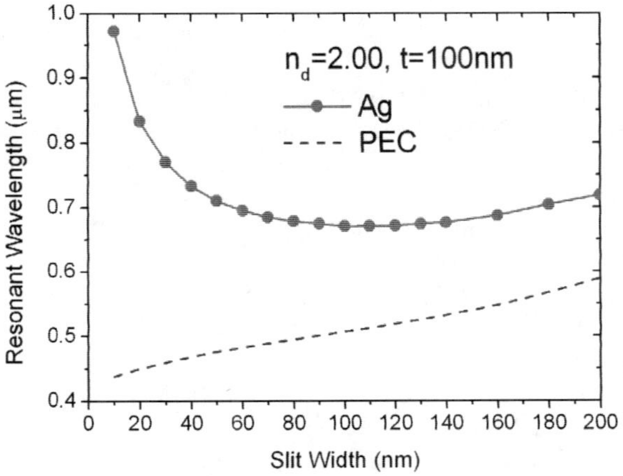

Figure 3. Dependence of the resonant wavelength with the slit width for the Ag and PEC cases. It shows that increasing the dielectirc constant of the filled insulator can weaken the impact of the phase of reflecton effectively.

A further increase of n_d to 3.45, which is the refractive index of GaAs or Silicon, makes the situation totally different. The results are shown in Fig. 4. The major difference between Fig. 4 and Fig. 3 is that $_0$ no longer increases with *w*. According to our simulation results, this trend does not change even when *w* becomes much larger than 200nm (not shown here). It is suggested that the effect of the phase of reflection in this case is negligible. Actually, as can be seen from the PEC result in Fig. 4, d /d*w* is very small and it reaches zero when *w* is larger than 160nm.

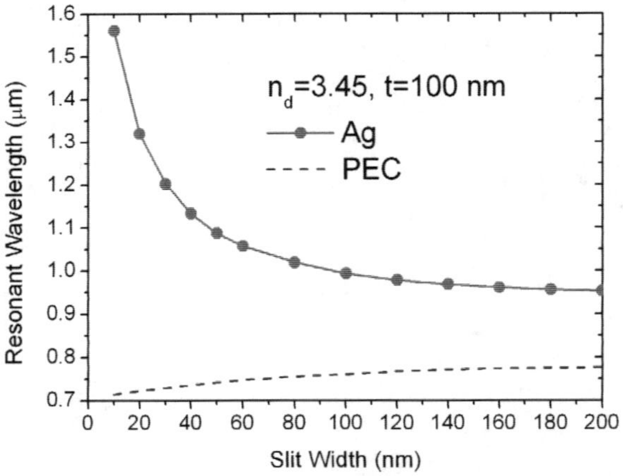

Figure 4. Dependence of the resonant wavelength with the slit width of the Ag and PEC cases. It shows that the effect of the phase of reflection is negligible.

3. Summary

In conclusion, the effective refractive index and the phase of reflection have opposite effects on the variation trend of resonant wavelength with respect to the metallic slit width. Minimum resonant wavelength can be obtained when the two effects balance. Further more, the resonant wavelength can decrease monotonously with the slit width if the slit is filled with an insulator with a large refractive index.

Acknowledgments
This work was supported by the National Natural Science Foundation of China under Grant Nos. 60778032 and 10974060.

References

[1] Barnes W L, Dereux A and Ebbesen T W 2003 *Nature* **424** 824
[2] Zia R, Selker M D, Catrysse P B and Brongersma M L 2004 *J. Opt. Soc. Am.* A **21** 2442
[3] Dionne J A, Sweatlock L, Polman A and Atwater H A 2006 *Phys. Rev.* B **73** 035407
[4] Pang Y, Genet C and Ebbesen T W 2007 *Opt. Comm.* **280** 10
[5] Chen L, Shakya J and Lipson M 2006 *Opt. Lett.* **31** 2133
[6] Veronis G and Fan S 2007 *Opt. Express* **15** 1211
[7] Veronis G and Fan S 2005 *Appl. Phys. Lett.* **87** 131102
[8] Lee T W and Gray S K 2005 *Opt. Express* **13** 9652
[9] Xiao S, Liu L and Qiu M 2006 *Opt. Express* **14** 2932
[10] Diest K, Dionne J A, Spain M and Atwater H A 2009 *Nano Lett.* **9** 2579
[11] Min C, Wang P and Chen C 2008 *Opt. Lett.* **33** 869
[12] Shen Y and Wang G P 2008 *Opt. Express* **16** 8421
[13] Yu Z, Veronis G, Fan S and Brongersma M L 2008 *Appl. Phys. Lett.* **92** 041117
[14] Min C and Veronis G 2009 *Opt. Express* **17** 10757
[15] Astilean S, Lalanne P and Palamaru M 2000 *Opt. Comm.* **175** 265
[16] Takakura Y 2001 *Phys. Rev. Lett.* **86** 5601
[17] Suckling J R, Hibbins A P and Lockyear M J 2004 *Phys. Rev. Lett.* **92** 147401
[18] Isaac T H, Rivas J G and Sambles J R 2008 *Phys. Rev.* B **77** 113411
[19] Gordon R 2006 *Phys. Rev.* B **73** 153405
[20] Miyazaki H T and Kurokawa Y 2006 *Phys. Rev. Lett.* **96** 097401
[21] Kurokawa Y and Miyazaki H T 2007 *Phys. Rev.* B **75** 035411
[22] Barnard E S, White J S, Chandran A and Brongersma M L 2008 *Opt. Express* **16** 16529

[23] Lindfors K, Lechner L and Kaivola M 2009 *Opt. Express* **17** 11026
[24] White J S, Veroins G and Yu Z 2009 *Opt. Lett.* **34** 686

Full 3D FDTD analysis of Electromagnetic Field in Photonic Crystal VCSEL

LIU Fa*, XU Chen, XIE Yi-Yang, ZHAO Zhen-Bo, ZHOU Kang, WANG Bao-Qiang, LIU Ying-Ming, SHEN Guang-Di
Key Laboratory of Opto-electronics Technology (Beijing University of Technology), Ministry of Education, Beijing University of Technology, 100 Ping Le Yuan, Chaoyang District, Beijing 100124, China

E-mail: liufa20719@126.com

Abstract. The effect of etch damage to the mode characteristics of photonic crystal vertical cavity surface emitting lasers was simulated in this paper. The devices simulated in this paper are 850-nm GaAs-based VCSELs with photonic crystal. And the devices were simulated by using finite difference time domain (FDTD) method. Limited to the computer resource, the top DBR was simulated only, and the traverse size was smaller than the real size. In order to highlight the impact of the etch damage, several kinds of light sources and photonic crystal structures were simulated separately, and each situation is calculated in the condition of ideal photonic crystal and photonic crystal with etch damage respectively. All parameters of device and light feature are referred to the real condition.

1. Introduction

The vertical-cavity surface-emitting lasers (VCSELs) are the key light sources used in telecommunication and the local networks [1-2]. During the fabrication of photonic crystal (PHC) lasers, the defects of cylindrical air holes are always generated in etching process. So, it's necessary for us to study the affection of defects to the PHC laser modes [3-4]. A group from MIT reported that the band gaps were not sensitive to the defects in the photonic crystals [5]. In this paper, finite difference time domain (FDTD) algorithm was developed and applied to investigate the affection of defects to the distribution of EM fields and laser mode characteristics. In order to contrast the results, the distribution of EM fields in the ideal structure and nonideal structure was simulated separately.

2. Methodology

Respecting the FDTD method is a classical numerical method [6-7], the introduction to it is omitted in this paper. Apparently, the mode characteristics rely on the reciprocity of the active region,

* Supported by the National High Technology Research and Development Program of China under Grant No 2008AA03Z402, and the Beijing Natural Science Foundation of Beijing under Grant No 4092007.
Email: liufa20719@126.com Tel: + 86 01067391641-834 Fax: +86 01067391641-822

DBR and photonic crystal structure, etc. In this paper, to simplify the simulation, we deal with the model as that the laser light projects vertically into photonic crystal and then detect the emergent light on the other side of photonic crystal. Via the comparison of the distribution of electromagnetic fields between the perfect device and the defective one, the influence of defect to the model could be estimated.

3. Device structure and Simulative domain

Owing to this paper is serviced for laser research in our lab, so photonic crystal structure and the defect in this paper imitate entirely the real device and etching defect. In this Letter, the VCSELs were fabricated from a 3-inch n-type wafer with epitaxial structure consisting of a bottom n-type (Si-doped) 34 pairs DBR, an undoped active region with three GaAs quantum wells, and a similar top (C-doped) 22 pairs DBR. The DBR are formed by graded heterojunction of AlxGa1-xAs, the unit period of bottom and top DBR mirrors consisting of 4 layers of $Al_xGa_{1-x}As$ (x: 0.9-0.12), $Al_{0.12}Ga_{0.88}As$, $Al_xGa_{1-x}As$ (x: 0.12-0.9), $Al_{0.9}Ga_{0.1}As$. In addition, a 30 nm $Al_{0.98}Ga_{0.02}As$ oxidization layer was inserted between the top DBR and the active region. The structure is shown in Figure 1(a). The operation wavelength of the device is designed at 850 nm. The top 22 pairs DBR is the disquisitive object, we assume that the laser is emitting from the active region into the top DBR, and we could detect the light field on the upper surface.

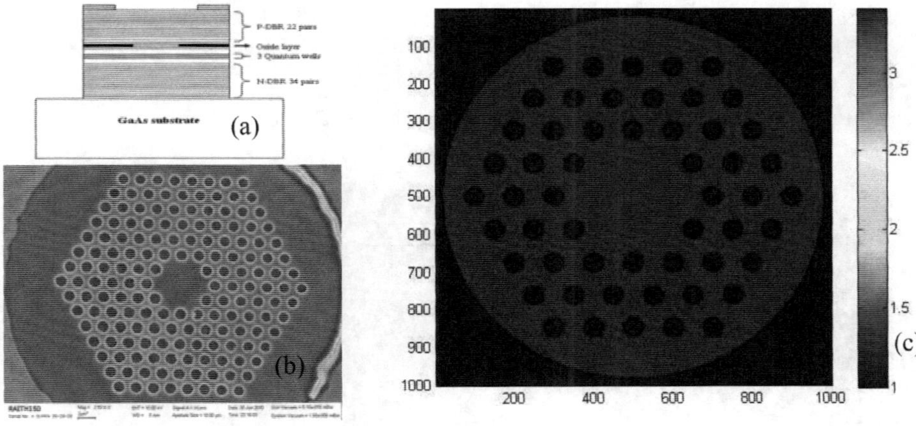

Figure 1. (a) Structure of the selectively oxidized VCSEL (b) The actual device photo (c) Simulating area on the xoy plane

The photonic crystal pattern is shown in Figure 1 (b). In this pattern, the diameter of the etched air circular holes is 0.01 μm; the distance of two adjacent circular centres is 0.02 μm. In the centre of the device, there is a region of 7 holes unetched. The light is well confined into or radiate out from this extent, we could see the area as the device aperture. In the region of far away from the aperture, the optical intensity decline rapidly, almost all energy is centralized in the aperture and the vicinity. So, it is enough to have a simulating region with three centre circle air holes. As shown in Figure1(c), the whole simulating district in XOY plane is 1000×1000 grids, represented 20 μm $\times$ 20 μm. The depth of the air holes is 1.6 μm, about 12 pairs of DBR. Figure 2 (a) shows a cross section of a real device with much of etching injury and Figure2 (b) shows the cut plane of ideal photonic crystal. For made the study object clearly, two kinds of defect are defined and researched in this paper. The first one is the diameter of air holes decreases gradually along with the increases of etching depth. This defect is caused by the bombarding rate of reacting ion decreases along with the increases of depth. In virtue of the unit period of DBR mirror consisting of four different layers, so the radius of etching holes on each layer is different and alters periodically. Therefore, the other sort of defect is the diameter of holes changed periodically along with the depth. Accordingly, the same photonic crystal structure with several state of defection as shown in Figure 3(a) ~ (e) was simulated. It needs to pay attention to the dimension, the abscissa scale direction is -10 μm to 10 μm, the vertical direction is only about 3 μm. For distinguishing the impact of DBR structure on the simulation result, the photonic crystal with DBR and without one is simulated respectively. Finally, a special defect configuration is simulated for much more approaching the shape in Figure 2(a). The affection of defect to the laser mode can be obtained through compare the distribution of near field between the ideal photonic crystal and the defective one.

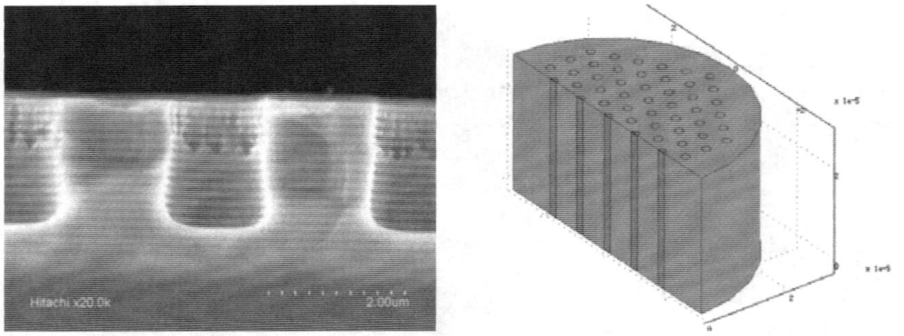

Figure 2. (a) Air holes with etching defect (b) The ideal photonic crystal

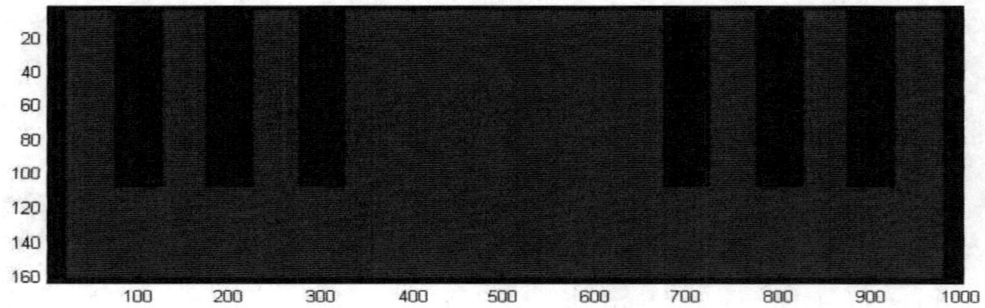

Figure 3. (a) The cut plane of simulative device without defect or DBR

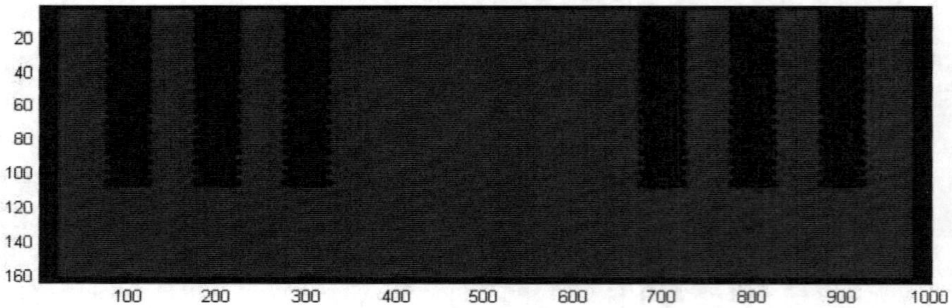

Figure 3. (b) The cut plane of simulative device with defect but no DBR

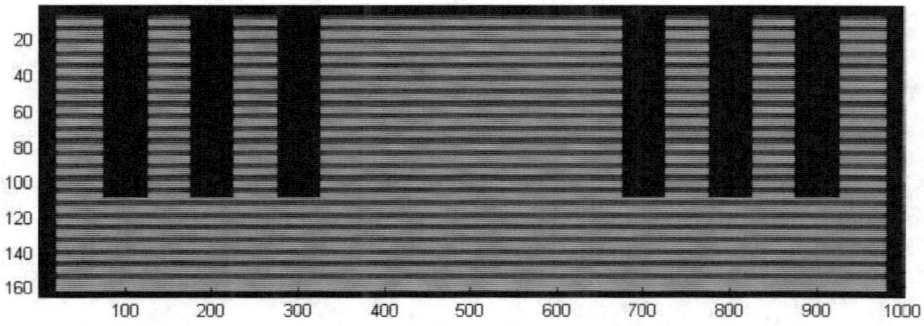

Figure 3. (c) The cut plane of simulative device with DBR but no defect

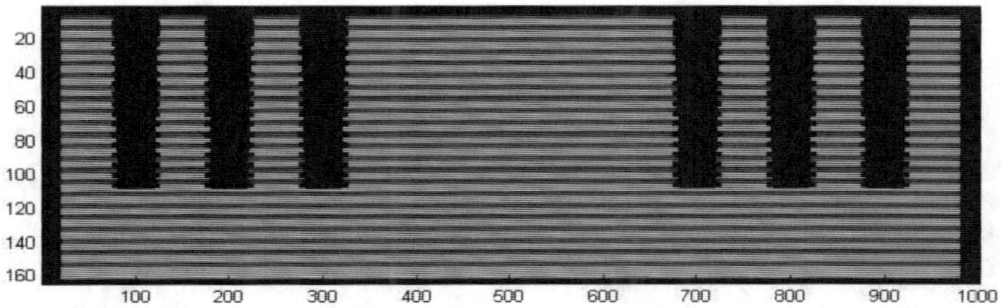

Figure 3. (d) The cut plane of simulative device with DBR and defect

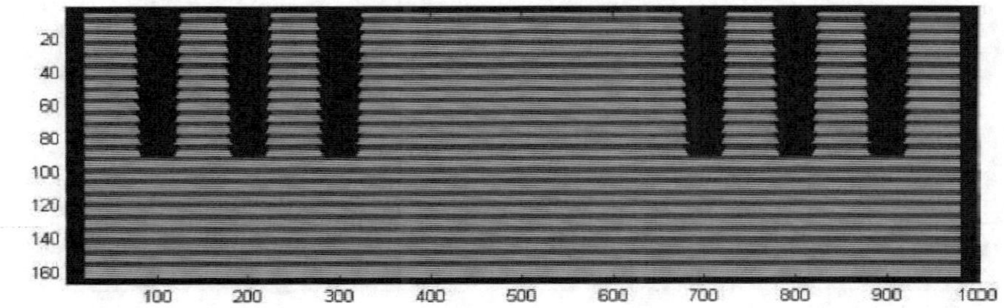

Figure 3. (e) A special etching defect follows the Figure 2(a)

4. Optical Source Selection

A field source must be selected in the method of FDTD simulation. Considering Gauss beam is the fundamental mode of many sorts of lasers, so this light source is used in this paper mainly. The size of Gauss light is consulted the real device completely, and selected several size for simulation. Due to the spot size in real device is confined by the device aperture and oxide-hole [8-9], the area of real light can not much larger than aperture.

Then the style of electromagnetic wave should be considered. In this paper transverse electric and magnetic field (TEM) is simulated. In the semiconductor laser the form of electromagnetic field is not strict TEM, but the transverse size of VCSEL device much greater that the area of emitting, so it can be treated approximately. Finally, continuous sine wave is used in simulation.

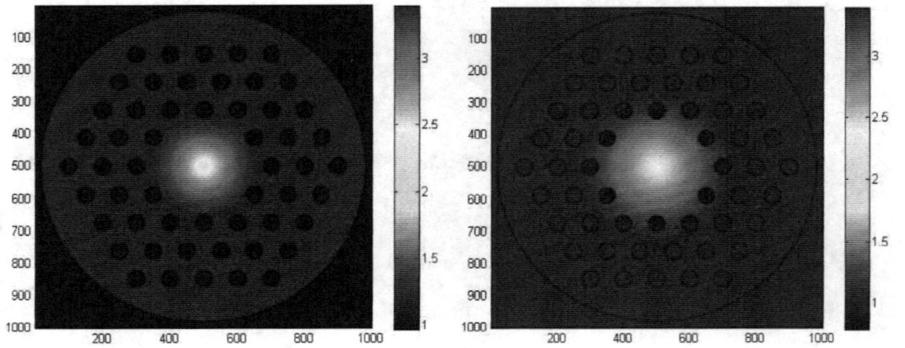

Figure 4. (a) The r_0 is 1.6 μm (b) The r_0 is 2.4 μm

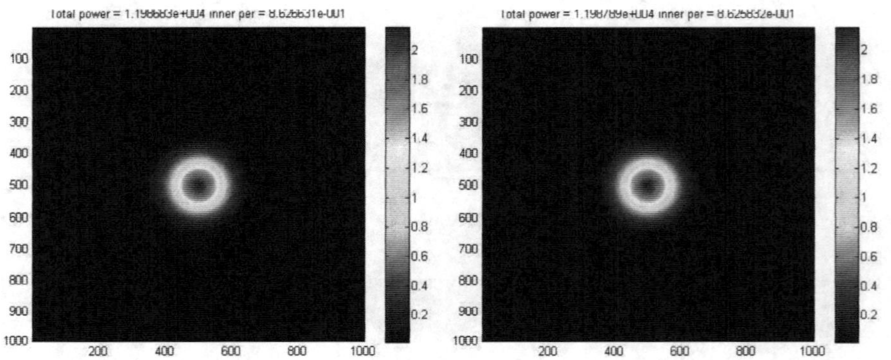

Figure 5. (a) Result in ideal device (b) Result in nonideal device

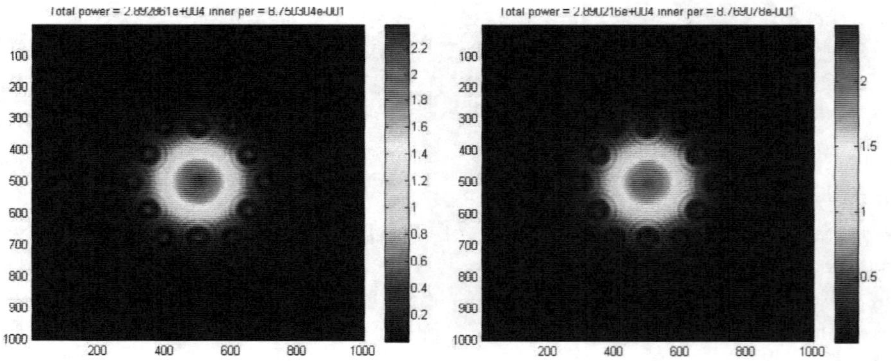

Figure 6. (a) Result in ideal device (b) Result in nonideal device

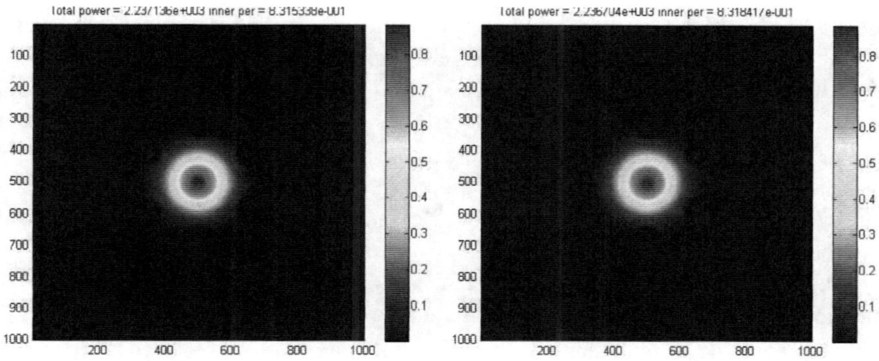

Figure 7. (a) Result in ideal device (b) Result in nonideal device

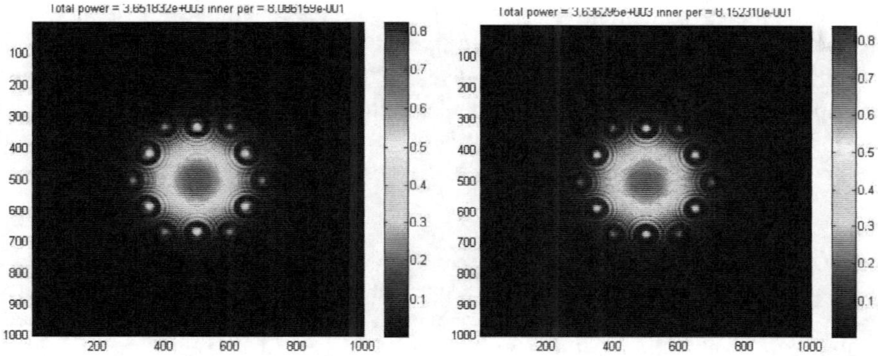

Figure 8. (a) Result in ideal device (b) Result in nonideal device

5. Results and Discussion

In this paper, we had used the FDTD method to simulate the performance of the photonic crystal VCSEL, by our FORTRAN code with 3000 time step and 4×10^8 grids. Once process needs memory 2.24 GB and 10 hours upwards. For the usage of symmetry in x axis and y axis, the analytical space was expanded to 20 $\mu m\times$20 $\mu m\times$3 μm and the DBR is arranged. They are still a little smaller than the actual device due to the limitation of computer resources. We considered, however, that origins of the fundamental mode could be estimated.

For evaluating the influence of defection to the light field, we define w_e as $\varepsilon\times E^2$ to describe the energy intensity on the given plane. If the distribution of output power in near field of device is more centralized, it is can be considered as more being close to fundamental mode. Apparently, the more percentage of energy in centre zone is closer to what we need. We define r_0 as the half bandwidth of Gaussian light, and the distance to the centre of device smaller than r_0 is defined as center zone. So, energy of any point in the centre area and in the whole plane is added separately, and the ratio (expressed as k) of centre power in the whole output power could express the modal characteristic. The figure on the finally time step of each simulative process is shown in figure 5~8.

Figure 5(a), (b) shows separately the simulative result of the ideal photonic crystal without DBR and the defective one, and the r0 equal to 1.6 μm as shown in figure 4(a). In the Figure 5(a), the total energy of relative magnitude 1.1988×10^4 is a little larger than 1.1987×10^4 which shown in figure 8, however, the ratio k 0.8750 in ideal photonic crystal is smaller than 0.8626 that in the defect.

Figure 6(a), (b) shows the similar simulative condition with the Figure 5(a) and (b), but the difference is that r0 is altered to 2.4 μm as shown in figure 4(b). The total power is 2.8929×10^4 and 2.8902×10^4; ratio k is 0.8750 and 0.8769. The varietal tendency is similar with that of the last 2 figures.

The two pairs of results simulated hereinbefore are the photonic crystal without DBR, now the results with DBR will be shown. As shows in Figure 11and 12, the total power in the ideal and nonideal is 2.2371×10^3 and 2.2367×10^3; the ratio k is 0.8315 and 0.8318. The half bandwidth r0 is 1.6um in this group of result.

The final group of result is shown in Figure 8(a) and (b). The structure of photonic crystal with DBR is shown in Figure 3(e), and r_0 is 2.4 μm. Figure 3 shows the etched defective style. The total power of relative magnitude in the ideal and the defective one is 3.6518×10^3 and 3.6363×10^3; and ratio k is 0.8086 and 0.8152. Apparently, as can be seen in this group, the total power in the ideal is little larger than in the nonideal, and the ratio k in the ideal is a little smaller.

An obvious law could be discovered from all the simulative data. The total output power of relative magnitude in the ideal photonic crystal is little higher than in nonideal, the ratio of the variety to the total is less than 1%. However, the ratio of k in the ideal is less after a sort, less than 1% also. The reason of these phenomena could be considered as the defect caused by etching technology enlarged the dispersion of EM wave, more power is scattered than in ideal structure; however, the centre aperture is unetched, so the fringe energy is more strongly suppressed than the centre [10]. Besides, the inconspicuous trend can be explained by the too small variety of defects.

6. Conclusion

We adopted FDTD method to analysis the 850-nm GaAs-based photonic crystal VCSELs. The impact of defect to the laser mode with varied light size and with and without DBR is simulated in this paper. Via these simulative data, it can be observed that, the defect in the photonic crystal should depress the output power and heighten the modal characteristic, but these trends are not obvious. The farther of the research is expanding the size of defect in simulation and comparing the results with experiment.

7.References

[1]. T Czyszanowski, M Dems and K Panajotov 2007 J. Phys. D: Appl. Phys. 40 2732–2735
[2]. Aaron J. Danner, James J. Raftery, Jr., Paul O. Leisher, and Kent D. Choquette 2006 APPLIED PHYSICS LETTERS 88, 091114
[3]. Xie Yi-Yang, Xu Chen and Kan Qiang 2010 Chin. Phys. Lett. 27 024206
[4]. Delai Zhou L. J. Mawst and Zheng Dai 2002 IEEE JOURNAL OF QUANTUM ELECTRONICS, 652-664, 38
[5]. Shan hui fan 1997 MIT Doctoral Dissertation 30
[6]. Jason S. Ayubi-Moak, Stephen M. Goodnick, Dan Stanzione, and Gil Speyer 2008 DoD HPCMP Users Group Conference of IEEE 319-326 85
[7]. K. Sheikhi · and N. Granpayeh 2008 Opt Quant Electron 991–1003
[8]. Anjin Liu, Mingxin Xing, Hongwei Qu, Wei Chen, Wenjun Zhou, and Wanhua Zheng 2009 APPLIED PHYSICS LETTERS 94, 191105
[9]. Tomasz Czyszanowski, Robert P. Sarzała, Łukasz Piskorski, Maciej Dems, Michał Wasiak, 2007 IEEE JOURNAL OF QUANTUM ELECTRONICS, 1041-1047 43
[10] Kosuke Morito, Daisuke Mori, Eichi Mizuta and Toshihiko Baba 2005 Proc. of SPIE Vol. 191-200 5722

The study of a kind of photonic crystal fiber grating based on structural change

Y Y Liu, C Q Xu, W H Bi and Y F Qi

School of Information Science and Engineering, Yanshan University, Qinhuangdao City 066004, China

E-mail:liuyanyan@ysu.edu.cn

Abstract. A kind of novel photonic crystal fiber grating based on structural change was studied, the effective refractive index induced by structural change was computed using Multipole Method, the relationship between the effective refractive index and the shrinkage of air-holes was got. Moreover, the fabrication principle was proved in the mathematics way. A model of photonic crystal fiber grating based on structural change was built, the transmission characteristics were analysed with Coupled Mode Theory. The emphasis of this study was the impacts of the rings of air-holes in the cladding, the rings of air-holes shrinkage and the degree of air-holes shrinkage on the transmission characteristics. The research results indicate that the periodic air-holes shrinkage can form grating, not only the degree of air-holes shrinkage, but also the shrinkage envelope has important effect on the magnitude and the distribution of the effective refractive index; the resonance wavelength, the resonance bandwidth and the intensity have relations to the rings of air-holes in the cladding, the rings of air-holes shrinkage, and the degree of air-holes shrinkage.

1. Introduction

Fiber grating (FG), as a kind of important passive optical device, has many advantages, such as compact size, light weight, low loss and easy to be integrated with the fiber system, so it is widely used in the fields of tunable lasers, optical add-drop multiplexers, fiber sensors, dispersion compensation [1-4]. The conventional fiber grating is written with the UV exposure method. Using the phase mask, the photosensitive fiber or the single mode fiber loaded hydrogen is written into the gratings. At present, the UV exposure method is a mature technology and easy to be operated, but the disadvantages of the fiber grating made by this method are obvious. For example, the gratings will undergo degeneration in the condition of long-term working and high temperature environment [5]. The instability of performance impedes the applications of fiber grating in many fields.

The invention and rapid improvement of photonic crystal fiber (PCF) present a new solution. After analyzing the advantages and disadvantages of the conventional fiber grating, this paper studied a kind of novel fiber grating, namely, photonic crystal fiber grating (PCFG). It need neither doping B/Ge in the core of PCF during the fiber producing process nor loading hydrogen before the fabrication of PCFG, it can be engendered by changing the structure of PCF. Thereby, this kind of PCFG based on structural change can overcome the shortcomings of long-term instability and thermal instability of the conventional fiber grating, it will have many potential applications in the long-term health estimation system or the special safety monitoring system of the big structure [6], such as the bridges, the dams and the oil storages.

2. The fabrication technique and principle analysis

2.1. The fabrication technique of PCFG based on structural change

PCF, generally called microstructured optical fiber (MOF) or holey fiber (HF) [7], is constructed by the homogeneous medium (pure SiO_2 or polymerized substance) and the air-holes orderly arranged or out-of-orderly arranged. In the core of PCF, there is a defect center (solid core or hollow core). According to the different principle of transmitting light, PCF is divided into two types, the total internal reflective PCF (TIR-PCF) and the photonic bandgap PCF. The transmission performance of PCF is determined by the size and arrangement of the air-holes in the cladding, the space between the air-holes and the type of the fiber core.

In 2003, Yinian Zhu of Nanyang Technological University of Singapore proposed the grating fabrication in PCF using CO_2 laser beam [8]. When a focused CO_2 laser beam notched periodically on the surface of PCF, the effective refractive index of the cladding would become periodic change, it was the reason of forming grating. The principle of this method was similar to the UV exposure method, which introduced the refractive index perturbation into the core of the fiber. In 2004, Lim J. H. presented the grating fabrication method in PCF using mechanical pressure. The equipment was composed of two plates, either of them had parallel plow grooves. The PCF was fixed between the two plates, using elastic-optic effect, the perturbation of refractive index was got at the pressure point, which could cause to form grating in the PCF.

As for the research of PCFG based on structural change, the present references just gave the description about the technique, and did not prove the principle from the standpoint of the mathematics [9]. Using Multipole Method and Coupled Mode Theory, with focus on the change of the effective refractive index brought by air-holes shrinkage, this paper analyzed the forming grating principle from the mathematic way.

2.2. The fabrication principle of PCFG based on structural change

When a CO_2 laser light beam heated a PCF with the properly designed period, spot, power and duration, the PCF would collapse with the certain period, depth of concavity and envelope profile. The deformation of the collapsed PCF in the longitudinal direction was such as the figure 1. *L* was the length of PCFG. was the period of the air-holes shrinkage, it was also the period of PCFG. *w* was the width of the collapse segment, which was determined by the size of the laser spot. *m* was the collapse depth, which was determined by the power and duration of the CO_2 laser beam. The collapsed envelope profile was depended on the shape of the laser spot.

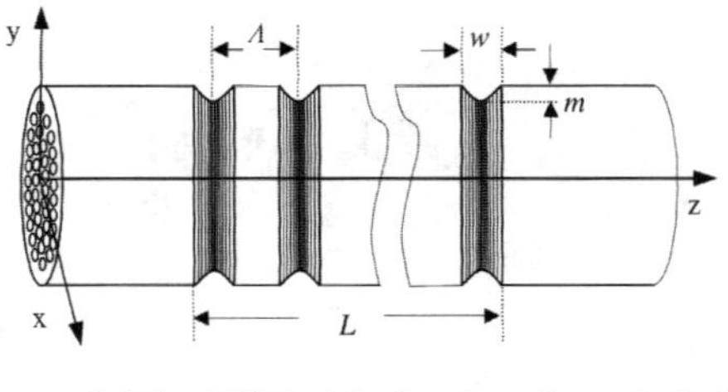

Figure 1. The PCF with the shrunken air-holes.

Using Multipole Method [10,11], this paper computed and compared the change of the effective refractive index induced by structural change in the collapse rules of sine curve and rectangular curve, and got the data of the effective refractive index. The figure 2 showed the relationship curves of the effective refractive index vs. the diameter of the shrunken air-holes. In the figure 2, the abscissa represented the diameter of the shrunken air-holes. The smaller the value was, the more seriously the air-holes shrunk. From left to right, the degree of the air-holes shrinkage decreased gradually, the effective refractive index showed a downtrend. Compared with the instance of the rectangular curve envelope function, the effective refractive index of sine curve envelope function descended smoothly and slowly. However, in the rule of rectangular curve, the effective refractive index sloped gently when the degree of air-holes shrinkage was serious, then it showed a rapid downtrend till the two n_{eff}

curves overlapped. The results of this study indicate that the more seriously the air-holes shrink, the bigger the effective refractive index is. Not only the degree of air-holes shrinkage, but also the shrinkage envelope has effect on the magnitude and the distribution of the effective refractive index.

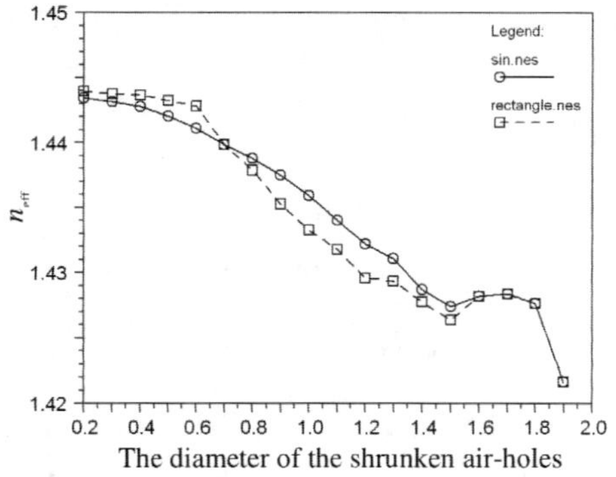

Figure 2. The curves of the effective refractive index vs. the diameter of the shrunken air-holes.

Using the data in the figure 2, the distribution function of the effective refractive index in the collapse segment could be obtained. Taking the typical sine curve envelope function as an example, this paper accomplished curve fitting with 5 orders for the effective refractive index, the distribution function expression along the longitudinal direction was

$$f(z)=\sum_{k=0}^{5} \boldsymbol{a}_k e^{kz} \quad \frac{w}{2} \le z \le \frac{w}{2} \tag{1}$$

Here $\boldsymbol{a}_k$, the coefficient of the series expansion, was determined by the following Matrix **A**.

$$\mathbf{A}=[1.138e+007, -5.581e+007, 1.636e+008, -3.187e+008, 4.331e+008, -4.189e+008] \tag{2}$$

If the effective refractive index of PCF was uniform distribution along the radial direction, the distribution model of sine curve envelope function was built such as the figure 3.

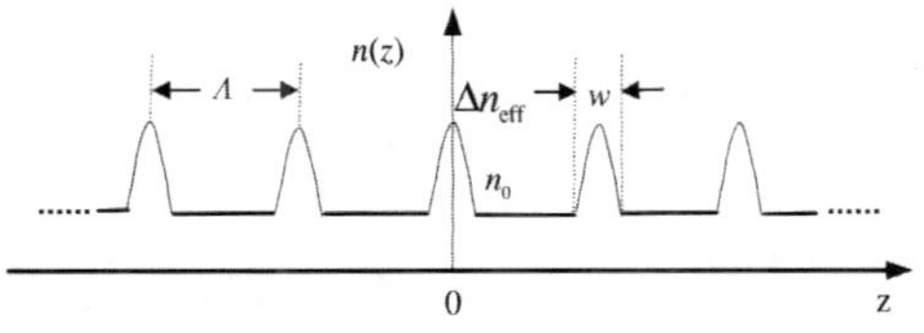

Figure 3. The distribution model of the effective refractive index in PCFG.

In the figure 3, n_0 was the effective refractive index of PCF. Δn_{eff} was the maximum of the effective refractive index in the collapse segment of PCFG, which was the modulation amplitude. Then $n(z)$ could be written as

$$n(z)=\begin{cases} n_0+f(z) & k\Lambda-\frac{w}{2}\le z\le k\Lambda+\frac{w}{2} \\ n_0 & k\Lambda-\frac{w}{2}> z, z> k\Lambda+\frac{w}{2} \end{cases} \tag{3}$$

Here, k was an integer. After putting the equation (1) into the equation (3) and doing Fourier series expansion, the expression including the top 5 orders was

$$n(z)=n_0+\sum_{k=0}^{5} \boldsymbol{b}_k \cos\left(k\frac{2\pi}{\Lambda}z\right) \tag{4}$$

In the expression, $\boldsymbol{b}_k$, the coefficient of the Fourier series expansion, was determined by the following Matrix **B**.

$$\mathbf{B} = [\ 1.435, -0.008822,\ -0.0002008, -0.1478305,\ -0.0001246, -0.0004898] \tag{5}$$

According to Coupled Mode Theory, the coupled coefficient C_q [12] between the q order mode and the fundamental mode was

$$C_q = \zeta_q(z) + 2k_q(z)\sum_{k=0}^{5} c_k \cos\left(k\frac{2\pi}{\Lambda}z\right) \tag{6}$$

In the equation (6), $\zeta_q(z)$ and $k_q(z)$ were defined as the self-coupling coefficient and the mutual-coupling coefficient respectively. They were determined by the following equations [12].

$$\begin{cases} \zeta_q(z) = \dfrac{\Delta n_{\text{eff}}\omega}{2}\iint\limits_{\Omega} e_q(x,y)\cdot e_{\text{core}}(x,y)dxdy \\ k_q(z) = \dfrac{s}{2}\zeta_q(z) \end{cases} \tag{7}$$

Here, $e_q(x,y)$ was electric field distribution of the q order mode in the cladding, $e_{\text{core}}(x,y)$ was electric field distribution of the fundamental mode. Ω was the overlap area of the both electric field. s was the visibility of grating fringe, this paper let s = 0.6. Putting the equation (6) and (7) into the coupled mode equations and solving them, the normalized amplitude and reflectivity could be obtained, the expression of the resonance wavelength [13] was

$$\lambda_{\text{clad},q} = (n_{\text{core}} + n_{\text{clad},q})P_{\text{FBG}} \tag{8}$$

In the equation (8), $n_{\text{clad},q}$ was the effective refractive index of the q order mode in the cladding, n_{core} was the effective refractive index of the fundamental mode, P_{FBG} was the period of Bragg fiber grating.

3. Numerical calculate and simulation results analysis

3.1. The model of PCFG based on structural change

Using the above analysis method, a model of C_{6v} TIR-PCFG was built and simulated. The PCFG model with 7 rings air-holes, it's cross-section and longitudinal structure were (a) and (b) in the figure 4.

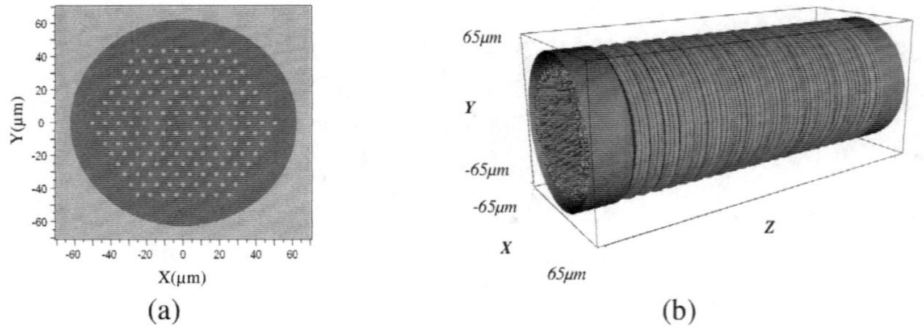

(a) (b)

Figure 4. The cross-section and longitudinal structure of PCFG.

From within to outside of PCFG, the rings number of air-holes was 1, 2, ······, 7. The diameter of the air-holes was $d = 2.16\mu\text{m}$, the space between the air-holes was $p = 7.2\mu\text{m}$, the outside diameter of PCFG was $D = 125\mu\text{m}$, the length of the fiber grating was $L = 20\text{mm}$, the diameter of the fiber core was $d_0 = 2p - d = 12.24\mu\text{m}$, the period of the fiber grating $\Lambda = 600\text{nm}$. The fiber material was SiO_2, it's effective refractive index was $n_{SiO_2} = 1.4466$. The maximum of the collapse depth of the outside diameter was $m = 4.5\mu\text{m}$. The collapse degree of air-holes was closely related to their position in PCFG. From the outermost ring to the innermost ring, the collapse depth was from 1.96 μm to 0.26 μm , the collapse period was grating period. With the above parameters, the reflectance spectrum of PCFG showed an obvious grating effect, the resonance wavelength was about 1.55μm, the resonance bandwidth was over 100nm, and the intensity of the peak was about 0.96.

3.2. The impact of the rings of air-holes on the performance of PCFG

The structure of PCF, especially the rings of air-holes, had an important impact on the transmission characteristics of PCFG. The table 1 was the data of the resonance wavelength and the resonance bandwidth varying with the rings of air-holes, whose curves were described in the figure 5.

Table 1. The data of the resonance wavelength and the resonance bandwidth with the different rings of air-holes.

The rings of air-holes	Resonance wavelength (μm)	Resonance bandwidth (nm)
1	1.5512	0.8
2	1.5511	1.6
3	1.5510	3.2
4	1.5506	8.0
5	1.5499	20.6
6	1.5479	61.8
7	1.5490	134.4

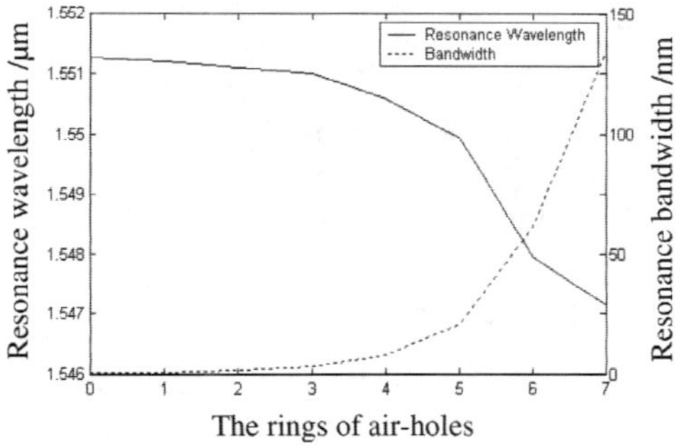

Figure 5. The curves of the resonance wavelength and the resonance bandwidth vs. the rings of air-holes.

From the Table 1 and the figure 5, as the rings of air-holes rose, the resonance wavelength showed blue-shift, the grating had a wider bandwidth. According to the equation (8), the resonance wavelength was determined by the sum of the effective refractive indexes of the core and the cladding. When the rings of air-holes in the cladding increased, the effective refractive index of the cladding would minish, this led to the sum of the effective refractive index of the core and the cladding reducing, so the resonance wavelength shifted right. At the same time, because the effective refractive index of the cladding minished, the energy was concentrated in the core, the fiber core could support more high-order modes, this was the reason of the bandwidth increasing.

3.3. The impact of the duty ratio of air-holes on the performance of PCFG

The duty ratio of air-holes (d/p) had an obvious impact on the performance of PCFG. The figure 6 disclosed the changing trends of the resonance wavelength and the resonance bandwidth vs. d/p . Furthermore, it was found that d/p increased from 0.3 to 0.5, the resonance wavelength and the resonance bandwidth showed a downtrend on the whole, that was because the effective refractive index of the cladding reduced with d/p increasing. According to the equation (8), the resonance wavelength would shift left. But in the simulation, it was supposed that the functional relation among the diameter of the fiber core, the space between the air-holes and the diameter of air-holes was $d_0 = 2p - d$. In this expression, if d held the current value all through, it could be deduced that d_0 would minish with d/p reducing; just the same, if p kept unchangingly, it could also be inferred that d_0 would minish with d/p increasing. These could cause the coupled area of the fundamental mode and the high-order modes to reduce, and the modes satisfied with the phase match condition would be cut down in quantity, so the reflectance spectrum had a narrower bandwidth. In the figure 6(b), while d/p was equal to 0.32, the bandwidth arrived at the peak value, it was about 140nm.

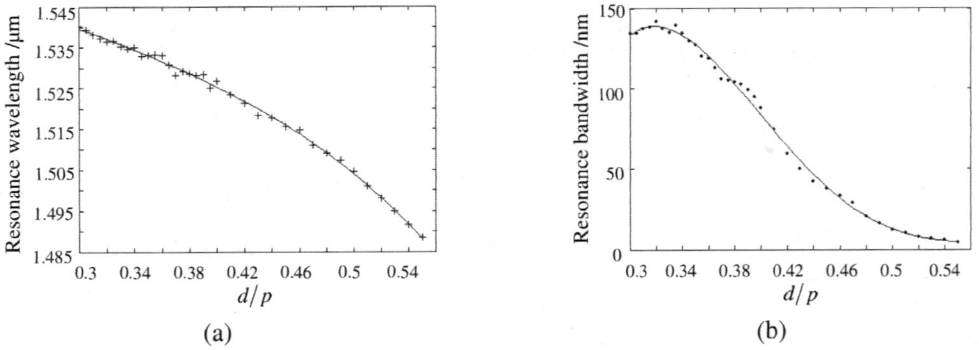

Figure 6. The curves of the resonance wavelength and the resonance bandwidth vs. d/p.

3.4. The impact of the rings and degree of air-holes shrinkage on the performance of PCFG

Not only the rings and the duty ratio of air-holes had the impact on the performance of PCFG, but also the rings and degree of air-holes shrinkage had the obvious effect on the transmission characteristics. In the simulation, it was assumed that the air-holes shrinkage was periodic and the rings of shrunken air-holes increased gradually. In addition, when the numbers of shrunken rings were same, the degree of air-holes shrinkage was divided into two grades.

The figure 7 was the changing curves of the effective refractive index varying with the rings of the shrunken air-holes. The real line showed the effective refractive index with the outer two rings air-holes shrinkage, the broken line showed the effective refractive index with the outer four rings air-holes shrinkage. From the figure 7, it could be drawn a conclusion that the effective refractive index went a downtrend with the wavelength increasing. Moreover, the more rings of shrunken air-holes were and the more seriously the air-holes shrunk, the bigger the effective refractive index of the cladding was. The figure 8 showed the curves of the resonance wavelength and the resonance bandwidth varying with the air-holes shrinkage. When the rings of shrunken air-hole increased, the effective refractive index rose. As the degree of air-holes shrinkage enhanced, the resonance wavelength showed red-shift and the grating had a narrower bandwidth. This coincided with the equation (8). Furthermore, because the effective refractive index of the cladding increased, it had the weak ability to confine the energy in the fiber core, the fiber could only support a few high-order modes, this led to the bandwidth reducing.

However, in the figure 8, when the rings numbers of air-holes shrinkage were same, it could be found that the resonance wavelength was unresponsive to the degree of air-holes shrinkage, the effective refractive index had not an obvious change in the condition of the same rings of shrunken air-holes. It was because of a small diameter of air-holes supposed in the simulation. If the diameter of the air-holes was bigger, the resonance wavelength would have a remarkable change.

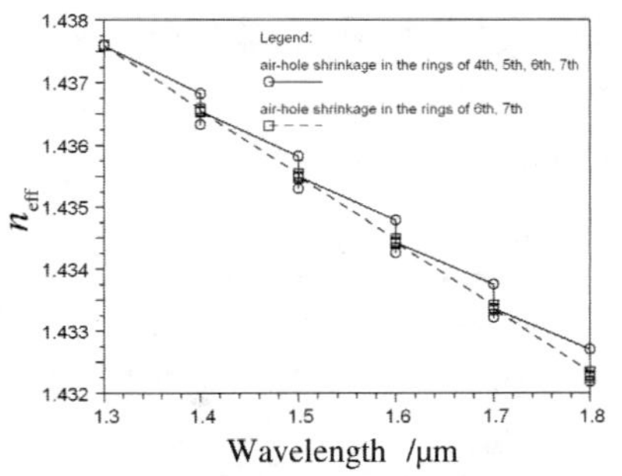

Figure 7. The n_{eff} vs. with the shrunken air-holes in the rings of 6th, 7th and the rings of 4th, 5th, 6th, 7th.

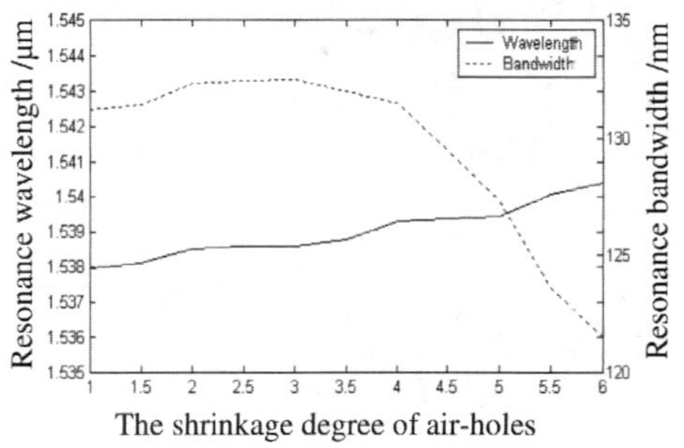

Figure 8. The curves of the resonance wavelength and the resonance bandwidth vs. the shrinkage degree of air-holes.

4. Conclusion

A kind of novel PCFG based on structural change was studied. It was the new progress that the fabrication principle was proved in the mathematics way. The relations between PCFG's performance and some important PCF's structure parameters were also discussed. The results of this study indicate that the periodic air-holes shrinkage can form grating, not only the degree of air-holes shrinkage, but also the shrinkage envelope has the very close relationships with the magnitude and the distribution of the effective refractive index; the structure parameters of PCF, such as the rings of air-holes in the cladding, the duty radio of air-holes, the rings and the degree of air-holes shrinkage, have the important impacts on the resonance wavelength, the resonance bandwidth and the intensity of resonance peak. Compared the PCFG based on structural change with one based on the UV exposure method, the former has a preferable stability in the high temperature environment or the longtime working system, it will have many potential applications in the field of optical fiber sensors.

References

[1] Ji H, Yang S, Zhang Y and Xie S 2008 *Acta Optica Sinica.* **28** 923
[2] Rao Y J, Zhu T and Mo Q J 2006 *Opt. Commun.* **266** 190
[3] Kivistö S, Herda R and Okhotnikov O G 2008 *Opt. Express* **16** 270
[4] Karim F and Seddiki O 2010 *Journal of Optical Communications* **31** 21
[5] Guan B O, Tam H Y, Tao X M and Dong X Y 2000 *IEEE Photon. Technol. Lett.* **12** 1351
[6] Sun L M and Cai H W 2006 *Chinese Journal of Lasers* **33** 99
[7] Liu D M 2008 Fiber Optics (Beijing:Science)
[8] Zhu Y, He Z and Du H 2003 *Sens. and Actuators* B **131** 269
[9] Tu Q, Zhang W, Jin L, Kai G, Yuan S and Dong X 2005 *Laser & Optoelectronics Progress* **42** 47
[10] White T P, Kuhlmey B T, McPhedran R C, Maystre D, Renversez G, De Sterke C M and Botten L C 2002 *J. Opt. Soc. Am.* B **19** 2330
[11] Kuhlmey B T, White T P, Renversez G, Maystre D, Botten L C, De Sterke C M and McPhedran R C 2002 *J. Opt. Soc. Am.* B **19** 2340
[12] Lázaro J M, Kuhlmey B T; Knight J C; Lopez-Higuera J M and Eggleton B J 2009 *Opt. Commun.* **282** 2361
[13] Li Y, Liu J G, KAI G Y, Xue L F, Liu Y G, Wang Z and Dong X Y 2007 *Laser & Infrared* **37** 150

Study of higher-order modes for waveguide grating coupler

J Liu [1], C Q Huang [2], P Tian [3], Z M Wan [1], Z M Luo [1] and M Chen [2]

[1] College of Information & Communication Engineering, Hunan Institute of Science and Technology, Yueyang 414006, China
[2] College of Physics & Electronics Information, Hunan Institute of Science and Technology, Yueyang 414006, China
[3] College of Optoelectronic Science and Engineering, Wuhan National Laboratory for Optoelectronics, Huazhong University of Science and Technology, Wuhan, 430074, China

E-mail: nameliujing@yahoo.com.cn

Abstract. The vector two-dimensional finite difference time domain method is applied to analyze the higher-order modes of waveguide grating couplers (WGCs). The phase matching function of WGC is deduced based on diffraction equation. The stable field distributions and mode patterns of TE_1、TE_2 and TE_3 in WGCs are obtained by using Gaussian beam as a driving source. It is found that the optical mode is determined by grating parameters and waveguide structures. According to the mode field distribution, the optimum coupling efficiency can be predicted, and thus highly-efficient WGCs which are of practical use in integrated optical circuits and optoelectronic devices can be designed conveniently.

1. Introduction

With the development of integrated optics and silicon photonics, waveguide grating couplers (WGC) have been gaining considerable interest and playing an important role in optical communication, optical compute, and optical information processing [1]. One of its applications is providing a means of coupling light into or out of waveguide between optoelectronic components, such as the coupling between lasers or detectors to waveguides [2]. This attractive coupling method greatly reduces the size and weight of optical systems, integrated waveguides and electronic components to form photoelectric systems, thus new application fields of WGC are being opened up. Some new communication apparatus, such as the optical interconnection devices, the integrated-optic disk pickup device, integrated-optic interferometer position sensor, integrated-optic scanning optical microscope and integrated-optic printer head are all employing the WGC [3].

There are two "special" problems in the use of WGC devices: cut-off frequency and higher-order mode. The widespread and thorough researches have been done on cut-off frequency in many fields, but comparatively little has been done in the work of higher-order mode presently. The main reasons for this case are follows: there are many advantages for basic-mode propagation, e.g., longer transmission distance, larger transmission capacity, bigger amplitude, wider transmission band width, stronger interaction between guided-mode and waveguide grating, and higher coupling efficiency, etc. Therefore, the research work for higher-order mode is ignored to some degree. However, in some specific situation, higher-order modes can be used to improve the optical quality for its fine performances, including bigger aperture, easier coupling, stronger capability of dispersion compensation and larger tuning range of wavelength, although the existence of intra-modal dispersion may restrict the bandwidth and increase the loss. It is very necessary to accurately know the field distributions of higher-order modes while dealing with the transition of different polarization modes or coupling of different order modes.

Journal of Physics: Conference Series **276** (2011) 012073 doi:10.1088/1742-6596/276/1/012073

2. Physical model and theoretical method

A schematic diagram of grating couplers which we discuss is shown in Figure 1. The refractive indices of air, grating base, coating, film, substrate are expressed by n_a (n_a =1), n_b, n_c, n_f and n_s ,respectively. While the incident beam is upon on the grating's surface at the incident angle θ from air, the relationship between the grating spacing and the angles of the incident and diffracted beams of light is expressed by the grating equation [4]

$$\sin\alpha_l = \sin\theta + l\frac{\lambda_0}{d} \qquad l = 0, \pm 1, \pm 2, \cdots \tag{1}$$

Where α、l、λ_0 and d are the number of the diffraction angle, the order of diffraction, the wavelength of incident wave and spatial period of grating, respectively.

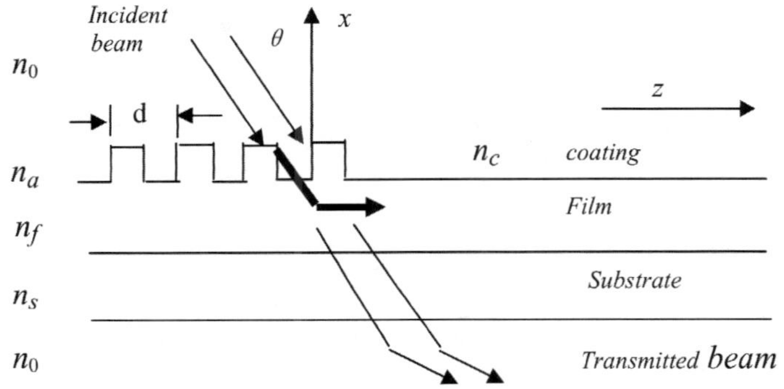

Figure 1. Basic structure of waveguid grating coupler

The phase displacement along the z-direction of different diffraction orders is determined by

$$\beta_l = \frac{2\pi}{\lambda_0} n_b \sin\alpha_l = \frac{2\pi}{d} l n_b + \frac{2\pi}{\lambda_0} n_b \sin\theta \tag{2}$$

The propagation constant of the m-order guided mode is named by β_m, and it must follow the condition of

$$\beta_m > \frac{2\pi}{\lambda_0} n_b \tag{3}$$

In order to reach the phase-matching between the grating and the waveguide mode, the following equation must be satisfied

$$\beta_l = \beta_m \tag{4}$$

This phase-matching condition can be satisfied in a suitable positive value of l [5]. As a result, the guided-mode can be stimulated inside the waveguide.

The finite difference time domain (FDTD) [6] technique with perfectly matched layer absorbing boundary condition [7] is used. From the FDTD numerical simulation, we can get the optical field distribution and stable mode pattern along the guide direction. There are many books and papers on the topic of FDTD, for a detailed and exhaustive description, please see Ref. [8] and references therein. In what follows, only essential steps are outlined.

In the passive and lossless region, the curl equations of Maxwell's equations are as follows

$$\nabla \times E = -\mu\frac{\partial H}{\partial t} = -\mu_0\mu_r\frac{\partial H}{\partial t} \tag{5}$$

$$\nabla \times \mathrm{H} = \varepsilon\frac{\partial E}{\partial t} = \varepsilon_0\varepsilon_r\frac{\partial E}{\partial t} \tag{6}$$

Where ∇ is Laplace operator, μ is magnetic permeability, μ_r is relative permeability, μ_0 is free-space permeability ($4\pi\times10^7 henrys/meter$), $\varepsilon(=\varepsilon_r\varepsilon_0)$ is electrical permittivity (farads/meter), ε_r is relative permittivity, and ε_0 is free-space permeability ($8.854\times10^{-12} farads/meter$).

It is obvious that the wave equations can be decomposed into six scalar equations for six vector field components. Only TE is considered in this paper. While waveguide structure has nothing to do with variable y, it is easily proved that there is only three field components (Ey, H_x, H_z), where H_x and H_z can be expressed by Ey. Thus, in our calculation, we choose vector FDTD to research the field distribution of Ey which can be used to represent the total field distribution. In rectangular coordinate system, the above expressions can be expanded as

$$\frac{\partial E_y}{\partial t}=\frac{1}{\varepsilon}\left(\frac{\partial H_x}{\partial z}-\frac{\partial H_z}{\partial x}\right) \tag{7}$$

$$\frac{\partial H_x}{\partial t}=\frac{1}{\mu}\left(\frac{\partial E_y}{\partial z}\right) \tag{8}$$

$$\frac{\partial H_z}{\partial t}=-\frac{1}{\mu}\left(\frac{\partial E_y}{\partial x}\right) \tag{9}$$

Using the central-difference formula, Equality (7) becomes

$$E_y^{N+1}(i,k)=E_y^N(i,k)+\frac{\Delta t}{\varepsilon}\left[\frac{H_x^{N+\frac{1}{2}}(i,k+1)-H_x^{N+\frac{1}{2}}(i,k-1)}{\Delta z}-\frac{H_z^{N+\frac{1}{2}}(i+1,k)-H_z^{N+\frac{1}{2}}(i-1,k)}{\Delta x}\right] \tag{10}$$

Where Δz stands for space step in both directions of the x and z, Δt stands for time step, N stands for mesh number, and $E_y^N(i,k)=E_y^N(i\Delta x,k\Delta z,N\Delta t)=E_y^N(x,z,t)$. In general, $\Delta x=\Delta z=\Delta s$ is chosen.

In order to satisfy the stability condition, the following relationship is usually taken

$$\Delta t=\Delta s/(2c_{max}) \tag{11}$$

Where c_{max} is the maximum velocity of light waves in the interaction space.To avoid numerical dispersion, $\Delta s=0.05\lambda_0$ is selected.

When the above conditions are satisfied, the Eq.10 becomes

$$E_y^{N+1}(i,k)=E_y^N(i,k)+A(i,k)[H_x^{N+\frac{1}{2}}(i,k+1)-H_x^{N+\frac{1}{2}}(i,k-1)-H_z^{N+\frac{1}{2}}(i+1,k)+H_z^{N+\frac{1}{2}}(i-1,k)] \tag{12}$$

Where $A(i,k)={}^{n_c}\!/_{4c\varepsilon n^2(i,k)}$

Similarly, the difference equations of Eqs.8 and 9 are as follows

$$H_x^{N+\frac{1}{2}}(i,k+1)=H_x^{N-\frac{1}{2}}((i,k+1)+B[E_y^N(i,k+2)-E_y^N(i,k)] \tag{13}$$

$$H_z^{N+\frac{1}{2}}(i+1,k)=H_z^{N-\frac{1}{2}}(i+1,k)+B\left[E_y^N(i,k)-E_y^N(i+2,k)\right] \tag{14}$$

Where $B={}^{n_c}\!/_{4c\mu_0}$

The perfectly matched layer absorbing boundary condition (derived by Mur) is used. The condition for Ey at $i=1$ (lower boundary), for example, can be written as

$$E_y^{N+1}(i,k)=-0.6E_y^{N+1}(i+2,k)+1.55[E_y^N(i,k)+E_y^N(i+2,k)]+0.025[E_y^N(i+2,k-2)$$
$$+E_y^N(i+2,k+2)+E_y^N(i,k-2)+E_y^N(i,k+2)]-E_y^{N-1}(i+2,k)-0.6*E_y^{N-1}(i,k) \tag{15}$$

3. Numerical results

The parameter values chosen in this paper are as follows: the thickness of waveguide is 1 μm, the incident wavelength is 1.3 μm, the grating spatial period is 0.4 μm, and n_c 、 n_s 、 n_f 、 Δs and Δt which could obtain stable optical field distributions of TE_1 、 TE_2 and TE_3 are listed in Table 1. All these values are base on the reference [4] which involves the complex physical principle and strict mathematical deduce. For a detailed derivation process, please see for it and do the needful.

The Gaussian beam is used as incident beam. The light field distribution profiles of TE_1 、 TE_2 and TE_3 are shown in Figure 2 (a), Figure 3 (a), and Figure 4 (a), respectively. It is obvious that, at the beginning of grating coupler, the stable mode is not formed, and therefore the case of field distribution is far from the real distribution of corresponding mode. With the increasing of propagation distance, the light field distributions stabilize gradually for each mode. It is shown that, by using grating coupler, the guided modes can be stimulated by radiation modes in waveguide. It is also found that the optical field distribution is related to the diffraction angle of each order. While the *l*-order is lacked, the coupling function is vanished. According to the mode field distribution profile, the optimum coupling efficiency can be predicted. It is worth mentioned that, the waveguide grating coupler could not only be used as the input coupler, but also the output coupler whose coupling effect could only occur in phase-matched *l*-order harmonic.

Table 1. refractive index distributions, space step and time step for different modes

mode / value	TE_1	TE_2	TE_3
n_c	1. 4	1. 25	1. 25
n_f	1. 563	1. 85	2. 35
n_s	1. 4	1. 25	1. 25
$\Delta s(\mu m)$	0. 0416	0. 0351	0. 0277
$\Delta t(ns)$	0. 097	0. 0732	0. 0576

The stable mode patterns of TE_1 、 TE_2 and TE_3 are illustrated in Figure 2 (b), Figure 3 (b), and Figure 4 (b), respectively. All of these can be seen that, while each mode propagates along the longitudinal direction, the mode pattern of cross direction maintains invariable. These results are in good agreement with analytic solutions. Thus the reliability and the accuracy of our simulation results by FDTD are guaranteed. The above results indicate that a stable light field mode at different order of *l* (*l* =1, 2, 3, …) can be coming into being as long as the grating parameters and waveguide structures and the incident wavelength are correctly chosen.

4. Conclusion

In this paper, the phase matching function of WGC is obtained based on diffraction equation, the difference equations and absorbing boundary conditions are deduced based on the vector FDTD according to Maxwell equations, and the field distributions and mode patterns of different modes in the WGC are calculated. The calculated results indicate that the different optical field distributions and stable mode patterns of different order modes can be obtained with suitable grating parameters and waveguide structures. According to the mode field distribution, the optimum coupling efficiency can be predicted. In this way, we are able to design highly-efficient waveguide grating couplers, which is of practical use in integrated optical circuits and optoelectronic devices. To the best authors' knowledge, the analysis of higher-order mode field distribution of WGC using the vector FDTD method has not been reported to date.

Acknowledgement. This work was supported by the Scientific Research Program of Education Bureau of Hunan Province (08C368) and Hunan Provincial Natural Science Foundation of China （10JJ6095）.

References

[1] Calvo M L and Lakshminarayanan V 2007 Wavelength dispersive planar waveguide devices: echelle and arrayed waveguide gratings in Optical Waveguides (London: CRC Press)
[2] Chaganti K, Salakhutdinov I, Avrutsky I and Auner G W 2006 Opt. Express **14** 4064-4072
[3] Papadopoulos A D and Glytsis E N 2009 Applied Optics **48** 5164-5175
[4]Ming H, Zhang G P and Xie J P 1998 Photoelectronic Technique (Hefei: China University of Science and Technology Press) pp 6-72
[5] Yuan Z H, Liu Y Z and Pang T 1997 Semiconductor Optoelectronics **18** 92-97
[6] Yee K S 1966 IEEE Trans. Antennas Propagat **14** 302-307
[7] Engquist B and Majda A 1975 Mathematics of Computation **31** 629-651
[8] Taflove A and Hagness S C 2000 Computational Electrodynamics: The Finite- Difference Time-Domain Method (Boston. London: Artech House)

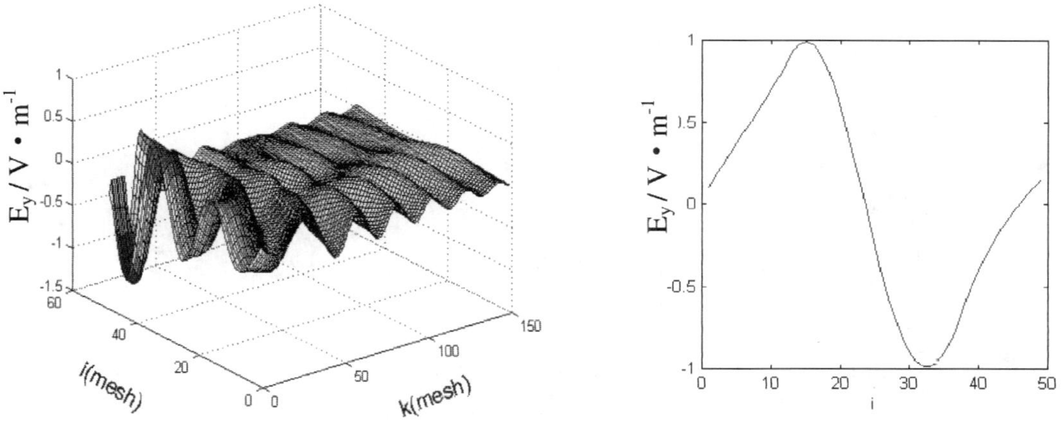

(a) (b)

Figure 2 (a) Field distribution of TE_1 (b) Mode pattern of TE_1

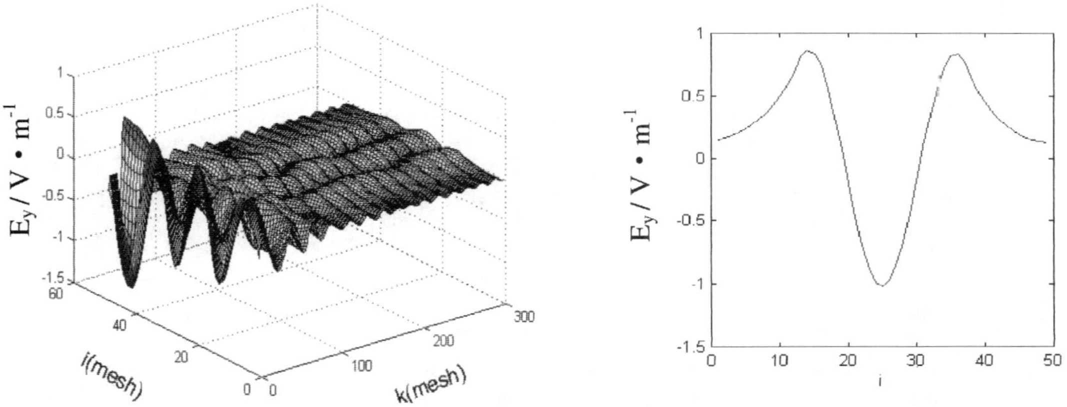

(a) (b)

Figure 3 (a) Field distribution of TE_2 (b) Mode pattern of TE_2

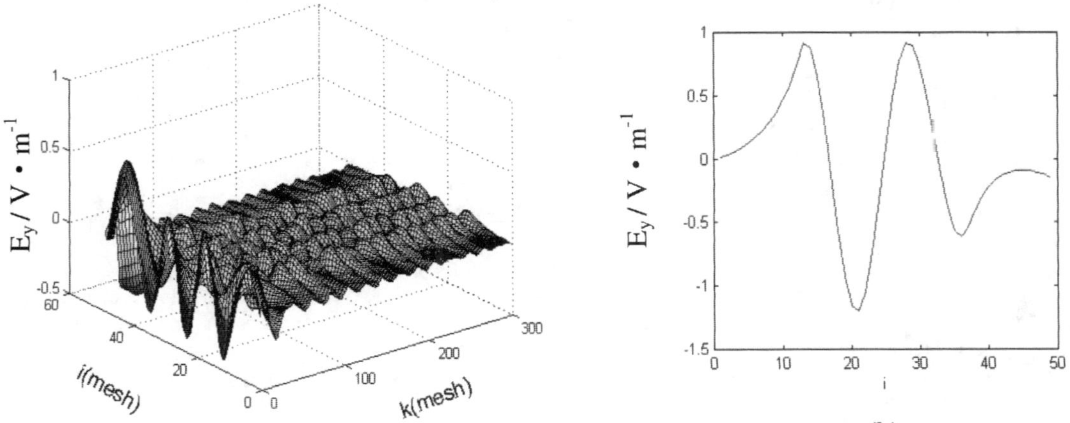

(a) (b)

Figure 4 (a) Field distribution of TE_3 (b) Mode pattern of TE_3

Controlling Extraordinary Optical Transmission by Subwavelength Hole Arrays -Bragg gratings structure

Bin Liu [1], Li Yu[2], Zhi Xin Lu, Ting Li and Gang Song
[1]School of Science, Beijing University of Posts and Telecommunications, Beijing, 100876
[2]Key Laboratory of Information Photonics and Optical Communications (BUPT), Ministry of Education, Beijing 100876

E-mail: jazz119@gmail.com

Abstract. We demonstrate a controlling of extraordinary optical transmission by using a novel structure composed of metal film with subwavelength hole arrays and Bragg structure. Extraordinary optical transmission through periodic arrays of subwavelength holes usually presents two obvious speaks. By controlling parameters of Bragg structure such as refractive indices or thickness, we can realize the switching of one of the peaks of extraordinary optical transmission. This structure might be used as tunable filters in future integrated optical circuit.

1. Introduction

Extraordinary Optical Transmission (EOT) has triggered tremendous attention focused to investigate the physical mechanism Since Ebbsen and co-workers discovered the phenomenon [1].Many researchers show that EOT is attributed to the surface plasmons polarization(SPP)resonance arising from periodic arrays of sub wavelength holes on metallic film [2, 3]. Transmission peaks are typically found at wavelengths close to the lattice periodicity. However some rearchers verified that varying the lattice periodicity, the change of EOT peaks is not enough to be explained by the SPP theory. In later study, wave guidance resonance has been raised up for another key factor to explain EOT peaks [4]. How to separate and control these two kinds of EOT peaks related to periodic arrays of sub wavelength holes on metallic film might exist some applications in future.
On the other hand, Bragg structure essentially is comprised of alternating layers of high and low-refractive dielectrics, which strongly reflects a band of wavelength depending on the thickness and refractive indices of the constituting layers [4, 5]. The band of wavelength which the reflectivity is significant is said to fall within the stop band of the Bragg structure [4]. Therefore, if the electromagnetic wave vector falls within the Bragg stop band of the periodically stratified medium, guidance wave could be achieved provided it is not radiating in the other medium [5]. This has been employed for designing electron accelerators and phase matching for nonlinear frequency conversion in Bragg reflection waveguides [6, 7].The polarization properties of the Bragg reflectors have been used for proposing polarization splitters and mechanically or optically tunable filters [8,9].
In this letter, we make use of the reflectivity character of Bragg structures to investigate the scheme of varying EOT originated from periodic arrays of subwavelength holes. We design a novel structure that is composed of an array of subwavelength holes perforated in a metal film and periodically stratified medium [10]. By using simulation method, we found that EOT can be controlled by varying parameters of Bragg structure. Our results might have an application in future optical integrated circuit

2. Simulation model

[1] School of Science, Beijing University of Posts and Telecommunications, Beijing 100876

Published under licence by IOP Publishing Ltd

We consider a free standing square Ag film with a thickness of h= 200nm and rectangular 675nm×150nm air holes. The rectangular holes array has a square lattice of a = 750 nm in the xy plane as shown schematically in Fig.1, where metal-Bragg interface formed by Ag film closed to a periodically stratified medium (Bragg structure). The refractive indices of stratified medium vary between n_1=3.5and n_2=2.04, and thickness of stratified medium has $d_1 = d_2$ both varying from 250nm to 350nm. The unit cell thickness for the Bragg structure is $\lambda = d_1 + d_2$ and the numbers of periodical unit is 6. Gaussian pump pulse, polarized TM with center frequency 2.5e+14Hz and width of spectrum 1200 nm, illuminated an array of holes in metallic film on substrate of Bragg structure. The direction of propagation light is chosen to be in the Z direction, which is also the one of refractive-index variation direction. EOT can be obtained when the pulse impinging on subwavelength hole arrays and transmission rate is calculated at the end of the Bragg structure by the full-vectorial three-dimensional (3D) finite difference time-domain (FDTD) method. Metal is modeled as Lorentz damping showed in Fig.1.

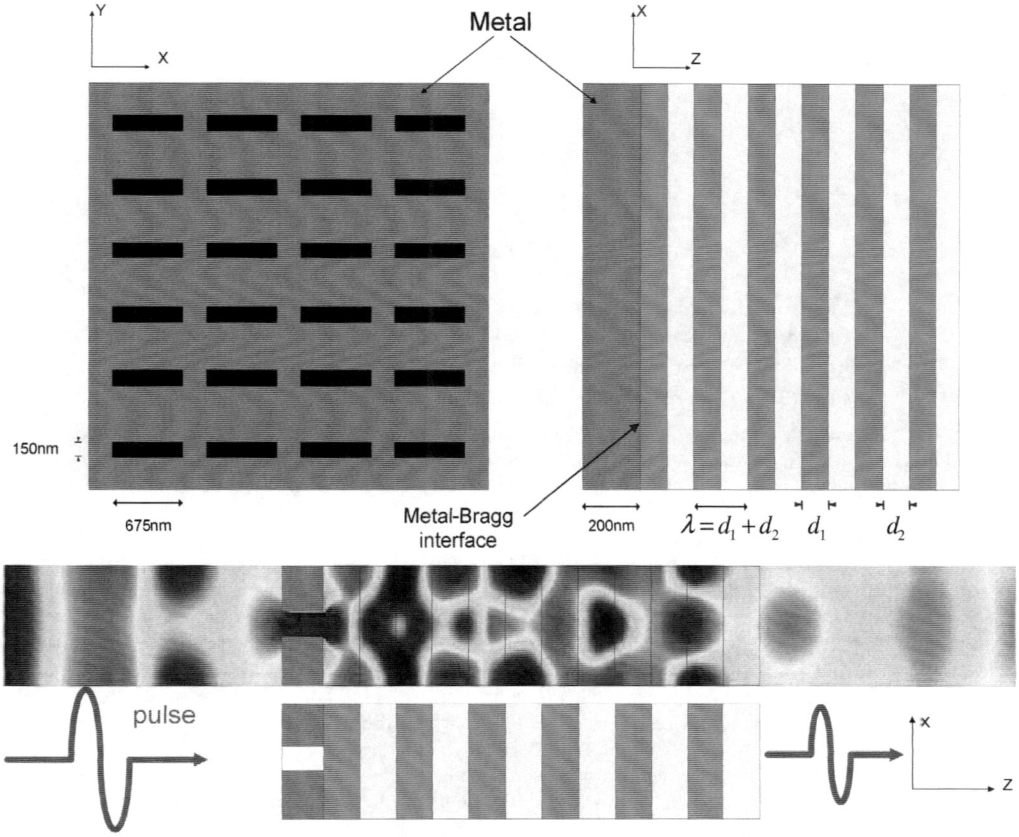

Fig. 1. (Color online) Schematic of the planar geometry showing the interface of a semi-infinite METAL and Bragg structure (periodically stratified region).(a) semi-infinite standing METAL film with a thickness 200nm at subwavelength, and square lattice of rectangular 675nm×150nm air holes.(b) periodically stratified medium (Bragg structure) of refractive index n1 and n2, having thickness d1 and d2, in a planar geometry. The unit cell thickness for the Bragg structure is $\lambda = d_1 + d_2$. A Gaussian pump pulse illuminates an array of holes in METAL film on substrate of Bragg structure. This unit is surrounded by periodic boundary. The rectangular holes array has a period of a = 750 nm in the xy plane.

3. Results and discussion.

In the optical regime, the EOT through metal films patterned with arrays of holes appears two obvious peaks at two different frequencies, which has been partially recognized as arising from the SPP. Some work has showed that except for SPP there are also other keys for this reason. If each air hole is considered as a rectangular waveguide with four metallic restrict and two sides opened to air, it forms a low quality factor cavity resonator, this resonant frequency is almost independent of the periodicity, and gives a flat peak in certain frequency band. Ruan referred to this kind of resonance as the localized waveguide resonance [4], and the flat peak is the signature of these resonances associated with each air hole. Additionally, the localized waveguide resonances always exist even if the metallic film is very thin (the thickness of the metallic films is less than the long axis of air holes).

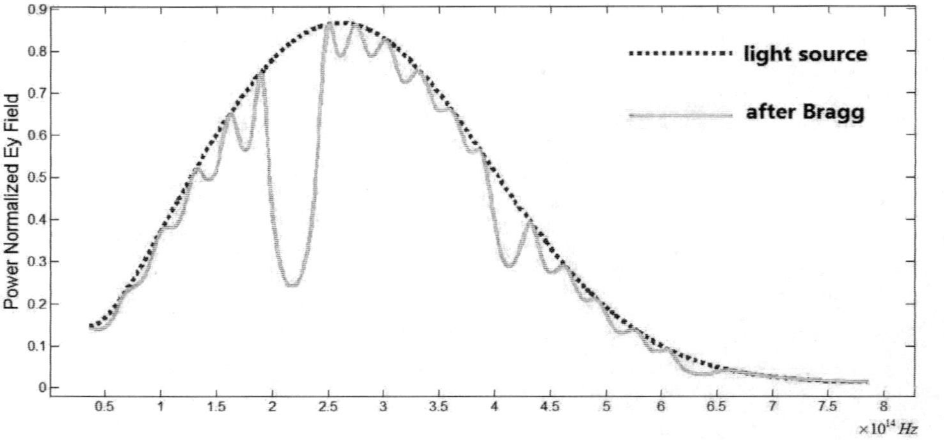

Fig.2. Diagram of the simulation measurement of Bragg

To investigate how to control the two peaks of EOT, we build the Bragg structure periodically stratified medium as mentioned above. We use the FDTD method to simulate the output Ey field of Bragg structures(n_1 =3.5and n_2 =2.04, d_1 =250nm and d_2 =250nm) as the function of incident frequency in the condition of the incident light illuminating directly to the Bragg structure without metal film, which the results are shown in Fig.2. Two different line shapes denote the spectrum of light of before and after Bragg structures respectively and the output Ey field is normalized. We see clearly that the field of incident light is reflected obviously at frequency of about 2.2e+14Hz. It is apparent that it must match with the stop band of Bragg structure. According to Bragg theory, two major parameters of both refractive dielectrics and unit cell thickness decide the frequency of stop band. The frequency tunable characteristic can be adjusted by carefully choosing the high and low-refractive dielectrics. So we can let anyone of these two peaks caused by wave guidance of a single hole or the SPP resonant of periodicity of rectangle holes array, matches the frequency of the stop band of Bragg structure. In this way, we can realize the controlling Extraordinary Optical Transmission by subwavelength hole arrays -Bragg gratings structure.

The transmission as function of frequency of metal film with subwavelength hole arrays are shown in Fig.3 (a), two obvious peaks of 2.2e+14Hz and 3.7e+14Hz respectively, where the transmissions are close to 1. Ruan's work shows the peak at frequency 2.2e+14Hz is caused by wave guidance of each single hole and the peak at 3.7e+14Hz is relative to the SPP of periodicity of rectangle holes array [4]. The transmission as function of frequency of composited structure of metal film with subwavelength hole arrays and Bragg structure are shown in Fig.3 (b), modify the parameter of $d_1 = d_2$=375nm, n_1 =3.5, n_2 =2.04, which make SPP peak at 2.2e+14Hz fall in the stop band of Bragg structure, the wave guidance resonate peak goes down nearly 60%. The reason why there is still an unapparent peak at the same frequency 2.2e+14Hz in Fig.3 (b) can be explained from two aspects. Firstly, in Fig.2 the reflective character is not perfect 100% due to limited numbers of periodicity layers of Bragg structure. In order to demonstrate the function of our structure, we only choose 6 units for simple consideration. Secondly, the bandwidth of Bragg stop band is less than the original wave guidance resonate peak at 2.2e+14Hz. Since the peak at 3.7e+14Hz relative to the periodicity of rectangle holes array does not fall into the stop band of Bragg, it almost has no change after the Bragg structure. In conclusion, with the help of Bragg reflector, a tunable pulse peak controller can be realized, which might have a bright future in nanooptical field.

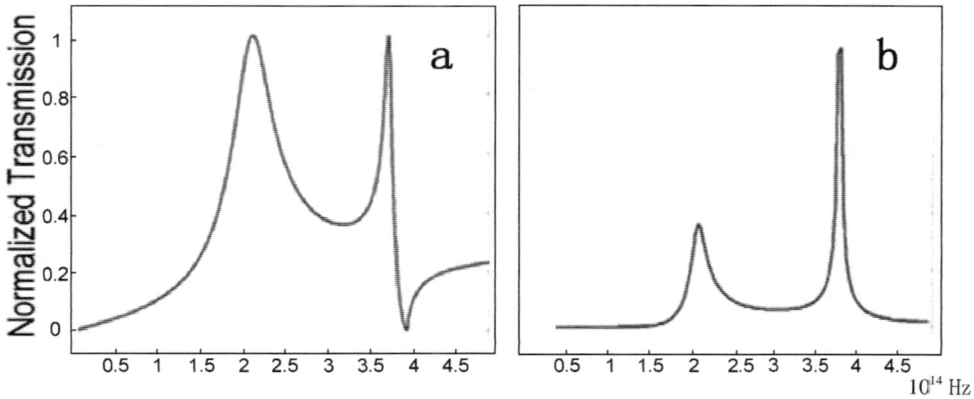

Fig.3. Diagram of the simulation measurement.

4. Conclusion

In conclusion, we have investigated and carried out an analysis of modes that are supported at the METAL–Bragg interface. The modes are essentially radiation modes in the dielectric region but fall within the stop band of the periodically stratified medium resulting in EOT control mode. Because of the presence of a strongly reflection periodically stratified element close to METAL film, these modes exhibit strong pulse control capability as compared to conventional METAL film with subwavelength air hole array modes. The strong EOT peak selected capability exhibited by the METAL–Bragg mode can appreciably improve the functionality of slow light devices and facilitate the tailoring of interacting modes in many linear and nonlinear optics applications. Moreover, there is significantly bigger propagation peak for this METAL–Bragg interface due to the presence of a Bragg structure, which can open up further new possibilities in photonic-circuit designing apart from its applications in integrated optic. Further, the study of modal characteristics of such a metal–Bragg structure can open up a new route for designing different types of ultra compact and precise optical sensors and designing electron accelerators and phase matching for nonlinear frequency conversion in Bragg reflection waveguides. In addition, the polarization properties of the Bragg reflectors have been used for proposing polarization splitters and mechanically or optically tunable filters.

5. Acknowledgments

This work is supported by the National Basic Research Program (973) of China under Project No.2010CB923202.

6. References

[1]. T.W. Ebbesen, H. J. Lezec, H. Ghaemi, T. Thio, and P. A.Wolf, Nature (London) 391, 667 (1998).
[2].H. F. Ghaemi, T. Thio, D. E. Grupp, T.W. Ebbesen, and H. J. Lezec, Phys. Rev. B 58, 6779 (1998).
[3].J. A. Porto, F. J. Garcia-Vidal, and J. B. Pendry, Phys. Rev. Lett. 83, 2845 (1999).
[4]. Z. Ruan and M. Qiu, Phys. Rev. Lett. 96, 233901 (2006).
[5]. D. Crouse, A. P. Hibbins, and M. J. Lockyear, Appl. Phys. Lett. 92, 191105 (2008).
[6].M. J. A. de Dood, E. F. C. Driessen, D. Stolwijk, and M. P. van Exter, Phys. Rev. B 77, 115437 (2008).
[7].J. Gómez Rivas, C. Schotsch, P. Haring Bolivar, and H. Kurz, Phys. Rev. B 68, 201306_R_ (2003).
[8]. E. Simova and I. Golub, Opt. Express 11, 3425–3430 (2003).
[9] .X. Hu, Q. Gong, S. Feng, B. Cheng, and D. Zhang, Opt. Commun. 253, 138–144 (2005).
[10]. Ritwick Das and Rajan Jha. APPLIED OPTICS / Vol. 48, No. 26(2009)

Synthesis and Fluorescence Switching of a New Photochromic Diarylethene

Ming Liu, Shouzhi Pu*,Gang Liu, Congbin Fan, Shiqiang Cui

Jiangxi Key Laboratory of Organic Chemistry Jiangxi Science and Technology Normal University Nanchang 330013, P. R. China

pushouzhi@tsinghua.org.cn

Abstract. A new photochromic diarylethene compound, 1-[2-methyl-5-(4-pyridine) -3-thienyl]-2-[2-methyl-5-(4-methoxylphenyl)-3-thienyl]perfluorocyclopentene(1a) was synthesized. its photochromism was investigated. It showed good photochromism both in solution and PMMA amorphous film upon photoirradiation. Its fluorescence properties in the solution also were reaearched.

1. Introduction

The design and synthesis of photochromic molecules is an area of intense research because of the widespread use in photonic device applications such as memory media and optical switches [1,2]. Diarylethene derivatives are the most promising candidates for the application among various thermally irreversible photochromic materials considering the excellent thermal stability in both isomers, fatigue resistance, rapid response, and high reactivity in solid state [3-5]. Photoinduced anisotropy is referred as a polarized-light anisotropic change for isotropic materials or anisotropic materials. This characteristic can be applied to optical recording, picture managing and spatial light modulating, etc. To date, photo-induced anisotropy of some materials, such as photochromic glass, color crystal, and dye-polymer, was investigated extensively. The photo-induced anisotropy of photochromic pyrrylfulgide has also been reported by Yao [6] et al..

In the present work, we synthesized a photochromic diraylethene 1-[2-methyl-5-(4-pyridine)-3-thienyl]-2-[2-methyl-5-(4-methoxylphenyl)-3-thienyl]perfluoro cyclopentene(1a). Its photochromic reactivity, fluorescence property and application for polarization holographic optical recording were investigated in detail. The photochromic reaction of diarylethene 1a is shown in Scheme 1.

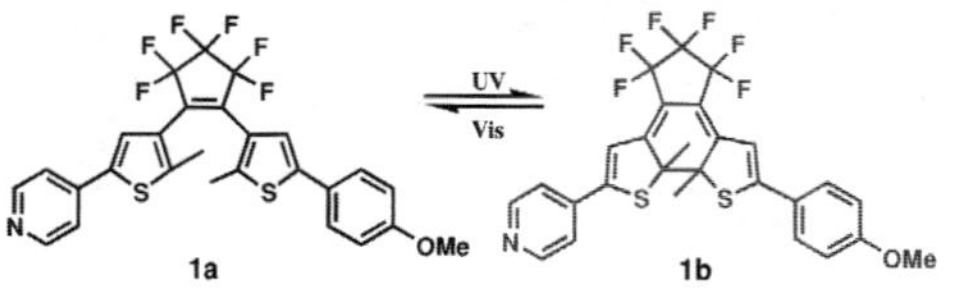

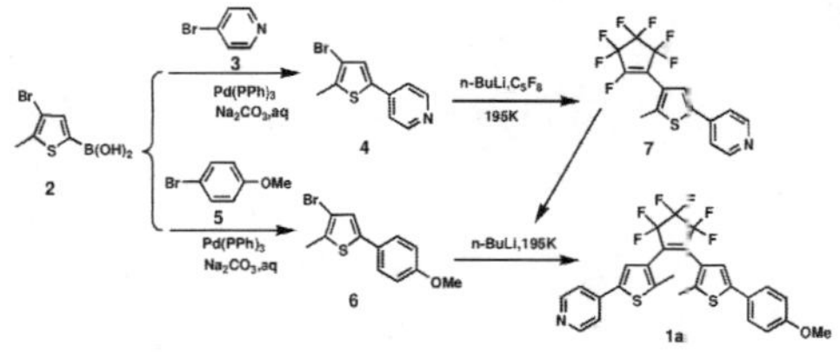

Scheme 1. Photochromism of diarylethene 1.**Scheme 2.** Synthetic route for the compound 1a.

2. Experimental

The diarylethene 1a of synthetic route is shown in Scheme 2. It was synthesized according to the similar procedure of Pu [7] et al.. Firstly, compound 4 and 6 was prepared by reacting compound 2 with 4-bromopyridine and 1-bromo-4-methoxybenzene in the presence of tetrakis(triphenylpho- sphine)palladium(0) and Na_2CO_3 in THF for 16 h at 343 K. Secondly, under an argon gas atmosphere, compound 4 was dissolved in THF and n-butyl lithium hexane solution was added at 195K. Stirring was continued for 30 min at this low temperature; octafluorocyclopentene was added and the mixture was stirred for 2 h at this temperature. The reaction mixture was extracted with diethylether, dried with anhydrous $MgSO_4$, filtered, and evaporated in vacuo, then purified by column chromatography to give compound 7 (2-methyl-5-(4-pyridine)-3-thienyl -perfluorocyclopenten) white solid. Finally, To a stirred THF solution of compounds 6 was slowly added a 2.5 mol/L n-BuLi/hexane solution at 195K under an argon atmosphere. After 30 min, compound 7 was added and the mixture was stirred for 2h at this temperature. The reaction mixture was extracted with diethylether, dried with anhydrous $MgSO_4$, filtered, and evaporated in vacuo, then purified by column chromatography to give 1a reddish brown oil liquid in 30% yield. The structure of 1a was characterized by ^{1}H NMR.

The structure of 1a was characterized by ^{1}H NMR. ^{1}H NMR (400 MHz, $CDCl_3$, ppm): 1.87 (s, 3H, -CH_3), 1.94 (s, 3H, -CH_3), 3.77 (s, 3H, -OCH_3), 6.84-6.86 (s, 2H, thiophene-H), 7.02-7.41 (d, 4H, benzene-H), 7.46-8.54 (d, 4H, pyridine-H).

3. Result and Discussion

3.1. Photochromism of diarylethene

The unsymmetric diarylethene 1a exhibits good photochromic properties and can be toggled between its colorless ring-open and colored ring-closed forms by alternate irradiation with appropriate wavelengths of light. Figure. 1 shows the absorption spectral changes of 1a in hexane (C =2.0 × 10^{-5} mol/L, Figure 1A) and in poly methyl methacrylate (PMMA) film (10% w/w, Figure 1B) by photoirradiation. In hexane, as shown in figure 1A, diarylethene 1a has the absorption maximum at 296 nm (= 5.8 × 10^4 L mol^{-1} cm^{-1}). Upon irradiation with 297 nm light, the colorless solution of 1a turned blue, in which absorption band was observed at 590 nm (= 1.97 × 10^4 L mol^{-1} cm^{-1}). The Orange-red color is due to the closed ring form 1b. The color disappeared by irradiation with visible light. Similar spectrum change of diarylethene 1a was observed in PMMA film (Figure 1B). Upon irradiation with 297 nm UV light, the colorless PMMA film turned blue with new broad absorption band centered at 604 nm for diarylethene 1a, which assigned to the open-ring isomer 1a changed to the closed-ring isomer 1b. The absorption maximum of 1a was longer than that in solvent, it can be ascribed to the stabilization of molecular arrangement in solid state [8].The colored PMMA film can also return to colorless upon irradiation with appropriate wavelength visible light.

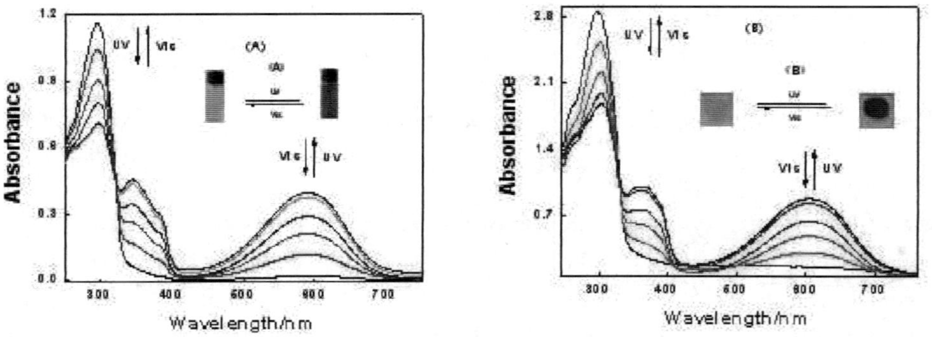

Figure 1. Absorption spectral changes of compound 1a in hexane (2.0 × 10^{-5} mol/L) and (B) in PMMA film (10%, w/w)

3.2. Fluorescence of diarylethenes 1a

The fluorescence properties in the solution of the title compound were measured using a Hitachi F-4500 spectrophotometer, and the breadths of excitation and emission slit were selected 10.0 nm and 20.0 nm. The fluorescence excitation and emission spectra of 1a in hexane (C = 5.0 × 10^{-4} mol/L) at room temperature was illustrated in figure 2. From this figure, we can clearly see that the hexane of diarylethene 1 exhibited relatively strong fluorescence at 434 nm when excited at 367 nm. The Stokes shift of the fluorescence was relatively large and the fluorescence spectral edge showed a red-shift in comparison with the absorption edge. This kind of large Stokes shift has already been discussed in detail by Sekiya and co-workers [9].

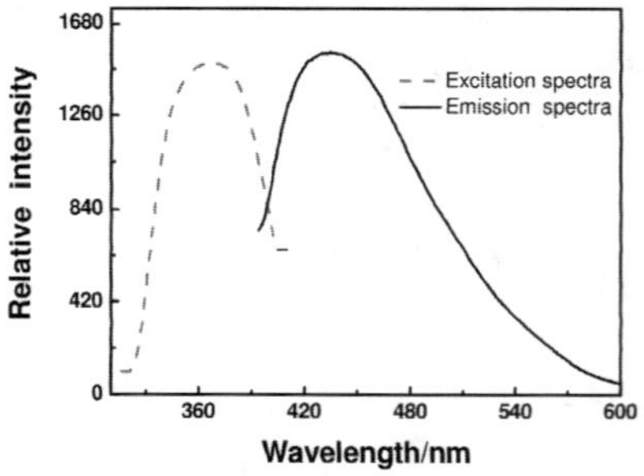

Figure 2. Excitation spectra (dash line) and fluorescence spectra (solid line) of 1a in hexane at room temperature.

The fluorescence intensity of diarylethene 1a decreased dramatically along with the photochromism from open-ring isomer to closed-ring isomer upon irradiation with 297 nm light in hexane. As shown in figure 3, when irradiated by light of 297 nm, the photocyclization reaction was carried out and the fluorescent closed-ring form of the compound became week. The back irradiation by appropriate wavelength visible light the open-ring isomer 1a and recovered the original emission spectra. The phenomena are useful for application as the fluorescence switches [10,11].

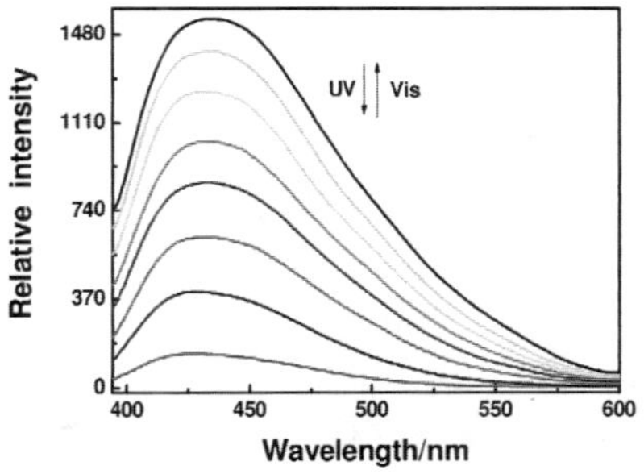

Figure 3. Emission intensity changes of diarylethene 1a in hexane (5.0×10^{-4} mol L^{-1}) upon irradiation with 297 nm UV light.

In conclusion, the diarylethene described here was synthesized and it showed good photochromic behavior both in solution and in PMMA amorphous film. In hexane, the open-ring isomer of diarylethene 1a exhibited relatively strong fluorescence intensity decreased along with the photochromism upon irradiation with 297 nm light and its closed-ring isomer showed almost no fluorescence.

Acknowledgment

This work was supported by the Project of National Natural Science Foundation of China (20564001), the Key Scientific Project from Education Ministry of China (208069), the Project of Jiangxi Youth Scientist, and the Science Funds of the Education Office of Jiangxi, China (GJJ08365).

References

[1] Pu S Z, Zhang F S, Xu J K, Shen L, Xiao Q and Chen B 2006 *Mate. Lett.* **60** 485
[2] Dulic D, Kudernac T, Puzys A, Feringa B L and Wees B J 2007 *Ad. Mater.* **19** 2898
[3] Pu S Z, Xu J K, Shen L, Xiao Q, Yang T S and Liu G 2005 *Tetrahedron Lett.* **46** 871
[4] Pu S Z, Yan L S, Wen Z D, Liu G and Shen L *2008 J. Photoch. Photobiol. A: Chem.* **196** 84
[5] Irie M 2000 *Chem. Rev.* **100** 1685
[6] Chen Y, Wang C M, Fan M G, Yao B L and Menke N 2004 *Opt. Mater.* **26** 75
[7] Li G, Zheng T, Pu S Z and Xu J K 2006 *Acta Cryst. E, International Union of Crystallography, Paris.* **62** o2877
[8] Kasatani K, Kambe S and Irie M, 1999 *J. Photochem. Photobiol. A: Chem.* **122** 11
[9] Tanaka N, Okabe C, Sakota K and Irie M 2002 *J. Mol. Struct.* **616** 113
[10] Chen B Z, Wang M Z and Tian H 2002 *Chem. Commun.* 1060
[11] Tian H, Chen B Z, Tu H Y and Müllen K 2002 *Adv. Mater.* **14** 918

Journal of Physics: Conference Series **276** (2011) 012076 doi:10.1088/1742-6596/276/1/012076

Optical and Electrical Properties of OLED with the Structure of ITO/ m-MTDATA/Meo-TPD/Alq3/LiF/Al

Gui T L, Dong X L, Wu F M, Jia W, Wen J Y and Liu Z Q

Applied Science College, Harbin University of Science and Technology, xue fu street, Harbin, China

guitl1952@yahoo.com.cn

Abstract. In this paper,the organic light emitting device (OLED) with the structure of ITO/m-MTDATA/Meo-TPD/Alq_3/LiF/Al has been fabricated using Alq_3 as a light-emitting layer, m-MTDATA(4,4',4''-tri(3-methyl-phenylphenylamino) tripheny- lamine) as a hole injection buffer layer and Meo-TPD (N, N, N ', N'-tetrakis (4-methoxyphenyl)-benzidine)as a hole transport layer, as well as luminous and electrical properties were studied. The turn on voltage of device is about 6v. Its maximum brightness reaches 14300cd/m^2 with applied voltage of 13v, sixteen times higher than that of the structure of ITO/Alq_3/LiF/Al. When the applied voltage is 8V, the electroluminescence spectra peak is at 540nm, showing green.

1. Introduction

In recent years, OLED has been widely used because of its low driving voltage, wide viewing angle, fast response, wide range of alternative materials, flexible displays, etc [1-2]. 8-hydroxyquinoline aluminum (Alq3) is first used by Kodak as light-emitting layer of EL device materials, as a result of its high glass transition temperature, good film forming, good carrier transport properties, high luminous efficiency and good stability, which has been widely used in organic light-emitting device. Its electron mobility was about $10^{-5}cm^2 \cdot V^{-1} \cdot s^{-1}$, which is lower by two orders of magnitude than hole mobility of the usual hole transport material TPD or NPB [3-4]. The LiF of appropriate thickness is added between the Alq_3 and the Al electrode to form composite electrodes, which can reduce Al diffusion from the metal electrode to organic light-emitting layer, making electrode and organic layer form a good interface characteristic, delay the photochemical and electrochemical aging of the device, and improve the device stability; which can reduce the effective potential barrier height of interface, and enhance the efficiency of carrier injection.

High heat stability (high glass transition temperature Tg), reducing interface energy potential barrier of the anode and hole transport layer, and good film-forming properties, which are the main requirements of hole injection layer [5-9]. In this paper, organic light-emitting device of multi-layer structure ITO/m-MTDATA/Meo-TPD /Alq3/LiF/Al has been fabricated. the hole injection layer m-MTDATA (4,4',4''-tri(3-methyl-phenylphenylamino)triphenylamine) was added between the ITO and hole transport layer Meo-TPD (N, N, N ', N'-tetrakis (4-methoxyphenyl)-benzidine), due to its relative stability, preventing effectively indium ions diffusing from the anode to hole transport layer, good filming, as well as hole injection capability, which greatly reduce the barrier height between the anode and hole transport layer, in favor of improving the brightness of light-emitting device and the balance of carrier injection, and reducing turn-on voltage.

Journal of Physics: Conference Series **276** (2011) 012076 doi:10.1088/1742-6596/276/1/012076

2. Experiment

2.1. Preparation of light-emitting device

First, selecting the size of 30mm×30mm ITO glass as the substrate, cleaning it for 10 minutes with acetone, ethanol, deionized water ultrasonic, respectively, and then drying it for 5 minutes, again using ultraviolet radiation raying it for 1 minute. Using 2001 multi-source automatic temperature control system of organic molecules to deposit films, when air pressure of the main evaporation chamber is 3.9×10-4Pa, evaporating thin films of m-MTDATA, Meo-TPD, Alq3, and LiF on the ITO surface sequentially, and finally depositing Al with DM-300B vacuum evaporation machine. deposition rate of m-MTDATA, Meo-TPD and Alq3 is 0.2nm / s, that of LiF and Al is 0.1nm / s and 1nm / s, deposition thickness of m-MTDATA, Meo - TPD, Alq3, LiF, and Al are respectively 45nm, 5nm, 50nm, 1nm and 100nm, with the FTM-V film thickness instrument monitoring thickness at the same time. The area of light-emitting device is 2×2mm2.

Structure of the device is ITO / m-MTDATA/Meo-TPD/Alq3/LiF/Al is shown in Figure 1. ITO as the anode, m-MTDATA as hole injection layer, Meo-TPD as hole transport layer, Alq_3 as the electron transport layer and light-emitting layer, LiF / Al as composite cathode.

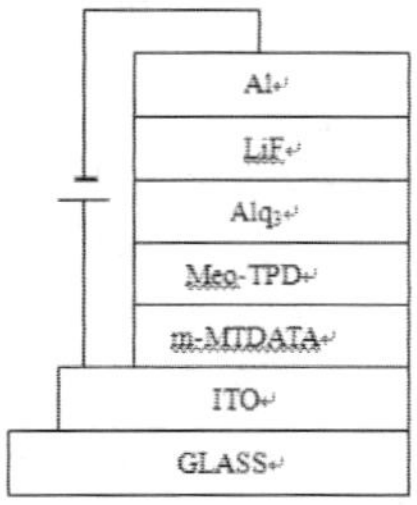

Figure 1. Schematic diagram of the device structure

2.2. Experimental material

As a hole injection layer, m-MTDATA namely 4,4 ', 4"-tri (3-methyl-phenylphenylamino) triphenylamine, star-type structure is shown in Figure 2, melting point is 203-207 ℃, the glass transition temperature is 75 ℃, ionization potential is 5.1ev, affinity energy is 1.9ev, its state is very similar to the glass material, so it can be deposited evenly on the ITO substrate.

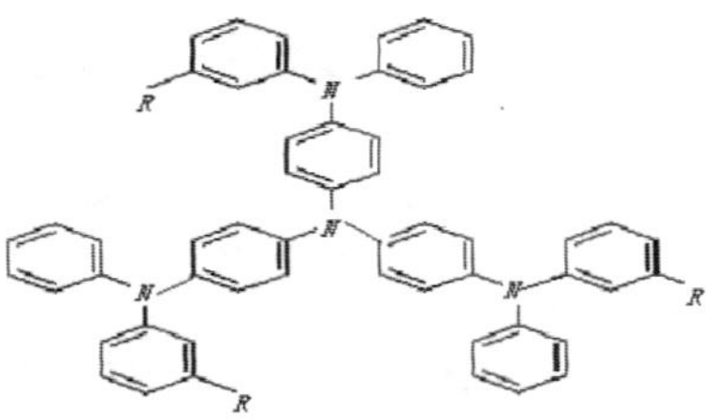

Figure 2. Star-type structure drawing of m-MTDATA

2.3. Experiment and analysis

After fabricating the device, testing its Current - voltage (IV) characteristics, the brightness - voltage (BV) characteristics, electroluminescence spectra immediately at room temperature under atmospheric environment, using the PR-650 spectral luminance meter. Figure 3 shows the current - voltage characteristic curve, which can be seen from the figure, that the current sharply jumps as the voltage exceeds 6V, the curve showes a typical diode characteristics.

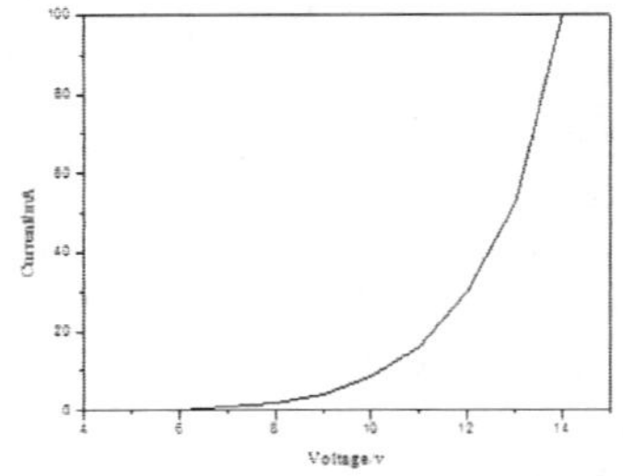

Figure 3. Current - voltage characteristics curve of device

Brightness - voltage curve of device is shown in Figure 4. turn on voltage of the device is 6V, when the applied voltage is 13V, the device reaches its maximum brightness 14300cd/m^2. owing to the Meo-TPD as hole transport layer and m-MTDATA as buffer layer, the light-emitting brightness of device is 16 times higher than that of simple structure of ITO / Alq_3/LiF (1nm) / Al, whose light-emitting brightness is about 850 cd/m^2 with applied voltage of 13V[10].

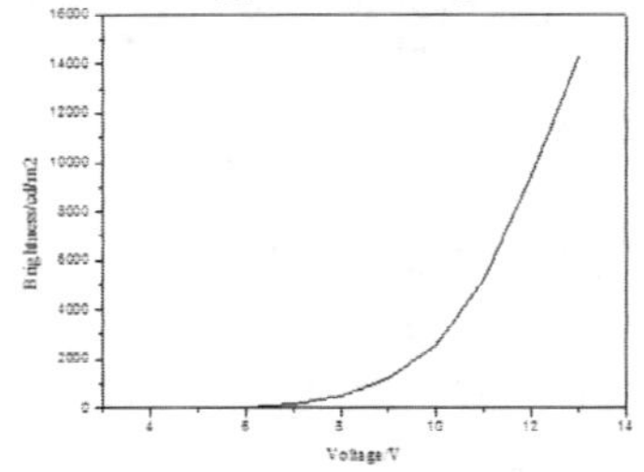

Figure 4. Brightness - voltage curve of device

Light-emitting mechanism of organic light-emitting device is that electron from cathode and hole from anode inject into the light-emitting layer respectively, under the influence of external electric field, hole and electron encounter to form exciton, exciton shows radioluminescence in the light-emitting layer, but because of the imbalance of the hole and electron injection, as well as the difference of the migration rate, carriers from both electrodes can not effectively be limited in light-emitting layer to form exciton, leading some of the excess carriers reach electrode, then making the light quenching at the electrodes, reducing the luminescence efficiency.

Considering about ionization energy, the lower the potential barrier of interface between the hole transport layer and the anode is, the easier hole injecting is. In this paper, using the m-MTDATA of 45 nm thickness as a hole injection buffer layer, the highest occupied molecular orbital energy (HOMO) of m-MTDATA is -5.1 eV, higher than the HOMO level of the Meo-TPD (-5. 7 eV). barrier height between ITO and m-MTDATA is 0.4 eV, but 1eV between ITO and Meo-TPD. The hole injection layer makes more holes inject into the transport layer, and holes and electrons combine more effectively in the light-emitting layer ,increasing the brightness of the current and lowering turn-on voltage.

Figure 5 shows the electroluminescence spectra as the applied voltage is 8V, the maximum peak of the light-emitting device is 540nm, CIExy = (0.3629,0.5309), light-emitting source is Alq3.

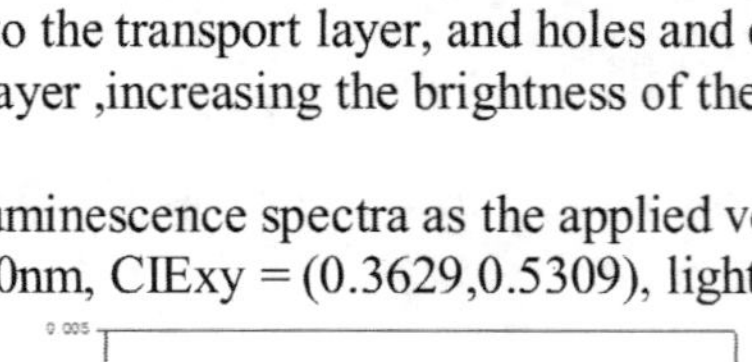

Figure 5. electroluminescence spectra of the device

3. Conclusion

m-MTDATA as a hole injection buffer layer improves the device's ability to enhance the hole injecting and boost the recombination rate of exciton, LiF / Al composite cathode enhances electron injection capacity of the device, thus the device current and brightness have been enhanced, as well as turn on voltage lowers. Owing to the Meo-TPD hole transport layer and m-MTDATA buffer layer, brightness of the device has been greatly improved, its maximum brightness reaches 14300cd/m^2 in the 13V, turn on voltage is about 6V, electroluminescence spectra peak is at 540nm when the applied voltage is 8V.

References

[1] Huang C H 2005 *Introduction to Organic Light-Emitting Materials and Devices*
[2] Zhou J L and Zhang F J 2008 *Journal of Optoelectronics ·Laser* **19** 12
[3] Wei F X, Fang L and Jiang X Y 2008 *Semiconductor Technology* **33** 12
[4] Yoshitaka N, Koji N, Hisayoshi F,Takeshi M and Takeshi O 2007 *Jpn. J. Appl. Phys.* 46
[5] Forsythea E W, Abkowitz M A and Gao Y L 2000 *J. Vac. Sci. Technol* **8** 4
[6] Cho H Y, Park E J, Kim J H and Park L S 2008 *Journal of Nanoscience and Nanotechnology* **8** 10
[7] Meng Y L, Xie W F and Ning Z 2008 *Microelectronics Journal* **39** 5
[8] Su S H, Hou C C, Tsai J S and Yokoyama M 2009 *Thin Solid Films* **517** 17
[9] Li S, Tong G P and George T F 2009 *Journal of Applied Physics* **106** 7
[10] Xu X M, Peng J C and Li H J 2004 *Optoelectronics.Laser.* **15** 5

Improved light extracion efficiency in AlGaInP-Based Diodes (LEDs) by applying a nanohole structure on GaP window layer

MA Li*, JIANG Wen-Jing, ZOU De-Shu, Meng Li-Li, SHEN Guang-Di
Key Laboratory of Opto-electronics Technology (Beijing University of Technology), Ministry of Education, Beijing University of Technology, 100 Ping Le Yuan, Chaoyang District, Beijing 100124, China

Email: mali@emails.bjut.edu.cn

Abstract: In this letter, to enhance light efficiency of AlGaInP-based LEDs, a nanohole structure was applied to GaP window layer by using self-assemble metal layer nano-masks and inductively coupled plasma (ICP). This method has the potential advantage because both the size and density of the nanoholes are controllable. The density of nanoholes varied from 5×10^8 to 2.8×10^8 cm^{-2} and size varied from 180-430 nm while increasing rapid thermal annealing temperature from 350-500℃. By using this surface texturing method, the light intensity and light output power of AlGaInP-based LEDs with textured surface (LED-Ⅱ) increases 27% and 15% comaparing to LEDs with flat surface (LED-Ⅰ), respectively.

1. Introduction

High-performance AlGaInP-based LEDs have been widely used for many applications such as traffic lights, full color-dispalys and genernal lighting[1,2]. As we know, the internal quantum efficiency of AlGaInP-based LEDs has closely approaches 99%, but light extraction efficiency is still restrained by the large difference in the refractive index between the GaP window layer(n=3.2) and

* Supported by the National High Technology Research and Development Program of China under Grant No 2008AA03Z402, and the Beijing Natural Science Foundation of Beijing under Grant No 4092007.
Email: mali@emails.bjut.edu.cn Tel: + 86 01067391641-842 Fax: +86 01067391641-822

air(n=1)[2]. Considering this large distinction, only about 2.5% of internally emitted light from the active region can escape from the surface of GaP window layer[3].

Roughening the top surface of an LED is one of the methods for improving the light extraction[4], lots of surface texturing techniques have been extensively reported. Jian-Jhong Chen et al fabricated LEDs with nano-mesh ZnO layers, where the nano-mesh ZnO were uniformly arranged on the top surface of the GaP window layer[5]. Liang-Jyi Yan applied a periodic texture surface using photolithography and a wet etching process, the exterior of the etched texture consist of a series of bowl-shaped recesses [6].

These surface texturing were all designed as the fixed periodic, but for GaP material itself, fixed sidewall angle is hard to fabricate and repeat, so random roughed surface profile is a much more attractive method for producing devices without the need for expensive lithography. So in this paper, we reported a method of fabricating GaP nanohole window layer in AlGaInP based LEDs using self-assemble metal nanomask and inductively coupled plasma (ICP).

2. Experiment

The schematic cross section graphs of the AlGaInP LEDs is shown in Figure 1. The wafers were epitaxially grown on (111) GaAs substrates by metal-organic chemical vapor deposition (MOCVD) system. The LED structure consist of AlGaAs- AlAs DBR, n-AlGa$In_{0.5}$P cladding layer, an undopted $(Al_{0.5}Ga_{0.5})In_{0.5}P$- $(Al_{0.1}Ga_{0.9})In_{0.5}P$ multiple-quantum well active region with dominate wavelength of 620 nm, a p-AlGaInP cladding layer and a 0.5 μ m-thick p-GaP layer.

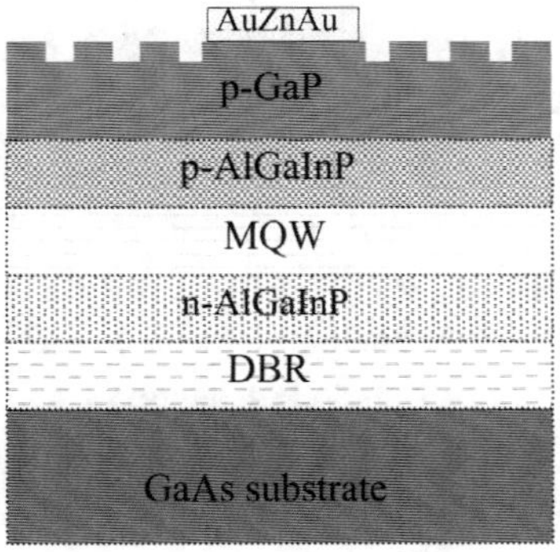

Figure 1. Schematic cross section graphs of AlGaInP LEDs

The process flowchart to form nanohole GaP window layer is shown in Figure 2(a). First, 30 nm SiO_2 film layer was deposited on the surface of GaP by Plasma Enhanced Chemical Vapor Deposition. The SiO_2 film layer can smooth the surface of GaP layer so that the nanomasks can form uniformly, and it also can prevent the metal atom from diffusing into GaP after rapid thermal annealing (RTA). Then, a 10 nm Au film layer was sputtered on the SiO_2 thin film. All the samples were subsequently sent to RTA under flowing N_2 at temperature 350-500 ℃ for 1 min. After RTA, the Au film would form self-assemble intermittent nano-partical because of the tension, as shown in Figure 2(b). Then, all the samples with nanomasks were etched by ICP system. The depth etched by ICP must control in about 100 nm so that surface texturing can give efficient light extraction without damaging the electrical or optical properties of the material [7]. After the etching progress, the residual metal on the surface was removed by diluted KI liquor, and SiO_2 was etched by HF liquor, remaining many nanoholes on the GaP window layer, shown in Figure 2(c), the size and density of the nanoholes were estimated by the scanning electron microscope (SEM).

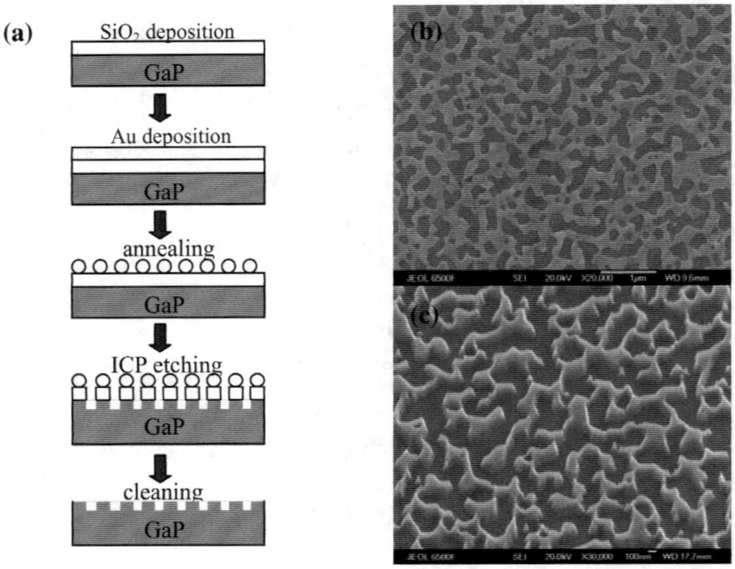

Figure 2. (a)Schematic illustration of the nanohole GaP window layer' s process; (b) SEM of surface morphology after RTA; (c) SEM of nanohole on GaP window layer

3. Results and Discussion

The rapid thermal annealing temperature need to be chosen above 300 ℃ to form Au self-assemble nano-masks. The samples were annealed at between 350 ℃ to 500 ℃ for 1 min. Figure 3(a) (b) (c) and (d) displays SEM images of GaP nanohole window layer of the samples annealed at 350 ℃, 400 ℃, 450 ℃ and 500 ℃, respectively. Figure 3(e) shows the dependence of density and mean size of the nanohole on annealing temperature. The mean size of the nanohole increases from about 220 nm to 430 nm when the temperature rises from 350 ℃ to 500 ℃. The densities decrease from 5×10^8 cm^{-2} to 2.8×10^8 cm^{-2} as well. At 350 ℃, the temperature is too low to produce enough tension for Au thin film forming self-assemble nano-masks, so nanohole formed at this temperature is much smaller than the others. When temperature rises, Au thin film can rupture adequately, the apertures between Au nano-masks become larger, so after ICP etching, size of the nanoholes increases with the rising of temperature.

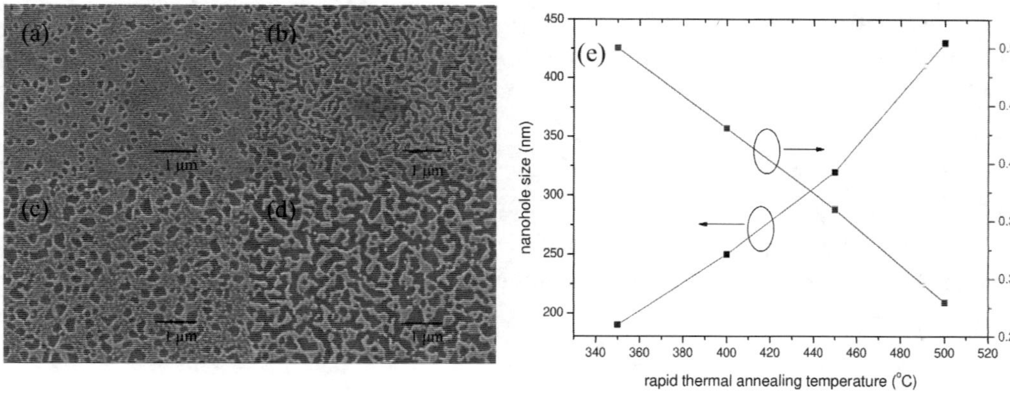

Figure 3. SEM images of GaP nanohole window layer of samples annealed at: (a) 350 ℃; (b) 400 ℃; (c) 450 ℃; (d) 500 ℃; (e) The dependence of density and mean size of the nanohole on annealing temperature

Jelena Vuckovic demonstrated that good dispersion appears on the textured surface while surface texturing dimension is half of the wavelength [8]. So we selected the sample which was annealed at 450 ℃ for 1 min to finish the process of the LED devices. This type of LED is named LED-Ⅱ. Meanwhile, conventional LEDs without surface texturing (named LED-Ⅰ) were also fabricated for comparison.

Comparison of measured current-voltage (I-V) of LED-Ⅰ and LED-Ⅱ is shown in Figure 4. Under an injection current (I_F) of 20 mA, the mesuread forward voltage (V_F) of LED-Ⅱ is 2.15 V, which is relatively high compared to 2.10 V of LED-Ⅰ. To understand the damage effect of ICP process, the dynamic resistances were extraced from the I-V characteristics. The series resistance for LED-Ⅰ and LED-Ⅱ were approximately 10.22 and 10.72 Ω at the injection current of 20 mA. The forward voltage and the series resistance of LED-Ⅱ were a little larger than that of LED-Ⅰ. It indicates that ICP etching may lead to some plasma damage on the surface of p-GaP wincow layer, but these differences are so tiny that they can not effect properties ot LED devices.

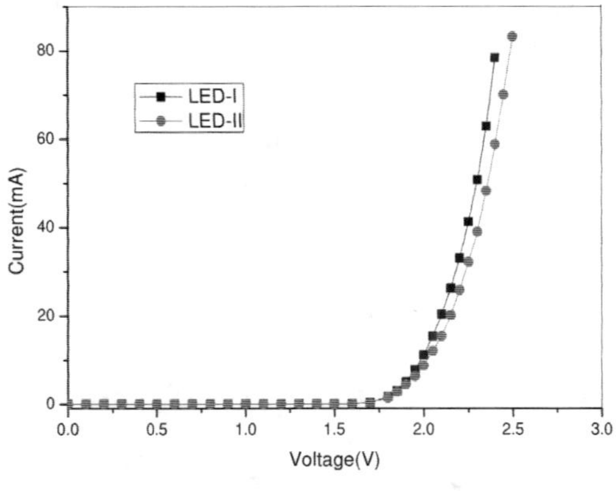

Figure 4. Typical I-V characterostics of LEDs with textured surface and LEDs with flat surface

Figure 5 (a) shows the possible photon paths at the interface between p-GaP and surrounding air for LEDs without surface texturing. For a LEDs with a microroughened p-GaP top surface, the angular randomization of photons can be achieved by surface scattering from the microroughened top surface of the LED, as shown in Figure 5(b). The light extraction efficiency of LEDs enhanced by surface roughness is shown in Figure 5(c). It shows the experimental light output power and light intensity versus the current for these two kinds LEDs. As Figure 5(c) shows, the light intensity and light output power of LED-Ⅱ increases about 27% and 15% compared to LED-Ⅰ at 20 mA, respectively. This indicates that the method of fabricating GaP nanohole window layer is powerful for light extraction enhancement. Moreover, it has the potential advantage because both the size and the density of the nanoholes are controllable.

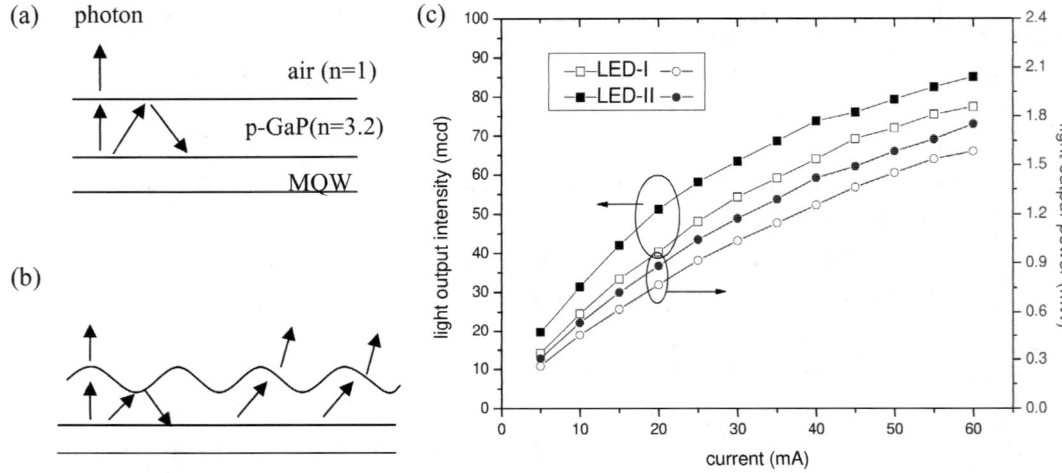

Figure 5. (a) Possible photon paths at the interface between the p-GaP surface and air; (b) The LED structure with a microroughened surface, the photons can be escaped out from the LED to air; (c) Light output power and light intensity versus the current in the flat surface and textured surface LEDs

4. Conclusion

In summary, GaP nanohole window layer was fabricated by using self-assemble metal nano-mask and inductively coupled plasma. Both the size and density of the nanoholes are controllable. By using this nanohole structure in AlGaInP-based LEDs, light intensity and light output power increases about 27% and 15% at 20 mA, respectively. To deep study the temperture and time of RTA, ICP etching condition, better result can be obtained.

5. References

[1] C P Kuo, R Fletcher, T Osentowski, M Lardizabal, M. Crafordm and V. Robbins 1990 Appl. Phys. Lett. **57** 2937

[2] Streubel K, Linder N and Wirth R 2002 IEEE J Sel Topics Quantum Electron **8** 321

[3] T. Gessmann and E. F. Schubert 2004 J. Appl. Phys. **95** 2203

[4] I. Schnitzer, E. Yablonovitch, C. Caneau, T. J. Gmitter and A. Scherer 1993 Appl. Phys. Lett. **63** 2714

[5] Jian-Jhong Chen and Yan-Kuin Su 2010 IEEE Photonics Technology Letters **22** 383

[6] Liang-Jyi Yan, J. K. Sheu, Wei-Chih Wen, Tien-Fu Liao, Ming-Jong Tsai and Chih-Sung Chang 2008 IEEE Photonics Thehnology Letters **20** 1724

[7] T. Fujii, Y. Gao, R. Sharma, E. L. Hu, S. P. DenBaars and S. Nakamuraa 2004 Appl. Phys. Lett. **84** 855

[8] Jelena Vuckovic, Marko Loncar and Axel Scherer 2000 IEEE Journal of Quantum Electronics **36** 1131

Flux based modeling and simulation of dry etching for fabrication of silicon deep trench structures

Rizwan Malik [1], Tielin Shi [2], Zirong Tang[2,]*, Shiyuan Liu[2]

[1] State Key Laboratory of Digital Manufacturing Equipment and technology, Huazhong University of Science and Technology, 1037 Luoyu road, Wuhan, China 43007 , rizwanmalik@smail.hust.edu.cn

[2] Wuhan National Laboratory for Optoelectronics, Huazhong University of Science and Technology, 1037 Luoyu road, Wuhan, China 430074, *zirong@mail.hust.edu.cn

Abstract

Deep reactive ion etching (DRIE) process is a key growth for fabrication of micro-electromechanical system (MEMS) devices. Due to complexity of this process, including interaction of the process steps, full analytical modeling is complex. Plasma process holds deficiency of understanding because it is very easy to measure the results empirically. However, as device parameters shrink, this issue is more critical. In this paper, our process was modeled qualitatively based on "High Density Plasma Etch Model". Deep trench solutions of etch rate based on continuity equation were successfully generated first time through mathematical analysis. It was also proved that the product of fluorine and gas phase concentration in SF_6 remains identical during both deposition and etching stages. The etching process was treated as a combination of isotropic, directional and angle-dependent component parts. It exploited a synergistic balance of chemical as well as physical etching for promoting silicon trenches and high aspect ratio structures. Simulations were performed for comprehensive analysis of fluxes coming towards the surface during chemical reaction of gas. It is observed that near the surface, the distribution of the arrival flux follows a cosine distribution. Our model is feasible to analyze various parameters like gas delivery, reactor volume and temperature that help to assert large scale effects and to optimize equipment design.

Keywords: DRIE, Plasma modeling, Silicon deep trench, MEMS

1. INTRODUCTION

The dry etching of deep trenches and contact holes in single crystal silicon is valuable for an ample variety of applications in micro-electromechanical systems (MEMS). Wet chemical etching was dually replaced in the late 1970; by a dry etching using reactive plasma in semiconductor processing technology field [1]. Reactive ion etching (RIE) that realized a directional etches profile without an undercut for excellent line delineation was established [2]. In the last 20 years , to endorse the auxiliary shrinkage and precision etching, low-pressure and higher density plasma generation were developed such as magnetron RIE [3], electron cyclotron resonance (ECR) and inductive-coupled plasma (ICP) sources. The plasma etching is especially beneficial in microfluidics because the hydrophilicity of the etched surfaces is regulated with O_2 plasma exposure after etching [4-5]. The design of the majority of micro-electro-mechanical systems (MEMS) requires silicon-etch step at some stage in production sequence because MEMS/ NEMS devices are being created commencing silicon due to sticking coefficient *Sc* inherent properties like material strength, miniaturization, integration, low electric power and low cost [6]. Therefore, polysilicon is used as a mature manufacturing material and to develop high aspect ratio and etch rate. As plasma etching of deep structures is complex because the etch rate depends on feature size, etch time and aspect ratio so dry etching of silicon oxide is one of the key steps not only in silicon integrated circuit (IC) fabrication but also in other applications such as the fabrication of optical wave guides [7-8]. Standard RIE of Si, used in the IC industry, utilizes a synergistic balance of chemical and physical etching to produce silicon trenches and structures with aspect ratios on the order of 5:1 [9-10]. Currently chlorine and fluorine based gases are being extensively used in plasma etching. To improve the device performance and yield, silicon etching requires a high etch rate, a vertical profile, surface morphology and low etch induced smash up. As the

MEMS devices entail extremely elevated structures or extremely deep trenches, now often on the order of 35:1 for which standard RIE is incapable of meeting this requirement. Only DRIE is handy for deep trenches. Previously, SF_6 and the $(c\text{-}C_4F_8)$ plasma reactions by themselves have been modeled extensively [11-12].

In this paper, we selected SF_6 gas for etching because it releases six fluorine atoms with more than 16 eV energy. These fluorine atoms then spontaneously retort with the *Si* material, forming volatile SiF_4. In this switching process, SF_6 etches the silicon substrate in the etch cycle and the sidewall will be coated by C_4F_8 in the passivation cycle to prevent undercutting. Modeling of the DRIE is to be involved into the integration of two models (deposition and etch models) and include a mechanism to remove the polymer from the trench bottom during each cycle. Our followed process is modeled on the bases of "High Density Plasma Etch Model" as this model is capable of specifying different etch rates or depositions for different materials. This model assumes that the etching is composed of isotropical and anisotropical conditions. Comprehensive analysis of fluxes towards the surface during chemical reaction of gas is obtained by using simulation software tool named as "Taurus Topography", in which numerical methods are used to describe fully the fluxes at the wafer surface for particular equipment and topographical configurations.

2. PROCESS MODELING

The deposition and etch steps are performed separately on *Si* wafer surface to analyze the percentage of fluorine and sulphur contents during chemical reaction of SF_6 with surface. Two steps are considered at this time for derivation. In the deposition step, plasma splits apart the strained cyclic hydrocarbon $(c\text{-}C_4F_8)$ into extremely excited fragments. The individual fragments react with one another on the exposed surface and build up a more or less strongly cross-linked layer of polymer. Although many neutral and ionic species are generated, the highest fluxes of species at the surface during deposition are measured to be CF_2 and C_2F_4. The general chemical reactions are

$$c-C*_4F_8+e^- \xrightarrow{gas-phase} C*, F, C*F, C*F_2, C*F_3+\xi^{SP}$$
$$C*F_2, C*_2F_4+\xi^{SP} \xrightarrow{adsorban\ \&\ polymerize\ at\ surface} (C*F)_n\xi^{SP}$$

Where the deposited polymer $(C*F)_n$ is Teflon and is ξ^{SP} the sum of neutral and ionic species. Here "*" sign is denoted to keep the concentration point of view closely in mind. Sidewalls are passivated by this layer of polymer during deposition step and then both polymer and silicon are etched from the base of the structure during etching step. Accurate control of anisotropy is achieved by a fine stability between etching and passivation [13].The deposition of the polymer film is affected by both etching and plasma characteristics [14-15]. Deposited polymer protects sidewalls as the etch step attacks on *Si* surface. The main chemical reactions occur in the gas phases of the SF_6 are

$$\begin{aligned} SF_6+e^- &\xrightarrow{k_1} SF_3+3F+e^-(PhaseI) \\ SF_6+e^- &\xrightarrow{k_2} SF_2+4F+e^-(PhaseII) \end{aligned} \quad (1)$$

Where k_1 and k_2 are phase constants in two phases (deposition and etching). At the wafer surface, *Si* is etched by the following reaction

$$\begin{aligned} F_{gas} &\xrightarrow{k_3} F_{ads} \\ Si_{solid}+4F_{ads} &\longrightarrow (SiF_4)_{gas} \end{aligned} \quad (2)$$

Where k_3 is the surface constant after two segments. Ion fluxes at the *Si* wafer surface are relatively low as compared to the *F* flux; therefore, SiF_6 is producing an isotropic profile. By continuity equation at the surface we obtain an analytical appearance for the etch rate of equation (1) by differentiating with ∂t, as follows

$$\frac{\partial SF_6}{\partial t} = \frac{FC^*_{SF6}}{V} - \frac{SC^*_{SF6}}{V} - k_1 n C^*_{SF6} - k_2 n C^*_{SF6} = 0 \quad (3)$$

$$\frac{\partial FC^*}{\partial t} = \frac{SC^*_F}{V} + 3k_1 n C^*_{SF6} + 4k_2 n C^*_{SF6} = 0 \quad (4)$$

Plasma consist of several reactions, but it is renowned that the one producing SF_3 according to equation (1) accounts for about *2/3* and that producing SF_2 accounts for *1/3* of the released etching species of *F* [16]. During chemical reaction it is very vital to keep an eye on gas phase concentration of SF_6 and *F* that can be obtained from equation (3) as $\frac{\partial SF_6}{\partial t} = 0$, so

$$C^*_{SF6} = \frac{\frac{FC^*_{SF6}}{V}}{\frac{S}{V} + k_1 n + k_2 n} \text{ and } n = \frac{FC^*_{SF6} - SC^*_{SF6}}{VC^*_{SF6}(k_1 + k_2)} \quad (5)$$

Where *n* and *V* denote the number of deposition / etch species and total volume of SF_6 respectively. The energy supplied to the surface during etch step in the form of fluorine concentration removes the passivation from the base of the features thus exposing silicon for spontaneous etching. It can be calculated as

$$C^*_F = -\frac{V}{S}(3k_1 n + 4k_2 n).\frac{\frac{FC^*_{SF6}}{V}}{\frac{S}{V} + \left(\frac{FC^*_{SF6} - SC^*_{SF6}}{VC^*_{SF6}}\right)} \quad (6)$$

Product of fluorine and gas phase concentration in SF_6 remains identical is derived by putting value of *n* in equation (6)

$$C^{*}_{SF_6} = \frac{\frac{FC^{*}_{SF_6}}{V}}{\frac{S}{V} + \frac{FC^{*}_{SF_6} - SC^{*}_{SF_6}}{VC^{*}_{SF_6}(k_1 + k_2)}.(k_1 + k_2)} \quad (7)$$

$$C^{*}_{SF_6} FC^{*}_{SF_6} = C^{*}_{SF_6} FC^{*}_{SF_6} \quad (8)$$

$$PhaseI(deposition) = PhaseII(etching)$$

On the basis of silicon surface resistance coefficient "Ω" and gas phase concentration of fluorine, etch rate is $\frac{\Omega_{Si}}{4} k_3 C^{*}_{F}$

$$Etch\ Rate = -\frac{\Omega_{Si}}{4} k_3.\frac{V}{S}(3k_1 n + 4k_2 n).\frac{\frac{FC^{*}_{SF_6}}{V}}{\frac{S}{V} + \left(\frac{FC^{*}_{SF_6} - SC^{*}_{SF_6}}{VC^{*}_{SF_6}}\right)} \quad (9)$$

Putting value of "n" from Eq. (5),

$$Etch\ Rate = -\frac{\Omega_{Si}}{4} k_3.\frac{V}{S}\left[\left(\frac{FC^{*}_{SF_6} - SC^{*}_{SF_6}}{VC^{*}_{SF_6}(k_1 + k_2)}\right)(3k_1 + 4k_2)\right].\frac{FC^{*}_{SF_6}}{C^{*}_{SF_6}} \quad (10)$$

Where Ω_{Si}, k_3 , V/S, FC^*_{SF6}, SC^*_{SF6} and VC^*_{SF6} symbolize resistance of Silicon material, silicon surface constant, ratio of total gas volume to sulphur contents, fluorine concentration in SF_6, sulphur concentration in SF_6,Volume concentration of both F and S in SF_6 . The negative sign in equation (10) presents evolving species during reaction.

3. SURFACE FLUX DISTRIBUTION MODELS

Etching process modeling of the trench bottom compared to its sides is divided into two models such as reactor as well as cosine distribution model explained beneath.

3.1. Reactor Modeling

A plasma-enhanced reactor model, neglecting the gas boundary layer near the surface is shown in Figure 1. The physical processes of deposition and etch are a function of the material fluxes of reactants at the wafer's surface.

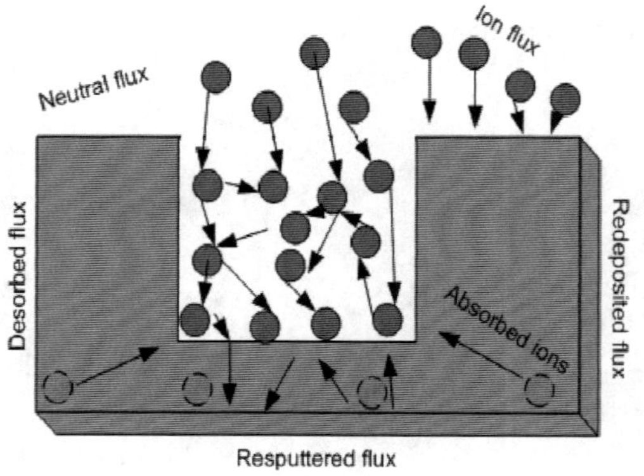

Fig. 1. Distribution of the arrival fluxes at the surface

Summation of all fluxes acting at the surface is

$$\gamma^{x^*}_{NET} = \gamma^{x^*}_{direct(neutrals)} + \gamma^{x^*}_{direct(ions)} - \gamma^{x^*}_{emitted} + \gamma^{x^*}_{redeposited} - \gamma^{x^*}_{sputtered} \quad (11)$$

Since plasma processing is performed at relatively low temperature, therefore, a diffusion flux expression is neglected here.

3.2. Cosine distribution model

The first two expressions in etch rate describe fluxes of species arriving directly from the reactor to the wafer surface. These can be calculated from the concentration and velocity of gas at the surface [9-10]. There are several models for describing the direct flux arrival with time [9]. A simple method with good results for most cases is the cosine distribution of flux model, shown in Figure 2.

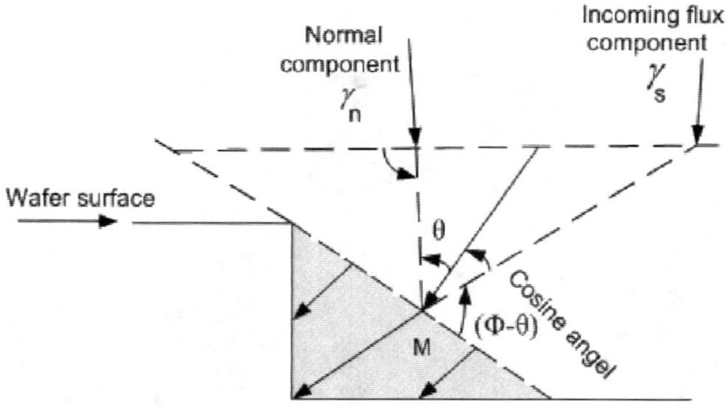

Fig. 2. Cosine distribution model

Surface modification rate at an arbitrary point *M* on the wafer surface depends on the normal component, or the cosine of the angle of incoming flux to the surface. Near the surface, the distribution of the arrival flux follows a cosine distribution. If γ_s is the flux toward the surface, and γ_n is the flux normal to the surface at point *M*, then

$$\gamma_s = \frac{\gamma_n}{2}(1+\cos\theta)*\begin{cases} \gamma_n, \theta = 0^o \\ \hline \frac{\gamma_n}{2}, \theta = 90^o \end{cases} \tag{12}$$

Therefore, point *M* seems to have maximum incoming flux at θ=0°. As θ increases, flux decreases until a minimum is observed at θ=90°. Actually in plasma systems, the incident flux becomes more directional than cosine distribution. So, the generalized flux expression by using $\cos^n(\theta)$ distribution becomes $\gamma_{direct}(\theta) = \gamma_n \cos^n(\theta)$ .The third expression in equation (11), the emitted flux expression, considers the fact that not all molecules stick where they arrive at the surface, and those which do not stick are re-emitted. The emitted flux expression is $\gamma_{emitted}^{x^*} = (1-Sc)*\gamma^{x^*}$, where *Sc* (modeled as a constant for a particular system) is the ratio of the number of incident atoms stick on the surface to the total number of incident atoms. Figure 3 illustrates the *Sc* concepts as,

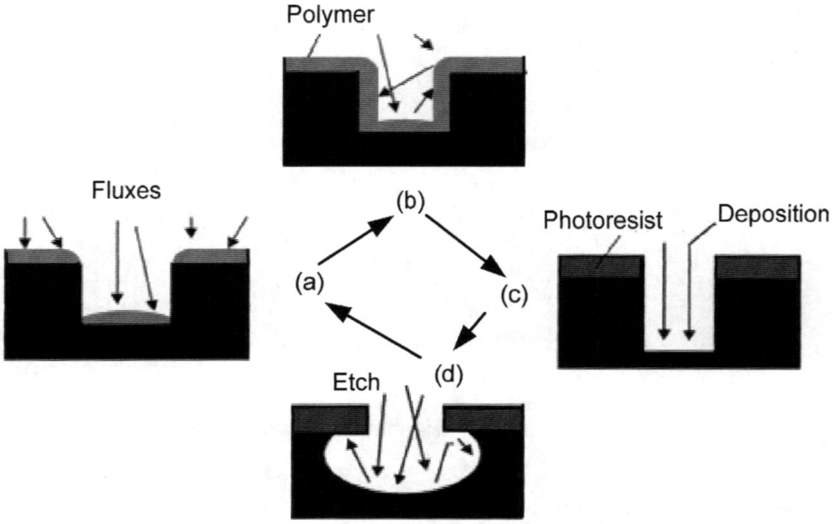

Fig. 3. Sticking coefficient effects on profile evolution (a) high Sc. (b) low Sc. (c) deposition. (d) etch.

Figure 3(a) presents that incident species with a high sticking coefficient react where they strike. In Figure 3(b), species with a low sticking coefficient don't "stick" where they first strike but re-emitted to re-deposit elsewhere. Emission/re-deposition occurs multiple times. Deposition and Emission are shown in Figure 3(c) And 3(d). Generally, ions stick at $Sc = 1$ and chemically reactive neutrals have $Sc < 1$. The fourth expression in Eq. (11) is the re-deposition flux, describes that species which do not stick and re-emit lands elsewhere on the surface. This is described with emitted flux expression $(1 - Sc)\gamma^{x^*}_{emitted}$ as

$$\gamma^{x^*}_{redeposited} = g^{x^*x'} * \gamma^{x^*}_{emitted} = g^{x^*x'} * (1 - Sc)\gamma^{x^*}_{emitted} \qquad (13)$$

x'

Where, the term $g^{x^*x'}$ represents the blocking effects of local topography, which limit the probability that the emitted flux from any point x' is able to be re-deposited at any point x^* on the surface. The redeposit flux also dependants on the sticking coefficient. The re-emitted flux is generally assumed to be re-emitted with a $\cos^n(\theta)$ distribution angle similar to that derived for direct fluxes, with no memory of the arrival angle at that point. The final term in equation (11) is the sputtered flux (energetic incoming ions). Y_s is the sputtering yield that is angle sensitive. The sputtered molecules re-deposited elsewhere, and enhance the deposition rate by supplying energy that modifies the surface or drives the chemical reaction to overcome a rate-limiting step given as

$$\gamma^{x^*}_{sputtered} = grind - rate * Y_s * \gamma^{x^*}_{ions} \tag{14}$$

4. SIMULATION RESULTS AND DISCUSSION

Comprehensive solutions of etch rate are obtained by using "Taurus Topography" software. In our simulations, the etch rate for photoresist is set much smaller than that for either polysilicon or polymer. The machine changes the polymer deposition at the same time as it changes polysilicon etching in any desired combination. The basic structure is presented by ideal depositions of polysilicon and photoresist on top of a silicon wafer, as shown in Figure 4(a). A photoresist coating of 1.5 μm over a small 3.5 μm polysilicon layer is used. We assume that each cycle will etch only about 0.6 μm to create a trench with a 35:1 aspect ratio. The photoresist layer is etched in the center to create opening in the mask of exactly 1.5 μm.

The first etch perspective used is the Dry Etch Model. Etching of photoresist, etching of polysilicon, and deposition of polymer are shown in Table l. It is observed that only polymer is re-deposited during etching and so needed a defined sticking coefficient and re-emission rate. This simulation shows anisotropic etch profile, as can be seen in Figure 4(b).The parameters are adjusted to produce an isotropic profile. The first two sputtering yield parameters, Sput.Y_1 and Sput.Y_2, are kept same throughout simulations. Reducing the sputter

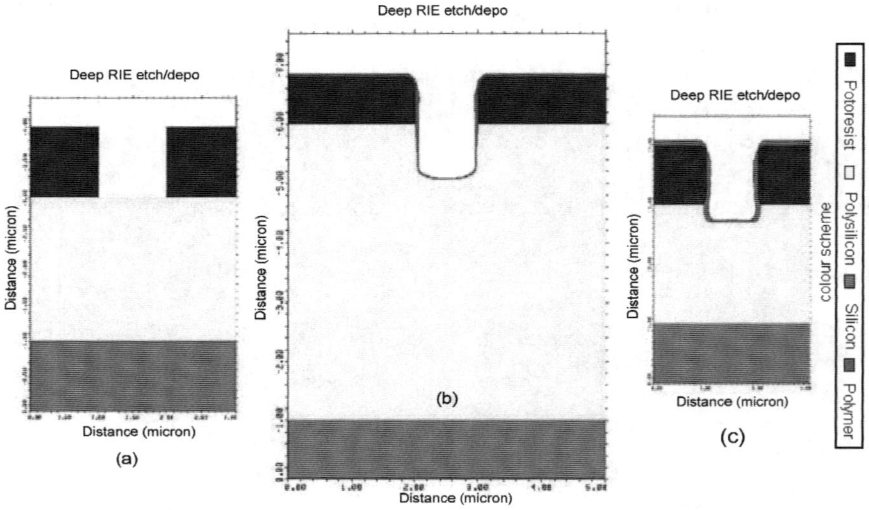

Fig. 4. (a) Basic simulation structure. (b) Dry etch model with polymer deposition. (c) Initial etch plus first deposition.

yield widens the view angle, reduces the overall etching rate and increases the deposition rate. Therefore, reducing the Sput.Y_4, sputtering yield for the polysilicon produces an anisotropic profile. Unfortunately, by increasing it does not produce any perceptible difference to the etch profile, no altering the view angle. Reducing the view angle for polymer deposition, however, gives result in more sidewall deposition. Reducing the etch rate of polysilicon results in a extra rounded bottom to the trench profile. Adding a lower rate for the polysilicon with a lesser sputtering term for the polymer greatly increases the overall etch rate.

Table I

Initial dry etch model and second etch cycle machine parameters

Initial Dry Etch Model Parameters				Second Etch Cycle Machine Parameters (Same etching properties)		
				Polymer Etch Machine		Polysilicon Etch Machine
Material	Photoresist	Polysilicon	Polymer	Polymer	Polysilicon	Polysilicon
Rate	0.16	1.2	0.1	2.0	0.1	1.0
Sput. Y_1	5.6	5.6	5.6	-1.0	-1.0	-1.0
Sput. Y_2	-6.1	-6.1	-6.1	1	1	1
Sput. Y_4	2.4	2.4	5.4	5.4	5.4	5.4
Sc	N/A	N/A	0.5	-6.0	-6.0	-6.0
Ra	N/A	N/A	0.1	1.4	2.4	1.4
Time(min)	1.1	1.1	0.15	0.16		0.15

Next step is to define polymer deposition perspective. In this model, the polymer is only defined for the deposition machine, not for the etch machine. The original parameters are given in Table ll. Figure 4(c) show that the initial deposition process results in more polymer coating on the sidewalls than on the horizontal surfaces. This is the result of a balance between the deposition and etching parameters chosen for the deposition machine. The etching parameters are the grind-rate and the sputtering yield coefficients, Y_1, Y_2, and Y_4. The deposition parameters are characterized by the flat, un-shadowed, surface rate, the re-deposition rate, the sticking coefficient and the anisotropy. Increasing the grind-rate and decreasing the re-deposition rate decreases the overall deposition on the horizontal surfaces. Initially, the re-deposition rate is set close to zero and the grind-rate close to one. This results that almost no polymer is deposited on the trench floor. Some deposition on the horizontal surfaces is needed for the next etch step and so the grind-rate is reduced and the re-deposition rate is increased until an unbroken layer of polymer covered the trench floor.

Table II

Initial HDP etch and deposition with second and third etch M/c parameters

Initial HDP Etch and Deposition M/c Parameters				Second Etch M/c Parameters		Third Etch M/c Parameters
	(left)		(right)			
Material	Photoresist	Polysilicon	Polymer	Polysilicon	Polymer	Polysilicon
Rate	0.1	1.15	1.0	1.2	2.0	1.0
Anisotropy	N/A	0.04	0.3	1.1	1.1	-1.1
Exponent	N/A	5		-1.0	-1.0	1
Sput. Y_1	5.6	5.6	5.6	5.6	5.6	5.6
Sput. Y_2	-6.1	-6.1	-6.1	-6.1	-6.1	-6.1
Sput. Y_4	1.1	1.4	1.4	2.4	1.4	2.4
Time (min)	0.4		0.3	0.2		0.5

The polymer on the sidewalls needs to stay, while the polysilicon below needs to be etched isotropically. This is accomplished by separating this step into two separate etch machines and etch times. The first etch machine etches the polymer more than the polysilicon, while the second machine etches only the polysilicon. The first machine removes the polymer from the entire floor of the trench, and leaves a good coating on the trench sidewalls. A set of time-lapse snapshots of this etch step are shown in Figure 5. Table ll outlines the parameters chosen for this etch machine. It is noted that the anisotropy is set to a negative value. This indicates that the polymer etches more laterally than vertically called a "super-isotropic" profile. This is necessary in order to remove the polymer from the bottom corners of the trench. Figure 5 presents that the polymer at the bottom of the trench is etched first. In reality, by 0.03 minutes, the polymer is almost completely gone. The corners of the trench, however, do not show appreciable etching until 0.09 minutes and the trench floor is not completely clear until 1.2 minutes.

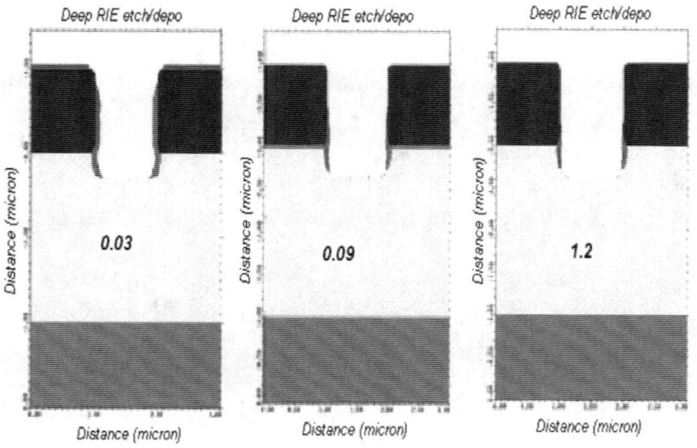

Fig. 5. Second Etch Profile Evolution over Time (0.03 to 1.2. minutes)

The second etch perspective is designed to perform a super-isotropic etch of polysilicon below the left-over polymer. The parameters chosen are shown in Table ll. The anisotropy is set to 1.1 to indicate that the polysilicon is etched more laterally than vertically. The second deposition machine parameters remained the same. The resulting profile is shown in Figure 6.

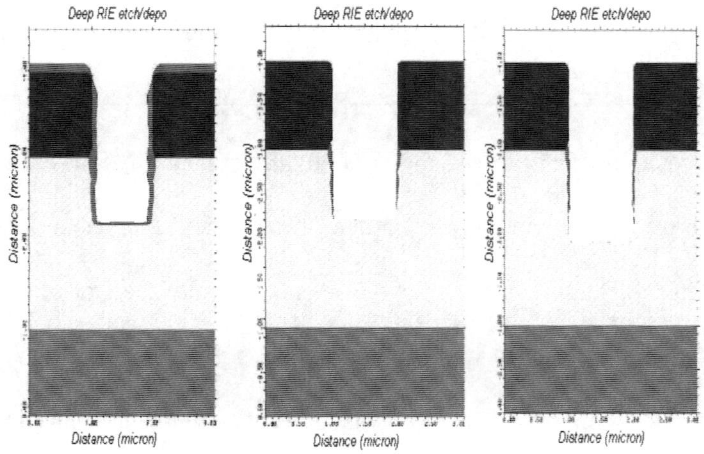

(a) (b) (c)

Fig. 6. Second Deposition/Etch Cycle (a) deposition. (b) Polymer Etch. (c) Polysilicon Etch.

The etch machine parameters are given in Table l. Two parameters are changed from the first deposition/etch cycle, the polysilicon etch rate and the polymer sputter yield. Reducing the polysilicon etch rate allows a corresponding increase in time and thus more control over the etching profile. We need the polysilicon sidewalls below the left-over polymer to bow out to allow room for the next polymer deposition. Leaving the polysilicon etch rate high for this etch machine does not give this profile. This machine is used primarily to etch through the polymer at the bottom of the trench so that the next machine could reach the polysilicon just beneath the polymer.

5. CONCLUSIONS

In summary, important chemical reactions are derived mathematically in which it is proved that the product of fluorine and gas phase concentration in SF_6 remains same during deposition and etching phases. First time, a simplified flux cosine distribution reaction rate model based on continuity equation is established. In this model, it is observed that silicon etch step requires a highly isotropic component and the polymer deposition process needs a simultaneous sputtering component. This is composed of equal parts of physical etch and sputtering re-deposition. We observed that a highly directional physical etch is also required to break through the thin polymer at the bottom of the trench in order to bare silicon for the fluorinated chemical etch. A fascinating correlation is realized between deposition and etching species presents that deposition species with a high *Sc*, generate less conformal films and vice versa while etching species with a high *Sc* generate more anisotropic etch and with a low Sc result in a more isotropic etch. It is also analyzed that ions are sticky when $Sc = 1$ and chemically reactive neutrals are originated at $Sc < 1$. Comprehensive clarification of etch rate is attained by using simulation software tool Taurus Topography. It predicts the feasibility of process changes, in which numerical methods are used to describe the fluxes at the wafer surface for particular equipment and topographical configurations.

Acknowledgements

The authors would like to thank Wuhan National Laboratory for Optoelectronics (WNLO) at HUST for preparation of samples. This work was partially supported by National Natural Science Foundation of China (Grant No.50875103 and 90923019).

References and notes

[1] H. Abe, Y. Sonobe, and T. Enomoto, Jpn. J. Appl. Phys. 12, 287 (1973).

[2] S. Franssila, John Wiley & Sons Ltd. England, 125 (2004).

[3] The MEMS Handbook, CRC Press, (2006).

[4] K. Kolari, J. Vac. Sci. Technol. A 24. 4,1005 -1011 (2006).

[5] L.-M. Lacroix, Surface. Sci. 592, 182-188 (2005).

[6] M. J. Madou, The Science of Miniaturization, 2nd ed., CRC Press, Boca Raton. (2001).

[7] I. Sakai, K. Tomioka, T. Ohiwa, M. Sekine, T. Mimura, K. Nageseki, Int. Symp. Dry Process, Tokoyo. 57 (2001).

[8] N. Takato, A. Sugita, K. Orose, H. Okazaki, M. Okuno, M. Kawachi, K. Oda, IEEE Photon. Technol. Lett. 2, 441 (1990).

[9] J.D. Plummer, M.D. Deal, P.B. Griffin, Prentice Hall Electronics and VLSI Series, Prentice Hall, Inc., New Jersey (2000).

[10] S. Wolf, R.N. Tauber, Lattice Press, Sunset Beach, CA (1986).

[11] S. Rauf, P. Ventzek, J. Vacuum Science Technology A 20 (1), American Vacuum Society. (2002).

[12] Zhang, D. Rauf, S. Sparks, Tl. Ventzek, J. Vacuum Science Technology B 21(2), American Vacuum Society. (2003).

[13] K. Bhardwaj, H. Ashraf, Proc. SPIE 224. 2639 (1995).

[14] M. Schaepkens, R.C.M. Bosch, T.E.F.M. Standert, G.S. Oehrlein, J.M. Cook, J. Vac. Sci. Technol. A. 16 , 2099 (1998).

[15] Y. Hikosaka, H. Sugai, Jpn. J. Appl. Phys. 32, 3040 (1993).

[16] S. Rauf, W. Dauksher, S. Clemens, K., Smith, J. Vacuum Science Technology A 20(4), American Vacuum Society. 12 (2002).

Study on the structure of bridge surface of the micro Fabry-Perot cavity tunable filter

Qinghua Meng[1], Huan Luo[2,3], Shiwei Bao[2,3], Yifan Zhou[2,3], Sihai Chen[2,3]
[1]College of Physics and Electronic Engineering, Guangxi Teachers Education University, Nanning, 530023, China
[2]College of Optoelectronic Science and Engineering, Huazhong University of Science and Technology, Wuhan, 430074, China
[3]Wuhan National Laboratory for Optoelectronics, Wuhan, 430074, China

Abstract. Micro Fabry-Perot cavity tunable filters are widely applied in the area of Pushbroom Hyperspectral imaging, DWDM optical communication system and self-adaptive optics. With small volume, lower consumption and cost, the Micro Fabry–Perot cavity tunable filter can realize superior response speed, large spectral range, high definition and high reliability. By deposition metal membrane on silicon chip by MEMS technology, the micro Fabry–Perot cavity has been achieved, which is actuated by electrostatic force and can realize the function of an optical filter. In this paper, the micro-bridge structure of the micro Fabry–Perot cavity tunable filter has been studied. Finite element analysis software COMSOL Multiphysics has been adopted to design the structure of the micro-bridge of the micro filter. In order to simulate the working mechanism of the micro Fabry–Perot cavity and study the electrical and mechanical characteristics of the micro tunable filter, the static and dynamic characteriastics are analyzed, such as stress, displacement, transient response, etc. The corresponding parameters of the structure are considered as well by optimizition the filter's sustain structure.

Keywords: MEMS, tunable filter arrays, FEA analysis, structure optimize

1. Introduction

In the field of satellite imaging with visible light, Pushbroom Hyperspectral Imager (PHI) is a new intensively researched subject in the spectral imaging field. It obtains hundreds of channels from the nearby wavelength and continuously sampling narrow bands [1]. It has wide application in commercial and military field.

Dispersive component is one of the important parts of Pushbroom Hyperspectral Imager. The traditional imaging spectrometers use prisms or gratings as dispersive components which are bulky and suffer low resolution. While the Fabry-Perot (F-P) cavity has high resolution and excellent frequency selectivity [2-4]. In this paper, MEMS F-P array filter is used as a dispersive component to replace the traditional one. It is compact, fast modulated, precise and easy to be integrated with other elements, such as photo-detectors.

1 mqhgx @163.com; phone: +86-13878868313; fax: +86-27-87792364

The authors have studied the structure of tunable micro F-P cavity filter integrated with the infrared Focal Plane. The development of the key MEMS device will make the Hyperspectral imager devices light and compact. In this paper, the structure of microbridge of the integrated tunable micro F-P cavity filter have been studied, and some proposals for further development are provided.

2. Design of the micro F-P filter

The hyperspectral imager chooses the wavelength by changing the F-P cavity length. As shown in Figure.1, based on multi-beam interference, if $\lambda = 2l/m$, $m = 1.2.3$ (λ: incidence wavelength, l: cavity length), the light (wavelength λ) can transmit at a vertical incidence angle [5]. When the incident light is a broadband source, F-P cavity can be used for frequency selecting.

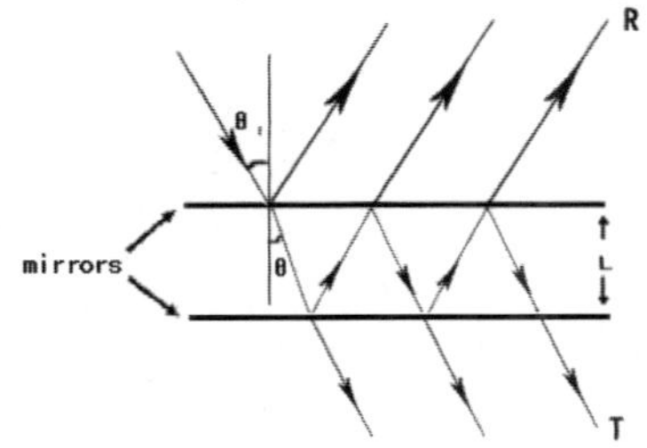

Figure 1. The principle of F-P cavity

The F–P cavity of MEMS hyperspectral imager is constituted by two dielectric mirrors, the top and the bottom mirrors, which are formed by alternately deposition high and low refractive index dielectric layers. The F–P cavity length can be tuned by applying electrostatic force to achieves the function of wavelength selectivity.

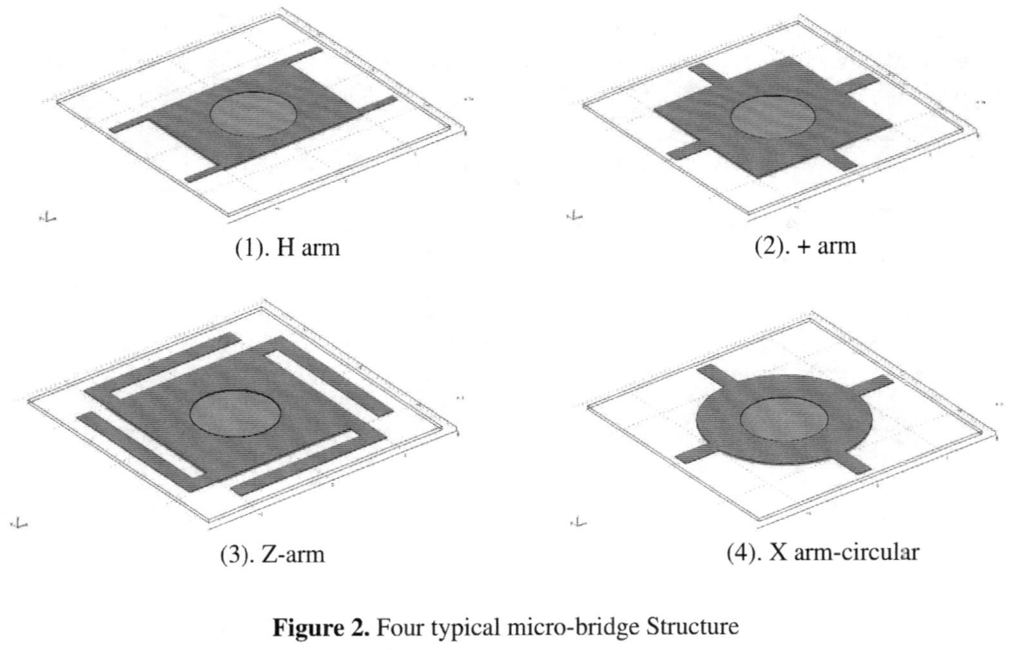

Figure 2. Four typical micro-bridge Structure

The four patters of F-P filters' bridge structure are shown in Fig. 2. They are H arm-square membrane, Cross arm-square membrane, Z-arm-square membrane and X arm-circular membrane structures, separately. The electrostatic force per unit area on the micro-bridge f_e is given by the formula

$$f_e = \frac{F}{A} = \frac{\varepsilon_0 U^2}{2(l-\Delta l)^2} \qquad (1)$$

where U is the voltage applied, L is the initial cavity length, L is the difference between the initial length and thickness of the air gap after the movement of top mirror, ε_0 is the vacuum dielectric constant.

3. Modeling and simulation analysis

The geometric parameter of the structures in Figure.2 is shown in the Table 1.

Table 1. Four typical micro-bridge structure, the set of their geometric parameter

type	Cantilever			Microbridge surface	Diameter of circular area for Microbridge surface
	Length	width	thickness		
(1).	50μm	10μm	1μm	200μm×150μm	100μm
(2).	50μm	20μm	1μm	200μm×200μm	100μm
(3).	(20+200) μm	20μm	1μm	200μm×200μm	100μm
(4).	50μm	20μm	1μm	(Outside diameter) 100μm	50μm

The circular areas in the center of the four structures are reserved for the mirror. The size of these areas are decided by the size of the microbridge. In the simulation, the material was chosen to be aluminum. The relationship between the displacement and voltage is shown in Table 2 and Figure 3.

Table 2. Four typical structures' simulation results

Type	Applied voltage/V	Maximum displacement /μm	Maximum stress/MPa
(1)	4.5	0.5063	9.22
(2)	8.5	0.4757	17.13
(3)	1.9	0.4802	4.615
(4)	9	0.5218	16.31

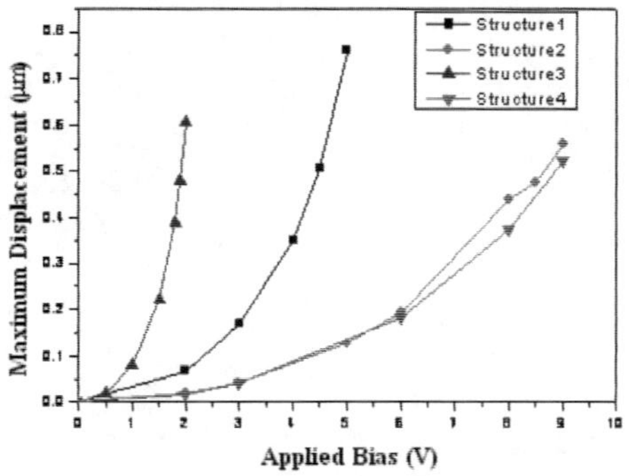

Figure 3. Relationship between the maximum displacement and the applied voltage

From Figure 3, we can make a conclusion that the Z-arm structure has lowest actuation voltage at same displacement of the mirror. It is important to keep two parallel mirrors flat which affects the micro-fine constants and the F-P cavity wavelength sensitivity. Mirror parallelism analysis is shown in Figure 4.

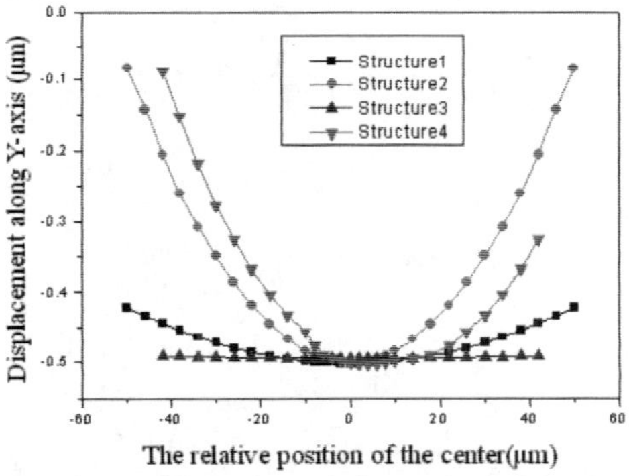

Figure 4. The comparison of the mirror deformation of the four typical structure

The simulation result in Fig.4 shows that Z-arm structure has the best parallelling performance. This structure with folded cantilever can balance the stress well, and the leg of the microbridge has a good flexibility. The above simulation shows that the Z-arm structure has advantages on actuation voltages and parallelism over the others.

4. The detailed analysis and simulation of Z-arm structure

There different Z-arm structures are shown in Figure 5. The parameters are listed in the table 3. In our simulation, the material of microbridge is Ni, the mirror is supposed to be made of SiO_2, the cavity space is filled up with air. Voltage is applied to the surface of micro-bridge, with the bottom electrode grounded.

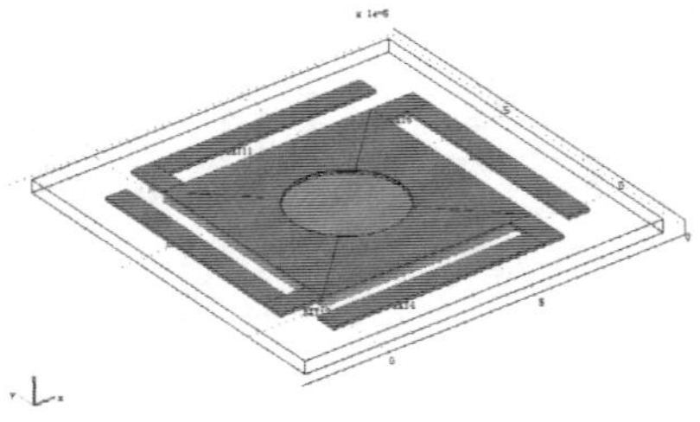

(a)

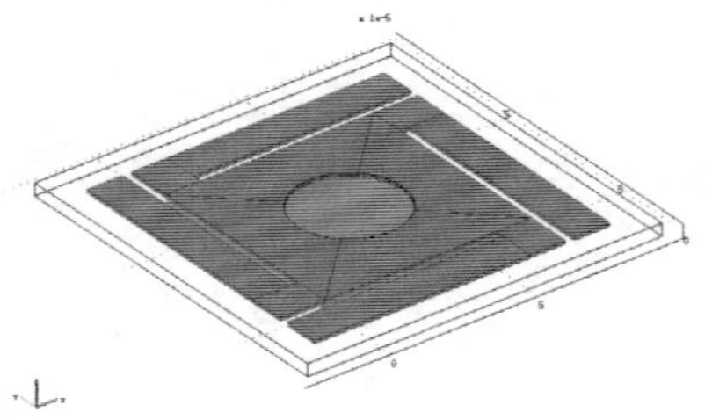

(b)

(c)

Figure. 5. Three geometrical structures with Z-arm

Table.3. The parameters of three Z-arm structures

Type	Cantilever			Bottom electrode	Diameter of circular area for top mirror	initial cavity length
	Length	width	thickness			
(a)	(7+70) μm	7μm	1μm	70μm×70μm	35μm	2μm
(b).Arm widenin g	(3+73) μm	12μm	1μm	70μm×70μm	35μm	2μm
(c).Arm extended	(19+63) μm	7μm	1μm	70μm×70μm	35μm	2μm

4.1. The relationship of Voltage, displacement, stress and deformation angle is shown in the table 4..

Table. 4. Comparison of the structure of the simulation parameters a.b.c

type	Applied voltage/V	Maximum displacement /μm	Maximum stress/MPa	Mirror deformation angle /°
(a)	29	542.9	57.53	0.00235
(b)	35.5	626.2	92.22	0.00504
(c)	29.1	645.9	51.93	0.0416

4.2. . Deformation of the three structures, displacement distribution are shown in theFigure. 6.

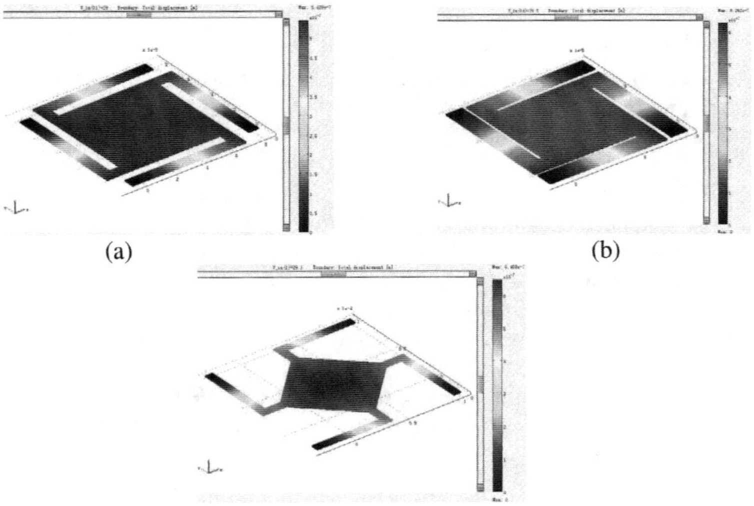

(a) (b)

(c)

Figure 6. a\b\c Displacement distribution of three structure

4.3. Stress distribution is shown in the Figure 7

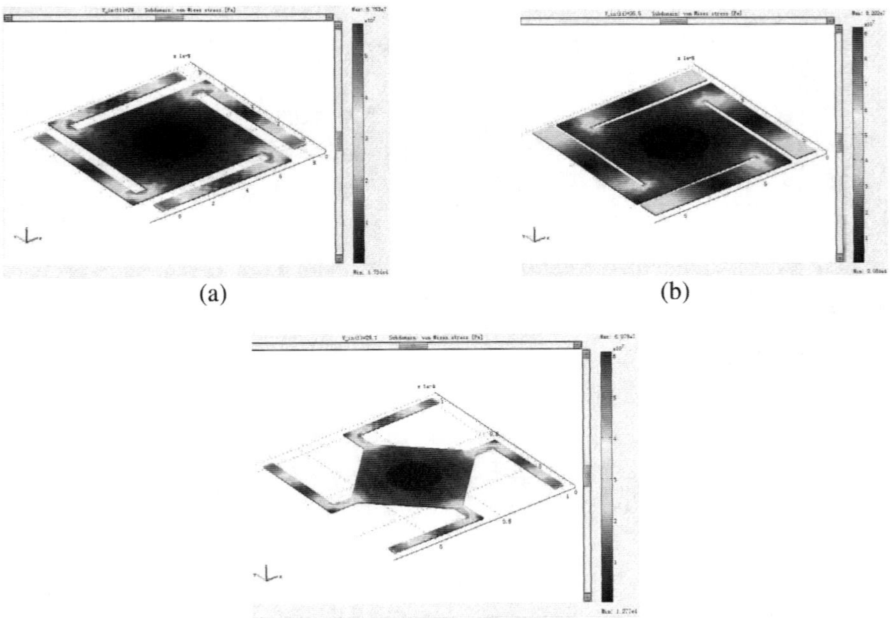

(a) (b)

(c)

Figure 7. Stress distribution on three structures

Analysis of structure (a):When the driving voltage is 29V, we get a maximum travel distance of 542.9nm, which meets our initial criteria. The distribution of displacement at the maximum driving voltage is shown in Fig.6(a). We can see from the figure that the larger deformation mainly lies in the cantilever. When the driving voltage V_in is 29V, the structure reaches the maximum moving distance, the stress at each point is shown in Fig7(a). We find out that the mirror part suffers the minimum force while the force on the sustain part and the middle part of the cantilever is medium, at cantilever's fixed end, and the folding site of the folding beam suffers a relatively large force.The maximum stress is up to 57.53Mpa, less than metal Ni's allowable stress.So theoretically, the phenomenon of bridge broken leg won't happen.

Analysis of structure (b): Increasing the width of cantilever can make it withstand greater stress while at the same time reduce the flexibility of the bridge legs. At maximum driving voltage, the displacement distribution is shown in Figure 6 (b).cantilever. When V_in = 35.5V, the structure reaches the maximum moving distance, the stress suffered on each point is shown in Figure 7 (b). Compared to structure (a), the maximum stress is relatively bigger. However,the remaining part's stress, especially stress on the cantilever are much smaller than structure(a).

Analysis of structure(c): Increasing the length of cantilever bridge legs can optimize the flexibility of eht cantilever. Figure 6 (c) shows the maximum displacement distribution. When reaching the maximum displacement, the bridge withstands a maximum stress of 51.93Mpa, within the allowance of Ni. The stress distribution on the microbridge is shown in Figure 7 (c).

5. Materials Influenes

The relationship of voltage, displacement, stress, deformation angle of the same mechanical structure with 6 different materials is shown in table 5. Relationship between the maximum displacement and applied voltage of these 6 models is shown in the Figure.8.

Table. 5. The simulation results comparing the six materials

	Applied voltage /V	Maximum displacement /μm	Maximum stress/MPa	Mirror deformation angle /°
Al	21	577.4	21.23	0.0057
Au	22.5	625.6	24.78	0.0087
Ni	37	645.4	74.28	0.0077
Cr	41	616.6	90.83	0.0081
Ti	16	593	12.49	0.0046
Pt	28.5	662.8	58.79	0.0115

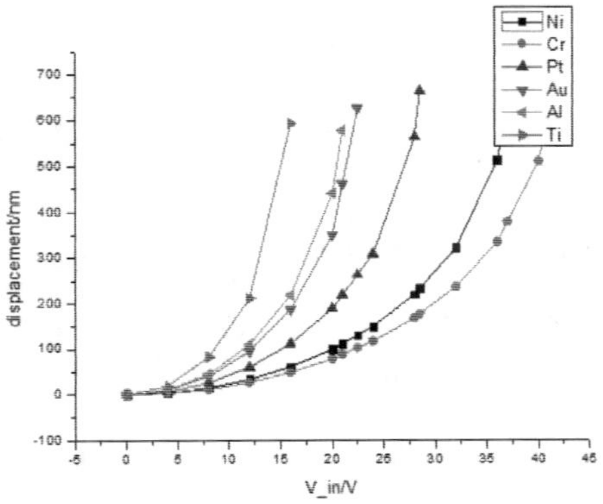

Figure 8. The voltage-displacement curves of six contrast material

Through the comparation of the voltage-displacement curves we find that Ti is the most sensitive, followed by Al and Au. Datum in Table 5 shows that, Pt has the greatest maximum displacement, followed by Ni, and then Au, the displacement of Ti is relatively small. When displacement occurs, bridges with the materials of Ti, Al, Ni can keep mirror parallelism better. As to bridge stress, Al, Au, and Cr structures are relatively small. In general, Ti and Al have the best performance in many aspects, and have achieved all the parameters required.

6. Conclusion

In this paper, different micro bridges structures of the F-P tunable filter is analyzed and simulated. The Z-arm structure has advantages on actuation voltage and parallelism over the others. We also make further analysis on the relationship of voltage, displacement, stress, deformation angle of the same mechanical structure with 6 different materials. The simulation results show that Ti is the most sensitive, Pt has the greatest maximum displacement and bridges with the materials of Ti, Al, Ni can keep mirror parallelism better. The simulation result in this paper will guide the device fabrication process in the near future.

Acknowledgments

This work is supported by the Program New Century Excellent Talents in University in China (No.NCET-07-0319), and also by the Program for Teacher Early basic research fund of Guangxi Teachers Education University(2010-3).

References

[1] Terry A Wilson, Steven K Rogers, Matthew Kabrisky. Perceptual-Based Image Fusion for Hyperspectral Data. IEEE Trans. on Geoscience and Remote Sensing, 1997, 35(4):1007~1017
[2] Thomae S. Pagano et . al. Moderate Resolution Imaging Spectroradiometer (MODIS) . SPIE, 1994, 1939:1~17
[3] Blechinger F et. al.. Optical Concepts for High Resolution Imaging Spectrometers, SPIE, 1995, 2408:165~179
[4] Hinnrichs M et. al.. Imaging Multispectral Sensing: a New and Innovative Instrument for Hyperspectral Imaging Using Dispersive Technology. SPIE, 1995, 2480:93~104
[5] Qijun Yao. Optical Guide[M]. Beijing: Higher Education Press, 1989. 64-71
[6] S.D. Senturia, Microsystem Design, 4th ed.(Kluwer, Norwell MA,2002)
[7] C. Prott, F. Romer, E. Ataro, J. Daleiden, S. Irmer, A. Tarraf and H. Hillmer,"Modeling of Ultrawidely Tunable Vertical Cavity Air-gap Filters and VCSELs", IEEE J. of. Quantum Electronics, 9(3) 918-928 (2003)

A new design of photonic crystal fiber with ultra-flattened dispersion to simultaneously minimize the dispersion and confinement loss

Saeed Olyaee and Fahimeh Taghipour

Nano-photonics and Optoelectronics Research Laboratory (NORLab), Faculty of Electrical and Computer Engineering, Shahid Rajaee Teacher Training University (SRTTU), Lavizan, 16788, Tehran, Iran

E-mail: s_olyaee@srttu.edu

Abstract. Photonic crystal fibers (PCFs) are highly suitable transmission media for wavelength-division-multiplexing (WDM) systems, in which low and ultra-flattened dispersion of PCFs is extremely desirable. It is also required to concurrently achieve both a low confinement loss as well as a large effective area in a wide range of wavelengths. Relatively low dispersion with negligible variation has become feasible in the wavelength range of 1.1 to 1.8μm through the proposed design in this paper. According to a new structure of PCF presented in this study, the dispersion slope is 6.8×10^{-4}ps/km.nm^2 and the confinement loss reaches below 10^{-6} dB/km in this range, while at the same time an effective area of more than 50μm^2 has been attained. For the analysis of this PCF, finite-difference time-domain (FDTD) method with the perfectly matched layers (PML) boundary conditions has been used.

1. Introduction

Nowadays, in WDM networks, the single mode fibers are widely used as optical data transmission media. Optical fibers which can transmit the data in the form of short optical pulses over long distances have revolutionized telecommunication industry in the last two decades. In WDM communication systems, it is essential to maintain a uniform response in the different wavelength channels, which requires that the transmission line approach the ideal state of ultra-flattened dispersion and ultra-low loss [1]. In the recent years, photonic crystal fibers (PCFs) have attracted much interest among researchers.

Photonic crystal fiber (PCF) is formed from a strand of silica glass with an array of microscopic air channels running along its length [2]. A defect in the center of this array is deliberately created to form the core of the fiber. In this paper, an index guiding PCF is introduced. In the index guiding PCF, the refractive index of the core is higher than that of the core's surrounding area; consequently the transmission in this type of PCFs is mainly resulting from total internal reflection (TIR).

Reducing dispersion and confinement loss are of the main concerns in designing PCFs. In conventional PCFs with uniform holes with equal diameters, proper control of dispersion and particularly its slope in a wide range of wavelengths is difficult to maintain. Today, on the other hand, exploiting multiple design parameters such as diameter and shape of the holes, the number of air hole rings and the spacing between these holes facilitates development of PCFs with improved properties.

Several designs for the PCFs have been proposed to achieve the nearly zero ultra-flattened chromatic dispersion properties and low confinement loss. Hansen proposed a hybrid-core photonic crystal fiber with three-fold symmetry, in which dispersion and its slope is well restrained, however, the confinement loss was shown to be about 10^{-3} [3]. In recent years, better results have been observed [3-11] as a result of different approaches adopted, e.g. modifying the form of the hole into ellipse [8], Gradually increasing the diameter of the hole from inner ring to the outer [4], selectively filling the PCF with liquids [6], application of double cladding [10], etc.

Most PCFs have hexagonal lattice structures. In reference [12], Razzak *et al.* introduced an octagonal structure for PCF with a better controllability of dispersion and confinement loss, yet a large effective area has not been realized in this pattern. Matsui *et al.* suggested a dual-core structure and numerically shown that the proposed PCF can achieve an ultralow dispersion variation of less than 0.8ps/nm·km in 1.3-1.65μm, with both a large effective area and a low confinement loss less than 0.01 dB/km [13].

In the present paper, a novel design for dispersion flattened PCF (DF-PCF) is put forth, by which both a very low confinement loss and a large effective area are acquired. Combining perfectly matched layer (PML) for the boundary treatment, an efficient compact two-dimensional finite-difference time-domain (2-D FDTD) method [14] is developed to model this photonic crystal fiber.

2. Theoretical discussion

In this paper, the three main properties of PCF, namely confinement loss, dispersion, and effective area are studied. Thus a brief introduction of these concepts will be presented in this section. Effective mode index of a guided mode for a given wavelength is obtained by solving an eigenvalue problem drawn from the Maxwell equations using the FDTD. Effective mode index, n_{eff}, can be obtained as:

$$n_{eff} = \frac{\beta}{k_0} \tag{1}$$

Here, β is the propagation constant and $k_0 = 2\pi/\lambda$ is the free apace wave number. The effective mode index has both real and imaginary parts.

The total dispersion is sum of the material dispersion and the waveguide dispersion. The chromatic dispersion, $D(\lambda)$ of a PCF is easily calculated from the second derivative of the real part of the effective mode index:

$$D(\lambda) = -\frac{\lambda}{C}\frac{d^2\,\mathrm{Re}[n_{eff}]}{d\lambda^2} \tag{2}$$

where, $\mathrm{Re}[n_{eff}]$ is the real part of n_{eff}, λ is the wavelength in units of μm, and C is the velocity of light in vacuum. The material dispersion given by Sellmeier formula is directly included in the calculation. Therefore, D in Eq. (2) corresponds to the total dispersion of the PCF [15].

The confinement losses, L_c, are due to the finite number of air-holes which can be made in the fiber cross section. As a consequence, all the PCF guided modes are leaky. In the solid-core PCF for small values of d/Λ, the resulting loss can be large unless a sufficiently large number of periods is used [16]. The confinement loss can be calculated from the imaginary part of the mode index as:

$$L_c = 8.686 k_0\,\mathrm{Im}[n_{eff}] \tag{3}$$

where, $\mathrm{Im}[n_{eff}]$ is the imaginary part of the effective mode index.

The effective mode area is another important parameter for fiber design. The effective area of fiber A_{eff} is calculated as:

$$A_{eff} = \frac{\left(\iint |E|^2\,dxdy\right)^2}{\iint |E|^4\,dxdy} \tag{4}$$

in μm^2, where E is the electric field distribution derived by solving an eigenvalue problem drawn from Maxwell's equations. By changing the geometric characteristics of the fiber cross-section, it is possible to design PCFs with completely different properties, which is with large effective area.

It is possible to significantly increase the effective area by narrowing the air-holes for a fixed Λ, or by enlarging the pitch for a fixed d/Λ value. Large mode area PCF can suppress undesirable nonlinear impairments [13]. On The other hand, low confinement losses and/or small effective mode areas are needed for some particular applications like nonlinear ones.

3. Design principles and results

The preliminary design of the proposed PCF is illustrated in Figure 1(a). This PCF comprises 6 air hole rings, embedded in pure silica with a refractive index of 1.45. The diameter of the holes in the inner two rings is chosen to be 0.6μm and that of the four outer rings equals 1.8μm. The lattice structure of the cladding in this PCF is hexagonal and the spacing between the centers of adjacent holes, Λ, is 2.3μm.

In index guiding PCF, the holes closer to the core have a stronger impact on dispersion, though their effect on confinement loss is quite negligible. A lower ratio of d/Λ in the cladding reduces the dispersion and its slope, therefore in this design the holes of the inner rings are chosen to be smaller. On the other hand, increasing the d/Λ results in reduction of the confinement loss, hence the choice of larger diameter for the holes in outer rings. By omitting the holes in the six corners of cladding in the hexagon structure, PCF will have even less dispersion.

Choice of unequal diameters for air holes provides better control of dispersion curve and also the confinement loss can be brought within the acceptable range with a less number of rings. Figure 1(b) shows the mode field distribution. As depicted in this figure, the field is confined well to the area proximately surrounding the center of the fiber. An effective area of 26.838μm^2 is achieved in the wavelength of 1.55μm.

The confinement loss of the PCF as a function of wavelength in the range of 1.35-1.8μm is illustrated in Figure 2(a). According to the figure, the confinement loss for wavelengths of less than 1.7μm drops to below 10^{-7}. Figure 2(b) illustrates the dispersion curve in the wavelength range 1.4-1.75μm. For a wavelength of 1.55μm, the dispersion is about -2.5ps/nm.km, which is fairly suitable and close to zero. Nevertheless we seek an even less slope of dispersion, which is not possible in this design. Thus for simultaneous improvement of dispersion slope, confinement loss and effective area, the preliminary design was modified which ultimately led to DF-PCF with the cross-section shown in Figure 3(a).

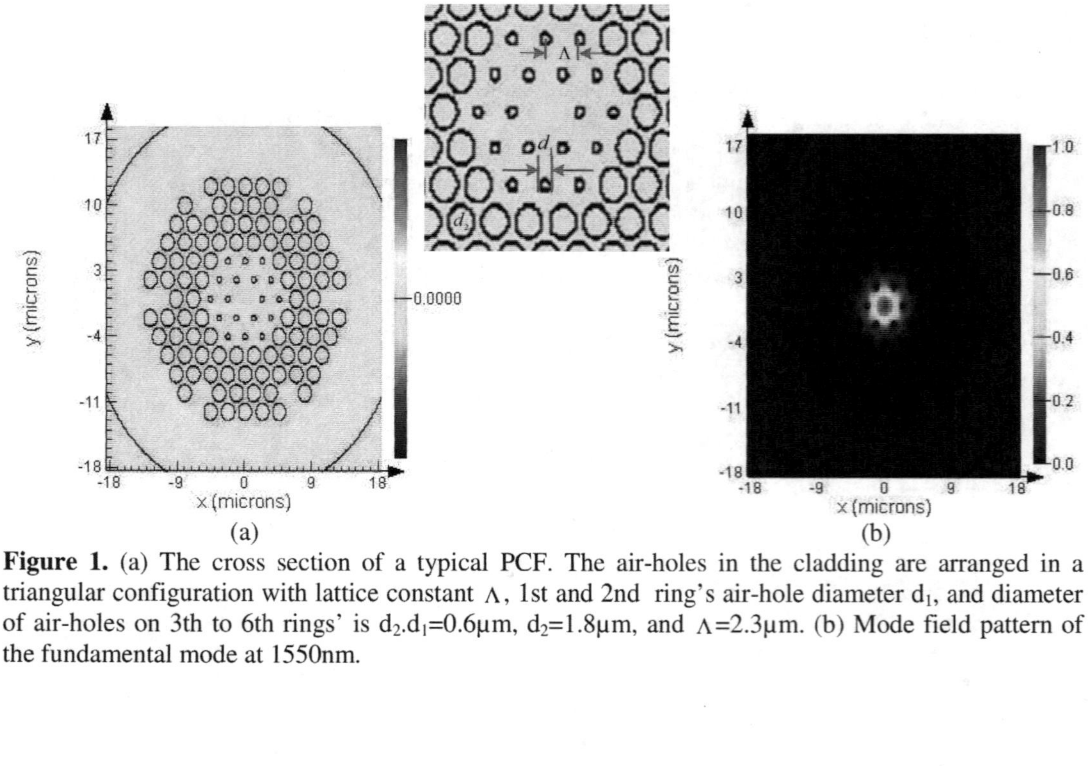

(a) (b)

Figure 1. (a) The cross section of a typical PCF. The air-holes in the cladding are arranged in a triangular configuration with lattice constant Λ, 1st and 2nd ring's air-hole diameter d_1, and diameter of air-holes on 3th to 6th rings' is d_2.d_1=0.6μm, d_2=1.8μm, and Λ=2.3μm. (b) Mode field pattern of the fundamental mode at 1550nm.

In the initial design (Figure 1), parameters of d_1, d_2, Λ and the six air hole rings provide the control of the properties of the PCF, while in the final design improved results are made possible by manipulating a variety of other parameters, though the multiplicity of the parameters makes the fabrication process slightly more elaborate. The values of these parameters in the final design are as follows:

d_1=0.4μm (diameter of the holes in the first ring)
d_2=0.76μm (diameter of the holes in the second ring)
d_3=1.8μm (diameter of the holes in the last five rings)
Λ_1=1.5Λ=3.45μm (spacing between the adjacent holes in the first ring)
Λ=2.3μm (spacing between the adjacent holes in other rings)

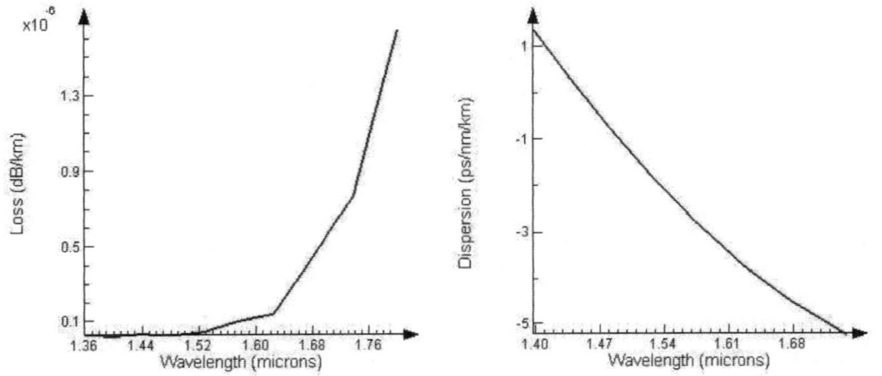

(a) (b)

Figure 2. (a) The confinement loss and (b) dispersion characteristics as a function of wavelength for the photonic crystal fiber of Figure 1.

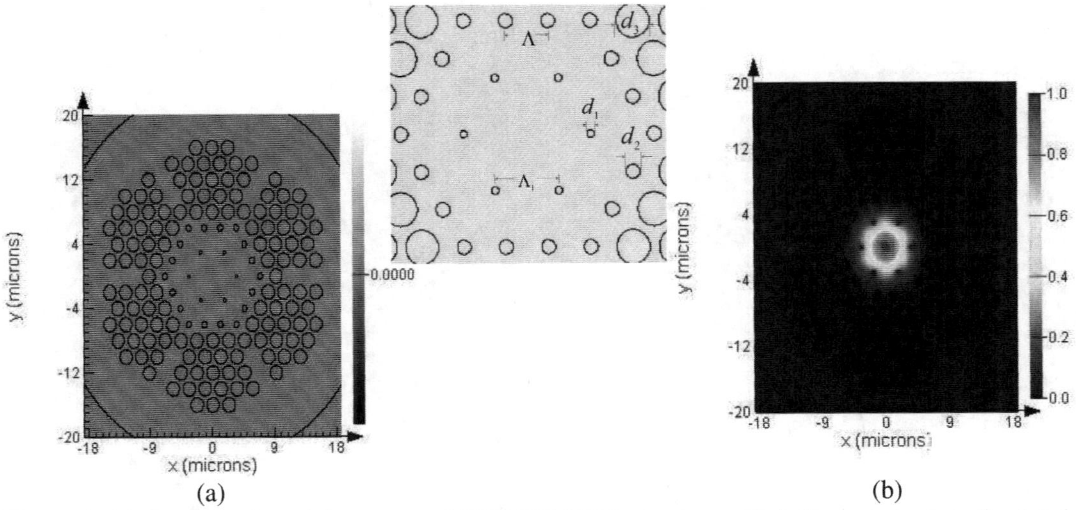

(a) (b)

Figure 3. (a) Cross section of the proposed PCF (DF-PCF) structure with lattice constant Λ_1 for 1st ring and Λ for 2nd to 7st rings and 1st ring's air-hole diameter d_1, 2nd rings' air-hole diameters d_2, and diameter of air-holes on 3th to 7th rings' is d_3.d_1=0.4μm, d_2=0.76μm, d_3=1.8μm, Λ=2.3μm, and Λ_1=3.45μm. This PCF structure can exhibit ultra-flattened dispersion characteristics with low confinement losses and large effective area. (b) The mode field pattern of the fundamental mode at 1550nm.

In order to reduce the dispersion even further, diameter of the holes in the first ring is decreased. Considering the fact that choosing smaller values for d and higher values for Λ in the inner rings leads to reduction of dispersion and its slope, the value of Λ is chosen as Λ_1=1.5Λ. Increasing the value of Λ and introducing the new parameter of Λ_1, as the lattice constant of the first ring, offer two main benefits. First, it contributes to reduction of the variations in dispersion in a broad range of wavelengths. Second, it increases the effective area in DF-PCF. However, choosing smaller values for d_1 and higher values for Λ_1 exacerbates the confinement loss; as a result an extra ring of holes with a diameter of d_3 is added to the final structure to offset this issue.

Figure 3(b) shows the mode field distribution of DF-PCF. As clearly shown in this figure, not only is the field still confined to the area around the core, but also the effective area is much larger compared to the initial design. In the proposed structure, the effective area is almost twice as large as in the preliminary design and reaches 53.279μm^2. Large mode area PCF can suppress undesirable nonlinear impairments [13].

Effective refractive index as a function of wavelength is presented in Figure 4(a). As previously mentioned in the second section, dispersion and confinement loss may then be calculated using the corresponding value of n_{eff} in a given wavelength. In Figure 4(b), the confinement loss curve as a function of wavelength is shown for a DF-PCF with 7 air-hole rings in the range of 1.2-1.7μm. Despite the large effective area in the whole wavelength range, loss has a value less than 10^{-7}. Figure 4(c) illustrates the dispersion of this PCF, and it is shown that in a wide range of wavelengths, both lower dispersion and very small variations of about 0.55ps/nm.km are obtained.

4. Conclusion

This paper shows that lower values of dispersion and confinement loss can be acquired by reducing the diameters of the holes in the inner rings and increasing the size of the holes in the outer rings. Moreover, increasing the spacing between adjacent holes in the first ring Λ_1 is shown to aid reaching ultra-low dispersion slope and large effective area, which are of our main concerns in designing the PCFs. In a broad wavelength range, dispersion rate of 6.8×10^{-4}ps/km.nm^2 is obtained, and concurrently confinement loss less than 10^{-7} and effective area larger than 50μm^2 are made feasible. Due to its low dispersion variations, small confinement loss and large effective area, this new design can be properly utilized in wideband transmission applications and can be suppress undesirable nonlinear impairments.

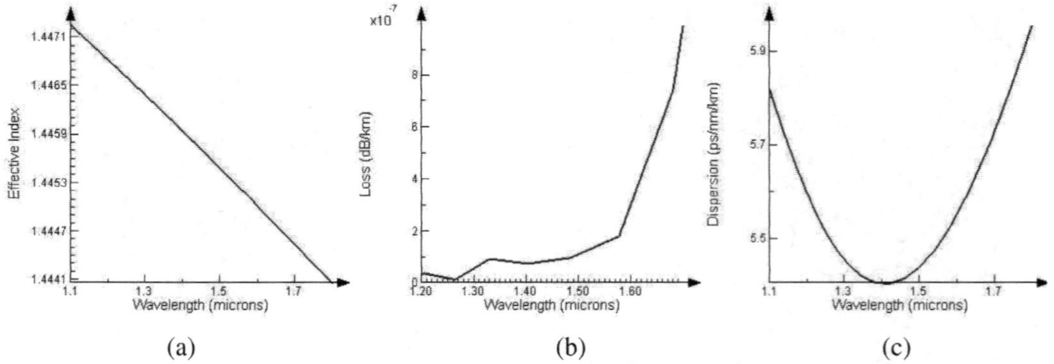

(a) (b) (c)

Figure 4. (a) The effective index, (b) confinement loss, and (c) dispersion characteristics as a function of wavelength for the designed photonic crystal fiber shown in Figure 3.

Acknowledgement
The authors would like to thank Iran Telecommunication Research Center (ITRC) for financial support.

References

[1] Hai N H, Namihiray Y, Kaijage S F, Kinjo T, Begum F, Razzak S M A, and Zou N 2008 *Multiple defect-core hexagonal photonic crystal fiber with flattened dispersion an polarization maintaining properties* Opt. Review **15**, 31

[2] Knight J C, Birks T A, Russell P S J, and Atkin D M 1996 *All-silica single-mode optical fiber with photonic crystal cladding*," Opt. Lett. **21**, 1547

[3] Hansen K P 2003 *Dispersion flattened hybrid-core nonlinear photonic crystal fiber* Opt. Express **11**, 1503

[4] Saitoh K and Koshiba M 2003 *Chromatic dispersion control in photonic crystal fibers: application to ultra-flattened dispersion* Opt. Express **11**, 843

[5] Saitoh K, Florous N, and Koshiba M 2005 *Ultra-flattened chromatic dispersion controllability using a defected-core photonic crystal fiber with low confinement losses* Opt. Express 13, 8365

[6] Gundu K M, Kolesik M, Moloney J V 2006 *Ultra-flattened-dispersion selectively liquid-filled photonic crystal fibers* Opt. Express **14**, 6870

[7] Chen M, Xie S 2008 *New nonlinear and dispersion flattened photonic crystal fiber with low confinement loss* Opt. Communications **281**, 2073

[8] Hai N H, Namihiray Y, Kaijage S F, Kinjo T, Begum F, Razzak S M A, and Zou N 2008 *A unique approach in ultra-flattened dispersion photonic crystal fibers containing elliptical air-holes* Opt. Rev.**15**, 91

[9] Uddin M J and Alam M S 2008 *Dispersion and confinement loss of photonic crystal fibers* Asian Journal of Information Technology **7(8)**, 344

[10] Wei W, Lan-Tian H, Jun-Jie S, and Gui-Yao Z 2009 *Design of nearly zero dispersion flattened photonic crystal fiber with double cladding* Chin. Phys. Lett. **26**, 0544

[11] Pourmahyabadi M and Nejad S M 2009 *Numerical analysis of photonic crystal fibers with low confinement loss and ultra-flattened dispersion by FDFD method* Iranian Journal of Electrical & Electronic Engineering **5(3)**, 170

[12] Razzak S M A, Namihiray Y, Miyaji K, Kaijage S F, Kinjo T, Hai N H, and Begum F 2008 *Simultaneous control of dispersion and confinement losses with octagonal photonic crystal fibers for communication systems* The International Conference on Electrical Engineering July 6-8, 2008

[13] Matsui T, Zhou J, Nakajima K, and Sankawa I 2005 *Dispersion-flattened photonic crystal fiber with large effective area and low confinement loss* Lightwave Tech. **23**, 4178

[14] Guo S, Wu F, Albin S, Tai H, and Rogowski R S 2004 *Loss and dispersion analysis of microstructured fibers by finitedifference method* Opt. Express **12**, 3341

[15] Razzak S M A, Khan M A G, Namihira Y, and Hussain M Y 2008 *Optimum design of a dispersion managed photonic crystal fiber for nonlinear optics applications in telecom systems* 5th International Conference on Electrical and Computer Engineering ICECE 2008, Bangladesh, IEEE 2008

[16] Reeves W H, Knight J C, Russell P S J, and Roberts P J 2002 *Demonstration of ultra-flattened dispersion in photonic crystal fibers* Opt. Express, **10**, 609

A novel phase noise suppression method for coherent detection by utilizing two adjacent longitudinal modes generated from a supercontinuum multi-wavelength source

D Pan[1,2,3], **C J Ke**[1,2,3,4] **,X H Zhu**[1,2,3], **D M Liu**[1,2,3]
[1] College of Optoelectronic Science and Engineering, Huazhong University of Science and Technology, 430074, Wuhan, China
[2] Wuhan National Laboratory for Optoelectronics, 430074, Wuhan, China
[3] National Engineering Laboratory for Next Generation Internet Access System, 430074, Wuhan, China

Abstract. A novel phase noise suppression method for coherent detection is proposed. It uses two adjacent longitudinal modes of a multi-wavelength source, as signal light and coherent light respectively, which are both launched into the optical fiber. And heterodyne synchronous demodulation structure is used in the receiver. The performance of a back-to-back system with 100 Gbit/s NRZ-QPSK utilizing this novel method is investigated by VPI. The simulation results show that phase noise in this coherent detection scheme is suppressed significantly, which is resulted from the correlating phase and fixed frequency spacing of adjacent longitudinal modes.

1. Introduction

In order to cope with the need for ultra high-speed, ultra high-capacity and ultra long-haul optical transmission, coherent detection has attracted significant attentions in research and development during the last several years [1-7]. Coherent receivers could linearly down-convert the whole optical signal to a baseband electrical signal by using heterodyne or homodyne detection, and have the following advantages against direct detection. First, its higher sensitivity makes coherent detection a promising technique for long-haul transmission. Second, the optoelectronic conversion process is linear, so that the whole optical signal information can be postprocessed in the electrical domain. Third, it is applicable to multilevel modulation formats such as multiple phase shift keying (mPSK), multiple quadrature amplitude modulation (mQAM) and orthogonal frequency division multiplexing (OFDM). However, coherent detection is sensitive to the phase noise associated with the transmitter laser and the local oscillator, which is the main impairment for the advanced lightwave systems [8-9].

Previously, phase noise suppression can be achieved by selecting narrow linewidth laser as local oscillator, or utilizing optical phase-locked loop (OPLL).And phase noise compensation can be realized by digital signal processing (DSP) [10]. Unfortunately，these solutions are relatively high-cost and complex for implementation.

In this paper, a novel method for phase noise suppression in coherent detection is proposed. Two adjacent longitudinal modes from a multi-wavelength source are employed as signal light and coherent

[4] Corresponding author, E-mail: cjke@mail.hust.edu.cn

light respectively, which are both launched into the optical fiber. And heterodyne coherent detection structure is used to demodulate the signal at the receiver. The simulation shows that this so-called "all-optical phase noise suppression" method would have the potential to replace the aforementioned solutions with simplified receiver structure.

2. System architecture and operation principle

2.1. System architecture

The architecture of the coherent optical communication system adopting the mentioned phase noise suppression techniques is shown in Figure 1. To distinguish from the conventional system which uses a local oscillator at the receiver for coherent detection, we call this NLO (non local oscillator) coherent optical communication system.

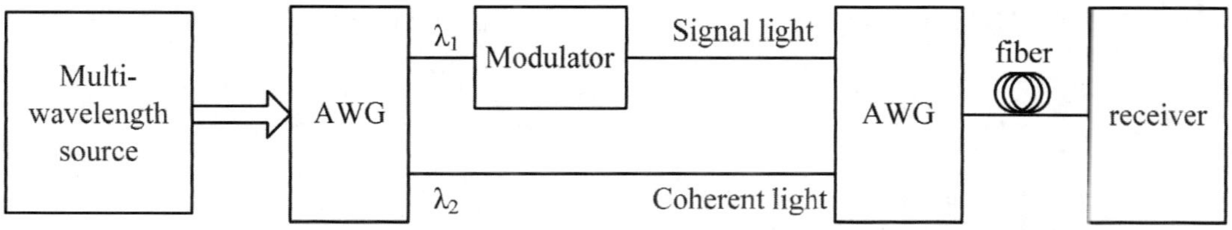

Figure 1. Schematic of NLO coherent optical communication system

In the transmitter, a supercontinuum multi-wavelength optical source is used to generate two wavelength simultaneously, or rather two adjacent longitudinal modes. One is modulated to carry the information as signal, and the other is continuous wave servicing as coherent light. Both of them are launched into optical fiber.

In the receiver, an optical filter is used to fetch out the CW coherence light from the fiber to play the role of local oscillator. Obviously, heterodyne detection is needed to demodulate the signal. By using phase diversities and balanced detection, we can obtain four outputs and detect differential in-phase and quadrature (IQ) components of the signal light. Synchronous demodulation is applied to restore the baseband in the end.

2.2. Operation principle

The AC component of photocurrent in heterodyne detection is given by Eq. (1) [11-12]:

$$I(t) = 2R\sqrt{P_S P_L}\cos(\omega_{IF}t + \phi) \tag{1}$$

where ω_S , ω_L are the angular frequency of the signal light and coherent light respectively, $\omega_{IF} = \omega_S - \omega_L$ is the intermediate frequency (IF). ϕ_S, ϕ_L are the phase of the signal light and coherent light respectively, $\phi = \phi_S - \phi_L$ is the phase of IF. R is the responsivity of photodiode. P_S, P_L is the optical power of signal light and coherent light respectively.

In coherent detection, phase fluctuations lead to current fluctuations and degrade the SNR. Both the signal phase ϕ_S and the local-oscillator phase ϕ_L should remain relatively stable to avoid the sensitivity degradation. A measure of the duration over which the laser phase remains relatively stable is provided by the coherence time. It is inversely related to the laser linewidth. In conventional system, the IF linewidth $\Delta\nu$ is actually the sum of the transmitter linewidth $\Delta\nu_S$ and the local oscillator linewidth $\Delta\nu_L$, i.e. $\Delta\nu = \Delta\nu_S + \Delta\nu_L$. This lies in the fact that random variables ϕ_S and ϕ_L fluctuate independently. While in the novel system mentioned in this paper, ϕ_S and ϕ_L fluctuate not independently stemming from the correlation between the two adjacent longitudinal modes of the same optical source. So σ_ϕ^2, variance of ϕ, can be described as the following,

$$\sigma_\phi^2 = \sigma_S^2 + \sigma_L^2 - 2\rho\sigma_S\sigma_L \tag{2}$$

where σ_S^2 and σ_L^2 are the variance of ϕ_S and ϕ_L respectively, ρ is the correlation coefficient between ϕ_S and ϕ_L. In our method, random variables ϕ_S and ϕ_L are nearly linear correlated, where $\rho \approx 1$. This results in the phase noise suppression intrinsically [13]. By this means, we can simplify the construction of the coherent receiver remarkably and will not need the OPLL & DSP anymore. Thus the novel "all-optical phase noise suppression" method comes into being.

3. Simulation results and discussions

VPI TransmissionMaker™ V8.5 is utilized to evaluate the performance of the aforementioned NLO system, by assuming that single channel 100Gb/s NRZ-QPSK back-to-back transmission. The related simulation parameters are given in table 1. We will emphasize on the performance affected by the linewidth of signal light and coherent light in conventional and NLO system, and analyse the noise limitation further.

Table 1. Simulation parameters

Parameters	**Values**
Average optical power of signal light	1 mW
Average optical power of coherent light	1 mW
Central frequency of signal light	193.1 THz
Central frequency of coherent light	193.15 THz
Linewidth of longitudinal mode	1MHz
Optical Signal Noise Ratio	35 dB
Channel spacing of AWG	50 GHz
3dB bandwidth of AWG	37 GHz
Passband type of AWG	Gaussian
Adjacent crosstalk of AWG	25 dB
Non-adjacent crosstalk of AWG	30 dB

3.1. Laser linewidth

Figure 2 shows the simulated symbol error rate (SER) performance of the NLO and conventional system under different laser (or longitudinal mode) linewidths. We can find that the narrower linewidth can lead to more excellent SER performance for both NLO and conventional system. Comparing the inset constellations, whose linewidth and received optical power (ROP) are 1MHz and -9dBm respectively, it can be seen that the NLO system, which uses two adjacent longitudinal modes from the same optical source to achieve coherent detection, is more effective for obtaining a noiseless accurate phase difference ϕ of IF than the conventional system on the same condition. It also shows that the NLO system can improve the tolerance of laser linewidth for coherent detection from 100 kHz to 10 MHz without any phase estimation techniques which is usually realized by DSP in the conventional system.

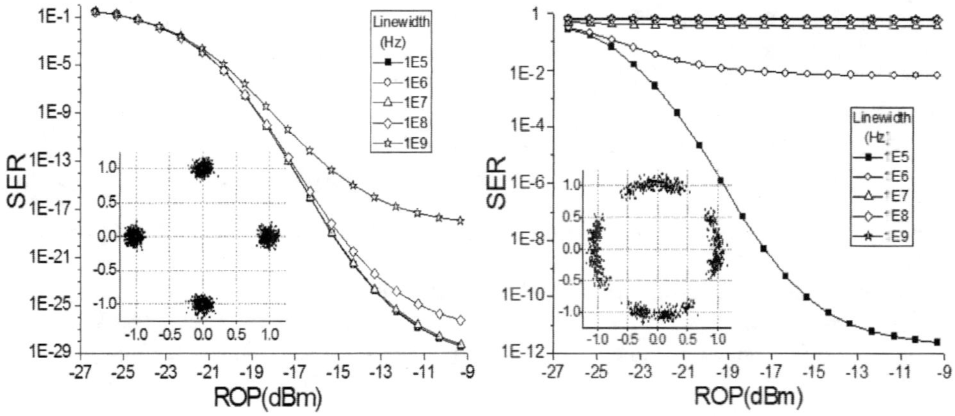

Figure 2. SER vs. ROP for different laser linewidths, where inset are constellations. (a) NLO system (b)conventional system

3.2. Noise limitation

Figure 3 shows the simulated SER performance of the NLO and conventional system with different optical signal to noise ratio (OSNR) when the laser linewidth for coherent detection is 1MHz. With the increasing of OSNR, SER performance will be enhanced to some extent for both of them, especially in the NLO system. And phase noise would overwhelm the intensity noise in the conventional system when the laser linewidth is about 1MHz order, which causes the SER floor. By the mentioned phase noise suppression method, the NLO system does exhibit better SER performance than the conventional one.

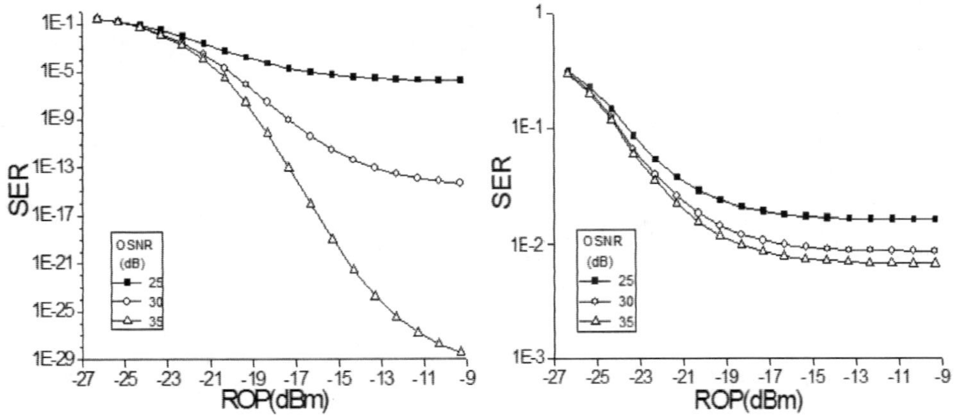

Figure 3. SER vs. ROP for different OSNR. (a) NLO system (b)conventional system

4. Conclusions

This novel method makes a contribution to practical applications of coherent detection in the future developing of optical transmission system. The performance of receiver can be effectively improved, because phase noise is suppressed when the accurate carrier phase estimation based on DSP is not used, so the cost has not increased. Another advantage is that system structure can be simplified. In conventional coherent optical system, it must be provided two sources as transmitter laser and local oscillator. In addition, two wavelengths should be controlled independently. Therefore, it is difficult to construct and control the sources, when wavelength channel number is more than one hundred.

However, in our method, supercontinuum generation is an effective way of obtaining more than one hundred optical longitudinal modes and the devices are easier to control.

Acknowledgement
This work was supported by the National Basic Research Program of China (No.2010CB328302)

References

[1] Rakefet W, Moshe N, Reinhold N and Isaac S 2009. *ECOC* (Vienna: Nexus Media Limited) Paper 2.3.3.
[2] Sano A, Yamada E and Masuda H 2008 *ECOC* (Brussels: IBBT) Th.3.E.1
[3] Winzer P J, Gnauck A H and Doerr C R 2010. *J. Lightwave Technol.* 2010, **28** 547
[4] Bertran-Pardo O, Renaudier J and Charlet G 2009 *ECOC* (Vienna: Nexus Media Limited) 2009. Paper 9.4.1
[5] Yu J J, Zhou X and Huang M F 2008 *ECOC* (Brussels: IBBT) Th.3.E.2
[6] Jansen S L, Morita I and Tanaka H 2008 *OFC/NFOEC.* (San Diego: OSA) OMU3
[7] Ammon Y 2004 *Optical Electronics in Modern Communication* (Beijing: Publishing House of Electronics Industry) p 302
[8] Franz J H and Jain V K 2002 *Optical Comunications Components and Systems* (Beijing: Publishing House of Electronics Industry) p 317
[9] Kazovsky L G, Benedetto S and Willner A 1999 *Optical Fiber Communication Systems* (Beijing: Publishing House of Electronics Industry) p 223
[10] Seb J S 2008 *Opt. Express* **16** 804
[11] Agrawal G P 2002 *Fiber-optic Communication Systems* (New York: John Wiley & Sons, Inc.) p 480
[12] Kazuro K 2008 Coherent optical communication systems Optical Fiber Telecommunications V B (System and Networks) ed Ivan P. Kaminow, Tingye Li and Alan E. Willner (New York: Academic Press) chapter 3 pp 95–129
[13] Zhu N H, Li W and Wang L X 2009 *J. Quant. Electron.* **45** 514

Journal of Physics: Conference Series **276** (2011) 012082 doi:10.1088/1742-6596/276/1/012082

A study of GaN-based LED structure etching using inductively coupled plasma

Pei Wang[1,2], Bin Cao[1], Zhiyin Gan[1,2], Sheng Liu[1,2]

[1]School of Mechanical Engineering, Huazhong University of Science & Technology, 1037 Luoyu Road, Wuhan, China, 430074

[2]Wuhan National Lab of Optoelectronics, Huazhong University of Science & Technology, 1037 Luoyu Road, Wuhan, China, 430074

E-mail: zhiyingan@yahoo.com.cn

Abstract. GaN as a wide band gap semiconductor has been employed to fabricate optoelectronic devices such as light-emitting diodes (LEDs) and laser diodes (LDs). Recently several different dry etching techniques for GaN-based materials have been developed. ICP etching is attractive because of its superior plasma uniformity and strong controllability. Most previous reports emphasized on the ICP etching characteristics of single GaN film. In this study dry etching of GaN-based LED structure was performed by inductively coupled plasmas (ICP) etching with Cl_2 as the base gas and BCl_3 as the additive gas. The effects of the key process parameters such as etching gases flow rate, ICP power, RF power and chamber pressure on the etching properties of GaN-based LED structure including etching rate, selectivity, etched surface morphology and sidewall was investigated. Etch depths were measured using a depth profilometer and used to calculate the etch rates. The etch profiles were observed with a scanning electron microscope (SEM).

1. Introduction

GaN as a wide band gap semiconductor has been employed to fabricate optoelectronic devices such as light-emitting diodes (LEDs) and laser diodes (LDs) [1, 2]. Etching is a required technique for patterning in the fabrication process of optoelectronic devices. Due to the chemical inertness and high thermal stabilities of GaN, a wet etching method without assistance is not suitable. Recently several different dry etching techniques for GaN-based materials including reactive ion etching (RIE), electron cyclotron resonance (ECR) etching and inductively coupled plasma (ICP) etching have been developed [3-7]. Among these dry etching methods ICP etching is attractive because of its superior plasma uniformity and strong controllability. Yeom et al. investigated the effects of process parameters on the plasma characteristics and GaN etch properties of inductively coupled Cl_2/Ar and Cl_2/BCl_3 plasmas [8, 9]. Im et al. presented a parametric study of ICP dry etching of undoped, n-type and p-type GaN films with Cl_2-based plasmas [10]. Lin et al. compared the etching characteristics and etching induced damages of n-type GaN films using Cl_2/He and Cl_2/Ar [11]. Kim et al. studied the GaN etching properties with HBr-based inductively coupled plasmas such as Cl_2/HBr, BCl_3/HBr and HCl/HBr [12]. Kao et al. applied the etching results of undoped, n-type and p-type GaN using Ni mask to fabricate highly smooth and anisotropic facets on InGaN laser structure and create the GaN-based material nanorods [13]. Most reports emphasized on the ICP etching characteristics of single GaN film. In this work, we focused on the etching properties of GaN-based LED structure including etching rate, selectivity, etched surface morphology and sidewall related to the key process parameters such as etching gases flow rate, ICP power, RF power and chamber pressure with Cl_2 as the base gas and BCl_3 as the additive gas. A detailed study was presented as below.

2. Experiment

In this work, GaN-based multiple quantum wells (MQWs) LEDs wafers were grown by metalorganic chemical vapor deposition (MOCVD). The LED structure comprised a 1.5μm thick undoped GaN layer, a 3μm thick n-GaN layer, a 0.15μm thick InGaN-GaN MQWs active layer, a 30nm thick p-AlGaN layer, and a 0.3μm thick p-GaN layer. A 480nm thick SiO_2 layer as etching mask was deposited on the surface of the GaN by plasma enhanced chemical vapor deposition (PECVD). A standard lithography process and wet etching were performed to form the patterns on the SiO_2 film as mask. Dry etching process was performed in an ICP system (Oxford Plasmalab System 100). Cl_2 was used as the base gas and BCl_3 was used as additive gases. The total gas flow rate was held at 50 standard cubic centimeters per minute (sccm). The ICP power was varied between 0 and 1200W, the RF power between 100 and 500W and the working pressure between 5 and 15mTorr. The temperature of the backside cooled sample chuck was kept at 23 ℃ during the etching process. Etch depths were measured using a depth profilometer after removing the SiO_2 layer in a buffered oxide etchant and used to calculate the etch rates. The etch profiles were observed with a scanning electron microscope (SEM).

3. Results and discussion

Figure 1 shows the etch rates of sample and the etch selectivity over SiO_2 with different gas combination at ICP power of 1000W, RF power of 100W and chamber pressure of 7mtorr. It can be seen from Figure 1 that the etch rate decreased with increased BCl_3 and the etch selectivity also decreased with increased BCl_3 percent in the gas mixture. The highest etch rate obtained was about 262 nm/min in pure Cl_2 gas. The etch selectivity decreased rapidly when BCl_3 was added into the pure Cl_2 gas and the selectivity was as high as about 66 in 100% Cl_2.

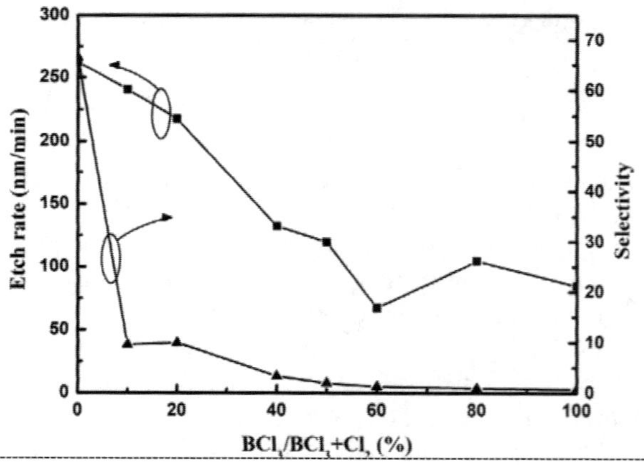

Figure 1. Etch rate and selectivity as a function of Cl_2/ BCl_3 gas ratio at 1000W ICP power, 100W RF power and 7mtorr chamber pressure.

The effect of ICP power on the etch rate and selectivity is shown in Figure 2. During the etching processes the gas ratio, RF power and chamber pressure were held at 20%BCl_3/80%Cl_2 and 7mtorr, respectively. The etch rate increased as the ICP power increased from 300W to 500W. The etch rate decreased when the ICP power increased further. The initial increase of etch rate could be resulted

Journal of Physics: Conference Series **276** (2011) 012082 doi:10.1088/1742-6596/276/1/012082

from the increase of ion density as the ICP power increased. The decrease of etch rate may be due to the lower ion energy. The etch selectivity decreased with increased ICP power.

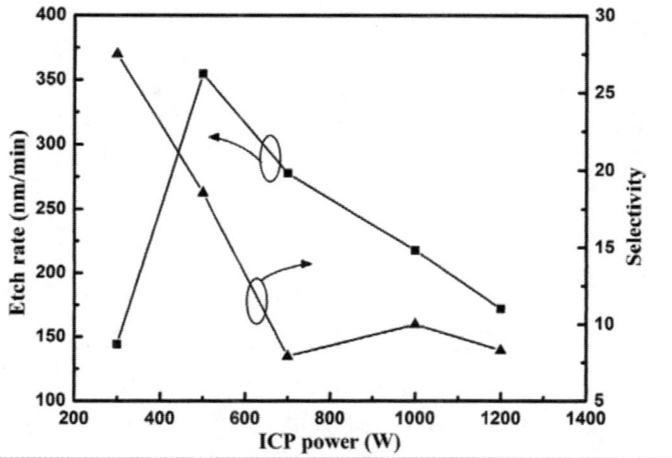

Figure 2. Etch rate and selectivity as a function of ICP power at 20% BCl_3, 100W RF power and 7mtorr chamber pressure.

It is shown in Figure 3 that the etch rate and selectivity as function of RF power at ICP power of 1000W, gas combination of 20%BCl_3/80%Cl_2 and chamber pressure of 7mtorr. It can be seen from Fig. 3 that the etch rate increased monotonically as the RF power increased. This result is consistent with some previous studies. The etch rate increase due to increase of ion energy which is controlled by RF power. The highest etch rate was 717 nm/min when the RF power was 500W. The etch selectivity decreased first and then increased when the RF power increased from 150W to 500W.

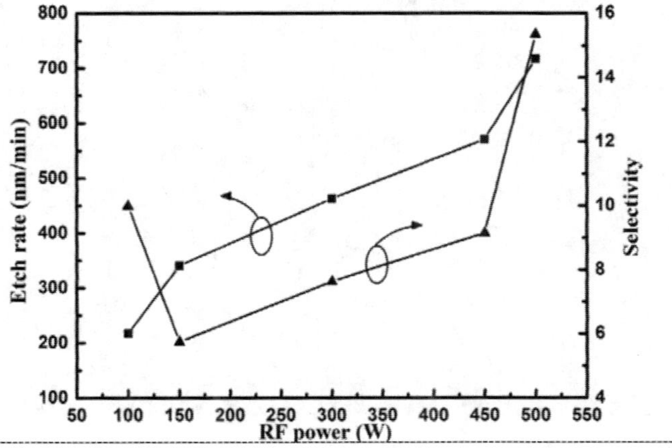

Figure 3. Etch rate and selectivity as a function of RF power at 20% BCl_3, 1000W ICP power and 7mtorr chamber pressure.

Figure 4 shows that effect of chamber pressure on the etch rate and selectivity at ICP power of 1000W, gas combination of 20%BCl_3/80%Cl_2 and RF power of 100W. The etch rate increased as the pressure increased and reached a maximum value, and then decreased with further increased chamber pressure. The highest etch rate obtained at chamber pressure of 10mtorr was about 253 nm/min. The etch selectivity increased monotonically as the chamber pressure increased.

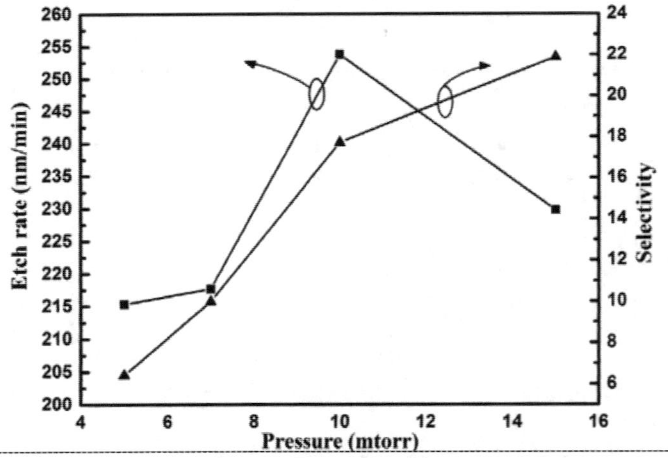

Figure 4. Etch rate and selectivity as a function of chamber pressure at 20% BCl_3, 1000W ICP power and 100W RF power.

Some etch profiles are shown in Figure 5. It can be seen that different etching parameters influenced the sidewall morphology. The increased BCl_3 percent in gas mixture reduced the etch anisotropy. The effect of ICP power on etch anisotropy is not obvious. Higher RF power could result in more perpendicular sidewall.

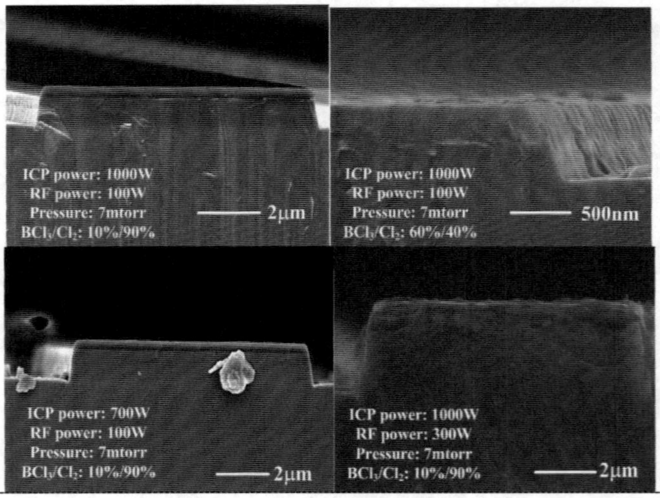

Figure 5. SEM photograph of some etch profiles with different etching parameters.

4. Conclusion

In summary, ICP etching of GaN-based LED structure was performed with Cl_2 as the base gas and BCl_3 as the additive gas. The effects of the key process parameters such as etching gas ratio, ICP power, RF power and chamber pressure on the etching properties of GaN-based LED structure was investigated. The highest etch rate was 717 nm/min at ICP power of 1000W, gas combination of

20%BCl_3/80%Cl_2, RF power of 500W and chamber pressure of 7mtorr. The etched sidewall morphology with different etching parameters was also investigated in this paper. It can be seen that the gas combination effect the etch anisotropy greatly.

References

[1] Nakamura S, Mokia S and Senoh M 1994 *Appl. Phys. Lett.* **64** 1689
[2] Nakamura S, Senoh M, Nagahama S, Iwasa N, Yamada T, Matsushita T, Sugimoto Y and Kiyodo H 1996 *Appl. Phys. Lett.* **70** 868
[3] Feng M S, Guo J D, Lu Y M and Chang E Y 1996 *Mate. Chem. Phys.* **45** 80
[4] Basak D, Nakanishi T and Sakai S 2000 Solid-State Electronics **44** 725
[5] Szczesny A, Sniecikowski P, Szmidt J and Werbowy A 2003 *Vacuum* **70** (2003) 249
[6] Shul R J, Kilcoyne S P, Hagerott Crawford M, Parmeter J E, Vartui C B, Abernathy C R and Pearton S J 1995 *Appl. Phys. Lett.* **66** 1761
[7] Shul R J, McClellan G B, Casalnuovo, S A, Rieger D J, Pearton S J, Constantine C, Barratt C, Karlicek R F, Tran C and Schurman M 1996 *Appl. Phys. Lett.* **69** 1196
[8] Kim H S, Yeom G Y, Lee J W and Kim T I 1999 *Thin Solid Films* **341** 180
[9] Kim H S, Yeom G Y, Lee J W and Tae-Il Kim T I 1999 *J. Vac. Sci. Technol. A* **17** 2214
[10] Im Y H, Park J, Hahn Y B, Nahm K S, Lee Y S, Cho B C, Lim K Y, Lee H J and Pearton S J 2000 *J. Vac. Sci. Technol. A* **18** 2169
[11] Lin Y C, Chang S J, Su Y K, Shei S C and Hsu S J 2003 *Mater. Sci. Eng. B* **98** 60
[12] Kim D W, Jeong C H, Lee H Y, Kim H S, Sung Y J and Yeom G Y 2003 *Sol. Stat. Elec.* **47** 549
[13] Kao C C, Huang H W, Tsai J Y, Yu C C, Lin C F, Kuo H C and Wang S C 2004 *Mater. Sci. Eng. B* **107** 283

Analysis of light extraction efficiency of GaN-based light-emitting diodes

Pei Wang[1,2], Bin Cao[1], Zhiyin Gan[1,2], Sheng Liu[1,2]

[1]School of Mechanical Engineering, Huazhong University of Science & Technology, 1037 Luoyu Road, Wuhan, China, 430074

[2]Wuhan National Lab of Optoelectronics, Huazhong University of Science & Technology, 1037 Luoyu Road, Wuhan, China, 430074

E-mail: zhiyingan@yahoo.com.cn

Abstract. GaN-based light emitting diodes (LEDs) as one of the most important light source in next-generation solid-state lighting have been extensively studied and remarkable progress has been obtained. However, the light extraction efficiency (LEE) is not sufficient to satisfy application requirements. Most of the photons generated in multiple quantum well (MQW) always are trapped inside the semiconductor because both the reflection index of GaN and InGaN are higher than that of air, which results in total internal reflection (TIR). Great efforts were made in enhancing light extraction of LEDs experimentally in previous investigation. However, detailed theoretical studies in predicting the LEE of different types of LEDs are not available. In this paper the light extraction efficiency(LEE) of conventional chip (CC), flip chip (FC) and vertical chip (VC) is investigated using Monte Carlo ray tracing method is presented in conventional chip (CC). Monte Carlo ray tracing simulation based on statistics is known to be one of the most suitable methods to analyze the dependence of the light extraction efficiency of LEDs on the variety of the device parameters. Diffused bottom surface was found to be better for improvement of LEE than the perfect mirror surface. When there is a material with higher refraction index on the top surface the VC structure has the highest LEE no matter the bottom surface is perfect mirror or diffuse surface, which is above 80%. It is indicated that VC is potential in high-power GaN-based LEDs from the results.

1. Introduction

GaN-based light emitting diodes (LEDs) as one of the most important light source in next-generation solid-state lighting have been extensively studied and remarkable progress has been obtained. However, the light extraction efficiency (LEE) is not sufficient to satisfy application requirements. Most of the photons generated in multiple quantum well (MQW) always are trapped inside the semiconductor because both the reflection index of GaN and InGaN are higher than that of air, which results in total internal reflection (TIR). Many methods have been proposed to improve the light extraction, such as shaping truncated-inverted-pyramid geometry, surface roughening, photonic crystal and so on [1-6]. Great efforts were made in enhancing light extraction of LEDs experimentally in previous investigation. However, detailed theoretical studies in predicting the LEE of different types of LEDs are not available. The light propagation effects of LEDs can be adequately described in terms of geometrical optics instead of the more rigorous wave optics [7]. Monte Carlo ray tracing simulation based on statistics is known to be one of the most suitable methods to analyze the dependence of the light extraction efficiency of LEDs on the variety of the device parameters. Recently this method has been employed to study the light output performance of LEDs with various structures [8, 9]. In this paper, comparison study is performed for the LEE of GaN-based LEDs of three main types using Monte Carlo ray tracing simulation and future LEE levels is estimated.

2. Simulation modeling

In a general way, GaN based LEDs comprise multiple films grown on sapphire substrates by metal-organic chemical vapor deposition (MOCVD) or molecular beam epitaxy (MBE) as shown in Figure 1. Figure 2 presents the structures of three main types of LED chips: conventional chip, flip chip and vertical chip [10, 11]. Figure 3 schematically describes the optical ray tracing model of the three different types of LEDs. The LED chip size is 1 mm×1mm. In this simulation we consider that the light rays are generated randomly within the active region and the light emission is isotropic. The side receiver and top receiver were introduced to collect rays emitting from the active region. The parameters of different materials were described for simulation as given in Table 1 [12]. The thickness of each layer is determined corresponding to our experimental condition. The absorption coefficient of sapphire substrate is assumed to be 0 since it is transparent to visible spectrum. The absorption coefficient of GaN is set to be 104 cm^{-1} according to previous studies [13, 14]. Silver is usually deposited on the bottom of the chips as a reflector in order to let more light escape upward. The reflectivity of silver to light of wavelength between 400nm and 500nm is about 96%. In this simulation we assume that the reflectivity is 100%. The following simulation results don't take the current spreading and the effects of the electrical pad into account and photon recycling is negligible.

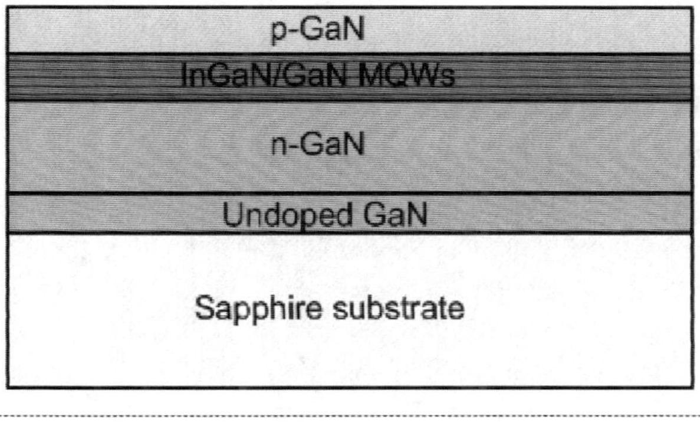

Figure 1. The schematics multi-layer structure of GaN-based LEDs.

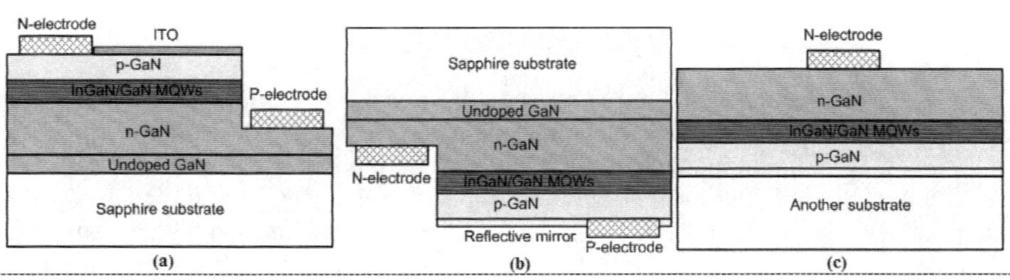

Figure 2. The schematic structures of (a) conventional chip, (b) flip chip and (c) vertical chip.

Assuming that the refractive index of GaN is 2.5, the critical angle of light escaping to air is 23° according to Snell's law. That is to say, the LEE is less than 20% and most of light would be trapped inside the semiconductor. As shown in Figure 3, total internal reflection happens when light propagates from one layer to another. In order to find out the upper limit of light extraction efficiency of each LED structure, we set a material which has higher reflection index on top of light-emitting surface instead of air to make all light rays reaching the interface escape outside.

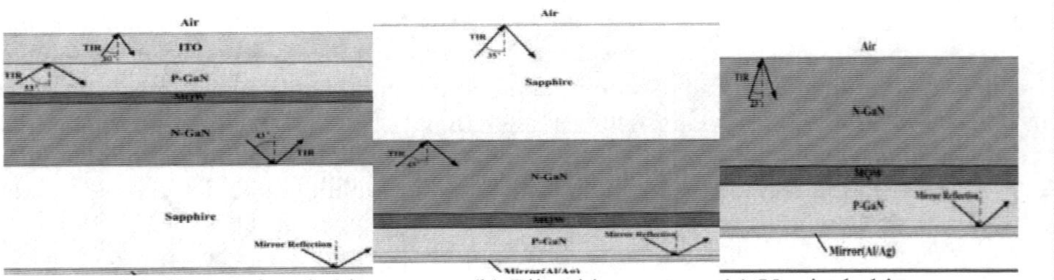

(a) Conventional chip (b) Flip chip (c) Vertical chip

Figure 3. Schematic diagrams of conventional chip (CC), flip chip (FC), and vertical chip (VC)

Table 1. Parameters of materials used in simulation.

Material	sapphire	n-GaN	MQW	p-GaN	ITO
Thickness(um)	90	3	0.2	0.3	0.39
Refraction index	1.78	2.45	2.54	2.4	2.0

3. Results and discussion

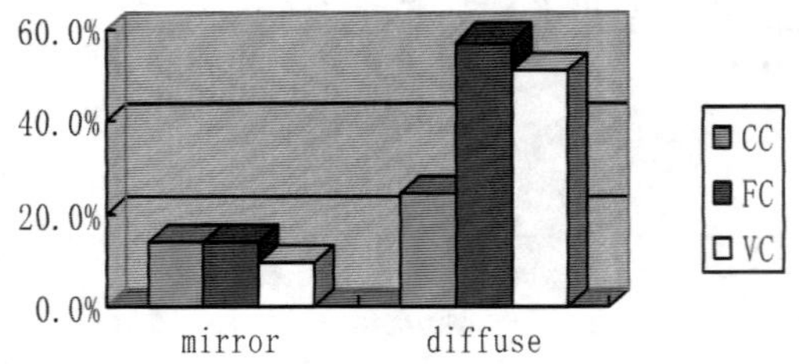

Figure 4. The normal LEEs of conventional chip (CC), flip chip (FC) and vertical chip(VC) with the bottom surface is mirror and diffuse reflection, respectively.

Figure 4 shows the LEEs of the three types of LED chips with bottom mirror reflection and diffuse reflection. It can be seen from Figure 4 when the bottom surface is perfect mirror VC have the lowest LEE 9.7% and CC have the highest LEE 14.7% in the three types. However, all of the LEEs of the three types of LED chips improve using diffuse reflection and LEE of FC and VC are above 50%.

The bottom surface with diffuse reflection is better for the improvement of LEE than the mirror reflection, especially in FC and VC. Perfect mirror surface can reflect light propagating to bottom back to the top surface, however, the incident angle of the light is not changed. Therefore most photons still could not escape to the air. The incident angle can be changed by diffused bottom surface. Roughing surface to improve LEE is to change the incident angle essentially.

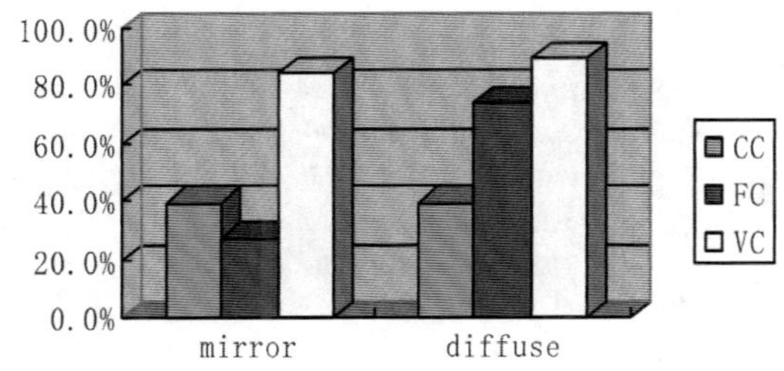

Figure 5. When there is a material with higher refraction index on the top surface of the LED chips, the LEEs of conventional chip(CC), flip chip(FC) and vertical chip(VC) with the bottom surface is mirror and diffuse reflection, respectively.

Figure 5 shows that the LEEs of the three types of LED chips with bottom mirror reflection and diffuse reflection when there is a material with higher refraction index on the top surface of the LED chips. It can be seen that the VC structure has the highest LEE no matter the bottom surface is mirror or diffuse reflection, which is above 80%.

The maximum LEE of CC with ITO is about 40% even all of photons reach the boundary between ITO and air can escape, which shows lots of photons are trapped in active layers. It isn't suitable for high-brightness LED except some textures or roughing could be made at the boundary between n-GaN and sapphire and/or p-GaN and ITO. The LEE of FC with diffuse reflection surface is about 57%, but there is no much room for it to enhance because the maximum LEE is 74%. VC has the greatest LEE. If the photons can't escape to air, they will be diffused at bottom surface and escape or come back at the boundary. Each photon has the chance to escape. If it is not absorbed by active layers, the photons will change the angle of incidence until it escape. Thin active layer is helpful to reduce absorption and the texture structure and/or roughing can be made at each layer. The higher material was introduced in the model in order to calculate the LEE in case that all light rays which came to the top surface could escape to air. The calculated LEE results help us estimate the potential of the three types of LED chips.

4. Conclusion

This study has investigated the LEEs of GaN-based LEDs of three main types using Monte Carlo ray tracing simulation. It was found that diffused bottom surface was better for improvement of LEE than the perfect mirror surface. It was also found that when there is a material with higher refraction index on the top surface the VC structure has the highest LEE no matter the bottom surface is perfect mirror or diffuse surface, which is above 80%. The results indicate that the VC could be potential in obtaining a high LEE.

References

[1] Krames M R, Ochiai-Holcomb M, Hofler G E, Carter-Coman C, Chen E I, Tan I H, Grillot P, Gardner N F, Chui H C, Huang J W, Stockman S A, Kish F A, Craford M G, Tan T S, Kocot C P, Hueschen M, Posselt J, Loh B, Sasser G and Collins D 1999 *Appl. Phys. Lett.* **75** 2365

[2] Wierer J J, Steigerwald D A, Krames M R, O'Shea J J, Ludowise M J, Christenson G, Shen Y C, Lowery C, Martin P S, Subramanya S, Gotz W, Gardner N F, Kem R S and Stockman S A 2001 *Appl. Phys. Lett.* **78** 3379

[3] Fujii T, Gao Y, Sharma R, Hu E L, DenBaars S P and Nakamura S 2004 *Appl. Phys. Lett.* **84** 855

[4] Horng R H, Zheng X, Hsieh C Y and Wuu D S 2008 *Appl. Phys. Lett.* **93** 021125

[5] Kim H G, Na M G, Kim H K, Kim H Y, Ryu J H, Cuong T V and Hong C H 2007 *Appl. Phys. Lett.* **90** 261117
[6] Kim K, Choi J, Jeon S C, Kim J S and Lee H M 2007 *Appl. Phys. Lett.* **90** 181115
[7] Shealy D L and Berg H M 1983 *Appl. Opt.* **22** 1722
[8] Tran N T and Shi F G 2006 *Microsystems Packaging Assembly Conference Taiwai* **18** 1
[9] T. X. Lee T X, K. F. Gao K F, Chien W T and Sun C C 2007 *Opt. Express.* **15** 6670
[10] Chang S J, Chen W S, Shei S C, Ko T K, Shen C F, Hsu Y P, Chang C S, Tsai J M, Lai W C and Lin A J 2007 *IEEE T Adv. Packaging* **30** 752
[11] Kim K, Choi K K, Kim K K, Cho J, Lee S N, Park Y, Kwak J S and Seong T Y 2008 *Opt. Lett.* **33** 1273
[12] Djurisic A B, Chan Y and Li E H 2001 Proc.SPIE **4283** 630
[13] Muth J F, Brown J D, Johnson M A L, Yu Z H, Kolbas R M, Cook R M and Schetzina J F, 1999 *MRS Internet J. Nitride Semicond. Res.* 4S1 G5.2
[14] Muth J F, Lee J H, Shmagin I K, Kolbas R M, Casey H C, Keller B P, Mishra U K and DenBaars S P 1997 *Appl. Phys. Lett.***71** 2572

Enhancement of evanescent waves in a planar waveguides with an anisotropic metamaterial layer

Wenqiang Qiu[1], Min Cheng[2], Rong Chen[1,*], Hongxin Lin[1], Zufang Huang[1],

[1]Key Laboratory of Optoelectronic Science and Technology for Medicine, Ministry of Education, Fujian Normal University, Fuzhou 350007, China

[2]Department of Physics and Electronic Information Engineering, Minjiang University, Fuzhou 350108, China

*Corresponding author: chenr@fjnu.edu.cn

Abstract. The enhancement of evanescent waves in waveguides consisting of an anisotropic metamaterials layer with partially negative permittivity and permeability is investigated for both TE and TM modes. The amplitudes of evanescent waves increase exponentially with the thickness of the metamaterials. The enhancement factor increases with the thickness of the metamaterial but reaches its maximum value when complete surface polaritons are established. Numerical results confirm our analysis. The enhancement effect may have potential applications in integrated optical devices.

1. Introduction

As early as 40 years ago artificial materials, called as the left-handed materials (LHMs) or metamaterials, were first analyzed theoretically by Veselago.[1] The novel materials exhibit simultaneous negative permittivity and permeability in certain frequencies and thus would possess blinking properties, such as negative refraction,[1,2] inverse Doppler shift,[1,2,3] reverse Goos-Hänchen shift[4] and reversed Cerenkov radiation[1,2,5]. Periodical arrays of metallic rods and split ring resonators (SRRs) were used to fabricate the LHMs by Shelby *et al.* in 2001. [6] Since that many potential applications of the LHMs have been proposed. Up to now almost all of the LHMs having been realized in experiments are actually anisotropic in nature and easier to implement in practical applications. Recently, the properties of evanescent field in the LHMs slab have attracted much attention [7-11]. The enhancement of evanescent waves in waveguides using isotropic metamaterials has been discussed [11]. However, the properties of the enhancement of evanescent waves using anisotropic metamaterials is sill not clear.

In this paper, we investigate the enhancement of evanescent waves in an anisotropic planar waveguides. The characteristic equations are derived. According to different combinations of parameters of the anisotropic medium, four distinct cases for oscillating and evanescent modes are analyzed. Then we present the transverse profiles for TE_0 and TE_1 modes and examine the enhancement factor in detail. This treatment reveals that the enhancement factor increases exponentially with the thickness of the metamaterial but saturates when complete surface polaritons mode are established at the interface between the metamaterial and the cladding. The numerical results confirm our analysis that the evanescent waves can be drastically enhanced by metamaterials, as long as we choose the appropriate parameters.

2. Method

We consider a structure of multilayer-integrated waveguides, adding an anisotropic metamaterials layer into the conventional dielectric three-layer waveguides, as illustrated in Fig.1. A substrate has positive permittivity ε_s and permeability μ_s. The width of the guiding layer with positive ε_g and μ_g is d_1. A cladding layer with positive ε_c and μ_c was coated onto the guiding layer. However, in the working region there is a metamaterials layer which is a biaxial anisotropic medium with permittivity tensor $\vec{\varepsilon} = \varepsilon_0\left(\varepsilon_x\hat{x}\hat{x}+\varepsilon_y\hat{y}\hat{y}+\varepsilon_z\hat{z}\hat{z}\right)$ and permeability tensor $\vec{\mu} = \mu_0\left(\mu_x\hat{x}\hat{x}+\mu_y\hat{y}\hat{y}+\mu_z\hat{z}\hat{z}\right)$. The thickness of metamaterials layer is d_2.

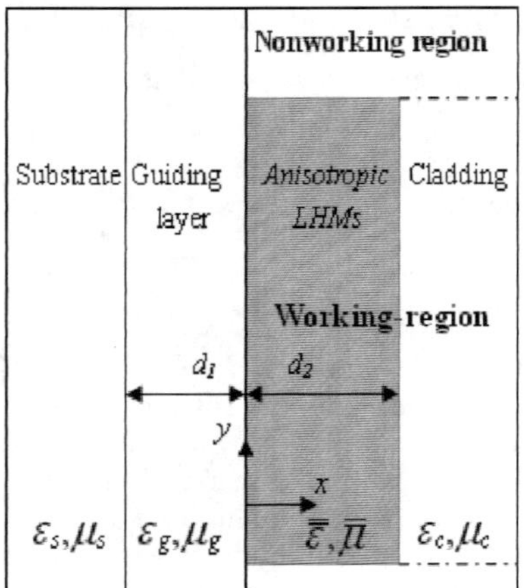

Figure 1. Schematic structure of the multilayer integrated waveguides with an anisotropic metamaterial layer.

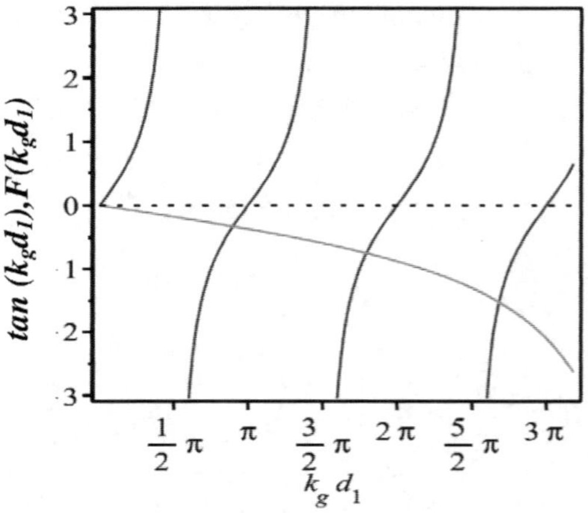

Figure 2. Graphical determination of $k_g d_1$ for guided modes corresponding to Eq.(11). The parameters are $\varepsilon_c\mu_c$=1.2 , $\varepsilon_g\mu_g$=1.55 , $\varepsilon_s\mu_s$=1.45, b=14 and c=25.

For simplicity, we assume all the materials are lossless and in order to analytically express the electromagnetic fields in such structures, our discussions are limited to planar structures. We begin by focusing on the transverse electric (TE) waves. An analogous result can be obtained for transverse magnetic (TM) waves by a dual analysis, i.e., by making the following corresponding transformations: $(\mu \rightleftharpoons \varepsilon)$ and $(E \rightarrow H)$. Suppose that the electromagnetic waves propagate along the x direction, and the electric field E is polarized along the z-axis (see Fig.1). The time dependence of the monochromatic field is expressed as $\exp(i\omega t)$, and hence the form of the solution is $E_Z(y)\exp[i(\beta x-\omega t)]$, where ω is the angular frequency of the field and β is the propagation constant. From Maxwell's equations, we can get the scalar wave equations of a waveguide with a coordination system shown in Fig.1 as follows:

$$\frac{d^2}{dy^2}E_z(y)+\frac{\mu_x}{\mu_y}\left(\frac{\omega^2}{c^2}\varepsilon_z(y)\mu_y(y)-\beta^2\right)E_z(y)=0, \tag{1}$$

$$\frac{d^2}{dy^2}E_z(y)+\left(\frac{\omega^2}{c^2}\varepsilon_i(y)\mu_i(y)-\beta^2\right)E_z(y)=0\,,(i=c,g,s). \tag{2}$$

Eq.(1) refers to the anisotropic metamaterials layer, and Eq.(2) corresponds to other regions, that is the cladding, the substrate and the guiding layer. The dispersion relations in normal isotropic media and anisotropic metamaterials can be written as

$$k_{iy}^2 = \frac{\omega^2}{c^2}\varepsilon_i(y)\mu_i(y) - \beta^2, (i=c,g,s), \tag{3}$$

$$k_{my}^2 = \frac{\mu_x}{\mu_y}\left(\frac{\omega^2}{c^2}\varepsilon_z(y)\mu_y(y) - \beta^2\right). \tag{4}$$

When β is large enough, there will be a case that only evanescent waves exist in the substrate, the cladding and the metamaterial layer while guided waves oscillate in the guiding layer. Under this condition, in the working region the electric field in each layer of the waveguides has the following solutions for TE modes:

$$E_z(y) = A\exp[k_s(y+d_1)], (-\infty < y < -d_1), \tag{5}$$

$$E_z(y) = B\cos(k_g y + \phi), (-d_1 < y < 0), \tag{6}$$

$$E_z(y) = C_1\exp(-k_m y) + C_2\exp(k_m y), (0 < y < d_2), \tag{7}$$

$$E_z(y) = D\exp[-k_c(y-d_2)], (d_2 < y < +\infty), \tag{8}$$

Where k_c , k_m and k_s correspond to the evanescent coefficients of the claddings, the metamaterial and the substrate, respectively, and k_g is the transverse wave number in the guiding layer.

Due to only evanescent waves existing in the metamaterial layer, k_{my} should be imaginary i.e., $k_{my} = ik_m$, where $k_m = \sqrt{(\beta^2 - \varepsilon_z\mu_y\omega^2/c^2)\mu_x/\mu_y}$, and the following inequality is valid:

$$\frac{\mu_x}{\mu_y}\left(\beta^2 - \frac{\omega^2}{c^2}\varepsilon_z\mu_y\right) > 0. \tag{9}$$

After matching the boundary conditions that require the tangential components $E_z(y)$ and $H_x(y)$ to be continuous at $y = -d_1, 0$ and d_2 , respectively, we get the characteristic equation of the multi-layer waveguides:

$$k_g d_1 = n\pi + \arctan\left(\frac{\mu_g k_s}{\mu_s k_g}\right) + \arctan\left(\frac{k_m\mu_g}{k_g\mu_x}\cdot\frac{(k_c\mu_x + k_m\mu_c) - (k_m\mu_c - k_c\mu_x)\cdot\exp(-2k_m d_2)}{(k_c\mu_x + k_m\mu_c) + (k_m\mu_c - k_c\mu_x)\cdot\exp(-2k_m d_2)}\right) \tag{10}$$

Where n is the mode number, and takes the value of 0, 1, 2, 3 and so on.

If we choose a metamaterial in a way that $k_c\mu_x + k_m\mu_c = 0$, according to Eqs.(7) and (8) and also the boundary conditions we get $C_1 = 0$, $D = C_2\exp(k_m d_2) = B\cos(\phi)\exp(k_m d_2)$. In the cladding of the conventional dielectric three-layer waveguides (the nonworking region, see Fig.1), it is easy to obtain that $D = B\cos(\phi)$. Here, to one's interest, it is found that the amplitude of evanescent wave is enhanced by $\exp(kd_2)$ due to the existence of metamaterials layer. Provided that $\mu_s = \mu_g = \mu_c = \mu_0$, Eq.(10) reduces to

$$k_g d_1 = n\times\pi + \arctan(k_s / k_g) + \arctan(k_c / k_g) \tag{11}$$

Equation (11) is exactly identical to the characteristic equation of the conventional dielectric three-layer waveguide without the metamaterials layer [12]. Therefore, the same characteristic equation holds for both the working region and the nonworking region, and consequently the propagation constants for them are identical [11].

3. Numerical analyses and results

In our coordinate system (see Fig.1), the time-averaged Poynting vector in the metamaterials layer can be developed as follows:

$$\vec{S} = \frac{\beta}{2\omega\mu_y}\mathrm{Re}\left[C_1^2 e^{-2ky} + C_2^2 e^{2ky} + 2C_1C_2\right]\hat{x}. \tag{12}$$

The boundary conditions require that the x component of the wave vectors in each layer should be continuous and equal to the same vector $\vec{\beta}$. Since the wave is propagating along the direction of x axis, parallel to the interface, we have $\beta > 0$. By further analyzing the time-averaged Poynting vector, we can easily find that the following inequality is satisfied: $\vec{\beta}\cdot\vec{S}<0$ when $\mu_y<0$. And thus the wave vector and the Poynting vector are antiparallel in the x direction for $\mu_y<0$. Similar conclusions were also achieved in Ref.10, 13 and 14.

The dispersion relation of the anisotropic medium, given in equation (4), may be elliptic or hyperbolic according to different combinations of parameters. Since the metamaterials exhibits the negative group refraction for $\mu_y<0$, in what follows we will discuss four distinct cases where μ_y is always negative.

Case (I): $\mu_x<0$ and $\varepsilon_z<0$. The tensor elements are just opposite in sign to those of the conventional anisotropic media. The metamaterials satisfy the elliptic dispersion relations, given in equation (4), and can support both oscillating and evanescent modes. The inequality (9) can be extended as follows: $\frac{\omega}{c}\sqrt{\varepsilon_g\mu_g}\geq\beta\geq\max\left\{\frac{\omega}{c}\sqrt{\varepsilon_z\mu_y},\frac{\omega}{c}\sqrt{\varepsilon_c\mu_c},\frac{\omega}{c}\sqrt{\varepsilon_s\mu_s}\right\}$.

Case (II): $\mu_x<0$ and $\varepsilon_z>0$. It is apparent from Eq.(4) that k_{my} is purely imaginary, i.e., the inequality (9) always holds, therefore the metamaterials only support the surface modes. No other addition condition is required upon the metamaterials region in this case, but the condition $\frac{\omega}{c}\sqrt{\varepsilon_g\mu_g}\geq\beta\geq\max\left\{\frac{\omega}{c}\sqrt{\varepsilon_c\mu_c},\frac{\omega}{c}\sqrt{\varepsilon_s\mu_s}\right\}$ is also needed.

Case (III): $\mu_x>0$ and $\varepsilon_z<0$. The dispersion relation of the metamaterial is a hyperbolic one. Such a metamaterial layer can support both oscillating and evanescent modes. The inequality (9) can be expressed as $\frac{\omega}{c}\sqrt{\varepsilon_z\mu_y}\geq\beta$, while $\frac{\omega}{c}\sqrt{\varepsilon_g\mu_g}\geq\beta\geq\max\left\{\frac{\omega}{c}\sqrt{\varepsilon_c\mu_c},\frac{\omega}{c}\sqrt{\varepsilon_s\mu_s}\right\}$ are the necessary conditions for such a waveguide. To sum up, in this case β must satisfy the conditions: $\min\left\{\frac{\omega}{c}\sqrt{\varepsilon_z\mu_y},\frac{\omega}{c}\sqrt{\varepsilon_g\mu_g}\right\}\geq\beta\geq\max\left\{\frac{\omega}{c}\sqrt{\varepsilon_c\mu_c},\frac{\omega}{c}\sqrt{\varepsilon_s\mu_s}\right\}$.

Case (IV): $\mu_x>0$ and $\varepsilon_z>0$. The metamaterials satisfy the hyperbolic dispersion relations. Due to the real component k_{my}, the inequality (9) is invalid, i.e., k_m is always imaginary, so it can only support the oscillating modes in the anisotropic media. Thus in this case it won't enhance the evanescent waves with the metamaterial layer.

Here, we take the case I as an example. The guidance condition curve and the dependence of the electric field distribution on the thickness are shown in Figures 2 and 3, respectively. For simplicity, we choose small refractive index contrast, namely, assume that $\varepsilon_g\mu_g=1.55$, $\varepsilon_c\mu_c=1.2$, $d_1=d_2$, $b=k_0d_1\sqrt{(\varepsilon_g\mu_g-\varepsilon_s\mu_s)}=14$, $c=k_0d_1\sqrt{(\varepsilon_g\mu_g-\varepsilon_c\mu_c)}=25$ and $\mu_x=-0.4$. We use a graphical method to determine the solution to k_gd_1 determined by Eq.(11), as illustrated in Fig.2, where the intersections show the existence of the guided modes. The transverse profiles corresponding to the two guided modes in Fig.2 are shown in Fig.3. As illustrated in Fig.3, the electric fields are oscillating in the guiding layer but evanescent outside the guiding layer. The fundamental mode TE_0 exists in this case, as shown in Fig.3(a). The field amplitudes have a critical enhancement in the metamaterial layer and decay exponentially in cladding layer for both TE_0 and TE_1 modes, while decaying slowly in the substrate, as Fig.3(a) and Fig.3(b) show. Both of these electric fields distribution make us believe that metamaterials can indeed enhance the evanescent wave in this multi-layer waveguides. However, as stressed in Ref.11, the enhancement effect is the result of redistribution of the electromagnetic field intensity profile due to the existence of the metamaterial, and does not contradict the energy-conservation law.

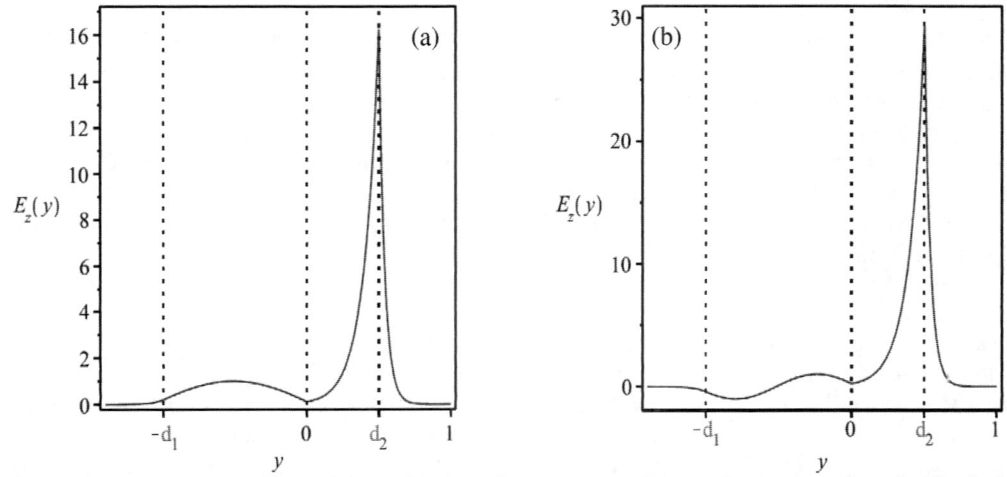

Figure 3 Transverse profiles of the guided modes corresponding to the intersections in Fig.2. (a) TE_0: $k_g d_1 = 2.82515$; (b) TE_1: $k_g d_1 = 5.64088$. Parameters are $\mu_x = -0.4$, $\varepsilon_c \mu_c = 1.2$, $\varepsilon_g \mu_g = 1.55$, $\varepsilon_s \mu_s = 1.45$ and $d_1 = 2d_2$. The vertical dot-lines denote the interface of two slabs. Amplitudes are in arbitrary units.

Secondly, comparing the power fraction in the cladding without and with a metamaterial will help us to evaluate the enhancement effect. Where no metamaterial exists (the nonworking region in Fig.1), the power fraction P_1 in the cladding for TE modes can be written as [12]

$$P_1 = \frac{k_g^2}{k_c(k_g^2 + k_c^2)\left(d_1 + 1/k_s + 1/k_c\right)} \tag{13}$$

Where there is a metamaterials layer, the power fraction in the cladding P_2 is

$$P_2 = \frac{k_g^2 \cdot \exp(2k_m d_2)}{k_c\left(k_g^2 + k_c^2\right)} \cdot \left(\frac{(k_m|\mu_y| + k_c) \cdot k_g^2 \cdot \exp(2k_m d_2) - k_c k_g^2 + k_m k_c^2 |\mu_y|}{|\mu_y| \cdot k_m \cdot k_c\left(k_g^2 + k_c^2\right)} + d_1 + 1/k_s\right)^{-1} . \tag{14}$$

The enhancement factor Q is defined by $Q = P_2/P_1$. Here, we take the case I as an example and plot how Q responds to d_2 and μ_x in Fig.4.

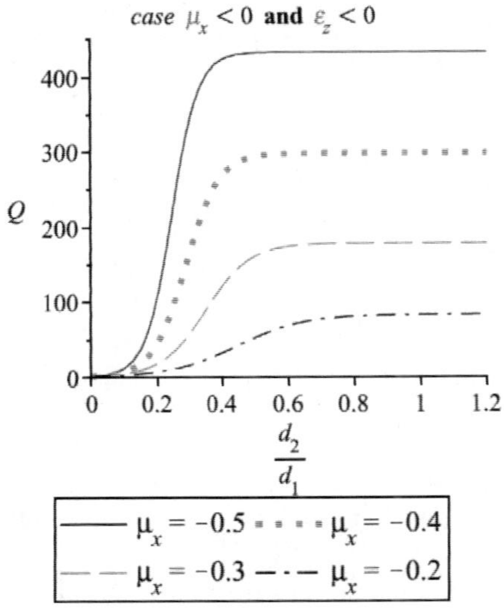

Figure 4. Enhancement factor Q as the function of the thickness of metamaterials d_2 for TE_0 (case I: $\mu_x < 0$, $\varepsilon_z < 0$). Parameters are the same as in Fig.2 except that μ_x changes for several other values.

As shown in Fig.4, one could see that for various parameters of μ_x, there are different values of enhancement factor. The enhancement factor for μ_x=−0.4 in Fig.4 corresponds to the case displayed by the electric field component in Figs.2.and 3. To one's interest, the enhancement factor increases with the thickness of metamaterials, but this does not last forever. Instead, there exists a saturation value of the enhancement factor. As *Qing et al.* have pointed out in Ref.11, this can be interpreted by surface polaritons effects [15] and is actually reasonable: when the thickness becomes large enough, a complete surface polaritons mode is established at the interface between the metamaterial and the cladding. Owing to the surface polaritons effect, all the energy flux is confined closely at the two sides of the interface, and thus the power fraction in the cladding will saturate. [11]

4. Conclusions

In this paper, we have demonstrated that metamaterials with partially negative permittivity and permeability can enhance the amplitude of the evanescent waves in waveguides for both TE and TM modes, provided that the surface polaritons conditions are satisfied. After the characteristic equations having been derived, we present the transverse profile for each layer. It is showed that the amplitude of evanescent waves is enhanced exponentially. We also examine the enhancement factor and find that it does not always increase with the thickness but saturates when complete surface polaritons have been established at interface between the metamaterial and the cladding. The enhancement effect may have potential applications in integrated optical devices, such as optical direction couplers and waveguide sensors. With the metamaterials, it may enhance the coupling effect and reduce the dimensions of these devices.

5. Acknowledgements

This work was supported by the National Natural Science Foundation of China (Nos. 60778046 and 60910106016), the Science and Technology Project of Fujian Province (Nos. 2008J0016 and 2009J01276), the Project of Science Foundation of Ministry of Health and United Fujian Provincial Health and Education Project for Tackling the Key Research (No.WKJ2008-2-046).

References

[1] V. G. Vesalago, The electrodynamics of substances with simultaneously negative values of ε and μ, *Sov. Phys. Usp.* **10(4),** 509-14

[2] D. R. Smith, Willie J. Padilla, D. C. Vier, S. C. Nemat-Nasser, and S. Schultz, Composite Medium with Simultaneously Negative Permeability and Permittivity, *Phys. Rev. Lett.* **84,** 4184-87

[3] N. Seddon, and T. Bearpark. Observation of the inverse Doppler effect, *Science*, **302,** 1537-40

[4] P. R. Berman, Goos-Hänchen shift in negatively refractive media, *Phys. Rev. E* **66,** 067603-05

[5] C. Luo, Mihai Ibanescu, Steven G. Johnson and J. D. Joannopoulos, Cerenkov Radiation in Photonic Crystals, *Science* **299,** 368-71

[6] R. A. Shelby, D. R. Smith, and S. Schultz, Experimental verification of a negative index of refraction, *Science* **292,** 77-79

[7] J. B. Pendry, Negative refraction makes a perfect lens, *Phys. Rev. Lett.* **85,** 3966-69

[8] N. Garcia and M. Nieto-Vesperinas, Left-Handed Materials Do Not Make a Perfect Lens, *Phys. Rev. Lett.* **88,** 207403(1)-(4)

[9] C. Liu, C. Yan, H. Chen, and Y. Liu, Evanescent field on the surface of a negative-index planar lens, *Appl. Phys. Lett.* **88**, 231102(1)-(2)

[10] Q. Cheng, and T. J. Cui, Electromagnetic interactions between a line source and anisotropic biaxial media, *J. Appl. Phy.* **98**, 074903(1)-(8)

[11] D.-K. Qing, G. Chen, Enhancement of evanescent waves in waveguides using metamaterials of negative permittivity and permeability, *Appl. Phy. Lett.* **84**, 669-71

[12] D.-K. Qing, X.-M. Chen, K. Itoh , and M. Murabayashi, A theoretical evaluation of the

absorption coefficient of the optical waveguide chemical or biological sensors by group index method, *J. Lightwave Technol.* **14**, 1907-16

[13] I.V. Lindell, S.A. Tretyakov, K.I. Nikoskinen, and S. Ilvonen, BW media-media with negative parameters, capable of supporting backward waves, *Microwave Opt. Technol. Lett.* **31**, 129-33

[14] L. Hu and S. T. Chui, Characteristics of electromagnetic wave propagation in uniaxially anisotropic left-handed materials, *Phys. Rev. B* **66**, 085108(1)-(7)

[15] R. Ruppin, Surface polaritons of a left-handed medium, *Phys. Lett. A* **277**, 61-64

Journal of Physics: Conference Series **276** (2011) 012085 doi:10.1088/1742-6596/276/1/012085

Novel all-optical dispersion monitoring technique for ultra-high-speed WDM networks

Sheng Cui[1,2], Li Li[1,2] and Deming Liu[1,2]
1 Wuhan National Laboratory for Optoelectronics, No.1037, Luoyu Road, Wuhan, Hubei, 430074, China
2 College of Optoelectronic Science and Engineering, Huazhong University of Science & Technology, No.1037, Luoyu Road, Wuhan, Hubei, 430074, China

Tel:086-27-61020961, Fax:086-27-87556188
E-mail: cuisheng@mail.hust.edu.cn

Abstract. This paper represents a novel all-optical dispersion monitoring technique based on fiber parametric amplifiers (FOPAs). The monitoring method is truly bit-rate transparent because it is enabled by the exponential power transfer function (PTF) provided by the FOPA gain. The slope of the PTF is increased from 2 to 3 by choosing appropriate phase-matching conditions. Due to the steeper PTF the monitoring sensitivity is greatly improved compared to the other PTF-based methods proposed before. The PTF obtained by numerical simulations agrees very well with the experimental results. Numerical simulations are then used to demonstrate that our method can be used to monitor signals in various modulation formats.

1. Introduction

Due to continued growth of the Internet and the introduction of new broadband services, it is expected that individual channel data rates will exceed 100Gb/s in the next 5 years. One of the major limitations to operating at such high data rates will be the amount of chromatic dispersion (CD) introduced [1]. As channel data rates increases, CD tolerance decreases precipitously as the square of the increase in data rate. Due to the severer CD tolerance CD fluctuations caused by seasonal temperature variations and mechanical stress become a serious problem [2]. In addition, in dynamic networks, each channel can experience a different amount of CD whenever the network is reconfigured. Thus the adaptive dispersion compensation technique becomes indispensable [1-5]. However, in order to realize the adaptive dispersion compensation, it is vital to develop a fast and accurate CD monitoring technique [3-5]. By now, a number of monitoring techniques have been proposed, including: RF tone measurement [1]; asynchronous histogram evaluation [3]; self-phase modulation [4] and cascaded four-wave mixing [5,6] in optical fibers. However each of these approaches tends to require either high-speed components (e.g. oscilloscope, photodetector and analog-to-digital converter) or high input power. Recently two-photo absorption in semiconductor detectors is also used in CD monitoring in high-speed, WDM networks [7,8]. In order to overcome the inefficiency associated with the nonlinear TPA process, a microcavity structure is employed. But the signal is only enhanced over a very narrow wavelength range. Thus to monitor different WDM channels mechanical movement components must be used to alter the incident angle of the input signal so that the resonance wavelength can be changed [8].

In this paper, we proposed a novel all-optical CD monitoring method based on a fiber optical parametric amplifier (FOPA). No fast electronics are needed due to the ultra-fast response time of FWM (<10fs). The pre-amplified signal is used as the pump of the FOPA. The pump and a cw probe wave are input into the FOPA. The accumulated dispersion of the input signal can be measured just by monitoring the average power of the idler wave at the output port of the FOPA using an optical power meter. The slope of the PTF is increased from 2 to 3 by choosing appropriate phase-matching conditions. Due to the steeper PTF the monitoring sensitivity is greatly improved compared to the other PTF-based methods proposed before. The dependence of sensitivity and operational bandwidth on phase matching conditions is analysed. The monitoring bandwidth can be tailored to cover any commonly used WDM bands. Numerical simulations are then used to demonstrate that our method can be used to monitor signals in various modulation formats.

2. Device concept

Because FWM is a quasi-instantaneous effect, the parametric gain actually depends on the instantaneous intensity of the pump. Thus if the signal serves as the pump and is input into the FOPA with a cw probe wave, the FOPA gain can map the CD induced instantaneous power variations of the signal pulse onto the average power of the output idler wave. The PTF, i.e. the gain, for one-pump FOPA can be derived analytically when pump depletion and fiber loss can be neglected and is given by [9].

$$G = (\gamma P_0 L)^2 [\sinh(gL)/gL]^2 , \tag{1}$$

where P_0 is pump power, γ is the fiber nonlinear coefficient, L is the fiber length. The output power of the idler wave is $P_{idler}(L) = GP_{probe}(0)$. In Eq. (1) g is the parametric gain defined as:

$$g = \sqrt{-\Delta\beta(\gamma P_0 + \frac{\Delta\beta}{4})} . \tag{2}$$

The linear phase mismatch $\Delta\beta$ is given by

$$\Delta\beta = \beta_{probe} + \beta_{idler} - 2\beta_{pump} , \tag{3}$$

where $\beta_{probe,idler,pump}$ are the respective propagation constants of the probe, idler and pump. Eqs. (1-2) show that when the condition

$$-4\gamma P_0 \le \Delta\beta \le 0 \tag{4}$$

is satisfied, g is real. In this case $P_{idler}(L)$ increases exponentially with P_0 [10-12] and in this way, an exponential PTF can be obtained. For ordinary fibers the effects of the fourth order dispersion can be neglected. Considering up to the third order dispersion and by Taylor expansion around the fiber zero dispersion frequency ω_0, $\Delta\beta$ can be expressed as [10]

$$\Delta\beta(\omega_p) = \beta_3(\omega_{pump} - \omega_0)(\omega_{pump} - \omega_{probe})^2 . \tag{5}$$

Here ω_{pump} and ω_{probe} are the pump and probe frequencies, respectively. Noting that $\beta_3 > 0$ in telecommunication bands for most optical fibers, to get the exponential gain, it is required that $\omega_{pump} \le \omega_0$ and $P_0 \ge P_{\min} = -\Delta\beta / 4\gamma$ [10]. The monitoring bandwidth depends on the allowable tuning range of the pump over which exponential gain can sustain. A detailed analysis represented before by us shows that when $\omega_{probe} = \omega_0 - 3(\gamma P_0 / \beta_3)^{1/3}$, the monitoring bandwidth can reach its maximum $\Delta\omega_{pump}^{\max} = 4(\gamma P_0 / \beta_3)^{1/3}$ for a given pump power P_0 and the operational band can be tailored by selecting fibers with appropriate ω_0 [11]. So the operational band of the monitor can be tailored according to the wavelength plan of the WDM system.

Journal of Physics: Conference Series **276** (2011) 012085 doi:10.1088/1742-6596/276/1/012085

3. Monitoring results

The monitoring is enabled by the nonlinear PTF of the FOPA. Once the PTF is obtained, the pulse shape can be mapped from the signal onto the idler wave by the PTF [6]. So we first use numerical simulations to calculate the PTF of the FOPA and compare it to the experimental results reported by [13,14] in order to validate the simulation results. The parameters used in the calculation and experiments are listed in Table.1. The setup of the experiment is showed in Fig.1. Fig.2 shows the PTFs for FOPA1 and FOPA2. For FOPA1, $\Delta\beta$ =1.98 $\times 10^{-4} m^{-1}$, thus $\Delta\beta \gg -4\gamma P_0$ for $2.5mW \le P_0 \le 5mW$. While for FOPA2 $\Delta\beta$ =-4e-3, thus $-4\gamma P_0 \le \Delta\beta \le 0$ for $100mW \le P_0 \le 200mW$. As we can see in Fig.2 (a) and (b) with the appropriate phase-matching conditions the slope is increased from 2 to 3. The results obtained by numerical simulations agree very well with the experimental results. Fig.3 shows the monitoring results utilized FOPA2. The output power of the idler wave is scaled by the one at CD=0ps/nm. As we can see it changes by about 5.9, 1.35, 0.39 and 4.6dB within the monitoring range for 33% RZ, CSRZ, NRZ and 33% RZ-DPSK signals. For comparison the quadratic PTF-based monitor relying on two photon absorption (TPA) has only 3.1, 0.45 and 0.088dB output dynamic range for 33% RZ, CSRZ and NRZ signals [7], while the quadratic PTF-based monitor relying on cascaded FWM has only 1.8dB dynamic range for a 40Gb/s RZ signals

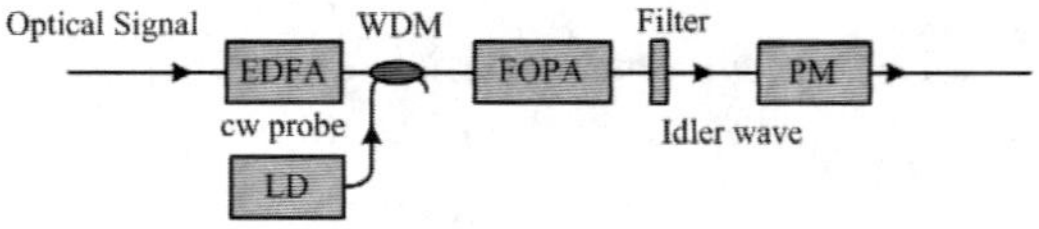

Fig.1 Configuration the monitor. EDFA: pre-amp erbium doped fiber amplifier, LD: laser diode, PM: power meter.

Table 1 Parameters used in calculation and experiments

	λ_p (nm)	λ_{probe} (nm)	λ_0 (nm)	S (ps/nm/km)	γ ($W^{1}km^{-1}$)	L (km)	α (dB/km)	P_0 (mW)
FOPA1	1559	1560.6	1557	0.06	2.1	17.5	0.21	2.5~5
FOPA2	1534.6	1545	1534	0.05	9.9	0.72	0.6	100~200

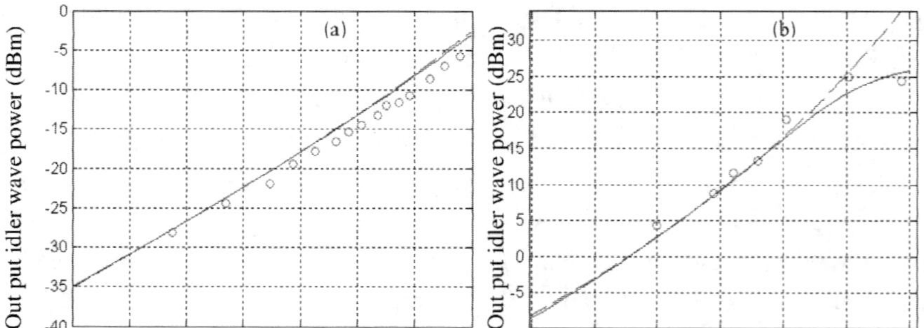

Fig.2 Comparison between experimental (circles) and calculated transfer functions of the idler wave power versus pump power for FOPA1 (a) and FOPA2 (b). The calculation was carried out analytically (dashed lines) and numerically (solid lines).

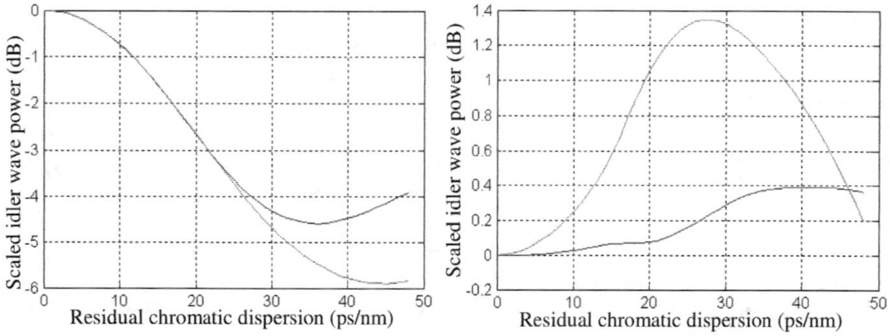

Fig. 3 Monitoring results utilizing FOPA2. (a): the modulation formats are 33% RZ (red) and 33% RZ-DPSK (blue). (b): the modulation formats are CSRZ (red) and NRZ (blue)

[6]. Thus The sensitivity and output dynamic range of the monitor are greatly improved compared to those PTF-based monitors proposed before. Due to the Talbot effect the signal pulse will reconstruct after the residual CD exceeds the Talbot period [7], for this reason the monitoring ranges are limited. For RZ, RZ-DPSK and NRZ signals the range is about 40ps/nm. While for CSRZ signals the range is about 30ps/nm. This underlying periodicity is fundamental to the data sequence itself, it is not a unique feature of the nonlinear monitor and will also manifest itself in many other dispersion-monitoring schemes such as RF spectral analysis [1,6,7]. Finally it is noteworthy that average input power required is actually about 50mW because P_0 is the peak power of the signal pulse. So the monitor based on the FOPA can work in a low power consumption mode because the FOPA gain can be very small (-10~10dB).

4. Conclusions

This paper presents a novel method to monitor CD in high-speed WDM networks. The method utilizes the fast response time and the broad bandwidth of FWM process. By choosing appropriate phase matching conditions the sensitivity of the monitor is greatly improved compared the other PTF-based methods proposed before. No fast electronics are needed and the required operational power is low. This monitor is very suitable for future ultra-high speed and reconfigurable WDM networks.

References

[1] Zhongqi Pan, Changyuan Yu, Alan E. Willner, "Optical performance monitoring for the next generation optical communication networks," *Optical Fiber Technol.* **16** 20-45

[2] Takatoshi Kato, Yasushi Koyano and Masayuki Nishimura, "Temperature Dependence of Chromatic Dispersion in Various Types of Optical Fibers," *OFC 2000, paper* TuG7.

[3] B.Kozicki, O. Takuya, and T. Hidehiko, "Optical performance monitoring of phase modulated signals using asynchronous amplitude histogram analysis," *J. Lightw. Technol.* **26** 1353-1361.

[4] Paul S. Westbrook, Benjamin J. Eggleton, Gregory Raybon, Stefan Hunsche, and Tsing Hua Her, "Measurement of Residual Chromatic Dispersion of a 40-Gb/s RZ Signal via Spectral Broadening," *IEEE Photon. Technol. Lett.* **14** 74-76.

[5] Trina T. Ng, Justin L. Blows, Joe T. Mok, Ross W. McKerracher, and Benjamin J. Eggleton, "Cascaded Four-Wave Mixing in Fiber Optical Parametric Amplifiers: Application to Residual Dispersion Monitoring," *J. Lightw. Technol.* **23** 818-826.

[6] T. T. Ng, J. L. Blows, M. Rochette, J. A. Bolger, I. Littler and B. J. Eggleton. "In-band OSNR and chromatic dispersion monitoring using a fiber optical parametric amplifier," *Optics Express* **13** 5542-5552 .

[7] Stephan Wielandy, M. Fishteyn and B. Zhu, "Optical performance monitoring using nonlinear detection," *J. Lightw. Technol.* **22** 784-793.

[8] K. Bondarczuk, P. J. Maguire and Douglas A. Reid, "Chromatic Dispersion Monitoring for High-Speed WDM Systems Using Two-Photon Absorption in a Semiconductor Microcavity," *J. Quantum Electron* **45** 223-232.

[9] G. P. Agrawal, *Nonlinear Fiber Optics*, (Academic Press, San Diego, 2001, Chap. 10.

[10] R. W. McKerracher, J. L. Blows and C. M de Sterke, "Wavelength conversion bandwidth in fiber based optical parametric amplifiers," *Optics Express* **11** 1002-1007.

[11] Sheng Cui, Deming Liu, Ying Wang, Feng Tu, "Method for effectively utilizing tunable one pump fiber parametric wavelength converters as an enabling device for WDM routers," *Optics Express* **17** 1454-1465.

[12] Sheng Cui, Deming Liu, Shaohua Yu, "Downhill simplex algorithm based approach to holey fiber design for tunable fiber parametric wavelength converters," *Optics Express* **18** 9831-9839.

[13] A. Bogris and D. Syvridis, "Regenerative Properties of a Pump-Modulated Four-Wave Mixing Scheme in Dispersion-Shifted Fibers," *J. Lightw. Technol.* **21** 1892-1902.

[14] G. A. Nowak, Y. -H. Kao, T. J. Xia and M. N. Islam, "Low-power high-efficiency wavelength conversion based modulational instability in high-nonlinearity fiber," *Optics Lett.* **23** 936-938.

Acknowledgements

This work is supported by the National Basic Research Program of China (973 Program, Grant No. 2010CB328300), China International Science and Technology Cooperation Project (Grant No. 2009DFA12640) and National Natural Science Foundation of China (Grant No. 61007043) .

Angular electromagnetic response of double-ring metamaterials for TE polarization

J.H. Shi[1, *1], N. Papasimakis[2], V.A. Fedotov[2], Z.P. Wang[1] and N.I. Zheludev[2]
[1]College of Science, Harbin Engineering University, Harbin 150001, China
[2]Optoelectronics Research Center, University of Southampton, Southampton SO171BJ, UK

[*] E-mail: shijinhui@hrbeu.edu.cn

Abstract We investigate transmission characteristics in symmetric planar double-ring metamaterials at oblique angles of incidence theoretically. The double-ring microwave metamaterials are used to study resonant response to incident TE electromagnetic wave. The trapped-mode resonant feature results from the excitation of antisymmetric current mode and the quality factor is associated to the angle of incidence.

1. Introduction

Metamaterials have attracted a tremendous amount of attention in recent years since it was fabricated to exhibit simultaneously negative values of effective permeability $\mu_{eff}(\omega)$ and permittivity $\varepsilon_{eff}(\omega)$ at microwave frequencies [1,2]. And planar metamaterials or metafilms are very thin metal films that are patterned on a subwavelength scale. Initially used as microwave frequency selective surfaces (FSS) [3,4], metafilms now attract attention due to their unusual functionalities ranging from invisibility [5], magnetic mirror [6] and optical magnetism [7] to asymmetric transmission [8] and optical activity [9], and have become a subject of intense research in the terahertz domain demonstrating application potential in wave manipulation and sensing [10-13]. The response of metafilms can also be engineered to mimic electromagnetically induced transparency (EIT) of atomic systems [14-19]. The spectral response of most FSSs, however, depends strongly on the polarization and, more importantly, on the angle of incidence of incident electromagnetic waves. Limited attempts were made to reduce their angular sensitivity [20-24], which led to a more complex structures exploiting bilayered compositions [25] and, more recently, fractal [26,27] and volume split-ring resonators [28]. Further, Metamaterials with polarization and direction insensitive resonant transmission response mimicked EIT [29]. However, the oblique incidence of electromagnetic wave provides some desirable properties of artificial composites. The oblique incidence resulted in optical activity of metamaterials without chirality [30] and split resonance [31]. The oblique response of a double negative metamaterial slab was investigated to show that the negative index characteristics remain nearly the same up to incidence angle of 45° [32]. One fishnet metamaterial was demonstrated tuning of the transmission resonance when the oblique incidence occurred [33].

[*] To whom any correspondence should be addressed.

In this letter, Fano-type resonances in the planar double-ring metamaterials was reported for TE polarization incidence theoretically that insensitively depend on angles of incidence, not resulting from breaking the symmetry of the metamaterial structure.

2. Double-ring metamaterial structure

The double-ring metamaterial structure has a simple design ideally suited for the existing manufacturing planar micro- and nanotechnologies, which was used to mimic electromagnetically induced transparency experimentally and show resonant features that weekly depend on polarization and angles of incidence [29]. Here, we provide theoretical support to the experimental results of double-ring trapped-mode metamaterial showing resonant transmission spectra. The metamaterial was formed by a planar array of concentric subwavelength ring resonators, which had ever been considered for achieving negative refraction in bulk composites. The double-ring pattern was etched from 35μm copper cladding covering FR4 PCB substrate of 1.6 mm thickness. The unit cell of the pattern contains one double-ring structure and is a square having size of 15×15 mm^2. The metamaterial structure is illustrated in figure1, showing geometrical parameters. The linewidth of metallic rings is 0.4mm and the mid-radii of inner and outer ring are 4.5mm and 5.45mm, respectively. In the experiment, the sample size was only 220μm×220μm that sufficiently generated strong electromagnetic response. Such a metamaterial did not diffract at frequencies below 10 GHz for any angle of incidence. However, for the simplification the overall size of the sample is infinite in the model, which enables us to conduct transmission for any incident angle ranging from 0° to 90° . The metal pattern was treated as a perfect conductor and the substrate was assumed to be a lossy dielectric with $\varepsilon = 4.05\text{-}i0.1$.

The origin of the narrow resonance can be traced to the excitation of the so-called trapped mode, a high-Q mode formed by counter-propagating current. Such a mode normally inaccessible but can be excited if, for example, the metamaterial symmetry is appropriately broken. As a result of the interference with continuum-like spectrum of the excitation the mode acquires an asymmetrically shaped resonance line characteristic to the well-known Fano resonance. The trapped mode nature of the observed pass-band was confirmed by numerical simulations based on method of moments.

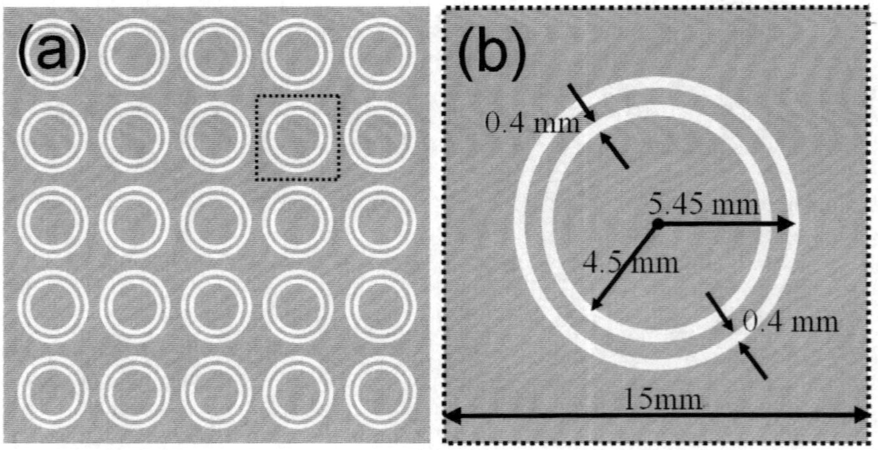

Figure 1 Double-ring metamaterial structure. Dashed box indicates its elementary unit cell.

3. Simulation results

The double-ring metamaterial transmission was measured in an anechoic chamber in the frequency range of 4-10GHz using broadband horn antennas (Schwarzbeck BBHA9120D) equipped with dielectric lens concentrators and a vector network analyzer (Agilent E8364B) [29]. Here, we present the theoretical transmission spectrum of double-ring metamaterial at angles of incidence ranging from 0°

to 80° for TE polarization. Figure 2 shows a typical response of the planar metamaterial to incident linearly polarized waves, which was recorded in transmission.

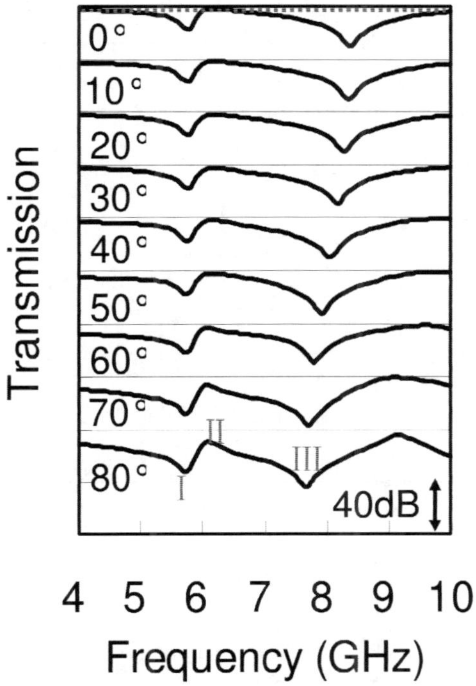

Figure 2 Transmission spectrum of double-ring metamaterial at angle of incidence for TE polarization. The red dashed line indicates zero dB level.

It should be noted that although the data are presented for TE polarization, the metamaterial also reveals weakly changed angular response for TM polarization (not shown here). The simulated spectra of double-ring metamaterial illustrated in figure 2 show a good agreement with the measured one in Ref.29. For the metamaterial, transmission pass-bands always happen at any angle of incidence and the electromagnetic response weakly depends on the angle of incidence. The peak transmission at the trapped-mode resonance gradually decreases with the increasing angle of incidence, but the transmission peaks are kept unmoved. Evidently, the spectrum of double-ring metamaterial reveals a sharp resonance near 6.2 GHz (marked as II), which is accompanied by two much weaker resonances (marked as I and III) corresponding to transmission dips. For the normal incidence, the sharp spectral response has the width of 0.8 GHz as measured at 3 dB below the maximum and the quality factor Q of such response is about 8. While the sharp spectral response has the width of 0.15 GHz as measured at 3 dB below the maximum and the quality factor Q of such response is up to 40 for the glazing incidence though a large reflection loss exists. The relation between the quality factor and the angle of incidence is shown in figure3. Only large angle of incidence, even to the glazing incidence, leads to the dramatic change of Q, where the metamaterial shows strong coupling with the magnetic component of incident electromagnetic.

The origin of the narrow resonance can be traced to the excitation of the so-called trapped mode, a high-Q mode formed by counter-propagating current [8]. Such a mode normally inaccessible but can be excited if, for example, the metamaterial symmetry is appropriately broken. As a result of the interference with continuum-like spectrum of the excitation the mode acquires an asymmetrically shaped resonance line characteristic to the well-known Fano resonance [34]. The absolute surface current distributions are shown in figure4, corresponding to I, II and III resonance. The strong antisymmetric current configuration of double metallic rings results in the electromagnetic energy confined to the metamaterial surface, revealing weak coupling to the free space and sharp resonance.

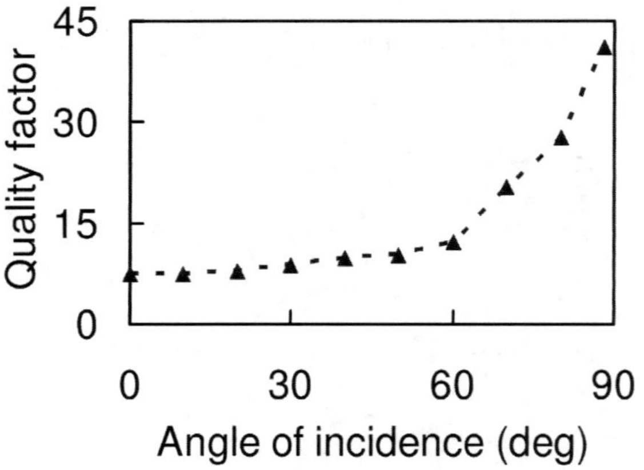

Figure 3 Angular response of the quality factor of double-ring metamaterial.

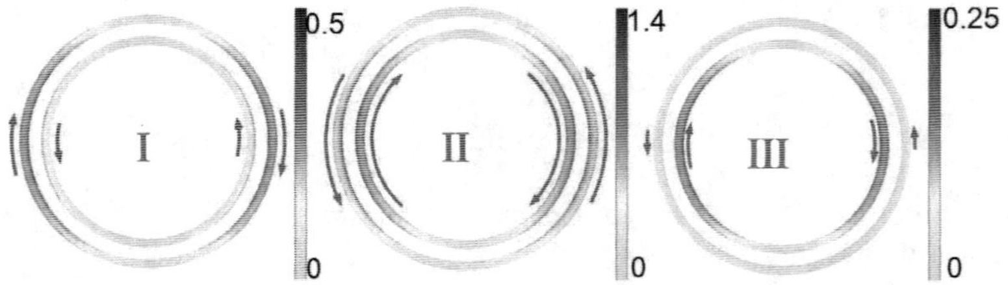

Figure 4 Absolute surface current distribution of double-ring metamaterial corresponding to the resonances.

4. Conclusions

In summary, the planar double-ring metamaterial exhibiting Fano-type resonance at oblique angles of incidence was demonstrated for TE polarization theoretically. The transmission resonances insensitively depend on the angles of incidence. The transmission response of double-ring metamaterials can show its unique functionality as wide-angular applications. Achieving the trapped-mode resonances will be especially important for low-loss metamaterials.

This work was supported by the Special Foundation for Harbin Young Scientists under grant 2008RFQXG031, and in part by the Special Foundation for Basic Scientific Research of Harbin Engineering University under grant HEUCF20101113.

Reference

[1] Veselago V G 1968 *Sov. Phys. Usp.* **10** 509
[2] Shelby R A, Smith D R, and Schultz S 2001 *Science* **292** 77
[3] Vardaxoglou J 1997 *Frequency Selective Surfaces* (Research Studies, England).
[4] Munk B A 2000 *Frequency Selective Surfaces: Theory and Design* (Wiley, New York)
[5] Fedotov V A, Mladyonov P L, Prosvirnin S L, and Zheludev N I 2005 *Phys. Rev. E* **72** 056613
[6] Sievenpiper D, Zhang L, Broas R, Alexopolous N, and Yablonovitch E 1999 *IEEE Trans. Microwave Theory Tech.* **47** 2059

[7] Linden S, Enkrich C, Dolling G, Klein M W, Zhou J, Koschny T, Soukoulis C M, Burger S, Schmidt F, and Wegener M 2006 *E J. Sel. Top. Quantum Electron.* **12** 1097
[8] Fedotov V A, Mladyonov P L, Prosvirnin S L, Rogacheva A V, Chen Y, and Zheludev N I 2006 *Phys. Rev. Lett.* **97** 167401
[9] Plum E, Fedotov V A, and Zheludev N I 2008 *Appl. Phys. Lett.* **93** 191911
[10] Chen H T, Padilla W J, Zide J M O, Gossard A C, Taylor A J, and Averitt R D 2006 *Nature* **444** 597
[11] Singh R, Smirnova E, Taylor A J, O'Hara J F, and Zhang W 2008 *Opt. Express* **16** 6537
[12] Singh R, Azad A K, O'Hara J F, Taylor A J, and Zhang W 2008 *Opt. Lett.* **33** 1506
[13] O'Hara J F, Singh R, Brener I, Smirnova E, Han J, Taylor A J, and Zhang W 2008 *Opt. Express* **16** 1786
[14] Fedotov V A, Rose M, Prosvirnin S L, Papasimakis N, and Zheludev N I 2007 *Phys. Rev. Lett.* **99** 147401
[15] Papasimakis N, Fedotov V A, Zheludev N I, and Prosvirnin S L 2008 *Phys. Rev. Lett.* **101** 253903
[16] Zhang S, Genov D A, Wang Y, Liu M, and Zhang X 2008 *Phys. Rev. Lett.* **101** 047401
[17] Liu N, Kaiser S, and Giessen H 2008 *Adv. Mater.* **20** 4521
[18] Tassin P, Zhang L, Koschny T, Economou E N, and Soukoulis C M 2009 *Phys. Rev. Lett.* **102** 053901
[19] Singh R, Rockstuhl C, Lederer F, and Zhang W 2009 *Phys. Rev. B* **79** 085111
[20] Roberts A and McPhedran R C 1988 *IEEE Trans. Antennas Propag.* **36** 607
[21] Shaker J and Shafai L 1993 *Electron. Lett.* **29** 1655
[22] Huang J, Wu T, and Lee S 1994 *IEEE Trans. Antennas Propag.* **42** 166
[23] Wu T K and Lee S W 1994 *IEEE Trans. Antennas Propag.* **42** 1484
[24] Van Labeke D, Gerard D, Guizal B, Baida F I, and Li L 2006 *Opt. Express* **14** 11945
[25] Shaker J and Shafai L 1995 *IEEE Microw. Guid. Wave Lett.* **5** 324
[26] Zhou L, Chan C T, and Sheng P 2004 *J. Phys. D* **37** 368
[27] Wen W, Zhou L, Hou B, Chan C T, and Sheng P 2005 *Phys. Rev. B* **72** 153406
[28] Baena J D, Jelinek L, Marques R, Mock J J, Gollub J, and Smith D R 2007 *Appl. Phys. Lett.* **91** 191105
[29] Papasimakis N, Fu Y H, Fedotov V A, Prosvirnin S L, Tsai D P, and Zheludev N I 2009 *Appl. Phys. Lett.* **94** 211902
[30] Plum E, Liu X X, Fedotov V A, Chen Y, Tsai D P, and Zheludev N I 2009 *Phys. Rev. Lett.* **102** 113902
[31] Enkrich C, Wegener M, Linden S, Burger S, Zschiedrich L, Schmidt F, Zhou J F, Koschny T, and Soukoulis C M 2005 *Phys. Rev. Lett.* **95** 203901
[32] Alici K B and Ozbay E 2009 *Opt. Lett.* **34** 2294
[33] Minovich A, Neshev D N, Powell D A, Shadrivov I V, Lapine M, McKerracher I, Hattori H T, Tan H H, Jagadish C, and Kivshar Y S 2010 *Phys. Rev. B* **81**15109
[34] Fano U 1961 *Phys. Rev.* **124** 1866
[35] Zheludex N, Prosvirnin S L, Papasimakis N, and Fedotov V A 2008 *Nat. Photonics* **2** 351

Improved the Surface Roughness of Silicon Nanophotonic Devices by Thermal Oxidation Method

Zujun Shi[1], Shiqian Shao[1], Yi Wang[1,*]

[1] *Wuhan National Laboratory for Optoelectronics, Huazhong University of Science and Technology,*
No. 1037, Luoyu Street, Wuhan 430074, China
Phone: 86-027-87792242
E-mail: ywangwnlo@mail.hust.edu.cn

Abstract- The transmission loss of the silicon-on-insulator (SOI) waveguide and the coupling loss of the SOI grating are determined to a large extent by the surface roughness. In order to obtain smaller loss, thermal oxidation is a good choice to reduce the surface roughness of the SOI waveguide and grating. Before the thermal oxidation, the root mean square of the surface roughness is over 11 nm. After the thermal oxidation, the SEM figure shows that the bottom of the grating is as smooth as quartz surface, while the AFM shows that the root mean square of the surface is less than 5 nm.

Key words- thermal oxidation, SOI, silicon waveguide, surface roughness

Ⅰ. INTRODUCTION

Silicon is an inexpensive and mature semiconductor material. Silicon waveguide is a basic component of silicon-based optoelectronics. SOI has met the needs of microelectronics technology development. Furthermore, it also has many advantages in the field of optoelectronic devices. Since silicon has a very low absorption loss in the infrared band. In the SOI waveguide, the refractive index difference between silicon and silica is nearly 2. The large index difference makes the optical field of the waveguide restricted more compactly, which will greatly reduce the propagation loss of the waveguide [1].

Fabrication of silicon gratings on an SOI wafer can be achieved by RIE (reactive ion etching) or ICP (inductively coupled plasma etching). The high-energy ion attacking will lead serious roughness of the etched surface of the silicon grating grooves and the silicon waveguide [2]. The transmission loss of the waveguide and the coupling loss of the grating are determined to a large extent by the surface roughness of silicon waveguide and grating [3]. In order to obtain smaller loss, using thermal oxidation to reduce surface roughness of the silicon grating is a good method [4].

II. PHYSICAL AND CHEMICAL MECHANISMS

The thermal oxidation method is categorized as dry oxygen oxidation and wet oxygen oxidation. Pure oxygen is used in the dry oxygen oxidation as the oxidizing ambient. During the dry oxidation, the silicon reacts with the ambient oxygen, forming a layer of silicon dioxide on its surface. In the wet oxidation, hydrogen gas and oxygen gas are introduced into a torch chamber where they react to form water molecules, which are then pumped into the reactor where they diffuse toward the wafers. The water molecules react with the silicon to generate the oxide. Since the reaction materials are a mixture of oxygen and water vapor, the oxidation rate of the wet oxygen oxidation is much faster than the dry oxygen oxidation.

The chemical equation of the dry oxygen oxidation is shown as following:

$$Si+O_2=SiO_2 \tag{1}$$

Besides the chemical equation (1), there is another chemical equation for the wet oxygen oxidation which is:

$$Si+2H_2O=SiO_2+2H_2 \tag{2}$$

The oxidizing ambient gas may also contain several percents of hydrochloricacid (HCl). The hydrochloricacid removes metal ions that may be contained in the oxide. The thermal oxide layer reacted with silicon consumed from the substrate and oxygen supplied from the ambient. Thus, it grows both down into the wafer and up out of it. For every unit thickness of consumed silicon, 2.27 unit thickness of the oxide will be engendered.

According to the commonly used Deal-Grove model, the time t required to grow an oxide thickness X_o, at a constant temperature, on a bare silicon surface, is:

$$t=\frac{X_0^2}{B}+\frac{X_0}{B/A} \tag{3}$$

where the constants A and B is determined by the characteristics of the reaction and the oxide layer, respectively. If a wafer that already contains oxide is placed in an oxidizing ambient, equation (3) must be modified by subtracting a corrective term τ at the right side of the equation, the time which would be required to grow the pre-existing oxide under current condition [5].

Solving the quadratic equation and obtaining X_o as below:

$$X_0(t)=A/2\cdot\left[\sqrt{1+\frac{4B}{A^2}(t+\tau)}-1\right] \tag{4}$$

Fig. 1 shows the kinetics of silicon thermal oxidation. In thermal oxidation of silicon, oxygen travels through the silica layer by diffusion. Oxygen reacts with silicon and then silica is formed around the silicon-silica interface. Silicon at the interface is consumed when oxidation takes place. As the oxide layer grows, the Si-SiO_2 interface moves into the silicon substrate. As a result, the Si-SiO_2 interface will always be below the original Si wafer surface. The SiO_2 surface, on the other hand, is always above the original Si surface. SiO_2 formation therefore proceeds in two directions relative to the original wafer surface. The shorter the diffusion distance is, the easier the diffusion is [6]. When the oxidation time is short, the thickness of the silica increased linearly with the oxidation time. When the oxide grows thicker, there is a parabolic relationship between the thickness of the silica and the oxidation time [7].

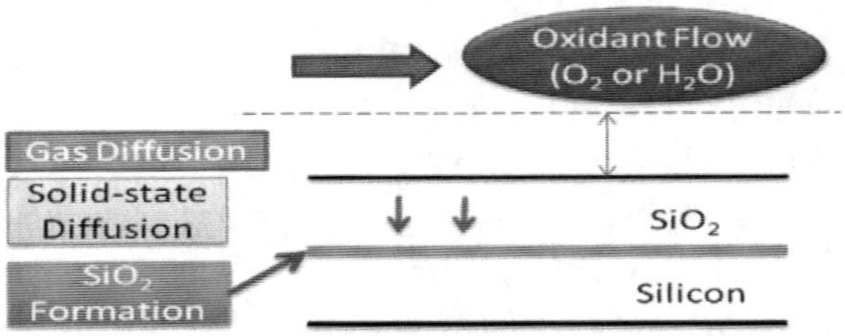

Fig. 1. The kinetics of silicon thermal oxidation

The compactness of the oxide formed by the dry oxygen oxidation is much better than that of the oxide formed by the wet oxygen oxidation, and the damage to the silicon by the dry oxygen oxidation is smaller than that by the wet oxygen oxidation. Furthermore, the thickness of silica is easier to control by the dry oxygen oxidation [8]. Therefore the dry oxygen oxidation is chosen for the further experiment of the thermal oxidation method.

The SOI wafer, on which the silicon waveguide and silicon grating is etched, is placed into a high-temperature furnace. The furnace is then vacuumized, heated to 1000 degrees Celsius and exposed in oxygen flow. The thickness of the oxide can be controlled by the oxidation time. It is important that all the materials put in the high-temperature furnace can withstand high temperature and will not import other impurities. The new impurities will increase transmission loss and coupling loss [9]. After the high temperature process, a thin oxide layer will be generated over the SOI grating surface, which can be removed by immersed in the hydrofluoric acid buffer. A smooth grating surface is realized.

III. RESULTS AND DISCUSSION

The silicon structures were etched by RIE first. The wafer is etched for 8 min in an air environment of 80% CHF_3, 3% O_2 and 20% SF_6 with an electrical power of 30 W. Then the thermal oxidation of the SOI devices is achieved by putting the wafer in the high-temperature furnace for one hour with 200 cc/min O_2 current under one thousand degrees.

The surface conditions of silicon grating before and after the thermal oxidation are observed by AFM (atomic force microscopy) firstly. The AFM testing results are shown in Fig. 2 and Fig. 3. Fig. 2 (a) is the AFM scanning result before the thermal oxidation and the right picture of Fig. 2 (a) is the sectional drawing of the back line in the left picture. Those sharp deep rags near the left sides of the grating bars are caused by measurement error of AFM. They are caused by the width of the scanning needle.

Before the thermal oxidation, the sidewalls are nearly perpendicular, and the needle scans from left to the right. When the right side of the needle reaches the left sidewall of the bars, the needle has to rise steeply to the top of the grating surface. A small gap is missed for the tip of the needle, and the tip touches nothing. In this case, the AFM considers it is a very deep gap as default. On the right sidewall, the needle scans from the top to the bottom and the needle has no encumbrance in its way. In this case, the top side and the bottom side are concatenated with a line. Fig. 2 (b) shows the AFM scanning result after the thermal oxidation and the red line and the green line in the right picture of Fig. 2 (b) are the sectional drawings of the red line and the green line in the left picture respectively. After the thermal oxidation, the sidewalls are less steep and thus the deep gaps caused by the measurement error of AFM are disappeared. Comparing the two figures, it can be seen that the roughness of the etched grooves of the grating has been greatly improved by the thermal oxidation method. Fig. 3 is the three dimension diagram of the grating after the thermal oxidation drawn by AFM, which shows that the bottom of the grating grooves is smooth and the roughness caused by the etching is removed by the thermal oxidation.

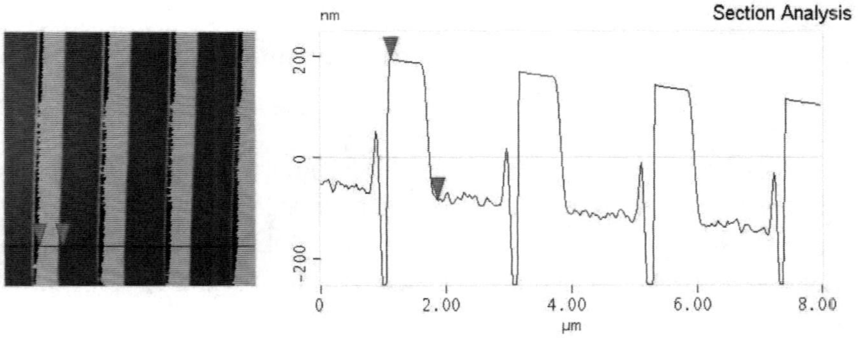

(a)

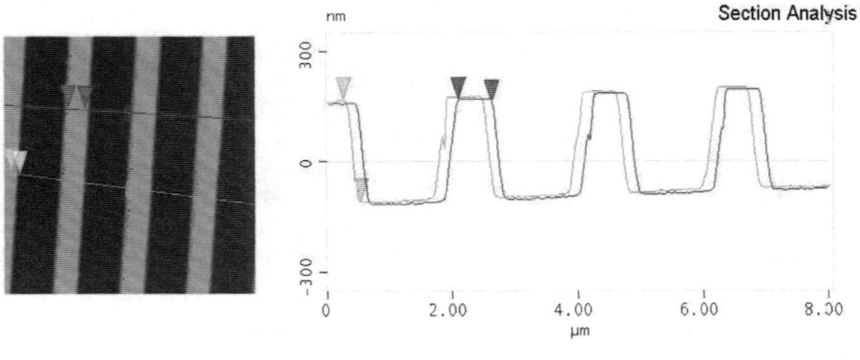

(b)

Fig. 2. (a) The surface conditions before the thermal oxidation by AFM. The right picture is the sectional drawing of the back line in the left picture. (b) The surface conditions after the thermal oxidation by AFM. The red line and the green line in the right picture are the sectional drawings of the red line and the green line in the left picture respectively.

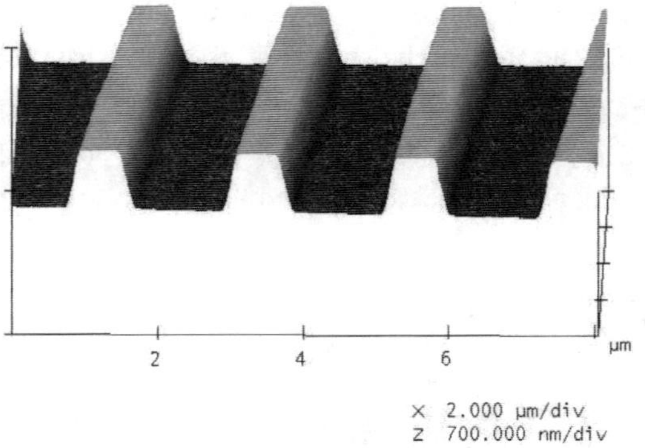

Fig. 3. 3D AFM diagram of the grating after the thermal oxidation.

The AFM data shows that before the thermal oxidation, the bottom of the etched grooves of the grating is very rough with many obvious protuberances and the root mean square of the surface roughness is over 11 nm. After the thermal oxidation, the peak protrusions have been eliminated mostly and the roughness of the bottom of the grating grooves has been approximately removed, while the root mean square of the surface is less than 5 nm.

The surface conditions of SOI grating before and after the thermal oxidation are also observed by SEM (scanning electron microscope). Fig 4 (a) is the SEM picture of a curved waveguide before the thermal oxidation. The etched surface is seriously rough according to the picture and there are many significant protuberances. Fig. 4 (b) is the SEM picture of the grating after the thermal oxidation. The roughness of the etched grooves of the grating is greatly improved and the notable protuberances are removed by the thermal oxidation, and the bottom of the grating is as smooth as quartz surface. Even the two objects are different, they are on the same wafer before and after the thermal oxidation, thus the results are comparable. The SEM results shows that the surface roughness have been greatly improved by the thermal oxidation method.

The oxide growth rate is affected by time, pressure and temperature. More specifically, oxide growth is accelerated by increasing oxidation time, pressure or temperature. When the thickness of oxide layer is small, the improvement of surface roughness is not obvious. Increasing oxidation time and oxygen flow can produce thicker oxide layer and make better improvement of the surface roughness, because the oxidation rate is the same at the beginning. With the oxidation time passing much longer, the oxidation rate of the vertex and the sidewall becomes smaller. Long-time thermal oxidation can reduce the surface roughness more significantly, but it also leads to more silicon consumption and thus narrower grating grooves. Proper dry etching and thermal oxidation parameters can lead to good surface-roughness improvement, which will reduce the scattering loss of the waveguide and guarantee the coupling efficiency of the grating.

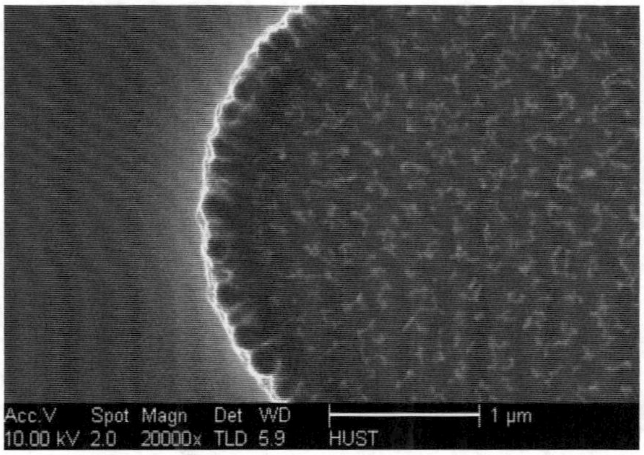

(a)

(b)

Fig 4. (a) Top view of the surface condition of a curved waveguide before the thermal oxidation by SEM. (b) Top view of the surface condition of the grating after the thermal oxidation by SEM.

IV. CONCLUSION

In this paper, the thermal oxidation method is used to improve the surface roughness of the etched grating grooves. Experimental results are shown by AFM and SEM. Before the thermal oxidation, the bottom of the etched grooves of the grating is very rough with many obvious protuberances. The root mean square of the surface roughness is over 11 nm. After the thermal oxidation, the peak protrusions have been eliminated mostly; the roughness of the bottom of the grating grooves has been approximately removed, and the SEM figure shows that the bottom of the grating is as smooth as quartz surface, while the AFM show that the root mean square of the surface is less than 5 nm. The surface condition of silicon grating is greatly improved. The thermal oxidation parameters can be properly chosen for good improvement of surface roughness of the grating.

REFERENCES

[1] R. J. Jaccodine and W. A. Schlegel, "Enhancement of silicon oxidation rate due to tensile mechanical stress," J. Appl. Phys. 37, 2429 (1966).

[2] J. P. R. Lacey and F. P. Payne, "Radiation loss from planar waveguides with random wall imperfections," IEEE Proc. Optoelectronics. 137, 282 (1990).

[3] K. K. Lee, D. R. Lim, H.-C. Luan, A. Agarwal, J. Foresi, and L. C. Kimerling, "Effect of size and roughness on light transmission in a Si/SiO_2 waveguide: Experiments and model," Appl. Phys. Lett. 77, 1617 (2000).

[4] M. A. Webster, R. M. Pafchek, A. Mitchell, and T. L. Koch, "Width dependence of inherent TM-mode lateral leakage loss in silicon-on-insulator ridge waveguides," IEEE Photon. Technol. Lett. 19, 429 (2007).

[5] R. Deaton and H. Z. Massoud, "Manufacturability of rapid-thermal oxidation of silicon: oxide thickness, oxide thickness variation, and system dependency," IEEE Trans. Semiconductor Manufacture, 5 (4), 347 (1992).

[6] H. Z. Massoud, "The onset of the thermal oxidation of silicon from room temperature to 1000°C," Microelectron. Eng. 28 (14), 109 (1995).

[7] H. Z. Massoud and J.D. Plummer, "Thermal oxidation of silicon in dry oxygen: growth-rate enhancement in the thin regime," J. Electrochem. Soc. 132, 2693(1985)

[8] D. B. Kao, J. P. McVittie, W. D. Nix, and K. C. Saraswat, "Modeling of stress effects in silicon oxidation," IEEE Trans. Electron Devices, 34, 1008 (1987).

[9] M. A. Webster, R. M. Pafchek, G. Sukumaran, and T. L. Koch, "Low-loss quasi-planar ridge waveguides formed on thin silicon- on- insulator," Appl. Phys. Lett. 87, 231108 (2005).

Modeling of Tunable DBRs Lasers Based on Transmission-Line Laser Model

Tan Shu and Yonglin Yu[1]

Wuhan National Laboratory for Optoelectronics, Huazhong University of Science and Technology,Wuhan, 430074, P. R. China

E-mail: yonglinyu@mail.hust.edu.cn

Abstract. The transmission-line laser model (TLLM) is extended to simulate the three-section distributed Bragg reflector (DBR) lasers, while a combined model of TLLM and digital filter approach is developed to study the sampled grating DBR (SGDBR) lasers. We integrate time-domain TLLM method and frequency-domain transfer matrix method (TMM) in the combined model. Characteristics of tunable DBR lasers and SGDBR lasers are all successfully simulated, such as L-I curve, and static tuning characteristics, which are in good agreement with the published experimental results. This new method is also used to study transient response and lasing spectrum of the devices, which are difficult to be obtained by the frequency domain models.

1. Introduction

Tunable lasers are important photonic integrated devices, and have potential wide applications in future high-capacity photonic networks based on dense wavelength division multiplexing and wavelength routing [1]. Among several structures, DBR based lasers seems more attractive due to their fast switching speed, such as distributed Bragg reflector (DBR) lasers [2], sampled grating DBR (SGDBR) lasers [3], superstructure grating DBR (SSGDBR) lasers [4], and digital supermode DBR (DSDBR) lasers [5] etc. In order to investigate the properties of these tunable DBRs lasers, powerful modeling tools are expected.

The TLLM [6]-[7] is a wide-bandwidth large signal dynamic laser model and developed by A J Lowery. It can take inhomogeneous effects, multiple longitudinal modes, carrier-induced refractive index change, and spontaneous emission noise into account. Moreover, it's a distributed-element circuit model, electrical parasitic and matching circuits can also be included in as a part of the model. TLLM has been successfully used to model Fabry-Perot lasers [8], distributed feedback semiconductor lasers [9], external-cavity DBR mode-locked lasers [10], and semiconductor optical amplifier [11]. In this paper, the TLLM is introduced to analyze tunable three-section DBR lasers and widely tunable four-section SGDBR lasers.

The traditional TLLM is extended to model three-section DBR lasers first. For four-section SGDBR lasers, a combined theory of TLLM and digital filter approach is developed. The active regions and the phase regions are easy to be described in time domain. However, the two reflectors of SGDBR are difficult to be described in time domain, due to the complex structure of sampled grating.

[1] To whom any correspondence should be addressed.

Published under licence by IOP Publishing Ltd

Here, the TMM, a frequency domain model, is used to characterize the sampled grating regions, and then is transformed into time domain via the digital filter approach. It has been verified the validity that time domain and frequency domain can be transformed into each other by using the finite impulse response (FIR) filter as a bridge [12]-[13].Based on the TLLM and digital filter approach, characteristics of tunable DBR lasers and SGDBR lasers, such as transient response, lasing spectrum, L-I curve, and static tuning characteristics are all successfully simulated.

The paper is organized as follows: In Section 2, a description of the TLLM used in the analysis is discussed. Different ways for the passive grating description of DBR laser and SGDBR laser, and the simulation results are given in Section 3. Finally, a brief conclusion is drawn in Section 4.

2. Model description

The basic principle of the TLLM is that the laser cavity length L is longitudinally devided into a number of equal length sections s, with a length ΔL, as shown in figure 1. In each section, a scattering matrix S is used to represent the optical process, including stimulation emission, spontaneous emission, and attenuation. The matrices of these sections are then connected by transmission-lines, which account for propagation delays of the optical wave. Each section includes a propagation delay of ΔT. This is related to the laser's cavity length and the group velocity v_g inside the cavity, given by

$$\Delta L = v_g \Delta T \tag{1}$$

The connecting processes can be represented by a connecting matrix C. From the iterations of scattering and connecting processes, the output optical field in the time domain can by obtained. Then, by applying fast Fourier transform (FFT), the laser output spectrum can be easily acquired. The independent carrier rate equation is used in each section to govern the carrier-photon interaction. In the following, we introduce the scattering and connecting matrices as well as the carrier density rate equation, separately.

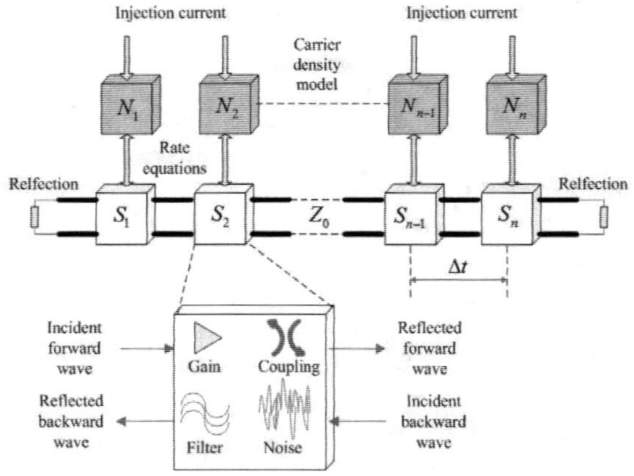

Figure 1. Schematic of the transmission line laser model

2.1. TLLM description for active section

2.1.1. Scattering Matrix. The scattering processes represent the optical process, including stimulated emission, spontaneous emission, and attenuation, in each section, which can be expressed as [14]

$$_k A(n)^r = S \cdot {}_k A(n)^i + {}_k A(n)^s \tag{2}$$

where ${}_{k}A(n)^{i}$ is the incident wave at the input of the n section at the k time step, ${}_{k}A(n)^{r}$ is the reflected wave from that section, and ${}_{k}A(n)^{s}$ is the wave due to the spontaneous emission. S respresents the scattering matrix, given by

$$S = \frac{1}{y(1+Z_s)}\begin{bmatrix} \gamma(G+y)(Z_s-1) & 2\gamma Y_C(Z_s-1) & 2\gamma Y_L(Z_s-1) & 2y \\ G(Z_s+1) & (2Y_C-y)(Z_s+1) & 2Y_L(Z_s+1) & 0 \\ G(Z_s+1) & 2Y_C(Z_s+1) & (2Y_L-y)(Z_s+1) & 0 \\ \gamma(G+y)2Z_s & 4\gamma Y_C Z_s & 4\gamma Y_L Z_s & y(1-Z_s) \end{bmatrix} \quad (3)$$

where G can be expressed as

$$G = \exp(\Gamma g \Delta L/2) - 1 \quad (4)$$

γ is the attenuation in each section, given by

$$\gamma = \exp(-\alpha_s \Delta L/2) \quad (5)$$

where Γ is the waveguide confinement factor, g is the material gain, α_s is the absorption coefficient.

Z_s is the impedance of the phase shift stub, Y_C is the capacitive reactance, Y_L is the inductive reactance, and $y = 1 + Y_C + Y_L$ is the total admittance of the RLC stub filter [7]. The phase shift stub is used to account the phase shift effect in the device. In fact, this phase shift is produced due to the impedance mismatching between the phase stub and the main transmission-line. The impedances of the phase shift stub are normalised to the cavity wave impedance and are given by

$$Z_s = \left|\cot(\pi f n_{eff} l/c)\right| \quad (6)$$

where f is the laser resonance frequency, n_{eff} is the effective refractive index, c is the velocity of light in a vacuum , l is the change in phase length across a section, given by

$$l = \frac{\Gamma \Delta L(N(n) - N_p)}{n_g}\frac{dn}{dN} \quad (7)$$

where N_p is an arbitrary carrier density for zero phase shift and is usually set to the threshold carrier density[15], n_g is the group effective refractive index, dn/dN is the refractive index carrier dependence coefficient. Therefore, the scattering process can be expressed as

$$ {}_{k}\begin{bmatrix} A(n) \\ A_C(n) \\ A_L(n) \\ A_P(n) \end{bmatrix}^{r} = S \cdot {}_{k}\begin{bmatrix} A(n) \\ A_C(n) \\ A_L(n) \\ A_P(n) \end{bmatrix}^{i} + {}_{k}\begin{bmatrix} I_s Z_C \gamma/2 \\ 0 \\ 0 \\ 0 \end{bmatrix}^{s} \quad (8)$$

where I_s is the noise current representing spontaneous emission [14], $A(n)$, $A_C(n)$, $A_L(n)$, and $A_P(n)$ are the traveling waves in the main transmission-line, the capacitive stub, the inductive stub, and the phase shift stub, respectively. Spontaneous emission sources in each section are generated by a Gaussian distributed random number generator. The mean-square value of these sources I_s is shown to be

$$I_s = mN(n)\sqrt{2\beta LhfB / Z_p} \tag{9}$$

where m is a unit constant with dimension of metres, $N(n)$ is the carrier density in section n, β is the spontaneous emission coupling coefficient, L is the laser cavity length, hf is the photon energy, B is the bimolecular recombination coefficients, and Z_p is the transverse-wave impedance of the waveguide [6].

2.1.2. *Connecting Matrix.* The connection matrix describes the cross coupling between forward and backward traveling waves. It provides the incident wave of the scattering section from the reflected wave of the previous section, which may be described by

$$_{k+1}A(n)^i = C \cdot {}_kA(n)^r \tag{10}$$

where C is the connecting matrix. For the active section and phase section of the tunable DBRs lasers, the connecting matrix of the main transmission-line can be written as:

$$_{k+1}\begin{bmatrix} A(n+1) \\ B(n) \end{bmatrix}^i = \begin{bmatrix} 1 & 0 \\ 0 & 1 \end{bmatrix}_k \begin{bmatrix} A(n) \\ B(n+1) \end{bmatrix}^r \tag{11}$$

For the inductive stubs in each section, the connection matrix is

$$_{k+1}\begin{bmatrix} A_L(n) \\ B_L(n) \end{bmatrix}^i = \begin{bmatrix} -1 & 0 \\ 0 & -1 \end{bmatrix}_k \begin{bmatrix} A_L(n) \\ B_L(n) \end{bmatrix}^r \tag{12}$$

For the capacitive stubs in each section, the connection matrix is

$$_{k+1}\begin{bmatrix} A_C(n) \\ B_C(n) \end{bmatrix}^i = \begin{bmatrix} 1 & 0 \\ 0 & 1 \end{bmatrix}_k \begin{bmatrix} A_C(n) \\ B_C(n) \end{bmatrix}^r \tag{13}$$

For the phase-adjusting stubs, which may be inductive or capacitive when the cotangent in eqn. 6 is negative, we have

$$_{k+1}\begin{bmatrix} A_p(n) \\ B_p(n) \end{bmatrix}^i = \begin{bmatrix} -1 & 0 \\ 0 & -1 \end{bmatrix}_k \begin{bmatrix} A_p(n) \\ B_p(n) \end{bmatrix}^r \tag{14}$$

and when it is positive,we have

$$\begin{bmatrix} A_p(n) \\ B_p(n) \end{bmatrix}_{k+1}^{i} = \begin{bmatrix} 1 & 0 \\ 0 & 1 \end{bmatrix}_k \begin{bmatrix} A_p(n) \\ B_p(n) \end{bmatrix}^{r} \tag{15}$$

2.1.3. *Carrier Density Rate Equation.* The time-dependent carrier density in the active region N satisfies

$$\frac{dN}{dt} = \eta \frac{I}{eV} - AN - BN^2 - CN^3 - 2gv_g S_p \tag{16}$$

where η is the injection efficiency, I is the injection current, e is the free electron charge, V is the volume of each section, A, B, and C are nonradiative linear, bimolecular and Auger recombination coefficients, respectively. S_p stands for the photon density, which is related to the incident waves from either side by[14]

$$S_p(n) = \frac{\left[(A^i(n))^2 + (B^i(n))^2\right] n_{eff}}{m^2 hfcZ_p} \tag{17}$$

The gain for multiple quantum well (MQWs) can be modeled by[16]

$$g = \frac{g_0 \ln(N/N_T)}{2(1+\varepsilon S_p)} \tag{18}$$

where g_0 is the gain coefficient, N_T is the transparent carrier density, and ε is the gain compression factor.

2.2. TLLM Description for Phase Section
The phase section has scattering and connecting processes, so the TLLM description for it is almost as same as that for the active section. There is no stimulated emission in the phase section, only attenuation is needed to be considered.

In the next section, different ways for the passive grating description of the DBR laser and SGDBR laser will be given.

3. Simulation results and disussion

3.1. Three-section tunable DBR lasers
In the DBR section, the refractive index n and the internal absorption α can be changed by injection currents of the DBR section, acccording to the free-carrier plasma effect [17]. The change of the carrier-induced index and internal absorption Δn and $\Delta \alpha$ can be described by

$$\Delta n = \Gamma \frac{dn}{dN} N \tag{19}$$

$$\Delta \alpha = \Gamma \frac{d\alpha}{dN} N \tag{20}$$

where dn/dN and $d\alpha/dN$ are material coefficients determined by the free-carrier plasma effect.

The carrier density related to the injection current of the passive section is given by

$$\frac{dN}{dt} = \eta \frac{I}{eV_p} - AN - BN^2 - C_p N^3 \tag{21}$$

where V_p, and C_p stand for waveguide volume, and Auger recombination coefficient in the passive waveguide, respectively.

For the Bragg grating regions of three-section DBR lasers, the coupling exists between forward and backward waves due to the modulation of the refrective index. As shown in figure 2, this effect can be described by the impedance transition of the transmission-line in the TLLM [10].

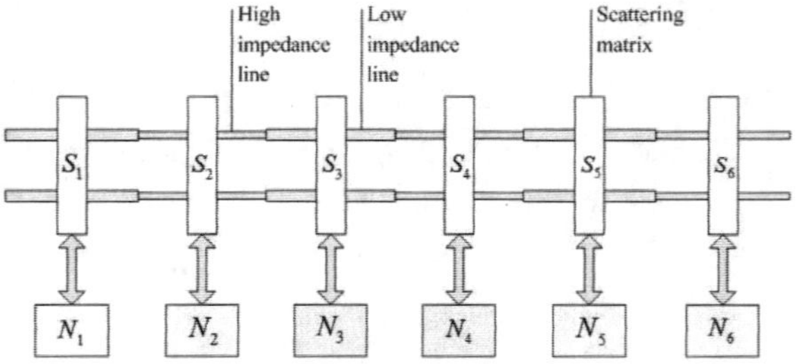

Figure 2. The TLLM for the DBR section

Two different matrixs are used alternately along the grating section. For a low-high impedance boundary, we have

$$\begin{bmatrix} A(n+1) \\ B(n) \end{bmatrix}_{k+1}^{i} = \begin{bmatrix} 1+k\Delta L & -k\Delta L \\ k\Delta L & 1-k\Delta L \end{bmatrix}_k \begin{bmatrix} A(n) \\ B(n+1) \end{bmatrix}^{r} \tag{22}$$

and for a high-low impedance boundary,we have

$$\begin{bmatrix} A(n+1) \\ B(n) \end{bmatrix}_{k+1}^{i} = \begin{bmatrix} 1-k\Delta L & k\Delta L \\ -k\Delta L & 1+k\Delta L \end{bmatrix}_k \begin{bmatrix} A(n) \\ B(n+1) \end{bmatrix}^{r} \tag{23}$$

k is the grating coupling per unit length.

The three-section tunable DBR laser for the simulation includes a 450-μm-long active section, a 150-μm-long phase section, and a DBR section with $\kappa L = 2$, Bragg wavelength at 1550nm, and the cavity length equal to 150μm. The other simulation parameters are given in Table 1.

Table 1. Simulation Parameters

Parameters		Symbol	Value
Waveguide width		w	2 μm
Waveguide thickness		d	0.05 μm
Waveguide loss	Active region	α_a	3000 m-1
	Passive region	α_p	200 m-1
Waveguide confinement factor	Active region	Γ_a	0.35
	Passive region	Γ_p	0.5
Effective refractive index		n_{eff}	3.23
Group refractive index		n_g	3.7

Transparent carrier density		N_T	1.5×1024 m-3
Initial carrier density		N_i	1.51×1024 m-3
Nonradiative linear recombination coefficient		A	1×108 s-1
Bimolecular recombination coefficient		B	1×10-16 m3 s-1
Auger recombination coefficient	Active region	C_a	2.5×10-41 m6 s-1
	Passive region	C_p	7×10-41 m6 s-1
Injection efficiency		η	0.8
Differentical gain coefficient		g_0	1.8×105 m-1
Nonlinear gain coefficient		ε	2×10-23 m3
Spontaneous coupling coefficient		β	5×10-5
Refractive Index derivative wrt carrier density		dn/dN	-2.34×10-26 m3
Absorption derivative wrt carrier density		$d\alpha/dN$	2.56×10-21 m2

The turn-on transient response and the output spectrum are simulated and shown in figure 3 and figure 4, respectively. The active section is initially biased at 0 mA, and then modulated with a step current of 100mA. The phase current and DBR current are stable at 0 mA. The optical spectrum is obtained by performing FFT of output optical field after the model running for some iteration. It should be noted that optical spectrum during the oscillation transient of the laser is somehow different from the one obtained when the laser is settled. Figure 4 gives a spectrum when the output field of the laser is stable.

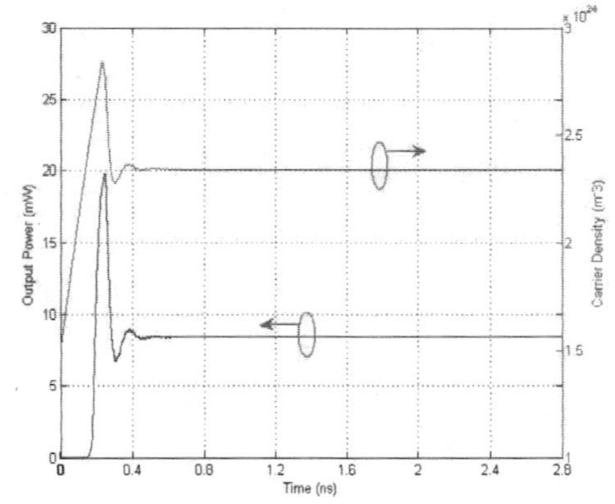

Figure 3. Turn-on transient response of DBR laser.

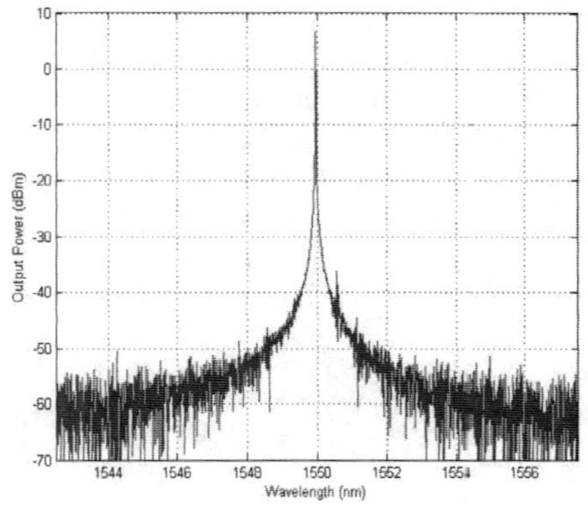

Figure 4. Output spectrum of DBR laser.

Figure 5 shows the output power versus bias current of the active section without application of the phase section and DBR section current. The simulations are performed with sufficient iterations for the turn-on transient to settle. As shown in Figure 5, the threshold current is about 18 mA, and the differential efficiency is about 0.104mW/mA.

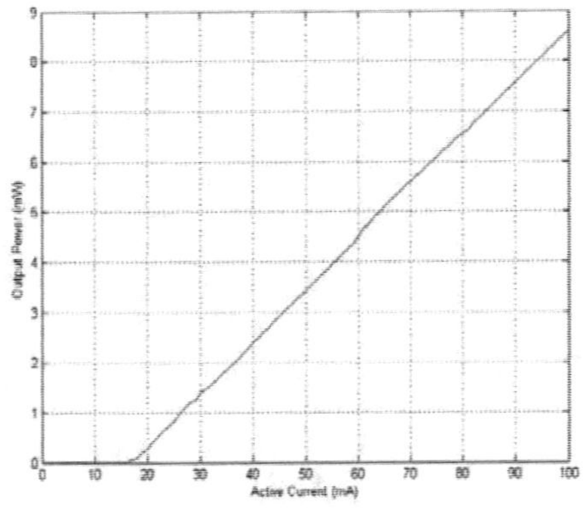

Figure 5. The output power versus bias current of the active section.

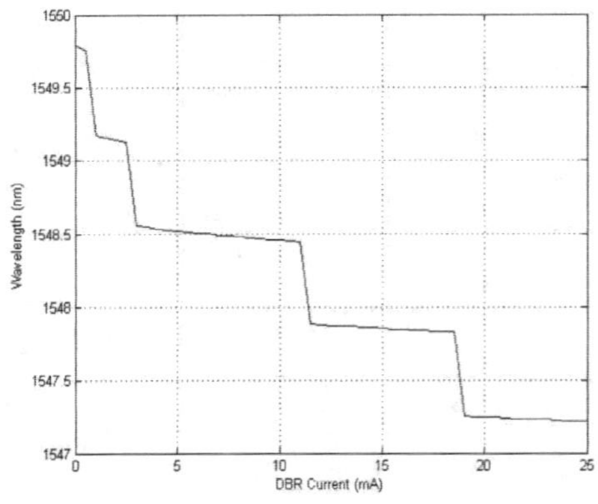

Figure 6. The lasing wavelength versus injection current of DBR section.

The DBR section can be seen as a frequency selective mirror. The refrective index changes with the injection current allows the peak wavelength to be tuned, thus selects one of the longitudinal mode of the active region as the dominate mode. The lasing wavelength versus injection current of DBR section is simulated and shown in Figure 6. The tuning curve is like a ladder, each step stands for a longitudinal mode. We can see that each step slants to short wavelength, since the longitudinal mode will shift toward the blue side with the increase of the DBR current.

3.2. Tunable SGDBR lasers

As mentioned above, the impedance transition of the transmission-line is used in the TLLM to model the DBR section of three-section DBR lasers. However, the two reflectors of SGDBR are difficult to be described in time domain, due to the complex structure of sampled grating. Here, the front sampled-grating (FSG) and rear sampled-grating (RSG) sections are first characterized by TMM in the frequency domain, and then they are transformed into time domain via the digital filter approach. As shown in figure 7, reflection and transmission filter are used to describe the reflectance and transmission characteristics of FSG and RSG sections, respectively.

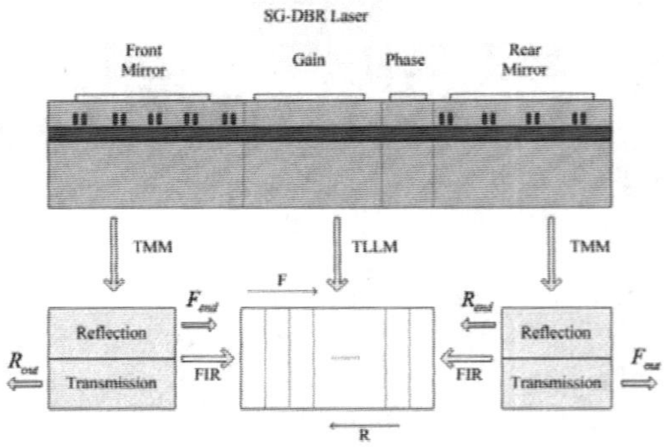

Figure 7. Schematic of the digital filter approach for the TLLM

FSR/RSG sections are first modeled by TMM [18]-[19] to get the reflectivity and transmission coefficient, then the FIR filter coefficients in the time domain $x(t)$ can be obtained easily by a reverse fast Fourier transform (FFT) transformation of the reflectivity and transmission coefficient function in the frequency domain, given by [12]

$$x(t) = \frac{1}{M}\sum_{k=0}^{M-1} X(f)e^{2\pi jkf\Delta t} \tag{24}$$

where $X(f)$ is reflectivity and transmission coefficient of FSR/RSG sections in the frequency domain, M is the FIR coefficient number.

The output optical field or the reflected field can be expressed by the M input field y^k and reflectivity and transmission FIR filter coefficients at the specific time k and earlier as [12]

$$F_{out}^{n} = \sum_{k=0}^{M} x^k(t) y^{n-k} \tag{25}$$

where $x^k(t)$ is the digital filter coefficients.

The SGDBR laser for the simulation includes a 450-μm-long active section, a 150-μm-long phase section, a 10-period front sampled-grating mirror with 6-μm-wide bursts using a 58.5-μm period and a 12-period rear sampled-grating mirror with 6-μm-wide bursts and 64.5-μm period. The other simulation parameters are given in Table 1.

Figure 8 is the turn-on transient response of SGDBR with a current step of 100 mA in the active section. The active section is initially biased at 0 mA. The FSG current, RSG current and phase current are stable at 3mA, 8mA, and 0mA, respectively. Figure 9 is the output spectrum obtained by FFT of the output optical field.

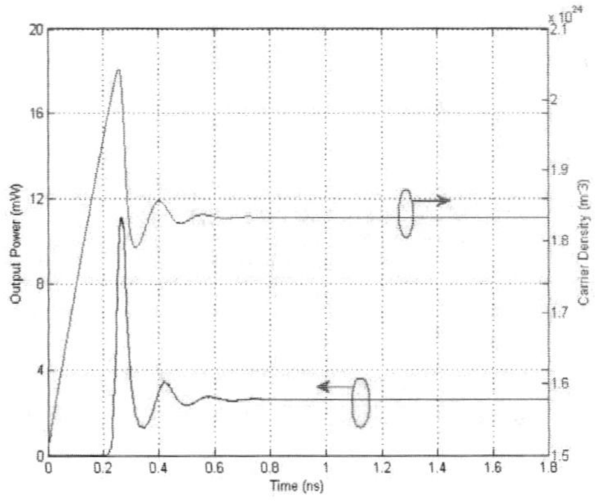

Figure 8. Turn-on transient response of the SGDBR laser.

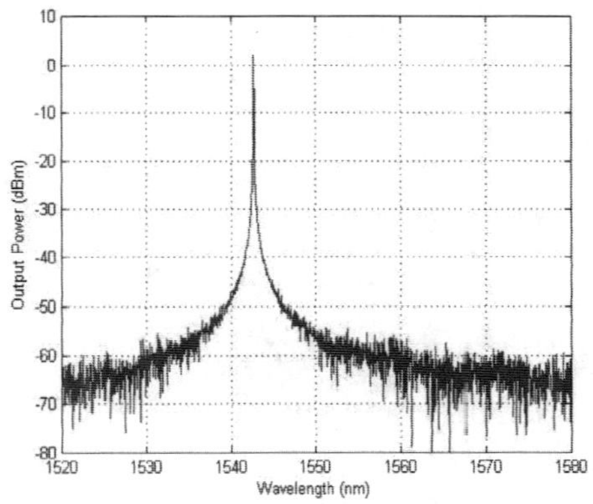

Figure 9. Output spectrum of the SGDBR laser.

Figure 10 shows the output power versus bias current in the active section for three different wavelength. It can be seen that the threshold currents are different for different lasing wavelength. These simulation results are agreed with the experimental report [20]. Figure 11 and figure 12 show the tuning curve of lasing wavelength versus current of FSG section and RSG section, respectively. We can see that these curves are alse like the ladder. Unlike the figure 6, each step stands for a super-mode, which represents a significant alignment of the comb-like reflection peaks of FSG and RSG, and the small steps are used to denote the change of different longitudinal modes. In figure 13, we show the output spectrum of different lasing wavelengths in the same graph. The tuning range is about 40 nm covered from 1533 to 1573 nm, which is about the same with the published experimental results [21].

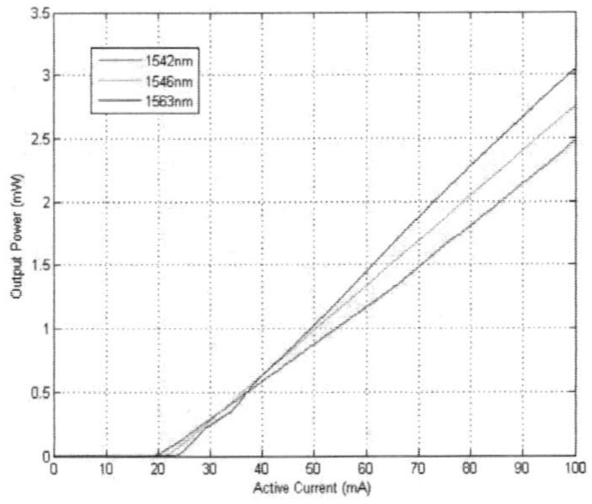

Figure 10. The output power versus bias current of the active section.

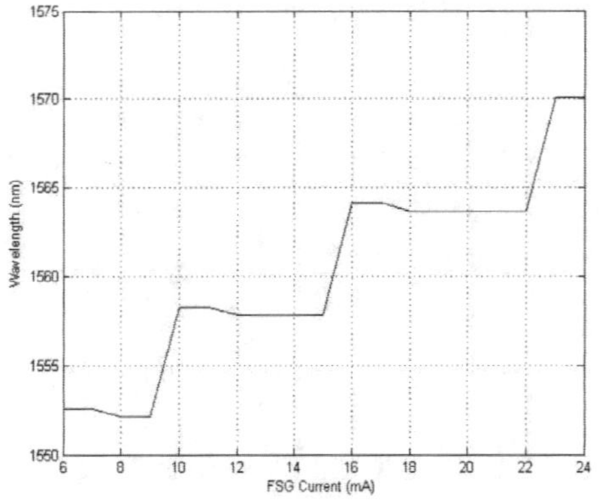

Figure 11. The lasing wavelength versus injection current of FSG section.

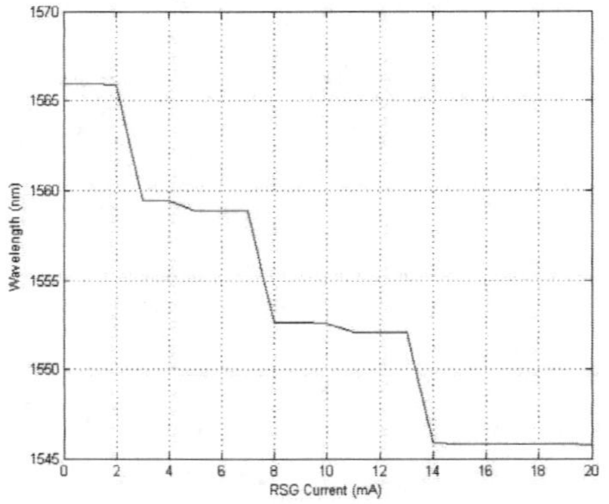

Figure 12. The lasing wavelength versus injection current of RSG section.

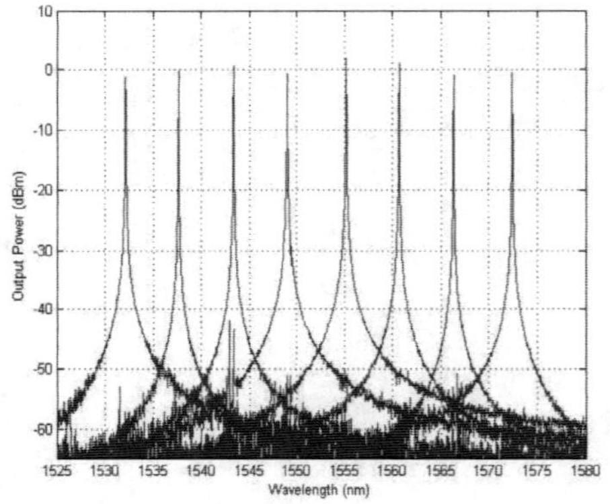

Figure 13. Output spectrum of different wavelengths in one graph.

4. Conclusion

The TLLM is used to model the tunable DBRs lasers, but different approaches are applied to simulate the passive grating section. The traditional TLLM developed by A J Lowery is used to simulate the three-section DBR lasers, while a combined theory of TLLM and digital filter approach is developed to model the SGDBR lasers. Characteristics of tunable DBR lasers and SGDBR lasers are all successfully simulated, including L-I curve, static tuning characteristics, lasing spectrum and transient response of the devices. Simulated results are in good agreement with the published experimental results. This method could be useful for for investigating the complex optoelectronic devices.

Acknowledgment

This work has been supported by the National High Technology Developing Program of China under Grant No. 2009AA03Z418, and the National Natural Science Foundation of China under Grant No. 60677024.

References

[1] Buus J and Murphy E J 2006 *J. Lightwave Technol.* **24** 5-11
[2] Zhang L and Cartledge J C 1995 *IEEE J. Quantum Electron.* **31** 75-81

[3] Jayaraman V, Chuang Z-M and Coldren L A 1993 *IEEE J. Quantum Electron.* **29** 1824-34
[4] Tomori Y, Yoshikuni Y, Ishii H, Kano F, Tamamura T, Kondo Y and Yamamoto M 1993 *IEEE J. Quantum Electron.* **29** 1817-22
[5] Ponnampalam L, Whitbread N D, Barlow R, Giacinto, Ward A J, Duck J P and Robbins D J 2006 *IEEE J. Quantum. Electron.* **42** 223-30
[6] Lowery A J 1987 *IEE Proceedings* **134** 281-89
[7] Lowery A J 1989 *International Journal of Numerical Modelling: Electronic Networks, Devices and Fields* **2** 249-65
[8] Nguyen L V T, Lowery A J, Gurney P C R and Novak D 1995 *Optical and Quantum Electron.* **27** 663-78
[9] Lowery A J 1989 *Electron. Lett.* **25**, 1307-08
[10] Lowery A J 1991 *IEE Proceedings* **138** 39-46
[11] Lowery A J 1988 *IEE Proceedings* **135** 242-50
[12] Li Wei, Huang Wei-Ping and Li Xun 2004 *IEEE J. Quantum Electron.* **40** 473-80
[13] Dong Lei, Zhang Ruikang, Wang Dingli, Zhao Shengzhi, Jiang Shan, Yu Yonglin and Liu Shuihua 2009 *J. Lightwave Technol.* **27** 3181-88
[14] Lowery A J 1990 *IEE Proceedings* **137** 293-300
[15] Lowery A J 1988 *IEE Proceedings* **135** 126-32
[16] Yu S F and Ngo N Q 2002 *IEEE J. Quantum Electron.* **38** 1062-74
[17] Buus J, Amann M C and Blumenthal D J 2005 *Tunable laser diodes and related optical sources*, Second Edition (New Jersey: John Wiley & Sons, Inc.)
[18] Shi Kai, Yu Yonglin, Zhang Ruikang, Liu Wen and Barry L P 2009 *Optics Communications* **282** 81-7
[19] Björk G and Nilsson O 1987 *J. Lightwave Technol.* **LT-5** 140-6
[20] Lee San-Liang, Heimbuch M E, Cohen D A, Coldren L A and DenBaars S P 1997 *IEEE J. Sel. Top. Quantum Electron.* **3** 615-27
[21] Mason B, Fish G A, DenBaars S P and Coldren L A 1999 *IEEE Photon. Technol. Lett.* **11**, 638-40

Journal of Physics: Conference Series **276** (2011) 012089 doi:10.1088/1742-6596/276/1/012089

Simple planar Bragg grating devices for photonic Hilbert transformer

C Sima, J C Gates, H L Rogers, B D Snow, C Holmes, M N Zervas, P G R Smith

Optoelectronics Research Centre, University of Southampton, Southampton SO17 1BJ, United Kingdom

E-mail: cs4g09@orc.soton.ac.uk

Abstract. In this work the theoretical design and experimental realisation of a photonic Hilbert transformer is investigated. Planar Bragg gratings are proposed as a route to realise photonic Hilbert transformers in terms of a finite bandwidth and temporal impulse response. With proper grating apodisation and π-phase shift, the Bragg grating will display the spectral features expected from a Hilbert transformer. The unique features of Direct UV grating writing technology are used to fabricate the proposed device in silica-on-silicon

1. Introduction

Hilbert transformers are important devices widely used in communication, information processing and signal analysis in the electronic domain [1]. A photonic Hilbert transformer (PHT) can be employed in a similar range of applications, which would enable the direct processing of optical signals at high speeds as well as operation bandwidths far beyond current electronic technologies.

Based on discrete free space photonic components, PHTs have been realised, e.g., multi-tap fiber-optics transversal filters and sampled fiber Bragg gratings [2–5]. Recently, single phase-shifted Fiber Bragg gratings (FBGs) with proper apodisation profile for PHT have also been presented [6,7]. However, most are about theoretical simulation or with multiple components; to date no simple PHT device has been experimentally reported. Here we present initial efforts to fabricate an integrated PHT, which provides a route to realise compact optical modules for applications such as single sideband (SSB) modulation.

In this paper, we present initial efforts to practically demonstrate a PHT using an appropriate apodised planar Bragg grating with a π-phase shift in the grating section. The Bragg grating is fabricated by direct UV grating writing (DGW) technology on a silica-on-silicon platform [8]. In the future, the integrated nature of planar geometry will allow these devices to be combined with other functions, such as liquid crystals and thermal tuning elements to generate reconfigurable devices.

2. Operation principle

The ideal frequency response of a Hilbert transform $\hat{H}(\omega)$ is defined as [1]:

$$\hat{H}(\omega) = \begin{cases} -i, & \omega > 0 \\ i, & \omega \le 0 \end{cases} \tag{1}$$

where ω is the baseband frequency. Figure 1 shows the ideal frequency response of a Hilbert transform with a constant amplitude and a π-phase shift at ω=0.

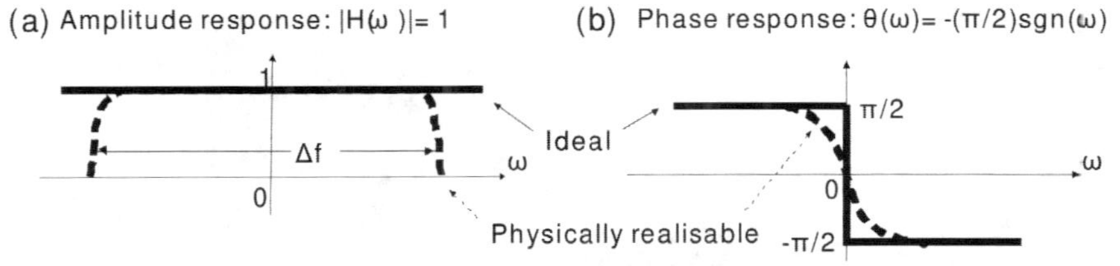

Figure 1. (a) PHT amplitude responses, (b) PHT phase responses (ideal in solid lines, physically realisable in dashed lines).

In practice a physically realisable device would have a limited frequency bandwidth and temporal impulse response, also shown in Figure 1. To realise such spectral features, the ideal practical Bragg device could incorporate an apodisation profile with the necessary π-phase shift, similar to the FBG derived from the first-order Born approximation (weak-coupling FBG) in reference [6]. The grating apodisation profile is given by the expression below:

$$\Delta n(z) \propto \frac{\sin^2\left(\pi n_{eff} \Delta f \left(z - z_0\right)/c\right)}{\left(z - z_0\right)} \tag{2}$$

In equation (2), z is the grating position along the waveguide with $0 \le z \le L$, while L is the total grating length, n_{eff} is the effective refractive index of the planar Bragg grating, Δf is the operating bandwidth, z_0 is the zero-crossing point in the apodisation function and c is the light speed in vacuum. z_0 is set to the centre of the grating length for simplicity. For the designed Bragg wavelength λ_B, the grating period is $\Lambda = \lambda_B / 2n_{eff}$.

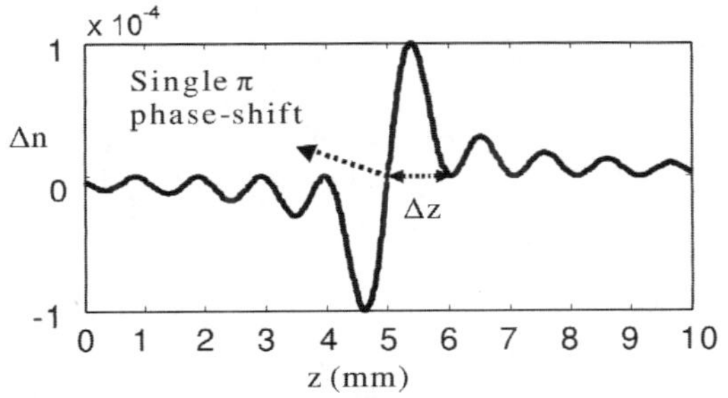

Figure 2. An example of the refractive index apodisation profile in the planar Bragg grating for a PHT.

Figure 2 shows the shape of the apodisation profile for the proposed grating. As anticipated, the resulting grating apodisation profile requires a single π-phase shift along the grating length. The zero-to-zero width of the side lobes Δz in the apodisation profile is directly related to the full operation spectral bandwidth Δf of the PHT, where $\Delta z = c/n_{eff}\Delta f$. Negative Δn relates to a π-phase shift in the apodisation profile.

3. Fabrication and experimental results

Figure 3 illustrates the direct UV grating writing (DGW) technology used to fabricate the proposed planar Bragg gratings [8]. This method involves focusing two crossed laser beams (λ=244nm) into the photosensitive core of a planar sample. UV irradiation increases the refractive index in this layer. Precise translation of the sample and modulation of the laser intensity both defines the channel waveguide and simultaneously the creation of the grating structures [8]. This spot size is approximately 6μm in diameter, providing the unique ability over traditional FBG to manipulate the grating structures at the micron level. This ability allows precise engineering of the amplitude and phase responses of the proposed gratings for PHTs.

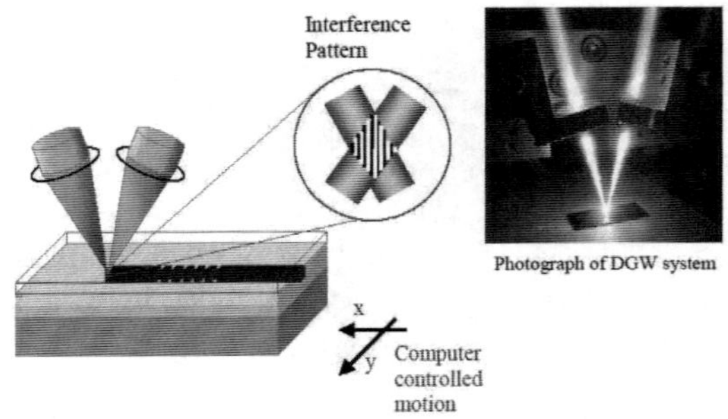

Figure 3. Illustration of the fabrication process with direct UV grating writing in silica-on-silicon.

The apodisation profile shown in equation 2 was used to fabricate a Bragg grating for application as a PHT. Modeling was used to ensure that grating parameters produced an apodisation profile which was truncated at points where $\Delta n = 0$. Thereby reduce rapid changes in the propagation constant of the waveguide and hence additional spectral features.

Figure 4 (a) shows the reflectivity of a Bragg grating that contained a π-phase shift and apodisation profile using the following parameters, L= 15mm, Λ=535.467nm, n_{eff}= 1.4478, Δn= 0.0003, and Δf= 140GHz. The reflectivity spectrum shows a rectangular spectral response, the result of the apodisation and a strong dip (~-20 dB) characteristic of the π-phase shift.

Figure 4(b) shows the relative group delay of the reflected signal. It demonstrates a comparatively flat group delay within the operating bandwidth with an abrupt change in the centre, a consequence of the phase shift. Both reflectivity and relative group delay were measured using the modulation phase-shift method. The poor dynamic range of the reflectivity measurement is a result of the limited sensitivity of the high bandwidth photoreceiver. According to Figure 4(a), the Bragg grating operates at 1550.5 nm and has a bandwidth of ~1.2nm (~150GHz) bandwidth, both values agreeing with the target parameters. It should be noted that these planar waveguide devices are birefrigent and are therefore polarization dependent.

Direct phase measurement of the Bragg grating will be required to qualify that this device performs as a Photonic Hilbert Transformer. However the asymmetric features on the reflectivity peak are characteristic of a small error in the π-phase shift. Modeling has shown that a ~5% error in the magnitude of the π-phase shift produces similar features as shown in figure 4(a). This analysis shows that further work is required to reduce such fabrication errors which are likely to be a result the stability of the laser source and nonuniformity of the planar sample. However this analysis also shows that the phase shift is within 5% of the target π-phase shift.

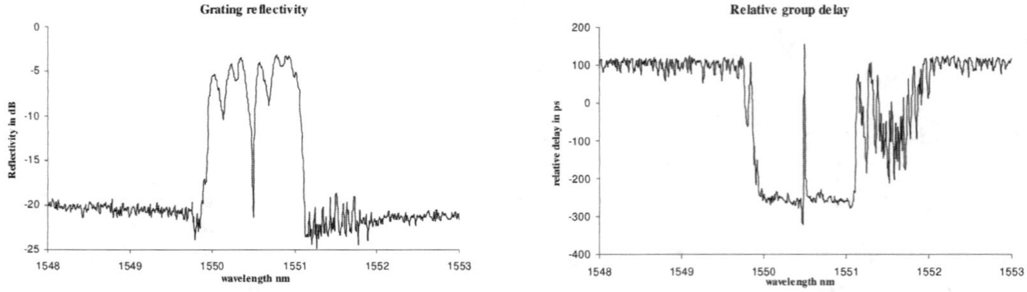

Figure 4. The measured reflection spectrum (a) and relative group delay (b) of the single π phase-shifted planar Bragg grating with the apodisation profile for Hilbert transforms.

4. Conclusions

We have presented data demonstrating π-phase shifted Bragg gratings with an apodisation profile, for applications as a PHT. The spectral results agree with the target parameters and suggest that the π-phase shift is correct to within 5%. Future work will investigate the phase response of the Bragg grating to qualify the device performance for operation as a Photonic Hilbert Transformer. The planar integrated format also offers a route to stable operation and allows high density integration.

References

[1] Hahn S L and Poularikas A D 2000 *the Transforms and Applications Handbook* 2nd Edition. (CRC Press) p 551.
[2] Takano K, Hanzawa N, Tanji S and Nakagawa K 2007 *Optical Fiber Communication Conference, OSA Technical Digest Series (CD)* **JThA48**.
[3] Emami H, Sarkhosh N, Bui L A and Mitchell A 2008 *Opt. Lett.* **33** 98–100.
[4] Emami H, Sarkhosh N, Bui L A and Mitchell A 2008 *Opt. Express* **16** 13707–13712.
[5] Hanawa M, Nakamura K, Takano K and Nakagawa K 2007 *Proceedings of the European Conference on Optical Communication (ECOC) (VDE-Verlag)* 01.5.6.
[6] Asghari M H and Azaña J 2009 *Opt. Lett.* **34** 334–336.
[7] Li M and Yao J 2010 *Opt. Lett.* **35** 223–225.
[8] Emmerson G D, Watts S P, Gawith C B E, Albanis V, Ibsen M, Williams R B and Smith P G R 2002 *Electron. Lett.* **38** 1531–1532.
[9] Hui R and O' Sullivan M 2009 *Fiber Optic Measurement Techniques* (UK: Elsevier Inc) pp 346–350.

Investigation of Surface plasmon microcavity for resonant transmission through a slit by using FP matrix method

M Afshari Bavil [1] ***, Xiu Dong Sun*** [1, *] ***(孙秀冬)***

[1] *Optical information processing group, Department of physics, Harbin Institute of Technology, Harbin 150001, China*

[*]corresponding author : Xiu Dong Sun xdsun@hit.edu.cn afsharibawili@gmail.com

Abstract. *Surface plasmon (SP) microcavity recently has been proposed to enhance transmission through a subwavelength slit. Here we employed Fabry-Perot matrix method to predict the transmission profile of the structure. The result show good agreement with simulation result based on finite difference time domain (FDTD) method. The calculation shows that the SP microcavity standing waves and the Fabry-Perot resonance mode are responsible for transmission enhancement. For layered slits FP matrix method is a powerful method for prediction transmission behaviors.*

1- Introduction

After discovering light transmission enhancement by tiny holes in a thin metal film [1], it has been applied to some important areas such as photolithography [2] and optical spectroscopy [3]. This effect consider as the interaction of the light with the electronic resonances in the interface between metal films and dielectric materials that are known as surface Plasmon polaritons (SPPs). For narrow slits in metal grating the same effect were reported [4]. Single slit aperture has been utilized in optical devices [5]. In earlier theoretical works the single slit was treated as a Fabry-Perot (FP) cavity with two open ends and light transmission through the slit was obtained [6]. Recently some deviated slits attract some interest [7]. Qiao Min et al. [8] proposed a compact reflector device by using the highly reflective gold as a simple mirror, which forms a compact surface plasmon (SP) microcavity. The SP microcavity was created by trench milling in a gold film, which trench was used simply to form mirror side-walls, and thereby to create a microcavity. They presented experimentally and numerically that the structure resonantly enhance the transmission of light through the slit [9].

In the case of microcavity the FP method is not valid. Here we employ Fabry-Perot matrix method to investigate the transmission property of microcavity slits. We show that even for microcavity slit the Fabry-Perot (FP) resonance still contribute in the light transmission. By defining effective refractive index for the microcavity and slit and calculating the refractive index numerically, transmission profile of the structure is obtained. The result shows good agreement with simulation and experimental results. The effect of microcavity Plasmon resonance explained as well.

2- Theoretical framework

Figure 1 shows the structure under study, the slit has 180nm width and 100nm thickness, the light impinges from up with wavelength 700nm. The thickness of microcavity considered as 200nm. In the experimental work different nanocavity widths has been investigated and it has been showed that the 1100nm width provides the maximum light transmission.

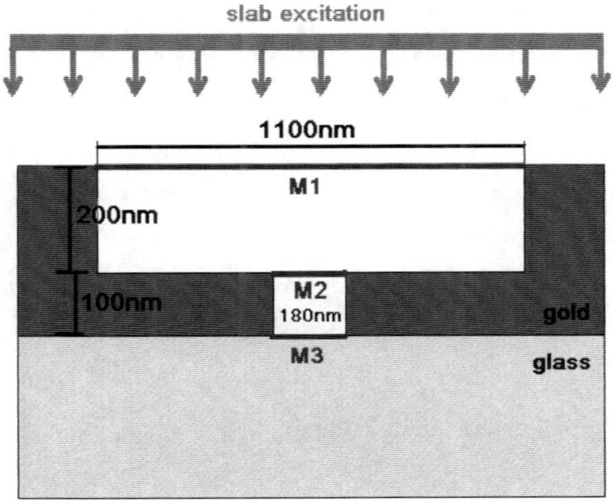

Figure 1. Schematic plot of the structure under study, width and thickness of the slit are w1=180nm and d1=100nm respectively. The microcavity has w2=1100nm width and d2=200nm thickness. The incident light impinges from up with wavelength 700nm. M1, M2, M3 are the monitors for detecting amplitude of the magnetic field.

For utilizing Fabry-Perot matrix method the refractive index should be defined. Here we employ finite difference time domain (FDTD) simulation to obtain refractive index numerically. The convolutional perfectly-matched-layer (CPML) boundary condition [9] is applied in the z and x-direction to absorb the back-reflected as well as transmitted radiation at the top and bottom boundaries of the mesh. The number of perfectly-matched-layer's cell is set to be 11. First we apply a normally incident periodic wave which impinge to the structure from up. The wave arrives monitors M1, M2, M3 at times t1, t2, t3 respectively. The position of monitors is presented in figure 1. C0*(t2-t1) equals to distance that light can travel in free space, which C_0 is the speed of light in free space. This distance is bigger than microcavity thickness that clarifies multireflection and squeezing effect inside the microcavity [10]. From classic optics, the refractive index of a medium can be defined by dividing the light speed in free space to light speed in the media. The light speed in the microcavity equals to the thickness over elapsed time to pass through the microcavity, so refractive index of the microcavity can be defined as

$$n_1 = \frac{C_0(t_2-t_1)}{d_1} \tag{1}$$

Which gives n_1= 1.17. By substituting the recorded time by monitor M3, the refractive index of the slit can be obtained as

$$n_2 = \frac{C_0(t_3-t_2)}{d_2} = 1.75 \tag{2}$$

By approximating the slit and nanocavity have identical widths but different thicknesses and refractive indexes, we are in position to utilize Fabry-Perot matrix method. Let consider the light impinges with amplitude E_0 and wavelength λ_0 from up. The refractive indexes of layers are determined and the reflection and transmission in the interfaces simply can be obtained by using Fresnel equations in the case of normal incidence. The transition matrix (T_{ij}) and propagation matrix (P_j) calculated excessively [11]. By using transfer matrix method (TMM), the electric field of incident layer associated to electric field at exit layer as

$$\begin{pmatrix} rE_0 \\ E_0 \end{pmatrix} = T_{12}P_1T_{23}P_2T_{34}P_3T_{45}\begin{pmatrix} 0 \\ tE_0 \end{pmatrix} \tag{3}$$

The matrixes combine together to give the stack matrix

$$\begin{pmatrix} rE_0 \\ E_0 \end{pmatrix} = S\begin{pmatrix} 0 \\ tE_0 \end{pmatrix} \tag{4}$$

The stack matrix S is a 2*2 dimensional matrix and the transmission can be calculated by $T=tt^*$ where $t=1/S_{22}$.

Now we are able to obtain the transmission property of the structure by Fabry-Perot matrix method.

3- Result and discussion

Figure 2 shows the transmission spectra of the structure for normal incidence and polarization based on FDTD simulation. The transmission was normalized to the incident power on the width of the slit. Due to the surface localization of the SPPs, the trench side-walls reflect the light very well and resonant enhancement of the optical transmission is expected from coupling of the bare slit to standing-wave resonances of the SP nanocavity. The SP microcavity had a calculated transmission peak at 730 nm, which has good agreement with experimentally measured peak.

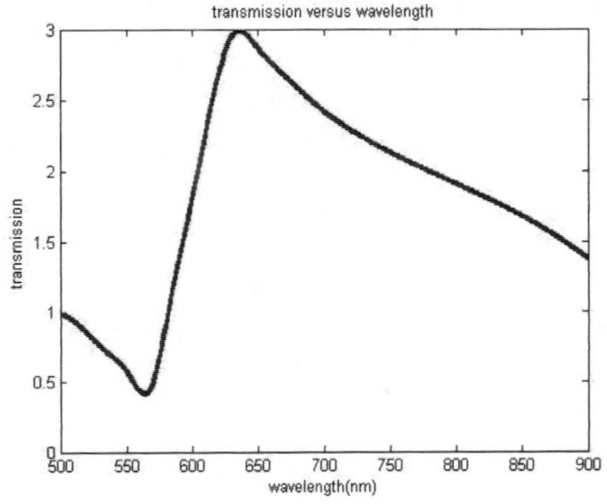

Figure 2. The transmission profile of the structure by utilizing FDTD simulation

Figure 3 shows the transmission profile by using Fabry-Perot matrix method. The map has a transmission peak at 750 nm. There is good quantitative agreement in the transmission resonance wavelength between simulation and our calculations. The small discrepancy can be attributed to approximating nanocavity and slit with identical widths. Because the absolute value of the amplitude of incident field consider as unit so the transmission can't exceed unit in this method. In other word matrix method is a useful method to define the position of peaks not the magnitude of the transmitted light. From the simulation and calculation result the FP matrix method provide a simple and accurate model to describe the transmission property of modified slits.

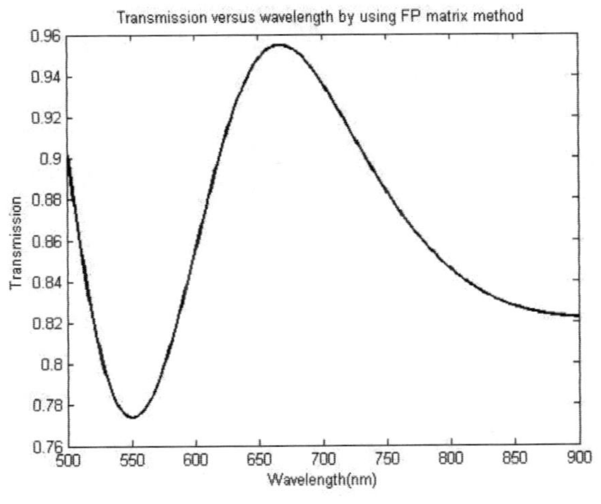

Figure 3. The transmission profile of the structure by utilizing Fabry-Perot matrix method

When light passes through the slit the surface Plasmon polaritons are excited and the standing wave inside the slit are created, that leads the light experiences multireflection and squeezing phenomena inside the slit. That causes the refractive index of the slit be bigger than the free space. As a result the slit behaves as a high refractive index media for the incident light. The refractive index of the slit is bigger than the microcavity that present the wider slits have weaker multireflection and squeezing effects and consequently lower refractive index. Due to higher refractive index of the slit, light needs much time to pass through the slit. As we see the light transmission enhancement can be explained by FP matrix method resonance, which prove our proposed method can analyze the transmission property of modified slits with good precision. The refractive index of slit originated from excitation of surface Plasmon polaritons. In the conclusion both surface plasmons and Fabry-Perot resonance contribute in the transmisison enhancement in an equal manner.

References

[1] Ebbesen T W, Lezec H J, Ghaemi H F, Thio T and Wolff P A 1998 *Nature* **391** 667

[2] Srituravanich W, Fang N, Sun C, Luo Q and Zhang X 2004 *Nano Lett.* **4** 1085

[3] Prikulis J, Hanarp P, Olofsson L, Sutherland D and Kall M 2004 *Nano Lett.* **4** 1003

[4] F. J. Garcı ´a-Vidal,and L. Martı ´n-Moreno, *PHYSICAL REVIEW B* **66**, 155412 (2002)

[5] T. Ribaudo, D. C. Adams, B. Passmore, E. A. Shaner, and D. Wasserman *APPLIED PHYSICS LETTERS* **94**, 201109 (2009).

[6] Y. Takakura, *PHYSICAL REVIEW LETTERS* **86**, 5601(2001).

[7] A. Battula and S. C. Chen, *APPLIED PHYSICS LETTERS* **89**, 131113 (2006).

[8] Qiao Min, Reuven Gordon, *OPTICS EXPRESS* **16**, 9708(2008).

[9] A.Taflove, S.C.Hagness, *Computational electrodynamics*, second edition, 2000.

[10] Chao Li, Yun-Song Zhou, Huai-Yu Wang, and Fu-He Wang, *J. Opt. Soc. Am. B,* **27**, 1(2010).

[11] N.Hodgson, H.Weber, *Laser resonators and beam propagation*, 2nd edition, Springer, 2005.

Journal of Physics: Conference Series **276** (2011) 012091 doi:10.1088/1742-6596/276/1/012091

Investigation of transmission performance of single sideband radio over fiber link

Daohai Tan, Xiaogang Chen*
College of Science, China Three Gorges University, Yichang, 443002, Hubei, China.

E-mail: cxg826@yahoo.com.cn

Abstract. We have presented a detailed investigation of the impact of group velocity dispersion (GVD) on the single sideband (SSB) radio over fiber (RoF) links. The theoretical and simulated results show that the transmission distance of the SSB RoF link is limited by the time shifting of the codes due to the GVD, and the code distortion could be suppressed if only one of optical carriers is modulated with baseband data. Moreover, the transmission performance of the SSB RoF link could be further improved by reducing the power of the optical carrier to optimize the carrier-to-sideband ratio.

1. Introduction

Radio over fiber (RoF) [1, 2] is a type of communication technology which combines millimeter wave with optical communication, which can avoid interference and also can meet the requirements of broad-band access network. RoF provides large bandwidth, low loss, and easy to install. It will play a pivotal role for the broad-band access network in the future communication.

The optical generation of the RF signal is the key technology to achieve radio over fiber link. Now, many approaches to generate millimeter waves have been proposed, mainly including external intensity modulation [3], optical heterodyne [4] and optical frequency multiplication [5]. The preferred technique for the optical generation of mm-wave is an intensity external modulation by using an external modulator [3, 6]. In order to overcome the problem of the power penalty, the use of single sideband (SSB) modulation has been proposed with a DD-MZM modulator [3]. Though the SSB mm-wave has good ability in anti-noise, the data signals carried by the optical carrier and its sideband will also get distortion because of the group velocity dispersion (GVD) [7] when they are transmitted along the fiber for a further transmission distance.

In the note, we perform theoretical analysis and simulation of the transmission performance of the SSB RoF links, and the effects of fiber chromatic dispersion on the SSB links are evaluated and discussed. In addition, the schemes to extend the transmission distance are put forward as well.

2. SSB modulation principle and the impact of fiber dispersion on the SSB RoF links

In ROF system, the output amplitude of DSB modulation is periodic fluctuation with the fiber length [8]. Now, theoretical analysis and experimental results have shown that SSB modulation could eliminate this effect. The principal diagram of SSB optical millimeter-wave generation and transmission is shown in Figure 1.

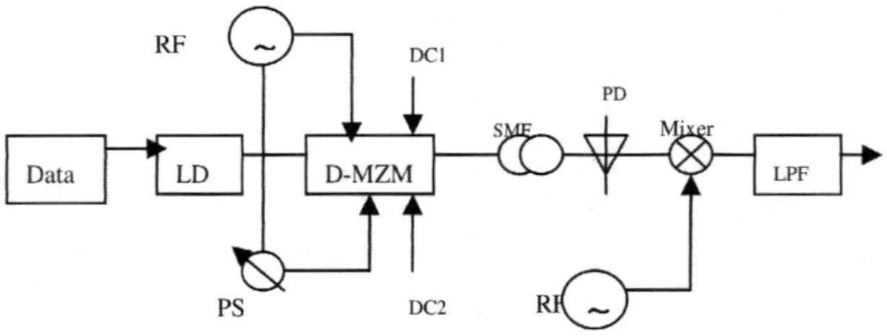

Figure1. The diagram of the optical mm-wave link. LD, directly modulated laser measured; D-MZM, LiNb Mach-Zehnder Modulator; RF, radio frequency source; PS, phase shifter; SMF,single mode fiber; PD, photo detector; Mixer, Electrical Multiplier; LPF, Low Pass Gaussian Filter.

As shown in Figure 1. The implementation of SSB modulation is to use a DD-MZM biased at quadrature point ($\gamma\pi = \frac{1}{2}\pi$), in which a RF signal is split into two branches, which drive the two arms of the MZM with $\varphi_2 = \varphi_1 + \frac{\pi}{2}$, and $V_{dc1} - V_{dc2} = \frac{V_\pi}{2}$. Then we can implement the SSB modulation. The output after DD-MZM can be written as [3]:

$$E(t) = \frac{E_0 X(t)}{2}\{J_0(\alpha\pi)\cos(\omega_c t) - J_0(\alpha\pi)\sin(\omega_c t) + 2J_1(\alpha\pi)\cos(\omega_c + \omega_{RF})t\} \tag{1}$$

The power-spectral density, $P(f)$:

$$P(f) = \frac{\pi E_0^2 X(t)}{4}\{J_0^2(\alpha\pi)\pi\delta(\omega + \omega_c) + 2J_1^2(\alpha\pi)\pi\delta[\omega + (\omega_c + \omega_{RF})]\} \tag{2}$$

In the optical receiver, only one sideband beat with the optical carrier, so, there is no superimposition destructively when the signal beats. The output power of RF signal after photoelectric detector is approximately equal as:

$$P_{f_{RF}} \propto m\sin(\omega_{RF} t + \frac{\pi DL}{c}\lambda_c^2 f_{RF}^2) \tag{3}$$

Where m is constant based on the system parameters. From (3), the amplitude of $P_{f_{RF}}$ is not periodic variation, and the dispersion effect in phase. Though the SSB mm-wave has good ability in eliminating the attenuation effect, the group velocity dispersion (GVD) cannot be ignored when they are transmitted along the fiber for a further transmission distance [9].After optical-millimeter wave is transmitted over a L-length fiber, we can get the following formula [9]:

$$E(L,t)=\frac{E_0X(t-\tau_0)}{2}\{J_0(\alpha\pi)\cos[\omega_c(t-\tau_0)]-J_0(\alpha\pi)\sin[\omega_c(t-\tau_0)]\}$$
$$+2E_0X(t-\tau_1)J_1(\alpha\pi)\cos[(\omega_c+\omega_{RF})(t-\tau_1)] \quad (4)$$

Where $\tau_0=\frac{\beta(\omega_c)L}{\omega_c}$, $\tau_1=\frac{\beta(\omega_c+\omega_{RF})L}{\omega_c+\omega_{RF}}$, is the time delay caused by the GVD after a L length fiber link. $\beta(\omega)$ is the propagation constant. From (4), the different time delay $\Delta\tau$ is existence after the data modulated by the carrier and the sideband is transmitting a L-length fiber. Ultimately cause serious distortion. $\Delta\tau$ can be approximated by:

$$\Delta\tau=\tau_1-\tau_0\approx|\omega_{RF}\beta''(\omega_c)L|=\frac{Df_{RF}Lc}{f_c^2} \quad (5)$$

To make SSB modulation get better application, the approach is that modulate signal after separation of carrier and sideband [10]. The principal diagram of optical millimeter-wave generation and transmission is shown in Figure 2. The data is only loaded into the sideband. From (4), after optical-millimeter wave is transmitted over a L-length fiber, we can get:

$$E'(L,t)=\frac{E_0}{2}\{J_0(\alpha\pi)\cos[\omega_c(t-\tau_0)]-J_0(\alpha\pi)\sin[\omega_c(t-\tau_0)]\}$$
$$+2E_0X(t-\tau_1)J_1(\alpha\pi)\cos[(\omega_c+\omega_{RF})(t-\tau_1)] \quad (6)$$

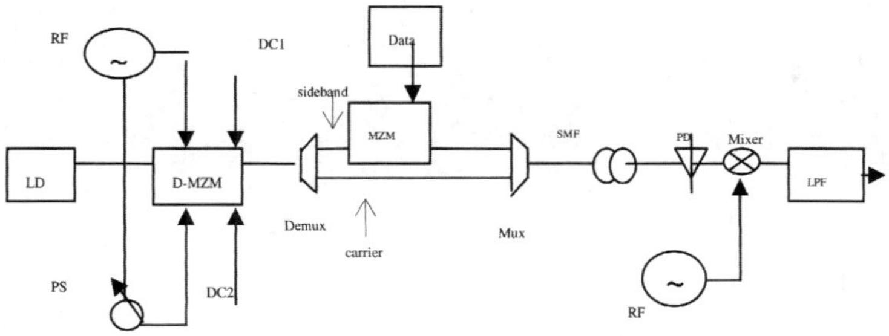

Figure 2. The diagram of scheme Ⅰ to improve the transmission performance of SSB RoF link.

Obviously, the signal still has a time delay τ_1, but the data has no relative time delay $\Delta\tau$. So the distortion caused by GVD can get suppression effectively.

3. Simulation

In order to verify the correctness of the theory, the SSB simulation calculation has been respectively performed for the following three different situations by using the commercial software package OptiSystem .

In the first situation, the principal diagram is shown in Figure 1. The system parameters are set as follows: The center frequency of LD is 193.1 THz, 10dB power and 10 MHz linewidth. The data rate is 2.5 Gbit/s and the RF frequency is 40GHz. We assume modulator is the ideal LiNb Mach-Zehnder Modulator and the two Modulation voltages are set to 1.02. Single-mode optical fiber attenuation can be compensation by amplifier, so the attenuation can be set to 0. The dispersion of the fiber is set to 16.75 ps(nm.km). The responsivity of photodetector is 1 A/W. The eye pattern of SSB electrical signal after the PD is shown as Figure 3. Under the requirement of BER of10^{9}, we can get that SSB electrical signal is able to transmit over 60 km as most of the literatures mentioned, which greatly exceeds the DSB modulation.

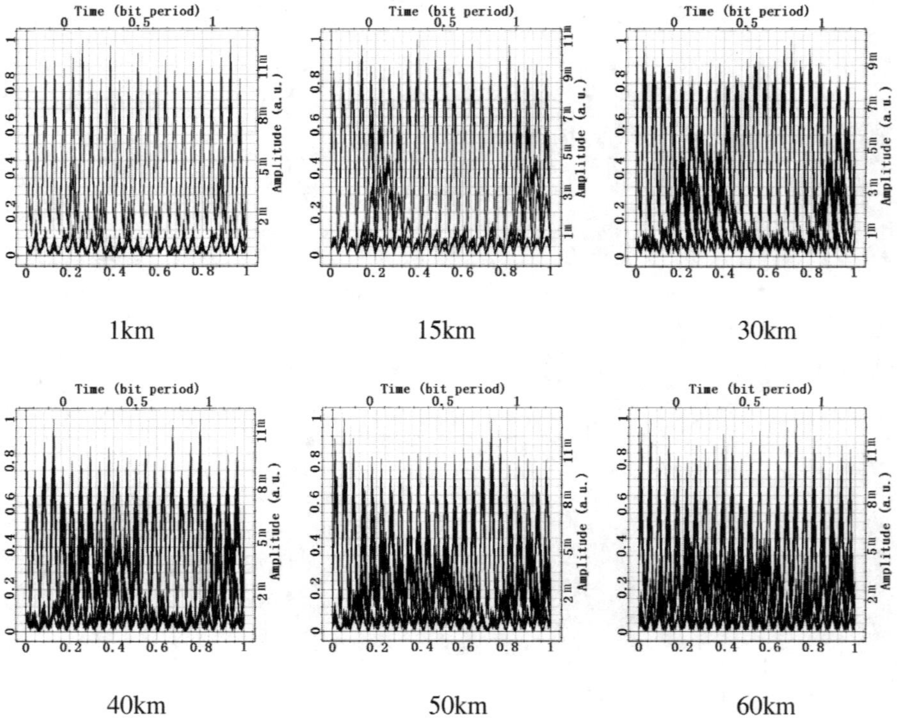

Figure 3. The eye pattern of SSB electrical signal after the PD

In the second situation, the principal diagram of scheme Ⅰ is shown in Figure 2. The data modulate sideband after the carrier and sideband is separated by using WDM Demux, and then transmits in fiber after coupled by WDM Mux. The light is using CW laser with 193.1 THz and the power is 0 dBm. The other parameters are all the same as the first situation. The eye pattern of SSB electrical signal after the PD is shown as Figure 4. Under the requirement of BER of10^{-9}, the SSB electrical signal is able to transmit over 90km. The transmission performance is improved obviously.

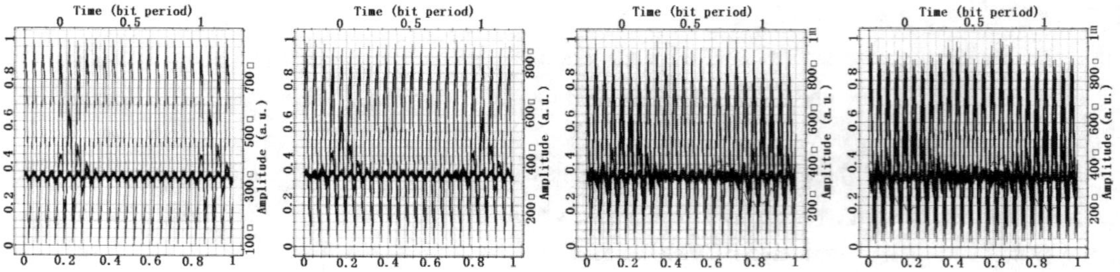

1km 30km 60km 90km

Figure 4. The eye pattern of SSB electrical signal after the PD of the scheme Ⅰ

In the third situation, we consider that the power difference of the carrier and sideband may also affect signal transmission distance. So we get a further improvement base on the second project. The diagram of scheme II is shown as Figure 5. The attenuation of optical attenuator is set to 10 dBm. The other parameters are all the same as the second situation (scheme I).

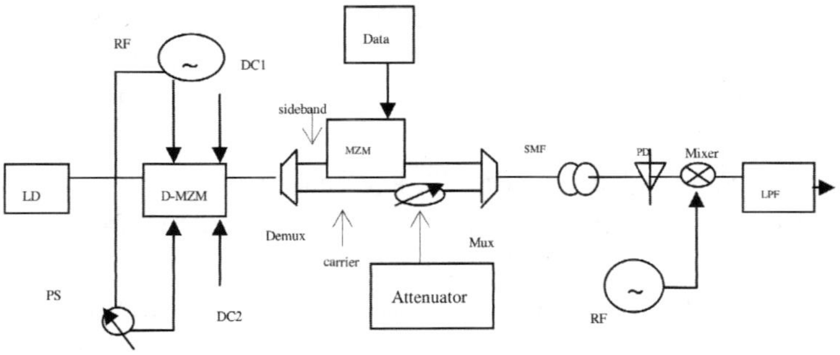

Figure 5. The diagram of scheme II to improve the transmission performance of SSB RoF link.

The eye pattern of SSB electrical signal after the PD of the third situation is shown as Figure 6. Under the requirement of BER of 10^{-9}, the SSB electrical signal is able to transmit over 180km. Our guess is verified, but also need to do further work.

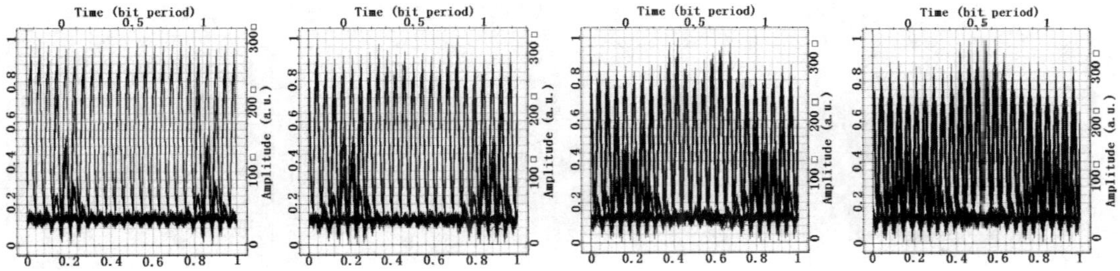

30km 80km 130km 180km

Figure 6. The eye pattern of SSB electrical signal after the PD of the scheme II

4. Conclusions

In conclusion, the impact of fiber dispersion on the transmission performance of the SSB RoF link is theoretical analyzed and simulated. The results show that it is able to suppress the performance degradation due to group velocity dispersion (GVD) if only sideband is modulated with baseband data, and the transmission distance could be further extended by optimizing the carrier-to-sideband ratio. In addition, the corresponding simulations were carried out using the commercial software package OptiSystem with the simulation parameters following the theoretical values to access the validity of the theoretical analysis.

References

[1] H. Ogawa, D. Polifko, and S. Banba 1992 Millimeter-wave fiber optics systems for personal radio communication *IEEE Transactions on Microwave Theory and Techniques* (vol 40) pp 2285–93

[2] A. J. Seeds 2002 Microwave photonics *IEEE Transactions on Microwave Theory and Techniques* (vol 50) pp 877–87

[3] G. H. Smith, D. Novak, and Z. Ahmed 1997 Overcoming chromatic dispersion effects in fiber-wireless systems incorporating external modulators *IEEE Transactions on Microwave Theory and Techniques* (vol 45) pp 1410–5

[4] G. J. Simonis and K. G. Purchase May 1990 Optical generation, distribution, and control of microwaves using laser heterodyne *IEEE Trans. Microwave Theory Tech* (vol 38) pp 667–9

[5] KoonenT., Ng'oma A., SmuldersP., et al 2003 In-house network using multimode polymer optical fiber for broadband wireless services *Photonics Network Communications* (Netherlands, Kluwer Academic Publisher vol 5) pp 177-87

[6] H. Schmuck 1995 Comparison of optical millimeter-wave system concepts with regard to chromatic dispersion *Electronics Letters* **31**(21) 1848–9

[7] J.X. Ma, J.J.Yu, Z.S. Jia, C.X.Yu and G.K. Chang 2007 Optical mm-wave generation based on phase modulation along with optical filtering *Microwave Opt. Technol. Lett* **49** 1787–93

[8] YANG Xiangyun, HU Weiwei and XU Anshi May 2006 Research on effects of dispersion of signal transmission on millimeter-wave ROF optical communication systems *Acta Scientiarum Naturalium Universititatis Pekinensis* **42**(3) 401-5

[9] JianGong Zhao, AiJun Wen, ZengJi Liu and XiangLing Liu 2009 An analytical model of the transmission capacity of optical mm-wave signal generated by single side band intensity-modulation *Journal of Optical Communications* **30** 53-6

[10] CHEN Chuan, JIN Ming-lu and HAO Yu-wen 2009 Research on technology of external modulation on millimeter-wave RoF optical communication systems *Optical Communication Technology* **33**(4) 59-61

Investigation on an optically-controlled phase-shifter based organic semiconductor poly-(3-hexylthiophene) (P3HT)

Hongyan Tang, Feifei Zhao, Fuyun Zhang

School of Electronic Engineering, University of Electronic Science and Technology of China, Chengdu, China, 610054

hongyan.tang@gmail.com

Abstract. In this paper, an investigation on optically-controlled phase-shifter based organic semiconductor P3HT is presented. The operation principle of optically-controllable phase-shifters and the applicability of organic semiconductor P3HT to such devices have been demonstrated. The reversible, optically-controlled phase-shifter employing P3HT as an active device layer, yielded differential phase shifts which varied with light intensity and signal frequency as well as device architecture. The results are non-optimized, and impedance matching needs to be improved in the future in order to obtain better signal transmission performance through the device.

1. Introduction

In the last couple of decades a significant amount of research has gone into optically controlling microwave and millimeter wave devices. Activity in the fields of lasers and photonics has led to such development. Research and development of photonic devices is driven by the need for dynamic control, fast response, immunity from electromagnetic interference and good isolation between the controlling and controlled devices [1]. Phase shifters, modulators, attenuators, slow wave structures etc, are some of the components of this important class of devices. The potential for the use of the optically-controlled phase-shifter in phased-array antenna systems lies in cost effectiveness, its low weight and low power consumption [2].

2. Organic semiconductor for optical techniques

Optical techniques offer the following advantages: 1) near perfect isolation, 2) low static and dynamic insertion losses, 3) fast response, 4) high power handling capability, 5) when nanosecond pulses are used, it is possible for very high density plasmas to be injected without damaging the material, 6) by a proper choice of semiconducting material and laser wavelength one can generate plasma with any desirable density distribution and at any desired time, 7) ultra fast switching and gating of microwave and millimeter-wave signals are possible, and 8) using nanosecond exciting-probing techniques, the dynamic evolution of the injected plasma can be studied in detail. Parameters related to transport properties of the carriers can be accurately determined.

The operating principle of these devices is photo-excitation. When a semiconductor substrate is being illuminated by the light source whose photon energy is greater than the bandgap energy of the semiconductor material, plasma (electron-hole pairs) is induced near the surface of the medium due to photo-absorption by the semiconductor [3]. The photo-excited carriers can change the propagation characteristics of the electromagnetic signals being propagated through the photo-illuminated semiconductor.

Organic semiconductors bring together the virtues of plastics and semiconductors in materials. This opens up new directions for electronic and optoelectronic materials. They have many advantages, such as easy fabrication, mechanical flexibility, tunable optical properties and low cost; therefore have outstanding potential for emerging technologies. There are a large number of applications for organic semiconductors, for instance organic light-emitting diodes (OLED), organic solar cells, organic field effect transistors (OFET). In the future, a number of new exciting developments are foreseeable.

3. Phase shifter employing P3HT/ITO

Organic semiconductor polymers have been the focus of many studies because they have shown promise for many applications such as organic solar cells, flexible display technology and radio frequency identification tags. Among various types of polymeric semiconductors, regioregular P3HT is the most explored due to its wide commercial availability and high carrier mobility [4]. The optically-controlled phase-shifters that have been reported so far were composed of conventional inorganic semiconductors like Si. In this section an optically-controlled phase-shifter comprised of organic semiconductor P3HT on an indium tin oxide (ITO) covered substrate is investigated.

3.1 Characteristics of Regioregular P3HT

The physical form of P3HT is dark brown to black power. Different processes and conditions for polymerization of 3-hexylthiophene may give rise to either regioregular P3HT (RR-P3HT) or regiorandom P3HT (RRa-P3HT) [5]. The degree of order in polymer films can have a drastic effect on both absorption properties and charge transport. This has fundamental effect on the absorption. The absorption peak for RR-P3HT is shifted to lower energies and shows more vibronic structure, all indicative of a higher degree of polymer chain ordering. It is clear that good ordering is essential for achieving high transport performance. It has also been demonstrated that the RRa-P3HT excitations consist primarily of intra-chain excitons, whereas RR-P3HT has a larger inter-chain component. Based on these advantages, RR-P3HT has much wider applications than RRa-P3HT. In this section, P3HT refers to RR-P3HT thereafter.

In recent years the potential application of P3HT in polymer electronics and optoelectronic applications, such as light emitting diodes, optical devices, solar cells etc, has gained significant attention [6]. The reason is its good spectral overlap with optical-wavelength irradiation, high charge-carrier mobility as well as having a low bandgap [7]. These features are relatively stable in ambient condition and it has excellent solubility in common organic solvents.

P3HT has a high absorption coefficient close to the maximum photon flux in the optical spectrum, peaking between the blue green (450nm) and green yellow (600nm). The UV-VIS absorption spectra for a thin film of pure P3HT indicates two peaks at 493 nm and 517 nm and one shoulder at 572 nm. These three bands can be attributed to the transition from valence band to conduction band. Therefore a general white-light source is sufficient for optical applications without particular equipment requirement.

In terms of literature, P3HT has the highest charge-carrier mobility among the conjugated polymers [7]. Mobility here can be referred to the agile movement of holes. Hole-mobility is typically $10^{-4} \sim 10^{-1}$ cm^2 /(Vs). The high mobility is mainly attributed to the formation of extended polaron states as a result of local self-organization, in contrast to the variable-range hopping of self-localized polarons found in more disordered polymers.

Another important characteristic of P3HT is its lower bandgap energy i.e. the energy difference between the bottom of the conduction band and the top of the valence band of the semiconductor material. The band gap of P3HT to be approximately 1.9 eV has been reported, with a conducting band of negative -5.4eV (1eV=1.6 x 10^{-19} J). These properties render P3HT a good candidate for the polymer in polymer-photovoltaic applications.

P3HT also has excellent solubility in common organic solvents, for instant Toluene, Chloroform, Chlorobenzene, Cyclohexylbenzene (CHB) or TriChloroBenzene (TCB). It has been reported that when different types of solutions were used with P3HT during the fabrication process, every solution revealed different mobility characteristics. It was discovered that using high boiling-point solvents such as Trichlorobenzene could lead to a high degree of crystallinity and enhanced mobility. So TCB is chosen to be the solvent for P3HT in our experiments.

3.2 Operating principle

Phase-shifting is a fundamental control operation. A general approach to this operation is to alter the phase-velocity along a fixed interval of a guided medium. In this case, the phase-shift per-unit-length is equal to the change in the propagation constant of the traveling wave. One well-know method of altering the dispersion of the signal is to introduce plasma into the medium.

According to the optical properties of organic semiconductor P3HT (as well as many other advantages), optically-controlled phase-shifters employing P3HT as an active device layer, has been investigated. The schematic cross-section of the proposed structure is shown in figure 1. Al microstrip is vapor-deposited onto a P3HT coated ITO substrate supported by glass. When illuminated with a white-light source, having component energy greater than the bandgap energy of P3HT, an electron-hole plasma region will be induced (in the active region, near the surface). This leads to a complex permittivity different from that of the complex permittivity of the non-illuminated region. As a result, electromagnetic waves will have different propagation factors through these activated regions. These yielded differential phase-shifts which varied with light intensity and signal frequency as well as device architecture.

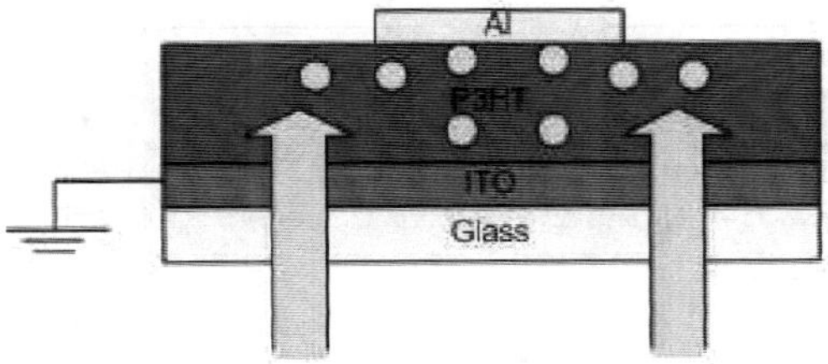

Figure 1: Schematic cross-section of the optically controlled phase shifter.

The plasma region is close to the region under the Al microstrip line as the microwave electromagnetic field is concentrated underneath the microstrip line and falls off sharply away from that region. The initial depth of plasma-injection is controlled by selecting an appropriate combination of irradiation wavelength and semiconductor absorption properties. At sufficiently small injection depths, the final thickness of the P3HT plasma is determined primarily by the processes of carrier diffusion and recombination. The effect of the plasma-occupied volume is to introduce a layer whose refractive index at the microwave frequency is larger than the remaining volume of P3HT. From simple physical arguments, one would expect the effect to become strong as one approaches and then exceeds the plasma density at which the value of the plasma frequency equals the frequency of the transmitted signal. However, the form and magnitude of the phase-shift versus the intensity of illumination (and hence the plasma density), depends in detail on the material and geometrical factors that characterize the optically-perturbed active structure. Their determination requires a detailed solution of the corresponding boundary value problem.

In the plasma region, the conductivity σ is a function of the mobility and the density of charge carriers:

$$\sigma = e(n\mu_e + p\mu_p) \tag{1}$$

where n is electron concentration, p is hole concentration, μ_e is electron-mobility and μ_p is hole-mobility.

If the conductor loss factor of the P3HT is neglected, and optically induced plasma is assumed to be of infinite extent in the transverse direction, then an equivalent dielectric constant of the plasma occupied region is expressed as [1]:

$$\varepsilon = \varepsilon_s - \sum_{i=e,h} \frac{\omega_{pi}^2}{\omega^2 + \upsilon_i^2}\left(1 + j\frac{\upsilon_i}{\omega}\right) = \varepsilon_R - j\varepsilon_I \tag{2}$$

where ε_s is the dielectric constant of the host lattice, υ_i is the collision frequency, ω is the radian frequency, and ω_{pi} is the plasma frequency given by

$$\omega_{\mathrm{pi}}^2 = \frac{N_p q^2}{\varepsilon_0 m_i^*} \tag{3}$$

where m_i^* is the effective mass of electron and hole, and q is the electronic charge. ε_0 is the free-space permittivity, and N_p is the density of the plasma-excited by optical illumination.

Since a quasi-TEM mode is the dominate mode of the microstrip line, the differential phase shift is given by:

$$\Delta\Phi = (\gamma_1 - \gamma_2)L \tag{4}$$

where L is the length of illuminated transmission line, γ_1, γ_2 are the propagation constant of the propagating wave during and before illumination. Here propagation constant is a function of both conductivity and equivalent dielectric constant.

3.3 Experiments and Results

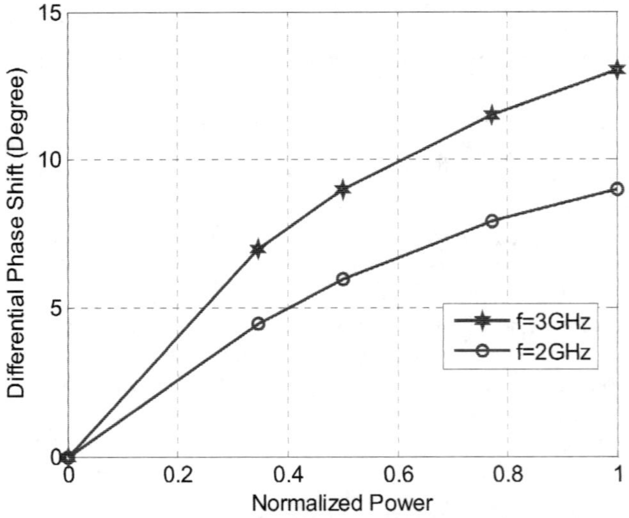

Figure 2: The measured differential phase shift for sample with different power lever at frequencies of 2GHz and 3GHz: 34.7%, 50.1%, 77.3% and 100%

Sample was fabricated for measurement. To measure the differential phase-shift with different power lever, three light filters were used to change the light intensity by 34.7%, 50.1% and 77.3%. Figure 2 gives the measured differential phase-shift for sample with different power lever at frequencies of 2GHz and 3GHz. Considering the 1V input signal at frequency of 2GHz, the amplitude of output is 109mV. The phase shift 3.2°/cm at 2GHz with energy of 90mW.

The reversible phase-shifts obtained by the two rudimentary devices presented here are comparable to phase shifts obtained using conventional silicon semiconductors and, given the lack of device optimization, very promising. They certainly demonstrate the principle of operation of optically controllable phase shifters and the applicability of organic semiconductors to such devices.

Qualitatively, one would expect a larger phase shift as one increases the signal dwell length/time over the illuminated dielectric and this is seen in the larger phase-shift obtained in sample A, which possessed the longer, meandering conductor, versus that obtained in sample B under otherwise identical conditions (figures 5.13b and 5.14b). Larger overall phase shifts can thus be obtained by increasing the length of the illuminated portion of the structure. As the charge carrier mobility of P3HT is currently less than 0.1 $cm^2/(Vs)$, which is much smaller than that of inorganic semiconductors (>100 cm^2/Vs), a relatively large power lever is needed to achieve the same phase-shift. Further development is needed to improve photo-coupling efficiency.

4. Conclusion

In this paper, an investigation on optically-controlled phase-shifter based organic semiconductor P3HT has been presented. The operation principle of optically-controllable phase-shifters and the applicability of organic semiconductor P3HT to such devices have been demonstrated. The reversible, optically-controlled phase-shifter employing P3HT as an active device layer, yielded differential phase shifts which varied with light intensity and signal frequency as well as device architecture. The results are non-optimized, and impedance matching needs to be improved in the future in order to obtain better signal transmission performance through the device.

Acknowledgments

Authors wishing to acknowledge Merck Chemicals Limited for supplying the P3HT and Chinese NSFC grant No. G0501020160801026 for financial support of this research.

References

[1] S. Shevgaonkar, A.N. Chandorkar, D.K. Sharma, "Optically controlled microwave phase shifter", Proc. Int. Conf., CODEC-98, 1998, pp. 53-56

[2] P.Herczfeld, "I Monolithic microwave-photonic integrated circuits: a possible follow-up to MIMIC", Microwave J., vol.35, 1992, pp.64-78

[3] C.H. lee, P.S. Mak, A.P.Defonzo, "Optical Control of Millimeter-Wave Propagation in Dielectric Waveguides", IEEE J. Quantum Electron.16, 277–288, 1980, pp. 277-288

[4] J.F. Chang, B. Sun, D. W. Breiby, M. M. Nielsen, T. I. Sölling, M. Giles, I. McCulloch, and H. Sirringhaus, "Enhanced Mobility of Poly(3-hexylthiophene) Transistors by Spin-Coating from High-Boiling-Point Solvents", Chem. Mater., Vol.16, 2004, pp. 4772-4776

[5] X.Jiang, R.Osterbacka, O.Korovyanko, "Spectroscopic studies of photoexcitations in regioregular and regiorandom polythiophene films", Advanced Functional Materials, 12, 2002, pp. 587-597

[6] W.U. Huynh, J. J. Dittmer, A. P. Alivisatos, "Hybrid Nanorod-Polymer Solar Cells", *Science,* Vol. 295. no.5564, 2002, pp. 2425 – 2427

[7] R.C.G. Naber, M. Mulder, B. de Boer, P.W.M. Blom, D.M. de Leeuw, "High charge density and mobility in poly(3-hexylthiophene) using a polarizable gate dielectric", Organic Electronics, 7 , 2006, pp. 132–136

Fabrication of the p-MgZnO/ZnO/n-MgZnO double-heterojunction by MOCVD

Xin Dong* , Hui Wang, Jin Wang, Wang Zhao, Long Zhao, Zhifeng Shi

State Key Laboratory on Integrated Optoelectronics, College of Electronic Science and Engineering, Jilin University, Qianjin Street 2699, ChangChun 130012, People's Republic of China

Email: dongx@jlu.edu.cn

Abstract

The p-MgZnO/i-ZnO/n-MgZnO double-heterojuntion was fabricated on GaAs (100) substrate by photon-assistant metal-organic chemical vapor deposition. We can found an obvious ultraviolet peak in the electroluminescence spectra and the causation has been discussed. The effects of the thickness of ZnO interlayer on the junctions properties was also analyzed. The conclusion indicated the intensity of the UV peak would increase by decreasing the thickness of ZnO interlayer. The thickness of ZnO layer had important impact on the optical qualities of the double-heterojunction.

Keywords: MgZnO; heterojunction; EL; MOCVD

1.Introduction

ZnO has recently attracted considerable attention due to its favorable properties such as the wider band gap (3.37eV) at room temperature, the large binding energy of excitons (60eV). ZnO has so good photoelectric and piezoelectric properties [1-4] that it has immensity space for developing at surface acoustic wave devices, light emitting diodes (LEDs) [5] , photodetectors [6], gas sensor and solar cells[7]etc. MgZnO has many similar properties to ZnO. Furthermore, the band gap of MgZnO is 3.3-4eV [8] due to the wider band gap of MgO (7.7eV [9]). Ternary MgZnO alloys can be utilized as barrier layers in ZnO/MgZnO multiple-quantum-well-structure [10-13].

In this paper, the p-MgZnO/i-ZnO/n-MgZnO double-heterojuntion was fabricated on *p*-GaAs substrates by metal-organic chemical vapor deposition (MOCVD). The *p*-type MgZnO layer was obtained by arsenic diffusion from the GaAs substrates. The optical properties of the heterojunction and its irradiance theory were analyzed detailed.

2. Experiment

The p-MgZnO films were grown on *p*-GaAs substrates (the hole concentration was about 10^{19} cm^{-3}) by low-pressure MOCVD. Photon-assistant was performed to reduce the mismatch of the lattices and improve the crystal quality. Diethyl zinc (DEZn), bis-cyclopentadienyl-Mg (cp_2Mg) and oxygen as the reactants were introduced into the reactor through separately injectors. 5N-purity argon was used as the carried gas of DEZn and cp_2Mg. The reactor pressure was kept at about 2 Torr during the growth. The growth was paused when the thickness of MgZnO achieved

500nm. The thickness of film was estimated by the known growth speed. Then the samples were annealed in oxygen for 1h to obtain p-type film by arsenic diffusion from the substrates. To test the impact of the diffusion, MgZnO film was also grown on semi-insulating GaAs substrate (the resistivity was about 10^8 Ω·cm) and the result indicated the carrier concentration of p-MgZnO film was about $+5.65\times10^{17}$ cm^{-3}. The insulating ZnO interlayer, about 5nm, 10nm, 15nm, 20nm were grown respectively in different samples with sufficient oxygen flux [14]. At last, the growth was performed under a lower flux of oxygen to obtain the *n*-MgZnO about 500nm according the former method [15]. The temperature was performed at 600℃ in the whole process and the content *x* of Mg in films determined by energy dispersive spectroscopy, was 0.12. To measure the properties of the double-heterojunction, the indium metalline electrode was deposited both on the top of *n*-$Mg_{0.12}Zn_{0.88}O$ and the GaAs substrate by conventional thermal evaporation. Then the samples were annealed at 300℃ for 4 min in N_2 ambient to reduce contact resistance. A schematic diagram of structure was shown in the Fig.1a. All the measurements were all operated at room temperature.

3. Results and discussions

The Fig.1b showed the I-V properties of the double-heterjunction with a 15nm thickness ZnO layer. The curve showed the typical rectifying behaviour of a p-i-n junction. The forward turn-on voltage appeared at about 6.0V and reverse break-down voltage was about -20V. The emission could be found by naked eye in the dark when the injection current attained 40mA. The inset shows the *I–V* properties of the indium metal contacts on n-MgZnO and p-GaAs, respectively. The results indicated that the rectifying characteristics do not result from the semiconductor/metal contacts, but from the double-heterojunction.

The SEM image (side view and top view) of the sample with 15nm thickness ZnO layer was shown in Fig.2. The Fig.2a showed the structure of the film was nanorods. The length and average diameters of the nanorods were 1000nm and in the range of 40-50nm respectively, which showed the typical 1-D grown mode. In comparison with bare ZnO nanorods, there is no significant change in the morphology, except that nanorod diameters decrease due to the smaller lattice parameter c of MgZnO alloy. The ZnO layer cannot be observed accurately due to its thinnest thickness. The top view image (Fig.2b) also indicated the fine crystal quality of the film.

Fig.3 showed the room temperature PL spectra of the *n*-type $Mg_{0.12}Zn_{0.88}O$ layer (A), *p*-type $Mg_{0.12}Zn_{0.88}O$ layer (B), ZnO interlayer (C) and EL spectra of the heterojunction (D) under the current of 60mA. The PL measurements were performed using a 325nm He-Cd laser. The curve A and B both exhibited major peaks at about 374nm attributed to the near-band-edge emission. The intensity of near-band-edge (NBE) peak in curve B was weaker than that of curve A due to the crystal quality degraded when the new defects were introduced by arsenic diffusion. A major peak positioned 388nm can be observed in curve C attributed to the same reason. The intensity ratios of the NBE to deep-level (DL) emission in curves C was almost 100 :

1, which implied the better crystal quality of ZnO film. The difference between the positions of the peaks was caused by the difference of the band gaps. From the curve D, a weak peak can be measured at the same positon of ZnO UV peak. It showed the radiation phenomenon occurred in ZnO interlayer and it was caused by the combination of carriers. The DL emission peaks, especially the ones at about 490nm, can be perceived in all curves. The phenomenon was consistent with our previous works [16, 17].

The effects of the ZnO layer thickness on the intensities of UV peaks and DL transition peaks in the EL spectra are shown in figure 4. Note that when the ZnO interlayer thickness was less than 10nm, the ZnO interlayer could be consider as a quantum well. The UV peak became much stronger with decreasing ZnO interlayer thickness. So the probability of recombination of the carriers increased due to the quantum size confinement effect. The thickness of ZnO layer had important effect on the optical qualities of the double heterojunction.

4. Conclusion

The p-MgZnO/i-ZnO/n-MgZnO double heterojunction structure was fabricated on GaAs substrate by MOCVD. The SEM images of the layers illustrated the fine crystal quality. An obvious ultraviolet peak can be observed in the electroluminescence spectra, which was result from the recombination of carriers in ZnO interlayer. The dependences of the characteristics on the thickness of ZnO interlayer were researched detailed. With decreasing the thickness of ZnO interlayer, the intensity of the UV peak increased remarkably due to the quantum size effects. The thickness of ZnO layer had important effect on the optical qualities of the double heterojunction.

Acknowledgments

This work was supported by National Natural Science Foundation of China under Grant No.61006006 and 60877020.

Reference:

[1] C.Klingshirn,Phys.Stat.Sol. (B) 71 (1975) 547.

[2] D.M.Bagnall, Y.F.Chen, M.Y.Shen, Z.Zhu, T.Goto, T.Yao, J.Crystal Growth 184 (1998) 605.

[3] P.Zu, Z.K. Tang, G.K.L. Wong, M. Kawasaki, H. Koinuma, Y.Segawa, Solid State Commun. 103 (1997) 459

[4] A.Ohtomo, M.Kawasaki, T.Koida, K.Matsubuchi, H.Koinuma, Y.Sakurai, Y.Yoshida, T.Yasuda, Y.Segawa, Appl.Phys.Lett. 72 (1998) 2466.

[5] T.Slki, Y.Hatanaka, D.C.Look, Appl.Phys.Lett.76 (2000) 3257.

[6] Y.Liu, C.R.Gorla, S.Liang,et al., J.Electron.Mater. 29 (2000) 60.

[7] U.Rau, M.Schmidt, thin Solid Films 387 (2001) 141.

[8] A.Ohtomo, R.Shiroki, I.Ohkubo, H.Koinuma, M.Kawasaki, Appl.Phys.Lett. 75 (1999) 980.

[9] Takashi Minemoto, Takayuki Negami, Shiro Nishiwaki, Hideyuki Takakura,

Yoshihiro Hamakawa, Thin Solid Films 372 (2000) 173.

[10] T. Makino, N. T. Tuan, H. D. Sun, C. H. Chia, Y. Segawa, M. Kawasaki, A. Ohtomo, K. Tamura, T. Suemoto, H. Akiyama, M. Baba, S. Saito, and T. Tomita, H. Koinuma, Appl.Phys.Lett.**78,** 1979 (2001).

[11] Jun Young Bae, Jinkyoung Yoo, Gyu-Chul Yi, Appl. Phys. Lett. **89,** 173114 (2006).

[12] Chinkyo Kim, Won □ Park, Gyu-Chul Yi, Miyoung Kim, Appl. Phys. Lett. **89,** 113106 (2006).

[13] Hiroaki Matsui, Hitoshi Tabata, Noriyuki Hasuike, Hiroshi Harima, J.Appl.Phys. **99,** 024902 (2006).

[14] Ma Y, Du G T, Yang T P, Qiu D L, Zhang X, Yang H J, Zhang Y T, Zhao B J, Yang X T and Liu D L J. Cryst. Growth **255,** 303 (2003).

[15] Ma Y, Du G T, Yang S R, Li Z T, Zhao B J, Yang X T,Yang T P, Zhang Y and Liu D L J. Appl. Phys. **95,** 6268 (2004).

[16] Du G T, Liu W F, Bian J M, Hu L Z, Liang H W, Wang X S and Liu A M Appl. Phys. Lett. **89,** 052113 (2006).

[17] Bian J, Liu W, Liang H, Hu L, Sun J, Luo Y and Du G Chem. Phys. Lett. **430,** 183–7 (2006).

List of captions:

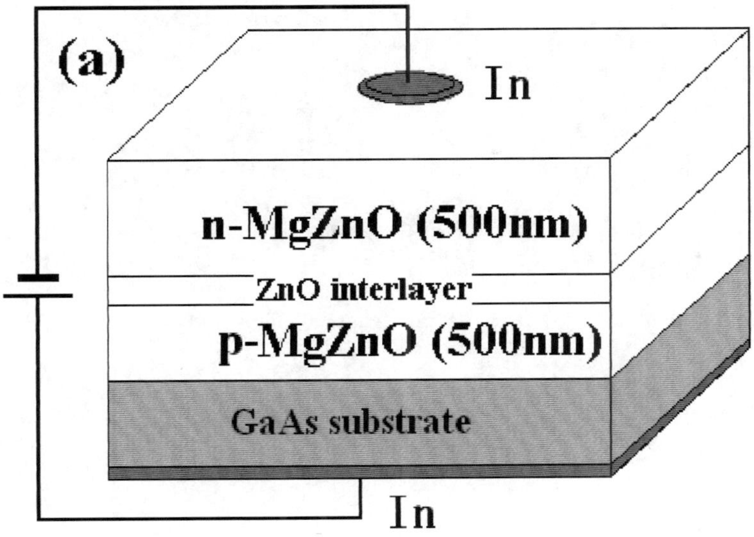

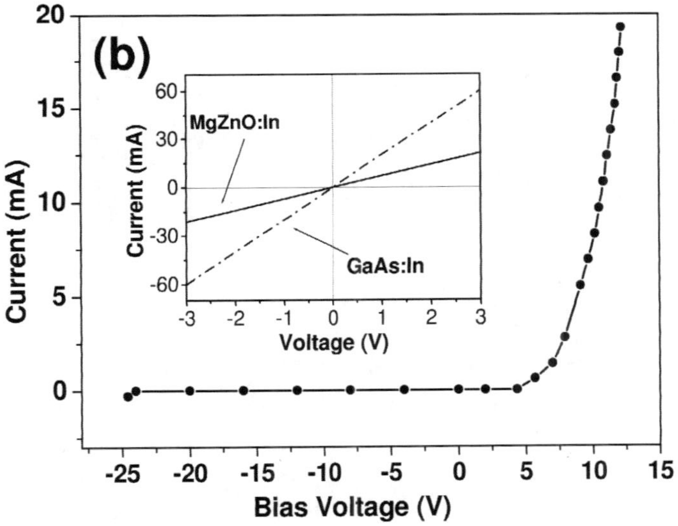

Figure 1. Schematic diagram of the double-heterojunction structure (*a*) and the *I–V* characteristics of the structure (*b*). The inset of fig.*b* shows the *I–V* dependences of indium contacts on the n-type MgZnO and the GaAs substrate.

Figure 2. The side-view (*a*) and the top-view (*b*) SEM images of the heterojunction with 15 nm-thick ZnO interlayer.

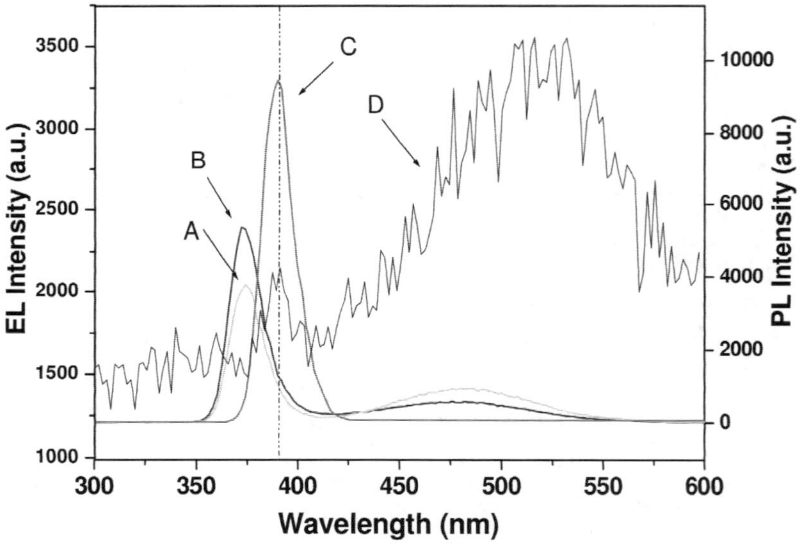

Figure 3. Room temperature PL spectra of n-type (A), p-type (B) MgZnO layer and ZnO interlayer (C). Curve D shows EL spectrum of the heterojunction with 15 nm-thick ZnO interlayer.

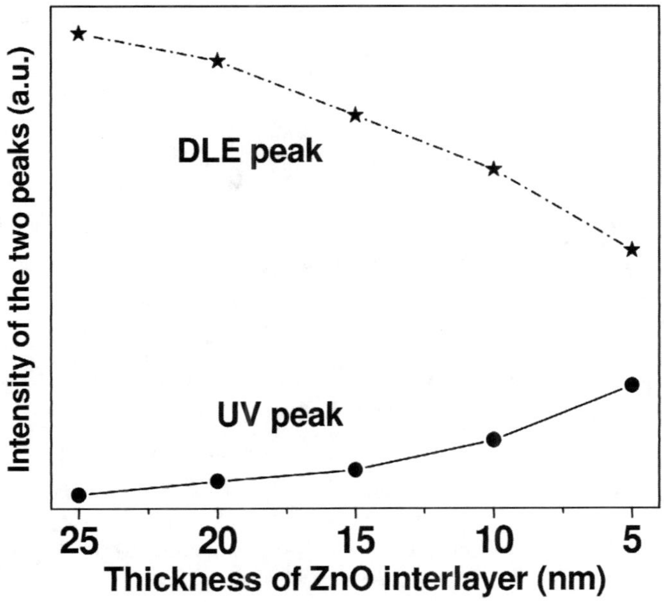

Figure 4. The intensities of the UV peaks and the DLE peaks in EL spectra as functions of the thickness of ZnO interlayer.

Influence of AlGaN Buffer Growth Temperature on GaN Epilayer based on Si(111) Substrate

Meng Wei[1,*], Xiaoliang Wang[1,2], Xu Pan[1], Hongling Xiao[1,2], Cuimei Wang[1,2], Minglan Zhang[1], Zhanguo Wang[2]

[1] Materials Science Center, Institute of Semiconductors, Chinese Academy of Sciences, P. O. Box 912, Beijing 100083, China
[2] Key Laboratory of Semiconductor Materials Science, Institute of Semiconductors, Chinese Academy of Sciences, P. O. Box 912, Beijing 100083,China

* Corresponding author: mengw@semi.ac.cn

Abstract. This paper investigated the influence of AlGaN buffer growth temperature on strain status and crystal quality of the GaN film on Si(111) sbustrates by metal organic chemical vapor deposition. It was demonstrated by the optical microscopy that AlGaN buffer gorwth temperature had a remarkable effect on compensating tensil stress in top GaN layer and preventing the formation of cracks. X-ray diffraction and atomic force microscopy analysis showed crystal quality and surface morphology of the GaN epilayer could be improved through increasing AlGaN buffer growth temperature. 1μm crack-free GaN epilayer on Si (111) substrates was obtained with graded AlGaN buffer layer at optimized temperature of 1050℃. Transmission electron microscopy analysis revealed that a significant reduction in threading dislocations was achieved in GaN epilayer.

1. Introduction

Gallium nitride (GaN) compound semiconductor material system has been extensively of great interest for the large band gap, high breakdown electric strength, high electron saturated drift velocity and good thermal stability. Furthermore, the particular spontaneous and piezoelectric polarization effects lead to high concentration of two-dimensional electron gas (2DEG) near the interface of AlGaN/GaN hetero-structure. Consequently, GaN material has a large potential application in the areas of high power and high frequency electronic devices, light emitting diodes (LEDs) and laser diodes [1-5]. Generally GaN is grown on sapphire, SiC and Si substrates. Sapphire substrates limit the improvement of the power density of the device because of the bad thermal conduction. Meanwhile, the high cost and small size of SiC substrates badly hinder the application of GaN material. Compared with sapphire and SiC, Si is the best alternative substrate for its low cost, good thermal conductivity and ability to be integrated with the mature Si-based processing techniques. However, the large lattice mismatch (17%) and coefficients of thermal expansion (CTE) mismatch (56%) between GaN epilayer and Si would induce large residual tensile stress in the GaN based epilayers [6,7], which lead to the formation of cracks and high defect density that seriously degrade the performance of GaN based devices on Si substrate. Thus the design of suitable buffer layers towards the reduction of residual stress is crucial. Several approaches, such as introduction of compressively strained layers or a bridging interlayer with similar CTE, epitaxial lateral overgrowth using patterned Si substrates, have been developed to compensate for the large thermal tensile stress and prevent the formation of cracks [8-14]. Among these high temperature (HT) AlN buffer is a simple and effective way to relieve tensile stress and further prevent cracks in GaN epilayer. However extra threading dislocations (TD) will be introduced into top GaN due to large cryatal lattice mismatch between AlN and GaN [15,16]. In our study, we grow an AlGaN buffer layer on HT AlN buffer prior to GaN growth. Compaired with a single AlN buffer, the AlN/AlGaN multibuffer facilitates the compensation for tensile stress in GaN film, as well as reduce the TD density, owing to the crystal lattice of AlGaN between those of AlN and GaN.

In this work, we studied the effect of AlGaN buffer growth temperature on the surface morphology, strain status and crystal quality of GaN epilayer on Si(111) substrates by comparing five samples with AlGaN buffer of different growth temperature. 1μm crack-free GaN epitaxy with graded AlGaN buffer at optimized temperature was presented. It was observed by transmission electron microscopy (TEM) to investigate the TD evolustion from through AlGaN buffer layer.

2. Experimental

In this work, the growth of samples was performed in a horizontal metal organic chemical vapor deposition (MOCVD) reactor on Si(111) substrates. Prior to loading in the growth chamber, the Si substrates were etched in boiling sulphuric acid, HCl:H2O2:H2O(1:1:5) solution, NH4OH:H2O2:H2O(1:1:5) solution and diluted HF solution in turn, which resulted in oxide-free hydrogen-terminated Si(111) surface. An in situ hydrogen exposure at 1100 ℃ was performed immediately after the substrates were loaded. Trimethylgallium(TMG), trimetheylaluminum(TMA) and ammonia(NH3) were used as Ga, Al and N precursors, respectively, while N2 and H2 as carrier gases. The entire GaN epitaxial layers containing of a HT AlN buffer and an AlGaN buffer followed by 1μm GaN film was shown in Fig. 1. Five samples with different AlGaN growth temperature were investigated, labled by samples A, B, C, D and E, respectively, as listed in Tab.1. For samples A, B, C and D, the Al composition of AlGaN buffer was fixed at 30% uniformly, with AlGaN growth temperature varied from 900, 1000, 1050, to 1080℃. For sample E, the AlGaN buffer growth temperature was 1050℃ while a graded AlGaN buffer layer instead of Al-composition-constant ones with the Al composition modulated gradually in the range of 30% to zero from the bottom to the top. Before AlN growth, the Si(111) sbustrates were exposed to TMA flow for several seconds without any ammonia, for the purpose of depositing an ultra Al layer to prevent the formation of a silicon nitride phase on the Si surfaces. The HT AlN buffer layer, and GaN film were grown at 1050℃ and 1060℃, respectively. The pressure was set to 100Tor for all deposition steps.

Optical microscopy (OM) and atomic force microscopy (AFM) were used to evaluate the crack density and surface morphology of GaN film. X-ray diffraction (XRD) was performed to measure the full width at half maximum (FWHM) of GaN(0002) rocking curve peaks to characterize the crystal quality. TEM was carried out to survey the TDs evolution with graded AlGaN buffer at optimized temperature.

3. Results and discussion

Crack density and surface morphology of the samples were investigated by OM. No white pits appeared to naked eyes on the surface of samples A, B, C and E. However some were presented in sample D. Under OM observation we found some curcular defects on the surface of sample D (shown in Fig. 2a) while mirror-like and specular surfaces were shown in other four samples (OM image of sample E was shown in Fig. 2b). Dadgar reported the curcular defects were caused by reaction of Ga and Si atoms outdiffusing from substrate during the high temperature growth process, and called it melt-back etching reaction [17]. The melt-back etching reaction of Ga and Si atoms is a well known problem of GaN epitaxy on Si, which resulted in circular defects and was severely destructive to the crystal quality of the GaN film [18]. The absent of circular defects in samples A, B, C and E proved the growth temperature (below 1050℃) of AlGaN buffer were proper to avoid melt-back etching reaction while the AlGaN growth temperature of sample D was so high that led to Si atoms diffusing out. A.M.Sanchez interpreted the circular defects origination by step theory [19].

The maximum crack-free ranges of the five samples were shown in Tab. 1. Samples A, B, D exhibited cracks across the entire wafer surface in three sets of parallel arrayed at 120° to one another, with the crack-free ranges of 9mm×10mm, 13mm×14mm, 11mm×11mm, respectively. In sample C the cracks were only found near the edges of the wafer while sample E revealed a crack-free surface on the overall wafer as shown in Fig. 2b. The results indicated AlGaN growth temperature influenced the crack density obviously and the optimized temperature was existed. It is known that the AlN interlayer will impose a compressive stress to the GaN epilayer which can partially offset the thermal tensile stress during the cooling down process. However, the GaN layer grown on the AlN buffer tends to relax due to the 2.4% lattice mismatch between GaN and AlN. The additional introduction of an AlGaN buffer layer with proper growth temperature, especially the grated AlGaN buffer, between GaN and AlN could lead to a soft transition in the lattice constants. Therefore, the compressive lattice mismatch stress may be more effectively maintained to compensate for the tensile stress induced by CTE mismatch. If the AlGaN growth temperature was not high enough the crystal quality was inferior, thus the compensation effect was limited. But if the growth temperature was too high, the thermal tensile stress became larger, and cracks emerged again.

The surface morphology of sample was characterized by AFM. The root mean square roughnesse (RMS) of sample was considerably decreased with AlGaN buffer growth temperature increasing. Meanwhile it was clearly seen the RMS of sample with graded AlGaN (sample E) was smaller than that of sample with Al-composition-constant ones (sample C) at the same growth teperature, due to the threading dislocations decreased by the employment of graded AlGaN. The results revealed that surface morphology of GaN film could be improved by increasing AlGaN growth temperature. Especially, it proved graded AlGaN buffer was more appropriate to perfect the surface morphology than the Al-composition-constant ones at the same growth temperature. The AFM image with 2μm×2μm scan scope of sample E was shown in fig. 3, which exhibited smooth surface with a RMS of 1.24nm.

The X-ray measurements for samples were crried out with symmetric GaN(0002) reflection using ω scan. GaN(0002) XRD FWHMs of rocking curve peaks were uesed to investigate the influence of AlGaN growth temperature on the crystal quality of GaN film. Results were shown in Fig. 4. It was clearly seen that FWHMs of samples decreased with AlGaN growth temperature increasing. It's suggested the AlGaN buffer crystal quality was improved with the growth temperature raising. Better buffer crystal quality led to better GaN epilayer crystal quality. Thus FWHM decreased with AlGaN growth temperature increasing.On the other hand, it was found FWHM of sample with graded AlGaN (sample E) was smaller than that of sample with Al-composition-constant AlGaN buffer (sample C) at the same growth temperature. Structure with graded AlGaN seemed more helpful to improve the crystal quality than that with the Al-composition-constant ones. These XRD results suggested that a graded AlGaN greatly improves the in-plane crystalline quality of GaN.The reason was that the gradual decrement of Al content resulted in a modest change in the lattice constants within AlGaN buffer layer and it could be expected to maintain a pseudo-2D growth mode for the successive GaN layer. Superior crystal quality of GaN film was presented due to the pseudo-2D growth mode. In another way, the decrease in FWHM could be attributed to a reduction in the dislocation density by the graded AlGaN buffer. Detailed discussions would be supplied by TEM analysis in the following.

In order to analyze the mechanism to achieve high crystal quality GaN on Si(111), cross-sectional TEM was performed. The electron beam is along the $[1\bar{1}00]$ zone axis so that all types of TDs can be revealed as shown in Fig. 5. A large number of TDs were clearly seen originating from the AlGaN/AlN interface due to the mismatch between AlGaN and AlN layers[20] and thread through the whole graded AlGaN buffer. Significant dislocation reduction took place at the AlGaN/GaN interface while few threading dislocations generated at the interface. Obviously the GaN layer owned lower TD density than the AlGaN buffer. Thus, we could declare the AlGaN graded buffer layer acted as a dislocation filter. Another dislocation reduction took place in the first 500nm GaN above the AlGaN buffer, where a large number of the residual TDs bent and interacted as shown in Fig. 5. It is clearly seen that TDs in the AlN buffer and the AlGaN buffer layer were normally perpendicular to the growth surface. Whereas TDs in GaN bent and deviated away from the growth direction in the region next to the AlGaN buffer layer. These inclined TDs interacted with those possessing opposite Burger vector and finally annihilated, resulted in the reduction of most TDs in GaN epilayer and the compressive strain relaxation[21]. Relaxation of compressive stress was produced by such bending due to the formation of in-plane misfit segments [22]. It was reported that TD inclination assisted stress relaxation in compressively strained structures[23,24]. By annihilation of bent dislocations, the TD density decreased during the growth of the GaN layer associated with compressive stress relaxation. Vice versa, the bend of dislocation symbolized the intense compressive stress exerted by the AlN/AlGaN mulitbuffer.

4. Conclusion

This paper was concentrated on the effect of AlGaN growth temperature on the crack density and cryatal quality of GaN epilayer on Si(111) substrates. We found an optimized temperature of 1050℃ was existed to maximize the compensation effect caused by the compressive stress from AlN buffer. Crystal quality seemed improved with AlGaN growth temperature increasing. 1μm crack-free GaN was obtained using graded AlGaN buffer at the optimized growth temperature with GaN(0002) XRD FWHM of 603.6arcsec. It was confirmed by TEM analysis that graded AlGaN with optimized temperature made a significant reduction of threading dislocations in GaN epilayer, thus improved the cryatal quality.

Acknowledgements:

This work has been supported by the Knowledge Innovation Engineering of Chinese Academy of Sciences (No. YYYJ-0701-02); the National Nature Sciences Foundation of China (No. 60890193, 60906006); the State Key Development Program for Basic Research of China (No. 2006CB604905, 2010CB327503), and the Knowledge Innovation Program of the Chinese Academy of Sciences (No. ISCAS2008T01, ISCAS2009L01, ISCAS2009L02).

References

[1] X. H. Wu, P. Fini, E. J. Tarsa, B. Heying, S. Keller, U. K.Mishra, S. P. DenBaars, and J. S. Speck, J. Cryst. Growth 189, 231 (1998).
[2] S. Tanaka, S. Iwai, and Y. Aoyagi, J. Cryst. Growth 170, 329 (1997).
[3] H. Lahre`che, M. Leroux, M. Lau¨gt, M.Vaille, B. Beaumont, and P. Gibart, J. Appl. Phys. 87, 577 (2000).
[4] T. Egawa, N. Nakada, H. Ishikawa, and M. Umeno, Electron. Lett. 36, 1816 (2000).
[5] S. A. Nikishin, N. N. Faleev, G. Antipov, S. Francoeur, G. A. Seryogin, H. Temkin, et al. Appl. Phys. Lett. 75, 2073 (1999).
[6] E. Arslan, M. K. Ozturk, A. Teke, S. Ozcelik, E. Ozbay, J. Phys. D: Appl. Phys. 41, 155317 (2008).
[7] H. Ishikawa, K. Yamamoto, T. Egawa, T. Soga, T. Jimbo, and M. Umeno, J. Cryst. Growth 189, 178 (1998).
[8] A. Krost, A. Dadgar, G. Strassburger, and R. Clos, Phys. Status Solidi A 200, 26 (2003).
[9] J. Blasing, A. Reiher, A. Dadgar, A. Diez, and A. Krost, Appl. Phys. Lett. 81, 2722 (2002).
[10] K. Y. Zang, Y. D. Wang, S. J. Chua, and L. S. Wang, Appl. Phys. Lett. 87, 193106 (2005).
[11] Y. Lu, X. Liu, D.-C. Lu, H. Yuan, G. Hu, X. Wang, Z. Wang, and X. Duan, J. Cryst. Growth 247, 91 (2003).
[12] A. Krost and A. Dadgar, Phys. Status Solidi A 194, 361 (2002).
[13] S. Raghavan and J. M. Redwing, J. Appl. Phys. 96, 2995 (2004).
[14] J. F. Wang, D. Z. Yao, J. Chen, et al, Appl. Phys. Lett. 89, 152105 (2006).
[15] W. Liu, J. J. Zhu, D. S. Jiang, et al, Appl. Phys. Lett. 90, 011914 (2007).
[16] A. Dadgar, A. Strittmatter, J. Blasing, et al. Phys. Stat. Sol. (C) 0, 1583 (2003).
[17] C.K. Shu, J. Ou, H.C. Lin, W.K. Chen, M.C. Lee, Appl. Phys. Lett. 73, 641(1998).
[18] A. M. Sanchez, G. P. Dimitrakopulos, P. Ruterana, et al. Appl. Phys. Lett. 86, 011917 (2005).
[19] Blasing J, Reiher A, Dadgar A, Diez A, and Krost A, Appl. Phys. Lett. 81, 2722 (2002).
[20] Weng X, Acord J D, Jain A, Dickey E C and Redwing J M , J. Electron. Mater. 36, 346 (2007).
[21] A. E. Romanov and J. S. Speck, Appl. Phys. Lett. 83, 2569 (2003).
[22] Wang J F, Yao D Z, Chen J, Zhu J J, Zhao D G, Jiang D S, Yang H and Liang J W, Appl. Phys. Lett. 89, 152105 (2006).
[23] P. Cantu, F. Wu, P. Waltereit, S. Keller, A. E. Romanov, U. K. Mishra, S. P. DenBaars, and J. S. Speck: Appl. Phys. Lett. 83 (2003) 674.
[24] A. E. Romanov and J. S. Speck: Appl. Phys. Lett. 83 (2003) 2569

Journal of Physics: Conference Series **276** (2011) 012094 doi:10.1088/1742-6596/276/1/012094

Figure Captions:

Fig. 1 Schematic diagram of sample structure

Fig. 2 OM images of sample D (a) and sample E (b)

Fig. 3 2μm×2μm scope AFM image of sample E

Fig. 4 GaN(0002) XRD FWHMs of samples vs AlGaN growth temperature

Fig. 5 Cross-sectional TEM image of sample E where the electron beam is along the $[1\bar{1}00]$ zone axis.

Table captions:

Tab.1 detailed description of the five samples

Fig. 1

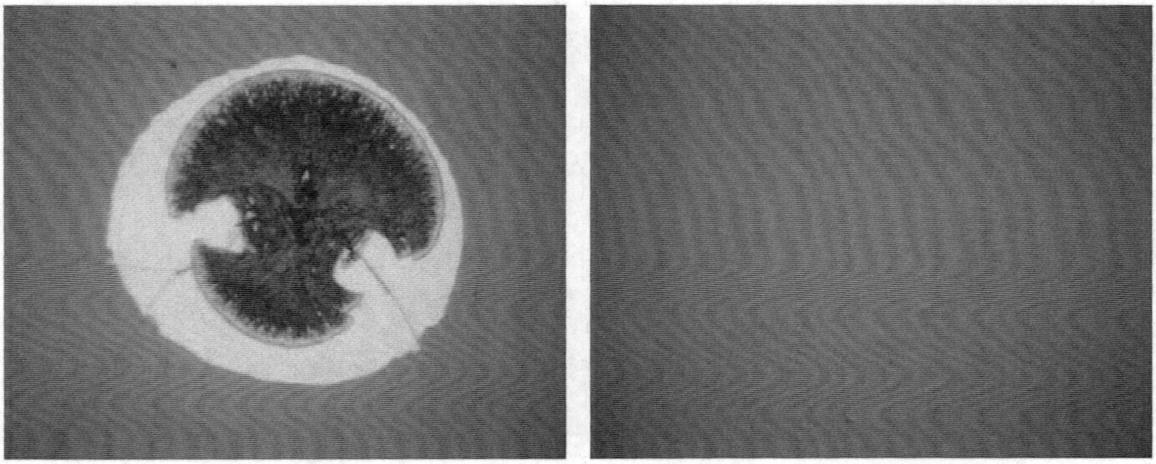

(a) (b)

Fig. 2

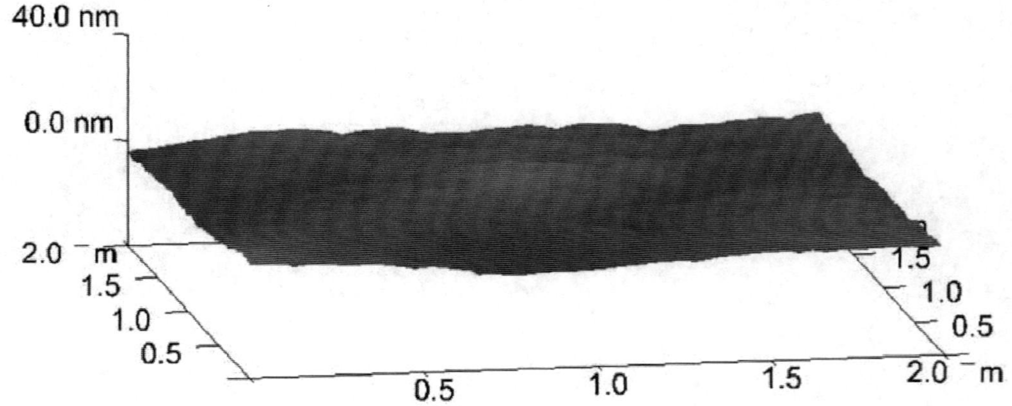

Fig. 3

Journal of Physics: Conference Series **276** (2011) 012094 doi:10.1088/1742-6596/276/1/012094

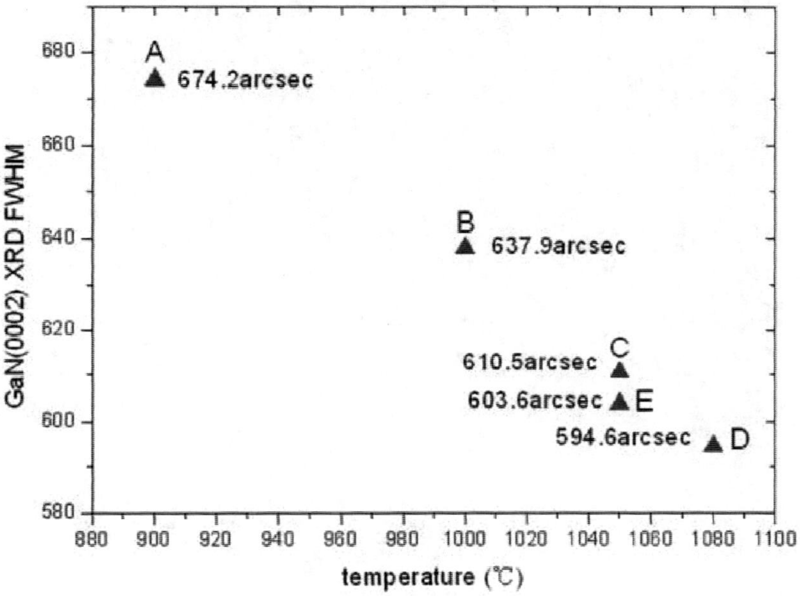

Fig. 4

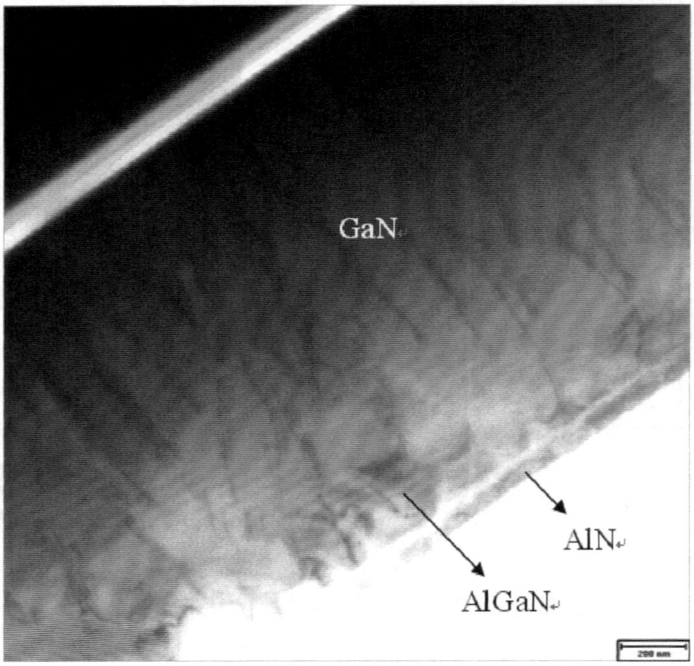

Fig. 5

Journal of Physics: Conference Series **276** (2011) 012094 doi:10.1088/1742-6596/276/1/012094

Tab. 1

Sample	A	B	C	D	E
AlGaN temperature (℃)	900	1000	1050	1080	1050
Crack-free range (mm^2)	9×10	13×15	Except the edge	11×11	All the area
XRD GaN(0002) FWHM(arcsec)	674.2	637.9	610.5	594.6	603.6

Journal of Physics: Conference Series **276** (2011) 012095 doi:10.1088/1742-6596/276/1/012095

High-speed silicon-based microring modulators and electro-optical switches integrated with grating couplers

Xi Xiao, Yingtao Hu, Haihua Xu, Zhiyong Li, Yuntao Li,

Yude Yu*, and Jinzhong Yu*

State Key Laboratory on Integrated Optoelectronics, Institute of Semiconductors, Chinese Academy of Sciences

*Corresponding author, E-mail: yudeyu@semi.ac.cn and jzyu@semi.ac.cn

Abstract. We demonstrate high-speed silicon-based microring modulators and electro-optical switches which are integrated with grating couplers. Wafer-level measurements are set up for the devices characterizations without wafer cleaving and facet polishing. 10 Gbit/s NRZ modulation is realized under pre-emphasis driving signals. Switch on / off time of 300 / 380ps are demonstrated in the asymmetrical 1×2 microring switch with an add-drop crosstalk of -23dB.

1. Introdution

With the help of advanced silicon processing techniques, silicon photonics is developing rapidly aiming at high operation speed, high integration density and CMOS-compatibility. Recently, great success has been made on the basic components of silicon photonics, especially for the silicon-based modulators and switches [1-5]. Compact, high speed silicon modulators and switches will play indispensable roles in the on-chip optical interconnects and the next generation communication networks.

However, electro-optical switching and modulation in silicon are not as easy as what in III-V materials. Pure silicon is a kind of centro-symmetric crystal which exhibits no linear (Pockels) effect and extremely weak Kerr and Franz-Keldysh effect. The mostly used fast modulation and switching method in silicon is the plasma dispersion effect, in which the concentration of free carriers in silicon changes the refractive index and the optical absorption [6]. So the modulation and switching speed is normally limited by the carrier injection, accumulation and depletion. Although the carrier mobility of single crystal silicon is much lower than that in III-V and Ge, the electro-optical response can still be greatly enhanced by adopting resonating optical structures which are very sensitive to the index change. Due to the high-index-contrast (HIC) offered by silicon-on-insulator (SOI) wafers, SOI microring resonators are able to provide very high quality factors (Q factors) while the footprint are small enough for dense photonic integrations. High bandwidth SOI microring switches have already realized nanosecond-scale switch time [7-9]. Ultra-high modulation bandwidth density of 100 Tbit/s· mm^2 has also been demonstrated based on the microring modulators [10]. Active microring devices are becoming the ideal building blocks of the optical networks-on-chip.

Journal of Physics: Conference Series **276** (2011) 012095 doi:10.1088/1742-6596/276/1/012095

In this talk, we review our recent work on the silicon-based 10 Gbit/s modulator and the sub-ns electro-optical switch based on the SOI microring resonators which are embedded with forward p-i-n diodes to fast manipulate the carrier concentrations [11, 12]. For the microring switch, an asymmetrical add-drop configuration is employed to reduce the add-drop in-band crosstalk. While for the modulator, we adopt the notch type microring structure to ensure the high Q factor and thus high modulation speed. Grating couplers have also been fabricated at all the input and output ports for wafer-scale online measurements without wafer cleaving and facet polishing. During the measurements, pre-emphasis signals were used to speedup the active microring devices, under which < 400 ps electro-optical switching and 10 Gbit/s NRZ modulation are respectively demonstrated.

2. Device fabrication and description

All the waveguide devices were fabricated on a SOI wafer with a 340 nm top silicon layer and a 2 μm buried oxide layer by electron-beam lithography (EBL) and inductively-coupled-plasma (ICP) processes. The waveguides are single mode rib waveguides designed for TE-mode transmission. The SEM images of the fabricated passive microrings are shown in figure 1(a) and (b). Both of the modulator and the switch consist of a 20 μm-radii microring with 600 nm wide bent waveguide. For the modulator, a notch-type microring was employed for the high Q factor. While for the switch, we chose the add-drop microring as the optical structure to increase the 3-dB bandwidth for high bit rate signals transmission. In order to obtain high extinction ratios, the coupling strength between the microring and the bus waveguide should be carefully optimized to achieve the critical coupling state.

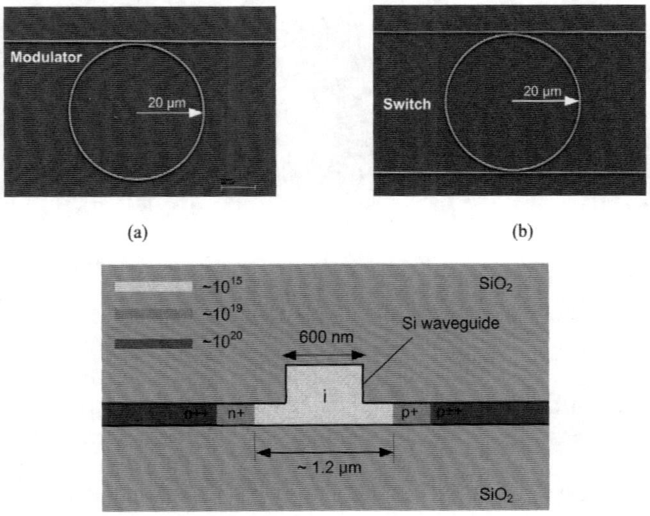

Figure 1. Top-view SEM images of (a) the notch type modulator and (b) the add-drop type switch's passive structures, respectively. (c) Schematic cross-sectional image of the microring waveguide embedded with the p-i-n diode.

Figure 1(c) shows the schematic cross-section of the ring waveguide surrounded by a lateral p-i-n diode. The manipulations of the waveguide's refractive index are implemented by the electrical signals forward applied to the p-i-n junction. As the operation speed of a p-i-n diode significantly depends on the carrier transport distance across the waveguide [13, 14], reducing the width of the intrinsic region becomes the essential way to decrease the diode's response time. We located both the p+ and n+ regions 300 nm away from the ring waveguide edges by EBL and ion-implantation processes. A location mismatch error of only ~ 40 nm was achieved by EBL. The p+ and n+ doping concentrations are both ~ 10^{19} cm^{-3} for an acceptable dopant absorption loss. In order to ensure good metal contacts, highly doping regions p++ and n++ with > 10^{20} cm^{-3} implantation concentrations are both defined 2 μm away from the waveguide ridge by optical lithography.

3. Measurement and results

3.1. Measurement setup

With the help of the grating couplers integrated at the devices' input and output ports, we have realized wafer-scale measurements without wafer cleaving and facet polishing. Before the fabrications, we used the software CAMFR to design the grating period Λ [15]. When the grating period is 620 nm and the duty cycle is 50%, simulation predicts an over 50% coupling efficiency and an 80nm 3-dB bandwidth for a single grating coupler. Figure 2(a) is the top-view SEM image of the grating coupler fabricated together with the waveguides. The coupler surface is about 14 × 19 μm^2 which is slightly larger than a single-mode fiber. The moderate surface size ensures a lot of tolerance to the fiber-coupler alignment. Figure 2(b) shows the detailed structure of the fabricated grating coupler. The measured grating period is 620 nm and the duty cycle agrees with the designed value very well.

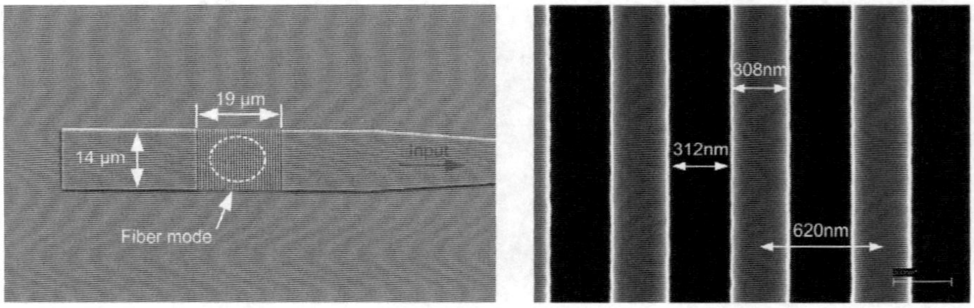

(a) (b)

Figure 2. (a) Over-view and (b) detail SEM image of the grating coupler. The grating coupler is 14 μm in width and 19 μm in length. The grating period is 620 nm and duty cycle is ~ 50%.

We set up the fiber-to-fiber transmission measurements to characterize the insertion loss of the microring devices integrated with grating couplers. During the experiment, light was coupled into and out of the chip by two single-mode fibers which were positioned at an angle of about 10° to the sample normal. The corresponding fiber-to-fiber transmission for TE-mode light measured as shown in figure 3. The 3-dB bandwidth is measured to be 60 nm (1527--1587 nm) fully covering the C-band. The minimal total insertion loss is measured to be 12.2 dB around 1550 nm which includes the waveguide propagation loss, alignment mismatch loss and polarization induced loss. As the maximal output of the laser source was 10 dBm and the output power from modulator was normally over -5 dBm, so the insertion loss was acceptable for the device characterizations even without erbium-doped fiber amplifier (EDFA).

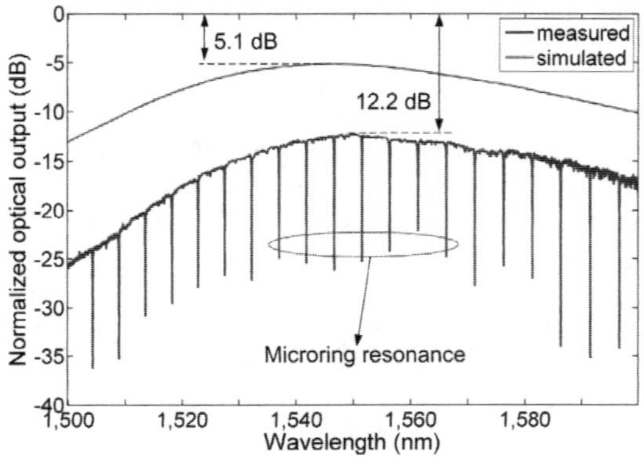

Figure 3. Measured (blue) and simulated (red) coupler-to-coupler transmission spectra. The spectral drops in the measured curve are the microring resonances.

3.2 Results of the microring switch

For the switch, an asymmetric dual-coupled microring was employed due to the characteristics of large through-port extinction and low crosstalk [16]. Figure 4(a) shows the top-view microscope image of the fabricated switch with electrodes. The gray lines highlight the 20 μm-radii microring and the straight bus waveguides. Under forward bias, carriers are injected into the intrinsic waveguide by the embedded p-i-n diode. The variation of the carrier density brings to the refractive index change due to the plasma dispersion effect in silicon. In this way, the driving electrical signal shifts the microring resonance and thus controls the on/off state of the optical signals outputting from the through-port and drop-port. Figure 4 shows the optical transmission spectra before and after a DC bias with the inset showing the current-voltage characteristics of the p-i-n diode. The forward differential resistance of the p-i-n diode is ~ 75 Ω, indicating a very good metal contact. A 1.4 V forward DC bias blue-shifted the resonant wavelength for ~ 0.23 nm which was sufficient for the switching due to the high Q factor of 11,000. Extinction ratios at the through-port and the drop-port were measured to be 25 dB and 13 dB, respectively, while the in-band add-drop crosstalk was less than -23 dB.

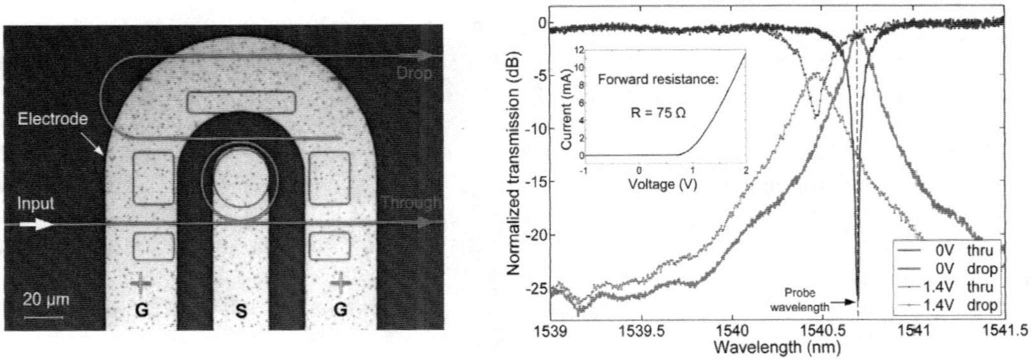

Figure 4. (a) Top-view microscope image of the add-drop microring switch. (b) Transmission spectra of the microring resonator at the bias voltages of 0 V and 1.4 V. Inset: Current-voltage curve of the microring switch.

Pre-emphasis signals have been demonstrated to greatly reduce the p-i-n diode's state transition time [3, 17]. In our case, the electrical pre-emphasis signal is adopted to accelerate the switch which is generated by combining two square wave signals. Figure 5(a) shows the waveform of the pre-emphasis driving signal with 500 ps wide emphasis pulses. We choose the switch on/off voltages to be 1.4/0.7 V, respectively. The forward and reverse pre-emphasized voltages are respectively 2.5 V and -0.5 V. Figure 5(b) and (c) are the corresponding normalized optical power measured from the through-port and drop-port. As the switch-off voltage is set close to the p-i-n junction's turn-on voltage (~ 0.7 V), the carrier density (as well as refractive index) would immediately change when the forward pre-emphasis pulse is applied and leads to a short switch-on time. When carrier density starts to stabilize, the driving voltage would then decrease to 1.4 V to maintain switch-on state with the least carrier densities. When the reverse emphasis pulse is applied to the switch, the carriers in the waveguide would be quickly swept by the reverse electric filed. According to the output optical waveforms, a 380 ps switch-on and 300 ps switch-off time are finally achieved by tailoring the driving signal's waveform.

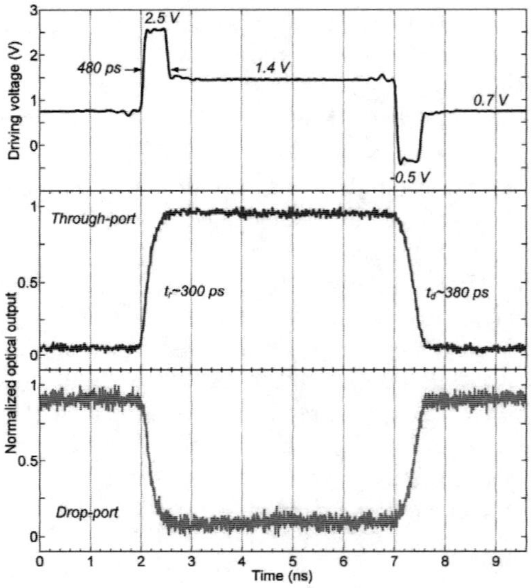

Figure 5. Dynamic responses of the microring switch driven by pre-emphasized square wave signals. (a) Measured waveform of the pre-emphasis square wave signal. (b) Measured optical output at the through-port. (c) Measured optical output at the drop-port.

3.3 Results of the microring modulator

For a high modulation speed, the Q factor has to be moderately increased. We adopt a notch type microring structure as shown in figure 6(a). Figure 6(b) shows the microring's transmission spectra curves under different bias voltages. When the modulator is zero-biased, the Q factor is measured to be ~ 23,000 ensuring fast electro-optical responses. By optimizing the bus-microring coupling to the critical coupling state, a 30 dB extinction ratio of is achieved under 1.2 V forward biased voltage.

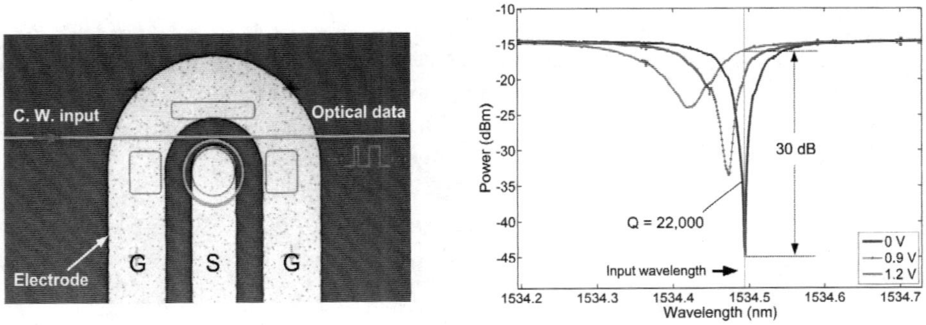

Figure 6. (a) Top-view microscope image of the notch microring modulator. (b) Transmission spectras of the microring modulator at the bias voltages of 0 V to 1.2 V.

In order to realize 10 Gbit/s modulation, we amplified the emphasis signals by a by a high bandwidth microwave amplifier. Figure 7(a) shows the 10 Gbit/s driving signals composed of an NRZ signal with V_{pp} ~ 2.2 V and pre-emphasis pulses with ~ 4 V peak amplitude occurring at each transition edges. Figure 7(b) illustrates the corresponding modulated output when the driving signals applied to the modulator. Actually, the electrical driving power would be greatly decreased if the modulation efficiency is improved which could be realized with the standard CMOS process.

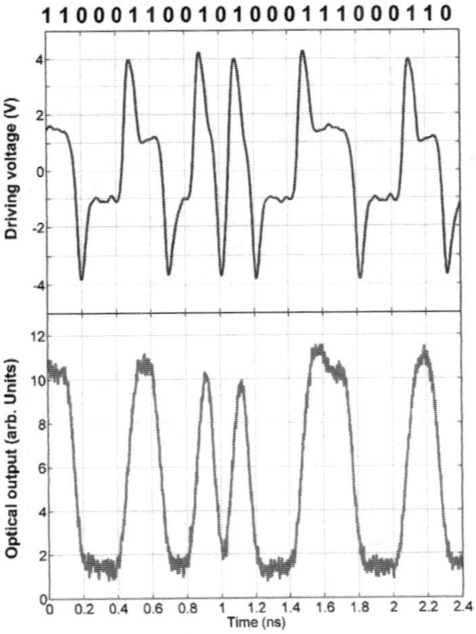

Figure 7. (a) Waveform of 24-bit electrical pre-emphasis signal at 10 Gbit/s. (b) Corresponding optical signal output from the modulator.

4. Conclusion

In conclusion, we have demonstrated a SOI microring switch and a microring modulator which are integrated with grating couplers. The grating couplers, which were exposed and etched together with the active microring devices, have enabled the wafer-level online measurements without wafer cleaving and facet polishing. In order to realize high speed operation of the forward p-i-n diode type devices, pre-emphasis driving signals were generated by combining two channels of NRZ signals. By optimizing the driving signals, switch on and off time of 300 ps and 380 ps are realized. Our measurements on the microring modulator demonstrated 10 Gbit/s NRZ modulation with the amplified pre-emphasis driving signals. This cost-efficient integration of active microring and grating couplers would have enormous potential in future photonic networks-on-chip.

References

[1] Liu A, Liao L, Rubin D, Basak J, Chetrit Y, Nguyen H, et al. 2008 *Semiconductor Science And Technology* **23**(6) 064001
[2] Gardes FY, Reed GT, Emerson NG and Png CE 2005 *Optics Expess* **13**(22) 8845
[3] Xu QF, Manipatruni S, Schmidt B, Shakya J and Lipson M 2007 *Opt. Express* **15**(2) 430
[4] Watts MR, Trotter DC and Young RW 2008 *National Fiber Optic Engineers Conference, OSA Technical Digest (CD) (Optical Society of America, 2008)* PDP14
[5] Sherwood-Droz N, Wang H, Chen L, Benjamin GL, Biberman A, Bergman K, and Lipson M 2008 *Opt. Express* **16** 15915
[6] Soref RA and Bennett BR 1987 *IEEE J. Quantum Electron.* **23** 123
[7] Luo X, Feng S, and Poon AW 2009 *The 14th OptoElectronics and Communications Conference (OECC 2009)* 13-17
[8] Lira HLR, Manipatruni S and Lipson M 2009 *Opt. Express* **17** 22271
[9] Watts MR, Trotter DC, Young RW and Lentine AL 2008 *5th IEEE Group IV Photonics* 4-6
[10] Manipatruni S, Chen L and Lipson M 2010 *Opt. Express* **18**(16) 16858
[11] Xiao X, Xu H, Zhou L, Li Z, Li Y, Yu Y and Yu J 2010 *Chinese Optics Letters* **8**(8) 757
[12] Xiao X, Zhu Y, Xu H, Zhou L, Hu Y, Li Z, et al. 2010 *Chinese Physics Letters* **27**(9) 094207
[13] Xu X, Chen S, Xu H, Sun Y, Yu Y, Yu J and Wang Q 2009 *Chinese Physcis B* **18** 3900
[14] Png CE, Chan SP, Lim ST and Reed GT 2004 *J. Lightwave Technol.* **22** 1573-1582
[15] Zhu Y, Xu X, Li Z, Zhou L, Han W, Fan Z, et al. 2010 *Chinese Physics B* **19** 014219-5
[16] Xiao X, Li Y, Yu Y and Yu J 2010 *Chinese Physics Letters* **27**(5) 054208
[17] Green WMJ, Rooks MJ, Sekaric L and Vlasov YA 2007 *Opt. Express* **15** 17106-17113

Flexible In-plane Photonic Devices Based on Transferrable Si Nanomembranes on Polyimide Film

Xiaochuan Xu[1], Harish Subbaraman[2], Daniel T. Pham[1], Amir Hosseini[1], Afshin Ghaffari[1], and Ray T. Chen[1]

1. Microelectronics Research Center, Electrical and Computer Engineering Department, the University of Texas at Austin, 10100 Burnet Rd., Bldg. 160, Austin, Texas, 78758
2. Omega Optics, Inc. 10306 Sausalito Dr. Austin, TX 78759

Email: xiaochuan.xu@mail.utexas.edu

Abstract. The flexible electronics and photonics have attracted a lot of attentions in past decade because of their potentially wide applications. A general method to fabricate nanomembrane flexible device involves two steps. At first, a nanomembrane on top of a sacrificial layer is patterned. Then, the nanomembrane is freed by high selective etching of the sacrificial layer and then weakly bonds to the handle substrate by Vander Waals forces. Later on, the released devices are transferred to the target flexible substrates. Notable successes include paper-like display, flexible silicon integrated circuits, photonic crystal filter and sensor skins, and so on. However, the progress in flexible photonic device is still limited to surface-normal devices such as photonic crystal filter. This type of devices usually has a large area and/or same dimensions which makes the transfer much easier than in-plane photonic device. This paper reports the transfer of in-plane Silicon nanomembrane photonic devices on polyimide flexible film. Employing a slightly modified transfer process, passive optical components, such as optical waveguide and multimode multimode interferometer, are successfully transferred on to Kapton polyimide film.

1. Introduction

Plenty of technologies require large-scale integration of different types of components fabricated separately, which cannot be realized through conventional microelectronic fabrication processes because of the different nature of materials. Examples of these systems that rely heavily on heterogeneous integration range from integrated photonic systems that contain III-V semiconductor lasers and silicon waveguides, to biomedical applications that usually involve organic tissues and inorganic sensors. Integrating electronic and photonic devices on flexible substrates provides a promising solution to these applications. Encouraging successes in this emerging field include but not limited to paper-like displays [1, 2], flexible silicon integrated circuits [3], photonic crystal filter [4, 5] and sensor skins [6] , and so on. Among all the transferable materials, Single crystal Si nanomembrane may be the most promising for both the electronic and photonic devices [4], because it not only has high carrier mobility and is mechanically robust, but also is transparent in the near infrared region. However, transferring in-plane photonic components on to flexible substrates are still a great challenge because of the dimension and shape sensitivity of the photonic devices [7].

Transfer printing technic is one of the prevalent method to integrate different components on flexible film. A typical transferring process includes two steps. First, a thin functional layer is deposit onto a sacrificial layer. In the following step, the sacrificial layer is removed and the functional layer is released and transferred to a flexible substrate [8]. The sacrificial layer needs to be very thin to let the functional layer bonded to the handle substrate through Van der Waals force. The typical value is ~200nm. If the sacrificial layer is too thick, special design is necessary to anchor the functional layer onto the handle substrate during the releasing procedure [2, 9]. In this paper, we introduce a new method to transfer photonic devices with 3 μm buried oxide layer as sacrificial layer.

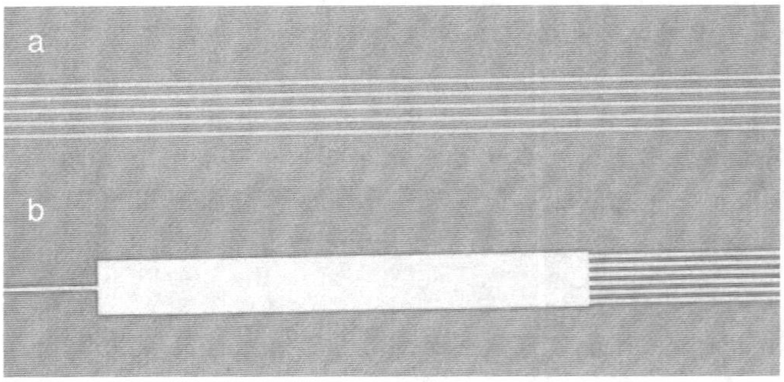

Figure 1. Silicon nanomembrane photonic components before transfer. a) 2 μm waveguide; b) 1×6 MMI

2. Fabrication Process

Photonic crystal waveguide devices are fabricated on commercially available SOI from SOITEC with 250nm top silicon layer and 3μm buried oxide. The silicon is first oxidized to create a 50nm thick top oxide layer which serves as a hard mask for the silicon etch. This oxidation consumes 20nm of silicon, leaving a final silicon thickness of 230nm. The photonic devices are patterned using electron beam lithography. After electron beam lithography and developing, a 20 nm nickle layer is deposited and a standard lift-off process is used to remove the extra nickel. The pattern is transferred to the silicon oxide hard mask by reactive ion etching (RIE). An HBr/Cl2 RIE etch is then used to transfer the pattern to the silicon layer. Finally, the metal is removed by piranha cleaning. The top oxide layer is removed by buffered hydrofluoric (BHF). Fig. 1 shows the pictures of the fabricated 2 μm waveguide and the 1×6 multimode interference (MMI) coupler, which will be used for transfer described in the following.

Unlike electronic devices, photonic devices usually have high length to width ratio which makes the transfer more difficult. The performance of photonic devices can be degraded by tiny shifting or breaking during the transfer procedure. Therefore, providing mechanical supporting during the transfer process becomes a necessary, especially when the buried oxide layer is as thick as 3 μm. Fig. 2 illustrates the transfer process. From top side down, the structure after previous steps is 230nm function layer, 3μm buried oxide layer and 0.5 handle silicon (Fig. 2a). Photoresist AZ 5209 is spun on (4000 rpm, 30sec) to provide mechanic support to the devices during the etching procedure after pretreating the chip by Hexamethyldisilazane (HMDS). After baking at 95℃, the photoresist is patterned to open several specially designed windows to let the etchant pass in. Hard bake at 110 ℃ is carried out after developing with AZ 726MIF (Fig.2b). Since the buried oxide layer is too thick, the

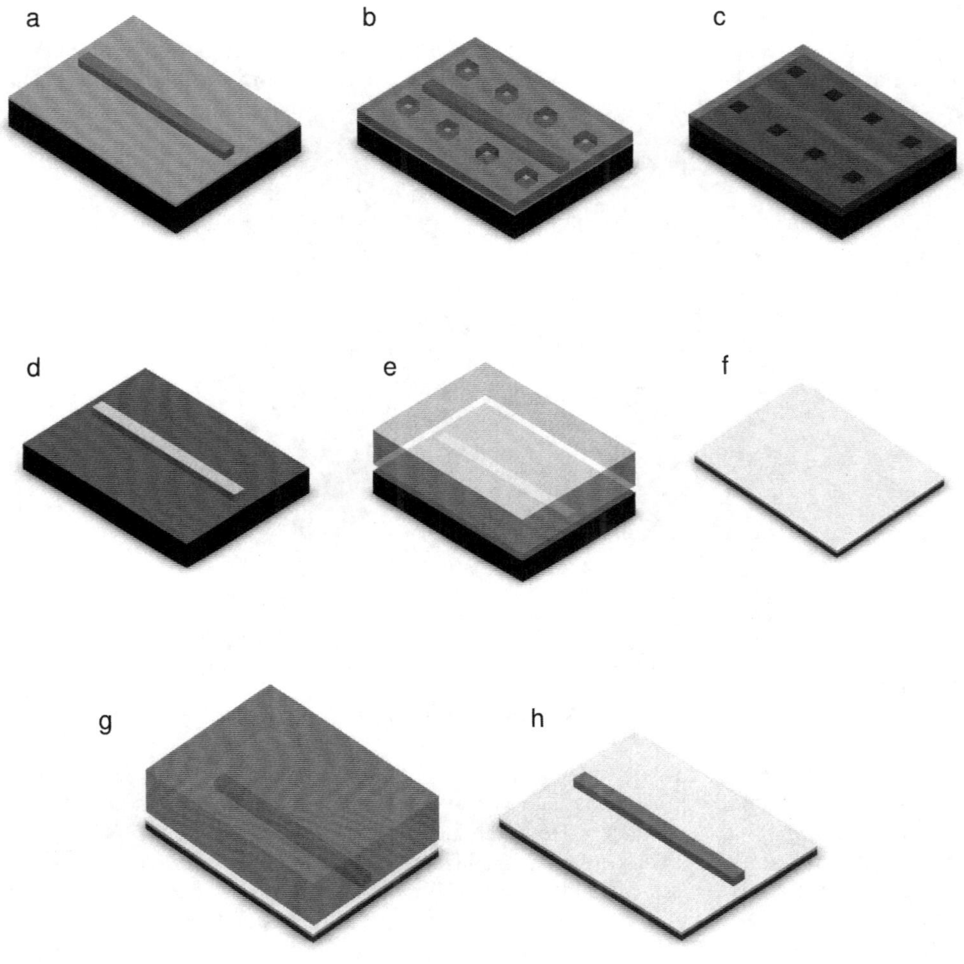

Figure 2. Transfer process flow chart. a. starting from SOI; b. form the supporting layer and open the etching window; c. release the function layer with HF vapor; d. strip the photoresist by oxygen plasma; e. using PDMS to peel up the device; f. prepare the target Kapton substrate; g. bring the inked PDMS into contact with the target substrate and remove PDMS slowly; f. the device is transferred to target substrate

function layer is weakly bonded to the handle silicon when either hydrofluoric (HF) or BHF is utilized to remove the sacrificial layer, which causes shifting and breaking. To address the critical issue, in our process, HF vapor is used instead of solution so that the shifting of the device can be minimized. HF vapor is generated simply by cover the beaker because HF is volatile. The etching rate is ~50 μm/h in our process. It can be adjusted to any desired value through opening holes on the cover. After the buried oxide layer is completely removed, the chip is moved to another beaker with acetone where the residue of HF can be cleaned thoroughly by the acetone vapor. Later on, the chip is baked at 65 ℃ to evaporate the water generated by the reaction of the HF and buried oxide so that the device can be tightly bonded to the silicon handle wafer (Fig.2c). Then, the photoresist is removed by oxygen plasma (200W, 13sccm, 10min), as shown in (Fig.2d). The PDMS is prepared by mixing the base and agent with a ratio of 8.5:1.5 and curing at 70 ℃ for 90 mins [2]. The optimized PDMS dimensions are 0.8 by 2 by 2cm. PDMS can be reused after cleaned by ethanol and tried by nitrogen. Bringing the prepared PDMS into conformal contact with the released photonic device and peel it up at high speed (Fig.2e) [10]. It has been demonstrated that at high peeling speed the adhesion between PDMS and silicon nanomembrane is larger than the Van der Waals force between silicon nanomembrane and the handle wafer [10]. The inked PDMS is shown in Fig. 3b.

Figure 3. a. the released device settles down to the silicon substrate; b. inked PDMS(the waveguide in this picture is 60μm).

Kapton film is widely accepted as a durable material. It has a unique combination of electrical, thermal, chemical and mechanical properties and retains these properties over a wide range of industrial environments and applications. However, because of the lack of the adhesions between Kapton and SU-8, the SU-8 layer can be delaminated from the Kapton during the printing process and lead to a false transfer. Therefore, after cleaning the film with isopropyl alcohol (IPA) and drying with nitrogen, we spin coat a thin layer of Omnicoat to promote the adhesion (3000rpm, 30sec, bake at 200 ℃ for 2 mins). Then SU-8 2002 is spun on at 3000 rpm and soft baked at 65 and 95 ℃ for 1 min and 2 min, respectively (Fig.2f). Later on, bring the inked PDMS into contact with the film and peel the device up at low speed after curing with UV lamp for 5 mins and baking at 110 ℃ for 10 mins. Fully cure the SU-8 at 150℃ for 30mins.

3. Results and Discussions

With the transfer process described in previous section, we transferred several classes of photonic components including 2 μm waveguide (Fig. 4a), multimode interference coupler (MMI), shown in Fig.4b and c. The SEM images are not clear because of the charging of the SU-8. To coupling light in, the transferred device is bonded to glass slides and cleaved by high speed dicing saw with resin blaze. Newport six-axis autoaligner system is used to couple the 1550nm TE polarized light in. An infrared camera is mounted up-down to examine the output end of the waveguide. Fig. 4d shows the output spot, illustrating the light passes through the waveguide.

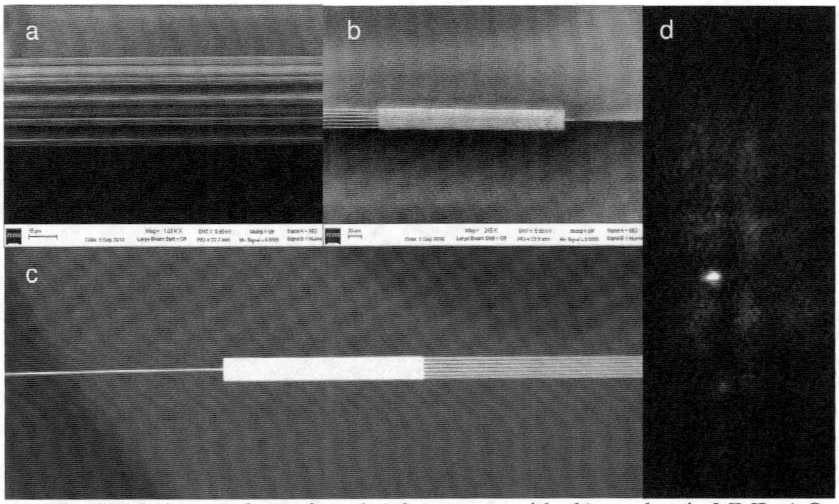

Figure 4. The SEM picture of transferred a) 2um waveguide; b) one by six MMI; c) Optical microscope picture of the transferred one by six MMI on Kapton film; d) the top down output infrared image when the waveguide is illuminated by 1550nm light

Nevertheless, how to couple light in and out is still a great challenge for the flexible photonic devices. The flexible substrate cannot be cleaved by ordinary methods used in silicon photonic devices based on natural cleavage. Surface normal coupling methods such as gratings will surely result in the increase difficulty of transfer.

4. Conclusion

In this paper, a method is proposed to transfer silicon nanomembrane photonic devices with thick buried oxide layer onto polyimide film. With this method, several types of devices are transferred successfully. Initial testing efforts show the promising future of this technique.

References

[1] Rogers JA, Bao Z, Baldwin K, Dodabalapur A, Crone B, Raju VR, et al. 2001 Paper-like electronic displays: Large-area rubber-stamped plastic sheets of electronics and microencapsulated electrophoretic inks P Natl Acad Sci USA **98** 4835-40

[2] Park SI, Xiong YJ, Kim RH, Elvikis P, Meitl M, Kim DH, et al. 2009 Printed Assemblies of Inorganic Light-Emitting Diodes for Deformable and Semitransparent Displays Science **325** 977-81

[3] Kim DH, Ahn JH, Choi WM, Kim HS, Kim TH, Song JZ, et al. 2008 Stretchable and foldable silicon integrated circuits Science **320** 507-11

[4] Zhou WD, Ma ZQ, Yang HJ, Qiang ZX, Qin GX, Pang HQ, et al. 2009 Flexible photonic-crystal Fano filters based on transferred semiconductor nanomembranes J Phys D Appl Phys **42** 234007-17

[5] Chen L, Qiang ZX, Yang HJ, Pang HQ, Ma ZQ, Zhou WD 2009 Polarization and angular dependent transmissions on transferred nanomembrane Fano filters Opt Express **17** 8396-406

[6] Someya T, Kato Y, Iba S, Noguchi Y, Sekitani T, Kawaguchi H, et al. 2005 Integration of organic FETs with organic photodiodes for a large area, flexible, and lightweight sheet image scanners Ieee T Electron Dev **52** 2502-11

[7] Zablocki MJ, Sharkawy AS, Ebil O, Prather DW 2010 Nanomembrane enabled nanophotonic devices Proceedings of the SPIE **7606** 9

[8] Cavallo F, Lagally MG 2010 Semiconductors turn soft: inorganic nanomembranes Soft Matter **6** 439-5

[9] Mack S, Meitl MA, Baca AJ, Zhu ZT, Rogers JA 2006 Mechanically flexible thin-film transistors that use ultrathin ribbons of silicon derived from bulk wafers Appl Phys Lett **88** 213101-3
[10] Meitl MA, Zhu ZT, Kumar V, Lee KJ, Feng X, Huang YY, et al. 2006 Transfer printing by kinetic control of adhesion to an elastomeric stamp Nat Mater **5** 33-8

Journal of Physics: Conference Series **276** (2011) 012097 doi:10.1088/1742-6596/276/1/012097

Diffraction properties study of reflection volume holographic grating in dispersive photorefractive material under ultra-short pulse readout

Yingyan Yi[1,2], Deming Liu[1], Hairong Liu[1]
[1] Wuhan National Laboratory for Optoelectronics, School of Optoelectronic Science and Engineering, Huazhong University of Science and Technology, Wuhan, 430074, China
[2] School of Science, Wuhan University of Technology, Wuhan, 430070, China

E-mail: yiyingyan0410@163.com

Abstract. Based on the modified Kogelnik diffraction efficiency equation, we study the diffraction intensity spectrum and the total diffraction efficiency of reflection volume holographic gratings in photorefractive media. Taking photorefractive $LiNbO_3$ crystal as an example, the effect of the grating parameters and the pulse width on the diffraction properties is presented under the influence of crystal material dispersion. Under the combined effects, the diffraction pulse profiles and the total diffraction efficiency are compared with and without crystal material dispersion. The results show that the dispersion will decrease the diffraction intensity. Moreover, when pulse width is smaller or the grating spacing and the grating thickness are larger, the influence of dispersion on diffraction is large. The results of our paper can be used in pulse shaping applications.

1. Introduction

Volume holographic gratings (VHGs) have a diverse range of applications, such as add/drop multiplexers, fiber and laser and beam coherent and incoherent combination etc [1-5]. Recently, the application of VHGs for manipulating laser output pulses to achieve functional waveforms with large bandwidths that can be utilized for a variety of applications has attracted a significant amount of attention because of the flexibility and the possibility of using VHGs to implement dynamic processing [6-10]. A most widely used basis for description of volume gratings is the theory of coupled waves developed by Kogelnik in 1969 [11]. However, the Kogelnik theory is from the plane wave discussion. In deduction, the effect of the material dispersion on diffraction is neglected, because there has been only one frequency component involved for plane wave; but for pulsed beam, the dispersion can not be neglected because of the large bandwidth of pulse beam, where each frequency component corresponds to a specific refractive index in the material, which will affect the diffraction greatly. Chen Yang and Xiaona Yan [11] investigated the diffraction properties of transmitting volume holographic grating (TVHG) in dispersion photorefractive material (InP:Fe crystal) for femtosecond pulse readout, but the work on reflection volume holographic grating (RVHG) has not been found yet.

In this paper, the diffraction properties of reflection volume holographic grating are investigated using the modified diffraction efficiency equation of Kogelnik. The analytical expressions of the normalized diffraction intensity spectrum and the total diffraction efficiency are derived by modified diffraction efficiency equation of Kogelnik in section 2. The numerical analysis of the normalized diffraction intensity spectrum and the total diffraction efficiency are presented in section 3. A summary of the results and conclusions are given in the final section.

2. Theory analysis

Our discussion is based on the diagram shown in Fig. 1. we consider an ultrashort pulsed beam incident on a lossless and unslanted RVHG, which is characterized by a refractive index in the modulated region($0 \le z \le d$) of the form

$$n = n_0 + n_1 \cos(\boldsymbol{K} \cdot \boldsymbol{x}) \qquad n_1 << n_0 \tag{1}$$

where n_0 is the refractive index of the grating region, n_1 is the amplitude of index modulation and $\boldsymbol{K}$ is the grating vector oriented perpendicular to the fringe planes with a module $|\boldsymbol{K}_G| = 2\pi / \Lambda$, where Λ is the period of the grating. d is the thickness of grating . θ_r is the incident angle (the angle inside crystal is θ_r').

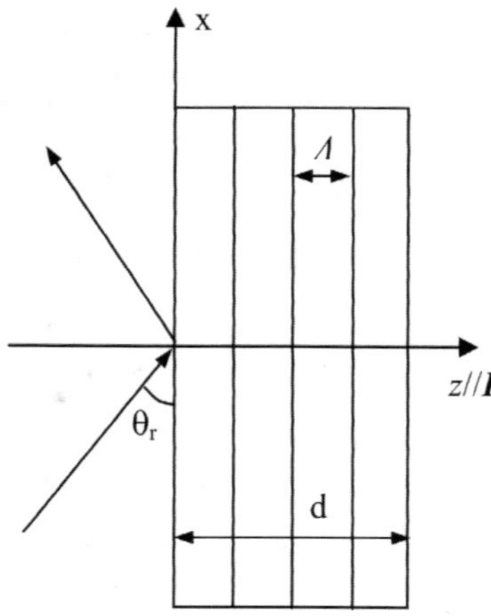

Figure1. Diffraction schematic of reflection volume holographic grating

The time variation of the electric field of the incident ultrashort pulse can be described as

$$E_r(t) = \exp(-i\omega_0 t - \frac{t^2}{T^2}) \tag{2}$$

where $\omega_0 = 2\pi c / \lambda_0$ is the central angular frequency corresponding to the central wavelength λ_0, parameter T is denoted by $T = \Delta\tau/(2\sqrt{\ln 2})$, with full width at half maximum(FWHM) $\Delta\tau$. The Fourier transform of $E_r(t)$ is defined as

$$E_r(\omega) = \frac{1}{2\pi}\int_{-\infty}^{\infty} E_r(t)\exp(i\omega t)dt = \frac{T}{2\sqrt{\pi}}\exp\left[-\frac{T^2(\omega-\omega_0)^2}{4}\right] \tag{3}$$

If the read out beam is a monochromatic plane wave, the diffraction efficiency is [12]

$$\eta = \left(1 + \frac{1-\xi^2/\nu^2}{\sinh^2\sqrt{\nu^2-\xi^2}}\right)^{-1} \tag{4}$$

where $\nu = j\pi n_1 d / \lambda (c_R c_S)^{1/2}$, $\xi = -\vartheta d / 2c_S$, c_R and c_S are obliquity factors, ϑ is the dephasing measure, and d is the grating thickness.

When the readout beam is pulse beam, parameter ν and ξ are frequency dependent, so η is frequency dependent. In our discussion, we assume all spectrum components incident at the same angle θ_r to readout. Due to the material dispersion, the frequency components in the ultra-short pulse will be separated according to the Snell refraction law $\sin\theta_r = n(\omega)\sin\theta_r'(\omega)$.Accordingly, the obliquity factors are changed to frequency-dependent, for a unslanted RVHG $c_R = -c_S = \cos\theta_r'(\omega)$, so

$$\nu(\omega) = \frac{\omega n_1 d}{2c\cos\theta_r'(\omega)} \tag{5}$$

As the central wavelength component of the incident pulse satisfies the Bragg diffraction condition, we have $\cos\theta'_{r0} = \lambda_0 / 2\Lambda n(\omega_0)$, then other wavelength components will not satisfy the Bragg condition, the dephasing measure can be redefined as $\vartheta(\omega) = K^2 c / 2n(\omega)(1/\omega_0 - 1/\omega)$, and thus

$$\xi(\omega) = \frac{\pi^2 cd}{\Lambda^2 n(\omega)\cos\theta_r'(\omega)}\left(\frac{1}{\omega_0} - \frac{1}{\omega}\right) \tag{6}$$

where c is the velocity of light in free space, $n(\omega)$ is the refractive index of the crystal to frequency ω.

In $LiNbO_3$ crystal, the wavelength dependence of the refractive index is given by [13]

$$n^2(\omega) = a_0 + \frac{a_1\omega^2 \times 10^{-12}}{(2\pi c)^2 - a_2\omega^2 \times 10^{-12}} - a_3(2\pi c/\omega)^2 \times 10^{12} \tag{7}$$

where a_0=4.9048, a_1=0.117680, a_2=0.047500, a_3=0.027169.

By substituting Eqs.(5) and (6) into (4) , the diffraction efficiency spectrum of the RVHG is obtained

$$\eta(\omega)=\left(1+\frac{1-\xi(\omega)^2/\nu(\omega)^2}{\sinh^2\sqrt{\nu(\omega)^2-\xi(\omega)^2}}\right)^{-1} \tag{8}$$

which is applicable to all readout components ω. In our discussion, the central frequency component satisfies the Bragg condition, so the diffraction efficiency comes to its maximum. However, the other readout components deviate from the central frequency, thus the diffraction efficiencies decrease; and the larger the deviation is, the smaller the diffraction efficiency becomes.

According to the definition of the diffraction efficiency, the diffraction intensity spectrum is obtained

$$I_d(\omega)=I_r(\omega)\cdot\eta(\omega)=|E_r(\omega)|^2\cdot\eta(\omega) \tag{9}$$

where $I_d(\omega)$ is the diffracted intensity corresponding to frequency ω.

Defining the total diffraction efficiency as the total intensity of diffracted beam expressed by Eq.(7) to that of incident intensity, and the representation is:

$$\eta=\frac{\int_{-\infty}^{+\infty}I_d(\omega)d\omega}{\int_{-\infty}^{+\infty}|E_r(\omega)|^2 d\omega} \tag{10}$$

From Eqs.(8) and (10), we can see that the diffraction intensity spectrum and total diffraction efficiency are affected by the grating parameters, readout pulse width and the material dispersion.

3. Numerical analysis and discussion of diffraction properties of RVHG illuminated by an ultrashort pulse

Assuming that the central wavelength of the incident pulse beam is 1.55μm, and the corresponding central angular frequency is $\omega_0=1.2161\times10^{15}$ Hz. Other parameters are $n_1=1\times10^{-4}$, d=1mm and Λ=1μm. In order to compare diffracted pulses with input pulses, according to Eq. (2), the input pulses at different duration **Δτ** versus angular frequency ω are shown in Fig. 2.

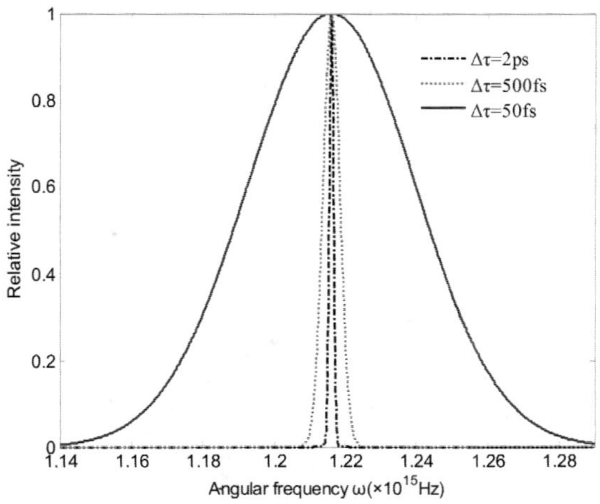

Figure 2. Spectral distribution of input Gaussian pulse at different pulse duration

3.1 Discussion of normalized diffraction intensity spectrum

First, we calculate the normalized diffraction intensity as a function of angular frequency. In order to facilitate compare, we only give a part frequency range of diffracted spectrum. The diagram is shown in fig.3. It's shown that the dispersion of the material has effect on the diffraction. When the readout pulse is 50fs, the influence of dispersion becomes obvious and cannot be neglected, while to the 2ps readout pulse, the effect of the dispersion on the diffraction can be neglected. The result is accordance with the TVHG, the reason is that the wider the temporal width of the pulse beam, the narrower the spectral width of it is, the smaller the dispersion of the volume holographic grating.

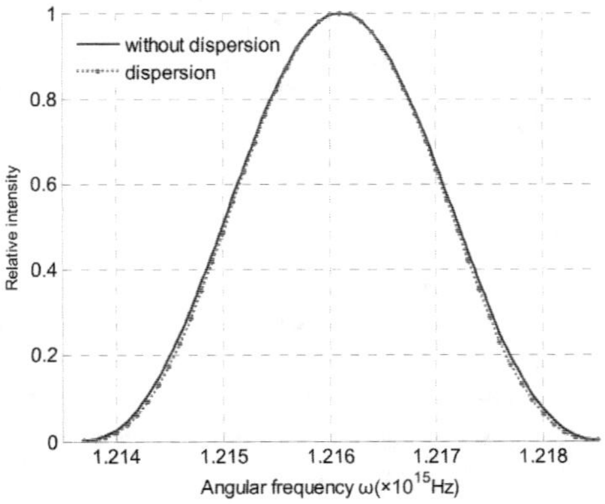

(a) Readout pulse 50fs

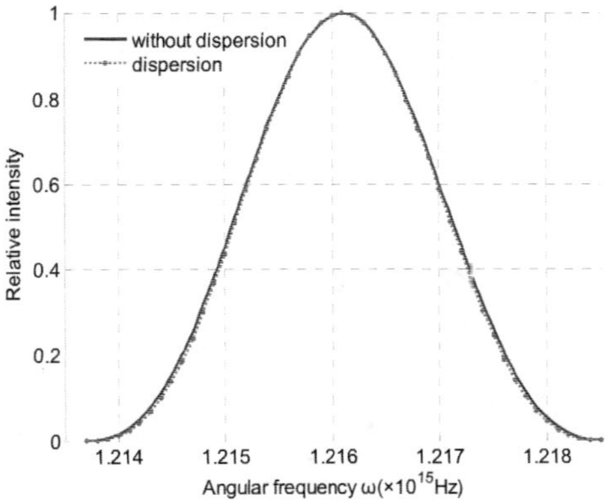

(b) Readout pulse 500fs

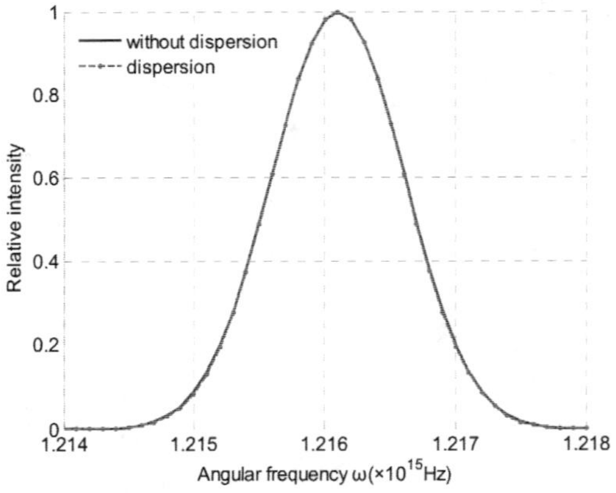

(c) Readout pulse 2ps

Figure 3. Normalized diffraction intensity spectrum against the angular frequency ω on conditions of different readout pulse width: (a) Δτ=50fs; (b) Δτ=500fs; (c) Δτ=2ps;.

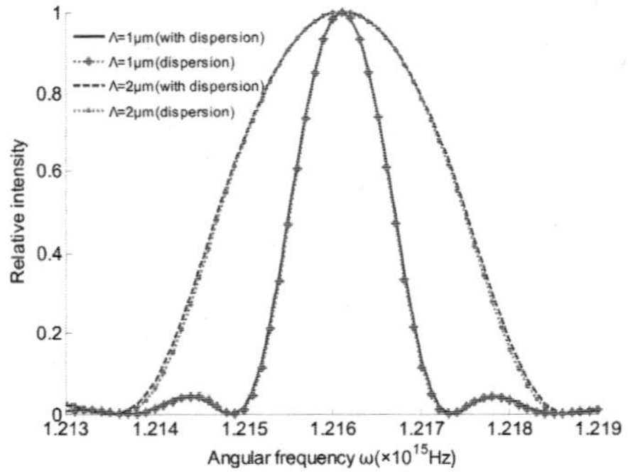

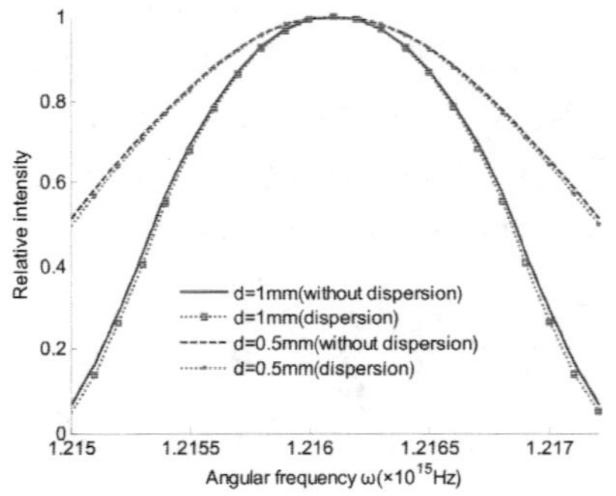

Figure 4. Normalized diffraction intensity spectrum against the frequency ω on condition of different grating thickness Λ=1μm and d=2μm, where d=1mm and Δτ=500fs

Figure 5. Normalized diffraction intensity spectrum against the frequency ω on condition of different grating thickness d=0.5mm and d=1mm, where Λ=1μm and Δτ=500fs

Fig.4 shows the relation between the diffraction spectrum and the grating periods when pulse duration is 500fs. We find that when the grating period increases, the decreasing speed of diffraction intensity becomes slower and the bandwidth of the diffracted pulse increases, the influence of the dispersion of diffraction becomes larger, which means that more frequency components have been diffracted out. With the increasing of the grating spacing, the diffraction angle of the readout frequency component becomes large, the angle deviates more from the Bragg angle, thus the more obviously of the dispersion effect on diffraction. This is the direct opposite to the TVHG.

Fig.5 is the diagram of the normalized diffraction intensity spectrum against ω on condition of different grating thicknesses. In order to survey the effect of dispersion, the figure only gives part of frequency range. It shows that the thicker the grating thickness is, the smaller the bandwidth of the diffracted pulse becomes and large influence of dispersion on diffraction. This can be explained as the following: when the thickness is large, the process between the grating and beam will become longer and more frequency components are filtered out. The larger grating thickness, the distance of diffraction is farther, the deflection angle becomes larger, this lead to the angle deviates more from the Bragg angle, so the dispersion is more obvious. The result is consistent with the TVHG.

From above discussion, we can know the function of the recorded grating can be acted as a band-pass filter, and the bandwidth can be changed by choosing different grating spacing or thickness. The effect of the dispersion on diffraction in these two diagrams can be interpreted as, with the decreasing of the grating spacing, the diffraction angle of the readout frequency component becomes larger; while to the larger grating thickness, the deflection angle becomes larger. The results of these two conditions make the angle deviate more from the Bragg angle, thus the diffraction decrease.

3.2 Discussion of the total diffraction efficiency

Fig.6 is the diagram of the total diffraction efficiency against the pulse width as the grating period is 0.5μm, 1μm and 2μm, the other parameter is the same as section 3.1, and Fig.7 show the relationships of the total diffraction efficiency to the grating spacing as pulse duration is 20fs, 100fs and 500fs. These two diagrams can be explained from Figs.3-5. For fixed grating spacing, Bragg selectivity is specified. To the larger duration pulse, more frequency components will be diffracted out and the total diffraction efficiency will increase. While to the fixed $\Delta\tau$, the smaller the grating spacing is, the stricter the Bragg condition becomes, thus fewer components will be diffracted out, the diffraction efficiency is decreasing. Moreover, the dispersion will reduce the total diffraction efficiency, and its influence on the diffraction is consistent with those in Figs3-5.

Fig.8 is the diagram of the total diffraction efficiency against the grating thickness as the pulse width is 100ps and grating spacings are 0.5,1 and 2μm. From Fig.8 we know the total diffraction efficiency increases when the grating thickness increases from 0.5mm to 2mm. It's also shown that the dispersion becomes more obvious with the increasing of grating thickness, which is consistent with the discussion in Fig.5.

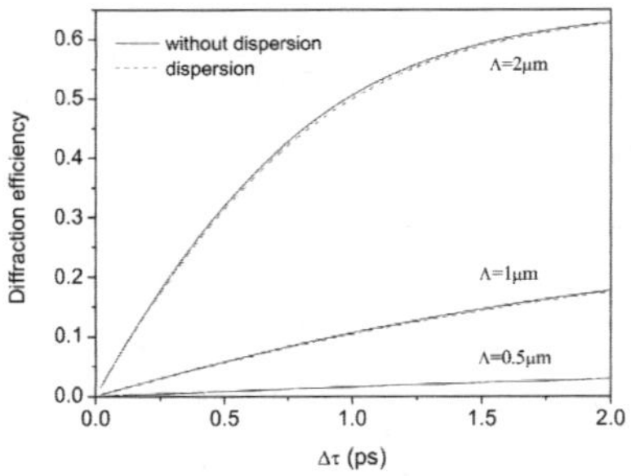

Figure 6. Diagram of the total diffraction efficiency as Δτ for both considering and neglecting the dispersion effect: with grating spacing of 0.5μm, 1μm and 2μm

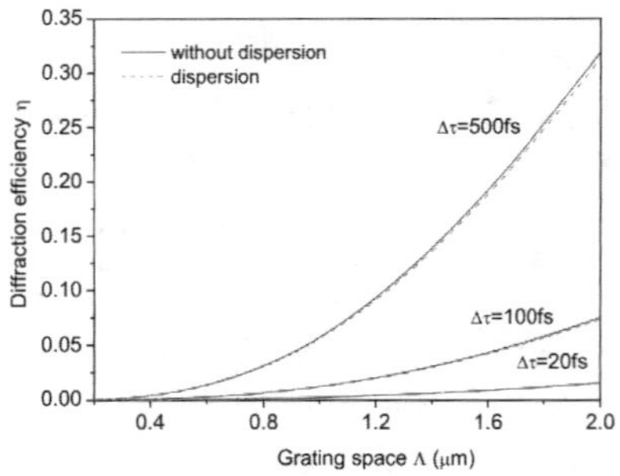

Figure 7. Diagram of the total diffraction efficiency against the grating spacing from 0.2 to 2um with readout pulse width of 20fs, 100fs, 500fs

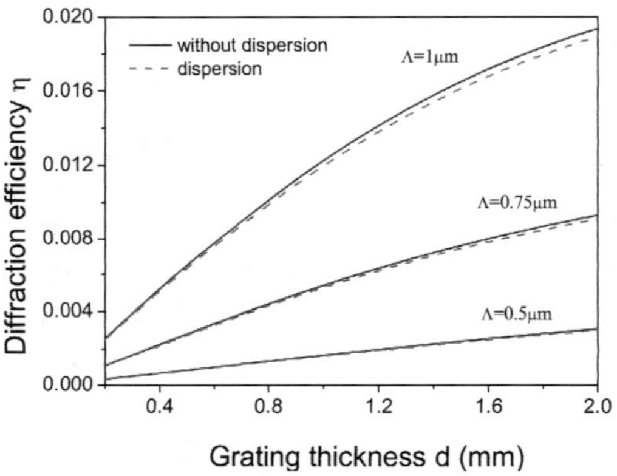

Figure 8. Diagram of the total diffraction efficiency against the grating thickness as the grating spacing is 0.5um, 0.75um and 1um.

4. Conclusion

Based on modified the diffraction efficiency equation of Kogelnik, the total diffraction efficiency and diffraction intensity spectrum of the reflection holographic volume grating under ultrashort pulse illumination are investigated in the dispersion material. The results show that by controlling the grating spacing, thickness and readout pulse width, the diffraction will have much difference. Moreover, the dispersion effect decreases the total relative intensity and the total diffraction efficiency. Especially when the readout pulse is small and the grating thickness is relatively large, the influence of the dispersion on diffraction becomes obvious, this result is consistent with the TVHG, but when the grating spacing is larger, the influence of the dispersion on diffraction is obvious, which is contrary to the TVHG. It's also show the effect of dispersion in $LiNbO_3$ crystal of RVHG is smaller than the TVHG in InP:Fe crystal. The results presented in this paper could be applied in pulse shaping and signal processing.

References

[1] Vikram Bhatia, K. P. Reddy, Marlene A. Marro, and David L. Weidman 1999 Proc. OSA Trends in Optics and Photonics, WDM Components **29** 12
[2] Pär Jelger, Fredrik Laurell 2007 Opt. Exp.**15** 11336
[3] Armen Sevian, Oleksiy Andrusyas and Igor Ciapurin et al., 2008 Opt. Lett. **33** 384
[4] George B.Venus. Armen Sevian, Vadim I. Smirnov. 2006 Opt. Lett. **31** 1453
[5] M. S. Shahriar, J. Riccobono, M. Kleinschmit and J. T. Shen 2003 Opt. Commun. **220** 75
[6]. G.A. Rakuljic, V. Leyva 1993 Opt. Lett. **18** 459
[7] X.Yan, B.Yang, B.Yu 2004 Optik **115** 512
[8] B. Yang, X. Yan, Y. Yang and H. Zhang 2008 Optics & Laser Technology **40** 906
[9] L. Guo, A. Yan 2009 Opt. Int .J. Light Electron.Opt. doi:10.1016/j.ijleo.2009.02.002
[10] C. Wang, L. Liu, A. Yan, D. Liu, D. Li, W. Qu 2006 J. Opt. Soc. Am. A **23** 3191
[11] Chen Yang, Xiaona Yan, Rong Zhu, Hua Zou and Feng Han 2009 Int .J. Light Electron. Opt.doi:10.1016/j.ijleo.2008.12.023
[12] H. Kogelink 1969 Bell Syst. Techn. J. **48** 2909
[13] D. S. Smith, H. D. Riccius, R. P. Edwin 1976 Opt. Commun. **17** 332

Fabrication of a MEMS micro-hotplate

Xian Yi[1, 2], Jianjun Lai[1*, 2], Huafeng Liang[1, 2] and Xiaofeng Zhai[1, 2]
[1]Wuhan National Laboratory for Optoelectronics, Wuhan, Hubei, 430074, China
[2]College of Optoelectronic Science and Engineering, Huazhong University of Science and Technology, Wuhan, 430074, China

E-mail: jjlai@mail.hust.edu.cn

Abstract. With the development of microelectronics and micromachining technology, micro-hotplate as a micro-heater is a very important component in micro-sensor. So far, it has wide applications in pressure detection, gas detection and so on. In this paper, a micro-hotplate based on micro-bridge structure was designed and fabricated. Here we present a micro-hotplate with SOI（silicon-on-insulator）as the substrate and metal Au as the heating electrode material. It shows low thermal conductivity coefficient, good electrical insulation, low power consumption, and excellent resistance to high temperature and low pressure. Some IC (integrated circuit) processes such as magnetron sputtering of Ni/Cr, ion beam sputtering of Au, dry and wet etching processes are employed to fabricate the device. Anisotropic wet etching of silicon and thin film deposition were mainly studied. To protect the front side micro structure during the wet etching process, a special apparatus was designed and fabricated. The optimization process for anisotropic wet etching was obtained for releasing the micro-bridge after a large number of experiments and the micro-hotplate was successfully fabricated. The electrical, thermal characteristics of the micro-hotplate were tested and approved it a well functional device.

1. Introduction

MEMS（micro electro-mechanical systems）is a high-tech and cross field to combine microelectronics with micromachining technology to integrate micro components, micro sensors, micro actuators, signal processing and control circuit etc[1]. In the common silicon substrate, micro-hotplates generally consist of a thin dielectric membrane suspended over an opening in a silicon substrate. In Microsystems technology, hotplates are mainly used for sensor applications where the sensing material is deposited onto the membrane and an electrical signal is also integrated in the hotplate [2]. Hotplates operate at temperatures from 200 to 400 degree. With the development of MEMS technology, micro-hotplate based on MEMS has gradually gained its wide applications on gas sensors [3], infrared detectors, infrared sources [4], infrared emitters, actuators etc [5]. By employing some thin films with good thermal insulation as the substrate, the micro-hotplate presents a series of advantages such as miniaturized size, low power consumption, fast response, high sensitivity, good reproduction and feasibility of sensor array[6] integration.

As the core component of MEMS sensors, structure, materials, mechanical and thermal design, processing technology of micro-hotplate varies for different applications. Here we present a

Published under licence by IOP Publishing Ltd

Journal of Physics: Conference Series **276** (2011) 012098 doi:10.1088/1742-6596/276/1/012098

micro-hotplate with silicon-on-insulator [7] (SOI) as the substrate and metal Au as the heating electrode material. It shows low thermal conductivity coefficient, good electrical insulation, low power consumption, and excellent resistance to high temperature and low pressure. Au has a good conductivity, small electromobility, stable TCR(temperature coefficient of resistance) and chemical properties. In accordance with the principle of structural design and the conditions of laboratory, the fabrication processes are designed for the micro-hotplate.

In this work, some IC processes such as magnetron sputtering of metal Ni, ion beam sputtering of Au, dry and wet etching[8] processes are employed to fabricate the device. Anisotropic wet etching of silicon and thin film deposition were mainly studied. Through the study of principle of the wet etching of silicon and a lot of corrosion experiments, the characteristics of wet etching were analyzed and the micro-bridge structures were designed to be compatible with the wet etching processes. TMAH [9] is used to be the etchant. To protect the front side micro structures during the wet etching process, a special apparatus with fixture was designed and fabricated. After doing many experiments, we found it works very well, no TMAH solution was found to corrode the front side structures. The optimization process for anisotropic wet etching was obtained for releasing the micro-bridge after a large number of experiments and the micro-hotplate was successfully fabricated. The electrical, thermal characteristics of the micro-hotplate were tested and approved it a well functional device.

2. Design of the micro-hotplate

There are two types of micro-hotplate structure: the closed-type membrane, where the membrane overlaps the silicon substrate along its periphery and the suspended-type membrane, where the membrane is supported on the Si substrate by means of supporting beams. In both cases, the membrane lies over a cavity etched in the silicon substrate. In the latter case, the thermal losses to the substrate take place only through the supporting beams, and thus they are minimized compared to the closed-type membrane. Micro-hotplate designed here belongs to the suspended-type membrane structure. It can be used as the public heating source platform of many different narrow-band infrared emitter structures, such as based on two-dimensional photonic crystal.

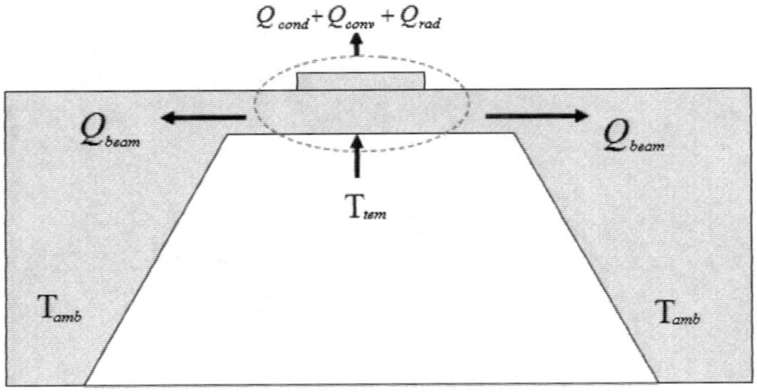

Figure 1. Schematic diagram of heat transfer of micro-hotplate.

The total heat loss of the micro-hotplate consists of conduction, convection and thermal radiation. As shown in figure 1, it includes: (1) heat conduction through the supporting beam to the surrounding frame region, (2) conduction, convection and thermal radiation through the high-temperature bridge surface(here assumed to be atmospheric environment, therefore the thermal conduction of air is also included).

The total heat loss can be expressed with the following formula (1), the first describe the heat conduction through the supporting beam, the second is conduction and convection of air, the third is thermal loss brought from the thermal radiation.

$$Q_{total} = G_m \lambda_m (T_{hot} - T_{amb}) + G_{air}(h_f + \lambda_{air})(T_{hot} - T_{amb}) + G_{rad}\sigma\varepsilon(T_{hot}^4 - T_{amb}^4) \quad (1)$$

Where λ_m and λ_{air} are thermal conductivity of membrane and gas respectively, h_f is the gas convection coefficient, G_m, G_{air} and G_{rad} are geometric factor related to the membrane structure, σ is the Stefan-Boltzman constant(5.67×10^{-8}W/(m^{2}T^2), ε is the surface emissivity, which range from 0 to 1, T_{hot} and T_{amb} are temperature of hotplate and ambient environment respectively.

We hope to obtain relatively high infrared emissivity, so the size of the micro-hotplate we design is larger, whose centre area is Mm order of magnitude. For such a large size, it is very difficult for us to use coating method to fabricate the non-closed membrane structure. In order to get the higher mechanical strength, improve device yield and simplify the production process, we adopt the SOI substrate to produce micro-hotplate, so the simulation of structure is also based on SOI substrate.

The micro-hotplate structure is shown in figure 2, to get the uniform heating effect, electrode use hole-like structure, as shown in Figure 3. For SOI we use, the thickness of underlying silicon is 675nm, the middle buried oxide layer is 500nm, and the top silicon is 10μm. The thickness of upper SiO_2 is 500nm, and gold resistance wire film is 100nm. The centre size of micro-hotplate is 2nm×2nm, the width of supporting beam is 50μm, and the length is 1400μm.

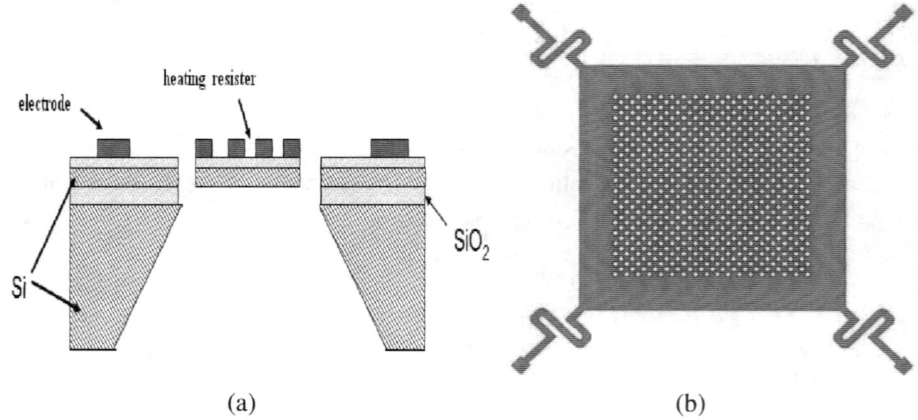

(a) (b)

Figure 2. Schematic diagram of micro-hotplate structure.
(a) Two-dimensional sectional view; (b) Heating electrode of front side.

3. Fabrication of the micro-hotplate

The fabrication process is based on standard technologies used in semiconductor industry; some IC processes such as magnetron sputtering of metal Ni/Cr, ion beam sputtering of Au, dry and wet etching processes are employed to fabricate the device. The process flow is shown schematically in figure 3. It starts with <100> orientation SOI wafers with a 500μm silicon oxide film grown by thermal oxidation both at front side and back side. The basic process steps are stated as follows:

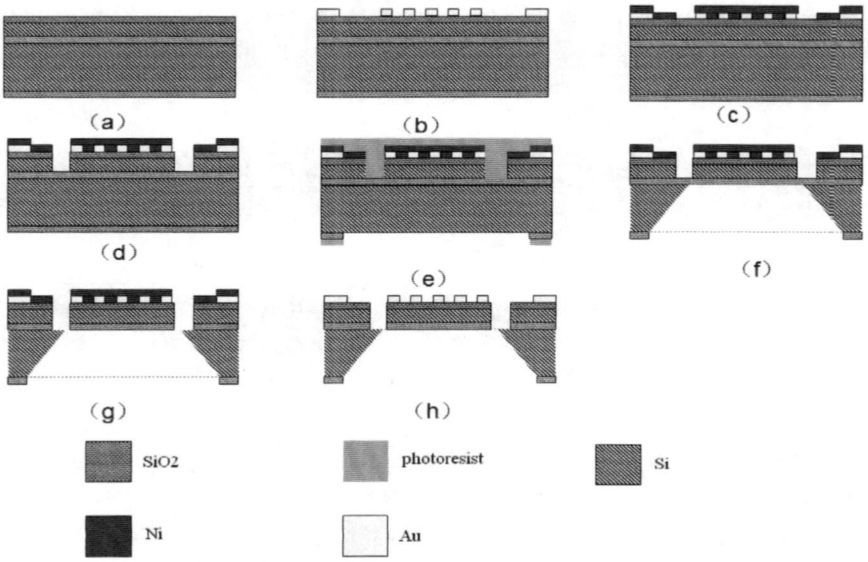

Figure 3. The process flow chart of micro-hotplate.

3.1. Fabrication of Au electrode

Metal film hotplate offer precise temperature control, which can also be used as thermometer, it greatly simplifies the device structure and fabrication technology. Due to metal gold some good electrical and thermal features, we select Au as the heating electrode. Here we adopt the lift-off process. First, the ENPI negative photoresist is spin coated on the front side of substrate, the inverse electrode pattern is formed by photolithography process, so the photoresist section comes into being. Then a RF magnetron sputtering of Ti layer is deposited on top of the formed oxide to improve adhesion between Au and SiO_2 substrate, the thickness of Ti is about 10nm. Au layer is deposited via ion beam sputtering, process parameter is shown in table 1, and its thickness is about 110nm. Metal layer covers the remaining photoresist as well as photoresist opening. The last step is to use stripper to remove the photoresist. Metal film attached to photoresist is stripped from the substrate, which deposited directly onto the substrate will be retained.

Table 1. Process parameters of ion beam sputtering Au film.

Parameter	Screen-grid voltage	Screen-grid current	Accelerating-grid voltage	Air pressure	Substrate temperature	Time
Value	500V	50mA	100V	3×10^{-2}Pa	80°C	20min

In older to strip effectively, the following requirements must be met: (a) the thickness of photoresist mask is greater than the thickness of metal layer; (b) after developing, the profile of the photoresist mask section goes to the next narrow width of Chinese character "八" shape, so the metal deposited on the substrate and metal on the photoresist film will be disconnected completely. When the photoresist layer is washed away, the metal film on it is lifted-off and washed together with ENPI photoresist layer. After this, the metal Au remains only in the region where it has a direct contact with substrate.

3.2. Fabrication of Ni/Cr mask and dry etching of Si

During the experiment, we found that Au layer will also be etched while RIE etching of Si, the surface of Au layer becomes rough. If etching time is long enough, Au layer will be etched completely. This is due to the hardness of Au is relatively low; the physical bombardment etches the Au as well as Si. So a Ni/Cr protective layer is added above Au layer. RIE etching rate of Ni/Cr is small, 100nm Ni/Cr protective layer can be enough.

Table 2. Process parameters of DC magnetic sputtering Ni/Cr film.

Parameter	Current	Base pressure	Air pressure	Substrate temperature	Time
Value	0.5A	3×10^{-3}Pa	3×10^{-2}Pa	Room temperature	5min

The etching mask of Ni/Cr is deposited via the DC magnetron sputtering, the process parameter is shown in table 2. The Au and the Ni/Cr layer are both less than 200 nm, so ordinary lift-off can fulfill the requirement. After that, front side etching window is opened. Subsequently, the front side silicon dioxide and the upper silicon of SOI are etched via RIE process. Dry etching of Si and SiO_2 is performed with fluorine-based chemistries, such as CHF_3 or SF_6, which break up into free fluorine atoms in the course of glow discharge, and then react with silicon to generate gaseous substance, finally achieve the purpose of etching. After several experiments, we have the best process flow. RIE dry etching process is as follows:

3.2.1. Remove the residual photoresist: O_2: 80SCCM, power: 100W, time: 1min;

$$C_xH_y + O_2 + e^- \rightarrow CO + H_2O \tag{3.2.1}$$

3.2.2. SiO2 etching: CHF_3:O_2 = 80SCCM:15SCCM, power: 100W, time: 3min;

$$CHF_3 + e^- \rightarrow CHF_2^+ + F + 2e^- \tag{3.2.2}$$

$$SiO_2 + 4F \rightarrow SiF_4 + O_2 \tag{3.2.3}$$

3.2.3. Si etching: SF_6:O_2 = 40SCCM:5SCCM, power: 100W, time: 15min.

$$SF_6 + e^- \rightarrow SF_5^+ + F + 2e^- \tag{3.2.4}$$

$$Si + 4F \rightarrow SiF_4 \tag{3.2.5}$$

Hence, the front side processes are achieved. The SEM photographs of structure after RIE etching is shown in figure 4.

Figure 4. The SEM photographs of structure after top silicon RIE etching.

3.3. SiO_2 wet etching of back side etching window

To release the micro-bridge structure, we will remove the underlying silicon. So it is necessary to etch the back side. Before doing this, the front side must be protected by spin coating a photoresist layer. Then the back side alignment is processed. Wet etching of SiO2 is carried out at room temperature, so photoresist can be used as mask layer. The adhesion of photoresist is to be very good, and it must be fully dried to prevent lateral corrosion. In older to enhance the adhesion of photoresist, we adopt two steps as following: (1) Place the substrate in the baking oven for 20～30 min to remove water on the surface thoroughly; (2) Spin coating a tackifier layer of HMDS (Hexamethyldisilizane) prior to photolithography.

BHF (buffered hydrofluoric acid) solution is used to etch the back side oxide layer; it is a mixture of 5 part 40% NH_4F: 1 part 49% HF by volume. To obtain stable etch rate, the prepared BHF solution must be sit at least 1 hour. In the etch process, we can clearly see the color change of substrate, the purple SiO_2 disappear gradually, exposing the Si color. After etching in BHF for 5 min, the back side etching window is opened. The chemical reaction as

$$SiO_2 + 4HF + 2NH_4F \rightarrow (NH_4)_2SiF_6 + 2H_2O \tag{3.3}$$

3.4. Anisotropic wet etching of Si

TMAH（tetramethyl ammonium hydroxide）is the most preferred ammonium hydroxide based silicon etchant, it is non-toxic to humans, no damage to integrated circuits, and compatible with CMOS technology. In addition, while using TMAH to etch silicon; thermal oxidation of silicon dioxide can be a good mask medium.

Schematic of equipment for silicon wet etching system is shown in figure 5, which is set up by thermostatic magnetic stirrer. It has function of temperature control and magnetic stirring to facilitate uniform etching. With the process of etching, due to chemical reaction and volatilize along with heat, solution concentration and components will change. In order to maintain the stability of TMAH solution concentration, we add a water cooled condensation-reflux system. Reflux can reduce the loss of volatile TMAH solution by the condenser. All experiments were carried out in a closed glass vessel with constant temperature bath.

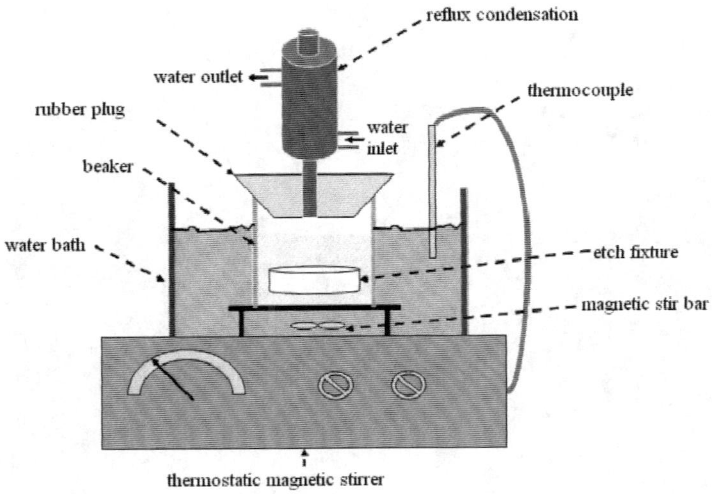

Figure 5. Schematic diagram of silicon anisotropic wet etching system.

The thickness of SOI wafer we used is about 675nm, so it takes 6～7 hours to etch fully. Substrate is immersed in TMAH solution so long that the front side structure will be destroyed, even the metal electrode will come off, and therefore it is necessary to take some measures to protect front side structure. A fixture is designed and made. The fixture material is Teflon, besides; all fluoride or Teflon coated O-seal ring is adopted. This fixture can be used repeatedly, just only replace the washer. The fixture we designed is shown in figure 6.

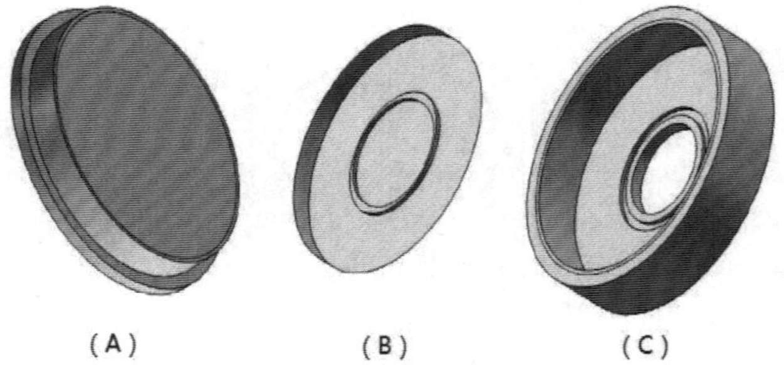

Figure 6. Structure diagram of Si TMAH wet etching fixture.

It mainly consists of three parts, as we can see from the figure 6, part A is a lid with screw thread, part B is a circular gasket, and part C is a sleeve with screw thread as well as a hole in the middle of it. The gasket and sleeve both have a groove which can place O-ring washer. Place the back side of substrate down in the middle of the sleeve, then cover the gasket on the substrate, finally tighten the lid. To prevent substrate from moving during the tightening process, we usually apply a small amount of glue-502 to the washer of sleeve to fix the substrate. In addition, packaged in the water can reduce the probability of breaking of substrate caused by the stress changes during the wet etching. After doing many experiments, we found it works very well, no TMAH solution was found to corrode the front side structure.

Wet anisotropic etching of silicon is accomplished by using 25% TMAH solution. After six-hour 90 degree water bath heating, the underlying silicon is etched entirely. The back side etching diagram is shown in figure 7. As we can see from the figure, the result of wet release is good. But due to etching time is too long, (111) plane is also be etched partly. Therefore, the etch window's size can be designed larger, so as not to effect the device function.

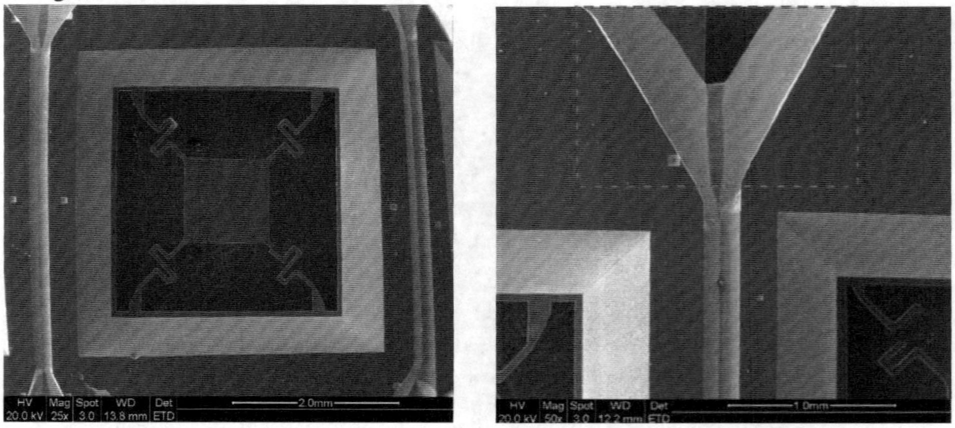

Figure 7. The SEM photographs of back side after wet etching.

3.5. Etching of the middle buried oxide layer and removal of Ni/Cr mask

The etching of buried oxide layer can be completed by wet etching as well as dry etching, specific methods has been introduced in the previous section. While Ni/Cr mask layer is removed by Ni/Cr etchant TFN [8], which is a mixture of 20% $(NH_4)_2Ce(NO_3)_6$, 6% HNO_3 and 74% H_2O. The reaction occurs at room temperature, etching rate is fast, is about 80nm/min. Because gold's color is yellow we can see clearly, there is no need to control etching time precisely to control the thickness of nichrome. Only observe the surface color of substrate in the etching process, if gold color emerges, prolong a few seconds, then clean the substrate carefully. Eventually, the micro-hotplate is successfully fabricated. The SEM graphic of micro-hotplate we make is shown in Figure 8.

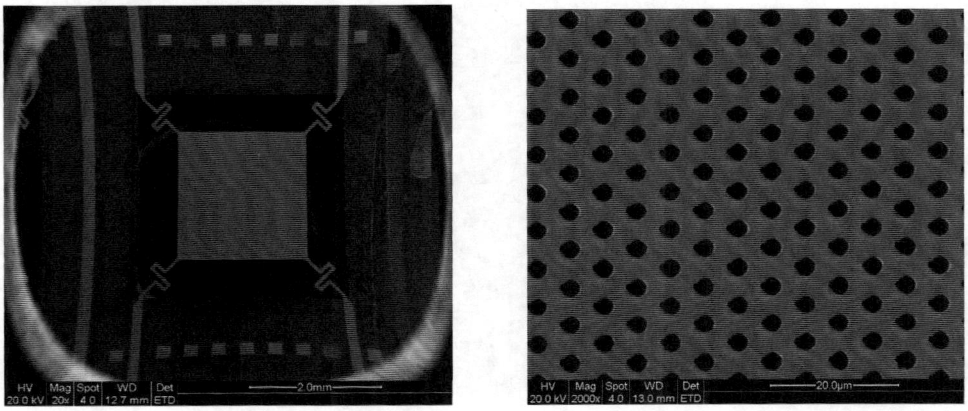

Figure 8. The SEM photographs of finished micro-hotplate.

4. Conclusion

A micro-hotplate with silicon-on-insulator (SOI) as the substrate and metal Au as the heating electrode material is designed and fabricated. Some IC processes such as magnetic sputtering, ion beam sputtering, dry and wet etching processes are employed to fabricate the devices. Anisotropic wet etching of silicon and thin film deposition were mainly studied. Through the study of principle of the wet etching of silicon and a lot of corrosion experiments, the characteristics of wet etching were analyzed and the micro-bridge structures were designed to be compatible with the wet etching processes. The optimization process for anisotropic wet etching was obtained for releasing the micro-bridge after a large number of experiments and the micro-hotplate was successfully fabricated.

References

[1] Gad-el-Hak M 2002 *The MEMS Handbook* (Boca Raton: CRC PRESS)

[2] Beckel D, Briand D, Anja Bieberle-Hutter, et al. 2007 Micro-hotplates—A platform for micro-solid oxide fuel cells *J. Power Sources* **166** 143-8

[3] Rettig F and Moos R 2004 Ceramic meso hot-plates for gas sensors *Sens. Actuators, B* **103** 91-7

[4] Konz W, Hlidenbrand J, Bauersfeld M et al. 2005 Micromachined IR source with excellent blackbody like behaviour Proc, SPIE. 5836 540-8

[5] Briand D, Colin S, Gangadharaiah A, et al. 2006 Micro-hotplates on polymide for sensors and actuators Sens. Actuators, A 1032 317-24

[6] Triantafyllopoulou R, Chatzandroulis S, Tsamis C and Tserepi A 2006 Alternative micro-hotplate design

for low power sensor arrays Microelectron. Eng. 83 1189-91

[7] Laconte J, Flandre D and Raskin J M 2006 Micromachined thin-film sensors for SOI-CMOS Co-Integration (Springer Ser. Chem. Phys.)

[8] Williams R, Gupta K and Matthew W 2003 Etch rates for micromachining—Part □ J. Microelectromech. Syst. 12 761-78

[9] Biswas K and Kal S 2006 Etch of KOH, TMAH and dual doped TMAH for bulk micromachining of silicon Microelectron. J. 37 519-25

Shaping device for high retangle factor tunable thin-film filter

K Yu[1], D X Huang[2], J Q Bao[1], J J Yin[1]
1: Huazhong University of Science and Technology Wenhua College, Wuhan, 430074, P.R.Chian
2: Wuahn National Laboratory for Optoelectronics, Wuhan, 430074, P.R.China

E-mail: onlyfish@126.com

Abstract. The beam divergence angle of the incident collimator will change the transmitting characteristics of the angle-tuned thin-film filter, which insert loss will increase and passband will decrease. With the incident angle is increasing, the phenomena will be more and more obviously. According to the Lagrange-Helmholtz principle, the beam expander has been designed to compress the divergence angle of the incident beam. The double filtering technology has also been used to enhance the retangle factor of the thin-film filter. The experiment results show that the shaping system can effectively eliminate the influence of the divergence angle to the thin-film filter.

Keywords: thin-film filter; divergence angle; angle-tuned; collimator

1. Introduction

Optical filter is a key component in optical communication system[1]. With the rapid development of the elimination polarization technology, more and more large tunable wavelength range angle-tuned thin-film filter are emerging[2,3]. With the low insert loss, high rectangle factor and good temperature stability,these tunable thin-film filters are widely used in the DWDM system. In their spatial coupling, the optical collimators are used to change the Gaussian beam to the approximately parallel beam. Due to the small divergence angle of the approximately parallel beam, the transmitting spectrum of the filter will worsen[4]. The insert loss will increase and the passband will decrease obviously, especially in the relatively large incident angle. In this paper we design and fabricate one kind of prisms to compress the divergence angle of the input beam. With a novel optical path structure the expanded beam can also revert to its original dimension. The theoretical analysis and experimental results show that the prisms can effectively compress the divergence angle, and it can greatly enhance the rectangle factor and passband stability of the transmitting spectrum.

2. Influence of the divergence angle

In our former work we design one kind of 100GHz channel spcing low-polarization four-cavity angle-tuned thin-film filter[5]. The stack of the thin-film filter is as follows:

[1] Dr. Kan Yu is the correspondence .

$$G/\left[\begin{matrix}(HL)^7 2L3H4L3H2L(LH)^7 L(HL)^8 2L3H4L3H2L(LH)^8 L \\ (HL)^8 2L3H4L3H2L(LH)^8 L(HL)^7 2L3H4L3H2L(LH)^7\end{matrix}\right]/A \tag{1}$$

In the stack, the high index material H is Ta_2O_5 and the low index material L is SiO_2. They are both quarter wavelength coatings[6]. The central wavelength is 1563nm when the thin-film filter is in normal incidence, and the central wavelength shift to 1528nm at the angle of 20° . In the device, a pair of C-lens collimators are positioning on the both sides of the interfergence thin-film filter. The C-lens collimator is widely use in the optical communication for spatial coupling. It can divert the Gaussian beam to the expanded parallel beam. However, the expanded beam is not the strict parallel beam, it also has a very small divergence angle. The C-lens is fabricated by the Accelink corporation in China, its divergence angle is at about 0.34° .

To the approximately parallel beam with the small divergence angle, the transmitting spectrum will not change much in normal incidence. However, the overlapped spectrum will separate obviously in oblique incidence because of the divergence angle[7]. So the insert loss of the thin-film filter will increase and its passband will decrease. These phenomena will become more and more obviously while the incident angle is increasing. The figure 1 shows the simulation result of the stack in 20° oblique incidence with the 0.34° divergence angle.

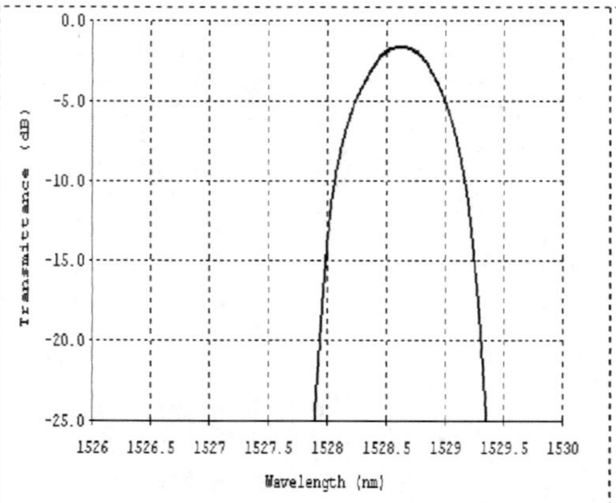

Figure 1. Transmittance curves of the stack(1) at the incident angle of 20° with the 0.34° divergence angle.

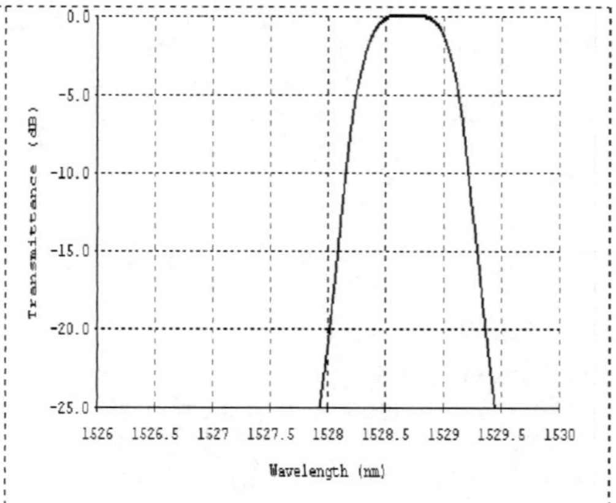

Figure 2. Transmittance curves of the stack(1) at the incident angle of 20° with 0.12° divergence angle.

We can see from these simulation results that the central wavelength of the figure 1 is in 1528.5nm. With the divergence angle of 0.34° , the passband is at about 0.2nm, the stopband is 1.4nm and the inset loss is more than 2dB.

3. Divergence angle compression

Due to the influence of the divergence angle, the transmitting curve of the thin-film filter will distort. The figure 2 shows the simulation result of the transmitting curve of the filter with the divergence of 0.12° . As shown in the figure 2, the passband is 0.36nm, the stopband is 1.24nm and the insert loss is less than 0.2dB. From the figure 1 and figure 2 we also can know that if the divergence is decrease, the phenomena of passband decrease and insert loss increase will be restrained. As the strict parallel beam of the C-lens is impossible, so we can search the way to compress the divergence angle to one third.

From the Lagrange-Helmholtz principle we can know that the object dimension y, the aperture angle u of the imaging beam and the medium refractive index n is a constant. It calls the Lagrange-Helmholtz invariant, which is shown as follows:

$$nuy = n'u'y' = J \tag{2}$$

From the equation we can see that in the same refractive index medium, the divergence angle will compress to one third if the input beam will expand triple dimension.

Due to the prism has a characteristic to deviate the light to its bottom, so we can use double right-angle prisms to change the dimension of the input beam. If the vertical angle of the right-angle prism is α and the refractive index of the prism medium is 1.52, the amplification ratio T of the prism is as follows[8]:

$$T = \frac{1}{\cos\alpha \cdot \sqrt{1-(1.52\sin\alpha)^2}} \tag{3}$$

As we use the double right-angle prisms to expand the beam triple, so the compress ratio of each prism is 0.57. According to the equation 3, we can get the vertical angle of each prism is 32° . Figure 3 shows the double prisms expander system. The divergence angle will be compressed to one third while the beam expand to triple through the prisms.

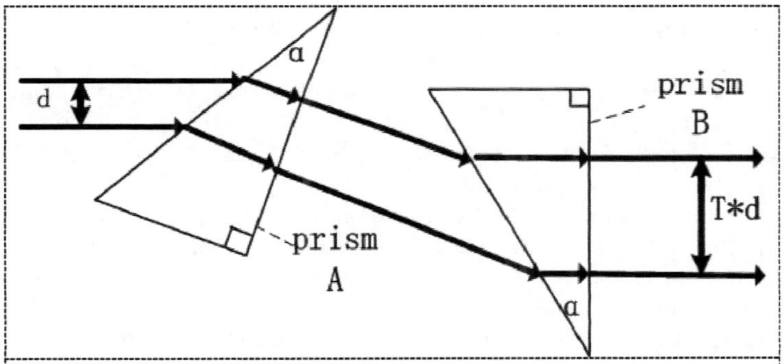

Figure 3. The double prisms expander system.

4. The structure of the device

Though the divergence angle can be compressed but the iuput beam dimension is alos expanded. So the couple efficiency of the C-lens collimator on the other side of the thin-film filter is very low. We should also compress the expanded beam to ite former dimension to the output collimator in the light path design. For the reason of device integration and low insert loss, we can reverse use the expand prisms to compress the beam dimension. Using a right-angle prism in the light path can solve the problem. Figure 4 shows the light path structure.

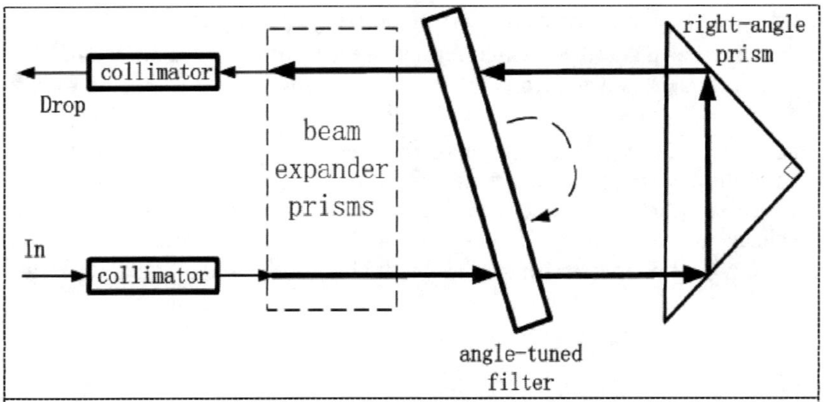

Figure 4. The structure of the tunable filter.

As shown in the figure 4, the expander prisms expand the beam triple from the input port and its divergence angle of the beam is compressed to one third. Due to the divergence angle of the beam is compressed, so the transmitting curve of the thin-film filter will be more stability. By using another right-angle prism the filter beam can be reflected to the thin-film filter again, the doubel transmitting beam with compress divergence angle will also pass the shaping prisms reversely. So at the drop port, the beam can be compress back to its former dimension and also its divergence angle will be triple. However, the relatively larger divergence will not influence the thin-film filter. In oblique incidence, the thickness of the thin-film filter will cause the light path displacement. However, the reverse double filtering can eliminate the light path displacement. So the collimator at the drop port can be fixed, which the tunable wavelength range and the angle modulation range will be greatly enlarged.

After the first filtering, the transmission degree beyond the passband has been decreased obviously. So the double filtering can greatly decrease the bandwidth of the stopband, which has little change to the bandwidth of the passband. Hence the rectangle factor of the thin-film filter will be increased, while the only cost is the double insert loss of the thin-film filter. Figure 6 shows the simulation result of the single and double filtering of the stack.

As shown in the figure 5, in the double filtering the passband is 0.37nm, and the stopband is 0.82nm. It will have better rectangle factor. Due to the insert loss of the angle-tuned filter we fabricated is less than 0.4dB, so the whole insert loss of the double filtering is less than 1dB.

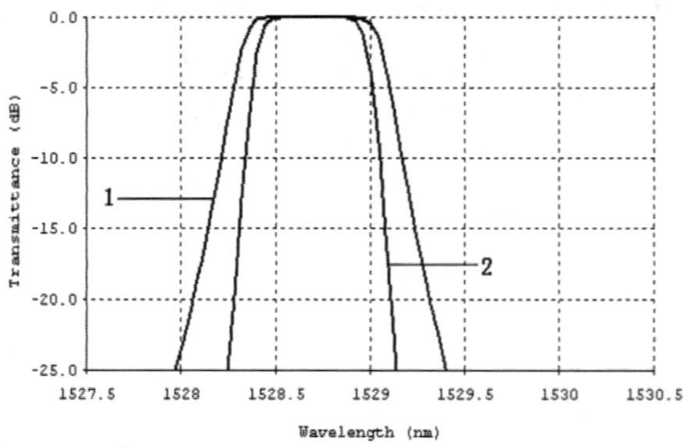

Figure 5. The transmittance curves of the stack(1) at the incident angle of 20° .1: single filtering; 2: double filtering.

5. Experiments

We first test the transmitting spectrum of the angle-tuned thin-film filter without the shaping device. As shown in the figure 6, the measured transmitting spectrum is in the 20° incidence. At the incident angle of 20° , the passband is 0.22nm, the stopband is 1.36nm and the insert loss is 4dB. The indexes in oblique incidence can not content the design demand of 100GHz DWDM channel spacing. Figure 7 shows the measured transmitting of the filter with the divergence angle shaping device in 20° oblique incidence. As shown in the figure 7, the passband is 0.35nm, the stopband is 1.0nm and the insert loss is 1dB. So these indexes content the demand of the 100GHz DWDM optical communication system, and the rectangle factor has been greatly enhanced.

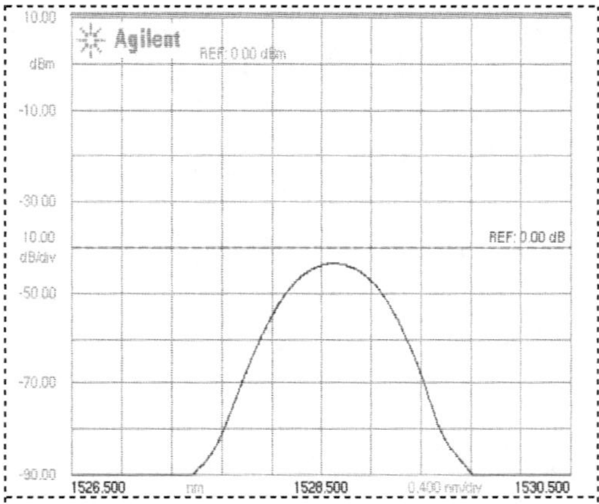

Figure 6. The measured spectrum of the stack(1) at the incident angle of 20° without the shaping device.

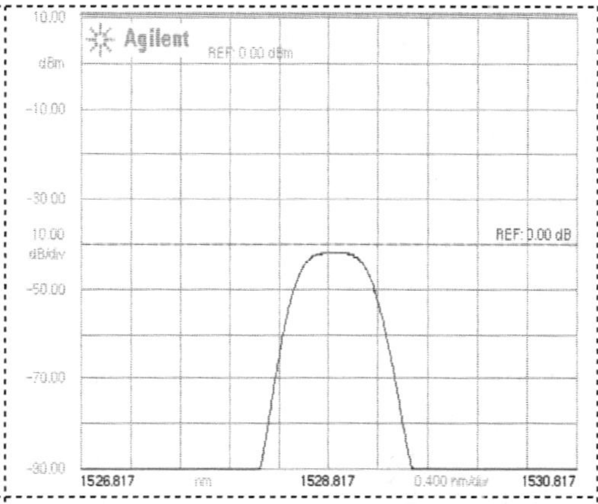

Figure 7. The measured spectrum of the double filtering at the incident angle of 20° with the shaping device.

6. Conclusion

In oblique incidence , the divergence angle of the incident beam will greatly change the transmitting spectrum stability. Due to the collimator C-lens can only input an approximately parallel beam which has a small divergence angle, we design and fabricate a shaping device to compress the divergence angle to one third. In the optical path structure, we use the double filtering technology to resume the dimension of the output beam and eliminate the light path displacement in the thin-film filter. So both the collimators can be fixed, the coupling degree and the rectangle factor can be also enhanced greatly. The simulation results and the experiments show the validity of the shaping device. On all accounts, the design has a bright application for wide tunable range angle-tuned thin-film filter.

Acknowledgments

Authors are grateful to engineer J Chang in the Accelink Technologies CO., LTD, for his help in the device fabrications and all the experiments.

References

[1] Junlchiro M, Yohji F. High performance band pass filter for WDM transmission. *Appl. Opt.*, 1984,**23**(2):193~194

[2] Yoshihiko S, Ahio T and Naoya H. An adaptive wavelength tunable optical filter employing an angle-tuned interference filter and an intelligent digital controller. *J. Lighttwave Technal.*, 1996,**14**(6):1048~1055

[3] Lawvence D, Ming W and Nikolay N. Tuanble and switchable multiple-cavity thin film filters.. *J. Lighttwave Technal.,* 2004,**22**(1):126~135
[4] Wu F, Lu W H and Ma B T. The beam propagation and transformation in axicon-lens system. *Acta Optica Sinica,* 2009, **29**(9):2557~2560
[5] Yu K, Liu W and Huang D X. C-band three-port tunable bandpass thin film optical filter with low polarization-sensitivity. *Optics Communications.,* 2008, **281**(14):3709~3714
[6] Luo Z Y, Liu X and Shen W D. Novel design method for initial structure of dispersion compensation filters. *Acta Optica Sinica,* 2009, **29**(9): 2619~2623
[7] Macleod H A. Thin-film optical filters. 3rd ed. *Bristol Pjoladelphia: Institute of Physics Publishing*, 2001:82~83
[8] Yu K, Liu W and Huang Design and research of facular-shaping on the angle-tuned filter. *Chinese J.Lasers.,* 2008, **35**(9):1375~1379

Fabrication of 1x3 Multimode Interference based Beam Splitter on Silicon Nanomembrane

Y Zhang, D Kwong, A Hosseini and R T Chen

Microelectronics Research Center, Department of Electrical and Computer Engineering, the University of Texas at Austin, Austin, TX 78758, USA

E-mail: magicloudy@gmail.com

Abstract. We present a 1x3 multimode interference based beam splitter on silicon-on-insulator wafer with very thin silicon layer. The device exhibits low cross-talk and high uniformity at 1550nm wavelength.

1. Introduction

Multimode Interference (MMI) Devices, based on self-imaging effect, are widely used as optical components in photonic integrated circuits (PICs) for beam splitting, combining and mode conversion [1]. Recently, silicon-on-insulator (SOI) technology, which allows the single mode waveguide core to be shrunk down to submicron level at 1550nm wavelength, is used as the PICs platform for increasing the integration density [2]. Conventionally, MMI devices on SOI use the ridge waveguide structure, which has large cross-section that limits their application in ultra-compact PICs [3]. In this paper, we report the design, fabrication and measurement of performance of 1x3 MMI based beam splitter on SOI wafer with very thin silicon layer using strip waveguide structure.

The self-imaging effect in multimode waveguides is that an input field profile is reproduces in single or multimode images at periodic intervals along the propagation direction of the guide [1, 4]. From figure 1, the multimode waveguide consists of a core with width W_{MMI}, length L_{MMI}, and N output ports. The multimode section can support a maximum of $N+1$ modes. From MMI theory, the propagation constant β_p can be approximated as $\beta_p \approx \beta_0 - p(p-2)/3L_\pi$, where L_π is defined as $L_\pi = \pi/(\beta_p - \beta_0) \approx 4n_{eff}W_e^2/3\lambda_0$, where n_{eff} is the effective refractive index of the fundamental mode in the multimode section, W_e is the effective width including the penetration depth due to the Goose-Hahnchen shift, and λ_0 is the wavelength, assumed to be 1550nm. Generally, an N-fold image of the input is formed at $L_{MMI}=3rL_\pi/N$, but in the case of a symmetric excitation to the center of the multimode waveguide, the length of the MMI becomes $L_{MMI}=3rL_\pi/4N$ where r is an integer [4].

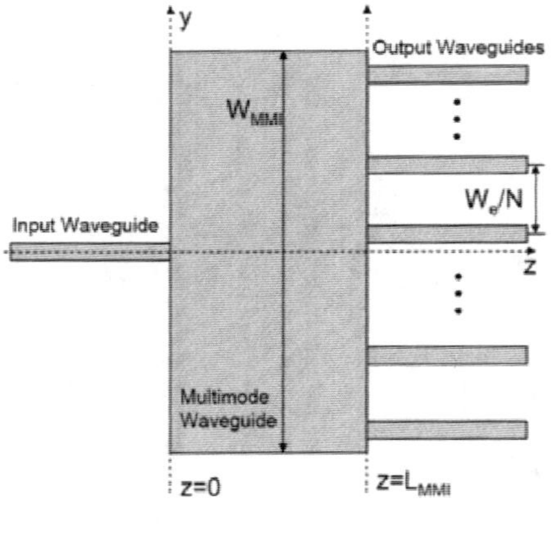

Figure 1. Schematic of a 1xN MMI beam splitter

2. Device design

The device was designed using the beam propagation method (BPM) module from RSoft. The length and width of the MMI region were designed as 54μm and 9.3μm respectively. The width of input and output waveguides were designed as 2.1μm, with the output waveguides tapered down to single mode waveguides, which have core cross-section of 500x230nm. Figure 2 shows the simulation of beam propagation in the MMI region.

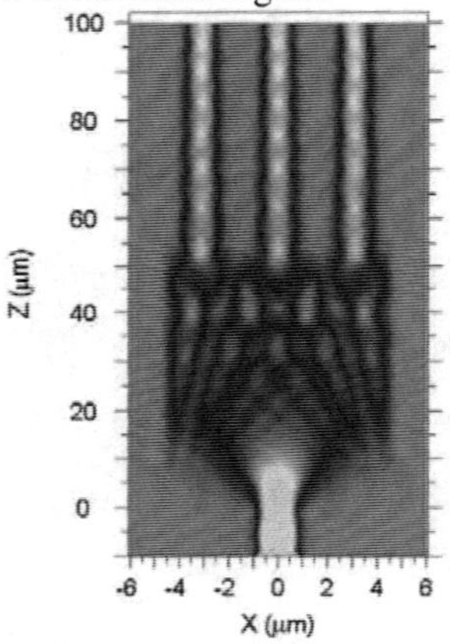

Figure 2. BPM simulation of beam propagation in the MMI region

The reason for not using single mode waveguides as input and output waveguides is to improve the insertion loss and the output uniformity of the device. The design rule for optimum input and output waveguides width can be found in ref. [5].

3. Device fabrication

The device was fabricated on a SOI wafer manufactured by SOITEC with 250nm of Silicon on a 3μm buried oxide (BOX) layer. The wafer was first oxidized to create a 40nm silicon dioxide layer as hard mask. The device was patterned using electronic beam lithography system and ZEP520A positive resist. The pattern was inversed by a nickel lift-off process and transferred to the silicon dioxide hard mask using CHF3/O2 based reactive ion etching (RIE). The nickel was then removed by piranha clean and the patterned silicon dioxide layer was used as a mask to transfer the pattern to the silicon layer via HBr/Cl2 based RIE. Finally, a 1μm silicon dioxide layer was deposited on the device as top cladding by plasma-enhanced chemical vapor deposition (PECVD) [6]. Figure 3 shows the scanning electron microscopy (SEM) picture of fabricated device.

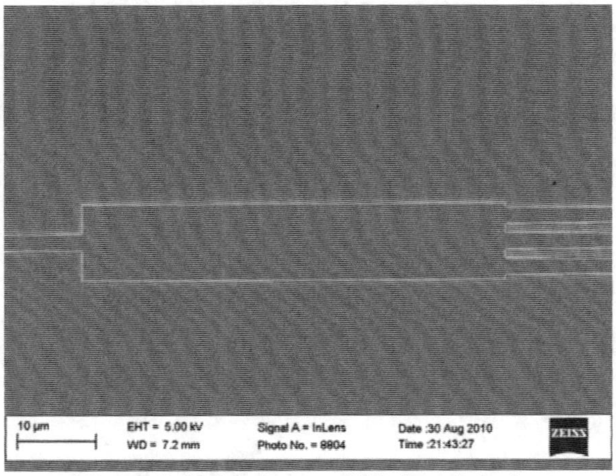

Figure 3. SEM picture of fabricated device

4. Device measurement and performance

A tunable semiconductor laser operating at a wavelength range of 1520-1620nm was used as the input light source. The input signal was amplified by a C-band erbium doped fiber amplifier (EDFA) and was made TE-polarized using a polarizer. A polarization maintaining fiber (PMF) and a single mode fiber (SMF) has been aligned to the input and output facets of the device precisely via a six-axis automated aligner. Both fibers are lensed fibers with a mode diameter about 2.5 microns. The output signal was collected by SMF, and fed to an InGaAs photo detector. A visible-infrared dual imaging system was set directly above the samples, including an objective lens with a magnification rate of 100X and a broad band CCD camera, to monitor the aligning process.

By scanning along the output facet of the device with SMF, we can get the intensity distribution at the device output, which is shown in figure 4. The result shows low cross-talk and high uniformity at 1550nm wavelength. Note that fabrication errors can cause the optimal working wavelength to deviate from 1550nm. We can further improve the output uniformity by fine tuning the wavelength.

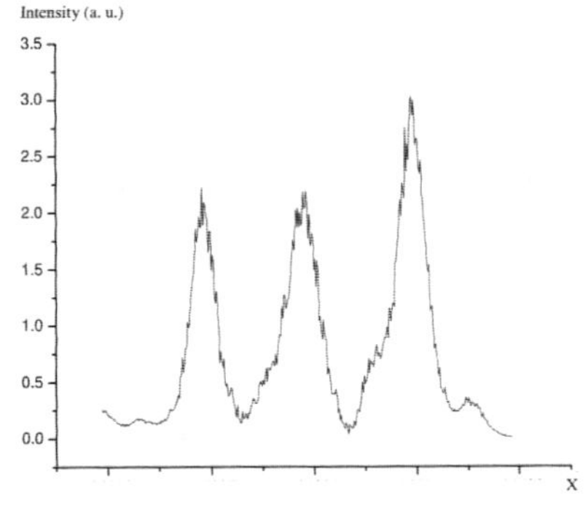

Figure 4. Intensity distribution at the device output

5. Conclusion

A 1x3 beam splitter based on MMI has been successfully demonstrated using strip waveguide structure on SOI.

References

[1] Soldano L B and Pennings E C M 1995 *Journal of Lightwave Technology* **13** 615-27
[2] Vlasov Y A and McNab S J 2004 *Optics Express* **12** 1622-31

[3] Lipson M 2005 *Journal of Lightwave Technology* **23** 4222-38
[4] Hosseini A, Kwong D N, Lin C Y, Lee B S and Chen R T 2010 *IEEE Journal of Selected Topics in Quantum Electronics* **16** 61-9
[5] Hosseini A, Subbaraman H, Kwong D N, Zhang Y and Chen R T 2010 *Optics Letters* **35** 2864-66
[6] Kwong D N, Zhang Y, Hosseini A and Chen R T, to appear 2010 *Electronics Letters*

Simultaneous Transmission of ASK, OCS-ASK and FSK Signals in A Radio over Fiber System Using A Single-drive Mach-Zehnder Modulator

Liang Zhang[1], Yanzhi Wu[1], Tong Ye[1], Junbo Xu[2], Xiaofeng Hu[1] and Yikai Su[1]

1. State Key Lab of Advanced Optical Communication Systems and Networks, Department of Electronic Engineering, Shanghai Jiao Tong University, Road Dongchuan 800#, Shanghai 200240, China

E-mail: yikaisu@sjtu.edu.cn

2. Fiberhome Technologies, Wuhan 430074, China

Abstract. We propose a bidirectional radio over fiber (RoF) system for simultaneous generation and transmission of amplitude shift keying (ASK), optical carrier suppressed amplitude shift keying (OCS-ASK) and frequency shift keying (FSK) signals based on one single-drive Mach-Zehnder modulator (MZM). The feasibility of the proposal is verified by experiments and error-free performances are achieved for all the data.

1. Introduction

With the development of optical and wireless technologies, it is attractive to simultaneously transmit baseband and radio-frequency (RF) signals using the same fiber infrastructure in an integrated platform [1-6]. In Ref. [1], individual light sources and modulators are required to generate wireline and wireless signals, which result in expensive transmitter architecture and complicated wavelength management. On the other hand, frequency shift keying (FSK) is gaining much more attentions duo to its high tolerance to fiber nonlinear impairment and chromatic dispersion (CD). Furthermore, frequency shift keying (FSK) balanced detection enables 3-dB receiver-sensitivity enhancement [7]. In Ref. [8], FSK signal is used for amplitude shift keying (ASK) re-modulation for its constant-intensity property, where the generation of FSK signal requires two light sources modulated by two inversed data, which significantly increase system complexity and cost.

In this paper, we propose and experimentally demonstrate a bidirectional radio over fiber (RoF) system for simultaneous transmission of downstream baseband-ASK, optical carrier suppression-amplitude shift keying (OCS-ASK) and FSK signals based on one single-drive Mach-Zehnder modulator MZM. At the central station (CS), a continuous-wave (CW) laser is launched into a MZM which is biased at its transmission null and driven by an electrical carrier-suppressed sub-carrier multiplexed (SCM) signal. With this modulation, the output of the MZM consists of two frequency components: the original optical carrier and the first-order sub-carriers, while the high-order sub-carriers are ignored due to the negligible optical power. After 12.5-km transmission of single mode fiber (SMF), at the base station (BS), the generated signal is split into two parts by a 50:50 optical coupler. One is separated by a fiber bragger grating (FBG) to obtain the baseband-ASK signal and 20-

Published under licence by IOP Publishing Ltd

GHz OCS-ASK signal, which are detected by individual receivers, respectively. For the other part, a band-pass filter is employed to filter out the original optical carrier and right tone of the first-order sub-carriers, to generate a FSK signal. To the best of our knowledge, this is the first time that FSK signal is realized using only a single MZM by both theoretical deviation and experimental demonstration. The generated FSK signal is re-modulated by upstream data, eliminating light sources and wavelength management in the BS. Compared with the previous works, the propose scheme processes two attractive advantages: 1) a cost-effective way to simultaneously generate wireline and wireless signals in a RoF system; 2) a novel method of producing FSK signal, which is used to realize ASK/FSK orthogonal modulation format, resulting in source-free ONUs.

2. Principles

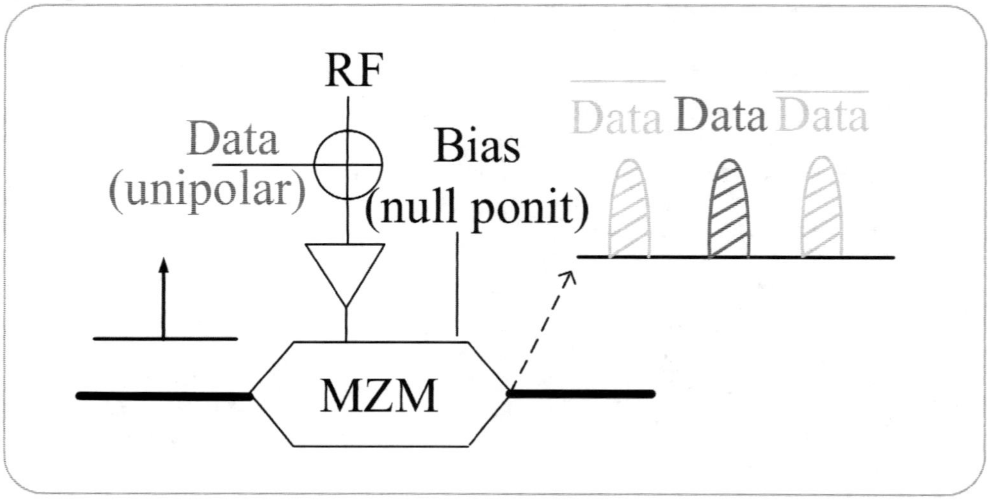

Figure 1. Principle of the generation of FSK signal.

The principle of the generation of FSK based on one single-drive MZM is shown in Fig. 1. An electrical unipolar non-return-to-zero (NRZ) Data and an RF clock are combined together and then fed to the RF input of the MZM, which is biased at the transmission null. The Data is $S_D = V_\pi \varepsilon(t)$, and the RF clock is $S_{RF} = \alpha V_\pi \cos(\omega_{RF}+\theta)$, where V_π is the half-wave voltage of the MZM, ω_{RF} and θ are the angle frequency and phase of the RF clock, respectively, and $\varepsilon(t) = 0$ or 1, depending on the Data. According to Ref. [9], the output field of the MZM can be expressed by:

$$E_{out} = E_0 J_0(\frac{\pi}{2}\alpha)\sin[\frac{\pi}{2}\varepsilon(t)]\cos(\omega_c t+\theta_0) + E_0 J_1(\frac{\pi}{2}\alpha)\cos[\frac{\pi}{2}\varepsilon(t)]\cos[(\omega_c \pm \omega_{RF})t+\theta_0 \pm \theta)] \tag{1}$$

where E_0, ω_c and θ_0 are the amplitude, angle frequency and phase of the input optical signal, respectively. In Eq. (1), only the baseband and the first-order components are considered, while the high-order components are ignored due to the negligible optical power.

When ε (t) is "1" (high electrical level), the first-order component are suppressed and the output signal only consists of the baseband carrier, which can be expressed by:

$$E_{out1} = E_0 \sin[\frac{\pi}{2}\varepsilon(t)] J_0(\frac{\pi}{2}\alpha)\cos(\omega_c t+\theta_0) \tag{2}$$

while, when ε (t) is set to "0" (low electrical level), the baseband carrier is suppressed and an optical carrier suppression (OCS) signal is generated which can be given by:

$$E_{out2} = E_0 \cos[\frac{\pi}{2}\varepsilon(t)]J_1(\frac{\pi}{2}\alpha)\cos[(\omega_c \pm \omega_{RF})t + \theta_0 \pm \theta)] \quad (3)$$

As a result, the baseband component carries Data, while, the first-order component carries inversed Data. By properly adjusting the amplitude and phase of the RF clock, the generated signal could have the same optical power in each band. If a filter is used to select the baseband and the right band of the first-order carriers, a FSK signal will be obtained duo to the inverse data on the selected two tones.

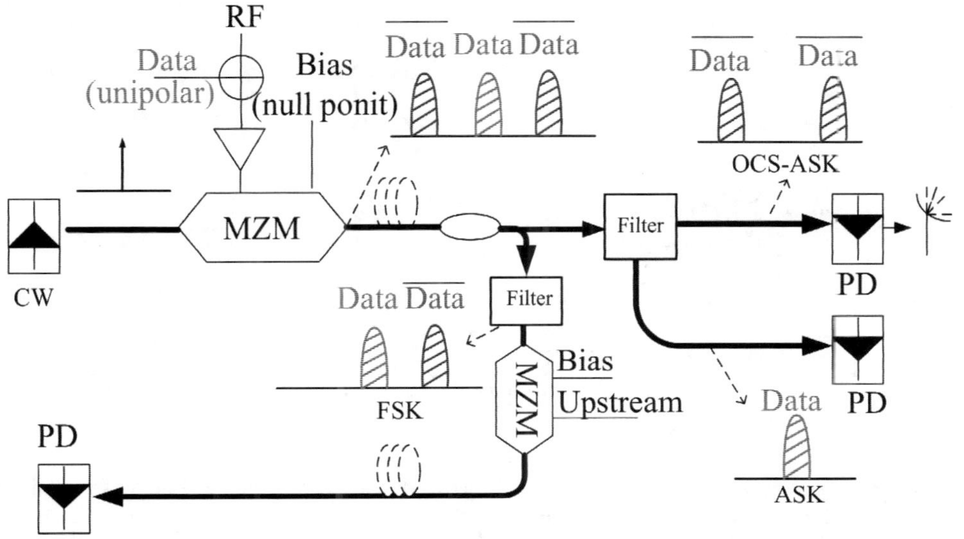

Figure 2. Schematic diagram of the proposed bidirectional RoF system.

The schematic diagram of the proposed bidirectional RoF system is depicted in Fig. 2. A continuous wave (CW) light is launched into a single-drive LiNbO3 MZM, which is modulated to simultaneously generate baseband ASK and OCS-ASK. After the transmission, at the BS, an optical filter is used to separate baseband ASK and OCS-ASK signals. A low-speed PD is employed to receive the baseband signal and a high-speed receiver is required for detection of the OCS-ASK signal. Thus the proposed scheme simultaneously provides wireline and wireless services for the end users. In order to save light sources at the BS, we use a pass-band filter to select the baseband and the right band of the OCS-ASK signal, to generate a frequency shifting keying (FSK) signal, which is used to be optical carrier for re-modulation of upstream data. The upstream signal is sent back to the CS and then detected by a low-speed receiver. Using this design, one can simultaneously deliver downstream wireline and wireless signals and upstream data with a single wavelength in a bidirectional RoF system.

3. Experiment and results

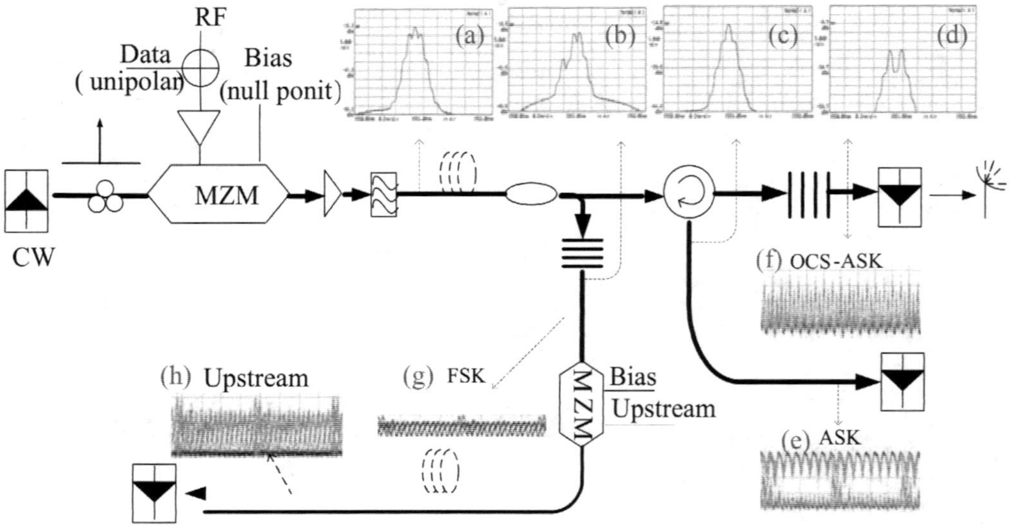

Figure 3. Experimental setup of the proposed bidirectional RoF system.

We perform an experiment to verify the feasibility of the proposed scheme, with its setup shown in Fig. 3. At the CS, a CW light from a tunable laser (Santec TSL-210F) at 1551.99 nm is launched into a single-drive MZM (JDSU OC-192). A 10-GHz clock is mixed with a 1.25-Gbps pseudorandom bit sequence (PRBS) data with a word length of 2^{31}-1 to generate SCM signal, which is employed to drive the MZM biased at the transmission null. The output of the MZM (inset (a) of Fig. 3) consists of two components: the original optical carrier with Data and the first-order sub-carriers with inversed Data. The generated signal is amplified to reach a power level of 6 dBm using an erbium-doped fiber amplifier (EDFA). Then a tunable optical filter (TOF) with a 3-dB bandwidth of 1.6 nm is used to suppress amplified spontaneous emission (ASE) noise.

After transmission of 25-km standard single-mode fiber (SMF), at the BS, the signal is divided into two parts by a 50:50 optical coupler. One part is separated by an FBG with a 3-dB bandwidth of 0.106 nm and a reflection ratio of 90%. The baseband signal is reflected and its spectrum and optical eye diagram are shown in insets (c) and (e) of Fig. 3. The residual clock signal on the eye of baseband ASK signal is due to non-ideal filtering effect. The passing signals are injected into a high-speed PD for detection, whose spectrum and optical eye are provided in insets (d) and (f) of Fig 3, respectively. The other part is filtered by a FBG to generate a FSK signal, whose spectrum and optical eye diagram are shown in insets (b) and (g) of Fig. 3. After passing through a polarization controller (PC), the generated FSK signal is ASK re-modulated using a MZM driven by a 1.25-Gbps upstream data with a PRBS length of 2^{31}-1. The optical eye diagram of the ASK/FSK orthogonal modulation signal is illustrated in inset (h) of Fig. 3. After transmission through 25-km SMF, at the CS, the upstream data is detected by a 2.5-GHz receiver.

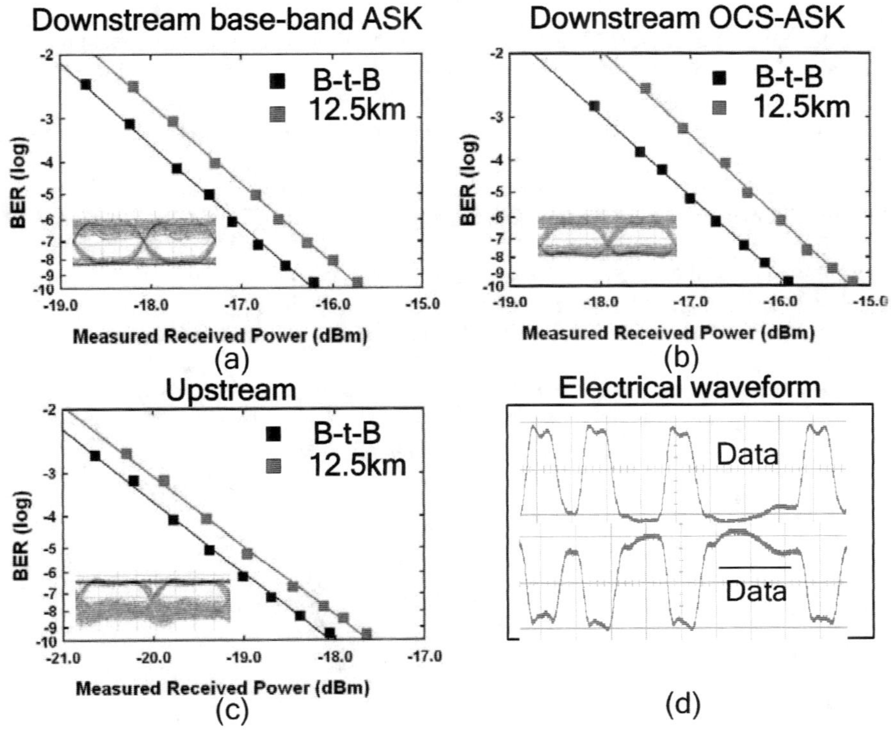

Figure 4. BER curves and electrical eye diagrams. (a) Downstream ASK signal, (b) downstream OCS-ASK signal, (c) upstream re-modulation signal, (d) electrical waveforms of the baseband and first-order component.

The BER performances and electrical eye diagrams of the downstream baseband ASK and OCS-ASK signals, and upstream data are shown in Fig. 4. Error-free performances are obtained for both bandband ASK and OCS-ASK signals with 0.3 dB and 0.8 dB power penalties, respectively. For the upstream data, the power penalty is ~ 0.2 dB. The electrical eye diagrams of the downstream signals and upstream data are provided in insets of Fig. 4. We also measure the electrical waveforms of the baseband and the first-order components, inversed data are achieved, which are depicted in Fig. 4 (d).

4. Conclusion

We propose a simple and novel way to generate FSK signal using only a single-drive MZM. Base on this technology, we experimentally demonstrated a bidirectional RoF system for simultaneous generation and transmission of amplitude shift keying (ASK), optical carrier suppressed amplitude shift keying (OCS-ASK) and frequency shift keying (FSK) signals, which is used as upstream optical carrier. The feasibility of the proposal is verified by experiments and error-free performances are achieved for all the data.

5. Acknowledgement

This work was supported by the 863 High-Tech program (2009AA01Z257).

6. References

[1] Chang G-K, Yu J and Z Jia 2006 "Novel optical-wireless access network architecture for simultaneously providing broadband wireless and wired services," in *Proc. OFC*, Los Angeles, CA, paper OFM1

[2] Lim C, Nirmalathas A, Novak D, Waterhouse R and Yoffe G 2000 *J. Lightw. Technol.* **18** 1355-63

[3] Martinez A, Polo V and Marti J 2001 *IEEE Trans. Microw. Theory Tech.* **49** 2018–24

[4] Jia Z, Yu J, Chowdhury A, Ellinas G and Chang G-K 2007 *IEEE Photon. Technol. Lett.* **19** 1691–93

[5] Yu J, Jia Z, Wang T and Chang G-K 2007 *IEEE Photon. Technol. Lett.* **19** 140–2
Lin C, Peng W and Peng P 2006 *IEEE Photon. Technol. Lett.* **18** 2481-83

[6] Kawanishi T, Higuma K, Fujita T, Ichikawa J, Sakamoto T, Shinada S, and Izutsu M, 2005 *J. Lightw. Technol.* **23** 87–94

[7] Tian Y, Ye T, Leng L, and Su Y, 2008 "A Scalable All-optical VPN in Multiple PONs with a Two-stage TDM-WDM Architecture," in *Proc.ECOC'*, paper Tu.1.F.3.

[8] Ye T, Yan C, Chang Q, and Su Y 2008 *Opt. Commum.* **281** 4648-52

Highly sensitive pressure-induced plasmon resonance birefringence in a silver-coated photonic crystal fiber

S Zhang[1,2], X Yu[1,5], P Shum[2], Y Zhang[1], H P Ho[3] and D Liu[4]

[1] Singapore Institute of Manufacturing Technology, 71 Nanyang Drive, S638075, Singapore

[2] School of Electrical and Electronic Engineering, Nanyang Technological University, Block S2, 50 Nanyang Avenue, S639798, Singapore

[3] Department of Electronic Engineering, The Chinese University of Hong Kong, Ho Sin Hang Engineering Building, CUHK Shatin, N.T. Hong Kong

[4] Department of Optical Communication and Optical Network Engineering, Huazhong University of Science and Technology, 1037 Luoyang Road, Wuhan, China

Email: xyu@simtech.a-star.edu.sg

Abstract. Using the finite element method, we have reported simulation and analysis results on the pressure-induced plasmonic loss in a photonic crystal fiber structure containing hexagonal silver-coated holes. We have also demonstrated that this configuration may offer highly sensitive pressure sensing capabilities. By measuring the resonance wavelength difference between two orthogonally polarized fundamental modes, a high sensitivity value of 32.89 nm/N has been achieved in the visible wavelength range. The effect of pitch size and hole diameter variation on the resonance wavelength have been studied in detail as well.

1. Introduction

A plasmon describes the quantized nature of energy of the surface electrons in the metal [1]. Surface plasmon resonance (SPR) refers to the excitation of surface plasmon polaritons, which are plasmons coupled with photons along the interface of the metal and dielectric materials. There are several methods for SPR to occur including prism coupling, waveguide coupling, fiber-optic coupling and grating coupling. Among all fiber-based plasmonics devices, there is a special class of optical fibers named photonic crystal fibers (PCFs) which possesses a periodic dielectric structure [2]. When metal inclusions are introduced into PCFs, new features which incorporate plasmonics effect have been the research highlight for the past decade, especially in the sensing technology field. PCF-based SPR sensors provide numerous advantages, such as miniaturization, a high degree of integration and remote sensing capabilities. Other applications such as optical circuitry, fiber-integrated optoelectronic components, electrochemistry, glass poling, liquid crystal devices and imaging on the sub-wavelength scale have also been proposed [3].

[5] To whom any correspondence should be addressed.

This paper mainly focuses on the pressure sensing application, which has also attracted many research groups worldwide. Bock *et al* have experimentally measured the pressure and temperature sensitivities of two PCFs and successfully developed a temperature-insensitive pressure sensor using a commercial polarization-maintaining PCF [4]. Subsequently, Gahir and Khanna have improved the model with simpler design configuration. They have theoretically and experimentally presented a temperature-compensated pressure sensor on PCFs operating at 1550 nm [5]. Similarly Fu *et al* have reported a large wavelength-pressure coefficient sensitivity using the Sagnac interferometer design with a polarization-maintaining PCF. The demonstrated measurement range was 0.3 MPa [6]. But until recently, the SPR effect in metal-based PCF with pressure sensing application has only been reported. Pone *et al* have demonstrated a metal-coated elliptical holey fiber based pressure sensor by measuring the splitting of the resonance wavelength of two orthogonally polarized fundamental modes in the infra-red wavelength region. The splitting value increases with the ellipticity of the air holes [7]. However, they only considered the deformation of holes while assuming that the pitch size, i.e. the distance between two adjacent holes remain the same. In fact, this does not reflect the actual situation when the fiber is exposed to an external force or pressure.

In the present paper, we report simulation results on pressure-induced plasmonic loss in a PCF structure containing hexagonal silver-coated holes taking into consideration of the mechanical deformation of both the pitch size (Λ) and the hole diameter (d) due to the external force applied on the fiber structure. Previously published results reveal that the plasmonic peak in the visible wavelength range exhibits higher sensitivity to changes in waveguide structural parameters and the analyte refractive index [8]. Thus our effort has been focused on the wavelength region of 400-800 nm, in order to achieve a high sensitivity and facilitate easy operation and detection. To our knowledge, the investigation on the PCF-based pressure sensor incorporating the SPR effect in the visible wavelength range has not been widely explored so far. We have numerically studied the polarization-dependent SPR condition by computing the difference between the resonance wavelengths of two degenerate fundamental modes in the x- and y-polarization direction. In our PCF device, both the Drude model and the Sellmeier equation have been adopted in the optical simulations to increase the accuracy. We report a large wavelength–force detection coefficient of 32.89 nm/N, which demonstrates the potential application of our device in highly sensitive pressure measurement.

2. Simulation structure

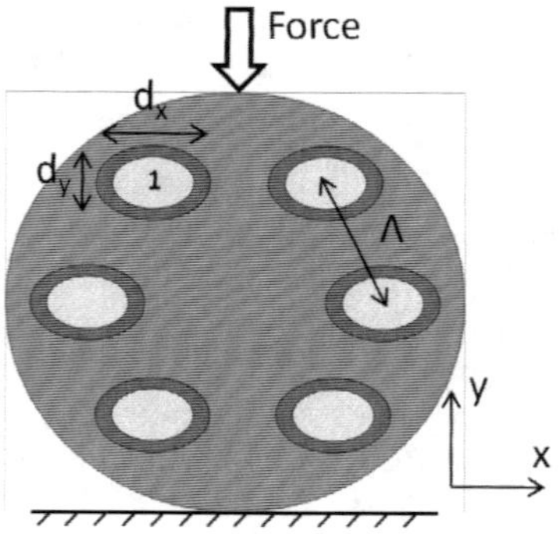

Figure 1. A PCF structure with silver-coated holes. "d_x" and "d_y" are the hole diameter in the x- and y-direction respectively. "Λ" is the pitch size indicating the distance between the center of two adjacent holes. The holes are numbered 1-6 clockwisely with "1" indicated in the figure (not drawn on scale).

The simulation model used in this paper is a PCF structure with six silver-coated holes as illustrated in figure 1. The hole pitch size is 1.5 μm and the diameter is 0.8 μm. The silver layer thickness shown in red is 50 nm. By compressing the fiber in the y-direction with an external force, the structure experiences stress and undergoes deformation. The simulation study is carried out in two parts. Firstly, the mechanical deformation due to the external force is studied using the finite element method based commercial software **ANSYS** (ANSYS, Inc. version 12.1). The mechanical property constants of the

Journal of Physics: Conference Series **276** (2011) 012102 doi:10.1088/1742-6596/276/1/012102

PCF materials are set as in table 1. The pressure-induced hole ellipticity δ is calculated according to the formula given in reference 6:

$$\delta = \frac{2 \times \left| d_x - d_y \right|}{d_x + d_y} \times 100\% \tag{1}$$

Table 1. Mechanical properties of silica glass and silver.

	Young's Modulus (GPa)	Poisson's Ratio	Yield Strength (MPa)	Ultimate Strength (MPa)
Silica glass[a]	73	0.17	-	69
Silver[b]	76	0.37	55	140

[a] Data are extracted from reference 9.
[b] Data are extracted from reference 10.

The second part of study is on the simulation of optical properties of the fiber device. **FemSIM** (RSoft Design Group, version 3.1), another finite element method based simulation tool with perfectly matched layer boundaries is adopted to find the complex propagation constants of coupled fundamental modes over a wide wavelength range based on the deformed fiber structure. The frequency-dependent dielectric constant of silver ε_{Ag} is approximated by the Drude model, together with the standard Sellmeier expansion for silica. In order to facilitate phase matching between a core-guided mode and a plasmonic mode in the visible wavelength range, the holes are assumed to be filled with analyte with a refractive index of n=1.41. The measurement parameter of interest is the resonance wavelength birefringence expressed as:

$$\Delta\lambda_r = \lambda_r^y - \lambda_r^x \tag{2}$$

where λ_r^x and λ_r^y are the resonance wavelength of the x- and y-polarized fundamental mode respectively.

It should be noted that several assumptions have been made in this paper to simplify the simulation model: (1) only the anamorphic effect caused by the external force is studied, the stress-optical effect is not considered at the current stage; (2) the silver layer thickness remains the same throughout the compression process; (3) the effect of the ambient temperature change on the structure deformation of the fiber is not considered.

3. Results and Discussion

The simulation experiment is first carried out with the **ANSYS** software to examine the mechanical deformation parameters, i.e. the Λ and d change. Then the results are fed to the **FemSIM** simulation to study the effect of the external force on the optical properties, i.e. fiber core confinement loss and $\Delta\lambda_r$ variation.

3.1. Mechanical deformation

An external force is applied uniformly in the y-direction on the surface of the PCF structure. The whole fiber becomes elliptical, with the x-direction as the major axis or transverse diameter as shown in figure 1. The deformation along the fiber length direction, i.e. the z-direction is ignored. Both the pitch size and hole diameter change in the x- and y-direction have been studied. The investigation range of the force value is 0-5 N, which is within the elastic deformation range for both silica glass and silver. It has been found in the simulation that the degree of hole diameter change is more significant than that of the pitch size change. The reason is that the Young's Modulus (E), or the slope of the linear deformation region of the stress-strain curve for silver is larger than that for silica glass as indicated in table 1, which means that under the same pressure condition, silver will deform more than materials made of silica glass. In addition, a video clip offering a zoom in view of the hole deformation process throughout the compression as well as the later recovery phase is available for the online version of this paper. The new Λ and d values after the deformation are recorded and fed to the later optical simulations to examine the resonance wavelength birefringence of the two fundamental modes.

3.2. Optical properties

The aforementioned **FemSIM** tool is used to find the complex effective mode index of the coupled fundamental mode, i.e. $n_{eff} = Re(n_{eff}) + jIm(n_{eff})$. The real part of the effective refractive index reflects the propagation constant while the imaginary part is proportional to the confinement loss.

The resonance wavelength λ_r for both x- and y-polarization state fundamental mode at different force values have been plotted in figure 2 (a) and (b) respectively. The peak in the loss spectrum signifies the coupling between a confined core mode and a plasmonic mode. The sharp increase in loss in the transmission corresponds to the matching of the frequency and momentum of the incident light with that of the surface plasmons, causing the plasmons to absorb the light energy and not reflected back. This core confinement loss is proportional to the imaginary part of the effective index by equation (3) below [11]. Thus the plasmon resonance wavelength can be identified by locating the peak of a loss spectrum against a wavelength plot [12].

$$\alpha(dB/cm) = \frac{40\pi}{\ln(10)} \frac{\mathrm{Im}(n_{eff})}{\lambda(cm)} \tag{3}$$

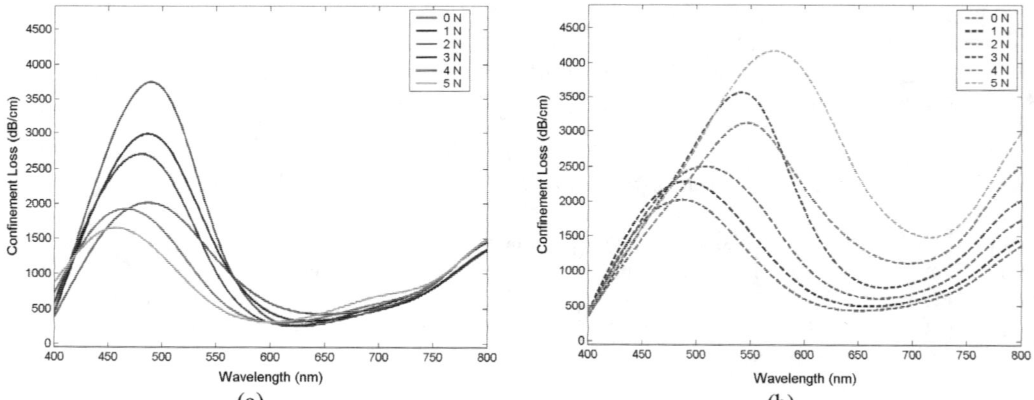

Figure 2. Confinement loss at different wavelengths for the (a) x-polarized and (b) y-polarized fundamental mode when an external force is applied ranging from 0 to 5 N.

From figure 2, it is observed that the resonance wavelength for the x-polarization state (λ_r^x) is generally blue-shifted as the force increases whereas λ_r^y is red-shifted. The amount of shift of λ_r^y is larger than that of λ_r^x since the force is applied in the y-direction. The splitting of the resonance wavelength $\Delta\lambda_r$ defined in equation (2) increases monotonically with the increase in force as represented in the blue curve in figure 3 below. Moreover, when there is no force applied and the holes are circular, the two fundamental modes are degenerate and $\lambda_r^x \approx \lambda_r^y$ or $\Delta\lambda_r \approx 0$. A cubic polynomial fitting result is obtained and indicated beside the curve with a close to 1 R-square value of 0.9882 and root mean square error (RMSE)=8.029. The highest resolution is calculated to be 32.89 nm/N when the force value is equal to 3.5 N. In addition, the green curve in figure 3 plots the average hole ellipticity δ variation defined in equation (1) with respect to the force. The ellipticity value of holes (1, 2, 4, 5) and (3, 6) as numbered according to figure 1 are different because they undergo different degree of deformation, i.e. the calculated δ_{1245} is smaller than δ_{36} because of the fact that the center two holes are further apart than the rest four. However, it is interesting to note that δ in average increases linearly with the external force. If we assume that 0.1 nm shift between two plasmonic peaks can be resolved, the ellipticity detection limit is estimated to be 2.89×10^{-4}, which shows a remarkable improvement than the results published previously. Our results also confirm the merits of a thinner metal coating layer (from 100 nm to 50 nm) and a visible wavelength investigation range (compared to the infra-red wavelength range) [7].

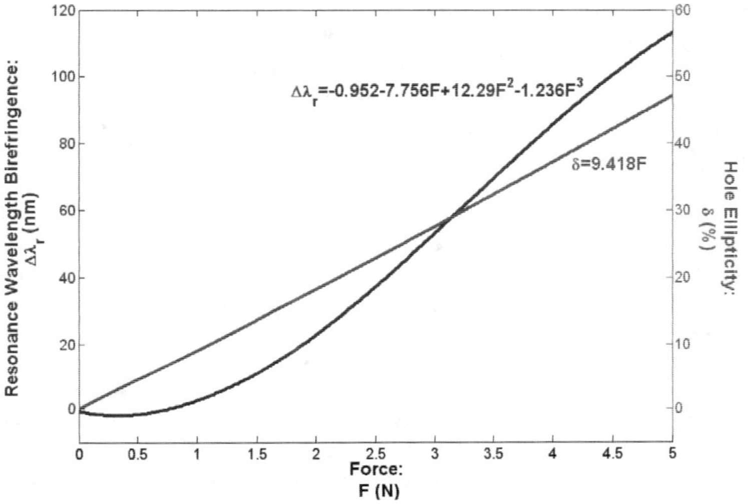

Figure 3. The resonance wavelength birefringence $\Delta\lambda_r$ and the average hole ellipticity δ variation with respect to the external force.

Table 2. Effect of Λ and d change on $\Delta\lambda_r$.

Force: **F** (N)	Λ change, d fixed: $\Delta\lambda_r^{\Lambda}$ (nm)	Λ fixed, d change: $\Delta\lambda_r^{d}$ (nm)	Λ & d change (actual): $\Delta\lambda_r$ (nm)
0	-0.95	-0.95	-0.95
1	2.10	6.45	3.70
2	-1.30	34.30	17.40
3	2.75	70.70	61.20
4	2.40	88.45	80.20
5	0.30	137.35	114.50

In order to examine in detail the parameters that contribute to the change of resonance wavelength birefringence $\Delta\lambda_r$ due to the external force, the effect of Λ and d change have been studied separately as named $\Delta\lambda_r^{\Lambda}$ and $\Delta\lambda_r^{d}$ respectively in table 2. It is evident that $\Delta\lambda_r$ change is more dependent on the hole diameter change which is in line with our previous discussion in section 3.1. As the force increases, d_x increases and d_y decreases. In the x-direction, the silver coating layer is closer to the core. Thus more energy in the core region is available to excite surface plasmons on the silver-silica interface, which results in stronger SPR coupling efficiency. The energy loss in the core is higher which translates to a larger value of the resonance wavelength in the x-direction for the y-polarized fundamental mode [13]. Similarly the change of d_y leads to a blue-shifted λ_r^{x} value. The results obtained agree well with our previously published conclusion [14]. In addition, it has been found that $\Delta\lambda_r$ variance of the pressure sensor, i.e. the last column in table 2, is actually the net effect of $\Delta\lambda_r^{\Lambda}$ and $\Delta\lambda_r^{d}$ change and can be estimated quantitatively using a polynomial equation below:

$$\Delta\lambda_r(\Lambda,d) = a + b(\Delta\lambda_r^{\Lambda})^1 + c(\Delta\lambda_r^{d})^1 + d(\Delta\lambda_r^{\Lambda})^1(\Delta\lambda_r^{d})^1 + e(\Delta\lambda_r^{\Lambda})^2 + f(\Delta\lambda_r^{\Lambda})^2(\Delta\lambda_r^{d})^1 \quad (4)$$

where a=1.201, b=0.595, c=0.785, d=0.142, e=-1.102, f=-0.030 with fitting parameters: R-square=1 and RMSE=4.11×10^{-14}. In fact, this curve fitting result is not unique and should be system specific. However, we have proven that such a correlation exists and can be expressed in a simple polynomial form depending on the number of data points available and the individual accuracy requirement. Further investigation shows that the last three nonlinear terms in equation (4) contributes a much more significant percentage to the final value than the first three linear terms. This indicates that the actual pressure sensor mechanism by measuring the resonance wavelength birefringence is not just a simple linear superposition of the individual effect of Λ and d change. However, the physical interpretations of the remaining nonlinear terms still require further exploration.

4. Conclusion

In this paper, we have analytically investigated the confinement loss and surface plasmon resonance condition of two orthogonally polarized fundamental modes in the context of a photonic crystal fiber with silver-coated elliptical holes. The relationship between the resonance wavelength birefringence $\Delta\lambda_r$ and the external force is studied in detail. The highest sensitivity demonstrated in the simulation is 32.89 nm/N, indicating that the device is capable for applications such as high sensitivity pressure measurement. In addition, a hole ellipticity detection limit value as low as 2.89×10^{-4} also shows a significant improvement over the current statistics reported. A polynomial equation which relates the effect of the pitch size and hole diameter change on the resonance wavelength birefringence is established. In addition, it should be highlighted that although the investigation is carried out for an external force applied in the y-direction, the results can be generalized to a force applied in any other directions. The stress-optical effect and the ambient temperature variance could be considered for further investigations on metal-based PCF pressure sensor devices.

References

[1] Davis J 1996 *Surface analytical techniques for probing biomaterial processes* (Boca Raton, FL: CRC Press)

[2] Lee B, Roh S and Park J 2009 Current status of micro- and nano-structured optical fiber sensors *Opt. Fiber Technol.* **15** 209-21

[3] Cerqueira A S Jr 2010 Recent progress and novel applications of photonic crystal fibers *Rep. Prog. Phys.* **73**, 024401

[4] Bock W J, Chen J, Eftimov T and Urbanczyk W 2006 A photonic crystal fiber sensor for pressure measurements *IEEE Tran. On Instru. And Meas.* **55** 1119-23

[5] Gahir H K and Khanna D 2007 Design and development of a temperature-compensated fiber optic polarimetric pressure sensor based on photonic crystal fiber at 1550 nm *Appl. Opt.* **46** 1184-89

[6] Fu H Y, Tam H Y, Shao L Y, Dong X, Wai P K A, Liu C and Khijwania S K 2008 Pressure sensor realized with polarization-maintaining photonic crystal fiber-based Sagnac interferometer *Appl. Opt.* **47** 2835-39

[7] Pone E, Hassani A, Lacroix S and Skorobogatiy M 2008 A pressure sensor based on the loss birefringence of a microstructured optical fiber containing metal coated elliptical inclusions *Conf. on Lasers and Electro-Optics/Quantum Electronics and Laser Science Conference and Photonic Applications Systems Technologies: OSA Technical Digest (CD) (Optical Society of America)* paper CMZ3

[8] Hassani A and Skorobogatiy M 2006 Design of the microstructured optical fiber-based surface plasmon resonance sensor with enhanced microfluidics *Opt. Express* **14** 11616-21

[9] Automation Creations, Inc. 2010 *Online Materials Information Resource* www.matweb.com

[10] ASM Handbooks Online 2002 *Design and engineering properties of glasses*

[11] Hassani A and Skorobogatiy M 2007 Design criteria for microstructured-optic-fiber based surface-plasmon- resonance sensors *J. Opt. Soc. Am.* B **24** 1423-29

[12] Yu X, Zhang Y, Pan S, Shum P, Yan M, Leviatan Y and Li C 2010 A selectively coated photonic crystal fiber based surface plasmon resonance sensor *J. Opt.* **12** 015005

[13] Sadiku M 2010 *Elements of Electromagnetics* (New York: Oxford University Press)

[14] Yu X, Zhang S, Zhang Y, Ho H P, Shum P, Liu H and Liu D 2010 An efficient approach for investigating surface plasmon resonance in asymmetric optical fibers based on birefringence analysis *Opt. Express* **18** 17950-57

Analytical Study on Characteristics of Dual Microring Resonators Combined with the Add-drop and All-pass Types

Xiaobei Zhang, Jinlong Li
Key Laboratory of Specialty Fiber Optics and Optical Access Networks, School of Communication and Information Engineering, Shanghai University, Shanghai, China

Email: xbzhang@shu.edu.cn

Abstract. This paper presents an analytical study on characteristics of dual microring resonators combined with the add-drop and all-pass types, which can provide novel filtering and delaying characteristics. Analytical expressions of transmission spectra and normalized delays of the dual microring resonators are developed using the transfer matrix method. The effects of coupling coefficients and loss on both transmission spectra and delays are then investigated. The analytical study shows that two resonance peaks of transmission spectra are existent. Under lossless coupling, the resonance frequency of the transmission spectra increases as the ring-ring coupling coefficient increases. However, it is independent on the ring-bus coupling coefficient. The valley amplitude of the transmission spectra increases as the ring-bus coupling coefficient increases, which is also independent on the ring-ring coupling coefficient. Generally, three delay peaks are existent. The corresponding frequencies of delay peaks on either side of the centre frequency are partially coincident with the normalized resonance frequencies of transmission spectra. The normalized delays can develop from three delay peaks to two delay peaks or single delay peak under specified coupling coefficients and loss.

1. Introduction

Microring resonators can be widely used as basic building blocks for future high density optical circuits, such as optical filters [1], wavelength converters [2], modulators [3], multiplexers [4], Lasers [5], switches [6] and sensors [7]. The spectral response and delays of a single ring resonator are typical periodical Lorenz shapes. Dual microring resonators are the simplest structure of multiple microring resonators, which have particular characteristics compared with the single microring resonator. Series-cascaded [8, 9] and parallel-cascaded [10, 11] dual microring resonators were studied in previous works. There are some other schemes to cascade two microring resonators. We have proposed a novel structure of dual microring resonators coupled via 3×3 couplers [12], which can be found applications in fields such as sensors and filters. Another novel configuration of dual microring resonators has been proposed and studied [13-15] and demonstrated [16]. Experimental results show the good agreement with the theoretical results. This structure is also a combination of the add-drop and all-pass types. Hence, the novel configuration of dual microring resonators is also named as dual microring resonators combined with the add-drop and all-pass types (DMRCAAT) in this paper. However, previous study on this novel structure is not completely. Characteristics of delays are also not analyzed and transmission spectra can be further studied.

In this paper, DMRCAAT is systematically investigated. First, analytical expressions of transmission spectra and delays of DMRCAAT are developed using the transfer matrix method. Then,

Published under licence by IOP Publishing Ltd

Journal of Physics: Conference Series **276** (2011) 012103 doi:10.1088/1742-6596/276/1/012103

characteristics of transmission spectra and delays of DMRCAAT are further investigated and discussed under different coupling coefficients and loss.

2. Theory

The DMRCAAT consists of a conventional add-drop microring resonator (ring1), with another all-pass microring (ring2) coupled to the add-drop microring, as shown in Fig.1.

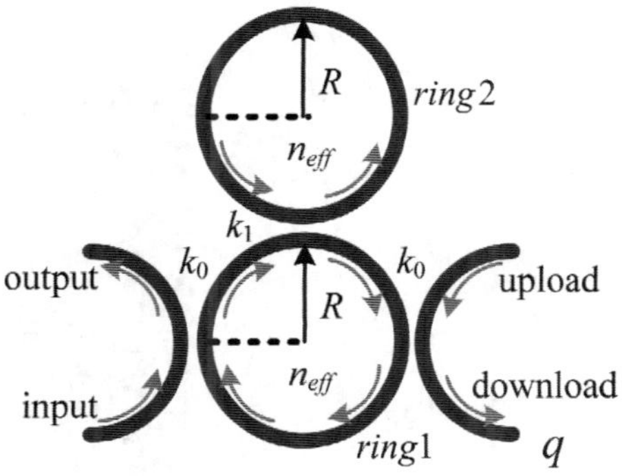

Figure 1. Schematic diagram of DMRCAAT.

The structure has four ports, which are input, output, upload and download respectively. The arrows in the Fig.1 indicate the light wave propagation directions. For simplicity, we assume the coupling coefficients at the ring-bus waveguides to be identical as k_0, and the coupling coefficient between two rings is k_1. Corresponding transmission coefficients are $t_0 = \sqrt{1-k_0^2}$ and $t_1 = \sqrt{1-k_1^2}$ respectively. The radius is R and hence the perimeter is $L = 2\pi R$. The effective index is n_{eff} and the optical wavelength is λ. The normalized frequency is denoted as $\theta = n_{eff}(2\pi/\lambda)L$ with τ as the resonator round-trip transmission coefficient. Using the transfer matrix method [17], the output complex optical field q, spectrum D, effective phase shift Φ and delay T at the download port are found as

$$q = \frac{-k_0^2 \tau^{\frac{1}{2}} \exp(i\theta/2)[t_1 - \tau\exp(i\theta)]}{1 - t_1\tau\exp(i\theta) - t_0^2 t_1 \tau \exp(i\theta) + t_0^2\tau^2\exp(i2\theta)} \tag{1}$$

$$D = |q|^2 = \frac{\sigma_1(\varepsilon_2 - \cos\theta)}{1 + \varepsilon_0(\varepsilon_1 - \cos\theta)^2} \tag{2}$$

$$\phi = \arg(q) = \frac{3\theta}{2} + a\tan\frac{t_1\sin\theta}{\tau - t_1\cos\theta} + a\tan\frac{t_1\tau(1+t_0^2)\sin\theta - t_0^2\tau^2\sin 2\theta}{1 - t_1\tau(1+t_0^2)\cos\theta + t_0^2\tau^2\cos 2\theta} \tag{3}$$

$$T = \frac{d\phi}{d\theta} = \frac{1}{\varepsilon_3(\varepsilon_2 - \cos\theta)} + \frac{\sigma_2(\varepsilon_4 - \cos\theta)}{1 + \varepsilon_0(\varepsilon_1 - \cos\theta)^2} \tag{4}$$

where $\sigma_1 = 2t_1\tau^2(1-t_0^2)^2/\{(1-t_0^2\tau^2)^2[1-(1+t_0^2)^2 t_1^2/4t_0^2]\}$, $\sigma_2 = t_1\tau(1+t_0^2)/\{(1-t_0^2\tau^2)[1-(1+t_0^2)^2 t_1^2/4t_0^2]\}$, $\varepsilon_0 = 4t_0^2\tau^2/\{(1-t_0^2\tau^2)^2[1-(1+t_0^2)^2 t_1^2/4t_0^2]\}$, $\varepsilon_1 = t_1(1+t_0^2)(1+t_0^2\tau^2)/4t_0^2\tau$, $\varepsilon_2 = (\tau^2+t_1^2)/2t_1\tau$, $\varepsilon_3 = 4t_1\tau/(\tau^2-t_1^2)$, $\varepsilon_4 = (1+t_0^2\tau^2)/t_1\tau(1+t_0^2)$. Hence the characteristics of transmission spectra and delays are related to parameters k_0, k_1 and τ.

3. Analysis

3.1. The effects of coupling coefficients on transmission spectra

Firstly we investigate the case of lossless resonators, i.e., τ =1. Fig.2 (a) shows the transmission spectra with different ring-bus coupling coefficient k_0 under a fixed ring-ring coupling coefficient k_1. Fig.2 (b) shows transmission spectra for k_0=0.1, 0.5 and 0.99 respectively. The transmission spectra split into two symmetrical resonance peaks near the centre frequency of $\theta = 0$, due to the mode splitting [18] between the coupled microring resonators.

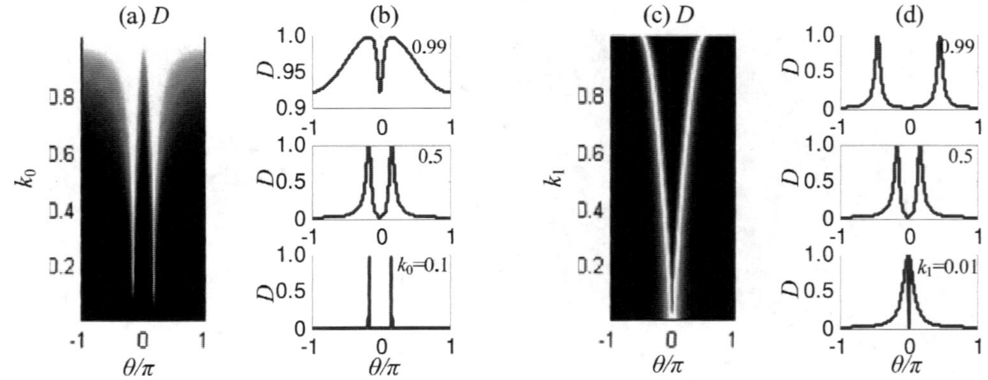

Figure 2. Transmission spectra D as functions of coupling coefficients. (a) Transmission spectra as a function of k_0 under k_1=0.5, (b) transmission spectra for k_0=0.1, 0.5 and 0.99 respectively under k_1=0.5, (c) transmission spectra as a function of k_1 under k_0=0.5, (d) transmission spectra for k_1=0.01, 0.5 and 0.99 respectively under k_0=0.5.

As shown in Fig.2 (a) and (b), both the full width at half maximum (FWHM) and the amplitude at the centre frequency ($\theta = 0$) of transmission spectra (D_1) increase as the ring-bus coupling coefficient k_0 increases under a fixed ring-ring coupling coefficient k_1. On the contrary, Fig.2 (c) and (d) show that both the FWHM and D_1 of transmission spectra are almost not changed, while the free spectral range increases as the ring-ring coupling coefficient k_1 increases under a fixed ring-bus coupling coefficient k_0.

From Eq. (2) and $dD/d\theta$ =0, the equation can be written as

$$\sin\theta[\cos^2\theta-2\varepsilon_2\cos\theta-(1/\varepsilon_0+\varepsilon_1^2-2\varepsilon_1\varepsilon_2)]=0 \qquad (5)$$

which suggests that there are off-resonance frequencies at $\sin\theta$=0 (i.e., $\theta=\pm\pi$ or θ=0) and resonance frequencies $\theta_1=\pm a\cos(\varepsilon_2-\sqrt{\varepsilon_2^2+1/\varepsilon_0+\varepsilon_1^2-2\varepsilon_1\varepsilon_2})=\pm a\cos(t_1)$. This means that resonance frequencies are only dependent on t_1 and independent on t_0. When θ_1 is very small, there is a near linear relationship $\theta_1\approx k_1$. θ_1 increases as k_1 increases in a near linear relationship as shown in the solid lines in Fig. 3. Then we analyze D_1 using numerical simulations. From Eq. (2), $D_1=[(1-t_0^2)/(1+t_0^2)]^2$, which means that D_1 is only dependent on t_0 and independent on the t_1. D_1 increases as k_0 increases as shown in the dashed lines in Fig. 3.

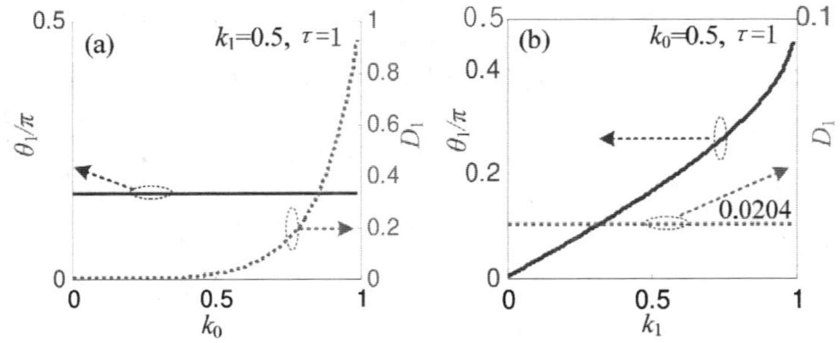

Figure 3. (a) θ_1/π (solid) and D_1 (dashed) as functions of k_0 under k_1=0.5 and τ =1, (b) θ_1/π (solid) and D_1 (dashed) as functions of k_1 under k_0=0.5 and τ =1.

3.2. The effects of coupling coefficients on delays

Under the lossless coupling, Fig.4 shows delays as functions of coupling coefficients. For simplicity, we make a normalized processing for delays as shown in Figs.4 (a) and (c). Moreover, we also limit the maximum delay as 15 in Figs.4 (b) and (d).

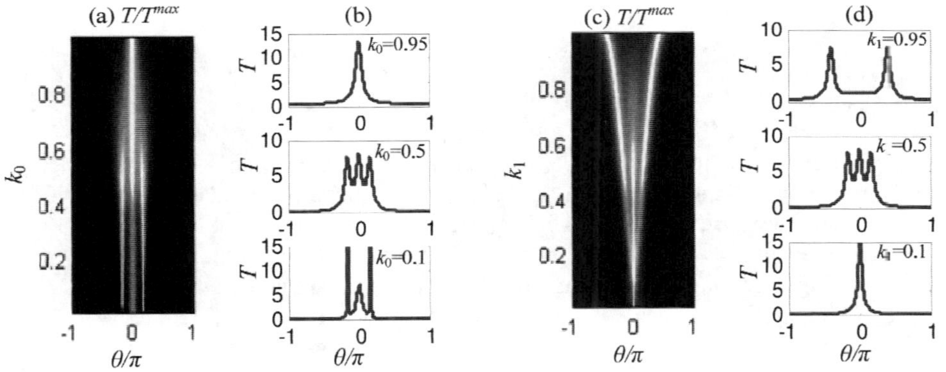

Figure 4. Delays as functions of coupling coefficients. (a) Normalized delays T/T^{max} as a function of k_0 under k_1 =0.5, (b) delays T for k_0=0.1， 0.5 and 0.95 respectively under k_1 =0.5, (c) normalized delays T/T^{max} as a function of k_1 under k_0 =0.5, (d) delays T for k_1=0.1， 0.5 and 0.95 respectively under k_0 =0.5.

Similar to the study of transmission spectra, there are three resonance peaks in delays due to the mode splitting and combinations of two optical paths. We define corresponding frequencies of resonance peaks on both sides of the centre frequency (θ = 0) as θ_2. Meanwhile, the corresponding amplitude of resonance peaks at the centre frequency (θ = 0) is denoted as T_1. As k_0 increases under a fixed k_1, T_1 increases and the amplitude of the resonance peak on both sides of the centre frequency decreases. Finally, three resonance peaks are degenerated into one peak at the centre frequency. On the contrary, T_1 decreases as the k_1 increases under a fixed k_0, and the single resonance peak splits into three peaks near the centre frequency, as shown in Fig. 4 (c) and (d).

From Eq. (4), the delays can be written as

$$T = \varsigma_1 + \varsigma_2 \tag{6}$$

where $\varsigma_1=1/\varepsilon_3\ (\varepsilon_2\text{-}\cos\theta)$, $\varsigma_2=\sigma_2(\varepsilon_4\text{-}\cos\theta)/[1+\varepsilon_0\ (\varepsilon_1\text{-}\cos\theta)^2]$, which suggests that $\varsigma_1=f(t_1, \tau, \theta)$ and $\varsigma_2= f(t_0,t_1, \tau, \theta)$. This means that T_1 is only determined by k_1 under the lossless case, i.e., by ring 2. The ring 2 confines most light when k_1 is small. The light confined by ring 2 decreases as k_1 increases. Thus the delay at the center frequency decreases gradually. The delay peaks on both sides of the centre frequency are mainly decided by ς_2. It is caused by the mode splitting. Fig. 5 (a) shows T_1 increases with k_0 increasing under a fixed k_1. θ_2 changes very small when k_0<0.5, and reduces to 0 rapidly near k_0=0.6. Fig. 5 (b) shows that T_1 decreases and θ_2 increases with k_1 increasing under a fixed k_0.

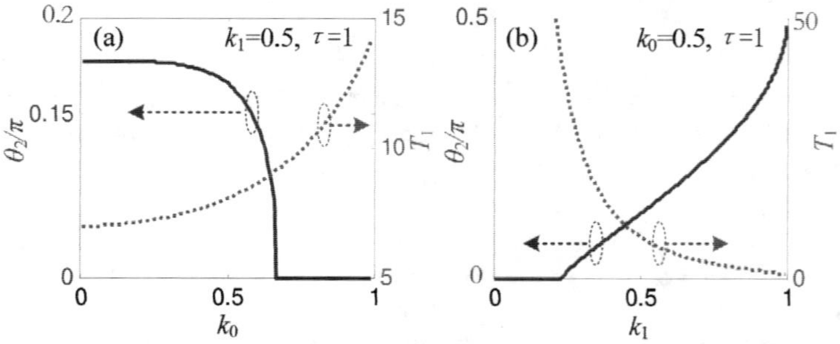

Figure 5. (a) θ_2/π (solid) and T_1 (dashed) as functions of k_0 under k_1=0.5 and τ =1; (b) θ_2/π (solid) and T_1 (dashed) as functions of k_1 under k_0=0.5 and τ =1.

3.3. Frequency detuning between spectra and delays

The optical path difference of forward and backward light waves induces interferes in optical resonators. Light field distributions in DMRCAAT decide the characteristics of transmission spectra and delays. However, there are two resonance peaks in both spectra and delays on both sides of the centre frequency, and the corresponding peak frequencies (i.e., θ_1 and θ_2) are different, also named as the peak frequency detuning. We obtain the relationship between $(\theta_1\text{-}\theta_2)/\pi$ and k_0 (or k_1) by numerical simulations. From Fig. 6 (a), θ_1 is coincident with θ_2 when k_0 is small under a fixed k_1, with θ_1 slightly higher than θ_2. After θ_2 decreases to 0 near k_0=0.6, the peak frequency difference will keep as a constant. Fig.6(b) shows that the peak frequency difference increases as k_1 increases to a maximum and then decreases under a fixed k_0.

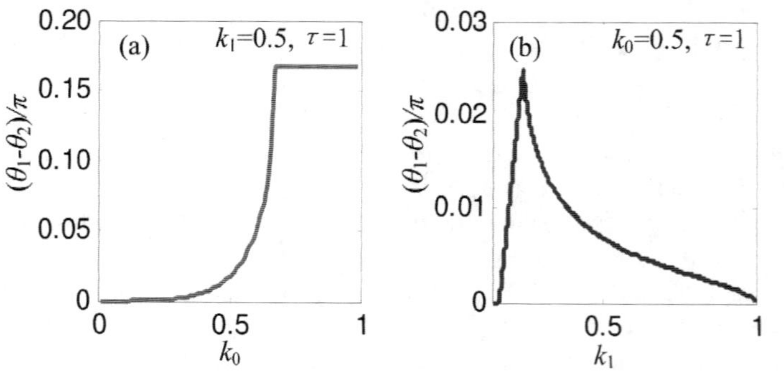

Figure 6. (a) $(\theta_1\text{- }\theta_2)/\pi$ as a function of k_0 under k_1=0.5 and τ =1, (b) $(\theta_1\text{- }\theta_2)/\pi$ as a function of k_1 under k_0=0.5 and τ =1.

3.4. The effects of loss on spectra and delays

To consider the case of loss, equal transmission coefficients $t_0=t_1$ are adopted. This discussion can be divided into three cases by different relations between τ and t_0. They are the critical coupling ($t_0= t_1= \tau$), over coupling ($t_0= t_1< \tau$), and under coupling ($t_0= t_1> \tau$) conditions. From Eqs. (2) and (4), the change regularities of D and T with τ can be obtained. Fig. 7 (a) shows that the amplitude of D decreases as τ decreases (i.e., the loss increases). As shown in Fig. 7 (b), θ_1 increases and then decreases to 0 as τ decreases. Meanwhile, D_1 decreases to 0 at the critical coupling condition. Then D_1 increases and decreases, as τ decreases.

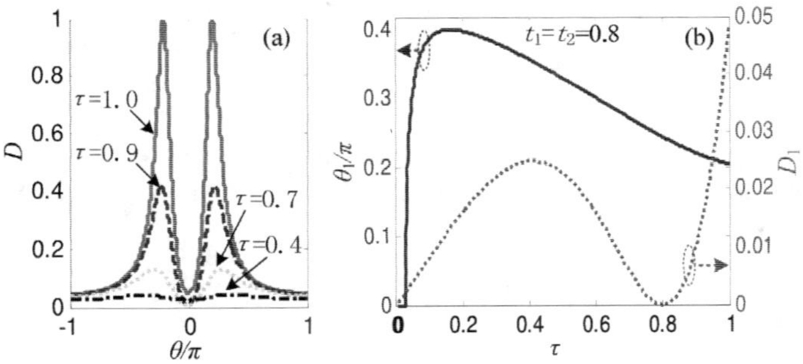

Figure 7. When $t_0= t_1=0.8$, (a) transmission spectra under τ adopted as 1.0, 0.9,0.7 and 0.4 respectively, (b) θ_1/π (solid) and D_2 (dashed) as functions of τ .

To analyze the delays, the normalized processing by $|T|^{max}$ is carried out because of negative delays existing. We can obtain the relations between normalized delays $T/|T|^{max}$ and τ, as shown in Fig. 8 (a). There are large negative and positive delays near the critical coupling region. As shown in Fig.8 (b), the delays are positive when τ is larger than 0.8, while the delays at the centre frequency are always negative when τ is smaller than 0.8. The details are also shown in Fig.8 (c). Fig.8 (d) shows θ_2 decreases slightly as τ decreases. Then θ_2 increases to a maximum and finally decreases to 0.

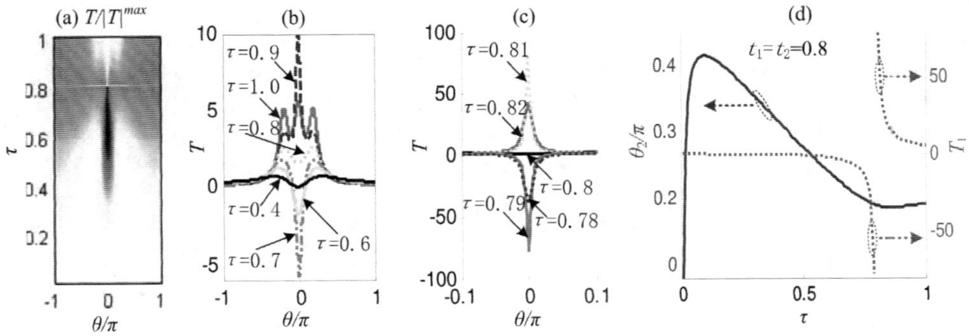

Figure 8. When $t_0= t_1=0.8$, (a) normalized delays $T/|T|^{max}$ as functions of τ; (b) delays T under τ adopted as 1.0,0.9,0.8,0.7,0.6 and 0.4 respectively; (c) delays T under τ adopted as 0.82,0.81,0.8,0.79 and 0.78 respectively, (d) θ_2/π (solid) and T_2 (dashed) as functions of τ .

4. Conclusions

In conclusion, we have presented a comprehensive study on characteristics of DMRCAAT, which can provide novel filtering and delaying characteristics. Analytical expressions of transmission spectra and delays of DMRCAAT are developed using the transfer matrix method. Then, the characteristics of transmission spectra and delays of DMRCAAT are investigated in detail.

Under lossless coupling, the transmission spectra and delays split into multiple symmetrical resonance peaks near the centre frequency because of the mode splitting induced by the two coupled-microring resonators. Corresponding resonance frequencies of the transmission spectra increase with the ring-ring coupling coefficient increasing, which are independent on the ring-bus coupling coefficients. The valley amplitude of the transmission spectra increases with the ring-bus coupling coefficients increasing, which is independent of the ring-ring coupling coefficient. Three delay peaks exist near the centre frequency, which can vary from three delay peaks to two delay peaks or single delay peak under specified coupling coefficients and loss. Corresponding frequencies of delay peaks on either side of the centre frequency are partially coincident with the resonance frequencies of transmission spectra. Therefore, there is a frequency detuning. The loss mainly affects the amplitude of the spectra and delays, with peak frequencies also changed. There is an interesting phenomenon at the critical coupling condition. The valley amplitude of the transmission spectra reaches to 0 at the critical coupling condition, while there are large negative and positive delays near the critical coupling region.

5. References

[1] T. Kato, and Y. Kokubun 2008 *Lightwave Technology* **26** 3694
[2] K. Vyrsokinos, L. Stampoulidis, Z. Sheng, E. Kehayas, P. Bakopoulos, D. Petrantonakis, C. Stamatiadis, Ch. Kouloumentas, P. Zakynthinos, R. Dekker, E. J. Klein, D. V. Thourhout, M. T. Korthorst and H. Avramopoulos 2009 *ECOC* '09
[3] Payam Rabiei, William H. Steier, Cheng Zhang, and Larry R. Dalton 2002 *Lightwave Technology* **20** 1968
[4] Andreas Vorckel, Mathias Monster, Wolfgang Henschel, Peter Haring Bolivar, and Heinrich Kurz 2003 *IEEE Photonics Technology Letters* **15** 921
[5] S. Matsuo, and T. Segawa 2009 *IEEE Selected Topics in Quantum Electronics* **15** 545
[6] Y. Tanushi, and S. Yokoyama 2005 *Group IV Photonics 2nd IEEE International Conference on* 165
[7] Lei Jin, Mingyu Li and Jian-Jun He 2009 *ACP* TuM4
[8] J. Yang, Q. Zhou, F. Zhao, X. Jiang, B. Howley, M. Wang, and R. T. Chen, 2003 *Optics Commuication.* **228**, 91
[9] Hryniewicz J. V., Absil P. P., Little B. E. 2000 IEEE Photonics Technology Letters **12** 320
[10] Xu Q., Sandhu S., Povinelli M. L. 2006 *Physical Review Letters* **96** 123901
[11] S. T. Chu, B. E. Little, W. Pan 1999 *IEEE Photonics Technol. Lett.* **11** 1426
[12] Xiaobei Zhang, Dexiu Huang and Xinliang Zhang 2007 *Optics Express* **15** 13557
[13] S. J. Emelett, R. A. Soref 2005 *Optics Express* **13** 4439
[14] L. Y. Mario, S. Darmawan, M. K. Chin 2006 *Optics Express* **14** 12770
[15] Landobasa Yosef Mario, Desmond C. S. Lim, and Mee-Koy Chin 2007 *IEEE Photonics Technol. Lett.* **19** 1688
[16] Landobasa Y. M. Tobing, Desmond C. S. Lim, Pieter Dumon, Roel Baets, and Mee-Koy Chin 2008 *Applied Physics Lett.* **92** 101122
[17] Yariv A. 2000 *Electronics Letters* **36** 321
[18] David D. Smith, Hongrok Chang, and Kirk A. Fuller 2003 *J. Opt. Soc. Am. B* **20** 1967

Acknowledgments

This work was supported by the National Natural Science Foundation of China (Grant No. 61007035), Shanghai Municipal Education Commission (11YZ01) and Shanghai Leading Academic Discipline Project (S30108).

High quality Bragg gratings fabricated by Nanoimprint Lithography

Yi Wen Zhang[1], Lei Wang[1], Yong Lin Yu[1*],Ding Li Wang[2], Ning Zhou[2] and Wen Liu[1,2]
[1] Wuhan National Laboratory for Optoelectronics, Huazhong Univeristy of Science and Technology, Wuhan, 430074, P.R.China
[2] Accelink Technologies Co. Wuhan, Hubei, China

E-mail: yonglinyu@mail.hust.edu.cn

Abstract. Nanoimprint lithography (NIL) is a very promising technology for nano structure fabrication, because of its high resolution, low cost and high throughput. In this paper, we have demonstrated high-resolution rectangular Bragg gratings for DFB LD by NIL technology. For the manufacture of high quality gratings, the stamp types and residual resist thickness are discussed, and an improved Simultaneous Thermal and UV (STU) imprint process with temperature-pressure variation is introduced.

1. Introduction

With narrow linewidth, high side mode suppression ration (SMSR) and wavelength stability, distributed feedback laser diodes (DFB LD) have been widely adopted in optical communication networks. Efforts to reduce cost of the device and simply its fabrication process have been continued since it invented.

Besides epitaxial process, performances of the devices are mainly depended on the quality of embedded waveguide Bragg gratings. Normally, there are two methods for fabricating waveguide Bragg gratings, interference exposure, electron beam lithography (EBL). Interference exposure can be used to perform uniform gratings, but not suitable to fabricate phase-shifted gratings. Electron-beam lithography (EBL) has sufficient resolution and the capability to fabricate irregular patterns, but it suffers form high cost and low throughput, which prevents this method from mass production. Nanoimprint lithography (NIL) [1] has been introduced in DFB fabrication as a new method to perform gratings with good manufacture quality [2]-[3], low cost and high throughput. Compared with conventional lithographic approaches, which define the patterns by using either photons or electrons to modify the properties of the photoresist, NIL is based on the mechanical embossing principle.

* To whom any correspondence should be addressed.

Published under licence by IOP Publishing Ltd

Journal of Physics: Conference Series **276** (2011) 012104 doi:10.1088/1742-6596/276/1/012104

It relies on direct mechanical deformation of the resist, and therefore can achieve high pattern resolutions beyond the limitations set by light diffractions or beam scattering. Actually, the pattern resolutions during the process are mainly limited by the stamp. NIL has the capability of patterning sub 10 nm features [4]-[5].

In order to enhance the fault tolerance capability for unexpected defects, such as particles or unflatness on the substrate, achieve large area, high uniformity imprint result and avoid damages to the stamp or substrate, usually, soft press process [6]-[7]with an flexible intermediate polymer stamp is introduced. The imprint process is shown in Figure 1.

For NIL technology, the stamps are usually made of some durable hard materials, such as silicon and silica, and the patterns on the stamps are usually of two types, positive and negative. For the first type, the top surface of the pattern is higher than the areas with no patterns, the patterns are anaglyphic; however, the negative patterns are invaginated into the stamp, they are beneath the stamp surface. These two types are both widely used in NIL technology, in this paper, experiments are carried out to make a comparison to find out the very type which is more suitable for NIL technology.

In addition, with the consideration of preventing the stamp and substrate from mechanical damages, stamp will not contact with the substrate during the imprint process, illustrated in Figure 1(b), they are separated by some residual resist. Too thick residual resist will hamper subsequent pattern etching process, as when the residual resist is removed by reactive ion etching (RIE) process, the pattern mask will be degraded. Continual efforts [8]-[9] have been trying to thin the residual resist, herein, an advanced process with temperature-pressure variation [10]-[11] is introduced.

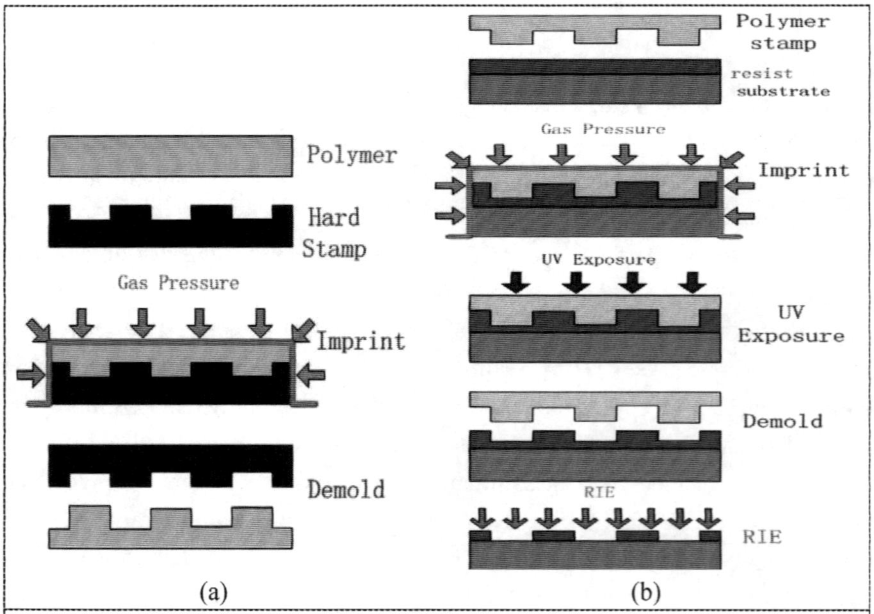

Figure 1. Schematic diagram process flow of nanoimprint with an intermediate polymer stamp: (a) fabrication of a polymer stamp, (b) imprint with the polymer stamp and STU process

2. Experiment

2.1. Stamps

Two fabricated stamps have nearly the same patterns, are both Bragg gratings for DFB LDs, however, one stamp is negative, the other is positive. On the stamps, there is a Bragg grating structure for one DFB LD in each 250*250 um^2 area. The gratings have the same period, and the depth of the negative patterns is about 124-130 nm, however, the height of the gratings is 60 nm on the positive stamp. Before the imprint process, an anti-adhesion layer coating is deposited on the stamps by immersing them in a tridecafluoro-1, 1, 2, 2-tetrahydrooctyl trichlorosilane (TFS) [12] solution.

To compare these two stamps, in this experiment their patterns are transferred onto intermediate polymer stamps first, secondly, polymer stamps are imprinted on thick resist with the process shown in Figure 2(a), imprint time varies from 1 to 30 minutes, and then the pattern quality is inspected by AFM (Veeco NanoScope MultiMode) and SEM (Quanta 200), the yield is computed under Olympus BX51 optical microscope.

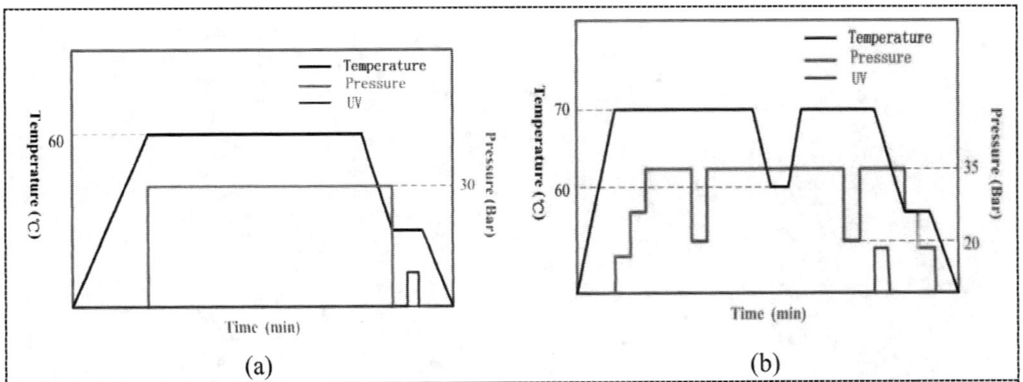

Figure 2. Schematic illustrations of (a) conventional NIL process and (b) NIL process with temperature-pressure variations

2.2. Residual resist

During the imprint process, to ensure full filling of the stamp and prevent the stamp or substrate from damages, the resist should have sufficient thickness, normally, the thicker the resist, the better the pattern quality. However, with the consideration of subsequent pattern transfer process, the resist should be as thin as enough to guarantee a thin layer of residual resist. Compared with conventional process, in this study we introduce an improved NIL process, shown in Figure 2(b), to obtain a very thin residual resist layer, and ensure good imprinted pattern quality at the same time.

The viscoelastic properties of the resist has been studied [13], and the relationship between temperature, pressure and the fractional free volume of the resist can be expressed in

$$f = f_g + \alpha(T - T_g) + \beta(P - P_g) \qquad (1)$$

Where f is fractional free volume at temperature T and pressure P, f_g represents fractional free volume at glass transition and demold pressure, α indicates the thermal expansion coefficient and β is compressibility coefficient. As shown in Figure 2(b), with the variations of the temperature and pressure, the fluidity and filling ability of the resist can be enhanced, thus thickness of the resist that needed during the imprint process can be thinned, and the residual resist thickness is reduced. Additionally, at the both ends of the process, the step variation of the pressure can reduce the distortion of the polymer stamp that may induced by large sudden pressure variations.

In this experiment, we compared the improved NIL process with a conventional one, the imprint time of the latter process is 3 minutes. The negative Bragg grating stamp is employed. The thickness of spun resist is varied from 110 nm to 160 nm, with 10 nm as an interval.

3. Results and Discussions

3.1. Stamps

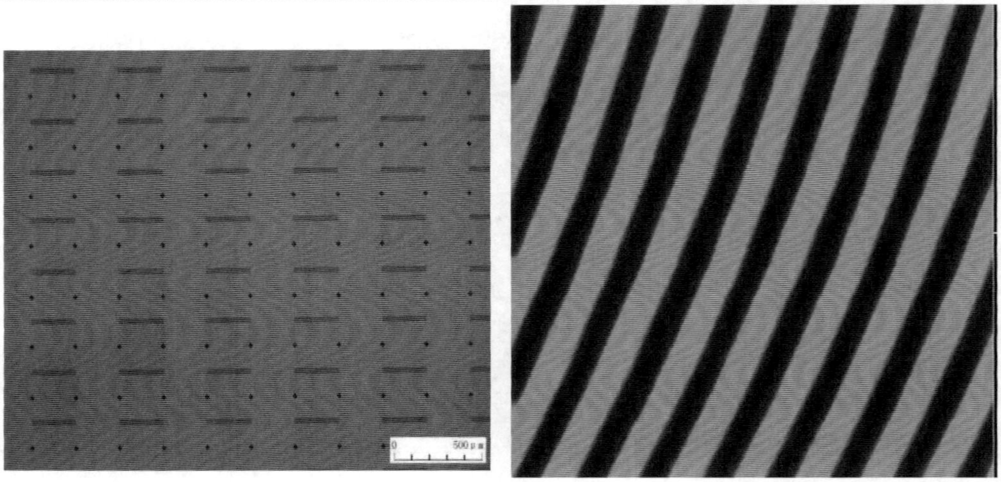

(a) (b)

Figure 3. (a) Microscope image of large area imprinted gratings, (b) AFM 2D image of Bragg grating structure. The stamp is negative, and the imprint time is 3 minutes.

As shown in the results illustrated in Figure 3, the negative stamp is easy to be filled by the resist, very good grating quality and high yield (higher than 80%) can be obtained in only 3 minutes. However, when it comes to the positive stamp, though the height of the patterns is nearly one half of the patterns in negative stamp, the filling behavior is very slow, indicated by Figure 4(a)-(c), the grating patterns can not be fully filled within 10 minutes, actually from Figure 4(d), it takes nearly 30 minutes to ensure a complete filling.

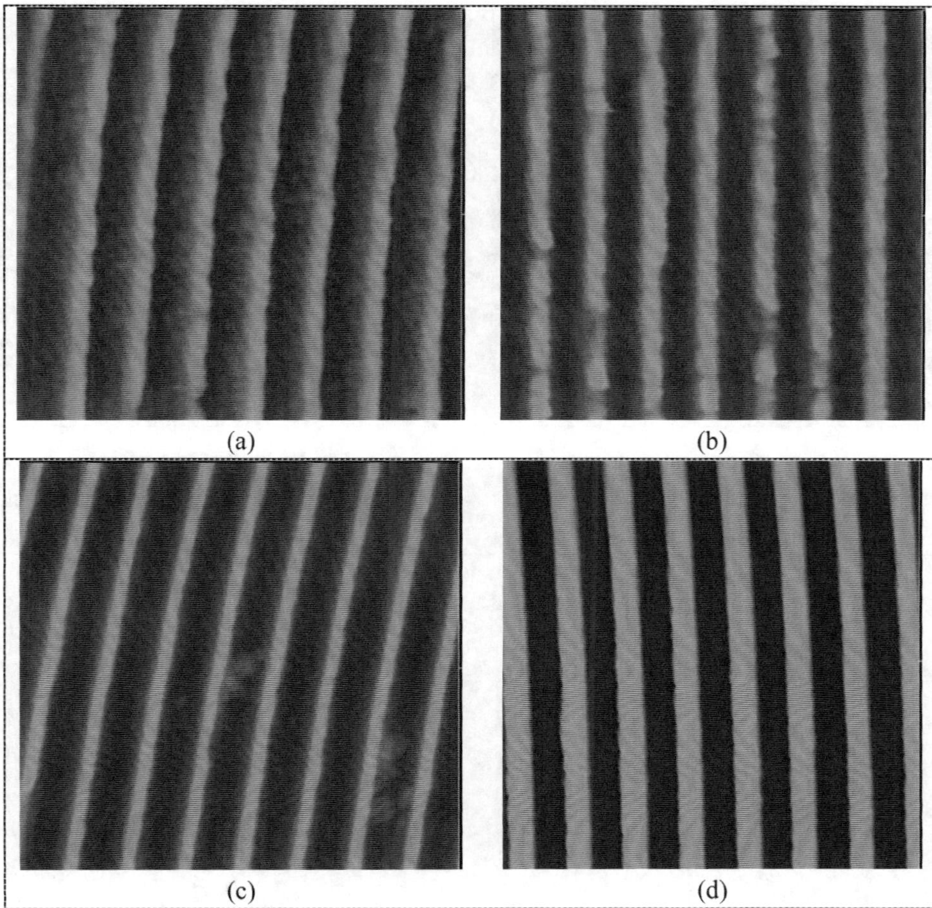

Figure 4. AFM images of imprint results under different time with positive stamp: (a) 3 min, (b) 5 min, (c) 10 min and (d) 30 min.

3.2. Residual resist

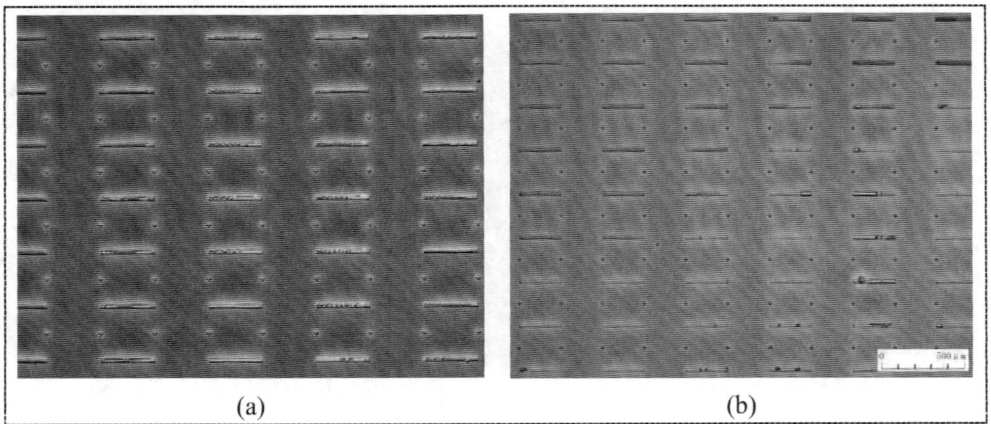

Journal of Physics: Conference Series **276** (2011) 012104 doi:10.1088/1742-6596/276/1/012104

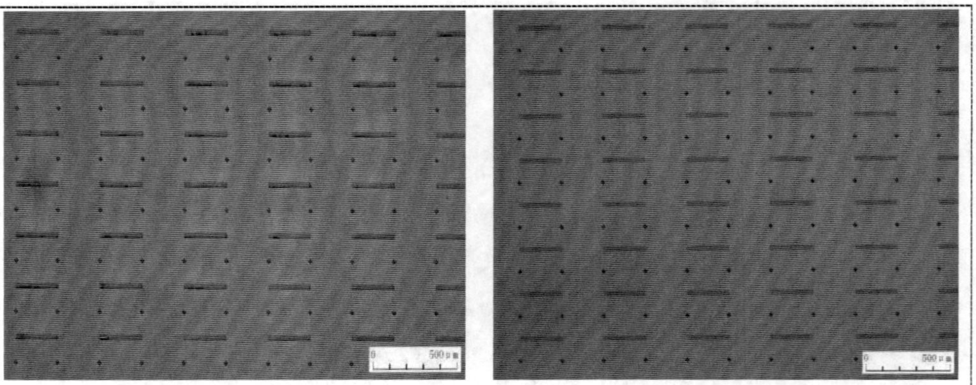

(c) (d)

Figure 5. Microscope images of large area gratings imprinted with varied resist thickness: (a) 120 nm, (b) 130 nm, (c) 140 nm and (d) 160 nm. The negative stamp is imprinted with conventional process, and the imprint time is 3 min.

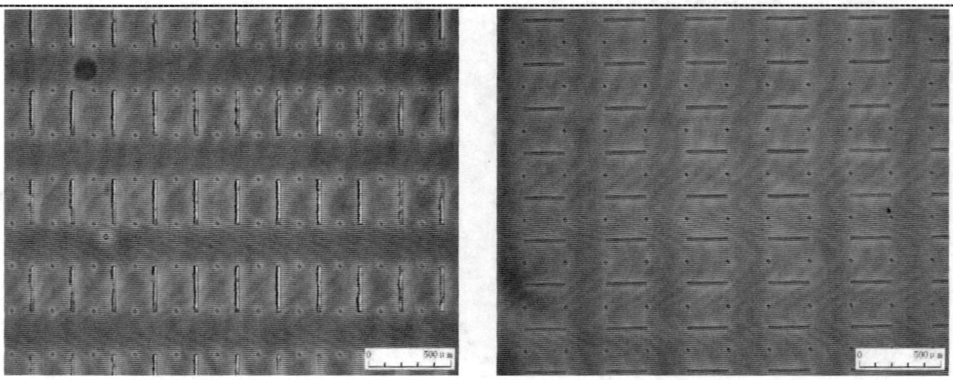

(a) (b)

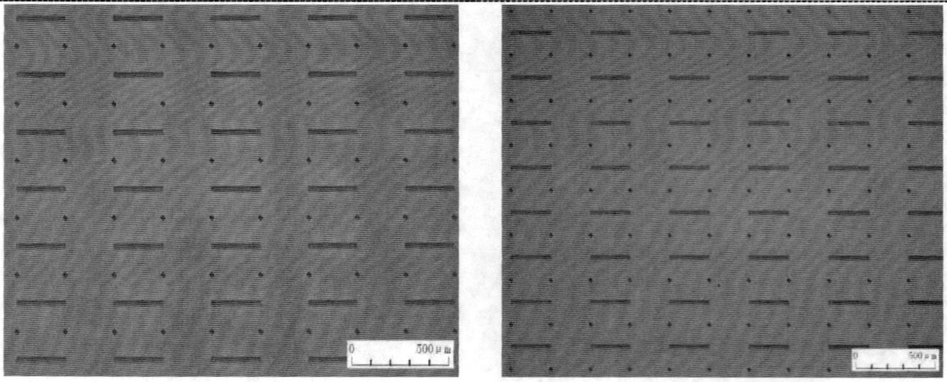

(c) (d)

Figure 6. Microscope images of large area gratings imprinted with varied resist thickness: (a) 110 nm, (b) 120 nm, (c) 130 nm and (d) 140 nm. The negative stamp is imprinted with improved process with temperature-pressure variations.

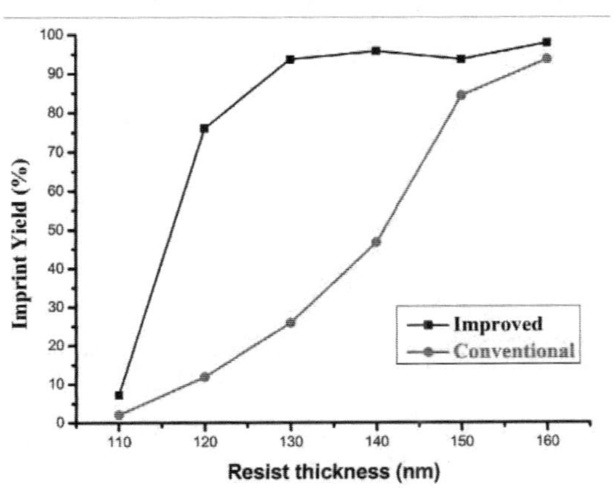

Figrue 7. Normalized statistical yield results of two different NIL process.

Shown in Figure 5 and Figure 7, when imprinted with conventional process, to obtain a good pattern quality and high yield, the resist should be thicker than 150 nm, which means the residual resist thickness is nearly 30 nm. For the improved process, illustrated in Figure 7 and Figure 7, nevertheless, despite the resist thickness is only 130 nm, the imprinted grating is already uniform and straight, the yield is higher than 90%. With the residual resist thinner than 10 nm, patterns can be easily transferred to the substrate by a reactive ion etching (RIE) process with high quality, shown in Figure 8. This result is owning to the temperature-pressure variations during the NIL process, which enhance the fluidity and filling effects of the resist, then the necessary resist thickness can be reduced, and so do the residual resist.

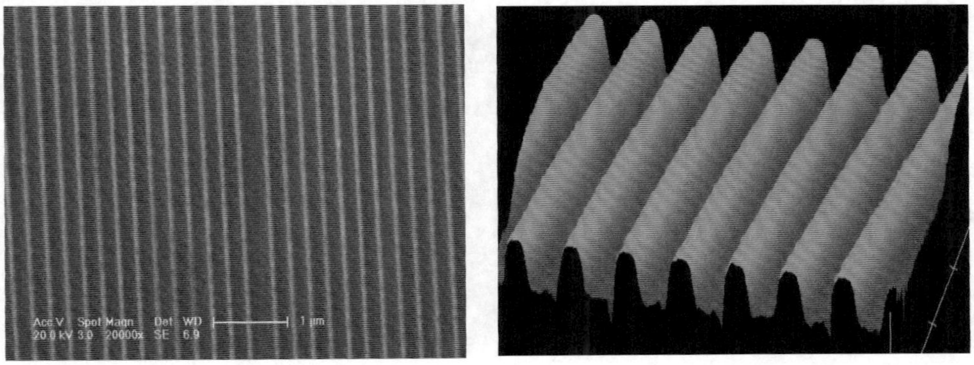

(a) (b)

Figure 8. Schemes of the gratings transferred onto substrates: (a)SEM, (b) AFM 3D.

4. Conclusions

In this paper, quality of Bragg gratings fabricated by Nanoimprint Lithograpthy are studied, with a focus on NIL stamp and residual resist thickness From the results, compared with positive stamps, negative ones are easier to be filled, the imprint time can be very short, consequently, negative stamps are more suitable for soft NIL process. In addition, an improved NIL process is introduced, the temperature-pressure variations can enhance the fluidity and filling effects of the resist, then reduce the necessary resist and subsequent residual resist thickness. With a layer of residual resist thinner than 10 nm, high quality rectangular Bragg gratings are transferred onto substrates by a reactive ion etching (RIE) process.

References

[1] Chou S Y, Krauss P R and Renstrom P J 1995 *Appl. Phys. Lett.* **67** 3114.

[2] Yanagisawa M, Tsuji Y, Yoshinaga H, Kono N and Hiratsuta K 2009 *Japan. J. Appl. Phys.* **48** 06FH11.

[3] Wang L, Zhang Y W, Qiu F, Zhou N, Wang D L, Xu Z M, Zhao Y L, Yu Y L and Liu W 2010 *Front. Opto. China* **3** 1674.

[4] Chou S Y, Krauss P R, Zhang W, Guo L J and Zhuang L 1997 *J.V.S.Technol.B* **15** 2897-2904.

[5] Schvartzman M and Wind S J 2009 *Nano Lett.* **9** 3629-3634.

[6] Gao H, Tan H, Zhang W, Morton K and Chou S Y 2006 *Nano Lett.* **6** 2438-2441.

[7] Obducat AB USA Patent US 7082876.

[8] Cheng X and Guo L J 2004 *Micro. Engineering* **71** 177-282.

[9] Liao W-C and Hsu S L-C 2007 *Nanotechnology* **18** 065303.

[10] Park H S, Shin H H, Sung M Y, Choi W B, Choi S W and Park S Y 2007 *T. S. Manufature*, **20** 13-19.

[11] Wang L, Liu W, Zhang Y W, Qiu F, Wang D L, Zhou N and Xu Z M, 2010 *A. Opti. Manu. Test. Tech.* 7657.

[12] Beck M, Graczyk M, Maximov I, Sarwe E-L, Ling T G I, Keil M and Montelius L 2002 *Micro. Engi.* **61-62** 441-448.

[13] Ferry J D, Viscoelastic Properties of Polymers, 3rd ed. New York: Wiley, 1980

Directed XOR/XNOR logic circuit implemented by microring resonators: simulation and demonstration

Lei Zhang, Ruiqiang Ji, Lianxi Jia, Lin Yang[1], Ping Zhou, Yonghui Tian, Ping Chen, Yangyang Lu, Zhenyu Jiang and Yuliang Liu

E-mail: oip@semi.ac.cn

Optoelectronic System Laboratory, Institute of Semiconductors, Chinese Academy of Sciences, Beijing 100083, China

Abstract. A directed logic circuit consisting of two silicon microring resonators which can perform XOR and XNOR operations has been proposed. These two operations are evaluated as the optical signal is directed along the circuit, and the results appear at some specific output ports of the circuit. A potential advantage of this new logical paradigm is that it has markedly less state delay than traditional logic circuits. Devices based on the above proposal are fabricated on an 8-inch silicon-on-insulator wafer. The microring resonators in the circuit are modulated through thermo-optic effect. Two electrical modulating signals applied to the microring resonators represent the two operands of the logical operation respectively. Bitwise XOR and XNOR operations at 20 kbit/s are demonstrated. The scattering matrix method is adopted to analyse the transmission spectra of the circuit. The symmetrical characteristics of the spectra are well described by this model. The potential applications of this device are given.

1. Introduction

Directed logic is a newly proposed logical paradigm in which the Boolean functions are calculated through the directed propagation of light [1–3]. Contrast to traditional implementation of optical logic, the directed logic is specially adapted to the features and promises of optics since it depends on the propagation of light other than the nonlinear interactions between light and materials. Directed logic circuit is a network consisting of several simple elements, each of which performs switching function on its input optical signal. Which operation to be performed is determined by a separate Boolean input applied to the corresponding element. The cumulative effect of the switching operations of all the elements yields the value of the function in question. In other words, the Boolean function is evaluated as the optical signal is directed along the switching network and the result appears at some specific output port. Compared to traditional digital logic circuit, directed logic has markedly less state delay because the inputs that determine the states of the switching elements do not pass through the preceding elements in the circuits. Therefore, all elements can perform their switching functions simultaneously and the results are given instantaneously [1, 2].

As pointed out in [1], the XOR operation is indispensable in all encoding and decoding schemes such as label processing, parity checking, data encryption/decryption, and pseudo-random binary sequence (PRBS) generation. Here we report a directed logic architecture consisting of two cascaded

[1] To whom any correspondence should be addressed.

microring resonators (MRRs) which can perform XOR and XNOR operations. The switching functions of the proposed directed logic circuit are performed by MRRs, which are widely used to function as filters [4], modulators [5], switches [6], and optical logic gates [7–9]. The XOR logic gate introduced here can be regarded as a newly-added item to the function library of MRRs and it can benefit from the potential high operation speed and low power consumption features of the MRR modulator [10–11].

2. Principle of the proposed logic circuit

The proposed architecture is schematically shown in figure 1. A continuous optical wave with the wavelength of λ is modulated by an electrical pulse train X using MRR 1 (we regard the electrical pulse train X as a sequence of Boolean *0s* and *1s*, taking no account of their actual amplitudes here for simplicity), the optical pulse trains that appear at the through and drop ports of MRR 1 are $\overline{X}$ and X respectively ($\overline{X}$ is the bitwise inverse of X). It should be noted that the input optical wave is assumed to be directed to the drop port of MRR 1 when the electrical modulating signal is *1*. In other words, the MRR 1 resonates at the wavelength of λ when the applied electrical signal is at high level. So does the MRR 2.

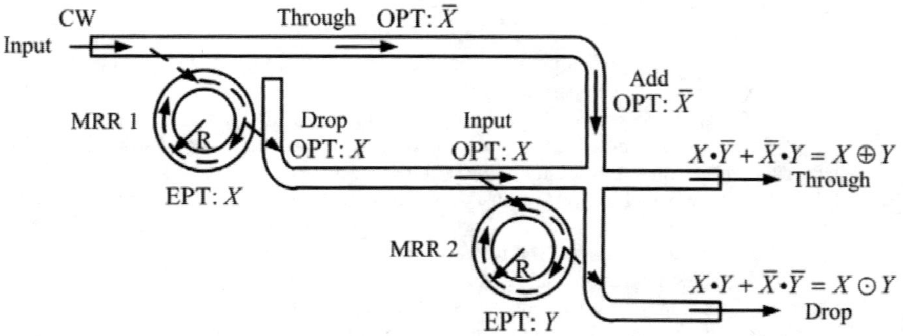

Figure 1. The proposed directed logic circuit as an XOR and XNOR calculator (CW: Continuous Wave; EPT: Electrical Pulse Train; OPT: Optical Pulse Train).

If an optical pulse train X is fed into MRR 2 from its input port solely and is modulated by an electrical pulse train Y bit-by-bit (here we assume that no signal is fed into MRR 2 from its add port), the optical pulse trains that appear at the through and drop ports of MRR 2 will be $X{\bullet}\overline{Y}$ and $X{\bullet}Y$, respectively, also taking the aforementioned assumption (the input optical wave is directed to the drop port of MRR 2 when the electrical modulating signal is *1*). The symbol $\bullet$ represents the logical operation *and* here.

If an optical pulse train $\overline{X}$ is fed into MRR 2 from its add port solely and is modulated by an electrical pulse train Y bit-by-bit (here we assume that no signal is fed into MRR 2 from its input port), the optical pulse trains that appear at the through and drop ports of MRR 2 will be $\overline{X}{\bullet}Y$ and $\overline{X}{\bullet}\overline{Y}$, respectively. If two optical pulse trains X and $\overline{X}$ are fed into MRR 2 from its input and add ports simultaneously and both of them are modulated by an electrical pulse train Y synchronously, we can obtain two optical pulse trains $X{\bullet}\overline{Y}+\overline{X}{\bullet}Y$ and $X{\bullet}Y+\overline{X}{\bullet}\overline{Y}$ at the through and drop ports of MRR 2, respectively, as shown in figure 1. The symbol $+$ represents the logical operation *or*, which is actually implemented by the multiplexing function of MRR 2.

It is well known in Boolean algebra that the XOR and XNOR operations of two logical variables X and Y can be calculated by the formulas $X\oplus Y=X{\bullet}\overline{Y}+\overline{X}{\bullet}Y$ and $X\odot Y=X{\bullet}Y+\overline{X}{\bullet}\overline{Y}$, where the symbols $\oplus$ and $\odot$ represent the XOR and XNOR operators respectively. Therefore, the proposed circuit shown in figure 1 can be used as an XOR/XNOR calculator.

3. Fabrication and characterization of the device

The working principle shown in the last section indicates that we can execute the XOR and XNOR operations simultaneously using the circuit in figure 1. Here in this part, we introduce the fabrication and characterization of the device. We adopt the silicon platform to fabricate the device due to its compact size, low power consumption and potential high speed [10–11]. The characterization procedure is composed of the static and the dynamic test. The former is for determining the working points, including the working wavelength and the tuning voltage representing logical *1* for each MRR. And the latter is for verifying the dynamic operation of the circuit as an XOR/XNOR calculator.

3.1. Fabrication of the device

The device schematically shown in figure 1 is fabricated on an 8-inch silicon-on-insulator wafer with 220-nm-thick top silicon (Si) layer and 2-μm-thick buried dioxide (BOX) layer. The cross-section of Si waveguides is 400×220 nm^2. The radii of the ring waveguides are both 10 μm. Gaps between the ring and the straight waveguides are 400 nm. Two 200-nm thick titanium micro-heaters are employed to tune the MRRs thermally, as shown in the inset of figure 2. Aluminium wires with the width of 50 μm and pads with the size of 150×150 μm^2 are fabricated after the heaters are done. The micrograph of the device is shown in figure 2, which has an effective area of about 1.5×0.6 mm^2. Details on the fabrication process are given in [12].

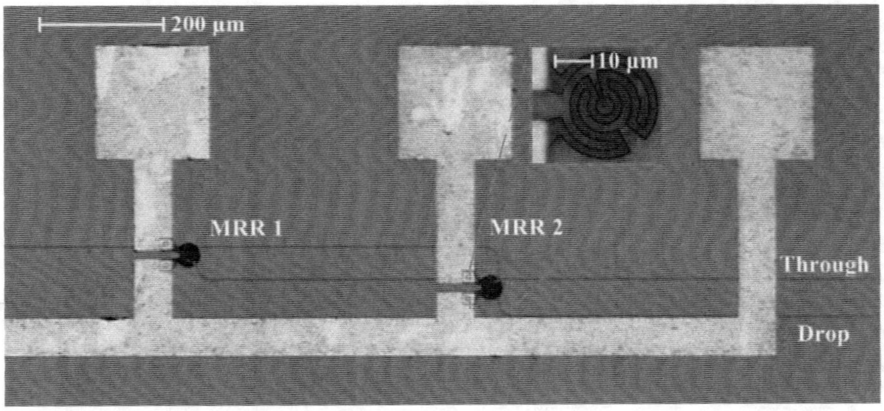

Figure 2. The micrograph of the fabricated device, with the heater shown in the inset.

3.2. Characterization of the device

The characterization of the device consists of static test and dynamic test. The former is for determining the working points and the latter is for verifying the dynamic operation of the circuit. A lensed single-mode fiber is used to couple light from an amplified spontaneous emission (ASE) source or a tunable laser source into the chip. Another lensed fiber is used to couple light from the chip to an optical spectrum analyser (OSA) or a photodiode for detection.

3.2.1. *Static test for determining the working points.* The broadband light is coupled into the device through the input lensed fiber. The output light from the through port of the circuit is collected by the output lensed fiber and fed into the OSA. The same procedure is used to obtain the spectrum of the drop port. Results of the foregoing tests are shown in figure 3. After obtaining the static spectra of the two output ports, two tunable voltage sources are adopted to provide voltages to the two heaters, respectively. When a MRR is heated up, the effective refractive index of the ring waveguide increases, and it results in a redshift of the resonant wavelength of the corresponding MRR.

As we get the highest extinction ratio in the region near 1536 nm when no electrical signals are applied (see figure 3), the working wavelength will be chosen in this region, which is enlarged to shown in figure 4(a) and figure 6(a). According to the principle aforementioned, a minimum (representing a logical *0* output) should be given at the through port when two applied electrical signals are both at low level (representing two logical *0* inputs) or high level (representing two logical *1* inputs). 1538.072 nm is chosen as the working wavelength because a minimum appears there when the electrical inputs are both 0 V (see figure 4(a)). With an appropriate voltage applied to corresponding MRR alone (4.70 V for MRR 1 and 5.04 V for MRR 2), it resonates at 1538.072 nm and a maximum appears there, as shown in figure 4(b) and 4(c). With two voltages of 4.70 V and 5.04 V applied to the corresponding MRRs simultaneously, a minimum appears at 1538.072 nm again, as shown in figure 4(d). In summary, voltages of 4.70 V and 5.04 V are applied to MRR 1 and MRR 2 respectively when the operands of the XOR operation are *1s*, and no voltages are applied to them when the operands are *0s*.

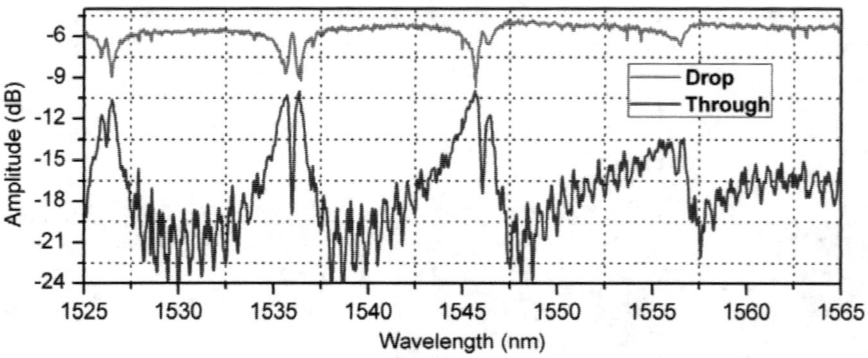

Figure 3. The transmission spectra obtained from the through and drop ports of the circuit.

We can find from figure 4(b) and 4(c) that the two MRRs without applied voltages have different resonant wavelengths (1536.188 nm for MRR 1 and 1535.898 nm for MRR 2). Such a slight difference mainly originates from the fabrication errors, which have no effect on the operation of the device except for the different tuning voltages required for two MRRs.

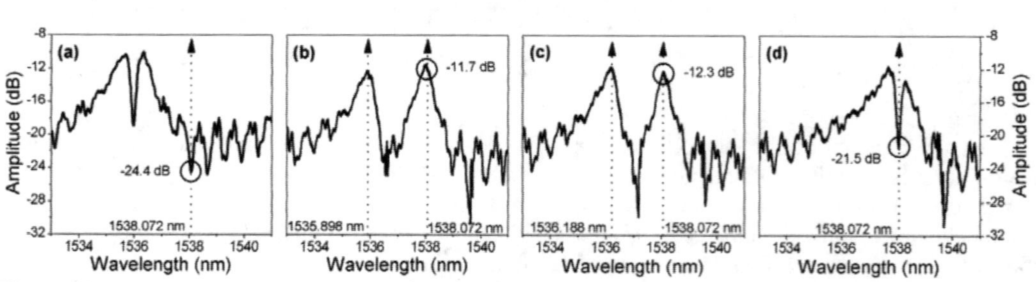

Figure 4. Response spectra at the through port of the circuit. Voltages applied to MRR 1 and MRR 2 are (a) both 0 V; (b) 4.70 V and 0 V; (c) 0 V and 5.04 V; (d) 4.70 V and 5.04 V, respectively.

3.2.2. *Dynamic XOR operation at the through port.* The dynamic performance of the device is shown in figure 5(a-c). Two pseudo-random binary sequence 2^4-1 non-return-to-zero signals with a data rate of 20 kbit/s are converted to two analog voltage signals bit-by-bit according to the rule presented above and then applied to the corresponding MRRs. Clearly, a logical *0* output is obtained when the two applied signals are both at high level or both at low level. And a logical *1* output is obtained otherwise. It indicates that the device performs the XOR operation correctly at the through port.

3.2.3. *Dynamic XNOR operation also at the through port.* According to the principle aforementioned, the XNOR operation can be achieved at the drop port at the same time when XOR operation is carried out at the through port. But due to the structure of waveguides we choose, the fabricated device has a low extinction ratio (about 3 dB as shown in figure 3) at the drop port, which makes the logical operations at the drop port inefficient. Nevertheless, we can achieve XNOR operation at the through port as well, just by choosing different working wavelength and tuning voltages corresponding to logical *1* inputs for two MRRs.

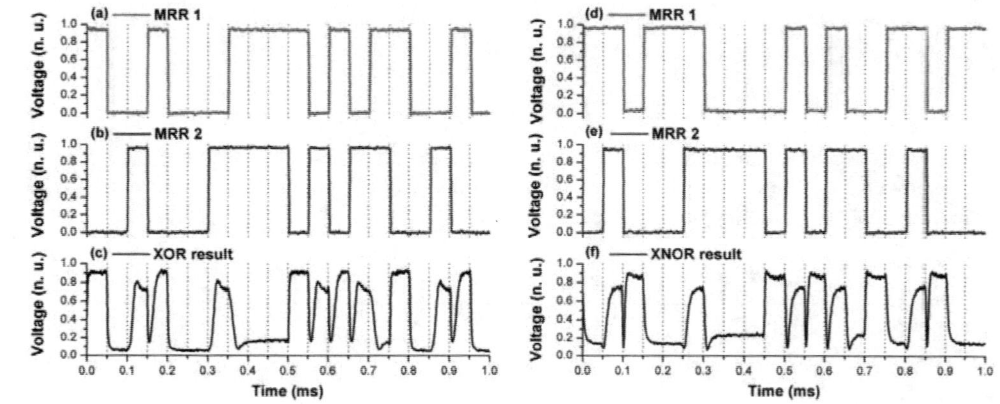

Figure 5. Signals applied to two MRRs and detected at the through port of MRR 2 when executing the XOR operation (a-c) and the XNOR operation (d-f). Signals applied to (a) MRR 1 and (b) MRR 2 and (c) the result of XOR operation at the through port. Signals applied to (d) MRR 1 and (e) MRR 2 and (f) the result of XNOR operation at the through port as well.

The working wavelength is still chosen from the region around 1536 nm. According to the principle of the XNOR operation, a maximum should be given when two operands are both *0s*.1536.335 nm is chosen as the working wavelength since a maximum appears there when the electrical inputs are both 0 V (see figure 6(a)). With a voltage of 3.58 V applied to MRR 1 alone, it is tuned to resonate far away from 1536.335 nm and a minimum appears there (see figure 6(b)). With a voltage of 2.58 V applied to MRR 2 alone, it resonates a little to the right of 1536.335 nm. The signal downloaded to the drop port of MRR 1 is further downloaded to the drop port of MRR 2 and therefore a minimum is achieved at the through port (see figure 6(c)). With two voltages of 3.58 V and 2.58 V applied to two corresponding MRRs simultaneously, a maximum appears at 1536.335 nm as shown in figure 6(d). The dynamic performance of the XNOR operation is shown in figure 5(d-f). It shows that the XNOR operation is carried out correctly at the through port as well.

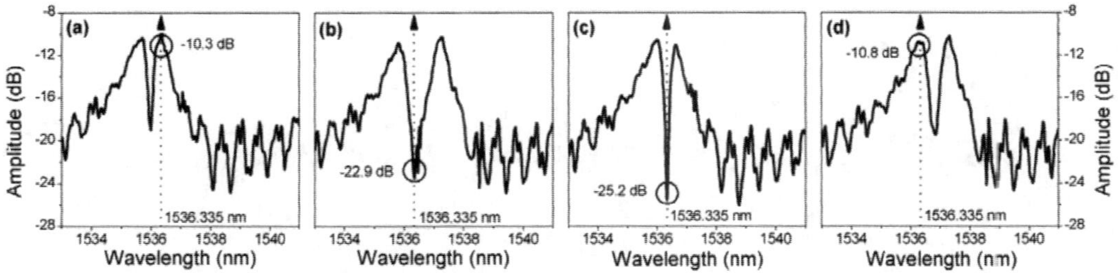

Figure 6. Response spectra at the through port of the circuit. Voltages applied to MRR 1 and MRR 2 are (a) both 0 V; (b) 3.58 V and 0 V; (c) 0 V and 2.58 V; (d) 3.58 V and 2.58 V respectively.

4. Scattering matrix model and spectrum evolution

The transmission spectra of the proposed structure are obtained using the scattering matrix method [13]. The whole device can be classified into three subsidiary sets according to their functions [3], as shown in figure 7. The first set consists of four coupling areas which cause energy exchanges between the straight waveguides and the ring waveguides. These coupling areas are described by four scattering matrixes, whose elements are assumed to be independent of wavelength [14]. The second set consists of two ring waveguides and two straight waveguides, which cause phase shifts and attenuations to the optical fields passing through them. The third set is the crossing of two straight waveguides, which possibly causes scattering when light passes through it. The scattering matrix describing the energy exchange in coupling area I is in the form of

$$\begin{pmatrix} E_{p1} \\ E_{r2} \end{pmatrix} = \begin{pmatrix} t_1 & k_1 \\ -k_1^* & t_1^* \end{pmatrix} \begin{pmatrix} 1 \\ E_{r1} \end{pmatrix} \tag{1}$$

where the complex mode amplitudes are normalized such that their squared magnitudes correspond to the modal powers. The input wave is chosen to be *1* so that all the field amplitudes are normalized to it. We assume that the four coupling areas are identical.

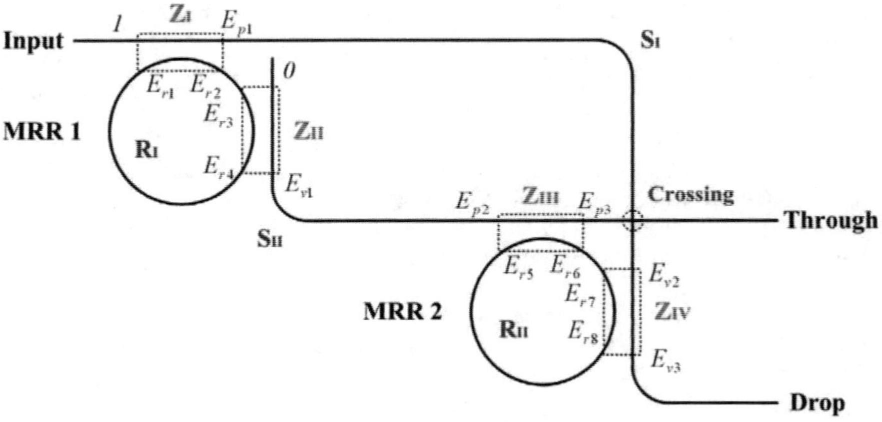

Figure 7. Scattering matrix model of the proposed circuit (Z_I~ Z_{IV}: coupling areas I~IV; S_I~S_{II}: straight waveguides I~II; R_I~ R_{II}: ring waveguides I~II).

The simulation result of the transmission spectra of the through and drop ports of the device based on the above model is shown in figure 8, which agrees well with the experimental result as shown in figure 3. The scattering at the crossing results in the ripples in the transmission spectra of the device.

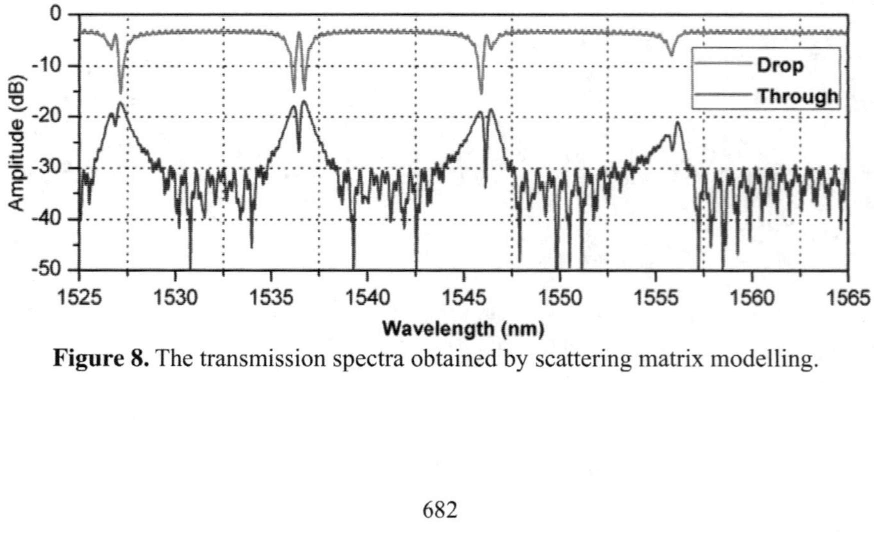

Figure 8. The transmission spectra obtained by scattering matrix modelling.

It can be noted in both figures 3 and 8 that the spectra show mirror symmetry, with the axis of symmetry located around 1536 nm. This is not caused by the radius deviation of the two MRRs, since we can still observe such kind of symmetry even when we adopt two identical radii in the numerical model.

Further investigation shows that such a property of mirror symmetry is related to the length difference between the two straight waveguides S_I and S_{II}.

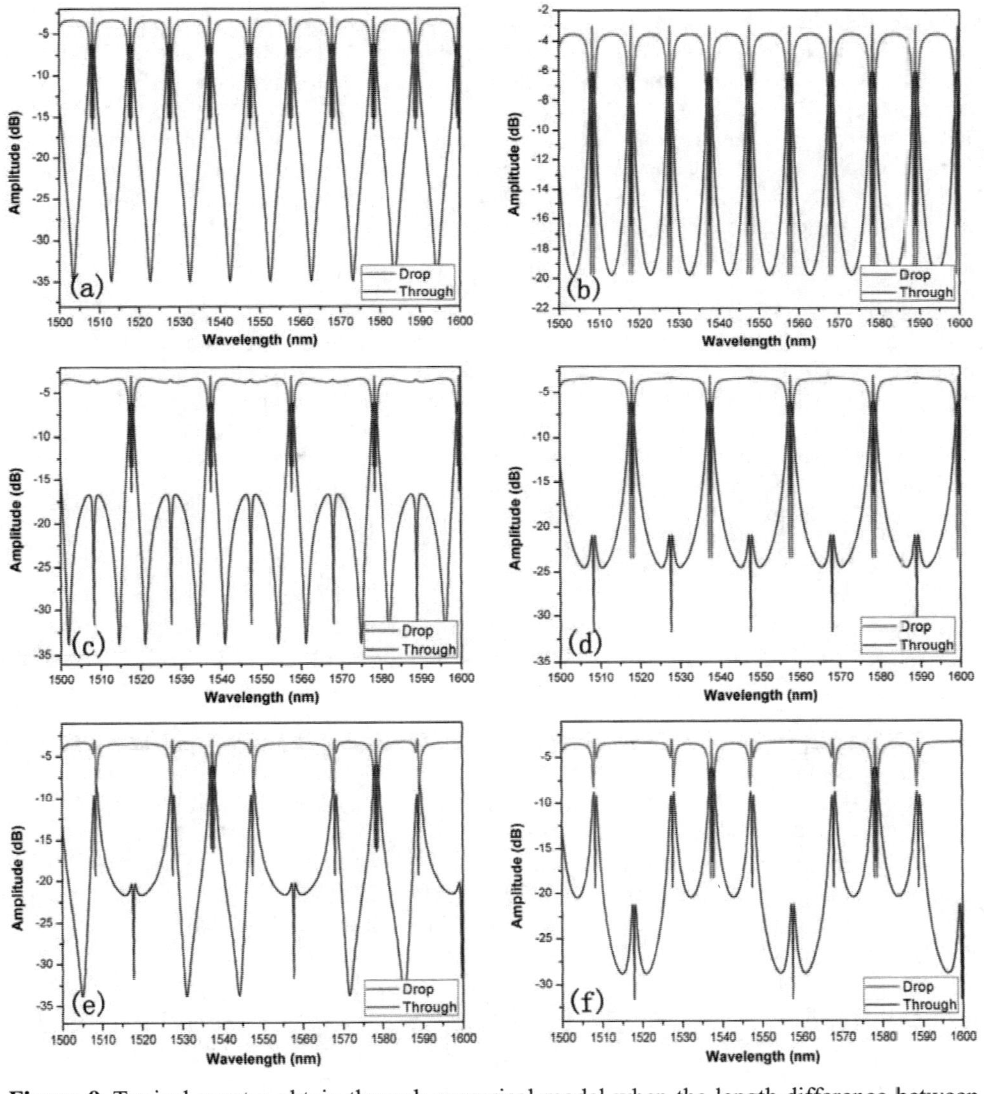

Figure 9. Typical spectra obtain through numerical model when the length difference between S_I and S_{II} are (a-b) odd multiple of one-half the circumference of the ring waveguide; (c-d) even multiple of one-half the circumference of the ring waveguide; (e-f) other values.

When the length difference between S_I and S_{II} is odd multiple of one-half the circumference of the ring waveguide, the spectra of the circuit show translational symmetry (see figure 9 (a) and (b)). In figure 9 (a) and (b), the length differences between S_I and S_{II} are $C/2$ and $-C/2$ respectively, where

C means the circumference of the ring waveguide. In such situations the periods of the spectra equal the free spectral range (FSR) of the MRR. When the length difference between S_I and S_{II} is even multiple of one-half the circumference of the ring waveguide, the spectra of the circuit show translational symmetry too (see figure 9 (c) and (d)), but with a period of 2*FSR. In figure 9 (c) and (d), the length differences between S_I and S_{II} are C and $-C$ respectively. In all other situations where the length difference between S_I and S_{II} is not integral multiple of one-half the circumference of the ring waveguide, the spectra show mirror symmetry (see figure 9 (e) and (f)), as demonstrated in figure 3 and figure 8. In figure 9 (e) and (f), the length differences between S_I and S_{II} are $C/4$ and $-C/4$ respectively. It should be noted here that the aforementioned mirror symmetry is a localized property. When we view the spectra in a wider wavelength region, translational symmetry will turn up again, with a much larger period than FSR.

Such a structure (see figure 1 and figure 7) can be regarded as a Mach-Zehnder interferometer (MZI) as a whole, while using the microring resonators as the splitting and combining units. In traditional MZI, only the behaviour of the two arms is wavelength-dependent. But for the structure presented in this paper, the splitting and combining units are also wavelength-dependent. That is why it shows such varied spectra in figure 9.

5. Conclusion

Directed logic circuit which can carry out the XOR and XNOR operations is proposed and experimentally demonstrated. The symmetry of the spectra is well described by the scattering matrix model. In order to achieve low latency of the directed logic circuit, the cavity photon lifetime should be minimized. An alternative way is to choose other kind of switching elements instead of resonators. This device can be used as a modulator with the function of scrambling to facilitate the timing recovery circuit of the receiver by eliminating long sequences of 0 or 1 only. It can also be regarded as a modulator with the function of XOR encryption.

References

[1] Hardy J and Shamir J 2007 *Opt. Express* **15** 150–165
[2] Zhang L, Ji R Q, Jia L X, Yang L, Zhou P, Tian Y H, Chen P, Lu Y Y, Jiang Z Y, Liu Y L, Fang Q and Yu M B 2010 *Opt. Lett.* **35** 1620–1622
[3] Zhang L, Ji R Q, Jia L X, Yang L, Zhou P, Tian Y H, Chen P, Lu Y Y, Jiang Z Y, Liu Y L, Fang Q and Yu M B 2010 *Proc. 7th Int. Conf. on Group IV Photonics (Beijing, China, 1–3 September)* p ThD1
[4] Little B E, Chu S T, Haus H A, Foresi J and Laine J P 1997 *J. Lightwave Technol.* **15** 998–1005
[5] Xu Q F, Schmidt B, Pradhan S and Lipson M 2005 *Nature* **435** 325–327
[6] Vlasov Y, Green W M J and Xia F N, 2008 *Nat. Photon.* **2** 242–246
[7] Van V, Ibrahim T A, Absil P P, Johnson F G, Grover R and Ho P T 2002 *IEEE J. Sel. Top. Quantum Electron.* **8** 705–713
[8] Caulfield H J, Soref R A and Vikram C S 2007 *Photon. Nanostr. Fundam. Appl.* **5** 14–20.
[9] Xu Q F and Lipson M 2007 *Opt. Express* **15** 924–929
[10] Zhang L, Li Y, Song M, Yang J Y, Beausoleil R G and Willner A E 2009 *Appl. Phys. A* **95** 1089–1100
[11] Dong P, Liao S, Feng D, Liang H, Zheng D, Shafiiha R, Kung C C, Qian W, Li G, Zheng X, Krishnamoorthy A V and Asghari M 2009 *Opt. Express* **17** 22484–22490
[12] Geng M M, Jia L X, Zhang L, Yang L, Chen P, Wang T and Liu Y L 2009 *Opt. Express* **17** 5502–5516
[13] Yariv A 2000 *Electron. Lett.* **36** 321–322
[14] Heebner J, Grover R and Ibrahim T 2008 *Optical microresonators: theory, fabrication, and applications* (London: Springer) p 75

Journal of Physics: Conference Series **276** (2011) 012106 doi:10.1088/1742-6596/276/1/012106

A novel burst-mode all-optical wavelength converter based on gain-clamping structure

B Zhao[1,2,3], C J Ke[1,2,3,4], Y Zhao[1,2,3] and D M Liu[1,2,3]
[1] College of Optoelectronic Science and Engineering, Huazhong University of Science and Technology, 430074, Wuhan, China
[2] Wuhan National Laboratory for Optoelectronics, 430074, Wuhan, China
[3] National Engineering Laboratory for Next Generation Internet Access System, 430074, Wuhan, China

Abstract. A novel gain-clamped wavelength converter (GCWC) is proposed to make the hybrid TDM-WDM PON architecture based on wavelength conversion feasible, and its operation principle is also demonstrated. The power budget of this network is measured to identify the architecture can support 4096 users who share 40Gbit/s accessing capacity on a single trunk fiber. And optical power equalization functionality and burst transparency characteristic of GCWC is experimentally evaluated, and the results show that the GCWC makes burst signals from different transmitters with up to 5 dB dynamic range equalized and is transparency for burst signals.

1. Introduction

Passive optical network (PON) is one of the most promising optical access network architectures in terms of cost-effectiveness. Time division multiplexing (TDM) PON offers low per-subscriber cost by sharing a single wavelength channel with multiple subscribers. However, it inevitably sacrifices per-subscriber bandwidth. The increasing demand for new high bandwidth services, such as VoIP, HDTV and cloud computing, will drive the enhancement of its cost-effective features. Wavelength division multiplexing (WDM) PON can provide large transmission capacity, network security and flexibility. However, the issue of the cost-effective optical source and arrayed waveguide grating (AWG) impedes a commercial WDM-PON deployed.

Recently, there has been increasing interest in hybrid PON (HPON) architecture that aggregates a number of TDM-PONs by WDM to share the same trunk fiber, since the costs is reduced by sharing the deployed infrastructure fiber among more customers[1-3]. Obviously, HPON is suitable for near future deployment in pragmatic migration from current TDM-PON to WDM-PON. All-optical wavelength converter (AOWC), as one of the key technologies employed in several reported HPON architectures, converts the various wavelengths of ONU transmitters to a stable wavelength allocated for upstream. In consequence, the colorless ONUs is feasible in the AOWC based HPON system, resulting in a reduced costs of operation, administration and maintenance. However, the reported AOWC scheme based on semiconductor optical amplifier (SOA) was complex in structure and the performance was poor for wavelength upconversion[4].

[4] Corresponding author, E-mail: cjke@mail.hust.edu.cn.

In this paper, we propose a novel semiconductor all-optical wavelength converter based on gain-clamping structure to ensure the hybrid TDM-WDM PON architecture feasible, and measure the power budget of the HPON. Then, the power equalization functionality and burst transparency characteristic of GCWC is experimentally investigated.

2. Network architecture and operation principle

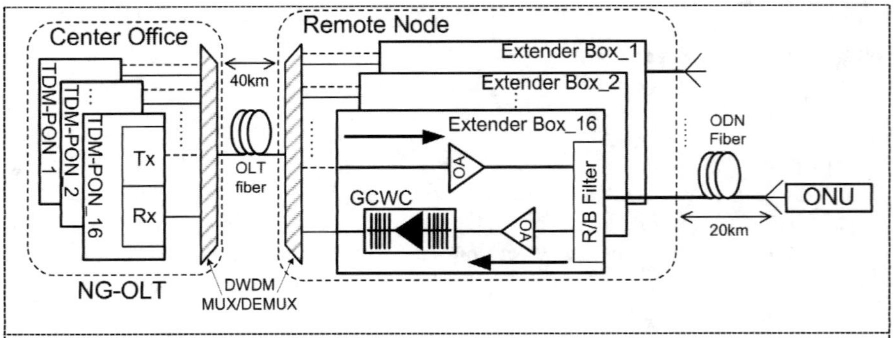

Figure 1. Proposed hybrid TDM-WDM PON architecture.

Figure 1 illustrates the architecture of the proposed hybrid TDM-WDM PON. The layout is divided into three locations, namely: central office (CO), remote node (RN) and colorless ONU. The fiber between CO and RN, referred to as the optical trunk line (OTL), is 40km. The fiber between RN and ONUs, referred to as the optical distribution network (ODN) fiber, is 20 km. A single fiber scheme is employed for the TDM-PONs, which feature a split of 256, 20km reach. Each TDM-PON is assigned an upstream (US) and a downstream (DS) wavelength, and both US and DS bands use 200GHz channel spacing. A guard band of 20nm is kept between the US and DS band, enabling the use of low-cost red/blue filters in the RN and ONU to separate US and DS channels. Reach extender box for each TDM-PON, consisting of two erbium-doped optical fiber amplifiers (EDFA) and a specific wavelength GCWC, is located at the RN to extend fiber length and splitting ratio.

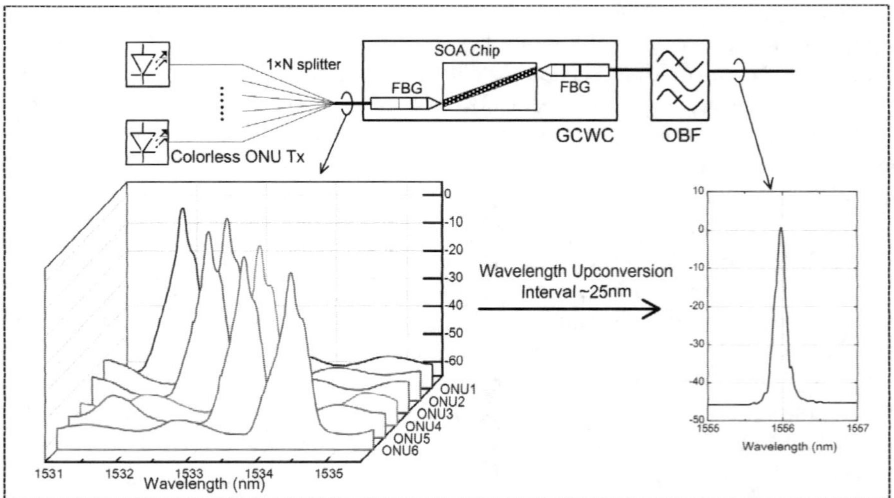

Figure 2. ONU colorless concept and the structure of GCWC.

The ONUs employed in the proposed HPON should be colorless (in other words, no ONU is wavelength specific) to decrease the costs, since mass production becomes possible with just one specification. However, WDM utilized in the architecture requires ONU wavelength strictly defined to ensure that is can pass through each optical filter. Many colorless proposals have been reported[5]. The concept of our system is shown in figure 2. We proposed a novel semiconductor optical wavelength converter based on gain-clamping structure, which consists of a tilted-waveguide SOA chip with anti-reflective coatings on facets and a pair of fiber Bragg gratings coupled to the SOA chip by conical fiber lenses. The optical spectrum of ONUs employed in the proposed HPON system, shown in the figure 2, is monitored by an optical spectrum analyzer [(OSA) Agilent 8614B]. The GCWC converts the wavelengths of the colorless ONUs to a stable wavelength with a 0.2nm linewidth@-20dB, which is strictly defined to ensure that is can pass through each optical filter of the WDM. Hence, the colorless ONU is feasible.

When a data signal is injected into the proposed GCWC, the internal laser power will vary inversely to the input signal power in order to obtain a constant cavity gain. When a "1" pulse is injected, the carriers are depleted and the gain of the internal laser decreases; therefore, the internal laser outputs a "0" pulse. For a "0" pulse, the gain is not affected; therefore, the internal laser outputs a "1" pulse. So, inverted wavelength conversion is achieved. Benefitting from the gain-clamped structure, GCWC also acts as an optical power equalizer[6].

3. Experimental demonstration and discussion

First, to evaluate available power budget in the HPON system utilizing the proposed GCWC, we measured the bit error rate (BER) performances of 2.5Gbps transceivers in the OTL between the CO and RN, and in the ODN between the RN and ONUs as a function of the inserted loss, respectively. From figure 3 and figure 4, we found that a power budget of more than 18.5 and 32.5 dB were achieved in the trunk span and ODN, respectively. In addition, we evaluated the dispersion penalties induced by 40km standard single mode fiber (SSMF) in the trunk span and 20km SSMF in the ODN. From figure 3, it was found that a dispersion penalty of 1.7 dB for 40km in the trunk, 1 dB for 20 km in the ODN and 2.0 dB for 60km was induced. From figure 4, it was found that no dispersion penalty for 20km in the ODN, a dispersion penalty of 1.8 dB for 40 km in the trunk and 60km in total was induced. These results indicated that by adopting the proposed hybrid TDM-WDM reach extender, a power budget in trunk span and ODN can be extended by more than 16.5 dB and 30.5 dB, respectively. The power budget in ODN ensure the 20 km ODN reach (4 dB) and 1:256 split ratio (24 dB) feasible, and the power budget in trunk span ensure the 40 km trunk fiber (8 dB) and 16 channels (WDM insert loss = 6 dB) feasible.

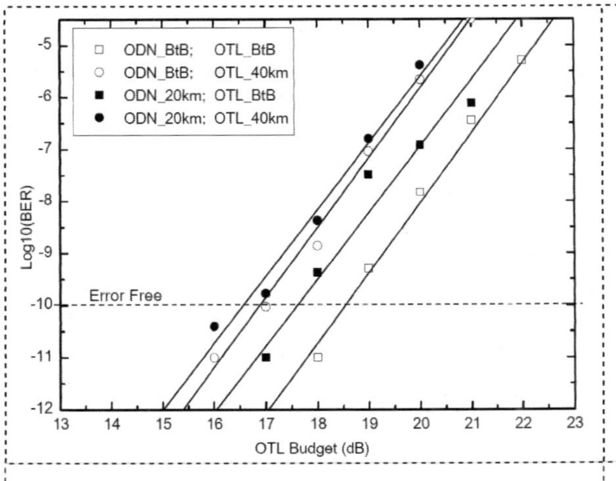

Figure 3. OTL budget performance for upstream.

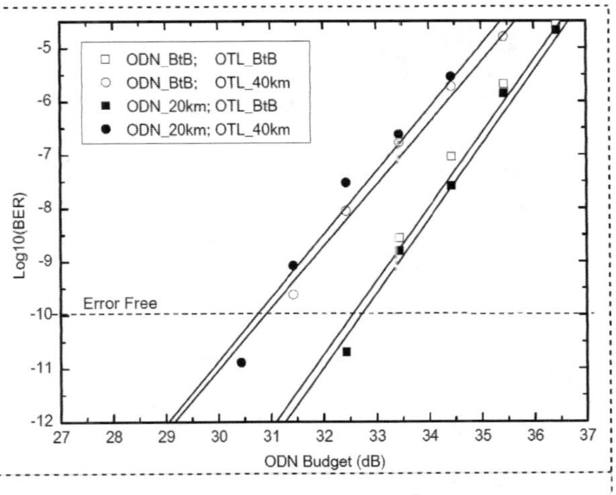

Figure 4. ODN budget performance for upstream.

Benefitting from the gain-clamped structure, GCWC also acts as an optical power equalizer to solve the near-far problem, which means that the burst-mode receiver used in this HPON system could be less complex in electronic circuits than the formerly reported one. To investigate the power equalization ability of GCWC, we monitored the input and output of GCWC through a communications signal analyzer [(CSA) Tektronix 7404B]. The waveforms obtained at input and output of the GCWC for 5dB power differences between burst frames are shown in figure 5. The results show that the GCWC could equalize burst signals with in excess of 5dB dynamic range. Figure 5 also shows that the time length of the burst frames and the time interval between them keep invariable, not as the optical electrical optical scheme[7]. The results show that the clock data recovery signals in the burst frames is not consumed and GCWC realized the transparent wavelength conversion for burst data.

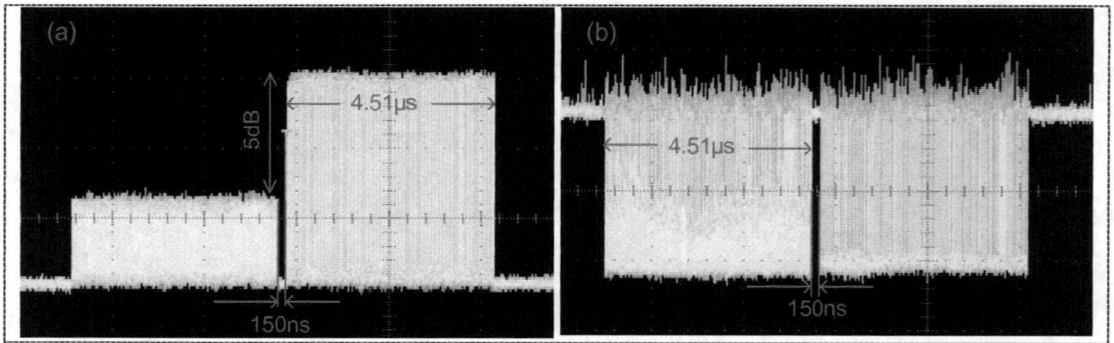

Figure 5. Waveforms obtained at the input (a) and output (b) of GCWC.

4. Conclusion

A novel semiconductor all-optical wavelength converter based on gain-clamping structure was proposed to meet the requirements of a hybrid TDM-WDM PON architecture, which has been adopted by ITU-T Rec. G.984.6 Amd.1[8].

We demonstrated that the HPON architecture utilizing the proposed GCWC is feasible and offers an extended power budget of in excess of 47dB, which reveals that the system can potentially support up to 16 TDM-PONs on a single fiber and accommodate 4096 users at 40Gbit/s in total. The power equalization functionality with at least 5dB dynamic range and the burst transparency characteristic of GCWC was experimentally confirmed.

Acknowledgements

This work was partially supported by the Hi-tech Research and Development Program of China (No. 2007AA01Z229) and the National Basic Research Program of China (No.2010CB328302).

References

[1] Lee S Mun S Kim M and Lee C 2007 *J. Lightwave Technol.* **25** 271
[2] Kazovsky L G Shaw W Gutierrez D Cheng N and Wong S 2007 *J. Lightwave Technol.* **25** 3428
[3] Miyazaki N Tanaka K and Horiuchi Y 2008 *OFC/NFOEC* p 1
[4] Shea D P and Mitchell J E 2009 *IEEE J. Selected Areas in Commun.* **27** 126
[5] Iwatsuki K and Kani J 2009 *J. Opti. Commun. and Net.* **1** C17
[6] Tran A V Chae C and Tucker R S 2006 *Opti. Commun.* **266** 500
[7] Deng L Liu D Zhang M and Qian Y 2009 *POEM (Wuhan)* p 751403
[8] ITU-T 2009 *ITU-T Recommendation G.984.6 Amendment 1*

An Equivalent Circuit Model for Separate Absorption Grading Charge Multiplication Avalanche Photodiode

Yanli Zhao[1*], Qiuyan Mo

Wuhan National Laboratory for Optoelectronics, School of Optoelectronics Science and Engineering, Huazhong University of Science and Technology, Wuhan 430074, China

E-mail: yanlizhao@mail.hust.edu.cn

Abstract. With the scale of microwave photonics system getting larger and the structure of it getting more complex, the CAD tools and accurate circuit models have become essential for designers, since difficulty for system design increases greatly. In this work, an equivalent circuit model for separate absorption grading charge multiplication avalanche photodiode (SAGCM APD) has been proposed. The model is applied to simulate the device performance and the simulating results show a well reasonable agreement with the physical model calculation and experimental data. A detailed comparison has been made for InP/InGaAs, InAlAs/InGaAs and Si/Ge APD with different structures. Our results show that Si/Ge APD is the most promising candidate for high performance optical communication system.

1. Introduction

SAGCM-APD has drawn a lot of attention in optical communication systems because of its internal gain, low noise and high bandwidth compared with APDs made in other structures [1-3]. For material used in APDs, ionization coefficient ratio k for InP is about 0.4-0.5[4], while for InAlAs that is about 0.2-0.3, which leads to a higher gain-bandwidth product (GBP) for InAlAs APD[5]. In recent years, Silicon has been considered to be the most promising multiplication material due to its favorable ionization coefficient ratio ($k<0.1$)[6]. In addition, compared with other semiconductors, Si/Ge APD has overwhelming advantages, such as high GBP (>200GHz), low excess noise, temperature independence, CMOS-compatible and feasible monolithic integration with Transimpedance amplifier (TIA).

As optical communication systems push toward multigigabit per-second transmission rates, the frequency response of these APDs has become a critical issue. The circuit model of APD is helpful to the optimization of APD[7-17]. In this work, an equivalent circuit model for separate absorption grading charge multiplication avalanche photodiode (SAGCM APD) has been proposed. The model is applied to simulate the device performance and the simulating results show a well reasonable agreement with the physical model calculation and experimental data. A detailed comparison has been made for InP/InGaAs, InAlAs/InGaAs and Si/Ge APD with different structures. Our results show that Si/Ge APD is the most promising candidate for high performance optical communication system.

2. Physical model

The two most important characteristics of a photodetector are its quantum efficiency and bandwidth. Increasing the thickness of absorption layer results in the collection of much more photons, and hence improves the quantum efficiency of the photodetector. However, the increase of the thickness of absorption layer leads to an increase of the transit time, and the 3-dB bandwidth decreases. According to the previous reports, WG-APDs or RCE-APDs can make a trade off between the bandwidth and the quantum efficiency of the photodetector. Based on Deen's theory, the frequency response of primary, second electrons and holes of RCE-SACM-APD are given in (1) -(4), respectively [7].

$$N(\omega)=\frac{P_i}{hv}\{(\eta_b+\eta_f)\frac{[1-\exp(-\alpha W_a)]}{j\omega}\times[1-\exp(-j\omega(x_t+x_m)/v_n)]+ \frac{1}{j\omega}\exp(-j\omega(x_t+x_m)/v_n)[1-\exp(-j\omega W_a/v_n)]\times[\eta_f-\eta_b\exp(-\alpha W_a)] + \exp(-j\omega(x_t+x_m)/v_n)\times\{\frac{\eta_f}{\alpha v_n-j\omega}[\exp(-\alpha W_a)-\exp(-j\omega W_a/v_n)] + \frac{\eta_b}{\alpha v_n+j\omega}[1-\exp(-\alpha W_a-j\omega W_a/v_n)]\}\} \quad (1)$$

$$P(\omega)=\frac{P_i}{hv}\{(\eta_b+\eta_f)\frac{[1-\exp(-\alpha W_a)]}{j\omega}\times[1-\exp(-j\omega x_{g1}/v_p)]+ \frac{1}{j\omega}\exp(-j\omega x_{g1}/v_p)[1-\exp(-j\omega W_a/v_p)]\times[\eta_b-\eta_f\exp(-\alpha W_a)] + \exp(-j\omega x_{g1}/v_p)\times\{\frac{\eta_b}{\alpha v_p-j\omega}[\exp(-\alpha W_a)-\exp(-j\omega W_a/v_p)] + \frac{\eta_f}{\alpha v_p+j\omega}[1-\exp(-\alpha W_a-j\omega W_a/v_p)]\}\} \quad (2)$$

$$P_s(\omega)=\frac{P_i(M-1)}{hv}\exp(-j\omega(x_t+\Delta u_m)/v_n)\times[1-\exp(-j\omega(x_{g1}+W_a+x_t+\Delta u_m)/v_p)] \times(\frac{1}{1+j\omega(M-1)\tau_m})\times\{\eta_f[\frac{1}{j\omega-\alpha v_n}-\frac{1}{j\omega}]\times[\exp(-W_a\alpha)-\exp(-j\omega W_a/v_n)] -\eta_b[\frac{1}{j\omega+\alpha v_n}-\frac{1}{j\omega}]\times[1-\exp(-W_a\alpha-j\omega W_a/v_n)]\} \quad (3)$$

$$N_s(\omega)=\frac{P_i(M-1)}{hv}\exp(-j\omega x_t/v_n)\times[\exp(-j\omega x_m/v_n)] \times(\frac{1}{1+j\omega(M-1)\tau_m})\times\{\eta_f[\frac{1}{j\omega}-\frac{1}{j\omega-\alpha v_n}]\times[\exp(-W_a\alpha)-\exp(-j\omega W_a/v_n)] -\eta_b[\frac{1}{j\omega}-\frac{1}{j\omega+\alpha v_n}]\times[1-\exp(-W_a\alpha-j\omega W_a/v_n)]\} \quad (4)$$

From (1)–(4), we can obtain the normalized frequency response function:

$$H(\omega)=\frac{hv}{P_0}\frac{\lfloor v_n(N(\omega)+N_s(\omega))+v_p(P(\omega)+P_s(\omega))\rfloor}{M(\eta_f+\eta_b)(1-\exp(-\alpha W_a))} \quad (5)$$

The effect of parasitic RLC elements are taken into account in the model, the output extrinsic current of APDs is presented as:

$$\frac{I_{ph}(\omega)}{I_{ph}(0)}=H(\omega)\cdot H_{ext}(\omega)=H(\omega)\frac{1}{1+j\omega R_tC_p-\omega^2C_pL_p} \quad (6)$$

$$\frac{1}{C_p}=\sum_i\frac{1}{C_i} \quad C_i=\frac{\varepsilon_0\varepsilon_iA}{x_i} \quad (7)$$

where $H(\omega)H_{ext}(\omega)$ reflects the frequency response of APDs system, and $P_0/(hv)$ denotes the input signal; R_t is equal to the series resistance (R_s) plus the load resistance (R_l), and L_p includes the inductance of the pad and bonding wire; C_i is the capacitance of the particular layer i with the thickness x_i and the relative permittivity ε_i in the depletion region, and A is the active area of the devices.

3. Circuit model

In signal and system theory, for a linear time-invariant system, the output signal equals the convolution of input signal and the impulse response of the system in time domain. On the other hand, the output signal is equal to the product of signal and the frequency response of the system in frequency domain. In time domain, the overall impulse response of the cascade of multiple linear time-invariant systems is the convolution of the impulse responses of individual systems. In frequency domain, the total frequency response of the cascade of multiple systems is the product of the individual frequency responses [18]. Base on this theory, $H(\omega)H_{ext}(\omega)$ is divided into small portions which are treated as frequency responses of sub-systems. From (1)-(4), five typical relationships of sub-system and the output signal are presented as follows (8)-(12):

$$I_{out}(\omega) = z \cdot I_{in}(\omega) \quad (8) \qquad I_{out}(\omega) = I_{in}(\omega) \cdot \exp(-j\omega \cdot z) \quad (9)$$

$$I_{out}(\omega) = I_{in}(\omega) \cdot \frac{1}{j\omega \cdot z} \quad (10) \qquad I_{out}(\omega) = I_{in}(\omega) \cdot \frac{1}{z + j\omega} \quad (11)$$

$$I_{out}(\omega) = I_{in}(\omega) \cdot \frac{1}{z - j\omega} \quad (12)$$

Based on JiaYin Wu's work [17], the equivalent circuit model of a RCE-SACM-APD is presented in Fig.1

4. Results and discussion

On basis of the former theory [2], the bandwidth of APD is mainly affected by transit time, RC time and avalanche buildup time. How these factors limit the bandwidth of APD will be discussed as follows.

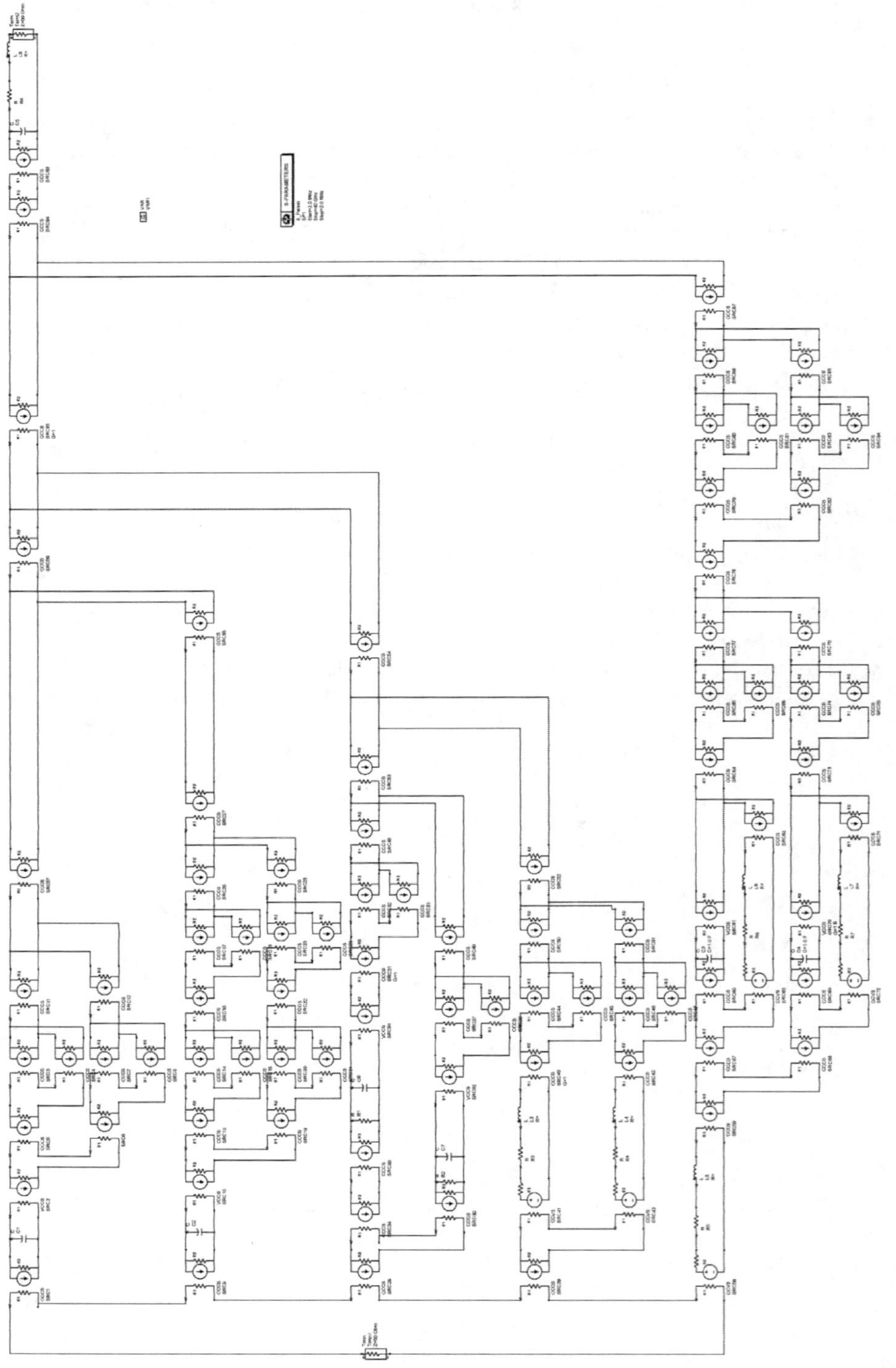

Fig.1 The equivalent circuit model of RCE-SAGCM-APD

In Fig.2, the gain-bandwidth (G-BW) characteristics of RCE-SACM-APD is simulated. The simulating results show that the circuit model simulation, physical model calculation and experimental data [19] are consistent with each other very well, which confirm the validity and application of circuit model for different structures.

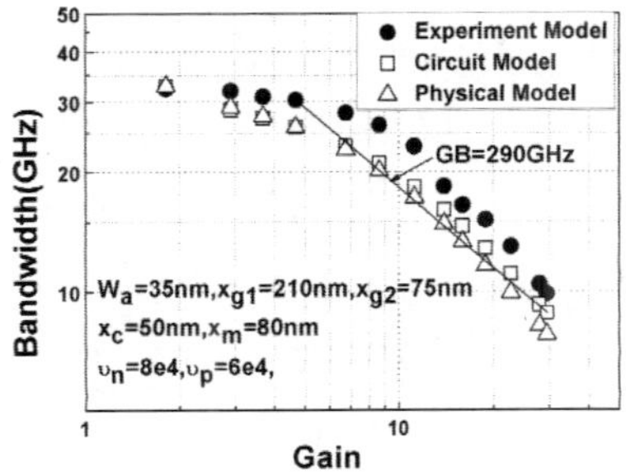

Fig.2 Circuit model simulated gain-bandwidth characteristics of RCE-SACM-APD compared with a physical model and experimental data from reference [19].

Fig. 3 illustrates G-BW characteristics for APD with different absorption layer's thickness and structures, where APD1, APD2 and APD3 represent REC-SAGCM-APD, WG-SAGCM-APD and normal incidence SAGCM-APD using InAlAs for multiplication material, respectively. As shown in Fig.3, at low gain, an increase of the thickness of absorption layer leads to a decrease of bandwidth of APDs. The increase of absorption layer's thickness results in an increase of transit time due to an increase of the distance which the photogenerated carriers in absorption layer have to travel, and hence the bandwidth decreases. At high gain, however, the bandwidth does not change obviously with the varying of the absorption layer's thickness, which proves at high gain, the change of the thickness of absorption layer does not result in an obvious change in the bandwidth of APDs, which may be ascribed to the avalanche buildup time determined APD's bandwidth. The avalanche buildup time is dependent on the multiplication layer's thickness, and the devices with the same thickness of multiplication layer have the same gain-bandwidth product. The simulated results show that the bandwidth of RCE-SAGCM-APD or WG-SAGCM-APD is higher than that normal incidence of the APD.

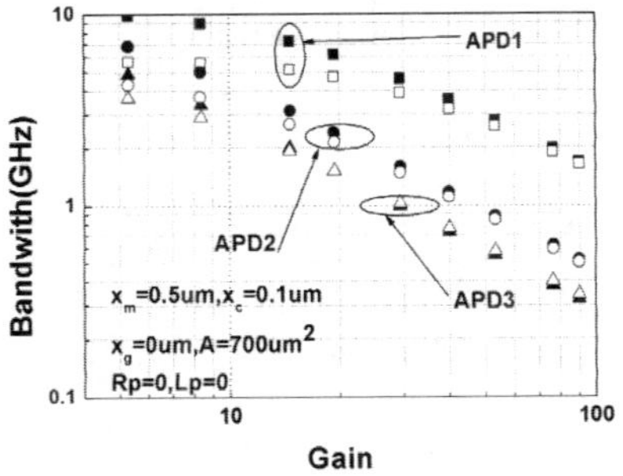

Fig.3 Circuit model simulated gain-bandwidth characteristics for APDs with different thickness of absorption layer and different structures (solid: x_a=1 um, empty: x_a=3 um)

According to (7), increasing the area of photosensitive region of APD leads to an increase of capacitance, and hence a decrease of the bandwidth of APD due to an increase of RC time. The G-BW characteristics for APDs with different the area of photosensitive region and different multiplication materials are simulated as shown in Fig.4. In this figure, APD4, APD5 and APD6 are Si/Ge, InAlAs/InGaAs and InP/InGaAs APD, respectively. The simulating results show that at low gain, the bandwidth of APD decreases with the increase of the area of photosensitive region. However, the effect of the area of photosensitive region for the bandwidth of APD is not significant at high gain, because the bandwidth is dominated by avalanche buildup time not RC time. In addition, APD using Si for multiplication region has a higher bandwidth compared with other materials due to its preferable ionization coefficience rate. It demonstrates that Si/Ge APD is the most promising candidate for high performance optical communication system.

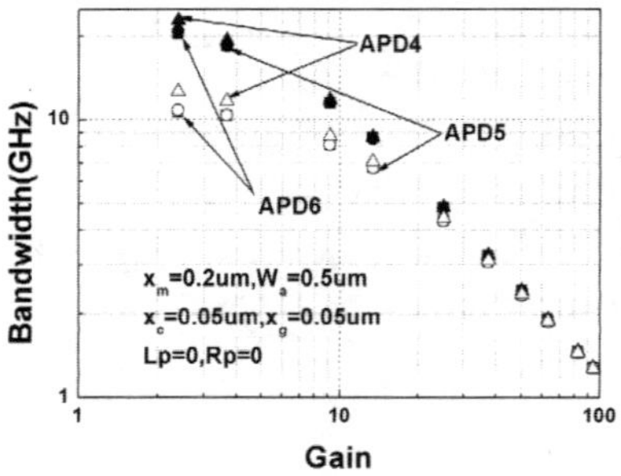

Fig.4 Circuit model simulated gain-bandwidth characteristics for APDs with different areas of photosensitive region and different multiplication materials
(solid: diameter is 10um;
empty: diameter is 20um).

In an APD's packaging module, the APD chip is connected usually to other device (like TIA) by bonding wires which can be equivalent to an inductance. The effect of varying of parasitic inductance are shown in Fig.5, where the frequency response of Si/Ge APD is simulated for different parasitic inductance at M=10. As shown in this figure, the bandwidth increases when the parasitic inductance increases from 0nHz to 1nHz. However, the bandwidth of APD decreases with further increase of inductance, while the resonant peak moves to lower frequency region, and the maximum S_{21} value increases. The resonant effect of parasitic capacitance and inductance demonstrates that the bandwidth can be compensated to some extent using the packaging technology according to the resonant effect [20].

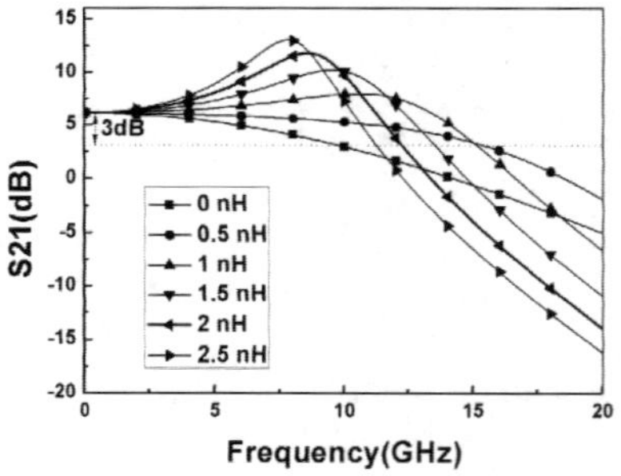

Fig.5 Circuit model simulated frequency response of Si/Ge APD chip for different inductance of bonding wires in the packaging state.

5. Conclusion

In this paper, an equivalent circuit model for Resonant Cavity Enhanced-Separated Absorption Graded Charge Multiplication-Avalanche Photodetector (RCE-SAGCM-APD) has been proposed, and the simulating results show a reasonable agreement with the physical model calculation and experimental data. A detailed comparison has been made for InP/InGaAs, InAlAs/InGaAs and Si/Ge APD with different structures. Our results show that Si/Ge APDs is the most promising candidate for high performance optical communication system. In addition, our circuit model can be applied for circuit simulation software, compared with simple circuit model of APDs, the circuit model is closed to the real APD's performance, which is helpful to the optimization of APDs.

Acknowledgements: This work was supported by the National High-Tech Research and Development Plan of China (Grant No. 2008AA01Z207) and Scientist Research Foundation for the Returned Overseas Chinese Scholars.

References

[1] Yagyu E, E. Ishimura M, Nakaji S, Ihara Y, Mikami H, Itamoto T, Aoyagi K Y, and Tokuda Y 2009 *J. Lightw. Technol.* **27** 1011
[2] Campbell J C, Holden W S, Qua G J 1985 *IEEE J. Quant. Electron.* QE-**21** 1743
[3] Campbell J C, Johnson B C, Qua G J, and Tsang W T 1989 *J. Lightw. Technol.* **7** 778
[4] Campbell J C 1988 *IEEE J. Quant. Electron.* **24** 496
[5] Lenox C. et al. 1999 *IEEE Photon. Tech Lett.* **11** 1162
[6] Kang Y M, Liu H D, Morse M et al. 2008 *Nature Photonics* **247** 1
[7] El-Batawy Y M and Deen M J 2003 *IEEE Trans.Electron Devices* **50** 790
[8] El-Batawy Y M and Deen M J 2005 *IEEE Trans Electron Devices* **52** 335
[9] Dai D X, Chen H W, Bowers J E et al. *Group IV Photonics, 2009 GFP '09 6th IEEE International Conference on* , $P_{3\text{-}15}$
[10] Dai D X, Chen H W and Bowers J E et al. 2009 *Optics Express* **17** 16549
[11] Banoushi A, Kardan M R, and Ataee M. 2005 *Solid State Electron.* **49** 871
[12] Wang G, Tokumitsu T, Hanawa I, Sato K, and Kobayashi M 2002 *IEEE Microw. Wireless Compon. Lett.* **12** 378
[13] Wang G, Tokumitsu T, Hanawa I, Yoneda Y, Sato K, and Kobayashi M 2003 *IEEE Trans. Microw. Theory Tech.* **51 1227**
[14] Chen W and Liu S 1996 *IEEE J. Quantum Electron.* **32** 2105
[15] Abbasi M R, Sheikhi M H 2007 IET-UK International Conference on Information and Communication Technology in Electrical Sciences 862-868
[16] Mai Y X and Wang G 2009 *J. Lightw. Technol.* **27** 1197
[17] Wu J Y and Wang G 2010 *J. Lightw. Technol.* **28** 784
[18] Oppenheim A V, Willsky A S, and Young I T 1983 Signals and Systems, 1st ed., Englewood Cliffs, NJ: Prentice Hall
[19] Nie H, Anselm K A, Lenox C, Yuan P et al. 1998 *IEEE Photon. Tech Lett.* **10** 409
[20] Zhu N H, Liu Y, Zhang S J, and Wen J M 2006 *Microwave Opt. Tech. Lett.* **48** 76

Appendix

NOMENCLATURE

v	Frequency of the incident light.
η_f, η_b	Quantum efficiency for APDs with forward and backward incidence, respectively.
h	Planck's constant.
w_a, x_g, x_c, x_m	Thickness of the absorption, the grading, the charge and the multiplication layers,
x_t, x_{g1}, x_{g2},	total thickness of the grading and the charge layers, and thickness of the two grading layers surrounding the absorption layer, respectively.
D	Width of the whole depletion region.
α	Absorption coefficient of the absorption region.
M	Multiplication gain.
τ_m	Avalanche buildup time.
v_n, v_p	Saturation velocities for the electrons and holes, respectively.
P_0	Power of Incident light
N, N_s	Photogenerated primary and secondary electrons, respectively.
P, P_s	Photogenerated primary and secondary holes, respectively.
$I_{ph}(\omega), I_{ph}(0)$	Output extrinsic current and dc photocurrent of APDs, respectively.
J, ω	Imaginary unit and angular frequency, respectively.

Journal of Physics: Conference Series **276** (2011) 012108 doi:10.1088/1742-6596/276/1/012108

Theoretical and experimental investigation on overshoot characteristic of gain-clamped wavelength converter

Y Zhao[1,2,3], C J Ke[1,2,3,4], B Zhao[1,2,3], D M Liu[1,2,3]
[1] College of Optoelectronic Science and Engineering, Huazhong University of Science and Technology, 430074, Wuhan, China
[2] Wuhan National Laboratory for Optoelectronics, 430074, Wuhan, China
[3] National Engineering Laboratory for Next Generation Internet Access System, 430074, Wuhan, China

Abstract. Gain-clamped wavelength converter is a key device for hybrid WDM/TDM passive optical network (HPON) based on all-optical wavelength conversion (AOWC). In this paper, the overshoot characteristic in converted signals of GCWC is numerically analyzed based on single-mode rate equation and large-signal analysis. The simulation shows that overshoot of converted signals could be decreased by increasing the bias current, length of active region, or reducing input signal power. And it also reveals that reflectivity of fiber Bragg grating has little effect on it. According to the above, structure parameters and operating parameters of GCWC are optimized. The experimental investigation shows great agreement with theoretical calculation.

1. Introduction

With the development of triple play service, passive optical network (PON), as a key technology to optical access network, tends to be long-reach access and large area covering to reduce the end-to-end cost of bandwidth and simplify the network. Current TDM-PON (Ethernet-PON and Gigabit-PON) architectures are economically feasible, but inevitably sacrifice per-subscriber bandwidth by sharing a single wavelength channel with multiple subscribers. WDM-PON offers great security and protocol transparency, and accommodates more subscribers without sacrificing bandwidth by adding wavelength channels, but the components in it are relatively expensive. A novel hybrid WDM/TDM passive optical network (HPON) based on all-optical wavelength conversion (AOWC) with high capacity and long reach and a single-fiber transmission is proposed. In AOWC-HPON architecture, wavelength conversion is a key technology to combine TDM and WDM. Several wavelength

[4] Corresponding author, E-mail: cjke@mail.hust.edu.cn

conversion mechanisms have therefore been presented in the past; e.g. cross-gain modulation, cross-phase modulation and Four-Wave Mixing in semiconductor optical amplifiers (SOAs). Conventional wavelength conversion techniques cannot meet the requirements in AOWC-HPON architecture due to the drawbacks of needing external CW-probe, low conversion efficiency, not implementing up-wavelength conversion efficiently, or having small dynamic range.

In this paper, a novel gain-clamped wavelength converter (GCWC) is proposed to make the AOWC-HPON architecture feasible. The overshoot characteristic in converted signals of GCWC is theoretical and experimental investigated.

2. Time domain model of GCWC

A GCWC consists of a titled-waveguide SOA chip (residual reflection $<10^{-3}$) and a pair of FBGs showned in figure 1. GCWC is modulated directly by input signals, implementing inverted wavelength conversion. The waveform of converted signals exhibits relaxation oscillations (overshoot) because of a time dependent energy exchange between laser field and carrier population in GCWC.

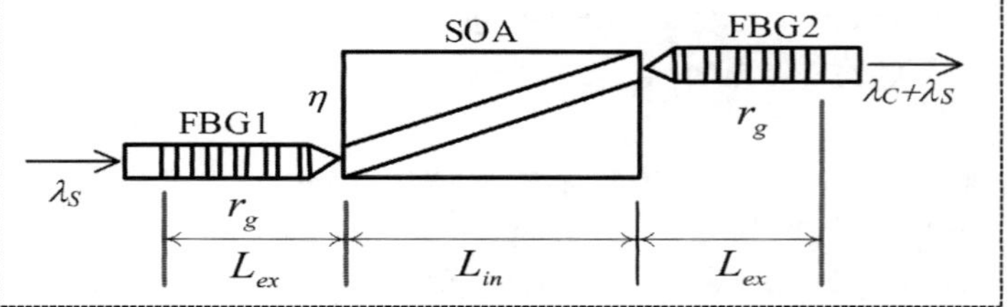

Figure 1. Schematic of gain-clamped wavelength converter.

We use the following rate equation model to study the overshoot characteristic of GCWC, taking into account the effects of nonlinear gain suppression:

$$\frac{\partial N}{\partial t}=\frac{I}{qV}-\frac{N}{\tau_e}-v_g G_1(1-\varepsilon_{11}S_1-\varepsilon_{12}S_{in})S_1-v_g G_2(1-\varepsilon_{21}S_1-\varepsilon_{22}S_{in})S_{in} \tag{1}$$

$$\frac{\partial S_1}{\partial t}=\Gamma v_g G_1(1-\varepsilon_{11}S_1-\varepsilon_{12}S_{in})S_1-\frac{S_1}{\tau_p}+\frac{\Gamma\beta_{sp}N}{\tau_e} \tag{2}$$

The effective photon lifetime of converted wavelength τ_p, which is important to gain-clamped wavelength conversion, is expressed as

$$\tau_p=\frac{n_1 L_{in}+2n_2 L_{ex}}{c(L_{in}+2L_{ex})}\cdot\left(\frac{L_{in}-2\eta L_{ex}(1+r_g{}^2)\ln(\eta r_g)/(1-\eta^2 r_g{}^2)}{\alpha L_{in}-2\ln(\eta r_g)}\right) \tag{3}$$

GCWC is typically biased close to threshold and modulated considerably above threshold to obtain optical pulses representing digital bits in HPON system. In this case of large-signals modulation, the rate equations should be solved numerically.

3. Numerical Simulation and Analysis

According to the previous model, we simulate the overshoot of converted signals using large-signal analysis. The parameters used in our simulation are shown in table 1.

Table 1. The parameters for numerical simulation.

Parameters	symbol	value	unit
Signals wavelength	λ_S	1530	nm
Extinction ratio of input signals	ER	10	dB
Active area width	w	2	μm
Active area thickness	d	0.2	μm
Effective group index of refraction in cavity	n_1	3.4	
peak wavelength of gain spectrum	λ_p	1550	nm
Lasing wavelength	λ_c	1557	nm
Transparency carrier density	N_t	0.8×10^{18}	cm^{-3}
Differential gain coefficient	dg/dN	4.0×10^{-16}	cm^2
Power confinement factor	Γ	0.49	
Internal wavelength losses	α	2×10^{-3}	μm^{-1}
Linear recombination coefficient	A	5.0×10^8	s^{-1}
Bimolecular recombination coefficient	B	870	$\mu m^3 s^{-1}$
Auger recombination coefficient	C	9.5×10^{-5}	$\mu m^6 s^{-1}$
Spontaneous emission factor	β_{sp}	0.001	
Nonlinear self gain suppression coefficient	$\varepsilon_{11},\varepsilon_{22}$	6.8×10^{-17}	cm^{-3}
Nonlinear cross gain suppression coefficient	$\varepsilon_{12},\varepsilon_{21}$	13.6×10^{-17}	cm^{-3}
Coupling efficiency	η	0.49	

3.1. The influence of operating parameters.

In this section, we will investigate the influence of some operating parameters on overshoot of converted signals. The peak value of normalized overshoot as a function of bias current and average input signal power is given in figure 2 and figure 3. It can be seen from figure 2 that the peak value of normalized overshoot of converted signals is decreased from 0.25 to 0.1 by increasing the bias current from 130 mA to 170 mA. This is directly related to the carrier variation in active region.

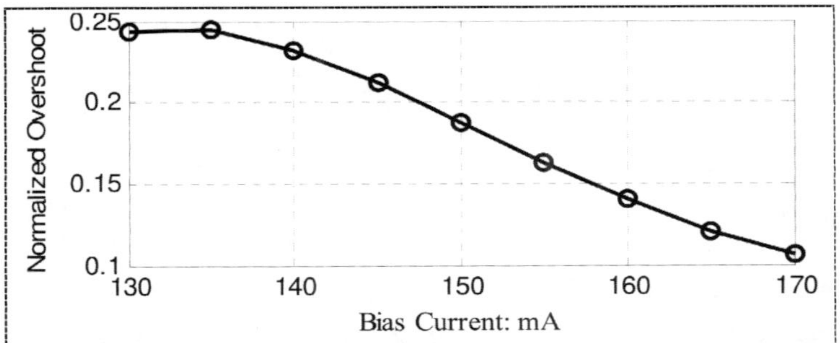

Figure 2. Overshoot of converted signals vs bias current (L_{in}=0.8 mm, r_g=0.5, P_{in}= -3 dBm)

Figure 3 shows that the peak value of normalized overshoot of converted signals increases from 0 to 0.9 when the input signal power changes from -10 dBm to 4 dBm. We can clearly see that lower input optical power causes the better performance of overshoot in converted signals. It should be noted that input signal power beyond a certain value will shut down the gain-clamping lasing in GCWC.

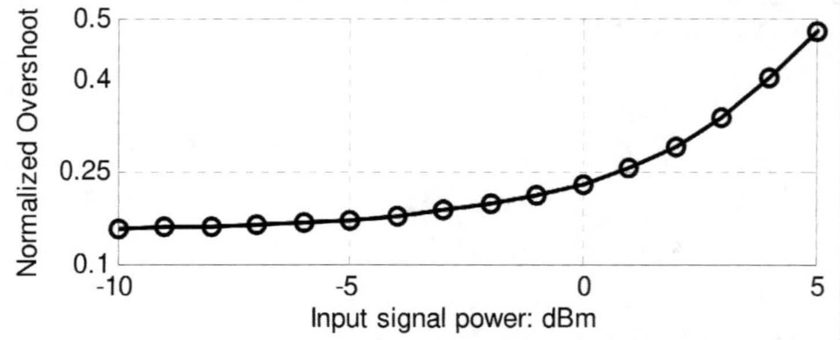

Figure 3. Overshoot of converted signals vs input signal power (L_{in}=0.8 mm, r_g=0.5, I_b=150 mA).

3.2. The influence of structure parameters.

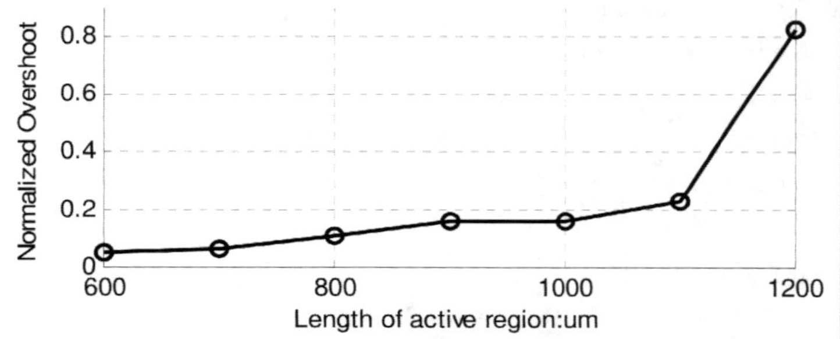

Figure 4. Overshoot of converted signals vs length of active region (r_g=0.5, I_b=150 mA, P_{in}=-3 dBm).

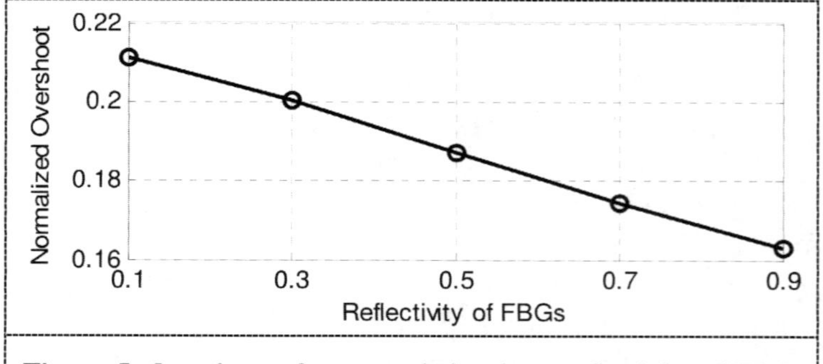

Figure 5. Overshoot of converted signals vs reflectivity of FBGs (L_{in}=0.8 mm, I_b=150 mA, P_{in}=-3 dBm).

The overshoot characteristic of GCWC is strongly dependent on the length of active region This dependence is illustrated in figure 4. From figure 4 we can conclude that the peak value of normalized overshoot of converted signals monotonely increases with the length of active region. Because the injected current density will decrease with increasing the length of active region. Furthermore, it should be paid attention in practical applications that decreasing the length of active region could lead to a serious degradation of extinction ratio of converted signal.

Figure 5 shows the influence of reflectivity of FBGs on overshoot of converted signals. As seen from the figure, overshoot is affected to some extent by reflectivity of FBGs, but not very seriously.

4. Experimental results

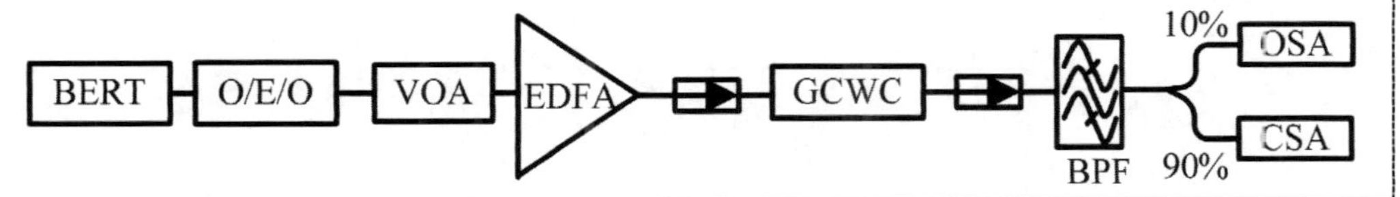

Figure 6. Experimental setup used for gain-clamped wavelength conversion.(BERT: Bit Error Rate Tester; VOA: Variable Optical Attenuator; OSA: Optical Spectrum Analyzer; CSA: Conmunications Signal Analyzer)

According to the simulation results, the optimized structure parameters are as the following: the length of active region operating at 1.550 μm is 800 μm, the reflectivity of FBGs is 0.5 with bandwidth of 0.3 nm @-3 dB. Based on the obove-mentioned structure, the optimized operating parameters are obtained as follows: the input signal power is -3 dBm, bais current is 150 mA. The experimental setup is presented in figure 6, which converts the signal wavelength at 1530 nm to the clamped-wavelength at 1557 nm. The converted signals is filtered out by a band-pass filter.

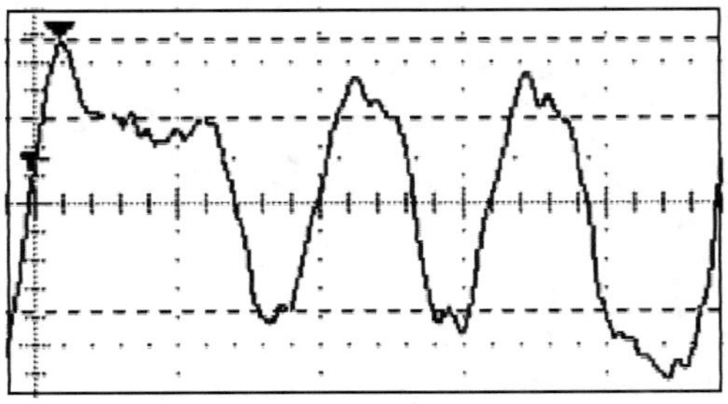

Figure 7. Waveform of converted signals for input signal power of -3dBm

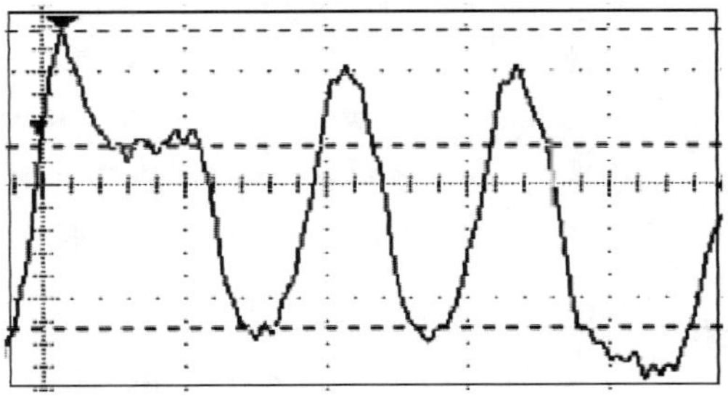

Figure 8. Waveform of converted signals for input signal power of 0 dBm

Figure 7 and figure 8 show the converted waveforms for different input optical power (-3 dBm and 0 dBm). The most remarkable notice is the wavelength converted signals undergoes significant overshoot on the rising edge while little undershoot on the falling edge. The results show that overshoot could be decreased by reducing input signals power. The experimental results agree well with the theoretical calculation.

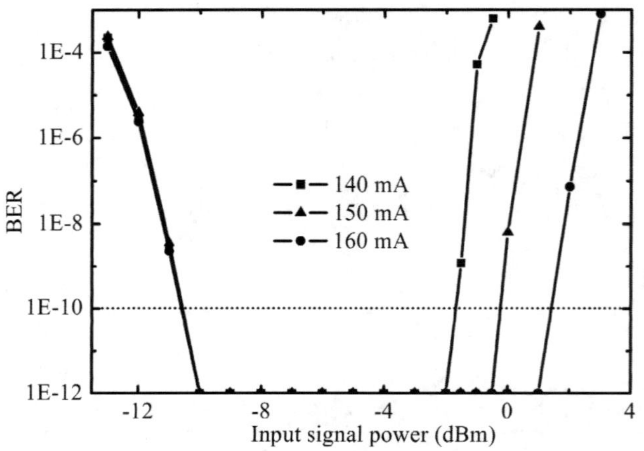

Figure 9. Input dynamic range of GCWC for HPON application

Figure 9 shows the BER performance for the converted wavelength under different bias current. The results confirmed that it is possible to achieve BER performance of 10^{-10} for AOWC-HPON application with an input dynamic rage up to 11 dB. This means GCWC of the obove-optimized parameters can implement wavelength conversion for burst signals with power different value of 11 dB. Although the overshoot of converted signals is monitored through a communications signal analyzer (CSA) when GCWC working at optimal operating point (P_{in}=-3 dBm, I_b=150 mA), according to figure 7, the BER performance of converted signals can still achieve 10^{-12}.

5. Conclusion

In this paper, the overshoot characteristic in converted signals of GCWC is numerically analyzed based on single-mode rate equation and large-signal analysis. The simulation shows that overshoot of converted signals could be decreased by increasing the bias current, length of active region, or reducing input signal power. And it also reveals that reflectivity of fiber Bragg grating has little effect on it. According to the above, structure parameters and operating parameters of GCWC are optimized. The experimental investigation shows great agreement with theoretical calculation.

Acknowledgements

This study was partially supported by the High Technology Research and Development Program of China (No. 2007AA01Z229) and the National Basic Research Program of China (No. 2010CB328302)

References

[1] Dong J J, Zhang X L, Huang D X 2005 *Acta Phys. Sin.* **59** 1021
[2] Sebastian W, Weng W C, Lukas C, Connie J C H 2006 *Quant. Electron.* **42** 552
[3] Jin Y S, Geert M, Roel B 1997 *Quant. Electron.* **3** 1162
[4] Farhan R, Paul G 2007 *Quant. Electron.* **43** 1109

Journal of Physics: Conference Series **276** (2011) 012108 doi:10.1088/1742-6596/276/1/012108

[5] D M Kane, Joshua P T 2009 *Quant. Electron.* **27** 2949
[6] Víctor T Company, Kamau P, Idelfonso T M 2008 *Photon. Technol. Lett.* **20** 1299
[7] Zhao T G, Ren J H, Li W, Zhao R H 2003 *Acta Optica Sinica* **23** 1071
[8] Govind P A 2001 *Fiber-Optic Communications Systems* (American: John Wiley & Sons, Inc.) P110
[9] Dong J J, Zhang X L 2007 *Chin. Phys. Lett.* **14** 990
[10] Dong J J, Zhang X L, Huang D X 2007 *Chin. Phys. Lett.* **24** 3450
[11] Larry A C and Scott W C 2006 *Diode Laaers and Photonic Integrated Circuits* (American: John Wiley & Sons, Inc.) P132

Journal of Physics: Conference Series **276** (2011) 012109 doi:10.1088/1742-6596/276/1/012109

Effect of SiO_2 Reflection Reducing Coating on the Vertical Cavity Surface Emitting Laser

ZHAO Zhen-Bo*, XU Chen, XIE Yi-Yang, ZHOU Kang, LIU Fa, WANG Bao-Qiang, LIU Ying-Ming, SHEN Guang-Di
Key Laboratory of Opto-electronics Technology (Beijing University of Technology), Ministry of Education, Beijing University of Technology, 100 Ping Le Yuan, Chaoyang District, Beijing 100124, China

E-mail: zhaozhenbo@emails.bjut.edu.cn

Abstract. The effects of epitaxis SiO_2 film on the characteristics of DBR reflectivity were simulated, and the reflectivity of graded interface DBR is lower than the abrupt one was studied. 850nm vertical-cavity surface-emitting lasers were fabricated, the DBR is formed by graded heterojunction of $Al_xGa_{1-x}As$, the top and bottom DBR have 22 and 34 pairs of mirror, and current was confined by oxide aperture, and then different thickness of SiO_2 film were grew on the top DBR. and found that a certain thickness ($\lambda/4$) of the SiO_2 reflection reducing coating could make the vertical-cavity surface-emitting lasers output power increased about 3-5 mW, the threshold current does not increase obviously, the reason is external quantum efficiency increases more than the threshold current. So the SiO_2 reflection reducing coating does not affect the lasing of lasers and threshold current obviously, but It can significantly improve the output power. The experimental results agree well with the theoretical expectation.

1. Introduction

Vertical cavity emitting laser (VCSEL) has a number of inherent advantages including low threshold current, single longitudinal mode operation, low-divergence circular output beams, broad modulation bandwidth, suitability for monolithic two-dimensional (2-D)integration and compatibility with on-wafer probe testing., and has many potential applications on data transmission in optical network, optical interconnect, optical storage and laser printing [1-2]. The single mode high output power is essential for the application of VCSEL. There have lots of studies on single mode high power VCSEL, such as photonic crystal, holey and surface relief VCSEL etc [3-5]. One critical issue for VCSEL is connected to the very short optical gain region as compared to edge-emitting lasers, hence requiring very high reflectivity(>99%) mirrors to achieve lasing action. This is realized by distributed Bragg reflector (DBR), whose design criteria are related to optical reflectivity, thermal and electrical conductivity, material index contrast and optical absorption. The analysis of DBR is critical in VCSEL design, because their reflectivity strongly affects all laser properties [6], traditional DBR is

* Supported by the National High Technology Research and Development Program of China under Grant No 2008AA03Z402, and the Beijing Natural Science Foundation of Beijing under Grant No 4092007.
Email: taikong906@163.com Tel: + 86 01067391641-834 Fax: +86 01067391641-822

GaAs/AlGaAs growed by alternately, but the abrupt interface of the heterojunction has a higher series resistance which will generate more heat, So recently graded interface heterojunction DBR was appeared, and through altered the reflectivity can affect the characteristics of the VCSEL [7-8]. In this paper, we simulated the reflectivity of abrupt and graded DBR, studied the effects of λ/4 reflection reducing coating on the reflectivity of DBR and the characteristics of VCSEL. The devices were produced and tested, and agreed well with the theoretical simulation.

2. Device structure and theory

In this Letter, the VCSELs were fabricated from a 3-inch n-type wafer with epitaxial structure consisting of a bottom n-type (Si-doped) 34 pairs DBR, an undoped active region with three GaAs quantum wells, and a similar top (C-doped) 22 pairs DBR. The DBR are formed by graded heterojunction of $Al_xGa_{1-x}As$, the unit period of bottom and top DBR mirrors consisting of 4 layers of $Al_xGa_{1-x}As$ (x: 0.9-0.12), $Al_{0.12}Ga_{0.88}As$, $Al_xGa_{1-x}As$ (x: 0.12-0.9), $Al_{0.9}Ga_{0.1}As$. In addition, a 30 nm $Al_{0.99}Ga_{0.02}As$ oxidization layer was inserted between the top DBR and the active region. The structure is shown in Figure 1 (a). The operation wavelength of the device is designed at 850 nm.

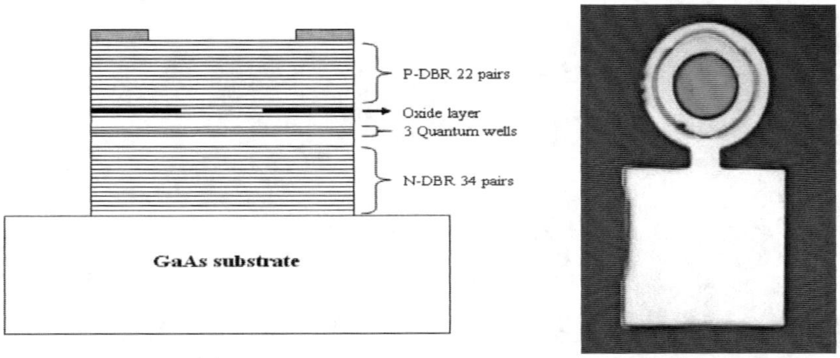

(a) (b)

Figure 1. Structure of the selectively oxidized VCSEL (a) and the actual device photo (b)

The circular mesa was formed by $SiCl_4/Ar/Cl_2$ inductively coupled plasma reaction ion etching (ICP- RIE) down close to the active region, using SiO_2 as the etching mask. Then the $Al_{0.99}Ga_{0.02}As$ layer was selectively oxidized to form the current aperture with diameter of 10 and 20 μm. After that, top ohmic (Ti/Au) ring contacts were patterned and bottom ohmic (AuGeNi/Au) contacts were formed. Then the device was annealed using rapid thermal annealing at 430℃ for 35s. The actual VCSEL device photo is shown in Figure 1 (b).

Considering the optical waves of the VCSELs were propagated along the vertical direction of the DBR, so we did not consider the entrance ray angle. According to E and H are continuously in the two side of the interface, we can obtain the transmission matrix method from Fresnel formula [9-10].

$$\begin{bmatrix} E_0 \\ H_0 \end{bmatrix} = \begin{bmatrix} \cos(kd+iad) & \frac{i}{n_1}\sin(kd+iad) \\ in_1\sin(kd+iad) & \cos(kd+iad) \end{bmatrix} \begin{bmatrix} E_2 \\ H_2 \end{bmatrix} = M \begin{bmatrix} E_2 \\ H_2 \end{bmatrix} \quad (1)$$

Where $k=2\pi n_1/\lambda$, E_0, E_2 are the electric-field intensity vector of the first film, H_0, H_2 are the intensity vector of magnetic field of the first film, λ is the wavelength in vacuo, n_1 is the index of refraction of the single dielectric film layer, d_i is the thickness of the single dielectric film layer, a is the absorption coefficient, the DBR multiplayer structure is shown in Figure 2.

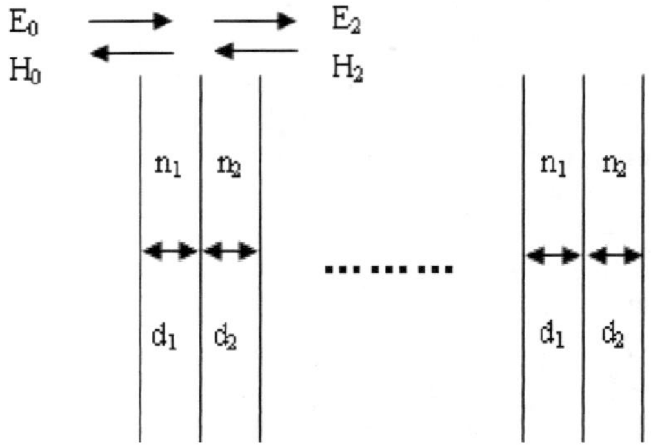

Figure 2. The DBR multiplayer structure consists of n_1 and n_2 mediums

The transmission matrix of every DBR layer is Mi, for the total layer of DBR, the overall transmission is described as equation (2), and the amplitude reflectivity of the DBR can be deduced as equation (3).

$$M = \prod_i M_i = \begin{bmatrix} m_{11} & m_{12} \\ m_{21} & m_{22} \end{bmatrix} \tag{2}$$

$$r = \frac{n_0 m_{11} + n_0 n_s m_{12} - m_{21} - n_s m_{22}}{n_0 m_{11} + n_0 n_s m_{12} + m_{21} + n_s m_{22}} \tag{3}$$

Where M_{ij} is the element of the matrix M, n_0 is the index of refraction in vacuo, n_s is the index of refraction of the substrate. According to the reflectivity of the amplitude r and the power R: $r = \sqrt{R} e^{j\varphi}$ (φ is the phase change on reflection), so the refractive ratio is described as $R = |r|^2$, then we can calculate the reflectance spectrum. Assume there are m pairs of λ/4 thickness multilayer DBR, the index of refraction of the two adjacent layers are n_1 and n_2, and absorbed coefficient is 0, so the wavelength reflectivity of DBR is described by equation (4). From that we can obtain the power reflectivity which is described by equation (5), and the more the change of index of refraction the higher reflection of power.

$$r_{wv} = \frac{1 - \frac{n_s}{n_o}\left(\frac{n_1}{n_2}\right)^{2m}}{1 + \frac{n_s}{n_o}\left(\frac{n_1}{n_2}\right)^{2m}} \tag{4}$$

$$R = \tanh 2\left[\frac{1}{2}\ln\left(\frac{n_o}{n_s}\right) + m\ln\left(\frac{n_1}{n_2}\right)\right] \tag{5}$$

Then the 850nm VCSEL structure which includes the refractive index with linearly graded Al composition from equation (6)-(11) was simulated. where n denotes the refractive of $Al_xGa_{1-x}As$, x is the composition of Al, E is optical energy, E_0 is effective oscillated energy, E_d is dispersion energy, E_g is band gap energy, the equations for E_0, E_d and E_g as the function of alloy composition yield [11].

$$n^2(E) = 1 + \frac{E_d}{E_0} + \frac{E_d}{E_0^3}E^2 + \frac{\eta}{\pi}E^4 \ln\left(\frac{2E_0^2 - E_g^2 - E^2}{E_g^2 - E^2}\right) \tag{6}$$

$$\eta = \frac{\pi E_d}{2E_0^3\left(E_0^2 - E_g^2\right)} \tag{7}$$

$$E = \hbar\omega \tag{8}$$

$$E_0 = 3.65 + 0.871x + 0.179x^2 \tag{9}$$

$$E_d = 36.1 - 2.45x \tag{10}$$

$$E_g = 1.424 + 1.266x + 0.26x^2 \tag{11}$$

The schematic diagrams of unit period in top and bottom DBR mirrors is mentioned in hereinbefore, the simulated results of top 22 pairs DBR is shown in Figure 3, we can see the reflectively of abrupt interface DBR is higher than graded one with the same pairs of DBR. That is because the index of refraction change between adjacent DBR layers of graded DBR is less than abrupt one, so the reflection of the whole DBR is lower.

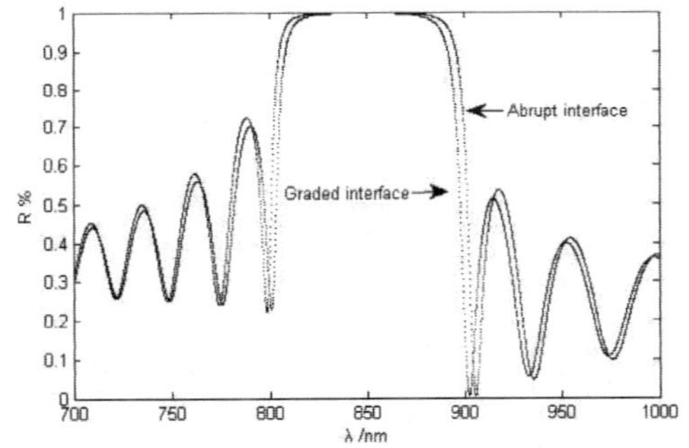

Figure 3. The spectra of top DBR of abrupt and grade interface

We calculated the effect of SiO_2 film on the DBR, so we had studied the reflection of the DBR with λ/2, λ/4 thickness of SiO_2, the simulated results are shown in Figure 4. At the high reflective zone, epitaxis λ/2 thickness of SiO_2 has the same reflectivity as with no SiO_2, but λ/4 thickness of SiO_2 can minish the reflectivity at the high reflectance zone.

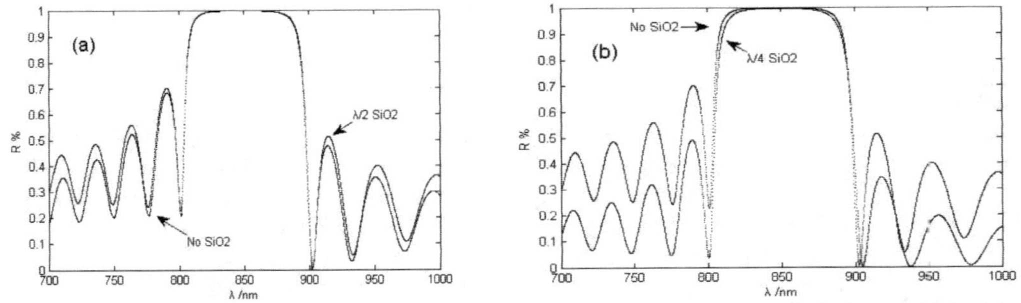

Figure 4. The reflectance spectrum of top DBR and with epitaxisλ/2 thickness of SiO_2 (a) The reflectance spectrum of top DBR and with epitaxisλ/4 thickness of SiO_2 (b)

3. Result and discussion

We fabricated the VCSELs, and epitaxis different thickness of SiO_2, we tested the output power and threshold current of the VCSELs, the results are shown in Figure 5 (a) and (b). we obtained that, when the oxide aperture is 10 μm, the output power of VCSEL with λ/4 SiO_2 is higher than the without one about 3 mW, and when the oxide aperture is 20μm, the output power of VCSEL with λ/4 SiO_2 is higher than the without one about 4.5 mW.

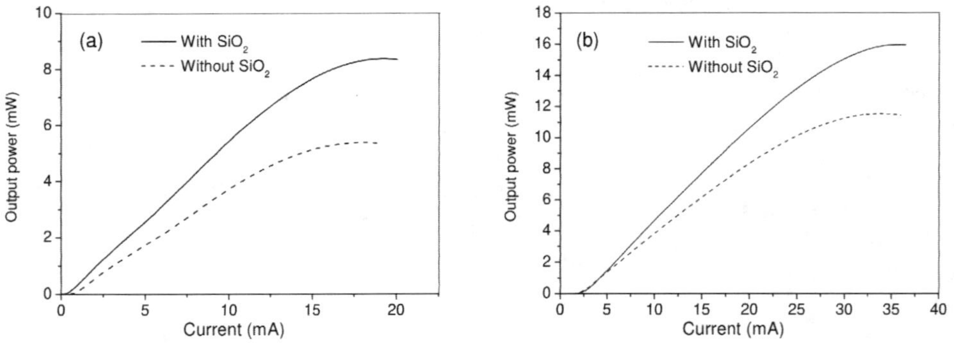

Figure 5. VCSEL oxide aperture 10μm, the output power and threshold current with SiO_2 is 8.3 mW and 0.5 mA, the without one is 5.3 mW and 0.5 mA (a) oxide aperture 20μm, the output power and threshold current with SiO_2 is 15.9 mW and 1.6 mA the without one is 11.4 mW and 1.5 mA (b)

The output power is increased with a small decreases of the DBR reflectivity, because the decreases reflectivity of DBR cause the increases of the loss of cavity face, the loss of top DBR cavity face is the output power, but the threshold current did not increase, from the equation (12) we know that, the increasing of the R_2 (the loss of top DBR cavity face) affects the threshold current weakly.

$$J_{th} = \frac{N_{\omega} J_0}{\eta_{\text{int}}} \exp\left(\frac{\alpha_{cav} + \frac{1}{2L_{cav}} \ln \frac{1}{R_1 R_2}}{N_{\omega} \Gamma_{\varpi} \Gamma_z g_0} \right) \tag{12}$$

R_1 and R_2 are the reflectivity of bottom (N = 34) and top (N = 22) DBR mirror, respectively, N_w is the number of active region quantum wells, and Γ_w and Γ_z are lateral and longitudinal confinement factors, respectively. Moreover, L_{cav} is the laser cavity length, α_{cav} is the optical cavity loss, g_0 is the material gain coefficient, J_0 is the transparency current density and η_{int} is the internal quantum efficiency.

4. Conclusion

The reflectivity of graded interface DBR is lower than the abrupt one was studied, and the effects of epitaxis SiO_2 film on the characteristics of DBR reflectivity were simulated. 850nm oxidation confined VCSELs were fabricated, then epitaxis different thickness of SiO_2 film on the top DBR. and found that a certain thickness (λ/4) of the SiO_2 reflection reducing coating can make the output power of VCSELs increases dramatically, the threshold current not increases obviously. From the theoretical simulation and experimental we can obtain that, the reflection reducing coating can increases the output power of VCSELs, then it can be used in the single mode VCSELs, such as photonic crystal VCSEL, holey VCSEL, to increase the output power.

5. References

[1] Yu, S.F. 2003 John Wiley & Sons **3**

[2] Da-Long Cheng, En-Chi Liu and Tsu-Chiang Yen 2004 IEEE Photonic Technology Letters **16** 278

[3] Xie Yi-Yang, Xu Chen and Kan Qiang 2010 Chin. Phys. Lett. **27** 024206

[4] Akio Furukawa, Mitsunari Hoshi and Satoshi Sasaki 2005 Physics and Simulation of Optoelectronic Devices XIII Proc. of SPIE **5722** 183

[5] A. Haglund, J. S. Gustavsson and J.Vukusic 2004 IEEE Photonics Technology Letters **16** 368

[6] Francesco De Leonardis, Vittorio M.N. Passaro and Francesca Magno 2007 J. Comput Electron **6** 289.

[7] A. D. Rakic, M. L. Majewski and A. B. Djurisic 2000, IEEE PROC. 22nd International Conference on Microelectronics **2** 607

[8] Wang X D, Wu X M and Wang Q 2006 Chin. Phys. Soc **55** 4983

[9] Qu Hongwei 2005 Beijing University of Technology Doctoral Dissertation **16**

[10] Hong-Mau Tsai, Shiang-Feng Tang and Shih-Tsang Su 2007 Proc. of SPIE **6484** 648401

[11] Bhattacharya Pallab 1996 INSPEC The Institution of Electrical Engineers London United Kingdom **283**

Design of a Real-time Signal Processing System for LIF Sensor

ZHAO Xian-de, ZHENG Wen-gang *, DONG Da-ming, SHEN Chang-jun, ZHANG Xin, ZHOU Jian-jun, YAN hua, WU Wen-biao

(National Engineering Research Center for Information Technology in Agriculture, P. R. China, 100097)

Abstract: Laser Induced Fluorescence (LIF) sensor is one of the most sensitive approaches available for a variety of analytical applications, such as determination of nitrogen content of plant leaves, detection of chlorophyll content in water, etal. As a core instrumental requirements of real time LIF sensor, signal processing system is used to store effective processing and identification algorithms in a short time. By analyzing the working principle of LIF sensor in detail, a novel platform of signal processing system used in LIF sensor is proposed in this paper. The design solutions and hardware architecture of the system are described in this paper, include Digital Signal Processor (DSP), data transmission block, and memory block. Several steps of signal processing methods are proposed, according to the characteristic of LIF sensor. At last, an application of using the signal processing system designed in this paper for measuring chlorophyll content in plant leaves is shown.

OCIS codes: 120.4640, 140.3450
Keywords: LIF sensor, real time, signal processing, chlorophyll

ZHAO Xian-de, male, born in1984, assistant engineer
ZHENG Wen-gang, male, associate professor E-mail: zhengwg@nercita.org.cn

1. INTRODUCTION

As a powerful spectroscopic technique, Laser Induced Fluorescence (LIF) is more and more commonly used for probing gas-phase atomic or molecular species, even in minute amounts, and determining their internal states distribution[1].

The basic principle of LIF is shown in Figure1. The species to be examined is excited with a laser. The wavelength is often selected to be the one at which the species has its largest cross section. The excited species will after some time, usually in the order of few nanoseconds to microseconds, de-excite and emit light at a wavelength larger than the excitation wavelength.

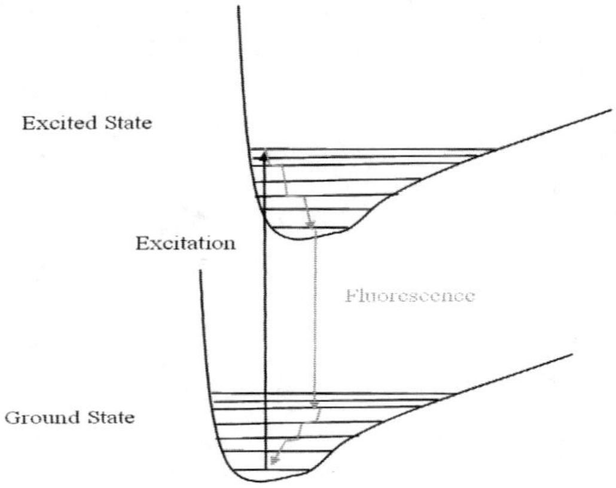

Figure1. The basic principle of Laser induced fluorescence

The relationship between fluorescence intensity and the material concentration can be described in following equation.

$$F = k * QE * Po * (1 - 10^{[-\alpha * L * c]}) \qquad (1)$$

F is measure of the fluorescence intensity, k is geometrical parameters of the instrument. QE is quantum efficiency (emission of photons / absorbtion of photons), Po is illuminant power, α is absorbed section associated with wavelength, L is wavelength, c is concentration. (the physics meaning of α , L, and c is same as the law of Beer-Lambert.)

Fluorescence analysis is a good method with the advantage of high selectivity, good reproducibility, simple, rapid, and easy sampling[2, 3].

The essence of signal processing is to extract information from the signal source with various noise and interference, and transform the extraction to a form which can be used by a man or a machine.

High-speed and real-time signal processing is a special branch of signal processing which is important to realization of real-time data processing and rapid response to get results and widely used in industrial and military, such as radar signal processing and communication signal processing, etc[4].

In this paper, a real-time signal processing for LIF sensor is designed. A new kind of spectral signal processing module, which uses a Digital Signal Processor (DSP) as its central processing unit, is designed for this system. Details on hardware and software design solutions of this module are discussed in later sections.

2. HARDWARE PART

The signal processing system designed for LIF sensor includes signal processing block, data transmission block, memory block and other devices. The hardware architecture is illustrated in Figure2.

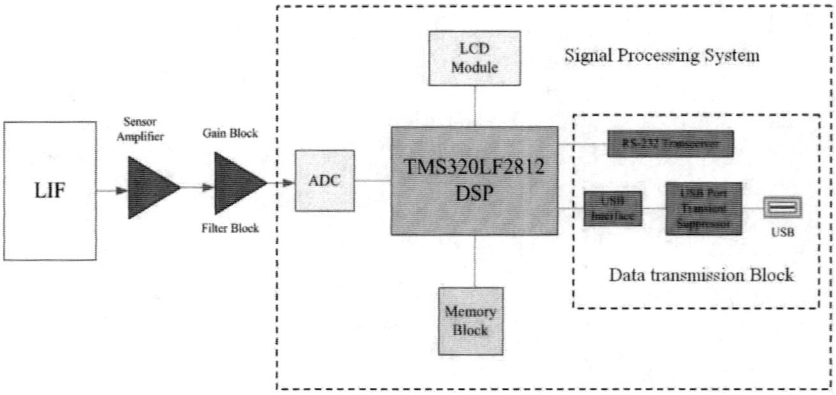

Figure2. The hardware architecture of the signal processing system

2.1 Signal Processing Block

The basic electronic module is based on a central processor, TMS320LF2812, a kind of digital signal processor made by Texas Instruments company. This chip is an advanced fixed-point 32-bit digital signal processor. It is based on 32-bit fixed-point DSPTMS320C28x™ CPU kernel and its working frequency is as high as 150MHz, very well suited for numerically intensive algorithms. On the chip, 128K×16 Flash and 56 independent programmable general-purpose input output interface (GPIO) are integrated. It has powerful capabilities of digital signal processing and event management, and embedded control function. It can be used in many fields such as industrial automation and control, application of power electronics technology, intelligent instrument and motor servo control system, etc[5].

TMS320LF2812 consists of a 12-bit analog-to-digital converter (ADC) to capture an input signal which is output from LIF sensor. The resulting digital representation of the captured signal is then processed by spectral signal processing software. All of the process above is implemented by the software. The software is running on DSP processor which is the central device of signal processing block.

2.2 Data Transmission Block

In spite of being almost completely physically embedded in the DSP, special reference should be made to the interface capabilities of the module. In this paper, two kinds of communication modes, Universal Serial Bus (USB) and the Universal Asynchronous Receiver/Transmitter (UART), are designed for the communication with the host computer.

In this design, serial communication based on RS-485 between the host PC and the slave device is realized. The product of MAXIM LLC, MAX485, is used to complete the standard interfaces of RS-485 communication between PC and TMS320LF2812. The method is shown in Figure3.

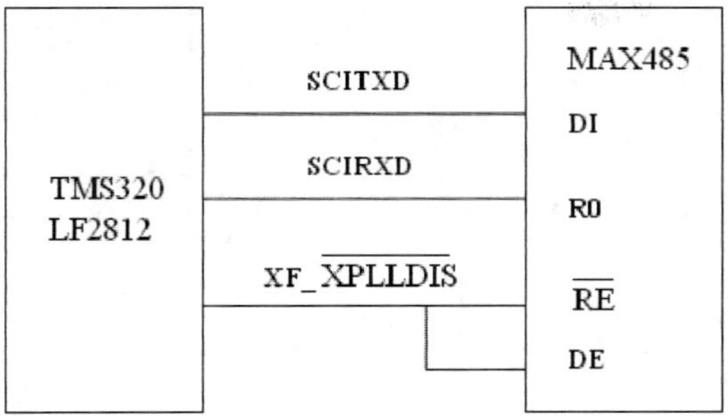

Figure3. Design solution of RS_485 standard by DSP

The ability to serial communication of TMS320F2812 is strong. Because relating to more control class registers, it makes system design more flexible. When design, firstly the control class registers are initialized, including data format definition, interrupt enable, interrupt priority, the baud rate setting and other parameters. After the initialization, sending or receiving data can be start. For sending data, the data should be written into SCITXBUF and then the SCI serial communication module will send the data automatically. For receiving data, the information from the pin SCIRXD / IO is automatically removed start bit, stop bit and parity bit by the SCI serial communication module and the data is written into SCIRXBUF.

AN2131Q is used for the USB interface which is a product of Crypress. There are four wires on the interface between the USB transceiver and PC: +5V, GND, D+, D-. Data is differential transmitted in the lines of D+ and D- as full-speed of 12Mb/s or low-speed of 1.5Mb/s. Transceiver is embedded in the chip, and there are not external circuits, but the pull-up resistor which used to distinguish the working mode of low-speed or full-speed are need to selected. The functional block diagram of AN2131Q is illustrated in figure4[6].

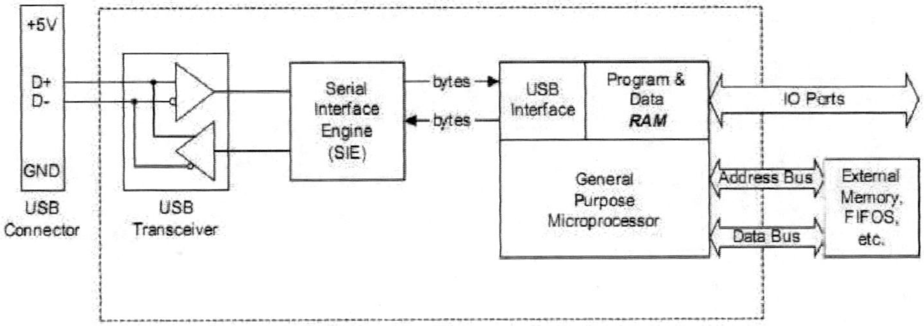

Figure4. The functional block diagram of AN2131Q

Two pieces of 8-bit FIFO (First in First out SRAM) chips are used to realize the two-way communication between the DSP and the USB interface. By this way, the data exchange between USB interface chip and DSP can be maximum faster than the speed of USB bus, so that the speed of data transfer between DSP and host PC is only limited by the USB protocol. Two pieces of IDT72V02 whose memory is 1K×9bit are used as the FIFO chips. If there is information to exchange between DSP and PC, the data is stored in FIFO, and then read by DSP or AN2131Q, in order to avoid data transmission congestion. The diagram of hardware connection is shown in Figure5.

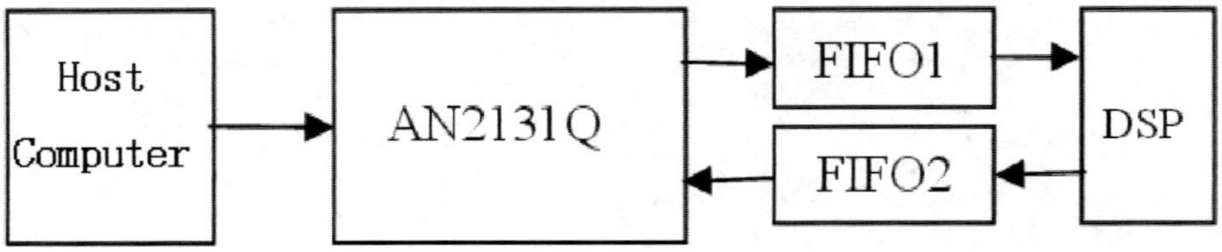

Figure5. The diagram of hardware connection

2.3 Memory Block

In some embedded applications based on DSP architectures, the limited on-chip memory capacity is often insufficient to meet the requirements of mass data throughput. So the strong support of high-speed and large-capacity storage space is needed. In this design, a high-speed synchronous dynamic RAM (SDRAM) chip, IS42SI6400, is selected to expand the storage space of embedded DSP systems. IS42SI6400 is a kind of high-speed SDRAM chip launched by the company of ISSI whose single-chip storage capacity is up to 64M bits.

TMS320LF2812 provides up to 1MB of external memory interface to access the external memory. The SDRAM must be initialized when it is powered. First every storage unit is charged, then SDRAM automatically enters idle state (IDLE). In this state, the SDRAM may enter auto-refresh state or access operation state. Once the refresh is executed, the control program will write the read/write parameter into register. If access to read or write operation, the controller will executes the command of ACTIVEROW, and writes the address line. When BANK is valid, according to the parameters in mode register, the command of READ/WRITE or READAP/WRITEAP is executed. If the operation is completed, the SDRAM automatically enters the idle state. Figure 6 describes the SDRAM status of metastasis.

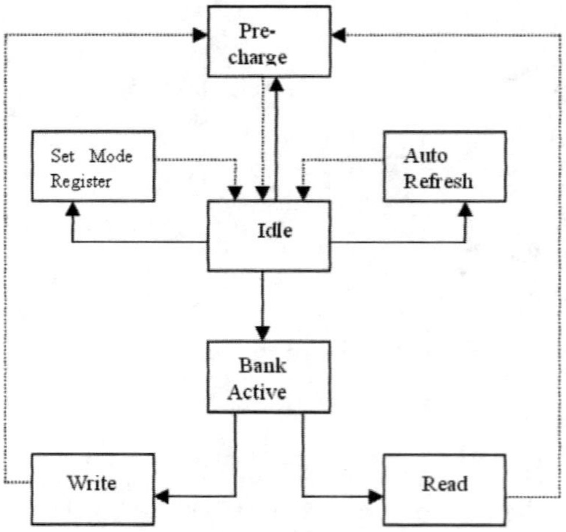

Figure6. The SDRAM status of metastasis

2.4 Other Devices

The system operates from a single +5V external power supply connected to the main power input. Internally, the +5V input is converted to +1.8V and +3.3V using separate voltage regulators. The +1.8V supply is used for the DSP core while the +3.3V supply is used for the I/O buffers of DSP and all other chips on the board.

A Complex Programmable Logic Device (CPLD) is used to implement functionality specific to the system. The CPLD has a register based user interface that lets the user configure the board by reading and writing to its registers.

3. SOFTWARE PART

3.1 Software Architecture

The majority of the software is written in C, using Code Composer Studio (CCS), which provides an integrated development environment to incorporate the software tools. CCS includes tools for code generation, such as a C compiler, an assembler, and a linker. For optimum performance, it is necessary to write some of the code in assembly language.

The C source program includes six modules shown in the figure8.

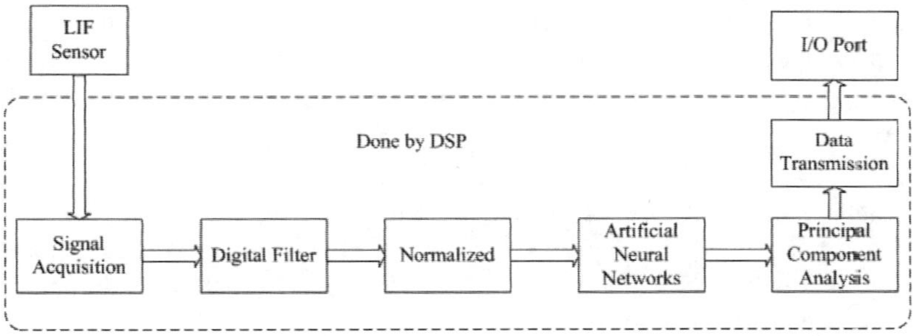

Figure8. Software architecture of the system

The signal from LIF sensor is firstly acquired and digitized by the signal acquisition module. Then the digital signal is clean though the digital filter, and normalized by the software. After the processing of artificial neural network algorithm, the target signal is analyzed by the principal component analysis module, then the result is got and transmitted by the data transmission module.

3.2 Software Implementation

The software mixed use of C language and Assembly Language, C language code is more easily and quickly achieved while Assembly Language has higher performance advantages. In certain cases, mainly control applications, C language code is sufficient. However, in most cases, the initial application written in C language can not meet the requirements of one or more goals. This usually means some assembly codes are needed. In this design, to improve the executive efficiency in the limited resources, and meet the real-time requirements, some steps are taken to ensure the program codes are efficient.

In this design, there are two kinds of communication port. They are used for the communication between this slave and host computer. A protocol is designed to ensure the data communication traffic effective and reliable. Figure9 illustrates the data frame structure of the protocol.

Start	Length of Data	Spectrum Data	Stop

Figure9. Data frame structure

4. IMPLEMENTATION

The signal processing system for LIF sensor is suit to real-time systems which need high-speed signal processing. It can be used to detect the existence of the target material and its concentration. An application for chlorophyll fluorescence sensor is implemented[7], and experiments were done with the algae artificially propagated. The experiments showed that the minimum detectable concentration of chlorophyll a is $1 \mu g/l$, and the signal processing system has good linearity, which prove that the design of the system is reasonable, feasible and practical.

5. CONCLUSION

In this paper, the design and implementation of a novel LIF sensor signal processing system is proposed. Both hardware and software are designed. The hardware is based on DSP TMS320LF2812, and expanded peripheral circuits, which to give it powerful capability of data processing and analysis. The Software is efficient and the algorithm is optimized. The applications show that the proposed signal processing system works quite well for LIF sensor.

REFERENCES

[1] N. BRAS. Laser Induced Fluorescence. Laser Chem, 1990. Vol. 10, pp. 405-412.
[2] M. Terzić, B.P. Marinković, D. Šević, J. Juretal, A.R. Milosavljević. Development Of Time-Resolved Laser-Induced Fluorescence Spectroscopic Technique For The Analysis Of Biomolecules. Physics, Chemistry and Technology, 2008, pp. 105 - 117.
[3] Sung Su Kim, Seul Jung. Hardware Implementation of a Real Time Neural Network Controller with a DSP and an FPGA. Proceedings of the 2004 IEEE International Conference on Robotics & Automation, 2004, pp. 4639-4644.
[4] Huang Wen-sheng, Bryan M. Reid, Timothy B. Smith, and Alec D. Gallimore. Laser-Induced Fluorescence of Singly-Charged Xenon in a 6-kW Hall Thruster Plume. 44th AIAA/ASME/SAE/ASEE Joint Propulsion Conference & Exhibit, 21 - 23 July 2008, Hartford, CT.
[5] Texas Instruments, TMS320F2810, TMS320F2811, TMS320F2812, TMS320C2810, TMS320C2811, TMS320C2812 Digital Signal Processors Data Manual, Literature Number: SPRS174M, Texas, USA, October 2005.
[6]Cypress Semiconductor Corporation. The EZ-USB™ Integrated Circuit Technical Reference Manual Cypress. Version 1.9. May 2000.
[7]Li Xiao, Feng Wei, Zeng Xiao-chun. Advances in Chlorophyll Fluorescence Analysis and Its Uses. Acta Botanica Boreali-Occidentalia Sinica. 2006, 26 (10):2186 - 2196.

Polarization Properties of an Acousto-optic Photonic Crystal Fiber Filter

C -L Zhao [1], W Jin *[2], J Ju [2], X Dong [1], L. Cheng [3]

[1] Institute of Optoelectronic Technology, China Jiliang University, Hangzhou, China
[2] Department of Electrical Engineering, Hong Kong Polytechnic University, Hong Kong, China
[3] Department of Mechanical Engineering, Hong Kong Polytechnic University, Hong Kong

E-mail: eewjin@polyu.edu.hk

Abstract. We investigate the polarization dependent characteristic of an acousto-optic filter in an endlessly-single-mode photonic crystal fiber and find that polarization dependence is high. The maximum wavelength shift due to polarization is ~ 2.25 nm and the polarization dependent loss is ~4.00 dB. The wavelength shift direction is different for different resonant dips in the acousto-optic filter when a polarized broadband light source is used as the input source, which is different largely compared with an acousto-optic filter based on a conventional single-mode fiber.

1. Introduction

Photonic crystal fibers (PCFs), also called microstructured optical fibers or holey fibers are a new class of optical fiber that emerged in recent years [1, 2]. Typically these fibers incorporate a number of air holes that run along the length of the fiber, and have a variety of different shapes, sizes, and distributions. A number of possible applications have been demonstrated because of their unique optical characteristics that cannot be achieved readily by conventional fibers [3, 4]. In recent years, many reports have been made for acousto-optic (AO) devices based on PCFs with interesting filter properties [5-10]. AO mode coupling has been observed in birefringent PCF [7], solid-core photonic bandgap fiber [8], and also hollow core PCF [9]. By utilizing a two-mode PCF that supports two core modes over a broad wavelength range, an all-fiber acousto-optic tunable filter (AOTF) with a wide tuning range as 1000 nm is achieved [11].

However, in a flexural-wave-induced fiber AO filter (AOF), only asymmetrical modes are excited because of the asymmetrical index perturbation profile. Several nearly degenerated constituent modes consist, which have different propagation constants. This leads to a polarization mode splitting effect, and the resonant wavelength of the filter shifts with the input state of polarization. In this letter, we investigate the polarization dependent characteristic of an AOF in an endlessly-single-mode photonic crystal fiber (ESM-PCF) in detail. Experimental results show that polarization properties of the AOF based on the ESM-PCF are different largely compared with that of an AOF based on a conventional single-mode fiber. The wavelength shift direction is opposite for two different resonant dip groups in the AOF based on the ESM-PCF when a polarized broadband light source is used as the input light. The maximum wavelength shift of the AOTF in the ESM-PCF at different polarized light as an input source is ~2.25 nm, and the polarization dependent loss (PDL) is as high as 4 dB.

Journal of Physics: Conference Series **276** (2011) 012111 doi:10.1088/1742-6596/276/1/012111

2. Experimental setup

The schematic diagram of the AOF in the ESM-PCF is shown in Figure 1. The AOF consists of an ESM-PCF, an acoustic horn and a thin piezoelectric transducer (PZT) attached to the horn. When driven a sinusoidal electric signal, the PZT vibration excites a shear mode, which energy is focused at the center and transferred to the fiber by the horn. The flexural acoustic wave propagates along the fiber and results in periodic micro-bends that cause coupling between the two modes at a phase matching wavelength. The period of the micro-bends equals to the acoustic wavelength of the flexural wave in the PCF. Thus, when we tune the acoustic wavelength by changing the frequency of the driven electrical signal, we get a tunable filter. The acousto-optic interaction takes place within length L defined from the tip of the acoustic horn to an acoustic damper where the acoustic wave is absorbed. Transmission spectra of the AOF are measured by using a broadband light source and an optical spectrum analyzer (OSA) with a spectral resolution of 0.05nm.

The ESM-PCF used in our experiments is purchased from Crystal Fiber A/S, and the cross-sectional scanning electron micrograph is also shown in Fig. 1. The mode field diameter is ~6.4 μ m, the spacing between the holes is ~7.78 μ m, and the diameter of the holes is ~3.55 μ m. The diameter of the entire holey region is ~ 60 μ m, and the outer cladding diameter of the PCF is 125 μ m. The total length of the PCF is about 25 cm, and both ends of the PCF are fusion spliced to single-mode fiber (SMF) by using a regular splicing machine (Erisson FSU975). The loss for each splice is about 0.5 dB. The acousto-optic interaction length L is about 16 cm.

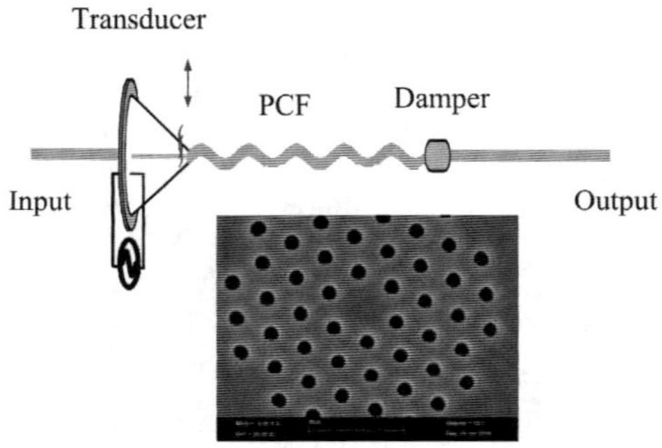

Figure 1. Experimental setup

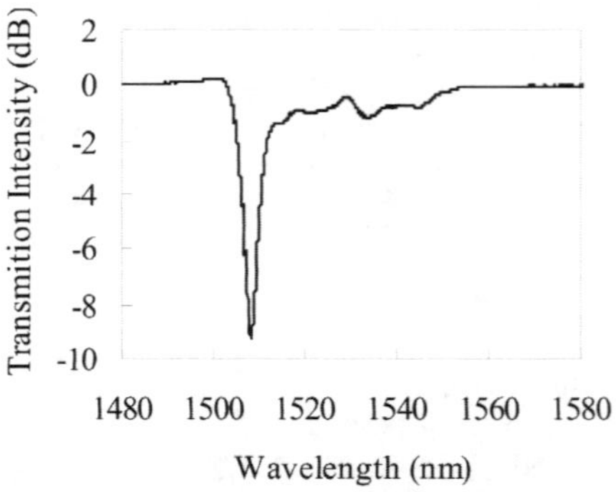

Figure 2. Transmission spectrum of the AOF when rotating the PCF

3. Experiment and results

Figure 2 shows the transmission spectrum of the AOF. The frequency and voltage driven on the PZT are 4100 kHz and 60 V, respectively. The transmission dips are about 1507 nm and about 1540 nm which are due to the fundamental core mode coupling to two different cladding modes. But the dip around 1540 nm is not clear. The acousto-optic coupling is affected strongly by a circumferential angle of initial acoustic vibration on the PCF cross-section [10]. When the vibration direction of the acoustic wave is adjusted by rotating the PCF, the transmission spectrum of the AOF is changed, as shown in Figure 3, in which the dip around 1540 nm becomes deep. It is clear that the dip around 1540 nm splits to two dips, 1535nm and 1541 nm, respectively. The splitting of transmission spectra is caused by the acoustic and optical birefringence and the angular misalignment between the two birefringent axes [10]. In our experiment, we observe the polarization properties of this state in detail.

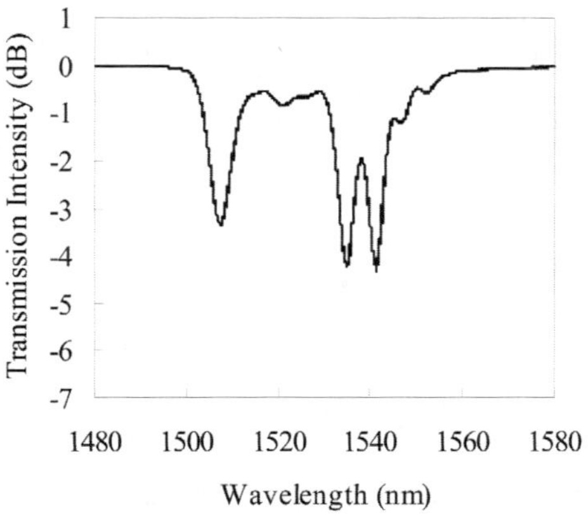

Figure 3. Transmission spectrum of the AOF based on the ESM-PCF

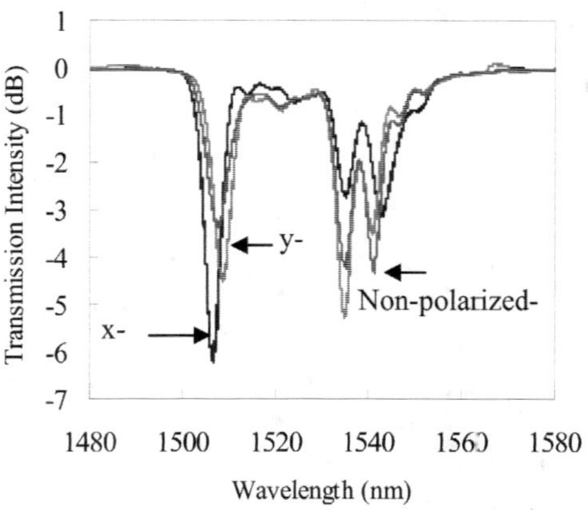

Figure 4. Transmission spectra of the AOF based on the ESM-PCF under non-polarized and polarized light sources

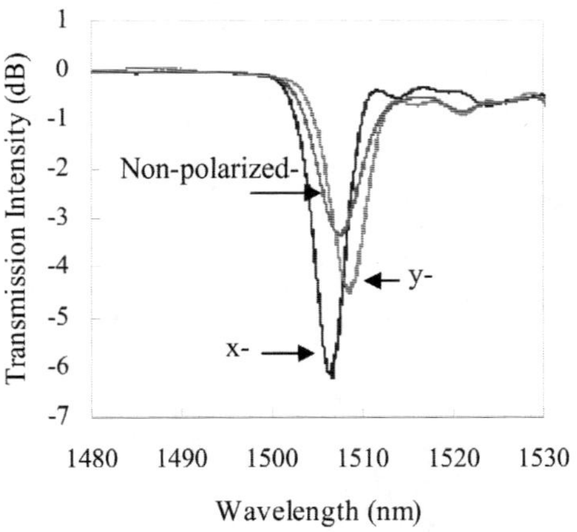

Figure 5. Transmission spectra of the dip at ~1507 nm under non-polarized and polarized light sources

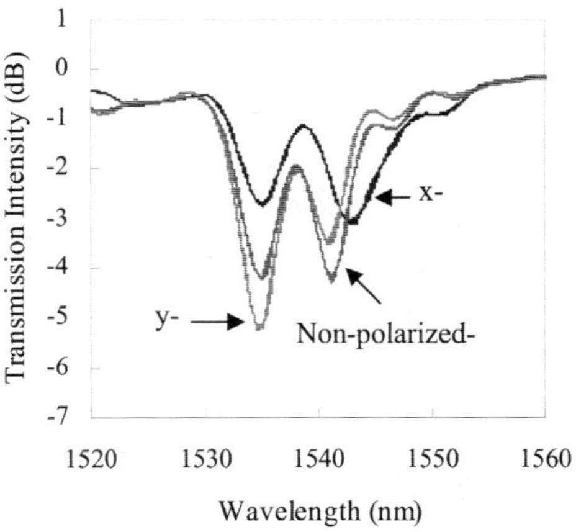

Figure 6. Transmission spectra of the dips at ~1540 nm under non-polarized and polarized light sources

To test the polarization dependent characteristics of the AOF based on the ESM-PCF, a broadband light source with a polarizer is used to provide a polarized broadband source. Figure 4 shows the transmission spectra of the AOF at horizontal and vertical polarization states of the broadband light source. As shown in Fig. 4, the AOF in the ESM-PCF is polarization dependent strongly. The resonant wavelength at around 1507 nm shifts to shorter wavelength when the polarized broadband light source is horizontal (x-polarized) with the vibration direction of the acoustic wave applied on the ESM-PCF, while the two splitting resonant wavelengths around 1535 nm and 1541 nm shift to longer wavelength. When the light source is vertically polarized (y-polarized) with the acoustic wave vibration, the phenomenon is opposite as shown in Figures 5 and 6. The wavelengths of three resonant dips in different light states are given in Table 1. For the resonant wavelength around 1507 nm, the wavelength shift is about -1.05 nm between the light states with x- polarization and non-polarization, and is about 1.2 nm between the y- polarization and the non-polarization light states, respectively. The maxim wavelength difference caused by the polarization is about 2.25 nm. This phenomenon is very different compared with an AOF based on a conventional single-mode fiber.

Journal of Physics: Conference Series **276** (2011) 012111 doi:10.1088/1742-6596/276/1/012111

Table 1. Wavelength of different resonant dips at different polarization states of the input light source.

	Resonant dip 1	Resonant dip 2	Resonant dip 3
Non-polarization state	1507.7 nm	1535.15 nm	1541.25 nm
x-polarization state	1506.75 nm	1535.36 nm	1543.15 nm
y-polarization state	1508.9 nm	1534.95 nm	1541.1 nm

The corresponding polarization dependent loss (PDL) versus wavelength for the AOF in the ESM-PCF is also measured by using an All-parameter-measurement equipment of Agilent Com.. Figure 7 shows the largest PDL is as high as 4.0 dB.

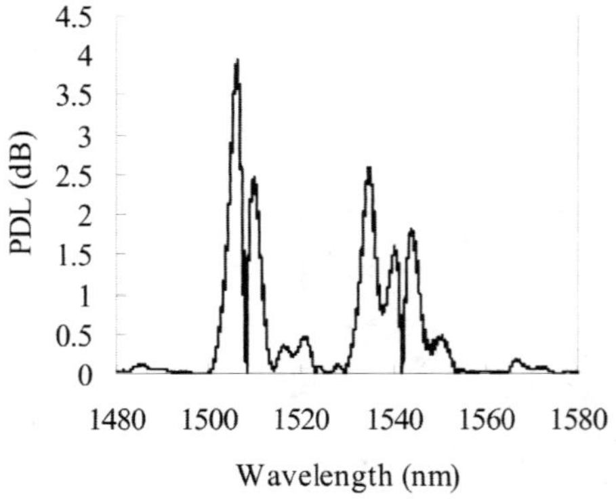

Figure 7. PDL of the AOF based on the ESM-PCF

In an AOF, periodic micro-bends formed by the flexural acoustic wave propagating along the fiber cause coupling between the fundamental core mode to some cladding modes and lead to some dips in the transmission spectrum at wavelengths that satisfy the resonant condition. The phase matching condition of an AOF can be expressed as [12]:

$$\lambda = (n_{co} - n_{cl})\Lambda \qquad (1),$$

where is the resonant wavelength, is the period of the micro-bends and equals to the acoustic wavelength in the ESM-PCF, and n_{co} and n_{cl} are the effective indices of the fundamental core mode, and one of the forward-propagating cladding modes, respectively. At a fixed acoustic wave frequency, the resonant wavelength is shorter for a coupling between the core mode and a lower-order cladding mode due to a larger effective index of the low-order cladding mode.

Since the asymmetrical index perturbation due to the flexural-wave leads to a polarization mode splitting, the resonant wavelength of the filter shifts with the input state of polarization. The resonant dip 1 (at around 1507 nm) shifts to shorter wavelength when an x-polarized light inputs, in the contrast, the resonant dip 1 shifts to longer when a y-polarized light inputs. This means that

$$(n_{co}^{y} - n_{cl1}^{y}) > (n_{co} - n_{cl1}) > (n_{co}^{x} - n_{cl1}^{x}) \qquad (2).$$

Where n_{co}, n_{co}^{y}, and n_{co}^{x} are the effective indices of the core mode, the x-polarization core mode, and the y-polarization core mode, respectively. And n_{cl1}, n_{cl1}^{y}, and n_{cl1}^{x} are the effective indices of the low-order cladding mode, and the corresponding x-polarization and y-polarization mode, respectively. Thus, we deduce (2) to

$$(n_{co}^{y} - n_{cl1}^{y}) - (n_{co}^{x} - n_{cl1}^{x}) > 0 \qquad (3)$$

Then, we can get

$$(n_{co}^{y} - n_{co}^{x}) > (n_{cl1}^{y} - n_{cl1}^{x}) \qquad (4)$$

Relation (4) shows that birefringence of the core mode is larger than that of the low-order cladding mode. By the same method, we can also get $(n_{co}^{y} - n_{co}^{x}) < (n_{cl2}^{y} - n_{cl2}^{x})$ for the resonant wavelength 2 and 3, where n_{cl2}, n_{cl2}^{y}, and n_{cl2}^{x} are the effective indices of the high-order cladding mode, and the corresponding x-polarization and y-polarization mode, respectively. Through the theoretical analysis, the experimental phenomenon will appear when birefringence of the core mode (Δn_{co}), the low-order cladding mode (corresponding to the resonant dip 1, $\Delta n_{cl,1}$), and the high-order cladding mode (corresponding to the resonant dips 2 and 3, $\Delta n_{cl,2}$) is satisfied with $\Delta n_{cl,2} > \Delta n_{co} > \Delta n_{cl,1}$. The origin of this polarization property of the AOF based on the ESM-PCF may be from the ellipticity in the PCF core and the irregularity in the air-hole lattice on the fiber cross-section that results birefringence of the core mode and two cladding modes is different. However, the further studies on the origin are needed.

4. Conclusion

The polarization properties of the AOF based on the ESM-PCF are different largely compared with that of an AOF based on a conventional single-mode fiber. The wavelength shift direction is opposite for two different resonant dip groups in the AOF based on the ESM-PCF when a polarized broadband light source is used as the input light. The reason may be due to the different birefringence of the core mode and two cladding modes of the ESM-PCF which is caused by the ellipticity in the fiber core and the irregularity in the air-hole lattice on the fiber cross-section. This work is very important to understanding the acousto-optic interaction in the PCF deeply, and the properties may be applied on an optical fiber sensor based on polarization.

Acknowledgments: This work was supported by the National Basic Research Program of China (973 Program) under grant No. 2010CB327804 and the project G-YF05 of the Hong Kong Polytechnic University.

[1] Birks T A, Knight J C, and Russell P St J 1997 Endlessly single-mode photonic crystal fiber, *Opt. Lett.* **22(13)** 961-963

[2] Knight J C, Broeng J, and Birks T A 1998 Photonic band gap guidance in optical fibers, *Science*, **282(5393)** 1476-1478

[3] Zhao C -L, Yang X, Lu C, Jin W, Demokan M S 2004 Temperature- insensitive interferometer using a highly berefringent photonic crystal fiber loop mirror, *IEEE Photon. Technol. Lett.*, **16(11)** 2535-2537

[4] Zhao C-L, Xiao L, Ju J, Demokan M S, Jin W 2008 Strain and temperature characteristics of a long- period grating written in a photonic crystal fiber and its application as a temperature-insensitive strain sensor, *J. Lightwave Technol.* **26(2)** 220-227

[5] Diez A, Birks T A, Reeves W H, Mangan B J, and Russell P St J 2000 Excition of cladding modes in photonic crystal fibers by flexural acoustic waves, *Opt. Lett.*, **25(20)** 1499-1501

[6] Haakestad M W and Engan H E 2006 Acoustooptic properties of a weakly multimode solid core photonic crystal fiber, *J. Lightwave Technol.* **24** 838-845

[7] Haakestad M W and Engan H E 2006 Acoustooptic characterization of a birefringent two-mode photonic crystal fiber, *Opt. Express* **14** 7319-7328

[8] Yeom D-I, Steinvurzel P, Eggleton B J, Lim S D, and Kim B Y 2007 Tunable acoustic gratings in solid core photonic bandgap fiber, *Opt. Express* **15** 3513-3518

[9] Yeom D-I, Park H C, Hwang I K, and Kim B Y 2009 Tunable gratings in a hollow-core photonic bandgap fiber based on acousto-optic interaction, *Opt. Express* **17** 9933

[10] Lim S D, Park H C, Hwang I K, and Kim B Y 2008 Combined effects of optical and acoustic birefringence on acousto-optic mode coupling in photonic crystal fiber, *Opt. Express* **16(9)** 6125-6133
[11] Hong K S, Park H C, and Kim B Y, Hwang I K, Jin W and Ju J, Yeom D I 2008 1000 nm tunable acousto-optic filter based on photonic crystal fiber, *Appl. Phys. Lett.* **92** 031110
[12] Bhatia V and Vengsarkar A M 1996 Optical fiber long-period grating sensor, *Opt. Lett.* **21(9)** 692-694

Journal of Physics: Conference Series **276** (2011) 012112 doi:10.1088/1742-6596/276/1/012112

A Single-Fundamental-Mode Photonic Crystal Vertical Cavity Surface Emitting Laser with Seven-point Defect

ZHOU Kang*, XU Chen, XIE Yi-Yang, ZHAO Zhen-Bo, Liu-Fa, LIU Ying-Ming, Wang Bao-Qiang, SHEN Guang-Di
Key Laboratory of Opto-electronics Technology (Beijing University of Technology), Ministry of Education, Beijing University of Technology, 100 Ping Le Yuan, Chaoyang District, Beijing 100124, China

E-mail: kangzhou1986@emails.bjut.edu.cn

Abstract. A single-fundamental-mode photonic crystal (PhC) vertical cavity surface emitting laser (VCSEL) with seven-point defect was fabricated by etching two-dimensional PhC on surface of the VCSEL's top distributed Bragg-reflector (DBR). The PhC was fabricated using electron beam lithography (EBL) and inductive coupled plasma reactive ion etching (ICP-RIE). The single-fundamental-mode performances were investigated and demonstrated. The device with output power 0.7mW under continuous-wave (CW) operate, half maximum of the lasing spectrum less than 0.06nm, far field divergence angle less than 10° and side mode suppression ratio over 35dB was obtained. The characteristics of the PhC VCSEL operating at pulse condition are compared with VCSEL with the similar aperture size, we found that the output power of PhC VCSEL increased more than conventional VCSEL with the impulse duty factor decreased, corresponding to 100% and 32%, respectively.

1. Introduction

The single-fundamental-mode vertical cavity surface emitting lasers (VCSELs) are important for a variety of applications, such as high-speed fiber optic data communication, free space optical interconnects and modulation spectroscopy [1-3]. For obtaining single mode the oxide aperture of conventional oxided VCSEL is restricted to 3 μm and the output power is limited to 5mW [4,5]. Getting large oxide aperture to achieve higher output power and at the same time operating at single fundamental mode is a real challenge while the VCSELs with large oxide aperture diameter usually operate at multiple transverse modes. Several methods have explored to solve this problem, such as

* Supported by the National High Technology Research and Development Program of China under Grant No 2008AA03Z402, and the Beijing Natural Science Foundation of Beijing under Grant No 4092007.
Email: kangzhou1986@emails.bjut.edu.cn, kangmeihou@163.com Tel: + 86 01067391641-834 Fax: +86 01067391641-822

etching surface relief, nonperiodic holes and photonic crystal on the surface of the VCSEL's top distributed Bragg-reflector (DBR) [6-10]. PhC VCSELs with a reproducible and simple fabrication process are extendible to any vertical cavity surface emitting device. Also the photonic crystal approach can provide single-fundamental-mode emission with a specific structure independent of wavelength [11].

PhC VCSEL operating at single-fundamental-mode can be explained by " endlessly single mode" PhC fiber model. In the year of 2003, Noriyuki Y and Danner A J introduced etching-depth dependence parameter to describer the influence of finite etching depth [12]. In this model the V_{eff} parameter is used to evaluate the laser modal properties:

$$v_{eff} = 2\pi \frac{a}{\lambda} \sqrt{n^2 - (n - \gamma \Delta n)^2} \qquad (1)$$

Where a is the lattice constant, λ is the emission wavelength, n is the average refractivity of the DBR, Δn is the reduction of the index after etching the air hole and λ is the etching-depth dependence factor. If V_{eff} was less than 2.405, the structure is considered to be a single-mode waveguide, but the V_{eff} should be big enough to overcome the other factors such as the index changing caused by thermal effect. Therefore, to obtain stable single fundamental mode the V_{eff} value of device should be less than but close to 2.405.

In this paper, a single-fundamental-mode PhC-VCSEL with seven-point defect was fabricated and characterized. The PhC structure was fabricated by using UV photolithography and inductive coupled plasma reactive ion etching on the top DBR. A laser with single mode output power 0.7mW, half maximum of the lasing spectrum less than 0.06nm, far field divergence angle less than 10° and side mode suppression ratio (SMSR) over 35dB was obtained. We compared the PhC-VCSEL with conventional VCSEL with the similar aperture size operating in different impulse duty factor when frequency is 1000Hz. We found that with the decreasing of the duty factor the output power of both PhC and conventional VCSEL are enhanced furthermore the PhC VCSEL's enhanced more quickly than conventional one.

2. Device structure and processing

The structure of device consist of a top p-type (C-doped) 22-period DBR of alternating layers of $Al_{0.12}Ga_{0.88}As$ and $Al_{0.90}Ga_{0.10}As$, a similar button n-type (Si-doped) 34-period DBR and an undoped three quantum-well active region sandwich between top and button DBR. A 5nm heavily doped p-type GaAs contact layer was grown above the top DBR. Also there is a 30nm $Al_{0.98}Ga_{0.02}As$ oxidization layer between top DBR and the active region. The VCSEL epitaxial structure was grown on n+-Gas (100) substrate.

We made the circular mesas by ICP-RIE with $SiCl_4/Ar/Cl_2$. Then the $Al_{0.98}Ga_{0.02}As$ oxide layer was oxided to form a oxide aperture with diameter of 10 μ m to confine the current. After then the top and button ohmic contact were formed respective with Ti/Au and Au/Ge/Ni/Au. Then we annealed the devices using rapid thermal annealing at 430 for 35s. With the EBL and ICP-RIE mask respective with ZEP and SiO_2 we etch the PhC in the top DBR. The depth of the PhC with lattice constant a=2 μ m, air hole diameter b=1 μ m and defect diameter 7 μ m is 1.7 μ m. The PhC structure and section of air hole are shown in Figure 1(a) and (b).

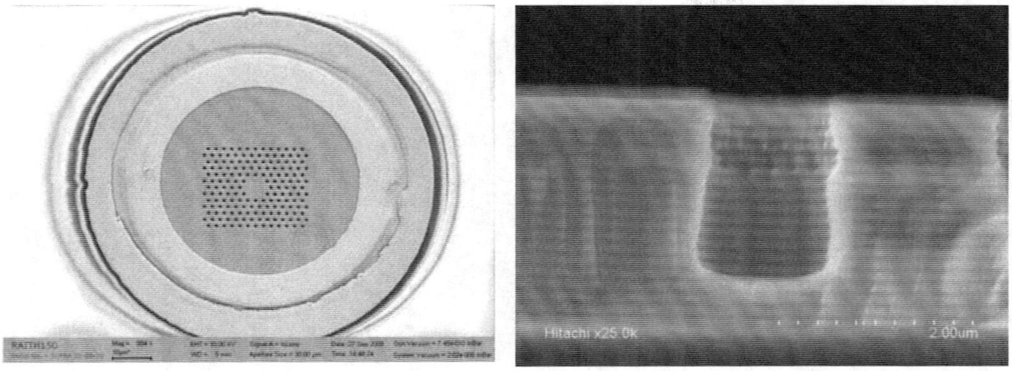

(a) (b)

Figure 1. (a)SEM image of the seven-point-defect PhC-VCSEL.(b)SEM image of section of an air hole

3. Device performances

Figure 2 (a) shows the measured CW L-I-V characteristic curves for the PhC VCSEL at room temperature. The threshold current of this device is 1mA, the slope efficiency is 0.08W/A and the series resistance is 67.32Ω. When the inject current is 11mA the output power of the PhC VCSEL is 0.7 mW. The CW L-I-V characteristic of VCSEL without PhC structure with the similar aperture size was also measured corresponding to the PhC VCSEL showed in Figure 2 (b). The VCSEL exhibits the threshold current of 0.4 mA, the slope efficiency of 0.415 W/A and the output power of 5.9 mW. As the figures show that the threshold current of the PhC VCSEL is higher than the VCSEL's while the output power is lower. The reason is that we etched air hole in the top of the DBR. Firstly, the air hole will reduce the refractive index of the top DBR so there must be higher inject current to laser the device. Secondly, the etch damage will be introduced which will increase the number of dangling bond, the increased dangling bonds will cause non-radiative recombination which will increase the threshold current and reduce the output power. Lastly the increased scattering loss introduced by air hole will also increase the threshold current.

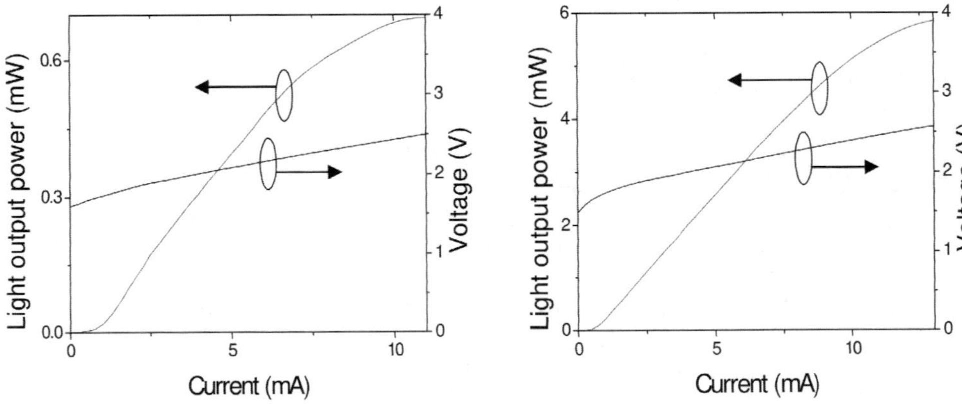

(a) (b)

Figure 2. (a) L-I-V characteristic of PhC VCSEL (b) L-I-V characteristic of VCSEL

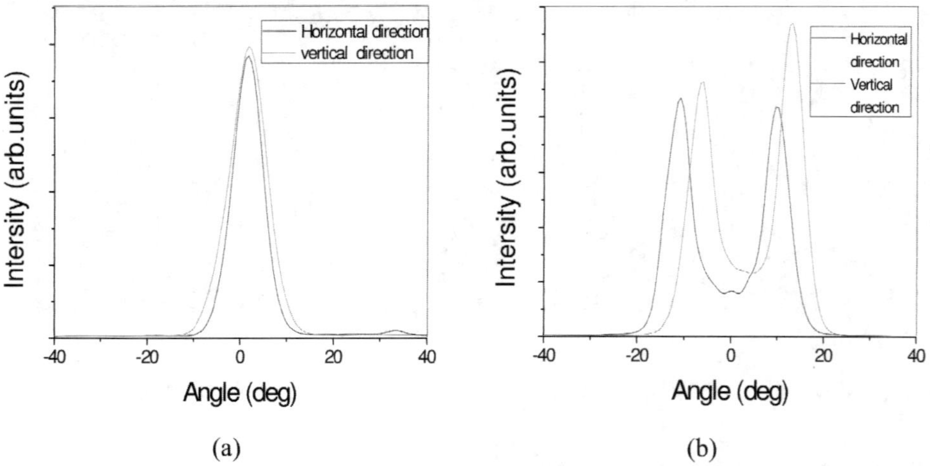

Figure 3. (a) the far field distribution of PhC VCSEL (b) the far field distribution of VCSEL

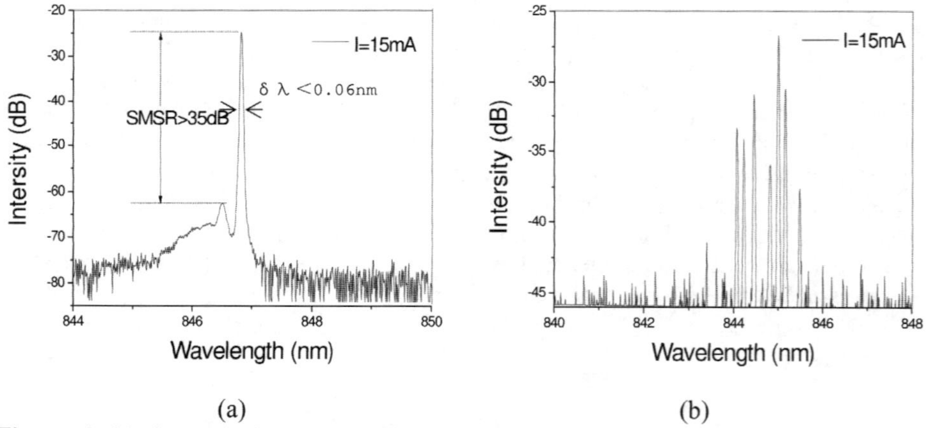

Figure 4. (a) the optical spectrum of the PhC VCSEL (b) the optical spectrum of VCSEL

Fig 3 (a) and (b) show the far filed distribution of PhC VCSEL and the conventional VCSEL without PhC, respectively. The divergence angles of PhC VCSEL with seven-point defect in both horizontal and vertical direction are less than 10° while the angles of conventional VCSEL are more than 20° . At the same time , the far filed distribution of PhC VCSEL in both horizontal and vertical direction have only one intensity peak while the conventional VCSEL have multiple intensity peaks. These indicate that the PhC VCSEL work in single-transverse-mode but the conventional VCSEL not. Fig 4 (a) and (b) show the optical spectrum of the PhC and conventional VCSEL at injection current I=15mA measured by optical spectrum analyzer, respectively. The optical spectrum of PhC VCSEL shows only on intercity at the wavelength λ=846.8nm with full width at half maximum (FWHM) less than 0.06nm and SMSR over 35dB. By contrast the optical spectrum of conventional VCSEL shows multiple intensity peaks corresponding to different wavelength. Therefore, by introducing PhC structure in the top DBR we can improve the quality of the beam and make VCSEL operating at single-fundamental-mode.

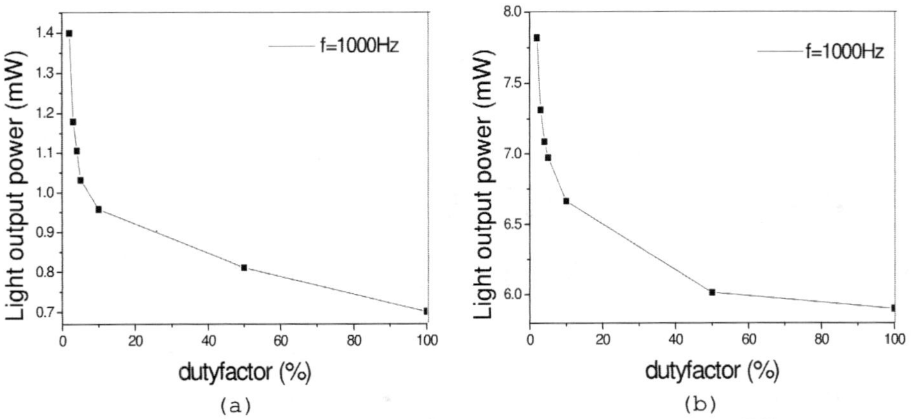

Figure 5. output power of (a) PhC and (b) conventional VCSEL in different dutyfactor

Fig 5 (a) and (b) show the output power of PhC and conventional VCSEL in different impulse duty factor when impulse frequency is 1000Hz. With impulse duty factor decreased from 100% to 2%, the output power of PhC and conventional VCSEL increased from 0.7 mW to 1.4 mW and 5.9 mW to 7.8 mW, corresponding to 100% and 32%, respectively. The reason is that we etched PhC pattern into the top DBR which will increase the resistance and operating current of device and that will result in generating more heat, at last the high junction temperatuer will depress the saturation power.

4. Conclusion

In this paper we report a PhC VCSEL with seven-point defect operating at single-fundamental-mode throughout the operating current range. This device with oxide aperture about 10 μ m has ultra-low threshold current about 1mA. By the lateral index confinement of seven-point 2D PhC pattern output light with SMSR over 35dB and divergence angles less than 10° was obtained. The experimental results agree with the modified effective index model.

5. Reference

[1] Fieldler U, Reiner G, Schnitzer P and Ebeling K J 1996 IEEE Photonics Technol. Lett. **8** 746
[2] Bristow J, Lehman J, Liu Y, Hibbs-Brenner M, Galarneau L and Morgan R A 1997 Proc. SPIE **3005** 112
[3] Mansuripur M and Sincerbox G 1997 Proc. IEEE **85** 1780
[4] Jung C, Jäger R, Grabherr M, Schnitzer P, Michalzik R, Weigl B, Muller S and Ebeling K J 1997 Electron. Lett. **33** 1790
[5] Dallesasse J M, Holonyak N, Sugg A R, Richard T A, and El-Zein N 1990 Appl. Phys. Lett. **57** 2844
[6] Martinsson H, Vukušic′ J A, Grabherr M, Michalzik R, Jäger R, Ebeling K J and Larsson A 1999 IEEE Photon. Technol. Lett. **11** 1536
[7] Haglund A, Gustavsson J S, Vukusic J, Modh P and Larsson A 2004 IEEE Photonics Technol. Lett. **16** 386
[8] Furukawa A, Sasaki S, Hoshi M and Matsuzono A 2004 Appl. Phys. Lett. **85** 5161
[9] Leisher P O, Danner A J, Raftery J J and Choquette K D 2005 Electron. Lett. **41** 1010–1011
[10] Song D S, Kim S H, Park H G, Kim C K and Lee Y H 2002 Appl. Phys. Lett. **80** 3901
[11] Birks T A, Knight J C and Russell P St J 1997 Opt. Lett. **22** 961
[12] Yokouchi N, Danner A J, and Choquette K D 2003 Appl. Phys. Lett. **82** 1344

Theoretical analysis and experiment performance of slow-light based on stimulated Brillouin scattering (SBS)

Hongyan Zhou[1]*, Kun Zhong[1], Ru Zhang[1] Peilin Lang[1]
1 School of Science, Beijing University of Posts and Telecommunications, Beijing 100876
E-mail: ziyan_zhou@sina.com.

Abstract. Slow light technology will play a key role in future all-optical communication. The slow-light technology based on stimulated Brillouin scattering has become a research highlight because of its additional advantages, such as compatibility of the devices with existing telecommunication systems, room-temperature operation, and tunable at arbitrary wavelengths. According to the propagation of a cw pulse through a Brillouin fiber amplifier, whose frequency is near the Stokes resonance, via three-wave coupling equations, both pump depletion and fiber losses taken into consideration, the principle of how slow-light effect based on stimulated Brillouin scattering produced and the mathematical expression of time delay are strictly deduced. A delay of 8 ns is obtained when the input Stokes pulse is 200ns and the SBS (stimulated Brillouin scattering) gain G is ~18 in our designed experiment of SBS slow-light system. Then the extent of transformation from pump waves to Stokes waves is measured using MATLAB numerical simulation according to the experiment dates, based on the relation between output pump light power and input pump light power and also the relation between output Stokes light power and input pump light power. And the relation between the input light power and propagation distance is discussed as well. Finally the relation between slow light pulse delay and SBS gain is also obtained.

1. INTRODUCTION

Slow light is widely used in many important sections in all-optical communication, such as optical buffering, optical storage, optical equilibrator and optical information processing, and it can also carry out all-optical route and all-optical swap. Slow light based on stimulated Brillouin scattering has become a hot topic. It is an important breakthrough in optical fiber communication that pulse time

* Corresponding author: Room 1214,Student Apartment 4,Beijing University of Posts and Telecommunications(post code:100876) TEL:(0)15210352628

delay is achieved based on stimulated Brillouin scattering—nonlinear effect in fiber.
Kwang Yong Song [1] got 30ns time delay for a 100ns pulse for the first time based on stimulated Brillouin scattering in fiber in 2005, and Zhu ZH M [2, 3] got 47 ns in 2007. Meanwhile, Kwang Yong Song research group improved wavelength to 25GHz, and got 10.9ps for a pulse of 37ps [4].
Zhu ZH M [6] made theoretical analysis on how slow light produced based on stimulated Brillouin scattering, according to three-wave equations in weak signal regime. Here we deduce the relations between emerged pump power and emerged Stokes power and incident puma power when fiber losses and pump depletion are both considered in the general case, and the relation between time delay and gain. A delay of 8 ns is obtained when the input Stokes pulse is 200ns and the SBS gain G is ~18 in the designed experiment of SBS.

2. THE PRINCIPLE OF SLOW-LIGHT BASED ON STIMULATED BRILLOUIN SCATTERING

The pump wave generates acoustic waves through the process of the electrostriction, which in turn causes a periodic modulation of the refractive index. Finally, the pump-induced refractive index grating scatters the light through Bragg diffraction [7]. Brillion scattering is then stimulated along the fiber by the electrostriction. Then the produce of slow light based stimulated Brillouin scattering can be realized as interaction of three waves: a backward strong pump wave (-z direction) , a forward weak Stokes wave (+z direction) ,and a backward acoustic wave, when special frequency condition achieved, that is $\omega_P = \omega_S + \Omega_B$.Because the coupling action among the three waves, then slow light emerges [6].

$$-\frac{\partial E_P}{\partial z}+\frac{n_{fg}}{c}\frac{\partial E_P}{\partial t}=-\frac{\alpha}{2}E_P+\frac{g_B}{2A_{eff}}\left|E_S\right|^2E_P+ig_2E_S\rho \tag{1a}$$

$$\frac{\partial E_S}{\partial z}+\frac{n_{fg}}{c}\frac{\partial E_S}{\partial t}=-\frac{\alpha}{2}E_S-\frac{g_B}{2A_{eff}}\left|E_P\right|^2E_S+ig_2E_P\rho \tag{1b}$$

$$\frac{\partial \rho}{\partial t}+(\frac{\Gamma_B}{2}-i\Delta\omega)\rho=i\frac{g_1}{\eta}E_PE_S^* \tag{1c}$$

Here E_P, E_S, ρ are the field amplitudes of pump, Stokes, and the acoustic, n_{fg} is group index of the fiber mode , α is fiber loss coefficient , $\Gamma_B/2\pi$ is the bandwith(FWTH) , $\Delta\omega=(\omega_{P0}-\Omega_B)-\omega_{S0}=\omega_0-\omega_{S0}$ is the detuning from SBS frequency shift, Ω_B is the SBS frequency. $\omega_{P0}(\omega_{S0})$ is the center angular frequency of the pump (Stokes)wave, $g_1=\gamma_e\varepsilon_0\Omega_B/(4v_a^2)$; $g_2=\omega_{p0}\gamma_e/(4cn_f\rho_0)$ $\eta=\frac{1}{2}c\varepsilon_0 n_f$; γ_e is electrostriction coefficient; v_a is the speed of the acoustic, n_f is mode index of the fiber mode, ρ_0 is the material density. $|E_P|^2(|E_S|^2)$ is optical intensity of the pump (Stokes) wave.

$$\frac{\partial \rho^*}{\partial t}+(\frac{\Gamma_B}{2}+i\Delta\omega)\rho^*=-i\frac{g_1}{\eta}E_P^*E_S \qquad (1c^*)$$

We use a Fourier-transform (FT) technique, $FT[E_S(z,t)]=\tilde{E}_S(z,\omega)$, $FT[E_P(z,t)]=\tilde{E}_P(z,\omega)$, $FT[\rho(z,t)]=\tilde{\rho}(z,\omega)$, transforming the equations into the frequency domain, then (1b)(1c) can be rewritten as follows:

$$\frac{\partial \tilde{E}_s}{\partial z}-i(\omega-\omega_{so})\frac{n_{fg}}{c}\tilde{E}_s=-\frac{\alpha}{2}\tilde{E}_s-\frac{g_B}{2A_{eff}}\left|E_P\right|^2\tilde{E}_s+ig_2\tilde{E}_P\tilde{\rho}^* \qquad (2b)$$

$$[\frac{\Gamma_B}{2}-i(\omega-\omega_{P0}+\Omega_B)]\tilde{\rho}^*=-i\frac{g_1}{\eta}\frac{1}{2\pi}\tilde{E}_P^*\tilde{E}_S \qquad (2c)$$

Then $$\tilde{\rho}^*=\frac{-i\frac{g_1}{\eta}\frac{1}{2\pi}\tilde{E}_P^*\tilde{E}_S}{\frac{\Gamma_B}{2}-i(\omega-\omega_{P0}+\Omega_B)} \qquad (3)$$

Then (2b) is

$$\frac{\partial \tilde{E}_S}{\partial z}-i(\omega-\omega_{SO})\tilde{E}_S=-\frac{\alpha}{2}\tilde{E}_S-\frac{g_B}{2A_{eff}}\left|\tilde{E}_P\right|^2\tilde{E}_S+ig_2\tilde{E}_P\frac{-i\frac{g_1}{\eta}\frac{1}{2\pi}\tilde{E}_P^*E_S}{\frac{\Gamma_B}{2}-i(\omega-\omega_{PO}-\omega_B)}$$

$$=-\frac{\alpha}{2}\tilde{E}_S-\frac{g_B}{2A_{eff}}I_P(z)\tilde{E}_S+\frac{\frac{1}{4\pi}g_0I_P(z)\tilde{E}_S}{1-i2\delta\omega/\Gamma_B} \qquad (4)$$

$$(g_0=\frac{4g_1g_2}{\eta/\Gamma_B},\delta\omega=\omega-\omega_{P0}+\Omega_B)$$

So $$\frac{\partial \tilde{E}_S}{\partial z}=i(\omega-\omega_{S0})\frac{n_{fg}}{c}\tilde{E}_S-\frac{\alpha}{2}\tilde{E}_S-\frac{g_B}{2A_{eff}}I_P(z)\tilde{E}_S+\frac{\frac{1}{4\pi}I_P(z)\tilde{E}_S}{1-i2\delta\omega/\Gamma_B} \qquad (5)$$

Comparing (5)with the SVEA wave equation in the frequency domain for Stokes pulse ,we can get :

$$(\frac{\omega}{c}-\frac{\omega_{S0}}{c})n_{fg}-i\frac{\frac{1}{4\pi}g_0I_P(z)}{1-i2\delta\omega/\Gamma_B}+i\frac{\alpha}{2}+i\frac{g_B}{2A_{eff}}=k_s(\omega)-k_s(\omega_{S0}) \qquad (6)$$

$$n_s=n_f-i\frac{c}{\omega}\frac{\frac{1}{4\pi}g_0I_P(z)}{1-i2\delta\omega/\Gamma_B}+i\frac{\alpha c}{\omega}+i\frac{g_Bc}{\omega A_{eff}} \qquad (7)$$

The new index of the medium is $$n_s=n_f-i\frac{c}{\omega}\frac{\frac{1}{4\pi}g_0I_P(z)}{1-i2\delta\omega/\Gamma_B}+i\frac{\alpha c}{\omega}+i\frac{g_Bc}{\omega A_{eff}} \qquad (8)$$

Group index is $n_g(\omega) = n_s + \omega(dn_s/d\omega) = n_{fg} + \frac{\frac{1}{2\pi} c g_0}{\Gamma_B} \frac{1-4\delta\omega^2/\Gamma_B^2}{(1+4\delta\omega^2/\Gamma_B^2)^2} I_P$ （9）

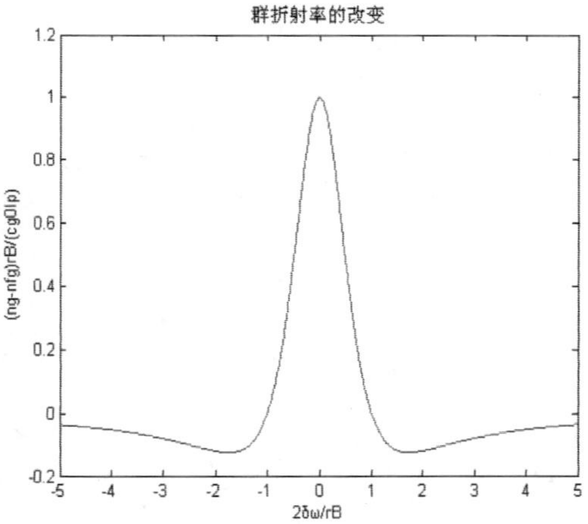

Figure1: the relation between the change of group index and the detuning of incident Stokes from resonance frequency

The propagation group velocity $v_g = \frac{d\omega}{dk} = \frac{c}{n_g} = \frac{c}{n+\omega\frac{dn}{d\omega}}$, （10）when the change of index $dn/d\omega$ is great, group velocity will change greatly accordingly. $dn/d\omega >> 1$, then group velocity is small, and slow light emerges.

3. OPTICAL POWER ANALYSIS IN A LOSSY SYSTEM

3.1 INTENSITY COUPLING EQUTION

When pump depletion is considered ,the interaction is governed by the following set of two coupled equations [8]:

$$\frac{dP_P}{dz} = -\gamma P_P P_S - \alpha P_P \qquad \frac{dP_S}{dz} = -\gamma P_P P_S + \alpha P_S \qquad (11)$$

Here, we have normalized length $\xi = \frac{z}{L}, 0 \le \xi \le 1$ [9], $w = \frac{P_P(z)}{P_P(0)}$, measurement of pump depletion, $u = \frac{P_S(z)}{P_P(0)}$, measurement of how much pump energy transformed to Stokes, $a = \alpha L$, measurement of the total linear absorption, $k = \gamma P_P(0)L$, $\gamma = \frac{g_B}{A_{ao}}$, $\varepsilon = \frac{P_S(L)}{P_P(0)}$.

$$\frac{dw}{d\xi} = -kuw - aw \qquad \frac{du}{d\xi} = -kuw + au \tag{12}$$

Boundary conditions: $w(0) = 1$ $u(1) = \varepsilon$

We recognize the unknowns through perturbations to the lossless systems,

$$w(\xi) = A(\xi)[1-\rho(\xi)] \qquad u(\xi) = B(\xi)[1-\mu(\xi)] \tag{13}$$

In the lossless systems: $\frac{dA}{d\xi} = -kAB$ $\frac{dB}{d\xi} = -kAB$ (14)

Boundary conditions: $A(0) = 1$ $B(0) = \varepsilon$

Then $A(\xi) = C_0[1-(1-C_0)\exp(-C_0 k\xi)]^{-1}$ $B(\xi) = C_0(1-C_0)[\exp(C_0 k\xi) - 1 + C_0]^{-1}$ (15)

$$C_0 = A(\xi) - B(\xi)$$

According to (13) (14) and (12) we obtain

$$\frac{d\rho}{d\xi} = (-kB\mu + a)(1-\rho) \tag{16}$$

$$\frac{d\mu}{d\xi} = -(kA\rho + a)(1-\mu) \tag{17}$$

Boundary conditions: $\rho(0) = 0$ $u(1) = 0$

For small perturbation, we discard perturbation to ρ, then $\mu(\xi) \approx 1 - e^{a(\xi-1)}$ (18)

We can get the follows from (18) and (16):

$$\frac{d\rho}{d\xi} = (-kB(\xi)1 - e^{a(\xi-1)} + a)(1-\rho) \Rightarrow \int \frac{-d\rho}{1-\rho} = \int_0^{\xi} kB(\xi)(1-e^{a(\xi-1)})d\xi - a\xi$$

$$In[1-\rho(\xi)] = k\int_0^{\xi} B(\xi)(1-e^{a(\xi-1)})d\xi - a\xi \approx k(1-e^{-a})\int_0^{\xi} B(\xi)d\xi - a\xi \tag{19}$$

Because $\frac{dA}{d\xi} = -kAB \Rightarrow k\int_0^{\xi} B(\xi)d\xi = -InA(\xi)$

Then $1-\rho(\xi) = A(\xi)^{\exp(-a)-1} e^{-a\xi}$ (20)

We can obtain the followings from (13) and (20)

$w(\xi) \approx A(\xi)^{\exp(-a)} \exp(-a\xi)$ (21) $u(\xi) \approx B(\xi)\exp[a(\xi-1)]$ (22)

Boundary conditions: $B(1) = \varepsilon = C_0(1-C_0)[\exp(C_0 k) - (1-C_0)]^{-1}$

C_0 can be written as $C_0 \approx \frac{1}{k}\{\Lambda + In[\Lambda(1-\frac{\Lambda}{k})]\}$ (23), $\Lambda = -In(\varepsilon k) = -In[\gamma P_s(L)L], \varepsilon k << 1$ [9]

Ⅰ.according to (15), (21) - (22), (23) ,we can obtain the simulated result of how pump light intensity $w(\xi)$ and Stokes light intensity $u(\xi)$ varies along the propagation distance.

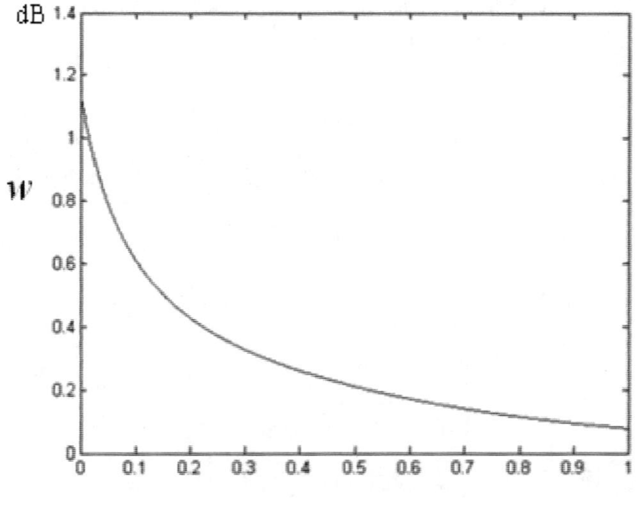

Figure 2: Normalized power of pump wave $w(\xi)$ versus propagation distance ξ

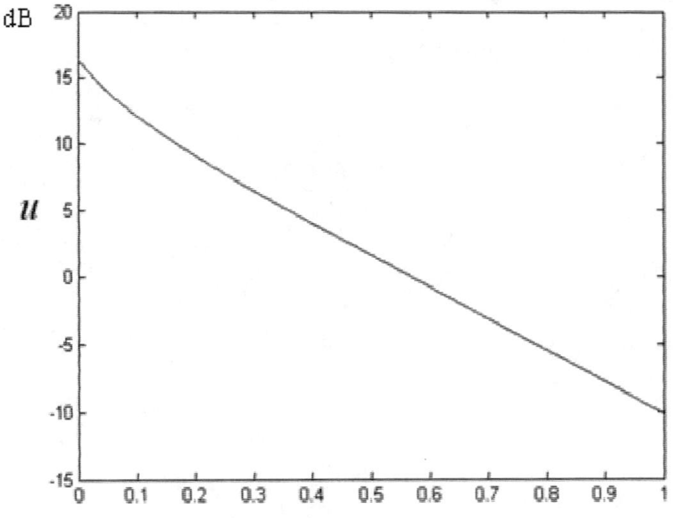

Figure 3: Normalized power of Stokes wave $u(\xi)$ versus propagation distance ξ

FIGURE 2 shows pump wave power attenuation constantly along the propagation distance ξ; FIGURE 3 shows Stokes wave power become stronger constantly along the propagation distance. We can come to the conclusion that pump wave translates energy largely to Stokes.

3.2 SBS experiment

3.2.1 composition of SBS experiment.

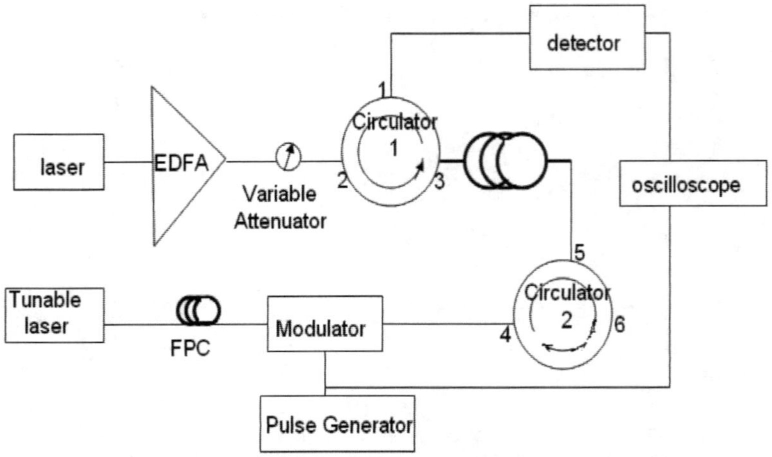

Experiment setup: FPC, fiber polarization controller ;EDFA, erbium-doped fiber amplifier

Figure 4: composition of SBS experiment

Altering different incident pump wave intensity, we get different curve according to the dates of spectrometer 1 and 2, showing the relation between output pump light power $I_p(L)$ and input pump light power $I_P(0)$ and also the relation between output Stokes light power $I_S(0)$ and input pump light power $I_P(0)$.

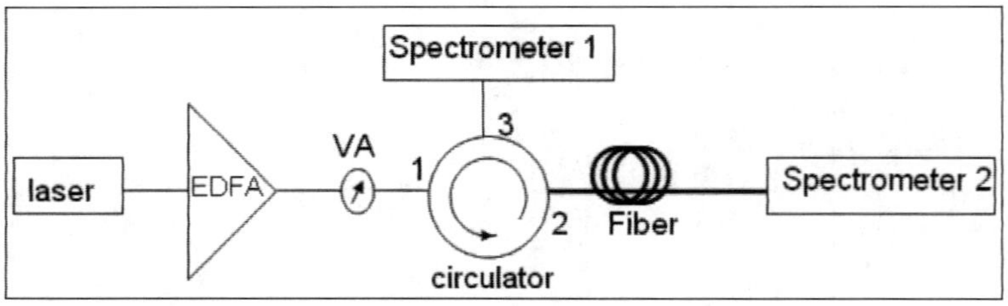

Figure 5: light intensity detecting

The following figures are the results according to the experiment dates

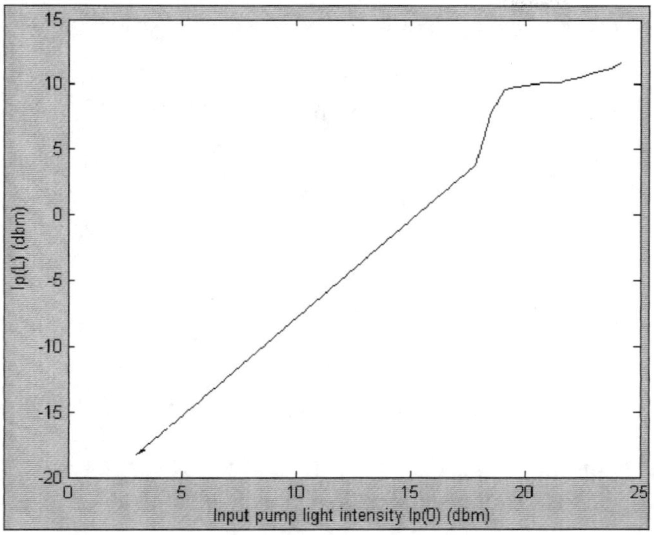

Figure 6: the relation between output pump light power $I_p(L)$ and input pump light power $I_P(0)$

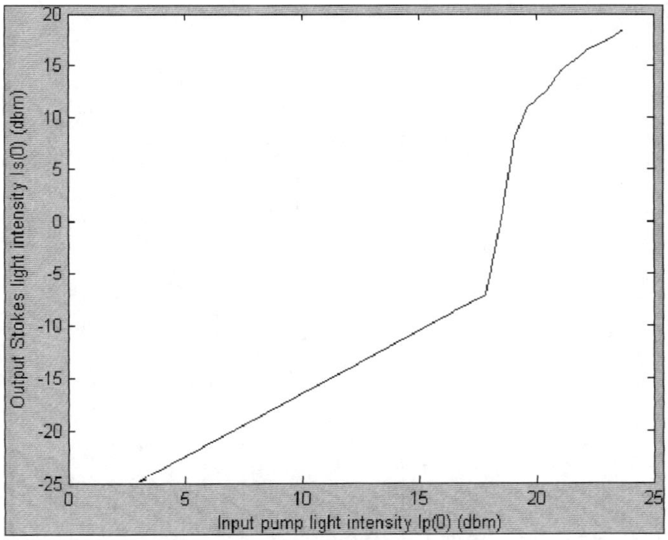

Figure7: the relation between output Stokes light power $I_S(0)$ and input pump light power $I_P(0)$.

We can come to the conclusion: ①when pump light intensity is weak, output pump light intensity and output Stokes light intensity become stronger because of input pump light intensity's becoming stronger, and nearly linear enlarged; ②only when input pump light intensity comes to a value, two curves change steeply, meaning both of output pump light intensity and output Stokes light intensity amplify rapidly. output pump light intensity is still large compared to be its depletion; while Stokes light intensity largely is amplified because of SBS gain; ③then input pump light intensity become stronger and accompanied is energy transferring largely to Stokes, output pump light intensity trends to be a gentle value due great depletion: output Stokes light intensity obtains gain, and turn into saturation regimes. So if the input pump light intensity is strong, its depletion is a problem that can't be ignored.

In our model we use the common single-mode fiber SMF-28 at 1550nm,and the value are as follows: $A_{eff}=50\mu m^2$, $\Gamma_B/2\pi=40MHz$, $g_0=5\times10^{-11}m/W$,L=50m, $\lambda=1550nm$, $n_{fg}=1.45$, $\alpha=0.2dB/km$。

Then the pulse time delay can be written [6]:

$$\Delta T_d=\frac{L}{\frac{c}{n_g}}-\frac{L}{\frac{c}{n_{fg}}}=\frac{L}{c}(n_g-n_{fg})=\frac{G}{\Gamma_B}\frac{1-4\delta\omega^2/\Gamma_B^2}{(1+4\delta\omega^2/\Gamma_B^2)^2}\frac{G}{\Gamma_B}(1-12\delta\omega^2/\Gamma_B^2) \qquad (24)$$

Gain parameter $G=g_0LI_p(0)$ [6].If detuning is ignored($\delta\omega=0$),then $\Delta T_d\propto G\propto I_p(0)$,and the time delay trend of the Stokes light is shown in fig 6

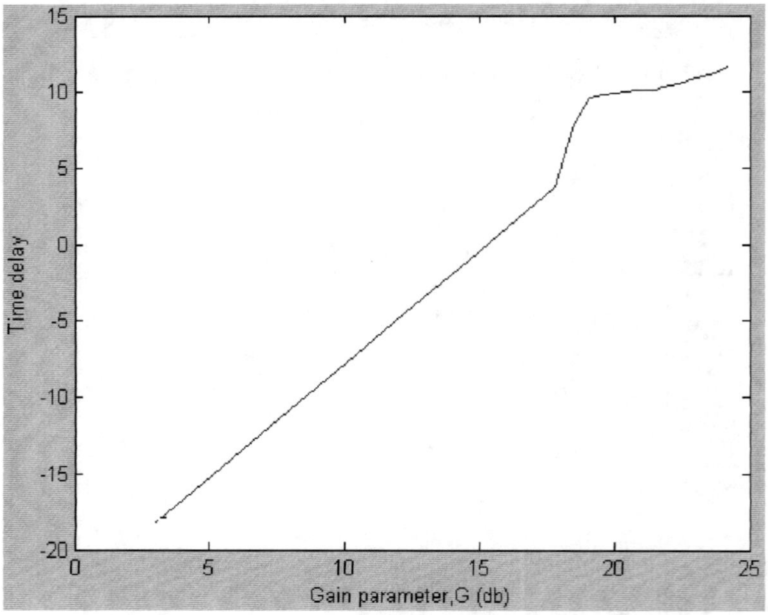

Figure 8: the relation between slow-light time delay and gain parameter G

3.2.2 the experiment performance of the SBS-slow light .In our experiment system, we get a delay of 8ns when the FWHM of input Stokes is 200ns, and G is 18.

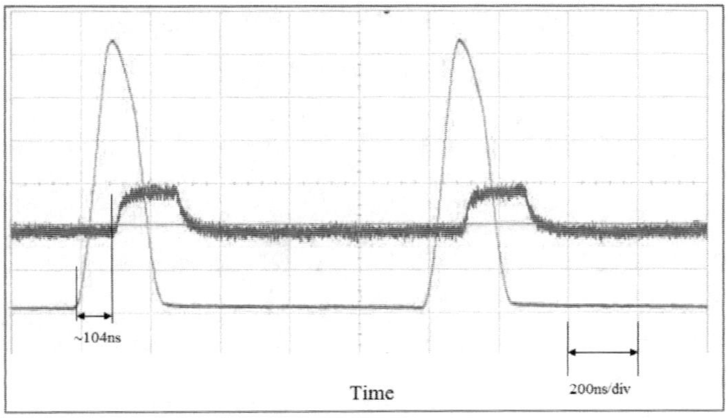

(a)

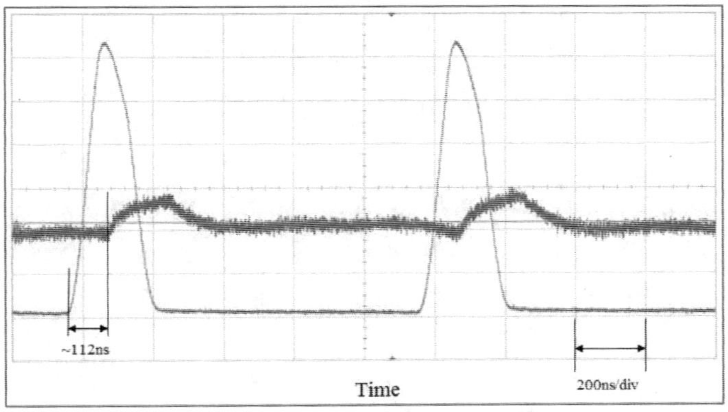

(b)

Figure 9: power spectrum of Stokes pulse time delay

FIGURE9 shows power spectrum of Stokes pulse time delay. The difference between the output and input light is 104ns when there is no pump as is shown in figure (a); and 112ns when pump light exists. So we can obtain a time delay of 8ns. But severe distortion emerged.

4. CONCLUTION

In this paper we consider both fiber absorption and pump depletion for light pulse propagating along the propagation distance, According to three-wave coupling equations, we deduce the principle of slow-light based on SBS, that is signal frequency causes resonance near Stokes frequency, and medium index become greater ,then group slow down. In our experiment system, we get a delay of 8ns when the FWHM of input Stokes is 200ns, and G is 18.And According to the experiment dates, we also get the qualitative relationship between pulse delay and gain parameter.

Time delay changes in line with pump power, that is pump depletion makes a big difference to time delay. So in order to get satisfactory results of time and gain, appropriate fiber length and input pump power must be attentively considered.

ACKNOWLEDGMENT

Project supported by the National Basic Research Program of China (Grant No. 2010CB923202) and by the National Natural Science Foundation of China (Grant No. 60871082).

REFERENCES

[1] SONG K Y, HERRAEZ M G, THEVENAZ L 2005 *Opt. Express.* **13** 82-88
[2] ZHU ZH M, DAWES A M C, GAUTHIER D J 2007 *Journal of Lightwave Technology* **25** 201-206
[3] ZHU ZH M, DAWES A M C, GAUTHIER D J 2006 *Optical Fiber Communication Conference (OFC)* PDP1
[4] SONG K Y, HOTATE K 2007 *Optical Fiber Communication Conference (OFC)*
[5] Yin J C, Xiao X S, Yang C X 2010 *ACTA PHYSICA SINICA* **59** 3986-91
[6] Zhu ZH M, Daniel J. Gauthier 2005 *J. Opt. Soc. Am. B* **22** 2378-84
[7] Aydin Yeniay, Jean-Marc Delavaux, Jean Toulouse 2002 *Journal of Lightwave Technology* **20** 1425-32
[8] G. P. Agrawal 2001Nonlinear Fiber Optics, 3rd ed chap**9** *ACADEMIC PRESS* 335-384
[9] Andrey Kobyakov, Sergey Darmanyan, Michael Sauer 2006 *OPTICS LETTERS* **31** 1960-62
[10] K. Y. Song, M. G. Herraez and L. Thevenaz 2005 Opt.Lett.**30** 1782
[11] Yoshitomo Okawachi, Jay E. Sharping, and Alexander L. Gaeta 2005 *Phys. Rev. Lett.* **94** 153902
[12] Miguel González Herráez, Kwang Yong Song, Luc Thévenaz 2006 *Optics Express.***14** 1395-1400
[13] David Dahan and Gadi Eisenstein, 2005 *Optics Express* **13** 6234-49
[14] M. G. Herraez ,K. Y. Song and L. Thevenaz 2005 *Appl.Phys.Lett.***87** 081113

AUTHOR INDEX

AUTHOR INDEX

AUTHOR INDEX

AUTHOR INDEX

AUTHOR INDEX

AUTHOR INDEX

AUTHOR INDEX

AUTHOR INDEX

Institute of Physics
Dirac House, Temple Back
Bristol BS1 6BE UK

CURRAN ASSOCIATES INC
proceedings.com

3rd International Photonics & OptoElectronics Meetings 2010

(POEM 2010)

Journal of Physics: Conference Series Volume 276

Wuhan, China
2-5 November 2010

Volume 2 of 2

ISBN: 978-1-61782-660-3
ISSN: 1742-6588

IOP Institute of Physics

3rd International Photonics & OptoElectronics Meetings 2010

(POEM 2010)

Journal of Physics: Conference Series Volume 276

Wuhan, China
2-5 November 2010

Volume 2 of 2

ISBN: 978-1-61782-660-3
ISSN: 1742-6588

Printed from e-media with permission by:

Curran Associates, Inc.
57 Morehouse Lane
Red Hook, NY 12571

Some format issues inherent in the e-media version may also appear in this print version.

Printed by Curran Associates, Inc. (2011)

For permission requests, please contact the Institute of Physics
at the address below.

Institute of Physics
Dirac House, Temple Back
Bristol BS1 6BE UK

Phone: 44 1 17 929 7481
Fax: 44 1 17 920 0979

techtracking@iop.org

Additional copies of this publication are available from:

Curran Associates, Inc.
57 Morehouse Lane
Red Hook, NY 12571 USA
Phone: 845-758-0400
Fax: 845-758-2634
Email: curran@proceedings.com
Web: www.proceedings.com

TABLE OF CONTENTS

Volume 1

INVITED PAPERS

LASER TECHNOLOGY AND APPLICATIONS (LTA)

NANO-ENABLED ENERGY TECHNOLOGIES AND MATERIALS (NETM)

OPTOELECTRONIC DEVICES AND INTEGRATION (OEDI)

Volume 2

OPTOELECTRONIC SENSING AND IMAGING (OSI)

TERA-HERTZ SCIENCE AND TECHNOLOGY (THST)

Journal of Physics: Conference Series **276** (2011) 012114 doi:10.1088/1742-6596/276/1/012114

Evolution of INO Uncooled Infrared Cameras Towards Very High Resolution Imaging

Alain Bergeron[1], Hubert Jerominek, Claude Chevalier, Loïc Le Noc, Bruno Tremblay, Christine Alain, Anne Martel, Nathalie Blanchard, Martin Morissette, Luc Mercier, Lucie Gagnon, Patrick Couture, Nichola Desnoyers, Mathieu Demers, Frédéric Lamontagne, Frédéric Lévesque, Sonia Verreault, François Duchesne, Julie Lambert, Marc Girard, Maxime Savard, François Châteauneuf

INO, 2740 rue Einstein, Quebec City, QC, Canada, G1P 4S4

E-mail: alain.bergeron@ino.ca

Abstract. Along the years INO has been involved in development of various uncooled infrared devices. Todays, the infrared imagers exhibit good resolutions and find their niche in numerous applications. Nevertheless, there is still a trend toward high resolution imaging for demanding applications. At the same time, low-resolution for mass market applications are sought for low-cost imaging solutions. These two opposite requirements reflect the evolution of infrared cameras from the origin, when only few pixel-count FPAs were available, to megapixel-count FPA of the recent years. This paper reviews the evolution of infrared camera technologies at INO from the uncooled bolometer detector capability up to the recent achievement of 1280x960 pixels infrared camera core using INO's patented microscan technology.

1. Introduction

INO (National Optics Institute), Canada's largest center of expertise in optics and photonics is the only Canadian developer and manufacturer of custom uncooled IR bolometric detector arrays [1] and detector-based instruments for industrial, military, and space applications [2-5]. Several types of camera modules have been developed at INO; this paper reviews the evolution of INO's infrared technology.

---

[1] To whom any correspondence should be addressed.

First, a brief overview of INO's uncooled bolometric technology is be presented including its1D and 2D Focal Plane Arrays. Then various electronics cores developed at INO are presented featuring its generic IRXCAM camera module offering outstanding image quality. The camera core generates a 16-bit raw signal output at 60 Hz, and the module gives complete access to the detector configuration parameters in order to ease the developer's integration process. Furthermore, the core can also drive the INO IRM160 FPA in random access mode. TECless operation minimizes module power consumption. This generic camera module can be configured for outdoor operation from -30°C to +55°C with 200°C scene dynamic range at maximum sensitivity. A high resolution camera module is then introduced. In today's fast evolving technology, there is an increasing demand for high resolution infrared modules that can be rapidly turned into a complete autonomous device in various fields of applications. To answer to this demand, INO has developed a 1280 x 960 pixel infrared imaging module that provides high resolution as well as wide field-of-view coverage. To achieve this high resolution, a custom optics was designed and manufactured.

2. Microbolometer Focal Plane Arrays

INO currently produces two standard microbolometer focal plane arrays, the IRM160A and IRL256B. In collaboration with the European and Canadian Space Agencies, INO has also developed a 512 x 3 pixel microbolometer focal plane array for satellite-based thermal pushbroom imaging.

The IRM160A (Figure 1) is a 160 x 120 pixel array of 50 x 50 µm2 microbolometer pixels on a 52 µm pitch. The active array area is 8.3 x 6.2 mm2 and pixel access may be performed by sequential scanning or random access mode. The achievable frame rate in full 160 x 120 mode is greater than 100 Hz. In random-access mode, subsections of the array may be run at higher frame rate or over-sampled to give improved thermal resolution. On-chip reference pixels are used to provide coarse offset correction as well as substrate temperature drift compensation. At the focal plane array level the achievable noise equivalent temperature difference (NETD) for f/0.86 at 300 K is below 60 mK (Figure 2). Next generation pixels currently implemented on prototype arrays will bring the NETD down to 30 mK. The IRM160A has already been integrated by INO's clients security cameras, unattended ground sensors, and airborne cameras.

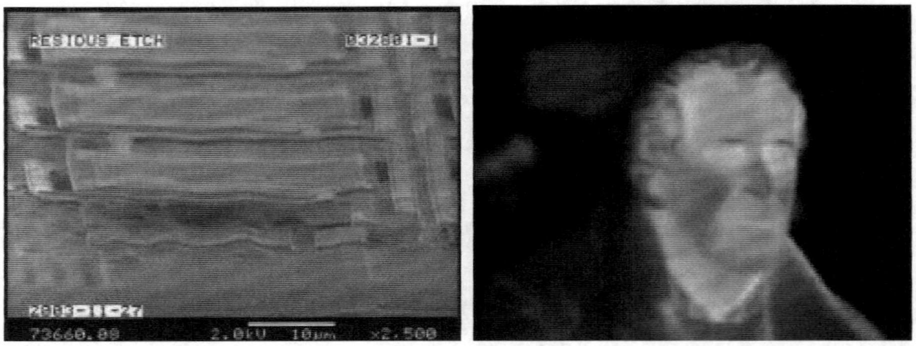

Figure 1: IRM160A: pixels (left) and thermal images taken with the 160x120 pixel FPA (right).

Journal of Physics: Conference Series **276** (2011) 012114 doi:10.1088/1742-6596/276/1/012114

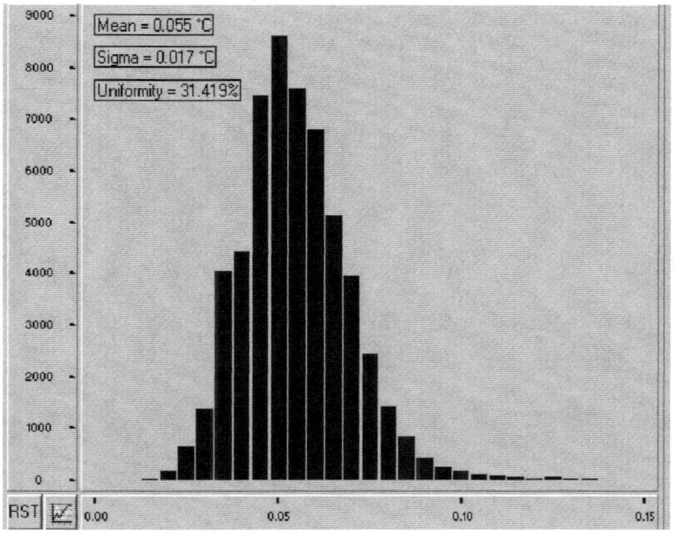

Figure 2: IRM160A sample NETD histogram

The IRL256B is a linear array of 256 pixels of 50 μm on a 52 μm pitch. A parallel array of reference detectors provides coarse offset correction and substrate temperature drift compensation. Pixel access may be in sequential scanning or random access mode. The pixel noise equivalent power (NEP) is on the order of 5 pW/Hz1/2 at 30 Hz. At the focal plane array level the single frame noise equivalent temperature difference (NETD) for f/0.86 at 300 K is below 100 mK. The IRL256 has already been integrated by INO's clients into pushbroom airborne cameras, spectrometers, and cameras for industrial temperature profile monitoring (Figure 3).

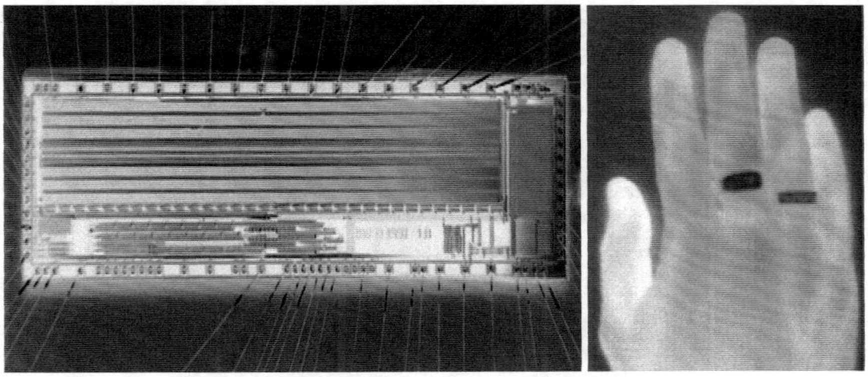

Figure 3: IRL256B: linear array (left); linescan thermal image obtained with the linear array (right)

The IRL512A (Figure 4) is designed specifically for satellite-based multispectral pushbroom imaging applications. It consists of 3 parallel lines of 512 pixels on a 39 μm pitch. Each pixel includes active and reference detectors to reduce pixel offset, eliminate common mode power supply noise and increase immunity to chip temperature drift. The chip integrates the signal of all the pixels in parallel and provides 14-bit digital output. Despite the small size of the pixels, the design target NETD with f/1 optics having 60% transmission and 300 K scene temperature is below 250 mK for any 1 μm wide spectral band between 8.3 and 12.5 μm.

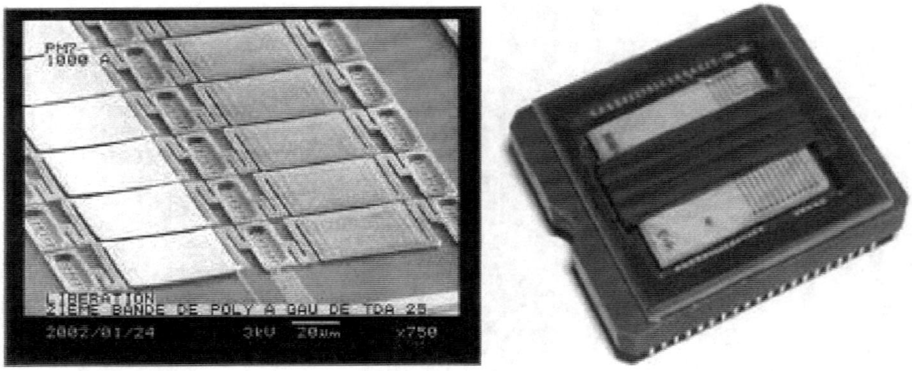

Figure 4: IRL512A: pixels (left) and FPA in a ceramic package (right).

3. Packaging and Environmental Testing

Packaging is a key element in miniaturization and cost reduction for microbolometer FPAs. All the parameters of the package, including its lateral dimensions, volume, weight, mechanical resistance and electrical, thermal and optical properties have influence on the properties of the packaged FPA as well as the properties and exploitation of complete instruments such as TWS making use of the bolometric FPAs. Thus, it is crucial to develop and optimize the package ensembles in order to achieve maximum performance of the FPAs as well as the complete cameras. Moreover, efficiency (i.e. yield) of the packaging procedure and material cost have an important influence on total cost of a particular FPA, often becoming a non-negligible factor determining the complete instrument cost. While the overall performance (maximum pixel count and minimum noise equivalent temperature difference NETD) is a driving factor behind various applications of bolometric FPAs, the cost issue becomes equally important limiting the proliferation of the technology. There are many applications where weight, volume and power consumption are particularly critical. In some application scenarios, these devices are being treated as disposable items which again emphasizes the importance of low-cost components and complete instruments. INO's standard ceramic FPA package is shown in Figure 5.

Figure 5: INO's standard ceramic package for microbolometer detector arrays.

4. MIMIC ii v2, uncooled thermal imaging development module with random access capabilities

The module MIMIC ii v2 was the first camera core based the INO IRM160A bolometric detector (160 x 120 pixels, 52 μm pitch). The unique random access capability makes it a unique tool for a device integrator. Windows as well as a completely random pattern can be read-out from the camera sensor. The camera operates in the 7-14 μm wavelength range. The main characteristics are presented in Table 1. The module consists of the detector, proximity electronics and a microshutter. The camera is operated from a portable computer equipped with an acquisition and control software. The link between the computer and the camera is of a Camera Link type. The module was characterized with various commercial or INO designed optics in terms of thermal sensitivity, and spatial resolution.

Figure 6: MIMIC ii v2 camera core module with INO IRM160A uncooled bolometer detector and microshutter

Table 1: Main characteristics of the MIMIC ii v2 module

Parameters	Specifications
# pixel	160 x 120 pixels, 52 μm pitch, interchangeable
NETD (F# 0.8, 300K)	< 10 mK
Vacuum package	INO Gen II, ceramic package
Detector weight	14 g
Module weight	< 50g
Microshutter	integrated with proximity electronics, removable
Frame rate	30 Hz
Reading mode	Sequential and random access
Operation mode	direct and differential
Thermal control accuracy	0.01 °C
Maximum current	500 mA
Data format	Digital 16 bits Camera LINK, 160 x 120 pixels
Dimensions	2" x 2" x 2"
Operating temperature	-20 à 40 °C
Supply	12 V +/- 3V
Software	Realtime display, offset and gain correction, automatic offset. The raw and corrected data can be saved in bitmap and ascii formats. Videos are saved in bitmap format.

5. INO IRL512 linear detector array

INO has developed the detector IRL512 composed of 3 linear arrays, and its proximity electronics The proximity electronics acquires the analogue signal at 2 Hz (to improve the thermal resolution of the detector, which otherwise operates at 50 Hz). The signal is digitized over 15 bits with a remaining electronic noise smaller than the detector noise and subsequently transferred via an RS-232 link to an external control electronics. This electronics is linked to a computer via a serial port operating at 115,000 bits/sec.

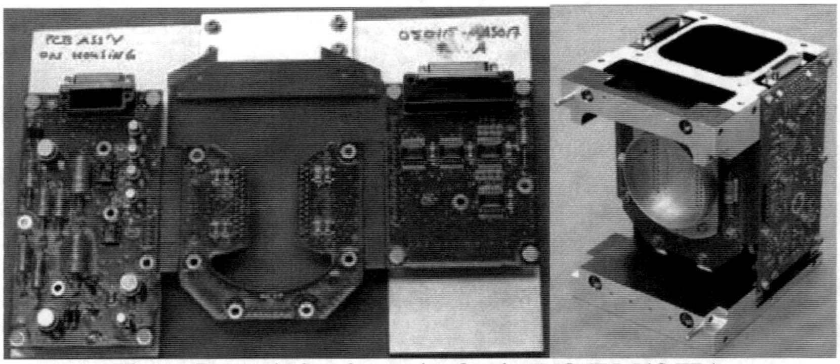

Figure 7: Proximity electronics for the INO IRL512 FPA.

6. IRXCAM camera core for fast prototyping

IRXCAM is a camera core designed to operate with several commercial infrared detectors, to acquire and postprocess the detector raw signal and to output the data/image in several formats. Figure 8 illustrates the concept of fast prototyping with IRXCAM. To obtain a prototype able to demonstrate the final imaging performance with a given detector, only optics has to be added in front of the detector. A microshutter is already integrated because it is critical to know the effect of the offset update on the image quality. The laptop is used to control IRXCAM and the detector parameters, to process the raw signal (offset and gain correction, pixel replacement, image improvement…), and to display the resulting image.

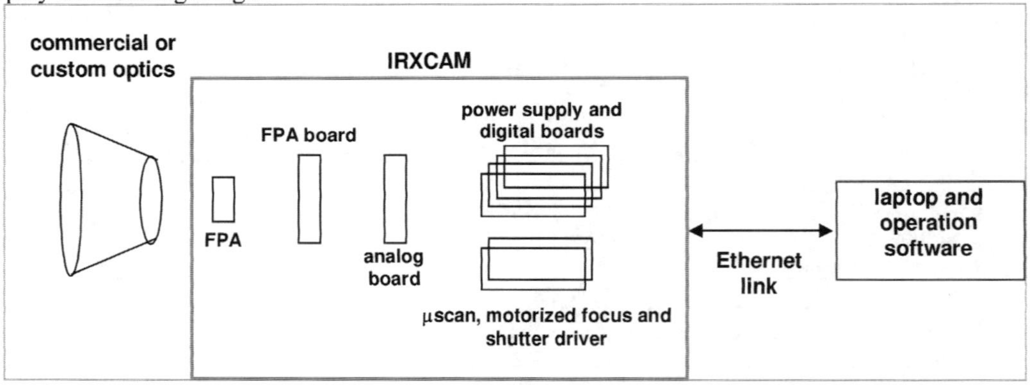

Figure 8 : Schematics of an imaging prototype rapidly achieved with IRXCAM..

To take advantage of the newest commercially available detectors, IRXCAM architecture is designed to minimize the effort necessary to implement a new detector. IRXCAM can currently support high data rate (more than 40 Mpixel/s, i.e. Ulis UL 05 25 1, 50 Hz, 1024 x 768 17 µm pitch pixel bolometric detector for example) and two detector output so that it is compatible with most currently and next available detectors.

The detector (FPA) of IRXCAM is interchangeable with minimum development. The FPA board, which is used to support the FPA, must be replaced to be compatible with the FPA pin out. The analog board, which realizes the signal digitization, is already designed to support at full frame rate of various currently available detectors. A version of the FPA and analog boards also exist for INO IRM160A (MWIR, LWIR and THz applications, 160 x 120 pixels, 52 μm pitch).

The permanent hardware part of IRXCAM consists in the power supply and digital boards that realize signal acquisition, postprocessing, and communication. Programs implemented in the integrated FPGA must be adapted to each detector but once the work has been realized for one detector, it is much less demanding for all detectors of the same family.

With its wide range of functionalities, IRXCAM is also designed for the development of prototypes which present most of or even all functionalities of a final system. Signal postprocessing and image enhancement capabilities are achieved thanks to the computing power of the DSP+FPGA pair (Digital Signal Processor). With the multiple format output (NSTC/PAL and VGA), IRXCAM can be connected to a large range of video displays. Through the video output and the Ethernet link, IRXCAM can also share information among multiple users.

IRXCAM is also designed to facilitate the minimization of the system weight and size and power consumption, the selection of the adequate components and the achievement of an optimal equilibrium of constraints on the submodules (optics, detector, data processing electronics and display submodules). The power consumption associated with a given functionality can be measured. It is therefore possible to select the functionalities, like algorithms of hyperresolution or edge enhancement, detector thermal regulation, microscanning, motorized focusing, Ethernet link, that can be implemented while maintaining the power consumption of the final system below a given value. For low power consumption, the electronic boards can be redesigned for the selected functionalities with a low level of risk. IRXCAM also allows the selection of the most adequate microdisplay for a given application because it can support multiple microdisplays through its NTSC/PAL and VGA output.

IRXCAM-640 and IRXCAM-160 have been developed. IRXCAM-640 is based on the Ulis UL 04 17 1 bolometric detector made of 640x480 25 μm pitch pixels. IRXCAM-160 is based on the INO IRM160A bolometric detector made of 160x120 52 μm pitch pixels. IRXCAM-640 is illustrated on Figure 9 and the specifications of IRXCAM-640 and IRXCAM-160 are shown in Table 2 [5]..

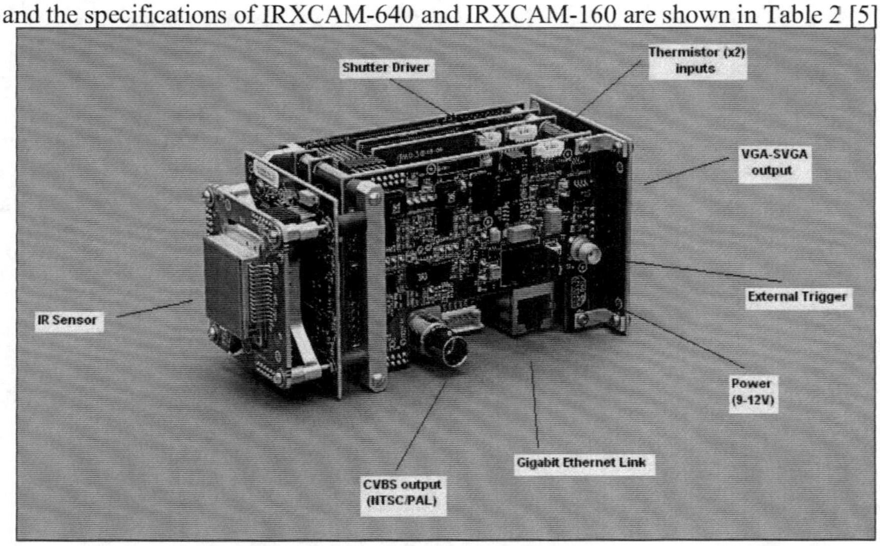

Figure 9: Picture of IRXCAM-640

Table 2: specifications of IRXCAM-640 and IRXCAM-160.

	IRXCAM-640	***IRXCAM-160***
Detector	640x480 VGA LWIR Uncooled, ULIS UL 04 17 1 Microbolometer	160x120 MWIR/LWIR/THz Uncooled Microbolometer
Resolution	640 x 480 pixels	160 x 120 pixels
Frame Rate	60 Hz	30 Hz
Video Output	1) GigE Link 2) CVBS (NTSC/PAL)	1) GigE Link
Control	GigE Link - System operation control - Loading of parameters & calibration tables - Windowing	GigE Link - System operation control - Loading of parameters & calibration tables - Random Access
Onboard Processing	Shutter Offset correction table Gain & Offset correction table Pixel replacement	
Other implemented function	External Trigger input TEC Driver Microscan Driver MicroShutter Driver	Ext. Trigger input TEC Driver Microscan Driver MicroShutter Driver
Power Supply	9-12V DC	9-12V DC
Dimension	65mm x 59mm x 125mm	65mm x 59mm x 105mm
Weight	250g	~230g
Temperature	Operating : -30 to 55°C Storage : -40 to 80°C	Operating : -30 to 55°C Storage : -40 to 80°C

7. HRXCAM high resolution camera module

HRXCAM, is an infrared device providing very high resolution of 1280 x 960 pixels, and a wide field of view of 22.6°. HRXCAM is build around IRXCAM detection electronics. The resolution was achieved by implementing a microscanning to double the infrared detector resolution, 640x480 pixels. The sensitivity is 60 mK (300 K, 10 Hz) and the frame frequency used was 10 Hz. When reducing the resolution to 640x480 pixels by stopping the microscanning, the sensitivity is unchanged while the frame frequency is increased to 58 Hz. HRXCAM imaging performance and resolution enhancement by microscanning are illustrated on Figure 10 and Figure 11. It can be clearly seen on Figure 10 that the level of details (number of windows, sharpness of the peaks) is improved. Also the aliasing appearing on the roof is clearly reduced.

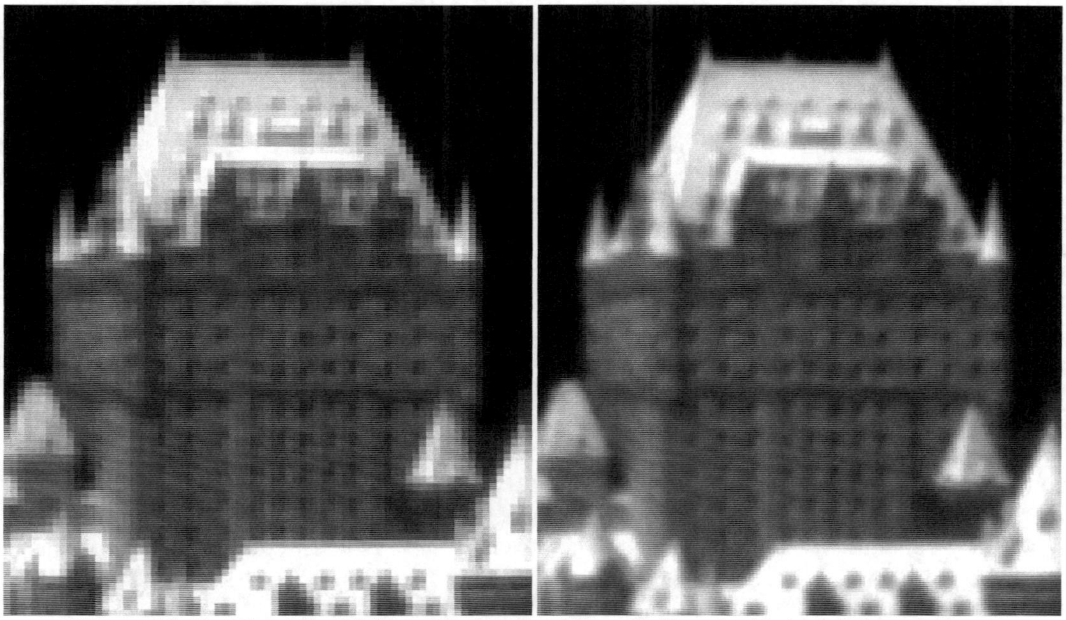

Figure 10: Details of the Château Frontenac without (left) and with (right) the microscan.

Figure 11: 1280 x 960 pixel image of Québec city with microscan.

8. NIRST, New IR Sensor Technology

This project was sponsored by the Argentinian Space Agency (CONAE). Aquarius/SAC-D is an international cooperative mission lead conjointly by NASA (National Aeronautics and Space Administration (NASA) of the United States of America (USA)) and CONAE. The mission objective is the global understanding of the terrestrial system and of the consequences of climatic changes, either from natural or human origin. Conjointly developed by CONAE and CSA (Canadian Space Agency), the instrument named NIRST (New IR Sensor Technology) will track high temperature events at the surface of the Earth, fires or volcanic activity, and will measure their physical parameters. In addition, NIRST will measure temperatures at the surface of oceans. The sampling on the ground is 534 µrad, or 350 m. A mirror allows scanning across +/- 500 km. An engineering prototype of the thermal infrared camera (TIR) has been built and its performances have been measured.

The system NIRST is a scanning multispectral imaging radiometer through pushbroom operation mode. It is made of two imagers operating over three wavelength bands chosen to optimize the detection of thermal events. The first imager is based on the INO IRL512 focal plan array made of 512 x 3 microbolometers. Two linear arrays are covered by pass-band filters for acquisition in two distinct wavelength bands at 10.85 µm (+/- 0.45 µm) and 11.85 µm (+/- 0.4 µm). These bands are denoted TIR1 and TIR2 respectively. The second imager is based on the same detector of 512 x 3 pixels, but this time covered by a pass-band filter at 3.8 µm (+/- 0.4 µm), denoted the MWIR band. Each imager is mounted with its dedicated optics. The acquisition is realized by a co-registration of the TIR1 and MWIR bands, while the TIR2 channel is spatially offset. The development of the radiometer has required the development of a read-out integrated circuit and of a control/acquisition software for the IRL512 detector and its electronics. The co-registration asks for the accurate measurement of the field of view, the magnification factor and the distortion.

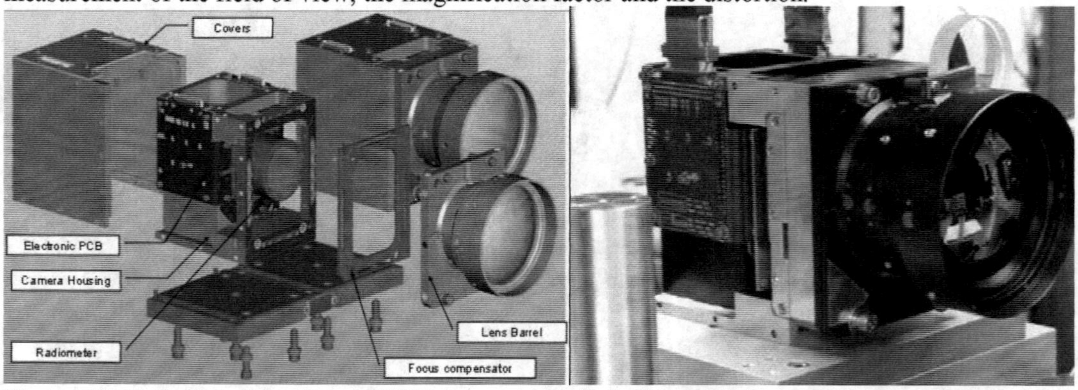

Figure 12: Exploded view of the NIRST camera module (left) and engineering prototype for the TIR1 and TIR2 bands (right).

9. Conclusion

Through the years INO has developed various IR systems . These systems are based on a wide expertise from the detector to the complete instrument integration. Moreover, the development of flexible cores open the way to faster prototyping minimizing the resources involved both for prototyping, and for product development.

References

[1] Bergeron, A., Jerominek, H, Laou, P., Doucet, M., Lagacé, F., Desnoyers, N., Bernier, S., Mercier, L., Boucher, M., Jacob, M., Alain, C, Pope, T.D., *Lightweight uncooled TWS equipped with catadioptric optics and microscan mechanism* (Invited Paper), Proc. SPIE 6206-59, (2006)

[2] Bergeron, A., Jerominek, H., Laou, P., Doucet, M., Lagacé, F., Desnoyers, N., Bernier, S., Mercier, L., Boucher, M., Jacob, M., Alain, C., Pope, T.D., *Dual-band dual field-of-view TWS prototype*, Proc. SPIE 6206-60 (2006)

[3] Bergeron, A., Jerominek, H., Laou, P., Lacoursière, J., Desnoyers, N., Alain, C., *Novel lightweight uncooled thermal weapon sight*, Proceeding SPIE Orlando [5406-43], 402-411 (2004).

[4] Smith, W.J.,[*Modern Lens Design*, McGraw-Hill Professional Engineering, SPIE Press, 41-45 (2005).

[5] Le Noc, Loïc; Tremblay, Bruno; Lagacé, François; Mercier, Luc; Duchesne, François; Bergeron, Alain; Jerominek, Hubert, *Modular infrared 640 x 480 pixel camera core for rapid device integration*, SPIE 7298-27, (2009).

Research on wavefront-sensing for improving receiver efficiency in turbulence atmosphere

Rui Bie, Xiuhua Yuan[1], Ming Zhao

Wuhan National Laboratory for Optoelectronics, Wuhan 430074 Hubei, P.R.China

bxacsq@gmail.com, yuanxh@mail.hust.edu.cn

Abstract. In the air-to-ground optical communication system, not only must we considerate the impact of beam pointing, but also the wavefront turbulence the atmosphere brings to the beam can not be ignored. This paper simply analyses the influence of the atmosphere turbulence on air-to-ground optical communication, and reforms the traditional adaptive optics(AO) system using the famous Zernike phase contrast method, which can makes it be applied to the air-to-ground optical communication system to correct the turbulence of the wavefront and improve the receiving efficiency and the system performance. In the end of this paper, the simulation of this method described above to detect phase is presented.

1. Introduction

In air-to-ground optical communication system, the refractive index of the air fluctuates randomly because of some factors such as temperature gradient, which results in the fluctuations of the intensity and phase[1], and increase of the BER(bit error rate). This paper reforms the traditional adaptive optics system, and applies it to FSOC (free space optics communication) to compensate the effect of atmosphere.

2. The application of adaptive optics[2] in FSOC

For a long time, adaptive optics has been applied to astronomy to overcome the impact of the atmosphere on the wavefront of the star light. Since the impact stays only in the near-field of telescope, the optical amplitude outside the receiver is approximately constant. So what we have to do is to compensate the phase distortion. But in FSOC, the atmospheric effect on optical signal stays in the far field, which will distort not only the phase but the optical amplitude.

The traditional AO system consists 3 main parts: wavefront sensor, wavefront controller, wavefront adjuster, which are corresponding to 3 stages of AO adjustment: wavefront sensing, wavefront reconstructing, and wavefront adjustment.

This AO system can compensate the distortion of the wavefront, but contains some components like wavefront sensor, which makes the system extremely complicated, so it cannot be applied on the satellite whose constructor and weight is strictly limited. On the other hand, this type of AO system cannot respond quickly, and will fail in the circumstance with strong scintillation.

[1] Corresponding author, Email: yuanxh@mail.hust.edu.cn

Published under licence by IOP Publishing Ltd

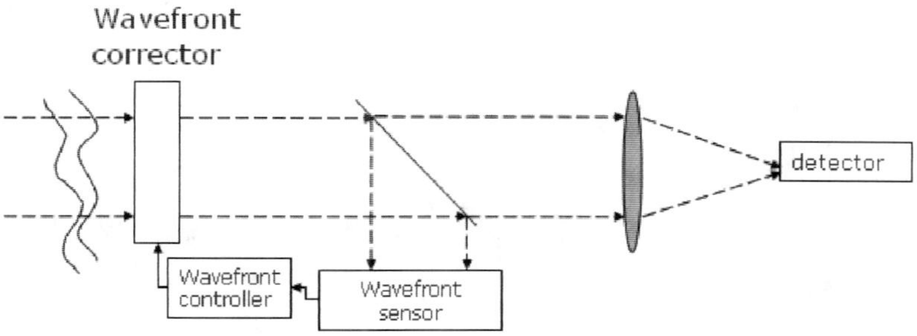

Figure 1. the diagram of traditional adaptive optics system

In order to overcome these disadvantages of traditional AO system, the novel AO system without wavefront sensor comes to existence[3].

This AO system adopts single estimation factor such as Strehl ratio, coupling efficiency, optical intensity etc. as the foundation for optical wave adjustment. It uses complex algorithm to control the wavefront adjuster in the close-loop recursively until the convergence point is reached. The algorithm adopted earlier are mountain-climbing method, multiple high-frequency vibration method[4], recently some heuristic algorithm like stochastic parallel gradient descent algorithm[5] and simulated annealing[6] is often used.

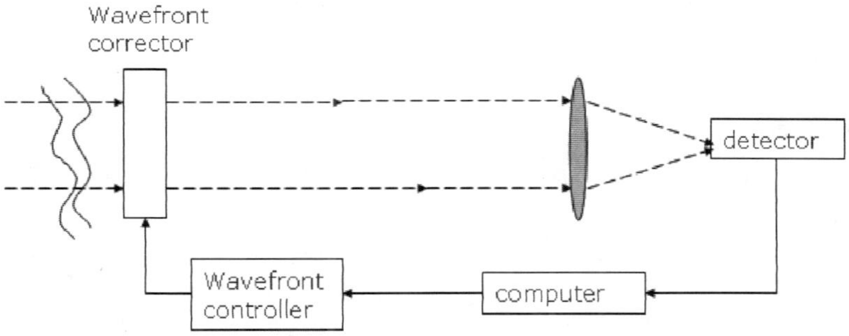

Figure 2. the diagram the common adaptive optics system without wavefront sensor

This type of AO system has lots of advantages such as the simplicity of system construction, the low cost. But the algorithm is complex and the convergence process is long.

This paper integrates both of the systems described above, and brings up a novel idea.

First of all, the complex wavefront sensor must be abandoned, but the constitution cannot be as simple as the AO system without wavefront which has so few estimation factors that the algorithm cannot converge as soon as possible. Second a CCD can be used to detect the intensity after the optical transformation to derive the phase distortion information. The specific system block diagram is shown in Fig.3.

In FSOC, of the optical distortion produced by the atmospheric turbulence, the tilt is the most serious aberration, but the deformable mirror deformation is not enough to compensate it, so it is necessary to add the steering mirror tilt to the system to deal with the tilt aberration, while the deformable mirror is to compensate for the higher-order aberrations.

This system has simple structure, and high optical efficiency because of the only one spectral. And we can also use the CCD of air-to-ground optical communication system designed for rough tracking to detect the optical intensity, which utilizes the existing resource and makes the structure simpler.

The key point of this method is to obtain phase information from the CCD, then generate the compensation signal through the algorithm to re-adjust the phase.

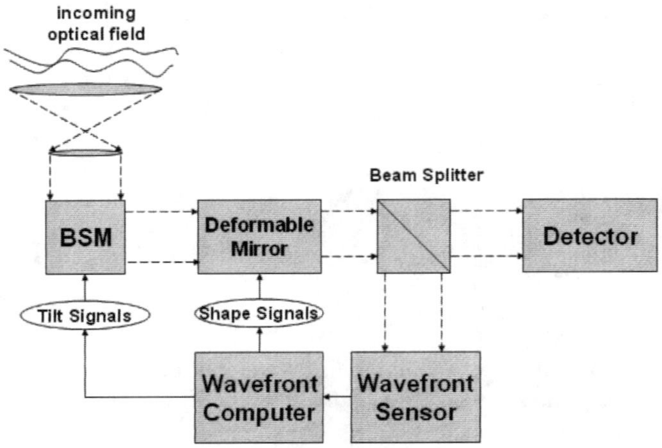

Figure 3. the diagram of adaptive optics system with CCD

3. Improved Zernike phase contrast method

In the theory of physical optics, there is a Zernike phase contrast method that can derive phase information from the light intensity distribution.

The theory of Zernike phase contrast method[7] is shown below.

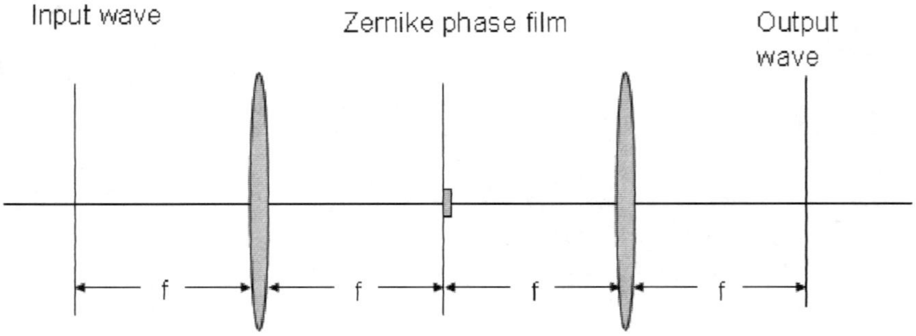

Figure 4. the diagram of Zernike phase contrast system

This figure actually shows a so-called 4f system. According to the theory of physical optics[2], the process that light waves spreads from the front focal plane of the lens to the rear focal plane can be seen as a two-dimensional spatial Fourier transform. So the FT(Fourier transform) plane(Zernike phase film) field shows the spatial frequency distribution of the incident wave. The structure of phase filter of the Zernike phase contrast method is very simple. What is needed is just a glass substrate with its central location coated with a membrane of certain thickness which makes the phase of zero-frequency light shift $\pi/2$ or $3\pi/2$ relative to other frequency light.

The transformation function is shown below:

$$H(f_x, f_y) = \begin{cases} \pm j & f_x = f_y = 0 \\ 1 & other \end{cases} \tag{1}$$

When this 4f system is illuminated by the coherent plane wave with unit amplitude, the incident wavefront distribution can be described as

$$f(x_1, y_1) = \exp[j\varphi(x_1, y_1)] \tag{2}$$

Assume the phase-shift is small, that is to say $\varphi(x_1, y_1) << 1rad$, and then the equation can be simplified as

$$f(x_1, y_1) \approx 1 + j\varphi(x_1, y_1) \tag{3}$$

That's the spatial frequency distribution

$$F(f_x, f_y) = \delta(f_x, f_y) + j\Phi(f_x, f_y) \tag{4}$$

after the transformation of Zernike phase film, it becomes

$$F(f_x, f_y)H(f_x, f_y) = \pm j\delta(f_x, f_y) + j\Phi(f_x, f_y) \tag{5}$$

Then the complex amplitude distribution of output plane is

$$g(x_3, y_3) = \pm j + j\varphi(x_3, y_3) \tag{6}$$

and the intensity distribution

$$I(x_3, y_3) \approx 1 \pm 2\varphi(x_3, y_3) \tag{7}$$

The fluctuation appears in the uniform background generated by φ, and their relationship is linear, so it is easy to derive the incident wave distribution form the intensity.

However, there is still a problem remained. The conclusions of Zernike phase contrast method are derived under the assumption that the amplitude of the incident wave is constant. But as described above, in FSOC, the atmospheric effect on optical signal stays in the far field, which will distort not only the phase but also the optical amplitude, so the light intensity of the receiving plane cannot be regarded as a constant parameter.

Supposing the phase and amplitude meet the requirement of weak fluctuation, the incident optical wave can be written as

$$u(x_1, y_1) = [1 + A(x_1, y_1)]\exp[i\varphi(x_1, y_1)]$$

In Eq.(8), $A(x_1, y_1)$ denotes the fluctuation of amplitude and $\varphi(x_1, y_1)$ the phase.

$$\begin{aligned} u(x_1, y_1) &= 1 + A(x_1, y_1) + j\varphi(x_1, y_1) + jA(x_1, y_1)\varphi(x_1, y_1) \\ &\approx 1 + A(x_1, y_1) + j\varphi(x_1, y_1) \end{aligned} \tag{9}$$

The spatial frequency is

$$F = \delta(f_x, f_y) + a(f_x, f_y) + j\Phi(f_x, f_y) \tag{10}$$

The system transforms the zero-frequency light with $\frac{\pi}{2}$ phase-shift. After the Fourier transformation of the second lens, the complex wave becomes

$$u(x_3, y_3) = j + A(x_3, y_3) + j\varphi(x_3, y_3) \tag{11}$$

And the intensity distribution is

$$\begin{aligned} I(x_3, y_3) &= |u(x_3, y_3)|^2 = [1 + \varphi(x_3, y_3)]^2 + A^2(x_3, y_3) \\ &= 1 + 2\varphi(x_3, y_3) + \varphi^2(x_3, y_3) + A^2(x_3, y_3) \\ &\approx 1 + 2\varphi(x_3, y_3) \end{aligned} \tag{12}$$

the Zernike phase contrast method is still suitable for this case.

4. Numerical simulation

Based on the theory described above, this chapter presents the simulation result of Zernike phase contrast method with weak perturbation.

Fig.(5) and Fig.(6) show the perturbations of phase and amplitude of the incident light field, respectively. The coordinates of all figures are relative values.

And Fig.(7) shows the intensity distribution after the transformation of the Zernike phase contrast method.

As shown in the figures, the output intensity information can accurately reflect the distribution of the incident light field phase, which can be used to adjust the wavefront phase.

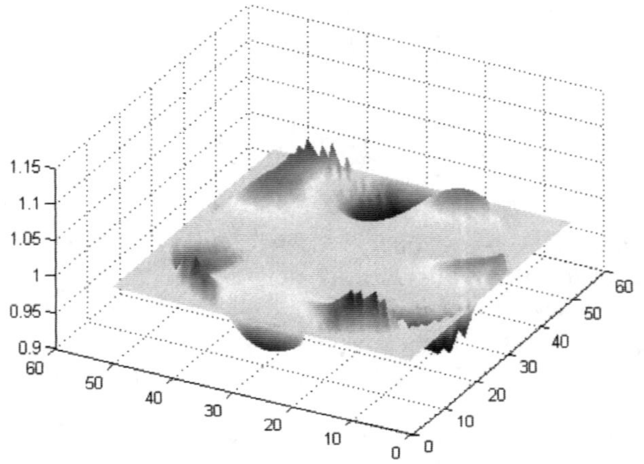

Figure 5. the amplitude perturbation of input optical field

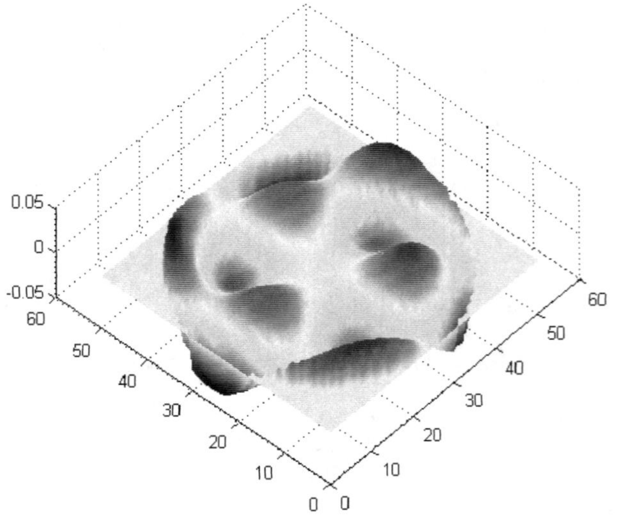

Figure 6. the phase perturbation of input optical field

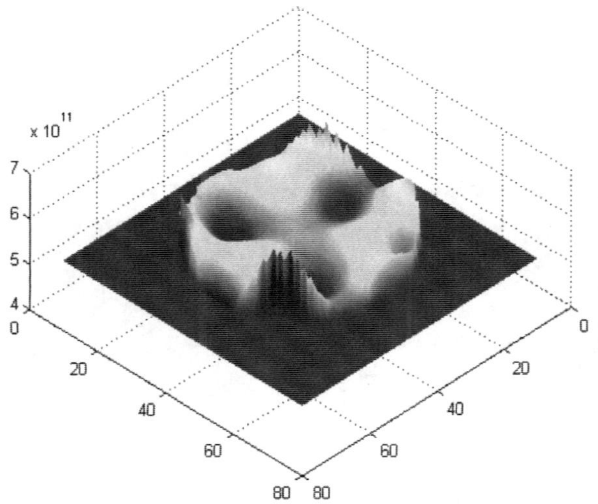

Figure 7. the intensity distribution of output optical field

5. Conclusion

This paper combines the characteristics of air-to-ground optical communication to improve the traditional adaptive optics system. In this paper, a phase adjustment device with simple structure is presented which has several advatages. First of all, it gets rid of the complex wave-front sensor, simplifying the system structure, which makes the system easy to install; second, the system utilizes the existing CCD in the ATP system of air-to-ground optical communication to derive the phase information through some algorithm. These features make it easier to be used in the air-to-ground optical communication systems.

Acknowledgments

This work is supported by the National High Technology Research and Development Program of China (Grant No. 2008AA01Z207).

References

[1] Wizen Ke, Xiaoli Xi, The introduction of wireless laser communication, Beijing University of Posts and Telecommunications Press. 2004
[2] Renzhong Zhou, The theory of adaptive optics, Beijing Institute of Technology Press.1996.1~3.
[3] HuiZhen Yang, Xinyang Li, Wenhan Jiang, Applications of Adaptive Optics Technology in Atmospheric Laser Communications System.2007, Vol44, No. 10: 61~68.
[4] Chunyi Chen, Huamin Yang, Huilin Jiang, Xin Feng, Hui Wang, Research Progress of Mitigation Technologies of Turbulence Effects in Atmospheric Optical Communication, Vol.30, No.6, 2009.6.
[5] Huizhen Yang, Xinyang Li, Wenhan Jiang, Comparison of several stochastic parallel optimization control algorithms for adaptive optics system, 2008,Vol.20, No.1: 11~16.
[6] R. El-Agmy, H. Bulte, A. H. Greenaway et al. Adaptive beam profile control using a simulated annealing algorithm.OPTICS EXPRESS, 2005, Vol. 13 Issue 16:6085~6091.
[7] Naiguang Lv, Fourier optics,2006.

Synthesis and Application for polarization Optical Storage of a New Photochromic compound

Shiqiang Cui, Shouzhi Pu*, Weijun Liu, Gang Liu

Jiangxi Key Lab of Organic Chemistry, Jiangxi Science and Technology Normal University, Nanchang 330013, China

pushouzhi@tsinghua.org.cn

Abstract. A new photochromic compound 1-[2-methyl-5-(4-methyl-phenyl)-3-thienyl]-2-[2-methyl-5-(4-fluor-phenyl)-3-thienyl]perfluoroyclopentene (1a) was synthesized, its photochromic properties were examined, the result indicated that the color of the compound 1a changed from colorless to a blue one upon irradiation with 297 nm UV light, a new absorption maxima were observed at 577 and 581 nm in hexane and PMMA film, respectively. This new photochromic compound also exhibited remarkable optical storage character.

1. Introduction

With the increasing requirements of the huge information storage, higher density and higher data transfer rate become the trends of optical data storage. Current commercial recordable and rewritable optical storage systems are heat-mode recording systems, which limits the size of recorded pits and writing speed[1]. To a great extent, the performance of holographic system depends on the properties of the recording media, and the lack of suitable recording materials has long been an obstacle to the development of holographic recording[2,3]. Photochromic compounds undergo a photo-induced reversible isomerization process between two isomers with different absorption spectra upon irradiation with light of appropriate wavelength[4,5]. Among various photochromic compounds, diarylethenes exhibit some excellent characteristics such as thermal stability of both isomers and fatigue-resistant property[1,6]. Such materials can be potentially used for application to optical memory media[7], switching devices[8], display materials[9,10], and photo-mechanical actuators [11,12], and so on. In particular, diarylethenes with different substituted ring systems have been of considerable interest [13-17], which different substituted ring could be resulted in different color change before and after light irradiation. In order to find out diarylethenes with excellent characters, a lot of molecules were designed and synthesized[18-20]. The solvent effect of the kinetics in polar solvent and the kinetics in solid state thin film have attracted more attention because most experiments are carried out in solution and the solid state thin film containing photochromic chromophore is the media for optical data storage, respectively.

In the present work, a new photochromic compound was designed and synthesized, the photochromic properties were examined, and its applied in optical storage was also demonstrated, respectively.

2. Experiments

2.1. Reagents and Instruments

Published under licence by IOP Publishing Ltd

^{1}H NMR spectra were recorded on Bruker AV400 (400 MHz) spectrometer with $CDCl_3$ as the solvent and tetramethylsilane as an internal standard. The absorption spectra were measured using Agilent 8453 UV/Vis spectrometer. Photoirradiation was carried out using SHG-200 UV lamp, CX-21 ultraviolet fluorescence analysis cabinet, and BMH-250 Visible lamp. Light of appropriate wavelengths was isolated by different light filters.

2.2. Synthesis of diarylethene 1a
Diarylethene 1a was prepared according to the method described in scheme 1. and its photochromism process shown in Scheme 2. Diarylethene 1a was characterized by ^{1}H NMR spectroscopy. ^{1}H NMR (400 MHz, $CDCl_3$): 1.87 (s, 6H, $-CH_3$), 2.30 (s, 3H, $-CH_3$), 6.99-7.03 (t, 2H, J = 8.0, thiophene-H), 7.11-7.13 (t, 4H, Ar-H), 7.36-7.38 (d, J = 8.0, 2H, Ar-H), 7.42-7.45 (q, 2H, Ar-H).

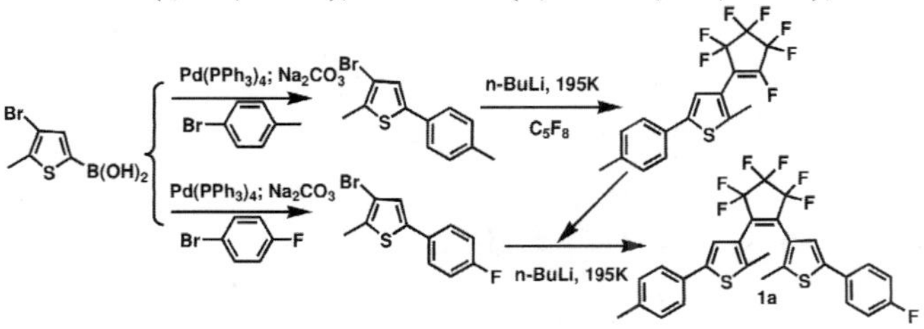

Scheme 1. Synthetic route for the compound 1a.

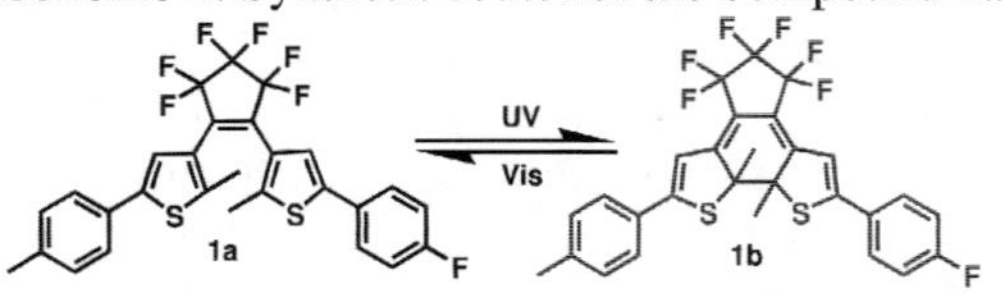

Scheme 2. Photochromism of diarylethene 1a.

3. Results and Discussions

3.1. Photochromism of diarylethene 1
The novel diarylethene 1a exhibits good photochromic properties and can be changed between its colorless ring-open and colored ring-closed forms by alternate irradiation with appropriate wavelengths of light. Figure 1 shows the absorption spectral changes and color changes of 1a in hexane(C = 2.0 × 10^{-5} mol/L), he absorption maximum of compound 1a was observed at 281 nm (= 3.34 × 10^4 L·mol^{-1}·cm^{-1}). Upon irradiation with 297 nm light, the color of the hexane solution turned blue, in which the absorption maximum was observed at 577 nm (=1.45 × 10^4 L·mol^{-1}·cm^{-1}). The blue color solution turned colorless, upon irradiation with visible light (> 500 nm). The quantum yields of cyclization and cycloreversion reaction of diarylethene 1 in hexane solution were 0.419 and 0.0013 at room temperature, respectively.

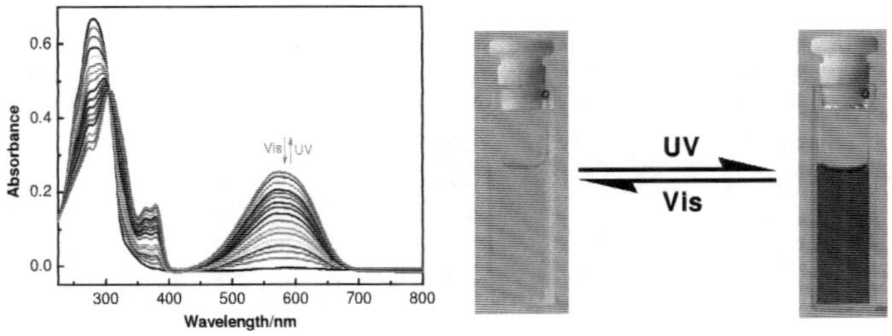

Figure1. Absorption spectral changes and color changes of compound 1 in hexane.

In PMMA film, the absorption spectral changes and color changes of 1a are showed in Figure 2. The absorption maxunum of compound 1a was observed at 282 nm. Upon irradiation with 297 nm light, the color of the PMMA film turned blue too, in which the absorption maximum was observed at 581 nm. The blue film turned colorless, upon irradiation with visible light (> 500 nm).

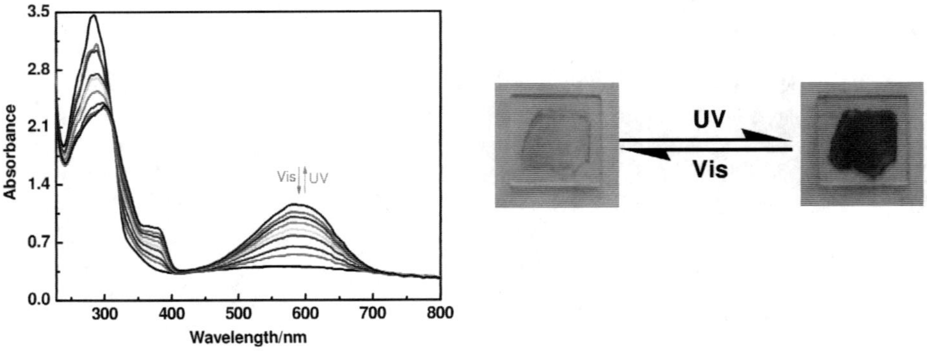

Figure 2. Absorption spectral changes and color changes of compound 1 in PMMA.

3.2. Optical storage property

With a He-Ne laser (wavelength: 650 nm, 105mW) in the experimental setup. the evaluation of potential of photochromic diarylethene as a polarization medium by recording and reading in a real-time operation was investigated. Using diarylethene 1b/PMMA film as recording medium, the polarization patterns were recorded and read on the film under the microscope. The polarizations of the patterns recorded on the film are shown in Figure 5. The exposure time for each pattern was 0.02 s and the wide of each pattern was about 20 μm. The result indicated that the new photochromic diarylethene can be applied in high capacity optical storage.

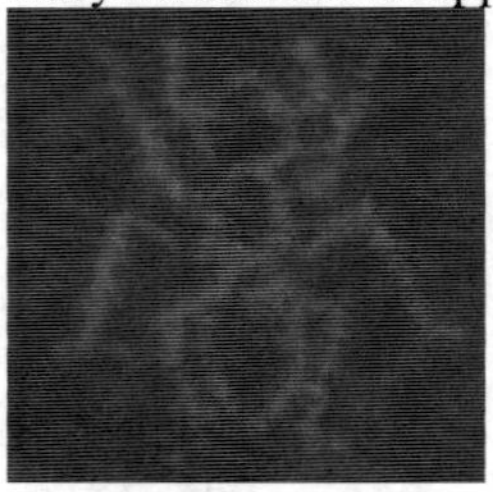

Figure 5. Readout patterns from the diarylethene 1b/PMMA film.

In Conclusion, a new photochromic compound was synthesized, the photochromic properties were studied, the results indicated that the compound has excellent photochromic properties. the evaluation of potential of photochromic diarylethene as a polarization medium by recording and reading in a real-time operation was investigated. The result indicated that the new photochromic compound can be applied in high capacity optical storage.

Acknowledgment

This work was supported by the the Youth Science Funds of the Education Office of Jiangxi, China (GJJ10241).

References

[1] Irie M 2000 *Chem. Rev.* **100** 1685
[2] Luo S J, Chen K X, Cao L C, Liu G D, He Q S, Jin G F, Zeng D X and Chen Y 2005 *Opt. Express* **13** 3123
[3] Pu S Z, Yang T S, Yao B L, Wang Y L, Lei M and Xu J K 2007 *Mater. Lett.* **61** 855
[4] Brown G H *Photochromism* Wiley-Interscience, New York, NY, 1971
[5] Dürr H and Bouas-Laurent H *Photochromism: Molecules and Systems* Elsevier, Amsterdam, 2003

[6] Kobatake S and Irie M 2004 *Bull. Chem. Soc. Jpn.* **77** 195
[7] Irie M, *High-Density Optical Memory and Ultrafine Photofabrication* Springer, Berlin, 2002
[8] Irie M *Photoswitchable Molecular Systems Based on Diarylethenes* Wiley-VCH, Weinheim, 2001.
[9] Yao J N, Hashimoto K and Fujishima A 1992 *Nature* **355** 624
[10] Fernandez-Acebes A and Lehn J-M 1999 *Adv. Mater.* **11** 910
[11] Irie M, Kobatake S and Horichi M 2001 *Science* **291** 1769
[12] Kobatake S, Takami S and Muto H 2007 *Nature* **446** 778
[13] Higashiguchi K, Matsuda K, Tanifuji N and Irie M 2005 *J. Am. Chem. Soc.* **127** 8922
[14] Pu S Z, Yang T S, Xu J K, Shen L and Liu G 2005 *Tetrahedron* **61** 6623
[15] Yamamoto S, Matsuda K and Irie M 2003 *Chem. Eur. J.* **9** 4878
[16] Fan C B, Pu S Z, Liu G and Yang T S 2008 *J. Photochem. Photobiol. A: Chem.* **194** 333
[17] Morimitsu K, Shibata K, Kobatake S and Irie M 2002 *J. Org. Chem.* **67** 4574
[18] Pu S Z, Yang T S and Xu J K 2006 *Tetrahedron Letters* **47** 6473
[19] Pu S Z, Yang T S and Li G Z 2006 *Tetrahedron Letters* **47** 3167
[20] Pu S Z, Liu G and Shen L 2007 *Organic Letters* **9** 2139

Journal of Physics: Conference Series **276** (2011) 012117 doi:10.1088/1742-6596/276/1/012117

Three-dimension reconstruction based on spatial light modulator

Xuejiao Deng, Nanyang Zhang, Yanan Zeng, Shiliang Yin, Weiyu Wang
Huazhong University of Science and Technology

daisydelring@yahoo.com.cn

Abstract. Three-dimension reconstruction, known as an important research direction of computer graphics, is widely used in the related field such as industrial design and manufacture, construction, aerospace, biology and so on. Via such technology we can obtain three-dimension digital point cloud from a two-dimension image, and then simulate the three-dimensional structure of the physical object for further study.

At present, the obtaining of three-dimension digital point cloud data is mainly based on the adaptive optics system with Shack-Hartmann sensor and phase-shifting digital holography. Referring to surface fitting, there are also many available methods such as iterated discrete fourier transform, convolution and image interpolation, linear phase retrieval.

The main problems we came across in three-dimension reconstruction are the extraction of feature points and arithmetic of curve fitting. To solve such problems, we can, first of all, calculate the relevant surface normal vector information of each pixel in the light source coordinate system, then these vectors are to be converted to the coordinates of image through the coordinate conversion, so the expectant 3D point cloud get arise. Secondly, after the following procedures of de-noising, repairing, the feature points can later be selected and fitted to get the fitting function of the surface topography by means of Zernike polynomial, so as to reconstruct the determinand's three-dimensional topography.

In this paper , a new kind of three-dimension reconstruction algorithm is proposed, with the assistance of which, the topography can be estimated from its grayscale at different sample points. Moreover, the previous stimulation and the experimental results prove that the new algorithm has a strong capability to fit, especially for large-scale objects .

1. Introduction

With the development of industry, the needs for improving quality and precision become increasingly urgent. Therefore, people work out some related technologies to solve this problem. Three-dimensional reconstruction, well known as an effectively solution and based on the measurement of analytes, can be divided into Contact and Non-contact. The contact measurement, which requires point-by-point scanning, has been gradually phased out considering its low efficiency. While, compared with the former, the non-contact measurement is now a hot topic in research field since it is simple and fast processing. Its theory can also simply described as following: via introducing various detectors, the system can obtain the information such as analytes's images so as to obtain digital point cloud in further processing. By now, there are mainly three non-contact measurement methods: one is based on the spatial light modulator, one wavefront instrument and the last polarization imaging technology. Among them, the system based on spatial light modulator is the most common way to obtain point cloud, so this method is chosen for further study in reconstruction algorithm.

2. Principle of reconstruction

Three-dimension reconstruction of objects is composed of two parts. One of them is how to obtain the point cloud in the measurement system while the other is surface fitting. Surface fitting, always with the disorderly scattered points, has four normal fitting ways to establish model: the least-squares fitting method, the model fitting method, the backbone method and the separate algorithm. And the point cloud which is used to reconstruct surface feature need to be the height or the surface vectors of object. In this case, we choose phase which is linear with height as the reconstruction data. Then, with the data, an arithmetic is presented and proved to be feasible with experiment demonstration.

2.1. The acquisiton of point cloud

Digital image reflecting the light intensity is discribed in the form of gray levels, so 8-bit digital figure correspond to the 256 grayscale pixel matrix. To make the gray levels change according to the cosine rule, the value in i row j column is expressed as follows:

$$G(i,j) = 255 \times [1+\cos(2\pi i/T+\phi_0)]/2 \tag{1}$$

where T is the change circle of pixel, and ϕ_0 is the initial phase.

When applying cosine modulation virtual grating in phase-shifting phase measurement, the intensity distribution of deformable virtual raster image recorded in the CCD camera is as the formula:

$$I(x,y) = A(x,y) + B(x,y)\cos[\varphi(x,y)+\theta(n)] \tag{2}$$

where A(x,y) represents background intensity, B(x,y) on behalf of the cosine field stripes contrast and (x,y) the phase function. Here (x,y) reflects the deformation of the stripes and the measured object height distribution. (n) is a controlled amount of phase modulation with the parameter n the number of changes in the amount of phase modulation, in other words, number of phase shift. When (n) changes, the intensity distribution of the deformable virtual raster images changes while (x,y) is equal to the initial phase of the cosine function. By means of changing (n) to obtain pieces of the deformable virtual raster images, next get its initial phase according to the characteristics of the cosine trigonometric function, this is quite the essence of shift phase demodulation.

Expand the intensity distribution showed above:

$$\varphi(x,y) = -\arctan[t(x,y)/s(x,y)] = -\arctan\frac{\sum_{i=1}^{Nq} I_i(x,y)\sin\theta(i)}{\sum_{i=1}^{Nq} I_i(x,y)\cos\theta(i)} \tag{3}$$

Through solving this equation, the phase map is gotten.

2.2. The arithmetic of reconstruction

As to the essence of three-dimensional reconstruction, we mainly take grid-class method, parameter-class of surface reconstruction and implicit surface fitting method to get the data point cloud after the surface fitting. Generally the way of forming parameters of polynomial fitting on the surface is most adopted for basis-function's simplicity. By changing the number of polynomial basis-function and the coefficient vector, rich and expressive shapes can be got. More than that, since basis-function is infinitely differentiable, the surface can be smooth sufficiently and it will be much easy to calculate the function value and the derivative values. For the most part, polynomial function are used as the basis-function, as well as polynomial's base. In addition, calculating the coefficients of basis-functions is very easy when carrying out surface fitting.

2.2.1. Bicubic spline interpolation arithmetic

The spatial light modulator process images by the method of phase shift, obtaining the phase distribution of continuous coordinates via unwrapping envelope. Phase value at each point (x, y) can be set as:

$$p(x,y)=\sum_{i=0}^{n} C_i f_i(x, y) \tag{4}$$

Bicubic interpolation is an extension of cubic interpolation for interpolating data points on a two dimensional regular grid. The interpolated surface is smoother than corresponding surfaces obtained by bilinear interpolation or nearest-neighbor interpolation.If typical value $\{r_{ij}\}$ is given in the coordinate $\{(x_i, y_j)\}$, (i=0,1,2…,n; j=1,2,…,m), the spline function s (x, y) can be written in the form of tensor produce:

$$s(x, y) = \sum_{i=-k}^{n-1} \sum_{j=-l}^{m-1} c_{ij} N_{i,k}(x) N_{j,l}(y) \tag{5}$$

when both k and l equal to 3, the equation is called bicubic spline interpolation arithmetic.

The continuation of data in spline surface is often required to be given at first with the boundary constraint condition. Bicubic spline surface usually refers to the subdivision cubic polynomial which belongs to $C^2(\Omega)$ in rectangular grid. In order to creat a smooth surface, different from the curce fitting, bicubic interpolation requires not only the intensity value, but also its derivative in any direction along x, y, or xy. So the interpolation condition is set as:

$$\begin{cases} s_3(x_i, y_j) = p(x_i, y_j), 0 \le i, j \le n \\ \dfrac{\partial s_3}{\partial x}(x_i, y_j) = \dfrac{\partial p}{\partial x}(x_i, y_j), 0 \le j \le n, i = 0, n \\ \dfrac{\partial s_3}{\partial y}(x_i, y_j) = \dfrac{\partial p}{\partial y}(x_i, y_j), 0 \le j \le n, i = 0, n \\ \dfrac{\partial^2 s_3}{\partial x \partial y}(x_i, y_j) = \dfrac{\partial^2 p}{\partial x \partial y}(x_i, y_j), i, j = 0, n \end{cases} \tag{6}$$

In general, The solution of bicubic spline function which meet the above conditons is unique, but when it comes to the condition that the number of equations great than unknown quantity, the solution of over-determined equation does not exist. So we can only seek the solution through approaching these points that is the closest to solution in some sense. Here we adopt the method called minimum 2 multiplication. It requires the square sum of deviation between the real value and the measurements minimum, that is:

$$E = \sum_{i=1}^{N} (p(u_i) - p_i)^2 \tag{7}$$

Where u_i represents the surface coordinates (x_i, y_j).

To obtain the minimum value E, it requires:

$$\frac{\partial E}{\partial c_{ij}} = 0 \tag{8}$$

According to the equation above, the value of c_{ij} (i=1,2,...n, j=1,2,...,n) can be obtained from selected feature points, so is the fitting function p(u), with which we can move forward to the phase of all the coordinates. Furthermore, because of the linear relationship between phase and height, the three-dimensional shape of analytes can be restored so as to achieve three-dimensional reconstruction.

3. Results of experiment

The five object phase shifting images modulated by spatial light modulator, after comparing with the conference images, are then demodulated to the phase map (Fig.1). Since the phase is discontinuous at the period of cosine function, the phase map is unwrapped with the threshold equivalent to 3. After that, the unwrapped phase map (Fig.2) which reflects the phase as gray scale is obtained.

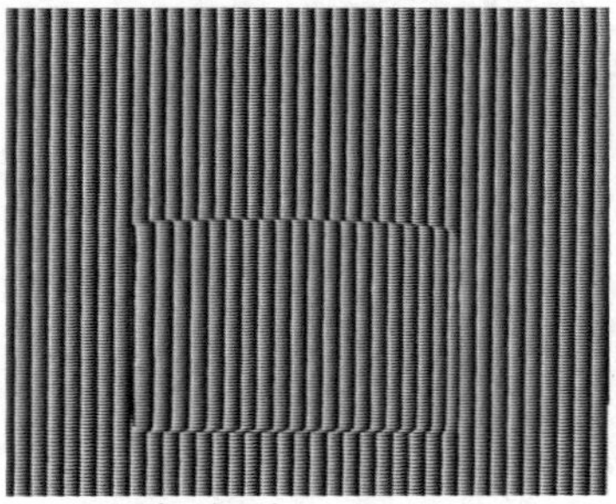

Fig.1. phase map

Fig.2. unwrapped phase map

As the unwrapped phase map has 1310720 pixels, to improve the computing speed and the accuracy of surface fitting, the pixels outside the target are cut out. After that, the data is used to compute a bicubic spline approximation.

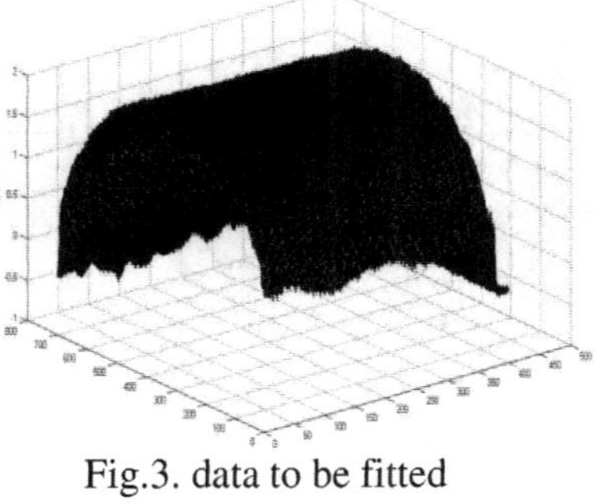

Fig.3. data to be fitted

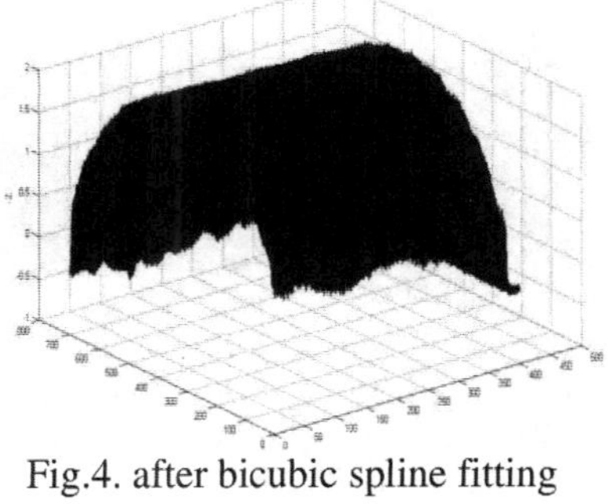

Fig.4. after bicubic spline fitting

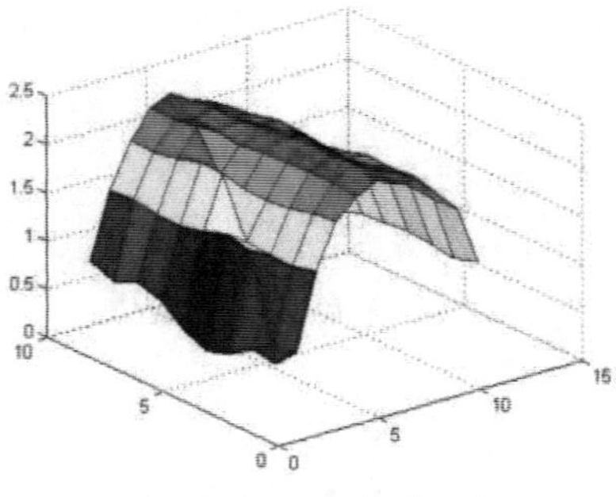

Fig.5. data to be fitted

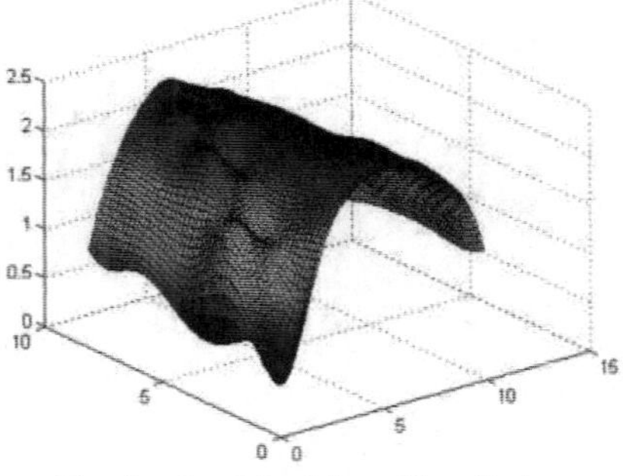

Fig.6. after bicubic spline fitting

In Fig.3 and Fig.4, 267000 point data is used to fit, while 99 point data is chosen in Fig.5 and Fig.6.

References

[1] Hu Yulan,Li Momo.Surface Fitting by Splines using Matlab[J]. Journal of Yunnan Nationalities University, 2005 , Vol.13 (No.2):168-171

[2] Li Ertao, Zhang Guoxuan, Zeng Hong. Algorithm of Surface Fitting Research Based on Least-squares Methods[J]. Jounal of Hangzhou Dianzi University, 2009, Vol.29(No.2):48-51

[3] Zhang Nanyangsheng.The technology of 3D measurement based on multiple spectral bands and spatial light modulation[D].Wuhan:Huazhong University of Sceien and Technology,2006

[4] Yuval Carmon,Erez N.Ribak. Phase retrieval by demodulation of a Hartmann-Shack sensor[J]. Optics Communications, 215(2003):285-288

[5] X.F. Meng, L.Z.Cai, etc. Wavefront reconstruction by two-step generalized phase-shifting interferometry[J]. Optics Communications, 281(2008):5701-5705

Analysis of image plane's Illumination in Image-forming System
Journal of Physics: Conference Series

Lihua Duan[1]**, Yan'an Zeng, Nanyangsheng Zhang, Zhiguo Wang, Shiliang Yin**

Huazhong University of Science and Technology, Wuhan, China

E-mail:victorydlh@163.com

Abstract. In the detection of optical radiation, the detecting accuracy is affected by optic power distribution of the detector's surface to a large extent. In addition, in the image-forming system, the quality of the image is greatly determined by the uniformity of the image's illumination distribution. However, in the practical optical system, affected by the factors such as field of view, false light and off axis and so on, the distribution of the image's illumination tends to be non uniform, so it is necessary to discuss the image plane's illumination in image-forming systems. In order to analyze the characteristics of the image-forming system at a full range, on the basis of photometry, the formulas to calculate the illumination of the imaging plane have been summarized by the numbers. Moreover, the relationship between the horizontal offset of the light source and the illumination of the image has been discussed in detail. After that, the influence of some key factors such as aperture angle, off-axis distance and horizontal offset on illumination of the image has been brought forward. Through numerical simulation, various theoretical curves of those key factors have been given. The results of the numerical simulation show that it is recommended to aggrandize the diameter of the exit pupil to increase the illumination of the image. The angle of view plays a negative role in the illumination distribution of the image, that is, the uniformity of the illumination distribution can be enhanced by compressing the angle of view. Lastly, it is proved that telecentric optical design is an effective way to advance the uniformity of the illumination distribution.

1. Introduction

In the detection of optical radiation, the detecting accuracy is affected by optic power distribution of the detector's surface to a large extent. In addition, in the image-forming system, the quality of the image is greatly determined by the uniformity of the image's illumination distribution. For example, as to cameras, the film's exposure is proportional to the amount of illumination. When the non-uniformity of the image's illumination overpass a certain level, the quality of the picture will be degraded sharply as a result of the uneven photosensibility[1]-[3]. What's more, in the practical optical system, affected by various factors such as field of view, false light and off axis and so on, the distribution of the image's illumination tends to be non uniform. Therefore, it is necessary to analyze the image plane's illumination theoretically.

Published under licence by IOP Publishing Ltd

In this paper, the calculation methods about image plane's illumination have been summarized, simultaneously, the mathematical model of out-of-focus light sources' illumination has been addressed. We take the emphasis on the effect of the angle of view, aperture angle, out-of-focus distance and off-axis distance on image plane's illumination. Finally, the results of the numerical results are given.

2. Mathematical analysis of image plane's illumination

2.1. Illumination of the central image

When the luminous surface is a Lambertian emitter and the area is small enough to be ignored, forming a clear image through the optical system. The model to calculate the illumination of the central image is given on figure 1, in which ds and ds' are the tiny areas of object and the image near the axis, y and y' are the tiny sizes of the object and the image, U and U' are the angles of the object view and image view, L and L' are the photometric brightness of the object and the image space, respectively[4]-[6].

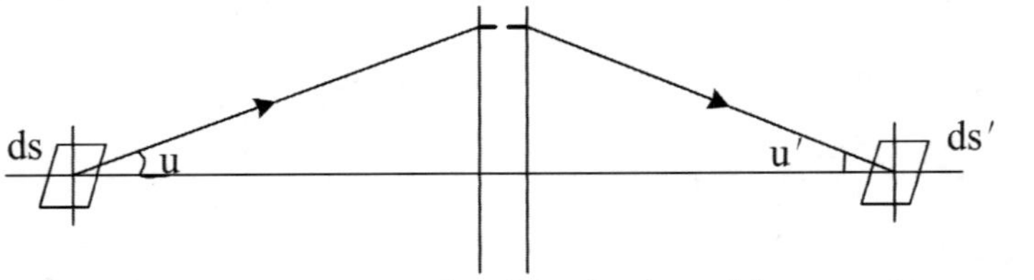

Figure 1. The model to calculate illumination of the central image.

The luminous flux of the tiny image emitted from the exit pupil is

$$\Phi' = \tau\pi L ds \sin^2 U \tag{1}$$

Where τ denotes luminous flux transmittance of the optical system, the illumination of the centre image is

$$E_0' = \frac{\Phi'}{ds'} = \tau\pi L \frac{ds}{ds'} \sin^2 U \tag{2}$$

In the ideal situation, the aperture angles meet the law of sines, that is

$$ny \sin U = n'y' \sin U' \tag{3}$$

Where n and n' denotes the refractive indexes respectively, from the formula (2) and (3), we have

$$E_0' = \frac{n'^2}{n^2} \tau\pi L \sin^2 U' \tag{4}$$

2.2. Image plane's illumination of the off-axis view

Figure 2 shows the model of calculating the image's illumination of off-axis view. A' and P' are the images which formed from object point A and P through optical system, respectively. ω' is the angle of image view from off-axis image point P'. Assuming the exit pupil is filled to the full with the image beam of image point P', in addition, the photometric brightness of object image is even and equivalent to all directions.

Ignoring the vignetting of inclined beams, if diameter of the exit pupil is not very large compared to the distance from the exit pupil to the image of off-axial view, the illumination of point P', the approximated illumination of point P' is

$$E_0' = \frac{I \cos^2 \omega'}{l'^2} \tag{5}$$

Where I_0 is the luminous intensity of the centre image, l' is the distance from the exit pupil to the image of off-axial view, l_0 is the horizontal distance between the image and exit pupil.

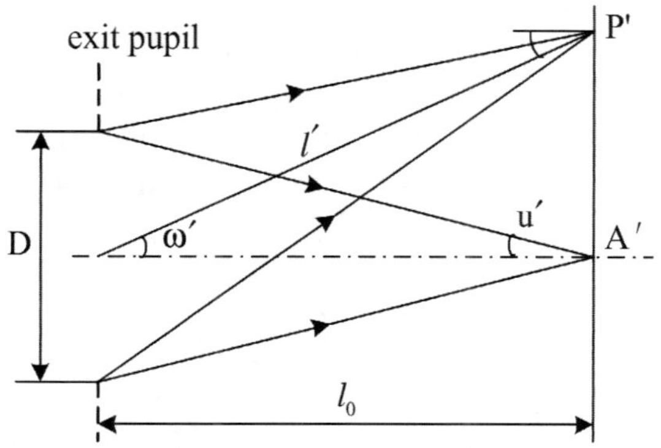

Figure 2. The model to calculate the image's illumination of off-axis view.

Because $I = I_0 \cos\omega'$ and $l' = l_0 / \cos\omega'$, thus

$$E' = I_0 \cos^4 \omega' / l_0^2 = E_0 \cos^4 \omega \tag{6}$$

Where E_0 is the illumination of the image's central point and $E_0 = I_0 / l^2$.

2.3. The relationship between image's illumination and the offset of light source under ideal optical system

Figure 2 shows the imaging model of ideal optical system. The vertical size of the object is y, forming a reversed image through the optical system. Assuming the optical system is symmetric, that is, the entrance pupil and the exit pupil coincide with the principal planes of the object space and the image space, respectively.

The vertical size of the image is y', focal distances of the object and the image are $-f$ and $-f'$, the object distance and image distance are $-l$ and l',respectively. Through Gauss formula, we have

$$\frac{f'}{l'} + \frac{f}{l} = 1 \tag{7}$$

When the medium of the object space is the same with the medium of the image space, we have

$$f' = -f \tag{8}$$

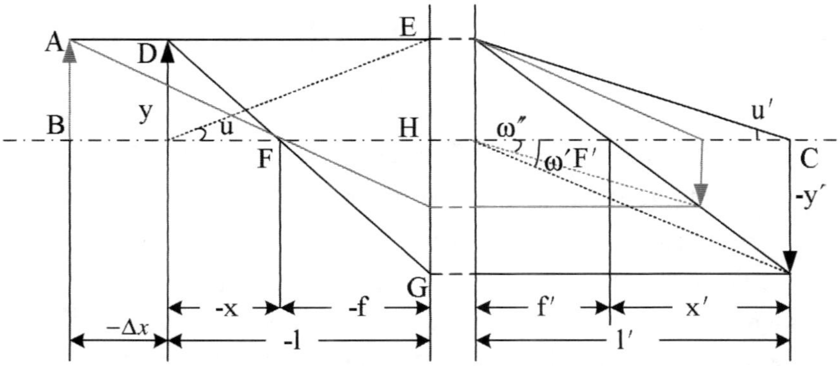

Figure 3. The model of the ideal optical system.

The line DE is parallel to the line FH, so FH/DE=GH/GE, that is

$$\frac{-f}{-l}=\frac{-y'}{y-y'} \tag{9}$$

When the object moves horizontally along the optical axis, the shifting distance is $-\Delta x$, the sign is negative if the object moves forward, on the other side, the sign is positive. The object distance changes into -（$\Delta x + l$）, substituting into the formula (3), combining the formula(4), the image distance l' changes into

$$l''=\frac{f'(l+\Delta x)}{f'+l+\Delta x} \tag{10}$$

y'' changes into

$$\frac{f}{l+\Delta x}=\frac{-y''}{y-y''} \tag{11}$$

After the object move, assuming the aperture angle of image space is U'' and the angle of image view is ω'', the radius of the exit pupil is r, we have

$$\sin U''=\frac{r}{\sqrt{l''^2+r^2}} \tag{12}$$

$$\cos\omega''=\frac{l'}{\sqrt{l''^2+y''^2}} \tag{13}$$

The formula to calculate the illumination of the image's center after the object moves is

$$E_0''=\tau\pi L\sin^2 U'' \tag{14}$$

The formula to calculate the illumination of the off axis changes into

$$E''=E_0''\cos^4\omega'' \tag{15}$$

3. Numerical results and discussions

In order to discuss the relationship among various parameters, numerical simulations have been performed. In the following calculations, the parameters used are: luminous flux transmittance $\tau = 0.6$, photometric brightness $L = 1W/(sr\cdot m)$. As to the object on the axis, we can assume a tiny area whose illumination distribution is uniform, ignoring the affect of aberration on the illumination of the image. Figure 4 shows the relation curve between the aperture angle of the image and the illumination of the centre image, it is demonstrated that the illumination of the image increases with the aperture angle of the image. Hence, we can increase the diameter of the exit pupil to increase the illumination of the image.

To analyze the illumination distribution of the off-axis image, we can use E'/E_0' to illustrate the relative illumination of the image, which can help research the issue about the uniformity of the illumination distribution. Figure 5 shows the effect of the angle of view on E'/E_0'. It is shown that E'/E_0' decreases as the angle of view increases. E'/E_0' get close to 1 as the angle of view reaches 0, which means the uniformity of the illumination is the best in this circumstance. In the practical optical design, designers tend to adopt the way of compressing the angle of view to improve the uniformity of the illumination distribution.

Moreover, we discuss the effect of horizontal offset of the object on image, respective parameter are as follow: the focal distance of the optical system $f'=10cm$, the initial object distance $l=-14cm$, other parameters are the same as figure 4. Assuming the optical system is ideal and the object is moving on the axis. Figure 6 shows the relation curve of horizontal offset of the object and E'/E_0', it is demonstrated that as the object get farther to the focal plane, E'/E_0' increases. As to telecentric optical system, the uniformity is the best. In addition, when the distance from the object to the focal plane is certain, as the object get closer to the axis, the uniformity decreases, which is an important factor that the illumination distribution is non-uniform.

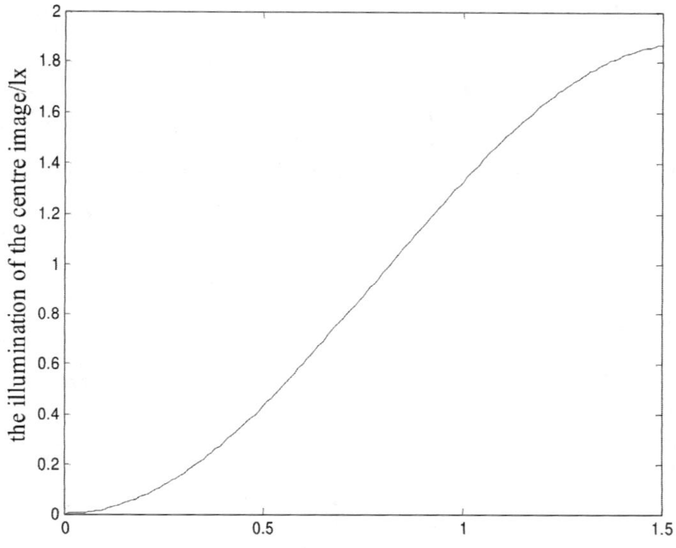

Figure 4. The relation curve between the aperture angle of the image and the illumination of the centre image

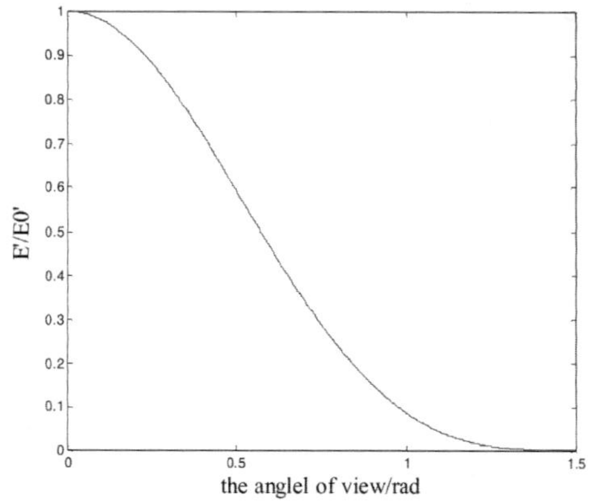

Figure 5. The relation curve between the angle of view and E'/E_0'.

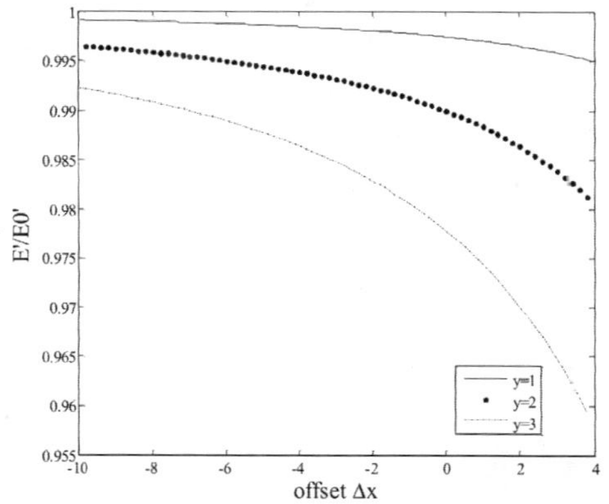

Figure 6. The relation curve between horizontal offset of the object and E'/E_0'

4. conclusions

As the aperture angle of the image increases, the illumination the centre image increases, so it is recommended to aggrandize the diameter of the exit pupil to increase the illumination of the image. The angle of view plays a negative role in the illumination distribution of the image, that is, we can try to compress the angle of view to improve the uniformity of the illumination distribution. Lastly, it is proved that telecentric optical design is an effective way to enhance the uniformity of the illumination distribution.

References

[1] Michael S. Brown, Peng S, Tat-Jen C 2006 *Image Pre-conditioning for Out-of-Focus Projector Blur IEEE. Vol 2.* p 1956-1963
[2] P. Vivirito S. Battiato, S. Curti 2002 *Restoration of out-of –focus images based on circle of confusion estimate SPIE. Vol 4790* 408
[3] Günter Z 1990 *Design and analysis of illumination system Proc. SPIE.* **1354** 617-626
[4] Cai H Y, Liu T G 2006 *Engineering Optics* (Beijing: Tsinghua University Press) p 90
[5] An L S 2006 *Applied Optics* (Beijing: China Machine Press) p 70-78
[6] Mark S 1997 *Analysis of illumination distribution in the plane of a thermal imager Proc. SPIE.* **3061** p662-672

Measurement of the velocities of dielectric particles in ER fluids with laser speckle

Fan Jijun[1*], Yu Nanhui[2], Zhai Chaowu[1], Du Zhenfei[1]
1 Department of Mathematics and Physics, Wuhan Polytechnic University, Wuhan 430023;
2 College of Mechanical Engineering, Wuhan Polytechnic University, Wuhan 430023
E-mail: jijun_fan@hotmail.com

Abstract: Usually, it is regarded that the movement of dielectric particles in ER fluids is related to the generation of electrorheological effect. In this paper, the principle of measuring infinitesimal displacement by laser speckle is proposed to measure the velocities of particles in ER fluids. It is verified that speckles can be formed by illuminating laser beam on the starch ER fluids. The speckle patterns was recorded with high speed CCD camera, and the speckle patterns sequences with short time intervals was obtained by intercepting laser speckle video. Displacements parallel to electric field and perpendicular to electric field were obtained by processing the two consecutive speckle patterns. The experimental results showed the velocities of particles exhibit the trend towards an oscillatory decreased with time, and a critical electric field and a critical sample concentration exist, when lower than this critical electric field (concentration), the mean particle velocity tends to increase; when high than that, decrease instead. Moreover, the horizontal velocity of particle is almost zero, there is only small oscillatory movement of particles in the vertical direction of electric field when the particle chain of ER fluid forms a relatively stable structure.

Keywords: ER fluids, velocity measurement, laser speckle, dielectric particles

1 Introduction

ER fluid is a kind of suspension by mixing the micron-sized particles of high dielectric constant with the insulating oil of low dielectric constant. When the applied electric field is changed from zero to high, ER fluid is usually considered to transform from Newton fluid under zero electric field to Bingham fluid which means the fluid can exhibit similar solid behavior and have certain yield stress under certain applied electric field[1-3]. In order to explain and quantitatively describe this singular behavior of ER fluid under applied electric field, many kinds of models were built to try to explain the ER effect. According to the current research results, it is generally believed that the ER effect is the result of the polarization of dielectric particles under the influence of an applied electric field[4].

It can be found from experiments that when the applied electric field is relatively low, the particles have certain movements but no ER effect is formed; as long as the applied electric field is increased to a critical range, the ER effect can be observed. Through the microscopic observation, it was found that the particles in ER fluid were arranged in nonuniform chain or column structure with the increase of the applied

* To whom any correspondence should be addressed

Published under licence by IOP Publishing Ltd

electric field, and the moving speed of the particles also changed along with the formation of the micro structure. Tamura and Doi[5] equated polarized particles with dipoles, simulated the Poiseuilleflow of ER fluid by molecular dynamics method. Gao[6,7], etc. considered the edge effect, simulated the Poiseuilleflow of ER fluid by using molecular dynamics of dipoles and found the movement velocity of particles not only had blocking area but also exhibited 'breath pattern' transition state in the contact zone. Liu Yulan, etc. studied the mechanism of the chain structure formation of particles in ER fluid, could realize the prediction of the formation rate and response time of the particle chain[8]. Yananda[9-11] studied coagulation of silica particles in oil by an a.c. electric field and the observation of the coagulation process under various mechanical conditions was made using a video-microscope with a CCD camera. Recently, Qian measured the velocities of nanoparticles in nanofluids by laser speckle velacimetry[12].

The characteristic and mechanism study on ER fluid involves: distribution, coagulation and structure of particles; shear stress of ER fluid; interaction among particles caused by applied electric field and synthesis of ER fluid. Only when the mechanism of ER fluid is investigated clearly, can the macroscopic properties of ER fluid be completely understood, so as to obtain the optimum synthesis plan of ER fluid and then develop more practical ones. So, whether from the point of practice or theory, the mechanism study of ER fluid is always of great importance.

When a laser beam passes through a transparent scatter, speckle will be formed. Much information of the light beam and the object passed through by it is carried by the speckle, it can be analyzed by using related calculation software, and it is known that starch ER fluid of low concentration is just regarded as transparent scatter. In the paper, different laser speckle were observed by laser speckle experimental facility in terms of different concentration of starch ER fluid under different applied electric field, and the movement velocities of starch particles were measured according to the dynamic situation of laser speckle. The relation between the formation of micro structure and the velocity of particles was also discussed.

2 Experiment

2.1 Sample preparation

Taking starch particles as dispersed phase, dimethyl silicone oil as carrier fluid, four different concentrations of samples were made, the mass percentage were 0.5%, 1%, 1.5%, 2%, 2.5%, 3%.

2.2 Experimental facility

Fig.1 is the schematic diagram of experimental facility. It includes: laser, double Polaroid, total reflection mirror, high voltage power supply and CCD camera. The laser's length is 25cm, its ideal wavelength is 632.8nm; the double Polaroid is used to adjust the intensity of light; the total reflection mirror is used to change the propagation path of light; CCD camera is to pickup image.

The sample container is made up of glass and copper electrode. Its length, width and height are 10.76mm, 2.5mm (inner distance) and 24.00mm. It should be noted that the length here means the distance between two electrodes.

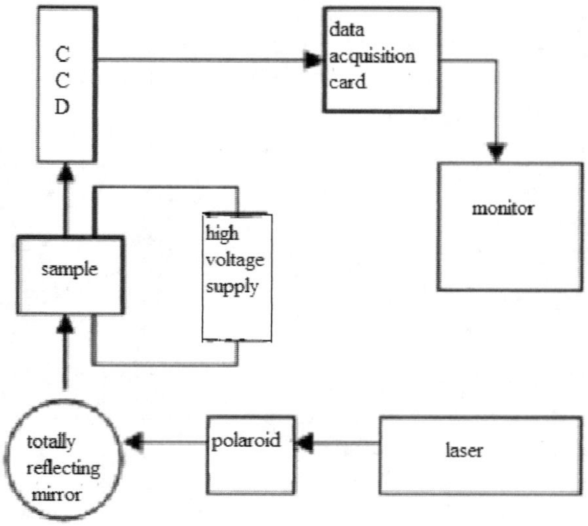

Fig.1 the schematic diagram of experimental facility

2.3 Experimental procedure

In this experiment, the distance from total reflection mirror to lens, from lens to sample, and from sample to CCD are fixed. Different electric fields were applied to each sample, a 30-second laserspeckle vedio was cut out for each electric field strength. Fig.2 is the laserspeckle pattern of ER fluid when the sample has 1%

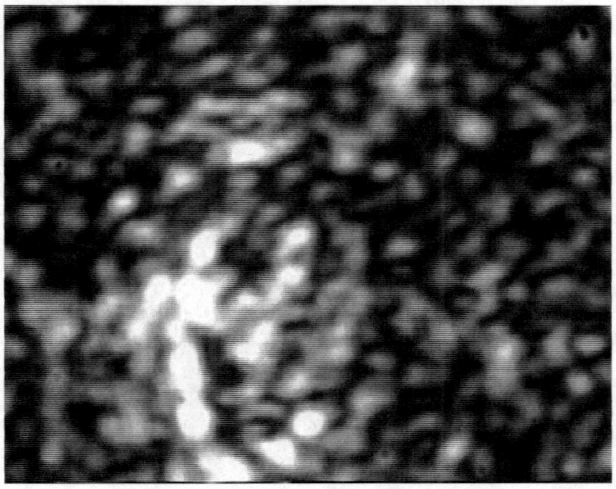

Fig.2 laserspeckle pattern of 1% starch ER fluid under 400v/mm electric field

concentration under 400v/mm electric field. Taking screenshots from the laserspeckle vedio by KMplayer software, a certain number of images were captured per second (in theory, the more images are got per second, the more accurate is the calculation for the velocity of starch particles, but the more calculated amount is needed). Make cross-correlation calculations for every two neighboring images, to obtain statistical displacements of particles per unit time interval in the direction of electric field and also in the vertical direction of it, so as to calculate the velocity. Thus velocity component in the direction of electric field (horizontal velocity for short) and that in the vertical direction of it (vertical velocity for short) could be obtained.

3 Experimental results and discussions

3.1 Particle velocity changes with time

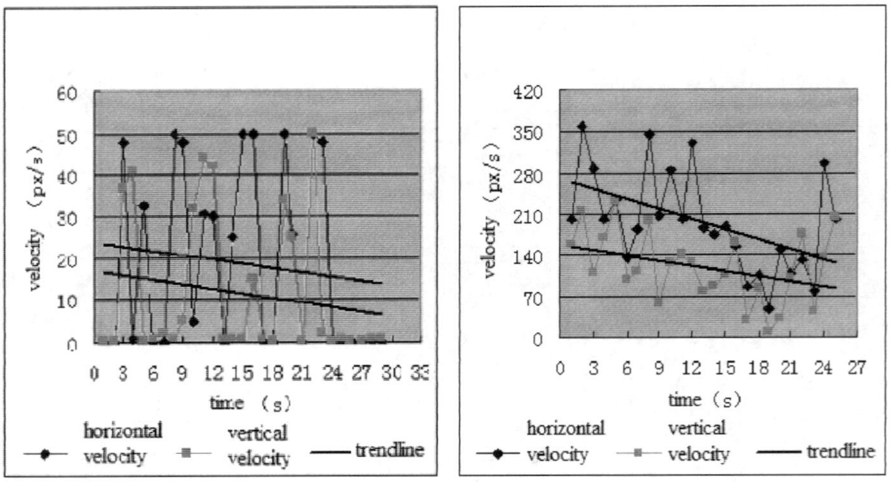

a. electric field of 400v/mm b. electric field of 700v/mm

Fig.3 for a given particle concentration (1%), particle velocity changes with time

Fig.3 gives the particle velocity variations in terms of time when the ER fluid with 1% particle concentration is applied in the electric field of 400v/m and 700v/m respectively. From the figures, it is known that horizontal and vertical velocity both show the trend towards an oscillatory decrease with time, moreover, the horizontal and vertical velocity in fig.3b are obviously larger than that in fig.3a, which illustrates that the horizontal and vertical velocity would increase with the increase of applied electric field.

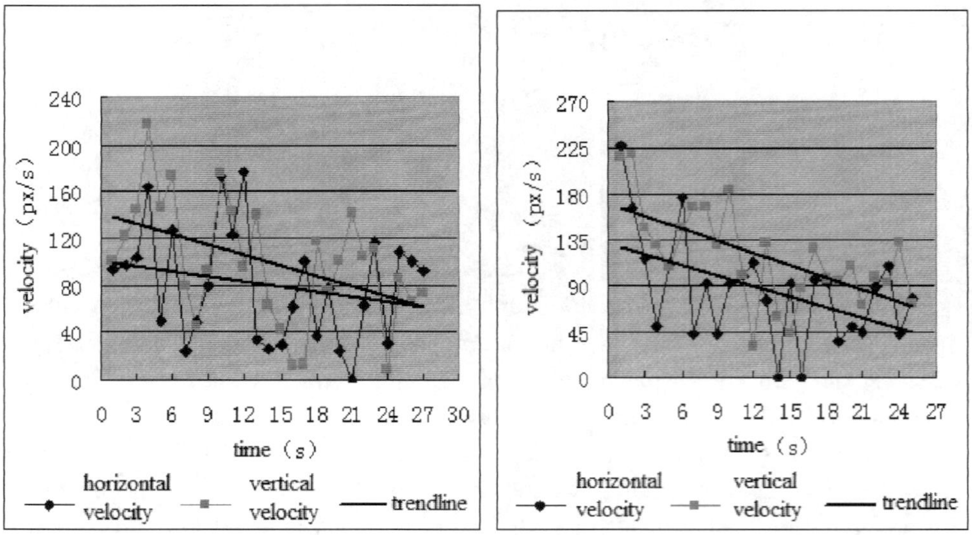

a. electric field of 500v/mm b. electric field of 800v/mm

Fig.4 for a given particle concentration (2%), particle velocity changes with time

Fig.4 shows the relation between velocity and time when the particle concentration of ER fluid is 2%. Fig.a and fig.b indicate two different applied electric fields of 500v/m and 800v/m respectively. It could be known from above figures that the horizontal and vertical velocity under two different electric fields both trend towards an oscillatory decrease, furthermore, the horizontal and vertical peak basically remain unchanged with the increase of electric field.

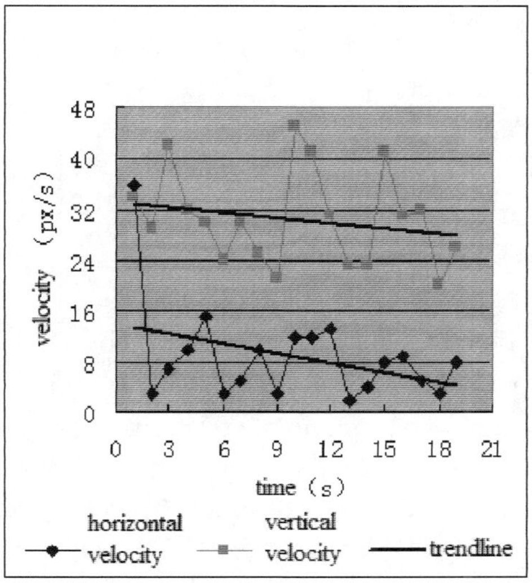

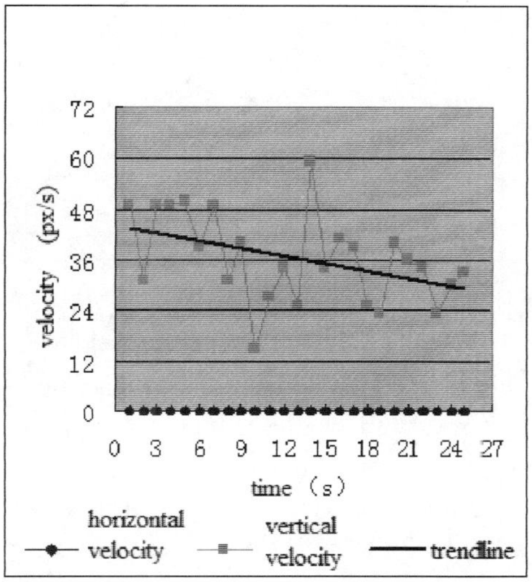

a. electric field of 300v/mm b. electric field of 600v/mm

Fig.5 for a given particle concentration (3%), particle velocity changes with time

Fig.5 gives the relation between velocity and time when the particle concentration of ER fluid is 3%. In fig.a, the applied electric field is 300v/m, it is obvious to see the particle velocity become smaller and the horizontal velocity is smaller than the vertical one. Fig.b shows, with the further increase of electric field and particle concentration, the horizontal velocity of particle is almost zero, there is only small oscillatory movement of particles in the vertical direction of electric field.

3.2 Particle velocity changes with sample concentration

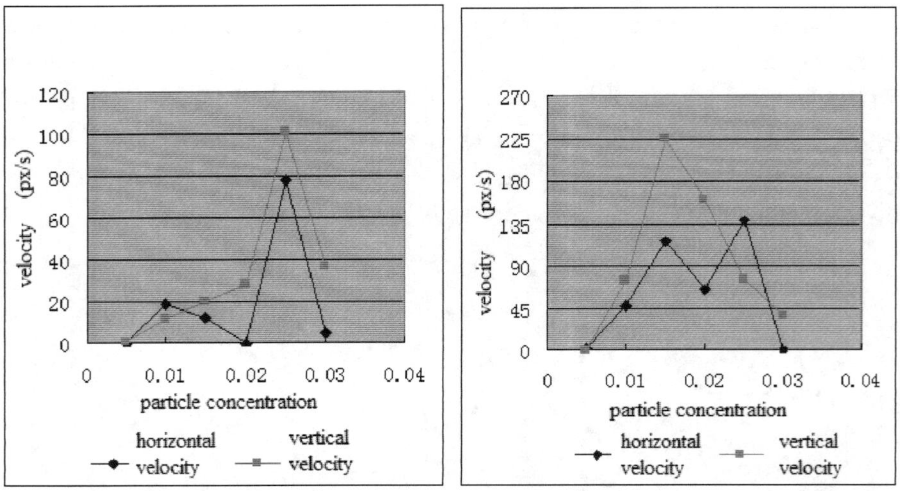

a. electric field of 400v/mm b. electric field of 600v/mm

Fig.6 for a given electric field, particle velocity changes with concentration

Fig.6 shows variation curves of mean particle velocity (that is, a mean value of velocity within 30 seconds) in term of sample concentration under two different electric fields (400v/mm and 800v/mm). Fig. a and fig.b both illustrate that a critical concentration exists when mean particle velocity changes with sample concentration, when lower than this concentration, mean velocity would increase with concentration, but when higher than that, it exhibits a trend of decrease. Comparing these two figures, it is also shown that in the condition of higher electric field, the critical concentration becomes smaller (that is, the peak of curve moves to left), which means when the sample concentration is low, provided the applied electric field is high enough, the particle velocity could also reach its peak.

3.3 Particle velocity changes with electric field

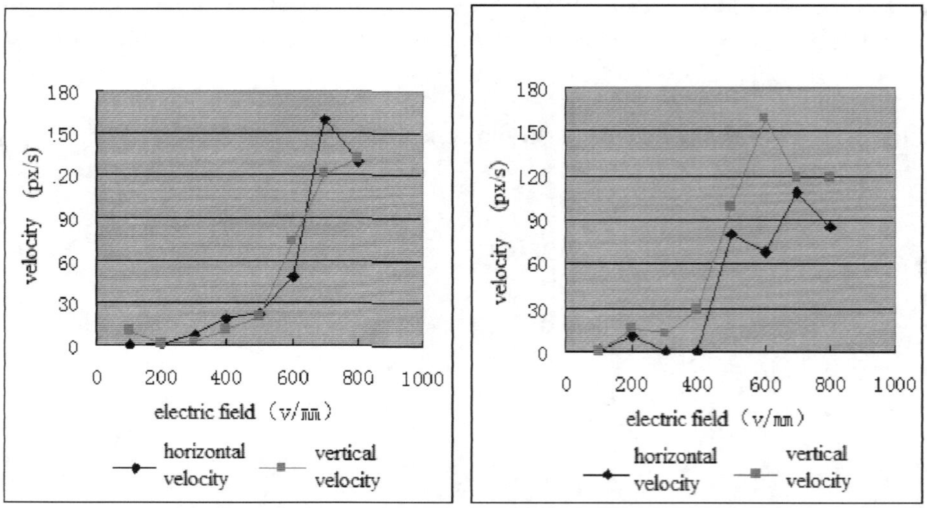

a. particle concentration of 1% b. particle concentration of 2%

Fig.7 for a given particle concentration, particle velocity changes with electric field

Fig.7 gives relation curves between mean particle velocity (that is, a mean value of velocity within 30 seconds) and applied electric field with two different sample concentrations (1% and 2%). Fig. a and fig.b both illustrate that a critical electric field exists when mean particle velocity changes with electric field, when lower than this electric field, mean velocity would increase with electric field, but when higher than that, it exhibits a trend of decrease. Comparing these two figures, it is also shown that in the condition of higher sample concentration, the critical concentration becomes smaller (that is, the peak of curve moves to left), which means lower electric field could be applied to a sample with high concentration, because the particle velocity would also reach its peak.

To sum up, the ER fluid can form a relatively stable structure only when the sample reaches a certain concentration and the applied electric field is high enough, at this time, the particle velocity of ER fluid is close to zero, as shown in fig.8.

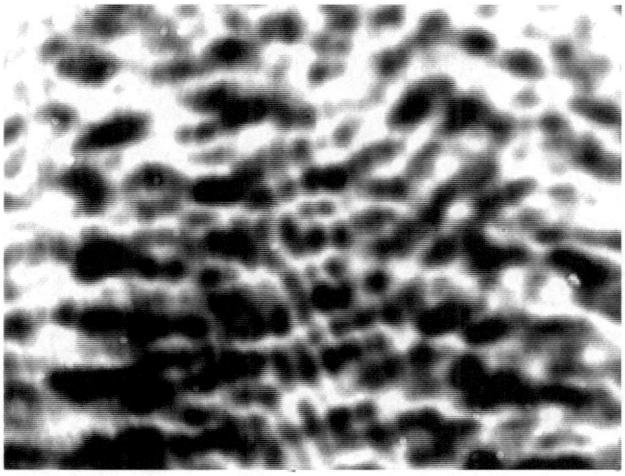

Fig.8 laserspeckle pattern of 3% ER fluid under 700v/mm electric field

4 Conclusion

In the paper, laser speckle measuring system was utilized to measure the velocity of micro particles in ER fluid, the influence of the applied electric field, sample concentration on the particle velocity was analyzed, and the relation between particle arrangement, coagulation and particle velocity was discussed when ER fluid is acted on by an electric field, the main conclusions are as follows:

1) Under certain electric field, the horizontal and vertical velocity both exhibit the trend towards an oscillatory decrease with time.
2) It is shown from the experiment that particle velocity is closely related to sample concentration and electric field, and a critical electric field and a critical sample concentration exist, when lower than this critical electric field (concentration), the mean particle velocity tends to increase; when high than that, decrease instead.
3) When the concentration reach a certain value, the particle chain of ER fluid

forms a relatively stable structure, so, under the applied electric field, the horizontal velocity of particle is almost zero, there is only small oscillatory movement of particles in the vertical direction of electric field.

Acknowledgments

Authors wishing to acknowledge assistance or encouragement from Dr. Guanghui Cheng and Dr. Guizhi Wen. The authors are grateful for the financial support from natural science foundation of Hubei Province(No. 2007ABA086).

References

[1] Block H and Kelly J P 1988 *J. Phys. D: Appl. Phys.* **21** 1661
[2] Atkin R J, Xiao Shi and Bullough W A 1991 *J. Rheol.* **35** 1441
[3] Lu K Q, Shen R and Wang X Z 2006 *J. Chinese Phys.* **15** 2476
[4] Wen W J and Huang X X 2003 *J. Nature Mater* **2** 727
[5] Tamura H and Doi M 1992 *J. Physical Society of Japan* **11** 3984
[6] Xiangyang G, Xiaopeng Z and Changqing Z 2000 *ACTA PHYSICA SINACA* **49** 272
[7] Xiaopeng Z and Danjun G 2001 *ACTA PHYSICA SINACA* **50** 1115
[8] Yulan L, Biao W and Dianfu W 2003 *Applied Mathematics and Mechanics* **24** 342
[9] Yananda H, Nishimura N and Imagawa T 2004 *Proc. Instn Mech. Engrs. Part C: J. Mechanical Engineering Science* **218** 317
[10] Yananda H and Okumura R 1997 *Proc Instn Mech Engrs. Part C: J. Mechanical Engineering Science* **211** 451
[11] Yananda H, Kikuchi M and Nishimura N 2001 *Proc Instn Mech Engrs. Part C: J. Mechanical Engineering Science* **215** 201
[12] Ming Q, Qin Y and Xiaowu N 2009 *ACTA OPTICA SINACA* **29** 290

Study on APD real time compensation methods of laser Detection system

FENG Ying, ZHANG He, ZHANG Xiangjin, LIU Kun

ZNDY of Ministerial Key Laboratory , Nanjing University of Science and Technology , Nanjing 210094 , China

E-mail: fy_caimi@163.com

Abstract. With the monochromatic and the directional capability of the laser, laser detection system is confidential in anti-jamming. Detection accuracy is improved significantly as the result of laser's good orientation ability. Sensitivity is enhanced as laser's high-brightness characteristic. With the development of laser technology and laser devices, laser detections are developed both in civilian and military areas. In the military field, laser detection system has been widely applied in various types of tactical missiles, the technique is more mature.

Because photo detectors receive the backscattering echo signal of target in laser detection system, they are required sensitive enough to weak signal. With APD's salient features of high sensitivity, rapid response speed, high response frequency and low noise equivalent power, etc.; PIN is replaced by APD to improve sensitivity of laser detection system in recent years. The signal magnification is inadequate in laser detection system, the detector output is usually amplified by multistage amplifiers. And then the system noise includes detector noise and latter amplifiers noise. With its high internal gain, APD becomes the primary noise source of receiving system. This point can be attested by analyzing the transfer function of laser detection system receiver. To ensure the system detecting ability, APD noise must be mitigated as low as possible. According to a large number of experiments, the power signal-to-noise (SNR) and the best multiplication factor of APD are mostly affected by background radiation and temperature. In order to make APD operate at state of the best multiplication factor, the optimum bias must be selected due to the actual operating circumstance. Therefore, APD real-time compensation must be adopted.

The existing APD compensation includes the constant false alarm rate compensation, the noise compensation and the temperature compensation. The features of these compensations are obtained by analyzing their operating principles. The constant false alarm rate compensation can't detect the pulse signal which comes randomly. Therefore real-time performance can't be realized. The noise compensation can meet the request of real-time performance. If it is used in the environment where background light is intense or changes acutely, there is a better effect. The temperature compensation can also achieve the real-time performance request. If it is used in the environment where temperature changes acutely, there is also a better effect. Aim at such problems, this paper presents that different APD real-time compensations should be adopt to adapt to different environments. The exiting temperature compensation adjusts output voltage by using variable resistance to regulate input voltage. Its structure is complex; the real-time performance is worse. In order to remedy these defects, a real-time temperature compensation which is based on switch on-off time of switching power supply is designed. Its feasibility and operating stability are confirmed by plate making and experiment.

At last, the comparison experiments between the real-time noise compensation and the real-time temperature compensation is carried out in the environments where temperature is almost invariant and background light acutely changes from5lux to150lux . The result shows that the operating effect of the real-time noise compensation is better here, the noise minifies to a sixth of original noise. The comparison experiments between the real-time noise compensation and the real-time temperature compensation is carried out in darkroom where background light is 5lux and temperature almost rapidly changes from -20℃ to 80℃. The result shows that the operating effect of the real-time temperature compensation is better here, the noise minifies to a seventh of original noise. Moreover, these methods can be applied to other type detection systems of weak photoelectric signal; they have high actual application value.

1. Instruction

Laser fuze has become one of the most important fuzes because of its high precision of detection, high sensitivity and strong resistibility. It is widely applied in various fields [1] [2]. Recently, in order to further improve the sensitivity, APD replaces PIN in laser detection system. Studies show that APD multiplication factor is the most influenced by the background radiation and the operating temperature. Selecting optimum bias based on the actual situation is its service point, so that APD always operates on the best multiplication factor state. To ensure that APD is always in the best state in laser detection system, appropriate compensatory measures must be taken. At present, APD compensation includes the constant false alarm rate compensation, the noise compensation and the temperature compensation. According to action principles of various methods, different methods have different environmental adaptability. If adopting an improper method, it cannot play its role; or even damages the device in given environment. To solve this problem, our paper presents that certain real-time measures must be taken to compensate laser detection system APD in given environment, ensuring that APD operates in the optimal gain state. In addition, the current temperature compensation adjusts output voltage by using variable resistance to regulate input voltage of switch power supply. The circuit structure is complex and the real-time performance of adjustment process is relatively poor. In order to remedy these deficiencies, a real-time temperature compensation which is based on switch on-/off- time of switching power supply is designed.

2. Analysis on APD noise model of laser detection system

The laser detection system amplifies signal inadequately. Generally, it amplifies the output of detector many times. Therefore, latter amplifiers make noise as well as detector does. If output signal is amplified n times, the overall noise coefficient is:

$$NF = NF_1 + \frac{NF_2 - 1}{K_{P1}} + \frac{NF_3 - 1}{K_{P1} \cdot K_{P2}} + \cdots + \frac{NF_n - 1}{K_{P1} \cdot K_{P2} \cdots K_{P(n-1)}} \tag{1}$$

Where, NF is noise coefficient; K_P is amplifier power transmission function.

APD is regarded as the first stage amplifier due to its high internal gain in detection system. The latter amplifiers are regarded as the second grade, the third grade and ……; then $K_{P1} = M^2$. If $M^2 \gg NF - 1$, $M^2 K_{P2} \gg NF_3 - 1$ and ……, equation (1) is approximated as:

$$NF = NF \tag{2}$$

Evidently, system noise coefficient equals with APD noise coefficient， and prime amplifier is not the main noise source which undesirably affects receiving system performances. It is the key to reduce system noise and enhance detection capabilities that restraint APD noise. APD noise equivalent model is shown in Figure 1.

Where, M is APD multiplication factor; i_s is signal photocurrent; $\overline{i_{nd}^2} = 2e(i_s + i_b + i_d)$ is shot noise power spectral density which is induced by signal light, background light and reverse saturation current; $\overline{i_{nT}^2} = 4TK_B / R_L$ is resistance thermal noise power spectral density. APD Power SNR is:

$$SNR = (i_s M)^2 / \left(2e(i_s + i_b + i_d) M^n \Delta f + 4TK_B \Delta f / R_L\right) \tag{3}$$

Selecting optimum bias accounts most in APD application. Doing this will make APD works at the best multiplication factor state, and then maximize the system SNR.
by:

$$\mathrm{d}SNR/\mathrm{d}M = 0 \tag{4}$$

At this point, the best multiplication factor is defined as:

$$M_{opt} = \left[4TK_B/(n-2)e(i_s + i_b + i_d)R_L\right]^{\frac{1}{n}} \tag{5}$$

Evidently, M_{opt} is related to i_s、i_b、i_d、T and R_L. Any change of each, M_{opt} will result in change of M. In which, i_b is inherent in the device. i_d is randomly introduced by signal, it can't be eliminated in application process.

We can see from the experiment that when R is fixed and background noise is constant, the optimum bias is slightly less than the V_B. V_B is closely related to temperature. The relationship is shown in Figure 2. At this point, it is the main factor that T affect M_{opt}. On the contrary, background brightness affects M_{opt} far beyond T when it is high or acutely change. Major determinate of M_{opt} is different in different environment; thus different compensations must be adapted to APD.

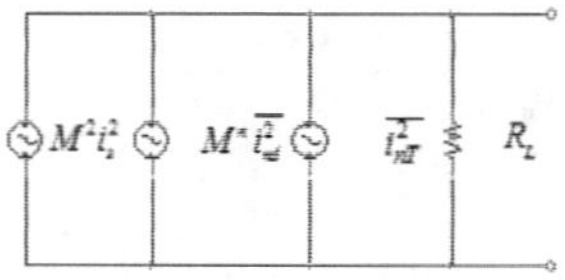

Figure.1 The signal noise equivalent model of APD

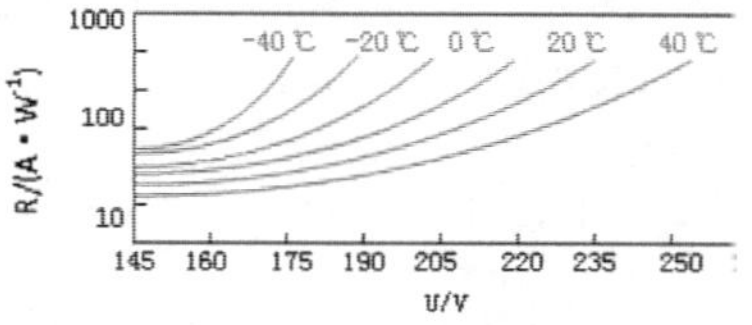

Figure2.APD breakdown voltage change curve with temperature

3. Laser detection system APD compensation

3.1. Laser detection system APD constant false alarm rate compensation

The operating principle of constant false alarm rate compensation is shown in Figure3

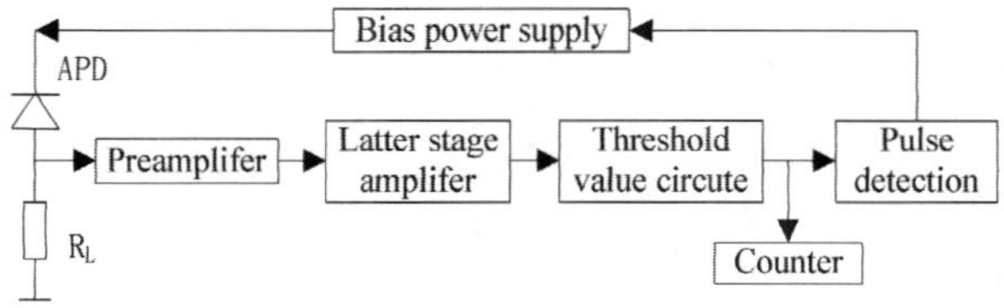

Figure3. APD CFAR Compensation schematic diagram

According to the operating principle of the constant false alarm rate compensation, the logic control circuit must reduce operating bias of APD before system operating. Although it is slightly influenced by background noise, temperature or other factors in external environment, and the application range is relatively wide; it can't detect laser pulse random signal. And then it can't meet the request of APD real-time compensation. Thus, it is seldom used in current laser detection system.

3.2. Laser detection system APD real-time noise compensation

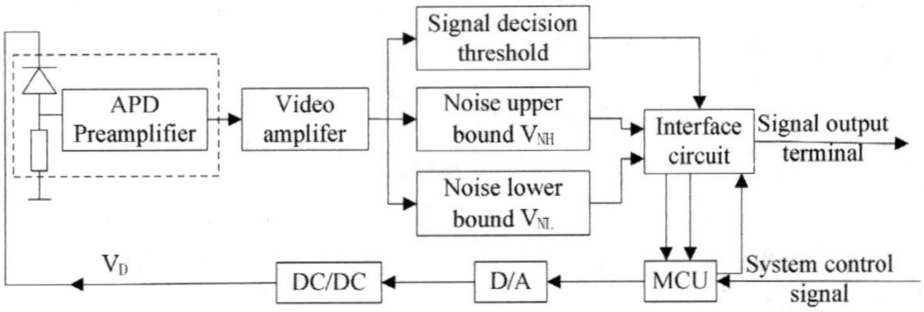

Figure4. APD real time noise compensation schematic diagram

The operating principle of real-time noise compensation [3] is illustrated in Figure 4. Decision point average noise power is:

$$N = eR/2\pi\left[1/\left(1-\left(-KV_gN_g/255V_B\right)^n\right)\right]^{2+X}\left[I_1/2\pi + P_bI_2\right]I_1 \tag{6}$$

Although N is related to P_b, T and N_B, the base of V_B is larger, $1-(-KV_gN_g/255V_B)^n$ change caused by the change of T is relatively small. If P_b greatly changes, the influence that V_B impacts on N is much smaller than the influence that P_b impacts on N. So noise compensation will achieve better results when background light is intense or greatly changes. As the noise compensation doesn't need to be adjusted to bias before detection, it can achieve APD real-time compensation.

3.3. Laser detection system APD real-time temperature compensation based on switching on-off time

If the background illuminance is low or constant in laser detection system operating environment, the influence that T change acts on M_{opt} will be highlighted. At this point, the noise compensation can't meet the M_{opt} requirements; or even PN junction is broken down. So the bias should be adjusted by temperature variation in relation to V_B. Considering the miniaturization of laser detection system, the non-isolated counterattack type transformer switch power is usually applied to supply high operating voltage for APD.

When $0 \le t < t_1$, the power switch turns on, the equivalent circuit is shown in Figure 5. R_1 is equivalent resistance of all components in the circuit. According to Kirchhoff's voltage theorem, when $0 \le t < t_1$, the voltage and current equations are:

$$\begin{cases} i_{L_2} = U\left(1-e^{-t/\tau}\right)/R_1 \\ u_{L_2} = Ue^{-t/\tau} \end{cases} \quad 0 \le t < t \tag{7}$$

Where, $\tau = L_2/R_1$ is the time constant. The magnetic flux Changes in inductor is ΔB_1, thus:

$$\Delta B_1 = Ut_1/NS \tag{8}$$

Where, N is the turn of L; S is the magnetic core sectional area of L.

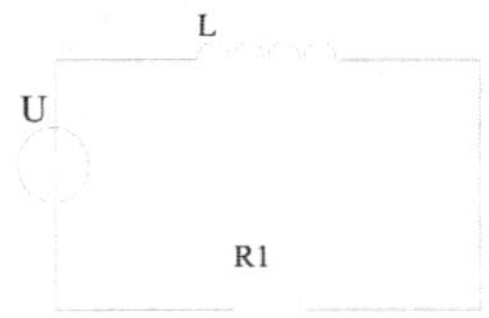

Figure5. The equivalent circuit of switching connecting

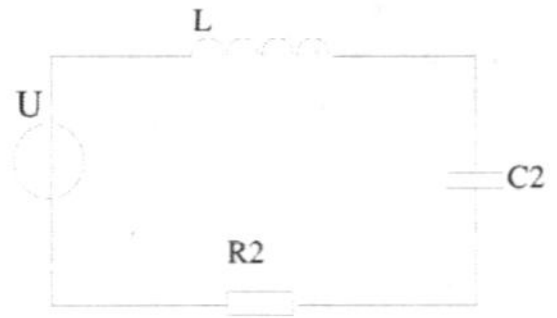

Figure6. The equivalent circuit of switching disconnecting

When $t_1 \le t \le t_2$, power switch cuts off, the equivalent circuit is shown in Figure 6. Where, R_2 and $C_2 (\approx C)$ are equivalent resistance and equivalent capacitance of all elements during the power switching off. According to Kirchhoff's voltage theorem, when $t_1 \le t \le t_2$, the voltage and current equations are:

$$\begin{cases} i_{L_2} = I_{L_2}(t_1)e^{-\delta t}(\cos\omega t + \frac{\delta}{\omega}\sin\omega t) \\ u_{L_2} = -\frac{I_{L_2}(t_1)}{\omega C_2}e^{-\delta t}\sin\omega t \qquad t_1 \le t \le t_2 \\ u_{C_2} = U - R_2 I_{L_2}(t_1)e^{-\delta t}(\cos\omega t + \frac{\delta}{\omega}\sin\omega t) + \frac{I_{L_2}(t_1)}{\omega C_2}e^{-\delta t}\sin\omega t \end{cases} \tag{9}$$

Where, $\delta = R_2/2L_2$, $\omega = \sqrt{\omega_0^2 - \delta^2}$, $\omega_0 = 1/\sqrt{L_2 C_2}$

The magnetic flux Changes in inductor is ΔB_2, thus:

$$\Delta B_2 = ((u_{OUT} - U)(t_2 - t_1))/NS \tag{10}$$

Because ΔB_1 and ΔB_2 are equal, the joint equation of (8) and (10) is:

$$u_{OUT} = U t_3/(t_3 - t_1) \tag{11}$$

According to equation (9) :

$$u_{C_2} = U - R_2 i_{L_2} - L_2 di_{L_2}/dt \tag{12}$$

When i_{L_2} is constant, u_{C_2} (namely u_{OUT}) is decided by Δt which is the off-time of switch element, $\Delta t = t_2 - t_1$.Once switch element is selected, and Δt is constant, thus, any u_{OUT} corresponds to a certain $I_{L_2}(t_1)$, and t_1 can be obtained, t_3 can be obtained by equation(11).

Evidently, if the input voltage is constant and the background radiation is low or constant. To complete switch on and off-time real-time adjustment can ensure that APD reverse bias is always close to V_B at any temperature of any. SNR is maximum. The laser detection system APD real-time temperature compensation principle which is based on the power switch on-off time is shown in Figure 7.

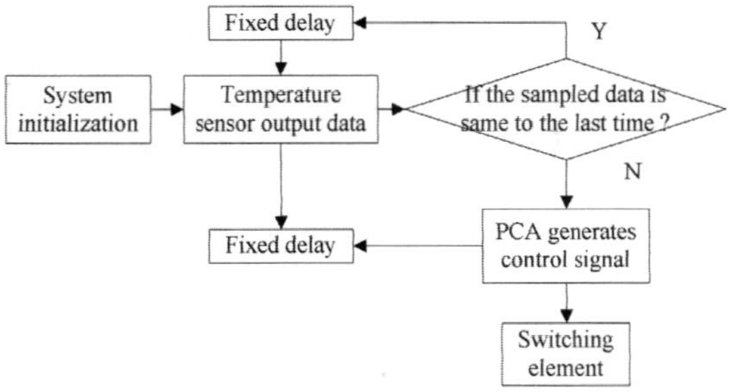

Figure7. APD real temperature compensation schematic diagram

4. Experiment results and analysis

4.1. Laser detection system real-time compensation experiment when the background light acutely changes

Based on the analysis of laser detection system real-time noise compensation principle, and according to Figure 4; the matching circuit board has been designed. AD7524 is used as 8-bit D/A converter, the reference voltage is -5V. AT89CXXXX is used as a control chip, the D/A circuit output is controlled by programming. C30707 is used as photosensitive tube. Its operating temperature range is -20 ℃ to

80 ℃, corresponding to the reverse bias voltage range of 120V to 200V, with a temperature coefficient of $0.6V\,/\,℃$.

In order to verify operating effect of laser detection system real-time noise compensation in the environment where background radiation is high or acutely changes, APD output signal is tested. The peak current of power transmitter is 15A. The rising edge is 8ns.The pulse width is 15ns.The frequency is 10 kHz. When the APD operating background illuminance abruptly changes from 5lux to 150lux, the output signal of receiver module without any compensation is showed in Figure 8 (a).The target echo signal is 4.5V, and then the noise is up to 2.5V. The output signal of receiver module with real-time temperature compensation is showed in Figure 8 (b).There is no obvious change. The output signal of receiver module with real-time noise compensation is showed in Figure 8 (c). The noise value is about 0.4V, reduced to one sixth of the original noise. Comprehensively, the APD output noise suppression effect of noise compensation is remarkable due to a drastic change in background light in the laser detection system where the temperature is almost constant. According to the measured value of APD reverse bias, the fitting curve is shown in Figure 9. And then theoretical K is 40. The Figure demonstrates that the measured value fairly well accords with the theoretical control curve. Calculating according to the measured bias, the real average K value is 39.8. When background illuminance changes from 3lux to 200lux, the APD reverse bias of each interval 50lux is shown in Table 1.Visibly, MCU output control signal changes with the background light changing, different output voltages are obtained.

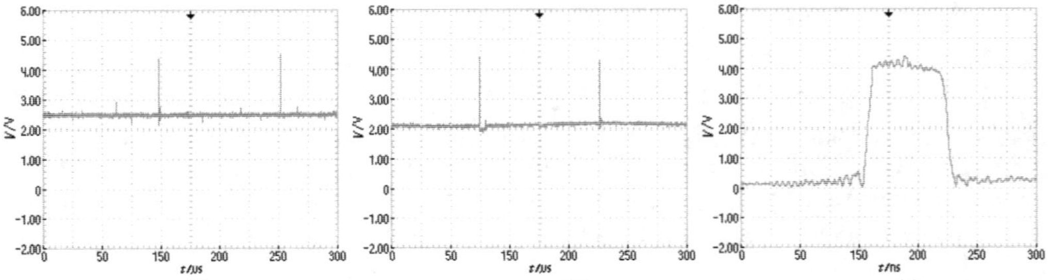

(a) (b) (c)

Figure8. Laser receiving module output

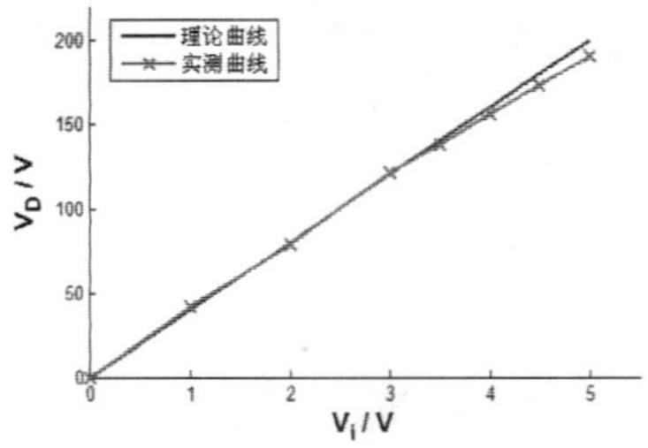

Figure9. Measured reverse bias fitting curve of APD schematic diagram

Table1. Output of D/A and bias of noise compensation under different illuminances

Illuminance	3	50	100	150	200
V_i / V	3.3	3.6	3.8	4.1	4.5
V_D / V	132	144	153	165	178

4.2. Laser detection system real-time compensation experiment when the environment temperature acutely changes

Based on the analysis of laser detection system real-time temperature compensation principle, and according to Figure 7, the matching circuit board has been designed. Temperature sensor used AD590, its rated temperature coefficient of 1μA / k. Power MOS as a switch, the maximum drain current of 6A; voltage is 200 V. C8051XXXX for the control chip. APD also use C30737.

In order to verify operating effect of Laser detection system real-time temperature compensation in the environment where temperature acutely changes, APD output signal is tested in a dark room (light intensity is 1.5lux).The peak current of power transmitter is 15A, the rising edge is 8ns; the pulse width is 15ns; the frequency is 10 kHz. The APD temperature rapidly changes from room temperature (25 ℃) to 60 ℃. The output signal of receiver module without any compensation is showed in Figure 10(a). The target echo signal is 4.4V, and then the noise reaches 2.2V. The output signal of receiver module with real-time noise compensation is showed in Figure 10(b). there is no significant change. The output signal of receiver module with real-time temperature compensation is showed in Figure 10(c). The noise is about 0.3V, falling to one seventh of the original noise. Comprehensively, when the background light is low and the environment temperature considerably changes, the APD output noise is better controlled by temperature compensation, but the effect of noise compensation is poor. When temperature changes from -20℃ to 80℃, the APD reverse bias of each interval 20℃ is shown in Table 2. Hence, MCU output control signals of different frequency and different duty cycle changes when the temperature changes, and different output voltages are obtained.

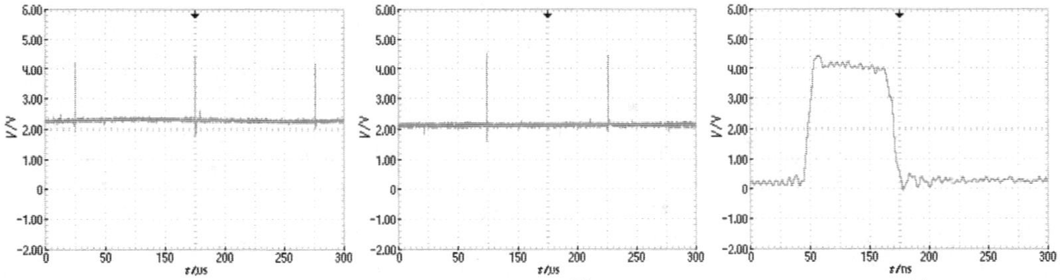

(a) (b) (c)

Figure10. Laser receiving module output

Table2. Changes of MCU controls Signal and output voltage temperature

Temperature /•	-20	0	25	40	60	80
Switching on-time	41.6	46.6	52.8	60.3	68.7	78.9
Switching off-time	16.6	17.2	17.3	18.7	19.7	21.0
Output voltage /V	131	144	160	173	185	198

5. Conclusion

By analyzing the operating noise characteristics of laser detection system transceiver and APD basic application circuit noise model; the relationships between best multiplication factor and dark current, signal power ,background power, temperature, load resistance are obtained. By analyzing principles of constant false alarm rate compensation, real-time noise compensation and real-time temperature compensation of laser detection system; this paper presents that laser detection APD system must take different compensations in different environments. A real-time temperature compensation which is based on switch on-/off- time of switching power supply is designed. The method can achieve real-time adjustment of APD operating bias, and the structure is relatively simple, stable and reliable; thus it can fill the shortages of current real-time temperature compensations. Meanwhile, the comparison experiments between laser detection system APD noise compensation and temperature compensation are carried out in different environments. Results show that adopting appropriate compensation in different environments can make APD operate in the optimal state. If the background illuminance and the temperature of the laser detection system operating environment change dramatically, the composite compensation of noise and temperature in the system can be taken into account. It is certain that the volume and complexity of the system will be correspondingly increased. Additionally, it is necessary to consider the power consumption and electromagnetic interference in the fuse system which is strictly limited in size. Moreover, in the basis of the application effect of APD real-time compensation in the laser detection system, these methods can be applied to other type detection systems of weak optical signal. Therefore, they are of high actual value.

References

[1] CAO Heng, ZHANG Xianjin,ZHANG He,WANG Xiaofeng. Laser Range-finding Technology Based on FPGA[J]. Journal of Nanjing University of Science and Technology (Natural Science) ,2008,32(5):561-564、636.

[2] KONG Jie, ZHANG Baomin. Analysis of Performance Moding on Underwater Imaging Detection System[J]. Journal of Nanjing University of Science and Technology (Natural Science), 2007,31(3): 332-336.

[3] Chen Dianren, Jiang Huilin, Yin Fuchang. Real Time Noise Compensation in Theapd Detection Technique [J].Acta Armamentaril,1999,20(2):127-131

[4] Dai Yang, Cui Xuesen, Wu Yumei, Cheng Xuewu, Gong Shun-sheng.Theoretical computation and experimental measurement on the properties of scattering irradiance in daytime[J].Laser Journal,2008,29(5): 39-41.

[5] Liu Xiaob, Li Li. Design of the optical system of flash lidar based on an APD array[J].Infrared and Laser Engineering,2009,38(5): 893-896.

[6] Combat identification system COMO integrated air defense model evaluation (CISE) study. AD - A206/ 515 - 942.461 PCA05/ MFA01

The design of preamplifier and ADC circuit base on weak e-optical signal

LENG Fen[1], YANG Ying-ping[1,2], YU Ya-nan[2], XU Xiao-ying[1,2]

（1 School of Information Engineering, WUT, Wuhan 430070, China;
2 School of Science, WUT, Wuhan 430070, China）

E-mail: fenleng@126.com

Abstract: Combined with the demand of the process of weak e-optical signal in QPD detection system, the article introduced the circuit principle of deigning preamplifier and ADC circuit with I/V conversion, instrumentation amplifier, low-pass filter and 16-bit A/D transformation. At the same time the article discussed the circuit's noise suppression and isolation according to the characteristics of the weak signal, and gave the method of software rectification. Finally, tested the weak signal with keithley2000, and got a good effect.

1. Introduction

Four-quadrant position detector (QPD) is a type of non-imaging detection that applied in position system widely. Because of its high detection sensitivity, simple signal processing & strong anti-interference, it is used widely in military, surveying, astronomy, communications, engineering surveys[1]. QPD detection system adopted 650nm dot lasers (Pmax≤5mW) as the emission source, QP20-6 as the receiver. As the signals which obtained from the four detector are very weak, the signal strength is generally between 10uA ~ 400 uA in natural light conditions. In order to use the signals more accurately and effectively, the signals must be processed before the further use.

The quality of an e-optical system performance is largely depended on its Suppression to the noise and interference. To solve the weak signal problem, the key is to solve the noise problem in order to improve the NRS. At present, the methods to amplified the weak signal included JFET amplification[2], instrument amplification[3], phase-locked amplification[4] and so on. But these are resolved with hardware, these may led to high cost of hardware, complexity circuit and large uncertainties. According to the characteristics of the QPD signal and noise, the paper designed preamplifier and ADC circuit base on CA3140, INA129 and MAX7705. It not only reduced the hardware interference, but also optimize the measurement signal by using software rectification method. The design has low cost, high accuracy, stable, easy operation, portability, and also has some practical value.

2. The design and implementation of the circuit

The signal obtained from the four-quadrant detector is generally weak、low frequency current signal. For the current signal, first, we should convert it into voltage signal, then after amplification, filtering and A/D conversion just can we send it into processor to do the further processing. In order to get the signal which can be identified by the processor, the whole system is mainly divided into four parts[5], they are I/V conversion, voltage amplification, low pass filtering and A/D transformation, just as Fig.1.

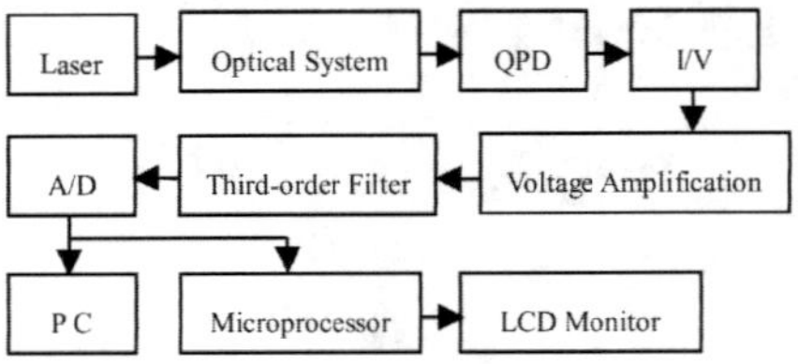

Fig.1 Structure of weak signal processing system

2.1. I/V conversion

The signal we got form the detector is weak current, so we always use high input impedance amplifier and take method of I/V conversion to achieve the weak current measurement. The main factor influencing the current measurement sensitivity is the bias current I_b of a amplifier, the smaller I_b is, the higher sensitivity are. Moreover, noise level and drift have some influence to a certain degree[6].

In the designing, we compared LF353、OPA2244 and CA3140 these there kinds operational amplifiers, Tablel is the measuring result ($R_f = 1k\Omega$).

Tablel Data of I/V conversion

	I(mA)	V(V)	ε (%)
LF 353	0.412	0.406	1.5
	0.393	0.379	3.56
	0.1984	0.2027	2.22
	I(mA)	V(V)	ε (%)
OPA 2244	0.3927	0.3871	1.43
	0.3382	0.3394	0.355
	0.1209	0.1192	1.41
	I(mA)	V(V)	ε (%)
CA 3140	0.4131	0.4148	0.4
	0.1483	0.1482	0.067
	0.0848	0.0857	1.06

According to the data in Tablel, we know that using CA3140 as the front I/V conversion can achieve good result. CA3140 is an integrated circuit operational amplifier that integrate the advantages of high-voltage PMOS and Bipolar process in one chip, developed by RCA, it has the 1.5TΩ high input impedance and input current low to -10pA(in the ±15V voltage circumstances), and its operating supply voltage is from 4V to 36V(either single or dual power supply), so CA3140 can meet the front I/V conversion need[7].

Fig.2 is the hardware circuit of I/V conversion with CA3140. In this circuit, R_4 play a role in adjusting bias, when the input current is 0mA, the output voltage is 0V, if R_4 is suitable. Meanwhile, there are two resistors R_5、R_6 in series at its tow ends to prevent one end of the current too large when adjusting.

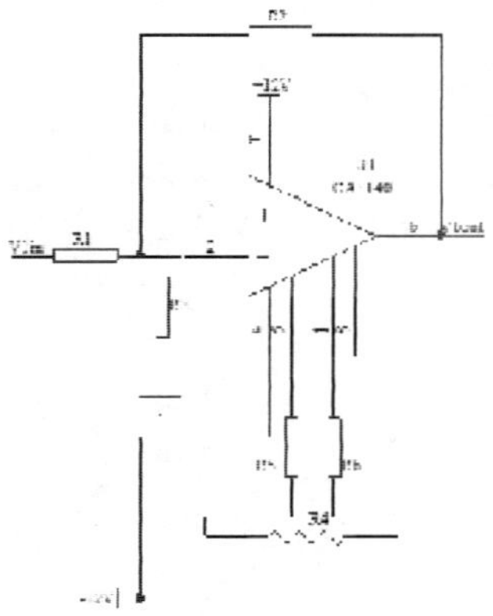

Fig.2 Circuit of I/V conversion with CA3140

2.2. Voltage amplification and filtering

As the signal obtained from the four-quadrant detector is very weak and easily influenced by noise, we select the precision、low power instrumentation amplifier INA129[8]. INA129 has a differential structure, has a strong inhibitory effect in common mode noise, and also has a high input impedance and low output impedance, so it is very suitable for the amplification of weak signals. Table2 show the measured data of voltage amplification, when the adjustment resistor $R = 5.18k\Omega$, $G = 9.5029$, we could get the desired results.

Table2 Data of voltage amplification with INA129

	V_i(V)	V_o(V)	ε (%)
INA	0.35	3.31	0.48
129	0.25	2.367	0.38
	0.15	1.420	0.35

In addition, in order to make the output voltage down faster at high frequencies and improve low-pass filter's ability to filter out noise, here we select the third-order low-pass filter[9].The low-pass filter is made up of dual operational amplifier LF353. LF353 has the feature of high precision、low offset and low power consumption, so could flexibility form various types of amplification and filter circuit, the hardware circuit is shown in Fig.3

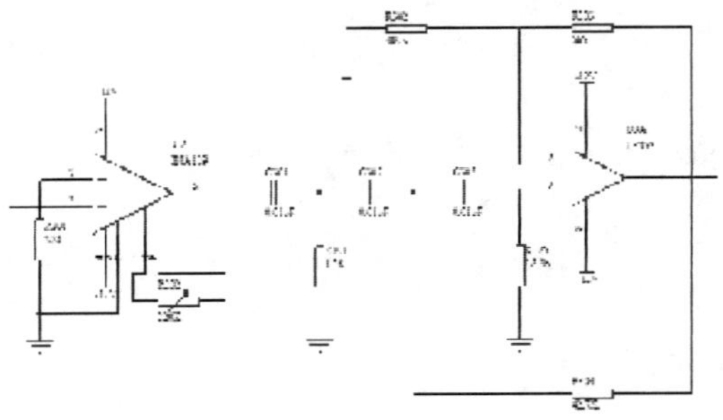

Fig.3 Circuit of vol amplification and wave filtering

In Fig.3, through adjusting sliding rheostat, we can adjust the voltage amplification on factor of 1 to 10000 of any value, to achieve a flexible adjustable magnification.

2.3. A/D transformation

The voltage values we got after amplified are analog signals, if you want to send them into microprocessor to do the further processing, here we should change them into digital signals. In order to ensure the conversion precision, we select 16 bit Σ-△ type AD converter MAX7705[10-12] (Fig.4). MAX7705 is a 2/3 channel analog fore-end device widely used in low frequency measurements, its typical features are high resolution、wide dynamic range、self-calibration and so on. In the program, we set the sampling frequency 250MHz, and use different rectification formulas to do software rectification according to the measuring range.

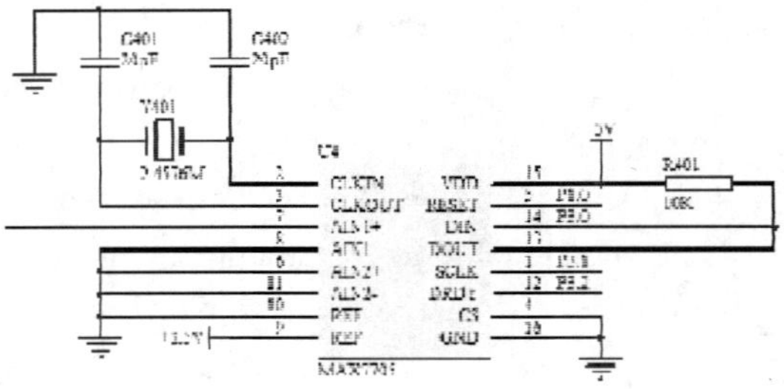

Fig.4 Circuit of A/D transformation

2.4. Circuit measurement

We make the PCB board according to Fig.2 to Fig.4, through debugging the system, we measured the weak signal that after I/V conversion、voltage amplification、low pass filtering and A/D transformation, the result is in Table3.

Table3 Some data of circuit testing

I_{in} (uA)	I/V (V)	A-F (V)	A/D (V)
10	0.01014	0.06069	0.0737
30	0.03036	0.18179	0.1892
50	0.05069	0.30326	0.3212
70	0.07102	0.42486	0.4378
90	0.09121	0.54586	0.5687
100	0.10137	0.60661	0.6345

150	0.15211	0.907211	0.9327
200	0.20276	1.210431	1.2462
250	0.25355	1.514021	1.5603
300	0.3043	1.817561	1.8745
350	0.35503	2.120551	2.1873
400	0.40576	2.424421	2.5022

In the production of circuit boards, we choose the basic components according to the output photocurrent's range of the four-quadrant position detector QP20-6 and reference voltage's range of the AD converter MAX7705. During I/V conversation, the feedback resistor of CA3140 $R_f \approx 1k\Omega$, so the actual magnification is 1.02379. During preamplifier, the adjustment resistor $R = 9.47k\Omega$, and magnification $G = 6.216473$.

3. Analysis and optimization of the system

In the actual test, we use constant current source keithley 2400 to output a constant current of 10uA ~ 400uA, and use accurate multimeter keithley 2000 with five and a half accuracy to measure data. But for the device noise and the interference between different devices, some error was existed in the measured data, so when designing the circuit, we can optimize the measured datas by means of software calibration while reducing the interference of hardware.

In the design of the system circuit, we use the tow-stage amplification in equal. Through the Friss formula[13] (1) based on multi-amplifier noise figure theory, we know that the size of multi-amplifier noise figure mainly depend on the first- stage amplifier noise figure. So as to reduce the noise figure of multi-stage amplifier, we should reduce the noise figure and improve the power magnification A_{p1} of the first-stage.

$$NF_{1,2\Lambda n} = NF_1 + \frac{NF_2 - 1}{A_{p1}} + \frac{NF_3 - 1}{A_{p1}A_{p2}} + \frac{NF_n - 1}{A_{p1} \cdot A_{p2} \Lambda A_{pn-1}} \quad (1)$$

In the design, we choose CA3140 as the first amplication, when processing their output signals, the most important factor determining the measurement is the bias current of CA3140 I_{os}, and then are insulating properties of the materials and wire connection method. The actual output voltage[14]is shown in formula (2).

$$V_o = -(I_p + I_{os})R_f + V_{os} \quad (2)$$

When load resistance R_f is too large, I_{os} will greatly affect the output result, so choosing a smaller bias current operational amplifier is quite needed. To CA3140, I_{os} is 1pA, and V_{os} is 5mV. Table4 show some result of primary amplification after been calibrated by linear function $y = 1.01007x + 2.12914 \times 10^{-6}$.

Table4 Some result of primary amplification after calibration

I_{in} (uA)	V_a (V)	V_t (V)	V_c (V)	ε(%)
10	0.01014	0.010238	0.010244	-0.06192
20	0.02028	0.020476	0.020486	-0.05152
30	0.03036	0.030714	0.030668	0.149268
40	0.04053	0.040952	0.04094	0.027676
50	0.05069	0.05119	0.051203	-0.02555
60	0.06085	0.061427	0.061465	-0.06103
70	0.07102	0.071665	0.071737	-0.10047
80	0.081	0.081903	0.081818	0.10427
90	0.09121	0.092141	0.092131	0.011381
100	0.10137	0.102379	0.102393	-0.0136

In order to improve the portability of the various modules, we use different functions to do software rectification according to different measurement range of each module, it does not list here. For the whole system, we divide the measurement range into 10uA~100uA and 100uA~400uA. Table5 show the result optimized by function $y = 1.04922x + 2.54155\times10^{-5}$ when the input current range of 10uA~100uA, and Table6 show the result optimized by function $y = 1.0494x + 0.00201$ when the input current range of 100uA~400uA.

Table5 Result 1 after optimization（10uA~100 uA）

I_{in} (uA)	V_a (V)	V_t (V)	V_c (V)	ε(%)
10	0.068062	0.063644	0.071437	-12.2455
20	0.116689	0.127287	0.122458	3.793731
30	0.179944	0.190931	0.188827	1.102157
40	0.243877	0.254575	0.255906	-0.5231
50	0.30781	0.318218	0.322986	-1.49825
60	0.356535	0.381862	0.374109	2.030343
70	0.420758	0.445505	0.441493	0.900588
80	0.483432	0.509149	0.507252	0.372663
90	0.547558	0.572793	0.574535	-0.30411
100	0.611297	0.636436	0.641411	-0.78164

Table6 Result 2 after optimization（100uA~400 uA）

I_{in} (uA)	V_a (V)	V_t (V)	V_c (V)	ε(%)
100	0.610551	0.636436	0.642722	-0.9877
125	0.755803	0.795545	0.79515	0.049699
150	0.900861	0.954654	0.947373	0.762701
175	1.061203	1.113764	1.115636	-0.16814
200	1.206066	1.272873	1.267655	0.409891
225	1.366408	1.431982	1.435918	-0.2749
250	1.511854	1.591091	1.58855	0.15968
275	1.672586	1.7502	1.757222	-0.4012
300	1.817741	1.909309	1.909547	-0.01248

325	1.962896	2.068418	2.061873	0.316443
350	2.122264	2.227527	2.229114	-0.07124
375	2.268392	2.386636	2.382461	0.174937
400	2.428832	2.545745	2.550826	-0.19958

In order to accurately and visually display the modified results, we use orgin8 to simulate the tested data, and Fig.5 is the result of data amended by linear curve fitting.

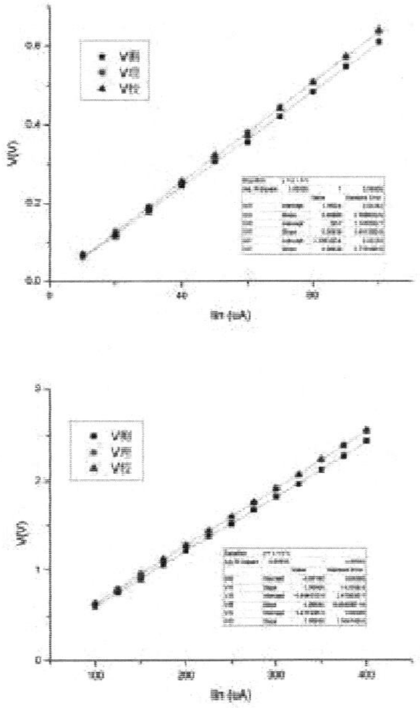

Fig.5 Result of data processing of the system

In the view of the data and curves above, although the circuit can be very good to restrain noise and interference, but the actual measured data slightly smaller than the theoretical value, this is mainly due to the loss of transmission. Through the corresponding liner fitting optimization, the output data is almost the same as the theoretical value. So the whole design can be good at pre-amplification and conversion of the weak e-optical signals, meet the requirement of quadrant detection system.

4. Conclusion

The design of preamplifier and ADC circuit which is discussed in the paper is applicable to QPD system, and each module has strong portability. The software rectification method which is adopted can reduced the hardware interference, and optimize the measurement signal too. Based on weak e-optical signal, through testing the preamplifier and ADC circuit which is mainly composed of low noise single dc amplifier CA3140, instrumentation amplifiers INA129 and 16 bit Σ-△ type AD converter MAX7705, the result was more satisfactory after software rectificating, and it indicated that the circuit could be applied in weak e-optical signal procession.

References:

[1] WANG Qing-you. Photoelectric technology[M]. Beijing: Electronic Industry Press, 2005. 4.

[2] JI Da-xiong, CHEN Xiao-zhen, et al. Optimal Design and Research of Small Signal Amplifier with Low Power Consumption[J]. Chinese Journal of Electron Devices, 2008, 31 (4): 1303-1306.

[3] LIU Hong-li, LI Chang-xi. Design of Lock-in Amplifier Circuit to Measure Weak Signal[J]. Journal of Wuhan University of Technology, 2002, 26(5): 619-621.

[4] ZHANG Shi-rui, ZHENG Wen-gang, et al. The design of preamplifier circuit base on weak signal detection[J]. CONTROL & AUTOMATION, 2009, 25 (8): 223-224.

[5] HOU Yue-xin, ZHOU Dong-liang, et al. The Amplifier Circuit for Weak Sianals[J]. Techniques of Automation & Application, 2008, 27 (11): 65-66.

[6] LI Yan, SONG Chang-qing, et al. Methods of improving performance of weak current integrated amplifier[J]. Nuclear Electronics & Detection Technology, 2007, 27 (5): 978-981.

[7] CA3140EZ Datasheet[S]. www.alldatasheet.com.

[8] INA128/INA129 Datasheet[S]. www.alldatasheet.com.

[9] MENG Li-xia, YU Lin-li, et al. Design of amplifying circuit for tiny signal[J]. Chinese Journal of Scientific Instrument, 2006, 27 (6): 1012-1013.

[10] MAX7705 Datasheet[S]. www.alldatasheet.com.

[11] MA Jing-yuan. Design of High-accuracy Temperature Measure Circuit Based On MX7705[J]. Automation Panorama, 2007. 6: 85-87.

[12] CHEN Yong-gang, WU Bo-nong. Implementation of high precision data acquisition based on AD7705[J]. Foreign Electronic Measurement Technology, 2006, 25(1): 38-40.

[13] HE Zhao-xiang. Photoelectric Signal Processing[M]. Wuhan: Huazhong University of Science and Technology Press, 2008. 1.

[14] KANG Hua-guang, CHEN Da-qing. Basic analog electronics(Fourth Edition)[M]. Beijing: Higher Education Press, 1999.

Application of Optical Fiber Sensing Technology to Ground Deformation Measurement

Feng Qian[1]**, Ding Wei, Wang Hao, Deng Tao**

Institute of Seismology, China Earthquake Administration, 40# Hongshan Ce Road, Wuchang District, Wuhan, China

E-mail: fengqian@eqhb.gov.cn

Abstract. A new method of applying optic fiber sensor to high accurate ground deformation measurement, which is based on the idea about the optic fiber interference combined with fiber grating and the utilization of the characteristics of Bragg wavelength absolute measurement and high accurate relative measurement with fiber interference, is proposed. This sensing system has many advantages, such as anti-interference, corrosion resistance, moisture proof, no zero drift and no need of calibration.

1. Foreword

Ground deformation measurement is the main method to acquire seismic precursor message and research seismicity problem. It has high accuracy to measure system. The methods to measure ground deformation such as hole body strain, borehole strain and GPS etc contain several disadvantages at present. While measuring at the pool environment condition, the first two are interfered by electromagnetic and light seriously, because of the sensing part is electronic device, which the capacity of anti lighting and ESD is weak, zero drift don't dismissed easily and it is difficult to distinguish between static strain. The GPS observation technology is mainly used to measure ground deformation for a long time. It couldn't meet the need of real-time observation about the local ground deformation. And the above several methods all want to take up larger space and capital cost. Therefore, new method to measure ground deformation is needed.

Optical fiber interference [1, 2, 3] measuring method is one of the most effective means in measure science. It is highly sensitive. The modern laser wavelength of light output line width could be very narrow, Fiber optic measurement system accuracy could be higher than l0 category if using this coherent light source. The physical quantities of interference measuring method are relative, cannot undertake absolute measurement, so it doesn't fit the need of long time observation of ground deformation. We use optical fiber grating to do apperceive physical quantity absolute measurement. The optical fiber grating is mature technology and is widely used in Engineering strain measurement. The precision reaches to 10-6 category, while it's difficult to improve measuring precision of the fiber grating sensor system. It couldn't meet the requirement of the ground deformation measurement precision.

The technology of optic fiber interference combined with fiber grating sensor used in ground deformation measurement is feasible. The advantage of high precision relative measurement of optic

[1] To whom any correspondence should be addressed.

Published under licence by IOP Publishing Ltd

fiber interference and absolute measurement of fiber grating is used combination. According to the practical application, it can meet the crystal deformation measurement accuracy requirements, and the demand of demodulation of Optical fiber and fiber grating is not high. It also could avoid stability which owns to the complexity of demodulating circuit and long-term continuous ground deformation measurement and provides low cost of fiber terrain change survey scheme for practical application.

2. Fiber sensing principle

2.1. The strain effects of optic fiber

When subjected to stress effect, the optical fiber length and refractive index of optical fiber core will change. These changes will lead to phase changes of lightwave. Lightwave through the fiber length for L, the output after the phase delay for:

$$\varphi = \frac{2\pi}{\lambda} L = \beta L \qquad (1)$$

In formula (1), β is the transmission constant of the light wave in the fiber. And λ is the transmission wavelength. The phase change of light wave can be written by:

$$\Delta\varphi = \beta\Delta L + L\Delta\beta = \beta L\frac{\Delta L}{L} + L\frac{\partial\beta}{\partial n}\Delta n + L\frac{\partial\beta}{\partial a}\Delta a \qquad (2)$$

In formula (2), a is the radius of the optical fiber core. $\beta L\frac{\Delta L}{L}$ is the phase delay caused by the fiber length change. $L\frac{\partial\beta}{\partial n}\Delta n$ is the phase delay caused by the refractive index change. $L\frac{\partial\beta}{\partial a}\Delta a$ is the phase delay caused by the fiber radius change.

2.2. Optical fiber interference measuring principle

Two columns of light wave, which have the same frequency and the same direction of vibration, In their overlapping areas, the light intensity of synthesis light wave is I. The light intensity of two columns of light wave respectively I_1 and I_2.

$$I = I_1 + I_2 + 2\sqrt{I_1 I_2}\cos\delta \qquad (3)$$

$\delta = \varphi_1 - \varphi_2$,it's the phase difference of the two columns of light wave. The intensity of two light beams is equal, which is emitted from the optical coupler. That is:

$$I = 2I_0 + 2I_0\cos\delta = 2I_0(1+\cos\delta) \qquad (4)$$

Virtually, interference measuring technology is converting the phase change to the light intensity change. At present, there are some forms of structure. Such as Mach-Zehnder, Michelson, Sagnac, Fabry-Pero. This paper adopts Mach-Zehnder interferometer structure.

2.3. Measuring principle of optical fiber bragg grating (FBG) sensor

Fiber grating belongs to the reflective work device. It can through the light wave which has the fixed wavelength. The center wavelength (Bragg wavelength) of the reflected back narrowband light is:

$$\lambda_B = 2n_{eff}A \qquad (5)$$

The Bragg wavelength is related to the grating period and effective refractive index of fiber core.

3. Sensing optical signal demodulation

As the sensor design [4, 5, 6], there should be signal output except sensitive element. The above analyze the design of the optical fiber sensing part, whose output is a beam of interference light, whereas the output of FBG is a beam of reflected light. Both of them are optical signal which carry deformation information, not physical quantity. Accordingly the output of the optical signal should be demodulated to physical quantity. A combination of the two sensing method is proposed in this paper and the method of demodulation is different, but can be combined in hardware implementation. Demodulation block diagram shown in Figure 1.

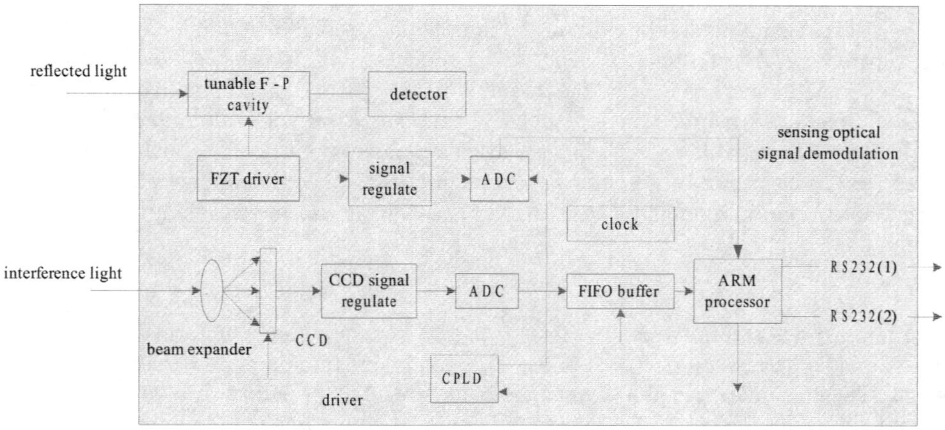

Fig.1 Demodulation of optical signals of sensor

3.1 The CCD demodulation of optical fiber interference

Linear CCD [7, 8] structure is photosensitive diode array. The TCD1501D is adopted, whose pixels value is 5340 and pixel pitch is 7μm. CCD is drive by CPLD, and the differential signal of its output is magnified by amplifier after filtered by low-pass filter. The magnified video signal is converted to digital signal by high-speed AD converter, through first in first out (FIFO) memory cache to memory and ARM processor to time-sharing read the sampling data. Then digital signal could be revert to physical quantity. It should be consider the evenness of CCD pixels photosensitive by CCD driver and avoid interference to adjacent pixels by well overflow.

Owning to the essence of CCD demodulation is identification a shining on the CCD by the surface of the light intensity, there are many factors influence the change of light strength such as the up and down of Light source power, decline of phase caused interference fringes contrast decline, and influence of environmental light etc. There should be adaptive function on CCD 0-level light stripe recognition, namely light intensity scan is needed before each measurement, and ensure the maximum and minimum of the light intensity. Then the zero-order light strip is positioned again based on above, hence the identification of stripe is relatively and irrelevant with Small fluctuations of light intensity.

It is necessary to discern the move of interference fringes on software, such known the current measurement data is adding or subtraction basis on the last time measurement data. The range of signal sensor is less than 2 π . The instantaneous frequency interference will often make stripe of mobile beyond a cycle, while causing measurement result error, using digital filtering method could filter the error. The measurement results are without influence, because ground deformation change is slow.

3.2 FBG optical fiber filter modems Fabry-Perot (F-P)

The reflected light of FBG enters into F, P antrum, which is changed by driving of piezoelectric ceramics. The wavelength of light is changed. If the transmission wavelength of F, P antrum is

Journal of Physics: Conference Series **276** (2011) 012122 doi:10.1088/1742-6596/276/1/012122

coincided with Reflection wavelength of FBG, detector could detect the biggest light intensity. Here the voltage V applied to PZT is corresponding to the FBG reflection wavelength. Small size and low price is the advantage of the demodulation scheme, which could directly output electronic signal changed with wavelength. It's a good demodulation scheme.

3.3 The measured results output and analysis

The demodulation of CCD and F-P is Independent of each other. The final measuring result relate with the both above. The measuring process is the FBG measurements VR being read when instrument boot, afterward, the CCD demodulation value V0 is read. The total measurement result is V=V0+VR. The value of VR accuracy requirement is not accurate. It is just for every time boot only for relative measurement of interfering method providing a deformation datum. If there is no VR, the output of interference instrument is from scratch. That is meaningless for asking for a continuous ground deformation observed value. The total output variation of the system could correctly reflect long-term changes of deformation. Sensitive demodulation transmission is placed in the mountain heading, as near as the sensor. The two demodulated signals convert to two channels digital signal output by RS232C, and send to instrument observed room through being isolated by opt coupler devices, which could process, memory and communicate to signal to complete the process of ground deformation observation.

4. Epilogue

Optical fiber interference and fiber grating sensor is used to measure ground deformation, with the prominent advantages as anti-interference ability, Corrosion, Moisture proof, non-zero drift, no calibration etc. The new method is proposed to solve the ground deformation observation of our earthquake precursory network.

References

[1] B.Y.Kim, J.N.Blake, H.E.Engan, and H.J.Shaw, Opt. Lett. 11,389(1986).
[2] W.V.Sorin, B.Y.Kim, and H.J.shaw, Opt. Lett.11,581(1986).
[3] H.G.Park, C.C.Pohalski, and B.Y.Kim, Opt> Lett.13,776(1988).
[4] T.A.Berkoff and A.D.Kersey, in Proceedings of 8th Optical Fiber Sensors Conf. 372(1992)
[5] K.A.Murphy,B.R.Fogg, and R.O.Claus, in Proceedings of 8th Optical Fiber Sensors Conf.129(1992).
[6] J.O.Askautrud and H.E.Engan, in Proceedings of 8th Optical Fiber Sensors Conf.101(1992).
[7] Y.Liao, Fiber Optics (in Chinese) (Tsinghua University Press, Beijing, 2000).
[8] M.A.Arbore,M.J.F.Digonner, and R.H.Pantell, Opt. Fiber Technol. 2, 400(1996).

Journal of Physics: Conference Series **276** (2011) 012123 doi:10.1088/1742-6596/276/1/012123

Design of High-Precision Frequency Measure System Based on CPLD Time Delay Unit

Feng Qian[1]**, Ding Wei, Wang Hao**

Institute of Seismology, China Earthquake Administration, 40# Hongshan Road, Wuchang District, Wuhan, China

E-mail: fengqian@eqhb.gov.cn

Abstract. Introduced a method for high-precision frequency measurement, which do do not need the complicated measuring control circumstance. CPLD is used for improving the precision of measurement by the method of quantization time-delay. High precision frequency adjustable module based on the method has been used on the photoelectricity data acquisition system. Frequency accuracy is $--8.306\times10^{-10}$, which meet the requirement of instrument.

1. Introduction

The measurement of frequency is actually the measurement of the time intervals. Direct count method [1], which using the quantity of known frequency, is the common way. The desynchronization of the rising edge of measuring frequency and the under test frequency may lead to ±1 pulse error. We can increase the frequency to avoid the problem. However, in most situations, clock frequency cannot change. How to get the high precision in the situation of stable clock frequency is a desiderate problem.

The appear of quantization time-delay [2-5] make it possible. We use it to design a high-precision frequency measure System Based on GPS second pulse, which is low cost and power consumption, without traditional complex frequency measure system. It meets the demand of high-precision frequency.

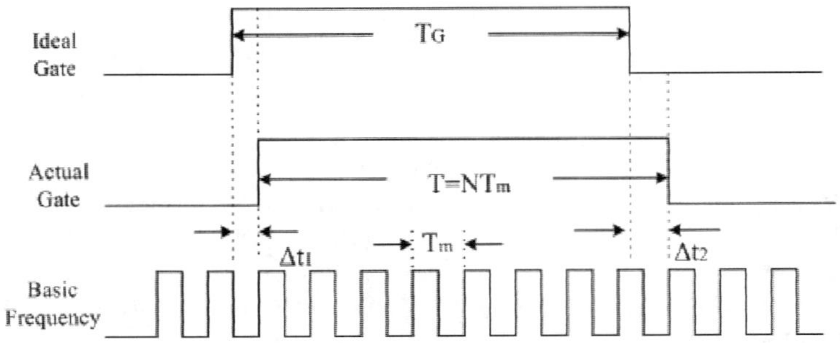

FIG.1 the phase of reference frequency and related frequency

[1] To whom any correspondence should be addressed.

In frequency measuring, such as GPS second pulse signal, phase equality and high precision of the under test frequency and basic frequency are most important. The core element is enhancing the resolution of the time interval. In direct counting method, 10Mhz Frequency can only lead to the measurement resolution of 10ns. Increasing or doubling the base frequency of the crystal oscillator, increases the resolution by a very limited level, it also increases the cost. Therefore, if we use the method of quantization time-delay, which is related with gate signal time and frequency of the basic signal, and is unconcerned with the resolution and frequency of the under test signal [6-8]. Such as FIG1, We can measure time interval Δt_1 and Δt_2 in order to improve the resolution without change the basic frequency.

Make use stability of signal delay, we can measure Δt_1 and Δt_2 quantitative the signal delay. The principle can explain below: let the signal pass by some delay unit, record the situation of the delay unit and calculate the relative difference of the frequency based on the basic frequency, then adjust the VCO sensitivity. Such as FIG1, $T_G=NT_m+\Delta t_1-\Delta t_2$, T_G is ideal gate signal. If we can measure the time interval Δt_1 and Δt_2, we can get T_G Accurate. Device of delay unit can use passive conductors, gate device or other way [9]. The delay time of conductor is too short and may cause complex circuit. The delay time of gate device is too long to control. In general, we choose LCELL logic unit in Complex Programmable Logic Device (CPLD) as delay unit [10], which can reduce conductor disturb, make system stable and simple.

2. Material and methods

In real design, using VCXO to generate 16.384M frequency standard, using GPS module Navman JUPITER 21 to generate standard second pulse, using MCU PIC18F452 control GPS module and adjust D/A device feedback voltage, which achieve using GPS second pulse lock high stable VXCO, and get adjusted second pulse output with long term stability.

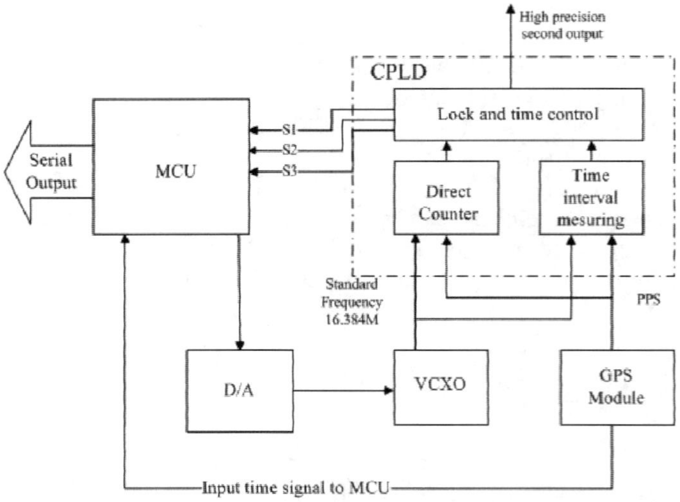

FIG.2 Functional block diagram of the high precision of measurement by the method of quantization time-delay based on CPLD

Such as FIG 2 structure, the key element of the design is the logic and time design, including direct counter, time interval measuring module, lock & time control module. The CPLD can use any Altera MAXII series with enough logic gates. We use LCELL as regular delay unit. In actual use[11], we should change SETTING option in QUARTUSII, turn ignore LCELL Buffer option to OFF, and turn Logic cell insertion to ON, which can avoid QuartusII synthesizing the LCELL delay unit when compiling. As shown in Fig 3, the input second pulse come through a group of LCELL, each LCELL buffers correspond a bit of 32 bits data bus sin[31..0] and connect the lock module. The signal in data bus will be locked at the raise edge of the clock frequency. Δt_1 can be measured by the quantity of LCELL which changed from 0 to 1, we also can measure Δt_2 by the quantity of LCELL which changed from 1 to 0. The key lock module is programmed by verilog HDL as below:

```
  module lock(s1,clk,sin,sout);
input s1,clk;
input[31:0] sin;//delay pulse input
output[31:0] sout;//lock signal output
reg flag=0;/*first pulse flag, flag=0 and S1=1show the first pulse coming in*/
reg[31:0] sout;
always @(posedge clk);
 if(s1)
   begin
     if(flag= =0)
     begin
        sout<=sin;//now input the value of time delay
          flag<=1;
       end
     else  //s1=1, not first pulse
           flag<=1;
         end
   else //s1=0,stop count, reset flag
       flag<=0;
```

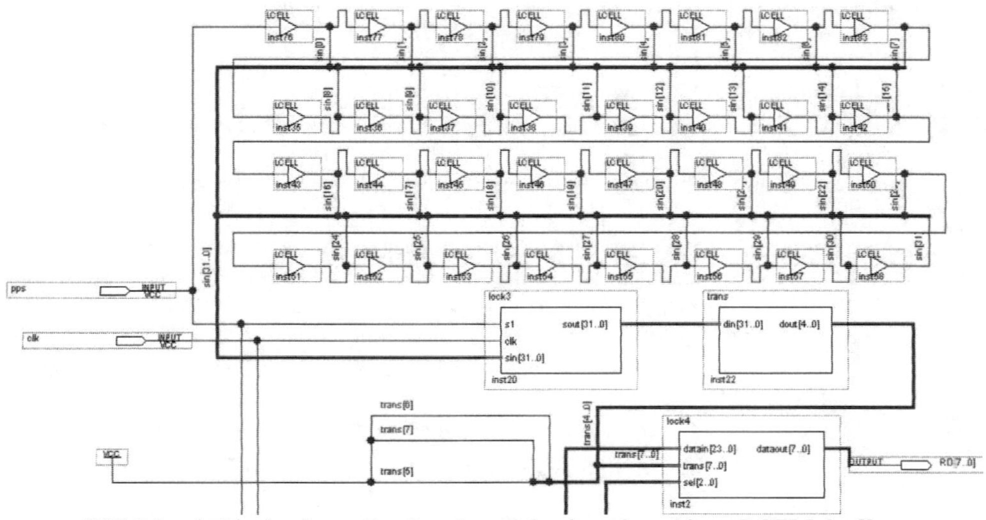

FIG 3 Logic block of quantization time-delay based on Altera LCELL buffer

The acquisition quantity LCELLs of Δt_1 and Δt_2 counted by CPLD transmit in MCU serial port. According to the change of Δt_1 and Δt_2,.We use The Kalman Filter Algorithm[12-13],which principle is minimum mean square error and using the previous estimate value and last observed value to correct standard value[14], deducing with state equation and recurrence method and reducing error by filtering the time difference. Standard frequency and GPS second pulse will output after adjusting D/A output.

3. Results & Discussion

The frequency correction system can adjust frequency accuracy to a higher resolution in 40 seconds when it correctly receive GPS second signal. Actual measurement may be disturbed by external factors such as field or weather, so the stability is decreased. Table 1 shows measured value of the crystal oscillator after correction. (Standard frequency is 16.384Mhz)

Time (h:m:s)	Frequency (Hz)	Time (h:m:s)	Frequency (Hz)
11:00:10	16383999.8772	11:00:15	16384000.1072
11:00:11	16383999.8792	11:00:16	16384000.1037
11:00:12	16383999.8766	11:00:17	16384000.0881
11:00:13	16383999.8875	11:00:18	16384000.0787
11:00:14	16383999.8873	11:00:19	16384000.0684

Table 1 Measured value of the corrected crystal oscillator

We can calculate the accuracy of the frequency is -0.8306×10^{-9}[15]. It's better than the initial frequency accuracy, which value is -4.963×10^{-8}. After being multiple tested, the system can reach 10^{-10} degree of accuracy in 5 minutes.

4. Conclusion

The method can enhance measurement accuracy without having to increase manufacturing cost and power consumption. Now it has worked well in the circuit of sampling photoelectricity system.

References

[1] Ryszard Szplet, Jozef Kalisz. Interpolating Time Counter with 100 ps Resolution on a Single FPGA Device[J]. IEEE Transactions on Instrumentation and Measurement. 2000,49(4):879-882

[2] This reference has two entries but the second one is not numbered (it uses the 'Reference (no number)' style.

[3] Zhou Wei, et.al. Some New Development of Precision Frequency Measurement Technique[J]. Proceedings of the 1995 IEEE International Frequency Control Symposium, 1995:354-359.

[4] ANG H,ZHOU W. Frequency measurement method based on delay chain[J].Chinese Journal of Scientific Instrument,2008,29(3):520-523.

[5] YU JIANGUO,CHEN MING. An Improved Method for Precision Time Interval Measurement[J].2003,23(3):15-19.

[6] ZHANG YING ZHOU WEI. Research of Techniques to Lock High-stablility Crystal Oscillators Based on GPS[J]. Journal of Astronautic Metrology and Measurement. 2005, 25(1):54-58.

[7] Qu Bayi, Zhou Wei. Time to digital method with good precision, linearity and temperature stablility[J]. Chinese Journal of Scientific Instrument. 2009, 30(10):2013-2016.

[8] LIU YILI, ZHOU WEI. A New Method of High Speed and Continuous Measurement of Frequency [J].Journal of Astronautic Metrology and Measurement. 2005, 20(1):32-38

[9] JIANG YUJIE,CHEN CHEN. A Study of New Methods about Frequency Measurement. Chinese Journal of Scientific Instrument. 2004,25(1):30-33.

[10] Zhou Wei et al. Some New Method for Precision Time Interval Measurement [J]. Proceedings of the 1997 IEEE International Frequency Control Symposium, 1997,418-421.

[11] ZHANG YAN LONG, CHU PENG. Design and Analysis of a Digitally Controlled Programmable Delay Element [J]. 2007, 24(8):142-144.

[12] WU JI HUA, WANG CHENG. Design of Altera FPGA/CPLD (Advance)[M].2005,160-164.

[13] LIU GUOHAI, LI QINXUE. Application of dynamic Kalman filtering in state estimation of navigation test [J]. Chinese Journal of Scientific Instrument. 2009,30(2):396-339.
[14] LI ZHAN,ZHANG YING. A Frequency Calibration System Based on Single Chip Processor and GPS Signal [J].Journal of Time and Frequency.2005,28(1):68-75.
[15] HAYTHAM Q, LEONHARD R. Unscented and extended Kalman estimators for non-linear indoor tracking using distance measurements[J]. Positioning Navigation and Communication, 2007(3):177-181.
[16] YAN XUE YI. Using Programmable Timer/Couter Improve Output Frequency Accuracy [J]. Microcontrollers & Embedded Systems. 2002, 2(6):14-16.

Journal of Physics: Conference Series **276** (2011) 012124 doi:10.1088/1742-6596/276/1/012124

Rigorous Three-dimension Electromagnetic Simulations and Optimization for CMOS Image Sensor pixels

Gao Tie-cheng ,Yao Su-ying,Xu Jiang-tao,Zhao Shi-bin

School of Electronic and Information Engineering; Tianjin University; Tianjin 300072

gaotiecheng@tjpu.edu.cn

Abstract: The size of pixels in complementary metal-oxide-semiconductor image sensor has steadily decreased during the last decade, which leads to a challenge on maintaining the pixels' optical performance. The diffraction effects result in increased spatial crosstalk and decreased optical efficiency, so rigorous simulation and optimization are important. FDTD method is used to simulate the optical performance and a diffuse-like source is used to reproduce real conditions. A commercial tool from Lumerical Solutions is used to do three-dimension FDTD simulation. The diffuse-like sources are simulated by the uniform sum of several focused beams characterized by a given f-number of the objective len. The typical 1.75μm, 1.45μm and 1.1μm pixels are simulated and analyzed. The results show that the optical efficiency of the pixels decreases dramatically when the pixels size scaling down. Several approaches to produce better device performance for sub-2um pixels are analyzed .Micro-lens optimization, dielectric stack height reduction can decrease the optical power loss and optical confinement realized by an air-gap can reduce the spatial crosstalk. The results show that the optical performances of the optimized 1.1μm pixel are comparable to those of conventional 1.75μm pixel.

1. Introduction

CMOS image sensors are experiencing exponential market growth due to the increasing demands of digital still and video cameras, security cameras, webcams, and mainly mobile cameras. complementary Metal Oxide Semiconductor (CMOS) technology has shown competitive performance but also many advantages in on-chip functionality, power consumption, pixel readout, and cost, so that it has become more and more important in the Image Sensor Industry.Market trend for higher pixel density leads to smaller and smaller pixel size. Thus the photodiode area shrink.On the other hand the thickness of interconnect layers scales less than the planar dimension, light has to travel through a narrower path to reach the photodiode which cause the problem of light collection. Thus optical simulation[1] is of great importance to characterize the sensor and optimize its performance such as Quantum Efficiency and Crosstalk. Figure 1 shows the structure of CMOS image sensor.

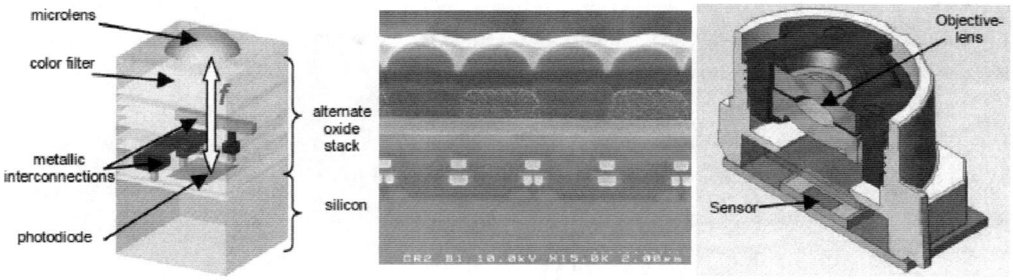

Fig. 1. CMOS image sensor: schematic (left) and SEM picture (center) of CMOS pixel, and the final module (right).

As diffraction effects can substantially affect light propagation when the pixel size is comparable to the wavelength of light waves , We chose an electromagnetic simulation tool based on Finite Difference Time Domain (FDTD)[4][5], available from Lumerical Solutions, to do 3-D simulation for light propagation and photon collection inside the pixels. Figure 2 shows the typical 3-D structure for cmos image sensor.

Fig.2. 3-D structure for cmos image sensor

The imaging performance of the 1.75-, 1.4-, and 1.1- μm pitch pixels is simulated by the FDTD methodology in this paper respectively. A rigorous light source is used to simulate the real light from the objective len.A set of new techniques for the pixel architecture optimization is proposed by analyzing the characteristic parameters. Finally, a comparison between the performance before and after optimization for the 1.1- μm pitch pixel is made and also a performance contrast between the optimized 1.1- μm pitch pixel and the conventional 1.75- μm pitch pixel is done.

2.Design and Methodology

2.1.pixel geometrical architecture.

Figure 2 shows the cross-section projection view of a typical conventional front-side-illuminated image sensor structure. It consist of the microlens ,the color filters,dielectric stacks,metal interconnection and the substrate. A gapless microlens array with an elliptic parabolic surface is used., and the CFA and the silicon substrate are simulated as absorptive materials. Ideal red–green–blue filters are used, and the metal in the dielectric stacks is treated as a perfect electrical conductor. The dielectric properties of materials are described by refractive indexes or dielectric constants obtained from Palik[2][3]. we assume a fill factor of 50% and the photo sensitive area is located at the center of the pixels.

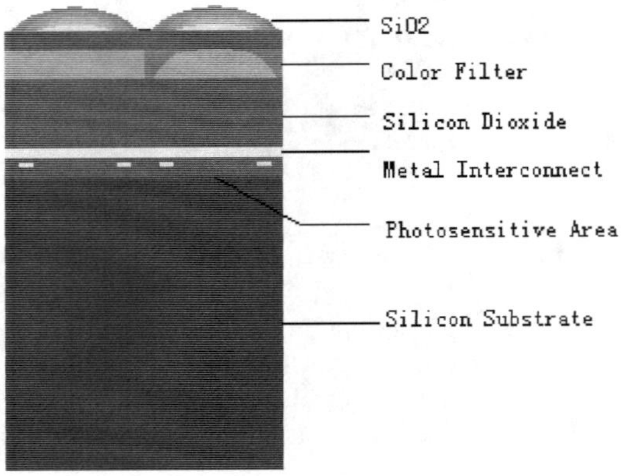

Fig. 3. Vertical structure projection of the four-pixel unit cell

2.2.FDTD

Perfectly matched layer conditions are used at the structure's boundaries. The simulation grid spacing is set to 25 nm for a fast simulation and approximately 10 nm for a detailed simulation. The z components of the Poynting vector available from the resulting fields, are integrated to obtain the total optical power delivered to the photodiode. Complete conversion of photons to electrons, i.e., quantum efficiency of 100%, is assumed when the light incidents on the photodiode area at the silicon interface. the plane-wave illumination with the wavelength of 650 nm is adopted as an incident light source located just above the microlens surface. The simulation area covers a square 2-by-2 pixel array .

2.3.Defuse like light source

In order to correctly evaluate pixels optical performances we must simulate a product-like Illumination. But the lens is hundreds of times bigger than the pixel. So it is difficult to simulate the objective-lens and pixels together because the computation will consume very large resources.

In Lumerical software, a source called "thin lens" allows the simulation of such an objectivelens with a given f-number and chief ray angle. Figure 4 below shows a schematic representation of the light source seen by the pixel array.

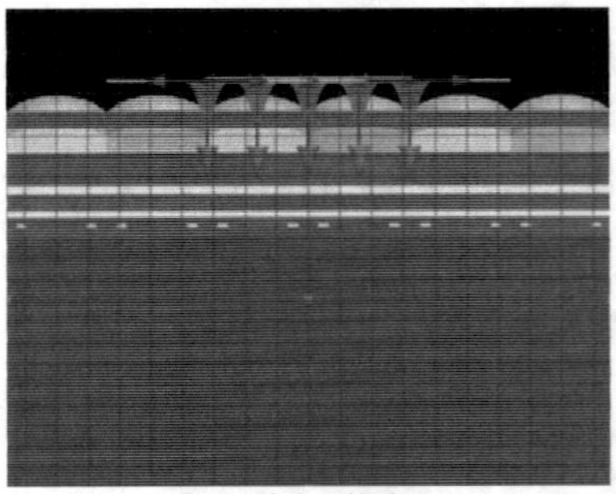

Fig.4. a schematic representation of the light source seen by the pixel array.

The f-numbe is assumed to be 2.8. Sources with a spacing of 0.5μm are employed to illuminate uniformly the microlens. The simulation result shows that this is sufficient to simulate the real conditions.

3.Result and optimization

3.1.simulation result

Optical efficiency and optical spatial crosstalk are introduced to characterize the pixel imaging performance. The pixel optical efficiency is defined as the ratio of optical power received by the photodiode at the silicon substrate to that incident on the pixel surface. Assuming that quantum efficiency of 100%, spatial crosstalk (CTK) is defined as the average ratio of optical power received by the photodiodes of the adjacent pixels (generally four-adjacent pixels) to that of the source directly incident on the pixel surface.

The electric field intensity through the middle of the Green - Blue pixels are shown in the simulation result. The figures show high transmission through the green pixels and low transmission through Blue pixels

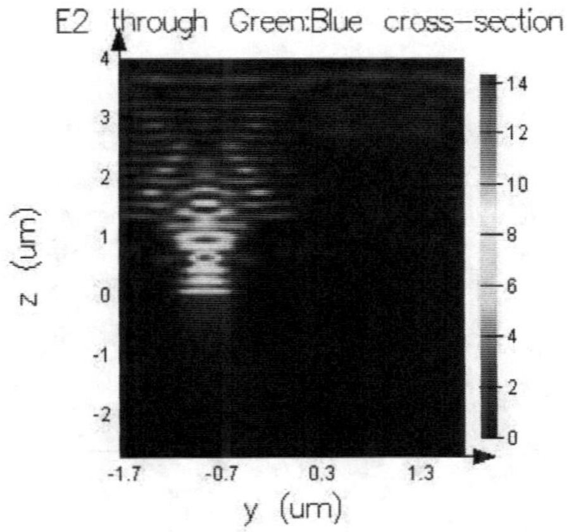

(a)

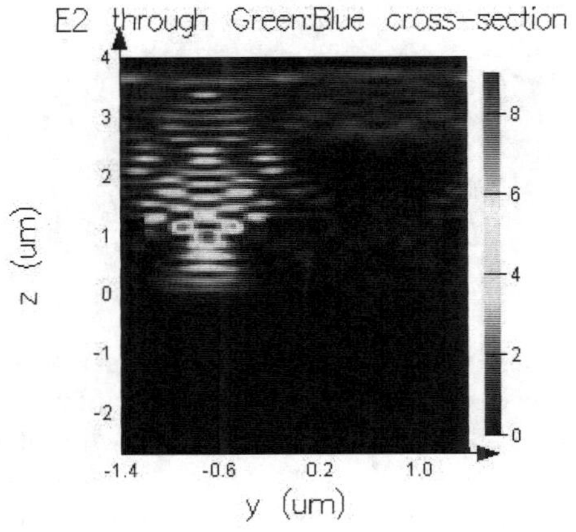

(b)

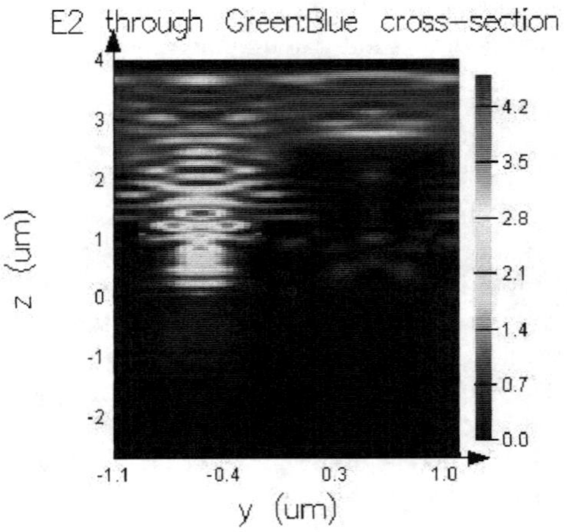

(c)

Fig.5. electric field intensity through the middle of the Green - Blue pixels

3.2.Optimization

*3.2.1 dielectric stack height reduction.*The pixel performance greatly depends on the height of the dielectric stack layer as the pixel size shrinks. A Cu process with 0.13- μ m design technology can be used to reduce the total pixel height by 20% through the thicker metals and interlayer materials compared with those of Al [6-9]. Figure 6 shows the simulation result of the pixel which the dielectric stack height was reduced by 1um.

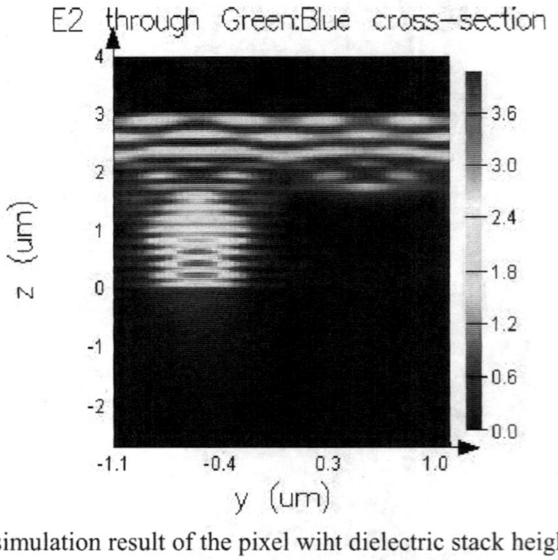

Fig.6. simulation result of the pixel wiht dielectric stack height reduced.

3.2.2 microlens optimization. In this case, the microlens is optimized for the1.05- μ m pixel. Effective optical efficiency at three locations with a microlens radius from 0.5 to 30 μ m is measured and curved utilizing the simulation results (shown in Fig. 7). The microlens height is kept constant at 500 nm, and the illumination is incident normal to the photodiode. Seen from Fig.7,the optimal microlens radius of curvature is between 1 and 1.5 μ m., a microlens radius of 1.1 μ m is used in later simulation for an optimal 1.1- μ m pixel to concentrate more light onto the photodiode.

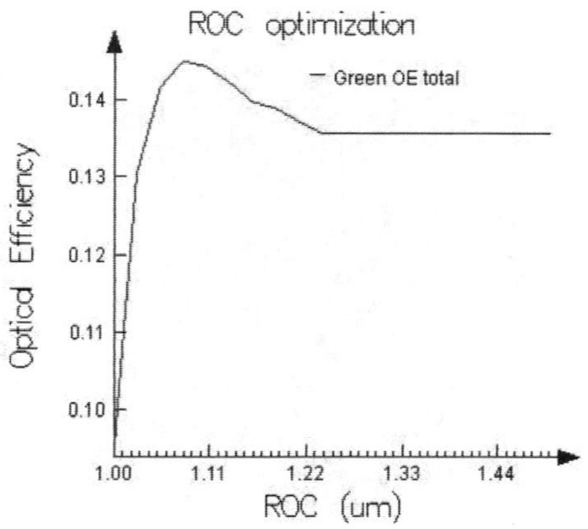

Fig.7. curves of optical efficiency versus the microlens radius

3.2.3 optical confinement method. Ddiffraction resulting in increased spatial optical crosstalk. Optical confinement methods for guiding incident light from the microlens down to the photodiode. These techniques rely on total internal reflection (TIR) at the boundary between dielectric media of different refractive indices[10-13].Figure 8 shows an example calculation for the 0.2 μm air gap thicknesses for light with nomal incidence angle.

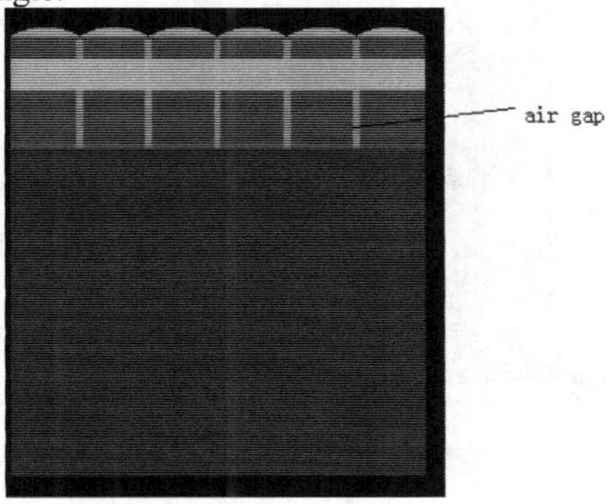

Fig.8.pixels with air gap

4.comparison

Figure 9 (a) shows the Poynting Vector on the active layer before optimization and Figure 9 (b) shows the Poynting Vector on the active layer after all the optimization is done. It is obvious that the optical crosstalk is reduced dramatically. The air gap play an important role in light confinement.

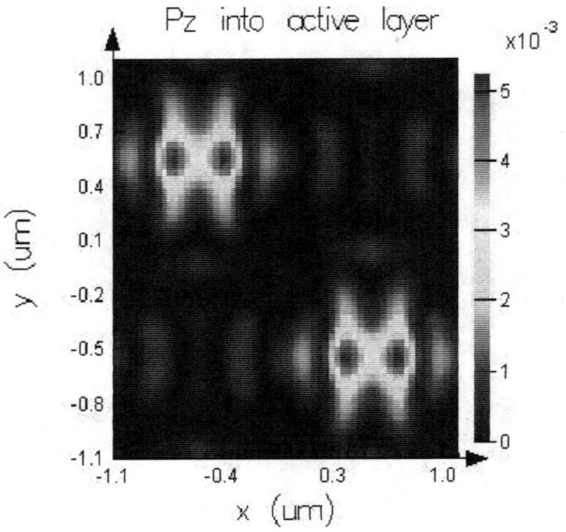

(a)

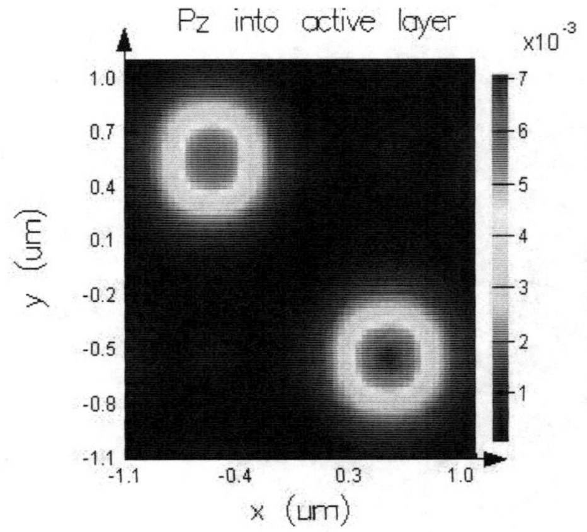

(b)

Fig.9. Poynting Vector on active layer:(a)before optimized (b) optimized

5. Concolusion

We have investigated the optical performance of sub-2-μm pixels using FDTD method. Three kinds of approaches are introduced to optimize the pixel architecture according to the simulation results. The performance improvements for the 1.1-μm pixel after optimization are shown. The optimization in microlens radius can concentrate more light to the photosensitive area. The combination of the dielectric stack height reduction and the air gap structure can maximize the optical efficiency and minimize the spatial crosstalk. By employing all these optimization methods, the optical performance of the 1.1-μm pixel after optimization are comparable to those of the larger 1.75-μm pixel.

References

[1] H. Mutoh, "3-D optical and electrical simulation for CMOS image ensors," IEEE Trans. Electron Devices, vol. 50, no. 1, pp. 19–25,Jan. 2003.
[2] E. D. Palik, *Handbook of Optical Constants of Solids*. Orlando, FL:Academic, 1985.
[3] M. J. Weber, *Handbook of Optical Materials*. New York: CRC Press,2003, sec. 1-2.
[4] F. Hirigoyen, A. Crocherie, J. Vaillant, and Y. Cazaux, "FDTD-based ptical simulations

methodology for CMOS image sensor pixels architecture and process optimization," in *Proc. SPIE*, San Jose, CA, Jan. 2008,vol. 6816, p. 681 609.
[5] Lumerical Solutions, Inc. [Online]. Available: http://www.lumerical.com
[6] P. B. Catrysse and B. A. Wandell, "Optical efficiency of image sensor pixels," *J. Opt. Soc. Amer. A, Opt. Image Sci.*, vol. 19, no. 8,pp. 1610–1620, Aug. 2002.
[7] G. Agranov, V. Berezin, and R. H. Tsai, "Crosstalk and microlens study in a color CMOS image sensor," *IEEE Trans. Electron Devices*, vol. 50,no. 1, pp. 4–11, Jan. 2003.
[8] Y. Kim, Y. Kim, S. Choi, H. Kong, S. Hwang, J. Ko, B. Kim,T. Asaba, S. Lim, J. Im, T. Oh, D. Yi, J. Lee, W. Yang, J. Ahn, E. Jung,Y. Lee, and J. Hahn, "1/2-inch 7.2 MPixel CMOS image sensor with 2.25 μm pixels using 4-shared pixel structure for pixel-level summation,"in *Proc. IEEE ISSCC Tech. Dig. Papers*, San Francisco, CA, Feb. 2006, pp. 494–495.
[9] C. C. Fesenmaier, Y. Huo, and P. B. Catrysse, "Optical confinement methods for continued scaling of CMOS image sensor pixels," *Opt. Exp.*,vol. 16, no. 25, pp. 20 457–20 470, Dec. 2008.
[10] C. C. Fesenmaier and P. B. Catrysse, "Mitigation of pixel scaling effects in CMOS image sensors," in *Proc. SPIE*, San Jose, CA, Jan. 2008, vol. 6817, p. 681 706.
[11] J. Vaillant, A. Crocherie, F. Hirigoyen, A. Cadien, and J. Pond, "Uniform illumination and rigorous electromagnetic simulations applied to CMOS image sensors," *Opt. Exp.*, vol. 15, no. 9, pp. 5494–5503,Apr. 2007.
[12] I. R. Çapoglu, A. Taflove, and V. Backman, "Generation of an incident focused light pulse in FDTD," *Opt. Exp.*, vol. 16, no. 23, pp. 19 208–19 220, Nov. 2008.
[13] A. Crocherie, J. Vaillant, and F. Hirigoyen, "Three-dimensional broadband FDTD optical simulations of CMOS image sensor," in *Proc. SPIE*, Bellingham, WA, Sep. 2008, vol. 7100, p. 710 02J.

Optical fiber ultrasonic sensor networks based on WDM and TDM

Zhenwu Guo[1,2] , Weixiang Li[2], Tiegen Liu[1]

[1]College of Precision Instrument & Opto-electronics Engineering,
Tianjin University, Tianjin 300072, China
[2]Binhai college, Nankai University, Tianjin 300270, China

E-mail: zw99@sohu.com

Abstract. An optical fiber sensor network for ultrasonic measurement based on wavelength division multiplexing (WDM) and time division multiplexing (TDM) technology is presented. Each of the sensor probes is an optical fiber extrinsic Fabry-Perot interferometer (EFPI) which is composed of the fiber's end face and the aluminum thin diaphragm. The sensors are arranged in different wavelength domains formed by a wavelength division multiplexer. Each wavelength division multiplexer, with a group of the sensors, is connected to one of the output ports of optical switch to realize TDM. The signal of each sensor is exported sequentially from a tunable narrowband optical filter (TNOF) that queries every sensor though scanning mode. The principle of the phenomenon of phase induced signal fade in interferometric fiber-optic sensors is also analyzed. Nicely, the detection method above implements the operation of anti-phase induced signal fade detection. The system is interrogated by broadband light source. The scanning range of TNOF is full of the bandwidth of the light source. The result of experiment in water show that the sensor sensitivity reaches -162dB(0dB=1rad/μPa), the frequency response range is from 10KHz to 5 MHz. The number of multiplexing sensors based on WDM and TDM reaches to 64.
Keywords: optical fiber sensor, WDM, TDM, Fabry-Perot interferometry, signal fade

1. Introduction

The detection of the ultrasound field pressure plays an important role in many fields. Optical fiber Fabry-Perot (F-P) interferometric sensor has a number of inherent advantages, including high sensitivity, immunity from electromagnetic interference, resistant to corrosion and small size, in application of ultrasound detection [1][2][3]. Array system based on various multiplexing technology is the current trend of research in ultrasound detection field [4], especially in optical fiber hydrophone field. At the same time, there is the phenomenon of phase induced signal fade in interferometric optical fiber sensors in measuring dynamic pressure signal. To implement anti-phase induced signal fade detection and multiplexing detection are the key points in measurement of ultrasound field. In this paper, we demonstrate an array structure of F-P interferometry sensors based on wavelength division multiplexing (WDM) and time division multiplexing (TDM) technology. And the anti-phase induced signal fade detection is implemented.

2. Structure of the networks

Figure 1 shows the total construction of the proposed array based on WDM and TDM methods. Sixty-four fiber F-P sensor probes are integrated in this trial. The input light for the system is a superradiance light emitting diode (SLD) broadband source (BBS) with 200nm full width at half maximum(FWHM). The injected light through the bidirectional optical switch (OS) is divided into eight beams by a coarse wavelength division multiplexer (CWDM) which has eight channels with 20nm interval. Thus, the every eight sensors in a group can all be interrogated within different optical spectrum, in which the sensors occupy different wavelength bands.

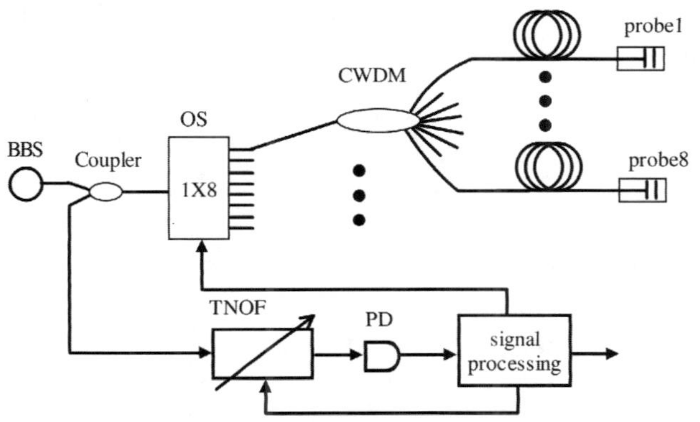

Figure 1. WDM and TDM scheme of the ultrasonic measurement system

All the reflected lights from the sensor elements are combined by a tunable narrowband optical filter(TNOF). The TNOF has a free spectrum range of 200nm, and the bandwidth is 0.1nm. A sawtooth-wave generated by processing circuit, as a linear-scanning voltage, is used to drive the TNOF. The output beam of the TNOF is a narrow band light with the information of interference signal come from sensor probes. And then the output beam detected by the PD is sampled and processed.

When the scan voltage is generated, the control signal is also exported to the optical switch from the signal processing circuit. These two signals together determine which probe the current signal waveform detected by PD corresponds to.

3. Principle of the probe

The optical fiber F-P sensor is composed of single mode fiber, silica glass tube and aluminum (Al) diaphragm, as shown in Figure 2. The inside diameter of the thick tube is 2mm, and its outside diameter is 4mm. The active diameter of Al diaphragm is equal to the inside diameter of the thick tube, 2mm obviously. The distance between Al diaphragm and the end face of single mode fiber ranges from several to several tens of micron, and it is the length of the F-P cavity.

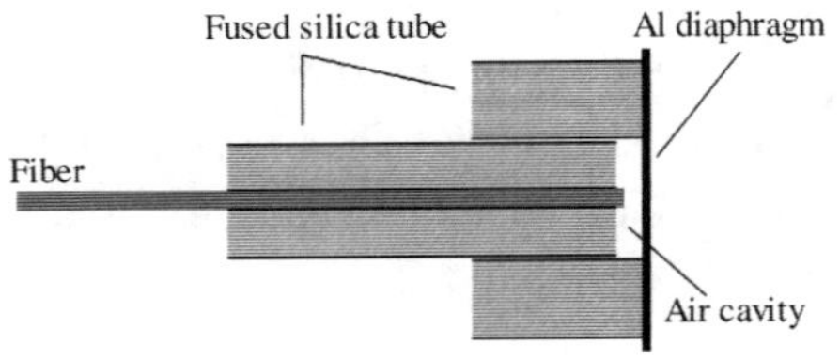

Figure 2. Structure of the sensor probe

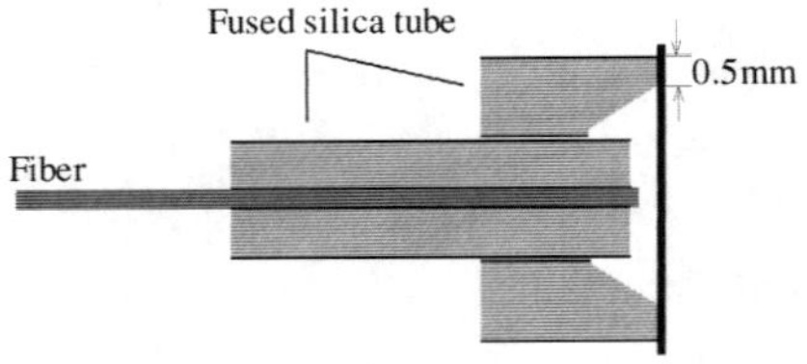

Figure 3. Higher sensitivity sensor

Light is injected into the optical fiber and partially reflected by the end face of the fiber as the first beam of the interference light. The other beam emitted from fiber is reflected by the inside surface of the Al diaphragm. Then the two reflected beams propagate back into the same fiber and generate interference fringes. Affected by ambient pressure, the Al diaphragm will deform, that is to say the cavity length is in response to ambient pressure variation, thus resulting in a slight change of the optical path difference of the two beams. And then the output intensity which generates output voltage signal will change with measurand.

The diaphragm will be deflected whenever there is a differential pressure ΔP between inside and outside of the cavity. Supposing the cavity length is d, Δd is the deflection of the diaphragm. With the assumption of uniform diaphragm thickness, small deflection, the deflection of Al diaphragm under pressure difference ΔP is expressed as [5]

$$\Delta d = \frac{3(1-\mu^2)(a^2-r^2)^2}{16Eh^3}\Delta P \tag{1}$$

Where a is the radius of the diaphragm defined by the inner diameter of the thick silica tube, r is any radial position leaving central point of the diaphragm, h is the thickness of the diaphragm, μ is Poisson's ratio of diaphragm material and μ=0.355 for Al material, E=7×10^{10}N/m^2 is modulus of elasticity of the aluminum material.

Based on the properties of aluminum diaphragm, the lowest natural frequency of the diaphragm can be theoretically calculated as follows [5]

$$f = \frac{\alpha}{4\pi}\sqrt{\frac{Eg}{3\omega(1-\mu^2)}}\left(\frac{h}{a^2}\right) \tag{2}$$

Where α is a constant related to the vibrating modes, and α=10.21 for the lowest natural frequency, $g = 9.815$ m/s^2 is the gravitational constant, ω is the mass density of the diaphragm material, $\omega = 2.7\times10^3$ kg/m^3 for Al material, and a, h, E, and μ are the same as the parameters in Eq.(1).

Can be seen from Eq.(1) and Eq.(2), the deflection of the diaphragm is proportional to pressure difference. And radius and thickness of the diaphragm are the key factors for sensitivity and frequency response of the sensor.

In order to further improve the sensitivity of the sensor, the thick silica tube is processed into tapered end, as shown in Figure 3. Then the efficient radius of the diaphragm increase to 1.5mm from 1mm, so the pressure sensitivity improves to 5 times of the original.

4. Anti-phase induce signal fading detection

If a sinusoidal phase modulated signal of the interferometric sensor with initial phase, φ_0, the interference intensity signal is

$$I = A + B\cos(\varphi_0 + \varphi_s + \varphi_n) \tag{3}$$

Where constants *A* and *B* are proportional to the input optical power, but *B* also depends on the mixing efficiency of the interferometer; and φ_s is the phase shift caused by the measured signal; φ_n is phase shift for environmental effects.

The phase shift φ_n caused by the environment is usually great signal of the low frequency and random fluctuations, but the measured signal is often weak signal. When there is a very small phase shift $\Delta\varphi_s$ caused by measured signal, the voltage converted from light intensity is

$$\begin{aligned} V &= kA + kB\cos\Delta\varphi_s\cos(\varphi_0+\varphi_n) - kB\sin\Delta\varphi_s\sin(\varphi_0+\varphi_n) \\ &\approx kA + kB\cos(\varphi_0+\varphi_n) - kB\Delta\varphi_s\sin(\varphi_0+\varphi_n) \end{aligned} \tag{4}$$

Where *k* is conversion factor. Since φ_n varied very slowly, it can be treated as DC signal. So the voltage changes ΔV can be expressed as

$$\Delta V \approx -kB\Delta\varphi_s\sin(\varphi_0+\varphi_n) \tag{5}$$

Therefore, when $\varphi_0 + \varphi_n = m\pi + \pi/2$ or $\varphi_0 + \varphi_n = m\pi + 3\pi/2$ (m is integer) with slowly drift of φ_n, the output signal will reach maximum. But, when $\varphi_0 + \varphi_n$ is close to $m\pi$, the output signal is about zero. This is the phenomena of phase induced signal fading of interferometric fiber-optic sensors.

In order to eliminate the signal fading and implement the signal stable examination, many methods of detecting signal have been proposed such as 3×3 coupler [6], phase generated carrier [7], phase-tracking technology [8].

In our design, it is to eliminate the signal fading automatically by introducing the TNOF and CWDM. The CWDM has the same operation principle as a coarse WDM that was used in fiber optic telecommunication. It is custom-made, and it is quite compatible with the optical fiber. The configuration of the CWDM and the spectra of the interference signal are shown in Figure 4.

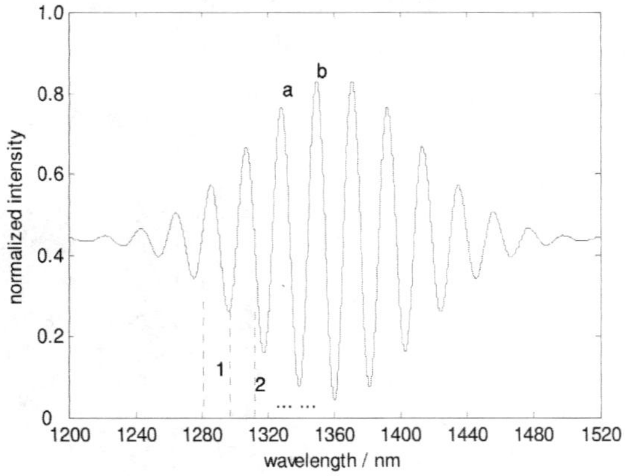

Figure 4. Interference spectrum and wavelength division interval

The interference signal waveform of one of the sensors is shown in Figure 4. The whole scope of source wavelength is divided into eight beams with the channel interval of the CWDM, $\Delta\lambda=\Delta\lambda_i(i=1\ldots8)$, as shown in Figure 4. Each beam is corresponding to a sensor probe respectively. If the TNOF is tuned in $\Delta\lambda_1$, probe 1 is activated while the other probes are not accessed, and the current of PD gives the output signal of probe 1. If the TNOF is moved into $\Delta\lambda_2$, the signal of probe 2 is detected, and so on.

The wavelength-period of the interference spectrum must be not less than 2 times of $\Delta\lambda$,

$$\lambda_b - \lambda_a \geq 2\Delta\lambda \quad (6)$$

Where λ_a and λ_b which are the peaks of the interference signal. When the TNOF scans within the scope of wavelength of the source, φ_0 is also to change continually. And then, the center wavelength of TNOF is always to achieve the orthogonal points to export the maximum of the signal when the TNOF scans in every channel of the CWDM. Even if there is disturbance from environment, it is also true. The voltage of orthogonal point is the value when $\varphi_0 + \varphi_n = m\pi + \pi/2$ or $\varphi_0 + \varphi_n = m\pi + 3\pi/2$.

The wavelength-period of the interference spectrum is inversely proportional to the cavity length of the sensor [9]. Therefore, we may control the period of interference signal by adjusting the cavity length of the F-P probe at first.

5. Experiments

The two wavelengths λ_a and λ_b are the peaks of the interference signal with 2π out of phase as shown in figure 4. Then the cavity length *d* is

$$d = \frac{\lambda_a \lambda_b}{2(\lambda_a - \lambda_b)} \quad (7)$$

The relationship of the differential variables, Δd and $\Delta\lambda$, is $\Delta d / d = \Delta\lambda_a / \lambda_a$.

First we test the static pressure characteristic of the F-P sensor. We may indirectly obtain cavity length *d* by the position of the peaks of the interference signal on the screen of spectroscopic instrument. The relationship between cavity length *d* and outside pressure is shown as Figure 5. The static pressure response curve has been fitted by line *y*=33.081-2.286*x*, and the correlation coefficient is 0.998. The sensitivity of the sensor is 1.96μm/KPa of change quantity of cavity length. We can also work out the phase sensitivity of the sensor is -162dB(0dB=1rad/μPa) at the wavelength, 1550nm.

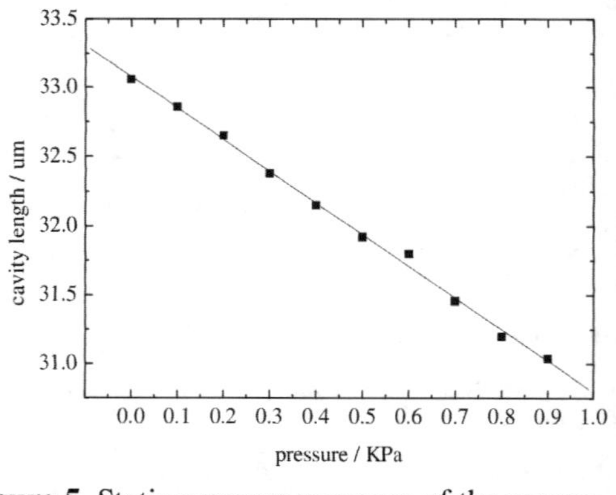

Figure 5. Static pressure response of the sensor

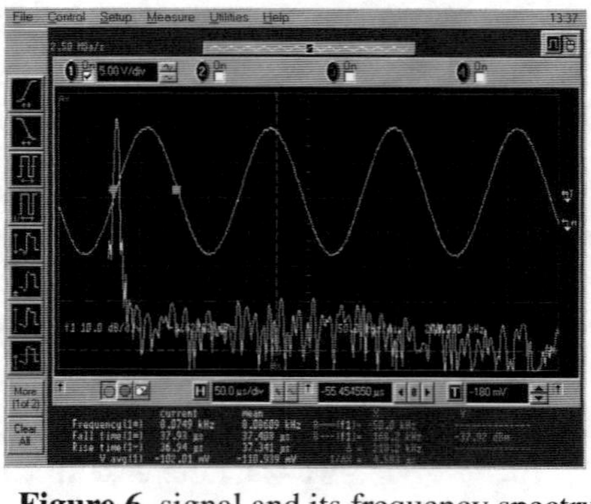

Figure 6. signal and its frequency spectrum

We set up another experimental system to test the performance of the array of F-P diaphragm-based sensors for ultrasonic wave signal detection in a water tank. The ultrasonic vibration signal is generated by a piezoelectric (PZT) transducer driven by an adjustable sine wave signal. The sensor probes receive underwater ultrasonic wave signal and transform the signal to light intensity signal, and the signal processing circuit transforms intensity signal to voltage signal. The output voltage signal waveform exported from one of the sensor and its frequency spectrum are shown in Figure 6.

Every sensor can receive the ultrasonic signal when the TNOF scans the corresponding passband of the CWDM. But the signal amplitude is variable with the center wavelength of the TNOF moving. When the center wavelength is at the orthogonal point of the interference signal, the amplitude is biggest. So the amplitude maximum should be treated as the normal amplitude of the signal detected by the sensor. The frequency range of the sensor sensing is from 10 KHz to 5MHz.

6. Conclusions

We have described a diaphragm-based miniature fiber-optic ultrasonic F-P sensor and the array detection system based on WDM and TDM technology. It has simultaneously implemented sensor multiplexing and anti-phase induced signal fade detection by using CWDM and a tunable narrowband optical filter. It is ideal for applications of acoustic and ultrasonic detection, such as hydrophone.

Acknowledgments

This work was supported by 973 Program of China (No. 2010CB327806), and Tianjin High Education Research Development Program of China (No. 20080715).

References

[1] Bing Yu, Dae Woong Kim, Jiangdong Deng, Hai Xiao and Anbo Wang 2003 Fiber Fabry–Perot sensors for detection of partial discharges in power transformers *Appl. Opt.* **42** 3241-50

[2] B T Cox, E Z Zhang, J G Laufer and P C Beard 2004 Fabry Perot polymer film fibre-optic hydrophones and arrays for ultrasound field characterisation *J. Phys.: Conf. Ser.* **1** 32-37

[3] M.Lamont and P.C.Beard 2006 2D imaging of ultrasound fields using CCD array to map output of Fabry-Perot polymer film sensor *Electronics Letters* **42** 187-89

[4] Yi Jiang and Caijie Tang 2009 A simultaneous interrogation system for intensity-type sensor / fiber Bragg grating / extrinsic Fabry–Perot interferometer *Meas. Sci. Technol.* **20** 025308

[5] Juncheng Xu, Xingwei Wang KristieL.Cooper and Anbo Wang 2005 Cooper Miniature all-silica fiber optic pressure and acoustic sensors *Opt. Lett.* **30** (24) 3269

[6] Todd M.D, Seaver M and Bucholtz F 2002 Improved operationally-passive interferometric demodulation method using 3×3 coupler *Electronics Letters* **38** 784 - 786

[7] ZHU Sun, HE Peikun and HUANG Changzhen 2004 A Design of Digital Demodulation System Using Phase Generated Carrier in Fiber Hydrophones Based on DSP *Proceedings of 3rd International Conference on Computational Electromagnetic and Its Applications* p 473

[8] K Toyama, K A Fesler, B Y Kim and H J Shaw 1991 Digital integrating fiber-optic gyroscope with electronic phase tracking *Opt. Lett.* **16** 1207-1209

[9] Guo Zhenwu, Sun Guiling, Li Weixiang 2008 Study on Temperature Characteristic of Optical Fiber Pressure Sensor Based on Fabry-Perot Interferometry *Semiconductor Optoelectronics* **29** 418

Laser Radar Range-Doppler Imaging and Simulation on High-speed Target

GUO Hua-ling[1, 2], DENG Jia-hao[1],CAI Ke-rong[1]

[1]National Key Laboratory for Mechatronics Engineering and Control, Beijing Institute of Technology,Beijing 100081,China

[2]National Key Laboratory for Electronic Measurement Technology, North University of China,Taiyuan 030051,China

E-mail： guohualing@nuc.edu.cn

Abstract. On the basic of the representation of the principle of Doppler imaging, an echo model of high-speed target illuminated by broadband linear frequency modulated (LFM) signal was established in this paper. For high-speed target, because of its nonlinear echo phase and time-varying Doppler frequency shift, the Doppler spectrum of the target acquired by the traditional Fourier method was ambiguous, and the same in the radar images of the target. Therefore, an adaptive Wigner-Ville time-frequency analysis method was presented in the paper, that is, the local fuzzy function could be obtained by applying the two-dimensional adding Window Fourier Transform to signal's Wigner-Ville distribution, which made the kernel function, not only be adaptive to time, but also to the frequency. Finally, the simulation results show that the method has a good time-frequency concentration, and effectively control the cross-term interference.

1. Introduction

In missile defense system, the detection and recognition of high speed missile target is one of the main research work. Through conducting accurate imaging on the target, features like size and motion parameters of the target can be obtained, then, these high resolution features can be used in the recognition of the target. In radar imaging field[1], the laser range-doppler imaging radar is a rapidly developing high and new technology, which is being extensively applied in the field of aerospace and military applications. At present, laser high resolution and range-doppler imaging system can measure the distance and set direction of the target by using the laser scattering features, and identify the target through the target location, radial velocity and reflection characteristics of objects. That range-doppler algorithm (RDA) [2-3]is a kind of algorithm to obtain the high resolution of longitudinal separation by broadband signals, and acquire the across high resolution by echo doppler bandwidth. When targets fly in high-speed maneuver flight, the echo of doppler frequency is time-varying during the observation, which is why we will get an ambiguous image if we use the traditional Fourier method to make observation on the targets. To conduct spectrum analysis on maneuvering targets, the method of time-frequency analysis not only can eliminate the fuzzy, but also can get target-instantaneous doppler imaging, which plays an important role in determining the flight attitude of the maneuvering target during observation. Therefore, the method has a great development potential.

2. Imaging principle of range-doppler

Laser range-doppler imaging is a kind of method to acquire the features of the locomotive information of targets, mainly based on the Doppler Effect resulted from the relative motion between radar and targets. The basic method of image is the range-doppler method, the longitudinal range resolution relies on broadband signals and the across resolution relies on doppler frequency of the echo[4-5]. Range-doppler imaging diagram is shown in figure.1

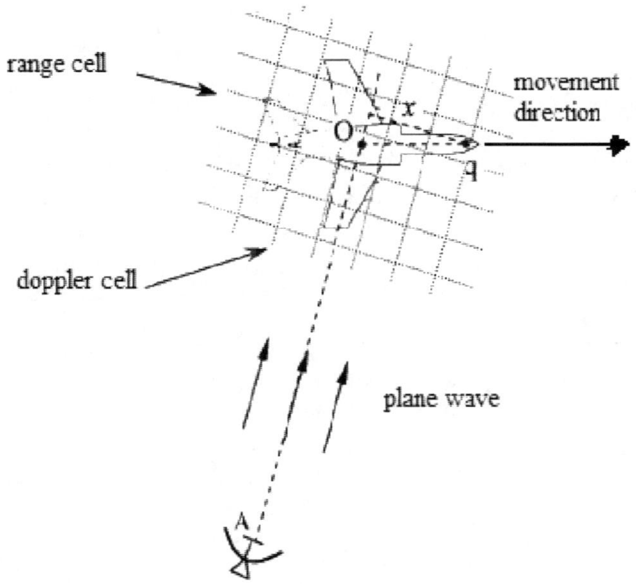

Figure. 1 Range-doppler imaging diagram

3. Echo model of high-speed moving target

Laser radar broadband chirp signal can be expressed as the following [6-7]:

$$S(\hat{t},t_m) = rect(\frac{\hat{t}}{T_p})\exp\left[j2\pi(f_c + \frac{1}{2}k\hat{t}^2)\right] \qquad (1)$$

where *rect(u)* represents rectangular pulse; f_c center frequency; T_p the pulse width; k frequency modulation ratio of launch signal. Launch time $t_m = mT$ represents slow time, $\hat{t} = t - mT$ represents fast time.

The distance of some point target to the radar is $R_n(\hat{t})$. Then, the echo signal of radar receiver can be expressed as follows:

$$S_r(\hat{t},t_m) = Arect(\frac{\hat{t} - 2R_n(\hat{t})/c}{T_p})\exp\left[j2\pi(f_c(t - \frac{R_n(\hat{t})}{c}) + \frac{1}{2}k(\hat{t} - \frac{2R_n(\hat{t})}{c})^2)\right] \qquad (2)$$

where A represents scattering intensity.

Through dechirp, the echo of target scattering point can be expressed as follows:

$$S_{if}(\hat{t},t_m) = Arect(\frac{\hat{t}-2R_n(\hat{t})/c}{T_p})\exp\left[-j\frac{4\pi}{c}k(\hat{t}-\frac{2R_{ref}}{c})\right]\left[R_n(\hat{t})-R_{ref}\right]\exp\left\{-j\frac{4\pi}{c}f_c\left[R_n(\hat{t})-R_{ref}\right]\right\}$$
$$\exp\left\{j\frac{4\pi}{c^2}k\left[R_n(\hat{t})-R_{ref}\right]^2\right\} = Arect(\frac{\hat{t}-2R_n(\hat{t})/c}{T_p})\exp\left\{-j\frac{4\pi}{c}\phi_n(\hat{t})\right\} \quad (3)$$

where R_{ref} represents reference distance;

$$\phi_n(\hat{t}) = k\hat{t}(R_n(\hat{t})-R_{ref}) + f_c(R_n(\hat{t})-R_{ref}) - \frac{k}{c}\left[R_n^2(\hat{t})-R_{ref}^2\right]$$

Assuming that target velocity is V, $R_n(\hat{t}) = R_{no} + V\hat{t}$, here R_{n0} represents the distance of the scattering point to radar at the starting time. Introducing into Eq.(3) , the precise model of high-speed moving target can be got:

$$S'_{if}(\hat{t},t_m) = S_{if}(\hat{t},t_m)\exp\left[-j2\pi\left(\frac{2f_c}{c}-\frac{4kR_{no}}{c^2}\right)V\hat{t}\right]\exp\left[-j2\pi\left(\frac{2}{c}kV-\frac{2}{c^2}kV^2\right)\hat{t}^2\right] \quad (4)$$

Eq.(4) is shown that the last two phase will make range imaging to produce displacement and stretch, which will effect feature extraction of target structure.
The signal distribution of radar transmission and reception in time-frequency is shown in figure.2[8].

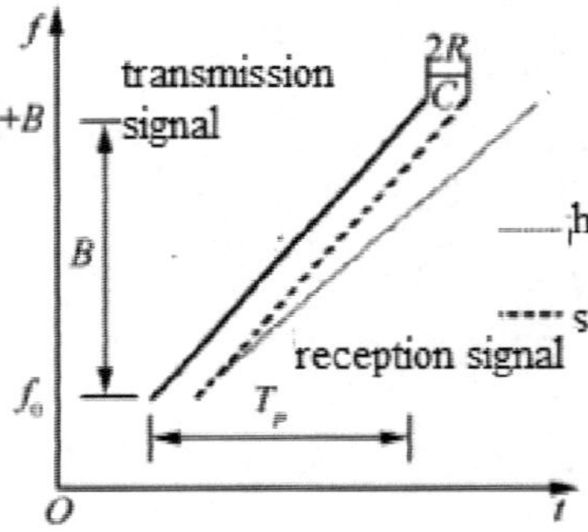

Figure.2 time-frequency properties of transmission and reception signal

Figure.2 illustrates that in pulse duration position, the static or low target scattering point relative to the radar is unchanged. For a stationary point, after a delay of 2R/c, the time-frequency curve of its echo signal is parallel to the time-frequency curve of radar launch signal. If this scattering point is in high-speed motion, in pulse duration position, because of the constant change of the position of the target, the echo signal delay and the slope of time-frequency curve will change greatly.

4. Time-frequency adaptive method

The time-frequency adaptive optimal kernel (TFAOK) time-frequency representation (TFR) [9] regards Wigner -Ville distribution of signal as a two-dimensional signal, and every bit in the time frequency plane optimizes its kernel function, which makes the kernel functions have adaptive characteristics in the time-frequency plane. The method of time frequency domain adding window can gain partial ambiguity function, and this local characteristic is able to restrain the cross-term interference. TFAOK TFR has a good cross inhibitory ability.
Wigner-Ville distribution is Fourier transform of the time-varying autocorrelation. It can be expressed as follows:

$$WVD(t,f)=\int s'_{if}\left(t+\frac{\tau}{2}\right)s'^{*}_{if}\left(t-\frac{\tau}{2}\right)e^{-2\pi f\tau}d\tau \quad (5)$$

The fuzzy function of signal is expressed as follows:

$$AF(\theta,\tau)=\int s'_{if}\left(t+\frac{\tau}{2}\right)s'^{*}_{if}\left(t-\frac{\tau}{2}\right)e^{-\theta t}dt \quad (6)$$

Reference to the thought of adding window Fourier transform, this paper put forward that, for Wigner-Ville distribution, the method of the two-dimensional adding Window Fourier Transform could make the fuzzy function have time characteristic and frequency local characteristic, and could define local ambiguity function. The fuzzy function with the characteristic of time-frequency localization can be expressed as follows:

$$AF(\theta,\tau,t,f)=\iint WVD(u,v)\cdot A(u-t,v-f)e^{-j\theta u+jv\tau}dudv \quad (7)$$

Variables t and f give the center of two-dimensional window function. Wigner-Ville distribution and window function can be multiplied, which can limit integral area in the neighborhood of ω and t. Thus, TFAOK TFR is expressed as follows:

$$p(t,f)=\iint AF(\theta,\tau;t,f)\phi_{opt}(\theta,\tau;t,f,s)e^{j\theta t-j2\pi f\tau}d\theta d\tau \quad (8)$$

Where $\phi_{opt}=\max_{\phi}\iint\left|AF(\theta,\tau)\phi(\theta,\tau;t,f,s)\right|^{2}d\theta d\tau$

5. The simulation analysis

This paper verified the correctness of the above analysis with simulation data, and got a clear 2-D image of the target. The following are the selected simulation:

the distance between radar and target is $R_0=20km$; the speed of the target is $v=10km/s$; the bandwidth is $B=800MHz$; the pulse width is $T_p=100\mu s$; and the pulse sampling rate is $f_s=100MHz$.

Assuming that the target model is an aircraft shape model consisting of 12 scattering points, and the scattering intensity are unitary to 1.

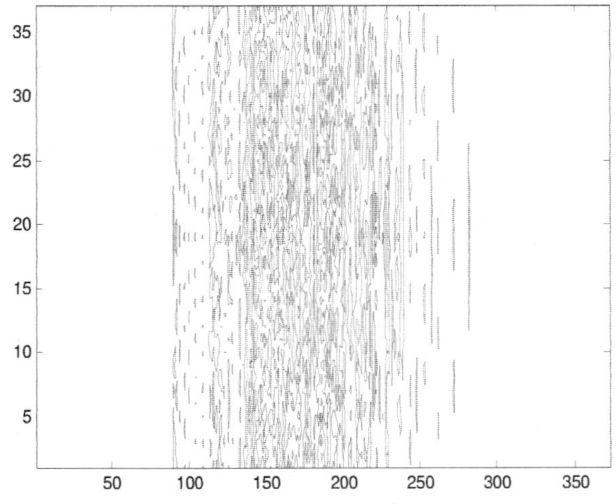

Fig.3 The range imaging of traditional method

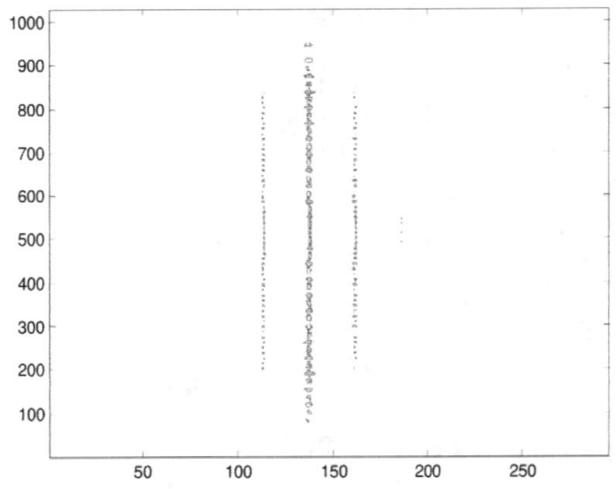

Figure.4 The 2-D imaging of traditional method

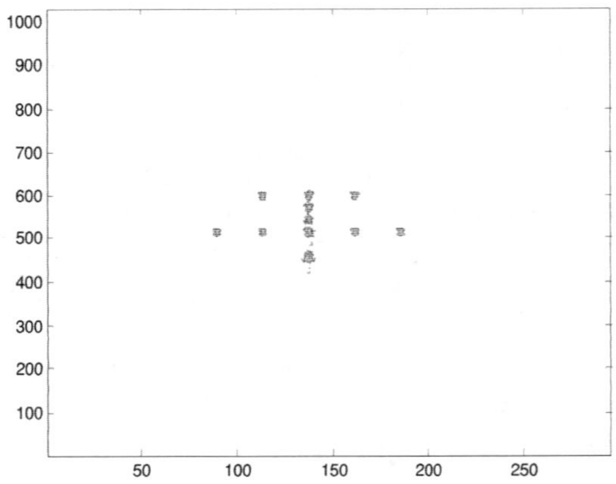

Figure.5 The2-D imaging of TFAOK TFR method

6. Conclusion

This paper mainly makes a study on the range-doppler imaging of LFM signal of laser radar. For high-speed target, because of its nonlinear echo phase and time-varying Doppler frequency shift, the Doppler spectrum of the target acquired by the traditional Fourier method was ambiguous, and the same in the radar images of the target. Therefore, this paper puts forward a new solution to the problem with the thought of speed supplement. In the consideration of the characteristics of the target in high-speed motion, the adaptive Wigner-Ville time-frequency analysis method can be used to eliminate fuzzy and coke caused by the high-speed radial motion. Finally, the simulation results show that this method has good time-frequency aggregation, and can effectively control the cross-term interference, and extract doppler spectrum characteristics and frequency spectrum characteristics from radar echo as well. The method not only can overcome the image blurring and the limit of the Doppler being constant in the traditional high speed target imaging, but also can get a clearer image, so as to provide new evidences for target identification.

References

[1] BAO Zheng 2005 *Technologies of radar imaging*（Beijing：Publishing House of Electronics Indust ry ）pp19-30 ,60-64 ,165-173
[2] J. M. Vaughan and P. A. Forrester 1989 *Wind Eng* **13** 1
[3] Shen Y Y, Liu Y T 1999 *Chinese J Electron* **18** 196 - 199
[4] BASHKANSKY M ,LUCLER L AND FUNKE 2002 *Opt Lett* **27** 1983-1985
[5] CHEN Fu-bin,SHI Zhi-guang,etc 2007 *Modern Radar* **29** 40-43
[6] ZHANG Yi-an ,YIN Zhi-ping, etc 2009 *JOURNAL OF UNIVERSITY OF SCIENCE AND TECHNOLO GY OF CHINA* **39** 954-960
[7] MA Liang,FENG De-jun, etc 2009 *SIGNAL PROCESSING* **25** 1051-1056
[8] YANG Zheng-long,LIU Ai-fang, etc 2007*Modern Radar* **29** 43-45
[9] Jones D L , Baraniuk R G 1995 *IEEE Trans on Acoustics Speech and Signal Processing* **43** 2361-2371

Journal of Physics: Conference Series **276** (2011) 012127 doi:10.1088/1742-6596/276/1/012127

The novel stable control scheme of the light source power in the closed-loop fiber optic gyroscope

Zhongxiao Ji[1,2] and Caiwen Ma[2]

[1] Graduate University of the Chinese Academy of Sciences, Beijing, P. R. China
[2] Xi'an Institute of Optics and Precision Mechanics of the Chinese Academy of Sciences, NO.17 Xinxi Road, New Industrial Park, Xi'an Hi-Tech Industrial Development Zone, Xi'an, Shaanxi, P.R.China

E-mail: jzx@opt.ac.cn

Abstract. The light source power stability of the Fiber-Optic Gyroscope (FOG) affects directly the scale factor and bias stability of FOG. The typical control scheme of the light source power employs an additional photodetector to detect the output power of the light source. When the fiber loss of FOG varied due to the temperature change, the light power in the additional photodetector did not indicate this change, which decreased the control effect. The spike pulse overlapping on the gyro signal denotes potentially the change of the light power and fiber loss. In the novel scheme, the spike pulse is extracted from the gyro signal, and is transformed into the square wave by the differential circuit. According to the change of the square wave amplitude, FOG adjusts the bias current of the light source to keep the stable light power in the signal photodetector. It is a simple and low-cost scheme without an additional photodetector.

1. Introduction

FOG is a kind of the new type all-solid-state inertial instrument. The interferometric FOG (I-FOG) which is one kind of FOG has been already used in the traditional application fields of the gyros successfully [1, 2]. The scale factor and bias stability of I-FOG is very important parameters for FOG performance [3], and the stability of the light source power affects directly these two parameters. So maintaining the light power stability is a vital way to enhance the gyro performance.

I-FOG uses generally the superluminescent diode (SLD). The major problem of this light source is its poor spectrum stability. The experiments showed that: SLD average wavelength drifts about 400ppm/K, and about 40ppm/mA with the bias current of the drive circuit [2]. The traditional scheme employs the temperature control circuit and constant current source to keep the stable output power and average wavelength of SLD. At the same time, an additional photodetector, which is connected to the dead end of the fiber coupler in FOG (Figure 1), is employed to detect the change of the light source power. And then adjusting the output power of SLD is done by the feedback control circuit. But the light wave achieving to the signal photodetector and the other achieving to the additional photodetector are different in the light path, as shown in Figure 1. When the fiber loss of the optical components and fiber junctures varied due to the environmental temperature change, the light power achieving to the signal photodetector became unstable. The light signal in the additional photodetector did not indicate this change, which affected the control effect finally. Moreover, the traditional scheme is an expensive scheme for an additional photodetector.

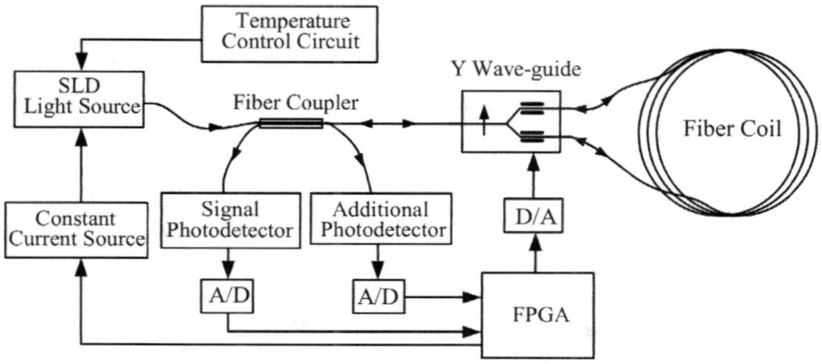

Figure 1. The traditional control scheme.

Due to using the unsymmetrical phase modulating square wave in I-FOG, the spike pulses periodically overlap on the gyro signal [2], which are unuseful signals for I-FOG. However, these signals potentially indicate the change of the light source power and fiber loss. In the novel scheme, the spike pulses are periodically extracted from the gyro signal, and are transformed into the square wave signal by the differential circuit. The change of SLD output power is known by measuring the amplitude of the square wave, and then adjusting the light source power by regulating the output current of the constant current source is to keep the stable light power in the signal photodetector.

2. The closed-loop I-FOG with the square wave biasing modulation

I-FOG is a kind of the inertial instrument based on Sagnac effect. When the gyro rotates, the light waves in the FOG produce interference, and then produce the Sagnac phase shift which is proportional to the rotation rate of the gyro [4].The response to the rotation rate of the gyro is:

$$P(t)=P_0[1+\cos(\Delta\phi_R)] \qquad (1)$$

Where P_0 is the light source power; $\Delta\phi_R$ is Sagnac phase shift. Equation (1) shows that the response is a raised cosine function, and symmetric about zero. At the zero point of the response, its slope is zero approximately. So when the gyro rotates slowly, the response is insensitive to the gyro rotation [2, 4].

To get high sensitivity, I-FOG employs the square wave biasing modulation technology so that I-FOG operates at the point with a non-zero response slope. A phase biasing modulation ϕ_b is applied periodically in the positive and negative half-periods of the modulating square wave. The square wave half-period $\Delta\tau$ is the time that the light wave spends on propagating a ring in the fiber coil, corresponding to the so-called eigen frequency $f_p=1/2\Delta\tau$ of the fiber coil [4]. The difference between the positive and negative modulating states is equal to $\Delta P(\Delta\phi_R,\phi_b)$.

$$\Delta P(\Delta\phi_R,\phi_b)=P_0[1+\cos(\Delta\phi_R-\phi_b)]-P_0[1+\cos(\Delta\phi_R+\phi_b)]=2P_0\sin\phi_b\sin\Delta\phi_R \qquad (2)$$

As ϕ_b is equal to $\pi/2$, the response to the gyro becomes a sine function with a stable zero point; while $\Delta\phi_R$ is in the small range, $\sin\Delta\phi_R\approx\Delta\phi_R$. So Equation (2) is simplified by Equation (3):

$$\Delta P=2P_0\cdot\Delta\phi_R \qquad (3)$$

Hence the difference ΔP between two modulating states is proportional to the rotation rate of the gyro.

However, the duty cycle of the modulating square wave is not absolutely equal to 1:1, which induces additional phase modulations. Within employing the unsymmetrical square wave modulation, the phase biasing modulation ϕ_b has four modulating states [2]:

$$\begin{cases} +\phi_b & t\in(1-\varepsilon)\Delta\tau \\ 0 & t\in\varepsilon\cdot\Delta\tau \\ -\phi_b & t\in(1-\varepsilon)\Delta\tau \\ 0 & t\in\varepsilon\cdot\Delta\tau \end{cases} \qquad (4)$$

Where ε is the error between the positive half-period and negative half-period of the square wave. During the time $\varepsilon \cdot \Delta\tau$, the gyro output in the signal photodetector is as shown in Equation (1). Since the I-FOG employs the closed-loop control technology, $\Delta\phi_R$ is regarded as the error signal. So it is limited at zero [2, 5]. Hence the gyro output in $\varepsilon \cdot \Delta\tau$ is:

$$P(t)=2P_0 \tag{5}$$

Corresponding to the four states of the phase biasing modulation, there are four different values of gyro output. Here ϕ_b is equal to $\pi/2$.

$$P(t)=\begin{cases} P_0\left[1-\sin\Delta\phi_R\right] & t\in(1-\varepsilon)\cdot\Delta\tau \\ 2P_0 & t\in\varepsilon\cdot\Delta\tau \\ P_0\left[1+\sin\Delta\phi_R\right] & t\in(1-\varepsilon)\cdot\Delta\tau \\ 2P_0 & t\in\varepsilon\cdot\Delta\tau \end{cases} \tag{6}$$

So it can be seen that these spike pulses overlap on the gyro signal, whose the frequency is two times the square wave frequency. When the square wave frequency f_m is equal to the eigen frequency f_p of the fiber coil, the width of all spike pulses is equal to $\varepsilon \cdot \Delta\tau$. If it not equal, one becomes wide and the other becomes thin, as shown in Figure 2 [2]. Equation (6) shows that: during the time $\varepsilon \cdot \Delta\tau$, the spike pulse amplitude is proportional to the output power of the light source, which provides a simple way to detect the change of the light source power.

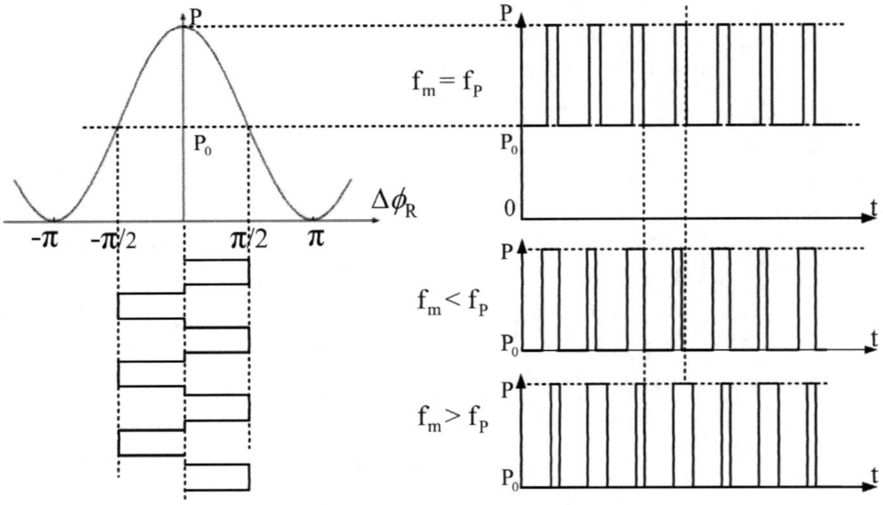

Figure.2 Spike pulses in the gyro signal.

3. The novel stable control scheme of the light source power

Spike pulses and signals are same in the light route and reach the signal photodetector synchronously. So the change of spike pulses directly indicates the change of the light source power and fiber loss. Because the spike pulse amplitude is proportional to the light source power. Considering the fiber loss and the photodetector responsivity, so the signal which is detected in the signal photodetector is:

$$P_t(t)=2\times10^{-L/10}\cdot K\cdot P_0 \qquad t\in\varepsilon\cdot\Delta\tau \tag{7}$$

where K is the photodetector responsivity and L is the fiber loss. Therefore, the light power in the signal photodetector results from the amplitude of spike pulses. However, the operational amplifier in the circuit, which drives the Y wave-guide, has a limited slew rate, so it has shorter rise time $\Delta\tau_1$ and fall time $\Delta\tau_2$. Therefore the actual square wave of the phase biasing modulation is trapezoidal wave, as shown in Figure 3.

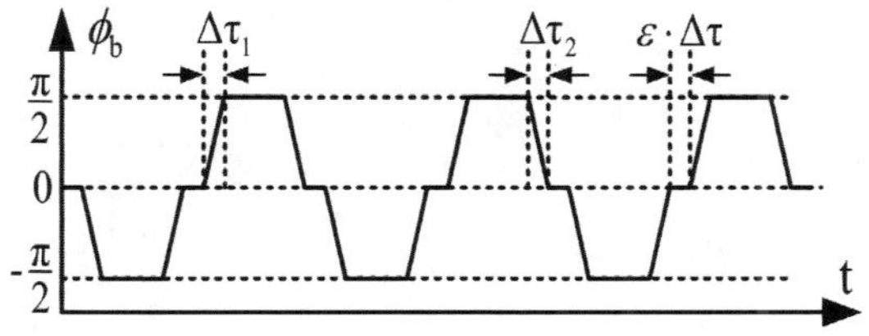

Figure 3. The phase biasing modulation of the trapezoidal wave.

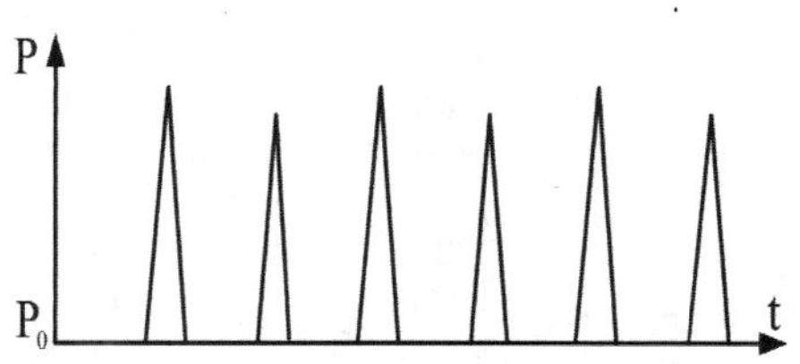

Figure 4. The response to the trapezoidal wave modulation and fiber loss.

Because the error ε is a very small value, the responses to the trapezoidal wave modulation and fiber loss are the triangular waves instead of spike pulses, and the tip of the triangular wave directly reflects the change of the light source power and fiber loss in I-FOG. Because the modulating frequency is not absolutely equal to the eigen frequency, one spike pulse is wide and the other is thin, corresponding to two triangular waves that one is bigger and the other is smaller, as shown in Figure 4.

Due to the disturbance of the circuit noise, sampling the tip of the triangular wave is a difficult thing. In Figure.4, it can be seen that the changes of the light source power and fiber loss will induce the change of the triangular wave amplitude, and then affect the slope of the rising edge and falling edge of the triangular wave. In the novel scheme, these triangular waves are transformed into the square waves by the differential circuit which consists of the operational amplifier, resistance R and capacitor C (Figure 5), as shown in Figure 6.

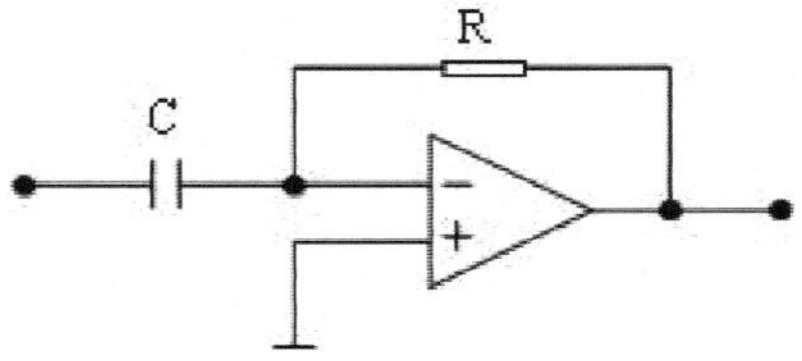

Figure 5. The differential circuit.

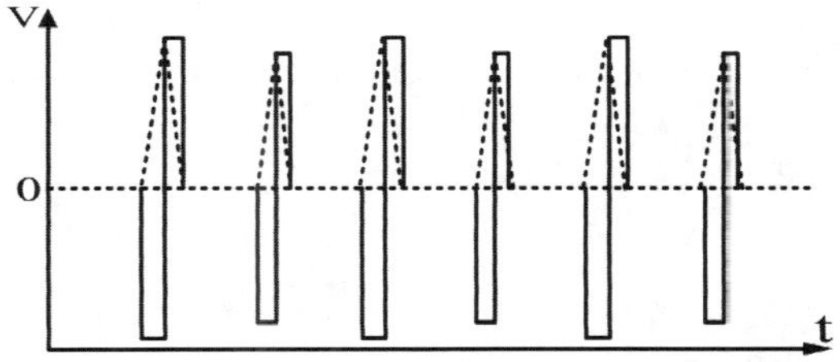

Figure 6. the differential transformation

The negative half-period amplitude P_n of the square wave is shown in Equation (8), and the positive half-period amplitude P_p is shown in Equation (9):

$$P_n = -\mathrm{RC}\cdot\frac{\mathrm{d}}{\mathrm{d}t}(K_r\cdot t) = -\frac{\mathrm{RC}\cdot P_t}{\Delta\tau_r} = -\frac{2\times10^{-L/10}\mathrm{KRC}\cdot P_0}{\Delta\tau_r} \tag{8}$$

$$P_p = -\mathrm{RC}\cdot\frac{\mathrm{d}}{\mathrm{d}t}(-K_f\cdot t) = \frac{\mathrm{RC}\cdot P_t}{\Delta\tau_f} = \frac{2\times10^{-L/10}\mathrm{KRC}\cdot P_0}{\Delta\tau_f} \tag{9}$$

Where K_r and K_f are the rising edge slope and the falling edge slope, P_t is the amplitude of the triangular wave, $\triangle\tau_r$ and $\triangle\tau_f$ are the rising time and falling time of the triangular wave, which are calculated by the sampling period of the A/D converter: the sampling values are negative in the time $\triangle\tau_r$, and the sampling values are positive in the time $\triangle\tau_f$. So $\triangle\tau_r$ is equal to the product of the number of the negative values and the sampling period, and $\triangle\tau_f$ is calculated by the same way. Equation (8) and Equation (9) show that P_n and P_p are all proportional to the light source power P_0. Therefore the change of the light source power is known by measuring the amplitude of P_n or P_p.

In the novel scheme, the bigger triangular wave is extracted from the gyro signal by the switch or multiplexer, and then it is transformed into the square wave by the differential circuit. According to the change of the square wave amplitude, FPGA adjusts the output current of the constant current source and keeps the stable light power in the signal photodetector. Moreover, the temperature control circuit is also employed in the novel scheme, which maintains the constant work temperature for the light source. The novel scheme is shown in Figure 8.

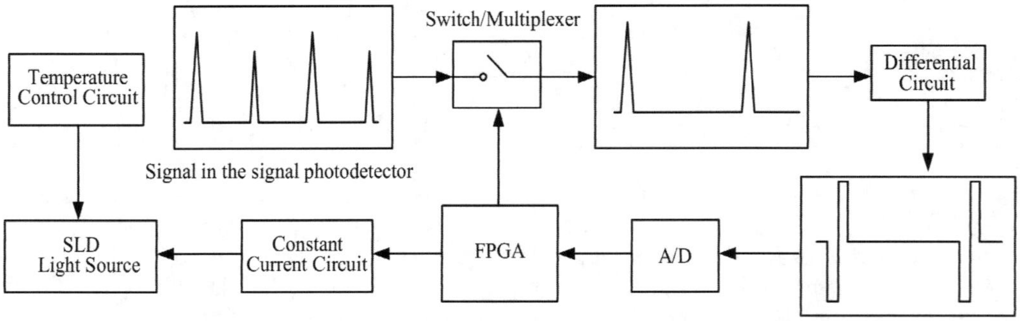

Figure.7 The novel stable control scheme of the light source Power.

4. Simulation

In the simulation, the eigen frequency of the fiber coil is 100KHz, which induces the frequency 200KHz of the triangular wave. In general, the width of the triangular wave is from one-eighth to one-tenth of its period. So supposing that the rising time of the triangular wave is 0.3us, and the falling time is 0.325us; the photodetector responsivity K is typically 8×10^5V/W, and RC 0.05us.

When the light source power changed, the square wave amplitude changed linearly (Figure 8). On the other hand, when the fiber loss in FOG increased due to the temperature change, the square wave amplitude decreased monotonously (Figure 9). The two figures showed that the square wave amplitude indicated directly the change of the light source power and the fiber loss in the I-FOG.

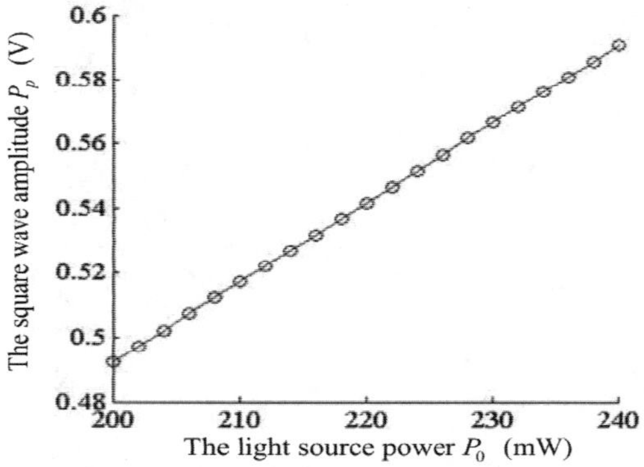

Figure 8. The square wave amplitude changes with the light source power. (L=20dB)

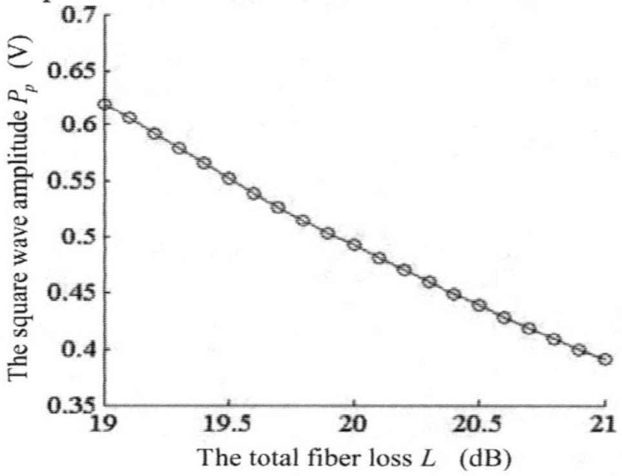

Figure 9. The square wave amplitude changes with the fiber loss. (P_0=200mW)

5. Conclusion

The spike pulses in the gyro signal could indicate the change of the light source power and fiber loss. The novel scheme makes use of this characteristic, and transforms the spike pulses into the square waves by the differential circuit. By measuring the square wave and adjusting the output current of the constant current source in I-FOG, the light power in the signal photodetector were kept at a stable value approximately. Comparing with the traditional scheme, it is a simple and low-cost scheme.

References

[1] Zhang G C 2008 *The Principles and Technologies of Fiber-Optic Gyroscope* (Beijing: National Defense Industry Press)
[2] Lefevre H C 1993 *The Fiber-Optic Gyroscope* (London: Artech House)
[3] Li Y, Ou P, Zhang C X and Xu H J 2006 *J. Beij. Univ. Aero. Astro* **32** 1439-42
[4] Lefevre H C 1997 *Optic. Rev.* **1A 4** 20-27
[5] Pavlath G A 1996 *Proc. Conf. Fiber Optic Gyros: 20th Anniversary* (Denver) **2837** 46

Theoretical analysis and simulation of fibre-top micro cantilever resonator excited optically

Jing Hua[1], Yueming Liu[1,2] and Weijian Tian[1]
[1] College of Optical and Electronic Technology ,China Jiliang University, Hangzhou 310018,China
[2] Corresponding Author, E-mail: liuym@cjlu.edu.cn

Abstract. Theoretical analysis and simulation of bi-layered optical fibre-end micro cantilever resonators was presented in this paper. First, corresponding theoretical model is setup and the resonating frequency is given in this case; Second, the typical characteristics of the micro resonator is simulated including deflection sensitivity, cantilever dimensions and metal coating optimization; Third, the working principle of this optimized micro cantilever is discussed based on optical interference theory of Fabry-Perot cavity. By detecting the optical output of Fabry-Perot cavity resulted from the cantilever deflection, the changes of environmental parameter (such as pressure or temperature) can be measured. Comparing with the traditional optical excited micro resonators, this fibre-top micro resonator has some favourable features, such as micro size, high sensitivity and optical integration and is more interesting and meaningful .

Keywords: micro cantilever, optical fibre, resonator sensors, simulation

1. Introduction

Micro cantilever resonator is widely used in micro-opto-electro-mechanical elements, such as micro-resonator sensors, micro-resonator switches, micro-modulators and etc. Normally the micro cantilever is fabricated on silicon wafer by micro machined process[1], and then should be excited firstly in related applications. The basic principle of these resonator sensors is that an environmental parameter (such as pressure or temperature) changes the first resonance frequency of the resonators, and then the environmental parameters can be detected through the change of the resonance frequency. Among all the exciting methods, the optical excitation is the most interesting method owing to its special advantages of non-contact and non-destructive[2]. For the traditional optically excited silicon cantilever resonator, there are some difficulties that baffle its practical applications, such as the accurate light coupled position between the micro cantilever and the excited optical radiation.

In this paper, a novel cantilever resonator is presented based on the optical fibre end fabricated by carving tiny mechanical cantilever beams directly on the cleaved edge of an optical fibre. For this optical fibre-end micro cantilever resonators, there are more great advantages comparing with the traditional optical excited micro silicon resonators, such as a monolithic structure comprising micro cantilever resonator being positioned easily in such a way that it covers the core of the optical fibre. By this way, the optical excited light coupled to the fibre core is thus put on the micro cantilever accurately and then partially reflected by the cantilever resonator back into the same optical fibre to the signal detector. A metal layer can be deposited atop of the cantilever in order to enhance the reflection of the incoming excited light. From the research of modelling of bi-layered structure frequency and temperature distribution by optical excitation, a more accurate sensor can be obtained by optimizing with better sensitivity and reasonable sensing mechanism.

2. Frequency of bi-layered optical fibre-end micro cantilever resonators

To the bi-layered optical fibre-end micro cantilever resonators in figure 1, the frequency can be deduced as follow. The intrinsic frequency of micro cantilever can be expressed as[3]

$$f_{ci} = 0.046\varphi_i \frac{t}{L^2}\sqrt{\frac{E}{\rho}} \qquad (1)$$

L、t、ρ、E are respectively the length, thickness, density and modulus of the micro cantilever. f_{ci} is the frequencies of different resonance modes, and φ_i are corresponding coefficients. When φ_1 equals 3.52, we can get the first mode frequencies as

$$f_{c1} = 0.162\frac{t}{L^2}\sqrt{\frac{E}{\rho}} \qquad (2)$$

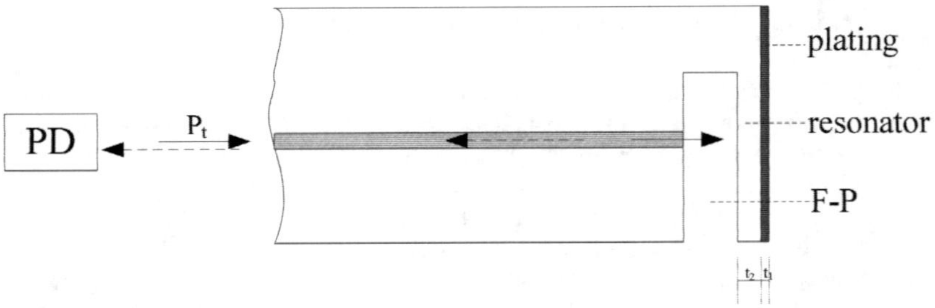

Figure 1. Setup of bi-layered optical fibre-end micro cantilever resonators

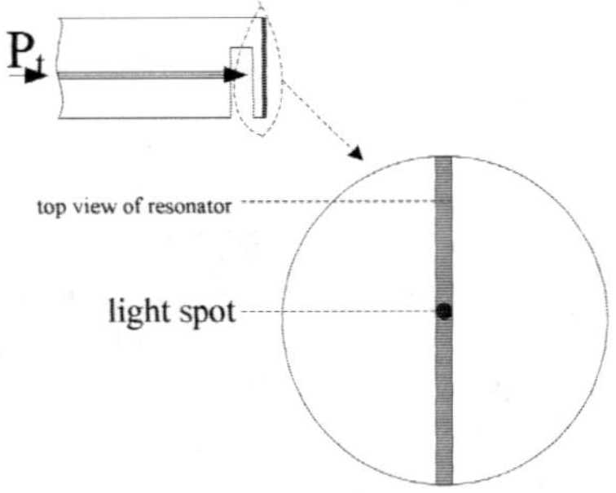

Figure 2. Top view of resonators

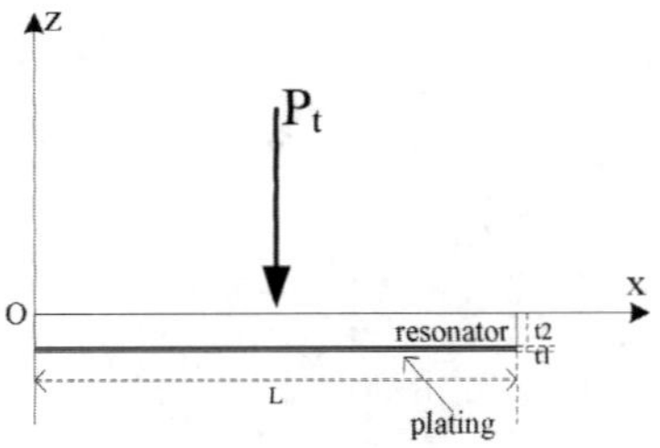

Figure 3. Simplified model of resonators

A better view of the resonator Structure is given in figure 2 and figure 3. The length, width of the resonator are respectively L and W, the thickness, modulus and density of the resonator and its coating are respectively t_2, E_2, $_2$ and t_1, E_1, $_1$. The excited light is guided onto its midpoint of the micro cantilever through the optic fibre. Based on the 'Equivalent methods of parameters' in vibration theory[4], the frequency of the micro resonators can be expressed as

$$f_c = 0.162\cdot\frac{t_2}{L^2}\sqrt{K'\frac{E_2}{\rho_2}} \qquad (3)$$

where K' is the equivalent coefficient which expressed as

$$K' = \frac{4+6\left(\frac{t_1}{t_2}\right)+4\left(\frac{t_1}{t_2}\right)^2+\frac{E_1}{E_2}\left(\frac{t_1}{t_2}\right)^3+\frac{E_2}{E_1}\left(\frac{t_2}{t_1}\right)}{\left(1+\frac{t_2E_2}{t_1E_1}\right)\left(1+\frac{\rho_1 t_1}{\rho_2 t_2}\right)}$$

3. Sensitivity analysis of micro resonator deflection and coating optimization

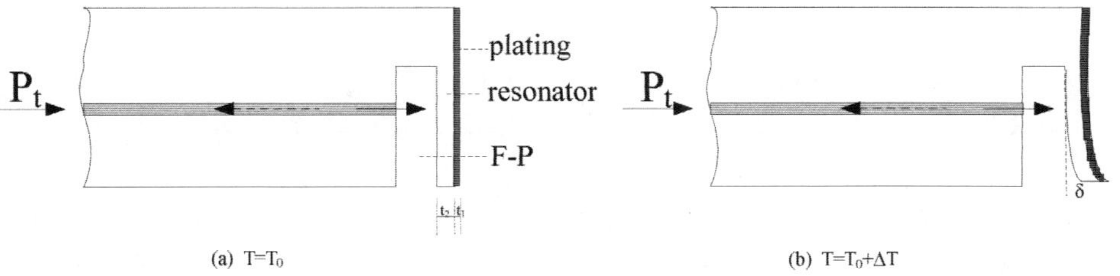

Figure 4. The comparison : (a) The normal structure (b) The deflection

Following the sensitivity of micro resonator shown in figure 4 must be discussed. Assuming that the thermal deflection of cantilever does not occur at the temperature T_0, but it appears when the temperature rises to T_0+ T. Taking thermal stress into account, the differential equation of deflection is setup by the thermal theory of elasticity as[5]

$$\frac{d^2 z}{dx^2} = 6\cdot(\alpha_1-\alpha_2)\cdot\left(\frac{t_1+t_2}{t_2^2\cdot K}\right)\cdot(T-T_0) \tag{4}$$

where K is the equivalent coefficient which is expressed as

$$K = 4+6\cdot\left(\frac{t_1}{t_2}\right)+4\cdot\left(\frac{t_1}{t_2}\right)^2+\frac{E_1}{E_2}\cdot\left(\frac{t_1}{t_2}\right)^3+\frac{E_2}{E_1}\cdot\left(\frac{t_2}{t_1}\right)$$

Ignoring the thermal convection between cantilever and outside, with this bi-layered optical fibre-end micro cantilever resonators as a temperature sensor, the temperature distribution in the resonator under optical excitation can be deduced based on the theory of heat conduction as[6]

$$(T-T_0)(x) = \frac{P\cdot\eta}{2LW(\lambda_1 t_1+\lambda_2 t_2)}\cdot x^2 \tag{5}$$

where W is the width of the cantilever, P is the power of optical excitation and η is optical absorption efficiency. After combined the equation (4) and equation (5), micro deflection at the free end of cantilever and the sensitivity of the optical excitation can be expressed as[7]

$$\delta_P = \frac{1}{4}\cdot(\alpha_1-\alpha_2)\left(\frac{t_1+t_2}{t_2^2 K}\right)\left(\frac{L^3}{W(\lambda_1 t_1+\lambda_2 t_2)}\right)\cdot P\cdot\eta \tag{6}$$

$$S_P = \frac{1}{4}\cdot(\alpha_1-\alpha_2)\left(\frac{t_1+t_2}{t_2^2 K}\right)\left(\frac{L^3}{W(\lambda_1 t_1+\lambda_2 t_2)}\right)\cdot\eta \tag{7}$$

The equation (7) can be changed into another form expressed as

$$S_P = \frac{1}{4}\eta\left(\frac{\alpha_2}{\lambda_2}\right)\left(\frac{L^3}{t_2^2 W}\right)\left[\frac{(\beta-1)(n+1)}{K(\gamma n+1)}\right] \tag{8}$$

where the relations of K, Φ, β, n and γ are as follows

$$K = 4 + 6n + 4n^2 + \Phi \cdot n^3 + \frac{1}{\Phi n}, \Phi = \frac{E_1}{E_2}, \beta = \frac{\alpha_1}{\alpha_2}, n = \frac{t_1}{t_2}, \gamma = \frac{\lambda_1}{\lambda_2}$$

Table 1. Relative property parameters of materials

The material	Density ρ $kg \cdot m^{-3}$	Young's Modulus E $10^{11} \cdot N \cdot m^{-2}$	Thermal conductivity λ $w \cdot m^{-1}k^{-1}$	Thermal expansion coefficient α $10^{-6} \cdot k^{-1}$	Specific heat C $J \cdot kg^{-1} \cdot k^{-1}$
Sio_2	2200	0.70	1.3	0.4	840
Au	19300	0.73	296	14.2	129
Cr	7100	0.25	96.5	8.5	438
Al	2702	0.8	237	23.6	908

It is essential to choose the suitable metal coating and its thickness to improve the detection sensitivity of micro cantilever resonators, normally there are three coating selections which are *Au, Al,* or *Cr*. In order to do coating optimization, according to equation (8), we take parameters of micro cantilever resonators as follows: length $L=200\mu m$, width $W=30\mu m$, thickness $t_2=2\mu m$ and absorption coefficient $\eta=0.1$, other material parameters were shown in table 1. Using the software MATLAB, the relationship between sensitivity and coating thickness ratio of three different metal materials are simulated in figure 5. From figure 5, it is clear that the best material as the metal coating is Aluminium, optimum coating thickness ratio is 0.039 and the maximum sensitivity is $1.43\times10^{-4}\mu m/\mu W$. Further simulation among the intensity of light, length and deflection of micro cantilever is shown in figure 6. We can know that when the light intensity changes from 0 to $400\mu W$, the maximum deflection changes from 0 to $0.46\mu m$.

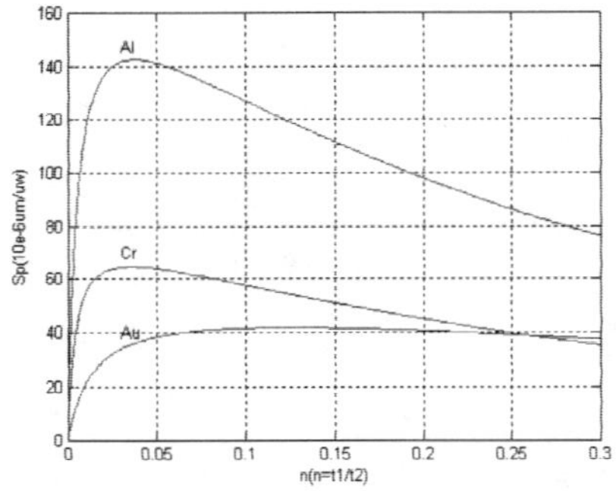

Figure 5. Comparison of the sensitivity in case of three different coating materials

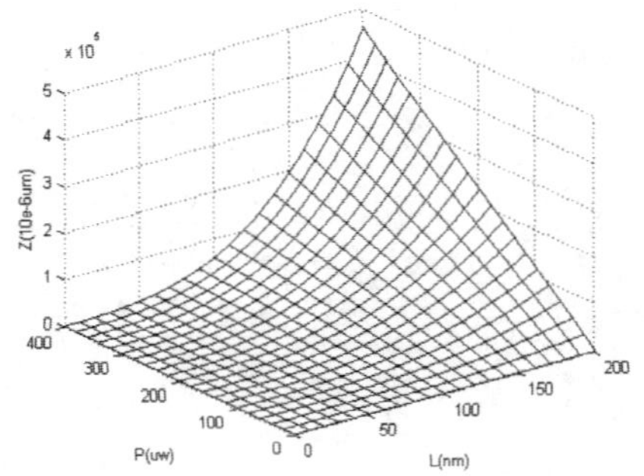

Figure 6. The relation among the intensity of light, length and deflection

4. Resonant sensing mechanism and theoretical analysis

When the periodically intensity-modulated light is guided onto the micro resonator, some of the optical energy is absorbed and the temperature distribution is formed along the resonator, then the resonator begins to vibrating via the photo thermal moment formed in the bi-layered materials which we named "bi-coating effect"[8]. If the frequency of the optical excitation is equal to the resonance frequency, the mechanical resonance occurs in the micro resonator.

A advantage in optical fibre-end micro resonators is that a Fabry-Perot Cavity formed between optical fibre end and micro cantilever which can be used to detect the resonant signals[9]. The reflected light signal is modulated by F-P interference cavity. Because micro cantilever resonators is made of bi-layer material, micro cantilever resonators will be deflected in response to changes of ambient temperature owing to different thermal expansion coefficient of two layer materials. Fabry-Perot cavity depth will be changed when the bi-layer material is deflected thermally, and this results in optical reflectance change of the Fabry-Perot basing on multiple optical interferences within the F-P cavity. The reflecting light from the Fabry-Perot cavity is carried back through the same fibre coupler and transfer to electricity by detector shown in figure 1.

The reflectance I_R/I_0 of Fabry-Perot cavity can be expressed as[10]

$$I_R / I_0 = \frac{a^2({r^2}_1 + {r^2}_2 - 2r_1r_2\cos\Psi)}{1+(r_1r_2)^2 - 2r_1r_2\cos\Psi} \tag{9}$$

where r_1 represents the reflectance of the fibre end face, r_2 represents the reflectance of micro cantilever resonators surface with coating, and Ψ, the round-trip propagation phase shift in the Fabry-Perot cavity, the value of Ψ is given by

$$\Psi = 4\pi h / \lambda_0 \tag{10}$$

where h is the cavity depth and $_0$ is the optical wavelength. Combined equation(9) and equation (10) above, the output light intensity varies periodically as a function of the cavity depth h, but on the other hand, the variety of the cavity h has particular relation with the environmental parameter. So when the output light intensity is measured, change of unknown parameter can be determined.

Next, we set some parameters and give the figure produced by software MATLAB to get a more clear understanding of relationship between cavity h and optical intensity ratio I_R/I_0. When input light wavelength is equal to *1310nm*, $r_1=0.5$, and r_2 is respectively equal to *0.2, 0.4, 0.6* or *0.8*, the performance curve of the cavity h versus intensity ratio is shown in figure 8. From figure 8, we can know two key points, one is that reflected light intensity changes periodically as the cavity length varies, by detecting the change of light intensity, F-P cavity length variation Δh can be calculated and then the environmental parameter can be determined; the other is that in the same circumstances, the higher the reflectivity r_2, the greater the reflectance I_R/I_0.

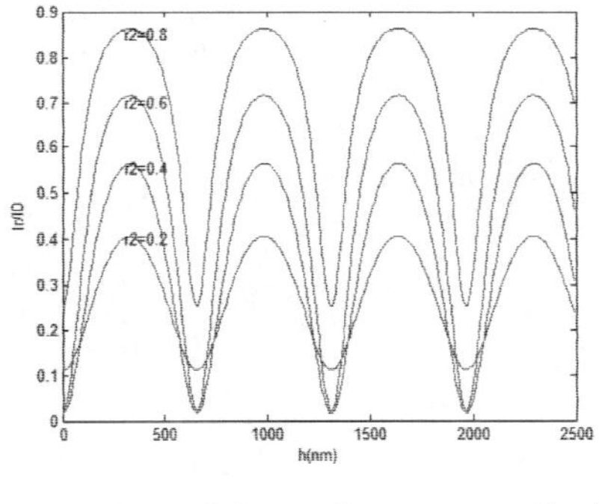

Figure 8. Performance curve of the cavity versus optical intensity ratio

5. Conclusions

Dimensions optimization and characteristics of bi-layered optical fibre-end micro cantilever resonators were described in this paper. The model of bi-layered structure frequency was setup and temperature distribution by excited light was analyzed firstly, then the sensitivity of the micro resonators was simulated and optimized by computer software MATLAB, in the end, resonant sensing mechanism and the character of Fabry-Perot cavity was concerned to detect the change of environmental parameters. The bi-layered optical fibre-end micro cantilever resonator which has greater sensitivity is a better choice to be used in further practical developments.

Acknowledgements

This paper was supported by Natural Science Foundation of Zhejiang Province of China (No. Y1091078) and Science Research Found of China Jiliang University.

Journal of Physics: Conference Series **276** (2011) 012128 doi:10.1088/1742-6596/276/1/012128

References

[1] Hua qingan. *Silicon micromachining technology*, 1996. (Beijing : Science Press)
[2] Li shaohui. *Optical fibre sensor*, 1997. (Wuhan: Huazhong Technology University Press)
[3] Langdon R M. Resonator sensors— a review, 1985. J. Phys. E: Sci. Instrum. **18**. 103–15
[4] Roard R J and Young W C. *Formulas for stress and strain*,1975. (New York: McGraw-Hill Book Company). p 112
[5] Barnes J R, Stephenson R J and Woodburn C N, 1994. A femtojoule calorimeter using micromechanical sensors. *Rev. Sci. Instrum.* **65 (12)**. 3793–98
[6] Holman J P. *Heat Transfer,* 1986. (New York: McGraw-Hill Book Company)
[7] Liu yueming. Theoretical model and techniques of silicon micro resonator under photo-thermal excitation (D), 2001. Xi'an Jiaotong University.
[8] Chen G Y, Thundat T and Wachter E A, 1995. Adsorptiob-induced surface stress and its effects on resonance frequency of micro cantilevers. *J. Appl. Phys.* **77(8)**. 3618–22
[9] Guo zhenhua. The inventors of the multibeam interferometer—C. Fabry and A. Perot, 2004. *Physics*. **4(33)**. 293–6
[10] Bi weihong. Mathematical model for fibre-optical non-symmetrical Fabry-Perot Interferometric cavity, 2000. *ACTA OPTICA SINICA*. **7(20)**. 874–6

MEMS integrated narrow band infrared emitter and detector for infrared gas sensor

J J Lai[1], **H F Liang, Z L Peng, X Yi and X F Zhai**
Wuhan National Laboratory for Optoelectronics, Wuhan 430074, P.R.China

E-mail: jjlai@mail.hust.edu.cn

Abstract. With the development of microfabrication technology, compact, lightweight, inexpensive and portable micro-spectrum detection and analyzing products are expected to be extensively applied in the future commercially. From the viewpoint of power consumption and compact volume, integration of various components on one substrate is desirable. In this work, integrated infrared gas sensor with micro infrared emitter and detector elements based on MEMS technology are developed. The integrated suspended infrared emitter and infrared micobolometers are designed and fabricated on a silicon substrate.

Introduction

Numerous techniques for gas concentration measurements have been developed over the years. These include metal-oxide gas senors, solid electrolyte gas sensors, capacitive gas sensor, gravimetric gas sensor and optical sensors. The first four methods rely on measuring the change of an electrical quantity of the sensor's active material. This change is caused by adsorption or absorption of chemical species at the surface of the active material. These methods have advantages such as easy and cheap fabrication by standard micro-electronic or micro-mechanical processes, tiny size, or high operating temperature. But the major drawback of the above sensor schemes is their lack of broad range selectivity due to a lack of sensitive materials for every gas and/or due to crosstalk. Another drawback of some of these sensors is short-lived due to the contact nature of active materials with specific gas which leads to the degradation of the active materials of the sensors. Infrared optical gas sensors, however, offer a variety of advantages compared to the sensors described above. The IR absorption detection mechanism is not influenced by the species used for detection so IR spectroscopy-based sensors do not rely on catalytic or electrochemical reactions; they are not prone to depletion or contamination of the surface and have a long period of operation life.

As can be seen, infrared absorption measurements have been considered more advanced than other sensors. The problem with optical devices has been their high price because of the needs of various infrared components like infrared source, infrared detector, two or more filters and gas absorption cell which is mostly with multiple optical elements and complex shape. However, as new technologies are developed and instrument volumes grow, these prices can be decreased by integration and miniaturization. MEMS/NEMS is one of these technologies which can promote these actions [1-4].

In this work, integrated chips with micro infrared emitter and detector elements based on MEMS technology are developed. The integrated suspended infrared emitter and infrared microbolometers are

[1] To whom any correspondence should be addressed.

Published under licence by IOP Publishing Ltd

designed and fabricated on a silicon substrate. In the integrated devices, the infrared emitting elements have good thermal isolation property and can reach 500℃ in tens of milliseconds and the microbolometer elements have one layer of infrared sensitive thin film of vanadium dioxide with high TCR and have measured D*(500K,1Hz) more than $5\times10^{8}cmHz^{1/2}W^{-1}$ at peak detection wavelength.

2. Narrow-band infrared devices

Infrared devices based on MEMS technology can achieve breakthrough in three aspects. First, thin film based structures replace the bulk material promote the ultrathin optical and mechanical structure and more easy for multifunction integration; Second, micro-fabrication of thermal isolation cavity makes the possibility of electrical modulation of infrared emission signal at higher frequency and low power consumption; Third, array devices can be batch fabricated and eventually lead to high product and low price. In fact, there are a mumble of researching institutes and companies have made commercial success in optical MEMS devices like DMD, micro-bolometer, infrared scene projector based on micro-emitter array and micro-spectrometry.

Narrow-band infrared devices, however, following the conventional MEMS technology to realize specific mechanical, electrical and thermal functions, must pay more attention to how to achieve narrowband filtering function. Without using individual optical filter, advanced schemes are proposed to pattern micro/nano structure on the optical surface to accomplish multiple optical functions. Compared to the realization of other structures for the MEMS devices, optical micro/nano structures are more challengeable.

Among various kinds of optical micro/nano structure, photonic crystal is the most popular structure used in spectral control. Photonic crystals of 1D to 3D have been designed for narrow band thermal emitter [5-9]. 2D photonic crystals with metal layer incorporating SPP effect were designed []. The SPP type microstructure is likely more effective due to field enhancement of coupled excitation and can achieve narrower emission or absorption. Coherent infrared emission can be obtained through SiC grating which supports surface phonon polariton, it can also be obtained through W grating supports surface plasmon polariton. However, the infrared spectrum angular emission distribution is like rainbow which means the spectrum is seriously angular dependent. The effect leads to the fact that the emission spectral width is mainly determined by the view of field of the optical system which possibly loss the narrow band advantage. So schemes for narrow band creation over wide angular range were proposed, such as MDM multilayer plate and MDM grating structures, where M represents metal or metallic like materials, D represents dielectric materials [10]. Thus spectral interference of these kinds of devices is greatly reduced.

3. Integrated structure

We propose an idea to construct infrared gas sensor with integrated sensor chip. The optical schematic diagram is shown in figure 1. The integrated sensor chip is composed of three on-chip elements, i.e. one emitter and two detectors. The working principle of this sensor is based on typical two-channel configuration, which means one channel is for measurement and another is for reference. The advantage of this configuration is that the fluctuation of source power and environment temperature is can be lessened. The emitter element emits infrared radiation with central wavelength λ_0 and bandwidth of $\Delta\lambda_0$, and the two detectors receive radiation with central wavelength λ_1 and λ_2 respectively, and with bandwidth of $\Delta\lambda_1$ and $\Delta\lambda_2$ respectively. The bandwidth of emitter is narrow but still larger than the sum of band-widths of the two detectors, that is $\Delta\lambda_0>\Delta\lambda_1+\Delta\lambda_2$.

The narrowband mechanism of the MDM structure is related to the electromagnetic modes excitation presented in curve as shown in figure 2, which is derived by transfer matrix method. The structure parameter of MDM is chosen as Au200nm/Silicon450nm/Au12nm. As shown in the figure, there are several kinds of narrow band modes excitation in the mid-infrared independent of incident angle of light. Since these modes are above to the light line, they can be directly excited by the radiation coming from the air side. This scheme reduces the processes of period structure fabrication adopted in conventional schemes. Moreover, if the metal layer is substituted by metal-like conducting ceramic such as TiN_x, emission or absorption spectrum width can be tuned by varying the electrical or optical properties of TiN_x.

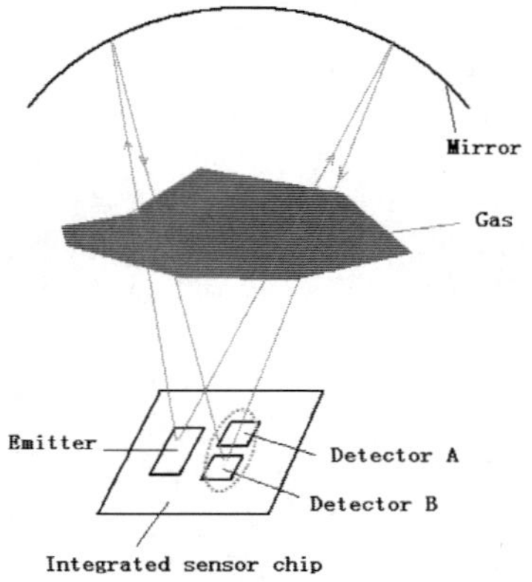

Figuire 1. Optical schematic diagram of infrared gas sensor using integrated chip.

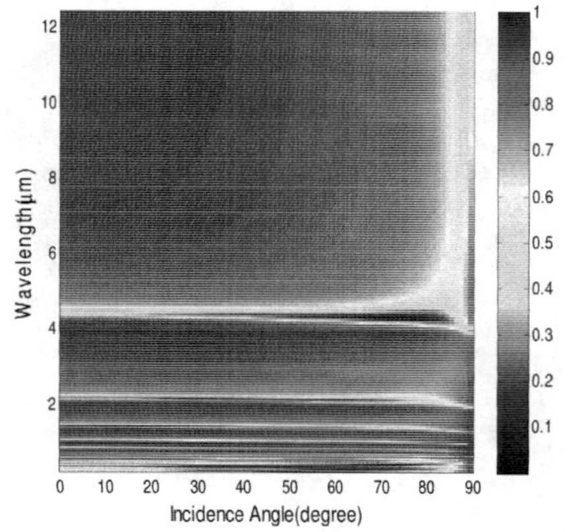

Figure 2. Reflectance spectrum of MDM structure.

4. FDTD simulation

To investigate the emission characteristics of the MDM structure, we calculate the emission spectrum in an indirect manner. First we use finite difference time domain(FDTD) method to calculate the reflectance of this structure, and then use Kirchhoff's law to deduce the emission spectrum from the relation: $\varepsilon_{\lambda,\theta} = A_{\lambda,\theta} = 1 - R_{\lambda,\theta} - T_{\lambda,\theta}$. In this specific structure, the transmittance $T_{\lambda,\theta}$ is zero (as the bottom metallic layer is optical thick), so the emittance can be calculated from the reflectance directly: $\varepsilon_{\lambda,\theta} = A_{\lambda,\theta} = 1 - R_{\lambda,\theta}$.

The metallic-like layer of the MDM structure is TiN_x, which has a resistivity of about 500μ ·cm, its optical constant is obtained through fit parameters by assuming material model of Drude–Lorentz-oscillator type to match the experimental reflectivity. Dielectric layer is a mixed multilayer of SiO_2/VO_x/silicon, which has a total thickness near to 0.52μm. The surface metal layer is a gold layer having a thickness of 0.012μm. The optical constants of other materials are all determined by proper model parameters fitting experimental reflectivity datum. The thin VO_x layer is need for infrared detection, but not necessary for emitter application.

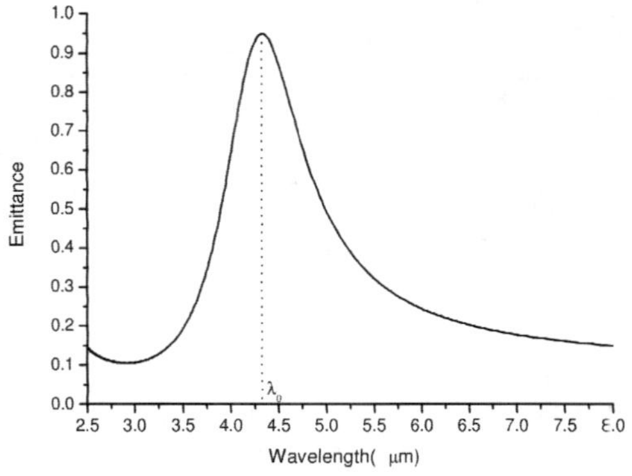

Figure 3. Simulated emittance of MDM at normal direction.

The simulated emittance of the MDM at normal direction is shown in figure 3. The central frequencies is at 4.25μm and the bandwidth is nearly 1μm. Increasing the conductivity of TiN_x can make the bandwidth narrower. The narrowest bandwidth we can reach is about 0.1μm by using high conductive TiN_x where its resistivity is about 40 μ Ω • cm very close to metal.

5. Protype of CO_2 gas sensor

Integrated chip composed of one emitter and two detectors are implemented by MEMS technology. Patterns are created by standard photolithography, thin film deposition and lift-off processes. Wet etching technique is used to make the patterned elements suspended on silicon supported by multiple long arms, with hollow cavity underneath of the elements. In the integrated devices, the infrared emitting elements have good thermal isolation property and can reach 500℃ in tens of milliseconds and the detection elements have one layer of infrared sensitive thin film of vanadium dioxide with high TCR and have measured D*(500K,1Hz) more than $5\times10^8 cmHz^{1/2}W^{-1}$ at peak detection wavelength. Two detection elements have different response, one has the largest response at CO_2 detection wavelength of 4.25μm, but another has response at the reference wavelength of 4.6μm. The integrated device is sealed in vacuum and packaged with leadless chip carrier (LCC) in the end.

Figure 4. Prototype of CO2 gas sensor module.

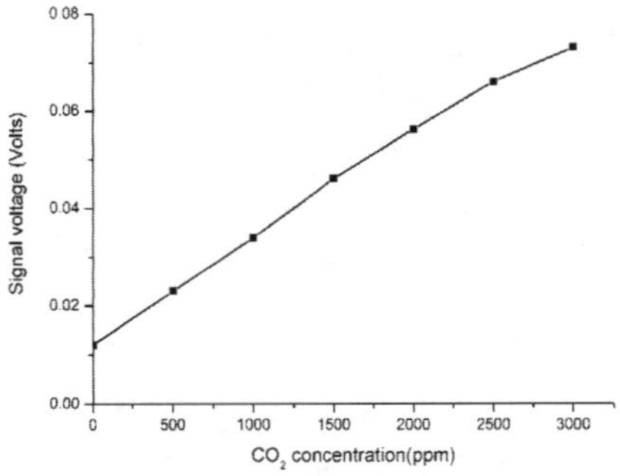

Figure 5. CO_2 concentration measurement from 0 ppm to 3000ppm

Having the integrated chip, CO_2 gas sensor is constructed by adding a gas absorption cell and an infrared reflective mirror as main components. The prototype of this sensor is shown in figure 4. The test was achieved in a vacuum chamber filled with N_2. The amount of N_2 and CO_2 gas flowing can be precisely controlled through two mass flow controllers..

After calibration and proper signal processing the concentration of gas can be obtained. We have measured standard CO_2 gas mixed in N_2 in the range between 0 ppm to 3000 ppm. The measurement results have demonstrated the capability of infrared microsystem in application of gas detection. Further research will extend to various kinds of gas or liquid sensing applications and also to establish advanced micro-spectrum analysis systems.

Reference

[1] Sai H, Yugami H, Kanamori Y 2003 *Microscale Thermophysical Engineering* **7** 101
[2] Mark P M, Nicholas M, Martin U P, Irma P, Lisa L, William H, Anton C G, James T D, Edward A J and Thomas G 2002 *Proceedings of SPIE* **4815** 30
[3] Simon T, Barsan N, Bauer M, *et al.* 2001 *Sensors and Actuators B-Chemical* **73** 1
[4] Moon B U, Lee J M and Shim C H 2005 *Sensors and Actuators B* **108** 271.
[5] Greffet J J, Carminati R, Joulain K, *et al.* 2002 *Nature* **416** 61
[6] Pralle M U, Moelders N, McNeal M P *et al.* 2002 *Applied Physics Letters,* **81** 4685
[7] Tsai M W, Chuang T H, Meng C Y, *et al.*2006 *Applied Physics Letters* **89** 173116
[8] Puscasu I and Schaich W L 2008 *Applied Physics Letters* **92** 233102
[9] Diem M, Koschny T and Soukoulis C M 2009 *Phys. Rev. B* **79** 033101
[10] Hocheol S, Mehmet F Y and Fan S H 2004 *Appl. Phys. Lett.* **48** 4421

300m optic fiber Bragg grating temperature sensing system for seawater measurement

Xingrong LI ,Yongqian Li and Zhengyang Wen
Dept. of Electronics and Communication Engineering, North China Electric Power University, Baoding 071003 ,China
li_xingrong@yahoo.cn

Abstract: Optic fiber grating sensor is a research hotspot.It has been used on many occasions,and how to use it for ocean detection is a new research directions. The paper introduced the calibration work of FBG temperature sensors. It confirmed that from being armored package,the sensors can eliminate the water pressure effect. From the calibration experiment and data processing,60 sensors has little error were screened out for experiment. 300m long optic fiber Bragg grating sensor array was designed.The marine experiments were achived in South China Sea with 300 meters long Bragg grating array and got the seawater profile temperature. Proposed the curve fitting method to process the data based on Levenberg-Marquardt algorithm.By curve fitting to the data acquired,the precision was better than 0.2℃,which verified the effectiveness of the method.This result has practical value.

Key words: optic fiber grating; sensing system; calibration experiment; curve fitting

1. INTRODUCTION

Our country has large ocean area,but we are not a marine power. China's marine surveying technology is relatively backward to developed country,and the proportion of output value of Chana's marine economy in GDP is not high.It's time to enhance marine technology research. China's ocean automatic monitoring equipment and information platform in the proportion of the marine economy is very little[1].

Currently,the detection system for marine environment has some urgent problem to be solved,such as the effectiveness of data acquisition,the network of monitoring sites,the real-time transmission of monitoring data,low cost and high reliability and so on[2]. Fiber-optic sensors can simultaneously measure water depth, temperature, currents, salinity and other parameters, so how to use a fiber optic sensor in marine detection has become a research focus at present. FBG sensor as a novel optic sensor which is easy to implement WDM,and easy to constitute a large number of sonsor arrays to facilitate distributed measurement[3,4]. The sensor array can be used in marine monitoring, and it's easy to get

Journal of Physics: Conference Series **276** (2011) 012130 doi:10.1088/1742-6596/276/1/012130

seawater temporature distribution,so as to provide reference material for marine research and development.We designed the 300-meter-long FBG sensor array, measured the marine water temperature distribution on South China Sea, and processed the measured data.At last,we got useful conclusions.Using fiber grating sensor array can reduce measurement costs, increase measurement flexibility. It proved that FBG sensor array practical value for the marine environment research.

2. EXPERIMENT SYSTEM DESIGN

In the experiment, a broadband light source was used, and its wavelength range was 40nm, the tunable Fabry-Perot cavity scanning frequency was 1Hz.Because each sensor wavelength occupied by about 2-3nm,so it may allow less than 20 sensors serial connected in a single optical fiber.In the experiment,60 temperature sensors were used,in order to prevent the wavelength overlap,four optical fiber were used,and 15 sensors were serial connected in each fiber. It was about five meters between two sensors,so all the length of the FBG grating array was 300m.The fiber Brag grating sensor was smartly packaged to eliminate the seawater pressure interference before the experiment. The vertical profile temperature of the seawater can be obtained by using such a sensing array. The FBG sensor array was shown as figure1[5].

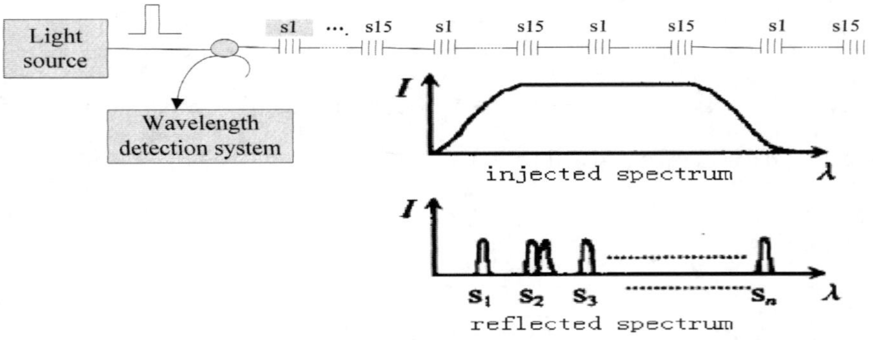

figure1. The chart of FBG sensor arrays and the reflection spectrum

3. GRATING TEMPERATURE SENSOR CALIBRATION

The center wavelength of FBG sensor changes with ambient temperature changes,while the internal characteristics of each sensor is different and the stability is different too,so the calibration work must be made for every FBG sensor before experiments[6-8].

The temperature calibration experiments were made in thermostat,and 2℃、5℃、10℃、15℃、20℃、25℃、30℃ were measured respectively.The data was very huge,some of them was listed as table1.In table1,it includes 10 seconds's data of one sensor.We can observed from table1 that the sensor's wavelength has small change at the same temperature,in order to improve the measurement accuracy,it need to measure a long time at each temperature,then we can calculate an average wavelength for each measured sensor.This average wavelength can be considered as the temperature sensor's center wavelength,through curve fitting,we can get the relationship between temperature and center wavelength.we can suppose that the relationship between the temperature and the center wavelength is y=ax2+bx+c,y represents the temperature,and x represents the center wavelength,a,b and c are the parameters to be determined.With Levenberg-Marquardt algorithm,we can get a,b and c's value from curve fitting.The sensor as table1 corresponding to the coefficients a,b and c's value are as below:

Table 1 Optic fiber grating center wavelength under different temperature

	2℃	*5℃*	*10℃*	*15℃*	*20℃*	*25℃*	**30℃**
1	1530.735	1530.761	1530.807	1530.852	1530.900	1530.947	1530.996
2	1530.735	1530.761	1530.807	1530.853	1530.899	1530.946	1530.995
3	1530.735	1530.761	1530.807	1530.853	1530.899	1530.947	1530.995
4	1530.735	1530.761	1530.807	1530.853	1530.900	1530.948	1530.994
5	1530.734	1530.761	1530.808	1530.852	1530.899	1530.947	1530.994
6	1530.735	1530.761	1530.807	1530.853	1530.900	1530.947	1530.995
7	1530.735	1530.761	1530.808	1530.852	1530.899	1530.946	1530.995
8	1530.735	1530.761	1530.807	1530.853	1530.900	1530.947	1530.995
9	1530.735	1530.761	1530.807	1530.853	1530.900	1530.946	1530.995
10	1530.735	1530.762	1530.807	1530.852	1530.899	1530.947	1530.994
average	1530.7349	1530.7611	1530.8072	1530.8526	1530.8995	1530.9468	1530.9948

$$\begin{cases} a = -17.0367091942 \quad 332 \\ b = 52269 \quad .4183137788 \\ c = -40091066 \quad .9862078 \end{cases}$$

All the sensors' coefficients a,b and c can be calculated with this method,then these values are stored into data processing system,so it can be used easily for practical measurements.The sensors which

Journal of Physics: Conference Series **276** (2011) 012130 doi:10.1088/1742-6596/276/1/012130

have relatively small temperature error and stable performance can be selected to construct sensing array from calculating temperature error.It can improve the system temperature measurement accuracy.

4. EXPERIMENTS AND DATA ANALYSE

The experiment has been completed on South China Sea.The 300 meter long sensor array was fell into seawater by pulley device,then started the wavelength demodulation system to acquire the reflected light by each FBG sensor.Part of the acquired data was listed as table2.

Table 2 Parts of the original data acquired(unit: nm)

sensor1	sensor 2	sensor 3	sensor 4	sensor 5	sensor 6
1527.034	1531.03	1533.974	1536.053	1538.067	1542.07
1527.034	1531.03	1533.974	1536.053	1538.068	1542.071
1527.036	1531.03	1533.974	1536.054	1538.068	1542.071
1527.034	1531.03	1533.975	1536.053	1538.067	1542.071
1527.036	1531.03	1533.977	1536.053	1538.067	1542.071
1527.035	1531.03	1533.975	1536.053	1538.067	1542.071
1527.035	1531.03	1533.974	1536.053	1538.068	1542.071
1527.035	1531.03	1533.975	1536.053	1538.067	1542.071
1527.035	1531.03	1533.975	1536.054	1538.067	1542.071
1527.035	1531.03	1533.975	1536.054	1538.067	1542.071

We can observe from table2 that some sensors were stable such as the 2nd sensor and the 6th sensor whose wavelengthes had almost no change over time,while other sensors' wavelength had changes in different degree,then we can calculate each sensor's center wavelength with average method as mentioned before.These calculated center wavelength was sent to background processing system,where we can substitude these wavelength into equation,such as y=ax2+bx+c,then the temperature can be calculated out.As these sensors'depth were different,so we can get the depth-temperature curve by curve fitting as figure2.

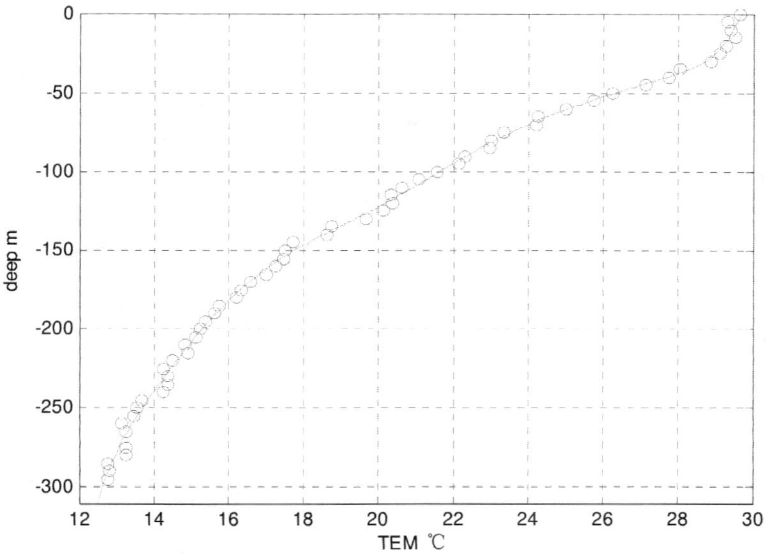

figure2. FBG temperature chain depth-temperature curve

As we can see from figure2 that in this region, from 0 to 40 meters depth, the temperature has small change with depth change,and from 40 meters to 140 meters depth,the temperature rapidly dropped about 10℃ with depth change nolinearly,and from 140 meters to 200 meters depth,the temperature changed slowly. In order to explore these phenomena,five sensors in different depth were analyzed with time change.Their depth were 17.8m, 44.6m, 71.3m, 98m, 124.7m respectively,and ten minutes'data was analyzed,their temperature-time curves can be got as figure3.

In figure3,we can see that at the depth of 71.3m and 124.7m,the seawater temperature changes greatly, and the other three depth,the seawater temperature has small change. Analyzed these phenomena,we can predict that in this area 1)seawater in different depth has different velocity 2)different ocean currents mix,these condition caused these difference.

In order to determine the measurement accuracy, contrastive test are needed,and the contrastive test between FBG chain and high precision themometer CTD, which was recommended by marine experiment experts,was completed. Three CTD themometers were tied beside three FBG sensors seperately,so it's easy to compare the measurement accuracy.The CTD themometer's measurement accuracy is 0.01℃,by curve fitting the raw data,we can get the relationship curve between temperature and time ,one of them was shown as figure4.

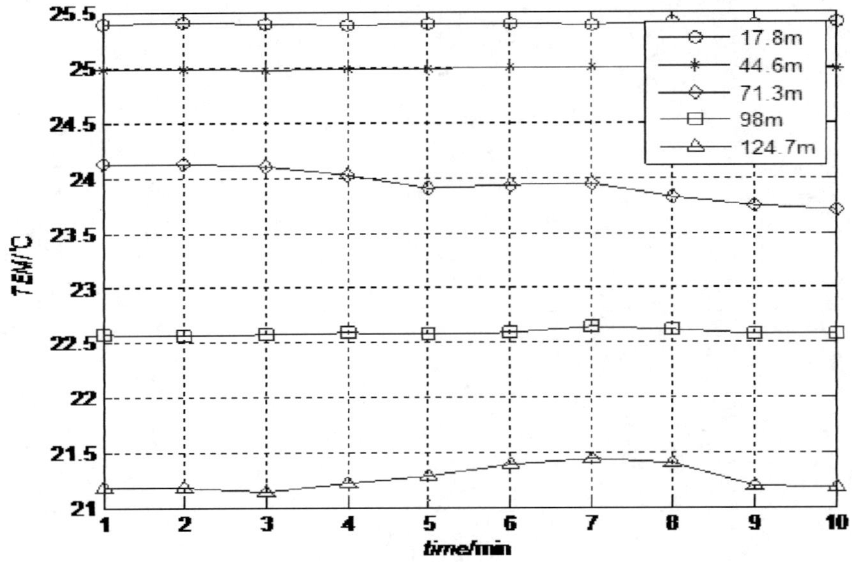

figure3 five different depth temperature-time curve

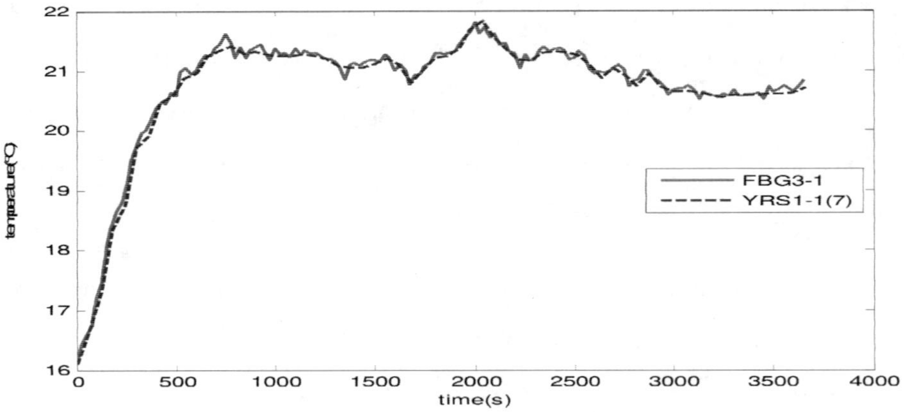

Figure 4 the contrastive curve between FBG3-1 sensor and CTD

In figure4,the dotted line is the temperature distribute curve for CTD named YRS1-1,and the solid line is the temperature sensor's distribution curve for FBG sensor named FBG3-1.As we can see from figure5,minor measurement errors was exist between the two kinds of sensors.We calculated the measurement error between them at three spots,which were listed as table3.

Table 3 the contrastive table between FBG sensors and CTD

depth(unit: m)	FBG sensor	CTD	errors(℃)
50	27.266	27.16	0.106
130	21.702	21.64	0.062
200	18.368	18.21	0.158

From the experiment result,we can conclude that the measurement accuracy of FBG chain is better than 0.2℃,which can meet the measurement requirement,and it proved that the FBG chain has practical value.

5. CONCLUSION

FBG sensor has been widely concerned because it's unique advantages,and has been applied in the area of building structures and other areas.We proposed the FBG temperature chains scheme for marine detection research,and designed 300-meter-long FBG temperature chain.The field test on South China Sea was accomplished.After the data processing,we got a satisfactory result.We believe that it will open a new situation for the application of FBG sensing system.

Acknowledgments

This work is supported by the National High Technology Research and Development Program of China (863 Program) NO.2006AA09Z141 and Young Teacher Funds of NCEPU.NO:200811006

Reference:

[1] Jin Xi-fan， Wang Shuo. 2009 Intelligence Monitoring Current Situationand and Key Technologies for Marine Environment MARINE FORECASTS.vol 26 pp：95-102

[2] Li Yan，Luo Xu-ye.2009 Research on Systems and Application Technology of Wireless Sensor Networks for Ocean Monitoring OCEAN TECHNOLOGY vol 28 pp:8-11

[3] Luiz Carlos Guedes Valente, A. M. B. Braga, Alexandre Sant'Anna Ribeiro,etc.2003.Combined Time and Wavelength Multiplexing Technique of Optical Fiber Grating Sensor Arrays Using Commercial OTDR Equipment IEEE SENSORS JOURNAL.vol 3 pp:31-35

[4] Shen Zhen-qiang,Zhao Jian-lin,Zhang Xiao-juan. 2007 Frequency-Division Multiplexing Technique of Fiber Grating Fabry-Perot Sensors ACTA OPTICA SINICA.vol 27 pp:1173-1177

[5] Li xingrong,Yang zhi,Li yongqian etal. 2009 Wavelength Division Multiplex Optic Fiber Brag Grating Temperature array for Ocean Detection.POEM

[6] Wang Xiao-na,Wang Qi ,Chen Le-hua etal. 2009 Swept Fiber Laser Based Fiber Optic Sensor Demodulator[J]. ACTA PHOTONICA SINICA.vol 38 pp:82-85

[7] Qian Xiang-zhong, Wang Xue-lei. 2007 FBG Based Sensing System for Temperature Monitoring of the High Voltage Apparatus Proceedings of the CSU-EPSA.vol 19 pp:49-51
[8] Fu Jian-wei,XIAO Li-zhi,Zhang Yuan-zhong etal. 2006 Wavelength calibration method for distributed f iber Bragg grating OPTICAL TECHNIQUE.vol 32 pp:903-905
[9] Yu You-long, ZHAO Hong-xia. 2005A novel demodulation scheme for fiber Bragg grating sensor system IEEE Photonics Technology Letters,vol 17 pp:166-168
[10] Jiang De-sheng, GAO Xue-qing. 2006 The Method of FBG Sensing for a Sort of Dense Distributed Measurement LASER & INFRARED.vol 39 pp: 960-962
[11] Wang Yu-tian, Cao Li-fang. 2007 Research of Stress Survey by Fiber Bragg Grating CHINESE JOURNAL OF SENSORS AND ACTUATORS.vol 20 pp:1193-1195
[12] Yi Hong-gang, Gong Xian-feng, Wang Chang-song. 2006 Research on high sensitivity fiber grating temperature Sensors Transducer andMicrosystem Technologies.vol 25 pp:10-11
[13] JIAO Sheng-jie,ZHU Jun,QU Bin. 2007 A Simple New Demodulating System for Fiber Grating Pressure Sensor ACTA PHOTONICA SINICA.vol 36 pp:18-20

The research of the accurate measure of static transfer function for the TDI CCD camera

LI Guo-ning[1], JIN Long-xu[1], REN Jian-yue[1], WANG Wen-hua[1], HAN Shuang-li[1]

[1] Changchun Institute of Optics, Fine Mechanics and Physics, Chinese Academy of Sciences, Changchun 130033, China

E-mail: Lgning@netease.com

Abstract. In the test course of static transfer function of TDI CCD camera, because of the influence that gets environmental and artificial etc. factor, the value of static transfer function measured at any time is between unceasing fluctuation, so, make accuracy reduce. To solve this problem, a kind of accurate measurement technique of static transfer function is put forward. First, before carrying out the measure of static quiet of transfer function, the best test point of transfer function of the TDI CCD camera must be determined, it is parallel to guarantee the rectangle target surface of parallel optical pipe and camera focal plane maintenance parallel, and again guarantee target strip in rectangle target and TDI CCD in camera focal plane maintenance vertical. TDI CCD catches rectangle target image, per 1000 lines of target mark image as a measures sample of static transfer function, exclude because of atmosphere tremble twisted, vague rectangle target mark image, retain 500 distinct and steady target mark image as measure sample set. Then, calculate the static transfer function of each measure sample respectively, take the average of all static quiet transfer function in measure sample set as the static transfer function of camera. Finally, the measure of the static transfer function for TDI CCD camera makes error analysis. Experimental results indicate that the value of the static transfer function of TDI CCD camera measured with this kind of method is 0.2923, with before measurement technique comparison, the value of static transfer function has raised 0.02 , makes the accuracy of the measure of static transfer function have gotten raising.

1 Introduction

The modern remote sensing digital camera of high resolution capacity adopt TDI CCD push sweep image pattern[1-3], MTF (Modulation Transfer Function) is to appraise a important index like image quality and application ability for this kind of camera, and it have reflected camera each space frequency weight in scenery respond property. The MTF of TDI CCD camera is including the static transfer function and the dynamic transfer function, after completing a series of works such as optical, mechanical and electronic installation and test in laboratory, the test of the static transfer function for TDI CCD camera must be carried out. In the test course of static transfer function of the TDI CCD camera, test environmental (atmosphere shake and environmental temperature), test method (target mark to carve line and transfer function calculation method), other aspects such as the space location of measure system can also affect as transfer function test produces. In which, major influencing factor is the space location of measure system and atmosphere tremble[4].

Now, the method of testing static transfer function of the TDI CCD camera has a lot, including the point source method, the line source method, edge side method, sine input method and rectangle input method, but in laboratory the method of major adopting is rectangle input method. This paper has put forward a new kind of method of accurate measure static transfer function on the foundation of rectangle input method, has raised accuracy and the test precision of static transfer function of the TDI CCD camera.

2 The form of the static transfer function measure system

The measure system of static transfer function of the TDI CCD camera forms as Figure 1 shows , including integrating sphere and rectangle target mark , parallel optical pipe, camera optical system, the TDI CCD focal plane, CCD signal handling, image collection card, computer, RS232 bus, the control system of revolving stage and target mark revolving stage. In which, parallel optical pipe includes spherical microscope mirror and plane microscope mirror, and integrating sphere includes silicon detector and illumination lamp group.

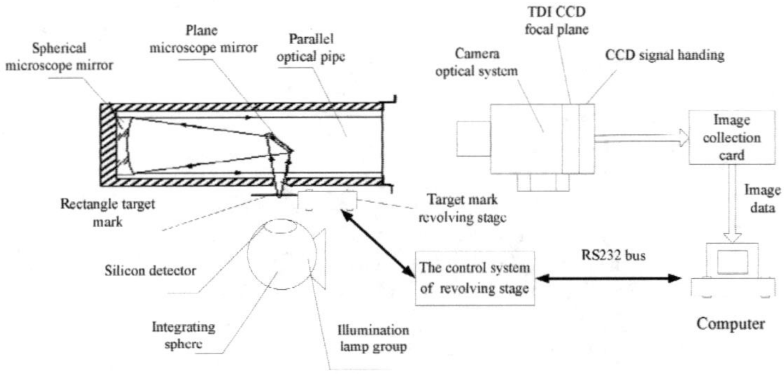

Figure 1. The form of the measure system of static transfer function of the TDI CCD camera

3 The accurate measure method of static transfer function

3.1 Initialize systematic location

Before measuring, the measure system of static transfer function of the TDI CCD camera must be initialized, the rectangle target mark is placed the entrance in parallel optical pipe, used for simulated infinite far scenery goal. The optical axis of parallel optical pipe and the optical axis of the TDI CCD camera are on same axle. TDI CCD camera has been put the export end in parallel optical pipe, and the image information gathered by camera is transmitted to computer through image collection card, computer communicated with the control system of revolving stage through RS232 bus. The control system of revolving stage controls the encoder and the motor of revolving stage, so realize target mark revolving stage turn with location information read.

3.2 The location relation that TDI CCD focal plane and rectangle target mark

The location relation that TDI CCD focal plane and rectangle target mark, and target mark revolving stage turn direction definition as Figure 2 show: the optical axis of parallel optical pipe and the optical axis of the TDI CCD camera are on same axle, it is Y direction to stipulate optical axis of parallel optical pipe, the positive Y direction points to camera optical system; X direction and Y direction are vertical and with ground parallel, the positive X direction points to left side visual field of parallel optical pipe; Z direction and Y direction are vertical and with ground also vertical, the positive Z direction points to the top of parallel optical pipe; θ_X Direction, θ_Y direction and θ_Z direction spin respectively to be axle with X direction, Y direction and Z direction, stipulate to obey the direction of hour hand as positive direction. In the course of searching for the best transfer function test point , it is standard with this coordinate department[5-8].

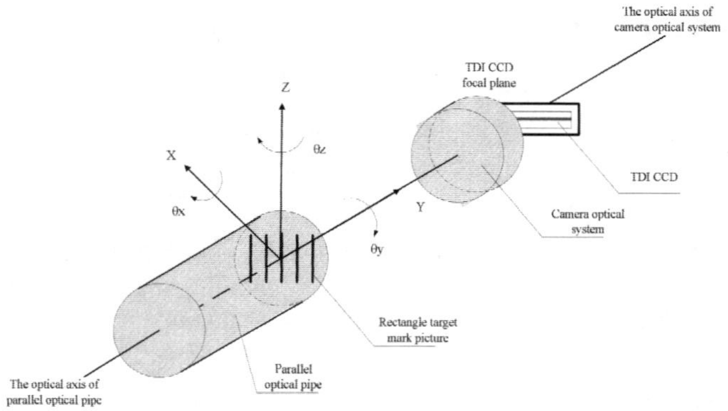

Figure 2. The location relation that TDI CCD focal plane and rectangle target mark, and target mark revolving stage turn direction definition

3.3 Search for the test point of the best transfer function

Make certain the relatively location relation of TDI CCD camera and parallel optical pipe[9-10], find out the best transfer function test point of TDI CCD camera, it can raise test precision of static transfer function, specific method is as follows:

a) The course of search is to adjust the location relation of target mark and TDI CCD focal plane. Guarantee target mark carve line and the horizontal direction of TDI CCD vertical, at the same time, it is parallel to again guarantee rectangle target mark surface and the TDI CCD focal plane;

b) Target mark revolving stage along X direction level to move, make the first image pixel of TDI CCD catch initial carve line of target marks image in X direction, in the course of moving along X direction, computer real time calculates and draws the transfer function change curve, the ideal transfer function change curve as Figure 3 shows, abscissa express the location of revolving stage, ordinate express the value of transfer function. Corresponding abscissa value of the curve highest point is the test point of the best transfer function on this direction. After testing, level revolving stage to move the test point of the best transfer function on X direction;

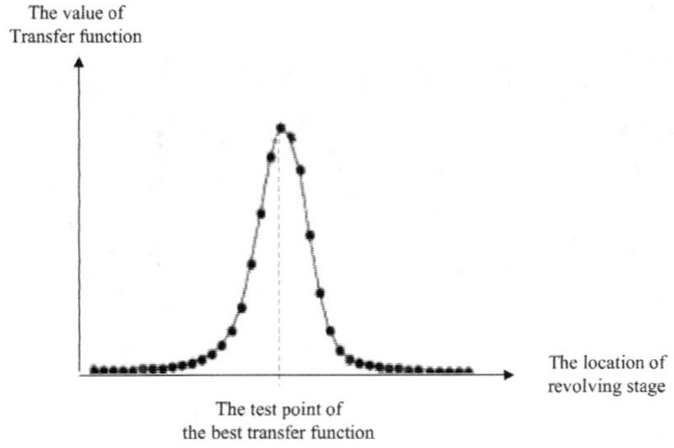

Figure 3. Transfer function change curve

c) Revolving stage is flat along Z direction to move, make TDI CCD in Z direction accurate alignment rectangle target mark, in the course of flat moving , computer real time calculates and draws the transfer function change curve of Z direction, after testing, level revolving stage to move the test point of the best transfer function on Z direction;

d) Revolving stage spins along θ_Y direction, make rectangle target mark carve line and the horizontal direction of TDI CCD vertical, in revolving course, computer real time calculates and draws the transfer function change curve, after testing revolving stage revolves to the test point of the best transfer function on θ_Y direction;

e) Revolving stage spins along θ_X direction and θ_Z direction, it is parallel to make camera focal plane and rectangle target mark surface, in revolving course, computer real time calculates and draws the transfer function change curve, after testing, revolving stage revolves to the test point of the best transfer function on θ_X direction and θ_Z direction;

f) The test point of the best transfer function of X direction, Z direction, θ_X direction, θ_Y direction, θ_Z direction are stored to the location information area. After establishing location information area, the test of static transfer function of TDI CCD camera can be carried out.

3.4 Calculate the static transfer function

Computer takes out the location information of revolving stage (the test point of the best transfer function) from location information area, and sends them to the control system of revolving stage. Revolving stage reaches the location intended, after confirmation, computer begins to calculate static transfer function. Revolving stage is beginning from the first pixel of TDI CCD, moves with regular step length along X direction. Computer calculates one time transfer function per move once. Fix step length is target mark image that occupies pixel number, and it is probably hundreds of pixel width.

The original target image that the collection of TDI CCD and the effective transfer function test area as Figure 4 show, image data in the frame of dotted line is effective transfer function test area. In Figure 4, per 1000 lines of target mark image as a measure sample, exclude because of atmosphere tremble twisted, vague rectangle target mark image (the image with low transfer function value), remain 500 no vague target mark images as measure sample set. Then, that calculates respectively static transfer function of each measure sample, takes the average of all measure samples in measure sample set as static transfer function of camera.

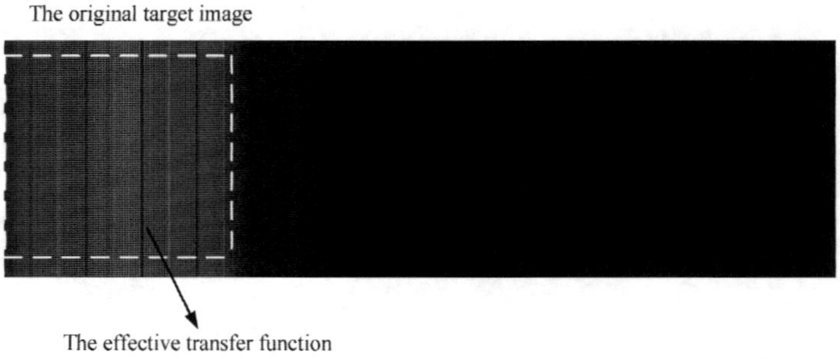

Figure 4. The original target image and the effective transfer function test area

The target mark image enlarged in effective transfer function test area as Figure 5 show, in the frame of dotted line to be adjacent target mark image, the first value of gray scale is low (dark stripe), the next value of gray scale is high (white stripe), through this two values of gray scale, transfer function can be calculated[11].

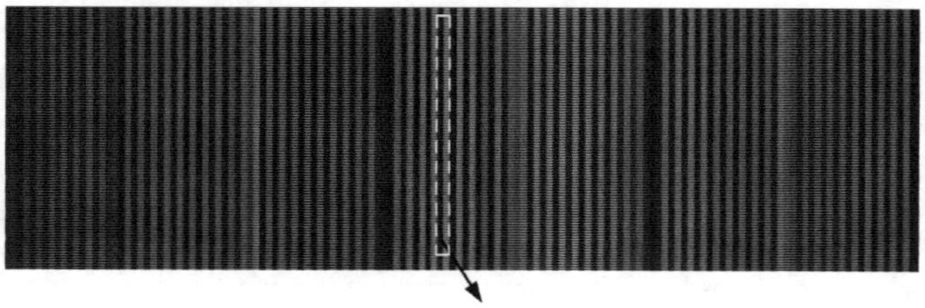

Figure 5. The target mark image enlarged in effective transfer function test area

The calculation formula of the average static transfer function for a marks image as Equation (1) show.

$$MTF_{avg} = \frac{\pi}{4} \times \frac{1}{mn} \times \sum_{i=0}^{m-1} \sum_{j=0}^{n-1} \frac{\left| I_{i,j} - I_{i+1,j} \right|}{I_{i,j} + I_{i+1,j} - 2I_{bg}} \quad , \qquad (1)$$

In Equation (1), MTF_{avg} expresses average static transfer function in effective transfer function test area in a target mark image. m and n express the row number and column number of target mark image in effective transfer function test area. $I_{i,j}$ expresses the gray value of target mark image in i row, j column, I_{bg} expresses the dark field value of target mark image.

According to average static transfer function MTF_{avg} calculated, exclude because of atmosphere tremble twisted, vague rectangle target mark image[12-13] (the MTF_{avg} is smaller than 0.1), select 500 no shaken mark image as measure sample set[14].

After measuring the static transfer function of the first step long, along X direction, move a step length, begin to test second step long and test method of static transfer function is identical, up to the final step length. After testing, the static transfer function test of entire visual field is completed. This kind of method has raised test accuracy and working efficiency.

4 Test result

In the course of searching for the test point of the best transfer function, revolving stage turns respectively along X direction, Z direction, θ_X direction, θ_Y direction, θ_Z direction, at the same time, computer real time records the value of static transfer function and draws the transfer function change curve. The highest point of the transfer function change curve is the test point of the best transfer function. After that, revolving stage is beginning from the first pixel of TDI CCD, moves with regular step length along X direction. Computer calculates one time transfer function per move once. Fix step length is target mark image that occupies pixel number, and it is six hundreds of pixel width. Because of each slice of TDI CCD effective pixel number is 2048, therefore the static transfer function that measures each slice of TDI CCD need to move 4 step lengths. In adjacent step the target mark image should have certain overlapping rate, so then can guarantee the continuity of the static transfer function in visual field of camera.

The test point of the best transfer function of revolving stage as table 1 shows.

Table 1. The test point of the best transfer function of revolving stage

	The direction of target mark revolving stage				
	Y direction	Z direction	θ_X direction	θ_Y direction	θ_Z direction
The 1st step	-179.75	-32.6	1.04	0.78	0.612
The 2nd step	-149.75	-22.1	1.076	0.78	0.576
The 3rd step	-119.75	-32.6	1.112	0.78	0.54
The 4th step	-89.75	-22.1	1.148	0.78	0.504

With same method, measure the static transfer function value of other step long of TDI CCD, and eventually draw the static transfer function curve of entire visual field of TDI CCD camera. The static transfer function curve of entire visual field tested using the before test method and test method of this paper as Figure 6 show , abscissa is expressed so as the visual field of TDI CCD camera with pixel partition, and ordinate expresses the static transfer function value. Dotted line expresses the static transfer function curve (drafted with discrete test points rear) tested using before method, the average of static transfer function is 0.2703. Solid line expresses the curve using test method of this paper, the average of static transfer function is 0.2923, and raised 0.02 compared with before test method.

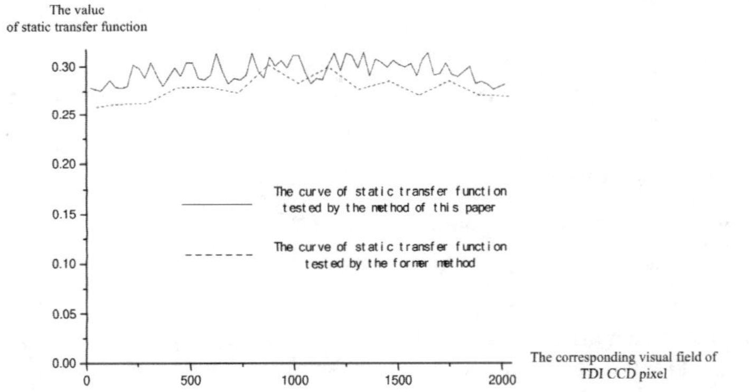

Figure 6. Static transfer function curve of the entire visual field tested using the method of the former and this paper

5 Conclusions

This paper has put forward a kind of accurate measurement technique of static transfer function, which applies to the TDI CCD camera. Computer real time analyzes the feature of target mark image, and draws the change curve of static transfer function, control revolving stage voluntarily to drive target mark to turn, search for the test point of the best transfer function. This measurement method solves the test error problem aroused by measure systematic space location and atmosphere shake in the course of testing static transfer function[15], raises the measure precision of static transfer function of TDI CCD camera. At the same time, this method has solved the test difficult, workload etc. problem of static transfer function, and has raised working efficiency. The average of static transfer function tested is 0.2923, and raised 0.02 compared with before test method.

References

[1] Li C H ,Wang Z X,Wu K Y. Optical assembly of CCD focal plane for space camera[J]. *Opt. Precision Eng*, 2000, 8 (3): 213-216

[2] Xu Q Y, Ye D, Che R S. On-line calibration of stereo vision measurement system based on optical reference bar[J]. *Acta Optica Sinica*, 2008, 28(1):81-86
[3] Zhou W. Study on Enhancing Dynamic Range of CCD Image Based on Digital Micro-Mirror Device[J]. *Acta Optica Sinica*, 2009, 29(3): 638-642
[4] Li H Z, Han C Y, Ma D M. Assessment Method of Modulation Transfer Function of On-Orbi t Space Optical Remote Sensor Using Neural Network [J] . *Acta Optica Sinica*, 2007, 27 (4): 631- 637
[5] Qian Y X, Liang W, Gao X D. Numerical Analysis of Dynamic Modulation Transfer Function for High-Resolution Aerial Camera[J]. *Acta Optica Sinica*, 2009, 29(1): 192-196
[6] Feng Z W,Cheng H B,Song Q,Shang Y Y. Modulation Transfer Function Measurement of Electron Multiplying CCD[J]. *Acta Optica Sinica*, 2008, 28(9): 1710-1716
[7] Guo Y, Liu X P, Yang H. MTF Analysis for CCD. *Spacecraft Recovery & Remote Sensing*, 2004, 25(3):25-28
[8] LAMBERT B M, HARBOLD J M. Experimental methods for measurement of the modulation transfer function (MTF) for time-delay-and-integrate (TDI) charge coupled device (CCD) image sensors. *SPIE*, 2009, 7405: 75-83
[9] Lu J, Chen W M , Cen J B. Effect of distance on modulation transfer function of imaging sensor using rectangle template contact method[J] . *Acta Optica Sinica*, 2006 , 26(7) :1021-1026
[10] Mao X F, Su X Y, Liu Y K, Chen W J. Analysis on Optical Coordinate Measurement Based on Phase Target[J]. *Acta Optica Sinica*, 2009, 29(9): 2452-2457
[11] Xue M Q, Shen W M. Optical Design of Light Weight and Compact High Resolution TDI CCD camera[M]. *Beijing: Science Publishing Company*, 2005 :158-163
[12] MAGNAN P, ESTRIBEAU M, ROLLAND G, *et al.*. Theoretical evaluation of MTF and charge collection efficiency in CCD and CMOS image sensors. *SPIE*, 2009, 7427: 23-34
[13] LI W J, QI C, ZHANG Y, *et al.*. Review on the theory and method of measuring MTF of CCD cameras. *SPIE*, 2005, 31:130-138
[14] Fei Y T. Error Theory and Data Processing[M] . *Beijing :China Machine Press* ,2007 :68-70.
[15] Han C Y. Study on optical system of high resolution space camera [J]. *Optics and Precision Engineering*, 2008, 16(11):2164-2172

Synthesis and properties of a new diarylethene having thiazole and biphenyl moieties

Hui Li, Shouzhi Pu*, Gang Liu, Weijun Liu

Jiangxi Key Laboratory of Organic Chemistry, Jiangxi Science and Technology Normal University, Nanchang 330013, P. R. China

pushouzhi@tsinghua.org.cn

Abstract. A novel photochromic diarylethene bearing thiazole and biphenyl moieties has been synthesized. Its properties, including photochromism and fluorescence were investigated in detail. It underwent reversible cyclization and cycloreversion reactions upon alternating irradiation with UV and visible light in hexane. The results showed that this compound exhibited reversible photochromism and good fluorescent switching in solution after irradiation with UV light and visible light.

1. Introduction

Due to photochromic diarylethenes have many excellent properties, notable thermally, irreversible photochromic behavior and remarkable fatigue resistance for instance [1, 2]. They were widespread applied to photonic devices such as high-density optical recording materials and photoswitches.[3–6] The design and synthesis of photochromic compounds has become an area of intense research. Among diarylethenes hitherto reported, the majority of the heteroaryl compounds bear thiophene or benzothiophene rings [7-12], with only a few reports concerning other five-membered heteroaryl moieties. In the case of six-membered, aryl ring moieties, only a few symmetrical diarylethene derivatives bearing two phenyl/naphthyl groups have been reported [13, 14], the majority of which are thermally reversible with poor photochromism. Recently, few publications concerned a new class of photochromic diarylethene derivatives bearing both five-membered and six-membered moieties reported by us [15-19]. They showed some novel characteristics different from diarylethenes containing five-membered aryl rings. In this study, we synthesized a new unsymmetrical photochromic diraylethene [1-(2, 4-dimethyl-5-thiazolyl)-2-(2-methyl-4-(3-trifluoromethylphenyl)-1-phenyl)]perflu orocyclopentene, Its photochromic reactivity , fluorescence were investigated in detail. The photochromic reaction of diarylethene 1a is shown in Scheme 1.

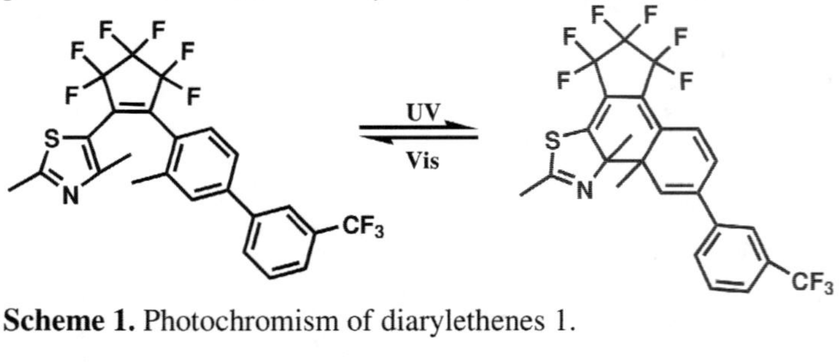

Scheme 1. Photochromism of diarylethenes 1.

2. Experiments

The synthetic route for diarylethene 1a is shown in Scheme 2. The structure of 1a was characterized by ^1H NMR spectroscopy. ^1H NMR (400 MHz, $CDCl_3$): 3.10 (s, 6H, $-CH_3$), 2.62 (m, 3H, $-CH_3$), 7.21 (s, 1H, Ar-H), 7.45 (s, 1H, Ar-H), 7.51-7.57 (m, 2H, Ar-H), 7.63 (d, 1H, J = 8.0 Hz, Ar-H), 7.76 (d, 1H, J = 8.0 Hz, Ar-H), 7.83 (s, 1H, Ar-H).

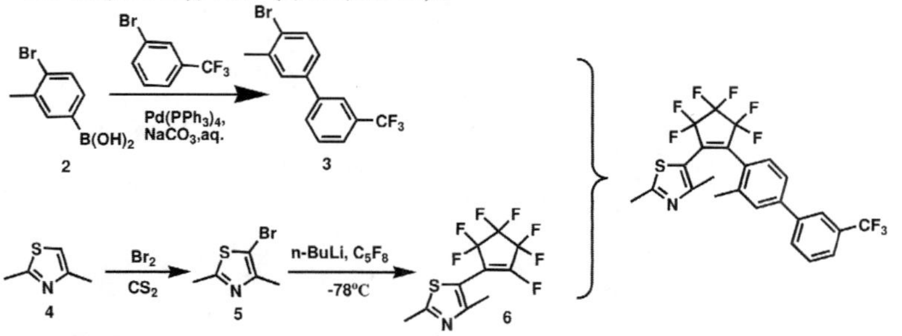

Scheme 2. Synthesis of diarylethene 1a.

3. Results and discussion

3.1. Photochromic behaviours

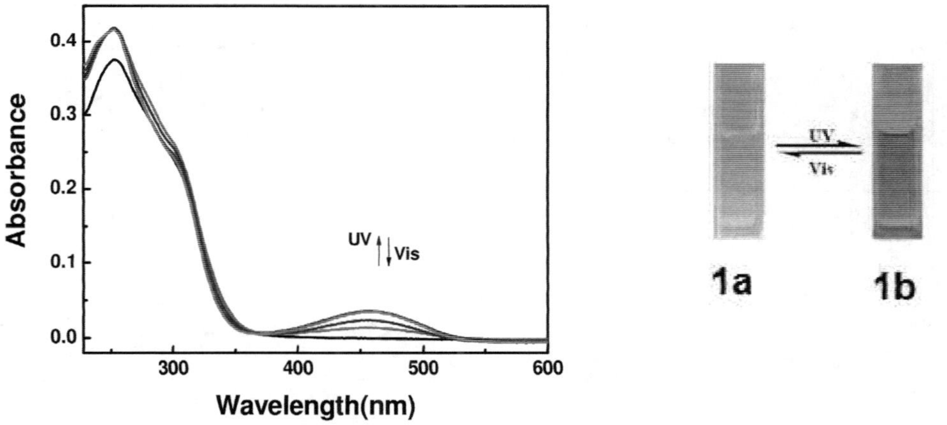

Figure 1. Absorption spectral changes and color change of compound 1 in hexane (2×10^{-5} mol/L).

Dithienylethene 1a underwent photochromism in hexane (2.0×10^{-5} mol/L) showed in Figure 1, the absorption spectral change and color change of 1a in hexane upon irradiation with 254 nm light. The colorless 1a has the absorption maxima at 253 nm (= 1.87×10^4 mol^{-1} L cm^{-1}). Upon irradiation with 254 nm light, the colorless solution of 1a turned yellow, in which absorption band was observed at 457 nm (= 1.80×10^3 mol^{-1} L cm^{-1}). The yellow color is due to the closed ring form 1b. The color disappeared by irradiation with visible light.

3.2. Fluorescence of diarylethene

The fluorescence properties in the hexane of the title compound (C = 1.0×10^{-4} mol/L) were measured using a Hitachi F-4500 spectrophotometer, and the breadths of excitation and emission slit were selected 10.0 nm. The hexane of diarylethene 1 exhibited relatively strong fluorescence at 453 nm when excited at 309 nm. The fluorescence intensity of diarylethene 1 decreased dramatically along with the photochromism from open-ring isomer to closed-ring isomer upon irradiation with 297 nm light in hexane. As shown in Figure 2, when irradiated by light of 297 nm, the photocyclization reaction was carried out and the low fluorescent closed-ring form of the compound was produced. The emission intensity of diarylethene 1 in a photostationary state was quenched to ca. 52%. The back irradiation by appropriate wavelength visible light regenerated the open-ring forms of diarylethene 1a and recovered the original emission spectra. The phenomena are useful for application as the fluorescence switches [20, 21].

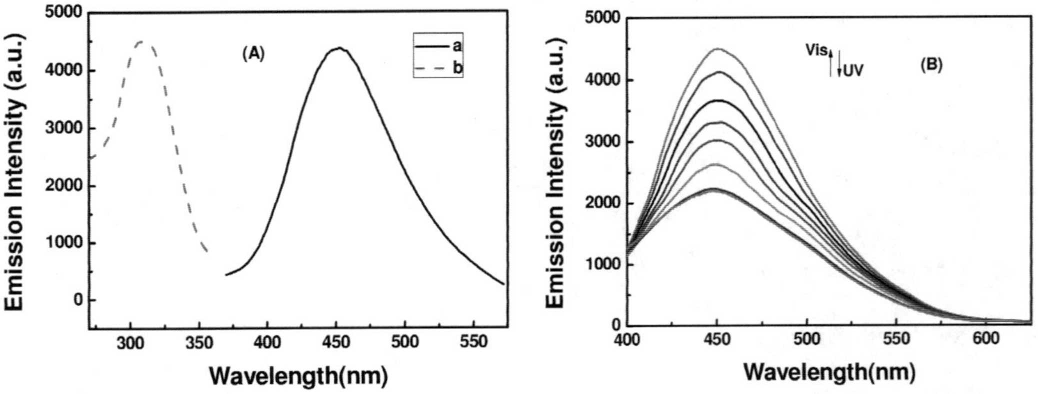

Figure 2. Fluorescence spectra (a) and Excitation spectra (b) of 1a (A), fluorescence spectral changes of (1.0×10^{-4} mol/L) in hexane upon irradiation with 297 nm UV light (B).

The solvent effect on the fluorescence spectra of diarylethene 1a was also investigated. Fluorescence spectra of 1a in different solvents at room temperature are shown as in Figure 3, and the fluorescent properties of the compound in various solvents at room temperature are summarized in Table 1. From these data, we can see that the fluorescence spectra depended on the polarity of the solvent. It gave a systematic red shift when the polarity of the solvent increased. And the relative fluorescence intensity increased with increasing the polarity of the solvent.

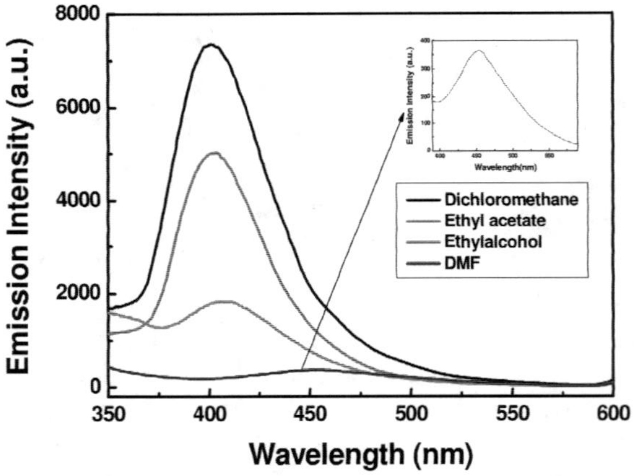

Figure 3. Fluorescence spectra of diarylethene 1a in various solvents in 1.0×10^{-4} mol/L at room temperature.

Table 1. Fluorescent properties of diarylethene 1a in various solutions at 1.0×10^{-4} mol/L.

Solvent	, max (nm)
Dichloromethane	401(7346)
Ethyl acetate	403(5023)
Ethylalcohol	411(1832)
DMF	453(364)

In conclusion, a new unsymmetrical dithienylethene was synthesized to investigate its photochromism, fluorescence and Polarization optical recording. The results showed that dithienylethene 1 exhibited reversible photochromism in hexane. In hexane, the open-ring isomer of diarylethene 1 exhibited relatively strong fluorescence at 453 nm when excited at 309 nm. The fluorescence intensity decreased along with the photochromism upon irradiation with 297 nm light. The fluorescence spectra of diarylethene 1a depended on the concentration and the polarity of the solvent.

Acknowledgments

This work was supported by the Youth Science Funds of the Education Office of Jiangxi, China (GJJ09572), and Graduated Research and Innovation Fund of Jiangxi Province.

References

[1] Irie M 2000 *Chem. Rev.* **100** 1685
[2] Tian H and Yang S J 2004 *Chem. Soc. Rev.* **33** 85
[3] Dürr H and Bouas-Laurent H 1990 *Photochromism: Molecules and System*
[4] Brown G H 1971 *Photochromism*
[5] Fukaminato T, Sasaki T, Kawai T, Tamai N and Irie M 2004 *J. Am. Chem. Soc.* **126** 14843
[6] Pu S Z, Zhang F S, Xu J K, Shen L, Xiao Q and Chen B 2006 *Mater. Lett.* **60** 485
[7] Higashiguchi K, Matsuda K, Tanifuji N and Irie M 2005 *J. Am. Chem. Soc.* **127** 8922
[8] Frigoli M, Welch C and Mehl G H 2004 *J. Am. Chem. Soc.* **126** 15382
[9] Moriyama Y, Matsuda K, Tanifuji N, Irie S and Irie M 2005 *Org. Lett.* **7** 3315
[10] Tanifuji N, Matsuda K and Irie M 2005 *Org. Lett.* **7** 3777
[11] Jeong Y C, Yang S I, Kim E and Ahn K H 2006 *Tetrahedron* **62** 5855
[12] Pu S Z, Li M, Fan C B, Liu G and Shen L 2009 *Journal of Molecular Structure* **919**, 100
[13] Takeshita M, Ogawa M, Keisuke M and Yamato T 2003 *J. Phys. Org. Chem.* **16** 148
[14] Uchida K, Nakamura S and Irie M 1995 *Research on Chemical Intermediates* **21** 861
[15] Pu S Z, Fan C B, Miao W J and Liu G 2008 *Tetrahedron* **64** 9464
[16] Pu S Z and Fan C B 2008 *Chinese Patent.* CNP101134753
[17] Pu S Z, Li H, Liu G and Liu W J 2010 *Tetrahedron Lett.* **51** 3575
[18] Li H, Pu S Z, Liu G and Liu W J 2009 *Proc. of SPIE* **7517** 751704-1
[19] Pu S Z, Miao W J, Cui S Q, Liu G and Liu W J 2010 *Dyes and Pigments* **87** 257
[20] Chen B Z, Wang M Z and Tian H 2002 *Chem. Commun.* 1060
[21] Tian H, Chen B Z and Tu H Y 2002 *Adv. Mater.* **14** 918

Influence of Atmospheric Propagation on Performance of Laser Active Imaging System

LI Ying-chun, ZHAO Yun [1], SUN Hua-yan, GUO Hui-chao [1]

Department of Optical and Electrical Equipment, the Academy of Equipment Command& Technology, Beijing 101416, China

1 Company of Postgraduate Management, the Academy of Equipment Command& Technology, Beijing 101416, China

E-mail: zhaoyunhappy@126.com

Abstract. Atmospheric propagation has serious influence on the performance of a good designed laser active imaging system. Atmospheric attenuation and turbulence are two main effects on laser atmospheric propagation. Imaging SNR (Signal-Noise-Ratio) and resolution are two key indexes to describe the performance of a laser active imaging system. Establishing the relation between system performance index and atmospheric propagation effect is significant. The paper analyzed the relation between imaging performance and atmospheric attenuation and turbulence through simulation. And also the experiments were done under different weather to validate the conclusion of simulation.

Key words: Laser active imaging system, Atmospheric propagation, Imaging performance

1. Introduction

Because of scattering and absorption of atmospheric aerosols and atmospheric elements, and also haze, fog, snow and rain, laser energy declines a lot through atmosphere. Atmospheric turbulence causes the laser beam deflection, jitter, intensity fluctuations (blinking) and other effects, so that laser beam quality gets worse[1]. All this make the performance of laser active imaging system worse. Laser image can reflect how atmosphere propagation influenced. Studying the relation between image quality and atmosphere was important. In this paper, the relation between atmospheric visibility and turbulence and image quality was studied by simulation. Experiments were done under different weather to validate the conclusion of simulation.

2. Laser Atmospheric Propagation

2.1. Laser Atmospheric attenuation

Horizontal propagation atmospheric attenuation can be described by transmittance as follow:

$$\tau(\lambda) = \exp\left[-\mu(\lambda)\cdot L\right] \tag{1}$$

[1] ZHAO Yun. Adress: Beijing Huairou 3380Mail-box NO.174, Huairou, Beijing. Post Code:101416

Here L is distance, $\mu(\lambda)$ is attenuation coefficient($\mu(\lambda) = \alpha(\lambda) + \beta(\lambda)$, $\alpha(\lambda)$ and $\beta(\lambda)$ are absorption coefficient and scattering coefficient.

In application, atmospheric transmittance can be calculated through experiential formula which described by visibility. Visibility was defined as follows

$$V_M = \frac{3.912}{\beta} \tag{2}$$

β is total extinction coefficient in 0.55 μm wavelength. The relation between Attenuation coefficient and V_M is as follow:

$$u = \frac{3.912}{V_M}\left[\frac{0.55}{\lambda}\right]^b \tag{3}$$

Here V_M is visibility, λ is wavelength, and

$$b = \begin{cases} 1.6 & (V_M \geq 50\text{km}) \\ 1.3 & (6\text{km} \leq V_M < 50\text{km}) \\ 0.16V_M + 0.34 & (1\text{km} \leq V_M < 6\text{km}) \\ V_M - 0.5 & (0.5\text{km} \leq V_M < 1\text{km}) \\ 0 & (V_M \leq 0.5\text{km}) \end{cases} \tag{4}$$

We can use this formula when $\lambda = 0.35 \sim 1.54\mu m$.

For computational purposes, we suppose target distance is 3km and 5km, the relation between transmittance and V_M is as follow:

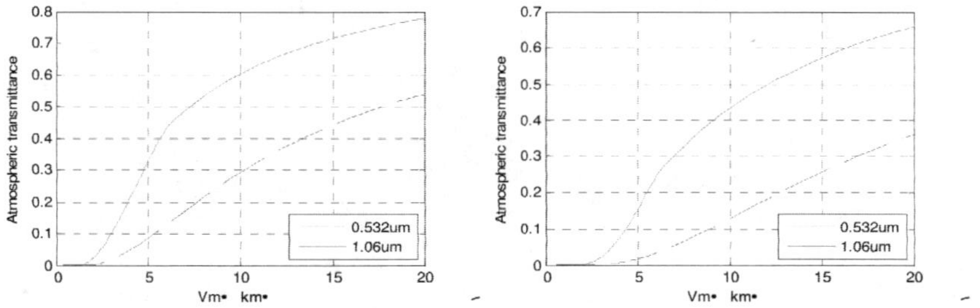

Figure-1.-Transmittance as a function of V_M for 3km and 5km

2.2. *Atmospheric turbulence*

Turbulence causes atmospheric refractive index stochastic changes. The value of change is 10^{-6}, and turbulence has a big influence on laser far distance propagation. Turbulence can be described by Kolmogorov model as usual. In local even area, atmospheric refractive index structure function $D_n(r)$ is as follow:

$$D_n(r) = C_n^2 r^{\frac{2}{3}} \qquad (l_0 << r << L_0) \tag{5}$$

r is the distance of two statistical points, l_0、L_0 is inner scale and outer scale, C_n^2 is the refractive index structure parameter, which is essential parameter of atmospheric optics. It can show turbulence intensity. The value of C_n^2 is little, and turbulence intensity is weak. The value of C_n^2 is changing with place, altitude, weather and season, etc.

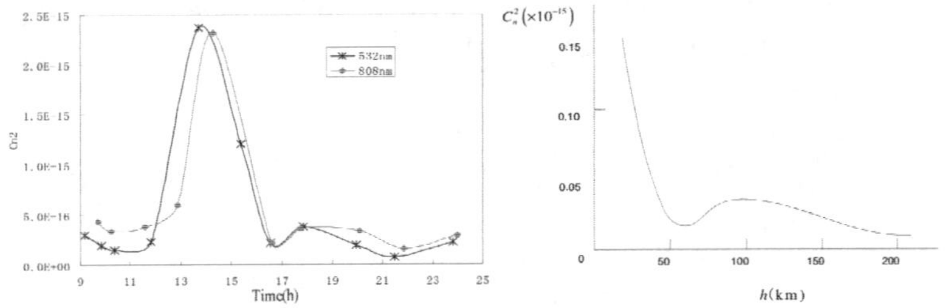

Figure 2. The value of C_n^2 is changing with time and altitude

Laser propagation through turbulence atmosphere, the intensity, phase, propagation direction, etc, get random changes. Here simulated the laser fluctuation in different turbulence using phase screen[2]. The propagation distance is 10km.

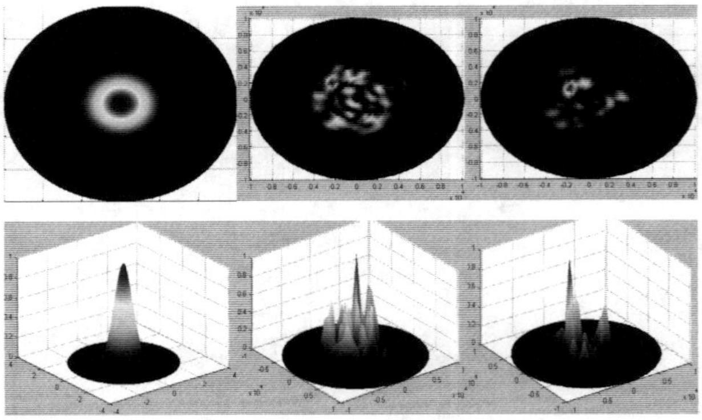

(a) (b) (c)

Figure 3. Laser shape and intensity through different turbulence ((a) Original, (b) $C_n^2=10^{-15}\ m^{-2/3}$, (c) $C_n^2=10^{-12}\ m^{-2/3}$)

3. Influence of laser atmospheric propagation on the performance of imaging

3.1. The decline of SNR

Imaging SNR is an important index to judge a system. SNR is defined as follow

$$SNR=\frac{Q_{target,pix}}{Q_{backgr,pix}+Q_{bsc,pix}+Q_{noise,pix}} \tag{6}$$

Here, $Q_{target,pix}$、$Q_{backgr,pix}$、$Q_{bsc,pix}$ are signal energy, background radiation energy and backscattered energy of every pixel. $Q_{noise,pix}$ is the equivalent noise energy of ICCD.

Based on the theory of laser active imaging, SNR can be further described as follow[3]

$$SNR=\frac{P_t\Delta\tau\cdot T_rT_e\frac{\rho}{\pi}\frac{\exp(-2\sigma_{ext}R)}{R^2}}{L_\lambda\Delta\lambda T_r\frac{\rho_b}{\pi}\Omega_r\Delta\tau+P_t t_p T_r T_e\frac{\sigma_{ext}}{8\pi}\int_{R1}^{R2}\frac{\exp(-2\sigma_{ext}R)}{R^2}dR+\frac{Q_{noise,pix}\times f^2N_p^2}{G\cdot A_r}} \tag{7}$$

Here P_t is laser emitting power, T_e 、T_r are emitting and receiving optical system transmittance, ρ is target diffuse reflection coefficient, $\sigma_{ext}(R)$ is total attenuation coefficient, R is distance, L_λ is background spectrum radiation brightness, $\Delta\lambda$ is optical filter bandwidth, Ω_r receiving system solid angle, ρ_b is background average reflection coefficient, R_1 and R_2 are gate range, A_r is entrance pupil area, f is focus, N_p is ICCD resolution, G is ICCD gain.

From (7), signal energy and backscattered energy lose because of attenuation. At night, f=800mm, Np=60lp/mm, $Q_{noise,pix}=6\times10^{-20}$ J/pix, R=5km, SNR is a function of V_M in different laser emitting power, graph as follow

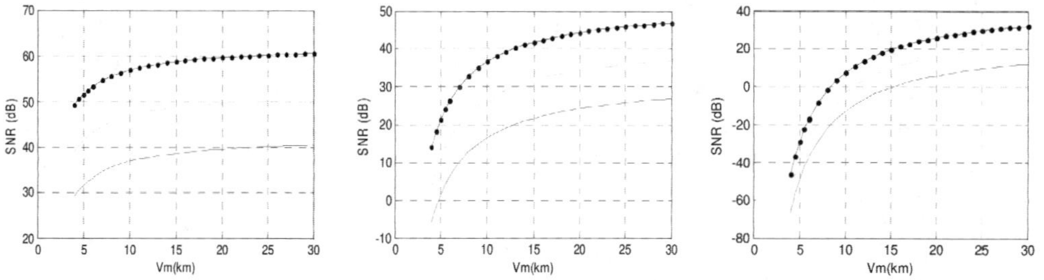

Figure 4. Relation between SNR and V_M(From left to right, imaging distance are 1.2km, 4km,10km. Power of laser are 0.1kw,1kw,10kw)

From the simulation, V_M<5km, imaging quality is bad, even imaging system cannot work. With visibility getting better, SNR is getting better. When V_M>20km, current of improving draws bit. Another, when laser power decuple, SNR improves about 10 dB.

3.2. Imaging resolution declines through turbulence

We use MTF to judge the resolution decline when studying laser active imaging system resolution. Without considering system itself, because the time of range-gated is short, imaging integral time is less than the time of atmospheric fluctuation when imaging in atmosphere. Intensity and phase are both important when Far-field propagation. Based on the studying of Fried, MTF_{tur} of Image blur which caused by atmospheric turbulence in short exposure is described as follow[4]:

$$MTF_{tur}(F)=\exp\{-3.44(\frac{\lambda f}{r}F)^{5/3}[1-0.5(\frac{\lambda f}{D}F)^{1/3}]\} \quad (8)$$

Here F is space frequency (*lp/mm*), λ is wavelength, f, D are focus and diameter of receiving optical system, r is atmospheric coherent length, described as follow

$$r=2.1\rho_0=2.1(1.46k^2C_n^2R)^{-3/5} \quad (9)$$

Where $k=2\pi/\lambda$, R is distance.

The strength of turbulence was classified in the following way[5]: Weak turbulence: $1.0\times10^{-15}(m^{-2/3})$, Medium turbulence: $1.0\times10^{-14}(m^{-2/3})$, Strong turbulence: $1.0\times10^{-13}(m^{-2/3})$. We simulate the graph of MTF_{tur} through three turbulence above. λ=0.532um, f=800mm,D=200mm, target distance R=5km.

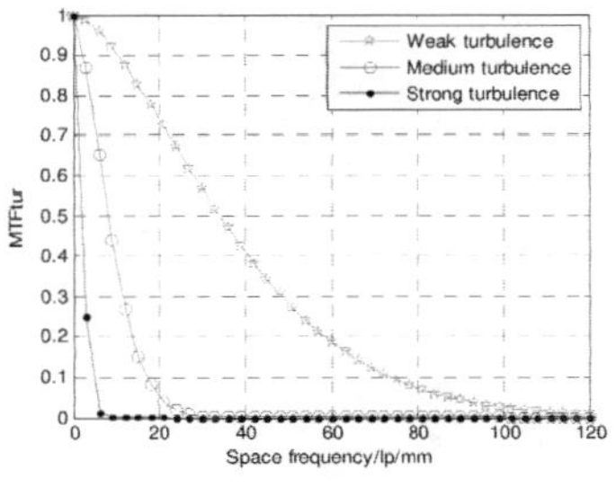

Figure 5. Influence on imaging resolution through three classes turbulence

It is evident that influence of atmospheric turbulence on imaging resolution is strong. For high frequency target, it is difficult to distinguish details from laser imaging through medium and strong turbulence. For low frequency target, Image resolution declines acutely. Imaging quality through medium turbulence is about 60% less than that weak turbulence. Imaging quality through strong turbulence is about 90% less than that weak turbulence.

4. Experiment and analysis

4.1. Calculation of image SNR

Based on the character of laser active imaging, we use local area standard deviation[6] to calculate image SNR. For digital image, signal intensity and image gray are linear relation in ICCD response range, so we could use gray to replace intensity to calculate. The ratio of average gray value and local area standard deviation is image SNR, steps as follow:

(1) Calculate the target signal area average gray value(M) of noise image;
(2) Divide image according 4×4, calculate every standard deviation of 4×4 image(*LSD*);
(3) Calculate average value of all standard deviation of 4×4 images(LSD_m);
(4) $SNR = 20\times\log_{10}\left(\frac{M}{LSD_m}\right)$.

4.2. Imaging experiment

The imaging system is designed and carried out by ourselves. It contains detector, lasers, optical system and control unit. Type of detector is 1XC18/18WHS-G range-gated ICCD. The spectrum response range is 450nm~900nm. Its resolution is more than ≥500TV-line. The least gate-time is 20ns. The laser we use to do far-distance imaging is Nd:YAG laser. Wavelength is 532nm and 1064nm. Single pulse energy is 80mJ. Pulse width is 10ns. The focus and diameter of receive optical system are 1800mm and 150mm.

Laser active imaging experiments were done at different time in different day. The target distance is about 4km. The values are in table 1. The data of visibility are from weather record web.

(a) (b) (c) (d) (e) (f)

Figure 6. Laser images of 3.7km distance target

Then, we calculate SNR of every image using the method in 3.1. The values are in table 1.

Table 1. The values of different images in Figure 5

Image	Time	humidity	Visibility/km	SNR
a	21:00~21:30,5th Aug	40%	30	17.1328
b	20:30~21:00,4th Aug	80%	25	14.8788
c	20:30~21:00,21th Jul	70%	20	11.6972
d	22:30~23:00,21th Jul	78%	8	10.9255

e	20:30~21:00,2nd Aug	50%	5	9.0718
f	21:00~21:30,22th Jul	79%	3	No image

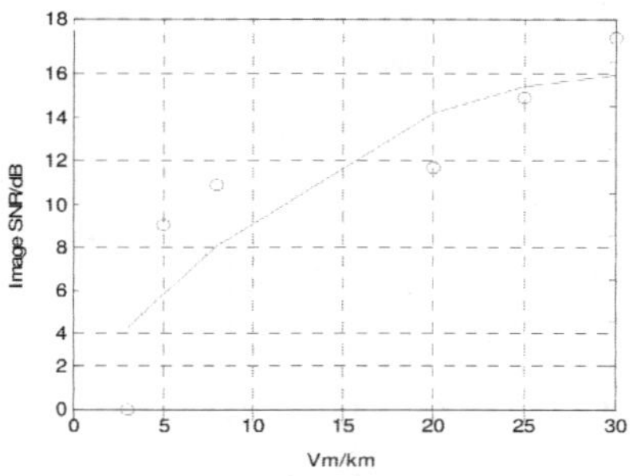

Figure 7. Graph of experimental images SNR

From the calculation results, visibility is so small that the laser energy declines too much because of atmospheric attenuation. So we can't see target through laser active imaging system. With visibility get far, we can see the target. And SNR of image get bigger. SNR of image in 30km visibility improves 36% more than in 8km. The result is close to the simulation result of 30%. Good visibility image has better details also. Because of different humidity and other reasons, experiment results have difference from theory results.

Then, we use blind deconvolution way[7] to calculate atmospheric MTF. From the calculation result of image (a)(b)(c)(d), we could qualitative see that value of MTF is getting small with Visibility getting worse. The imaging resolution is getting worse.

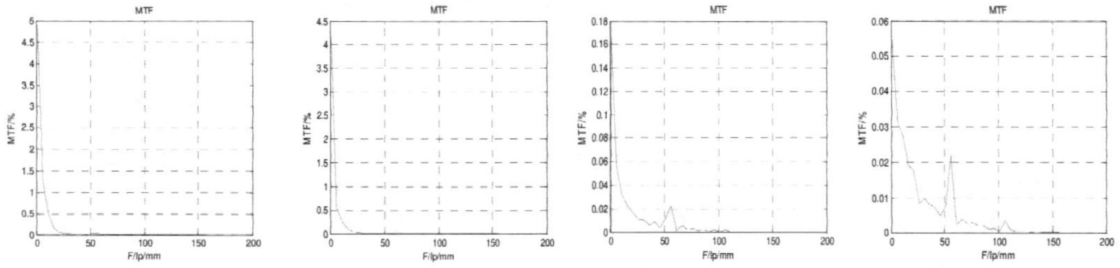

(a) (b) (c) (d)

Figure 8. **MTF of image** (a)(b)(c)(d)

5. Conclusion

The influence on imaging quality of atmospheric propagation is complex. This paper analyzed that how imaging SNR and resolution declines with visibility getting worse and turbulence getting strong by simulation way. And we did experiments to analyze the influence.

The performance of laser active imaging system in sunshine day (V_M>20km) can improve 20%~30% more than in light fog day (V_M is about 10km). If laser energy is not enough, the system can't see far target in strong fog day (V_M<5km). And power of laser decuple, SNR improves about 10dB. Atmospheric turbulence has strong influence on imaging resolution. It can make target image detail blurred. Imaging quality through strong turbulence is about 90% less than that weak turbulence.

6. References

[1] YIN Jun-yan, YIN Fu-chang, CHEN Ming, YU Hao, "Impact on laser transmission in atmosphere"[J], Infrared and Laser Engineering, 2008, Vol.37 Supplement:399-401.

[2] Christopher J. Subich, Stephen Watson, Cynthia Y. Young, **Simulation of Wave Structure Function Through Atmospheric Propagation** [J], SPIE 2005 Vol. 5793: 231-236.
[3] PANG Chun-ying, ZHANG Tao, "Signal-to-noise ratio model of laser active imaging system" [J], Optics and Precision Engineering, 2008, Vol 16 NO 2: 319-323.
[4] YANG Rui-ke, LIU Qi, MA Chun-lin, "Analysis of the influence of atmospheric turbulence on lidar direct detection"[J], Infrared and Laser Engineering, 2008, Vol.37 Supplement: 137-140.
[5] Karin R. Weiss-Wrana, "Influence of atmospheric turbulence on imaging quality of electro-optical sensors in different climates"[J], SPIE 2004 Vol. 5237 1-12.
[6] Dai De-de, "The study of appraising method on performance of laser active imaging system"[D], the Academy of Equipment Command& Technology, 2009: 36-38.
[7] Hong Hanyu, "Research on Image Restoration Algorithms in Imaging Detection System"[D], Huazhong University of Science and Technology,2004: 91-103.

Journal of Physics: Conference Series **276** (2011) 012134 doi:10.1088/1742-6596/276/1/012134

Realization of 16-channel digital PGC demodulator for fiber laser sensor array

Lin Wang, Jun He, Fang Li, and Yu-Liang Liu

(Optoelectronics System Laboratory, Institute of Semiconductors, Chinese Academy of Sciences, P.O. Box 912, Beijing 100083, P.R. China)

E-mail:wlcas@semi.ac.cn

Abstract. This paper describes a 16-element DFB FL (distributed feedback fiber laser) sensor array system interrogated by NI-PXI-based (PCI extensions for Instrumentation) digital PGC (phase generated carrier) technique. The lasing wavelengths of the DFB FLs are changed by the external strains or temperatures, and hence they can be used as sensors by detecting the wavelength shifts. An unbalanced MI (Michelson interferometer) is employed in the sensor array system to amplify the wavelength shifts of DFB FL sensors to detectable phase shifts. The output phase signals of the MI are separated into different channels by a DWDM, and then detected by a low-noise photodiode array. The digital PGC algorithm is realized on a PXI platform (NI, National Instruments), which consists of three FPGA (Field Programmable Gate Array) modules and a high performance system controller. The normalization of the interference fringe is proposed and realized in this paper to reduce the influence of the light intensity fluctuations, and a trigger mechanism is introduced into the digital multi-channel PGC demodulation scheme to synchronize the date among different channels. A 16-element DFB fiber laser sensor array system has been set up in the experiment and the demodulated results have demonstrated a minimum detectable wavelength shift of 1×10^{-6} pm/√Hz, a linearity of as high as 0.9994, and a dynamic range of 110dB@100Hz.

1. Introduction

With the development of optical fiber sensor techniques, fiber sensor networks have been widely studied for the applications in the fields of petroleum seismic exploration, earthquake prediction, and security. The DFB FL sensor has attracted considerable research interests for its many unique advantages, including small size, light weight, high sensitivity, and especially the intrinsic ability to be wavelength division multiplexed along a single fiber [1]. In 1995, K P Koo and A D Kersey achieved a two fiber laser sensors array system using phase-locked demodulating technique [2]. In 2007, a four-element fiber laser hydrophone array was reported with the demodulation by a 3×3 coupler [3].

Among the various demodulation schemes adopted in optical fiber sensor system, the PGC technique has been widely used for its advantages of high resolution, large dynamic range and high stability. However, the application of the traditional analog-circuit-based PGC technique is limited by the characteristics of the electronic device, and it has the disadvantages as the difficulty of changing parameters, the sensitivity to environmental noise, and the instability. Compared with the analog-PGC, the digital PGC technique is much easier to be realized, and it is insensitive to the environmental interference, so it is more suitable for demodulating large scale WDM fiber laser sensors array. But on the other hand, the digital PGC algorithm needs relatively complex calculations to recover the phase signals [4]. It requires a high performance demodulation system to carry out the great amount of real-time calculations. In fact, it is difficult to realize the PGC algorithm on a personal computer for the simultaneous demodulation for a DFB FL sensor array system.

The NI-PXI system is a commercialized digital signal processing system produced by the National Instrument Corporation. It is based on the FPGA modules and a powerful system controller, which provides high performance and reliability. A 16-element fiber laser sensor array system is set up using the digital PGC demodulation technique based on the NI-PXI system and corresponding LabVIEW software in this paper. The system has achieved a high performance demodulation of 16-channel PGC signals in a real-time manner. And the simultaneous demodulation of the 16-element DFB fiber laser sensor array system with a high wavelength resolution is also confirmed by the experiments in this paper.

2. Principle

2.1. The principles of PGC algorithm

The lasing wavelength of a DFB FL sensor will be changed by the external strains or temperatures. An unbalance Michelson interferometer is employed to amplify the wavelength shift of DFB FL sensor into the detectable phase shift of the MI [5].

$$\Delta\varphi = -\frac{2\pi n \Delta L}{\lambda^2}\Delta\lambda \tag{1}$$

where $\Delta\varphi$ is the phase shift of the MI, $\Delta\lambda$ is the wavelength shift of laser, n is the refractive index of fiber, and ΔL is the path difference of the two arms of the unbalance MI. In the PGC technique a large amplitude phase modulation carrier at a higher frequency out of the signal band is introduced into one arm of the interferometer, and hence eliminates the signal-fading caused by the initial phase drifts of the MI. The output signal of the MI can be expressed as:

$$I = A + B\cos\left(C\cos\omega_c t + \varphi_s(t) + \varphi_{env}\right) \tag{2}$$

where the constant A and B are proportional to the incident light intensity, B also correlates with the visibility of the interferometric signal, C is the modulation depth, ω_c is the angular frequency of the carrier, $\varphi_s(t)$ is the signal of interest, and φ_{env} is the initial phase of the MI with the external disturbance.

The output phase signal of the MI is mixed with $G\cos\omega_c t$ and $H\cos(2\omega_c)t$ to produce two channels of signals, respectively. After the following operations as shown in figure 1, including LPF (low-pass filtering), differential, cross-multiplying, integrating, and HPF (high-pass filtering), the recovered sensing signal of interest can be obtained as:

$$B^2 GHJ_1(C)J_2(C)\varphi(t) \tag{3}$$

where $J_1(C)$ and $J_2(C)$ are Bessel functions.

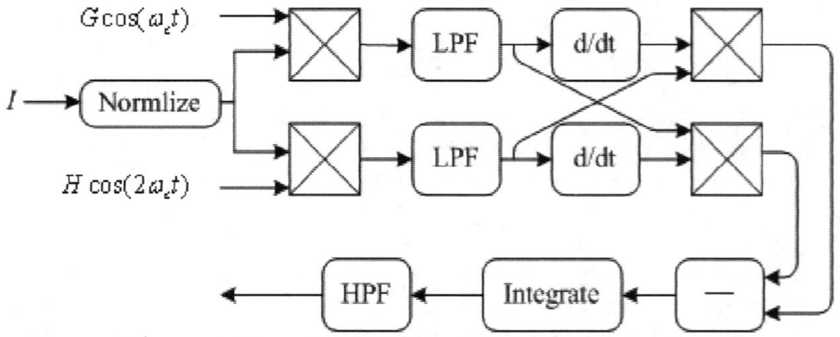

Figure 1. The schematic algorithm of PGC technique.

Because the final demodulated result is proportional to B^2. B is decided by the light intensity and the visibility of the MI, the recovered signal will change with the fluctuation of the light and external disturbance. So the system cannot get the desired signal of interest. To eliminate the influence of the variations of light intensity, the peak-to-peak normalization is introduced into the digital PGC algorithm. The basic principle of the normalization is shown as follows:

$$y = \frac{x - x_{Min}}{x_{Max} - x_{Min}} \tag{4}$$

where y is the result of the normalization of x, x_{Max} and x_{Min} is the maximal value and the minimum value of the sampled date. By adopting the normalization scheme, the demodulation result can be expressed as $GHJ_1(C)J_2(C)\varphi(t)$, so the influence of light intensity has been eliminated.

2.2. The realization of PGC on NI-PXI

PXI is a modular instrumentation platform originally introduced in 1997 by National Instruments as a basis for building electronic test equipment or automation systems. And the platform always combines a signal system controller and various kinds of modules to set up a real measurement system. On account that the PC-based (personal computer) PGC demodulation scheme can not afford the great amount of calculations, three high performance FPGA modules are selected to build up the PXI platform to realize the simultaneous PGC demodulation for the 16-element DFB fiber laser sensor array system in this paper. The PGC algorithm is mainly programmed on the FPGA except for the band-pass filtering on the duo-core system controller, as shown in figure 2. And the programme is achieved by a graphical development environment: NI LabVIEW, which can help to save developing time.

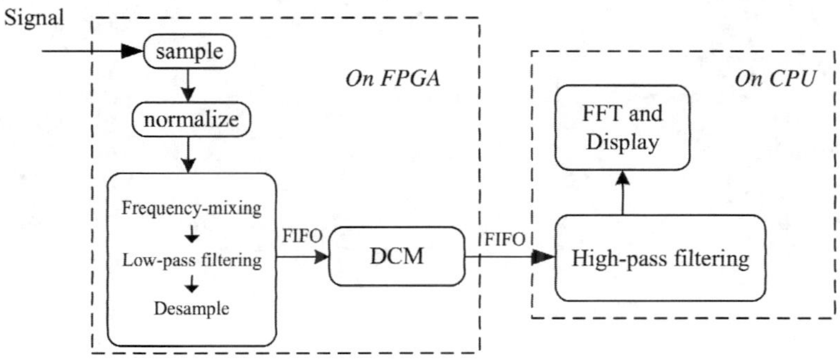

Figure 2. The structure of the program of PGC algorithm realized on the FPGA modules and the system controller.

In order to realize the PGC technique on FPGA, three key techniques have been adopted in the program. Firstly, the program should be developed as parallel execution to make sure that the PGC technique could be carried out simultaneously for different channels, and the PXI could demodulate the signals from the multiplexed DFB laser sensor system. Secondly, the algorithm is processed digitally one point by one point on the FPGA module. It means that the program will not allow new samples until the processing of a former point has been finished. Therefore, the PGC technique should be separated into several sections, as shown in figure 2, so as to ensure a high sampling rate [6]. At last, a trigger mechanism, as shown in figure 3, must be employed in the program for the synchronization between different channels to ensure the validity of the demodulated results. When the demodulation system starts to work, a trigger signal is sent by a master FPGA module at first, and then the other two FPGA modules receive the trigger, which is an order as to begin to sample. Such a trigger mechanism can assure the synchronization of all the FPGA modules.

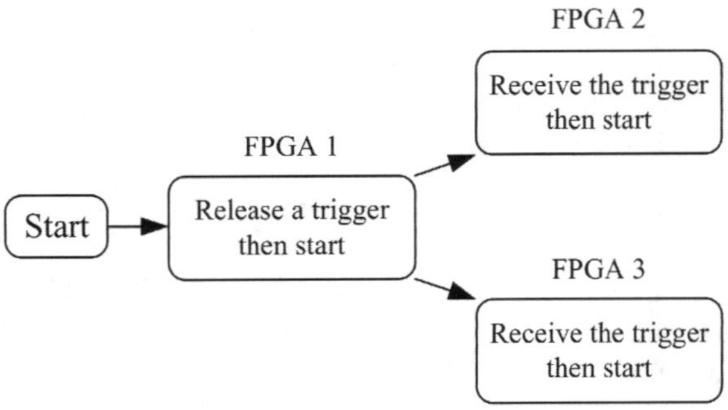

Figure 3. The trigger mechanism adopted in the digital multi-channel PGC demodulation scheme.

The program on the system controller is also completed using LabVIEW. Its main function is to carry out the process of band-pass filtering and the fast Fourier transform (FFT), and to display the demodulated results both in the time domain and in the frequency domain. The interface of the program is shown in figure 4.

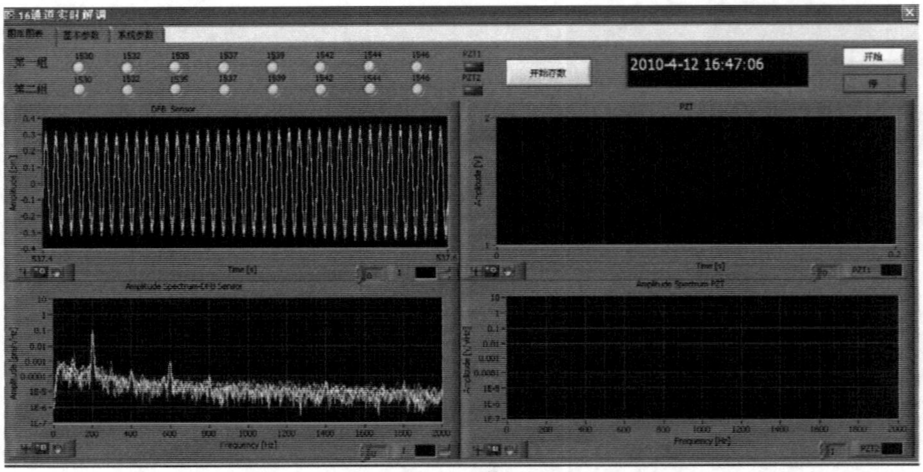

Figure 4. The interface of the demodulation program on the system controller.

3. Experiment and result

The schematic diagram of the proposed 16-element DFB fiber laser sensor array system is shown in figure 5.

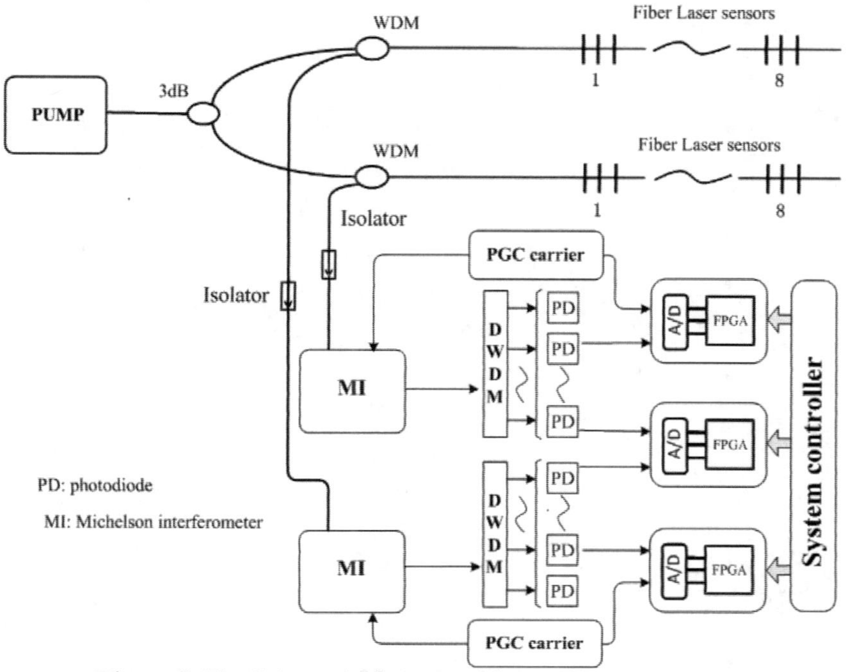

Figure 5. The Schematic diagram of the proposed 16-element DFB FL sensors array system.

Sixteen DFB fiber lasers were pumped by a 980nm semiconductor laser. They were stuck parallelly on a rectangular piezoelectric transducer (PZT). A sinusoidal test signal was given to the sensors by electrical modulating the PZT. The path difference of the MI was 5 m. Two faraday rotator mirrors were incorporated at the end of the MI arms to make the interferometer polarization insensitive. The output phase signals of the MI were separated into different wavelength channels by DWDM, and the detected analog electronic signals were converted into digital signals. Then the PGC algorithm was processed on the three I/O FPGA modules. The sampling rate of A/D converter is 100 kS/s and the frequency of the carrier is 10 kHz. The test signal is 100Hz and the demodulated results of sixteen sensors are demonstrated in time domain as shown in figure 6. The minimum detectable wavelength shift of the system is determined by the noise level, which is 1×10^{-6} pm/√Hz, as shown in figure 7. The linearity between the test signal and the demodulated results can reach as high as 0.9994 at each test frequency. as the linearity at 100Hz shown in figure 8. A dynamic range of 110dB@100Hz has been achieved by the PGC demodulation scheme in this system, as shown in figure 9.

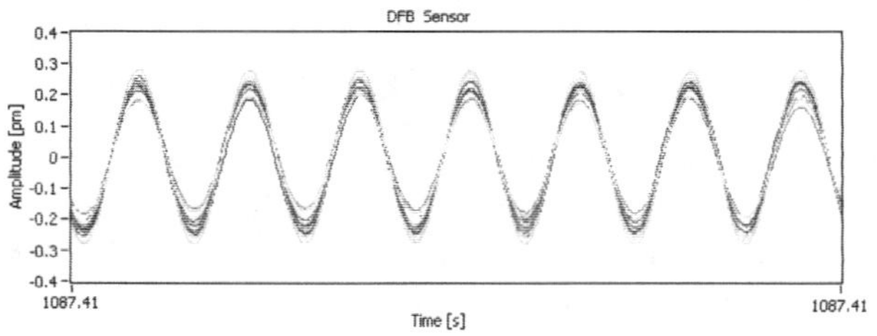

Figure 6. The demodulated results of 16 channels in time domain.

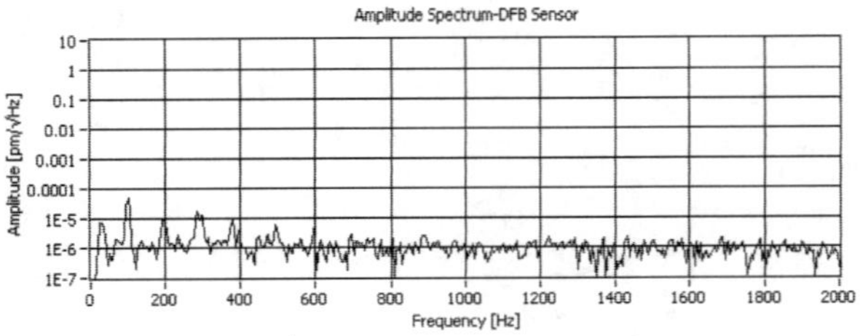

Figure 7. The figure of the noise level.

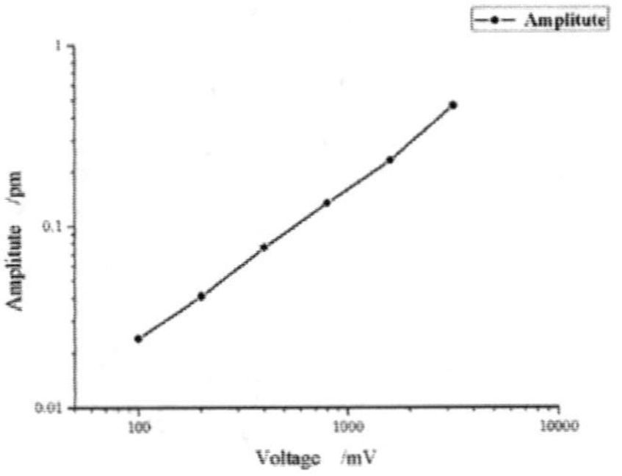

Figure 8. The linearity of the system at the frequency of 100Hz.

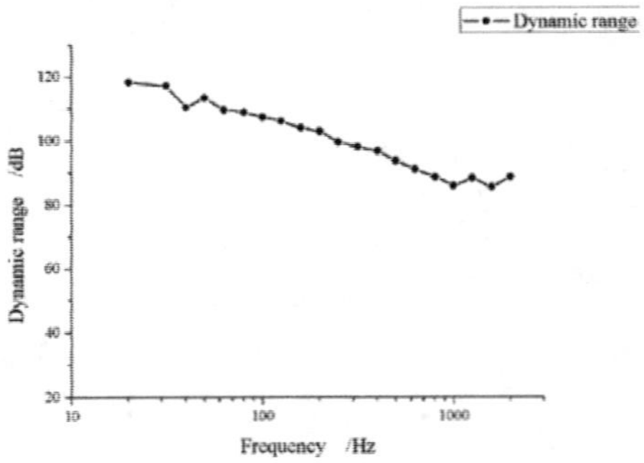

Figure 9. The dynamic range of the system.

4. Conclusion

In this paper, a 16-element DFB fiber laser sensor array system is demonstrated using PGC demodulation scheme. The key techniques of the high performance real-time 16-channel PGC demodulation based on a NI-PXI platform have been introduced. Experimental system has been set up in the laboratory. And the demodulation results show that a minimum detectable wavelength shift of 1×10^{-6} pm/√Hz, a linearity of as high as 0.9994, and a dynamic range of 110dB@100Hz.

References

[1] Anthony Dandridge, Alan B Tveten and Thomas G Giallorenzi 1982 Homodyne demodulation scheme for fiber optic sensors using phase generated carrier *IEEE Journal of Quantum Electronics.* **18**

[2] K P Koo, A D Kersey 1996 Fiber laser sensor with ultrahigh strain resolution using interferometric interrogation *Electron. Lett.* **31(4)** 1180-82

[3] Yi Jiang 2007 Wavelength division multiplexing addressed four-element fiber optical laser hydrophone array *Applied Optics.* **46(15)** 2939-48

[4] Jun Zhu, Pei-Kun He and Chang-Zhen Huang 2004 A design of digital demodulation system using phase generated carrier in fiber hydrophones based on DSP *the 3rd Int. Conf. ICCEA*

[5] Hao Xiao, Fang Li, and Yu-Liang Liu 2008 Crosstalk analysis of a fiber laser sensor array system based on digital phase-generated carrier scheme *Journal of Lightwave Technology.* **26**

[6] Lei Feng, Jun He, Jing-Yuan Duan, Fang Li, and Yu-Liang Liu 2008 Implementation of Phase Generated Carrier Technique for FBG Laser Sensor Multiplexed System Based on Compact RIO *JESTC.* **6** 385-8

Journal of Physics: Conference Series **276** (2011) 012135 doi:10.1088/1742-6596/276/1/012135

Research on the characteristics of the Q-switched ytterbium-doped fiber lasers

Lin xiang, Yuan xiuhua, Xiong jie, Zhao ming

Huazhong University of Science and Technology, Wuhan National Lab for Optoelectronics, Hubei Wuhan 430074

E-mail: yuanxh@mail.hust.edu.cn

Abstract. Fiber lasers with pulse duration and repetition tuned have great application in space probe. On the basis of typical Q-switched ytterbium-doped double-clad fiber lasers under forward pump, the ion population and propagation equations for both pumping and signal light are presented ,and the steady-state solution of inversion ion population, pumping power distribution and signal power distribution due to ASE are studied by the traveling-wave method. Moreover, the effects of fiber-laser parameters such as Yb-doped fiber length, pump power and fiber core diameter on pulse characteristics are analyzed.

1. Introduction

Lasers have great application in space exploration owing to their high angle, long distance, high velocity solution, wide speed range and excellent anti-interference ability, etc. The common lasers used in this area are mostly Nd:YAG lasers and CO_2 gas lasers, which are bulky and heavy, sensitive to vibration and temperature changes. In addition, the attached cooling devices increase the size and quality of the lasers. In the past two decades, the rare earth doped fiber lasers have undergone intense research due to their small size, reliability and high efficiency[1-3]. Fiber lasers in advantage of double-clad ytterbium-doped structures have been extensively studied[4,5] . The most appropriate laser source for space debris detection requires low output pulse repetition rate and pulse duration tunable properties. The pulse repetition frequency generally decided by the processing capacity of the system and the detection range. Moreover, the influence of the atmosphere should be considered. Therefore, studying the output characteristics of the fiber laser has further significance for space exploration.

2. Model

A typical end-pumped Q-switched fiber laser is shown in Fig 1, the pump output power and the signal output separate through the beam splitter(SP) in the left. Acousto-optic Q switch(AOQS) is in the right. The left mirror M1 is cut fiber facet with a reflection about 4% in the signal frequency. Suppose the Yb-doped fiber is much longer than the other fiber. The modulation of the right mirror reflection R2 is achieved by the AOQS. R2 is zero when the switch off while 0.45 when the switch on. Suppose all the mirrors are transparent for the pump light. Fig 2 shows the energy lever diagram of ytterbium ions.

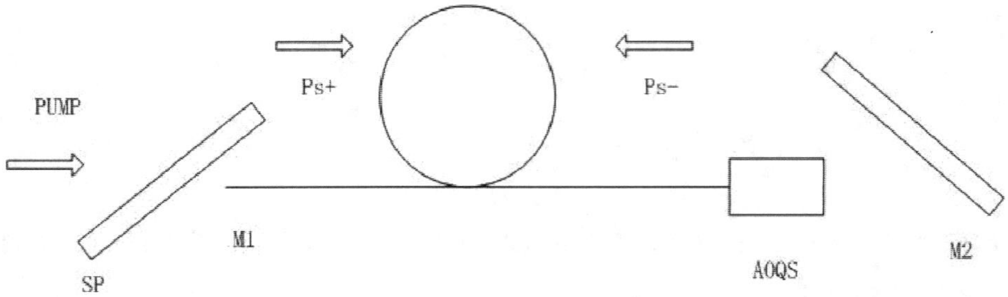

Figure 1. Set up of the Q-switched fiber laser

When analyzing the process of the laser generation it is necessary to derive the value of the pump power Pp(x), the right-moving signal power $Ps^+(x)$, the left-moving signal power $Pp^-(x)$ and the ion population inversion. The upper lever ion population N_2 (x) and the lower lever ion population N_1 (x) satisfy the equation $N_2(x)+N_1(x)=N$, where N is the total ion population. Those can be derived from the following steady-state equations[6]:

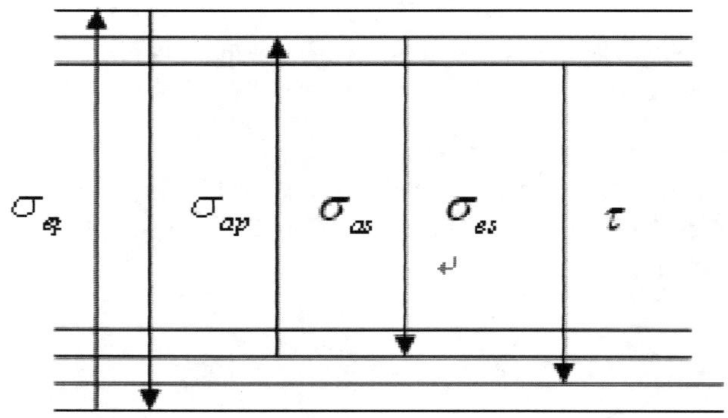

Figure 2. Energy band diagram of Yb3+ with the cross sections and lifetime as indicated

$$[\frac{Pp(x)\sigma_{ap}.\Gamma p}{h\cdot v_p\cdot A}+\frac{(Ps^+(x)+Ps^+(x))\sigma_{as}.\Gamma p}{h\cdot v_s\cdot A}]N_1(x)-[\frac{Pp(x)\sigma_{ep}.\Gamma p}{h\cdot v_p\cdot A}+ \frac{(Ps^+(x)+Ps^+(x))\sigma_{es}.\Gamma p}{h\cdot v_s\cdot A}+\frac{1}{\tau}]=0 \quad (1)$$

$$\frac{dPp(x)}{dx}=-\Gamma p\{\sigma_{ep}.N_2(x)-\sigma_{ap}N_1(x)\}Pp(x)-\alpha_p Pp(x) \quad (2)$$

$$\pm\frac{dPs^{\pm}(x)}{dz}=\Gamma s(\sigma_{es}N_2(x)-\sigma_{as}N_1(x))\cdot Ps^{\pm}(x)+ \Gamma s\cdot N_2(x)\sigma_{es}\cdot 2n_m hc^2\Delta\lambda_s/\lambda_s^3-\alpha_s Ps^{\pm}(x) \quad (3)$$

In the above equations, σ_{ap} , σ_{ep} are pump power absorption and emission cross section , respectively; σ_{as} , σ_{es} are signal power absorption and emission cross section, respectively; Γp , Γs are the pump power and the signal power filling factors, respectively; A is the core area; h is Planck's constant; τ is the upper level ion lifetime; $h.Vp$ is the pump power photon energy; α_p , α_s are the pump and signal loss factors, respectively. In order to effectively derive the approximate analytical solutions, it is necessary to make some simplification. Suppose that pump power is uniformly distributed in the cross section, Γp can be expressed as $\Gamma p = A / A_{cl}$, A_{cl} is the inner cladding cross section area. Take the boundary conditions as follow into account:

$$Ps^{+}(0) = R1 \times Ps^{-}(0) \tag{4}$$

$$Ps^{-}(L) = R2 \times Ps^{+}(L) \tag{5}$$

$$Pp(0) = W_0 \tag{6}$$

Where W_0 is the end pump power, L is the fiber length. When the Q switch is on, R2 becomes 0.45 from 0 in about 300 nanoseconds.

3. Simulations and analysis

According to the mathematical model mentioned above, we use iterative algorithm to simulate the output characteristics of Q-switched fiber laser. Moreover, the laser structure and parameters' influence on the output are analyzed. The related parameters are shown in Table 1.The initial steady state value of the inversion population, pump power and signal power are shown in Figure 3.

Table 1. The parameter used in the simulation

Parameter	value
σ_{ap}	$3\times10^{-21} cm^2$
σ_{as}	$1.4\times10^{-23} cm^2$
n	1.45
Γs	0.75
τ	1ms
R	$8\mu m$
N	$4\times10^{19} cm^{-3}$
σ_{es}	$2.5\times10^{-21} cm^2$
λs	*940nm*
λ p	*1080nm*
L	*5 m*
W_0	*5W*
Rcl	$200\mu m$
α_s	$5\times10^{-5} cm^{-1}$

It is seen that in Fig.3(a) the inversion populations distribute nearly uniform when the pump power are 15W and 25W, whereas in the 35W case it has a dramatic decline at the right end. The reason for those can be seen in Fig.3(c), which indicates the spontaneous emission noise is extremely high at the right end. From Fig.3(b) it is obvious that the pump powers are not completely absorbed by the fiber laser.

Take it into consideration that the laser operates in a single transverse mode, when the Q switch process starts, Fig 4 shows the variation of the pulse duration and energy with pump power. It is seen

Journal of Physics: Conference Series **276** (2011) 012135 doi:10.1088/1742-6596/276/1/012135

that the pulse energy changes linear with the pump power; whereas the variation relationship between the pulse duration and pump power is more complicated, when the pump power is larger than the threshold, the pulse duration increased due to the fluorescence, and pump power increased, the net cavity gain coefficient becomes larger and larger, the raise of photon and the decrease of the inversion populations are much faster, the front and rear of the pulse contract ,making the pulse duration reduced and eventually to be a constant that is the photon lifetime of the cavity.

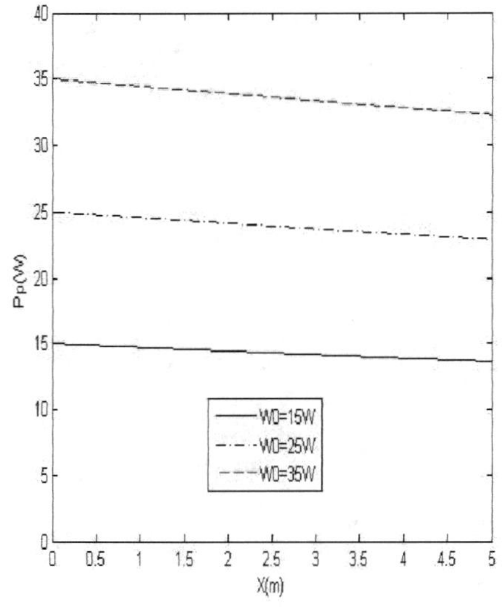

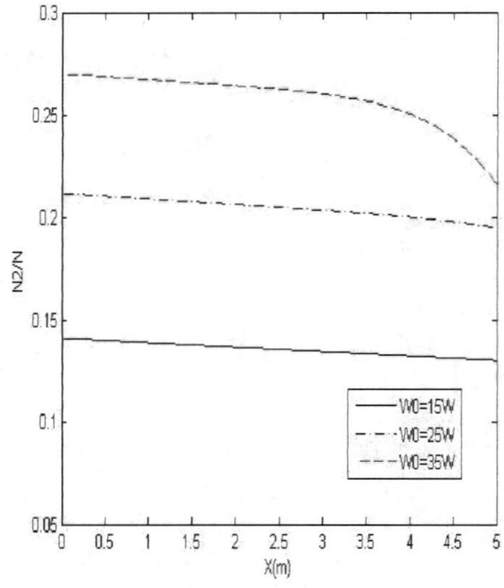

(a) population N_2/N (b)pumping power distribution

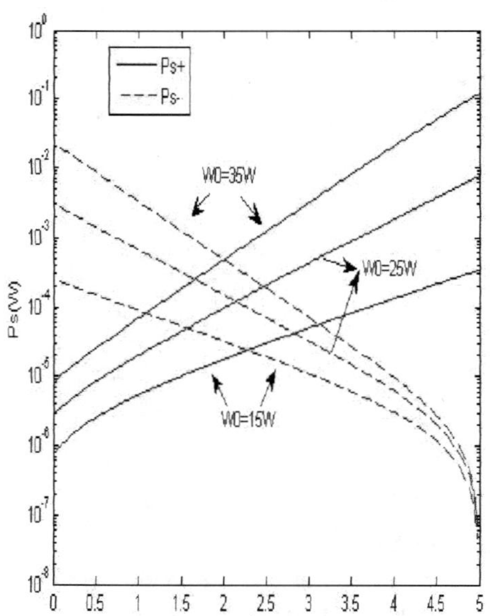

(c) signal power distribution due to ASE

Figure 3 Steady-state solution of a fiber laser at three pump power levels at switch off

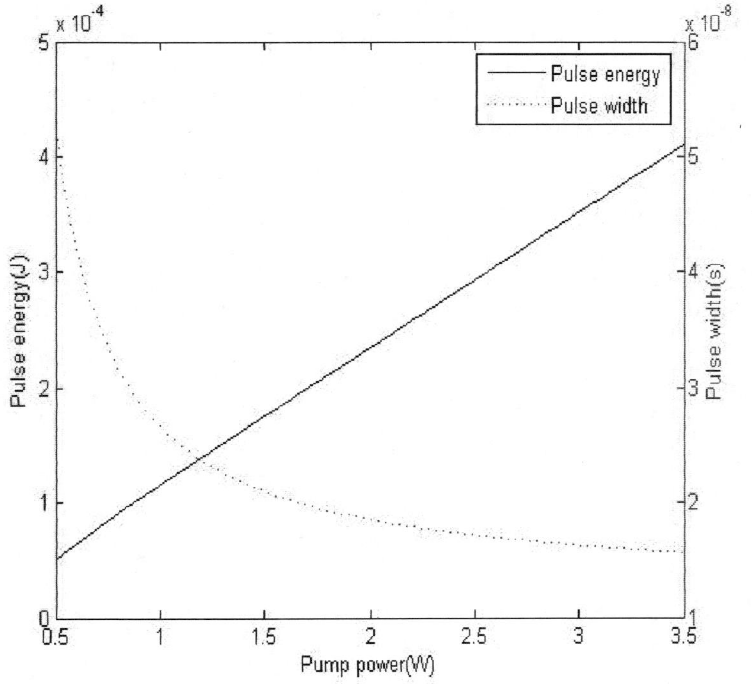

Figure 4. Pulse duration versus pump power

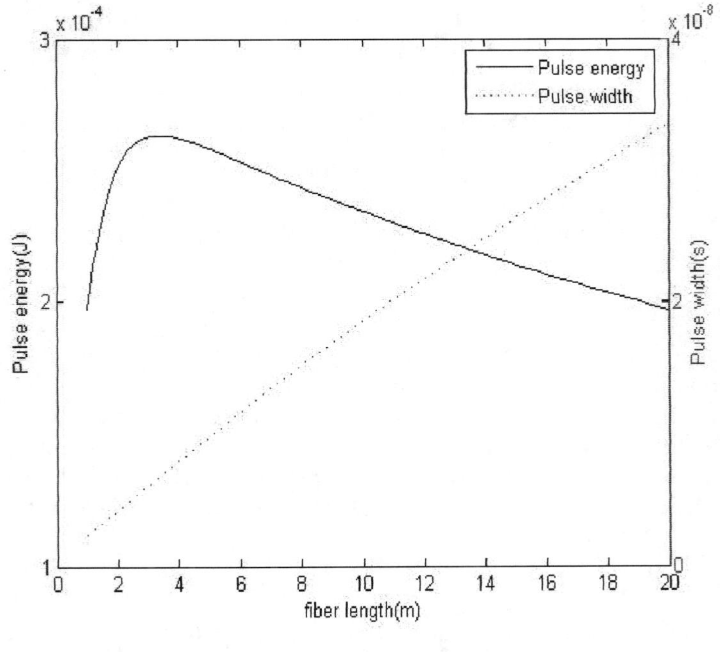

Figure 5.Pulse duration versus Yb-doped fiber length

Fig 5 shows the influence of the fiber length on the output pulse. The distribution of the pump power becomes uneven because the different distance from the pump source as the fiber length increases. So the pulses generate at times along the different position in the Yb-doped fiber, resulting in the multiple pulse overlap, and at last broaden. The pulse energy are decided by the absorption of the pump and signal power, as a result there is an optimal length, making the maximum output pulse energy.

Fig 6 illustrates the pulse duration and energy vary with the core diameter. As the core diameter increases, the uniform distribution of the pump power spoils and broaden the duration. The higher the absorption efficiency of the core, the greater signal power loss in the fiber core, so the variation of the pulse energy depends on the interaction of both.

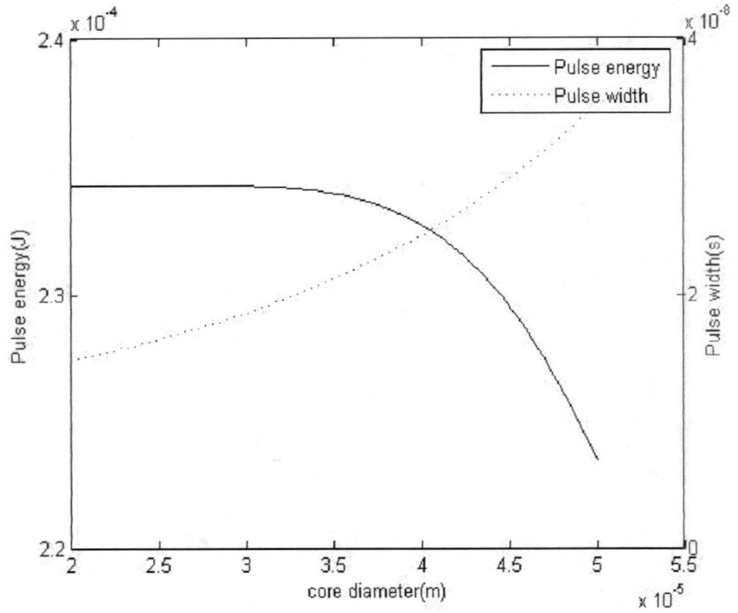

Figure 6. Pulse width versus core diameter of fiber

4. Conclusion

The improvement of output characteristics of the Q-switched fiber laser mainly focus on the pulse duration and energy. Output pulse characteristics relate with the initial parameters once the fiber laser running at its best, therefore the choice of parameters is completely crucial. Increasing pump power, reducing the rare-earth doped fiber length and decreasing the fiber core diameter are beneficial for short pulse duration. Unfortunately, those above are negative for improving the pulse energy. Therefore, it should be integrated to measure the impact of each parameter to determine the whole system.

Reference

[1] J.Kirchhof, T.Sandrock, A.Harschak． 1.3kW Yb-doped fiber laser with excellent beam quality． CLEO,2004,CPDD2

[2] Gapontsev ,V Gapontsev D,Platonav N,et a1． 2kW CW ytterbium fiber laser with record diffraction-limited brightness． CLEO，2005，508

[3] DingWei Huang,WenFung Liu,Yang C.C.Q-switched all-fiber laser with an acoustically

modulated fiber attenuator. IEEE Photonics Technology
[4] Letters,2000,12(9):1153-1155
[5] H.M.Pask,J.L.Archambault,D.C.Hanna,et al. Operation of cladding-pumped Yb3+-doped silica fiber lasers in the 1um region. Electron. Lett. ,1994,30,11:863-864
[6] H. Po, J. D. Cao, B. M. Laliberte, R. A. Minns, R. F. Robinson, B. H. Rockney, R. R. Tricca, and Y. H. Zhang, "High power neodymium-doped single transverse mode fibre laser," Electron Lett.29, 1500-1501
[7] Yanming Huo, Robert T. Brown, George G. King, et al. Kinetic modeling of Q-switched high-power ytterbium-doped fiber lasers. APPLIED OPTICS , Vol. 43, No. 6:1404-1411

Optoelectronic System for Measuring Warhead Fragments Velocity

Ji Liu[1,2,3], Donge Zhao[1,3], Yangjun Li[1] and Hanchang Zhou[1,2]
[1]National Key Laboratory on Electronic Testing Technology, Taiyuan, 030051, China
[2]Education Ministry Key Laboratory on Instrumentation Science and Dynamic Testing, Taiyuan, 030051, China
[3]School of Information and Communication Engineering North University of China, Taiyuan, 030051, China

E-mail:gdcs2009@sina.com

Abstract. High-speed warhead fragments velocity measurement is one of the key technologies in investigating damage efficiency of warhead. We have designed and constructed a system to accurately determine the velocity of warhead fragments by measuring the time of flight between two parallel laser screens is presented. Each screen is formed by a laser source, a large photodetector and a retro-reflector. Optical output of the laser source is a collimated beam. The beam passes through cylindrical lens and the slit of photodetector reach to retro-reflector .The energy, reflected by retro-reflector; focus on the active area of photodetector. The system utilizes reflected ray by scotchlite retro-reflector as the start and end signal. And utilizes wideband circuit and data acquiring system to condition and sample signals. Experimental results show the system can measurement velocity are within the range from 20m/s to 2000m/s on target area of $1m^2$ and can perform satisfactorily with a wide range from 2000 lx to100,000lx. The measurement system also can be used to test velocity of projectile.

Keywords: laser screen; non-contact measurement; fragment velocity; retro-reflector; external ballistics

1. Introduction

Various means have been used in the prior art to determine the velocity of fragments undergoing development. The technique of measurement of the time of flight (ToF)of a fragment between two fixed points or planes is routinely employed to determine the speed of fast-moving fragment s .The passage of the fragment creates a field perturbation that can be sensed in terms of some measurable quantity such as the pressure, electric or magnetic field, or radiation field. The Doppler frequency-shift measurement technique is also used for measurement of the velocity of projectiles, but it is not suitable for small-caliber fragments. Speed measurement apparatus using screens based on infrared (IR)or optical radiation fields offer very high resolution and sensitivity while being immune to electromagnetic interference and incompatibility, but the measurement schemes of the prior art typical exhibit one or more of the following disadvantages:

(1)They require larger optical components to construct larger screens, whose sensitivity is not uniform over the entire screen area.

(2) They need careful alignment during installation.

(3) They are unreliable due to excessive false triggering, particularly when ambient lighting is used.

(4) Systems employing sunlight as the source cannot be used indoors, and the precision and accuracy of such systems are very much dependent on the intensity fluctuations of the sunlight.

Foundation items: Project supported by the National Key Laboratory on Electronic Testing Technology ,China (Grant No.9140C1204041009) and National Key Laboratory on Electronic Testing Technology Fund for Young Scholars ,China

Journal of Physics: Conference Series **276** (2011) 012136 doi:10.1088/1742-6596/276/1/012136

This paper will be described a laser-based system, employ two parallel laser screens for measurement of the speed of fragments moving up to 2000 m/ s. The optical screens are spaced apart at a know or fixed distance for providing time measurement stat and stop signals. Each screen is constructed from a single source and does not require any expensive aspheric optical components. The system ensures accurate performance over a wide range of ambient light, temperature, and other environment conditions, including the high ballistic pressure and shock waves generated during measurements on high-speed projectiles.

2. Design and description of system

2.1. Basic Principle of System

The system is comprised of an optical unit, electronic unit, and a data acquisition & processing unit. Complete configuration, as shown in Fig.1. Whenever a fragment crosses either of the screens, the corresponding photodetector senses the event, due to partial or full obscuration of the incident energy. The change of light flux is transformed by photodiode in to electronic signal. The weak signal amplified and filtered by signal conditioning, then, sampled by data acquiring devices. Data processing is finished by professional software in computer. The distance between the screens being known, the velocity is displayed on a computer screen.

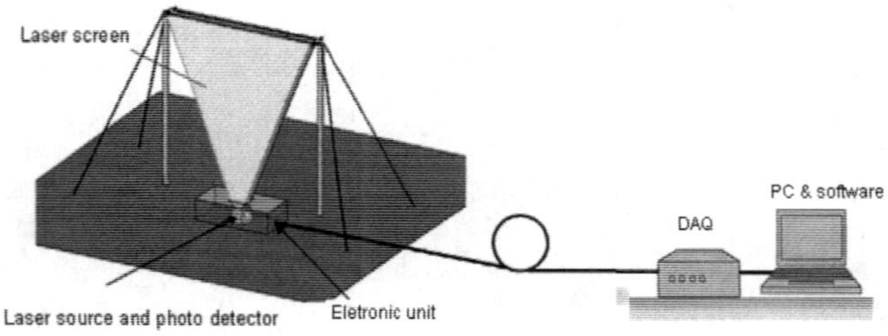

Figure 1. Composition of system

2.2. Large laser screen

The novel laser screen measurement system include two parallel laser screens, each screen is formed by a laser diode, photo detector array and a retro-reflector. Laser diode emitting visible radiation at 650nm wavelength and having suitable power, which may vary from 5 to 50 mW depending on the size of the laser screen. Photo detector array assembled with lots of PIN photodiode which response time less than 10 ns, responsivity 0.4 A/W, typical at 635 nm. A slit of size 4×4mm in the center of photo detector. Optical output of the laser source is a collimated beam. The beam passes through cylindrical lens and the slit of photo detector reach to retro-reflector (a non-metalized micro-prismatic lens reflective sheeting).Based on the theory of remained divergence angle, the energy of laser source focus on photo detector. As shown in Fig.2.

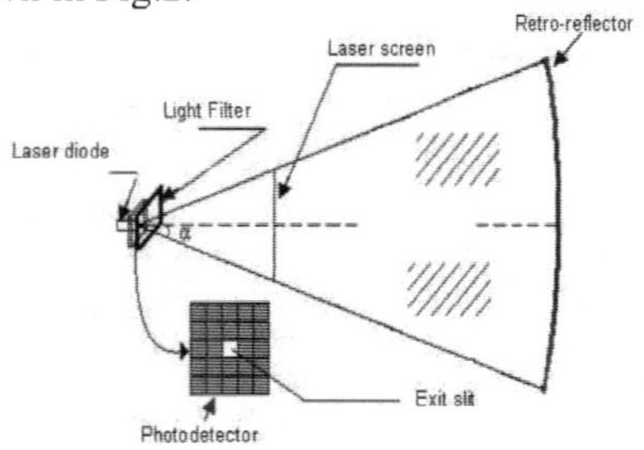

Figure 2. Optical system schematic

2.3. Scotchlite Retro-reflector

Scotchlite is reflective sheeting has the property of directing a large fraction of the light reflected by it back toward the vicinity of light source. The optical elements of Scotchlite consist of small beads of transparent material with a metallic reflective coating, and are attached on their coated side to a backing sheet. In the Scotchlite sheeting, the beads are packed closely so as to effectively provide a continuous surface having the reflective characteristics of the beads which were used. The structure of Scotchlite, as shown in Fig.3.It provides long-term reflectivity and durability.

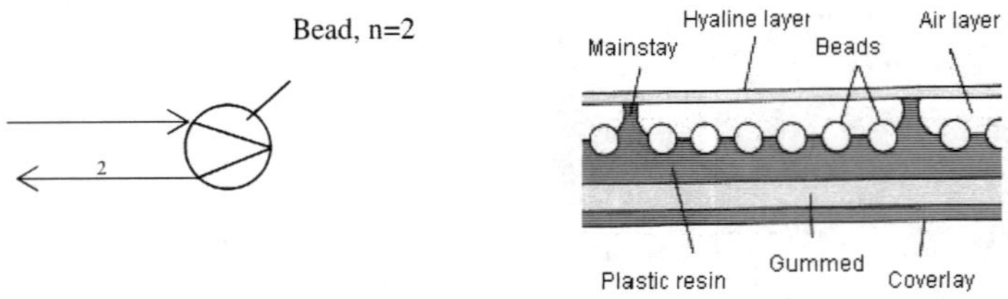

Figure 3 Micro-prismatic lens schematic

3. Electronic Circuit and Data acquisition system

Optical module connects two analogue channels, which contains a preamplifier, a conditioning circuit and DAQ system. The preamplifier amplifies the signal coming from the photodiode. For high-speed components, the output noise of the photodetector increases exponentially with the speed of the projectile whereas the signal amplitude decreases, which reduces the signal-to-noise ratio(S/N). The preamplifier and signal-conditioning circuits are designed to minimize the noise, such as insects flying through the light screen. To increase the sensitivity of measurement, a variable dc reference voltage corresponding to ambient noise, such as flicker of other light sources nearby, is subtracted from the signal. As a preamplifier we used an AD8066 (Analog Devices), which is characterized by high speed (145 MHz), low input voltage noise (7 nV/Hz$^{1/2}$), and Wide supply voltage range (5 V to 24 V). Conditioning circuit consist of a main amplifier. We use the AD8027,which is a high speed amplifier with rail-torail input and output that operates on low supply voltages (2.7V to12V) and is optimized for high performance and wide dynamic signal range and low noise (4.3 nV/Hz1/2) and low distortion (120 dB @ 1 MHz). The signal, which output of conditioning circuit, proceeds towards the DAQ system, which consist of 4 analog channels, every channels sampling rate at 10 Mbit/s, input resolution at 12 bits and with 32M memory. DAQ system based on the open industry-standard PXI (PCI eXtensions for instrumentation) architecture deliver high-performance. Output digital signals can then be transmitted to the remote computer through 1000 base Ethernet. Velocity data can be generated instantaneously by the computer based addition of appropriate software and displayed on a computer screen. The electronic circuit is sensitive enough to detect even1%change in intensity. The system is an active system capable of detecting advancing objects, and does not require synchronization with the firing signal.

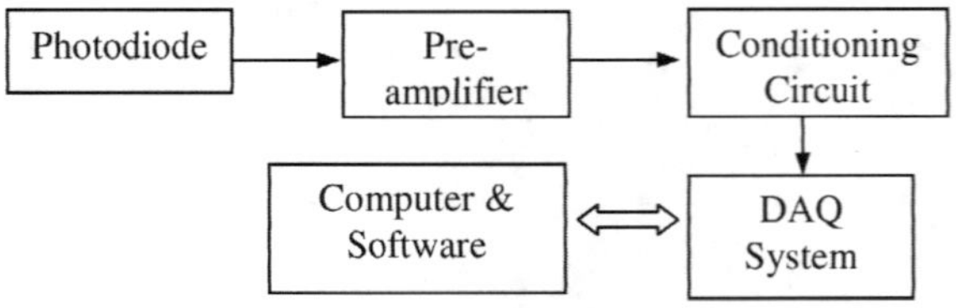

Figure 4. Block diagram of electronic circuit

4. Experimental results

4.1. Test 1

A number of experiments were performed using different warhead fragments. The warheads were placed at a distance of about 10m from the system. The laser screen, properly aligned, with each laser screen of size 1000×1000mm and a distance between two screens of 250mm. The main parts, include laser source and photo detector, are underground. Thus, equipment to avoid being hit by fragments. After the experiment, apparatus is intact, although covered by some dust, shown in Fig.5. The measured typical results are shown in Fig.6 and Table1.The results are within the ranges supplied by the manufacturers, which validate the measurement.

Figure 5. Apparatus after the experiment

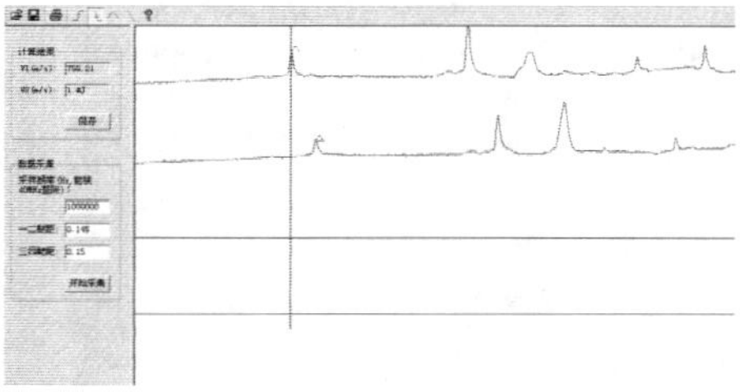

Figure 6. Signal waveform of fragments.

Table1. Velocity data of warhead fragments.

	Velocity(ms^{-1})				
	v_1	v_2	v_3	v_4	v_5
1	1332.7	1292.4	1192.6	1192.4	1091.9
2	740.74	714.29	719.42	600.24	543.46

System sampling frequency at 5 MHz,
distance between two screens at 250 (mm)

4.2. Test 2

Laser-based apparatus can be use to measure velocity of repeating rifle projectiles. The rifls were rigidly mounted on a table placed at a distance of about 20m from the system. The measured typical results are shown in Table 2. The results show that the system working properly, non-occurrence of false triggering.

Table 2. Results of test 2 speed data of AK-47 rifle（7.62mm）

Velocity(ms^{-1})

	v_1	v_2	v_3	v_4	v_5	$\bar{v}$
1	715.5	717.3	720.5	714.8	721.2	717.90
2	718.4	720.4	716.3			718.40
3	717.4	711.3	713.3	711.4		713.28
4	711.3	714.3	718.7	714.6	716.1	715.00

4.3. Test 3

The system has been tested in different ambient light conditions using different-caliber guns. A number of experiments have been performed in an outdoor range with ambient light of about 100,000 lx, in an indoor range with an artificial floodlight of about 2000 lx directed towards the system, and in a darkroom. The results show that the system yields consistent readings, which ensure that the system can perform satisfactorily in outdoor and indoor ranges with a wide range of ambient light conditions.

5. Conclusions

Large amounts of experiments have been performed to measure. All of experiments acquired effective data. The results show that the system working properly, non-occurrence of false triggering. The main advantages of the reported system in comparison with commercially available systems are as follows:
(1) Optical screens can be used at indoor or outdoor ranges equally effectively.
(2) A lot of advantages such as larger effective light screen area, quick response speed, high sensitivity, low uncertainty, strong repetition and reliability, simple and easy to set up etc.
(3) Its speed range covers subsonic to supersonic.
(4) The measurement system also can be used to test velocity of projectile.

References

[1] Zhao Donge 2003 Laser screen measuring system for gun velocity, *ISTM 5*, Bei Jing.
[2] Lu, S.T., Chlou, C., Lee, M.C. and Wu, Y.P 1993 Electro-optical target system for position and speed measurement, *IEE Proceedings-A*,140(4), 252-256.
[3] C. C. Chang, H.C. Chang, L.C. Tang, W.K. Young, J.C. Wang, K.L 2005 Huang Hybrid-integrated prism array optoelectronic targeting system, *Opt Laser Technol*;"37:591–596.
[4] Han Feng, Liu QunHua, Sun GuoBin,2008 Study on Measurement Method for Projectile Location Based on Light Screen, *Proc. SPIE*, 7155, 71552K.
[5] J. M. Sánchez-Pena, C. Marcos, and M. Y. Fernández 2007Cost-effective optoelectronic system to measure the projectile velocity in high velocity impact testing of aircraft and spacecraft structural elements, *Opt. Eng.*; 46(5), 051014.
[6] R. C. Kalonia, G. Mitra, and A. Kumar 2007 Laser-based projectile speed measurement system, *Opt. Eng.*, 46(4) 044303 .
[7] Q. Liu, H. Shi, B. Yan, F. Han, and R. Cai 2004 The infrared light screen system and accuracy analysis, *Acta Photonica Sin* 33(11), 1409–1411.
[8] Zhexin Jia, Ke Gong, Yujing Huo 2009 Optoelectronic system for high-speed flier velocity measurement based on laser scattering, *Opt. Eng.*; 48(4), 043601 .

Acknowledgments

The authors would like to thank National Key Laboratory on Electronic Testing Technology (Grant No.9140C1204041009) and National Key Laboratory on Electronic Testing Technology Fund for Young Scholars for providing funds for this project.

A new diarylethene and its application in optical storage

Weijun Liu, Duohua Jiang, Shouzhi Pu*, Shiqiang Cui

Jiangxi Key Laboratory of Organic Chemistry, Jiangxi Science and Technology Normal University, Nanchang 330013, P. R. China

pushouzhi@tsinghua.org.cn

Abstract. A new blue photochromic diarylethene compound, 1-[5-(3-methoxyphenyl)-2-methyl-3-thienyl]-2-[5-(3-cyanophenyl)-2-methyl-3-thienyl]perfluoroclopentene (1a), was synthesized and its photochemical properties were investigated in detail. The measurement results demonstrated that the compound has good photochromism in solution and in PMMA amorphous film, fluorescence and optical storage properties.

1. Introduction

Among various photochromic compounds, diarylethene with heterocyclic rings has been developed as a new type of photochromic derivative. In particular, these diarylethenes derivatives bearing two thiophene or benzothiophene rings are regarded as the best candidates for photonics application such as optical memory [1,2] and switch [3], because of their excellent thermally irreversibility, high fatigue resistance, and rapid response [4]. Therefore, the molecular design and synthesis are the crucial factor for optical characterization. Especially, diarylethenes with different substituted ring systems have been of considerable concentration [5-9], which different substituted ring could be resulted in different photochromism properties before and after light irradiation. In this work, a new blue photochromic diarylethene compound, 1-[5-(3-methoxyphenyl)-2-methyl-3-thienyl]-2-[5-(3-cyanophenyl)-2-methyl-3-thienyl]perfluoroclopentene (1a), was synthesized and its photochemical properties, such as photochromism in solution and in PMMA film, fluorescecne and optical storage properties were investigated in detail.

2. Experiments

2.1. Synthesis of diarylethene 1a

The photochromic compound was prepared by the method [10] and the synthetic route was shown in Scheme 1. Diarylethene 1a was characterized by ^{1}H NMR spectroscopy. ^{1}H NMR ($CDCl_3$, 400 M Hz, TMS): 1.88 (d, 6H, J = 6.2 Hz, $-CH_3$), 3.75 (s, 3H, $-OCH_3$), 6.76 (d, 1H, J = 6.2 Hz, thiophene-H), 6.97 (s, 1H, thiophene-H), 7.05 (d, 1H, J = 8.0 Hz, phenyl-H), 7.17-7.23 (m, 2H, phenyl-H), 7.26 (s, 1H, phenyl-H), 7.39-7.49 (m, 2H, J = 8.0 Hz, phenyl-H), 7.68 (t, 2H, J = 7.8 Hz, phenyl-H); ^{13}C NMR (100 MHz, $CDCl_3$): 14.06, 14.10, 54.85, 110.53, 110.95, 112.68, 115.62, 117.70, 117.83, 122.04, 123.42, 125.11, 125.80, 128.44, 129.16, 129.41, 129.61, 130.54, 134.01, 134.09, 135.13, 139.03, 140.83, 141.84, 142.35.

2.2. Preparation of PMMA film

One hundred milligram of PMMA was first dissolved in 1 mL chloroform, which was stirred ultrasonically to make it homogeneous. Samples were obtained by dissolving ultrasonically 10 mg of compounds 1 with 1 mL of the PMMA solution, respectively, and the solution was spin-coated on a glass substrate with a spin rotation speed of 1500 rpm.

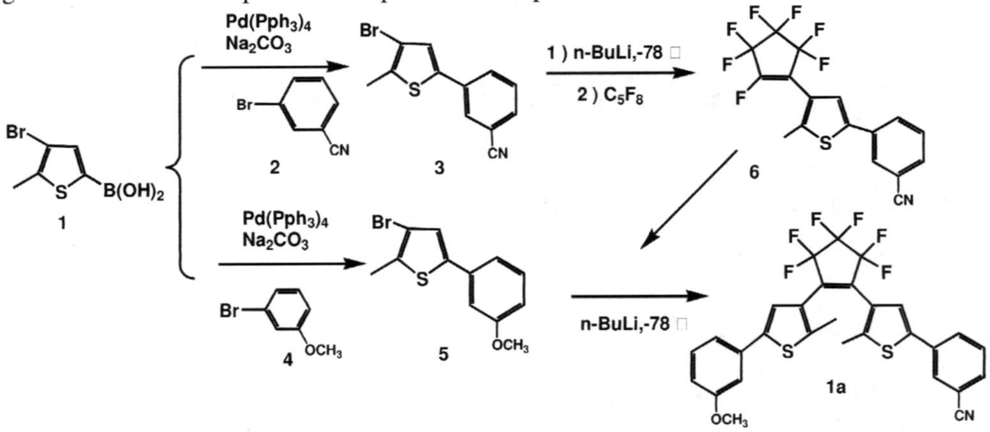

Scheme 1. Synthesis of diarylethene 1a

3. Results and Discussions

3.1. Photochromism of diarylethene 1a

Figure 1(A) shows the absorption spectral changes of diarylethene 1a in hexane (2.0 × 10-5 mol/L) by photoirradiation. It can be seen that diarylethene 1a has the absorption maximum at 275 nm (= 4.34 × 104 L mol-1 cm-1, o-c = 0.39). Upon irradiation with 297nm UV light, the colorless solution of 1a turned blue (Figure 1(B)), in which absorption maximum was observed at 583 nm (= 1.93 × 104 L mol-1 cm-1, c-o = 0.015). The blue color is due to the formation of closed-ring isomer 1b. When the blue solution was irradiated with visible light (λ > 450 nm), the blue color returned colorless and its spectrum become the same as that of original one.

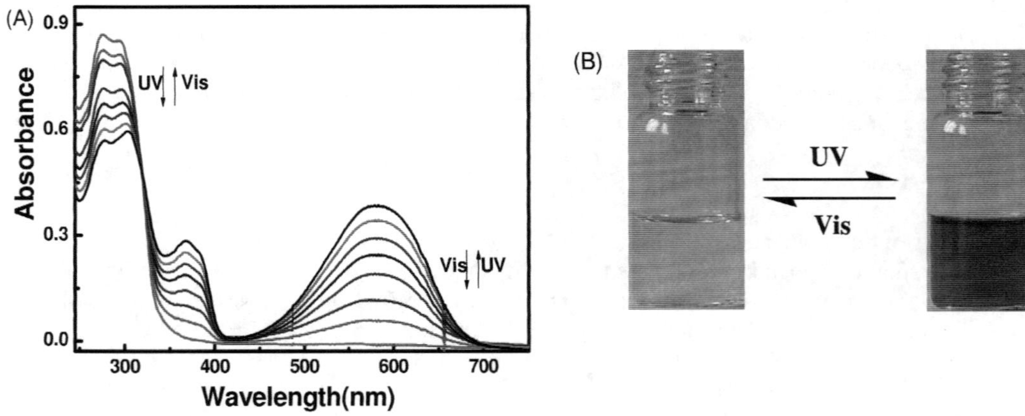

Figure 1. Absorption spectral changes of compound (A) and color changes (B) 1a in hexane (2.0×10^{-5} mol/L).

The further experiments revealed that the solution can keep similar absorption spectrum under several cycles irradiation with UV and visible light, indicated the high fatigue resistance property. Similar spectrum change of diarylethene 1a was observed in PMMA film (10% w/w). Upon irradiation with

297 nm UV light, the colorless PMMA film turned blue (Figure 2(B)) with new broad absorption band centered at 590 nm (Figure 2(A)), which assigned to the open-ring isomer 1a changed to the closed-ring isomer 1b. The colored PMMA film can also return to colorless upon irradiation with appropriate wavelength visible light.

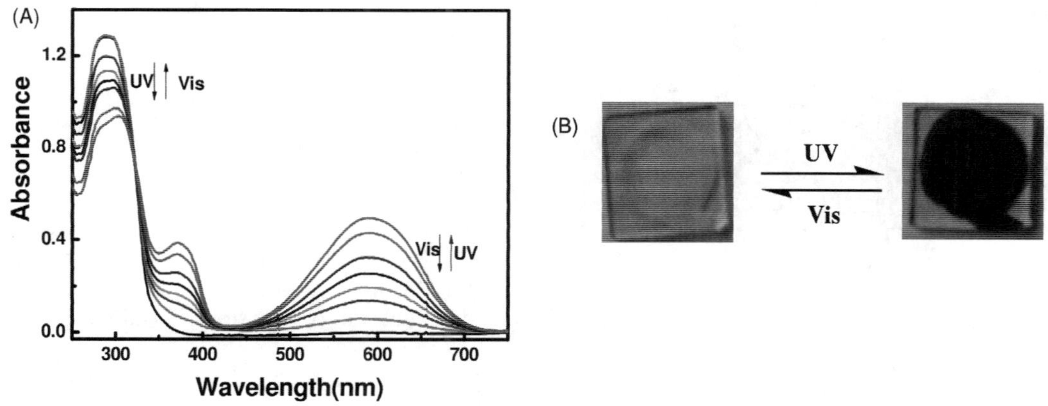

Figure 2. Absorption spectral changes of compound (A) and color changes (B) 1a in PMMA film.

3.2. Fluorescence switch properties

Figure 3 showed the fluorescence switch properties of 1 in hexane (Figure 3(A))and in PMMA film (Figure 3(B)) at room temperature. It exhibited a strong fluorescence switch properties excited at 300 nm along with photochromism from open-ring isomer to closed-ring isomer upon irradiation with 297 nm UV light. When arrived at photostationary state, the emission intensities of diarylethenes 1 in hexane was quenched to ca. 72% and in PMMA film was quenched to ca. 45%. The back irradiation with visible light of wavelength longer than 450 nm regenerated the open-ring forms and recovered the original emission spectra. The phenomena are useful for potential application as the fluorescence switches [11,12].

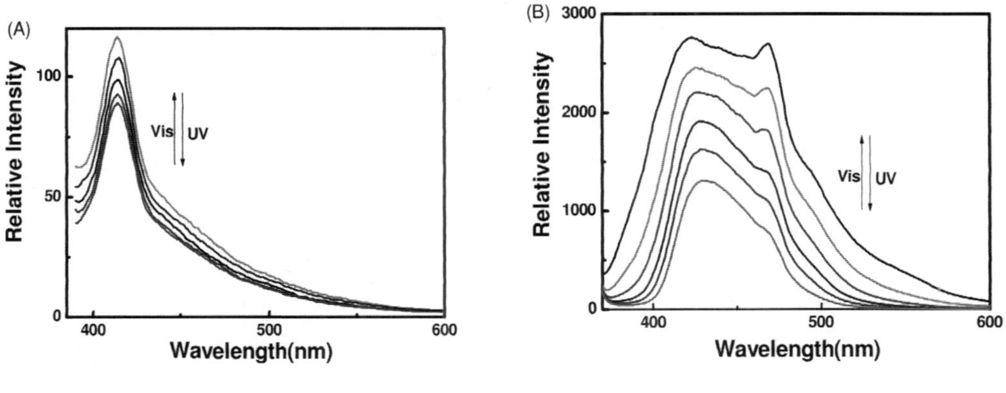

Figure 3. Emission intensity changes of diarylethene 1 in hexane (A) and in PMMA film (B) upon irradiation with 313 nm UV light at room temperature, excited at 300 nm.

3.3. Optical storage properties

With a He-Ne laser (wavelength: 650 nm, 105mW) in the experimental setup. the evaluation of potential of photochromic diarylethene as a polarization medium by recording and reading in a real-time operation was investigated. Using diarylethene 1b/PMMA film as recording medium, the polarization patterns were recorded and read on the film under the microscope. The polarizations of the patterns recorded on the film are shown in Figure 4, The exposure time for each spot was 0.04 s and the patterns were about 20 μm. The result indicated that the new diarylethene compound can be applied in high capacity optical storage [13].

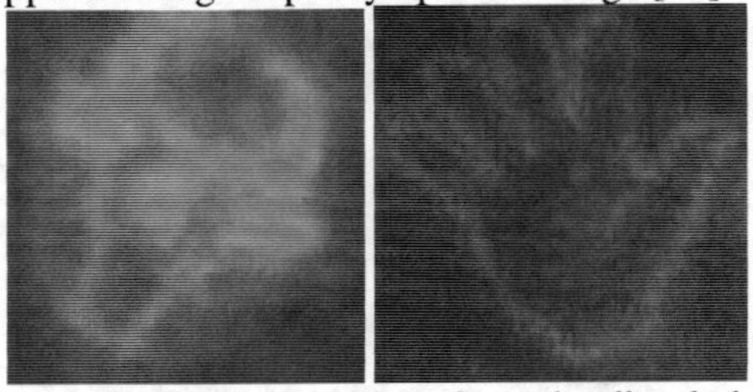

Figure 4. Readout patterns from the diarylethene 1b/PMMA film.

In conclusion, a new blue photochromic diarylethene was synthesized and the photochromic properties such as photochromism in solution and in PMMA amorphous film, fluorescence and optical storage properties, were investigated in detail. The results showed that it can be useful in fluorescence molecule switch, scale optoelectronics and potential use for data storage, and so on.

Acknowledgment

This work was supported by the Youth Science Funds of the Education Office of Jiangxi, China (GJJ09572).

References

[1] Pu S Z, Zhang F S, Xu J K, Shen L, Xiao Q and Chen B 2006 *Mater. Lett.* **60** 485
[2] Pu S Z, Yan L S, Wen Z D, Liu G and Shen L 2008 *J. Photochem. Photobiol. A: Chem.* **196** 84
[3] Tian H and Yang S J 2004 *Chem. Soc. Rev.* **35** 85
[4] Dulic D, Kudernac T, Puzys A, Feringa B L and Wees B J, 2007 *Adv. Mater.* **196** 2898
[5] Higashiguchi K, Matsuda K, Tanifuji N and Irie M 2005 *J. Am. Chem. Soc.* **127** 8922
[6] Pu S Z, Yang T S, Xu J K, Shen L and Liu G 2005 *Tetrahedron* **61** 6623
[7] Fan C B, Pu S Z, Liu G and Yang T S 2008 *J. Photochem. Photobiol. A: Chem.* **194** 333
[8] Morimitsu K, Shibata K, Kobatake S and Irie M 2002 *J. Org. Chem.* **67** 4574
[9] Yamamoto S, Matsuda K and Irie M 2003 *Chem. Eur. J.* **9** 4878
[10] Pu S Z, Liu G, Shen L and Xu J K 2007 *Org. Lett.* **9** 2139
[11] Pu S Z, Yang T S and Li G Z 2006 *Tetrahedron Lett.* **47** 3167
[12] Fan C B, Pu S Z, Liu G and Yang T S 2008 *J. Photochem. Photobiol. A: Chem* **197** 415
[13] Yao B L, Wang Y L and Lei M 2005 *Optics Express* **13** 20

Journal of Physics: Conference Series **276** (2011) 012138 doi:10.1088/1742-6596/276/1/012138

High-resolution Measurement of Refractive Index Based on Resonant Tunneling Effect

J Liu [1], C Q Huang [2], P Tian [3], Z M Wan [1], Z M Luo [1] and M Chen [2]

[1] College of Information & Communication Engineering, Hunan Institute of Science and Technology, Yueyang 414006, China
[2] College of Physics & Electronics Information, Hunan Institute of Science and Technology, Yueyang 414006, China
[3] College of Optoelectronic Science and Engineering, Wuhan National Laboratory for Optoelectronics, Huazhong University of Science and Technology, Wuhan, 430074, China

E-mail: nameliujing@yahoo.com.cn

Abstract. A high-resolution measurement device of refractive index with high transmissivity and Q factor based on resonant tunneling effect is presented. The device is constructed by one dimensional photonic crystal (PC) with a defect which would produce different resonant peaks in PC band gap. These peaks are sensitive to the change of the refractive index for different mediums which are filled in the defect. It is found that with a spectrometer that can resolve a resonant peak shift of 0.5nm, at the resonant wavelength of about 1550*nm*, the resolution of refractive index should be better than $\Delta n = 0.0008$, and the transmissivity could be accessible to 1. Both the plane wave method and the finite difference time domain method are employed in numerical analysis. The device is suitable in a variety of applications of practical interest, since many physical, chemical, and biological parameters can be known through measurement of the refractive index.

1. Introduction

Refractive index is one of the fundamental material properties, because it is closely related to the electronic polarizability of ions and the local field inside the material [1]. The accurate measuring, monitoring or sensing of refractive index is of scientific and technological importance in many cases [2]. In physics, by accurate measuring the refractive index, many physical features may be perceived, such as light spectrum structure, optical properties, purity and concentration of gas or liquid, and so on [3]. In chemical, pharmaceutical, beverage, or food industries, the monitoring of refractive index is part of the quality control. In biosening, with the changing of refractive index, molecular bindings, chemical, or biochemical reactions are carried out continuously, thus the detection of minute refractive index changes is crucial [4]. In the fabrication of integrated optical devices, such as filters, switches and modulators, etc., where the refractive index of material is the key parameter for device design, the sensing of tiny refractive index changes is decisive for monitoring the production process.

Owing to the broad range of applications just mentioned, new measuring principles and methods of refractive index are constantly appearing. Many devices based on different principles have been proposed, such as angular surveying instrument, interferometric testing apparatus, F-P interference sensor, all-fiber interferometers or resonators, and biochemical sensors [5]. However, they have also some drawbacks. Some measuring devices, for example, have complex design with high price, are fragile, instable, or incapable of on-line monitoring. In most cases, it is difficult to reach relevant request of sensitivity, precision and measuring range simultaneously.

In this paper, we report a novel high-resolution measuring method which detects the value of refractive index based on resonant tunneling effect in photonic crystal (PC) [6, 7]. One dimensional

PC is used because there have more methods and more mature technology in its fabricating compared with two and three dimensional PC. The paper is organized as follow. Firstly, we will define our model system and describe the theoretical method. Secondly, we will try to perform the optimal design with geometric and material parameters, and the policy of design is expounded. Thirdly, numerical results will be presented together with an improved discussion. Finally, some remarks will be given. In our study, both the plane wave method [8] and the finite difference time domain method [9, 10] are used for numerical analysis.

2. Physical model and theoretical method

The schematic diagram of our test device researched in this work is composed by stacking T_iO_2 and S_iO_2 layers with a defect which is used to fill different materials, as shown in Figure 1. For convenience, we denote the structural configuration as $(HL)_m(D)_n(LH)_m$, where *m* is the repeated number of the unit "*HL*", *n* is the periods of the unit "*D*" which represents the thickness of the defect. The refractive index and thickness of layer *H* (*L*) are n_H=2.2 (n_L=1.44) and $d_H = 0.4a$ ($d_L = 0.6a$), respectively, where the parameter *a* is the lattice constant of the PC. The layer *D* is chosen with the same thickness as layer *H* ($d_D = 0.4a$) but different refractive index n_D. Obviously, one kind of PC is formed by $(HL)_m$ (marked as PC $(HL)_m$). It is not difficult to imagine that this structure is just similar to semiconductor quantum well system. The PC $(HL)_m$ can be regarded as a potential barrier for the propagation of electromagnetic waves and be utilized to quantize the continuous states of a suitably chosen quantum well material into a series of photonic bound states, much like the way potential barriers are used to quantize the electronic states of a semiconductor; while "$(D)_n$" may be thought of as potential well, whose material may be chosen according to ours needs.

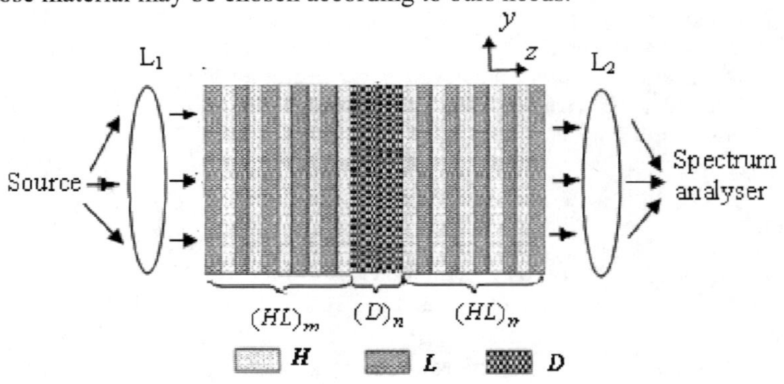

Figure 1. Schematic of measuring device of refractive

In the case of the same thickness $(D)_n$ with the invariable geometric and dielectric parameters of PC $(HL)_m$, the transmission spectrum is only determined by refractive index of n_D. Thus we can call this device as "refractive index sensor". The different transmission spectra with different n_D are calculated based on the plane wave method and the finite difference time domain method. To minimize back reflection, the computational region is bounded by the PC-based perfectly matched layers. The parallel light originated from broadband optical source and transformed from convex lens L_2 is normally launched upon the surface of PC $(HL)_m$. In the design of the device, the coaxiality and the parallelization of the convex lens should be insured, so as to improve the coupling efficiency of light. The other request is that the input and output faces of PC $(HL)_m$ should have good temperature stability and should not receive effects of shearing stress, so as to insure the stability of optical property. In favor of the observation, in the calculated diagrammatic curves of transmission spectra, the Y-coordinate value is calculated as the ratio of transmitted power to the incident power; the frequencies described in the X-coordinate are in the normalized frequencies which take $(\omega a/2\pi c)$ as their units, c is the propagation speed of light in vacuum, and ω is the angular frequency.

3. Numerical results

The PC $(HL)_m$ provides a photonic band gap (PBG), in which the propagation of electromagnetic (EM) waves or photons is strongly inhibited. Although there are no photon modes in the PBG, the defect modes (sharp peaks) could be induced by breaking its periodicity due to change of the interference behavior of light. Figure 2 (a) and 2(b) show the transmission spectra of $(HL)_m$ and $(HL)_3(D)_4(LH)_3$, respectively. It can be seen that defect modes would emerge in the PBG and could transmit through the well region. We interpret this phenomenon as the effect of resonant tunneling, not in a usual way. Obviously, such a PBG may be viewed as a perfect mirror and is particularly useful for constructing the resonant cavities in the optical region, and only the resonant frequencies can be captured by this cavity. We can conjecture that the resonant frequency of the sharp peaks would shift when the refractive index of defect layer $(D)_n$ changes. Therefore the change in the refractive index of the defect layer $(D)_n$ can be obtained measured by measuring the output spectrum.

It is worth mentioning that both the transmissivity and the quality factor Q are important parameters for the measurement of refractive index, which are closely related with the sensitivity and the precision. The Q factor is defined as $\omega_0/\Delta\omega$, where ω_0 is the center frequency of the defect modes, and $\Delta\omega$ is half-height width of the resonance peak. The higher the transmissivity is, the lower energy loss is, the lower the temperature rise is, and thus the simpler the instrument's maintenance is. The larger the Q factor, the higher the sensitivity can be achieved. Figure 3 show the transmission spectra of $(HL)_m(D)_n(LH)_m$ with the same n=4 but different m=4, 8, 10, respectively. These results demonstrate that, with the increasing of the thickness of PC $(HL)_m$, the Q factor is increased, but the transmissivity is reduced. It can be explain that, the wider the thickness of $(HL)_m$ is, the stronger the reflection is, and therefore the lower the transmissivity is; simultaneously, the less the overlapping of wave functions is, the weaker the coupling effect between the input and output PC is, the stronger the filterability of this device is, the sharper the resonant peaks is, and thus the higher Q factor is. From the above discussions, we should suitably choose the value of m. In our next design, m= 8, n=4 are chosen with a transmissivity of more than 0.9 and a Q factor of more than 750.

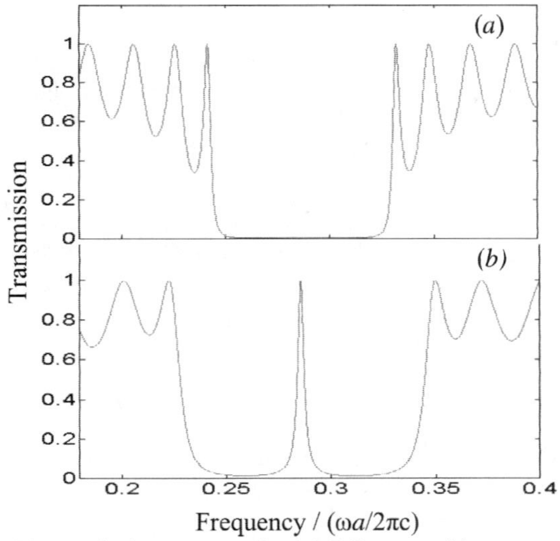

Figure 2. Transmission spectra for (a) PC $(HL)_m$ (b) $(HL)_3(D)_4(LH)_3$

Figure 4 shows the transmission spectra for different refractive indexes in an increment of $\Delta n = 0.0008$. Please noted that the refractive index of the middle spectrum (labeled the mark of "*") is $n_D = 2.2$. While the refractive index increases, the resonance hump shifts toward left, i.e., red shift. The expression of refractive index n_D can be expressed as $n_D = 2.2 + p * 0.0008$ ($p = \pm1, \pm2, \pm3, \cdots$), where p is the number offset away from the middle spectrum for researched resonance hump; while the researched resonance hump is on the left of the middle spectrum, p takes the sign of "+"; on the contrary, takes the sign of "-". It is not difficult to see there is direct proportion relation between the normalized frequency of the resonance hump and the refractive index n_D, and the normalized frequency between two neighboring resonance peaks is about 0.00006. According to the relation: $\Delta\lambda = (\Delta f \cdot \lambda_0^2)/c$,where λ_0 is the center wavelength of resonance peak, Δf is frequency change value, and $\Delta\lambda$ is wavelength change corresponding to Δf. From the above relation, in the case of $\lambda_0 = 1550$, it can be obtained that the resonant wavelength shifts 0.5*nm* with $\Delta n = 0.0008$ (corresponding to normalized frequency $\Delta f = 0.00006$). Because of the high Q factor, the full width at half-maximum is only approximately 0.48*nm*, thus the peak values of the spectra can be clearly resolved and the measurement accuracy is assured. As a result, a relative measurement error is less than 0.5%.

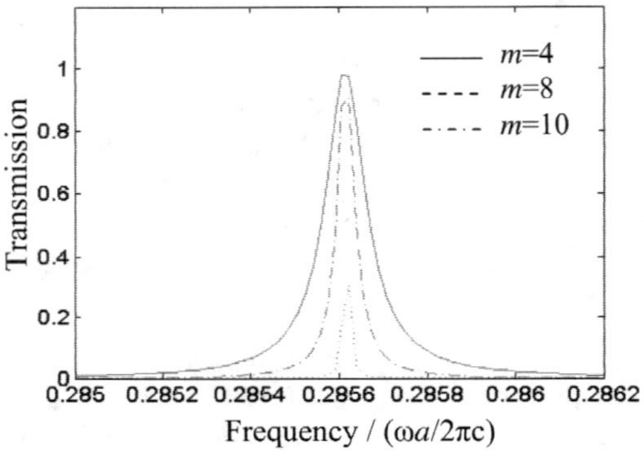

Figure 3. Transmission spectra for different m

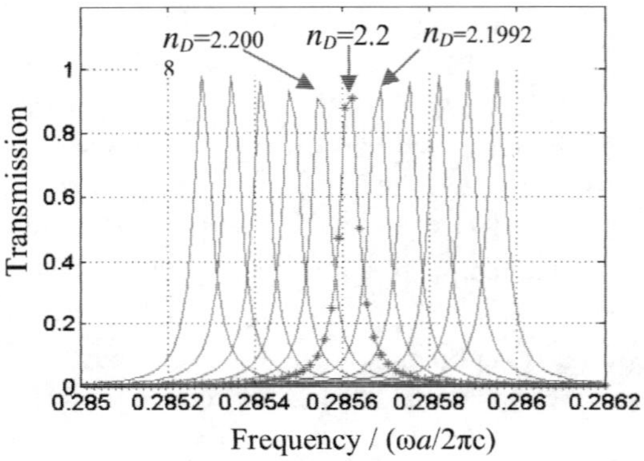

Figure 4. Transmission spectra for different n_D

4. Conclusion

We have presented a measurement device of refractive index with high transmissivity and Q factor based on resonant tunneling effect. Our device is constructed by one dimensional PC with a defect which gives rise to different resonant peaks in PC band gap. These peaks are sensitive to the medium which is filled in the defect. One of the features that make the device proposed here is the fact that all the peaks can be used simultaneously to monitor the refractive index with high resolution. With a spectrometer that can resolve a resonant peak shift of 0.5nm, at the resonant wavelength of about 1550*nm*, the resolution of refractive index should be better than $\Delta n = 0.0008$, and the transmissivity could be accessible to 1. The device is suitable in a variety of applications of practical interest, since many physical, chemical, and biological parameters can be known through measurement of the refractive index. In addition, this device is compact, robust, and highly stable over time, and thus can be widely applied in the fields of optical sensor and integrated optics.

Acknowledgement. This work was supported by the Scientific Research Program of Education Bureau of Hunan Province (08C368) and Hunan Provincial Natural Science Foundation of China （10JJ6095）.

References

[1] White I M and Fan X D 2008 Optics Express **16** 1020-1028
[2] Ksendzov A and Lin Y 2005 Opt. Lett. **30** 3344-3346
[3] Armani A M and Vahala K J 2006 Opt. Lett. **31** 1896-1898
[4] Chao C Y, Fung W and Guo L J 2006 IEEE J. Sel. Top. Quantum Electron **12** 134-142
[5] Yalcin A, Popat K C, Aldridge O C and Desai T A 2006 IEEE J. Sel. Top. Quantum Electron.**12** 148-155
[6] Yablonovitch E 1987 Phys. Rev.Lett. **58** 2059-2062
[7] Jone S 1987 Phys. Rev. Lett. **58** 2486-2489
[8] Plihal M and Maradudin A A 1991 Phys. Rev. B **44** 8565-8571
[9] Yee K S 1966 IEEE Trans. Antennas Propagat **14** 302-307
[10] Taflove A and Hagness S C 2000 Computational Electrodynamics: The Finite- Difference Time-Domain Method (Boston. London: Artech House)

A surface plasmon biosensor based on sub-wavelength metal slit arrays

Weitao[1] Ma, Jun Zhou[1], Mingyang Yang [1], Ruiqin Tan [2] and Hongwei Yuan [3]

[1] Institute of Photonics, Faculty of Science, Ningbo University, Ningbo, Zhejiang 315211, China
[2] Department of Electronic and Electric Engineering, School of Information Science and Engineering, Ningbo University, Ningbo 315211, China
[3] Department of Optical Engineering, School of Mechanical Engineering, Jiangsu University, Zhenjiang 212013, China

E-mail: ejzhou@yahoo.com.cn

Abstract. A novel surface plasmon sensor which is constructed with sub-wavelength metal slit arrays deposited on a buffer media layer is proposed. To obtain the maximum resonant transmission, the buffer layer is coated on a glass substrate and the refractive index of the buffer layer is selected close to that of the water environment. Based on rigorous coupled wave analysis, the numerical simulation shows that the resonant transmission is enhanced and the peak mediated by surface plasmon resonance is much sharper due to the buffer layer. Therefore, this surface plasmon sensor with a buffer layer provides an improved measurement precision in biosensing application.

1. Introduction

The surface plasmon resonance (SPR), excited in periodic metal nonastrutures, causes an enhanced transmission of light which is accompanied by strong field localization and has great potential applications in biosensors [1-5]. The wavelength of enhanced transmission depends on the refractive index of a thin layer of solution near metal-solution interface. Thereby, the binding of biomolecules to the metal nonastrutures surface causes a resonant wavelength shift due to their refractive index change. Moreover, Using the resonant wavelength shift in transmission spectrum as sensing signal results in a miniature bioanalytical instrument due to the simple optical setup, the metallic nanostrutures are expected to be applied for high-throughput and chip-based detections, such as DNA microarrays and protein microarrays [6, 7].

Two types of structure of periodic nanohoes and nanoslits are used as sensing elements [8-9]. Nanoslit-based structure is demonstrated with higher sensitivity than that of nanohole-based structure and is good candidates for sensing designs. In general, sub-wavelength metal slit arrays are directly deposited on the substrate such as silicon, glass. Such structure was demonstrated with high sensitivity, but its wavelength resolution was poor due to large bandwidth of transmission resonance spectrum. To get sharp transmission resonance peak, short slit gap and large metal layer thickness are needed, however, which would result in the resonance peak attenuated due to strengthening of absorption phenomena and different resonant phase matching conditions between the refractive indices of substrate and outside water environment.

In this paper, we propose a BK7 glass/Cytop/sub-wavelength Ag slit arrays /water environment multilayer structure where Cytop is an amorphous fluoropolymer with a refractive index very close to that of water. High resonant transmission peak intensity is achieved for the Cytop buffer layer and outside water environment have the same resonant phase matching conditions.

2. Principle and design

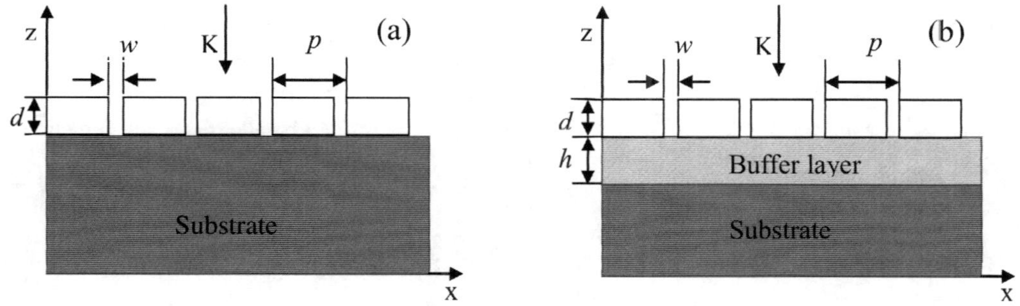

Figure 1. Schematic of sub-wavelength metal slit arrays without (a) and with (b) the buffer layer.

The schematic of sub-wavelength metal slit arrays is shown as figure 1 and all the components are treated as semi-infinite along y-direction. In figure 1 (a), the metal layer is directly deposited on the substrate without the buffer layer, while in figure 1 (b), a buffer layer is deposited on the substrate first and then the metal layer is deposited on the buffer layer. The material of metal slit array is Ag and its dielectric constant is described by the Lorenze-Drude model [10]. The substrate is selected as BK7 glass whose refractive index (n_S) is 1.51, the material of the buffer layer is Cytop whose refractive index is 1.31, close to that of pure water (1.3335).

For sub-wavelength metal slit arrays, two types of SPR excitations are shown to be possible, where one at the upper grating-analyte interface and the other one at lower grating-substrate interface which is affected by the multiple interference or cavity-type effects. The transmission of normally incident light through sub-wavelength metal slit arrays is enhanced at the wavelengths that satisfy the SPR conditions as

$$\lambda = p\sqrt{\frac{\varepsilon_m \varepsilon_d}{\varepsilon_m + \varepsilon_d}} \tag{1}$$

where λ is the resonant wavelength, p is the period of array, ε_m and ε_d is the dielectric constant of metal and the adjacent medium, respectively. The enhanced transmission peak mediated by SPR at the upper grating-analyte interface is related to the outside surface condition and using this enhanced peak wavelength as signals, high density sensing arrays have been demonstrated.

It is well known that small bandwidth of the enhanced transmission peak will result in large measurement precision. In sub-wavelength slit arrays, the resonant bandwidth is dependent on slit width and metal layer thickness. The bandwidth of resonant peak can be narrowed with slit width decreasing and array thickness increasing, however, for the structure as shown in figure 1 (a), the resonance peak becomes attenuated due to strengthening of absorption phenomena. To improve this problem, the Cytop is chosen as buffer layer material whose refractive index is close to that of the water environment as shown in figure 1(b). From equation (1), it is known that the SPRs at upper grating-analyte interface and at lower grating-buffer interface are almost the same, therefore, they could couple with each other and more energy can be transmitted at the resonance wavelength.

3. Simulation and discussion

The rigorous coupled wave analysis (RCWA) [11] is adopted to analyze sub-wavelength metal slit arrays. In most SPR biosensors, water environment is preferred; therefore all the simulation is under the water environment condition. To achieve narrow bandwidth, the dimensions of sub-wavelength metal slit arrays are chosen as: the silt width w=40 nm, the Ag array thickness d=150 nm and the grating period p=600 nm, and TM incident light (with the magnetic field parallel to the slit direction) is only considered.

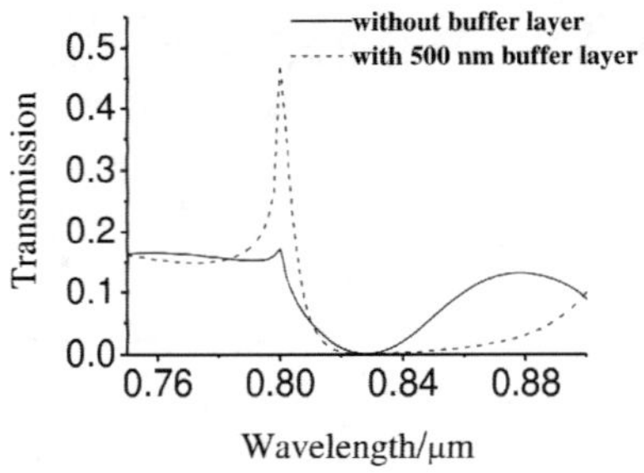

Figure 2. The transmission spectrum of sub-wavelength metal slit arrays with and without the Cytop buffer layer.

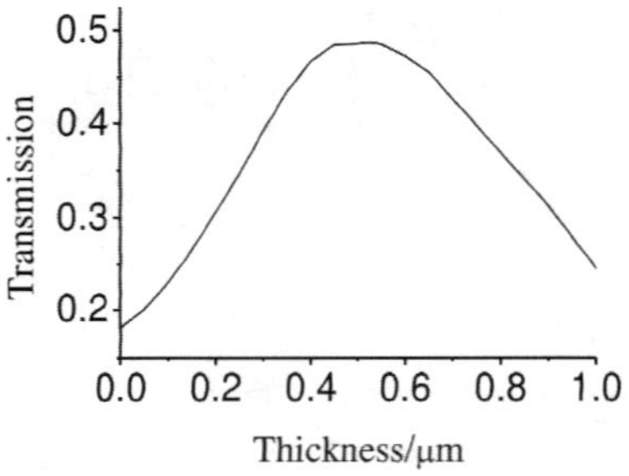

Figure 3. The dependence of the enhanced peak intensity on the buffer layer thickness.

The calculated transmission spectrum of sub-wavelength metal slit arrays with and without the Cytop buffer layer is shown in figure 2. The enhanced transmission peaks for the two cases are both at 800 nm, which are induced by the SPR at the upper grating-analyte interface. The intensity of the enhanced transmission peak when without the buffer layer is too small, just about 1.67 to identify. However, with 500 nm thick buffer layer, the intensity of the enhanced transmission is almost three times larger than that without the buffer layer and the bandwidth of peak is as narrow as 5 nm.

The SPR mode localized on the lower grating-buffer interface is influenced by the substrate, if the buffer layer thickness is less than the SPR field penetration depth. In this case, the buffer layer and the substrate can be considered as an equivalent semi-infinite medium with an effective dielectric constant ε_{eff}, where $\varepsilon_{eff}^{-1} = \varepsilon_{buf}^{-1}\left[1-\exp(-h/g)\right] + \varepsilon_{sub}^{-1}\exp(-h/g)$ under lowest order approximation [12], ε_{buf} and ε_{sub} are the dielectric constants of the buffer layer and the substrate, respectively, g is the penetration depth of the SPR field localized on the lower grating-buffer interface and h is the thickness of the buffer layer. The effective refractive index n_{eff} of the equivalent semi-infinite medium can be calculated from the effective dielectric constant of ε_{eff}. So the effective refractive index n_{eff} depends on the buffer layer thickness and is in the range of the buffer refractive n_{buf} index (1.31) to the substrate refractive index n_s (1.51). When the effective refractive index n_{eff} is equal to the water environment refractive index, the resonant peak obtains the maximum transmission.

To get the optimal thickness range of the buffer layer, various values of the buffer layer thickness are used in the calculation. Figure 3 shows the dependence of the enhanced peak intensity on the buffer layer thickness. It can be clearly seen that there is a maximum transmission value about 0.487 at depth of 500 nm, when the depth is less than 500 nm, the transmission increase with depth; when the depth is more than 500 nm, it is in verse. Moreover, when the thickness is in the range of 370-670 nm, the transmission is beyond 0.45, which is considered as the optimal thickness range of the buffer layer due to high resonant transmission.

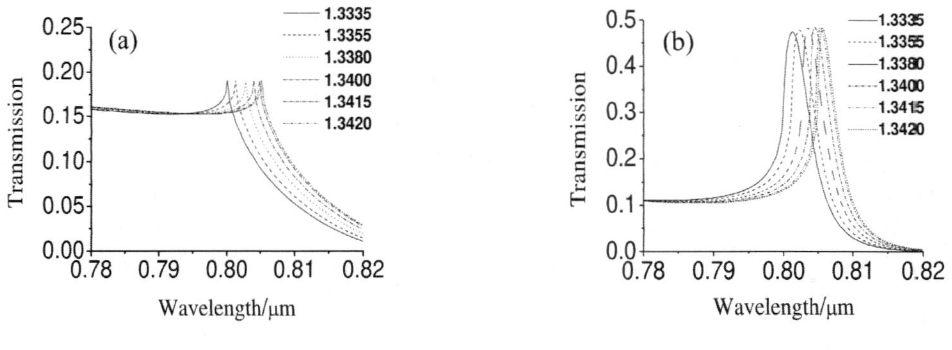

Figure 4. Transmission spectrum against the refractive index of the water environment (a) Without the buffer layer. (b) With the buffer layer.

Figure 4 shows the transmission spectrum in water and different methanol/water mixtures under the conditions without and with the buffer layer. The refractive indexes of pure water and different methanol concentrations of 10%, 20%, 30%, 40%, 50% are 1.3335, 1.3355, 1.3380, 1.3400, 1.3415 and 1.3320, respectively. The buffer layer thickness is 500 nm. It is noted that both the enhanced resonant peaks in figure 4 (a) and (b) move toward long wavelength with increasing refractive index of the outside medium, which is predicated by the equation (1). The resonant peaks in figure 4 (b) have better spectral resolution than that in figure 4 (a) due to higher resonant peak intensity and narrower bandwidth.

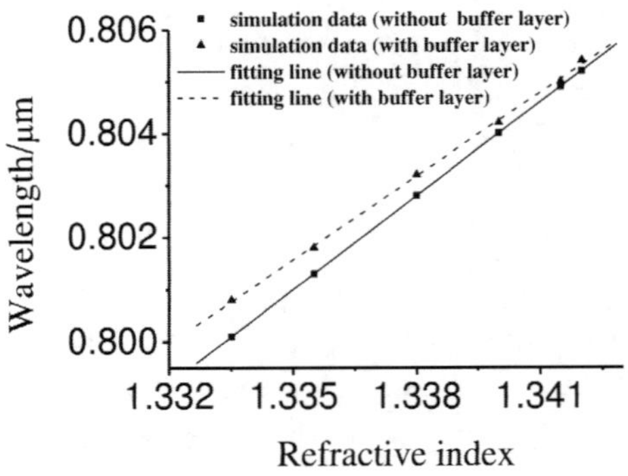

Figure 5. Resonance peak wavelengths against the refractive index of the outside medium and linear fitting curves.

Figure 5 shows the dependence of the resonant peak wavelengths on the refractive index of the outside medium and the corresponding linear fitting curves. The slope of the fitting curves shows the sensitivity to bulk refractive index change of the outside medium. When the structure without the buffer layer, the sensitivity to bulk refractive index is as high as 600 nm/RUI while 535 nm/RUI for the structure with the buffer layer. So the sensitivity is affected by the buffer layer. Meanwhile, the bandwidth of resonant transmission peak is narrowed and become sharper, greater accuracy on the measurement of the resonant wavelength provides better refractive index resolution.

4. Conclusion

In summary, the transmission spectrum of sub-wavelength metal slit arrays in the water environment is investigated based on the rigorous coupled wave analysis. The numerical simulations show that the resonant transmission can be enhanced by inducing the Cytop buffer layer. Choosing the optimum buffer layer thickness, the resonant transmission peak intensity becomes largest and the sensitivity of sub-wavelength metal slit arrays remains high although which is affected by the buffer layer. Therefore adopting the buffer layer of the optimum thickness and the periodic array of narrow slit and thick metal layer, the SPR peak with narrow bandwidth and high transmission peak intensity would be achieved, which guarantees accurate determination of peak location.

Acknowledgments

This work is supported by the Nature Science Foundation of China (60977048), Qianjiang Talent Program of Zhejiang (2009R10064), the Nature Science of Ningbo (2008A610001) and the K. C. Wong Magna Fund in Ningbo University, China.

References

[1] Ebbesen T W, Lezec H J, Ghaemi H F, Thio T and Wol P A 1998 *Nature* **391** 667
[2] Kawata S 2001 *Near-Field Optics and Surface Plasmon Polaritons* (New York: Springer)
[3] Lezec H J, Degiron A, Devaux E, Linke R A, Moreno L M, J F, Vidal G and Ebbsen T W 2002 Science **297** 820

[4] Brolo A G, Gordon R, Leathem B and Kavanagh K L 2004 *Langmuir* **20** 4813
[5] Gordon R, Sinton D, Kavanagh K L and Brolo A G 2008 *Acc. Chem. Res.* **41(8)** 1049-57
[6] Schena M, Shalon D, Davis R and Brown P 1995 *Science* **270** 467
[7] Gavin M B and Stuart L S, 2009 *Science* **289** 1760
[8] Murray-Methot M P, Ratel M and Masson J F 2010 *J. Phys. Chem. C* **114** 8268
[9] Karabchevsky A, Krasnykov O, Auslender M, Hadad B, Goldner A and Abdulhalim I 2009 Plasmonics **4** 281
[10] Rakic A D, Djurisic A B, Elazar J M and Majewski M L 1998 *Appl. Opt.* **37** 5271
[11] Moharam M G, Grann E B and Pommet D A 1995 *J. Opt. Soc. Am. A* **12** 1068
[12] Fiore A, Berger V, Rosencher E, Bravetti P and Nagle J 1998 *Natrue* **391** 463

Journal of Physics: Conference Series **276** (2011) 012140 doi:10.1088/1742-6596/276/1/012140

Analysis of the optimized low-nonlinearity lateral effect sensing detector

Saeed Olyaee, Ali Bidrang, Mahdie Izadpanah

Nano-photonics and Optoelectronics Research Laboratory (NORLab), Faculty of Electrical and Computer Engineering, Shahid Rajaee Teacher Training University (SRTTU), Lavizan, 16788-15811, Tehran, Iran

E-mail: s_olyaee@srttu.edu

Abstract- In this paper, the current time delay (TD) and the voltage difference (VD) modes for some practical lateral effect position sensing detectors are investigated, simulated, and compared. Some advantages of the current time delay mode are more linear response, independent position from magnitude and frequency, background illumination-independent response, and no need to signal normalization.

Key words: Lateral effect position sensing detectors, voltage delay, time delay.

1. Introduction

Position-sensing detectors are known as simple photodiodes capable of detecting centroid of light beam position. There is an important feature for PSDs due to no need to have contact with specimen being inspected. Position information is calculated from photocurrent signal magnitude by these detectors. For designing these detectors with high resolution, the most effective method is to create Schottky barrier with surface diffusion method [1].

When an optical spot incidents on the detector surface, electric charge is generated proportional to the light intensity at the collision region. Electric charge passes from resistive layer and collected as the photocurrent by electrodes. It is divided between two terminals with an inverse relation to the distance of optical spot from electrodes.

Applying full reverse bias to LEP, the thickness of depletion region increases, diffusion current decreases to zero and surface regeneration is stopped. To achieve best performance quality, the photodiode should be able to absorb maximum photons in depletion region. It means photons should not be absorbed anywhere except depletion region before passing it.

So, when a p-n junction is illuminated by the optical spot, little energy is saved in space charge region. Under reverse bias, this light energy will move horizontally to the electrodes mounted next to the junction. This phenomenon is called lateral effect first presented by Schottky in 1930, but has not been developed much [2]. In 1959, Wallmark used this theory practically [3]. He designed a device including a p-n junction. By illuminating light beam to the surface, a position-dependent voltage difference is generated between the contacts [3].

The position sensing detectors have many applications in distance measurement, machine vision, atomic force microscope, surface shaping, robots, tracking laser beam, etc. They are classified into two types: lateral effect position sensing detectors and four-quadrant position sensing detectors [4], with some similarities and differences to each others, physically and practically [1].

Voltage delay mode has been widely used to determine the position of optical spot until now. As shown in Figure 1, if optical spot incidents on the detector surface, I_1 and I_2 photocurrents are generated at electrodes and then convert to voltage, so the optical spot position can be obtained by calculating voltage difference between two electrodes. The disadvantage of this method is diameter-dependent voltage in optical spot led to detector nonlinearity. In this paper, current time delay mode is proposed to make detector more linear.

In this method, as the optical spot incidents on the detector surface, a current has been generated. The current reaching time, for each electrode is different, depending on the optical spot position. So, there is a time difference between two signals reached to terminals. Here, the diameter of optical spot has no effect on signals arrival time. For example, if reaching signals oscillate synchronous, the optical spot will mount at the center of detector surface.

The rest of the paper is organized as follows. In section 2, a brief investigation is presented about LEPs, their dynamic equations and electronic circuit based on the [7]. The simulation results for three practical LEPs are discussed in section 3 and the conclusions are drawn in the final section.

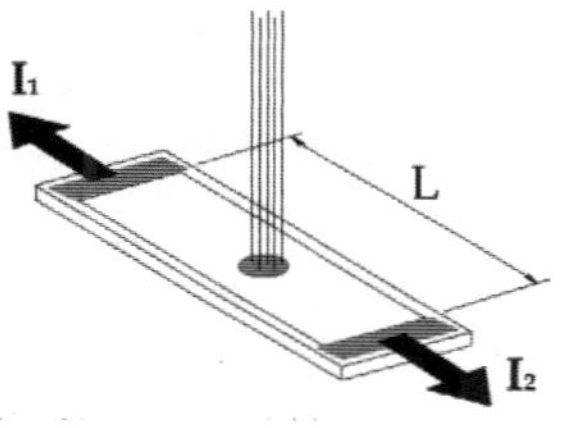

Figure 1. One-dimensional detector structure.

2. Lateral effect position sensing detectors (LEPSDs)

Lateral effect position sensing detectors include a uniform resistive layer mounted in one or two sides of semiconductor sub-layer with high resistivity (Figure 2).

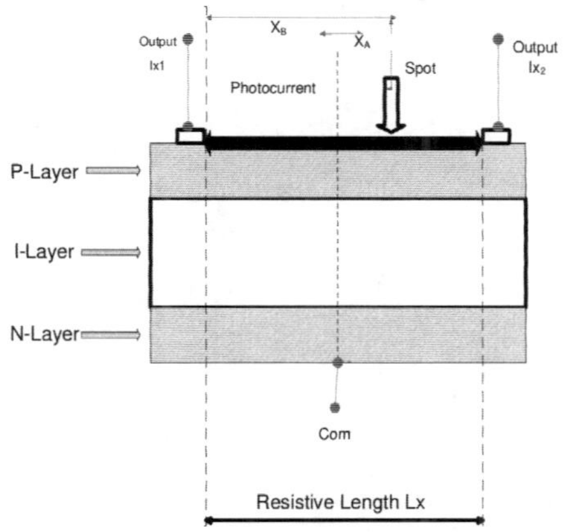

Figure 2. Schematic of lateral effect position sensing detectors (LEPs) [5].

In one-dimensional case, there is a pair of electrodes to obtain position signals at the two ends of resistive layer, and in two-dimensional case, two pairs of electrodes mount on the edge of the detectors mutually. The active area (or resistive layer) includes a p-n junction generated photocurrent due to light collision to the surface (photovoltaic effect).

LEPs have uniform big structures with an anode and a cathode. Because these detectors do not depend on optical spot diameter magnitude, detectivity quality is independence of their distance from the object. There are two models for LEPs: one-dimensional and two-dimensional models, i.e. the ability to detect 1-D or 2-D positions. Usually in detector analysis, first 1-D case is investigated and then it can be generalized to 2-D case. When an optical spot mounts on the detector surface, photocurrent is generated by each electrode, depending on the distance of beam centroid from electrodes. So, with measuring output, the beam position can be also measured [6]. The equation of 1-D position detectivity, based on Figure 1 is given as [7]:

$$\Delta X = S_1 \frac{I_2 - I_1}{I_2 + I_1} \tag{1}$$

where I_1 and I_2 are the output currents of electrodes. After calculating currents, we can write them in the form of voltage difference between two electrodes. As shown in Figure 3, Eq. 1 can be rewritten as:

$$\Delta X = S_1 \frac{V_2 - V_1}{V_2 + V_1} \tag{2}$$

and the time delay between two signals can be obtained by Eq. 3:

$$\Delta X = S_2 \Delta t \tag{3}$$

where Δt is the time delay between the two signals.

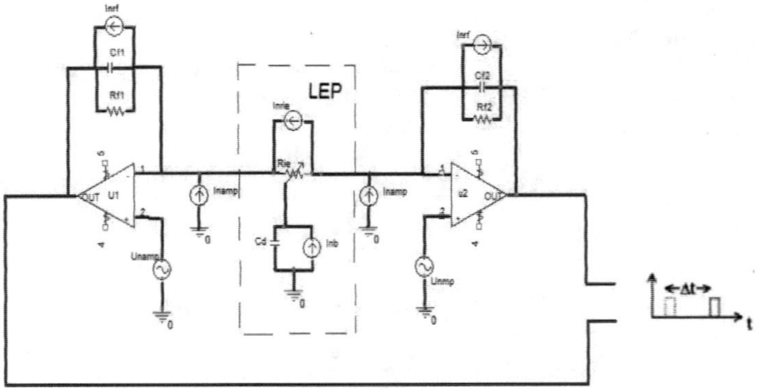

Figure 3. Converting output currents to voltage.

When the optical spot incidents on the LEP surface, there will be a carrier distribution led to a gradient in excess carrier concentration. The optical spot will start to spread in a Gaussian manner controlled by the diffusion coefficient and the recombination lifetime without any lateral or transverse external electric fields. So, the obtained current difference can be in the form of voltage difference between electrodes. This idea was presented first by Walmark. In this method, current difference led to voltage difference mode while time delay causes time difference and more linear response. To simulate LEP, we used Lucovsky Equation [6-7]:

$$\frac{\partial^2 V}{\partial x^2}-\frac{J_s\rho_P}{w_P}[\exp(\frac{q_e V}{k_B T})-1]-\frac{\rho_P c}{w_P}\frac{\partial V}{\partial t}=-\frac{q_e\rho_P}{w_P}g(x,t) \tag{4}$$

where V is the lateral photo-voltage, J_s is the transverse saturation current density, ρ_P is the resistivity of the p-doped region and w_p is the thickness of the p-doped region.

In small signal analysis, because $q_e V << k_B$, it can be ignored and the Eq. 4 can be rewritten as:

$$\frac{\partial^2 V}{\partial x^2}-\frac{\rho_P c}{w_P}\frac{\partial V}{\partial t}=-\frac{q_e\rho_P}{w_P}g(x,t) \tag{5}$$

With normalizing relation 5 (to the $V_{G_{max}}$ and by the time constant RC) it can be written [7]:

$$\frac{\partial V}{\partial' t}=\frac{\partial^2 V_r}{\partial' x^2}-\frac{R_{se}}{R_{sh}}V_r+g_r(x',t') \tag{6}$$

where $V_r=\frac{V}{V_G}$ is normalized to $V_{G_{max}}$ (represented by $V_G(x',t')$) and $g_r(x',t')=\frac{V_G(x',t')}{V_{G_m}}$ is normalized photon flux distribution in the detector surface [5]. Equation 5 is valid for any LEPs with a similar ratio $\frac{R_{se}}{R_{sh}}$. We can replace a similar relation for current distribution not voltage instead of Eq. 5. For this reason, the initial and boundary conditions should be considered as:

$$V_r(x',0)=V_r(0,t')=V_r(1,t')=0 \tag{7}$$

And for generated photocurrent in two electrodes, we have:

$$i_{r1,r2}=\frac{\partial V_r}{\partial x'}\downarrow\ x'=0, x'=1 \tag{8}$$

By expanding Eq. 5 in Fourier series form and by converting voltage to current, the currents will be:

$$i_{r_o}=I(1-x'_p)\sin(\omega' t'+\varphi_0) \tag{9}$$

$$i_{r1}=-I(x'_p)\sin(\omega' t'+\varphi_1) \tag{10}$$

x' is the normalized position for detector length and φ_0 and φ_1 denote phase delays for signals.

By expanding Fourier series around mid point, the slope of linear region is given [7]:

$$S_2=\frac{\omega L}{2}\left\{\frac{\cosh(\varepsilon_r)-\cos(\varepsilon_i)}{\varepsilon_r\sin(\varepsilon_i)-\varepsilon_i\sinh(\varepsilon_r)}\right\} \tag{11}$$

Eq. 11 shows that S_2 is linearly proportional to frequency and detector length. For linear region slope, in voltage difference mode, we have [7]:

$$S_1\approx L\left\{\frac{\cosh(\varepsilon_i)-\cos(\varepsilon_r)}{\varepsilon_i\sin(\varepsilon_i)+\varepsilon_r\sinh(\varepsilon_r)}\right\} \tag{12}$$

S_1 and S_2 are the slopes of time delay and voltage difference modes [7].

3. Simulation results

The functions resulted from Eqs. 11 and 12 are shown in Figure 4. The linearity of time delay mode can be easily inferred in comparison to voltage difference mode .With R_{se} and R_{sh}, we can use these relations for any photodetectors. We know when an optical spot approaches to detector terminals, voltage difference decreases. As shown in Figure 4, over a dynamic frequency range, time delay mode is more linear than voltage difference mode. Figure 5 shows the simulation results for S3931 detector. This detector is more linear in time delay mode than S3979 detector over a larger frequency range. The simulation results for S3932 detector are also shown in Figure 6. It can be concluded that this detector has better performance comparing two other detectors.

Figure 7 is a 3-dimensional graph for voltage difference and time delay mode signals. As shown in this figure, in voltage difference mode, signals will saturate at the edges.

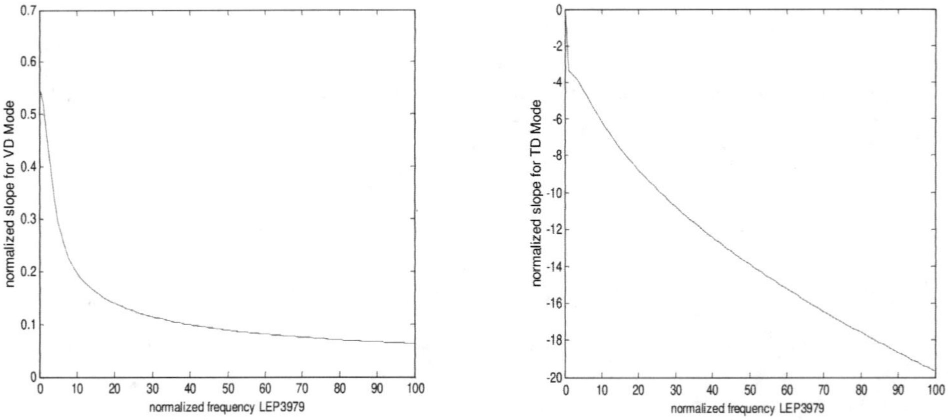

(a) (b)

Figure 4. Normalized slope with respect to the normalized frequency for (a) VD and (b) TD modes in S3979 detector.

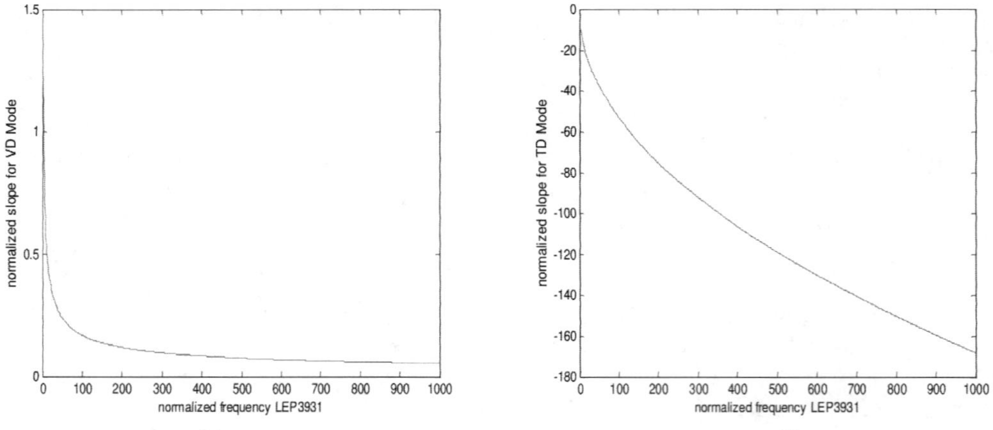

(a) (b)

Figure 5. Normalized slope with respect to the normalized frequency for (a) VD and (b) TD modes in S3931 detector.

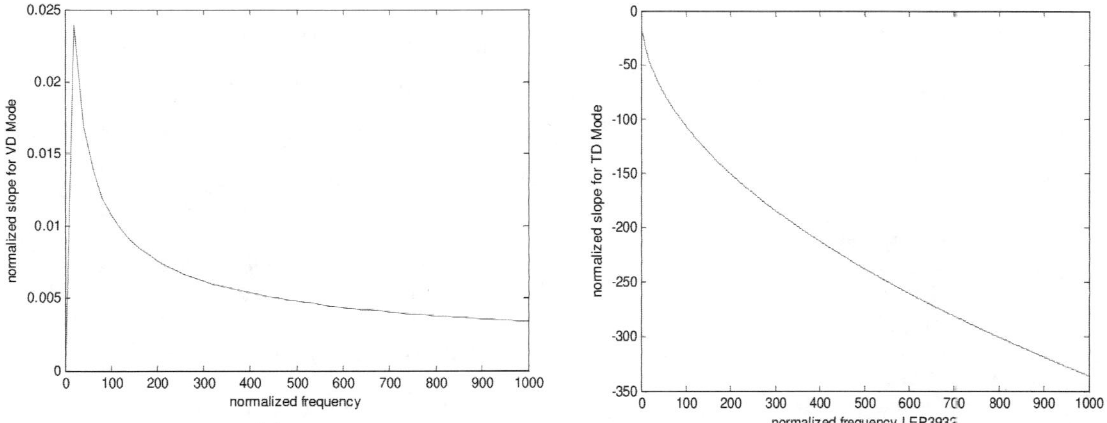

(a) (b)

Figure 6. Normalized slope with respect to the normalized frequency for (a) VD and (b) TD modes in S3932 detector.

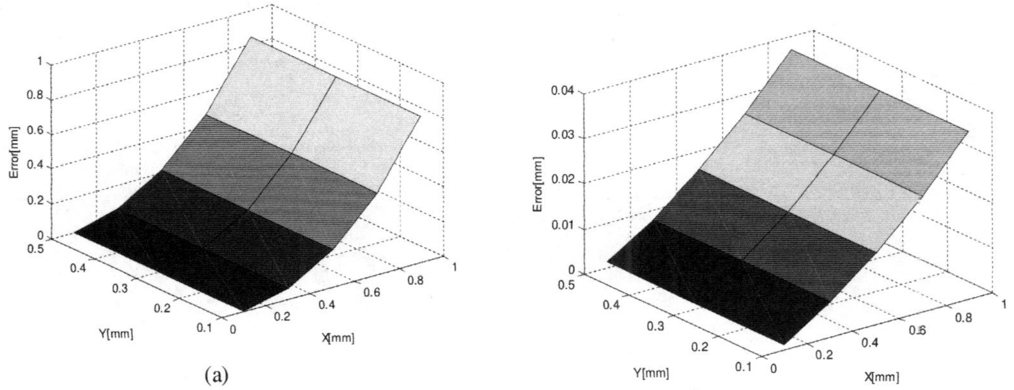

(a) (b)

Figure 7. The error of position detection in X- and Y-axes for (a) VD and (b) TD modes in S3931 detector.

4. Conclusion

In this paper we have compared three practical LEPSDs based on the analysis reported by Abdullhalim in 2004 [7]. According to the obtained results, time delay mode has more advantages comparing to the voltage difference mode. Also with having the ratio R_{se} / R_{sh}, for any lateral effect position sensing detectors, TD and VD modes relations can be written and investigated. Here, S3932 has more linearity in the response comparing to the S3931 and S3979.

References

[1] Petersson G P, Lindholm L, 1998 ,"Position sensitive light detectors with high linearity", *IEEE Journal of Solid State Circuits*,Vol.sc-13, No.**3**.

[2] Schottky W,1930, "On the origins of photoelectrons in Cu_2O-Cu photocells", *Phys. Z.*, **31**, pp.913-925.

[3] Wallmark J T, 1957, "A new semiconductor photocell using lateral photoeffect", *Proc. IRE*, pp. 474 - 483.

[4] Makynen A, 2000,"Position-sensitive devices and sensor systems for optical tracking and displacement sensing applications", Department of Electrical and informational Engineering university of Oulu.

[5] H X Song, X D Wang, L Q Ma, M Z Cai and T Z Cao "Design and performance analysis of laser displacement sensor based on position sensitive detector (PSD)" *J. Phys. D: Appl. Phys.* 48 (2006) 217–222

[6] Lucovsky G, 1960, "Photoeffects in nonuniformly irradiated p–n junctions,"*J. Appl. Phys.*, pp. 1088–1095.

[7] Abdulhalim I, 2004, Device physics of the time-delay mode of operation of optoelectronic lateral effect position sensors, *J. Phys. D: Appl. Phys.* 37 (2004) 1376–1384.

Neural network approximation of nonlinearity in laser nano-metrology system based on TLMI

Saeed Olyaee and Samaneh Hamedi

Nano-photonics and Optoelectronics Research Laboratory (NORLab), Faculty of Electrical and Computer Engineering, Shahid Rajaee Teacher Training University (SRTTU), Lavizan, 16788, Tehran, Iran

E-mail: s_olyaee@srttu.edu

Abstract. In this paper, an approach based on neural network (NN) for nonlinearity modeling in a nano-metrology system using three-longitudinal-mode laser heterodyne interferometer (TLMI) for length and displacement measurements is presented. We model nonlinearity errors that arise from elliptically and non-orthogonally polarized laser beams, rotational error in the alignment of laser head with respect to the polarizing beam splitter, rotational error in the alignment of the mixing polarizer, and unequal transmission coefficients in the polarizing beam splitter. Here we use a neural network algorithm based on the multi-layer perceptron (MLP) network. The simulation results show that multi-layer feed forward perceptron network is successfully applicable to real noisy interferometer signals.

1. Introduction

A nano-metrology system can include length or size measurements, (where dimensions are typically given in nanometers and the measurement uncertainty is often less than 1nm), as well as measurement of force, mass, electrical and other properties like thin films layers and surfaces, carbon nano-tubes, inorganic nano-tubes, and nano-wires [1-3]. Precise control of dimensions of objects is the key issue of nanotechnology and the science of nano-objects. The measurement techniques developed for conventional materials in many cases cannot be simply applied to nanostructures. Nano- metrology has a crucial role in order to produce nanomaterials and devices with a high degree of accuracy and reliability. With development in semiconductor industry to make devices and circuits in nanometer size such as nanoelectro-mechanical systems (NEMS), semiconductor nanosensors, and nanoelectronics is asking for improved resolution and reached the high accuracy of nano-metrology systems [1, 4]. Non-contact, absolute-distance measurement system with high resolution and an extending range is essential for a number of applications like semiconductor manufacture, calibration, precision cutting, lattice constant measuring, robotic systems, and controlling the gap of liquid crystal display (LCD) panels [1-4]. The laser heterodyne interferometer has special benefits for accurate length/speed measurements because it allows very precise measurements with nanometer resolution based on phase measurement [5-7].

Interferometers are typically influenced by polarization ellipticity or non-orthogonality of the laser beams. Even perfect optical components can also contribute to this nonlinearity. Furthermore, the misalignment of the interferometer seriously affects the nonlinearity of a heterodyne interferometer. This cyclic nonlinearity exists in all laser interferometers. This error can be worse under certain conditions. The nonlinearity errors happen mainly from two factors: polarization-mixing and frequency-mixing [8]. In a three-longitudinal-mode laser heterodyne Doppler interferometer (TLMI), the phase is detected with some electronic sections [9]. The periodic nonlinearity in this nano-metrology system is arising from polarization effects, frequency mixing, intermodulation distortion, phase detection uncertainty, and electrical cross-talk. However, the periodical nonlinearity property is an obstacle to improve the high measurement accuracy in nanometer scale [9]. In order to minimize the nonlinearity error of the heterodyne interferometer, it is important to model the nonlinearity and distinguish it from actual phase.

Quenelle [10] and Sutton [11] performed investigations of periodic error in heterodyne Michelson interferometers. Specific efforts have included to measure periodic error under various conditions, frequency domain analyses [12], analytical modeling techniques [13], and Jones calculus modeling methods [14]. Artificial Neural Networks are a programming paradigm that seek to emulate the microstructure of the brain, and are used extensively in artificial intelligence problems from simple pattern-recognition tasks, to advanced symbolic manipulation. Li *et al.* have first modeled and corrected the nonlinearity in the homodyne interferometers based on the multi-layer perceptrons neural networks in 2003 [15]. Then, Heo *et al.* employed a combined algorithm, which adopts recursive least square error (RLS) method and a neural network back-propagation algorithm to compensate the nonlinearity in the heterodyne interferometer in 2007 [16]. Olyaee *et al.* have modeled the nonlinearity resulting from non-ideal polarized light of a modified high-resolution laser heterodyne interferometer by using multi-layer perceptrons and radial basis function as single neural networks, and stacked generalization method by combination of neural networks [17].

By replacing a three-mode laser heterodyne interferometer the requirement for high resolution about less than 1 nanometer is achieved [14]. Yokoyama *et al* have first analyzed periodic nonlinearity error in TLMI [18]. Then Olyaee *et al* used Jones matrices and plane wave methods to predict and simulate nonlinearity originating from errors in optical alignment and polarization states of the components of nano-metrology system based on TLMI [7-9, 14]. But all of these methods need a lot of time and computational analysis.

The polarization-mixing in the TLMI commonly happens within elliptically polarized laser beams, non-orthogonally polarized laser beams, rotational error in the alignment of laser and beam splitter, rotational error in the alignment of the mixing polarizer, and unequal transmission coefficients in the polarizing beam splitter [14].

In this paper, we propose a simple algorithm using an artificial intelligence method which is set up in front of the detectors in the interferometer. The ability to learn and generalize is fundamental to any learning machine. Artificial neural networks (ANNs) use inductive learning to find general concepts from their concrete examples [19]. The feed forward neural network trained by back-propagation predicts the nonlinearity error and difference with the reference signal. In this paper, sections 2 and 3 introduce the principals of the optical setup and analysis of nonlinearity error, respectively. In section 4, the simulation results are presented. And finally conclusions are drawn in section 5.

2. Principals

In the TLMI, three orthogonally polarized beams with a split frequency of approximately 500MHz generated within the He-Ne laser first pass through a non-polarizing beam splitter (BS). One part is led to a linear polarizer (LP) in the base arm, the orthogonally polarized fields of the three modes are combined with the polarizer, and then, it is focused on the avalanche photodiode (APD_b) to generate a reference signal. The other part of emerged light is led to the measurement system (interferometer). There are two paths namely reference and target paths, which the incident light is separated in two orthogonal orientations with the polarizing beam splitter (PBS). In a three-longitudinal-mode laser heterodyne Doppler interferometer (TLMI) the phase is detected with some electronic sections [14].

By using the super-heterodyne phase detection circuit in the following of base and measurement avalanche photodiodes (APD_b) and (APD_m), the base and measurement signals are respectively concluded as [14]:

$$V_{o_b} = g\cos\left(\omega_S t + \Phi_{nl_b}\right) \tag{1}$$

$$V_{o_m} = g\cos\left(\omega_S t - 2\Delta\Phi + \Phi_{nl_m}\right) \tag{2}$$

where g is the total transfer gain of the signal conditioner, $\omega_S = |\omega_H - \omega_L|$ is the secondary angular beat frequency, $\omega_H = \omega_3 - \omega_2$ and $\omega_L = \omega_2 - \omega_1$ are the higher and lower intermode angular beat frequencies, respectively. Here Φ_{nl_b} and Φ_{nl_m} are the phase nonlinearities in the base and measurement arms, respectively.

In the above formula, $\Delta\Phi = (4n\pi / \lambda_2)\Delta z$ is the phase shift due to the optical path difference (Δz) [8], where n is the refractive index and λ_2 is the wavelength of the central mode. Finally total nonlinearity phase is concluded as:

$$\psi_{nl-total} = \Phi_{nl_m} - \Phi_{nl_b} \tag{3}$$

3. Neural network approximation of nonlinearity

The block diagram of nonlinearity modeling in TLMI based on the neural networks is shown in Figure 1. Using artificial neural networks, it is possible to approximate the full complexity computational problems in a high dimensional space. As shown in Figure 2, ANNs will consist of at most a few hundred neurons and very limited connections between them. Generally, artificial neural networks are basic input and output devices with the neurons organized into layers [19]. In this paper, input layers are output voltage of base and measurement arms (v_{om} and v_{ob}) and output layer is nonlinearity error ($\psi_{nl-total}$). The learning process consists of finding the correct values for the weights between the input and output layer. In the metrology industry, it is increasingly led to small (nano) dimensions and of course we need more precise measurement system with higher resolution. The main goal of using neural network is its fast feasibility.

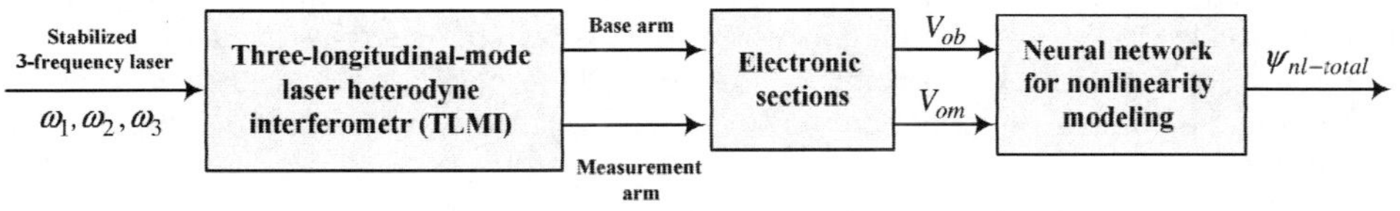

Figure 1. The block diagram of nonlinearity modeling in TLMI.

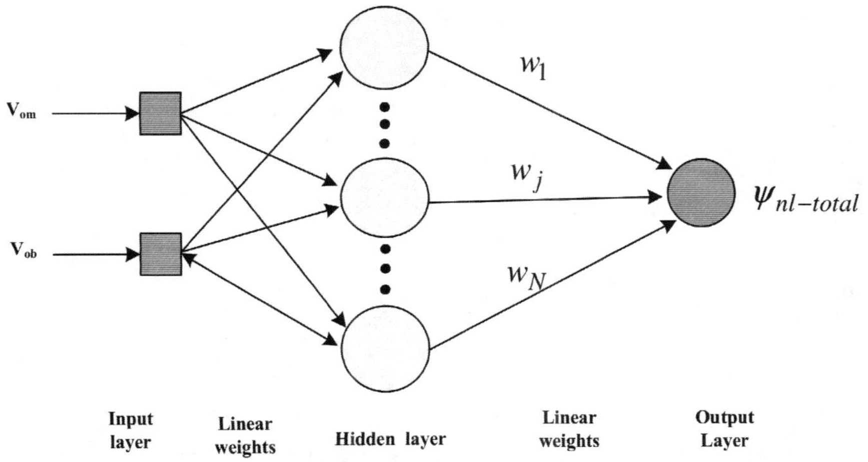

Figure 2. The schematic of MLPN for nonlinearity modeling in the TLMI.

The Multilayer perceptron is an artificial neural network that is used extensively for the solution of a number of different problems, including pattern recognition and interpolation [20]. The network consist of a layer of input neurons, coupled with a layer of output neurons, and one or more layers between them, as shown in Figure 2. The back propagation algorithm is most famous to layered feed forward networks, or multilayer perceptron is used generally to train neural networks. In this paper, we used MLP network with input and output layers and one hidden layer.

4. Simulation Results

In this section, the performance of the MLP network is studied in terms of modeling nonlinearity accuracy and training time. The results obtained are compared with actual nonlinearity. Input data for networks are output voltage of electronic circuit from base and measurement arms. The input vector and output vector are respectively as [19]:

$$\mathbf{P} = [\mathbf{V}_{om} \,, \mathbf{V}_{ob}] \qquad\qquad (4)$$
$$\mathbf{T} = [\psi_{nl-total}]$$

The full data are about 6000 data, which 4000 is for training phase and the rest is for test phase. The training input vector is obtained by inserting constant nonlinearity parameters in Eqs. (1) and (2) and the target vector is generated from Eq. (3). Table 1 represents properties of MLP network for nonlinearity modeling and time consuming for some deviation parameters.

According to Table 1, multi-layer perceptron with one hidden layer is used for all deviation parameters. There are 6 neurons with 60 epochs and learning rate about 0.01 for the best mean square error. The time consuming for this result is 3.09s. The result of nonlinearity modeled with MLP network according to wavelength is plotted in Figure 3 for parameters of Table 1.

Comparison of mean square error as the number of training samples for nonlinearity modeling for deviation parameters $\delta_t = 0.2$, $\delta_r = 0.28$ (ellipticity), $\tau_t = 0.90$, $\tau_r = 0.95$ (transmission coefficients of the polarizing-beam splitter), and $\alpha = 2$ (rotation angle) is summarized in Table 2. According to the Figure 4, by increasing number of training data, the neural network can model nonlinearity better. In conclusion, the number of training samples is affect on the results. Table 3 represents the performance of nonlinearity modeling as number of epochs for same deviation parameters by MLP network. Results are plotted in Figure 5. It is concluded that the optimum mean square error is with 60 epochs and by increasing the number of epochs from 60, the performance is not modified, although the time consuming is increased.

Table 1. Properties of multi-layer perceptrons (MLP) network for nonlinearity modeling for some deviation parameters.

Curve	Condition					Properties of nonlinearity modeling by MLPs network			Results	
	δ_t	δ_r	τ_t	τ_r	α	Number of neurons in the hidden layer	Number of epochs	Learning rate	consuming time for neural network (seconds)	Mean square error (dB)
a	0.2	0.28	0.90	0.95	2	5	50	0.01	3. 935	-66.07
b	0.2	0.28	1	1	2	5	50	0.01	3.15	-76.61
c	0.2	0.2	1	1	2	5	50	0.01	3.072	-100.90
d	0.2	0.2	0.9	0.95	0	6	60	0.01	3.09	-105.83
e	0.2	0.28	1	1	0	6	50	0.01	3.243	-83.15

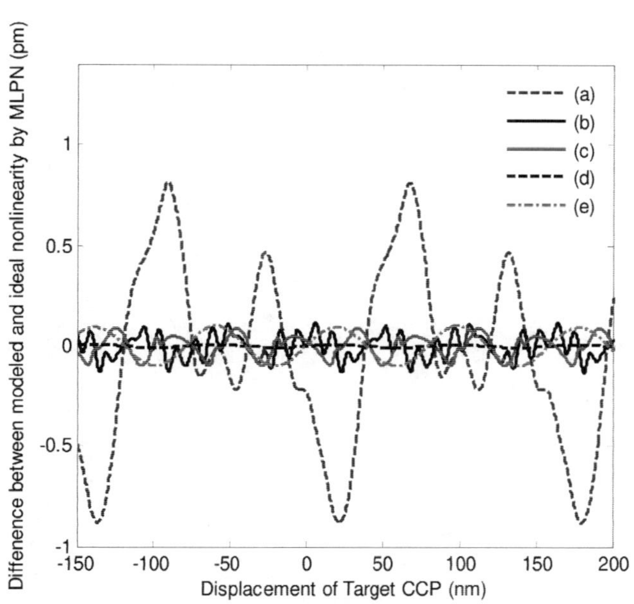

Figure 3. The difference between expected nonlinearity and modeled nonlinearity with MLP network for deviation parameters listed in Table 1.

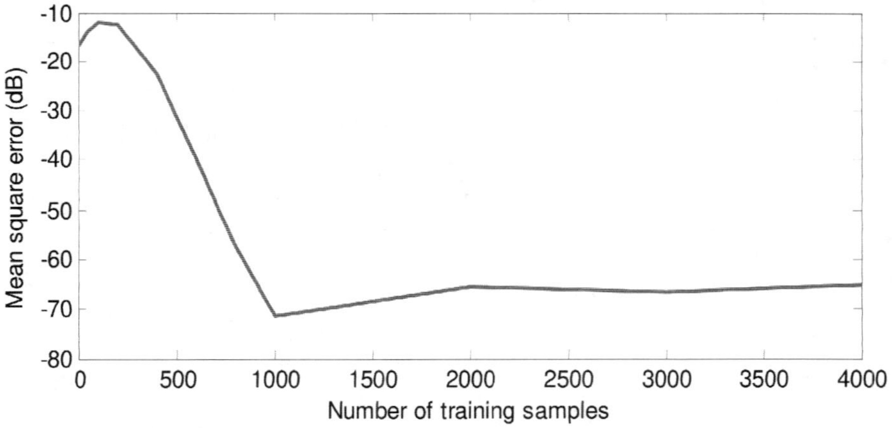

Figure 4. Mean square error as increasing the number of training samples of MLP network for nonlinearity modeling in the TLMI.

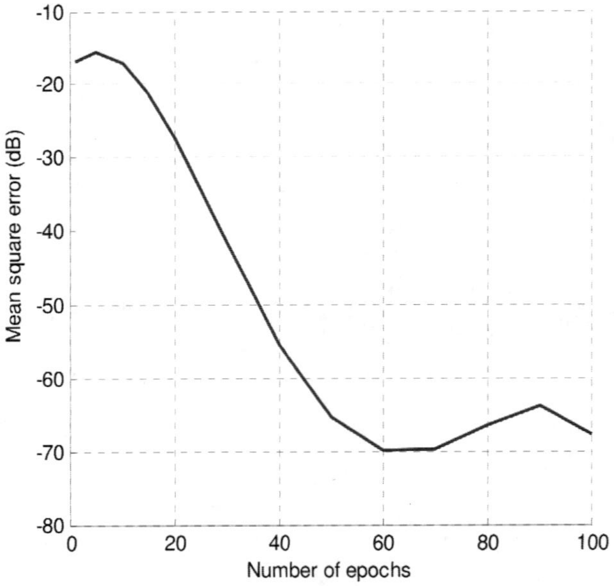

Figure 5. Mean square error as increasing the number of epochs of MLPN for nonlinearity modeling in the TLMI.

Table 2. Results of mean square error as the number of training samples for nonlinearity modeling with MLP network.

Parameter	Number of training samples									
	1	50	100	200	400	800	1000	2000	3000	4000
MSE (dB)	-11.91	-16.18	-17.05	-14.55	-13.97	-66.09	-66.49	-65.65	-66.53	-65.06
Time (seconds)	1.18	1.62	1.995	2.1715	1.995	2.126	2.066	2.282	2.562	2.574

Table 3. Results of mean square error as the number of epochs for nonlinearity modeling with MLP network.

Parameter	Number of epochs												
	1	5	10	15	20	30	40	50	60	70	80	90	100
MSE (dB)	-14.79	-17.64	-19.37	-22.34	-25.35	-33.36	-64.72	-66.77	-67.07	-67.07	-66.80	-66.61	-66.15
Time (seconds)	2.52	1.99	2.06	2.48	2.30	2.72	2.93	3.33	3.05	4.04	4.25	5.07	5.45

5. Conclusion

In this paper, an approach based on neural network for modeling nonlinearity in a three-longitudinal-mode laser heterodyne interferometer (TLMI) has been introduced. We modeled nonlinearity errors that arise from elliptically and non-orthogonally polarized laser beams, rotational error in the alignment of laser and beam splitter, rotational error in the alignment of the mixing polarizer, and unequal transmission coefficients in the polarizing beam splitter. The MLP neural network algorithm was used. The simulation results have shown that this network has successfully modeled nonlinearity errors with simplest and fast structure, including 5 neurons in the hidden layer. It was also confirmed that the system with the neural network modeling part could respond at a speed about that exceeds only to 3 seconds. We note that the method presented here is applicable to all of nano-metrology systems like conventional and extended two- or three-frequency heterodyne interferometers to extend the integrated measurable system of the heterodyne frequency.

References

[1] Schattenburg M L and Smith H I 2001 *The critical role of metrology in nanotechnology* Vol 4608 (SPIE) pp 116–124

[2] Pieter B. 1999 *Optical lithography to 2000 and beyond* Vol 3 (Solid State Technol) pp 31-41

[3] Hou W 2006 *Optical parts and the nonlinearity in heterodyne interferometers* Vol 30 (*Prec. Eng)* pp 337–346

[4] *The international technology roadmap for semiconductors* 2008 (accessed 21 July 2008) http://public.itrs .net

[5] Lawall J and Kessler E 2000 *Michelson interferometry with 10 pm accuracy* Vol 71 (Review of Scientific Instruments) pp 2669–2676

[6] Demarest F C 1998 *High-resolution, high-speed, low data age uncertainty, heterodyne displacement measuring interferometer electronics* Vol 9 (Meas. Sci. Technol) pp 1024–1030

[7] Olyaee S and Nejad S M 2007 *Design and simulation of velocity and displacement measurement system with subnanometer uncertainty based on a new stabilized laser Doppler-interferometer* Vol 32 (The Arabian J. Sci. Eng) pp 89-99

[8] Olyaee S and Nejad S M 2007 *Nonlinearity and frequency-path modelling of three-longitudinal-mode nanometric displacement measurement system* Vol 1 (IET Optoelectron) pp 211-220

[9] Olyaee S and Nejad S M 2007 *Error analysis, design and modeling of an improved heterodyne nano-displacement interferometer* Vol 3 (Iranian J. Electrical and Electronic Eng) pp 53-63

[10] Quenelle R C 1983 *Nonlinearity in interferometric measurements* Vol 3410 (Hewlett-Parkard J)

[11] Sutton C M 1987 *Nonlinearity in the length measurement using hetrodyne laser Michelson interferometery* Vol 20 (J. Phys. E: Sci. Instrum) pp 1290-1292

[12] Cosijns S J A G, Haitjema H and Schellekens P H J 2002 *Modeling and verifying non-linearities in heterodyne displacement interferometry* Vol 26 (Prec. Eng. and Nanotechnol) pp 448–455

[13] Badami V G and Patterson S R 2000 *A frequency domain method for the measurement of nonlinearity in heterodyne interferometry* Vol 24 (Prec. Eng) pp 41–49

[14] Olyaee S, Yoon T H and Hamedi S 2009 *Jones matrix analysis of frequency mixing error in three-longitudinal-mode laser heterodyne interferometer* Vol 3 (IET Optoelectronics) pp 215-224

[15] Li Z, Herrmann K and Pohlenz F 2003 *A neural network approach to correcting nonlinearity in optical interferometers* Vol 14 (Meas. Sci. Technol) pp 376-381

[16] Heo G, Lee W, Choi S, Lee J, and You K 2007 *Adaptive neural network approach for nonlinearity compensation in laser interferometer* (Springer Press) pp. 251-258

[17] Olyaee S, Ebrahimpour R and Hamedi S 2010 *Modeling and compensation of the periodic nonlinearity in two-mode laser heterodyne interferometer by using stacked generalization method* Vol 56 (IETE Journal of Research) pp 102-110

[18] Yokoyama T, Araki T, Yokoyama S, and Suzuki N 2001 *A subnanometre heterodyne interferometric system with improved phase sensitivity using a three-longitudinal-mode He–Ne laser* Vol. 12 (Meas. Sci. Technol) pp 157–162

[19] Haykin S 1999 *Neural networks: a computer foundation* (Prentice Hall)
[20] Luger G F 2005 *Artificial Intelligence* (Addison Wesley)

Compressive sampling and gated viewing three-dimensional laser radar

Ximing Ren[1], Li Li, and Ersheng Dang
School of Electronics and Information Engineering, Beihang University, Beijing 100191, P. R. China

E-mail: rxmbit@gmail.com

Abstract. By a new framework named compressed sensing (CS) for the compressive sampling achieving simultaneous sampling and compression of signals, it is possible for us to perform laser imaging using a single detector rather than focal plane arrays having millions of pixels. Moreover, based on acquiring a serial of time-slicing images through combining compressive sampling with gated viewing, the three-dimensional(3D) scene can be reconstructed, wherein a single-pixel gated viewing 3D imaging LADAR system prototype is proposed in this paper; numerical experiments using the proposed 3D image formation model evaluate the system preliminarily and draw the main result that higher range accuracy for 3D scene as the increase of pseudorandom projections by qualitative analysis from Lissajous figures.

1. Introduction

Laser active imaging systems have the advantages of high reliability and spatial resolution, in particular the three-dimensional (3D) laser detecting and ranging (LADAR) systems which obtain full space information [1]. Recently, 3D imaging LADAR systems have drawn a great deal of attention on the use of target detection, recognition and classification [2,3,4,5]. It is common and efficient for most 3D imaging LADAR systems to acquire object's 3D data through focal plane arrays sensing. However, adapting to image at wavelengths, especially low-light, near-IR or short-wavelength IR imaging, requires expensive sensors, most of which are still in the experimental research stage. Additionally, massive data acquisition and storage in the receivers are required in 3D imaging LADAR systems. To reduce the quantity of imaging data and lessen power consumption, in two dimensional imaging, there is a related research, such as Ma in[6], proposing a single-pixel imaging system for remote sensing.

---

[1] School of Electronics and Information Engineering, Beihang University, New Main Building F1013, 37 Xue Yuan Road, Hai Dian District, Beijing 100191, P. R. China

Likewise, in order to minimize the 3D imaging LADAR systems, get rid of the dependence on focal plane arrays with special wavelengths, and reduce data storage overhead and energy consumption, as an alternative, we propose a single-pixel gated viewing 3D imaging LADAR system in this paper, based on a new sensing mechanism named compressed imaging[7,8],which is a revolutionary technology of image generation created by Rice University. And as known from gated viewing [9,10,11,12], it is used for laser ranging and for overcoming backscatter brought by obscurants by camouflage, fog or water has matured into a robust and efficient method for recovering the surfaces of objects.

The rest of the paper is organized as follows. In the next section, we describe the single-pixel gated viewing 3D imaging LADAR system prototype. In Section 3, we describe the detailed process of 3D reconstruction from time-slicing 2D images. In Section 4, we present our simulations generated by the use 3D image formation model. Section 5 concludes the paper.

2. SYSTEM DESCRIPTION

Our single-pixel gated viewing 3D imaging LADAR system is illustrated in **Figure 1**. Its sampling mechanism resembles traditional gated viewing. It means that during the pulsed laser travels, the detector gate remains closed, thereby eliminating backscattered light from the light scattering media and improving the system's SNR. But its distinct feature is that the incident light field from the scene is modulated by the digital micro-mirror device (DMD) [7] before passing the ultrafast shutter, then reaching the single detector. Note that the principle of the gated viewing technology is based on the ability to control the laser beam and the detector shutter in a very accurate way in order to limit the detection to a specific range of space [12], so the synchronous control technology is applied in this system.

There are two choices in keeping with the synchronization triggered: internal trigger and external trigger. On the one hand, internal trigger mode need delay pulse generator to generate trigger signal source. Then the trigger signal is separately imported into the pulsed laser and the detector shutter from the two delay channel. By control the delay time of two delayed channel respectively. Therefore, pulsed laser synchronize with detector has been achieved. One of the necessary conditions for internal trigger is that the pulsed laser can be controlled by external trigger. In other words, laser which is enabled depends on external trigger signal. In addition, laser light frontier jitter of the pulse laser must be small, and the delay time from receiving trigger signal to producing laser output must be fixed, which means that it is overcritical for the high performance laser. So, this mode is not a good choice. On the other hand, it is the external trigger mode. As depicted in **Figure 1**, unlike the internal trigger mode, here, the pulsed laser works autonomously. At the same time, the emitted laser is partially into the photodiode, followed converting light to electric signal. Then the electric signal imports into delay pulse generator. The delay pulse generator output trigger delay and the pulse signal width to detector shutter and DMD. Delay time and duty cycle can be regulated by delay pulse generator. Therefore, the detector shutter and DMD can be controlled to open when the emergent light arrived which reflected by the target. It sets delay time based on the range of target.

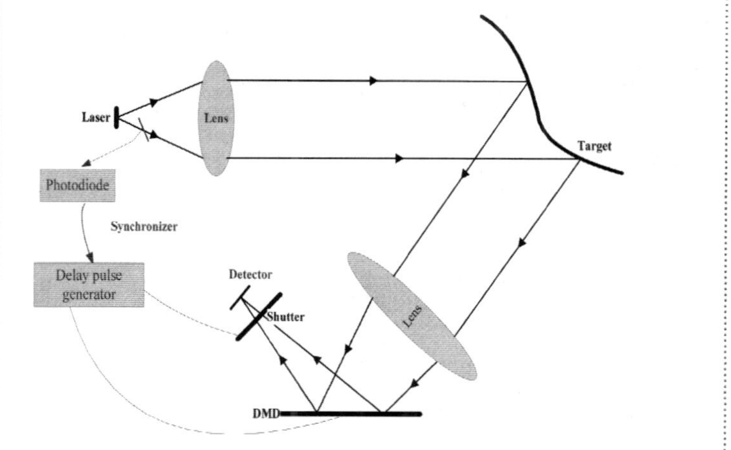

Figure 1. Sketch of single-pixel gated viewing 3D imaging LADAR system.

3. 3D Image Formation Model

The proposed imaging LADAR system can be operated in a specific way in order to produce a set of slicing images of the observed scene which are still reconstructed by compressing imaging. On the basis of three slicing images, a 3D scene reconstruction can be achieved by the algorithm mentioned in[12].

3.1 Time-Slicing Technique based on compressed sampling

Specifically, the mechanism of the proposed imaging LADAR system has two different parts: 1) sensing "hardware", and 2) reconstructing "software". The "hardware" is comprised of a synchronizer, an ultrafast shutter, a DMD and a single detector. And the sampling scheme is as same as the one described in[7] and [8]. However, the measurement basis matrix applies a Walsh Hadamard matrix. Meanwhile, there is a synchronized mechanism to ensure acquiring the ranges for the interest scene. Additionally, for the "software", the reconstruction algorithm uses "Your Algorithms for L1 Optimization (YALL1)" [13].

To be more exact, a permutated Walsh Hadamard matrix A (M*N) generated by the digital DMD is programmed. Here, M is the number of rows and N is number of columns, note that $M \ll N$. In fact, to correctly implement in the DMD, the matrix entries −1 and 1 are shifted to 0 and 1[14].

In terms of gated viewing active imaging, it is based on synchronization of illumination and image sensor gate .**Figure 2** shows how $P(t)$ and $G(t)$ come together to form I. It can be depict [12]as :

$$I_i = \int P(t - 2r / c)G(t - t_i)dt \tag{1}$$

where the pulse is delayed by the round-trip travel time 2r/c at a range of r. The detector gate is delay by the time $t_i, i \in \{1,2,3...\}$ sequentially. The sequence of time slices or range Gates I_i is record for all (x, y) pixel, resulting in a data set:

$$I_i = \{p_{i1}, p_{i2}, p_{i3}..., p_{in}\}, n \in \{1,2,3...N\}, i \in \{1,2,3...\}.$$

Then in the certain t_i, it is modulated by DMD and measurements in matrix notation can be described as:

$$Y = AI + \varepsilon \quad (2)$$

where ε denotes the possible measurement errors or noises. Y is an M×1 column vector, denoting in a data set:

$$Y_i = \{y_{i1}, y_{i2}, y_{i3}..., y_{im}\}, m \in \{1,2,3...M\}, i \in \{1,2,3...\}.$$

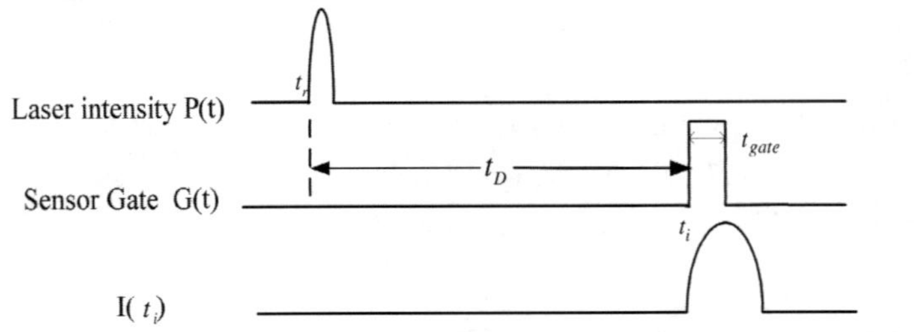

Figure 2. Time sequence of emitted light pulse, sensor gate and received light pulse.

Based on the theory of compressive sensing introduced recently in [15], through a reconstruction algorithm YALL1, I can be recovered from the measurements Y, and the result corresponding to reconstructed gated images is denoted in a set as:

$I_i' = \{p_{i1}', p_{i2}', p_{i3}'...p_{in}'\}, n \in \{1,2,3...N\}, i \in \{1,2,3...G\}$, where G is the total number of the reconstructed gated images.

3.2 Three Dimensional Reconstruction

With the appropriate gate control, we can obtain a series of images from a gated viewing system in special delay time $t' = \{t_1', t_2', ...t_n'\}$. These images can be recorded as: $I' = \{I_1', I_2', I_3', ...I_n'\}$.

The task in 3D image is to determine a range value for each pixel. And the round-trip time t_r

($t_r = 2r/c$) of the laser pulse is proportional to range. So we can use the data set $\{I', t_i\}$ to determine the t_r. As a result, then the range corresponding to each pixel can be achieved. In the standard setup, the detector is gated such that it is open for a time t_{gate} and with a time delay after pulse emission of $t_i = t_0 + i * \Delta t$. Here t_0 is the initial delay, Δt is delay step, the sequence of delays runs $i = 1, 2, ..., T_s$, where T_s is time slices in the sequence.

$I_i'(x, y)$ is the gray value corresponding to (x, y) in the ith image in a sequence of 2D images obtained from CS-reconstruction. By summing the 2D images $i = 1, 2, 3, ...G$, so we have follows equation.

$$I_r = \sum_{i=1}^{G} I_i'(x, y) \tag{3}$$

Therefore we can calculate the average round-trip time $\langle t \rangle$ .to an object point within the field of view of a pixel.

$$\langle t \rangle = \frac{2r}{c} = I_r^{-1} \sum_i I_i' t_i = I_r^{-1} \sum_i I_i'(t_0 + i\Delta t) = t_0 + \Delta t * I_r^{-1} \sum_i i * I_i' \tag{4}$$

Finally, we obtained the distance $z(x, y)$ to a target point in any pixel.

$$z(x, y) = \frac{c}{2}\langle t \rangle = \frac{c}{2}(t_0 + \Delta t * I_r^{-1} \sum_i i * I_i') \tag{5}$$

Then 3D reconstruction of the scene can be achieved based on the equation (5).

4. Validation via Simulation

To validate the proposed 3D image formation model, we performed simulations based on four gated viewing images from[16], which are shown in **Figures 3**. During the simulations, the size of images was chosen to be 256×256(hence, N=65536). Using M=16384 and 32768 pseudorandom projections ($\frac{N}{4}$ and $\frac{N}{2}$, respectively), we reconstructed the 2D images using YALL1 shown in **Figures 5(i)** and **Figures 6(i)** as well as the 3D views shown in **Figures 5(ii)** and **Figures 6(ii).** In contrast, the original reconstruction was shown in **Figures 4**. From these results, it was clear that the recognizable features of the Tank can be recovered. The reconstruction quality was also better as the higher M, according to the PSNR of the reconstructed 2D images using YALL1. However, it was not clear to find the slight range difference among these three 3D reconstructions.

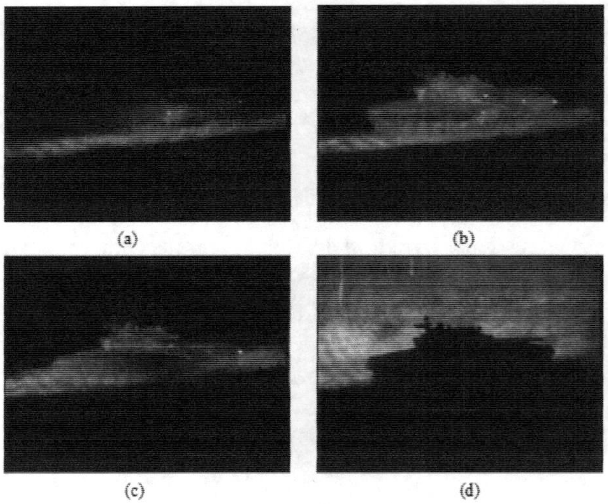

Figure 3. Images of a scene captured using four different settings of the gate delay time, resulting in the laser reflections from the front (a),entire(b), back(c) , and background(d) of the tank being imaged. The range to the target is 0.9km and the gate width is 40ns [16] .

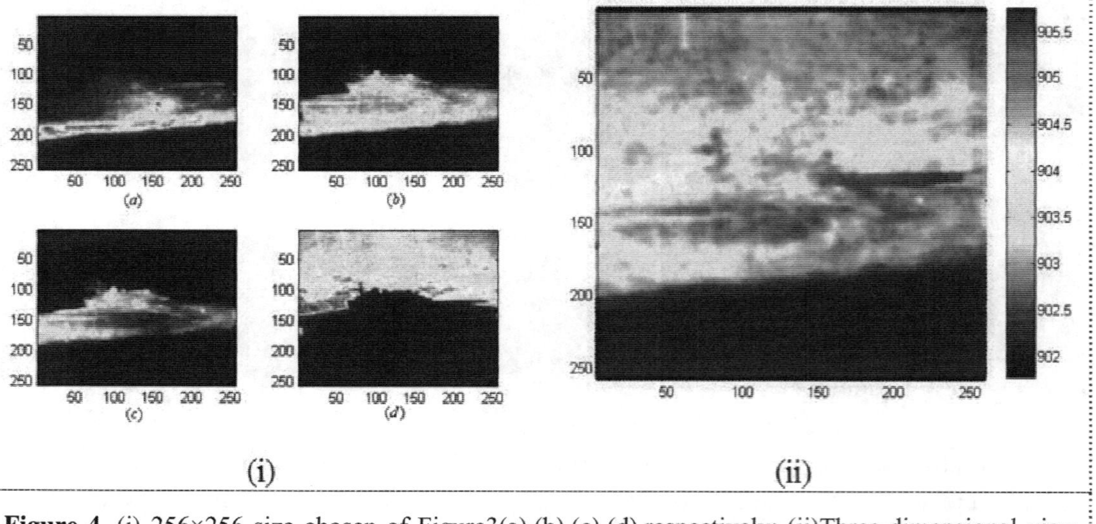

Figure 4. (i) 256×256 size chosen of Figure3(a),(b),(c),(d),respectively; (ii)Three-dimensional view of Tank reconstructed from the original gated image sequence simulated by Matlab.

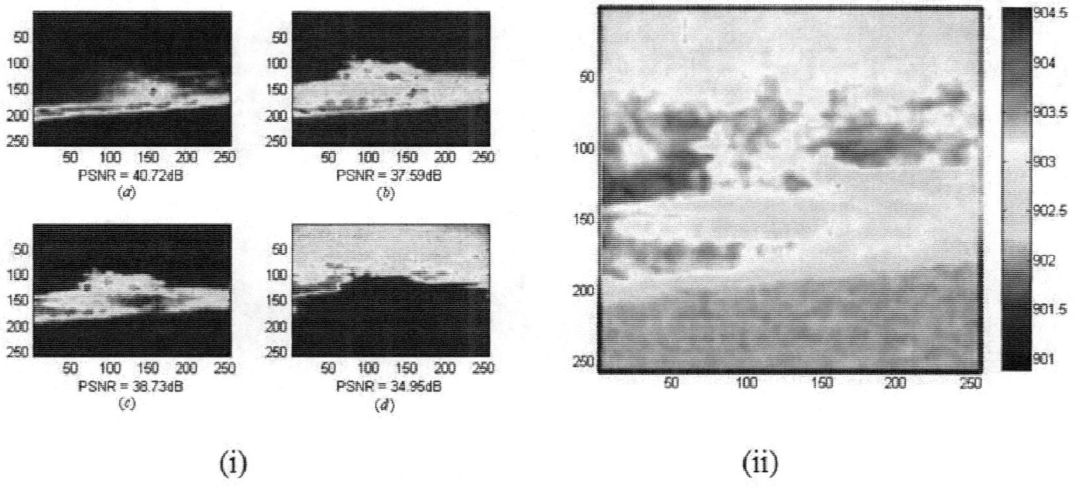

Figure 5. (i)Wherein (a),(b),(c),(d) represent the CS-reconstructed gated image (PSNR=40.72dB, 37.59dB, 36.73dB, 34.95dB), respectively; (ii)Three-dimensional view of Tank reconstructed from the CS-reconstructed gated image sequence based on 16384 measurements equal to $\frac{N}{4}$ simulated by Matlab.

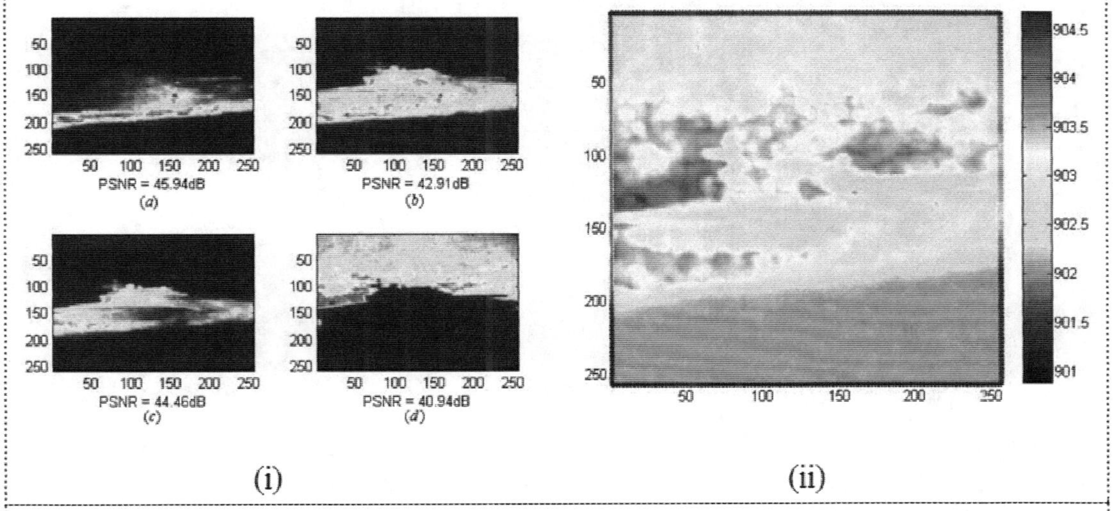

Figure 6. (i)Wherein (a),(b),(c),(d) represent CS-reconstructed gated image(PSNR=45.94dB, 42.91dB, 44.46dB, 40.94dB), respectively; (ii)Three-dimensional view of Tank reconstructed from the CS-reconstructed gated image sequence based on 32768 measurements equal to $\frac{N}{2}$ simulated by Matlab.

So, to evaluate the range differences between the original 3D reconstruction and the 3D reconstruction based on the proposed 3D image formation model of the scene with the tank, Lissajous figures were utilized. As known from Lissajous figures [17], if no optical path difference, equaling to no range difference, appears in the signals, one observes a single straight line; while with an increasing optical

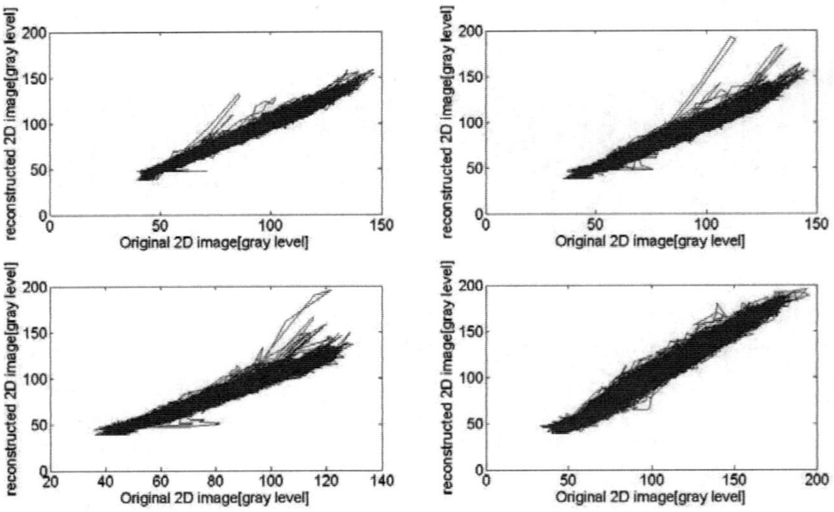

Figure 7. Eye pattern Lissajous figure (M = 16384).

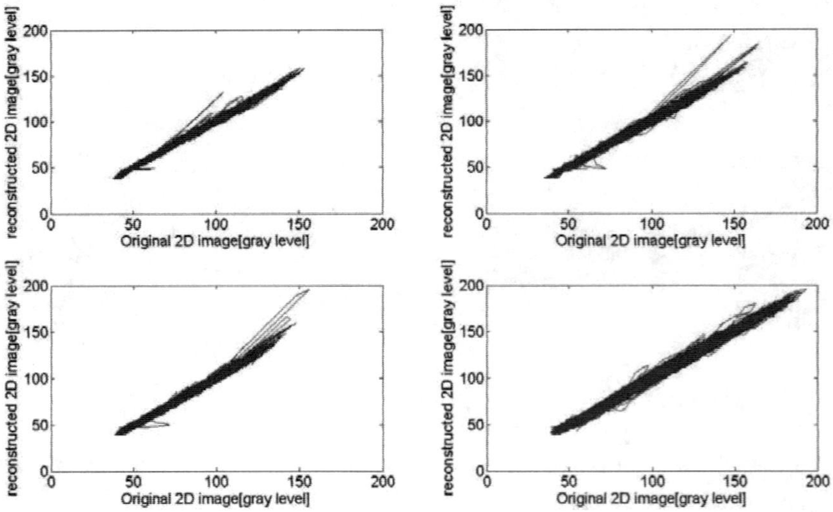

Figure 8. Eye pattern Lissajous figure (M = 32768)

path difference, the loop of the Lissajous figure opens and widens. Therefore, in terms of small optical path differences, the data representation in a Lissajous-type eye pattern is very sensitive. Then from the **Figure 7** and **Figure 8**, the eye patterns was similar to a single straight line, reflecting that the range difference between original and reconstructed ones was not marked, but the eye pattern based on using M=16384 pseudorandom projections opened wider than that based on using M=32768 pseudorandom projections, which meant that the range accuracy was better as the increase of pseudorandom projections.

5. Conclusions

In this paper, through numerical experiments, we successfully applied YALL1 algorithm to estimate 2D measurements and generate 3D reconstruction of the scene with the tank by 3D image formation model based on the proposed prototype (a single-pixel gated viewing 3D imaging LADAR system). To minimize noise, the synchronous control mechanism in the prototype has been proposed. And the Lissajous figures based on Lissajous-type eye-pattern method was utilized with the aim of analyzing 3D reconstructions from different pseudorandom projections, which is an enhanced analysis in the means of range differences. However, the proposed system is limited due to the capabilities of the DMD arrays. The current DMD arrays can change their geometric configuration approximately 10K to 40K times per second [18]. For some applications, the system must keep the characteristic of high real-time at the expense of resolution, even at a rate of 40K times per second. To beyond the simulations, in terms of a single-pixel gated viewing 3D imaging LADAR system, building a physically realizable hardware optical scheme for practical application will be investigated in a future work.

Acknowledgment

Thanks to Yin Zhang and Wotao Yin of Department of Computational and Applied Mathematics, Rice University for providing the MATLAB package called YALL1. Thanks also to Chengbo Li of Department of Computational and Applied Mathematics, Rice University for enlightening discussions.

References

[1]R L Espinola, E L Jacobs, C E Halford, R Vollmerhausen,and D H Tofsted 2007 "Modeling the target acquisition performance of active imaging systems," *Opt. Express* **15** 3816–3832

[2] O Steinvall, H Olsson, G Bolander, C Carlsson, and D Letalick 1999 "Gated viewing for target detection and target recognition,"*Proc. SPIE* **3707** 432-448

[3] R Marino, T Stephens, R Hatch, et a1 2003 "A Compact 3D Imaging Laser Radar System Using Geiger-Mode APD Arrays:System and Measurements" *Proc.SPIE* **5086** 1-15

[4] Brian F Aull, Andrew H Loomis, Douglas J Young,et a1 2002 "Geiger-Mode Avalanche Photodiodes for Three-Dimensional Imaging" *Lincoln Laboratory Journal* **13** (2):335-350.

[5]R G Driggers, R H Vollmerhausen, N Devitt, C Halford,and K J Barnard 2003 "Impact of speckle on laser range-gated shortwave infrared imaging system target identification performance," *Opt. Eng.* **42** 738-746

[6]Ma J W 2009 "A Single-Pixel Imaging System for Remote Sensing by Two-Step Iterative Curvelet

Thresholding," *IEEE Geosci. Remote Sensing Lett.* **6** 676-680
[7] D Takhar, J N Laska, M B Wakin 2006 "A New Compressive Imaging Camera Architecture using Optical-Domain Compression" *Proc. SPIE* **6065** 1-10
[8]M F Duarte, M A Davenport,D Takhar, J N Laska, T Sun, K F Kelly, and R G Baraniuk 2008 "Single-pixel imaging via compressive sampling," *IEEE Signal Proc. Mag.* **25** 83-91.
[9] E Belin, F Christnacher, F Taillade, and M Laurenzis 2008 "Display of an analytical model for backscattered luminance and a full-field range gated imaging system for vision in fog," *Proc. SPIE* **7088** 708-800
[10]G R Fournier, D Bonnier, J L Forand, and P W Pace 1993 "Range-gated underwater laser imaging-system," *Opt. Eng.***32** 2185-2190
[11] J Busck 2005 "Underwater 3-D optical imaging with a gated viewing laser radar," *Opt. Eng.* **44** 116001
[12]J Busck and H Heiselberg 2004 "Gated viewing and high-accuracy three dimensional laser radar,"*Appl. Opt.* **43** 4705-4710
[13]J F Yang and Y Zhang **2010** *Alternating direction algorithms for l1-problems in compressive sensing* (TR09-37, CAAM, Rice University)
[14] C B Li 2009 *An Efficient Algorithm For Total Variation Regularization with Applications to the Single Pixel Camera and Compressive Sensing* (Houston, Texas: Rice University)
[15] D Donoho 2006 "Compressed sensing," *IEEE Trans. Inf. Theory* **52** 1289-1306
[16]P Andersson 2006 "Long-range three-dimensional imaging using range-gated laser radar images", *Opt. Eng.***45** 1-10
[17] M Laurenzis 2010 "Evaluation metrics for range-gated active imaging systems using a Lissajous-type eye pattern" *Appl. Opt.* **49** 2271-2276
[18] D Dudley, W Duncan, J Slaughter 2003 "Emerging Digital Micromirror Device (DMD) Applications," *Proc. SPIE* **4985** 14-25

Experimental Investigation of Supersonic Mixing Mechanisms of HYLTE Nozzle for DF Chemical Laser

SHAO Yan[1], ZHOU Jin, ZHANG Sunping and LAI Lin

College of Aerospace and Material Engineering, National University of Defense Technology, Changsha 410073, China

E-mail: freebird_18@163.com

Abstract. Utilizing experimental techniques of Nano-particle based Planar Laser Scattering (NPLS) and schlieren photography, the flow patterns and mixing characteristics of a designed HYLTE (HYpersonic Low TEmperature) nozzle were investigated in this paper. In order to visualize the non-reacting flowfield of supersonic angled jets into a supersonic crossflow in the HYLTE nozzle, a testing section with windows was designed and manufactured. The effects of different total pressure ratio of the twin jets to the freestream and different injectants on supersonic mixing are examined. Instantaneous side- and end-view NPLS images provide transverse penetration and lateral spread information for the secondary twin jets. As an assistant method, schlieren photos display the shock patterns that exist in the HYLTE nozzle.

1. Introduction

HYLTE nozzle is an advanced nozzle concept which is widely used in DF chemical laser for its high mixing efficiency. In order to optimize the nozzle design and the laser performance, it is necessary to get a clear understanding of how the gaseous injections interact with the primary flow.

Non-intrusive visualization techniques, such as schlieren and NPLS are powerful and commonly used optical methods at present. It is considered that streamwise vortices produced by the interaction of injections and mainstream dominate and accelerate mixing. NPLS technique provides convenient conditions of the observation of streamwise vortices development of supersonic transverse injection. Schlieren photography is a convenient experimental method in non-reacting flowfield visualization, but its images are not clear enough to study the mixing mechanism in the supersonic crossflow of HYLTE nozzle.

NPLS is an advanced method for supersonic flow visualization, which is newly developed based on Particle Image Velocimetry (PIV) technology by YI Shihe and ZHAO Yuxin of National University of Defense Technology in 2005. NPLS uses nanometer TiO_2 and pulsed planar laser as tracer particles and illumination respectively, not only solving the problem of particles flow-following but also improving the Signal-to-Noise of supersonic flow imaging. Due to the good flow-following ability of nanoparticles, if the concentration distribution of particles on the inlet of flow field is uniform, the variation of particle will reflect the density variation and mixing structure of the flow field. Thus the NPLS method is appropriate for the study of transverse injection of the HYLTE nozzle in which the density and concentration gradient are rather large.

2. Experimental equipments and methods

2.1. NPLS technology

The NPLS system consists of a computer, synchronizer, CCD camera, pulse laser and nanoparticle generator, the structure of which is schematically shown in figure 1. In NPLS, the computer controls the collaboration of the components and receives the experimental images. The input and output parameters of the synchronizer are controlled by software, collaboration of other components is controlled by signal of the synchronizer. The timing diagrams of exposure of CCD and laser output of pulse laser can be adjusted according to the purpose of measurement. The laser beam is transformed to a sheet with cylindrical lens. The nanoparticle generator is driven by high pressure gas, and the output particles concentration can be adjusted precisely by the driving pressure. While measuring the flow field with NPLS, the nanoparticles are injected into and mixed with the inflow of the flow field, while the flow is established in observing widow, the synchronizer controls the laser pulse and CCD to ensure synchronization of the emission of scattering laser by nanoparticles and the exposure of CCD[1].

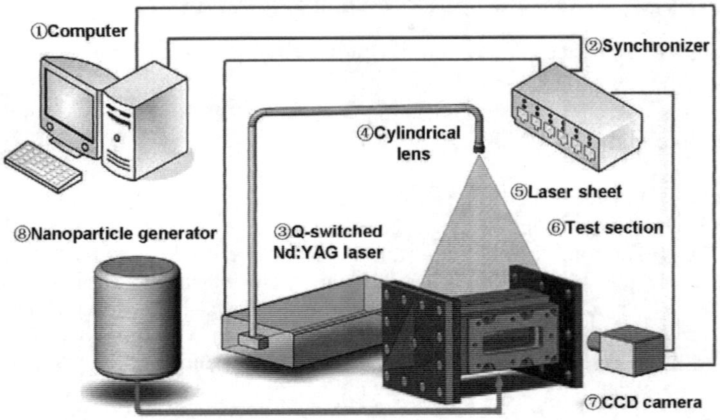

Figure 1. Schematic diagram of NPLS system.

The NPLS experimental system used in this paper is shown in figure 2. The synchronizer has eight output ports with temporal precision at 0.25 ns. The CCD is an interline transfer CCD equipped with micro-lens, of which the shortest frame straddle time is 200 ns, the number of CCD array is 2000× 2000 pixels with 4096 grayscale grades. The pulse laser is a double pulse Nd:YAG laser, the wavelength of the laser is 532 nm with duration of 6 ns and 350 mJ per pulse. The location of light sheet lens, polarization angle of laser, and particle flow rate of nanoparticle generator can be adjusted according to experimental requirements.

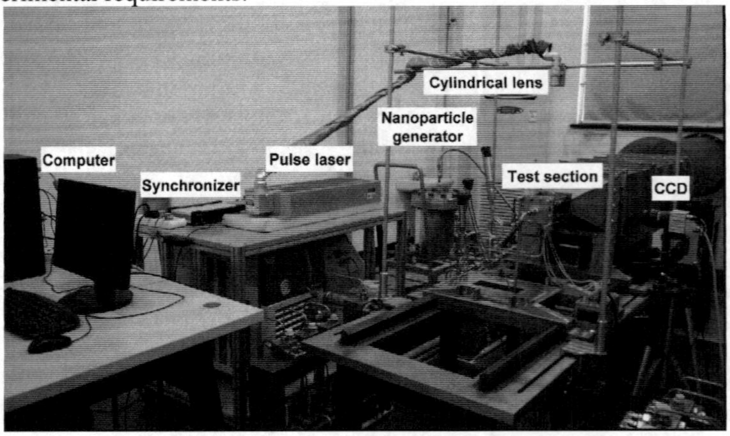

Figure 2. Supersonic NPLS experimental system.

2.2. High speed schlieren system

Journal of Physics: Conference Series **276** (2011) 012143 doi:10.1088/1742-6596/276/1/012143

The mirror-based, Z-arrangement schlieren system is used in this study, which is displayed in figure 3. The point light source is a halogen bulb. The slit sits at the focal point of the condenser lens of the lamp. Two 120-cm focal length concave mirrors are used to collimate the light through the test section and then refocus it onto a knife edge. The knife edge at the focal point of the second schlieren mirror is used to partially cut off the deflected rays for observing the schlieren effect (visualization of the density gradients). Blocking more of the light by moving the knife edge transverse to the optical axis makes the system more sensitive, showing more features of the jet. The knife edge, oriented horizontal or vertical with respect to the focused light, will emphasize the density gradients in the horizontal or vertical directions respectively[2]. The test object is imaged with a zoom lens onto a high-speed framing camera. Frame rate of the camera can be up to 120,000 frames per second (fps).

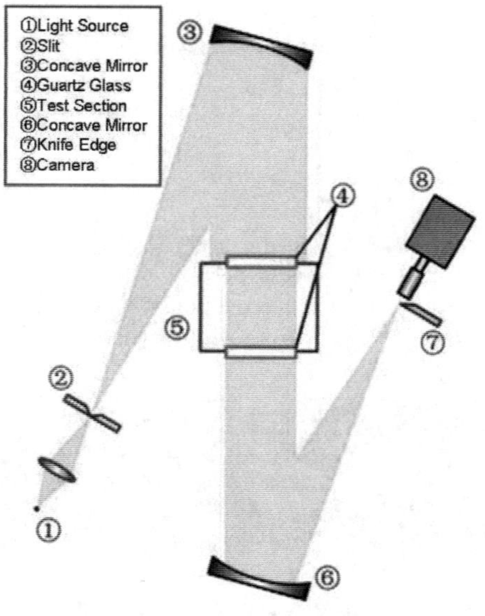

Figure 3. Schlieren system.

2.3. HYLTE nozzle testing section

In order to visualize the non-reacting flowfield of supersonic angled jets into a supersonic crossflow in the HYLTE nozzle, a testing section with observation windows is designed and manufactured, which is shown in figure 4. Three pieces of quartz glass are embedded in the top and side observation windows respectively for the convenience of capturing the NPLS images of centerline plane and planes in several spanwise locations. The injection angles and the secondary fluids orifice separation can be changed by replacing the forepart of the nozzle component.

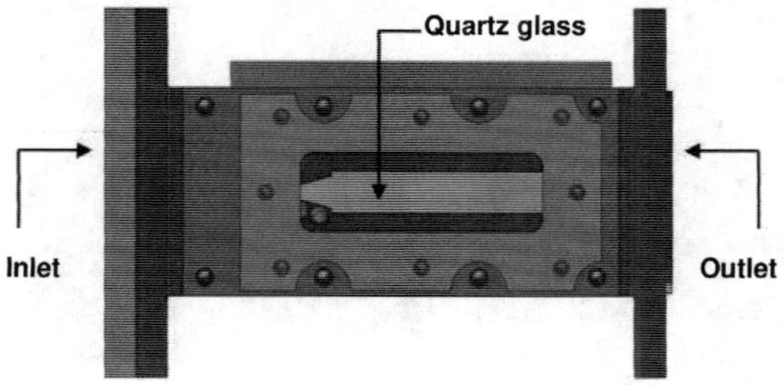

(a) Front view

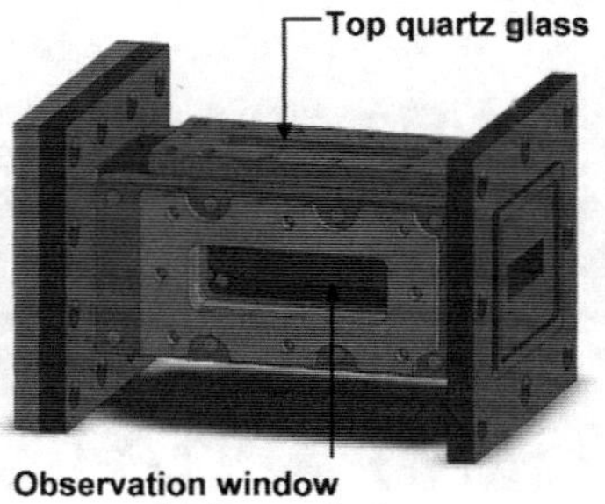

(b) Lateral view

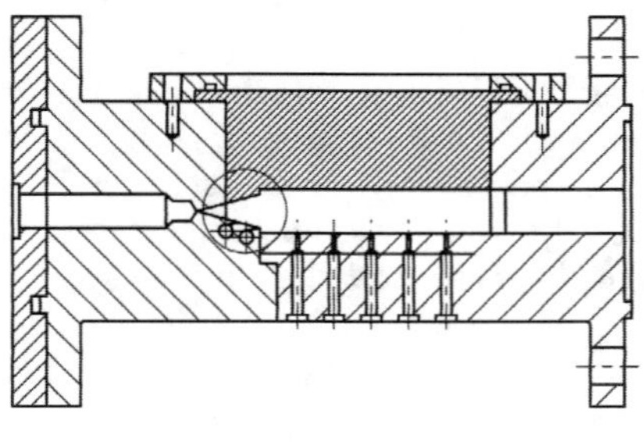

(c) Full section view

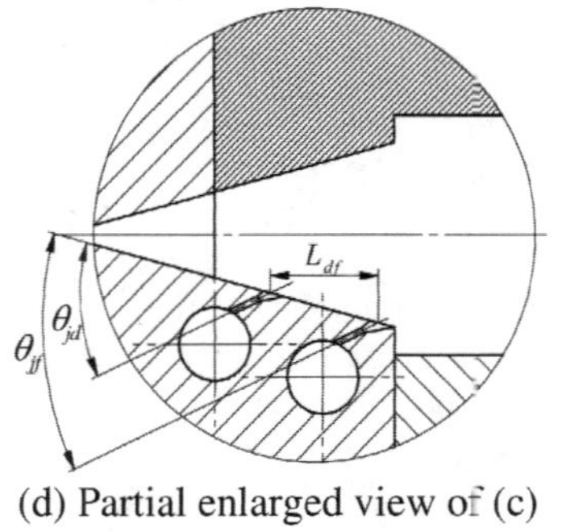

(d) Partial enlarged view of (c)

Figure 4. HYLTE nozzle testing model.

Gas feeding and pressure measuring orifices are placed at the bottom of the testing section. The diffuser is installed at the end of the testing section which connects with vacuum tank and provides equivalent pressure between them. As for cold experiment, gaseous nitrogen or helium is used to simulate the jet fluids of secondary nozzle, while the crossflow of primary nozzle is always nitrogen. Because secondary jets are 0.3mm in diameter, the tiny gaseous flow is dominated by mass flow controller (MFC).

3. Results and discussions

By visualizing the supersonic transverse injection flowfield in the HYLTE nozzle, streamwise and spanwise images of NPLS are taken, the flow structures of these twin supersonic secondary injections are identifiable. Owing to the block effect of the first injector to the main flow, the second injector has a stronger injection and deeper penetration than the former injector. For the Mach number of the mainstream is a little too high (Mach 4-5), shear-induced mixing is poor[3], the mixing layer tends to become stable, as a result, some large-scale structures of the flowfield have been restrained.

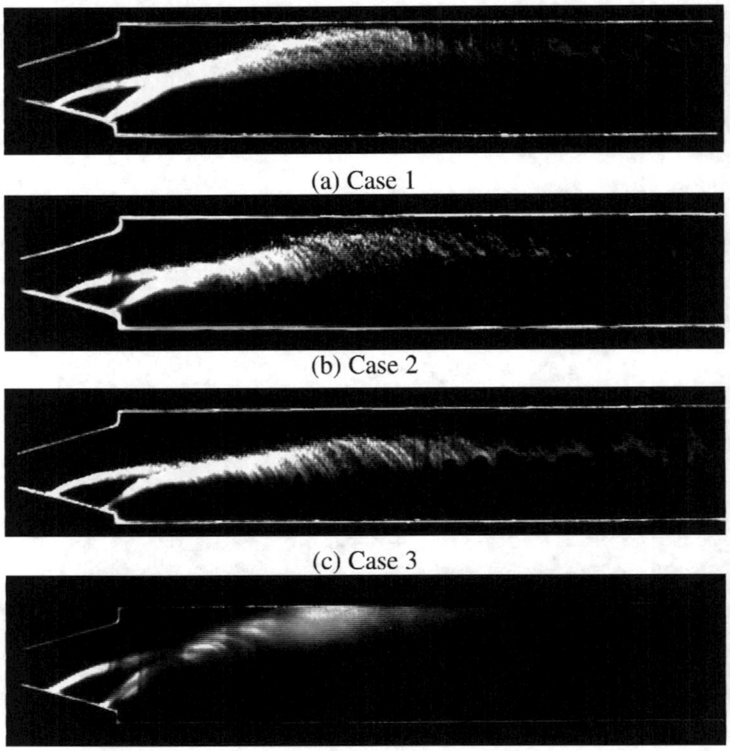

(a) Case 1

(b) Case 2

(c) Case 3

(d) Case 4

Figure 5. Instantaneous side-view NPLS images.

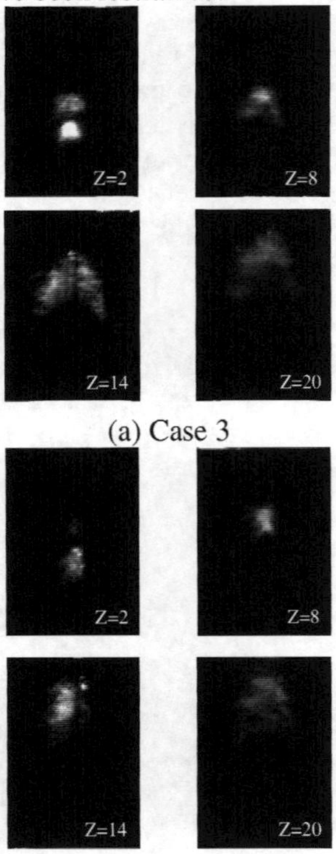

(a) Case 3

(b) Case 4

Figure 6. Instantaneous end-view NPLS images.

Figure 5 presents instantaneous side-view NPLS images which offer a view of the jet/crossflow interaction at the spanwise centerline of the twin injectors. From case 1 to case 3, there is a increase in total pressure of the primary flow progressively, while the pressure values of secondary jets are the same. The penetration of secondary injections is depressed obviously with the increasing crossflow pressure. Results also show that the mixing quality is greatly enhanced by jet surface strentching and entraining the crossflow fluid continuously, which is clearly visualized in figure 5(c). That is to say, higher momentum jets entrain less crossflow fluid than lower momentum jets. Case 3 and case 4 are designed to operate at the same condition, while case 4 is supplied with helium to the secondary injections. Because of the smaller molecular weight of helium, the jets penetrate farther vertically in case 4 than in case 3.

The cross-sectional structure of the twin jets is observed in figure 6. The image plane moves downstream from the base to 2mm, 8mm, 14mm and 20mm respectively, the near-field development of the jet/crossflow interaction may be examined. Counter-rotating vortex pairs emerge near the base, which is very close to the injectors. The jet cross sections change from ellipse-shaped to kidney-shaped consequently. Further downstream the two jets attract each other and start to merge, at the same time, the strength of the counter-rotating vortices becomes weaker.

As can be seen from figure 7, two curved-shaped bow shock waves appear in the vicinity of the twin injectors. The injected gases act as obstructions to the freestream and thus produce interaction bow shock waves upstream of the injector locations. In addition, it is pointed that angled injection will result in weaker bow shock than normal injection[4], that is why the injector bow shock waves in all obtained schlieren images are not strong enough. The shock waves coming from the HYLTE throat and from the joint of base and wall have been marked together. It is obvious that the shock inclination angle from the neck depends on the nozzle's expansion degree[5].

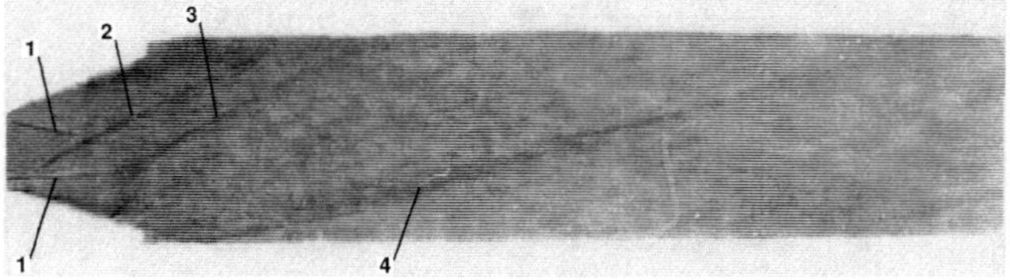

Figure 7. Schlieren photos of flow: 1–Shock waves from the nozzle throat; 2–first injector bow shock; 3–second injector bow shock; 4–shock wave from the joint of base with wall.

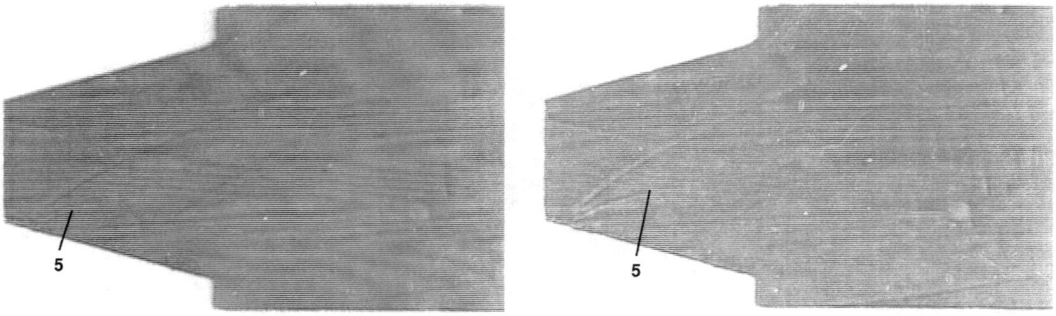

(a) Case 5 (b) Case 6

Figure 8. Partial enlarged view of Schlieren images: 5–Barrel shock.

From figure 7, we can see the whole system but not the details of shock waves presented to the flow in HYLTE nozzle. Figure 8 displays the partial enlarged view of schlieren images. The difference between case 5 and case 6 is the injection pressure. The increasing pressure (case 6) makes the jet momentum larger which would increase the oblique angle of the bow shock waves. This means that the total pressure losses created from the injectors may increase with the momentum flux ratios and mass fluxes at which the injectors are operated due to increased blockage effects. Furthermore, the barrel shock could be discerned downstream of the first injector bow shock[6, 7]. Another barrel shock behind the second injector bow shock and Mach disks are not visible in schlieren images for low resolution.

4. Conclusion

In order to visualize the supersonic flow field of transverse injection in the HYLTE nozzle, NPLS and schlieren photography was used in this paper. The objective was to examine the influence of different total pressure ratio of the twin jets to the freestream and different injectants may exert on the flow structure and mixing. It is clarified that in consequence of the shielding effect of the front jet, the rear jet penetrated further into the mainstream. The mixing characteristics of these tandem jets were dominated by the contact surface strentching, entraining and counter-rotating vortices generated beside the jets.

As for HYLTE nozzle, mixing enhancement is mostly characterized for the variation of several flow control and geometrical parameters including the effective jet-to-crossflow velocity ratio, jet injection angle and secondary fluids orifice separation. Further research should focus on examining the impact of the secondary fluids injection angle and orifice separation on the mixing mechanism of transverse injection in HYLTE nozzle.

References

[1] ZHAO Yuxin, YI Shihe, TIAN Lifeng, et al 2009 Supersonic flow imaging via nanoparticles *Sci China Ser E-Tech Sci* **52** 3640
[2] Ben-Yakar A 2000 Experimental investigation of mixing and ignition of transverse jets in supersonic crossflows Ph.D. thesis Stanford Uiversity
[3] Waitz I A, Marble F E, Zukoski E E 1993 Investigation of a contoured wall injector for hypervelocity mixing augmentatio *AIAA Journal* **31**
[4] Abdehafez A, Gupta A K, Balar R, et al 2007 Evaluation of oblique and traverse fuel injection in a supersonic combustor AIAA 2007-5026
[5] Endo M, Walter R F 2006 *Gas lasers* (New York: CRC Press) pp 67–68
[6] Srinivasan R, Bowersox R D W 2005 Simulation of transverse gaseous injection through a diamond port into a Mach 5.0 freestream AIAA 2005-894
[7] Dickmann D A, Lu F k 2006 Jet in supersonic crossflow on a flat plate AIAA 2006-3451

A Rapid Method for Determining the Concentration of Recombinant Protein Secreted from Pichia pastoris.

L W Sun[1] *, Y Zhao[1], L P Niu[2], R Jiang[1], Y Song[1], H Feng[1], K feng[1] and C Qi[2]*

1 Life Science Research Center of Beihua University, Jilin, China

2 College of Life Sciences, Huazhong Normal University, Wuhan, China.

Abstract Pichia secretive expression system is one of powerful eukaryotic expression systems in genetic engineering, which is especially suitable for industrial utilization. Because of the low concentration of the target protein in initial experiment, the methods and conditions for expression of the target protein should be optimized according to the protein yield repetitively. It is necessary to set up a rapid, simple and convenient analysis method for protein expression levels instead of the generally used method such as ultrafiltration, purification, dialysis, lyophilization and so on. In this paper, acetone precipitation method was chosen to concentrate the recombinant protein firstly after comparing with four different protein precipitation methods systematically, and then the protein was analyzed by SDS-Polyacrylamide Gel Electrophoresis. The recombinant protein was determined with the feature of protein band by the Automated Image Capture and 1-D Analysis Software directly. With this method, the optimized expression conditions of basic fibroblast growth factor secreted from pichia were obtained, which is as the same as using traditional methods. Hence, a convenient tool to determine the optimized conditions for the expression of recombinant proteins in Pichia was established.

1. Introduction

Pichia expression system is a kind of widely used eukaryotic expression system for recombinant proteins[1,2].It is of advantages of higher eukaryotic expression system such as protein processing, protein folding, and posttranslational modification, higher expression levels, higher density fermentation and suitable for secretive protein expressions [3]. Because of few native proteins secreted, it is also easy to seperate and purify the heterologous proteins.As the secreted heterolgous protein comprises the vast majority of the total protein in the medium, it will reduce the procedures of purification therefore reduce the loss of the target protein during purification. Therefore, P. pastoris was widely used as a tool for the seccetive expression of recombinant proteins [4-6].

Published under licence by IOP Publishing Ltd

However, protein expression level in P. pastoris is generally low in the initial expression experiments. Only after the expression conditions were optimized, protein yeilds will reach the peak level. Moreover, usually protein amount should be determined through a series of protocols such as ultrafiltration, concentration, seperation and purification by chromatography, dialysis, lyophilization, and quantitation analysis, So it is necessary to set up a rapid analysis method for the secretive heterologous protein during the optimizing process of protein expression.

Hence, combining with the Automated Image Capture and 1-D Analysis Software using bovine serum albumin (BSA) as a control, an optimized method of protein precipitation and concentration was set up after systematically comparing with other four protein precipitaion methods. It is a rapid, convenient and effective method for low concentration protein assay. Then the supernatant protein assay of basic fibroblast growth factor (bFGF, FGF2) which was secretively expressed in Pichia pastoris, was determined by this optimized assay method. The results were compared with the traditional method. bFGF has a lot of biological effection in different cells and organs, and has been applied in the thearapeutics of wound healing[7-8]. So it is important to improve the recombinant bFGF yeild for both clinical application and mechanistic investigations. The optimized assay method in this paper has solved a difficult analysis problem and may be also suitable for other secretive expression system of heterologous protein expression.

2. Materials and Methods

2.1. Materials

Peptone and yeast extract was purchased from OXOID. Yeast Nitrogen Base, Amino acid Mix, D-Biotin, and Dextrose (D-Glucose) were obtained from Invitrogen. Standard hbFGF was purchased from national institute for the control of pharmaceutical and biological products. BSA was got from BBI. Tris base，Tricine ，and SDS were purchased from BioRad. Acrylamide, N,N'-Methylenebisacrylamide, Glycine, Ammonium persulfate, and TEMED were purchased from Amresco. Methanol, enthanol, and acetone were all obtained from Beijin Chemistry Reagent Factory. Pichia pastoris strain which expresssed bFGF was constructed by our lab. Heparin affnity column was purchased from GE healthcare. All other chemicals and solvents used in the study were analytical grade.

2.2. Methods

2.2.1. Protein precipitation Methods.

2.2.1.1. 10% TCA/acetone precipitation.

10% TCA/acetone precipitation was based on the method described before Lu [9].1ml of sample protein (0.5mg/ml BSA) was added into 8-10 volume of ice-cold acetone containing 10% (w/v) TCA. After vortexed the mixture was incubated at -20℃for 2 h. The sample was centrifuged at 15,000 rpm at 4℃for 15min. The protein pellet was washed three times with 3-4 volume of ice-cold acetone, dried at room temperature for about 30min， and dissolved in 100ul double distilled water.

2.1.2.2. TCA precipitation.

Journal of Physics: Conference Series **276** (2011) 012144 doi:10.1088/1742-6596/276/1/012144

TCA protein concentration was carried out accordingly Sambrook J *et al* [10].100% (w/v) TCA was added to 1ml of sample (0.5mg/ml BSA) and the final TCA concentration was 20%. The mixture was stored at -20℃for 2 h. The protein pellet was collected by centrifugation at 15,000 rpm at 4℃ for 15min and washed with 3-4 volume of ice-cold acetone for three times, dried at room temperature for about 30min and then dissolved in 100ul double distilled water.

2.1.2.3. Alcohol precipitation.
Alcohol protein concentration method was derived from Xu *et al* [11].1ml of sample protein (0.5mg/ml BSA) was mixed with 8-10 volume of ice-cold alcohol, proteins was left to precipitate at -20℃for 2 h, then centrifuged at 15,000 rpm at 4℃ for 15min. Pellet was washed three times with 3-4 volume of ice-cold alcohol, dried at room temperature for about 30min and then dissolved in 100ul double distilled water.

2.2.1.4. Acetone precipitation.
Acetone precipitation method was based on the method described before Lodish H *et al*[12].1ml of sample protein (0.5mg/ml BSA) was added in 8-10 volume of ice-cold acetone, after vortexed the mixture was stored for at -20℃ 2 h.. It was then centrifuged at 15,000 rpm at 4℃for 15min. The final protein pellet was washed with 3-4 volume of ice-cold acetone three times, dried at room temperature for about 30min and then dissolved in 100ul double distilled water.

2.2.1.5. Determination of protein concentration.
Protein concentrations of all samples were determined according to Bradford assay using a commercial dye reagent with BSA as a standard and MiliiQ water as blank and each assay repeated three times. Protein yield was calculated using the followed formula: Protein yield=Protein concentration×Dilution multiples×Sample volumes.

2.2.1.6. SDS-PAGE and Tricine-SDS-PAGE.
SDS-PAGE analysis was performed using 12.5% gel according to the method of Laemmli[13], and the method of Simpson R *et al*[14]. Proteins obtained with different precipitation methods were separated by polyacrylamide gel electrophoresis (SDS-PAGE) or tricine sodium dodecyl sulfate polyacrylamide gel electrophoresis (Tricine-SDS-PAGE), and stained with Coomassie Blue G-250.

2.2.1.7. Image scan and analysis.
The gels were scanned by Image Master Scanner 6.0 and the image was analyzed with BIO-RAD Quantity One 4.3.1 software.

2.2.2. Verification of the bFGF amount in the fermentation supernatant of P.pastoris.
The fermentation supernatant was harvested by centrifugation at 15000 r/min for 20 min at 4 ℃. 500mL of supernatant was precipitated by the above methods (2.2.1.2, 2.2.1.3, 2.2.1.4) respectively. Then protein pellets were resuspended and analyzed by Tricine-SDS-PAGE. The gels were scanned and analyzed according to 2.2.1.7.

2.2.3. Comparison with affinity chromatography and acetone precipitation methods on the effects of bFGF yields induced by methanol at different concentration.

Fermentation supernatants of bFGF which was induced by 0.4% and 0.8% methanol were purified with heparin-Sepharose Fast Flow column by Fast Purification Lipid Chromotography (FPLC). The column was washed with 50 ml of buffer A (20mM Tris-HCl, pH 7.4, 1mM EDTA) and then with buffer A containing 0.5 M NaCl. The recombinant 18-kDa human bFGF was eluted with buffer A containing 3 M NaCl. The elution was dialyzed against 4 liters of buffer B (20mM Tris-HCl, pH 7.4, 150 mM NaCl), and lyophilized. The protein concentration was determined with Bradford protein assay (2.2.1.5). The same sample was precipitated by acetone and the protein pellet was resuspended and analyzed by Tricine-SDS-PAGE. The gels were scanned and analyzed according to 2.2.1.7.

3. Results and Discussion

3.1 Screen and verification of the protein concentration methods

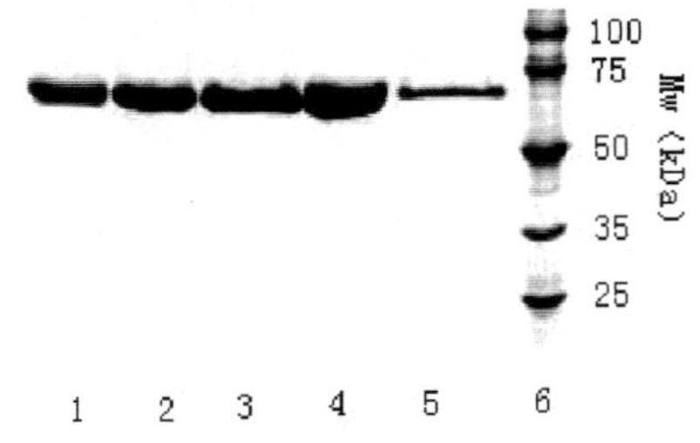

Figure1. Trincine-SDS-PAGE of four protein concentration methods.
1. 10% TCA/acetone precipitation method; 2. TCA precipitation method; 3. Ethanol precipitation method; 4. Acetone precipitation method; 5. BSA; 6. Marker.

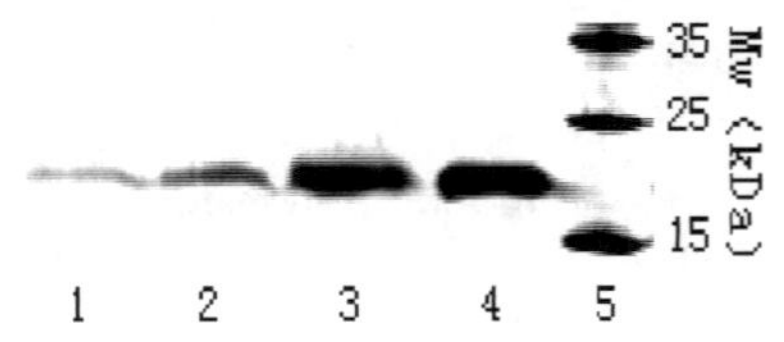

Figure 2. Tricine -SDS -PAGE of three kinds of protein concentration methods.
1. TCA precipitation method; 2. Ethanol precipitation method; 3. Acetone precipitation method; 4. Pure bFGF obtained by HPLC; 5. Marker.

Table1. Protein yields from four protein concentration methods (n=3).

Concentration methods	10% TCA/acetone precipitation	TCA precipitation	Ethanol precipitation	Acetone precipitation
Protein yield(mg)	242±2.18	368±3.45	393±2.94	440±5.89

Table2. Protein yields from different protein concentration methods (n=3).

Concentration methods	TCA precipitation	Ethanol precipitation	Acetone precipitation
Protein yield(mg)	26±0.33	40±0.83	78±1.41

Protein bands using four protein concentration methods were obtained by SDS-PAGE. The densities of protein bands were assayed using BIO-RAD Quantity One 4.3.1 software (Figure1, Table1). The protein band from acetone precipitation was the thickest and its yield is always the highest. This indicated that protein could be precipitated completely by acetone. Although the protein yield from alcohol precipitation was also higher, it still has significant difference compared with acetone precipitation. As a superior polarity organic solvent, alcohol has better water-solubility and it could not completely precipitate proteins, which resulted in certain amount loss of protein. The protein band and yield obtained using 10% TCA/acetone was the weakest and lowest. There are two reasons for this result: Firstly, 10% TCA/acetone may not completely precipitate all the proteins, and secondly 10% TCA/acetone destroyed the protein structure and reduced the resolution. Although the concentration of TCA in TCA precipitation was increased to 20%, no obviously improvement was obtained according to Figure1 and Table1. Hence, acetone precipitation method got the highest protein yield and followed with the alcohol, and TCA precipitation. However, 10% TCA/acetone revealed the lowest protein yield. The conclusion was verified on the supernatant of bFGF which was secreted from P. pastoris, in which higher protein yield methods (acetone precipitation, alcohol precipitation and TCA precipitation method) were applied. Data were shown in Figure2 and Table2. The protein band obtained from acetone precipitation was the thickest, which further illustrated that as a low polarity dissolvent, acetone can precipitate the water-solubility protein commendably. In addition, this method needs less reagents and the operation is simple with better stability. It is an adaptive protein precipitation method for the low concentration protein, especially for secretive recombinant proteins.

3.2 Effects of concentration methods on bFGF yields from different methanol induction

Table3. Protein yields from different methanol induction using affinity chromatography and acetone precipitation （n=3）.

Methanol Concentration (%)	Protein yield（mg）	
	affinity chromatography	acetone precipitation
0.4	440±4.55	418±5.23
0.8	780±5.82	753±2.04

The fermentation supernatants of bFGF which was induced by different concentration of methanol, were purified by traditional affinity chromotography, followed by dialysis and lyophilization. Proteins were assayed, with yields at 440±4.55mg and 780±5.82 mg, respectively (Data shown in Table 3). Likewise, the samples were analyzed by acetone precipitation and image analysis, resulting in protein yields at 418±5.23mg and 753±2.04 mg, respectively. These data indicated that 0.8% methanol is better than 0.4% during inducing expression process, which means that the optimized assay (acetone precipitation and image analysis) can reach the same analysis result with the traditional one. In addition, the combination of acetone precipitation and image analysis is a rapid, time saving and less consuming assay.

4. Conclusion

The levels of majority target proteins at initial period in secretive expression systems were low, so how to determine the target protein level rapidly and accurately remains a difficult problem. Acetone precipitation combined with image analysis was used in this study which revealed a rapid analysis of protein amount. Moreover, the conclusion that acetone precipitation assay has the same results as the traditional chromatography assay on the analysis of bFGF yields under different inducing conditions in P. pastoris indicated that optimized assay is a rapid, convinent, accurate, and less consuming method, which will provide a powerful tool for all secretive expression systems including P. pastoris.

5. Acknowledgements
This research was financially supported by national key technology R&D program (No. 2007BAI38B02), Jilin Provincial scientific and technologic development project (No. 200705403 and 20070923).

References

[1] Cregg J M, Tolstorukov I, Kusari A, Sunga J, Madden K and Chappell T 2009 Expression in the yeast Pichia pastoris J. Methods Enzymol.463 169-89

[2] Weidner M, Taupp M and Hallam S J. 2010 Expression of recombinant proteins in the methylotrophic yeast Pichia pastoris J. Vis Exp. 25 36

[3] Patrick S M, Fazenda M L, Neil B M and Harvey L M 2005 Heterologous protein production using the Pichia pastoris expression system J. Yeast. 22 249-270

[4] Li Z, Moy A, Sohal K, Dam C, Kuo P, Whittaker J, Whittaker M, Düzgünes N, Konopka K, Franz A H, Lin-Cereghino J and Lin-Cereghino GP 2009 Expression and characterization of recombinant human secretory leukocyte protease inhibitor (SLPI) protein from Pichia pastoris J .Protein Expr Purif. 67 175-81

[5] Buensanteai N, Mukherjee P K, Horwitz B A, Cheng C, Dangott L J and Kenerley C M 2010 Expression and purification of biologically active Trichoderma virens proteinaceous elicitor Sm1 in Pichia pastoris J . Protein Expr Purif. 72 131-8.

[6] Gach J S, Maurer M, Hahn R, Gasser B, Mattanovich D, Katinger H and Kunert R 2007 High level expression of a promising anti-idiotypic antibody fragment vaccine against HIV-1 in Pichia pastoris J. Biotechnol. 128 735-46

[7] Bikfalvi A, Klein S and Pintucci G 1997 Biological roles of fibroblast growth factor-2 J. Endocr Rev.18 26-45

[8] Przybylski M A 2009 review of the current research on the role of bFGF and VEGF in angiogenesis J. Wound Care. 18 516-9. Review.

[9] Lu S D 1999 Current Protocols for Molecular Biology second ed（Beijing:Peking Union Medical College Press）

[10] Sambrook J and Russell D W 2001 M Molecular Cloning: A Laboratory Mannual, 3 rd ed (New York: Cold Spring Harbor Laboratory Press)

[11] Xu R A, Chen L and Xiao W D 2008 Molecular Gene Medicine (Beijing:Peking University Medical Press)

[12] Lodish H, Berk A, Zipursky S L, Matsudaira P, Baltimore D and Darnell J 2000 Molecular Cell Biology 4th ed (New York: W.H. Freeman and Company)

[13] Laemmli U K 1970 Cleavage of structural proteins during assembly of head of bacteriophage T4 J. Nature. 277 680-685

[14] Richard J S 2003 Protein and Proteomis:A Laboratory Manual(NewYork: Cold Spring Harbor Laboratory Press)

Simulation and Experimental Investigations on the Strain Measurement of the Uniform Strength Beam Using a FBG sensor

Yumeng Tu[1], Huaping Gong, Jixuan Chen, Yongxing Jin
Institute of Optoelectronic Technology, China Jiliang University, Hangzhou, 310018, China
E-mail:lovetymabc@126.com

Abstract The model of force analysis on a uniform strength beam is built by the general finite element program (ANSYS software). The flexivity profile produced by uniform strength beam with different forces is simulated by ANSYS software. In experiment, a fiber Bragg grating sensor is fixed on the uniform strength beam with modified acrylate. The flexivity and strain are varied by changing the load on the end of the beam. The strain of the uniform strength beam is measured with FBG when applied force is varied from 2.45N to14.7N with a step of 2.45N. Both the simulated and experimental results show that, the strain induced by the uniform strength beam is linear with the load force. The sensitivity is 18.32με/N for experimental measurement, and 19.72με/N for simulation. The experimental results are consistent with the simulation results, with the maximum measurement error of strain being 7.4%. It indicates that, the FBG sensor fixed with modified acrylate is proved to be effectively and reliably in the applications of civil engineering.

Key words fiber Bragg grating sensor; simulation; uniform strength beam; strain measurement; flexivity

1. Introduction

The safety assessment of civil engineering becomes more important with the development of the society. Detecting the strain in the civil engineering correctly can realize structural health monitoring for a long time. Sensor technology is the foundation of the engineer security. In the field of fiber-optic sensors, FBG sensors are the most outstanding. Its sensing principle is based on the sensitive characteristic of its wavelength to the measurand, such as temperature and strain. FBG sensors have several inherent advantages over conventional electrical sensors, such as small size, light weight, non-conductivity, fast response, resistance to corrosion, higher temperature capability, and immunity to electromagnetic noise and radio frequency interference[1-5]. When civil engineering is monitored by FBG sensor, sensing characteristics of FBG sensor are affected by different fixing methods. Therefore, it is necessary to investigate the difference between measurement value and true value.

in this paper, finite element method is used, this is to say that, uniform strength beam is divided into a number of finite elements to analyze. The strain produced by uniform strength is simulated based on ANSYS 10.0 software, and measured by FBG. Finally, the difference of the measured value and the

[1] Foundation item: 浙江省科技厅重大科技专项（编号 2008C13041-1）
Corresponding author:Yumeng Tu(1986-),male,Tel 13675854196

simulated value is investigated [6].

2. Operating principle

The Bragg wavelength related to the grating period Λ, and the effective refractive index n_{eff} , which is given by

$$\lambda_B = 2n_{eff}\Lambda \tag{1}$$

When the temperature is constant, both n_{eff} and Λ vary with the changes of the strain in the fiber. The Bragg wavelength shift caused by variation of the strain is given by is given by

$$\frac{\Delta\lambda}{\lambda} = (1-P)\Delta\varepsilon \tag{2}$$

Where P is the valid elastic-optic constant, $\Delta\varepsilon$ is the variation of the strain[7]. The strain caused by uniform strength beam is given by

$$\varepsilon = \frac{hX}{l^2} \tag{3}$$

Where h is the thickness, l is the length, and X is the maximal flexivity of the uniform strength beam. From the above equations, it is easy to find that the Bragg wavelength shift is linear with the strain caused by the uniform strength beam [8, 9].

3. ANSYS simulation

The ANSYS program is composed of PREP7, SOLUTION, and POST1. In the PREP7, the model of the uniform strength beam is built and meshed. The element type is defined soulid95, which is a higher order version of the 3-D 8-node solid element solid45. It can tolerate irregular shapes without as much loss of accuracy, so it can be used to analyze the strain caused by uniform strength beam. The result is shown through the cloud picture.

The material of the uniform strength beam is iron, whose elastic modulus is 1.6e11pa. The left end of the beam is fixed, and the right end is applied some force, as shown in Fig.1. The flexivity of the uniform strength beam is simulated by ANSYS software, and the could pictures are acquired when applied force 2.45N、4.9N、7.35N、9.8N、12.25N、14.7N, as shown in Fig.2. The strain can be calculated by equation (3), as shown in Fig.3. The results show that the strain caused by the uniform strength beam is linear with the applied force, which is consistent with the theory.

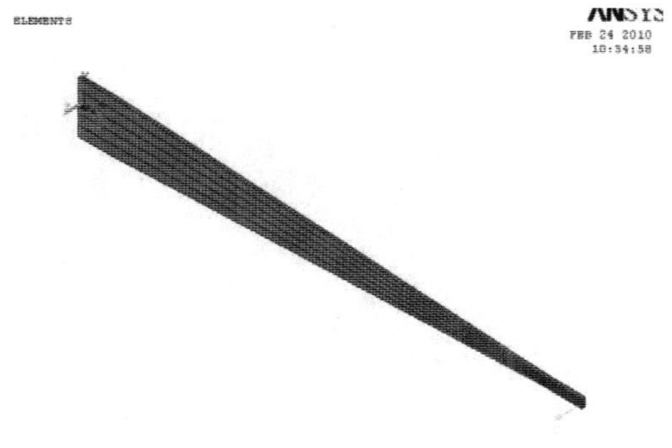

Figure.1 The model of the uniform strength beam

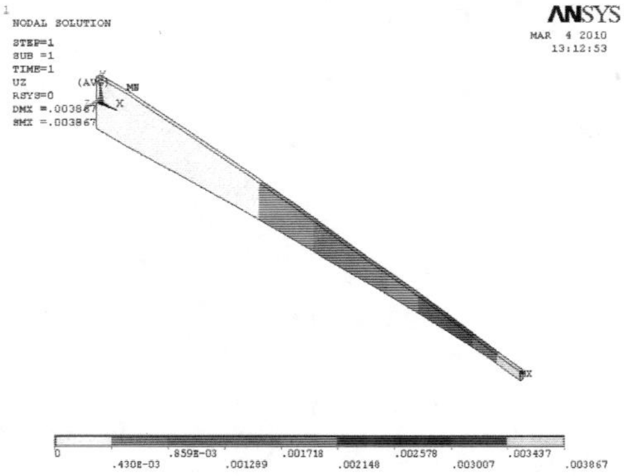

（a）applied force 2.45N

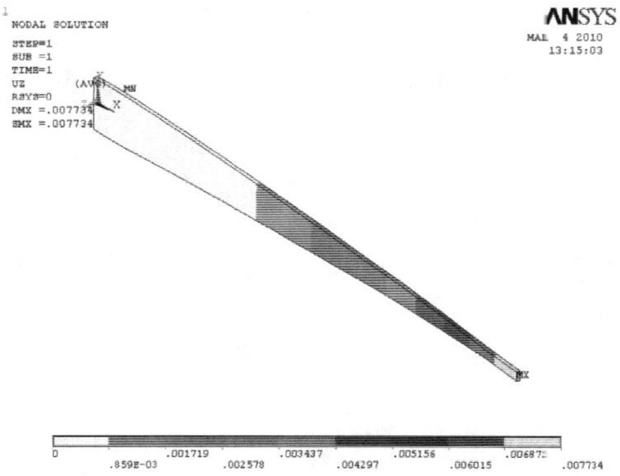

(b) applied force 4.9N

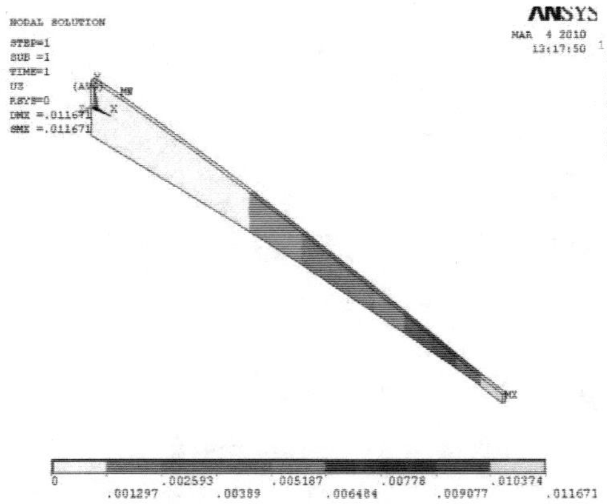

(c) applied force 7.35N

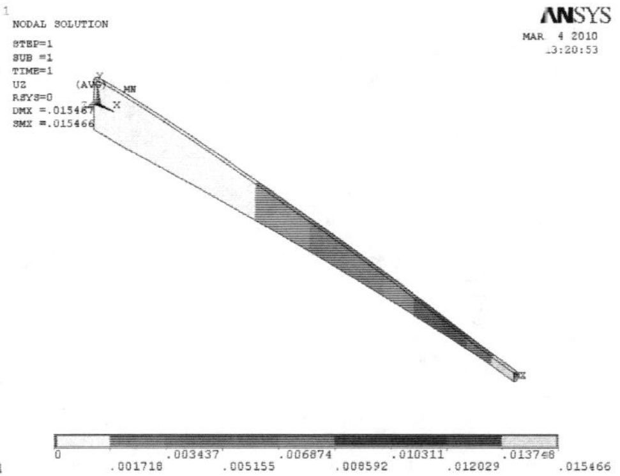

(d) applied force 9.8N

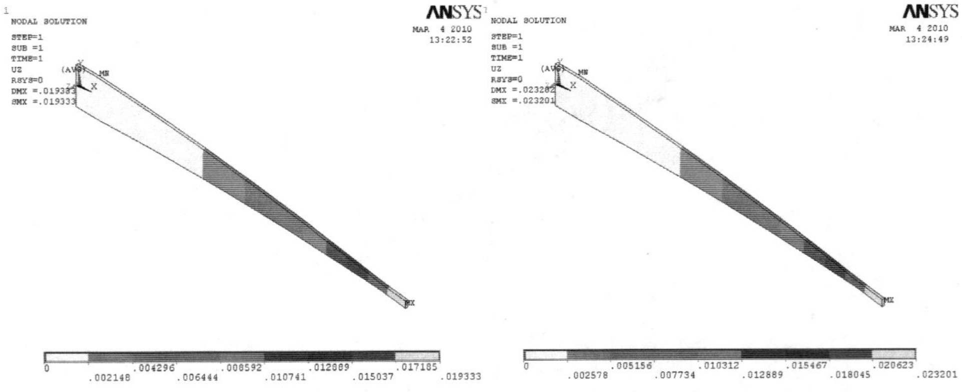

(e) applied force 12.25N (f) applied force 14.7N

Figure.2 The flexivity of the uniform strength beam when applied different forces, the colors represent the values of the flexivity profiles.

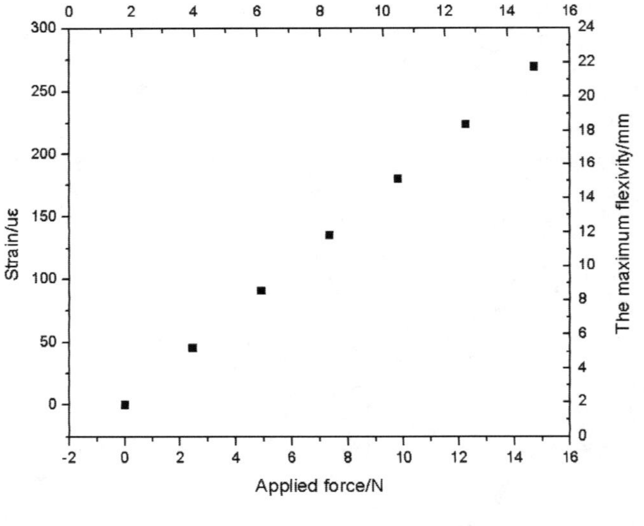

Figure.3 The maximum flexivity and the strain obtained from simulation

4. Experiment

4.1 Experimental setup

We use the SM130 FBG demodulation produced by MICRON OPTICS Ltd, whose resolution is 2pm. The FBG sensor is produced by NINGGBO SANGONG Ltd. Its center wavelength is 1547.799nm at temperature 17℃. The experimental setup is shown in Fig.4. The probing light is transmitted to the FBG sensor from the broadband source which is integrated in the demodulator. And then the reflected signal of the FBG sensor is transmitted back to the demodulator, which will be display on the computer. A fiber Bragg grating sensor was fixed on the uniform strength beam with modified acrylate.

In order to change the strain in the FBG sensor, the weight at the end of the uniform strength beam is varied.

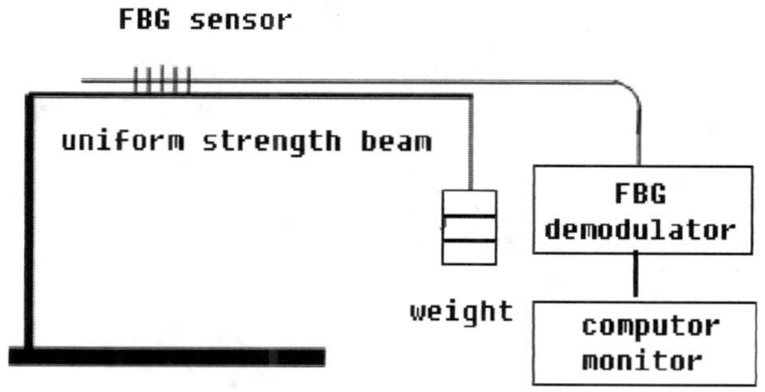

Figure.4 Experimental setup

4.2 Experimental results and discussion

The wavelength shifts of FBG sensor fixed with modified acrylate are measured at the force 0、2.45N、4.9N、7.35N、9.8N、12.25N、14.7N when the temperature is 25℃. The strain can be obtained through the equation (2), as shown in the Fig.5. It shows that the strain caused by the uniform strength beam is linear with the load force. Also, in the three repetitive measured experiments, the error of the wavelength shift is less than 5pm, which shows that the temperature is constant.

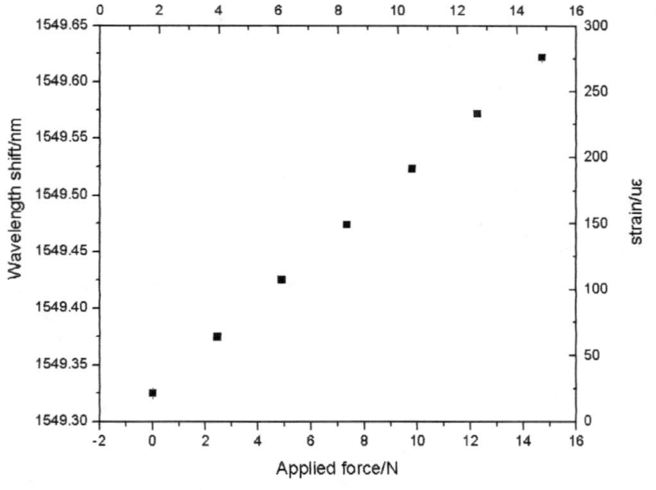

Figure.5 FBG sensor wavelength and strain response under different applied forces

The strain comparison between the simulated and measured is shown in Fig.6. The strain of the uniform strength beam measured with FBG when applied force is 2.45N, 4.9N, 7.35N, 9.8N, 12.25N, and 14.7N. The corresponding strain is 48.46 με, 96.92 με, 143.38 με, 192.64 με, 242.20 με. 290.70 με by ANSYS simulation, and 45.33 με, 90.66 με, 135.08 με, 179.50 με, 223.92 με, 269.25 με by FBG demodulation. The strain obtained by simulation and measurement both has excellent linearity. The sensitivity is 19.72με/N for ANSYS simulation and 18.32με/N for measurement by FBG, respectively.

The linearly dependent coefficient is 0.99997 for ANSYS simulation and 0.99999 for measurement by FBG, respectively, which proves that the strain is excellent linear with the applied force. The experimental results are consistent with the simulation results, with the maximum measurement error of strain being 7.4%. The error is caused by the error of the elastic modulus between the parameter value and the real value, and by the stress transmission error caused by FBG fixing method.

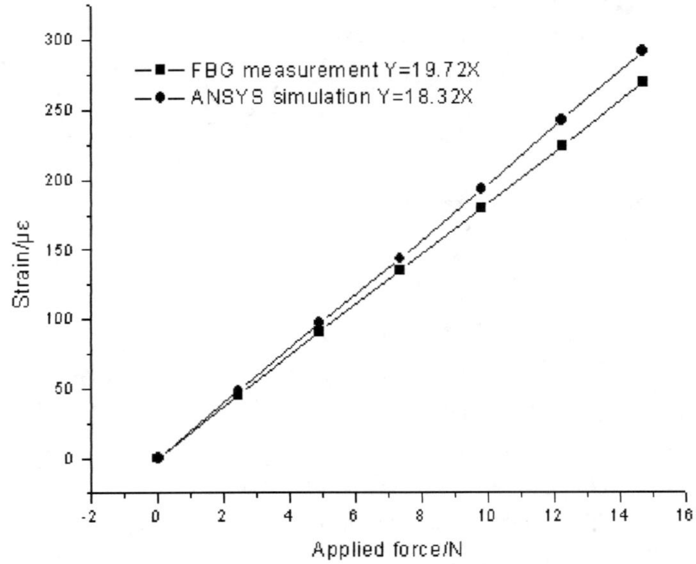

Figure.6 Strain of simulation and experimental measurement versus applied forces.

5. Conclusions

The flexivity profile produced by uniform strength beam with different forces is simulated by ANSYS software. In experiment, a fiber Bragg grating sensor is fixed on the uniform strength beam with modified acrylate. The strain is measured by FBG. Both the simulation and experimental results show that, the strain induced by the uniform strength beam is linear with the load force. And the experimental results are consistent with the simulation results, with the maximum measurement error of strain being 7.4%, which is acceptable to the applications of the civil engineering. This paper indicates that, the FBG sensor fixed with modified acrylate is proved to be effective and reliable in the applications of civil engineering.

References

[1] T.H.T. Chan, L. Yu, H.Y. Tam. Fiber Bragg grating sensors for structural health monitoring of Tsing Ma bridge: Background and experimental observation. Engineering Structures, 2006, 28:684-659

[2] LU Zhe-an, JIANG Zhi-xue, SHI Yu-hua, WANG Ji. Application Study of FBG Sensing Technology in Bridge Monitoring [J]. Journal of Wuhan University of Technology, 2003, 25(11): 57-59

[3] HUANG Rui, CAI Hai-wen, QU Rong-hui. Separately Measurement of Strain and Temperature Using a Single Fiber Bragg Grating [J]. Chinese Journal of Lasers, 2005, 32(2): 233-235

[4] WANG Jian-ping. Application of Fiber Grating sensors to Civil Engineering Structures Health Monitoring[J].Journal of Guizhou University of Technology, 2004, 33(1): 78-80

[5] SU Mu-biao, DU Yan-liang, SUN Bao-chen. Study on the Long-term Health Monitoring and Alaming

System for the Wuhu Yangtze River Bridge[J].Journal of the China Railway Society, 2007, 29(2): 71-76

[6] JIANG Yong, ZHANG Bo. Analysis on ANSYS7.0 examples [M]. Press of Qsinghua university, 2004 Jan, pp110-130

[7] FU Hai-wei, QIAO Xue-guang, JIA Zhen-an. Stress Sensitivity Enhanced in-Fiber Bragg Grating Pressure Sensor [J].Chinese Journal of Lasers, 2004, 31(4): 474-476

[8] LUO Ying-xiang,JIANG Xing-da.Research on Tuning Characteristics of the Equi-intensity Cantilever Beam Fiber Optic Grating[J]. Journal of Sichuan Normal University, 2006, 29(1): 86-88

[9] ZHANG Wei-gang,KAI Gui-yun,ZHAO Qi-da.Novel Sensing Device of Fiber Bragg Grating with Temperature Active Compensation[J]. Acta Optica Sinica, 2002, 22(8): 999-1002

Journal of Physics: Conference Series **276** (2011) 012146 doi:10.1088/1742-6596/276/1/012146

A novel temperature self-compensation FBG vibration sensor

Wang Tao, He Dawei[1], Wang Ziqian, Wang Yongsheng
Key Laboratory of Luminescence and Optical Information, Ministry of Education, Institute of Optoelectronic Technology
Beijing Jiaotong University, Beijing 100044, China

E-mail: dwhe@bjtu.edu.cn

Abstract. There has been considerable interest recently in the development of fiber optic sensors based on Fiber Bragg gratings（FBGs）and FBGs are being used in various applications. In the measuring of the FBG, temperature compensation is one of the most important aspects. This paper describes a new FBG sensors use two FBGs and packaged them with novel structure, and then they can achieve self-compensating of temperature. Two FBGs with 0.2nm different wavelengths are made in one optic fiber, and they are fixed on symmetrical position of a cantilever girder's surface. Wavelength shift of two FBG caused by the temperature is equal and can be eliminated. To inspect and verify the FBG vibration sensor, we fixed it on the standard vibration system, and set the vibration frequency from 5Hz to 30Hz, record the wavelengths of two FBGs over time. The experimental value caught by this novel sensor and setting value is agreement. This sensor have a good frequency response characteristic, it can measure vibration frequency accurately.

1. Introduction

There has been considerable interest in the development of fiber optic sensors based on Fiber Bragg gratings（FBGs）over the last few years, and FBGs are being used in various applications[1]. FBG sensors have many inherent advantages over conventional electrical sensors such as fast response, non-conductivity, small size, light weight, and resistance to corrosion. FBG sensors have large potential for applications where quasi-distributed measurements of physical parameters such as strain, pressure, vibration and temperature.
In the measuring of the FBG, temperature compensation is one of the most important aspects.
The FBG were produced using the phase mask method. Part of the incident light was reflected by FBG in a narrowband, it is the Bragg wavelength, $_{B.}$

$$_B=2ne \qquad (1)$$

And wavelength shift, $_B$.

$$_B = {}_B[(\ +\ )\ T+(1-\ e)\ \ ] = {}_T+\ {}_S \qquad (2)$$

ne is the effective refractive index of the fiber core; is the grating period; is the fiber linear thermal coefficient (0.55×10-6/℃); is the thermo-optic coefficient (8.3×10-6/℃); T is the change in temperature; is the change in strain; $_e$ is the effective photo-elastic coefficient (0.26).

---

1 dwhe@bjtu.edu.cn; phone +86-10-51688018, fax +86-10-51683933

When there are external disturbances in the grating part, the grating period is changed, which leads the Bragg wavelength changes linearly with temperature (10 pm/°C) and strain (1.2 pm/μ). The change of the period length is caused by the change of temperature ($_T$), strain ($_S$), and other physical influences, so physical quantity can be known by analysis the wavelength.

If the temperature have a change when we measuring the strain, We have to tell the wavelength shift caused by temperature apart strain. At present, there are two main ways to achieve temperature compensation when we used FBG vibration sensors. The first way is active, which means control temperature by external circuit. But this way is complex, difficult to achieve and with high failure rate. The second and better way is to achieve temperature compensation by using special packaged FBG, or add another FBG in the same area to elimination the influence of temperature. We follow the second way with two FBGs

2. Principle and Structure of FBG vibration sensor

As is shown in Figure1, A cantilever girder was set in the internal steel shell, two FBGs are fixed on it, they are made in one optic fiber and there are 0.2nm different between their wavelengths. FBG 1 was pasted on the surface of the cantilever girder with resene, FBG 2 was pasted on symmetrical position under the surface, they are justify align and on the center line of the cantilever girder. A mass was set on the free end of the cantilever girder; it can oscillate freely on the plane which is vertical with the cantilever girder. Vibration frequency of the cantilever girder can be adjusted by changing the mass.

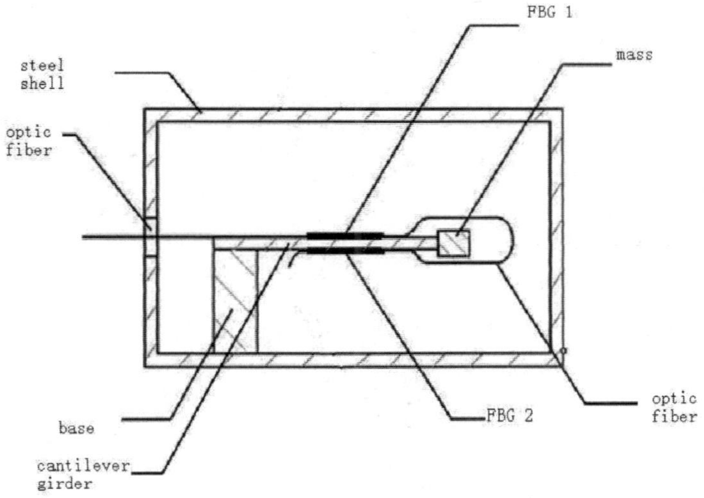

Figure1. Structure of FBG vibration sensor

When the vibration monitoring system is working, the sensor will be fixed on the detected objects, and its surface parallel the surface of detected objects. When sensor is forced to vibrate, deflection of cantilever girder will have a change, and the distance between the base and the mass will change with the same frequency of vibration, then the axial strain of FBG will follow the frequency of vibration.

When there is only temperature shift in the area, the wavelengths of FBG 1 and FBG 2 will change synchronously. Two FBGs are written on the same optic fiber, set in the same temperature field, so they have the same temperature sensitivity, and their wavelength shift caused by temperature shift must be equal:

$$ {}_{B1} = {}_{T1} = {}_{T2} = {}_{B2} \qquad (3) $$

When there is no vibration on the measured structure, the difference of wavelengths between two FBGs is constant. Once there is vibration on the detected objects, it will cause forced vibration of the cantilever girder. Because cantilever girder is constant strength, the position of two FBGs are justify align and on the center line of the cantilever girder, then the wavelengths of two FBGs will shift in the opposite direction, and their variation must be equal.

$$_{B1}= {}_{T1}+ {}_{S1} \tag{4}$$

$$_{B2}= {}_{T2}+ {}_{S2} \tag{5}$$

$$_{T1}= {}_{T2} \tag{6}$$

$$_{S2}=- {}_{S1} \tag{7}$$

Then

$$_{B1}- {}_{B2}=2 {}_{S1} \tag{8}$$

The difference of wavelengths between two FBGs can reflects the vibration information of detected objects, and wavelength shift caused by the temperature factor is eliminated. So this FBG vibration sensor achieves temperature compensation by itself.

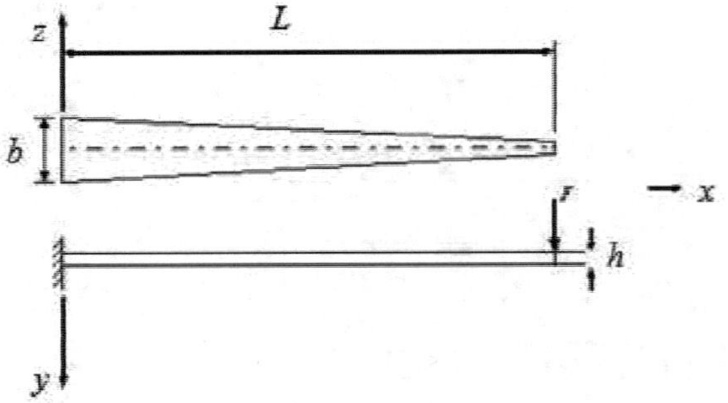

Figure2. Structure of cantilever girder

L is the length of cantilever girder, h is the thickness of cantilever girder.
Frequency of the system:

$$\omega_0 = 2\pi f = \sqrt{\frac{k}{m}} = \sqrt{\frac{bh^3E}{4L^3m}} \tag{9}$$

Wavelength shift of each FBG:

$$\frac{\Delta\lambda_B}{\lambda_B} = -0.78\frac{6Lma}{bh^2E} \tag{10}$$

Sensitivity of the system:

$$S = \frac{\Delta\lambda_B}{a} = -0.78\frac{6Lm\lambda_B}{bh^2E} \tag{11}$$

the material of cantilever girder is glucinfum bronze， with young's modulus E=128GPa, m=20g,L=20mm, b=8mm, h=1mm. Centre wavelength of FBGs is 1557.997nm.
Then, ω_0=1206Hz， f=$\omega_0/2\pi$=191Hz, S=2.845*10^{-12}m/m.s^{-2}

3. Measurements used FBG vibration sensor

To inspect and verify the FBG vibration sensor, we fixed it in the standard vibration system. The vibration system consists of standard vibrator JZ-10, Signal Generator and station amplifier. The FBG vibration sensor was fit on the standard vibrator JZ-10. FBG interrogate system which we used has 4 channels and its maximum sampling rate of FBG interrogate system is 500 Hz. The wavelength data will be recorded and reappeared by the analyses software.

Figure3. FBG vibration sensor with two FBG

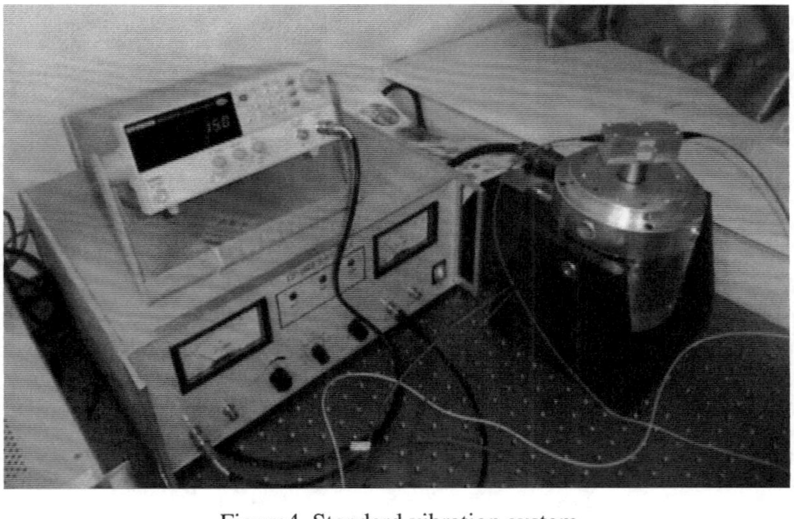

Figure4. Standard vibration system

We set the vibration frequency from 5Hz to 20Hz, and the system can record the wavelengths of two FBGs over time. The experimental value caught by the FBG vibration sensor and setting value is agreement

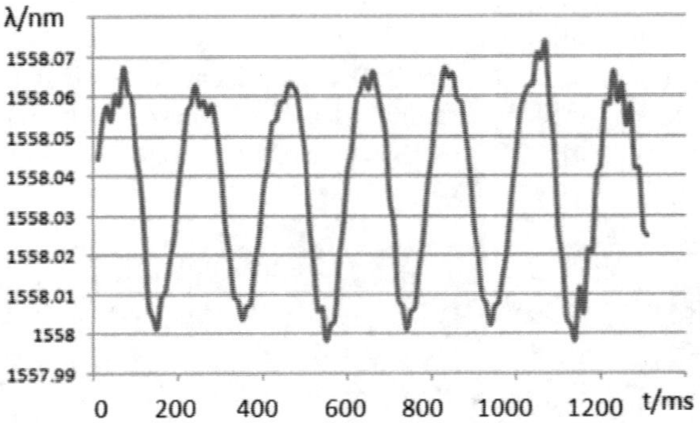

Figure5. Vibration frequency 5Hz

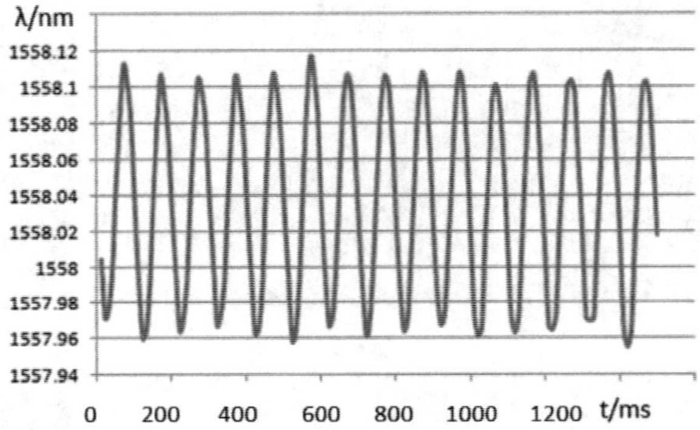

Figure6. Vibration frequency 10Hz

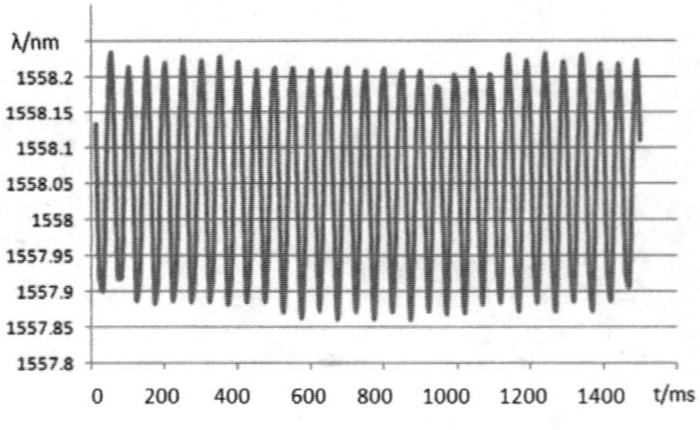

Figure7. Vibration frequency 20Hz

4. Conclusion

We set two FBG on cantilever girder to make a novel vibration sensor. The frequency of the sensor is ω_0=1206Hz, f=$\omega_0/2\pi$=191Hz, Sensitivity of the sensor is S=2.845*10^{-12}m/m.s^{-2}. This FBG vibration sensor can achieves temperature compensation by itself with two FBG. The experimental value caught by this novel sensor and setting value is agreement. This sensor have a good frequency response characteristic, it can measure vibration frequency accurately.

Acknowledgments
The authors would like to acknowledge support via [2009J006-K] funded by Ministry of Railways of PR China.

References

[1] K.T.V.Grattan and B.T. Meggitt 2000 *Optical Fiber Sensor Technology: Advanced Applications* (The Netherlands：Kluwer Academic Publishers)
[2] Berkoff, T.A,Kersy and A.D. Phot. 1996 *Tech.Lett.* **12** 1677
[3] Todd M.D., Johnson G.A. and Althouse B.A. 1998 *Phot.Tech,Lett,* **10** 1605
[4] Mitia A. 2000 *Fiber Bragg grating-based acceleration sensors for civil and building structures,Interntional workshop on present and future health monitoring* (Weimar: Bauhans-University Weimar)

Journal of Physics: Conference Series **276** (2011) 012147 doi:10.1088/1742-6596/276/1/012147

Application of fiber Bragg grating sensors for the micro-strain measurement of optical components surface

Z G Wang, J X Li, R H Zhu

Department of Optical Engineering, School of Electronic Engineering and Optoelectronic Technique, Nanjing University of Science and Technology, Nanjing 210094, China

E-mail: zhurihong@mail.njust.edu.cn

Abstract. Fiber Bragg Grating Sensors (FBGS) are gaining increasing attention in the fields of micro-stress analysis, showing high accuracy and sensitivity. In this paper, FBGS which wavelength variations can be converted to strains, are used to measure the micro-strain variation of a plane mirror where the forces acting upon. Both the self-gravity and extrusion forces are applied to the surface of the optical elements, generating inevitable deformation and decreasing the test accuracy of the whole optical system. First, a micro-strain model is built using finite element analysis to obtain the qualitative value of the micro-strain variation. Next, a series experiments are performed to validate the model is effective. A detailed analysis of the micro-strain variation is acquired from the phase generated carrier homodyne demodulation and then the three-step phase-shifting algorithm based on an unbalanced Mach-Zehnder interferometer system. The experimental results match well with the theory. Both the modeling and measurement results indicate that the sensitivity of this method in measuring the micro-strain can be reached as small as $10\mu\varepsilon$.

1. Introduction

Fiber Bragg gratings (FBGs) have recently been applied in a growing range of sensing applications, such as aerospace, military, petrochemical, structural monitoring, chemical and biomedical sectors [1].Their many advantages include their compact size which allows for embedding in advanced structures, easily distributed sensing, high precision and sensitivity, wavelength-encoded operation immune to light power fluctuations and system consumption and, in particular, their quasi-linear spectral response to external influences [2,3].Of these, strain measurement is a major area of interest, and a number of configurations of FBG strain sensors have been demonstrated. The key point of the FBGS sensors systems is to attain the Bragg wavelength shifts caused by the strain. Various demodulation methods for the detection of the Bragg wavelength shift have been conceived. These approaches include techniques based on direct wavelength analysis, phase delay response, and optical intensity measurement [4], most of which involved some form of optical filtering [5]. Filters consisting of tunable filters incorporating matched FBGS, have been widely used [6,7]. Using interference technology can implement real-time measurement. Unbalanced Mach-Zehnder interferometer changes the wavelength shifts into phase change, has extremely high sensitivity and accuracy. The strain on the surface of the optical component is slight and can not be easily discerned, but it can change the surface shape and is harm to the test accuracy of the whole optical system. So it is essential to measure the strain change of an optical component, better combined with the interferometer.

In this paper, we measure the micro strain on a plane mirror surface due to the extern pressure based on the unbalanced Mach-Zehnder interferometer system based on a 3×3 fiber coupler. Experimental results are well matched with the results from finite element analysis.

2. Principle

From the coupled wave theory, when the phase matching conditions are satisfied, FBG's Bragg wavelength is (1)

$$\lambda_B = 2n_{eff}\Lambda \tag{1}$$

Where λ_B is Bragg wavelength; n_{eff} is effective refractive index of optical fiber spread abroad mode; Λ is optical grating period. Here we assume the temperature is constant.

The relation between the shift of the Bragg wavelength of FBG and the axial strain applied to a fiber grating is (2)

$$\frac{\Delta\lambda_B}{\lambda_B} = (1-P_e)\varepsilon = k\varepsilon \tag{2}$$

where $P_e = n_{eff}^2[P_{12} - \nu(P_{11}+P_{12})]/2$ is the effective photoelastic coefficient of the glass fiber with Poisson ratio ν, P_{11}, and P_{12} denote the photoelastic coefficients, and n_{eff} represents the effective refractive index of the guide mode. For a typical fused-silica fiber, $\nu = 0.16$, $n_{eff} = 1.46$, $P_{11} = 0.12$, $P_{12} = 0.27$, and thus, $P_e = 0.22$, $k = 0.78$.

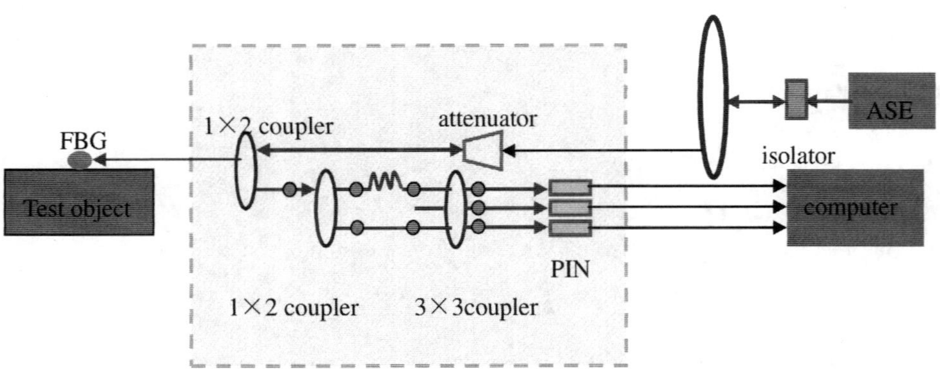

Figure 1 Scheme of measuring FBG system

The measuring scheme is shown in figure 1. Take one single fiber-optic interferometer for example, other three interferometers are the same.

Using unbalanced Mach-Zehnder interferometer system, the changes of the reflection wavelength cause the diversification of the light intensity periodically,

as equation(3) shows, $$\Delta\varphi = \frac{2\pi nl}{\lambda^2}\Delta\lambda = \frac{2\pi nl}{\lambda}k\varepsilon \tag{3}$$

So can measure the strain via the phase information as (4)

$$\varepsilon = \frac{\lambda}{2\pi nlk}\Delta\varphi = G\Delta\varphi \tag{4}$$

Where $G = \lambda / 2\pi nlk$ is the slope coefficient.

Thus ε and $\Delta\varphi$ have a linear relationship. Gauge the three output light intensity we can get the change of the phase $\Delta\varphi$ via (5)

$$\Delta\varphi = \arctan(\frac{\sqrt{3}(I_1 - I_3)}{2I_2 - I_1 - I_3}) \tag{5}$$

The three outer has a 120° phase error between each other.

In practical, $\Delta\varphi$ is represented the phase difference between strained and zero strain. When the strain ε is affected, the phase value corresponded with the inteference signal is

$$\varphi_\varepsilon = \frac{2\pi n\Delta l}{\lambda + \Delta\lambda} = \frac{2\pi n\Delta l}{\lambda(1 + k\varepsilon)} \tag{6}$$

Thus the phase difference between it and zero strain is

$$\begin{aligned}\Delta\varphi = \varphi_\varepsilon - \varphi_0 &= \frac{2\pi n\Delta l}{\lambda(1+k\varepsilon)} - \frac{2\pi n\Delta l}{\lambda} \\ &= 2\pi n\Delta l \frac{-k\varepsilon}{\lambda(1+k\varepsilon)}\end{aligned} \tag{7}$$

In the test range, the value of strain ε is between $0 \sim 1.8\times10^{-3}$, obviously $k\varepsilon << 1$, so $1 + k\varepsilon \approx 1$.Hence

$$\Delta\varphi = -\frac{2\pi n\Delta l}{\lambda}k\varepsilon \tag{8}$$

The form of equation (8) is the same as the result from formula (3).

3. Finite element model

A round plane mirror is choosed as the experiment object. And a circular groove is used to fix the mirror to avoid shifting.

Firstly, a finite element model is created. The mirror is made of glass, with diameter $d = 66mm$ and height $h = 10mm$. Its density is $\rho_1 = 2.6\times10^3 kg/m^3$, the Young's modulus is $E_1 = 7.2GPa$, the Poisson's ratio is $\nu_1 = 0.2$. The material of the groove is steel. Its outer diameter is $120mm$, inner diameter is $66mm$, height is $66mm$, the depth of the groove is $5mm$. The density of the groove is $\rho_2 = 7.8\times10^3 kg/m^3$, the Young's modulus is $E = 200GPa$, the Poisson's ratio is $\nu = 0.3$. The created model is shown as figure 2.

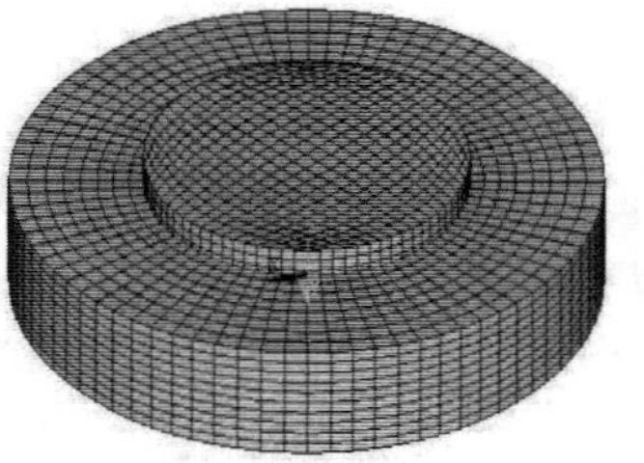

Figure 2 Finite element model of the mirror

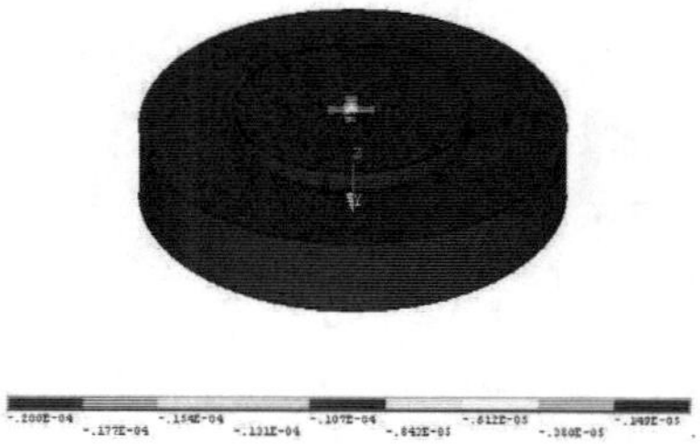

Figure 3 Deformation of the mirror

Pressure of $1N$ is exerted on the middle point of the mirror surface, and deformation is generated to the mirror, as shown in figure 3.

From the theoretical results, the deformation of the mirror is $1.49\mu m$. So the strain of the mirror is $75\mu\varepsilon$, due to $\varepsilon = \frac{\Delta l}{l}$, where Δl is the changed length, and l is the former length, $l = 20mm$.

4. Experimental results

To measure strain the FBGS must be fixed to the plane mirror, typically by gluing. Traditionally there are three kinds of adhesives, which were modified acrylate, glass glue and epoxy resin. The modified acrylate is choosed because the FBG sensor fixed by the modified acrylate has a higher linearity and sensitivity than others [8].

The experiment is carried, using the first interferometer, with an instantaneous pressure of $1N$ applied in the surface of the mirror. The three outer signal is shown in figure 4 , we calculate the changed phase is $\Delta\varphi = 0.6541$, and then $\varepsilon = 64\mu\varepsilon$. In practical, the relationship between the strain and the changed phase is not strictly linear.

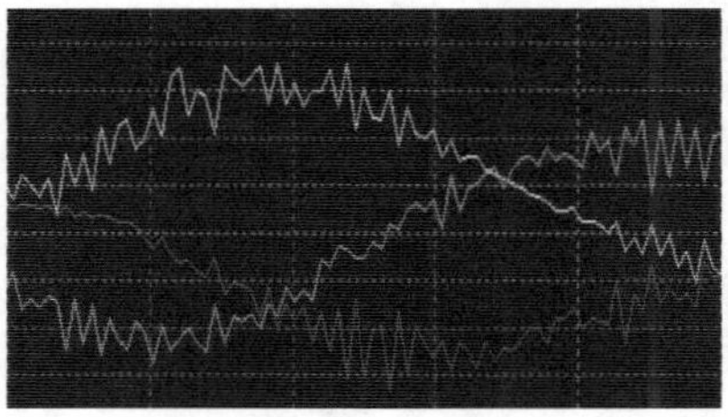

Figure 4 Three outputs of the first interferometer

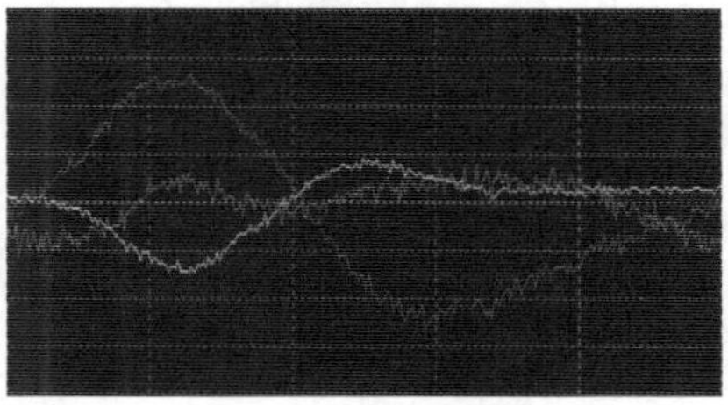

Figure 5 Three outputs of the second interferometer

So under the effect of $1N$ pressure, the strain changed of the mirror is $64\mu\varepsilon$.

The same experiment is done using the second interferometer, the output is shown in figure 5. It is calculated $\Delta\varphi = 0.4164$, then $\varepsilon = 69\mu\varepsilon$.

Another experiment using the first interferometer is done again, and the result is $\Delta\varphi = 0.6654$ and $\varepsilon = 65\mu\varepsilon$. These experimental results are well matched with the result from finite element model.

5. Conclusion

We have proposed and demonstrated a FBG strain sensor based on an unbalanced Mach-Zehnder interferometer system. The demodulation that translates the Bragg wavelength of FBGS caused by the strain into the phase information. A linear relationship between the strain and the phase variation is built up. The phase variation is obtained through the three-step phase-shifting algorithm. This method is applied in the micro-strain measurement of a plane mirror. A finite element model is created and a series of experiments are carried up. The results show that this method can be applied in measuring plane mirrors' surface micro-strain and has a high sensitivity and accuracy. The error between the experiment and the theory is less than $10\mu\varepsilon$. It can be extended to measure optical components' surface micro-strain and can be applied in real-time monitoring of interferometers. However, this method has limitation that it is applied to transient measurement, not suitable to gauge strain functioned under long time. This will be our further research.

References

[1] A. D. Kersey, M. A. Davis, H. J. Patrick, M. LeBlanc, K. P. Koo, C. G. Askins, M. A. Putnam, and E. J. Friebele 1997 *J.Lightwave Technol.* **15** 1442–57.

[2] K. O. Hill and G. Metz 1997 *J.Lightwave Technol.* **15** 1263–64.

[3] Y. Zhao and Y. Liao 1999 *Opt. Laser Eng.* **31** 297–324.

[4] Xiufeng Yang , Shaojun Luo, Zhihao Chen , Jun Hong Ng and Chao Lu 2007 *Opt. Commun.* **271** 203–206.

[5] M.Suleiman, H.C.Seat, and T.Bosch 2008 *IEEE Sens.J.* **8** 1317-23.

[6] Y. N. Ning, A. Meldrum, W. J. Shi, B. T. Meggitt, A. W. Palmer, K.T. V. Grattan, and L. Li 1998 *Meas.Sci.Technol.* **9** 599–606.

[7] L. A. Ferreira, J. L. Santos, and F. Farahi 1997 *IEEE Photon.Technol.Lett.* **9** 487–9.

[8] J X Chen, H P Gong, S Z Jin and S H Li 2009 *SPIE* **7514** 751419-1-7.

Journal of Physics: Conference Series **276** (2011) 012148 doi:10.1088/1742-6596/276/1/012148

Study on HDPE-PCF evanescent wave sensor

Haixia Cui[123], Jianquan Yao[2], Ying Lu[2], Chunming Wan[1]

[1]Changchun University of Science and Technology,Changchun, Jilin 130022, China

[2]Institute of Laser & Optoelectronics, Tianjin University, Tianjin 300072, China

[3]Key Laboratory of Opto-electronics Information Science and Technology, Ministry of Education, Tianjin University, Tianjin 300072, China

Email: meihao3698@163.com

Abstract. The photonic crystal fiber (PCF) evanescent sensor has unique superior characteristics among the optical sensors, especially in terahertz (THz) wavelength range, PCF evanescent sensing will have huge application potentials. PCF characteristics and THz propagation characteristics in high-density polyethylene (HDPE) were combined. Finite element method was used to simulate and calculate THz-HDPE-PCF some parameters, at last the design considerations were given.

1. Introduction

As a new type of optical fiber, photonic crystal fiber (PCF) shows some of the unique characteristics compare to the traditional optical fiber (OF), such as unlimited single-mode, dispersion controllability, birefringence, can be designed for special mode field area, high non-linear, etc [1-6]. With the development of THz technology, THz play an important function in gas detecting and substance identify. Thus, in this paper we try to combine the PCF characteristics with THz characteristics in order to study and explore PCF sensing for finding a superior sensor [7-9].

2. The principle of THz-PCF evanescent sensing

The principle of THz-PCF evanescent sensing is that the part of THz frequency radiation is absorbed through the reaction of being measured substance and evanescent in PCF, then analyze substance by detected the THz radiation.

Total internal reflection-type PCF is characterized by guided-mode optical power concentrated at the center of the solid core defects, only a very small part of the optical energy (evanescent wave field) distribute in the surrounding air holes. The evanescent wave was absorbed by gas medium in the air, according to a Lambert Beer's law, the relationship between gas concentration and the optical intensity can be expressed as

$$I(\lambda) = I_0(\lambda)\exp[-r\alpha(\lambda)lC] \quad (1)$$

To facilitate the analysis, the equation will be rewritten as

$$C = -\frac{1}{r\alpha(\lambda)l}\ln\frac{I(\lambda)}{I_0(\lambda)} \quad (2)$$

Where I and I_0 denote output light intensity and input light intensity of being tested gas respectively. $\alpha(\lambda)$ for the gas absorption coefficient, l for the length of the PCF as a probe, C is gas concentration, n_r for the gas refractive index, which is approximately equal to 1; n_e for guide-mode effective refractive index, r is the relative sensitivity coefficient, defined as:

$$r = (n_r / n_e) f \quad (3)$$

For the model to be determined, f is the ratio for the optical power to focus on the air holes divides the total power, through the Poynting's theorem:

$$f = \frac{\int_{holes}(E_x H_y - E_y H_x)dxdy}{\int_{total}(E_x H_y - E_y H_x)dxdy} \quad (4)$$

Equation, Ex, Ey, Hx, Hy, express transverse electric and magnetic field model components.

By equation (2), we can see, if l, r, and $\alpha(\lambda)$ were known, by detecting the I and can get the gas concentration C in PCF holes. And the longer of PCF, the better of the test , in general, for fiber-optic gas sensor, air chamber can be made longer. This PCF is used for gas sensing has a great advantage. In addition, you can change the PCF fiber mode field distribution the by changing structure , thus changing the relative sensitivity r.

3. THz-PCF structure shape of the clad holes influence to propagation constant

Using finite element method to simulate total internal reflection-type transmission characteristics of THz-PCF evanescent wave, analyse the influence change of THz-PCF structural parameters on the evanescent wave.The background material is optical fiber of high-density polyethylene (HDPE), because of THz transmission losses are very small in plastic,also plastic is flexible comparing with silicon,It can be used hash environments.

So the HDPE material is selected to analyse THz–PCF evanescent wave sensing, the conventional silicon optical fiber material is not to be chosen. [10-16]

In the whole, supposed the refractive index of HDPE n =1.5 in the THz band, a triangular arrangement air holes, and the whole calculation region size is 3.4mm × 3 mm, in the process of meshing if a Yee's unit cell contains two kinds of different refractive index materials, use $\varepsilon = \varepsilon_a f + \varepsilon_b(1-f)$ to average them, ε_a, ε_b is the dielectric constant of air and HDPE respectively. PML absorbing boundary conditions was placed around the calculation region, the outermost boundary is zero, The thickness of PML regions is 10% calculating region. Figure 1 is cross-sectional structure diagram of THz photonic crystal fiber, which the clad hole structure is circle.

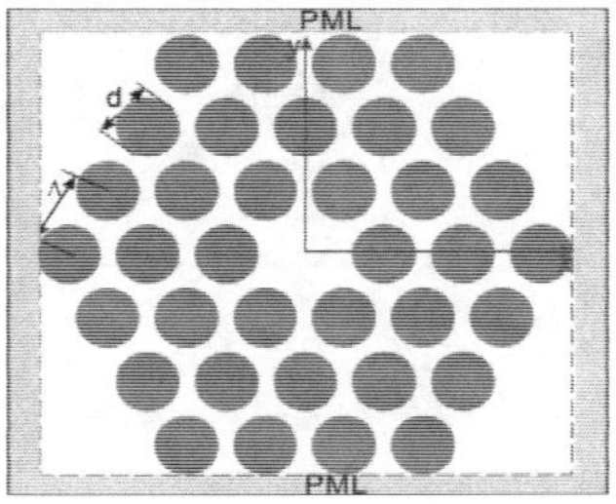

Fig. 1 Cross-sectional structure diagram of THz photonic crystal fiber

People used D-fiber optical fiber as evanescent wave gas sensors before[17], but the relative sensitivity is very low, only about 0.2% of the model energy can be used as a probe. The evanescent wave in PCF cladding air holes interact with the gaseous materials, it provides a new way. Monro, who verified the special PCF structure can be used as gas sensors, and pointed out from the theory that people can obtain high sensitivity gas sensor through the selection of suitable parameters [18].

Numerical simulation using perfectly matched layer (PML) structure, when applied PML absorbing boundary conditions, the propagation constant will be a complex number, in fact, the real part for a phase constant, the imaginary part is related to the loss amount. Thus, the relationship of the mode effective refractive index will be [19-21]

$$n_{eff} = \frac{\beta}{k_0} = n + \tilde{n} i \qquad (5)$$

$$\alpha(\text{dB/m}) = \frac{20}{\ln 10}\text{Im}(\beta) == 8.686\tilde{n}k_0 \qquad (6)$$

$$\alpha(\text{cm}^{-1}) = 2\tilde{n}k_0 \qquad (7)$$

Where n is real part of mode effective refractive index, $\tilde{n}$ for the effective refractive imaginary part, k_0=2/λ for the vacuum wave number (where k_0 unit is m^{-1}), λ is the wavelength in vacuum, the α unit is dB/m, if the unit is cm^{-1}, formula (6) express the loss of expression (which k_0 unit of cm^{-1}).

Figure 2 shows the results by calculations in the THz band, the condition is the same area but different clad hole structure, we can be see from the figure, when the frequency is high, the effective refractive index changes slowly with frequency, indicating PCF cladding air column structure do not influence the effective refractive index. Under the same value in

0.1THz-2.5THz round, the relation is $n_{ellipse} > n_{square} > n_{triangular} > n_{circle}$, in the relation n represents real part effective refractive index.

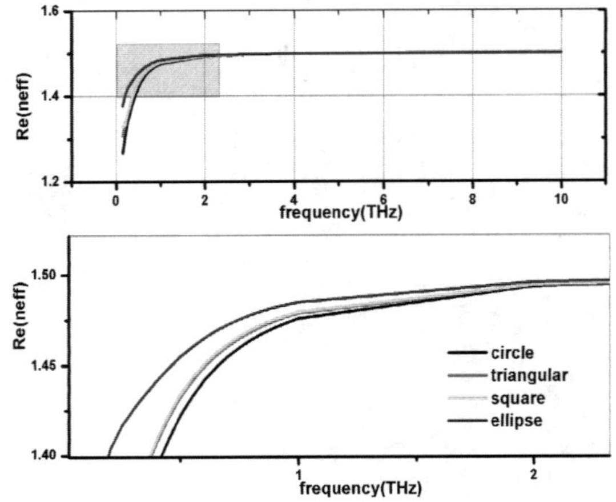

Fig. 2 The relationship between the fundamental mode effective index real part and the frequency with different clad hole structure

Figure 3 shows the relationship between the imaginary part of effective refractive index and the THz frequency as different clad hole structure with same area. The relationship are $n_{ellipse} > n_{square} > n_{triangular} > n_{circle}$, in the relation n represents imaginary part effective refractive index.

According to equation 6, the imaginary part of effective refractive index is related to the loss amount of PCF, so the circular hole structure denotes the smallest loss, ellipse hole structure denotes the biggest loss.

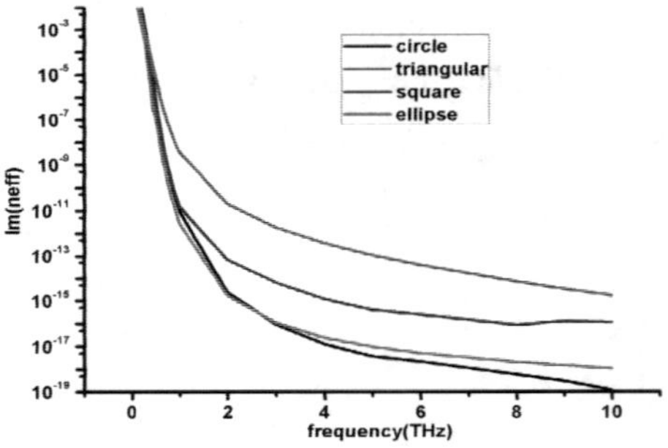

Fig. 3 The relationship between the imaginary part of fundamental mode effective index and the terahertz frequency

4. The relationship between the structure parameters of clad hole in THz-PCF and the evanescent wave energy

Structure shapes of holes are changed, respectively circular, square, triangle and ellipse, under the conditions of holes distances are constant and hole areas are the same. Fig.4 shows evanescent wave energy of PCF changes with THz frequency (0.1THz-10THz). With the frequency rising (2THz-10THz), fiber clad hole structure the THz-PCF sensor performance, while with the frequency decreasing (0.1THz-2THz), it is clear that the different clad hole structure of PCF has larger influence to transmission performance. Following relation were obtained: ellipse($E_{evanescent} / E_{total}$)<square($E_{evanescent} / E_{total}$)<triangular($E_{evanescent} / E_{total}$) <circle ($E_{evanescent} / E_{total}$) .

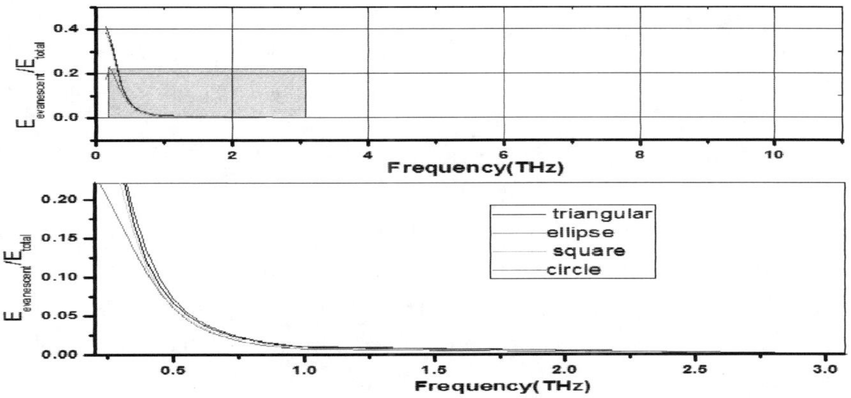

Fig. 4 The evanescent wave energy with THz frequency with four different hole structure Shapes

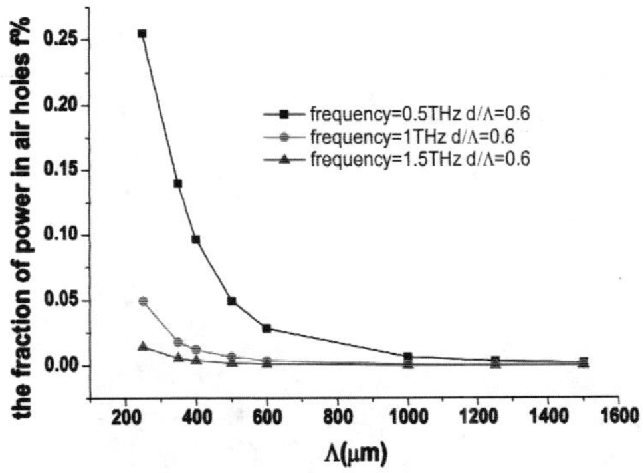

Fig.5 the fraction of power in air hole,as a function of d/Λ=0.6

From Fig.5 describes that d/Λ equal constant, the fraction of power in air hole will increase with the frequency, at the same time, the bigger of hole diameter, the better of the result.

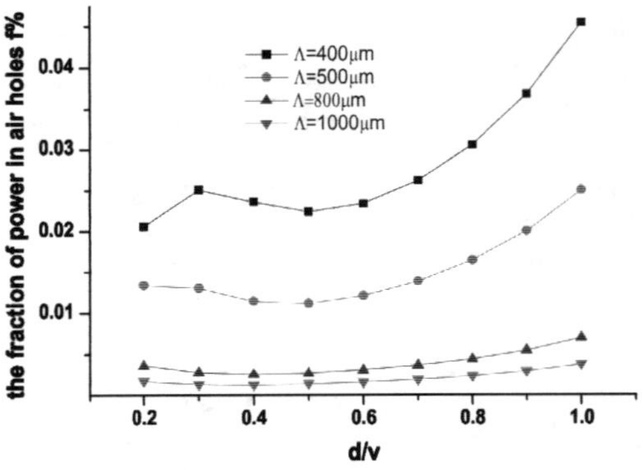

Fig.6 the fraction of power in air hole f as a function of d/Λ,at frequency=1THz

Choosing circle hole shape PCF, taking d/Λ=0.4, d/Λ=0.6 and d/Λ=0.8(Fig.6), the relationship between the fraction of power in air holes f % and d/Λ is simulated by finite element method. From the simulation results it can be seen that increase the hole diameter with the same hole structure shape, evanescent wave energy as the sensing will be significantly increased, so as a PCF sensor, large diameter structure should be used.

5. The relationship between THz-PCF effective area of the fundamental mode and the evanescent wave energy

(a) *f*=0.4THz (b) *f*=0.8THz (c) *f*=3THz

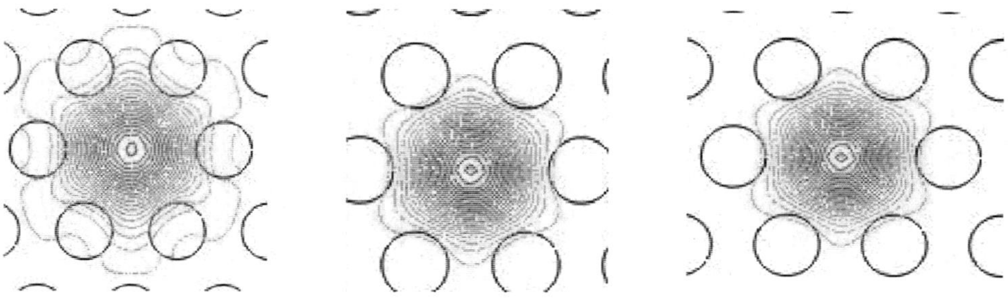

(d) f=0.45THz (e) f=0.85THz (d) f=5THZ

Fig. 7 Fundamental mode field distributions with corresponding frequency are 0.4THz、0.45THz、0.8THz、0.85THz、3THz、5THZ

Taking f = 0.4THz, f = 0.8THz, f = 3THz with the three-dimensional mode field distribution, f = 0.45THz, f = 0.85THz, f = 5THz with two-dimensional mode field distribution. From Fig 7, we can see that the mode field diameter will increase when the frequency decrease, the reason is that the restrictive influence of air holes decreases to mode field along with the frequency decrease. Without taking into account the case of dispersion, SMF fiber core and the cladding refractive index does not change with the frequency changing, so the mode field diameter increases with the frequency reduced.

Mode effective area is defined as

$$A_{eff} = \frac{\left(\iint_{-\infty \to +\infty} |E(x,y)|^2 \, dxdy \right)^2}{\iint_{NLR} |E(x,y)|^4 \, dxdy} \tag{8}$$

In the equation, E (x, y) is the electric field distribution of mode field; NLR is non-linear waveguide region, for this thesis, NLR is the fiber background region besides the air-hole. Fig. 8 shows the relationship between effective area of fundamental model and frequency with four different clad hole structure shape.

The relationship between fundamental effective area and the THz frequency at four different clad hole structure shape is: affective area (ellipse)> affective area (square)> affective area (triangular)> affective area (circle). Because the effective area and the energy density distribution of evanescent wave inverse, so the evanescent wave energy density distribution should be: ellipse（$E_{evanescent}/E_{total}$）<square（$E_{evanescent}/E_{total}$）<triangular（$E_{evanescent}/E_{total}$）<circle（$E_{evanescent}/E_{total}$）.

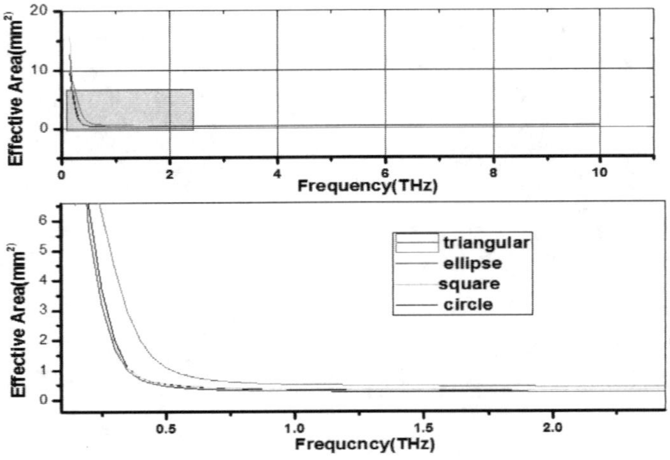

Fig. 8 The fundamental mode effective area with THz frequency at four different hole structure shape

6. The relationship between mode field distribution and evanescent wave energy in THz-HDPE-PCF

The circle clad hole structure shape is selected with hole period 500μm, d/Λ=0.6, Fig.9 shows the relation between mode field distribution of the fundamental mode and second-order mode with frequency. The changes of mode effective area with a circular clad hole at the THz frequency range is shown in figure 10. The proportion is shown in figure 11, which is the energy of fundamental mode and second-order model with frequency changing.

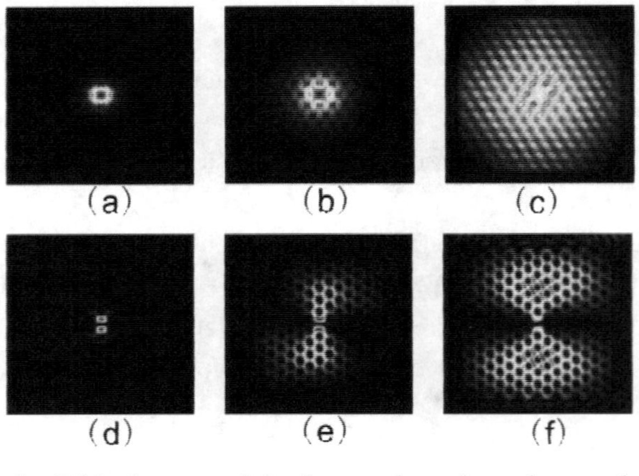

Fig9 Mode field changes of fundamental mode and second-order mode with the THz frequency

（a）-（c）represent frequency at 0.33THz、0.25THz and 0.15THz，corresponding area: 1.70 mm^2，9.67 mm^2 和 29.2mm^2；（d）-（f）represent frequency 1.2THz、1THz and 0.93THz, corresponding area :0.56mm^2，8.6 mm^2 and 12.1mm^2；

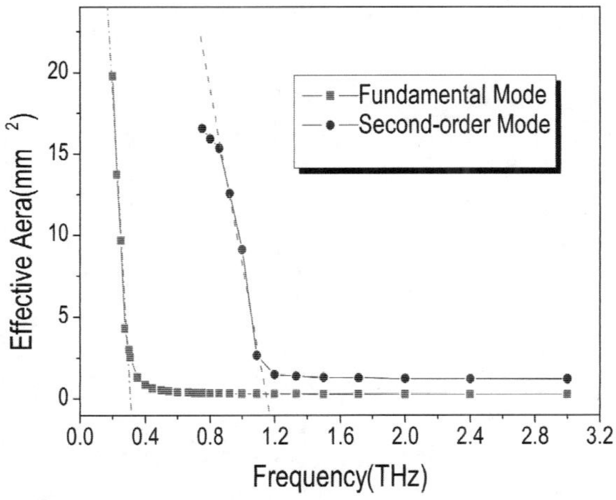

Fig.10 The mode effective area of a circular hole structure shape with the THz frequency

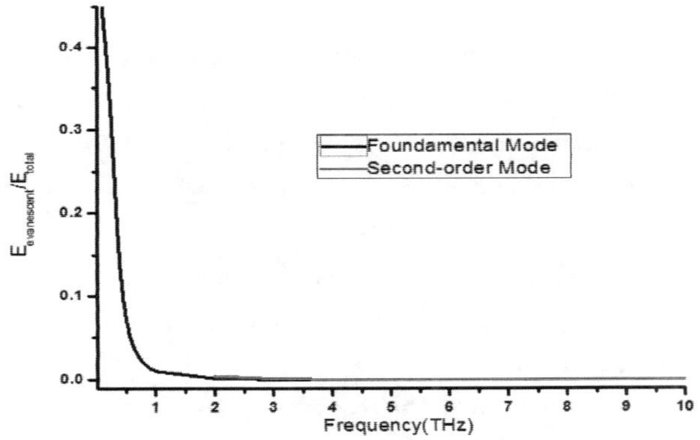

Fig. 11 The proportion of the energy of fundamental mode and second-order mode with THz frequency

The two ratio curve is almost coincident that the evanescent wave field energy of fundamental mode HE11 and second-order mode TE01 in the THz region. The second-order mode can not find in the low frequency. Further more fundamental model can get relatively high power density, so in evanescent wave sensing, it is much better using the fundamental mode as sensing than second-order mode. Single-mode operation was shown in Figure 12. The conclusions of figure 11 and figure 12 consistent with reference literature [22].

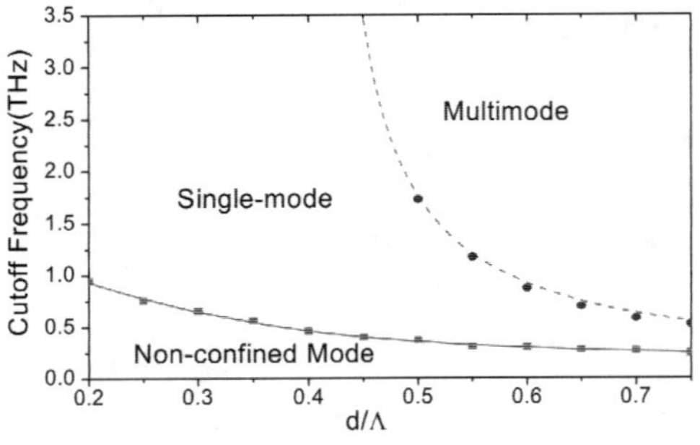

Fig. 12 Phase diagram of single-mode operation

In figure12, the dotted line represents cutoff frequencies of the second-order mode, solid line represents boundaries with the Non-confined mode, so three parts can be obtained: multi-mode area, single-mode area and the non-restricted-mode area, Only the structure in the middle ,can achieve single-mode optical fiber transmission.

7. Conclusion

THz-HDPE-PCF evanescent wave sensor is a very important direction of development, this sensor play an important function in gas detecting and substance identify. At the same time, THz transmission losses in plastic small, flexibility is good, it can work in harsh environments, THz-HDPE-PCF evanescent wave sensors is a superior sensor. Following conclusions can be obtained.

a、The principle of THz-PCF evanescent sensing is that part of THz frequency radiation is absorbed through the reaction of being measured substance and evanescent in PCF, then analyse substance by detected the THz radiation.

b、THz transmission losses in plastic is very small; also plastic is flexible comparing with silicon. It can be used at harsh environments. So the HDPE material is selected to analyse THz–PCF evanescent wave sensing, the conventional silicon optical fiber material is not to be chosen.

c、Corresponding propagation constant relationship is Re（neff）（ellipse）> Re（neff）(square) > Re（neff）triangular> Re（neff）(circle) with the same area, different clad hole shape structure. THz-PCF effective index is different with different propagation constant, result different electromagnetic fields transmission in PCF.

d、At low frequency (0.1THz-2THz) with four different clad holes structure shape, the evanescent wave energy ratio in total energy is shown: ellipse（$E_{evanescent} / E_{total}$）<square（$E_{evanescent} / E_{total}$）<triangular（$E_{evanescent} / E_{total}$）<circle（$E_{evanescent} / E_{total}$）. In 0.1THz-2THz sensing, sensing performance with a circular clad hole structure shape is superior to the other three shape of THz-PCF sensor.

e、To get a better sensing performance in THz-HDPE-PCF, to use large aperture single-mode operation PCF to be used.

References

[1] P. Y. Han, G. C. Cho and X. C. Zhang, Time-domain transillumination of biomedical tissue with terahertz pulses, Opt.Lett., 2000,25, 242 .

[2] T. Luffler, T. Bauer, K. J. Siebert et al., Terahertz dark-field imaging of biomedical tissue, Opt. Express, 2001, 9, 616.

[3] M. K. Seongsin, F. Hatami and J. S. Harris, Biomedical terahertz imaging with a quantum cascade laser, Appl. Phys. Lett., 2006, 88, 153903.

[4] T. Loffler, K. Siebert, S. Czasch et al., Visualization and classification in biomedical terahertz pulsed imaging, Phys. Med. Biol., 2002, 47, 3847.

[5] N. A. Mortensen, M. D. Nielsen, J. R. Folkenberg, A. Petersson, and H. R. Simonsen, Improved large-mode-area endlessly single-mode photonic crystal fibers, Opt. Lett., 2003, 28(6):393-395

[6] H. B. Liu, Y. Q. Chen, Detection and identification of explosive RDX by THz diffuse reflection spectroscopy, Opt. Express, 2006, 14, 415.

[7] Y. C. Shen, T. Lo, P. F. Taday et al., Detection and identification of explosives using terahertz pulsed spectroscopic imaging, Appl. Phys. Lett., 2005, 86, 241116.

[8] J. F. Robert, R. E. Allman, E. Zuckerman, Terahertz absorption measurement for gas-phase 2,4-dinitrotoluene from 0.05 THz to 2.7 THz, Chem. Phys. Lett., 2006, 431, 34.

[9] J. F. Federici, B. Schulkin, F. Huang et al., THz imaging and sensing for security applications-explosives, weapons and drugs, Semicond., Sci. Technol. 2005,20, S266.

[10] T. Idehara, H. Tsuchiya, O. Watanabe et al., The first experiment of a THz gyrotron with a pulse magnet, Int. J. Infrared and Millimeter Waves, 2006,27, 319.

[11] T. Idehara, I. Ogawa, T. Saito et al., Development of THz gyrotrons and application to high power THz technologies, Terahertz Science and Technology, 2008, 1, 100.

[12] E. Grosse, THz radiation from free electron lasers and its potential for cell and tissue studies, Phys. Med.Biol. 2002, 47, 3755.

[13] R. Kohler, et al., Terahertz semiconductor-heterostructure laser. Nature, 2002, 417, 156.

[14] M. S. Vitiello et al., Electron-lattice coupling in bound-to-continuum THz quantum-cascade lasers, Appl. Phys. Lett., 2006, 88, 241109.

[15] A. Straub et al., Threshold reduction in quantum cascade lasers with partially undoped, dual-wavelength interdigitated cascades, Appl. Phys. Lett., 2002, 80, 2845.

[16] G Wang, J Shen, et al.. Vibrational spectra of ketamine hydrochloride and 3, 4-methylenedioxymethamphetamine in terahertz range[J]. J. Appl. Phys., 2007, 102(1): 013106/1~013106/4

[17] O. Abdi, K.C. Wong, T. Hassan, K.J. Peters, M.J. Kowalsky, Cleaving of solid single mode polymer optical fiber for strain sensor applications, Optics Communications, 2009, 282:856～861

[18] Yanjie Zhu, Xiaopeng Dong, Yingchao Chen, The implication and development of optical waveguide analysis by finite difference time domain methods(Chinese), optics technology, 2001. 21: 25 1-258.

[19] T.M. Monro, D.J.Richardson,P.J.Bennett,Developing holey fibers for evanesceot field

devices, Electron.Lett,1999, 35: 1188～1189

[20] Y. Li, Ch. Wang, N. Zhang, Analysis and design of terahertz photonic crystal fibers by an effective-index method, Appl.Opt., 2006,45:8462

[21] Sandra Börner, Rozalia Orghici, Siegfried R. Waldvogel, Ulrike Willer, and Wolfgang Schade Evanescent field sensors and the implementation of waveguiding nanostructures, Aplied Optics, 2009, Vol. 48:4

[22] Youfu Geng. Study on Difference-freqency Generation of Terahertz Waves and Low Loss Waveguides for Terahertz Transmission [D] (Chinese). Tianjin, Tianjing University, 2009.

Acknowledgement

Supported by **973 Plan "Fundamental research on new generation optical fiber intelligent sensing network and key devices": Sub-Project: "Fundamental research on new PCF Sensor（2010CB327801）"**

Synthesis and application for optical recording of a new diarylethene having a benzene unit

Peijian Yan, Shouzhi Pu*, Weijun Liu, Gang Liu

Jiangxi Key Laboratory of Organic Chemistry, Jiangxi Science and Technology Normal University, Nanchang 330013, P. R. China

pushouzhi@tsinghua.org.cn

Abstract. A novel photochromic diarylethene was synthesized and its photochromism, fluorescence properties have been investigated. This compound exhibited good photochromism and functioned as a fluorescence switch upon alternating irradiation with UV and visible light both in solution and in PMMA film. When diarylethene was used as recording medium, the results demonstrated that the diarylethene compound had attractive properties for high density optical storage.

1. Introduction

The demand for higher storage capacity in optical storage is rapidly increasing. So, many research activities have focus on the development of new storage materials [1] and innovative technologies [2] to simultaneously increase the storage density and data-transfer rates. Organic photochromic compounds that undergo reversible color changes upon illumination have long been viewed as potential media for high-density optical memory devices. Upon photoirradiation photochromic compounds reversibly change not only the absorption spectra but also other physicochemical properties, such as refractive index, electrical moment, dielectric constants, chelate formation, ion dissociation, oxidation/reduction potentials and phase transitions during the photochromic reaction [3]. There are some advantages of photochromic materials for optical recording: (1) fast response time, (2) high spatial resolution, (3) re-writability, and (4) without special fixing information. Moreover, to the best of our knowledge, diarylethene is regarded as one of the most promising candidates because of their good thermal stability of two isomers, notable fatigue resistance, high quantum yields, rapid response and high reactivity in solid state [4].

To date, there are many reports on diarylethene derivatives with heteroaromatic rings as both thermally irreversible and reversible photochromic compounds. Among diarylethene derivatives so far synthesized, most of the hetero-aryl moieties have been thiophene or benzothiophene rings, and other hetero-aryl moieties, such as such as furan [5], thiazole [6], indole [7], benzofuran [8], pyrrole [9], and indene [10], have also been reported partially. Recently, photochromic hybrid diarylethene derivatives bearing both five-membered and six-membered moieties have been reported by Pu et al. [11]. They are the first class example of photochromic diarylethene derivatives bearing both five-membered and six-membered moieties, and all of them showed good reversible photochromism both in solution and in PMMA amorphous film. In this paper, we also have designed and prepared a new photochromic diarylethene derivative bearing benzene and thiophene moieties, i.e., 1-(2-trifluoromethylphenyl)-2-[2-methyl-5-(4-cyanophenyl)-3-thienyl]perfluorocyclopentene (1o). Its photochromic reactivity, fluorescence switch property and application for optical recording were investigated. The photochromic reaction of diarylethene 1 is shown in Scheme 1.

2. Experiments

The synthetic route for diarylethene 1o is shown in Scheme 2. Firstly, 2-bromobenzotrifluoride (2) was lithiated and then coupled with octafluorocyclopentene to give 2-trifluoromethylphenyl-perfluorocyclopentene (3); Then, Suzuki coupling of 3-bromo-4-cyano-benzene with 3-bromo-2-methyl-5-thienylboronic acid (4) [12] gave 3-bromo-2-methyl-5-(4-cyanophenyl)thiophene (5). Finally, compound 5 was lithiated and then coupled with compound 3 to give the target diarylethene 1o. The structure of diarylethene 1o was confirmed by ^{1}H NMR and ^{13}C NMR spectroscopy. ^{1}H NMR ($CDCl_3$, 400 MHz, TMS): 2.22 (s, 3H, $-CH_3$), 7.07 (s, 1H, thienyl-H), 7.47 (t, 3H, *J*=9.1 Hz, phenyl-H), 7.58 (t, 1H, *J*=7.7 Hz, phenyl-H), 7.64-7.72 (m, 2H, phenyl-H), 7.92 (d, 2H, *J*=8.3 Hz, phenyl-H). ^{13}C NMR ($CDCl_3$, 100 MHz, TMS) 13.92, 121.92, 124.64, 124.77, 125.38, 125.59, 125.63, 127.58, 127.63, 127.68, 127.72, 128.89, 128.99, 130.43, 130.99, 131.81, 136.07, 137.47, 140.29, 142.28.

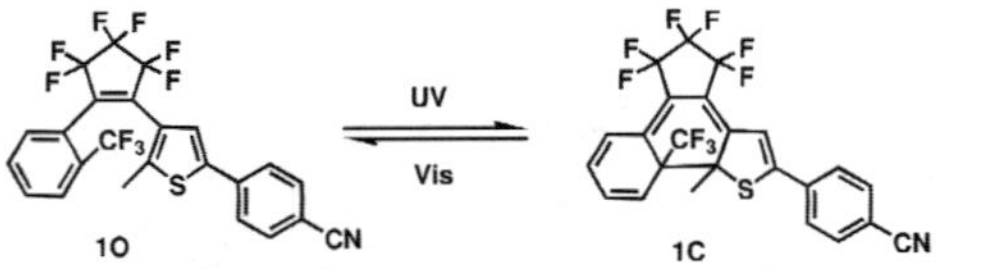

Scheme 1. Photoisomerization of diarylethene 1.

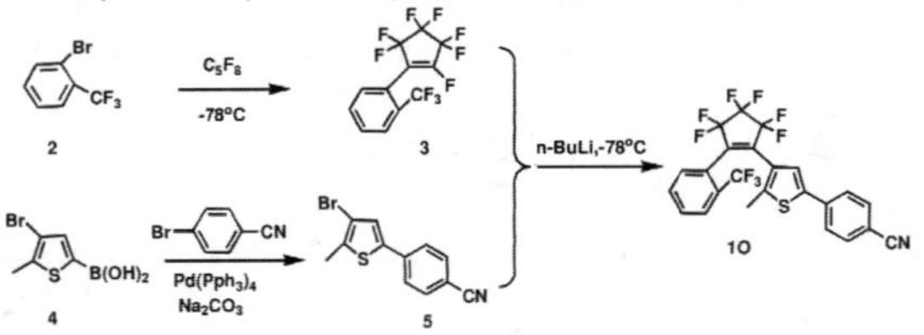

Scheme 2. Synthetic route for the diarylethene (1o).

3. Results and Discussion

3.1. Photochromism of diarylethene

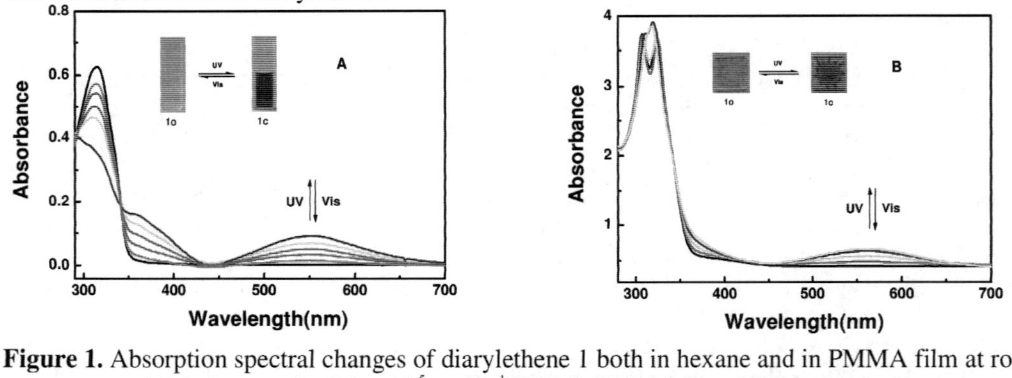

Figure 1. Absorption spectral changes of diarylethene 1 both in hexane and in PMMA film at room temperature: (A) in hexane (c=2.0 × 10^{-5} mol L^{-1}); (B) in PMMA film (10%, w/w).

The diarylethene 1o exhibits good photochromism both in hexane solution and in PMMA amorphous film at room temperature. This diarylethene could be toggled between its colorless open-ring isomers (1o) and colored closed-ring isomers (1c) by alternating irradiation with UV light and appropriate wavelength visible light, and its absorption spectral changes in hexane upon photoirradiation were shown in Figure 1A. As shown in Figure 1A, the absorption maximum of colorless open-ring isomer 1o was observed at 315 nm (= 3.1 × 10^{4} L mol^{-1} cm^{-1}) in hexane. Upon irradiation with UV light (=297 nm), the colorless solution turned purple with a new visible absorption band centered at 551 nm (= 4.6 × 10^{3} L $mol^{-1}cm^{-1}$) emerging, while the original peak at 315 nm decreased, indicating the formation of the closed-ring isomer 1c. Correspondingly, The purple colored solution returned to colorless upon irradiation with visible light (> 500 nm), indicating that 1c returned to the initial open-ring isomer 1o. Similarly, as shown in Figure. 1B, upon irradiation 297 nm light, the color of dithienylethene 1o /PMMA film ($_{max}$ = 320 nm) changed from colorless to purple with the appearance of a new broad absorption band at $_{max}$= 561 nm, which was assigned to the formation of the closed form 1c. The colored PMMA film can invert to colorless, on irradiation of appropriate visible light (> 500 nm). As has been observed for most of the reported diarylethenes [13,14], the maximum absorption peaks of both the open-ring and the closed-ring isomers in PMMA film are longer than those in hexane solution. The red shift phenomena may be ascribed to the stabilization of molecular arrangement in solid state.

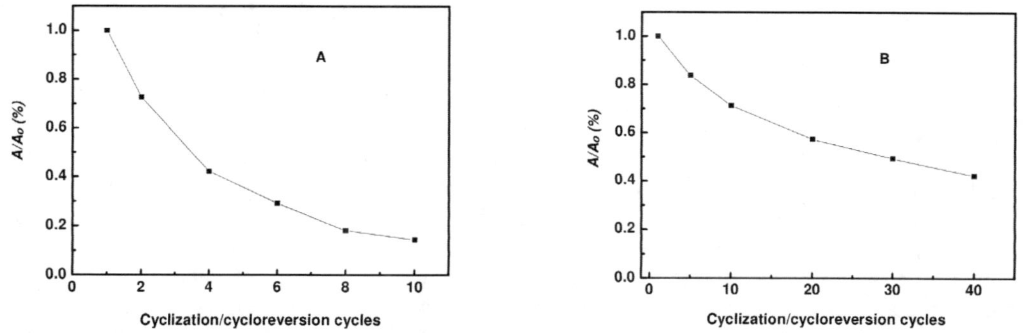

Figure 2. Fatigue resistance of diarylethene 1 in hexane and in PMMA film in air atmosphere at room temperature. Initial absorbance of the sample was fixed to 1.0; (A) in hexane; (B) in PMMA film

Fatigue-resistant property, i.e., how may times photocyclization and cycloreversion reaction cycles can be repeated without loss of performance, is an very important factor for practical applications in optical devices. The fatigue resistances of diarylethene 1 were examined both in hexane and in PMMA amorphous film at room temperature, as shown in Figure 2. In hexane, diarylethene 1 were irradiated alternatively with 297 nm and visible light (> 500 nm), respectively. The irradiation time was long enough for coloration to reach the photostationary state and for the color to be completely bleached. As shown in Figure 2A, the fatigueresistant characteristic of 1 in solution indicated that ~85% of 1c was destroyed after 10 repeat cycles. This may be ascribed to degradation resulting from the formation of an epoxide.[15] Similarly, the fatigue resistances of diarylethene 1 were tested in PMMA film by alternatively irradiating with 313 nm and visible light, respectively. As shown in Figure 2B, the fatigue resistance of diarylethene 1 in the solid state (PMMA amorphous film) are much stronger than those in solution. After 40 repeated cycles, this compound still showed good photochromism with ~58% degradation of 1c. This remarkable improvement may result from effectively suppressing the oxygen diffusion.[16] As described above, it can be easily concluded that the fatigue resistance of diarylethene 1 is weakness than those of report by Pu et al. [11].

3.2. Fluorescence switch properties

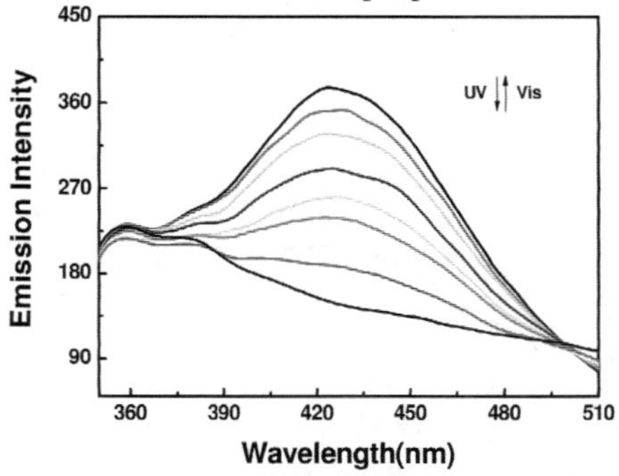

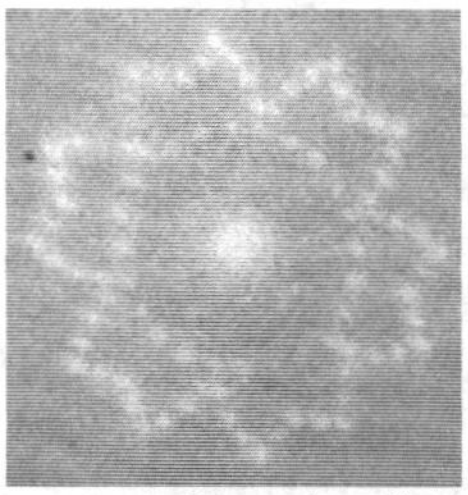

Figure 3. Fluorescent intensity changes of diarylethene 1 in hexane($c=2.0\times10^{-5}$ mol L^{-1}).

Figure 4. Readout patterns from the diarylethene 1c/PMMA film.

The fluorescence properties in hexane solution of the compound 1o were measured using a Hitachi F-4500 spectrophotometer, and the breadths of excitation and emission slit were selected 10.0 nm. Figure. 3 shows the emission intensity changes of diarylethene 1 during the process of photoisomerization in hexane. Upon irradiation with 313 nm light at room temperature, the emission intensity of compound 1 decreased remarkably along with the photoisomerization from open-ring isomers to closed-ring isomers when excited at 318 nm. When arrived at the photostationary state, the emission intensity was quenched to ca. 18%. The incomplete cyclization reaction and the existence of parallel conformations may be the main cause for the moderate change in fluorescence induced by photoirradiation in hexane. The back irradiation by appropriate wavelength visible light regenerated its open-ring isomer and recovered the original emission intensity. The reversible changes of the emission intensity of diarylethene 1 are useful for application as optical memory with fluorescence readout method and photo-modulation switches.

3.3. Optical recording

The evaluation of potential of photochromic material as an erasable storage medium by recording, reading and erasing the optical image in a real-time operation was investigated. For the sake of long-term stability and environmental durability of storage devices, photoresponsive materials must be processed into large-area, high quality solid films. Usually, photochromic materials are doped into a polymer matrix for investigation of applications, and PMMA was chosen in this paper to be used as polymer matrix for optical recording. When this diarylethene was used as recording medium, optical storage was performed by a 650 nm He-Ne laser. With a He-Ne laser in the experimental setup, the exposure time for each spot was 0.08 s and the patterns were about 20 μm. The polarizations of the patterns recorded on dithienylethene 1c/PMMA film is shown in Figure 4. The result indicated that the new photochromic diarylethene can be applied in high capacity optical storage. As shown in Figure 4, the quality of optical recorded in 1c/PMMA film is also fairly good.

In conclusion, a novel photochromic diarylethene 1o was synthesized and its photochromic and fluorescent properties were also investigated. The results showed that this compound exhibited good reversible photochromism both in solution and in PMMA amorphous film. It also exhibited remarkable fluorescence switching in hexane solution. When this diarylethene was used as recording medium, the results demonstrated that the diarylethene compound had attractive properties for high capacity optical storage.

Acknowledgment

This work was supported by the Project of National Natural Science Foundation of China (20564001), the Key Scientific Project from Education Ministry of China (208069), the Project of Jiangxi Youth Scientist, and the Science Funds of the Education Office of Jiangxi, China (GJJ08365).

References

[1] Kawata S and Kawata Y. 2000 *Chem. Rev.* **100** 1777
[2] Chen Y, Li T, Fan M, Mai X, Zhao H and Xu D. 2005 *Mater. Sci. Eng. B.* **123** 53
[3] Chen Y, Wang C M, Fan M G, Yao B L and Menke N. 2004 *Opt. Mater.* **26** 75
[4] Matsuda K and Irie M. 2004 *J. Photoch. Photobiol.* **5** 169
[5] Deng X H, Lanny S and Liebeskind. 2001 *J. Am. Chem. Soc.* **123** 7703
[6] Takami S and Irie M. 2004 *Tetrahedron.* **60** 6155
[7] Yagi K, Soong C F and Irie M. 2001 *J. Org. Chem.* **66** 5419
[8] Yamaguchi T and Irie M. 2005 *J. Org. Chem.* **70** 10323
[9] Pu S Z, Yang T S, Xu J K and Chen B. 2006 *Tetrahedron Lett.* **47** 6473
[10] Yamaguchi T and Irie M. 2006 *Tetrahedron Lett.* **47** 1267
[11] Pu S Z, Fan C B, Miao W J and Liu G. 2008 *Tetrahedron.* **64** 9464
[12] Pu S Z, Liu G, Li G Z, Wang R J.and Yang T S. 2007 *J. Mol. Struct.* **833** 23
[13] Fan C B, Pu S Z, Liu G and Yang T S. 2008 *J. Photochem. Photobiol., A: Chem.* **194** 333
[14] Tsujioka T, Kume M and Irie M. 1997 *J. Photochem. Photobiol. A: Chem.* **104** 203
[15] Higashiguchi K, Matsuda K, Yamada T, Kawai T and Irie M. 2000 *Chem. Lett.* 1358
[16] Irie M. 2000 *Chem. Rev.* **100** 1685

Based on MDCE to simulate bell-shaped interferograms produced by a capillary tube interferometer

Yang Ailing [1] and Tang Mingming [2]

[1] Department of Physics, Ocean University of China, No.238 Songling Road, Qingdao 266100, Shandong, China
[2] Resource College, Beijing Normal University, Beijing, 100875, China
ailingy@ouc.edu.cn

Abstract. The vector theory to calculate the optical path length (OPL) of the rays passing through a capillary tube was given and the OPLs including the collimated lens in 3D space were calculated, the interferograms produced by the capillary tube were simulated. A pinhole localized at the focal point of the collimated lens was seen as a point source of light. Based on ray-tracing, the OPL distribution from the point source of light, passing by the collimated lens, a cylindrical lens, the capillary tube and a screen was calculated. Because the huge calculation in the ray-tracing, a distributed calculating net based on MATLAB Distributed Computing Engine (MDCE) was established to save time. In order to get the simulated interferograms from the envelope of the OPLs, Delaunay triangulation algorithm and cubic spline interpolation were used. We record the interferograms of the capillary tube filled with liquids with different refractive indices and compared with the simulated interferograms. The results showed the two coincided well.

1. Introduction

Capillary tube interferometer is a new method to measure the refractive indices of transparent [1,2] or low transparent liquids[3,4]. Because fiber is a kind of solid core capillary tube, this method was also used to measure the refractive index of fiber or testing fiber [6-8]. Ramadan called this method lens–fiber interference method [8]. In above references, the experimental setup was similar, the mechanism of the interferograms was proved to be two beam interference in experiment and theoretical analysis. The simulated linear fringe patterns were the extended results in 2D space, but the experimental fringes were bell-shaped. The main reasons were the authors omitted the influence of the collimated lens and thought that the beam after the collimated lens was perfect parallel. In 2008, Yang Ailing [9] *et al* took account the influence of the collimated lens and simulated the interferograms in 3D space by Delaunay triangulation and sector boundary algorithm. The calculated interferograms were curved, which shows this method was effect. But Ref. 9 was lack of theoretical analysis and the calculating procedure took long time, so this problem needs a further probing.

In this paper, the vector theory to calculate the OPLs of the rays passing through the capillary tube was given and the OPLs including the collimated lens in 3D space were calculated. Because the rays will come across many interfaces with different forms, and at every interface, the refraction law [10] has to be matched. For simplicity, a 3D ray-tracing procedure on a single interface was complied. As a ray passing through an interface, one only needs to modify the procedure slightly. Although 324×30 rays were chose to make ray tracing, the number was still limited, which caused the simulated optical path surface was not so smooth. By using Delaunay triangulation [9,11-16], sector boundary algorithm [17-19] and cubic spline interpolation method [20-21], a smooth optical path surface was obtained. For the huge calculation, distributed computing method based on MDCE was used to make 3D ray-tracing, which saved the calculating time greatly. The simulated curved interferograms coincided well with the experiment results.

2. Expreimental setup

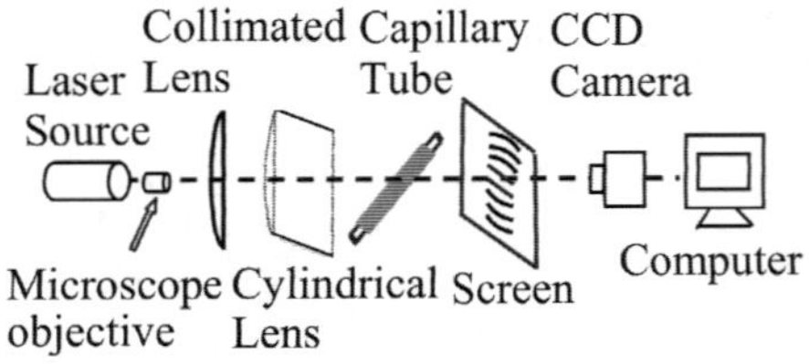

Figure 1. The experimental set up.

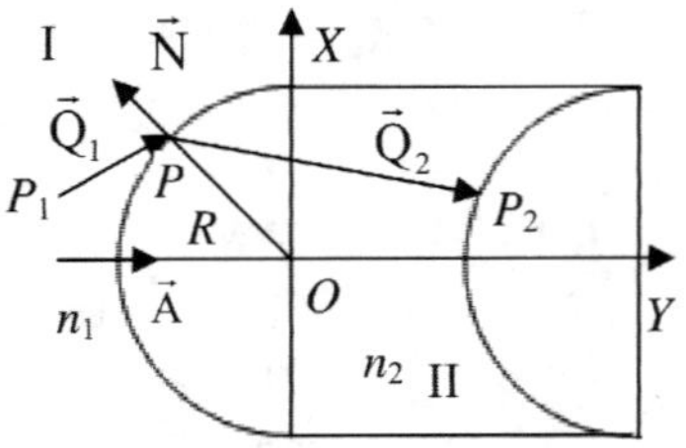

Figure 2. Ray tracing on the spherical interface.

Figure 1 shows the experimental setup. A He–Ne laser with 632.8 nm wavelength was used as a light source. After a spatial filter and a collimated lens, a parallel beam with nearly 10cm diameter was obtained. The cylindrical lens was made of K-9 glass of refractive index 1.5163, whose focal length is 31.2 cm. The inner and outer diameters of the capillary tube were 3 and 4 mm, respectively. The width of the focused beam is about 0.75mm, which is very small compared with the diameter of the capillary tube. The capillary tube is parallel to and within the focal lines of the cylindrical lens. The rays passing through the capillary tube formed an interference pattern on the screen. A high performance CCD camera recorded the interference pattern, which was stored in a computer for later digital processing.

3. Theory

When a ray passes through the elements of the capillary tube then arrives to a screen, it will intersect with spherical, cylindrical or plane interfaces. However, cylindrical and plane interfaces can be seen as special spherical interfaces, so one only needs to compile a procedure about how a ray passes through a spherical interface. For other interfaces, one only needs to modify the procedure slightly, then repeat the procedure, and can simulate the rays of the whole optical system.

Consider the refraction at a sphere interface (see figure 2). Here O is the centre of the sphere interface, R its radius, n_1 and n_2 are refractive indices of the surrounding medium and the lens respectively. Let P_1P be the incident ray upon the sphere interface. And PP_2 is refracted ray. P_2 is the point intersection of the refracted ray and the successive surface. $\vec{N}$ is the unit normal vector of the spherical surface. $\vec{Q}_1$ and $\vec{Q}_2$ are the unit vectors of the incident and refracted rays. Assuming $\boldsymbol{P}_1$ and $\vec{Q}_1$ are known, it is necessary to calculate $\boldsymbol{P}$, $\boldsymbol{N}$, and $\boldsymbol{Q}_2$. Assuming R is finite, we could calculate $\boldsymbol{P}$ and $\boldsymbol{N}$ from incident light ray equation and spherical equation:

$$\begin{aligned} (\vec{P}-\vec{P}_1) &= k\vec{Q}_1 \\ \left|\overrightarrow{PO}-\vec{O}\right| &= R. \end{aligned} \tag{1}$$

Since there are two solutions for the spherical quadratic equation, so it is necessary to select one solution based on the selected region of the surface. If the left region of the surface is selected, $\vec{N}\bullet\vec{A}\le 0$, otherwise $\vec{N}\bullet\vec{A}\ge 0$. Here $\vec{A}$ is the unit vector of Y axis.

By the law of refraction [10],

$$\begin{aligned} n_1\vec{Q}_1 \times \vec{N} &= n_2\vec{Q}_2 \times \vec{N} \\ \left|\vec{Q}_2\right| &= 1. \end{aligned} \tag{2}$$

$\boldsymbol{Q}_2$ can be obtained. For the second interface, PP_2 is incident ray. Taking into the equation of the second interface, and repeating above procedure, one can easily determine $\boldsymbol{P}_2$ and the unit direction vector of the refracted rays. By repeated applying refraction law and the ray equation, the intersection points and the unit direction vectors can be obtained gradually.

4. MDCE

The calculating results showed that one light ray tracing on a computer (WinXP, AMD Athlon 64 Processor) cost 0.35s. 324×30 light rays will cost about 1.5h. From section 3, one can see that each light ray tracing is independent. For saving calculating time, one can take advantage of distributed computing engine to execute a lot of independent ray tracing on a group of computers. We used MDCE to realize distributed ray tracing on local area network.

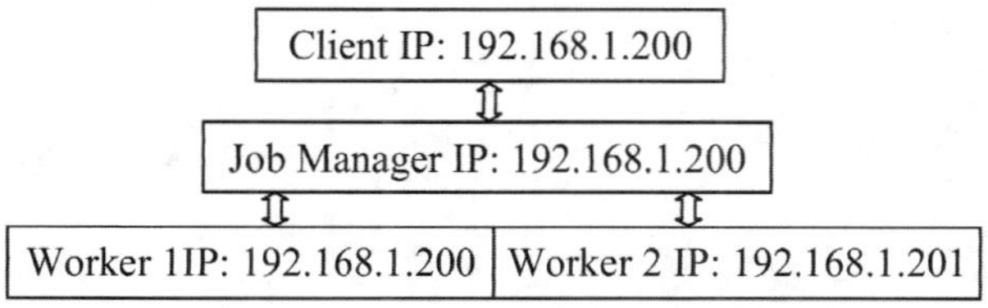

Figure 3. The distributed ray tracing diagram.

Figure 3 shows the distributed ray tracing diagram. In the MDCE, two computers were used. One was assigned client (IP address: 192.168.1.200), job manager and the first worker. The other (IP address: 192.168.1.201) was as the second worker. Each worker executed tasks simultaneously. The whole ray tracing MATLAB codes were put on the client computer. The definition of ray tracing job and tasks were executed by distributed computing toolbox. When all of 324×30 rays were assigned to the two workers, each one executed 324×15 rays at the same time. The calculation results showed that the distributed ray tracing method cost about 0.8h, which saved about half time.

5. Experimental and simulated results

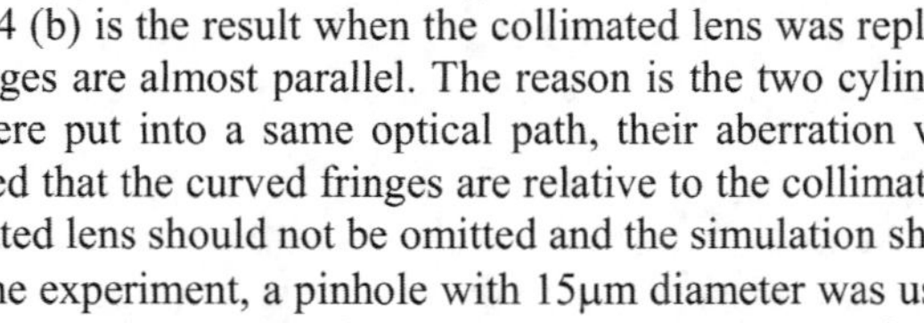

Figure 4. The interferogroms of ethanol (a) and (b) the collimated lens was replaced by a cylindrical lens.

Figure 4 (a) shows the interferograms of ethanol when the collimated lens localized like figure 1. Figure 4 (b) is the result when the collimated lens was replaced by a cylindrical lens. One can see that the fringes are almost parallel. The reason is the two cylindrical lenses has opposite aberration, when they were put into a same optical path, their aberration will be cancelled out. Figure 4 (a) and (b) indicated that the curved fringes are relative to the collimated lens. When one calculates the OPLs, the collimated lens should not be omitted and the simulation should be in 3D space.

In the experiment, a pinhole with 15μm diameter was used as a space filter. Because it is small, we saw it as a point source of light. The OPL distribution from the point source of light, passing by the collimated lens, the cylindrical lens, the capillary tube and the screen was calculated. Although 324×30 rays were chose to make ray tracing, the number was still limited, which caused the simulated optical path surface was not so smooth. By using Delaunay triangulation, sector boundary algorithm [17-19] and cubic spline interpolation method [20-21], a smooth optical path surface was obtained (see figure 5(b)). For the huge calculation, distributed computing method based on MDCE was used to make 3D ray-tracing, which saved the calculating time greatly. Figure 5 (a) and (c) are the experimental and simulated interferograms of auto gearing oil. The two interferograms coincides well. This oil is semi-transparent and refractive index is 1.4885. Figure 5 (d) and (e) are the experimental and simulated interferograms of ethanol, whose refractive index is 1.3510. The number of fringes of the simulated curved interferograms is less than the experimental result.

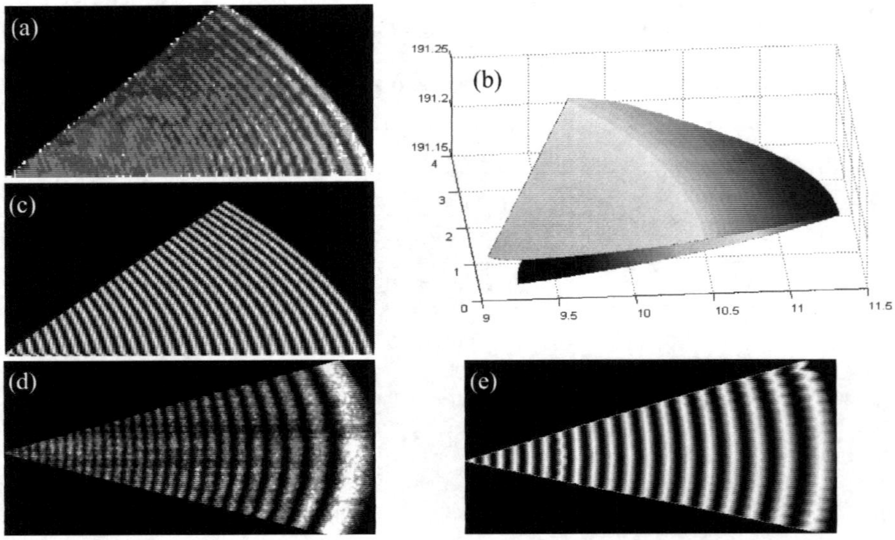

Figure 4. The experimental ((a) and (d)) interferograms, the simulated optical path surface (b) and the simulated interferograms ((c) and (e)). In which, (a), (b) and (c) are for auto gearing oil with 1.4785 refractive index and (d), (e) for ethanol with 1.3510.

6. Conclusion

A pinhole localized at the focal point of the collimated lens was seen as a point source of light to trace the OPLs of the rays in a capillary tube interferometer. Because the rays will come across sphere, cylindrical and plane interfaces from the pinhole to the screen. And the cylindrical and plane interfaces can be seen as special sphere interfaces. A procedure was compiled to trace a ray passing through a sphere interface. Other interfaces are only repeating of the sphere interface, which simplified the calculation. A distributed calculating net based on MDCE was established and saved about half time. Delaunay triangulation algorithm, section boundary algorithm and cubic spline interpolation were used to obtain a smooth optical path surface. Two beam interference theory was used to simulate the interferograms. The results showed the above methods were effect to simulate the curved interferograms produced by the capillary tube interferometer.

Acknowledgements

This research was supported by Chinese National 863 High Technology Plan (grant number: 2004AA616230). The authors are very appreciate to discuss with Dr. Li Wendong.

Journal of Physics: Conference Series **276** (2011) 012150 doi:10.1088/1742-6596/276/1/012150

References

[1] Ramadan W A and Wahba H H 2006 *J. Meas. Sci. Techno.* **17** 215-20
[2] Yang Ailing, Li Wengdong, Yuan Guang, Dong Junyu and Jinliang Zhang 2006 *J. Appl.Opt.* **45** 7993-7
[3] El Ghandoor H, Hegazi E, Nasser Ibraheem and Behery G M 2003 *J. Opt. Laser Techno.* **35**, 361-7
[4] Yang Ailing, Li Wendong, Dong Junyu, Yuan Guang, Sun Guimei, Bi Decang and Zhang Jinliang 2007 *Proc.SPIE* **6279** 62794w-1-8.
[5] El-Ghandoor H, El-Ghafar E Abd and Hassan R 1999 *J. Opt. Laser Techno.* **31** 481-8
[6] Eldoor H, Nasser I, Abdel-Hady Afaf and Al-Shukri A 2002 *J. Opt. Lasers Engrg.* **42** 277-87
[7] Hamza A A, Mabrouk M A, Ramadan W A and Wahba H H 2004 *J. Opt. Lasers Engrg.* 42 121-30
[8] Ramadan W A 2000 *J. Opt. A: Pure and Appl. Opt.* **2** 234-8
[9] Yang Ailing, Tang Mingming and Zhang Jinliang 2008 *Proc., Int., Conf., 1st on Modelling and Simulation in Electronics, Computing and Bio-Medicine* (Nanjing,China) Vol Ⅲ p 86-90
[10] Born and Wolf 1959 *Principles of Optics* (New York:Pergamon).
[11] Lemaire C and Moreau J M 2000 *J. Comput. Geom.* **17** 69–96
[12] Song Young Joon 1997 *J. Comput. Methods Appl. Mech. Engrg.* **150** 57-64
[13] Mulchrone Kieran F 2002 *J. Struct. Geolo.* **25** 689-702D
[14] Lasser 1996 *J. Comput. Math. Appli.* **32** 43-58
[15] Jayachandra M Reddy, George M Turkiyyah 1995 *J. Computer-Aided Design* **27** 677-94
[16] Golias N A and Dutton R W 1997 *J. Finite Elements in Analysis and Design* **25** 331-41
[17] Chen Tao and Li Guangyao 2004 *J. Comput. Sim.* **21**, 21-63
[18] Fan Guangquan, Ma Liping and Yang Bingru 2006 *J.Geogra. and Geo-Info. Sci.* **22**, 38-41
[19] Wu Wenring and Wu Yiring 2007 *J. ACTA PHOTONICA SINICA* **36** 1561-64
[20] Li Xianhong 2006 *MATLAB 7.x Application* (Beijing:Electronics Industry)
[21] Wang Jiawen 2006 *MATLAB 7.0 Image Processing* (Beijing:National Defense Industry)

Fiber Temperature Sensor Based on Micro-mechanical Membranes and Optical Interference Structure

Yueming Liu[1], Weijian Tian[1,2], Jing Hua[1]
[1] College of Optical and Electronic Technology ,China Jiliang University, Hangzhou 310018,China
[2] Xi'an Institute of Optics and Precision Mechanics of CAS, Xi'an 710068, China

Email：liuym@cjlu.edu.cn

Abstract. A novel fiber temperature sensor is presented theoretically and experimentally in this paper. Its working principle is based on Optical Fabry-Perot interference structure that is formed between a polished optical fiber end and micro-mechanical Bi-layered membranes. When ambient temperature is varying, Bi-layered membranes will be deflected and the length of Fabry-Perot cavity will be changed correspondingly. By detecting the reflecting optical intensity from the Fabry-Perot cavity, the ambient temperature can be measured. Using finite element software ANSYS, the sensor structure was optimized based on optical Interference theory and Bi-layered membranes thermal expansion theory, and theoretical characteristics was simulated by computer software. In the end, using optical fiber 2×2 coupler and photo-electrical detector, the fabricated sample sensor was tested successfully by experiment that demonstrating above theoretical analysis and simulation results. This sensor has some favorable features, such as: micro size owing to its micro-mechanical structure, high sensitivity owing to its working Fabry-Perot interference cavity structure, and optical integration character by using optical fiber techniques.

1. Introduction

Temperature is one of the most important measurement parameters in industrial and agricultural productions, so it is meaningful to research and develop high-accuracy and miniature temperature sensors that can be applied in such situations. Because of their simple construction, anti-electromagnetism interference and high accuracy, the Fabry-Perot optical fiber sensors get widespread research in the word [1-4]. One of the earliest Fabry-Perot temperature sensors was designed by forming Fabry-Perot cavity between a thin silicon membrane and the end of a multimode fiber [5], but its sensitivity is low and its size is large. Another Fabry-Perot temperature sensor utilized a cavity formed between two cleaved ends of two fibers fixed in a capillary [6], and its temperature sensitivity in this case is due to the thermal expansion of capillary that changes the length of

Fabry-Perot cavity, but its structure is complicated and not easy to use. In this paper, a novel Fabry-Perot temperature sensor is presented which is based on Fabry-Perot cavity and Bi-layered membranes. Because this temperature sensor has some features, such as: small size, high accuracy, and integrating with optical fibers, this Fabry-Perot temperature sensor would be suitable for accurate temperature measurements where the traditional Thermometer can not work, such as: remote place, narrow position, and strong electromagnetism interference working area.

2. Basic structure and principles

2.1. Basic Structure of Sensor Head

The configuration of the Fabry-Perot temperature sensor head consists of a silicon plate with a shallow square cavity and bi-layer membranes that covers the cavity by bonding process, as shown schematically in Figure1. Fabry-Perot cavity's upper mirror is bi-layer membranes and lower mirror is formed by polished fiber end, so the Fabry-Perot cavity is formed between bi-layer membranes and polished fiber end, and in this structure the parallel between upper and lower mirror must be guaranteed. Micro-fabrication methods were used to fabricate the temperature sensor head with micro-mechanical process.

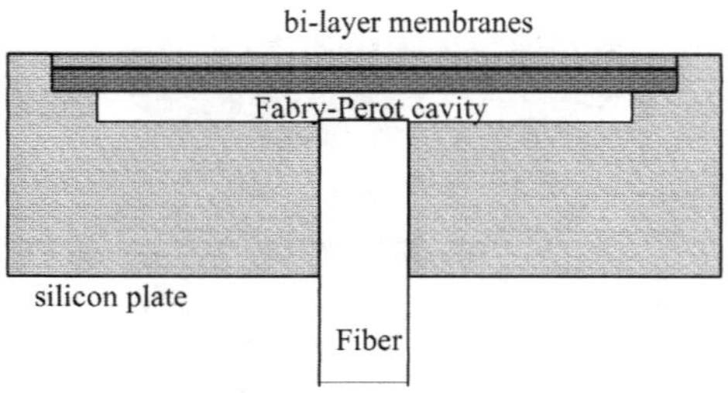

Figure 1. Configuration of the Fabry-Perot sensor heard

2.2. Operating Principle of Fabry-Perot Sensor

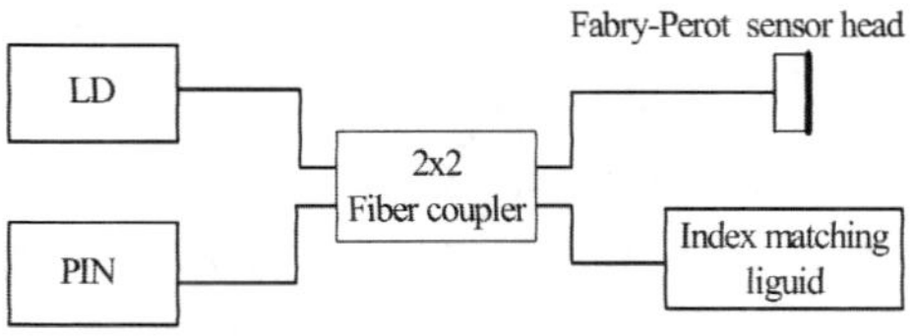

Figure 2. Configuration of the sensor system

Figure 2 shows the configuration of the sensor system, the system consists of four parts: a laser diode(LD) light source, a photo-electrical detector PIN, a 2×2 multi-mode fiber coupler, and the Fabry-Perot sensor head. The output of LD source is coupled into 2×2 single mode fiber and propagates through the fiber coupler to the Fabry-Perot sensor head. Because Fabry-Perot cavity's upper mirror is made up of bi-layer membranes, bi-layer membranes will be deflected in response to ambient temperature change owing to different thermal expansion coefficient of this two layer materials, this is so-called 'Bi-coating Effect'. Fabry-Perot cavity depth will be changed when the bi-layer membrane is deflected thermally, and this results in optical reflecting power change of the Fabry-Perot basing on multiple optical interferences. The reflecting light from the Fabry-Perot cavity is returning back through the same fiber coupler and converted to electricity by detector. The reflectance R of Fabry-Perot cavity can be expressed by [7]

$$R = \frac{R_1 + R_2 - 2\sqrt{R_1 R_2}\cos\phi}{1 + R_1 R_2 - 2\sqrt{R_1 R_2}\cos\phi} \tag{1}$$

where R_1 represents the reflectance of the fiber face, R_2 represents the reflectance of membranes surface, and ϕ,the round-trip propagation phase shift in the Fabry-Perot cavity, The value of ϕ is given by

$$\phi = 4\pi h / \lambda_0 \tag{2}$$

where h is the cavity depth, and λ_0 is the optical wavelength. As seen from equ.(1) and equ.(2), the reflectance R varies periodically as a function of the cavity depth h. If the input light intensity is defined as I_i, output light intensity is defined as I_o, and we can get the equation

$$I_o = KRI_i \tag{3}$$

where K is constant. Combined equations above, the output light intensity varies periodically as a function of the cavity depth h, and so the variety of the cavity h has particular relation with measured temperature accordingly. when the output light intensity is measured, measured temperature can be determined.

3. Thermodynamics characteristics and optimization

3.1. Theoretical Model Based on Thermodynamics Characteristics

The basic construction of bi-layer membranes is shown schematically in Figure3.

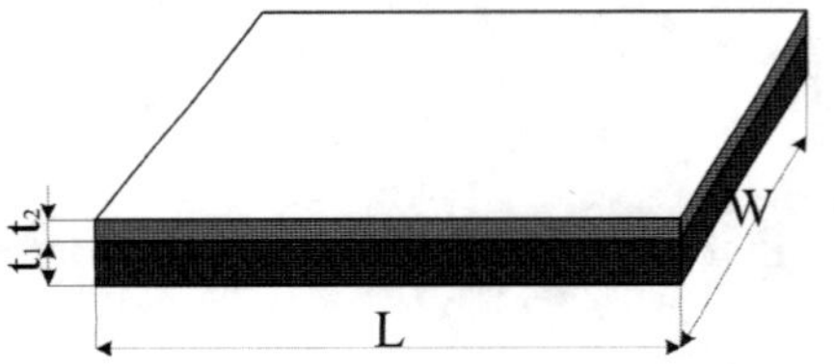

Figure 3. Basic construction of bi-layer membranes

The bottom layer material's thickness, thermal expansion coefficient, Poisson ratio and Young's modulus are t_1, α_1, ν_1 and E_1 respectively, they are t_2, α_2, ν_2 and E_1 respectively for the upper layer material. According to thermal mechanics theory, maximum temperature deflection of the membranes is given by [8]

$$\delta_{\max} = 3K_y LW(\alpha_2 - \alpha_1)\Delta T / 4t_1 K_D = 3K_Y LW\Delta TM / 4t_1 \tag{4}$$

where L is the length of the rectangle diaphragm, and W is the width of the rectangle diaphragm, ΔT is the temperature's variety, K_Y is the constant and determined by boundary conditions, M is equal to $(\alpha_2 - \alpha_1)/\mathrm{K}_D$ and K_D is relative rigidity of bi-layer membranes, which is expressed

$$K_D = \frac{1}{n+1}(4 + 6n + 4n^2 + \omega\phi n^3 + \frac{1}{\omega\phi n}) \tag{5}$$

where $n = \frac{t_2}{t_1}, \phi = \frac{E_2}{E_1}, \omega = \frac{1-\nu_1}{1-\nu_2}$

3.2. Selection of Materials for Bi-layer Membranes

As can be seen from the equ. (4), the maximum deflection of bi-layer membranes $\delta_{\max}$ is in proportion to the temperature's variety ΔT and the difference of bi-layer material's thermal expansion coefficient. It implies that bi-layer membranes' deflection becomes larger when the difference of bi-layer material's thermal expansion coefficient increases. Bottom layer of bi-layer membranes can be made of silicon because silicon's thermal expansion coefficient is relative small and it is easy to bond with silicon plate, aluminum can be selected for the upper layer material of bi-layer membranes because its difference of thermal expansion coefficient with silicon is more obvious than that of silicon and others material. From the equ. (4), It can be deduced that M gets maximum when the thickness ratio of bi-layer membranes is 0.55 for silicon and aluminum coating, thus it implies that $\delta_{\max}$ gets maximum.

3.3. Determination of Membranes Dimensions

Combined equation (1), equation (2) and equation (3), the reflected light intensity varies periodically as a function of the cavity depth h, and variation period of h is $\lambda_0/2$. To avoid ambiguous results from the period variation of the reflected light intensity, the maximum diaphragm deflection is typically restricted to guarantee reflected light intensity have a single definite value relation with the cavity depth h, so the maximum diaphragm deflection should be restricted to less than $\lambda_0/4$ for a single valued relation between reflected light intensity and measured temperature, where λ_0 is the optical wavelength. Furthermore, in order to let the sensor have the best sensitive work region and have good linearity, the maximum deflection of bi-layer membranes should be restricted to $\lambda_0/8$.

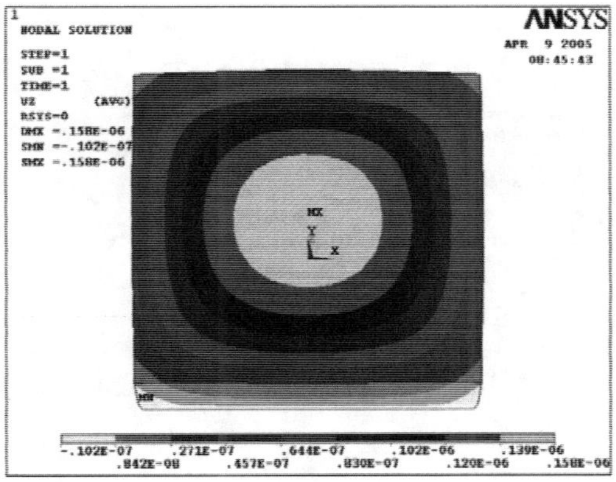

Figure 4. The deformation of square membranes

The thickness and the length of the square membranes are crucial parameters for sensor design. We initially designed temperature sensors with the temperature range 0℃ to 100℃, to operate at λ_0=1310nm, so the thickness and the length of the square membranes should be chosen to guarantee that the range of cavity's variety is less than 164 nm. Using software ANSYS to analyze the bi-layer membranes, Figure4 shows the deformation of square membranes when measured temperature is 100℃ for selecting parameters, and different gray color represents different deformation scope in Figure4. As can be seen from Figure 4, the white region's area is far larger than fiber end face, therefore the centre of square membranes have a region that can guarantee parallel motion of lower mirror when it is deflected. When square shape is selected for bi-layer membranes' geometrical shape, considering the optimization of sensor's sensitivity and structure, optimized parameters were finally determined as: L=W=500μm，t1 =40μm, t2 =22μm. For getting coefficient relation between measured temperature and central maximum deflection, central maximum deflection versus measured temperature have been calculated by means of software ANSYS given in Table 1.

Table 1. Central deflection o versus measured temperature

Measured temperature（℃）	0	20	40	60	80	100
Central max-deflection（nm）	0	32	63	95	126	158

4. Fabrication and experiment

4.1. Fabrications of Temperature Sensor

Using optical fiber 2×2 coupler and photo-electrical detector PIN, the sensor head was fabricated successfully according to the illustration shown in Figure 5, multi-mode telecommunication fibers with diameter 125 um were used to fabricate the sensor head. Firstly, the protective layers of fibers were stripped off at one end, and inserted into ceramic ferrules; secondly, the end face of the fiber were polished with the ferrules; thirdly, one fiber was fixed with bi-layer metal membranes, and two facet-polished fibers were inserted into a holey sleeve, and an air cavity was formed between two fiber facets by careful controlling. As can be seen from Figure 5, when bi-layer membranes were deflected in response to detected temperature changes, the fiber facet fixed with membranes would move relatively from another fiber facets, then Fabry-Perot cavity depth would be changed .

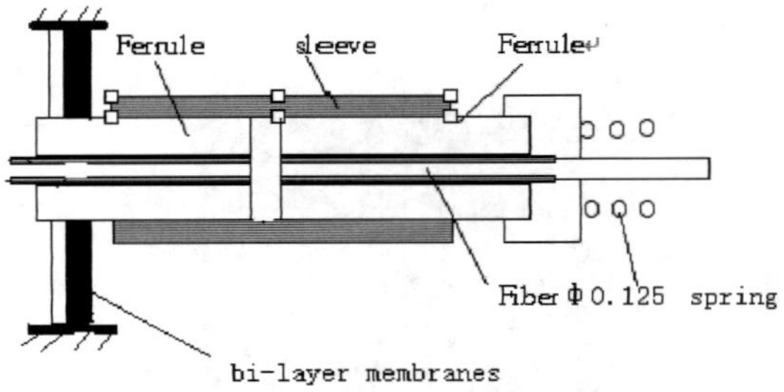

Figure 5. Illustration of sensor with membranes

4.2. Experimental Results

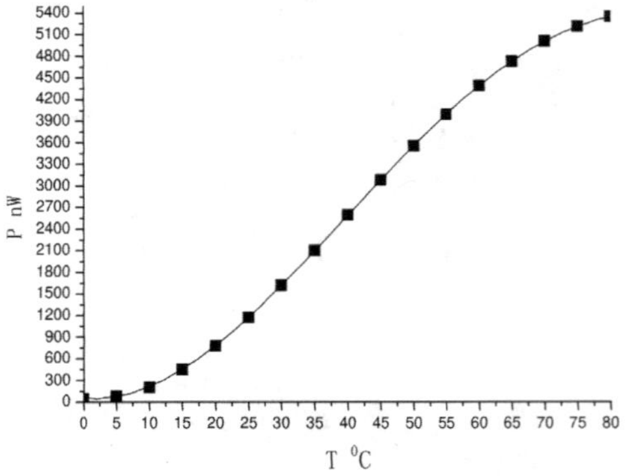

Figure 7. Output vs. detected temperature of the sample temperature sensor.

The temperature sensor based on the extrinsic fiber Fabry-Perot cavity and bi-layer metal membranes was done to verify the feasibility of the sensor model descried in Section 2. A laser diode ($\lambda = 1310\ nm$) illuminates the Fabry-Perot temperature through a 2×2 coupler. The reflected light power was measured from one input ports of the coupler by opto-electrical detector. One output port is connected to the temperature sensor placed in the temperature-controlled sample chamber. The other end of the output port of the coupler is dumped into index matching liquid to avoid any disturbing reflected light from the unused output. The experimental curve was fitted by experimental data points shown in Figure 7, The full testing scope is from 0 to 80℃, a average sensitivity of 62.82nw/℃ and accuracy of $\pm0.7\%$ were obtained over the full temperature range.

5. Conclusion

The paper has successfully demonstrated a novel micro-mechanical optical temperature sensor capable of temperature measurements in related applications. Construction of the all-optical miniature temperature sensor has been described and was optimized theoretically. The fabricated sample sensor was tested successfully by experiment that demonstrated above theoretical analysis and theoretical simulation results. The advantages of this temperature sensor are small size, high sensitivity, immunity to electromagnetic interference and available for remote sensing. This type of optical fiber sensors can offer its applications in harsh and narrow environments.

Acknowledgements

This work was supported by Natural Science Foundation of Zhejiang Province of China (No. Y1091078) and Science Research Found of China Jiliang University.

References

[1] S H Aref, H Latifi, M I Zibaii, et al. Fiber optic Fabry–Perot pressure sensor with low sensitivityto temperature changes for downhole application ,Optics Communications,V. 269, 2007: 322–330

[2] G Z Xiao, A Adnet, Z Zhang. Monitoring Changes in the Refractive index of Gases by Means of a Fiber Optic Fabry-Perot Interferometer Sensor. Sensors and Actuators A, 2005, 118:177-182

[3] K Smith, S. de Man, H Zeijlemakeret, al. Fiber-top atomic force microscope: a worthwhile challenge, Authorized licensed use limited to: China Jiliang University. Downloaded on February 1, 2009, at 22:44 from IEEE Xplore. Restrictions apply

[4] J Fujii, T Nakamura, T Tsuda, et al. Comparison of Winds Measured by Mu Radar and Fabry-Perot Interferometer and Effect Of OI1577 Airglow Height Variations. Journal of Atmospheric and Solar-Terrestrial, 2004, 66: 573-583

[5] C E Lee, H F Taylor. Sensors for Smart Structures Based upon the Fabry-Perot interferometer. Fiber Optic Smart Structures, 1995, 1(1): 249-269

[6] W H Tsai, C J Lin. A Novel Structure for the Intrinsic Fiber-Perot Temperature Sensor, Journal of Light Technology, 2001, 19(5): 682-686

[7] M Jiang, E Gerhard. A Simple Strain Sensor Using a Thin Film As a Low-Finesse Fiber-Optic Fabry-Perot Interferometer. Sensors and Actuators A, 2001, 88: 41-46

[8] J R Barnes, R J Stephenson, C N Woodburn. A Calorimeter Using Micro-mechanical Sensors. Rev. Science Instrument, 1994, 65(12): 3793-3798

Research of High Sensitivity Uncooled Infrared Detector Array

Ping-chuan Zhang

HUAZHONG University of Science& Technology， WUHAN China 430074

LUOHE Vocational Technology College, LUOHE China 462002

E-mail: redmoon123456@126.com

Bo Zhang

LUOHE Vocational Technology College, LUOHE China 462002

E-mail: lhzyzb@126.com

Abstract---The infrared thermal imaging technology has been widely used in military and civilian fields and the field of the infrared detection and infrared thermal imaging technology has been of concern for a long time. On infrared thermal imaging, its core components for the infrared focal plane arrays, how to develop a high sensitivity of the multi-focal plane infrared detector is a key issue. Although the Common focal plane array of quantum has high sensitivity, but it requires low temperature cooling work environment and led to complexity and high cost, difficult to compact. Conventional uncooled infrared focal plane array is contrast to the quantum focal plane arrays. Therefore, this article preceded by the uncooled infrared detector array to improve the wide temperature sensitivity in examining the feasibility PMN composite film, materials composition, structure design and preparation process technology.

Keywords: Pyroelectric effect; Pyroelectric detectors; Infrared radiation; Etching process

1. Introduction

Pyroelectric detectors[1-5] is the light receiving device using the pyroelectric effect to detect infrared radiation, its advantages are: able to work at room temperature; a broad spectrum of responses, can be used to detect any radiation caused by body temperature changes in the pyroelectric; tight structure, simple operation, high mechanical strength; no external bias field: only response to the temperature change rate, can detect transient signals and so on.

The pyroelectric detector can be divided into two categories, the photon detector of required cooling and thermal detector without cooling . Photon detector response rates, response time, have a wide range of military applications, but the range of applications was affected by the wavelength selectivity, to work at low temperature, high cost and large power consumption, etc,. the thermal detector, although its response rate and the sensitivity of devices is less than the high-photon detectors, but with no wavelength selectivity, without cooling, simple structure, low price and therefore have more extensive application prospects [6]. The pyroelectric detector is a kind of hot detectors, its structure mainly consists of five parts, radiation modulation system, absorbing layer, thermal insulation layer, sensitive element and reading circuit, in which the sensitive element is the core [7,8].

Using the both broadening effects of inorganic phase change materials and the multiple- multiphase of composites, to design a new type of wide temperature range composite response pyroelectric materials[9], to get a good overall performance 0-3 composite pyroelectric materials by casting, hot pressing and simultaneous electric field induced polarization techniques, also studied the broadening mechanisms of the film composite pyroelectric effect in the temperature zone;

On the other hand, a new micro-bridge with thermal insulation and good conductive properties of Si3N4 insulation core / Pt thermal conductivity shell was achieved by etching micro-channel - steamed Pt electrode-filled organic material- membrane- a new type of corrosion process by filling.

And designed distributed-temperature human identification system using the development of 16*16 Element Array PMNST / PVDF composite uncooled infrared detector array. This would have important theoretical and practical value in the promotion of research and development for uncooled infrared detector array and raising its application level.

2. the pyroelectric mechanism and materials analysis

2.1 Pyroelectric Effect

Pyroelectric effect refers to the dielectric polarization phenomena change with temperature [10]. When some of the strong dielectric material of the surface received infrared radiation, its surface effect of thermal change, as the temperature up or down, in which generating electric charges on the material surface, a phenomenon known as the pyroelectric effect, is a thermoelectric phenomenon, a phenomenon in Barium ferrite of dielectric material was particularly notable. Imagine a single domain of a ferroelectric material, which are arranged so close to the polarization vector across the surface near the bound charge. In thermal equilibrium, the bound charge is from the body of the equivalent number of free charge against the shield, so ferroelectrics on the role of the outside world does not show power. When the temperature changes, the polarization changes, the original charge can not be completely free screen bound charge, then the surface free charge, which formed in the space near electric field on charged particles are attracted or repulsion. If it's connected with the external circuit, the circuit can be observed in the current, heating and cooling in both cases the current in the opposite direction.

Since coating electrodes on the two surface perpendicular to the crystal (the direction of spontaneous polarization Ps) of Pyroelectric detector in the polar axis, the thermal sensor is similar to the plate capacitor. The working mechanism [3] shown in Figure 1.

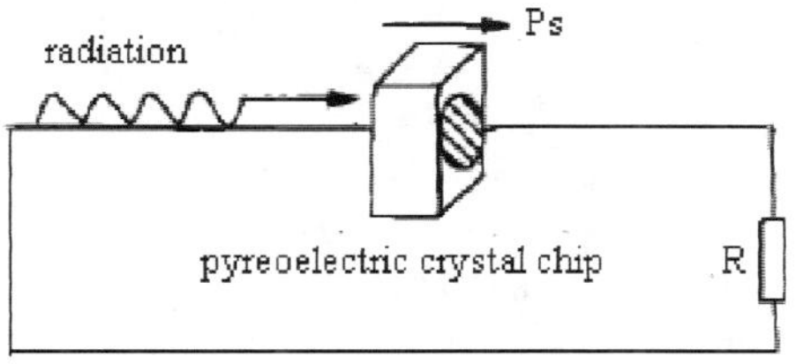

Fig,1 the theory of pyroelectric detectors

Under the assumption that the pyroelectric infrared film temperature distribution is uniform, square wave pulse function modulated infrared radiation is given by:

$$T(t)=\begin{cases}\dfrac{aP_o}{H}(1-e^{-t/H})+T_O & 0\le t\le \tau_o\\ \dfrac{aP_o}{H}(1-e^{-t/\tau_1})e^{\frac{t-\tau_0}{\tau_1}}+T_O & t\succ\tau_0\end{cases} \quad (1)$$

H for the heat capacity; $\tau=\varepsilon/\sigma$, $\varepsilon$for dielectric constant, $\sigma$the conductivity of the crystal; a for heat detector for the absorption rate; set the ambient temperature is consistent with T0 said.

2.2 pyroelectric material

Polyvinylidene fluoride [11]. PVDF resin mainly refers to the fluoride homopolymer or a small amount of fluoride and other fluorinated vinyl monomer copolymer. PVDF resin, fluorine resin, and both General of the resin, in addition to good chemical resistance, high temperature resistance, oxidation resistance, weather resistance, radiation properties, also with piezoelectricity, dielectric property, thermal resistance, and other special properties.

It's a crystal materials with the spontaneous polarization. Spontaneous polarization is intrinsic polarization of the material itself, because the centers of structure of positive and negative charge in one direction do not overlap. Under normal circumstances, the bound charges on the crystal surface produced by the spontaneous polarization are adsorbed. The spontaneous polarization changes when the temperature changes, thus freeing up some of the Adsorption of charge. It's opposite for the polarity charge when crystal cooling and heating. Pyroelectric material is a piezoelectric material, is not a centeral symmetry of the crystal. There are thousands of materials with pyroelectric properties, but only a dozen widely used, mainly such as, sulfuric acid peptide in three glycosides, lead lanthanum zirconate titanate transparent ceramics, and polymer film. Pyroelectric materials can be used in industrial infrared detectors, thermal camera tube has some special purpose in the military defense. Its advantage is not low-temperature cooling, but the sensitivity lower than the corresponding semiconductor devices.

3. Process implementation

The proposed schematic design of the infrared detector structure shown in Figure 2.

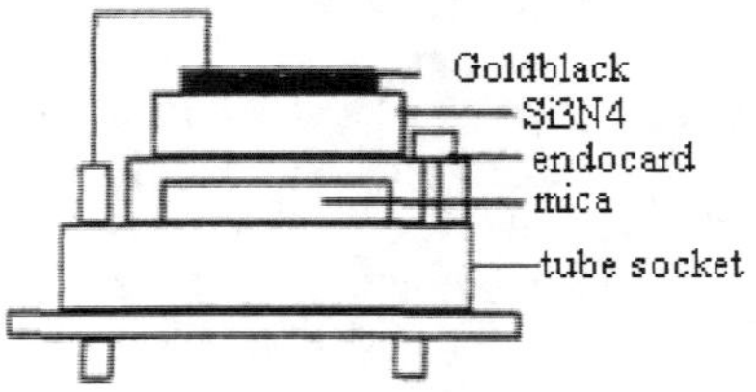

Fig. 2 Si3N4 thin film IR detector schematic

Conventional lithography employs UV of wavelength from 2000 to 4500 angstrom as the image information carrier and photo resist as intermediate (image record) media to achieve the transformation graph, transfer and processing of image data ultimately delivered to the chip (mainly refers to silicon), or dielectric layer on a process (Figure 3). In a broad sense, this includes two main aspects of light copying and etching.

- Optical copying process: The exposure system accurately passes the prefabricated components in the mask or the circuit on the graphic to the desired location, which is the pre-coated photo resist layer on the chip surface or a dielectric.
- etching process: using the chemical or physical methods, to remove surface resisting mask wafer of thin dielectric layer or the chip surface of a dielectric material on the resist layer, on which to get a graphic exactly the same as the copied. The integrated circuits functional layer is overlapped in three-dimension, and therefore always lithography process repeated several times. For example, large scale integrated circuit lithography have to go through about 10 times to complete the full transfer.

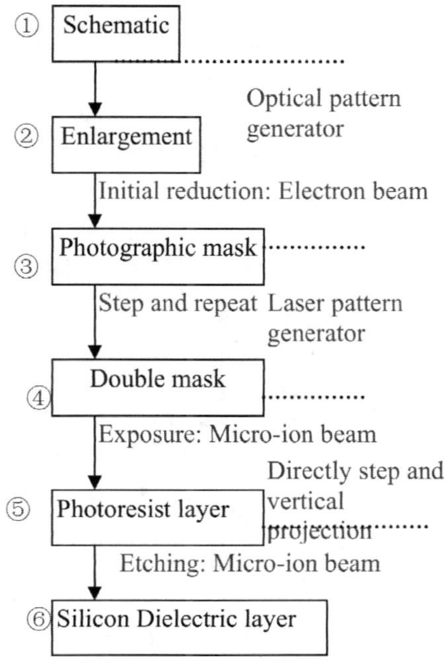

Fig. 3 Flow chart of image transfer pyroelectric devices

In the narrow sense, the optical lithography process refers only to copying process, that is, from ③ to ⑤ or ④ to ⑤ of the process in Figure 3.

The electrode under the microscope after graphic photos shown in Figure 4. Response curve shown in Figure 5.

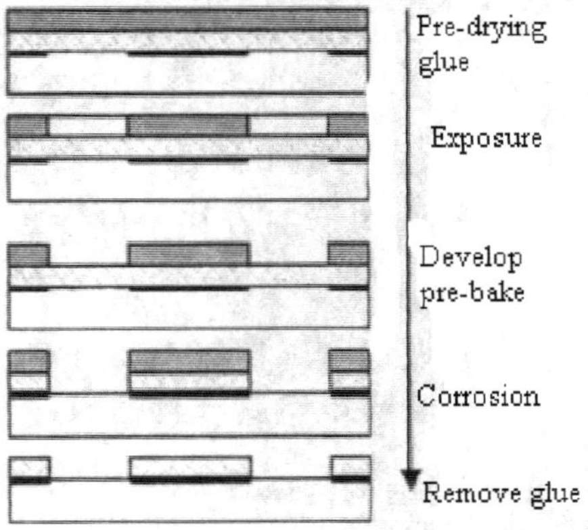

Fig. 4 electrode microscope after graphic photos

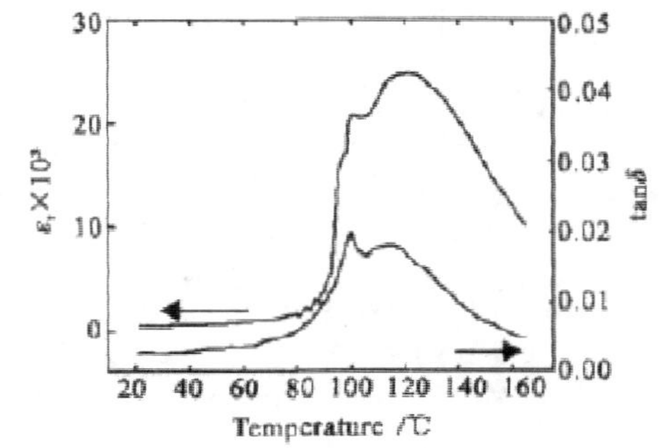

Fig.5 Dielectric characteristics for PMN single crystal chip

Exposure is the most critical lithography technology processes, it is directly related to the resolution of the resist pattern, leaving film rate, line width control and register accuracy. In certain light, the effect is due to the exposure time. Exposure time is too short, then the photoresist sensitive enough not to fully resist chemical reactions, poor graphics resolution; if exposure time is too long, will resist the edge of a weak non-sensitive part of the light-sensitive, have a "Halo" phenomenon is also made worse resolution of the resist pattern. In the experiment using contact exposure, exposure time is in 45s, the best graphic lithography. Graphic imaging is the mask to the photoresist to achieve the transfer of graphics. In the developer solution unchanged, must control a good visualization of the time. Experiment is impregnated with visualization, imaging time control at 1 minute 30 seconds.

Generated by sputtering on silicon base to get Si3N4, and then etching the Pt film, MgO single crystal, MgO film deposited Pt substrate, and the mica substrate film, about 20μm thick deposited Si3N4 film, the Si3N4 film on evaporated A1 electrode, then evaporating a layer of gold black as electrodes for the infrared absorption, with 20V voltage for polarization 20min. The heat capacity Si3N4 film is small, so its sensitivity can be very high, its response is proportional to the input infrared light intensity, its voltage responsivity Rv is 465V/W, detectivity D* is $1.7\times 10^{8}cmHz^{0.5}$ / W.

4. Conclusions

A typical semiconductor lithography and metal electrode stripping technology are combined with DC sputtering method, to achieve a graphic of infrared pyroelectric detector sensitive element for the lower electrode. Achieved the preparation for the uncooled infrared detector ferroelectric thin film composite structure (absorbing layer of top electrode / ferroelectric film / bottom electrode / insulating layer / Si substrate), including the insulation layer, bottom electrode, ferroelectric thin films, semi-permeable membrane electrode and absorption layer of the preparation and performance optimization, cell detection rate was 2.8×108cmHz0.5 / W.

references

[1] LANG S B.Pyroelectricity:from ancient curiosity to modern imaging tool[J].*Physics Today*, 2005, 58(8): 31-36.

[2] ROGER W W. Pyroelectric arrays:ceramics and thin films[J].*J Eleetroceram*, 2004, 13(1-3): 139-147.

[3] ZHONG Wei-lie.*Ferroelectric Physics* [M]. Beijing. Science Publishing House,1998.

[4] KULWICKI B M,AMIN A,BERATAN H R, et al. Pyroelectric imaging[C].Proceedings of IEEE, *Inter Sympos Appl Ferroelectric*,1992,90,CH3080-9:1-10.

[5] Cooper A W, Lentz W J, Walker P L,et al. Infrared Polarization Measurement of Ship Signature and Background Contrast[C]. *SPIE,* 1994, 2223:301-309.

[6] WHATMORE R W, AINGER F W. Pyroelectric ceramic materials for uncooled IR detectors[C].Proceedings of SPIE, *Advances in Infra-Red Sensor Technology*,1983.395: 261-26.

[7] WHATMORE R W, BELL A J. Pyroelectric ceramics in the lead zirconate-lead titanate -lead iron niobate system[J].*Ferroelectrics*,1981(35):155-160.

[8] SHAW C P,GUPTA S,STRINGFELLOW S B, et al. Pyroelectric properties of Mn-doped lead zirconate-lead titenatelead magnesium niobate ceramics[J]. *J Euro Ceram Soc*,2002,22(13): 21-23.

[9] WATTON R.Ferroelectric materials and IR bolometcr arrays from hybrid arrays towards integration[J]. Integr Ferroelectr,19914, 4(22): 175-186.

[10] WHATMORE R W,WATTON R.Pymelectric ceramics and thin films for uncooled thermal imaging[J].*Ferroelectrics*,2000,236(11):259-279.

[11] Shoucair F, Hwaug W, Jain P. Electrical characteristics of Large scale integration (LSI) MOSFETs at verv high temperatures[J]. *Micro electron reliability*, 1984, 24(3):46-485

Real-time Infrared Characterization Simulation and Research of Wind-roughened Sea Surface

Shicheng Zhang[1], Zhen Yang, Li Yang

College of Naval Architecture and Power, Naval University of Engineering, Wuhan 430033, China

Email: schzhang@yahoo.com.cn

Abstract. The sea surface is an important background of sea surface targets and its infrared radiation characterization is the basis of infrared detection and recognition of sea surface targets. The model of linear wave theory and JONSWAP spectrum was used to achieve shape and display of ocean wave. Real-time infrared characterization simulation model was established for the sea surface, considering the solar radiance, the sky background radiance and the effects of atmospheric transmission. The sea surface was divided into many wavelet facets and the radiance of every facet was calculated on the basis of infrared simulation model. Three typical time, 9:00, 12:00, 15:00 in one day were selected and the radiance of every facet about 3~5μm and 8~12μm waveband was respectively calculated. Real-time infrared simulation images of sea surface were generated by converting wavelet facets radiance value to image gray value. The results show that the sea surface infrared images of 3~5μm waveband display the bright spots of reflected solar irradiance, and the spots appear the same direction with the solar azimuth while the impact of solar irradiance in 8~12μm waveband isn't usually required, and the method can effectively simulate the sea surface infrared characterization and provide the basis for infrared detection and recognition of sea surface targets.
Keywords: Infrared characterization; JONSWAP spectrum; Sea surface simulation; Radiance

1. Introduction

With the development of infrared track and guidance technology, different kinds of infrared precision-guided weapons are increasingly serious threats against sea surface military targets. The sea surface is a decisive background of sea targets, and its infrared radiation characterization is the basis of infrared detection and recognition of sea targets, which can directly influence the results of detection. Meanwhile, due to the impacts of sea surface waves and sea clutter effects existence, there is a certain roughness of the sea itself and inconsistent reflection of radiation on the surrounding environment. The detection of comparatively weak radiant objects in the thermal band becomes much more difficult. Therefore, analysis and simulating sea surface infrared characterization are significant for studying missile tracking guidance, feature recognition and infrared stealth technology of ship.

A widely previous work has been done about the infrared characterization of the sea. In 1954, C. Cox and W. Munk established the classic Cox-Munk model[1]. Based on this model, other scholars established the calculation model of sea surface emissivity and reflectivity[2][3], which had a wide range of application in infrared remote sensing and simulation fields[4]. In the construction of the ocean wave, using the wave spectrum analysis method, currently used mainly focus on Bretschneider spectrum, PM spectrum, Phillips spectrum, JONSWAP spectrum, Wallops spectrum, etc.. Each model has its scope of application and can be selected according to actual needs. In 1968-1969, scientists from NATO countries built the first international spectrum, called JONSWAP[5] spectrum, which is widely used by many researchers over the world as a useful tool for analyzing the optical, infrared, and other spectrum characteristics of the sea for its authority and practicality. In terms of infrared imaging simulation, I. Wilf and Y. Manor[6], F. Schwenger and E. Repasi[7] did the effective work of excellence. In this paper, we focus on the real-time infrared characterization of the sea surface. The ocean waves are simalted with JONSWAP spectrum model. Then the radiance of sea surface is calculated, considering the solar radiance, sky background radiance, atmospheric path transmission and surface structure of the wavelet facets' impacts on emission and reflection of the sea. In the end, infrared radiation characterization of the sea of different times in one day under the 3~5μm and 8~12μm waveband is simulated.

2. Modelling the sea surface pattern

Linear wave theory[8] indicates that the ocean waves can be superimposed by different periods and different random initial phases of irregular cosine waves. We use the method based on wave spectrum to achieve the wave simulation and display. The international generalduty JONSWAP spectrum is used, which can be expressed as:

$$S(\omega,\mu)=S(\omega)G(\omega,\mu)$$

$$S(\omega)=\frac{\alpha g^2}{\omega^5}\exp\left[-\frac{5}{4}\left(\frac{\omega_m}{\omega}\right)^4\right]\gamma^{\exp[-(\omega-\omega_m)^2/(2\sigma^2\omega_m^2)]} \qquad \bar{L}=\frac{Lg}{v^2}$$

$$\alpha=0.076\bar{L}^{-0.22} \qquad \sigma=\begin{cases}0.07 & (\omega\le\omega_m)\\ 0.09 & (\omega>\omega_m)\end{cases}$$

$$\omega_m=2\pi\times3.5\frac{g}{v}\bar{L}^{-0.33} \qquad \gamma=0.33$$

$$G(\omega,\mu)=\begin{cases}\frac{2}{\pi}\cos^2(\mu-\psi) & -\pi/2+\psi\le\mu\le\pi/2+\psi\\ 0 & others\end{cases} \tag{1}$$

Where, ω is the spatial angular frequency; θ is the wave vector direction; g is the acceleration of gravity; ω_m is the spectrum peak frequency; v is the wind velocity; L is the wind fetch; $G(\omega,\mu)$ is directional distribution function; ψ is the angel of wind direction and X-axis.

The 3D grids image of ocean wave is presented according to the linear wave theory and JONSWAP spectrum model (Figure 1).

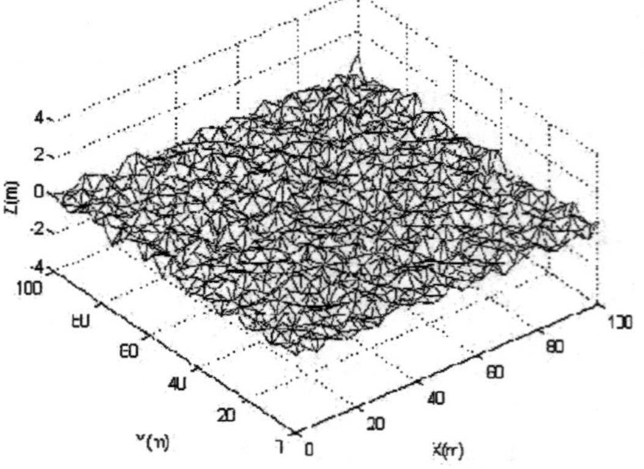

Fig.1. 3D grids of ocean waves

3. Modelling the infrared radiation of sea surface

Generally, the sea surface can be approximated as a surface of a non-black body. When the sea surface is watched by a camera, then the radiance reaching the detector from an element of the sea can be divided into three parts (Figure 2): (1) sea itself infrared radiance; (2) sky background radiance, solar irradiance and reflection radiance; (3) atmospheric path radiance. In addition, atmospheric absorption and scattering effect should be considered in the calculation of radiance.

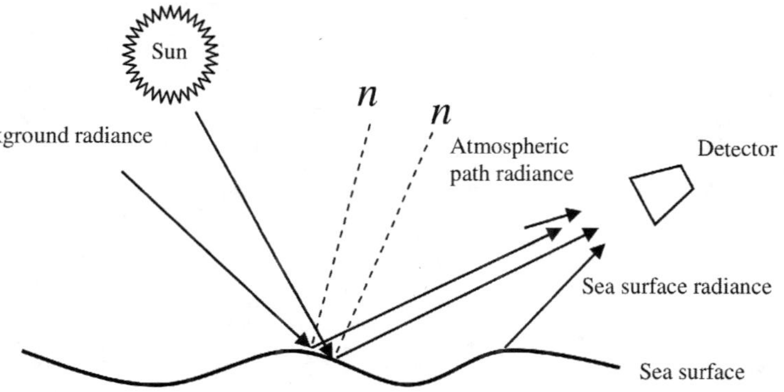

Fig.2. Infrared radiance schematic diagram of sea surface

Therefore, the total radiance $L_{sea}^{\Delta\lambda}$ of one waveband at the detector from the sea surface in the direction of (θ_e, φ_e) can be approximated as:

$$\begin{aligned} L_{sea}^{\Delta\lambda} = \left(1-\tau_{air}^{\Delta\lambda}\right)\cdot L_{air}^{\Delta\lambda} + s_e \cdot \tau_{air}^{\Delta\lambda} \cdot \Big(\varepsilon_{sea}^{\Delta\lambda}(\theta_e,\varphi_e)\cdot L_{bb}^{\Delta\lambda}(T_{sea}) + \rho_{sea}^{\Delta\lambda}(\theta_e,\varphi_e)\cdot(L_{sky}^{\Delta\lambda} + L_{sun,s}^{\Delta\lambda}) \\ + \rho_{sea}'^{\Delta\lambda}\cdot(\theta_e,\varphi_e,\theta_{sun},\varphi_{sun})L_{sun,d}^{\Delta\lambda}\Big) \end{aligned} \quad (2)$$

Where, $\tau_{air}^{\Delta\lambda}$ is radiant transmittance of the path between a point on the sea and the sensor; s_e is the slope-shadowing factor; $\varepsilon_{sea}^{\Delta\lambda}(\theta_e,\varphi_e), \rho_{sea}^{\Delta\lambda}(\theta_e,\varphi_e), \rho_{sea}'^{\Delta\lambda}(\theta_e,\varphi_e,\theta_{sun},\varphi_{sun})$ are respectively sea surface emissivity, hemisphere reflectivity and bidirectional reflectivity in the direction of (θ_e,φ_e); $\theta_{sun}, \varphi_{sun}$ are respectively the zenith angle and azimuth angle of the sun; $L_{air}^{\Delta\lambda}$ is the radiance of the atmospheric path; $L_{bb}^{\Delta\lambda}(T_{sea})$ is the radiance of a black body of T_{sea}; $L_{sky}^{\Delta\lambda}$ is the radiance of the sky measured at sea level; $L_{sun,s}^{\Delta\lambda}, L_{sun,d}^{\Delta\lambda}$ are respectively the radiance of the scattered and direct sun energy.

Because of the influence of wind-roughened waves, sea surface itself is rough. In the calculation of infrared radiation, we can not simply take it as a flat surface. However, it can be seen as a combination of different slopes of wavelet facets. Reflection schematic diagram wavelet facet is shown in Figure 3.

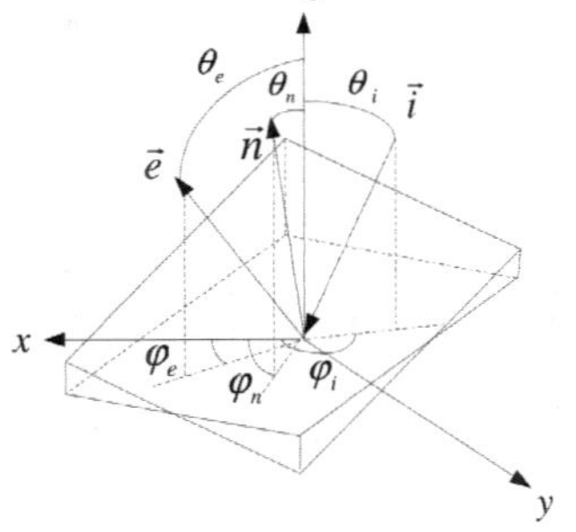

Fig.3. Reflection schematic diagram of wavelet facet

The bidirectional reflectivity of a wavelet facet can be expressed as[1]:

$$\rho_{sea}'^{\Delta\lambda} = \rho(n,\chi)\cdot(\pi\sigma^2)^{-1}\exp(-\frac{s_x^2+s_y^2}{\sigma^2})\cdot\frac{1+\cos\theta_e\cos\theta_i+\sin\theta_e\sin\theta_i\cos(\varphi_e-\varphi_i)}{(\cos\theta_i+\cos\theta_e)^2} \quad (3)$$

Where, $\rho(n,\chi)$ is the reflectivity of calm sea, which can be calculated from the Fresnel formula[9]; χ is the incidence angle, n is complex refractive index of seawater; $\theta_i,\varphi_i,\theta_e,\varphi_e$ are respectively zenith angle and azimuth angle in the direction of incidence and reflection; σ is the mean-square slope.

$$\sigma^2 = 0.003 + 0.00512\upsilon \pm 0.004 \quad (4)$$

Where, υ is the wind velocity above the sea.

Therefore, $\rho_{sea}^{\Delta\lambda}(\theta_e,\varphi_e)$, $\varepsilon_{sea}^{\Delta\lambda}(\theta_e,\varphi_e)$ can be approximated as:

$$\rho_{sea}^{\Delta\lambda} = \int_0^{2\pi}\int_0^{\pi/2}\rho_{sea}'^{\Delta\lambda}(\theta_i,\varphi_i,\theta_e,\varphi_e)\sin\theta_i d\theta_i d\varphi_i \quad (5)$$

$$\varepsilon_{sea}^{\Delta\lambda} = 1-\rho_{sea}^{\Delta\lambda}(\theta_e,\varphi_e) \quad (6)$$

Sky background radiance is diffused reflection, and it can be acquired without considering the issue of location of radiation sources. But the reflection of solar direct irradiance is concerned with the surface roughness. For the incident direct radiance of sun, the surface only to meet the specific conditions of the wavelet facets can reflect to the detector. Suppose incident direction of the sun **i**, wavelet facet normal to **n**, the direction of detector **e**. If the surface solar radiation can be reflected by the wavelet facet into the detector, then the sun, the detector and wavelet facet must satisfy:

$$\begin{cases}\mathbf{n}\bullet\mathbf{i}=\mathbf{n}\bullet\mathbf{e}\\ \mathbf{e}=2\mathbf{n}(\mathbf{n}\bullet\mathbf{i})-\mathbf{i}\end{cases} \quad (7)$$

In the calculation of the sea infrared radiance, firstly, it need to generate the surface shape of sea surface to calculate the wavelet facets slope distribution of all area; secondly, according to the slope size of each wavelet facet to judge whether the sensor can receive the reflection of solar irradiance from the surface; finally, on the combination of sea itself and its surrounding radiance, and considering the impact of atmospheric transmission, acquire the value of the infrared radiance of the entire surface. After calculation of radiance of facets, we also have to know how many facets correspond to a pix of the sensor. The number of wavelet facets in the perpendicular and parallel direction N_H, N_V can be approximated as[6]:

$$\begin{aligned}N_H &= R\Delta\theta/\Delta y\\ N_V &= R\Delta\theta/(\Delta x\sin\phi)\end{aligned} \quad (8)$$

Where, R is the path length between the sensor and the facet; $\Delta\theta$ is the instantaneous field of view; $\Delta x,\Delta y$ are the horizontal and vertical size of a facet; ϕ is the viewing angle. Therefore, a pix of the sensor can be the average of $N_H\times N_V$ facets' radiance.

4. Infrared characterization simulation of sea surface

After infrared radiance value of each wavelet facet based on sea surface infrared radiance model, convert wavelet facets radiance value to image gray value according to some algorithm, then the infrared radiation imaging will be generated. The algorithm can be expressed as:

$$G_{mi} = \frac{L_{mi}^{\Delta\lambda}-L_{\min}}{L_{\max}-L_{\min}}\cdot 255 \quad (9)$$

Where, G_{mi} is the grey value of mi wavelet facet; $L_{\max}$, $L_{\min}$ are respectively the maximum value and minimum value of radiance of entire surface.

Three typical time, 9:00, 12:00, 15:00 in one day are selected and the radiance of every facet about 3~5μm and 8~12μm waveband is respectively calculated on the basis of infrared simulation model. The following parameters are used. The direction of detector is north, detecting elevation angle is 40°, and the height is 5 km. The wind velocity is 6 m/s and wind fetch is 100 km. The seawater temperature is 20 □. The highest temperature of local atmosphere is 25 □, and the lowest is 18 □. The field of the sensor is 1.8°×3.5°.

(a) 09:00

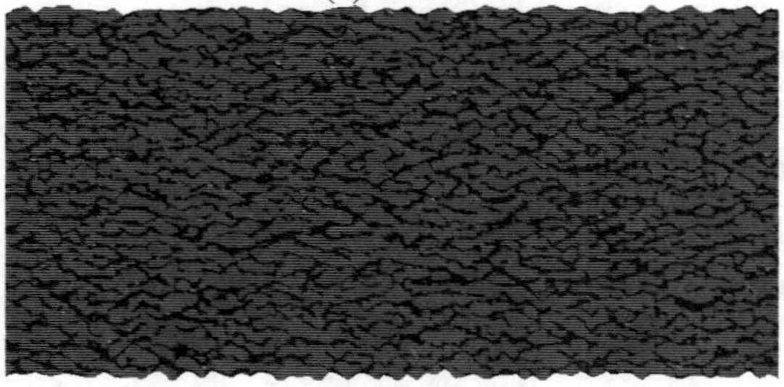

(b) 12:00

(c) 15:00

Fig.4. Infrared image of sea surface (3~5μm)

In figure 4, for the 3~5μm waveband, we can find that, as the sun changes position, the sea surface infrared image shows difference corresponding to the changes. At 9:00 in the morning, when the sun's in the eastern direction, then only the specific wavelet facets of eastern direction can reflect the incident radiation of direct sunlight, so that part of the eastern wavelet facets appear bright spots for the whole surface of the sea. At 12:00 in the daytime, the sun direction points to south, and this time the incident direction of the sun is almost perpendicular to the sea. There are less wavelet facets directly reflecting sunlight to the sensor and less difference in the overall brightness of the sea surface. At 15:00 in the afternoon, the sun's direction is western. Sea surface infrared image is the corresponding features with 9:00 pm. Part of the western wavelet facets appear bright spots. Comparison of three images of figure 4, it is found that at 9:00 and 15:00, the sun's elevation angle is small, so that more solar irradiance can be reflected and bright spots appeared more.

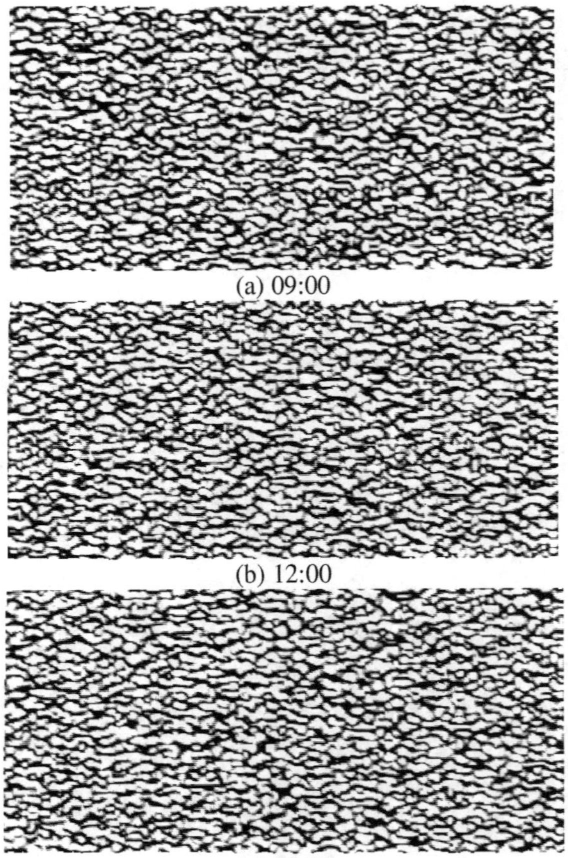

(a) 09:00

(b) 12:00

(c) 15:00

Fig.5. Infrared image of sea surface (8~12μm)

In Figure 5, for the 8~12μm waveband, it can be seen, with the changes of sun orientation in the day, the sea surface infrared characterization change little. Compared with 3~5μm waveband, the solar irradiance is very low, and it also indicates that solar irradiance in 8~12μm waveband is usually negligible.

To further discuss the impacts of solar irradiance in the 3~5μm and 8~12μm two wavebands on sea surface infrared images, the results of considering and not considering the solar irradiance in both wavebands are compared, then the difference and derivation of two results in each waveband are calculated, setting sun elevation angle 14°, the viewer elevation angle changing from 0° to 40°.

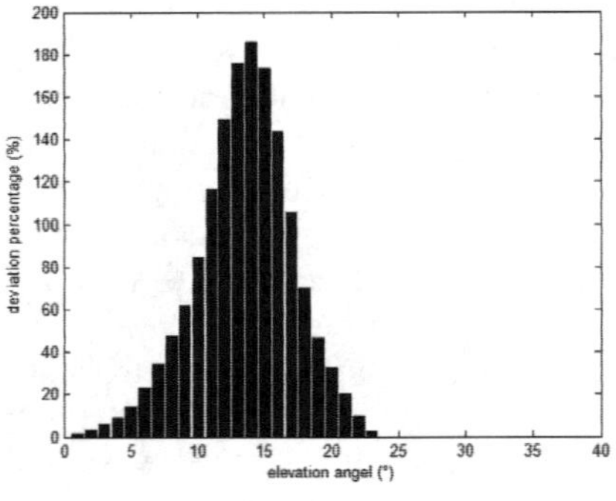

Fig.6. The radiance comparison of 3~5μm

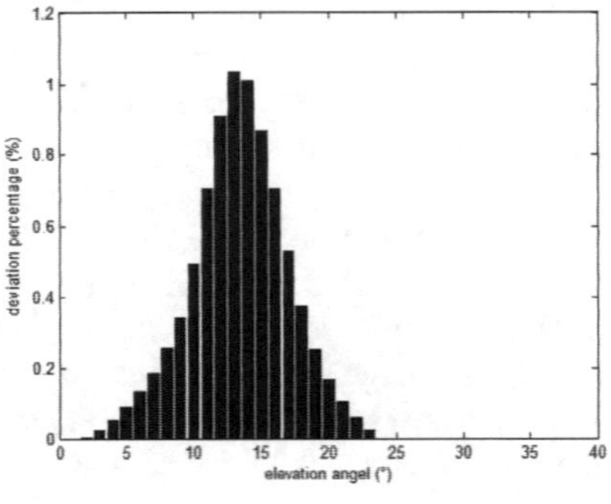

Fig.7. The radiance comparison of 8~12μm

From Figure 6 and Figure 7, it can be found that, in the 3~5μm waveband, solar irradiance is very apparent. The reflection intensity is largest when the elevation angle of both sun and sensor is approximated, and the deviation percentage of solar irradiance is even approaching to 200%. In the 8~12μm waveband, solar irradiance has little effect, usually negligible, which is consistent with well-known results.

5. Conclusions

In this paper, the model of linear wave theory and JONSWAP spectrum is used and the sea surface is simulated. The sea surface infrared radiation model is established, considering the radiance of sea surface itself, surrounding environment and effects of atmospheric transmission path. Infrared radiation images of 3~5μm and 8~12μm waveband of different day times are generated. The results are consistent with the actual situation and the difference of whether or not considering solar irradiation is compared. The conclusions are as follows:
(1) Infrared characterization of sea surface in 3~5μm waveband changes over time, and bright spots of the sea and the sun appear bearing the same orientation, while infrared characterization change little over time in 8~12μm waveband;
(2) When the elevation angle of detector and sun is similar, the wavelet facets can reflect more solar irradiance to the sensor;
(3) In 8~12μm waveband, solar irradiance has little impact on the sea surface infrared characterization, often negligible.

References

[1] Cox C and Munk W H 1954 Measurement of the roughness of the sea surface from photographs of the sun's glitter *J. Opt. Soc. Am,* **44** 838-50.
[2] Masuda K, Takashima T and Takayama Y 1988 Emissivity of pure and sea waters for the model sea surface in the infrared windows regions *Remote Sensing of Environment* **24** 313-329
[3] Wu X and Smith W L 1997 Emissivity of rough sea surface for 8~13μm: modeling and verification *Appl. Optics* **36** 2609-19
[4] Nalli N R, Minnett P J, Delstc P, Barnetd C D and Goldberg M D 2005 Developments in ocean infrared emissivity/reflection modelling *Remote Sensing of the Ocean, Sea Ice, and Large Water Region: Proc. SPIE* (vol 5977) ed Bostater C R and Santoleri R
[5] Hasselmann D E, Dunckel M and Ewing J A 1980 Directional wave spectra observed during JONSWAP 1973 *J. of Phy. Oceanographty* **10** 1264-80
[6] Wilf I and Manor Y 1984 Simulation of sea surface images in the infrared *Appl. Optics* **23** 3174-80
[7] Schwenger F and Repasi E 2006 Sea surface simulation in the infrared modeling and validation *Targets and Backgrounds XII: Characterization and Representation, Proc. of SPIE* (vol 6239) ed Watkins W R and Clement D
[8] Yu Y X 2003 *Random wave and its applications to engineering* (Dalian: Dalian university of technology press) pp 139-42
[9] Born M and Wolf E 1989 *Principles of Optics: Electromagnetic Theory of Propagation, Interference, and Diffraction of Light* 6th ed (New York: Pergamon) p 38

Study on time-varying velocity measurement with self-mixing laser diode based on Discrete Chirp-Fourier Transform

ZHANG Zhao-yun *[a,b], GAO Yang [a,c], ZHAO Xing-hai [a], ZHAO Xiang [a]
[a] Institute of Electronic Engineering, China Academy of Engineering Physics, Mianyang 621900, China; [b] Department of Graduate Student, China Academy of Engineering Physics, Beijing 100088, China; [c] Key Laboratory of Optoelectronic Technology and System, Chongqing University, Chongqing 400044, China

ABSTRACT

Laser's optical output power and frequency are modulated when the optical beam is back-scattered into the active cavity of the laser. By signal processing, the Doppler frequency can be acquired, and the target's velocity can be calculated. Based on these properties, an interferometry velocity sensor can be designed. When target move in time-varying velocity mode， it is difficult to extract the target's velocity. Time-varying velocity measurement by self-mixing laser diode is explored. A mathematics model was proposed for the time-varying velocity （invariable acceleration） measurement by self-mixing laser diode. Based on this model， a Discrete Chirp-Fourier Transform （DCFT） method was applied， DCFT is analogous to DFT. We show that when the signal length N is prime， the magnitudes of all the side lobes are 1 ， whereas the magnitudes of the main lobe is $\sqrt{N}$ ， And the coordinates of the main lobe shows the target's velocity and acceleration information. The simulation results prove the validity of the algorithm even in the situation of low SNR when N is prime.

Keywords: Velocity measurement, interferometry, laser diode, Self-mixing

1. INTRODUCTION

Laser Diode (LD) Self-Mixing (SM) interferometry is an emerging precision photoelectric measuring technology. When a small fraction of the light emitted by LD is backscattered or reflected by an external target and re-enter into the laser active cavity, the laser's output power and frequency are modulated. Because LD output signal's property is similar to the property of conventional two-beam interferometers, so it is called "Self-Mixing". The self-mixing interference measurement system has the merits of simple architecture, naturally self-aligned optical characteristics, compact size and low cost. So, it can replace conventional interferometers in many situations for physical measurements, such as displacement [1-3], distance [4], velocity [5], and vibration [6].

In 1970s and 1980s, LDSM had been already used to measure velocity. For example, Shigenobu Shinohara reported a laser Doppler velocimeter using self-mixing in 1989 [7], which can discern the movement direction of the target , and the measurement range is 23 mm/s - 23 m/s. The measured results for a reciprocally moving target as well as a rotating target agrees well with the theoretical value. But all LDSM velocimeter documents we retrieved are aim at low-speed and invariable velocity motion target, variable velocity measurement is not covered. In many practical applications, the movement of objects is uniform, and therefore needs to solve the problems of variable velocity measurement by self-mixing interference, such as the signal processing method.

The method of FFT is usually adopted in laser self-mixing velocity measurement technology[8]. But FFT is used to analyze stationary signal. When the target moves with acceleration, the output signal is time-varying. If the method of FFT is still used to analyze the output signal, some problems will appear as follows: 1) lose the output SNR, leading to detection performance becoming bad, 2) the inherent resolution of Doppler frequency detection is down [9]. Therefore, this paper puts forward a method based on the Discrete Chirp-Fourier Transform (DCFT) to analyze laser self-mixing variable velocity measurement.

*E-mail: zzy_caep@163.com

2. MATHEMATICS MODEL OF SELF-MIXING VARIABLE VELOCITY MEASUREMENT

Fabry-Perot (F-P) cavity is used to analyze the self-mixing interference because of its simple and intuitionistic. Fig. 1 introduces a model of a solitary semiconductor laser diode in which laser facets and the external reflector form a three-mirror compound cavity. r_1 and r_2 form the laser's inner cavity, r_3 and the external reflector form the laser's external cavity. Length of the inner cavity and external cavity are L and L_{ext} respectively. Reflection coefficient of the three mirrors are r_1, r_2 and r_3.

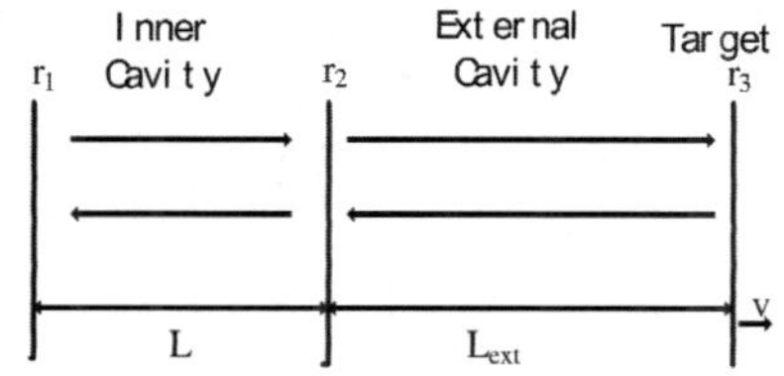

Fig. 1 Schematic diagram of a three-mirror cavity self-mixing interference

From three-mirror cavity model, the change of laser's output gain and frequency are [8]

$$\begin{cases} g_{th} = g_o - \dfrac{\xi}{L}\cos(2\pi f \tau_{ext}) \\ f = f_o - \dfrac{\xi}{2\pi\tau}\sqrt{1+\alpha^2}\sin(2\pi f \tau_{ext} + \arctan\alpha) \end{cases} \quad (1)$$

Where, g_o is initial gain, g_{th} is threshold gain, $f = f_o + \Delta f$, f_o is laser's initial frequency, Δf is the change of frequency, $\tau = 2L/c$ is optical round-trip time in inner cavity, $\tau_{ext} = \dfrac{2L_{ext}}{c}$ is optical round-trip time in external cavity, $\xi = \dfrac{r_3}{r_2}\left(1-|r_2|^2\right)$ is the coupling coefficient of the laser coupled from external cavity to inner cavity, α is the laser's linewidth enhancement factor.

From Eq. (1), the laser's threshold gain g_{th} is modulated by feedback light. Because the laser's output power is proportion to the laser's threshold gain, so the laser's output power is also modulated by feedback light. The output power of LD can be expressed as

$$p = p_o\left[1 + m\cos(2\pi f \tau_{ext})\right] \quad (2)$$

Where, m is the undulation coefficient.

From Eq. (2), the laser's output power changes periodically when the length L_{ext} of external cavity changes. When the target moves away from laser in velocity v, the angle between target moving direction and light axial is θ, the external cavity length can be expressed as $L_{ext}^{'} = L_{ext} + vt\cos\theta$ (when move converse $-v$), then $\tau_{ext} = 2\dfrac{L_{ext} + vt\cos\theta}{c}$, the Eq.(2) can be expressed as

$$p = p_o\left[1 + m\cos\left(4\pi f\frac{L_{ext} + vt\cos\theta}{c}\right)\right] \quad (3)$$

From Eq. (3), the laser output power's frequency fluctuations can be expressed as

$$\Delta f = \frac{2v}{\lambda}\cos\theta = f_D \tag{4}$$

Namely, self-mixing frequency is equal to the target's Doppler frequency. So, based on the change of laser output power, the change of frequency Δf can be attained, then the target's velocity can be acquired. This is the principle of LD SM velocity measurement.

From Eq. (3) and Eq. (4) , when the target moves in invariable velocity, the target's Doppler can be extracted by FFT, and then obtain the target's velocity through Eq. (4). But, when the object moves in variable velocity, this method is not feasible.

If the target moves in variable velocity and the acceleration is a and invariable, $v = v_o + at$, Eq. (3) can be expressed as

$$p = p_o\left[1 + m\exp i2\pi\left(\begin{matrix} 2f\frac{L_{ext}}{c} + \frac{2fv_o\cos\theta}{c}t \\ + \frac{2fa\cos\theta}{c}t^2 \end{matrix}\right)\right] \tag{5}$$

If the sampling frequency is f_s, sampling number is N, then $t = n/f_s$, Eq. (5) can be expressed as

$$p = p_o + E\exp\frac{i2\pi}{N}\left(l_o n^2 + k_o n\right) \quad ,n = 0,\cdots,N-1 \tag{6}$$

Where

$$E = p_o m\exp\left(\frac{i4\pi fL_{ext}}{c}\right) \tag{7}$$

$$l_o = \frac{2fa\cos\theta N}{cf_s^2} \tag{8}$$

$$k_o = \frac{2fv_o\cos\theta N}{cf_s} \tag{9}$$

In order to facilitate subsequent signal processing, Eq. (6) can be expressed as

$$s(n) = p - p_o = E\exp\frac{i2\pi}{N}\left(l_o n^2 + k_o n\right) \tag{10}$$

Consider the output signal noise, the output signal is

$$y(n) = s(n) + \omega(n) \tag{11}$$

Where $\omega(n)$ is white noise with Zero mean and variance σ^2. So, the mathematical model of self-mixing variable velocity measurement is established. And then, we will use the signal processing method of Chirp-Fourier transform to get the object velocity and acceleration information.

3. SIGNAL PROCESSING OF CHIRP-FOURIER TRANSFORM

For a signal $y(n)$ with length N , its N-point DCFT is defined as

$$Y(l,k) = \frac{1}{\sqrt{N}}\sum_{n}^{N-1} y(n)W_N^{l\cdot n^2 + k\cdot n} \quad ,0 \le l,k \le N-1 \tag{12}$$

Where $W_N = \exp\left(-j2\pi/N\right)$. When $l = 0$, the DCFT is transformed to DFT, this indicates that DFT belongs to DCFT. Let's consider following lemma.

Theorem 1: When N is a prime, $0 \le l, k \le N-1$, we have the following equation:

$$\left|\sum_{n=0}^{N-1} W_N^{l\cdot n^2 + k\cdot n}\right| = \begin{cases} N & l = 0, k = 0 \\ \sqrt{N} & l \ne 0 \\ 0 & l = 0, k \ne 0 \end{cases} \tag{13}$$

Theorem 2: Let $x(n) = \sum_{n}^{N-1} W_N^{-(l_o\cdot n^2 + k_o\cdot n)}$, for some integers l_o and k_o with $0 \le l_o, k_o \le N-1$. If the N is a prime, its DCFT magnitude has the following form:

$$|X(k,l)| = \begin{cases} \sqrt{N} & l = l_o, k = k_o \\ 1 & l \ne l_o \\ 0 & l = l_o, k \ne k_o \end{cases} \tag{14}$$

Theorem 2 tells us that for the signal $x(n)$, the peak or the mainlobe of its DCFT appears at (l_o, k_o), and its value is $\sqrt{N}$.

Based on the theorem 2, for the signal $y(n)$ in the equation (11), the peak of its DCFT appears at (l_o, k_o), and its value is $|E|\sqrt{N}$. So for the output signal $y(n)$, has its DCFT, and the value l_o and k_o where its mainlobe appears can be obtained, and then through the equations (8) and (9), the target's velocity and acceleration can be acquired.

4. NUMERICAL SIMULATION

When the object moves in variable speed, $v_o = 1103 m/s$ and $a = 4.27\times10^{11} m/s^2$, if use conventional FFT to analyze the output signal, the result is as figure 2.

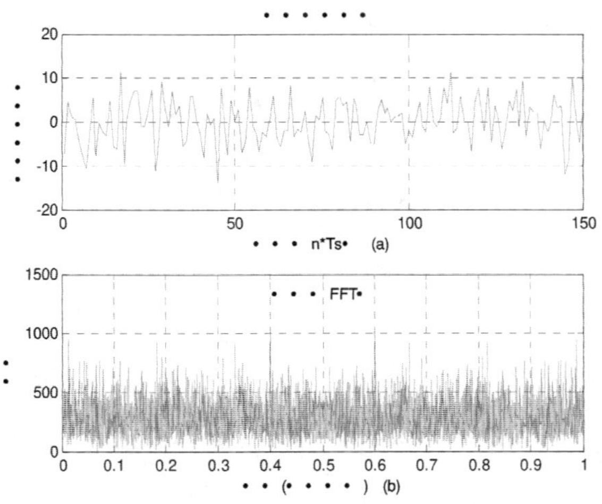

Fig. 2 Simulation result by FFT method

From the simulation result, the FFT can not be used to extract target's movement information when the target moves in variable speed. Let's use DCFT to analyze the output signal in the following.

Set the simulation parameters: $N = 127$, $v_o = 1103 m/s$, $a = 4.27 \times 10^{11} m/s^2$, $E = 1$, $f_s = 3 \times 10^9$, then $l_o = 3.9987$, $k_o = 30.9877$. When $SNR = 0$ dB, and $SNR = 7$ dB, the simulation result is $l_o^{'} = 4$, and $k_o^{'} = 31$, just as figure 3. The simulation result tells us that under low SNR conditions, through the DCFT method can get object's velocity and acceleration information.

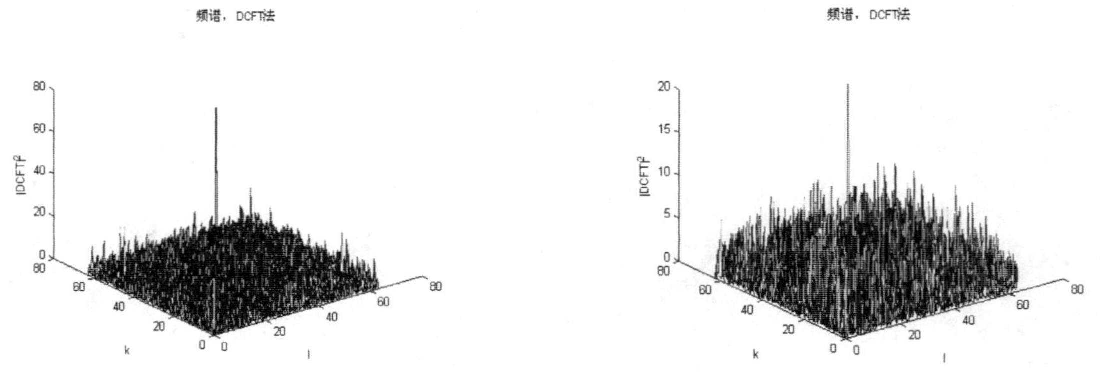

（a） $SNR = 7dB$ （b） $SNR = 0dB$

Fig. 3 Simulation result by DCFT method

This method has a big advantage when used to analyze high-speed and high acceleration movement. But this method has some limitations: 1) require the sampling points N is prime, otherwise the measurement accuracy becomes low, 2) measurement results are only integer solutions, so the measured parameters l_o and k_o should be approximate integers, otherwise the measurement error become much bigger. So, in order to make this method is more effective and adapt much broader field, it needs further improvement

5. CONCLUSION

This paper discusses the problem of laser self-mixing variable velocity measurement firstly. On the basis of self-mixing variable velocity measurement model, the method of Discrete Chirp-Fourier Transform is proposed to analyze self-mixing variable velocity measurement signal. Through the simulation analysis, this method can effectively obtain object's velocity and acceleration information, and has good noise performance. This paper is a useful exploration for self-mixing variable velocity measurement, promoting its practical process. And the corresponding experimental study will be carried out.

REFERENCES

[1] Kato, J., Kikuchi, N., "Optical feedback displacement sensor using a laser diode and its performance improvement," Measurement and Technology, 6(1)： 45-52 (1995).

[2] Servagent, N.， Bosch, T.， Lescure, M., "A laser displacement sensor using the self-mixing effect for modal analysis and defect detection," Instrumentation and Measurement, 46(4)： 847-850 (1997).

[3] Yu, Y. G.， Qiang, F., "Displacement measurement by differential type of self-mixing interference system," Optical Journal, 37(28)： 1269-1273 (1998).

[4] Servagent, N.， Bosch, T.， Flore, G., "Absolute distance measurement with an optical feedback interferometer," Applied Optics, 37(28)： 6684-6689 (1998).

[5] Servagent, N.， Bosch, T.， "PLANRIER G., "A low-cost optical feedback interferometer for real-time velocity measurement". Instrumentation and Measurement Technology Conference," 36(27)： 738-743 (2000).

[6] Roos, P. A., Stephens, M., Wieman, C. E., "Laser vibrometer based on optical feedback-induced frequency modulation of a single-mode laser diode," Applied Optics, 35(34): 6754-6761 (1996).
[7] Shinohara, S., "Compact and versatile self-mixing type semiconductor laser doppler velocimeters with direction discrimination circuit," Instrumentation and Measurement, 38(2): 574-577 (1989).
[8] Fu S Z, Han D F, Ma L. Reseach of laser Doppler velocimetry based on the Self-mixing effect [J] .Journal of Nanchang University (Natural Science, 31(3): 264-267 (2007)
[9] Zhao H Z, Fu Q. Analyze of radar signal's acceleration resolving capability [J] .Science in China (Series E), 7(23): 638-646 (2003).
[10] SUN X M, MA J S, QIANG X F. Theoretical analysis of self-mixing laser diode interference [J] .China Journal of Laser, 25(11). 1018-1022 (1998).
[11] Xia Xiang-Gen. Discrete chirp-Fourier transform and its application to chirp rate estimation [J] .IEEE Trans. SP, 48(11): 3122-3133 (2000).

The influence of radio frequency power on the characteristics of carbon-rich hydrogenated amorphous silicon carbide films

Juncao Bian, Zhe Li, Haiyan He, Xiwen Zhang[1] , Jiang Lin, Qian Gao, Zhongdong Chen and Gaorong Han

[1] State Key Laboratory of Silicon materials, Zhejiang University, Hangzhou 310027, PR China

E-mail: zhangxw@zju.edu.cn

Abstract. A series of carbon-rich hydrogenated amorphous silicon carbide (a-$Si_{1-x}C_x$:H) films were prepared at different radio frequency (RF) powers from silane-ethylene-hydrogen plasma. The effect of the RF power on the bonding configurations and microstructures has been investigated. The grown films were characterized by a collection of techniques including Scanning Electron Microscope, Fourier transform infrared(FTIR) spectroscopy, Raman scattering and photoluminescence spectroscopy. The deposition rate increases upon RF power due to the enhancement of chemical reactivity of plasma. The carbon to silicon ratio increases, for more C_2H_4 molecules decompose with the enhancement of RF power and more carbon atoms are bonded into the films. Raman G peak position shifts to a higher wavenumber, which indicates that the size and concentration of sp^2 carbon clusters increase as the RF power becomes stronger. Blue-green photoluminescence is detected at room temperature. The PL band can be attributed to the existence of the amorphous carbon clusters in films with high carbon concentrations.

1. Introduction

Hydrogenated amorphous silicon carbide (a-$Si_{1-x}C_x$:H) attracts much attention for use in solar cells, light emitting diodes, microelectronic dielectric layers, colour sensors and high temperature engineering materials etc [1-5]. The structural and optoelectronic properties of this material are known to be considerably sensitive to the nature of bonding of their constituent atoms. For instance, carbon can have two different probable bonding configurations characterized by the sp^2 and sp^3 hybridizations which exibit very different physical properties. The relative proportions of the sp^2 and sp^3 components in these films are therefore believed to control the quality of the films and are expected to be

[1] To whom any correspondence should be addressed.

influenced to a large extent by the deposition parameters such as the substrate temperature [6], radio frequency (RF) power [7, 8], gas flow rates [1, 9] and hydrogen dilution [10]. Even the nature of the source gases (CH_4 or C_2H_4) can affect the bonding properties [11]. Therefore, from the viewpoint of actual applications it is important to clarify the relationship between the deposition parameters and the film microstructures and demonstrate luminescence mechanism. In the work reported here, a-$Si_{1-x}C_x$:H thin films were fabricated by PECVD technique on Si (100) substrates at room temperature, and the influence of RF power on the characteristics of a-$Si_{1-x}C_x$:H thin films was investigated in details.

2. Experimental details

The a-$Si_{1-x}C_x$:H thin films were prepared in an RF-PECVD system equipped with a radio frequency source operated at a frequency of 13.56 MHz using SiH_4 (diluted to 10% in H_2) and C_2H_4 as gas sources. The deposition process was carried out at room temperature (RT) and the RF power was changed from 20 to 140 W increase by 30 W, named as P1~P6. The H_2-diluted 10% SiH_4 and C_2H_4 gas were fixed at 20 and 10 sccm respectively, and the chamber pressure was maintained at 50 Pa. P-type Si (100) wafers were used as substrates.

The chemical bonding within the films was determined by using Fourier transform infrared spectrometry (FTIR). RT micro Raman spectroscopy analyses were performed on a Labor Raman HR800 spectrometer at a laser power of 30 mW. The surface morphology and film thicknesses were studied by Hitachi S4800 field-emission scanning electron microscopy (FE-SEM). The RT PL spectra were obtained with a FLS920 (Edinburgh Instruments Ltd.) fluorescence spectrometer using a 325 nm UV xenon lamp as the excitation source.

3. Results and discussion

The cross-sectional FE-SEM image depicted in figure 1 clearly indicates the smooth interface between the a-$Si_{1-x}C_x$:H layer and the Si substrate, which is attributed to the enough density of reactants and a good dissociation of the gases [12]. Figure 2 shows the deposition rate as a function of RF power, which is calculated by dividing the film thickness by the deposition time. It is observed that the deposition rate increases as the RF power becomes higher and exhibits a quasi-linear relationship. The results are in good agreement with previously reported data obtained by Ciprian I et al [12]. This phenomenon can be explained as follows. The higher power enhances the plasma in the chamber, which subsequently yields higher energy electrons. The increased energy of the electrons enhances the dissociation of the main gases and, as a result, increases the deposition rate [13]. It is also found from figure 2 that when the RF power is more than 100W, the increase becomes slow. It most probably stems from the erosion effect of the plasmas and the limit of mass transport at high RF power.

Figure 3 shows the FTIR absorption spectra of all the samples. In order to compare the content of each bond among the films, FTIR spectra were normalized by the film thicknesses. They are characterized by 3 regions: a) 600-1200 cm^{-1}, b) 2000-2300 cm^{-1} and c) 2800-3000 cm^{-1}.

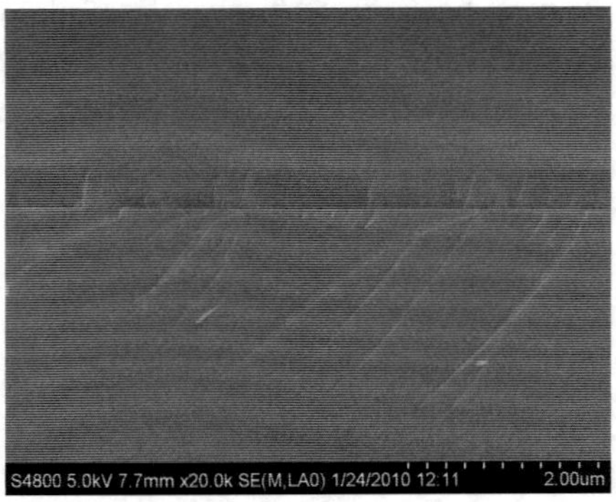

Figure 1. The cross-sectional SEM image of the film P2.

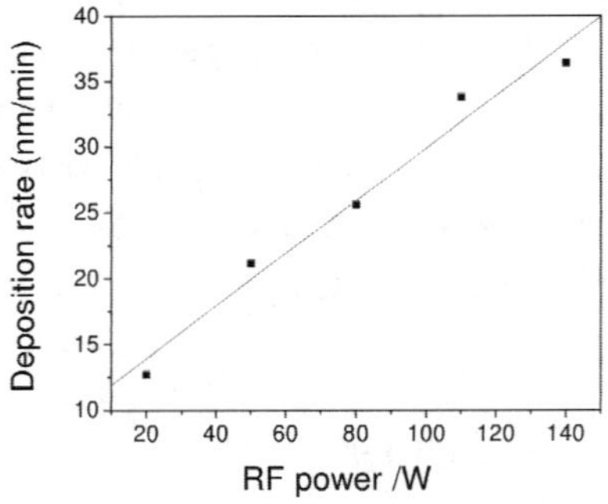

Figure 2. The variation of deposition rates as a function of RF power.

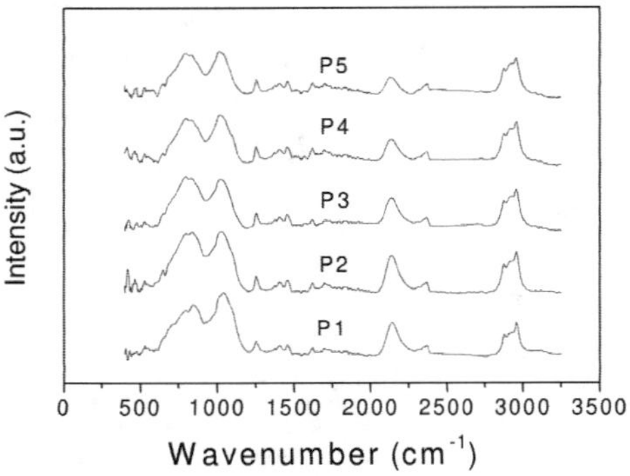

Figure 3. FTIR absorption spectrum of a-$Si_{1-x}C_x$:H films deposited at different RF powers.

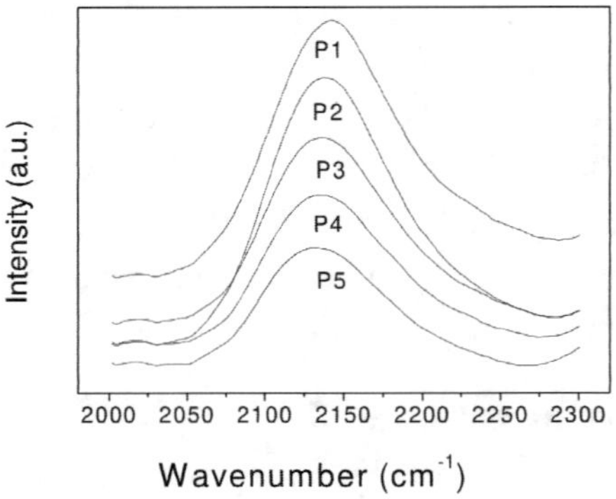

Figure 4. 2000-2300 cm^{-1} FTIR spectrum of a-$Si_{1-x}C_x$:H films deposited at different RF powers.

The mode at 790 cm^{-1} is due to Si-CH_3 rocking/wagging or Si-C stretching modes [2, 5, 8-11]. The peak associated with a Si-C stretching mode should be at 670 cm^{-1} which is covered by the strong 790 cm^{-1} peak and thus indicates the formation of a-$Si_{1-x}C_x$:H film. A strong feature at 1040 cm^{-1} is assigned both to the C-H wagging mode around 1000 cm^{-1} [3-6, 8] and the Si-O stretching vibration at 1100 cm^{-1} [4, 5]. The Si-O stretching mode is due to oxygen contamination during deposition process and sample preservation.

In the region between 2000-2300 cm^{-1} displayed in figure 4, the band appearing at 2080 cm^{-1} in the spectra corresponds to the shifted stretching mode of Si-H_n vibrations [14]. This vibrational mode has a frequency of 2000 cm^{-1} in pure a-Si:H while it shifts to higher frequencies in a-$Si_{1-x}C_x$:H films. This chemical shift is due to the bonding of C atoms with Si atoms bonded with H atoms.

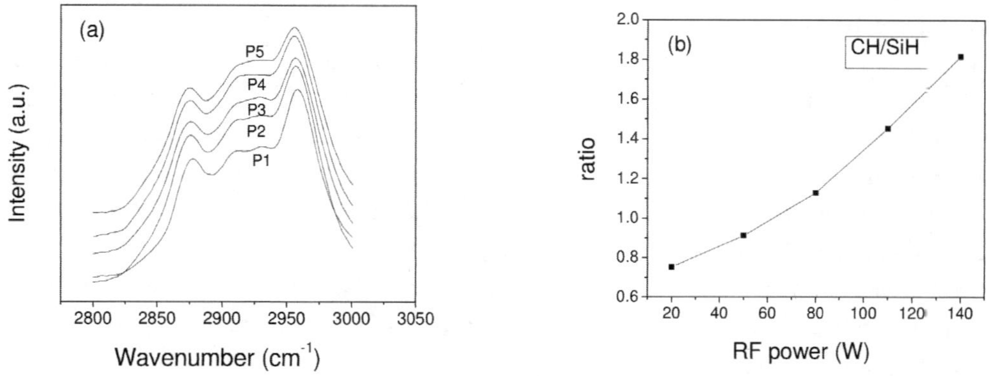

Figure 5. (a) 2800-3000 cm^{-1} FTIR absorption spectrum of a-$Si_{1-x}C_x$:H films deposited at different RF powers, **(b)** The variation of C/Si ratio as a function of RF power.

The third absorption band around 2800-3000 cm^{-1}, as shown in figure 5(a), is due to sp^3 C-H_n stretching modes [8, 9]. The sp^2 C-H_n absorption band is around 2975-3085 cm^{-1}, which is so weak that it is covered by the broad sp^3 C-H_n absorption band. The peak at 2956 cm^{-1} is attributed to sp^3 C-H_3 and sp^2 C-H_n. And the peak at 2875 cm^{-1} is attributed to sp^3 C-H_2 and C-H_3. The relative intensity of C-H_n stretching modes (2000-2300 cm^{-1}) to that of Si-H_n stretching mode, displayed in figure 5(b), decreases with the increase of RF power. It indicates that the C/Si ratio of the films increases upon RF power. As the RF power increases, the energy of the electrons increases, which enhances the dissociation of the C_2H_4, and hence more C atoms are cooperated into the films.

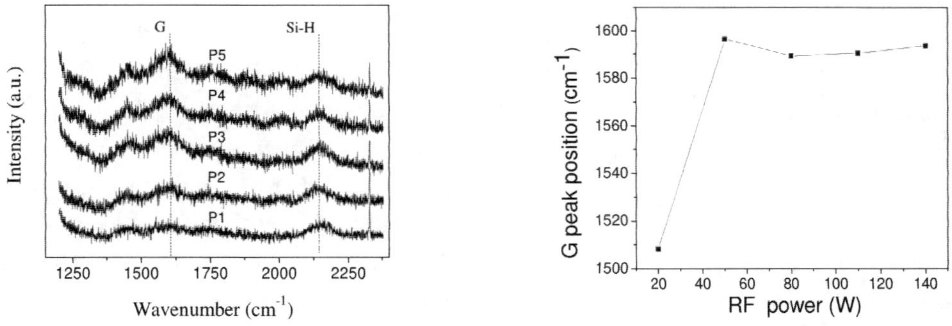

Figure 6. (a) Raman spectra of a-$Si_{1-x}C_x$:H films deposited at different RF powers, **(b)** The variation of G peak position as a function of RF power.

Figure 6a shows Raman spectra of various a-$Si_{1-x}C_x$:H films deposited at different RF powers. No amorphous Si-C band at around 780 cm^{-1} are observed, which is due to their extremely low scattering cross-section and low absorption coefficient in the amorphous state [15]. Si-H_n Raman modes at 2000-2300 cm^{-1} [9] are also observed, which exhibit a redshift with increasing RF power. The carbon-rich a-$Si_{1-x}C_x$:H basically has a structure similar to pure a-C:H, so the Raman spectrum can be analyzed relative to amorphous carbon [8, 12]. The broad band near 1400-1500 cm^{-1} is attributed to the C-C intrinsic vibrations [15]. The vibrations of graphitic carbon have the characteristics of a high scattering cross-section and high absorption coefficient for the excitation frequency. Hence, the presence of even a small fraction of graphitic carbon can be detected. The G peak positions are obtained by fitting the region of 1500-1700 cm^{-1} into Gaussian curves since the poor signal to noise ratio of Raman results. And the G peak positions are plotted in figure 6b, from which it is observed that the position of the G peak shifts to the higher wavenumber as the RF power increases. This shift implies that a fraction of sp^2 bonded carbon atoms are increased upon RF power. Based on the trend in the Raman spectra, it appears that high RF power promotes an increase in the size and number of sp^2 clusters in a-$Si_{1-x}C_x$:H.

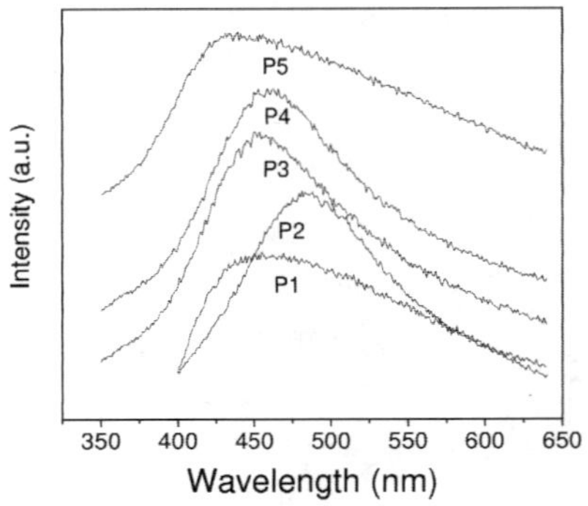

Figure 7. PL spectrum of a-$Si_{1-x}C_x$:H films deposited at different RF powers.

Figure 7 presents the room temperature PL spectra of samples deposited at various RF power, from which no clear discipline is found. The PL mechanism of carbon-rich a-$Si_{1-x}C_x$:H films can be interpreted by Robertson's model [16]. Photoexcited carriers are believed to be confined inside sp^2 clusters. The origin of the PL is the geminate recombination of electron-hole pairs within sp^2 bonded carbon clusters in a sp^3 bonded amorphous matrix. The π states of the sp^2 sites form the valence-band and conduction-band edges and lie closest to the band gap. Hence optical excitation creates electron hole pairs in the π and π^* states. The narrow π-π^* gap of sp^2 sites surrounded by the wider gap of sp^3 sites creates large band-edge fluctuations which tend to strongly localize tail states. PL occurs by the radiative recombination of photo-excited electrons and holes in localized tail states within sp^2 clusters after loosing their excess energy through the thermalization processes. The sp^2 clusters contain different structural units with double conjugated C=C bonds, which act as recombination centers.

4. Conclusion

The carbon-rich a-$Si_{1-x}C_x$:H films were deposited by PECVD technique with SiH_4 and C_2H_4 as as reactant gases. The influence of RF power on the configurations and microstructures of carbon-rich a-$Si_{1-x}C_x$:H films was explored. It is found that the RF power is an important parameter which affects the carbon content, the structure and the bonding configurations of the films. The carbon to silicon ratio increases due to more C_2H_4 molecules decomposed with the enhancement of RF power and more carbon atoms are bonded into the films. Raman G peak position shifts to a higher wavenumber, which indicates that the size and concentration of sp^2 carbon clusters increases as the RF power became stronger. Blue-green photoluminescence is detected at room temperature on the as-deposited samples with UV excitation. The PL band can be attributed to the existence of the amorphous carbon clusters in films with high carbon concentrations.

Acknowledgements
This work was supported by the National Basic Research Program of China (973 Program) "2007CB613403" and Foundation of the scientific research base development (Engineering Research Center of the Education Ministry for the Surface and Structure Modification of Inorganic functional Materials) "KYJD09014".

References
[1] Ihsanul A Y, Kenji H, Shinsuke M, Akira Y and Makoto K 2009 *Sol. Energ. Mat. Sol. C.* **93** 1056
[2] Ray S C, Okpalugo T I T, Pao C W, Tsai H M, Chiou J W, Jan J C, Pong W F, Papakonstantinou P, McLaughlin J A and Wang W J 2005 *Mater. Res. Bul.* **40** 1757
[3] Swain B P and Dusane R O 2007 *Mater. Lett.* **61** 4731
[4] Wang Y H, Lin J and Huan C H A 2002 *Mater. Sci. Eng.* B-*Solid State Mater. Adv. Technol.* **95** 43
[5] Xu J, Yang L, Rui Y J, Mei J X, Zhang X, Li W, Ma Z Y, Xu L, Huang X F and Chen K J 2005 *Solid State Commun.* **133** 565
[6] Bibhu P S and Rajiv O D 2007 *Mater. Lett.* **61** 4731
[7] Iliescu C, Chen B T, Poenar D P and Lee Y Y 2008 *Sensor. Actuat.* B **129** 404
[8] Bhusari D M and Kshirsagar S T 1993 *J. Appl. Phys.* **73** 1743
[9] Wang L, Xu J, Ma T F, Li W, Huang X F and Chen K J 1999 *J. Alloy. Compd.* **290** 273
[10] Yoon S F and Ahn J 1997 *J. Vac. Sci. Technol.* A **15** 1832
[11] Tawada Y, Tsuge K, Kondo M, Okamoto H and Hamakawa Y 1982 *J. Appl. Phys.* **53** 5273
[12] Ciprian I, Bangtao C, Daniel P P and Yong Y L 2008 *Sensor. Actuat.* B **129** 404
[13] Ciprian I, Francis E H T and Jiashen W 2006 *J. Micromech. Microeng.* **16** 869
[14] Wieder H, Cardona M and Guamieri C R 1979 *Phys. Status Solidi* B. **99** 92
[15] Inoue, Nakashima S, Mitsnishi A, Tabdta S and Tsuboi S 1981 *Solid State Commun.* **48** 1071
[16] Rusli and Robertson J 1996 *J. Appl. Phys.* **80** 2998

Optimizing interdigitated contact scheme of GaN-based light-emitting diodes

Bin Cao[1], Pei Wang[2], Zhiyin Gan[1,2], Wei Wei [1] and Sheng Liu[1,2*]

1)Wuhan National Lab for Optoelectronics, 1037 Luoyu Road, Wuhan 430074, China
2)School of Mechanical Science and Engineering, Huazhong University of Science&Technology, 1037 Luoyu Road, Wuhan 430074, China

E-mail: victor_liu63@126.com

Abstract. The unapt design of electron layout involved in the fabrication of light emitting diode(LED) chip, especially for conventional mesa-etch structure, will cause a series of problems including electrical, optical, and thermal issues. The effects of the figure number of p- and n- electrodes as well as their spacing and width on the performance of the mesh-etched GaN-based LED chip with interdigitated contact scheme electrode had been studied and optimized by numerical simulation. The results showed that operation voltage decreased with increasing the numbers or/and the width of electrode, meanwhile the current became more uniform. However, the wall-plug efficiency droop due to the decrease of active layer and the shading for emitted light. A uniform current distribution was attained and local overheating alleviated when the distance between p- and n- electrode was about 250um using a small LED chip with 2 and 1 of n- and p-electrode, respectively. At the same time the output power reached its maximum.

1. Introduction

Light emitting diodes(LEDs) have attracted much attention as the next generation solid-state lighting source for its lower power consumption, high reliability, long lifetime, and so on[1]. GaN-based LEDs have been studied comprehensively and much progress has been made. However, there are still many problems that need to be solved in order to satisfy various application requirements, one of which is the current spreading [2].

The cathode and anode of the conventional chip are on the same side due to the insulating sapphire substrate, thus the current flows laterally which leads to non-uniform current spreading and subsequently asymmetric distributions of electron-hole pairs, photons and heat. These distributions result in low efficiency and poor reliability of GaN-based LED. Its geometric structure also has a significant influence on the light extraction efficiency. Kim *et al.*[3] analyzed theoretically the lateral current transport path length and the relationship with material parameters when uniform current spreading was achieved. Indium tin oxide (ITO) current spreading layer was applied to improve the electrical and optical performance of GaN-based LEDs [4, 5]. Guo *et al.* [6] proved that the optical emission of the chip with interdigitated patterning was better than that with traditional square-shaped pattern. Sheu *et al.*[7,8] did lots of work on the effect of electrode pattern on current spreading by simulation.

However, there are rarely reports about optimizing the figure number of p- and n-electrode as well as their spacing in a GaN-based LED with interdigitated scheme by comprehensive consideration of electrical and thermal performance of the chip. In this paper, this issue will be analyzed through a three-dimensional numerical model of mesa-etch GaN-based LED chip in order to achieve an optimized electrode scheme.

2 Model of GaN-based LED

The emission spectra and the current density-voltage characteristics of the epitaxial structure were solved via the self-consistent solution of the Schrodinger-Poisson equation coupled with the continuity equations for electrons and holes based on a one-dimension drift-diffusion model. With the help of the parameter relationship obtained from the one-dimension model, the distribution of the current density and temperature was obtained in a three-dimension model with boundary conditions and electrostatic field equation, the heat conduction equation was considered. More details were described in[9].

A conventional GaN-based LED chip based on a simple model was modeled in this paper. The area of 1000um*1000um LED chip model consists of a 200um-thick sapphire substrate on the bottom, a 3um-thick n-type GaN layer which doped Si has a concentration of $4*10^{18}cm^{-3}$, a thin InGaN/GaN active layer and a 0.3um thickness p-type GaN doped with Mg whose concentration is $2*10^{19}cm^{-3}$. The active layer is a multiple-quantum-well region including 5nm thick $Al_{0.1}Ga_{0.9}N$ wells and 15nm thick GaN barriers, which has 10 periods together.

Interdigitated electrode included many couples of n- and p-electrode and a structure with two stripes of p-electrode and three stripes of n-electrode (here called 2p3n for short in the following) was shown in Figure 1, in which n-electrode formed in the mesa etched into the LED n-type layer. Different figure numbers of electrode were simulated and then the effect of the distance D between n- and p-electrode based on a simplified structure of 2n1p electrode was analyzed. The p-electrode was set at the center of the chip and n-electrodes were set at both sides of the p-electrode symmetrically. The specific contact resistance of the n- and p-contact electrode were set to be of $10^{-5}\,\Omega\,cm^2$ and $10^{-4}\,\Omega\,cm^2$, respectively. The heat conductivities of the electrode, ITO transparent conductive layer, and GaN were set to be 280, 11, and 130W/mK, respectively.

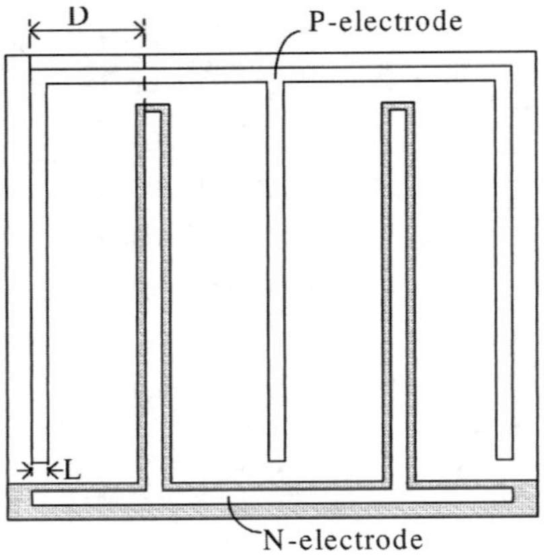

Figure 1. Top view of a GaN-based LED model with interdigitated contact scheme with 3 stripes n-electrode and 2 stripes p-electrode (2n3p)

3 Results and discussion

Figure 2 showed voltage and current density at active layer under different figure number. At the same injection current density (set as 35A/cm^2), operation voltage greatly drooped with the increasing number of electrode, from 3.97V to 3.39V, as can be seen from Figure 2. This result indicated that increasing electrode number was beneficial to improving the current-voltage (*I-V*) characteristics and decrease the series resistance, which affected LED's electrical performance significance[10]. At the same time, the maximum current density of active layer decreased with increasing electrode number, while the minimum current density at active layer increased, which illustrated that current injection of active layer was more uniformly distributed and therefore, photon generated was more uniform, so that the internal quantum efficiency (IQE) was enhanced. Current crowding was released due to multiple-electrode interdigitated structure and localized heat concentration was alleviated, which resulted in the reliability enhanced and lifetime becoming longer.

The effect of different electron width on the performance of GaN-based LED was investigated based on a 2n1p LED model, with the size of the chip changed to be 350um*350um for simplifying computation here. The characteristic of the voltage versus current was shown in Figure 3. The operation voltage was decreasing with increasing the electron width at different current density, which suggested that lager contact area could improve electricity property and then improve the performance of the chip. Moreover, the improvement was more significant under big current density.

However, it is unpractical to reduce voltage and improve current uniformity by adding the number of electrode or/and increasing the electron width, because there is smaller area of active layer for a conventional chip with larger area of n-electrode. Moreover, light emitting to the air would be hold back by n- and p-electrode at the interface, which resulted in a poor light extraction efficiency. On the other hand, it was seen from Figure 2 that the minimum current density tended to be stable and the drop of the voltage became slower, suggesting that the effect of increasing electrode on improving LED's performance becoming weaker.

Figure 4 showed the simulation result of the structures of 2p1n electrode with different spacing. The number of electrode was fewer than that of actual applied LED chip, therefore, it had poor electrical properties, which was proven by the results above. From the results it could be seen that the distance of p- and n-electrode had greatly influence on distribution of the current density and temperature at the active layer. The maximum value of current density decreased from 1153 A/cm^2 to 909.4 A/cm^2 when the spacing increased from 30um to 200um and then increased to 955.3 A/cm^2 with the spacing increased continuously to 450um, while the minimum value increased and then decreased with the spacing increased. It had a peak value at 250um. These data showed that the distribution of current at active layer was more uniform at 250um. The distance between the p- and n- electrode being far smaller or longer than 250um would result in the difference of current density at different region increased. The current crowding produced remarkable temperature non-uniform in the active layer. The maximum temperature at p-n junction was shown in Figure 5, which revealed the LED chip had a lower temperature and smaller thermal when the distance of p- and n-electron was about 250um. The unapt distance between p- and n- electrode of LED chip with Interdigitated pattern would cause severe local heating, reducing the reliability of the GaN-based LED chip.

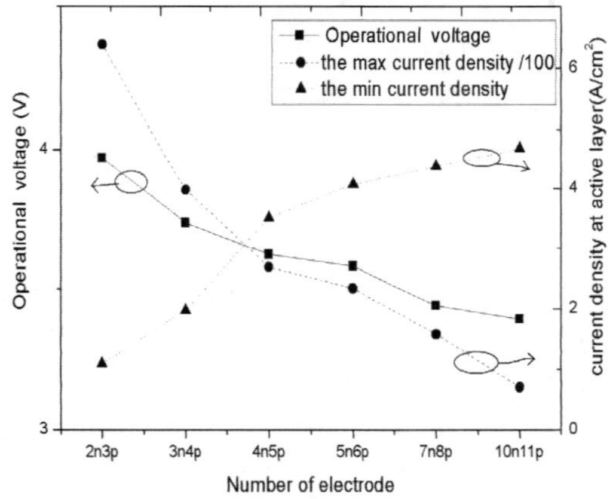

Figure 2. Operation voltage and current density with different number of cathode and anode.

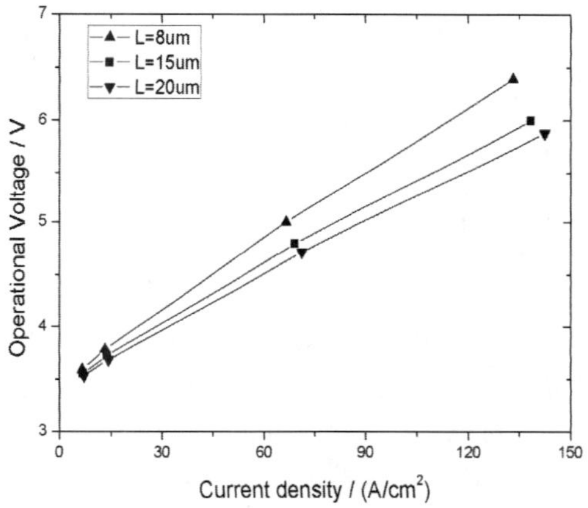

Figure 3. Operation voltage versus current density for LED die with different electrode width (L).

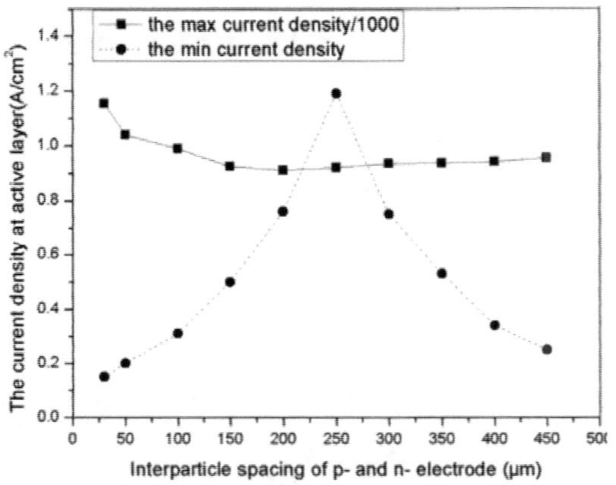

Figure 4. Current density at active layer versus different spacing of cathode and anode.

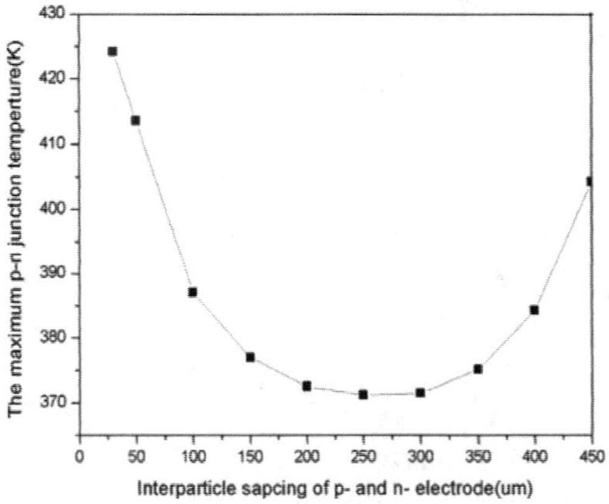

Figure 5. The maximum p-n junction temperature versus different spacing of cathode and anode.

Figure 6 showed the relationship between the wall-plug efficiency (WPE) and output power and the distance of p- and n-electrode. The WPE was almost the same from 50um to 400um, however, the output power reached the highest value at 250um, up to about 99.8mW. Although the output power at the spacing from 150um to 350um wasvery close, the spacing of n- and p- electrode from 200um to 300um was appropriate by comprehensive consideration of electrical and thermal performance.

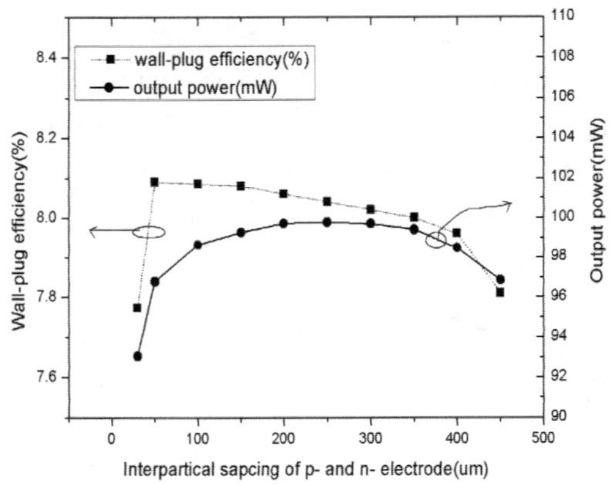

Figure 6. Wall-plug efficiency and output power of GaN-based LED with different spacing of cathode and anode

4 Conclusion

Current spreading at active layer and corresponding self-heating as well as output power were analyzed by simulation based on a conventional GaN-based LED with interdigitated contact scheme electron. Although the current spreads more uniformly with larger contact area through increasing the number of electrode or the width of electron, the LED chip with too many electrode is unpractical. Under the condition of GaN-based LED with 2n1p electrode, the optimizing spacing between p- and n-electrode was about 250um, where the more uniform current was attained and self-heating relaxed. The electrode scheme should be consider edincluding current spreading, heat, IQE and EQE.

Acknowledgment

This work was supported by a key project of the National Natural Science Foundation of China under Project 50835005.

Reference

[1] Schubert E Fred 2003. *Light-Emitting Diodes*(New York, Cambridge)

[2] Eliashevich I, Li Y and Osinsky A, 1999 *Proc. SPIE* 3621, 28

[3] Kim H, Park S J and Hwang H, 2001 *IEEE TRANSACTIONS ON ELECTRON DEVICES*, V**48**, 1065

[4] Liu Z Q, Guo D, Yi X Y, Chen Y, Wang L B, Wang L C, Wang G H and Li J M. 2007 *Proc. SPIE* **V6841** 68410N

[5] Margalith T, Buchinsky O, Cohen D A, Abare A C, Hansen M, DenBaars S P and Coldren L A. 1999 *App.l Phys. Lett.* V**74** 3930

[6] Guo X, Li Y L and Schubert E F. 2001 *Appl. Phys. Lett.*V**79** 1936

[7] Sheu G J, Hwu F S, Chen J C, Sheu J K and Lai W C. 2008 *J. Electrochem. Soc*, **155** H836
[8] Hwu F S, Chen J C, Tu S H, Sheu G J, Chen H and Sheu J K. 2010 *J. Electrochem.Soc*, **15** H31
[9] Bulashevich K A, Mymrin V F, Yu S, Karpov I A and Zhmakin A I. 2006 *J. Comput. Phys* **213** 214
[10] Chakraborty A, Shen L, Masui H, DenBaars S P and Mishra U K. 2006 *Appl. Phys. Lett.* V**88** 181120

Suppression of the formation of interstitial Li through (F, Li) codoping ZnO

Lanli Chen, Zhihua Xiong*, Qixin Wan, Dongmei Li

Key Laboratory for Optoelectronics and Communication of Jiangxi Province, Jiangxi Science & Technology Normal University, Nanchang 330013, China

E-mail: xiong_zhihua@126.com

Abstract. Based on the first-principles calculations with density functional theory, the formation energy and electronic structure of (F, Li) codoping ZnO has been investigated, where the configurations including related defects for the ratio of 1:1 of (F, Li) codoping ZnO, and the ratio of 1:2 of (F, Li) codoping ZnO are studied. The calculated results show that the formation energy of F_O-Li_{Zn} is lower than that of F_O-Li_i when the Fermi level approaches the valence band under the O-rich condition, which indicates that the (F, Li) codoping can suppress the Li interstitial. Furthermore, it is found the formation energy of F_O-$2Li_{Zn}$ reduces more dramatically than that of F_O-Li_{Zn}. Moreover, the ionization energy of F_O-$2Li_{Zn}$ is also lower than that of F_O-Li_{Zn}. It is found, for F_O-$2Li_{Zn}$ complex, an occupied state at the top of valence band is formed, meaning that the presence of F, not only enhances the acceptor concentration, but also leads to shallower acceptor level. Therefore, we draw a conclusion that the presence of F atom could suppress the formation of interstitial Li, and F_O-$2Li_{Zn}$ may form good p-type tendency in ZnO. Our results are in good agreement with the experiment results.

1. Introduction

Because of wider band gap (3.4 eV) and larger binding energy (60 meV) [1], ZnO is suitable for short wavelength device applications, such as blue and ultraviolet lasers, light-emitting diodes, and solar cells. To realize ZnO-based optoelectronic devices, n-type and p-type are essential. However, it is difficult to dope p-type in ZnO, because of a low solubility of the dopants and high defect ionization energy, and intrinsic defects, such as Zn interstitial, O vacancy, which may act as acceptor compensation centers, make it even more difficult to achieve p-type ZnO [2]. The doping asymmetry problems have hindered potential applications of ZnO materials. So, many groups have employed monodoping ways to realize p-type conductivity ZnO, such as N [3], Li [4], and Ag [5]. C.H. Park et al. [6] pointed out: although N is considered as a good p-type dopant, because of the smallest atomic size and lowest p-orbital energy, it is difficult to obtain better quality p-type ZnO due to the deep acceptor level of 0.4 eV relative to the valence band maximum (VBM). For Li doping ZnO, despite shallow Li acceptor levels, the hole density of these compounds was fairly low. The reason is that interstitial Li acting as donors compensate Li acceptors. So the compensation of Li acceptors by interstitial Li is a serious problem to realize p-type ZnO by Li-doping [7]. Later, Tetsuya Yamamoto et al. [8] have reported p-type doping using Li, N or As species causes an increase in Madelung energy,

* Corresponding author. E-mail: xiong_zhihua@126.com, Tel: (86)791-3831968.

n-type doping using B, Al, Ga, In or F species gives rise to a decrease in Madelung energy. Simultaneously, Tetsuya Yamamoto et al. have found that codoping using reactive codopants, Al, Ga, or In, enhances the incorporation of N acceptors in p-type codoped ZnO, and found the delocalized states of N for p-type ZnO codoped with N and Al [9]. So many groups have managed to gain p-type ZnO by codoped ZnO, such as N-Li [10], N-Ag [11], N-Mg [12], N-Zr [13], N-Al [14], and N-P [15]. Lately, Young et al. [16] have reported that the effects of co-doping with fluorine on properties of ZnO thin films grown by pulsed-laser deposition. They have pointed that an ambiguous carrier type was observed in Li-F codoped ZnO films grown at a temperature of 500℃ with the oxygen pressures of 20 and 200 mTorr. K.Kobayashi et al. [17] have proposed that Li, F-codoped ZnO films have been prepared by simultaneous evaporation of LiF at 780℃ during ZnO growth. They found that the codoped ZnO film has the resistivity of 79 cm and the electron density of $6.8\times10^{16}cm^{-3}$. However, there are few reports about the mechanism of (F, Li) codoping ZnO from theoretical calculations based on the first principles.

In this paper, we proposed codoping method for fabricating low resistivity p-type ZnO with F and Li, based on first-principles calculations, where the formation energy, electronic structure have been investigated. The formation energy of F_O-Li_{Zn} is lower than that of F_O-Li_i under O-rich condition, which indicates that F_O-Li_{Zn} can suppress the formation of F_O-Li_i. It is also found that F_O-Li_{Zn} form the passivated complex, which demonstrates as an intrinsic state, and could not realize p-type ZnO. So we proposed a new method of one F atom and two Li atoms codoping in ZnO, which can promote good p-type ZnO. The calculated results support and help recent experiment results [16, 17].

2. Calculation models and methods

2.1. Calculation models

In order to investigate the possibility of achieving p-type ZnO using F and Li codoping, we have constructed the configuration with a 2x2x2 supercell of ZnO. The F atom prefers to occupy the O site. The Li atom can choose two sites: one is occupying the Zn site, and the other is occupying the interstitial. The interstitial coordinate is (0.5, 0.5, 0.625). So, the models can be listed: model a: pure-ZnO (32 atoms), model b: one F atom substitutes O site and Li substitutes Zn site (F_O-Li_{Zn}), model c: F substitutes O site and Li is located in interstitial (F_O-Li_i), model d: F substitutes O site and two Li substitutes Zn sites (F_O-$2Li_{Zn}$). All models can be seen in Fig.1c-1d.

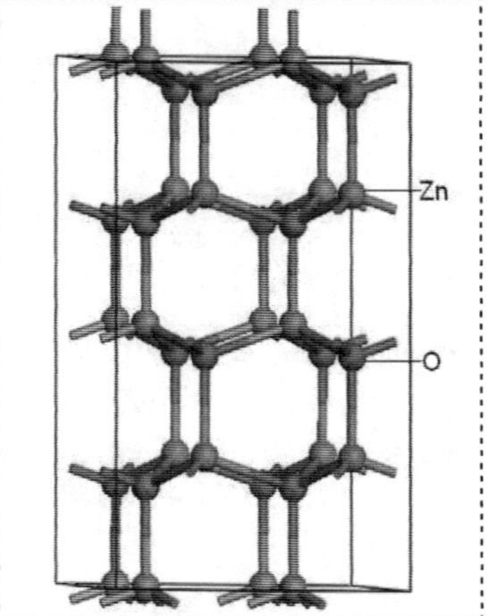

Figure 1a. pure-ZnO

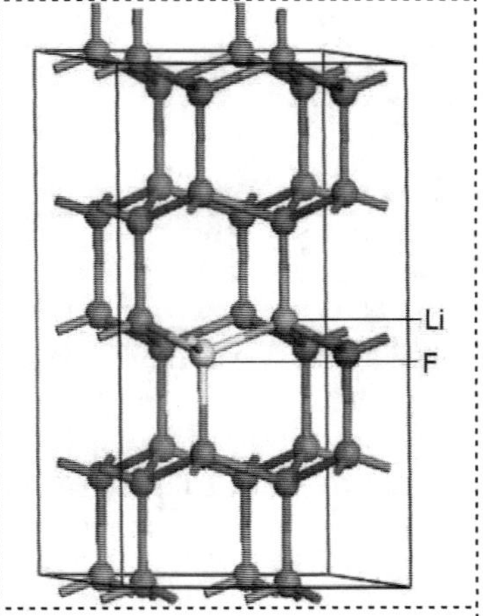

Figure 1b. F_O-Li_{Zn}

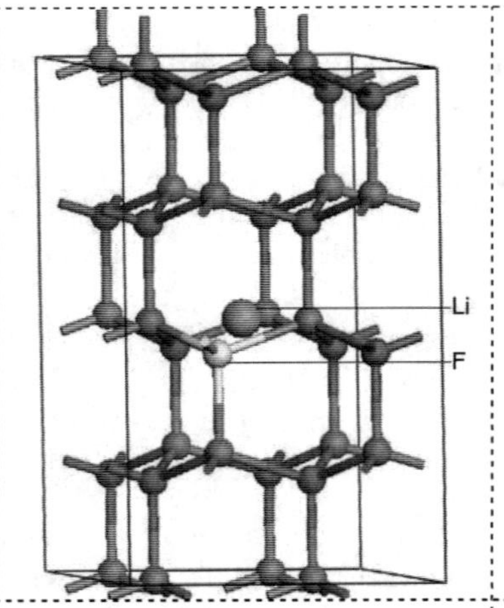

Figure 1c. F_O-Li_i

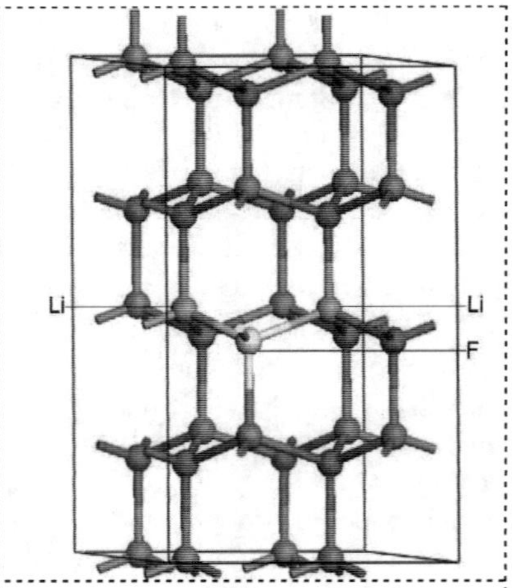

Figure 1d. F_O-$2Li_{Zn}$

2.2. Calculation methods

All calculations were carried out using the density functional theory with the generalized gradient approximation (GGA) and the projector augmented wave (PAW) pseudo-potentials, as implemented in the VASP codes [18][19]. A gamma centred 5×5×3 k-point mesh for the Brillouin zone was employed. The energy cutoff is 450 eV. We determined the crystal structure of the models by minimizing the total energy. The total energy calculations show that the formation of the F and Li atoms occupied the nearest neighbour sites is favourable. All atoms have been fully relaxed until the force on each atom was below 0.01 eV/Å. By optimizing the geometry of the pure-ZnO, we got the lattice parameters as follows: a= 3.281 Å, b= 3.281 Å, c= 5.277 Å, c/a= 1.608. They are in good agreement with the experimental results [20], which imply that our calculation methods are reasonable, and our calculated results should be authentic.

The formation energy ($E^f(D,q)$), with a defect D at charge state q, can be calculated as [21]

$$E^f(D,q) = [E_{tot}(D,q) - E_{tot}(ZnO)] + \sum_i n_i \mu_i + q(E_F + E_{VBM}) \tag{1}$$

$$\mu_i = \Delta\mu_i + \mu_i(bulk) \tag{2}$$

where $E^f(D, q)$ is the defect formation energy for the studied complex. $E_{tot}(D, q)$ is the total energy of a supercell containing the defect D at the charge state q. $E_{tot}(ZnO)$ is the energy of the pure ZnO in the same supercell. E_F is the Fermi energy level with respect to the valence band maximum (VBM) of the host, which varies up to the experimental value of 3.4 eV. E_{VBM} is the VBM of the pure ZnO. μ_i is the chemical potential of the elements. μ_i is the chemical potential of atom i referenced to the element. μ_i (bulk) is the energy of the bulk material. From the equation: $2\mu_O + \mu_{Zn} = \Delta H_f(ZnO)$, we can get: The formation energy ($H_f(ZnO)$) of ZnO is calculated to be -3.449 eV, as compared to the other calculated value of -3.525 eV [22]. To calculate the formation energies, we use Li_2O, LiF as the Li and F sources, respectively. The maximum chemical potentials of Li and F satisfy the following two relations:

$$2\mu_{Li} + \mu_O = \Delta\mu_{Li_2O} \tag{3}$$

$$\mu_{Li} + \mu_F = \Delta\mu_{LiF} \tag{4}$$

3. Results and discussion

3.1. Formation energy of F-Li codoping ZnO

Fig.2 is the formation energies of various dopants systems. Under Zn-rich condition (as shown in Fig.2a), the formation energy of F_O-$2Li_{Zn}$ is higher than that of F_O-Li_{Zn} when the Fermi level is close to the VBM, indicating that it is not available for p-type for F-2Li codoping ZnO under this condition. Simultaneously, when $E_F<0.6$ eV, the formation energy of F_O-Li_i is lower than that of F_O-Li_{Zn}. However, when $E_F>0.6$ eV, the formation energy of F_O-Li_i is higher than that of F_O-Li_{Zn}. There is a crossing point between the formation energies of F_O-Li_{Zn} and F_O-Li_i, thus it is unlikely to play important role in p-type conductivity. So, Zn-rich condition is not suitable for the p-type doping. These results are also consistent with the experiment results [17].

On the contrary, under O-rich condition (as shown in Fig.2b), the formation energy of F_O-$2Li_{Zn}$ is lower than that of others when the Fermi level is close to the VBM. That is to say, F_O-$2Li_{Zn}$ is easy to form, and it could be effective to incorporate the donors and acceptors into the system, Furthermore, F_O-$2Li_{Zn}$ bring about the lowering the acceptor level, which indicates that the acceptor could be activated easily. Simultaneously, it also indicates that O-rich condition is better for p-type doping than Zn-rich condition. Therefore, we can conclude that codoping with F atom and two Li atoms is more favorable to realize good p-type ZnO under O-rich condition, which suppresses the formation of interstitial Li.

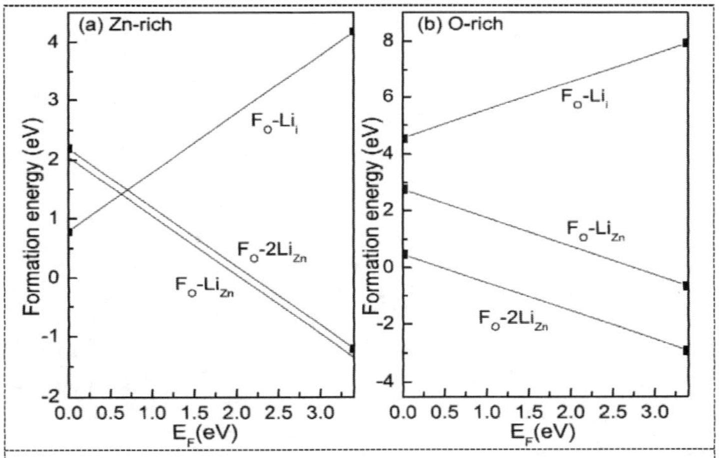

Figure 2. Formation energies of various dopants systems.

3.2. Electronic structure

Firstly, we performed a calculation on the total density of states (TDOS), partial density of states (PDOS), and band structure of F_O-$2Li_{Zn}$ complex, as shown in Fig.3. All energies are relative to the respective Fermi energies. It is found that the conduction bands are mainly introduced by Zn-s, F-p and O-p states. The valence bands are divided into two parts: the lower valence bands are mainly determined by the F-s and O-s states, and the upper valence bands are mainly introduced by the F-p, O-p and Zn-d states. The hybridization between Zn-d, Li-p and F-p, O-p states, which are near the top of the valence band, leads to the shallower acceptor level. We can see that the Fermi level is located on the valence band (as shown in Fig.3a). It can be observed that some additional peaks appear near the VBM. F atom and Li atoms introduced in ZnO can contribute to the additional peak near the Fermi level. So, F_O-$2Li_{Zn}$ complex could generate a fully occupied impurity band above the VBM, which can increase the hole concentration, implying the characteristic of p-type ZnO. In Fig.3b, a new occupied impurity band is generated at the top of the valence band. The acceptor level is expected to decrease when additional Li atoms are introduced into the system for (F, 2Li) codoping ZnO, since electrons may transit from the impurity band. So it indicates that F_O-$2Li_{Zn}$ complex is favorable for fabricating p-type ZnO.

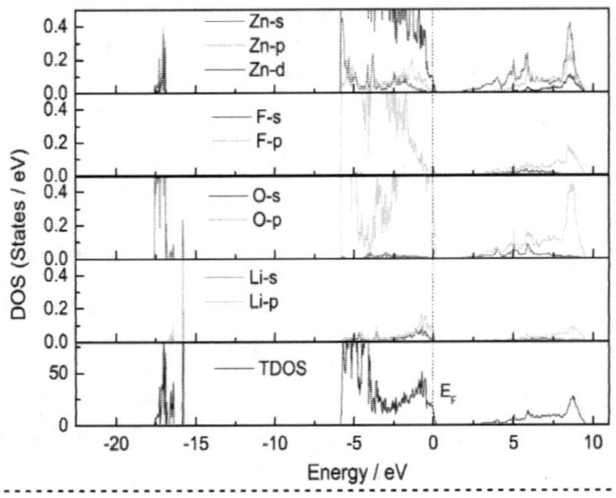

Figure 3a. PDOS of F_O-$2Li_{Zn}$.

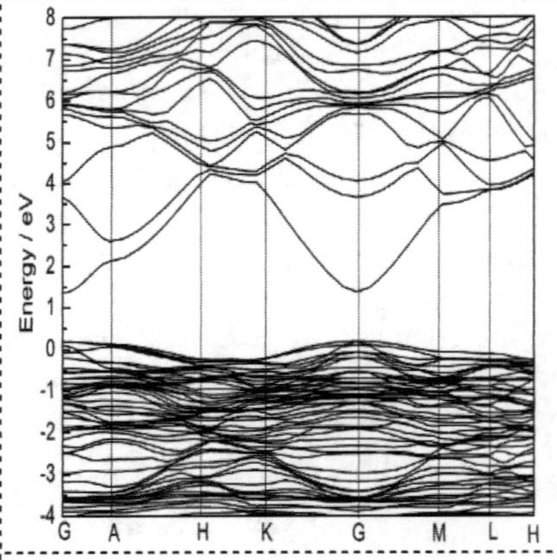

Figure 3b. band structure of F_O-$2Li_{Zn}$.

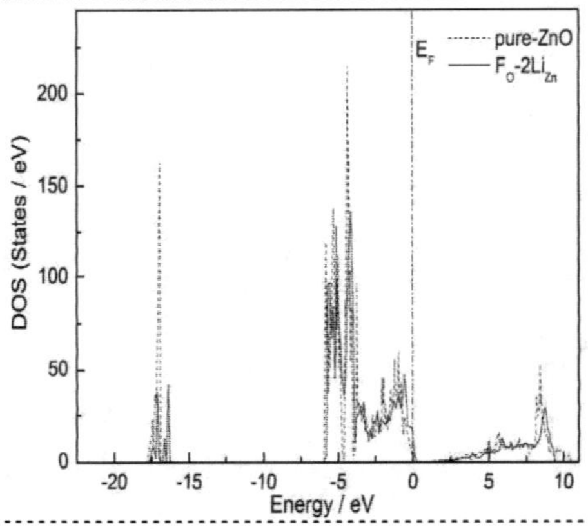

Figure 4a. Comparison of TDOSs between pure-ZnO and F_O-$2Li_{Zn}$.

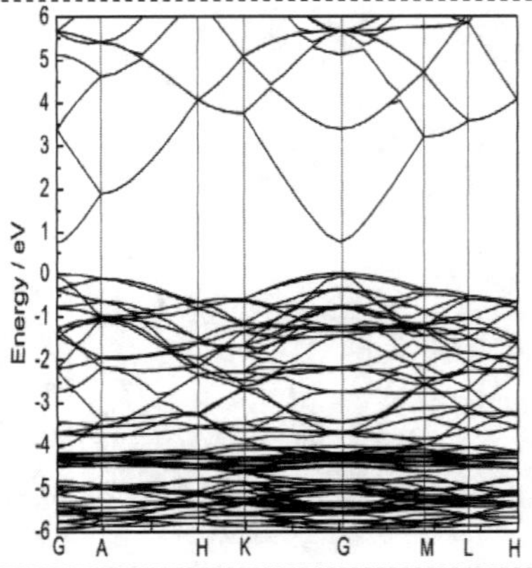

Figure 4b. Band structure of pure-ZnO.

Then, we have a detail comparison of TDOSs between the pure-ZnO and F_O-$2Li_{Zn}$ complex as shown in Fig.4a. Compared with the TDOS of the pure-ZnO, it is found that there is an additional peak near the Fermi level for F-2Li codoping ZnO. It demonstrates that F_O-$2Li_{Zn}$ complex does not change the basic electronic structure, which is similar to the pure-ZnO, but generates an additional fully occupied impurity band above the VBM, which is consistent with the cases of N-Zr [13] and N-Ga [23]. From Fig.3b and Fig.4b, we also can see that the basic structures have not been changed, but a new occupied band generates at the top of valence band, indicating that F-2Li codoping could obtain p-type conductivity ZnO.

Next, we investigated the influence of F-Li codoping related defect upon the p-type ZnO. The PDOSs of F_O-Li_{Zn} complex and F_O-Li_i complex are shown in Fig.5a. It is observed that F_O-Li_{Zn} demonstrates as an intrinsic state, because F-Li codoping forms the passive complex. So, there is no peak value at the top of the valence band, which is similar to that of pure-ZnO. That indicates F-Li codoping could not promote p-type ZnO. Fig.5b shows that the Fermi level enters the conduction band, which forms the donor level. The donor level is mainly introduced by Zn-s and O-p states. Therefore, self-compensation would be expected to generate, if the two complexes are coexisting in the system under Zn-rich condition. However, has been discussed earlier, the formation energy of F_O-Li_{Zn} is lower than that of F_O-Li_i under O-rich condition. So, if we can control the growth condition available, F_O-Li_{Zn} complex could suppress the formation of Li interstitial.

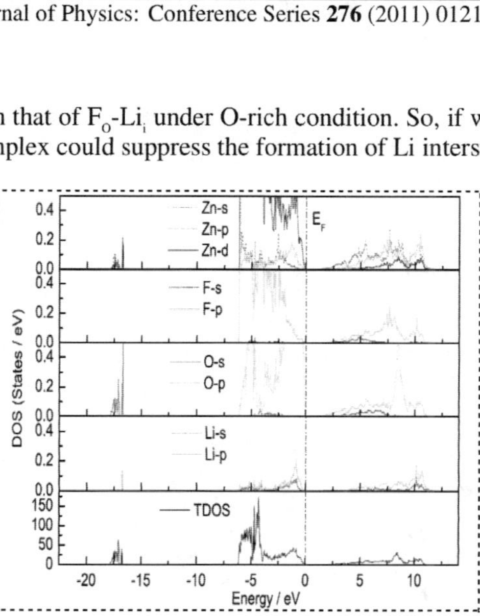

Figure 5a. PDOS of F_O-Li_{Zn}.

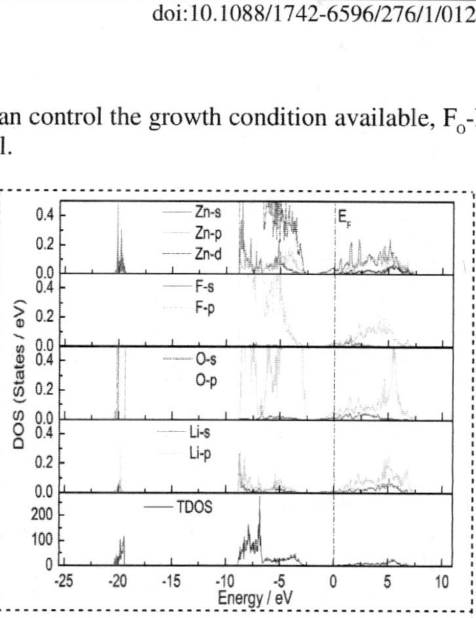

Figure 5b. PDOS of F_O-Li_i.

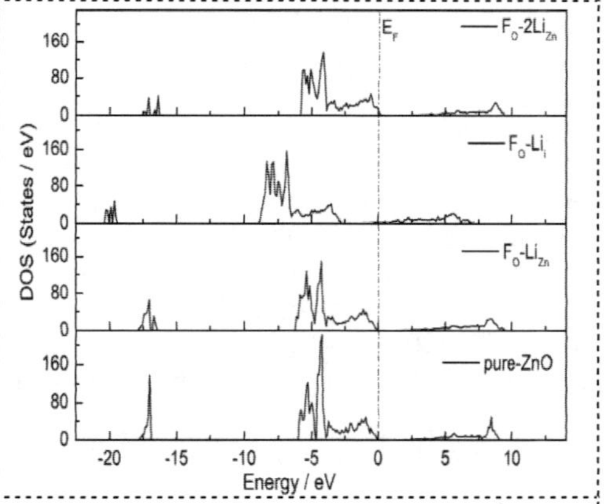

Figure 6a. TDOSs of various dopants systems.

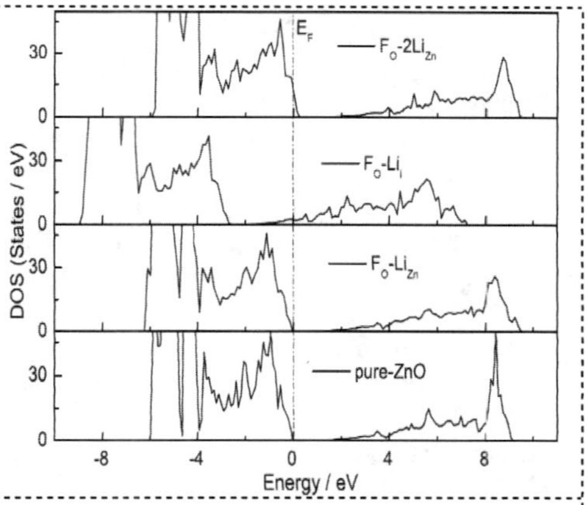

Figure 6b. Amplified curve near the band gap.

In order to compare with the electronic structures of the system about the related defects in F-Li codoping ZnO, the TDOSs have been plotted (as shown in Fig.6). In the graphs, the Fermi levels have been set as zero. For F_O-Li_{Zn}, it demonstrates as an intrinsic state, because F-Li forms the fully passive complex. However, for F_O-Li_i, the Fermi level is located at the conduction band minimum, indicating that behave as a donor. By contrast, for (F, 2Li) codoping in ZnO, it can be seen that the Fermi level is moved into the valence band, which implies good transfer characteristics of charge carriers. And the impurity peak is generated, which possibly demonstrate the characteristic of p-type ZnO. Therefore, we have proposed a new method of F-2Li codoping ZnO. It could be concluded that p-type conductivity is easier to achieve under O-rich condition for F_O-$2Li_{Zn}$ complex when we control the growth condition availably.

4. Conclusions

The calculations, including the structure, formation energy and electronic structure of F-Li codoping in ZnO, have been performed. The calculated results indicate: F_O-Li_{Zn} complex and F_O-Li_i complex have the effect of compensation under Zn-rich condition, while the formation energy of F_O-Li_{Zn}

complex is lower than that of F_O-Li_i complex under O-rich condition. Interestingly, it was found that the presence of F atom could suppress the formation of interstitial Li and F_O-$2Li_{Zn}$ forms as a shallower acceptor, which is benefit for gaining good p-type ZnO.

Acknowledgments
This work was supported by Science and Technology Staff Service Enterprise Project of the Ministry of Science and Technology of China (No. 2009GJC50021), National Natural Science Foundation of China (No.51062003), Key Science and Technology Project of Jiangxi Provincial Department of Education (No.GJJ10030), and Natural Science Foundation of Jiangxi Province (No. 2008GZW0014).

References
[1] Y.Segawam A.Ohtomo, M.Kawasaki, H.Koinuma, Z.K.Tang et al. Phys. Status. Solid. (b) 202 (1997) 669.
[2] Bei Deng, Zhiyou Guo, and Huiqing Sun, Appl. Phys. Lett. 96 (2010) 172106.
[3] Jingbo Li, Su-Huai Wei, Shu-Shen Li, Jian-Bai Xia, Phys. Rev. B 74 (2006) 081201(R).
[4] A. Allenic, X. Q. Pan, Y. Che, Z. D. Hu, and B. Liu, Appl. Phys. Lett. 92 (2008) 022107.
[5] Qixin Wan, Zhihua Xiong, Jiangnan Dai, Jianping Rao, Fengyi Jiang, Opt. Mater. 30 (2008) 817.
[6] C. H. Park, S. B. Zhang, and S.-H. Wei, Phys. Rev. B 66, 073202 (2002).
[7] Y.J. Zeng, Z.Z. Ye, J.G. Lu, W.Z. Xu, L.P. Zhu, B.H. Zhao, and S. Limpijumnong, Appl. Phys. Lett. 89, 042106 (2006).
[8] Tetsuya Yamamoto and Hiroshi Katayama-Yoshida, Physica B 302-303, 155-162 (2001).
[9] Tetsuya Yamamoto,and Hiroshi Katayama-Yoshida, Appl. Phys. 38 (1999) pp. L166-L169.
[10] X. –Y. Duan, R. –H.Yao, Y. –J. Zhao, Appl. Phys. A 91(2008) 467-472.
[11] Zhihua Xiong, Lanli Chen, Qixin Wan, Dongmei Li, the 5th SPIE international Symposium on Advanced Optical Manufacturing and Testing Technologies, Dalian (2010).
[12] Lanli Chen, Zhihua Xiong, Qixin Wan, Dongmei Li, Opt. Mater. 32 (2010) 1216.
[13] X. –Y. Duan, Yu-Jun Zhao, R. -H. Yao, Solid State Communications 147 (2008)194.
[14] X.M. Duan, C.Stanmpfl, M.M.M.Bilek, and D.R.Mckenzie, Phys. Rev. B 79 (2009) 235208.
[15] Ren-Yu Tian, Yu-Jun Zhao, Journal of Applied Physics 106 (2009) 043707.
[16] Young-Woo Heo and D.P.Norton, Journal of the Korean Ceramic Society, 43, No.11,738-742 (2006).
[17] K.Kobayashi, Y.Tomita, Y.Maeda, and H.Haneda, Phys.stat.sol.(c) 5, No.9,3122-3124 (2008).
[18] G. Kresse, and J. Hafner, Phys. Rev. B 47, 558 (1994).
[19] G. Kresse, and J. Furthermuller, Phys. Rev. B 54, 11169 (1996).
[20] W. Z. Xu, Z. Z.Ye, T. Zhou, B. H. Zhao, and J. Y. Huang, et al., J. Cryst. Growth. 265,133 (2004).
[21] Woo-Jin Lee, Joongoo Kang, Phys. Rev. B, 73:024117 (1-6) (2006).
[22] Ren –Yu Tian and Yu –Jun Zhao, Journal of Applied Physics 106, 043707 (2009).
[23] Y. F. Yan, J. B. Li, S. H. Wei, and M. M. Al-Jassim, Phys. Rev. Lett. 98, 135506 (2007).

Design Secondary Optical System Applied in White-LED General Illumination

CHEN Chen[1], ZHANG Xianhui[1]
(1.Dept. of Weaponry Engineering, Naval University of Engineering, Wuhan Hubei 430033, China)

Abstract: As a new generation of light source, LED has many advantages that other light sources do not have. However, due to the nonuniform lighting of LED, secondary LED optical system design is particularly important. Freeform surface tailoring method, an important method of lighting design, establishes a light intensity change model after smooth surface refraction (reflection) of the light and simplifies the solution process for more complex issues of solution using the free surface tailoring method. Based on this method, secondary LED optical system is designed, and the light intensity distribution is simulated after LED light passes through the secondary optical system. The results show that the method has not only simplified the calculation process of the free surface tailoring method, but also the designed LED secondary optical system has achieved the purpose of uniform lighting to a certain degree.
Keywords: Freeform surface tailoring method, LED lighting, Geometrical optics, Secondary optical system

Introduction

As a new solid cold light source, LED, unmatched by traditional light sources, has many advantages such as energy saving, resistance to impact, resistance to moisture, longer service life and so on. In recent years, with the continuous advancement of semiconductor lighting technology, luminous efficiency of LED has already reached $100 lm/W$, so that LED lighting lamps have gradually replaced traditional light sources and been applied to general lighting field.

Luminescent properties of LED lamp are similar to Lambert, which results in its nonuniform illumination light field. When it is applied to lighting, its light intensity must be converted using a secondary optical system. If LED is regarded as a point light source, its secondary optical system can be designed using the free-form surface tailoring method. Free surface tailoring method is a design method that partial differential equations are solved using numerical values so as to construct the surface figure [1] of free-form surface in lighting optics, it is very widely applied in the field of lighting design. One of the most important steps in designing tailored free-form surface is to calculate [2] the light intensity after light passes through the optical surface with light wavefront as the carrier. As shape of the light wavefront has changed after passing through the optical surface, therefore, changes [3] of the light wavefront curvature tensor and optical surface tensor must also be considered, so that complexity of the differential equation that is required to be ultimately solved is increased. In this paper, a light intensity change model that does not take the light wavefront as the light intensity carrier is put forward, which reduces the complexity of optical surface to be solved and designs the secondary optical system for LED lamps using an improved method. The simulation results have proved the efficiency of this method.

1 Improved theoretical model

It is hypothesized that both the optical surface $S_1: \vec{r_1}=\vec{r_1}(u_1,v_1)\in R^3$, $(u_1,v_1)\in D_1\subset R^2$ and target surface for illumination $S_2: \vec{r_2}=\vec{r_2}(u_2,v_2)\in R^3$, $(u_2,v_2)\in D_2\subset R^2$ are regular surfaces of R^3, that is, there are definite unit normal vectors at various points on S_1 and S_2, which are recorded as $\overrightarrow{N_1}(u_1,v_1)$ and $\overrightarrow{N_2}(u_2,v_2)$ respectively. Point light source T: $\vec{r}=\vec{r_0}$ is outside S_1, any incident ray $\overrightarrow{TP_1},(P_1\in S_1)$ is reflected via S_1 to obtain its reflected ray $\overrightarrow{P_1P_2}(P_2\in S_2)$. Due to the smoothness and regularity of curved surface, a continuous mapping $\sigma:\Delta S_1\to\Delta S_2$ can be defined between sufficiently small neighborhoods $\Delta S_1\in S_1$ of the reflection point P_1 and $\Delta S_2\in S_2$ of the receiver point P_2, and this mapping can be known as one-to-one correspondence by the free surface tailoring method. P_1 is solely determined by $\vec{r_1}(u_1,v_1)$, and P_2 is solely determined by $\vec{r_2}(u_2,v_2)$ and $\sigma:\Delta S_1\to\Delta S_2$, therefore, the mapping can be recorded as

$$\sigma:\begin{pmatrix}u_2\\ v_2\end{pmatrix}=\begin{pmatrix}f(u_1,v_1)\\ g(u_1,v_1)\end{pmatrix} \qquad (1)$$

Wherein, $(u_1,v_1)\in\Delta D_1\subset D_1$, $(u_2,v_2)\in\Delta D_2\subset D_2$, this mapping is only related to the light source $T:\vec{r}=\vec{r_0}$ and curved surfaces S_1 and S_2. When the reflection point changes in a tiny neighborhood ΔS_1 of P_1 on S_1, a tiny neighborhood ΔS_2 of P_2 on S_2 is obtained according to the continuity of σ and under the action of mapping. Thereupon, lights shined into ΔS_1 can all be projected onto ΔS_2. If illumination intensity at P_1 is E_1, light intensity at P_2 is E_2, area of ΔS is recorded as $S(\Delta S)$, and diameter of ΔS is recorded as $d(\Delta S)$, then there must be $\lim\limits_{d(\Delta S_1)\to 0} E_1\cdot S(\Delta S_1)=\lim\limits_{d(\Delta S_2)\to 0} E_2\cdot S(\Delta S_2)$ in the light of the conservation of energy, that is, the following formula can be obtained:

$$E_2 / E_1 = (\lim_{d(\Delta S_1)\to 0} S(\Delta S_1))/(\lim_{d(\Delta S_2)\to 0} S(\Delta S_2)) \quad (2)$$

Let $I_1 / I_2 = \lambda$, then

$$\begin{aligned}\lambda &= \frac{\lim_{d(\Delta S_2)\to 0} S(\Delta S_2)}{\lim_{d(\Delta S_1)\to 0} S(\Delta S_1)} = \frac{\left|\overrightarrow{r_{2u_1}} du_1 \times \overrightarrow{r_{2v_1}} dv_1\right|}{\left|\overrightarrow{r_{1u_1}} du_1 \times \overrightarrow{r_{1v_1}} dv_1\right|} \\ &= \frac{\left|(\overrightarrow{r_{2u_2}} \cdot f_{u_1} + \overrightarrow{r_{2v_2}} \cdot g_{u1}) du_1 \times (\overrightarrow{r_{2u_2}} \cdot f_{v_1} + \overrightarrow{r_{2v_2}} \cdot g_{v_1}) dv_1\right|}{\left|\overrightarrow{r_{1u_1}} du_1 \times \overrightarrow{r_{1v_1}} dv_1\right|} \\ &= \frac{\left|\overrightarrow{r_{2u_2}} \times \overrightarrow{r_{2v_2}}\right| \cdot \left|\frac{\partial(f,g)}{\partial(u_1,v_1)}\right|}{\left|\overrightarrow{r_{1u_1}} du_1 \times \overrightarrow{r_{1v_1}} dv_1\right|}\end{aligned} \quad (3)$$

Unit direction vector of the incident ray $\overrightarrow{TP_1}$ is recorded as $\vec{\rho}$, if $\vec{n_1}$ is the unit normal vector of S_1 at P_1, then direction vector of the emergent ray can be obtained as $\vec{A} = \frac{n_1}{n_2}\vec{\rho} + \left(-(\vec{\rho}\cdot\overrightarrow{N_1})\cdot\frac{n_1}{n_2} + \sqrt{1-\left(\frac{n_1}{n_2}\right)^2 [1-(\vec{\rho}\cdot\overrightarrow{N_1})^2]}\right)\overrightarrow{N_1}$ [4] according to the Snell's law. In addition, direction of the emergent ray can also be expressed as $\overrightarrow{P_1P_2}$, which is recorded as $\vec{b}$, that is, $\vec{b} = \overrightarrow{P_1P_2} = \vec{r_2} - \vec{r_1}$. Then $\vec{b} = \vec{r_2}(f(u_1,v_1), g(u_1,v_1)) - \vec{r_1}(u_1,v_1)$ is obtained by the definition of the mapping σ.

Both $\vec{A}$ and $\vec{b}$ represent the direction of emergent ray, therefore, $\vec{A} // \vec{b}$, that is

$$\vec{A} \times \vec{b}(f,g) = 0 \quad (4)$$

Formula (4) is the implicit definition of the mapping σ. Both sides of formula (4) are derived by u_1 to obtain

$$\vec{A} \times [\overrightarrow{r_{2u_2}} \cdot f_{u_1} + \overrightarrow{r_{2v_2}} \cdot g_{u_1}] = \vec{A} \times \overrightarrow{r_{1u_1}} - \overrightarrow{A_{u_1}} \times \vec{b} \quad (5)$$

Likewise, there is

$$\vec{A} \times [\overrightarrow{r_{2u_2}} \cdot f_{v_1} + \overrightarrow{r_{2v_2}} \cdot g_{v_1}] = \vec{A} \times \overrightarrow{r_{1v_1}} - \overrightarrow{A_{v_1}} \times \vec{b} \quad (6)$$

Formula (7) can be obtained by mutual cross product of both sides of formulae (5) and (6) as follows:

$$(\vec{A}\times\overrightarrow{r_{2u_2}})\times(\vec{A}\times\overrightarrow{r_{2v_2}})\cdot\frac{\partial(f,g)}{\partial(u_1,v_1)}=(\overrightarrow{A_{u_1}}\times\vec{b}-\vec{A}\times\overrightarrow{r_{1u_1}})\times(\overrightarrow{A_{v_1}}\times\vec{b}-\vec{A}\times\overrightarrow{r_{1v_1}}) \tag{7}$$

Moreover,

$$\begin{aligned}(\vec{A}\times\overrightarrow{r_{2u_2}})\times(\vec{A}\times\overrightarrow{r_{2v_2}})&=\overrightarrow{r_{2u_2}}\cdot(\vec{A}\cdot(\vec{A}\times\overrightarrow{r_{2v_2}}))-\vec{A}\cdot(\overrightarrow{r_{2u_2}}\cdot(\vec{A}\times\overrightarrow{r_{2v_2}}))\\&=0-\vec{A}\cdot(\vec{A}\cdot(\overrightarrow{r_{2v_2}}\times\overrightarrow{r_{2u_2}}))=\vec{A}\cdot(\vec{A}\cdot(\overrightarrow{r_{2u_2}}\times\overrightarrow{r_{2v_2}}))\\&=\vec{A}\cdot(\vec{A}\cdot\vec{n_2})\cdot\left|\overrightarrow{r_{2u_2}}\times\overrightarrow{r_{2v_2}}\right|\end{aligned} \tag{8}$$

Formula (8) is substituted into formula (7) to obtain

$$\vec{A}\cdot(\vec{A}\cdot\overrightarrow{n_2})\cdot\left|\overrightarrow{r_{2u_2}}\times\overrightarrow{r_{2v_2}}\right|\cdot\frac{\partial(f,g)}{\partial(u_1,v_1)}=(\overrightarrow{A_{u_1}}\times\vec{b}-\vec{A}\times\overrightarrow{r_{1u_1}})\times(\overrightarrow{A_{v_1}}\times\vec{b}-\vec{A}\times\overrightarrow{r_{1v_1}}) \tag{9}$$

Numerator of λ is present in the left side of formula (9), which can be converted into

$$\lambda=\left|\frac{(\vec{b}\cdot\vec{n_1})}{(\vec{b}\cdot\overrightarrow{n_2})}\cdot(4K_1\cdot|\vec{b}|^2+\frac{|\vec{b}|^2}{|\vec{A}|^2}+\frac{2\cdot|\vec{b}|}{|\vec{A}|}+1)\right.$$
$$\left.-\frac{2\cdot|\vec{b}|^2\cdot\left|1+\frac{|\vec{b}|}{|\vec{A}|}\right|}{(\vec{b}\cdot\overrightarrow{n_2})(E_1G_1-F_1^2)}\cdot(\overrightarrow{r_{1u_1}}\cdot\vec{A},\overrightarrow{r_{1v_1}}\cdot\vec{A})\begin{pmatrix}-N_1 & M_1\\ M_1 & -L_1\end{pmatrix}\begin{pmatrix}\overrightarrow{r_{1u_1}}\cdot\vec{A}\\ \overrightarrow{r_{1v_1}}\cdot\vec{A}\end{pmatrix}-4H_1\cdot|\vec{b}|^2\frac{\left|1+\frac{|\vec{b}|}{|\vec{A}|}\right|}{(\vec{b}\cdot\overrightarrow{n_2})}\right| \tag{10}$$

Wherein, E_1, F_1 and G_1 are coefficients of the first basic formula of the curved surface S_1, and M_1, N_1 and L_1 are coefficients of the second basic formula of the curved surface S_1.

2 Improved free surface tailoring method

It is hypothesized that light source is located at point s, direction vector of the incident ray at point $\vec{p}$ on the optical surface is $\vec{In}$, direction vector of the reflected (refracted) ray is $\overrightarrow{Out}$, unit normal vector is $\vec{N}$, and refractive indexes at both sides of the optical surface are n_1 and n_2, formula (11) can be obtained according to the Snell's law as follows:

$$[1+\frac{n_1^2}{n_2^2}-2\frac{n_1}{n_2}(\overrightarrow{Out}\cdot\vec{In})]^{1/2}\vec{N}=\overrightarrow{Out}-\vec{In} \tag{11}$$

The incident ray reaches the point $\dot{t}$ on target surface after optical surface reflection (refraction), illumination intensity at point t can be obtained according to the derivation in Section 1 as follows:

$$E(t) = \lambda \frac{I(\overrightarrow{In})}{(\vec{p}-\vec{s})(\vec{p}-\vec{s})} \tag{12}$$

It is hypothesized that the second-order equation is integrated and normal of the optical surface is determined, then necessary and sufficient condition of existence of a curved surface that is perpendicular to the vector field $\vec{N}$ everywhere is that the vector field satisfies the following conditions [5]:

$$\vec{N} \cdot curl(\vec{N}) = 0 \tag{13}$$

Surface figure of the curved surface can be solved according to the above 3 equations.

3 Secondary LED optical system design

Luminescent properties of LED are similar to Lambert, when it is applied to lighting, its light intensity must be converted using a secondary optical system. LED is regarded as a point light source, and then its secondary optical system can be designed using the free-form surface tailoring method. During solution, a spherical coordinate system and an equation of the optical surface and target surface are established to solve the surface figure data of the optical surface by computer programming using a numerical method. Here, only the lens face in case of rotational symmetry is calculated to simplify the solution process. Figure (1) is the finally obtained lens. Figure (2) is the light intensity distribution on target surface.

It can be seen from Figure (2) that the purpose of uniform lighting is basically achieved on target surface, but light intensity at some points still fluctuates, causes of which are analyzed as follows: (1) Insufficient calculation precision; (2) Insufficiently precise LED simulation. But it still shows the efficiency of this method.

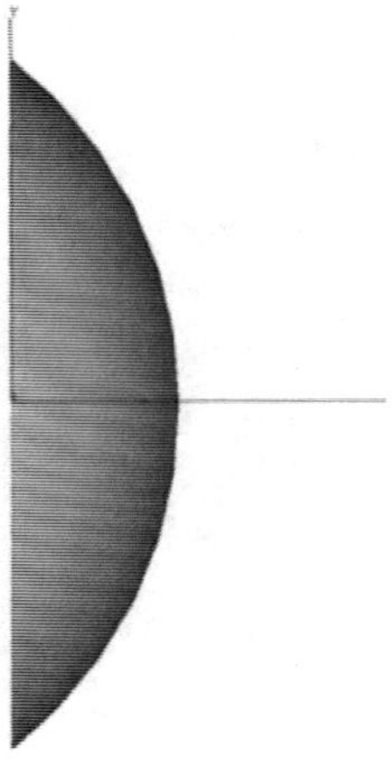

Figure (1) LED secondary lens

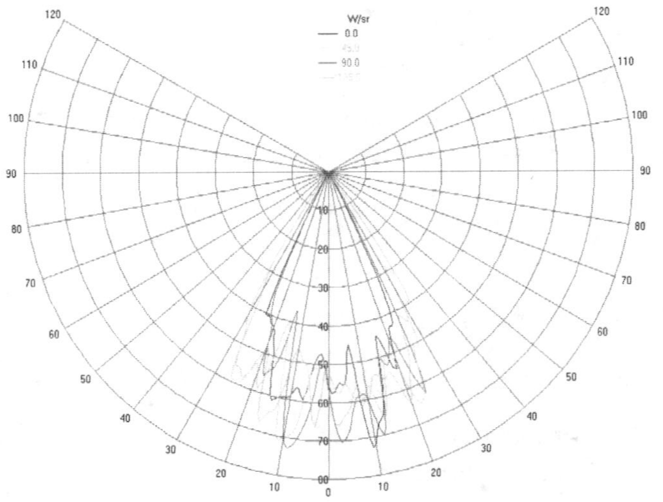

Figure (2) Light intensity distribution on target surface

4 Conclusion

Free surface tailoring method, an important method of illumination optical design, is improved in this paper to simplify the calculation process of solving the optical surface figure. Secondary LED optical system is designed using an improved tailoring method, which has proved the efficiency of this method.

References

1 Ries, H. and J. Muschaweck, Tailored freeform optical surfaces, *Journal of the Optical Society of America*, 590-595(2002).

2 Ries, H and J. Muschaweck, Tailored freeform optical surfaces for illumination, in *Proc. SPIE. San Diego*, Ca. 43-50(2001)

3 A. Timinger, J. Muschaweck and Ries, H, Designing Tailored Free-Form Surfaces for General Illumination, in *Proc. SPIE. Bellingham*, WA. 128-132(2003)

4 J. Chaves, Introduction to Nonimaging Optics, (CRC Press, 2007)

5 W. A. Parkyn, The design of illumination lenses via extrinsic differential geometry, in *International Optical Design* Conference, Proc. SPIE, 191–193(1998).

The influence of using different substrates on the structural and optical characteristics of $Bi_2VO_{5.5}$ thin films

F Chen[1], H M Deng[2], D J Huang[1], P X Yang[1*] and J H Chu[1]
[1] Key Laboratory of Polar Materials and Devices, Ministry of Education, Department of electronics, East China Normal University, 500 Dongchuan Road, Shanghai 200241, China
[2] Key Laboratory for Material Microstructures of Shanghai University, Shanghai University, 99 Shangda Rd, Shanghai 200444, China
[1]E-mail: pxyang@ee.ecnu.edu.cn

Abstract. The BVO thin films have been successfully grown on Si (100) and SiO_2/Si (100) substrates via sol-gel technique followed by rapid thermal annealing at 650 °C. The crystalline nature of the films has been studied by X-ray diffraction and indicates all the films are well-crystallized. Atomic force microscopy was used to study the surface morphology of the BVO thin films which shows the films are compact and homogeneous. The BVO film with SiO_2 buffer layer has a larger grain size and lower root mean square roughness than that directly grown on Si. The Raman spectra measurement in the range of 300-1500 cm^{-1} reveals that the buffer SiO_2 layer definitely influences the structure of BVO thin films. The optical properties of the BVO thin films are investigated using spectroscopic ellipsometric in the wavelength range of 400–900 nm. In the visible range, the refractive index of the film is about 2.42 and the extinction coefficient is 0.036.

1. Introduction

Ferroelectric thin films have been studied for many years owing to their interesting electronic and optical properties, which have led to a wide range of technical applications. In recent years, there has been much significative progress in the applications of ferroelectric thin films in practical devices such as nonvolatile ferroelectric memories, infrared sensors, and electro-optic devices [1-4]. $Bi_2VO_{5.5}$ (BVO), as a typical ferroelectric material with a layered perovskite structure, is promising and has attracted much attention [5-7]. It is an interesting member with n=1 of Aurivillius family $(Bi_2O_2)^{2+}$ $(A_{n-1}B_nO_{3n+1})^{2-}$ of Bi-based layered structured oxides consisting of $(Bi_2O_2)^{2+}$ layers interleaved with pervoskite blocks like sheets of $(VO_{3.5})^{2-}$ octahedra [8].

Metal/ferroelectric/insulator/semiconductor (MFIS) structure is a key part of ferroelectric field effect transistor (FeMFET) [9]. Therefore it is necessary to investigate the influence of insulating buffer layers on the structural characteristics of the BVO films. In our works, Sol-gel technique, which is good for grown large area of high quality thin films, is the method we have used to prepare the BVO thin films on Si (100) and SiO_2/Si (100) substrates, while SiO_2 is the insulating buffer layer. Furthermore, the rapid thermal annealing (RTA) has also been adopted to avoid the interface diffusion

* Corresponding author. Tel.: +86 21 54345157; fax: +86 21 54345119. E-mail address: pxyang@ee.ecnu.edu.cn (PX. Yang).

and the formation of interfacial layer between the BVO films and the substrates [10]. In addition, because of the applications in electro-optic devices, optical sensors and waveguide, the optical characteristics of $Bi_2VO_{5.5}$ thin films grown on Si substrate are obtained.

2. Experimental

The ferroelectric BVO thin films were prepared on p-Si (100) and the configuration SiO_2/Si (100) substrates by the sol-gel process with rapid thermal annealing techniques. In order to obtain the precursor solution, ammonium metavanadate and bismuth nitrate were selected as starting materials. The amount of bismuth nitrate is excessive by 5% for compensating the loss of Bi. The solvent was acetic acid. The details about the process of preparing the precursor solution are described in our previous work [11]. The resultant precursor solution was coated on Si and SiO_2/Si substrates by spin coating at 4000 rmp for 20s, followed by heating at 380 °C for 3 min to remove organic solvent and other volatile materials. Then, the films were annealed at 650 °C for 5 min. The deposition as well as the heating and annealing procedure was repeated for eight times.

The crystalline nature of the BVO films was characterized by X-ray diffraction (XRD; D/max 2200 VPC) using CuKα radiation. Atomic force microscope (AFM; Dimension 3100) was used to investigate the surface morphology and microstructure. Raman scattering spectra in the range of 300-1500 cm^{-1} were obtained by Jobin-Yvon LabRAM HR 800UV micro-Raman spectrometer. And the optical properties of the BVO films deposited on Si were studied via Spectroscopic ellipsometry (SC630) in the wavelength range of 400-900 nm.

3. Results and discussion

Figure 1 represents the X-ray diffraction (XRD) patterns of the BVO films deposited on Si and SiO_2/Si substrates, respectively. Both of the films are annealed at 650°C. As it can be seen, the film deposited on Si exhibits a highly c-preferred orientation structure and all of the diffraction peaks match well with those of typical BVO layered perovskite polycrystalline structure as well as on SiO_2/Si. There are no impure and secondary phases. And strong and sharp peaks mean the good crystallinity of films. The BVO films grown on SiO_2/Si show stronger and sharper diffraction peaks than those directly deposited on Si substrate. The reason must be that SiO_2 buffer layer decreases the mismatch of lattice constant between BVO and Si. So, the existence of the SiO_2 layer results in better crystallized BVO films.

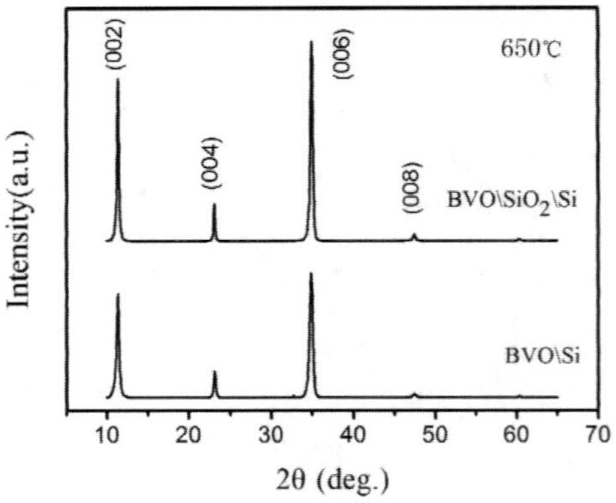

Figure 1. XRD patterns of the BVO films deposited on Si and SiO_2/Si substrates.

The surface morphology and microstructure of the BVO films for a 10×10μm^2 area have been studied by AFM. Figure 2a and 2c show the two-dimensional AFM images of the BVO thin films grown on Si and SiO_2/Si substrates, respectively. It had been proved that both of them exist no cracks and flaws, which suggests that the films compact and homogeneous. But the grain size of films including SiO_2 buffer layer is larger than that directly grown on Si. Figure 2b and 2d show the three-dimensional micrographs of BVO on Si and SiO_2/Si substrates, respectively. The root mean square (rms) roughness of the BVO film on SiO_2/Si is 8.242 nm, which is better than 9.361 nm derived from the BVO films deposited on Si.

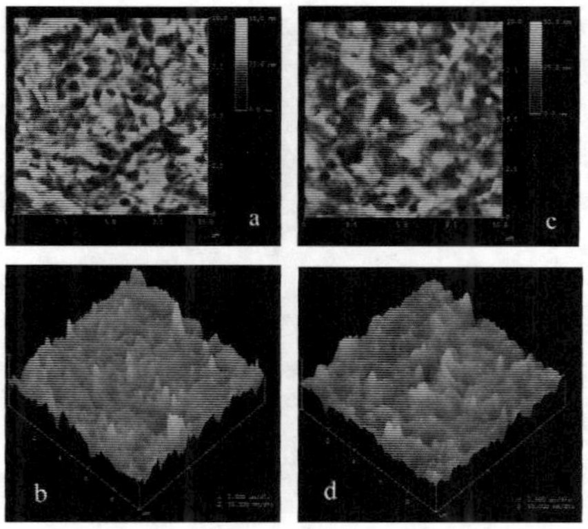

Figure 2. The two-dimensional (a) and three-dimensional (b) AFM images of the BVO thin film grown on Si substrate, the two-dimensional (c) and three-dimensional (d) AFM images of the BVO thin film grown on SiO_2/Si substrate.

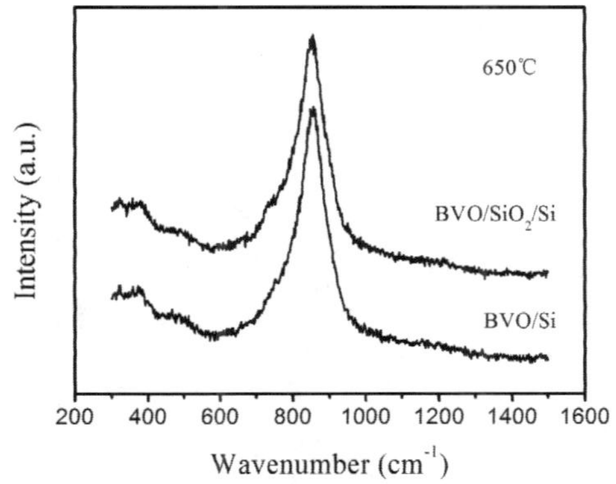

Figure 3. Raman scattering spectra of BVO deposited on Si and SiO_2/Si substrates.

Raman spectra of BVO thin films are depicted in Figure 3. We can find the Raman peaks around 324, 372, 512, 752 and 858 cm^{-1} for the BVO/Si film. Because of existence of SiO_2 layer, the strongest peak at 858 cm^{-1} becomes weaker. The symmetric stretching of $VO_{3.5}$ octahedra which the band at 858 cm^{-1} represents, shows large change [12-13]. Hence, the buffer SiO_2 layer definitely influences the structure of BVO thin films. However, the evident changes of other Raman peaks have not been noticed.

The optical constants of the BVO films deposited on Si substrate are detected by spectroscopic ellipsometry at room temperature. Figure 4 gives the refractive index and the extinction coefficient of the BVO film at 400-900 nm wavelength range, respectively. Initially, the refractive index decreases sharply with wavelength increasing up to 550 nm and then keeps almost constant. On the other hand, the extinction coefficient drops rapidly in the range of 400-650 nm, and reduces to 0.017 when the wavelength is larger than 650 nm. In the visible range (600 nm), the refractive index is about 2.42 and the extinction coefficient is 0.036, which is almost the same as the optical constants of the BVO films obtained by pulsed laser ablation technique [14].

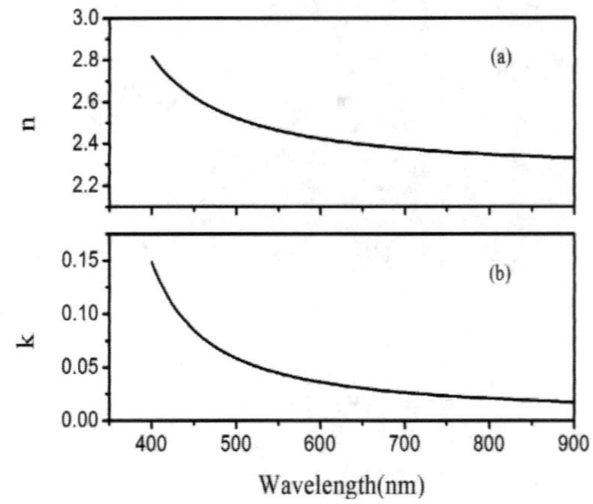

Figure 4. The refractive index n (a) and the extinction coefficient k (b) of BVO film deposited on Si substrate.

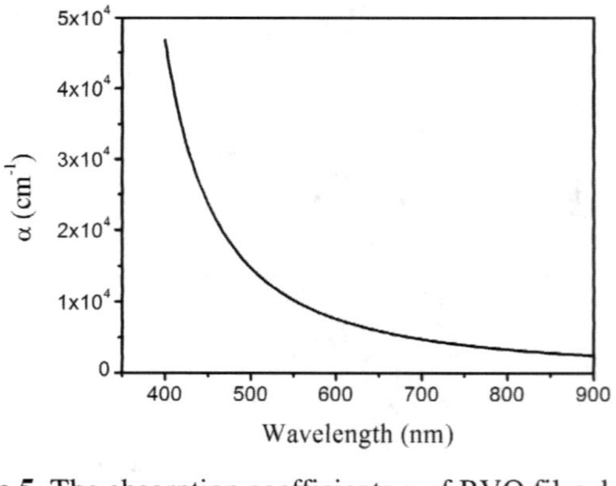

Figure 5. The absorption coefficients α of BVO film deposited on Si substrate.

Figure 5 shows the absorption coefficients α of BVO films in the range of 400-900 nm. It is observed that the absorption coefficients is quite high at 400nm and then decreases dramatically with wavelength increasing up to 650 nm. After that, the change goes gently.

4. Conclusion

The high quality BVO thin films were successfully grown on Si and SiO_2/Si substrates via sol-gel method. All films are c-axis orientation. The existence of SiO_2 buffer layer brings in better crystallized BVO films and better surface morphology and microstructure with lager grain size. Besides, the buffer SiO_2 layer definitely influences the internal structure of BVO thin films. The refractive index of BVO film is about 2.42 and the extinction coefficient is 0.036 in the visible range.

Acknowledgements

This work was supported by the National Natural Science Foundation of China (No. 60990312), the State Key Basic Research Program of China (2007CB924902), Shanghai Leading Academic Discipline Project (B411) and Science and Technology Commission of Shanghai Municipality (10JC1404600).

References

[1] Scott J F and Paz de Araujo C A 1989 *Science* **246** 1400.
[2] Scott J F 2000 *Ferroelectrics* **236** 247.
[3] Paz de Araujo C A, Cuchiaro J D, McMillan L D, Scott M C, and Scott J F 1995 *Nature (London)* **374** 627.
[4] Yang P X, Carroll D L, and Ballato J 2002 *Appl. Phys. Lett.***81** 4583-4585.
[5] Kumari N, Krupanidhi S B and Varma K B R, 2007 *Mater. Sci. Eng.* B **138** 22-30.
[6] Kumari N, Parui J, Varma K B R and Krupanidhi S B 2006 *Solid State Communications* **137** 566-569.
[7] Guo M, Deng H M, Yang P X and Chu J H 2009 *Mater.Lett.* **63** 1535-1537.
[8] B.Aurivillius 1950 *nature (Lond)* **2** 519.
[9] Rost T A, Lin H and Rabson T A 1991 *Appl. Phys. Lett.* **59** 3654.
[10] Wang H 2004 *Materials Science and Engineering* B **111** 64-68.
[11] Zhang Z L, Liu Z P, Guo M and Yang P X 2009 *Proceedings of the SPIE*, 7518.
[12] Wang J, Cheng G X, Zhang S T, Cheng H Wand Chen Y F 2004 *Physica* B **344** 368–371.
[13] Anwar H, Akishige Y and Kojima S 2005 *Materials Science and Engineering* B **120** 45–49.
[14] Kumari N, Krupanidhi S B and Varma K B R 2008 *Appl. Phys.* A. **91** 693-699.

The I-V Measurement System for Solar Cells Based on MCU

Chen Fengxiang[1], Ai Yu, Wang Jiafu, Wang Lisheng
Department of physics science and technology, Wuhan University of Technology, Wuhan city, Hubei Province,China,430070

E-mail: phonixchen79@yahoo.com.cn

Abstract. In this paper, an I-V measurement system for solar cells based on Single-chip Microcomputer (MCU) is presented. According to the test principles of solar cells, this measurement system mainly comprises of two parts—data collecting, data processing and displaying. The MCU mainly used as to acquire data, then the collecting results is sent to the computer by serial port. The I-V measurement results of our test system are shown in the human-computer interaction interface based on our hardware circuit. By comparing the test results of our I-V tester and the results of other commercial I-V tester, we found errors for most parameters are less than 5%, which shows our I-V test result is reliable. Because the MCU can be applied in many fields, this I-V measurement system offers a simple prototype for portable I-V tester for solar cells.

1. Introduction

Booming world economy's increases the demand for energy, with the traditional fossil energy sources dwindling, while the use of fossil energy has also caused serious environmental pollution [1]. In order to solve the resulting energy crisis and environmental problems, more attention orientates to renewable energy, in which development and utilization of solar energy is a viable option. Now solar photovoltaic technology is the highly regarded, and the PV as the core components of solar technology has become the focus of many researchers. In solar cell research and industrial production process, it is necessary to test the performance of solar cells and sort solar cells.

In this paper, a new solar cell I-V tester based on MCU is presented. This measurement system mainly comprises of two parts-- data gathering, data processing and displaying. After the development is finished, this system was used to test solar cells. Comparing our results and the test result from other tester, we find the test results coincide with each other, which shows the reliability of our test system.

2. Design of Data Collecting

The optical and electrical properties of solar cells are mainly calibrated by its I-V curve. Because the standard measurement condition is under spectrum AM1.5, the irradiance is $100mW/cm^2$ and the temperature is fixed at 25 degree. Both the variation of temperature, spectrum and irradiance will affect the results of the I-V curve, so the goals of data collecting are to gather the information of temperature, irradiance condition, current and voltage, then send these information to the computer. The computer was used to deal with the above data, and convert the data gathered from non-standard condition to the data under standard test condition.

[1] This paper is supported by "The Fundamental Research Funds for the Central Universities"(Project number:2010-la-019)

In our system, DS18B20 was used to acquire the temperature information, for this sensor can directly output the binary data. The gathering of irradiance information relies on the principles that the short-circuit current is always proportional to the irradiance. So a very small standard solar cell was used as an irradiance detector. By measuring the voltage upon a known resistance, we can deduce the irradiance. The current and voltage measurement depends on the adjustable resistance linked to the detected solar cell. MCU can control the stepping motor to change the adjustable resistance to one fixed value, then measure the voltage and current at that time. This process was repeated many times, and then current and voltage in the loop were collected. The schematic of current and voltage collecting was shown in figure 1 [2]:

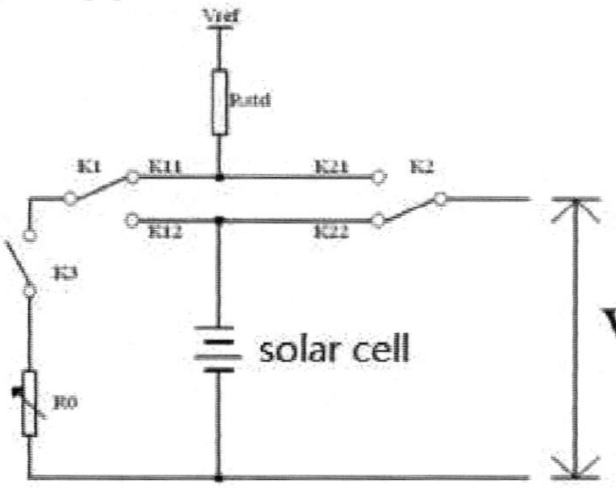

Figure1 The current and voltage collecting principle diagram.

In figure1, R_0 is the load resistance, R_{std} is the standard resistance, V_{ref} is the reference voltage source. The switch K_1, K_2 and K_3 all are solid state relay. It is the MCU who control the conduction and interrupt of the solid state relay, and then different loops were formed to gather the current and voltage signal. Through the following procedures, we can acquire the voltage and current values of solar cell under different load conditions.

1) The irradiance is fixed, then the load resistance was tuned to a unique value, K_3 is close. K_1 links to K_{11}, K_2 connects to K_{21}, The voltage between R_0 is

$$V = \frac{R_0}{R_0 + R_{std}} V_{ref} \quad (1)$$

From the Eq.(1), the expression of resistance R_0 is:

$$R_0 = \frac{V}{V_{ref} - V} R_{std} \quad (2)$$

2) Keep the irradiance and R_0 unchangeable, the switch K_3 is close, K_1 connects to K_{12}, K_2 connects to K_{22}. Then this electric circuit only contains solar cell and R_0. At this time the port voltage is named as V_c. Through the procedure 1, the value of R_0 is deduced, then the current in the loop is:

$$I_c = \frac{V_c}{R_0} = \frac{V_{ref} - V}{V} \frac{V_c}{R_{std}} \quad (3)$$

3) In this step, the MCU is used to control the stepper motor driver and the rotate of the stepping motor, then the value of R_0 is changed. Repeat the procedure 1 and 2, the current and voltage under different load resistance can be obtained.
4) Change the R_0 to the minimum as possible, so the resistance of load is much smaller than the inner resistance of solar cell. According to the procedure 1 and 2, the short-circuit current I_{sc} is obtained.
5) The switch K_3 is off, K_1 connects to K_{12}, K_2 connects to K_{22}. Under these conditions, the loop which contains the solar cell and load resistance is open, so the voltage at this time is the open-

circuit voltage V_{oc} of the solar cell.

The hardware diagram of data sampling circuit is shown in figure2.

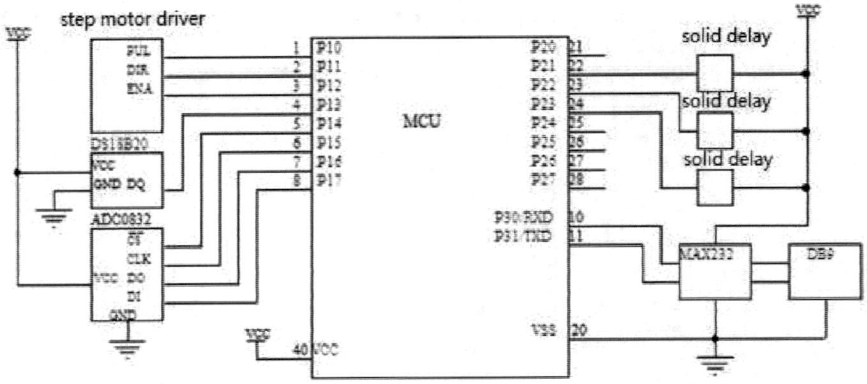

Figure2 Data sampling hardware circuit diagram.

Firstly the collecting data was saved in the memory of MCU. After the sampling process ends, the LED lights and informs the user that the data collecting process is finished. Then the data was send to the PC by the serial port. All the follow-up data processing was completed by computer.

3. Design of data processing and result displaying

The work mode adopted in our system is upper computer and lower computer mode. PC is the upper computer, MCU is the lower computer. The data sending, receiving and control communication between two computers are through the RS-232 serial port. The human-computer interface is developed by VC++6.0 software [3]. Any operator can monitor the process of data collecting through the human-computer interface and acquire the I-V curve and the related parameters of the detected solar cell.

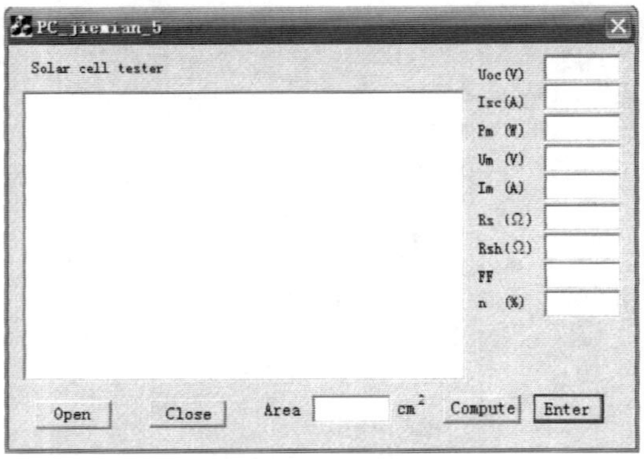

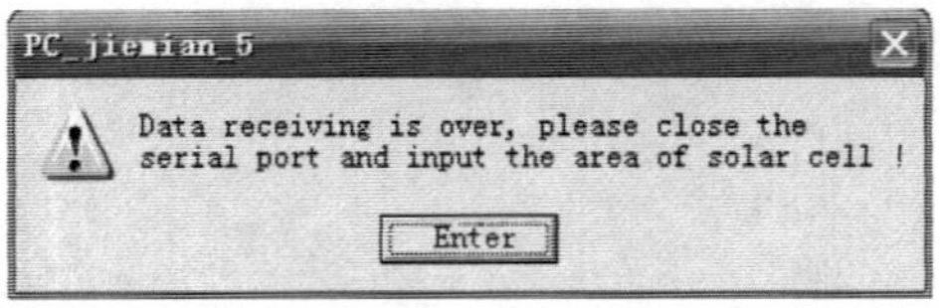

Figure4 The dialog box after data transmission is over.

Figure3 Human-computer interaction interface.

The finished human-computer interaction interface is shown in figure 3. The interface comprises of three push buttons, namely "Open", "Close" and "Compute", respectively. When the user received the signal which shows the collecting process was finished, clicks "Open", the serial port of computer began to receive the information from the lower computer. After the data transmission is over, the dialog box as shown in figure 4 will pop, which informs the user the data receiving process is over. At that time the area of the detected solar cell should be input. Operator inputs the area of solar cell to the "Area", click "Compute " button, then the I-V curve and all related parameters will display in the

Journal of Physics: Conference Series **276** (2011) 012161 doi:10.1088/1742-6596/276/1/012161

interface.

4. The experimental results

After the test system is developed, three solar cells connected by series connection with area 7.5*7.5cm^2 were test by this system and the test result is shown in figure 5.

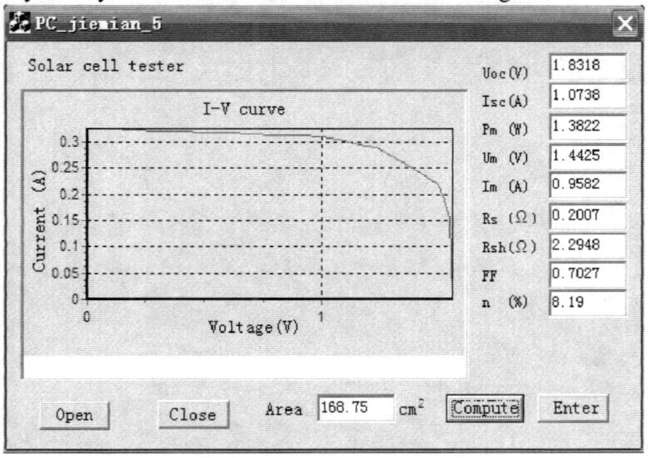

Figure5 I-V test results.

It can be seen from figure 5 that both the I-V curve and the related parameters shown in the interface are very straightforward. The same solar cell test in Jiawei Solar Cell(Wuhan) Cor. reports the results that the Voc is 1.797V, the maximum power is 1.349W, the work current is 1.008A and the conversion efficiency is 7.98%. By comparing two test results, we find the errors of most parameters are less than 5%, which shows the development of our system is successful.

5. Conclusions

In this paper, an I-V tester based on MCU is introduced. It can be seen from the experimental results that the operation of this measurement system is simple and the related characteristic of solar cell can be displayed in human-computer interface by graphical and numerical methods. Under the condition that the spectrum derivation was not included, most parameters error for solar cell were less than 5%, which shows the experimental results for our test system is reliable. For MCU is small bulk, this tester can be applied to portal I-V tester, which is helpful to diagnose the large-scale PV array.

References

[1] Zhang S K, Xu X G 2004 *Energy Research and Information.***20 211**

[2] Geng H X, Tang Z Y 2005 *Journal of Chengdu University(Natural Science).* **2 118**

[3] Liu Y,Shi X M and Cheng X L 2007 *Chinese Journal of Engineering Geophysics.* **4 455**

Optical properties of $BiCr_xFe_{1-x}O_3$ films grown by sol-gel method

H Deng[1], H M Deng[2], D J Huang[1], L P Zhu[1] and P X Yang[1*]

[1]Key Laboratory of Polar Materials and Devices, Ministry of Education, Department of Electronics, East China Normal University, 500 Dongchuan Road, Shanghai 200241, China
[2]Key Laboratory for Material Microstructures of Shanghai University, Shanghai University, 99 Shangda Rd, Shanghai 200444, China

Abstract. The $BiCr_xFe_{1-x}O_3$ thin films were fabricated on Si substrate by sol–gel method with *x* range from 0 to 0.05. X-ray diffraction analysis shows that both pure $BiFeO_3$ thin film and Cr-doped $BiFeO_3$ thin films have rhombohedral structure with space group R3c. Moreover, the influence of different values of *x* on the extinction coefficient and refractive index were investigated in order to exploit their possible application in optoelectronic devices. The optical constants of the Cr-doped $BiFeO_3$ thin films in the wavelength range 500-1100 nm were obtained by spectroscopic ellipsometry, and the band-gap energy of $BiCr_{0.03}Fe_{0.97}O_3$ was found to be 2.40 ±0.02 eV, and that of $BiCr_{0.05}Fe_{0.95}O_3$ was 2.30 ±0.02 eV.

1. Introduction

Multiferroic materials are currently attracting extensive attentions due to their potential applications and fascinating physical phenomena. However, single-phase multiferroic materials are rare because of contradictory requirement of d-electron occupancy in the A and B metal sites of ABO_3 type perovskite structure[1]. $BiFeO_3$ (BFO) is a multiferroic material in which ferroelectricity is due to the presence of $6s^2$ lone pair of electrons present in Bi^{3+} ions and (anti) ferromagnetism is due to the occupancy of unpaired *d*-orbital electrons[2]. BFO is a rare single phase material which gives multiferroic characteristics at room temperature. BFO was discovered to show ferroelectric order with high Curie temperature (T_C~850 °C) and G-type antiferromagnetic order with Neel temperature (T_N~370 °C)[3]. The structure of BFO is rhombohedrally distorted perovskite, which belongs to the space group *R3c*. However, the optical properties of the Cr-doped BFO materials were seldom investigated.

In this paper, The $BiCr_xFe_{1-x}O_3$ (BCFO) thin films were fabricated on silicon substrates by sol–gel method, and the optical properties of the films were investigated.

2. Experimental details

$Fe(NO_3)_3{\cdot}9H_2O$, $Bi(NO_3)_3{\cdot}5H_2O$ and $Cr(NO_3)_3{\cdot}9H_2O$ were adopted as raw materials. $Cr(NO_3)_3{\cdot}9H_2O$ was dissolved into the ready precursor solution, which was prepared by mixing $Bi(NO_3)_3{\cdot}5H_2O$ solution with $Fe(NO_3)_3{\cdot}9H_2O$ solution, and then the catalyst acetic acid was added. The solutions with Cr^{3+} concentrations of 0 %, 3 %, and 5 % were prepared. These solutions was stirred for 2 hrs at room

* Corresponding author. Tel.: +86 21 54345157; fax: +86 21 54345119. E-mail address: pxyang@ee.ecnu.edu.cn (P. Yang)

Published under licence by IOP Publishing Ltd

temperature. All solutions were prepared with excess 5 % Bi to compensate the bismuth loss during annealing process. The concentration of the precursor solution was adjusted to approximately 0.2 mol·L^{-1}. The BCFO thin films were spin-coated on Si substrate at 4000 rpm for 25 s. Each layer was dried at 200 °C for 2 min and then pyrolyzed at 350 °C for 3 min in a rapid thermal annealing in air. After deposited two layers, the amorphous films were annealed at 600 °C for 10 min. The above process was repeated four times to obtain the desired thickness.

The structural properties were examined by x-ray diffraction (XRD, Rigaku D/max-2200), and the influence of different values of *x* to the extinction coefficient and refractive index were investigated using spectroscopic ellipsometry.

3. Results and discussions

Figure 1(a) shows XRD patterns of pure BFO film, 3 % Cr-doped BFO film (Cr 3 %, BCFO-3 %), and 5 % Cr-doped BFO film (Cr 5 %, BCFO-5 %). The diffraction peaks were identified by PDF cards[4]. As can be seen in figure 1a, there are typical peaks of BFO in the XRD pattern, such as [012], [104], [110], [202], [024], [116], [300]. However, with increasing the concentration of Cr some additional peaks were observed in BCFO that might be due to some impurity phase. It is clear that impurity phases around 32° are observed in the films with Cr doping concentration varying from 3 % to 5 %, it might belong to Fe_2O_3. There are also some other impurity phase of Cr_2O_3 and $Bi_2O_4Fe_9$, as shown in figure 1(a).

Figure 1(b) shows the diffraction peaks corresponding to (104) and (110) of the pure and Cr-doped BFO films. In the present case, the prominent splitting around 32° for the 104/110 reflections were clearly visible for the pure and Cr-substituted BFO films, suggesting invariance of structural change upon Cr substitution. The XRD patterns indicate the rhombohedral structure of BFO with space group R3c, that is because the radii of Cr^{3+} (0.0615 nm) is similar to Fe^{3+} (0.0645 nm)[5] and the BFO structure could be less affected by light doping.

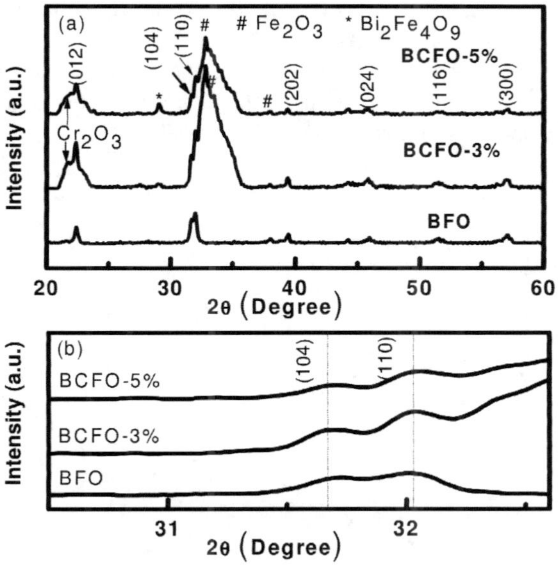

Figure 1.(a) XRD patterns of Cr 0%, Cr 3%, and Cr 5% doped BFO films, (b) the diffraction peaks corresponding to (104) and (110) of Cr 0%, Cr 3%, and Cr 5% doped BFO films.

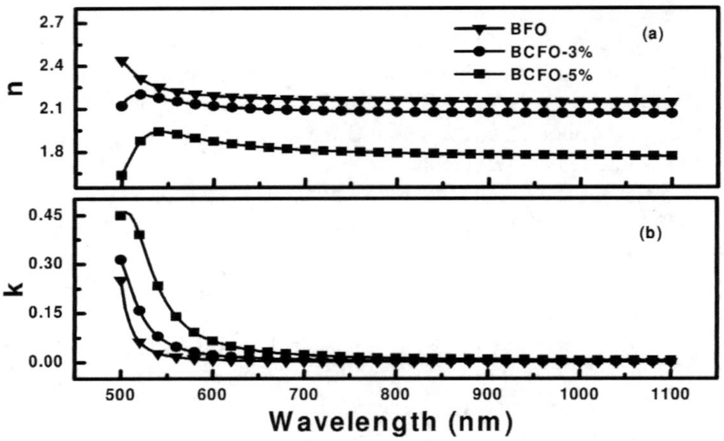

Figure 2. The refractive index *n* (a) and the extinction coefficient *k* (b) of Cr 0%, Cr 3% and Cr 5% doped BFO thin films deposited on Si.

Spectroscopic ellipsometry (SE) is a powerful technique to investigate the optical constants of the materials. It has been used to measure the refractive index *n* and extinction coefficient *k*. Figure 2 presents the optical constants (*n*, *k*) of the samples with the wavelength in the range 500~1100 nm. The refractive index of the pure BFO thin films reduces over 500 nm, however, the curves of BCFO-3 % and BCFO-5 % rise during short wavelength. The maximum refractive index of BCFO-3 % and BCFO-5% is 2.20 at 515 nm (2.41 eV) and 1.94 at 534 nm (2.32 eV), respectively. Moreover, with increasing the amount of Cr, the refractive index decreases and the extinction coefficient increases, and the extinction coefficient curves showing a bunching behavior towards higher wavelengths.

Figure 3 displays absorption coefficient of Cr 0 %, Cr 3 % and Cr 5 % doped BFO thin films deposited on Si. We assigned the observed excitations based on recent first-principles calculations[6-8]. Although the overall absorption profiles are similar, the absorption coefficient increases with the Cr doping from 0 to 5 % at the same photon energy.

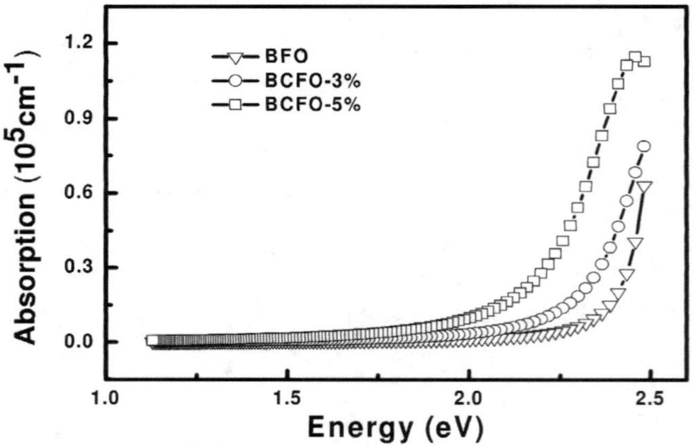

Figure 3. Absorption coefficient of Cr 0%, Cr 3% and Cr 5% doped BFO thin films deposited on Si.

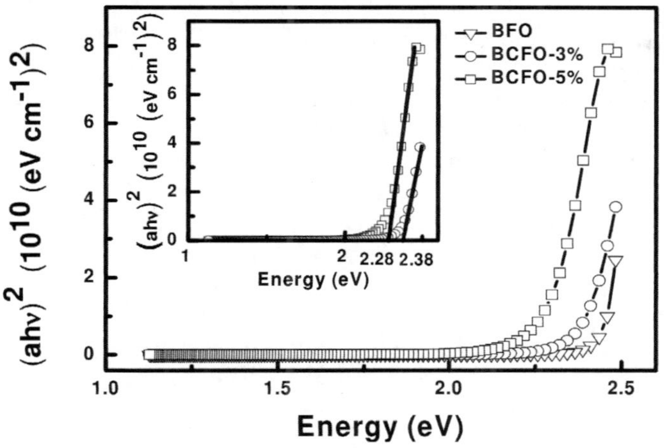

Figure 4. The plot of ( )^2 versus photon energy for Cr 0%, Cr 3% and Cr 5% doped BFO thin films deposited on Si, and direct band gap analysis of Cr 3% and Cr 5% doped BFO thin film (inset).

The direct optical band gap energy E_g of the BFO thin films was deduced from the spectral dependence of the absorption coefficient[9]. The energy E_g of the BCFO-3 % is about 2.40±0.02 eV and that of BCFO-5 % is 2.30±0.02 eV, which were estimated from the inset of figure 4 by extrapolating the linear portion of the plot to $(\alpha\hbar v)^2 = 0$. The values are smaller than that of pure BFO films[10], which was 2.67±0.02 eV.

4. Conclusions

In summary, the prominent splitting around 32° for the 104/110 reflections were clearly visible for the pure and Cr-substituted BFO films, suggesting the rhombohedral structure of BFO with space group R3c. The maximum refractive index of BCFO-3% and BCFO-5% is 2.20 at 515 nm and 1.94 at 534 nm, respectively. Moreover, the absorption coefficient increases with the Cr doping from 0 to 5% at the same photon energy, and the extinction coefficient curves show a bunching behavior towards higher wavelengths. The energy E_g of the BCFO-3% is about 2.40±0.02 eV and that of BCFO-5% is 2.30±0.02 eV, whereas that of pure BFO thin films was 2.67±0.02 eV.

Acknowledgments

We wish to thank Kai Jiang and Jinzhong Zhang for their great help in the spectroscopic ellipsometry measurement. This work was supported by the National Natural Science Foundation of China (No. 60990312), the State Key Basic Research Program of China (2007CB924902), Shanghai Leading Academic Discipline Project (B411) and Science and Technology Commission of Shanghai Municipality (10JC1404600).

References

[1] Hill N A 2000 *J. Phys. Chem.* **B 104** 6694-6709
[2] Khomskii D I 2006 *J. Magn, Magn. Mater.* **306** 1-8
[3] Fischer P, Połomska M, Sosnowska I and Szymański M 1980 *J. Phys. C: Solid State Phys.* **13** 1931-40
[4] Tomashpol'sky Y Y, Venevtsev Y N, Zhdanov G S and Kristallografiya 1967 PDF Card 74-2016
[5] Shannon R D 1976 *Acta Cryst.* **A32** 751–767
[6] Tütüncü H M and Srivastava G P 2008 *Phys. Rev.* **B78** 235209

[7] Ricinschi D, Yun K Y and Okuyama M 2006 *J. Phys.: Condens. Matter* **18** 97-105
[8] Neaton J B, Ederer C, Waghmare U V, Spaldin N A and Rabe K M 2005 *Phys. Rev.* **B71** 014113
[9] Kamalasanan M N, Chandra S, Joshi P C and Mansingh A 1991 *Appl. Phys. Lett.* **59** 3547-49
[10] Chen P and Podraza N J 2010 *Appl. Phys. Lett.* **96** 131907

Microstructure and optical properties of Zn-doped $BaTiO_3$ thin films

W L Gao[1], H M Deng[2], D J Huang[1], P X Yang[1*], J H Chu[1]

[1] Key Laboratory of Polar Materials and Devices, Ministry of Education, Department of electronics, East China Normal University, 500 Dongchuan Road, Shanghai 200241, China

[2] Instrumental Analysis & Research Center, Institute of Materials, Shanghai University, 99 Shangda Rd, Shanghai 200444, China

E-mail: pxyang@ee.ecnu.edu.cn

Abstract. The high-quality $BaTiO_3$ and Zn-doped $BaTiO_3$ thin films were prepared on Si substrates by choosing the suitable raw materials and optimizing the preparation of thin films and the post process of annealing. Through the characterization technique of x-ray diffraction (XRD) and atomic force microscope (AFM), the influences of the different doping levels was discussed on the sizes, shapes and uniformities of $BaTiO_3$ and Zn-doped $BaTiO_3$ thin films respectively. Raman spectra measurements were carried out to study the lattice vibration modes of thin films. Meanwhile, the type of dopant and the occupied site in the $BaTiO_3$ lattice were also discussed in detail. For investigating of the mechanism of the doping effects, the optical properties of thin films were studied by spectroscopic ellipsometry furthermore. And the optical constant refractive index and extinction coefficient have been obtained in the wavelength from 400nm to 700nm.

1. Introduction

Barium titanate is a kind of perovskite ferroelectric materials with the earlier investigation. It owns the excellent piezoelectric, ferroelectric, dielectric and pyroelectric qualities. $BaTiO_3$ (BTO) thin films can be used to develop microelectronic devices. With the combination of superconducting thin films, they can also be used to develop ferroelectric superconducting devices[3-6]. In order to develop ultra large scale integrated devices, the new demands for the quality of the films have been put forward. The Sol-gel method is simple and easy to manipulate, which can be used to prepare $BaTiO_3$- based

* Corresponding author. Tel.: +86 21 54345157; fax: +86 21 54345119.E-mail address: pxyang@ee.ecnu.edu.cn (P. Yang).

ferroelectric thin films with low cost and high quality[1,2]. With more in-deep research, metallic ion doping was also considered to be a good way to optimize the properties of $BaTiO_3$ thin films[1,10,15].

In spit of the facts that zinc compounds show excellent photoelectric and piezoelectric properties [10], few reports are available on the use of Zn as a doping in barium titanate. Therefore, this paper deals with the preparation and characterization of Zn-doped $BaTiO_3$ (Zn-BTO) thin films and its influence on the structural as well as the optical properties.

2. Experiment

The precursor solution of the BTO thin films were obtained by barium acetate [$Ba(CH_3COO)_2$] and titanium tetrabutoxide [$Ti(OC_4H_9)_4$]. Firstly, barium acetate and glacial acetic acid were mixed by constant stirring at 40 °C for 60 min. After cool to room temperature, the chelating agent acetyl acetone and absolute ethyl alcohol were added successively. Secondly, tianium tetrabutoxide in equimolar ratio was added. This solution which was at constant volume of 40 ml was dispersed on the substrates and then spin-coated at 4200 rev. /min for 45 s. Lastly, the as-spun films were annealed at 700 °C for 5 min. In order to prepare thicker BTO films, a multilayer deposition procedure was followed and the spin-coating/thermal treatments cycle was repeated up to 8 times.

The chemical solution of Zn-BTO was prepared by the same way with adding zinc acetate, at different molar ratio of Zn: Ti between 2% to 6% at a step of 2% on silicon substrates [2,14].

The microstructures of the films were characterized by X-ray diffraction (XRD; Rigaku D/max-2200) using CuK radiation and atomic force microscope (AFM; Dimension 3100). Raman spectras were investigated by Jobin-Yvon LabRAM HR 800UV micro-Raman spectrometer. The optical properties were studied by spectroscopic ellipsometry (SE) in the 400-1000 nm wavelength range using the angle of 65° incidence. All measurements were carried out at room temperature.

3. Result and discussion

3.1. Structure analysis-XRD

Figure 1 shows the X-ray diffraction patterns of BTO and Zn-BTO thin films with different doping levels grown on Si substrates by sol-gel method, respectively. From this graph, it can be seen that all the films were perovskite structures. The observation of the (110) peaks is in agreement with the formation of a BTO perovskite phase [10,15].

Compared with the BTO film, the intensity of the diffraction peaks for the Zn-BTO thin film are relatively strong. Moreover, no other phases are observed. It can be concluded that a minute amount of Zn dopants can improve the crystal quality of the BTO film without changing perovskite structure. According the parallel slight movements of the curves of Zn-BTO thin films, it can be deduced that the changes of lattice constantly caused by dopant are existed but inconspicuous. The a(110) cell parameter is found to be 39.3nm taking account of the residual peak at 31.48°.

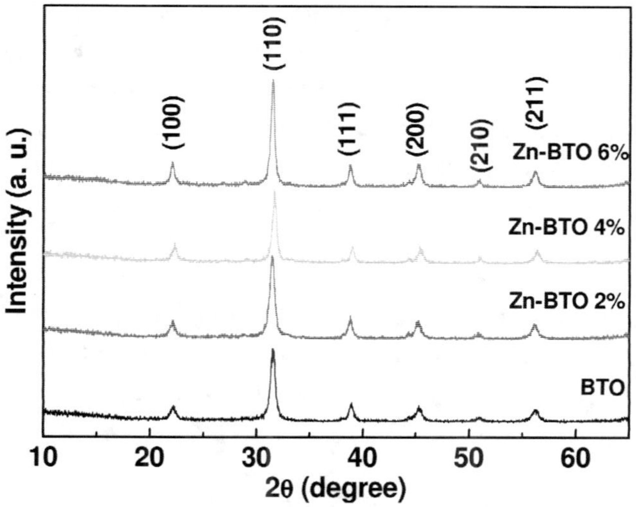

Figure 1. XRD patterns of BTO and Zn-BTO thin films grown on Si substrates by sol-gel method.

3.2. Surface analysis-AFM

The surface conditions of BTO and Zn-BTO films grown on Si substrates are observed by AFM micrograph, as shown in Figure 2. The results can be seen that the surfaces of all these films are uniform, dense and free of voids. However, the average size of the BTO grains estimated is much bigger than critical size of 49 nm for BTO thin films. It can be inferred that the growth properties of these BTO films present columnar with its particular orientation figure at high annealing temperature [9,10].

From images of Figure 2 it can also be included that Zn dopant can improve the crystal quality of BTO thin films on Si.

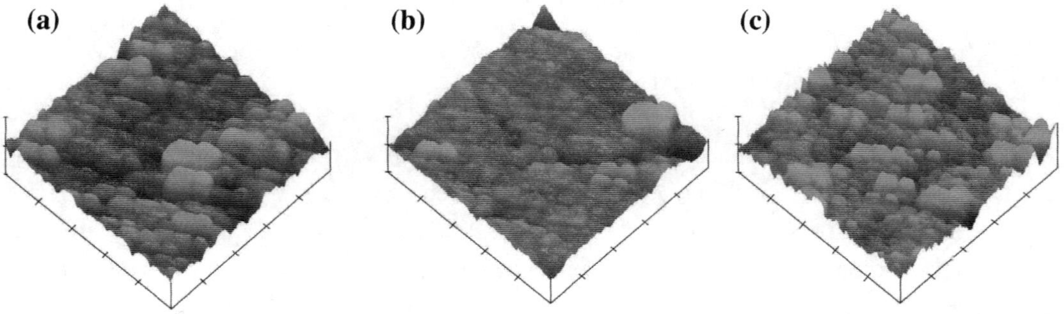

Figure 2. AFM images (10μm×10μm) (a) BTO/Si, (b) Zn-BTO/Si 2%, (c) Zn-BTO/Si 6% of thin films by sol-gel method.

3.3. Optical properties analysis

Since Raman selection rules are strictly associated with the orientation of the polar axis, the presence of only one kind of domains with polarization normal to the substrate (c-domains) or parallel to the substrate (a-domains) can be easily determined from polarized Raman studies.

In the tetragonal ferroelectric BTO one expects $3A_1+4E+B_1$ Raman active optical phonons. Here, A_1 is a nondegenerate mode, E is the double degenerate mode, and B_1 is the silent mode. Due to the action of the long-range electrostatic potential field, the polar A_1 and E modes split into transverse (TO) and longitudinal (LO) components [11-13].

Figure 3 shows the peaks at ~305 cm^{-1} ($E(TO_2+LO_2)$) and ~719 cm^{-1} ($A_1(LO_3)$) are specific to the tetragonal phase of BTO and the peak at ~504 cm^{-1} ($A_1(TO_2)$) which occurs both in the tetragonal and cubic phase of BTO. It, increasing the amount of doping, leads to the enhancement of the specific peaks of BTO. These observations indicate that Zn dopant improves the density and the crystal quality of BTO films.

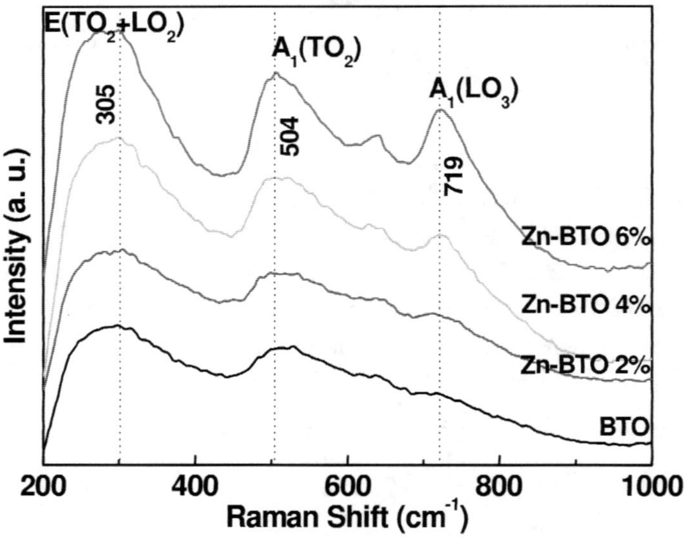

Figure 3. Raman spectra of the Zn-BTO thin films on Si substructures at different doping levels.

In order to investigate the optical properties of the Zn-BTO thin films deposited on Si substrate, spectroscopic ellipsometry is used. Figure 4 shows the refractive index, *n* and extinction coefficient, *k* of the Zn-BTO thin films. Comparing with the films of different doping levels, it can be clearly seen that both *n* and *k* decrease obviously with the doping addition increases. Moreover, *k* decreases rapidly with the increasing wavelength [7].

With the molar ratio of Zn rises from 2% to 6%, the refractive index of the films declines from 2.15 to 1.98 at 400 nm. Meanwhile the extinction coefficient of them decreases rapidly from 0.025 to 0.0135 at 400 nm and to zero at 700 nm. The implausible conclusion is that higher Zn dopant level leads to an increase in the band gap, due to Zn irons can compensate for the oxygen vacancy formed. Take the changes of deterioration " " for account, it presents the same tendency as the refractive index, which means the crystal structure and atomic arrangement is experiencing the transitions from disorder to order with adding the amount of dopant.

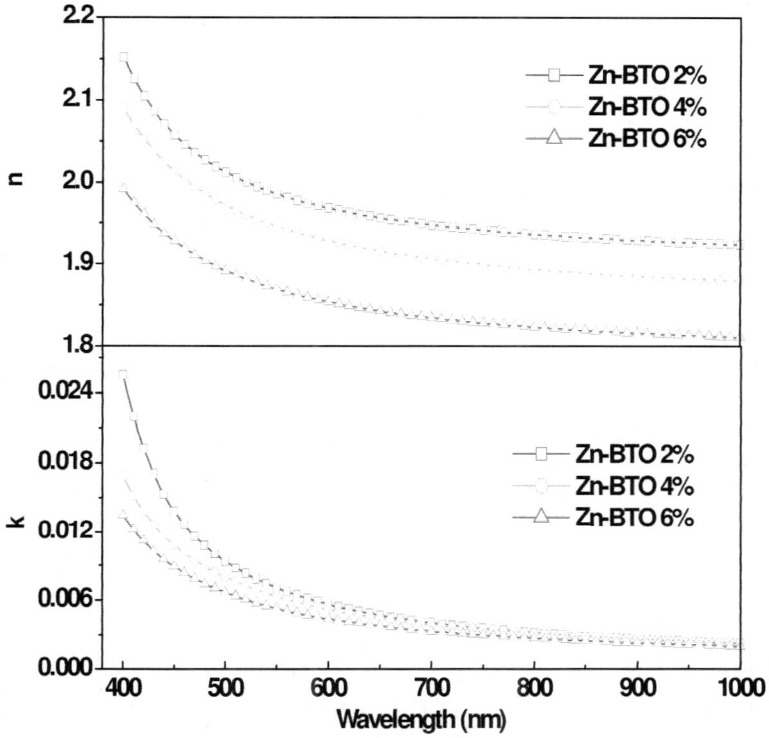

Figure 4. Refractive index and extinction coefficient curves of the Zn-BTO thin films on Si substructures at different doping levels

4. Conclusion

In this work, the high-quality ferroelectric BTO and Zn-BTO thin films were successfully fabricated on Si substrates by sol-gel method. The principle of the chemical solution for preparing Zn-BTO thin films was analyzed, and the theory showed that the best way to produce Zn-BTO thin films on Si substrates was by using zinc acetate, titanium tetrabutoxide and barium acetate as raw materials, acetic acid as the solution mixed with the chelating agents, with multi-times repeats of spin-coating and thermal treatments.

According the X-ray diffraction patterns, the addition of Zn did not modify the perovskite structure of BTO, which remained tetragonal, but led to lattice expansion a little which was caused by microstructure changes. It could be seen from the AFM images that the surfaces of Zn-doped films are more uniform and denser than un-doped ones. There was an increase of the optical band gap with the amount of Zn addition increased, which is due to the improvement of crystal quality and the changes of deterioration by dopants.

Acknowledgements

Thanks to Mr. Xu Wenfei for AFM investigation, Mr. Jiang Kai and Mr. Chen Xiangui for optical investigations. This work was supported by the National Natural Science Foundation of China (No. 60990312), the State Key Basic Research Program of China (2007CB924902), Shanghai Leading Academic Discipline Project (B411) and Science and Technology Commission of Shanghai Municipality (10JC1404600).

References

[1] Liu Z P, Deng H M, Sheng G, Yang P X and Chu J H 2009 *J. Mater. Lett.* **63** 2622-24
[2] Manso-Silvan M, Fuentes-Cobas L, Martın-Palma R J, Hernandez-Velez M and Martınez-Duart J M 2002 *Surf. Coat. Technol.* **151-152** 118-121
[3] Yang P X, Shi M R, Qin S M, Deng H M 2007 *Mater. Lett.* **61** 2687-2691
[4] Fong D D, Stephenson G B, Streiffer S K, Eastman J A, Auciello O, Fuoss P H and Thompson C 2004 *Sci.* **304** 1650-53
[5] Choi K J, Biegalski M,Li Y L, Sharan A, Schubert J, Uecker R, Reiche P, Chen Y B, Pan X Q, Gopalan V, Chen L Q, Schlom D G and Eom C B 2004 *Sci.* **306** 1005-09
[6] Scott J F 2007 *Sci.* **315** 954-9
[7] Clark N A and Lagerwall S T 1980 *App1. Phys. Lett.* **36** 899-901
[8] Hu Z G, Li Y W, Zhu M, Zhu Z Q and Chu J H 2008 *Appl. Phys. Lett.* **92** 081904
[9] Dong S S, Sang T P, Hoon S C, In H C and Jeong Y L 1999 *Thin Solid Films* **354** 251-255
[10] Fasasi A Y, Maaza M, Rohwer E G, Knoessen D, Theron Ch, Leitch A and Buttner U 2008 *Thin solid films* **516** 6222-32
[11] Xu J B, Shen B and Zhai J W 2009 *Appl. Surf. Sci.* **255**, 5922-25
[12] Guo H Z, Chen Z H, Cheng B L, Lu H B, Liu L F and Zhou Y L 2005 *J. Eur. Ceram. Soc.* **25** 2347-52
[13] Katiyar S Ram, Yuzyuk I Yu, Das R R, Bhattacharya P and Gupta V 2005 *Ferroelectrics* **329** 3–12
[14] Chen M S, Shen Z X, Tang S H, Shi W S, Cui D F and Chen Z H 2000 *J. Phys.: Condens. Matter* **12** 7013-23

Energy-Saving Tunnel Illumination System Based on LED's Intelligent Control

Shanshan Guo[1], Hanting Gu[2], Lan Wu[1] and Shuixiu Jiang[3]

[1]State Key Laboratory of Modern Optical Instrument, Department of Optical Engineering, Zhejiang University, Hangzhou 310027, China
[2] Zhejiang Jinji Electronics Co., Ltd, Hangzhou 310013, China
[3]Hangzhou Vocational and Technical College, Hangzhou 310018, China

E-mail:guopingguoshan@163.com,
zjght@163.com,
moimaster@zju.edu.cn,
jsxiu@yahoo.com.cn

Abstract. At present there is a lot of electric energy wastage in tunnel illumination, whose design is based on the maximum brightness outside and the maximum vehicle speed all year round. LED's energy consumption is low, and the control of its brightness is simple and effective. It can be quickly adjusted between 0-100% of its maximum brightness, and will not affect the service life. Therefore, using LED as tunnel's illumination source, we can achieve a good energy saving effect. According to real-time data acquisition of vehicle speed, traffic flow and brightness outside the tunnel, the auto real-time control of tunnel illumination can be achieved. And the system regulated the LED luminance by means of combination of LED power module and intelligent control module. The tunnel information was detected by inspection equipments, which included luminometer, vehicle detector, and received by RTU(Remote Terminal Unit), then synchronously transmitted to PC. After data processing, RTU emitted the dimming signal to the LED driver to adjust the brightness of LED. Despite the relatively high cost of high-power LED lights, the enormous energy-saving effect and the well-behaved controllability is beyond compare to other lighting devices.

1. The present condition of tunnel lighting

The tunnel includes entrance section, transition section, middle section and exit section[1], lighting requirement of different tunnel section is different, even though the same section in different condition has different lighting requirement, which concerns outside brightness, traffic flow, vehicle speed, and so on[2].

Now most of tunnels adopt high pressure sodium lamp(HPS) to light, since HPS starts up slow and can't dim light[3], still using it can't guarantee the tunnel brightness uniformity and smooth transition between the different sections. Despite LED tunnel lamp has been applied in tunnel, the controlling methods have no difference with high pressure sodium lamp's, which almost adopt simple stepping control or loop control. These methods have not considered the factors that concern tunnel lighting such as outside brightness, outside weather, vehicle speed, traffic flow, and this often leads to the brightness exceeds actual demand, and results in energy wastage[4]. Therefore the important problem that needs to be solved in tunnel lighting control is real-time tracking the traffic flow and the outside brightness, adopting different intensity light to illuminate, and then improving the tunnel lighting and its distribution.

1.1. The features of LED lamp as lighting source
LED 's luminous efficiency is better compared with traditional HPS and fluorescent lamp, in the case of the same illumination, energy consumption of LED is 25%-30% of HPS'. LED's brightness can be quickly adjusted between 0-100% of the maximum[4], it will not affect the service life and can achieve a good energy saving effect. At the same time, LED has a long life and firm structure, and its start time is short[5], all these features make LED be preferred light source of tunnel lighting[6].

2. The architecture of the system

Tunnel lighting system requires to effectively monitor all of the LED lamps in the tunnel, it can remotely find fault and control the LED's on, off and brightness at any time as requirement. The centralized monitoring approach will greatly ease the labor intensity of workers, reduce equipment consumption, and improve work efficiency, furthermore the system's real-time shorten the time of fault finding and maintenance period, thus passive management can be changed to active management.

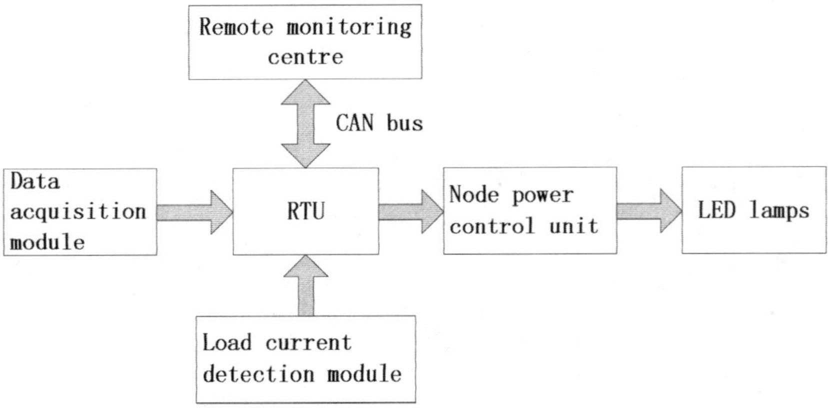

Figure 1. the control system

The figure 1 shows the components included in the control system, and then we simply explain each component.

1) Data acquisition module
   Equipments used in data acquisition module mainly include brightness detector, illumination detector, vehicle detector and outside weather detector, which are respectively used to collect the information of outside brightness, illumination inside the tunnel, traffic flow, wind speed and visibility.
2) RTU(Remote Terminal Unit)
   RTU is responsible for monitoring and controlling on-site signals. After the acquisition data received, RTU analysed these data, and then conveyed the processing results to monitoring centre, simultaneously adjusted the LED drive current through the node power control unit.
3) Remote monitoring centre
   The monitoring centre monitored all kinds of information within the tunnel, it received data from RTU, then saved them in the database, and according to the corresponding program output light-dimming value to RTU that controls the node control unit.
4) Node power control unit
   The unit received the light-dimming signal of RTU, then output PWM(Pulse-Width Modulation) signal of corresponding duty cycle to LED power.

5) Load current detection module
   Adopted dual operational amplifiers to detect the load current of LED tunnel lamp and then determined the operating condition.

3. The specific implementation of the system

3.1. The introduction of operating mode

The system controlled the brightness of all the LED lamps in the tunnel by RTU. All of the tunnel lamps were connected into a complete control network through the communication cables, and some detectors were accessed in the network, such as vehicle detectors, outside brightness detectors, illumination detectors. The system could automatically track the real-time changes of outside brightness and traffic flow, and real-time dim all the LED tunnel lamps steplessly[4], and achieve real "on-demand lighting" and substantial energy savings.

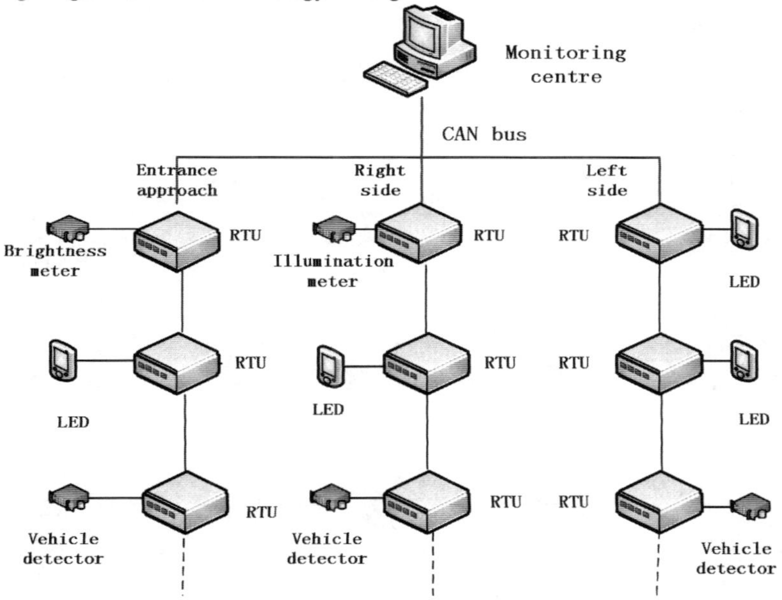

Figure 2. the illumination system

The structure of the illumination system is shown in figure 2. In the system, every RTU controls one or several neighboring LED tunnel lamps[5], which were connected through CAN bus, and then cascaded to CAN bus network, which could communicate with monitoring centre and formed local area network. The tunnel local area network automatically adjusted the brightness of every LED lamp according to predetermined program based on actual traffic flow and environmental brightness, and it can be controlled by administrator. The monitoring centre can obtain and store the power consumption and working state information of every LED, and can turn on or turn off every LED lamp in the tunnel or steplessly adjust their brightness. The system monitored the running state of every lamp in the network through load current detection module, and it facilitated maintenance and repair. In order to wire conveniently, the tunnel entrance approach, the left side and right side of tunnel, can be formed separate subnet, and then be connected into the monitoring centre. When network interruption detected, each RTU adjusted the brightness of all the tunnel lamps to the maximum according to the original setting in order to ensure traffic safety.

3.2. Build tunnel simulation model

Since it is complicated to determine the dimming basis in real tunnel through actual test, so we choose LabVIEW software to analyze and calculate the tunnel parameters, and then use DIAlux lighting design software to build the tunnel simulation model.

When analyzing the tunnel parameters in LabVIEW, the vehicle speed and traffic mode must be determined firstly, and then call the traffic flow and outside brightness information of actual measurement, then calculate the required illumination of tunnel entrance section, transitional section, middle section and exit section at different times according to highway tunnel ventilation and lighting specifications(JTJ 026.1-1999)[1]. Here, the brightness of entrance section and transitional section were adjusted according to outside brightness, and the brightness of middle section and exit section were adjusted according to traffic flow[6].

Here we give some data that were collected in the summer solstice, which show the average traffic flow and the outside brightness between 6:00-8:00 in the day. Import these data into the LabVIEW and then calculate the length and the illumination of each section. In table 1, we just list the calculation results of entrance section and the transition section.

Table 1. An example of calculating the illumination and the length of each section in the tunnel.

The time	Traffic flow (/h)	Outside brightness (cd/m^2)	Entrance section Illuminance (lux)	Transition section illuminance TR1(lux)	TR2(lux)	TR3(lux)
6:00-7:00	213	179.3	67.23	30.00	30.00	30.00
7:00-8:00	298	731.6	274.35	82.31	30.00	30.00
8:00-9:00	386	1723.1	653.31	195.99	65.33	30.00
9:00-10:00	493	2873.9	1138.71	341.61	113.87	39.85
10:00-11:00	576	3965.4	1623.70	487.11	162.37	56.82
11:00-12:00	478	4865.8	1916.29	574.89	191.63	67.07
12:00-13:00	518	5503.6	2202.61	660.78	220.26	77.09
13:00-14:00	638	5833.7	2446.43	733.93	244.64	85.62
14:00-15:00	617	5839.9	2429.46	728.84	242.95	85.03
15:00-16:00	652	5521.5	2327.84	698.35	232.78	81.47
16:00-17:00	618	4895	2037.15	611.15	203.71	71.30
17:00-18:00	600	4004	1654.84	496.45	165.49	57.92

In accordance with the calculated length and illuminance in different tunnel sections by LabVIEW, we determine the dimension of tunnel model in DIAlux, and select the material and attribute of space components (mainly include the material and the reflectance of road surface, wall and ceiling), and then choose suitable LED lamp with desired power, and determine the layout mode of lamps, then set the calculation points on the working surface, at last calculate and generate report. According to the report and compared to the highway tunnel ventilation and lighting specifications (JTJ 026.1-1999), check whether the report result meet the specifications, if not, adjust the power or the layout mode of LED lamps. When simulating real-time adjustment of each section's brightness, regulate the power of LED in the tunnel simulation model, and finally make the calculation results consistent with the lighting specifications. The dimming basis is just the illumination information of each section and the power of LED lamp in the simulation model.

At present the lighting brightness curves of most tunnels render step, furthermore when reducing the lighting level, as the lamps were light up alternately or every few, this will lead to the deterioration of road surface illumination uniformity, and make the drivers feel visual fatigue, and even bring about security risk.

When simulating and calculating by DIAlux, place the calculation points of the appropriate number, finally according the illumination values of each point in the calculated report, draw the illumination curve, which The illumination curve obtained according to DIAlux shown in Figure 3.The outside brightness is 5000cd/m^2, traffic flow is 600 per hour, and the length of the tunnel is 600m.

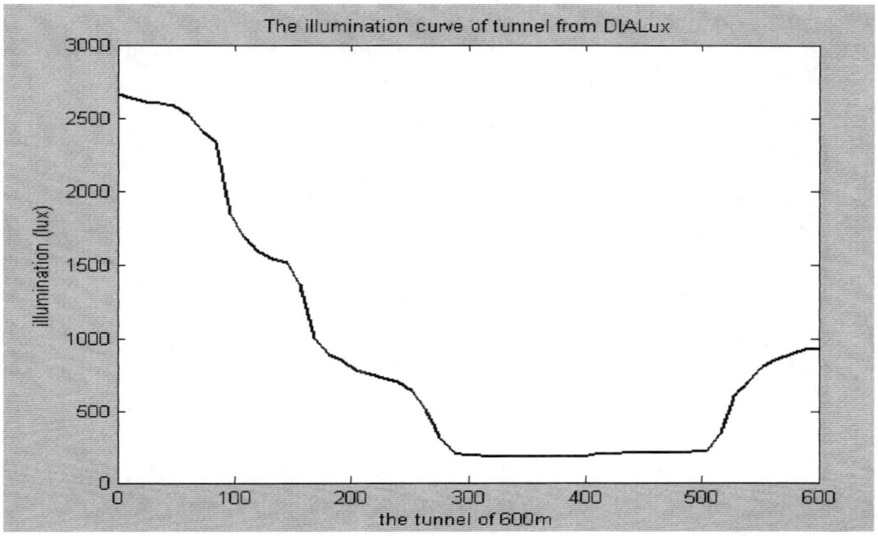

Figure 3. In the illumination curve, the x axis shows the numbers of the calculation points, and the y axis shows the illumination value.

The curve fits the demand of CIE illumination degression curve recommended by International Commission on Illumination. And using LED intelligent control system to dim light on-line and steplessly, we can not only reduce the lighting level but also keep the illumination uniformity in a good level, which also meet the physiological characteristics of human vision[7].

3.3. The features of the tunnel illumination system

The greatest feature of the system is that it can steplessly dim all of the LED lamps on-line between 0-100% of rated power. The system can either dim one lamp, and also can dim all the lamps simultaneously, it also can divide all the lamps into groups and then dim the different groups in different degrees, all these are important to tunnel energy-saving illumination. The system monitors on-line the working current of each lamp, when one lamp is broken or not working properly, the alarm will show in the monitoring centre and it will remind staff to maintenance and replacement. It can avoid the workers to inspect on the spot thus economize the human resources. We build the tunnel simulation model by DIALux lighting design software, which can not only simulate the real tunnel environment, and also calculate the illumination in the tunnel according to the selected lamps and the layout, which provide the basis for the dimming of the system.

4. The conclusion

The article describes the theory and the working methods of the tunnel illumination system, despite the relatively high cost of high-power LED lights, the enormous energy-saving effect and the well-behaved controllability is beyond compare to other lighting devices.This project is the key research topic of Zhejiang Jinji Electronics Co., Ltd. The system can ensure the safety of the drivers without reducing the brightness within the tunnel, and it can produce good energy saving effect, this is also the main direction of the tunnel lighting.

References

[1] Ministry of Communications Highway Research Institute in Chongqing 2001 *Professional Standards Compilation Group of People's Republic of China, Specifications for Design of Ventilation and Lighting of Highway Tunnel (JTJ 026.1-1999)* (Beijing:China Communications Press)p 41
[2] Hua Xiao, Fuhai Li and Cheng Dong 2009 *Application Of Electronic Technique* **35** 128
[3] Yu Zhu, Tibo Zhao and Dingjiu Zhou 2006 *China Light & Lighting* **9** 1
[4] Maohua Li 2008 *Transportation Information Industry* **5** 107
[5] Yaqiong Wang, Yongli Xie, Jinxing Lai 2000 *Chinese Journal Of Underground Space and Engineering* **5** 505
[6] Zhongliang Chen 2009 *Modern Tunnelling Technology* **46** 105
[7] Bertoldi, Paolo, Atanasiu and Boddan 2008 *International Journal of Green Energy* **5** 15-34

Journal of Physics: Conference Series **276** (2011) 012165 doi:10.1088/1742-6596/276/1/012165

Structure, optical and electrical properties of $Bi_2VO_{5.5}$ films deposited on ITO/glass substrates by chemical solution method

Ming Guo[1,2], Yude Shen[2], Changxin Liu[1], Pingxiong Yang[2*]

[1] College of Electronics and Information Engineering, Jingganshan University, ji'an, jiangxi 343009, China

[2] Key Laboratory of Polar Materials and Devices, Department of Electronics, East China Normal University, 500 Dongchuan Rd, Shanghai 200241, China

Abstract

Bismuth vanadate ($Bi_2VO_{5.5}$) thin films were fabricated on indium-doped tin oxide (ITO)-coated glass substrates by chemical solution method combined with a rapid thermal annealing process. The structure of the films was characterized with X-ray diffraction and atomic force microscopy. The $Bi_2VO_{5.5}$ films annealed at 600 °C showed a good match with the ITO coated glass substrates and had a desired perovskite structure with high (00*l*) preferred orientation. The spherical grains with a homogeneous distribution of high crystallinity and packing density were observed. Optical properties of the $Bi_2VO_{5.5}$ thin films were studied by Raman spectra and the lattice vibration modes of the films were obtained. A low frequency dielectric dispersion was observed in the films. Dielectric constant and loss was about 75 and 0.076 at 10kHz, respectively. The ac conductivity obeyed Jonscher's universal power law, which may be originated a possible hopping mechanism for $Bi_2VO_{5.5}$ thin films. The complex impedance traces revealed material dielectric dispersion nature and the presence of grain effects in the films.

BVO; chemical solution method; (00*l*) preferred orientation; optical properties; electrical properties

1. Introduction

Recently, the ferroelectric materials belonging to the family of the Aurivillius phases have attracted great attention for a variety of integrated device applications such as nonvolatile memories, optical memories, piezoelectric and electro-optic devices[1, 2]. $Bi_2VO_{5.5}$ (BVO), a member of bismuth layer family, has a relatively low crystallization temperature and low dielectric constant, and thus BVO thin films have many potential and demonstrated applications, such as integrated ferroelectric devices[3]. The BVO films have been fabricated on metallic as well as oxide electrodes and their structural and electrical properties were investigated[3-7]. Indium-doped tin oxide (ITO) is the most extensively used material as a transparent electrode to fabricate oxide thin films. However, few studies have involved the optical and electrical properties of BVO thin films deposited ITO/glass (ITO/glass) substrates by chemical solution deposition (CSD) processing. In this paper, we grew BVO thin films on ITO/glass substrates by CSD process annealing at 600 °C, and studied their structural, optical and electric properties.

2. Experimental

The BVO films were prepared on ITO/glass substrates by a CSD method. $Bi(NO_3)_3{\cdot}5H_2O$ and NH_4VO_3 were used as raw materials. All the chemical agents were analytical grade purity. Transparent green precursor solutions were obtained by dissolving corresponding stoichiometric reactants into the

* Corresponding author. E-mail: pxyang@ee.ecnu.edu.cn. Tel: 86-21-54345157, Fax: 86-21-54345119

acetic acid, respectively. Using the aged solution, BVO thin films were deposited on ITO/glass substrates. The spin-coated films were annealed at 120 °C for 180 s, 380 °C for 180 s and crystallized at 600 °C for 240 s in a rapid thermal process (RTP) furnace. These coating and annealing process were repeated several times to achieve desired film thickness. For the electric measurements, the platinum top electrodes of 9×10^{-4} cm^2 area were deposited on the surface of the BVO films through a shadow mask and thermal evaporation technique.

Phase identification of the samples were performed by X-ray diffraction (XRD; D/max2200VPC, Cu Ka, λ=1.54056Å). The surface features of the films were examined by means of a tapping mode atomic force microscope (AFM) (Veeco CP-II). The Raman spectra in the 1000–200 cm^{-1} range were measured using a Jobin-Yvon LabRAM HR 800UV micro-Raman spectrometer. The electric properties were measured by a HP 4194A impedance analyzer in the frequency range of 100 Hz–10 MHz with an applied voltage of 0.1 V. All measurements were carried out at room temperature.

3. Results and discussion

Fig. 1 shows the XRD patterns of BVO/ITO thin films annealed at 600 °C. XRD results showed that the films in our study had a high degree of crystallinity and a desired perovskite structure with preferred (00*l*) preferred orientation for the selective peaks at (002) and (006) are strong and sharp[3, 7]. It indicates that the BVO films match well with the ITO/glass substrates and have a lower crystallinity sintering temperature in the CSD process[8].

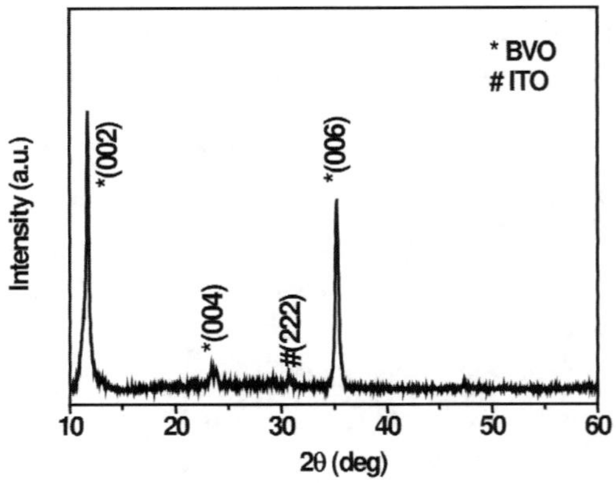

Fig. 1. XRD patterns of the BVO/ITO/glass thin films annealed at 600 °C.

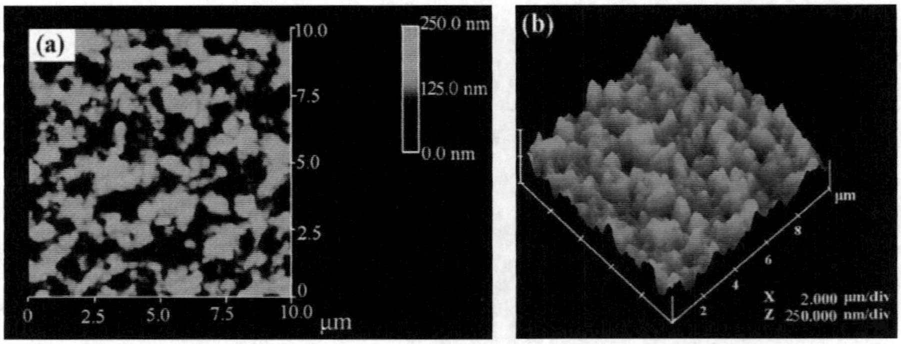

Fig.2 The AFM images of BVO/ITO/glass films annealed at 600 °C.
(a) surface morphology and (b) three-dimensional topography

The surface morphology and three-dimensional of BVO/ITO/glass thin film were characterized through AFM (Fig. 2). The spherical grains with a homogeneous distribution of high crystallinity and packing density were observed. The homogeneous crack-free appearance and dense structure in all scanned areas of the sample indicated the BVO thin film have a good match with the ITO/glass substrates. The surface showed a mean grain size about 200 *nm* and the root-mean-square (*RMS*) value of the surface roughness is about 10-12 *nm*.

Fig.3 showed Raman spectra of BVO/ITO/glass thin films annealed at 600 °C. In n=1 perovskite materials, the Raman and IR modes can be described as following[9],

$$2A_{1g}(\mathrm{R}) + B_{1g}(\mathrm{R}) + 3E_g(\mathrm{R}) + 5A_{2u}(\mathrm{IR}) + B_{2u}(\mathrm{IR}) + 6E_u(\mathrm{IR}) \quad (1)$$

where A_{1g}, B_{1g} and E_g are Raman active modes. From the spectra, Raman band around 853 cm^{-1} was the strongest of all the Raman peaks which represented the symmetric stretching of $VO_{3.5}\square_{0.5}$ octahedra for A_{1g} vibrational stretching mode. The band 200-500 cm^{-1} can be assigned the reversion and vibration of $VO_{3.5}\square_{0.5}$ oxygen octahedra corresponding vibrational modes. These B_{2g} and B_{3g} mode peaks were very close and the degenerate vibrational modes make the peaks merge.

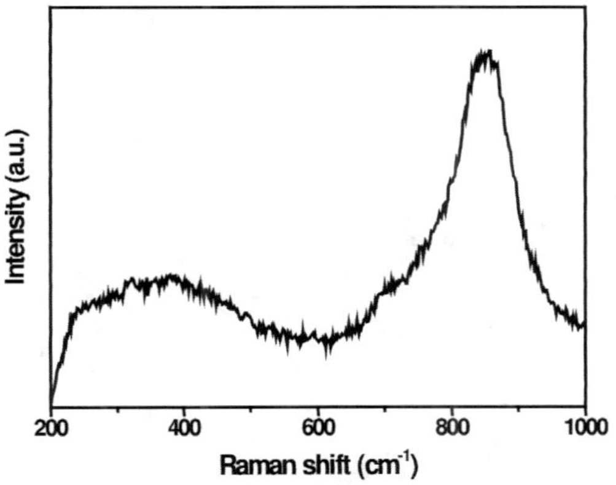

Fig.3 Raman spectra of BVO/ITO/glass thin films annealed at 600 °C.

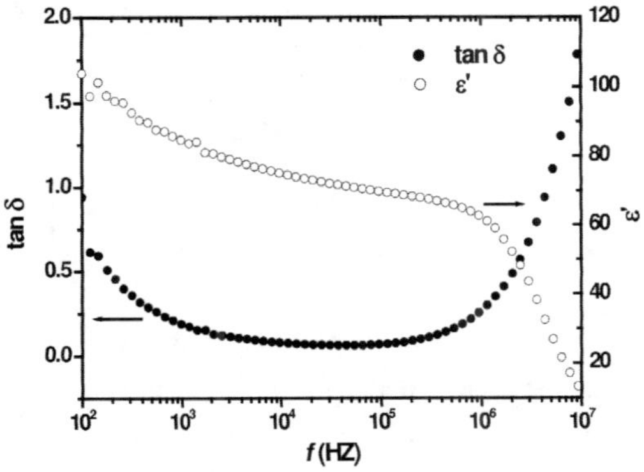

Fig.4 Relative dielectric constant (ε ') and dielectric loss (tanδ)as a function of frequency of BVO/ITO/glass films annealed at 600 °C.

The electrical properties such as dielectric constant (ε'), dielectric loss (tanδ)and ac conductivity as a function of frequency of the annealed BVO/ITO/glass were also investigated. The frequency variation (100 Hz–10 MHz) of both the dielectric constant and dielectric loss of BVO thin films were shown in Fig. 4. The ε′ was 75 with the tanδ of 0.076 at 10 kHz. A low frequency dielectric dispersion was observed for ε′ was found to decrease gradually with the increase of frequency. It originated the ferroelectric nature of BVO material and was closely related to the grain boundaries, free charges and oxide ion vacancies[4]. The dielectric constant of as deposited BVO film in our experiments was higher than that of the BVO films on ITO/glass substrates (ε′~52 at 10 kHz) owing to the increased crystallinity and packing density, as reflected in AFM[10].

The *ac* conductivity measurements by impedance spectroscopy are a useful technique for investigating the charge carriers' nature and the ac conductivity response. The frequency dependence of ac conductivity for BVO films was shown in Fig. 4. It was clear that the *ac* conductivity can be classified two different process of ionic movement: low frequency independent region (*dc* conductivity) was arising from inter-well hopping domination and plateau at higher frequencies dependent region (*ac* conductivity) contributing from intra-well hopping domination of this system [11]. This behavior suggested that electrical conduction in the films was a hopping mechanism and can be defined as conventional Jonscher's power law[11],

$$\sigma_{ac} = \sigma_{dc}[1 + (\frac{f}{f_p})^s] \tag{2}$$

where σ_{dc} is the dc conductivity, f_p is a crossover frequency and s is a fractional exponent related to the degree of correlation among ions. The term $(f/f_p)^s$ comprises the ac dependence and characterizes all dispersion phenomena. The change in the values of s is related to the charge carriers. A well-fitted solid line using Eq. (2) for sample film was also shown in the Fig. 4.

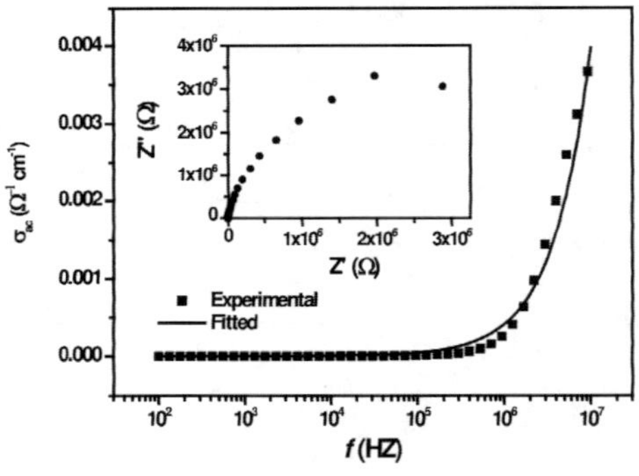

Fig. 5. The experimental and theoretical fit for *ac* conductivity as a function of frequency of BVO/ITO/glass thin films at different frequencies. The inset is the complex impedance spectra.

The inset of Fig. 4 showed the complex impedance spectra (Nyquist plots) for the BVO thin films. Only a arc was present, exhibiting material dielectric dispersion nature[12]. Single semicircle nature predicted the grain interior charge transfer mechanism for BVO films, with higher resistance. In fact, the complex plots can be modeled with an equivalent circuit containing a parallel resistance (R_g) and capacitance (C_g) elements. When frequency ω equals to the relaxation frequency $\omega_\tau = 1/R_gC_g$, the imaginary part (Z″) achieves maximum.

4. Conclusion

High-quality BVO thin films with (00*l*)-preferred orientation have been successfully prepared on ITO/glass substrates by CSD method. A homogeneous distribution of gains were observed in the films from the AFM microscopy. The lattice vibration modes of BVO thin films were obtained by Raman spectra, which mainly originated from the reversion and vibration of $VO_{3.5}\square_{0.5}$ oxygen octahedra. Dielectric constant and loss was about 75 and 0.076 at 10kHz, respectively. The *ac* conductivity obeyed Jonscher's universal power law, which may be assigned a possible hopping mechanism for BVO thin films. The complex impedance traces revealed material dielectric dispersion nature and the presence of grain effects in the films.

Acknowledgements

This work was supported by the National Natural Science Foundation of China (No. 60677022), Science Foundation (Research Fund) for Excellent Youth Scholars of Shanghai Municipal Education Commission, Science and Technology Commission of Shanghai Municipality (07JC14018), the State Key Basic Research Program of China (2007CB924902) and Science and Technology Foundation of Jiangxi Municipal Education Commission (GJJ08417).

References

[1] B.H. Park, B.S. Kang, S.D. Bu, T.W. Noh, J. Lee and W. Jo, Lanthanum-substituted bismuth titanate for use in non-volatile memories [J]. *Nature*, *401*, 682-684 (1999).

[2] P.X. Yang, M. Guo, M.R. Shi, X.J. Meng, Z.M. Huang and J.H. Chu, Spectroscopic ellipsometry of $SrBi_2Ta_{2-x}Nb_xO_9$ ferroelectric thin films [J]. *J. Appl. Phys.*, *97*, 106106 (2005).

[3] M. Guo, H.M. Deng, P.X. Yang and J.H. Chu, The influence of annealing temperatures on the properties of $Bi_2VO_{5.5}/LaNiO_3/Si$ thin films [J]. *Mater. Lett. 63(17)*, 1535-1537 (2009).

[4] K.V.R. Prasad, K.B.R. Varma, A.R. Raju, K.M. Satyalakshmi, R.M. Mallya and M.S. Hegde, Growth and ferroelectric properties of $Bi_2VO_{5.5}$ thin-films with metallic $LaNiO_3$ electrodes [J]. *Appl. Phys. Lett.*, *63*, 1898-1900 (1993).

[5] M. Joseph, H.Y. Lee, H. Tabata and T. Kawai, Ferroelectric behavior of epitaxial $Bi_2VO_{5.5}$ thin films on Si(100) formed by pulsed-laser deposition [J]. *J. Appl. Phys.*, *88*, 1193-1195 (2000).

[6] K. Tsukada, T. Nagahama, M. Sohma, I. Yamaguchi, T. Manabe, T. Tsuchiya, S. Suzuki, T. Shimizu, S. Mizuta and T. Kumagai, Characterization of epitaxial thin films of $Bi_2VO_{5.5}$ on La-doped $SrTiO_3$ substrates prepared by coating-pyrolysis process [J]. *Thin Solid Films*, *42*, 73-792 (2002).

[7] N. Kumari, S.B. Krupanidhi and K.B.R. Varma, Dielectric, impedance and ferroelectric characteristics of c-oriented bismuth vanadate films grown by pulsed laser deposition [J]. *Mater. Sci. Eng. B*, *138*, 22-30 (2007).

[8] M. Guo, H.M. Deng and P.X. Yang, Synthesis and characterization of nanoscale $Bi_2Cu_{0.1}V_{0.9}O_{5.35}$ powders by solution-based chemical methods [J]. *Mater. Sci.-Poland.*, 27, 1055-1066 (2009).

[9] P.R. Graves, G. Hua, S. Myhra and J.G. Thompson, The raman modes of the aurivillius phases - temperature and polarization dependence [J]. *J. Solid State Chem.*, *114*, 112-22 (1995).

[10] S. B. K. N. Kumari, and K. B. R. Varma,, "Structural, optical and electrical characteristics of transparent bismuth vanadate films deposited on indium tin oxide coated glass substrates," *J Mater Sci: Mater Electron,* (2009).

[11] V. Pena, A. Rivera, C. Leon, J. Santamaria, E. Garcia-Gonzalez and J.M. Gonzalez-Calbet, Correlated oxygen diffusion in BIFEVOX [J]. *Chem.Mater.*, *14*, 1606-1609 (2002).

[12] J.R. Dygas, Dielectric function of ionic conductors studied by impedance spectroscopy [J]. *Solid State Ionics*, *176*, 2065-2078 (2005).

Test of battery internal resistance

Hou Luan, Li Tianze, Zhang Xia, Jiang Chuan

School of Electric and Electronic Engineering, Shandong University of Technology

No.12, Zhang Zhou Road, Zhangdian District, Zibo 255049, China

E-mail: ltzwang@163.com

Abstract. Responding the problem in the measurement of battery internal resistance that it is difficult to improve the accuracy because of the power supply ripple and many other factors, a method used chaotic oscillator is studied. Using the principle that chaotic oscillator is sensitive to initial value this method simplifies the measuring circuit and ignores the anti-interference circuit. To combine the good inhibitory effect on noise to AC internal resistance detection method, the on-line detection is realized.

1. Introduction

In early 1992, Birx showed chaotic oscillator and complex mappings prior to the feedback network (CMFFNS) used in weak signal detection. In recent years, Wang Guanyu showed chaotic oscillator array method of detection of weak signal frequency. As the solar energy conversion backup, battery is the important guarantee of power output when the sun light is lack. So it becomes the important work that the effective maintenance quality of battery. It can be a very good mastery of the current work capacity by measuring the internal resistance of storage battery. But the battery resistance is not a constant. It changed with the working state of battery and the environmental conditions, especially when the battery remaining capacity is less than 40%. Battery resistance is very small(generally speaking, the battery capacity greater, the resistance smaller.). So it is difficult to measure the internal resistance technically with the outside interference and noise.

Under certain conditions chaotic system has noise immunity and it is sensitive with small signals. Even if the signal amplitude is lesser, it also can be made that the phase transition of system essentially. Then the weak signal can be detected from the noise. Because of the good properties of chaotic systems, it can reduce and even ignore the circuit design anti-interference using chaos detection system to realize the battery resistance testing. Then impedance measurements are simple, rapid and low cost.

2. Chaotic oscillator theory

2.1 Chaotic oscillator Duffing equation

The movement in chaos system is random, but the equation describing the movement is determined. The Duffing equation is chosen as the chaotic system equations detecting of weak periodic small signal under the background of strong noise.

The Duffing equation

$$\ddot{x}+k\dot{x}+\alpha x+\beta x^3=0 \tag{1}$$

is a motion equation of description of the nonlinear elastic system. Where k is the damping ratio, $\alpha x+\beta x^3$ is non-elastic restoring force, $\ddot{x}$ is the second derivative of x and $\dot{x}$ is the first derivative of x. When external force γ cos(ω t) in Duffing equation, the equation turns into

$$\ddot{x}+k\dot{x}+\alpha x+\beta x^3=\gamma\cos(\varpi t) \tag{2}$$

The analysis shows that the system characteristics mainly depends on the elastic recovery force when the signal in equation (1) is determined. So α and β are the keys of Duffing equation detecting weak periodic small signal. Through the research Duffing equation turns to be that

$$\ddot{x}+k\dot{x}-x+x^3=\gamma\cos(\varpi t) \tag{3}$$

when α =-1and β =1. Then the weak periodic small signal can be realized exactly.

2.2 Detection principle of chaos Duffing oscillator equation

The parameters value of k and γ is the key of that the chaotic Duffing oscillator is in the state of chaotic or periodic. In order to the better calculation and simulation, equation (3) turns to that:

$$\begin{cases}\dot{x}=y\\ \dot{y}=x-ky+y\cos(\varpi t)\end{cases} \tag{4}$$

According to the melnikov equation, the chaos threshold of Duffing oscillator equation is that:

$$\frac{\gamma_c}{k}\geq R=\frac{J_0}{J_1}=1.6769 \tag{5}$$

In this equation, γ_c is small signal threshold amplitude cycle, k=0.5, ω =1rad/s, $J_0=\int_0^T \dot{x}^2(t)\,dt$, and $J_1=\int_0^T x(t)\cos(t)dt$. There is the only determined corresponding threshold with the changing k, therefore, this paper study the change of state of chaos in the condition of k=0.5. For chaotic system, it is in the movement that the threshold is γ_c, where in the critical point of cycle mode and chaos state. Adding weak periodic small signal x(t)to the testing strong noise, chaos Duffing oscillator equation turns into

$$\begin{cases}\dot{x}=y\\ \dot{y}=x-ky+x^3+\gamma\cos(\varpi t)+xn(t)\end{cases} \tag{6}$$

In this equation, $xn(t)=a\cos(t)+k(randn)$, $randn$ is the generated function of matlab gaussian white noise. The equation is getted according the equation (5)

$$\dot{y}=x-ky+x^3+\gamma_c\cos(t)+xn(t)$$

$$= x - ky + x^3 + \gamma_c \cos(t) + a\cos(t) + k(randn)$$

$$= x - ky + x^3 + (\gamma_c + a)\cos(t) + k(randn)$$

According to the characteristics of chaotic systems, chaos system will be periodic from the original state when $\gamma > \gamma_c$. And it will be chaotic from the periodic state when $\gamma < \gamma_c$. So the mixing system will be periodic state when γ_c +a>γ_c in equation (6). Then the detection of weak small cycle signal will be realized in the strong noise conditions.

3. Solar battery essential resistance measurements

Solar battery essential resistance measurement is a complicated process. The common methods mainly are the density, opening voltage, dc discharging and communication.

The methods of density estimate the essential resistance mainly through measuring of the density of battery electrolyte. The method usually is used to measure the impedance of lead-acid batteries in open style, apparently not the sealed lead-acid batteries.

The methods of open circuit voltage estimate the essential resistance by measuring terminal voltage of battery with poor precision even the wrong conclusion. Because even in the condition of a small capacity battery, the terminal voltage filling can still be very normal in the floating state.

The voltage changes when step current I flows in batteries with the method of DC discharge. The changes of battery voltage Δ U decreases with the shorter time measurement. This shows that Δ U value contains the concentration polarization composition reduced which indicates that the concentration polarization composition the shorter containing in Δ U reduced. The measured value Δ U is unchanged when the measuring time less than 1ms.

The method of DC discharge is that the change voltage Δ U between 0.5ms-1.0ms is measured during constant current I discharging. And then calculate the ohm resistance by $r = \Delta U / I$. The main defects of the methods are as following. Friet this measurement must be realized in the condition of static or offline. So the online measurement can't be really realized. Then this will inevitably bring security hidden danger of equipment operation. It is still unknown whether so huge current affects the dc system when it is static. The system will be in more safe hidden trouble during weaning if it is a long time measurement. Second the battery is damaged with huge discharge current and some even hundreds of ampere. Communication is one of the important electrochemical measurements. The feedback voltage is measured through exerting a small low-frequency current to the battery.

Calculate the essential resistance through the ratio between voltage and their phase separation. The battery electrodes are porous and connected by more pieces of electrodes in parallel, so its equivalent circuit of exchange impedance is extremely complex. Therefore treating porous battery electrodes is based on the theoretic analysis results of planar electrode.

4. Battery essential resistance testing principle based on chaotic system

4.1 Alternating method of battery essential resistance testing

When using controlled current $\Delta I = I_{\max} \sin(2\pi ft)$, $\Delta U = U_{\max} \sin(2\pi ft + \phi)$ the impedance is $Z(f) = (U_{max} / I_{max})e^{j\phi}$. It is in connection with the frequency and |Z|=U/I with photo corner ϕ . In theoretic field, essential resistance can be measured measuring the voltage change from the alternating current signal fed into battery.

$$R = U_{AV} / I_{AV} \tag{7}$$

In this equation, U_{AV} is the average of ac voltage detection. I_{AV} is the present average of alternating current signal. Alternating method is good at that it doesn't need electric discharge during the measuring process, and repeated the measurement conveniently. But because the signal amplitude fed into is limited and the resistance is weak, response signal of amplitude the battery is small. Then it is easy interfered, especially in the online measurement disturbance. Therefore the anti-interference ability of alternating method is weak, and is vulnerable to external noise. So alternating method needs to eliminate power supply ripple and other sources. Chaotic oscillator model testing techniques is used to eliminate the influence factors. Specific means are that the battery ends on I_s constant current source, through the data acquisition card collecting the battery voltage U_O into the computer, and get into chaotic oscillator model. Then test the battery ends voltage and get the essential resistance by $Z = U_o / I_s$.

4.2 Design structure of essential resistance based on chaos

Study design of excitation device, and special to be fed into the battery is controlled by alternating current signal sampling card by measuring the battery voltage signal. The internal structure of the test design is shown in figure 1.

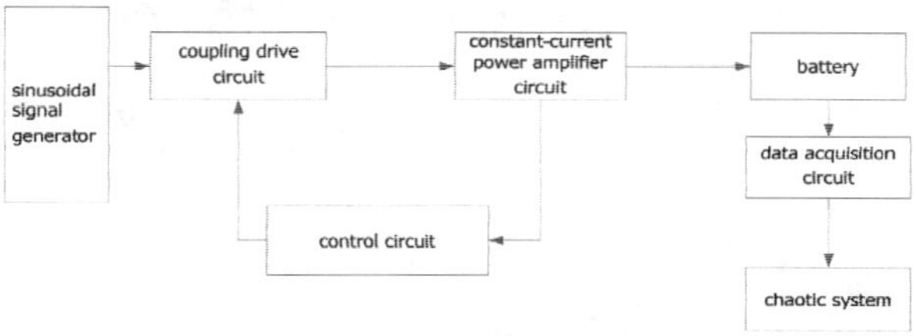

Figure 1. System diagram

This structure is mainly by the sinusoidal signal generator, coupling drive circuit, constant-current power amplifier circuit, control circuit, constant flow data acquisition circuit and chaotic system simulation model.

4.3 Battery resistance measurement principle

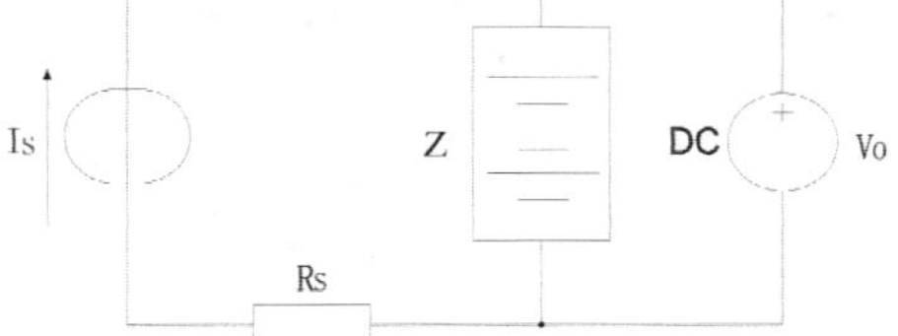

Figure 2. Measuring principle diagram of resistance four-wire method

Normally solar battery resistance is very small, at about milliohm, so the impedance measurement line can ignore. So using four-wire method for measuring the driving current loop, and induced voltage circuit. After four line method measuring principle shows in figure 2, such as R_S sampling resistance.

4.4 Chaotic system

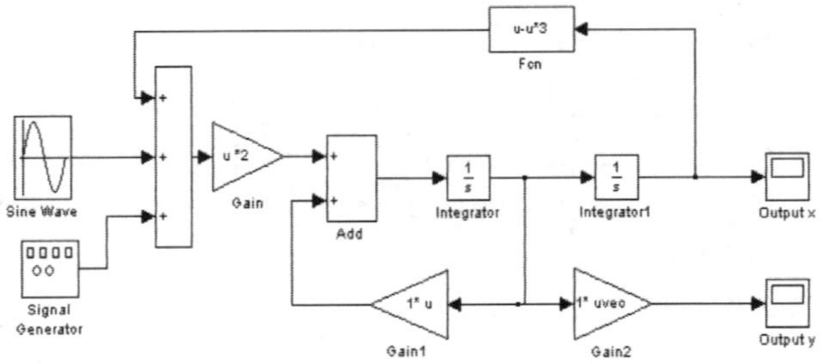

Figure 3. Duffing oscillator simulation model

Duffing oscillator simulation model in the system of chaotic systems shows in figure 3.

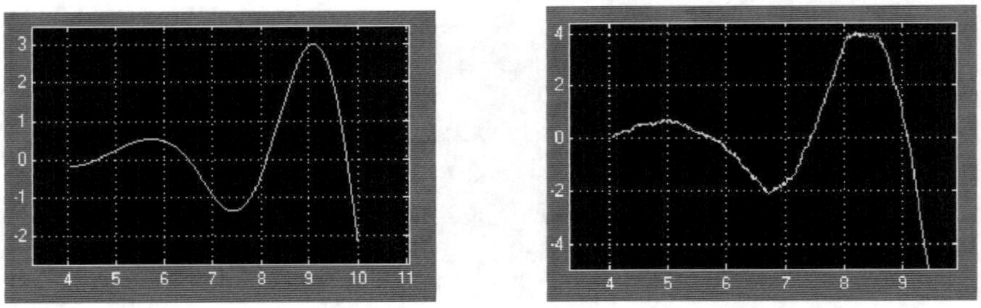

（a） waveform of output x （b）waveform of output y

Figure 4. Output waveform

The output of output x is in the cycle state of the Duffing system, and the output of output y is in the chaos state of the Duffing system. Therefore, the waveforms of output x and output y are extracted. Then the state of Duffing system is judged according to the relationships between the waveforms and their phase track. So select appropriate system parameter, and get the testing voltage value according to the figure 4.

5. Discussion and conclusion

An effective essential resistance detection method is achieved using ac impedance method and Duffing chaos system. Simulation results show that this method not only can more easily detect internal storage battery essential resistance and can accomplish the on-line detection.

References:

[1] Yu Peiliang，Chen Tixian. VRLA battery essential resistance[J]. Battery, 1995 (3) : p.325.

[2] Dong Chaohui. The behavior change rule of VRLA battery essential resistance under diffirent conditions[c]. The seventh conference on lead-acid batteries China electrotechnical society，2000.p.73-79.

[3] Ma Guangwei, Jing Youquan, Gan Linchuan. Battery resistance measurement and its application[J]. Communication power technology, 2007, 24(2):p.63-64.

[4] Liu Li. Study of weak signal detection approaches based on chaos theory [D] NCEPU, College of communication and information. 2006.1.17

[5] Cao Jianping, Jiang Guodong. Practical research of Dc charging method in the analysis of battery essential resistance. 2008.34(12):p.37-39.

The investigation of preparation by magnetron sputtering and application in amorphous silicon solar cells of pyramid-like textured ZnO:Al thin film

Y H Hu, H J Xu, H Gao, Y C Chen, L F Wang and W H Jiang
Department of Mechanical and Electronic Engineering, Jingdezhen Ceramics Institution, Jingdezhen, 333403, China

E-mail: 8489023@163.com

Abstract. In this paper, the pyramid-like texture aluminum doped zinc oxide (ZnO:Al) thin films was prepared by adding negative bias voltage and controlling the sputtering parameters properly, and the lowest resistivity of 7.8×10^{-4} ·cm, light transmittance above 83% was achieved, respectively. Instead of the traditional tin oxide doped with fluorine (SnO_2:F), using the ZnO:Al film as a front electrode transparent conducting oxide (TCO) layer for hydrogenated amorphous silicon (a-Si:H) solar cells, the higher contact potential barrier of ZnO:Al /p-a-SiC:H have to be overcome. In this paper, we attempt to resolve the contact potential barrier by inserting a buffer layer of microcrystalline silicon (μc-Si:H) between ZnO:Al layer and p-SiC: H windows layer, and its efficiency increases from 7.6% for the SnO_2:F TCO to 8.3% for the ZnO:Al /μc-Si:H TCO.

1. Introduction

As a front electrode transparent conducting oxide (TCO) layer for hydrogenated amorphous silicon (a-Si:H) solar cells, the electrical properties of tin oxide doped with fluorine (SnO_2:F) can be degraded seriously at the hydrogen plasma atmosphere. Compared with SnO_2:F, the textured aluminum doped zinc oxide (ZnO:Al) thin-film not only has the equivalent electrical properties, but also has a lot of advantages, such as the high electrical properties stability under the hydrogen plasma environment, the effective light trapping action. Those are the potential advantages in improving the a-Si:H solar cell performance when using it as a front electrode, and it has attracted a great deal of attention[1-3]. However, there are still two problems as follows: (1) the preparing technology of pyramid-like textured ZnO:Al thin-film is not easy to grasp by magnetron sputtering; (2) the problem of higher contact potential barrier between ZnO:Al layer and p-SiC: H windows layer need to be overcome when used as front electrode in solar cells.

In this paper, the pyramid-like texture ZnO:Al films was prepared by adding negative bias voltage and controlling the sputtering parameters properly. In order to use the ZnO:Al thin film as a front electrode in solar cells well, we attempt to resolve the contact potential barrier by inserting a buffer layer of microcrystalline silicon (μc-Si:H) between ZnO:Al layer and p-SiC: H windows layer, namely ZnO:Al /μc-Si:H TCO/pin solar cell was fabricated.

2. Experiment

The ZnO:Al films were deposited on the quartz substrates by RF magnetron sputtering. A sintered ceramic ZnO:Al target with 90 mm in diameter and 5mm of thickness was used. Before each deposition, the sputtering chamber was pumped down to below 4.2×10^{-4} Pa, and then working pressure was maintained at 1 Pa with Ar/O_2 (30/3) ambient gas and with the sputtering power of 150 W. The sample was prepared at 120 ℃ for 60min with negative bias voltage showing in table 1.

Table 1. Parameters of magnetron sputtering.

Bias voltage (V)	0V	20V	50V	100V	130V
Samples No.	A1	A2	A3	A4	A5

In order to study the light-scattering abilities of the textured surfaces for solar cell applications, two a-Si:H solar cells were deposited in an RF PECVD system. The each solar cell structure was as follows: the cell A with structure of glass/SnO_2:F (0.7 μm) / p-SiC:H, p-a-Si:H (~30 nm) / i-a-Si:H (~0.4 μm) / n-a-Si:H (~35 nm) /Ag; the cell B with structure of glass /textured ZnO:Al (0.8 μm) / μc-Si:H (~90 nm) / p-SiC:H, p-a-Si:H (~30 nm) / i-a-Si:H (~0.4 μm) / n-a-Si:H (~35 nm) / Ag.
The resistivity of the sample was measured by four-probe method. A X-ray instrument (XRD-Rigaku D/ max 2500) was used to characterize crystallization orientation of sample, SEM (SEM-JSM 6700F) was used to observe surface morphology of the film. For the thin film light-transmission, A UV2Vis-NIR Spectrophotometer (Backman-Du 8B Spectrophotometer) whose wavelength is in the range of 400~900 nm was carried out. All the sample measurements were performed at room temperature. The solar cell illuminated current-voltage characteristics were measured using a solar simulator (Wacom WXS-140S-Super) at standard test conditions (AM 1.5, 100 mW/cm^2, 25 ℃).

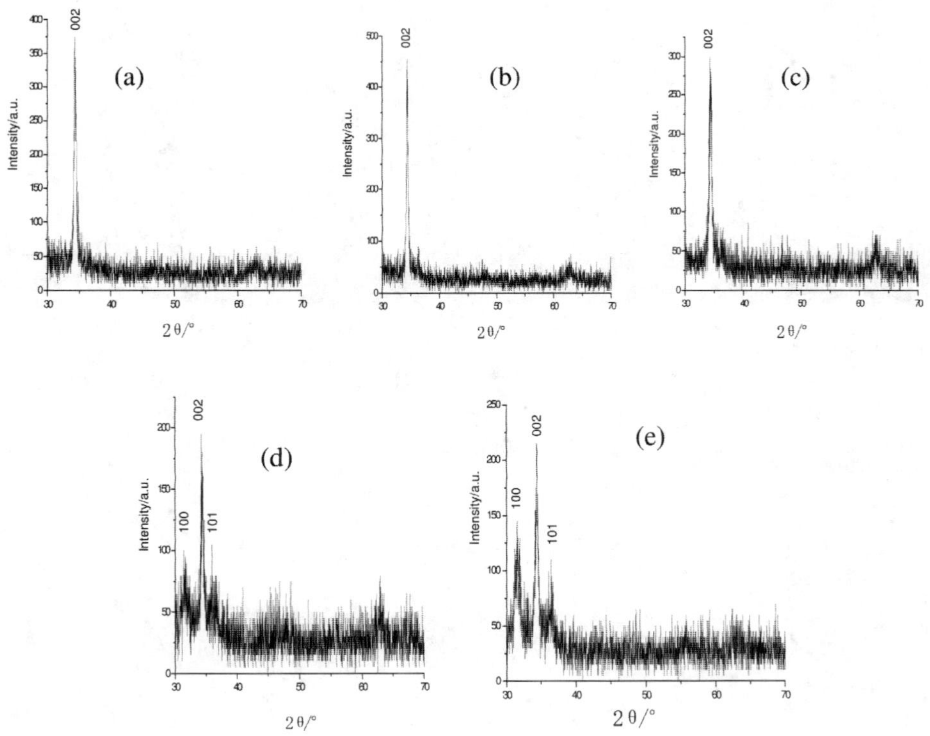

Figure 1. The XRD patterns of samples A1, A2, A3, A4 and A5.

3. Results and Discussion

3.1. The influence of the bias on the structure of ZnO:Al films

The XRD patterns of samples with different bias are shown in Fig.1. As shown in Fig.1, the sample A1 has a preferred orientation with the c axis perpendicular to the substrate, which agreed with a large number of research results of the ZnO:Al films fabricated by magnetron sputtering. From the XRD pattern of the sample A2 whose bias is 20V in Fig.1(b), the peak intensities of (002) preferred orientation is somewhat increase. It is maybe that the larger bias increase particles energy, and the particles have enough energy to reach to the lowest energy place after reaching the substrate. The XRD pattern of sample A3 is given in Fig.1(c). It can be seen the value of (002) peak decrease and the preferred orientation is inhibited with the bias increasing to 50V. The XRD patterns of samples A4 and A5 are shown in Fig.1(d) and Fig.1(e), respectively. The value of (002) peak are getting smaller and smaller with bias increasing, and the peaks (100) and (101) appear when the bias is higher than 100V. These suggest that the growth of ZnO: Al thin film and crystal quality may be changed. The result can be interpreted as follows: the bigger of bias voltage is, the higher of particles` energy is. Therefore, there are large of particles reaching the substrate, and the growth of ZnO: Al thin film has the chance to change from the lowest energy (002) crystal plane to the higher energy (101) and (100) crystal plane.

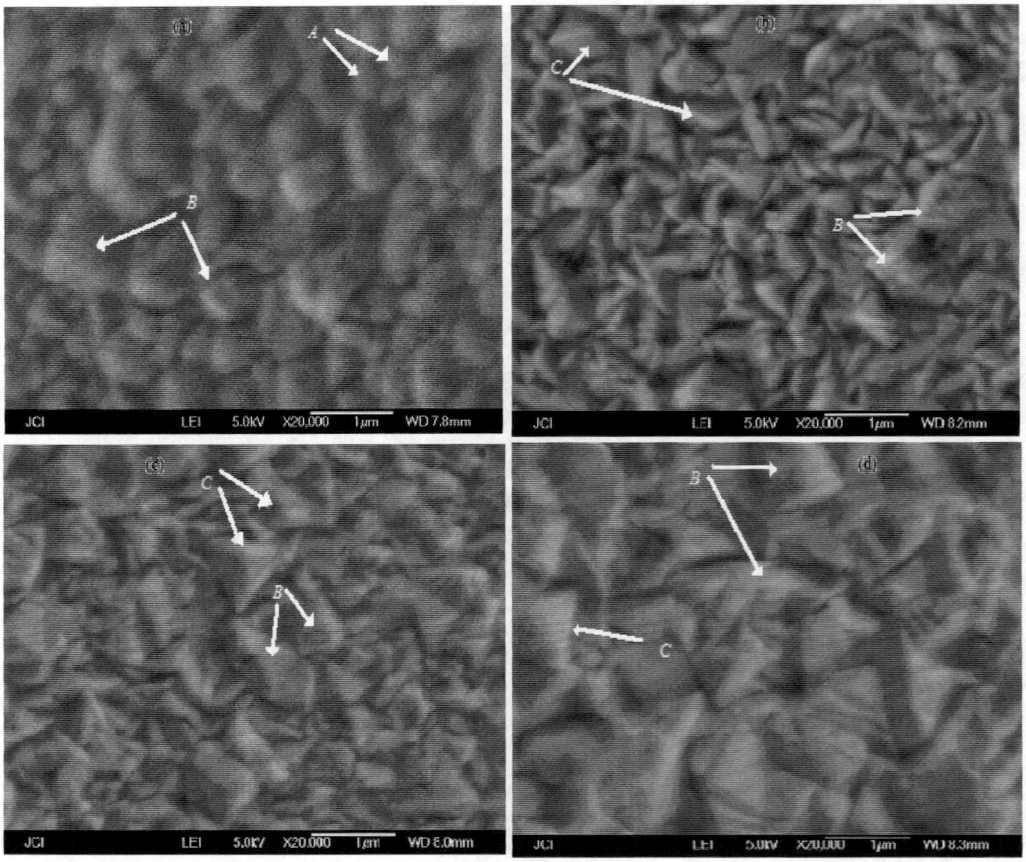

Figure 2. The SEM patterns of samples A2, A3, A4 and A5. **A** represented columnar structure, **B** represented polygon structure, **C** represented pyramid-like structure.

3.2. The influence of the bias on the surface morphology of ZnO:Al films

The SEM pictures of samples A2, A3, A4 and A5 are shown in Figure 2 (a), (b), (c) and (d), respectively. It can be seen that surface morphology of A2 sample shows the mainly columnar structure with relative small grain size, containing some parts of the polygon structure. Samples A3 and A4 possess obvious pyramid-shaped surface structure with larger grain size, also including some parts of the polygon structure. While sample A5 shows the polygon surface structure with the largest grain size, containing some parts of the pyramid-like structure. There is a possible reason to explain the phenomena above. As negative bias increase, not only most of ions grow along the lowest surface energy (002) face, but also some higher energy ions can grow along other crystal face. According to the literature [4], ZnO: Al film grains exhibit an obvious columnar growth, when it appears the preferred orientation peak (002). When (100) peak of thin film is observed, the grain growth along rod or polygon structure, while the appearance of (101) peak suggests that the grain growth exhibits pyramid-like structure, and the film surfaces are roughness with a certain textured structure.

3.3. The influence of the bias on the optical and electrical properties of ZnO:Al films

Fig. 3 shows the UV transmission spectrum of samples A1, A4, A5. The visible light transmission rate was above 83%. Among all the samples, sample A4 has the lowest light transmittance, and sample A1 has the highest one. According to the literature [5], sample A4 is relatively thinner compared with the other two samples, and samples A1 and A5 have the same thickness. However, the light transmittance of sample A4 is the lowest, and the light transmittance of sample A5 is lower than sample A1. The reason for sample A4 and A5 with abnormal light transmittance phenomenon is that rough surfaces of samples A4 and A5 with like-pyramid structure morphology can enhance scattering effect of short-wavelength light, and light vertical transmittance decreases obviously, which can be proved by SEM images in figure 2 (c) and (d).

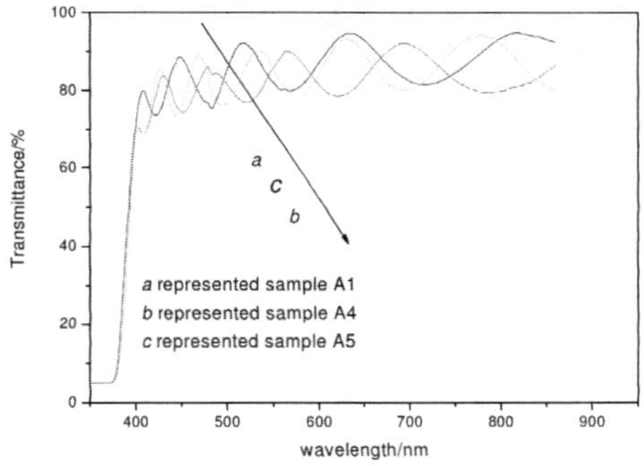

Figure 3. Optical transmission spectra of spectrum of samples A1, A4 and A5.

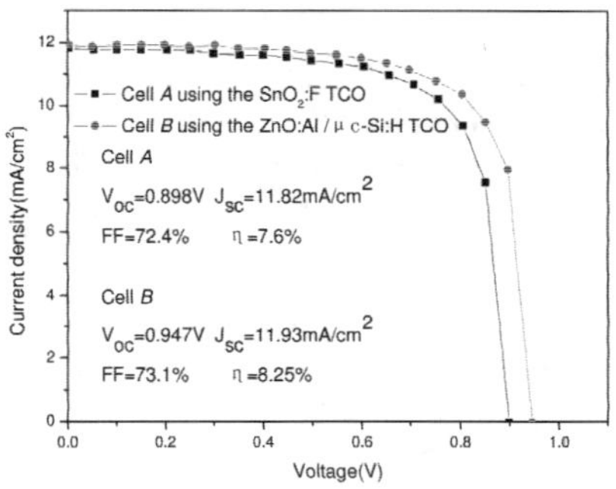

Figure 4. The I-V curves for solar cell A and B

The resistivity of sample A1, A2, A3, A4 and A5 was measured. For sample A1, the resistance is high which can not detect by four-probe method, and for A2, A3, A4 and A5, the resistance is 30 •cm, 0.146 •cm, 0.083 •cm, and 0.00078 •cm, respectively. From the electrical properties measurement, it is shown that the resistivity decreases with the increasing of the bias. This result may be relative to grain size, and it can be proved by SEM image in figure 2. The increase of the ZnO:Al grain size can decrease the mounts of grain boundary, and reduce the carries scattering effect, which can improve the electron mobility.

The synthetic analysis of electrical properties and the light trapping properties indicated that the sample A4 is an ideal transparent conductive film for amorphous silicon solar cells.

3.4. The application of textured ZnO:Al thin films on amorphous silicon solar cells

For comparison, cell A using SnO_2:F and cell B with ZnO:Al films of sample A4 as transparent conductive electrodes is fabricated, respectively. As buffer layer a μc-Si:H layer is inserted between ZnO:Al layer and p-SiC:H layer for cell B. The I-V curves of the two type solar cells are shown in figure 4. It can be seen that efficiency increases from 7.6% for the SnO_2:F TCO to 8.3% for the ZnO:Al /μc-Si:H TCO. The results indicated that the ZnO:Al /μc-Si:H films with pyramid-shaped textured structure as front electrode can improve the cell performance compared with that of SnO_2:F. Meanwhile, μc-Si:H, as a buffer layer, can overcome the higher contact potential barrier of of ZnO:Al /p-a-SiC:H.

4. Conclusions

The following results have been obtained: (1) the pyramid-like textured ZnO:Al films were prepared directly by magnetron sputtering through adding negative bias voltage and controlling the sputtering parameters properly. (2)when the pyramid-like textured ZnO:Al film was used as a front electrode transparent conducting oxide layer for a-Si:H solar cells, a μc-Si:H layer was inserted between ZnO:Al and p-a-SiC:H, and it is in favor to the improvement of solar cell performance.

Acknowledgment

This work was financially supported by the following funds: (1) the Natural Science Foundation of China (No. 60940016); (2) the Education Bureau of Jiangxi Province, China (No. GJJ08317, GJJ08318, GJJ09540, GJJ10567, GJJ10568); and (3) the Natural Science Foundation of Jiangxi Province (No. 2007Gzw0787, 2009GZW0013).

References

[1] Berginski M, Hüpkes J, Reetz W, Rech B and Wuttig M 2008 *Thin Solid Films* **516** 5836
[2] Groenen R, Linden J L and van Lierop H R M 2001 *Applied Surface Science* **173** 40
[3] Schropp R E I, Li H and Franken R H J 2009 *Solar Energy Materials & Solar Cells* **93** 1129
[4] Hu Y H, Wang L F and Xu H J 2010 *Proc. Int. Conf. on Advanced Optical Manufacturing and Testing Technologies 2010 (Dalian)* vol. **7658** (Dalian: SPIE proc) p. 7658E-1
[5] Swanepoel R 1983 *J. Phys. E: Sci. Instrum.* **16** 1214

Journal of Physics: Conference Series **276** (2011) 012168 doi:10.1088/1742-6596/276/1/012168

Optical properties of $BiFeO_3$ and $Bi_{0.9}La_{0.1}$ FeO_3 films on silicon substrates

D J Huang[1], H M Deng[2], F Chen[1], H Deng[1], P X Yang[1*] and J H Chu[1]

[1] Key Laboratory of Polar Materials and Devices, Ministry of Education, Department of electronics, East China Normal University, 500 Dongchuan Road, Shanghai 200241, China
[2] Key Laboratory for Material Microstructures of Shanghai University, Shanghai University, 99 Shangda Rd, Shanghai 200444, China

Abstract. $BiFeO_3$ (BFO) and $Bi_{0.9}La_{0.1}FeO_3$ (BLFO) thin films were deposited on Si (100) wafer by sol-gel technique. The microstructure and surface morphology of the films were characterized using X-ray diffraction and atomic force microscope. After annealing at 600 °C, all the films were well-crystallized, and the diffraction peaks corresponded to BFO standard pattern quite well. The optical properties of BFO and BLFO film were studied by spectroscopic ellipsometry in the wavelength range 420 to1700 nm. Compared with the BFO film, the BLFO film has lower refraction index and higher extinction coefficient in the wavelength range 600 to 1700 nm. The absorption coefficient of the BLFO film is greater than 1700cm^{-1} in total measured wavelength region. The results suggest that the potential role of La dopant improved the optical properties of BFO thin film and the BLFO film is a promising material for photovoltaic devices.

1. Introduction

In recent years, multiferroic materials have attracted extensive attention due to their tremendous potential for applications in multiple state memory elements and other integrated ferroelectric devices. Among all multiferroic materials studied so far, ABO_3-type perovskite structure $BiFeO_3$ (BFO) is a known Pb-free and environmentally friendly material, which possesses a high Curie temperature (TC~1103 K), and a high Néel temperature (TN~643 K) at room temperature[1]. The BFO film also has other excellent applications such as photocatalytic compound[2], infrared detector and ultrafast optoelectronic device due to its small bandgap (Eg~2.5 eV)[3]. Recent researches indicate that, a substantial visible-light photovoltaic effect was observed in BFO diode structures, which has demonstrated the photovoltaic application of BFO films[4]. In this paper, the BFO and BLFO films were deposited on Si (100) wafer by sol-gel technique, the microstructure, surface morphology and the optical properties of the films were studied.

2. Experiment

Bismuth nitrate $[Bi(NO_3)_3{\cdot}5H_2O]$, iron nitrate $[Fe(NO_3)_3{\cdot}9H_2O]$, and lanthanum nitrate $[La(NO_3)_3{\cdot}6H_2O]$ were dissolved in glacial acetic acid under magnetic stirring for 30 min, and bismuth

* Corresponding author. Tel.: +86 21 54345157; fax: +86 21 54345119. E-mail address: pxyang@ee.ecnu.edu.cn (PX. Yang).

nitrate should be 5% excess to compensate the evaporation of bismuth during the high temperature annealing. Then ethylene glycol was added to adjust the stickiness of the precursor. After the solution was stirred for another 30 min, the sol was obtained and then was aged for 72 h at room temperature. The samples were fabricated by spin-coating technique on silicon wafer with the velocity 4000~5000 rpm for 20 seconds. After each coating, the coated layer was annealed using rapid thermal processor (RTP) in air. The layer was annealed at 200 °C for 2 min, 350 °C for 3 min, and 600 °C for 4 min. These coating and annealing processes were repeated for several times to achieve desired film thickness.

The microstructures analysis of the films was carried out by X-ray diffraction (XRD, D/MAX 2550, with Cu Kα radiation). The surface morphology was detected by atomic force microscope (AFM, DEMENSION 3100). The optical properties were investigated via spectroscopic ellipsometry (SE) with an angle of incidence of 70° in the wavelength range of 420–1700 nm. All measurements were carried out at room temperature.

3. Results and discussion

The XRD patterns of the BFO and BLFO thin films are shown in Figure 1. The BFO film and BLFO film being well crystallized with (012), (110), (202), (024), (122) and (300) peaks can be observed. The BFO film with only a small second-phase of $Bi_2Fe_4O_9$ can be detected [5], while the BLFO film shows almost single-phase and without other phase. As a volatile element Bi will evaporate during the high temperature annealing, the stable nature of lanthanum compared to the volatile bismuth might have played a role in the suppression of secondary and impure phases [6]. It is obvious that the dopant of La has not changed the structure of the BFO film. The fine crystallite size of 13.1 nm and 17.8 nm were obtained from the dominant diffraction peak at (110) of the BFO and BLFO films using Scherrer's formula, respectively.

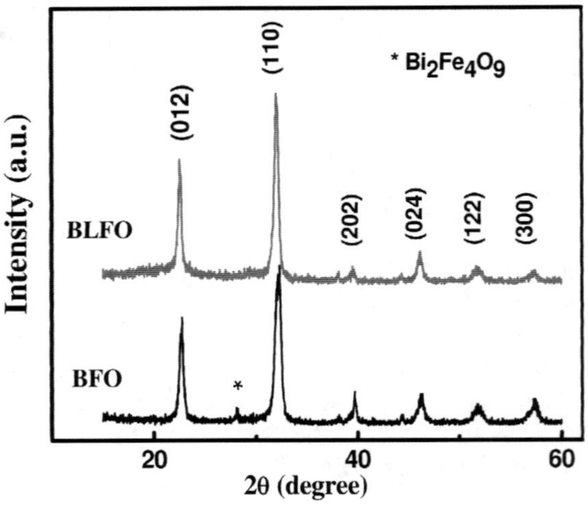

Figure 1. The XRD patterns of BFO and BLFO thin films grown on Si substrates by sol-gel method.

The microstructure and surface morphology of the thin films are shown in the Figure 2. Both of the films have a continuous surface microstructure without presence of particulates and cracks. It can be seen that the surface of BFO films are uniform and dense, the grains of the film are small and the grain boundary is clear. The BLFO film exhibits less dense but larger grain size, and the film surface is not as uniform as BFO, indicating that the doping La influences the microstructure and the surface morphology of the films.

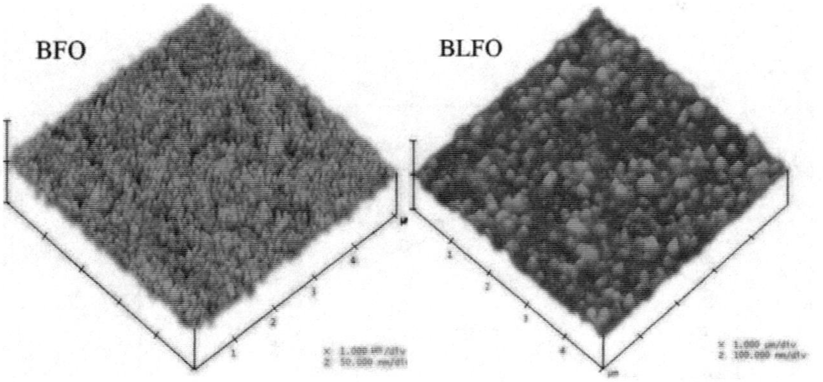

Figure 2. AFM images (5μm×5μm) of BFO /Si and BLFO/ Si thin films by sol-gel method.

Figure 3 demonstrates the refractive index n and extinction coefficient k of the BFO and BLFO thin films. These optical coefficients of films were obtained from the Lorentz oscillator model [7] using optical thin film software, namely SC630UVN (Shanghai Sanco Instrument Co. Ltd)

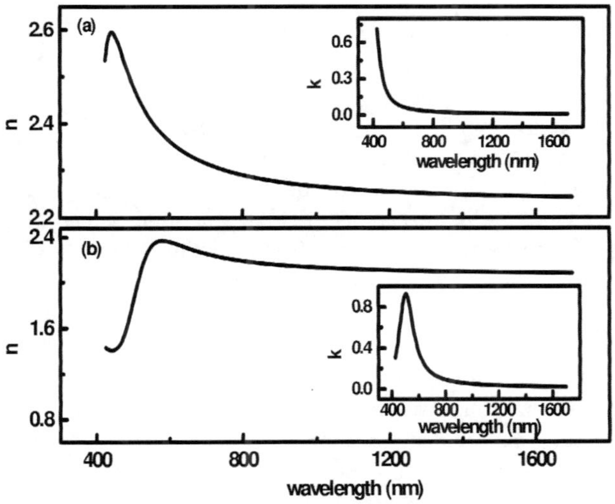

Figure 3. Refractive index (n) and extinction coefficient (k) (inset) curves of the BFO film (a) and the BLFO film (b).

As shown in Figure 3(a), a peak of the refractive index of the BFO film estimated to be approximately 2.6 appears at 445 nm. Then the refractive index slightly declines with wavelength increasing, and is smaller than 2.265 when the wavelength is larger than 1000 nm. The extinction coefficient drops sharply from 420 to 600 nm and tends to smooth when the wavelength is beyond 800 nm. On the other hand, the refraction index of the BLFO film increases sharply and shows a maximum at about 577 nm in Figure 3(b). Then the refractive index decreases smoothly with the wavelength increasing. Compared with that of the BFO film, the refractive index of the BLFO film is smaller. The extinction coefficient of the BLFO film reaches its maximum of 0.92 at 512 nm, and then decreases rapidly with the wavelength increasing.

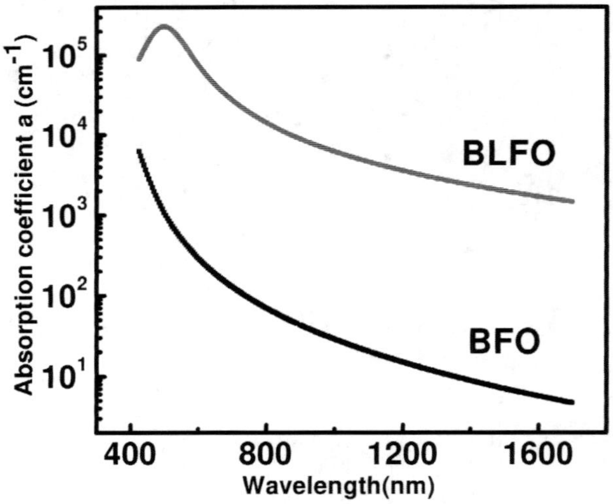

Figure 4. The absorption coefficient α of BFO and BLFO thin films.

The absorption coefficient α of the films, deduced from the extinction coefficient by the formula: $\alpha=4\pi k/\lambda$, are shown in Figure 4. It is seen that the absorption coefficient of the BFO film declines when the wavelength increases. The value is 6240 cm^{-1} at 420 nm and less than 10 cm^{-1} when the wavelength is lager than 1300 nm. While the absorption coefficient of BLFO is greater than 1489 cm^{-1} in the measured wavelength region and reaches its maximum of 230969 cm^{-1} at 500 nm, which means a great absorption of blue and green light. This absorption coefficient of the BLFO film is greater than that of the BFO films reported by other researchers [8] but still smaller than that of the single crystal BFO films [9], which is resulted from the dopant of La, indicating a favorable light absorption of the BLFO film.

4. Conclusion

The BFO and BLFO thin films were successfully prepared on Si substrates by sol-gel method. The XRD patterns showed that all the films were rhombohedral perovskite structure and well crystallized. The films were uniform and without crack from the AFM analysis. The La dopant leads to a larger grain structure but less dense film. The maximum absorption coefficient of the BLFO film reaches order of magnitude 10^5 cm^{-1}. These results could be attributed to the changes of microstructure caused by the dopants. The potential role of La dopant improving the optical properties of the BFO thin film and the obtained BLFO film is a promising material for photovoltaic devices.

Acknowledgements

The authors wish to thank Mr. Xu Wengfei for AFM investigation, Mr. Zhang Jingzhong for optical investigation. This work was supported by the National Natural Science Foundation of China (No. 60990312), the State Key Basic Research Program of China (2007CB924902), Shanghai Leading Academic Discipline Project (B411) and Science and Technology Commission of Shanghai Municipality (10JC1404600).

Reference

[1] Fruth V, Popa M, Calderon-Moreno J M, Anghel E M , D. Berger , Gartner M , Anastasescu M, Osiceanu P, Zaharescu M 2007 *J. Eur. Ceram. Soc* **27** 4417–4420
[2] Li Shun, Lin Yuan-Hua, Zhang Bo-Ping, Nan Ce-Wen, and Wang Yao 2009 *J. Appl. Phys.* **105** 056105
[3] Quan Zuci, Liu Wei, Hu Hao, Sheng Xu, Sebo Bobby, Fang Guojia, Li Meiya, and Zhao Xingzhong 2008 *J. Appl. Phys.* **104** 084106
[4] Choi T, Lee S, Choi Y J, Kiryukhin V, Cheong S W 2009 *Science* **324** 63-66
[5] Carvalho T T, Tavares P B 2008 *Mater Lett* **62** 3984–3986
[6] Kazhugasalamoorthy S, Jegatheesan P, Mohandoss R, Giridharan N V, Karthikeyan B, Justin Joseyphus R , Dhanuskodi S 2010 *J. alloys compd* **493** 569–572
[7] Ding L, Chen T P, Liu Y, Ng C Y 2006 *J. Cryst. Growth* **288** 87–91
[8] Gu Bing , Wang Yang , Wang John , and Ji Wei 2009 *Opt. Express* **17** 10970- 10975
[9] Xu X S, Brinzari T V, Lee S, Chu Y H, Martin L W, Kumar A, McGill S, Rai R C, Ramesh R, Gopalan V, Cheong S W, and Musfeldt J L 2009 *Phys. Rev. B* **79** 134425

Research on the Characteristics of Organic Solar Cells

Jiang Chuan, Li Tianze, Hou Luan, Zhang Xia

School of Electric and Electronic Engineering, Shandong University of Technology

No.12, Zhang Zhou Road, Zhangdian District, Zibo 255049, China

E-mail: ltzwang@163.com

Abstract. Recently, organic solar cells are attracting a lot of attention due to flexibility, low cost, light weight and large-area applications, and significant improvement in the power conversion efficiency. But the efficiencies of organic solar cells are still lower than crystal silicon solar cells. By studying the photovoltaic properties of organic solar cells it is possible to influence the short circuit current (I_{sc}), the open circuit voltage (V_{oc}), the fill factor (FF), hence the cells efficiencies can be improved. First of all, the paper describes the characteristics of organic solar cell. Several characteristics impact on the conversion efficiency of organic solar cells, including the series resistance (R_s), the shunt resistance (R_{sh}) and so on. A circuit model of the solar cell provides a quantitative estimate for losses in the solar cell to interpret the characteristics of the solar cell. The conventional circuit model which is suitable for inorganic solar cells has emerged. However, compared with inorganic solar cells, organic solar cells are lack of a three-dimensional crystal lattice, different intramolecular and intermolecular interactions, local structural disorders, amorphous and crystalline regions, and chemical impurities. Therefore, an appropriate circuit model is represented in the paper.

1. Introduction

The worldwide consumption of energy has increased every year by several percentages over the last thirty years [1]. Nowadays, most of energy is supplied by fossil fuels on the one hand and by nuclear energy on the other hand. However, these resources are limited and their use has a serious environmental impact, which extends probably over several future generations. Solar light is the most important source of regenerative energy and represents mankind's only inexhaustible energy source. One of today's most promising tools to make use of solar energy is its direct conversion into electrical energy in photovoltaic cells [2]. The applied photovoltaic cells mainly are silicon-based solar cells in most cases. However, compared with traditional silicon-based solar cells, Organic solar cells are a promising way towards large-area and low-price photovoltaic systems. The main advantages are the easy preparation, low process temperatures, low-cost materials and processing technology and the possibility to produce flexible devices on plastic substrates [3].When the power conversion efficiency of organic solar cells is close to 10%, it can be served as an excellent candidate.

The low power conversion efficiency of organic solar cells confines its application. The power conversion efficiency was only 10-5% about thirty years ago. It is reported that the highest power conversion efficiency of polymer-based bulk-heterojunction solar cells is approach to 5% and the efficiency of small-molecule-based solar cells is up to 6.0% [4]. It is gratifying that there's been a significant progress. It can encourage more and more people to go on research.

When researching organic solar cells, we should figure out the characteristics of organic solar cells and elements which impact on the power conversion efficiency of organic solar cells. An equivalent circuit of organic solar cells contributes to understand and analyze the characteristics of organic solar cells. This paper elaborates the characteristics of organic solar cells by analyzing equivalent circuits of organic solar cells.

2. Characteristics of organic solar cells

Organic solar cells and inorganic solar cells are based on photovoltaic effect. There are some well-known differences between organic solar cells and inorganic solar cells. Basically, compared with inorganic solar cells, organic solar cells are lack of a three-dimensional crystal lattice, different intramolecular and intermolecular interactions, local structural disorders, amorphous and crystalline regions, and chemical impurities. Obviously, when light absorption in inorganic solar cells, it leads to the formation of more free carriers, but when light absorption in organic solar cells, it leads to the formation of more excitons. Excitons have more binding energy in organic solar cells, and the mobilities of charge carriers are far lower in organic solar cells than in inorganic solar cells. As one of the many consequences of this, the power conversion efficiency of organic solar cells is much less than inorganic solar cells.

3. Equivalent circuit

3.1 Conventional equivalent circuit

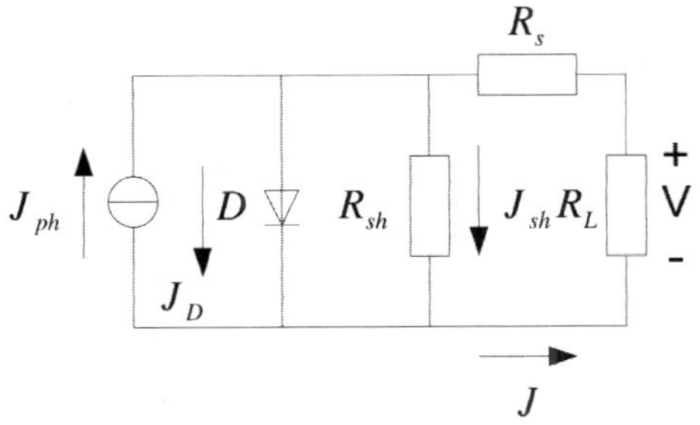

Figure 1. Conventional equivalent circuit of solar cells

Conventional equivalent circuit is usually adequate for describing characteristics of inorganic solar cells. Figure 1 shows conventional equivalent circuit of solar cells. J_{ph} represents the photo-generated current, D is a diode which expresses current-voltage characteristics under dark condition, R_s is the parasitic series resistance, R_{sh} is the shunt resistance, R_L is the load resistance, and V expresses the voltage of R_L. The following equations which express conventional equivalent circuit can be derived from the Kirchhoff's Voltage Law and the Kirchhoff's Current Law.

$$V = J_{sh}R_{sh} + JR_s \tag{1}$$

$$J + J_{ph} = J_D + J_{sh} \tag{2}$$

$$J_D = J_0(e^{(V - JR_s)/nV_T} - 1) \tag{3}$$

Solving above equations, we obtained the following equation:

$$\ln(\frac{J + J_{ph}}{J_0} - \frac{V - JR_s}{J_0 R_{sh}} + 1) = \frac{V - JR_s}{nV_T} \tag{4}$$

Where, J_0 is the saturation current density, n is the diode ideality factor, and V_T is the temperature potential.

The fourth equation is a transcendental equation. It is very difficult to solve the equation. Jain and Kapoor used Lambert W-function to solve the equation. It is proved that using Lambert W-function can make solving more previously [5].

3.2 Two-diode equivalent circuit

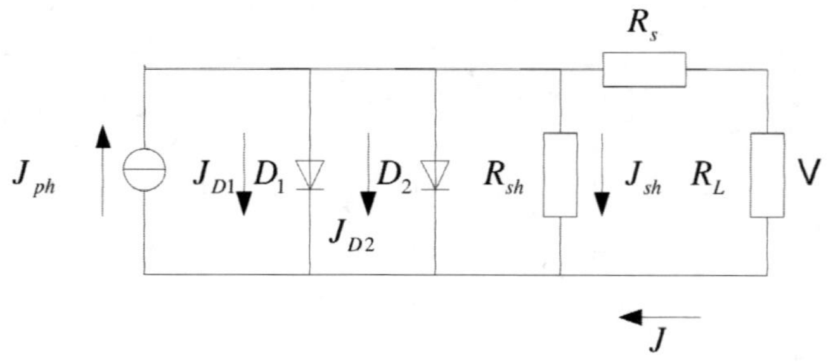

Figure 2. Two-diode equivalent circuit of solar cells

Two-diode equivalent circuit has been used to describe the characteristics of solar cells under illumination condition for thirty years. Figure 2 shows the two-diode equivalent circuit. We got the next equation from the Kirchhoff's Current Law.

$$J = J_{ph} - \frac{V + JR_s}{R_{sh}} - J_{D1}[e^{B_1(V+JR_s)} - 1] - J_{D2}[e^{B2(V+JR_s)} - 1] \tag{5}$$

Where, $B_1 = q/kT$, $B_2 = B_1/n$.

D_1 diode stands for the electronic conduction phenomena in the quasi neutral region of the junction, such as diffusion, recombination, and the drift effect. D_2 diode represents the carrier recombination via deep levels in the space-charge region of the junction, with n = 2 for the approximation corresponding to the Shockley–Read–Hall recombination current density in the space-charge region. This model is consistent with the underlying physics of the electrical characteristics and should give an accurate description of the real solar cell for low-level-injection conditions.

3.3 A simplified two-diode equivalent circuit

A central assumption in the conventional model is that photo-generated current Jph is constant for given incident light intensity and is independent of voltage. Mazhari put forward a simplified two-diode equivalent circuit. The central assumption of his viewpoint is photo-generated current Jph is a function of voltage. Figure 3 shows a simplified two-diode equivalent circuit. R_{sh}^{int} represents loss due to polaron recombination and R_s^{int} models charge extraction to the electrodes. Diodes D_1 and D_2 are ideal diodes that are like short circuit under forward bias and open circuit under reverse bias, respectively. Diode D_{dark} represents the current-voltage characteristics of organic solar cells under dark condition. Vint expresses the internal voltage.

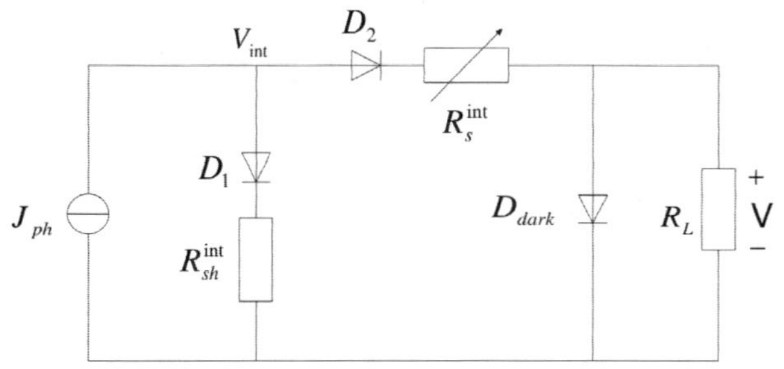

Figure 3. A simplified two-diode equivalent circuit of organic solar cells

4. Discussion and conclusion

The single diode equivalent circuit usually is used to illustrate the characteristics of solar cells. It is a kind of simple understanding about characteristics of solar cells. Compared other kinds of solar cells, organic solar cells have unique properties. The two-diode equivalent circuit has occurred for thirty years. It considers low-level-injection conditions in short. The injection level of organic solar cells is low, which is in favor of more precise illustrating the characteristics of organic solar cells. The third circuit considers more factors. B. Mazhari has proven that his method is adequate to depict the characteristics of organic solar cells.

When measuring solar cells, temperature is 25℃ and light intensity is $100mW/cm^2$, which are International measuring standards. Ali Cheknane et al made a measuring experiment. They chose an MEH-PPV. It is proven that the single diode equivalent circuit is not adequate to depict the characteristics of organic solar cells. When light intensity is $100mW/cm^2$, the third equivalent circuit stands for elaborating the characteristics of organic solar cells.

References

[1] Brabec, C., et al., *Production Aspects.* Mrs Bulletin, 2005. 30: p. 51.

[2] Wöhrle, D. and D. Meissner, *Organic solar cells.* Advanced Materials, 1991. 3(3): p. 129-138.

[3] Maennig, B., et al., *Organic p-i-n solar cells.* Applied Physics A: Materials Science & Processing, 2004. 79(1): p. 1-14.

[4] Shrotriya, V., et al., *Accurate measurement and characterization of organic solar cells.* Advanced Functional Materials, 2006. 16(15): p. 2016-2023.

[5] Jain, A. and A. Kapoor, *A new approach to study organic solar cell using Lambert W-function.* Solar Energy Materials and Solar Cells, 2005. 86(2): p. 197-205.

[6] Mazhari, B. (2006). "An improved solar cell circuit model for organic solar cells." Solar Energy Materials and Solar Cells 90(7-8): 1021-1033.

[7] Charles, J., I. Mekkaoui-Alaoui, et al. (1985). "A critical study of the effectiveness of the single and double exponential models for IV characterization of solar cells." Solid-State Electronics 28(8): 807-820.

[8] Cheknane, A., H. Hilal, et al. (2008). "An equivalent circuit approach to organic solar cell modelling." Microelectronics Journal 39(10): 1173-1180.

3rd International Photonics & OptoElectronics Meetings (POEM 2010) IOP Publishing
Journal of Physics: Conference Series **276** (2011) 012170 doi:10.1088/1742-6596/276/1/012170

Processing Parameters and Property of AZO Thin Film Prepared by Magnetron Sputtering

Hui Kong [1,2], Pingxiong Yang [1*], Junhao. Chu [1]

[1] College of Information Science & Technology, East China Normal University, 500 Dong chuan Road,Shanghai 200241, People's Republic of China

[2] Shanghai Topsolar Green Energy Co.,Ltd , No.951 Jianchuan Road , Shanghai 200240, People's Republic of China

E-mail: pxyang@ee.ecnu.edu.cn

Abstract. As we all know, transparent and conductive Al:ZnO (AZO) films are being considered for manufacturing transparent electrodes in flat panel displays, solar cells and organic light emitting diodes due to their high electro optical quality, high material availability and low material cost for large area applications. In this experiment, AZO thin films were prepared on glass substrate by Magnetron Sputtering method with a 2wt % Al –doped ceramic target .The processing parameters on electrical properties of AZO thin film were investigated. When processing pressure is 0.8Pa,the sputtering power is 225W， the temperature of substrate is 225℃, the AZO thin films have a lowest resistivity of 9.81×10^{-4} Ω.cm and visible light transmittance of more than 85%..

1. Introduction

As we all know, transparent and conductive Al:ZnO (AZO) films are being considered for manufacturing transparent electrodes in flat panel displays, solar cells and organic light emitting diodes due to their high electro optical quality, high material availability and low material cost for large area applications.[1-2] There are several deposition techniques to grow AZO films such as chemical vapor deposition (CVD) [3], spray pyrolysis [4], pulsed laser deposition [5], magnetron sputtering [6] and sol–gel [7].Among the processes used to prepare AZO coatings, magnetron sputtering is considered to be a suitable technique due to its inherent characteristics such as high deposition rate, good Controllability and scalability to large areas.

In this experiment, AZO thin films were prepared on glass substrate by Magnetron Sputtering method with a 2wt % Al –doped ceramic target .The processing parameters on electrical properties of AZO thin film were investigated.

2. Experimental details

The AZO films were deposited onto glass substrates in a homemade radio- frequency magnetron sputtering system. A 2wt % Al –doped ceramic target was used. Glass substrates were ultrasonically and chemically cleaned in acetone， alcohol and deionized water, and dried. Then, the substrates were mounted on a 50 mm diameter holder. The target-substrate distance was 40 mm. The chamber was evacuated to a base pressure of < 1×10^{-4} Torr by a turbomolecular pump. The RF sputtering power was 125-225 W and the sputtering time was 20min. Before deposition, the target was presputtered for 10 min prior to the film deposition. During the deposition, the substrate holder continuously rotated to enhance film uniformity.

The morphology, electrical and optical properties of AZO films were examined and characterized. The influence of sputter power, working pressure on the electrical and optical properties of the deposited film was investigated in detail. The surface roughness was measured by an atomic force microscope (AFM). Optical transmittance of the films was measured by using a UV-VIS-NIR spectrophotometer. The resistivity of AZO thin films were determined by a Hall Effect measurement and the sheet resistance measured by a four-point probe measurement.

3.Results and discussion

The two criteria of TCO (Transparent Conductive Oxide) thin films are highly transparent in the visible wavelength range where the solar cell is operating to minimize the photon absorption and high conductivity to reduce the resistive losses. The electrical properties of AZO films are essential to adjust the sputtering parameters for its change obviously, so in this experiment sheet resistance and the resistivity were investigated in detail.

There is an equation between the sheet resistance and the resistivity:

$$Rs = \frac{\rho}{t}$$

where ρ = resistivity and t = thickness of films. Rs = sheet resistance

Figure1 shows the sheet resistance and resistivity with different O2 flow rate at room temperature under conditions of the sputtering power of 225w, process pressure of 0.8Pa and thinkness of 4780-4800nm. With an increase in O2 flow rate from 0 to 20 sccm, the resistivity goes from 3.88×10^{-2} Ω.cm to 4.28×10^{-1} Ω.cm and the sheet resistance from 81Ω to 895 Ω . The conduction mechanism of TCO investigated by many researchers can be classified into two large groups: one was the occurrence of intrinsic carrier by oxygen vacancy, and the other was the introduction of extrinsic carrier by impurity doping. The occurrence of the carrier caused by the development of resistivity was greatly affected by impurity doping than oxygen vacancy when the film was deposited by the doped target. The behaviour was consistent with the other authors in previous works [8].

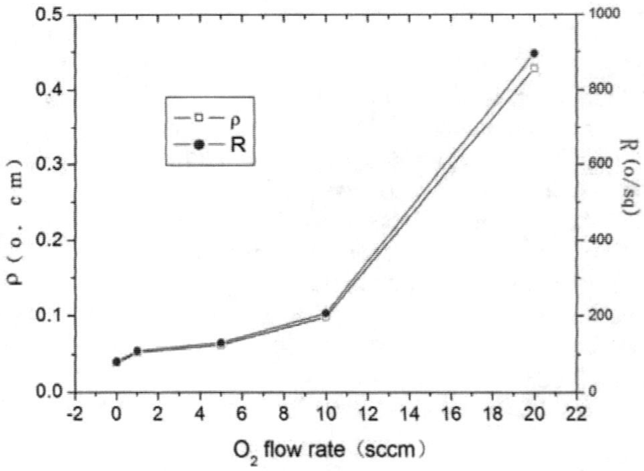

Figure .1 Sheet resistance and resistivity with different O2 flow rate at room temperature

Several researches on AZO thin films indicate that increasing the substrate temperature in the 200-250℃ range enhanced the crystallite size and electrical conductivity, but the resistivity increased when the substrate temperature was above that range[9-11].Figure2 shows the sheet resistance and resistivity of AZO thin films with different substrate temperature under conditions of the sputtering power of 225w , process pressure of 0.8Pa, thinkness of 725-732nm and the Ar flow rate of 100 sccm . The lowest resistivity is 9.81×10^{-4}Ω.cm and the lowest sheet resistance is 14Ω at 250 ℃.

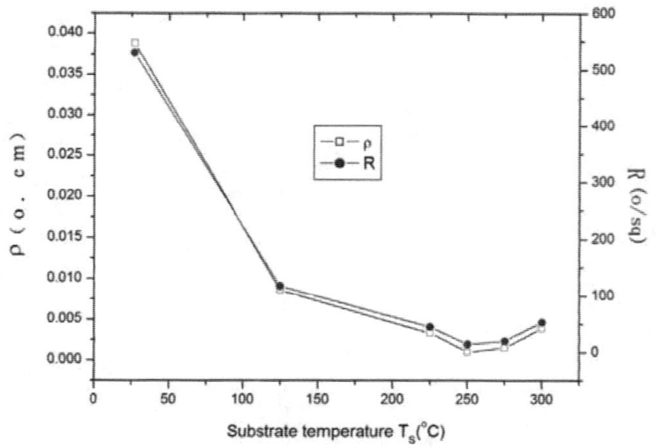

Figure2. Sheet resistance and resistivity with different substrate temperature

Figure 3 shows the sheet resistance and resistivity at different deposition pressure for AZO films under conditions of the sputtering power of 225w, temperature of 250℃, thinkness of 700-780nm and the Ar flow rate of 100 sccm. From the figure,the sheet resistance and resistivity of the film deposited at process pressure of 0.8Pa were lowest.

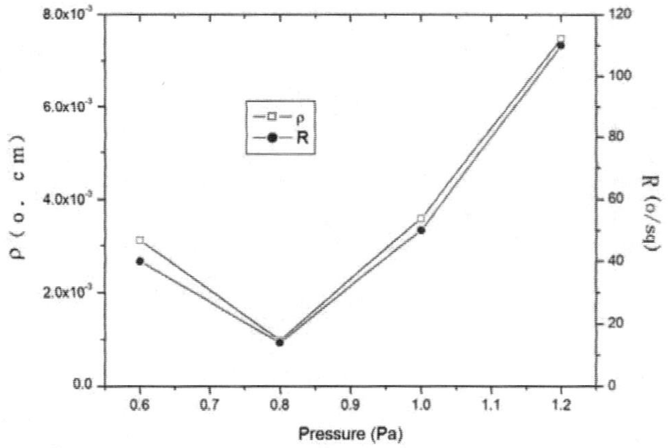

Figure 3.Sheet resistance and resistivity with different process pressure at 250℃

Figure 4 shows the sheet resistance and resistivity at different sputtering power for AZO films under conditions of the process pressure of 0.8Pa, temperature of 250℃, thinkness of 500-730nm and the Ar flow rate of 100 sccm . From the figure,the sheet resistance and resistivity of the film decrease quickly with an increase of sputtering power. Take the stability of sputtering power into consideration, the sputtering power was choose as 225 W.

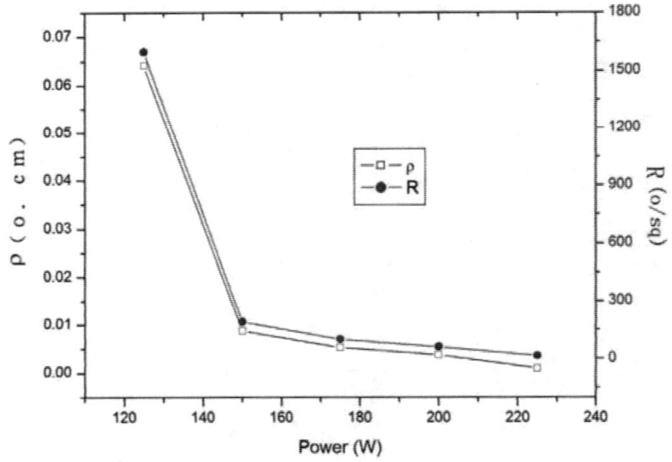

Figure 4.Sheet resistance and resistivity with different sputtering power at 250℃

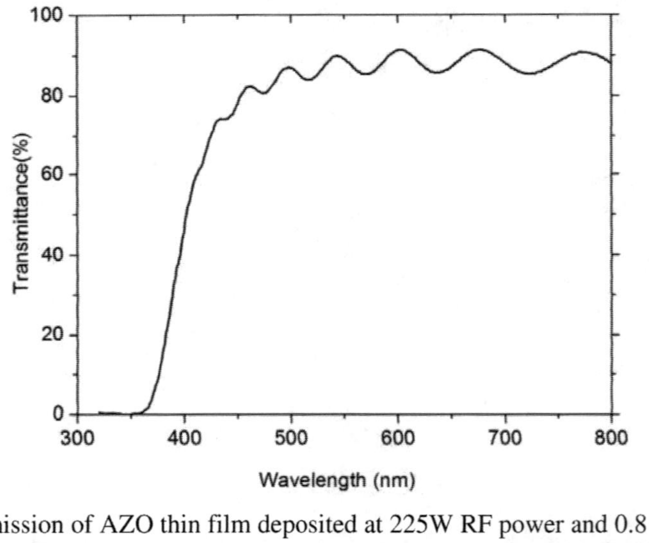

Figure5. The transmission of AZO thin film deposited at 225W RF power and 0.8 Pa working pressure

The transmittance of the AZO thin film is another critical factor for TCO applications. Figure 5 shows the transmission of AZO thin film deposited at 225W RF power and 0.8 Pa working pressure. From Figure5, the AZO thin film are highly transparent in the visible region of the electromagnetic spectrum.The average transmission in the visible range (305 - 800 nm) was greater than 85% ,even 90% excluding the transmission of glass background .The The AFM image of the AZO thin film deposited at 225W RF power and 0.8 Pa working pressure is shown in Figure 6. From the AFM image, the average grains are 28.93 nm in size.

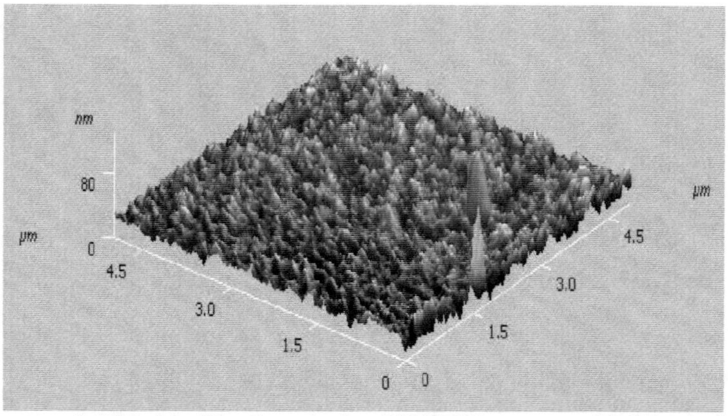

Figure6. The AFM image of AZO thin film deposited at 225W RF power and 0.8 Pa working pressure.

4. Conclusions

AZO thin films were prepared on glass substrate by Magnetron Sputtering method with a 2wt % Al –doped ceramic target .The processing parameters on electrical properties of AZO thin film were investigated. When processing pressure is 0.8Pa,the sputtering power is 225W， the temperature of substrate is 225℃, the AZO thin films have a lowest resistivity of 9.81×10^{-4} Ω.cm and visible light transmittance of more than 85%.The results obtained are very important to the selection of the technical conditions.

References

[1] Minami T.2008 *Thin Solid Films* **516** 5822
[2] Granqvist CG. 2007 *Sol Energ Mater Sol Cell* **91** 1529
[3] T. Terasako, M. Yagi, M. Ishizaki, Y. Senda, H. Matsuura, S. Shirakata, 2007 *Surf. Coat. Technology* **201** 8924
[4] W.T. Seeber, M.O. Abou-Helal, S. Barth, D. Beil, T. H ¨ oche, H.H. Afify, S.E. Demian, 1999*Material Science. Semicond. Process* **2** 45.
[5] F.K. Shan, Y.S. Yu, 2003*Thin Solid Films* **435** 74
[6] K. Prabacar, C. Kim, C. Lee, UV, 2005 *Cryst. Res. Technol.* **12** 1150
[7]. P. Nunes, E. Fortunato, P. Tonillo, F.B. Fernandes, P. Vilarinho, 2002 *Vacuum* **64** 281
[8] Minami T. 2005 *Semicond Sci Techno* **20**:35
[9] Chang JF, Hon MH. 2001 *Thin Solid Films* **386** 79.
[10] Haug FJ, Geller Z, Zogg H, Tiwari AN, Vignali C. 2001 *J Vac Sci Technol A* **19** 171
[11] Li W, Sun Y, Wang Y, Cai H, Liu F, He Q. 2007 *Sol Energy Mater Sol Cells* **91** 659

Journal of Physics: Conference Series **276** (2011) 012171 doi:10.1088/1742-6596/276/1/012171

πShaper - Refractive Beam Shaping Optics for Advanced Laser Technologies

Alexander Laskin[1], **Vadim Laskin**

Molecular Technology (MolTech) GmbH
Rudower Chaussee 29-31, 12489 Berlin, Germany

alex@mt-berlin.com

Abstract. Laser beam shaping brings to various industrial and scientific laser techniques effects that improve their performance comparing to what can be achieved with using Gaussian or Gaussian-like laser beams: more stability, less tough positioning tolerances make the technologies easier to use, higher efficiency of using of costly laser energy, etc. The task of such a transformation is solved by series of refractive beam shaping optics of field mapping type. This solution is important in irradiating the cathode of Free Electron Lasers, confocal microscopy, biomedical fluorescence techniques, many industrial technologies like welding, cladding, hardening, various laser techniques in photovoltaics, homogenizing of pump radiation by building powerful femtosecond lasers, etc. The refractive beam shapers can be used with TEM_{00} and multimode laser beams, achromatic design provides the same conditions of beam shaping for several lasers of a certain spectrum range simultaneously, low inherent losses allow to use them with powerful laser sources, particular models can be implemented as Galilean Telescope without internal focusing or as Collimators. This paper will describe the principles of operation, design features of the achromatic refractive beam shapers; there will be presented examples of beam intensity transformation and effects on material processing achieved in several industrial applications.

1. Introduction

The lasers are widely used in various applications in science and industry and their effective using is very important. Typically the intensity distribution of laser sources is described by the Gaussian function provided by physics of creating the laser radiation. From one side, this Gaussian profile provides high energy concentration, however, from another side for many industrial, scientific and life science applications it is not an optimum one because of non-uniform intensity distribution within the laser beam. In plenty of laser techniques a homogenized laser beam is most preferable from the point of view of saving the energy and providing the same conditions of illumination or material treatment in the beam area. Therefore, the task of re-distribution of energy within the laser beam to provide uniform intensity profile is an actual scientific and industrial task. In many cases a solution can be realized on the base of a family of refractive beam shaping systems ***πShaper***.

Unlike other beam shaping solutions, like truncation of a beam by an aperture or attenuation by apodizing filters, integration systems based on arrays of microlenses, micromirrors, prisms or DOE, the ***πShaper*** realizes so called field mapping approach featuring by the intensity profile transformation

[1] To whom any correspondence should be addressed.

in a controlled manner by introducing a pre-determined wave aberration with, after the intensity re-distribution, its compensation. As result accurate, theoretically almost lossless transformation is provided and a collimated, speckle-free beam is created.

2. Principles of operation

2.1. Motivation – increasing efficiency of using laser energy
All laser applications have specific features of interaction of material and laser radiation. There is, however, something common for all single mode (TEM_{00}) lasers — the Gaussian function of intensity distribution. Therefore approximate evaluation of the efficiency of using laser energy can be done by considering just the geometrical features of the Gaussian function, without taking into account effects accompanying laser treatment of materials like burning, melting, etc.

The three-dimension intensity profile can be interpreted as a geometrical figure bounded by a horizontal plane and a surface of the Gaussian function $I(r)$

$$I(r) = I_0 \exp(-2 \cdot r^2/\omega^2) \tag{1}$$

where r designates a variable beam radius in polar coordinates,
ω designates a waist radius of the Gaussian laser beam,
I_0 is constant.

Fig.1 shows a section of such a figure, its volume has physical sense of energy of the laser beam. Let's denominate by variables E_1, E_2, E_3 different parts of that figure:

- E_1 - an "apex" of Gauss function is an excess of intensity over the working level I_h, very often this is a loss of energy or a source of overheating the central portion of a zone under treatment,
- E_2 - "tails" of Gaussian distribution, almost always this is a loss of energy or a source of unwished effects like heat affected zone (HAZ), and
- E_3 - "effective cylinder" of energy.

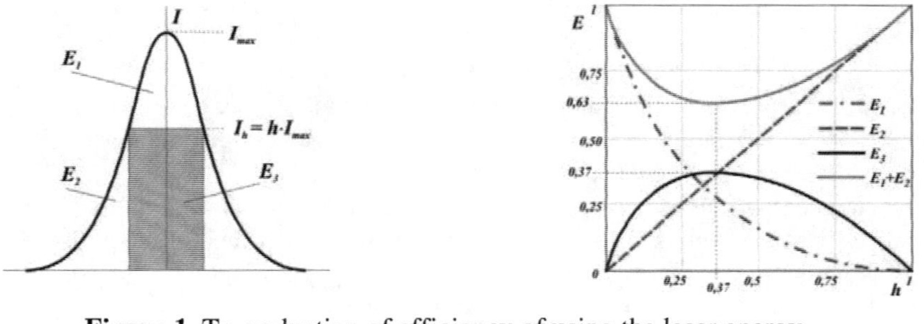

Figure 1. To evaluation of efficiency of using the laser energy

By mathematical transformation one can get following formulae to calculate the energy parts E_1, E_2 and E_3:

$$\left\{ \begin{aligned} E_1 &= 1 - h + \ln h \cdot h \\ E_2 &= h \\ E_3 &= -\ln h \cdot h \\ E_1 + E_2 &= 1 + \ln h \cdot h \end{aligned} \right\} \tag{2}$$

where $h = I_h / I_{max}$.

The results of calculations are presented on right diagram in Fig.1.

The unconditional energy loss E_2, "tails", can reach a very high level - for example, if a working energy level I_h is half of maximum (h = 0.5, very often just this level is considered as a working one) the energy losses are 50 percent of full laser beam energy!

In the case of laser treatment of thin films the energy part E_1, "apex" of Gauss, is also considered as a loss of energy because this part exceeds the working energy level I_h. Thus both energy parts E_1 and E_2 are losses, the sum $E_1 + E_2$ has sense of combined losses and minimum of this function is 0.63! In other words, when treating thin films, in the best case, "only" 63 percent of energy is lost and 37 percent is effective!

No doubt, transformation of the original Gaussian shape to "an effective cylinder", called as a flattop or top hat profile would help to save laser energy and improve those technologies where uniform intensity is most desirable. This transformation is a main function of the ***πShaper*** systems.

2.2. *Design features of the* ***πShaper***

An idea of the ***πShaper*** operation is illustrated with Fig. 2. Gaussian (or close to Gaussian) intensity distribution of a TEM_{00} or multimode laser beam is converted to a flat-top distribution (similar to a Greek letter π) that stays invariable over long distance after the ***πShaper***.

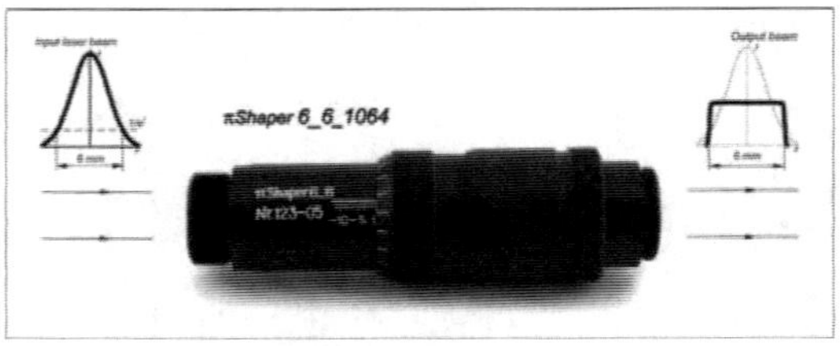

Figure 2. Principle of the ***πShaper*** operation

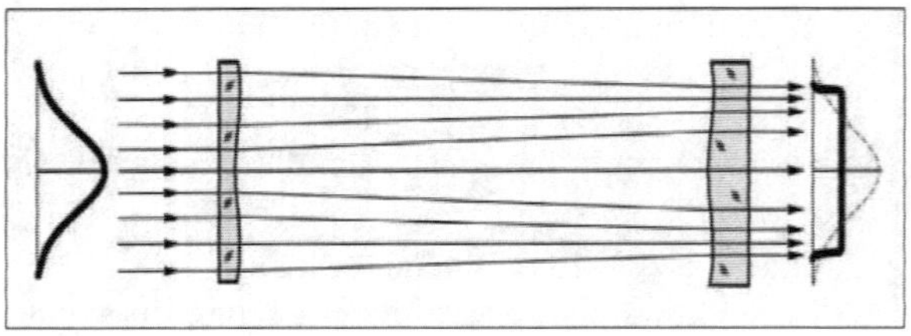

Figure 3. Example of optical layout

This transformation is realized through distortion of the beam wave front inside the optical system under the condition of conservation of energy, this effect is illustrated with a picture in Fig. 3. Mathematically this condition is formulated as follows

$$\int_0^{r_{in}} I_{in}(r)\cdot r\cdot dr = \int_0^{R_{out}} I_{out}(R)\cdot R\cdot dR\,, \tag{3}$$

where r is an input beam radius in polar coordinates,
$I_{in}(r)$ is a function of intensity distribution of the input laser beam section,
r_{in} is a max. radius of the input beam subjected to considered intensity distribution,
R is an output beam radius in polar coordinates,
$I_{out}(R)$ is a function of intensity distribution of the output beam,
R_{out} is a max. radius of the output beam resulting after the intensity redistribution.

Another basic principle of a beam shaper is zero wave aberration; this means the aberration introduced by first optical component is then compensated by the second one. Other details of refractive beam shapers of field mapping type like ***πShaper*** can be found in publications [1,2,3,4,5], here we will emphasize on some features important for their practical using.

An essential design feature of the ***πShaper*** optical system is in consisting of two optical components and controlled wave front transformation in the space between them due to applying of special optical surfaces, as result necessary intensity redistribution is achieved. Another important feature of the ***πShaper*** optical design is zero or negligible for practical applications residual wave aberration, this means equal path lengths for all rays of input beam passing through the optical system; this condition is very important for practice since guarantees flat output wave front and avoidance of appearing undesirable interference fringes, this provides also low output divergence and keeping the result intensity profile over long distance after the ***πShaper***.

One of important features of the ***πShaper*** optical systems over other approaches is in their achromatic design that guarantees simultaneous fulfillment of conditions of intensity redistribution and zero or negligible wave aberration for a certain spectral range, as result the achromatic optical systems provide the same operation at each wavelength of this spectral range. This feature is realized through using materials with different dispersion characteristics.

Providing of the same operation of the achromatic ***πShaper*** optical systems at each wavelength of the certain spectral range makes these systems very useful in the applications where several laser sources are applied simultaneously, for example in spectroscopy, fluorescence life science technologies, confocal microscopy. This feature is very important, also, in material processing technologies where various lasers are applied in one technology cycle, for example by manufacturing the solar cells. Another important application example is ultrashort pulse lasers, like femtosecond Ti:Sapphire lasers, where small pulse duration is achieved due to wide spectral bandwidth of a laser source.

To prevent internal focusing of a laser beam all ***πShaper*** optical systems realize so called Galilean type telescope, this is a critical point for high power and short pulse laser applications.

The ***πShaper*** are capable to work with TEM_{00} as well as multimode lasers.

Summarizing, the most important features and basic principles of ***πShaper*** are:

- telescopic or collimating refractive optical systems that transform Gaussian, or close to Gaussian intensity distribution of source laser beam to a flattop (or top-hat, or uniform) one;
- The initial laser beam can be either a TEM_{00} or a multimode one;
- The uniform intensity is kept after the ***πShaper*** over a large distance;
- There are available telescope models, as well as systems with optical power like collimator;
- The transformation is provided for a certain spectral range, thus ***πShaper*** is an achromatic system;
- Galilean design, thus there is no internal focusing of a beam.

Currently the ***πShaper*** model line includes many systems capable to work with laser beams of various input size as well as of various wavelengths, Fig.4.

Most of models are implemented as telescopes, thus collimated input and collimated output. However, there are also available collimator type models with a divergent input beam and, hence, combining functions of beam shaping and collimation, this feature might be of interest for fiber laser sources.

Figure 4. Outlook of some ***πShaper*** models

An example of a collimating beam shaper is ***πShaper 37_34_1064,*** Fig.5, this system is intended to work with powerful, up to 6 kW, industrial lasers like fiber-coupled multimode solid state lasers or diode lasers, fiber lasers.

(a)

(b)

Figure 5. ***πShaper 37_34_1064***
(a) outlook,
(b) mounted on industrial robot.
Courtesy of Daimler AG

Due to their features the ***πShaper***'s are useful beam shaping tools in various laser techniques where working beam sizes are orders of magnitude greater than an operating wavelength, usually they are recommended to be applied when required laser beams sizes are more than 0.2 mm. Low divergence of output beam makes possible further beam size and spot shape manipulation by applying additional lenses, beam-expanders or anamorphic optics.

3. Examples of optical layouts on the base of *πShaper*

3.1. Beam Size manipulation with beam-expanders

According to basic design the output beam after refractive field mapping beam shapers is round and has a pre-determined size, for example, in case of ***πShaper 6_6*** the resulting beam diameter is about 6 mm. For some tasks that's enough, however most often it is necessary to change the beam diameter. For example, the applications, where it is necessary to illuminate a Spatial Light Modulator or a mask with a collimated laser beam of uniform intensity, usually require expansion of a beam after the refractive field mapping beam shaper. At the same time many scientific and industrial tasks can be successfully solved when a collimated beam of uniform intensity and diameter of about 1 mm is provided, some of them are laser welding, hardening, cladding, etc.

Common features of these beam transformations are varying the beam size and leaving the laser beam collimated. Evidently, this optical task can be easily solved by applying a beam-expander or a beam-reducer of telescopic type with an appropriate magnification factor; the principle optical layouts are shown on Fig.6.

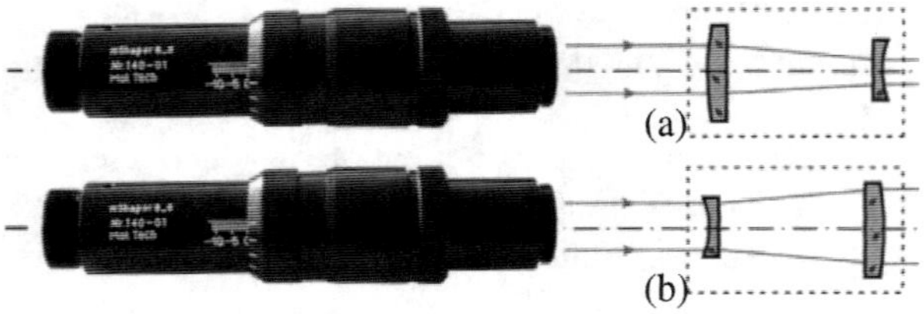

Figure 6. Examples of (a) reduction and (b) expansion of a beam after a ***πShaper***.

When variable final beam size is required one can apply a zoom beam-expander.

The beam expansion leads to extending of the space where a resulting beam profile is kept stable because the diffraction effects influencing on the beam profile transformation become less strong; another reason is in reduction of residual wave aberration always existing in real optical systems. Therefore, expansion factor is limited rather by capabilities of applied beam-expanders.

In case of beam demagnification it is recommended don't exceed a factor 10, since too much beam reduction would lead to increasing the residual wave aberrations and, hence, quicker beam profile deterioration when its propagation in space.

3.2. Generation of linear spots with using anamorphotic optics

There are many industrial applications which performance can be seriously improved by applying a linear shape of laser spot, some examples of them are laser cleaning, annealing, hardening, cladding. Therefore, the task of generation of a "laser line" is very important and refractive field mapping beam shapers in combination with special anamorphotic optics can be successfully used as a solution. An example of a layout of such a combined system is shown on the below Fig. 7.

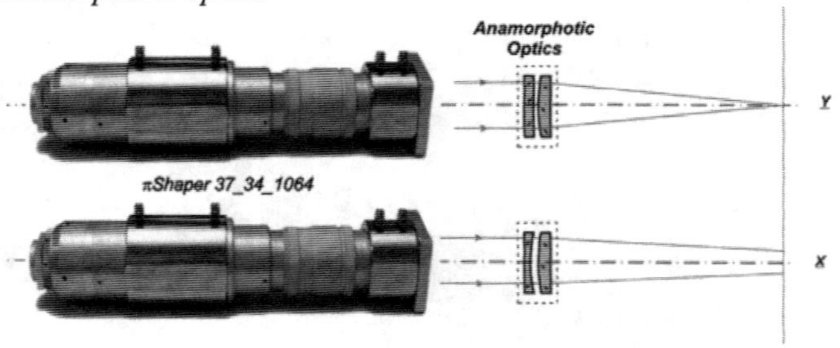

Figure 7. Generation of a "Laser Line" by attaching anamorphic optics to the ***πShaper***.

The collimated beam of uniform intensity is emerging from a refractive field mapping beam shaper and is then focused by an anamorphotic optics that is implemented as a pair of lenses, one of them is an ordinary spherical lens and another one is a negative cylinder lens. Due to the inherent astigmatism of the anamorphotic optics the beam is focused in one plane, Y in Fig. 7, but stays unfocused in the perpendicular plane X, hence a spot of linear shape is created. The length and the width of the line are defined by parameters of anamorphotic optics and aspect ratio can be up to 1:1000! The above layout was realized for the task of metal hardening with using radiation of high power fiber laser and a line of

10 mm length and about 0,5 mm width was realized, more detailed description and results achieved are discussed later in the chapter of experimental results.

4. Examples of operation of the refractive field mappers

Essential features of operation of the refractive field mapping beam shapers can be seen in the example of beam intensity re-distribution of a TEM_{00} laser beam on Fig. 8, here there are presented beam profiles measured with a camera based beam profiler.

In spite of deviation of intensity distribution of the input beam from the perfect Gaussian function the beam shaper provides quite good quality of re-distribution. Due to operational principle a field mapping beam shaper cannot suppress high frequency intensity modulation of original beam; that is why that modulation presents in resulting profile as well. The bright rim in the output spot is result of too much energy in "tails" of that non-perfect Gaussian original beam.

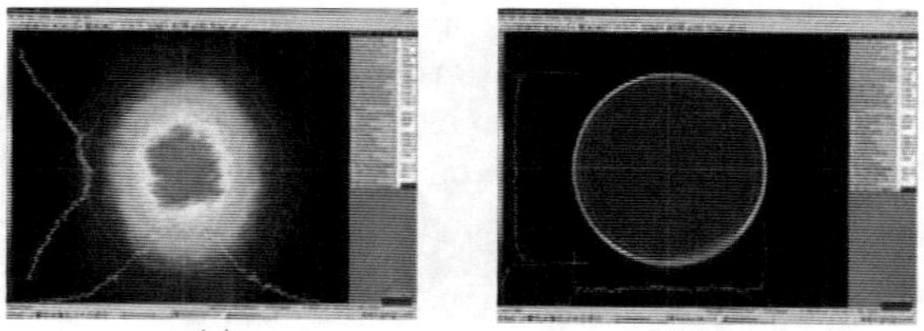

(a) (b)

Figure 8. Intensity profiles:
(a) source beam; (b) after the ***πShaper***

Since the original beam is rather elliptic the horizontal and vertical output profiles have minor difference – there exists some arising of intensity in periphery of vertical profile; that difference is, however, quite weak, and this shows the achromatic beam shaper is capable to smooth away the ellipticity of original beam.

An illustration of the beam shaper influence on the results of material treating is presented in Fig. 9. Here one can see a comparison of engraving of a circular hollow in material used in photovoltaic: with a TEM_{00} Nd:YAG laser and the same laser with the refractive field mapper.

The difference of engraving results is evident - irregular shape of the hollow with a ragged unwished hole in the center in case of original TEM_{00} laser, and the good shaped round hollow with a controlled depth when applying the field mapping beam shaper.

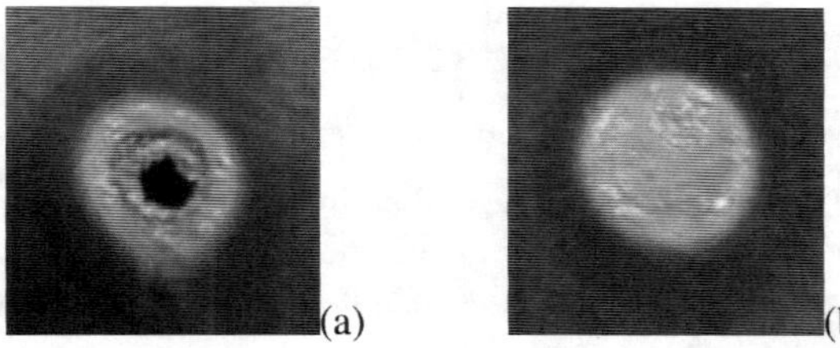

(a) (b)

Figure 9. Comparison of material processing:
a) TEM_{00} laser beam, (b) after the ***πShaper***.
Courtesy of EO Technics

Fig. 10 presents results of intensity re-distribution of a multimode beam of 2 kW fiber coupled solid-state laser with using the collimating refractive beam shaper shown in Fig. 5.

Evidently the collimating refractive beam shaper can successfully operate with multimode beams and is capable to manipulate the resulting beam intensity distribution by varying internal settings (distance between components), - either the flattop profile, or a profile with steep edges and intensity downing in the beam centre can be realized. In other words, under certain conditions the beam shapers of field mapping type can realize not only the flattop but also some other profiles, this gives users freedom in choosing an optimum for a particular application profile.

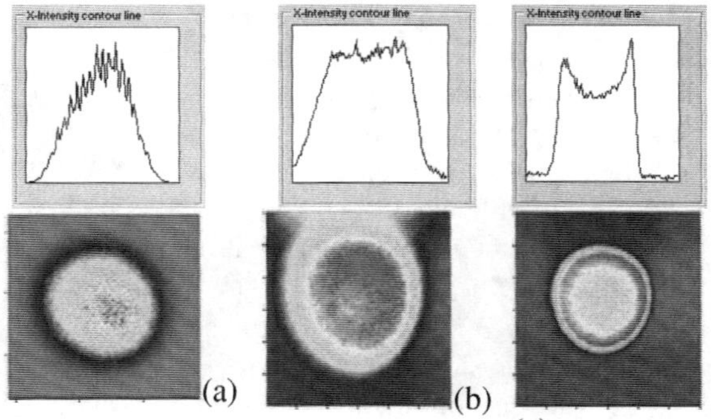

(a) (b) (c)

Figure 10. Intensity profiles of fiber-coupled multimode laser:
(a) initial beam; (b) after the beam shaper, 1st setting,;
(c) after the beam shaper, 2nd setting.
Courtesy of Daimler AG

Focusing of a multimode beam after a refractive field mapping beam shaper leads to some useful effects. The Fig. 11 shows a comparison of view of caustics for multimode 2 kW solid state laser, fiber-coupled to 600 micron fiber, by focusing with using the same lens in 2 cases: (a) with ordinary collimator and (b) with the ***πShaper 37_34_1064*** as a collimator. Evidently, there is a difference in the beam profile behaviour in zone of image plane of a fiber end – in case of the beam shaper the caustic is characterized by an extended zone along the optical axis where the intensity profile is uniform and beam size stays almost invariable, this is something like a "pipe of uniform intensity". This feature makes it possible to achieve more stable and predictable results of welding, brazing, and many other applications where a uniform intensity profile is advisable, while the extended "pipe of uniform intensity" increases a tolerance extent on position of a tool with respect to a workpiece that simplifies applying of a laser technology.

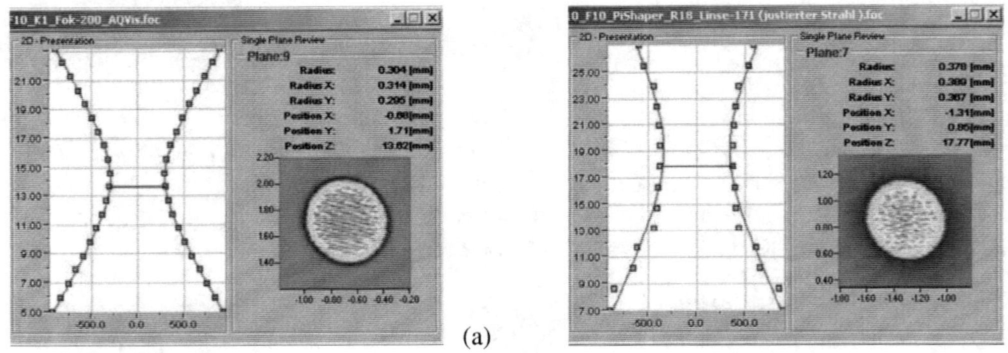

(a) (b)

Figure 11. Caustic view: (a) Ordinary Collimator and Lens; (b) Collimating beam shaper and Lens.

The optical system described in section 3.2 and presented in Fig. 7 was realised in IPG Photonics to generate a linear shape of the spot of multimode 3 kW fiber laser for laser hardening of metal parts, sure, this approach can be applied in many other technologies where a linear spot can improve their performance.

The uniform intensity was provided over the long axis which length was about 10 mm, the line width was about 0.5 mm. Since it was planned to move the linear spot over a workpiece in direction perpendicular to the long axis the intensity profile in short axis wasn't specified, a main aim was to achieve as narrow as possible line.

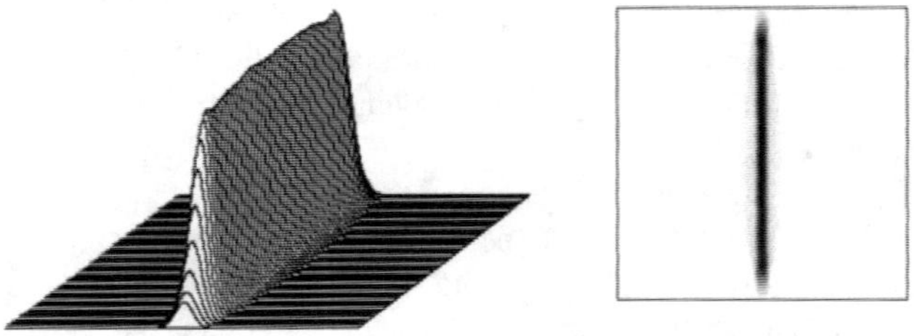

Figure 12. Computer simulation of beam profile created in optical system described in Fig. 7.

Results of numerical calculations for the optical system, Fig. 7, implemented as a ***πShaper 37_34_1064*** with an anamorphotic system are shown in Fig. 12. Results of intensity profile measurements in area of working plane are presented in Fig. 13.

Evidently, there exists good correspondence between theoretical and experimental results.

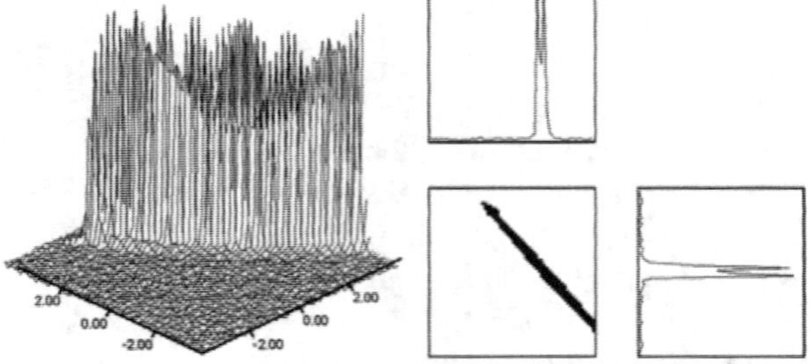

Figure 13. Results of measurements with using a beam profiler: On left - 3D view of the spot, On right – profiles in sections. *Courtesy of IPG Photonics*

Let us note one important feature of the field mapping beam shapers that is very good seen in this example –

capability to create not only uniform resulting profiles but some other beam shapes like so called "inverse Gauss" characterized by steep edges and downing of intensity in the centre. Just this "inverse Gauss" profile was achieved in short axis of the final linear spot while focusing the laser radiation of multimode fiber laser, Fig. 13 on right. Thus, depending on settings the field mapping beam shapers provide various beam profiles, and this flexibility is very useful and important in various laser technologies since gives the possibility to choose an optimum intensity distribution for a particular application.

5. Conclusions

Growing complexity and variety of laser applications demand various designs of beam shaping optics. Laser technologies and techniques, where a required final laser beam should have other than Gaussian profile, low divergence and the size orders of magnitude larger than a wavelength, can be improved by applying refractive beam shapers of field mapping type, like ***πShaper***, providing collimated beams of uniform intensity. The specific features of refractive field mapping beam shapers provide flexibility in manipulation of the size and the spot shape by applying additional optical components. Low divergence of output collimated beam, near the same like in input beam, leads to extended space after a beam shaper where a resulting beam profile is kept stable, which, in turn, guarantees the long depth of field of a combined optical system. At the same time very high factors of expansion, demagnification as well as high aspect ratios of linear spot shapes become available. These features in combination with inherent capability of field mappers to create not only flattop but also other beam profiles ("inverse Gauss", etc.) make these devices a convenient tool to build beam shaping optics for various industrial and scientific applications.

6. References

[1] Dickey F.M., Holswade S.C. (eds) 2000 *Laser Beam Shaping: Theory and Techniques*, (New York Marcel Dekker)
[2] Laskin A.V. 2009 *Laser Beam Shaping X Proceedings of SPIE* **7430** 743003,
[3] Laskin A.V. 2008 *Achromatic optical system for beam shaping*, European Patent Application EP1 998 215 A1
[4] Hoffnagle J.A., Jefferson C.M. 2000 *Appl. Opt.* **39** 5488
[5] Kreuzer J. 1969 *Coherent light optical system yielding an output beam of desired intensity distribution at a desired equiphase surface*, US Patent No. 3 476 463

Numerical Investigations on a Distributed Fiber-Optic Lighting System with an End Reflector

Shuhua Li, Huaping Gong, Yumeng Tu, Ying Meng
College of Optical and Electronic Technology, China Jiliang University, Hangzhou, China 310018

E-mail：lishuhua1010@163.com

Abstract A novel distributed fiber-optic decorative lighting system with the reflection coating on the extremity of fiber-optic is designed, which used the multi-mold optical fibe made up of large core diameter(Diameter of core and cladding is 105μm and 125μm, respectly).After introducing the distributional optical fiber decorative lighting system briefly, the ralationship between corrosion depth of the optical fiber core and the leakage of fiber-optic has been analyzed with the Rsoft, and then the relationship of the lighting power and the uniformity of lighting power with the leakage rate of optical fiber lamp, the reflective of reflection coating has been discussed.The simulation analysis shows that, when the core diameter is corroded to 80~85 μm, the leakage rate of optical fiber may achieve 5.0%, which suits the optical fiber decorative lighting. Considering all kinds of factors, when optical fiber lamp's quantity is 20, the coating index of reflection is 95%, optical fiber lamp's leakage of light rate is 5.0%, and the optical fiber lamp's distance is 1 meter, the quite high illuminating power may be achieved, as well as the good lighting uniformity.Finally the experimental study of decorative lighting system is given. And the experimental result is in keeping well with the theory simulation conclusion.
Keywords Fiber-Optic Lighting, Reflector, Numerical Simulation

1. Introduction

Fiber-optic lighting has been developed for 30 years in 20th century. In recent years, it has been changed from theory into a practical stage gradually, and it provides a new way for modern decorating and lighting [1~3]. Fiber-optic lighting offers a unique new way for the decoration and lighting with pro-environment, safety, energy saving, convenient and high efficiency [1~4]. At the same time, fiber-optic lighting has some advantages that general lighting can not provide. For example, the spread of light can be flexible, and light and electric is separated. It only has the visible light and no infrared or ultraviolet light, and it is colorful and versatile. Because of these virtues, fiber optic decorative lighting is widely used in advertising, landscape design, entertainment, industrial field, medical treatment, museums, mines and other special lighting occasions[4~7]. It is prospective in the field of decorative lighting in the 21st century.

2. Constitution of fiber-optic lighting system

Fiber-optic lighting system is mainly constituted of light source, couple and fiber, as shown in Fig.1. Light source provides energy for the whole lighting system, and couplers could couple light into the fiber to transmit. Light transmitting in optical fiber leaks out at the fiber-optic loops by the certain ratio for lighting, because the cladding of fiber loop is removed partially. In this way, we could control the leakage rate of the fiber-optic loops, the reflectivity of the reflector to obtain a good effect in decoration lighting, and satisfy the requirements of modern decorative lighting.

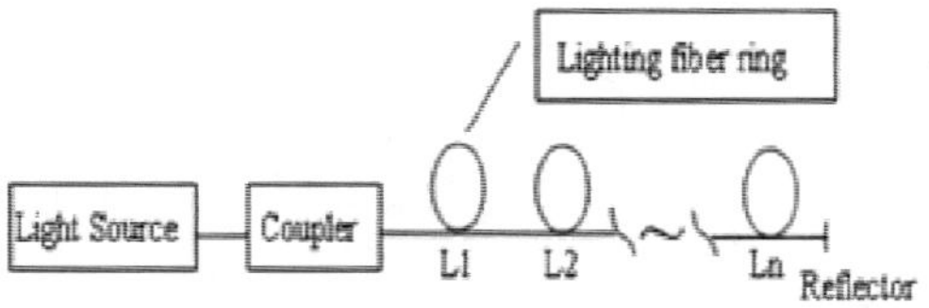

Figure 1. The configuration of the fiber lighting system

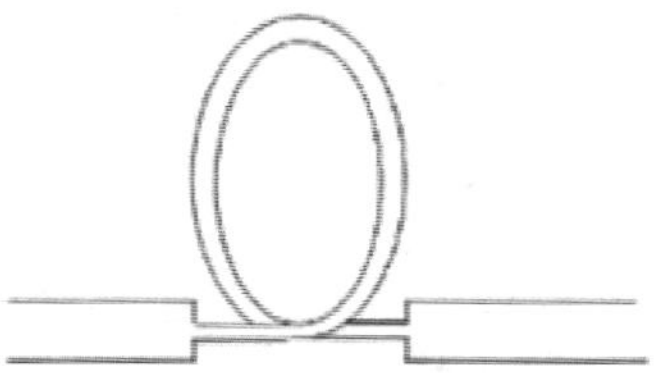

Figure 2. amplificatory picture of fiber loop

The configuration of the fiber-optic lighting system is shown as Figure 1.It contains one light source at initial of fiber, one coupler, fiber and one reflector at end of fiber. Light generating from the light sources is coupled into the optical fiber with coupler, which leaks out at the fiber loop Ln (n = 1, 2, 3 ...) for lighting and decoration [7]. Then the leftover part light at the end is reflcted back to fiber to reuse by reflector. Fig. 2 is an enlarged view of the optical fiber loop. The loop which the cladding layer is removed partially is made up of a circle fiber. The paper analyzes the key factors affecting the lighting power and uniformity of the distributed fiber lighting system, such as light leakage rate of fiber loops, and the reflectivity of reflector. The above parameters are studied with MATLAB7.0 software.

3. Simulating the performance of with Rsoft software

The multimode step index fiber is used and the transmission and loss of the visible light in fiber whose core layer is corroded is simulated and analyzed. The parameters of the multimode step-index fiber are as follows: core layer diameter is 105 μ m, cladding layer diameter is 125 μ m, coating layer is 250 μ m, core index is 1.4662, and cladding index is 1.45. The incident light is He-Ne laser, and wavelength is 632.8nm.

The simulation for different diameter and length of corrosive fiber shows: holding the corrosive diameter, the power loss in fiber core layer is not same for different corrosive length. And the power has loss at ends of corrosive part. According to the theory of optical waveguide [8], the loss at the initial part of corrosion is light leak that the light transforms from the larger diameter of fiber core into smaller one, which belongs to the lighting power, but the end does not. According to the optical mode theory, the power ratio of fiber core can be expressed as [8]

$$\eta_{co}(m) = 1 - \frac{U^2}{V^2}(1-\kappa) \tag{1}$$

Where V is the normalized frequency, $V = k_0 a\left(n_{co}^2 - n_{cl}^2\right)^{\frac{1}{2}}$, U is normalized phase constants, $U = a\left(k_0^2 n_{co}^2 - \beta^2\right)^{\frac{1}{2}}$.

Based on the simulation results, the loss is not linear relationship. Holding the length of corrosion, the loss increases with decreasing of diameter of core layer. The curve is obtained as figure 3 for loss of initial point.

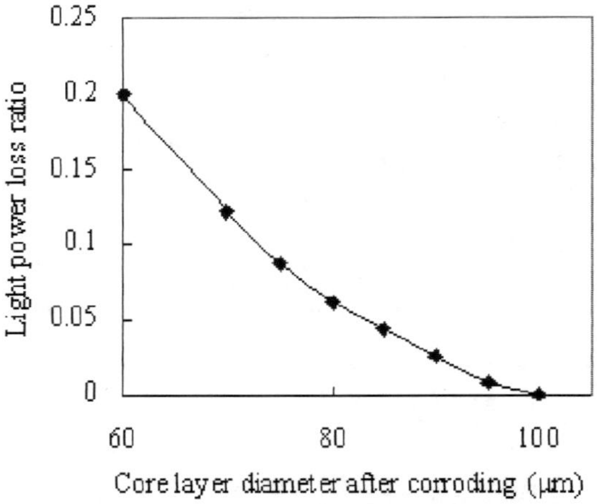

Figure 3. Simulation results with Rsoft

4. Simulation and analysis

We assume that the optical power coupled into the fiber is p_0 (unit: W), the distance between light source and the nearest loop is a (unit: m), the adjacent distance is b (unit: m), the loss of optical fiber is α (units: /m), the leakage rate of the fiber-optic loops is μ, the reflectivity of reflector is r, and the number of optical fiber loops is n (n = 1,2,3 ...). Then we can get lighting power W_{1k} (unit: W) of the *k* (k=1,2,3,···, n)fiber loop by calculating and stimulation without refector as follows:

$$W_{1k} = p_0 e^{-\alpha\,[a+(n-1)b]}(1-\mu)^{n-1}\mu \tag{2}$$

The lighting power W_{2k}(unit: W) of the *k*(k=1,2,3,···, n) fiber loop by calculating and stimulation with refector as follows:

$$W_{2k} = p_0 e^{-\alpha[a+(k-1)\,b]}(1-\mu)^{k-1}\mu + p_0 r e^{-\alpha[a+(2n-k-1)\,b]}(1-\mu)^{2n-k}\mu \tag{3}$$

From the above equations, we can obtain lighting power of each loop.

In this fiber-optic lighting system, μ and r are varied parameters. Other parameters can be set according to the actual conditions as follows. The light source power coupled into the fiber is p_0 = 200W; the distance between light source and the nearest loops is a=3m; the number of fiber loops is n=20; the adjacent distance is b=1 meters. Range of the visible light, the minimal loss of POF made by PMMA-d8 is19.4dB/km at the 680nm[9], so we can set the attenuation of fiber-optic as 40dB/km, and it could be converted into attenuation coefficient $\alpha = 9.21 \times 10^{-3}$ / m. Based on the above formula (1), we use MATLAB 7.0 software to analyze the effects of these three variable parameters and illumination uniformity of lighting system.

4.1 Analysis on the light leakage rate μ of fiber loops

Holding r = 0.95 and other parameters ,and varying μ, we can obtain variable power curve of fiber loops through simulation and calculation with MATLAB when μ=0.080, μ=0.050, μ= 0.035, as figure 3.

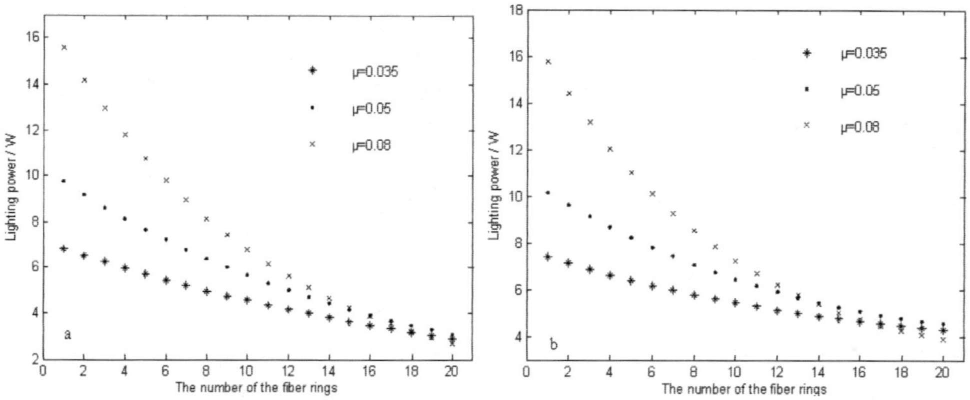

Figure 4. Optical power curve without (a) reflector and with (b) refector for different μ .

It is shown from Fig 4, , the difference of each loop's lighting power is large with large μ, regardless of whether there is reflector at the end of fiber. The lighting power curve bends obviously, then the uniformity of lighting power is poor, but the average power is larger. When the value of μ is small, the difference of each loop's lighting power decreases. So the curve is close to a flat line, and the uniformity is very well. As it is shown in the fig.3, the curve signed by " • " indicates the lighting power with μ=0.05. Although the average power corresponding to the two curves above is very large, the difference of the variable points' power is large too, and the uniformity is poor. The curve below is close to a flat line, but the power of loops is too small. In the other words, the average power is small and the lighting effect is poor. At the same time, we can see from fig.3, that the power of loops decreases seriously with μ reducing. Thus, we should contrive to select a larger μ and ensure the effective power and average power is better.

4.2 Analysis on the reflectivity r of the end reflector

The function of reflector is to reuse the leftover power continuely in fiber. Because of various technology, the reflectivity of reflector has difference, which would lead to different effects to the lighting power and uniformity. Using advanced precision technology, the reflectivity of reflector can achieve 99.9% above in domestic [10]. So we can set r = 0.95, r = 0.90 and r = 0.80, then we carry out the numerical simulation and obtain the results as figure 4.

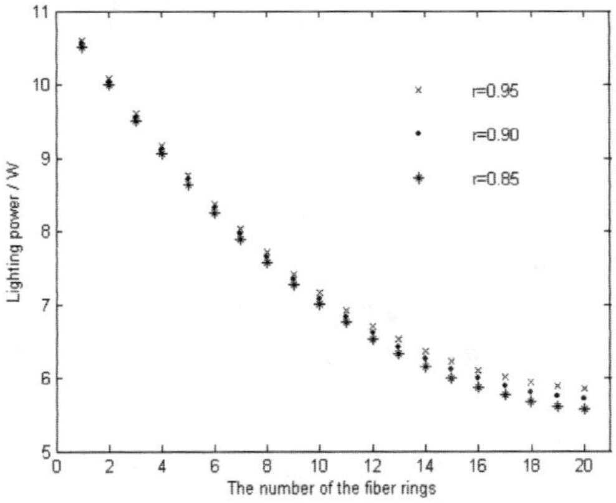

Figure 5. Optical power curve with different reflectivity

In figure 5, the lighting power of each loop increases after Coating film reflector, especially at the end. The increased lighting power has some difference with various reflectivity. Namely, the larger the reflectivity is, the more the increasing is. Conversely, the power increases gradually. Therefore, we should choose larger reflectivity to get better lighting power and uniformity,

5. Conclusions

The lighting power and lighting uniformity of a distributed fiber-optic lighting system with reflector are simulated and analyzed with MATLAB 7.0 and Rsoft software. During the numerical simulation, we study the parameters one by one with varying a parameter and holding other parameters. First, the relationships of the lighting power and uniformity of the distributed fiber lighting system with light leakage rate of fiber loops is analyzed. The next is reflectivity of the end reflector. The simulation results show that, the light leakage rate μ of fiber loops, the reflectivity of reflector in the fiber-optic lighting system have an obvious effect on the lighting power and uniformity, and the high lighting power and good uniformity are conflictive objects. The compromising effect is obtained on the condition that the light leakage rate is 0.05 and the reflectivity of reflector is 95%.

References

[1] Yuan Guoliang. "Linear Optical Fiber Illuminator and Its Application of the Expressway Traffic Arrow ". Electrotechnical Application.Vol, 27:44- 45(2008)

[2] Geng jin, Xu Xiaoxing. "In the Development of Fiber Optic Lighting". Optics,Mechanics & Electronics Information.Vol,02.(2003)

[3] Jiang Yuan, Li Zecai. "Design of Decorative lighting with Polymer Optical Fiber Ceiling". China Illuminating Engineeloop Journal.Vol, 16:44 - 51(2005)

[4] Han Qingshi, Wang Xiaozhen. "The Application of Optical Fiber in Adornment Illumination". Energy Conservation and Environmental Protection.Vol, 2:54- 55(2004)

[5] Wei Shancheng, Cao Junfeng, Han Xueyun. "Ramble on fiber optic lighting". Modern Physics.Vol, 06:45-46(2006)

[6] WANG Liu-ling, Zheng Qing-hong, Li Ming. "Designing and Realization of the Illuminating Device Based on Optical Fiber Transmitting". Journal of Yunnan Normal University (Natural Sciences Edition.Vol, 28:47-49(2008)

[7] ZHU Yun-qing, JIANG Yuan, CHEN Li. "Application of optical fiber to medical illumination". Laser Journal. Vol,6:66~68(2005).

[8] Li Yuquan, Cui Min. The Theory and Technology of Waveguides[M]. Beijing: Posts and Telecom Press, 2002:112~120

[9] TOSHIKUNI KAINO. "Preparation of lastic optical Fibers for Near-IR region transmission". Jounal of Polymer Science.Vol, 25:37-46(1987)

[10] Hu Jiangchuan, Wang Wanlu, Feng Qing. "Progress in Research on Protection of Highly Reflective Coatings from Laser Induced Damage". Materials Review.Vol, 17:238-241.(2003)

Journal of Physics: Conference Series **276** (2011) 012173 doi:10.1088/1742-6596/276/1/012173

The effect of relatively low hydrogen dilution on the properties of carbon-rich hydrogenated amorphous silicon carbide films

Zhe Li, Juncao Bian, Haiyan He, Xiwen Zhang[1] and Gaorong Han

State Key Laboratory of Silicon materials, Zhejiang University, Hangzhou 310027, PR China

E-mail: zhangxw@zju.edu.cn

Abstract. Carbon-rich hydrogenated amorphous silicon carbide (a-$Si_{1-x}C_x$:H) films were deposited by plasma enhanced chemical vapor deposition (PECVD) using silane, ethylene and hydrogen as gas sources. The effect of relatively low hydrogen dilution on the properties of as-deposited samples was investigated. A variety of techniques including Scanning Electron Microscope (SEM), Fourier transform infrared spectroscopy (FTIR), Raman scattering (RS), UV-VIS spectrophotometer and photoluminescence (PL) spectroscopy were used to characterize the grown films. The deposition rate decreases with hydrogen dilution. The silicon to carbon ratio increases slightly with the addition of hydrogen. The phenomenon can be attributed to the dissipation of power density caused by hydrogen dilution. Raman G peak position shifting to a lower wave number indicates that hydrogen dilution reduces the size and concentration of sp^2 carbon clusters, which is caused by the etching effect by atomic hydrogen. The optical band gap, which is controlled by the sp^2 carbon clusters and Si/C ratio, changes unmonotonously. The as-deposited samples exhibited a blue-green room-temperature (RT) PL well visible to the naked eye with UV excitation. The PL band can be attributed to the radiative recombination of electron-hole pairs within small sp^2 clusters containing C=C and C-H units in a sp^3 amorphous matrix.

1. Introduction

The hydrogenated amorphous silicon carbide (a-$Si_{1-x}C_x$:H) films have been extensively studied in recent years due to its optoelectronic and electronic applications in solar cells, light emitting diodes, microelectronic dielectric layers, color sensors and high-temperature engineering materials etc [1-5]. The structural, optical and electrical properties of this material can be controlled by the carbon, silicon and hydrogen composition of the films. However, the chemical bonding configurations and film microstructures of a-$Si_{1-x}C_x$:H materials are very complicated due to the incorporation of carbon and hydrogen since carbon atom can be in a form of sp^2 or sp^3 configuration when it is bonded to Si and H [5]. It is well accepted that the film structures and in turn the physical properties of a-$Si_{1-x}C_x$:H materials are strongly influenced by the preparation techniques and conditions [5]. a-$Si_{1-x}C_x$:H films with low carbon content have been studied for many years and are presently used in many fields. However, a-$Si_{1-x}C_x$:H films with high carbon content emitting visible photoluminescence (PL) at room

[1] To whom any correspondence should be addressed.

temperature has led to studies of this material for application as the active layer in large-area electroluminescent devices [1, 6]. a-$Si_{1-x}C_x$:H films were usually prepared by the plasma enhanced chemical vapor deposition (PECVD) [1, 2, 4, 5], hot wire CVD [3], electron-cyclotron resonance CVD [6] or r.f. sputtering techniques [7] etc. It was reported that the use of reactive plasmas is an effective way to obtain very high deposition rates for a-$Si_{1-x}C_x$:H films [1]. And the films prepared by plasma deposition from hydrogen diluted source gases had superior photoelectric properties due to fewer defects, a much less clustered hydrogen phase and more ordered network [4]. Moreover, uniform films with good electronic properties could be easily deposited over a large area by PECVD, so a-$Si_{1-x}C_x$:H is quite suitable to be used in flat-screen display devices. Considering its compatibility with semiconductor integrated circuit technology, a-$Si_{1-x}C_x$:H is also a good candidate for application in Si-based electroluminescent or optoelectronic devices [8].

The effect of high hydrogen dilution on the structural and optical properties of a-$Si_{1-x}C_x$:H has been investigated[9]. However, the influence of relatively low hydrogen dilution on this material has not been systematically studied. In this investigation, we have explored the role and effect of hydrogen dilution on the optical and structural properties of carbon-rich a-$Si_{1-x}C_x$:H thin films grown by PECVD technique. A variety of techniques, including FTIR spectroscopy, Raman scattering, SEM and PL spectroscopy were used to characterize the grown films.

2. Experimental details

The a-$Si_{1-x}C_x$:H thin films were prepared in an r.f.-PECVD system using SiH_4 (diluted to 10% in H_2), C_2H_4 and H_2 as gas sources. The reactor was capacitively coupled, consisting of two parallel electrodes, powered electrode and grounded electrode, within a stainless steel chamber (grounded), operated at a radio frequency of 13.56 MHz. During deposition the C_2H_4 and H_2-diluted 10% SiH_4 gas were fixed at 10 and 20 sccm respectively. The total flow rates of H_2 (with 18 sccm in H_2-diluted 10% SiH_4) were controlled at 18, 20, 43 and 88 sccm, and the samples were marked as H1~H4 respectively. The total gas flow rate of all precursors was in the range of 30-100 sccm. P-type silicon (100) wafers and glass slides (7101) were used as the substrates. The chamber pressure and plasma power were maintained at 80 Pa and 50 W, respectively. The growth temperature was fixed at room temperature.

The film thicknesses were obtained by cross-sectional SEM (Hitachi S4800) images of the samples. The FTIR measurements were carried out in transmission mode by using a Nicolet 380 FTIR Spectrometer in the mid-IR region with a resolution of 4 cm^{-1}. RT micro Raman spectroscopy analyses were performed on a Labor Raman HR800 spectrometer. A laser line 514.5 nm was used as the excitation source. The RT PL spectra were obtained with a FLS920 (Edinburgh Instruments Ltd.) fluorescence spectrometer using a 325 nm UV xenon lamp as the excitation source.

3. Results and discussion

The cross-sectional SEM of film H2 is shown in figure 1.The surface of the film and the interface between the Si substrate and the a-$Si_{1-x}C_x$:H layer is smooth. The film is homogeneous, and no crystallites are observed in figure 1.

The thicknesses of all films were obtained from SEM images, and the film deposition rate is calculated by dividing the film thickness by the deposition time. Figure 2 shows the deposition rate as a function of hydrogen flow rate. The deposition rate of the a-$Si_{1-x}C_x$:H film decreases as the hydrogen flow rate increases because of the decreasing partial pressure of silane and ethylene and an etching effect of hydrogen plasma[10]. With the increase in hydrogen flow rate, the deposition rate decreases from 26.1 nm/min at the hydrogen flow rate of 18 sccm to the minimum value of 13.8 nm/min at 88 sccm.

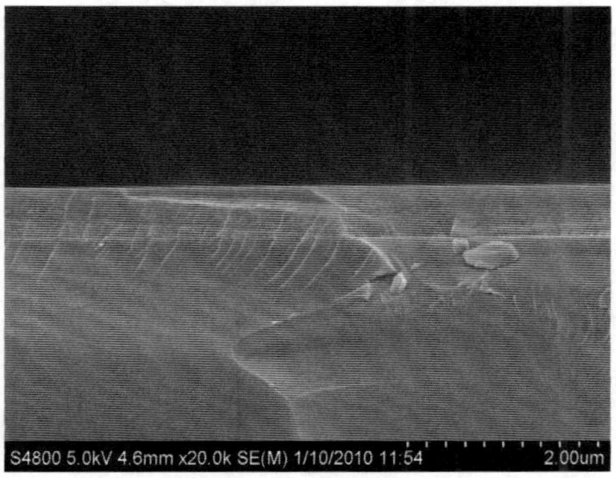

Figure 1. The cross-sectional SEM image of film H2

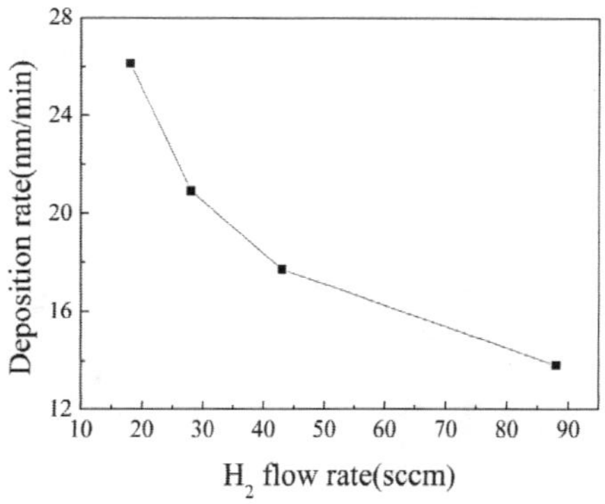

Figure 2. The variation of deposition rate as a function of total H_2 flow rate

Figure 3(a) shows the FTIR transmission spectra of the a-$Si_{1-x}C_x$:H films as a function of hydrogen flow rate. In order to compare the content of each bond between the films, FTIR spectra were normalized by the film thicknesses. The spectra are characterized by 3 regions:

a) 500-1200 cm^{-1}
b) 2000-2300 cm^{-1}
c) 2800-3000 cm^{-1}

The spectra between 500-1200 cm^{-1} cannot be fitted into several Gaussian peaks precisely because of too many IR-characteristic peaks in this region. We had to pick peaks directly from the absorption spectra of this region (figure 3(b)), and plotted them in figure 3(c) as a function of hydrogen flow rate. Although the peak positions cannot indicate corresponding IR vibration modes accurately, we can obtain some bonding information from their shifts. The peak at 790 cm^{-1} is due to Si-CH_3 rocking/wagging or Si-C stretching mode [1, 2, 5, 8, 11, 12]. The peak associated with a Si-C stretching mode should be at 670 cm^{-1}, which is covered by the strong 790 cm^{-1} peak. The absorption band at 845 and 890 cm^{-1} is attributed to $(SiH_2)_n$ bending [9]. The relative intensity of $(SiH_2)_n$ bending modes to that of Si-C stretching mode decreases with the increase of hydrogen flow rate, shown in figure 3(b). This result indicates that the SiH_2 radicals of the films decrease upon hydrogen dilution. A strong feature at 1040 cm^{-1} is assigned both to the C-H wagging mode around 1000 cm^{-1} [1, 3-5, 9, 11, 13] and the Si-O stretching vibration at 1100 cm^{-1} [2, 11]. The Si-O stretching mode is due to oxygen contamination during deposition process and sample preservation. The absorption band at around 1250 cm^{-1} is attributed to the bending mode of Si-CH_n bonds [1, 4, 5, 8, 11]. From figure 3(c), it was found that the 790 cm^{-1} and 840 cm^{-1} peak position tend to shift to a lower wavenumber upon hydrogen dilution. The 1040 cm^{-1} peak has the same trend except for sample H2 whose blue-shift may be caused by a stronger Si-O peak. As will be discussed later, the downward shift of these peaks is caused by the decreasing carbon content.

Figure 3(a) also shows the band of Si-H_n stretching between 2000-2300 cm^{-1} [1, 2, 4, 5, 8]. Si-H and Si-H_2 stretching mode are supposed to be located at 2000 and 2090 cm^{-1} respectively. The two peaks in our experiment shift to a higher wavenumber than those of a-Si:H. This is caused by the increase in electronegativity due to C substitution when C is bonded to Si, which shortens the Si-H distance and increases the vibrational frequency of Si-H bonds. The broad Si-H_n absorption peak was fitted with two Gaussians and a linear background. From this curve fitting, it was found that the positions of these two peaks tended to shift to a lower wavenumber (figure 3(d)), which means less carbon content upon hydrogen dilution. Meanwhile we plotted the ratio of integrated intensities of the Si-H and Si-H_2 stretching mode as a function of hydrogen flow rate. The decrease of Si-H_2/Si-H ratio indicates that SiH_2 radicals in the films decreases upon hydrogen dilution. The result is consistent with the discussion of figure 3(b).

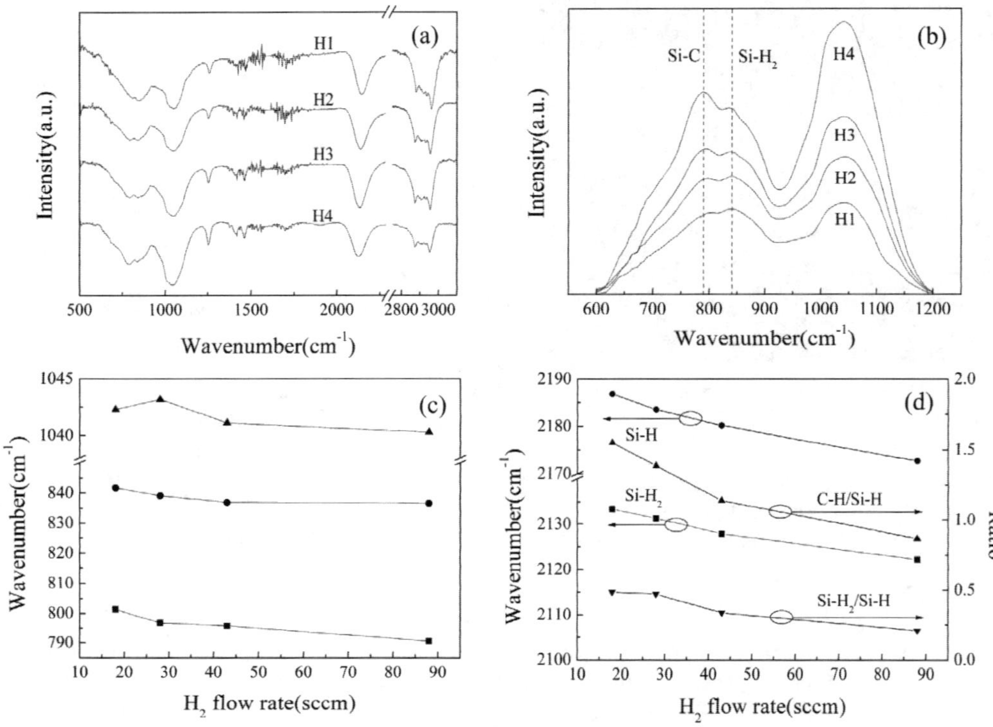

Figure 3. (a) FTIR spectra of a-$Si_{1-x}C_x$:H films deposited at different H_2 flow rates, (b) FTIR absorption spectra between 600-1200 cm^{-1}, (c) FTIR peak positions at 800, 840, and 1040 cm^{-1}, (d) FTIR peak positions of Si-H and Si-H_2 stretching mode, and the ratio of integrated intensities of Si-H/Si-H_2 peak and C-H/Si-H peak.

The absorption band around 2800-3000 cm^{-1} [1, 2, 5, 8, 14] is attributed to sp^3 C-H_n stretching modes. The sp^2 C-H_n absorption band is around 2975-3085 cm^{-1}, which is so weak that it is covered by the broad sp^3 C-H_n absorption band. Within the 2800-3000 cm^{-1} region, notably the peak at 2956 cm^{-1} (attributed to sp^3 C-H_3 and sp^2 C-H_n) becomes weaker compared with 2875 cm^{-1} peak (sp^3 C-H_2 and C-H_3). It indicates the reduction of sp^2 C-H_n upon H_2 dilution, and this is supported by the following Raman spectra analysis. Sp^1 C-H_n absorption band at 3300 cm^{-1} is not observed [8, 14]. The relative intensity of C-H_n band (2000-2300 cm^{-1}) to that of Si-H_n band (2000-2300 cm^{-1}) decreases with the increase of H_2 flow rate (shown in figure 3(d)). This indicates that the Si/C ratio of the films increases upon H_2 dilution. With increasing H_2 flow rate, less C atoms are cooperated into the films. And the decrease of carbon content leads to all the FTIR peaks shifting to lower wavenumbers.

In this experiment, the flow rates of SiH_4 and C_2H_4 is 40% of total flow rate to 12% for sample H1 to H4. Hence, the power density distributed to SiH_4 and C_2H_4 decreases notably from 40% of the total input power to 12% with the increase of H_2 flow rate. It has been reported that r.f. power density is an important parameter which affects the carbon content of the films [15]. SiH_4 and C_2H_4 contribute to the formation of a-$Si_{1-x}C_x$:H films. As the addition of H_2 leads to less power density distributing to SiH_4 and C_2H_4, the Si/C ratio increases with increasing H_2 flow rate. In the case of high H_2 dilution, the Si/C ratio dose not change with the addition of H_2. It can be explained that the power density distributed to the C and Si related precursors is not affected evidently, since their content is very low compared with H_2. For example, in the report of Park et al. [9], the tetramethylsilane (TMS) is no more than 3% of the total flow rate. So the change of power density with H_2 dilution is less than 3% of the total input power. Thus the Si/C ratio kept almost constant.

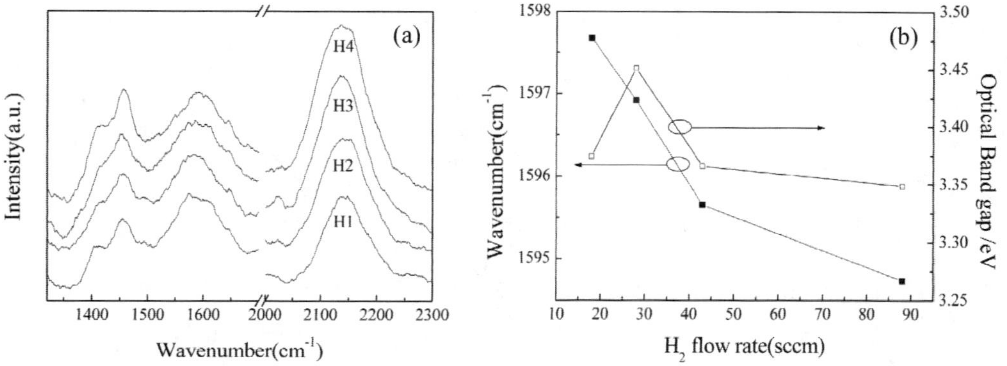

Figure 4. (a) Raman spectra a-$Si_{1-x}C_x$:H films deposited at different H_2 flow rates, (b) The variation of G peak position and optical band gap as a function of H_2 flow rate.

Figure 4(a) shows Raman spectra of the as-deposited a-$Si_{1-x}C_x$:H films as a function of hydrogen flow rate under visible excitation (514.5 nm). In accordance with FTIR, Si-H_n Raman modes at 2000-2300 cm^{-1} [4] are also observed, which exhibit a redshift with increasing hydrogen flow rate. No amorphous Si-Si Raman mode at 480 cm^{-1} was detected, indicating no amorphous silicon clusters exist in our samples. The Si-C band, which is expected to be at around 780 cm^{-1}, is not observed since the Raman efficiency of the Si-C band is much smaller than that of C-C and Si-Si bands [1, 4].

The carbon-rich a-$Si_{1-x}C_x$:H basically has a structure similar to that of pure a-C:H, so the Raman spectra can be analyzed relatively to that of amorphous carbon [9, 16]. The typical Raman spectra of a-C:H consists of two broad bands: G band (G for graphite) at about 1580 cm^{-1} and D band (D for disordered) at approximately 1350 cm^{-1}. The G band is the stretching vibration of any pair of sp^2 sites, whether in C=C chains or in aromatic rings, while the D band is breathing mode of those sp^2 sites only in rings, not chains [14]. G band is observed at about 1590 cm^{-1} in our experiment. The result indicates that the samples contain amorphous carbon clusters consisting of sp^2 C=C chains. And the sp^2 sites are organized in very small clusters because D band is absent in all samples [14]. Since the signal to noise ratio of Raman results is poor, the G peak positions are obtained by fitting the region of 1500-1700 cm^{-1} into Gaussian curves then plotted in figure 4(b). In a typical Raman spectra for amorphous carbon, the position of the G band shifts significantly if the sp^2/sp^3 carbon ratio of the film varies [9]. As is seen from figure 4(b), the position of the G peak shifts to the lower wave number region as the flow rate of hydrogen increases. This shift implies that hydrogen dilution promotes a reduction in the size and number of sp^2 carbon clusters in a-$Si_{1-x}C_x$:H [9]. The peak around 1400 cm^{-1} is attributed to carbon network attached with Si [12]. The peak at 1450 cm^{-1} assigned by Ferrari and Robertson [17] arises from vibrations of trans-polyacetylene like structural units.

The optical band gap derived from optical transmission spectra for a-$Si_{1-x}C_x$:H films deposited at different hydrogen flow rates are shown in figure 4(b). The optical band gap first slightly increases, then decreases as the hydrogen flow rate increases. The result is different from the report of high hydrogen dilution [9]. The optical band gap behavior as a function of hydrogen dilution is rather complex because it is controlled by two factors: the sp^2 carbon clusters and the Si/C ratio. As revealed by Raman spectra, carbon-rich a-$Si_{1-x}C_x$:H films contain sp^2 carbon clusters embedded in sp^3-bonded matrix, and hydrogen dilution promotes a reduction in the size and number of sp^2 clusters in a-$Si_{1-x}C_x$:H. The σ-bonds of sp^3 sites have a wider band gap than that of the sp^2 carbon cluster sites. Therefore, the optical band gap is partly determined by the size and distribution of sp^2 carbon clusters. A reduction in relative concentration of sp^2 carbon results in increase in optical band gap [9]. However, in this experiment, Si/C ratio increases upon hydrogen dilution. More silicon content leads to lower optical band gap. The two factors contributing to the optical band gap simultaneously causes the non-monotonous change shown in figure 4(b).

The room temperature PL spectra from samples deposited at various hydrogen flow rates are shown in figure 5. The as-deposited samples exhibited a blue-green room-temperature PL well visible to the naked eye. Using the framework of a-C:H, the PL mechanism of carbon-rich a-$Si_{1-x}C_x$:H films can be interpreted by Robertson's "cluster" model[1, 9]. Photoexcited carriers are believed to be confined inside sp^2 clusters in the films. The origin of the PL is the geminate recombination of electron hole pairs within sp^2 bonded carbon clusters in a sp^3 bonded amorphous matrix. The π states of the sp^2 sites form the valence- and conduction-band edges and lie closest to the band gap, so optical excitation creates electron hole pairs in the π and π^* states. The narrow π-π^* gap of sp^2 sites surrounded by the wider gap of sp^3 sites creates large band-edge fluctuations which tend to strongly localize tail states. PL occurs by the radiative recombination of photo-excited electrons and holes in localized tail states within sp^2 clusters after loosing their excess energy through the thermalization processes [18, 19]. The sp^2 clusters contain different structural units with double conjugated C=C bonds, which act as recombination centers [10].

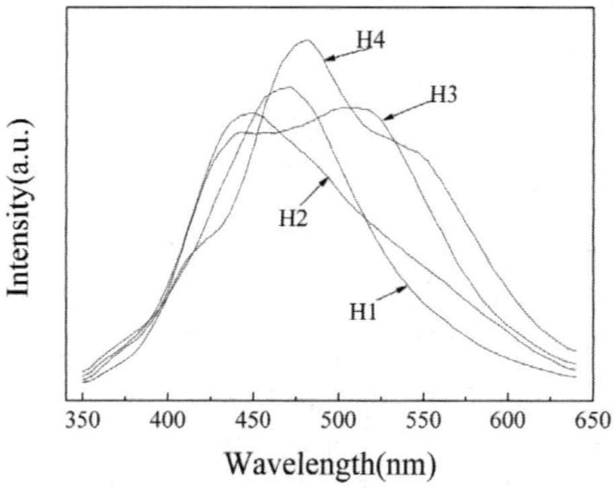

Figure 5. Room temperature photoluminescence spectra of a-$Si_{1-x}C_x$:H films at different H_2 flow rates.

4. Conclusion

The properties of carbon-rich a-$Si_{1-x}C_x$:H films prepared by PECVD was investigated. The samples were deposited using silane, ethylene and hydrogen as gas sources at a relatively low hydrogen dilution. The flow rates of silane and ethylene and other deposition parameters were kept constant, meanwhile the concentration of hydrogen was controlled between 60%-88%. A variety of techniques including SEM, FTIR, RS, UV-VIS spectrophotometer and PL were used to characterize the grown films. The deposition rate decreases with hydrogen dilution due to the decreasing partial pressure of silane and ethylene and an etching effect of hydrogen plasma. The Si/C ratio slightly increases with the addition of hydrogen, which is different from that of high hydrogen dilution. The phenomenon can be attributed to the dissipation of power density caused by hydrogen dilution. And less power density leads to higher Si/C ratio. The shift of the Raman G peak position to a lower wave number indicates that hydrogen dilution reduces the size and concentration of sp^2 carbon clusters, caused by the etching of sp^2 clusters by atomic hydrogen. The optical band gap, which is controlled by the sp^2 carbon clusters and the Si/C ratio, changes unmonotonously. The as-deposited samples exhibited a blue-green room-temperature PL well visible to the naked eye with UV excitation. The PL band can be attributed to the radiative recombination of electron-hole pairs within small sp^2 clusters containing C=C and C-H units in a sp^3 bonded amorphous matrix.

Acknowledgements
This work was supported by the National Basic Research Program of China (973 Program) "2007CB613403" and Foundation of the scientific research base development (Engineering Research Center of the Education Ministry for the Surface and Structure Modification of Inorganic functional Materials) "KYJD09014".

References

[1] Cheng Q J, Xu S, Long J D, Ni Z H, Rider A E and Ostrikov K 2008 *J. Phys.* D. **41** 055406
[2] Ray S C, Okpalugo T I T, Pao C W, Tsai H M, Chiou J W, Jan J C, Pong W F, Papakonstantinou P, McLaughlin J A and Wang W J 2005 *Mater. Res. Bull.* **40** 1757
[3] Swain B P and Dusane R O 2007 *Mater. Lett.* **61** 4731
[4] Wang Y H, Lin J and Huan C H A 2002 *Mater. Sci. Eng.* B. **95** 43
[5] Xu J, Yang L, Rui Y J, Mei J X, Zhang X, Li W, Ma Z Y, Xu L, Huang X F and Chen K J 2005 *Solid State Commun.* **133** 565
[6] Conde J P *et al.* 1999 *J. Appl. Phys.* **85** 3327
[7] Sha Z D, Wu X M and Zhuge L J 2005 *Vacuum* **79** 250
[8] Ma T F, Xu J, Du J F, Li W, Huang X F and Chen K J 2000 *J. Appl. Phys.* **88** 6408
[9] Park M, Teng C W, Sakhrani V, McLaurin M B, Kolbas R M, Sanwald R C, Nemanich R J, Hren J J and Cuomo J J 2001 *J. Appl. Phys.* **89** 1130
[10] Xu J, Mei J X, Huang X H, Li W, Li Z F, Li X and Chen K J 2003 The change of photoluminescence characteristics of amorphous carbon films due to hydrogen dilution *20th International Conference on Amorphous and Microcrystalline Semiconductors* (Campos do Jordao, BRAZIL: Elsevier Science Bv): pp 481
[11] Vasin A V, Kolesnik S P, Konchits A A, Rusavsky A V, Lysenko V S, Nazarov A N, Ishikawa Y and Koshka Y 2008 *J. Appl. Phys.* **103** 123710
[12] Swain B P 2006 *Surf. Coat. Technol.* **201** 1589
[13] Coscia U, Ambrosone G, Lettieri S, Maddalena P and Ferrero S 2005 Microcrystalline silicon-carbon films deposited by silane-methane mixture highly diluted in hydrogen *Symposium on Thin Film and Nanostructured Materials for Photovoltaics held at the 2005 EMRS Meeting* (Strasbourg, FRANCE: Elsevier Science Sa): pp 399
[14] Robertson J 2002 *Mater. Sci. Eng.* R. **37** 129
[15] Wang L, Xu J, Ma T F, Li W, Huang X F and Chen K J 1999 *J. Alloy. Compd.* **290** 273
[16] Zhang X, Weber W H, Vassell W C, Potter T J and Tamor M A 1998 *J. Appl. Phys.* **83** 2820
[17] Ferrari A C and Robertson J 2001 *Phys. Rev.* B **63** 121405
[18] Demichelis F, Schreiter S and Tagliaferro A 1995 *Phys. Rev.* B **51** 2143
[19] Rusli, Robertson J and Amaratunga G A J 1996 *J. Appl. Phys.* **80** 2998

Journal of Physics: Conference Series **276** (2011) 012174 doi:10.1088/1742-6596/276/1/012174

Microstructures and properties of the multiferroic Bi_xFeO_3 thin films

X X Li[1], H M Deng[2], D J Huang[1], P X Yang[1*] and J H Chu[1]

[1] Key Laboratory of Polar Materials and Devices, Ministry of Education, Department of electronics, East China Normal University, 500 Dongchuan Road, Shanghai 200241, China
[2] Key Laboratory for Material Microstructures of Shanghai University, Shanghai University, 99 Shangda Rd, Shanghai 200444, China

Abstract. The BixFeO3 thin films were grown on both Si and LaNiO3-coated Si substrates via sol-gel technique. The x-ray diffraction patterns were used to check the preliminary structures and crystallinities. The P-E curves which suggests the electrical characterizations of the films were analyzed. Furthermore, Raman spectra were also exploited to evaluate their optical properties.

1. Introduction

During past few years there is a tremendous flurry of research interests in materials which show simultaneous ferromagnetic order and ferroelectric ordering. As a lead-free multiferroic material with large remnant polarization (Pr~100 μC/cm^2), $BiFeO_3$ (BFO) draws extraordinary interests due to its particular properties such as high Curie temperature (T_C~850 °C) and antiferromagnetic Néel temperature (T_N~370 °C) well above the room temperature. Therefore, the $BiFeO_3$ thin films have many potential applications such as high-density ferroelectric random access memories, sensors and spintronics devices.

Although numerous of researches have been done since its first discovery in 1960s, the electrical properties of pure BFO have rarely been reported due to its high conductivity, which may be originated from uncertain oxygen stoichiometry, high defect density and poor sample quality. The volatility of bismuth usually leads to an uncertain proportion among the Bi-Fe-O system. If the proportion cannot be controlled accurately during crystallization of the BFO film, there will always be other impurity phases in the $BiFeO_3$ films such as $Bi_2Fe_4O_9$, $Bi_2O_{2.75}$, and $Bi_{46}Fe_2O_{72}$, etc.[1][2]

In this paper, we changed the index x (x = 0.95, 1, 1.05 and 1.10) in Bi_xFeO_3 thin films on purpose and prepared several Bi_xFeO_3 films with different contents of bismuth, and investigate the microstructures and properties of multiferroic Bi_xFeO_3 thin films.

2. Experimental details

In order to obtain Bi_xFeO_3 thin films (x = 0.95, 1, 1.05 and 1.10), $Bi(NO_3)_3 \cdot 5H_2O$ and $Fe(NO_3)_3 \cdot 9H_2O$ with different molar ratio were dissolved in the mixtures of ethylene glycol and acetic acid by constant stirring at room temperature, respectively. Then the transparent solutions of $Bi(NO_3)_3$ and $Fe(NO_3)_3$

* Corresponding author. Tel.: +86 21 54345157; fax: +86 21 54345119. E-mail address: pxyang@ee.ecnu.edu.cn (P. Yang).

were mixed together by stirring at room temperature for another hour. The resultant solutions were transparent, blackish red, and clear. [2]

The films were fabricated on both Si (100) and LNO/Si substrates. The precursors were dropped on substrates at a low rotating speed of 600 rpm, then spin coated at a high speed of 4000 rpm for 20 s. The films were pre-fired at 350 °C for 3 minutes and annealed at 650 °C for 4 minutes in air. This process was repeated six times for each sample to obtain the film of desired thickness. [3]

The crystalline structures of the films were investigated using an x-ray diffractometer. Raman spectra were exploited to evaluate their optical properties. The P-E curves which suggests the electrical characterizations of the films were analyzed.

3. Results and discussion

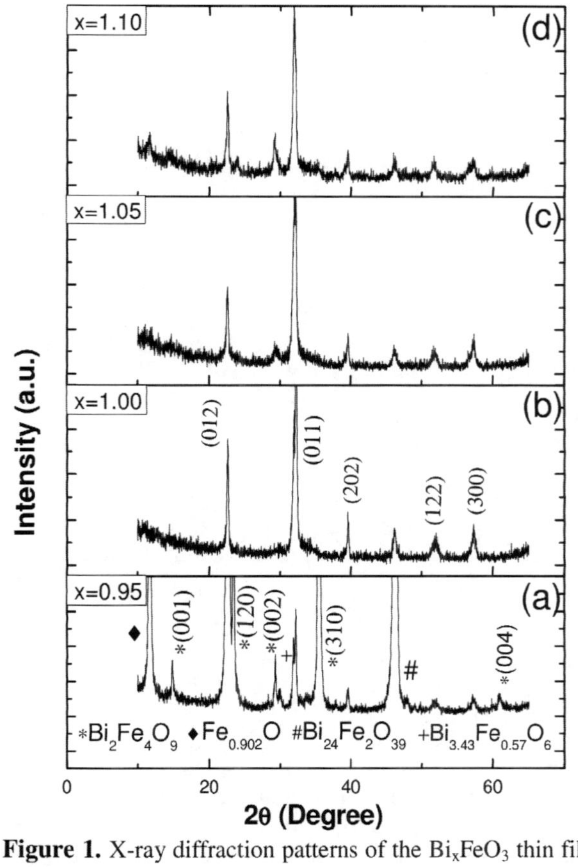

Figure 1. X-ray diffraction patterns of the Bi_xFeO_3 thin films on Si substrates

Figure.1 compares the XRD patterns of the Bi_xFeO_3 films on Si substrates with different Bi contents in the precursors. The patterns indicate that polycrystalline Bi_xFeO_3 thin films with a perovskite structure are obtained, and the influence of bismuth loss on crystallization and purity of the films is obvious. Besides the peaks associated with the perovskite structure of the BFO thin films, additional peaks are detected in samples (a). These peaks may result from the presence of $Bi_2Fe_4O_9$ and $Fe_{0.902}O$ impurities since bismuth is insufficient in the film. Other impurity phases such as $Bi_{3.43}Fe_{0.57}O_6$ and $Bi_{24}Fe_2O_{39}$ are observed at the same time. As shown in the patterns of the samples (b), (c) and (d), the non-perovskite phases are suppressed by compensating Bi atoms using the Bi-excess precursor solutions. But when *x* grows to 1.10, there is a very small amount of non-perovskite $Bi_2Fe_4O_9$ phase.[1][2][4][5]

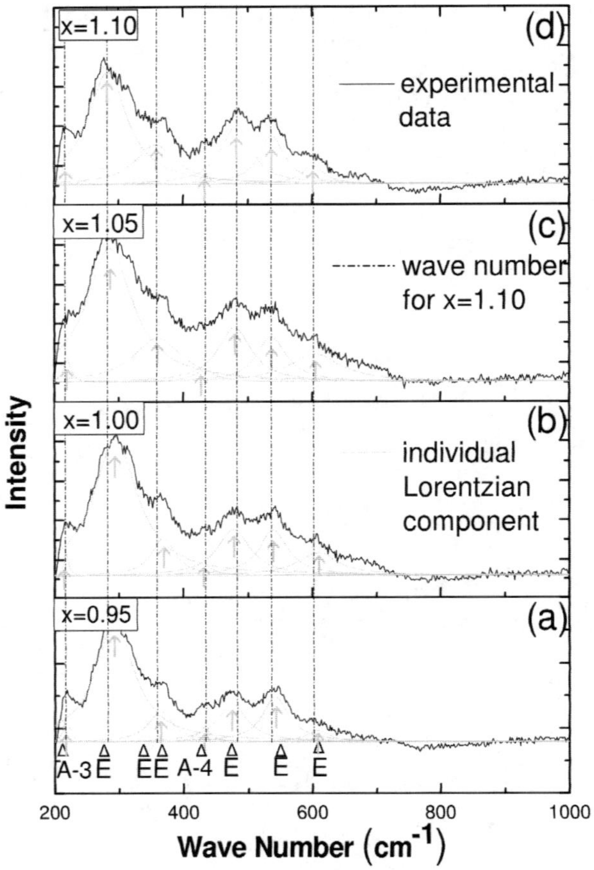

Figure 2. Raman spectra of the Bi_xFeO_3 thin films on Si substrates

The structural phase change in BFO can also be investigated by the Raman spectra with the range of 200~1800 cm^{-1}. In order to clarify the Raman active modes in the samples, the measured spectra are fitted and decomposed into individual Lorentzian components. As can be seen in Figure 2, it has been reported that the Raman active modes of BFO with rhombohedral R3c symmetry can be summarized using the following irreducible representation: $\Gamma = 4A_1 + 9E$. These ten normal modes, except for A-1 and A-2 blow 200 cm^{-1}, are also included for comparison with the decomposed active modes of our Bi_xFeO_3 samples. The first E mode, which has been demonstrated governed by Bi–O covalent bonds, shows red shift with increasing *x* from 0.95 to 1.10, indicating a gradual change of Bi–O covalent bonds.[5][6]

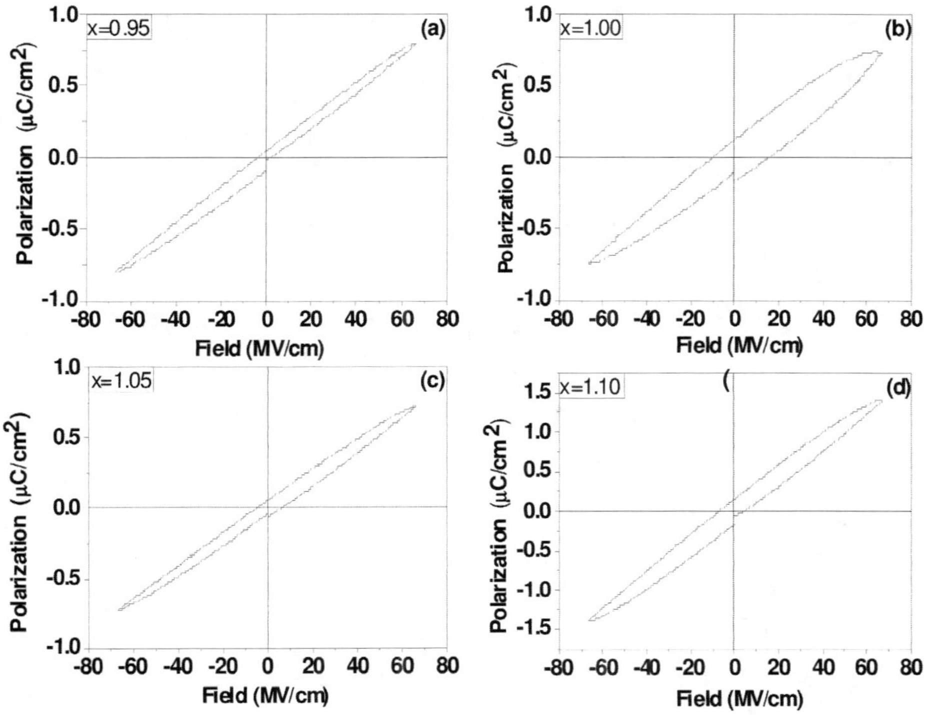

Figure 3. Polarization-electric field hysteresis loops of BixFeO3 thin films on LNO/Si substrates.

The difference of the ferroelectric properties among the four samples can be confirmed by polarization-electric field (P-E) measurement. Figure 3 shows the polarization hysteresis curves of the samples at 1 kHz. The largest remnant polarization appears when x grows to 1.10, but still much smaller than that of the pure BFO film without second phase. The existence of Bi_2O_3 and other phases may be the reason for the difference in electrical properties. The other phases such as Bi_2O_3 and Fe^{2+}, Fe^{3+} mixed-valent system with oxygen vacancies can strongly influence the BFO film on ferroelectricity.[7]

4. Conclusions

In summary, the Bi_xFeO_3 thin films with different contents of bismuth are fabricated on both Si and LNO/Si substrates by sol-gel technique. The crystalline BFO phase was observed in the films using Bi-excess solutions. The fraction of the secondary phases strongly depended upon the contents of bismuth. Secondary phases emerge and hysteresis loop is degraded as a result of insufficient bismuth. The first E mode, which has been demonstrated governed by Bi–O covalent bonds, shows red shift with increasing x from 0.95 to 1.10, indicating a gradual change of Bi–O covalent bonds.

Acknowledgements

This work was supported by the National Natural Science Foundation of China (No. 60990312), the State Key Basic Research Program of China (2007CB924902), Shanghai Leading Academic Discipline Project (B411) and Science and Technology Commission of Shanghai Municipality (10JC1404600).

References

[1] Jong Kuk Kim, Sang Su Kim, and Won-jeong Kim 2005 *Mater Lett* **59** 4006-4009

[2] Singh S K，Funakuba H, Uchida H，and Ishiwara H 2005 *Integrated Ferroelectrics* **76**: 139-146
[3] Singh S K, Kim Y K, Funakubo H, and Ishiwara H 2006 *Appl. Phys.Lett.* **88**, 162904
[4] Shannigrahi S R, Huang A, and Chandrasekhar N 2007 *Appl. Phys.Lett.* **90**, 022901
[5] Yuan G L, Wing Siu Or, Lai Helen, and Chan Wa 2007 *J. Appl. Phys.* **101**, 024106
[6] Liu Jian, Fang Liang, Zheng Fengang, Ju Sheng, and Shen Mingrong 2009 *Appl. Phys.Lett.* **95**, 022511
[7] Luo W B, J Zhu, Li Y R, Wang X P, Zhao D, Xiong J, and Zhang Y, *Appl. Phys.Lett.* **91**, 082501

Application and design of solar photovoltaic system

Li Tianze, Lu Hengwei, Jiang Chuan, Hou Luan, Zhang Xia
School of Electric and Electronic Engineering, Shandong University of Technology, China
No.12, Zhang Zhou Road, Zhangdian District, Zibo 255049, China
E-mail: ltzwang@163.com

Abstract. Solar modules, power electronic equipments which include the charge-discharge controller, the inverter, the test instrumentation and the computer monitoring, and the storage battery or the other energy storage and auxiliary generating plant make up of the photovoltaic system which is shown in the thesis. PV system design should follow to meet the load supply requirements, make system low cost, seriously consider the design of software and hardware, and make general software design prior to hardware design in the paper. To take the design of PV system for an example, the paper gives the analysis of the design of system software and system hardware, economic benefit, and basic ideas and steps of the installation and the connection of the system. It elaborates on the information acquisition, the software and hardware design of the system, the evaluation and optimization of the system. Finally, it shows the analysis and prospect of the application of photovoltaic technology in outer space, solar lamps, freeways and communications.

1. Introduction

Energy is material base of the economic development. To ensure the sustainable development of national economy, there must be the energy which can be continually provided for support. In the past 200 years, the energy system based on coal, oil, natural gas and other fossil fuel has greatly promoted the development of human society. However, material life and spiritual life is increasing, the awareness of serious consequences brought from the large-scale use of fossil fuels is increasing at the same time: depletion of resources, deteriorating environment, in addition to all of the above, it induce political and economic disputes of a number of nations and regions, and even conflict and war. After in-depth reflection of the development process of the past, human advance seriously the future path of sustainable development. Today in the 21st century, there is no a problem as important as a sustainable energy supply, especially for the benefit of solar energy development and has been highly concerned by all mankind. Around the world are faced with limited fossil fuel resources and higher environmental challenges, it is particularly important to adhere to energy conservation, improve energy efficiency, optimize energy structure, and rely on scientific and technological progress, development and utilization of new and renewable sources. In this paper, starting from the composition of the solar photovoltaic system, working principle and the photovoltaic power generation system design approach and design elements expatiate the steps and the idea of photovoltaic system design. Instructions for the applications of solar photovoltaic system in various aspects.

2. Solar photovoltaic system and operation principle

Solar photovoltaic system is power generation system of translating solar radiation into electrical energy directly using solar cell based on photovoltaic effect. Solar energy resources are dispersive and available everywhere, so solar energy photovoltaic power generation system is particularly suitable for use as an independent power supply. Solar photovoltaic system mainly includes three parts: solar components; power electronic equipment such as charge-discharge controller, inverter, test instrumentation and computer control; battery or other energy storage and auxiliary power generation equipment, as shown in Fig1. The working principle of photovoltaic power supply system is that the electricity produced from the solar component will charge for the battery controlled by the controller, directly to the load power supply in the context of meeting the load demand under the sunlight. The battery supply power for the DC load under the control of the controller if the sunlight is lack or at night. The converter is need to translate AC into DC for the photovoltaic system with AC load.

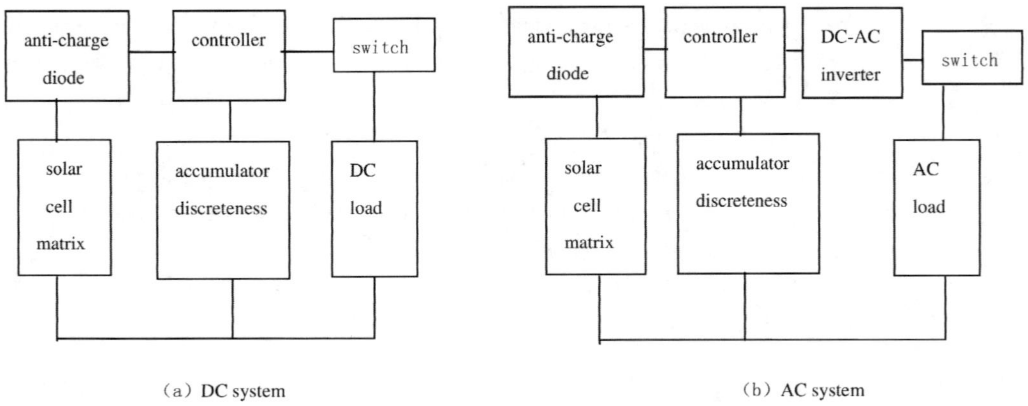

Fig.1 Solar photovoltaic system

2.1 Solar photovoltaic components

Consisting of solar components by serious-parallel connection according to requirement converting solar radiation into electrical energy under the light, it is the core component.

2.2 Anti-recoil diode

Choose the appropriate rectifier diodes as anti-anti-charge in diode Solar photovoltaic system. Its role is to ensure the battery can discharge through the solar cell matrix when the solar cell matrix don't generate electricity or appear short-circuit fault in the rainy days and nights

2.3 Accumulator discreteness

Electric energy generated by solar power is stored, and the stored energy will release to meet the energy needs of the load when the light is insufficient, night, or the load demand is greater than the solar power.

2.4 Controller and inverter

Controller is the equipment of control and management solar photovoltaic system, its control has the two ways of logic control and computer control. The main complete function: First, detection of a variety of PV systems installations and the state and parameters of all modules for the system, provide the basis for judgement, control and protection; Second, optimal charging control for battery, the controller determine the optimal charging method based on the current status of battery in the state of solar energy resources, in order to achieve efficient, rapid charging, and fully consider the battery life with the charge method; Third, the management of battery discharge process; Forth, provide protection for electrical equipment connected to photovoltaic electricity system to prevent damage of the PV system or electrical equipment; Fifth, fault diagnosis positioning and operation instructions.

Inverter is the equipment turned DC into AC. Because the output of solar cell and battery is direct current, the inverter is indispensable when the load is AC load. The technical requirements of the inverter are: steady output voltage and frequency, adjustable in a certain range; a certain over loading capability; output voltage waveform with the smaller harmonic components.

2.5 Measuring equipment

According to the scale of solar photovoltaic systems, there are different detection and measurement of photovoltaic power generation system. Only simple measurements are carried for the small-scale solar photovoltaic systems, but more measurement parameters are called for large and medium-sized solar photovoltaic power plants, and industrial power systems, such as: solar power system communication, pipeline cathodic protection systems.

3. Design principles of solar photovoltaic system

3.1 Design principles of solar photovoltaic system

The overall principles of designing of solar photovoltaic system are: under the premise of electricity supplying meeting the load, the economy of the system is the best. Photovoltaic power generation system can be divided into software design and hardware design, usually software design is earlier than the general hardware. Software design includes the investigation of the load and estimating of load power consumption, calculation of radiation of solar cell surface matrix, the calculation of solar cell components and battery capacity, and the optimization match between them, the best calculation of square angle, the prediction of system performance and the analysis of cost-effective. Hardware design include the selection and design of the load, the choice of solar cells and batteries, the design of components and support of the array, the selection and design of inverter, as well as the selection and design of the control and measurement system.

3.2 Design methods and procedures of software and hardware of solar photovoltaic system

Photovoltaic system's designing is based on understanding and knowing many relevant information and carry out the necessary economic analysis. Mainly it has the following methods and steps:

The first is do the detailed inspection to the location of the photovoltaic system installation getting detailed information, including geographical location, weather data, on-site situation, load conditions, the user requirements and so on; the second is the calculation and design of software; the third is the design of system hardware. After the software is designed, it is necessary to consider the requirements of the performance, and a relatively good economy; the fourth is the installation and connection of the system; the fifth is to do the monitor, evaluation, optimization to the operation of the system.

4. System design explain

4.1 For the detailed information

To accomplish the design task better ,you must acquire the accurate information before design of the software .Mainly do the follow works: The first is to do the detailed inspection to the location getting the geographic information including the longitude latitude devation.The second is to analyze the weather data deeply, mainly including the all energy through radiation ,amount of direct dadiation ,amount of discrete dadiation, amount of anti-discretness radiation, maximum and minimum temperature ,the average and maximum speed of wind ,the hailstone ,snow and so on. Fig2 gives the curve of power consumption of every month in different seasons.

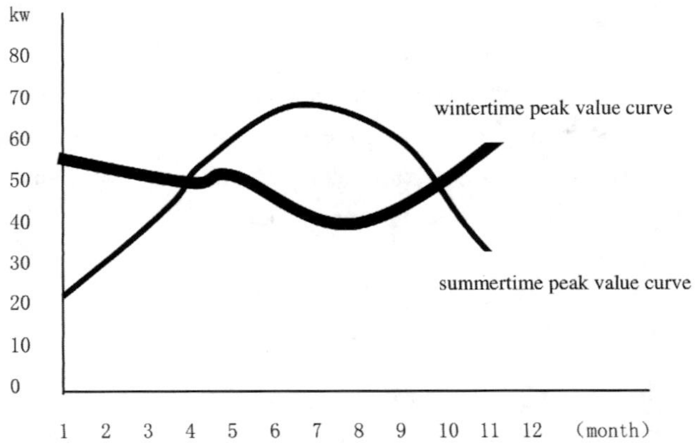

Fig.2 Curve of power consumption of every month in different seasons

4. 2 Design of software

On the base of acquiring detailed information you can get down to the design of software for the solar photovoltaic system. The principles the you should always follow in the design of software are: under the premise of electricity supplymeeting the load, the calculation of solar cell components and battery capacity should math the load power consumption furthest. This important principle decides the selection of solar cell component, the selection of battery capacity, the selection of best angle and so forth.

In the design, under the presime of considering the investment cost and the working condition of the load, you can consider using the mixed generating system to have a complementary system.

4. 3 Design of hardware

Make software design prior to the hardware design .we should select on the base of detailed related information and the design of software before ,considering not only meeting requirement of function but also making it more economical .The details should be considered :the selection of diode, the design of electric cable ,the design of prop stand ,the selection of controller and inverter ,considering installing following ,measuring and data collecting facilities with biggest power ,protection from thunder and lighting with ground contact ,square design field, the selection of auxiliary power source ,the design of transmitting and distributing electricity system and so on.

4. 4 Evaluation and optimization to the system

It is necessary to analyze the function of the solar and photovoltaic system for the solar and photovoltaic already built. the main purpose of function analysis is to know the working condition of the solar photovoltaic system ,seeing whether it can work normally. Find out the main factors that affect the system function through the analysis of various parameters, accumulating more experience data for the future solar and photovoltaic system.

5. Solar photovoltaic power and technology application

5.1 Solar light

The solar light is a solar-powered light, which is composed of solar components, batteries, charge-discharge controller, lighting circuits and poles, etc. Light, electricity, machinery, control technologies etc. That the light is gathering in integral whole and often integrates with the surrounding scenic environment. As long as sunny is enough it can install in situ, the light is a green environmentally friendly product and free from the effects of power lines, without ditching and embedding, non- consumption of conventional energy, and attracted a wide spread attention and application.

5.2 Solar car

With the arrival of the 21st century, the automotive industry advanced countries are researching and developing in energy saving and environment-friendly electric car. Solar-powered electric car developed rapidly in some developed countries as a result of advances in technology, especially the improvement of the cell and control technology. The primary users of solar-powered electric car are urban and rural middle-income residents, individual traders, and the tourism sector.

5.3 The application of PV in the communications

The most familiar application of solar photovoltaic power system is communications in the industrial field. Solar power used in unmanned microwave relay station, cable maintenance station, electricity / radio / communications / paging power systems, rural telephone carrier photovoltaic systems, small communication equipment, and soldier GPS-powered, etc.

5.4 The application of PV in the highway

Because of their unique characteristics of the highway, it is one of the solar photovoltaic place. Power supply system of highway plays a crucial role in the safety of the highway. In the urban areas of less electricity, if you use mains as power supply, the cost of pull-based power grid is very expensive. If using solar energy photovoltaic power generation on the highway to supply power to necessary electrical facilities, it is energy saving , environmental protection and economic security. Its applications is in the following areas: First, the service area on the highway which is away from the city power can build photovoltaic power station or photovoltaic-diesel hybrid systems, to supply area lighting, catering and other power needs to the service; The second is the emergency telephone system.

the highway pass through many remote areas, in order to deal with emergency incidents, an emergency call must be provided as a means of communication. Using the solar-powered long-distance transmission distribution equipment is not necessary, there is no transmission loss, safe and reliable when operate.

5.5 Applications in space

The first application field of photovoltaic technology is in space as a human satellite power, and later prevalence to the ground application. Solar cell can work in a wide range of sun intensity and temperature for a long period of time, with high reliability, high efficiency, long life and good anti-radiation properties, etc. making it obtain a wide range of application as a ideal space power. So far the vast majority of all types of aircraft launched into outer space by humanity are using solar cells as power supply.

References

[1] Xing yunmin ,Tao yonghong . Modern energy and power generation technology. Xi an :xi an electric science and techonology university press,2007
[2] Prince M B. Silicon solar energy converters.J Appl Phys,26:534-540(1955)
[3] Kallmann H,Pope M.Photovoltaic effect in organic crystals.J Chem Phys,30:585-586(1959)
[4] Wu zhijian,Ye zhiquan,Shen hui. The use of new and renewable energy.Beijing: Machinery Industry Press,2006
[5] Zhao yuwen,Wu dacheng,Li xudong,Song shuang. Development of photovoltaic industry and market in China. Solar Energy,(3):7-10(2007)
[6] Chapin D M, Fuller C S, Pearson G L. A new Silicon p-n junction photocell for converting solar radiation into electrical power.J Applphys, 25:676-677(1954)
[7] Li jianhua ,Li jingfeng. The new energy material and application technology Beijing : Qinghua university press,2005)
[8] Wang gehua .New energy and sustainable development .Beijing :chemical industry press,2005
[9] Xi wenhua etc Solar practical engineering techonology.Lanzhou: Lanzhou university press,2002

Journal of Physics: Conference Series **276** (2011) 012176 doi:10.1088/1742-6596/276/1/012176

Methods and analysis of factors impact on the efficiency of the photovoltaic generation

Li Tianze, Zhang Xia, Jiang Chuan, Hou Luan
School of Electric and Electronic Engineering, Shandong University of Technology, China
No.12, Zhang zhou Road, Zhangdian District, Zibo 255049, China
E-mail: ltzwang@163.com

Abstract. First of all, the thesis elaborates two important breakthroughs which happened In the field of the application of solar energy in the 1950s.The 21st century the development of solar photovoltaic power generation will have the following characteristics: the continued high growth of industrial development, the significantly reducing cost of the solar cell, the large-scale high-tech development of photovoltaic industries, the breakthroughs of the film battery technology, the rapid development of solar PV buildings integration and combined to the grids. The paper makes principles of solar cells the theoretical analysis. On the basis, we study the conversion efficiency of solar cells, find the factors impact on the efficiency of the photovoltaic generation, solve solar cell conversion efficiency of technical problems through the development of new technology, and open up new ways to improve the solar cell conversion efficiency. Finally, the paper connecting with the practice establishes policies and legislation to the use of encourage renewable energy, development strategy, basic applied research etc.

1. Introduction

Solar energy is the most important basic energy in all sorts of renewable energy, biomass, wind, and water are all from the oceanic solar energy etc. Through converting equipment change solar radiant energy into electricity, the photoelectric convert device is using photoelectric principle of semiconductor devices, to achieve the conversition between the sun light and the electric power. In the field of solar energy utilization, in the 1950s, there were two big technology breakthrough: one is the bell LABS developed 6% practical monocrystalline cells in 1954, the second is Tabor in Israel put forward the selective absorption surface theories and successfully developed the selective absorption coating the 1955. These two breakthroughs established technology base for solar energy utilization to entere the modern development period. With the development of the material and spiritual in people's life, the energy resources is exhausing and environment is worsening, in addition to induce political and economic disputes in a lot of national and regional, even conflicts and war. In the 21st century, the world is facing the challenges of limited fossil fuels resources and higher requirements of environmental protection, so adhering the priority of energy conservation, improving energy efficiency, optimizing energy structure, relying on scientific and technological progress, the exploitation and utilization of new energy are particularly important. Based on the analysis the energy crisis of the world is facing in 21st century, the influence factors of improving photovoltaic energy efficiency, this paper research the new technology in the efficiency of solar cells converting, and opening a new way to improve the efficiency of solar cells.

2. The development characteristics of solar photovoltaic power generation in 21 century

Solar power has the following features in 21st century: the first one is the industry will be developed with a high speed. The development of photovoltaic industry is one of the highest growth rate and most stable for many years, the prospection of photovoltaic energy has been recognised by more and more governments and finance circles, many developed countries and regions have set pv development planning, and photovoltaic energy is expected to be mankind's basic energy the middle of this century, the second one is the component costs will decrease, photovoltaic energy system's installation cost is decreasing by 9-10% rate per year, by expanding scale, improving the level of technology and automation technology, improving the efficiency of the cells approach to reduce the cost. The third one is photovoltaic industry develop to larger scale and higher technical level. The fourth is film battery technology will obtain breakthrough. The fifth is the rapid development of solar building integrated and grid-connected generation system.

3. Analusis of working principle and producting material of solar cells

3.1 The principle of solar cells

Solar cells work principle is pv effect of p-n knot in semiconductor. When the object is irradiated by light, the charge distribution state is changed and produce emf and current. When the sun or other illumination irradiate semiconductor p - n knot, both sides of p - n knot will appeare the voltage. Every atom outer electron have fixed position, and is constrainted by the nucleus, after stimulated by the energy in the outside,it get rid of the nucleus, and become a free electron,and at the same time , its original place will leave a space (named cavity), electron is negative electricity, while cavity is positive electricity. Combine p–type and n-type semiconductor together which are mixed with two different type impurities. Because the diffusion of electron and cavity in the interface, p-n knot will be formed, and electric field will be formed too. When the sun light to p-n knot, the electron in semiconductor will releasing electron. Accordingly to produce electron - cavity, and under the potential field electron is drived to n area, the cavity is drived to p area, thereby n area have excess electron, p area have exceed hole, then, the Photo-electric is born near p-n knot which is opposite to the barrier electric field. Part of the Photo-electric offset potential field, the rest let p area have positively charged , n zones have negatively charged, which makes EMF on the thin layer between n area and p area. This is named the pv emf, when switching the external circuit there is electricity output. If dozens or hundreds of solar battery series connected or parallel connected to composite solar modules, in the sun radiation, considerable power output of electric power can be gained. Figure 1 gives a solar cells energy level.

3.2 Analysis of solar cells production materials

In principle, almost all the materials have pv effect, can be used as the base material of solar cells. So, in theory, all the semiconductor materials should be solar photoelectric material preparation. But because of the following three aspects, not all of the semiconductor materials can be used solar photoelectric material in actual:the first one is the physical properties of materials, such as forbidden band width , carriers mobility, and light absorption coefficient, which makes some materials prepared for solar cell's theory transform is very low, have no development and application value. The Second one is the preparation and purification of the material are difficulty, in the present technical conditions, not all of the semiconductor materials are used for preparation of solar cells which can achieved high purity. The third one is the cost of materials and the preparation of solar battery, if the relevant costs is too high, the development and application will have no meaning. Therefore, although there are many types of semiconductor materials, but the real actual application semiconductor materials in solar cell industry is short. In the 19th century, the researchers had found the semiconductor solar effect in selenium, that is to say in sunshine illuminate, semiconductor materials would appear current, but has not been widely studied and applied. Until the 1950s, due to the invention of silicon and germanium ,the transistor of solar energy conversion is applicated. In 1954 monocrystalline silicon solar cells are developed, the photoelectric conversion efficiency soon more than 10%. Then, amorphous silicon, cast polysilicon, film polysilicon are all researched and applicated as solar photovoltaic material. In this period, compound semiconductor film materials, since it has appropriate forbidden band width and light absorption coefficient, therefore it is the important solar photoelectric material and is researched extensive.

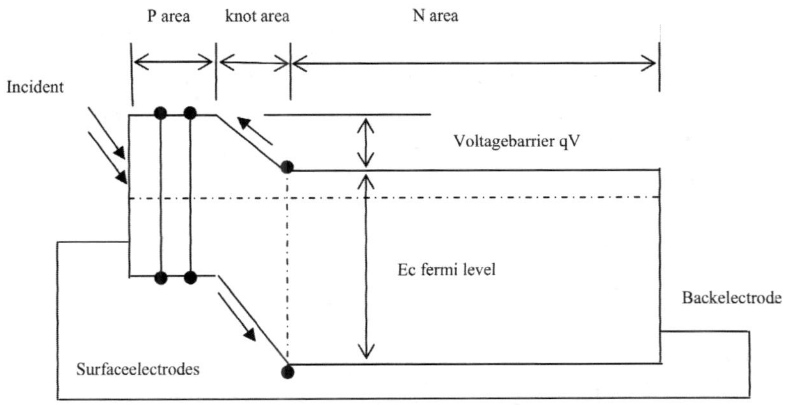

Figure 1. Solar cells energy level

Organic materials have strong light absorption ability, absorpt spectra is narrow, material energy gap can decreases with the increasing of conjugate length. When radiation of sunlight is AM1.5, after comparision between different maximum absorption curve of different energy gap, we know that narrow gap material (absorption peaks 600nm)is good to the absorption of sun energy. Organic materials' absorption is usually in the visible region, most materials absorptiorate on the sunlight is no more than 40%, therefore improving molar absorption coefficient of the active material, reducing energy gap, letting material have a wide absorpt spectrum is the useful method to improve the sunlight absorption. To increase the problem of sun absorption, except through changing the molecular structure function design to improve the active material's absoptipn, but also strong absorpt characteristics materials can be introducing the device structure. Using them to absorb part solar energy, then through exciton diffusion to transferred it to active material, in active material occur exciton disintegrate process and induced current.

3.3 Performance analysis of solar cell devices

Solar cells have three important output characteristic parameters: the first one is open voltage, that is to say,when there is no current in the circuate, the produced voltage after lighting. The second one is the short-circuit current density (Jsc), that is to say when the extra electric field is zero, the current in the circuate. The third one is the fill factor (FF), it refers to the specific value between maximum output power (U_mJ_m) and the product of open voltage short-circuit current.

When the load resistance of solar cells is zero (RL$\rightarrow$0), can detect short-circuit current of solar cells, When the load resistance of solar battery is infinite (RL$\rightarrow\infty$), can detect the open voltage. Solar battery open voltage is relative to the spectral intensity, have no relation with the cells size. Solar cells open voltage is proportional to the incident spectral irradiance. An ideal solar battery have small series resistance Rs, while the leakage resistance Rsh is large. Because Rsh and Rs is respective tandem and parallel in the circuit, so in the ideal circuit, they are negligible, the current through load is

$$I_L = I_{SC} - I_D \quad (1)$$

Ideal p-n junction characteristic curve equation is

$$I_L = I_{SC} - I_0(e^{qV/Ak_BT} - 1) \quad (2)$$

wherein, I0 saturation current when solar cells in the absence of light; q is electron charge; kB is boltzmann constant; A is constant factor. When IL=0, U is U_{OC}, and can be expressed as

$$U_{OC} = \frac{A_{k_B}T}{Q}\ln\left(\frac{I_{SC}}{I_O} + 1\right) \quad (3)$$

So the resulting fill factor (FF) is

$$FF = \frac{U_m J_m}{U_{oc} Jsc} \quad (4)$$

In the single-layer organic solar cells, the open circuit voltage value does not exceed the difference value between the two electrodes' work function. Figure 2 shows the solar cells output three important characteristic parameters.

In the heterojunction between donor and receptors, open voltage have relation with LUMO and HOMO, also have relation with interface between the active material and the electrodes, light current is linear to absorpt light ntensity. Through a series study the characteristics of heterojunction which is formed by different levels of LUMO small molecular materials and donor MEH – PPV. We found that the receptor LUMO significantly affect the open voltage, the lower of LUMO is linear to open-circuit voltage.

Solar cells energy conversion efficiency is have three representations: external quantum efficiency(η_{EQE}) also is named input photo conversion efficiency(η_{IPCE}); the second one is internal quantum efficiency， η_{IQE}, which is named light conversion efficiency (η_{APCE}); the third one is η, the total

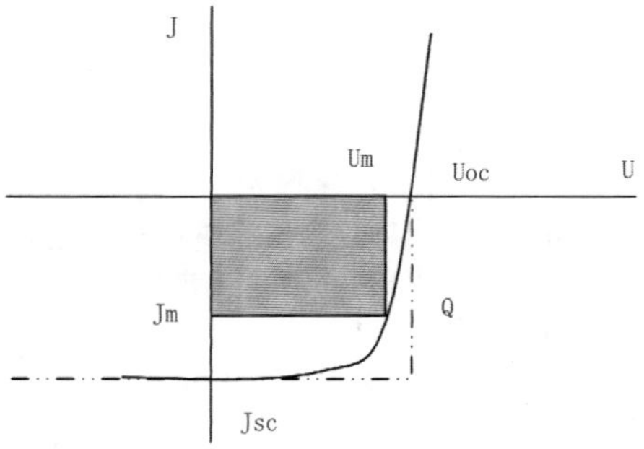

Figure 2.The solar cellsJ-U curve under light

conversion efficiency of the device,which is named （power efficiency η_P）. the solar cells which is based on the dissociation mechanism of donor and receiver. η_{EQE} is signified as

$$\eta_{EQE} = \eta_A \eta_{IQE} = \eta_A \eta_{ED} \eta_{ET} \eta_{CT} \eta_{CC} \tag{5}$$

wherein, η_{IQE} is internal quantum efficiency, defined as the ratio between the number of carriers and the number of absorbed photons by the device. η_A is photon absorption efficiency; η_{ED} is exciton diffusion efficiency; η_{ET} is Exciton dissociation efficiency; η_{CT} is free charge transport efficiency; η_{CC} is charge collection efficiency.Cells power conversion efficiency is η_P

$$\eta_P = \frac{P_m}{P_{in}} = \frac{U_{OC} J_{SC} FF}{P_{in}} \tag{6}$$

Defined as the ratio between the maximum output power Pm and radiation power Pin.

4. Factors of affecting the efficiency of photovoltaic power generation and new technologies to improve efficiency

4.1 Factors of affecting the efficiency of energy conversion and the two major bottlenecks

In recent years,there are many problems in the application of solar energy technologies, especially the conversion efficiency of the battery technology has encountered many obstacles: the shading effects of grating on the surface of solar cell; the loss of surface reflection; the loss of optical transmission; the loss of internal composite; the loss of surface composite. The core of solar energy appliction is generating electricity, generating electricity system is like a mini power station, which absorbs solar light, and directly changes solar into electrical energy. Not only satisfying their own electricity demand, excess electricity can also be incorporated into the power grid. The key bottleneck of limited solar energy application is technical environment and policy environment. Interconnection model with solar power equipment based on urban and rural power grid, far higher power grid companies may not buying grid type household solar power equipment redundant power. Even if the government subsidies, still quite difficult. To solve the key technical problems, is to reduce the cost of electricity can, in the true sense of the large-scale promotion.

4.2 New technologies to improve conversion efficiency

To solve these problems, the following new technologies are developed: Single-layer antireflection coating;laser carving groove buried grating technology; rongmian technology; back contact electrode, efficient back reflect technology; light absorption technology. With the application of these new technologies, issusing a series of studies which is focus on solar cell materials, conversion efficiency and stability, inventing many new types of cell, which is greatly enhancing the conversion efficiency of solar cells. In organic solar cells, three primary factors which is decide photoelectric conversion efficiency is the absorption of solar spectrum, exciton dissociation, free charge transfer. So when designing and choosing active material should pay attention to these characteristic to obtain high-performance battery device.

Improving the conversion efficiency of solar cells and reducing the cost of photovoltaic technology are the key. At present the impact of major obstacle to large-scale application of photovoltaic cells is its high manufacturing costs. In many power generation technologies, solar photovoltaic is still one of the highest cost, therefore, the main objective to the development of solar power generation technology is to design new battery structure by improving the existing manufacturing processes, to develop new battery materials, so the manufacturing cost is lower and improve the efficiency of photoelectric conversion., there are two main factors to improve the efficiency of photovoltaic solar cells:the first is how to make electron donor and receptor molecules in the mixture in the separate state; the second is how to expand the contact area between the two. In order to meet these two conditions, after studying the simulated new materials of molecular structure of the biofilm, hydrophobic side-chain and hydrophilic side-chain can be added, then to combine the two types of material to study the arrangement of the molecular structure, in order to form a liquid crystal state layered structure to improve the photoelectric conversion efficiency.

5. Policy support

- To strengthen publicity, improve the consciousness of developing and using solar energy. The potential crisis degree of China's energy and the ecological environment is more serious than the world average level. strengthening propaganda, improving energy and environmental crisis consciousness and developing renewable energy sources, such as solar industry is necessity and urgency.
- Making the long-term development industry plan of photovoltaic. Organizing pv bounded experts of technical and economic aspects to make the plan, which make photovoltaic technology and industries developed coordinative and healthy.
- Encourage photovoltaic companies actively explore the international market. To reduce the company dilemma due to domestic policy delay, countries should give more capital and technical support. To export enterprises in formulating a policy.

- Develop standards for the production of photovoltaic and policies, regulations of PV industry development.
- Government support policies. Learn from the developed countries started to rely on government support market experience, through the Government's policy orientation and demonstration projects, in a given period, giving appropriate subsidies to the users who using solar photovoltaic products, gradually guiding solar PV products to be used in more fields, cultivating the domestic consumer market.

References

[1] Xing Yunmin,Tao Yonghong. Modern energy and technology of power generation. Xian: Press of Xidian University, 2007
[2] Prince M B. Silicon solar energy converters.J Appl Phys,26:534-540(1955)
[3] Kallmann H,Pope M.Photovoltaic effect in organic crystals.J Chem Phys,30:585-586(1959)
[4] Wu Zhijian, Ye Zhiquan, Shen Hui.The utilization of New energy and renewable energy. Beijing: Press of Mechanical Industry, 2006
[5] Zhao Yuwen, Wu Dacheng, Li Xudong, Song Shuang. Development of China's Photovoltaic industry and market. Solar (SOLAR ENERGY),(3):7-10(2007)
[6] Chapin D M,Fuller C S,Pearson G L. A new Silicon p-n junction photocell for converting solar radiation into electrical power.J Applphys, 25:676-677(1954)
[7] Li Jianbao, Li jingFeng. The new energy material and application technology. Beijing: Tsinghua university press, 2005
[8] Wang Gehua. Energy and sustainable development. Beijing: Chemical technology press, 2005
[9] Yang Deren. Solar battery materials. Beijing: Chemical technology press, 2007
[10] Gao Baoxiu,Gao Zhiqiang,Deng Xianyu,Huang Wei. materials and device research based on the organic film solar cells. Science in China Series B, 38（11）：957-975（2008）
[11] Belcher W J,Wagner K I,Dastoor P C.The effect of porphyrin inclusion on the spectral response of ternary P3HT:porphyrin:PCBM bulk heterojunction solar cells.Sol Energy Mater Sol Cells,91:447-452(2007)
[12] Bundgaard E,Krebs F C.Large-area photovoltaics based on low band gap copolymers of thiophene and benzothiadiazole or benzobis. Sol Energy Mater Sol Cells, 91:1019-1025(2007)
[13] Chan M Y,Lai S L,Fung M K,Lee C S,Lee S T. Doping-induced efficiency enhancement in organic photovoltaic devices.Appl Phys Lett,90:023504-1~0235504-3(2007)

Simulation of High Efficiency Heterojunction Solar Cells with AFORS-HET

Wang Lisheng[1], Chen Fengxiang and Ai Yu

Department of physics science and technology, Wuhan University of Technology, Wuhan city, Hubei province, P.R.China. 430070

E-mail:wang_lesson@yahoo.com.cn

Abstract: In this paper, the high efficiency TCO/a-Si:H (n)/a-Si:H(i)/c-Si(p)/uc-Si(p^+)/Al HIT (heterojunction with intrinsic thin-layer) solar cells was analyzed and designed by AFORS-HET software. The influences of emitter, intrinsic layer and back surface field (BSF) on the photovoltaic characteristics of solar cell were discussed. The simulation results show that the key role of the intrinsic layer inserted between the a-Si:H and crystalline silicon substrate is to decrease the interface states density. If the interface states density is lower than $10^{10}cm^{-2}V^{-1}$, thinner intrinsic layer is better than thicker one. The increase of the thickness of the emitter will decrease the short-current density and affect the conversion efficiency. Microcrystalline BSF can increase conversion efficiency more than 2 percentage points compared with HIT solar cell with no BSF. But this BSF requires the doping concentration must exceed $10^{20}cm^{-3}$. Considered the band mismatch between crystalline silicon and microcrystalline silicon, the optimal band gap of microcrystalline silicon BSF is about 1.4-1.6eV.

1. Introduction

Silicon HIT solar cell is now attracting much interest, because very high efficiencies (above 22%) have been demonstrated [1]. This technology is based on Sanyo's successful technologies for fabricating high-quality a-Si film and solar cells with low plasma damage processes. It combines many advantages, such as (1) it simultaneously enables an excellent surface passivation and p-n junction, resulting in high efficiency, (2) its low-temperature processed (<200℃) can prevent any degradation of bulk quality that happen with high-temperature cycling processes in low-quality silicon materials such as solar grade Czochralski (CZ) Si, and (3) compared with conventional diffused cells, a much

[1] This paper is supported by "The Fundamental Research Funds for the Central Universities"(Project number:2010-la-019)

better temperature coefficient can be obtained with high V_{oc} cells [2]. Though SANYO's original design used an n-type substrate as the absorber for the HIT solar cell, current researches concentrate on developing the HIT solar cell on a p-type substrate, because of its popularity in the photovoltaic industry [3]. However, inferior performance was observed for devices fabricated on c-Si(p) as compared with those on c-Si(n) except SANYO, which suggests that a further investigation is necessary to fully understand the factors that affect the performance of the HIT solar cells.

Owing to the large numbers of processing variables, such as the thickness of the emitter, the interface states density, the band gap of the back surface field. etc, it is a formidable task to scrutinize the effect of each variable on the performance of the solar cell experimentally. Fortunately, numerical simulation software AFORS-HET provides a convenient way to evaluate the role of the various parameters present in the fabrication processing of HIT solar cells, and its reliability is proved by many references [3-4]. In this paper, the HIT solar cell TCO/a-Si:H (n)/a-Si:H(i)/c-Si(p)/uc-Si(p^+)/Al is designed and optimized by AFORS-HET.

2. Simulation

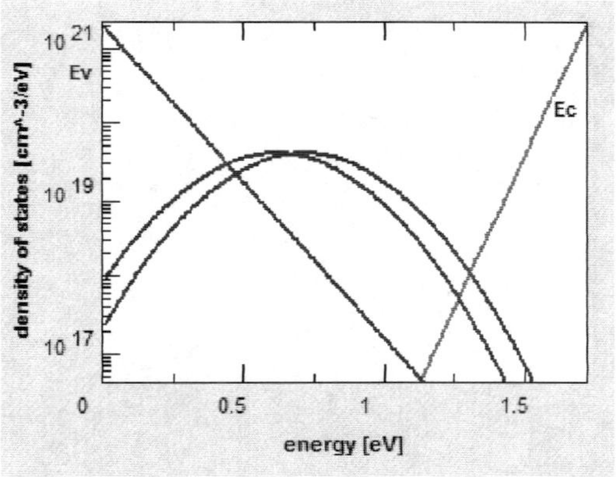

Figure 1 The gap state distribution in n-type a-Si:H layer.

AFORS-HET is a very general computer simulation code for analyzing and designing the HIT solar cells, device operation may be in dark or under light illumination. In our simulation process, the solar AM1.5 radiation was adopted as the illumination source with a power density of 100mW/cm^2. The front and back contacts are assumed as flat band ones to neglect the contact potential influence. The surface recombination velocities of both electrons and holes are set as 10^7cm/s. For the density of localized states in the band gap of amorphous silicon it has been assumed that there are both acceptor-like states and donor-like states modeled by exponential band tails (Urbach tails) and Gaussian mid-gap states (associated to silicon dangling bonds) [5]. The detailed localized states distributions are shown in figure 1. Other parameters used in the simulation are given in Table 1. During the simulations, all the parameters were adopted as the above setting values except for the specific declared one.

Table 1 The parameters used for HIT solar cell simulation

Parameter and units e,h for electrons and holes, respectively	a-Si(n)	a-Si(i)	c-Si(p)	uc-Si(p+)
Thickness(nm)	variable	variable	30000	variable
Dielectric constanct	11.9	11.9	11.9	11.9
Electron affinity (eV)	3.9	4	4.05	4
Bandgap(eV)	1.72	1.72	1.12	variable
Effective conduction band density (cm^{-3})	10^{20}	10^{20}	$2.8 \cdot 10^{19}$	10^{20}
Effective valence band density (cm^{-3})	10^{20}	10^{20}	$1.04 \cdot 10^{19}$	10^{20}
Electron mobility ($cm^2V^{-1}s^{-1}$)	5	5	1041	10
Hole mobility ($cm^2V^{-1}s^{-1}$)	1	1	412	3
Acceptor concentration (cm^{-3})	0	0	$1.5 \cdot 10^{16}$	variable
Donor concentration (cm^{-3})	$2.49 \cdot 10^{19}$	100	0	0
Thermal velocity of electrons($cm.s^{-1}$)	10^7	10^7	10^7	10^7
Thermal velocity of holes($cm.s^{-1}$)	10^7	10^7	10^7	10^7
Layer density ($g.cm^{-3}$)	2.328	2.328	2.328	2.328
Auger recombination coefficient for electron($cm^6.s^{-1}$)	0	0	$2.2 \cdot 10^{-31}$	0
Auger recombination coefficient for hole($cm^6.s^{-1}$)	0	0	$9.9 \cdot 10^{-32}$	0
Direct band-to-band recombination coefficient ($cm^3.s^{-1}$)	0	0	0	0

3. Results and Discussion

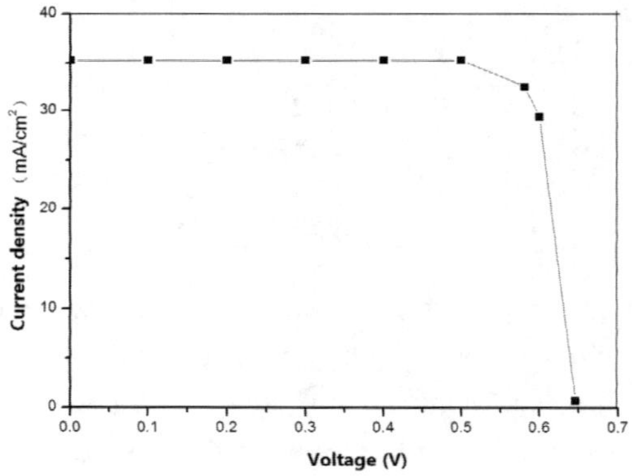

Figure 2 The J-V curve of TCO/a-Si:H(n)/c-Si(p)/Al cell.

Firstly, the HIT solar cell with structure TCO/a-Si:H(n)/c-Si(p)/Al is considered. The emitter thickness is 5nm and the related J-V curve is shown in figure 2. The detailed photovoltaic properties are V_{oc}=646.1mV, J_{sc}=35.32mA/cm^2, FF=82.68% and E_{ff}=18.87%.

3.1. Optimization of the intrinsic layer

While SANYO develops HIT solar cells with a very thin intrinsic a-Si:H(i) layer inserted between different type a-Si:H and c-Si, it should be noted that there is a controversy on the need of such a layer

[1]. Some authors claim that it is beneficial, while others get good results without it and do not see significant improvements if they introduce it. One reasonable explanation for the benefit of this undoped buffer layer is that the density of states in undoped a-Si:H is weaker than in doped a-Si:H, so we can expect to have less interface defects when the heterointerface is formed with undoped rather than doped a-Si:H.

3.1.1. The influence of the intrinsic layer thickness. Figure3 shows the photovoltaic characteristics of HIT solar cell varies with the thickness of the intrinsic layer. As can be seen from figure 3, the conversion efficiency of HIT is up to 19.27% with 1nm intrinsic layer is inserted, which is 0.4% higher compared with no intrinsic layer solar cell. But as I layer thickness increases, the conversion efficiency decreases. When the thickness of intrinsic layer reaches 5nm, the conversion efficiency of HIT solar cell with intrinsic layer is equivalent to solar cell with no intrinsic layer. Compromising the production processes and the conversion efficiency, the optimal intrinsic layer thickness should be set at 3nm. In addition, in figure 3, we can see that with the increase of intrinsic layer thickness, the open circuit voltage keeps almost unchanged, while the short circuit density decreases. This is because, as I layer thickness increases, the electric field strength of space charge region decreases and the short-spectrum absorption of amorphous silicon increases. The corresponding light-induced carriers cannot be effectively collected, resulting in the decrease of short circuit current density.

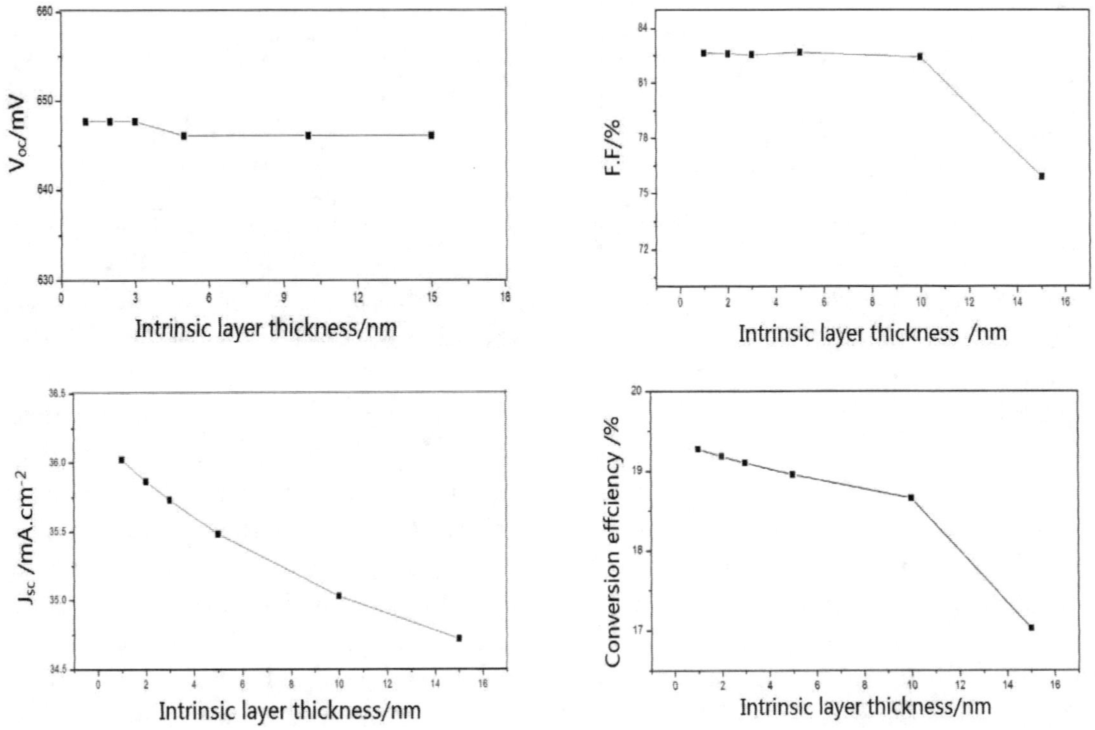

Figure 3 The performances of TCO/a-Si:H(n)/a-Si:H(i)/c-Si(p)/Al solar cell vary with the intrinsic layer thickness.

3.1.2. The influence of the interface states density. In the actual solar cell production processes, the impact of interface states density can not be ignored. In this paper, the interface defect states is assumed to insert between the I-type a-Si:H layer and the p-type crystalline silicon substrate layer. The interface states is assumed as continuous donor-like states and acceptor-like states with average distribution in the band gap, with the capture cross sections of electron and hole both are $10^{-14}cm^2$. The interface state density of a-Si:H/c-Si interface varies from $10^9cm^{-2}.eV^{-1}$ to $10^{13}cm^{-2}.eV^{-1}$, and its effects on the photovoltaic performances of the solar cell are shown in figure 4.

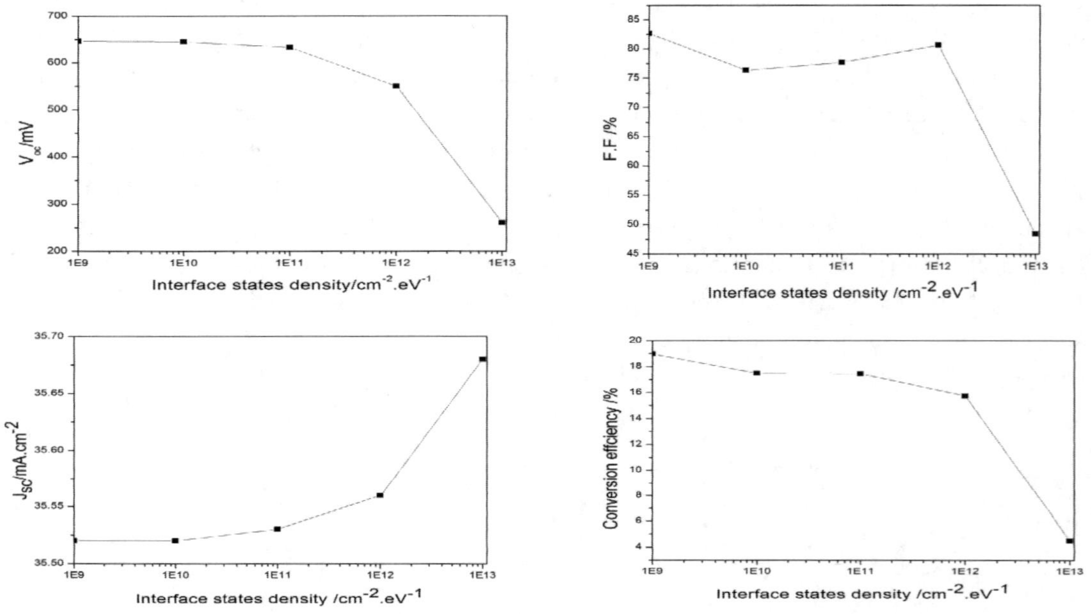

Figure 4 The impact of interface states density on the photovoltaic performances of HIT solar cell.

As shown in figure 4, when the interface state density D_{it} is lower than$10^{10}cm^{-2}.V^{-1}$, the solar cell performance is almost unaffected by the interface states. Under this condition, the efficiency is still maintained at 18.97 %, which is very close to the one without interface states. But when the interface state density D_{it} is higher than $10^{10}cm^{-2}.V^{-1}$, the open circuit voltage decreased rapidly, while short-circuit current remained almost unchanged, even increased slightly. When the interface state density varies from $10^{10}cm^{-2}.V^{-1}$ to $10^{12}cm^{-2}.V^{-1}$, the open circuit voltage decrease from 644.5mV down to 549.2mV and the short-circuit current density increased from $35.52mA/cm^2$ to $35.56mA/cm^2$. The open circuit voltage reduction results from the increase of p-n junction reverse saturation current. As the interface defect density increases, the carrier recombination probability at interface increases, leading to the increase in reverse saturation current and reducing the open circuit voltage and fill factor. In order to obtain high-efficiency HIT solar cells, surface passivation methods is essential, such as plasma-assisted H passivation, in order to control interface state defect density as down to $10^{10}cm^{-2}.V^{-1}$ as possible.

3.2. Optimization of the emitter

Because of the structural disorder and high doping concentration, the diffusion length of carriers in amorphous emitter is shorter. And there is only drift current and no diffusion current in emitter. Furthermore, since the amorphous silicon layer is a higher-doping layer, the space charge region of p-n junction has fallen mainly on crystalline silicon side, and even there is no electric field in amorphous silicon area. Therefore, emitter should be prepared as thin as possible with higher-doping. Figure 5 shows the influence of emitter thickness on the photovoltaic properties of solar cell.

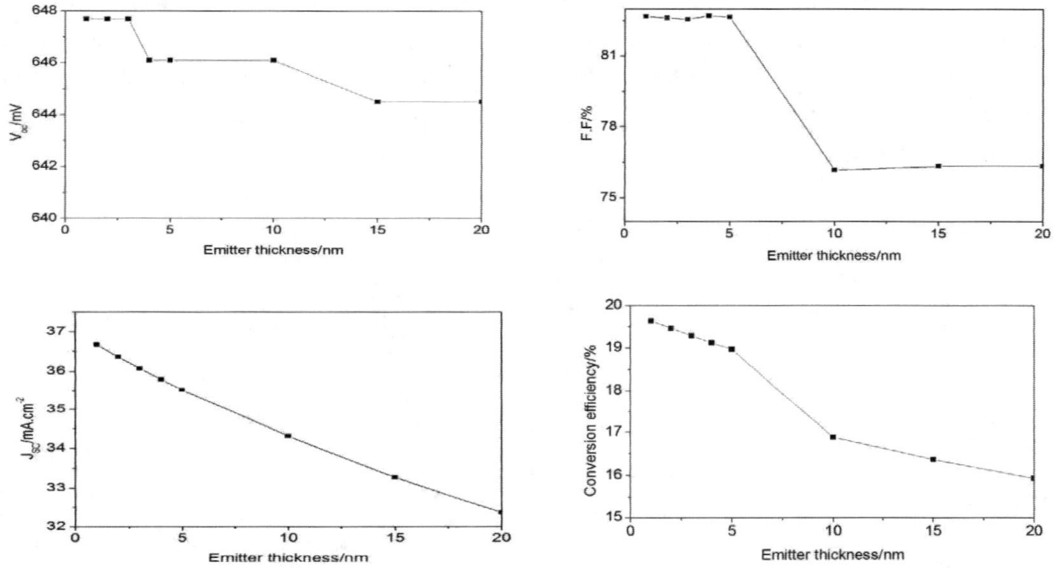

Figure 5 The impact of emitter thickness on the photovoltaic properties of HIT solar cell.

As can be seen from figure 5, with the emitter thickness increases, the open circuit voltage changed little, while the short circuit current is dramatically reduced. This is because with the emitter thickness increases, the absorption of the photon in emitter has increased. Considering the large amounts of recombination centers and the feature of no electric field in emitter, the photo-induced carriers are impossible to reach the edge of space charge region and contribute to light current. On the contrary, they will be recombined in the region and disappeared, resulting in the reduced short-spectrum response and short-circuit current. Fill factor also decreases with the n region thickness increases, for as the n layer thickness increases, series resistance will increases, which will reduce the fill factor [6]. Taking into account the technical production and conversion efficiency, the emitter thickness should be chosen about 5nm.

3.3. Optimization of the back surface field

The back surface field refers to a region which will be a barrier of minority carrier, thereby reducing the recombination rates of photo-induced carriers in the back surface. This barrier can not only enhance the photocurrent, but also increase photo-voltage to some extent [7]. Similar with the crystalline silicon solar cells, the back surface field can be achieved by a layer doped with the same type, but with higher doping concentration. Considering the microcrystalline silicon is a coexistence mixed phase of nano-crystalline silicon, amorphous silicon and grain boundaries, and its band gap can be adjusted by crystallization phase, so microcrystalline silicon was used to simulate the back-field structure [8], and the corresponding parameters were listed in Table 1.

3.3.1. The effect of the BSF thickness. Table 2 lists the effect of the BSF thickness on the J-V characteristics of HIT solar cell. We can see that when the thickness increases, the open circuit voltage and short circuit current are nearly unchanged, and only a slight change in fill factor and cell efficiency. So the efficiency can be considered as independent with the thickness [7], which is very beneficial. In view of technology, the difficulty in fabricating microcrystalline silicon thin films is that its growth rate is very low. If the required thickness is very thin, production time can be dramatically saved. Therefore, the BSF thickness can be set to 5nm.

Table 2 The J-V characteristics of solar cell with the variation of the BSF thickness

thickness/nm	V_{oc}/mV	J_{sc}/mA.cm^{-2}	F.F/%	□/%
2	646.1	35.53	82.65	18.97
5	646.1	35.53	82.67	18.98
10	646.1	35.54	82.69	18.98
15	646.1	35.54	82.71	18.98
20	646.1	35.55	82.73	19

3.3.2. The effect of the BSF doping concentration. Figure 6 shows the influence of BSF doping concentration on the photovoltaic characteristics of solar cell. In figure 6, we can see that the doping

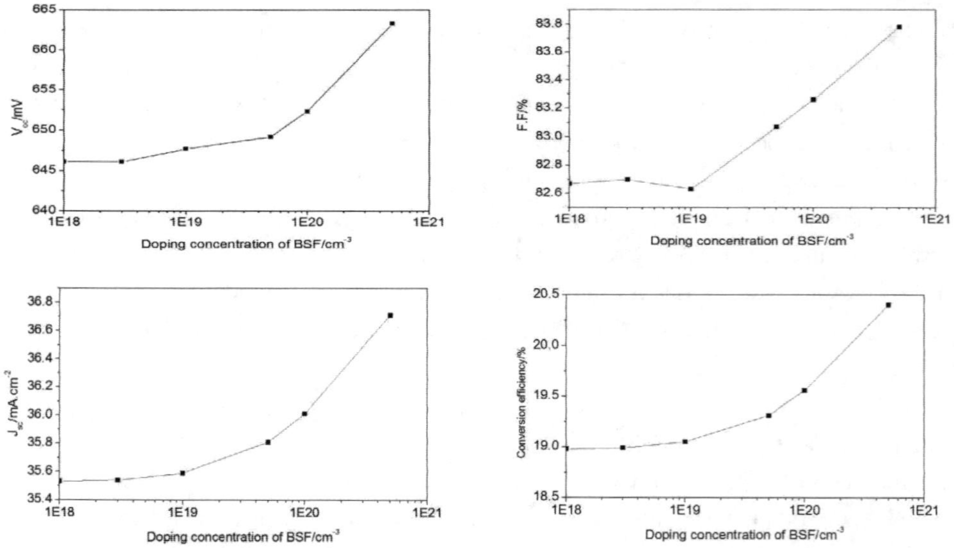

Figure 6 The influence of the doping concentration of BSF.

concentration must reach a certain value, preferably more than 10^{20}cm^{-3} before the conversion efficiency increased by 0.5 percentage points. With the doping concentration increases, the open circuit voltage, short circuit current, fill factor and so on all are increasing. This can attributed to the BSF band structure. When the doping concentration is low, the reflection role of BSF is not clear, and the barrier on the carrier transport can be reduced by increasing the doping concentration, which explains the high doping concentration is guarantee of good BSF. Microcrystalline silicon, due to its special crystallization phase and possibility of high doping, is selected as the BSF here.

3.3.3. The effect of the BSF band gap. With the development of thin film deposition process, silicon films with different band gap can be prepared under different deposition temperature and crystallization phase. In general, the band gap of nano-crystalline silicon is larger, up to 1.8eV and above; followed by amorphous silicon, the band gap of crystalline silicon is smaller. Therefore, we supposed that the band gap of microcrystalline silicon varies between 1.12eV-2eV. Figure 7 shows the parameters of HIT solar cells vary with the band gap of microcrystalline silicon BSF.

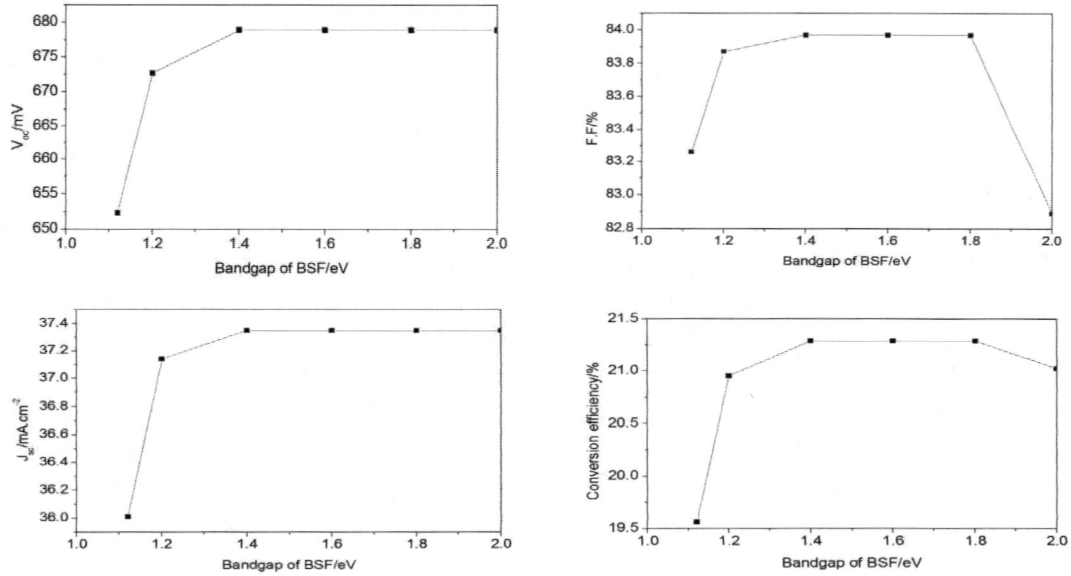

Figure 7 The effect of BSF band-gap.

As can be seen from figure 7, with the band gap increases, the open circuit voltage increased firstly and then saturated; when the band gap is above 1.8eV, the fill factor began to decrease, leading to decreased conversion efficiency. The reason is that with the band gap broader, although the BSF provides barrier for the minority carriers, but also will impede the transport of majority carriers. So an optimum band gap should be selected in the 1.4-1.6 eV.

3.3.4. The comparison between HIT solar cells with and without BSF. Figure 8 shows the effciency comparison of HIT solar cells before and after the BSF. In figure 8, we can see that with the BSF is added, all parameters of solar cells are improved, indicating a reasonable design of the back field, especially the optimization of doping concentration and the band gap, will increase the conversion efficiency by 2 percentage points. The detailed information also shown in figure.

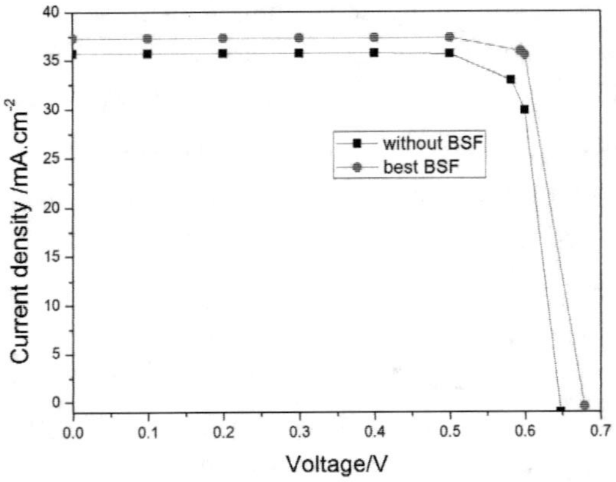

Figure 8 The J-V curves for HIT solar cells before and after BSF.

Type	V_{oc}	J_{sc}	FF	□
n/i/p	647.7	35.73	82.55	19.1
n/i/p/p^+	678.9	37.35	83.97	21.29

As shown in figure 8, when a 5nm thick microcrystalline silicon layer with doping $10^{20}cm^{-3}$ and band gap 1.5eV is added as BSF, it can improve V_{oc} 30mV and raise J_{sc} 1.5mA/cm^2.

4. Conclusion

In this paper, the TCO / a-Si: H (n) / a-Si: H (i) / c-Si (p) / uc-Si (p^+) / Al HIT solar cell is simulated with AFORS-HET software. It was found that with the increase of the n layer thickness, the short-spectrum response and the short-circuit current density of solar cell will decrease; the introduction of intrinsic amorphous silicon layer is mainly to reduce the interface defect state density. If the interface states density is lower, the better intrinsic thickness is no more than 5nm. The thickness of BSF has little effect on the conversion efficiency, while reasonable doping and band gap design can improve efficiency more than 2 percentage points. At that time, the photovoltaic parameters of HIT solar cells are V_{oc}=678.9mV, J_{sc}=37.35mA/cm^2,F.F=83.97%, □=21.29%.

References:

[1] Kleider J P, Chouffot R, Gudovskikh A S, Cabarrocas P R and Labrune M 2009 *Thin Solid Films* **517** 6386-6391

[2] Mikio T, Hitoshi S, Yukihiro Y, Eiji M, Akira T and Makoto T 2005 *PVSEC* 501

[3] Zhao L, Zhou C L, Li H L, Diao H W and Wang W J 2008 *Solar Energy Materials & Solar cells.* **92** 673-681

[4] Ren B Y, Zhang Y, Guo B, Zhang B, Li H Y, Xu Y and Wang W J 2008 *Acta Energy Solaris Sinica* **29(9)** 1112-1116

[5] Norberto H C and Arturo M A 2010 *Solar Energy Materials & Solar cells* **94** 62-67

[6] Ren B Y, Wang M H, Liu X P and Li Y L 2008 *Acta Energy Solaris Sinica* **29(2)** 126-129

[7] Zhao L, Zhou C L, Li H L, Diao H W and Wang W J 2008 *Acta Physcia Sinica* **57(5)** 3212-3217

[8] Li L M, Zhou B Q, Chen X, Han B and Hao L Y 2009 *Information Recording Materials* **10(3)** 18-21

Journal of Physics: Conference Series **276** (2011) 012178 doi:10.1088/1742-6596/276/1/012178

Characteristics of GaN-based p-i-n photodetector at low temperature and performances of MCT/GaN-based infrared/ultraviolet dual-colour detector

Fuhao Liu[1,3], Jintong Xu[2], Ping Wang[2], Chao Li[1], Dahan Qian[1], Nili Wang[2] and Xiangyang Li[2]

[1] State Key Laboratories of Transducer Technology, Shanghai Institute of Technical Physics, Chinese Academy of Sciences, Shanghai 200083, China
[2] Key Laboratory of Infrared Imaging Materials and Detectors, Shanghai Institute of Technical Physics, Chinese Academy of Sciences, Shanghai 200083, China

E-mail: liufuhao@mail.ustc.edu.cn

Abstract. The performances of GaN-based p-i-n photodiode at low temperature are studied. The current-voltage characteristics at different temperatures were measured. The turn-on voltage of GaN-based p-i-n detector decreases from 3.7 V to 3.0 V as temperature increases from 91 K to 286 K. The series resistance and ideality factor of the GaN-based detector is calculated. Values of series resistance are 1508 Ω, 453 Ω, 353 Ω and 295 Ω, at 91 K, 174 K, 224 K and 286 K, respectively .The values of ideality factors are 43.7, 15.4, 12.6 and 11.9, respectively. The spectral response of GaN p-i-n photodiode was measured at different temperature .The results show that with the temperature decreasing, the peak response wavelength gets shorter, and the responsibility gets smaller. The structure of ultraviolet/infrared MCT (mercury cadmium tellurium)/ GaN-based dual-colour detector is designed. MCT medium wave photovoltage photodetector is employed in dual-colour detector. The responsivities of GaN-based detector and MCT-based detector were measured. The dual-colour device can realize the detection of the light between 240 nm and 330 nm, and light between 3.2 μm and 5.8 μm.

1. Introduction

Recently, gallium nitride (GaN) based detectors have been investigated intensively because of their wide direct band gap, high breakdown fields, and high-temperature operation. They are applicable in flame detection, chemical/biological sensing, optical communication and scientific research [1-3]. However, gallium nitride (GaN) based detectors can only get information in the range of 200-365 nm. To collect more information, many dual-colour and multi-colour detectors are investigated these years. A. Rogalski etc. reported infrared dual-colour detector, sensitive to short-wavelength (SW) and long-wavelength (LW) infrared light [4], offering good performance. AlGaN/PZT ultraviolet/infrared dual-colour detector has also been presented [5]. MCT-based detector covers the whole infrared wavelength, and the fabrication of the device develops very well, the performance of MCT-based detector is

[3] Fuhao Liu, Email: liufuhao@mail.ustc.edu.cn

excellent [6-8]. Hitherto, as far as we know, there have been no reports about MCT/GaN-based dual-colour detector.

In this paper, MCT/GaN-based medium-wavelength (MW) infrared and ultraviolet detector is presented. MCT-based infrared detectors have excellent performance at the temperature of liquid nitride, GaN-based ultraviolet detectors commonly work at room temperature, and the performance of GaN-based device of this dual-colour detector at low temperature proves to be important. The current-voltage characteristics and spectrum of GaN-based device from 90 K to 300 K are investigated intensively. Series resistance and ideality factor of the GaN-based detector at different temperature are calculated. The reason why as temperature decrease the responsivity becomes smaller is discussed. Spectrums of GaN-based detector and MCT-based detector in the MCT/GaN-based detector are given, respectively. The dual-colour device can detect the ultraviolet light between 240 nm and 330 nm, and infrared light between 3.2 μm and 5.8 μm.

2. Experiment

GaN materials were grown in a metal-organic chemical vapor deposition (MOCVD) reactor on double-side polished sapphire substrates. The template consisted of a 500 nm AlN buffer to improve the crystal quality. The structure consisted of a 0.6 um Si-doped n type $Al_{0.2}Ga_{0.8}N$ layer, a 0.2 μm unintentionally doped GaN layer, and a 0.2 μm Mg-doped p type GaN layer. Carrier concentrations in these layer were estimated by Hall-effect measurement, with the results of 2×10^{18} cm^{-3}, 1×10^{16} cm^{-3} for electron concentration in the n-$Al_{0.2}Ga_{0.8}N$ and i-GaN layers, respectively, and 2×10^{17} cm^{-3} for hole concentration in the p-GaN layer. A Ni/Au/Ni/Au (20 nm/20 nm/20 nm/20 nm) p-type contact was evaporated on surface of p-GaN, and annealed in air atmosphere to form ohmic contact. Circle mesas were defined via standard photolighography technology and inducted coupled etching (ICP) under Cl_2, BCl_3 and Ar plasmas. Ti/Al/Ti/Au (30 nm/30 nm /50 nm/50 nm) was deposited via thermal evaporation. Devices were passivated with SiO_2 of 300 nm. Finally, 20 nm/300 nm Cr/Au metal pads were deposited.

Epoxy adhesive was plastered on the MCT-based infrared devices, and then followed by GaN-based ultraviolet detector adhering on it. Stabilization of epoxy was much better than other glue at low temperature. Before that, the filter film which allows 3-5 μm wavelength light transmits through was deposited on the substrate side of GaN-based detectors. The profile of GaN-MCT dual-colour detector is given in figure 1.

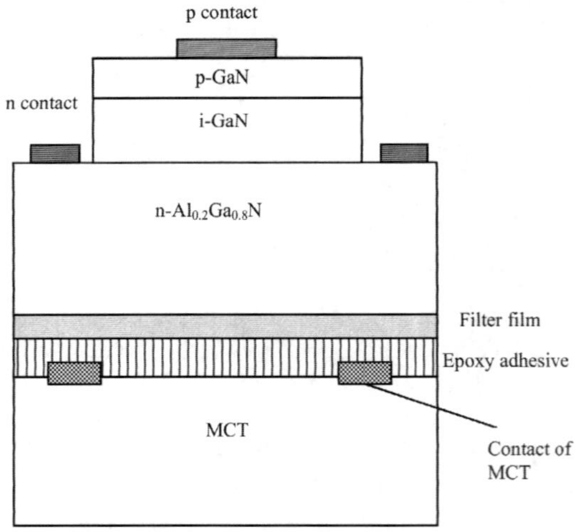

Figure 1. The cross structure of MCT/GaN-based dual-colour detector.

3. Results and discussion

3.1. I-V characteristics of GaN-based detector at different temperature from 91 K to 286 K

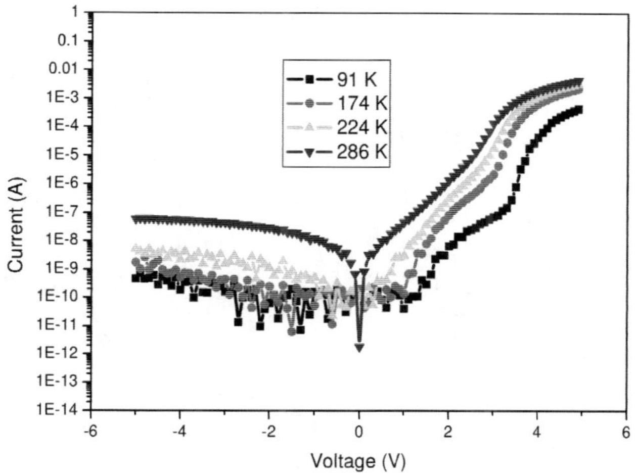

Figure 2. Current-voltage characteristics at different temperatures.

Current-voltage characteristics under bias voltage from -5 V to 5 V were measured with a Keithley236 semiconductor parameter analyzer in the temperature range from 91 K to 286 K. Figure 2 shows that current at forward bias becomes higher as the temperature goes higher. Table 1 shows the electric parameters calculated from I-V curves. As shown in table 1, the turn-on voltage of GaN-based p-i-n detector decreases from 3.7 V to 3.0 V as temperature increases from 91 K to 286 K. It is clearly seen that the current at reverse bias becomes higher as temperature goes higher. The current under -5 V bias at 286 K is about 2 orders of magnitude larger than that at 91 K. The curve at 91 K between 0 V and 5 V is easily separated into two parts, the slope of the first part between 0 V and 3.7 V is lower, than that of the second part between 3.7 V and 5 V. The slope of the first part at 174 K is higher than that at 91 K, and lower than that at 286 K. This is mainly due to the variable series resistance at different temperature. The series resistance of the diode is calculated from the current-voltage characteristics under forward bias according to

$$\frac{dV}{dI}I = RsI + \frac{nkT}{q} \tag{1}$$

As discussed in reference [9-10]. Values of series resistance are1508 Ω, 453 Ω, 353 Ω, 295 Ω at 91 K, 174 K, 224 K and 286 K, respectively. It is an exponential decay with temperature increasing. The mechanism of this phenomenon will be discussed in another paper. In addition, the ideality factor is extracted [10] in order to gain insight into the conduction mechanisms operating in the GaN-based p-i-n detector. The values are 43.7, 15.4, 12.6 and 11.9 for device at 91 K, 174 K, 224 K and 286 K, respectively. R. McClintock [10] reported that the ideality factor is expected to have a value between 1 and 2, if n is closer to 2, recombination current dominates, if closer to 1, diffusion current dominates. The ideality factor they calculated was 2.89, and it fell well outside of this range. The explanation they gave was that the weak Schottky-like nature of the p contact created a second diode in series with the p-i-n junction.

Table 1. The turn-on voltage, series resistance and ideality factor of GaN-based detector at different temperatures.

Temperature (K)	91	174	224	286
Turn-on voltage (V)	3.7	3.4	3.2	3.0
Series resistance (Ω)	1508	453	353	295
Ideality factor	43.7	15.4	12.6	11.9

3.2. The spectrum of GaN-based detector at different temperature

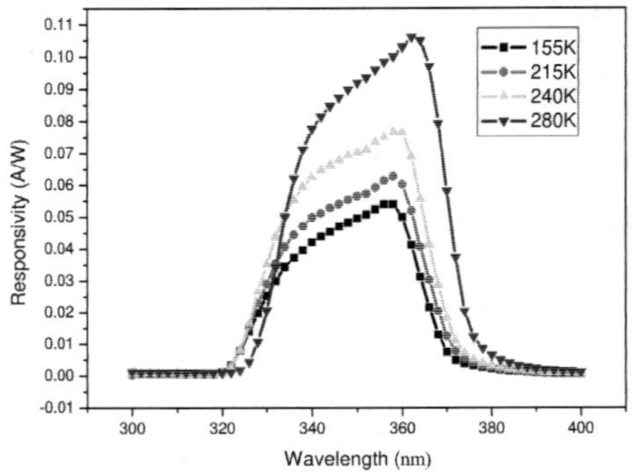

Figure 3. Responsivity of GaN-based detector at different temperature.

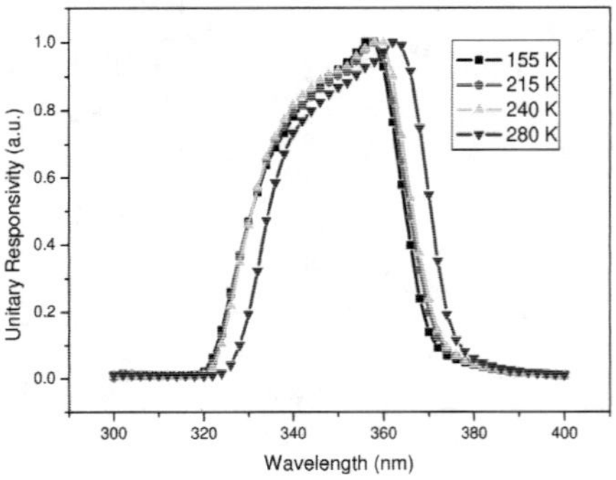

Figure 4. The unitary responsivity of GaN-based detector at different temperature.

The spectral dependence of the photoresponse was measured with a Xe-arc lamp source coupled to a grating monochromator. The relative response was calibrated with a large square area UV-enhanced Si detector. Figure 3 shows the response of the GaN-based p-i-n photodiode at different temperatures. We observe that as temperature becomes higher, the responsivity of the device becomes larger, and the largest peak responsivity at 280 K is about 0.106 A/W at 362 nm. This is because the depletion gets wider as temperature increases. From the fit of quantum efficiency, it is observed that the width of the depletion affects the quantum efficiency severely, and with wider depletion width, quantum efficiency is larger. It is consistent with the experiment result: as temperature becomes higher, the width of depletion region gets wider, and the quantum efficiency becomes larger [11].The short wavelength edge of spectral response lies at 300 nm, corresponding to the band gap of $Al_{0.2}Ga_{0.8}N$, and the long wavelength edge is at 366 nm, corresponding to the band gap of GaN. The quantum efficiency of the photodiode at 280 K is calculated from spectral responsivity, and the largest value is 36.4%. The unitary spectral response curves at different temperature from 155 K to 280 K are shown in figure 4. The peak response wavelength changes from 356 nm to 362 nm as the temperature goes from 155 K to 280 K.

The shape of the spectrum at different temperatures is nearly the same, only with the peak wavelength shifting. The difference is attributed to the decrease of GaN band gap when the temperature becomes higher. The band gap of GaN versus temperature is given as follow [12-13]:

$$E_g(T) = E_g(0) - \frac{\alpha T^2}{(\beta + T)} \qquad (2)$$

Where E_0 (0) is the band gap of GaN at 0 K, α is the temperature coefficient dE_g/dT, and β is the character temperature related to Debye temperature. It is obvious from the formula that as temperature increases the band gap of GaN becomes smaller, consistent with the trend of our experiments.

3.3. MCT/GaN-based dual-colour detector

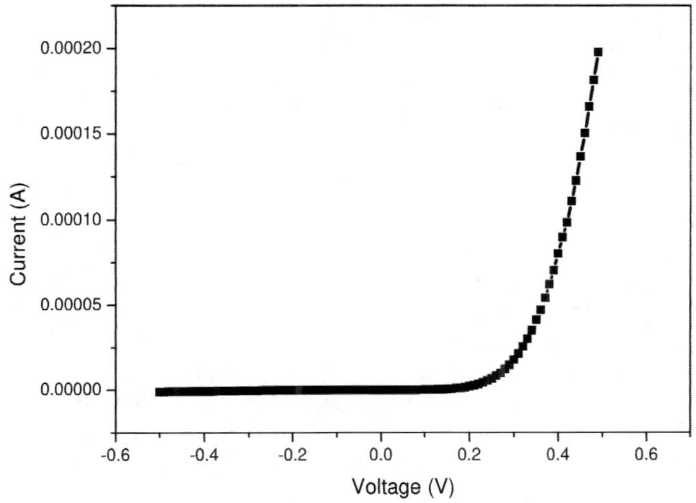

Figure 5. Current-voltage characteristics of GaN-based detector in MCT/GaN-based dual-colour detector.

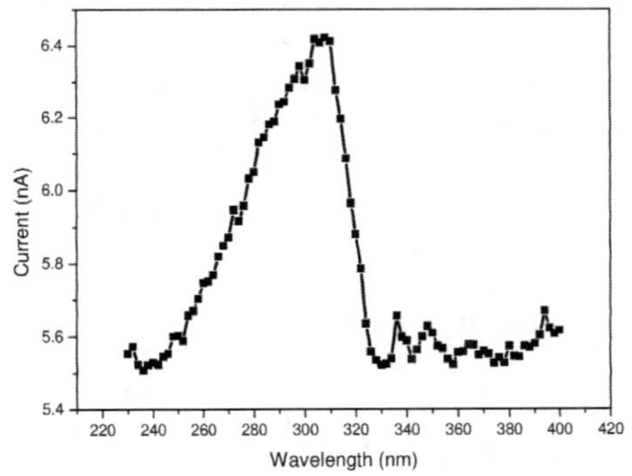

Figure 6. The responsivity of GaN-based detector in the MCT/GaN-based dual-colour detector.

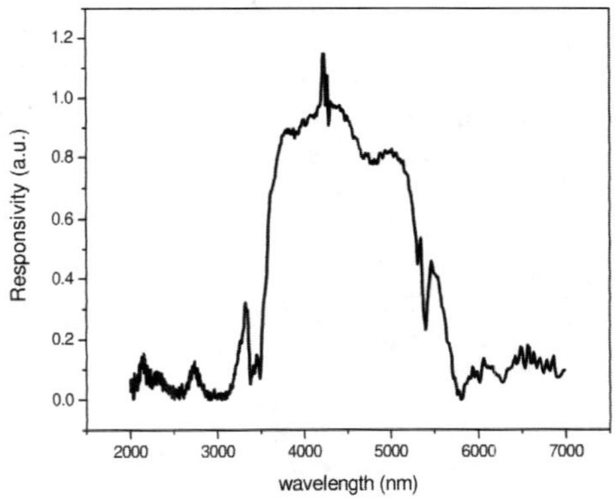

Figure 7. The responsivity of MCT-based detector in the MCT/GaN-based dual-colour detector.

Current-voltage characteristics under bias voltage from -0.5 V to 0.5 V were measured with a Keithley236 semiconductor parameter analyzer at 90 K. The spectral responsivity of GaN-based detector in MCT/GaN-based dual-colour device was measured with a Xe-arc lamp source coupled to a grating monochromator, and it is not calibrated. The response of MW MCT-based detector of dual-colour device was measured with Varian 3100 FT-IR at 90 K.

It is observed from figure 5 that the leakage current of GaN-based detector at -0.5 V is fairly larger than the GaN-based device we show before. This is mainly because the quality of GaN crystal is worse than before. As observed in figure 6, the peak response wavelength is at about 308 nm. The dark current at zero bias is about 10^{-11} A; as a result, the response current at the wavelength which should not create photocurrent cannot be neglected.

There are two windows for the dual-colour detector, which is 240 nm-330 nm for GaN-based device, and 3.2 μm-5.8 μm for MCT-based photovoltage device. Photocurrent measurement was performed under front illumination. The ultraviolet light between 240 nm and 330 nm is absorbed by the GaN-based detector, and the infrared light transmits through the GaN-based detector. Because of the filter film adhering on the sapphire side of GaN, the light between 3.2 μm-5.8 μm can transmit into the MCT-based detector. The MCT/GaN-based dual-colour detector shows that it can realize both ultraviolet and infrared radiation detection.

4. Conclusions

The fabrication and measurement of GaN-based detectors are reported. From the current-voltage characteristics with different temperature, it is observed that the current under both forward and reverse bias voltage becomes higher as the temperature increases. The resistance and ideality factor get smaller, the responsivity of GaN-based device becomes larger, and the peak response wavelength becomes longer as the temperature becomes higher. The peak responsivity is about 0.106 A/W at 362 nm at 280 K, corresponding to the quantum efficiency of 36.4%. The MCT/GaN-based dual-colour detector is presented. The leakage current at 0V of GaN-based detector in dual-colour device is quite large due to the bad quality of material, resulting in the low signal-to-noise ratio. The spectrum of MCT-based detector is shown. Finally, the MCT/GaN-based ultraviolet and infrared dual-colour detector is realized, and it can detect the light between 240 nm and 330 nm, and light between 3.2 μm and 5.8 μm.

Acknowledgements

The authors would like to thank the invaluable contributions to Professor Chen Changqing, Dr. Dai Jiangnan, and Dr. Wu Zhihao for offering GaN-based p-i-n material in National Photonics Laboratory, Wuhan, Hubei province of China.

We also thank Zhu Longyuan, Liu Shijia, Ye Liping, Jiang Peilu, Zhang Lianmei, Jiang Ke, Jia Jia, Qiao Hui, and Liu Xiangyang for their technical assistance. We gratefully acknowledge the stimulating discussions with Liu Fei, Ye Baisong and Zhang Liyao.

References

[1] Minder K, Pau J L, McClintock R, Kung P, Bayram C, and Razeghi M, *Appl. Phys.Lett.* **91**,073513 (2007).
[2] Wang L, Xu J T, Yuan Y G, Jiang P L, Zhang Y and Li X Y, *Proc. Of SPIE* **Vol. 7381** (2009).
[3] Lin J C, Su Y K, and Chang S J, *Appl. Phys. Lett.* **91**, 173502 (2007).
[4] Rogalski A, Antoszewski J and Faraone L, *J. Appl. Phys.* **105**, 091101 (2009).
[5] Zhang Y，Wang N L，Sun J L，Han L，Liu X Y，Li X Y，Meng X J, *Infared and Laser Engineering*, **Vol 38** No.2 (2009).
[6] Bratt P R, Johnson S M, Rhiger D R, Tung T, Kalisher M H, Radford W A, Garwood G A, and Cockrum C A, *Proc. of SPIE* **Vol. 7298** (2009).
[7] Reine M B, Proc. of SPIE **Vol. 7298** (2009).
[8] Lu W, He L, Chen X S, Ding L, Sun S L, Tang J, Su J H, *Proc. of SPIE* **Vol. 7298** (2009).
[9] Collins C J, Li T, Beck A L, Dupuis R D, and Campbell J C, *Appl. Phys. Lett.* **Vol 75**, No.14 (1999).
[10] McClintock R, UV Photodetectors, Focal Plane Arrays, and Avalanche Photodiodes, Doctor of Philosophy (2007).
[11] Kang Y, Fabrication of GaN-based p-i-n UV array photodetector and study of its opto-electronic characteristics, Doctor of Philosophy (2006).
[12] Li X, The MS contact of GaN-bsed semiconductor and investigation of p-i-n ultraviolet

photodetector, Doctor of Philosophy (2006).
[13] Brunner D, Angerer H, Bustarret E, Freudenberg F, *J. Appl. Phys.* **82** (10), (1997).

Journal of Physics: Conference Series **276** (2011) 012179 doi:10.1088/1742-6596/276/1/012179

Synthesis, photochromism and holographic optical recording of a novel diarylethene bearing a pyrrole unit

Gang Liu, Ming Liu, Shouzhi Pu*, Congbin Fan

Jiangxi Key Laboratory of Organic Chemistry, Jiangxi Science and Technology Normal University, Nanchang 330013, P. R. China

pushouzhi@tsinghua.org.cn

Abstract. A unsymmetrical photochromic diarylethene, 1-[2-methyl-5-(3-fluorophenyl)-3-thienyl]-2 -(2-cyano-1,5-dimethyl-4-pyrryl)hexafluorocyclopentene (**1o**), was synthesized. The photoconversion ratios from open-ring to closed-ring isomers at photostationary state under UV irradiation in hexane solution was measured by HPLC, and the properties, including photochromism and fluorescence properties were also investigated systematically. The results showed that this compound had good thermal stability and exhibited reversible photochromism, changing from colorless to blue after irradiation with UV light both in solution and in PMMA amorphous film. The open-ring isomer of the diarylethene 1 exhibited relatively strong fluorescence at 427 nm in hexane solution (5×10^{-5} mol/L) when excited at 340 nm. The fluorescence intensity decreased along with the photochromism upon irradiation with 297 nm light and its closed-ring isomer showed almost no fluorescence. Using diarylethene 1 as recording medium, polarization holographic optical recording was carried out successfully.

1. Introduction

Photochromic molecules attract much attention from fundamental as well as practical points of view because of their potential for applications to optical devices, such as optical memories and switches [1,2]. During the past decades, there have been important achievements in the synthesis of new families of organic photochromic molecules [2-4]. Among the photochromic compounds, Diarylethenes are regarded as the best candidates for such devices, owing to its excellent thermal stability, fatigue resistance, and easy monitoring using UV-vis spectroscopy [5,6]. As components of molecular electronics, it is desirable to develop various types of molecular switch devices with diarylethene compounds.

Photochromism is a light-triggered reversible reaction between two isomers. Most diarylethene derivatives show very large spectral shifts upon photoisomerization from the open-ring to closed-ring isomers [1]. The open-ring isomer is colorless in most cases, while closed-ring isomer shows some color, such as yellow, red, or blue. The open and closed isomers differ from each other not only in their absorption but also in various physical and chemical properties, such as -conjugation length [7], refractive indices [8,9], geometrical structures [10,11], and chiral properties [12,13]. The property changes and the photochromic reactivity strongly depend on the kind of the heteroaryl moieties and the electron donor/acceptor substituents. Formerly, most of the heteroaryl moieties have been thiophene or benzothiophene rings, with just a few reports on other heteroaryl moieties such as thiazole [14], oxazole [15], and so on. Pyrrole is an attractive aryl unit due to its biological characteristics and has been applied in biosensors widely. However, diarylethene derivatives with a pyrrole unit hitherto reported is very rare. In this study, a photochromic diarylethene compound, 1-[2-methyl-5-(3-fluorophenyl)-3-thienyl]-2 -(2-cyano-1,5-dimethyl-4-pyrryl)hexafluorocyclopentene (1o), was synthesized. Its photochromic reactivity, fluorescence property and application for polarization holographic optical recording were investigated in detail. The photochromic reaction of diarylethene 1a is shown in Scheme 1.

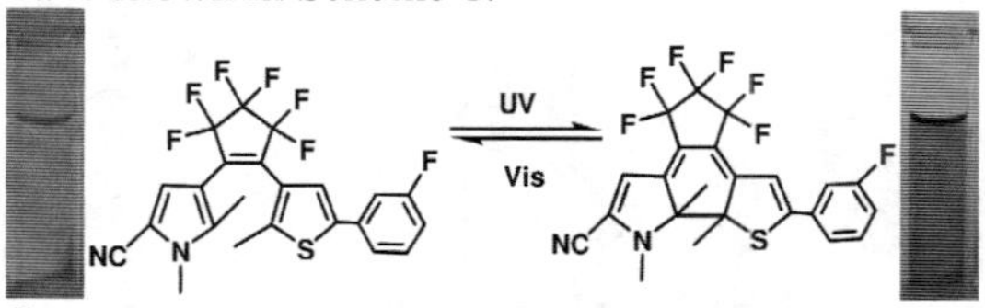

Scheme 1. Photochromism of diarylethene 1o.

2. Experiments

The synthetic route for diarylethene 1o is shown in Scheme 2. It was synthesized according to the similar procedure of Pu et al. [16]. The structure of compound 1o was confirmed by IR, ^{1}H NMR and ^{13}C NMR spectroscopy: Anal. Calcd for $C_{23}H_{15}F_7N_2S$ (%): C, 57.02; H, 3.12; N, 5.78. Found C, 57.09; H, 3.07; N, 5.85; ^{1}H NMR (400 MHz , $CDCl_3$): δ 1.77 (s, 3H, $-CH_3$), 1.97 (s, 3H, $-CH_3$), 3.60 (s, 3H, $-CH_3$), 6.85 (d, 1H, J = 7.6 Hz, phenyl-H), 6.92 (s, 1H, pyrrole-H), 7.05 (s, 1H, thiophene-H), 7.12 (d, 1H, J = 7.6 Hz, phenyl-H), 7.21 (s, 1H, phenyl-H), 7.28 (t, 1H, J = 8.4 Hz, phenyl-H); 13C NMR (100 MHz, $CDCl_3$): δ 11.3. 14.5, 33.1, 55.4, 105.5, 109.9, 111.4, 113.3, 118.1, 118.7, 122.8, 125.9, 130.1, 134.6, 135.9, 140.4, 142.3; IR (ν, KBr, cm^{-1}): 442, 525, 682, 740, 777, 900, 983, 1032, 1056, 1114, 1137, 1185, 1274, 1335, 1384, 1441, 1488, 1561, 1585, 1612, 2222, 2345, 2370, 2852, 2921.

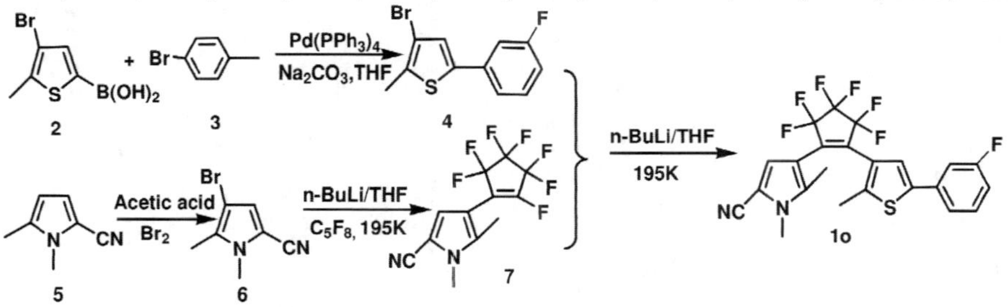

Scheme 2. Synthetic route for the compound 1o.

3. Results and Discussions

3.1. Photochromism of diarylethene 1o

The unsymmetric diarylethene 1o exhibits good thermal stability and photochromic properties and can be toggled between its colorless ring-open and colored ring-closed forms by alternate irradiation with appropriate wavelengths of light. Figure 1 shows the absorption spectral changes of 1o in hexane (2.0×10^{-5} mol/L) and in PMMA film by photoirradiation, respectively. In hexane solution, the absorption maximum of compound 1o was observed at 253 nm ($\varepsilon = 3.56 \times 10^4$ $L\cdot mol^{-1}\cdot cm^{-1}$). Upon irradiation with 297 nm light, the color of the hexane solution turned blue, in which the absorption maximum was observed at 589 nm ($\varepsilon = 1.08 \times 10^4$ $L\cdot mol^{-1}\cdot cm^{-1}$). The blue solution turned colorless, upon irradiation with visible light ($\lambda > 500$ nm). In the PMMA amorphous film, diarylethene 1o also showed good photochromic property. Upon irradiation 297 nm light, the color of 1o/PMMA film (λmax =255 nm) changed from colorless to blue with the appearance of a new broad absorption band at 614 nm, which was assigned to the formation of the closed form 1c. The colored 1c/PMMA film can invert to colorless, on irradiation of appropriate visible light ($\lambda > 500$ nm). The red shift of the ring-closed form of diarylethene 1 in PMMA film in comparison with that in hexane solution can be ascribed to the polar effect of the polymer matrix and the stabilization of molecular arrangement in solid state [17]. In addition, the photoconversion ratio from open-ring to closed-ring isomer of compound 1 at photostationary state was analyzed under UV irradiation in hexane solution by HPLC, which is shown in Figure 2. The values is 94%.

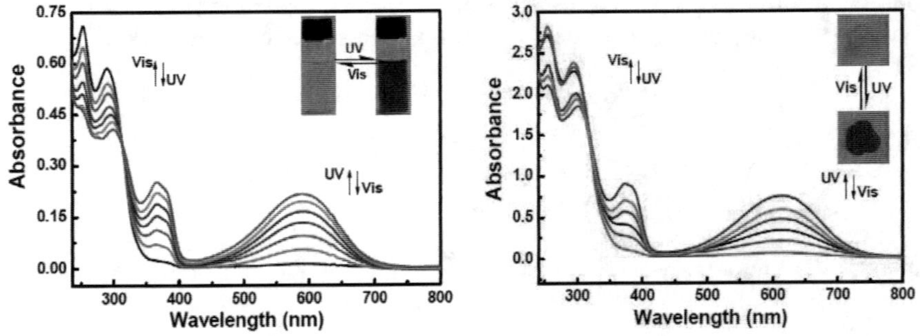

Figure 1. Absorption spectral changes of compound 1 at room temperature (A) in hexane (2.0×10^{-5} mol/L) and (B) in PMMA film.

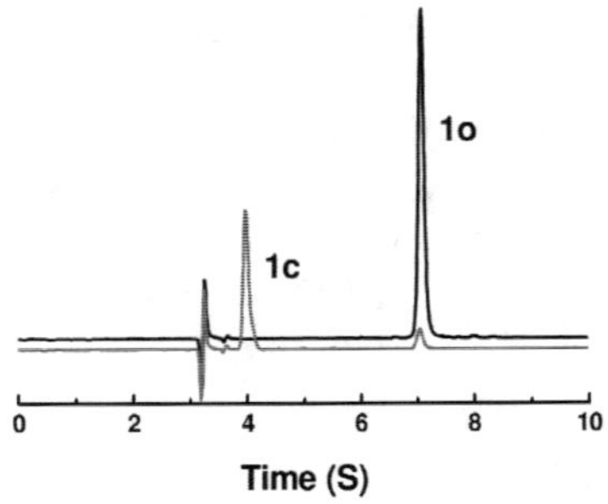

Figure 2. the photoconversion ratios of diarylethenes 1 at photostationary state under UV irradiation in hexane solution by HPLC.

3.2. Fluorescence switch properties

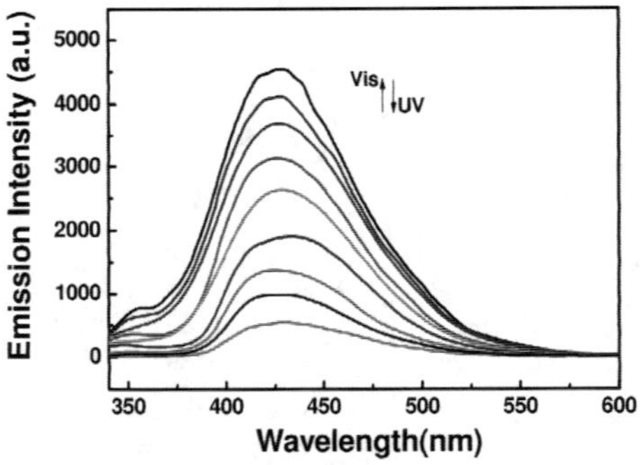

Figure 3. Fluorescence spectral changes of 1o upon irradiation in hexane (5×10^{-5} mol/L).

Fluorescent properties can be useful not only in molecularscale optoelectronics, but for digital photoswitching of fluorescence [18]. In this work, the fluorescence properties of the diarylethene 1 in solution was measured using a Hitachi F-4500 spectrophotometer. The result showed that 1o exhibited good fluorescence in hexane solution. By using anthracene (0.27 in acetonitrile) as the reference, the fluorescence quantum yields of 1o was determined to be 0.035. The fluorescence intensity of diarylethene 1 decreased dramatically along with the photochromism from open-ring isomer to closed-ring isomer upon irradiation with 297 nm light. As shown in Figure 3, when irradiated by light of 297 nm, the photocyclization reaction was carried out and the non-fluorescent closed-ring form of the compound were produced. The back irradiation by appropriate wavelength visible light regenerated the open-ring forms of diarylethene 1o and recovered the original emission spectra. The emission intensity of diarylethene 1 in a photostationary state was quenched to ca. 12% in hexane. The incomplete cyclization reaction and the existence of parallel conformations may be the main cause for the moderate change in fluorescence induced by photoirradiation. The reversible changes of the emission intensity of diarylethene 1 are useful for application as the fluorescence switches [19].

3.3. Holographic optical recording

Based on the photochromic and photo-induced anisotropic properties of diarylethene 1o, we proposed an application as polarization pattern display. We have accomplished three different types of polarization holographic optical recording using diarylethene 1c/PMMA film as recording medium, i.e., copolarization, cross-polarization and 45°-polarization holographic recording. The exposure time for holograms recording was 6 seconds. Readout patterns with the three types of polarization holographic optical recording are shown in Figure 4. As shown in Figure 4(a), we can see that the copolarization holographic recording obtains high diffraction efficiency as well as high scattering noise when the intensities of the recording wave and the readout wave are fixed. Figure 4(b) shows the observed pattern by just using cross-polarization light for readout. The result indicated that cross-polarization holographic recording obtains low diffraction efficiency as well as low scattering noise when the intensities of the recording wave and the readout wave are fixed. If the polarization of P is 45°, from the horizontal direction, only the vertical polarization light patterns can pass through the A-diarylethene-P system. Thus, the readout pattern is like Figure 4(c). By this way, two polarization patterns can be hidden in the pattern and recorded in the diarylethene 1c/PMMA film based on its photochromic and photoinduced anisotropic properties, and only readout with two orthogonal polarizers at certain polarization angles can pick up the patterns separately [20].

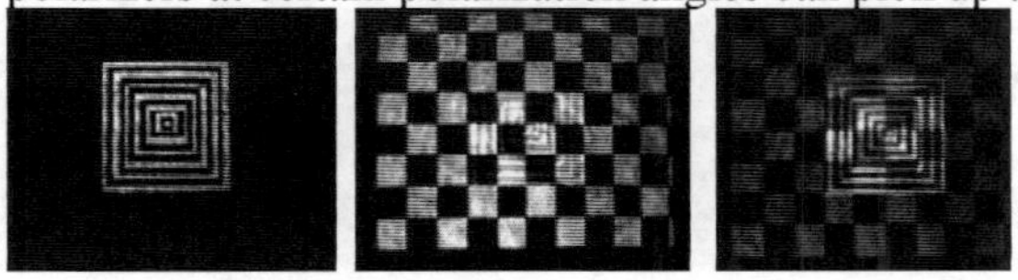

(a) (b) (c)

Figure 4. Readout patterns from the diarylethene 1c/PMMA film with different types of holographic optical recording. (a). The polarization of P is 90°; (b). 0°; (c). 45°.

In conclusion, A symmetrical photochromic diarylethene was synthesized and its photochromic and fluorescent properties were also investigated. The results showed that this compound exhibited reversible photochromism both in solution and in PMMA amorphous film. The photoconversion ratio from open-ring to closed-ring isomer at photostationary state was analyzed under UV irradiation in hexane solution by HPLC, and the values is 94%. In hexane solution, the open-ring isomer of diarylethene 1 exhibited relatively strong fluorescence at 426 nm when excited at 340 nm. The fluorescence intensity decreased along with the photochromism upon irradiation with 297 nm light and its closed-ring isomer showed almost no fluorescence. Using diarylethene 1 as recording medium, polarization holographic optical recording was carried out successfully. The results demonstrated that the compound had attractive properties for polarization holographic optical recording.

Acknowledgment

This work was supported by the Project of National Natural Science Foundation of Jiangxi Province (2009GZH0034), the Key Scientific Project from Education Ministry of China (208069), and the Science Funds of the Education Office of Jiangxi, China (GJJ09302).

References

[1] Irie M 2000 *Chem. Rev.* **100** 1685
[2] Tian H and Yang S 2004 *Chem. Soc. Rev.* **33** 85
[3] Matsuda K and Irie M 2006 *Chem. Lett.* 1204
[4] Tian H and Wang S 2007 *Chem. Commun.* 78
[5] Tian H, Chen B, Tu H and Müllen K 2002 *Adv. Mater.* **14** 918
[6] Raymo F M and Tomasulo M 2005 *Chem. Soc. Rev.* **34** 327

[7] Bens A T, Frewert D, Kodatis K, Kryschi C, Martin H D and Trommsdorf H P 1998 *Eur. J. Org. Chem.* 2333
[8] Bertarelli C, Bianco A, D'Amore F, Gallazzi M C and Zerbi G 2004 *Adv. Funct. Mater.* **14** 357
[9] Kim E, Choi Y K and Lee M H 1999 *Macromolecules* **32** 4855
[10] Norikane Y and Tamaoki N 2004 *Org. Lett.* **6** 2595
[11] Kocer A, Walko M, Meijberg W and Feringa B L 2005 *Science* **309** 755
[12] Okuyama T, Tani Y, Miyake K and Yokoyama Y 2007 *J. Org. Chem.* **72** 1634
[13] Tani Y, Ubukata T and Yokoyama Y 2007 *J. Org. Chem.* **72** 1639
[14] Uchida K, Ishikawa T, Takeshita M and M Irie 1998 *Tetrahedron* **54** 6627
[15] Takami S, Kuroki L and Irie M 2007 *J. Am. Chem. Soc.* **129** 7319
[16] Pu S Z, Liu G, and Xu J K 2007 *Org. Lett.* **9** 2139
[17] Kasatani K, Kambe S and Irie M 1999 *J. Photoch. Photobiol. A* **122** 11
[18] Fukaminato T, Sasaki T, Kawai T, Tamai N and Irie M 2004 *J. Am. Chem. Soc.* **126** 14843
[19] Chen B Z, Wang M Z and Tian H 2002 *Chem. Commun.* 1060
[20] Yao B L, Wang Y L, Lei M, Menke N and Chen G F 2005 *Opt. Express* **13** 20

Study on XRDK Optical and Electrical performance of Transparent Conducting ZnO:Al(ZAO) Thin Films

LU Feng[1], XU Cheng-hai[2], WEN Li-shi[3]
1.School of Transportation and Mechanical Engineering, Shenyang Jianzhu Uinversity, Shenyang,110168,China; 2. School of Mechanical Engineering & Automation, Northeastern University,Shenyang 110004, China; 3. Institute of Metal Research, The Chinese Academy of Sciences, Shenyang 110015, China.

E-mail: lufeng72@126.com

Abstract: The ZAO (ZnO: Al) thin films were prepared by DC reactive magnetron sputtering technique. The XRD、electrical and optical properties of films are particular investigated。The results show that ZAO films are polycrystalline hexagonal wurtzite structure, and we are not find Al2O3 crystal phase。At the same time, we gained the high quality ZAO films with the minimum resistivuty of 4.5x10-4 Ω • cm, the transmittance in visible region above 80% and the reflectivity in IR region above 70%。

1.Introduction

The transparent conductive oxide (TCO) thin films have been extensively used as transparent electrode for liquid crystal displays (LCDS) , organic light emitting diodes(OLEDS) and solar cells. Recently ZnO:Al (ZAO) thin film has attracted interests as a transparent conductive film material because of its low cost, stable performance, simple preparation. Therefore, ZAO thin film has been actively investigated as an alternative to ITO by the scientific and research workers.

ZAO thin film was studied mainly on the organizational structure, optical and electrical properties as basic theories research [1,2]. To prepare high quality ZAO thin films for application, it is importance to research the preparation of ZAO film processing technology. ZAO thin film can be prepared by RF magnetron sputtering [3,4] , chemical vapor deposition [5] , pulsed laser [6] , sol-gel method [7] etc.

In this work, highly conductive and transparent ZAO thin films onto glass substrates were prepared by DC magnetron reactive sputtering technique. The structural ,electrical and optical properties of the films are investigated in detail.

2.Experimental

ZAO thin film was prepared by DC planar target magnetron sputtering device. The target was a composite plate consisting of mixture of Zn and Al. The glass was used as substrates which were mounted parallel to the target surface. During deposition, argon was used as sputtering gas and the oxygen was used as reactant gas. The substrates were cleaned in ultrasonic bath with acetone and dried in hot air before deposition. The substrates could be heated before deposition. The deposition parameter are summarized in Table.1.The ratio of oxygen and argon must be well regulated before deposition. The structure of the sample was determined with D/max-rA X-ray diffraction instrument. The carrier concentration and the Hall coefficient were measured by the van der Pauw method. The double-beam spectrophotometer was used to measure the optical properties of the films. The optical and electrical properties were characterized and analyzed by resistivity, transmissivity and reflectivity, respectively.

Table 1. Deposition conditions of ZAO by DC reactive magnetron sputtering

Parameter	value
Base vacuum/Pa	$1x10^{-3}$
Working gas pressure/Pa	0.1-1
Sputtering power/W	75~250
Sputtering gas	Ar
Reactive gas	O_2
Deposition temperature/℃	50~350
Deposition time/min	30,45,60,90
Target-substrate distance/cm	2-20

3.Results and Discussion

3.1 The Analysis of XRD、 SEM

The prepared films on optimum conditions on glass substrates were physically stable and had very good adherence to the substrates. The X-ray diffractometer measurements revealed that all of the obtained films were polycrystalline with the hexagonal wurtzite structure and had a preferred orientation with the c axis perpendicular to the substrate. The X-ray diffraction patterns of ZAO thin films are shown in Fig 1.The only diffraction peak observed is the (002) peak and the diffraction angle is very close to that of the standard ZnO crystal (34.45°). This implies that Al replaces zinc substitutionally in the hexagonal lattice and Al2O3 crystal phase isn't found. At the same time, the grain size of ZAO thin films are increased and the crystalline quality can be improved with increasing depositing temperature, because it can provide sufficient energy in the crystallization process.

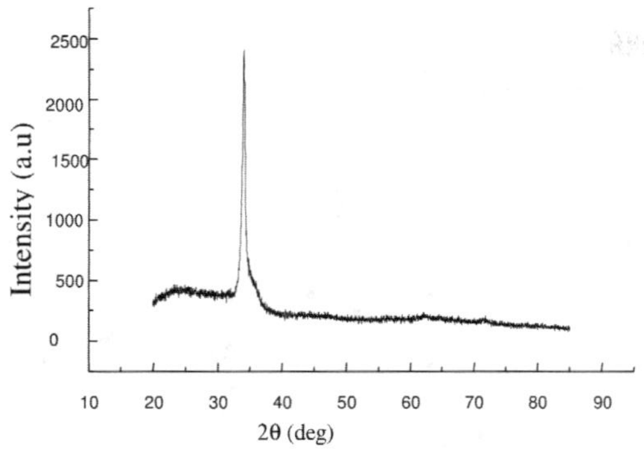

Fig 1 XRD spectra of ZAO thin films

Based on the XRD results, the average crystallite sizes were estimated by Scherrer formula as follows[8]:

$$D=0.9\ \lambda / \beta \cos \theta \qquad (1)$$

Where D is the grain size, λ is the X-ray wavelength, θ is the Bragg angle, and β is the full width at half maximum(FWHM) of ZAO (002) diffraction peak. The corresponding parameters of the (002) peak are listed in Table 2.

Table 2 Parameters of the X-ray diffraction patterns of ZAO films

sample	2 θ (°)	FWHM	D(nm)
1	34.02	0.399	20.82
2	34.24	0.353	23.55
3	34.41	0.327	25.43

Fig 2 shows the SEM morphologies of ZAO films deposited by DC magnetron reactive sputtering. The ZAO thin films are composed of many crystallite, which is cylindrical and spherical. The grain size is about 30nm,. And the grain boundary is clear. The Crystal defects aren't found in the surface of ZAO thin films by the internal stress and micro-cracks. This result is consistent with X-ray diffraction.

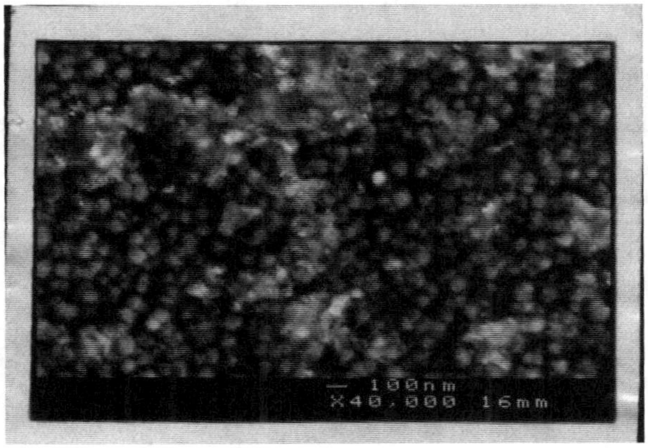

Fig 2 SEM picture of ZAO films

3.2 The Analysis of Electrical Properties

All transparent conducting ZAO thin films are n-type conductivity. The resource of carriers was mainly attributed to oxygen absence and Al-doped, and the carrier mobility was determined by all kinds of scattering mechanisms.

Carrier concentration depends on with Al content. The proportion of different Al content can affect the ratio of substitutional donor and oxygen absence. It can Produce the different oxygen adsorption in the process of film, and cause the accumulation of miscellaneous grain boundaries of different quality, and affect the grain size and orientation. As the role of Al doping for the position change, which will affect the production of thin film carrier

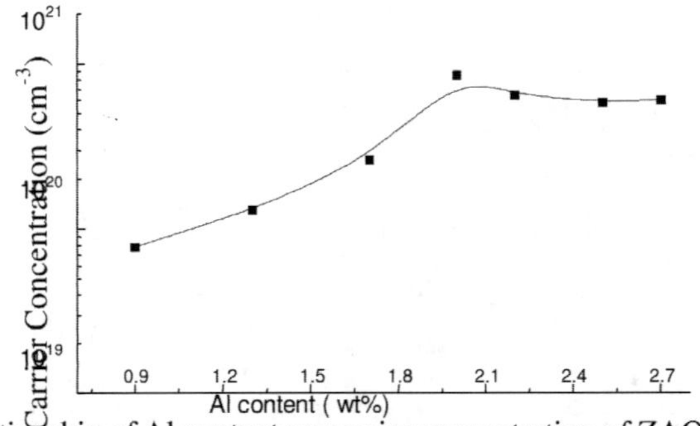

Fig 3 The relationship of Al content on carrier concentration of ZAO thin films

Figure 3 shows the relationship between Al content and carrier concentration of ZAO thin films. It can be seen, when the Al content is 0.9 ~ 2.7wt%, carrier concentration is 1019 ~ 1021/cm3, and the maximum is 9.74x1020. The reason is, with the Al content increasing, Al doping effect more obvious, and the carrier concentration will increase. But when the Al content up to a certain extent, some Al atoms combine with oxygen into Al2O3 crystals, the substitutional doping effect will became weaken, the ratio of oxygen vacancy will also be reduced, therefore the carrier concentration will be weakened and became a saturation value.

For ZAO thin films, the carrier mobility was determined by all kinds of scattering mechanisms, which are ionized impurity scattering, neutral impurity scattering and lattice vibration scattering. Therefore, the films scattering mechanisms may be expressed:

$$\frac{1}{\mu} = \frac{1}{\mu_i} + \frac{1}{\mu_g} + \frac{1}{\mu_l} \quad (2)$$

Where μ i is for ionized impurity scattering mobility, μ g is for grain boundary scattering mobility, μ l is for lattice vibration scattering mobility. For TCO thin films, grain boundary scattering isn't main scattering mechanism, and it may be neglected.

According to Conwell-Weissk formula, the carrier mobility can be expressed:

$$\mu_i = \frac{e}{m^*}\tau_i(E_F) \tag{3}$$

Where $\tau_i(E_F)$ is relaxation time, e is the ion charge and m* is the effective mass of electrons.

From formula (3), it can be shown that ionized impurity scattering is independent of temperature. Ionized impurity concentration is usually greater than 1020cm-3, so these ionized impurities will be the carrier scattering center. The ionized impurity scattering is main scattering mechanism in lower temperature range (<250K), and mobility basically keeps constant.

With increasing temperature, lattice vibration scattering becomes the major scattering mechanism. The relation of mobility and temperature may be briefly expressed:

$$\mu_l \propto \frac{1}{T} \tag{4}$$

For formula (4), it is seen that mobility is inversely proportion to temperature. So Through the above analysis, formula (2) can be rewritten as:

$$\mu = \frac{\mu_i \mu_l}{\mu_i + \mu_l} \tag{5}$$

By theoretical analysis of Hall mobility, it can be drawn: ionized impurity scattering is dominate in lower temperatures, lattice vibration scattering has little influence, ionized impurity concentration is a certain value, therefore, mobility is independent of temperature and can be expressed as μ_i, in higher temperature conditions (>300k), the lattice vibration scattering becomes a major scattering mechanism, mobility can be expressed as μ_l, it is inverse relationship between mobility and temperature. This can be proved from Figure 4.

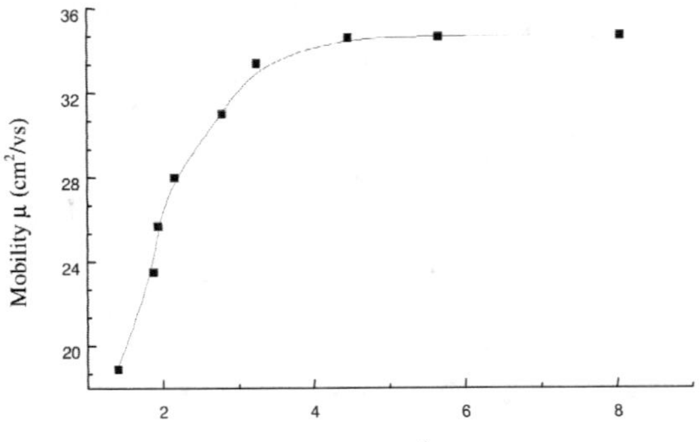

Fig 4 The reciprocal dependence of mobility of ZAO thin films on temperature

Improving the electrical conductivity of ZAO thin films is to maximize the films in the carrier concentration, reduce the scattering center of the films and improve the carrier mobility. The electrical resistivity ρ, Hall mobility μ and carrier concentration n of ZAO thin films are summarized in Table 3. With carrier concentration increase, which means an increase of scattering centers, it led to Hall mobility decreased. So in a certain range, with the increase of carrier concentration, resistivity decreased. And over this range, since the value of Hall mobility is relatively small, the resistivity will increase. Therefore, in order to get the best conductivity, Carrier concentration and Hall mobility must be met the larger value at same time, the lowest resistivity is $4.5x10^{-4}$ Ω • cm

Table 3 The result of resistivity, carrier concentration and Hall mobility for ZAO films

Sample	resistivity ρ (Ω • cm)	carrier concentration n (cm^{-3})	Hall mobility μ(cm^2/vs)
1	$5.21x10^{-3}$	$3.65x10^{19}$	10.23
2	$1.08 x10^{-3}$	$1.31 x10^{20}$	14.36
3	$8.73 x10^{-4}$	$2.60 x10^{20}$	25.23
4	$4.5 x10^{-4}$	$5.539 x10^{20}$	27.81
5	$7.97 x10^{-4}$	$6.93 x10^{20}$	34.35
6	$6.35 x10^{-3}$	$7.47 x10^{20}$	21.66
7	$2.14 x10^{-3}$	$9.74 x10^{20}$	12.78

3.3 The Analysis of Optical Properties

Optical properties of ZAO thin films include the transmittance in visible region and the reflectivity in IR region. Fig 5 shown the ZAO thin film transmission and reflection profiles, we can know some optical properties of ZAO thin films from Fig 5.

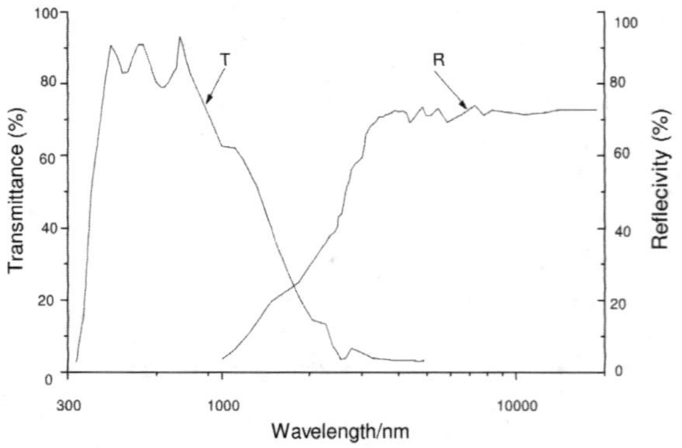

Fig 5 Transmittance and Reflectivity of ZAO films

The band gap of ZAO thin films is about 3.4ev, it means the intrinsic absorption of ZAO thin films on electromagnetic waves is about 360nm, which is in the ultraviolet region. That is UV cut-off performance of ZAO thin films. In the visible range (380-760nm), transmission of ZAO thin film is above 80%. Because the wavelength is greater than the intrinsic absorption limit (360nm), the electromagnetic wave absorption coefficient of ZAO thin films decrease rapidly, most of the energy of visible light can be transmitted, only a very small number of energy is reflected and absorbed. Therefore, high transmission can be gained in the visible range.

Most energy of the electromagnetic wave belong the infrared band (760-20000nm). The significant features of ZAO thin films is high reflectivity in IR region, which can be reflected away most of the thermal radiation energy. It can prevent the absorption of energy and the higher temperature in the application of electronic devices. The reflectivity of ZAO thin films is more than 70% in IR region.

4.Conclusion

Good transparent conducting ZAO thin films can be prepared on glass substrate by DC reactive magnetron sputtering. XRD analysis indicates that the ZAO thin films are polycrystalline with a hexagonal wurtzite structure and have a preferred orientation (002) along the c axis. Carrier concentration depends on with Al content. Hall mobility was determined by all kinds of scattering mechanisms, ionized impurity scattering dominates the hall mobility of the ZAO thin films in lower temperature, and in the higher temperature range, the lattice vibration scattering will become major scattering mechanism. Therefore, the carrier concentration and mobility must be met the larger value at same time ,the lowest resistivity (4.5x10-4 Ω • cm) can be obtained. It was showed that the transmittance and the reflectance in visible region and IR region was above 80% and 70%, respectively.

References

[1]. Wei Shao, Ruixin Ma, and Bin Liu. Fabrication and properties of ZAO powder, sputtering target materials and the related films, Journal of University of Science and Technology Beijing, 2006,Vol 13, No 4, 346-349.

[2]. Mass J., Bhattacharya P., and Katiyar R.S., Effect of high substrate temperature on Al-doped ZnO thin films grown by pulsed laser deposition, Mater. Sci. Eng. B, 2003, 103 (1): 9.

[3]. J.Ma, X.T.Hao, S.Y.Zhang,et al. Characterization of ZnO:Al films deposited on organic substrate by RF magnetron sputtering, Mater.Sci.Technol,2003,Vol.19 No.4:363–365

[4]. Guillen C , Herrero J. High conductivity and transparent ZnO :Al films prepared at low temperature by DC and MF magnetron sputtering. Thin Solid Films , 2006 , 515 :640 - 6431

[5].Groenen R,Sommeing P M,Linden J L, et al. Surface Textured ZnO Films for Thin Film Solar Cell Application by Expanding Thermal Plasma CVD, Thin Solid Films,2001,392:226-230.

[6].M.Rusop,K.Uma,T.Sog, et al, Post-growth annealing of zinc oxide thin films pulsed laser deposited under enhanced oxygen pressure on quartz and silicon substrates, Materials Science and Engineering B,2006,127:150-153

[7].Srinivasan G,Kumar R T R , Kumar J . Influence of Al dopant on microstructure and optical properties of ZnO thin films prepared by sol-gel spin coating method, Optical Materials , 2007 , 30 : 3142317.

[8]. Cullity B., Elements of X-ray Diffraction, Addison-Wesley, London, 1959: 99.

The influence of the conical incidence on the waveguide-type colour-separating backlight for liquid crystal display

Ji Luo, Wenfeng Wang
Faculty of Physics and Electronics, Hubei University, Wuhan, China

E-mail: w.wang@hubu.edu.cn

Abstract. The lightguide assisted by diffraction gratings is used to implement colour-separating backlight for Liquid Crystal Display (LCD). In the configuration, the light incidence angle is required to be a precise designed one. However, the light incidence is conical always. In this paper, by rigorous coupled wave theory, the influence of conical input on the colour-separating backlight is investigated. It shows that, when the azimuthal angle of the light increases, the diffraction efficiency could be enhanced by a factor of 2, which means that the conical input will benefit the diffraction efficiency of the lightguide. However, it also leads to a larger output spot area at the same time, which could result in the mixing of RGB pix. When the azimuthal angle is smaller than 12°, the maximum diffraction angles of RGB light can be completely separated, which means that, with a designed distance or with an assisted micro-lens layer between the grating and the pix layer, the RGB light can reach their corresponding pix. The influence of the conical input on the polarized light is also investigated. The conical input will result in the output with composite polarization, which, in principle, disables the use of polarized light source in LCD.

1. Introduction

Liquid crystal display (LCD) technology has been wildly used in the TV, computer and various mobile terminals. However, the light management in present LCD is not efficient. Only around 5% of the light generated in the backlight of the LCD reaches the eyes of the user [1]. It has been suggested that, by a diffraction grating, the energy efficiency of the LCD can be greatly enhanced since the grating can separate the colours of the backlight to RGB light [2], and thus avoid using the colour filter, whose absorption is estimated to be over 66% [1]. This backlight requires that the light is guided in a lightguide and captured inside due to the total internal reflection. When the grating is applied on the surface of the lightguide, the light can be partially diffracted out the lightguide [3] [4] [5], at the same time, the light is separated according to the colour due to the dispersion of the grating [6]. To implement this idea, the light should be input in a design angle [7]. However, the light incidents are conical always, i.e. the plane of the incidence is not exactly perpendicular to the grating, therefore, it is important to investigate the influence of the conical incidence on the colour separated backlight.

By rigorous coupled wave analysis theory [8], the influence of conical input on the colour-separating backlight is investigated. The results show when the azimuthal angle of the incidence light increases, the diffraction efficiency could be enhanced by a factor of 2, which means that the conical input can even benefit the diffraction efficiency of the lightguide. However, it also leads to a larger output spot area at the same time, which could result in the mixing of RGB pix. When the azimuthal angle is smaller than 12°, the maximum diffraction angles of RGB light can be completely separated, which means, with a designed distance or with an assisted micro-lens layer between the grating and the pix layer, the RGB light can reach their corresponding pix. The influence of the conical input on the polarized light is also investigated. The azimuthal angle will result in output light with composite polarization, which, in principle, disables the use of polarized light source in LCD.

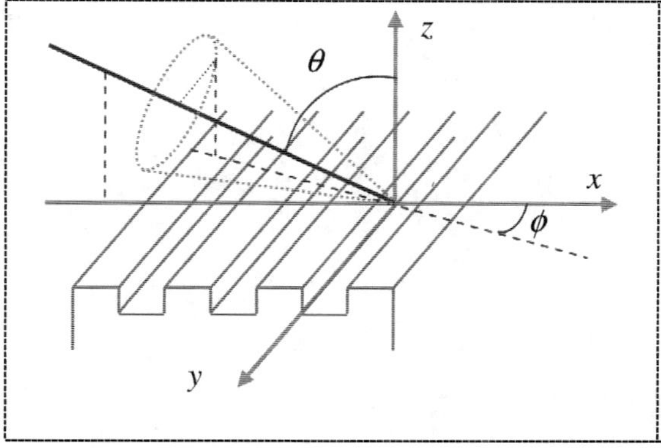

Figure 1 Geometry for the grating and the conical incidence.

2. Theory

The study is based on the rigorous coupled wave analysis (RCWA) [8]. The incident electric field is given by

$$E_{inc} = U\exp[-\mathrm{i}k_0 n_1(\sin\theta\cos\phi xk + \sin\theta\sin\phi y + \cos\theta z],$$

Where $U = (\cos\psi\cos\theta\cos\phi - \sin\psi\sin\phi)\vec{x} + (\cos\psi\cos\theta\sin\phi + \sin\psi\cos\phi)\vec{y} + \cos\psi\sin\theta\vec{z}$.

Here ψ is the angle between the incident plane and the electric field vector, θ and ϕ as shown in Figure 1. The electrical fields in region 1 and region 2 are

$$E_1 = E_{inc} + \sum_j R_j \exp[-\mathrm{i}(k_{xj}x + k_y y - k_{1zj}z)] \ , \quad \text{and} \quad E_2 = \sum_j T_j \exp\{-i[k_{xj}x + k_y y - k_{1zj}(d-z)]\} \ ,$$

where $k_{xj} = k_0[n_1\sin\theta\cos\phi - j(\lambda_0/\Lambda)]$, $k_y = k_0 n_1\sin\theta\sin\phi$, and

$$k_{1zj} = \begin{cases} [(k_0n_1)^2 - k_{xj}^2 - k_y^2]^{\frac{1}{2}} & (k_{xj}^2 + k_y^2)^{\frac{1}{2}} < k_0 n_1 \\ -[(k_0n_1)^2 - k_{xj}^2 - k_y^2]^{\frac{1}{2}} & (k_{xj}^2 + k_y^2)^{\frac{1}{2}} > k_0 n_1 \end{cases}, \quad k_{2zj} = \begin{cases} [(k_0n_2)^2 - k_{xj}^2 - k_y^2]^{\frac{1}{2}} & (k_{xj}^2 + k_y^2)^{\frac{1}{2}} < k_0 n_2 \\ -[(k_0n_2)^2 - k_{xj}^2 - k_y^2]^{\frac{1}{2}} & (k_{xj}^2 + k_y^2)^{\frac{1}{2}} > k_0 n_2 \end{cases}.$$

Here R_j is the normalized vector electric-field amplitude of the j-th backward-diffracted wave in region 1. T_j is the normalized vector electric-field amplitude of the j-th forward-diffracted wave in region 2. We will get the normalized vector electric and magnetic field amplitude of the j-th backward-diffracted and forward-diffracted wave in region 1 and region 2 by Maxwell equation. The normalized vector electric and magnetic field amplitude of the j-th backward-diffracted and forward-diffracted wave in the grating are given by:

$$E = \sum_j [D_{xj}(z)\vec{x} + D_{yi}(z)\vec{y} + D_{zi}(z)\vec{z}]\exp[-\mathrm{i}(k_{xi}x + k_y y)], \text{ and}$$

$$H = -\mathrm{i}\sqrt{\varepsilon_0/\mu_0}\sum_j [C_{xj}(z)\vec{x} + C_{yi}(z)\vec{y} + C_{zi}(z)\vec{z}]\exp[-\mathrm{i}(k_{xi}x + k_y y)].$$

Here D_j and C_j are the amplitude of electric and magnetic field vector. They satisfy the Maxwell's equation in the grating region: $\nabla\times E = -i\omega\mu_0 H$, $\nabla\times H = -\mathrm{i}\omega\mu_0 E$. Thus a set of coupled-wave equations in a matrix form is obtained:

$$\begin{bmatrix} \partial D_x / \partial(k_0 z) \\ \partial D_y / \partial(k_0 z) \\ \partial CK_x / \partial(k_0 z) \\ \partial C_y / \partial(k_0 z) \end{bmatrix} = \begin{bmatrix} 0 & 0 & K_y E^{-1} K_x - I & -K_x E^{-1} K_y \\ 0 & 0 & K_y E^{-1} K_x & I - K_y E^{-1} K_y \\ K_x^2 - E & -K_x K_y & 0 & 0 \\ K_x K_y & E - K_y^2 & 0 & 0 \end{bmatrix} \times \begin{bmatrix} D_x \\ D_y \\ C_x \\ C_y \end{bmatrix}.$$

We can obtain that

$$\partial D_x / \partial(k_0 z) = (K_y E^{-1} K_x - I) C_x - K_x E^{-1} K_y C_y,\ \partial D_y / \partial(k_0 z) = K_y E^{-1} K_x C_x + (I - K_y E^{-1} K_y) C_y,$$

$$\partial C_x / \partial(k_0 z) = (K_x^2 - E) D_x - K_x K_y D_y,\ \partial C_y / \partial(k_0 z) = K_x K_y D_x - (K_y^2 - E) D_y,$$

It can be written by

$$\begin{bmatrix} \partial^2 D_x / \partial(k_0 z)^2 \\ \partial^2 D_y / \partial(k_0 z)^2 \end{bmatrix} = \begin{bmatrix} K_y^2 + (K_x E^{-1} K_x - I) E & (E^{-1} K_y E - K_y) K_x \\ (E^{-1} K_x E - K_x) K_y & K_x^2 + (K_y E^{-1} K_y - I) E \end{bmatrix} \begin{bmatrix} D_x \\ D_y \end{bmatrix},$$

$$\begin{bmatrix} \partial^2 C_x / \partial(k_0 z)^2 \\ \partial^2 C_y / \partial(k_0 z)^2 \end{bmatrix} = \begin{bmatrix} K_x^2 + E(K_y E^{-1} K_y - I) & (K_y - E K_y E^{-1}) K_x \\ (K_y - E K_y E^{-1}) K_x & K_y^2 + E(K_x E^{-1} K_x - I) \end{bmatrix} \begin{bmatrix} C_x \\ C_y \end{bmatrix},$$

where K_y is a constant matrix, then it can be

$$\left[\partial^2 C_x / \partial(k_0 z)^2\right] = \left[\left(k_y / k_0\right)^2 I + K_x^2 - E\right]\left[C_x\right],$$

$$\left[\partial^2 D_x / \partial(k_0 z)^2\right] = \left[\left(k_y / k_0\right)^2 I + (K_x E^{-1} K_x - I) E\right]\left[D_x\right].$$

Thus the electric and the megnitic fields are given by

$$D_{xj}(z) = \sum_{m=1}^{n} a_{1,j,m} \left\{-c_{1,m} \exp(-k_0 q_{1,m} z) + t_{1,m} \exp\left[-k_0 q_{1,m}(z-d)\right]\right\},$$

$$C_{xj}(z) = \sum_{m=1}^{n} a_{2,j,m} \left\{c_{2,m} \exp(-k_0 q_{2,m} z) + t_{2,m} \exp\left[-k_0 q_{2,m}(z-d)\right]\right\},$$

$$C_{xj}(z) = \sum_{m=1}^{n} g_{21,j,m} \left\{c_{1,m} \exp(-k_0 q_{1,m} z) + t_{1,m} \exp\left[k_0 q_{1,m}(z-d)\right]\right\} + \sum_{m=1}^{n} g_{22,j,m} \left\{c_{2,m} \exp(-k_0 q_{2,m} z) + t_{2,m} \exp\left[k_0 q_{2,m}(z-d)\right]\right\},$$

$$D_{xj}(z) = \sum_{m=1}^{n} g_{11,j,m} \left\{-c_{1,m} \exp(-k_0 q_{1,m} z) + t_{1,m} \exp\left[k_0 q_{1,m}(z-d)\right]\right\} + \sum_{m=1}^{n} g_{12,j,m} \left\{-c_{2,m} \exp(-k_0 q_{2,m} z) + t_{2,m} \exp\left[k_0 q_{2,m}(z-d)\right]\right\},$$

where $a_{1,j,m}$ and $q_{1,m}$ are the element of the eigenvector matrix and the positive square root of the eigenvalues of the matrix $\left[k_y I + K_x^2 - E\right]$, $a_{2,j,m}$ and $q_{2,m}$ the element of the eigenvector matrix and the positive square root of the eigenvalues of the matrix $\left[k_y^2 I + (K_x E^{-1} K_x - I) E\right]$, $c_{1,m}$, $t_{1,m}$, $c_{2,m}$ and $t_{2,m}$ is the unknown constant, g_{11}, g_{12}, g_{21} and g_{22} given as

$$g_{11} = \left[K_x^2 - E\right]^{-1} W_1 Q_1,\ g_{12} = (k_y / k_0)\left[K_x^2 - E\right]^{-1} K_x W_2,$$

$$g_{21} = (k_y / k_0)\left[K_x E^{-1} K_x - I\right]^{-1} K_x E^{-1} W_1,\ g_{22} = \left[K_x E^{-1} K_x - I\right]^{-1} W_2 Q_2.$$

Here Q_1 and Q_2 are the diagonal matrices with the diagonal elements $q_{1,m}$ and $q_{2,m}$.

At the boundary $z = 0$,

$$\sin\psi\delta_{j0} + R_{s,j} = \cos\varphi_j D_{yj}(0) - \sin\varphi_j D_{xj}(0),$$

$$i\left[\sin\psi n_1 \cos\theta - (k_{1,zj} / k_0)R_{sj}\right] = \sin\varphi_j C_{yj}(0) - \cos\varphi_j C_{xj}(0),$$

$$-in_1 \cos\psi + R_{p,j} = \sin\varphi_j C_{yj}(0) - \cos\varphi_j C_{xj}(0), \text{ and}$$

$$\cos\varphi\cos\theta - i\left[k_{1,zj} / \left(k_0 n_1^2\right)\right] R_{pi} = \cos\varphi_j D_{xj}(0) - \sin\varphi_j D_{yj}(0).$$

It can be written as

$$\begin{bmatrix} \sin\psi\delta_{j0} \\ \sin\psi n_1 \cos\theta\delta_{j0} \\ -in_1\cos\psi\delta_{j0} \\ \cos\varphi\cos\theta\delta_{j0} \end{bmatrix} + \begin{bmatrix} I & 0 \\ -\mathrm{i} & 0 \\ 0 & \mathrm{I} \\ 0 & -\mathrm{i} \end{bmatrix} \begin{bmatrix} R_s \\ R_p \end{bmatrix} = \begin{bmatrix} F_c g_{11} & F_c g_{11} X_1 & F_c g_{12} - F_s W_2 & (F_c g_{12} - F_s W_2) X_1 \\ F_c W_1 + F_s g_{21} & -(F_c W_1 + F_s g_{21}) X_1 & F_s g_{22} & -F_s g_{22} X_1 \\ F_c g_{21} - F_s W_1 & -(F_c g_{21} - F_s W_1) X_2 & F_c g_{22} & -F_c g_{22} X_2 \\ F_s g_{11} & F_s g_{11} X_1 & F_c W_2 + F_s g_{12} & (F_c W_2 + F_s g_{12}) X_1 \end{bmatrix} \begin{bmatrix} c_{1m} \\ t_{1m} \\ c_{2m} \\ t_{2m} \end{bmatrix}$$

When $z = d$

$$T_{s,j} = \cos\varphi_j D_{yj}(d) - \sin\varphi_j D_{xj}(d),$$

$$R_{sj} = -\sin\varphi_j C_{yj}(d) - \cos\varphi_j C_{xj}(d),$$

$$T_{p,j} = \sin\varphi_j C_{yj}(d) - \cos\varphi_j C_{xj}(d), \text{ and}$$

$$i\left[k_{1,zj} / \left(k_0 n_1^2\right)\right] T_{pi} = \cos\varphi_j D_{xj}(0) + \sin\varphi_j D_{yj}(0).$$

It can be written as

$$\begin{bmatrix} I & 0 \\ -\mathrm{i} & 0 \\ 0 & I \\ 0 & -\mathrm{i} \end{bmatrix} \begin{bmatrix} T_s \\ T_p \end{bmatrix} = \begin{bmatrix} F_c g_{11} X_1 & F_c g_{11} & (F_c g_{12} - F_s W_2) X_1 & (F_c g_{12} - F_s W_2) \\ (F_c W_1 + F_s g_{21}) X_1 & -(F_c W_1 + F_s g_{21}) & F_s g_{22} X_1 & -F_s g_{22} \\ (F_c g_{21} - F_s W_1) X_2 & -(F_c g_{21} - F_s W_1) & F_c g_{22} X_2 & -F_c g_{22} \\ F_s g_{11} X_1 & F_s g_{11} & (F_c W_2 + F_s g_{12}) X_1 & (F_c W_2 + F_s g_{12}) \end{bmatrix} \begin{bmatrix} c_{1m} \\ t_{1m} \\ c_{2m} \\ t_{2m} \end{bmatrix}.$$

The diffraction efficiency of the grating can be obtained by

$$DE_{ri} = |R_{si}|^2 \operatorname{Re}\left[k_{1,zj} / \left(k_0 n_1 \cos\theta\right)\right] + |R_{pi}|^2 \operatorname{Re}\left[k_{1,zj} / \left(k_0 n_1^3 \cos\theta\right)\right], \text{ and}$$

$$DE_{ti} = |T_{si}|^2 \operatorname{Re}\left[k_{2,zj} / \left(k_0 n_1 \cos\theta\right)\right] + |T_{pi}|^2 \operatorname{Re}\left[k_{2,zj} / \left(k_0 n_1 n_2^2 \cos\theta\right)\right].$$

3. Simulation and results

The grating adopted in the study has following parameters: grating period $\Lambda = 400$ nm, depth $d = 180$ nm, fraction of the period $f = 0.5$, refractive index $n = 1.5$ [9]. The wavelengths of the incident light are: 450nm (B), 550nm (G) and 650nm (R). When incidence angle $\theta = 66°$ and azimuthal angle $\phi = 0$, only -1T order diffraction light can output from the light guide. The diffraction efficiencies for RGB lights are 5.18%, 4.70%, and 5.70%. The diffraction angles are: -14.2 °, 0 ° and 14.8 °, respectively. They get well separated symmetrically. By using the configuration in Ref[4],colour-separating backlight can be implemented for the LCD. However, the light source, like LED, normally has a field angle. This means the light incidences are conical always, which leads to azimuthal angle $\phi \neq 0$. In this case, for LED light source, the light incidence angles span a range, rather than a single designed one. Correspondingly, the diffraction light will not output as expected.

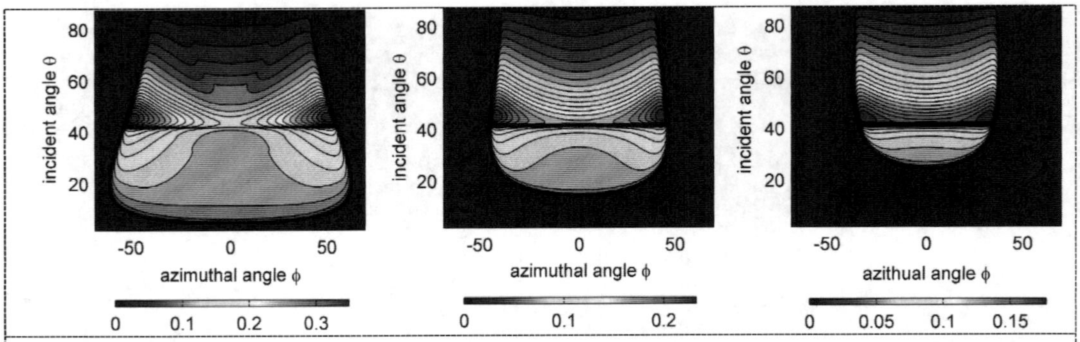

Figure 2 Diffraction efficiencies as functions of incident angle θ and azimuthal angle ϕ for un-polarized light. The -1T order light can output when the incidence angle lies in the bright region. As the azimuthal angle ϕ increasing, the diffraction efficiency could be enhanced by a factor 2.

The diffraction efficiency as the function of the light incidence angle (θ and ϕ) are presented in Figure 2. One could notice that the -1T order diffraction light can output when the incidence angle lies in the bright region. When incident angle $\theta \approx 45°$, with the increasing of azimuthal angle ϕ, the diffraction efficiency could be enhanced by a factor 2, and the maximum efficiency can be obtained.

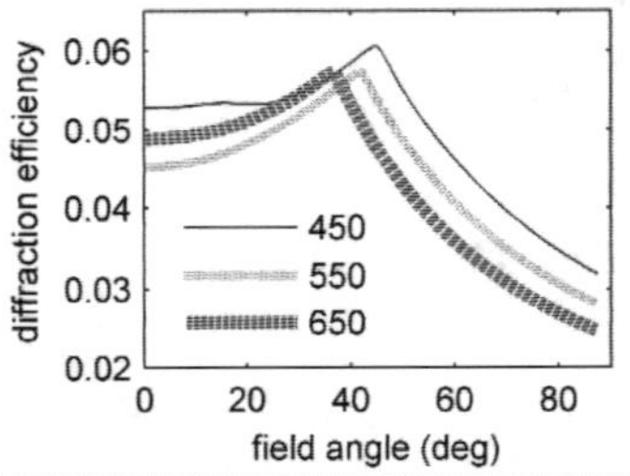

Figure 3. The diffraction efficiencies as the functions of the field angles of the light source. A small field angle can enhance the diffraction efficiency.

In Figure 3, the diffraction efficiencies as the functions of the field angles of the light source are shown. When the light source has a small filed angle, the diffraction efficiency is even larger. This can also be found from Figure 2, since larger azimuthal angle have higher diffraction efficiency. However, when the field angle is larger than a critical angle, where the -1T order diffraction cannot output, the diffraction efficiency decrease quickly. Thus, if one considers diffraction efficiency solely, a light source with moderate field angle is desirable.

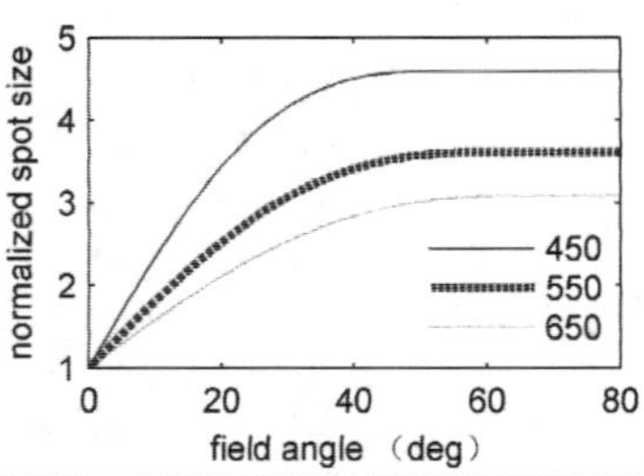

Figure 4. The normalized output spot size as a function of the field angle of the light source. The spot size increase quickly as the filed angle increase, which could lead to the mixing of RGB pix.

However, although the conical incidence can enhance the diffraction efficiency, at the same time, we should consider its influence on the angle of -1T order diffraction light, which determines the spot size

Journal of Physics: Conference Series **276** (2011) 012181 doi:10.1088/1742-6596/276/1/012181

of the output light. In Figure 4, the influence of the field angle of the light source on the spot size is shown. The incident angle is $\theta = 66°$. Solid, dotted, and dashed lines are for R, G, and B lights. Obviously the normalized spot area enlarges as the field angles increasing, which could results in the mixing of RGB pix. The spot size of the B-light enlarges much fast than those of the R- and G-light as shown in Figure 4.

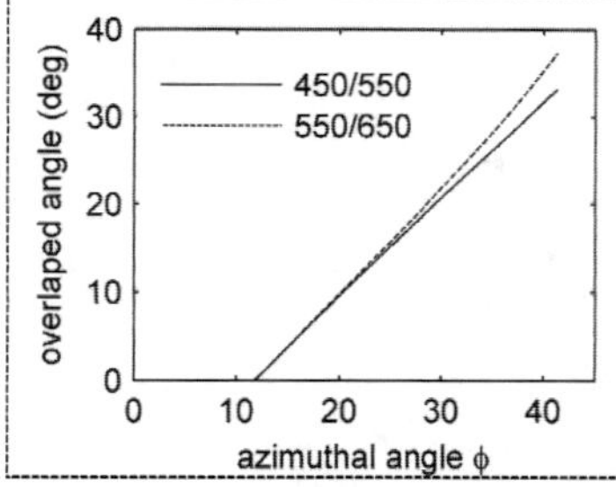

Figure 5. The overlap angle between the RGB diffraction light as a function of azimuthal angle. When $\phi < 12°$, RGB light are completely separated

To show the RGB mixing problem clearly, the overlap of the maximum diffraction angle of RGB light as a function of azimuthal angle ϕ is presented in Figure 5. One could notice, when $\phi < 12°$, the diffraction angle of RGB light are completely separated. This means, with a designed distance or with an assisted micro-lens layer between the grating and the pix layer, RGB lights can reach their corresponding pix without mixing. Thus a conical light source, used in the lightguide, must have a field angle smaller than 12 °.

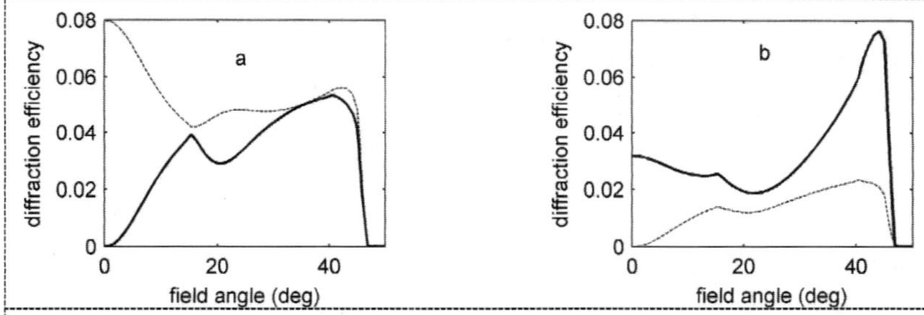

Figure 6. The influence of the conical input on the polarized light. **(a)**TE input, (b) TM input: Dotted line(TE), solid line(TM). As the field angle increase, the output light has a composite polarization.

The influence of the conical input on the polarized light is also investigated as in Figure 6. Figure 6a and 6b show the TE and TM linearly polarized inputs. One can find, in both cases, as the field angle of the light source increases, the output light becomes composite polarized, which, in principle, disables the use of polarized light source in LCD. Or the light field of the light source is extremely small.

4. Conclusion

By rigorous coupled wave theory, we have studied the influence of conical input on the colour-separating of LCD backlight assisted by diffraction gratings. The results show that, when the azimuthal angle of the light incidence increases, the diffraction efficiency could be enhanced by a factor of 2, which means that the conical input can benefit the diffraction efficiency of the lightguide. However, azimuthal angle also leads to a larger output spot area at the same time. When the azimuthal angle is smaller than 12°, the maximum diffraction angles of RGB light can be completely separated. This means the conical light source must have a field angle smaller than 12 °.The influence of the conical input on the polarized light is also investigated. The azimuthal angle will result in output light with composite polarization, which, in principle, disables the use of polarized light source in LCD.

Acknowledgment

This work is supported by NSF (No. ZRY0847) of Hubei Province, Education commission of Hubei Province (No. D20081001) and NSF (No.608080001) of China

References

[1] Roberto Caputo, Luciano De, Sio Martin J J Jak 2007 Short period holographic structures forbacklight display applications *Opt. Exp.* **15** 10540
[2] M Xu, H P Urbach and D K G de Boer 2007 Simulations of birefringent gratings as polarizing color separator in backlight for flat-panel displays *Opt Exp.* **15** 5789
[3] C M VAN HEESCH, H Jagt and C SÁNCHEZ 2005 Polarized Light Out-coupling from Lightguides for LCDs *Chem. Rec.* **5** 59
[4] Ben Layet, Iain G and Cormack 1999 Stripe color separation with diffractive optics *Appl. Phys. Lett.* **38** 7193
[5] Ko-Wei Chien and Han-Ping D Shieh 2004 Design and fabrication of an integrated polarizedlight guide for liquid-crystal-display illumination *Appl. Opt.* **43** 20
[6] Dick K G de Boer and Roberto Caputo 2006 Diffractive grating structures for colour-separating backlights *Proc. of SPIE* **6196** 61960R
[7] Yan Ye and Donglin Pu 2007 Diffraction characteristics of a submicrometer grating for a light guide plate *Appl. Opt.* **46** 3396
[8] M G Moharam and T K Gaylord 1981 Rigorous coupled-wave analysis of planar-grating diffraction *J. Opt. Soc. Am.* **71** 811
[9] L L Doskolovich and N L Kazanskiy 2007 Design and investigation of color separation diffraction Gratings *Appl. Opt.* **46** 20

Journal of Physics: Conference Series **276** (2011) 012182 doi:10.1088/1742-6596/276/1/012182

MEMS-based flexible reflective analog modulators (FRAM) for projection displays: a technology review and scale-down study

Francis Picard[1,3], Samir Ilias[1], Daniel Asselin[1], Marc-André Boucher[1], François Duchesne[1], Michel Jacob[1], Carl Larouche[1], Carl Vachon[1], Keith K Niall[2] and Hubert Jerominek[1]

[1] INO, 2740 Einstein street, Sainte-Foy, Québec, Canada, G1P 4S4
[2] DRDC Toronto, 1133 Sheppard Ave. W., PO Box 2000, Toronto, Ontario, Canada, M3M 3B9

E-mail: Francis.Picard@ino.ca

Abstract. A MEMS based technology for projection display is reviewed. This technology relies on mechanically flexible and reflective microbridges made of aluminum alloy. A linear array of such micromirrors is combined with illumination and Schlieren optics to produce a pixels line. Each microbridge in the array is individually controlled using electrostatic actuation to adjust the pixels intensities. Results of the simulation, fabrication and characterization of these microdevices are presented. Activation voltages below 250 V with response times below 10 μs were obtained for 25 μm x 25 μm micromirrors. With appropriate actuation voltage waveforms, response times of 5 μs and less are achievable. A damage threshold of the mirrors above 8 kW/cm^2 has been evaluated. Development of the technology has produced projector engines demonstrating this light modulation principle. The most recent of these engines is DVI compatible and displays VGA video streams at 60 Hz. Recently applications have emerged that impose more stringent requirements on the dimensions of the MEMS array and associated optical system. This triggered a scale down study to evaluate the minimum micromirror size achievable, the impact of this reduced size on the damage threshold and the achievable minimum size of the associated optical system. Preliminary results of this scale down study are reported. FRAM with active surface as small as 5 μm x 5 μm have been investigated. Simulations have shown that such micromirrors could be activated with 107 V to achieve f-number of 1.25. The damage threshold has been estimated for various FRAM sizes. Finally, design of a conceptual miniaturized projector based on 1000x1 array of 5 μm x 5 μm micromirrors is presented. The volume of this projector concept is about 12 cm^3.

1. Introduction

More than ever, displays are omnipresent in our societies. HD televisions, cell phones, MP3 players, projectors, all rely on advanced display technologies based on LCD, PDP, OLED or MEMS to list just a few. Many applications make use of projection displays and associated technologies have received particular attention over the years. This paper reviews the effort devoted by INO and partners for the development of one such technology for projection display based on MEMS. The active element of

[3] To whom any correspondence should be addressed.

this technology is a linear array of microbridges acting as flexible micromirrors. This mirror array is illuminated with a laser source and produces an image line at the output of an optical relay. Each pixel of the image line corresponds to one flexible micromirror which modulates the pixel intensity in an analog fashion with switching times in the range of 5 μs. The light modulating scheme shown in figure 1 employs Schlieren optics [1] which translate the micromirror curvature into light intensity at the Schlieren relay output. Micromirror curvature is controlled by electrostatic actuation. A complete 2-D image is obtained by a scanning mechanism that displays each image column sequentially. Projection optics is used to tailor the final image.

One force driving the development of many display technologies is miniaturization. Such requirement for low weight and compactness is obviously essential for many applications. The projection display platform described herein is no exception. Results describing a preliminary study aiming at decreasing the size of the developed prototypes will be presented. In addition, this paper first provides a review of the technology including simulation results, a description of the micromirror fabrication process and characterization results for these microdevices and associated projection engines developed for principle demonstration.

2. FRAM simulation

In order to limit the parameter design space for the micromirrors, electromechanical static simulations and dynamic simulations were performed with the MEMS simulation software Intellisuite™. The software ANSYS™ was also used for thermal simulations to estimate the micromirror damage threshold. The simulated micromirror structure (see figure 2) consists of a reflecting flexible area supported by two legs a few microns above an electrode located on the substrate underneath the mirror. Aluminum with a low residual tensile stress of about 10 MPa was set as the micromirror material.

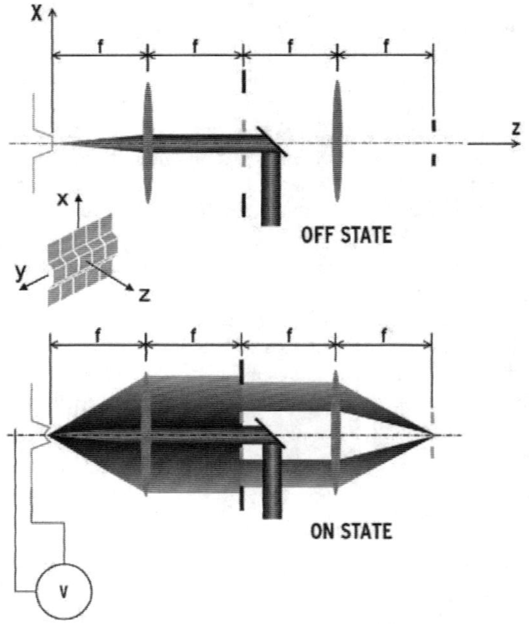

Figure 1. Light modulation approach using flexible micromirrors.

As an additional requirement, the optical design of the display engine assumes a maximum micromirror curvature corresponding to f-number of 2. The maximum mirror sag or deflection is therefore approximated by the following expression:

$$\delta_{max} \approx L / 16f\# \approx L / 32 \quad (1)$$

where δ_{max} is the maximum mirror sag, f# is the micromirror f-number at maximum curvature and L is the mirror length as defined in figure 2.

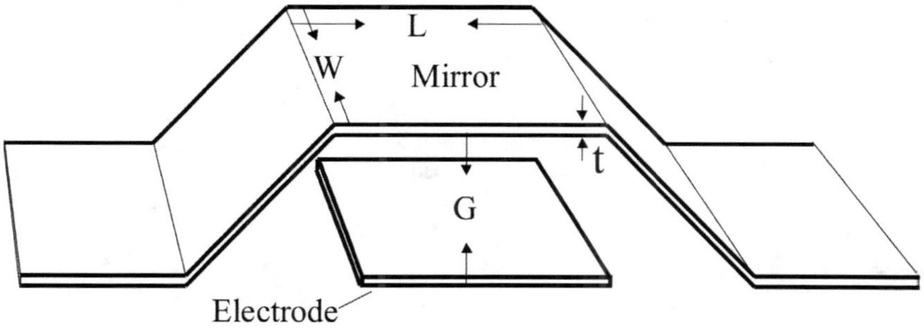

Figure 2. Flexible micromirror structure (W: mirror width, L: mirror length, G: mirror/substrate gap size, t: membrane thickness).

2.1. Static simulations

Electromechanical static simulations have been performed to evaluate the micromirror deflection as a function of the applied voltage for various FRAM configurations. These response curves are of particular interest to estimate actuation voltage required to reach the mirror sag corresponding to f# of 2 and to verify that a particular design presents a safe margin between the maximum actuation voltage and the pull-in voltage. Pull-in phenomenon is well-known [2] and occurs for a wide variety of electrostatic MEMS. In the particular case considered here, pull-in is caused by nonlinear variation of electrostatic force with mirror curvature. At some point, the restoring force caused by the membrane curvature cannot counterbalance the electrostatic force; the mirror either stops as it touches the substrate or it collapses. It is therefore critical that the designed micromirror exhibits control range sufficient to reach the required maximum curvature without approaching too closely the pull-in threshold. Another important design issue is minimizing the activation voltage while maintaining the feasibility of the structure by surface micromachining. This was also considered within this simulation phase.

The influence of various parameters including membrane length (L), membrane width (W), membrane thickness (t) and gap size between the membrane and the underlying electrode (G) (see figure 2 for definitions) was studied. Preliminary fabrication tests indicated that the membrane thickness should be 0.1 μm or more to keep the structure reliable and rugged. Gap size had to be a few micrometers only (1.5 to 4.5 μm typically) to maintain a reasonably low maximum activation voltage. Due to this small gap size, membranes with length of 30 μm or more exhibited a maximum activation voltage very close to the pull-in threshold. For this reason, the study considered mainly membrane lengths shorter than 30 μm. Membrane width was not found to have a significant impact on static performance.

Detailed results of these simulations have been presented elsewhere [3]. Typical results are shown in figure 3 for 25 μm x 25 μm micromirror with a thickness of 0.15 μm. The horizontal line in this figure indicates the required deflection corresponding to f# 2 for a mirror length of 25 μm. As expected, the actuation voltage increases with the gap value. The margin between actuation voltage and pull-in threshold (the voltage at which the curve becomes vertical) also increases with the gap size. Similar results were obtained for thinner membranes with t= 0.1 µm but with the actuation voltages shifted toward lower values. Simulations also investigated the behavior of shorter membranes with L= 15 µm. Shorter membranes are stiffer but require smaller deflection to achieve a f-number of 2. Simulations showed that, for the selected parameters, these effects combine to result in actuation voltages similar to the ones for 25 µm long membranes. However, due to the reduced gap, activation of these structures produces electric field higher than for the 25 µm long membranes. These microstructures are therefore more prone to electrostatic discharges which must also be taken into account when optimizing the device design.

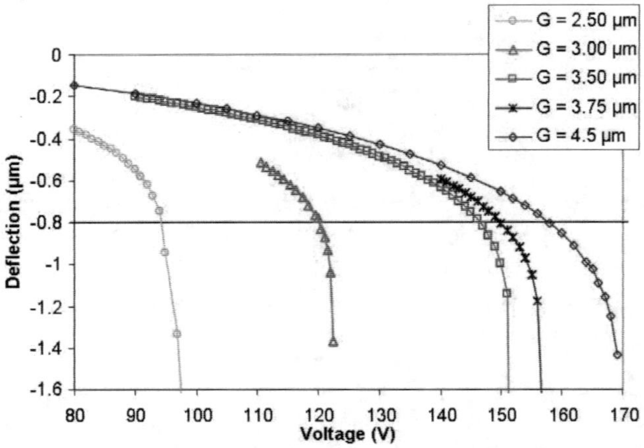

Figure 3. Simulation of the micromirror static response for a membrane thickness of 0.15 µm. Structure dimensions: L: 25 µm, W: 25 µm, t: 0.15 µm, G: 2.5 to 4.5 µm

2.2. Dynamic simulations

Since resolution of a projection engine based on linear FRAM array depends critically on the fast response of the micromirrors, dynamic simulation of these microdevices were performed to verify that response times within the required range could be achieved. Dynamic simulations included structural and squeeze film damping effects [4]. The MEMS simulation software used for these simulations required that the non uniform electrostatic pressure field acting on the membrane and varying in time with the mirror deflection to be converted into a uniform pressure field. This uniform field was a function of time but did not include the influence of the mirror curvature variations. The uniform pressure waveforms used for these simulations stabilized at a value corresponding to the required final micromirror deflection. The correspondence between voltage and pressure was obtained by determining what uniform pressure would produce the same static deflection as a given voltage. This was considered an acceptable simplification for this preliminary study of dynamic performance of the microdevice. The dynamic simulations were performed for various FRAM structures and different pressure variations with time. Of particular practical interest were pressure waveforms corresponding to a voltage step function or to an exponentially varying voltage. For performance characterization, the time required by the micromirror to oscillate within ± 5 % of the maximum deflection value was used as a definition of response time.

As mentioned above, various FRAM configurations were considered with different gap, thickness, length and width. All studied structures typically exhibited response times well below 10 µs when actuated with the proper pressure waveform. Detailed results were discussed elsewhere [3]. This necessity of using an appropriate waveform is particularly critical and one of the most important information resulting from this dynamic study. Using step functions to actuate the micromirrors caused large overshoots and made the device oscillate for relatively long time. These microsystems were clearly underdamped. It was therefore necessary to modify the actuation waveforms to minimize the mirror oscillations. Good results were obtained with pressure waveforms corresponding to exponentially varying voltages. Figure 4 shows an example of such improvement in response time obtained when passing from a step function to an exponential voltage waveform to activate a given micromirror structure.

Another interesting observation obtained from the dynamic simulations is related to the effect of the mirror width. As mentioned in section 2.1., micromirror width is not an important parameter when considering the static membrane performance. However, the situation changes significantly when dynamic response is considered. In this case, for FRAM with different widths and all other parameters being the same, simulations showed that response time can be strongly increased for narrower structures. This demonstrates, as expected, that squeeze film damping is significantly reduced for narrow membrane. This is another important information that must be taken into account when designing this type of micromirror.

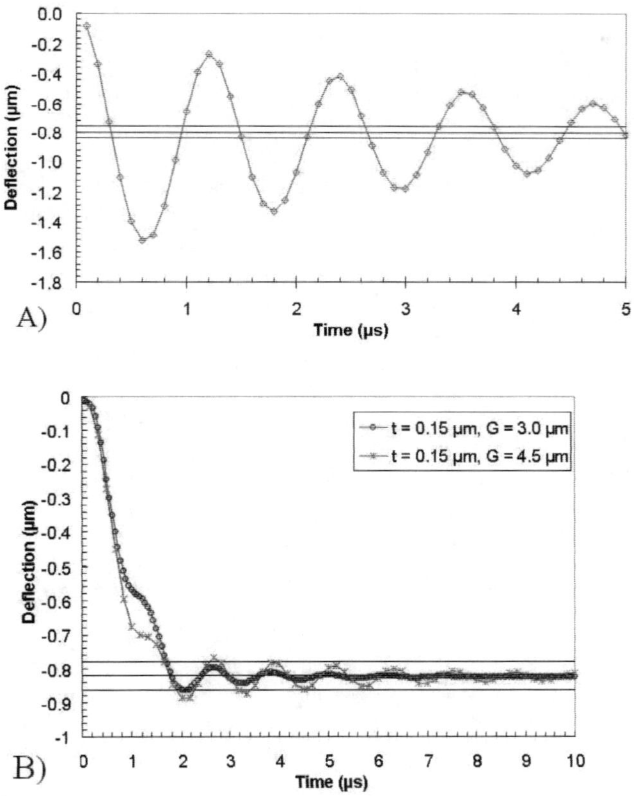

Figure 4. A) Simulation of the micromirror dynamic response to a 158 V step function. Structure dimensions: L: 25 µm, W: 25 µm, t: 0.15 µm, G: 4.5 µm B) Simulation of the micromirror dynamic response to an exponential voltage variation with a time constant of 0.43 µs. Structure dimensions: L: 25 µm, W: 25 µm, t: 0.15 µm, G: 3 and 4.5 µm.

2.3. Damage threshold estimate

Thermal and stress analyses have been performed with ANSYS™ software to estimate the damage threshold for 25 x 25 µm micromembranes. These simulations indicate that a membrane initially at 20°C could absorb up to 6750 W/cm^2 before reaching 450°C that is about 75% of the melting point of the Al alloy. Assuming optical absorption of 10 % for the alloy, this means that an optical power density of 67.5 kW/cm^2 could illuminate the membrane before its temperature reached 450 °C. However, stress analysis revealed that the Al alloy yield – about 190 MPa – is reached in some points of the structure when absorbed power density is 885 W/cm^2. Still assuming absorption of 10%, membrane damage threshold is then estimated to correspond to an incident optical power density of about 8850 W/cm^2. Most importantly, the damage threshold for the studied micromirror structure is not limited by the Al alloy melting temperature but by the Al alloy mechanical yield.

3. Micromirror array fabrication

Over the years, micromirror linear arrays with various sizes have been fabricated within development activities, going from 480 x1 mirrors for VGA format to the demonstration of 4000x1 FRAM. These arrays have been fabricated using surface micromachining techniques. The micromirror design rules resulted either from simulations described briefly above or from practical considerations linked to the developed fabrication process. As part of the technology development, micromirrors presenting various dimensions and structures have been built. This allowed investigation of the actual microdevice performance variation within the design parameter space. Table 1 below presents a representative set of parameters that has been used over the years for the fabrication of FRAM.

3.1. Fabrication process

The developed fabrication process uses 3 to 4 photolithographic masks. The devices are built on silicon substrates on which a silicon nitride (SiN) film has been deposited for electrical insulation. A first metallic layer is deposited and patterned to produce the bottom electrodes (see figure 5). This metallic layer material is typically aluminum or aluminium alloy. The electrodes are then passivated using a

Table 1. Fabrication parameters

Mirror parameters	Dimensions (µm)
Mirror length	20, 25 and 30
Mirror width	10 and 25
Gap size	3.5 and 4.5
Distance between mirrors	2, 3 and 5
Mirror thickness	0.1 to 0.3

SiN layer and windows are opened in the SiN to produce bonding pads. This passivation step is optional, i.e., it is not absolutely required to achieve functional micromembranes. The membrane fabrication begins with the definition of the gap in a relatively thick sacrificial polyimide layer. The thickness of this polyimide determines the gap size between the mirror and the electrode. The next step comprises the deposition and patterning of an aluminum alloy thin film. The aluminum alloy layer is deposited using Physical Vapor Deposition (PVD). This aluminum alloy film is patterned using a photoresist mask and reactive ion etching based on chlorine chemistry. This produces the micromembrane itself. Finally, the sacrificial polyimide layer is isotropically removed using a plasma asher which releases the microstructures by generating a gap between membrane and substrate.

3.2. Some fabrication results

Many micromirror array fabrication runs have been completed using the process described above. The obtained FRAM exhibit a smooth reflecting surface and are relatively flat (see figure 6). At the mirror level, the structures show typically a membrane width which is narrower than the nominal width. Moreover, the legs supporting the mirror are found to be slightly wider at the substrate level than at the mirror level. These small defects come from the fact that the photoresist used to mask the Al alloy for the etching step does not cover the wafer topography in a perfectly conformal manner. This results in a photoresist mask which does not match perfectly the photomask pattern. This difference,

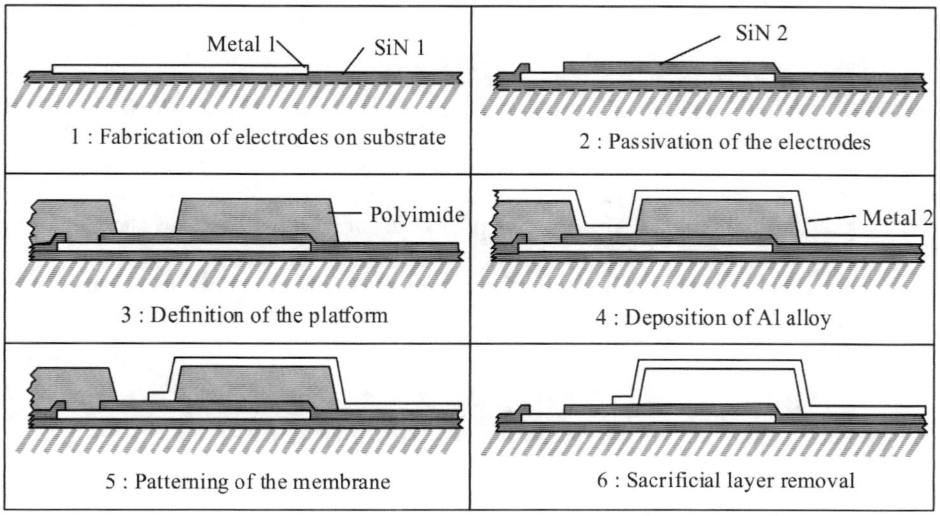

Figure 5. Flexible micromirror fabrication process

combined with some slight undercut occurring during the Al alloy etching, caused the observed geometrical imperfections. The texture on the leg surface reproduces surface topography of the sacrificial polyimide which supported the leg before being finally etched away. In addition, small defects can be observed at the membrane edges. The characterization performed on the membranes indicates that these defects do not distort the mirror profile significantly.

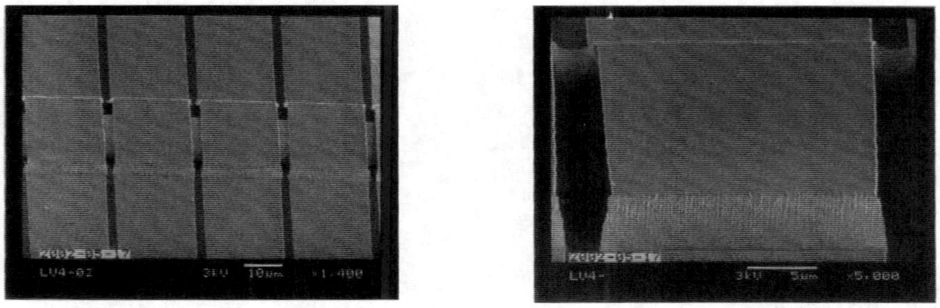

Figure 6. Flexible micromirrors. Left side: nominal dimensions are L: 30 μm, W: 25 μm, t: 0.15 μm, G: 4.7 μm, spacing between mirrors: 2 μm. Right side: nominal dimensions are L: 20 μm, W: 25 μm, t: 0.15 μm, G: 4.7 μm, spacing between mirrors: 3 μm.

4. Micromirror characteristics

Three main aspects characterizing the micromirror performance have been measured: the static response, the dynamic response and the damage threshold. The static response was characterized by measuring the mirror deflection as a function of the voltage applied to the FRAM. This was done using an interferometric microscope equipped with a 10 x Mirau objective with a numerical aperture of 0.25. The wavelength of the light illuminating the mirror under test was 548 nm. This produced an interference pattern in which two consecutive dark or light fringes are generated by mirror regions separated by a vertical distance of 0.27 μm. The voltage was increased gradually across the FRAM and the corresponding mirror deflection was measured by recording the number of fringes between the mirror centre and the mirror edges.

An optical method (see figure 7) has been used to characterize the micromirror dynamic response. The micromirror under test was illuminated with a laser beam and the reflected diffraction pattern was observed through a microscope. When the micromirror curvature was changed, the observed diffraction pattern was modified and some regions of the intensity profile showed an important intensity variation. A photodetector was positioned at the microscope output to measure the intensity corresponding to such high contrast regions. The photodetector response time was short enough (about 10 ns) to provide intensity measurements resolved in time. The voltage waveform applied to the micromembrane and the photodetector signal were recorded simultaneously using an oscilloscope. The analysis of recorded signals allowed the evaluation of the micromirror response time and overall dynamic performance.

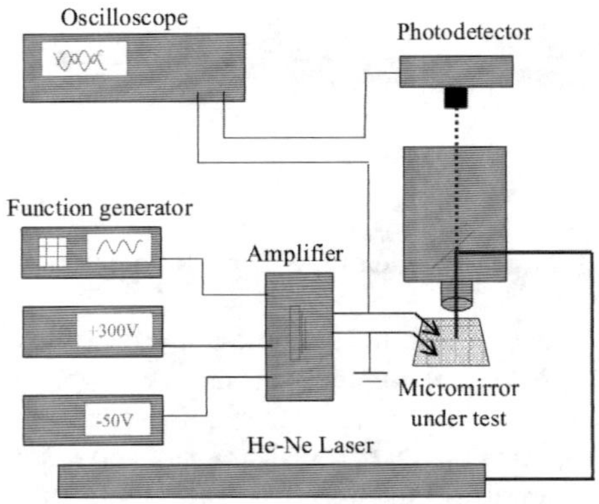

Figure 7. Set-up for the micromirror dynamic response characterization

Finally, a doubled YAG laser source was focussed on groups of micromirrors to evaluate damage threshold. The micromembranes were illuminated with a fixed laser intensity. After each illumination session of a few minutes, the micromirrors were examined with a microscope to verify if visible damage had occurred. Laser intensity was then increased and the procedure repeated until the mirrors were damaged.

4.1. Static response characterization

The micromirror curvature at rest (no voltage applied) depends on the residual stress in the membrane. A compressive residual stress causes a convex initial curvature while a tensile residual stress results in a downward membrane deflection. The residual stress can be modified, to some extent, by varying the aluminum alloy deposition parameters. Using such deposition parameter variation, it has been possible to fabricate micromirrors exhibiting various initial curvatures.

Typical static responses for micromirrors exhibiting residual compressive and tensile stresses are presented in figure 8. In the case of the membrane with a compressive stress, a voltage bias is necessary to bring back the mirror curvature to a minimum value corresponding to a deflection of 0 μm. In the present example, this offset voltage is about 136 V and a deflection of 0.8 μm corresponding to f/2 is reached with 233 V. The membrane with a tensile residual stress exhibits a downward deflection at 0 V. In the example shown, this deflection is 0.25 µm which represents a significant portion of the required deflection range. A deflection of 0.8 µm corresponding to f/2 is obtained with 158 V.

From these examples, it is clear that a membrane with a tensile residual stress has the advantage of reaching the maximum required deflection at a lower voltage than a membrane with a compressive residual stress. However, this reduced actuation voltage for membranes with a residual tensile stress is obtained at the expense of a significant decrease in the useful deflection range. This is an important drawback. The larger deflection range available for membranes with a compressive residual stress and the corresponding larger pixel intensity change are key features for projection applications. Although they typically require a higher activation voltage, these micromirrors seem a better choice for projection engines than FRAM with tensile stress.

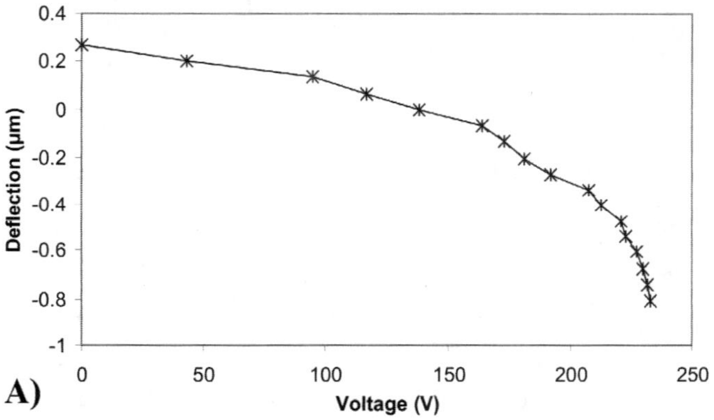

A)

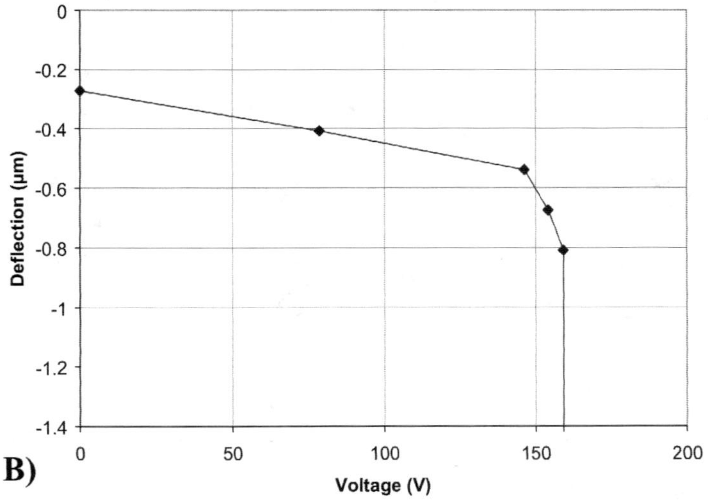

B)

Figure 8. A) Static response of a micromirror with compressive residual stress B) Static response of a micromirror with tensile residual stress. Nominal micromirror dimensions are L: 25 µm, W: 25 µm, t: 0.15 µm and G: 4.7 µm for both cases.

4.2. FRAM dynamic response

The dynamic responses for micromirrors with a compressive or a tensile residual stress were measured. Results are presented in figure 9 for membranes moved from the minimum to the maximum deflection and then back to the minimum deflection. The voltage waveforms used to control the membranes are also shown. In the compressive residual stress case, the activation voltage waveform is added to the voltage offset which is required to achieve an initial membrane deflection of 0 μm. In this case, the membrane stabilizes in about 10 μs. With the voltage waveform used, the stabilization time is even longer when the micromembrane is deactivated. These relatively long settling times indicate that the transient times of the voltage waveform (less than 2 μs) were not correctly adjusted for activation of the micromirror under test. Based on the simulation results presented above and on the observed response, it is clear that the settling times of the tested mirror could be improved by increasing the voltage transient times to 3 or 4 μs.

In the case of a membrane with a tensile residual stress, no voltage offset is applied to the membrane under test. A voltage waveform with a transient time of 3 μs has been used to control the micromirror. When activated, the micromembrane stabilizes in 4 μs and the settling time required to bring it back to its minimum deflection position is less than 5 μs.

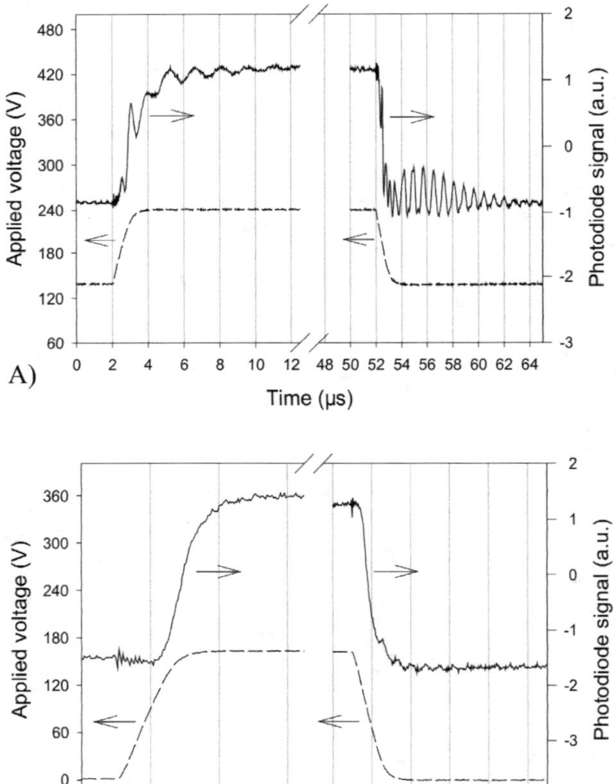

Figure 9. A) Dynamic response of a micromirror with compressive residual stress. B) Dynamic response of a micromirror with tensile residual stress. Nominal micromirror dimensions are L: 25 μm, W: 25 μm, t: 0.15 μm, and G: 4.7 μm in both cases.

4.3. Damage threshold evaluation
The tests performed to estimate the membrane damage threshold demonstrated that no visible damage is produced on the micromirror for incident laser intensities up to 8000 W/cm^2. However, permanent membrane deformations are observed for a laser intensity of 16000 W/cm^2. Although the actual damage threshold was not precisely pinpointed, these experimental results still agree with the simulations reported in section 2.3 which evaluated a damage threshold higher than 8000 W/cm^2. These simulations also predicted that thermally generated stresses exceeding the material yield would cause the device failure.

5. Projection engine prototypes
As part of the technology development, two projection engines have been built to demonstrate the principle. These engines include control software and electronics driving packaged FRAM linear arrays. They were integrated with illumination systems based on doubled YAG lasers. A line of light produced by these systems illuminates the micromirror arrays. For each setup, the light reflected by the micromirror array is collected by an optical relay which produces an image of the modulator array. A stop is positioned in the Fourier plan of the relay to block part of the collected light. The intensity of a given modulator image at the optical relay output depends on the amount of light coming from this modulator and not blocked by the stop (see figure 1). This amount of light is controlled by the curvature of the micromirrors. A galvanometer and projection optics located at the optical relay output produce the final 2-D image.

Characterization of these prototypes has revealed contrast values up to 300. The contrast is defined as the ratio between the intensity for a fully on pixel and the intensity for a fully off pixel, the intensity being measured at the optical relay output. For these measurements, a CCD camera combined with a commercial beam analyser were used. A 10 X objective magnified the image at the relay output prior recording by the CCD camera. In addition, the temporal response of pixels has been studied with a high speed photodetector recording pixel intensity variations at the relay output. It has been verified that the speed at which the pixel intensity can be modulated is in good agreement with response time measurements described in section 4.2.

Figure 10 shows the most recently built projection engine. This engine can display monochrome VGA video streams at the rate of 60 frames/sec. The software developed for this engine allows image data transfer compatible with DVI protocol. It also features a built-in look-up table (LUT). The default LUT used by the software includes 14-bit voltage input data varying linearly with the grey level data for all FRAM of the array. Using this LUT, the voltages applied to the FRAM are proportional to the grey levels of the original images to be displayed by the projection engine. Alternatively, the user can modify the LUT to adjust the Pixel Intensity vs Grey Level relation for each FRAM. This feature is very useful to optimize the Pixel Intensity vs Grey Level curve to match the eye response. It also allows correction of the pixel-to-pixel non uniformity if necessary. In addition, contrast, brightness and gamma correction can be performed on the images before data conversion by the LUT. It is also possible for the user to manually adjust frame synchronization and the number of displayed columns.

The architecture of the controls electronics built for the projection engine of figure 10 includes a FPGA board, a high-voltage digital-to-analog converter (HV DAC) board and a high-voltage DC power supply board. The FPGA board is dedicated to locking to the incoming video stream received from the host PC through a DVI port, management of the line-wise/column-wise addressing scheme for read/write accesses in a double-buffer memory and generation of frame synchronization for the DAC board. These tasks are controlled by a VHDL application programmed in the FPGA (Xilinx Virtex-2 Pro). Images are sent from the FPGA board to the DAC board through a CameraLink port.

The CameraLink port provides 24 independent data channels and each channel is used as a serial data link to access the serial input port of each of 24 multi-DAC chips installed on the DAC board. The DAC board makes use of 480 individual 14-bits DAC from 24 AD5535 Analog Devices chips populating a printed circuit board. A common voltage reference is connected to each AD5535 chip. This signal allows adjustment of the maximum output voltage from 50 to 200 V. As mentioned above, the number of 480-pixel columns that is written into the DAC for each image frame is defined by the user and can vary from 480 to 640. If desired, this allows the adjustment of the time available for the HV DAC to display each image column. The DAC board also provides a signal available to drive a galvanometer which is the scanning device typically used to produce a 2-D image. This signal is a periodic triangular waveform synchronized with the frame rate of the incoming video stream. In addition, the rising time, falling time and amplitude of the triangular waveform can be set with potentiometers part of the galvanometer driver circuit located on the DAC board. More details on this control electronics can be found in reference [5].

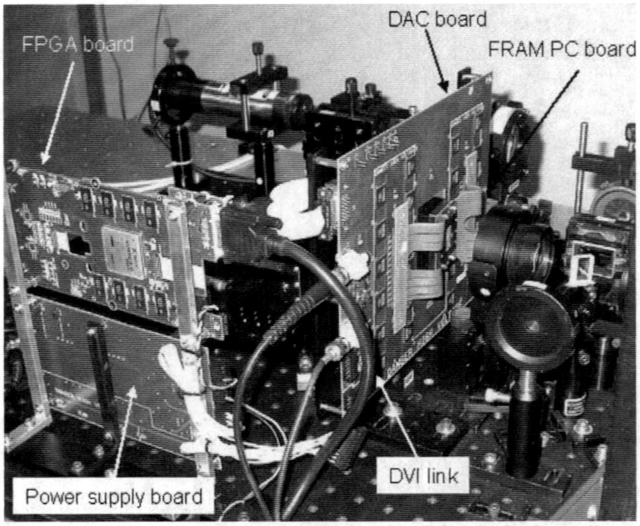

Figure 10. The projector engine and part of the optical test bench.

The developed engine relies on a 480x1 FRAM array for light modulation. Due to the large number of inputs for this chip, a custom package and packaging procedure were developed to connect all 480 micromirrors of the array to the control electronics. The selected packaging approach is a modified die-on-board packaging method. A flex board exhibiting rigid central and external portions is used for this approach. Standard high pin count connectors are soldered to the external rigid portions of the board. A hermetic enclosure with an optical window is clamped to the central portion of the board. The micromirror array is mounted in this enclosure filled with a dry nitrogen atmosphere. The central rigid portion of the flex board exhibits bonding pads to which the FRAM are connected. The hermetic enclosure has been designed to allow easy rework of the package and regeneration of the dry nitrogen atmosphere, if necessary. The main purpose of the flexible portions of the board is to provide some degree of freedom during optical alignment of the FRAM array. Resulting packaging assembly is shown in figure 11. As can be seen in figure 10, this assembly is connected directly to the DAC board which completes the projection engine. All components of the projector engine have been assembled within an existing test bench (figure 10). This test bench provides the required optical setup to produce images with the fabricated projector engine. Tests confirmed functionality of all features of the engine and projection of VGA video streams at a frame rate of 60 Hz.

6. Preliminary scale down study

Initial development of the technology was triggered by identified needs linked to flight simulation for jet fighter pilots. For such application, size and weight of the projector were not such a concern. Therefore, until recently, no effort had been devoted to minimizing the size of the projection engine based on FRAM. This resulted in the test bench shown in figure 10 which has a volume of about 0.2 m^3. However, for some applications, it is essential to strongly reduce this size. As a preliminary target, the possibility of decreasing the FRAM active area from 25 μm x 25 μm to 5 μm x 5 μm has been investigated. As a starting point, electromechanical simulations have been performed with Intellisuite™ to assess the performance of such micromirror. For these simulations, the FRAM material was an aluminum alloy with a density of 2.7 g/cm^3, Young's modulus of 70 GPa and Poisson's ratio of 0.36. The membrane surface (active area) and the actuation electrode dimensions were set to 5 μm x 5 μm. The mirror/substrate gap was fixed to 1.5 μm. For a membrane thickness of 40 nm, a mirror deflection of 0.25 μm corresponding to f/1.25 was obtained of an actuation voltage of 107 V. Due to the small mirror size and optical design requirements, the f-number at maximum deflection was set at 1.25 instead of 2 (see section 2 above) to insure sufficient reflected beam divergence. For this mirror thickness, activation at full deflection generates a relatively high stress above 250 MPa in the structure. From these preliminary results, feasibility of a 5 μm x 5 μm FRAM seems possible although optimization of the structure is still required to reduce the stress at maximum deflection.

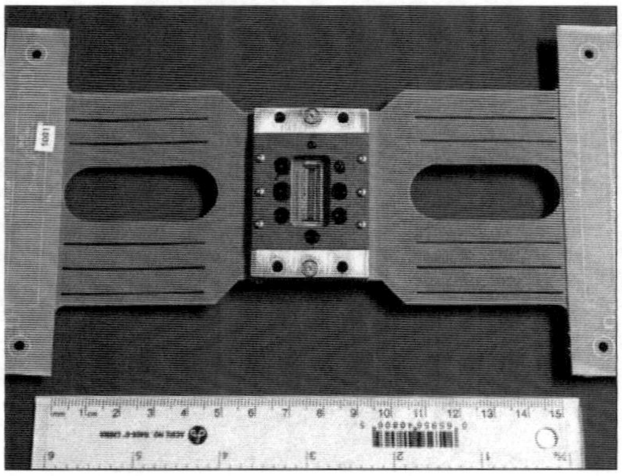

Figure 11. Packaging assembly for a 480x1 FRAM array.

Another interesting characteristic of these micromirrors is their capability to withstand high incident power density. This is of particular interest for various applications and it was important to estimate the damage threshold for FRAM with reduced active area. The approach used for this evaluation is the same that produced the results presented in section 2.3. As previously, the damage threshold was defined as the incident power density generating maximum stress in the structure corresponding to the yield stress of aluminum alloy. Results of this study are presented in table 2 below.

As can be seen in table 2, the damage threshold increases for smaller micromirrors since the effective thermal resistance is lower for these shorter structures and heat is more readily evacuated toward the substrate. Within the assumption underlying these simulations, the damage threshold is driven by the stress generated by the temperature increase. As can be seen, aluminum alloy yield stress is reached for fairly low temperatures. This is due to the relatively high structure stiffness and to the mismatch of coefficient of thermal expansion (CTE) between the mirror and the silicon wafer covered with a silicon nitride layer used as substrate. The maximum total power illuminating a 1000x1 FRAM array decreases for smaller micromirrors. This is due to the decrease in total array area which cannot be fully compensated by the damage threshold increase. For applications requiring higher incident power, approaches have been identified to further improve the micromirror damage threshold. One possibility would be the addition of buffer layers of appropriate material at strategic locations to decrease the stress due to CTE mismatch. Another approach would involve the improvement of mirror reflectivity to decrease the amount of light absorbed by the mirror. As for previous simulations, the results of table 2 were obtained assuming mirror absorption of 10 %. It is believed that this absorption could be reduced to 5 % or less by using proper reflective coating on the FRAM. This would result in significant improvement of the damage threshold and substantial increase of the maximum incident power.

Table 2. Thermal simulation estimates for damage threshold

Mirror size (μm)	Gap size (μm)	Thickness (μm)	Damage threshold (W/cm^2)	Total power[a] (W)	Maximum temperature[b] (°C)
25 x 25	4.5	0.15	8 850	55	76
15 x 15	2.8	0.1	13 400	30	69
5 x 5	1.5	0.04	25 600	6.4	54

[a]Total power: power incident on a 1000x1 mirror array at estimated damage threshold.
[b]Maximum temperature reached at damage threshold, substrate kept at 20 °C.

The results presented above give preliminary indications that functional micromirrors exhibiting an active surface as small as 5 μm x 5 μm should be feasible. Another interesting aspect of this scale down study was to determine the minimum dimensions for a projector based on FRAM arrays presenting a reduced size. The optical design of this concept of miniaturized projector is based on a 1000x1 array of 5 μm x 5 μm micromirrors. The maximum deflection for these mirrors is assumed to be 0.25 μm which corresponds to f/1.25. Illumination of the array is performed with a focused laser line with a width of 2 μm. Generation of this line is based on f/3 cylindrical optics and requires a source with good optical quality to achieve the required line narrowness. The targeted throw distance is 15 cm. The resulting design is depicted in figure 12.

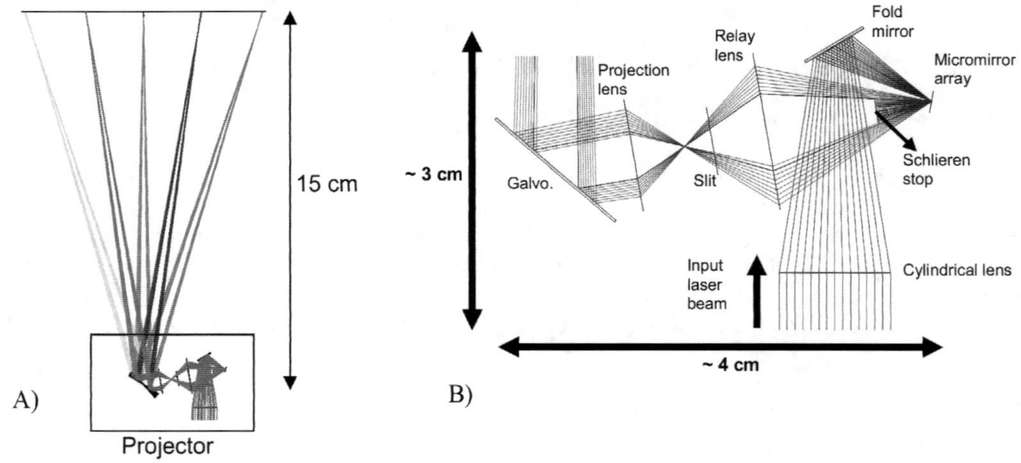

Figure 12. Conceptual design of a miniaturized projector based on a 1000x1 array of 5 μm x 5 μm micromirrors. A) General view of the system showing the throw distance. B) Enlarged view of the projector showing details of the optical system.

This projector concept exhibits a volume of about 12 cm^3. This volume does not take into account the generation of the collimated beam at the input of the system. Preliminary evaluation of this non optimized design shows some image distortion. The estimated resolution is about 40 μm at the centre of the image plane and about 400 μm in periphery. In practice, aberrations will limit the contrast of the designed system. It is likely that correction of these aberrations will be difficult due to the limited number of optical components that can be used for this miniaturized projector. Custom aspherical components would possibly be needed.

7. Conclusions

INO and partners have been working on a MEMS based technology for light modulation and projection display. This technology relying on flexible reflective analog modulators has been reviewed in this paper. Typical results have been provided regarding simulations of these microdevices and their fabrication. Characterization of the fabricated micromirrors has revealed activation voltages below 250 V with response times below 10 μs. With appropriate actuation voltage waveforms, response times of 5 μs and less are achievable. A damage threshold of the mirrors above 8 kW/cm^2 has been evaluated. Projector engines based on the FRAM have been built to demonstrate the technology principle. The most recent of these engines is DVI compatible and displays VGA video streams.

In addition, preliminary results of a technology scale down study have been reported. FRAM with active surface as small as 5 μm x 5 μm have been investigated. Electromechanical simulations have shown that such micromirrors could be activated with 107 V to achieve f-number of 1.25. However, the stress level generated in the structure at full actuation is high and further optimization is necessary to decrease this stress. The damage threshold has been estimated for various FRAM sizes. The damage threshold increases with decreasing active areas due to improved thermal coupling with the substrate. However, this increase cannot fully compensate for the decrease in active area when smaller micromirrors are used. Therefore, the maximum incident power allowed for illuminating an array of a given size increases with the micromirror dimension. Augmenting the available incident power is possible by further reducing the damage threshold. Possible approaches to do so have been suggested. Finally, design of a conceptual miniaturized projector based on FRAM array has been presented. The optical design is based on a 1000x1 array of 5 μm x 5 μm micromirrors for which the f-number reaches 1.25 at maximum deflection The throw distance specified for the projector is 15 cm. This projector concept exhibits a volume of about 12 cm^3. Although not optimized and still exhibiting some performance problems, this preliminary design and other results related to the scale down study provide interesting insights and suggest directions for a significant miniaturization of the technology. This is encouraging considering the importance of such miniaturization for most display and light modulation technologies.

References

[1] Van Raalte J A 1970 A new schlieren light valve for television projection Applied Optics **20** 2225-30

[2] Hung E S and Senturia S D 1999 Extending the travel range of analog-tuned electrostatic actuator Journal of Microelectromechanical Systems **8** 497-505

[3] Picard F, Campillo C, Niall K K, Larouche C and Jerominek H 2002 MEMS-based light valves for ultra-high resolution projection displays *Interservice/Industry Training, Simulation and Education Conference proc.*

[4] Yang Y-J and Senturia S. D 1996 Numerical simulation of compressible squeezed-film damping Technical Digest of the Solid State Sensor and Actuator Workshop (Hilton Head Island, USA) pp 76-80

[5] Picard F, Duchesne F, Jacob M, Larouche C, Vachon C and Niall K K 2010 Development of a DVI-compatible VGA projector engine based on flexible reflective analog modulators *Vision and Displays for Military and Security Applications* ed K K Niall (New York: Springer) chapter 2 pp 13-26

Journal of Physics: Conference Series **276** (2011) 012183 doi:10.1088/1742-6596/276/1/012183

Influence of Ni-dopant on structural and optical properties of $Ba_{0.6}Sr_{0.4}TiO_3$ thin film by sol–gel technique

Linqin Qian[a], Hongmei Deng[b], Liping Zhu[a], Pingxiong Yang[a1], Junhao Chu[a]
[a]Key Laboratory of Polar Materials and Devices, Ministry of Education, Department of Electronics, East China Normal University, 500 Dongchuan Road, Shanghai 200241, China
[b]Key Laboratory for Material Microstructures of Shanghai University, Shanghai University, 99 Shangda Rd, Shanghai 200444, China

E-mail: pxyang@ee.ecnu.edu.cn

Abstract. The pure and Ni-doped $Ba_{0.6}Sr_{0.4}TiO_3$ thin films were prepared on Si and $LaNiO_3$/Si substrates via sol-gel technique, respectively. The microstructure and surface morphology of the films were characterized using x-ray diffraction and atomic force microscope. The films on $LaNiO_3$/Si substrate were well-crystallized of the perovskite structure, and the diffraction peaks corresponded to BST standard pattern quite well. The Raman spectra showed the Ni dopant could improve the crystal quality of the BST films on Si substrate. The optical properties of the pure and Ni-doped films were studied by spectroscopic ellipsometry at room temperature.

1. Introduction

Barium strontium titanate, $Ba_xSr_{1-x}TiO_3$ (BST), has been investigated as a high-dielectric constant material for various applications in advanced electronic devices, such as storage capacitors in gigabit dynamic random-access memory, microwave tunable devices, and ferroelectric random-access memories. Recently, the ferroelectric thin-film optical devices are especially attractive because of their predominance for monolithic integration with electronic and optical electronic devices. As a typical ferroelectric material, the BST thin films have high optical transparency in the visible wavelength region, high refractive index, and low extinction coefficient, which have been used for integrated optic devices, such as optical memory, optical storage and low-voltage electro-optic switching.[1,2] As a result, it is worthwhile to investigate the optical properties of the BST thin films.

In this paper, the structural and optical properties of the pure and Ni-droped BST thin films were investigated, which were deposited on Si and $LaNiO_3$-coated Si (LNO) substrates by sol-gel technique, respectively.

2. Experimental Details

The precursor solution of the BST thin films were obtained by barium acetate [$Ba(CH_3COO)_2$], strontium acetate [$Sr(CH_3COO)_2\cdot 1.5\ H_2O$], and tetra butyl titanate [$Ti(OC_4H_9)_4$]. Firstly, Barium acetate, strontium acetate and nickel acetate were mixed in a mole ratio of Ba:Sr = 6:4, and then dissolved in heated glacial acetic acid with constant stirring. After cooling to room temperature, the

[1] Corresponding author. Tel.: +86 21 54345157; fax: +86 21 54345119. E-mail address: pxyang@ee.ecnu.edu.cn (P. Yang)

mixture was added with tetra butyl titanate drop by drop (the mole ratio of Ni:Ti is 2:98). Then, an appropriate amount of acetylacetone was added to stabilize the solution. Lastly, glacial acetic acid was added to make sure the concentration of the final precursor solutions was 0.3 M in Ti.

The crystallization of the pure and Ni-doped BST (Ni-BST) thin films was achieved via a rapid thermal annealing (RTA) process. These thin films were grown on Si and LNO substrates by spin-coating of the above precursor solutions at 4000 rpm for 25 s. After the spin-coating of each layer, the films were dried at 200 °C, prefired at 400 °C to remove the residual organic groups, and then annealed at 750 °C for 300 s in air. Finally, the six-layered BST thin films were post-annealed at 750 °C for 10 min. [3] The average thickness of a single-annealed layer can be obtained by spectroscopic ellipsometry, found to be about 65 nm. The desired films with a thickness of about 400 nm were achieved by repeating the spin-coating and heat treatment cycles.

The structure of the BST and Ni-BST films was analyzed by x-ray diffraction (XRD;D/max2200VPC), using CuKa radiation. The surface morphology of the films was examined using atomic force microscopy (AFM; Dimension 3100). Raman spectra were investigated by Jobin-Yvon LabRAM HR 800UV micro-Raman spectrometer. The thickness and the optical constants were determined by spectroscopic ellipsometry (SE). All the measurements were carried out at room temperature.

3. Results and Discussions

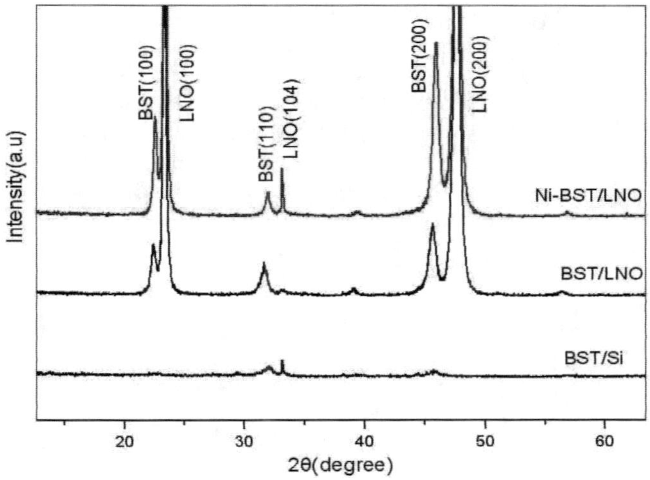

Figure 1. XRD patterns of BST and Ni-BST films deposited on Si and LNO substrates.

Fig. 1 shows the XRD patterns of the BST and Ni-BST thin films annealed at 750 °C on Si and LNO substrates, respectively. It can be seen that the film directly on Si substrate was not well crystallized, because the lattice constants of BST and Si do not match well. As shown in Fig. 1, the films deposited on LNO substrate were well crystallized to be perovskite structure, and the dominant orientations of the film were (100) and (200). $LaNO_3$ used as buffer layers can lead to a relatively high quality of the BST thin films. Compared with the BST film, the intensity of the diffraction peaks for the Ni-BST film is relatively strong, and no other phase is observed. It means that a small amount of Ni dopant can improve the crystal quality of the BST film without changing perovskite structure.

Fig. 2 presents the 2D and 3D AFM images of the BST and Ni-BST thin films grown on LNO substrates by sol-gel technique. The surfaces of the films are, in general, very smooth. The grains are densely and regularly packed without cracks or voids. The surface roughness is below 4 nm for both of them via AFM measurement. The average grain sizes on the surface of the BST and Ni-BST thin films

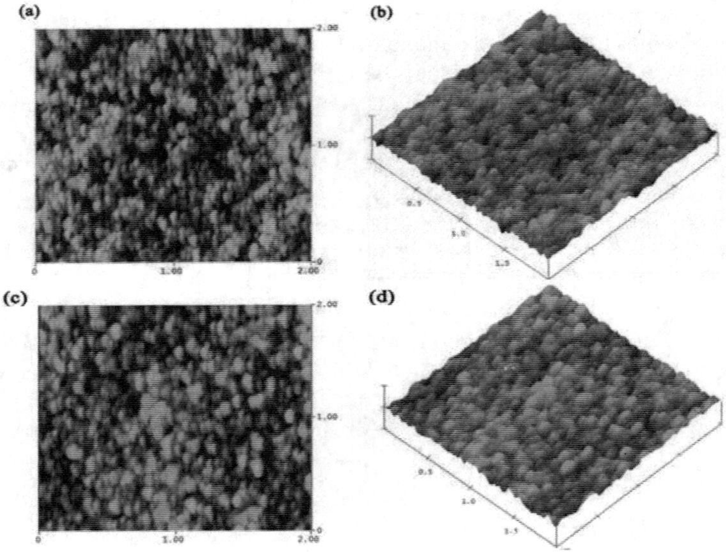

Figure 2. AFM images (2μm×2μm) of (a) 2D BST, (b) 3D BST, (c) 2D Ni-BST and (d) 3D Ni-BST thin films grown on LNO.

are nearly the same, about 60–100 nm, indicating that Ni dopant has little influence on grain size of the BST thin film. According to Chen's article,[4] the crystal structures of the BST thin films are very close to cubic phase when grain sizes are smaller than 20 nm at the room temperature, while, they are tetragonal phase when grain sizes are > 20 nm. Namely, in this paper the crystal structures of the BST and Ni-BST thin films are tetragonal phase.

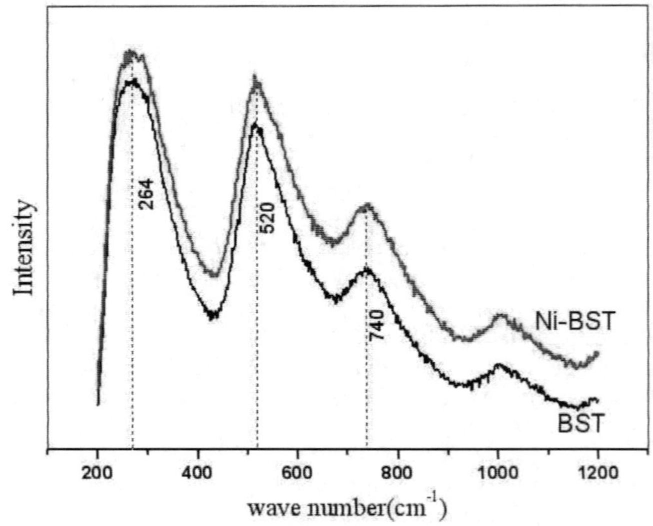

Figure 3. Raman spectra of the BST and Ni-BST thin films grown on Si.

Fig. 3 shows the Raman spectra of the BST and Ni-BST films grown on Si substrate by sol-gel method, annealed at 750 °C. From the figure, we found that a broadband centre at 264 cm^{-1} corresponds to the $A_1(TO_2)$ phonon mode, the peak at ~520 cm^{-1} corresponds to $A_1(TO_3)$ modes, and the peak at ~740 cm^{-1} is attributed to $A_1(TO_3)$. [3,5] Anyway, the characteristic Raman peaks persist for all BST films in the tetragonal phase. The Raman spectra confirm that these films prepared in the present work have an excellent crystallinity with a tetragonal crystal structure. Compared to the BST film, no other Raman peaks appear in the Ni-BST film. However, Ni dopant enhances the specific peaks of the BST film. It means that Ni dopant improves the crystal quality of the BST film. This result is consistent with what was analyzed above.

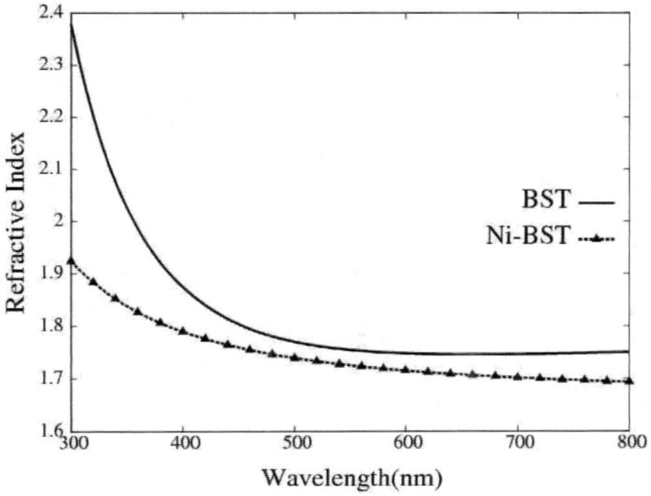

Figure 4. Refractive index of the BST and Ni-BST films deposited on Si.

Spectroscopic ellipsometry (SE) is a powerful technique to investigate the optical constants of the materials. [3] The refractive index and the extinction coefficient of the BST films can be obtained by fitting the model function to the experimental spectroscopic ellipsometric data in expression.

The refraction index of the BST and Ni-BST thin films as a function of wavelength which range from 300 to 800 nm are shown in Fig. 4. As can be seen, the curve of the refractive index of the BST film is fairly flat above 500 nm, rises in short wavelength range, and increases quickly when the wavelength arrives at ultraviolet range (380 nm). The refractive index of the Ni-BST film increases slightly from 1.70 to 1.92 as wavelength decreases from 800 nm to 430 nm. It is obvious that the refractive index of the Ni-BST film is lower than that of the BST film in the metrical wavelength range. The difference is due to the Ni dopant in the BST thin films. The refractive index of the BST film decreases abruptly in comparison with that of the Ni-BST film. This means that the Ni dopant can change the optical properties of the BST films.

4. Conclusions

In this paper, the BST and Ni-BST thin films were successfully prepared on Si and LNO substrates by sol-gel method, annealed at 750 °C. The x-ray diffraction showed that all the films were perovskite structure. However, the films deposited on LNO substrate were crystallized much better, suggesting that the use of buffer layers strongly affected the crystallization and structural orientation. According to the AFM images, the surfaces of these thin films are smooth, and the grains are dense. All the grain sizes are over 20 nm, indicating that the crystal structures of the present thin films are tetragonal phase. The Raman spectra can also confirm the conclusion. From the XRD patterns and Raman spectra, it can be found that the crystal quality of the films with Ni dopant was improved, without changing perovskite structure. The effects of Ni dopant on the refractive index of the BST films have been studied by spectroscopic ellipsometry in the visible region. The result shows that the refractive index of the Ni-BST film is lower than that of the BST film in the metrical wavelength range, and increases more slowly as wavelength decreases in short wavelength range. This means that the Ni dopant changes the optical properties of the BST films.

Acknowledgements
The authors thank Mr. Wenfei Xu for AFM investigation, Mr. Xiangui Chen and Mr. Jie Ge for optical investigation. This work was supported by the National Natural Science Foundation of China (No. 60990312), the State Key Basic Research Program of China (2007CB924902), Shanghai Leading Academic Discipline Project (B411) and Science and Technology Commission of Shanghai Municipality (10JC1404600).

References
[1] Kingon A. I, Maria J. P, and Streiffer S. K 2000 Nature (London) 406, 1032
[2] Seo, H, Kim Y.B 2010 *J. Appl. Phys.* 107(2): 024109
[3] Wang G. S, Yu J, and Wang Q 2002 phys. stat. sol. (a) 194, No. 1, 56–63
[4] Chen, H, Yang C 2009 Journal of Materials Science: Materials in Electronics 21(3): 236-240
[5] Kuo S.Y. and Hsieh W.F 2005 Journal of Vacuum Science & Technology A: Vacuum, Surfaces, and Films 23(4): 768
[6] Leng, W. J, Yang C. R. 2006 *J. Appl. Phys.* 99(11): 114904
[7] Catalan A.B, Mantese J.V, Micheli A.L 1994 *J. Appl. Phys.* 76. 2541
[8] Li, S, Yang Y 2008 Physica B: Condensed Matter 403(17): 2618-2623
[9] Zhu J.S, Lu X.M, and Jiang W. 1997 *J. Appl. Phys.* 81
[10] S. Saha and S.B. Krupanidhi.2000 *J. Appl. Phys.* 87 p. 849
[11] Zhou Q.F, Chan H.L.W, and Choy C.L 2000 *Appl. Phys.* A 70 p. 293

Journal of Physics: Conference Series **276** (2011) 012184 doi:10.1088/1742-6596/276/1/012184

White LED based on $CaAl_2Si_2O_8:Eu^{2+}$, Mn^{2+} phosphor and CdS/ZnS quantum dots

Changyu Shen*, Chuan Zhong, Qianglong Hou, Ke Li
Institute of Optoelectronic Technology, China Jiliang University, Hangzhou, China

E-mail: shenchangyu@gmail.com

Abstract. Core/shell CdS/ZnS quantum dots (QDs) with the emission wavelength of 610nm, was synthesized by thermal deposition using cadmium oxide and selenium as precursors in a hot lauric acid and hexadecylamine trioctylphosphine oxide hybrid. $CaAl_2Si_2O_8:Eu^{2+}$, Mn^{2+} phosphor was synthesized by high-temperature solid state reaction at 1290 °C for 2 hours under the H_2 reducing atmosphere, and X-ray powder diffraction analysis confirmed the formation of it. It has two emission bands peaking at 420 nm and 580nm originated from the transition 5d to 4f of Eu^{2+} and $^4T_1$-$^6A_1$ of Mn^{2+}, respectively. Blends of $CaAl_2Si_2O_8:Eu^{2+},Mn^{2+}$ phosphor and CdS/ZnS QDs exhibited the prominent spectral evolution with an increasing content of QDs. A hybrid white LED, which combines a blue LED with the blend of $CaAl_2Si_2O_8:Eu^{2+}$, Mn^{2+} phosphor and QDs with a weight ratio of 2:1, with the CIE coordinate of (0.3183, 0.3036) and CRI of 85 was obtained.

1. Introduction

White LEDs have the potential to replace traditional incandescent and fluorescent lamps for their characteristics of long lifetimes, energy saving and environmental protection. However, the commercialized white LED combining a blue LED with YAG: Ce^{3+} suffers from poor color rendering property caused by lack of red component in the spectrum. For an effort to improve color rendering properties of white LED, a novel means to obtain high color rendering index white LED by using semi-conductive colloidal quantum dots (QDs) blending with YAG phosphor was developed, recently[1-3]. Semi conductive colloidal quantum dots with exhibit quantum confinement effect (QCE) have drawn significant attention over the past decade [1-7]. QDs have been regarded as promising materials for white LEDs due to their size-dependent emission tunability and high quantum yield. The emission peak of CdS can be shifted from red to blue by reducing particle size. However, because the emitting spectral of YAG phosphor is not wide enough, the white LED blending of QDs and YAG suffers from the lower CRI, too. Recently, co-doping of the Eu^{2+}-Mn^{2+} pair in the silicate host has been adopted as a means to obtain white emission such as $Ba_3MgSi_2O_8$: Eu^{2+}, Mn^{2+}[8,9] and a-$Ca_2P_2O_7$:Eu^{2+}, Mn^{2+} [10]. Eu^{2+} can transfer its absorbed energy fully or partially to Mn^{2+}, consequently red (orange-reddish) or single-phase-full-color phosphors can be obtained. In this paper, highly luminescent red emitting core/shell CdS/ZnS QDs and $CaAl_2Si_2O_8:xEu^{2+},yMn^{2+}$ phosphor were synthesized, and the high CRI white LED based on the hybrid of them was obtained.

* shenchangyu@gmail.com

2. Experiments

The materials for CdS/ZnS QDs were as following: $Na_2S(9H_2O)$, RSH, Cadmium acetate, Zinc acetate and lauric acid, de-ionized water and hexadecylamine were analytically pure. The Cd precursor was prepared by dissolving 5 mmol of CdO (99.9%) in a mixture of 25 mmol lauric acid (LA, 99%) and 8m mol hexadecylamine (HDA, 95%) at 300 °C under argon flow. After that, $Na_2S(9H_2O)$ was injected into the hot Cd precursor, CdS nuclei formed and the temperature dropped to ~280 °C. The reduction of temperature stops the nucleation but allows the nuclei to continue growing. The CdS nanocrystal sizes were controlled by different growth time. The product was washed by hot methanol and collected by a size-selective method. For the in-situ ZnS overcoating process, Zinc acetate and $Na_2S(9H_2O)$ were directly injected into the CdSe nanocrystal solution alternately, and keeping the mixture's back flow 100min at the temperature of ~40 °C, the CdS/ZnS nanocrystal was obtained. The raw materials of $CaAl_2Si_2O_8$:xEu^{2+},yMn^{2+} were as follows: $CaCO_3$ (AR, 99.99%), SiO_2 (AR, 99.9%), $MnCO_3$ (AR, 99.999%), Al_2O_3 (AR, 99.9%), and Eu_2O_3 (AR, 99.999%). The concentration of $CaCO_3$, SiO_2 and Al_2O_3 was fixed at 1mol, 2mol and 1mol, while the Eu^{2+} and Mn^{2+} concentrations were varied from 0.02 to 0.4mol. After the ingredients were mixed thoroughly using ethanol as a solvent for mixing, the mixture was pestled for 1 h in a crucible with a lid. The crucible was preheated at 300 °C for 1 h, cooled down, and then sintered at 1290°C for 2 hours under the H_2 reducing atmosphere. The phase purity of the sample was checked by powder X-ray diffraction analysis with a Bruker AXS D8 advanced automatic diffractometer with Cu Kα radiation (λ=0.15418 nm) operating at 40 kV and 20mA. The measurements of photoluminescence spectra were performed using a TIAX550 fluorescent spectrometer (J-Y Co., France).

3. Results and discussion

The shell material of the quantum dots is important for the core/shell quantum dot's luminescence. The band gap energy and the crystal lattice mismatch between the core and the shell semiconductor were considered. Because of the larger crystal lattice mismatch between the core and shell material, the core/shell CdS/ZnS quantum dots has the characteristics of the structure defect and the photoluminescence quenching effect. Fig.1 is PL excitation and emission spectra of the CdS/ZnS quantum dots. For the QDs of CdS/ZnS, the excitation peaking wavelengths are 385nm and 450nm. The peak energies of CdS/ZnS QDs exceed the bulk CdS's and depend on the particle size [2-4]. Photoluminescence of CdS/ZnS quantum dots demonstrated high photoluminescence efficiency and size-tunable emission wavelengths from 400 to 620nm. In our experiments, core/shell CdS/ZnS quantum dots(5.6nm) has the emission wavelength of 618nm.

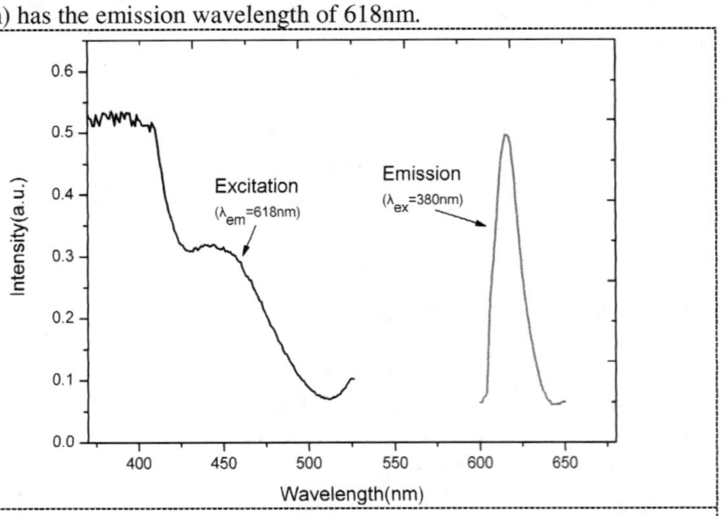

Figure 1. PL excitation and emission spectra of core/shell CdS/ZnS quantum dots

Journal of Physics: Conference Series **276** (2011) 012184 doi:10.1088/1742-6596/276/1/012184

The emission spectra of $CaAl_2Si_2O_8$: Eu^{2+},Mn^{2+} phosphor were shown in Fig. 2. As shown in this figure, the $CaAl_2Si_2O_8$: Eu^{2+},Mn^{2+} phosphor had a wide emission band ranging from blue to yellow peaking at 420 nm and 580nm respectively under excitation of 380nm irradiation.
The emission peaks at 420nm originate from the transition 5d to 4f of Eu^{2+} that occupy Ca^{2+} sites in the crystal. Compare to the emission spectra of the Eu^{2+} singly doped sample, the Mn^{2+} co-doped $CaAl_2Si_2O_8$: Eu^{2+} has two emission peak wavelengths of 420nm and 580nm. The emission band peaking at 580nm originated from the $^4T_1$-$^6A_1$ transition of Mn^{2+}. Corresponding to forbidden transitions from the ground state $^4A_1(^4G)$ to $^4E(^4G)$ of Mn^{2+}. The PLE spectrum of $CaAl_2Si_2O_8$: xEu^{2+},yMn^{2+} sample reveals an effective energy transfer from Eu^{2+} to Mn^{2+} in $CaAl_2Si_2O_8$. The energy transfer effect increasing with the concentration of the Mn^{2+} increases. As the x and y has the value of 0.05 and 0.35mol, the sample has the maximum emission intensity. When the Mn^{2+} and Eu^{2+} concentration exceeded 0.35mol and 0.05mol, concentration quenching occurred.

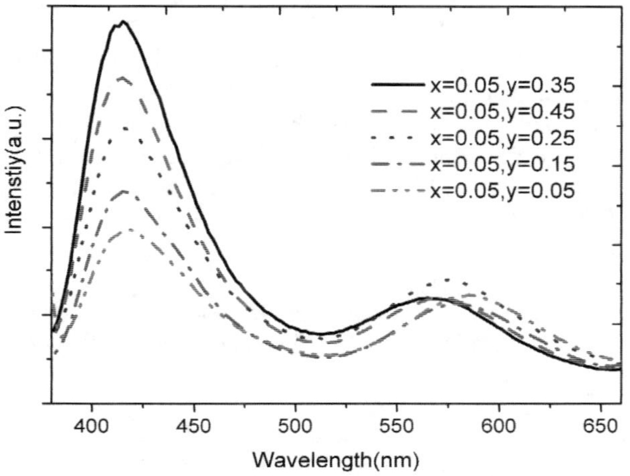

Figure 2. Emission spectra of $CaAl_2Si_2O_8$: xEu^{2+},yMn^{2+}

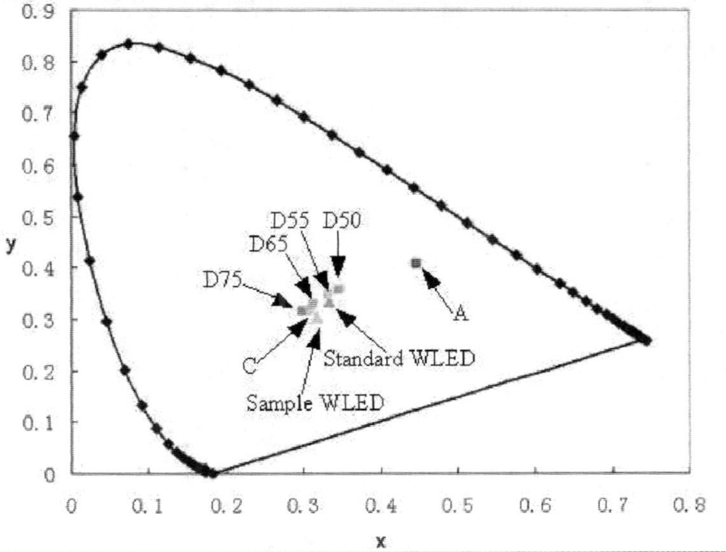

Figure 3. Coordinates of the white LED combining GaN chip (380nm) with $CaAl_2Si_2O_8$:$0.05Eu^{2+}$, $0.35Mn^{2+}$ and CdS/ZnS quantum dots

White LEDs by combining GaN chip (380nm) with the $CaAl_2Si_2O_8$: $0.05Eu^{2+}$,$0.35Mn^{2+}$ phosphor and core/shell CdS/ZnS quantum dots were obtained. With an increase of QDs content in the blends the white LEDs from 10:1, 10:3, and 10:5 blends showed the CRI values of 80, 81, and 85, respectively. However, as the hybrid ratio of the phosphor and the QDs larger than 2:1, the CRI value does not increase, yet. Fig.3 were the coordinates of the white LED by combining GaN chip (380nm) with the blends under the forward-bias current of 20 mA and standard illuminant A, C, D50, D55, D65 and D75 in the CIE 1931 chromaticity diagram, where the blend ratio of the $CaAl_2Si_2O_8$: $0.05Eu^{2+}$,$0.35Mn^{2+}$ phosphor and core/shell CdS/ZnS quantum dots is 2:1. The calculated CIE color coordinates of sample white LED were (0.3183, 0.3036) and the color rendering index (CRI) was 85, which was slightly larger than the CRI of the YAG:Ce white LEDs. The calculated color coordinates of the sample WLED were located in that of six standard illumination areas of A, C, D50, D55, D65 and D75. And it was very close to that of the standard white light of (0.3333, 0.3333).

4. Conclusion

In summary, $CaAl_2Si_2O_8$: $0.05Eu^{2+}$, $0.35Mn^{2+}$ with a wide emission band peaking at 420nm and 580nm was prepared by solid-state reaction at 1290°C. Core/shell CdS/ZnS quantum dots with the emission wavelength of 618nm was synthesized by thermal deposition. The hybrid $CaAl_2Si_2O_8$: $0.05Eu^{2+}$, $0.35Mn^{2+}$and CdS/ZnS quantum dots with blue LED chip were fabricated into a novel WLED. As the ratio of $CaAl_2Si_2O_8$: $0.05Eu^{2+}$, $0.35Mn^{2+}$ and CdS/ZnS QDs was 2:1, the white LED with color coordinates of (0.3183, 0.3036) and CRI of 85 was obtained.

Acknowledgements
This work was supported by Zhejiang Provincial Natural Science Foundation of China (Grant No. Y1100792), Science and Technology Department of Zhejiang Province (Grant No. 2008C21172).

5. References

[1] H-S Chen, C-K Hsu, and H-Y Hong, IEEE Photonics Technology Letters, 2006,18(1), 193
[2] C-Y. Shen, Ke Li, Q-L Hou, H-J Feng, and X-Y Dong, IEEE Photonics Technology Letters, 2010, 22(12), 884
[3], Y. Yang, S. Jin, J. Ming, Proc.SPIE, 2008,7135, 71353E
[4] A. Tang, F. Teng, Y. Gao, Journal of Luminescence, 2007,122,123,649
[5] C. Shen, H. Feng, Z. Xu etc, Chinese Optical Letters, 2008,6, 152
[6] H-S Chen, B Lo, J-Y Hwang, J. Phys. Chem. B, 2004,108, 17119
[7] F. Boxberg and J. Tulkki, Rep. Prog. Phys. 2007,70, 1425
[8] J. S. Kim, P. E. Jeon and J. C. Choi, et al, Appl. Phys. Lett. 2004,84, 2931
[9] J. S. Kim, K. T. Lim and Y. Seok, et al, Sol. Stat. Commun. 2005,135, 21
[10] Z. Hao, J. Zhang and X. Zhang, et al, Appl. Phys. Lett. 2007,90, 261113

Journal of Physics: Conference Series **276** (2011) 012185 doi:10.1088/1742-6596/276/1/012185

The effects of growth temperature of AlN buffer layers on *a*-plane GaN grown on *r*-plane sapphire by MOCVD

Y Q Sun[1], Z H Wu[1*], J Yin[1], Y Y Fang[1], H Wang[1], J N Dai[1], J Zhang[1], C H Yu[1], C Feng[1] and C Q Chen[1*]

[1]Wuhan National Laboratory for Optoelectronics, College of Optoelectronic Science and Engineering, Huazhong University of Science and Technology, Wuhan 430074, China

E-mail: cqchen@mail.hust.edu.cn; zhihao.wu@mail.hust.edu.cn

Abstract. To grow high quality *a*-GaN films on *r*-sapphire substrates, it is essential to optimize the buffer layers (BLs) growth parameters such as temperature, thickness, and V/III ratio. In this work, we investigated the effects of growth temperature for AlN BLs on the surface morphology and the crystal quality of *a*-GaN films. The films studied here were grown epitaxially by metalorganic chemical-vapor deposition (MOCVD) with different AlN BLs grown at 690 °C, 720 °C, 750 °C, respectively. The properties of *a*-GaN films were comprehensively studied via high resolution x-ray diffraction (HRXRD), optical microscope (OM) and atomic force microscope (AFM). It is found that the crystal quality and the surface morphology of *a*-GaN films have been remarkably affected by the growth temperature of the AlN BLs. At the optimum AlN BLs growth temperature of 720 °C, the *a*-GaN films have the best crystal quality and the smoothest surface morphology with the $(11\bar{2}0)$ HRXRD FWHM of 864 arcsec and the root mean square (RMS) surface roughness of 1.2 nm, respectively.

Keywords: *a*-plane GaN; AlN buffer layers; MOCVD; AFM; XRD

Introduction

*Corresponding author. Tel: +86-27-87793024, Fax: +86-27-87793024

Optoelectronic and microelectronic devices with a wurtzite GaN underlying layer heteroepitaxially grown on *c*-plane sapphire substrate have been widely developed by inserting low-temperature (LT) BLs to achieve high quality *c*-GaN single crystals [1-4]. However, the internal quantum efficiency of the *c*-plane GaN-based light emitting diodes (LEDs) is severely affected by the unavoidable strong polarization fields in quantum wells along the [0001] growth direction [5-7]. To circumvent the strong internal electric field in quantum wells, epitaxial growth of GaN on nonpolar planes have attracted much attention for recent years [8-10]. However, so far, the light emitting efficiencies of the non-polar GaN-based light-emitting diodes are still lower than those based on *c*-plane GaN due to the presence of high density of stacking faults and threading dislocations in non-polar GaN films. To improve the light emitting efficiencies, it is critical to improve the crystal quality of non-polar GaN. As the non-polar GaN bulk substrates are not readily available, many kinds of substrates have been used to heteroepitaxially grow nonpolar GaN, such as *r*-plane sapphire, *m*-plane sapphire, *a*-plane sapphire, *a*-plane SiC, *m*-plane SiC, and $LiAlO_2$ [11-15]. Similar to the growth of *c*-plane GaN, LT BLs are also necessary here, as they can provide adequate nucleation sites for subsequent GaN epitaxial growth and directly affect the crystalline quality of the epilayers. Therefore, it is essential to optimize the growth parameters of BLs such as temperature, thickness, and V/III ratio for high quality non-polar GaN growth.

In this work, *a*-plane GaN films have been heteroepitaxially grown on *r*-sapphire by metalorganic chemical-vapor deposition with AlN BLs grown at different temperatures. The surface morphology and the crystalline quality of the nonpolar *a*-plane GaN films have been investigated. The results showed that the growth temperature of LT AlN BLs has significant effects on the crystalline quality of *a*-plane GaN films and the temperature of 720 °C is found to be the most suitable condition to achieve nonpolar *a*-plane GaN films with flat surface and high crystalline quality.

Experiment Details

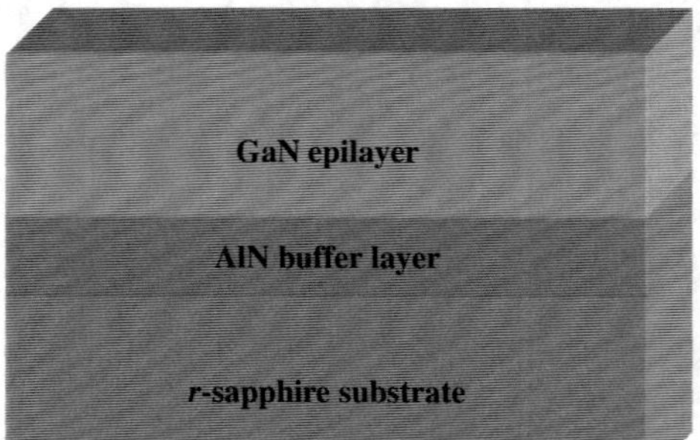

Figure 1. The schematic structures of the three samples.

Nonpolar $(11\bar{2}0)$ *a*-plane GaN films were grown on 2-inch $(1\bar{1}02)$ *r*-plane sapphire in a low-pressure and vertical showerhead MOCVD reactor. Figure 1 shows the schematic structure of these samples. Triethylgallium (TEGa), trimethylaluminum (TMAl) and ammonia (NH_3) gases were used as the precursors for Ga, Al and N, respectively. First, the substrate was thermally cleaned and nitridized under NH_3/H_2 mixture at a temperature of 960 °C. Then a thin AlN buffer layer was deposited at different growth temperature of 690 °C, 720 °C and 750 °C, respectively. The subsequent growth of GaN layers was conducted at 1020 °C with a V/III ratio of 1250. For all the above processes the reactor pressure was kept constant at 40 Torr. The samples with AlN BLs deposited at growth temperature of 690 °C, 720 °C and 750 °C, were denoted as sample A, B and C, respectively. The orientation and crystal quality of *a*-plane GaN films were investigated by high resolution X-ray diffraction (HRXRD) (Panalytical X'pert PRO MRD, Holand) using CuKα1 radiation (λ = 1.54056Å) as source. The surface morphology of the samples was examined by a differential interference contrast (DIC) optical microscopy (OM) (Olympus-BX51) and an atomic force microscopy (AFM) (Veeco NanoScope MultiMode) in contact mode.

Results and Discussion

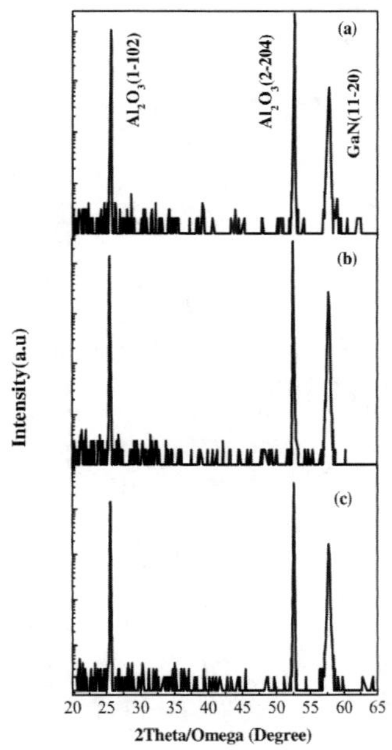

Figure 2. HRXRD 2θ-ω scan for (a)sample A, (b)sample B and (c)sample C.

Figure 2 shows the XRD 2θ-ω scan profile of the *a*-plane GaN films for samples A, B and C. The sharp peaks at 25.5° and 52.6° correspond to $(1\bar{1}02)$ and $(2\bar{2}04)$ diffractions from *r*-plane sapphire substrate, respectively. The $(11\bar{2}0)$ peak from the a-plane GaN can be observed at 57.6° and no other diffraction peaks are observed, which indicates that the *a*-plane GaN film is solely grown on the $(1\bar{1}02)$ *r*-plane sapphire.

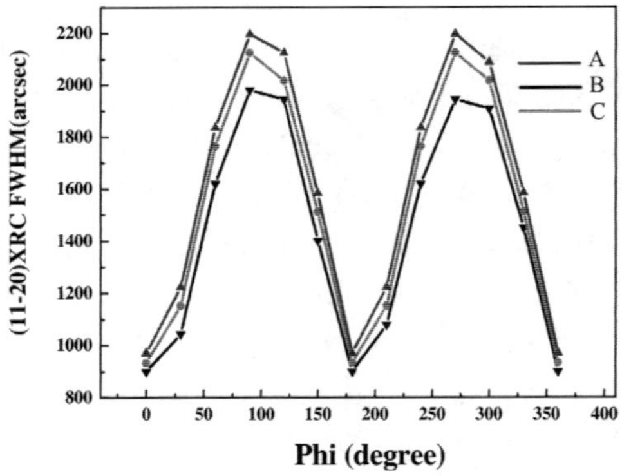

Figure 3. XRD $(11\bar{2}0)$ *a*-plane GaN FWHM anisotropy with respect to the in-plane beam orientation of all samples.

Figure 3 shows that the $(11\bar{2}0)$ *a*-GaN XRD FWHM value is a function of the in-plane azimuth angle for all the samples, exhibiting a "M" profile similar to what has been reported by Wang et al. [16]. The azimuth angle is defined with respect to the [0001] direction. At azimuth 0° the projection of

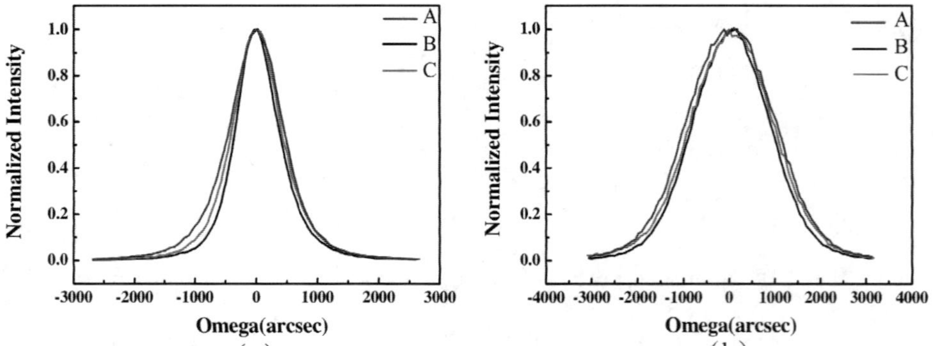

(a) (b)

Figure 4. Rocking curves of a-plane GaN film on $[11\bar{2}0]$ axis reflection with the incident beam parallel to the GaN (a) [0001] direction and (b) $[1\bar{1}00]$ direction.

the incident x-ray beam is parallel to the [0001] direction, and at azimuth 90° the projection of the incident X-ray beam is parallel to the $[1\bar{1}00]$ direction. Strong anisotropy of the $(11\bar{2}0)$ *a*-GaN XRD FWHM values with respect to the in-plane beam orientation for all the samples can be seen from the large difference between the maximum and the minimum points of the "M" profile. The sample B has an overall smaller FWHM value as compared to the other two samples. Figure 4 further gives the $(11\bar{2}0)$ X-ray rocking curves (XRC) of the *a*-plane GaN films measured along the [0001] direction

Table 1. The summary of the HRXRD FWHM and AFM measurement results of *a*-GaN films for the three different samples.

Samples	c-axis XRC FWHM (arcsec)	m-axis XRC FWHM (arcsec)	RMS roughness (nm)
A	936	2268	3.4
B	864	1980	1.2
C	900	2124	2.0

and the $[1\bar{1}00]$ direction within the scattering plane of the diffractometer. The FWHM values of $(11\bar{2}0)$ XRCs along the $[0001]_{GaN}$ and $[1\bar{1}00]_{GaN}$ directions were listed in table 1. The FWHM for $(11\bar{2}0)$ XRCs of sample B were the smallest with values of 864 arcsec and 1980 arcsec along the $[0001]_{GaN}$ and the $[1\bar{1}00]_{GaN}$, respectively. The crystalline quality is mainly affected by the mosaic structure of slightly misoriented sub-grains in the epilayer, which is caused by the large lattice mismatch between the epitaxial films and the substrates, as well as the epilayer growth conditions. The XRCs results show that the AlN BLs with the optimum growth temperature of 720 °C can remarkably decrease the mosaic structure by providing adequate and uniform nucleation sites for the subsequen a-GaN films.

The 1080 arcsec difference between the maximum and minimum FWHM values for the sample B was also the smallest among the three samples (sample A: 1224 arcsec; samples C: 1189 arcsec), indicating the anisotropy of the $(11\bar{2}0)$ *a*-GaN is the smallest at the optimum AlN BLs growth temperature of 720 °C. The large anisotropy of the $(11\bar{2}0)$ *a*-GaN is mainly due to the atomic arrangements on the $(11\bar{2}0)$ *a*-plane GaN surface. There are both Ga and N atoms at the step edge in the [0001] direction, whereas there are either Ga or N atoms at the step edge in the $[1\bar{1}00]$ direction, which induces a smaller density of the nitridation bonding in per unit area of $(1\bar{1}00)$ facet at the step edge in the [0001] direction. The different atomic arrangements results in that the migration length of gallium adatoms at the step edge with smaller density of the nitridization bonding in the [0001] direction is much longer than that in the $[1\bar{1}00]$ direction. The quality of epilayers will be better with longer distance order in the [0001] direction in comparison with the $[1\bar{1}00]$ direction. So the FWHM is smaller when the incident X-ray beam is parallel to the [0001] direction. The AlN buffer layers grown at 720 °C can make the growth rates of the subsequent *a*-GaN films in the [0001] direction and the $[1\bar{1}00]$ direction more homogenous, which can decrease the crystalline anisotropy with respect to the in-plane orientation.

(a) (b) (c)

Figure 5. Differential interference contrast OM images for sample A (a) , sample B (b), and sample C (c), taken with the same magnification.

Figure 5 shows differential interference contrast OM images of the three samples. The surface of all the samples exhibits streak features along the [0001] direction and undulations along the $[1\bar{1}00]$ direction, but the surface morphology of sample B grown with the AlN buffer growth temperature of 720 °C appears to be smoother than the other two samples. In addition, sample A and sample C have pits spreading over the surface, which originates from the island growth and coalescing at the early stage of films growth. By contrast, such surface pits disappeared almost completely in sample B, which benefits from the optimum AlN BLs grown at the temperature of 720 °C.

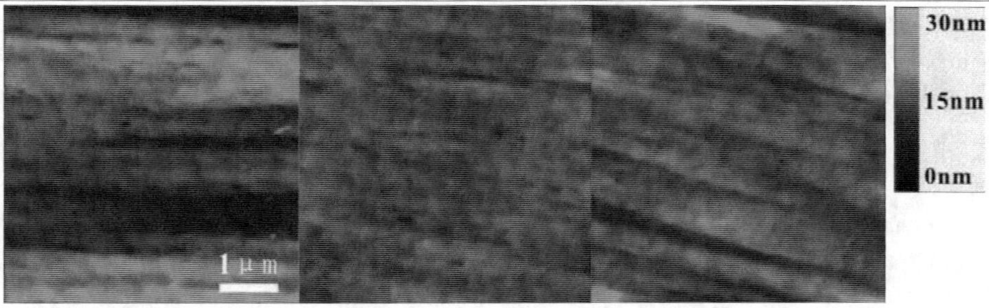

(a) (b) (c)

Figure 6. 5×5 μm^2 AFM images of the surface morphologies of the $(11\bar{2}0)$ *a*-plane GaN films grown on $(1\bar{1}02)$ *r*-plane sapphire substrates with different AlN buffer growth temperature (a) 690 °C, (b) 720 °C, and (C) 750 °C.

The surface morphologies of wafers with different AlN BLs growth temperature were further characterized by AFM, as shown in figure 6, and the measured root mean square of surface roughness over an area of 5×5 μm^2 was listed in table 1. The sample B had the smallest surface undulations compared with sample A and sample C, and the RMS of sample A and sample C with AlN BLs growth temperature of 690 °C and 750 °C, was 3.4 nm and 2.0 nm, respectively, while the value was reduced to only 1.2 nm for sample B with AlN BLs growth temperature of 720 °C.

Conclusion

In summary, we have succeeded in performing the $(11\bar{2}0)$ *a*-plane GaN films growth on $(1\bar{1}02)$ r-sapphire substrates by MOCVD. The HRXRD characterization results shows that all the samples have a solely $[11\bar{2}0]_{GaN}$ epitaxial orientation, and the optimum FWHM values for $(11\bar{2}0)$ *a*-GaN in the azimuths $[0001]_{GaN}$ and $[1\bar{1}00]_{GaN}$ are 864 and 1980 arcsec, respectively, at the AlN BLs growth temperature of 720 °C. The OM and the AFM measurements reveal that the surface of all the samples have the streak features along the [0001] direction and undulations along the $[1\bar{1}00]$ direction, but sample B with the optimum growth temperature of 720 °C has the slightest surface undulations, and the root mean square of surface roughness over an area of 5×5 μm^2 is reduced to 1.2 nm. All the results demonstrate that the optimum AlN BLs growth temperature is 720 °C, which remarkably improves the crystalline quality and the surface morphology of the obtained nonpolar $(11\bar{2}0)$ *a*-plane GaN films.

Acknowledgments

This work was surported by the Doctoral Program Foundation of Institutions of Higher Education of China (Grant No. 200804871144), the Program for New Century Excellent Talents in University (NCET-08-0214), the Hubei Province Science Fund for Distinguished Young Scholars (Grant No.2008CDB334), the Key Programs of the Natural Science Foundation of Huazhong University of Science and Technology (Grant No. 20072008B), the National Natural Science Foundation of China (Grant No.60976042), the Open Project of State Key Laboratory of Functional Materials for Informatics, and the National Natural Science Foundation of China (Grant No. 60906023), the Major Program of National Natural Science Foundation of China (Grant No. 10990100), and the National Basic Research Program of China (973 Program: 2010CB923204).

References

[1] H. Amano, N. Sawaki, I. Akasaki, and Y. Toyoda 1986 Appl. Phys. Lett. **48** 353
[2] S. Nakamura 1991 Jpn. J. Appl. Phys. **30** L1705
[3] F.A. Ponce 1997 Mater. Res. Soc. Bull. **22** 51
[4] I. Akasaki, H. Amano, Y. Koide, K. Hiramatsu, and N. Sawaki 1989 J. Cryst. Growth. **98** 209
[5] T. Takeuchi, S. Sota, M. Katsuragawa, M. Komori, H. Takeuchi, H. Amano, and I. Akasaki 1997 Jpn. J. Appl. Phys. **36** L382
[6] T. Takeuchi, C. Wetzel, S. Yamaguchi, H. Sakai, H. Amano, I. Akasaki, Y. Kaneko, S. Nakagawa, Y. Yamaoka and N. Yamada 1998 Appl. Phys. Lett. **73** 1691
[7] S. F. Chichibu, A. C. Abare, M. S. Minsky, S. Keller, S. B. Fleischer, J. E. Bowers, E. Hu, U. K. Mishra, L. A. Coldren, S. P. DenBaars and T. Sota 1998 Appl. Phys. Lett. **73** 2006
[8] T. Takeuchi, H. Amano, and I. Akasaki 2000 Jpn. J. Appl. Phys. **39** 413
[9] K. Kojima, M. Funato, Y. Kawakami, S. Masui, S. Nagahama, and T. Mukai 2007 Appl. Phys. Lett. **91** 251107
[10] M. Kubota, K. Okamoto, T. Tanaka, and H. Ohta 2008 Appl. Phys. Express **1** 011102
[11] M. D. Craven, S. H. Lim, F. Wu, J. S. Speck, and S. P. DenBaars 2002 Appl. Phys. Lett. **81** 469
[12] R. Armitage and H. Hirayama 2008 Appl. Phys. Lett. **92** 092121
[13] M. D. Craven, F. Wu, A. Chakraborty, B. Imer, U. K. Mishra, S. P. DenBaars, and J. S. Speck 2004 Appl. Phys. Lett. **84** 1281
[14] T. Kawashima, T. Nagai, D. Iida, A. Miura, Y. Okadome, Y. Tsuchiya, M. Iwaya, S. Kamiyama, H. Amano, and I. Akasaki 2007 J. Cryst. Growth **298** 261
[15] K. Xu, J. Xu, P. Deng, R. Qiu, and Z. Fang 1999 Phys. Status Solidi A **176** 589
[16] H. M. Wang, C. Q. Chen, Z. Gong, J. P. Zhang, M. Gaevski, M. Su, J. W. Yang, M. Asif Kahn 2004 Appl. Phys. Lett. **84** 499

Journal of Physics: Conference Series **276** (2011) 012186 doi:10.1088/1742-6596/276/1/012186

The influence of La content on properties of PLZT thin films

Qian Sun[1], Hongmei Deng[2], Xiaoxi Li[1], Pingxiong Yang[1], Junhao Chu[1]
[1] Key Laboratory of Polar Materials and Devices, Ministry of Education, Department of electronics, East China Normal University, 500 Dongchuan Road, Shanghai 200241, China
[2] Key Laboratory for Material Microstructures of Shanghai University, Shanghai University, 99 Shangda Rd, Shanghai 200444, China
E-mail: pxyang@ee.ecnu.edu.cn

Abstract. The polycrystalline complex compounds of PLZT, the formula $(Pb_{1-x}La_x)(Zr_{0.52}Ti_{0.48})O_3$ (x = 0.10, 0.15, 0.20 and 0.25), were fabricated by sol-gel method. the influence of La content on the microstructure, optical properties and electrical properties of PLZT materials were studied. Phase characterization and crystal orientation of the PLZT films was investigated by X-ray diffraction analysis. The extinction coefficient k of the films drops dramatically in the range of 300-500 nm, and reduces to 0.0355, 0.0357 and 0.0451 with the La concentration of 15%, 20% and 25%, respectively, when the wavelength is larger than 550 nm. The remnant polarization of the films decreases with the La content increasing.

1. Introduction

Since the discovery of ferroelectricity in single-crystal materials (Rochelle salt) in 1921, ferroelectric thin films have attracted a great amount of attention because of its excellent ferroelectric, pyroelectric and electrooptic properties. The field of ferroelectrics have been dominated by the materials based on barium titanate ($BaTiO_3$) and lead zirconate titanate (PZT). It is well known that a suitable ionic substitution ABO_3 type material has a dramatic effect on their physical properties [1]. Lanthanum-doped lead zirconate titanate (PLZT) material is one of the perovskite-type ferroelectric materials based on PLZT which is formed by doping La^{3+} ions at the A-site of PZT. PLZT has excellent dielectric, ferroelectric, piezoelectric and pyroelectric properties. It has varieties of applications which depend on composition ranging from piezoelectric applications to pyroelectric devices. Ferroelectric property of PLZT is sensitive to the concentration of La, and La content affects the dielectric property obviously. Recently, it is discovered that PLZT has great photoelectric characteristics and PLZT materials have great potential for applications of photoelectric devices [2].

In this paper, the polycrystalline complex compounds of PLZT were fabricated with the formula $(Pb_{1-x}La_x)(Zr_{0.52}Ti_{0.48})O_3$ (x = 0.10, 0.15, 0.20 and 0.25) by sol-gel technique. The microstructures, optical properties and electric properties of ferroelectric PLZT thin films were investigated.

2. Experimental

The ferroelectric PLZT thin films containing 10%, 15%, 20% and 25% La were prepared on Si (100) and $LaNiO_3$(LNO)/Si substrates by the sol–gel method, respectively. An excess of 10 wt.% of PbO was added to the initial solutions to compensate the loss of PbO from evaporation during the rapid thermal annealing process. Lead acetate trihydrate $Pb(CH_3COO)_2 \cdot 3H_2O$, lanthanum nitrate hexahydrate $La(NO_3)_3 \cdot 6H_2O$, zirconium nitrate pentahydrate $Zr(NO_3)_4 \cdot 5H_2O$ and tetrabutyl titanate

$Ti(OC_4H_9)_4$ were used as precursors and metoxyethanol $C_3H_8O_2$ was selected as solvent. Acetyl acetone $CH_3COCH_2COCH_3$ and acetic acid CH_3COOH were used as chelating agents. Lead acetate trihydrate, lanthanum nitrate hexahydrate and zirconium nitrate pentahydrate were first dissolved in methoxyetanol at 60 °C under stirring. Then the solution was cooled down until the room temperature and tetrabutyl titanate was added and dissolved under stirring. At last acetyl acetone and acetic acid were added to the solution of tetrabutyl titanate, the ratios were 1.4:1 and 4:1, respectively. The concentration of solution was about 0.3 M and the precursor solutions were transparent. Before the spin-coated, the solutions were filtered to avoid particulate contamination. The wet PLZT films were prepared on substrates by spinning at a rotation rate of 4000 rpm for 20 s at room temperature. Every layer was pre-annealed at 200 °C for 200 s to dry, and pyrolyzed at 380 °C for 200 s to remove residual organic compounds, then the polycrystalline PLZT films were obtained by annealed at 600 °C for 300 s in air by a rapid thermal annealing (RTA) process. The procedures were repeated for six times to obtain PLZT thin films with desired thickness.

Phase characterization and crystal orientation of the PLZT films were investigated by X-ray diffraction analysis (XRD). The optical properties of the PLZT films deposited on Si were studied via spectroscopic ellipsometry (SC630) in the wavelength range of 300-900 nm. The hysteresis loops of the PLZT films on LNO/Si substrates were measured to study the electrical properties.

3. Results and discussion

Fig.1 shows the XRD patterns of the PLZT films on Si substrates with different La contents. For the pattern (a), a broad peak observed at $2\theta = 29^\circ$ may be due to $(Pb,La)O{:}3(Zr,Ti)O_2$ (PLZT3) phase. According to other studies, the formation of PLZT3 is caused by the deficiency of Pb, and it will be converted to perovskite phase by the solution containing excess of Pb (>10%) in the system [3]. The pattern (b) exhibits peaks at $2\theta = 33^\circ$, 38° and 44° corresponding to the (220), (111) and (002), respectively [4, 5]. Moreover, the phase at $2\theta = 29^\circ$ diminishes as the intensity of the peak reduces when the content of La increases. When the content of La reaches 20%, as shown in pattern (c) of Fig.1, perovskite phase starts to appear at $2\theta = 31^\circ$ and 64° corresponding to the perovskite (110) and (202) reflections, respectively. The relatively high diffraction peaks of (110), (111) and (200) suggest that the grain is highly oriented growth. In pattern (d) of Fig.1, another peak at $2\theta = 22^\circ$ corresponding to (100) clearly emerges. The film shows a highly (110) oriented profile with a narrower peak. We can get the conclusion that La doping can improve the crystallinity of the samples.

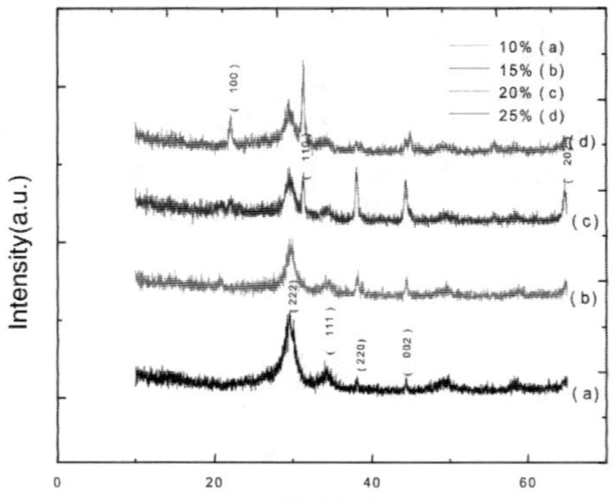

Fig. 1 X-ray diffraction patterns of the PLZT thin films on Si substrates.

Fig. 2 shows the refractive index (n) and the extinction coefficient (k) at 300-900 nm wavelength range derived from spectroscopic ellipsometry data for PLZT thin films on Si substrates. The optical constants of the PLZT thin films decrease with the wavelength increasing, and then approach unity. As the profiles show, the refractive index of the PLZT thin films decreases with La concentrations increasing at 390-900 nm wavelength range, in agreement with the conclusion that La reduces the light scattering[6]. The extinction coefficient k drops dramatically in the range of 300-500 nm, and reduces to 0.0355, 0.0357 and 0.0451 with the La concentration of 15%, 20% and 25%, respectively, when the wavelength is larger than 550 nm.

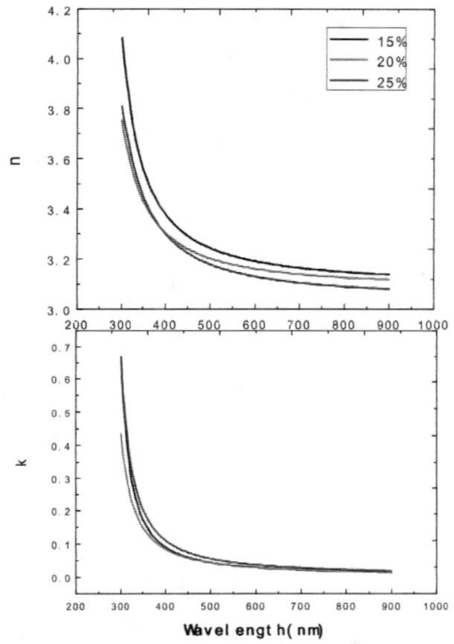

Fig. 2 The refractive index (n) and the extinction coefficient (k) of PLZT films deposited on Si substrates.

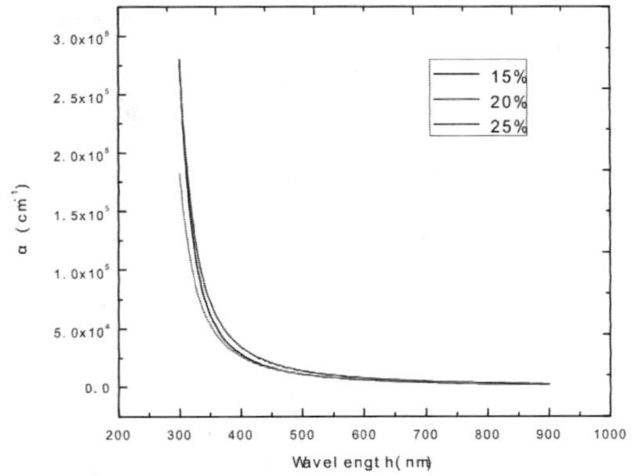

Fig3. The absorption coefficients α of the PLZT films deposited on Si substrate.

The absorption coefficients α of the PLZT films in the range of 300-900 nm are shown in Figure 3. It is obvious that the absorption coefficient drops dramatically in the range of 300-500 nm, and then the changes of the absorption coefficients go gently when the wavelength is larger than 500 nm. It can be seen that absorption coefficient and extinction coefficient have the same trend.

Fig.4 illustrates the room temperature P–E hysteresis loops of the PLZT thin films on LNO/Si substrates with varying applied voltage (10, 20, 30 and 40 V) to characterize the ferroelectric properties. As the applied voltage increases, hysteresis loops gradually broaden and the remnant polarization and coercive field of the film increase. The asymmetry of the hysteresis loops is induced by the different upper (Pt) and lower (LNO) electrodes[7,8]. Fig.5 presents the P－E hysteresis loops of the PLZT thin films on LNO/Si substrates with different La contents (x = 0.1, 0.15, 0.20 and 0.25), and the applied voltage is 40 V. It is obvious that the remnant polarization decreases with the La content increasing, and inversely the values of coercive field decrease. Moreover, the PLZT thin films containing 15% La have the highest breakdown voltage, about 65 V.

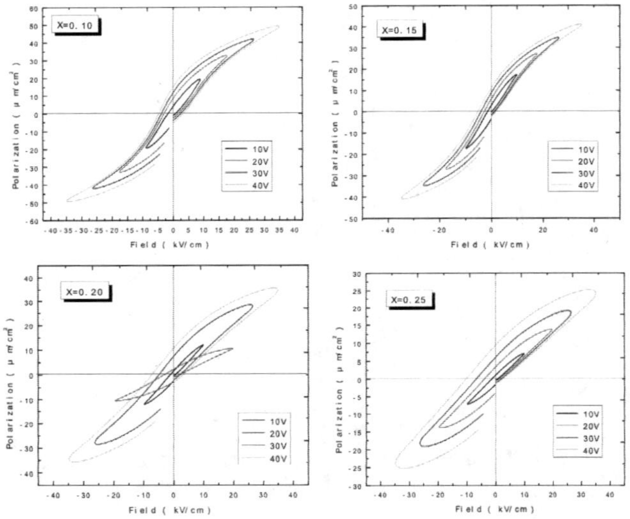

Fig.4 P-E hysteresis loops of the PLZT thin films with different applied voltages on LNO/Si substrates.

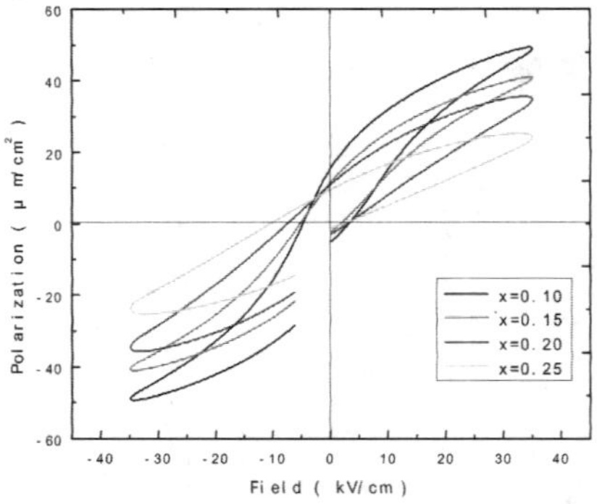

Fig.5 P-E hysteresis loops of the PLZT thin films with different La contents on LNO/Si substrates.

4. Conclusion

In summary, the PLZT thin films with different contents of La were prepared on Si and LNO/Si substrates, respectively, using the sol-gel method. La doping can improve the crystallinity of the PLZT thin films. La reduces the light scattering based on the analysis of the refractive index. The change of the extinction coefficient and the absorption coefficient have the same trend with La concentration, and decrease with the La concentration increasing. Moreover, hysteresis loop is degraded because of more La content.

Acknowledgements

This work was supported by the National Natural Science Foundation of China (No. 60990312), the State Key Basic Research Program of China (2007CB924902), Shanghai Leading Academic Discipline Project (B411) and Science and Technology Commission of Shanghai Municipality (10JC1404600).

References

[1] Miao Bin-bin, Wang Jun, Chen Jiang-tao and Yan Peng-xun, 2006 *Jurnal of Synthetic Crystals* **35** 540
[2] Meng Qin, Kui Yao and Yung C. Liang 2009 *Applied Physics Letters* **95** 022912
[3] Reji Thomas, Shoichi Mochizuki, Toshiyuki Mihara and Tadashi Ishida 2003 *Thin Solid Films* **443** 18
[4] Hong-Jin Zhao, Tian-Ling Ren, Ning-Xin Zhang, Ru-Zhong Zuo, Xiao-Hui Wang, Li-Tian Liu, Zhi-Jian Li, Zhi-Lun Gui and Long-Tu Li 2003 *Materials Science and Engineering* **B99** 197
[5] N.W. Jang, S.B. Mah, D.S. Paik, H.W. Choi and C.Y. Park 1999 *Materials Research Bulletin* **34** 1465-1466
[6] Z.G. Hu, F.W. Shi, T. Lin, Z.M. Huang, G.S. Wang, Y.N. Wu and J.H. Chu 2004 *Physics Letters A* **320** 483
[7] L.B. Kong, J. Ma, W. Zhu and O.K. Tan 2001 *Journal of Alloys and Compounds* **322** 295
[8] Hyeong-Ho Park, Woo Sik Kim, Jun-Kyu Yang, Hyung-Ho Park and Ross H. Hill 2004 *Microelectronic Engineering* **71** 218–219

Structural and optical analysis of CdS thin films grown by magnetron sputtering technique

Y H Sun[1*], Y J Ge[2], W W Li[1], D J Huang[1], F Chen[1], L Y Shang[1], P X Yang[1*], and J H Chu[1]

[1] Key Laboratory of Polar Materials and Devices, Ministry of Education, Department of Electronic Engineering, East China Normal University, Shanghai 200241, People's Republic of China

[2] Shanghai Center for Photovoltaics, Shanghai 201201, People's Republic of China

E-mails: yhsun@mail.sitp.ac.cn and pxyang@ee.ecnu.edu.cn

Abstract. Structural and optical characterizations were performed on CdS polycrystalline thin films grown by magnetron sputtering method. The thicknesses of the films were estimated to be 200-550 nm by scanning electron microscopy (SEM) measurements. X-ray diffraction (XRD) data show that CdS films are in the stable hexagonal crystalline structure. Using Debye-Scherrer's formula, the average grain size for the samples was found to be 40 ~ 45 nm. X-ray photoelectron spectroscopic (XPS) experiments indicate that the CdS samples are Cd rich in the surface region and S rich in the bulk region of the film. Photoluminescence (PL) bands, including green emission band, yellow emission band, orange emission band, and red emission band, were observed at room temperature and analyzed for the CdS films. The energy gap of the samples was also estimated and discussed from the room temperature optical transmittance spectra.

1. Introduction

Cadmium Sulphide (CdS) material has attracted much attention due to its applications in optical-electro devices, such as solar cells and light emitting diodes [1, 2]. Owing to the wide band gap of about 2.4 ~ 2.5 eV, CdS polycrystalline thin films are widely used as the window material in heterojunction solar cells [1]. CdS can crystallize in hexagonal (h) wurtzite and cubic (c) zinc blende structures, in which the hexagonal structure is more stable than the cubic structure. Since sulphur is a volatile element, CdS is prone to generate lattice defects during fabrication procedures. For CdS, interstitial Cd atoms and S vacancies act as donors, and Cd vacancies and interstitial S atoms behave like acceptors. These native defects introduce trapping states into the energy gap, which can significantly influence the optical-electric properties of CdS material.

CdS polycrystalline films have been grown by a variety of film deposition techniques such as thermal evaporation [3], sputtering [4], pulsed laser deposition [5], and chemical bath deposition (CBD) [6]. Meanwhile, various structural and optical characterizations have been done to CdS films, as widely reported in literatures. However, since the crystal structure and optical-electric properties of CdS films are sensitive to the specific methods and conditions employed during fabrication, investigations on structural and optical properties of CdS films are still underway to enrich the knowledge in this area [6, 7]. For instance, although photoluminescence (PL) measurements have been extensively used to obtain the information of band-edge transitions and defect-associated transitions in CdS films, the unambiguous origins of some PL emission peaks are still unclear. In this work, CdS thin films were prepared by magnetron sputtering method. Room temperature (RT) structural and optical properties of the CdS films were analyzed by scanning electron microscopy (SEM), X-ray diffraction (XRD), X-ray photoelectron spectroscopic (XPS), PL spectra, and optical transmittance spectra.

2. Experimental

The studied CdS thin films were grown by magnetron sputtering method with the deposition temperature of about 300℃. After the deposition procedure, CdS films were annealed in the Ar + S_2 mixture of gases flow for two hours at about 500℃. Substrates used in this work include glass slices coated with F-doped transparent conductive tin oxide (FTO) layer and quartz slices. The film thickness was controlled by varying the sputtering time and measured by SEM. The crystalline structure of CdS samples was analyzed through the XRD measurement by using the Cu Kα radiation with the wavelength of about 0.154 nm. The XPS experiments were performed at RT with a Perkin-Elmer PHI 550 spectrometer to investigate the elemental composition in the samples. The PL spectra at RT were recorded in the range of 1.4 – 2.75 eV by using a He-Cd laser with the wavelength of 325 nm (3.82 eV) as the excited light source. The RT optical transmittance measurements were carried out with a double beam ultraviolet-infrared spectrophotometer (PerkinElmer Lambda 950) in the photon energy range from 1.5 eV to 3.1 eV.

3. Results and discussion

A series of four CdS thin films with different thicknesses and substrates were studied in this work, as listed in Table 1. The typical surface micrograph obtained by SEM for CdS sample C was given in figure 1(a). By SEM measurements, the thicknesses of the four films were estimated to be 200-550 nm. Meanwhile, columnar shape growth in the CdS films is visible from the cross section view, as shown in figure 1 (b). From XRD measurements, it is found that CdS films grown on the same substrate possess the same XRD pattern. Therefore, only the XRD results for samples A and C were presented in figure 2. The XRD spectra of the substrates are also presented in figure 2 for comparison. Diffraction peaks with the Bragg angle 2θ = 26.7° were due to the (002) reflections in the hexagonal phase for the CdS films. CdS films grown on quartz substrates exhibit only the {002} reflection peaks, indicating a single preferential orientation and better crystal quality in these samples. As for CdS films grown on FTO substrates, apart from {002} reflection peaks, (101), (102), and (103) reflection peaks of the hexagonal phase were also observed in the XRD patterns, as designated in figure 2. The average grain size of the CdS was estimated by using Debye-Scherrer's formula [5], $D = 0.9\lambda/\beta\cos\theta$, in which D is the average grain size, λ is the X-ray wavelength, and β is the full-width at half-maxima (FWHM) in radians and θ is the diffraction angle. The grain size for all the CdS samples was found to be 40 ~ 50 nm, as given in Table 1. Since all the CdS films were prepared by employing nearly the same growth procedure as described previously, XPS measurements were performed only on CdS sample B to obtain the information of the elemental composition. XPS results reveal the CdS sample are Cd rich with the molar ratio of S:Cd = 0.73:1 in the surface region. Whereas, the molar ratio of S:Cd is 1.16:1 at the depths of about 20 nm after a etching procedure, indicating the fact of S rich in the bulk of the film.

Table 1. Substrates, thicknesses, average grain sizes, and values of energy gap for the four CdS thin films.

CdS Samples	Substrate	Thickness (nm)	Grain size (nm)	E_g (eV)

A	FTO	550	43.9	2.41
B	FTO	300	44.6	2.41
C	quartz	450	44.1	2.40
D	quartz	200	40.6	2.44

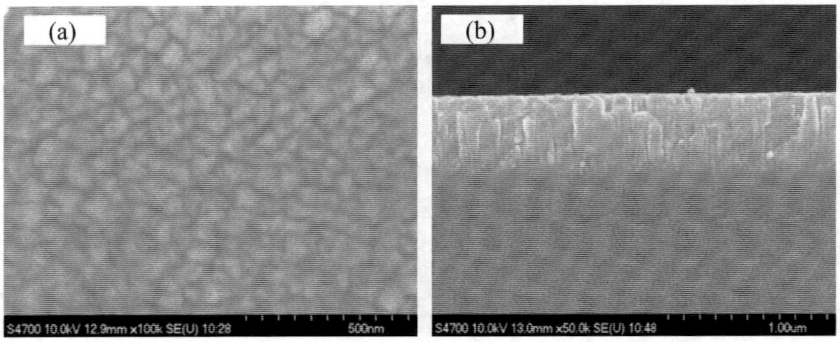

Figure 1. The surface (a) and the cross section (b) micrograph for CdS sample C by SEM.

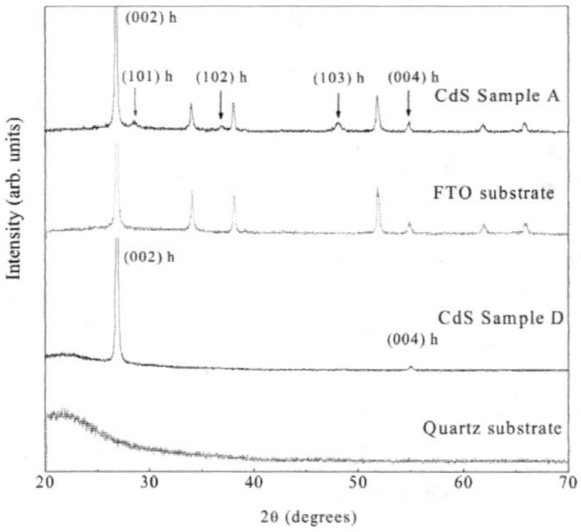

Figure 2. X-ray diffraction patterns of CdS films. The XRD spectra of the substrates are also given for comparison purpose.

As known, PL spectra reveal radiative recombination processes, which are usually related to the variety of the defects, in the material. Therefore, the analysis of PL spectra could be a complicated issue since the types of defects are related to the specific sample preparation procedures. Figure 3 then shows the room temperature PL spectra of the studied CdS thin films as a function of the emission energy. One can see from figure 3 that there are two well resolved emission structures in the PL spectra for each CdS sample: the emission band centered at the photon energy of about 2.37-2.43 eV and the broad band with the photon emission energy approximately ranging from 1.5 to 2.2 eV. As for the latter band, it is a superposition of several bands, which is evidenced by the peaks and weak shoulders in the PL features.

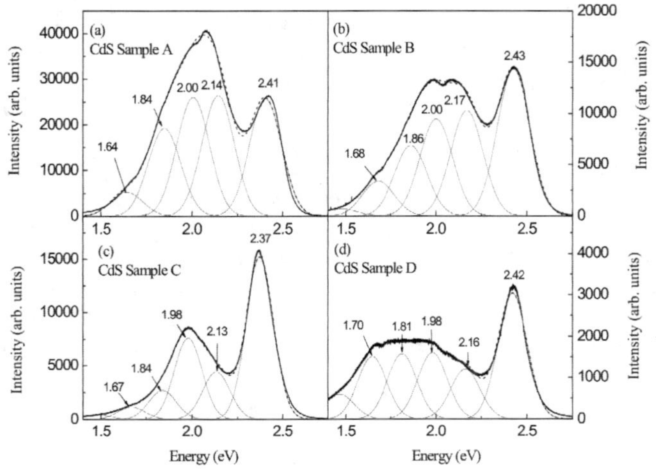

Figure 3. Experimental photoluminescence spectra and deconvoluted emission bands for CdS films.

We try to decompose the experimental PL spectra into several Gaussian curves for each CdS film, as given in figure 3. Emission bands centered at 2.37-2.43 eV are attributed to the green emission (GE) bands. At RT, GE bands centered at about 2.4 eV or above might origin from band to band transitions and/or recombinations associated with shallow trapping states [8, 9]. Note that a second GE peak located at about 2.3 eV has been observed and attributed to the transitions of donor-acceptor pairs (DAP) in CdS by some authors [3, 9]. The second GE band is also considered as the sign of the presence of cubic to hexagonal phase transition in CdS [10]. The lower GE peak energy of 2.37 eV for CdS sample C could be ascribed to the band tailing effect. Emission bands centered at 2.13-2.17 eV are yellow emission (YE) bands, which are ascribed to the transitions from interstitial Cd to the valence band [3] or Cd related DAP transitions [8, 11].

Emission bands centered at 1.98-2.00 eV in figure 3 are assigned to be orange emission (OE) bands. The well-resolved OE bands were only observed in CdS samples grown by CBD and microwave-assisted hydrolysis methods to our knowledge [6, 7]. OE bands might originated from deep DAP associated raddiative recombinations [6] or transitions from S vacancies to the valence band [7]. Emission peaks centered at 1.81-1.86 eV and 1.64-1.70 eV are attributed to red emission (RE) bands in this work [4, 6]. Generally speaking, RE bands are related to the transitions from surface states to the valence band [4]. O. Vigil et al. have argued that the emission peaks centered at 1.83 eV peak might be related to the transitions from S vacancies at the surface to the valence band [12]. Meanwhile, the more than one RE band might be due to the coupling effects with local vibration modes as reported in a recent work [6].

From figure 3, it is seen that the dominate emission band for the CdS films grown on quartz substrates is GE bands. Whereas, the intensities of the YE and OE bands increase relative to the GE bands in the films grown on FTO substrates when compared to those grown on quartz substrates. Combining with the XRD data, it is inferred that CdS films grown on FTO substrates possess a higher structural disorder (lattice defects). The intensity of RE band for CdS sample D increases due to the less thickness of the film, and therefore, radiative transitions from surface states become more important.

Figure 4 shows the optical transmission spectra of CdS films, in which the transmittances have been divided by the ones of substrates. For CdS films grown on the same substrates, the absorption edge shifts towards the higher photon energy when the film thickness decreases. At the photon energy higher than the absorption edge, the transmittances increase in thinner CdS films when compared to the thicker ones which grown on the same substrates, as shown in figure 4, indicating the less absorption to the incident light in these thinner samples [13]. By using the well known Tauc plot of $(\alpha h\nu)^2$ versus photon energy $h\nu$, the energy gap of CdS films was determined, as shown in the inset of figure 4. The absorption coefficient α were calculated by the relation $\alpha = \ln(1/T)/d$, where d is the film thickness, and T is the transmission coefficient. The energy gap can then be obtained by extrapolating the linear region of the plot $(\alpha h\nu)^2$ versus photon energy hν to $\alpha = 0$. The obtained values of the energy gap for CdS films are listed in Table 1. It is found that CdS films grown on FTO substrates have nearly the same value of energy gap, and the energy gap for CdS films grown on quartz substrates shows an increase for the thinner sample. We also found the energy gap determined by transmission spectra is close to the values of GE peak for each CdS film, indicating the GE bands are associated with the radiative recombination near the band edge.

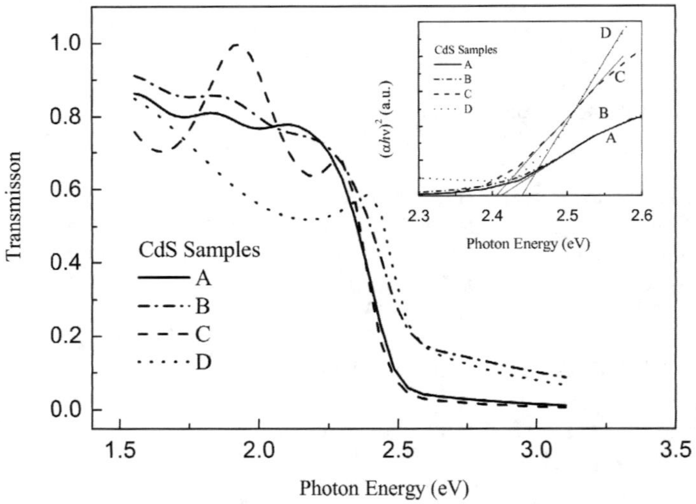

Figure 4. Transmission spectra of CdS thin films. The inset shows the Tauc plot of $(\alpha h\nu)^2$ versus photon energy $h\nu$ to determine the energy gap.

4. Conclusions

In the present study, structural and optical properties were investigated on CdS hexagonal polycrystalline films which are grown on the FTO and quartz substrates by magnetron sputtering method. The thicknesses of the films were in the range 200-550 nm and the average grain size was found to be 40 ~ 45 nm. It is also inferred that CdS samples are Cd rich in the surface region and S rich in the bulk of the film from the XPS data. The observation of PL emission bands, including green emission band, yellow emission band, orange emission band, and red emission band, allows one to predict the possible defects in the CdS samples. Meanwhile, the energy gap for the CdS thin films was also estimated and discussed from the room temperature optical transmittance spectra.

5. References

[1] Das S K and Morris G C 1993 *J. Appl. Phys.* **73** 782
[2] Wagner Sigurd 1974 *J. Appl. Phys.* **45** 246
[3] Agata M, Kurase H, Hayashi S and Yamamoto K 1990 *Solid State Commun.* **76** 1061
[4] Tsai C T, Chuu D S, Chen G L and Yang S L 1996 *J. Appl. Phys.* **79** 9105
[5] Perna G, Capozzi V, Ambrico M, Augelli V, Ligonzo T, Minafra A, Schiavulli L and Pallara M 2004 *Thin Solid Films* **453** 187
[6] Abken Anke E, Halliday D P and Durose Ken 2009 *J. Appl. Phys.* **105** 064515
[7] Podborska Agnieszka, Gaweł Bartłomiej, Pietrzak Łukasz, Szyma^{n}ska Iwona B, Jeszka Jeremiasz K, Łasocha Wiesław and Szaciłowski Konrad 2009 *J. Phys. Chem.* C **113** 6774
[8] Lozada-Morales R, Zelaya-Angel O, and Torres-Delgado G 2001 *Applied Physics A: Materials Science & Processing* **73** 61
[9] Lee Jaehyeong 2004 *Thin Solid Films* **451** 170
[10] Ariza-Calderon H, Lozada-Morales R, Zelaya-Angel O, Mendoza- Alvarez J G and Baños L 1996 *J. Vac. Sci. Technol.* A **14** 2480
[11] Mochizuki Katsumi, Satoh Masaaki and Igaki Kenzo 1983 *Jpn. J. Appl. Phys.* **22** 1414
[12] Vigil O, Riech I, Garcia-Rocha M and Zelaya-Angel O 1997 *J. Vac. Sci. Technol.* A **15** 2282
[13] Ebothe Jean 1995 *J. Appl. Phys.* **77** 233

Acknowledgements
This work was supported by the Major State Basic Research Project of China (Grant No. 2007CB924902), the National Natural Science Foundation of China (Grant No. 60990312), Science and Technology Commission of Shanghai Municipality (Grant No. 10JC1404600), and Shanghai Postdoctoral Scientific Program (Grant No. 10R21412600).

Analysis on dye-sensitized solar cell's efficiency improvement

Hanmin Tian [1,2], Jiyuan Zhang[1,3], Yangjing[4], Tao Yu[1,3*], Zhigang Zou[1,3]

1. Eco-materials and Renewable Energy Research Center (ERERC), Department of Physics, Nanjing University, Nanjing 210093, P.R. China;

2. Department of Electronic Science and Technology,School of Information,Hebei University of Technology,Tianjin 300401, P.R. China;

3. National Laboratory of Solid State Microstructures, Nanjing 210093, P.R. China

4. College of Chemistry and Chemical Engineering, Nanjing University, Nanjing 210093, P.R. China;

[*]Corresponding Author, E-mail: yutao@nju.edu.cn (Yu Tao)

Tel.: +86-25-83686304; Fax: +86-25-83686632

Abstract

The influence of I_{ph}, I_0 , R_s , R_{sh} , n on the I-V curve, which are the equivalent circuit parameters of a dye-sensitized solar cell, was presented. A series of TiO_2-based dye-sensitized solar cells were prepared, and experiment results consisted with our theoretical deduction that the increase of R_s would change the *FF* of DSSC while none influence on short-circuit current I_{sc}, but I_{ph} changed I_{sc} greatly by increasing the thickness of the TiO_2 layer of a set of cells from 8 um (120 um in wet paste layer state) to 27 um (220 um in wet paste layer state) gradually. These factors that affect efficiency were analyzed on the basis of equivalent circuit for further analysis of experimental results. Such equivalent-circuit-based electron transmission analysis of dye-sensitized solar cells is very useful for the establishment of the link between cells' electrical properties and its physical and chemical fabrication processes.

Keywords: Dye-sensitized solar cells; Equivalent circuit; Initial current; Resistance; Photo current;

Journal of Physics: Conference Series **276** (2011) 012188 doi:10.1088/1742-6596/276/1/012188

1. Introduction

The dye-sensitized solar cell (DSSC) is an attractive candidate for a new renewable energy source because of the low-cost materials and the facile manufacturing used in its production[1]. The photoelectrochemical process of DSSC originates at the interface between a redox electrolyte containing iodide and triiodide (I^-/I_3^-) ions and a dye-derivatized mesoscopic TiO_2 electrode[2]. Local changes in DSSC's complex physical and chemical compositions often give rise to interactions with other parts.

For example, an increase of T-BuPy in electrolyte, increase open-circuit voltage (V_{oc}) by inhibiting the electron-hole compound[3] [4], but also caused reduction of the photocurrent generated [5]. Introduction of MgO coating layer on TiO_2 improved the V_{oc} but break up the fill factor (*FF*)[6]. Our experiments also showed that, the actual unstableness of experimental materials and experimental processes made the law of electrical properties of cells even more untraceable. So, the analysis of DSSC's efficiency should draw out a concise abstraction from the DSSC's complex systems by mathematical epagoge.

Equivalent circuit of DSSCs precisely is the mathematical abstract of electronic transmission in cells already used in some research [4] [5, 7, 8], while most of them are for qualitative discussion [5, 7, 8]. Here presented the quantitative calculation of the relationship of I_{ph}, I_0 , R_s , R_{sh} , n with the cell's electrical properties, with well confirmed experiment results. According to the study of Masaki [4], decreasing the series resistance R_s of cells would increase the short-circuit current I_{sc} by acetic acid treatment, and increasing shunts resistant R_{sh} of the cell by 4t-butylpyridine would increase open-circuit voltage V_{oc}. However both our theoretical deduction and experiments showed the changed series resistance of DSSC, only would change the *FF* of DSSC while none influence on short-circuit current.

Further way, the thickness of mesoscopic TiO_2 layer is thought important as photon-generated electrons originate from the dye-derivative TiO_2 layer [9-11]. ITO et al summarized an analysis about TiO_2 layer thickness on the fill factor, short circuit current, open circuit voltage of DSSC I-V property from experiments, but was lack confirmed by theoretical deduction and his experiment results were also slightly confusing [12]. One of his two groups experiment results, which is AcCN-based DSSC, showed the thickness incrassation was followed by an increase in short-circuit current and a steady in fill factor (*FF*), while his another ionic-liquid-based DSSC experiment showed thickness incrassation caused short-circuit current (I_{sc}) decline slightly and the drop in *FF* [12]. Our theoretical deduction and experiments showed the changed I_{ph} of DSSC by increasing the thickness of the TiO_2 layer of a set of cells from 8 um to 27 um gradually, mainly resulted in the change of I_{sc} . Changes of *FF* would be due to changes in R_{sh} and R_s when the thickness of the TiO_2 layer was changed.

2. Experiment

The dye-sensitized solar cells were prepared as following[13]. TiO_2 film was made by extruding a precursor paste onto a F:SnO_2 conductive glass substrate and heating it at 500 C for 30 min. Dye absorption was carried out by dipping TiO_2 electrode in a $4x10^{-4}$ M ethanolic solution of Ru (2,2-bipyridine-4,4-dicarboxylic acid)$_2$(NCS)$_2$. The

Journal of Physics: Conference Series **276** (2011) 012188 doi:10.1088/1742-6596/276/1/012188

electrolyte is composed of 0.6 M PMII, 0.1 M I_2, and 0.45 M NMBI in MPN solvent.

The I-V curves were obtained using a source measure unit (Model 236, Keithley Instruments Inc.) under irradiation using a Solar Simulator (92251A ,ASTM class A solar simulate, Oriel Instruments, Inc.).

The thickness of sintered TiO_2 films was measured by a surface topography instruments (Dektak6M, VEECO) and the SEM photo was obtain by LEO 1530, LEO Inc. The DSSC I-V curve is displayed in Fig 1, in which η is 5.3%. Fig 2 shows the SEM photograph of TiO_2 film of the DSSC.

Further experiments confirmed well with the theoretical results about R_s. First, we measured the I-V curve of a dye sensitized solar cell. And then by series-connecting a 10 ohm resistor and a 20 ohm resistor in the outside circuit of the cell respectively so that to augment the R_s of this cell, we measured the corresponding I-V curves. As shown in Fig 3, when the R_s increases, no obvious change is observed in I_{sc} , V_{oc} of I-V curve, and *FF* drops distinctly. The experiments result confirmed that R_s does not influence I_{sc} and V_{oc} evidently but influence *FF* markedly.

For photon-generated electrons originate from the dye-derivative mesoscopic TiO_2 layer[9-11], we increased the thickness of the TiO_2 layer of a set of cells from 8 um (120 um in wet paste layer state) in a to 27 um in g (220 um in wet paste layer state) gradually, as measured results shown in Fig 4, so that to augment the amount of photon-inspired electrons I_{ph} of Cells. The TiO_2 precursor paste layers were coated onto F:SnO_2 conductive glass substrates by auto-coating machine (ZAA2300, Zehntner, Swiss), which could adjusts the thickness of the coating layer with the precision tolerance of 5 um. Other fabrication conditions and components of these cells are same.

As shown in Fig 5, the measured I-V curves of Cells, which have the enhanced I_{ph}, show significantly increased values of I_{sc} . V_{oc} of Cells were slightly improved, while no obvious change was observed in FF and the shape of I-V curve. From the measured I-V curves from Cell a to g, the analysis about I_{ph} to the I-V curve was confirmed.

3. Results and discussion

3.1 The estimation of equivalent circuit parameter

The generic solar cell may be described by a lumped parameter equivalent circuit model consisting of a single exponential-type ideal junction, a constant photo-generated current source, a series parasitic resistance (R_s) and a parallel parasitic conductance (R_{sh}) [14, 15], and recently some researchers confirmed the existence of capacitance characteristic in DSSCs's equivalent circuit by EIS, which is time-dependent and not significant in the Si cell [16-19].

The process to calculate equivalent circuit parameters, which could fit to the measured I-V curve, is derived from the equivalent circuit (Fig 6) as below.

According to Kirchhoff's law[20] , equation 1, 2 and 3 are deduced as follows:

$$I(t) = I_{Cs}(t) + I_{Rs}(t) = C_S \cdot \frac{d(V_{Cs}(t))}{d(t)} + \frac{V_{Cs}(t)}{R_S}; \qquad (1)$$

$$I_{ph} - I_{Rsh} - I_d - I_{Csh} - I_{Rs} - I_{Cs} = 0; \qquad (2)$$

$$V_d = V_{Rsh} = V_C = V_{Rs} + V(t) = I_{Rs} \cdot R_S + V(t); \qquad (3)$$

The currents of diode, capacitance and resistance could be obtained via equations

4, 5, 6 and 7, according to their physical definitions.

$$I_d = I_0 \cdot (\exp(\frac{q \cdot V_d}{K \cdot T \cdot n}) - 1); \quad (4)$$

$$I_{Rsh} = \frac{V_{Rsh}}{R_{sh}} = \frac{V_d}{R_{sh}}; \quad (5)$$

$$I_{Csh} = -C_{sh} \cdot \frac{d(V_d)}{d(t)}; \quad (6)$$

$$I_{Rs} = \frac{(V_d - V(t))}{R_s}; \quad (7)$$

I_0 initial current; R_s series resistance; R_{sh} parallel (shunt) resistance;

T temperature;

n diode factor; q elementary electric charge; K Boltzmann constant;

C_s series capacitance; C_{sh} parallel (shunt) capacitance; I_{ph} photo current;

The current expression has been derived as equation 8, according to equations 1-7.

$$I = I_{ph} - I_0 \cdot (\exp(\frac{q}{K \cdot T \cdot n} \cdot (V(t) + I \cdot R_s)) - 1) - \frac{(V(t) + I \cdot R_s)}{R_{sh}} + (C_{sh} + C_{Cs}) \cdot R_s \cdot \frac{d(I)}{d(t)} + C_{sh} \cdot \frac{dV(t)}{d(t)}; \quad (8)$$

For the I-V curve measured in stable state, the $\frac{dV(t)}{d(t)}$ and $\frac{dI(t)}{d(t)}$ impact come up to absence [13, 21]. Therefore, equation 9 is get from equation 8.

$$I = I_{ph} - I_o \left\{ \exp(q \frac{V + IR_s}{nKT}) - 1 \right\} - \frac{V + IR_s}{R_{sh}}; \quad (9)$$

According to the measured I-V data, as shown in Fig 1, the values of equivalent circuit parameters are obtained by mathematics estimation. The estimation method is described as following. Firstly, we assign a set of initial values optionally to the equivalent circuit parameters in equation 9, and then assess the difference between I

and measured data. If the difference is obvious, the next group of new parameters' values will be optimized by Newton–Raphson's method, and further evaluation will be performed until the most suitable parameters obtained. Taking into account the possibility of the existence of multiple solutions, we determined the only appropriate solution in accordance with the published improved method[22]. The obtained values of the parameters are listed below. The calculated I-V curve by these parameters fits the measured I-V curve appropriately, as shown in Fig 1.

R_s *3.5 Ω;* n *1.83;* I_{ph} *10.5mA* I_0 *1.01e-7 mA;* R_{sh} *1461 Ω*

3.2 Analysis on the dye-sensitized solar cells (DSSCs)'s equivalent circuit parameter

Based on these parameters gotten above, the trends of the influences of each parameter are obtained and perspicuous illustrated in Fig 7.

The red-dot-line curves in Fig 7 are the measured I-V curve of the cell. The blue-constant-line curves are the calculated I-V curves, which are obtained by stepwise changing the value of one parameter in equation 9 and fixing the values of the other equivalent circuit parameters. The influence of this parameter on the I-V curve is therefore quantificational revealed by ranking these curves together. By this method, the influences of I_{ph}, I_0 , R_s , R_{sh} , n on the I-V curve are gained respectively(Fig 7.a-e).

I_{ph} significantly influences the I_{sc} of the I-V curve, which is co-directionally moved following the variety of I_{ph}. That is I_{sc} will be improved following the increase of I_{ph} . V_{oc} is also improved slightly when I_{ph} increase, while no obvious change is observed in *FF* and the shape of I-V curve(Fig 7.a).

I_0 significantly influences the V_{oc} of the I-V curve, which is co-directionally moved following the variety of I_0. When I_0 increases, no obvious change is observed in I_{sc} , *FF* and the shape of I-V curve (Fig 7.b).

As shown in Fig 7.c, R_{sh} significantly influences *FF*, which is co-directionally moved following the variety of R_{sh}. The I-V curve between the I_{sc} point and the maximum power point becomes more declining when R_{sh} decreases, so that *FF* is diminished. No obvious change is observed in I_{sc} and V_{oc} when R_{sh} varies, which is different with the conclusion of some study[4].

As shown in Fig 7.d, R_s contrary-directionally changes the *FF* of the I-V curve. The I-V curve between the maximum power point and the V_{oc} point becomes more upright when R_s decreases, so that *FF* is enhanced. When R_s increases, no obvious change is observed in I_{sc} , V_{oc} of I-V curve, which is also different with the conclusion of some study[4] while is consistent with experiment result (Fig 3). As shown in Fig 7.e, n significantly influences the V_{oc} of the I-V curve, which is co-directionally moved following the variety of n. When n increases, no obvious change is observed in I_{sc}, *FF* and the shape of I-V curve.

3.3. Strategy for raising efficiency and experiments verification

On the base of the above atlas of I-V curves, some strategy on the energy conversion efficiency improvement of DSSCs can be deduced as following:

1) The improvement of *FF* could be obtained from the diminution of R_s and the aggrandizement of R_{sh} of the DSSCs equivalent circuit. Masik conjecturethed $R_{sh} \setminus R_s$

of change will lead to changes in open circuit voltage [4]. However, both experiments and calculation shows that, R_{sh} increase or decrease of R_s both the increase in fill factor, but not caused by short-circuit current, open circuit voltage of the significant changes. The charge-transfer process and impedance in DSSC has been studied in various parts respectively, such as porous TiO_2 electrode[23-25], counter electrode[26-28], electrolyte[27]. In addition to the internal resistance of conductive glass, resistance also related with transfer kinetics of electrolyte restrictions, such as the proliferation speed limit of redox in electrolyte[5].

2) The improvement of I_{sc} could be obtained from the aggrandizement of I_{ph} of the DSSCs equivalent circuit. For $I_{ph} = qAI_a$, q is electronic charge, A is coefficient of the optical absorption I_a. I_a is proportional to incident light intensity P_0. Therefore, the increase in optical absorption coefficient will lead to short-circuit current enhancement of DSSC. Moreover, studies have shown that in 0.1AM, I_{sc} changes linearly with light intensity [3, 29]. Increased sunlight absorption of DSSC is another way to increase I_{ph}. At present, the DSSC with highest reported efficiency which band gap is 1.8eV, has been basically absorbed sunlight below 800nm[30].

3) The improvement of V_{oc} could be obtained from the diminution of I_o and the aggrandizement of n of the DSSCs equivalent circuit. For $I_0 = qK_{et}C_{ox}{}^{m}n_o{}^{ua}$[31], I_0 depends on the electron back-transfer constant K_{et}, redox concentration C_{ox}, none-light semiconductor concentration n_0. Therefore, we think that to improve K_{et}, C_{ox}, n_0 will increase the DSSC's open circuit voltage. There have experiments show that components of electrolyte greatly impact on DSSC performance. For example, LiI or

t-BuPy in electrolyte led lower V_{oc} [5]. Lithium-ion which adsorbed on the TiO_2 surface led the surface of TiO_2 neutral so that the injected electron transfers inside TiO_2 more easily [23, 32, 33]. T-BuPy reduced the TiO_2 surface state so that suppressed the compound reactions of electronics with I_3^- [3].

In short, both experiments and our analysis shows that, by increasing the redox electrolyte C_{ox}, the increased I_0 will help improve the open circuit voltage of DSSC.

More direct benefit from Fig 7 is that, for the two different shapes of the I-V curves, we can directly distinguish in principle the differences of these DSSCs' equivalent circuit parameters which illuminate the internal physical mechanisms of them.

4. Conclusions

In summary, we derived the influences of all parameters in DSSC equivalent circuit to I-V curve and perspicuous illustrated. The strategy to improve DSSCs' efficiency according to the analysis on the trends to I-V curve is given. At last, the analysis is confirmed in experiments.

Acknowledgements

The authors would like to thank the National Natural Science Foundation of China (No 10874077) as well as the Jiangsu Provincial High Technology Research Program (Nos BG2006030 and BK2008252). The National Basic Research Program of China (No. 2007CB613301) is also gratefully acknowledged. We would like to thank the support by the Scientific Research Foundation of Graduate School of Nanjing University and Grant CX08B_009 from Jiangsu Province Innovation for Ph.D. candidate. Professor Zou and Yu would like to thank the Jiangsu Provincial Talent Scholars Program.

References

1. Kerr, R.A. and R.F. Service, *What Can Replace Cheap Oil--and When?* Science, 2005. **309**(5731): p. 101.
2. Oregan, B. and M. Gratzel, *A Low-Cost, High-Efficiency Solar-Cell Based on Dye-Sensitized Colloidal TiO_2 Films.* Nature, 1991. **353**(6346): p. 737-740.
3. Nazeeruddin, M.K., et al., *Conversion of light to electricity by cis-X_2bis(2,2'-bipyridyl-4,4'-dicarboxylate)ruthenium(II) charge-transfer sensitizers (X = Cl^-, Br^-, I^-, CN^-, and SCN^-) on nanocrystalline titanium dioxide electrodes.* Journal of the American Chemical Society, 1993. **115**(14): p. 6382-6390.
4. Murayama, M. and T. Mori, *Evaluation of treatment effects for high-performance dye-sensitized solar cells using equivalent circuit analysis.* Thin Solid Films, 2006. **509**(1-2): p. 123-126.
5. Hoshikawa, T., R. Kikuchi, and K. Eguchi, *Impedance analysis for dye-sensitized solar cells with a reference electrode.* Journal of Electroanalytical Chemistry, 2006. **588**(1): p. 59-67.
6. Kumara, G.R.A., et al., *Dye-sensitized solid-state solar cells made from magnesiumoxide-coated nanocrystalline titanium dioxide films: enhancement of the efficiency.* Journal of Photochemistry and Photobiology A: Chemistry, 2004. **164**(1-3): p. 183-185.
7. Virginia, Y., H. Seng-Tiong, and P.H.C. Robert, *Modeling and simulation for dye-sensitized solar cells.* Applied Physics Letters, 2008. **92**(14): p. 143506.
8. Tornow, J. and K. Schwarzburg, *Transient electrical response of dye-sensitized ZnO nanorod solar cells.* Journal of Physical Chemistry C, 2007. **111**(24): p. 8692-8698.
9. O'Regan, B. and M. Gratzel, *A low-cost, high-efficiency solar cell based on dye-sensitized colloidal TiO_2 films.* Nature, 1991. **353**(6346): p. 737-740.
10. Bach, U., et al., *Solid-state dye-sensitized mesoporous TiO_2 solar cells with high photon-to-electron conversion efficiencies.* Nature, 1998. **395**(6702): p. 583-585.
11. Gratzel, M., *Photoelectrochemical cells.* Nature, 2001. **414**(6861): p. 338-344.
12. Ito, S., et al., *High-efficiency organic-dye-sensitized solar cells controlled by nanocrystalline-TiO_2 electrode thickness.* Advanced Materials, 2006. **18**(9): p. 1202-1205.
13. Ito, S., et al., *Fabrication of thin film dye sensitized solar cells with solar to electric power conversion efficiency over 10%.* Thin Solid Films, 2008. **516**(14): p. 4613-4619.
14. Nogueira, A.F., C. Longo, and M.A. De Paoli, *Polymers in dye sensitized solar cells: overview and perspectives.* Coordination Chemistry Reviews, 2004. **248**(13-14): p. 1455-1468.
15. Mora-Sero, I., et al., *Implications of the negative capacitance observed at forward bias in nanocomposite and polycrystalline solar cells.* Nano Letters, 2006. **6**(4): p. 640-650.
16. Ruhle, S. and D. Cahen, *Electron tunneling at the TiO_2/substrate interface can determine dye-sensitized solar cell performance.* Journal of Physical Chemistry B, 2004. **108**(46): p. 17946-17951.
17. Asano, T., T. Kubo, and Y. Nishikitani, *Electrochemical properties of dye-sensitized solar cells fabricated with PVDF-type polymeric solid electrolytes.* Journal of Photochemistry and Photobiology A: Chemistry, 2004. **164**(1-3): p. 111-115.
18. Zhang Jing, et al., *Conductive carbon nanoparticles hybrid PEO/P(VDF-HFP)/SiO_2 nanocomposite polymer electrolyte type dye sensitized solar cells.* Solid State Ionics, 2007.

178(29-30): p. 1595-1601.

19. Koide, N., et al. *Improvement of efficiency of dye-sensitized solar cells based on analysis of equivalent circuit.* in *7th International Symposium on Photoreaction Control and Photofunctional Materials.* 2006. Tsukuba, JAPAN: Elsevier Science Sa.
20. Johnson, D.H., *Origins of the equivalent circuit concept: the current-source equivalent.* Proceedings of the Ieee, 2003. **91**(5): p. 817-821.
21. Koide, N. and L.Y. Han, *Measuring methods of cell performance of dye-sensitized solar cells.* Review of Scientific Instruments, 2004. **75**(9): p. 2828-2831.
22. Hanmin, T., et al., *An improved method to estimate the equivalent circuit parameters in DSSCs.* Solar Energy, 2009. **83**(5): p. 715-720.
23. Zaban, A., A. Meier, and B.A. Gregg, *Electric Potential Distribution and Short-Range Screening in Nanoporous TiO_2 Electrodes.* The Journal of Physical Chemistry B, 1997. **101**(40): p. 7985-7990.
24. Bisquert, J., et al., *Doubling Exponent Models for the Analysis of Porous Film Electrodes by Impedance. Relaxation of TiO_2 Nanoporous in Aqueous Solution.* The Journal of Physical Chemistry B, 2000. **104**(10): p. 2287-2298.
25. Bisquert, J. and V.S. Vikhrenko, *Interpretation of the Time Constants Measured by Kinetic Techniques in Nanostructured Semiconductor Electrodes and Dye-Sensitized Solar Cells.* The Journal of Physical Chemistry B, 2004. **108**(7): p. 2313-2322.
26. Papageorgiou, N., W.F. Maier, and M. Gratzel, *An iodine/triiodide reduction electrocatalyst for aqueous and organic media.* 1997.
27. Hauch, A. and A. Georg, *Diffusion in the electrolyte and charge-transfer reaction at the platinum electrode in dye-sensitized solar cells.* Electrochimica Acta, 2001. **46**(22): p. 3457-3466.
28. Fang, X., et al., *Effect of the thickness of the Pt film coated on a counter electrode on the performance of a dye-sensitized solar cell.* Journal of Electroanalytical Chemistry, 2004. **570**(2): p. 257-263.
29. Papageorgiou, N., et al., *The Performance and Stability of Ambient Temperature Molten Salts for Solar Cell Applications.* Journal of the Electrochemical Society, 1996. **143**(10): p. 3099-3108.
30. Sauve, G., et al., *High quantum yield sensitization of nanocrystalline titanium dioxide photoelectrodes with cis-dicyanobis(4,4 '-dicarboxy-2,2 '-bipyridine)osmium(II) ortris(4,4 '-dicarboxy-2,2 '-bipyridine)osmium(II) complexes.* Journal of Physical Chemistry B, 2000. **104**(15): p. 3488-3491.
31. Huang, S.Y., et al., *Charge recombination in dye-sensitized nanocrystalline TiO_2 solar cells.* Journal of Physical Chemistry B, 1997. **101**(14): p. 2576-2582.
32. Nakade, S., et al., *Effects of Lithium Ion Density on Electron Transport in Nanoporous TiO_2 Electrodes.* The Journal of Physical Chemistry B, 2001. **105**(38): p. 9150-9152.
33. Kambe, S., et al., *Influence of the Electrolytes on Electron Transport in Mesoporous TiO_2 Electrolyte Systems.* The Journal of Physical Chemistry B, 2002. **106**(11): p. 2967-2972.

Figure captions:

Fig.1 Measured I-V curve of a DSSC and the calculated current-voltage curve by equivalent circuit parameters

Fig.2 SEM of the TiO_2 film.

Fig.3 Measured I-V curves of cell and cell with series-connected resistor

Fig.4 The thicknesses of the sintered TiO_2 film of cells a to g.

Fig.5 Measured I-V curves of cells a and g which have different thicknesses

Fig.6 The equivalent circuits of DSSC.

Fig.7.a the calculated I-V curves obtained by stepwise changing the value of I_{ph} which revealed the influence of Iph to I-V curve.

Fig.7.b the calculated I-V curves obtained by stepwise changing the value of I_0 which revealed the influence of I_0 to I-V curve.

Fig.7.c the calculated I-V curves obtained by stepwise changing the value of Rsh which revealed the influence of R_{sh} to I-V curve.

Fig.7.d the calculated I-V curves obtained by stepwise changing the value of Rs which revealed the influence of Rs to I-V curve.

Fig.7.e the calculated I-V curves obtained by stepwise changing the value of n which revealed the influence of n to I-V curve.

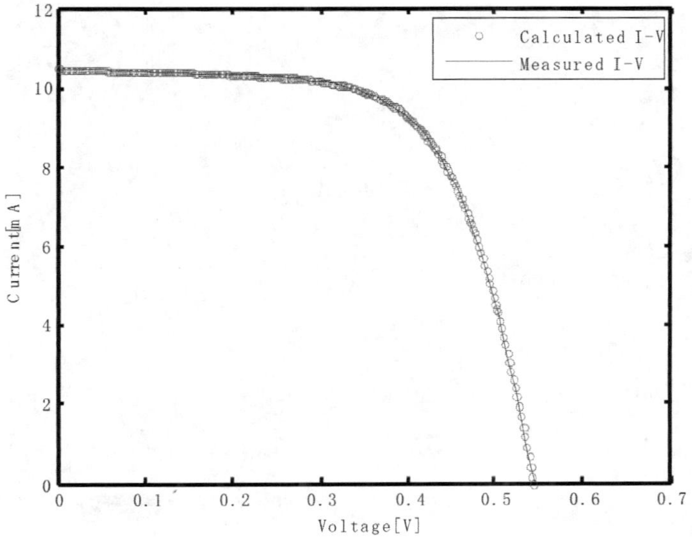

Fig 1 By Tian HM et al

Fig 2 By Tian HM et al

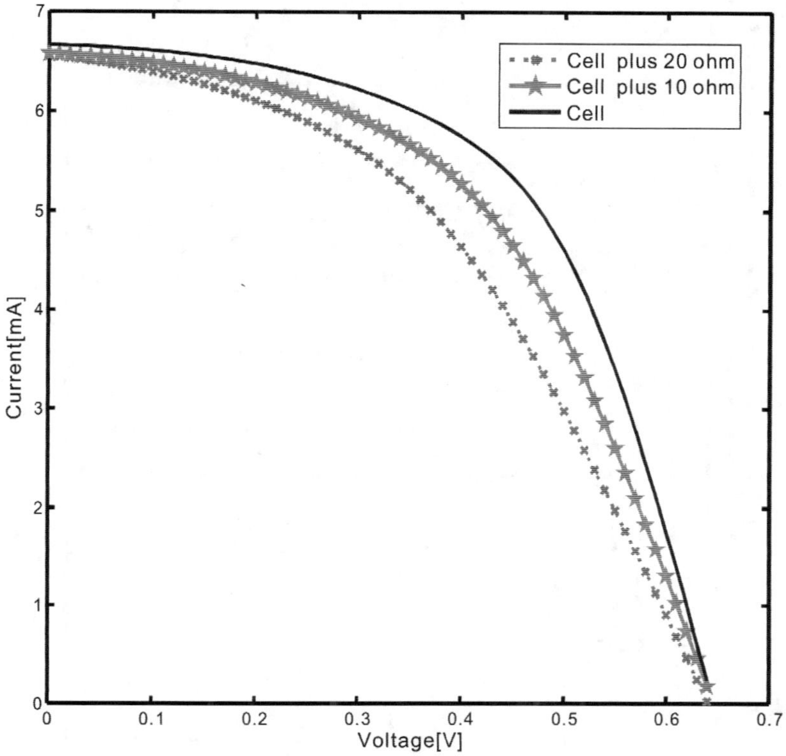

Fig 3 By Tian HM et al

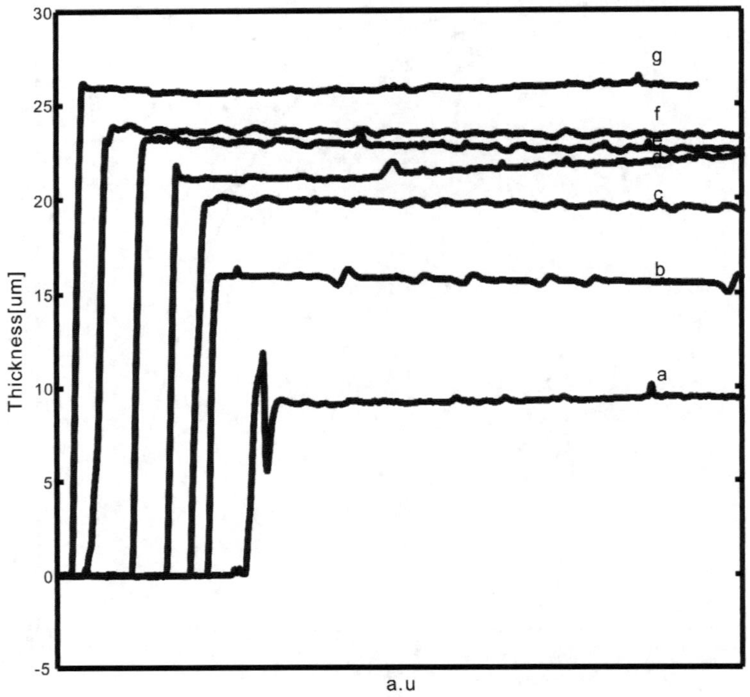

Fig 4 By Tian HM et al

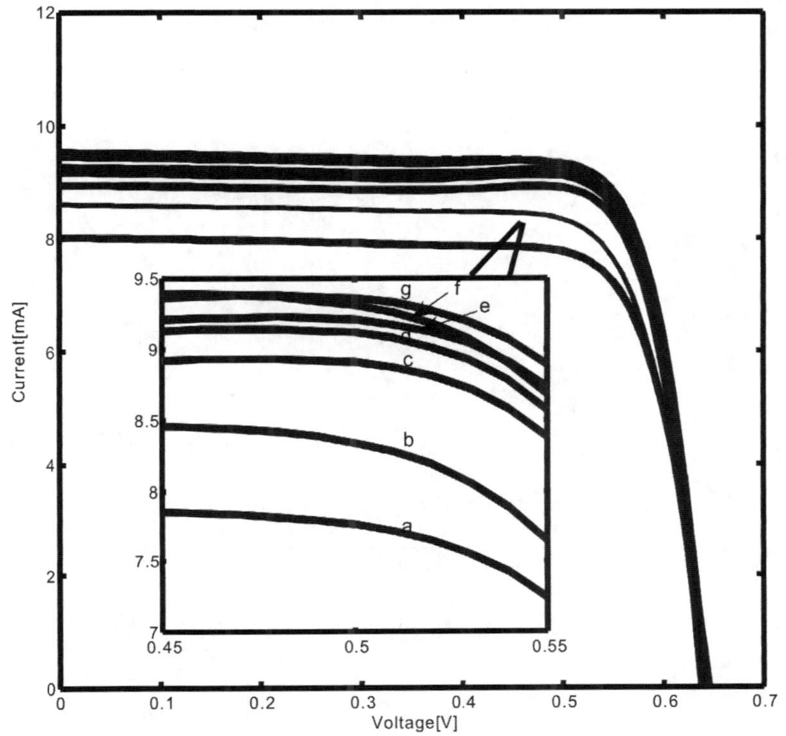

Fig 5 By Tian HM et al

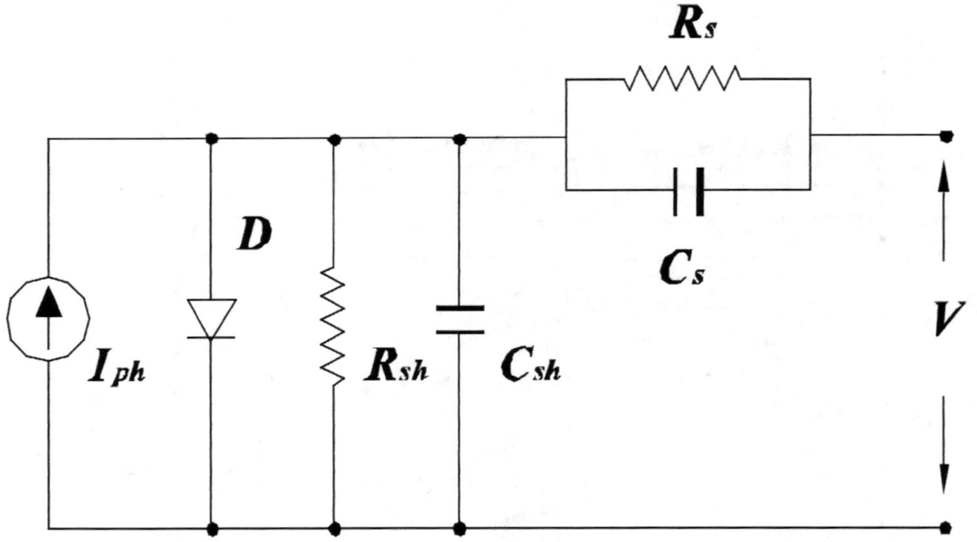

Fig 6 By Tian HM et al

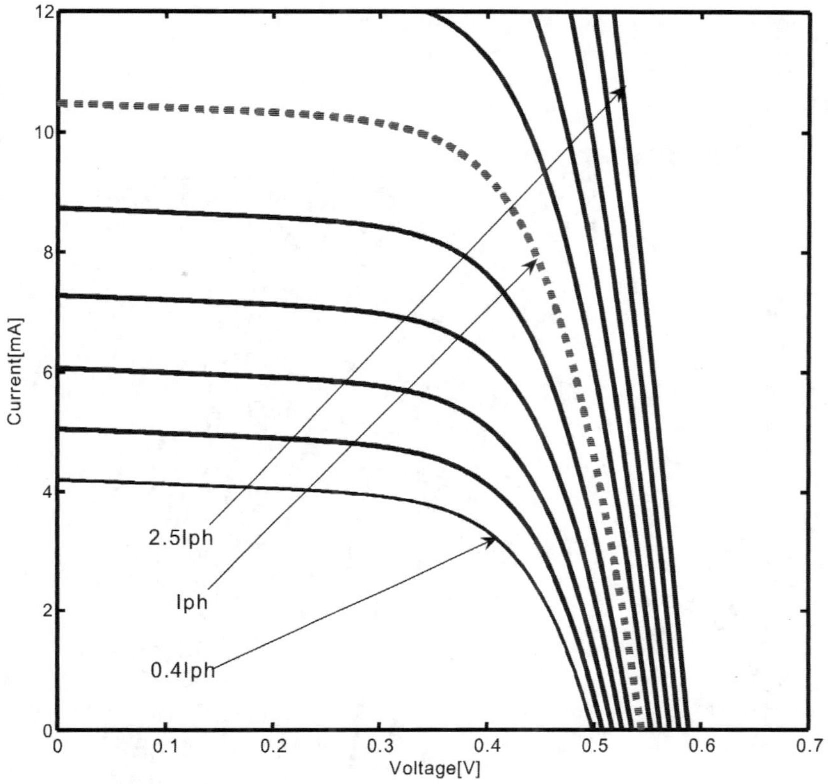

Fig 7 a By Tian HM et al

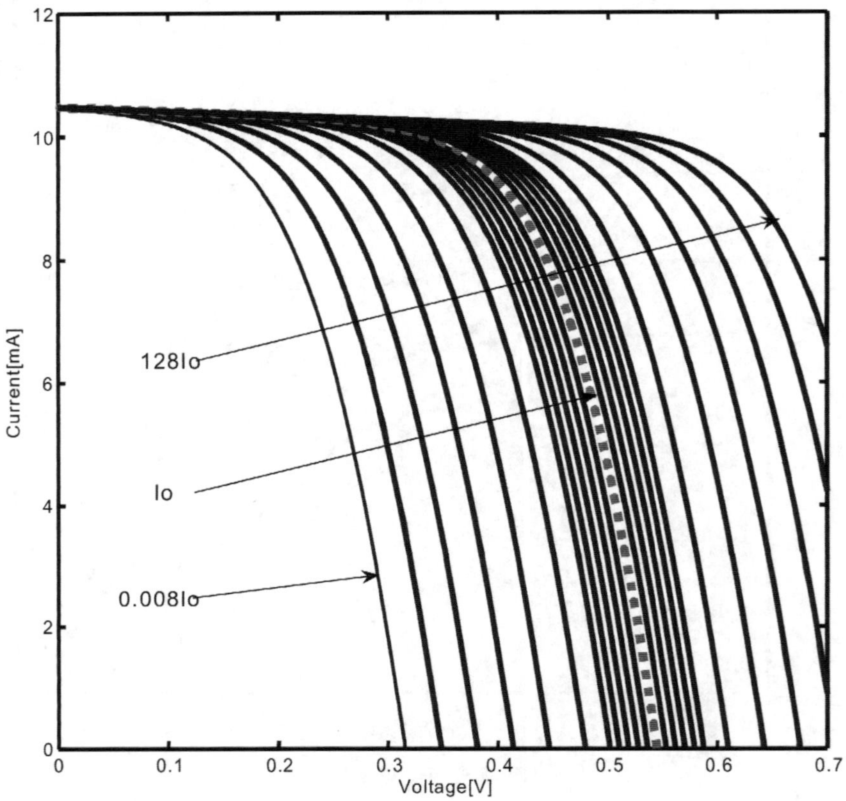

Fig 7 b By Tian HM et al

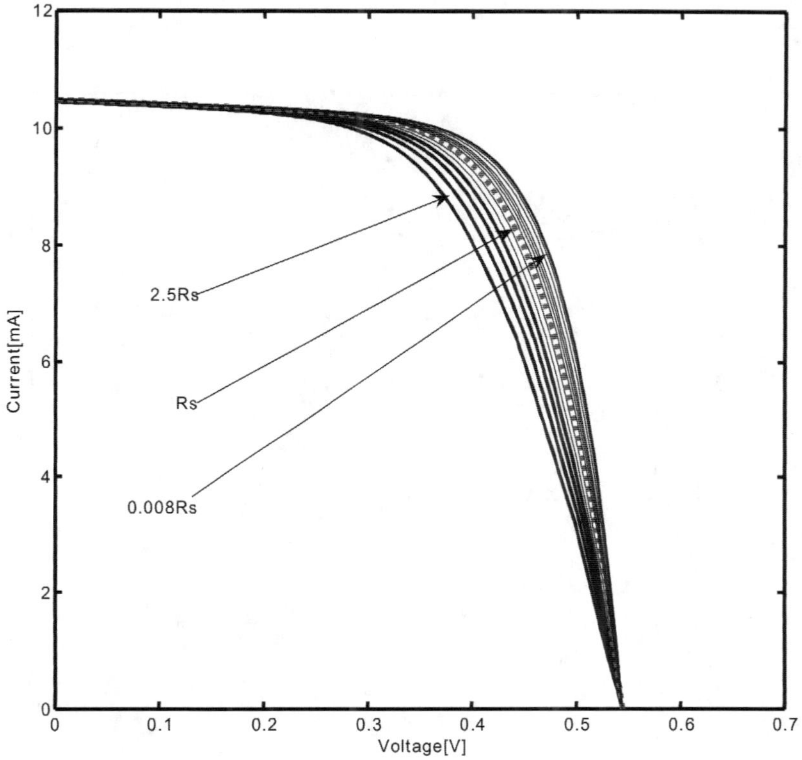

Fig 7c By Tian HM et al

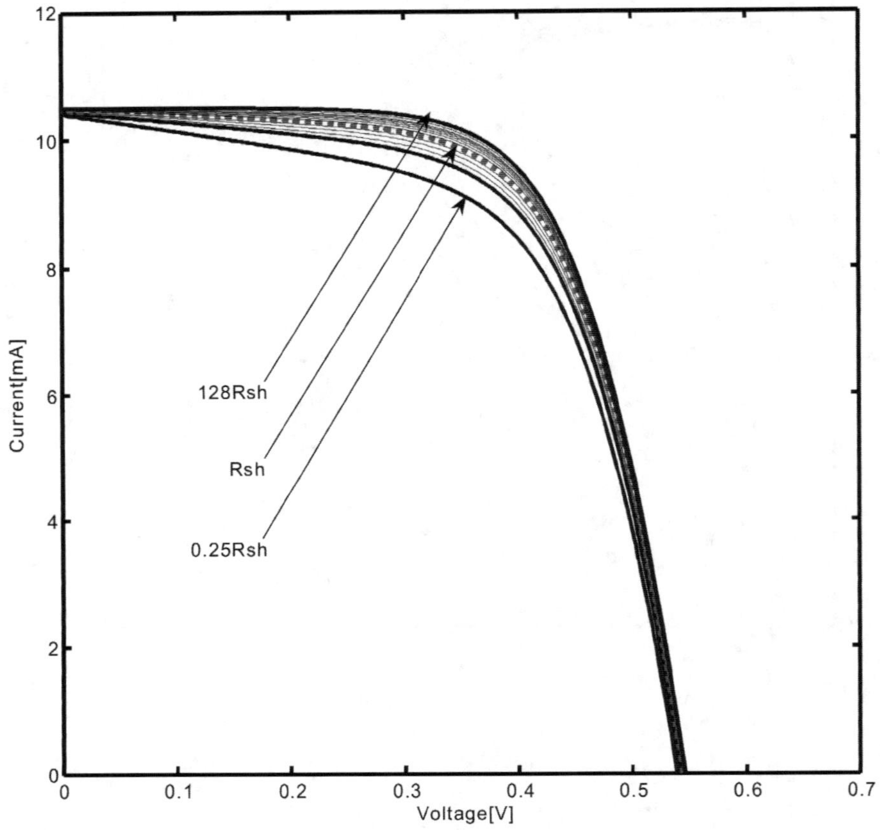

Fig 7d By Tian HM et al

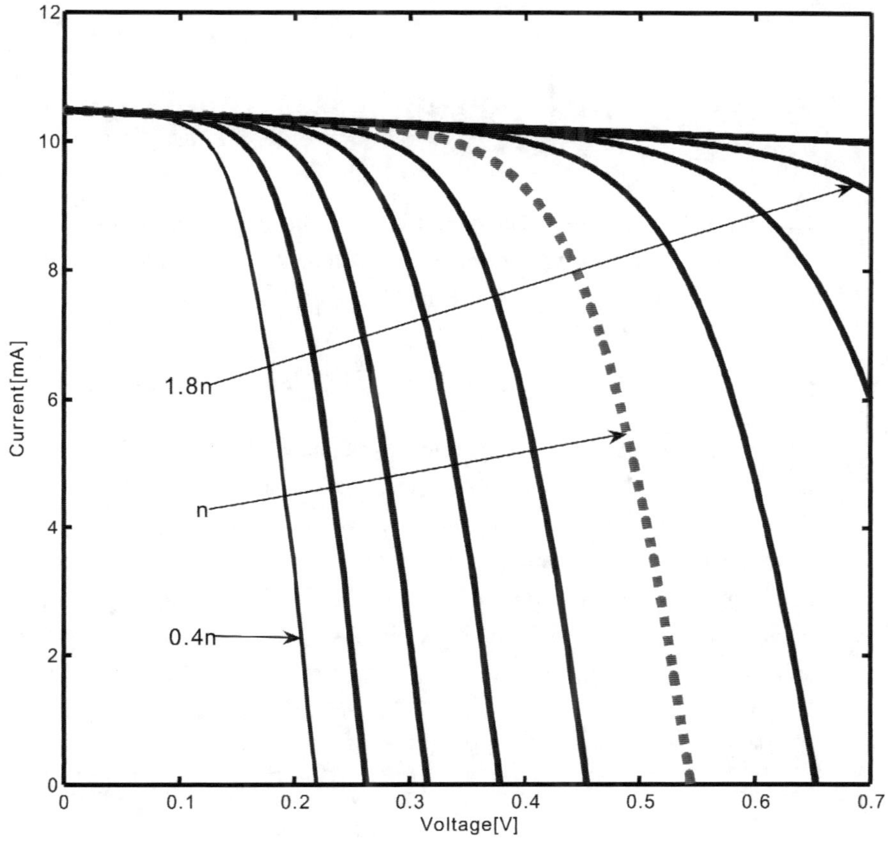

Fig 7e By Tian HM et al

Theory study of Ag-S codoping in ZnO

Qixin Wan[1]*, Lanli Chen[1], Guodong Liu[1], Dongmei Li[1], Zhihua Xiong[1]*
[1] Key Laboratory for Optoelectronics and Communication of Jiangxi Province,
Jiangxi Science & Technology Normal University, 330038, P. R. China

E-mail: wanqixin@yahoo.com.cn; xiong_zhihua@126.com

Abstract: All calculations, regarding S-monodoped and Ag-S codoped ZnO, have been performed by first-principles density-functional calculations. In S-monodoped ZnO, the formation energies show that S_{Zn} and S_O is easier to form in O-rich conditions while S dopant prefers to occupy the substitutional O site in Zn-rich conditions. In Ag-S codoped ZnO, the electrical structures show that $2Ag_{Zn}+S_O$ complex is better for *p*-type doping while $2Ag_{Zn}+S_{Zn}$ and $2Ag_{Zn}+S_i$ complexes are not suitable for *p*-doping. Moreover, $2Ag_{Zn}+S_O$ complex is easier to form than the other two complexes according to the formation energy. Therefore, in 2Ag-S codoped ZnO, the Zn-rich condition is helpful to gain *p*-type ZnO.

Keywords: ZnO, Ag, S, Density function theory, *p*-type

Journal of Physics: Conference Series **276** (2011) 012189 doi:10.1088/1742-6596/276/1/012189

Introduction

ZnO potentially offers many complementary and competitive advantages to GaN, which has a similar band gap and lattice constant. Wurtzite-ZnO, a wide-gap semiconductor, is considered as the most imaginable material substituting for GaN material, which is extensively used in optoelectronic application. However, possible applications for ZnO are currently limited by the difficulty of achieving *p*-type doping of this semiconductor, which as-grown is always n-type. There have been many methods to obtain the *p*-type ZnO by using different forms of dopant sources. In these methods, the Codoping method is popular to imply to achieve *p*-type ZnO. Yamamoto et al. have firstly proposed a new acceptor doping method, donor－acceptor codoping, to fabricate low-resistivity *p*-type ZnO [1]. Consequently, Joseph et al. recently reported that *p*-type ZnO films with low resistivities could be fabricated by pulsed laser deposition using Ga and N (Ga+N) codoping [2]. At the same time, Vlasenflin et al [3] observed that both N and P were used as the dual-acceptor doping to produce *p*-type conduction in ZnO by secondary ion mass spectroscopy. The theory has been studied by Tian et al [4]. Many theoretical and experimental workers both suggested that the codoping methods were significant way to overcome the difficulty to fabricate *p*-type ZnO. Simultaneity,

Previously, we reported on the Ag monodoping in ZnO by first principles methods [5], implied that single doping Ag is difficult to gain the *p*-type ZnO. Simultaneously, Sun [6] suggested that Ag-S codoping is promising method for the realization of *p*-type ZnO. In this paper, based on the first principles calculations, the study of S monodoped ZnO and Ag-S codoped ZnO was done.

Computation Method

The density functional theory was used to carry out calculations with the generalized gradient approximation (GGA) and the projector augmented wave (PAW) pseudopotentials, as implemented in the VASP codes [7]. 32 atoms were in the supercell which corresponded to a 2×2×2 supercell of ZnO. Moreover, until the force on each atom was below 0.01eV/Å, all atoms were allowed to relax. The calculations were found to be converged to be better than 0.002eV/atom with these parameters.
As we know, the formation energy was calculated to reflect the bond strength between the impurity and the host atom around the impurity. The formation energy E_f (D,q) was found for the complexes. The formation energy was calculated by assuming Ag atom reservoir was bulk Ag. μ_S was calculated by the expression $\mu_S = E_{tot}(ZnS) - \mu_{Zn}$, where $E_{tot}(ZnS)$ is the total energy of a two-atom unit of bulk ZnS. Thus the complex defect D formation energies were obtained from the following expression [8]:

$$E_f(D,q) = E_{tot}(D,q) - n_{Zn}\mu_{Zn} - n_O\mu_O - n_{Ag}\mu_{Ag} - n_S\mu_S$$

Where E_{tot} (D,q) is the total energy of a supercell containing the complex defect. Here n_i and μ_i are the number of species *i* (*i* =Zn, O, Ag, H) in the supercell and the corresponding reservoir chemical potential, respectively.

Moreover, the chemical potentials depend on the experimental growth conditions, which can be either zinc-rich (Zn-rich) or oxygen-rich (O-rich). For Zn-rich conditions, we assumed the ZnO is in equilibrium with Zn-metal, whereas for O-rich conditions ZnO is in equilibrium with oxygen-gas. For example, in Zn-rich conditions, μ_{Zn} is subject to an upper bound given by the energy of bulk Zn. Based on the expression $\mu_O = E_{tot}(ZnO) - \mu_{Zn}$, where $E_{tot}(ZnO)$ is the total energy of a two-atom unit of bulk ZnO, it results in a lower limit on μ_O. Similar, in O-rich conditions, the conditions place an upper limit on μ_O given by $\mu_{O[O2]}$ (the energy of O in an O_2 molecule). And the upper limit on μ_O results in a lower limit on μ_{Zn}: $\mu_{Zn} = E_{tot}(ZnO) - \mu_O$ [9].

Results and Discussion

1. S monodoped ZnO

Tab.1 Formation energies of Ag-monodoped or S-monodoped ZnO（unit in eV）

		Ag_{Zn}	Ag_O	Ag_i	S_{Zn}	S_O	S_i
E_f	Zn-rich	3.451	2.729	4.303	6.038	0.907	7.223
	O-rich	0.0139	6.167	4.303	-0.83772	0.907	3.7857

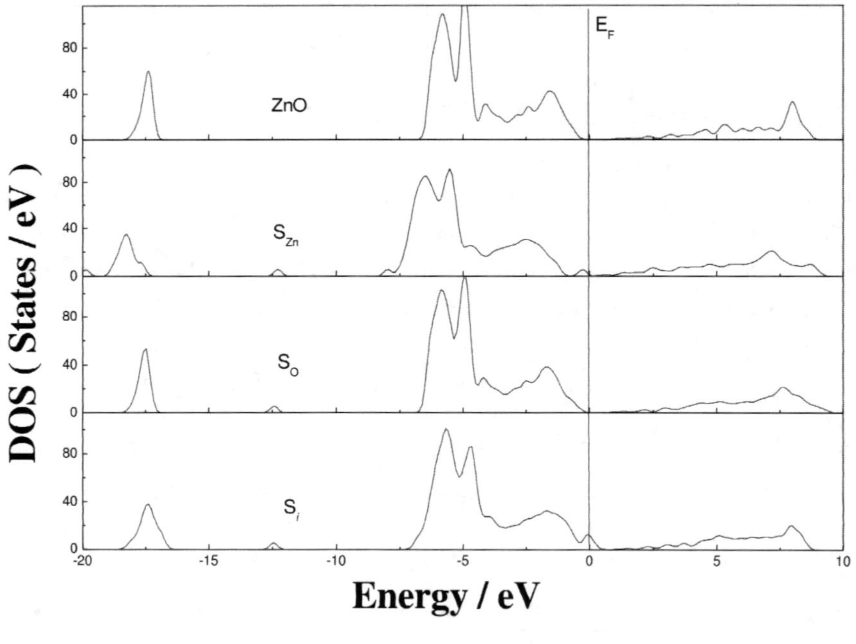

Fig.1 DOS of ZnO, S_{Zn}, S_O and S_i

The Ag monodoped ZnO and the S monodoped ZnO were calculated respectively. In Ag monodoped ZnO, there are three possible existence for Ag which are on the Zn site（Ag_{Zn}）, the O site（Ag_O）and the interstitial site（Ag_i）. Similarly, in S monodoped ZnO, there are also three possible existence for S which are on the Zn site（S_{Zn}）, the O site（S_O）and the interstitial site（S_i）. Here, we calculated the formation energies of Ag_{Zn}, Ag_O, Ag_i, S_{Zn}, S_O and S_i. Table 1 shows the calculated formation energies. In Zn-rich conditions, compared with the formation energy of the other defects, the formation energy of S_O is smaller. In O-rich conditions, Ag_{Zn}, S_{Zn} and S_O are smaller. Therefore, Ag_{Zn}, S_{Zn} and S_O is easier to form in O-rich conditions. Nominally, Ag, as a group IB element, which could act as an acceptor in ZnO, if incorporated on substitutional Zn sites [10]. Consequently, the electrical structure of S_{Zn} and S_O were calculated respectively. The density of states (DOS) for a crystal with S_{Zn} and S_O is shown in Fig.1. In the graph, Fermi level indicated by a solid line is set as zero which is the highest occupied state. It is showed that S_{Zn} results in unoccupied states near the top of the valence band. Simultaneously, S_i behaves as acceptor in S monodoped ZnO. However, the formation energy of S_i is too high as compared with the other defects in S-monodoped ZnO. By contrast, S_O does not evidently results in new occupied levels near the top of the valence band. DOS near the band gap of S_O is close to the states of undoped ZnO.

2. Ag-S codoped ZnO

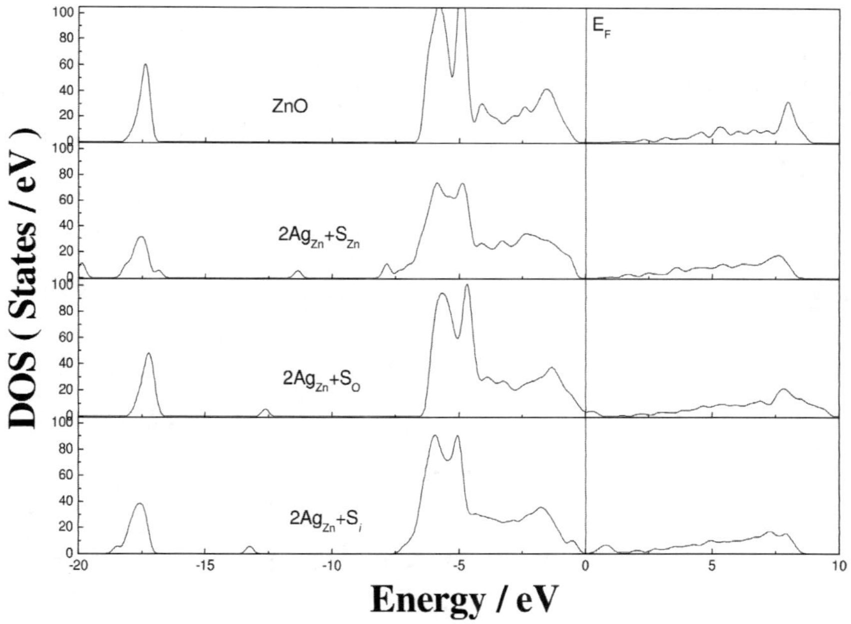

Fig.2 DOS of complexes in 2Ag-S codoped ZnO

In Ag-S codoped ZnO, two Ag and one S was doped in ZnO. In the monodoped ZnO, Ag_{Zn}, S_{Zn} and S_O is easier to form in O-rich conditions. Accordingly, two Ag both incorporate on substitutional Zn sites while one S incorporate on substitutional Zn sites ($2Ag_{Zn}+S_{Zn}$) or O sites ($2Ag_{Zn}+S_O$) in Ag-S codoped ZnO. At the same time, one S on interstitial site was calculated too ($2Ag_{Zn}+S_i$). Tab.2 shows the formation energy of three complexes in 2Ag-S codoped ZnO. In Zn-rich conditions, the formation energy of $2Ag_{Zn}$+ S_O is smaller. The Zn-rich condition is helpful to gain $2Ag_{Zn}$+ S_O. On the contrary, In O-rich conditions, the formation energy of $2Ag_{Zn}$+ S_{Zn} is smaller than that of the other two complexes. It is illuminated that $2Ag_{Zn}$+ S_{Zn} is easy to form in O-rich conditions.

Tab.2 Formation energies of complexes in 2Ag-S codoped ZnO（**unit in eV**）

		$2Ag_{Zn}$+ S_{Zn}	$2Ag_{Zn}$+ S_O	$2Ag_{Zn}$+ S_i
E_f	Zn-rich	7.8967	6.053	10.092
	O-rich	-5.8545	-0.823	-0.222

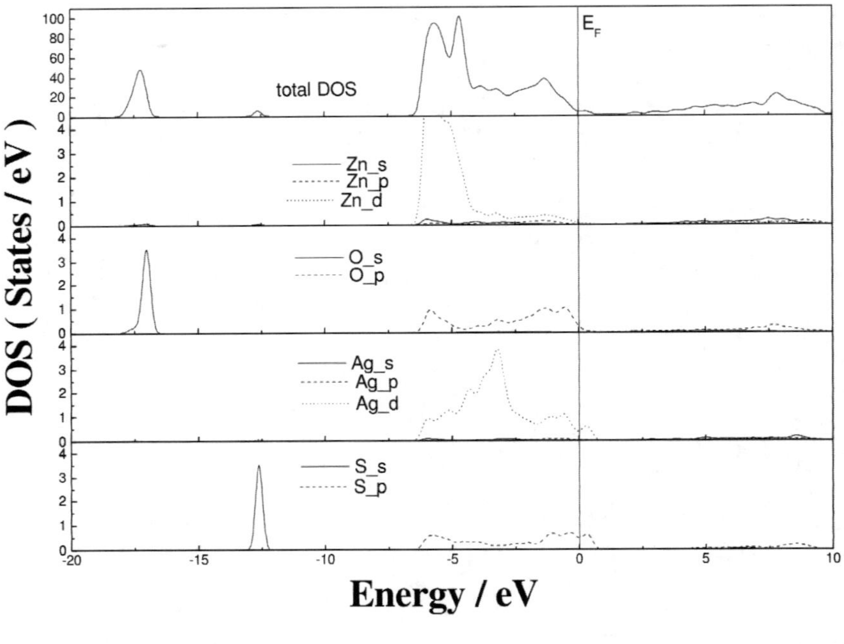

Fig.3 PDOS of ZnO ($2Ag_{Zn}$ + S_O)

The density of states (DOS) for a crystal with the three complexes is shown in Fig.2. In Fig.2 the DOS were ZnO, ZnO with $2Ag_{Zn}$+S_{Zn}, $2Ag_{Zn}$+S_O and $2Ag_{Zn}$+S_i respectively. It is showed that $2Ag_{Zn}$+S_O results in unoccupied states near the top of the valence band while Femi level is in the band gap of DOS of $2Ag_{Zn}$+S_{Zn} and $2Ag_{Zn}$+S_i. It is illuminated that $2Ag_{Zn}$+S_O results in acceptor in Ag-S codoped ZnO. Therefore, in this Ag-S codoped ZnO, $2Ag_{Zn}$+S_O is helpful to fabricate *p*-type ZnO. Accordingly, the partial density of states (PDOS) of ZnO with $2Ag_{Zn}$+S_O complex is plotted in Fig.3. From Fig. 3 we found that the lower valence band (-7.37~ -4.20eV) consists of a mixture of Zn (3d), Ag (4d) and O (2p) states. The upper valence band (-4.20~-0.61eV) is composed mainly of O (2p), Ag

(5s) and S (3p) states with a small contribution from Zn (3d) states. The conduction band is mainly a mixture of Zn (4s), O (2p), Ag (5s) and S (3p) states. It is showed that the contribution to the acceptor level is mainly from Ag (4d) states, O (2p) states and S (2p) states. It is therefore illuminated that the hybridization between Ag (4d) states, (2p) states and S (2p) states which are near the top of the valence band leads to the acceptor level. As a result, in the three complexes, $2Ag_{Zn}$+S_O complex is better for *p*-type doping while $2Ag_{Zn}$+S_{Zn} and $2Ag_{Zn}$+S_i complexes are not suitable for *p*-doping.

Conclusion

In summary, all calculations, regarding S-monodoped ZnO and Ag-S codoped ZnO, have been performed. In S-monodoped ZnO, S_{Zn} results in unoccupied states near the top of the valence band. Simultaneously, S_i behaves as acceptor in S monodoped ZnO. By contrast, S_O does not evidently results in new occupied levels near the top of the valence band. DOS near the band gap of S_O is close to the states of undoped ZnO. However, the formation energies show that S_{Zn} and S_O is easier to form in O-rich conditions while S dopant prefers to occupy the substitutional O site in Zn-rich conditions. In Ag-S codoped ZnO, there are two Ag and one S was used to dope in ZnO. Two Ag both incorporate on substitutional Zn sites while S incorporates on substitutional Zn sites ($2Ag_{Zn}$+S_{Zn}), O sites ($2Ag_{Zn}$+S_O) or interstitial sites ($2Ag_{Zn}$+S_i). The electrical structures show that $2Ag_{Zn}$+S_O complex is better for *p*-type doping while $2Ag_{Zn}$+S_{Zn} and $2Ag_{Zn}$+S_i complexes are not suitable for *p*-doping. Moreover, $2Ag_{Zn}$+S_O complex is easier to form than the other two complexes according to the formation energies in Tab.2. Therefore, in this Ag-S codoped ZnO, the Zn-rich condition is helpful to gain *p*-type ZnO.

Acknowledgments

This work was supported by National Natural Science Foundation of China (No. 51062003), Natural Science Foundation of Jiangxi Province (No. 2008GZW0014) and Technology Project of Jiangxi Provincial Education Office (No. GJJ10245).

References

[1] T. Yamamoto, H. Yoshida, Jpn. J. Appl. Phys. 38 (1999) L166.
[2] M. Joseph, H. Tabata, T. Kawai, *Jpn. J. Appl. Phys.* 38(1999) L1205.
[3] T. H. Vlasenflin, M. Tanaka, *Solid State Communications* 142, 292-294 (2007)
[4] R. Y. Tian, Y. J. Zhao, *Journal of Applied Physics* 106**,** 043707 (2009)
[5] Q. X. Wan, Z. H. Xiong, J. N. Dai, J. P. Rao, and F. Y. Jiang, *Opt. Mater.* 30, 817-821 (2008)
[6] L.J.Sun, J.Hu, H.Y.He, et.al. *Solid State Communications*, 149, 1663-1665(2009)
[7] G. Kresse and J. Furthmüller, *Comput. Mater. Sci.* **6** (1996) 15.
[8] W. J. Lee, J. Kang, K. J. Chang, *Phys. Rev. B.* **73** (2006) 024117.
[9] A. Janotti, C. G. Van de Walle, *J. Cryst. Growth.* **287** (2006) 58.
[10] C. F. Klingshirn, Semiconductor Optics (Springer, Berlin, 2006).

Properties and Application of a Novel Unsymmetrical Photochromic Diarylethene

Liqin Wang[a,b] Shouzhi Pu[a,*], Gang Liu[a], Congbin Fan[a], Shiqiang Cui[a]

[a]Jiangxi Key Lab of Organic Chemistry,Jiangxi Science & Technology Normal University 330013, P.R. China

[b]College of Environmental & Chemical Engineering, Applied Chemistry, Nanchang Hangkong university 330034, P.R. China.

pushouzhi@tsinghua.org.cn

Abstract. A new unsymmetrical photochromic diarylethene, 1-(2-methyl-5-phenyl-3-thienyl)-2-(2-n-butyl-5-(1,3-dioxolane)-3-thienyl)perfluoroncyclopentene (1a), was synthesized and its optoelectronic properties, such as photochromism in solution as well as in poly-methylmethacrylate (PMMA) amorphous films and fluorescences were investigated. The photochromic reaction kinetics was also studied. Using diarylethene 1b/PMMA film as recording medium and a linearly polarized 633 nm laser diode for recording and readout, polarization multiplexed image recording can be carried out in this film, which illustrated that the diarylethene can be potentially used as holographic optical recording medium.

1. Introduction

Photochromic diarylethenes exhibit reversible variations in their physicochemical properties when stimulated by light and are amongst the most promising candidates for photoelectronic applications such memories and switches because of their thermao-irreversible and fatigue-resistant photoisomerization performances[1]. This photoinduced change is not limited to the color or absorption spectra, as the change occurs between two different molecules with different physical and chemical properties[2,3]. Derivatives of diarylethenes, especially those with thiophene rings, are currently being investigated as candidates for switches in molecular electronic devices, or for other applications, because of their photochromic properties[4]. The important features of this class of molecules are the good fatigue resistance, fairly high photocyclizationbquantum yields, high reversibility of the forward and reversed photo processes triggered by irradiation with light of different wavelengths and thermal stability of the products[5]. The majority of the research work has been devoted to the development of new molecules and investigative studies into their fundamental properties, and the results obtained have contributed to a broad understanding of the effects of substituents on the photochromism of diarylethenes[6]. Up to date, design and synthesis of new Photochromic compounds which is an active area of research[7]. In this paper, we synthesized a new diarylethene 1-(2-methyl-5-phenylthiophene-3-yl)-2-(2-nbutyl-5-(1,3-dioxolane) thiophene-3-yl) perfluoroncyclopentene. Its photochromic reactivity and fluorescence were investigated in detail.

Journal of Physics: Conference Series **276** (2011) 012190 doi:10.1088/1742-6596/276/1/012190

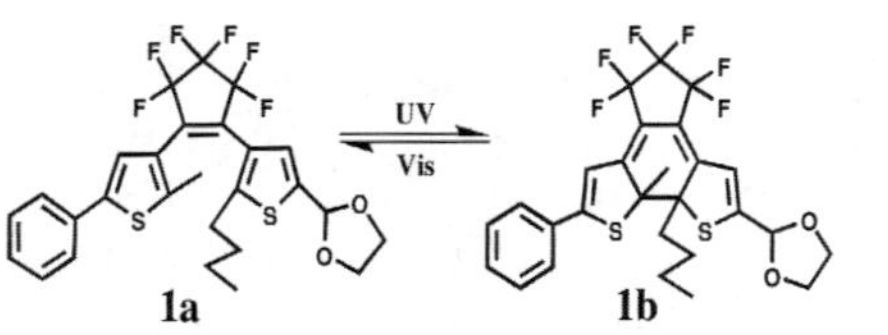

Scheme 1. Photochromism of diarylethene 1a

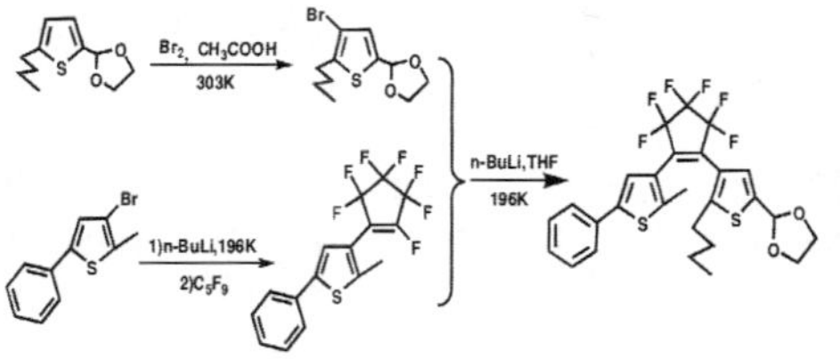

Scheme 2. Reagents and conditions.

2. Experiments

2.1. Synthsis of diarylethene 1a

The synthetic route for diarylethene 1a was shown in Scheme 2 and its synthetic steps based on the similar procedure of Pu et al[8,9]. It optoelectronic properties, such as UV–Vis absorption spectra, fluorescence spectra were investigated in detail. Diarylethene 1a was characterized by ^{1}H NMR spectroscopy. NMR spectra were recorded on Bruker AV400 (400MHz) spectrometer with $CDCl_3$ as the solvent. ^{1}H NMR (400MHz, $CDCl_3$): 1.14(d, 3H, $-CH_3$), 1.31 (s, 2H, $-CH_2$),1.53 (s, 2H,$-CH_2$), 2.48(d, 3H, $-CH_3$),2.68 (s, 2H, $-CH_2$), 4.01 (s, 4H, $-CH_2$-CH_2-),6.89 (d, 1H, Ar-H),7.25((d, 1H, Ar-H),7.32 (d, 1H, Ar-H),7.39 (s, 2H, Ar-H), 7.53(s, 2H, Ar-H).

2.2. Preparation of PMMA film

First of all, one hundred milligram of PMMA was dissolved in 1 ml chloroform, which was stirred ultrasonically to make it homogeneous. Samples were obtained by dissolving ultrasonically 10 mg of compounds 1 with 1 ml of the PMMA solution, respectively, and the solution was spin-coated on a glass substrate with a spin rotation speed of 1500 rpm.

3. Results and Discussion

3.1. Photochromic in solution and in PMMA film

Figure 1a shows absorption spectral changes of 1 in hexane solution (2×10^{-5} mol · L^{-1}). The hexane solution of the open-ring isomer 1a was colorless without absorption in the visible region. The absorption maximum of its colorless open-ring isomer 1a was observed at 285nm ($= 1.43 \times 10^4$ mol · L^{-1} · cm^{-1}). Upon irradiation with UV light (= 297 nm), the colorless solution turned violet. Isomer 1b has an absorption maximum at 522 nm ($= 4.45 \times 103$ mol · L^{-1} · cm^{-1}) in the visible region. Upon irradiation with visible light (> 500 nm), the violet solution of 1b was completely bleached, and the absorption spectrum returned to that of 1a[10]. The quantum yields of cyclization and cycloreversion reaction of diarylethene 1 were 0.19 and 0.000086 at room temperature. Similarly, the color of dithienylethene 1a in PMMA film (10% w/w) changed from colorless to violet with the appearance of a new broad absorption band at $_{max}$ = 568 nm upon irradiation 297 nm light. The colored PMMA film can invert to colorless, upon irradiation appropriate visible light (> 500 nm)[11].

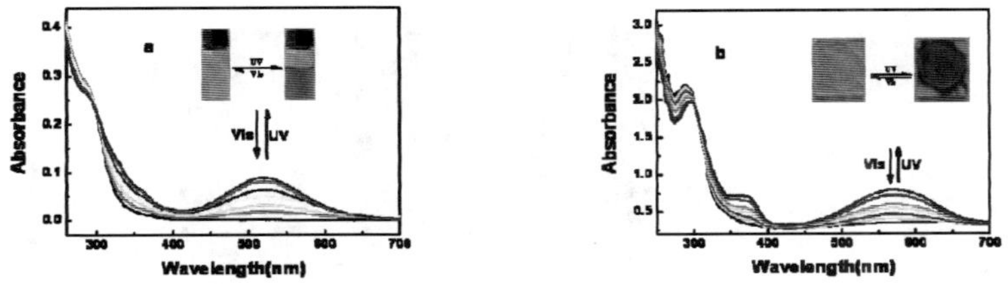

Figure 1. Absorption spectral changes of compound 1a :(a) in hexane and (b) in PMMA.

3.2. Photochromic reaction kinetics in hexane solution

As shown as in Figure 2, the photochromic cyclization/cycloreversion kinetics of diarylethene 1a in hexane solution was determined by UV-Vis spectra upon alternating irradiation with UV and appropriate wavelength visible light at room temperature. It can be seen that the relationships between the absorbance and exposal time have good linearity upon irradiation with 297 nm UV light suggesting that the cyclization processes of 1a belong to the zeroth order reaction when open-ring isomer changed to closed-ring isomer. The slope of every line in Figure 2a represents the reaction rate constant (k) of diarylethene 1a in hexane solution. So all k of cyclization process (k_{o-c}, 10^{-3}) of diarylethene 1a can be easily obtained, which is 2.01 s^{-1} in hexane solution. As shown in Figure 2b, during the cycloreversion of 1b, the relationship between –Log(Abs) and exposal time also behave perfect linearity, indicating that the cycloreversion process belong to the first order reaction. The k of cycloreversion (k_{c-o}, 10^{-3}) process of 1b can be easily obtained, which is 16.69 s^{-1} in hexane solution.

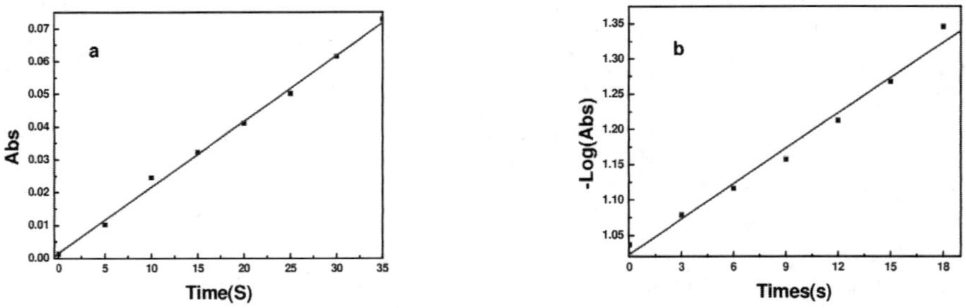

Figure 2. The cyclization/cycloreversion kinetics of compound 1a in hexane solution

3.3. Fluorescence of diarylethene

The fluorescence spectra of diarylethenes 1a in hexane at room temperature is illustrated in Figure 3. We could clearly see that the fluorescent emissions of 1a was at 356 nm when excited at 297 nm. When it arrived at photostationary state, the emission intensities of diarylethenes 1a was quenched to ca 51%. The fluorescence was efficiently quenched when the diarylethene unit converts from the open- to closed-ring form. The fluorescence intensity gradually decreased in proportion to the conversion from 1a to 1b[12].The back irradiation with visible light of wavelength longer than 450 nm regenerated the open-ring forms and recovered the original emission spectra. Otherwise, the average times of "on" and "off" state shortened in proportion to the reciprocal power of radiated light by changing the power of the UV and visible light, indicating that the switching effect is perfect in hexane solution.

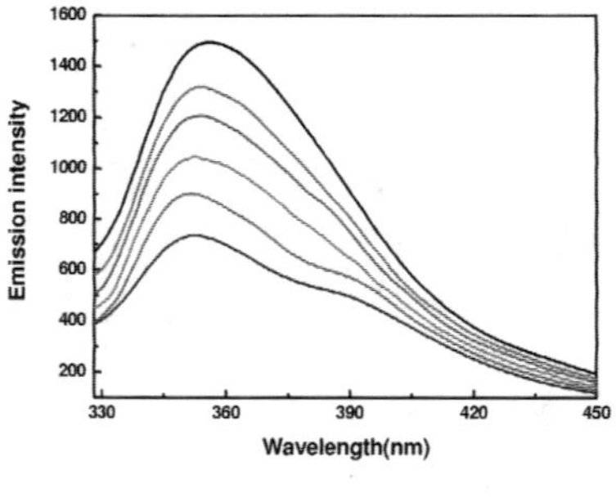

Figure3. Fluorescence spectral of 1a in hexane

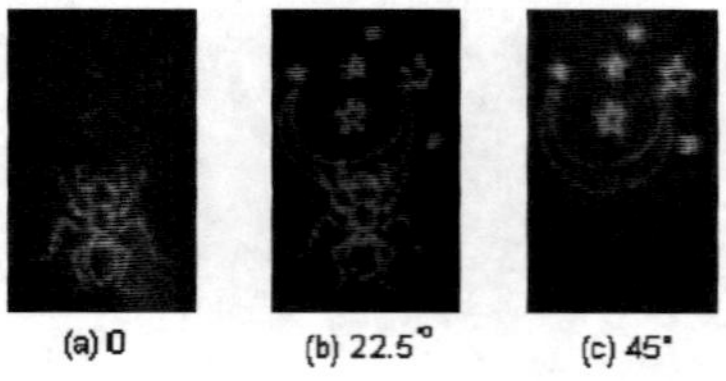

Figure4. Readout patterns from the film

3.4. Photoinduced anisotropy and polarization multiplexed image recording

Witn a linearly polarized 633 nm laser diode in the experimental setup[13,14], the evaluation of potential of photochromic diarylethene as a polarization holographic medium by recording and reading the holograms in a real-time operation is investigated. Using diarylethene 1b/PMMA film as recording medium. Two patterns are written on and readout successfully. To observe and measure the photoinduced anisotropy, diarylethene 1b/PMMA film is placed between two orthogonal polarizes P and A. Two patterns are written on 1b/PMMA film by a linearly polarized 633 nm laser diode. Readout patterns with the three types of polarization holographic optical recording are shown in Figure 5. The first pattern is written with the beam polarization at 0 deg (Figure 5a), while the second was written with the beam polarization at 45 deg (Figure 5c). After writing, the patterns are readout with the sample placed between two orthogonal polarizers P and A. If the polarization of P is 22.5 deg from the horizontal direction, both the vertical polarization light and horizontal polarization light patterns can pass through the A-diarylethene-P system. Thus, the readout pattern is like Figure 5b. The results demonstrate that the diarylethene compound has attractive properties for polarization holographic optical recording.

In conclusion, diarylethene 1a changed the color from colorless to violet upon irradiation with 297 nm UV light, in which absorption maxima were observed at 522 nm in hexane and at 568nm inPMMA. The cyclization processes of 1a belong to the zeroth order reaction and the cycloreversion process belong to the first order reaction. The k of cyclization process (k_{o-c}, 10^{-3})/cycloreversion (k_{c-o}, 10^{-3}) process of 1 can be easily obtained, which are 2.01 s^{-1}/16.67 s^{-1} in hexane solution. Diarylethene 1a showed relatively strong fluorescence intensity in hexane solution. Polarization multiplexed images was performed successfully in diarylethene 1b/PMMA film, the results demonstrated that the compound was very sensitive responding to 633 nm recording laser, and the recorded holographic images have good contrast, which illustrated that the diarylethene can be potentially used as holographic optical recording medium.

Acknowledgment

This work was supported by the Project of National Natural Science Foundation of Jiangxi Province (2008GZH0020), the Key Scientific Project from Education Ministry of China (208069), the Project of Jiangxi Youth Scientist, and the Science Funds of the Education Office of Jiangxi, China (GJJ08365, GJJ09306).

References

[1] Liu H H and Chen Y 2009 *J. Mater. Chem.* **19** 706
[2] Yokoyama Y, Shiozawa T, Tani Y and Ubukata T 2009 *Angew. Chem.* **121** 4591
[3] Zhang Z, Liu X, Li Z Y, Chen Z H, Zhao F Q, Zhang F S and Tung C H 2008 *Adv. Funct. Mater.* **18** 302
[4] Sauer P and Allen R E 2007 *Chemical Physics Letters* **434** 260
[5] Kobatake S and Yamashita I 2008 *Tetrahedron* **64** 7611
[6] Pu S Z, Fan C B, Miao W J and Liu G 2009 *Dyes and Pigments* **84** 25
[7] Pu S Z, Sang T, Li G Z, Xu J K and Chen B 2005 *Tetrahedron* **61** 6623
[8] Fan C B, Pu S Z, Liu G and Yang T S 2008 *Journal of Photochemistry and Photobiology A: Chemistry* **19** 333.
[9] Pu S Z, Li M, Liu G and Le Z G 2009 *Aust. J. Chem.* **62** 1
[10] Morimoto M and Irie M 2006 *Chem. Eur. J.* **12** 4275
[11] Pu S Z, Fan C B, Miao W J and Liu G 2008 *Tetrahedron* **64** 9464
[12] Pu S Z, Li W J and Miao W J 2009 *J. Phys. Org. Chem.* **22** 954
[13] Chen Y, Wang C M, Fan M G, Yao B L and Menke N M 2004 *Opt. Mater.* **26** 75
[14] Yao B L, Wang Y L, Lei M, Menke N M and Chen G F, 2005 *Opt. Express* **13** 20

Synthesis, kinetics and application for optical recording of a new asymmetrical photochromic diarylethene

Renjie Wang, Hongying Xia, Shouzhi Pu *, Gang Liu, Shiqiang Cui

Jiangxi Key Laboratory of Organic Chemistry, Jiangxi Science and Technology Normal University, Nanchang 330013, P. R. China

pushouzhi@tsinghua.org.cn

Abstract. A novel photochromic diarylethene was synthesized and its photochromic, fluorescence and kinetics properties were investigated detail. This compound exhibited remakable photochromism, changing from colorless to blue after irradiation with UV light both in solution and in PMMA amorphous film. In hexane, the kinetic experiments showed that the cyclization /cycloreversion process of the compound was determined to be the zeroth/first reaction, respectively. Finally, The results demonstrated that the diarylethene compound had attractive properties for high density optical storage.

1. Introduction

Organic photochromic molecules have raised considerable interest because of their potential applications to optical memory and molecular switches [1]. Among the various photochromic molecular systems, diarylethene with derivatives bearing thiophene or benzothiophene ring systems are regarded as the best candidates for applications because of their good thermal stability of both forms, remarkable fatigue resistance, rapid response and high reactivity in the solid state [2-5]. Recently, many investgations have been carried out on photochromic diarylethens in various optical memory media [6], such as switch devices [7], and photomechanical actuators [8, 9], etc. In order to research the diarylethene, which have the excellent properties, a larger number of molecules have been designed and synthesized. In particular, diarylethenes with thiophene ring systems bearing terminal phenyl groups have been of considerable interest, because the hydrogen atoms attached to the terminal phenyl groups can be substituted by many other electron-donating groups or electro-withdrawing groups, which inevitably influence the optoelectronic properties of corresponding diarylethenes [10]. In this paper, we have designed and synthesized a new unsymmetrical photochromic diarylethene compound 1-[2-methyl-5-(3-methylphenyl)-3-thienyl]-2-[2-methyl-5-(2,4-difluorophenyl)-3-thienyl]perfluorocyclopentene (1a), and its photochromic reactivity, fluorescence, kinetics and application for holographic optical recording medium was investigated in detail. The photochromism of diarylethene 1a is shown in Scheme 1.

2. Experiments

The target diarylethenes were synthesized by the method as shown in Scheme 2. Similar detailed synthetic method was described in references [11]. First, 3-bromo-2-methyl-5-methylphenyl-thiophene (3) and 3-bromo-2-methyl-5-(2,4-difluorophenyl)thiophene (5) was prepared by reacting 3-bromo-2-methyl-5-thenylboronic acid (1) with 1-bromo-3-methylbenzene and 1-bromo-2,4-difluorobenzene in the presence of tetrakis(triphenylphosphine)palladium(0) and Na_2CO_3 in THF for 16 h at 378K. Under an argon gas atmosphere, compound 3 was dissolved in THF and n-butyl lithium hexane was added at 195K. Stirring was continued for 30 min at this low temperature; octafluorocyclopentene was added and the mixture was stirred for 2 h at this temperature. After extracting with diethyl ether and evaporation in vacuum, the residue was purified by column chromatography on silica gel (petroleum ether) to obtain 6. Compound 1a was prepared by reacting compound 5 in dry THF at 195K, then through purified by column chromatography on silica gel (petroleum ether), and 1a was obtained. The structure of compound diarylethene 1a was characterized by ^{1}H NMR spectroscopy. ^{1}H NMR (400 MHz, $CDCl_3$): ^{1}H NMR (400 MHz, $CDCl_3$, ppm): δ 2.28 (s, 3H, $-CH_3$), δ 2.32 (s, 3H, $-CH_3$), δ 2.47 (s, 3H, $-CH_3$), δ 6.65 (d, 1H, thiophene-H), 6.72 (d, 1H, thiophene-H), δ 6.75(d, 1H, benzene − H), δ 7.06(s, 1H, benzene − H), δ 7.10(m, 2H, benzene − H), δ 7.19(d, 2H, benzene − H), δ 7.3(s, 1H, benzene − H).

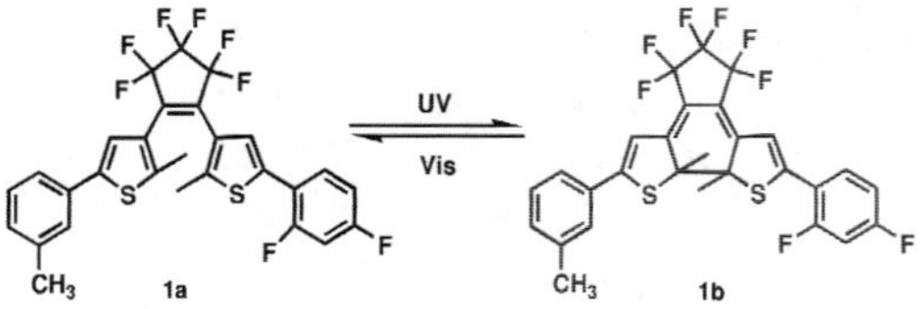

Scheme . Photochromism of diarylethene 1a.

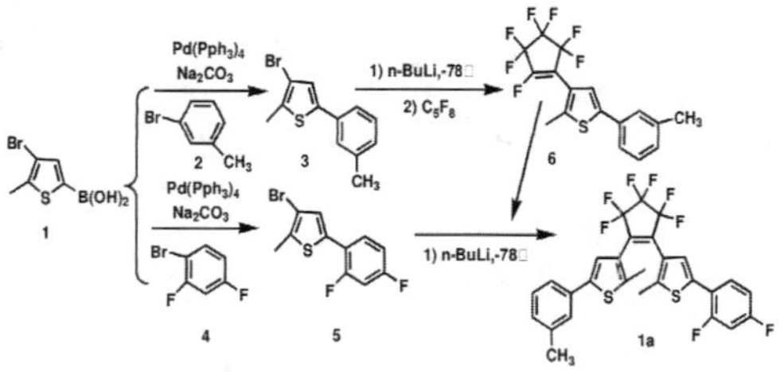

Scheme 2. Synthetic route for the compound 1a.

3. Results and Discussions

3.1. Photochromism of diarylethene 1a

The photochromic behaviors of diarylethene1a in hexane (2.0×10^{-5}mol/L) and in PMMA film (10% w/w) induced by photo-irradiation at room temperature are shown in Figure 1A and 1B. From Figure 1A, the colorless diarylethene 1a show the sharp absorption peak at 284 nm in hexane solution. The colorless solution 1a turned blue and a new absorption band centered at 575 nm emerged while the original peak at 284 nm decreased when irradiation with 313 nm UV light. The results indicated that the closed-ring isomer 1b was formed. Upon irradiation with visible light of wavelength longer than 500 nm, the blue color turned to colorless and the maximum absorption returned to the initial state, which demonstrated that the closed-ring 1b was returned to open-ring 1a, and a clear isosbestic point was observed at 303 nm.

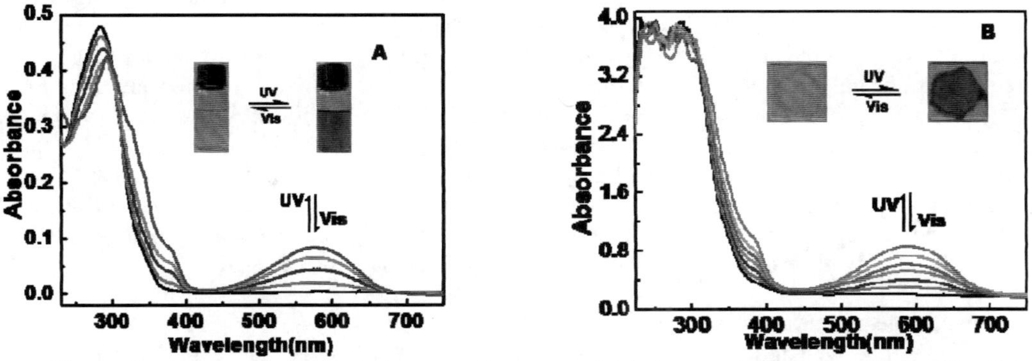

Figure 1.. Absorption spectral changes of compound 1 in hexane (A) and in PMMA film (B)

In PMMA amorphous film, diarylethene 1a has also shown good photochromism. Upon irradiation with 313 nm UV light, the colorless PMMA film changed to blue, and a new absorption band appeared at 587 nm, which was assigned to the formation of the closed-ring isomer 1b, the colored PMMA film can invert to colorless on irradiation of appropriate wavelength visible light (>500 nm). The red shift of the maximum absorption of 1b in PMMA film in comparison with those in hexane solution could be mainly ascribed to the molecular arrangement in solid state is more stable than in solution [12]. The quantum yields of cyclization and cycloreversion reaction of diarylethene 1 in hexane solution were 0.27 and 0.20 at room temperature, respectively. In addition, it was also found that both open-ring isomer and closed-ring isomer of diarylethene 1 were stable at room temperature in darkness.

3.2. Photochromic reaction kinetics in hexane solution

The photochromic cyclization/cycloreversion kinetics of 1a in hexane were determined by UV-Vis spectra upon alternating irradiation with UV and appropriate wavelength visible light at room temperature. The cyclization and cycloreversion curves of 1a were shown in Figure 3. It can be seen that the relationships between the absorbance and exposal time have good linearity upon irradiation with 298 nm UV light suggesting that the cyclization processes of 1a belong to the zeroth order reaction when open-ring isomer changed to closed-ring isomer. The slope of every line in Figure 3(A) and 3(B) represents the reaction rate constant (k) of diarylethene 1a in hexane. So all k of cyclization/cycloreversion process ($k_{o\text{-}c}$/ $k_{c\text{-}o}$, 10^{-3}) of diarylethene 1a can be easily obtained, which are 1.07 s^{-1} and 7.02 s^{-1} in solution, respectively. As shown in Figure 3(B), during the cycloreversion of 1b, the relationship between –log(Abs) and exposal time also behave perfect linearity, indicating that the cycloreversion process belong to the first order reaction.

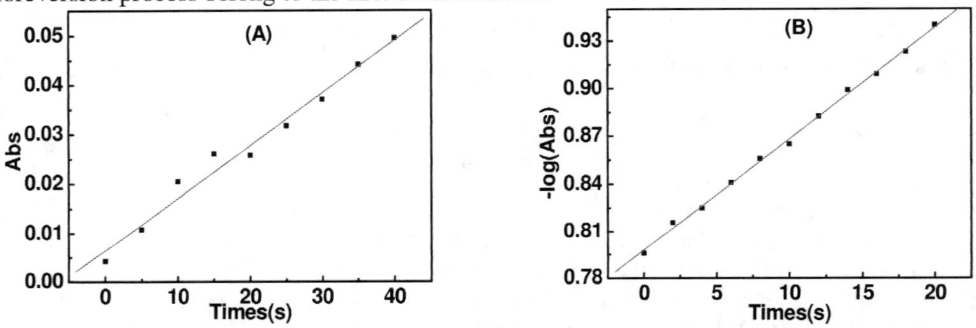

Figure 3. The cyclization/cycloreversion kinetics of compounds 1a in hexane

3.3. Fluorescence of diarylethene

The fluorescence properties in the hexane solution of the compound were measured using a Hitachi F-4500 spectrophotometer, and the breadths of excitation and emission slit were selected 10.0 nm. From Figure 4, it could be clearly seen that the solution of 1a exhibited relatively strong fluorescence at 388nm when excited at 328 nm. In order to investigate the relationship between fluorescence and concentration, we measured the fluorescence in different concentration of hexane solution at room temperature. As shown in Figure 4, when the concentration of diarylethene 1a in hexane increased from 2×10^{-6} mol/L to 1×10^{-4} mol/L the emission intensity is significantly increased, and the fluorescence peak gave a minor red shift from 366 nm to 388 nm when irradiated by light of 328 nm. However, when the concentration increased 2×10^{-4} mol/L the fluorescence intensity decreased, and the fluorescence appeared self-quench when the concentration at 1×10^{-3} mol/L. This self-quench phenomenon maybe resulted from the formation of excimers and or exciplexes in high concentration solutions so that deactivation effect can occur during the excited-state lifetime [13].

3.4. Optical storage properties

With a He-Ne laser (wavelength: 650nm, 105mW) in the experimental setup. The experimental system for recording was described and shown in [14]. Using diarylethene 1b as the recording medium, polarization recording was performed. The images recorded on diarylethene 1b/PMMA film is shown in Figure 5. The size of recording marks was about 20μm, the interval distance between two spots was 80μm. As shown in Figure 5, the quality of images recorded in 1b/PMMA film showed a clear contrast and without cross talk, which are important in rewritable data storage. The recorded images can be erased completely along with irradiated with UV light. The result demonstrated that the photochromic diarylethene can be used in high capacity optical storage.

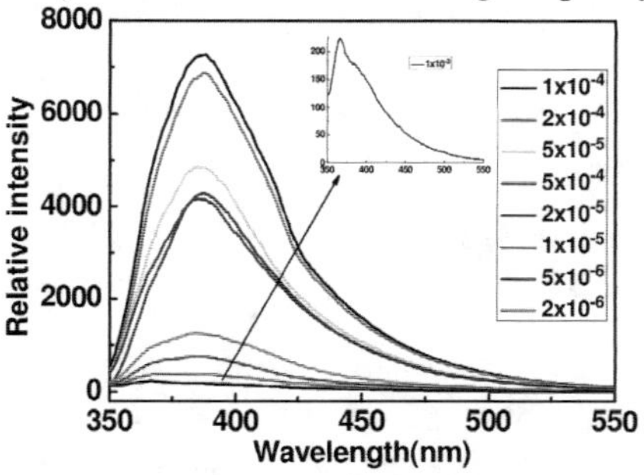

Figure 4. Fluorescence spectra of diarylethene 1a in various concentrations in hexane excited at 328 nm.

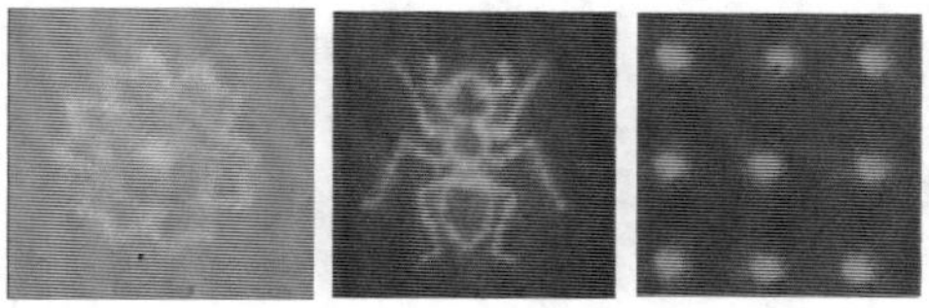

Figure 5. Rewritable optical images on diarylethene 1b/PMMA film.

In conclusion, a new unsymmetrical photochromic diarylethene was synthesized and its photochromic, fluorescence and kinetics properties were investigated. The results indicated that the compound exhibited good reversible photochromic behaveior both in solution and in PMMA film. In hexane solution, The fluorescence spectra of diarylethene 1a depended on the concentration, when the concentration at 1×10^{-3} mol/L the fluorescence self-quench was observed. The kinetics properties indicated that the cyclization/cycloreversion processes belong to zeroth/first order reaction in hexane. Rewritable holographic recording experiments were carried out on 1b/PMMA film successfully. The results demonstrated that the diarylethene compound had attractive properties for polarization holographic optical recording.

Acknowledgment

This wok was supported by the Project of National Natural Science Foundation of Jiangxi Province (2008GZH0020), The Key Scientific Project from Education Ministry of China (208069), the Project of Jiangxi Youth Scientist, and the Youth Science Funds of the Education Office of Jiangxi, China (GJJ09572, GJJ09567).

References

[1] J C Crano, R J Guglielmetti (Eds.), Organic Photochromic and ThermochromicCompounds 1999 *Main Photochromic Families*. **1** 315
[2] Irie M. 2000 *Chem. Rev.* **100** 1685
[3] Tian H, Yang S J. 2004 *Chem. Soc. Rev.* **33** 85
[4] Chen Y, Xie N J. 2005 *Mater. Chem.* **15** 3229
[5] Morimoto M, Irie M. 2005 *Chem. Commun.* 3895
[6] Masahiro Irie. High-Density Optical Memory and Ultrafine Photofabrication. 2002 *Springer, Berlin.* 137
[7] Masahiro Irie, Tuyoshi Fukaminato, Takatoshi Sasaki, Naoto Tamai, Tsuyoshi Kawai. 2002 *Nature.* **420** 759
[8] J N Yao, K Hashimoto, A Fujishima. 1992 *Nature.* **355** 624
[9] Alvaro Fernandez-Acebes, Jean-Marie Lehn. 1999 *Adv. Mater.* **11** 910
[10] Pu S Z, Yang T S, Xu J K, Shen L, Li G Z, Xiao Q, Chen B. 2005 *Tetrahedron* **61** 6623.
[11] Liu G, Zheng T, Pu S Z, Xu J K. 2006 *Acta Cryst.* **E62** o2877
[12] Kazuo Kasatani, Shingo Kambe, Masahiro Irie, 1999 *J. Photoch. Photobiol. A. Chem.* **122** 11
[13] T Fukaminato, T Kawai, S Kobatake, M Irie. 2003 *J. Phys. Chem. B.* **107** 8372
[14] Pu S Z, Yang T S, Yao B L, Wang Y, Lei M, and Xu J K. 2007 *Mater. Lett. Papers.* **61** 855

Synthesis of a new Diarylethene for Optical Recording

Taofeng Wang, Shouzhi Pu*, Shiqiang Cui, Gang Liu

Jiangxi Key Laboratory of Organic Chemistry, Jiangxi Science and Technology Normal University, Nanchang 330013, P. R. China

pushouzhi@tsinghua.org.cn

Abstract. A novel photochromic diarylethene bearing five and six-membered aryl units 1-[(2-methyl-5-(3-pyridinyl)-4-thienyl)]-2-[(2-methoxylphenyl)]perfluorocyclopentene was synthesized and its photochromic, fluorescent electrochemical propertieswere investigated in detail. using diarylethene (1a) as recording medium, polarization holographic optical recording was carried out successfully.

1. Introduction

With the increasing requirements of the huge information storage, higher density and higher data transfer rate become the trends of optical data storage. Current commercial recordable and rewritable optical storage systems are heat-mode recording systems, which limits the size of recorded pits and writing speed. During the past decades, there have been important achievements in the synthesis of new families of organic photochromic molecules [1,2]. Among the various photochromic molecular systems, diarylethene espec-ially diarylethenes with heterocyclic aryl rings, is one of the most promising organic photochromic molecules because of its high fatigue resistance.These molecules can play a key role in materials for photonic switches, memories, sensors, or actuators, and this wide range of applications involve large-size devices (e.g., sunglasses) as well as miniaturized ones (e.g., memory media) [3]. In order to find out diarylethenes with excellent characters, a lot of molecules were designed and synthesized [4-6]. Up to date, design and synthesis of new photo-chromic compounds is an very active area of research, and many publications concerning synthesis and investigation of the properties of diarylethenes with the heterocyclic aryl rings have been reported. However, there are few publications that contribute to holographic recording using diarylethenes as recording media, except for several papers reported by us [7, 9].

The photochromic reaction of diarylethene 1a is shown in Scheme 1. In this paper, we synthesized a photochromic diraylethene 1-[(2-methyl-5-(4-pyridinyl)-3-thienyl)]-2-[(2-methoxylphenyl)] perfluorocyclopentene. Its recording media properties was also investigated in detail.

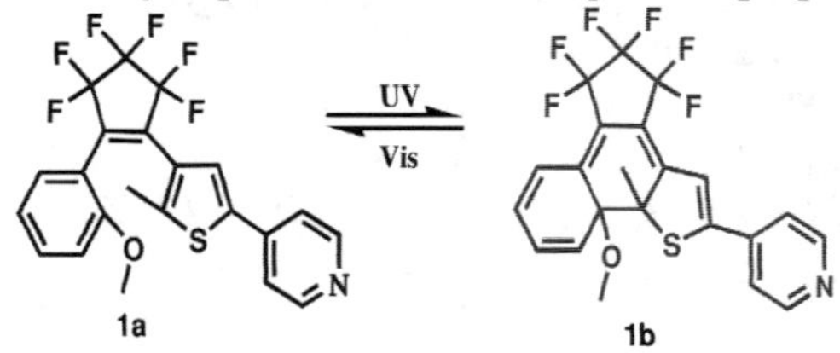

Scheme 1. Photochromism of diarylethene 1a.

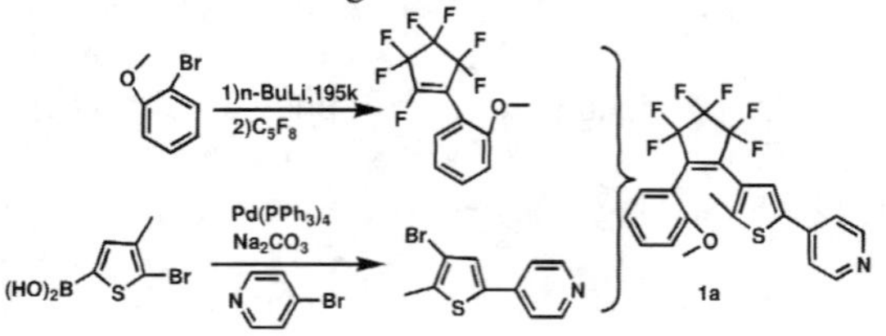

Scheme 2. Synthetic route for the compound 1a

Journal of Physics: Conference Series **276** (2011) 012192 doi:10.1088/1742-6596/276/1/012192

2. Experimental

The synthetic route for diarylethene 1a is shown in Scheme 2. It was synthesized according to the similar procedure of [10]. NMR spectra were recorded on Bruker AV400 (400 MHz) spectrometer with $CDCl_3$ as the solvent and tetramethylsilane as an internal standard. The absorption spectra were measured using Agilent 8453 UV/Vis spectrometer. Photoirradiation was carried out using SHG-200 UV lamp, CX-21 ultraviolet fluorescence analysis cabinet, and BMH-250 Visible lamp. Light of appropriate wavelengths was isolated by different light filters. The fluorescent property was measured using a Hitachi F-4500 spectrophotometer. Electrochemical examinations were performed in a one-compartment cell with the use of a Model 263 potentiostat–galvanostat (EG&G Princeton Applied Research), The structure of compound 1a was confirmed by 1H NMR. NMR spectra were recorded on Bruker AV400 (400 MHz) spectrometer with $CDCl_3$ as the solvent. 1H NMR (400 MHz, $CDCl_3$): 1.96(s, 3H,-CH_3), 3.51(s, 3H, -OCH_3), 6.84(d, 1H, J=8.0), 7.00(t, 1H), 7.38(t, 3H, J=7.8), 7.52(s, 1H), 7.21(d, 1H), 8.58(d, 2H).

3. Result ant discussion

3.1. Photochromism of diarylethene 1a

The photochromic properties of diarylethene 1a irradiation by photoirradiation at room temperature were measured both in hexane and in PMMA amorphous film. As shown in Figure 1A, compound 1a exhibited an absorption peak at 297 nm (= 1.7×10^4 L mol^{-1} cm^{-1}) in hexane, which was arisen from the • * transition [11] Upon irradiation with 313 nm light, a new visible absorption band at 584 nm(= 3.51×10^3 L mol^{-1} cm^{-1}) appeared while the original peak at 297 nm decreased, indicating the new formation 1b produced. Correspondingly, the colorless solution of 1a turned blue. The blue colored solution turned colorless by irradiation with visible light (> 500 nm), indicating that 1b returned to the initial state 1a.

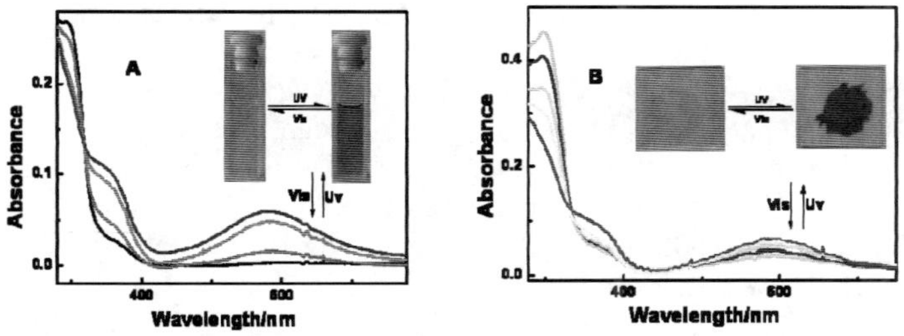

Figure 1. Absorption spectral changes of compound 1a in hexane (A) (2.0×10^{-5}mol/L) and in PMMA film (B) (10%, w/w)

For practical applications in optical devices, it is very important that photochromic materials can keep good photochromism in a polymer film, such as the PMMA film [12-14]. In PMMA amorphous film, diarylethene 1a also showed good photochromism (Figure 1B) as similar to that in solution. In PMMA amorphous film, diarylethene 1a also showed good photochromism (Figure 1B) as similar to that in solution. The colorless diarylethene 1a/PMMA film turned blue for which the maximal absorption were observed at 585 nm, as the closed-ring isomer 1b was generated, Upon irradiation with 313 nm light. The colored diarylethene/PMMA films can return to colorless upon irradiation with visible light (> 500 nm). Figure 1 show the color changes of diarylethene upon photoirradiation in hexane and in PMMA respectively. The quantum yields of cyclization and cycloreversion reaction of diarylethene in hexane solution were 0.36 and 0.07 at room temperature, respectively.

3.2. fluorescence of diarylethene 1a

The fluorescence intensity of diarylethene 1a decreased along with the photochromism from open-ring isomer to closed-ring isomer upon irradiation with 265 nm light in hexane(5.0×10^{-5}mol/L). As shown in Figure 2, the photocyclization reaction was carried out and the fluorescent closed-ring form of the compound was produced. The irradiation by appropriate wavelength visible light regenerated the open-ring forms of diarylethene and return the original emission spectra. The phenomena are useful for application as the fluorescence switches [15].

The concentration dependence on the florescence spectrum of diarylethenes 1a was measured in hexane at room temperature, as shown in Figure 3.

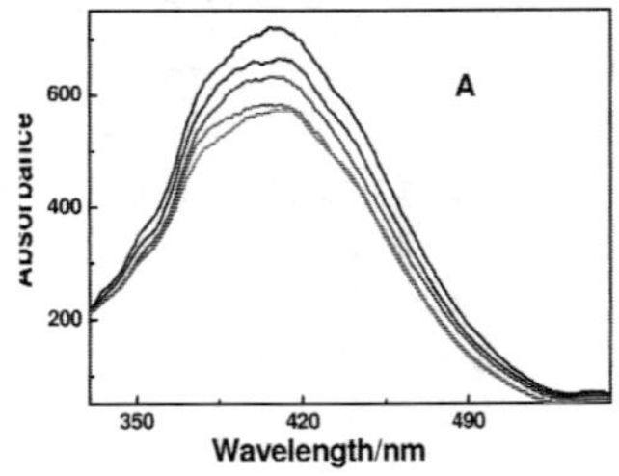

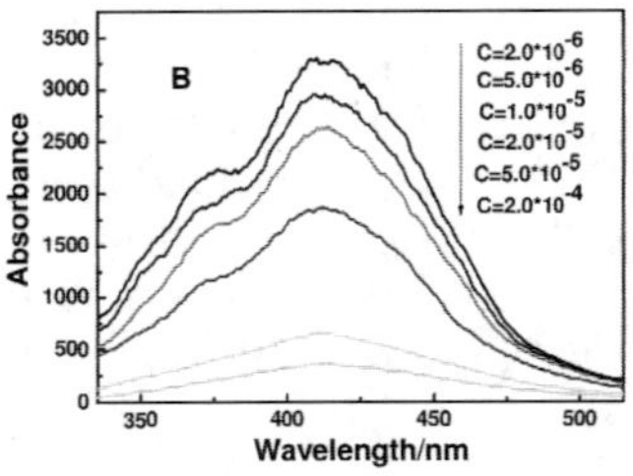

Figure 2. Fluorescence spectral changes (A) of spectra(B) of diarylethene1a in hexane (5.0×10^{-5}mol/L).

Figure 3. Fluorescence emission 1a in various concentrations in hexane.

When the solutions at 2.0×10^{-6}, 5.0×10^{-6}, 1.0×10^{-5}, 2.0×10^{-5}, 5.0×10^{-5}, and 2×10^{-4} mol/L were excited at 305 nm, the relative fluorescence intensity decreased the increase in the concentration. when the concentration decreased to 2.0×10^{-5} mol/L, the fluorescence intensity remarkably.

3.3. Electrochemical properties of diarylethenes 1a

Reversible modulation of electrochemical properties by photoirradiation is of basic importance for the development of molecular electronic devices. To date, many diarylethene derivatives and their electrochemical properties have been reported [16-18].

In this work, the electrochemical properties of 1a was investigated. The typical electrolyte was acetonitrile(5 mL) containing 0.10 mol/L tetrabutylammonium tetrafluoroborate ((TBA)BF4) and 2.0×10^{-5} mol/L diarylethenes 1a. Figure 4 shows the CV curves of diarylethene 1a with the scanning rate of 50 mV/s.

With regard to the energy level of the ferrocene reference(4.8 eV below the vacuum level), the HOMO and LUMO energy levels can also be estimated. As shown in Figure 5, it can be clearly seen that the onset potentials of oxidation and reduction for 1a were observed at +1.44V and -0.68V, and those for 1b were observed at +1.57 V and -0.79V, respectively. Therefore, the values of IP and EA were calculated to be +6.24 and 5.56eV. Based on the HOMO and LUMO energy levels, the band gap (Eg, Eg = LUMO-HOMO) of 1a can be calculated as 2.12 eV, and that of 1b is 2.36 eV.

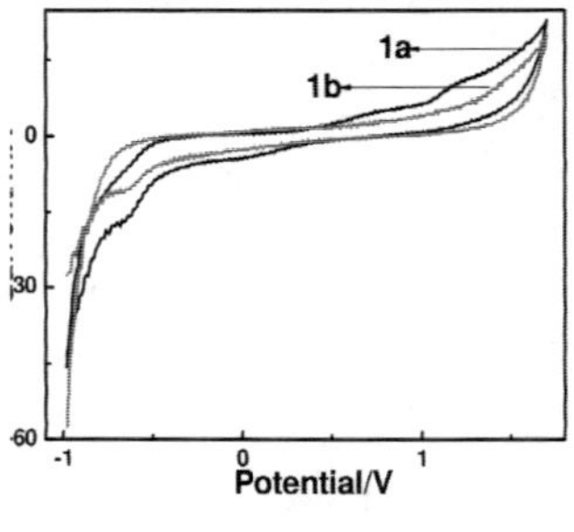

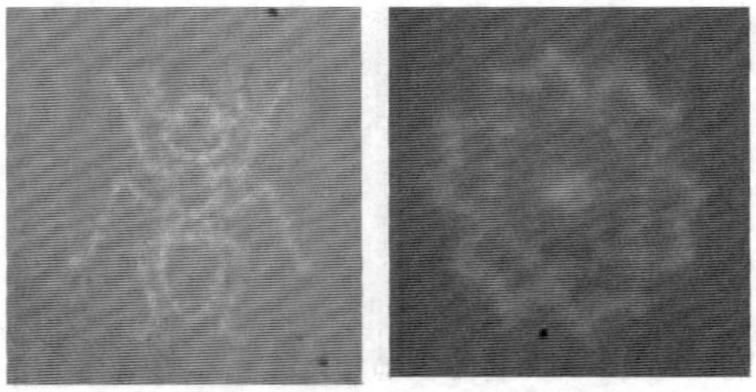

Figure 4. CV of 1a in 0.1 mol/L (($(TBA)BF_4$)

Figure 5. Readout patterns from 1b/PMMA film

3.4. Polarization multiplexing holographic recording

With a He-Ne laser (wavelength: 650 nm, 105mW) in the experimental setup. the evaluation of potential of photochromic diarylethene as a polarization medium by recording and reading in a real-time operation was investigated. Using diarylethene 1b/PMMA film as recording medium. The polarizations of the patterns recorded on the film are shown in Figure 5. The exposure time for each pattern was 0.04 s and the wide of each pattern was about 20 μm. The result indicated that the new photochromic diarylethene can be applied in high capacity optical storage.

In conclusion, An unsymmetrical dithienylethene was synthesized to investigate its photochromism, fluorescence and electrochemical properties. Dithienylethene 1a showed good photochromism both in solution and in PMMA amorphous film. The diarylethene exhibited a relatively strong fluorescence switches along with the photochromism from open-ring isomers to closed-ring isomers in hexance. The oxidation potential differences of the open-ring and closed-ring forms of compound 1 were 0.24 V, because the -conjugation length of 1b was get longer.The present results are useful for the designation of efficient photoactive and excellent characteristic diarylethene compounds.

Acknowledgments

This work was supported by the Project of National Natural Science Foundation of Jiangxi Province (2008GZH0020), the Key Scientific Project from Education Ministry of China (208069), the Project of Jiangxi Youth Scientist, and the Youth Science Funds of the Education Office of Jiangxi, China (GJJ09572, GJJ09567).

References

[1] Irie M 2000 *Chem. Rev.* **100** 1685
[2] Tian H and Yang S J 2004 *Chem. Soc. Rev.* **33** 85
[3] Irie M and Kingo U 1998 *Bull Chem. Soc.* **5** 985
[4] Pu S Z, Yang T S and Xu J K 2006 *Tetrahedron Lett.* **47** 6473
[5] Pu S Z, Yang T S and Li G Z 2006 *Tetrahedron Lett.* **47** 3167
[6] Liu G and Shen L 2007 *Org. Lett.* **9** 2139
[7] Yang T S, Shen L, Xu J K, Xiao Q and Chen B 2005 *Proc. PICS* 266
[8] Pu S Z, Yao B L, Liu G,and Wang Y L 2008 *Opt. Eng. SPIE* 030502 1
[9] Pu S Z, Fan C B , Miao W J and Liu G 2008 *Tetrahedron Lett.***64** 9464
[10] Liu G, Zheng T, Pu S Z and Xu J K 2006 *Acta Cryst E* **62** o2877
[11] Li Z X , Liao L Y , Sun W, Xu C H , Zhang C and Yan C H 2008 *J. Phys. Chem. C* **112** 5190
[12] Fan C B, Pu S Z, Liu G and Yang T S 2008 *J. Photochem Photobiol A: Chem.* **194** 333
[13] Fan C B, Pu S Z, Liu G and Yang T S 2008 *J. Photochem Photobiol A: Chem.* **197** 415
[14] Pu S Z, Liu G,Yang T S, Xu J k, Shen L and Li G Z 2005 *Tetrahedron Lett.* **61** 6623
[15] Tsujioka T, Kume M and Irie M 1997 *J. Photochem Photobiol A: Chem.* **104** 203
[16] Pu S Z, Yang T S, Li G Z, Xu J K and Chen B 2006 *Tetrahedron Lett.* **47** 3167
[17] Moriyama Y, Matsuda K, Tanifuji N, Irie S and Irie M 2005 *Org. Lett.* **7** 3315
[18] Peters A and Branda N R 2003 *J. Am. Chem. Soc.* **125** 3404

A compact signal transmitter for UV communication system

Rongyang Wang[1, 3], Yonggang Yuan[2], Chao Li[1], Yan Zhang[2] and Xiangyang Li[2]

[1] State Key Laboratories of Transducer Technology, Shanghai Institute of Technical Physics, Chinese Academy of Sciences, Shanghai 200083, China
[2] Key Laboratory of Infrared Imaging Materials and Detectors, Shanghai Institute of Technical Physics, Chinese Academy of Sciences, Shanghai 200083, China

E-mail: rywang@mail.ustc.edu.cn

Abstract. Three LED-based signal transmitting schemes were presented to build the transmitter of UV communication system. Microprogrammed Control Unit (MCU) was taken to generate digital driving signal and to accomplish modulation with Keil C51 development tool. The modulated signal with a certain pulse width was sent to a LED driving circuit. High power 365nm UV LED was adopted according to the trade-off between its scattering coefficient and price. Utilization of high power UV LED improved the integration density of printed circuit board (PCB), which was beneficial to assemble a beam shaping system. The whole transmitter was installed on a four-dimensional optical stage for system adjustments and measurements. The radiant optical power of UV LED was measured. Some issues about the PCB design were also discussed. The MCU output waveforms were demonstrated on an oscilloscope and the UV LED output optical pulse were verified by a Hamamatsu APD module. Different modulation data rates were investigated. Experimental result showed that this signal transmitter for UV communication system could transmit optical signal well and greatly shorten the system development cycle.

1. Introduction

Nowadays, research on UV communication systems are greatly stimulated by rapid progresses of semiconductor emitters and detectors operating in ultraviolet (UV) region. Successful UV communication experiments and systems have been achieved by many research groups [1, 2]. Generally, a typical UV communication system consists of a digital data source, a modulator, a light source that can be directly modulated in intensity at fairly data rates, a beam shaping and angular adjustment optical system; a receiving optical system to collect and focus the UV light onto a detector, a signal pre-amplifier and demodulator to recover the data. The acquisition of modulated data is often complicated, because the developer has to consume a large amount of time to make clear how to implement the architecture of hardware. However, the essence of UV communication system is to deliver the UV light pulse at a certain rate corresponding to the required communication bandwidth. So it is essential to find a convenient way to generate the digital signal with required pulse width and modulation scheme, in order to reduce system development cycle and simplify the system construction.

On the other hand, the choice of light source is also essential for a UV communication system. UV gas lamp, UV laser and UV LED are three types of light source currently available. LED is widely

[3] Rongyang Wang, Email: rywang@mail.ustc.edu.cn

used due to its small size, low power consumption, high lifetime, high bandwidth, and modulating facility. Meanwhile, for the radiant power of a solar blind region (240 nm-280 nm), UV LED is unique at the moment. A 365 nm UV LED was taken as the UV source because it has a relatively high optical power and an appreciable scattering coefficient.

The scope of this paper will cover an introduction to the transmitter schemes, design, and the hardware required to construct a working transmitter prototype of UV digital communication system. Also, a compact signal transmitter will be described and its modulation performance will be observed.

2. Three digital signal processing schemes

The subject of digital communications involves the movement of information from the source equipment to the terminal equipment. At the transmitter side, the information is processed into digital bit stream through software or coder chips. Since the LED is peak-power limited, the data must be compressed and transmitted at low rate in order to make achievable energy per pulse larger. Then the bit stream is modulated by the embedded system and sent to an analog circuit to drive the LED. We put forward three transmitter schemes to implement modulation and compressing of input information.

The first transmitter scheme utilizes a computer as an information processing and delivering carrier, which can handle text, speech or even image information. Figure 1 describes the block diagram of this transmitter with a clear delineation between hardware and software.

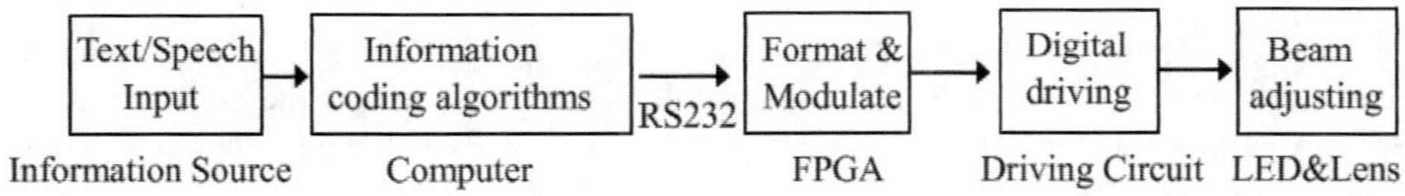

Figure 1. Block diagram of computer-based transmitter scheme.

The information is fed into a computer through the peripheral equipment at the front end. Then the software based on corresponding coding algorithms is used to transform the information into formatted or compressed data. This low rate digital bit stream is transferred into FPGA-based modulation hardware with an RS232 interface. After channel coding by FPGA, the data are taken to drive LED or LED array. As described above, this scheme is neither portable nor scalable, because a computer and certain software are needed during the communication process both at the emitter and receiver sides. It is especially discommodious to conduct outdoor field testing.

The second transmitter scheme based on a special vocoder chip can handle UV speech communication well. Its block diagram is shown in figure 2.

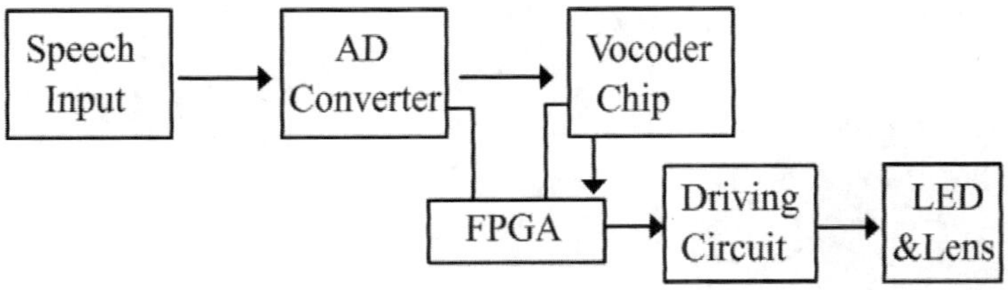

Figure 2. Block diagram of vocoder chip-based transmitter scheme.

The arrows in figure 2 indicate the data transfer direction. A microphone or an mp3 can be used to generate the analog audio signal. An analog-to-digital converter, integrated with anti-aliasing filter, digital high-pass filter, and an output low-pass filter, collects the audio signal and converts it into digital bit stream. Then the digital bit stream is input into a vocoder chip, such as AMBE-2000. The vocoder chip provides a real-time implementation of the standard-setting voice compression software algorithm. It maintains natural voice quality and speech intelligibility at extremely low data rate. At last, the data can be utilized to drive the light source through driving circuits. This transmitter scheme is compact and easy to be performed outside. However, it is time consuming to get familiar with the characteristics of the chips and the FPGA digital communication system, which will lengthen theUV system construction cycle.

In the UV non line of sight (NLOS) communication system, the channel path loss and bit error rate (BER) are the most important properties. To analyze them, a series of digital modulated optical pulses should be sent out at the transmitter side, and the received optical pulses are measured and recorded at the receiver side. According to comparison with the source binary codes, the path loss and BER characteristics can be deduced. Figure 3 gives another workable transmitter scheme. Different with the other two schemes, all the digital processing work is managed by one MCU chip.

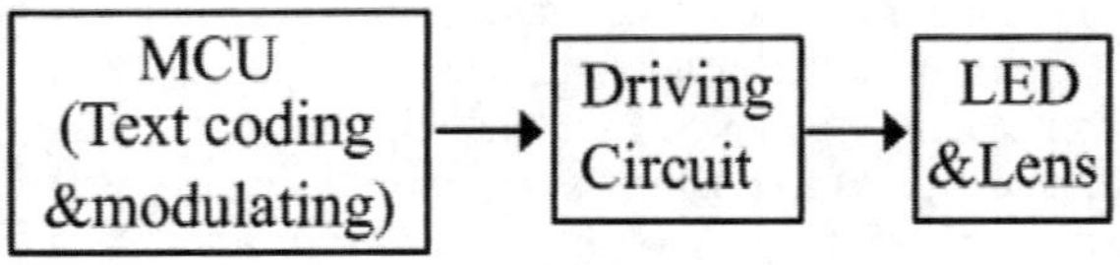

Figure 3. Block diagram of MCU-based transmitter scheme.

The principle of the MCU-based transmitter scheme is as follows: first, the text information to be sent is input into the MCU; then, a program of C stored in the MCU changes the source information from ASCII format to binary format; after that, the binary data stream is modulated by the same program and sent to the IO ports of MCU; finally, the IO ports are connected to the control pins of the driving circuit, making the light source emit modulated optical pulses.

All the above UV transmitters use LED or LED array as light sources. But from the complexity perspective, the third scheme can be implemented easily and can process text information well. So a demonstration system is constructed to verify its performance.

3. The MCU-based UV transmitter system

A picture of our MCU-based UV transmitter is presented in figure 4. It consisted of three components: a printed circuit board (PCB), an angular adjustment stage, an LED and lens. The PCB was made up of a high constant current LED driver, a manual "transmit" button and an external 5V power. Signal integrity was not concern much because this design operated at relatively low speed. Power decoupling capacitors were placed around the power sockets and each digital component. MCU and LED were designed on two PCBs. The LED board was vertically stacked above the MCU board to provide enough space for optical system design.

Figure 4. MCU-based UV transmitter system. This transmitter scheme emphasizes flexibility and modularity, providing the capability to transmit text information at adjustable data rate.

The PCBs were mounted on an optical stage to adjust the pitching angle. The lens was installed on a cross-post adaptor to achieve the regulation of beam divergence angle.

SETI now offers deep UV LEDs with peak emission wavelengths in the range of 240 to 400 nm [3]. To implement outdoor UV communication, the peak wavelength of UV LED should be in the solar blind region. But in this region, a single commercial LED typically consumes electrical power of 100 mW and only radiates an optical power of less than 1 mW. Meanwhile, a solar blind UV LED is extremely expensive. We have taken Nichia's high power 365 nm UV LED as the UV source since it has a comparable scattering coefficient with solar blind region UV LED. For instance, the scattering coefficient of 275 nm UV radiance is approximately 0.47 km^{-1}, while the scattering coefficient of 365 nm UV radiance is 0.27 km^{-1}[1]. The electrical and optical characteristics of Nichia 365 nm UV LED is shown in figure 5[4].

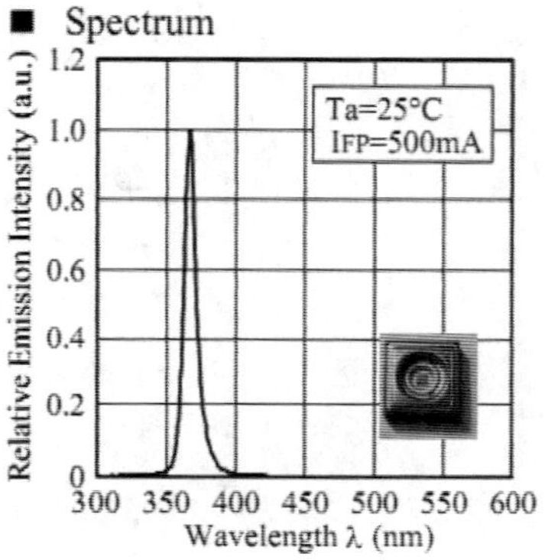

Parameters	Value(Typ.)	Unit
Power Dissipation	3.3	W
Forward Voltage	3.8	V
Forward current	500	mA
Peak Wavelength	365	nm
Spectrum Half Width	9	nm
Radiant Flux	210	mW

Figure 5. Specifications for Nichia high power 365nm UV LED.

4. Experimental result

In our experiments, the text information "Shanghai EXPO" was stored in the MCU and modulated by several popular modulation schemes at different data rates. We examined the performance of the transmitter and checked out the communication data rates from 1k to 125k bps.

4.1. The optical power of LED

Received optical power for various transmitter and receiver （Tx-Rx） separation distances under the condition of point to point line of sight (LOS) link were examined. Using a LabMax_TOP laser power meter, we recorded the power at different Tx-Rx separation distances ranging from 0.02 m to 5 m. The LED was driven at a current of 500 mA, 3.6 V DC. The optical sensor was OP-2 UV type, having a spectral response range of 250 to 400 nm and an active area of 28 mm^2. Figure 6 shows the relationship between the optical power arrived at the photosensitive surface and the Tx-Rx separation distance. It is noticed that the power density decays sharply as separation distance increases. At the range of 5 m, the received optical power by the OP-2 UV sensor is only 81 nW.

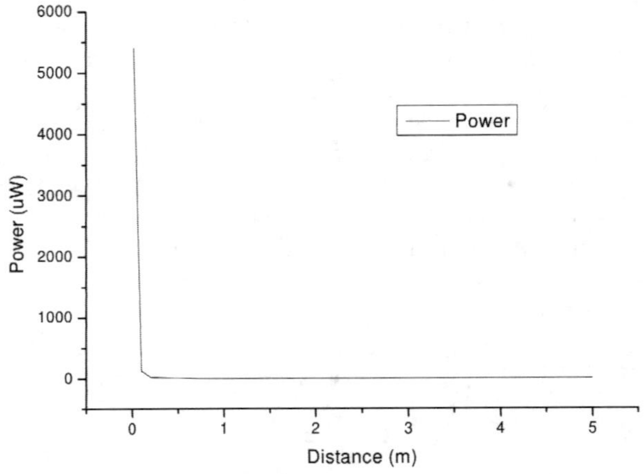

Figure 6. The relationship between the received optical power and Tx-Rx separation distance.

4.2. The results of different modulation schemes

Modulation schemes are judged on the following criteria: power efficiency, flexibility, bandwidth efficiency, and complexity. Baseband modulation schemes, including on-off keying (OOK) and pulse position modulation (PPM), are commonly used in NLOS UV communication. They can offer good power efficiency and bandwidth efficiency. In this section, the performances of OOK and 4PPM modulation are investigated. We modulated the text information "Shanghai EXPO" by both OOK and 4PPM schemes. The output waveforms of MCU were shown in figure 7. Since the LED is peak power limited, employing OOK modulation for a fixed data rate will upgrade the bandwidth efficiency.

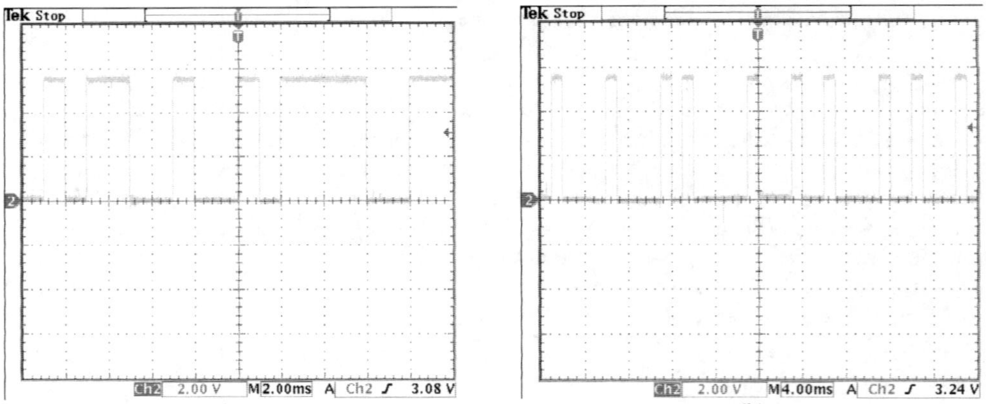

(a) (b)

Figure 7. Text information "Shanghai EXPO" modulated by OOK (a) and 4PPM (b) with a pulse width of 1 ms, corresponding to data rates of 1k bps and 500 bps, respectively.

4.3. Data communication rates test

A Hamamatsu APD module C5331-12 was used to examine the optical pulse emitted by the Nichia 365 nm LED. The optical pulse width was set to 8 μs, with the OOK data rate of 125k bps and the 4PPM data rate of 62.5k bps. APD module was put in front of the LED and the output waveforms were probed with an oscilloscope. Pictures observed from the oscilloscope are presented in figure 8. We can see that the digital signal can be recovered well by the APD module. But the transmitting distance is only within 0.5 m, because the module has a low gain at 365 nm wavelength.

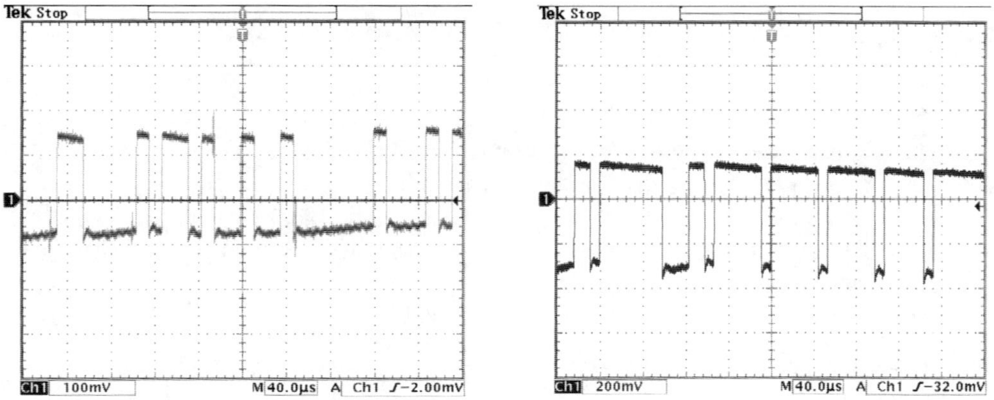

(a) (b)

Figure 8. The output waveforms of APD module. (a)OOK, data rates 125k bps. (b) 4PPM, data rates 62.5k bps. Note that the 4PPM waveform was reversed.

5. Conclusions

In this article, three transmitter schemes of UV communication system were summarized. An MCU-based UV communication system was built up using Nichia high power 365 nm UV LED. Different UV communication modulation schemes were also achieved on this transmitter system. Data transmission bit rate can reach 125k bps using OOK modulation. However, this system would certainly be further developed in the near future. The receiver with high sensitivity and proper bandwidth should be built in order to detect the weak scattering signal. Also, the BER property should be investigated with a BER analyzer. Finally, the optical concentration system would be embedded into the receiver system to realize the scattering communication, which is the ultimate goal of UV communication.

Acknowledgments

This project is supported in part by National Science funds of China PR, and the serial number is 60807037, 60708028.

References

[1] Shaw G A, Siegel A M and Model J 2006 *Proc. of SPIE* **6231C** 1-12
[2] Xu Z Y, Chen G, Abou-Galala F, and Leonardi M 2007 *Proc. of SPIE* **6709Y** 1-12
[3] Sensor Electronic Technology, http://www.s-et.com
[4] Nichia Corporation, http://www.nichia.com

Ab initio studies of the Sn-doped ZnO transparent conductive oxide

Zhihua Xiong[1]*, Lanli Chen[1], Cuiqun He[2], Zhijun Liang[3]

[1] Key Laboratory for Optoelectronics and Communication of Jiangxi Province, Jiangxi Science & Technology Normal University, Nanchang 330013, China
[2] Department of Mechanical Electrical, Nanchang Institute of Technology, Nanchang 330099, China
[3] Naval Equipment Department, Xi'an 710000, China

E-mail: xiong_zhihua@126.com

Abstract. Based on density functional theory within the generalized gradient approximation, Sn-doped wurtzite ZnO transparent conducting oxide (TCO) was investigated, where structures relaxations, formation energies, and electronic structure of Sn doped in ZnO are studied. It was found that the lattice parameters of the Sn-doped ZnO increased with increasing Sn concentration. The substitutional Sn (Sn_{Zn}) atom has low formation energy and introduces impurity levels below the conduction band minimum. Also, substitutional Sn atoms contribute significantly delocalized s orbitals in the conduction band states, which are expected to increase the conductivity of the material. Additionally, with increase of Sn concentration, the band gap is narrowed. We further identify that the lower concentration of heavily doped of Sn, the conductivity of ZnO is stronger. The calculated results are helpful to gain a systematic understanding of structure, electrical properties of Sn-doped ZnO transparent conducting oxide.

1. Introduction

Znic oxide (ZnO) has owned many features, including optical, semiconducting, piezoelectric, electro-optic, and optoelectronic properties. Because of wide band gap (3.4 eV) and larger binding energy (60 meV), ZnO can be used in many fields, such as light appliance, solar cell, sensors, actuators, flat panel liquid crystal monitor, and infrared reflector [1]. It is well known that pure-ZnO is polar semiconductor, demonstrating as n-type, and it has usually low conductivity. In order to gain transparent conducting oxide materials, it should employ doping donor elements to improve the conductivity such as Al [2-4], Ga [5], In [6]. For Al doped ZnO, on experiment, Al doping ZnO could lower the resistivity of ZnO [2-4], lower Madelung energy [7], improve the electrical conductivity, and its optical absorption edge blue-shifted [8], which can gain good quality TCO. In theory, Zhang et al. [9] have calculated the electronic structure and optical properties of ZnO in different Al doping, showing that AZO materials can used as excellent transparent conducting films. Recent, many groups have chosen Sn as the dopant in ZnO. Seval Aksoy et al. [10] have reported that the undoped and Sn doped ZnO films were deposited by a spray pyrolysis method onto the glass substrates, and suggested that the absorption edge shifts to the lower wavelengths with Sn dopant. Benelmadjat et al. [11] proposed sol-gel method to investigate the effects of Sn dopant on structural and optical properties. It shows an important effect of Sn dopant on optical band gap. Saliha et al. [12] reported that Sn doped ZnO films have been deposited by sol gel spin coating method. They investigated the effect of Sn

* Corresponding author. E-mail: xiong_zhihua@126.com, Tel: (86)791-3831968.

incorporation on structural and electro-optical properties of ZnO films. They observed that the shift of absorption edge is associated with shrinkage effect and the electrical conductivity of the ZnO film enhanced with the Sn dopant. Li et al. [13] reported that Sn-doped ZnO microrods have been fabricated by a thermal evaporation method. They investigated the effect of Sn dopant on the microstructure, morphological and composition of as-prepared Sn-doped ZnO microrods. Sn doping of ZnO films have been widely reported to investigate the influence of doping on the structural and optical properties. K.C.Yung et al. [14] have proposed that the optical band gap in free standing transparent ZnO films using the sol-gel method can narrow with the dopant of Sn. They found that the optical band gap of Sn-doped ZnO is red-shifted due to the deep states in the band gap. It is well known that using the higher valence of dopants is mainly to provide more free electrons, which leads to lower the impurity doping and improve the quality and optical properties of thin films [15]. However, there are few studies about the electronic structure and the effects of Sn concentration of ZnO from the theoretical calculation using the first principles. In this work, the paper is organized as follows: Firstly, the computational methods are introduced. Then, the formation energy and electronic structure of $Sn_{0.0625}Zn_{0.9375}O$ are calculated, and finally, the effects of concentration of Sn dopant on electrical properties of ZnO are considered by the first principles.

2. Computational methods

All calculations were carried out using the density functional theory with the generalized gradient approximation (GGA) and the projector augmented wave (PAW) pseudo-potentials, as implemented in the VASP codes [16,17]. A gamma centred 5×5×3 k-point mesh for the Brillouin zone was employed. The energy cutoff is 450 eV. We have constructed the four configurations, which are listed as following: Configuration (a): 2×2×2 (pure-ZnO), Configuration (b): 2×2×2 ($Sn_{0.0625}Zn_{0.9375}O$), and Configuration (c): 2×2×3 ($Sn_{0.0417}Zn_{0.9583}O$), Configuration (d): 3×3×2 ($Sn_{0.0278}Zn_{0.9722}O$). All the atoms have been relaxed until the force on each atom was below 0.01 eV/Å. By optimizing the geometry of the pure-ZnO, the calculated lattice parameters were got as follows: a= 0.328 nm, b= 0.328 nm, c= 0.527 nm, c/a= 1.608. They were consistent with experimental results [18]: a= 0.324 nm, b= 0.324 nm, c= 0.519 nm, c/a= 1.602. The calculated parameters are in excellent with the experiment results, indicating that our calculation methods are reasonable. We have mainly considered the formation energy ($E_f(D,q)$) for a complex with a defect *D* at charge state *q*. Calculated the formation energy of Sn substituted Zn site and Sn substituted O site. The calculation equation can be calculated as [19]

$$E^f(D,q) = [E_{tot}(D,q) - E_{tot}(ZnO)] + \sum_i n_i\mu_i + q(E_F + E_{VBM}) \tag{1}$$

$$E_{tot}(ZnO) = \mu_{Zn} + \mu_O \tag{2}$$

where $E_f(D, q)$ is the defect formation energy for the studied supercell. $E_{tot}(D, q)$ is the total energy of a supercell containing the defect *D* at the charge state q. $E_{tot}(ZnO)$ is the energy of the pure ZnO in the same supercell. EF is the Fermi energy level with respect to the VBM of the host, which varies up to the experimental value of 3.4 eV. E_{VBM} is the valence band maximum of the pure ZnO. μ_i is the chemical potential of the elements. μ_i is the chemical potential of atom i referenced to the element. μ_i *(bulk)* is the energy of the bulk material. The chemical potentials of Zn and O satisfy the following equations:

$$\mu_O^{min} = E_{tot}(ZnO) - \mu_{Zn}(bulk) \tag{3}$$

$$\mu_{Zn}^{min} = E_{tot}(ZnO) - \mu_O(O_2)\ . \tag{4}$$

And the chemical potential of Zn is the energy of Zn bulk. The chemical potential of O is the half energy of oxygen.

3. Results and discussion

3.1. Electronic structure of $Sn_{0.0625}Zn_{0.9375}O$

Firstly, we have calculated the formation energies of Sn occupied Zn site (Sn_{Zn}) and Sn occupied O site (Sn_{O}). The results are listed as Table 1, from which it can be found that the formation energy of Sn_{Zn} is less than that of Sn_{O}. It means that Sn occupied Zn may be the most stable doping way among all the related defects. This is also in excellent consistent with the experiment [12].

Table 1. Formation energy of ZnO for Sn_{Zn} and Sn_{O}

Formation energy /eV	E_f /eV
Zn site	5.2421
O site	11.6061

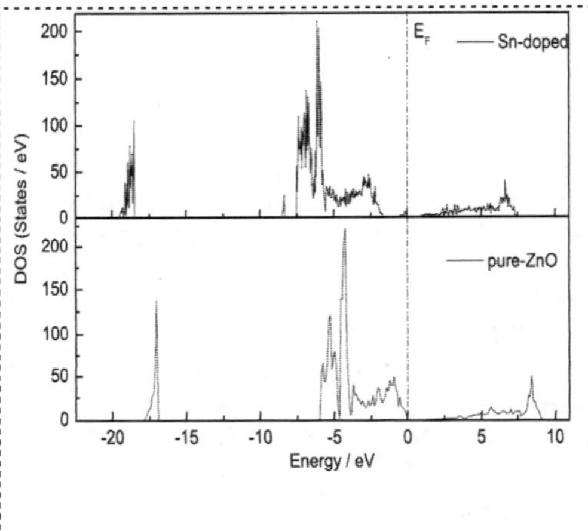

Figure 1. TDOSs of pure-ZnO and $Sn_{0.0625}Zn_{0.9375}O$

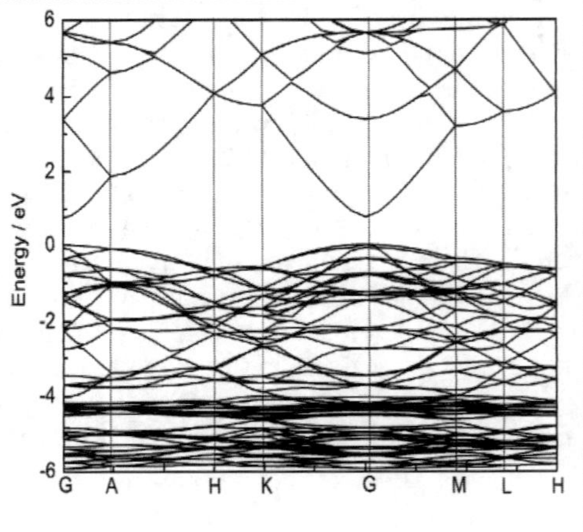

Figure 2. Band structure of pure-ZnO

Sequentially, we can make a clear comparison of pure-ZnO and Sn doped ZnO with $Sn_{0.0625}Zn_{0.9375}O$ (as shown in Fig. 1). All energies are relative to the respective Fermi energies. It shows that the Fermi level is located at the bottom of conduction band for Sn doped ZnO. Moreover, the valence bands and the conduction bands are both shifted into the lower energy. In addition, as the Sn is doping in ZnO, Sn supplies four electrons, while Zn provides two electrons. That is to say more free electrons are introduced into the doped systems. Therefore, it can be concluded that Sn is a good candidate for enhancing the electrical conductivity. The band gap of pure-ZnO is 0.76 eV (as shown in Fig. 2), which is lower than that of experiment value (3.4 eV), because of the well-known of GGA. Fig.3 is the band structure of $Sn_{0.0625}Zn_{0.9375}O$, from which we can see the band gap narrowed to 0.15 eV. It can be observed that Sn doped ZnO can narrow the band gap, which is demonstrates as red-shifted. This result agrees well with the experiment [14].

From PDOS of $Sn_{0.0625}Zn_{0.9375}O$ show in Fig. 4, it can be seen an impurity peak near the Fermi level is formed, which is favorable for improve conductivity of ZnO. The upper conduction band within -0.96 to 0 eV is mainly contributed by Sn-s states near the Fermi level. However, the lower conduction band is constituted by Sn-p and Zn-s states. The valence bands are divided into two parts: the upper valence band within -19.73 to -18.30 eV is introduced by O-s, and the lower valence band within -8.71 to -8.31 eV is introduced by Zn-d.

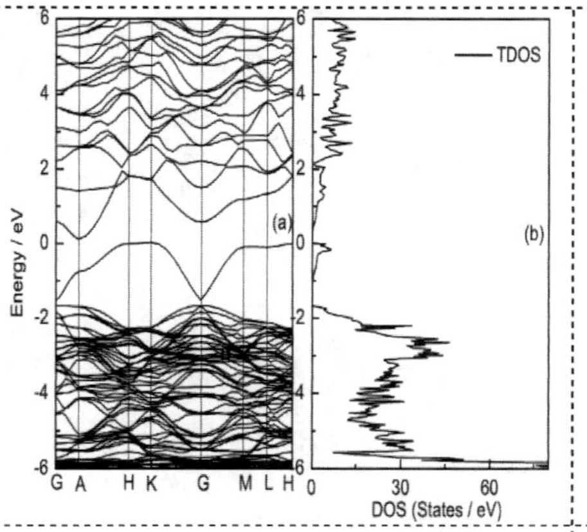

Figure 3. Band structure and DOS of $Sn_{0.0625}Zn_{0.9375}O$

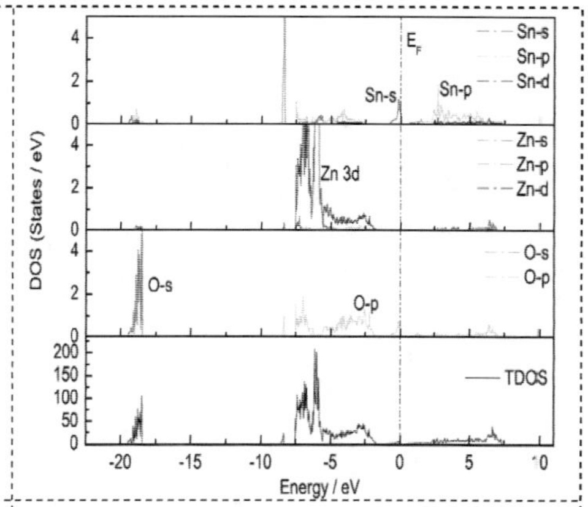

Figure 4. (Color online) PDOS of $Sn_{0.0625}Zn_{0.9375}O$

3.2. Effects of Sn concentration on conductivity of ZnO

From the Table 2, we can see that as increase of x, the lattice parameters of a and c are both increased. The calculated result agrees well with the experimental results [11]. It is well known that Zn ionic radius has almost the same as the Sn ion ($r_{Zn}^{2+} = 0.74$ Å and $r_{Sn}^{4+} = 0.69$ Å) [20]. When Sn is incorporated to the ZnO film, Sn^{4+} substitutes Zn^{2+} site leads to two more free electrons to contribute to the electrical conductivity. So, Sn doping ZnO possibly can realize good conductivity. In Fig.5, we can see that the Fermi levels are all located at the bottom of conduction band of $Sn_xZn_{1-x}O$. Moreover, the peak values of the impurity states are increased with the decrease of x near the Fermi level. In addition, it can be observed that the band gap becomes narrow with increase of x, due to the bottom of conduction band of gradually shifted to lower energy.

Table 2. Lattice parameters and band gap energy of $Sn_xZn_{1-x}O$.

$Sn_xZn_{1-x}O$	a/Å	c/Å	c/a	band gap energy / eV
0	3.281	5.277	1.608	0.76
0.0278	3.285	5.309	1.616	0.46
0.0417	3.290	5.311	1.614	0.27
0.0625	3.294	5.354	1.625	0.15

To further discuss the effects of Sn concentration on electronic structure of ZnO, PDOSs of $Sn_xZn_{1-x}O$ are plotted in Fig.6a to 6d. It can be seen that the location of the valence band maximum is determined by O-p and Sn-s. Furthermore, with increase of x, the proportion of Zn is reduced, the peak value of O-p are increased near the Fermi level. It also can be observed that the conduction band is mainly depended on Sn-s, which plays an important role for ZnO high conductivity. As increase of x, the Sn-s shifts to the lower energy, which promote the band gap is narrow. And then red-shifted occurs.

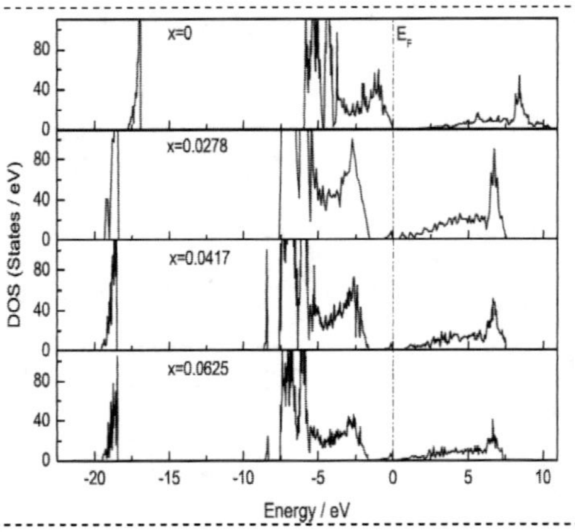

Figure 5a. TDOS of $Sn_xZn_{1-x}O$.

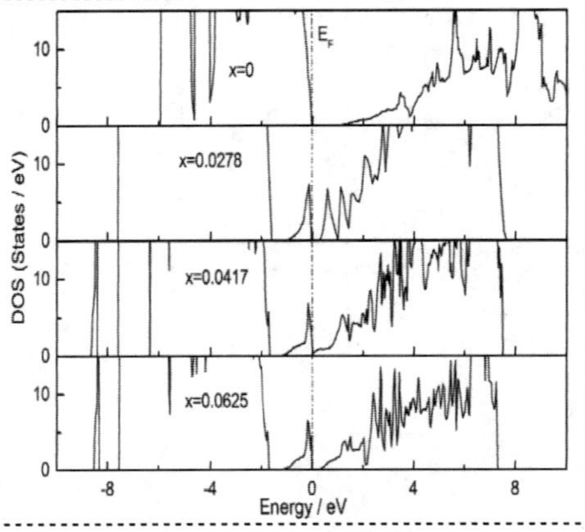

Figure 5b. TDOS of $Sn_xZn_{1-x}O$ for amplified figure.

However, the conductivity property of a semiconductor is not only dominated by the number of electrons, but also depends on the mean free time. According to the solid theory [21], mean free time $\tau_i \bullet N_i^{-1}T^{3/2}$. Here, the calculated values of dopant concentration are $2.48404\times10^{21}cm^{-3}$, $1.6691147\times10^{21}cm^{-3}$, and $1.11957\times10^{21}cm^{-3}$ for $Sn_{0.0625}Zn_{0.9375}O$, $Sn_{0.0417}Zn_{0.9583}O$, and $Sn_{0.0278}Zn_{0.9722}O$, respectively. Obviously, for our first-principle calculations, Sn atoms are heavily doped in ZnO compared with experiment. The calculated numbers of electrons are 2.01842, 1.88004, and 1.69589 for $Sn_{0.0625}Zn_{0.9375}O$, $Sn_{0.0417}Zn_{0.9583}O$, and $Sn_{0.0278}Zn_{0.9722}O$. Due to the configurations are n-type semiconductor, the electronic conductivity can be computed as

$$\sigma_i = \frac{n_i q^2 \tau_i}{m_e^*} \tag{5}$$

where m_e^* is the electron effect mass. n_i is the number of electrons. q is the charge state, and $_i$ is the free time on average. $_i$ is the conductivity of the dopant system. Thus, the conductivity ratios of various complexes should be calculated. The conductivity of $Sn_{0.0625}Zn_{0.9375}O$, $Sn_{0.0417}Zn_{0.9583}O$, and $Sn_{0.0278}Zn_{0.9722}O$ was set to be $_1$, $_2$, $_3$, respectively, and then the ratios are as follows:

$$\frac{\sigma_3}{\sigma_2} = \frac{n_3 q^2 \tau_3 / m_e^*}{n_2 q^2 \tau_2 / m_e^*} = \frac{1.69589\times1.669114}{1.88004\times1.11957} = 1.3448,$$

$$\frac{\sigma_2}{\sigma_1} = \frac{n_2 q^2 \tau_2 / m_e^*}{n_1 q^2 \tau_1 / m_e^*} = \frac{2.48404\times1.88004}{1.6691147\times2.01842} = 1.3862.$$

Obviously, we can get: $_3 > _2 > _1$. It can be seen that the conductivity of $Sn_{0.0278}Zn_{0.9722}O$ is larger than that of others. From the calculated results, we can see that the conductivity of Sn-doped ZnO is closely related to the concentration of impurities and the average number of electrons in the conduction band. And the concentration of impurities is depending on the free time on average. Therefore, we can draw a conclusion that the lower concentration of heavily doped of Sn, the conductivity of ZnO is stronger.

Journal of Physics: Conference Series **276** (2011) 012194 doi:10.1088/1742-6596/276/1/012194

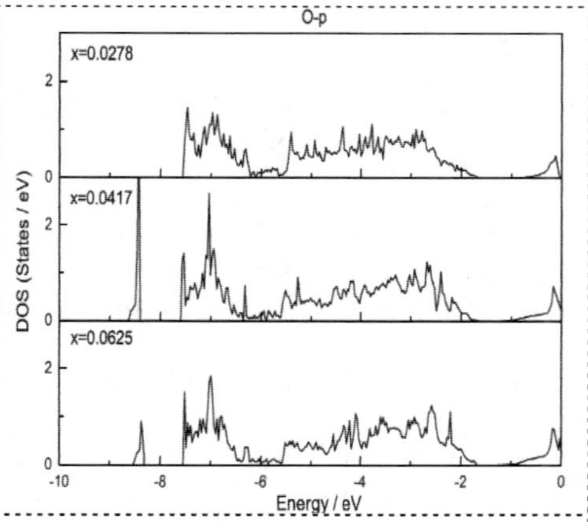

Figure 6a. PDOS of $Sn_xZn_{1-x}O$ for O-p wthin -10 to 0 eV

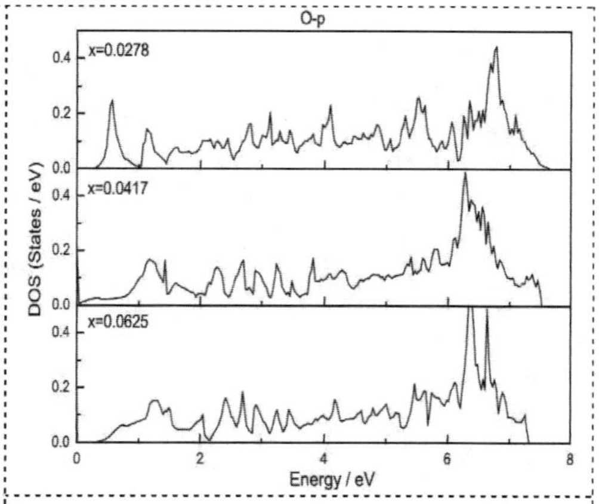

Figure 6b. PDOS of $Sn_xZn_{1-x}O$ for O-p within 0 to 8 eV

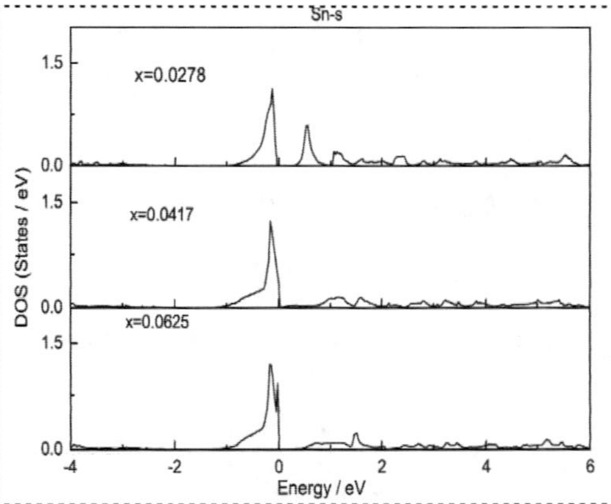

Figure 6c. PDOS of $Sn_xZn_{1-x}O$ for Sn-s

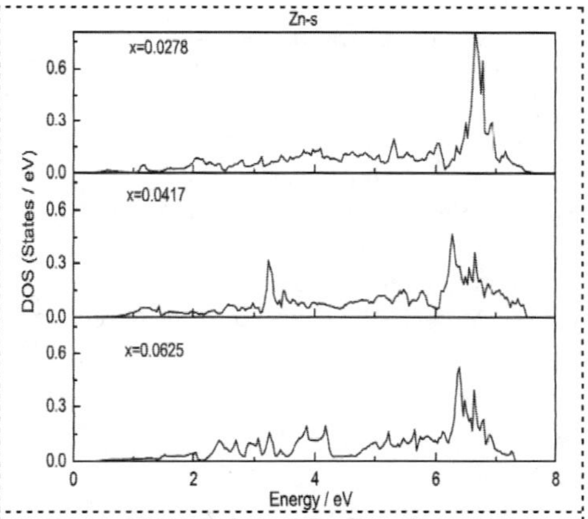

Figure 6d. PDOS of $Sn_xZn_{1-x}O$ for Zn-s

4. Conclusions

The formation energy, geometry structure, and electronic structures of $Sn_xZn_{1-x}O$ have been calculated by the first principle method. The calculated results show that Sn_{Zn} has lower formation energy than that of Sn_O, and the lattice parameters are both increased with the increase of x. We also found that the conduction band near the Fermi level is mainly determined by the Sn-s states near the Fermi level, and the band gap of $Sn_xZn_{1-x}O$ narrows, as the increase of x. Furthermore, the conductivity have been employing to compared, we can draw a conclusion that the conductivity of $Sn_xZn_{1-x}O$ decrease with the concentration of Sn increasing. The results are helpful to gain a systematic understanding of structure, electrical properties of Sn-doped ZnO TCO, and also can support and explain the experimental results.

Acknowledgments

This work was supported by Science and Technology Staff Service Enterprise Project of the Ministry of Science and Technology of China (No. 2009GJC50021), National Natural Science Foundation of China (No.51062003), Key Science and Technology Project of Jiangxi Provincial Department of Education (No.GJJ10030), and Natural Science Foundation of Jiangxi Province (No. 2008GZW0014).

References

[1] D C. Look, Mater.Sci. Eng., B 80, 383 (2001).

[2] Minami T, Nanto H, Takata S, Jpn. J. Appl. Phys. Part I 23 L280 (1984).
[3] Hu J, Gordon R G J., Appl. Phys. 71 880 (1992).
[4] Sato H, Minami T, Takata S, Miyata T, Ishii M, Thin Solid Films 236 14 (1993).
[5] Q.B. Ma, Z.Z. Y, H. P. He, L.P. Zhu, W. C. Liu, Y.F. Yang et al., J. Phys. D: Appl. Phys. 41, 055302 (2008).
[6] N. Ito, Y. Sato, P. K. Song, A.Kaijio, K. Inoue, Y. Shigesato. Thin Solid Film 496, 99 (2006).
[7] Yamamoto T, Katayamae Yoshida H. Physics B 302-303 155 (2001).
[8] Xu Z Q, Deng H, Xie J, Li Y. Chin. J. of Liquid Crystals and Displays 20 503 (2005).
[9] Zhang F.C, Zhang Z.Y, Zhang W.W, Yan J.F, and Yun J.N, Acta Optica Sinica 0253-2239, 04-1025-07 (2009).
[10] Seval Aksoy, Yasemin Caglar, Aaliha Ilican, Mujdat Caglar, Optica Applicata, Vol.XL,No.1 (2010)
[11] H. Benelmadjat, B. Boudine, O. Halimi, and M. Sebais, Optics & Laser Technology 41 630-633 (2009).
[12] Saliha Ilican, Mujdat Caglar, Yasemin Caglar, Applied Surface Science doi:10.1016/j.apsusc.2010.05.052 (2010).
[13] Lijun Li, Ke Yu, Yang Wang, and Ziqiang Zhu, Applied Surface Science, 256 3361-3364 (2010).
[14] K.C.Yung, H.Liem, and H.S.Choy, J. Phys.D: Apll.Phys.42 (2009) 185002.
[15] B.M.Ataev, A.M.Bagamadova, V.V.Mamedov, A.K.Omaev, M.R.Rabadanov. J. Cryst. Growth 198-199, 1222 (1999).
[16] G. Kresse, and J. Hafner, Phys. Rev. B 47, 558 (1994).
[17] G. Kresse, and J. Furthermuller, Phys. Rev. B 54, 11169 (1996).
[18] S. J. Pearton, D. P. Norton, K. Ip, Y. W. Heo, and T. Steiner, J. Vac. Sci. Technol. B 22, 932 (2004).
[19] Woo-Jin Lee, Joongoo Kang, Phys. Rev. B, 73:024117(1-6) (2006).
[20] R. Deng, X. T. Zhang, J. Luminescence 128 1442 (2008).
[21] Kun Huang, and Nvqi Han, Solid State Physics, Higher Education Press, p300,344-345 (1985).

Large Enhancement in short-circuit current densities of dye-sensitized solar cells via photon conversion

Jie Xu[1,*], Hui Zhang[1], Qi Xiong[1], Guijie Liang[2], Lei Wang[1], Xiaolin Shen[1], Li Liu[1], and Weilin Xu[1]

[1] Key Lab of Green Processing & Functional Textiles of New Textile Materials, Ministry of Education, Wuhan University of Science & Engineering, Wuhan 430073, China

[2] College of Electronics & Information Engineering, Wuhan University of Science & Engineering, Wuhan 430073, China

Abstract. Converting the solar spectrum to match the absorption properties of dye-sensitized solar cells (DSSCs) can reduce spectral losses and thus enhance the photovoltaic performance. In this paper, liquid electrolytes containing Fluorescent Yellow 8G, which could absorb light at around 458 nm and down-convert it to the absorption band of *cis*-bis(isothiocyanate)-bis(2,2`-bipyridine-4,4`-dicarboxylate) ruthenium (N3), are prepared and applied in DSSCs. The short-circuit current density of the DSSC with the optimal concentration of 8G shows an increase of 110.7%, leading to an increase of 88.5% in the overall conversion efficiency, compared to that of the referenced DSSC assembled with the blank electrolyte, which is attributed to the improvement of light harvesting from the solar spectrum range.

* Corresponding author. Tel/Fax: 86-27-87426559; *E-mail: xujie0@ustc.edu*

Published under licence by IOP Publishing Ltd

1. Introduction

Dye-sensitized solar cells (DSSCs) have attracted much attention as promising alternatives to conventional solid-state photovoltaic devices because of their low cost production, environmentally friendly components, and relatively high conversion efficiency [1-5]. In these cells, the charge-transfer sensitizer is one of the key elements in achieving high solar-to-electric conversion efficiency. The most successful sensitizers used in DSSCs are ruthenium polypyridyl complexes, which yield conversion efficiencies of 10-11% with simulated sunlight [6]. The majority of ruthenium complexes reported to date show absorption in the visible region at around 535 nm. Essential for efficient conversion of solar energy by DSSCs is the spectral match of the sensitizer absorption to the solar radiation, and in this regard, the ruthenium complexes are inadequate. Therefore, developments of sensitizers with extended absorption and spectral sensitivity in the broader wavelength region is imperative. The simultaneous adsorption of multiple sensitizers which have complementary spectral absorption on the TiO_2 electrode, the so called co-sensitization, was widely employed to broaden the spectral response. Various combinations of sensitizers have been investigated in the context of co-sensitization; however, only limited success is achieved to date [7-10]. In most cases the dye component having the most potent sensitization ability in the dye combination behaved less efficiently in the co-sensitization than in the single-dye sensitization due to unfavorable interactions between the dyes [10].

Journal of Physics: Conference Series **276** (2011) 012195 doi:10.1088/1742-6596/276/1/012195

The concept of photo conversion is fundamentally different from the co-sensitization. Photo conversion processes aim at converting via luminescence the solar spectrum to match the absorption properties of the sensitizers. This is in contrast to the co-sensitization which all focuses on developing combinations of sensitizers better able to match the polychromatic solar spectrum. Consequently, photo conversion processes have a number of advantages over the co-sensitization. Photo conversion can be treated and modeled as an optical process alone, dissociated from the actual operating procedure of the solar cell. There are two photon conversion processes of interest, namely, down-conversion and up-conversion. However, no efficient up-conversion materials are available to shift the infrared photons into the visible spectrum at low light intensity conditions (solar radiation). Therefore, in this work we focused on the down-conversion of liquid electrolytes containing an organic fluorescent dye and improvements in the short-circuit current density and the overall conversion efficiency compared to the blank electrolytes were investigated. Fig. 1 shows the structure of DSSCs based on electrolytes containing fluorescent dye.

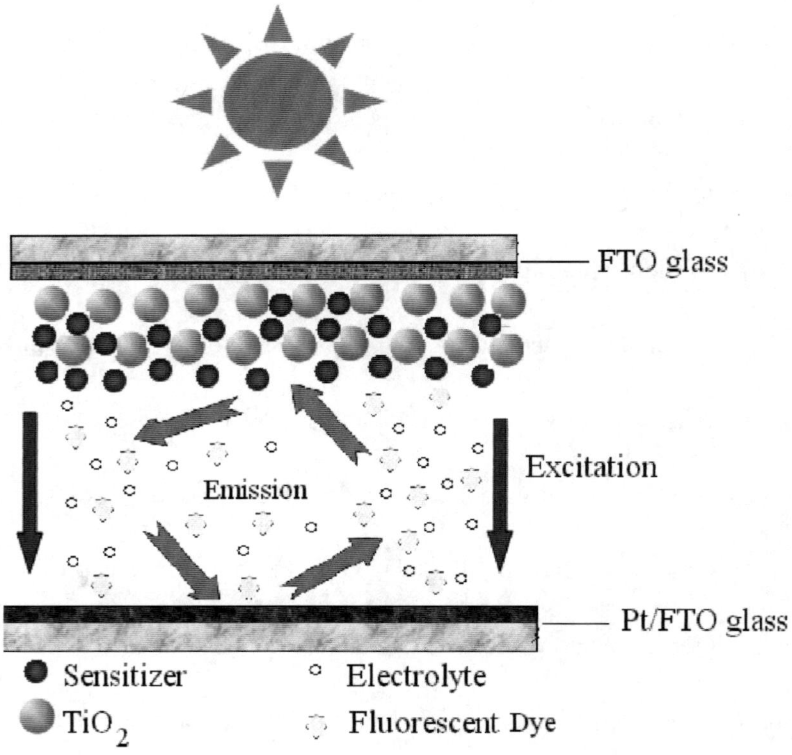

Fig. 1. Structure of DSSCs based on electrolytes containing fluorescent dye

2. Experimental

The TiO_2 (P25, Degussa AG, Germany) paste was coated on a conducting glass substrate (F-doped SnO_2, $15\Omega cm^{-2}$, Libbey Owens Ford Industries, USA) by a doctor-blading technique. After drying, the coated substrate was sintered at 450°C in air for 0.5 h. The TiO_2 film (10μm thick) was cooled to 80°C and then immersed in the 0.3 mM ethanol solution of N3 (*cis*-bis(isothiocyanate)-bis(2,2`-bipyridine -4,4`-dicarboxylate) ruthenium, Solaronix SA, Switzerland) for 12 h. The dye-coated TiO_2 electrode and platinized counter electrode were assembled into a sandwich type cell. The electrolyte solutions (0.3 M iodine, 0.03 M LiI in acetonitrile) with various concentrations of Fluorescent Yellow 8G (FY 8G, the structure was shown in the inset of Fig. 2) were used. It should be noted that no 4-*tert*butylpyridine was added into the electrolytes. UV/Vis spectra were recorded on a Shimadzu UV-2550PC (Japan) at room temperature. Emission spectra were analyzed with an Acton SpectraPro 500i spectrometer (USA). The photocurrent and photovoltage of the cells were measured on an electrochemical workstation analyzer (model LK9805) by a two-electrode arrangement with A 500 W Xenon lamp (100 $mWcm^{-2}$) as light source.

3. Results and discussion

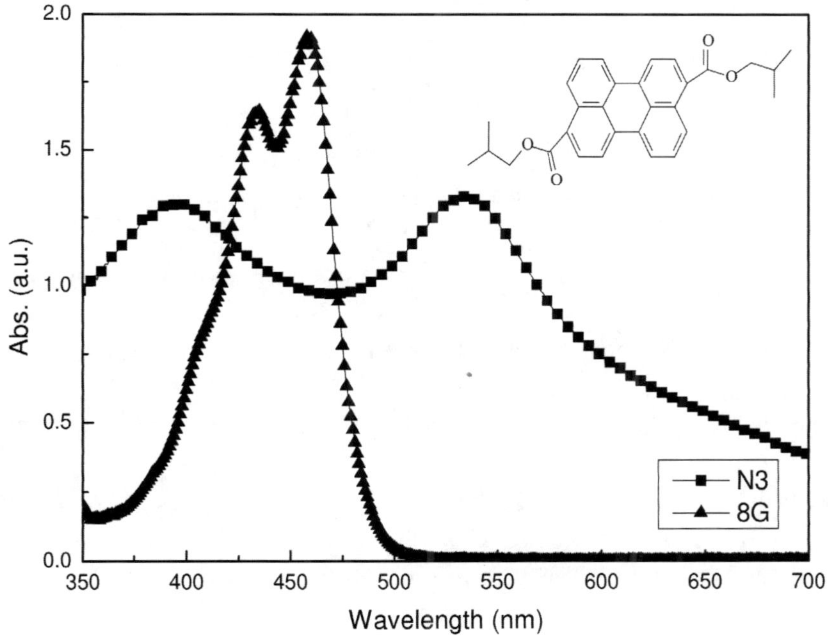

Fig. 2. UV/Vis absorption spectra of N3 (solid line, 300 μM) and Fluorescent Yellow 8G in ethanol (dotted line, 25 μM). The inset is the chemical structure of Fluorescent Yellow 8G.

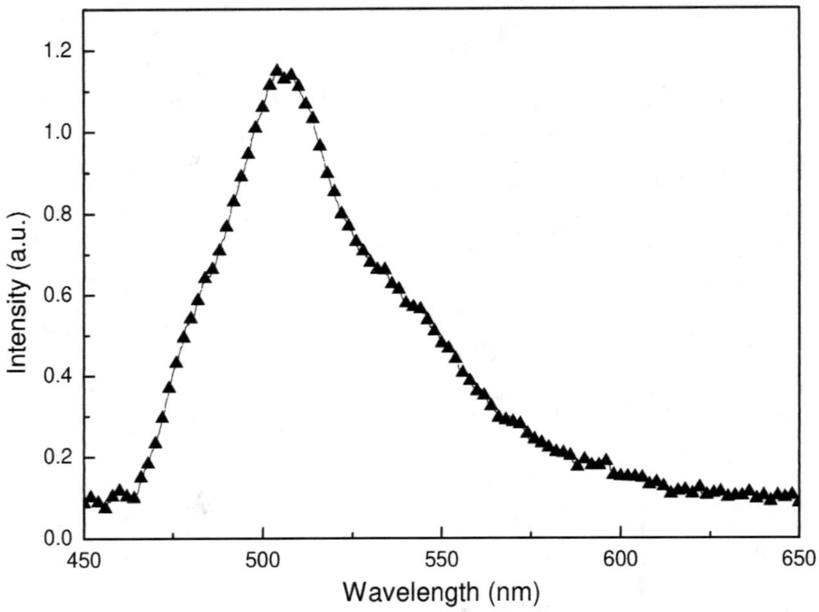

Fig. 3. Emission spectra of Fluorescent Yellow 8G in ethanol solutions (25 μM)

The absorption spectra of N3 and FY 8G in ethanol are given in Fig. 2. The N3 solution has two peaks at around 533 and 395 nm and a trough at around 466 nm. The absorption spectrum of FY 8G shows a maximum at 458 nm, which is complementary to that of N3. The visible light absorption of FY 8G even at much lower concentration was much stronger than that of N3 because the molar absorption coefficient of FY 8G is significantly greater than that of N3. Therefore, introduction of FY 8G into the DSSCs would result in an efficient light-harvesting over a wide spectrum region. Fig. 3 shows the emission spectra of FY 8G in ethanol solutions (25 μM) excited with 469 nm at 298 K. There is one intense and broad band in the range of 450-600 nm, centered at 504 nm, just remarkably overlapping the absorption band of N3. Thus, the emitted light could be absorbed by N3 dye to generate excitons, and considerably increase the light harvesting and the photocurrent of DSSCs.

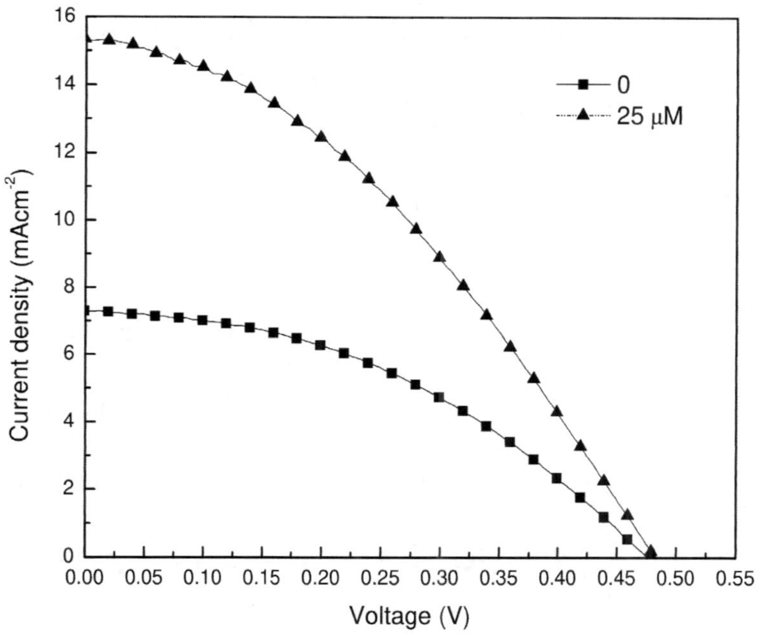

Fig. 4. Photocurrent-voltage characteristics of DSSCs based on the electrolyte with and without FY 8G under AM 1.5 irradiation

Fig. 4 shows the photocurrent-voltage characteristics of the DSSCs based on the electrolyte with and without FY 8G under AM 1.5 irradiation. The reference DSSC (without FY 8G) gives a short-circuit current density (I_{SC}) of 7.28 mAcm^{-2}, an open-circuit voltage (V_{OC}) of 0.476 V, and a filling factor (ff) of 0.412. The V_{OC} and ff is relatively low due to the absence of 4-*tert*butylpyridine in the electrolyte. After adding the FY 8G into the electrolyte, the DSSC exhibits an I_{SC} of 15.34 mAcm^{-2}, a V_{OC} of 0.483 V, and an ff of 0.368. Compared to that of the reference DSSC, the I_{SC} value shows an increase of 110.7%, corresponding to an increase of 88.5% in the overall conversion efficiency. The improvement in photovoltaic performances of DSSCs has confirmed that the light harvesting is effective through the photon conversion via luminescence. The decrease in ff is attributed to the increase in series resistance caused by a higher density of photon flux.

Liu et al. [11] had introduced the Dy^{3+} doped $LaVO_4$ luminescent film onto back of the DSSC, and the overall conversion efficiency was improved by 23.3% relative to the referenced cell coated with an undoped $LaVO_4$ film. The much higher improvements of the present work can be explained by taking the different effects that contribute to the photocurrent as photon conversion used. Firstly, the absorption coefficient and the quantum yield of FY 8G are much higher than those of the inorganic Dy^{3+} doped $LaVO_4$. Secondly, the most intense emission of the Dy^{3+} doped $LaVO_4$ film located in the range of 560-580 nm, which is not a close match for the absorption of N3. Finally, the Dy^{3+} doped $LaVO_4$ luminescent film was cast onto back of the DSSC, resulting in a relatively lower fraction of the emitting light directed toward to the sensitized TiO_2 film; while here the fluorescent dye was directly integrated inside the DSSC, thus the harvesting of the emitting light could be effectively improved.

4. Conclusion

The present work demonstrates that enhanced photovoltaic performance of the DSSCs can be achieved by simply adding an organic fluorescent dye into the liquid electrolytes. The introduction of the fluorescent dye could reduce the spectral losses by transferring the unabsorbed light to the absorption band of the sensitizer through a down-conversion process. With the optimal concentration of the fluorescent dye, the short-circuit current density shows an increase of 110.7%, leading to an increase of 88.5% in the overall conversion efficiency, compared to that of the referenced DSSC assembled with the blank electrolyte. Further improvements are expected from the use of a more appropriate fluorescent dye with a better match of absorption band and emission band.

Acknowledgements

This work was supported by the Natural Science Foundation of China (No. 51003082), the Key Project of Science and Technology Research of Ministry of Education (No.208089), the Natural Foundation of Hubei Province (No.2008CDB261) and Educational Commission of Hubei Province (Q20101606).

References

[1] Heimer T A and Heilweil E J 2000 *J. Phys. Chem. A* **104**, 4256
[2] Park N-G, Kang M G, Kim K M, Ryu K S and Chang S H 2004 *Langmuir* **20**, 4246
[3] Gao F, Wang Y, Shi D, Zhang J, Wang M, Jing X, Humphry-Baker R, Wang P, Zakeeruddin S M and Grätzel M 2008 *J. Am. Chem. Soc.* **130**, 10720
[4] Yang Y, Zhou C-H, Xu S, Hu H, Chen B-L, Zhang J, Wu S-J, Liu W and Zhao X-Z 2008 *J. Power Sources* **185**, 1492
[5] Shen X, Xu W, Xu J, Liang G, Yang H and Yao M 2008 *Solid State Ionics* **179**, 2027
[6] Nazeeruddin M K, Angelis F D, Fantacci S, Selloni A, Viscardi G, Liska P, Ito S, Takeru B and Grätzel M 2005 *J. Am. Chem. Soc* **127**, 16835
[7] Ehret A, Stuhl L and Spitler M T 2001 *J. Phys. Chem. B* **105**, 9960
[8] Chen Y, Zeng Z, Li C, Wang W, Wang X and Zhang B 2005 *New J. Chem.* **29**, 773
[9] Yum J-H, Jang S-R, Walter P, Geiger T, Nüesch F, Kim S, Ko J, Grätzel M and Nazeeruddin M K 2007 *Chem. Commun.* 4680
[10] Inakazu F, Noma Y, Ogomi Y and Hayase S 2008 *Appl. Phys. Lett.* **93**, 093304
[11] Liu J, Yao Q and Li Y 2006 *Appl. Phys. Lett.* **88**, 173119

TiO_2 Nanowire dye-sensitized solar cells Fabricated by Hydrothermal Method

Yang Xu Haoshuang Gu Huating Xia Mingzhe Hu*

Faculty of Physics and Electronic technology, State Key Laboratory of Ferro- & Piezoelectric Materials and Device of Hubei Province, Hubei University, Wuhan 430062, Hubei, P.R. China

Abstract. TiO_2 nanowire dye-sensitized solar cells were fabricated by using hydrothermal method. The synthesizing of TiO_2 nanowire/nanorod arrays directly on FTO substrate would cause a high conducting loss to solar cells. Through laminating a compact layer between FTO substrate and photonic anode layer, the conducting loss could be effectively prevented. Results indicated that using different concentration of titanium tetrabutoxide would affect the photoelectric conversion efficiency and different producing methods of the compact layer also played an important role to the conversion efficiency.

1.Introduction

Nowadays, energy crisis and environmental pollution are two major challenges in 21st century. Developing renewable green energy, especially solar cells, has become major issues of human society facing. Up to now, the highest efficiency of the solar cells prepared by single crystal silicon or polycrystalline silicon films could be up to 20% or more, but the cost was too expensive. Developing low-cost photovoltaic active materials as well as fabrication process are important. Currently ZnO, CdS, CdSe, CdTe, Fe_2O_3, SnO_2, TiO_2 and many other compounds have been used for dye-sensitized solar cells (DSCs)[1-4]. Although CdSe, CdTe and other solar cells have some advantages, they are highly toxic and harmful to environment. Dye-sensitized TiO_2 powder cell possessed the advantages of both low cost and simple fabrication process, it propels great amount of investigation all over the world and becomes the forefront of chemistry and material science fields. Up to now, the highest photoelectric conversion efficiency of TiO_2 DSC surpassed 10% and further improvement is also underway to reach the theoretical prediction. However, the lastest theoretical analysis indicated that TiO_2 powder had a lower photoelectric conversion efficiency compared with that of TiO_2 nanowire[5, 6].

The typical composition of a TiO_2 DSC included the dye-sensitized semiconductor anode, the electrolyte and the counterpart electrode. The TiO_2 photonic anode was produced by absorbing the photosensitizers on the surface of TiO_2 nanowire, the counterpart electrode was produced by sputtering a certain thickness Pt thin film on FTO substrate and the electrolyte was permeated in between. The dye molecular can absorb the visible photo energy and transit from groud state to excited state. Due to the unstable of excited state, the electronic will fast return to the lower energy conducting band of TiO_2 through the interaction of dye molecular and the TiO_2 suface. And subsequently the electronic will inject into the FTO conducting thin film and finally generate photoelectric current through outside circuit. The oxided dye molecular will be reduced by the electrolyte, which will then diffuse to the counterpart electrode to be charged. The complete circulation could be described as[7-10]:

Ground state Dye molecular (S)+hv→Excited dye molecular (S3)

Excited dye molecular (S3)+TiO_2 → e- (conducting band of TiO_2) + oxided dye molecular (S+) (generating photoelectric current)

Oxided dye molecular (S+) + reduced electrolyte(R-)→Ground State dye(S)+ Oxided electrolyte (R) (Dye reduction)

Oxided electrolyte (R)+e- (cathode)→Reduced electrolyte(R-) (electrolyte reduction)

Oxided electrolyte (R)+e- (TiO_2 conducting band)→Reduced electrolyte (R-) (Dark current)

In present paper, well-aligned TiO_2 nanowire arrays were successfully prepared on the FTO substrate by hydrothermal method with different amount of titanium tetrabutoxide ($Ti(OC_4H_9)_4$) in starting materials. The as-prepared TiO_2 nanowire arrays with a rutile structure were about 10μ m in length and 200nm in width. The photoelectric conversion efficiency of the DSCs fabricated by the as-synthesized TiO_2 nanowire arrays was detaily investigated.

2. Experiment

The TiO_2 nanowire/nanorod arrays were synthesized by hydrothermal method. The FTO conducting glass was cleaned by soaking in the deioned water for 20min as well as by ultrasonic cleaning for 20min. The titanium tetrabutoxide,high concentration hydrochloric acid and deioned water were chosen as the precursors, which were hydrothermally synthesized in the reaction kettle at 150°C for 20hrs. The TiO_2 photonic anodes were then soaked in the N97 dye for 12hrs before the final encapsulation.

3.Results and discussion

Fig. 1 illustrated the J-V curve of TiO_2 DSC as a function of concentration of titanium tetrabutoxide. As the concentration of titanium tetrabutoxide increased, the conversion efficiency increased firstly and when the concentration reached 0.75 M, the photoelectric conversion efficiency became the highest of 0.31%, with the current density and short circuit voltage being 3.27mA/cm^2 and 0.465 mV respectively. As the concentration further increased to 1.0M the conversion efficiency decreased instead.

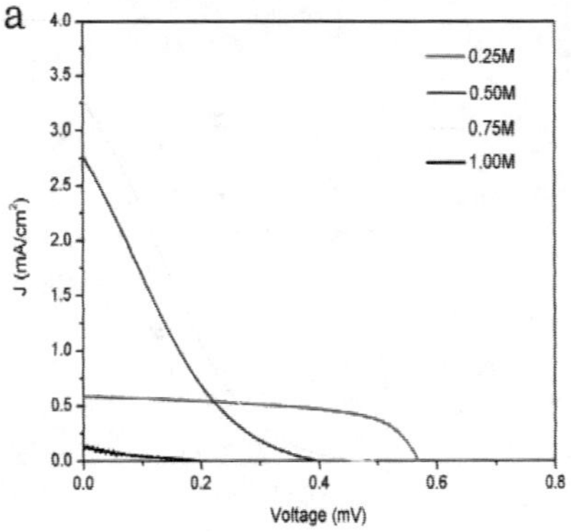

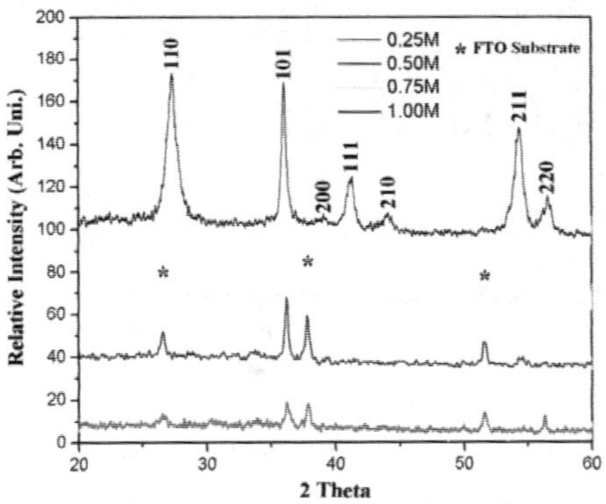

Fig.1 The photoelectric conversion efficiency of TiO_2 DSC as a function of concentration of titanium tetrabutoxide. Fig. 2 XRD patterns of TiO_2 photonic anode varied with concentration of titanium tetrabutoxide.

Fig.2 illustrated the as-prepared XRD patterns of TiO_2 photonic anode varied with concentration of titanium tetrabutoxide. The diffraction peaks could be indexed based on rutile TiO_2 phase (JCPDF card file No. 881175) with tetragonal $P4_2/mnm$ (No.136) space group, which indicated TiO_2 composition was successfully synthesized under present parameters. And the higher the concentration of titanium tetrabutoxide was, the stronger the diffraction peaks of TiO_2 phase exhibited. The lower concentration samples also exhibited some diffraction peaks of FTO substrate, which mightbe arose from the thin size of the TiO_2 layer or the very loose surface structure of TiO_2 layer under the low concentration.

Fig.3 indicated the SEM photos of the DSC with different concentration of titanium tetrabutoxide. As could be seen, the concentration affected the density of the nanowire arrays. As the concentration increased, the nanoarray became denser and the distribution of TiO_2 nanowire became more uniform. The density had a close relationship to the surface area of the photonic anode, and the photoelectric conversion efficiency increased with the increase of surface area and reached the maximum when the concentration was 0.75M. The cross section photograph (d) indicated a thickness of about 10 microns TiO_2 photonic anode layer.

Fig. 3 SEM photographs of the TiO_2 photonic anode fabricated with the titanium tetrabutoxide concentration being (a) 0.5 M titanium, (b) 0.75M and (c)1.0 M. (d) was the cross section of picture b.

SEM pictures also indicated that there existed some area without the growth of TiO_2 nanowire, where became the place of current leakage in DSCs, and thus lowered the photoelectric conversion efficiency. In order to avoid this problem, we employed a buffer layer between FTO substrate and TiO_2 photonic anode to improve the efficiency. The buffer layer was produced by the following two methods: rotary gelatinizing method using titanium tetrabutoxide and directly deposited TiO_2 buffer layer on FTO by magnetron sputtering method, using sputtering time to control the film thickness.

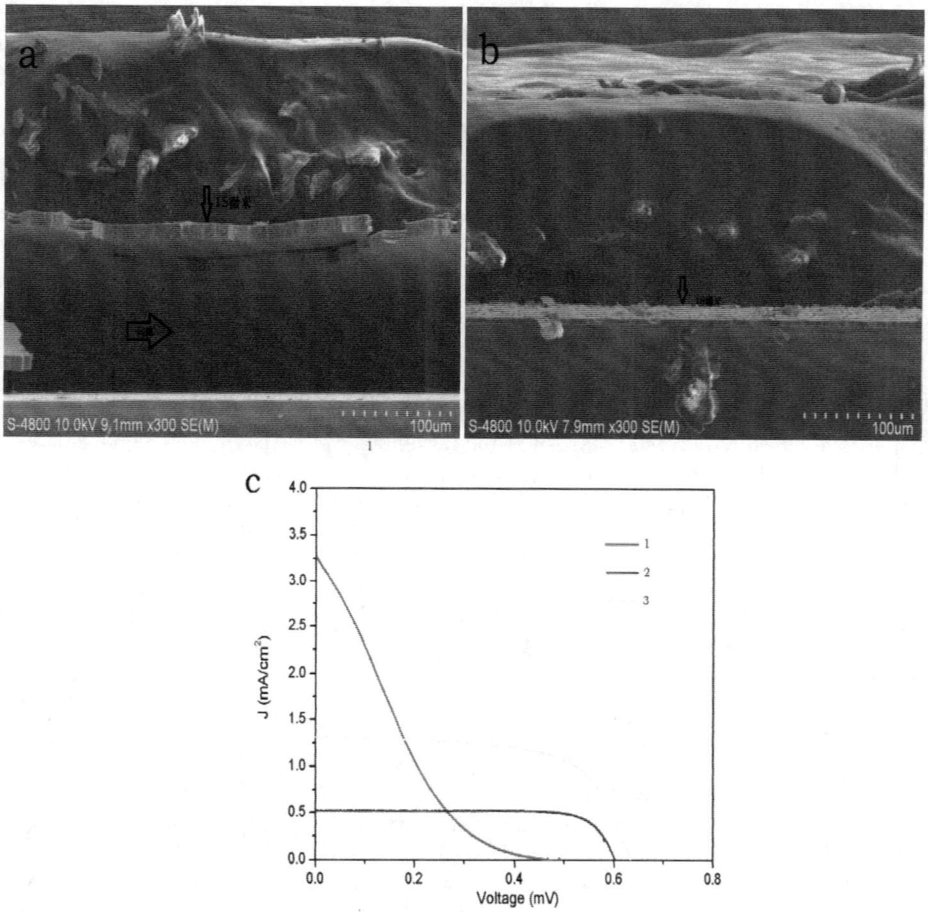

Fig.4 SEM cross-section images of the buffer layer produced by (a) rotary gelatinizing method, (b) magnetron sputtering medhod. (c) J-V curves of the three type battery with titanium tetrabutoxide being the same concentration of 0.75 M, (1 was the DSC without buffer layer, 2 represented the DSC with magnetron sputtered buffer layer, 3 was the DSC with rotary gelatinized buffer layer).

Fig4 illustrated the cross-section SEM images of the DSCs produced with buffer layer. The buffer layer could be clearly observed between FTO substrate and TiO_2 photonic anode. The one produced by rotary gelatinizing method was about 15um thickness, while that produced by magnetron sputtering method was about 18 um. Fig4(c) indicated that among the three DSCs, the one with buffer layer produced by rotary gelatinizing method possessed the highest cell efficiency. The open circuit voltage and fill factor were improved by about and respectively, compared with the DSC without buffer layer. However, when the baffer layer was produced by magnetron sputtering method, the short-circuit current decreased. The possible reasons for this phenomenon lied in that the crystallization in the buffer layer made by magnetron sputtering method was worse than that in and the lattice defects inside acted as traps for electronic transmission and lowered the short-circuit current. Moreover, one could expect a higher conversion efficiency in DSC-3 if the gap existed beween the buffer layer and conductive glass could be eliminated.

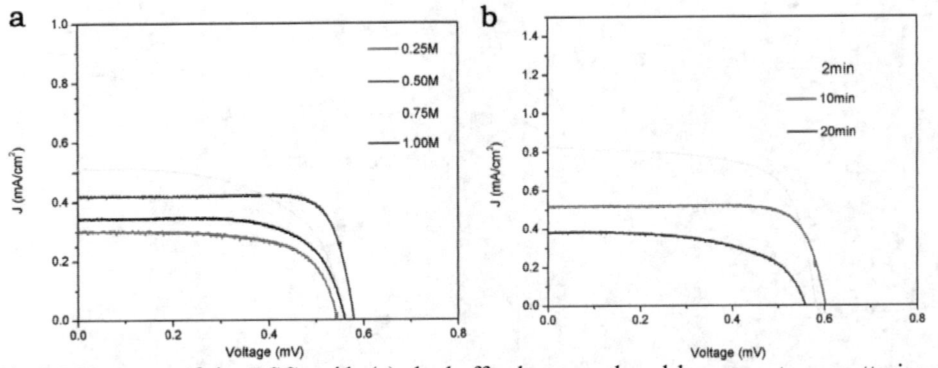

Fig.5 J-V curves of the DSCs with (a) the buffer layer produced by magnetron sputtering for 10 minutes, (b)different thickness of buffer layer fabricated by different sputtering time.

Fig.5(a) illustrated the J-V curves of DSCs with the buffer layer produced by magnetron sputtering for 20 minutes. The highest conversion efficiency was still obtained in the 0.75M sample, which corresoponded very well with the conclusion in Fig1, however, the efficiency was improved because of the buffer layer. The thickness of the buffer layer prepared by magnetron sputtering also had a significant influence on the photoelectric conversion efficiency of DSCs. As could be seen in Fig5(b), when the sputtering time was two minutes for the DSC produced with 0.75M titanium tetrabutoxide, the highest conversion efficiency of 0.43% was obtained, with the current density and open circuit voltage being 0.83mA/cm2 and 0.58mV respectively.

4. Conclusions

TiO_2 nanowire/rod array DSCs were prepared by hydrothermal method in the present research, results showed that:

1、The concentration of tetrabutyl titanate played an important role on the conversion efficiency of the DSCs. The highest conversion efficiency of 0.31% was obtained when the concentration of tetrabutyl titanate was 0.75M, with the current density and open-circuit voltage being 3.27mA/cm^2 and 0.47mV respectively.

2. Using buffer layer between FTO substrate and TiO_2 photonic anode could effectively improve the conversion efficiency. When the rotary gelatinizing method prepared buffer layer had a higher conversion efficiency than the magnetron sputtering prepared one. When the concentration of tetrabutyl titanate was 0.75M, the conversion efficiency of the former was improved to 0.71%, with the current density and open-circuit voltage being 1.36mA/cm^2 and 0.63mV respectively.

3. The thickness of the buffer layer also affected the battery conversion efficiency. When magnetron sputtering time was 2 minutes, the conversion efficiency of DSC reached the highest of 0.43%, with the current density and open-circuit voltage being 0.83mA/cm^2 and respectively 0.58mV respectively.

The optimal photoelectric conversion property was obtained when the amount of $Ti(OC_4H_9)_4$ was 0.75M. The open-circuit voltage and short circuit current density were 0.47 V and 3.27 mA/cm^2, respectively. The photoelectric conversion efficiency was about 0.31%. However, when the TiO_2 buffer layer were prepared on the FTO substrate by spin coating before hydrothermal treatment, the open circuit voltage was improved to 0.63V and the current density was 1.32mA/cm^2. The conversion efficiency was increased to 0.7%.

Acknowledgements

The authors would like to thank the National Science Foundation of China (No.50572026, No. 61076049) for the financial support of the present work. The same gratification is also debted by the author to the financial support from Science Foundation of Guangdong Province, under granted No. 2009B090300368.

Reference

[1] Guo P and Aegerter M A 1999 *Thin Solid Films* **351** 290-4
[2] Turkovic A and Cmjak Orel Z 1997 *Solar Energy Materials and Solar Cells* **45** 275-81
[3] Ferrere S, Zaban A and Gregg B A 1997 *The Journal of Physical Chemistry B* **101** 4490-3
[4] Keis K, Lindgren J, Lindquist S-E and Hagfeldt A 2000 *Langmuir* **16** 4688-94
[5] Grätzel M 2004 *Journal of Photochemistry and Photobiology A: Chemistry* **164** 3-14
[6] Ghannam M, Sivoththaman S, Poortmans J, Szlufcik J, Nijs J, Mertens R and Van Overstraeten R *Solar Energy* **59** 101-10
[7] O'Regan B and Gratzel M 1991 *Nature* **353** 737-40
[8] Wang Z-S, Huang C-H, Li F-Y, Weng S-F, Ibrahim K and Liu F-Q 2001 *The Journal of Physical Chemistry B* **105** 4230-4
[9] Hu L, Dai S, Weng J, Xiao S, Sui Y, Huang Y, Chen S, Kong F, Pan X, Liang L and Wang K 2006 *The Journal of Physical Chemistry B* **111** 358-62
[10] Ogbonna J C and Tanaka H 2000 *Journal of Applied Phycology* **12** 207-18

The Study on the Reliability of High Power LED Streetlights

YAO Dong-Ge, CHEN Jian-Xin,
Key Laboratory of Opto-electronics Technology (Beijing University of Technology), Ministry of Education, Beijing University of Technology, 100 Ping Le Yuan, Chaoyang District, Beijing 100124, China

E-mail: yaodongge@emails.bjut.edu.cn

Abstract. This paper was about a reliable research on high-power LED lighting. Based on the samples of the self-developed high-power LED streetlights, an electrical stress ageing test was carried out and thermocouple method was used in the temperature te**st. T**he ageing test showed that the initial flux reduction was mainly due to the absorption of the light lamp or the block by some parts of the lighting. And the late light decling was mainly caused by the decay of the high-power LED light source itself. Some suggestions on improving the design of streetlights will be given according to my research.

1. Introduction

In recent years, light emitting diode (LED) has begun to play a more and more important role in many applications including back lighting for cell phones, LCD displays, interior and exterior automotive lighting such as headlights, large signs and displays, signals and illumination[1]. LED will soon be used in general illumination because of its distinctive advantages including high efficiency, good reliability, long life, variable colors and low power consumption. An expectation about high power LED is that it will be the dominant lighting technology by 2025.[2] But in the actual application process, LED's performance is not as expected as shown, but the emergence of various problems. For example, System life quoted by manufacturers, commonly 100,000 hours, is often based on the average lifespan of a single LED[3]. But in actual use conditions, only a few thousand hours. Therefore, if we want the energy-saving products for the LED lights to obtain a long-term development, and ultimately to replace the traditional street lamps, the study on its reliability is essential and urgent.

The study on the reliability of LED lighting belongs to the study on systemic reliabitlity, LED light source which restricted the reliability of the overall reliability of lamps is an important part of the system. In this paper, we mainly study the LED's ageing characteristics through the electrical stress on LED light source so that the reliability of the LED streetlights could be known and helpful advice could be given on the design of LED streetlights .

2. Experimental Design

Based on the self-developed LED streetlight, an ageing test was carried out to study the reliability of the LED source in the application. This LED lamp (patented) was made up of many article lamps from different angles. The amount of the LED light source needed depended on the photometric

Published under licence by IOP Publishing Ltd

requirements of article lamps. And then the secondary distribution of light on LED was done by reflective bowl. The article lamps with a structure of groove were made from the aluminium with suppression and shape. And the LED was fixed to the article lamps by reflective bowl directly. Then, thermal grease was coated on LED heat sink evenly, thermal conductivity coefficient of> 1.5W / m ℃, working temperature -60 ℃ ~ 250 ℃. Upon the experimental condition available of the laboratory. The test on some article lamps was made to give a reference for studying the reliability of the LED streetlights.

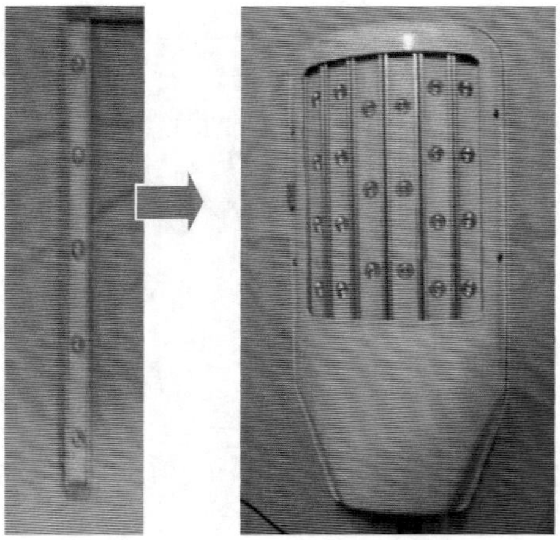

Fig.1 Independent research of LED lamp

10 LED lamps, 1w with white light, from the same factory were chosen as test samples. The samples were numbered from 1 to 10. And the initial test on the 10 samples showed the quality of samples can meet the requirements of the experiment. making NO. 1-2-3-4 LED lamps in series into LED article lamps; then installing the lamps on the cooling plate; and NO.9-10 LED lamps serving as a backup device. Before the test, light-colour-electricity experiment was performed on the 1-2-3-4 LED strings and the 5-6-7-8 LED strings respectively. And the ageing test was an electrical stress test under room temperature. To ageing the two LED in-series devices with 700mA DC accessed. From the comparison of the results, we can know whether the cooling effect of the article lamps can ensure the LED light resource had a high reliable quality. The ageing test on article lamps was easy to perform and cost little. And it can truly reflect the characteristics of LED in its aging process when used. This test was meaningfully creative.

3. Results and Discussion

(1) The curve of 1-2-3-4, it is parameter comparison of the light fitting which fore and after assembled.

Table.1 Test data before and after installing

	1-2-3-4# (before-stalling)	1-2-3-4# (after-stalling)
Voltage(v)	14.89	14.8
luminous flux(lm)	498.88	386.88
colour temperature(K)	5174	5048
Luminous Efficacy(lm/w)	48.9	37.3

Due to table 1, the LED voltage did not change seriously fore and after assembled, but the luminous flux and luminous efficiency change clearly, before assemble, luminous flux is 498.88 lm, the after assemble, it is 386.88lm, decreased about 22.5%, and the luminous efficiency arrive to 23.72%, the main reason is the absorb of the heating panel and the reflecting shade to the LED output light.

Test system is the Three-color 8716S Digital Optical parametric test systems, including: IT6800 Series Power, CCD fiber spectrometer, 1.5m integrating sphere and test software.

(2)The aging of assembled light fitting

In the process of aging, we tested the light, colour, electricity characters of assembled light fitting and 5-6-7-8 LED. From the figure 2, 3 and 4, we analyzed the change of luminous flux, colour temperature and luminous efficiency, and obtained the result as follow:

1. Due to the LED of assembled light fitting and 5-6-7-8 are produced by the same manufacturer, and they have better uniformity after initial filter. So in the experiment, they had the uniformity performance change, it is agree well with the theoretical.

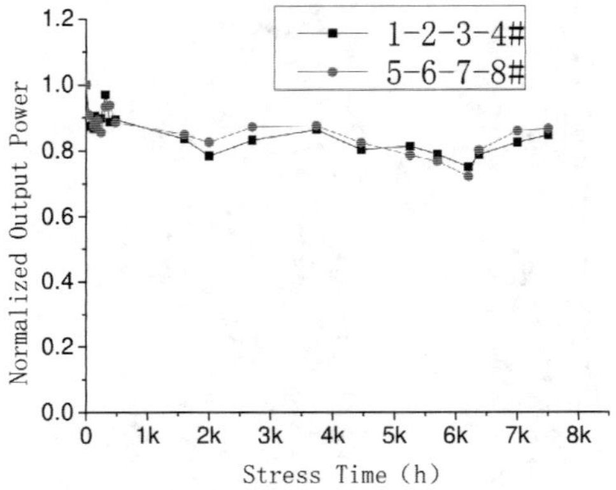

Fig.2 Normalized light output vs. Stress time

2. Toward luminous flux, assembled light fitting and 5-6-7-8 LED have the same tendency of change, and it can maintain above 80% of the initial after 7500 hours later, so these LED have higher reliability. In another side, it indicate that the heating panel and meet the needed of LED, and keep the LED work normally, and in the process of aging, it can emanate the heating of LED timely, decrease the decay of luminescence of LED.

3. During the process of aging, the colour temperature increase gradually, the change of assembled light fitting and 5-6-7-8 LED are uniformity, it is clearly that the light fitting itself and reflecting shade have less effect on the colour temperature. It is caused by the LED itself; the main reason is the decay of fluorescent powder.

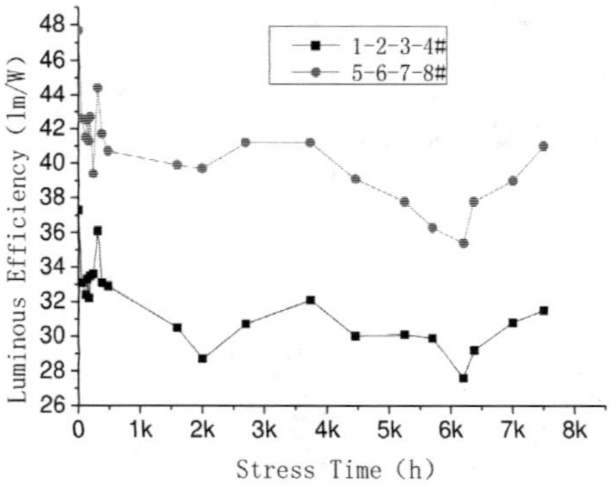

Fig.3 Colour temperature vs. Stress time

4. The ability of LED invert electricity into light is luminous efficiency, due to the light absorb of the light fitting, as table 1, the luminous efficiency of it is 3/4 of the 5-6-7-8 LED, in the process of aging, it has the same result (figure 2).

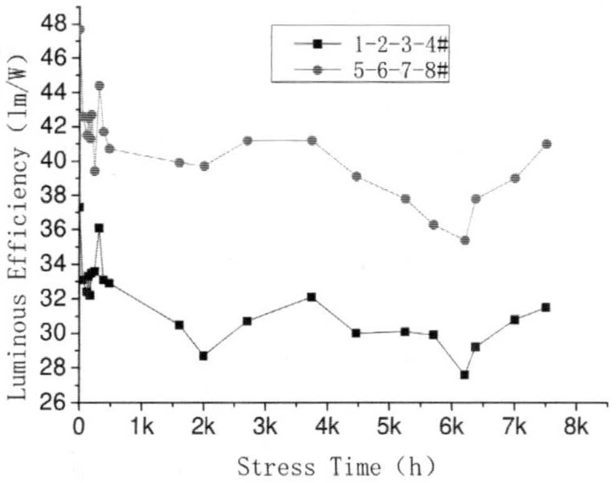

Fig.4 Luminous efficiency vs. Stress time

5. From the exterior of the assembled light fitting, the main change is reflecting shade, the colour change from cream white to pale white and the shape is changed, and it shows that the reliability of reflecting shade is not good, it need amelioration.

LED chip junction and heat sink temperature is linear, the higher temperature heat sink, LED light the greater the decline[4].Then, in the process of aging, we tested the different places of assembled light fitting using thermocouple, the result shown that, the average temperature of assembled light fitting is between 17℃ and 36℃. So, this type of assembled light fitting can meet the needed of LED radiating, and can maintain the LED work stable and long-term.

4. Summary

Based on the equipments available of the laboratory, the ageing test on LED streetlights was designed. The results showed that the initial flux reduction was mainly due to the absorption of the light lamp or the block by some parts of the lighting. And the late light declining was mainly caused by the decay of the high-power LED light source itself. There was a direct relationship between the degree of the attenuation and the quality of LED products. So we should choose feasible high-power LED light resource when designing LED streetlights. Moreover, we should try to avoid existence of light blocked. And materials with high thermal conductivity should be used on the contacts of the LED light resource and the cooling plate of lighting. Because LED lighting system to add thermal coating can improve the cooling efficiency of LED systems, the higher thermal conductivity material, the better heat dissipation[5].In addition, the design of article lamps and the use of the reflective bow should be improved . The structure should be optima led. And we should try to lessen the energy loss due to the absorption of the light lamps so that the efficiency could be improved. Furthermore, although the reflective bowl could meet the requirements of light distribution, the light lamp was less reliable. It needed for further improvements to ensure that the whole LED circuit was reliable for a long time.

To sum up , to design a good LED streetlights products , designer's knowing the quality of LED products well and having a good knowledge on the differences between the unique nature of LED light source and the nature of traditional LED source are not enough. Furthermore, the cooperation and communication between the workers from LED manufacturing industry and lighting-made industry should be strengthened and the standards for road lighting, corresponding to the performance of LED source, should be drawn as soon as possible.

5. References

[1] M. Alan, "Solid state lighting-a world of expanding opportunities at LED 2002,"III-Vs Review, Vol 16 (1) (2003), pp.30-33.

[2] M.Alan, "Lighting: The progress & promise of LEDs,"III-Vs review, Vol 17 (4) (2004), pp. 39-41.
[3] Eugene Hong Nadarajah Narendran Third International Conference on Solid State Lighting (SPIE, Bellingham, WA, 2004), p.93-95, (2004).
[4] ZHANG D. Automated Biometrics-Technologies and Systems.USA: KluwerAcademic Publishers, 2000: 102.
[5] Shufang Zhang, Liang Fang, Guangzong Fu. Effect of Thermal conductive coating on Thermolysis Property of LED. Semiconductor Optoelectronics, 2007,28（6）：793-786

Photochromic and Field Emission Properties of Ag-TCNQ Micro/Nanostructures

Chunnuan Ye [1], Guanying Cao [2], Kaibo Zheng [3] and Guorong Chen [3]

[1]College of Chemistry, Chemical Engineering and Materials Science, Soochow University, Suzhou 215123, China;

[2]Research Institute of Photonics, Dalian Polytechnic University, Dalian 160034, China;

[3]Department of Materials Science, Fudan University, Shanghai 200433, China

Corresponding author, E-mail: cnye@suda.edu.cn; grchen@fudan.edu.cn

Abstract. To study the opto/electrical properties of metal organic complex Ag-TCNQ mico/nano structure, they were grown on Si, glass and silver plate substrates by a solution reaction and a vapor-transport reaction, respectively. They were firstly characterized with scanning electron microscopy (SEM), X-ray diffraction (XRD), and Raman spectroscopy. Then Raman observation of single microwire and field emission display of nanowires array were studied experimently.

1. Introduction

The morphology, structure and phase control is a long standing problem for understanding the switching effect of charge transfer complex M-TCNQ composed of one metal and organic molecular tetracyanoquinodimethane(TCNQ). Among of them, Ag-TCNQ and Cu-TCNQ are both reported with two phases in form of bulk materials [1,2]. Xiao K et. al have reported the synthesis of Ag-TCNQ phase II as a three dimensional nanowires array by a vapor–solid chemical reaction growth method in the temperature region of 150-180^oC at "atmosphere" pressure with the assistance of argon flow[3]. In our previous work, Ag-TCNQ nanostructured arrays were grown on single crystal silicon substrate by a "vacuum" vapor-transport reaction to study the influence of growth conditions on the structure, morphology and field emission properties. And the field emision from Ag-TCNQ nanowires has also been reported by other work[4, 5], but in fact the phase of them is not clear or

Published under licence by IOP Publishing Ltd

doubtful. In our study, the nanowires synthesied at different tempertures are in different phases and the phase change from II to I happens with increasing the reaction temperature[6].

2. Experimental details

Metal organic complex Ag-TCNQ mico/nano structure were grown on Si, glass and silver plate substrates by a solution and a vapor-transport reaction, respectively. They were characterized with X-ray diffraction (XRD), scanning electron microscopy (SEM) and Raman spectroscopy. Then Raman observation of single microwire was carried out to study the photochromic property and a field emission current density-field curve and display image of nanowires array were tested in a two-teminate device by a home-made setup.

Raman measurements were carried out on LabRam-1B microscope Raman spectrum instrument (Dilor, France). The source exciting laser wavelength is 632.8 nm and the power is 3 mW. The photochromic measurement was carried out at room temperature in air. Using the CCD system attached to the measurement system, Ag-TCNQ microwires can be easily identified from the as-prepared sample and the light spot was adjusted on one single crystal.

3. Results and discussion

The XRD analysis showed they were crystals in form of phase II and phase I by different method, respectively. The SEM results showed that Ag-TCNQ wire or tube synthesized by the solution reaction was in a micro/nano scale. Ag-TCNQ nanowires by the vapor reaction between Ag droplets and TCNQ molecules were grown vertical to the silver plate substrate, while those were mussy on the glass substrate by the same vapor method. Raman data showed they were both incomplete charge-transfer product.

After laser irradiation toward a micro single Ag-TCNQ wire, the dark blue colored wire changed into a yellow one. Moreover，a characteristic peak of neutral TCNQ was observed in the Raman spectrum, but vanished after 30 s or longer laser irradiation. So the single Ag-TCNQ quasi-one-dimensional micro/nano wire displays a reversiable photochromic property which can be used to make optical memory elements and optical switches.

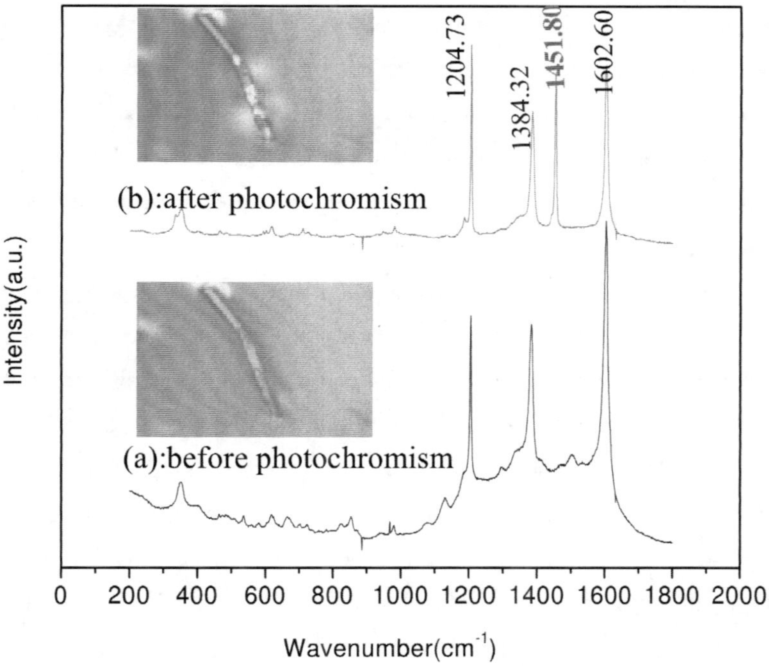

Fig.1. Raman spectra of a single Ag-TCNQ crystal before and after photochromism.

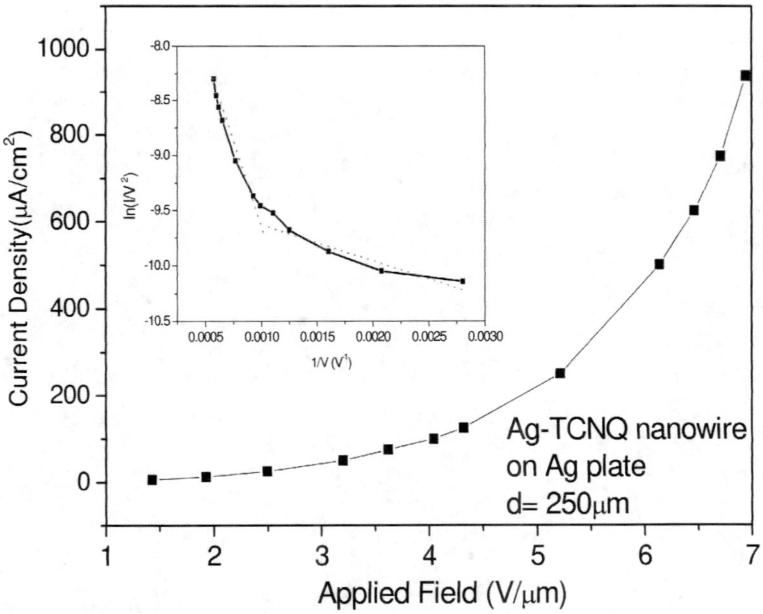

Fig. 2 *I-E* curve and F-N plot(inset) of field emission for Ag-TCNQ nanowires on Ag plate

Journal of Physics: Conference Series **276** (2011) 012198 doi:10.1088/1742-6596/276/1/012198

Direct experiment for field emission of Ag(TCNQ) nanowires array on Ag plate is carried out here. Results are shown in Fig. 2. It shows the *I-E* curve and its corresponding F-N plots(inset) for field emission from Ag-TCNQ nanowires reacted with 30 nm-thick Ag film on Ag plate. The *I-E* curve has a seemingly feature of the exponential relation. The turn-on field for emission is about 1.50 V/μm, which is much lower than those nanowires on Si substrate. The F-N plot is seperated with low and high field regions for the whole emission process, not in complete agreement with the exponential relation especially in the low field region. The slopes and intercedes of these two dotted lines showed that the effective emission area and the work function for the Ag-TCNQ nanowires are not constant under different fields.

As far as the enhancement of field emission from the special structure signed by ITO/Ag/Ag(TCNQ) to ITO/Ag(TCNQ), the contribution of the buffer layer of metal Ag film is considered to be predominant [4]. Here, the as-grown Ag-TCNQ nanowires generating from 30 nm-thick Ag film on Ag plate and under other same conditions have the same morphology and phase structure with that from the buffer layer. XRD results show it belongs to phase I. So the field emission is in the same or alike order.

Further experiment was performed to fabricate field emitter in order to evaluate the properties of Ag-TCNQ nanowires with form of phase I in the device application. The following Fig.3 showed the photographs of prototype device and lighting display image during the emission at the field of 3.0 V/μm, They suggested a promising reliability in FE device application.

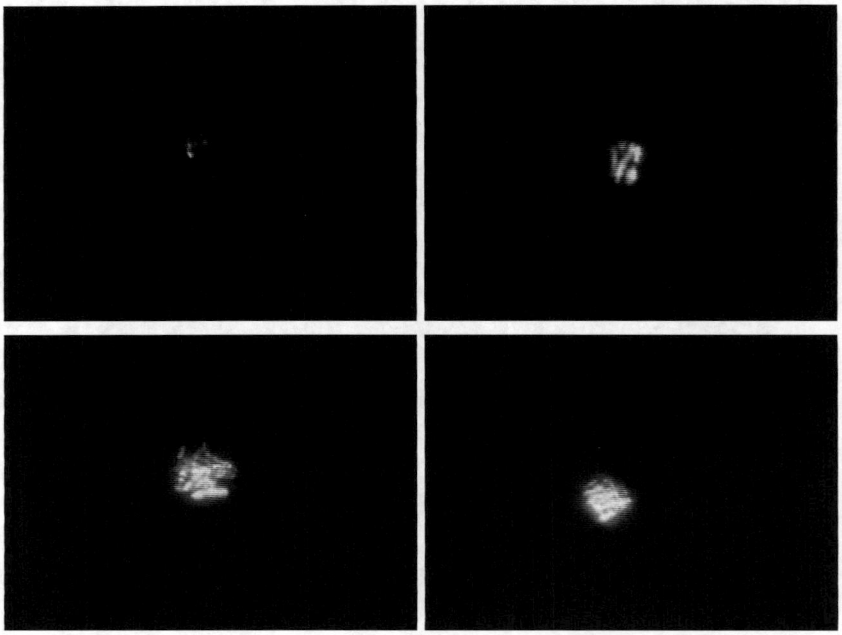

Fig. 3 Photographs of display image from prototype device at the field of 3.0 V/μm

4. Conclusion

Ag-TCNQ single micro/nano wire and nanowires array, in form of phase II and I, were successfully grown on Si wafer and on Ag plate or glass substrate without any template. The reversiable photochromic property from micro/nano wires can be used to make optical memory elements and optical switches. Field emission enhancement were gained by adding a μm-thick metal buffer layer between the glass or silver plate substrate and the nanowires. This makes it convenient to further study their fundamental problem and potential application in optical storage, ultrahigh-density nanoscale memory, and logic devices and display devices as well.

References:

1. S. A. O'Kane, R. Clérac, H. Zhao, X. Ouyang, J. R. Galán-Mascarós, R. Heintz, K. R. Dunbar. New Crystalline Polymers of Ag(TCNQ) and Ag($TCNQF_4$): Structures and Magnetic Properties. Journal of Solid State Chemistry. 2000, 152(1): 159-173
2. R. A. Heintz, H. Zhao, X. Ouyang, G. Grandinetti, J. Cowen, K. R. Dunbar. New Insight Into the Nature of Cu(TCNQ): Solution Routes to Two Distinct Polymorphs and their Relationship to Crystalline Films that Display Bistable Switching Behavior. Inorg. Chem. 1999, 38(1): 144-156
3. Xiao, K. Tao, J. Puretzky, A A, Ivanov, I N, Retterer, S T, Pennycook, S J, Geohegan, D B. Selective Patterned Growth of Single-Crystal Ag-TCNQ Nanowires for Devices by Vapor-Solid Chemical Reaction. Adv. Funct. Mater. 2008, 18: 3043–3048.
4. Zheng, K, Li, X, Mo, X, Chen, G, Wang, Z, Chen, G. Applied Surface Science. 2010, 256: 2764-2768
5. H. Liu, Q. Zhao, Y. Li, Y. Liu, F. Lu, J. Zhuang, S. Wang, L. Jiang, D. Zhu, D. Yu. Field Emission Properties of Large-Area Nanowires of Organic Charge-Transfer Complexes. J. Am. Chem. Soc. 2005, 127(4): 1120-1121
6. Chunnuan Ye, Kaibo Zheng, Wenlong You, Guorong Chen. Nanoscale Research Letters. 2010, DOI 10.1007/s11671-010-9643-9

The effects of V/III ratio on nonpolar *a*-plane GaN films grown on *r*-plane sapphire by MOCVD

J Yin[1], Z H Wu[1]*, Y Q Sun[1], Y Y Fang[1], H Wang[1], C Feng[1], J Zhang[1], J N Dai[1], and C Q Chen[1]

[1]Wuhan National Laboratory for Optoelectronics, College of Optoelectronic Science and Engineering, Huazhong University of Science and Technology, Wuhan 430074, China

E-mail: wuzhihao.sd@gmail.com

Abstract. It was critical to optimize V/III ratio for growing high quality GaN epilayers, because a suitable V/III ratio can promote the two-dimensional growth mode and decrease the crystalline anisotropy. In this paper, in order to investigate the influence of the V/III ratio on the surface morphology and the crystal quality of nonpolar *a*-GaN, *a*-plane GaN thin films with different V/III ratios were grown on *r*-plane sapphire substrates with low-temperature (LT) GaN buffer layers (BLs) by metal-organic chemical-vapor deposition (MOCVD). The surface morphology of these *a*-GaN thin films was characterized by optical microscopy (OM) and atomic force microscopy (AFM). The crystal quality of the *a*-GaN thin films was investigated by high resolution X-ray diffraction (HRXRD). The results showed that the *a*-GaN thin films with a V/III ratio of 1200 had the smoothest surface morphology with a root mean square (RMS) surface roughness of 1.6 nm and the best crystal quality with a full-width-at-half-maximum (FWHM) of $(11\bar{2}0)$ x-ray rocking curve of 864 arcsec.

1. Introduction

Currently, III -nitrides have been widely used in the field of optoelectronics and microelectronics [1, 2]. Most of the commercial GaN-based blue/green light-emitting diode (LED) materials are grown along the polar *c*-axis on *c*-plane sapphire substrates. Due to noncentrosymmetric crystal structure, the positive and negative charge centers do not coincide and which induces a bulk spontaneous polarization along the *c*-axis. The in-plane strain in the *c*-plane nitrides will also induces piezoelectric polarization along the polar *c*-axis. The spontaneous and piezoelectric polarization fields existed in this kind of epitaxial structure would build large internal electrostatic fields of the order of ~MV/cm in the active layers. The internal electrostatic field would reduce the overlap of the electron-hole wave functions, and consequently lead to a decrease in the rate of radiative recombination which brings low internal quantum efficiency in the active layers [3]. On the other hand, GaN grown along nonpolar directions, such as *a*-plane $(11\bar{2}0)$ and *m*-plane $(1\bar{1}00)$, are free of spontaneous and piezoelectric polarization along the growth direction [4-6]. Therefore, these nonpolar GaN epilayers are expected to have a distinct advantage for fabricating high-efficiency InGaN-based LEDs over conventional *c*-plane counterpart [7]. In spite of of great efforts to develop nonpolar GaN based LEDs, the internal quantum efficiency is still lower than those based on *c*-plane sapphire substrate [8-14]. The main

* Corresponding author Tel: +86-27-87793024, Fax: +86-27-87793024

reason for the low internal quantum efficiency of nonpolar GaN based LEDs is the presence of high density of threading dislocation and stacking faults [15]. Therefore, the emission efficiency is still low because the high density of defects acts as the efficient non-radiative recombination centers. Hence, the improvement in crystal quality of non-polar nitrides still needs more efforts.

In this paper, the effects of V/III ratio on nonpolar *a*-plane GaN films grown on *r*-plane sapphire by MOCVD had been investigated. The surface morphology and the crystalline quality of the nonpolar *a*-plane GaN films have been found to be sensitive to V/III ratio, and a suitable V/III ratio can improve the crystal quality and surface morphology.

2. Experiments

All the three *a*-plane ($11\bar{2}0$) GaN films were grown on *r*-plane ($1\bar{1}02$) sapphire substrates by low-pressure (LP) MOCVD. Figure 1 shows the schematic structure of these samples. Before the growth, the *r*-plane sapphire substrates were annealed at 1100 ℃ for 5 minutes in hydrogen (H_2) ambient. Following depositing low-temperature (LT) GaN buffer layers (BLs) at 960 ℃ for 6 minutes, thick GaN layers about 2.5 μm were deposited at 1020 ℃ with different V/III ratio. The different V/III ratio from 800 to 2400 was changed by flow rate modulation of TMGa from 100 to 300 sccm. The approximate thickness of the GaN films with different V/III ratio was controlled by growth time from 60 to 30 minutes. The chamber pressure of MOCVD system kept at 40 Torr when in the growth proceeds. Trimethylgallium (TMGa) and ammonia were used as the Ga and N sources, respectively. The surface morphology of these *a*-GaN thin films was characterized by optical microscopy (OM) and atomic force microscopy (AFM). The crystal quality of the *a*-GaN thin films was investigated by high resolution X-ray diffraction (HRXRD).

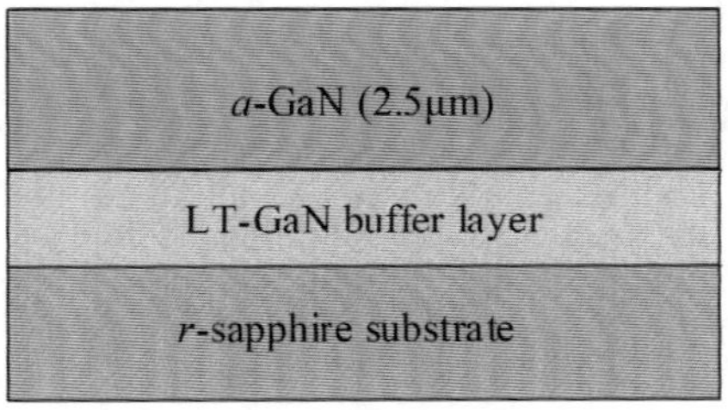

Figure 1. The schematic structures of the samples.

3. Results and discussions

The grown samples were firstly examined with high-resolution XRD 2theta-omega (2θ-ω) scans with the X-ray incident plane parallel to the *c*-axis of the *a*-plane GaN films. Figure 2 shows the scan pattern of sample B. The diffraction peak of ($11\bar{2}0$) plane GaN at 2theta 57.7 ° indicated that the grown film was *a*-plane GaN. Table 1 shows the FWHM value of X-ray Rocking Curve (XRC) of *a*-plane ($11\bar{2}0$) GaN films parallel/perpendicular to *c*-axis. The FWHM value parallel to *c*-axis varies significantly with the change in V/III ratio, and the sample B has the smallest value of 864 arcsec. On the other hand, for the XRC perpendicular to *c*-axis, the FWHM value is almost the same in all three samples.

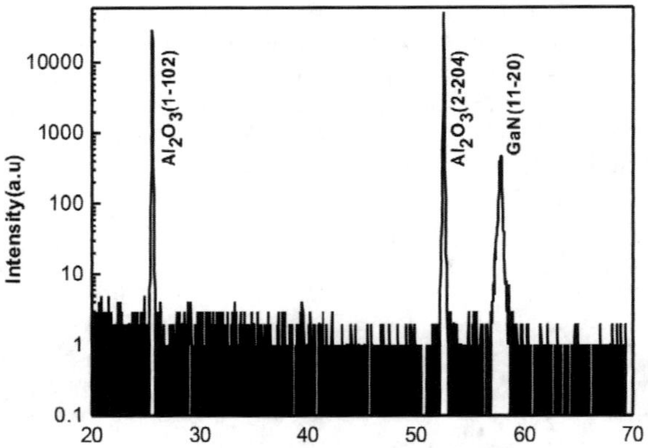

Fig. 2. XRD 2θ-ω scan pattern of the *a*-plane GaN film sample B.

Table 1 FWHM value of X-ray Rocking Curve of *a*-plane ($11\bar{2}0$) parallel/perpendicular to *c*-axis for the *a*-plane GaN samples grown with different V/III ratios

Sample No.	V/III ratio	Rocking curve FWHM // *c*-axis (arcsec)	Rocking curve FWHM ⊥ *c*-axis (arcsec)
A	2400	936	2156
B	1200	864	2120
C	800	972	2168

The surface morphology of the grown samples was then examined by Optical-Microscope and AFM. Figure 3 shows the optical images of the sample A, B and C with different V/III ratios. Although all the three *a*-plane GaN films showed a fully coalesced surface morphology with a few occasional pits, the *a*-plane GaN film grown with the V/III ratio at 1200 showed a smoother surface, indicating a significant improvement in surface morphology.

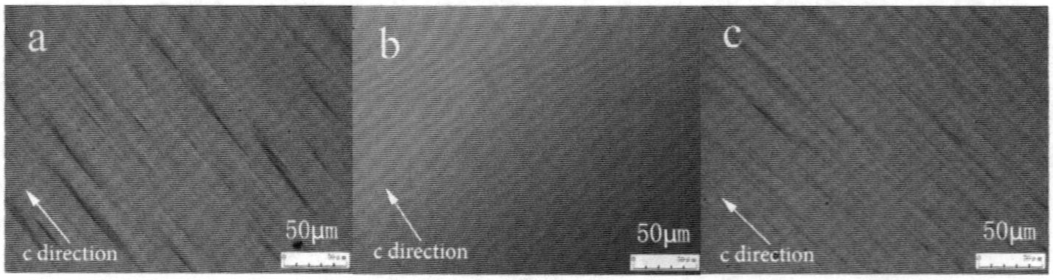

Figure 3. Optical images of samples A (a), B (b) and C (c) under 500× magnification.

The 5×5 μm^2 AFM images of all the *a*-plane GaN films were shown in Figure 4. The AFM image of a fully coalesced and pit free *a*-plane GaN film with a V/III ratio at 1200 exhibits a smoother surface (see Fig. 4 (b)). Stripe-like rise and fall in sample B with a V/III ratio at 1200 became not so obvious compared with sample A and C. The RMS roughness over a scanned area of 5×5 μm^2 for sample A, B and C was 2.4 nm, 1.6 nm and 3.6 nm respectively. The dramatic reduction of RMS roughness to 1.6 nm indicates a significant improvement in the surface morphology after optimizing the V/III ratio.

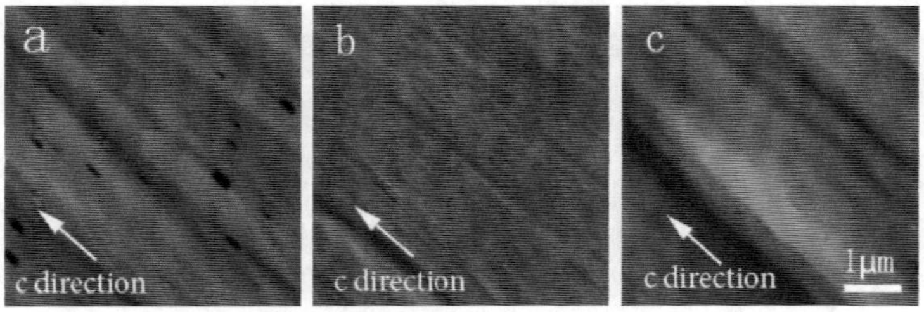

Fig. 4. 5×5 μm^2 AFM images of samples A (a), B (b) and C(c).

4. Conclusion

Fully coalesced and relatively higher crystalline-quality *a*-plane GaN thin films with different V/III ratios were grown on *r*-plane sapphire substrates with low-temperature (LT) GaN buffer layers (BLs) by MOCVD. The V/III ratio had a significant effect on the crystalline-quality and especially on the surface morphology of nonpolar *a*-GaN films grown on *r*-plane sapphire substrate. A suitable V/III ratio could be used to grow higher crystalline-quality *a*-plane GaN thin films with a smoother and pit-free surface morphology.

5. Acknowledgement

This work was supported by the Doctoral Program Foundation of Institutions of Higher Education of China (Grant No. 200804871144), the Hubei Province Science Fund for Distinguished Young Scholars (Grant No.2008CDB334), the National Natural Science Foundation of China (Grant No. 51002058), and the National Natural Science Foundation of China (Grant No.60976042).

References

[1] H. Xing, M. D. Craven 2008 *J. Appl. Phys.* **101** 023506
[2] L. Shen, S. Keller, A. Chakraborty, S. Heikman, S. P. DenBaars, U. K. Mishra, J. Liberis, O. Kiprijanovic, A. Matulionis 2006 *Appl. Phys. Lett.* **89** 073508
[3] M. Kuroda, H. Ishida 2007 *J. Appl. Phys.* **102** 093703
[4] V. Darakchiev, T. Paskov, M. Schubert 2007 *Phys. Rev.* **75** 195217
[5] M. D. Craven,S. H. Lim, F. Wu, J. S. Speck, S. P. DenBaars 2002 *Appl. Phys. Lett.* **81** 1201
[6] F. Wu, M. D. Craven, S. H. Lim, J. S. Speck 2003 *J. Appl. Phys.* **94** 166
[7] P. Waltereit, O. Brandt, A. Trampert, H. T. Grahn, J. Menniger, M. Ramsteiner, M. Reiche, and K. H. Ploog 2000 *Nature* **406** 865
[8] T. Detchprohm, M. Zhu, Y. Li, Y. Xia, C. Wetzel, E. A. Preble, L. Liu, T. Paskova, and D. Hanser 2008 *Appl. Phys. Lett.* **92** 241109
[9] S. M. Hwang, Y. G. Seo, K. H. Baik, I. S. Cho, J. H. Baek, S. Jung, T. G. Kim, and M. Cho, 2009 *Appl. Phys. Lett.* **95** 071101
[10] A. Chakraborty, B. A. Haskell, S. Keller, J. S. Speck, S. P. Denbaars, S. Nakamura, and U. K. Mishra 2005 *Jpn. J. Appl. Phys.* **44** 173
[11] M. Funato, M. Ueda, Y. Kawakami, Y. Narukawa, T. Kosugi, M. Takahashi, and T. Mukai, 2006 *Jpn. J. Appl. Phys.* **45** 659
[12] B. Neubert, T. Wunderer, P. Bruckner, F. Scholz, M. Feneberg, F. Lipski, M. Schirra, and K. Thonke 2007 *J. Cryst. Growth* **298** 706

[13] M. Schmidt, K. C. Kim, H. Sato, N. Fellows, H. Masui, S. Nakamura, S. P. DenBaars, and J. S. Speck 2007 *Jpn. J. Appl. Phys.* **46** 126
[14] T. Fujii, Y. Gao, R. Sharma, E. L. Hu, S. P. DenBaars, and S. Nakamura 2004 *Appl. Phys. Lett.* **84** 855
[15] M. D. Craven, S. H. Lim, F. Wu, J. S. Speck, and S. P. DenBaars 2002 *Appl. Phys. Lett.* **81**, 469

Optical properties of $LuFeO_3$ thin films prepared by pulsed laser deposition

L P Zhu[1], H M Deng[2], X X Li[1], P X Yang[1*] and J H Chu[1]
[1] Key Laboratory of Polar Materials and Devices, Ministry of Education, Department of Electronics, East China Normal University, 500 Dongchuan Road, Shanghai 200241, China
[2] Key Laboratory for Material Microstructures of Shanghai University, Shanghai University, 99 Shangda Rd, Shanghai 200444, China

Abstract. The $LuFeO_3$ thin films were successfully prepared by pulsed laser deposition on silicon and $LaNiO_3$-coated silicon substrates, respectively. The crystalline structure of the films has been studied by x-ray diffraction and indicated that all the films were well-crystallized. Atomic force microscope was used to characterize the surface morphology of the films and the surface was uniform and dense. Raman spectra measurements were carried out to study the lattice vibration modes of the films in the 200-1000 cm^{-1} wave number range. Polarization-electric field hysteresis loops of the $LuFeO_3$ thin films were measured at applied electric field.

1. Introduction

Ferroelectric materials are used in modern electric devices, such as memory elements, filtering devices and high-performance insulators [1-4]. $RFeO_3$ compounds, belonging to the rare-earth orthoferrite, are a big family, where R is a trivalent rare earth cation. They have orthorhombic distorted perovskite structures (space group Pnma) [5]. They have been found a host of fascinating magnetic properties including canted anti-ferromagnetism and spin reorientation transitions [6]. $LuFeO_3$ (LFO) is one of the $RFeO_3$ families.

Recently, the charge–frustrated system $LuFe_2O_4$ and its dielectric and multiferroic characteristics have attracted much attention of the scientists [7]. $LuFeO_3$ can be seen the giant dielectric constant step [8]. Kumar [9] has prepared hexagonal metastable phase from undercooled $LuFeO_3$ melting with low oxygen partial pressure.

In this work, the $LuFeO_3$ thin films were prepared on silicon and $LaNiO_3$/Si substrates by pulse laser deposition. The structural, electrical and optical properties of the films were investigated.

2. Experimental

The $LuFeO_3$ thin films were prepared by pulsed laser deposition (PLD) using the KrF (λ_{ex} = 248 nm) pulsed excimer laser ablating a $LuFeO_3$ target. The films were deposited on Si substrate and $LaNiO_3$-coated Si (LNO) substrate, respectively. The base pressure of the deposition chamber was 7×10^{-4} Pa. The films were deposited in oxygen or argon atmosphere for 30 minutes. The substrate temperature is 750-800 °C. After deposition, the films were in situ heat preservation in oxygen or argon atmosphere.

* Corresponding author. Tel.: +86 21 54345157; fax: +86 21 54345119. E-mail address: pxyang@ee.ecnu.edu.cn (P. Yang).

The structure of the films was investigated by x-ray diffraction (XRD) using D/max2200. The surface morphology of the films was examined by tapping mode atomic force microscopy (AFM) (Veeco Dimension 3100). For the investigations of the electrical properties of the films, top Pt electrodes with area of 9×10^{-4} cm^2 were sputtered on the films. Ferroelectric property was measured by Precision Materials Analyzer (Precision Premier II) at 1 KHz with different applied electric field. The optical properties of the films were studied by Raman spectra using Jobin-Yvon LabRAM HR 800UV micro-Raman spectrometer equipped with He–Cd 325 nm (3.82eV) laser as an exciting source.

3. Results and discussion

Figure 1 shows the XRD patterns of the LFO thin films deposited on LNO substrate. The crystal structure of the LFO film at room temperature was determined to be perovskite structures. All the samples have been well-crystallized and give x-ray peaks corresponding to LFO except for the impure phase, which turns out to be Si. The peaks at (110) and (220) are strong and sharp, which indicate that the PLD method has grown the films with preferred well orientation at 750 °C.

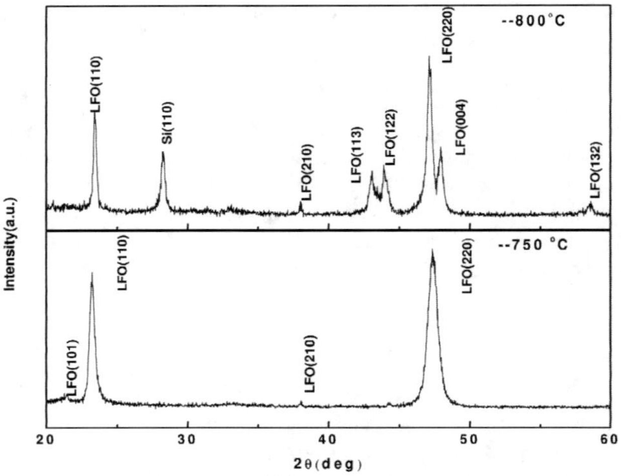

Figure 1. XRD patterns for the $LuFeO_3$/LNO thin films deposited at 750 °C and 800 °C, respectively

Microstructure and surface morphology of the LFO thin films deposited at 750 °C were characterized by AFM. The surfaces of the films were uniform and dense, as shown in Figure 2. The grain size and root-mean-square roughness were obtained from the analysis of the AFM images. The grain size was estimated to be 50 nm. The root-mean-square roughness was calculated to be 1.231 nm.

Figure 2. AFM images (1×1 μm^2) of the films. (a) two dimensional and (b) three dimensional

Figure 3 shows the room-temperature polarization-electric field (P-E) hysteresis loops of the LFO thin films measured at applied electric field. The remnant polarization (P_r) was detected is 1.38 μC/cm^2, and the coercive field (E_c) was 74.8 kV/cm. The loop was not saturated, indicating that the leakage current of the films is high.

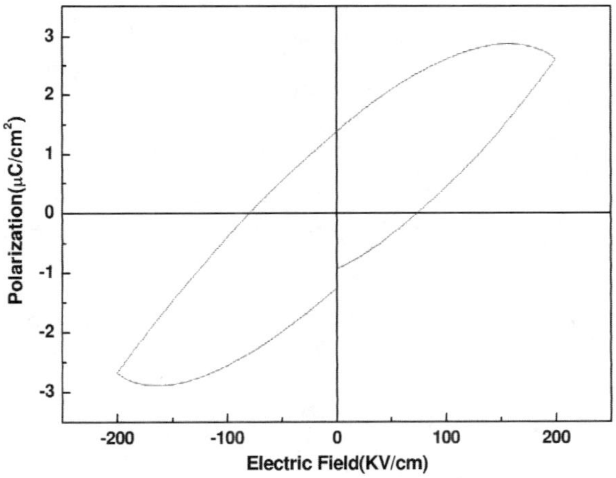

Figure 3. Ferroelectric P-E hysteresis loops of $LuFeO_3$ films

Figure 4 shows the Raman spectra of the $LuFeO_3$ films grown on Si substrate by PLD. For single crystal, Venugopalan [6] proved that there were sixty phonon modes expected for $RFeO_3$ crystals. At the center of the Brillouin zone, the irreproducible representation is given as $7A_g+5B_{1g}+7B_{2g}+5B_{3g}+8A_u+10B_{1u}+8B_{2u}+10B_{3u}$, where A_g, B_{1g}, B_{2g} and B_{3g} were Raman-active modes of species, three infrared modes are species B_{1u}, B_{2u} and B_{3u}, and A_u as the inactive mode.

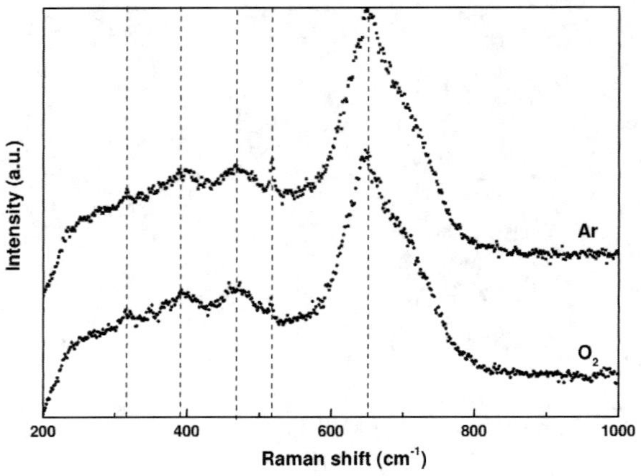

Figure 4. Raman spectra of the $LuFeO_3$ films deposited on Si substrate by PLD

As shown in Figure 4, the Raman peaks are almost the same whatever the films deposited in oxygen and argon atmosphere at 800 °C, respectively. But the intensity of the Raman peaks on the films deposited in argon is stronger than that in oxygen. The peaks appear at 315, 393, 465, 516, 650 cm^{-1}, respectively. There are assignments and symmetries of A_g and B_{1g} spectra, and comparison from present results and the literature [6] is in Table 1. The positions of Raman peaks at present results are slightly different from the literature, which can be attributed to be crystalline state. The films of the present study are polycrystal, and that of the literature was single crystal. In other words, the preparation process changes the oxygen bonding and disorder that is reflected in the vibration frequencies of mode involving oxygen [10].

Table 1 Comparison of Raman peaks from present results and literature

Present results (cm^{-1})	A_g [6] (cm^{-1})	B_{1g} [6] (cm^{-1})
315	278	279
393	350	350
/	427	425
465	450	/
516	516	518
650	654	/

4. Conclusions

The lutetium ferrite thin films were prepared using pulsed laser deposition. The x-ray diffraction results indicate that the $LuFeO_3$ thin films on LNO at 750 °C show a highly orientation. The polarization–electric field hysteresis loops of the film can prove that the LFO film has ferroelectric property. Uniform and dense surface morphology is observed in the AFM micrograph. The first-order Raman scattering gave the peak positions at 315, 393, 465, 516 and 650 cm^{-1}, respectively.

Acknowledgments

The author thanks to Mr. Wenfei Xu for AFM investigation, Mr. Xiangui Chen for optical investigation, Dr. Lin Sun and Mr. Nuofan Ding for experimental work. This work was supported by the National Natural Science Foundation of China (No. 60990312), the State Key Basic Research

Program of China (2007CB924902), Shanghai Leading Academic Discipline Project (B411) and Science and Technology Commission of Shanghai Municipality (10JC1404600).

References

[1] N. Ikeda, H. Ohsumi, K. Ohwada, K. Ishii, T. Inami, K. Kakurai, Y. Murakami, K. Yoshii, S. Mori, Y. Horibe, H. Kito, *Nature* **436** (2005) 1136-1138.
[2] Y. Yamada, K. Kitsuda, S. Nohdo, N. Ikeda, *Physical Review B* **62** (2000) 12167-12174.
[3] M. A. Subramanian, T. He, J. Z. Chen, N. S. Rogado, T. G. Calvarese, A. W. Sleight, *Advanced Materials* **18** (2006) 1737-1739.
[4] N. Ikeda, K. Kohn, N. Myouga, E. Takahashi, H. Kitoh, S. Takeawa, *Journal of the Physical Society of Japan* **69** (2000) 1526-1532.
[5] A. A. Bossak, I. E. Graboy, O. Y. Gorbenko, A. R. Kaul, M. S. Kartavtseva, V. L. Svetchnikov, H. W. Zandbergen, *Chemistry of Materials* **16** (2004) 1751-1755.
[6] S. Venugopalan, M. M. Becker, *Journal of Chemical Physics* **93** (1990) 3833-3836.
[7] Y. Zhang, H. X. Yang, C. Ma, H. F. Tian, J. Q. Li, *Phys Rev Lett* **98** (2007) 247602.
[8] L. Zhang, X. M. Chen, Solid State *Communications* **149** (2009) 1317-1321.
[9] M. S. V. Kumar, K. Nagashio, T. Hibiya, K. Kuribayashi, *Journal of the American Ceramic Society* **91** (2008) 806-812.
[10] D. Kothari, V. R. Reddy, V. G. Sathe, A. Gupta, A. Banerjee, A. M. Awasthi, *Journal of Magnetism and Magnetic Materials* **320** (2008) 548-552.

An optical design and simulation of LED low-beam headlamps

Xiangbing Zhu, Jian Ni[1] and Qiaoyun Chen

The Photo-Electronic Technique Research Center, Anhui Normal University, Wuhu 241000, People's Republic of China

E-mail: nij126@yahoo.com

Abstract. The low-beam headlamp is an important component for the automobile safety. With the improvement of optical efficiency and heat dissipation' technology of white LEDs, it becomes feasible to design low-beam headlamps with LEDs. The principle of B-spline surfaces is used to construct the free-form surface reflector meeting the requirement. First, the initial B-spline surface reflector is established on the basis of the light source structure, emitting features and capability of light distribution. Optical simulation is carried out according to the principle of ray tracing. And then the simulation results will be compared with the standard of photometric characteristics. The segmented surfaces fine-tuning method and the method of trial and error are used to trim the part that failed to meet requirements gradually. The vector groups of surfaces are obtained. Finally, the desired free-form surface reflector meeting the ECE regulations is got. The experimental results can meet the standard of photometric characteristics. The impact of the technique showed in this paper in the field of LED illumination design seems to be a very promising topic.

1. Introduction

The safety of car-driving relates to automotive headlamps' quality. When we are driving at night, headlamps not only illuminate the road ahead to provide drivers with a bright and broad vision, but also avoid causing head-on car drivers glare. With the development of automobile technology, customers have increasingly high quality demands for headlamps.

LEDs (Light Emitting Diodes) have advantages of long service life, low power consumption, small packaging, very short switching time, and anti-vibration, compared with conventional light sources of headlamps [1,2]. In addition, LEDs have the mounting depth on the order of millimeter. It agrees with modern cars' styling demands that car front is low and streamlined. A more usable space inside the car is obvious [3].

Early, manufacturers produced some vehicle lamps of low brightness with LED light sources, for example, CHMSLs (center high mount stop lamps), position lights, turn signals, and daytime running lights [4]. Thanks to the rapid increase in power in LEDs and the development of manufacturing technology, applications in headlamps can now also be possible. In the summer of 2008, Cadillac Escalade Platinum was the first high-volume vehicle in the world to be equipped with Hellas's full-LED headlamps. Subsequently, new Audi A8 also used the LED headlamps.

As LEDs have unconventional structure and light emitting characteristics, use of only LEDs array is difficult to satisfy the illumination distribution in most cases. Therefore, using multiple LEDs in

[1] Author to whom any correspondence should be addressed.

combination with optical elements is necessary. One of the keys is the optical design of low-beam lights in the whole optical design process of LED headlamps.

There are three forms to design optical systems for low-beam lights, namely, refraction systems, reflection systems and hybrid systems. In 2008, André Domhardt etc. designed a projection-type low-beam lamp which satisfied the standard with VRS (virtual reflection or refraction of surface) technology [5]. Lexus LS600h equipped with a full headlamp system with all functions in LEDs. Three projection units and a reflection unit constituted its low-beam part. In China, Jilin Dongguang Group Corporation developed the LED-headlight sample at research funding of the 863 Program. Its low-beam light was composed of three projection units.

For low-beam headlamps using ordinary light sources, free-form segmented surfaces have been applied to design the reflector. Almost no paper introduced optical design methods of LED low-beam lamps within publishing literature, whose design theory and methods are not perfect. The studies are concerned with a new technique for designing LED-based automotive low-beam lamps. To solve the problems above, this article will introduce B-spline surfaces theory into optical design of LED low-beam headlamps. According to the simulation results, the beam pattern met the standard of photometric characteristics of headlamps. We expect to improve the efficiency of designing LED low-beam lamps.

2. ECE legal regulations for low-beam light distributions

The light performance of automotive headlamps is greatly related to driving safety. So there are many regulations and standards about light distributions. For example, there are more than 30 regulations which are directly related to vehicle lighting and signal lights. ECE regulations for low-beam light distributions are very strict. China's standards about light distributions of automobile headlamps mainly refer to the ECE legal regulation. Figure 1 is the low-beam light distributions. André Domhardta etc. drew it in terms of ECE.

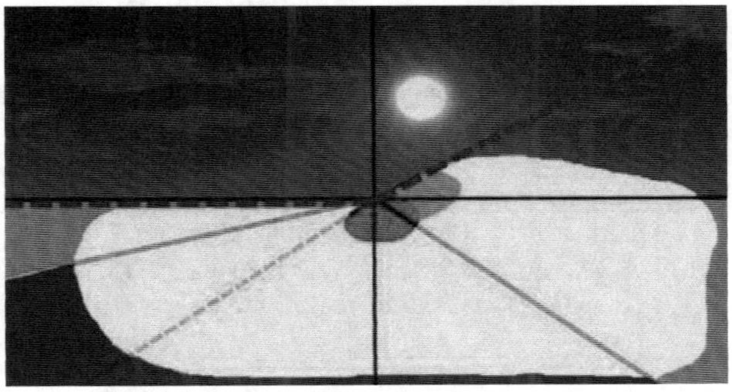

Figure 1. Typical low-beam light distributions.

There is the night sky on the top of the figure, and the road below. The red zones with high brightness, the hot spot, illuminate the road far ahead. A broad yellow distribution illuminating the road surface in front of the car has low illumination. Obviously, there is a blue dotted line in the figure 1. It is called the cut-off line to avoid glaring of the oncoming drivers. In the top of the horizontal portion of the dotted line, light intensity is strictly very weak. The fifteen degrees tilted line is on the right. It mainly provides better lighting for oneself.

As can be seen in figure 1, the desired low-beam light distribution is uneven and asymmetrical. To illuminate both sides of the road ahead, light distributions in horizontal directions need to have a certain expansion. So drivers can acquire a broad and secure view. On the left side of the vertical direction, light intensity above is weak, and the one below is strong. For illuminating distant road surface and traffic signs, the maximum light intensity is in the centre and weakens along the cut-off line of fifteen degrees. As LED light sources compared with traditional ones have small volume and very different optical characteristics, conventional methods for designing the optics of LED low-beam headlamps are hardly practical. In order to meet requirements, new mathematical methods should be introduced.

3. Mathematical expressions of B-spline surfaces

3.1. B-spline curve and its derivative vector

B-spline surfaces are constructed from basic B-spline curves. It has some beneficial properties for designing surfaces: 1.local support; 2. differentiability; 3. convex hull properties. According to the theory of B-spline, B-spline curves can be expressed as:

$$P(u)=\sum_{i=0}^{n} d_i N_{i,k}(u) \tag{1}$$

Where $d_i, i=0,1\cdots,n$ are the control points and $N_{i,k}(u), i=0,1\cdots,n$ are the kth orders B-spline basis functions. $U=[u_0,u_1,\ldots u_{n+k+1}]$ is a knot vector which specifies the distribution of the parameter u on the curve.

There are many different equivalent definitions for B-spline basis functions. The following describes an algorithm and it is defined by

$$\begin{cases} N_{i,0}(u)=1 & u_i \le u \le u_{i+1} \\ N_{i,0}(u)=0 & u<u_i \, or \, u>u_{i+1} \\ N_{i,k}(u)=\dfrac{u-u_i}{u_{i+k}-u_i}N_{i,k-1}(u)+\dfrac{u_{i+k+1}-u}{u_{i+k+1}-u_{i+1}}N_{i+1,k-1}(u) \end{cases} \tag{2}$$

In above formulas， we define $\frac{0}{0}=0$.

According to the above definitions, we can determine the B-spline curves after having given control points and the parameter-interval u. The de Boor algorithm is used to calculate the derivative vector of the B-spline curve. To solve normal vector groups of surfaces, we only introduce an algorithm for obtaining curves' first-order derivative vector (tangent vector). It can be calculated as

$$\left.\frac{\partial p(u,v_0)}{\partial u}\right|_{u=u_0} = \sum_{j=i-k+1}^{i} d_j^1 N_{j,k-1}(u), u\in[u_i,u_{i+1}] \tag{3}$$

$$d_j^1 = k\frac{d_j - d_{j-1}}{u_{j+k}-u_j} \qquad j=i-k+1,\cdots,i \tag{4}$$

3.2. Construction of B-spline surfaces

A B-spline surface in three-dimensional space is defined by

$$P(u,v)=\sum_{i=0}^{m}\sum_{j=0}^{n} d_{i,j} N_{i,k(u)} N_{j,L(v)} \tag{5}$$

Where $d_{i,j}$, $i=0,1\cdots m; j=0,1\cdots n$ are control points, and $N_{i,k(u)}$ *and* $N_{j,L(v)}$ are the Kth-degree and Lth-degree B-spline basis functions, respectively. Assume that we have all control points, the degree of parameters u *and* v and knot vectors. Then the B-spline surface can be defined.

The calculation for normal vectors of B-spline surfaces can refer to B-spline curves' one. When a control point $d_{i,j}$ in the definition domain is given, parameter values (u, v) can be calculated. According to the calculation for normal vectors of B-spline curves, the normal vector of the point $\left.\frac{\partial p(u,v)}{\partial u \partial v}\right|_{u=u_0, v=v_0}$ can be similarly obtained.

4. Construction of free-form reflectors

4.1. Establishment of initial surfaces

In order to construct B-spline surfaces, control points $d_{i,j}$, $i = 0,1 \cdots m; j = 0,1 \cdots n$ firstly should be defined. Let us assume that there are $(m+1) \times (n+1)$ control points located on the paraboloid (see figure 2).

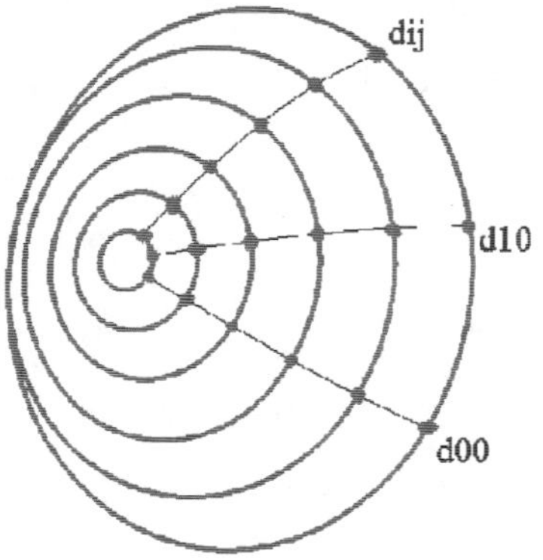

Figure 2. Some controls points located on the paraboloid.

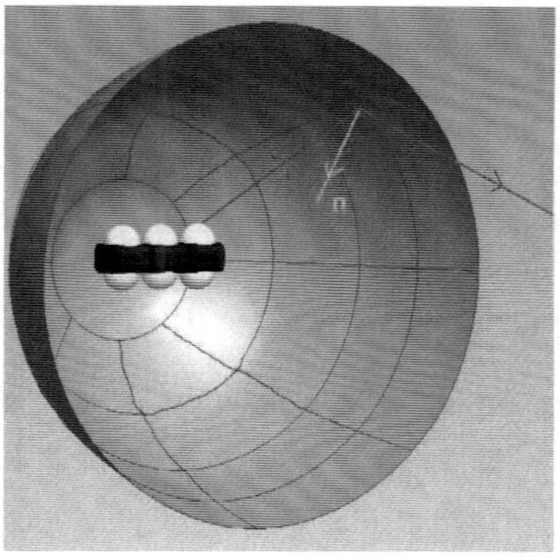

Figure 3. The rudiment of optical system.

The red line-group presents closed B-spline curves of u cluster, the other line segment group presents B-spline curves of v cluster. There are some control points distributed on curves. The density of control points are determined by the size of light sources, the initial reflector and desired light distributions. Control points are calculated according to parabolic equations.

After defining control points, we can fit a $K \times L$-order B-spline surface which is similar to paraboloid. As figure 3 shows, the center of LED array will be placed in the B-spline surface and longitudinal axis of the array and optical axis are made in a vertical.

The size of LED array is slightly smaller than the one of the reflector, so it can not be simplified as point light sources. We need to describe near-field photometric data of LED array. Good manufacturers will provide accurate LEDs' parameters used for modeling and analysis.

4.2. A method for fine-tuning segmented surfaces

As figure 4 shows, the basic reflector is divided into fifteen segmented surfaces. The beam pattern can be got from the contribution of all segments of the reflector. A1, A2, B1, B2, C1 and C2 are part of reflector, respectively. A1 and A2 are mainly used to illuminate central zones. C1 and C2 are provided for forming the level expansion of light distribution. B1 and B2 are for the transition of light beam. Shallow gray area is used to form the cut-off line. Dark gray area is aimed to strengthen the effect of the tilted part of the cut-off line.

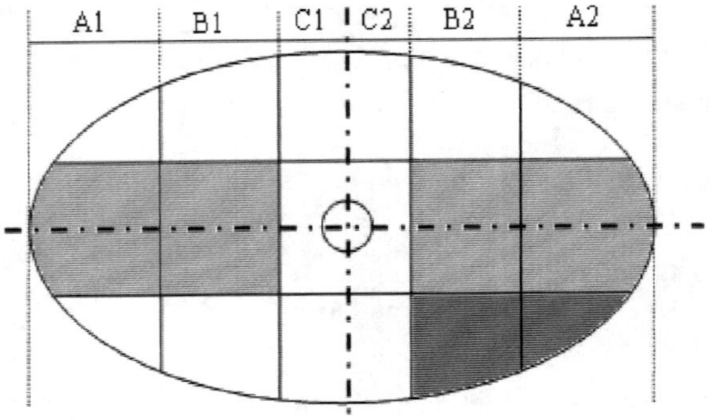

Figure 4. The segmentation of basic reflector.

The fifteen segmented surfaces will be more segmented in the detailed design process to obtain more B-spline segmented surfaces, and then fine-tune them. The function of each segmented surface must refer to the corresponding area on the basis reflector, especially in the center of each segmented surface.

Figure 5 shows the illuminated area corresponding with reflected area on the light-match screen.

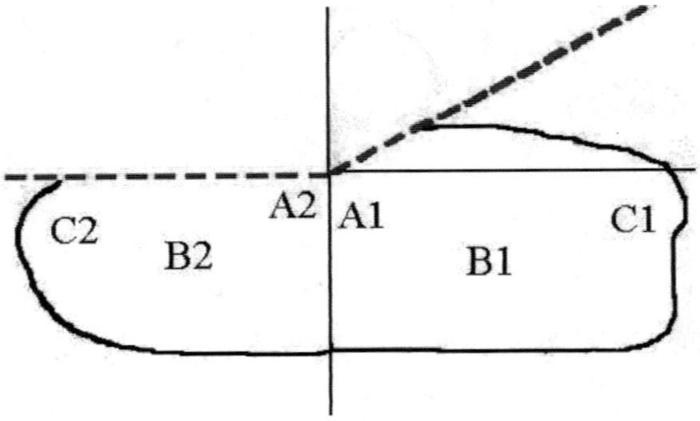

Figure 5. The illuminated area of fine-tuning.

With near-field photometric data of light sources, we can adjust the normal vector of B-spline surfaces by the ray-tracing theory. The principle is shown in figure 6.

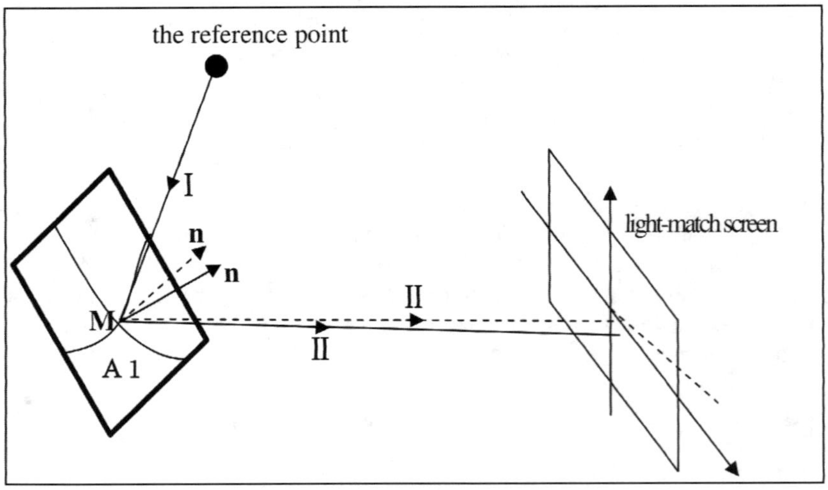

Figure 6. The schematic diagram of fine-tuning.

We choose any intersection of B-spline curves, $d_{i,j}$, in field A1 of reflector, and name the bin with the center of $d_{i,j}$ M. We divide the extended light source into several pieces. Each small piece of light source emits light to the bin M. With received luminous flux of the bin M and position of each piece, we can assume that all rays illuminating the bin M come from a reference point which is obtained by weighting method. The incident ray I is from the reference point to the center $d_{i,j}$ of the bin M. After obtaining the incident ray I, the ray intersection point $d_{i,j}$ and the normal vector $\left.\frac{\partial p(u,v)}{\partial u \partial v}\right|_{u=u_i, v=v_j}$ of the bin M, it is easy to calculate reflected ray II by the ray-tracing theory. Because the above reflector is only the initial surface, the reflected ray II obtained by ray tracing may be diverse on the light-match screen. For example, the intersection point of the reflected ray II may be located on the left of the desired light distribution, and may also be located on the right. To adjust surfaces, we can assume that the reflected ray II deviates from the center region and illuminates field B1. The position of the ideal reflected ray II' is marked in the figure 6. Finally, the required normal vector $\boldsymbol{n}'$ can be easily obtained.

The light distributions are uneven and asymmetrical, so the entire fine-tuning process is relatively complicated. The requirements of light distributions and light intensity are strictly prescribed. The marginal area of the reflector, A1and A2, and the middle regions, C1and C2, are key areas, we must subtly divide them. More and complete desired normal vectors $\boldsymbol{n}'$ are obtained by calculating. For transitional area, B1 and B2, we can adopt an approach of dividing large region, and then gain less normal vectors. To obtain a vector grid, all normal vectors are arranged by the position of control points. New B-spline surfaces are built according to the position of control points and normal vector groups, $[\boldsymbol{n}']$. We use optical software for ray tracing. According to simulation results, we must adopt amendment until the results are consistent with the standard by above design idea. However, it is very difficult to manufacture some of surfaces in the engineering process, and then we need to do re-calculation and simulation to acquire new positions of control points, $d_{i,j}$.

5. Optical simulation of LED low-beam headlamps

We select three LED light sources whose total luminous flux meets the requirements of the low beam lamp. The luminous flux of single LED is 425 lumens. For complete product specifications, please see table 1.

Table 1. Philips LUXEON Altilon product specifications.

Part Number	Form Factor	Min.LuminousFlux @1A	Typical Vf @1A	Typical CCT
LAFL-C2*-0425	1×2	425 lm	7.0	5600K

The initial reflector of B-spline surfaces is established based on the design method above, the structure of LEDs and photometric characteristics. Figures 7 and 8 show the section diagram of the model in adjusting process.

Figure 7. The section diagram of the reflector 1.

Figure 8. The section diagram of the reflector 2.

According to the desired light distributions and the ray-tracing theory, we repeatedly adjust the model. Finally, the distribution curve flux is gained (see figure 9).

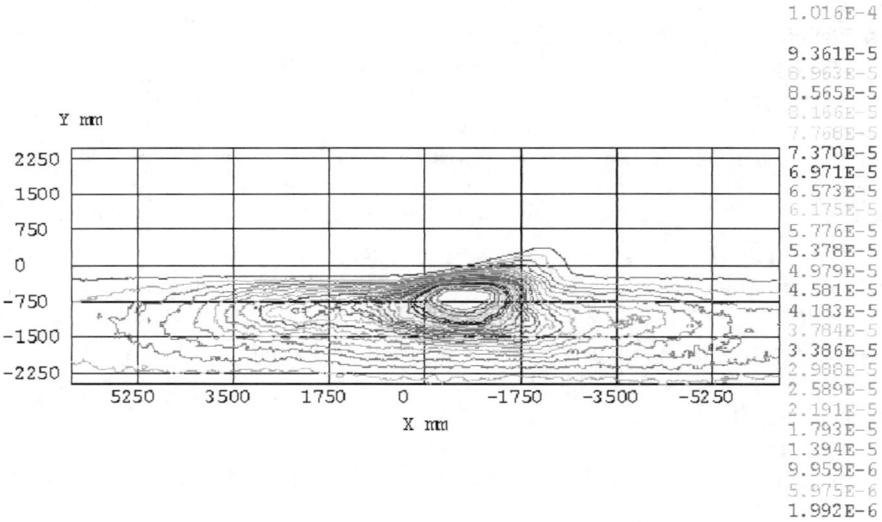

Figure 9. The final simulation results of light distributions.

As can be seen in figure 9, the low-beam light distributions can well fit the regulation. If LEDs can be coupled with lens, light distributions would be better.

6. Conclusion

In this paper, we introduced B-spline surfaces and designed an optical system of LED low-beam headlamps. Finally, the optics satisfying the regulations was obtained. The simulation results show that the method is feasible. In recent work, we will apply this method to design a LED low-beam headlamp for a kind of car. The method can solve optical design of LED low-beam lightings, but the calculation amount is large. We still need further improvement in theory.

Acknowledgments

The authors gratefully acknowledge the financial support provided to this study by the Natural Science Foundation of Anhui Province of China, under grant no. 090414173 and the Science & Research Program of Wuhu of China.

References

[1] Su A, Liu Y C and Chen C Y 2007 *Heat Transfer-Asian Research* **36** p 449-458
[2] Eichhorn K 2006 *Proc. of SPIE* Vol 6134 Streubel K P Yao H W and Schubert E F p 613405-1-6
[3] Wrdenweber B, Wallaschek J, Boyce P and Hoffman D D 2007 *Automotive Lighting and Human Vision* (New York: Springer Publishing Company)
[4] Eichhorn K 2005 *Proc. of SPIE* Vol 5663 T P Pearsall p 7-14
[5] Domhardta A, Rohlfingb U, Weingaertnera S, Klinger K, Kooß D, Manz K and Lemmer U 2008 *Proc. of SPIE* Vol 7003 F Berghmans, A G Mignani, A Cutolo, P P Meyrueis and T P Pearsall p 70032C-1-10

Journal of Physics: Conference Series **276** (2011) 012202 doi:10.1088/1742-6596/276/1/012202

Study on characteristics of terahertz radiation in the surface and far field generated by large-aperture photoconductive antennas

G Chen, T J Cui, Z J Jiang and J Zhang
Institute of Target Characteristics and Identification and State Key Laboratory of Millimeter Waves, Department of Radio Engineering, Southeast University, Nanjing 210096, People's Republic of China

E-mail: gangchen@emfield.org

Abstract. The time-domain characteristics of terahertz radiations emitted from a biased large-aperture photoconductive antenna trigged by an ultra-short optical pulse are studied. Succinct explicit expressions in the surface field and in the far field for the emitted THz radiation fields are deduced based on the well-known current-surge model. The expressions indicate that the THz radiations in the surface and far fields are mainly affected by the incident light intensity, the incident light wavelength, the carrier relaxation time, the reflectivity illuminated at the photoconductor, the carrier lifetime, and the incident light pulse width. The characteristics of THz radiation in the surface and far fields are discussed, and the phenomena of changes in the radiation peak values, the radiation pulse width, and the rising edges are explained in details. Especially the changes in surface fields are explained with the increasing of the ratio of luminous flux. The numerical results indicate that the emitted THz radiation intensity can be increased by enhancing the ratio of luminous flux, prolonging the carrier lifetime, or reducing the incident light pulse width. To avoid the shielding effect of the radiation field, an effectively method is presented that the intense of the incident laser pulse could be divided into multiple beams with low energy to trigger the photoconductive antenna, thus the THz radiation power can be further increased significantly.

1. Introduction

Terahertz (THz) waves occupy a large portion of electromagnetic spectrum between the infrared and microwave bands ranging from 0.1 to 10 THz, having wide potential applications in the field of material science, astronomy, and safety guard and so on. Being lack of efficiently generation and detection methods, the terahertz waves are called "THz gap" before the mid-80s, in the 20th century. The terahertz source is an important part in the terahertz technology and it has become the goal pursued by the scientists in terahertz fields in the 21st century [1].

With the rapid development of ultra-short laser pulse technology over the last decades, the terahertz pulse can be obtained by impelling the semiconductor material or non-linear optical crystal through the ultra-short laser pulse. A lot of terahertz radiation generation and detection methods are proposed such as photoconductive antenna, nonlinear optical rectification, Bloch oscillations and semiconductor lasers [1]. The most applied terahertz generation method propagated in free space is the use of femto-second laser pulse to trigger the biased photoconductive antenna [2]. The distance in the two electrodes of the large-aperture photoconductive antenna is greater than the centre wavelength of electromagnetic radiation. Experimental results show that the use of large-aperture photoconductive antenna can generate relatively strong THz radiation. The mechanism can be depicted by the famous current surge model [3]. The model deems that the THz electromagnetic waves are generated by transient current in photoconductive antenna. Justin T. Darrow [3], A.J.Taylor [4], Sang-Gyu Park [5] and Zhang tongyi [6] discussed the far-field radiation characteristics of the large-aperture THz photoconductive antenna and derived time-domain expression of the THz radiation field. On this basis, the paper further considers the influence of the ratio of luminous flux, carrier lifetime and the incident THz pulses width on the radiation THz pulses in the surface field and in the far field. An effectively method is presented to avoid the shielding effect of the radiation field.

2. Theoretical analysis

Based on the photoconductive incentive mechanism with the large-aperture photoconductive dipole antenna, the conductivity of GaAs is modulated by electron hole pairs generated in GaAs material by the use of ultra-short laser pulse. Suppose that $\boldsymbol{E}_b$ is the DC (direct current) bias electric field between the two electrodes in the chip of photoconductive dipole. When the chip is not illuminated, it lies in high block state. When the chip is illuminated by ultra-short laser pulse, the photo-induced carriers in the chip of GaAs are accelerated and form transient current $\boldsymbol{J}_s$. The accelerated motion of carriers will generate THz radiation pulse. When the incident laser pulse illuminates the chip, the GaAs photoconductive dipole antenna will radiate THz electromagnetic pulse, as shown in Figure 1 [7].

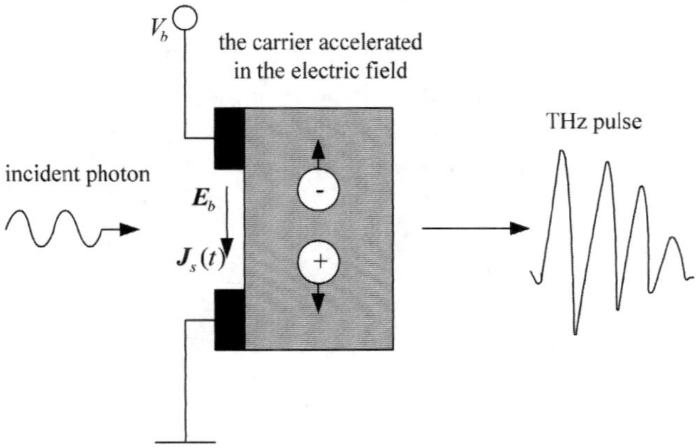

Figure 1. Schematic of THz pulse radiation emitted from GaAs photoconductive antenna.

Through the current surge model [3], the surface current $\boldsymbol{J}_s(t)$ can be expressed as

$$\boldsymbol{J}_s(t) = \sigma_s(t)[\boldsymbol{E}_b + \boldsymbol{E}_s(t)] \quad (1)$$

where $\sigma_s(t)$ is the time-dependent surface conductivity of photoconductive materials, $\boldsymbol{E}_b$ is the static bias field, and $\boldsymbol{E}_s(t)$ is the generated radiation electric field on the photoconductor surface. According to boundary conditions of the Maxwell's equations, the relationship between $\boldsymbol{J}_s(t)$ and $\boldsymbol{E}_s(t)$ is given by [3]

$$\boldsymbol{E}_s(t) = -\frac{\eta_0}{1+\sqrt{\varepsilon}}\boldsymbol{J}_s(t) \qquad (2)$$

where $\eta_0 = 120\pi\ \Omega$ is the free-space wave impendence, and ε is the relative permittivity of the photoconductor.

Substituting equation (2) into equation (1), $\boldsymbol{E}_s(t)$ can be written as

$$\boldsymbol{E}_s(t) = -\frac{\eta_0\sigma_s(t)}{\eta_0\sigma_s(t)+(1+\sqrt{\varepsilon})}\boldsymbol{E}_b\,. \qquad (3)$$

Equation (3) is the radiation surface field by the photoconductive antenna.

Substituting equation (3) into equation (2), the relationship between $\boldsymbol{J}_s(t)$ and $\boldsymbol{E}_b$ can be expressed as

$$\boldsymbol{J}_s(t) = \frac{(1+\sqrt{\varepsilon})\sigma_s(t)}{\eta_0\sigma_s(t)+(1+\sqrt{\varepsilon})}\boldsymbol{E}_b\,. \qquad (4)$$

In the far field, the THz radiation fields can be denoted as [8]

$$\boldsymbol{E}_{far}(r,t) = -\frac{1}{4\pi\varepsilon_0 c^2}\frac{A}{(x^2+y^2+z^2)^{1/2}}\frac{d}{dt}\boldsymbol{J}_s(t-\frac{r}{c}) \qquad (5)$$

in which A is the effective illumination area of photoconductor electrode, (x, y, z) is the coordinate of observation point, and r is the distance from the radiation centre to the observation point.

Suppose that one detects the THz radiation fields along the photoconductive axis (i.e., $x = y = 0$) and $t \to t-(z/c)$, then equation (5) can be reduced as

$$\boldsymbol{E}_{far}(r,t) \approx -\frac{1}{4\pi\varepsilon_0 c^2}\frac{A}{z}\frac{d}{dt}\boldsymbol{J}_s(t)\,. \qquad (6)$$

Substituting equation (4) into equation (6), $\boldsymbol{E}_{far}(r,t)$ is further written as

$$\boldsymbol{E}_{far}(r,t) \approx -\boldsymbol{E}_b\frac{1}{4\pi\varepsilon_0 c^2}\frac{A}{z}(1+\sqrt{\varepsilon})^2\frac{1}{\left[\eta_0\sigma_s(t)+(1+\sqrt{\varepsilon})\right]^2}\frac{d\sigma_s(t)}{dt}\,. \qquad (7)$$

From reference [3], the conductivity of the semiconductor is given by

$$\sigma_s(t) = \frac{q(1-R)}{h\nu}\int_{-\infty}^{t}\mu(t-t')I_{opt}(t')\exp\left(-\frac{t-t'}{\tau_c}\right)dt' \qquad (8)$$

in which q is the electric charge, R is the photoconductor reflectivity, $h\nu$ is the photon energy, I_{opt} is the energy of laser single pulse, τ_c is the carrier lifetime, and μ is the transient mobility of photo-induced carriers.

When the bias voltage is low, the photo-induced carriers are not transferred from the centre valley to the satellite valley. The transient mobility of photo-induced carriers can be expressed as [6]

$$\mu(t) = \frac{q\tau_s}{m^*}\left[1-\exp\left(-\frac{t}{\tau_s}\right)\right] \qquad (9)$$

in which m^* is the electron effective mass and τ_s is the relaxation time of photo-induced carriers.

To analyze the influence of the incident light pulse, I_{opt} is supposed to be Gaussian function [4]

$$I_{opt}(t) = \frac{F}{2\sqrt{\pi}\Delta t}\exp\left(-\frac{t^2}{\Delta t^2}\right) \tag{10}$$

where F is the total luminous flux, and $\sqrt{\ln 2}\Delta t$ is the full width at half maximum of excitation light pulse. Substituting equation (9) and equation (10) into equation (8), we have:

$$\sigma_s(t) = \frac{q^2\tau_s(1-R)F}{4h\nu m^*}X(t) \tag{11}$$

$$\frac{d\sigma_s(t)}{dt} = \frac{q^2\tau_s(1-R)F}{4h\nu m^*}Y(t) \tag{12}$$

where

$$X(t) = \left\{\exp\left(-\frac{t}{\tau_c}\right)\exp\left(\frac{\Delta t}{2\tau_c}\right)^2\left[1+erf\left(\frac{t}{\Delta t}-\frac{\Delta t}{2\tau_c}\right)\right] - \exp\left[-t\left(\frac{1}{\tau_c}+\frac{1}{\tau_s}\right)\right]\exp\left[\frac{\Delta t}{2}\left(\frac{1}{\tau_c}+\frac{1}{\tau_s}\right)^2\right]\right.$$
$$\left.\times\left[1+erf\left(\frac{t}{\Delta t}-\frac{\Delta t}{2}\left(\frac{1}{\tau_c}+\frac{1}{\tau_s}\right)\right)\right]\right\} \tag{13}$$

$$Y(t) = \left\{\exp\left(\frac{1}{\tau_c}+\frac{1}{\tau_s}\right)\exp\left(-t\left(\frac{1}{\tau_c}+\frac{1}{\tau_s}\right)\right)\exp\left(\frac{\Delta t}{2}\left(\frac{1}{\tau_c}+\frac{1}{\tau_s}\right)\right)^2\left[1+erf\left(\frac{t}{\Delta t}-\frac{\Delta t}{2}\left(\frac{1}{\tau_c}+\frac{1}{\tau_s}\right)\right)\right]\right.$$
$$\left.-\frac{1}{\tau_c}\exp\left(-\frac{t}{\tau_c}\right)\exp\left(\frac{\Delta t}{2\tau_c}\right)^2\left[1+erf\left(\frac{t}{\Delta t}-\frac{\Delta t}{2\tau_c}\right)\right]\right\}. \tag{14}$$

Substituting equations (11) and (12) into equations (3) and (7), the surface field and far field of the terahertz radiation can be written as

$$\boldsymbol{E}_s(t) = -\frac{(F/F_{sat})X(t)}{1+(F/F_{sat})X(t)}\boldsymbol{E}_b \tag{15}$$

$$\boldsymbol{E}_{far}(t) = -\frac{A(1+\sqrt{\varepsilon})}{4\pi\cdot z\cdot P}\frac{(F/F_{sat})Y(t)}{\left[1+(F/F_{sat})X(t)\right]^2}\boldsymbol{E}_b \tag{16}$$

where $4h\nu(1+\sqrt{\varepsilon})m^*\left(q^2\tau_s(1-R)\eta_0\right)^{-1} = F_{sat}$ denotes the saturated luminous flux, $P = \left(\varepsilon_0\eta_0 c^2\right)^{-1}$.

3. Analysis and discussion

3.1. The influence of the ratio of luminous flux $F(F_{sat})^{-1}$ on the THz radiation waveform

Suppose that $A = 1cm^2$, $Z = 1cm$, $\tau_c = 1ps$, $\tau_s = 0.5ps$, $\Delta t = 0.1ps$ and F / F_{sat}=1.0, 2.0, 3.0, 4.0, 5.0, we can obtain the THz radiation waveform in the surface and in the far field, as shown in Figures 2(a) and 2(b). The abscissa in Figure 2 and the following figures denotes the delay time, and the ordinate denotes the ratio of the THz radiation field and the incident field.

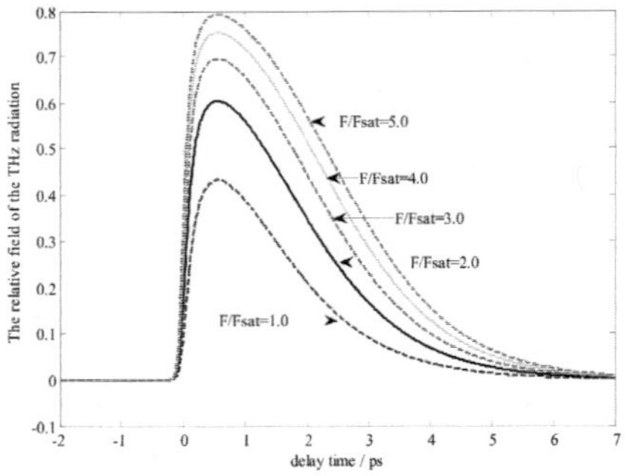

Figure 2(a). The surface field waveform of the THz radiation

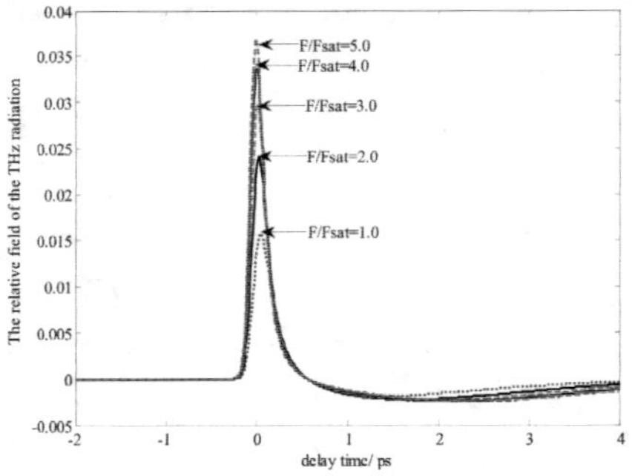

Figure 2(b). The far field waveform of the THz radiation

From Figures 2(a) and 2(b), the peak values of the THz radiation in the surface field and in the far field are on rise and the rising edges of radiation pulses become more and more steep with the increasing of $F(F_{sat})^{-1}$. Meanwhile, the electron hole pairs generated in GaAs become more and more, the carrier concentration becomes higher, and the current change rate versus time becomes larger, which makes the THz radiation intensity in the surface field and in the far field larger and the radiation pulse width smaller in the far fields. Owing to the characteristics of similar steady-state field in the surface field, the transient current on the surface prolongs longer and the radiation pulse width becomes wider and wider. According to the expression of F / F_{sat}, the incident pulse intensity, the incident wavelength, the carrier relaxation time, and the reflectivity in the effective photoconductor illumination area can change the radiation waveform. Therefore, the terahertz radiation field of photoconductive antenna can be improved by increasing $F(F_{sat})^{-1}$.

3.2. The influence of the carrier lifetime τ_c on the THz radiation waveform

Suppose that $A = 1cm^2$, $Z = 1cm$, $F / Fsat = 1.0$, $\tau_s = 0.5ps$, $\Delta t = 0.1ps$ and τ_c=0.1ps, 0.2ps, 0.3ps, 0.4ps, 0.5ps, we can obtain the THz radiation waveforms in the surface field and in the far field, as shown in Figures 3(a) and 3(b).

From Figures 3(a) and 3(b), with the more longer life of the photon-induced carrier, the peak values of THz radiation in the surface field and in the far field are on rise and the rising edges of THz radiation pulses become more and more steep. It is due to fact that with the extension of photon-induced carrier lifetime, more and more photon-induced carriers accumulate in the photoconductor. Thus more current can be obtained under the effect of static bias field and the radiation peak value becomes larger. Hence the transient current on the surface lasts longer and the THz radiation pulse width becomes wider and wider. Therefore, the terahertz radiation field of photoconductive antenna can be improved by selecting photoconductive materials with long lifetime.

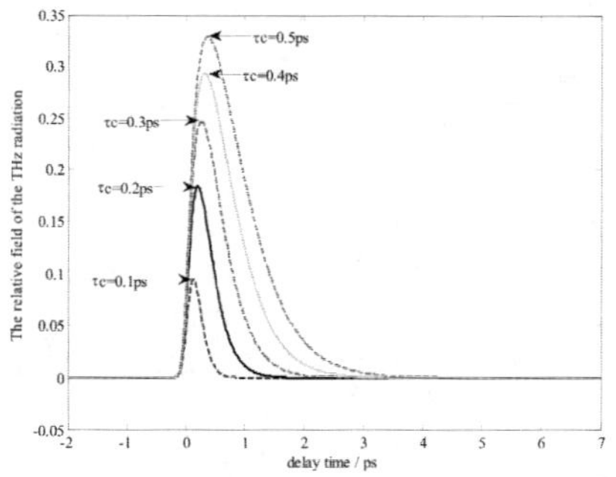

Figure 3(a). The surface field waveform of the THz radiation

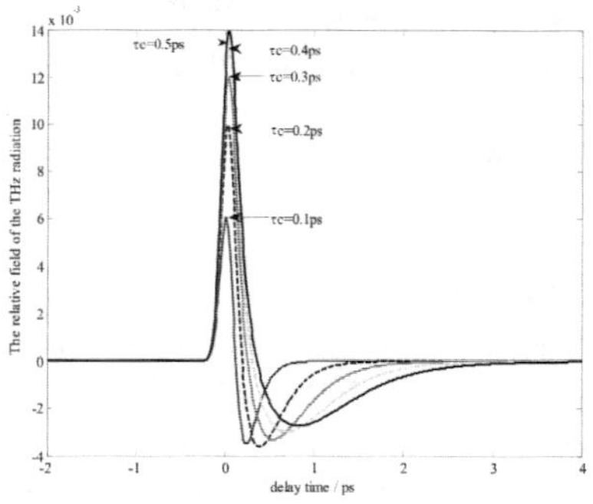

Figure 3(b). The far field waveform of the THz radiation

3.3. The influence of the full width at half maximum $\sqrt{\ln 2}\Delta t$ of exciting pulse on the THz radiation waveform

Suppose that $A = 1cm^2$, $Z = 1cm$, $F / Fsat = 1.0$, $\tau_s = 0.5ps$ and Δt=0.1ps,0.2ps,0.3ps,0.4ps,0.5ps, we can obtain the THz radiation waveforms in the surface field and in the far field, as shown in Figures 4(a) and 4(b).

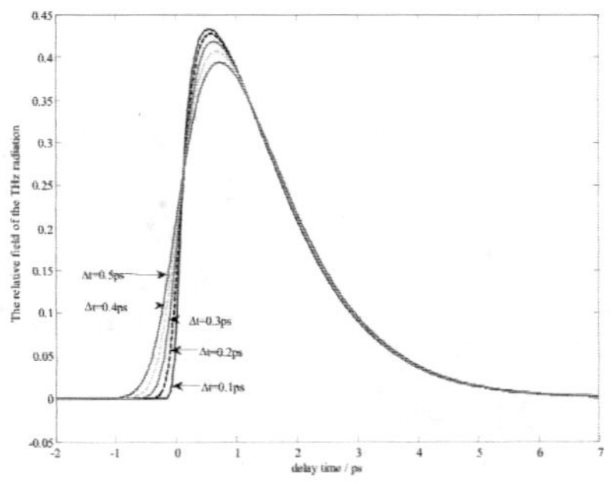

Figure 4(a). The surface field waveform of the THz radiation

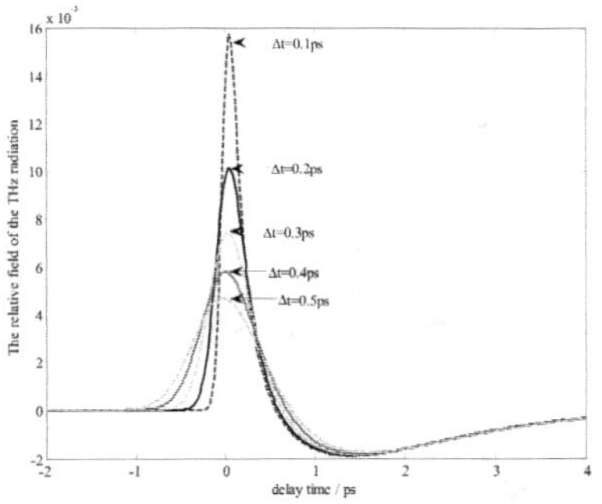

Figure 4(b). The far field waveform of the THz radiation

From Figures 4(a) and 4(b), with the increasing of the full width at half maximum of drive pulse, the peak values of THz radiation in surface field and far field are decreased continuously , the rising edges of radiation pulses become more slower and the radiation pulse width becomes increasingly wider. It is due to the fact that with the Δt lasts longer, the photocurrent change excited in the photoconductor is more slower, the average speed of carriers gets smaller and the conductivity of photoconductive materials gets smaller. Thus the excited THz radiation peak value gets lower and lower, the transient current lasts longer and the excited THz radiation pulse width becomes increasingly wider. Therefore, the terahertz radiation field of photoconductive antenna can be improved by reducing the incident pulse width.

When the incident radiation illuminates on the surface of photoconductor, the number of photo-induced carriers' increases rapidly in one pulse and the photo-induced carriers are accelerated under the effect of bias electric field. Hence the transient current density increases rapidly, which corresponds to the positive part in the far fields of THz radiation waveform. Then the photo-induced carriers experience an attenuation process rapidly which makes the decrease rate with time of the photo-induced carriers be greater than the increase rate with time of current density caused by the acceleration under the effect of bias field. Thus the current density shows the phenomenon of attenuation and makes the growth rate is negative, which corresponds to negative part in the far fields of THz radiation waveform. While the surface field does not appear negative values is due to the radiation field on the photoconductive surface has nothing to do with the time rate of change.

4. Conclusions

In this paper, the THz radiation analytical expressions in the surface field and in the far field are obtained through the famous current-surge model by using large-aperture photoconductive antennas. The time-domain waveforms of THz radiation fields are numerical investigated. The results show that the THz radiation in the surface field and in the far field is affected by many factors, such as the incident intensity, the incident wavelength, the carrier relaxation time, the reflectivity illuminated at photoconductor, the carrier lifetime and the incident light pulse width. The THz radiation intensity can be increased by enhancing the ration of luminous flux, prolonging the photo-induced carrier lifetime or reducing the incident light pulse width. However, due to the shielding effect of space charge and the shielding effect of radiation field, the THz radiation field can appear in the saturated phenomenon. In order to improve the THz radiation power, the intense incident laser pulse can be divided into multi-beam pulse with low energy to trigger photoconductive antenna. Thus the shielding effect of the THz radiation field can be avoided effectively to increase the THz radiation power.

Acknowledgements

The work was supported by innovation fund of the State Key Laboratory of Millimeter Waves, Southeast University.

References

[1] Zhang C L 2008 *Terahertz Sensing and Imaging* (Beijing: National Defense and Industry Press) pp 10-45
[2] Xu J Z and Zhang X C 2007 *Terahertz Science and Technology* (Beijing: Pecking University Press) pp 27-31
[3] Darrow J T, Zhang X C, Auston D H and Morse J D 1992 *IEEE J. Quantum Electron.* **28** 1606-1607
[4] Taylor A J, Benicewicz P K and Young S M 1993 *Opt. Lett.* **18** 1340-42
[5] Sang G P, Weiner A M and Michael R M 1999 *IEEE J. Quantum Electron.* **35** 1257-68
[6] Zhang T Y and Cao J C 2004 *J.Rare Materials.* **28** 588-590 (in Chinese)
[7] Shi W and Zhang X B 2004 *J.Semiconductors.* **25** 1735-38 (in Chinese)
[8] Benicewicz P K, Roberts J P and Taylor A J 1994 *J.Opt.Soc.Am.* B **11** 2533-56

Theoretical and Experimental studies on CH_3OH THz Laser Pumped by Pulse Carbon Dioxide Laser

Fei Fei Wang Jing Tian Zhaoshuo* Zhang Yanchao Fu Shiyou Wang Qi

Information Optoelectronics Research Institute, Harbin Institute of Technology at Weihai, Sandong, Weihai, 264209, China

E-mail: tianzhaoshuo@126.com

Abstract. In this paper, according to the molecular structure and vibration mode of micro-asymmetric gyroscope CH_3OH molecule, dynamic process of optically pumped Terahertz laser is analyzed theoretically. The rate equation models based on three level systems are given according to the theory of typical laser rate equation. The output THz pulsed laser waveform is obtained by solving the rate equation model. An all-metal Terahertz laser pumped by RF waveguide carbon dioxide laser is designed with CH_3OH as its working gas. The pulsed Terahertz laser output is obtained. The waveform and repetition frequency of the optically pumped laser are measured in the experiments. The Terahertz laser designed does not need water cooling system. It also has the advantages of simple structure and small size.

1. Introduction

High repetition frequency pulse optically pumped Terahertz lasers can be applied in high-speed Terahertz imaging, Terahertz laser radar, medical diagnosis and other fields. The optically pumped THz laser can be continuous or pulse operation [1~4]. Currently, most pulse optically pumped THz lasers adopt $TEACO_2$ laser as their pump sources. The wave lengths range is from mm to 40μm. Although the peak power of the THz lasers with $TEACO_2$ laser as their pump source are high, its repetition frequencies are low (usually 10 times per second), which greatly limits their application. To solve this problem, J.Bae et al. developed the Q-switched carbon dioxide laser with gain and mechanical switching. When the repetition frequency is 1 KHz, the peak power of the output laser is more than 1KW. When pumped with 500Hz repetition frequency by this laser, the CH_3F gas can generate a 496μm laser, whose peak power is 6.5W and the pulse width is 100ns. R.C.Viscovini et. al. reported the CD_3OD THz laser pumped by adjustable wideband waveguide carbon dioxide laser, which can worked either continuously or with long pulses, whose the peak power is 100W, the repetition frequency ranges from 800Hz to 1200 Hz and the pulse widths are 30μm to 150μm [5]. In recent years, international researchers have been constantly explored new THz pump laser light spectrum lines and have some new findings of optical pumped THz laser spectral lines [6~8].

In China, researchers at Sun Yat-sen University and Huazhong University of Science and Technology have carried out a series of theoretical and experimental researches in THz lasers pumped by $TEACO_2$ laser. The focus of their researches is the NH_3 gas THz laser. They have performed a series of profound and systematic researches on the mechanisms and the spectrum characteristics of the laser, as well as improving the working parameters and miniaturizing the size of the laser [9]. He

Zhihong et. al. in Tianjin University studied the $TEACO_2$ laser pumped heavy water gas theoretically and experimentally and obtained the pulsed THz laser output of 385μm.

The Pump source of most THz lasers are the water cooling glass DC-exciting carbon dioxide laser or $TEACO_2$ laser, which are bulky and complicated [10,11]. The highest pulse repetition frequency is about 1000Hz. Based on the result of the previous studies, we designed an all-metal RF waveguide carbon dioxide laser pumping THz laser, whose pulse repetition frequency can reach 5KHz. The pump source is the all-metal RF waveguide carbon dioxide laser. The THz generator is also of all-metal structure. Therefore, the laser can work with air cooling system instead of water cooling. Under the low pulsing frequencies, it can work without cooling. The size of the THz laser system can thus be designed smaller and the structure less complicated. The laser working gas is CH_3OH. THz laser wave form and the features of the output THz lasers under different repetition frequencies of the pumping laser are measured. These researches might serve as a reference to other relevant researches and as the foundation of later researches.

2. Dynamic Analysis for Pumping Progress

Up to now, the rate equation (R.E) models have been one of the main quantitative methods used to evaluate the power efficiency of pulse optically pumped Terahertz lasers [12~14].

Figure 1 shows the energy level transfer process of CH_3OH when pumped by $CO_2 9P(36)$. The THz laser generated by stimulated absorption transition $(0,16,0,1,8) \rightarrow (1,16,0,1,8)$ and stimulated emission transition $(1,16,0,1,8) \rightarrow (1,15,0,2,7)$ which are main parts of the energy transition is chosen as the main research object [15]. The three-energy-level theoretical model consists of $(0,16,0,1,8)$、$(1,16,0,1,8)$、$(1,15,0,2,7)$ is shown in Figure 2.

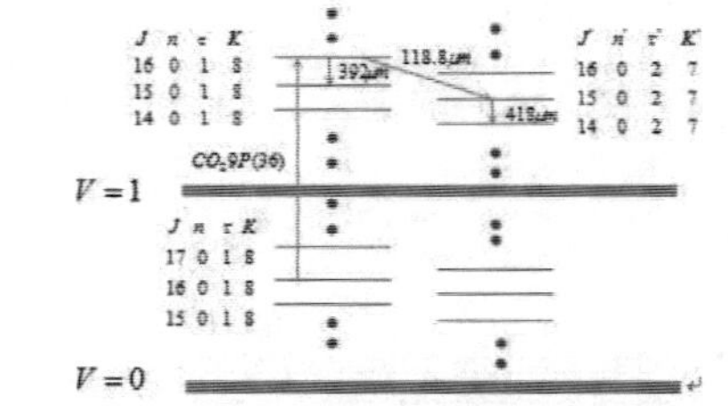

Fig.1 The relevant energy transition of CH_3OH pumped by $CO_2 9P(36)$

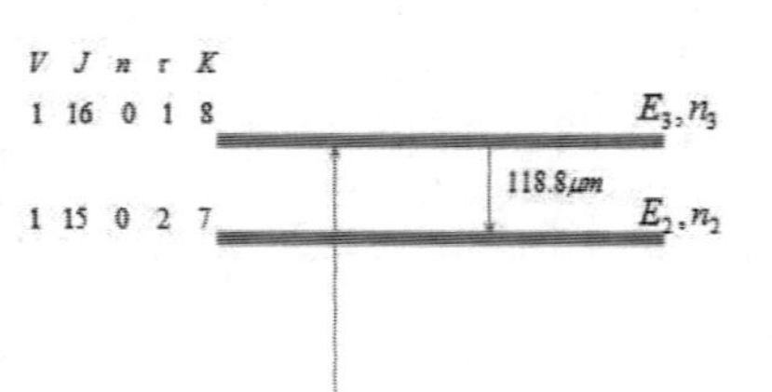

Fig.2 The Simplified CH_3OH three-energy-level system model

E_1 shows the ground state of the energy level. Under the action of pump light $CO_2 9P(36)$, particles in the E_1 ground state are pumped to E_3, the upper laser level, which will lead to the particles in E_1 exploded in number. Before the population inversion, particles in E_3 transit to E_2 and E_1, mainly by means of spontaneous transition and radiationless transition. However this transition is incredibly rare, so particles in E_3 has a longer residence time (a longer life span). If the pump speed of particles to E_3 is high enough, there will easily form a distribution for population inversion between E_2 and E_1, Thus the stimulated radiation between E_3 and E_2 will take absolutely advantage, and further generate THz laser radiation with its wavelength of 118.8μm.

Through analysis of above experiment, we can establish the rate equation model of optically pumped CH_3OH molecular laser as follows:

$$\frac{dn_3}{dt}=n_1\left(\frac{g_3}{g_1}\right)B_{31}\rho_b(t)+\frac{1}{hv}\left[\left(\frac{g_3}{g_2}\right)n_2-n_3\right]\sigma_{32}(v,v_0)I_T-n_3\omega_{31}$$

$$\frac{dn_2}{dt}=\frac{1}{hv}\left[n_3-\left(\frac{g_3}{g_2}\right)n_2\right]\sigma_{32}(v,v_0)I_T-n_2\omega_{21}$$

$$\frac{dN_l}{dt}=\frac{1}{hv}\left[n_3-\left(\frac{g_3}{g_2}\right)n_2\right]\sigma_{32}(v,v_0)I_T-\frac{N_l}{\tau}$$

$$n_1+n_2+n_3=n$$

The $\rho_b(t)$ stands for energy density of the carbon dioxide laser pump light; I_T stands for intensity of THz optical wave output; n is the total number of per unit volume particles in the three-energy-level while n_1、n_2 and n_3 represent the particles per unit volume of the ground state E_1, the lower laser level E_2 and the upper laser level E_3. The $\sigma_{32}(v,v_0)$ represents the emission cross section of THz laser . $g_i=2J_i+1$ represents the statistical weight of each energy level and N_l represents the density of photon in laser resonant.

3. Calculation Results

For convenience, pump pulse is some what similar to the Gaussian pulse on the basis of the experimental results, which can be expressed as:

$$\rho_p(t)=\rho_{p0}\exp\left[-(t-1)^2/\sigma^2\right] \quad 0<t<1.5\mu s$$

Where ρ_{p0} represents the photon density of the pump pulse peak and σ is the width of pulse laser. The parameters for the rate equations' calculation are as follows:

Table 1. Parameters of CH_3OH gas molecular

Physical Parameter	Sign	Value	Unit
Electric dipole moment	μ_{32}	2.23×10^{-30}	C·m
	μ_{31}	1.79×10^{-30}	C·m
THz spectrum line homogeneous broadening line width	Δv_H	2.8×10^6	Hz
THz spectrum line Doppler broadening line width	Δv_D	7.9×10^6	Hz
Upper energy level particle number before pumping	$n_3(0)$	$0.00183n_0$	m^{-3}
Lower energy level particle number before pumping	$n_2(0)$	$0.00264n_0$	m^{-3}
Each statistical weight level ratio	g_3/g_2	1.065	——
	g_3/g_1	1	——

By using Matlab program, we have numerically solved the rate equations. Fig. 3 shows that the particle distribution is in accordance with Boltzmann distribution in thermal equilibrium. The particle number of lower energy level is more than the particles in upper energy level in the beginning. The particles of both the lower and upper energy level begin to increase slowly under the action of pump light. However, there is a rapid increase after 0.8μs. The THz laser output is obtained as a result of particles in upper energy level increase rapidly under the action of pump pulse and form the population inversion. Due to the stimulation, the particles in lower energy level also increase and reach to the peak after 1μs. Then the particles of the lower and upper energy level decrease gradually with pump light decreasing until it resorts to the particles distribution in the thermal equilibrium. From Fig. 4, we can get a conclusion that the output THz wave reaches its peak around 1.22μs and the decline stage after the peak is relevantly slow which is due to the relaxation from lower energy level to the ground state. In the pumping process, the relaxation of particles transiting from upper energy level to the lower level and from the lower level to the ground state makes some particles still gather in the lower energy level which will influence the population inversion and further reduce the speed of photon generation, thus makes the pulse wave relatively slow. From Fig. 5, we can get a conclusion that the output THz pulse wave has some time delay compared with the input pumping pulse wave. Its peak power is lower and pulse width is shorter.

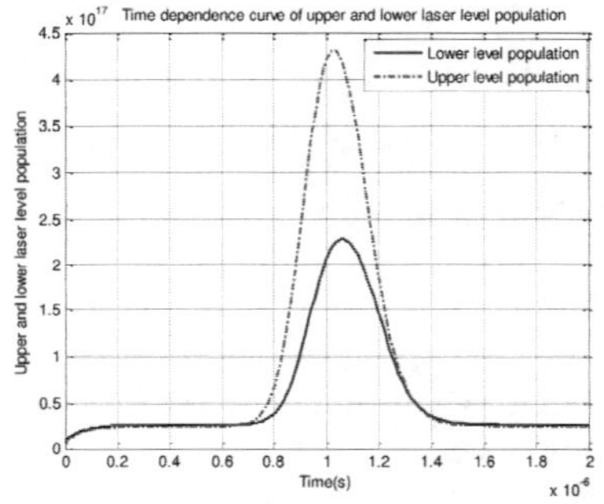

Fig. 3 Time dependence curve of upper and lower laser level population

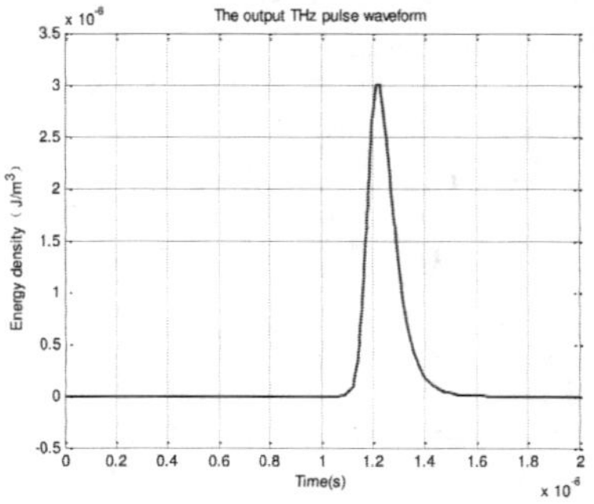

Fig. 4 The output THz pulse waveform

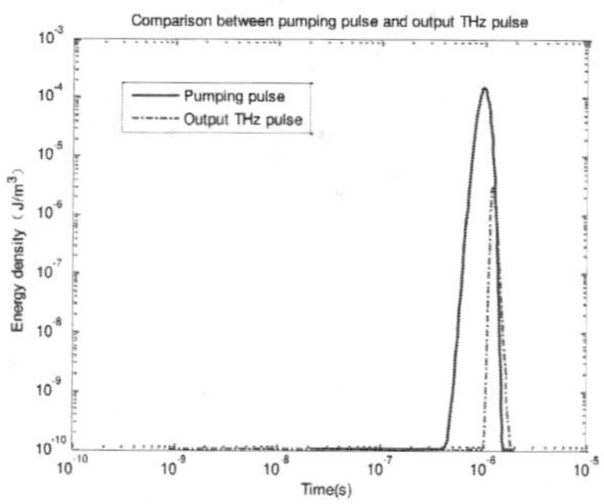

Fig. 5 Comparison between pumping pulse and output THz pulse

4. Experimental results

The THz laser consists of two parts: One is the pump source [16]; the other is the THz laser generator. The pump source is grate tuning all-metal RF waveguide carbon dioxide laser [17]. The CH_3OH gas is inflated into the THz laser through the air supply system. According to the references, the laser can output 118.8μm THz laser with the 9P (36) line pump laser [18]. In the beginning, the pump laser was operated in low repetition frequency (90Hz). The voltage on the piezoelectric ceramics was then adjusted, so as to tune the frequency of the pump laser and maximize the intensity of the THz output. The waveform and intensity of the generated THz were observed via oscilloscope. As was shown in Fig. 6, the THz laser resembled the triangular wave, the reason of which is that the band width of our detector is 300KHz, which is only enough to detect the laser signal and is not capable of showing the real waveform of the pulsed THz laser. Theoretical analyses above show that the output waveform of the THz laser should be similar to the pump laser, with only a lower intensity.

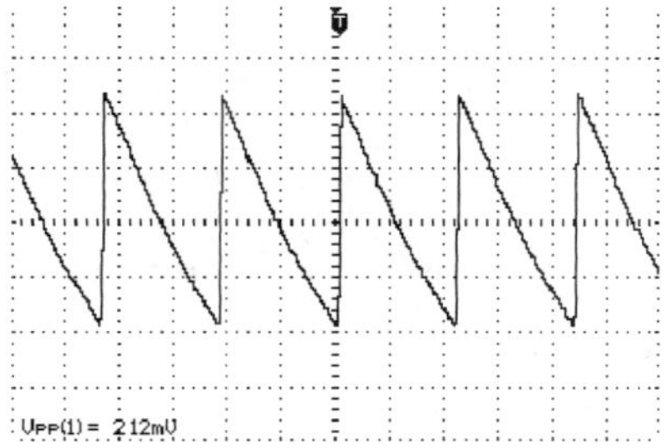

Fig. 6 Measured THz laser output (5ms/div.)

The repetition frequency of the pump laser can be as high as 5KHz. Theoretically, the frequency of the THz output should also be 5 KHz. Thus, we measured the waveform of the THz output while increasing the repetition frequency of the pump laser. In the experiment, the oscilloscope shows that the intensity of THz signal decreased as the repetition frequency increased. The obtained signal is shown in Fig.7 at the pump laser repetition frequency 200Hz. If further improving the repetition frequency of the pump laser to 300Hz, the signal would scarcely be observable. We believe the phenomenon is due to the band width of the detector being narrow. If a detector with high band width being adopted, the true waveform of THz output could be obtained.

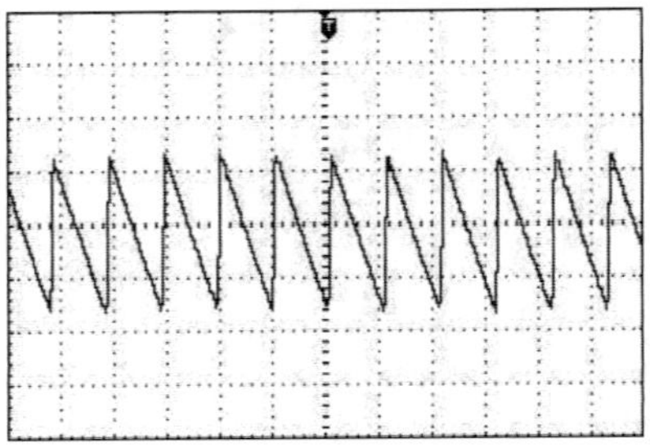

Fig. 7 THz laser output at the Pulse Repetition Frequency of 200Hz (5ms/div.)

5. Conclusion

In this paper, the dynamic process of CH_3OH THz laser pumped by carbon dioxide laser is studied. Through the energy-level analysis of CH_3OH, a three-level system model of energy level transition of CH_3OH is set up. Based on the typical theory of laser rate equation, the rate equation of the three-level system model is also given. The dynamic process of pulse THz laser is analyzed theoretically. Through the numerical calculation, we have got the transitions of particles on the different energy levels.

On the experimental side, we have studied the THz laser pumped by all-metal Radio-frequency waveguide carbon dioxide laser with grating line selected. The whole laser including pump source and THz generator is all-metal structure. It has the advantages of a simple structure and small volume without water cooling. The laser can work either on continuous or pulse modes. In this experiment, we measured the parameters for the THz pulse laser.

On further experiment, the following points need to be modified or improved: (1) The CH_3OH used in this experiment is common in the market with a purity of only 99.5%. The laser should have high efficiency by using spectroscopically pure CH_3OH. (2) Presently, we use manually control the voltage on the piezoelectric ceramic to change the pump laser frequency to get the strongest output of THz laser, which result in a poor stability of the output of laser. On further experiment, we consider employing an automatic control system to improve the output stability. (3) Limited by the detector bandwidth, the true THz pulse laser wave form couldn't be obtained, and it is also not able to measure THz laser at high-repetition rate. So on further study we will consider purchasing a broadband detector or cooperate with qualified organizations.

References

[1] E.R.Mueller, R.Henschke, W.E.Robotham Terahertz local oscillator for the Microwave Limb Sounder on the Aurasatellite. APP.OPT. 2007,46(22):4907-4915
[2] J.FFederici, B.Schulkin, F.Huang, D.eGary ,THz imaging and sensing for security applications-explosives, weapons and drugs. Semicond.Sci.Technol. 2005,20: S266-S280
[3] Fu Shiyou, Tian Zhaoshuo, Yi Fulong et al. Development and prospect for optically pumped THz laser[J] JOURNAL OF HARB IN STITUTE OF TECHNOLOGY2008,40(3): 435-439(in Chinese)
[4] Baej, Nozokedo.T, Shirai.H et al. An EMQ-Switched CO2 laser as a pump source for a far-infrared laser with a high peak power and a high repetition. IEEE.J.Q.E., 1989,25(7), 1591-1594.
[5] Viscovini.R.C,Scalabrin.A, and Pereira.D. CD3OD Optically pumped by a waveguide CO2 laser: New FIR laser lines. IEEE.J.Q.E. 1997,33(6): 916-918.
[6] Viscovini.R.C, Cruz.F.C, Scalabrin.A, et al. CHD2OH optically pumped by a Waveguide CO2 laser: New far-infrared laser lines from the CD2 wagging vibrational mode. IEEE.J.Q.E. 2002,38(8): 1029-1030.

[7] Michele.A.D, Bousbahi.K, Carelli.G, et al. The 13CH3OH far-infrared laser: new lines and assignments. Infrared Physics&Technology. 2004,45: 243-248.
[8] A.Moretti,G.Moruzzi,F.Strumia,J.C.S.Moraes,and G.Carelli. New Terahertz Laser Lines From 13CD3OH Pumped by Regular and HotBands CO2 Laser. IEEE J.QUANT.ELECT., 2008,44(11):1104-1106
[9] Qin.J.Y, Huang. X, Luo X Z, and Zheng X S. Spectral chararcteristics comparasion of pulsed optically pumped NH3 and CH3OH far-infrared laser. Int.J.IR&MMW, 1999,20(8): 1515-1524.
[10] Zhang Ping ,Zhang Xun ,Huang Xiao et al. The Study of Two L ong itudinal M ode Optically Pum ped Super Radiatn t THz Laser and Cavity THz Laser[J] Journal of Optoelectronics L aser.2002,13(4): 332-335(in Chinese)
[11] QI Chun-Chao,ZUO Du-Luo,LU Yan-Zhao,etal.An Eicient Photon Conversion Eiciency Ammonia Terahertz Cavity Laser. CHIN.PHYS.LETT. 2009,26 (12):124201
[12] He Zhihong,Yao Jianquan,Shi Huafeng et al.Semicla s sical theory of optically pumped D2O gas tera-Hz laser[J] Chin. Phys. Soc. 2007,56(7):5802-5807(in Chinese)
[13] HE Zhi-hong, YAO Jian-quan,REN Xia etal. Experimental and theoretical research on a compact superradiant optically pumped D2O gas terahertz laser. [J] Journal of Optoelectronics Laser. 2008,19(1):35-37. (in Chinese)
[14] Sun Y D, Fu S Y, Wang J, Zhenghe Sun, Yanchao Zhang, Zhaoshuo Tian, Qi Wang. Optically pumped terahertz lasers with high pulse repetition frequency: theory and design. Chin. Opt. Lett.2009,7(2)127_129
[15] Bao Yanxiang , Huang Xiao, Luo Xizhang. Choice of Three-Energy-Level Groups on Optically Pumped Sub-Millimeter Wave Laser Medium[J] Journal of Optoelectronics L aser 2004,15(11):1320-1323. (in Chinese)
[16] Sun Y D, Tian Z S, Fu S Y, et al. Zhenghe Sun. Study of active frequency stabilized CO2 laser for THz radiation. POEM, 2008,7727-44
[17] SUN Zhenghe, WANG Jing,FU Shiyou et al. A tunable RF excited waveguide CO2 laser with dual-channel ,high synchronism and coelectrode[J] Journal of Optoelectronics L aser. 2009,20(8):997_999(in Chinese)
[18] Tian Zhaoshuo,Wang Jing,Fei Fei,Yang Junguo,Zhang Yanchao,Fu Shiyou,Wang Qi. Study of Optically Pumped All-Metal Terahertz Laser. Chinese Journal of Lasers, 2010, 37(9): 2323～2327 (in chinese)

Theoretical Analysis and Numerical Simulation of Terahertz Wave Generation and Modulation based on GaAs:O

Jian HE[1*], Yinli JIANG[1], Qingyuan MIAO[2], Jiayuan HE[3], Jinsong LIU[1], Jianquan YAO[1, 4]

1. Wuhan National Lab for Optoelectronics, Huazhong University of Science and Technology, Wuhan 430074, Hubei China
2. College of Electronics and Information Engineering, Wuhan University, Wuhan 430071, Hubei, P. R. China
3. Department of Electronics & Information Engineering, Huazhong Univeristy of Science and Technology, Wuhan 430074, China
4. Institute of Laser and Optolectronics, College of Precision Instrument and Optoelectronic Engineering, Tianjin University, Tianjin 300072, China

[*]E-mail: jianhe@mail.hust.edu.cn

Abstract. In this paper, the generation and modulation of terahertz wave based on Oxygen-implanted GaAs (GaAs:O) is proposed and discussed. The theoretical analyses and numerical simulation results about the modulation characteristics of the terahertz wave using the proposed scheme have been obtained. Through comparing the numerical simulation results we can find that the information carried on the optical light can be transferred onto the generated terahertz wave and the power of the modulation side bands in terahertz frequency can keep stable in the whole modulation bandwidth. The relationship between the terahertz power and carrier lifetime has also been studied.

1. Introduction

Terahertz wave has numerous benefits, such as high bandwidth, good anti-interference and small antenna size, which make it suitable for short-distance high-speed wireless communications[1-11]. However, the generation and modulation of terahertz wave with high stability and small size is still a big problem need to be resolved for the future terahertz wireless communication systems.

Terahertz photomixer based on low-temperature-grown GaAs(LTG GaAs) is one of the possible solutions for the terahertz genearton with small size and compact structure[10-11]. Campared with LTG GaAs, ion-implanted material not only retains the advantages of high carrier mobility and short carrier lifetime, but also can overcome the temperature sensitivity problem encountered in LTG GaAs growth[1]. So far, many ion-implanted material based terahertz sources were proposed, such as: GaAs:As, GaAs:H, GaAs:N, GaAs:O, InGaAs:Au[2-9]. Among these new materials, the energy level of oxygen ions in GaAs is close to the Fermi level, it makes GaAs:O material almost electrically neutral, and hence a comparatively higher resistance. This property is benefit for signal to noise ratio of the photomixer.

On the other hand, the modulation of terahertz wave is another key problem need to be resolved in the future terahertz communication system. In this paper, two Mach Zender Modulators (MZM) are used as the external modulator to modulate the injected laser light into the photomixer based on oxygen-implanted GaAs (GaAs:O). Theoretical analyses and numerical simulaiton results about the terahertz modulation have been obtained and presented.

2. Theoretical Analysis and Numerical Simulation

Terahertz photomixing process occurs when the photoconductive device is illumiated by two single mode lasers, which having a frequency differency in terahertz range. Thus the carrier density in the active area will change with the modulation signals when the modulated optical signals are focused onto the photomixer.

The injected instantaneous optical power on the active area can be expressed as:

$$P_i = P_0[1+\cos(\Omega t)+\cos(\omega t)+\frac{1}{2}\cos(\omega+\Omega)t+\frac{1}{2}\cos(\omega-\Omega)t] \tag{1}$$

Here the ω is the frequency difference between the two different optical frequencies corresponding to ω_1, ω_2 respectively. The Ω is the modulation frequency applied to the two MZMs. The P_0 is the power of the optical signals. Thus the change of the photocarrier density in the semiconductor active area can be expressed using the dynamic equation as follows[10-11]

$$\frac{dn(\omega,t)}{dt} = \frac{\eta_e P_i}{hfAL} - \frac{n(\omega,t)}{\tau} \tag{2}$$

Where, A and η_e are the effective active area and the external quantum efficiency, respectively.

The change of the instantaneous value of the carrier density, which can be obtained from the Equation 2, will induce change of the photoconductance of the active area. Thus the photoconductance can be expressed as follows:

$$G_p = G_0\left[1+\beta_1\sin\omega t+\beta_2\sin\Omega t+\beta_3\sin(\omega+\Omega)t+\beta_4\sin(\omega-\Omega)t\right] \tag{3}$$

Where, $G_0 = \frac{e\mu\eta_e N_g L\tau P_0}{hfAW_g}$, $\beta_1 = \frac{1}{\sqrt{1+\tau^2\omega^2}}$, $\beta_2 = \frac{1}{\sqrt{1+\tau^2\Omega^2}}$, $\beta_{3-4} = \frac{1}{2\sqrt{1+\tau^2(\omega\pm\Omega)^2}}$.

The photoconductance density is an instantaneous value, which will produce a voltage v across the photogap. The change of the voltage can be described using the following differential equation:

$$\frac{dv}{dt} = \frac{v_B - v}{R_L C} - \frac{v}{C}G_p(\omega,t) \tag{4}$$

Generally the genearted terahertz wave power can be expressed using the equation of $P = \frac{v_{ac}^2}{2R_L}$, where, v_{ac} is the ac part of v. Under the small signal codition, $G_0R_L \ll 1$, v_BG_0 can be regarded as the dc photocurrent $I_{DC,ph} = v_BG_0$, thus the power corresponding to the different frequency components of $\omega, \Omega, \omega+\Omega$ and $\omega-\Omega$ can be obtained as follows

$$P_\omega = \frac{\frac{1}{2}I_{DC,ph}^2 R_L}{\left(1+\tau^2\omega^2\right)\left[1+(\omega R_L C)^2\right]} \tag{5}$$

$$P_\Omega = \frac{\frac{1}{2}I_{DC,ph}^2 R_L}{\left(1+\tau^2\Omega^2\right)\left[1+(\Omega R_L C)^2\right]} \tag{6}$$

$$P_{\omega-\Omega} = \frac{\frac{1}{8}I_{DC,ph}^2 R_L}{\left[1+\tau^2(\omega-\Omega)^2\right]\left[1+\left((\omega-\Omega)R_L C\right)^2\right]} \tag{7}$$

$$P_{\omega+\Omega} = \frac{\frac{1}{8}I_{DC,ph}^2 R_L}{\left[1+\tau^2(\omega+\Omega)^2\right]\left[1+\left((\omega+\Omega)R_L C\right)^2\right]} \tag{8}$$

The frequency components of $\Omega, \omega+\Omega$ and $\omega-\Omega$ are present because of the modulation. .

In our simulaion, the central wave length of the two input laser beams is assumed to be 850nm, and the frequency difference between the two wavelengths is 1.0THz. The photomixer have 10 electrodes, and 9 photogaps, the width of the electrodes and photogaps are 0.2μm and 1.8 μm, respectively. The length of electrodes and photogaps are both 20μm. Some other parameters used in the simulatin are shown in the following table 1.

Table 1. Parameters used in the simulation

Parameters	Symbol	Value
Central wavelength	λ	850 nm
Power of Laser beams	P_0	40 mW
Internal quantum efficiency	η_S	0.8
Bias voltage	V_B	1.5 V
Resistance of antenna	R_L	50 Ω
Coupling capacitor	C	$3.0341*10^{-15}$ F

Figure 1 shows the simulation results when the modulation rate is 20GHz and the carrier lifetime is 0.3 ps. When there are no modulation signals applied to the optical modualtors, there exist no modulation sidebands near the central frequency. On the other hand, when the modulation signals are applied to the optical modualtor, we can find easily that the optcial modulation signals have been transferred onto the generated terahertz wave.

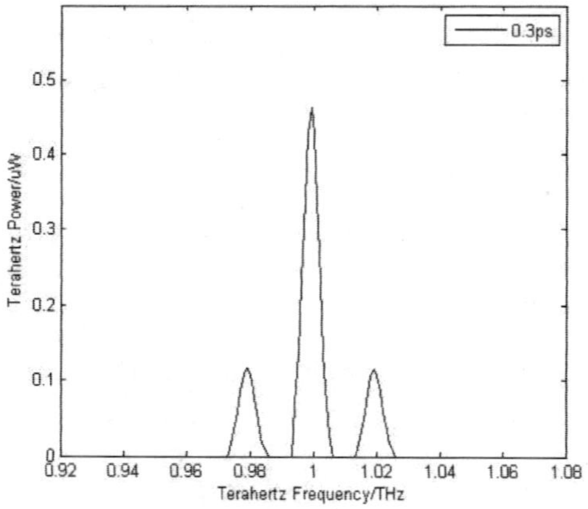

Fig.1. Generation and modulation of Terahertz wave corresponding to 20GHz modulation frequency

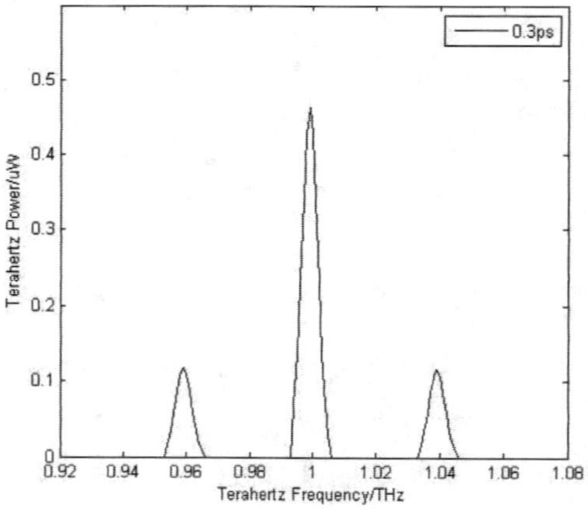

Fig.2. Generation and modulation of Terahertz wave corresponding to 40 GHz modulation frequency

Figure 2 shows the results when modulating at the frequencies 40GHz. It can be seen that the modulation sidebands move away from the central frequency when the modulation frequency increases. What's more, the generated terahertz wave power kept stable. That's to say that this modulation method have a very good power stability and can keep the pace of the modulation bandwidth in the optical transport networks.

Firure 3 shows the simulation results about the modulation characteristics for four different materials with different carrier lifetimes 0.2ps, 0.3ps, and 0.4ps. For the four different materials, the simulation results shows that the output power of generated terahertz grows up with the increasing of carrier lifetime. On the other hand, some simulation results of the generated terahertz power to the terahertz frequency are shown in the Figure 4.

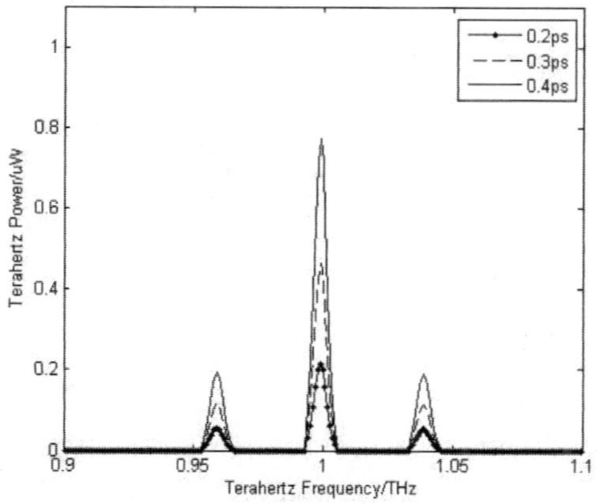

Fig.3. Modulation characteristics corresponding to different carrier lifetime

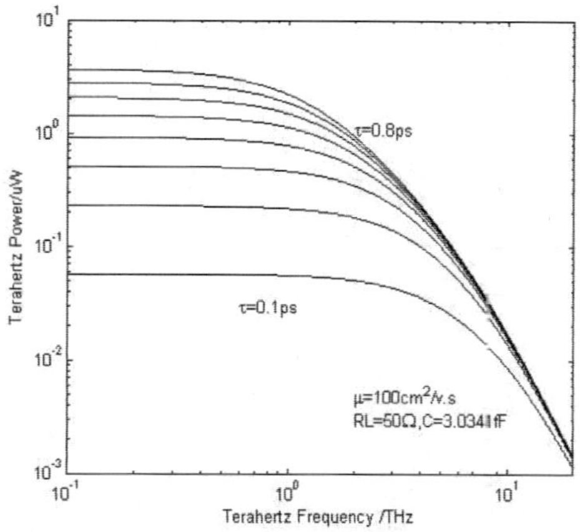

Fig.4. Power of the generated terahertz wave corresponding to different carrier lifetime

3. Conclusion

A highspeed terahertz modulation scheme based on photomixer is proposed and discussed. The optical modulaton signals can be easily transferred onto the generated terahertz wave. The terahertz modulation power is quite stable in the whole modulation bandwidth,. Further more, the power of the generated modulation signals keeps stable when the modulation rate changes from 20 GHz to 40 GHz. Comparing results of the four kinds of selected material we can find that selecting the bigger lifetime material under meeting the requirements of bandwidth will be good for improving terahertz power.

Acknowledgement

This work was supported in part by the National Natural Science Foundation of China under Grant No.60877059. The authors would like to sincerely thank Mr. Zuan Li and Yinghao Yuan at Huazhong University of Science & Technology, Wuhan, China, for their discussions.

References

[1] Chen KJ, Li YT 2009 *Optics Letters* **34** 935
[2] T. A. Liu, M. Tani, and C. L. Pan 2003 *J. Appl. Phys.* **93** 2996
[3] B. Salem, D. Morris, V. Aimez, J. Beauvais, and D. Houde 2006 *Semiconductor Science and Technology* **21** 283
[4] M. Mikulics, E. A. Michael, M. Marso, M. Lepsa, A. van der Hart, H. Luth, A. Dewald, S. Stancek, M. Mozolik, P. Kordos 2006 *Appl. Phys. Lett.* **89** 071103
[5] B. Salem, D. Morris, Y. Salissou, V. Aimez, S. Charlebois, M. Chicoine, F. Schiettekatte, 2006 *Journal of Vacuum Science & Technology A.* **24** 774
[6] I. C. Mayorga, E. A. Michael 2007 *Appl. Phys. Lett.* **91** 031107
[7] Chimot N, Mangeney J, Crozat P 2006 *Opt. Exp.* **14** 1856
[8] M. Martin, J. Mangeney, P. Crozat, Y. Chassagneux, R. Colombelli, N. Zerounian, L. Vivien, K. Blary 2008 *Appl. Phys. Lett.* **93** 131112
[9] Jian He and et al 2008 *Proceedings of the 33st International Conference on Infrared, Millimeter, and Terahertz Waves ,Pasadena, CA, SEP 15-19*
[10] E. R. Brown 1999 *Appl. Phys. Lett.* **75** 769
[11] E. R. Brown, F. W. Smith, and K. A. Mcintosh 1993 *J. Appl. Phys.* **73** 1480

Terahertz wave attenuation analysis of silicon photonic crystal fiber

He yun-hong[1], **Li Jiu-sheng**[1, 2]

[1]Centre for THz Research, China Jiliang University, Hangzhou 310018, China

[2]College of Precision Instrument and Opto-Electronics Engineering, Institute of Laser and Optoelectronics, Tianjin University, Tianjin 300072, China

E-mail: lijsh2008@126.com

Abstract. Finite element method is employed to analyze terahertz wave transmission characteristic in hexagonal photonic crystal fiber. We compared the terahertz field distribution and attenuation of the proposed photonic crystal fiber for different air-hole diameter. When d/ equals to 0.95, the lowest of transmission loss is 0.0012dB/m. The simulated results show that the performance of the presented photonic crystal fiber meets with the demand of terahertz transmission.

1. Introduction

Terahertz wave frequency range is 0.1THz~10THz, located midway between the microwaves infrared light. Due to the lack of effective means of THz generation and detection, there is the THz gap segment in the electromagnetic spectrum. In recent years, with developing THz wave sources and THz detection, THz has increasingly gained wide attention. THz wave technology has widely application in object imaging, environmental monitoring, medical diagnostics, radio astronomy, mobile communications and broadband .So the problem of effective transmission of THz become a very pressing issue. Since transmission loss of THz in free space is large, whether terahertz widely used lies in the availability of low transmission loss waveguide. the relevant research on THz waveguide is carried out at home and abroad, such as the Rice University, M. Mittleman [1] achieve the stainless steel waveguide which has a low dispersion . The average loss coefficient of this waveguide is less than 0.03cm-1. Hidaka et al [2] designed hollow waveguide using ferroelectric polymer polyvinylidene fluoride ethylene material as cladding. Transmission loss of the waveguide is only 1.5m-1 in 1~2THz band. R. Mendis et al [3-4] has produced THz plastic photonic crystal fiber. Jamison et al [5] designed sapphire fiber for terahertz transmission. How to achieve THz waveguide of low-loss is the current research focus.

In this paper, a kind of photonic crystal fiber with point defect has put forward. Using finite element analysis method, the attenuation properties of terahertz transmission waveguide based on photonic crystal fiber was analyzed. The results show that the THz transmission waveguide has some advantages, such as transmission performance, small size and ease of integration, etc.

2. Design of silicon photonic crystal fiber

The terahertz photonic crystal fiber structure is shown in figure 1. It is the hexagonal photonic crystal fiber. The substrate material is silicon with resistivity of 10k .cm. Cross-section is divided into cladding and core areas. Cladding is periodic structure which is formed by hexagonal air. Air hole defect in the center of core region formed fiber core. The geometric parameters of photonic crystal fiber as follows: the air hole period , the diameter of air hole d, the relative aperture d/ , the layer of air-hole cladding N, the refractive index of air n_1, the refractive index of silicon n_2.

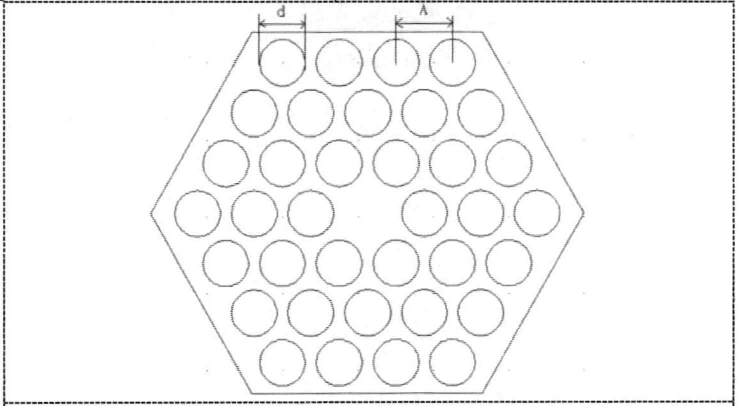

Figure 1. Cross-section photonic crystal fiber of THz waveguide

3. Simulation results

In this papar, the number of air-hole layers N=7, the air hole period =75μm, the air refractive index n_1=1, the refractive index of silicon n_2=3.48. Fixed air-hole pitch , changing the air hole diameter d, the terahertz wave transmission characteristic in a photonic crystal fiber is analyzed when d/ =0.4, and d/ =0.8. Figure 2 is the distribution of THz energy in the photonic crystal fiber when d/ =0.4. Figure 2 (a), (b), (c) show that most of the energy concentrated in the fiber core in photonic crystal fiber, but also a part of the energy distributes round of the photonic crystal fiber core. The results show that the transmission attenuation of THz is great in the photonic crystal fiber when d/ =0.4. The energy distribution of THz in photonic crystal fiber shown in figure 3, when d/ = 0.8. Compared with d/ =0.4, when d/ =0.8, the energy of THz is more concentrated in the core of fiber photonic crystal fiber and the energy leakage reduced significantly from figure 3 (a), (b), (c).

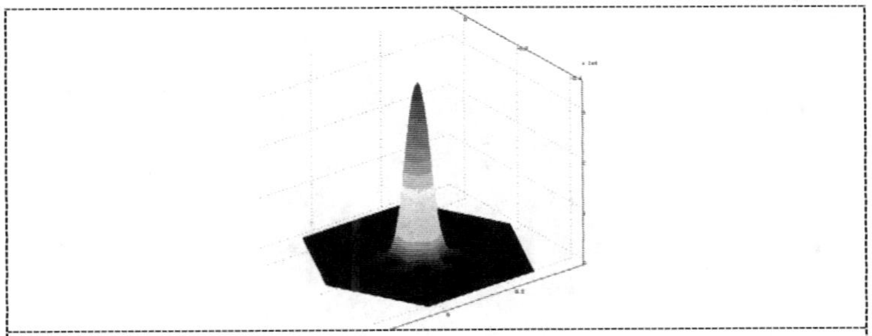

(a)

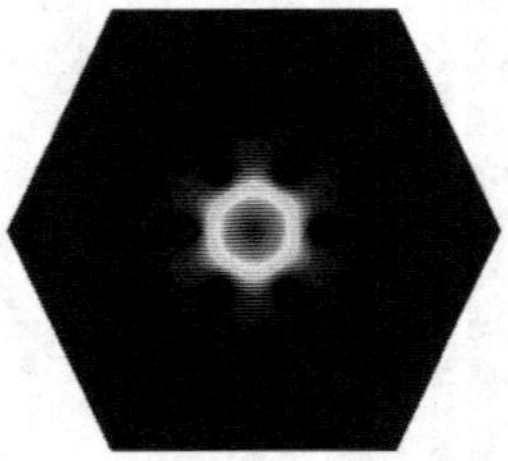

(b)

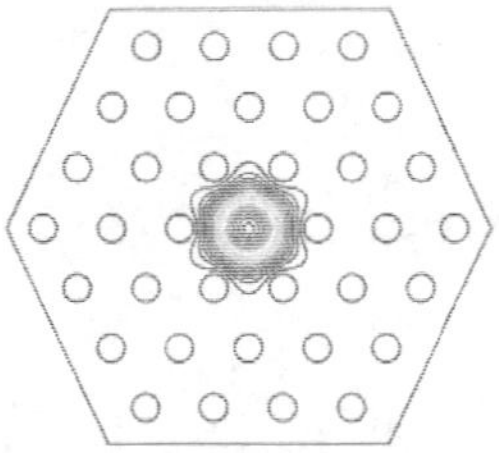

(c)

Figure 2. The distribution of THz in the photonic crystal fiber（ d/ =0.4）

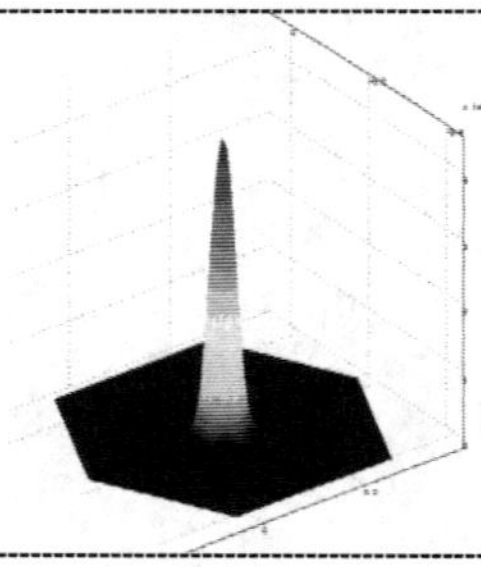

(a)

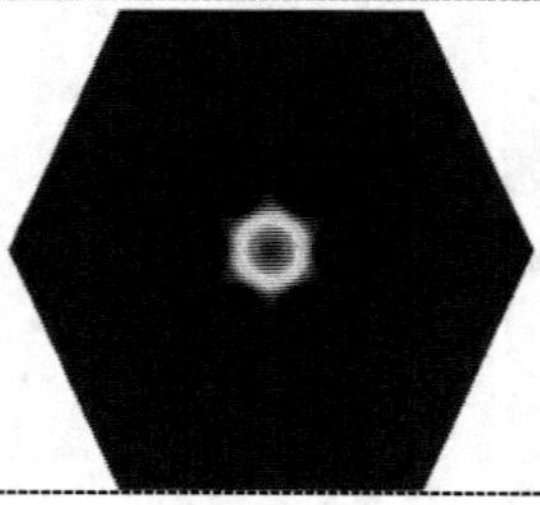

(b)

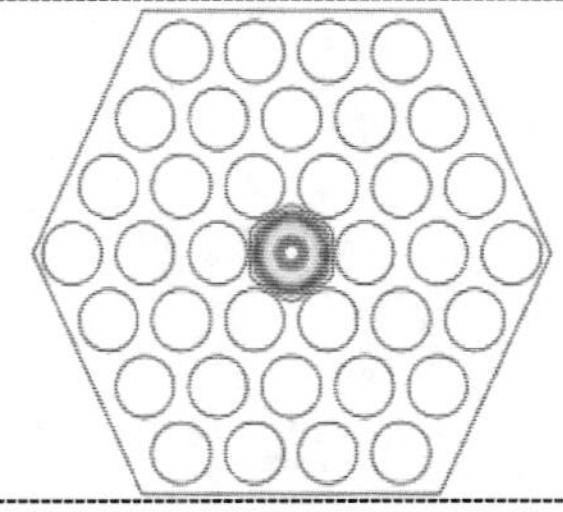

(c)

Figure 3. The distribution of THz in the photonic crystal fiber（d/ =0.8）

4. Conclusion

In this paper, THz photonic crystal fiber based on the silicon substrate, the periodic air holes formed hexagonal and fiber core with point defect. With increase the air-hole diameter, leakage loss reduced significantly.When the transmission frequency f=0.1THz, duty cycle d/ =0.8, the air-hole spacing =75μm, the minimum transmission loss is 0.0012dB/m. The results show that terahertz photonic crystal waveguide with a low transmission loss can be designed if there are suitable parameters.

Acknowledgments

This research was partially supported by the National Natural Science Foundation of China (No.60971027).

References

[1] Wang K L and Mittleman D M 2004 *Nature* **432** 376
[2] Hidaka T, Minamide H, Ito H, Nishizawa J, Tamura K and Ichikawa S 2005 *J. Lightwave Technol.* **23** 2469
[3] Mendis R amd Grischkowsky D 2000 *J.Appl Phys* **88** 4449
[4] Goto M, Quema A, Takahashi H, Ono S and Sarukura N 2004 *Jpn J Appl Phys Part2* **43** L317
[5] Jamison S P, McGowan R Q and Grischkowskv K 2000 *Appl.Phys. Lett* **76** 1987

Journal of Physics: Conference Series **276** (2011) 012206 doi:10.1088/1742-6596/276/1/012206

Terahertz spectroscopic investigation of magnetic materials

He yun-hong[1], **Li Jiu-sheng**[1, 2]

[1] Centre for THz Research, China Jiliang University, Hangzhou 310018, China

[2] College of Precision Instrument and Opto-Electronics Engineering, Institute of Laser and Optoelectronics, Tianjin University, Tianjin 300072, China

E-mail: lijsh2008@126.com

Abstract. The absorption spectra of three kinds of magnetic materials in the frequency range of 0~2.0THz has been measured with terahertz time-domain spectroscopyat room temperature in nitrogen atmosphere. Simultaneously, the terahertz refraction, transmission and absorption spectrum of the three kinds of magnetic material are obtained through the experiment. Comparing the spectroscopic characteristics of the samples of the three kinds of magnetic materials, the difference can be obviously observed. The results provided in this paper will help us to study the terahertz application to magnetic materials quality evaluation and the development of rare earth resources.

1. Introduction

Terahertz wave is usually refers to the frequency of 0.1 ~10THz of the electromagnetic wave, of which the photon energy is about 1~10 meV, coinciding with the energy between the molecular rotation vibration and energy transition. THz radiation can penetrate through dielectric and non-polar substances such as ceramic, fat, carbon board, plastic, cloth, etc. Globus and Shen have made terahertz spectrum identification on various categories of biological molecules such as nucleotide, carbohydrate and protein [1-2]. There is high correlation between biological molecules and the overall structure of the macromolecules and their terahertz spectrum properties. In recent years, terahertz spectrum technology attracts more and more attentions [3-4].

In this paper, terahertz time domain spectroscopy (THz-TDS) technique is introduced to test the spectrum of three kinds of magnetic materials at room temperature in nitrogen atmosphere. And the terahertz refraction spectrum and absorption spectrum of the three samples are obtained in the 0.1~2.0THz-band. Through the spectroscopic characteristics, the difference can be obviously observed and which can be used to distinguish the three kinds of samples magnetic materials. The results show that the THz-TDS detection technique has an important meaning on magnetic materials quality evaluation.

2. Theory analysis

The schematic diagram of the THz-TDS setup used in this experiment is shown in figure 1. A repetition rate of 80MHz, diode-pump mode-locked Ti:sapphire laser (MaiTai, Spectra Physics) provided the femtosecond pulses with duration of 100fs and center wavelength of 810nm. A p-type InAs wafer with <100> orientation was used as the THz emitter and a 2.8 mm-thick <110> ZnTe was employed as the sensor. A standard lock-in technology was used in this system. A femtosecond laser pulse was split into two beams. The pump beam was used to generate THz radiation and the probe

Journal of Physics: Conference Series **276** (2011) 012206 doi:10.1088/1742-6596/276/1/012206

beam acted as a gated detector to monitor the temporal waveform of THz field. A silicon lens and parabolic mirrors were used to collimate and focus the THz beam through free space onto the detector. A balanced photodiode detector detected the probe beam, and the signal was amplified by a lock-in amplifier and sent to the computer for processing. The THz beam path was purged with dry nitrogen to minimize the absorption of water vapor and enhance the signal to noise ratio (SNR). The humidity was kept less than 1% and the temperature was kept at 298K. The dynamic range was about 4000:1, and the spectral resolution was better than 60 GHz in the 0.2–1.8THz region.

The THz-TDS can be used to extract optical parameters such as the absorption coefficient and refractive index of the sample. The transmitted field, $H = E_{sam}(\nu) / E_{ref}(\nu)$ is related to the complex refractive index $N = n + ik$ by

$$H = \frac{4N}{(1+N)^2} e^{i2\pi(N-1)d\omega/c} = Ae^{-i\varphi} \tag{1}$$

where ω is the frequency of the THz wave, A the amplitude of the THz wave, φ the phase difference between the sample and reference waveforms, and d the thickness of the sample. The multiple internal reflection is ignored in this equation under the condition $K \ll n$, where K is the extinction coefficient. Hence the Fabry–Perot effect can be neglected. Thus, we can obtain the refractive index n, absorption coefficient α, respectively:

$$n = \frac{c\varphi}{2\pi d\omega} + 1 \tag{2}$$

$$\alpha = \frac{4\pi k\omega}{c} = \frac{2}{d} \ln\left[\frac{4n}{A(n+1)^2}\right] \tag{3}$$

3. Experiment results

The measured THz time-domain spectroscopy of the three magnetic materials is shown in figure 2. The experimental results were redone more than 3 times, the results provided here are repeatable and credible. Compared to the first peak of the reference signal, one sees that the time delays of the magnetic materials are 2.61ps, 2.60ps, and 2.49ps, respectively.

Figure 3 indicates the absorption coefficient of three magnetic materials. It can be noted that the three magnetic materials have some obvious characteristic absorption peaks around 1THz. The absorption coefficient for the first sample is 52.68cm^{-1} at 0.72THz, 51.75cm^{-1} at 0.91THz, 50.43cm^{-1} at 1.07THz and 89.72cm^{-1} at 1.29THz and for the second sample is 31.99cm^{-1} at 0.68THz, 47.46cm^{-1} at 1.02THz, and 68.68cm^{-1} at 1.31THz. The absorption coefficient for the third sample is a little weak, and is about 26.24cm^{-1} at 0.69THz, 32.57cm^{-1} at 1.01THz, and 40.19cm^{-1} at 1.32THz.

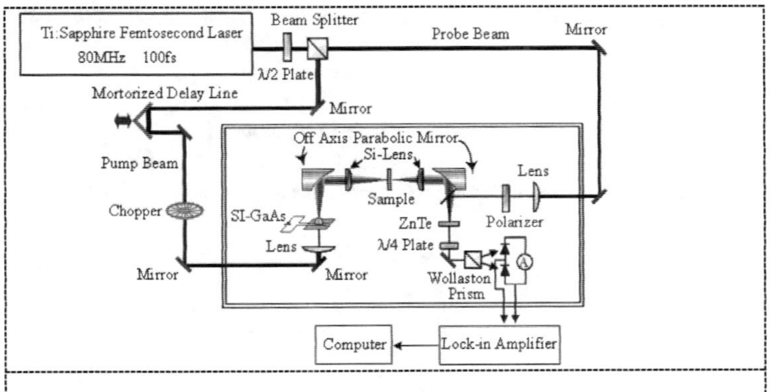

Figure 1. A schematical THz-TDS system

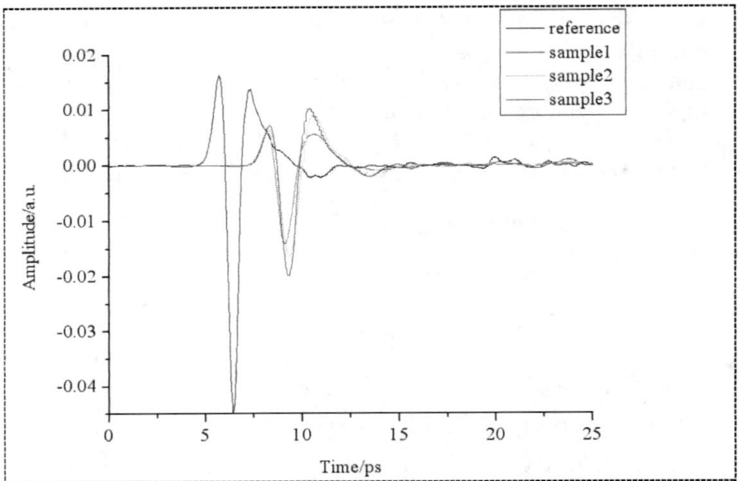

Figure 2. Measured terahertz time-domain spectra of reference and samples

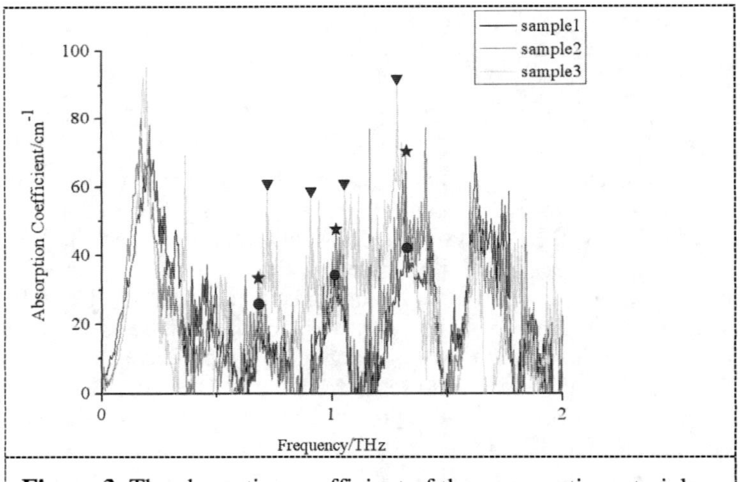

Figure 3. The absorption coefficient of three magnetic materials

The refractive index of three magnetic materials is illustrated by figure 4. In the frequency band from 0.2 to 1.8THz, the refractive index decline with the increased frequency. The refractive index of three magnetic materials is 1.066, 1.060 and 1.054 at 1THz respectively. There are large refractive index differences between the three samples, which can obviously identify the three magnetic materials.

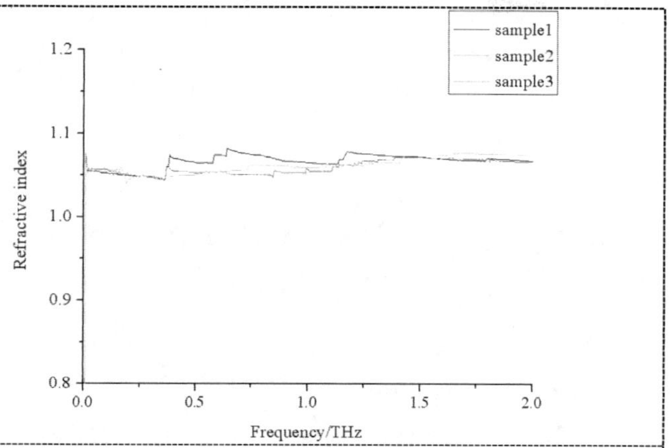

Figure 4. The refractive index of three magnetic materials

4. Conclusion

The THz spectral transmission of three kinds of magnetic materials in 0.1~2.0THz range has been measured and analyzed. Simultaneously, the terahertz refraction, transmission and absorption spectrum of the three kinds of magnetic material are obtained through the experiment. Differences between refractive index are obvious enough to identify the three magnetic materials. The results provided that the THz-TDS detection technique has an important meaning on magnetic materials quality evaluation and the development of rare earth resources.

Acknowledgments

This research was partially supported by the National Natural Science Foundation of China (No.60971027).

References

[1] Daniel M, Martin C N and Vicki L C 1997 *Chemical physic Letters* **275** 332
[2] Globus T R, Wooland D L and Khromova T 2003 *Journal of biological physics* **29** 89
[3] Shen Y C, Updhya P C and Linfield E H 2004 *Vibrational spectrocopy* **35** 111
[4] M.Ge, Zhao H and Ji T 2006 *Science in China Series B: Chemistry* **49** 204

Transmission properties of terahertz pulses through semiconductor split-ring resonators

He yun-hong[1], **Li Jiu-sheng**[1, 2]

[1] Centre for THz Research, China Jiliang University, Hangzhou 310018, China

[2] College of Precision Instrument and Opto-Electronics Engineering, Institute of Laser and Optoelectronics, Tianjin University, Tianjin 300072, China

E-mail: lijsh2008@126.com

Abstract. In this paper, two novel planar terahertz semiconductor split-ring resonators are successfully constructed and measured using the commercial software CST Microwave Studio. They exhibit a duel-band and a triple-band transmission property within the frequencies ranging from 0.1THz to 3THz. We have simulated the dual-band planar metamaterial with two distinct electric resonances at 0.81THz and 1.818THz, and triple-band planar metamaterial with three distinct electric resonances at 0.543THz, 1.044THz, and 1.506THz. These developments are further steps towards the development of broadband terahertz devices.

1. Introduction

Recently, artificially structured electromagnetic materials, which are often called metamaterials, have increasingly become an attractive research area because of the possibility of their novel electromagnetic responses not available in natural materials, such as negative refractive index. Metamaterials are a new class of composite artificial materials that acquire their properties from internal subwavelength structure rather than the materials of which they are composed [1-4]. Metamaterials are built from the bottom up, at the unit cell level, to produce independently tailored electric and magnetic responses to the incident electromagnetic waves over a significant portion of the spectrum including the THz region. Planar THz metamaterials fabricated on suitable substrates have shown electric resonant response, which causes the band-pass or band-stop property in THz transmission and reflection. The advent of metamaterials has given rise to numerous electromagnetic functionalities previously unimaged, including filters, modulators, amplifiers, transistors, resonators, sub-diffraction- limited superlenses, and cloaks.

In this paper, we have employed InAs to fabricate semiconductor split-ring resonators instead of conventional metal SRRs. Two novel planar THz semiconductor metamaterials are successfully constructed and measured using the commercial software CST Microwave Studio, exhibiting a duel-band and a triple-band transmission property, respectively.

2. Device configuration and simulation results

The proposed novel planar structures are depicted in figure 1 and figure 2. To be specific, the essence of the metamaterial is a meta-surface with a thickness of 200nm printed on an isotropic dielectric substrate covering a period of 60 μm. The commercial software CST Microwave Studio was employed to simulate the designed planar array SRRs. During the simulation of the designed structure, we supposed the planar SRR array was surrounded by air and we set the boundary conditions as open along the propagation direction. We calculated the transmittance spectrum of the planar array SRRs when illuminated by a plane wave with its wave vector oriented parallel to the z axis and its electric field oriented parallel to the y axis. The transmittance is the ratio of the power density of the transmitted terahertz pulse to that of the incident terahertz pulse.

Figure 1 (c1) shows that a transmittance dip exists at 0.81THz and 1.818THz frequency. Compared to figure 1 (f1), we can attribute the dip at 0.81THz to a resonance of the SRRs: as shown in figure 1 (d1), the electric field is concentrated in both gaps of every SRR at this frequency. Figure 1 (e1) displays the vector plot of the current density at 0.81 THz. If both gaps of every SRR were closed, the transmittance dip at 0.81 THz would disappear, as can be gathered from the transmittance spectrum presented in figure 1 (f1).

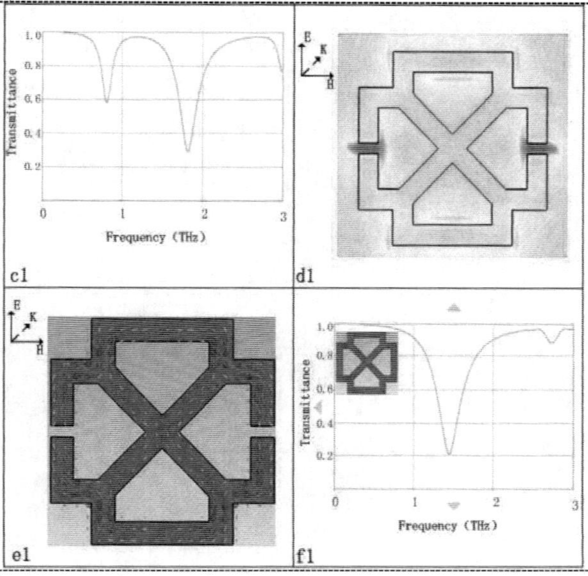

Figure 1. Transmittance spectrum of the single SRR illuminated by a terahertz pulse with electric field oriented perpendicular to the gaps in the SRR. (d1) Map of the magnitude of the electric field in an SRR at 0.81THz. (e1) Map of the current density in an SRR at 0.81THz. (f1) The same as (c1) but when both gaps are closed in every SRR.

Figure 2 (c2) shows that a transmittance dip exists at 0.543THz, 1.044THz and 1.506THz frequency. Compared to figure 2 (f2), we can attribute the dips at 1.044THz and 1.056THz to the resonance of the SRRs: as shown in figure 2 (d2), the electric field is concentrated in both gaps of every SRR at these frequencies. figure 2 (e2) displays the vector plot of the current density at 1.044 THz. If all gaps of every SRR were closed, the transmittance dips at 1.044THz and 1.056THz would disappear, as can be gathered from the transmittance spectrum presented in figure 2 (f2).

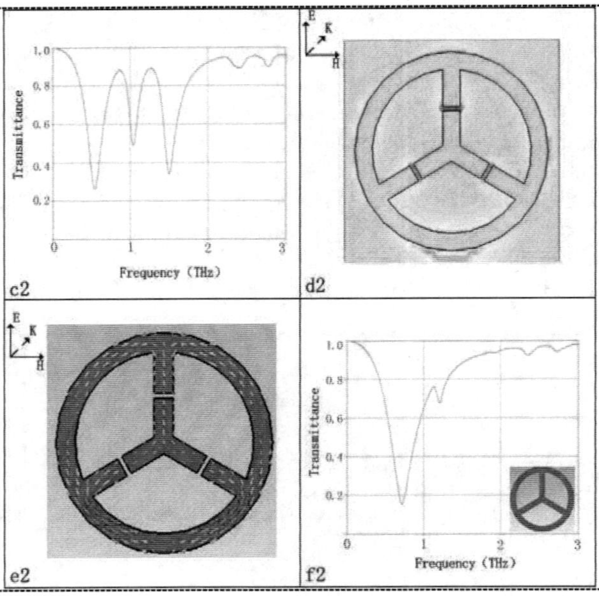

Figure 2. (c2) Transmittance spectrum of the single SRR illuminated by a terahertz pulse with electric field oriented perpendicular to the gaps in the SRR. (d2) Map of the magnitude of the electric field in an SRR at 0.81THz. (e2) Map of the current density in an SRR at 0.81THz. (f2) The same as (c2) but when both gaps are closed in every SRR.

3. Conclusion

In summary, we present our recent developments in terahertz metamaterials. Two novel planar THz semiconductor metamaterials are successfully constructed and measured using the commercial software CST Microwave Studio, respectively exhibiting a duel-band and a triple-band transmission property within the frequencies ranging from 0.1THz to 3THz. We have measured the dual-band planar metamaterial with two distinct electric resonances at 0.81THz and 1.818THz, and triple-band planar metamaterial with three distinct electric resonances at 0.543THz, 1.044THz, and 1.506THz. These developments are further steps towards the development of frequency agile or broadband THz materials and devices.

Acknowledgments

This research was partially supported by the National Natural Science Foundation of China (No.60971027).

References

[1] Tao H, Landy N, Bingham C M, Zhang X, Averitt R D and Padilla W J 2008 *Opt. Exp.* **16** 7181
[2] Smith D R, Padilla W J, Vier D C and Nemat-Nasser S C 2000 *Phys. Rev. Lett.* **84** 4184
[3] Shelby R A, Smith D R and Schultz S 2001 *Science* **292** 77
[4] Smith D R, Pendry J B and Wiltshire M C K 2004 *Science* **305** 788

Journal of Physics: Conference Series **276** (2011) 012208 doi:10.1088/1742-6596/276/1/012208

Intense terahertz radiation from μm-gap GaAs photoconductive antenna

L Hou[1], **W Shi**[1,2]

[1] Department of Applied Physics, Xi'an University of Technology, No.5, South Jinhua Road, Shaanxi 710048, China
[2] State key Laboratory of Electrical Insulation for Power Equipment, Xian Jiaotong University, No.28, Xianning West Road, Shaanxi 710049, China

E-mail: swshi@mail.xaut.edu.cn

Abstract. The terahertz pulse energy from photoconductive antenna is mainly derived from that stored in the static bias field. To obtain high intense THz radiation, the distribution of electrical field in semi-insulating GaAs photoconductive antennas was studied at square DC bias voltage. Due to the trap-enhanced space-charge, the electrical field is enhanced at anode and extent to cathode. So the highest effective terahertz radiation is obtained when the laser beam just overlaps the gap. A larger laser beam can reduce the Coulomb screening and radiation screening caused by a focused smaller beam, so the larger beam can increase the efficiency of antenna. At the same electrical field and laser power, an antenna with a larger gap can generate more intensive terahertz radiation than that with a smaller gap.

1. Introduction

The development of terahertz (THz) source is of great importance for a large variety of scientific and technological application [1]. Recently, various methods for generating pulsed THz waves have attracted considerable attention. However, the main methods are photoconductive emitters, surface emitters (such as GaAs, InP, InAs, InSb) [2-6], and electro-optic crystals (optical rectification) [7, 8]. Among these, photoconductive emitters are considered to be the most efficient THz emitters. Various antenna geometries have been used, including Hertzian dipoles, resonant dipole antennas, spiral antennas, stripline antennas, antenna arrays, and so on [9]. Dipolar antennas and stripline dipole antennas are the most common structures.

The shortage of high-power THz sources is the bottleneck of many THz applications. Improving the power of THz waves is important [10].The radiation mechanism of a photoconductive antenna is attributed to a time-varying current, which results when the photo excited carriers are accelerated across the surface of the photoconductor in the presence of the applied electric field. The terahertz pulse energy is derived from that stored in the static bias field [11]. But the electrical field is not evenly distributed in the electrode gap of antenna [12], so the terahertz amplitude will be different with the laser position [13, 14]. It is important to study the distribution of electrical field in the antenna. The efficiency of antenna also is related to the beam size of antenna. In this paper, we studied

[1] W Shi, Department of Applied Physics, Xi'an University of Technology, No.5, South Jinhua Road, Shaanxi 710048, China. Email: swshi@mail.xaut.edu.cn

the different electrical field distribution on stripline semi-insulting (SI) GaAs antennas and optimized the THz generation efficiency of stripline antennas.

2. Experiment

In our experiment, the substrate of antenna is a commercial high-resistivity (>107 Ω·cm) liquid-encapsulated-Czochralsky grown, (100)-oriented, SI-GaAs wafer. The width of electrodes is 100 μm, and the gaps between the two electrodes are 50, 100, 150 and 200μm, respectively. The electrode pattern was made using a conventional photolithography technique. Ohmic contacts were obtained by using a standard mixture of Ni/Au-Ge/Au for the metallization and the thickness is 700 nm.

Those antennas were tested by a THz time domain spectroscope (THz-TDS) system. A mode-locked laser based on Er doped fiber oscillator (Femtolite-100 IMRA) was used to generate excitation laser pulses with the center wavelength of 800 nm with the repetition rate of 75 MHz. The width of the pulses was less than 150 fs. The pump beam has a power of 90 mW. The bias power supply provided a square DC voltage, and they were modulated by a lock-in amplifier, the frequency was 75 kHz. Ultrafast switching of the biased emitter by the ultrafast laser generated transient current and produced broadband THz radiation. The THz beam from the emitter was collimated and focused on a 1-mm-thick ZnTe crystal by a pair of off-axis parabolic mirrors. A probe pulse of 10 mW traveled collinear with the THz pulse through the ZnTe crystal. The THz pulse modified the optical refractive index of the probe pulse by means of the linear electro-optic effect and thereby induced an ellipticity of the probe polarization. By measuring this ellipticity as a function of delay with respect to the THz pulse, we can retrieve the time dependence of the electric field of the THz pulse.

3. Distribution of electrical field

To find the laser position dependence of the THz intensity, the laser beam was focused into an about 20 μm spot, the laser spot can illuminate different position of the antenna by precisely adjusting the antenna, and the position of laser spot can be accurately recorded.

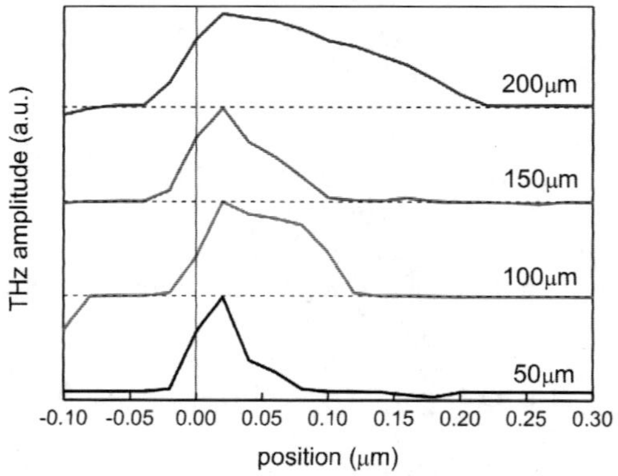

Figure 1. Laser position dependence of THz amplitude of SI-GaAs stripline antennas with different gaps biased by square DC voltage.

Figure 1 shows the laser position dependence of the THz amplitude of the 4 antennas biased by the square DC power supply. The THz generation process become more efficient when the excitation is carried out closed to the anode electrodes due to trap-enhanced fields (TEF) in proximity of the anode when the DC bias voltage is used. This TEF results from a number of physical mechanisms including the quasi-static nonlinear field distribution, electron velocity saturation, space charge of uncompensated residual acceptors resulting from the increasing occupancy of deep trap, and field-enhanced carrier injection [15]. Because the THz intensity is proportional to the bias electrical field [16], the distribution of the laser position dependence of electrical field is same to figure 1. The right edge of the anode was at zero. The maximum THz amplitude locates at the right of zero where the laser spot can entirely illuminate the nearest region close to anode, because the laser spot is not small enough. And in these cases, the electrical fields also distribute in a large region, not are limited in a several microns zone near the anode. This result is in accord with the reference 17. The distribution of electrical field is not like reference 15 reports that up to 90% of the applied potential may exist within 5 μm of the anode, but the considerably intensive electrical field permeated the entire or mostly whole gap.

4. THz generation efficiency

In our experiment, we found the most efficient terahertz radiation generated by SI-GaAs antenna was not obtained by a small focused laser spot. However, when the laser spot with the same power just overlaps the entire gap, the most intensive THz radiation will be generated and the THz amplitude is 3 to 8 times larger than that generated by a small focused spot illuminating the region close to anode, which is shown in figure 2.

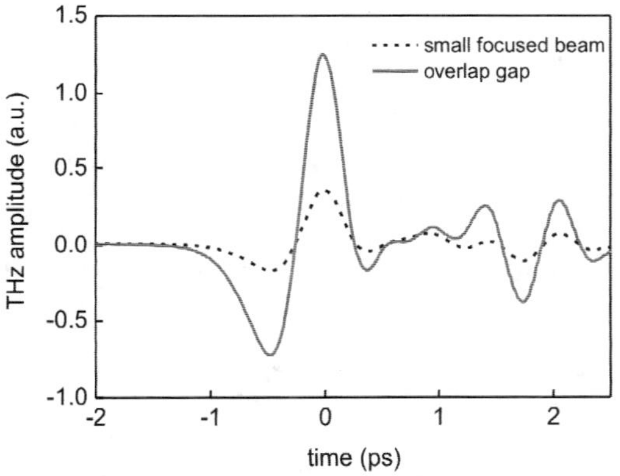

Figure 2. THz waveforms generated by a 200-μm-gap antenna illuminated by small focused beam (with a 20 μm diameter) and the beam (with a 200 μm diameter) just overlapping the gap.

In this figure, the antenna with 200 μm gap was biased by a square DC voltage with the peak to peak value of 120 V. The THz intensity by a laser spot (with 200 μm diameter) overlapping gap is 3.5 times larger than that from a small focused spot (with 20 μm diameter) with the same power illuminating the region close to anode. This is caused by the screening of photo excited carriers. When the femtosecond laser illuminates the photoconductive antenna, the accelerated photo excited carriers by the bias field will emit THz radiation, but at the same time, as the carriers undergo their spatial dynamics in the bias field, they partially screen out the bias field. The origins of the screening consist of the radiation field and the space-charge (Coulomb) field, which contribute to the collapse of the total electric field acting on the carriers at high carrier density. D. Kim and D. S. Citrin studied the effects of the excitation-spot size and excitation level on the emitted THz radiation by the Monte Carlo method [18]. In the simulation, the laser has a fixed optical power, and the spot size can be changed by focusing optics. In this case, as the excitation-spot size increases, the Coulomb screening decreases rapidly because the distance between the carriers increases due to the decrease of carrier density. In addition, as the carrier density becomes sparse, the carrier numbers contributes the Coulomb and radiation screening field decrease. As a result, the radiation screening field also decreases as the excitation-aperture size increases. Another shortcoming lies in generating THz radiation by small focused spot illumination, the small focused beam can easily damage the antenna, because the current passes through a narrow passage and the heat caused by current easily damages the material of the narrow passage.

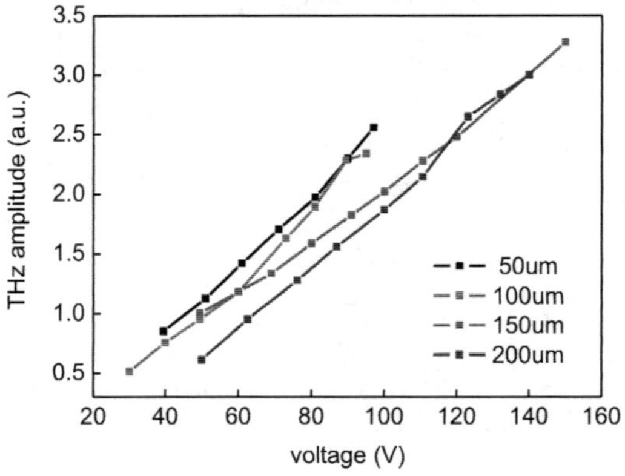

Figure 3. Voltage dependence of the THz intensity of antennas with the exciting laser overlapping the gap.

Then we compared the THz radiation capacity of the four antennas with different gaps. Figure 3 shows the bias voltage dependence of the THz intensity measured for the four antennas with the exciting laser overlapping the gaps. In all cases, the THz intensity linearly increases with the bias voltage and there is no saturation. At the same bias voltage, the antenna with smaller gap can emit a more intensive THz radiation because the electrical field of smaller gap antenna is larger than that of the larger one. Then we investigated the electrical field dependence of the THz intensity signals measured for the four antennas with the exciting laser overlapping the gaps, which is shown in figure 4. At the same electrical field, the THz intensity of larger gap antenna is much higher than the smaller gap antenna. In the test, the THz intensity of 200 μm antenna is more than 4 times larger than that of 50 μm antenna at about 7.5 kV/cm. This is also caused by screening [19]. Large gap antenna can reduce screening because its large gap needs a large laser spot to illuminate, and the distance of photo exciting carriers increase and the density of carriers reduce. So, large gap antennas have high electrical to THz and optical to THz conversion efficiency.

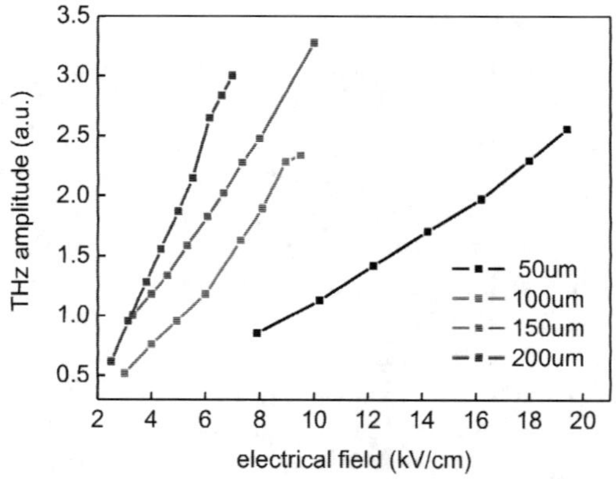

Figure 4. Electrical field dependence of the THz intensity of antennas with the exciting laser overlapping the gap.

5. Conclusions

We have investigated the distribution of electrical field in SI-GaAs stripline antennas biased by a square DC power supply, respectively. And we find the electrical field is increased at anode for DC square bias voltage due to a trap-enhanced space-charge region. But the electrical field is not limited in an about several microns region near anode, considerably intensive electrical field permeates the entire or mostly whole gap. The effective THz radiation is obtained when the laser beam just overlaps the gap, because the larger beam can reduce the Coulomb screening and radiation screening caused by a focused small beam with same power. And at the same electrical field, large gap antenna can generate more intensive THz radiation than small gap antenna also because the large gap antenna can reduce the screening. THz radiation can be improved by using laser beam cover the gap of a large gap antenna.

6. Acknowledgements

This work is supported by the National Natural Science Foundation of China (Grant No. 61007060, No.60906029, No.50837005), the National Basic Research Program of China (Grant No.2007CB310406).

7. References

[1] Dreyhaupt A, Winnerl S, Dekorsy T, and Helm M 2005 *Appl. Phys. Lett.* **86** 121114
[2] Darmo J, Strasser G, Müller T, Bratschitsch R, and Unterrainer K 2002 *Appl. Phys. Lett.* **81** 871
[3] Jagadish and Johnston M B 2007 *Opt. Express* **15** 7047
[4] Heyman J N, Neocleous P, Hebert D, Crowell P A, Müller T, and Unterrainer K 2001 *Phys. Rev. B* **64** 085202
[5] Reid M and Fedosejevsa R 2005 *Appl. Phys. Lett.* **86** 011906
[6] Welsh G H, Hunt N T, and Wynne K 2007 *Phys. Rev. Lett.* **98** 026803
[7] Rice A, Jin Y, Ma X F, and Zhang X.-C. 1994 *Appl. Phys. Lett.* **64** 1324
[8] Kadlec F, Kuzel P, and Coutaz J 2005 *Opt. Lett.* **30** 1402
[9] Zhang X.-C. 1996 *J. Lumin.* **66-67** 488
[10] Dreyhaupt A, Winnerl S, Dekorsy T, and Helm M 2005 *Appl. Phys. Lett.* **86** 121114
[11] Darrow J T, Zhang X.-C., and Auston D. H 1991 *Appl. Phys. Lett.* **58** 25
[12] Krökel D, Grischkowsky D and Ketchen M B 1989 *Appl. Phys. Lett.* **54** 1046
[13] Katzenellenbogen N and Grischkowsky D 1991 *Appl. Phys. Lett.* **58** 222
[14] Gregory I S, Baker C, Tribe W R, Bradley I V, Evans M J, Linfield E H, Davies A G and Missous 2005 *IEEE J. Quan. Elec.* **41** 717
[15] Ralph S E and Grischkowsky D 1991 *Appl. Phys. Lett.* **59** 1972
[16] Benicewicz P K, Roberts J P, and Taylor A J 1994 *Opt. Lett.* **11** 2533
[17] Salem B, Morris D, Aimez V, Beerens J, Beauvais J and Houde D 2005 *J. Phys.: Condens. Matter.* **17** 7327
[18] Kim D S and Citrin D S 2006 *Appl. Phys. Lett.* **88** 161117
[19] Darrow J T, Zhang X.-C., Auston D H 1992 *IEEE J. Quan. Elec.* **28** 1607

Spectrum characteristics of nitrofen by terahertz time-domain spectroscopy

Huali WANG and Qiang WANG*
Dept. of Quality & Safety Engineering, China Jiliang University, Hang-Zhou 310018, China

*Corresponding author. E-mail: qiangwang@cjlu.edu.cn

Abstract. This paper firstly presents the preliminary research on nitrofen with terahertz time-domain spectroscopy (THz-TDS). Different weight ratios of nitrofen in nitrofen-polyethylene combinations were measured. We employed linear fit between not only absorption coefficient but also average refractive index and the concentration of nitrofen. In the frequency region from 0.4 THz to 2.2 THz, seven characteristic absorption peaks of pure sample had been observed, four of which were assigned by using density function theory. Our study demonstrates that THz-TDS is an effective tool to distinguish nitrofen.

1. Introduction

Pesticide residue is one of the prominent factors on influencing the food safety, which has increasingly been paid attention by consumers. The existing detection methods of pesticide residues can be roughly classified into spectrometry, enzyme inhibition and chromatography etc. Spectrometry is generally acted as a qualitative method because of low sensitivity. The sensitivity and reproducibility of enzyme inhibition is poor due to the instability of the sources and types of enzymes. It is difficult to intuitively detect the pesticide residues using chromatography owing to the complex preparation technique of samples, expensive equipments and professional staff [1-2]. Nitrofen, as a kind of selective contact ether herbicides, is widely used on various kinds of vegetables. Research shows that nitrofen will cause hemolytic anemia, toxic hepatitis or various dermatitis while act on the human body [3]. Therefore, it is greatly significant in developing a kind of easy-to-use, quick and sensitive detection method.

This paper achieved the study on the nitrofen molecule by using a relatively novel detection technology of far-infrared spectral -terahertz time-domain spectroscopy technology. Terahertz (THz) wave is the electromagnetic wave of 0.1 ~ 10 THz (1THz = 10^{12} Hz) range, corresponding to a wavelength of 3000 ~ 30 μ m. Terahertz radiation lies between the microwave and infrared regions of the electromagnetic spectrum, and THz spectrums of substances contain considerable physical and chemical information [4]. In recent years, many scholars at home and abroad have made investigation on considerable substances based on THz - TDS system. For instance, Mittleman D M et al studied polar gases and made identification and detection for gas mixtures [5]. Hua Y F et al quantitatively detected cyfluthrin content in n-hexane solvent by utilizing THz-TDS [6]. THz spectrum of Polycrystalline RDX was acquired by Allis D G et al [7]. Yan Z G et al studied on vibration mode of the organophosphate pesticide acephate in the THz range [8]. The nitrofen detection by utilizing THz-TDS technique has not yet found at home.

We reported the response characteristics and transmission mechanism of nitrofen in the terahertz range by using THz-TDS technology at the room temperature and nitrogen environment. Varieties of nitrofen-polyethylene combinations with differing nitrofen contents were measured to obtain absorption and refractive index spectra. We employed linear fit between not only absorption coefficient but also average refractive index and the concentration. Basing on the experiment, simulation calculation of quantum chemistry software Gaussian 03 was used to analyze experiment data.

2. Experimental methods

2.1. Sample preparation and experimental apparatus

Nitrofen with the purity of 99% was purchased from J&K SCIENTIFIC LTD, Shanghai city, China and was used without further purification or treatment. Firstly, the sample and polyethylene (PE) powder were dried at 310 K for 2 hours. Secondly, they were mixed well with several different concentrations, as shown in table 1, and grinded with a agate mortar to reduce baseline offsets. Furthermore, laboratory tablet press was employed to compress the circular discs with diameter of 13 mm and thickness ranged from 1 to 2 mm at 30 MPa pressure.

Table 1. Parameters of nitrofen pellets and their upper limit of significant frequency

No.	date	thickness(d) / mm	approx weight percentage of sample / %	upper limit THz
1	1st	1.52	66%	1.5
2	1st	1.20	75%	2.2
3	2nd	1.16	100%	2.4.
4	2nd	1.60	50%	1.5
5	2nd	1.57	33%	1.5

THz-TDS spectroscopy experiment setup refers to literature [8]. Experiment was carried out at room temperature (about 294 K), with the optical path of THz Filled with nitrogen and the relative humidity of about 3.0%.

2.2. Data acquisition

The real refractive index $n(\omega)$ and absorption coefficient $\alpha(\omega)$ respectively describing the dispersion and absorption characters of substance are generally used to represent macroscopic optical properties of substance, and calculated as follows:

$$n(\omega) = \frac{\phi(\omega)c}{\omega d} + 1 \quad (1)$$

$$\alpha(\omega) = \frac{2\kappa(\omega)\omega}{c} = \frac{2}{d}\ln\frac{4n(\omega)}{A(\omega)(n(\omega)+1)^2} \quad (2)$$

Where c represents the light velocity, d is the sample thickness, $A(\omega)$ is the amplitude ratio of the sample to reference signal, and $\phi(\omega)$ is the relative phase difference between sample and reference signal. The above are based on the processing model proposed by Dorney T D et al [9] and Duvillaret L et al [10-11].

3. Result and discussion

Molecular formula of nitrofen is $C_{12}H_7C_{12}NO_3$. The molecular structure is shown in Fig. 1, the hydrogen bonds and hydrogen atoms had been omitted. The structure includes a nitrobenzophenone and a 2,4-dichlorophenyl, which are related to vibration modes.

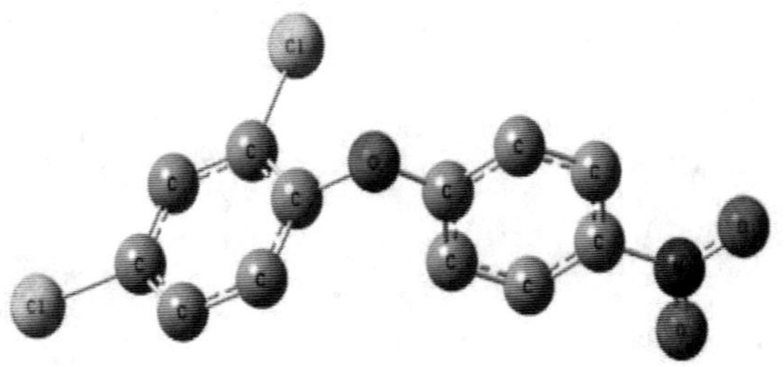

Fig. 1. Nitrofen molecular structure diagram.

THz time-domain waveforms of nitrofen and reference signal are shown in Fig. 2. A certain attenuation of the sample signal relative to the reference signal is due to sample absorption, scattering and other reasons. A certain delay on time is due to the different refractive index of sample and nitrogen.

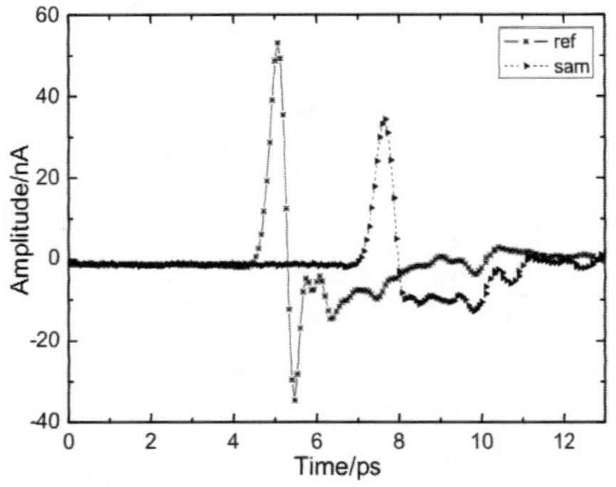

Fig. 2. THz time-domain wave of nitrofen.

THz absorption spectra of nitrofen without baselines corrected, shown in Fig. 3(a), are calculated by equation (2). The samples with different thickness have the distinct upper limit THz. That have been assumed to result from different signal-to-noise ratio (SNR).The thinner the sample is, the higher the upper limit THz is. The upper limit THz is shown in Table 1. From Fig. 3(a), we can see that the 3th sample exhibits 7 absorption peaks in the effective frequency range from 0.4 THz to 2.4 THz. The test absorption peaks locates respectively at 0.5412, 0.6592, 0.98145, 1.3330, 1.6260, 1.9775 and 2.2412 THz. With frequency increasing, the baseline of the absorption spectrum has a slowly rising. The phenomenon may be caused by light scattering or sample surface reflection.

The frequencies of the test characteristic absorption peaks a_1 and a_2 are within effective frequency range of all the samples. The weight content can be regressed linearly against the absorption coefficient at a_1 and a_2 THz, with correlation coefficients for the linear fit of 0.99049 and 0.99967, respectively. The results of linear fitting are shown in Fig. 4(c) and Fig. 4(a). The solid line is the linear fit. The solid triangles and stars represent the absorption coefficient.

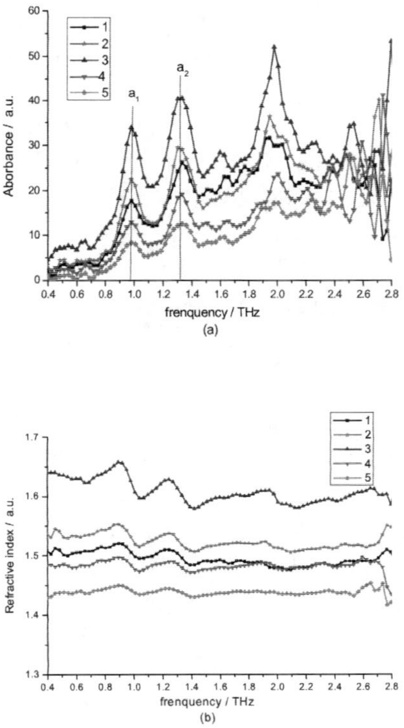

Fig. 3. (a) Absorption spectra of nitrofen and (b) refractive index spectra of nitrofen.

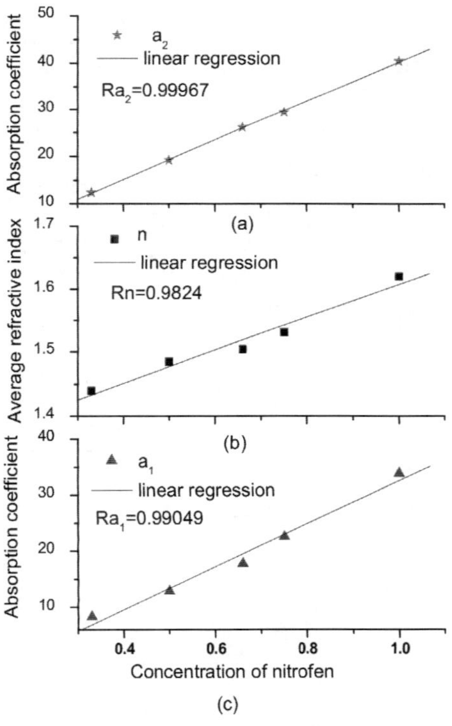

Fig. 4. the results of the linear fit by utilizing (a) the absorption coefficient of a_2, (b) average refractive index, (c) the absorption coefficient of a_1.

Refractive index spectra of nitrofen shown in Fig. 3(b) are calculated by equation (1). Refractive index decreases with the frequency gradually increasing. In the effective frequency range, refractive index has a significant change associated with each characteristic absorption peak due to anomalous dispersion. From 0.5 to 1.5 THz, we can learn the average refractive index is 1.43975, 1.48451, 1.50413, 1.53043, 1.62056 with corresponding ascending concentration of nitrofen. We employed linear fit between average refractive index and the concentration of nitrofen and obtained the correlation coefficient 0.9824. In Fig. 4(b), the solid line is the linear fit and the solid square is average refractive index.

We employed linear fit between not only absorption coefficient but also average refractive index and the concentration of nitrofen. In terms of the correlation coefficient, the fit line with absorption coefficient was superior to that with average refractive index. Knowing the fit linear parameters, it is easy to allow a reliable determination of weight content. Nevertheless, the linear fit has been supposed an empiric approximation for the test date. The precision of determining unknown weight content depends on the data used for fitting, while relating to SNR.

The Becke-3-Lee-Yang-Parr (B3LYP) method of DFT at 6-31G (d) basis sets as implemented in Gaussian03 was used to geometrically optimize nitrofen molecules [12]. The frequency calculation results had no imaginary frequency utilizing the same method, indicating that molecular optimization get the minimum energy molecular structure. Four of the characteristic absorption peaks were assigned theoretically, and the intermolecular modes are supposed to dominate the remaining spectrum.

Theoretical simulation results showed that: the four characteristic absorption peaks of nitrofen molecules at this range are corresponding to the different vibration modes of nitrobenzene and 2, 4 - dichlorophenyl. With the visual function of Gaussian View 3.09, the vibration modes of characteristic absorption peaks were assigned in Table 2.

Table 2. Dominant motions at the frequencies of theory absorption peaks

serial number	theory / THz	test / THz	ascription
1	0.5343	0.5412	swing vibration of nitrobenzophenone and torsional vibration of nitrobenzophenone
2	0.9975	0.9815	torsional vibration of nitrobenzophenone and bending vibration of 2,4 – dichlorophenyl
3	1.1746	1.3330	torsional vibration of nitrobenzophenone and swing vibration of 2,4 – dichlorophenyl
4	2.1220	2.2412	torsional vibration of nitrobenzophenone

There are some deviations between the experimental frequencies of characteristic absorption peaks and the theoretical values. The reasons causing the difference between the theoretical and experimental values are as follows: Gaussian can calculate the intermolecular and intramolecular chemical properties on the condition of gas or liquid, and the simulative model is a single molecule, while the nitrofen is crystal, involving intramolecular or phonon interactions. Simulation theory often neglects many practical factors and the experimental results are affected by the absorption of system and environment. To sum up, absorption of nitrofen in the terahertz range contains not only the intramolecular interactions, but also the collective -vibration modes or lattice vibrations associated with intermolecular hydrogen bonds.

4. Conclusion

This paper studied on absorption and refractive index spectra of nitrofen based on THz-TDS technology and density functional theory. In the range from 0.4 to 2.4 THz, there are 7 obvious characteristic absorption peaks which respectively locate at 0.5412, 0.6592, 0.98145, 1.3330, 1.6260, 1.9775 and 2.2412 THz, and four of them were assigned theoretically to the different vibration modes of nitrobenzene and 2, 4 - dichlorophenyl. The experimental results were in good agreement with the theory simulation data. Absorption coefficient at a_1 and a_2 THz and average refractive index raging from 0.5 to 1.5 THz have optimum linear relations with the concentration of nitrofen, with correlation coefficients of 0.99049, 0.99967 and 0.9824, respectively. Knowing the fit linear parameters, it is easy to get a reliable determination of concentration. This study provides qualitative and quantitative analysis technology for pesticide molecular detection by using THz-TDS.

Acknowledgments

This work is supported by the National Quality Supervision and Inspection Public Welfare Project of China under Grant 200910181 and in part by the National Natural Science Foundation of China under Grant 60902095.

5. References

[1] Cai X X, Zhang Z L Liang S Q and Mu X M 2009 *Cereals and Oils Processing* **11** 105

[2] Niu W Q, Wang H Y, Li J Q, Han H Y, Shen C Y and Zhou S K 2009 *Journal of Anhui Agri Sci* **37** 14005

[3] Wang N, Gu Z W and Zhang S N 1996 *Modern Occupational Medicine* (Beijing: People's Medical Publishing House) p 501

[4] Wang W N 2009 *Chin. Phys. Soc.* **11** 7640

[5] Mittleman D M, Jacobsen R H, Neelamani R, Baraniuk R G and Nuss M C 1998 *J. Appl. Phys.* **B 67** 379

[6] Hua Y F, Zhang H J and Zhou H L 2010 *IEEE transactions on instrumentation and Measurement* **59** 1414

[7] Allis D G, Zeitler J A and Taday P F 2008 *Chem. Phys. Lett.* **463** 84

[8] Yan Z G, Hou D B, Cao B H, Zhang G X and Zhou Z K 2008 *Journal of Zhejiang University* **42** 2207

[9] Dorney T D, Baraniuk R G and Mittleman D M 2001 *J. Opt. Soc. Am.* A **18** 1562

[10] Duvillaret L, Great F and Coutaz J L 1996 *IEEE J. Sel. Top. in Quantum Electron. 2 739*

[11] Duvillaret L, Great F and Coutaz J L 1999 *Appl. Opt.* **38** 409

[12] Frisch M J, Trucks G W and Schlegel H B et al 2004 *Gaussian 03 Revision 9C.02* Gaussian Inc. Wallingford CT.

A 3-Dimensional display and process software for THz spectrum

Xiaowen Jiang[1], Zhaohui Zhang[1], Xiaoyan Zhao[1], Yixin Yin[1], Katsuhiro Ajito[2] and Hojin Song[2]
1 Information Engineering School, University of Science and Technology Beijing, Beijing 100083, China
2 NTT Basic Research Laboratories, Kanagawa243-0198, Japan

E-mail: zhangzhaohui@ustb.edu.cn

Abstract. An underpinning software is devoted to THz spectrum analyzing and 3-D imaging. The paper describes the software's outline, structure, functions and some of considerations. Users in LAN (local area network) can access it and implement some basic and advanced works such as files operation, echoes cutting, spectrum calculation, baseline cancelling, peak fitting, qualitative and quantitative measuring of solid-state samples.

1. Introduction

Terahertz sensing and imaging technologies [1] push massive data to us, and we hardly make use of them without specific software. This paper describes a software package for THz spectrum process, in which the primary data can be transferred to absorbance or transmittance spectra at any pixel of a pellet sample surface, and 3D images can be shown with messages of concentration distributions. Several algorithms, such as baseline fitting, peak fitting, and substance identifying and quantitative measuring of mixture are involved.

2. Remote login

The software bases on Java Web. Figure 1 shows that users in LAN can login, and upload files to or download files from a Tomcat server where the software runs.

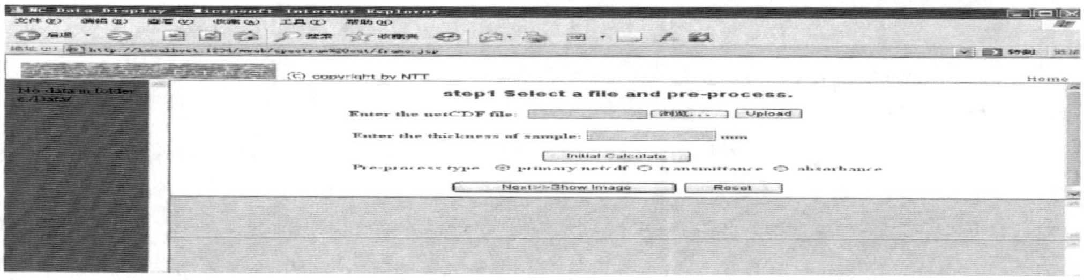

Figure 1. Interface after login

3. Data process for a pixel

Published under licence by IOP Publishing Ltd

3.1. Echo cutting

As both sides of a solid-state sample resulting in incident reflection, the acquired waveform data need preprocessing to eliminate echoes and extract the essential part of them. The schematic diagram of cutting echoes from Esam(t) and Eref(t) which are time-resolved data of the sample and the reference correspondingly[2, 3], is shown in figure 2.

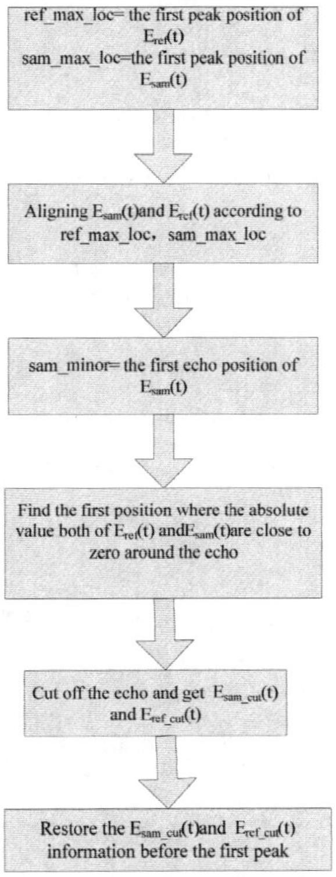

Figure 2. Procedures of echo cutting

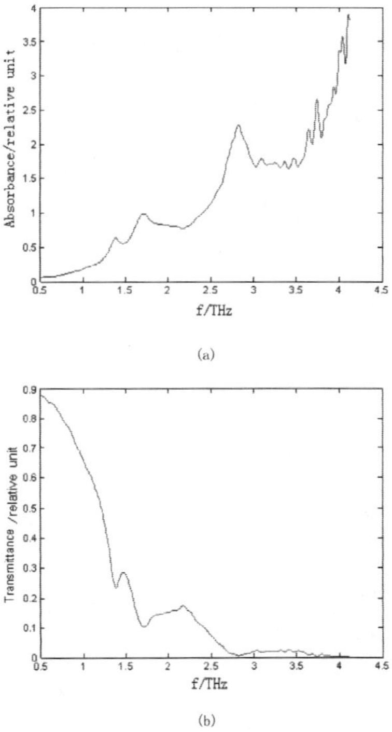

Figure 3. Spectra carried out from formulas
(a) absorbance spectra; (b) transmittance spectra

3.2. Absorbance and transmittance spectra

Denote the time-resolved electric fields of a sample and a reference as $E_{sam\_cut}(t)$ and $E_{air\_cut}(t)$ after echo cutting, and their Fourier transformations as $E_{sam}(\omega)$ and $E_{ref}(\omega)$. Based on Duvillare's theory [2], the absorbance spectra $\alpha(\omega)$ is calculated as following.

$$\alpha(\omega) = -\frac{2}{d}\ln\left\{\frac{|E_{sam}(\omega)|}{|E_{ref}(\omega)|}\frac{[n_s(\omega)+1]^2}{4n_s(\omega)}\right\} \tag{1}$$

Where d is the sample's thickness; $n_s(\omega)$ is its refractive index. Similarly, the transmittance spectrum is as following:

$$T(\omega) = \left\{ \frac{|E_{ref}(\omega)|}{|E_{sam}(\omega)|} \frac{4n_s(\omega)}{[n_s(\omega)+1]^2} \right\}^{\frac{2}{d}} \tag{2}$$

See figures 3(a) and 3(b). Both of them exhibit specific peaks [4], but the former baseline is rising while the later is declining as increasing frequencies. That's reasonable because absorbance is always against transmittance.

3.3. Spectrum baseline fitting and removing

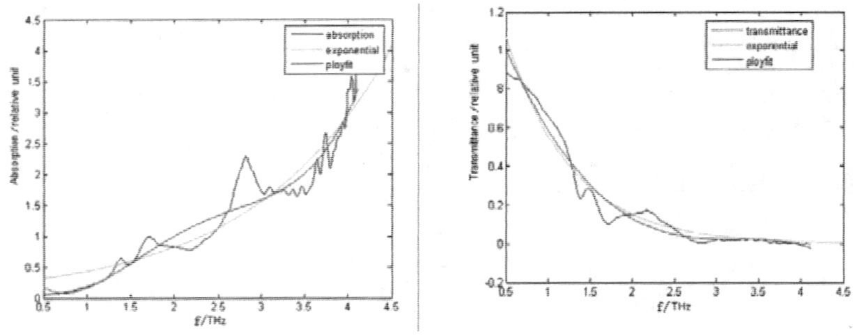

(a) (b)

Figure 4. Baseline fitting. (a) absorbance baseline fitting; (b) transmittance baseline fitting

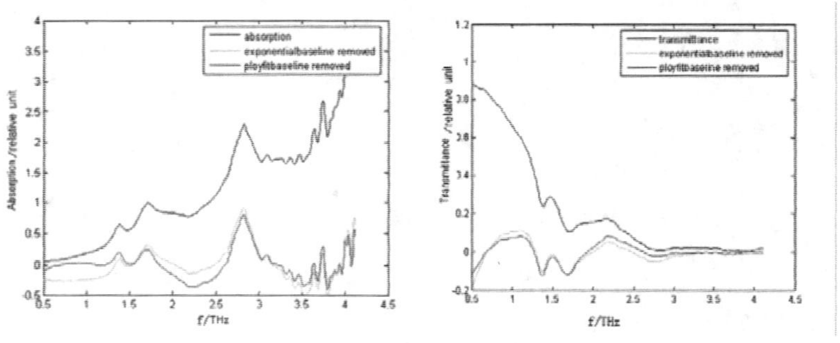

(a) (b)

Figure 5. Spectra with removing baselines. (a) absorbance spectra; (b) transmittance spectra

Spectrum baselines come from scattering of incident light by sample's granules. In order to analyze absorbing or transmitting phenomena without disturbance, the baselines should be removed. In figure 3, the baseline can be regarded as an exponential (blue curve) or polynomial (red curve) function of frequency. The baseline then should be fit by an exponential or polynomial function as following[5,6]:

Exponential model

$$\alpha'(\omega) = b_0 + b_1 \exp(-b_2\omega) \tag{3}$$

Polynomial model

$$\alpha'(\omega) = a_0 + a_1\omega^1 + a_2\omega^2 + \cdots + a_m\omega^m \tag{4}$$

Here m is an integer. Figure 4 shows the baseline fitting where m=4 in polynomial model. Then the spectra with removing baselines are depicted in figure 5.

3.4. Peaks fitting

Each substance has its inherent peaks of spectra. And the peaks are generally overlapped or disturbed by noises. We try to recovery them by peak fitting of a Gaussian or Lorentzian function.

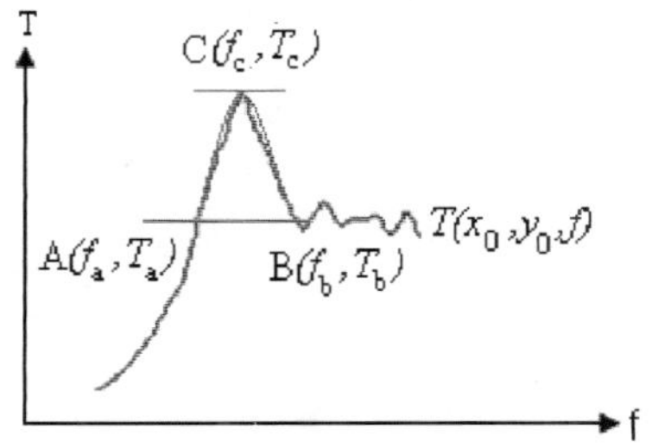

Figure 6. Peak fitting

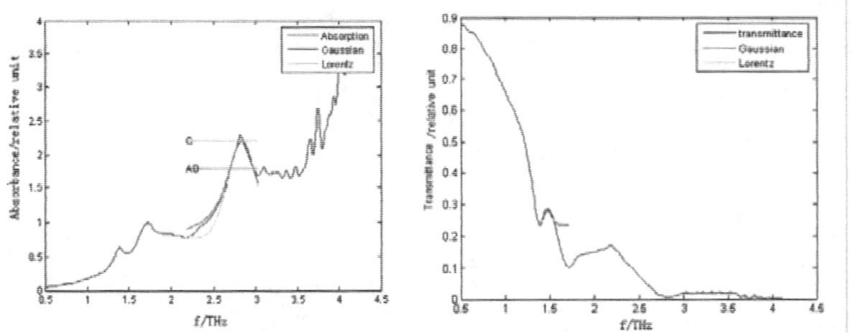

(a) (b)

Figure 7. Peak fitting. (a) absorbance peak fitting; (b) transmittance peak fitting

In figure 6, $T(x_0, y_0, f)$ is a peak of transmittance spectrum on pixel (x_0, y_0). Make a top mark C and a cross mark AB where the amplitude is as low as possible without other peaks overlap. Then, the height T_c, the intersection points $(f_a, T_a), (f_b, T_b)$ and the width AB are calculated by the software. Hence, the Gaussian model is

$$T = T_0 + \frac{D}{w\sqrt{\pi}} e^{-(\frac{f-f_c}{w})^2} \tag{5}$$

and the Lorentzian model is

$$T = T_0 + \frac{D}{2\pi} \frac{w}{(f-f_c)^2 + w^2} \tag{6}$$

Where T_0 is the baseline, w is the width of peak, D is the area under the peak, and $f_c = (f_a + f_b)/2$ is a symmetric axis. The value of D is derived from two points at different frequencies, $(f_a, (T_a + T_b)/2)$ and $((f_a + f_b)/2, T_c)$. After all, w and D in Gaussian model are as following.

$$w = (f_a - f_c)[-0.5\lg((T_a - T_0)/(T_c - T_0))]^{-0.5} \tag{7}$$

$$D = (T_c - T_0)w\sqrt{\pi/2} \tag{8}$$

And in Lorentzian model, they are as following.

$$w = 2(f_a - f_c)[(T_a - T_0)/(T_c - T_a)]^{0.5} \tag{9}$$

$$D = (T_c - T_0)w\pi/2 \tag{10}$$

In figure 7, the peaks fit with Gaussian and Lorentzian models are correspondingly depicted on red and green curves.

4. 3-D display of spectra

The spectrum process for a single pixel is represented above. One by one, huge mount of data form a series of 3-dimension images.

4.1. 3-D display of absorption and transmittance spectra

Figures 8 (a) and (b) represent the transmittance and absorbance spectra at specific frequency 1.1577 THz. The spectrum amplitudes and concentrations of substances are colorized according to a colour scale at the right corner.

By changing frequencies and identifying their peaks, substances included in the sample could be confirmed qualitatively. The 3-D images are shown in figure 9 at a series of frequencies.

4.2. Some of considerations

In figure 8, some substances out of our interesting are completely transparent in THz scope. Extremely big transmittance would take place on those pixels, and corresponding spectra go beyond the regular range. We take them to be noises and wipe them off by an adaptive threshold.

In order to process massive data in format of NetCDF (Network Common Data Form), some tricks must be taken in case of overflow of memories and to ensure that all tasks implement in an acceptable period of time. These tricks include multiplex of variables, dynamic relocation of memories, and recall of temporal results.

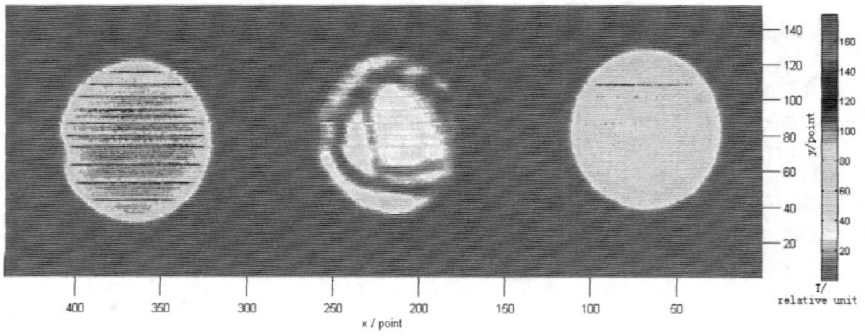

(a)

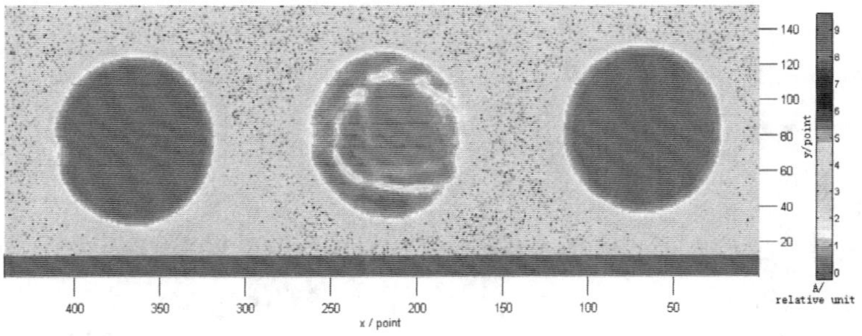

(b)

Figure 8. Reconstructed THz images at 1.1577 THz. (a) transmittance (b) absorbance

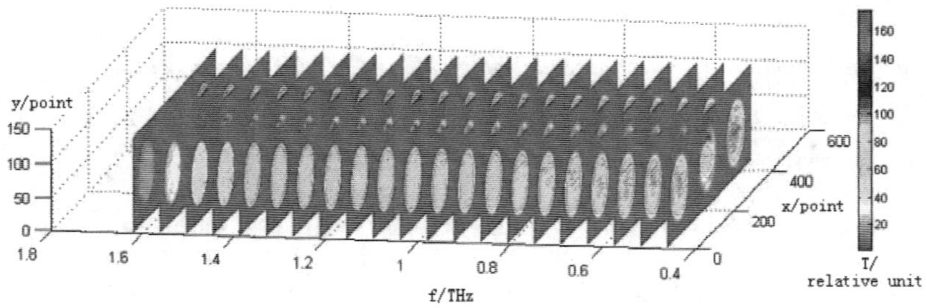

Figure 9. Reconstructed 3D THz images of transmittance at 1.1577 THz

5. Conclusion

The primary distribution of THz experiment data can be transferred to absorbance or transmittance spectra in this software. Due to be able to treat echoes, noises, baselines and peaks, it becomes powerful to qualitative and quantitative analyses of solid-state samples.

Acknowledgments

This work was supported by the National Nature Science Foundation of China (No60977065).

References

[1] X.-C. Zhang 2002 *Physics in Medicine and Biology* 47 3667
[2] Peter Uhd Jepsen and Bernd M Fischer 2005 *Optics Letters* 30 29-31
[3] Zhaohui Zhang, Zuxing Zhang, Yixin Yin, Yuko Ueno, Rakchanok Rungsawang and Katsuhiro Ajito 2009 *Frontiers of Optoelectronics in China* 2 239-243
[4] Taday PF, Bradley IV and Arnone DD 2003 *Journal of Biological Physics* 29 109-115
[5] Zhaohui Zhang, Luodong Lv, Yixin Yin, Yuko Ueno, Rakchanok Rungsawang and Katsuhiro Ajito 2007 *Journal of Analytical Science* 23 651-654 (in Chinese)
[6] Yamamoto K, Kabir M H and Tominaga K J 2005 *Opt. Soc. Am.* B 22 2417-26.

Synthesis and characterization of octahedral PbS nanocrystals assisted with mixed cationic/anionic surfactants

Shengyu Jing [a,*], **Zhihua Sun** [b], **Hai Liu**[a]

[a,*] *School of Information and Electrical Engineering, China University of Mining and Technology, Xuzhou 221000, People's Republic of China*

[b] *Discipline Development Office, China University of Mining and Technology, Xuzhou 221000, People's Republic of China*

jingshengyu1@yahoo.com.cn

Abstract. Octahedral PbS nanocrystals were synthesized in aqueous solution assisted with mixed cationic/ anionic surfactants. X-ray diffraction, transmission electron microscopy were used to characterize the products. The formation mechanism of octahedral PbS nanocrystals was also studied. The study result would be a huge contribution to the terahertz research.

1. Introduction

In recent years, many efforts have devoted to the morphology control of semiconductor materials, because the shape and size of inorganic nanocrystals control their widely varying optical and electric properties [1]. Among the familiar semiconductor materials, PbS is an important binary IV–VI semiconductor material which have novel semiconducting and optical properties [2]. It has attracted considerable attention in the field of materials science due to its specially small direct band-gap energy (0.41 eV, for bulk at room temperature) and a large exciton Bohr radius (18 nm) [3-6]. PbS nanostructures may be one of the most potential candidates for optical devices such as light-emitting diodes and optical switches due to its exceptional third-order nonlinear optical properties [7, 8]. Up to now, various shapes of PbS nanocrystals such as nanospheres, nanobelts, nanorods, nanotubes, nanowires, dendritic nanostructures, star-shapes and flower-shapes have been prepared [9-17]. Moreover, many methods have been developed to fabricate PbS nanostructures. Among them, the hydrothermal method and thermal decomposition provides a more promising way for the synthesis of crystals due to its low cost, high efficiency and potential for large-scale production [18].

Qian et al. reported a one-step route to crystalline PbS by a solvothermal reaction between lead oxalates and elemental sulfur in organic solvents such as ethylenediamine and pyridine [19]. Cheon et al. have synthesized PbS nanocrystals with various rod-based structures including highly faceted star shapes, truncated octahedrons and cubes from the thermal decomposition of a molecursor [16]. Zou et al. have synthesized PbS microcrystals including dendrites, flowers, multipods and cubes from the reaction of dissoluble lead salt (lead nitrate or lead acetate) and thiourea through a simple

* Corresponding author: jingshengyu1@yahoo.com.cn

hydrothermal process [5]. Warner has synthesized PbS nanorods by lead acetate, elemental sulfur and oleic acid as surfactant [20].

On the other hand, the surfactant-assisted synthesis of nanostructured materials has attracted considerable attention for its soft-template effect, reproducibility and simple maneuverability [21, 22]. Additional, it is to be noted that mixtures of various surfactants, including alkyl amines, alkyl acids, alkylphosphonic acids, and trioctyl phosphine oxide, are frequently used as capping agents to tailor the crystal shape in high-temperature solution-phase synthesis. The mixed cationic/ anionic surfactants reverse-micelle solutions were also used to synthesis of morphology-controlled one-dimensional nanostructures and hierarchical architectures at low temperature [23-25]. Herein, we synthesize octahedral PbS nanocrystals in aqueous solution assisted with mixed cationic/ anionic surfactants.

2. Experimental section

2.1. Synthesis of PbS nanocrystals with octahedral morphology

All of the chemical reagents used in this experiment were analytical grade. PbS nanocrystals were synthesized as follows: 3.0 mL of water, 0.5 mL of 0.05 M Cetyltrimethylammonium bromide (CTAB), 0.1 mL of 0.05 M dodecyl sulfonie acid sodium salt (SDSA), and 0.4 mL of 1 M acetic acid were mixed at room temperature, followed by the addition of 0.2 mL of 0.5 M $Pb(OAc)_2$ and 0.2 mL of 0.5 M TAA under stirring, giving concentrations of 5.7 mM and 1.1 mM for CTAB and SDSA, respectively. The resultant solution was thermo-stated at 80• for 6 h. The resultant colloidal particles were collected by centrifugation, washed with water, and dried in air.

2.2. Characterizations

The X-ray diffraction (XRD) patterns were recorded using a Shimadzhu XRD 6000 X-ray diffractometer. Transmission electron microscopic (TEM) image were recorded using a HITACHI H-8100 electron microscope. For TEM characterization, the products were ultrasonically dispersed in ethanol, and then a drop of the resulting suspension was placed on a Cu grid coated with carbon film.

3. Results and discussion

3.1. XRD pattern

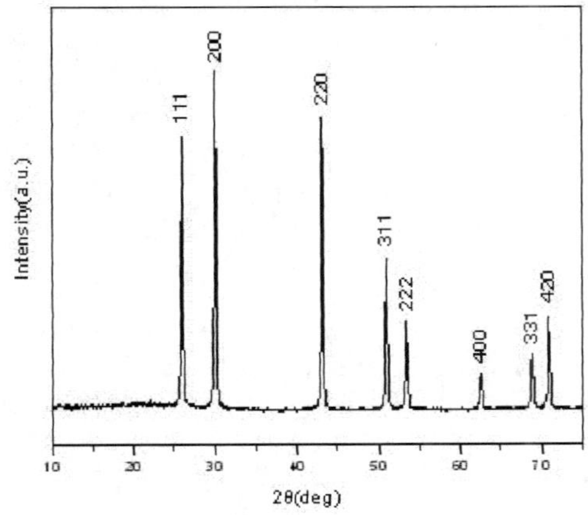

Figure 1. The XRD pattern of the octahedral PbS nanocrystal.

Figure 1 shows an XRD pattern of the octahedral PbS nanocrystal. The XRD pattern of the octahedral PbS nanocrystal contains eight peaks that are clearly distinguishable. All the diffraction peaks can be indexed according to the cubic rock-salt structured PbS, which is in agreement with the value in the literature (JCPDS card No. 5-592). The peaks with 2θ values of 26.03°, 30.18°, 43.12°, 51.03°, 53.55°,

62.61°, 68.91° and 71.04° correspond to the crystal planes of 111, 200, 220, 311, 222, 400, 311 and 420 of crystalline PbS, respectively. The strong and sharp peaks indicate that PbS nanocrystals are highly crystalline.

3.2. TEM image

Figure 2. TEM image of the octahedral PbS nanocrystals. The inset shows the electron diffraction (ED) pattern of the octahedral PbS nanocrystals.

Figure 2 shows a typical TEM image of the octahedral PbS nanocrystals. The products exhibit high uniform, regular octahedra with an average edge length of 70 nm. The selected area electron diffraction (SAED) pattern indicates that the octahedral nanocrystal is a single crystal of PbS.

3.3. Mechanical anylysis

The presence of the cationic/ anionic mixture in the reaction system plays a key role in the formation of octahedral PbS nanocrystals. Cetyltrimethylammonium bromide (CTAB), the cationic surfactant and dodecyl sulfonie acid sodium salt (SDSA), the anionic surfactant were used in the system. As we known, CTAB is often used as "capping reagent" in nanomaterial fabrication because it can bond to solid surface and selectively adsorb on some specific facets of crystals to control crystal growth direction consequently. The presence of CTAB will bring an increase in the growth rate of {100} faces, which favors the formation of octahedrons with eight {111} faces. In the present case, when CTAB was used in reaction system, it maybe preferentially adsorb on the faces parallel to (100) axis direction of PbS nanocrystals, leading to preferential growth along the (100) direction, resulting in the formation of octahedral morphology.

We expect that the formation of these octahedral nanocrystals should be generally applicable to many other metal or semiconductor systems. When the amount of surfactant used exceeded a certain value, it probably not only limited the diameter but also the length, in other words, under the condition of too much CTAB in reaction system, a small part of CTAB maybe adsorb on (100) crystal face, resulting in the decrease of length with the increment of surfactant. At the same time, we found the obtained octahedral PbS nanocrystals will form a 3D well-packed octahedral PbS nanocrystal arrays when the reaction time was prolonged. We think that it may be probably determined by the assistant of SDSA.

4. Conclusion

In summary, we demonstrated a novel method for the synthesis of octahedral PbS nanocrystals. This octahedral PbS nanocrystal may become an advanced material for its special optical and electric properties determined by its special structure. And it will play an important applying role in the optical and electrical fields in the future. Meantime, this study result would be a huge contribution to the terahertz research.

References

[1] Yang P D and Lieber C M 1996 *Science* **273** 1836
[2] Qin A M, Fang Y P, Zhao W X, Liu H Q and Su C Y 2005 *J. Cryst. Growth* **283** 230
[3] Xu L, Zhang W, Ding Y, Yu W, Xing J, Li F and Qian Y T 2004 *J. Cryst. Growth* **273** 213
[4] Zhang Y C, Qiao T, Hu X Y, Wang G Y and Wu X 2005 *J. Cryst. Growth* **277** 518
[5] Wang S, Pan A, Yin H, He Y, Lei Y, Xu Z and Zou B 2006 *Mater. Lett.* **60** 1242
[6] Zhou S, Feng Y and Zhang L 2003 *J. Mater. Res.* **18** 1188
[7] Colvin V L, Schlamp M C and Alivisatos A P 1994 *Nature* **370** 354
[8] Dementjev A, Gulbinas V and Valkunas L 2004 *Phys. Stat. Sol.* B **241** 945
[9] O'Brien P 1997 *J. Mater. Chem.* **7** 1011
[10] Wang S and Yang S 2000 *Langmuir* **16** 389
[11] Liveri V T, Rossi M D, Arrigo G, Manno D and Micocci G 1999 *Appl. Phys.* A **69** 369
[12] Wang S F, Gu F, Lü M K, Zhou G J and Zhang A Y 2006 *J. Cryst. Growth* **289** 621
[13] Yu D, Wang D, Meng Z, Lu J and Qian Y 2002 *J. Mater. Chem.* **12** 403
[14] Kuang D, Xu A, Fang Y, Liu H, Frommen C and Fenske D 2003 *Adv. Mater.* **15** 1747
[15] Ma Y, Qi L, Ma J and Cheng H 2004 *Cryst. Growth Des.* **4** 351
[16] Lee S M, Jun Y W, Cho S N and Cheon J W 2002 *J. Am. Chem. Soc.* **124** 11244
[17] Ni Y, Liu H, Wang F, Liang Y, Hong J, Ma X and Xu Z 2004 *Cryst. Growth Des.* **4** 759
[18] Jana S, Goswami S, Nandy S and Chattopadhyay K K 2009 *J. Alloy. Compd.* **481** 806
[19] Yu S H, Yang J, Wu Y S, Han Z H, Lu J, Xie Y and Qian Y T 1998 *J. Mater. Chem.* **8** 1949
[20] Warner J H 2008 *Adv. Mater.* **20** 784
[21] Liu Y, Hou D and Wang G 2003 *Chem. Phys. Lett.* **379** 67
[22] Yu Y, Du F P, Yu J C, Zhuang Y Yand Wong P K 2004 *J. Solid State Chem.* **177** 4640
[23] Shi H, Qi L, Ma J and Cheng H 2003 *J. Am. Chem. Soc.* **125** 3450
[24] Shi H, Qi L, Ma J, Cheng H and Zhu B 2003 *Adv. Mater.* **15** 1647
[25] Shi H, Qi L, Ma J and Wu N 2005 *Adv. Funct. Mater.* **15** 442

Pulsed CH_3OH terahertz laser radiation pumped by 9P(36) CO_2 lasers

Jiu Zhi-Xian [1], Zuo Du-Luo [2], Miao Liang [2], and Cheng Zu-Hai [2]

1. Department of Mathematics and Physics, Wuhan Polytechnic University, Wuhan 430023 China

2. Wuhan National Laboratory for Optoelectronics, College of Optoelectronic Science and Engineering, Huazhong University of Science and Technology, Wuhan 430074 China

E-mail: jiuzhixian@163.com

Abstract An efficient pulsed CH_3OH terahertz (THz) laser pumped by a TEA CO_2 laser was investigated experimentally. A simple terahertz cavity and a TEA CO_2 laser for the optically pumped THz radiation were studied experimentally. To improve THz laser energy and photon conversion efficiency, two different TEA CO_2 lasers were developed to pump CH_3OH. When CH_3OH was pumped by the 9P(36) line with different powers of the CO_2 laser, the generation of terahertz radiation with energy as high as 0.307mJ and 23.75mJ were obtained, respectively. The corresponding photon conversion efficiencies were 0.29% and 2.4%. The photon conversion efficiency increases by a factor of about 8. Meanwhile, higher peak power of pump laser effectively improves the photon conversion efficiency. And the optimum THz laser pressure increases with narrower pulse width of pump laser because of increasing absorptive gases molecules of CH_3OH with higher peak power of pump laser.

1. Introduction

With the development of the applications of terahertz (THz) (30 μm ~ 3 mm) laser, the THz laser source technology is becoming increasingly important. Since T. Y. Chang et. al. obtained THz laser by using a CO_2 laser to pump CH_3OH and other molecules in 1970 [1], methanol has become the THz laser-active molecule with the richest spectral lines. Just at that time, TEA CO_2 lasers have been employed as the optical pumping sources generally. The most powerful and efficient THz laser line was the 118.8 μm of CH_3OH pumped by the 9P(36) CO_2 laser line [2]. Most research has been dedicated on the experimental apparatus of THz laser pumped by CW CO_2 lasers or chopped CO_2 lasers with 9P(36) CO_2 laser line [3-10]. For example, THz power with the addition of buffer gases has been measured to be 1.25 W with 125 W of CO_2

pump power. The corresponding photon conversion efficiency has been 24.51% [6]. However, relatively less attention has been concentrated on energy of pulsed optically pumped CH_3OH THz lasers.

In this paper, a simple optically pumped THz laser system was designed. Two tunable TEA CO_2 lasers were developed to pump CH_3OH active medium, respectively. The corresponding generation of terahertz radiation with energy as high as 0.307mJ and 23.75mJ was obtained, with the photon conversion efficiency of 0.29% and 2.4%. All of these prove our experimental system is efficient.

2. Experimental system and results

The schematic of the experimental apparatus was consists of three main parts: the TEA CO_2 laser tuned by a 150 gmm^{-1} grating, the THz cavity, and the detection system.

Both home-made TEA CO_2 laser served as the THz laser pump source, respectively. The first TEA CO_2 laser was that the discharge chamber was terminated with an antireflection-coated GaAs window and a semitransparent-coated GaAs window on both sides spacing 1.4 m apart. Dispersive element was a metallic reflective grating (150 gmm^{-1}) mounted on a rotating platform and blazed at 10.6 μm. Two flat gold-plated Si mirrors and a GaAs lens with focal length 2 m were used to transmit the output CO_2 laser radiation and compress its beam size. The second TEA CO_2 laser was that the discharge chamber was terminated with an antireflection-coated GaAs window and a ZnSe output coupler with only one surface antireflection-coated on both sides spacing 2.64 m apart. The dispersive element in the CO_2 laser cavity was a metallic reflective grating (150 gmm^{-1}) blazed at 10.6 μm and mounted on a rotating platform. The grating was placed in the position of 4.15 m away from the output coupler. Two flat gold-plated Si mirrors were used to transmit the output CO_2 laser radiation.

The THz cavity was composed of a 2 m long and 50 mm i. d. quartz tube, an input window and an output window. The input window of the THz laser was a 6 mm thick GaAs plate coated antireflection film, which fully transmits CO_2 laser. The output window was a 2 mm thick crystal quartz. It is opaque to CO_2 laser, and transparent THz radiation, which ensures detector only receives the latter. A tsurupica lens of 180 mm focal length (Microtech Instruments) is used for focusing the THz emission. It's highly transparent to THz emission. The THz pulse energy was detected by a high responsive pyroelectric detector (Spectrum Detector SPJ-A-8-OB) which can detect an energy range from 60 nJ to 600 μJ with a wavelength range from 0.1 to 1000 μm. Finally, the signal was recorded by a digital oscillograph (Agilent DSO7034A) with a 350 MHz bandwidth.

The photon conversion efficiency is one of the conversion efficiency of the THz laser. The photon conversion efficiency is $\eta=(E_T/E_p)\cdot(2\lambda_T/\lambda_p)$, where E_T (E_p) is THz output energy (CO_2 laser pump energy) and λ_T (λ_p) is THz wavelength (CO_2 laser pump wavelength). $\lambda_p/(2\lambda_T)$ is the energy conversion efficiency in quantum limit [9, 11].

When the first TEA CO_2 laser served as the THz laser pump source, the pressure ratio of the gas mixture in the cavity of the TEA CO_2 laser was CO_2: N_2: He = 3: 1: 6 in the total pressure of about 50 kPa. The corresponding CO_2 energy was 2.6 J. The generation of terahertz radiation with energy of 307 μJ was obtained [12]. The photon conversion efficiency at 118.8 μm was 0.29%. The THz energy versus the CH_3OH pressure is shown in figure 1. An optimum THz laser pressure of about 260 Pa was observed. For low gas pressures, the pumping pulses are sufficiently intense to saturate the pumping transition so that the output is limited only by the density of the gas. On the other hand, at higher pressures the laser transition becomes collision broadened, the upper laser level is no longer saturated so that the THz output eventually decreases to zero [13].

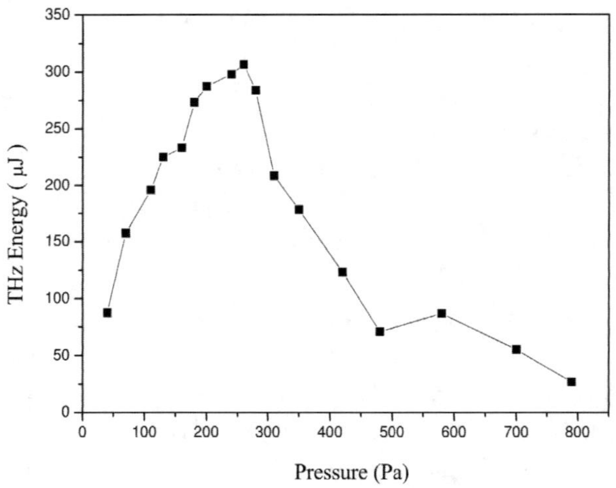

Figure 1. The THz energy vs. the CH_3OH pressure pumped by the first TEA CO_2 laser.

When the second TEA CO_2 laser served as the THz laser pump source, the pressure ratio of the gas mixture in the cavity of the TEA CO_2 laser was CO_2: N_2: He = 3: 1: 12.7 in the total pressure about 100 kPa. Shown in figure 2 are the laser pulse waveforms detected by a photo drag detector (Hamamatsu B749). The corresponding pump intensities and the full width at half-maximum (FWHM) were about 15.75MW/cm^2 and 42 ns, indicating that the pump intensity is greatly improved. The CO_2 energy was 24.2 J. The generation of terahertz radiation with energy of 23.75 mJ was obtained. The photon conversion efficiency was 2.4%. In figure 3, the optimum THz laser pressure of about 820 Pa increases by three-fold, compared with the former of 260 Pa.

In the results, the photon conversion efficiency increases by a factor of about 8. Meanwhile, The THz laser energy increases with the pump intensity, because the CH_3OH transition gain is dependent on the pump intensity [14]. Higher peak power of pump laser effectively improves the photon conversion efficiency. And the optimum THz laser pressure increases with narrower pulse width of pump laser because of increasing absorptive gases molecules of CH_3OH with higher peak power of pump laser [15].

3. Conclusions

The efficient pulsed CH_3OH terahertz (THz) laser pumped by two different TEA CO_2 lasers was studied experimentally. . When CH_3OH was pumped by the 9P(36) line with different powers of the CO_2 laser, the generation of terahertz radiation with energy as high as 0.307mJ and 23.75mJ were obtained, respectively. The corresponding photon conversion efficiencies were 0.29% and 2.4%. The photon conversion efficiency increases by a factor of about 8. The results prove that higher peak power of pump laser effectively improves THz laser energy and photon conversion efficiency. And the optimum THz laser pressure increases with narrower pulse width of pump laser because of increasing absorptive gases molecules of CH_3OH with higher peak power of pump laser.

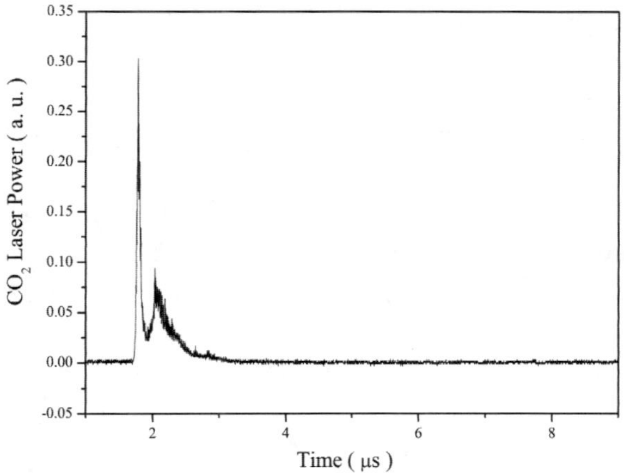

Figure 2. The TEA CO_2 laser pulse temporal waveform.

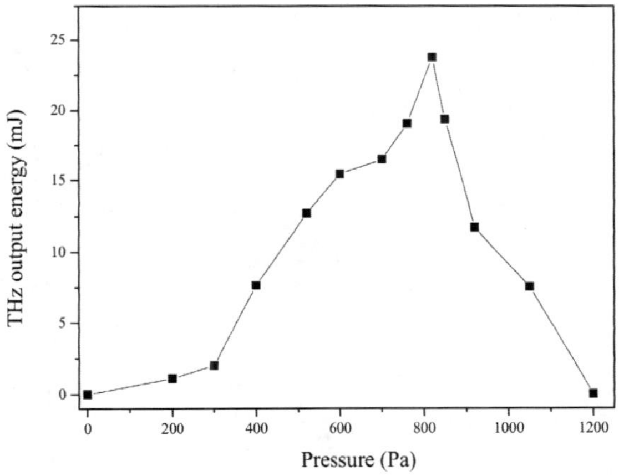

Figure 3. The THz energy vs. the CH_3OH pressure pumped by the second TEA CO_2 laser.

References

[1] Chang T Y, Bridges T J and Buihardt E G 1970 *Appl. Phys. Lett.* **17** 249
[2] Moruzzi G, Silos Moraes J C and Strumia F 1992
[3] Devoy D and Walker B 1984 *J. Phys. E: Sci. Instrum.* **17** 1132
[4] Boscolo L 1988 *Appl. Opt.* **27** 3325
[5] Mansfield D K, Horlbeck E, Bennett C L and Chouinard R 1985 *Int. J. IR & MM Waves.* **6** 867

[6] Farhoomand J and Pickett H M 1987 *Int. J. IR & MM Waves.* **8** 441

[7] Pellemans H P M, Burghoorn J, Klaassen T O and Wenckebach W Th 1996 *Infra. Phys. & Tech.* **37** 635

[8] Heppner J, Weiss C O, Hubner U and Schinn G 1980 *IEEE J. Quantum Electron.* **16** 392

[9] Hodges D T, Foote F B and Reel R D 1976 *Appl. Phys. Lett.* **29** 662

[10] Bakos J S, Mandula K and Sorlei Z 1991 *IEEE J. Quantum Electron.* **27** 1094

[11] Marchetti S, Martinelli M, Simili R, Fantoni R and Giorgi M 2000 *Infra. Phys.& Tech.* **41** 197

[12] Jiu Z X, Zuo D L, Miao L, Cheng Z H and Qi C C 2010 *J. Infra. Milli. Terahz. Waves.* **31** 885

[13] Jiu Z X, Zuo D L, Miao L, Cheng Z H and Qi C C 2010 *Chin. Phys. Lett.* **27** 024211

[14] Heppnet J, Welss C O, Hubner U and Schinn G 1980 *IEEE J. Quantum Electron.* **16** 392

[15] Bae J, Nozokido T and Shirai H 1989 *IEEE J. Quantum Electron.* **25** 1591

Journal of Physics: Conference Series **276** (2011) 012213 doi:10.1088/1742-6596/276/1/012213

Terahertz wave modulator based on metamaterial

Li Jiu-sheng[1,2], **Zhao Xiao-li**[1]

[1] Centre for THz Research, China Jiliang University, Hangzhou 310018, China

[2] College of Precision Instrument and Opto-Electronics Engineering, Institute of Laser and Optoelectronics, Tianjin University, Tianjin 300072, China

E-mail: jshli@126.com

Abstract. We demonstrated experimentally a terahertz wave modulator based on microstrip resonator structure. The signal modulation mechanism of the presented terahertz wave modulator is based on the resonance characteristic of resonator microstrip structure controlled by without or with light excitation. A modulated semiconductor laser with 808nm wavelength is employed to light the microstrip resonator structure. The experimental results show that the modulation speed and attenuation of the proposed terahertz modulator are of 0.1Kb/s and 25dB at frequency of 0.32THz, respectively

1. Introduction

The terahertz (THz) frequency range 0.1–10 THz, located midway between microwaves and infrared light, presents a new frontier containing numerous technical applications and fundamental research problems. With the realization of the terahertz generator and detector, terahertz wave has attracted significant attention and has been extensively investigated. Due to their special prosperities, within the past few years, many potential applications of terahertz waves have been dramatically explored in many fields such as biomedical diagnostics, security screening, military detection, radio astronomy, atmospheric studies, high speed communication, quality control of packaged goods, and moisture analysis for agriculture [1]. With wide bandwidth and high data transmission bit rates, terahertz wave wireless communications have great potential in the future short range wireless communications [2]. As an important device in terahertz communication for signal processing, terahertz wave modulator has been attracted significant attention. In 2004, T.K.Ostmann [3] designed an electrically driven terahertz modulator. In 2007, using a one dimensional photonic crystal with a GaAs defect, L. Fekete et al. [4] demonstrated the possibility of ultrafast modulation of THz radiation. At this year, J. Li [5] analyzed a terahertz wave modulator using photonic crystals. However, there are still relatively undeveloped with only a few examples of cryogenically cooled and room temperature modulators. Quantitative studies on terahertz wave modulator are still very limited. Robust terahertz wave modulators are still needed that can be easily implemented and integrated into a chip-scale platform. Therefore, it is valuable to investigate the design of modulator in the terahertz range.

In this letter, we have proposed and demonstrated a novel terahertz wave modulator using microstrip resonator structure. The terahertz wave modulation mechanism of the proposed terahertz wave modulator is based on the resonance characteristic of resonator microstrip structure which is controlled by using a modulated laser. Experimental results show that the presented terahertz wave modulator has modulation speed of 0.1Kb/s, extinction ratio of 25dB, and simplicity.

2. Device fabrication

The resonator microstrip structure unit are optimized and obtained through simulation using commercial CST software. We transferred resonator microstrip structure pattern to mask. 210nm of aluminum on a 400μm-thick intrinsic silicon wafer was deposited by radio frequency magnetron sputtering method. About 1μm AZ-601 photoresist be coated on the wafer by GKF-411 gluing purifying machine. To evaporate the solvent and densify the AZ-601 film, put the wafers in hotplate at 70°C persistence 5 minutes, and cool the wafer for 5 minutes. Mount the mask and the wafer onto the mask aligner, the AZ-601 film be exposed by URE-2000S deep ultraviolet lithography exposure plane. Bake the AZ-601 on a hotplate at 70°C persistence 5 minutes for acid-initiated, thermally driven expoxy cross-linking. Develop with RZX-3038 positive photoresists developer with agitation until the pattern is clear, rinse with deionized water. To evaporate deionized water and densify the AZ-601 film, put the wafers in hotplate at 70°C persistence 5 minutes, following with 120°C oven 10 min, and cool the wafer for 15 minutes. The wafer is etched by MNL/DIII reactive ion etcher, rinse with deionized water and bake with 120°C oven 10 minutes. Strip photoresist with ORNIII5532 plasma stripping photoresist system. The sample is evaluated by optical microscope and shown in figure 1.

Figure 1. Photography of the modulator

3. Experiment results

A backward-wave oscillator (BWO) system is used to test the terahertz-wave modulator which is placed at the focal point of the terahertz radiation. Both excitation and terahertz wave spot sizes while the carrier versus terahertz spot size overlap was determined by sweeping the optical excitation. The backward-wave oscillator is set to be 0.32THz and used as the continuous terahertz wave source. A modulated semiconductor laser at 808 nm with incident optical intensities 100mW is employed to light the microstrip resonator structure based on intrinsic silicon. The resonance of the microstrip resonator structure is controlled by the photo-excited carriers. Measured terahertz wave transmittance characteristic of the novel terahertz wave modulator with the modulated laser shows in Figure 2. Applied square wave signal with 0.1Kb/s for laser modulation is shown in Figure 2 (a). Figure 2 (b) shows the detected signal from terahertz wave pyroelectric detector with 0.1Kb/s. From figure, it can be noted that a better than 25dB attenuation of the novel terahertz wave modulator is obtained at terahertz frequency of 0.32THz. The modulation speed of the proposed modulator is of 0.1Kb/s. Being the modulation speed is limited by the response time of the detector, and further studies will be conducted with a rapid response terahertz detector.

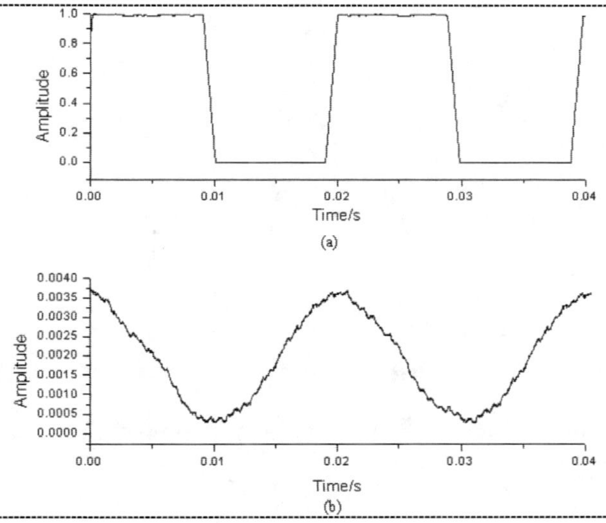

Figure 2. Measured modulation characteristic of the proposed terahertz wave modulator (a) Applied signal, (b) Detected signal

4. Conclusion

We experimentally demonstrated that the terahertz wave modulator based on microstrip resonator structure is able to achieve terahertz wave signal modulation. The terahertz wave modulator presented here, is based on the modulated semiconductor laser excitation at 100mW average power at 808nm to light the microstrip resonator structure, demonstrating terahertz wave transmission modulation that has a modulation speed of 0.1Kb/s. The extinction ratio of the presented terahertz wave modulator is more than 25dB at frequency of 0.32THz.

Acknowledgments

This research was partially supported by the National Natural Science Foundation of China (No.60971027).

References

[1] Li J and Li X 2009 *Chem. Phys. Lett.* **476** 92
[2] Piesiewicz R, Ostmann T K, Krumbholz N, Mittleman D and Koch M 2007 *IEEE Antennas and Propag. Mag.* **49** 24
[3] Ostmann T K, Pierz K, Hein G, Dawson P and Koch M 2004 *Electron. Lett.* **40** 124
[4] Fekete L, Kadlec F, Kuˇzel P and Nˇemec H 2007 *Opt.Lett.* **32** 680
[5] Li J 2007 *Optics Commun.* **269** 98

Omnidirectional mirror for terahertz wave

Li Jiu-sheng[1,2], He yun-hong[1]

[1]Centre for THz Research, China Jiliang University, Hangzhou 310018, China

[2]College of Precision Instrument and Opto-Electronics Engineering, Institute of Laser and Optoelectronics, Tianjin University, Tianjin 300072, China

E-mail: jshli@126.com

Abstract. We design an omnidirectional dielectric mirror that operates over a wide terahertz wavelength range and is based on a periodic bilayer (tellurium and polystyrene) structure. The structure is characterized by using transfer matrix calculations. Results of simulation show that the presented dielectric mirror is highly reflecting for all incidence angles and TE as well as TM polarization in the frequency band between 0.451THz and 0.821THz (wavelength range 365.5μm~665.5μm), when the incidence angle is less than 45°. Tolerance analysis reveals a large tolerance to fabrication errors.

1. Introduction

With the realization of THz generator and detector, electromagnetic terahertz frequency range (0.1THz~10THz) attracted significant attention and has been extensively investigated [1-3]. Terahertz applications include imaging techniques for quality control of industrial goods, medical diagnosis, security screening, radio astronomy, atmospheric studies, short-range indoor communication, chemical, biological sensing, and security checks of people, mail, or luggage. For all these applications efficient, inexpensive, and compact terahertz sources and detectors need to be developed. In addition, passive functional devices such as beam splitter, switches, filters, and terahertz wave mirrors are required. Mirrors which are highly reflective for all incidence angles are called omnidirectional mirrors. Some approaches to achieving terahertz wave mirrors have been reported in the literature. However, quantitative studies on terahertz wave omnidirectional mirrors are still very limited. Therefore, it is valuable to investigate the design of omnidirectional mirrors in the terahertz range.

In this letter, we present the numerical design and analysis of an omnidirectional dielectric mirror for the terahertz range made from periodic bilayer tellurium and polystyrene structure. The novel terahertz wave omnidirectional dielectric mirror has been designed and calculated through transfer matrix calculations. Simulation results show that the omnidirectional dielectric mirror has a highly reflecting for all incidence angles, TE as well as TM polarization and wide bandwidth. Moreover, we show that its performance accommodates reasonable fabrication tolerances.

2. Device design and analysis

Here, we use thin slices of tellurium and polystyrene to build an omnidirectional terahertz mirror. The structure consisted of thirty alternating layers of tellurium and polystyrene on a teflon substrate. The sketch of the high-efficiency terahertz wave omnidirectional dielectric mirror is shown in figure 1.

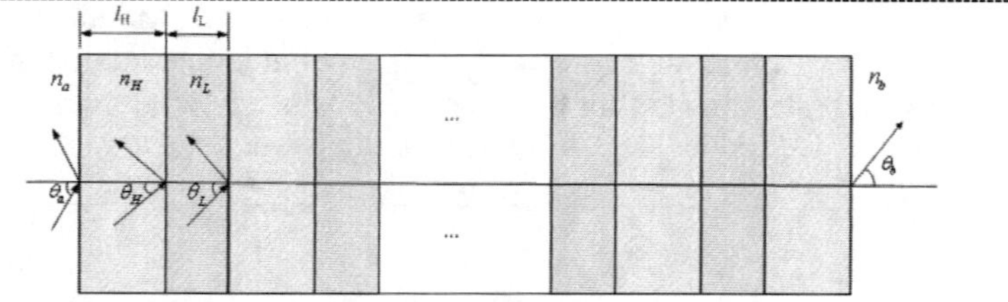

Figure 1. Schematic diagram of the terahertz wave omnidirectional dielectric mirror

3. Simulation results

Here, the physical length of both tellurium and polystyrene is $l_H= l_L= {}_0/4$. The number of periodic bilayers is N=30. The refactive index of tellurium and polystyrene are n_H = 4.6 and n_L=1.6, respectively. The simulation is performed with the parameters given above. The incidence angles ( ) is equal to 0°, 30°, and 45°, respectively. The spectral response of this structure for both TE and TM-polarized incident terahertz wave is shown in Fig. 2 as calculated by the transfer matrix calculations. Figures 2 show the terahertz reflection spectra through the structure for TE and TM polarizations. As the incidence angle increases this reflection band blueshifts for both TE and TM polarizations. For both polarizations a frequency band can be found for which the mirror is highly reflecting for all angles. According to analysis as above, when the incidence angle is less than 45°, TE and TM-polarized terahertz wave modes are reflected with 100% for a wide wavelength range 365.5μm and 665.5μm.

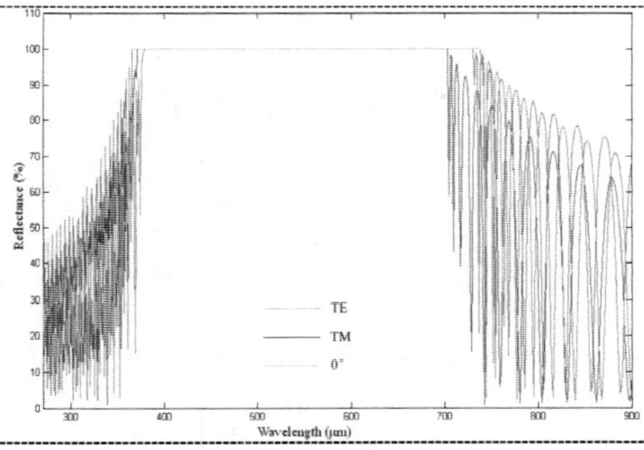

(a)

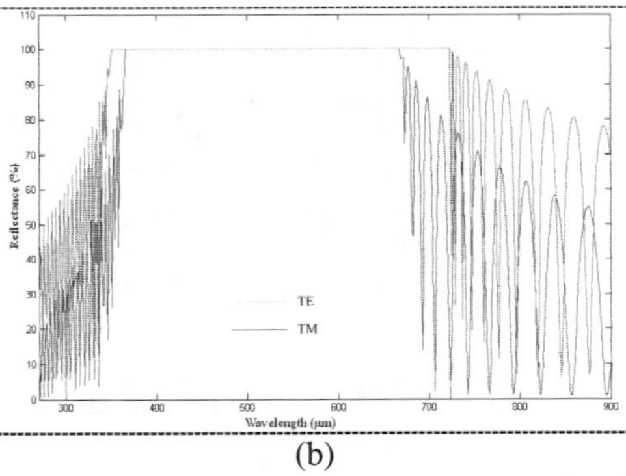

(b)

Fig.2 TE and TM reflectances (a) incidence angle =30°(b) incidence angle =45°

In addition, we examined the performance of the omnidirectional dielectric mirror as a function of potential fabrication errors that affect dielectric layer length. Simulation results (sees Fig.3) for a variation in dielectric layer length from 135μm to 115μm (20μm range for the design dielectric layer length) show that both TE and TM modes reflectivity bandwidth has lettle change.

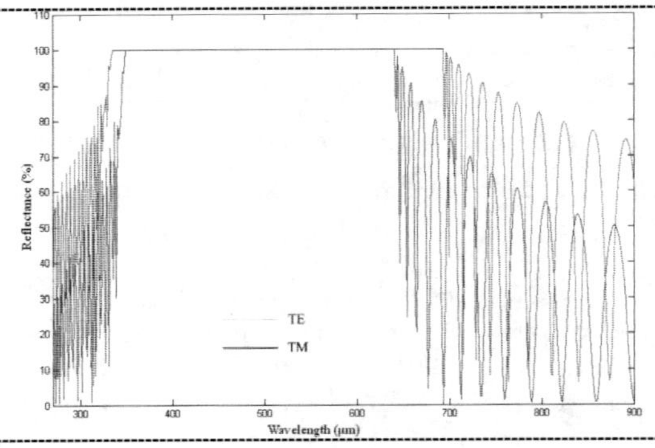

(a)

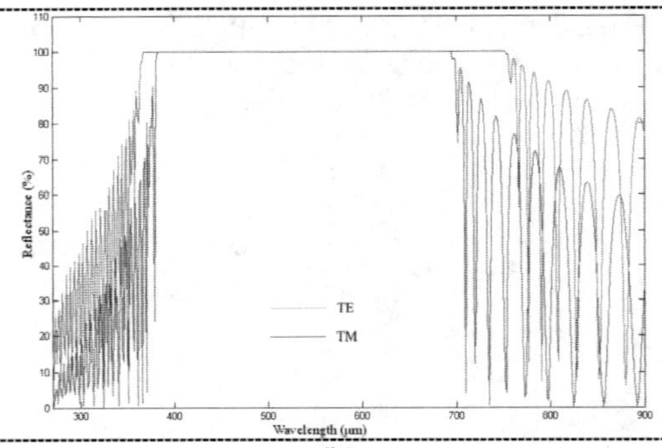

(b)

Fig.2 TE and TM reflection spectrum of the proposed mirror structure for incidence angle =45° (a) dielectric layer length of 115μm, (b) dielectric layer length of 135μm

4. Conclusion

We theoretically demonstrated an omnidirectional mirror for the terahertz range made from polystyrene and thin slices of tellurium. This offers the opportunity to achieve a high-efficiency omnidirectional dielectric mirror to reflect TE and TM polarized modes. For frequencies between 0.451THz and 0.821THz, the structure is highly reflecting for all incidence angles and TE as well as TM polarization. The present mirror is mechanically flexible and could be used for in-door terahertz wave communication systems. The proposed terahertz wave omnidirectional mirror has reasonable tolerances to fabrication errors.

Acknowledgments
This research was partially supported by the National Natural Science Foundation of China (No.60971027).

References

[1] Kohler R, Tredicucci A, Beltram F, Beere H E, Linfield E H, Davies A G, Ritchie D A, Iotti R C and Rossi F 2002 *Nature.* **417** 156
[2] Li J, He J and Hong Z 2007 *Appl. Opt.* **46** 5034
[3] Kawano Y and Ishibashi K 2008 *Nature Photon.* **2** 618

Terahertz wave polarizer

Li Jiu-sheng[1,2], He yun-hong[1]

[1] Centre for THz Research, China Jiliang University, Hangzhou 310018, China

[2] College of Precision Instrument and Opto-Electronics Engineering, Institute of Laser and Optoelectronics, Tianjin University, Tianjin 300072, China

E-mail: jshli@126.com

Abstract. We design terahertz wave reflective polarizer that operates over a wide terahertz wavelength range and is based on a periodic bilayer structure. The structure is characterized by transfer matrix calculations. Results of simulations show that the mirror is highly reflecting for all incidence angles and s as well as p polarization in the frequency band between 319 and 375 GHz. Tolerance analysis reveals a large tolerance to fabrication errors.

1. Introduction

With the realization of THz generator and detector, electromagnetic terahertz frequency range (0.1THz~10THz) attracted significant attention and has been extensively investigated. Terahertz applications include imaging techniques for quality control of industrial goods, medical diagnosis, security screening, radio astronomy, atmospheric studies, short-range indoor communication, chemical, biological sensing, and security checks of people, mail, or luggage [1-3]. For all these applications efficient, inexpensive, and compact terahertz sources and detectors need to be developed. In addition, passive functional devices such as switches, filters, modulator and polarizer are required. Some approaches to achieving polarizer of optical wave have been reported in the literature. However, quantitative studies on terahertz wave polarizer are still very limited. Therefore, it is valuable to investigate the design of polarizer in the terahertz range.

In this letter, we present a polarizer for the terahertz range made from periodic bilayer structure. The novel terahertz wave polarizer has been designed and calculated through transfer matrix calculations. In the x-direction indices are different and the structure will act as a mirror for the TM polarization. The TE polarization will be reflected only by the air-high interface. In the case, the materials are matched in their y-direction indices, the structure becomes a mirror for the TE polarization. The present mirror is mechanically flexible and could be used for in-door terahertz wave communication systems. The proposed terahertz wave reflective polarizer has reasonable tolerances to fabrication errors.

2. Device design and analysis

In polarizer, we are interested in separating the TE and TM polarizations. This can be accomplished with a periodic bilayer structure of the type shown in figure 1. Figure 2 show the terahertz reflection spectra through the structure for s and p polarizations, respectively. It is highly reflecting only for TE and highly transmitting for TM polarizations. The spectral response of this structure for both TE and TM-polarized incident terahertz wave is shown in figures 2 as calculated by the transfer matrix

Journal of Physics: Conference Series **276** (2011) 012215 doi:10.1088/1742-6596/276/1/012215

calculations. Figures 2 show the terahertz reflection spectra through the structure for TE and TM polarizations.

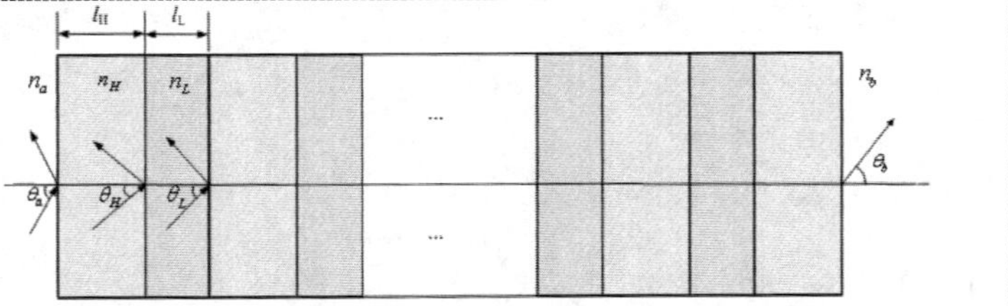

Figure 1. Schematic diagram of the terahertz wave polarizer

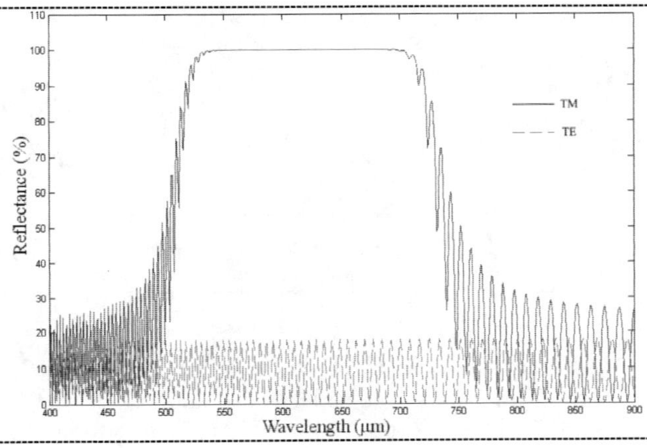

(a)

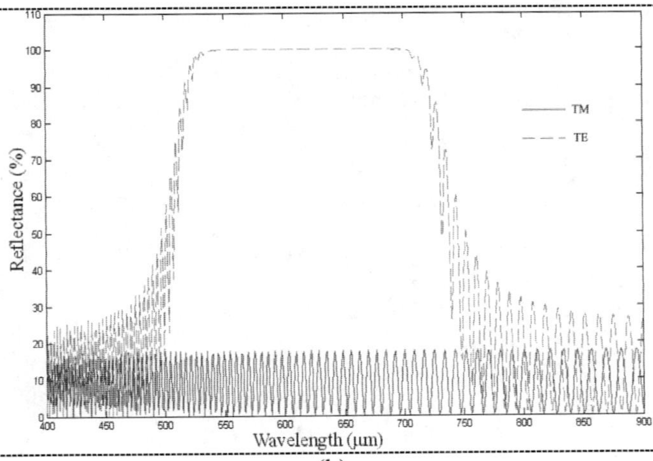

(b)

Figure 2 TM and TE mirror polarizer (a) TM polarizer (b) TE polarizer

In addition, we examined the performance of the omnidirectional dielectric mirror as a function of potential fabrication errors that affect dielectric layer length. Results (sees figure 3) for a variation in dielectric layer length from 135μm to 115μm (20μm range for the design dielectric layer length) show that both TE and TM modes reflectivity efficiency are not changed, whereas the corresponding TE and TM-polarized modes reflection wavelength range varies from 410~645μm.

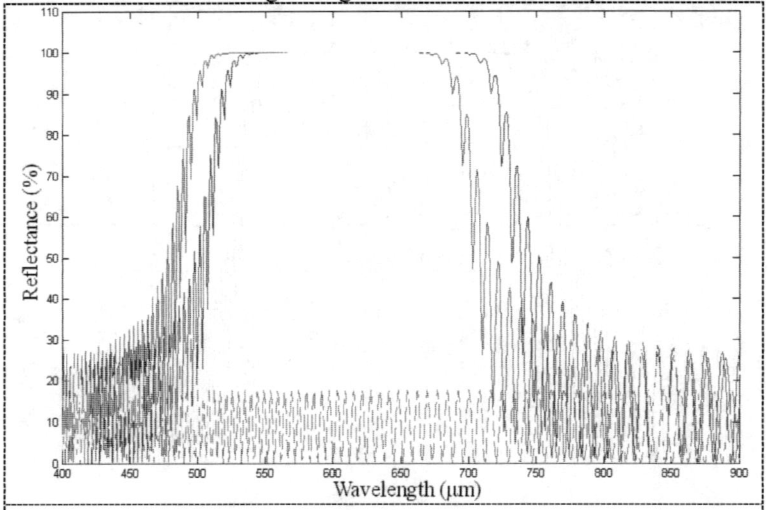

Figure 3 TM and TE mirror polarizer (a) TM polarizer (b) TE polarizer

3. Conclusion

In summary we theoretically demonstrated a reflective polarizer for the terahertz range. For frequencies between 319 and 375 GHz the structure is highly reflecting for all incidence angles and s as well as p polarization. The present mirror is mechanically flexible and could be used for in-door terahertz wave communication systems. The proposed terahertz wave reflective polarizer has reasonable tolerances to fabrication errors.

Acknowledgments

This research was partially supported by the National Natural Science Foundation of China (No.60971027).

References

[1] Kohler R, Tredicucci A, Beltram F, Beere H E, Linfield E H, Davies A G, Ritchie D A, Iotti R C and Rossi F 2002 *Nature.* **417** 156
[2] Li J and Li X 2009 *Chem. Phys. Lett* **476** 92
[3] Li J 2007 *Optics Commun.* **269** 98

Journal of Physics: Conference Series **276** (2011) 012216 doi:10.1088/1742-6596/276/1/012216

Terahertz quantum cascade lasers operating above liquid nitrogen temperature

J Q Liu[1], J Y Chen, L Li, F Q Liu, L J Wang, Z G Wang
Key Laboratory of Semiconductor Materials Science, Institute of Semiconductors, Chinese Academy of Sciences, Beijing 100083, People's Republic of China

E-mail: jqliu@semi.ac.cn

Abstract. The development of quantum cascade lasers at 2.94 -3.10 THz are reported. The materials were prepared by Veeco GEN-II solid source molecular beam epitaxy and devices were processed using standard photolithography and wet chemical etching. Lasing is observed up to a heat-sink temperature of 80 K in pulsed mode with peak light power of 14 mW at 10 K and 2 mW at 80 K.

1. Introduction

The invention of quantum cascade laser (QCL) blazed a new way in the development of coherent laser sources in THz frequencies[1]. Since the first demonstration, significant improvements have been made over the past few years. Recently, the available highest operating temperature is 186K in pulse mode [2] and 117K in continuous-wave (CW) mode [3], the highest output peak power is 248 mW in pulse mode and 138 mW in CW mode at 10 K [4], the lowest threshold current density is 110A/cm^2 at 10 K [5], and the longest lasing wavelength is 250 μm [6]. However, for the final sake of continuous wave operation at room temperature, there still much to be desired. Nowadays, a practical way is to stably operate above liquid nitrogen temperature. Here, we report the development of terahertz quantum cascade lasers about 3 THz. The chirped superlattice active region based on GaAs/$Al_{0.15}Ga_{0.85}As$ builds bound-to-continuum transition. The materials were prepared by Veeco GEN-II solid source molecular beam epitaxy (MBE) and devices were processed using standard photolithography and wet chemical etching. The quantum cascade lasers operate about 2.94~3.10 THz due to the growth uncertainties for different wafers. For the 3.10 THz quantum cascade laser, optical powers of 14mW at 10 K and 1.5mW at 80 K are observed from a 180-μ m-wide and 2-mm-long device with a pulsed mode of 1% duty cycle. These results are an encouraging step towards a widely applicable solid-state terahertz source operating above liquid nitrogen temperature.

2. Page Device structure and fabrication processing

The structure labeled G355 and G394 were grown by EPI Gen II solid source molecular beam epitaxy on a semi-insulating (SI) GaAs substrate. The growth started with 0.8 μm highly-doped (Si, 2.5×10^{18}

[1] To whom any correspondence should be addressed.

cm^{-3}) GaAs cladding, followed by 120 periods of active regions unit, which is demonstrated in reference [7]. Finally, a 0.2 μm highly-doped (Si, 5×10^{18} cm^{-3}) GaAs was grown as the top contact layer. The highly-doped GaAs cladding could provide a strongly confined low loss TM mode bound to that layer and make it easy to use the n+ layer for electrical contacting. During the growth, a substrate temperature of 580 ℃, an arsenic overpressure of 3×10^{-6} Torr and a V/III ratio of as large as about 12 are selected.

After growth, samples were processed into 180-μm-wide ridge stripes by optical lithography and wet chemical etching （H_3PO_4:H_2O_2:H_2O=1:1:6, 22℃） to the bottom contact layer. The top electrical contact was provided by two 20-μm-wide stripes of Ge/Au/Ni/Au (26/54/15/100nm) deposited along the edges of each ridge. Two bottom 220-μm-wide contacts were evaporated on both sides of the ridge stripes with 55-μm-wide separations between the metal on the side and the edge of the ridge. The top and bottom contacts were formed by a standard lift-off deposition at the same time. Thermal annealing for 20s at 390℃ under nitrogen atmosphere was followed to provide Ohmic contact. After that, Ti/Au (8.6/200 nm) was deposited completely covering both the bottom contact and the top of the ridge to provide the top confinement and allow for wire bonding. Then, the substrate was thinned down to 200μm and Ti/Au (25/250 nm) contact was deposited on the backside allowing for easy soldering. Finally, the sample was cleaved into laser bars with different length and Al_2O_3/Ti/Au layers (100/10/100nm) were evaporated on the back facets to form a high reflectivity (HR) coating.

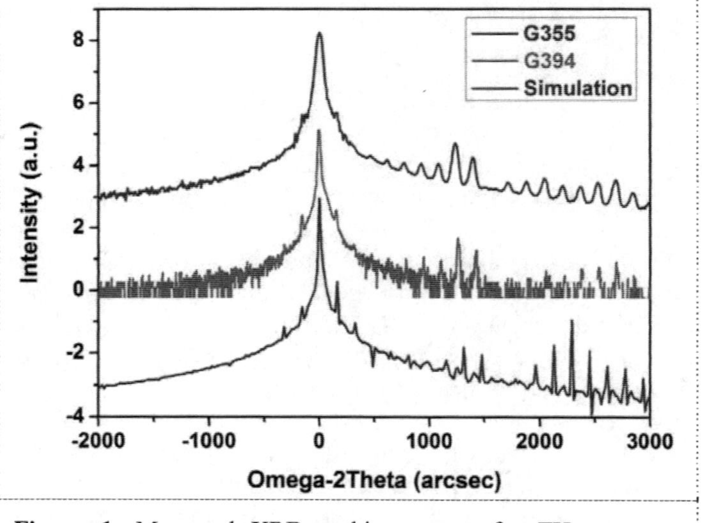

Figure 1. Measured XRD rocking curves for THz quantum cascade laser structures of G355 and G394, and the simulated result.

Double crystal X-ray diffraction (XRD) is used to analyze the periodicity of the wave-guide core [8]. Fig.1 shows the measured XRD rocking curve of G355 and G394 together with the simulated result. The actual period extracted from the XRD results are 120.79nm and 119.66 nm for G355 and G394 respectively, which show deviations about 3.95% and 2.98% from the design value. A FWHM of about 68 arcsec is shown for the satellite peak of G355 and 40 arcsec for G394.

For a comprehensive understanding about the device processing, the scanning electron microscopic image of device structure and its cross-section are shown in Fig. 2 (a) and Fig. 2(b) respectively.

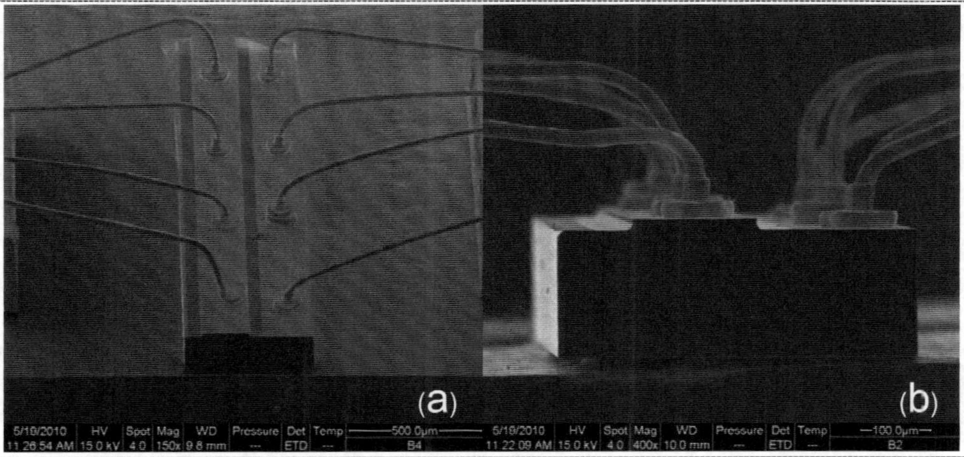

Figure 2. The scanning electron microscopic image of a laser bar: (a) the surface and (b) the cross-section.

3. Experiments and results

For device characterization, the samples were soldered epilayer up onto Cu heat-sinks with In and wire bonded. The packaged laser bars were then mounted to the cold finger of a He flow cryostat with polyethylene windows. The pulsed light power was measured by a calibrated thermopile powermeter placed directly near the polyethylene window without any accessorial optical component in the optical path. The collection efficiency was estimated to 20% in view of the beam divergence and the transmittance of the polyethylene window. For the spectral measurement, the emission was collected by two sequential off-axis parabolic mirrors, passed through a Fourier-transform infrared spectrometer and focused again onto a room temperature DTGS detector. All laser spectra were collected in rapid scan mode with a resolution of 0.5 cm^{-1}.

Fig. 3 shows the lasing spectra of G355 and G394 at a heat-sink temperature of 10 K. The laser spectra were measured with current pulses of 1-μs duration and 100-kHz frequency at about 1.2 times threshold current. The lasing frequency is 2.94 THz for G355 and 3.10 THz for G394 at 10 K, which is in accordance with the XRD results.

Fig. 4 displays the optical power versus drive current (*L-I*) curves for G355 at various heat-sink temperatures with current pulses of 1-μs duration and 20-kHz frequency. Lasing is observed up to 70 K with power level of 1 mW, compared to 4.75 mW observed at 10 K. The slope efficiencies are 4.72 and 1.25 mW/A for 10 K and 70 K respectively. The threshold current density as a function of temperature for the same device is plot in Fig. 4. The threshold current densities are J_{th}=296.5 A/cm^2 at 10 K and J_{th} =428 A/cm^2 at 70 K, and the characteristic temperature is extracted to be T_0=57.5 K.

Fig. 5 displays the optical power versus drive current (*L-I*) curves for G394 at various heat-sink temperatures with current pulses of 1-μs duration and 10-kHz frequency. Fig.5 (a) is *L-I* characteristics for a 1.5-mm-long device and Fig. 5(b) for a 2-mm-long one. The two devices are all lasing up to 80 K with power level of 2 mW and 1.5 mW at 80 K. The peak powers at 10 K are 9 mW and 14 mW for the 1.5-mm-long device and the 2-mm-long one respectively. The threshold current densities J_{th} =370 A/cm^2 at 10 K and J_{th} =667 A/cm^2 at 80 K for the 1.5-mm-long device, and J_{th} =685 A/cm^2 at 10 K and J_{th} =1260 A/cm^2 at 80 K for the 2-mm-long one. The characteristic temperatures are extracted to be T_0=31.5 K and 35.9 K respectively.

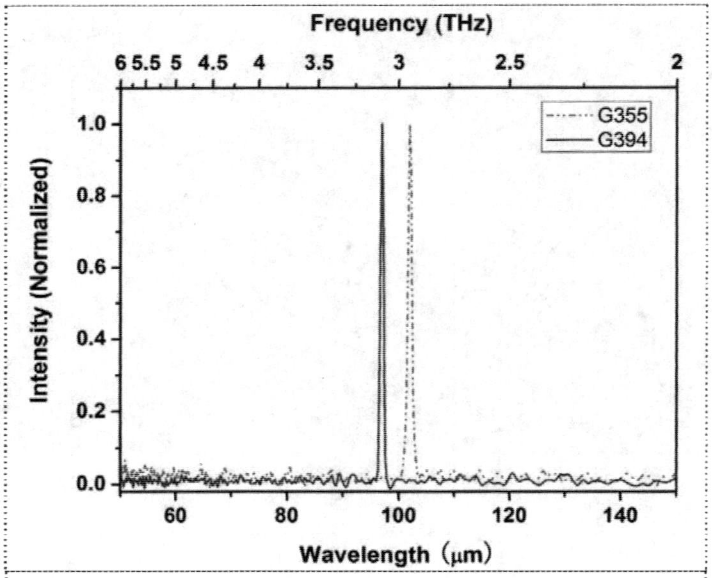

Figure 3. Lasing spectra for G355 and G394 at 10 K under current pulses of 1-μs duration and 100-kHz frequency at about 1.2 times threshold current.

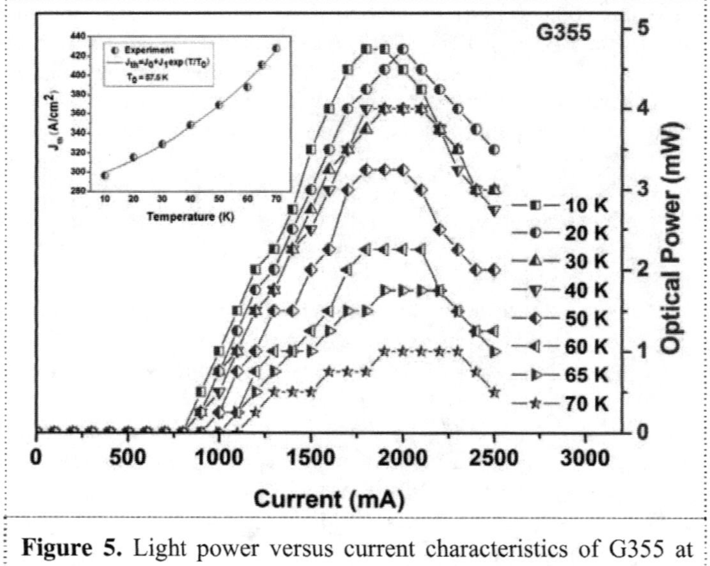

Figure 5. Light power versus current characteristics of G355 at various heat-sink temperatures measured using 1-μs pulses repeated at 20 kHz. The inset is the variation of threshold current density (J_{th}) with heat-sink temperature (T) along with a fit to the phenomenological relation $J_{th}=J_1\exp(T/T_0)$.

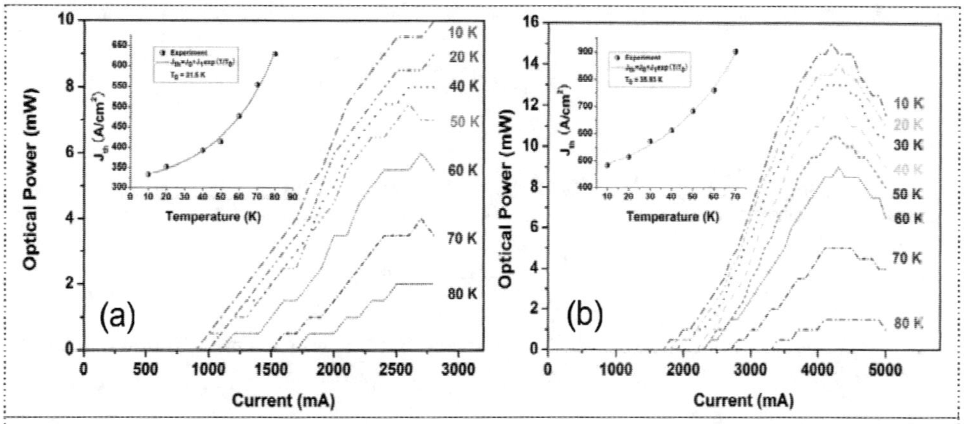

Figure 5. Light power versus current characteristics of G394 at various heat-sink temperatures measured using 1-μ s pulses repeated at 10 kHz for lasers with different length: (a) 1.5-mm-long, (b) 2-mm-long. The relations between Jth and T is also included respectively.

4. Conclusions

We have demonstrated THz quantum cascade lasers based on the bound-to-continuum active region and the semi-insulating surface-plasmon waveguide. Due to the growth uncertainties for different wafers, the quantum cascade lasers can operate from 2.94~3.10 THz. For the 3.10 THz quantum cascade laser, lasing is observed up to a heat-sink temperature of 80 K with a peak power of 2 mW. These devices exhibit a potential of higher temperature operation by more well-defined device structure and optimized fabrication processing.

5. References

[1] Köhler R, Tredicucci A, Beltram F, Beere H E, Linfield E H, Davies A G, Ritchie D A, Iotti R C and Rossi F 2002 *Nature* **417** 156
[2] Kumar S, Hu Q and Reno J L 2009 *Appl. Phys. Lett.* **94** 131105
[3] Williams B S, Kumar S, Hu Q and Reno J L 2005 *Opt. Express* **13** 3331
[4] Williams B S, Kumar S, Hu Q and Reno J L 2006 *Electron. Lett.* **42** 89
[5] Scalari G, Amanti M I, Fischer M, Terazzi R, Walther C, Beck M and Faist J 2009 *Appl. Phys. Lett.* **94** 041114
[6] Walther C, Fischer M, Scalari G, Terazzi R, Hoyler N and Faist J 2007 *Appl. Phys. Lett.* **91** 131122
[7] Liu J Q, Chen J Y, Liu F Q, Li Lu, Wang L J and Wang Z G 2010 *Chin. Phys. Lett.* **27** 104203
[8] Liu J Q, Kong N, Li Lu, Liu F Q, Wang L J, Chen J Y and Wang Z G 2010 *Semicond. Sci. Technol.* **25** 075011

Acknowledgments

This work is supported by the National Program of China (Grant Nos. 2006CB604903, 2007AA03Z446 and 2009AA03Z403). The authors would like to thank P. Liang, Y. Hu and L. Jiang for their help in processing.

The study on THz light-waves in 0ne-dimensional random media

H Liu[1*], S Y Jing [1], X J Zhang[1], J T Lv[2] and Z H Ren[1]

[1] School of Information and Electrical Engineering, China University of Mining and Technology, Xuzhou 221116, China;
[2] Wuhan National Laboratory for Optoelectronics, School of Optoelectronic Science and Engineering, Huazhong University of Science and Technology, Wuhan 430074, China

Abstract. The research on optical properties of the light-waves in random media has become research hotspot in laser physics. Many theoretical models have been developed to explain the experimental results. With the further research on the optical properties of the light-waves in different random media, the great importance of THz section is attached. We research on the THz light-waves in 0ne-dimensional random media, and mainly focus on the Frequency spectrum and threshold property. The results show that each of these THz modes also has a typical threshold gain behavior and different threshold pumping rate .

1. Introduction

The research on optical properties of the light-waves in random media has become research hotspot in laser physics [1-11]. In the past decade, several experiments on these two types of random lasers have been reported, and the output wavelengths almost all concentrated in a few hundred nanometer band [1-2]. Many theoretical models have been developed to explain the experimental results [8-10]. In our prior research [12-14], we create a numerical model of random media by simultaneously solving Maxwell's equations and rate equations. Based on the model, the electromagnetic field inside the disorder system can be researched particularly. And we calculate and investigate the particular characters of the light-waves existed in the random media.

With the further research on the optical properties of the light-waves in different random media, the great importance of THz section is attached. Terahertz radiation is a newly developed coherent far infrared source, which has been applied in many research fields. In this paper, we improved the theoretic model and changed the parameters in order to find some difference of THz light-waves in one-dimensional random media. As the peak frequencies in the power spectrum reflect the frequency character of the localized modes， we can study the property of THz light-waves in disordered media

*Corresponding author: lhai_hust@hotmail.com

Supported by the National Natural Science Foundation of China (Grant No. 10876080) and China University of Mining and Technology Development Funds (Grant No. 2009A021).

from the spectrum. We research on the THz light-waves in the random media, and mainly focus on the frequency spectrum and threshold property. In the paper, we mainly study the spectrum property of THz modes in random gain media. And the semi-classical model equations of transverse electric of random laser are presented. For scattering particles with same spatial distribute, the properties of random system are computed numerically. It is denoted that the emission frequency and spatial pattern of modes are dependent on the parameters of the random media heavily. The results are similar as before.

In addition, the pumping rate dependence of the peak intensity of the individual lasing THz mode in two-dimensional active media is investigated. The results show that each of these modes has a typical threshold gain behavior and different threshold pumping rate. The newly research on the THz light-waves in random media have important impact on the application of THz bandwidth. And the new-style THz device enjoys easy fabrication and low cost.

2. Theoretical Model

In isotropic, non-magnetic, homogeneous dielectric random media, the Maxwell equations shown as following:

$$\nabla \times \mathbf{H} = \varepsilon_0 \varepsilon \frac{\partial \mathbf{E}}{\partial t} + \frac{\partial \mathbf{P}}{\partial t}, \qquad (1.1)$$

$$\nabla \times \mathbf{E} = -\mu_0 \frac{\partial \mathbf{H}}{\partial t}, \qquad (1.2)$$

In one-dimensional case, for that propagates along the z direction, the equation (1) simplified as:

$$-\frac{\partial H_y}{\partial z} = \varepsilon_0 \varepsilon \frac{\partial E_x}{\partial t} + \frac{\partial P_x}{\partial t}, \qquad (2.1)$$

$$\frac{\partial E_x}{\partial z} = -\mu_0 \frac{\partial H_y}{\partial t}, \qquad (2.2)$$

Eventually, $\vec{P}$ is described as obeying the quantum population equation of motion. There-into,

$\Delta N = N_2 - N_3$, $\Delta\omega_l = 1/\tau_{32} + 2/T_2$, $\kappa = 6\pi\varepsilon_0 c^3/(\omega_l^2 \tau_{32})$

$$d^2 P_x / dt^2 + \Delta\omega_l dP_x / dt + \omega_l^2 P_x = \kappa \Delta N(t) E_x, \qquad (3)$$

The three-level rate equations read as following:

$$dN_1/dt = N_2/\tau_{21} + N_3/\tau_{31} - N_1 W_p, \qquad (4.1)$$

$$N_1 = N_T - N_2 - N_3 \qquad (4.2)$$

$$dN_2 / dt = N_3 / \tau_{32} - N_2 / \tau_{21} + (E_x / \hbar\omega_l) dP_x / dt, \qquad (4.3)$$

$$dN_3 / dt = N_1 W_p - N_3 / \tau_{31} - N_3 / \tau_{32} - (E_x / \hbar\omega_l) dP_x / dt \,, \tag{4.4}$$

Where W_p is the pump rate; N_i is the population density in level i (i=1 to 3), τ_{21}, τ_{32}, and τ_{31} are the lifetime of corresponding level, respectively; ω_l is the transition frequency between levels 3 and 2.

The values of those parameters in above equations that will be used in simulating the active part in the following numerical calculations are taken as: $l = 5000\mu m$, $d = 50\mu m$, $T_2 = 10^{-11} s$, $\tau_{32} = 10^{-10} s$, $\tau_{31} = \tau_{21} = 3\times10^{-3} s$, $N_T = \sum_{j=1}^{3} N_j = 10^{27} m^{-3}$. We use a broadband Gaussian pulse with random amplitude as an excitation source, and the electromagnetic fields can be calculated by use of the FDTD methods and PML absorbing boundary manner to solve equations above [13].

3. Results and Discussions

First, we obtained the spectrum which is shown in figure.1, in which we select the THz mode marked by λ_1 with λ_1 =0.47 μ m , λ_2 =0.50 μ m and λ_3 =0.52 μ m. Obviously, the mode λ_2 can be stimulated with the pumping rate $W_p = 10^{16} s^{-1}$.

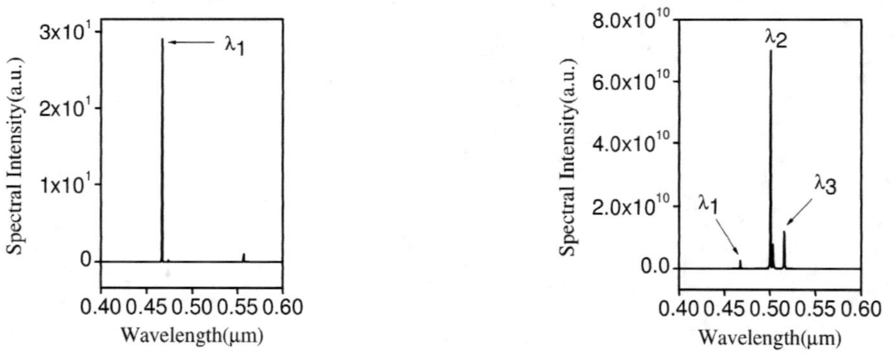

Figure 1 The spectrum of THz modes (a)Passive mode (b) Active mode and $W_p = 10^{16} s^{-1}$

As can be seen, the peak intensity of the THz mode with pumping is much stronger than that of the THz mode in passive case. It indicates that THz modes also can be excited. In fact, in our foregoing studies, we have demonstrated the point. In order to have a further research on the threshold property of THz mode, we trace out the evolution curves of marked THz modes by recording the peak intensities of them at different pumping rate which are shown as figure.2. There-into, the thresholds of the marked THz modes are $W_{pth2} = 6.8\times10^{14}s^{-1}$, $W_{pth3} = 7.1\times10^{14}s^{-1}$ and $W_{pth1} = 8.35\times10^{14}s^{-1}$, respectively.

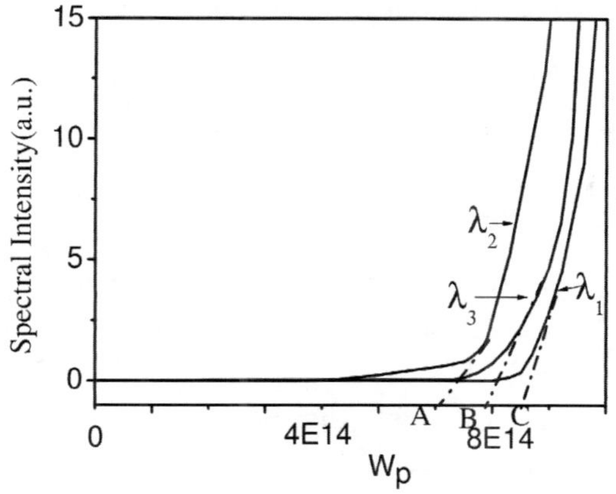

Figure. 2 The pumping rate dependence of the power of the three marked modes in the active system

The evolution process indicates the threshold property of THz modes distinctly. Therefore, we can study the property of THz modes in two-dimensional random media with the similar method. As we know each lasing mode has its own spatial region in a 2D medium, pumping this region locally would exalt the mode preferentially, thus offering an effective method to obtain some special THz modes we want.

4. Conclusion

In the paper, we studied the threshold property and spectral property of THz light-waves in the random media. The results show that each of these modes has a typical threshold gain behavior and different threshold pumping rate. It is denoted that the emission frequency and spatial pattern of modes are similar as forgoing study. The THz modes have similar property. Simultaneity, we can introduce the fabrication and testing of the THz random media film as the common case we studied before. The newly research on the THz light-waves in random media have important impact on the application of THz bandwidth. And the new-style THz device enjoys easy fabrication and low cost.

References

1. Cao H, Zhao Y G, Ong H C, Ho S T, Wu J Y and Chang R P H 1998 Ultraviolet lasing in resonator formed by scattering in semiconductor polycrystalline films *Appl. Phys. Lett.* **73** 3656-8.
2. Cao H, Zhao Y G, Ong H C, Ho S T, Seelig E W, Wang Q H and Chang R P H 1999 Random Laser Action in Semiconductor Powder *Phys. Rev. Lett.* **82** 2278-81.
3. Anglos D, Stassinopoulos A, Das R N, Zacharakis G, Psyllaki M, Jakubiak R, Vaia R A, Giannelis E P and Anastasiadis S H 2004 Random laser action in organic–inorganic nanocomposites *J. Opt. Soc. Am. B* **21** 208-13.

4. Soest G V, Tomita M, Lagendijk A 1999 Amplifying volume in scattering media *Opt. Lett.* **24** 306-8.
5. Burin A L, Ratner M A, Cao H and Chang R P H 2001 Model for a Random Laser *Phys. Rev. Lett.* **87** 215503.
6. Burin A L, Ratner M A, Cao H and Chang R P H 2002 Random Laser in One Dimension *Phys. Rev. Lett.* **88** 093904.
7. Patra M 2003 Influence of spatial correlations on the lasing threshold of random lasers *Phys. Rev. E* **67** 065603 .
8. Jiang X and Soukoulis C M 2000 Time Dependent Theory for Random Lasers *Phys. Rev. Lett.* **85** 70-3.
9. Sebbah P and Vanneste C 2002 Random laser in the localized regime *Phys. Rev. B* **66** 144202.
10. Wang C and Liu J S 2006 Polarization dependence of lasing modes in two-dimensional random lasers *Phys. Lett. A* **353** 269–72.
11. Balachandran M. and Lawandy N. M 1996 Understanding bichromatic emission from scattering gain media *Opt. Lett.* **21** 1603-6.
12. Liu Hai, Liu Jinsong, Lv Jiantao, et al. 2009 Spectral time evolution of polarized modes under local pumping in two-dimensional random medium. *J Opt A: Pure Appl Opt.*, 11: 065202
13. Liu Hai, Liu Jinsong, Feng Bo, et al. 2008 The competition between two polarization states in two-dimensional random medium. *Optics Communications*, 281: 2964
14. Liu Hai, Liu Jinsong, Lv Jiantao , et al. 2009 The research on polarization states with local pumping in two-dimensional active random medium. *Optics Communications*, 282: 1004-1008.

Analysis on characteristic and application of THz frequency comb and THz sub-comb

Pengxiang Liu, Degang Xu*, Jianquan Yao

The Institute of Laser & Optoelectronics, The college of precision instruments and opto-electronic engineering, Tianjin University, Tianjin, 300072, P. R. China

Key laboratory of Opto-electronic Information Science and Technology, Ministry of Education, Tianjin, 300072, P. R. China

*Corresponding author: xudegang@tju.edu.cn

Abstract: In this paper, we proposed a method for THz sub-comb generation based on spectral interference. The result of our calculation indicated that the THz pulse train, generated by surface-emitted optical rectification of femtosecond (fs) laser pulse in periodically poled lithium niobate (PPLN), has a comb-like spectrum. The characteristic of this THz sub-comb was analyzed both in frequency and time domain. Compared with the THz frequency comb emitted by a photoconductive antenna (PCA), THz sub-comb has a lower spectral resolution and wider free spectral range. Thus it could be an ideal source for wavelength division multiplexing (WDM) in THz wireless communication system.

Key words: terahertz wave, frequency comb, optical rectification

1. introduction

Optical frequency comb is a precision ruler in frequency metrology [1] and an ideal source in an optical communication system with wavelength division multiplexing [2]. In terahertz region, the output of a photoconductive antenna, triggered by a periodical train of fs laser pulses (fs optical comb), has a spectrum with a regular comb of sharp lines [3, 4]. The temporal form is an infinite series of periodic THz pulse train. As a THz frequency comb, it can be used to measure the frequency of a continuous-wave THz with high precision.

Similar to the THz comb mentioned above, a finite series of periodic THz pulse train also has a comb-like spectrum, but with the tooth (spectral components) coarser. Thus it is called coarse THz comb or THz sub-comb. It has been reported that the transmitted spectrum of a THz pulse through a multilayer filter is a sub-comb structure [5]. This filter has a one-dimensional dielectric multilayer structure composed of alternating materials with high and low index of refraction. The main characteristic of this THz spectrum is attributed to THz pulses bouncing back and forth via multilayer structures.

The result of our calculation indicated that the surface-emitted THz wave, by optical rectification of femtosecond laser pulse in a periodically poled lithium niobate, can also be used as a THz sub-comb. The temporal form of radiated THz wave is a train of THz pulses with one single cycle. The polarities of every two adjacent pulses are opposite, and the number of pulses is proportional to the number of poled periods of the crystal. The physical mechanism of THz sub-comb generation was explained by spectral interference between early and late pulses.

The characteristic of THz sub-comb was compared with the THz frequency comb emitted by a photoconductive antenna. As a finite series of THz pulses, the spectral resolution of this sub-comb is not high enough for precision measurement. But with a wide free spectrum range, it is suitable for the source of THz wireless communication with wavelength division multiplexing.

2. characteristic of THz sub-comb in frequency and time domain

The principle of narrow-band THz wave generation by surface-emitted optical rectification of fs-pulse in PPLN has already been discussed in detail by Avetisyan *et al.* [6, 7]. A sketch of surface-emitted geometry is shown in figure 1. In previous works [6, 7], the output THz wave is narrowband with one center frequency. The corresponding temporal form is quasi-harmonic.

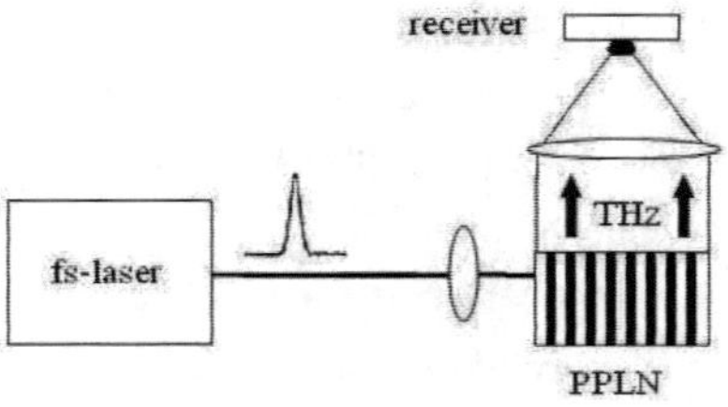

Figure 1 Surface-emitted geometry

If we change the structure of the PPLN and use larger poling period (e.g. 520μm), there would be multiple frequency components in the spectrum (figure 2). It is a comb-like spectrum, with the interval between every two adjacent components the same. The interval is inversely proportional to the poling period.

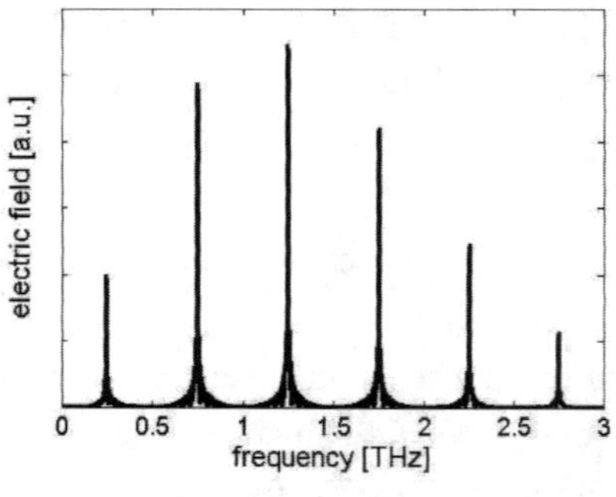

Figure 2 Spectrum of THz sub-comb for the case poling period Λ = 520 μm

The corresponding waveform is presented in figure 3. Figure 4 is the expansion of the temporal region of -6~6 ps. Similar to the temporal form of mode-locked fs optical comb, which is a train of periodical fs laser pulses, the waveform of this THz radiation consists of a train of single cycle pulses. Here the number of pulses is 66, which equals to the number of interfaces between individual domains in the crystal (for the number of domains in PPLN is $2N+1 = 65$) and two surfaces of the crystal. As seen from figure 6, the polarities of every two adjacent pulses are opposite to each other.

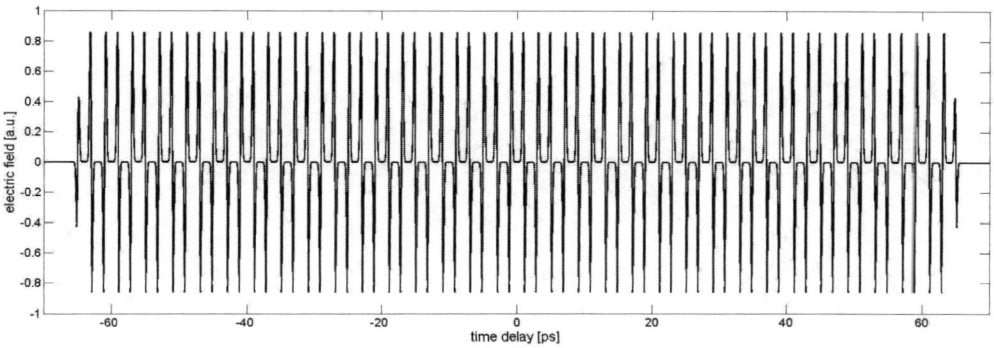

Figure 3 Calculated temporal form of THz sub-comb

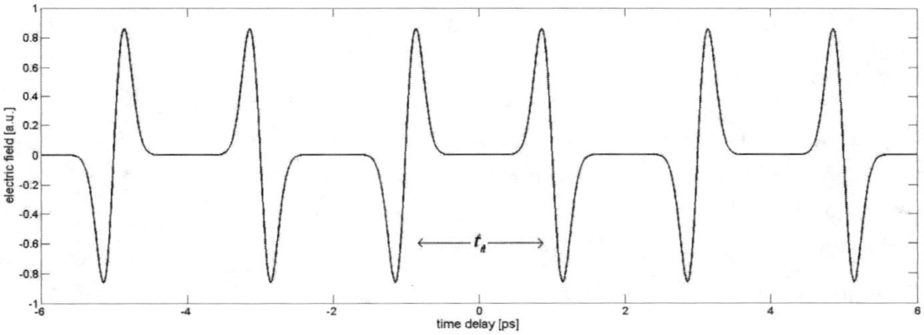

Figure 4 the expansion of the temporal region of -6~6 ps in Figure 3

As a train with 66 pulses, this THz radiation has a lower spectral resolution, compared with a frequency comb, which is an infinite series of pulses. Rigorously, it is not a THz frequency comb. But with a comb-like spectrum, we call it "THz sub-comb". The spectral resolution (bandwidth of each component) in the comb-like spectrum is defined as: • $1.77c/Nl_c\ n$ [6]. For the case $N = 32$, l_c = /2 = 260μm and the mismatch of refractive index n is 2.3, the bandwidth of one frequency component is 27.75GHz (circular frequency). The free spectral range (interval between two adjacent components) is 2 $c/l_c\ n$ = 3.14THz (the corresponding frequency, divided by 2 , is 0.5THz).

3. mechanism of THz sub-comb generation

The mechanism of this surface-emitted THz sub-comb generation can be explained by spectral interference. According to figure 3, THz wave generated from a PPLN is a series of THz pulses. When a pump laser pulse travels through an interface between two adjacent domains of PPLN, one THz pulse is radiated, which has a wideband spectrum. Pulses, radiated from different domains, "interfere" with each other. This behavior is spectral interference between early and late pulses [8]. The interference effect has accumulated by increasing the number of poling periods in LN crystal. As a periodical continuation of one THz pulse, the THz pulse train emitted from a PPLN has a quasi-discrete spectrum, which is on the analogy of mode-locked fs optical frequency comb [1].

Only at the condition of constructive interference, the certain frequency components can arise in the output spectrum (peak values in figure 2). The condition can be expressed as:

$$\omega = (2k+1)\frac{\pi c}{l_c \Delta n}, k \in \mathbf{N} \quad (1)$$

Not only the fundamental frequency (k = 0), but also high-order harmonics (k = 1, 2 ...) satisfy this condition. Thus, there could be multiple frequency components in the output spectrum (comb-like spectrum).

4. comparison with THz frequency comb

For the THz frequency comb emitted by a PCA, when a mode-locked femtosecond laser pulse train is incident on the antenna gap of a photoconductive film, a periodical THz pulse train is radiated. In the frequency domain, the optical frequency comb is down-converted to the THz region without any change in its frequency spacing. As an infinite series of periodic THz pulse train, it has a high spectral resolution ~11MHz (the corresponding free spectral range is 81.8MHz) [3]. This THz frequency comb is quite suitable for measurement of the frequency of a continuous-wave THz with high precision.

As calculated above, the spectral resolution of THz sub-comb is = 4.42GHz (=27.75GHz/2) and the free spectral range is 0.5THz, which is much wider than . Compared with the THz frequency comb , THz sub-comb has a lower spectral resolution and wider free spectral range. The THz sub-comb will be an ideal source for wavelength division multiplexing in THz wireless communication system, for it can be separated into its individual components easily for demultiplexing and modulation of each channel.

5. conclusion

In this paper, we proposed a method for THz sub-comb generation based on spectral interference by surface-emitted optical rectification of fs laser pulse in PPLN. The result of our calculation indicated that the surface-emitted THz wave has a comb-like spectrum and the corresponding temporal form is a train of THz pulses, if the structure of PPLN is designed properly. The physical mechanism of this phenomenon was explained by spectral interference between early and late pulses.

The characteristic of THz sub-comb was compared with the THz frequency comb emitted by a photoconductive antenna. As a finite series of THz pulses, this calculated THz radiation has a lower spectral resolution. However, with a wide free spectrum range, this sub-comb can be easily separated into its individual components for demultiplexing and modulation of each channel. So it could be an ideal source for wavelength division multiplexing in THz wireless communication system.

Acknowledgments

This paper is supported by the National Basic Research Program of China (2007CB310403) and National Natural Science Foundation (NSFC) (60801017).

Journal of Physics: Conference Series **276** (2011) 012218 doi:10.1088/1742-6596/276/1/012218

Reference

[1] Th. Udem, R. Holzwarth, and T. W. Hansch, "Optical frequency metrology," Nature (London) 416, 233-237 (2002).

[2] P. J. Delfyett, et al, "Optical frequency combs from semiconductor lasers and applications in ultrawideband signal processing and communications," J. Lightw. Technol. 24(7), 2701–2719 (2006).

[3] T. Yasui, Y. Kabetani, E. Saneyoshi, S. Yokoyama, and T. Araki, "Terahertz frequency comb by multifrequency-heterodyning photoconductive detection for high-accuracy, high-resolution terahertz spectroscopy," Appl. Phys. Lett. 88, 241104 (2006).

[4] S. Yokoyama, R. Nakamura, M. Nose, T. Araki and T.Yasui, "Terahertz spectrum analyzer based on a terahertz frequency comb," Opt. Express 16, 13052-13061 (2008).

[5] Min-Woo Yi, Young-Chan Kim, Dae-Su Yee, and Jae-Wook Ahn, "Terahertz Frequency Spreading Filter via One-dimensional Dielectric Multilayer Structures," J. Opt. Soc. Korea 13, 398-402 (2009)

[6] Y. Avetisyan, C. Weiss, G. Torosyan, R. Beigang, "Generation of narrow-band THz radiation by surface-emitted optical rectification in periodically poled lithium niobate," Proc. SPIE 4490, 134-142 (2001).

[7] J. A. L'huillier, G. Torosyan, M. Theuer, Y. Avetisyan, R. Beigang, "Generation of THz radiation using bulk, periodically and aperiodically poled lithium niobate – Part 1: Theory" Appl. Phys. B 86, 185-196 (2007).

[8] Y. -S. Lee, T. Meade, V. Perlin, H. Winful, T. B. Norris, "Generation of narrow-band terahertz radiation via optical rectification of femtosecond pulses in periodically poled lithium niobate," Appl. Phys. Lett. 76(18), 2505-2507 (2000).

Journal of Physics: Conference Series **276** (2011) 012219 doi:10.1088/1742-6596/276/1/012219

The new diagnosing method basing on the frequency analysis of human's nails in T-RAY

Pengcheng Ma*, Long Xiao

Wuhan National Laboratory for Optoelectronics, College of Optoelectronic Science and Engineering, Huazhong University of Science and Technology, Wuhan 430074, China

*pengchengma@smail.hust.edu.cn

Abstract. It has been reported that many kinds of sickness can be reflected in one's nails. And because of the special composition and structure of nails, they are very suitable for T-ray detecting. Therefore we want to see whether we can diagnose some sickness by only analyzing the transmission of the nails. The experiments consist of three parts. In the first part, we took 200 samples which came from about 20 years old people. By that way, we got that the peak transmission wavelengths of human's nail is about 1.7THz and we almost acknowledged the characteristics of nails. Then we changed our samples coming from different ages, and some interesting phenomenon were coming out. Lastly, by detecting the nails which are taken from special patients, we discovered some sickness can be detected using T-ray.

1.Introduction

Lying between the microware and the infrared, the terahertz (THz) frequency radiation has been proven to be a fertile region in the electromagnetic spectrum and a powerful tool in diagnosing sickness, scientific research and many applications. As the advance of femtosecond laser and the realization of travelling wave electro-optic detection technique, [1] terahertz time-domain spectroscopy (THz-TDS) has been widely used in many field, such as, condensed matter physics,[2,3] biological detection,[4,5] material characterization,[6,7] and near-field imaging,[8] etc. As THz-TDS is different from traditional optical measures, it measures the THz electric field $E(t)$ directly, and its Fourier transform $E(\omega)$ directly yields both amplitude and phase information at the same time. This feature makes THz-TDS a very convenient and valuable tool to study material.[9,10,11]

The Theoretical figure of THz-TDS is introduced just as Fig.1. Photoconductor antenna, illuminated by ultrashort laser pulses, is used for THz radiation and detection. The pumping laser should provide the pulses of 760-840 nm wavelength, 50-150 fs pulse duration and 50-500 mW output power at approx 80 MHz pulse repetition rate. For more efficient collimation and focusing of THz radiation, a substrate lens fabricated from high resistance Silicon is attached to the backside of each antenna.

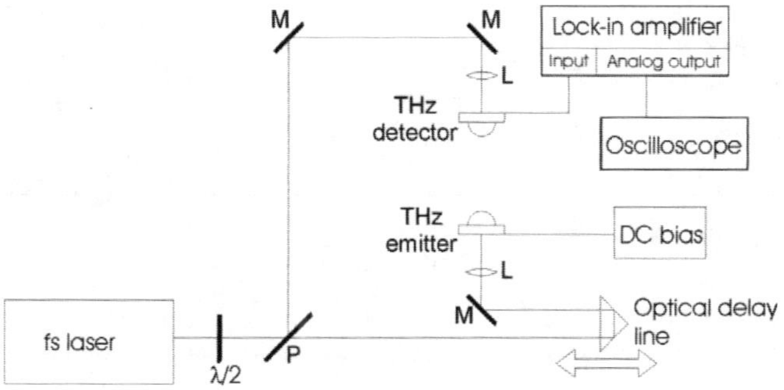

Fig.1
The theroy of THz-TDS

2.The designing of experiments

It has been reported that many kinds of sickness can be reflected in one's nails. And the nails are very suitable for T-ray detecting. Therefore we want to see whether we can diagnose some sickness by only analyzing the transmission of the nails. If it could be achieved, those kinds of sickness could be checked cheaply, harmlessly and conveniently.

In order to get better results, we designed the following instrument to fix the nails(just as show in the Fig.2). The instrument play an important role in this experiment.

Fig.2
The instrument to fix the nails

3.The analysis of results

In the following Fig.3, the different colorful lines represent different transmission of the samples. We can get that the peaks of transmission are delay. The reason is obviously due to the index of nails. The different delays may cause by the thickness and the index of the different nails.

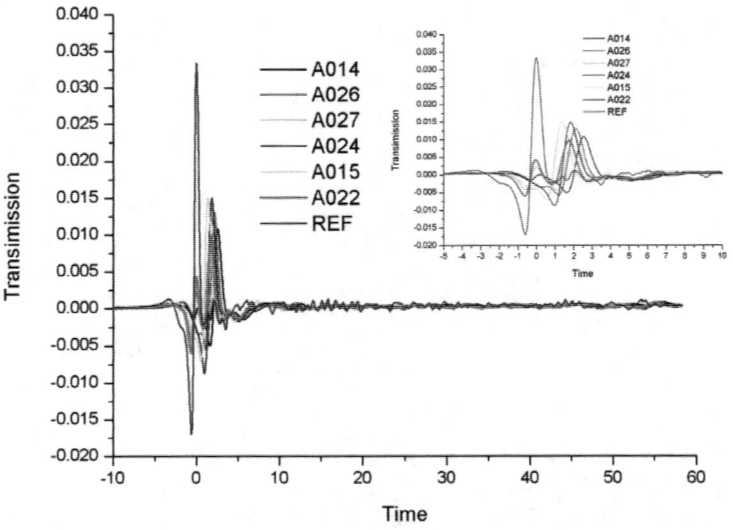

Fig.3

The intensity of transmission varies with time

When we translated the Fig.3 to Fourier transform, we got the Fig.4, the intensity of transmission varies with frequency. And we discover there is a peak nearby 1.7THz. We guess this may be the characteristic peak of human's nails. In that way we should concentrate our mind at the analysis of this point. By analyzing this point, we can get the key information. Fig.4 is the range of 1.6THz to 1.8THz. We found almost all the samples have a peak at 1.67THz, except the sample A024' peak is near 1.72THz. This may be caused by the experiments' miss. So we conformed the peak is the human's characteristic one.

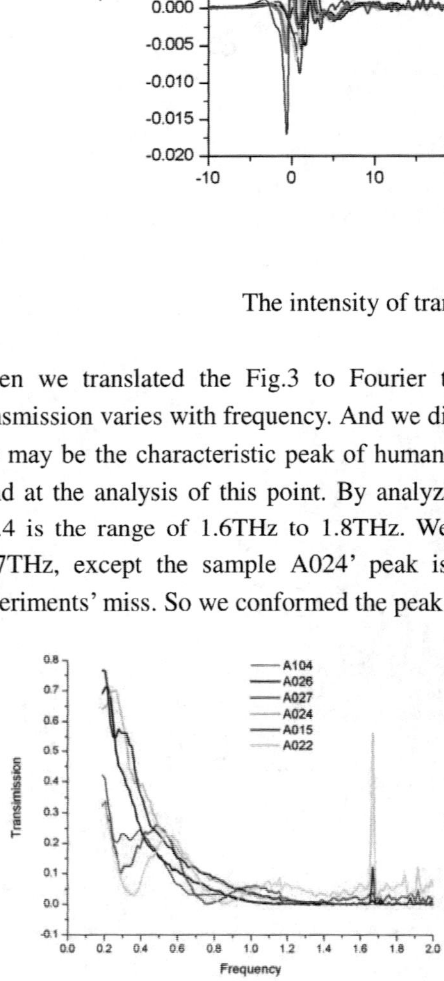

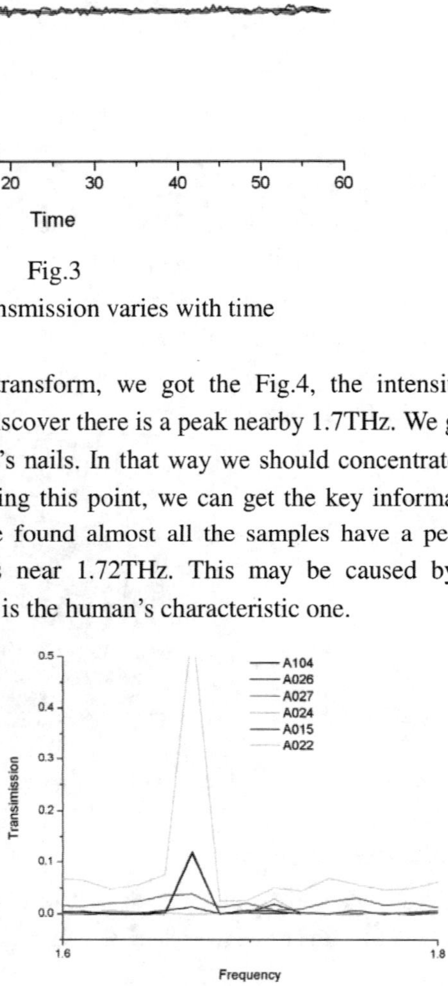

Fig.4(a) Fig.4(b)

The intensity of transmission varies with frequency (THz)

If we have the picture enlarged (as Fig.4(b)), we could notice there surely existing a peak at 1.67THz. But to my surprise the intensity of different people are various. Then we checked the basic information of those samples, we have found that both A014 and A015 prefer to drinking tea ordinarily. And sample A024 enjoys coffee very much. From the above-mentioned materials, we inferred that the Caffeine may take a great place in the intensity of the peak. However we discover a fun phenomenon that the intensity has a strong relationship with one's GPA. As we can see that the GPA is higher when there exit a higher intensity.

Table.1 The GPA vary with the Intensity

NUM	A104	A026	A027	A024	A015	A022
GPA(by 2010/3/2)	83.53	79.63	81.03	80.72	83.44	87.32
Intensity	0.114	0.016	0.045	0.019	0.115	0.562

As the water plays a great place in the intensity of the peak, so we guess if a student drinks more water, he may get a higher GPA. In that reason, we may use this method to take the place of exams to test the ability of the students. Certainly, some works should be taken to prove our idea.

4.Conclusion

we have got the giant birefringence of human's nails in T-RAY. We discovered that the characteristic peak of human's nails is at 1.67THz. And the intensity of the peak may vary with one's eating habits, Caffeine and the scores. Maybe, by the analysis of one's nails, can we get his score. And our students can be liberated from examinations. However, some additional experiments should be taken to confirm our ideas.

References

[1] Wu Q, Litz M and Zhang X C1996 *Appl. Phys. Lett.* **68** 2924

[2] Exter M and Grischkowsky D 1990 *Appl. Phys. Lett.* **56** 1694

[3] Thzmizhmani L, Azad A K, Dai J and Zhang W 2005 *Appl. Phys. Lett.* **86** 131111

[4] Ferguson B, Wang S, Gray D, Abbott D and Zhang X C 2001 *Proc. SPICE* **4591** 172

[5] Brucherseifer M, Nagel M, Bolivar P H et al 2000 *Appl. Phys. Lett.***77** 4049

[6] Linden S, Enkrich C, Wegener M et al 2004 *Science* **306** 1351

[7] Moser H O, Casse B D F, Wilhelmi O and Saw B T 2005 *Phys. Rev. Lett.* **94** 063901

[8] Yang Y P, Yan W, Xu X L, Shi Y L and Wang L 2005 *Chin. Phys. Lett.* **22** 1401

[9] SUN Y M, MAO Z L, HOU B H, LIU G Q and WANG Li 2007 *Chin. Phys. Lett.* **24** 2414

[10] BRADLEY FERGUSON and XI-CHENG ZHANG nature materials VOL 1SEPTEMBER 2002

[11] Lang L Y, Xing Q X, Li S X, Mao F L, Chai Lu, and Wang Q Y 2004 *Chin. Phys. Lett.* **11** 0677

Detection and identification of selected alcohols using terahertz time-domain spectroscopy

Qing Miao[1], Lu Tian[1], Kun Zhao[1,2,4,5], Qingli Zhou[3], Yulei Shi[3], Dongmei Zhao[3], Songqing Zhao[1] and Cunlin Zhang[3]

[1] Laboratory of Optic Sensing and Detecting Technology, College of Science, China University of Petroleum, Beijing 102249, People's Republic of China
[2] State Key Laboratory of Heavy Oil Processing, China University of Petroleum, Beijing 102249, People's Republic of China
[3] Beijing Key Laboratory for Terahertz Spectroscopy and Imaging, Key Laboratory of Terahertz Optoelectronics, Ministry of Education, Department of Physics, Capital Normal University, Beijing 100048, People's Republic of China
[4] International Center for Materials Physics, Chinese Academy of Sciences, Shenyang 110016, People's Republic of China

E-mail: zhk@cup.edu.cn (Kun Zhao)

Abstract. The optical property and spectroscopy of selected kinds of alcohols are studied based on the terahertz time-domain spectroscopy (THz-TDS) in the spectral range of 0.5-2.5 THz. The B3LYP method in conjunction with the 6-31G (d) basis set was used to optimize the geometric structure of samples and simulate the alcohols spectrum. No absorption peak occurs in THz wave band based on our simulation, suggesting that the present experimental absorption peaks resulted from the intermolecular interaction.

1. Introduction

The alcohols such as 2-propanol, 1,2-popanediol and glycerol, as solvents and raw materials for synthetic resins, esters and other substances, are widely used in pharmaceutical, cosmetics, plastics, fragrances, paint and electronics industries and so on. Since the spectrometers have played a great role in chemical analysis, types of spectroscopy including raman scattering, mass spectroscopy, emission spectroscopy, chromatogram, ultraviolet absorption, infrared absorption, and nuclear magnetic resonance came into widespread in chemical area. At present much attention has been paid to the terahertz (THz) spectroscopic studies of chemical products. THz spectroscopy can serve as a potential tool in petroleum products identification, explosives and contaminations inspection, quality control, pharmaceutical analysis, medical diagnostics and security inspection or illegal drug detection [1-5].

THz time-domain spectroscopy (THz-TDS) as a new detection technology can effectively obtain amplitude, phase and structural information of the materials. Many rotational and vibrational spectra of organic molecules and related compounds lie in the THz range. In addition, THz-TDS provides absorption coefficient of a sample with high signal-to-noise ratio without using the Kramers-Kroning relation, which is ideal for the measurement of optical properties.

[5] Author to whom any correspondence should be addressed.

In this paper, we studied the optical property and spectroscopy of selected alcohols in terms of their different kinds, based on the THz-TDS. By applying a numerical fast Fourier transform (FFT), the optical constants of selected alcohol have been calculated. Furthermore, the simulation has been carried out using the GAUSSIAN 03 software package, where the B3LYP method was used in conjunction with the 6-31G (d) basis set.

2. Experimental procedure

The THz spectrometer in the transmission mode is used to investigate the optical properties of samples. A spectra physics regenerative amplifier system produces 800 nm pulses of 100 fs duration with 1 kHz repetition rate. The source beam is split into two portions, corresponding to THz generation and probe beams, respectively. THz-TDS is based on a ZnTe emitter for THz generation and electro-optic sampling [6]. The THz radiation is detected by free-space electro-optic sampling in a <110> ZnTe crystal. Then, the signal is collected by a lock-in amplifier with phase locked to an optical chopper [7]. The path with THz radiation is enclosed and purged with dry nitrogen. THz-TDS allows us to measure both the phase and the amplitude of the THz pulses propagating through the sample and reference, respectively. The frequency-dependent refractive index and absorption coefficient are extracted from the derivation of Fresnel law. By comparison with reference and sample pulse, and use of a numerical FFT, the absorption coefficient can be calculated [8].

Figure 1 shows the different molecular structures of alcohols due to the OH group numbers, which affect the formation of hydrogen bonds in the liquid state. All these chemicals were of analytical reagent and were used without further purification. The liquid samples were put into polyethylene cells with a side thickness less than 0.5 mm, which is transparent for visible light and has a low refractive index and a low THz absorption. And the sample thickness was fixed at 0.2 cm.

3. Results and discussion

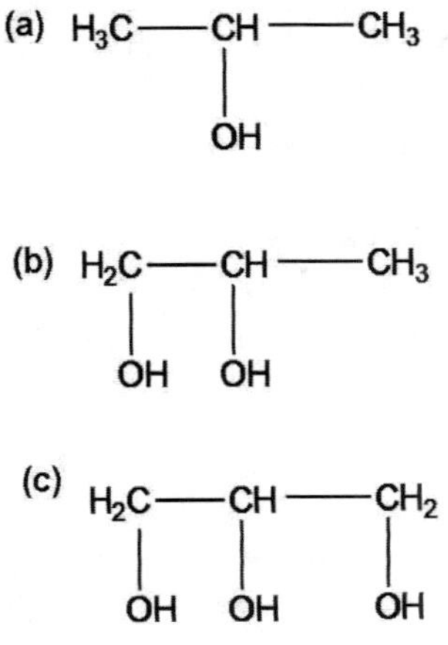

Figure 1. Molecular structures of (a) 2-propanol, (b) 1, 2-popanediol and (c) glycerol.

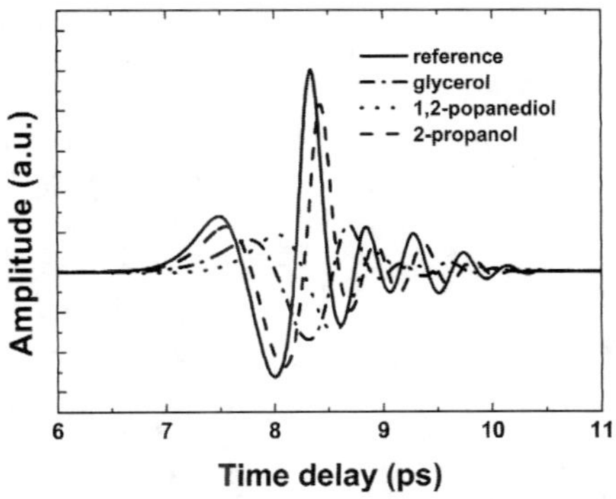

Figure 2. The THz pulses of reference in free cell and the signals of alcohol.

Figure 2 displays the typical temporal profiles of THz waves. Compared with that of the reference THz pulse of free cell the amplitude of the THz wave in the time domain is attenuated

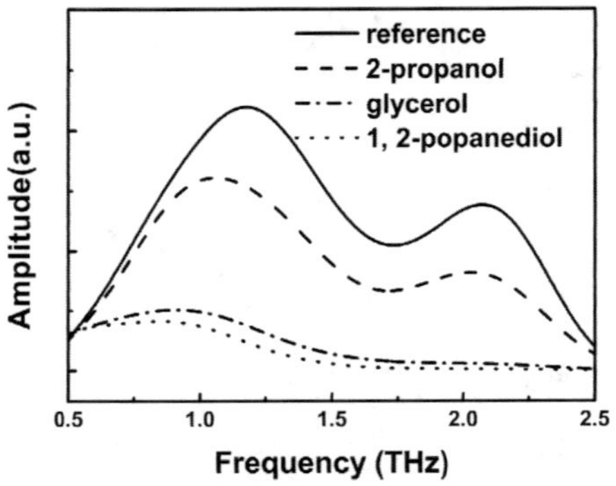

Figure 3. The frequency-dependent spectral amplitudes of reference and the alcohols obtained by FFT.

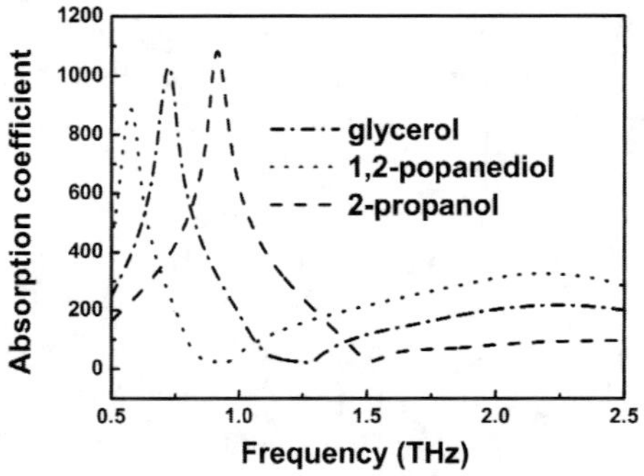

Figure 4. The calculated absorption coefficient of different alcohol in the 0.5-2.5 THz region.

when the THz wave passes through the alcohol samples. At the same time, the pulses of the samples relative to the reference pulse show time delays, 0.14 ps, 0.70 ps, and 0.27 ps for 2-propanol, 1,2-popanediol and glycerol, suggesting that the physical and chemical properties of the samples differ from each other [9]. In addition, the scattering of sample cause the time-domain wave broaden.

Subsequent FFT of the waveforms yields the frequency-dependent amplitude as shown in figure 3. The differences in absorption and surface reflectivity resulted in the dropped amplitude for three samples. The different frequency-domain waveform indicates that the alcohols have different properties in some aspects such as refractive index, absorption coefficient and dielectric constant.

The absorption coefficients of the alcohol in 0.5-2.5 THz were shown in figure 4. The absorption peaks were found at 0.9, 0.58 and 0.73 THz with intensity of 1143, 927 and 1096 for 2-propanol, 1,2-popanediol and glycerol, respectively, which were due to rotation and vibration of the molecule. To further reveal the origins of those characteristic spectra, the simulation has been carried out using the GAUSSIAN 03 software package. The B3LYP method was used in conjunction with the 6-31G (d) basis set. Simulated result did not show significant absorptions in THz wave band, suggesting that the present experimental absorption peaks resulted from the intermolecular interaction. The alcohols we used form a network-like local structure in the liquid through intermolecular hydrogen bonds. The hydrogen bonds often impact on the location and intensity of peak, leading to stretching vibration to move to low-frequency. And a hydrogen bond X-H$\cdots$Y can be formed between a group X-H with a proton H and a proton accepted group Y. Y is usually electronegative oxygen atoms or larger electronegative atoms. This effect can average the electron density, thus reducing bond force constant, then frequency decreased. And the hydrogen bond is stronger, the more frequency reduces. In our paper, the absorption peaks can be clearly seen below 1.0 THz, so this result indicated that the low-frequency vibration of sample is sensitive to the structure in hydrogen-bonded liquids.

4. Conclusions

In summary, the different kinds of selected alcohols have been studied using the THz-TDS. The obtained results indicated that the alcohols were sensitive in the THz range, and theoretical studies on the vibration spectra suggested that the hydrogen bond resulted in the absorption peaks of the samples. THz-TDS technology applied to alcohols analysis has a potentially significant impact on the field of chemicals, chemical products, raw materials and chemical engineering.

Acknowledgments
This work has been supported by Program for New Century Excellent Talents in University, National Natural Science Foundation of China, Research Fund for the Doctoral Program of Higher Education and State Key Laboratory of Heavy Oil Processing, China University of Petroleum.

References

[1] Zhou Q L, Zhang C L, Mu K J, Jin B, Zhang L L, Li W W and Feng R S 2008 *Appl. Phys. Lett.* **92** 101106
[2] Wang W N, Yan H T, Yue W W, Zhao G Z, Zhang C L and Zhang X C 2005 *Sci. China Ser. G: Phys. Mech. Astron.* **48** 585
[3] Lu M H, Shen J L, Li N, Zhang Y, Zhang C L, Liang L S and Xu X Y 2006 *J. Appl. Phys.* **100** 103104
[4] Tian L, Zhou Q L, Jin B, Zhao K, Zhao S Q, Shi Y L and Zhang C L 2009 *Sci. China Ser. G: Phys. Mech. Astron.* **52** 1938
[5] Bao R M, *et al* 2010 *Scientia Sinica Phys. Mech. Astron.* **40** 1
[6] Zhou Q L, Shi Y L, Jin B and Zhang C L 2008 *Appl. Phys. Lett.* **93** 102103
[7] Zhang C L, Zhang Y and Zhou Q L 2008 *National Defense Industry Press* **64**
[8] Hu Y, Wang X H, Guo L T, Zhang C L, Liu H B and Zhang X C 2005 *Acta. Phys. Sin.* **54** 4124
[9] Al-Douseri F M, Chen Y Q and Zhang X C 2006 *Int. J. Infrared Millimeter Waves* **27** 481

Journal of Physics: Conference Series **276** (2011) 012221 doi:10.1088/1742-6596/276/1/012221

Design of an on-line measuring system for 0.14THz high-power terahertz pulse

Guangqiang Wang[1], Jianguo Wang[2,3], Xiaoze Li[2], Changjiang Tong[2] and Xuefeng Wang[2]

[1] Department of Engineering Physics, Tsinghua University, Beijing 100084, P.R.China;
[2] Northwest Institute of Nuclear Technology, Xi'an 710024, P.R.China;
[3] School of Electronics and Information Engineering, Xi'an Jiaotong University, Xi'an 710049, P.R.China

E-mail: wgq02@mails.tsinghua.edu.cn

Abstract. An on-line measuring system, including an aperture-coupling structure and a novel high-power pulse detector, is proposed in this paper to measure the output pulses from high-power 0.14THz surface wave oscillator (SWO). At first a T-type coupling structure between the TM_{01} mode of circular waveguide with radius of 6mm and TE_{10} mode of rectangular waveguide WR6 is designed. Based on loose coupling theory, the coupling degree of this structure is derived and calculated, reaching about -47dB with the aperture radius of 0.4mm and length of 0.5mm. The reasonable coincidence is found between the theoretical computation and numerical simulation employing the three-dimensional finite difference time domain method. Then a novel high-power terahertz pulse detector based on hot electron effect in semiconductors is developed for the detection of output pulses from T-type coupling structure. With hot electron theory, the working principle of the detector is elucidated, also its sensitivity is simply analyzed, showing that this detector is capable of handling the pulse power as high as 2kW. The present 0.14THz on-line measuring system would be convenient to monitor the terahertz pulse shape and pulse power during the application researches of SWO besides increasing the accuracy of its pulse power measurement.

1. Introduction

With the developments of terahertz technologies, there have been considerable demands on the compact high-power terahertz radiation sources due to their unique applications in the military security, fundamental science, communications, and so on [1]. However, typical photonic based and solid-state sources have rather low energy conversion rate, yielding output power with the maximum magnitude of only tens of watts, while the free electron laser based on vacuum electronics has the largest output power but also system size. So the other two kinds of vacuum electronic devices, namely relativistic Cerenkov radiation generators and gyrotrons, offer the promising candidates and become hot researches for their high-power output and compact configurations [2,3].

Recently, our research group has successfully designed and tested a 0.14THz high-power surface wave oscillator (SWO) [4], a type of relativistic Cerenkov radiation generators. According to the preliminary experimental results, the output pulses from SWO have the characteristics of high frequency, high-power, and nanosecond pulse duration, thus, their measuring methods and systems are needed. To directly detect the output pulses, this paper presents the design of an on-line measuring system for our 0.14THz SWO, which is composed of a T-type aperture-coupling structure and a novel high-power pulse detector. Using the loose coupling theory, the coupling degree of the designed T-type aperture-coupling structure is analyzed and calculated. Then a novel 0.14THz high-power pulse detector based on hot electron effect in n-Si is introduced with detailed illumination on the design of the sensing element and working principle of detector.

2. Aperture-coupling system

2.1. Coupling formulas of a single cylindrical hole

The small apertures are usually used to partially couple electromagnetic (EM) wave energy from output waveguides of sources for following measurements in the on-line measuring systems [5,6]. In general, the apertures are positioned on the common waveguide walls between different waveguides, schematically shown in figure 1. When the source wave in waveguide below encounters the aperture, wave diffraction occurs, exciting positive and negative waves simultaneously in waveguide above and changing the EM field distribution in waveguide below. According to the loose coupling theory, the aperture-coupling coefficient A of electromagnetic field in excited waveguide can be determined by the expression [7]

$$A^{\pm} = (-1)^{\pm}\frac{j\pi f}{2}(\mu p_u^* H_{2u}^{0\mp} H_{1u}^{0+} + \mu p_v^* H_{2v}^{0\mp} H_{1v}^{0+} + \varepsilon p E_{2n}^{0\mp} E_{1n}^{0+}) \quad (1)$$

where the superscripts "+" and "-" denote positive wave and negative wave, respectively, the subscripts 1 and 2 separately represent parameters in source waveguide and excited waveguide, f is the frequency of EM wave, ε is the permittivity of medium inside waveguide, μ is the permeability, (u,v,n) is the local coordinates at the aperture where n is normal direction of the cross-section of the aperture, p^* are the tangential magnetic polarizability of the aperture, p is the normal electrical polarizability, H^0 and E^0 are normalized magnetic and electric fields at the aperture neglecting the effect of the hole on the EM field in waveguides, respectively. For a cylindrical hole with radius r and length h, the p^* and p are [7]

$$p^* = 4r^3\left|10^{-0.8\sqrt{1-(3.41rf/c)}h/r}\right|/3, \quad p = 2r^3\left|10^{-1.045\sqrt{1-(2.61rf/c)}h/r}\right|/3 \quad (2)$$

where c is the speed of light. Finally, the coupling degree C of the hole is calculated by

$$C^{\pm} = 10\lg\left|A^{\pm}\right|^2 = 20\lg\left|A^{\pm}\right| \quad (3)$$

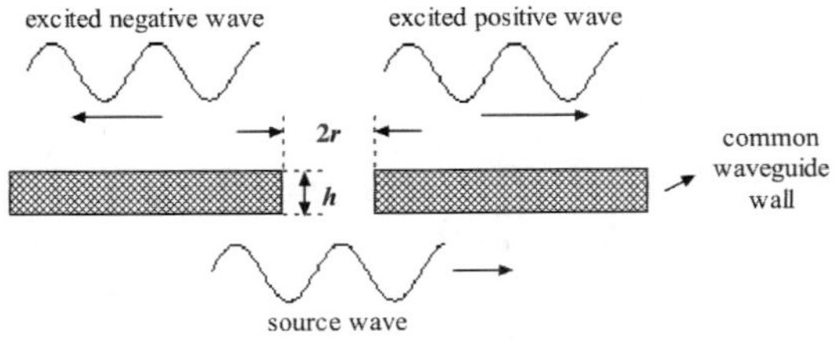

Figure 1. Sketch map of aperture-coupling on the common waveguide wall

2.2. Design and analysis of a T-type aperture-coupling structure

Suppose that the 0.14THz SWO is operated near the π mode of TM_{01} wave and has an overmoded cylinder waveguide output [4], the aperture-coupling structure should be able to effectively couple the energy from TM_{01} mode of circular waveguide to the TE_{10} mode of rectangular waveguide WR6. Through the analysis of EM field distribution at both sides of the aperture, a T-type aperture-coupling structure, where one port of waveguide WR6 is on the wall of cylinder waveguide and a cylindrical hole in the center of the interface penetrates it, is designed for the sake of easy fabrication and installation, schematically shown in figure 2. Actually the angle between the direction along the narrow wall of rectangular waveguide and the length of the cylinder waveguide, which is denoted as θ in the following analysis, would be adjustable to acquire optimal coupling degree.

The selected coordinates in each waveguide and at the aperture are also displayed in figure 2. Obviously, the local coordinates (u,v,n) correspond to the coordinates (φ,z,ρ) in cylinder waveguide while coordinates (u,v) and (x,y) lie at an angle of θ with n and z' coinciding in rectangular waveguide. Consequently, the EM field components in Eq. (1) can be replaced, yielding that,

$$A^{\pm}=(-1)^{\pm}\frac{j\pi f}{2}[\mu p_u^*(H_x^{0\mp}\sin\theta-H_y^{0\mp}\cos\theta)H_\varphi^{0+}+\mu p_v^*(H_x^{0\mp}\sin\theta+H_y^{0\mp}\cos\theta)H_z^{0+}+\varepsilon pE_{z'}^{0\mp}E_\rho^{0+}] \quad (4)$$

For TM_{01} mode in cylinder waveguide, the EM field components normalized to the power flow are derived as follows:

$$\begin{cases} E_\rho=-\dfrac{j\sqrt{2\beta_{01}}}{\sqrt{\omega\varepsilon\pi}R}\dfrac{J_0'(x_{01}\rho/R)}{J_0'(x_{01})}e^{-j\beta_{01}z} \\ E_z=\dfrac{\sqrt{2}x_{01}}{\sqrt{\omega\beta_{01}\varepsilon\pi R^2}}\dfrac{J_0(x_{01}\rho/R)}{J_0'(x_{01})}e^{-j\beta_{01}z} \\ H_\phi=-\dfrac{j\sqrt{2\omega\varepsilon}}{\sqrt{\beta_{01}\pi}R}\dfrac{J_0'(x_{01}\rho/R)}{J_0'(x_{01})}e^{-j\beta_{01}z} \\ E_\varphi=H_\rho=H_z=0 \end{cases} \quad (5)$$

where ω is the angular frequency of EM wave, β_{01} is the longitudinal wavenumber of TM_{01} mode, R is the radius of the cylinder waveguide, J_0 is zero order Bessel function, x_{01} is the first root of $J_0=0$, and J_0' is its first differential. The normalized EM field components for TE_{10} mode in rectangular waveguide are

$$\begin{cases} E_y=-\sqrt{\dfrac{\beta_{10}'}{\omega\mu}}\dfrac{2j\eta_{TE_{10}}}{\sqrt{ab}}\sin(\dfrac{\pi}{a}x)e^{-j\beta_{10}'z'} \\ H_x=\sqrt{\dfrac{\beta_{10}'}{\omega\mu}}\dfrac{2j}{\sqrt{ab}}\sin(\dfrac{\pi}{a}x)e^{-j\beta_{10}'z'} \\ H_{z'}=\sqrt{\dfrac{\beta_{10}'}{\omega\mu}}\dfrac{2\pi}{\omega\mu a\sqrt{ab}}\cos(\dfrac{\pi}{a}x)e^{-j\beta_{10}'z'} \\ E_x=E_{z'}=H_y=0 \end{cases} \quad (6)$$

where a and b are the length of the wide wall and narrow wall of rectangular waveguide, respectively, β_{10}' is the longitudinal wavenumber of TE_{10} mode, and η_{TE10} is the wave impedance. Substituting Eqs. (5) and (6) to (4), one can get

$$\left|A^{\pm}\right|=(-1)^{\pm}\frac{j\pi f\mu p_u^*}{2}H_x^{0\mp}H_\varphi^{0+}\sin\theta=\frac{2r^3\left|10^{-0.8\sqrt{1-(3.41rf/c)}h/r}\right|\omega\left|\cos\theta\right|}{3Rc}\sqrt{\frac{2}{ab\pi}}\sqrt{\frac{\beta_{10}'}{\beta_{01}}} \quad (7)$$

In terms of the structural characters of the T-type aperture-coupling, the negative wave excited in rectangular waveguide would be fully reflected by the metal wall of the cylinder waveguide, then superimposed on the positive wave. Thus, the total field coupling coefficient is

$$|A| = \frac{4r^3 \left|10^{-0.8\sqrt{1-(3.41rf/c)}h/r}\right| \omega |\cos\theta|}{3Rc} \sqrt{\frac{2}{ab\pi}} \sqrt{\frac{\beta_{10}^{'}}{\beta_{01}}} \tag{8}$$

Evidently there is only magnetic coupling in the designed T-type aperture-coupling structure, and the maximum coupling degree is achieved when θ is 0, namely the narrow wall of the rectangular waveguide parallels the longitudinal direction of cylinder waveguide. In this case, the coupling degree is obtained

$$C = 20\lg\left(\frac{4r^3 \left|10^{-0.8\sqrt{1-(3.41rf/c)}h/r}\right| \omega}{3Rc} \sqrt{\frac{2}{ab\pi}} \sqrt{\frac{\beta_{10}^{'}}{\beta_{01}}}\right) \tag{9}$$

In our design, the following geometric parameters of the waveguides and aperture are chosen: a=1.651mm, b=0.8255mm, R=6mm, r=0.4mm, and h=0.5mm. The coupling degree calculated by Eq. (9) in the frequency range 0.113-0.173THz is shown in figure 3, represented by dotted line. Along with the increase of the frequency the coupling degree increases approximately linearly, ranging from -52dB to -42dB. To validate the theory analysis above, a three-dimensional EM finite-difference time-domain method [8] is used to numerically simulate the T-type aperture-coupling structure. The simulation result is also shown in figure 3 corresponding to the solid line, agreeing with the theory calculation fairly well. A discrepancy no more than 1.5dB is found probably caused by the following two reasons: 1)the attenuation of polarizability along the longitudinal direction of the hole in Eq. (2) is an approximate estimation [7]; 2)the cylinder waveguide with radius of 6mm belongs to overmoded waveguide and has more than 200 propagation modes in 0.14THz frequency range, leading to other modes besides TM_{01} mode excited in cylinder waveguide by the effect of the hole on the waveguide wall.

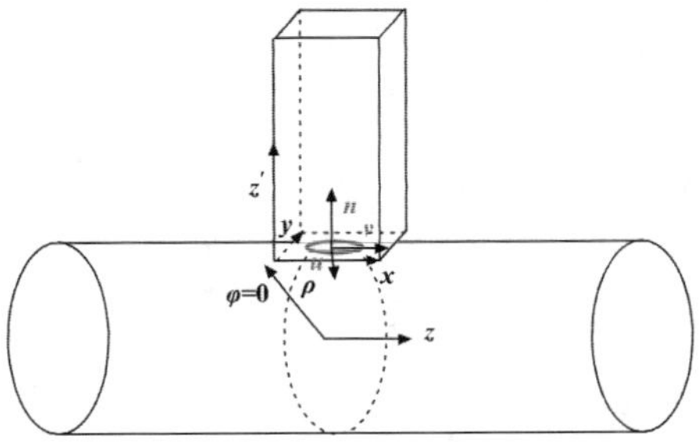

Figure 2. Schematic diagram of T-type aperture-coupling structure

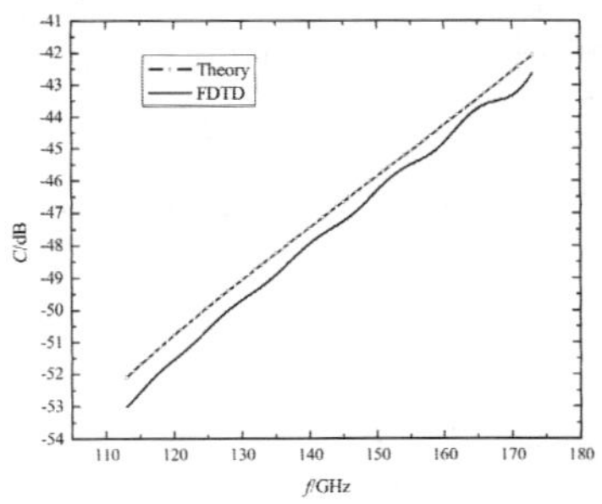

Figure 3. Coupling degree of T-type aperture-coupling structure

3. Novel high-power terahertz pulse detector

Based on the simulation results of our SWO, it has an output power of at least several megawatts around the frequency of 0.147THz [4]. Accordingly, the coupled energy from SWO employing the foregoing T-type aperture-coupling structure would get the power level of tens of watts, much bigger than the bearing power level of conventional diode detector. So large attenuation is needed, resulting in the accuracy decrease of terahertz pulse power measurement.

To solve this problem, a novel high-power terahertz pulse detector based on hot-electron effect in semiconductors gives an alternative device, illustrated in figure 4(a). Referred to the resistance sensor in W-band [9], the developed detector in 0.14THz frequency band comprises a standard waveguide WR6, a sensing element made of n-type silicon, and a bias constant current source. The sensing element, key component of this detector, is mounted on the center of wide wall of the waveguide WR6, the cross-section of which is shown in figure 4(b). Two same n-Si cubes with ohmic contact are shorted with copper foil on the upper contact, while the bottom contact of one cube is grounded with wide wall and the other is isolated from it. Connected with a constant current source, the feeding and output signal measurement of the sensing sample are fulfilled.

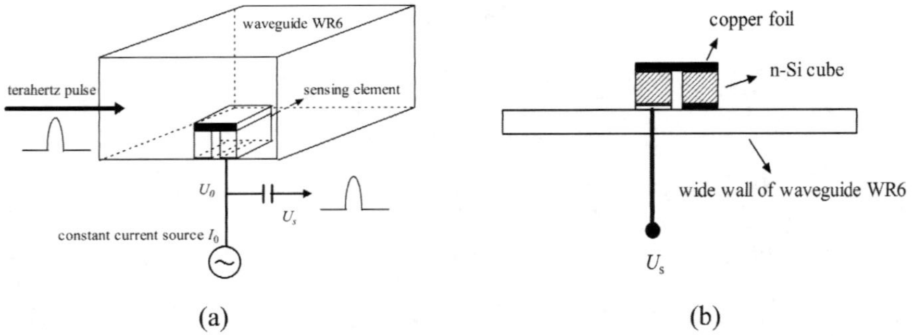

Figure 4. (a)The schematic diagram of the novel detector and (b) cross-section of the sensing element

At first, the constant current I_0 is applied on the sensing element, forming a DC voltage drop U_0. When a high-power 0.14THz pulse propagates within the waveguide WR6, the TE_{10} mode is excited, where the electric field parallels the direction along the narrow wall. According to the hot-electron effect in semiconductors [10], the high electric field heats the electrons in sensing element, producing so-called hot electrons and resulting in the increase of its resistance. Simultaneously, the voltage drop on the sensing element changes with a shape the same as the envelop of the terahertz pulse, namely a voltage pulse with amplitude of U_s is superposed on initial voltage U_0. Therefore, so long as the Joule heat effect on the sensing element caused by measured pulse is ignorable, this voltage pulse fully indicates the envelop shape and amplitude of the 0.14THz high-power pulse. With careful calibration in 0.14THz frequency band, the designed detector can be used to measure the power of 0.14THz high-power pulses.

Through simple circuit analysis of the detector, the relation between U_s and U_0 is written as

$$\frac{U_s}{U_0} = \frac{I_0 \Delta R_0}{I_0 R_0} = \frac{\Delta R_0}{R_0} = \xi P \tag{10}$$

where R_0 is the initial resistance of the sensing element, P is the power of the measured pulse, and ξ is the relative sensitivity of the detector, which is the function of the frequency and power of the measured pulse. When P is rather small, the heated electrons are called warm electrons. In this case, the relative sensitivity is independent of P [9], just like the diode detector working in the region of square law.

The researches on the hot-electron effect in n-Si show that the saturation field strength is about 20kV/cm and the field strength in warm electron region is less than 1kV/cm [10]. However, there is atmospherical breakdown in waveguide WR6, and the maximum electric field strength is about 18kV/cm considering the standing wave and atmospheric humidity in waveguide [11]. Thus the detector here could bear the electric field with maximum strength of 18kV/cm while its linear working region is that the electric field strength is no more than 1kV/cm. For TE_{10} mode in an empty rectangular waveguide, the maximum electric field is described by

$$E_{\max} = \frac{480\pi}{ab\sqrt{1-(f_c/f)^2}} P \tag{11}$$

where f_c is the cut-off frequency of TE_{10} mode in waveguide. Substituting the corresponding electric field to Eq. (11), we obtain that the maximum power that the detector can handle and works in the linear region is about 2kW and 7W, respectively. As the distributed capacitance and inductance are very small in the sensing element, the response time of this detector is mainly decided by the pickup circuit of the output voltage pulse. With appropriate design on the detecting circuit, the novel detector could directly measure the nanosecond pulses with power as high as 2kW in 0.14THz frequency band. The manufacture and calibration technologies of the detector are under research now.

4. Conclusions

In this work, an on-line measuring system for high-power 0.14 THz SWO is designed, which is composed of the T-type aperture-coupling structure and novel high-power pulse detector for 0.14THz frequency band. With aperture radius of 0.4mm and length of 0.5mm, the T-type aperture-coupling structure has the coupling degree from -52dB to -42dB, attested by the three-dimensional EM finite difference time domain method. Then a novel detector based on hot-electron effect in n-Si is proposed. The capability of handling the maximum power of about 2kW is highlighted with detailed analysis on the working principle of the detector. Combining the novel detector with T-type aperture-coupling structure, the designed 0.14THz on-line measuring system can directly measure the output pulses from our megawatt-class SWO without attenuation, increasing the accuracy of the power measurements, and also it is very suitable to monitor the terahertz pulse shape and pulse power during the applications of SWO.

References

[1] Siegel H P 2002 *IEEE Trans. Micro. Theory Tech.* **50** 910-928
[2] Bratman V L, Denisov G G, Ofitserov M M, Korovin S D, Polevin S D and Rostov V V 1987 *IEEE Trans. Plasma Sci.* **PS-15** 2-15
[3] Bratman V, Glyavin M, Idehara T, Kalynov Y, Luchinin A, Manuilov V, Mitsudo S, Ogawa I Saito T, Tatematus Y and Zapevalov V 2009 *IEEE Trans. Plasma Sci.* **37** 36-43
[4] Tong C J, Li X Z, Wang J G, Wang X Z and Wang G Q 2009 *Proc. SPIE* **7385** 101-106
[5] Yuan N C and Hao C J 1997 *Acta Electronica Sinica* **25** 114-115
[6] Cao N S, Luo Y and Wang J X 2008 *High Power Laser and Particle Beams* **20** 637-640
[7] Huang H J 1963 *Microwave Principles* (Beijing: Science Press) pp 211-306
[8] Bing W H and Lu W D 1997 *J. Nature Sci. Heilongjiang Univ.* **14** 60-64
[9] Kancleris Ž, Simniškis R, Dagys M and Tamošiūnas V 2007 *IET Microw. Antennas Propag.* **1** 757-762
[10] Liu E K, Zhu B S and Luo J S 1994 *Physics of Semiconductors* (Beijing: Publishing House of National Defence Industry) pp 105-110
[11] Ying J N, Gu M Z and Zhang K Q 1994 *Microwave and Optical Guided Wave Technology* (Bejing: Publishing House of National Defence Industry) pp 89-91

Journal of Physics: Conference Series **276** (2011) 012222 doi:10.1088/1742-6596/276/1/012222

Surface profile measurement by terahertz interferometric phase imaging

Yingxin Wang[1,2], Ziran Zhao[2], Zhiqiang Chen[2], Li Zhang[2] and Jingkang Deng[1]

[1] Department of Physics, Tsinghua University, Beijing, 100084, China
[2] Key Laboratory of Particle & Radiation Imaging (Tsinghua University), Ministry of Education, Department of Engineering Physics, Tsinghua University, Beijing, 100084, China

E-mail: wangyingxin2000@tsinghua.org.cn

Abstract. We demonstrate an interferometric phase imaging technique by using a continuous-wave terahertz system with a far-infrared gas laser as the radiation source. A terahertz Michelson interferometer is constructed based on a reflection imaging system. By analyzing the measured fringe patterns, the phase distribution information can be obtained and then used to reconstruct the profile of the object surface under test. We also show that, for imaging a nearly transparent object with terahertz radiation, our technique can provide a high contrast image which would be quite valuable for accurate identification of the examined object.

1. Introduction

Non-contact and non-destructive characterization of the surface topography of an object plays an important role in a wide range of fields, such as industrial inspection, quality control, optical component testing and biomedical engineering. One common means to achieve this task involves determining the wavefront distortion of an illumination light beam induced by the surface under test. In visible and infrared regions, many interferometric techniques have been developed to detect the spatial phase variation of the optical beam and have been widely used for surface profile measurement, including holographic [1], moiré [2], speckle [3], and fringe projection [4] interferometry. However, these optical techniques encounter problems for the surface with roughness on the order of the wavelength or more. Due to the longer wavelength of terahertz radiation, it is insensitive to this roughness. Also terahertz radiation can penetrate most non-metallic and non-polar materials, which enables the surface of a concealed object to be measured. Consequently, sensing with terahertz waves can serve as a complementary tool to conventional optical techniques for surface profilometry.

At terahertz frequencies, phase-shift interferometry for few-cycle optical pulses has been demonstrated to enhance the depth resolution dramatically in tomographic measurements [5]. Recently, continuous-wave (CW) terahertz heterodyne profilometry [6] and fiber-endoscopy [7] based on interferometric imaging systems have been reported to reconstruct the object surface profile. Using CW terahertz interferometer, noncontact sensing vibrations of macroscopic objects behind optically opaque barriers is also possible [8]. These investigations indicate that terahertz interferometric imaging has a variety of valuable applications. But the above-mentioned CW systems worked in the submillimeter region (below 1 THz). This limits the lateral spatial resolution of imaging. In this report, we present a terahertz interferometric phase imaging (TIPI) technique that can be used to determine the object surface profile as well as image a weak absorption or nearly transparent object. The phase measurement is performed by a Michelson interferometer that is constructed from a CW reflection imaging system employing a far-infrared gas laser as the source of 2.52-THz radiation. We show that TIPI could reveal the surface profile with a high sensitivity and give a good contrast for the weak absorption object compared with the conventional terahertz absorption imaging so as to obtain its true material distribution.

2. Experimental setup

The schematic diagram of the TIPI experimental system is shown in figure 1(a). The interferometer operates in a Michelson configuration for reflection imaging. An optically-pumped far-infrared laser (FIRL 100, Edinburgh Instruments Ltd., UK) is employed for terahertz generation. Its output power can reach approximately 150 mW at 2.52 THz (corresponding to a wavelength of λ=118.8 μm). To address the problem of beam divergence, the terahertz beam is first focused by a gold-coated off-axis parabolic mirror and then collimated by a high-density polyethylene lens. After a beam splitter (a high resistivity silicon wafer), one part of the terahertz radiation is incident on a flat mirror M1 mounted on a linear translation stage, serving as the reference beam. Another part is guided to the sample under test at normal incidence by a focusing element. The reference beam and the object beam are recombined to produce interference and the interference signal is recorded via a Golay cell detector. The optical path difference (OPD) between these two beams is adjusted by translating the flat mirror in the reference arm. In order to obtain the two-dimensional surface profile of the sample, it is located at the focus of the object beam and raster scanned in a plane perpendicular to the beam axis. For measuring transparent object, an additional flat mirror M2 is placed behind the object to retroreflect the transmitted terahertz radiation. An optical chopper and a lock-in amplifier are used to improve the signal-to-noise ratio (SNR) of the system. Meanwhile, the chopper creates a modulated radiation input (with a frequency of 25 Hz) for the Golay cell. In addition, an attenuator set with four transmission levels (30%, 10%, 3%, and 1%) is inserted into the beam path to avoid saturation of the detector.

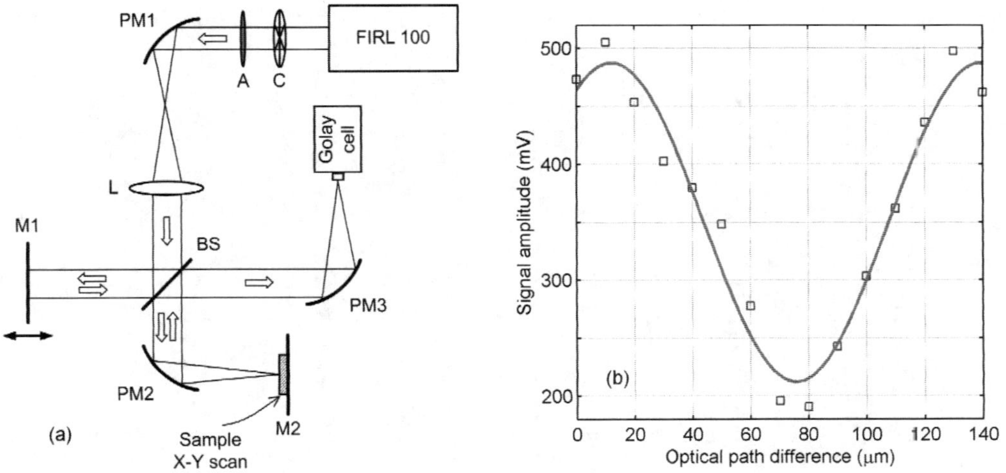

Figure 1. (a) Schematic diagram of the TIPI system. C, chopper; A, attenuator; PM, parabolic mirror; L, lens; BS, beam splitter; M, mirror. (b) Measurement of the interference intensity variation with respect to the OPD (square points) and the corresponding cosine fit curve (solid line).

By blocking the terahertz radiation in the reference and sample arm, respectively, we evaluated the light intensity ratio of the reference and sample beam. It was measured to be 1.24, close to unity, which satisfies the basic requirement that the interferometer has a high fringe contrast (i.e., high sensitivity). Figure 1(b) displays the measurement result of the interference intensity variation with respect to the OPD of the two beams. The measurement was performed by scanning the reference mirror M1 with a step size of 5 μm (corresponding to an OPD of 10 μm). Fitting the raw data points to a cosine function will yield an estimate of the wavelength of the CW terahertz radiation. It was obtained to be 127.5 μm, having a relative error of 7.3% compared to the theoretical value (118.8 μm). This slight inconsistency may arise from the position error of M1 and misalignment of the optical elements. The fringe contrast was calculated to be about 40%, which is sufficient for further analysis. Consequently, performance evaluation for the interferometer demonstrates the successful construction of our TIPI system.

3. Results and discussion

We select a plastic sample and a reflective metal surface as the test objects to illustrate how to image the weak absorption material as well as reconstruct the surface profile by using the TIPI technique. The weak absorption sample is a plastic sheet cut into the shape of character "T" [see region A in Figure 2(a)] and has a generally uniform thickness of 0.6 mm. The sample was held in a rectangular aperture [not shown in Figure2(a) except for one boundary marked with C] by a layer of adhesive tape and then mounted on a metal plate (region B). In the imaging experiment, we first blocked the reference beam and performed the conventional terahertz reflection imaging for the sample. The time constant of the lock-in amplifier was set to be 30 ms. The sample was raster scanned over an area of 37×45 mm^2 with a step size of 0.5 mm and a speed of approximately 10 pixels per second. Mirror M2 was fixed during the scan. Figure 2(a) shows the resulting terahertz image of the plastic sample and the metal plate. We can see that only the shape of the plastic sample is visible owing to the scattering effect occurring at the sample edge. Because of the high transmittance of plastic to terahertz radiation, no obvious intensity difference emerges between the sample region and the background region. Thus obtaining the true material distribution information for transparent object is impossible through such common absorption imaging. In other words, one could not distinguish which region in the image is plastic. Since the metal plate is covered with a layer of adhesive, it exhibits nonuniform gray level. The weak signal intensity around the aperture boundary originates from the strong attenuation of the sample holder.

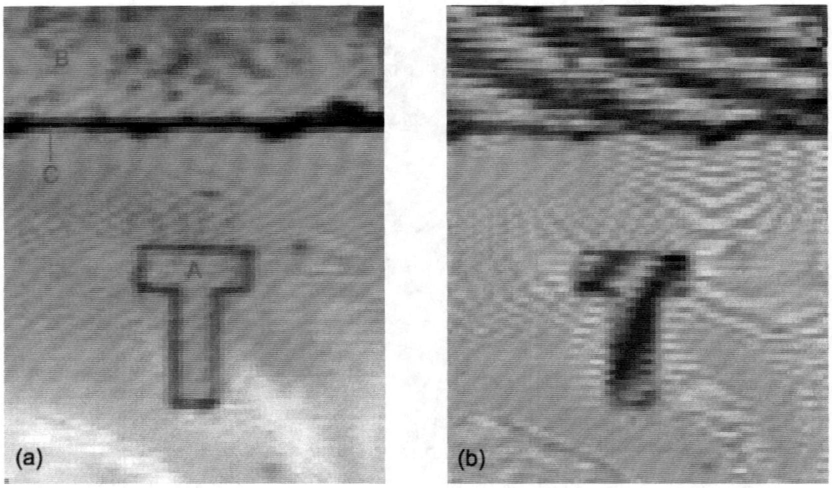

Figure 2. (a) Conventional terahertz reflection imaging result. A represents a plastic sheet; B represents a metal plate; C represents the edge of the sample holder. (b) Terahertz interferometric imaging result. The red line indicates the data adopted for surface profile reconstruction.

With the reference beam block removed, we acquired an interferometric image, as shown in Figure 2(b). The scanning parameters were the same as above. It is clear that fringe patterns appear both in the plastic sheet and the metal plate region. Note that if the plastic sheet has an absolutely uniform thickness, it will exhibit an almost constant difference in intensity compared with the background signal, which could help recognize the sample distribution. This difference is determined by the phase shift $\Delta\varphi$ of terahertz radiation induced by the sample and becomes more notable when $\Delta\varphi$ approaches π, considering the 2π ambiguity of $\Delta\varphi$. Assuming the refractive index and thickness of the sample are n and d, respectively, the phase shift can be expressed as $\Delta\varphi = 4\pi(n-1)d/\lambda$. We substitute the measured values, $n = 1.5$, and $d = 0.6$ mm, into the above relation and get $\Delta\varphi \approx 10\pi$. However, the existence of fringes shows that the plastic sheet has a varied thickness. In another aspect, reflections from the sample interfaces may also lead to such interference fringes [9]. It is difficult to separate these two contributions accurately. Anyway, compared with the conventional terahertz imaging, interferometric imaging for weak absorption materials can provide an image with a high contrast, from which we can obtain the information about the exact material distribution.

The fringe patterns in the metal plate region are generated from inclination of the plate plane, i.e., not perpendicular to the incident beam. The degree of inclination or, equivalently, the surface profile can be evaluated by demodulating this fringe pattern. We used the Fourier transform technique [10] for this purpose. The data along a horizontal line [as indicated by the red line in Figure 2(b)] were extracted for analysis. Figure 3(a) plots the signal amplitude variation along this line. After a filtering procedure in the spatial frequency domain, the phase distribution was calculated, as shown in Figure 3(b). Finally, the object surface profile could be formed with a phase unwrapping algorithm, as shown in Figure 3(c). Owing to the low SNR of the system, the measured profile looks less smooth. From this reconstruction result, we know that the plate surface has a maximum height of about one wavelength in the observation direction. In other words, the inclination angle of the metal plate along this direction is only about 3.2×10^{-3} rad.

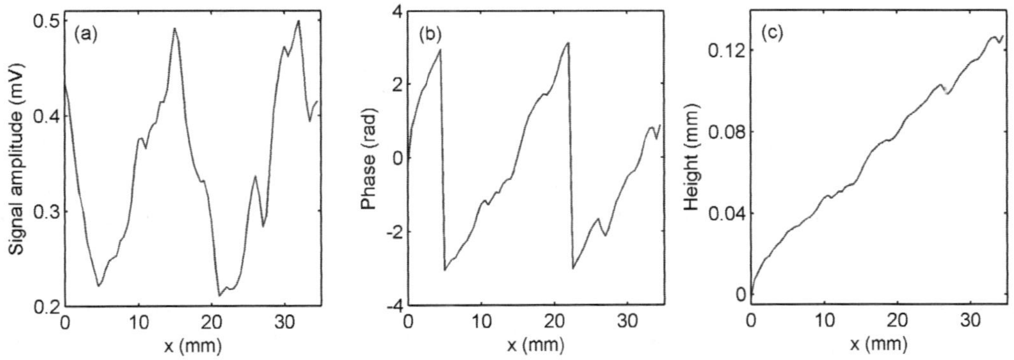

Figure 3. (a) Signal amplitude variation of the fringe pattern in the metal plate region along a horizontal direction indicated by the red line in Figure 2(b). (b) Phase distribution obtained by the Fourier transform method. (c) Reconstructed surface profile along the observation line.

4. Conclusion

In conclusion, we have presented an interferometric phase imaging technique at terahertz frequencies. An interferometric imaging setup based on a CW terahertz system has been constructed successfully. Experimental results show that our method can not only produce an image with a higher contrast for transparent objects compared to that acquired from the conventional absorption imaging, but also provide the surface profile information of the sample. Therefore, our method has a promising application potential in the fields of industrial inspection, quality control, etc.

Acknowledgments
This work is supported by China Postdoctoral Foundation (No. 20100470017).

References

[1] Cuche E, Bevilacqua F and Depeursinge C 1999 *Opt. Lett.* **24** 291
[2] Takasaki H 1973 *Appl. Opt.* **12** 845
[3] Creath K 1985 *Appl. Opt.* **24** 3053
[4] Indebetouw G 1978 *Appl. Opt.* **17** 2930
[5] Johnson J L, Dorney T D and Mittleman D M 2001 *Appl. Phys. Lett.* **78** 835
[6] Hils B, Thomson M D, Löffler T, Spiegel W, Weg C, Roskos H G, Maagt P, Doyle D and Geckeler R D 2008 *Opt. Express* **16** 11289
[7] Lu J Y, Kuo C C, Chiu C M, Chen H W, Hwang Y J, Pan C L and Sun C K 2008 *Opt. Express* **16** 2494
[8] Chen J C and Kaushik S 2007 *IEEE Photon. Technol. Lett.* **19** 486
[9] Song Q, Yu F, Zhao Y J, Zhang C L, Yu Q, Liu M and Liu X H 2009 *Proc. SPIE* **7443** 74432G
[10] Takeda M, Ina H and Kobayashi S 1982 *J. Opt. Soc. Am.* **72** 156

Journal of Physics: Conference Series **276** (2011) 012223 doi:10.1088/1742-6596/276/1/012223

The physical theory and propagation model of THz atmospheric propagation

R Wang[1,2,3], J Q Yao[1,2], D G Xu[1,2], J L Wang[1,2] and P Wang[1,2]

[1] College of Precision Instrument and Opto-electronics Engineering, Institute of Laser and Opto-electronics, Tianjin University, Tianjin 300072, China

[2] Key Laboratory of Opto-electronics Information Science and Technology, Ministry of Education, Tianjin University, Tianjin 300072, China

[3] Email:wangran19861014@163.com

Abstract. Terahertz (THz) radiation is extensively applied in diverse fields, such as space communication, Earth environment observation, atmosphere science, remote sensing and so on. And the research on propagation features of THz wave in the atmosphere becomes more and more important. This paper firstly illuminates the advantages and outlook of THz in space technology. Then it introduces the theoretical framework of THz atmospheric propagation, including some fundamental physical concepts and processes. The attenuation effect (especially the absorption of water vapor), the scattering of aerosol particles and the effect of turbulent flow mainly influence THz atmosphere propagation. Fundamental physical laws are illuminated as well, such as Lamber-beer law, Mie scattering theory and radiative transfer equation. The last part comprises the demonstration and comparison of THz atmosphere propagation models like Moliere(V5), SARTre and AMATERASU. The essential problems are the deep analysis of physical mechanism of this process, the construction of atmospheric propagation model and databases of every kind of material in the atmosphere, and the standardization of measurement procedures.

1. Introduction

Terahertz radiation is the electromagnetic wave in the frequency interval from 0.1 to 10 THz ($1THz=10^{12}Hz$). It lies in the so-called "unexplored" range between visible light and radio wave, due to the difficulty of generating and detecting techniques in this region [1]. Recently, basic research and development in terahertz technologies are now underway for an increasing wide variety of applications, including information and communication technology (ICT), biomedical field, safety and security, non-destructive examination, quality control of food and agricultural products, and global environmental monitoring [1,2]. The rapid development can be attributed to the nature of terahertz radiation, which offers the advantages of both microwave and light wave.

The atmospheric transmittance of terahertz wave now ranks among the most critical issues in the principal application of space communication and atmospheric remote sensing [2].

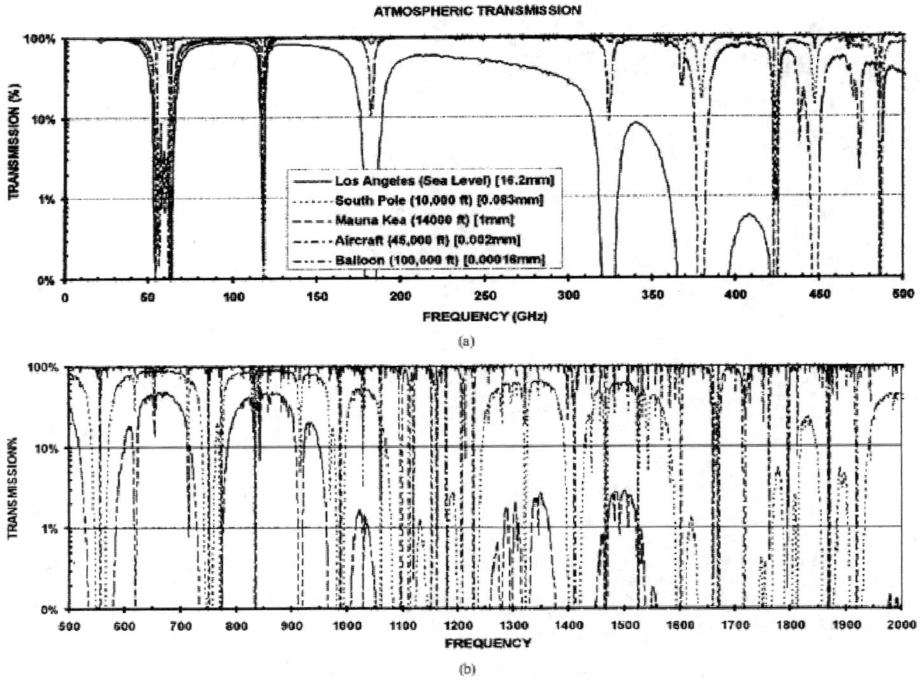

Figure 1. Atmospheric transmission in the terahertz region at various locations and altitudes (a) 0-500 GHz, (b) 500-2000GHz.

The atmospheric opacity severely limits the communication application at terahertz range, which has been shown in figure 1[3]. Terahertz communication will benefit from the high-bit-rate wireless technology which takes advantage of higher frequency and broader information bandwidth allowed in this range than microwave [4]. Some THz frequency bands can be used for communications between locations separated several kilometers, and knowing accurately the location and width of atmospheric transmission windows will be of great importance. Other terahertz waves propagate less than 1m due to the significant attenuation of rotational transitions of molecules in the air and can be used for certain applications such as short-range secure communication.

As for the atmospheric remote sensing, THz technology is important in the environmental monitoring of atmospheric chemical species (water, oxygen, ozone, chlorine and nitrogen compounds, etc.), cloud and ice as well as taking temperature and other measurement [2]. A special emphasis has been put on the troposphere and lower stratosphere to learn climate evolution and the influence of pollution or biomass burning on the atmospheric composition. The knowledge about atmospheric attenuation will illustrate the optimum bands for sensing systems while the material database will discriminated atmospheric components. The overview of the THz remote sensing from the National Institute of Information and Communications Technology (NICT) in Japan is given in figure 2[5].

Figure 2. Overview of NICT THz remote sensing.

With respect to the research on terahertz atmospheric propagation, there are three fundamental problems as follow: [6]

(1)To confirm the atmospheric transparency in the THz range and find out the air transmission windows for communicating and sensing system.

(2)To collect the spectroscopic fingerprinting of atmospheric molecules for Terahertz atmospheric monitoring.

(3)To improve the signal to noise ratio and restore the original signal from the observed signal by the process of deconvolution [7].

Based on these considerations, for many types of practical application, it is essential to understand the actual effects on the amplitude and phase of THz wave propagating through the atmosphere, which depends on the frequency of incident wave, gas components, and ambient temperature or barometric pressure in different atmospheric conditions.

This investigation aims to construct a terahertz radiation propagation model on the basis of the theoretical analysis and material database. It comprises the construction of an atmospheric radiative transfer equation as well as the collection of accurate spectroscopic parameters including standardization of the measurement protocol [8]. Section 1 outlines the significance and application of THz atmospheric propagation, including space communication and atmospheric remote sensing. Section 2 describes the fundamental theory in the THz atmospheric propagation, with the physical process of Lamber-beer law, Mie scattering theory and so on. The atmospheric absorption, emission and scattering are taken into account, and a special focus is put on the detailed derivation and physical significance of radiative transfer equation. Several THz atmospheric propagation model are introduced in Section 4, including Moliere, SARTre and AMATERASU. The last one has a strong heritage from the former two, respectively in the non-scattering case and scattering version. The conclusions are drawn in section 5 by giving the future evolutions and suggestions of further study in this region.

2. The fundamental physical theories

2.1. The framework of atmospheric propagation characteristics of THz-Wave

In the process of terahertz atmospheric propagation, there are several fundamental physical concepts and theories as figure 3:

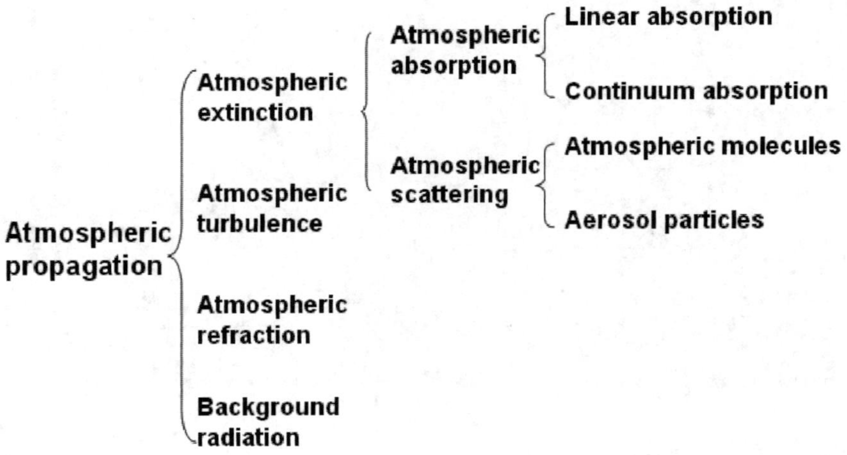

Figure 3. The fundamental physical concepts and theories.

(1)Atmospheric extinction

The atmospheric extinction consists of the absorption and scattering effects, described by Lamber-beer law and mainly causing the energy attenuation of incident wave. The differential and integral forms of the mathematical expression is

$$dI(v) = -\alpha_v(z)I(v)dz \qquad I_{r_1}(v) = I_{r_0}(v)\exp[-\int_{r_0}^{r_1}\alpha_v(z)dz] \tag{1}$$

$\alpha_v(z)$ is the extinction coefficient and can be expressed as the summation of the absorption and scattering coefficient. But in accordance with most literatures at present, $\alpha_v(z)$ only represents the absorption coefficient while the scattering effect has been discussed separately, so did we in this paper. $I_{r_0}(v)$ denotes the incident radiance entering the optical path (r_0, r_1) at the frequency v and $I_{r_1}(v)$ is the outgoing radiance. The opacity or optical thickness is defined as $\tau_v(r_0, r_1) = \int_{r_0}^{r_1}\alpha_v(z)dz$ and the transmission is $\eta_{r_0,r_1} = \frac{I_{r_1}}{I_{r_0}} = \exp[-\tau_v(r_0, r_1)]$. The atmospheric absorption involves the linear absorption and continuum absorption, which will be described particularly in the next part.

The scattering effect comprises the molecular Rayleigh scattering and the Mie scattering by aerosols. The former can be neglected with the wavelength of THz radiation being in the order of aerosols. Aerosol particles mainly refer to the solid and liquid particles suspending in the atmosphere, for example, dusts, salts, ice particles and water droplets, and the Mie scattering effect mainly depends on their size-distribution, complex refractive index and the wavelength of incident radiation.

(2)Atmospheric turbulence

The essence of turbulence effect is the impact of medium disturbance on the transmission of incoming wave, due to the fluctuation of atmospheric refractive index. Fully coherent light beams are sensitive to the properties of the medium through which they are propagating and the turbulence-induced spatial broadening is the major limiting factor in most applications. Partially coherent beams are less affected by atmospheric turbulence than fully ones [9].

(3)Atmospheric refraction

The atmospheric refraction results from the uneven distribution of air in horizontal and vertical directions. When passing through the atmosphere, the line of sight is refracted and bended towards the surface of the planets. Taking refraction into account will correct and promote the radiative transfer path with some elementary geometrical relationships, as figure 4 shows.

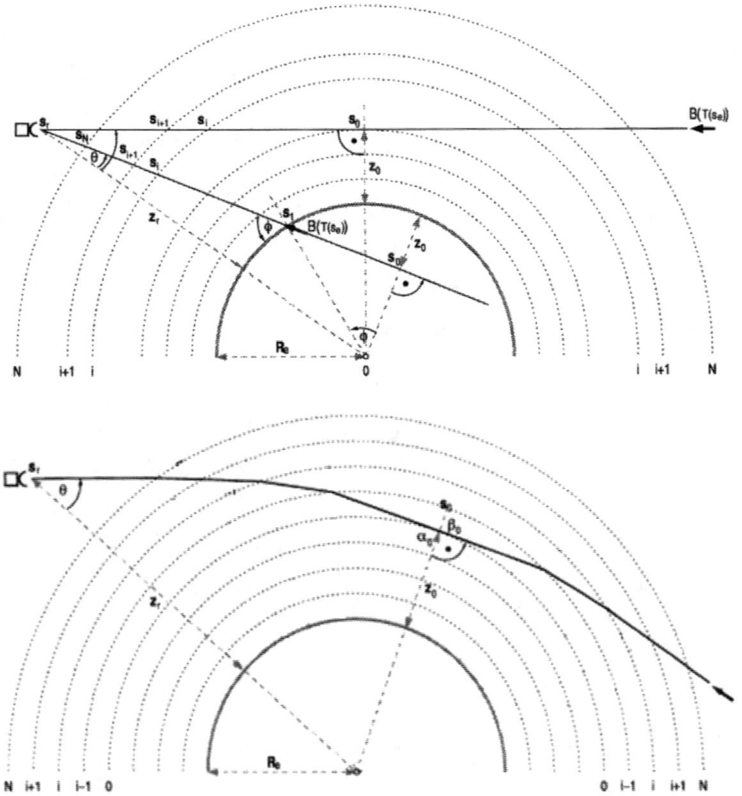

Figure 4. The radiation path and its modification due to atmospheric refraction.

(4)Background radiation

The background radiation of THz wave in atmosphere, depending on the geometries, mainly results from many kinds of electromagnetic radiation in the interstellar space or from the planet surface. For limb-sounding and up-looking, it is the cosmologic radiation at 3K, and for nadir-sounding (or down-looking), it is the earth surface emission.

In conclusion, the general idea to solve these problems above is to study the various effects independently and superpose them. Currently, most researches are mainly focused on the atmospheric extinction and the establishment of radiative transfer model, which will be introduced in the following text.

2.2. The absorption of water vapor

The linear and continuum absorption constitutes the atmospheric absorption of terahertz waves. The former is comprised most of the water vapor absorption lines in the air, which is due to the molecular rotational transitions. Their spectroscopic parameters, including the center frequency, oscillator intensity, and pressure broadening coefficient, should be measured by laboratory experiments precisely, as they will directly influence the accuracy of atmospheric propagation model [10]. There are several databases, such as JPL [11] and HITRAN [12], to stimulate the line by line absorption.

The continuum absorption is arbitrary because it is what remains after subtraction of linear contributions from the total absorption that can be measured directly [13]. Its generating mechanism is not sufficiently understood while several theories have proposed various causes, including anomalous far-wing absorption [14], absorption by dimmers and larger clusters of water vapor, and absorption by collisions between atmospheric molecules [15]. An empirical CKD model [16], proposed by Clough, Kneizys, and Davis, is applicable in a wide frequency range and has been proven successful in some aspects. For the simulation at frequencies below 400GHz, Liebe model [17] should be used for dry air and water vapor continua. Figure 5 [10] illustrates the discrepancy between radio-wave and infrared wave propagation models. The radio-wave model is calculated with JPL line catalog and Liebe model for continuum absorption while the infrared model is on the basis of HITRAN line catalog and CKD continuum model.

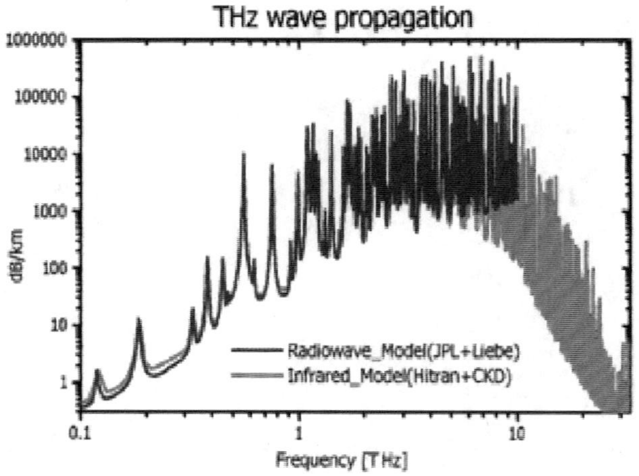

Figure 5. The linear and continuum absorption of THz wave from NICT.

It is difficult to measure spectroscopic parameters accurately in THz region due to the limitations of measurement instruments in the past. Currently, Terahertz Time-domain Spectroscopy (THz-TDS) technology and Fourier-transform Infrared Spectroscopy (FT-IR) have attracted a great deal of attention in precise measurement of spectroscopic parameters. The former is advantageous at low frequencies less than 3 THz, while Fourier transform spectroscopy works better at frequencies above 5 THz [18].

2.3. The derivation of radiative transfer equation

The radiative transfer equation is determined by absorption and emission processes occurring along the line-of-sight. It is the foundation of atmospheric propagation model, and the derivation is as follow [19,20]:

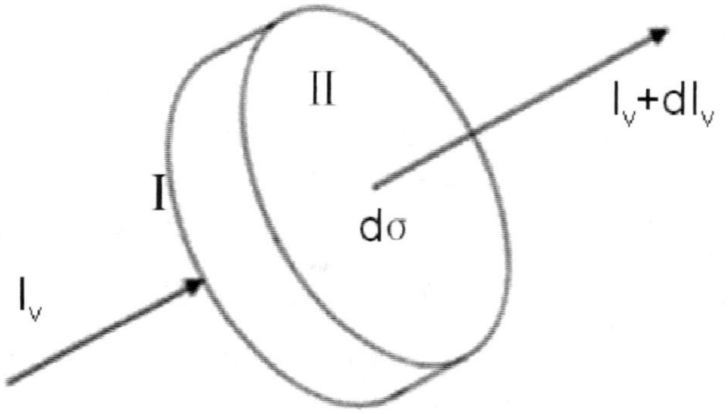

Figure 6. The input and output optical intensity.

As the figure above, the radiant energy of incident light in the surface I of an infinitesimal volume is:

$$dE^{in} = I_v d\omega d v d\sigma dt \quad (2)$$

where I_v is radiant intensity, $d\omega$ solid angle, dv frequency interval, $d\sigma$ basal area, and dt denotes the time of radiation. And the emergent radiant energy from surface II is:

$$dE^{out} = (I_v + dI_v) d\omega d v d\sigma dt \quad (3)$$

According to the Lamber-beer law, with the absorption coefficient α_v, the radiant energy absorbed by the medium is:

$$dE_\alpha = -\alpha_v dE^{in} dr = -\alpha_v I_v d\omega d v d\sigma dt dr \quad (4)$$

With the emission coefficient j_v, the radiant energy of medium emission is:

$$dE_e = j_v d\omega d v d\sigma dt dr \quad (5)$$

In accordance with energy conservation law, we get:

$$dE^{out} = dE^{in} + dE_e + dE_\alpha \quad (6)$$

Substituting equation (2)~(5) into equation (6):

$$dI_\nu d\omega d\nu d\sigma dt = j_\nu d\omega d\nu d\sigma dt dr + (-\alpha_\nu I_\nu) d\omega d\nu d\sigma dt dr \tag{7}$$

The source function is defined as $S_\nu \equiv j_\nu / \alpha_\nu$, and assuming the local thermodynamic equilibrium (LTE), it equals the Planck function, according to Kirchhoff's law:

$$S_\nu \equiv j_\nu / \alpha_\nu = B_\nu(T) \tag{8}$$

Given the definition of opacity or optical thickness: $d\tau_\nu = \alpha_\nu dr$, we get the differential form of radiative transfer equation from equation (7):

$$\frac{dI_\nu}{d\tau_\nu} = S_\nu - I_\nu \tag{9}$$

To solve this single-order partial differential equation along integral path (r_0, r_1), with the integral variable *r*, we get the integral form of radiative transfer equation:

$$I_\nu(r_1) = I_\nu(r_0)\exp[-\int_{r_0}^{r_1}\alpha_\nu(r)dr] + \int_{r_0}^{r_1}\exp[-\int_{r}^{r_1}\alpha_\nu(r')dr']\ S_\nu(r)\alpha_\nu(r)dr \tag{10}$$

Under the assumption of local thermodynamic equilibrium, the equation can be written as:

$$I_\nu(r_1) = I_\nu(r_0)\exp[-\int_{r_0}^{r_1}\alpha_\nu(r)dr] + \int_{r_0}^{r_1} B_\nu(T)\alpha_\nu(r)\exp[-\int_{r}^{r_1}\alpha_\nu(r')dr']dr \tag{11}$$

The physical significance of radiative equation lies in the processes of absorption and emission of atmosphere at the position *r* along a given optical path (r_0, r_1), with the first term on the right side describing the background radiation attenuated by atmosphere while the second one standing for atmospheric emission and absorption. $I_\nu(r_1)$ is the outgoing radiance reaching the sensor at the frequency ν and $I_\nu(r_0)$ corresponds to the background radiance entering the optical path, either the cosmologic radiation at 3K or the Earth surface emission, depending on geometries mentioned above. $B_\nu(T)$ denotes the atmospheric source function which is given by Planck's function describing the radiation of a black-body at temperature *T*:

$$B_\nu(T) = \frac{2h\nu^3}{c^2}\frac{1}{\exp[h\nu / k_B T] - 1} \tag{12}$$

where *h* is Planck's constant, *c* the speed of light, and k_B denotes Boltzmann's constant.

Note that the radiative transfer equation results from energy conservation law, so it is applicable to the whole electromagnetic spectrum, from radio wave to visible light including THz wave.

For the special case when the optical path goes through a medium at a constant temperature *T*, the equation above simply becomes:

$$\begin{aligned} I_\nu(s_r) &= I_\nu(s_e)\exp[-\textstyle\int_{s_e}^{s_r}\alpha_\nu(s)ds] + \int_{s_e}^{s_r} B_\nu(T)\alpha_\nu(s)\exp[-\textstyle\int_{s}^{s_r}\alpha_\nu(s')ds']ds \\ &= I_\nu(s_e)\cdot\eta_{s_e,s_r} + B_\nu(T)\int_{s_e}^{s_r}\alpha_\nu \exp[-\alpha_\nu(s_r - s)]ds \\ &= I_\nu(s_e)\cdot\eta_{s_e,s_r} + B_\nu(T)\cdot(1-\eta_{s_e,s_r}) \end{aligned} \quad (13)$$

3. THz atmospheric propagation model

3.1. Moliere

Microwave Observation Line Estimation and Retrieval (Moliere) is the versatile forward and inversion model for millimeter and sub-millimeter wavelength observations on board the Odin satellite, including a non-scattering radiative transfer model, a receiver simulator and an inversion code. The forward models comprise spectroscopic parameters, atmospheric radiative transfer model, and instrument characteristics in order to model and compute the searched atmospheric quantities. In parallel, inversion techniques have been developed to retrieve geophysical parameters such as temperature and trace gas mixing ratios from the remotely measured spectra [21].

Moliere is presently applied to data analysis for ground-based and space-borne heterodyne instruments and definition studies for future limb sensors dedicated to Earth observation and Mars exploration. However, this code can not be used when both up-looking and down-looking geometries should be considered together, and for limb geometry if the receiver is inside the atmosphere, such as balloon and airplane.

3.2. SARTre

The new radiative transfer model [Approximate] Spherical Atmospheric Radiative Transfer model (SARTre) has been developed to provide a consistent model that accounts for the influence of aerosols and clouds, e.g. water droplets or ice particles. It includes emission and absorption as well as scattering as sources/sinks of radiation from both solar and terrestrial sources in the spherical shell atmosphere and is able to analyze data measured over the spectral range from ultraviolet to microwaves [24]. SARTre is designed for monochromatic, high spectral resolution forward modeling of arbitrary observing geometries, especially for the limb observation technique.

The line-by-line calculation of molecular absorption cross sections has been adapted from the radiative transfer package MIRART (Modular Infrared Atmospheric Radiative Transfer). And the DISORT (Discrete Ordinate Radiative Transfer Model) package is used for the calculation of the incident radiation field when taking multiple scattering into account, under the assumption of a locally plane-parallel atmosphere [22].

3.3. AMATERASU

The Advanced Model for Atmospheric Terahertz Radiation Analysis and Simulation (AMATERASU) is developed by the National Institute of Information and Communications Technology (NICT) THz project. This project aims to develop THz technology for various applications concerning the telecommunications, atmospheric remote sensing and the study of the thermal atmospheric emission in the Earth energy budget.

The first stage of this model concerns a non-scattering and homogeneous atmosphere, based on the original Moliere receiver simulator and retrieval codes. The absorption coefficient module has been extent to THz region and a more general radiative transfer module has been implemented to handle different geometries of optical paths and any location for the receiver [23].

The advanced version has a strong heritage from the SARTre model, considering the scattering effect. Modules related to optical properties of atmospheric particles and to scattering have been adapted from SARTre. The complex refractive index data of aerosols in THz region should be emphasized as a crucial parameter for single scattering properties of monodispersions and bulk optical properties, which are the necessary input of radiative transfer algorithms [24].

4. Conclusion

In this paper, we have discussed the fundamental theory in the process of THz atmospheric propagation, including Lamber-beer law, Mie scattering theory and the radiative transfer equation in detail. The key problems lie in the construction of radiative transfer algorithm, the collection of accurate spectral parameters, such as linear and continuum absorption and complex refractive index in THz region, and the standardization of measurement procedures. However, due to the lack in terahertz source and receiver technology, few experimental measurements have been made within this special region. Several kinds of THz atmospheric propagation model are introduced as well. There is a close inheritance relationship between them and these models should be compared with each other and validated against real laboratory measurements in order to check the correctness of the data accuracy and the algorithm hypothesis. The ultimate aim is to construct the atmospheric propagation model under different kinds of climatic conditions on the basis of the theoretical analysis and the material database.

5. Reference

[1] Zhang X C and Xu J Z 2009 *Introduction to THz Wave Photonics* (New York: Springer Science+Business Media) pp 1-20

[2] Tonouchi M 2007 Cutting-edge terahertz technology *Nat. Photonics* **1** 97-105

[3] Siegel P H 2002 Terahertz Technology *IEEE T. on microw. Theory* **50** 910-28

[4] Lee Y S 2009 *Principles of Terahertz Science and Technology* (New York: Springer Science+Business Media) pp 1-47

[5] Yasuko K 2008 Terahertz-Wave Remote Sensing *Journal of the National Institute of Information and Communication Technology* **55** 79-81

[6] Foltynowicz R J, Wanke M C, *et al* 2005 *Atmospheric Propagation of THz Radiation* (Albuquerque: Sandia National Laboratories) pp 1-21

[7] Ryu C and Kong S G 2010 Atmospheric degradation correction of terahertz beams using multiscale signal restoration *Appl. Opt.* **49** 927-35

[8] Hosako I, Seking N, *et al* 2007 At the Dawn of a New Era in Terahertz Technology *Proc. of the IEEE* **95** 1611-23

[9] Shirai T 2003 Mode analysis of spreading of partially coherent beams propagating through atmospheric turbulence *J. Opt. Soc. Am.* A **20** 1094-102

[10] Yasuko K and Takamasa S 2008 Atmospheric Propagation Model of Terhertz-Wave *Journal of the National Institute of Information and Communications Technology* **55** 73-77
[11] JPL database,http://spec.jpl.nasa.gov/
[12] Rothman L S, Gordon I E, *et al* 2009 The HITRAN 2008 molecular spectroscopic database *J. Quant. Spectrosc. & Radiat. Transfer* **110** 533-72
[13] Rosenkranz P W 1998 Water vapor microwave continuum absorption: a comparison of measurements and models *Radio Sci.* **33** 919-28
[14] Ma Q and Tipping R H 1992 A far wing line shape theory and its application to the foreign-broadened water continuum absorption *J. Chem. Phys.* **97** 818-28
[15] Ma Q and Tipping R H 1999 The averaged density matrix in the coordinate represesntation: application to the calculation of the far-wing line shapes for H_2O *J. Chem. Phys.* **111** 5909-21
[16] Clough S A, Kneizys F X, *et al* 1989 Line shape and the water vapor continuum *Atmos. Res.* **23** 229-41
[17] Liebe H J 1989 MPM-An atmospheric millimeter-wave propagation model *Int. J. Infrared Milli* **10** 631-50
[18] Han P Y, Tani M, *et al* 2001 A direct comparison between terahertz time-domain spectroscopy and far-infrared Fourier transform spectroscopy *J. Appl. Phys.* **89** 2357-59
[19] Chandrasekhr S 1960 *Radiative Transfer* (New York: Dover Publications) pp 1-44
[20] Thomas G E, Stamnes K 2002 *Radiative Transfer in the Atmosphere and Ocean* (Cambridge: Press Syndicate of the University of Cambridge) pp 34-83
[21] Urban J, Baron P, *et al* 2004 Moliere(v5): a versatile forward-and inversion model for the millimeter and sub-millimeter wavelength range *J. Quant. Spectrosc. Ra.* **83** 529-54
[22] Mendrok J 2006 *The SARTre Model for Radiative Transfer in Spherical Atmospheres and its application to the Derivation of Cirrus Cloud Properties* (Germany: Freie Universität Berlin) pp 1-109.
[23] Baron P, Mendrok J, *et al* 2008 AMATERASU: Model for Atmospheric TeraHertz Radiation Analysis and Simulation *Journal of the National Institute of Information and Communications Technology* **55** 109-21
[24] Mendrok J, Baron P, Yasuko K. The AMATERASU Scattering Module *Journal of the National Institute of Information and Communications Technology* **55** 123-32

Acknowledgement
This program is supported by the National Basic Research Program of China under Grant No. 2007CB310403.

The Vibrational Spectra of Bactericide molecules: Terahertz Spectroscopy and Density Functional Theory Calculations

Xiaowei WANG and Qiang WANG*
Department of Quality & Safety Engineering, China Jiliang University, Hangzhou, 310018, China

*Corresponding author. E-mail: qiangwang@cjlu. edu. cn

Abstract. In the room temperature and nitrogen conditions, we presented well-resolved absorption spectra and indexes of refraction of bactericide molecules in the far infrared radiation (FIR) spectral region recorded by terahertz time-domain spectroscopy (THz-TDS). As illustrative examples we discussed the absorption spectra of captan and folpet in THz region. The absorption coefficient and index of refraction of them were obtained. Meanwhile, density functional theory (DFT) with software package Gaussian 03 using B3LYP theory was employed for optimization and vibration analysis. With the help of Gaussian View 3.09, the distinct absorption peaks of those molecules were assigned with reliable accuracy. They were caused by intermolecular hydrogen-bonding, molecular torsion or vibration modes, absorption of water molecules, etc. As the absorption spectra are highly sensitive to the overall structure and configuration of the molecules, the THz-TDS procedure can provide a direct fingerprint of the molecular structure or conformational state of a compound.

1. Introduction

Pesticides, which are defined by the UN Food and Agricultural Organization (FAO) as substances or mixtures that are used to prevent, destroy, repel or mitigate any pest, including insects, rodents and weeds, have been used in crude forms since antiquity [1]. The expenditure around the world on pesticides totaled more than $32.5 billion in 2000 and nearly $32.0 billion in 2001, and the amount of pesticide used around the world exceeded 5.0 billion pounds in 2000 and 2001 [2]. Food safety is becoming more and more concerned, especially pesticide residual threats today due to the mass use of pesticides in agriculture [3]. Conventionally, HPLC [4], Gas chromatography-mass [5], HPLC-MSMS [6], enzymatic inhibition method [7] are employed for pesticide detection, which are labor intensive and sample and time consuming.

Terahertz time-domain spectroscopy (THz-TDS) is a technique for measuring the complex refractive index of materials over a wide range of frequencies, usually spanning from a few tens of

GHz to several THz. Terahertz spectroscopy causes little damage to the target material due to its low photon energy, can provide rich intermolecular and low-frequency intra-molecular modes of the chemicals, is less sensitive to the thermal background radiation and the scattering effect in the sample, has a stable radiation source, and more especially, can give the amplitude and phase information of the sample simultaneously, proving to be a promising tool for qualitative analyses with the signal-to-noise ratio (SNR) up to 80 dB [8,9]. Due to its uniqueness, this technique has been used in many fields, e.g. chemistry, biology, medical science, homeland security, information and community technology [10-13], etc. In the field of food safety control, especially pesticide detection, Y. Zhang et al. qualitatively studied acephate based on both first principles calculations and experimental investigations [14]; Khan et al. investigated the dielectric property, as well as its relationship to the density of bacillus thuringiensis, along with other household powders [15]. Few papers have been found concerning the analysis of bactericide pesticide as of yet, which will be a meaningful progress.

In this paper, solid-form pesticides were analyzed. We studied two bactericide molecules of captan and folpet, which are similar molecular structure and chemical properties. We tried to obtain the THz temporal waveforms of bactericides from the experiment and acquire their absorbance coefficient and index of refraction. Moreover, we obtained their theory spectra by density functional theory (DFT) calculations, and with the help of Gaussian View 3.09 [16], we account for their different distinctive absorption peaks.

2. Experimental methods and materials

2.1. Sample preparation

The standard folpet and captan were obtained from AccuStandard, Inc., New Haven, USA, with the purity of 97.5% and 99%, respectively. And the polyethylene was bought from the Sigma-Aldrich Company, St.Louis, MO. All the powder-form samples were used without further purification, and were dried at 423 K for at least 1 hour to remove water. The molecular structure of folpet and captan is shown in figure 1(a) and figure 1(b), respectively.

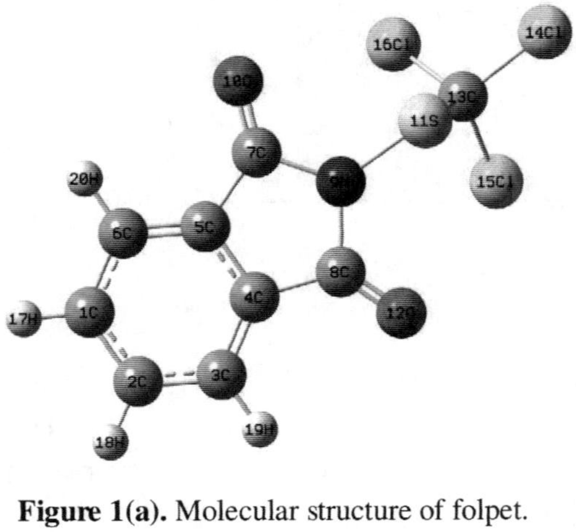

Figure 1(a). Molecular structure of folpet.

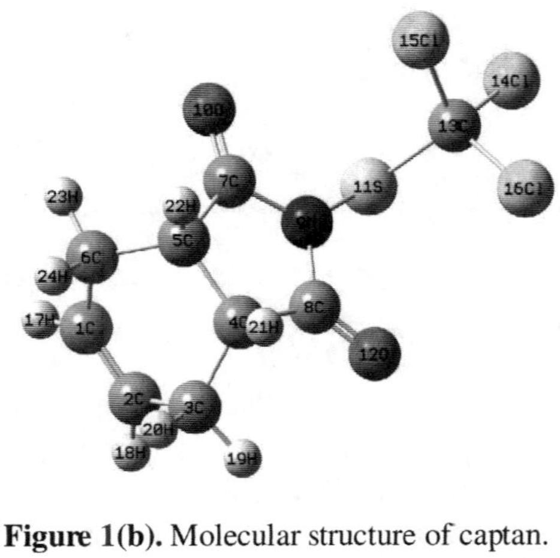

Figure 1(b). Molecular structure of captan.

We prepared material and polyethylene mixtures with the material weight ratio of about 50%. All the mixtures were carefully mixed and ground using a pestle and mortar, and then were pressed into

1-2-mm-thick disks with a aerostatic press under the pressure of approximately 30 MPa.

2.2. Terahertz time-domain spectroscopy (THz-TDS)

A THz-TDS system is employed, which is a THz wave generation and detection setup as illustrated in figure 2. The heart of our setup is a femtosecond titanium sapphire laser. It produces less 100 fs laser pulses with a repetition rate of 80 MHz, and the power of nearly 960 mW is split into the pump beam (I) and probe beam (II) at the splitter for terahertz generation and detection, respectively. After transmitting through the delay line, the pump beam elicits the terahertz beam at the emitter, the photoconductive dipole antenna under the 17-kHz sinusoidal voltage bias. The generated terahertz beam is focused onto the sample by a pair of gold-coated parabolic mirrors, carries sample characteristics, and meets the probe beam at the ZnTe crystal detector, where the probe beam is modulated by the terahertz radiation through the electro-optic effect in which a birefringence is induced in the ZnTe crystal and the polarization of the probe beam is rotated by the terahertz beam. After transmitting through the $\lambda/4$ wave plate and the Wollaston prism, the modulated probe beam is then detected by a set of balanced photodiodes, and the generated photocurrent via a lock-in amplifier is collected by a computer through a Labview program. To avoid moisture absorption in the air, the whole setup is enclosed in a nitrogen-purged box. For further details see reference [17]. The effective frequency range of this research is 20 – 80 cm^{-1} (0.6 – 2.4 THz), with the measurable maximum frequency greatly reduced because of the strong absorption of the pesticide sample.

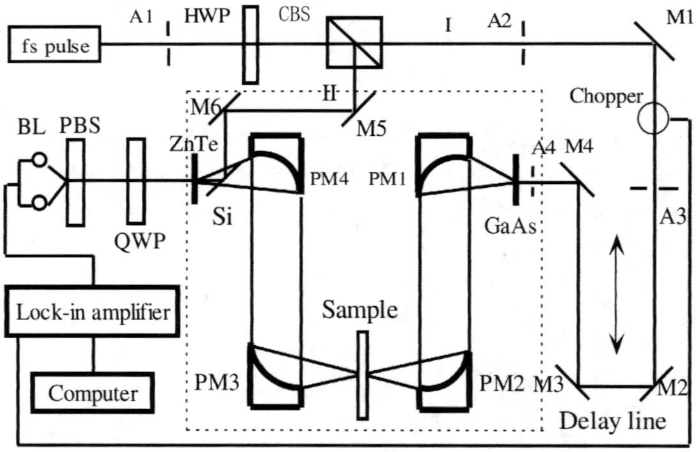

Figure 2. Experimental set-up for transmission THz-TDS.

2.3. Data analysis method

In the last decade several authors presented material parameter extraction algorithms to determine the complex refractive indexes of nearly homogenous solid samples with THz-TDS [18,19]. Commonly, for material parameter extraction, a THz pulse which has propagated through a sample is compared to the other THz pulse without the sample in its propagation path. This is achieved by tracing the temporal shape of the electric field with sample $E_{sam}(t)$ and without sample $E_{ref}(t)$, where t is the optical delay time. These two pulses are transformed into the frequency domain using fast Fourier transform (FFT). The frequency-domain reference and sample spectrum denoted as $E_{ref}(v)$ and $E_{sam}(v)$, respectively. The ratio of these fields are related to the absorption coefficient $\alpha(v)$ and index of refraction $n(v)$ of the sample as follows

$$\begin{aligned} E_{sam}(v)/E_{ref}(v) &= 4n(1+n)^{-2}\exp\left[-jvdc^{-1}(n-1)\right] \\ &= A(v)\exp\left[-j\phi(v)\right] \end{aligned} \quad (1)$$

where v represents the frequency, c is the speed of light in vacuum, d is the sample thickness, $A(v)$ is the amplitude ratio between the sample and reference spectrum, and $\phi(v)$ is the relative phase difference. Thus, we can obtain

$$n(v) = 1 + (2\pi vd)^{-1} c\phi(v) \quad (2)$$

$$\alpha(v) = -2d^{-1}\ln\left\{[4n(v)]^{-1} A(v)[1+n(v)]^2\right\} \quad (3)$$

3. Results and discussion

As shown in figure 3, the time-domain waveforms of captan and folpet are obtained, respectively. Where solid line represents reference and dashed line represents sample. The reduction in amplitude is attributed to the reflective loss at both surfaces and the absorption during passage through the sample. And the delay in time derives from the different refractive index of the different samples relative to the nitrogen. Meanwhile, owing to the different thickness of captan and folpet, the delay time is different.

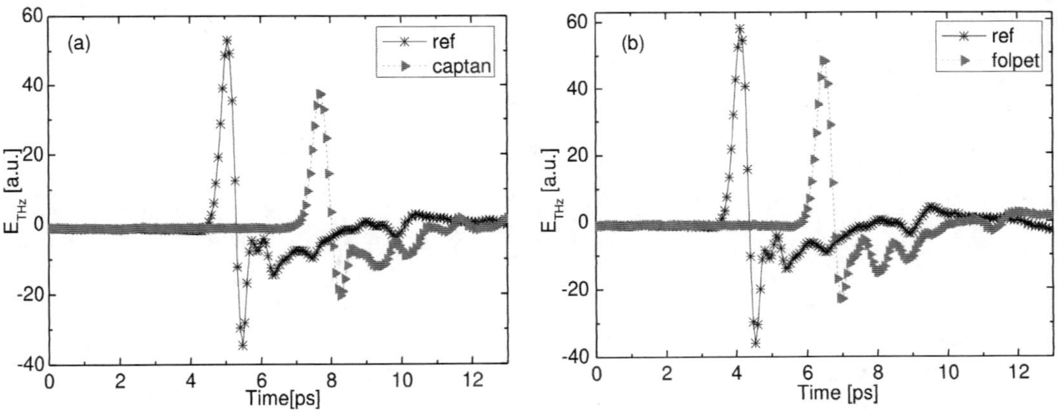

Figure 3. The THz time-domain waveform of reference (solid curves) and samples (dashed curves).

Figure 4 depicts the frequency-domain waveform of reference and samples (captan and folpet). Below 20 cm^{-1} oscillatory feature occur due to multiple reflection, and above 80 cm^{-1} the noise increases significantly and excessive attenuation of the thick sample. Therefore, our reliable data of captan and folpet are from 20 cm^{-1} to 60 cm^{-1} and from 20 cm^{-1} to 80 cm^{-1}, respectively.

Spectra of the absorption coefficient and the index of refraction of the different samples were obtained in the figure 5. We can see that the both bactericide samples exhibit distinctive absorption peaks, and can be easily discriminated from each other with the peaks positions of captan at 39.3 cm^{-1},

46.9 cm^{-1} and 58.1 cm^{-1}, and folpet at 42.7 cm^{-1} , 67.4 cm^{-1} and 72.6 cm^{-1}. There are obviously different in position of peaks and absorption intensity.

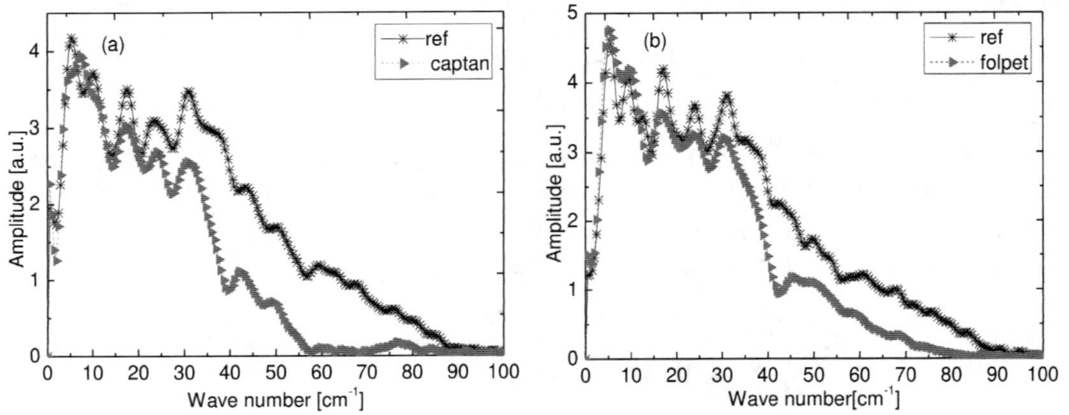

Figure 4. (a) and (b) are the THz frequency-domain waveform of reference (solid curves) and samples (dashed curves).

We can also obtain that the average index of refraction of captan and folpet is 1.49 and 1.43 from figure 5. All absorption peaks are accompanied by a characteristic change of the index of refraction. Comparing the absorption coefficient and index of refraction between captan and folpet, the former is more smooth and distinct than the latter; this is very possibly caused by the unsatisfactory sample and unsuitable thickness.

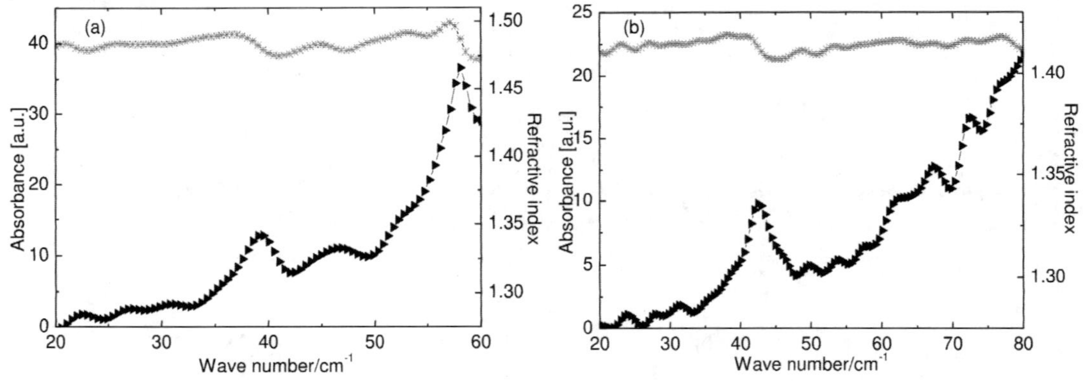

Figure 5. Absorption coefficient and index of refraction of captan (a) and folpet (b).

In order to back up the above conclusions we performed a theory analysis based on density functional theory (DFT) for captan and folpet molecule, performed with the software package Gaussian 03 using B3LYP theory and the 6-31G basis set. In figure 6 we compared the experimental and calculated absorption spectra of captan and folpet in the range 20 – 60 cm^{-1} and 20 – 80 cm^{-1}, respectively. The experimental spectrum of captan shows obvious absorption peak at 39.3 cm^{-1}, but the peak of calculated spectrum indicates at about 43.0 cm^{-1}. And the obvious absorption peaks of folpet at 42.7 cm^{-1} and 67.4 cm^{-1} in the experimental spectrum and at 38.2 cm^{-1} and 61.3 cm^{-1} in the calculated spectrum. We can see that two different bactericide molecules have different positions and absorption intensity in their distinctive spectra. So these characteristic absorption peaks as fingerprint can be used to detect and identify this substance.

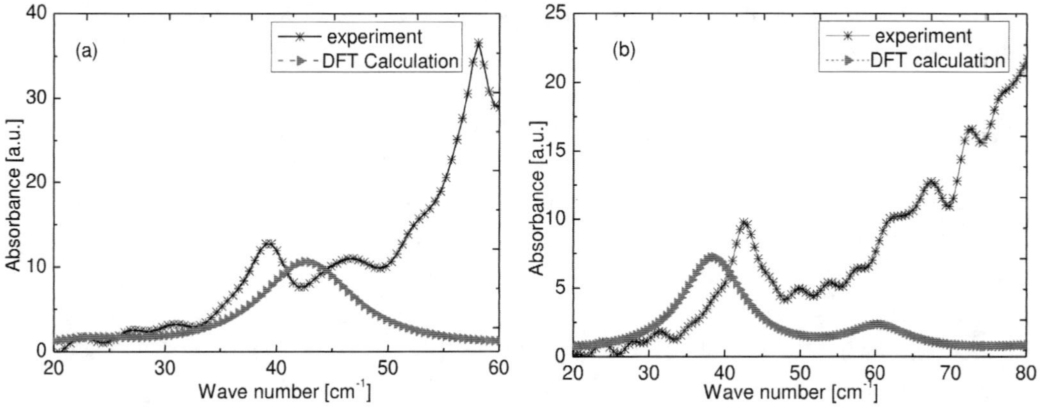

Figure 6. Comparison between a DFT calculation and experimental determination of absorption spectrum of (a) captan and (b) folpet.

With the help of Gaussian View 3.09, the peak of captan at 39.3 cm^{-1} and the peak of folpet at 42.7 cm^{-1} all correspond to the trichloromethyl-sulphur vibration, but the peak of folpet at 57.4 cm^{-1} corresponds to the phthalimide torsion. While the peaks of captan at 46.9 cm^{-1}, 58.1 cm^{-1} and folpet at 72.6 cm^{-1} perhaps correspond to the intermolecular hydrogen-bonding, absorption of water molecules, etc. The concrete reasons for those peaks are yet unknown and will be the subject of further studies.

4. Conclusions

In this article we report measurements of the FIR absorption spectra of captan and folpet by THz-TDS. The results show that the similar molecules have obviously absorption peaks and the positions of peaks are totally different. We observe well-resolved absorption features due to collective intermolecular motion and show their strong sensitivity to the molecular structure. We calculate the theory spectra of bactericide molecules based on DFT. And with the help of Gaussian View 3.09, we illustrates reasons of the distinctive absorption peaks. This illustrates how THz-TDS provides a fingerprint of the molecular conformation, and it must be pointed out that THz-TDS is an optical technique that can detect and distinguish the molecular structure and bactericide residues. It not only offers references to the application of other bactericide, but also is significant for the detection of pesticide residues.

Acknowledgments

This work is supported by the National Quality Supervision and Inspection Public Welfare Project of China under Grant 200910181 and in part by the National Natural Science Foundation of China under Grant 60902095.

References

[1] UN FAO 2002 International code of conduct on the distribution and use of pesticides *UN Food and Agricultural Organization* Rome
[2] US EPA 2004 *US Environmental Protection Agency* Washington DC

[3] Bai N B 2005 *Pesticide • Health • Environment* Beijing: China Environment Science Publication
[4] Konstantinos P P, Dimitris A P and Alexandra S 2004 *Anal. Bioanal. Chem.* **379** 458
[5] Lacassie E, Dreyfuss M F, Daguet J L Vignaud M, Marquet P and Lachatre G 1998 *J. Chromatography* A **805** 319
[6] Qian S, Jun H, Bao D S and Gang Y 2009 *Chinese Science Bulletin* **54** 4633
[7] Campanella L, Bonanni A, Martini E, Todini N and Tomassetti M 2005 *Sensors and Actuators* B **111–112** 505
[8] Han P Y, Tani M, Usami M, Kono S, Kersting R and Zhang X C 2001 *J. Appl. Phys.* **89** 2357
[9] Ferguson B and Zhang X C 2002 *Nature Materials* **1** 26
[10] Peter H and Siegel 2004 *IEEE Transactions Microwave Theory and Techniques* **52** 2438
[11] Dragoman D and Dragoman M 2004 *Progress in Quantum Electronics* **28** 1
[12] Nishikiori R, Yamaguchi M, Takano K, Enatsu T, Tani M, Silva U C, Kawashita N, Takagi T, Morimoto S, Hangyo M and Kawase M 2008 *Chem. Pharm. Bull.* **56** 305
[13] Watanabe Y, Kawase K, Ikari T, Ito H, Ishikaw Y and Minamide H 2004 *Optics Communications* **234** 125
[14] Zhang Y, Peng X H, Chen Y, Chen J, Curioni A, Andreoni W, Nayak S K and Zhang X C 2008 *Chem. Phys. Lett.* **452** 59
[15] Khan U A, Nguyen N and Mohammed N A 2008 *IEEE Transactions on Instrumentation and Measurement* **57** 286
[16] Frisch M J, Trucks G W, Schlegel H B, Scuseria G E, Robb M A, Cheeseman J R, Montgomery J A, et al 2004 *Gaussian 03, Revision9C.02* Gaussian, Inc., Wallingford, CT.
[17] Grischkowsky D, Keiding S, Exter M V and Fattinger C 1990 *J. Opt. Soc. Am.* B **7** 2006
[18] Duvillaret L, Great F and Coutaz J L 1996 *IEEE J. Select. Topics in Quantum Electron.* **2** 739
[19] Dorney T D, Richard G B and Daniel M M 2001 *J. Opt. Soc. Am.* A **18** 1562

The study on THz wave propagation feature in atmosphere

Haixia Cui[1,2,3], Jianquan Yao[2,3], Chunming Wan [1]

[1]Changchun University of Science and Technology, Changchun, Jilin 130022, China

[2]Institute of Laser and Optoelectronics, Tianjin Uiversity, Tianjin 300072, China

[3]Key Laboratory of Opto-electronics Information Science and Technology, Ministry of Education, Tianjin University, Tianjin 300072, China

e-mail:meihao3698@163.com

Abstract. THz wave has many applications, such as space communication, Earth environment observation, atmosphere science, remote sensing and so on. The atmospheric propagation of terahertz waves now rank among the most critical issues in the societal implementation of terahertz technology. In this paper we introduce the constitution of THz wave atmosphere propagation system, decay, then analyze the decay, turbulent and enhancement effects about THz wave atmosphere propagation, then carry out atmospheric propagation measurement system based on THz-TDS.

1.Importance of terahertz-wave propagation in atmosphere

Terahertz (THz) region (0.1-10 THz; 3mm-30um) plays important role for the Earth's radiation budget, the atmospheric transmittance of terahertz waves now rank among the most critical issues in the societal implementation of terahertz technology[1]. THz atmospheric propagation technology is expected to open new opportunities for the accurate prediction of the natural disaster, such as localized torrential rain and extraordinarily heavy snowfall,due to the climate change, THz atmospheric technology is suitable to observe humidity, ice cloud, and the temperature simultaneously with high spatial resolution. Despite its importance, due to the difficulty of generating and detecting waves in this region, we did not yet reach a good understanding of the behaviour of THz-wave in the atmosphere. Moreover, there still exist large difficulties in developing appropriate observation technology for THz-wave[2].

2. THz wave atmosphere propagation system

THz wave atmosphere propagation system includes emission, propagation, detector, amplifier and processing systems(Fig.1).

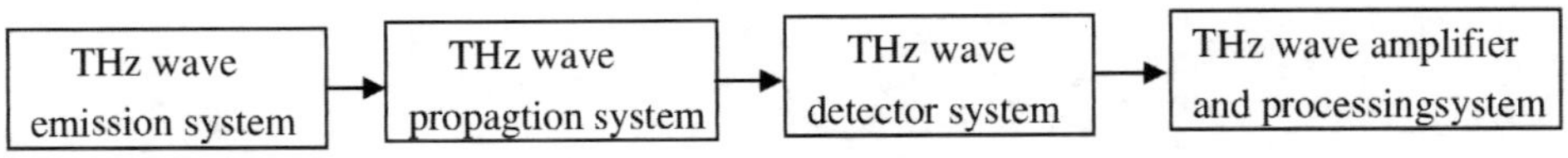

Fig.1 THz wave atmosphere propagation system

THz propagation is an important application technology, THz wave has a very wide unallocated band, which transfer rate is high, direction is good, and security is high, and diffused antlers is small. In the 90's, because of lacking effective THz source and detection technology, people's knowledge on THz is very limited, so called "unexplored" range between visible light and radio wave. Recently, with the reliable source and detection technology developing, THz will be used to short distance, wireline and wireless communication[4]. In communication and propagation ,THz wave transmission is a bond, which associates wave THz launch and receiving two parts, it is of vital importance on the whole THz applications. During propagation, various media and amplifier technologies can be adopted to overcome the absorbtion, for example, water absorption.

3.THz wave atmosphere propagation effects

There are three main effects in THz wave atmosphere propagation, including decay, turbulent and enhancement effects[5].

3.1 The decay effect of THz propagation in atmospheric

3.1.1 Absorption process about THz propagation in atmospheric

About the THz radiation, the absorption and the scattering are the basic decay effects in atmosphere propagation. When THz radiation goes through atmosphere, its total or part energy transfer to the atmosphere, this process is called absorption. Some terahertz wave-lengths can be used for communications between locations separated several hundred kilometers on the ground, other terahertz waves propagate less than 1m due to significant, it is essential to know the applicable attenuation characteristics under different atmospheric condition.

Atmospheric absorption of THz waves is commonly divided into two components. One is the line contribution, which arises mainly from the resonance absorption of water vapour in atmosphere. One of the absorption line is comprised mostly of the absorption lines of water vapour present in air. The absorption lines of water vapour are characterized by spectroscopic parameters, including the center frequency of the absorption line, oscillator intensity, and the pressure broadening coefficient. These spectroscopic parameters are generally determined through precise, careful measurement in laboratory experiments. Common spectrum parameter in the atmosphere may find in existing database (for instance JPL[6], GEISA database, HITRAN[7]and so on), at present the most commonly used database is HITRAN the database. With the spectrum parameter precision's developing, the atmospheric propagation model's accuracy will enhance[7].

The other component of atmospheric absorption is the "continuum absorption"(Fig.2), which in practice is defined as the excess of absorption unaccounted for by the resonance water vapour spectrum. The major contribution to the THz-wave propagation is in total the water vapour lines and their wing absorption. It is well known that the results of simulation under the assumption that the atmospheric absorption spectrum corresponds to the accumulation of water vapour absorption lines do not agree with experimental observation of atmospheric absorption.continuum absorption cannot be described by water vapour absorption lined and it may be observed in very wide electromagnetism wave frequency scope (from microwave to infrared)[8].

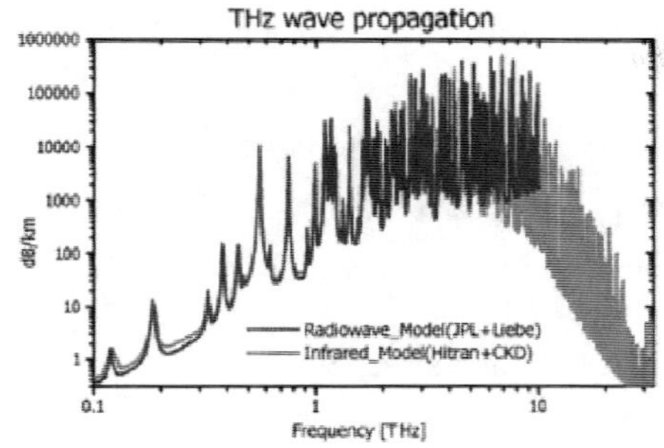

Figure 2. The linear and continuum absorption of THz wave from NICT

Absorption coefficients strongly depend on pressure broadening of water vapour lines, which is caused by collision between a water molecule itself and the other atmospheric molecule like N_2, O_2, or H_2O. In Fig.1 atmospheric H_2O line-by-line and continuum absorption coefficients calculated on the basis of several spectroscopic models are shown. It demonstrates that there are significant discrepancies among the models. One of the reasons of the discrepancy is due to the small number of laboratory spectroscopic measurements made so far. The accuracy of atmospheric parameters derived based on the model directly depends on the accuracy of the spectroscopic parameters[8].

But so far, we can not solve completely the continuous absorption physical machine-made question. The water vapour continuous absorption is the biggest question in THz radiation, Until now people find three kinds theories, which can explained the water vapour continuous absorption: Far wing absorption, collision induced absorption and double syndrome absorption. The corresponding processing method includes: CKD [9]continuous absorption method; M&T continuous absorption theory method; MT-CKD continuous absorption theory; Semiempirical processing methods and so on. Without a doubt, the water vapour continuous absorption originates interaction between the syndrome, but how to describe these interactions, is still the question which waits for solving. We must have experimental data to verify these predictions. At same time, we can adopt various media and amplify technologies to overcome the absorption shortcoming in space applications.

3.1.2 Scattering process about THz propagation in atmospheric

It is well known, scatters including the Rayleigh scattering and the Mie scattering.To THz wave band, because its wave length is much bigger than the wave length of visible light and the ultraviolet ray, therefore in general we do not consider the Rayleigh scattering[10].

During THz propagation process, the main scattering granule are aerosol and the water vapour coagulum. The aerosol (refers to multi-disperse systems from small granule constitution) and the water vapour coagulum (cloud, rain, snow and so on). Aerosol radiative transfer is a difficulty question because of its big change in the space and time. The criterion distribution is an important concept in describing aerosol, the different criterion's granule has different scattering wave length. At present, the aerosol criterion spectrum pattern commonly include[11]:

1) Revision spectrum:

$$\frac{dN(r)}{dr} = ar^{\alpha} \exp(-b^{\gamma}) \qquad (1)$$

Where N is granule number in the unit volume, r is the radius of particle, a, b, Alpha, Gamma is the constant which depends the resource. But the origin of aerosol can be divided into: Mainland (Haze L), sea (Haze M) and high stereotype (Haze H)[12]

2）Junge spectrum

$$\frac{dN}{d\log r} = cr^{-v} \tag{2}$$

In the formula (2), v is the spectrum parameter, v takes 2~4, c related to the total density in general.

3.2 The turbulent effect of THZ propagation in atmospheric

The atmospheric physical property is dissimilar in different position and different height. Even if at one place, the property can have the change with different time.The turbulent motion is the most important characteristic usually, the turbulent motion causes the atmospheric refractive index to fluctuate stochastic[13].

The atmospheric turbulent cause the atmospheric refractive index to fluctuate unceasingly, thus causes the THz wave amplitude and the phase to fluctuate stochastic, so the THz beam will twinkle, curve, fission, expand, polarization condition will fluctuate[14], even the spatial coherence is reduced and so on.

The influence may divide into three kinds according to the turbulent criterion: When the turbulent criterion is bigger than the THz beam diameter, the THz beam deflects stochastic, causing THz beam drifting on receiving end[15]; When the turbulent criterion is equal to the THz beam diameter nearly, the light beam will also have the deflection stochastic, main performance is arrival angle fluctuating, image point vibrating[16]; When the turbulent criterion is smaller than the THz beam diameter, the light beam has the diffraction, the coherence character dropping, the main performance is expanding, the luminous intensity fluctuating and the luminous intensity reducing. The atmospheric turbulent has critical affection to THz wave propagation, this will also be our research direction later.

3.3 The enhancement effect of THz propagation in atmospheric

Atmospheric and the THz radiation interact during THz wave propagation in atmospheric, it will experience the enhancement process. Usually we introduce the source function J item to describe THz radiation enhancement. The scattering and the heat radiation cause the enhancement. These two machine relative importances depend on the difference wave length . The expression of source function J is defined:

$$J = J_B + J_{SS} + J_{MS} \tag{3}$$

J_B described the heat radiation, the definition as follows[17]:

$$J_B = (1-\omega_0)B(T) \tag{4}$$

Where B is the Planck function, represents blackbody heat radiation at temperature T. The mono frequent Planck function can be written:

$$B_v(T) = \frac{2hv^3c^2}{e^{hvc/kT}-1} \tag{5}$$

Where v is a wave number, c is the speed of light, k and h on behalf of the Boerziman constant and the Planck constant separately

J_{ss} comes from the source (e.g.: Sun) sole scattering on direct radiation condition

$$J_{SS} = \omega_0 \frac{1}{4\pi} P(\Theta_\Theta) F_\Theta T_\Theta \tag{6}$$

Where P is the phase function, F represents sun radiance from the exterior land , F =exp (-) is transmissibility on condition of the crown atmosphere to the scattering perpendicular way[18], $_0$ is the sole scattering albedo:

$$\omega_0 = \frac{\beta_S^{par}}{\beta_S^{par} + \beta_a^{Par} + \beta_S^{mol}} \qquad (7)$$

s and a is the scattering and the absorption coefficient, the superscript "mol" and "par" represents molecular and the granule material characteristics.

J_{MS} item of multiple scattering source comes from multiple scattering radiative[19]

$$J_{MS} = \omega_0 \frac{1}{4\pi} \int_0^{4\pi} P(\Theta) I(\Omega') d\Omega' \qquad (8)$$

Where I (') representes the incident scattering radiation field.

4.Atmospheric propagation measurement scheme based on THz-TDS

THz-TDS spectroscopic system includes Pumping laser, optical delay, THz emitter, detector sample, THz detector locking-amplifier and computer, and so on. Sub-ps THz pulse penetrate sample, the detector receive the signal after free space propagation.THz-TDS is a pulsed terahertz wave to the sample, measures the time variation of the electric field of the transmitted wave, and obtains the amplitude magnitude and phase information of the electric field of the terahertz wave at each wavelength by Fourier transformation(Fig.3)[20]. The time-domain spectrum measured transform Fourier transformed into a frequency spectrum and then modified to an absorbency spectrum using the background spectrum measured directly prior to the main measurement.

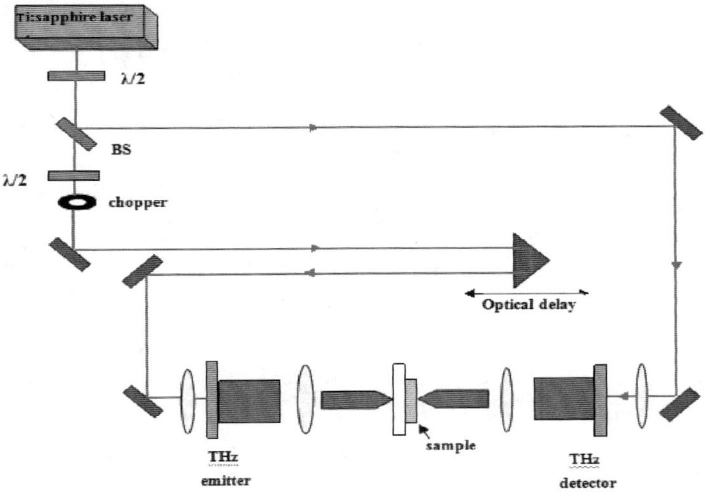

Fig.3 THz-TDS spectroscopic system

Fig.4[20] shows the measurement results of atmosphere propagation characteristics based on THz-TDS, the time-domain spectrum and frequency spectrum by Fourier transformation, as an spectroscopic parameters required to construct an appropriate model.

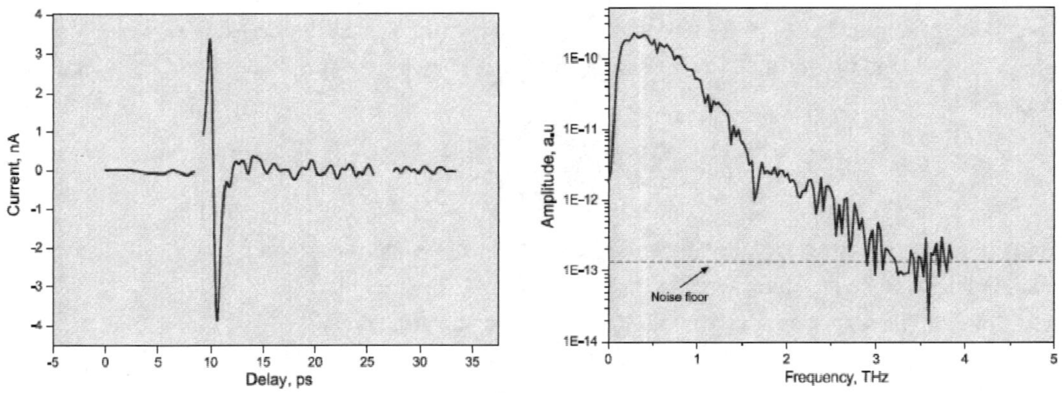

Fig. 4time-domain spectrum and frequency spectrum

4. Conclusion

Terahertz (THz) radiation is extensively applied in diverse fields, such as space communication, earth environment observation, atmosphere science, remote sensing and so on. And the research on propagation features of THz wave in the atmosphere becomes more and more important.In this paper we introduce the constitution of THz wave atmosphere propagation system, decay, then analyze the decay, turbulent and enhancement effects about THz wave atmosphere propagation, then carry out atmospheric propagation measurement system based on THz-TDS.

References

[1] Albert Redo Sanchez, Xi Cheng Zhang. Terahertz Science and Technology Trends[J].IEEE J.Sel.Topics Quantun Electron., 2008, 14(2): 260-269.

[2] R.Appleby, H.B.Wallance. Standoff detection of weapons and contraband in the 100GHz to 1THz region[J]. IEEE Trans. Antennas Propag., 2007,55(11):2944-2956.

[3] Radoslaw Piesiewicz, Martin Jacob, Martin Koch, et al. Performance analysis of future multi-gigabit wireless communication systems at THz frequencies with highly directive antennas in realistic indoor environments [J].IEEE Electron, 2008, 14(2): 421-430.

[4] Nagel M, Bolivar P H, Brucherseifer M, et al. Integrated THz technology for label-free genetic diagnostics [J].Appl.Phys.Lett., 2002, 80(1):154-156.

[5] Shen Y C, Lo T, Taday P F, et al. Detection and identification of explosives using terahertz pulsed spectroscopic imaging[J]. Appl. Phys.Lett., 2005,86:241116-241116-3.

[6] JPL database,http://spec.jpl.nasa.gov/

[7] Rothman L S, Gordon I E, *et al* The HITRAN molecular spectroscopic database *J. Quant. Spectrosc. & Radiat. Transfer* 2009, 110 **:**533-72

[8] Jiang Zhiping, Xu X G, Zhang X-C. Improvement of terahertz imaging with a dynamic subtraction technique [J]. Appl. Opt., 2000, 39(17):2982-2987.

[9] John F Federici, Brian Schulkin, Feng Huang, et al. THz imaging and sensing for security applications - explosives, weapons and drugs[J].Semicond. Sci. Technol., 2005, 20:S266-S280.

[10] K.S.Yasuko and S.TA Takamasa Atmospheric Propagation Model of Terahertz-Wave[J].NICT.2008,5(1).73-77

[11] J.R.Pardo, E.R.Serabyn, J.Cernicharo. Submillimeter atmospheric transmission measurements on Mauna Kea during El Nino conditions: implications for broadband opacity

contributions[J].J.Quant.Spectrosc.&Radiat. Transfer, 2001, 68:419-433.

[12] D.A.Naylor, G.R.Davis, B.G.Gom,et al. Atmospheric transmission at submillimetre wavelengths from Mauna Kea[J].Monthly Notes of the Royal Astronomical Society, 2000,315:622-628.

[13] E.Serabyn, E.W.Weisstein, D.C.Lis, et al. Submillimeter Fourier-transform spectrometer measurements of atmospheric opacity above Mauna Kea[J]. Applied Optics, 1998, 37:2185-2198.

[14] R.J.Emery, P.Moffat, R.A.Bohlander,et al. Measurements of anomalous atmospheric absorption in the wavenumber range $4cm^{-1}$-15 cm^{-1} [J].Journal of Atmospheric and Terrestrial Physics, 1975,37:587-594.

[15] R.J.Emery, A.M.Zavody. Atmospheric propagation in the frequency range 100-1000GHz[J].The Radio and Electronic Engineer, 1979 49,(7/8):370-380.

[16] D.L.Woolard, E.R.Brown, A.C.Samuels, et al. Terahertz-frequency spectroscopy as a technique for the remote detection of biological warfare agents[A].23rd Army Science Conf. Orlando, FL, 2002.

[17] W.A.Traub, M.T.Stier. Theoretical atmospheric transmission in the mid- and far-infrared at four altitudes[J]. Applied Optics, 1976, 15(2):364-377.

[18] Jeremy Pearce, Daniel M.Mileman. Using terahertz pulses to study light scattering[J].Physica B,2003,338:92-96.

[19] A. A. Danylov, J. Waldman. THz Laboratory Measurements of Atmospheric Absorption Between 6% and 52% Relative Humidity[R]. Internal STL Report, 2006,1-7.

[20] www.oplanchina.com www.dpiv.cn

Acknowledgement

Supported by 973 Plan "Fundamental research on new generation optical fiber intelligent sensing network and key devices"： Sub-Project（2010CB327801）"

The impact of MgO-doped near-stoichiometric lithium niobate crystals on the THz wave output characteristics

Zhang Xianbin[1], Li Yunfeng[1], Ma lijuan[1], Yuan ke[1], Shi Wei[1*]

[1] Department of Applied Physics , Xi'an university of technology, Xi'an 710048, China

[1]E-mail: Zhangxianbin@xaut.edu.cn
[*]E-mail: Swshi@mail.xaut.edu.cn

Abstract: The control experimental study on the THz wave parametric oscillator (TPO) output characteristics based on the congruent $LiNbO_3$ crystal (CLN) and stoichiometric MgO-doped lithium niobate (SLN) crystal is performed. As a nonlinear medium, in the aspect of the THz wave output, experiments show that the congruent $LiNbO_3$ crystal is more stable than the SLN crystal. Compared with the CLN crystal, SLN showed significant photorefractive effect, which adversely the stability of the THz wave output. Experiments indicated that different molar concentration of MgO doped, can significantly change the photorefractive properties of SLN crystal. The results showed that with the increase of MgO doping concentration, the photorefractive of SLN gradually become weaker, and THz wave output stability has the significantly increase. The output stability of mol 5.0% MgO droped SLN crystal, has not significantly different with the CLN. In the contrast experiment of TPO with the 160mm cavity length and 65mm crystal length, the pump laser threshold of the 5% mol MgO: SLN crystal decreased by 23% than the CLN crystal, while the peak THz energy output increased 28%.

1. Introduction

In the middle of the last century , Y R.Shen[1], B.C.Johnson and J.Nishizawa[2] have proposed the theory that sub-millimeter electromagnetic wave can be generated by nonlinear optical crystal. In 1996, the Ito lab of Japan first reported the research work about THz wave parametric oscillator based on nonlinear $LiNbO_3$ crystal[3]. This THz wave source has the characteristics of frequency tunable, narrow line width (high monochromatic), high spatial and temporal coherence, high energy output, and operation at room temperature. The threshold of TPO is closely related with the parameter of resonant cavity itself. In addition, experiments show that the characteristics of the crystal material had made a great impact on the operation threshold of oscillator. With the rapid development of $LiNbO_3$ research, choosing the appropriate crystal to reduce the TPO-threshold and to optimize the output characteristics is not only necessary, but also has the exploration significance in the research.

$LiNbO_3$ crystal has the important feature and application prospects, primarily due to its elements doping[4,5]. In recent years[6], compared with congruent $LiNbO_3$ crystal (CLN), stoichiometric MgO-doped lithium niobate (SLN) crystal is more attracting attention, many researchers have achieved the significant progress in this area. The $LiNbO_3$ crystal is the typical non-stoichiometric ratio crystal, the $LiNbO_3$ crystal which is widely used is the same ingredient allocated proportion ([Li]/[Li]=48.3~48.6/51.7~51.4) at present. The $LiNbO_3$ crystal (CLN) growed in the same ingredient allocated proportion has obviously superiority in the growth technology, but it bing the large number of intrinsic defects at the same time, mainly due to the antisite defect of (NbLi) and vacancies of (VLi) . The intrinsic defects in CLN reduced the many properties of crystal. The growth intrinsic defects of stoichiometric ([Nb]/[Li]=1:1) $LiNbO_3$ crystal (SLN) was reduced obviously. Compared with the CLN crystal, the results show that the SLN crystal performance has shows considerable improvement, only in terms of nonlinear coefficient and electro-optic coefficient enhanced about 30%. The nonlinear coefficient of CLN is 13.8pm / V, the SLN is 17.5 pm / V. The photo-damage resistance of CLN is 100kW/cm^2, and the SLN photo-damage resistance is 104kW/cm2.

From the application view of TPO, we hope that crystal non-linear coefficient has a large nonlinear coefficient and the strong ability of anti-photorefractive. The big nonlinear optics coefficient is not only advantageous in reduces the oscillation threshold of TPO, but also can enhances the THz wave energy output; meanwhile, the photorefractive resistance ability is beneficial to the stability of THz wave radiation. In this paper, 5% mol MgO doped SLN crystal was used as a nonlinear optical medium of TPO. The proposal of this experiment is to enhance the THz wave output by using the large nonlinear coefficient of SLN, and reducecd the threshold of TPO. Simultaneously, the photorefractive effect is expected to reduce due to the MgO droped.

2.The impact of SLN and CLN on the output stability of the TPO

Only from the level of crystal growth and crystal structure can make a fundamental interpretation for the physical performance differences among CLN and SLN, and different molar concentration of MgO doped SLN crystal. This paper showed the qualitative analysis by the contrast experiment. In the contrast experiment used four types $LiNbO_3$ crystal: (1) the congruent $LiNbO_3$ crystal (CLN); (2) near-stoichiometric 0.89% mol MgO: SLN crystal; (3) 1.7% mol MgO: SLN; (4) 5.0% mol MgO: SLN crystal. The crystal geometry of four types $LiNbO_3$ is same, all for the 65mm × 10mm × 10mm. Experimenting , in the exact same experimental condition, all types crystal as the nonlinear medium of a TPO with the 154mm cavity length, respectively, the THz wave colleted in a vacuum calorimeter, and then the calorimeter signal inputed oscilloscope , to record the electrical signals amplitude corresponding to THz wave.

Figure 1 is the THz wave output with time of 0.89% mol MgO: SLN in 27 minutes. Labview control platform automatically recorded the output voltage in every 10 seconds. You can see, the signal amplitude of THz wave first reached a peak in four minutes, then gradually attenuated to a

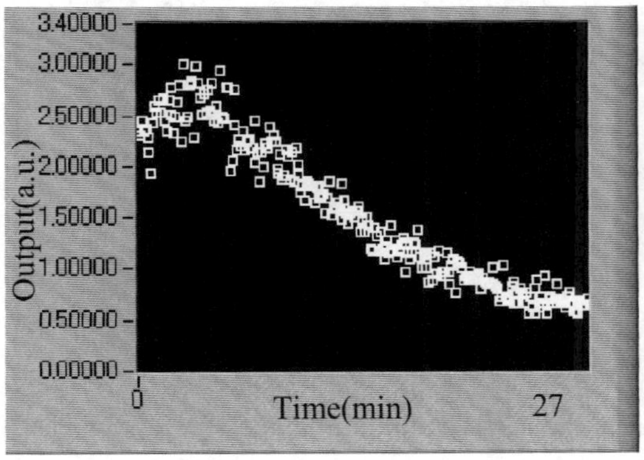

Figure 1. Output curve of 0.89mol%MgO:SLN in 27 minutes

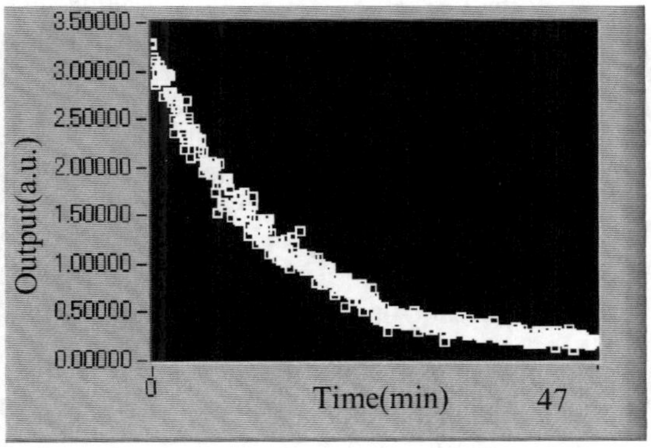

Figure 2. Output curve of 0.89mol% MgO:SLN in 47 minutes.

certain value. In the initial stage THz wave rised because of the YAG laser device was turned on, there is a gradual process of laser energy rised and stabilized , THz wave signal also tended to increase with the laser intensity increased; When the laser output has stabilized, the output of the THz wave reflected attenuation. This showed that physical property of the 0.89% mol MgO: SLN is not stable,

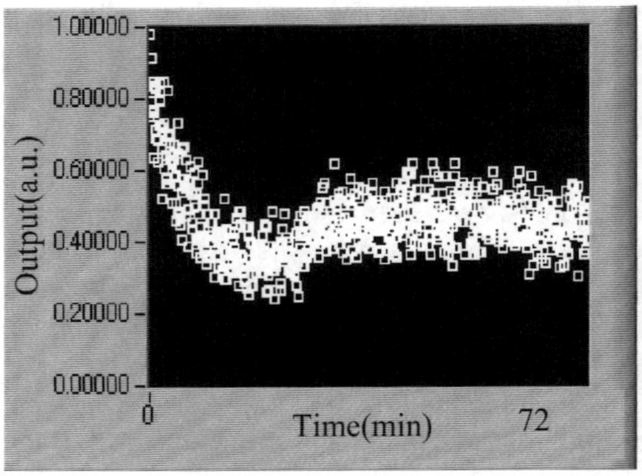

Figure 3. Output curve of 1.7mol% MgO:SLN in 72 minutes

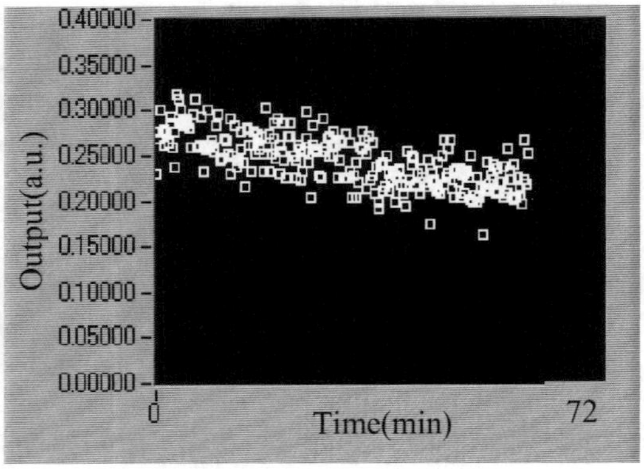

Figure 4. Output curve of 5.0mol% MgO:SLN in 72 minutes.

will change in the light. Figure 2 is the output curve of 0.89% mol MgO: SLN in 47 minutes in the case of laser output stabilized. The curve also reflected the instability characteristics of SLN, and with continuing of the illumination time, THz wave output even attenuated to zero. Figure 3 is the output curve of 1.7% mol MgO: SLN crystal when the laser energy output stabilized. The curve showed the THz wave output still attenuated when the pump incidence, then tended to stabilize at a lower average level. Compared with 0.89% mol MgO: SLN, the crystal output stability has improved. Figure 4 is the output curve of 5.0% mol MgO: SLN crystal under the same conditions. Compared with the above three types crystal,the THz wave output of crystal stabilized at a average level, reflected the stability of the physical properties of crystal.

Figure 5 is the output curve of the congruent $LiNbO_3$ crystal (CLN) under the same conditions. Compared with the SLN, output of CLN is stable, reflected the stability of its physical properties better than near-stoichiometric SLN crystal. The contrast experiments showed the instability of the THz wave output was due to photorefractive effect of $LiNbO_3$ crystal.

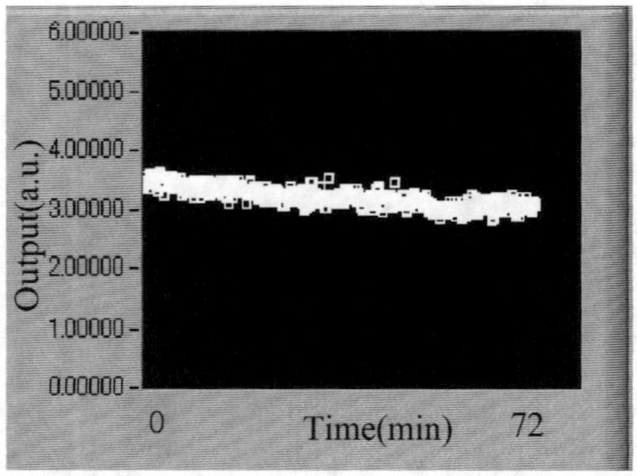

Figure5. Output curve of CLN in 72 minutes

3. Comparison of the oscillation threshold and energy output of CLN and 5% mol MgO: SLN

For THz wave parametric oscillator, the adverse impact of the photorefractive are : (1) phase mismatch caused by photorefractive effect would affect the efficiency of THz wave radiation; (2 the loss of light in the crystal become very serious; (3) photo-damage caused by photorefractive would seriously influences the performance of TPO.

Journal of Physics: Conference Series **276** (2011) 012226 doi:10.1088/1742-6596/276/1/012226

Figure 6 is the threshold comparison of the SCN-TPO with that the CLN-TPO at 160mm cavity length. To ensure the comparability in experimental test, the frequency of output THZ wave both are 1.2 Thz, and the crystal length are 65mm.

From the curve, the pump laser threshold of SCN-TPO approximately is 23mJ/pulse, and MgO:SLN approximately is 18mJ/pulse. That is to say: 5%mol MgO:SLN crystal used in the THz wave parametric oscillator make the TPO threshold reduced 21.7%. Just like front introduces, the cause is that 5%mol MgO:SLN has the big nonlinear optics coefficient, and dopes the appropriate mole fraction MgO, which suppressed the photorefractive effect. The vibration threshold value reduces means: at the same level condition of the incident pump luminous energy invariable, the output THz wave in all frequency tuning range is to increase. Figure 7 is under the same pump energy is 20mJ/pulse, the comparison of THz output of 65mm 5% mol MgO: SLN and CLN crystal with equal length. The peak energy of SLN is 90pJ/pulse, while CLN is 65pJ/pulse, the peak energy of SLN has increased 28%.

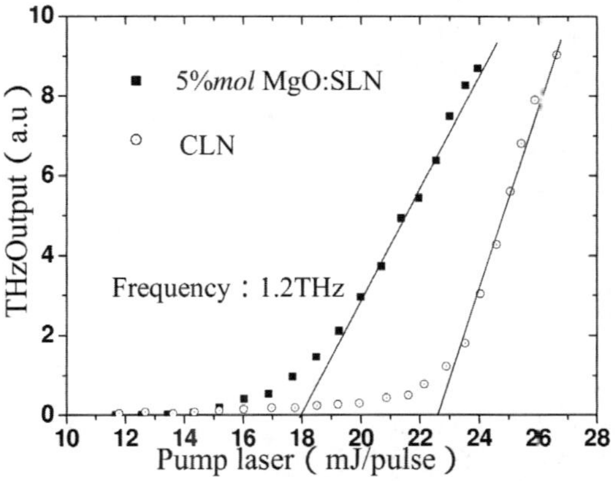

Figure 6. Comparison of threshold of SCN-TPO with CLN at 160mm cavity length

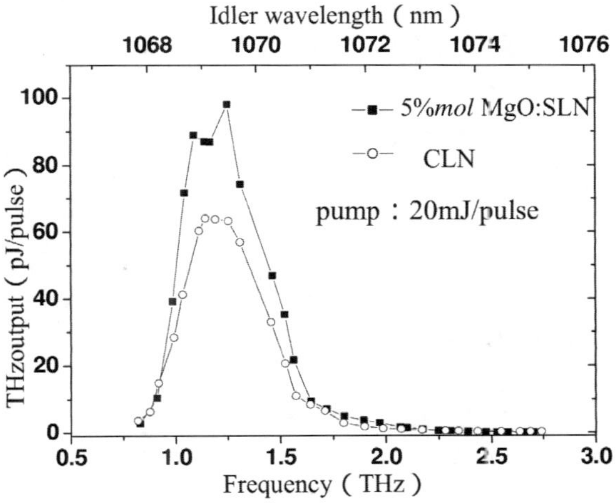

Figure 7. Compare THz Output of SCN-TPO with CLN at 160mm cavity length

4. CONCLUSIONS

As a nonlinear medium which generated THz wave on the output stability of the device, the CLN crystal of congruent ratio growth is more stable than the near-stoichiometric SLN crystal.

The experiment results indicated that compared with the CLN crystal, SLN showed significantly photorefractive effect; But can significantly change the photorefractive properties of SLN crystal by different molar concentration of MgO-doped. With the increase of MgO doping concentration, the photorefractive of SLN gradually becomes weaker , the output stability has the increase. Experiments pointed out that the THz wave output stability of mol 5.0% MgO:SLN crystal, has not significantly different with the CLN. In the contrast experiment of TPO with the 160mm cavity length and 65mm crystal length, the pump laser threshold of the 5% mol MgO: SLN crystal decreased by 21.7% than the CLN crystal, while the peak energy output increased 28%.

This paper only analysed some differences on the THz-wave output of the MgO: SLN and CLN crystal used in THz wave parametric oscillator, relevant experimental phenomenon can be used as a reference.

References

[1] Shen Y R 1977 *Nonlinear Infrared Generation*(New York:Springer-Verlag)

[2] Nishizawa J and Suto K 1980 *J.Appl.Phys* **51** 2429
[3] Kawase K, Manabu Sand Ito H 1996 *Appl. Phys. Let* **68** (18) 2483
[4] Ionita I, Jaque F 1998 *Optical Materials* **10** 171
[5] Xihe Z , Rui W a Wusheng X.2002 *Optical Materials* **19** 427
[6] Fontana M and Chah K 2001 **16** 111

Acknowledgments
Authors wishing to acknowledge the financial support from the National Natural Science Foundation of China (Grant No. 60971015, 60771053), Natural Science Foundation of Shaanxi Province (Grant No:8JK402), Science Foundation of Xi'an University of Technology (Grant No. 108-210703) and Well-educated Staff Science and Technology Start-up Foundation of Xi'an University of Technology (Grant No. 108-210801). The authors especially thanks the Ito-lab of RIKEN for their support.

THz-TDS Signal Analysis and Substance Identification

Weixin Xie[1,2] **Jing Li**[1,2] **Jihong Pei**[2]

[1] School of Electronic Engineering, Xidian University, Xi'an 710071, P. R. China;
[2] Intelligent Information Institute, Shenzhen University, Shenzhen, 518060, P. R. China

E-mail: jhpei@szu.edu.cn

Abstract. The terahertz time-domain spectroscopy (THz-TDS) imaging system can obtain high-dimensional signals of the substance fingerprint information. It is necessary to process properly to use some signal processing techniques especially for high dimensional signals. As a mathematical description language, geometric algebra (GA) provides not only a powerful algebraic framework for the multi-dimensional vector analysis and computing, but also a unified measurement and geometrical description for different geometric models. On the basis of the GA theories, a new signal analysis method of the THz-TDS is presented. Based on the characteristics of THz-TDS signals, signals are mapped into vectors in the high-dimensional real vector space. The vectors are represented with hyper-complex numbers. We can analyze the vectors using theories of GA. Based on the physical mechanism of the THz-TDS signal analysis, geometric distribution properties and algebraic relationships of THz-TDS signals are deduced. It is demonstrated that every complex refractive index of the sample relates to a unique 2-blade $\boldsymbol{B}_2$, all vectors corresponding to the samples of the same substance are collinear and belong to the intrinsic 2-blade of the substance. In projective interpretation, the 2-blade $\boldsymbol{B}_2$ represents a fixed line and all vectors related to the same substance are along that line. Accordingly, a novel substance identification method based on the relative THz-TDS is presented. In the method, two THz-TDS signals through the samples of the same substance but different thickness are measured. The intrinsic 2-blade $\boldsymbol{B}_2$ of the substance is then determined by the outer product of these two corresponding vectors. Using the conformal split by the fixed bivector $\boldsymbol{B}_2$, each vector corresponding to THz-TDS signals in the vector space V_n can be linearly splitted into vectors in vector spaces V_2 and V_{n-2}. Since that 2-dimensional subspace V_2 is the support of $\boldsymbol{B}_2$, the subspace is also a label for substances. So substances of samples can be identified on the magnitudes of projection vectors in that subspace. This method can contribute to the accurate classification and identification, and facilitate the feature extraction. Finally, experiments are presented and show that the substance identification method is feasible and effective.

1. Introduction

THz time-domain spectroscopy (THz-TDS) is one of the important detecting techniques in the THz research field, by which the time-dependent electric field of THz signals are measured [1-3]. Each THz-TDS signal records a temporal response of the THz reference pulse and can be represented by a pulse waveform, from which the THz spectrum including the amplitude and phase information is obtained by the Fourier transform [4]. Accordingly, the technique of terahertz spectroscopy is very attractive because many materials have an absorption band, which is called a fingerprint. THz-TDS image signal provides not only the spatial information but also the spectral information of the material at each pixel. One category of Terahertz spectroscopy studies is aiming to identify or differentiate substances on the basis of their terahertz time-domain spectroscopy (THz-TDS) signals and has been applied in the security checks, drug analysis, genetic engineering and chemistry and other fields.

Current processing and analysis of THz-TDS signals are generally using methods of the scalar data. Usually substances are identified using the amplitude waveform or the phase waveform in the frequency domain, focusing on the waveform properties, such as peaks, slopes and shifts [5,6]. Data processing techniques, such as neural network, wavelet, are mainly on the noise reduction of THz-TDS signals[7-9]. However, these scalar data processing methods have limitations. In practical applications, the analysis methods using waveform characteristics require high signal to noise ratio, and are not suitable to deal with signals of similar waveforms. Recently, the component spatial pattern analysis (CSPA) is proposed by Fukunaga, etc., [10,11]. Substances are identified using absorption spectra measured with a tunable THz wave source. Stability of the THz imaging system is required for high performance. And the substance identification is based only on amplitudes of THz waves regardless of their phase information.

One THz-TDS signal usually contains 1024 samples or more. Such high dimension would cause the data redundant and the high-dimensional disaster. It is necessary to process properly and effectively using some further signal processing techniques especially for high dimensional signals. As a mathematical description language, geometric algebra (GA) provides not only a powerful algebraic framework for the multi-dimensional vector analysis and computing, but also a unified measurement and geometrical description for different geometric models[12,13]. In recent years, geometric algebra has been successfully applied in the computer vision, color image processing, spectral image processing and high-dimensional data processing.

THz-TDS signals can be represented as hyper complex numbers [14-16]. Based on the physical mechanism of the THz-TDS signal analysis, geometric distribution properties and algebraic relationships of THz-TDS signals are analyzed under the framework of GA. From that, signals are demonstrated to belong to the intrinsic blade for each substance. Accordingly, a substance identification method via the conformal split is proposed. In this method, intrinsic blades for each substance are obtained using the normalized outer product of corresponding vectors to the substance with two different thicknesses. Substance of a sample is identified using the maximum magnitude of the corresponding vectors projected onto these blades. Finally, experiments are presented and show that the analysis in this paper is feasible and effective.

2. Physical Model of the THz-TDS Transmission System

The complex refractive index, $\hat{n}(\omega)=n(\omega)-j\kappa(\omega)$, comprises the optical constants: the index of refraction $n(\omega)$ and the extinction coefficient $\kappa(\omega)$. These constants of the sample can be calculated from the phase and amplitude of the transmitted THz waveforms.

A T-ray signal that passes through a parallel surfaced dielectric sample at the normal angle of incidence can be expressed as a function of the frequency, assuming no reflection, by $\hat{E}_{\text{sam}}(\omega)$. The signal measured with the same settings but without the sample is expressed by $\hat{E}_{\text{ref}}(\omega)$. Then the sample response normalized by the reference yields the complex transfer function of a material in the frequency domain [4]

$$\hat{h}(\omega)=\frac{\hat{E}_{\text{sam}}(\omega)}{\hat{E}_{\text{ref}}(\omega)}=\frac{4\hat{n}(\omega)n_0}{\left[\hat{n}(\omega)+n_0\right]^2}\exp\left\{-\kappa(\omega)\frac{\omega l}{c}\right\}\exp\left\{-j\left[n(\omega)-n_0\right]\frac{\omega l}{c}\right\} \qquad (1).$$

Taking the argument and logarithm of the transfer function gives, respectively,

$$\angle\hat{h}(\omega)=\angle\hat{f}(\omega)-\left[n(\omega)-n_0\right]\omega l/c \qquad (2a)$$

$$\ln\left|\hat{h}(\omega)\right|=\ln\left|\hat{f}(\omega)\right|-\kappa(\omega)\omega l/c \qquad (2b)$$

where n_0 is the refractive index of free air, l is the sample thickness and $\hat{f}(\omega) = 4\hat{n}(\omega)n_0 / [\hat{n}(\omega) + n_0]^2$ is a consequence of the Fresnel loss at the sample-air interfaces.

3. THz-TDS Signals Analysis and Substance Identification via the Conformal Split

Throughout this article, real vectors are denoted by boldface lowercase letters, real scalars are denoted by italic lowercase letters, complex vectors or scalars are denoted by the corresponding ones with a superscript ^, r-blades are denoted by boldface uppercase letters with the subscript r, while labels of substances are denoted as uppercase letters in superscripts.

Let any data vector $\mathbf{x} = [x_i]_{1\times m}$ be the discrete representation of a continuous signal $x(\omega)$ sampled at m frequencies $\{\omega_i, i = 1,2,\ldots,m\}$. Then discrete representations of the complex refractive index $\hat{n}(\omega)$, the transfer function $\hat{h}(\omega)$ and the Fresnel loss $\hat{f}(\omega)$ are respectively $\hat{\mathbf{n}} = [\hat{n}_i]_{1\times m}$, $\hat{\mathbf{h}} = [\hat{h}_i]_{1\times m}$ and $\hat{\mathbf{f}} = [\hat{f}_i]_{1\times m}$.

Note that, these discrete representations are complex vectors in an m-dimensional complex space. It is difficult to analysis THz-TDS signals or perform substance identification directly in such a high dimensional space. In this section, THz-TDS signals are first analyzed using GA. Two functions are defined mapping those complex vectors into real vectors. The linear relationship among those vectors is deduced on the basis of the physical model. From that, three theorems are proved and discussed. Following the fundamental decomposition theorem – the conformal split, the decomposition of vectors corresponding to THz-TDS signals are obtained. Consequently, a novel method of the substance identification based on THz-TDS signals is proposed using the conformal split.

3.1. Signal Analysis

To analysis THz-TDS signals using GA, two functions are defined mapping complex vectors into real vectors. Let C_m be an m-dimensional complex space, and V_n be a n-dimensional (n=2m) vector space over real numbers with the orthonormal basis of n space hyper-imaginary units $\mathbf{e}_i, i = 1,2,\ldots,n$, denoting the corresponding geometric algebra by $\mathrm{G}(\mathrm{V}_n)$. For any $\hat{\mathbf{c}} = [\hat{c}_i]_{1\times m} \in \mathrm{C}_m$, functions are defined by

$$g_1 : \mathrm{C}_m \to \mathrm{V}_n , \quad \hat{\mathbf{c}} \mapsto \mathbf{c} = g_1(\hat{\mathbf{c}}) = \Sigma_{i=1}^m \mathrm{Re}(\hat{c}_i)\mathbf{e}_i + \Sigma_{i=1}^m \mathrm{Im}(\hat{c}_i)\mathbf{e}_{m+i} \tag{3a}$$

$$g_2 : \mathrm{C}_m \to \mathrm{V}_n , \quad \hat{\mathbf{c}} \mapsto \mathbf{c} = g_2(\hat{\mathbf{c}}) = \Sigma_{i=1}^m \angle\hat{c}_i\mathbf{e}_i + \Sigma_{i=1}^m \ln|\hat{c}_i|\mathbf{e}_{m+i} \tag{3b}$$

where $\mathrm{Re}(\hat{c}_i)$ and $\mathrm{Im}(\hat{c}_i)$ are real part and imaginary part of respectively.

Using the function (3b), the vector relates to the transfer function $\hat{\mathbf{h}}$ is

$$\mathbf{h} = g_2(\hat{\mathbf{h}}) = \Sigma_{i=1}^m \angle\hat{h}_i\mathbf{e}_i + \Sigma_{i=1}^m \ln|\hat{h}_i|\mathbf{e}_{m+i} \tag{4}$$

Let $\hat{\alpha}_i = -(n_i - n_0 + j\kappa_i)\omega_i / c$ and $\hat{\boldsymbol{\alpha}} = [\hat{\alpha}_i]_{1\times m}$, from the equation (2a) and (2b), we have

$$\begin{aligned}\mathbf{h} = g_2(\hat{\mathbf{h}}) &= \Sigma_{i=1}^m \{\angle\hat{f}_i - [n_i - n_0]\omega_i l / c\}\mathbf{e}_i + \Sigma_{i=1}^m [\ln|\hat{f}_i| - \kappa_i\omega_i l / c]\mathbf{e}_{m+i} \\ &= \Sigma_{i=1}^m \angle\hat{f}_i\mathbf{e}_i + \Sigma_{i=1}^m \ln|\hat{f}_i|\mathbf{e}_{m+i} + \Sigma_{i=1}^m \{-[n_i - n_0]\omega_i l / c\}\mathbf{e}_i + \Sigma_{i=1}^m [-\kappa_i\omega_i l / c]\mathbf{e}_{m+i} \\ &= \Sigma_{i=1}^m \angle\hat{f}_i\mathbf{e}_i + \Sigma_{i=1}^m \ln|\hat{f}_i|\mathbf{e}_{m+i} + \Sigma_{i=1}^m \mathrm{Re}(\hat{\alpha}_i)\mathbf{e}_i + \Sigma_{i=1}^m \mathrm{Im}(\hat{\alpha}_i)\mathbf{e}_{m+i}\end{aligned}$$

that is $g_2(\hat{\mathbf{h}}) = g_2(\hat{\mathbf{f}}) + lg_1(\hat{\boldsymbol{\alpha}})$. Denoting $\boldsymbol{\alpha} = g_1(\hat{\boldsymbol{\alpha}})$ and $\mathbf{f} = g_2(\hat{\mathbf{f}})$, that equation can be simplified as:

$$\mathbf{h} = \mathbf{f} + l\boldsymbol{\alpha} \,. \tag{5}$$

The equation (5) reveals a linear relationship among real vectors correspond to the measurement of THz-TDS signals of the sample and the optical constants of the substance in the sample. The thickness of the sample is exactly the proportional scalar. To explicit deduce the following theorems, some notations and assumptions are clarified first.

For two substances, such as substance A and substance B, suppose $\hat{\mathbf{n}}^{\mathrm{A}} = \left[\hat{n}_i^{\mathrm{A}}\right]_{1\times m}$ and $\hat{\mathbf{n}}^{\mathrm{B}} = \left[\hat{n}_i^{\mathrm{B}}\right]_{1\times m}$ are their complex refractive indexes, l^{A} and l^{B} are their thickness values respectively, $\hat{\mathbf{h}}^{\mathrm{A}}$ and $\hat{\mathbf{h}}^{\mathrm{B}}$ are the corresponding measurements obtained from the THz-TDS transmission system, we made up the following assumptions.

Assumption1. Substances could be identified on basis of their THz-TDS signals. That is the vector, $\mathbf{h} = g_2\left(\hat{\mathbf{h}}\right)$, which relates to the measurement from the THz-TDS system via the equation (4), corresponds to the unique l and $\hat{\mathbf{n}}$ in the T-ray range.

We do not consider other sources' effecting on the measurements from the THz-TDS system in this paper.

Assumption2. It is assumed that there is no such substances A and B coexist that: their distinct complex refractive indexes, $\hat{n}_i^{\mathrm{A}} \neq \hat{n}_i^{\mathrm{B}}$, could satisfy any of the following relations for $\forall i = 1, 2, ..., m$:

(a) $\exists \lambda \in R$, $\hat{n}_i^{\mathrm{A}} = \lambda \hat{n}_i^{\mathrm{B}}$;

(b) $\exists \lambda \in R$, $\left(\hat{n}_i^{\mathrm{A}} - n_0\right) = \lambda\left(\hat{n}_i^{\mathrm{B}} - n_0\right)$;

(c) $\exists \lambda \in R$, $\hat{n}_i^{\mathrm{A}} = 4\lambda \hat{n}_i^{\mathrm{B}} n_0 / \left[\hat{n}_i^{\mathrm{B}} + n_0\right]^2$;

(d) $\exists \lambda \in R$, $\dfrac{\hat{n}_i^{\mathrm{A}}}{\left(n_0 + \hat{n}_i^{\mathrm{A}}\right)^2} = \lambda \dfrac{\hat{n}_i^{\mathrm{B}}}{\left(n_0 + \hat{n}_i^{\mathrm{B}}\right)^2}$;

(e) $\hat{n}_i^{\mathrm{A}} \hat{n}_i^{\mathrm{B}} = n_0^2$;

That is they are neither linearly dependent nor complex conjugate of each other. Then for substances with distinct complex refractive indexes, their corresponding vectors $\boldsymbol{\alpha}$ and $\mathbf{f}$ are linearly independent of each other, that is

$$\hat{n}_i^{\mathrm{A}} \neq \hat{n}_i^{\mathrm{B}} \; (\forall i = 1, 2, ..., m) \Leftrightarrow \boldsymbol{\alpha}^{\mathrm{A}} \wedge \mathbf{f}^{\mathrm{A}} \wedge \boldsymbol{\alpha}^{\mathrm{B}} \wedge \mathbf{f}^{\mathrm{B}} \neq 0 \,.$$

Based on the assumptions and the equation (5), following properties of the vectors can be deduced.

Theorem1. Under the assumptions, every sequence of the complex refractive index, $\hat{\mathbf{n}}$, relates to a unique $\hat{\boldsymbol{\alpha}}$ and a unique $\hat{\mathbf{f}}$ in C_m, consequently, every distinct $\hat{\mathbf{n}}$ determines a unique vector $\boldsymbol{\alpha}$ and a unique vector $\mathbf{f}$ in V_n, and vice versa.

Proof For the substances A and B, which are with exactly the same complex refractive indexes

$$\hat{\mathbf{n}}^{\mathrm{A}} = \hat{\mathbf{n}}^{\mathrm{B}}, \quad \text{i.e.} \quad \forall i, n_i^{\mathrm{A}} = n_i^{\mathrm{B}} \text{ and } \kappa_i^{\mathrm{A}} = \kappa_i^{\mathrm{B}} \,.$$

From the definition of $\hat{\alpha}_i = -\left(n_i - n_0 + j\kappa_i\right)\omega_i / c$, it is obviously $\forall i, \hat{\alpha}_i^{\mathrm{A}} = \hat{\alpha}_i^{\mathrm{B}}$. That is

$$\hat{\mathbf{n}}^{\mathrm{A}} = \hat{\mathbf{n}}^{\mathrm{B}} \Leftrightarrow \hat{\boldsymbol{\alpha}}^{\mathrm{A}} = \hat{\boldsymbol{\alpha}}^{\mathrm{B}} \Leftrightarrow \boldsymbol{\alpha}^{\mathrm{A}} = \boldsymbol{\alpha}^{\mathrm{B}} \,.$$

$$\because \quad \forall i, \quad \frac{\hat{n}_i^{\mathrm{A}}}{\left(n_0 + \hat{n}_i^{\mathrm{A}}\right)^2} = \frac{\hat{n}_i^{\mathrm{B}}}{\left(n_0 + \hat{n}_i^{\mathrm{B}}\right)^2} \Leftrightarrow \hat{n}_i^{\mathrm{A}} = \hat{n}_i^{\mathrm{B}} \text{ or } \hat{n}_i^{\mathrm{A}} \hat{n}_i^{\mathrm{B}} = n_0^2 \,.$$

Based on our assumption 2(e), such substances A and B would not be considered coincidently that their complex refractive indexes could satisfy $\hat{n}_i^{A}\hat{n}_i^{B} = n_0^2 \quad (\forall i = 1, 2, ..., m)$.

$$\therefore \quad \hat{\mathbf{n}}^{A} = \hat{\mathbf{n}}^{B} \Leftrightarrow \hat{\mathbf{f}}^{A} = \hat{\mathbf{f}}^{B} \Leftrightarrow \mathbf{f}^{A} = \mathbf{f}^{B} \blacksquare$$

So, for distinct complex refractive indexes in the space C_m, each determines a unique vector $\boldsymbol{\alpha}$ and a unique vector $\mathbf{f}$ in V_n. Conversely, each distinct vector $\boldsymbol{\alpha}$ in V_n is corresponding to a unique $\hat{\mathbf{n}}$ in C_m and so is $\mathbf{f}$.

Theorem2. Each complex refractive indexes $\hat{\mathbf{n}}$ relates to a unique 2-blade $\mathbf{B}_2$ in V_n, where $\mathbf{B}_2 \; \dfrac{\boldsymbol{\alpha} \wedge \mathbf{f}}{|\boldsymbol{\alpha} \wedge \mathbf{f}|}$. And it determines a unique 2-dimensional subspace V_2 in V_n which is the solution set of the equation $\mathbf{x} \wedge \mathbf{B}_2 = 0$.

Proof From our assumption 2, it can be deduced that

$$\hat{\mathbf{n}}^{A} = \hat{\mathbf{n}}^{B} \Leftrightarrow \boldsymbol{\alpha}^{A} \wedge \boldsymbol{\alpha}^{B} = 0, \mathbf{f}^{A} \wedge \mathbf{f}^{B} = 0.$$

And so, $\hat{\mathbf{n}}^{A} = \hat{\mathbf{n}}^{B} \Leftrightarrow \exists \lambda_1, \lambda_2 \in R$, and $\lambda_1\lambda_2 \neq 0, \ni \boldsymbol{\alpha}^{A} = \lambda_1 \boldsymbol{\alpha}^{B}$, and $\mathbf{f}^{A} = \lambda_2 \mathbf{f}^{B}$

$$\because \frac{\boldsymbol{\alpha}^{A} \wedge \mathbf{f}^{A}}{|\boldsymbol{\alpha}^{A} \wedge \mathbf{f}^{A}|} = \frac{\lambda_1 \boldsymbol{\alpha}^{B} \wedge \lambda_2 \mathbf{f}^{B}}{|\lambda_1 \boldsymbol{\alpha}^{B} \wedge \lambda_2 \mathbf{f}^{B}|} = \frac{\boldsymbol{\alpha}^{B} \wedge \mathbf{f}^{B}}{|\boldsymbol{\alpha}^{B} \wedge \mathbf{f}^{B}|}$$

$$\therefore \hat{\mathbf{n}}^{A} = \hat{\mathbf{n}}^{B} \Leftrightarrow \mathbf{B}_2^{A} = \mathbf{B}_2^{B} \blacksquare$$

Theorem3. All vectors, $\mathbf{h}_l = \mathbf{f} + l\boldsymbol{\alpha} \; (\forall l \in R)$ are collinear and belong to $\mathbf{B}_2$.

Proof Using properties of the outer product, it is obviously true that

$$\mathbf{h}_l \wedge \mathbf{A}_2 = (\mathbf{f} + l\boldsymbol{\alpha}) \wedge \boldsymbol{\alpha} \wedge \mathbf{f} = \mathbf{f} \wedge \boldsymbol{\alpha} \wedge \mathbf{f} + l\boldsymbol{\alpha} \wedge \boldsymbol{\alpha} \wedge \mathbf{f} = 0 \; (\forall l \in R) \blacksquare$$

These theorems are important and usable in the substance identification. Theorem 2 demonstrates a unique 2-blade $\mathbf{B}_2$ relates to each substance. And theorem 3 reveals that vectors corresponding to THz-TDS signals from the same substances but with different thickness belong to the same blade. The geometric meaning of these theorems is interpreted in the figure 1. From the theorem 2, as the complex refractive indexes are different, the blades $\mathbf{B}_2^{A}$ and $\mathbf{B}_2^{B}$ are linearly independent $\mathbf{B}_2^{A} \wedge \mathbf{B}_2^{B} \neq 0$. That in the geometric interpretation is they are not lying on the same plane. Since h_{l1}^{A} and h_{l2}^{A} are corresponding to the same substance A, they are collinear and on the same plane of $\mathbf{B}_2^{A}$, regardless of their different thickness values.

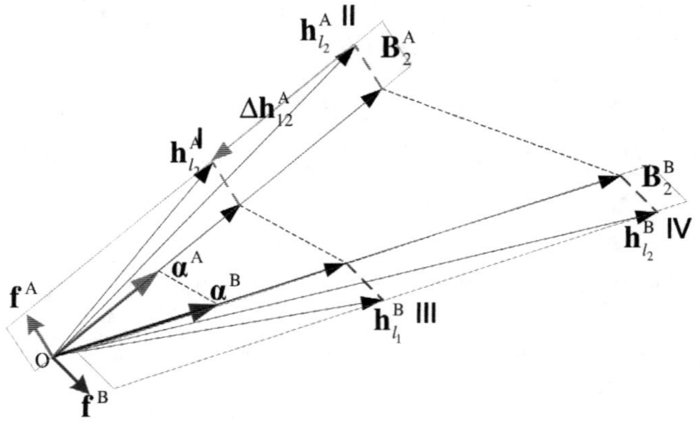

Figure 1. Distribution of vectors. Four different samples: samples I and II are homogeneity of substance A but different thickness l_1 and l_2, samples III and IV are homogeneity of substance B but different thickness l_1 and l_2, and $\tilde{\mathbf{n}}^{A} \neq \tilde{\mathbf{n}}^{B} 0, l_1 \neq l_2 \neq 0$.

3.2. Substance Identification

As in GA, to every r-dimensional subspace V_r in V_n in there corresponds an r-blade $\mathbf{B}_r$ such that V_r is the solution set of the equation $\mathbf{x} \wedge \mathbf{B}_r = 0$. $\mathbf{B}_r$ is called the blade of V_r while V_r is called the support of $\mathbf{B}_r$ [12]. The 2-blade $\mathbf{B}_2$ defined in the theorem 2 determines a unique subspace V_2 for each substance, which is independent of thickness values of samples. So, the substance identification can be performed by mapping n-dimensional vectors $\mathbf{h}_l$ into each 2-dimensional subspace V_2 (the support of $\mathbf{B}_2$). Accordingly, we proposed a substance identification method via the conformal split (CS).

In the study of D. Hestenes, geometric algebra can be factored multiplicatively according to the fundamental theorem: $G_n = G_2 \otimes G_{n-2}$, (The Kronecker product $\otimes$) [12]. A conformal split by a bivector determines a geometrically significant relation between the geometric algebras G_n and G_{n-2}. Let $\mathbf{B}_2$ be a fixed unit 2-blade in G_{n-2} and let $\boldsymbol{x}$ be a generic vector in $V_{n-2} = G_{n-2}$. A linear split of V_n into vector spaces V_2 and V_{n-2} is determined by the equation

$$\mathbf{x}\mathbf{B}_2 = \mathbf{x} \cdot \mathbf{B}_2 + \mathbf{x} \wedge \mathbf{B}_2 = \mathbf{x}_0 + \rho\mathbf{X} \tag{6}$$

where $V_2 = \{\mathbf{x}_0 = \mathbf{x} \cdot \mathbf{B}_2\}$ and $V_{n-2} = \{\rho\mathbf{X} = \mathbf{x} \wedge \mathbf{B}_2\}$ [13].

Given a fixed 2-blade $\mathbf{B}_2$ of one substance, the vector $\mathbf{h}_l$ in V_n corresponding to THz-TDS signals from one sample, it can be linearly mapped into one 2-D subspace V_2 and the other subspace V_{n-2} using the conformal split (6):

$$\mathbf{h}_l\mathbf{B}_2 = \mathbf{h}_l \cdot \mathbf{B}_2 + \mathbf{h}_l \wedge \mathbf{B}_2 = \mathbf{h}_0 + \rho\mathbf{H} \tag{7}$$

where $\mathbf{h}_0 = \mathbf{h}_l \cdot \mathbf{B}_2$. Using the above equation, high dimensional vectors of THz-TDS signals is mapped into 2-D vectors.

In the process of mapping the vector $\mathbf{h}_l$ into the 2-D subspace intrinsic to one substance, it is necessary to obtain the unique 2-blade $\mathbf{B}_2$ of that substance first. In the condition the complex refractive index $\hat{\mathbf{n}}$ of that substance is already known, the 2-blade $\mathbf{B}_2$ can be easily obtained through its definition in the theorem 2. In the general cases the complex refractive indexes of substances are unknown, the 2-blade $\mathbf{B}_2$ of each substance can be evaluated using experimental measurements, which is discussed in the following.

Intrinsic blades for "known" substances can be experimental computed. Given two THz-TDS signals measured respectively from samples of the same substance "A" but with thickness l_1 and l_2, they can be expressed by hypercomplex numbers, $\mathbf{h}_{l1}$ and $\mathbf{h}_{l2}$, in form of the equation (4). From the equation (5), it is deduced that

$$\begin{aligned}\mathbf{h}_{l1}\wedge\mathbf{h}_{l2} &= (\mathbf{f}+l_1\boldsymbol{\alpha})\wedge(\mathbf{f}+l_2\boldsymbol{\alpha})\\ &= \mathbf{f}\wedge\mathbf{f}+l_2\mathbf{f}\wedge\boldsymbol{\alpha}+l_1\boldsymbol{\alpha}\wedge\mathbf{f}+l_1\boldsymbol{\alpha}\wedge l_2\boldsymbol{\alpha}\\ &= (l_1-l_2)\boldsymbol{\alpha}\wedge\mathbf{f}\end{aligned}$$

As in the definition of $\mathbf{B}_2$ in the theorem 2, $\mathbf{B}_2$ is a unit 2-blade, so

$$\mathbf{h}_{l1}\wedge\mathbf{h}_{l2} = \lambda\mathbf{B}_2, \qquad (\lambda\in \mathrm{R})$$

Consequently the computation of $\mathbf{B}_2$ is as:

$$\mathbf{B}_2 = \frac{\mathbf{h}_{l1}\wedge\mathbf{h}_{l2}}{\left|\mathbf{h}_{l1}\wedge\mathbf{h}_{l2}\right|}. \tag{8}$$

Blades of some substances to be identified can be obtained using the above equation. For the ith substance A_i , i=1,2,…, it is measured first THz-TDS signals from the samples of the same substance but with two different thicknesses. With vectors corresponding to those signals, $\mathbf{h}_{l1}^{\mathrm{A}_i}$ and $\mathbf{h}_{l2}^{\mathrm{A}_i}$, the blade of that substance can be evaluated using the equation (8).

For the vector $\mathbf{h}_l^{\mathrm{M}}$ corresponding to signal from one sample to be identified, it can be linearly decomposed into one 2-D subspace of and the other subspace of 2 dimension lower with respect to the blade $\mathbf{B}_2^{\mathrm{A}_i}$ of each substance using the equation (7):

$$\mathbf{h}_l^{\mathrm{M}}\mathbf{B}_2^{\mathrm{A}_i} = \mathbf{h}_l^{\mathrm{M}}\cdot\mathbf{B}_2^{\mathrm{A}_i} + \mathbf{h}_l^{\mathrm{M}}\wedge\mathbf{B}_2^{\mathrm{A}_i} = \mathbf{h}_0^{\mathrm{MA}i} + \rho\mathbf{H}^{\mathrm{MA}i}.$$

Magnitudes of the projection of $\mathbf{h}_l^{\mathrm{M}}$ onto the blades can be obtained, which is used as the criterion in the substance identification. The sample is then identified as the k-th substance based on the maximum magnitude:

$$k = \arg\max_i \left|d_{\mathrm{A}i}^{\mathrm{M}}\right|, \tag{9}$$

where $d_{\mathrm{A}i}^{\mathrm{M}} = \left|\mathbf{h}_0^{\mathrm{MA}i}\right| = \left|\mathbf{h}_l^{\mathrm{M}}\cdot\mathbf{B}_2^{\mathrm{A}_i}\right|$

Consequently, the THz-TDS signal of an unknown substance can be identified by comparing its mapping vectors in these blades intrinsic of "known" substances. The figure 2 illustrates the process of the substance identification (CS).

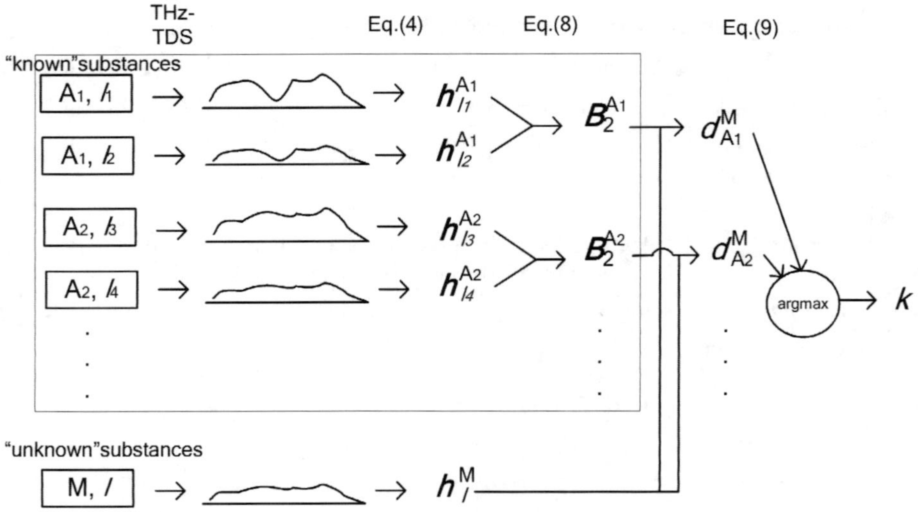

Figure 2. Illustration of the substance identification (CS).

4. Results

4.1. Experiment I

Ten THz-TDS signals are obtained and also their related reference signals from five substance samples using our THz-TDS system in the transmission model. The substances are marked as CARD, OSA, PTFE, RB, and WB. Figure 2 (a) shows the amplitudes of the transfer functions for these signals. These waveforms are quite noisy and fluctuated. It is difficult to classify three substances, which are WB, RB and OSA, from each other as their waveforms are quite similar. Blades intrinsic to these five substances are computed as the outer product of the vector pairs corresponding to the signals from the same substance.

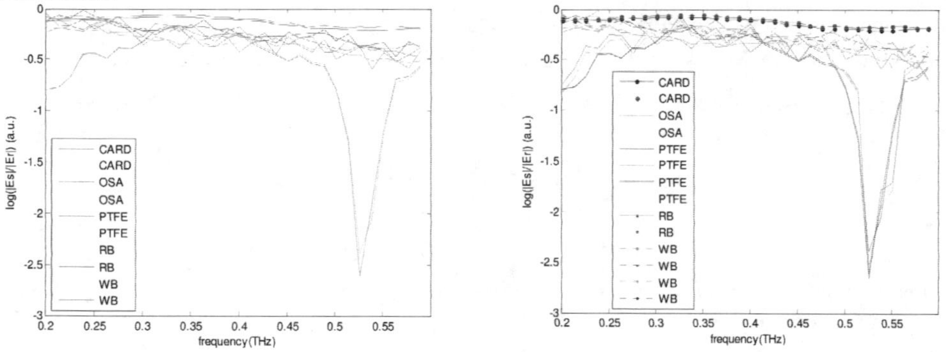

(a) signal pairs from five distinct substances (b) fourteen signals from five samples

Figure 3. THz-TDS signals

Fourteen raw THz-TDS signals (shown in the figure 3(b)) are also measured from these five distinct samples. Noisy signals simulated on these fourteen signals are identified using our method. Simulations are performed by adding Gaussian white noises to signals with zero mean and different variances (listed in the table 1). Each simulation is run 500 times and 500 noisy signals are obtained for each raw signal. The substances identification method is performed on these noisy signals. Table 1 shows the accurate rate of identification over 500 runs for each simulation. The average of the signal noise ratios (SNR) in each simulation are also calculated and listed out in the table 1. Simulation experiments show the high performance of the method on the identification of substances on the basis of their THz signals. This experiment demonstrates the feasibility of our method.

Table 1. Accurate rate of identification

The standard deviation ( )	SNR	Accurate rate of identification (%)
0.001	51.0786	100%
0.01	31.1131	100%
0.05	17.2672	98.10%
0.1	11.6075	86.50%

4.2. Experiment II

A THz-TDS image was simulated. It was composed of 192*265 pixels and imaged at 32 frequencies (0.2THz~0.6THz). Experimental THz-TDS signals of substances 'CARD', 'OSA' and 'PTFE' were used. Each substance was simulated as being placed in a small block (a region of interest, ROI). Gaussian noises with zero mean and different variances (listed in the table 2) were added to the THz-TDS image. Such that, four simulated noisy THz-TDS images were obtained.

Substances identification on these THz-TDS images was performed using our method (CS). The visualized results are shown in the figure 4. For compare, substances identification was also performed using the component spatial pattern analysis (CSPA) [10,11]. Accurate rates (listed out in the table 2) of both method in each experiment are calculated. In the noisy conditions σ =0.5 and 0.1, the accurate rates of our method (CS) are higher than those of the CSPA method. That shows the advantage of our method (CS).

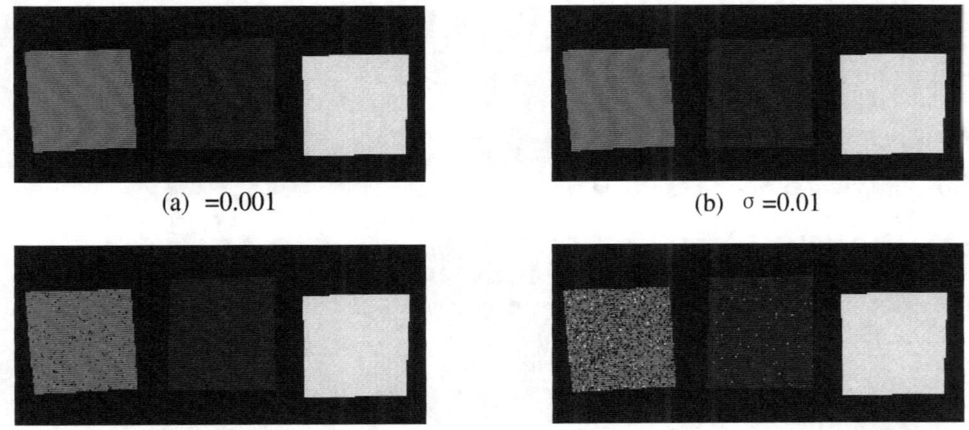

(a) =0.001 (b) σ =0.01

(c) σ =0.05 (d) σ =0.1

Figure 4. results of the substance identification (CS)

Table 2. Accurate rate of identification

The standard deviation ( )	SNR	Accurate rate of identification (%)	
		CS	CSPA
0.001	50.6070	100%	100%
0.01	30.6097	100%	99.99%
0.05	16.7254	98.83%	91.34%
0.1	10.9694	90.78%	83.05%

5. Conclusion

In conclusion, a novel analysis of the THz-TDS signals using GA is presented. THz-TDS signals, described in the high dimensional vector space, are mapped into the hyper-complex numbers in GA. Geometric properties of vectors corresponding to signals from the same substance but different thickness are deduced using the language of GA. From that, signals are demonstrated to belong to the intrinsic blade for each substance.

A substance identification method via the conformal split is proposed. In this method, intrinsic blades for each substance are obtained using the normalized outer product of vectors corresponding to the THz-TDS signals from the same substance but two different thicknesses. And for each sample to be identified, the corresponding vector is projected onto each blade. The maximum magnitude of the projected vector is used to identify the sample's substance.

Experiments demonstrate the feasibility and high accuracy of our method on the applications of identify substances on the basis of their THz transmission spectra.

Acknowledgment
This study is supported by the National Natural Science Funds (No. 61071206, P.R.China).

References

[1] D. M. Mittleman, S. Hunsche, etc., T-ray tomography [J]. Optics letters, 1997, Vol.22(12):904-906
[2] W. X. Xie, J. H. Pei, "Review of Terahertz signal processing and analysis", ACTA Electronica sinica, 35(10) , 1973-1978, (2007)
[3] Pei Ji-hong, Hu Yong, Xie Wei-xin, PCA-based visualization of terahertz time-domain spectroscopy image[A], ,MIPPR2007: Multispectral Image Processing[C], Wuhan, China, Proc. of SPIE, 2007 Vol. 6787, 67871M-1: 67871M-7Nov 15-17
[4] W. Withawat, B.M. Fischer, D. Abbott, "Material thickness optimization for transmission-mode terahertz time-domain spectroscopy", Optics express, 16(10), 7382-7396, (2008)
[5] H. B. Liu, X. C. Zhang, Dehydration kinetics of D-glucose monohydrate studied using THz time-domain spectroscopy [J]. Chemical Physics Letters, 2006, Vol.429:229–233
[6] Pupeza, R. Wilk, M. Koch, Highly accurate optical material parameter determination with THz time-domain spectroscopy [J]. Optics express,2007, Vol.15(7):4335-4350
[7] J.H. Pei•, P.L. Ye and W.X. Xie, Optimal wavelet analysis for THz-TDS pulse signals[A], Photonics and Optoelectronics Meetings, POEM 2008[C], Wuhan, China
[8] Sillas Hadjiloucas, Roberto K. H. Galvão, John W. Bowen, Analysis of spectroscopic measurements of leaf water content at terahertz frequencies using linear transforms [J]. Opt. Soc. Am. A, Vol. 19(12), 2002: 2495-2509
[9] Roberto K. H. Galvão, Sillas Hadjiloucas and John W. Bowen, Clarimar J. Coelho, Optimal discrimination and classification of THz spectra in the wavelet domain[J]. OPTICS EXPRESS, OSA, 2003, Vol. 11(12),:1462-1472
[10] K. Fukunaga, Y. Ogawa, etc, Application of terahertz spectroscopy for character recognition in a medieval manuscript[J]., IEICE Electronics Express, 2008, Vol.5(7):223-228
[11] K. Kawase, Y. Ogawa, Y. Watanabe, Non-destructive terahertz imaging of illicit drugs using spectral fingerprints, OPTICS EXPRESS, OSA, 2003, Vol.11(20):2549-2554
[12] D. Hestenes, Universal geometric algebra, [J], A quarterly J. of Pure and Applied Mathematics, 1988, Vol. 62(3-4)
[13] D. Hestenes, The design of linear algebra and geometry, [J], Acta Applicandae Mathematicae, 1991, Vol. 23, pp. 65-93
[14] W. Xie, J. Li, J.H. Pei, A Clifford Algebra Analysis Method of THz-TDS Images [A]. 9th International Conference on Signal Processing (ICSP 2008 Proceedings) [C]. Beijing, China, IEEE Press, 2008.128-131
[15] W. Xie, J. Li, J.H. Pei, An Analysis of THz-TDS Signals using Geometric Algebra [A]. Photonics and Optoelectronics Meetings, POEM 2008, Wuhan, China

[16] W. Xie, J.H. Pei, J. Li, A Dimensionality Reduction Method for THz-TDS Signals via the Recursive Projective Splits based on PCA, Terahertz Science and Technology, the On-line International Journal, JTST, Vol.2(2), 2009

Electrically controlled broadband THz switch based on liquid-crystal-filled multi-layer metallic grating structures

Yinghao Yuan[1], Jian He[1], Jinsong Liu[1,3] and Jianquan Yao[1,2]

[1] Wuhan National Laboratory for Optoelectronics, School of Optoelectronic Science and Engineering, Huazhong University of Science and Technology, Wuhan 430074, China
[2] College of Precision Instrument and Optoelectronics Engineering, Tianjin University, Tianjin 300072, China

E-mail: jsliu4508@vip.sina.com

Abstract. We propose an electrically controlled broadband THz switch based on the liquid-crystal-filled multilayer metallic grating (MMG) structures. During the THz wave passing thought the device, it is multiple reflected among the MMG layers and suffers loss due to the absorption of the LC. The on and off state of the switch is realized by the different absorption intensity of the LC when its complex refractive index is switched between ordinary n_o and extraordinary n_e value in the presence of an electric field. Simulation results show the proposed device exhibits a respectable switching effect for the THz wave over a wide frequency range.

1. Introduction

Terahertz has been proved greatly potential applications in many domains including imaging, environment monitoring, medical diagnosing and broadband mobile communications [1]. However, until now there are few reports on the commercial THz products, despite the THz generation and detection techniques being available for many years. This is partly due to the lack of components to guide and manipulate THz waves. In the field of building tunable THz devices, the liquid crystal (LC) materials are potentially suitable, as they have relatively large birefringence and low absorption in the THz range [2]. Up to now, several LC-based THz devices have been reported [3-14], such as phase shifters [4-7, 11], filters [14, 15], switches [9, 12] and polarizers [13]. At THz region, the thickness of the LC must be the order of wavelength or longer in order to achieve significant tunable range. However, it is difficult to align the thick LC cell, especially when the thickness is of the order of hundred microns [8, 14]. Although this problem can be overcome by using strong external magnetic fields, it makes them inconvenient for practical use [4, 7]. For a compact design, another approach to control the orientation of the LC is by applying a high electric field. But it is not easy to fabricate the transparency conductive electrodes, as even metal layers as thin as tens of nanometers are almost completely opaque to THz waves. This issue makes it more difficult to design an effective electrical controlled THz device based on LC.

Recently, we proposed an electrically controlled THz switch based on the metallic grating-liquid crystal-metallic grating (MG-LC-MG) structures [16]. The switching mechanism is realized by

[3] To whom any correspondence should be addressed.

modifying the effective refractive of the LC using different bias electric fields. In this novel design, the MGs not only support the super transmittance at certain frequencies, but also act as the electrodes. Therefore, the proposed device has the potential advantage of avoiding the insertion loss due to the indium tin oxide (ITO) electrodes used in [14, 17]. However, the operate frequencies should be carefully chose at the Fabry-Perot (FP) resonance peaks of the MG-LC-MG structures, that is, it is a narrowband THz switch.

Here, we propose an electrically controlled broadband THz switch based on the liquid-crystal-filled multilayer metallic grating (MMG) structures. During the THz wave passing thought the device, it is multiple reflected among the MMG layers and suffers loss due to the absorption of the LC. The on and off state of the switch is realized by the different absorption intensity of the LC when its complex refractive index is switched between ordinary n_o and extraordinary n_e value in the presence of an electric field. Simulation results show the proposed device exhibits a respectable switching effect for the THz wave over a wide frequency range.

2. Theory model and simulation method

Figure 1 shows the schematic diagram of our proposed MMG device. The device comprises multilayer identical MGs parallel to each other with the same spacing (S) filled with liquid crystal. The MG periodicity and the thickness of the metal stripe are p = 100 μm and h = 0.32 μm, respectively. The incident THz beam is p-polarized.

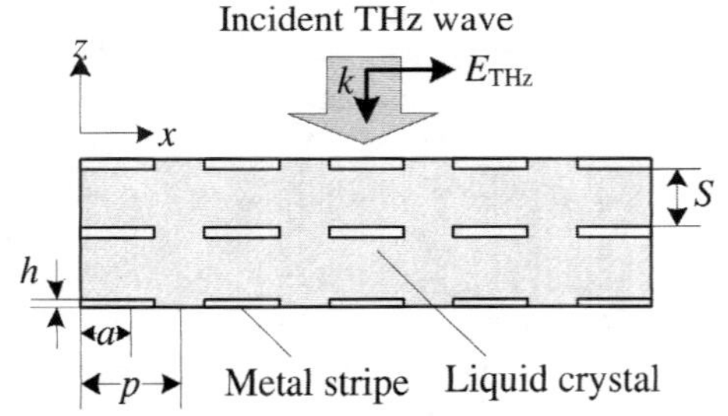

Figure 1. Schematic of the multi-layer metallic grating structures. The spacing of the MGs is filled with LC.

When the wavelength of the incident THz wave is much greater than the period of the MG, the MG can be regarded as homogeneous lamella. Each MG can provide a long-wavelength reflectance with negligible absorption losses [18]. For the normal incidence, the effective refractive index for the p-polarized wave is [19]

$$n_{\mathrm{m}} = \left(\frac{f}{\epsilon_{\mathrm{m}}} + \frac{1-f}{\epsilon_{\mathrm{LC}}} \right)^{-\frac{1}{2}} \tag{1}$$

where the fill factor $f = \frac{a}{p}$ is defined as the ratio of the width of metal stripe a to the MG period p. ϵ_{m} and ϵ_{LC} are the relative permittivity of metal and the LC, respectively. If the spacing S is sufficiently great, there will no evanescent coupling among the MGs, the proposed device shown in Figure 1 acts as a multilayer mirror. The transmission spectrum exhibit reflection bands around the frequencies:

$$f_{\mathrm{B}} = \frac{cm}{2n_{\mathrm{LC}}S} \qquad (m = 1, 2, 3 \cdots) \tag{2}$$

here, c is the speed of light in vacuum, n_{LC} is the effective refractive index of the LC, which can be switched between its ordinary n_o and extraordinary n_e value in the presence of an electric filed. This results in a change of the transmission properties of the multilayer structure.

In order to study the detail transmission characteristics of the device, especially the influence of changing the LC refractive index on the transmission spectra, we carried out rigorous coupled wave analysis (RCWA) method in the in the next section. For simplicity, we simulate the model in 2D space, and the metal stripes are considered infinitely along the horizontal direction. In the calculation, the LC is chosen as 7CB, whose n_o and n_e are approximate to 1.61+i0.028 and 1.73+i0.016, respectively [17]. The relative permittivity of metal is $\epsilon_m = -3\times10^4 + i10^5$ [20].

3. Results and discussions

Under the normal incidence, over the frequency range from 0.3THz to 2 THz, the transmission spectra of the p-polarized wave for different MG layers and LC refractive indices are shown in Figure 2. In the transmission spectra of the dual-layer MGs structure (L=2), large transmission peaks are observed, they correspond to the FP multiple reflections between the two MGs. When the layers of the MG is greater than two, serials of small oscillations arise around the FP resonance peaks, which are equal to the number of the LC layers (L-1) in quantity. As the MG layers increases, theses small oscillations superimposed together, forming a broad transmission band, as the one seen at the range of 0.5 THz to 0.6 THz in Figure 2(a). The MMG (L>2) shows the transmission characteristics of a Bragg structure, narrow reflection bands appear near the frequencies indicated by (2). The width of the reflection band is narrowed with the increasing of the MG layers. For the MMG with eight layers, the width of the reflection bands already reduced to less than 50 GHz.

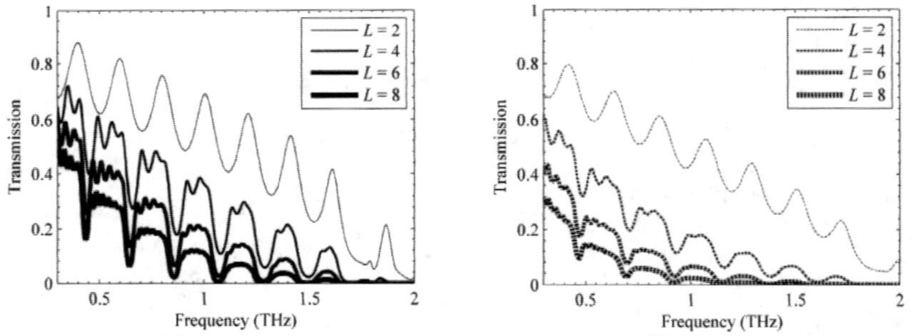

(a) $n_{LC} = n_e$. (b) $n_{LC} = n_o$.

Figure 2 Calculated transmissions of the MMG structures versus frequency for different MG layers and refractive indices of LC. The spacing of the MGs is 400 μm. The width of the metal stripe is 50 μm.

Comparing Figure 2(a) with Figure 2(b), one can see that the positions of the narrow reflection bands are shifted when the refractive index of the LC changes form n_e to n_o. This feature can be used to design a narrow band THz switch. However, the operate frequencies should be accurately chosen in the narrow reflection bands. This may be inconvenient in practical applications. On the other hand, we can notice that the transmission rates of the transmission bands are also strongly influenced. For example, the average transmission rates in the frequency range from 0.5 THz to 0.6 THz shown in Figure 2(a) and Figure 2(b) for L = 6 are changed from 0.4 to 0.2. The changing of the transmission rate is mainly due to the different absorption loss of the LC when its imaginary part of the complex refractive index is switched from Im(n_e) to Im(n_o). This conclusion can be confirmed by Figure 3. In Figure 3, we plot the transmission spectra of a six-layer MMG structure with the same calculation parameters used in Figure 2, except the absorption of the LC, i.e., the imaginary part of its complex refractive index is neglected. In this case, when the LC refractive index changes, the shift of the narrow reflection band is still observed, but there is no significant transmittance modulation phenomenon. In addition, the transmission spectra in the range above 1.8 THz show different characteristics, this is because the incident frequency has beyond the diffraction limit, part of the incoming THz beam is coupled into guided mode in the LC layer [21]. Such phenomenon is not seen in Figure 2, as most of the THz wave has been attenuated due to the loss of the LC.

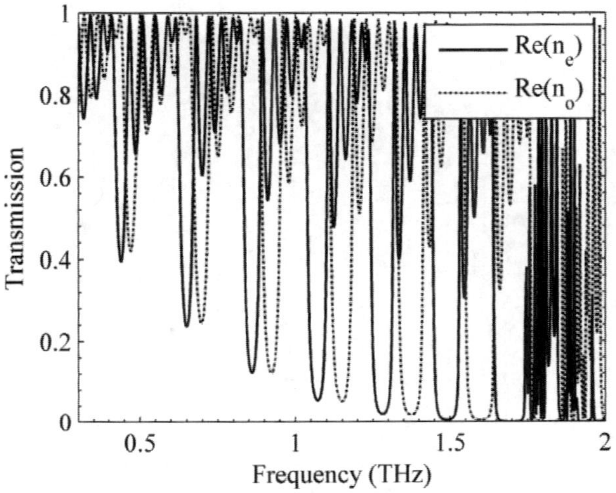

Figure 3. Calculated transmissions of a six-layer MMG structure versus frequency for different refractive indices of LC. The absorption losses of the LC are neglected.

The transmittance modulation phenomenon described above is very suitable to design a broadband THz wave switch. As the modulation mechanism is the result of changes of the LC absorption, the variation of the imaginary part of the LC refractive index is essential. For the LC material 7CB, such value is quite small: $\Delta\mathrm{Im}(n_{LC}) = 0.012$ [17]. In order to obtain more significant modulation effect, it is necessary to increase the THz wave propagate path in the LC. This can be directly achieved by increase the thickness of LC layer. However, it becomes difficult to control the orientation of LC molecules, especially when the thickness is of the order of hundred microns [8]. Another solution is to increase the number of MG layers, as more MG layers means that the THz wave will be reflected more times when passing through the device. We define the extinction ratio $\mathrm{EXT} = 10\log(T_{ON}/T_{OFF})$, where T_{ON} and T_{OFF} are the transmission rates within the frequency range from 0.5 THz to 0.6 THz in the case of $n_{LC} = n_e$ and $n_{LC} = n_o$, respectively. From Figure 2, it can be seen that the EXT is approximately increasing from 4 dB to 10 dB when the number of MG layers changes from 4 to 8. If the metal layers continue to increase, it is possible to get a higher EXT. However, this will lead to serious insertion loss (IL). So there is a trade-off between the EXT and IL.

In certain applications, it is expected that the THz switch can be operated over a broad frequency range. One can easily observe from Figure 2 that the operating bandwidth of the THz switch is affected by the width of the reflection band. Figure 4 shows a comparison of the calculated transmissions of a six-layer MMG structure with different metal strip widths and refractive indices of LC. It is clear that the width of the reflection band can be adjusted by changing the width of metal stripe under the condition of constant MG layers. The reflection band becomes more and more narrow with decreasing width of the metal stripe. When the width of the metal stripe decreased to 20 μm, the reflection band is almost disappeared in the transmission spectra. In this case, the THz switch can be worked over the whole frequency range shown in Figure 4(a). However, this ideal situation is impossible in practice. In the design of this paper, a bias voltage is applied on the metal stripes of the top and bottom layers of the MGs to provide the electric field to orient the LC molecules along the z-axis. If the metal stripes are too narrow, the electric field distribution will no longer be quasi-longitudinal. In other words, the reflection of the LC cannot be effectively changed from n_e to n_o. As a result, the EXT will become worse. Therefore, there is also a trade-off between the EXT and working bandwidth.

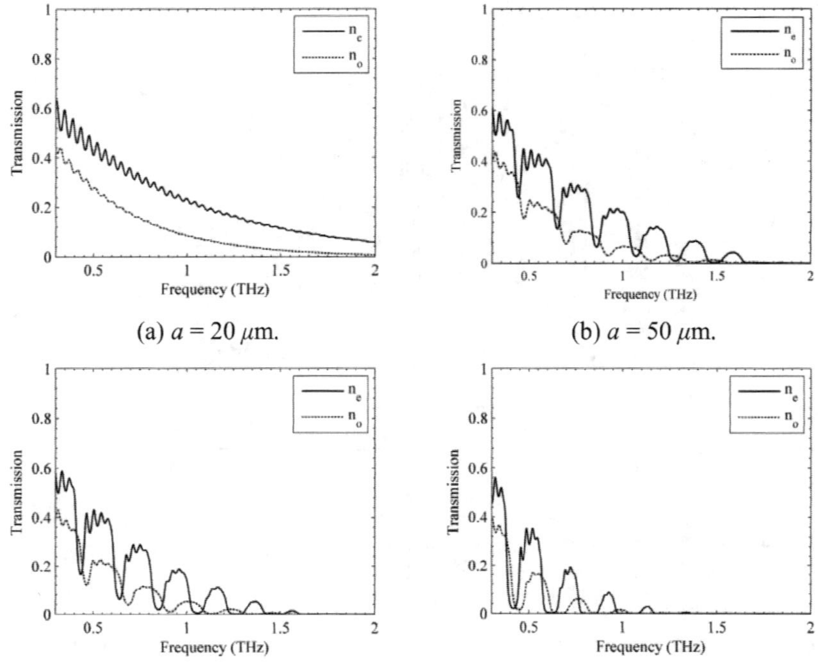

(a) $a = 20\ \mu$m. (b) $a = 50\ \mu$m.

(a) $a = 60\ \mu$m. (a) $a = 80\ \mu$m.

Figure 4. Calculated transmissions of a six-layer MMG structure versus frequency for different metal strip widths and refractive indices of LC.

4. Conclusion

In conclusion, we have proposed an electronically controlled broadband THz switch based on the liquid-crystal-filled multilayer metallic grating structures. During the THz wave passing thought the device, it is multiple reflected among the metallic grating layers and suffers loss due to the absorption of the LC. The on and off state of the switch is realized by the different absorption intensity of the LC when its complex refractive index is switched between ordinary n_o and extraordinary n_e by external electric field. The relationship between the performance parameters of the THz switch and the device design parameters are studied by RCWA method. Simulation results show that: (1) the extinction ratio is increasing with the number of the metallic grating layers, but accompanying with the deterioration of the insertion loss; (2) a wide operating bandwidth can be obtained with relatively narrow metal strips, however, this will lead to a decline in the extinction ratio. There are trade-off among the extinction ratio, insertion loss and operating bandwidth. As a demonstration, a THz switch using six-layer metallic gratings exhibit a 7 dB extinction ratio and 100 GHz operating bandwidth with an acceptable low insertion loss. Although the switching performance is relatively poor, it is just the simulation results without any further optimization. There is considerable room for performance improvement. Even though there remain challenges to carry out experimental verification, the principle of the device presented here provides a novel strategy to design broadband THz switch.

This work is supported by the National Natural Science Foundation of China under Grant No. 10974063, the Research Foundation of Wuhan National Laboratory under Grant No. P080008, and the National "973" Project under Grant No. 2007CB310403.

References

[1] Horiuchi N 2010 Terahertz technology: Endless applications *Nat Photon* **4** 140-
[2] Wilk R, Vieweg N, Kopschinski O, Hasek T and Koch M 2009 THz Spectroscopy of Liquid Crystals from the CB Family *J. Infrared Millim. Terahertz Waves* **30** 1139-47
[3] Pan C-L and Pan R-P 2006 Recent progress in liquid crystal THz optics. In: *Liquid Crystal Materials, Devices, and Applications XI,* (San Jose, CA, USA: SPIE) pp 61350D-13
[4] Chen C-Y, Hsieh C-F, Lin Y-F, Pan R-P and Pan C-L 2004 Magnetically tunable room-temperature 2 pi liquid crystal terahertz phase shifter *Opt. Express* **12** 2625-30
[5] Tsong-Ru T, Chao-Yuan C, Ru-Pin P, Ci-Ling P and Xi-Cheng Z 2004 Electrically controlled room temperature terahertz phase shifter with liquid crystal *Microwave and Wireless Components Letters, IEEE* **14** 77-9
[6] Lin C J, Lin C H, Li Y T, Pan R P and Pan C L 2009 Electrically Controlled Liquid Crystal Phase Grating for Terahertz Waves *Ieee Photonics Technology Letters* **21** 730-2
[7] Lin C J, Li Y T, Hsieh C F, Pan R P and Pan C L 2008 Manipulating terahertz wave by a magnetically tunable liquid crystal phase grating *Opt. Express* **16** 2995-3001
[8] Pan C-L, Hsieh C-F, Pan R-P, Tanaka M, Miyamaru F, Tani M and Hangyo M 2005 Control of enhanced THz transmission through metallic hole arrays using nematic liquid crystal *Opt. Express* **13** 3921-30
[9] Zhang H, Guo P, Chen P, Chang S J and Yuan J H 2009 Liquid-crystal-filled photonic crystal for terahertz switch and filter *Journal of the Optical Society of America B-Optical Physics* **26** 101-6
[10] Scheller M, Jordens C and Koch M 2010 Terahertz form birefringence *Opt. Express* **18** 10137-42
[11] Hsieh C-F, Pan R-P, Tang T-T, Chen H-L and Pan C-L 2006 Voltage-controlled liquid-crystal terahertz phase shifter and quarter-wave plate *Opt. Lett.* **31** 1112-4
[12] Ghattan Z, Hasek T, Wilk R, Shahabadi M and Koch M 2008 Sub-terahertz on-off switch based on a two-dimensional photonic crystal infiltrated by liquid crystals *Optics Communications* **281** 4623-5
[13] Hsieh C-F, Lai Y-C, Pan R-P and Pan C-L 2008 Polarizing terahertz waves with nematic liquid crystals *Opt. Lett.* **33** 1174-6
[14] Jewell S A, Hendry E, Isaac T H and Sambles J R 2008 Tuneable Fabry-Perot etalon for terahertz radiation *New J. Phys.* **10** 6
[15] Chen C-Y, Pan C-L, Hsieh C-F, Lin Y-F and Pan R-P 2006 Liquid-crystal-based terahertz tunable Lyot filter *Applied Physics Letters* **88** 101107
[16] Yuan Y, He J, Liu J and Yao J 2010 Propose of an electrically controlled THz switch based on liquid-crystal-filled dual-metallic grating structures *Appl. Optics* **49** (to be published on 20 Nov 2010)
[17] Wilk R, Vieweg N, Kopschinski O and Koch M 2009 Liquid crystal based electrically switchable Bragg structure for THz waves *Opt. Express* **17** 7377-82
[18] Lalanne P and Lemercier-lalanne D 1996 On the effective medium theory of subwavelength periodic structures *J. Mod. Opt.* **43** 2063 - 85
[19] Flory F, Escoubas L and Lazarides B 2002 Artificial anisotropy and polarizing filters *Appl. Optics* **41** 3332-5
[20] Lee J W, Seo M A, Park D J, Jeoung S C, Park Q H, Lienau C and Kim D S 2006 Terahertz transparency at Fabry-Perot resonances of periodic slit arrays in a metal plate: experiment and theory *Opt. Express* **14** 12637-43

[21] Armand D, Todorov Y, Garet F, Minot C and Coutaz J L 2008 Study of the transmission of subwavelength metallic grids in the THz frequency range *IEEE J. Sel. Top. Quantum Electron.* **14** 513-20

Propagation characteristics of THz radiation in hollow rectangle metal waveguide

Y Zhang[1, 3], Y Z Zhang[1], D M Lu[1], D Y Wang[1], M Liu[1], S Y WANG[1] ,P D Zhao[1],Z D ZHANG[1] and J Q Yao[2]
[1]School of Science, Hebei University of Technology, 300130, Tianjin PRC
[2]College of Precision Instrument and Opto-Electronics Engineering, Tianjin University, 300072, Tianjin PRC

E-mail: zhyong1970@163.com

Abstract. The secular equation of medium coating were obtained by considering the particularity of THz, which had some reference value in designing of THz waveguides. The attenuation of medium coating metal hallow waveguides were theoretically calculated, the conclusion is that the power loss is less than the metal waveguides. Also we obtained the mode characters and attenuation in metal waveguides, and we noted that there exists an absorption peak. And we studied the affection of shape and size on propagation and got some useful conclusion. The split rectangular waveguide (SRW) is suitable for THz transmission which is confirmed by experiment. Our secular equation can be considered to be a theoretically discussion on it.

1. Introduction

The generation of THz signals at high power levels and the efficient transmission of THz signals using wave guiding structures are two major challenges researchers are currently facing while trying to satisfy the strong demand for compact, Practical THz waveguides with losses below the level of 1 dB/m only recently low-cost and robust THz systems have been realized[1-5]. Certain key applications anticipated at microwave or optical frequencies (e.g. integrated circuit devices or fiber-based lasers, respectively) seem to be impossible at THz frequencies due to the lack of sufficiently efficient waveguides. Transmission properties of dielectric single-mode flexible rod and tube waveguides made of plastic have been investigated recently at millimeter-wave frequencies between 33 GHz and 50 GHz [6]. Here we investigate loss, dispersion and mode confinement properties of the proposed structures in the THz range and compare the results with state-of-the-art THz waveguides.

2. Transmission characteristics of THz pulse in dielectric coated metal pipe waveguide

The structure of the waveguide is shown in figure 1. Suppose the wave transmits along z axis, the refractive index of the inner and outer is n_1 and n_2 respectively, and $n_1 > n_2$.

[3] To whom any correspondence should be addressed.

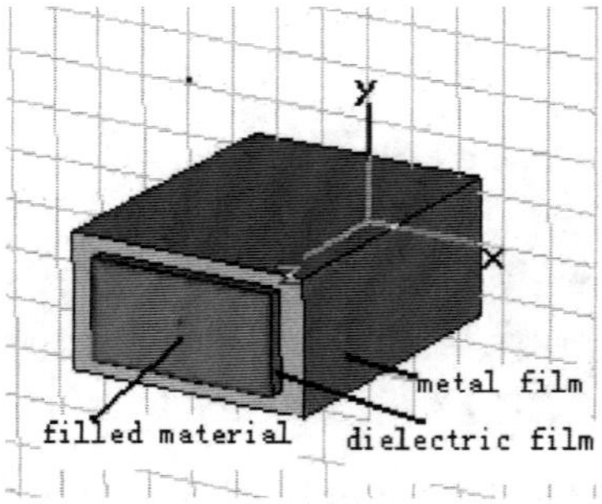

Figure 1. Structure of rectangle dielectric coated metal-pipe waveguides.

2.1 Mode analysis of THz transmit in the waveguide

Assume that the refractive index of the waveguide film have symmetry distribution ,and the differential coefficient equation concerning to mode field only have one variable x, partial differential coefficient equation can be expressed as [7]:

$$\frac{d^2E}{d^2x}+(k^2n^2-\beta^2)E=0 \tag{1}$$

$$\frac{d^2H}{d^2x}+(k^2n^2-\beta^2)H=0 \tag{2}$$

Take TE mode for example, under boundary condition, E_y is limited in the core film, when film x goes to unlimited, E_y tend to 0. We have expression of TE even mode as:

$$E_y=\begin{cases} b_1\cos(\sqrt{k^2n_1^2-\beta^2}\,x) & (|x|<a) \\ b_2\exp(-\sqrt{\beta^2-k^2n_2^2}\,|x|) & (|x|>a) \end{cases} \tag{3}$$

$$H_z=\begin{cases} \dfrac{b_1\sqrt{k^2n_1^2-\beta^2}}{\omega\mu_0}\sin(\sqrt{k^2n_1^2-\beta^2}\,x) & (|x|<a) \\ \dfrac{b_2\sqrt{\beta^2-k^2n_2^2}}{\omega\mu_0}\exp(-\sqrt{\beta^2-k^2n_2^2}\,|x|) & (|x|>a) \end{cases} \tag{4}$$

Where, b_1 and b_2 are integral constants.

2.2 Attenuation characteristics of THz transmit in the waveguide

The attenuation of the dielectric film can be expressed as [8, 9]:

$$2a(\theta)=\frac{1-R(\theta)}{2b\cot(\theta)} \tag{5}$$

Where θ the incident angle, b is is the distance between layers, $R(\theta)$ is the refractive index of different polarized directions,

$$R(\theta)=\frac{R_S(\theta)+R_P(\theta)}{2} \tag{6}$$

The power of transmitting distance of z is:

$$P(z)=\int_0^{\theta_{\max}} p_0(\theta)\exp\left[-2a(\theta)z\right]\sin\theta d\theta=\int_0^{\theta_{\max}} p_0(\theta)\exp\left[-\frac{1-R(\theta)}{2T\cot\theta}z\right]\sin\theta d\theta \tag{7}$$

Take the roughness into consideration and the refractive index of power loss is:

$$R'(\theta) = R(\theta)\exp[-(\frac{4\pi n_i \sigma \sin\theta}{\lambda})^2] \quad (8)$$

For mental waveguide coated with dielectric film, the refractive index for inner layer can be expressed as [9]

$$r_{1p} = \frac{n_F \cos\psi_1 - \cos\psi_2}{n_F \cos\psi_1 + \cos\psi_2} \quad (9)$$

$$r_{1s} = \frac{\cos\psi_1 - n_F \cos\psi_2}{\cos\psi_1 + n_F \cos\psi_2} \quad (10)$$

For outer layer, the refractive index can be expressed by two parameters:

$$r_2 = \rho_2 \exp(j\varphi_2) \quad (11)$$

The expression of two parameters is as fallows:

$$\rho_{2p}^2 = \frac{[(n^2-\kappa^2)\cos\psi_2 - n_F u]^2 + (2n\kappa\cos\psi_2 - n_F \upsilon)^2}{[(n^2-\kappa^2)\cos\psi_2 + n_F u]^2 + (2n\kappa\cos\psi_2 + n_F \upsilon)^2} \quad (12)$$

$$\tan\varphi_{2p} = 2n_F \cos\psi_2 \frac{-2n\kappa u - (n^2-\kappa^2)\upsilon}{(n^2+\kappa^2)^2\cos^2\psi_2 - n_F^2(u^2+\upsilon^2)} \quad (13)$$

$$\rho_{2s}^2 = \frac{(n_F\cos\psi_2 - u)^2 + \upsilon^2}{(n_F\cos\psi_2 + u)^2 + \upsilon^2} \quad (14)$$

$$\tan\varphi_{2s} = \frac{-2\upsilon n_F \cos\psi_2}{u^2+\upsilon^2 - n_F^2\cos^2\psi_2} \quad (15)$$

Where n and κ are the image and real part of the refractive index:

$$u^2 = \frac{1}{2}\left\{ n^2-\kappa^2-n_F^2\sin^2\psi_2 + [(n^2-\kappa^2-n_F^2\sin^2\psi_2)^2 + 4n^2\kappa^2]^{\frac{1}{2}} \right\} \quad (16)$$

$$\upsilon^2 = \frac{1}{2}\left\{ -(n^2-\kappa^2-n_F^2\sin^2\psi_2) + [(n^2-\kappa^2-n_F^2\sin^2\psi_2)^2 + 4n^2\kappa^2]^{\frac{1}{2}} \right\} \quad (17)$$

Thus, we get the refractive index of dielectric film

$$r = \frac{r_1 + r_2\exp(-j2\Gamma)}{1 + r_1 r_2 \exp(-j2\Gamma)} = \frac{r_1 + \rho_2\exp[j(\phi_2 - 2\Gamma)]}{1 + r_1\rho_2\exp[j(\phi_2 - 2\Gamma)]} \quad (18)$$

where $\Gamma = k_0 n_F d\cos\psi_2$. Power refractive index $R(\theta)$ can be expressed as:

$$|r|^2 = \frac{r_1^2 + \rho_2^2 + 2r_1\rho_2\cos(\phi_2 - 2\Gamma)}{1 + r_1^2\rho_2^2 + 2r_1\rho_2\cos(\phi_2 - 2\Gamma)} \quad (19)$$

3. Transmitting character of THz of THz pulse in metal-coated pipe waveguide

3.1 The cut-off frequency of THz radiation in the waveguide

The cut-off frequency and wavelength of TE (or TM) modes are:

$$f_c = \frac{k_c}{2\pi\sqrt{\mu\varepsilon}} = v\sqrt{(m/a)^2 + (n/b)^2}/2, \quad \lambda_c = v/f_c = \frac{2}{\sqrt{(m/a)^2 + (n/b)^2}}$$

3.2 Mode analysis of THz radiation in the waveguide

Using right angle coordinate system, Helmholtz equation can be expressed as:

$$\frac{\partial^2 H_z}{\partial x^2}+\frac{\partial^2 H_z}{\partial y^2}=-k_c^2 H_z,\ \frac{\partial^2 E_z}{\partial x^2}+\frac{\partial^2 E_z}{\partial y^2}=-k_c^2 E_z \tag{20}$$

The boundary condition of inner wall is:

$$E_y\Big|_{\substack{x=0\\x=a}}=0, E_x\Big|_{\substack{y=0\\y=a}}=0\ ; H_z=H_{mn}\cos(\frac{m\pi}{a}x)\cos(\frac{n\pi}{b}y)e^{-j\beta_{mn}z} \tag{21}$$

By solving the equation, we get all the field values (take TE mode for example):

$$\left.\begin{aligned}
E_x&=\frac{j\omega\mu n\pi}{k_c^2 b}H_{mn}\cos(\frac{m\pi}{a}x)\sin(\frac{n\pi}{b}y)e^{-j\beta_{mn}z}\\
E_y&=\frac{-j\omega\mu m\pi}{k_c^2 a}H_{mn}\sin(\frac{m\pi}{a}x)\cos(\frac{n\pi}{b}y)e^{-j\beta_{mn}z}\\
E_z&=0\\
H_x&=\frac{j\beta m\pi}{k_c^2 a}H_{mn}\sin(\frac{m\pi}{a}x)\cos(\frac{n\pi}{b}y)e^{-j\beta_{mn}z}\\
H_y&=\frac{j\beta n\pi}{k_c^2 b}H_{mn}\cos(\frac{m\pi}{a}x)\sin(\frac{n\pi}{b}y)e^{-j\beta_{mn}z}
\end{aligned}\right\} \tag{22}$$

3.3 Attenuation characteristics

Commonly, loss contains the loss of waveguide wall and the filled. Thus, transmission constant is $\gamma=\alpha+j\beta$. When wave transmits in waveguide, its electromagnetic field decrease by index rule. Suppose wave transmits from z=0 to z, the relation of power is $P(z)=P_0\exp(-2\alpha z)$ .Obviously, the power loss in transmitting unit length is $P_1=P_0[1-\exp(-2\alpha)]\approx 2\alpha P_0$.

Therefore, the attenuation coefficient of the waveguide should be:

$$\alpha=\frac{P_1}{2P_0}=\alpha_c+\alpha_d \tag{23}$$

where α_c is attenuation constant of conductivity, α_d is attenuation constant of filled medium. α_c is caused by the current on the wall. The power loss of conductivity of unit length is

$$P_l=0.5R_s\oint_l |J|^2 dl=0.5R_s\oint_l |H_t|^2 dl \tag{24}$$

where $R_s=(\sigma\delta)^{-1}$ is the surface resistance of waveguide wall in unit width, and s is the conductance rate, $\delta=2^{\frac{1}{2}}(\omega\mu\sigma)^{-\frac{1}{2}}$ is the skin depth. Integral trace l is the closed line along the cross section of inner wall. H_t is the tangential component of magnetic field at the inner wall of waveguide.

Using the equations above, we can easily derive the expressions for attenuation of TE and TM mode:

$$a_{cTE_{mn}}=\frac{2R_s}{\eta b\sqrt{1-(\lambda/\lambda_c)^2}}\left\{\frac{\frac{b}{a}(\frac{b}{a}m^2+n^2)}{(\frac{b}{a})^2m^2+n^2}[\frac{\varepsilon_{0n}}{2}-(\lambda/\lambda_c)^2]+(1+\frac{b}{a})(\lambda/\lambda_c)^2\right\}(Np/m) \tag{25}$$

$$a_{cTM_{mn}} = \frac{2R_s}{\eta b\sqrt{1-(\lambda/\lambda_c)^2}} \frac{(\frac{b}{a})^3 m^2+n^2}{(\frac{b}{a})^2 m^2+n^2}(Np/m) \tag{26}$$

where $\varepsilon_{0n} = \begin{cases} 1, n=0 \\ 2, n\neq 0 \end{cases}$. When loss of dielectric exists, its dielectric constant takes complex number as $\varepsilon' = \varepsilon(1+j\sigma/(\omega\varepsilon))$. Then the longitudinal transmission constant of waveguide changes to:

$$\begin{aligned} \gamma &= \sqrt{k_c^2-k^2} = j\omega\sqrt{\mu\varepsilon}\sqrt{1-j\frac{\sigma}{\omega\varepsilon}-(\lambda/\lambda_c)^2} \\ &\approx j\omega\sqrt{\mu\varepsilon}\sqrt{1-(\lambda/\lambda_c)^2}\left\{1-j\frac{\sigma/\omega\varepsilon}{2[1-(\lambda/\lambda_c)^2]}\right\} \\ &= a_d + f\beta \end{aligned} \tag{27}$$

where $\alpha_d = \frac{\sigma\eta}{2\sqrt{1-(\lambda/\lambda_c)^2}} = \frac{\pi\tan\delta}{\lambda\sqrt{1-(\lambda/\lambda_c)^2}}(Np/m)$ is the attenuation of medium, $\tan\delta = \sigma/(\omega\varepsilon)$ is tangent of dielectric loss angle.

4. Numerical simulations

Taking aluminum for example, we calculated the attenuation constant using Matlab programs, as shown in figure 3. We can see from figure 3 that, at higher value of b/a, higher mode has bigger loss.

Figure 4 shows that attenuation changes almost exponentially with wavelength, the attenuations of TE_{1X} modes are far below those of TE_{2X} modes.

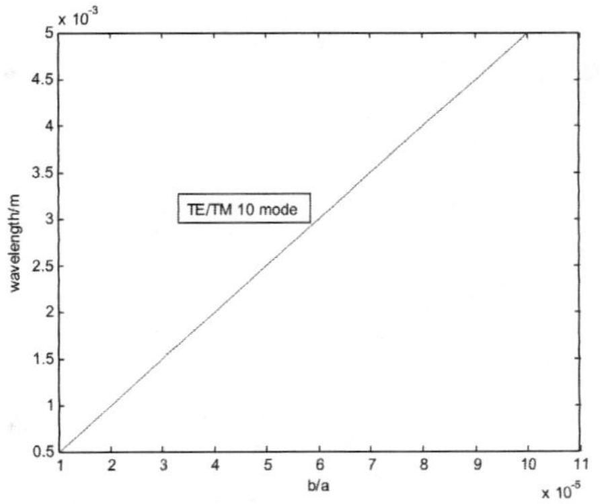

Figure 2. Dependence of cut-off frequency on b/a.

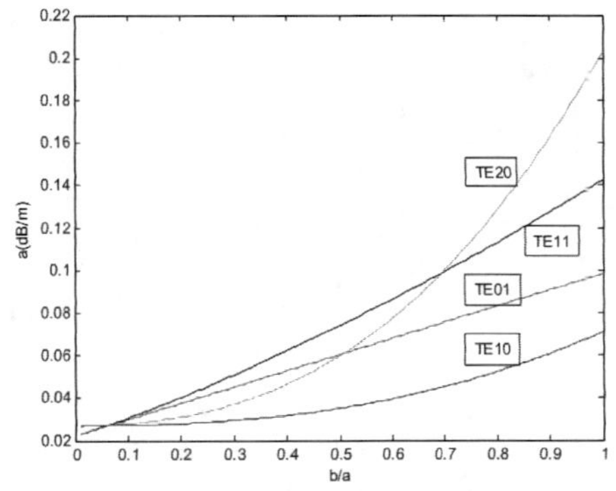

Figure 3. Attenuation changes with b/a of TE modes @ 0.1THz, b=0.01m, aluminium.

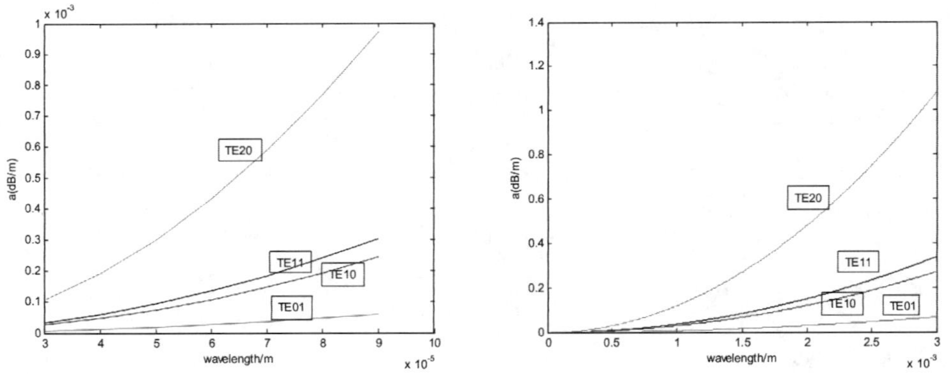

Figure 4. Dependence of attenuation on wavelength for TE modes
@ b/a=2 , b=0.01m, aluminium-coated waveguide.

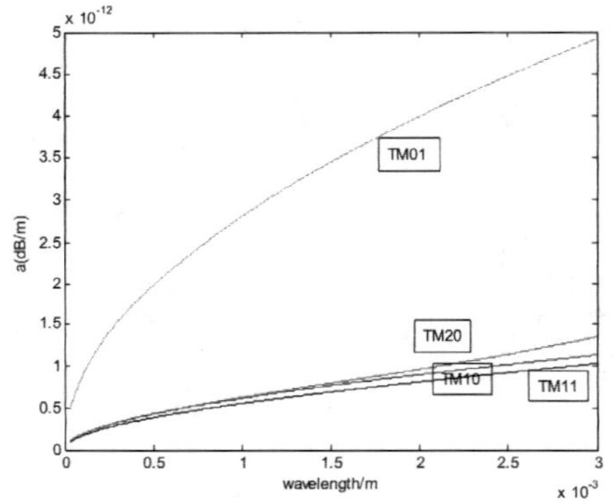

Figure 5. Dependence of attenuation for TM modes on wavelength@ b/a=2, b =0.01m.

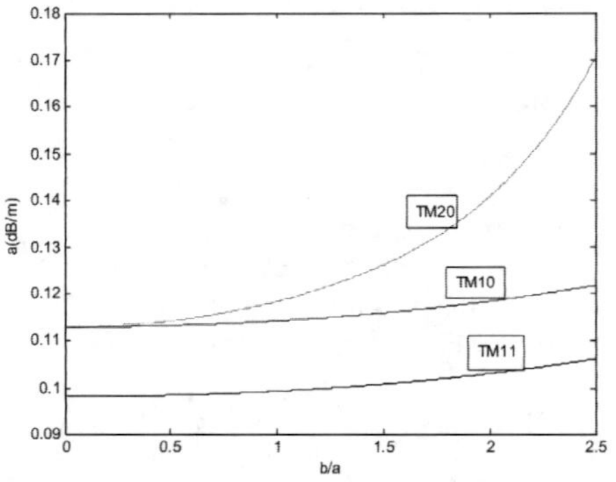

Figure 6. Loss characteristics of TM modes @ 0.1THz, b=0.01m.

The results show that TM_{01} mode has a bigger loss than that of other modes. The results can be used in designing THz devices, and TM_{01} mode should to be avoided.

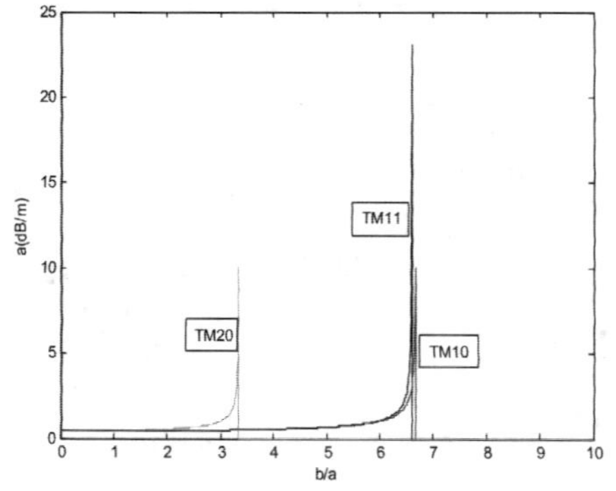

Figure 7. Loss characteristics of TM modes @0.1THz, b=0.01m, aluminum waveguide.

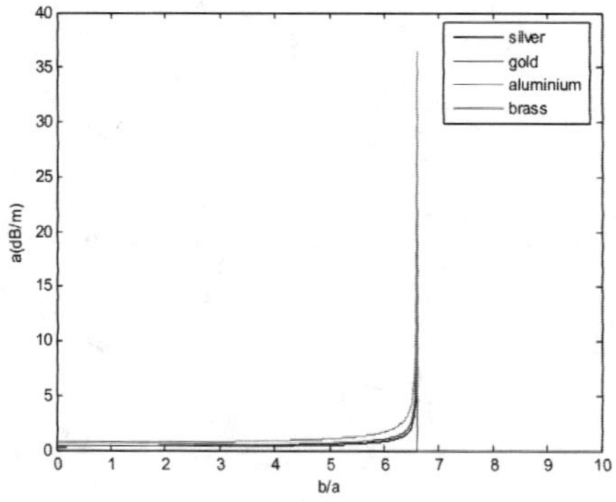

Figure 8. Loss characteristics of TM modes @ 0.1THz, b=0.01m, several metals.

Figure 7 and figure 8 show that the waveguide has a bigger loss when b/a is close to 3 or 7 for different modes. That is to say, when designing waveguides we should avoid the two situations. Figure 8 shows that silver is better than other metals in transmitting THz wave. While brass is not suitable for coating waveguides.

5. Conclusions

The results show that for TM_{20} mode, it is a good choice when b/a is less than 3 for transmitting THz waves. While for TM_{11} mode and TM_{10} mode, it is good when b/a is less than 6. THz pulse transmits in metal waveguide suffering more loss of current on the surface of metal wall. Dielectric film can lower this loss by reducing its power loss on the metal. The loss of rectangle waveguide is lower than that of circular hollow waveguide as discussed in reference [10]. The attenuation is bellow 0.2dB/m when b/a is bellow 1.The TM_{01} mode is at a high loss than other modes, so the TM mode should be avoided in application.

Acknowledgements

Project supported by the National Basic Research Program of China (973Program) (Grant No. 2007CB310403) and Key Subject Construction of Hebei Provincial University and the Doctoral Foundation of Hebei University of Technology.

References

[1] Mitrofanov1Oleg and Harrington James A 2010 *Opt. Express* 18**(3)** 1898
[2] Bowden B, Harrington J A and Mitrofanov O 2007 *Opt. Lett.* **32(20)** 2945
[3] Matsuura Y and Takeda E 2008 *J. Opt. Soc. Am. B* **25(12)** 1949
[4] Nielsen K, Rasmussen H K , Adam A J , Planken P C , Bang O and Jepsen P U 2009 *Opt. Express* **17(10)** 8592
[5] Lai C H, Hsueh Y C, Chen H W, Huang Y J, Chang H C and Sun C K 2009 *Opt. Lett.* **34(21)** 3457
[6] Nagel Michael, Marchewka Astrid and Kurz Heinrich 2006 *Opt. Express* **14(21)** 9944
[7] Musin Roman R, Xing Qirong, Li Yanfeng, Hu Minglie, Chai Lu, Wang Qingyue, Mikhailova Yuliya M, Nazarov Maksim M, Shkurinov Alexander P and Zheltikov1Aleksei M 2008 *Appl. Opt.* **47(4)** 489
[8] Yuji Matsuura,Mitsunori Saito and Mitsunobu Miyagi 1989 *J. Opt. Soc. Am. A* 6**(3),** 423
[9] Mitsunori Saito, Yuji Matsuura, Masashi Kawamura and Mitsunobu Miyagi 1990 *J. Opt. Soc. Am. A* 7**(11)** 2063
[10 Yuping Zhang, Huiyun Zhang, You fu Geng, Xiaoling Tan and Jianquan Yao 2009 *ACTA Physica Sinica* **58** 7030 (in Chinese)

Journal of Physics: Conference Series **276** (2011) 012230 doi:10.1088/1742-6596/276/1/012230

Optimal design of hollow elliptical waveguide for THz radiation

Y Zhang[1,3], Z Y Zhang[1], D M Lu[1], D Y Wang[1], M Liu[1], P D Zhao[1] and J Q Yao[2]
[1]School of Science, Hebei University of Technology, 300130, Tianjin PRC
[2]College of Precision Instrument and Opto-Electronics Engineering, Tianjin University, 300072, Tianjin PRC

E-mail: zhyong1970@163.com

Abstract. Based on propagation equations of and metal-coated elliptical waveguide, we derived the characteristic equations of the waveguides. And the relationship between effective refractive index and ellipticity for two-layer media waveguide. Also obtained the relationships between attenuation coefficient and metal conductivity, elliptical eccentricity and ellipse semi-major axis for the latter waveguide. We got the corresponding ellipse eccentricities of different frequency in the THz band. The results showed that the elliptical waveguide was more suitable than the circular waveguide for transmitting THz wave. Our results were helpful in optimal designing of elliptical waveguide.

1. Introduction

Applications of terahertz wave in many fields had highlighted the importance of terahertz researches [1-4]. Study on transmission of terahertz wave was one of the important parts of THz field [5]. Although terahertz technology had developed greatly, both basic research and applied basic research were still to be done. As no mature waveguide in this band had been found to the present, it usually transferred in open space. And this undoubtedly increased costs and operating difficulties and limited the application of terahertz technology. So, it had become a hot to find practical waveguides to control the propagation of terahertz wave [6-8]. This paper analyzed the transmission characteristics of terahertz wave in the medium and metal-coated layer hollow elliptical waveguide.

2. Fundamental theories

2.1 Propagation characteristics of terahertz radiation in two-layer elliptical waveguide

The structure of the waveguide was shown in figure 1. The dielectric constants and magnetic permeability of inner and outer media were supposed to be ε_i 、 ε_o and μ_i 、 μ_o respectively, and $\varepsilon_i > \varepsilon_o$. The conductivities of two media were zeros.

[3] To whom any correspondence should be addressed.

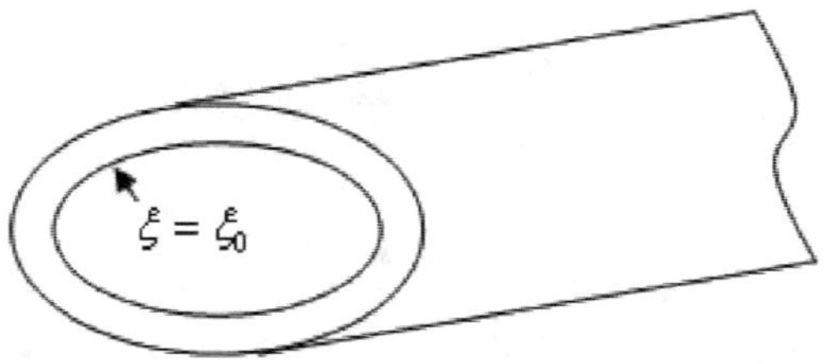

Figure 1. Structure of elliptical dielectric waveguide.

According to our former work [9] , characteristic equations were as follows:

$$\frac{k_0^2}{\beta^2}\left(\frac{a'_m}{a_m}+\frac{\gamma_c^2}{\gamma_d^2}\frac{l'_m}{l_m}\right)\left(n_c^2\frac{b'_m}{b_m}+n_d^2\frac{\gamma_c^2}{\gamma_d^2}\frac{p'_m}{p_m}\right)+\left(X_{mm}+\frac{\gamma_c^2}{\gamma_d^2}\frac{W_{mm}}{\alpha_{mm}}\right)\left(V_{mm}+\frac{\gamma_c^2}{\gamma_d^2}\frac{\theta_{mm}}{\beta_{mm}}\right)=0 \quad \text{(even modes)} \quad (1)$$

$$\frac{k_0^2}{\beta^2}\left(\frac{b'_m}{b_m}+\frac{\gamma_c^2}{\gamma_d^2}\frac{p'_m}{p_m}\right)\left(n_c^2\frac{a'_m}{a_m}+n_d^2\frac{\gamma_c^2}{\gamma_d^2}\frac{l'_m}{l_m}\right)+\left(X_{mm}+\frac{\gamma_c^2}{\gamma_d^2}\frac{W_{mm}}{\alpha_{mm}}\right)\left(V_{mm}+\frac{\gamma_c^2}{\gamma_d^2}\frac{\theta_{mm}}{\beta_{mm}}\right)=0 \quad \text{(odd modes)} \quad (2)$$

Where, m=1 stood for fundamental mode, m>1 stood for higher-order modes. k_0 was wave number in vacuum.

2.2 Propagation characteristics of THz radiation in metal-coated elliptical waveguide

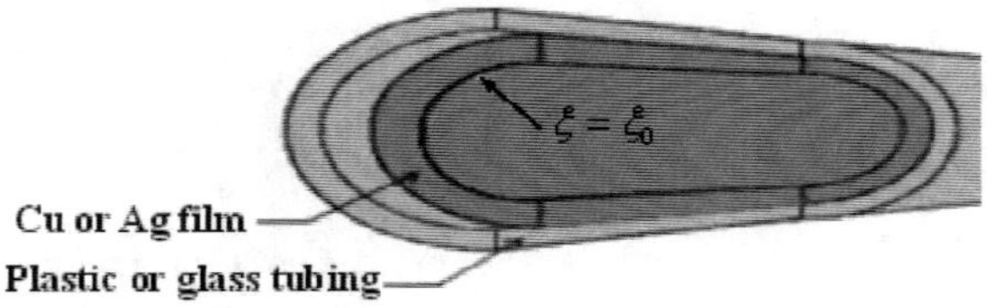

Figure 2. Structure of metal coating hollow elliptical waveguide.

Waveguide structure shown in figure 2, set 2h to be the distance between the two focus, a and b, the semi-major axis and semi-minor axis, respectively; $\xi = \xi_0$, internal border of confocal elliptical waveguide; elliptic cylinder $\xi = \xi_0$ was perfectly conducting wall, thus the space of $\xi < \xi_0$ constitutes a hollow waveguide, which is air-filled or vacuum, its magnetic conductivity and dielectric constant were μ and ε , respectively. Assume that electromagnetic field in the waveguide changes along the axis of propagation by $\exp(j\omega t - \gamma z)$, where ω was the terahertz frequency, γ was the propagation constant. $e = 1/coh\zeta_0$ was ellipse eccentricity.

Similar to the previous analysis, we got:

$$\begin{bmatrix} E_z \\ H_z \end{bmatrix} = \begin{bmatrix} C_m Ce_m(\xi,q)ce_m(\eta,q) \\ S_m Se_m(\xi,q)se_m(\eta,q) \end{bmatrix} \exp(j\omega t - j\beta z) \quad (3)$$

For TM field, $H_z \equiv 0$, equation (3) changed to

$$E_z = \begin{bmatrix} C_m Ce_m(\xi,q)ce_m(\eta,q) \\ S_m Se_m(\xi,q)se_m(\eta,q) \end{bmatrix} \exp(j\omega t - j\beta z) \quad (4)$$

Then, we had the components as follows

$$E_\xi = -\frac{j\beta}{q_1 k^2}\begin{bmatrix} C_m Ce'_m(\xi,q)ce_m(\eta,q) \\ S_m Se'_m(\xi,q)se_m(\eta,q)\end{bmatrix}\exp(j\omega t - j\beta z) \quad (5)$$

$$E_\eta = -\frac{j\beta}{q_1 k^2}\begin{bmatrix} C_m Ce_m(\xi,q)ce'_m(\eta,q) \\ S_m Se_m(\xi,q)se'_m(\eta,q)\end{bmatrix}\exp(j\omega t - j\beta z) \quad (6)$$

$$H_\xi = \frac{j\omega\varepsilon}{q_1 k^2}\begin{bmatrix} C_m Ce_m(\xi,q)ce'_m(\eta,q) \\ S_m Se_m(\xi,q)se'_m(\eta,q)\end{bmatrix}\exp(j\omega t - j\beta z) \quad (7)$$

$$H_\eta = -\frac{j\omega\varepsilon}{q_1 k^2}\begin{bmatrix} C_m Ce'_m(\xi,q)ce_m(\eta,q) \\ S_m Se'_m(\xi,q)se_m(\eta,q)\end{bmatrix}\exp(j\omega t - j\beta z) \quad (8)$$

where C_m and S_m were constants to be determined. For even modes, at the border, that is when $\xi = \xi_0$, $Ce_m(\xi_0,q)=0$. Based on Umov-Poynting vector, we got the time rate of change of energy flow per unit area in the waveguide: $\vec{p} = \frac{1}{2}\vec{E}\times\vec{H}^*$ Where $\vec{H}^*$ was the conjugate value of $\vec{H}$, $\frac{1}{2}$ was the time factor, then:

$$p_z = \frac{1}{2}\left(E_\xi H_\eta^* - E_\eta H_\xi^*\right) \quad (9)$$

For TM even modes, considering the normalization condition, and ignore the time factor, the total average power flew through each section can be obtained by integral of equation (9).

$$P_z = \iint p_z q_1^2 d\xi d\eta = \frac{\pi}{2}|C_m|^2\frac{\beta\omega\varepsilon}{k^4}\int_0^{\xi_0}\left(Ce'^2_m(\xi,q)+\theta_m Ce^2_m(\xi,q)\right)d\xi \quad (10)$$

where C_m was relevant to the amplitude of E_z. The amplitude of electromagnetic components attenuated by $C_m e^{-\alpha z}$. Transmission power attenuated by $P_z\cdot e^{-2\alpha z}$. The attenuation constant $\alpha = \frac{1}{2}\frac{-\frac{\partial P_z}{\partial z}}{P_z} = \frac{L}{2P_z}$. Take $C_m = 1$ in equation (10), inside the waveguide (that is $\xi \le \xi_0$), we got

$$E_z = Ce_m(\xi,q)ce_m(\eta,q) \quad (11\text{-}1)$$

$$H_\eta = -\frac{j\omega\varepsilon}{k^2 h_1}Ce'_m(\xi,q)ce_m(\eta,q) \quad (11\text{-}2)$$

Field in the metal (according to skin effect) [11] could be written as

$$E_z = \sum_{p=0}^{\infty} B_p Ce_p(\xi,q')ce_p(\eta,q') \quad (12)$$

$$H_\eta = -\frac{\sigma + j\omega\varepsilon}{q_1 k'^2}\frac{\partial E_z}{\partial\xi} \quad (13)$$

Change $Ce_p(\xi,q')$ in equation (12) to $Ce_p(\xi,q') = C\cosh(2\xi)^{-1/4}e^{\pm j\int_0^{\xi_0}\sqrt{2q'\cosh(2\xi)}d\xi} = y(\xi,q')$, then equation (12) became

$$E_z = \left[\sum_{p=0}^{\infty} B_p ce_p(\eta, q')\right] y(\xi, q') \tag{14}$$

Considering boundary conditions, equation (14) was rewritten as

$E_z = BCe'_m(\xi_0, q)\frac{y(\xi, q')}{y(\xi_0, q')}ce_m(\eta, q)$. Accordingly, equation (13) became

$$H_\eta = -\frac{\sigma + j\omega\varepsilon}{q_1 k'^2} BCe'_m(\xi_0, q)\frac{y'(\xi, q')}{y(\xi_0, q')}ce_m(\eta, q) \tag{15}$$

At the border of $\xi = \xi_0$, $H_\eta = E_z$, using the following equation $\frac{\frac{d}{dx}y(x,q)}{y(x,q)} = \pm j2q^{\frac{1}{2}}\sqrt{\frac{\cosh(2x)}{2}}$,

we could obtain normal component of Umov-Poynting vector as follow:

$$p_\xi = -\frac{1}{2}E_z H_\eta^* = Ce'^2_m(\xi_0, q)ce^2_m(\eta, q)\left(-\frac{j}{2}\frac{\sqrt{\omega\mu\sigma}e^{j\frac{3\pi}{4}}}{k^4\sigma h q_1}\frac{\omega^2\varepsilon^2}{\sqrt{\frac{\cosh(2\xi_0)}{2}}}\right) \tag{16}$$

Thus, the power loss through surface of the waveguide into the metal per unit length to be:

$$L = P_\xi = \int_0^{2\pi} \mathrm{Re}(p_\xi)h_1 d\eta = \frac{\pi}{2}Ce'^2_m(\xi_0, q)\left(\frac{1}{\sqrt{2}}\sqrt{\frac{\omega\mu}{\sigma}}\frac{\omega^2\varepsilon^2}{k^4 h}\frac{1}{\sqrt{\frac{\cosh(2\xi_0)}{2}}}\right) \tag{17}$$

Introduce L and P_z we have obtained into $\alpha = \frac{1}{2}\frac{-\frac{\partial P_z}{\partial z}}{P_z} = \frac{L}{2P_z}$ (the attenuation constant), we got:

$$\alpha_{TM} = \left(\pi\sqrt{\frac{\varepsilon}{\mu}}\right)^{1/2}\frac{1}{\sqrt{\sigma}}\left(\frac{1/\lambda}{1-(\lambda/\lambda_c)^2}\right)^{1/2}\frac{Ce'^2_m(\xi_0, q)}{2h\sqrt{\cosh(2\xi_0)/2}\int_0^{\xi_0}\left(Ce'^2_m(\xi, q) + \theta_m Ce^2_m(\xi, q)\right)d\xi} \tag{18}$$

Similarly, we could derive the attenuation coefficient of even mode for TE field:

$$\alpha_{TE} = \left(\pi\sqrt{\frac{\varepsilon}{\mu}}\right)^{1/2}\frac{1}{\sqrt{\sigma}}\left(\frac{1/\lambda}{1-(\lambda/\lambda_c)^2}\right)^{1/2}\frac{Ce^2_m(\xi_0, q)\times 2q\cosh(2\xi_0)(\lambda/\lambda_c)^2 + \left[1-(\lambda/\lambda_c)^2\right]\vartheta_m}{2h\sqrt{\cosh(2\xi_0)/2}\int_0^{\xi_0}\left(Ce'^2_m(\xi, q) + \vartheta_m Ce^2_m(\xi, q)\right)d\xi} \tag{19}$$

Where f_c and λ_c were cut-off frequency and cut-off wavelength of the waveguide.

3. Numerical simulations

3.1 two-layer media Elliptical waveguide

By solving characteristic equation of even mode of fundamental mode, we gave relationship between effective refractive index ($N = \beta/k_0$) and normalized frequency ($V_y = bk_0\sqrt{n_i^2 - n_0^2}$) under different elliptical ratio, as shown in figure 3. The horizontal axis corresponding to frequencies in the range of 0.8-4THz.We could see from figure 3 that, the effective refractive index increased with the ellipticity at same normalized frequency. When $V_y > 20$ (that is when $f > 4THz$), the effective refractive index tended to be closer to 1.48 and no longer increased with the frequency. The results also showed that under same normalized frequency, the change of effective refractive index (ΔN) decreased with the increase of ellipticity.

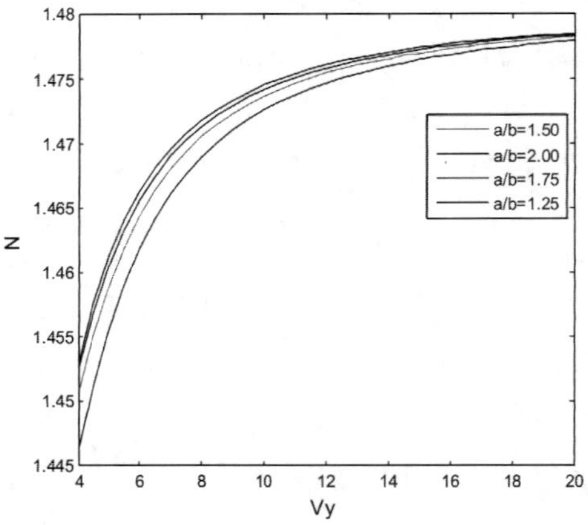

Figure 3. Curve of propagation characteristic of even mode.

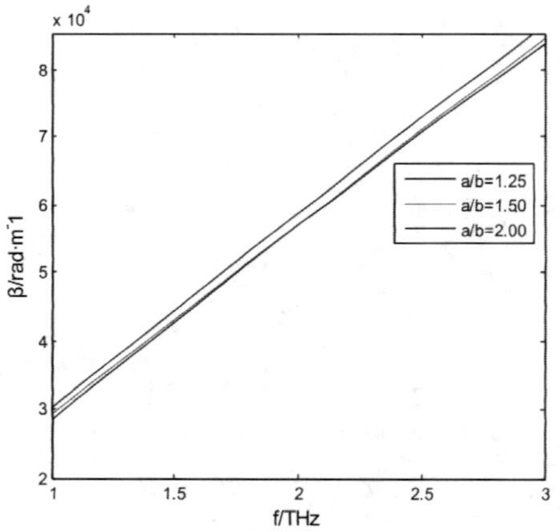

Figure 4. Curve of propagation constant changing with frequency.

Take refractive index ratio of the waveguide $noi = n_i / n_o = 0.9$, refractive index of inner waveguide $n_i = 1.48$; semi-major axis $a = 1mm$. And the elliptical ratio a / b =1.25, 1.5 and 2.0, respectively. We could see that the propagation constant was proportional to frequency. Propagation constant increased with the elliptical ratio under same frequency. The results indicated that bigger elliptical ratio is in favour of propagating terahertz wave. That is to say elliptical waveguide is more easily than circular waveguide to propagate terahertz wave.

3.2 Metal-coated hollow elliptical waveguide

We calculated transmission loss of terahertz wave in silver-coated hollow elliptical waveguide. Took the conductivity $\sigma = 6.1\times10^7 (\Omega \cdot m)^{-1}$, elliptical eccentricity e = 0.955, 0.9, 0.8 respectively, and kept the semi-major axis remain $a = 1mm$ unchanged, the results were shown in figure 5. We could see from figure 5 that there was a minimum attenuation coefficient for each curve. This means that each elliptical waveguide parameters were corresponding to an optimal transmission frequency. For example, when e = 0.955, the smallest attenuation coefficient occurred at the frequency of approximately 1THz, which is most suitable for this frequency of transmission in this waveguide. We could also see from figure 5 that, under same frequency, the smaller e the smaller α . This indicated that smaller e is more suitable for transmitting THz wave.

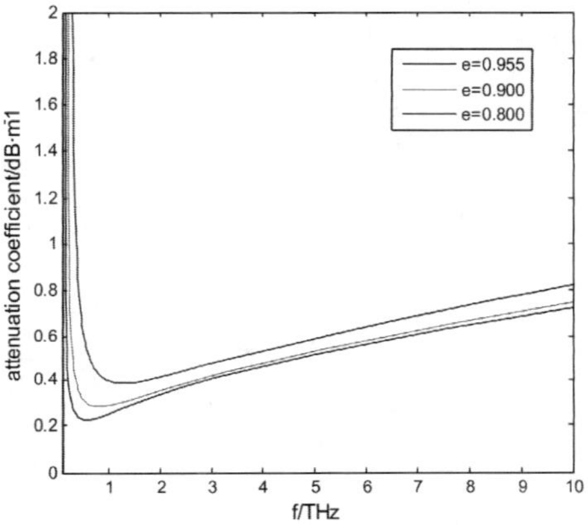

Figure 5. Curves of attenuation coefficient changing with the frequency.

4. Conclusions

We derived the characteristic equations of terahertz in the two-layered dielectric waveguide .We obtained the relationship between normalized frequency and the effective refractive index. The normalized frequency increased with the effective refractive index. We got the relationship between transmission constant and the propagation frequency. In this waveguide, the propagation constant increased with transmission frequency in terahertz band. Based on the Maxwell equations, we derived transmission equation of THz wave in the metal-coated hollow elliptical waveguide. And we derived the transmission attenuation coefficient equation of THz wave in the waveguide. Then we simulated using mat lab programs, we got the relationship between attenuation coefficient and other parameters of the waveguide, especially the elliptical eccentricity of the waveguide.

Acknowledgements

Project supported by the National Basic Research Program of China (973Program) (Grant No. 2007CB310403) and Key Subject Construction of Hebei Provincial University and the Doctoral Foundation of Hebei University of Technology.

References

[1] Chen Q, Jiang Z P, Xu G X and Zhang X C 2000 *Opt. Lett.* **25** 1122
[2] Yao Jianquan,Lu Yang, Zhang Baigang and Wang Peng 2005 *J. Optoelectr. Laser* **16** 503(in Chinese)
[3] Leahy Hoppa M R, Fiteh M J and Zheng X 2007 Chem. Phys. Lett. **434** 227

[4] Yoshida Shigeki, Suizu Koji, Kato Eiji, Nakagomi Yoichi, Ogawa Yuichi and Kawase Kodo 2009 *J. Mol. Spectrosc.* **256** 146
[5] Krumbholz N, Gerlach K and Rutz F 2006 *Appl. Phys. Lett.* **88** 112
[6] Lu Dongmei, Yao Jianquan, Zheng Y, Geng Youfu, Li Zhongyang and Wang Peng 2007 *Laser & Infrared* **37** 1287(in Chinese)
[7] Liu Xiaojie, Lei Hong, Yu Tian, Feng Jinjun and Liao Fujiang 2008 *Opt.Comm.* **281** 102
[8] Zhang Yuping, Zhang Huiyun, Geng You fu, Tan Xiaoling and Yao Jianquan 2009 *ACTA Physica Sinica* **58** 7030 (in Chinese)
[9] Zhang Yong, Zhang Zhongya, Wang Dayun, Bao Wenxia, Liu Ming, Lu Dongmei, Zhao Peide, Zhang Zhidong and Yao Jianquan 2008 *Photonics and Optoelectronics Meetings (POEM) 2008:Terahertz Science and Technology， Proc. Of SPIE* **7277** O-1-8

Detection of pesticides residues in rapeseed oil by terahertz time domain spectroscopy

Zhao Xiao-li[1], Li Jiu-sheng[1, 2]

[1] Centre for THz Research, China Jiliang University, Hangzhou 310018, China

[2] College of Precision Instrument and Opto-Electronics Engineering, Institute of Laser and Optoelectronics, Tianjin University, Tianjin 300072, China

E-mail: forever-li@126.com

Abstract. The absorption spectra of panax notoginseng and glycyrrhiza in the frequency range of 0.2~1.6THz has been measured with terahertz time-domin spectroscopy at room temperature. Simultaneously, the corresponding theoretical spectra were given by using density functional theory methods. It was found that the absorption peaks of the two molecules obtained by theoretical were in good agreement with the experimental results.

1. Introduction

Terahertz wave usually refers to the frequency range of 0.1 ~ 10THz in electromagnetic waves, which coincides with the energy between the molecular vibration and rotational energy transitions. Because of the macromolecules highly relevant nature of the spectroscopic properties in the scope of the THz band, the terahertz time-domain spectroscopy has received extensive attention in recent years [1-3].

In recent years, food safety problems occur frequently. Edible oil, which is one of the three major nutrients, is closely related to human life. People begin to pay attention to the safety of edible oil with the improvement of living standard. At present, edible oil inspection and detection depend largely on gas chromatography [4], and HPLC [5]. In order to guarantee food safety, there is an urgent need to develop a user-friendly, fast and effective test to identify the quality of edible oil. Although some of the methods mentioned above have high detection precision, the preprocessing technology such as purification and extraction is very complex. In this paper, we proposed a new detection method employing terahertz time-domain spectroscopy to investigate the pesticides residues in oil.

2. Theory analysis

The schematic diagram of the THz-TDS setup used in this experiment is shown in figure 1. A repetition rate of 80MHz, diode-pump mode-locked Ti:sapphire laser (MaiTai, Spectra Physics) provided the femtosecond pulses with duration of 100fs and center wavelength of 810nm. A p-type InAs wafer with <100> orientation was used as the THz emitter and a 2.8 mm-thick <110> ZnTe was employed as the sensor. A standard lock-in technology was used in this system. A femtosecond laser pulse was split into two beams. The pump beam was used to generate THz radiation and the probe beam acted as a gated detector to monitor the temporal waveform of THz field. A silicon lens and parabolic mirrors were used to collimate and focus the THz beam through free space onto the detector. A balanced photodiode detector detected the probe beam, and the signal was amplified by a lock-in amplifier and sent to the computer for processing. The THz beam path was purged with dry nitrogen to minimize the absorption of water vapor and enhance the signal to noise ratio (SNR). The humidity was kept less than 1% and the temperature was kept at 298K.

According to the figure, we can describe the Fresnel reflection (R_{12}) and transmission (T_{12}) coefficients of a THz wave at the interface of media 1 and media 2 in the time domain as follows :

$$T_{12} = \frac{2\times\tilde{n}_1}{\tilde{n}_1+\tilde{n}_2}$$
$$R_{12} = \frac{\tilde{n}_2-\tilde{n}_1}{\tilde{n}_2+\tilde{n}_1} \quad (1)$$

The transmission function of the first order Fabry-Perot can be given by

$$H_{\text{theory}} = (E^1_{\text{sam}}(1)+E^1_{\text{sam}}(2)+E^1_{\text{sam}}(3))/(E^1_{\text{ref}}(1)+E^1_{\text{ref}}(2)+E^1_{\text{ref}}(3))$$
$$H_{\text{experiment}} = E_{\text{sam}}/E_{\text{ref}} \quad (2)$$

We define the following error function

$$\mathrm{Re}(erf) = \mathrm{Re}(H_{\text{theory}}) - \mathrm{Re}(H_{\text{experiment}})$$
$$\mathrm{Im}(erf) = \mathrm{Im}(H_{\text{theory}}) - \mathrm{Im}(H_{\text{experiment}}) \quad (3)$$
$$Total(erf) = \sum\left|\mathrm{Re}(erf)\right| + \left|\mathrm{Im}(erf)\right|$$

The optical path and thickness of the cuvette equal to 5mm and 1.2mm, respectively. The optical path of THz wave in cuvette is shown in figure 2.

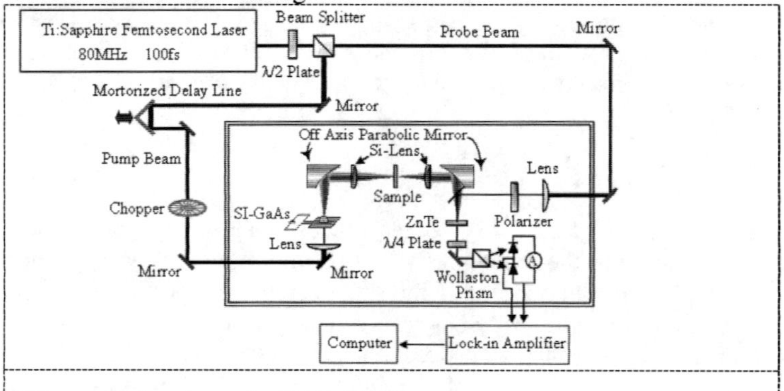

Figure 1. A schematical THz-TDS system

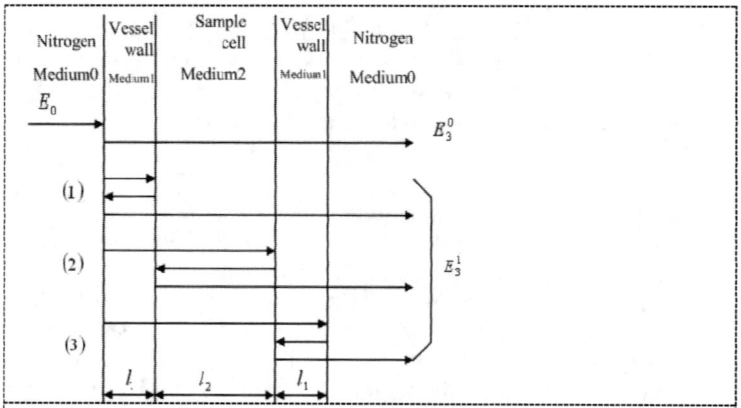

Figure 2. Optical path figure of THz in cuvette

Journal of Physics: Conference Series **276** (2011) 012231 doi:10.1088/1742-6596/276/1/012231

3. Experiment results

In order to detect pesticide ingredient in oil, we choose two samples as follows: Sample 1(rapeseed oil), and Sample2 (rapeseed oil +Imidacloprid). Figure 3 shows the terahertz time-domain spectra of the THz pulse traveling through reference and the pulse traveling through the small quartz cuvette with the two samples, respectively.

The refractive indices and the power absorption coefficients of the investigated two samples are calculated from the measured data by using Eqs. (1) ~ (3). There have three absorption peaks of rapeseed oil at 1.56THz, 1.45THz, and 1.35THz. For rapeseed oil/imidacloprid mixture, there have three absorption peaks at 1.58THz, 1.47THz, and 1.37THz. The refractive index decrease as the frequency increases.

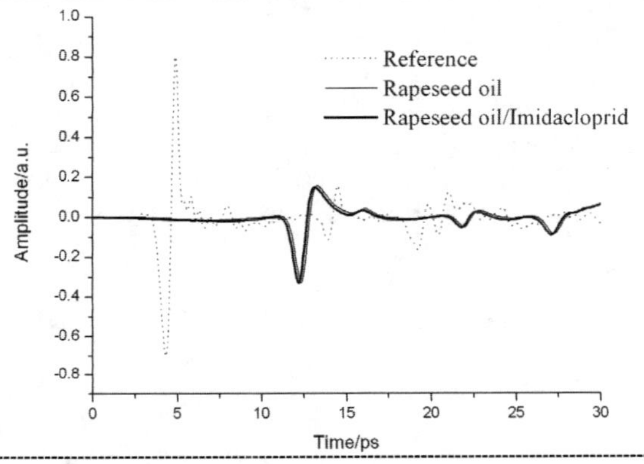

Figure 3. The time-domain spectra of reference and two sample

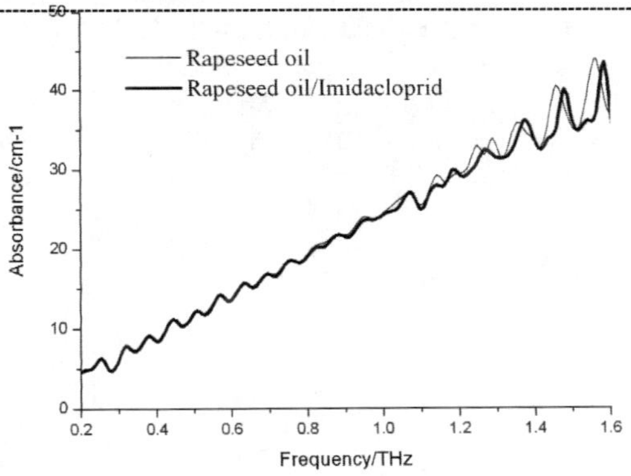

Figure 4. Absorption spectra

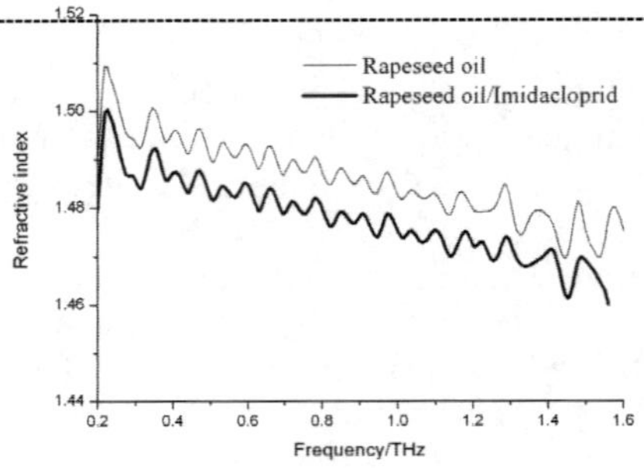

Figure 5. Refractive index of two samples

4. Conclusion

We analyzed the absorption coefficient and refractive index of both the rapeseed oil and rapeseed oil/imidacloprid mixture in the frequency range from 0.2 to 1.6THz respectively. The results show that it is feasible to employ THz-TDS to identify imidacloprid in oil.

Acknowledgments

This research was partially supported by the National Natural Science Foundation of China (No.60971027).

References

[1] Taday P, Bradley I V, Arnone D and Pepper M 2003 *J. Pharm. Sci.* **92** 831
[2] Nagai N, Imai T, Fukasawa R, Kato Kand Yamauchi K 2004 *Appl. Phys. Lett.* **85** 4010
[3] Federici J F, Schulkin B, Huang F, Gary D, Barat R, Oliveira F and Zimdars D 2005 *emicond.Sci.Technol.* **20** S266
[4] Jun H, Hua X and Qing X 2008 *Food science* **29** 568
[5] Dorney T, Baraniuk R and Mittleman D 2001 *J. Opt. Soc. Amer. A* **18** 1562

Broadband Terahertz Emission Based on the Femtosecond Laser Pulses

Guozhong Zhao [*,(1)], Haiyan Wang [(1)], Xinqiang Wang [, (2)]**

(1) Key laboratory of THz Spectroscopy and Imaging, Departments of Physics, Capital Normal University, Beijing 100048, China
[*] Email: guozhong-zhao@mail.cnu.edu.cn

(2) Affiliation, State Key Laboratory of Artificial Microstructure and Mesoscopic Physics, School of Physics, Peking University, Beijing 100871, China
[**] Email: wangshi@pku.edu.cn

ABSTRACT: The characteristics of terahertz (THz) radiation from the surface of InN excited by the femtosecond laser pulses are presented. The effect of intensity and central wavelength of the excited laser pulse on THz emission are studied experimentally. THz radiations from the surface emission of Mg-doped InN and InAs are compared. The results show the stronger dependence of THz radiation on the different intensity and central wavelength of the excitation pulses. THz radiation from InAs than InN has higher emission efficiency. However, the strong enhancement of THz emission is observed from InN with appropriate Mg concentrations. The mechanism of THz emission from InN is found to be the photo-Dember effect and the emission intensity is inversely proportional to the conductivity, which is beneficial to investigate THz source based on InN.

INTRODUCTION

The terahertz (THz) region of the electromagnetic spectrum (0.1-10THz) has potential applications in many fields of science and technology, including imaging, time-domain spectroscopy, and nondestructive materials identification and so on [1-3]. It was found by Zhang *et al.* in 1990 that ultrashort THz pulses can be generated by illuminating semiconductor surfaces with femtosecond laser pulses [4].Due to the low effective mass and the high mobility and saturation velocities of III nitrides [5], InN is considered to be an exciting source of terahertz radiation. The emission of THz pulses from InN thin films was observed firstly by Ascazubi *et al.* [6].After that, several groups reported the THz emission from InN films with different growth orientations [7-11]. Others reported Mg-doped InN samples can greatly enhance the THz radiation intension [12-14]. Recently, Ahn *et al.* reported that THz emission from InN nanorods can be three times stronger than THz emission from InN thin films [15]. Chern *et al.* reported excitation wavelength dependence of THz emission from InN and InAs [16]. In that letter, they used a ultrafast laser pulses at wavelengths tuned between 800 and 1500nm.The THz amplitude, normalized to pump and probe power, from both narrow bandgap semiconductors remains relatively constant over the excitation wavelength range.

Semiconductor surface irradiated with femtosecond fs laser pulses generates terahertz THz radiation pulses which are of great importance for time-resolved spectroscopy in condensed matter, chemical spectroscopy, imaging in medicine and material science. Owing to strong photo-absorption and high electron mobility, narrow band gap semiconductors are promising sources of optically excited THz irradiation. Among them, indium nitride (InN) is considered as one of the best candidates due to the large difference between the diffusion lengths of electrons and holes and very low probability of intervalley scattering.[2,6] Several groups have reported THz emission from InN films and nanostructures until now.[5-8,17] It has been reported recently that the emission intensity from InN is even higher than that from InAs which was found to show the highest emission efficiency.[18] Since the native InN layer is seriously defective and shows high electron concentration ranging from 10^{17}–10^{20} cm^{-3}, which degrades the emission efficiency, the THz emission can be further enhanced by compensating the donors through doping.

In this paper, we present THz emission from InN excited by the ultrafast laser pulses at central wavelengths from 750nm to 840nm in the same pump and probe power. THz radiation generated from InN exhibits greatly dependence on the excitation wavelength and intensity. Its amplitude, radiation efficiency and the effective spectral width are greatly different. On the other hand, the strong enhancement of THz emission from InN with appropriate Mg concentrations is shown. The mechanism of THz emission from InN is found to be the photo-Dember effect and the emission intensity is inversely proportional to the conductivity, which is beneficial to investigate THz source based on InN.

EXPERIMENT AND SAMPLES

Our measurements are performed using a Ti:sapphire laser, which operates at a center wavelength of 800 nm, and has a pulse duration of 100 fs and a repetition rate of 82 MHz. Its average output power is about 930 mW. The laser is split into two beams.The stronger beam used as pump beam is focused on the sample surface with a diameter of 2 mm and incident angle of 45° from normal.The pump power is ~250wM while the detect power is ~25mW for all excitation wavelengths. The THz signal was detected by free-space electro-optic sampling in a 2 mm thick ZnTe crystal.

InN films with both In- and N-polarities were grown by molecular beam epitaxy. [12,13] For the series of InN:Mg films in In-/N-polarity regimes, the typical thickness is 450/600 nm with Mg concentration [Mg] at 10^{16}–10^{22} cm^{-3}.[13] Other samples are either undoped or Mg-doped with the thickness up to 3 μm. Some InN:Mg layers showed *p*-type conduction in electrolyte capacitance-voltage measurement.[13,19] Hall effect measurement was performed to investigate the conductivity, carrier concentrations and mobility. Here, it should be noticed that the measured carrier concentrations and mobility are apparent ones. Since the surface electron accumulation layer with the density as high as $2–5\times10^{13}$ cm^{-2} exists in both *n*- and *p*-type InN due to the surface Fermi level pinning above the conduction band minimum,[20] the apparent conduction is *n*-type even for *p*-type InN and this is the reason why it is called the "buried *p*-type."

Time-domain THz emission measurements were performed by using a Ti:sapphire fs laser with a center wavelength of 800 nm, a pulse duration of 100 fs and a repetition rate of 82 MHz.[21] THz radiation is generated by the pump beam focused on the InN surface in *p*-polarization with a diameter of 0.2 mm and an incident angle of 45°. The energy of the pump beam was 4 nJ per pulse. THz signal was detected by the free-space electro-optic sampling in a 1 mm thickness of ZnTe crystal.

RESULTS AND DISCUSSION

The wavelength and intensity dependence of THz emission from InN on the excitation laser pulse are studied experimentally. By keeping at the intensity of pump and probe laser pulses and adjusting the central wavelength of excitation laser pulse, the THz emission efficiency of InN by the different central wavelength of excitation pulses is measured. At the same way, the intensity dependence of THz emission is investigated at 800 nm of central wavelength of excitation pulse. Finally, THz emission from the different InN with undoped and Mg-doped samples are compared and analyzed in detail.

Figure 1 shows the THz peak-peak amplitude as a function of wavelength of excitation laser pulse.from 750 nm to 840 nm at the same pump and probe power. The obvious change of THz emission efficiency for the different wavelength of excitation laser pulse is observed experimentally. It is clear that there exist a minimum of THz emission at 770 nm of excitation wavelength. THz emission fron InN excited by the femtosecond laser pulse at more than 800 nm of wavelength shows a great enhancement of radiation efficiency.

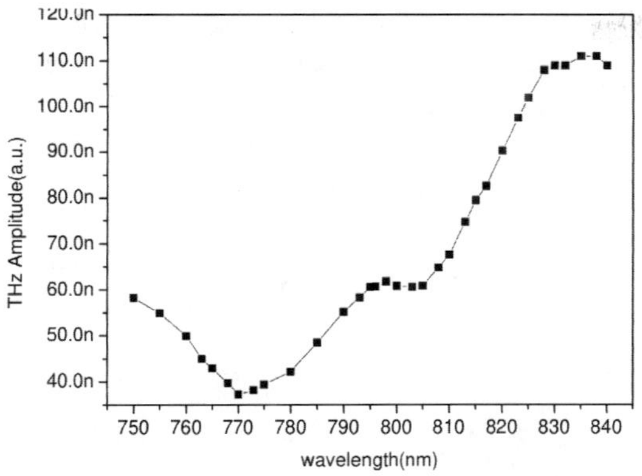

Fig. 1 THz peak-peak amplitude as a function of wavelength of excitation pulse

Figure 2 illuminates the relationship of THz emission spectra of InN and the pump intensity of femtosecond laser pulse. In the experiment, we kept the probe intensity as a constant and measured THz emission spectra at different pump power of excitation laser pulse. With increasing of pump power, THz emission intensity shows an obvious increase which is almost linear proportional to the pump power. At the same time, it is noticed that the half-maximum-full width of THz spectra increased firstly and reached saturation at a critical pump power.

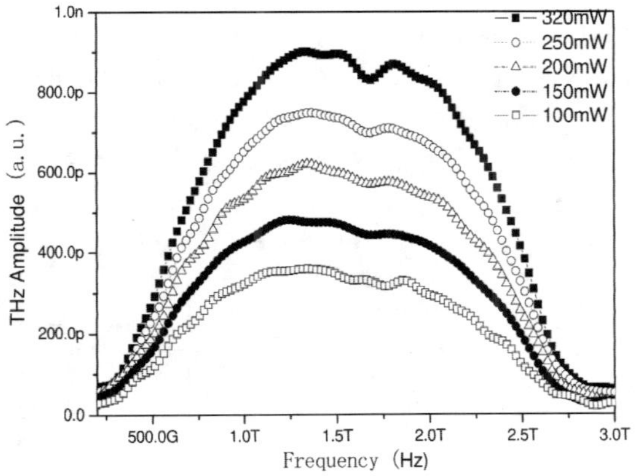

Fig. 2 THz emission spectra for the different pump power of excitation pulses

Figure 3 shows the typical time-domain waveforms of THz emission from nondoped and Mg-doped InN layers with In- and N-polarities and *p*-InAs (p=1×10^{16} cm^{-3}). In our measurement, more than 40 samples including undoped and Mg-doped ones in different lattice polarity regimes were investigated. No azimuthal angle dependence of THz radiation was observed in all these samples, indicating that THz emission is not related to the optical rectification effect. As shown in Fig.1, time-domain waveforms of THz emission from all InN layers show the same polarity as *p*-type InAs, which is independent of conduction and lattice polarities. It is known that the electric field of the THz waveform in the far field is given by $E_{THz} \sim \partial J / \partial t$, where *J* is the transient current induced by photoexcitation and originated from the photo-Dember effect due to the difference between the diffusion lengths of electrons and holes and the acceleration of photoexcited carriers by the surface depletion or accumulation field. At lower carrier densities, the optical rectification contribution is no longer the main THz generation mechanism in InN due to the effects of the difference of the electron and hole diffusion coefficients, which is so called photo-Dember effect. The same THz wave polarity of all InN layers and *p*-type InAs contradicts the second origin. It has been reported that *p*-type InAs showed the same THz wave polarity as the *n*-type GaAs, indicating the same wave polarity of InN and *n*-type GaAs. Since the surface electron accumulation layer exists in both *n*- and *p*-type InN, the surface band bending is downward, which is opposite to that of *n*-type GaAs, where the surface band bending is upward. InN would show opposite wave polarity to the *n*-type GaAs if the second origin is correct, which does not coincide with our experimental observation. Thus, the dominant mechanism for THz emission from InN is the photo-Dember effect.

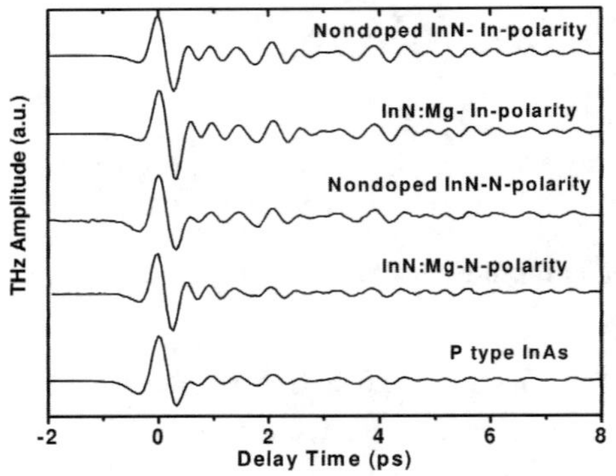

Fig. 3 The typical THz time-domain waveforms generated from nondoped InN and Mg-doped InN in In- and N-polarity, respectively. The THz waveform generated from *p*-type InAs (*p*=1_1016 cm^{-3}) is also shown for comparison.

Figure 4 shows the magnitude of THz emission from InN layers as a function of [Mg] in both In- and N-polarity regimes. It is obvious that the intensity of THz emission is greatly affected by Mg doping. With increasing [Mg], the intensity first increases and then decreases with the maximum intensity at [Mg] of ~1×10^{18} and 6×10^{18} cm^{-3} for In- and N-polar InN films, respectively. In addition, the p-type InN samples generally shows stronger THz emission than the n-type ones.

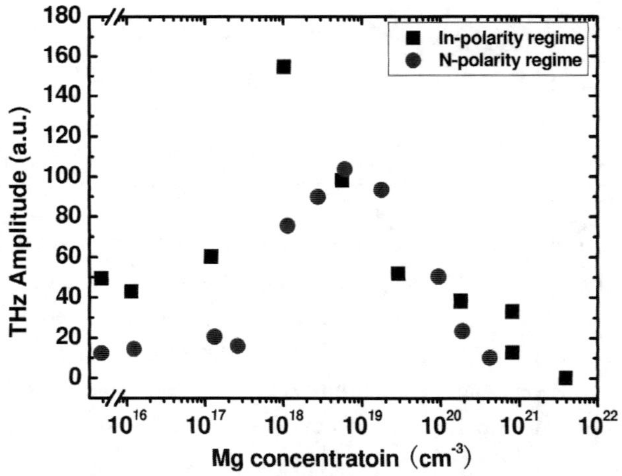

FIG. 4 THz emission intensity and resistivity of InN layers with N-polarity as a function of Mg doping concentrations.

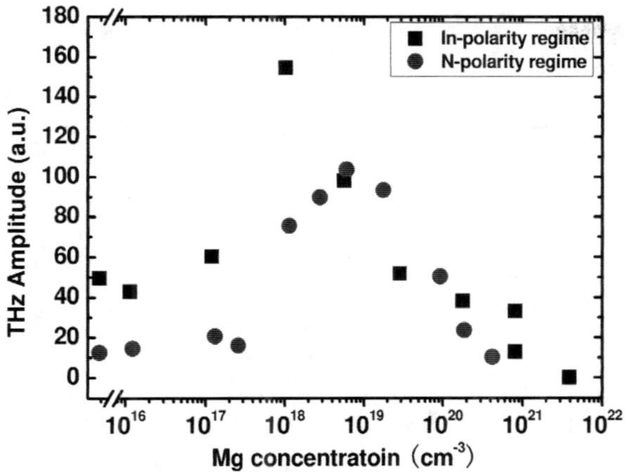

FIG. 5 THz emission intensity and resistivity of InN layers with N-polarity as a function of [Mg]. The sample shown in front of the break at the *x*-axis is undoped one.

It is known that origins of transient currents in semiconductors induced by photo-excitation come from (1) the photo-Dember effect originating from the difference between the diffusion constants of electrons and holes and (2) the acceleration of photo-excited carriers by a surface depletion field or accumulation field. In the case of InN, which is similar as InAs, a surface electron accumulation layer exists. This surface electron accumulation layer is independent of conduction polarity and thus it's electric field has the same direction for both n- and p-type InN layers. However, this electric field should lead to negative polarity time-domain wave forms, which is different from the experimental observation shown in Fig. 3. Therefore, the contribution from the surface field should not be the dominant one. This is probably because the thickness of surface accumulation layer is less than 10 nm, which is much smaller than the penetration depth of ~200 nm for a 800 nm laser beam. As for the depletion layer, its contribution is also small due the small thickness as reported previously. In addition, it has been reported by Lin *et al* that the dominant radiation mechanism is the drift current induced by the internal electric field due to spontaneous polarization in InN.[22] However, this is only correct at low density excitation and the polarity of time-domain wave forms should be different for InN layers with different lattice polarities, which does not coincide with our experimental results as shown in Fig. 3. Thus, it is suggested that that the dominant mechanism of THz emission from our InN samples is the photo-Dember effect. In this case, the intensity of THz emission greatly depends on the concentration and mobility of carriers. However, for InN:Mg, it is difficult to precisely determine the carrier concentration due to the surface electron accumulation, in particular for p-type sample. In our measurements, it was found that the dependence of THz emission intensity on [Mg] shows almost the same tendency as that of the resistivity (inverse of conductivity), as shown in figure 5. This is reasonable since the dominant mechanism for THz emission is the photo-Dember effect, where drift and diffusion currents of electrons J_n and hole J_p in an optical pumped semiconductor are described by the following equations [23]:

$$J_n = eE(\mu_n n + \mu_n \Delta n) + \mu_n k_B T(\partial n/\partial z + \partial \Delta n/\partial z) \quad (1)$$

$$J_p = eE(\mu_p p + \mu_p \Delta p) - \mu_p k_B T(\partial p/\partial z + \partial \Delta p/\partial z) \quad (2)$$

Where the E is electric field, T is carrier temperature, μ_n and μ_p are mobilities of electrons and holes, $n/\Delta n$ and $p/\Delta p$ are bulk/photo-generated electron and hole concentrations, $\partial n/\partial z$, $\partial \Delta n/\partial z$, $\partial p/\partial z$, $\partial \Delta p/\partial z$ are the gradients of bulk carrier concentrations and photocarrier concentrations. Thus, the photo-Dember field can be deduced according to the Dember hypothesis that the total current equals to zero:

$$E_D = \frac{k_B T}{e} \bullet \frac{\partial \Delta\sigma_p/\partial z - \partial \Delta\sigma_n/\partial z}{\sigma_n + \sigma_p + \Delta\sigma_p + \Delta\sigma_n} = \frac{k_B T}{e} \bullet \frac{\partial \Delta\sigma_p/\partial z - \partial \Delta\sigma_n/\partial z}{\sigma_{total} + \Delta\sigma_{total}} \quad (3)$$

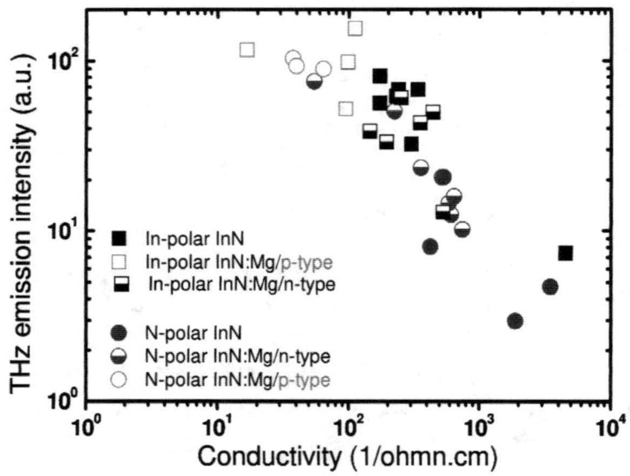

FIG. 6 The THz emission intensity generated from undoped and Mg-doped InN layers at both In- and N-polarities as a function of conductivity.

Where $\sigma_n = ne\mu_n$, $\sigma_p = pe\mu_p$, $\sigma_{total} = \sigma_n + \sigma_p$ are conductivities contributed from electrons and holes, and the both carriers, the $\Delta\sigma_n$, $\Delta\sigma_p$ are photoconductivity. The $\partial\sigma_n/\partial z$ and $\partial\sigma_p/\partial z$ are negligible. It is obvious from Eq. (3) that the intensity of THz emission is strong for InN layer with low conductivity since the $\Delta\sigma_{total}$ is smaller than the σ_{total} in our measurement condition. This relationship is true for both Mg-doped and undoped InN samples as shown in figure 6. It can be seen that the intensity of THz emission decreases with increasing conductivity, which is independent of conduction polarity. The p-type samples exhibit stronger intensity due to its lower apparent conductivity. Meanwhile, the undoped InN samples with low conductivity also show stronger THz emission. Thus it is suggested that it is more reasonable to correlate the THz intensity with the conductivity than with carrier concentration since (1) the conductivity includes contribution from electrons and holes and (2) it is very difficult to precisely determine the bulk carrier concentration for InN, in particular for p-type InN. The dependence of THz intensity on conductivity also confirms that the dominant mechanism of THz emission of InN is photo-Dember effect under our measurement condition. It can be suggested from Fig. 6 that it is necessary to decrease the conductivity of InN samples by improving crystalline quality or carrier compensation.

SUMMARY

In summary, the THz emission from InN excited by the femtosecond laser pulse is studied experimentally. THz radiations from the surface emission of Mg-doped InN and InAs are compared. The effect of intensity and central wavelength of the excited laser pulse on THz emission are studied experimentally. The results show the stronger dependence of THz radiation on the different intensity and central wavelength of the excitation pulses. THz radiation from InAs than InN has higher emission efficiency. However, the strong enhancement of THz emission is observed from InN with appropriate Mg concentrations. The mechanism of THz emission from InN is found to be the photo-

Dember effect and the emission intensity is inversely proportional to the conductivity, which is beneficial to investigate THz source based on InN.

ACKNOLEDGEMENT

This work is supported by National Natural Science Foundation of China under Grant No. 50971094, Beijing Natural Science Foundation under Grant No. 1092007. It is also supported by Beijing Ren Cai Program for Education.

REFERENCES

[1] D Dragoman, M Dragoman, *Progress in Quantum Electronics,* 28, 1, 2004
[2] I. Wilke, Proc. SPIE 6772, 67720N, 2007
[3] C L Zhang, Y Zhang, G Z Zhao, *"Terahertz Sensing and Imaging".* Beijing: National Defense Industry Press (in Chinese), pp.3-5, 2008.
[4] X C Zhang, B B Hu, D Darrow, D H Auston, *Appl. Phys. Lett.* 56, 1011,1990
[5] V. Cimalla, B. Pradarutti, G. Matthaus, C. Bruckner, S. Riehemann, G.Notni, S. Nolte, A. Tunnermann, V. Lebedev, and O. Ambacher, Phys.Status Solidi B 244, 1829, 2007
[6] R. Ascázubi, C. Shneider, I. Wilke, R. Pino, and P. S. Dutta, Phys. Rev. B 72, 045328, 2005
[7] K Liu, J Xu, T Yuan , X C Zhang, *Phys. Rev. B* 73, 155330, 2006
[8] H Ahn, Y P Ku, C H Chuang, C L Pan, *Appl. Phys. Lett.* 92**,** 102103, 2008
[9] V M Polyakov, F Schwierz, *Semiconductor science and technology* 22, 1016, 2008
[10] H Q Sun, G Z Zhao, C L Zhang, G Z Yang, *Acta Phys. Sin.* (in Chinese) 57, 790, 2008
[11] H Chen，L Wang, *Acta Phys. Sin.* (in Chinese).58, 4605, 2009
[12] X. Wang, S. Che, Y. Ishitani, and A. Yoshikawa, Jpn. J. Appl. Phys., Part 2, 45, L730, 2006
[13] X. Wang, S. Che, Y. Ishitani, and A. Yoshikawa, J. Appl. Phys. **99**, 073512, 2006
[14] X. Wang, S. Che, Y. Ishitani, and A. Yoshikawa, Appl. Phys. Lett. 91,242111, 2007
[15] H. Ahn, Y. P. Ku, Y. C. Wang, C. H. Chuang, S. Gwo, and C. L. Pan, Appl. Phys. Lett. 91, 163105, 2007
[16] G D Chern, E E Readinger, H Shen, *Appl. Phys. Lett.* 89, 141115, 2006
[17] I. Wilke, R. Ascazubi, H. Lu, and W. J. Schaff, Appl. Phys. Lett. 93, 221113, 2008
[18] B. Pradarutti, G. Matthäus, C. Brückner, S. Riehemann, G. Notni, S. Nolte, V. Cimalla, V. Lebedev, O. Ambacher, and A. Tünnermann, Proc.SPIE 6194, 61940I, 2006
[19] X. Wang, S. Che, Y. Ishitani, and A. Yoshikawa, Appl. Phys. Lett. 92,132108, 2008
[20] I. Mahboob, T. D. Veal, C. F. McConville, H. Lu, and W. J. Schaff, Phys.Rev. Lett. 92, 036804, 2004
[21] G. Zhao, R. N. Schouten, N. van der Valk, W. Th. Wenckebach, and P. C. M. Planken, Rev. Sci. Instrum. 73, 1715, 2002
[22] L. Yang, G. Zhao, W. Li, Y. Liu, X. Shi, X. Jia, K. Zhao, X. Lu, Y. Xu, D.Xie, J. Wu, and J. Chen, Spectrochim. Acta, Part A 73, 884, 2009
[23] X. C. Zhang, Y. Jin, and X. Ma, Appl. Phys. Lett. 61, 2764 , 1992

Journal of Physics: Conference Series **276** (2011) 012233 doi:10.1088/1742-6596/276/1/012233

Terahertz spectroscopic investigation of Chinese herbal medicine

Zhao Xiao-li[1], **Li Jiu-sheng**[1, 2]

[1] Centre for THz Research, China Jiliang University, Hangzhou 310018, China

[2] College of Precision Instrument and Opto-Electronics Engineering, Institute of Laser and Optoelectronics, Tianjin University, Tianjin 300072, China

E-mail: forever-li@126.com

Abstract. The absorption spectra of panax notoginseng and glycyrrhiza in the frequency range of 0.2~1.6THz has been measured with terahertz time-domin spectroscopy at room temperature. Simultaneously, the corresponding theoretical spectra were given by using density functional theory methods. It was found that the absorption peaks of the two molecules obtained by theoretical were in good agreement with the experimental results.

1. Introduction

Terahertz wave usually refers to the frequency range of 0.1 ~ 10THz in electromagnetic waves, which coincides with the energy between the molecular vibration and rotational energy transitions. Because of the macromolecules highly relevant nature of the spectroscopic properties in the scope of the THz band, the terahertz time-domain spectroscopy has received extensive attention in recent years [1-4].

Drug ingredients identification and quality control is the key for the widely used Chinese herbal medicine. In this paper, we investigate the terahertz spectra of two kinds of Chinese herbal medicines using terahertz time-domain spectroscopy (THz-TDS) as well as computationally using the B3LYP density functional theory (DFT) method. The research shows that THz-TDS technique can be used for the identification of panax notoginseng and glycyrrhiza, which provides guidance significance to further research using THz time-domain spectroscopy to identify other herbs.

2. Theory analysis

The schematic diagram of the THz-TDS setup used in this experiment is shown in figure 1. A repetition rate of 80MHz, diode-pump mode-locked Ti:sapphire laser (MaiTai, Spectra Physics) provided the femtosecond pulses with duration of 100fs and center wavelength of 810nm. A p-type InAs wafer with <100> orientation was used as the THz emitter and a 2.8 mm-thick <110> ZnTe was employed as the sensor. A standard lock-in technology was used in this system. A femtosecond laser pulse was split into two beams. The pump beam was used to generate THz radiation and the probe beam acted as a gated detector to monitor the temporal waveform of THz field. A silicon lens and parabolic mirrors were used to collimate and focus the THz beam through free space onto the detector. A balanced photodiode detector detected the probe beam, and the signal was amplified by a lock-in amplifier and sent to the computer for processing. The THz beam path was purged with dry nitrogen to minimize the absorption of water vapor and enhance the signal to noise ratio (SNR). The humidity was kept less than 1% and the temperature was kept at 298K. The dynamic range was about 4000:1, and the spectral resolution was better than 60 GHz in the 0.2–1.8THz region.

The THz-TDS can be used to extract optical parameters such as the absorption coefficient and refractive index of the sample. The transmitted field, $H = E_{sam}(\nu) / E_{ref}(\nu)$ is related to the complex refractive index $N = n + ik$ by

$$H = \frac{4N}{(1+N)^2} e^{i2\pi(N-1)d\omega/c} = Ae^{-i\varphi} \tag{1}$$

where ω is the frequency of the THz wave, A the amplitude of the THz wave, φ the phase difference between the sample and reference waveforms, and d the thickness of the sample. The multiple internal reflection is ignored in this equation under the condition $K \ll n$, where K is the extinction coefficient. Hence the Fabry–Perot effect can be neglected. Thus, we can obtain the refractive index n, absorption coefficient α, respectively:

$$n = \frac{c\varphi}{2\pi d\omega} + 1 \tag{2}$$

$$\alpha = \frac{4\pi k\omega}{c} = \frac{2}{d}\ln\left[\frac{4n}{A(n+1)^2}\right] \tag{3}$$

Here, the panax notoginseng is named sample 1, and glycyrrhiza is named sample2. To reduce the scattering effect of the sample, these samples were ground into fine particles and then pressed into pellets. The approximate pressure is 4500 PSI. The THz spectra of three kinds of POPs were measured with free space electro-optic sampling. We calculated the THz spectrum of the panax notoginseng and glycyrrhiza by using the method of DFT method. Optimization of the molecular structure was shown in figure 2.

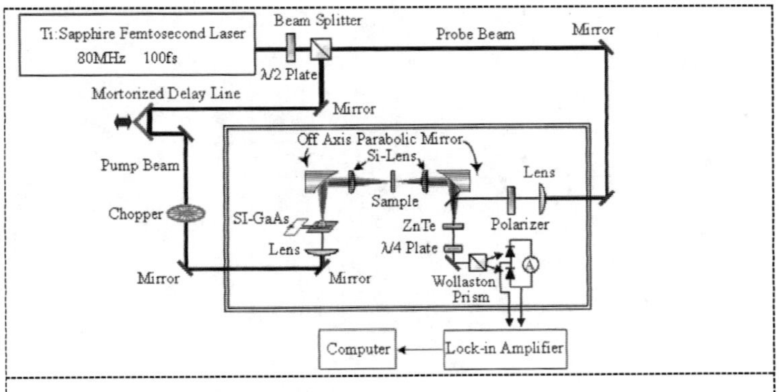

Figure 1. A schematical THz-TDS system

3. Experiment results

The terahertz time-domain spectra and frequency domain spectrum of the terahertz pulse traveling through air and the pulse traveling through the four kinds of soil samples are shown in figure 3 and figure 4. Every pulse curve is the average of three individual measurements in order to increase the signal-to-noise ratio. The experimental results were redone more than 3 times, the results provided here are repeatable and credible. Figure 5 shows the absorption curves of panax notoginseng and glycyrrhiza. Panax notoginseng has three characteristic absorption peaks which are located at 1.29THz, 1.40THz and 1.51THz, while glycyrrhiza in the frequency of 1.56THz has a characteristic absorption peak. The theoretical calculation has a good accordance corresponding to each other with experimental spectra characteristic absorption peaks.

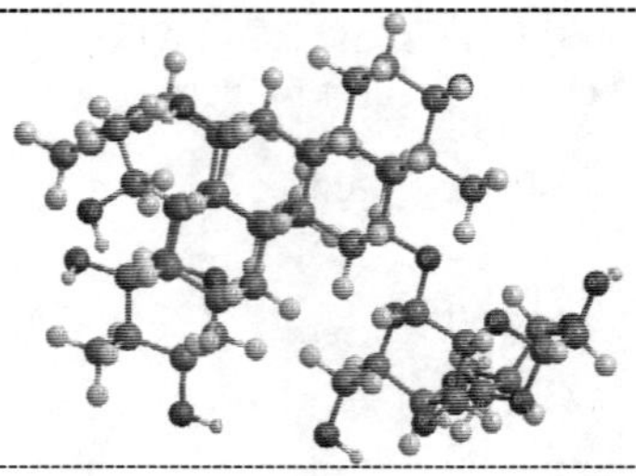

(a)

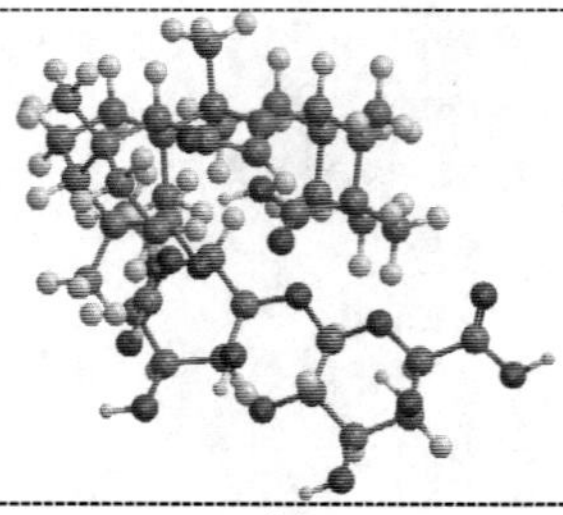

(b)

Figure 2. Molecular structure of panax notoginseng and glycyrrhiza

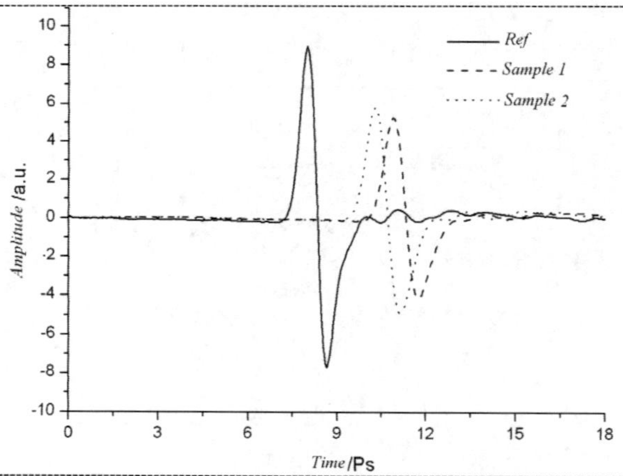

Figure 3. Measured terahertz time-domain spectra of reference and samples

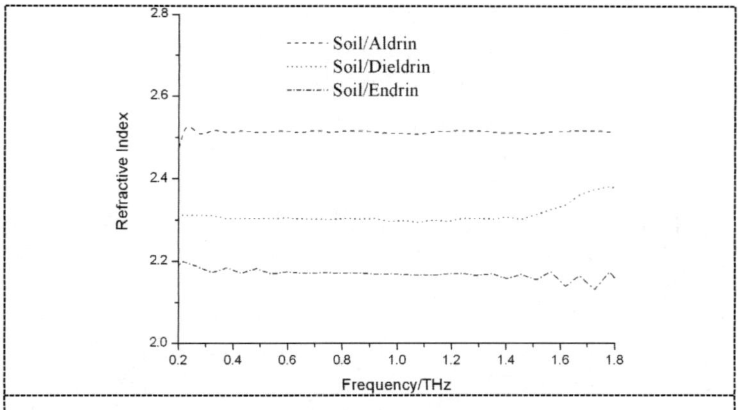

Figure 4. THz Frequency-domain spectra

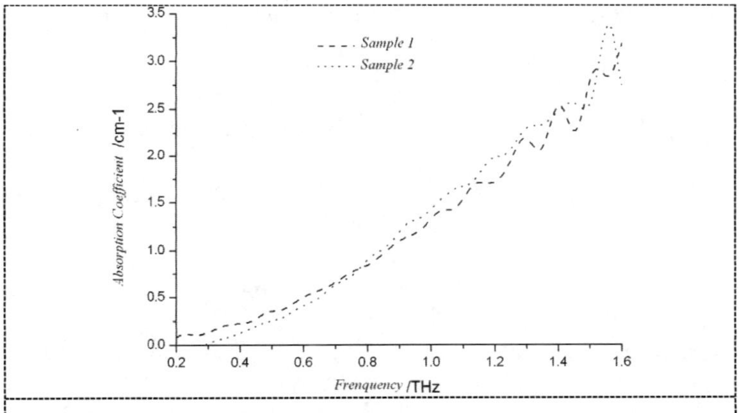

Figure 5. Absorption spectra

4. Conclusion

We tested panax notoginseng and glycyrrhiza in frequency of 0.2 ~ 1.6THz. The experimental results and theoretical calculations are basically in accordance with each other. The results of this study provide a practical application of THz-TDS in the detection quality of Chinese herbal medicine.

Acknowledgments

This research was partially supported by the National Natural Science Foundation of China (No.60971027).

References

[1] Walther M, Fischer B and Uhd Jepsen P 2003 *Chem. Phys.* **288** 261
[2] Taday P F, Bradley I V, Arnone D D and Pepper M 2003 *J. Pharm. Sci.* **92** 831
[3] Nagai N, Imai T, Fukasawa R, Kato K andYamauchi K 2004 *Appl. Phys. Lett.* **85** 4010
[4] Federici J F, Schulkin B, Huang F, Gary D, Barat R, Oliveira F and Zimdars D 2005 *Semicond.Sci.Technol.* **20** S266

Journal of Physics: Conference Series **276** (2011) 012234 doi:10.1088/1742-6596/276/1/012234

Diagnostic techniques of talc powder in flour based on the THz spectroscopy

Zhao Xiao-li[1], **Li Jiu-sheng**[1, 2]

[1] Centre for THz Research, China Jiliang University, Hangzhou 310018, China

[2] College of Precision Instrument and Opto-Electronics Engineering, Institute of Laser and Optoelectronics, Tianjin University, Tianjin 300072, China

E-mail: forever-li@126.com

Abstract. The terahertz time-domain spectroscopy technique is used to test flour/talc powder mixture samples. It is found that the two samples have obvious absorption peaks and different refractive index. The results provided in this paper will help us to study the THz application to the field of food safety.

1. Introduction

Over the past 10 years, electromagnetic terahertz (THz) frequencies region from 100 GHz to 10 THz have received extensive attention and investigation. Many potential applications are also explored such as radio astronomy, atmospheric studies, remote sensing, imaging, broadband communications, spectroscopy, and plasma diagnostics. In recent years, there has been a remarkable effort in employing terahertz spectroscopy for investigating material properties [1-4].

In this article, the THz-TDS is employed to characterize the THz optical properties and the complex dielectric function of five kinds of flour/talc powder mixture samples. The refractive index, the power absorption coefficient of flour/talc powder mixture in the THz region are tested and analyzed.

2. Experiment and theory analysis

A standard terahertz time domain spectroscopy system is employed and is illustrated schematically in figure 1. A repetition rate of 80MHz, diode-pump mode-locked Ti:sapphire laser (MaiTai, Spectra Physics) provided the femtosecond pulses with duration of 100fs and center wavelength of 810nm. A p-type InAs wafer with <100> orientation was used as the THz emitter and a 2.8 mm-thick <110> ZnTe was employed as the sensor. A standard lock-in technology was used in this system. A femtosecond laser pulse was split into two beams. The pump beam was used to generate THz radiation and the probe beam acted as a gated detector to monitor the temporal waveform of THz field. A silicon lens and parabolic mirrors were used to collimate and focus the THz beam through free space onto the detector. A balanced photodiode detector detected the probe beam, and the signal was amplified by a lock-in amplifier and sent to the computer for processing. The THz beam path was purged with dry nitrogen to minimize the absorption of water vapor and enhance the signal to noise ratio (SNR). The humidity was kept less than 1% and the temperature was kept at 298K. The dynamic range was about 4000:1, and the spectral resolution was better than 60 GHz in the 0.2–1.8THz region. The flour/talc powder mixture samples are prepared as follows: pure flour, flour/talc powder (0.5%)

Journal of Physics: Conference Series **276** (2011) 012234 doi:10.1088/1742-6596/276/1/012234

mixture, flour/talc powder (1%) mixture, flour/talc powder (3%) mixture, and flour/talc powder (5%) mixture. To reduce the scattering effect of the sample, these samples were ground into fine particles and then pressed into pellets. The approximate pressure is 4500 PSI.

In the THz-TDS system, the mathematical relationships connecting these measurable quantities with the dielectric parameters of GaAs material $n(\omega)$ and $\alpha(\omega)$ are as follows:

$$n(\omega) = \frac{\phi(\omega)}{\omega d}c + 1 \qquad (1)$$

$$a(\omega) = \frac{2\kappa(\omega)\omega}{c} = \frac{2}{d}\ln[\frac{4n(\omega)}{\rho(\omega)(n(\omega)+1)^2}] \qquad (2)$$

where $\phi(\omega)$ is the phase shift of the transmitted wave, $\rho(\omega)$ is the magnitude the transmitted wave, d is the sample thickness, $n(\omega)$ is the refractive index, $\alpha(\omega)$ is the absorption coefficient.

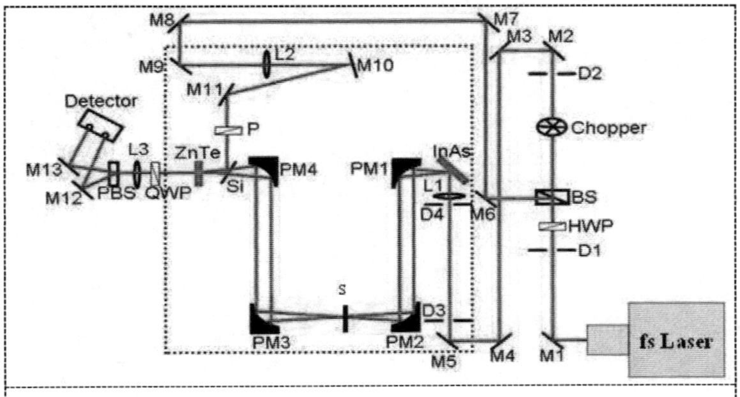

Figure 1. A schematical THz-TDS system

3. Results and discussion

Figure 2 (a) and (b) show the time-domain and frequency-domain THz pulse traveling through air and the samples. The refractive indices and the power absorption coefficients of the measured samples are calculated from the measured data. The calculation results are plotted in figure 3 (a) and (b). It can be noted that all the samples have three absorption peaks at the frequencies of 0.9557THz, 1.364THz, and 1.568THz. Absorption peak may be mainly from macromolecular organic of flour and talcum powder containing heavy metal ions. The absorption spectra of the samples increase slowly as the frequency increases.

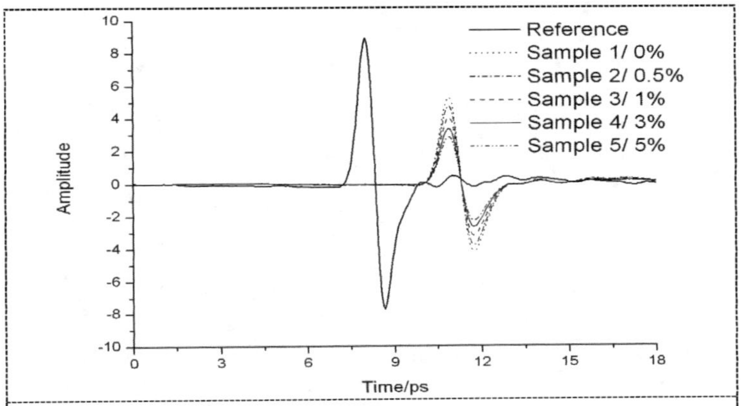

(a)

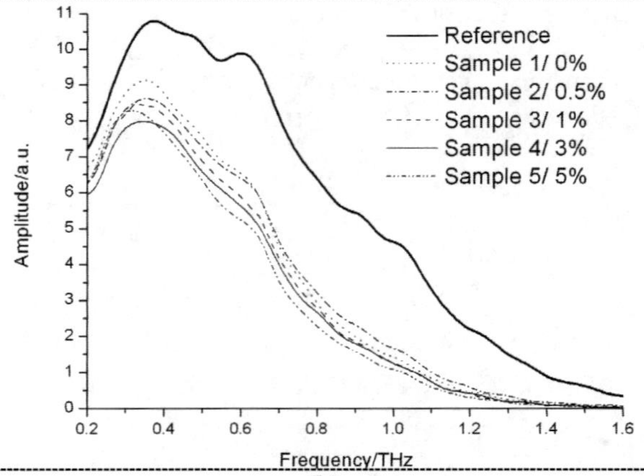

(b)

Figure 2. Time-domain (a), and frequency domain (b) terahertz spectra of reference and five samples

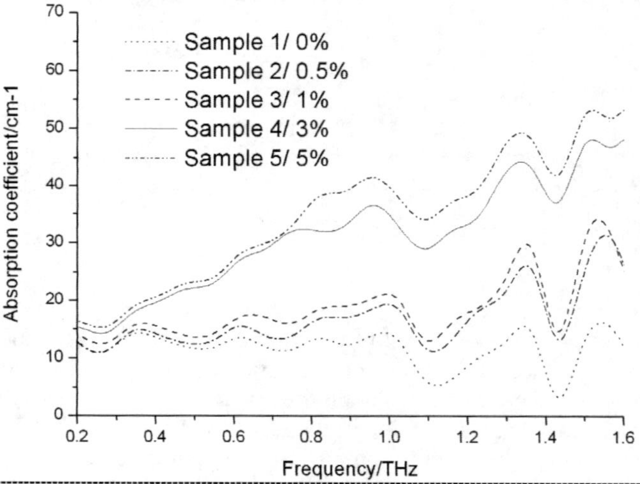

(a)

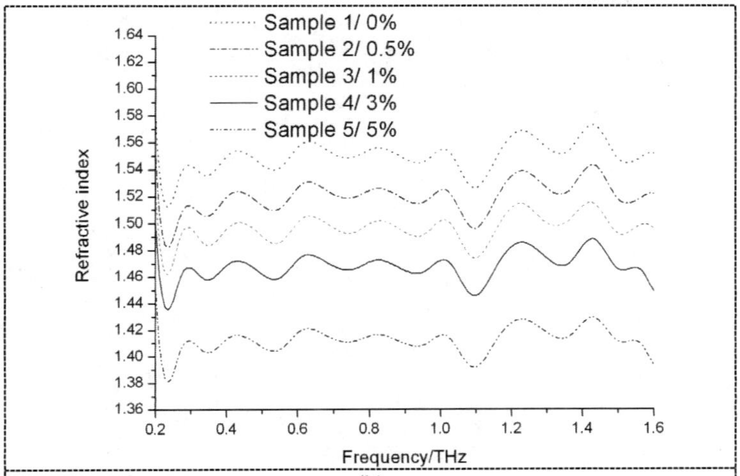

(b)

Figure 3. Absorption coefficient (a) and refractive index (b) of samples

4. Conclusion

In this paper, the spectral characteristics of five kinds of flour/talc powder mixture samples have been measured with terahertz time-domain spectroscopy (THz-TDS) in the range of 0.2~1.6THz. We can accurately extracted refractive index and absorption coefficient of samples by this method. The results provided in this paper will help us to study the THz application to the field of food safety.

Acknowledgments
This research was partially supported by the National Natural Science Foundation of China (No.60971027).

References

[1] Taday P F, Bradley I V, Arnone D and Pepper M 2003 *J. Pharm. Sci.* **92** 831
[2] Nagai N, Imai T, Fukasawa R, Kato K and Yamauchi K 2004 *Appl. Phys. Lett.* **85** 4010
[3] Chen J, Chen Y and Zhang X C 2007 *Optics Exp.* **15** 12060
[4] Laman N, Harsha S S and Grischkowsk D 2008 *Optics Exp.* **16** 4094

AUTHOR INDEX

AUTHOR INDEX

AUTHOR INDEX

AUTHOR INDEX

AUTHOR INDEX

AUTHOR INDEX

AUTHOR INDEX

AUTHOR INDEX

Institute of Physics
Dirac House, Temple Back
Bristol BS1 6BE UK

CURRAN ASSOCIATES INC
proceedings.com